(10)	1B (11)	2B (12)	3A (13)	4A (14)	5A (15)	(16)		
							H 1.0079	He 4.0026
			Boron 5 **B** 10.811	Carbon 6 **C** 12.011	Nitrogen 7 **N** 14.0067	Oxygen 8 **O** 15.9994	Fluorine 9 **F** 18.9984	Neon 10 **Ne** 20.1797
			Aluminum 13 **Al** 26.9815	Silicon 14 **Si** 28.0855	Phosphorus 15 **P** 30.9738	Sulfur 16 **S** 32.066	Chlorine 17 **Cl** 35.4527	Argon 18 **Ar** 39.948
Nickel 28 **Ni** 58.69	Copper 29 **Cu** 63.546	Zinc 30 **Zn** 65.39	Gallium 31 **Ga** 69.723	Germanium 32 **Ge** 72.61	Arsenic 33 **As** 74.9216	Selenium 34 **Se** 78.96	Bromine 35 **Br** 79.904	Krypton 36 **Kr** 83.80
Palladium 46 **Pd** 106.42	Silver 47 **Ag** 107.8682	Cadmium 48 **Cd** 112.411	Indium 49 **In** 114.82	Tin 50 **Sn** 118.710	Antimony 51 **Sb** 121.75	Tellurium 52 **Te** 127.60	Iodine 53 **I** 126.9045	Xenon 54 **Xe** 131.29
Platinum 78 **Pt** 195.08	Gold 79 **Au** 196.9665	Mercury 80 **Hg** 200.59	Thallium 81 **Tl** 204.3833	Lead 82 **Pb** 207.2	Bismuth 83 **Bi** 208.9804	Polonium 84 **Po** (209)	Astatine 85 **At** (210)	Radon 86 **Rn** (222)

Europium 63 **Eu** 151.965	Gadolinium 64 **Gd** 157.25	Terbium 65 **Tb** 158.9253	Dysprosium 66 **Dy** 162.50	Holmium 67 **Ho** 164.9303	Erbium 68 **Er** 167.26	Thulium 69 **Tm** 168.9342	Ytterbium 70 **Yb** 173.04	Lutetium 71 **Lu** 174.967

Americium 95 **Am** (243)	Curium 96 **Cm** (247)	Berkelium 97 **Bk** (247)	Californium 98 **Cf** (251)	Einsteinium 99 **Es** (252)	Fermium 100 **Fm** (257)	Mendelevium 101 **Md** (258)	Nobelium 102 **No** (259)	Lawrencium 103 **Lr** (260)

The Chemical World

CONCEPTS AND APPLICATIONS

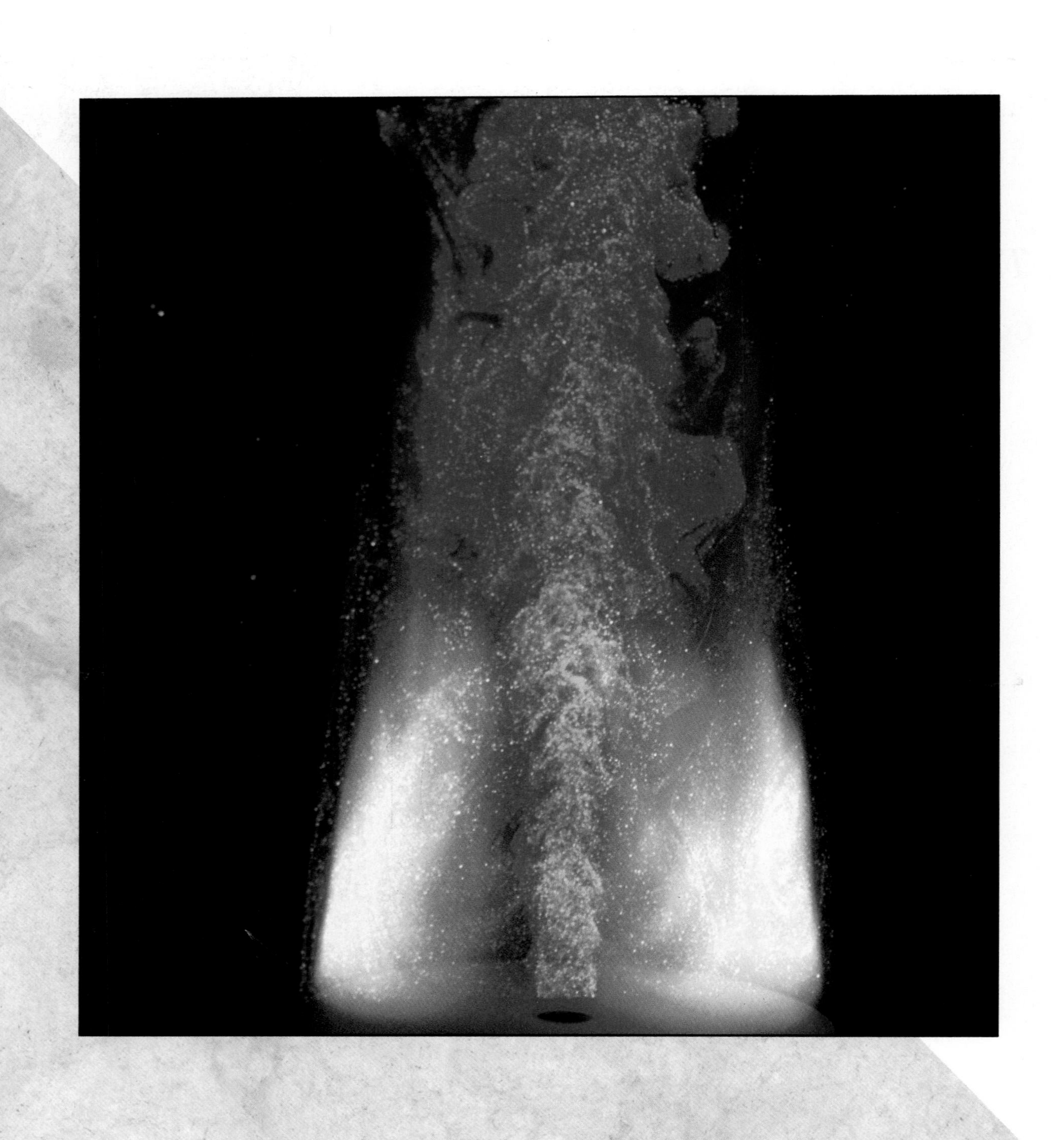

The Chemical World

CONCEPTS AND APPLICATIONS

JOHN C. KOTZ
SUNY Distinguished Teaching Professor
State University of New York
College at Oneonta
Oneonta, New York

MELVIN D. JOESTEN
Professor of Chemistry
Vanderbilt University
Nashville, Tennessee

JAMES L. WOOD
Resource Consultants, Inc.
Brentwood, Tennessee

JOHN W. MOORE
Professor of Chemistry
Director, Institute for Chemical Education
University of Wisconsin-Madison
Madison, Wisconsin

Illustrated by George V. Kelvin

Saunders Golden Sunburst Series
SAUNDERS COLLEGE PUBLISHING
Harcourt Brace College Publishers

Fort Worth Philadelphia San Diego New York Orlando Austin
San Antonio Toronto Montreal London Sydney Tokyo

Text Typeface: Times Roman
Compositor: York Graphic Services
Publisher: John Vondeling
Developmental Editors: Jennifer Bortel and Mary Castellion
Managing Editor & Project Editor: Carol Field
Copy Editor: Jay Freedman
Manager of Art & Design: Carol Bleistine
Text Design: Rebecca Lloyd Lemna
Cover Design: Lawrence R. Didona
Text Artwork: George Kelvin
Layout Artist: Dorothy Chattin
Photo Editor: Kathrine Kotz
Director of EDP: Tim Frelick
Production Manager: Charlene Squibb
Marketing Manager: Marjorie Waldron

Cover: Turbulence of flame-combustion. © YOAV LEVY/PHOTOTAKE

Printed in the United States of America

THE CHEMICAL WORLD: CONCEPTS AND APPLICATIONS

0-03-094659-X

Library of Congress Catalog Card Number: 93-085366
3456 032 987654321

Preface

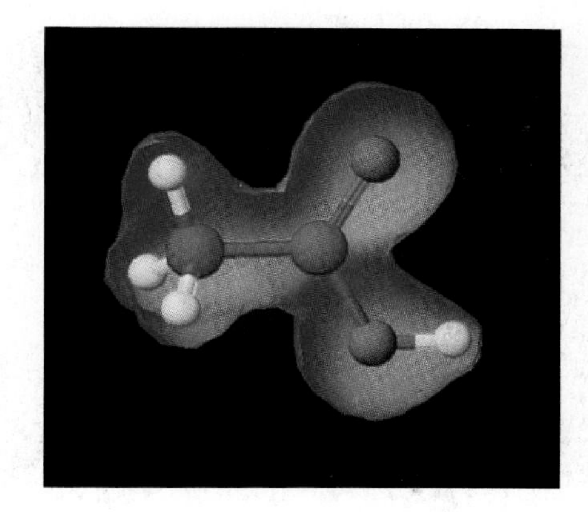

As authors of a chemistry textbook, we are often asked why books are revised every few years. "There really isn't much new in the theory of chemistry. Don't you just move words around?" The situation is just the opposite. There is a great deal that is new every day in chemistry, and much of it can be understood and appreciated by beginning students. *So, many new developments are often included in new editions, while still describing all the main topics in the theory of chemistry in order to serve the broadest possible audience.* In addition, because chemistry courses in the United States currently emphasize numerical calculations, there are a large number of solved examples, drill problems, and study questions. The result is that books have become longer and longer, and more and more complex.

Many people concerned with chemical education believe that it is time to reverse this trend, that re-examination of the goals of introductory college and university chemistry courses is long overdue, and that several challenges must be addressed:

- Lectures should be fewer.
- Examinations must require critical thinking in addition to numerical problem solving skills.
- Laboratories should be more flexible.
- Concepts and applications should be better integrated into the courses.
- Conceptual problem solving should be emphasized.
- Numerical problem solving should be de-emphasized.

GOALS OF THIS BOOK

Our response to these and other concerns in science education takes the form of this book. The title, THE CHEMICAL WORLD: CONCEPTS AND APPLICATIONS, conveys its principal themes: a broad overview of the concepts of

chemistry and their applications in the world around us. Our goals in this book have been

- To cover the truly important concepts of chemistry.
- To enable students to see how chemistry is done.
- To demonstrate how to solve conceptual as well as numerical problems.
- To convey the sense that chemistry is a dynamic field, a subject with both a fascinating history and continually unfolding new developments.
- To provide an understanding of everyday phenomena.
- To illustrate where chemistry fits among other sciences and technologies.
- To begin to integrate information technology appropriately into the learning process.
- To prepare students to evaluate the risks and benefits of science and technology in modern society.

AUDIENCE

THE CHEMICAL WORLD: CONCEPTS AND APPLICATIONS is a textbook for the introductory course in chemistry for students pursuing further study in science, whether that science is chemistry, biology, engineering, geology, physics, or related subjects. Our assumption is that students beginning this course have had a basic foundation in algebra and in general science. Although undeniably helpful, a previous exposure to chemistry is neither assumed nor required.

PHILOSOPHY AND APPROACH

Only concepts that we believe are truly fundamental to understanding basic chemistry have been included, so that those concepts can be seen more clearly. Topics such as stoichiometry, acid–base chemistry, basic theories of chemical bonding, equilibrium, thermodynamics, kinetics, and electrochemistry are covered. However, when discussing bonding, for example, orbital hybridization and molecular orbital theory are excluded. These theories are of undeniable importance to students in higher level courses in chemistry, and they are introduced in such courses when they are needed. The chemistry that provides a background for biology, geology, nutrition, and engineering can be understood without them.

Many topics in current books are just "there," and there is never enough time to cover them all. So much time is spent on the details of chemical equilibria, kinetics, and thermodynamics, bonding, electrochemistry, and other subjects that the applications of chemical concepts cannot be included. By eliminating some of the details, we have created a book that can be used in its entirety. In order to better integrate the concepts of chemistry with their applications, we have not introduced a principle in this book unless it is needed later in the book and unless it is placed in the context of its applications. For example, the chapter on solids (Chapter 14) covers not only crystal structure, but also cement, ceramics, superconductors, and the use of silicon in semiconductor devices. A course based on this book will truly be a "general chemistry" course—an introduction to the field of chemistry that will leave students well prepared for further studies in science.

In order to teach both numerical and conceptual problem solving, we have included a number of conceptual problems involving critical thinking skills in addition to plenty of the usual numerical problems.

It is our belief that books will indeed become thinner as some of the functions of textbooks are taken over by computers, video, and related technologies that better show the dynamic nature of chemistry. For this reason we have included in a number of chapters some questions based on the use of a computer program or videodisc.

Finally, we believe that it is important to involve the reader in "doing" chemistry and to show that chemistry can be done with everyday things. For this reason, each chapter includes several "take-home/dorm-room" experiments called ***Chemistry You Can Do.*** These illustrate the topic of the chapter and can be done with simple and familiar chemicals and equipment found at home or on a college campus.

ORGANIZATION

The first chapter, "The Nature of Chemistry," is a description of the science of chemistry and how it is done in principle and in practice. The question of risks and benefits is addressed.

"The Nature of Matter," Chapter 2, introduces the kinetic theory of matter, states of matter, and classification of matter as mixtures or compounds. Some units and numerical quantities important to chemistry are discussed in this context.

The next four chapters—"Chemical Elements" (Chapter 3), "Chemical Compounds" (Chapter 4), "Chemical Reactions," (Chapter 5) and "Stoichiometry" (Chapter 6)—are the basis of all that follows. The fundamental concepts of element, compound, atom, molecule, ion, and chemical change are developed here and placed in the context of applications and their practical importance in chemistry. Simple organic molecules are introduced so that their role in applications can be meaningfully included.

Chapters 7 and 8 describe the most important principles of chemical reactivity. The first of these chapters deals with thermodynamics and the second with kinetics and chemical equilibrium. THE CHEMICAL WORLD differs from other books by introducing these key concepts early so that they can be applied to the chemistry that is discussed subsequently.

The next three chapters—"Electron Configurations, Periodicity and Properties of Elements" (Chapter 9), "Covalent Bonding" (Chapter 10), and "Molecular Geometry and Isomerism" (Chapter 11)—return to the topic of atomic and molecular structure and to the ways that atoms are assembled into molecules. We have not included the theories of orbital hybridization or of molecular orbitals. In addition, some details of atomic structure are excluded to make room for a description of the role of atomic and molecular structure in determining the chemistry of elements and compounds.

Chapters 12 ("Gases, Intermolecular Forces, and Liquids"), 13 ("Energy, Organic Chemicals, and Polymers"), and 14 ("Solids") are concerned broadly with states of matter. Chapter 13 is unique in a textbook at this level because it covers organic chemicals and their uses early in the text and not as a peripheral subject.

The next three chapters, "Solution Chemistry" (Chapter 15), "The Importance of Acids and Bases" (Chapter 16), and "Electrochemistry" (Chapter 17) include many examples and exercises that focus on practical applications.

Chapters 18 through 21 close the book with topics that provide abundant opportunities for the integration of concepts and applications. First, "Elements from the Land, Sea, and Air" (Chapter 18) shows how electrochemistry and other techniques are used to produce elements and compounds that are important in our economy. Next, we discuss two topics—"Nuclear Chemistry" (Chapter 19) and "Chemistry of the Atmosphere" (Chapter 20)—that are constantly in the news and should be understood by scientifically literate citizens. In Chapter 20, the chemical concepts that bear on air pollution and smog formation, the greenhouse effect, and ozone depletion are fully explored. Finally, Chapter 21 on "Chemistry of Life" describes important and newsworthy topics such as genetic engineering.

FEATURES

Each chapter of the book includes an **introductory paragraph** to give an overview of the chapter and to provide practical reasons why the topics are worth studying. In addition, there are worked **Examples** throughout the text to illustrate important principles and to serve as models for solving Exercises and end-of-chapter problems. In many cases the examples are integrated into the text. There are also **Exercises,** both quantitative and qualitative, that follow many of the examples. Answers to the Exercises are found in Appendix L.

Each chapter also contains at least one ***Chemistry You Can Do*** experiment. This feature encourages students to perform chemistry experiments on their own, using materials readily available at home or on campus. Thought-provoking questions at the conclusion of the open-ended experiments encourage students to consider the reasons for the outcome of the experiment. The results of the experiments and answers to the questions are provided in the instructor's manual.

All chapters have a number of boxed features. ***Portrait of a Scientist*** boxes provide a glimpse into the lives of scientists, both those who helped create the science of chemistry in the past and those who continue the tradition today. ***Deeper Look*** boxes furnish more detailed information on a particular principle or the application of a specific concept. These include "The Wave-Particle Dual Nature of Light" or "What General Conditions Produce Smog?"

Another type of box, ***The Chemical World,*** features excerpts by the late Isidore Adler of the University of Maryland and by Nava Ben-Zvi, of the Hebrew University of Jerusalem, from the popular Annenberg/CPB telecourse series "The World of Chemistry." These describe such topics as "A Better Aluminum Foil" and "Unraveling the Protein Structure." Each box identifies "The World of Chemistry" videotape in which the material appears.

Finally, the *News Feature* boxes show what is happening on the cutting edge of chemistry research. Topics include creating fake fats and isolating potential drugs from frog skin.

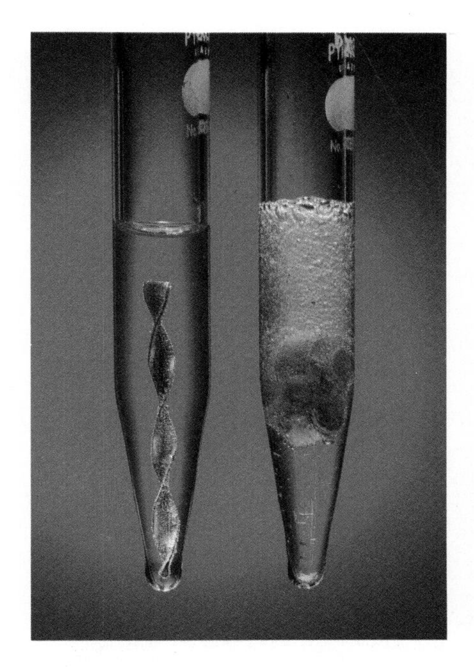

This book also contains a number of **full-color original photographs** by Charles D. Winters to create a visual accompaniment to the explanation of principles and applications. Photos are also used in some end-of-chapter Study Questions. In addition, all the art has been drawn by premier scientific illustrator George Kelvin. His use of color and eye for composition and detail make the illustrations and diagrams even more understandable.

We have used **margin notes** to highlight important points and interesting facts, to note topics to be discussed later in the text, and to help locate material. **Boldfaced terms** within the text indicate the introduction of new concepts, and most of these are also defined in the **Glossary/Index.**

At the end of each chapter you will find a section called "In Closing," which lists **goals** and concepts that should be understood after reading the chapter. Each goal is followed by the relevant section number. There are also a number of end-of-chapter **Study Questions,** some illustrated with photographs. They have been chosen to provide a balance between numerical problem-solving questions and conceptual "thought" questions. Question types include review questions, questions classified by category, general questions, and questions that require using the computer programs and videodiscs that are available to adopters of the text. Each chapter may also contain one or more summary questions, which tie in concepts from previous chapters. Answers to questions numbered in boldface are given in Appendix M.

Finally, there are numerous **appendices,** which include problem-solving skills, basic math operations; common units, equivalences and conversion factors; physical constants; naming simple organic compounds and coordination compounds; ionization and solubility constants; reduction potentials; hybridization solutions to in-text exercises; and answers to selected study questions.

SUPPORTING MATERIALS

Written Materials for the Student and Instructor

The **Student Study Guide** by Dean Nelson of the University of Wisconsin-Eau Claire has been designed around the key objectives of the book. Each chapter of the study guide includes study hints, true/false questions, a list of symbols presented in the chapter, a list of terminology, concept maps of the chapter topics, a concept test, a discussion of how the chapter material relates to other chapters, crossword puzzles, and supplementary reading of current interest.

The **Student Solutions Manual** by John Moore and Greg Steinke of the University of Wisconsin-Madison gives detailed solutions to designated end-of-chapter study questions. The manual also contains strategies for problem solving.

The **Student Lecture Outline** by Ronald O. Ragsdale of the University of Utah helps students organize material in the text and serves as a helpful classroom note-taking supplement.

A **Problem Solving Workbook** by Ronald O. Ragsdale provides many multiple-choice questions, most with complete solutions, to help students practice their test-taking and problem-solving skills.

An **Instructor's Resource Manual** by the authors provides solutions to the end-of-chapter study questions. The manual also lists the important objectives for each chapter, gives suggestions for possible ways to organize the course, suggests classroom demonstrations, and explains the results of the "Chemistry You Can Do" experiments and answers the questions about them.

A **Test Bank** provides hundreds of conceptual and numerical problems for use by the instructor. It is available in written, Macintosh, and IBM PC versions.

Overhead Transparencies of 150 figures from the book are available to adopters of the book. In addition, these transparencies as well as those from all other Saunders College Publishing chemistry textbooks are available on the Shakhashiri Chemical Demonstrations videodisc mentioned below. Both the transparencies and the videodisc are available at no charge to the adopters of this book.

Laboratory Manuals

Saunders offers many excellent general chemistry laboratory manuals, all of which can be used in conjunction with THE CHEMICAL WORLD. An instructor's manual is available for each title. In the event that none of these laboratory manuals meets your needs individually, selected experiments can be custom published as a separate laboratory manual, in accordance with Saunders' custom publishing policy. Please contact your local sales representative for more information on custom publishing.

Laboratory Experiments for General Chemistry, second edition, by Harold R. Hunt and Toby F. Block of Georgia Institute of Technology contains experiments that are designed to minimize waste of materials and stress safety. Pre-lab exercises and post-lab questions are included. The manual includes 42 experiments.

Experimentation and Analysis in the Chemistry Laboratory by Daniel Reger of the University of South Carolina, Eugene R. Weiner of the University of Denver, and William Gilkerson of the University of South Carolina presents clear, precise instructions in pre-lab exercises and emphasizes safety throughout. This manual includes 40 experiments.

Chemical Principles in the Laboratory, fifth edition, by Emil J. Slowinski and Wayne Wolsey of Macalaster College and William J. Masterton of the University of Connecticut provides detailed directions and advance study assignments. These thoroughly class-tested experiments were chosen with regard to cost and safety. The manual includes 43 experiments.

Experiments in General Chemistry by Frank Milio and Nordulf Debye of Towson State University and Clyde Metz of the College of Charleston, is the result of many years of collaboration on the development of laboratory

tested, and attention has been given to cost and safety. This manual contains 44 experiments.

Experimental General Chemistry, second edition, by Carl B. Bishop and Muriel B. Bishop, both of Clemson University, Kenneth Gailey, late of University of Georgia, Athens, and Kenneth Whitten of the University of Georgia, Athens, contains descriptive, quantitative, and instrumental experiments. This manual provides detailed instructions and incorporates helpful notes into the experiments to help students gain confidence. The manual includes 27 experiments.

MULTIMEDIA MATERIALS

Many data in this book are from an extensive computerized collection of information on the chemical elements known as the **KC? Discoverer** database. Data were carefully assembled by a team of chemists, and references are provided, making this an excellent "official" database. Many of the tables herein reflect the information from *KC? Discoverer,* a program and database available for the four different platforms described below. Each is a significant and independent program in its own right but is built around the complete database.

- "*KC? Discoverer* with Knowledgeable Counselor" by Daniel Cabrol, John W. Moore, and Robert C. Rittenhouse for IBM PS/2 or PC-compatibles with at least 640K memory and a hard disk.
- "*KC? Discoverer*" by Aw Feng and Moore, for IBM PC and PC-compatibles.
- "*KC? Discoverer* for the Apple II" by Michael Liebl, for Apple II machines with 128K RAM.
- "The Periodic Table Stack" by Michael Farris for Apple Macintosh, with HyperCard.

These programs are available at no charge to the adopters of the text, or they may be purchased directly from *Journal of Chemical Education: Software (JCE: Software):* Department of Chemistry, University of Wisconsin—Madison, 1101 University Avenue, Madison, WI 53706. For further information on *KC? Discoverer* and its use in the classroom setting, see J. Kotz, *J. Chem. Educ.,* **1989**, *66,* 750.

The Periodic Table Videodisc: Reaction of the Elements by Alton J. Banks is also available in accordance with Saunders' adoption policy or may be purchased from *JCE: Software.* This videodisc is a *visual* database; it shows still and motion images of the elements, their uses, and their reactions with air, water, acids, and bases. It is a particularly useful way to demonstrate chemical reactions in a large lecture room. The videodisc can be operated from a videodisc player by a hand-controlled keypad, a barcode reader, or an interface to a computer running any of the four programs above. When the videodisc is run from a computer, the many additional features of each program become available to users.

Shakhashiri Chemical Demonstration Videotapes contains a unique set of 50 three- to five-minute chemical experiments performed by Bassam

Shakhashiri. These videos bring the drama of chemistry into the classroom. An instructor's manual describes each experiment and includes questions for discussion. The videotapes are available free to adopters of the text. **Saunders General Chemistry Videodisc** contains most of the Shakhashiri demonstrations. In addition to this live-action footage, almost 3000 still images taken from eight of Saunders' general chemistry textbooks, including THE CHEMICAL WORLD, are included on the two-disc set. This videodisc is free upon adoption of any of Saunders' general chemistry texts.

The World of Chemistry Videotapes are based on the television series and telecourse, "The World of Chemistry," with Roald Hoffmann as host. These 26 tapes are each about 30 minutes long and provide introductory material on the principles and applications of chemistry. The tapes may be ordered through the Annenberg Foundation at 1–800–LEARNER ($350).

Demonstrations and animations from The World of Chemistry Videotapes are offered in a streamlined two-videodisc version called **The World of Chemistry: Selected Demonstrations and Animations I and II.** These are available in accordance with Saunders' adoption policy or may be purchased from *JCE: Software.*

Reviewers

The success of any book is due in no small way to the quality of the reviewers of the manuscript. All of the authors of this book have written books before. We all believe that the reviewers listed below are the best we have ever worked with. Many of them believe, as we do, that it is time to change how we teach introductory chemistry, so many went well beyond their charge and produced highly detailed reviews that were invaluable in developing our new approach to the subject. We would especially like to thank Greg Steinke, who provided solutions to the end-of-chapter problems, and Clyde Metz and Donald Kleinfelter, who took extreme care under great time constraints to review the galleys and page proofs for accuracy. A special thanks also to John DeKorte, who wrote the appendix on hybridization. They and all of our other reviewers will find themselves in this book, and it is a pleasure to acknowledge their efforts.

Jon M. Bellama
University of Maryland

Rathindra N. Bose
Kent State University

Mary K. Campbell
Mount Holyoke College

Dewey Carpenter
Louisiana State University

Coran L. Cluff
Brigham Young University

John DeKorte
Northern Arizona University

Gary Edvenson
Moorhead State University

John Fortman
Wright State University

Ron Furstenau
U.S. Air Force Academy

Roy Garvey
North Dakota State University

Tom Greenbowe
Iowa State University of Science & Technology

Henry Heikkinen
University of Northern Colorado

Paul Hladky
University of Wisconsin, Stevens Point

John Hostettler
San Jose State University

James House
Illinois State University

Dave Johnson
University of Oregon

Stanley Johnson
Orange Coast College

Paul Karol
Carnegie Mellon University

Doris Kimbrough
University of Colorado, Denver

Donald Kleinfelter
University of Tennessee, Knoxville

Anna McKenna
College of St. Benedict

Clyde Metz
College of Charleston

Patricia Metz
Texas Tech University

Dean Nelson
University of Wisconsin, Eau Claire

Ronald O. Ragsdale
University of Utah

Dave Robson
Chem Matters Magazine

Tom Rowland
University of Puget Sound

Arlene Russell
University of California, Los Angeles

Barbara Sawrey
University of California, San Diego

Conrad Stanitski
University of Central Arkansas

George Stanley
Louisiana State University

Gail Steehler
Roanoke College

Tamar Y. Susskind
Oakland Community College

Donald Titus
Temple University

Raymond Trautman
San Francisco State University

Deborah Wiegand
University of Washington

Donald Williams
Hope College

Stanley Williamson
University of California, Santa Cruz

Susan Weiner
West Valley College

R. Terrell Wilson
Virginia Military Institute

Acknowledgments

Preparing this book was an intricate process that took over two years. However, we have had the support and encouragement of family and of some wonderful friends, colleagues, and students.

The editorial staff of Saunders College Publishing has once again been extraordinarily helpful. The project has been enjoyable because of their good humor, friendship, and dedication. Much of the credit goes to our Publisher, John Vondeling. John believes, as we do, that the time has come for a different book for introductory chemistry, and he has been instrumental in bringing this team together to produce just such a book. His support and confidence are greatly appreciated.

This book was put together in a relatively short time for such a major project, and it simply could not have been done without our two developmental editors, Mary E. Castellion and Jennifer Bortel. Mary is an old hand at editing textbooks, and she worked extraordinarily hard on this project. We believe this book will enjoy a measure of success, and its success is in no small way due to Mary's efforts. Thanks from all of us!

Jennifer Bortel is just beginning her editorial career, which we expect will be a promising one because she has done a wonderful job. She has kept this complicated project organized and helped us meet our deadlines. Her hard work, warmth, and good humor were invaluable in helping us all through the project.

We have worked again with Jay Freedman, who, simply put, is the best copy editor in the business. He was meticulous in helping us prepare this text, and we are again grateful for his efforts.

Carol Field has been our managing editor, and she has been as patient as ever. Her attention to detail is extraordinary, and she keeps all of us calm as deadlines approach. We are pleased that Carol Bleistine was once again the manager of art and design. She has a wonderful sense of color, design, and layout, as this and other books produced by Saunders make clear. Finally, our team is completed by Charlene Squibb, production manager, and Tim Frelick, the director of editing, design, and production, who have kept all of this effort organized.

Many of the color photographs for this edition are the product of the creative eye and mind of Charles D. Winters of Oneonta. He spends countless hours in his studio to get a photograph just right. This book was created in record time, so he has produced excellent work under considerable pres-

sure. It is always a great pleasure to work with Charlie, and one of the authors was especially pleased and honored to be asked to be the best man at Charlie's recent wedding.

The phenomenal drawings in this book were done by George Kelvin, perhaps the finest science illustrator in the United States. His drawings not only illustrate the principles of chemistry, but also are truly works of art.

One goal of this book was to show the many applications of chemistry. This required many color photographs, and meant, therefore, that many photographs had to be found from all over the world. This task was done by Kathrine Kotz. Katie met all of our requests and kept the entire illustrations program on track. We are grateful for her efforts.

Even if a book is very well written and edited, and beautifully illustrated, it may not be successful without effective marketing. Therefore, we are also very pleased to work again with Margie Waldron, one of the most creative and effective—not to mention most pleasant—marketing managers in the publishing world. Margie was assisted by Randi Misher, Marketing Coordinator, and Laura Coaty, Field Product Manager. The marketing team did an impeccable job of conducting market research, helping us refine and test ideas, and getting the book into the right hands.

This book is illustrated with a number of computer-constructed molecular models. All were done using software from CAChe Scientific, Inc. We wish to thank them for their generous grant of this software to JCK.

We wish to express our gratitude to the students and staff of the chemistry department at the University of Wisconsin at Madison for class testing the entire manuscript in 1992 and 1993. Students in Chemistry 103–104 also tried nearly all of the Chemistry You Can Do experiments. Their opinions were invaluable in developing this book.

Many of the Chemistry You Can Do experiments were adapted from activities published by the Institute for Chemical Education as "Fun with Chemistry" volumes I and II. These were originally collected for ICE by Mickey and Jerry Sarquis of Miami University, Ohio. Thomas Kim was responsible for coordinating student comments and suggestions regarding *Chemistry You Can Do.*

Finally, thanks to the many unnamed people and sources from which we received information and advice. The responsibility for the contents of the text, however, rests with us. In spite of our many efforts to produce a perfect book, we are not infallible. Therefore, if you find a better way to explain a concept, a more relevant example or demonstration, or errors in calculations or discussions, please do not hesitate to write us or our editors at Saunders College Publishing.

We hope you enjoy teaching or learning from this book as much as we enjoyed writing it.

John C. Kotz
Oneonta, New York

James L. Wood
Nashville, Tennessee

Melvin D. Joesten
Nashville, Tennessee

John W. Moore
Madison, Wisconsin

August 1993

Contents Overview

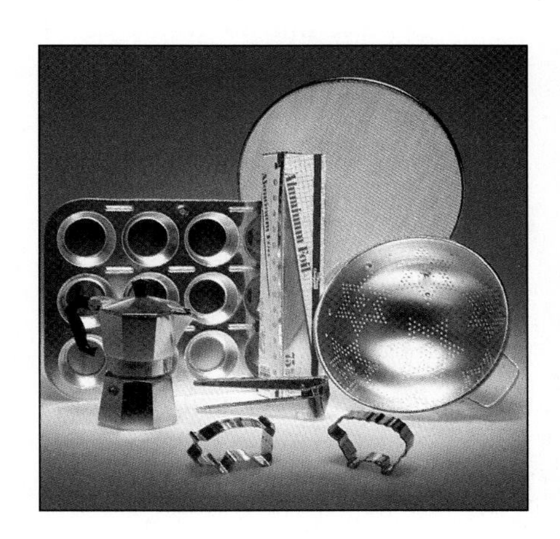

Contents

A Fresh Box for Baking
ARM & HAMMER
PURE BAKING SODA
DISTILLED
WHITE
VINEGAR

Some common chemical elements and chemical compounds.
(C.D. Winters)

The Nature of Chemistry

CHAPTER

1

Why study chemistry? Each person probably has a different answer. Some find the subject itself fascinating, but many are in a college chemistry course because someone else has decided it is useful as part of the background for a particular career. But why should it be so useful? Because chemistry is central to our understanding of biology, geology, materials science, medicine, and branches of engineering, among others. In addition, chemistry plays a major role in our economy; chemistry and chemicals affect our daily lives in a variety of ways. Furthermore, a college course in chemistry can help you see how a scientist thinks about the world and how to solve problems. The knowledge and skills developed in such a course will benefit you in many career paths and will help you become a better informed citizen in a world that is becoming technologically more and more complex—and interesting. Therefore, to begin your study of chemisty, this first chapter discusses some of the most fundamental of the ideas used by practicing chemists. It also introduces you to a problem confronting chemists, and indeed all citizens, almost every day—the weighing of the benefits of a discovery or a practice against its risks to individuals and society.

Figure 1.1
Photosynthesis is a process that uses sunlight to convert carbon dioxide and water to oxygen and glucose, as well as other carbon-containing chemical compounds. *(Steve Wilson, ENTHEOS)*

"Chemistry is the science of molecules and their transformations. It is the science not so much of the . . . elements, but of the variety of molecules that can be built from them." This statement is by Roald Hoffmann, one of the recipients of the 1981 Nobel Prize in chemistry. It reflects the theme of this book and of another book, *The Periodic Table,* the beautifully written autobiography of the chemist Primo Levi. Different periods of Levi's life reminded him of various chemical elements and their compounds, so he constructed chapters of *The Periodic Table* around elements such as nitrogen, phosphorus, gold, and chromium. In the final chapter of his book he imagines the journey of a single carbon atom.

"Our character lies for hundreds of millions of years, bound to three atoms of oxygen and one of calcium, in the form of limestone: it already had a long cosmic history behind it . . . " The limestone was removed from the ground and *"sent on its way to the lime kiln, plunging it into a world of things that change. It was roasted until it separated from the calcium, which remained so to speak with its feet on the ground and went on to meet a less brilliant destiny. . . . Still firmly clinging to two of its three former oxygen companions, [the carbon atom] issued from the chimney and took the path of the air. Its story, which once was immobile, now turned tumultuous."*

"Our atom of carbon enters [a] leaf, colliding with other innumerable (but here useless) molecules of nitrogen and oxygen. It adheres to a large and complicated molecule that activates it, and simultaneously receives the decisive message from the sky, in the flashing form of a packet of solar light: in an instant, like an insect caught by a spider, it is separated from its oxygen, combined with hydrogen and (one thinks) phosphorus, and finally inserted in a chain, whether short or long does not matter, but it is the chain of life."

The carbon dioxide molecule of Levi's story has formed part of a molecule within a green plant by the process of photosynthesis (Figure 1.1). One molecule formed this way is *taxol,* which was very recently discovered to be useful in the treatment of certain kinds of cancers. The story of taxol's discovery is not only interesting but also provides some insight into how science works.

Taxol—A New Drug from Nature

Taxol was discovered as a result of a program sponsored by the U.S. National Cancer Institute to test plants for anticancer activity. More than 35,000 plant species were tested, including a concentrate of material from the bark of the Pacific yew tree, which contains taxol (Figure 1.2). After early clinical trials with taxol gave promising results for treating certain types of cancer, more of the compound was required. This meant in turn that huge quantities of yew tree bark were required, since the bark from about three trees is needed to give enough taxol to treat one cancer patient. Therefore, a program began in 1991 in the Pacific Northwest to collect the bark from about 38,000 trees, which should give about 25 kilograms of taxol or enough to treat nearly 12,000 patients.

The problem with harvesting thousands of yew trees in one year just for their bark is that the source of taxol is ultimately destroyed and the damage done to the environment in the process can be considerable. How can a

Portrait of a Scientist

Primo Levi, 1919–1987

Primo Levi. (Giansanti/Sygma)

Levi was born in Turin, Italy in 1919. He was trained as a chemist, and worked in the field of paint chemistry for 30 years. His knowledge and love of chemistry not only saved his life but made him the kind of writer he was.

During World War II Levi was a member of the anti-Fascist resistance. He was arrested and sent to a death camp, Auschwitz, in 1944. He survived the camp because the Nazis found his skills as a chemist useful. The experience, however, left him with "an absolute need to write." His memoirs of life in the death camps "are considered a triumph of lucid intelligence over modern barbarism." In addition to *The Periodic Table* (Schocken Books, New York, 1984) Levi also wrote *Survival in Auschwitz, The Reawakening, The Drowned and the Saved, The Monkey's Wrench,* and *Other People's Trades.*

---

valuable compound be made available inexpensively and safely while preserving natural resources? Finding the solution to just such challenging and interesting problems is the reason that men and women become chemists, plant scientists, or biologists.

One solution to the taxol dilemma would be to assemble much simpler, more readily available chemical compounds into taxol. Chemists call this a "total synthesis" approach. Unfortunately, taxol is a very complex molecule, as you can appreciate by looking at its structure in Figure 1.2(b). This means

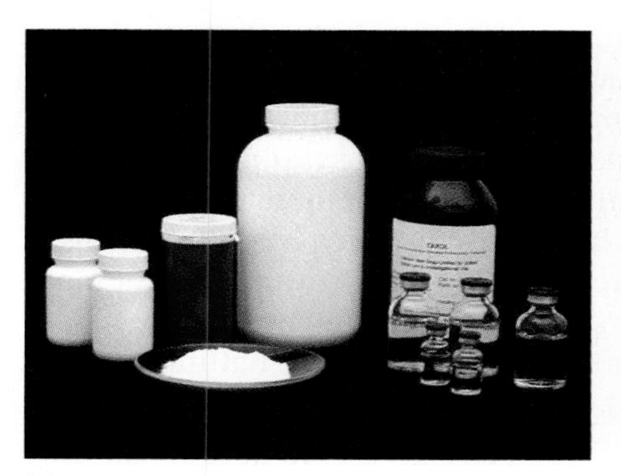

(a)

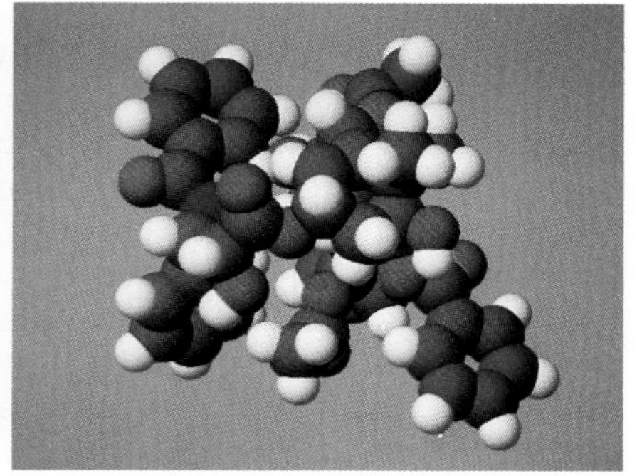

(b)

Figure 1.2
(a) A sample of taxol, a compound recently approved by the U.S. Food and Drug Administration for use in treatment of ovarian cancer. (b) Taxol is a complex molecule composed of the chemical elements carbon, hydrogen, oxygen, and nitrogen. *(a, NaPro; b, C.D. Winters)*

Figure 1.3
The metallic element sodium reacts very vigorously with water. *(C.D. Winters)*

the "total synthesis" approach is very difficult, although it may be solved eventually.

In the meantime, taxol's potential as a cancer chemotherapy agent is too great to wait for chemists to solve the "total synthesis" problem. Other scientists began a search for a close relative or "precursor" of the taxol molecule that could be converted easily and inexpensively to the active molecule. Indeed, this "semisynthesis" approach is the one now being used. Biologists have found just such a precursor molecule in the twigs and needles of certain species of yews. The twigs and needles are a renewable resource, since it is necessary only to clip the plants periodically, just as you would trim a hedge along a sidewalk. After extracting the precursor from the twigs and needles, chemists convert it to taxol or to a very similar but also effective compound called taxotere.

No matter what approach is used, the pace of work in taxol research can only be described as feverish. In fact, it has progressed far enough that it seems certain the bark of the Pacific yew tree will no longer be needed for taxol production by the end of 1995. But there is still a concern about the supply of the drug because early clinical studies indicate that taxol may be useful in treating a broad range of cancers such as those of the breast, lung, head, and neck. Its main use now is the treatment of advanced ovarian cancer, and the United States Food and Drug Administration (FDA) approved taxol for this use in December, 1992. Indeed, the FDA felt that the drug was so important that it gave its approval after only five months of investigation, instead of the years that have normally been required.

Science and Its Methods

Gould's definition of science is found in his "Essay on a Pig Roast" in Bully for the Brontosaurus, *W.W. Norton, New York, 1991. This book also includes essays on textbooks ("The Case of the Creeping Fox Terrier"), on statistics ("The Median Isn't the Message"), and on the great chemist Lavoisier ("The Passion of Antoine Lavoisier").*

The taxol story illustrates how chemistry and other sciences work to solve a specific problem. But how can we define science and its methods more broadly? It is difficult to find a better definition than one given by Stephen Jay Gould, a noted paleontologist. *"Science is a method for testing claims about the natural world, not an immutable compendium of absolute truths."* What does this mean? And why should you care?

As scientists we study questions of our own choosing or ones that someone else poses—such as the synthesis of taxol—in the hope of finding an answer or of discovering some useful information. Although it may not seem so to you now, it is easy to conceive of an idea to study. The goal is to clearly state a problem that is worth studying and that is also narrow enough in scope that there is a realistic chance of coming to a useful conclusion. It is reasonable to ask, for example, how one might cure or prevent AIDS, but it is certainly not reasonable to expect an answer in the next year or so. The answer will surely come only as the result of thousands of experiments done by hundreds of chemists, biologists, and physicians in dozens of laboratories. On the other hand, it is reasonable to ask how one might make taxol cheaply, easily, and safely from readily available chemicals.

Having posed a reasonable question—a claim about the natural world—you would first look at the work that others have already done in the field so that you have some notion of the best direction to take. After forming a hypothesis—a tentative explanation or prediction of experimental observations—you perform experiments designed to give results that may confirm your hypothesis or eliminate wrong explanations, that is, to test the claim. In chemistry this will surely require that you collect both qualitative and quantitative information. *Quantitative* information usually means numerical data, such as the temperature at which a chemical substance melts. *Qualitative* information consists of nonnumerical observations, such as the color of a substance or its physical appearance. With these results, you then revise and extend the original hypothesis and test it with new experiments. After you have done a number of experiments and continually checked to make sure your results are truly *reproducible,* a pattern may begin to emerge. At this point, you may be able to summarize your observations in the form of a rule, a concise verbal or mathematical statement of a relation that seems always to be the same under the same conditions.

You will see many laws in chemistry, and we base much of what we do on them because they help us predict what may occur under a new set of circumstances. For example, we know from experience that if we allow the chemical element sodium to react with the compound water, the result will be a very violent reaction and the production of several new substances (Figure 1.3). But sometimes the result of an experiment is different from what is expected on the basis of a law or general rule. Then a chemist gets excited, because experiments that do not follow the laws of chemistry are the most interesting. We know that understanding the exceptions almost invariably gives new insights.

Once you have done enough reproducible experiments to lead to a law, you may be able to formulate a theory to explain the observation. A *theory* is a unifying principle that explains a body of facts and the laws based on them. It is capable of suggesting new hypotheses. Theories abound to account for the way our economy or society works, for example. In chemistry, excellent examples of theories are those developed to account for chemical bonding (Chapter 10). It is a fact that atoms are held together or bonded to one another, as in a molecule such as taxol (Figure 1.2). But how and why? Several theories are currently used in chemistry to answer these questions, but are they correct? What are their limits? Can the theories be improved or are completely new theories necessary? Laws summarize the facts of nature and rarely change, but theories are inventions of the human mind. *Theories can and do change* as new facts are uncovered.

People outside of science usually have the idea that science is an intensely logical field. They picture a white-coated chemist moving from hypothesis to experiment and then to laws and theories without human emotion or foibles. Nothing could be further from the truth! Scientific results and understanding often arise quite by accident. Creativity and insight are needed to transform a fortunate accident into useful and exciting results, and a wonderful example is the discovery of a cancer drug by Barnett Rosenberg (Figure 1.4).

Figure 1.4
Dr. Barnett Rosenberg moved from New York City, where he had studied physics, to Michigan State University in East Lansing, Michigan, to establish a department of biophysics in 1961. He is now the head of Barros Research Institute near East Lansing. Dr. Rosenberg has received many awards for his work, including the Galileo Medal from the University of Padua (Italy) and the Charles F. Kettering Prize from the General Motors Cancer Foundation. *(Doug Elbinger)*

Dr. Rosenberg and his collaborators discovered that a chemical compound that had been known for a century or more, a compound composed of the elements platinum, chlorine, nitrogen, and hydrogen, was effective in the treatment of certain types of cancers. The discovery of the utility of this simple compound, now commonly known as *cisplatin,* in cancer therapy is an example of serendipity in science.

The dictionary defines serendipity as "the faculty of making fortunate and unexpected discoveries by accident." In this case Rosenberg had set out to study a problem that had interested him for some time—the effect of electric fields on living cells—but the results of the experiment were quite different than expected. He and his students had passed a suspension of a live culture of *Escherichia coli* bacteria in water through an electric field between supposedly inert platinum plates. Much to their surprise they found that the growth of the cells was affected; cell division was no longer occurring. After careful experimentation, they traced the effect on cell division to tiny amounts of cisplatin produced by reaction of platinum with electrically charged chemical species in the water.

Much to the benefit of all of us, Rosenberg recognized that these laboratory results had implications in cancer chemotherapy, and subsequent experiments led to compounds now used by cancer patients. He recently said that the use of cisplatin has meant that "Testicular cancer went from a disease that normally killed about 80 percent of the patients, to one which is close to 95 percent curable. This is probably the most exciting development in the treatment of cancers that we have had in the past 20 years. It is now the treatment of first choice in ovarian, bladder, and osteogenic sarcoma [bone] cancers as well."

Simple But Elegant Experiments

The remarks by Richard Feynman (1919–1988), and the story of his role in the Challenger disaster, are found in his book What Do You Care What Other People Think? *(W.W. Norton and Company, New York, 1988). See also his autobiography* Surely You're Joking Mr. Feynman *(W.W. Norton and Company, New York, 1985), in which he tells about his life in physics and his passion for the bongo drums.*

Another person who understood very well how science works was Richard Feynman, the recipient of the Nobel Prize in Physics in 1965 and one of the most original thinkers of this century. Feynman said that "Scientific knowledge is a body of statements of varying degrees of certainty—some most unsure, some nearly sure, but none *absolutely* certain. . . . Now, we scientists are used to this, and we take it for granted that it is perfectly consistent to be unsure, that it is possible to live and *not* know." It is well to remember Feynman's observation as you read statements about our environment, about health care and medicine, and about nutrition.

Richard Feynman gave the United States a lesson in how science is done when he used a simple experiment to uncover the reason for the disastrous explosion of the space shuttle *Challenger*. The *Challenger* was launched on Tuesday, January 28, 1986. The day was unusually cold for Florida—the temperature at the time of launch was 29 °F. The world watched in horror when, after a minute or so of flight, the shuttle and its rockets exploded in a fireball, killing the astronauts on board.

To understand the reason for the explosion, and the importance of Feynman's experiment, requires knowing something of the design of the shuttle. The main engine is fueled by liquid hydrogen and oxygen, which are contained in the large tank strapped onto the belly of the shuttle. The hydrogen and oxygen combine to give water, and the energy released by this reaction provides the thrust to boost the shuttle into orbit. However, to provide additional thrust at the time of launch, there are solid-fuel booster rockets strapped on each side of the shuttle. The solid-fuel rockets are made in sections, and the sections are shipped from the factory to The Kennedy Space Center where they are joined to make the completed booster rocket. The joint between the sections was designed so that hot gases from the

(a)

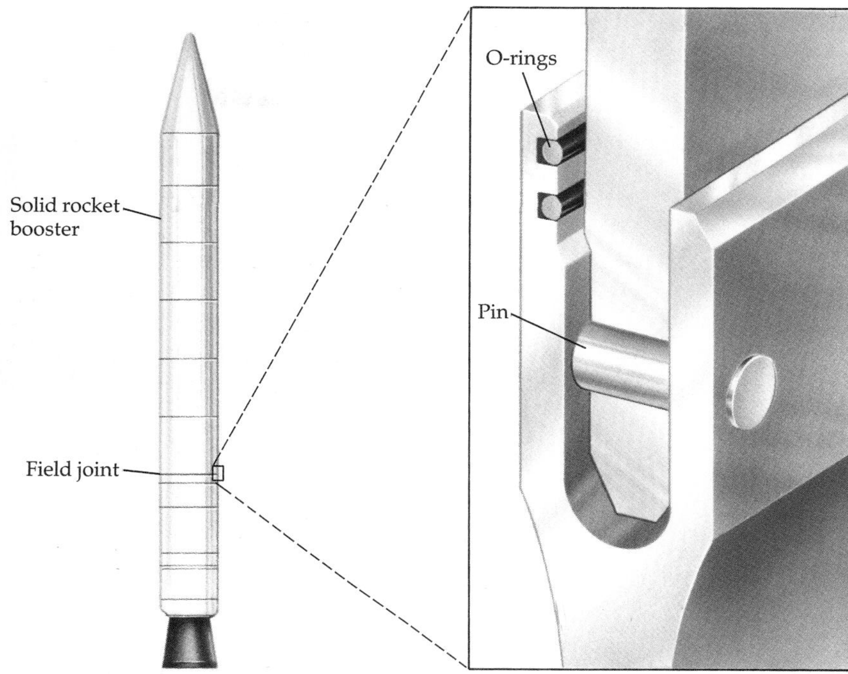

(b)

Figure 1.5
(a) The space shuttle. The large orange tank on the belly of the orbiter contains hydrogen and oxygen, which are burned in the orbiter's main engines. The long tubes on either side are the solid fuel boosters. (b) Each solid fuel booster rocket is made of a number of sections, bolted together. The joints between sections are complicated and involve, among other things, O-rings made of a special rubber. The function of the O-rings is to seal the joints and prevent hot gases from leaking out of the sides of the rocket. *(a, NASA)*

burning solid fuel would not leak through the walls of the rocket. Part of the design to close the joint, and yet make it somewhat flexible, was the use of a thin O-ring made of a special rubber (Figure 1.5). From the beginning Feynman and others thought that a possible cause of the *Challenger* explosion was that the solid fuel had burned through the wall of the booster rocket and then burned into the tank holding the liquid hydrogen, thus exploding the hydrogen. But how did this happen?

When the shuttle is launched, and the solid fuel begins to burn, the walls of the rocket casing move slightly outward. If this movement caused a joint between sections to open up, then the fuel would burn through the joint. But the O-rings were supposed to prevent this. Based on information from engineers involved in the design of the solid rocket boosters, Feynman's hypothesis was that, due to the unusually cold weather, the rubber O-rings had not expanded properly, and flame burned through the joint. To prove his point, Feynman did a dramatic—but very simple—experiment. During a public hearing Feynman held a sample of the rubber O-ring tightly in a C-clamp, and put it into a glass of ice water (Figure 1.6). Everyone could make the qualitative observation that the rubber did not spring back to its original shape! The poor resilience of the rubber at low temperatures had almost certainly doomed the *Challenger.* Feynman's hypothesis was supported by his elegantly simple experiment.

Not all experiments are as simple as Feynman's, or so dramatically illustrate the point. Some take days or even months to complete and involve complex and expensive instruments. Nonetheless, it is often true that the

Figure 1.6
Richard Feynman showing experimentally that a cold O-ring, of the type used in the space shuttle *Challenger*, does not snap back quickly to its original shape.
(Marilynn K. Yee/NYT Pictures)

very best experiments, the ones that produce the most useful and persuasive results, are the simplest.

Why Care?

The taxol problem, the *Challenger* disaster, and Feynman's experiment bring us to the reasons you should care about science and the way it is done. Many in the United States were saddened by the *Challenger* explosion, not only because of the deaths of the astronauts but also because, unless solved, it could clearly affect the future of our space program and its role in our national goals. Further, such engineering failures can lead uninformed citizens to lose faith in all technology, to the detriment of our economy and society. Professor Feynman and his common-sense approach uncovered the reasons for the *Challenger* explosion, and the space program is moving cautiously ahead. However, there are hundreds of problems confronting our society that depend in some way on science for their resolution. Almost every day we read statements in newspapers or magazines, or hear them on television, that

- we could develop an AIDS vaccine if we just moved more quickly.
- asbestos must be removed from all public buildings.
- all pesticides and herbicides should be banned.
- the Antarctic ozone hole is becoming larger.
- burning tropical forests is an ecological disaster.
- global warming is a reality.
- global warming is not a reality.
- homeopathic "medicines" are a sure cure for whatever ails us.
- you should not eat butter but substitute margarine.
- there are health risks in eating margarine.
- you should check the basement of your house for dangerous levels of radon.

- eating foods with high levels of selenium will prevent cancer.
- incineration of wastes is highly damaging to the local ecology.

The list goes on and on. What are you to believe? How do you analyze the problem? We believe that some knowledge of chemistry and the methods used by chemists and other scientists can be of great benefit. Such knowledge can help you at least begin to analyze the risks and benefits of public policies.

Risks and Benefits

The problems listed above—all related directly to chemistry and chemicals—will be with us for many years, and others will surely be added. No matter where you live or what your occupation, you are exposed every day to chemicals, some of which may be hazardous. Some you have decided you can tolerate because the risk they pose seems to be outweighed by the benefits they offer. Examples may include having an occasional beer, which contains toxic alcohol, or using an herbicide on your garden or lawn. In other cases, however, it is difficult to weigh the risks and benefits. This means that you will hear more about the assessment and management of risks.

"Risk assessment" is a process that brings together the scientific disciplines of chemistry, biology, toxicology, epidemiology, and statistics and that attempts to ascertain the severity of a health risk when people are exposed to a particular chemical. The first step is to try to identify the hazard by doing animal tests or by examining data for human exposure, and to find the relation between the amount of exposure and the chances of experiencing an adverse health effect. Also involved is an attempt to determine how people are exposed and to how much. Once these things are known, one can then try to make an estimate of the overall risk.

In recent years a new discipline called "risk communication" has grown up to promote the accurate transfer of information between the scientists who assess risk, the government agencies that manage risk, and the public. Those who study risk communication have found that the response of people to risks depends on a number of interesting factors. For example, people accept voluntary risks, such as smoking or playing in a high school football game, much more readily than involuntary risks such as being in a building that contains asbestos (Figure 1.7) Objectively, while the annual death rate per million people is 10 for high school football or 40 for a long-term smoker, the death rate from asbestos in schools is estimated to be in the range from 0.005 to 0.093.

It is also interesting that people often conclude that anything synthetic is "bad," while anything natural is "good." However, this is not always confirmed by risk assessment, as you can see in Table 1.1.

"Risk management" involves ethics, equity, and economics and other matters that are part of government and politics. The removal of asbestos from public buildings is an example of the management of a risk that many scientists believe to be quite low for the general public. The public has not been willing to tolerate the risk, so the political system has responded.

Figure 1.7
There are many voluntary risks that we all take on in everyday living.
(Image Enterprises, James Cowlin)

Table **1.1**

Estimates of Risk: Activities that Produce One Additional Death per One Million People Exposed to the Risk

Activity	Cause of Death
Smoking 1.4 cigarettes	Cancer, lung disease
Living 2 months with a cigarette smoker	Cancer, lung disease
Eating 40 tablespoons of peanut butter	Liver cancer caused by the natural carcinogen aflatoxin B
Drinking 40 cans of saccharin-sweetened soda	Cancer
Eating 100 charcoal-broiled steaks	Cancer
Traveling 6 minutes by canoe	Accident
Traveling 10 minutes by bicycle	Accident
Traveling 300 miles by car	Accident
Traveling 1000 miles by jet aircraft	Accident
Drinking Miami tap water for 1 year	Cancer from chloroform
Living 2 months in Denver	Cancer caused by cosmic radiation
Having one chest x-ray in a good hospital	Cancer
Living 5 years at the boundary of a typical nuclear power plant	Cancer

Chem Matters, April 1993.

Another example of risk management is the ban on chlorofluorocarbons (CFCs) in many nations because of the damage CFCs are thought to cause to the earth's ozone layer (see Section 20.6 for a complete discussion). Here ethics and economics clearly clash. After so many years of development, refrigeration equipment that uses CFCs is efficient, relatively inexpensive, and widespread. Its use is spreading to Third World countries such as the nations of Africa and many in South America, and its availability has a profound effect on their economies. Now, just as these countries are beginning to develop, the countries of the so-called developed world tell them that CFC-based refrigeration equipment should no longer be used. What is better for the greater good: a ban on equipment and processes using CFCs, which will apparently have a long-term effect on the ozone layer, or some limited, reasonable use of CFCs in the refrigeration equipment that is vital to the development of Third World countries? If alternatives to CFCs can be found, are they economic and without damaging environmental effects? Can the technology be transferred readily and efficiently to Third World countries? To return to where this chapter began—is it reasonable to sacrifice 38,000 yew trees per year in our forests in order to extract enough taxol to treat 12,000 cancer patients? Do the benefits to relatively few people outweigh the forest and habitat destruction that would occur?

Why study chemistry? The reasons are clear. You will be called upon to make many decisions in your life for your own good, or for the good of those in your community—whether that is your local community or the global community. An understanding of science in general, and chemistry in particular, can only serve to help in these decisions.

Chemistry is the science of change. Here a fuel is burning, and the energy derived produces a change in a chemical compound.
(C.D. Winters)

The Nature of Matter

CHAPTER

2

The first sentences in a chemistry textbook published in 1879 state that "The material objects surrounding us present striking and infinite differences. Sulphur [an old way of spelling sulfur] is readily distinguished from charcoal, rock-crystal from flint, iron from copper, water from spirit of wine, and wood from ivory: It is known to all that these bodies differ not only in form, density, and structure, but also in their proper substance." These sentences, written so many years ago, convey an idea of what this chapter is about: the properties of substances and how understanding those properties helps us adapt the material world to our needs. Substances can be separated, identified, and classified according to their physical and chemical properties. Some of these properties are measured quantitatively, using carefully defined units. To help explain the properties of substances, people have constructed models which assume that matter is made of atoms and molecules—particles that are too tiny to see with the human eye. An ability to imagine how atoms and molecules behave is fundamental to understanding modern chemistry.

You put about half a pound of sugar (whose chemical name is sucrose), along with half a cup of water and half a cup of corn syrup (a glucose solution), in a pan and heat the mixture while stirring steadily. You continue to boil the syrup and watch as it slowly turns brown; a vapor rises above the bubbling mixture. When the temperature reaches 140 °C, you toss in a handful of peanuts and a pinch of sodium bicarbonate and pour the liquid mixture quickly onto a piece of aluminum foil. When it cools, you smash the solid into small pieces and pop some into your mouth! You have just made peanut brittle—and have carried out a wonderful kitchen experiment, in the course of which you have observed many of the things to which chemists pay attention every day. You have seen chemical and physical changes and different states of matter, have made quantitative and qualitative observations, and have made measurements. Such activities are what this chapter is all about—making sense of the material world through quantitative and qualitative observations, and using the knowledge so gained to adapt the material world to our own interests.

For more information on making peanut brittle and the chemistry of the process, see "Peanut Brittle" by E. Catelli in Chem Matters, *December, 1991, page 4.*

Making peanut brittle. *(C.D. Winters)*

2.1 PHYSICAL PROPERTIES

Your friends can recognize you by your physical appearance: your height and weight and the color of your eyes and hair. The same is true of chemical substances. While making peanut brittle you could distinguish sugar from

(a)

(b)

Figure 2.1
The chemical elements differ from one another in their physical properties. Here are four elements, three of them solid and one a liquid. (a) Solid iron (*left*) and sulfur (*right*) are clearly different in color. (b) Bromine (*left*) and iodine (*right*) differ in color and in their physical state. Notice that bromine is a liquid, but it is volatile as indicated by the red-orange color of bromine vapor in the flask. Iodine is a purple solid, but it too is volatile; iodine vapor fills the flask. *(C.D. Winters)*

water because you know that sugar consists of small white particles of solid while water is a colorless liquid. Corn syrup is also a liquid, but it comes in light and dark colors and is much thicker (pours more slowly) than water. Properties such as these, which can be observed and measured without changing the composition of a substance, are called **physical properties.** As you can see in Figure 2.1, the solids iron and sulfur clearly differ in color, while liquid bromine and solid iodine differ in their physical states as well as color. Physical properties allow us to classify and identify substances.

States of Matter

An easily observed and very useful property of matter is its physical state. Is it a solid, liquid, or gas? A **solid** can be recognized because it has a rigid shape and a fixed volume that changes very little as temperature and pressure change (Figure 2.2). Like a solid, a **liquid** has a fixed volume, but a liquid is fluid—it takes on the shape of its container and has no definite form of its own. **Gases** are fluid also, but gases expand to fill whatever container they occupy, and their volume varies considerably with temperature and pressure (Figure 2.3). For most substances, the volume of the solid is slightly less than the volume of the same mass of liquid, but the volume of the same mass of gas is much, much larger. Solids occur at low temperatures, and as the temperature is raised they melt to form liquids; eventually, if the temperature is raised enough, liquids boil to form gases.

There is a fourth state of matter, plasmas, which are gases composed of electrically charged particles. Plasmas exist naturally only in the outer portion of the earth's atmosphere, the atmosphere of stars, much of the material in nebular space, and a beautiful aurora borealis or "northern lights."

All of the physical properties just described can be observed by the unaided human senses and refer to samples of matter large enough to be seen, measured, and handled. Such samples are called **macroscopic** sam-

Figure 2.2
Solid "ice cubes" have a definite shape and do not fill the container evenly. After melting, the liquid water assumes the shape of the container. *(C.D. Winters)*

Figure 2.3
The brown gas nitrogen dioxide, NO_2, completely fills the container. *(C.D. Winters)*

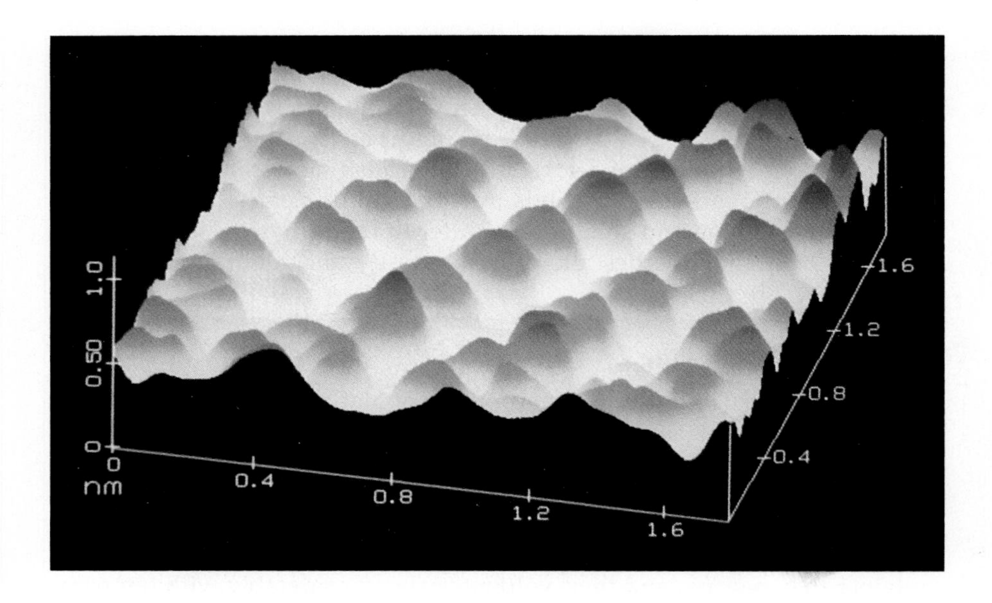

Figure 2.4
Scanning tunneling microscopy (STM) and atomic force microscopy (AFM) are powerful methods of probing the surfaces of solids with atomic resolution. This figure shows an AFM topographic image of the copper atoms on the surface of silica. The image is 170 nm square, and the rows of atoms are separated by about 0.44 nm. [X. Xu, S.M. Vesecky, and D.W. Goodman, *Science*, **258**, 788 (1992)]

The world of biological cells and bacteria is at the microscopic scale.

ples, in contrast to **microscopic** samples, which are so small that they have to be viewed with a microscope. The structure of matter that really interests chemists, however, is at the submicroscopic scale of atoms and molecules (Figure 2.4). It is a fundamental idea of chemistry that the properties of a sample of matter are determined by the nature of its parts, and those parts are very, very tiny. Therefore we need to use imagination to develop useful theories that connect the behavior of those tiny parts to the observed behavior of matter in the macroscopic world. Chemistry enables us to "see" in the things all around us submicroscopic structure that cannot be seen with our unaided eyes.

Kinetic-Molecular Theory

One theory that helps us interpret the physical properties of solids, liquids, and gases is the **kinetic-molecular theory of matter.** According to this theory, all matter consists of extremely tiny particles (atoms and molecules) that are in constant motion. In a solid these particles are packed closely together in a regular array as shown in Figure 2.5. The particles vibrate back and forth about their average positions, but seldom does a particle in a solid squeeze past its immediate neighbors to come into contact with a new set of particles. Because the particles are packed so tightly and in such a regular arrangement, a solid is rigid, its volume is fixed, and the volume of a given mass is small. The external shape of a solid often reflects the internal arrangement of its particles. For example, a piece of "fool's gold" or iron pyrite (Figure 2.6) is made up of iron and sulfur, and the cubic solid you can hold in your hand is simply a large version of the cubic arrangement of iron and sulfur atoms in the solid crystal. This relation between the observable structure of the solid and the internal arrangement of its particles is one reason scientists have long been fascinated by the shapes of crystals and minerals.

Many solids such as "fool's gold," table salt, and some minerals have very distinctive shapes and because of this are called crystalline solids. By contrast, peanut brittle is an amorphous solid, which means "without regular shape."

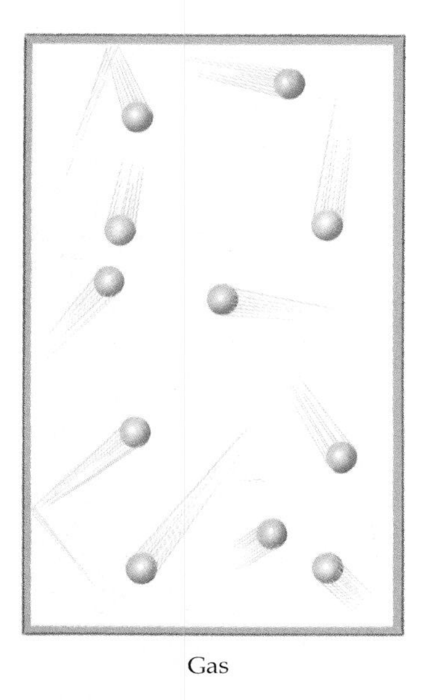

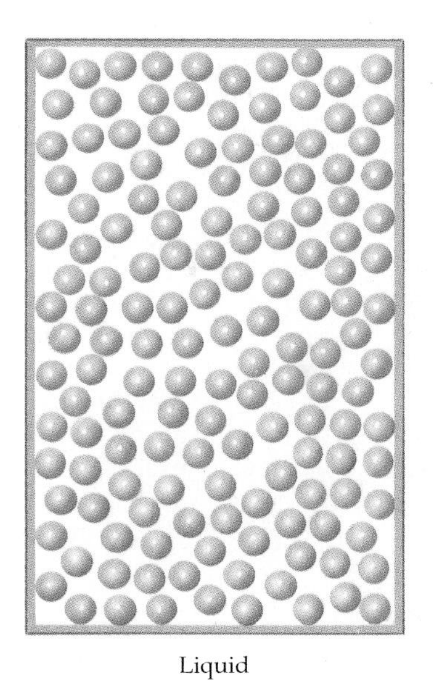

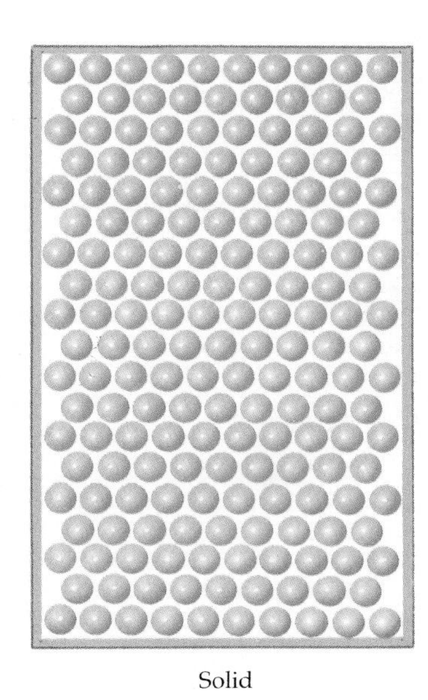

Figure 2.5
The three states of matter. In the gas phase, atoms or molecules move rapidly over distances larger than the sizes of the atoms or molecules themselves. There is little interaction between them. Cooling, increasing the pressure, or both, converts gases to liquids. The atoms or molecules are now much closer together, and they interact with one another. Motion of the particles is still very evident, although the particles move over only very small distances. Cooling still further converts a liquid to a solid. The particles are even closer together and almost totally restricted to specific locations.

The kinetic-molecular theory of matter can also be used to interpret the properties of liquids and gases, as shown in Figure 2.5. Liquids and gases are fluid because the atoms or molecules are arranged more randomly than in solids. They are not confined to specific locations but rather can move past one another. Because the particles are a little farther apart in a liquid than in the corresponding solid, the volume is a little bigger. No particle goes very far without bumping into another—the particles in a liquid interact with each other constantly. In a gas the particles are very far from one another and moving quite rapidly. (In air at room temperature, for example, the average molecule is going faster than 1000 miles per hour.) A particle hits another particle every so often, but most of the time each is quite independent of the others. The particles fly about to fill any container they are in; hence a gas has no fixed shape or volume.

Figure 2.6
Iron pyrite can form large cubic crystals that reflect the arrangement of the atoms deep inside the crystal. *(C.D. Winters)*

In addition, the kinetic-molecular theory states that the higher the temperature the more active the particles are. A solid melts when the temperature of the solid is raised to the point at which the particles vibrate fast enough and far enough to push each other out of the way and move out of their regularly spaced positions. The substance becomes a liquid because the particles are now behaving as they do in a liquid. As the temperature goes even higher, the particles move even faster until finally they escape the clutches of their comrades and become independent; the substance becomes

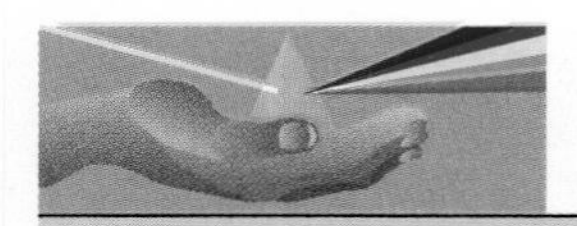

CHEMISTRY YOU CAN DO

Crystalline Solids

This experiment will allow you to observe properties of crystals of ordinary table salt. Other solids may be included as well. You will need:

- a small quantity of salt—one or two servings from a fast-food restaurant should be enough
- a small quantity of water
- a small container, such as a film canister, medicine cup, or plastic glass
- a mirror, or something else with a smooth, flat surface that can be washed clean
- a towel or tissue and some soap or detergent
- a magnifying glass (optional)

Place about two thirds of the salt into the bottom of the container and just cover it with water. Swirl the water around for a minute or so. Allow the container to stand for five minutes. Repeat the swirling or stirring every five minutes or so for half an hour. If all of the salt dissolves in the water, add more salt. Allow the mixture to stand for at least five minutes so that all the solid salt falls to the bottom of the container.

While the salt is dissolving, wash the mirror with soap, dry it with a towel, and allow it to stand and dry completely. Place the mirror on a horizontal surface in a place where it can remain undisturbed for several hours. Carefully pour a little salt water onto the mirror without allowing any of the solid to fall onto the mirror. Allow the liquid to stand, and observe it every 15 minutes or so for several hours; then let it stand overnight. If crystals form, observe them carefully. What shape are they? Do crystals of different shapes form from the same solution? Compare your crystals with those other students have grown.

You can repeat the experiment with several other solids (sugar, Epsom salts, sodium bicarbonate, instant coffee, and others that will dissolve). In each case it is important that the quantity of water added is not enough to dissolve all of the solid. Observe the results as the water slowly evaporates from the solution on the mirror. Do all of the solids form crystals as they come out of solution? What shapes are the crystals? It is handy to keep notes.

Equipment needed for Chemistry You Can Do. (C.D. Winters)

In each chapter of the book there will be at least one Chemistry You Can Do, *which describes an experiment that illustrates the themes of the chapter. Although the experiment may be done easily with inexpensive equipment, it can lead to useful conclusions and give you valuable insight into the world of chemistry.*

a gas. Increasing temperature corresponds to faster and faster motions of atoms and molecules, a concept you will find useful in understanding chemistry.

Because the particles in a gas are independent of each other, most gases respond in similar ways to temperature and pressure changes. Their behavior can be described quantitatively by a single mathematical equation. The pressure, volume, amount of gas, and temperature are related by the ideal gas equation,

$$PV = nRT$$

(pressure)(volume) = (amount)(universal gas constant)(temperature)

This equation is important because it can be used to predict, for example, the pressure exerted by a given amount of gas when the temperature of the gas and the container volume are known. In designing a tank to hold oxygen gas for use in scuba diving, this would be a crucial piece of information. The equation would not apply equally well for all gases if the gas particles were as close together as the particles are in solids or liquids. When particles are close together they can interact in different ways depending on the exact nature of the particles; when they are far apart, as in a gas, those interactions are almost negligible and the characteristics of the particles matter hardly at all.

Scientific Models

The qualitative descriptions of microscopic structure in solids, liquids, and gases, and the quantitative predictions made by the ideal gas law are both examples of scientific models. Scientists often take an approach to understanding nature that is exemplified by the title of a recent book, ''Consider a Spherical Cow'' (J. Harte, University Science Books, 1988). The title comes from a joke the author heard. ''Milk production at a dairy farm was low so the farmer wrote to the local university, asking for help from academia. A multidisciplinary team of professors was assembled, headed by a theoretical physicist, and two weeks of intensive on-site investigation took place. The scholars then returned to the university, notebooks crammed with data, where the task of writing the report was left to the team leader. Shortly thereafter the farmer received the write up, and opened it to read on the first line: 'Consider a spherical cow . . .' ''

The ''spherical cow'' approach is to consider only the essentials of a problem, to strip away all of the unnecessary details, and to build a simple model, even if we know that model does not describe a situation (or a cow) exactly. If the simple model answers the questions we want to know about, fine—we use it. If we need more detail, then often we can add a correction to the simple model or extend it, thereby expanding its ability to make predictions about nature. The equation $PV = nRT$ is an example of a mathematical ''spherical cow'' model of gases. To keep things simple, we assume that the particles in a gas do not interact with each other at all, and so the ideal gas law applies to all gases with a reasonable degree of accuracy. However, there are cases where the interactions among gas particles are big enough to affect the accuracy of predictions. These cases are said to be nonideal, and force us to extend the model to accommodate them, or to scrap it entirely.

Chemistry involves a variety of extensions of the kinetic-molecular theory of matter—extensions that describe the behavior and structure of atoms and molecules in more detail. Our senses have no direct access to the submicroscopic world of atoms and molecules, but our imaginations do. It is quite useful to draw conclusions about the submicroscopic world on the basis of macroscopic and microscopic observations. Often these observations are made with the aid of electronic instruments. For example, experiments show that the atomic-level structure of ice is reflected in (and helps explain) the

Figure 2.7
A snow flake is invariably based on a hexagon, a reflection of the way that the water molecules in the snow are arranged within the solid.
(E. Grave/Phototake)

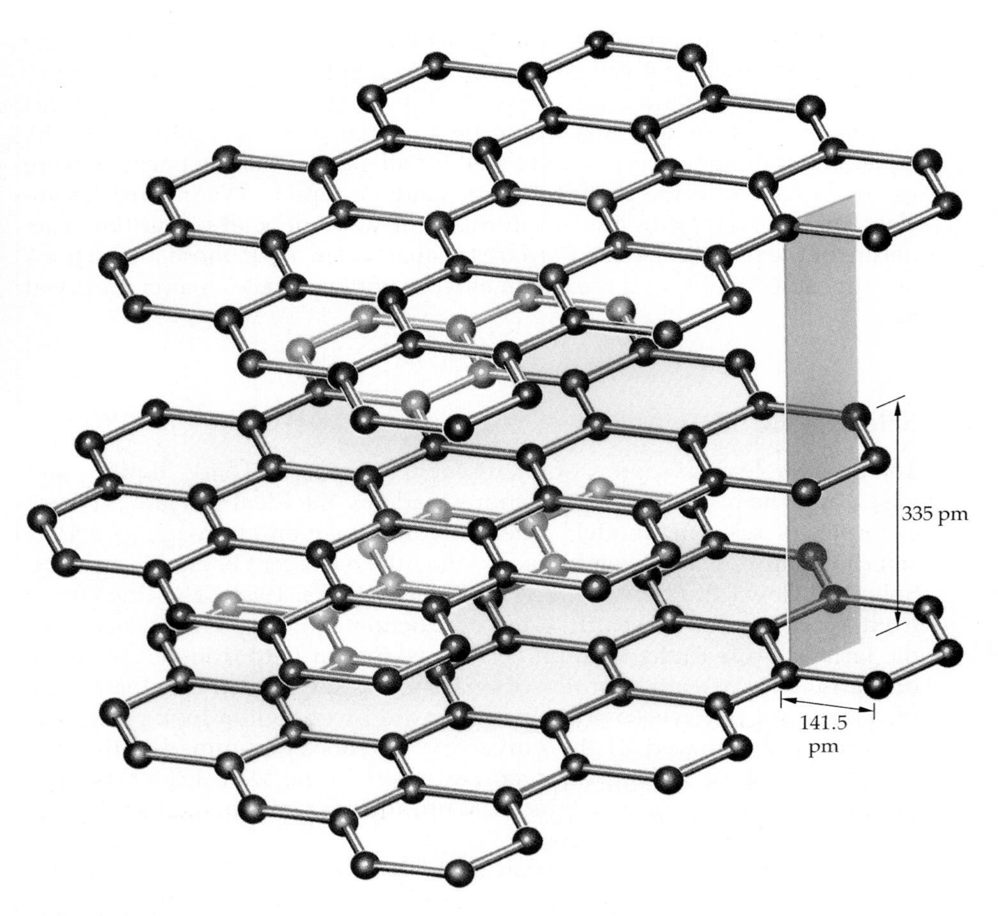

Figure 2.8
Graphite consists of layers of carbon atoms bonded firmly to one another. Between layers, however, the forces of attraction are weaker, and the layers can slide over one another quite readily. This accounts for the lubricating properties of the solid.

A "book" of transparent sheets of mica. *(C.D. Winters)*

six-sided shape of a snow flake (Figure 2.7). The slippery feel of graphite (the chief component of pencil lead and some lubricants) reflects the weakness of attraction between flat sheets of carbon atoms (Figure 2.8) that can move past one another easily.

EXERCISE 2.1 • *Structure and Properties*

Mica, which is seen in the photo in the margin, is a common mineral. It is often found as "books": stacks of sheets of the mineral that look like the pages of a book. Indeed, one can peel off very thin, transparent layers. Because of this property, mica was used to make crude windows many years ago. More recently, it has been used as windows in furnaces, since the mineral is heat resistant as well as transparent. Finally, mica appears as the shiny flecks in "metallic" paint on cars. What might these macroscopic properties tell you about the submicroscopic structure of the material?

EXERCISE 2.2 • ***States of Matter***

Identify the solids, liquids, and gases in the photo in the margin.

Pouring a carbonated soft drink into a glass. *(C.D. Winters)*

Answers to the Exercises in this book are given in Appendix K.

Density

Density, the ratio of the mass of an object to its volume, is another physical property that is useful for identifying substances.

$$\text{density} = \frac{\text{mass}}{\text{volume}}$$

Your brain unconsciously uses the density of an object you want to pick up by estimating its volume visually and preparing your muscles to lift the expected mass. A metallic object is denser than a wooden one, for example, and so your muscles are prepared to work harder.

Density can be determined more precisely by measuring the mass of an object and its volume. Nickel, the element that gave its name to the U.S. 5-cent coin, has a density of 8.90 g/cm^3, for example. That is, 1.00 cm^3 of nickel has a mass of 8.90 g. In contrast, titanium (a metal used for its resistance to corrosion) has a much lower density, only 4.5 g/cm^3. Pieces of both of these metals are shiny and gray, but you could tell the difference between them by measuring their densities or by selecting pieces of each metal of nearly identical volume and then weighing them.

If any two of the three quantities—mass, volume, and density—are known for a sample of matter, they can be used to calculate the third quantity. For example,

$$\text{volume} \times \text{density} = \text{volume } (\cancel{\text{cm}^3}) \cdot \frac{\text{mass (g)}}{\text{volume } (\cancel{\text{cm}^3})} = \text{mass (g)}$$

You could use this approach to find the mass of 25 cm^3 (or 25 milliliters) of mercury (Figure 2.9). A database of information for chemistry lists the density of mercury as 13.534 g/cm^3 (at 20 °C). Therefore,

$$25\ \cancel{\text{cm}^3} \cdot \frac{13.534 \text{ g mercury}}{1\ \cancel{\text{cm}^3} \text{ mercury}} = 340 \text{ g mercury}$$

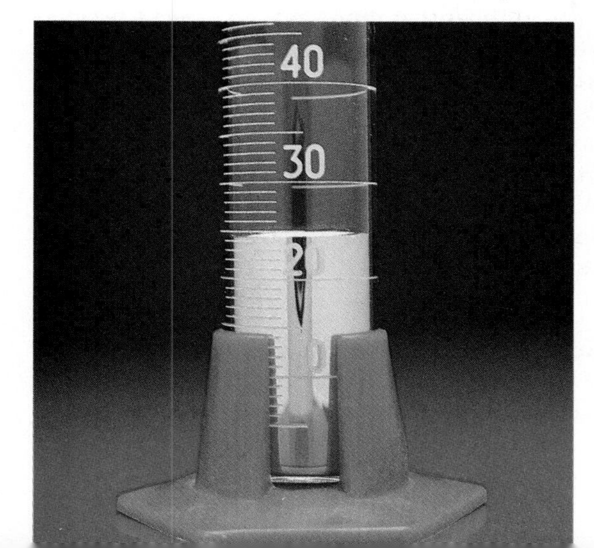

Figure 2.9
Mercury, 25 mL (or 25 cm^3), in a graduated cylinder. *(C.D. Winters)*

The data used in numerical calculations in this book, such as the density of mercury, must be found in tables of information. These are usually in various handbooks available in most libraries. One of the best sources is the Handbook of Chemistry and Physics *published by the CRC Press. The authors of this book also relied upon* KC? Discoverer, *a computer database of information on the elements. See the Preface for more information.*

This very small volume of mercury has a large mass (about 3/4 of a pound using the English system of measurements).

EXAMPLE 2.1

Density

Suppose you have 250. cm^3 (that is, 250. milliliters) of ethyl alcohol (the common "alcohol" in alcoholic beverages). If the density of ethyl alcohol is 0.789 g/cm^3 (at 20 °C), what mass, in grams, of alcohol does this represent?

SOLUTION You know the density and volume of the sample. Since density is the ratio of the mass of a sample to its volume (density = mass/volume), then "mass = density × volume." Therefore, to find the sample's mass, multiply the volume by the density; units of cm^3 cancel.

$$250.\,\cancel{\text{cm}^3} \cdot \frac{0.789\text{ g ethyl alcohol}}{1\,\cancel{\text{cm}^3}\text{ ethyl alcohol}} = 197\text{ g ethyl alcohol}$$

By simply recognizing the units of the given information, of the desired answer, and of the multiplying factor, you can avoid memorizing equations like "$m = d \times v$" and then forgetting them at a time of crisis—such as in the middle of an examination.

EXERCISE 2.3 • ***Density***

The density of dry air is 1.12×10^{-3} g/cm^3. What volume of air, in cubic centimeters, will have a mass of 1.00×10^3 g (that is, 1.00 kilogram)?

Melting Point, Boiling Point, and Temperature

There is more about temperature and heat in Chapter 7.

Another very useful physical property of pure elements and compounds is the temperature at which the solid melts (its **melting point**) or the liquid boils (its **boiling point**). **Temperature** is the property of matter that determines whether there can be heat (energy) transfer from one body to another, and the direction of that transfer: heat transfers spontaneously *only* from a hotter object to a cooler one. The number that represents an object's temperature depends on the unit chosen for the measurement. (The number representing the distance between two points depends on its unit of measurement in the same way. The length of a football field could be expressed as 100 yards or 91.4 meters, for example.) Unfortunately, there are three units of temperature measurement in common use today: Fahrenheit, Celsius, and kelvin units (Figure 2.10). When calculations are done using temperature data, it is usually best to use kelvin units. In the laboratory, however, the Celsius scale is used for most measurements.

The Celsius Temperature Scale In the United States everyday temperatures are reported using the Fahrenheit scale, but the Celsius scale is used in most other countries and in science. Both scales are based on the properties

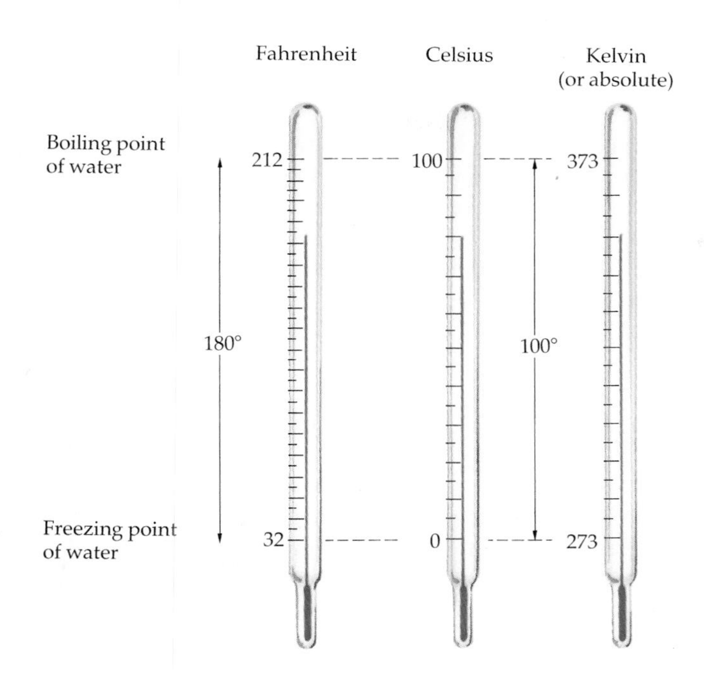

Figure 2.10
A comparison of Fahrenheit, Celsius, and Kelvin temperature scales. The reference or starting point for the Kelvin scale is absolute zero (0 K = −273.15 °C), which has been shown theoretically to be the lowest possible temperature. Note that the abbreviation K for the kelvin unit is used without the degree sign (°). Also note that 1 °C = 1 K = (9/5) °F.

of water. The size of the Celsius degree is defined by assigning zero as the freezing point of pure water (0 °C) and 100 as its boiling point (100 °C). The size of the Fahrenheit degree is equally arbitrary. Gabriel Fahrenheit defined 0 °F as the freezing point of a solution in which he had dissolved the maximum amount of salt (because this was the lowest temperature he could reproduce reliably), and he intended 100 °F to be the normal human body temperature (but this turned out to be 98.6 °F). Today, the reference points are set at 32 °F and 212 °F, the freezing and boiling points of pure water, respectively. The number of units between these points is 180 Fahrenheit degrees (Figure 2.10). Comparing the two units, the Celsius degree is almost twice as large as the Fahrenheit degree; it takes only 5 Celsius degrees to cover the same temperature range as 9 Fahrenheit degrees, and this relationship can be used to convert a temperature on one scale to a temperature on the other if one wishes to do so. However, since laboratory work is almost always done using Celsius units, there is really no need to make conversions to and from the Fahrenheit scale.

To be entirely correct, we must specify that water boils at 100 °C and freezes at 0 °C only when the pressure of the surrounding atmosphere is 1 standard atmosphere. We shall discuss pressure and its effect on boiling point in Chapter 12.

To help you think in terms of the Celsius scale, it is useful to know that water freezes at 0 °C and boils at 100 °C, a comfortable room temperature is about 22 °C, average body temperature is 37 °C, and the hottest water you could put your hand into without serious burns is about 60 °C.

The Kelvin Unit Winter temperatures in many places can easily drop below 0 °C, that is, to temperatures given by negative numbers on the Celsius scale. In the laboratory, even colder temperatures can be achieved easily, and the temperatures are given by even larger negative numbers. However, there is a limit to how low the temperature can go, and hundreds of experiments have found that this limiting temperature is −273.15 °C (or −459.67 °F).

The degree symbol (°) is not used with Kelvin temperatures, and the unit is called a kelvin *(not capitalized).*

William Thomson, known as Lord Kelvin (1824–1907), first suggested a temperature scale that does not use negative numbers. Kelvin's scale, now adopted as the international standard for science, uses the same size degree as the Celsius scale, but it takes the lowest possible temperature as its zero, a point called **absolute zero.** Because kelvin and Celsius units are the same size, the freezing point of water is reached 273.15 degrees *above* the starting point; that is, 0 °C is the same as 273.15 kelvins or 273.15 K. Temperatures in Celsius units are readily converted to kelvins, and vice versa, using the relation

$$T(\text{K}) = t(°\text{C}) + 273.15$$

Thus, a common room temperature of 23.5 °C would be (23.5 °C + 273.15) or 296.7 K.

Chemistry deals with changes in matter, and one type of change is a **physical change.** In a physical change a pure substance is preserved even though it may have changed its physical state or the gross size and shape of its pieces. An example of a physical change is the melting of a solid, and the temperature at which this occurs is often so characteristic of the solid that it can be used to identify the substance. For example, tin and lead resemble one another in outward appearance, but tin melts at 231.8 °C,

Note the use of significant figures when adding 273.15 in the calculation of kelvin temperatures. See Appendix B for a discussion of the use of significant figures.

$$\text{melting point of tin} = 231.8\ °\text{C} + 273.15 = 505.0\ \text{K}$$

while lead melts almost 100 Celsius degrees (or kelvins) higher (327.5 °C) (Figure 2.11).

$$\text{melting point of lead} = 327.5\ °\text{C} + 273.15 = 600.7\ \text{K}$$

EXERCISE 2.4 • *Temperature Conversions*

Liquefied nitrogen boils at 77 K. What is this temperature in Celsius degrees?

Figure 2.11
Lead melts—changes from the solid to the liquid state—at 327.5 °C or 600.7 K (*left*). The molten lead is poured into an antique mold where it solidifies into a musket ball (*right*). These are examples of a physical change. *(C.D. Winters)*

Portrait of a Scientist

Anders Celsius (1701–1744)

Until 1948 scientists around the world worked with temperatures in "degrees centigrade." In that year, however, it was agreed to rename the scale for Anders Celsius, an 18th-century Swedish astronomer. Celsius, who was from a family of scientists, studied mathematics, astronomy, and physics at Uppsala University and was a professor of astronomy there beginning in 1730. He published important papers on the aurora borealis (the "northern lights") and the brightness of stars. Sometime in the 1730s, Celsius described a thermometer scale based on dividing the interval between the freezing and boiling points of water into 100 equal parts. Whether he actually devised the scale, though, is in some dispute. One person who claimed to have invented it was a protégé and friend of Celsius, the botanist Linnaeus.

Celsius apparently used melting snow to fix the lower temperature of his scale, but curiously he marked it as 100. The upper point, the boiling point of water, was marked as the 0 point. Beginning in June 1743, the "Swedish thermometer" with this scale was used for regular observations at the observatory at Uppsala, Sweden. In 1747, a few years after Celsius's death, Linnaeus reversed the scale so that the freezing point of water was 0 °C.

Anders Celsius. (Bettmann Archive)

2.2 MIXTURES AND PURE SUBSTANCES

Most natural samples of matter consist of two or more substances; that is, they are mixtures (Figure 2.12). Peanut brittle is obviously a mixture, because one can see that the peanuts are different from the surrounding mate-

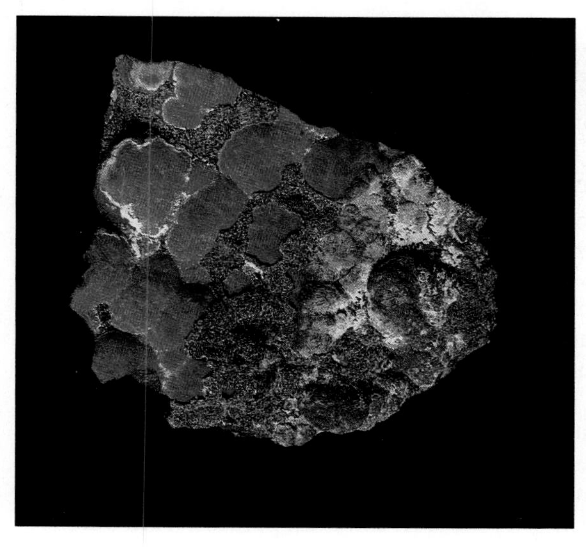

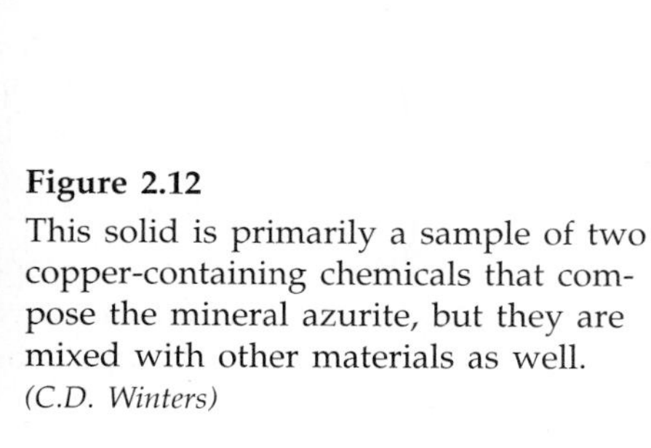

Figure 2.12
This solid is primarily a sample of two copper-containing chemicals that compose the mineral azurite, but they are mixed with other materials as well.
(C.D. Winters)

Figure 2.13
The air in a forest would seem to be homogeneous, but it is really a heterogeneous mixture. That is, in the proper light, one can see particles of dust floating in the air. *(C.D. Winters)*

Figure 2.14
A homogeneous mixture or solution. A solid yellow chemical compound called potassium chromate (*at the right*) is stirred into water, where it dissolves to form an aqueous solution. *(C.D. Winters)*

If you put a handful of fine sand into water and shook it vigorously, you would have a ***suspension.*** *Sand particles would still be clearly visible, and the particles would gradually settle to the bottom of the bottle.*

A colloidal dispersion such as milk represents a situation intermediate between a solution and a suspension. See Chapter 15.

rial. A mixture in which the uneven texture of the material is visible is called a **heterogeneous mixture.** Some heterogeneous mixtures may appear completely uniform, but on closer examination are not. For example, the air in your room may appear the same everywhere until a beam of light enters the room, revealing floating dust particles (Figure 2.13). Milk appears smooth in texture to the unaided eye, but magnification reveals fat and protein globules within the liquid. In a heterogeneous mixture the properties in one region are different from the properties in another region.

A **homogeneous mixture** or **solution** is completely uniform at the microscopic level and consists of two or more substances in the same state (Figure 2.14). No amount of optical magnification will reveal a solution to have different properties in one region than in another, because heterogeneity exists in a solution only at the atomic or molecular scale, where the individual particles are too small to be seen with ordinary light. Examples of solutions are clear air (mostly a mixture of nitrogen and oxygen gases), sugar water, and some brass alloys (which are homogeneous mixtures of copper and zinc). The properties of a homogeneous mixture are the same everywhere in a sample, but they can vary from one sample to another depending on how much of one component is present relative to another component.

When a mixture is separated into its components, the components are said to be purified. However, most efforts at separation are not complete in a single step, and repetition is almost always necessary to give an increasingly pure substance. For example, the iron can be separated from a hetero-

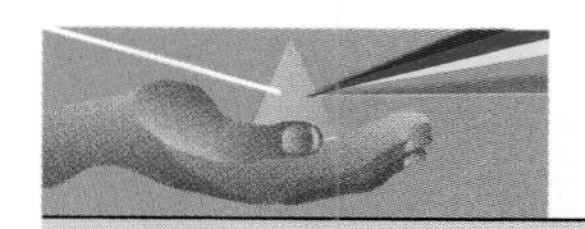

CHEMISTRY YOU CAN DO

Separating a *Mixture*

Breakfast cereal is a heterogeneous mixture. Many cereals are fortified with iron, but did you know what form the iron may be in?

To do this experiment you will first need to find a reasonably strong magnet. You might find a good one holding things to a refrigerator or in a toy or hobby store. A really strong magnet can be found in an agricultural supply store—ask the salesperson for a "cow magnet." (A dairy farmer feeds a magnet to a cow, and the magnet lodges in the first stomach of the animal. If the cow swallows pieces of wire fence or old nails, the magnet stops the junk from getting into the cow's second stomach and doing damage.) You will also need a package of Instant Cream of Wheat® and a plastic sandwich bag.

Place the magnet and all the cereal from an individual serving packet into a plastic sandwich bag, and stir the contents for several minutes. Remove the magnet from the bag and observe it carefully. Is anything attached to the magnet that was not there before?

Here are some interesting questions for you to think about:

- On the basis of this experiment and your experience, does metallic iron dissolve in water?
- Essential elements in your diet, such as iron and zinc, are carried around in your body dissolved in blood, which is mostly water. If you ingest pieces of metallic iron, what do you suppose happens to the iron after you eat it?
- What do the labels of several iron-fortified cereals or other foods say about the form of iron in the food?

geneous mixture of iron and sulfur by repeatedly stirring the mixture with a magnet (Figure 2.15). When the mixture is stirred the first time and the magnet is removed, much of the iron is removed with it, leaving the sulfur in a higher state of purity. After just one stirring, however, the sulfur may still have a dirty appearance due to a small amount of iron that remains. Repeated stirrings with the magnet, or perhaps the use of a very strong magnet, will finally leave a bright yellow sample of sulfur that apparently cannot be purified further, at least by this technique. In this purification process a property of the mixture, its color, is a measure of the extent of purification. (The color depends on the relative quantities of iron and sulfur in the mixture.) When the bright yellow color is obtained, it is assumed that all the iron has been removed and that the sulfur is purifed.

Figure 2.15
The iron chips in a mixture of iron and sulfur may be removed by stirring the heterogenous mixture with a magnet. *(C.D. Winters)*

Drawing a conclusion based on one property of the mixture may be misleading, however, because other methods of purification might change some other properties of the sample. It is safe to call sulfur pure only when a variety of methods of purification fail to change its properties. Each pure substance has a set of physical properties by which it can be recognized, just as you can be recognized by a set of characteristics such as your hair or eye color or your height. We define a **pure substance** as a sample of matter with properties that cannot be changed by further purification.

There are a few naturally occurring pure substances. Gold, diamonds, and sulfur occur naturally in very pure form (Figure 2.16), but these substances are special cases. We live in a world of mixtures—all living things, the air and food we depend on, and many products of technology are mixtures.

See Chapter 1 for a discussion of taxol and how scientists are solving the problem of finding enough of this promising new drug that has been separated from yew tree bark.

Although naturally occurring pure substances are not common, it is possible to separate many of them from naturally occurring mixtures by

(a)

(b)

Figure 2.16
Pure substances from the earth. (a) Clockwise from the bottom the substances shown are sulfur, diamonds, and gold. (b) A sulfur vent on the Kilauea volcano in Hawaii. *(a, C.D. Winters; b, D. Cavagnaro)*

modern purification techniques. Familiar pure substances obtained in this manner include refined sugar, table salt (sodium chloride), copper, nitrogen gas, and carbon dioxide, to mention just a few. However, there are other, less familiar but equally important ones such as taxol, a promising drug derived from the bark of the yew tree and used for the treatment of advanced cases of ovarian, breast, and other cancers. In all, more than 12 million pure substances have been obtained from natural mixtures or created by chemists in the laboratory.

Homogeneous and heterogeneous mixtures. *(C.D. Winters)*

EXERCISE 2.5 • *Mixtures and Pure Substances*

The photo at the left shows two mixtures. Which one is homogeneous and which is heterogeneous? Which one is a solution?

2.3 ELEMENTS AND ATOMS

Pure table sugar, when heated, will decompose in a complex series of chemical changes (caramelization) that produces the brown color and flavor of peanut brittle or caramel candy. If heated for a longer time at a high enough temperature, however, sucrose can be converted completely to two other pure substances, carbon and water (Figure 2.17). Furthermore, if the water is collected, it can be decomposed still further to pure hydrogen and oxygen by passing an electric current through it. Pure substances like carbon, hydrogen, and oxygen that are composed of only one type of atom are classified as **elements.** Only 109 elements are known at this time. Of these, only 90 are found in nature while the other 19 have been created by human beings. Each element has a *name* and a *symbol,* which are listed in the table at the front of the book. The nine elements listed in the margin were known in relatively pure form to the early Greeks and Romans and to the alchemists of ancient China, the Arab world, and medieval Europe. However, many others, such as aluminum (Al), silicon (Si), and iodine (I), were not discovered in the minerals of the earth or the atmosphere until the 18th and 19th centuries.

The elements known since ancient times are carbon (C), sulfur (S), iron (Fe), copper (Cu), silver (Ag), tin (Sn), gold (Au), mercury (Hg), and lead (Pb).

Figure 2.17
Sucrose can be decomposed using sulfuric acid to form carbon (the black solid) and water (seen as steam emerging from the beaker). *(C.D. Winters)*

Finally, artificial elements such as technetium (Tc), plutonium (Pu), and americium (Am) were not made until the 20th century when the techniques of modern physics became available.

Many elements have names and symbols with Latin or Greek origins, but more recently discovered elements have been named for their place of discovery or for a person or place of significance (Table 2.1). The table at the

Table 2.1

The Names of Some Chemical Elements

Element	Symbol	Date of Discovery	Discoverer	Derivation of Name or Symbol
berkelium	Bk	1950	G.T. Seaborg S.G. Thompson A. Ghiorso (U.S.)	Berkeley, California was the site of Seaborg's laboratory
copper	Cu	Ancient		Latin, *cuprum*, copper, or *cyprium*, from Cyprus
lead	Pb	Ancient		Latin, *plumbum*, lead, meaning heavy
oxygen	O	1774	J. Priestley (G.B.) K.W. Scheele (Sweden)	French, *oxygene*, generator of acid, derived from the Greek, *oxy* and *genes* meaning acid-forming (because oxygen was thought to be part of all acids)

V. Ringnes, "The Origin of the Names of Chemical Elements," *Journal of Chemical Education*, **1989**, 66, 731.

Gold Bugs

FROM OLD WESTERN MOVIES ON television, your vision of a gold miner is a grizzled old prospector panning for gold in a stream. He scooped up some sand and gravel from the bottom of the stream and swirled it in a shallow pan. Since gold is much denser than gravel, the gold settled to the bottom of the pan and could be picked out as tiny flecks and even larger nuggets. This is called "placer" gold, material long thought to have been eroded from an ore bed by wind and water and washed into streams.

John R. Watterson of the U.S. Geological Survey in Denver has a new theory about the way this gold found its way to the streams. He has found evidence that stream-borne gold particles in Alaska, and presumably elsewhere, were formed by bacteria that produced a thin skin of pure gold! The photograph at the right is a mass of bacteria (of the genus *Pedomicrobium*) that accumulated gold around themselves; the piece is about 0.1 millimeter wide. These coatings are extraordinary in that they are almost pure, 24-karat gold. Apparently the bacteria attract gold compounds selectively from ore beds and convert them into a metallic gold coating. When the bacteria die, the coating remains and can be washed into streams as a fleck of gold.

Watterson and others are far from understanding these "gold bugs," which are also known to form coatings of other minerals around themselves. When asked why the bacteria use gold, Watterson replied, "That's still an absolute mystery to me." (See *Scientific American*, September, 1992, page 27.)

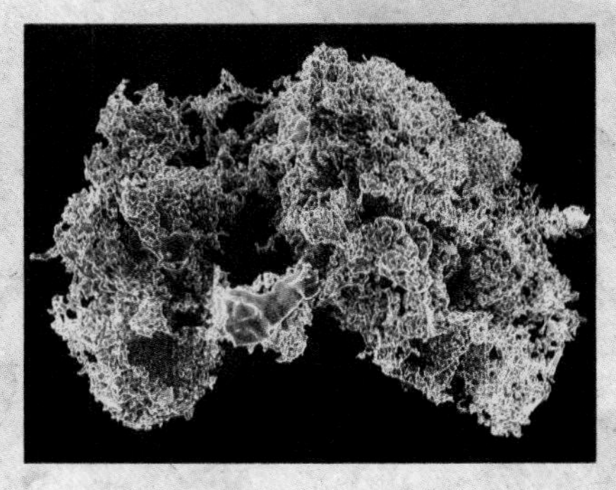

The shell from a microbe that accumulates gold. (John Watterson, U.S. Geological Survey, *Geology*, volume 20, pp. 315–318, 1992)

Be sure to notice that only the first letter of an element's symbol is capitalized. For example, cobalt is Co and not CO. The notation CO would mean the combination of carbon (C) and oxygen (O), the formula for carbon monoxide.

front of the book, in which the symbol and other information for each element are enclosed in a box, is called the **periodic table.** We shall describe this important tool of chemistry in more detail beginning in Chapter 3.

An **atom** is the smallest particle of an element that retains the chemical properties of that element. Modern chemistry is based on an understanding and exploration of nature at the atomic level, so we will have much more to say about atoms in Chapters 3 and 9, in particular.

EXERCISE 2.6 • *Elements*

Using the periodic table in the front of the book,

a. find the names of the elements with symbols Na, Cl, and Cr.

b. find the symbols for the elements zinc, aluminum, and silicon.

2.4 COMPOUNDS AND MOLECULES

Even though only 109 elements are known, there appears to be no practical limit to the number of compounds that can be made from those elements.

Pure substances such as sucrose and water that can be decomposed into two or more different pure substances are referred to as **chemical compounds.** Compounds can be decomposed because they are made of atoms of more than one element combined in a manner distinctive to that compound.

When elements are part of a compound, their original characteristic properties (such as color, hardness, and melting point) are replaced by the characteristic properties of the compound. Consider ordinary table sugar, sucrose, which is composed of three elements:

- carbon, which is usually a black powder, but is also found in the form of diamonds
- hydrogen, the lightest gas known
- oxygen, a gas necessary for human and animal respiration

Sucrose, a white, crystalline powder, has properties completely unlike any of these three elements.

It is important to make a careful distinction between a compound of two or more elements and a mixture of the same elements. Iron and sulfur can be mixed in varying proportions (Figure 2.15). However, in the chemical compound known as iron pyrite (Figure 2.6), this kind of variation cannot occur. Not only does iron pyrite exhibit properties peculiar to itself and different from those of iron and sulfur, but it has a definite percentage composition by mass (46.55% Fe and 53.45% S). Thus, there are two major differences between pure compounds and mixtures: compounds have distinctly different properties from their parent elements, and *compounds have a definite percentage composition by mass of their combining elements.*

Some compounds—such as salt—are composed of **ions,** electrically charged atoms or groups of atoms, and we shall describe such compounds beginning in Chapter 4. Other compounds—such as water and sugar—consist of **molecules,** which are the smallest discrete units that retain the chemical characteristics of the compound. The composition of any compound can be represented by its **formula.**

You know already that the formula for water is H_2O. The symbol for hydrogen, H, is followed by a subscript 2 to indicate that there are two atoms of this type in the water molecule. Similarly, there is one O atom, since the appearance of the symbol for oxygen without a subscript signals a single atom of that type in the molecule. Water's formula also indicates that, if you have a glassful of water, there are twice as many H atoms in this quantity of water as there are oxygen atoms. The other compound we have already mentioned frequently, sucrose, has a molecular formula of $C_{12}H_{22}O_{11}$. Therefore, a single molecule is composed of 12 C atoms, 22 H atoms, and 11 O atoms. Similarly, a bowl of sugar has a ratio of C atoms to H atoms to O atoms of 12/22/11.

Salt, sodium chloride, is a good example of an ionic compound, since it is composed of sodium ions and chloride ions. In salt there is no such thing as a simple molecule of NaCl. Its formula, NaCl, reflects the fact that a salt crystal has as many sodium ions as chloride ions.

Many properties of elements and compounds depend on the arrangement of the atoms and molecules within the substance. For this reason a variety of models are used to represent those atoms and molecules. For example, a detailed model of the simple sugar glucose, $C_6H_{12}O_6$, may be drawn on paper, a computer can draw the model (top of next page), or we can use wood and plastic to make a model we can hold and manipulate.

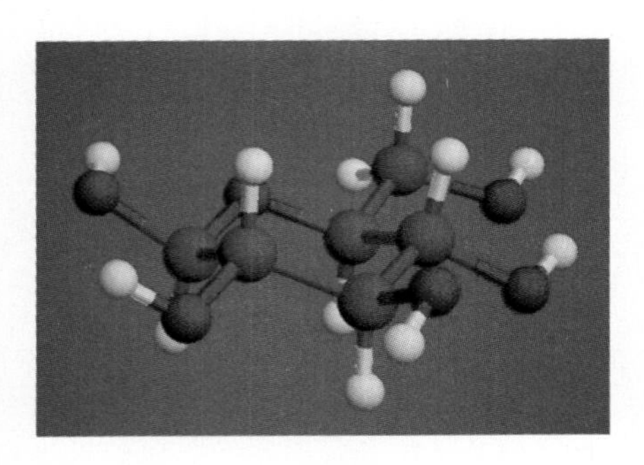

The structure of glucose. Carbon atoms are gray, oxygen atoms are red, and hydrogen atoms are white. *(CAChe Scientific Co.)*

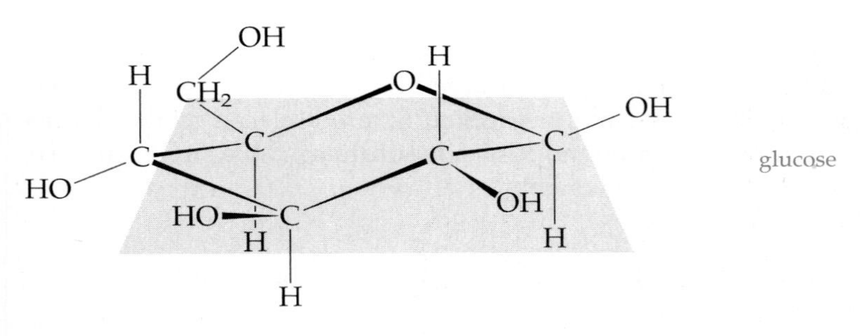

glucose

2.5 CHEMICAL CHANGE

Heating sucrose produces a tan color and then eventually a brown color as the compound decomposes to give water and other new compounds. Among these is $C_6H_6O_3$, which is partly responsible for the brown color in caramel. The formation of this compound from sugar is an example of a **chemical change** or **chemical reaction** because one or more substances (the **reactants**) are transformed into one or more different substances (the **products**). We say that sucrose reacts to form the brown substance and water, and this can be represented as:

$$\underbrace{1 \text{ molecule } C_{12}H_{22}O_{11}}_{\text{reactants}} \rightarrow \underbrace{5 \text{ molecules } H_2O + 2 \text{ molecules } C_6H_6O_3}_{\text{products}}$$

At the molecular level the chemical change produces a new arrangement of atoms without a gain or loss in the number of atoms. The structures of the molecules present after the reaction are different from the structures of those present before the reaction. In a physical change the molecules present before and after are the same, but their arrangement relative to one another (farther apart in a gas, closer together in a solid) is different.

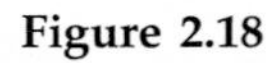

Figure 2.18
The chemical change that occurs when the element aluminum is mixed with the compound iron oxide. This reaction, known as the "thermite reaction," produces a great deal of heat and light. *(C.D. Winters)*

Figure 2.19
A "light stick" produces light by a chemical reaction.
(C.D. Winters)

Physical changes, and to a greater extent chemical reactions, are accompanied by transfers of energy. The thermite reaction (Figure 2.18) transfers tremendous quantities of heat and light energy to its surroundings when an oxide of iron donates its oxygen to the element aluminum. The reaction in a commercial "light stick" evolves light energy and little heat (Figure 2.19), and a battery makes a calculator work because a chemical reaction forces electric current to flow through the circuits.

Energy transfer in the course of chemical and physical changes is discussed in more detail in Chapter 7.

EXERCISE 2.7 • *Chemical Reactions and Physical Changes*

In the photo at the right, you see a candle burning. The burning wick gives off the heat that melts some of the wax, which then travels up the wick and also burns. Name the chemical and physical changes you see. Is energy involved in the process? If so, how?

Wax melting from burning candles.
(C.D. Winters)

2.6 SEPARATION OF MIXTURES INTO PURE SUBSTANCES

The separation of mixtures is usually more difficult than the magnetic separation of iron and sulfur described earlier (see Figure 2.15). Many trained chemists would find it difficult to separate the copper-containing mineral shown in Figure 2.12 into pure substances. Nevertheless, since each pure substance has a set of properties unlike those of any other, it should be possible to use these properties to separate the pure substances in this mineral, just as we described for the separation of iron and sulfur.

Many different methods have been developed to separate mixtures into pure substances. In each case, different *physical* properties of the pure substances—such as solubility in water, density, melting point or boiling point—

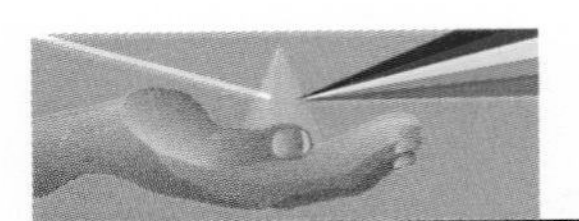

CHEMISTRY YOU CAN DO

Separation of Dyes

Find a piece of absorbent paper such as a coffee filter or paper towel. Cut it into a circle and fold it in half and then in half again. Put the paper in front of you with the point toward you. Make a horizontal mark with a water-soluble marker pen about 2 cm from the point.

Pour about an inch of water into a glass. Poke a pencil through the paper near the curved top so that you can rest the pencil on top of the glass and suspend the paper down inside the glass. The pointed tip should just enter the water so that the lowest 1 cm is below the surface. Make careful observations every minute or two, depending on how fast things are happening. What do you observe? Before the water reaches and wets the top of the paper, remove the paper from the glass and stop the experiment.

Look around for more pens of different colors and try them. Do one experiment in which you allow the paper to remain in the glass overnight. How does this affect your results? What other variations besides pens and length of time can you try?

Consider the following questions and indicate how your results apply to answering them:

Set up for separation of dyes. (C.D. Winters)

1. Is the ink in your pen a pure substance?
2. Assuming that water is made of molecules, do you think these molecules are "sticky"? That is, is there some force of attraction between water molecules?
3. Is there a force of attraction between water molecules and absorbent paper?
4. What are some possible uses for the type of experiment you have just done? What could you do with the paper you have made?
5. Draw a picture of what the molecules might be doing as the experiment proceeds. Label the ink, water, and paper molecules. Try to make your picture as accurate as possible; it should not contradict any of the observations you have made.

are exploited to make the separation. A practical example is the separation of a metal from its ore. In an **ore,** a mineral (a compound containing the metal combined chemically with other elements) is usually mixed with dirt and other, unwanted minerals. The first step in obtaining the metal is separation of the desired mineral from the ore. The production of copper, the metal used in electrical wiring, is an excellent illustration. The most abundant copper-bearing mineral, chalcopyrite ($CuFeS_2$), can be separated from the unwanted material by a process called *flotation.* Here the powdered ore is mixed with oil and agitated with soapy water in a large tank (Figure 2.20). Compressed air is forced through the mixture, and lightweight, oil-covered particles of nearly pure chalcopyrite are carried to the top and float on the froth. The other, heavier material settles to the bottom of the tank, and the copper-laden froth is skimmed off.

Nitrogen, N_2, and oxygen, O_2, are annually among the top five chemicals produced in the United States. Both are separated from air by using the fact that the boiling points of liquid nitrogen and liquid oxygen are slightly different from one another and are different from those of other components of air (Section 18.3).

Figure 2.20
"Winning" copper from its ores. The copper-containing ore (a copper sulfide) is enriched in the flotation process. The lighter particles containing the copper compound are trapped in soap bubbles and float on the water. The heavier "gangue" or waste material settles to the bottom.

Figure 2.21
Chlorine gas can be obtained by the electrolysis of brine, a concentrated solution of salt. *(Oxy Tech Systems)*

Pure substances have fixed compositions and cannot be purified further by methods such as those we have mentioned. However, by using the *chemical properties* of the substance, it may be possible to decompose a pure substance into other pure substances, which may be chemical elements or compounds. Once again the production of a commonly used chemical is an illustration of this technique. Chlorine, Cl_2, is tenth on the list of chemicals produced in the United States, with 22.28 billion pounds produced in 1992. The element is produced from salt, NaCl, which is usually obtained by mining salt beds deep in the earth or by forcing water into the beds and then pumping out the dissolved salt. Electricity is passed through the solution of NaCl in water, and Cl_2 is formed (Figure 2.21).

The consideration of separations leads to a useful way to classify matter (Figure 2.22). Heterogeneous mixtures, such as sand in water, can be separated by simple manipulation, for example, by allowing the sand to settle and pouring off the water. Homogeneous mixtures are somewhat more difficult to separate, but physical processes will serve. For example, salt water can be purified by heating to evaporate the water, collecting the salt crystals, and eventually condensing the water vapor back to liquid. Most difficult of all is separation of the elements that form a compound. This requires one or more chemical reactions and may involve input of other substances or of energy.

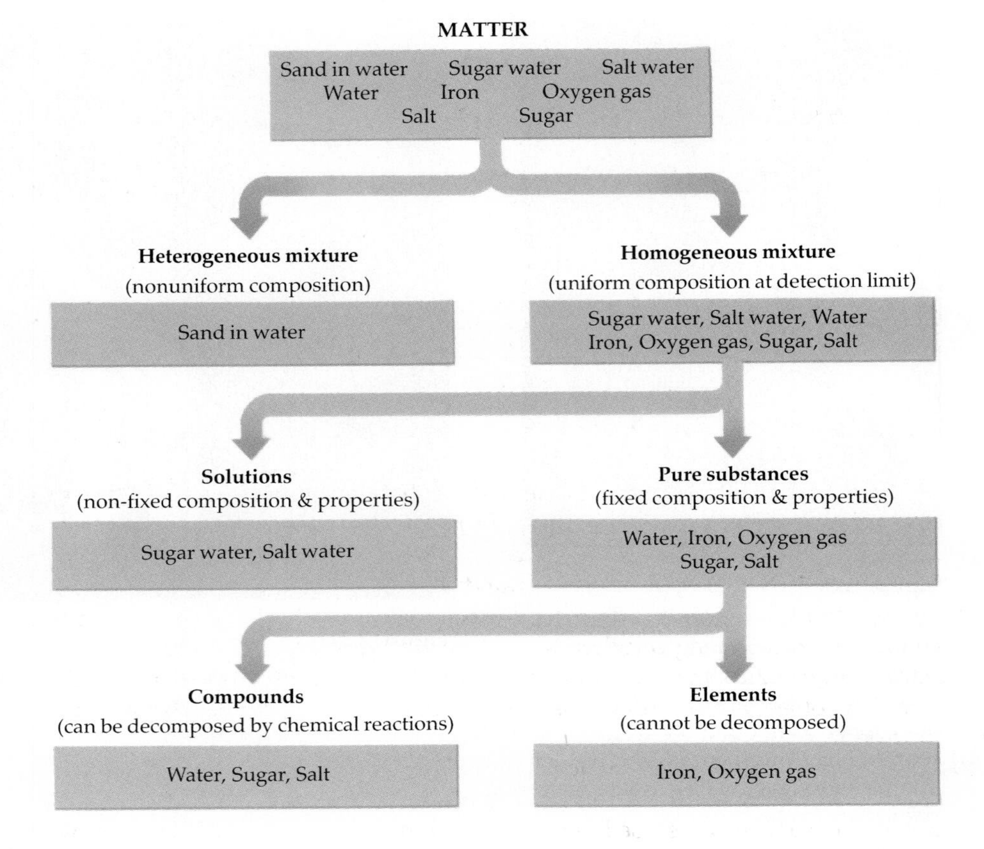

Figure 2.22
A scheme for the classification of matter.

2.7 UNITS OF MEASUREMENT

"Scales and magnitudes are part of the stuff that scientists love. Cosmology and megascales on the one hand, and atoms . . . and microscales on the other, give us a sense of how grand nature is and how consistent our physical pictures are." J.I. Brauman, Science, *Vol. 254, p. 1277, 1991.*

Doing chemistry requires observing chemical reactions and physical changes. When making peanut brittle, for example, you see the color turn brown and watch the evolution of steam as water escapes when the mixture is heated. These are **qualitative observations.** No measurements and numbers are involved, but something obviously happened, probably a chemical reaction, and it can be described in words.

To understand more precisely a reaction such as the caramelization reaction in making peanut brittle, people usually make **quantitative measurements.** For example, if two compounds react with one another, will some of one of them be left over? How much new substance will form? Do we have to heat the substances to make the change occur, and if so, how much heat and for how long? If light is given off, is it red, blue, green, or is it not in the visible range at all? If a new molecule is formed, what size is it? What is the distance between two atoms in the molecule?

Such questions require us to think in terms of physical quantities such as length, volume, temperature, and time. Because each such quantity consists of a number and a unit of measurement, making a measurement consists of counting the number of units (how many centimeters, for example) that correspond to the measured item.

Table 2.2

*SI Base Units**		
Physical Quantity	**Name of Unit**	**Abbreviation**
mass	kilogram	kg
length	meter	m
time	second	s
temperature	kelvin	K
amount of substance	mole	mol
electric current	ampere	A

*A seventh base unit, the candela for luminous intensity, is not needed in chemistry.

Figure 2.23
Nanotechnology: a scanning electron microscope view of a micromotor. The rotor of the motor is only 150 μm across. This dimension (150 micrometers) is equal to 150×10^{-6} meters or 0.015 cm or 0.0059 inches. See *Science*, Volume 254, page 1340, November 29, 1991.

One of the frontier areas of chemistry where an understanding of measurements and their units is crucial is *nanotechnology*, the construction of devices (Figure 2.23) that have dimensions of 1 to 100 nanometers. A nanometer (nm) is 10^{-9} meters, a very small dimension in the world of engineers, a large one to chemists, but a common one to biologists—a bacterium is about 1000 nm in length, whereas a typical molecule is only about 0.1 nm across. Figure 2.24 shows a nanostructure of xenon atoms on a nickel surface. This object is 5 nm on a side or only about 2×10^{-7} inches square, and the xenon atoms are about 0.5 nm apart.

To know more about nanostructures—or about much of science and engineering—it is essential to be familiar with the dimensions of the objects to be studied, the units used, and the relation between them. The scientific community has chosen a modified version of the **metric system** as the standard system for recording and reporting measurements. This is a decimal system, in which all of the units are expressed as powers of 10 times some basic unit. The resulting system, as applied internationally in science, is called the *Systeme International d'Unités* (International System of Units), abbreviated **SI**.

SI Units

In SI, all units are derived from seven base units (listed in Table 2.2), and larger and smaller quantities are expressed by using an appropriate prefix (Table 2.3) with the base unit. For instance, highway distances are given in

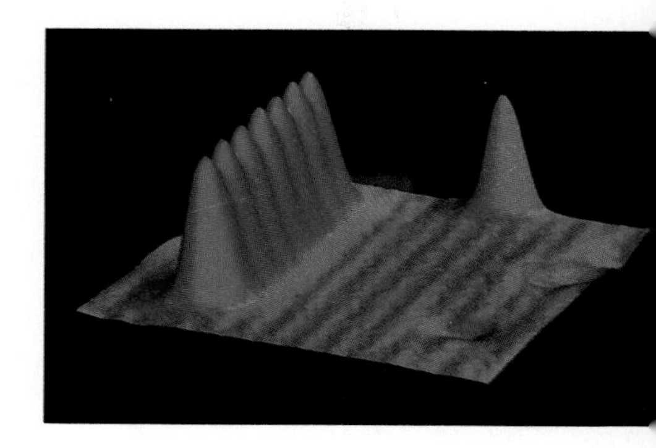

Figure 2.24
The first "hand-built" atomic structure. Scientists at IBM's Almaden Research Center in San Jose, California, were able to move individual atoms across a nickel surface and position them at will. Here seven xenon (Xe) atoms (each represented by a peak) form a linear chain. The image is 5 nm on each side. This is 5×10^{-9} m or 5×10^{-7} cm (2×10^{-7} inches). The atoms are 6.4 billionths of an inch (0.16 nm) high and are spaced 20 billionths of an inch (0.5 nm) apart. See *Science*, Volume 254, page 1324, November 29, 1991. *(IBM Corporation, Research Division, Almaden Research Center)*

Table 2.3

Selected Prefixes Used in the Metric System

Prefix	Abbreviation	Meaning	Example
mega-	M	10^6	1 megaton = 1×10^6 tons
kilo-	k	10^3	1 kilogram (kg) = 1×10^3 g
deci-	d	10^{-1}	1 decimeter (dm) = 0.1 m
centi-	c	10^{-2}	1 centimeter (cm) = 0.01 m
milli-	m	10^{-3}	1 millimeter (mm) = 0.001 m
micro-	μ*	10^{-6}	1 micrometer (μm) = 1×10^{-6} m
nano-	n	10^{-9}	1 nanometer (nm) = 1×10^{-9} m
pico-†	p	10^{-12}	1 picometer (pm) = 1×10^{-12} m

*This is the Greek letter mu (pronounced "mew").
†This prefix is pronounced "peako."

Table 2.4

Some Common Conversion Factors

Length: SI Unit, meter (m)	
1 kilometer	= 1000 meters = 0.62137 mile
1 meter	= 100 centimeters
1 centimeter	= 10 millimeters
1 nanometer	= 1×10^{-9} meter
1 picometer	= 1×10^{-12} meter
1 inch	= 2.54 centimeters (exactly)
1 Ångstrom	= 1×10^{-10} meter
Volume: SI Unit, cubic meter (m^3)	
1 liter (L)	= 1×10^{-3} m^3
	= 1000 cm^3 = 1.056710 quarts
1 gallon	= 4.00 quarts
Mass: SI Unit, kilogram (kg)	
1 kilogram	= 1000 grams
1 gram	= 1000 milligrams
1 pound	= 453.59237 grams = 16 ounces
1 ton	= 2000 pounds
Temperature: SI Unit, kelvin (K)	
0 K	= −273.15 °C
K	= °C + 273.15 °C
? °C	$= \frac{5\ °C}{9\ °F}(°F - 32\ °F)$
? °F	$= \frac{9\ °F}{5\ °C}(°C) + 32\ °F$

*kilo*meters, where 1 kilometer (km) is exactly 1000 or 10^3 meters (m). In chemistry, length is most often given in subdivisions of the meter, such as *centi*meters (cm) or *milli*meters (mm). The prefix *centi-* means 1/100, so 1 centimeter is 1/100 of a meter (1 cm = 1×10^{-2} m); 1 millimeter is 1/1000 of a meter (1 mm = 1×10^{-3} m). On the atomic scale, dimensions are often given in nanometers (nm) (1 nm = 1×10^{-9} m) or picometers (pm) (1 pm = 1×10^{-12} m).

The quantity 0.001 or 1/1000 is written as 1×10^{-3}. This notation—called scientific or exponential notation—is used throughout this book and is explained in Appendix B.

Several **conversion factors** that enable you to convert between SI and non-SI units are given in Table 2.4 and in Appendix C.

Units and Numerical Problem Solving

As an example of the use of units and conversion factors, consider first the **meter,** which is equivalent to 3.281 feet or 39.37 inches. Since the meter is the standard unit of length used in science, you should be able to convert measurements into this unit and convert between meters and centimeters or other metric units. As an example of the approach to such conversions, let us find out how long a football field is in centimeters and meters. The field for U.S. football is 300. feet long, a distance readily converted to inches and from there to centimeters.

The meter is a convenient unit in human terms since a human leg is roughly a meter long, and you can hold a meter stick lengthwise comfortably between your outstretched hands.

$$300.\ \cancel{\text{feet}} \cdot \frac{12\ \text{inches}}{1\ \cancel{\text{foot}}} = 3.60 \times 10^3\ \text{inches}$$

You may have done the conversion between feet and inches many times, but this time it is set it up carefully to show how problems should be solved in chemistry by using **dimensional analysis,** an approach explained in more detail in Appendix A. Notice that one unit is converted to another by multiplying by a "factor" (in this case 12 inches/1 foot) that expresses the equivalence of a quantity in two different units (1 foot = 12 inches). Since the numerator and denominator describe the same quantity, the factor is equivalent to the number 1. Therefore, multiplication of the original quantity (300. feet) by this factor does not change the magnitude of the original quantity, only its units. The factor is always written so that the units in the denominator cancel the original units, leaving the desired units. Here units of feet cancel, and you are left with units of inches.

This example, and all others in this book, observe the rules of using significant figures. Be sure to see Appendix A for a discussion of these important rules.

With the distance known in inches, you can now make the conversion to the metric scale, since you know that there are 2.54 centimeters in one inch and 100 centimeters in 1 meter.

$$3.60 \times 10^3\ \cancel{\text{inches}} \cdot \frac{2.54\ \text{cm}}{1\ \cancel{\text{inch}}} = 9.14 \times 10^3\ \text{cm}$$

$$9.14 \times 10^3\ \cancel{\text{cm}} \cdot \frac{1\ \text{meter}}{100\ \cancel{\text{cm}}} = 91.4\ \text{m}$$

As a final check when doing a conversion, or any other calculation, you should ask yourself whether the answer is reasonable. Here you can see that 91.4 meters is about right: since a meter is a little more than 3 feet, the distance should be a little less than 100 meters.

EXAMPLE 2.2

Distances on the Molecular Level

The distance between the oxygen atom and a hydrogen atom in a water molecule is 95.7 pm. What is this distance in meters? In nanometers?

SOLUTION In each case, you must first know the relation between the picometer (pm) and the desired unit: 1 pm is exactly 1×10^{-12} m (see Table 2.3). Then, multiply the distance in pm by a factor that has the form (units for answer/units of number to be converted). As you work the problems in this and other chapters, notice that this is always the way problems are solved. Thus, here you multiply the distance in pm by the factor (10^{-12} m/1 pm) so that the units of pm cancel, leaving an answer in meters.

$$95.7\ \cancel{\text{pm}} \cdot \frac{10^{-12}\ \text{m}}{1\ \cancel{\text{pm}}} = 95.7 \times 10^{-12}\ \text{m} = 9.57 \times 10^{-11}\ \text{m}$$

To relate picometers and nanometers, you must know that 1 nm = 10^{-9} m (Table 2.2). Therefore, you can multiply the distance in meters by the factor (1 nm/10^{-9} m).

$$9.57 \times 10^{-11}\ \cancel{\text{m}} \cdot \frac{1\ \text{nm}}{10^{-9}\ \cancel{\text{m}}} = 9.57 \times 10^{-2}\ \text{nm}$$

EXERCISE 2.8 • *Interconverting Units of Length*

The pages of a typical textbook are 10.0 inches long and 8.00 inches wide. What is each length in centimeters? In meters? In millimeters?

Figure 2.25
Some common laboratory glassware.

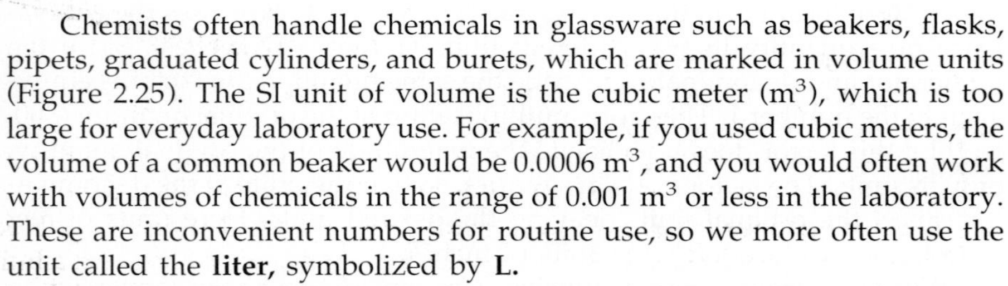

Chemists often handle chemicals in glassware such as beakers, flasks, pipets, graduated cylinders, and burets, which are marked in volume units (Figure 2.25). The SI unit of volume is the cubic meter (m^3), which is too large for everyday laboratory use. For example, if you used cubic meters, the volume of a common beaker would be 0.0006 m^3, and you would often work with volumes of chemicals in the range of 0.001 m^3 or less in the laboratory. These are inconvenient numbers for routine use, so we more often use the unit called the **liter,** symbolized by **L.**

A cube with sides equal to 10 cm (0.1 m) has a volume of 10 cm × 10 cm × 10 cm = 1000 cm^3 (or 0.001 m^3). This is defined as 1 liter. Thus,

$$1\ \text{liter}\ (1\ \text{L}) = 1000\ \text{cm}^3$$

The liter is a convenient unit to use in the laboratory, as is the **milliliter (mL).** Since there are 1000 mL and 1000 cm^3 in a liter, this means that

$$1\ \text{cm}^3 = 0.001\ \text{L} = 1\ \text{milliliter}\ (1\ \text{mL})$$

Chemists often use the terms milliliter and cubic centimeter (or "cc") interchangeably. Therefore, a flask that contains exactly 250. mL has a volume of 250. cm^3

$$250.\ \cancel{cm^3} \cdot \frac{1\ L}{1000\ \cancel{cm^3}} = 0.250\ L$$

or one fourth of a liter.

10 cm is called a decimeter (dm), since it is 1/10 of a meter. Therefore, since a cube 10 cm on a side contains a liter, a liter is equivalent to a cubic decimeter: $1\ L = 1\ dm^3$. Although not widely used in the United States, the cubic decimeter is used in the rest of the world.

EXAMPLE 2.3

Units of Volume

A laboratory beaker has a volume of 0.6 L. What is its volume in cm^3 and mL?

SOLUTION The relation between the given unit and the desired unit of cm^3 is $1\ L = 1000.\ cm^3$. Therefore, multiply by the factor ($1000.\ cm^3/1\ L$) so that units of L cancel to leave the answer in cm^3.

$$0.6\ \cancel{L} \cdot \frac{1000\ cm^3}{1\ \cancel{L}} = 600\ cm^3$$

Since $1\ cm^3$ and 1 mL are equivalent, we can say the volume of the beaker is also 600 mL.

$$600\ \cancel{cm^3} \cdot \frac{1\ mL}{1\ \cancel{cm^3}} = 600\ mL$$

EXERCISE 2.9 • *Volume*

a. A standard wine bottle has a volume of 750 mL. How many liters does this represent?
b. One U.S. gallon is equivalent to 3.7865 L. How many liters are there in a 2.0-quart carton of milk?

The **mass** of a body is the fundamental measure of the quantity of matter in that body, and the SI unit of mass is the **kilogram (kg)**. Smaller masses are expressed in **grams (g)** or **milligrams (mg)**. Again, the prefixes in Table 2.3 apply.

$$1\ kg = 1000\ g$$
$$1\ g = 1000\ mg$$

In the United States we are used to the English system of mass measurement, so most masses are given in pounds,* where *1 pound is equivalent to 453.59237 grams*. This means that a mass of 1.00 kg is equivalent to 2.20 pounds, or about the mass of a quart of milk. But let us say you are living in Portugal and you have bought 750 grams of strawberries. How many kilograms does this represent and how many pounds?

*Strictly speaking, the pound is not a unit of mass, but rather of *weight*. The weight of an object depends on the local force of gravity. If you took the object into space or to another planet, its weight would be different from its terrestrial value although its mass would remain the same. For measurements made on the earth's surface, however, this distinction is not generally useful.

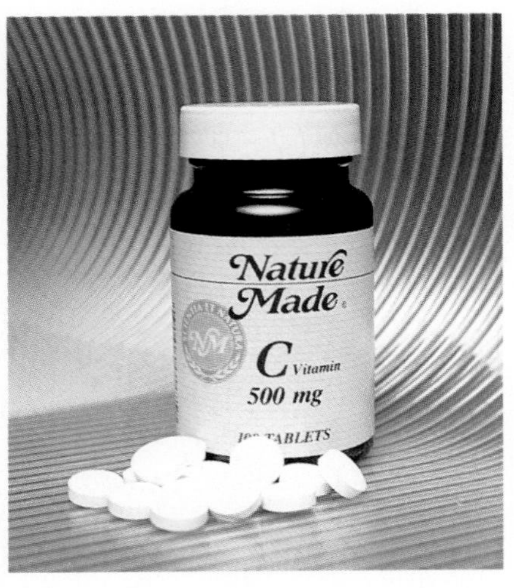

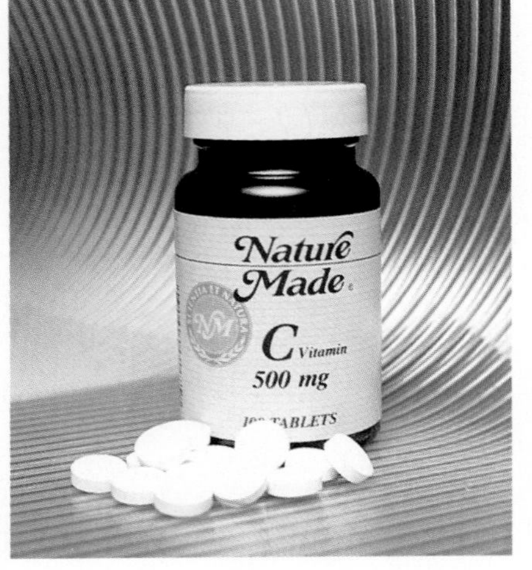

Vitamin C tablets. *(C.D. Winters)*

$$750\ \text{g} \cdot \frac{1\ \text{kg}}{1000\ \text{g}} = 0.75\ \text{kg}$$

$$750\ \text{g} \cdot \frac{1\ \text{pound}}{454\ \text{g}} = 1.7\ \text{pounds}$$

EXAMPLE 2.4

Mass in Kilograms and Grams

A new U.S. penny has a mass of 2.49 g. Express this mass in kilograms and milligrams.

SOLUTION Here the relation between the unit of the desired answer and the given unit is 1 kg = 1000 g or 1000 mg = 1 g. Therefore, multiply the mass in grams by a factor that has the form (units for answer/units of number to be converted).

a. $2.49\ \text{g} \cdot \dfrac{1\ \text{kg}}{1000\ \text{g}} = 0.00249\ \text{kg}$

b. $2.49\ \text{g} \cdot \dfrac{1000\ \text{mg}}{1\ \text{g}} = 2.49 \times 10^3\ \text{mg}$

EXERCISE 2.10 • *Mass*

a. The amount of vitamin C in a tablet is 500. mg. How many grams is this?

b. One pound is equivalent to 453.6 g. How many kilograms are equivalent to a 2.00-pound package of hamburger meat?

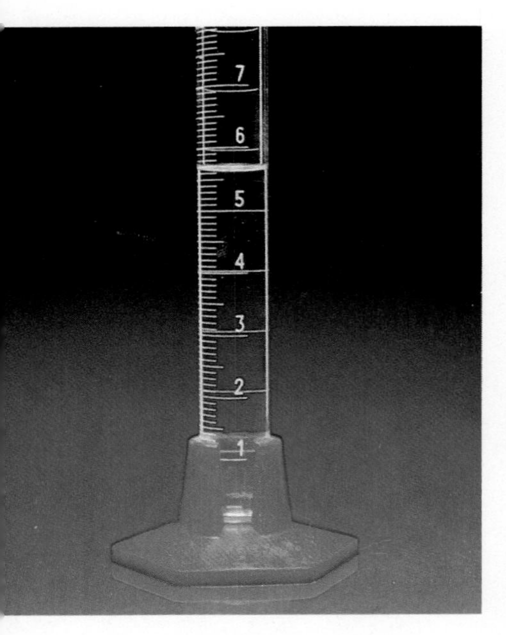

Carbon tetrachloride in a graduated cylinder. *(C.D. Winters)*

You learned earlier that density is an important physical property, one that enables a chemist to identify a substance or separate it from another one. Therefore, extensive tables of densities are available, and the determination of the density of a sample can be done fairly readily. Now that we have discussed the units of mass and volume in more detail, suppose you are asked to approximate the density of a liquid commonly used to dissolve grease in industrial processes, the carbon- and chlorine-containing compound CCl_4. You first weigh a clean, dry graduated cylinder. Then you place some CCl_4 in the cylinder, and, reading from the scale on the cylinder, you would find the volume of the liquid. Suppose it is 5.6 mL.

Next, you would find the mass of the liquid by weighing the cylinder with the liquid and then subtracting the mass of the empty cylinder. In this experiment, let us say you find the liquid mass to be 8.90 g. Therefore, remembering that milliliters and cubic centimeters are the same thing, you would calculate the density of the liquid compound as

$$\text{density} = \frac{8.90\ \text{g}}{5.6\ \text{cm}^3} = 1.6\ \text{g/cm}^3$$

EXERCISE 2.11 • *Density*

Some chemical reactions are made to occur by passing electricity through a compound, and this is often done by immersing a thin sheet of platinum in a solution of the compound. Suppose that the platinum sheet you are using is 2.50 cm square and has a mass of 1.656 g. The density of platinum is 21.45 g/cm^3. What is the thickness of the platinum sheet, in millimeters?

The Concept of "Percent"

Chemists often express the composition of matter in terms of **percent.** For example, we know that 88.81% of a given mass of water is oxygen, that sucrose is 42.11% carbon, or that a 5¢ coin, a nickel, is only about 25% nickel (the rest is copper). Since percent is a widely used concept in chemistry, it is worth taking a moment to think about it while we are discussing units.

You are familiar with "percent" from looking for a bargain at the local mall or from paying sales taxes. For example, paying a sales tax of 6.0% means that you pay 6.00 dollars per 100 dollars of things purchased. Therefore, if you buy a new shirt for \$31.50, the tax is \$1.89.

$$31.50\ \cancel{\$\ \text{spent}} \cdot \frac{6.00\ \$\ \text{tax}}{100\ \cancel{\$\ \text{spent}}} = 1.89\ \$\ \text{tax}$$

There are two important points to make here. First, notice that the problem—one that you have probably done in your head many times—was solved using units. We used the conversion factor "\$ tax/\$100 spent," and the unit "\$ spent" cancels out and leaves units of "\$ tax." Second, the word "percent" tells us what the conversion factor must be, since *per cent* literally means *per 100*. The conversion factor in a percent calculation will always be the value of the percent divided by 100.

EXAMPLE 2.5

Using Percent

Battery plates in lead storage batteries (the type used in automobiles) are made from a mixture of two chemical elements: lead (Pb, 94.0% by mass) and antimony (Sb, 6.0% by mass). If a battery plate has a mass of 25.0 g, what masses of lead and of antimony (in grams) are present?

SOLUTION Let us first solve for the mass of lead, using the known percentage of lead. The plate is 94.0% lead, which means there are 94.0 g of lead in every 100. g of plate.

$$25.0\ \cancel{\text{g battery plate}}\ \frac{94.0\ \text{g lead}}{100.\ \cancel{\text{g battery plate}}} = 23.5\ \text{g lead}$$

We know that the plate contains only lead and antimony, so

$$25.0\ \text{g plate} = 23.5\ \text{g lead} + X\ \text{g antimony}$$

Solving for X, we find that the mass of antimony is 25.0 g − 23.5 g = 1.5 g of antimony. Of course, we could obtain the same result from the calculation

$$25.0\ \cancel{\text{g battery plate}}\ \frac{6.0\ \text{g antimony}}{100.\ \cancel{\text{g battery plate}}} = 1.5\ \text{g antimony}$$

EXERCISE 2.12 • *Percent*

What is the mass of gold in a 15.0-g earring if it is made of 14-karat gold? Fourteen-karat gold is 58% gold, the remainder being copper and silver.

IN CLOSING

Having studied this chapter, you should be able to

- define physical properties of matter and give some examples (Section 2.1).
- explain the difference between chemical and physical change (Sections 2.1 and 2.5).
- identify the name or symbol for an element, given its symbol or name (Section 2.3).
- describe some common-sense ways of separating mixtures (Sections 2.2 and 2.6).
- explain the difference between elements and compounds, and homogeneous and heterogeneous mixtures (Section 2.6).
- recognize and know how to use the prefixes that modify the sizes of metric units (Section 2.7).
- convert between temperatures on the Celsius and kelvin scales (Section 2.1).
- use density as a way to connect the volume and mass of a substance (Sections 2.1 and 2.7).
- use the concept of percent in calculations in chemistry (Section 2.7).
- use dimensional analysis to carry out unit conversions and other calculations.

STUDY QUESTIONS

Review Questions

Questions whose numbers appear in boldface have answers in Appendix L.

1. In the photo (*right*) you see tiny crystals of the mineral vanadinite, a mineral containing the elements lead, vanadium, chlorine, and oxygen. What are the symbols for these elements? Are these crystals in the macroscopic or submicroscopic world? How would you describe the shape of these crystals? What does this tell you about the arrangement of the atoms inside the crystal?
2. Galena, pictured on the next page, is a black mineral that contains lead and sulfur and that shares its name with a number of towns in the United States; they are located in Alaska, Illinois, Kansas, Maryland, Missouri, and Ohio. What are the symbols of the elements in the mineral? How would you describe the shape of the

The mineral vanadinite. *(C.D. Winters)*

Galena is the black mineral in a matrix of quartz. *(C.D. Winters)*

galena crystals? What does this tell you about the arrangement of the atoms inside the crystal?

3. What are the states of matter and how do they differ from one another?
4. The photo below shows some copper balls, immersed in water, floating on top of mercury. What are the symbols of the elements copper and mercury? What are the liquids and solids in this photo? What does this photo tell you about the relative densities of these materials?

Water, copper balls, and liquid mercury. *(C.D. Winters)*

5. Small chips of iron are mixed with sand (see at top of next column). Is this a homogeneous or heterogeneous mixture? Suggest a way to separate the iron and sand from each other.

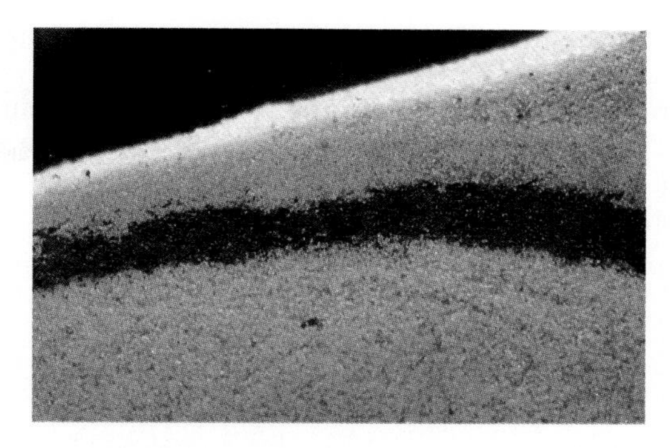

Chips of iron mixed with sand. *(C.D. Winters)*

6. The photo below shows the label from a bottle of cranberry juice. Is this a homogeneous or heterogeneous mixture? Is the juice a pure substance or not?

Cranberry juice. *(C.D. Winters)*

7. The photo below shows an element that can occur in the elemental form in nature. From the color and metallic luster can you tell what this element might be? What are its name and symbol?

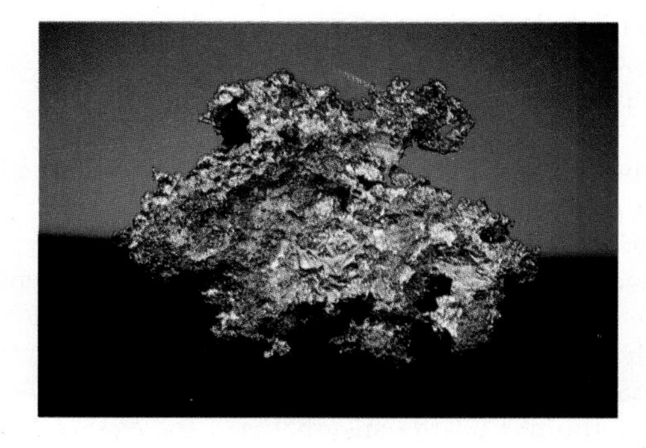

The natural form of an element. *(C.D. Winters)*

8. Describe the difference between a compound and a molecule.
9. In the photo below, you see a crystal of the mineral calcite surrounded by piles of calcium and carbon, two of the elements that combine to make the mineral. (The other element combined in calcite is oxygen.)

Calcite is the clear crystal, the calcium chips are white, and the pile of carbon (as graphite) is at the right. *(C.D. Winters)*

   (a) What are the symbols of the three elements that combine to make calcite?
   (b) Based on the photo, describe some of the physical properties of the elements and the mineral. Are any the same? Are any properties different?
10. In each case, tell whether the underlined property is a physical or chemical property:
   (a) The normal color of bromine is red-orange.
   (b) Iron is transformed into rust in the presence of air and water.
   (c) Dynamite can explode.
   (d) The density of uranium metal is 19.07 g/cm^3.
   (e) Aluminum metal, the "foil" you use in the kitchen, melts at 660 K.
11. In each case, tell whether the change is a chemical or physical change.
   (a) A cup of household bleach changes the color of your favorite T-shirt from purple to pink.
   (b) The fuels in the space shuttle (hydrogen and oxygen) combine to give water and provide the energy to lift the shuttle into space.
   (c) An ice cube in your glass of lemonade melts.
12. While camping in the mountains you build a small fire out of tree limbs you find on the ground near your campsite. The dry wood crackles and burns brightly and warms you. Before slipping into your sleeping bag for the night, you put the fire out by dousing it with cold water from a nearby stream. Steam rises when the water hits the hot coals. Describe the physical and chemical changes in this scene.
13. In the photo below you see a crystal of quartz (often called a Herkimer diamond because it comes from mineral deposits near Herkimer, NY). If you studied the crystal, you might observe that it is colorless and clear, and it has a mass of 2.5 grams and a length of 4.6 cm. Which of these observations are qualitative and which are quantitative?

A sample of quartz called a Herkimer "diamond." *(C.D. Winters)*

Elements and Atoms

14. Give the name of each of the following elements:
   (a) C (c) Cl (e) Mg
   (b) Na (d) P (f) Ca
15. Give the name of each of the following elements:
   (a) Mn (c) K (e) As (g) Cu
   (b) F (d) Fe (f) Kr (h) V
16. Give the symbol for each of the following elements:
   (a) lithium (d) silicon
   (b) titanium (e) cobalt
   (c) iron (f) zinc
17. Give the symbol for each of the following elements:
   (a) silver (e) tin
   (b) aluminum (f) barium
   (c) plutonium (g) krypton
   (d) cadmium (h) palladium

Density

18. Ethylene glycol, $C_2H_6O_2$, is a liquid that is the base of the antifreeze you use in the radiator of your car. It has a density of 1.1135 g/cm^3 at 20 °C. If you need 500. mL of this liquid, what mass of the compound, in grams, is required?

19. The piece of silver metal in the photo below has a mass of 2.365 grams. If the density of silver is 10.5 g/cm^3, what is the volume of the silver?

Silver nuggets and a silver belt buckle. *(C.D. Winters)*

20. Water has a density at 25 °C of 0.997 g/cm^3. If you have 500. mL of water, what is its mass in grams? In pounds?

21. A chemist needs 2.00 g of a liquid compound. (a) What volume of the compound is necessary if the density of the liquid is 0.718 g/cm^3? (b) If the compound costs $2.41 per milliliter, what is the cost of the compound?

22. The "cup" is a volume widely used by cooks in the United States. One cup is equivalent to 225 mL. If 1 cup of olive oil has a mass of 205 g, what is the density of the oil?

23. Peanut oil has a density of 0.92 g/cm^3. If a recipe calls for 1 cup of peanut oil (1 cup = 225 mL), what mass of peanut oil (in grams) are you using?

24. A sample of unknown metal is placed in a graduated cylinder containing water. If the mass of the sample is 37.5 g, and the water levels before and after adding the sample to the cylinder are as shown in the figure, which metal listed below is most likely the sample? (*d* is density)

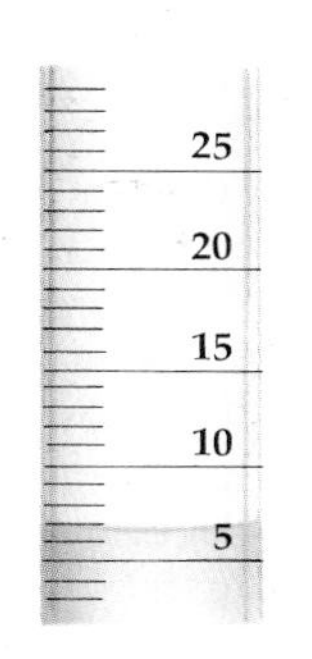

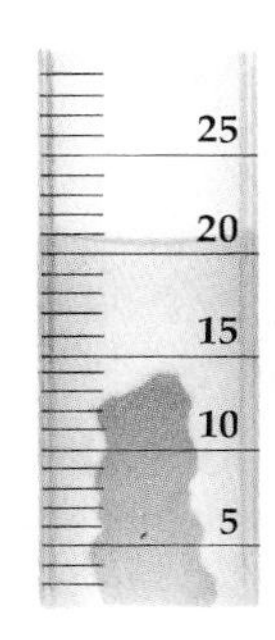

(a) Mg, $d = 1.74\ g/cm^3$
(b) Fe, $d = 7.87\ g/cm^3$
(c) Ag, $d = 10.5\ g/cm^3$
(d) Al, $d = 2.70\ g/cm^3$
(e) Cu, $d = 8.96\ g/cm^3$
(f) Pb, $d = 11.3\ g/cm^3$

25. Fool's gold, sometimes called iron pyrite, can look very much like gold (see Figure 2.6). Assume that you have a piece of a solid that looks as though it may be gold, but you believe it to be fool's gold. The sample has a mass of 23.5 g. When placed in some water in a graduated cylinder (see Study Question 24), the water level increases from 47.5 mL to 52.2 mL. Is the sample fool's gold ($d = 5.00\ g/cm^3$) or is it gold ($d = 19.3\ g/cm^3$)?

Temperature

26. Many laboratories use 25 °C as a standard temperature. What is this temperature in kelvins?

27. The temperature on the surface of the sun is 5.50×10^3 °C. What is this temperature in kelvins?

28. Make the following temperature conversions:

	°C	K
(a)	16	______
(b)	37	______
(c)	−40	______

29. Make the following temperature conversions:

	°C	K
(a)	______	77
(b)	60.	______
(c)	______	1450

30. Solid gallium has a melting point of 29.8 °C. If you hold this metal in your hand, what will be its physical state? That is, will it be a solid or a liquid? Explain briefly.

31. Neon, a gaseous element used in signs, has a melting point of −248.6 °C and a boiling point of −246.1 °C. Express these temperatures in kelvins.

Units and Unit Conversions

32. The average lead pencil, new and unused, is 19 cm long. What is its length in millimeters? In meters? In inches?

33. An excellent height in the pole vault is 18 feet, 11.5 inches. What is this height in meters?

34. The maximum speed limit in the United States is 65 miles per hour. What is this speed in kilometers/hour?

35. A sailboat has a length of 36 feet 7 inches; its beam (the widest point of the hull) is 12 feet. What are these distances in meters? In centimeters?

36. A standard sheet of notebook paper is $8\frac{1}{2} \times 11$ inches. What are these dimensions in centimeters? What is the area of the paper in cm^2?

37. A standard U.S. postage stamp is 2.5 cm long and 2.1 cm wide. What is the area of the sample in cm^2? In m^2? In $inches^2$?

38. A typical laboratory beaker has a volume of 800. mL. What is its volume in cm^3? In liters? In m^3?

39. A Volkswagen engine has a displacement of 120. cubic inches. What is this volume in liters? In cm^3?

40. An automobile engine has a displacement of 250. cubic inches. What is this volume in liters? In cm^3?

41. Suppose your bedroom is 18 feet long, 15 feet wide, and the distance from floor to ceiling is 8 feet, 6 inches. You need to know the volume of the room in metric units for some scientific calculations. What is the room's volume in cubic meters? In liters?

42. A new U.S. quarter has a mass of 5.63 g. What is its mass in kilograms? In milligrams?

43. The fluorite crystal (which contains the elements calcium and fluorine) in the photo (*above, right*) has a mass of 2.83 grams. What is this mass in kilograms? In pounds? Give the symbols for the elements in this crystal.

A fluorite crystal. *(C.D. Winters)*

44. Complete the following table of masses.

Milligrams	*Grams*	*Kilograms*
______	0.693	______
156	______	______
______	______	2.23

45. Complete the following table of masses.

Milligrams	*Grams*	*Kilograms*
10.2	______	______
______	16.56	______
______	______	0.545

Percent

46. Silver jewelry is actually a mixture of silver and copper. If a bracelet with a mass of 17.6 g contains 14.1 g of silver, what is the percentage of silver? Of copper?

47. The solder once used by plumbers to fasten copper pipes together consists of 67% lead and 33% tin. What is the mass of lead (in grams) in a 1.00 pound block of solder? What is the mass of tin?

48. Automobile batteries are filled with sulfuric acid. What is the mass of the acid (in grams) in 500. mL of the battery acid solution if the density of the solution is 1.285 g/cm^3 and if the solution is 38.08% sulfuric acid by mass?

49. The density of a solution of sulfuric acid is 1.285 g/cm^3, and it is 38.08% acid by mass. What volume of the acid solution (in mL) do you need to supply 125 g of sulfuric acid?

General Questions

50. Molecular distances are usually given in nanometers ($1\ nm = 1 \times 10^{-9}\ m$) or in picometers ($1\ pm = 1 \times 10^{-12}\ m$). However, a commonly used unit has been the Ångstrom, where $1Å = 1 \times 10^{-10}m$. (The Ångstrom unit is not an SI unit.) If the distance between the Pt atom and the N atom in the cancer chemotherapy drug cisplatin is 1.97 Å, what is the distance in nm? In pm?

51. The separation between carbon atoms in diamond is 0.154 nm. (a) What is their separation in meters? (b) What is the carbon atom separation in Ångstrom units (where 1 Å=10^{-10} m)?

52. The smallest repeating unit of a crystal of common salt is a cube with an edge length of 0.563 nm. What is the volume of this cube in nm^3? In cm^3?

53. The mass of a gemstone is often measured in "carats" where 1 carat = 0.200 g. If the annual worldwide production of diamonds is 12.5 million carats, how many grams does this represent?

54. Metals such as gold and platinum are sold in units called "troy ounces," where 1 troy ounce has a mass of 31.103 g. Is this larger or smaller than an ounce in the usual scale used in the United States? [Recall that there are 16 ounces in 1 pound (called an "avoirdupois" pound) and that 1 pound has a mass of 453.59237 grams.]

55. As discussed in Chapter 9, light and other forms of radiation can be described as waves, and the distance between adjacent crests of a wave is called the wavelength. If a radio wave has a wavelength of 13 cm, what is its wavelength in meters? In inches? In feet?

56. The platinum-containing cancer drug cisplatin contains 65.0% platinum. If you have 1.53 g of the compound, how many grams of platinum can be recovered from this sample?

57. At 25 °C the density of water is 0.997 g/cm^3, whereas the density of ice at −10 °C is 0.917 g/cm^3. (a) If a soft-drink can (volume = 250. mL) is filled completely with pure water and then frozen at −10 °C, what volume will the solid occupy? (b) Could the ice be contained within the can?

58. When you heat popcorn, it pops because it loses water explosively. Assume a kernel of corn, weighing 0.125 g, weighs only 0.106 g after popping. What percent of its mass did the kernel lose on popping? Popcorn is sold by the pound. Using 0.125 g as the average mass of a popcorn kernel, how many kernels are there in a pound of popcorn?

59. An ancient gold coin is 2.2 cm in diameter and 3.0 mm thick. It is a cylinder for which volume = (π) $(\text{radius})^2$ (thickness). If the density of gold is 19.3 g/cm^3, what is the mass of the coin in grams? Assume a price of gold of $410 per Troy ounce. How much is the coin worth? (1 Troy ounce = 31.10 g)

60. You have a 100.0-mL graduated cylinder (*above, right*) containing 50.0 mL of water. You drop a 154-g piece of pure brass (density = 8.56 g/cm^3) into the water. How high will the water rise in the graduated cylinder?

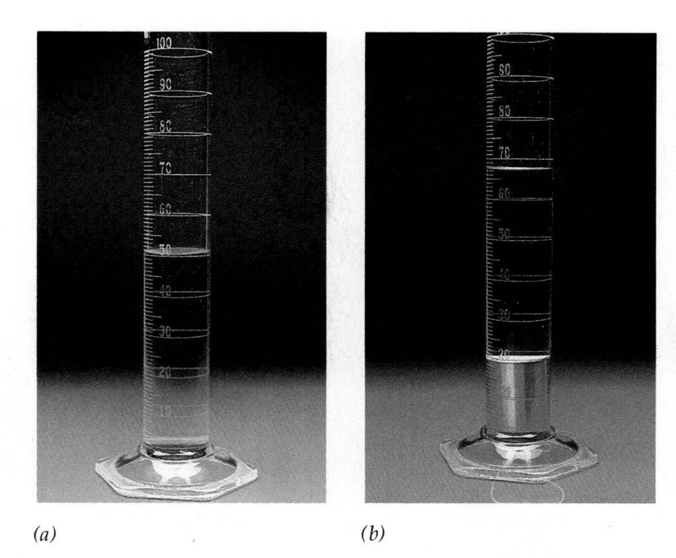

(a) (b)

(a) A graduated cylinder with 50.0 mL of water. (b) A piece of brass is added to the cylinder. *(C.D. Winters)*

61. A common fertilizer used on lawns is designated as "16-4-8" (see photo below). These numbers mean that the fertilizer contains 16% nitrogen-containing compounds, 4.0% of a phosphorus-containing compound, and 8.0% of a potassium-containing compound. If you buy a 40.0-pound bag of this fertilizer, how many grams of the phosphorus-containing compound are you putting on your lawn? If the phosphorus-containing compound consists of 43.64% phosphorus (the rest is oxygen), how many grams of phosphorus are there in 40.0 pounds of fertilizer?

A bag of lawn fertilizer. *(C.D. Winters)*

62. The aluminum in a package containing 75 square feet of kitchen foil weighs approximately 12 ounces. Aluminum has a density of 2.70 g/cm^3. What is the approximate thickness of the aluminum foil in millimeters? (1 ounce = 28.4 g)

A package of aluminum foil. *(C.D. Winters)*

63. The fluoridation of city water supplies has been practiced in the United States for several decades, because it is believed that fluoride prevents tooth decay, especially in young children. This is done by continuously adding sodium fluoride to water as it comes from a reservoir. Assume you live in a medium-sized city of 150,000 people and that each person uses 175 gallons of water per day. How many tons of sodium fluoride would you have to add to the water supply each year (365 days) in order to have the required fluoride concentration of 1 part per million (that is, 1 ton of fluoride per million tons of water)? (Sodium fluoride is 45.0% fluoride, and one U.S. gallon of water has a mass of 8.34 pounds.)

64. Copper has a density of 8.94 g/cm^3. If a factory has an ingot of copper that has a mass of 125 pounds, and the ingot is drawn into wire with a diameter of 9.50 mm, how many feet of wire can be produced?

65. Measure the length of your foot in inches. What is the length in centimeters? In meters?

66. Which occupies a larger volume, 600 g of water (with a density of 0.995 g/cm^3) or 600 g of lead (with a density of 11.34 g/cm^3)?

67. You can identify a metal by carefully determining its density. An unknown piece of metal, with a mass of 29.454 g, is 2.35 cm long, 1.34 cm wide, and 1.05 cm thick. Which of the following is the element?

(a) nickel, 8.91 g/cm^3
(b) titanium, 4.50 g/cm^3
(c) zinc, 7.14 g/cm^3
(d) tin, 7.23 g/cm^3

68. About two centuries ago, Benjamin Franklin showed that 1 teaspoon of oil would cover the surface of about 0.5 acre of still water. If you know that 1.0×10^4 m^2 = 2.47 acres, and that there are approximately 5 cm^3 in a teaspoon, what would be the thickness of the layer of oil? How might this thickness be related to the sizes of molecules?

69. Lead is sold as tiny spherical beads called "shot." What is the average diameter (in millimeters) of a piece of lead shot if 25 beads have a total mass of 2.31 g? (Recall that the volume of a sphere is given by the formula $(4/3)(\pi r^3)$ where r is the radius of the sphere. The density of lead is 11.3 g/cm^3.) Outline another, simple way to find the diameter.

70. What SI base unit, accompanied by what prefix, would be most convenient for describing each of the quantities that follows?
(a) the mass of the earth
(b) the distance from New York City to London, England
(c) the mass of a single atom
(d) the mass of an orange
(e) the thickness of a human hair

71. It has been proposed that dinosaurs and many other organisms became extinct 65 million years ago because the earth was struck by a large asteroid. (See *Science*, August 14, 1992.) The idea is that dust from the impact was lofted into the upper atmosphere all around the globe and blocked the sunlight reaching the earth's surface. On the dark and cold earth that temporarily resulted, many forms of life become extinct. Available evidence suggests that about 20% of the asteroid's mass ended up as dust spread uniformly over the earth after eventually settling out of the upper atmosphere. This dust amounted to about 0.02 g/cm^2 of the earth's surface. The asteroid likely had a density of about 2 g/cm^3. What was the mass of the asteroid? If the asteroid is considered to have been spherical, what was its diameter? (The earth has a surface area of 5.1×10^{14} m^2.) (From *Consider a Spherical Cow—A Course in Environmental Problem Solving* by J. Harte, University Science Books, Mill Valley, CA 1988. Used with permission.)

72. The coatings of M&M™ candies contain several dyes (food colorings) that are water soluble. They are Red #3, Red #40, Yellow #5, and Blue #1. (Red #3 and Red #40 differ in that one of them is more orange than the other.) Each of the colors on the surface of the M&Ms (yellow, orange, red, green, brown, and tan) is made up of one of the dyes or a combination of two or more dyes. It has been reported that some people are allergic to Yellow #5. Suppose you know such a per-

son, and that person likes M&Ms. Presumably you would tell her not to eat yellow M&Ms, since these almost certainly contain the yellow dyes. However, since M&Ms of other colors might also contain Yellow #5, you would like to find out which ones are safe for your friend to eat. Outline a procedure for a simple experiment you could do to find out what dyes are in each kind of M&Ms. Your experiment should not involve expensive lab equipment, but rather should be done with the things your friend could find around the house.

73. A dry black powder is placed in a clean, dry glass tube. Pure hydrogen gas ($H_2(g)$) is passed down the tube in contact with the powder and drives out all the air. With $H_2(g)$ still flowing, the powder is heated. The powder turns red, and water vapor can be detected coming out the end of the tube. Was the original black powder an element or a compound? Explain your reasoning.
74. Make a drawing, based on the kinetic-molecular theory and the ideas about atoms and molecules presented in this chapter, of a submicroscopic model of the arrangement of particles in each of the cases listed below. For each case draw ten submicroscopic particles of each substance. It is OK for your diagram to be two-dimensional—it need not be fancy. Represent each atom as a circle and distinguish each different kind of atom by shading its circle.
    (a) A sample of solid aluminum (which consists of aluminum atoms).
    (b) A sample of liquid water (which consists of H_2O molecules).
    (c) A sample of pure water vapor.
    (d) A homogeneous mixture of water vapor and helium gas (which consists of He atoms).
    (e) A heterogeneous mixture consisting of liquid water and solid aluminum; show a region of the sample that includes both substances.
    (f) A sample of brass, which is a homogeneous mixture of copper and zinc.

Using Computer Programs and Videodiscs

As outlined in the Preface, a computer program—*KC? Discoverer*—and two videodiscs—*The Periodic Table* and *Chemical Demonstrations*—are available to be used with this book. There are questions at the end of each chapter that make use of these materials as a way to further explore the chemical and physical principles outlined in the chapter.

75. Use *KC? Discoverer* to find elements that fulfill the following criteria at room temperature:
    (a) An element that is a gas.
    (b) An element that is a liquid.
    (c) An element that is a solid.
    (d) An element that is yellow.
76. Find the image of bromine on the *Periodic Table Videodisc.* (See frame number 14228.) What qualitative observations can you make about this element? What are some uses of this element? (See frame number 36735.)
77. Find the images of the reaction of magnesium and air on the *Periodic Table Videodisc.* (See frames 7535–7856.)
    (a) Is magnesium a solid, liquid, or gas? What is its color?
    (b) Is this a chemical or physical change? What qualitative observations can you make about the changes you see? What happens to the magnesium as it burns in air?
78. The *Chemical Demonstrations* videodisc describes the formation of fog (frames 26635–29530 on Disc 2, Side 2). (See Demonstration 40 on the videotape version.)
    (a) What observations can you make as you watch this demonstration? Are you seeing chemical or physical changes?
    (b) The demonstrator adds chunks of Dry Ice to hot water. What is the chemical name of Dry Ice? What is the temperature of Dry Ice in Celsius degrees?
    (c) Why does the fog "sink"? What does this tell you about the density of cold air?

When elemental sulfur, a yellow solid, is heated it turns to red, plastic sulfur.
(C.D. Winters)

Chemical Elements

CHAPTER

3

A sample of matter can be classified according to its properties and composition as either a mixture or a pure substance. Of the more than 12 million substances, only about 100 are elements—the building blocks of all the others. At the submicroscopic level elements consist of atoms, and so an understanding of the properties of atoms provides a variety of insights into the properties and behavior of the elements. Atoms are made up of even smaller particles (protons, electrons, neutrons), and the masses of atoms differ from one element to another. When the elements are listed in order of increasing number of protons in their atoms, there is a repeating pattern of similar properties. This is the basis for the periodic table, which summarizes a great deal of information about the chemical behavior of the elements.

Hydrogen is the most abundant element in the universe, and there is an enormous amount of it at the surface of the earth, mostly in water in the oceans. Nevertheless, it was not until 1660 that Robert Boyle (1627–1691) prepared hydrogen by the reaction of iron with oil of vitriol (sulfuric acid), and not until 1766 that Henry Cavendish (1731–1810) separated a pure sample and distinguished it from other gases, calling it "inflammable air." Cavendish found that hydrogen did not dissolve in water, had a smaller density than any other gas, formed an explosive mixture with normal air, and produced water when burned. Cavendish did not claim to have discovered a new element, but shortly after he reported his experiments hydrogen was recognized as an element and given its name, which means "water former."

An early use of hydrogen depended on its low density. In 1783 Jacques Charles filled a large balloon with hydrogen and flew over the French countryside in a basket suspended from the balloon. This use carried over to World War I, when hydrogen-filled balloons carried observers aloft to report on troop movements at the battle front. In 1928 Germany built the *Graf Zeppelin,* a rigid airship that was lighter than air because it was filled with hydrogen. It carried more than 13,000 people between Germany and the United States until 1937, when it was replaced by the *Hindenburg.* The *Hindenburg* was designed to be filled with another gaseous element, helium, which does not react with air and presents no explosion hazard, but Germany had no source of helium. While landing at Lakehurst, New Jersey in

Portrait of a Scientist

Henry Cavendish (1731–1810)

Cavendish was an English chemist and physicist. He was the first to distinguish hydrogen from other gases (1776) and independently discovered nitrogen. He also established the composition of water (1784) and synthesized nitric acid (1784).

One of Cavendish's greatest contributions to chemistry was his reliance on precise experiments to develop his ideas, something not done by earlier scientists. Not only did he apply this to chemistry, but to other fields as well. Based on a simple experiment, for example, Cavendish determined the density of the earth. His value, 5.48 times greater than the density of water, proved that the earth was solid. Further, since ordinary rocks on the earth's surface are not that dense, this showed that the earth must have a very dense core (which we know now to be largely molten iron).

Although he was a man of great wealth and social position, Cavendish devoted himself almost totally to scientific studies. In fact, he so disliked wasting time and words that he worked out a system of signals with his servants so that they did not need to speak. He also lived a very regulated and reclusive life, and so the sketch of Cavendish shown here is a copy of the only picture of him in existence; it was made by an artist when Cavendish was not aware of his presence.

The great English scientist Sir Humphry Davy (1778–1829) said of Cavendish that "His name will be an immortal honor to his house, his age and to his country."

Henry Cavendish. (Oesper Collection in the History of Chemistry, University of Cincinnati)

Figure 3.1
The disastrous end of the hydrogen-filled dirigible Hindenburg in May 1937. The accident occurred on landing in Lakehurst, New Jersey. *(The Bettmann Archive)*

May, 1937, the airship exploded and burned (Figure 3.1), and hydrogen gas earned a reputation as a very dangerous substance.

Actually, with proper techniques, hydrogen can be handled safely in large quantities. It is the principal fuel in the space shuttle, for example, and until about 1950, when pipelines brought natural gas to homes throughout the United States, hydrogen was a major component of the gas used in kitchen ranges. Currently the most important use of hydrogen is in manufacturing ammonia, which is used as fertilizer. Approximately three million tons of hydrogen are combined with nitrogen each year in the United States to make this valuable chemical. Hydrogen is also used to manufacture methyl alcohol, which is used as a gasoline additive and in windshield washer antifreeze, and to hydrogenate oils such as those found in peanut butter. Most of the hydrogen we use today is obtained from hydrocarbons—compounds of hydrogen and carbon.

The uses of hydrogen depend on its physical and chemical properties, and on how much it costs to manufacture the pure element by separating it from compounds in which it occurs. Those physical and chemical properties can be interpreted in terms of the properties of the atoms of hydrogen. For example, hydrogen's low density is understandable when it is known that its atoms are the lightest of all atoms. We turn now to a discussion of atoms and their structure as the basis for understanding the properties of the elements.

3.1 ORIGINS OF ATOMIC THEORY

The Greek Philosophers

Chapter 2 presented a picture of atoms that was not very detailed. Tiny spheres in constant motion were adequate to explain the macroscopic properties (such as differences among solids, liquids, and gases) that were described. This simple picture of atoms goes back a long way in history—to the Greek philosopher Leucippus and his student, Democritus (460–370 BC). Democritus reasoned that if a piece of matter such as gold were divided into smaller and smaller pieces, one would ultimately arrive at a tiny particle of gold that could not be divided further but would still retain the properties of gold. He used the word **atom**, which literally means "uncuttable," to describe this undividable, ultimate particle of matter. According to Aristotle (384–322 BC), Democritus taught that atoms are hard, move spontaneously, and link to one another by some hook-and-eye connection. Epicurus (341–270 BC) attributed mass to atoms, and about 100 BC Asklepiades introduced the idea of clusters of atoms, corresponding to what we would now call molecules.

Democritus used his concept of atoms to explain the properties of substances. For example, the high density and softness of lead could be interpreted if lead atoms were packed very closely together like marbles in a box and moved easily past one another. Iron, on the other hand, was less dense and harder than lead, and Democritus argued that iron atoms might be shaped like corkscrews so that they would entangle in a rigid but relatively lightweight structure. Democritus was able to explain in a simple way other well-known phenomena, such as drying of clothes, the appearance of moisture on the outside of a vessel of cold water, how an odor moves through a room, and how crystals grow from a solution. He imagined scattering or collecting together of atoms as needed to explain macroscopic events. All atomic theory has been built on the assumption of Leucippus and Democritus: the properties of matter that we can see are explained by the properties and behavior of atoms that we cannot see.

Jacob Bronowski, in a television series and book titled The Ascent of Man, *had this to say about the importance of imagination: "There are many gifts that are unique in man; but at the center of them all, the root from which all knowledge grows, lies the ability to draw conclusions from what we see to what we do not see."*

Plato (427–347 BC) and Aristotle argued against atoms, and their ideas prevailed. For centuries most of those in the mainstream of enlightened thought rejected or remained ignorant of the atomic theory proposed by Democritus, though a few well-known scientists did refer to atoms. Galileo Galilei (1564–1624) reasoned that the appearance of a new substance through chemical change involved a rearrangement of parts too small to be seen, and Francis Bacon (1561–1626) speculated that heat might be a form of motion of small particles. Robert Boyle (1627–1691) and Isaac Newton (1642–1727) used atomic concepts to interpret physical phenomena. However, none of these people provided detailed, quantitative explanations of physical and chemical facts in terms of atomic theory.

EXERCISE 3.1 • *Atomic Theory*

Use the idea that matter consists of atoms or molecules to interpret each observation. Describe what the atoms or molecules are doing and how that explains what happens. (a) Wet clothes hung on a line eventually become dry. (b) Moisture appears on the outside of a glass of ice water. (c) Crystals of solid sugar dissolve in water. (d) Sugar dissolves faster in hot water than in cold water.

John Dalton and His Atomic Theory

In 1803 John Dalton (1766–1844) forcefully revived the idea of atoms. Dalton linked the existence of elements, which cannot be decomposed chemically, to the idea of atoms, which are indivisible. Compounds, which can be broken down into two or more new substances, must contain two or more different kinds of atoms. Dalton went further to say that each kind of atom must have its own properties—in particular, a characteristic mass. This idea allowed his theory to account quantitatively for the masses of different elements that combine chemically to form compounds. Thus, unlike earlier ideas about atoms, Dalton's ideas could be used to interpret known chemical facts, and to do so quantitatively.

The postulates of Dalton's atomic theory are

- All matter is made of atoms. These indivisible and indestructible objects are the ultimate chemical particles.
- All atoms of a given element are identical, both in mass and in properties. Atoms of different elements have different masses and different properties.
- Compounds are formed by combination of two or more different kinds of atoms. Atoms combine in the ratio of small whole numbers, for example, one atom of A with one atom of B, or two atoms of A with one atom of B.
- Atoms are the units of chemical change. A chemical reaction involves only combination, separation, or rearrangement of atoms, but atoms are not created, destroyed, divided into parts, or converted into other kinds of atoms during a chemical reaction.

John Dalton's ideas were accepted by the scientific community because they could be used to explain several general rules or scientific laws that were already known when he proposed the atomic theory. Some years earlier Antoine Lavoisier (1743–1794) had carried out a series of experiments in which the reactants were carefully weighed before a chemical reaction and the products were carefully weighed afterward. He found no change in mass when a reaction occurred, proposed that this was true for every reaction, and called his proposal the **law of conservation of matter.** Others verified his results, and the law became accepted. Dalton's second and fourth postulates imply the same thing. If each kind of atom has a particular characteristic mass, and if there are exactly the same number of each kind of atom before and after a reaction, the masses before and after must also be the same. Consequently, Dalton's theory is able to explain the quantitative chemical fact of conservation of matter.

The contributions of Lavoisier, the great 18th century French chemist, are discussed in Chapter 5.

Another chemical law known in Dalton's time had been proposed by Joseph Louis Proust (1754–1826) as a result of his analyses of minerals. Proust found that a particular compound, once purified, always contained the same elements in the same ratio by mass. As an example, one of the compounds formed from carbon and oxygen always has one third more oxygen than carbon. This was called the **law of constant composition** or the **law of definite proportions,** because a given compound always has the same proportion of each of its constituent elements. Dalton suggested that the carbon-oxygen compound just described always had one third greater mass of oxygen than carbon because the compound contained one carbon atom for each oxygen atom, giving the formula CO, and the mass of one oxygen atom was one third greater than the mass of a carbon atom. Using this kind of reasoning, Dalton was able to determine which atoms were

Portrait of a Scientist

John Dalton (1766–1844)

John Dalton. (Oesper Collection in the History of Chemistry, University of Cincinnati)

John Dalton was born about the fifth of September, 1766, in the village of Eaglesfield in Cumberland, England. His family was quite poor, and his formal schooling ended at age 11. However, he was clearly a bright young man, and with the help of influential patrons he began a teaching career at the age of 12. Shortly thereafter he made his first attempts at scientific investigation, observations of the weather. This was a study that was to last his lifetime. In fact, he made more than 200,000 observations of weather conditions by the end of his life.

In 1793 Dalton moved to Manchester, England, where he took up a post as tutor at the New College, but he left there in 1799 to pursue scientific inquiry on a full-time basis. It was not long afterward, on October 21, 1803, that he read a paper introducing his "Chemical Atomic Theory" to the Literary and Philosophical Society of Manchester. His presentation was followed by lectures in London and in other cities in England and Scotland, and his reputation as a scientist rapidly increased. He was first proposed for membership in the top scientific society in Britain, the Royal Society, in 1810, and many other honors followed over his lifetime.

---

heavier than others and how much of one element would be expected to combine with another element in a compound.

Dalton's theory was valuable because it could explain existing facts, but Dalton went further and proposed a new law on the basis of his theory and a few experiments. To see how his reasoning worked, suppose that another compound can form in which there are two oxygen atoms per carbon atom; that is, the molecules have the formula CO_2. This compound will have twice as great a mass of oxygen per gram of carbon as did the compound CO, because CO_2 has twice as many oxygen atoms per carbon atom. If an oxygen atom weighs one third more than a carbon atom, then in CO the ratio (mass of O)/(mass of C) would be $1.33/1 = 1.33$; in the compound CO_2, with 2 O atoms per C atom, this ratio would be $2 \times 1.33/1 = 2.66$, which is what the proportions turned out to be when they were measured. Dalton stated that when two elements form two different compounds, the mass ratio in one compound will be a small whole number times the mass ratio in the other. (In the case of CO and CO_2, the small whole number is 2.) This law is called the **law of multiple proportions.**

This aspect of Dalton's atomic theory—that it suggested a new law—stimulated Dalton and his contemporaries to do a great deal more research, thereby contributing to scientific progress. A successful theory not only accounts for existing knowledge but also stimulates the search for new knowledge. Though it was not until the 1860s that a consistent set of relative masses of the atoms was agreed upon, Dalton's idea that the masses of atoms are crucial to quantitative chemistry was accepted from the early 1800s on.

3.2 ATOMIC STRUCTURE

Dalton's atomic theory said nothing about one of Democritus' ideas—the idea that atoms have structure. Knowledge of atomic structure is important because it gives us insights into how and why atoms bond together to form molecules. Though we now disagree with Democritus' idea that atoms were held together by hooks and eyes, his fundamental notion that we can better understand how elements behave if we know about the structures of atoms turned out to be correct. The next few sections describe how people arrived at our current understanding of the structure of atoms.

Evidence that Atoms Are Divisible

Electricity is involved in many of the experiments from which the theory of atomic structure was derived. Electric charge was first observed and recorded by the ancient Egyptians, who noted that amber, when rubbed with wool or silk, attracted small objects. You can observe the same thing when you comb your hair on a dry day—your hair is attracted to the comb. A bolt of lightning or a shock upon touching a doorknob results from an electric charge moving from one place to another. Two types of electric charge had been discovered by the time of Benjamin Franklin (1706–1790). He named them positive (+) and negative (−), because they appear as opposites and can neutralize each other. Experiments with an electroscope (Figure 3.2) show that *like charges repel* one another and *unlike charges attract* one another.

(a)

(b)

Figure 3.2
An electroscope demonstrates electric charge. (a) With no electrical influence, the foil leaf hangs straight down. (b) A rubber rod that has been rubbed with fur is brought up to the bulb of the electroscope. The electric charge that has built up on the rod flows onto the electroscope, and the movable leaf diverges from the stationary leaf. The reason for this observation is that the same charge has flowed onto both leaves. Since like charges repel, the leaves repel one another. *(C.D. Winters)*

Franklin also concluded that charge is conserved: if a negative charge appears somewhere, a positive charge of the same size must appear somewhere else. The fact that a charge builds up when one substance is rubbed over another implies that the rubbing separates positive and negative charges. Apparently positive and negative charges are somehow associated with matter—perhaps with atoms.

Radioactivity

In 1896 Henri Becquerel (1852–1908) discovered that a uranium ore emitted rays that could expose a photographic plate, even though the plate was covered by black paper to protect it from being exposed by light rays. In 1898 Marie Curie and co-workers isolated polonium and radium, which also emitted the same kind of rays, and in 1899 she suggested that atoms of radioactive substances disintegrate when they emit these unusual rays. She named this phenomenon **radioactivity.** Whether in a compound or uncombined, a radioactive element gives off exactly the same rays; about 25 elements exist only in radioactive forms.

Radioactive elements spontaneously emit three kinds of radiation: alpha, beta, and gamma rays. These behave differently when passed between electrically charged plates, as shown in Figure 3.3. Alpha and beta rays are deflected while gamma rays pass straight through. This implies that alpha and beta rays are electrically charged particles, since particles with a charge would be attracted or repelled by the charged plates. Even though an α particle has an electrical charge (+2) twice as large as a β particle (−1), α particles are deflected less; hence α particles must be heavier than β particles. Gamma rays have no detectable charge or mass—they behave like light rays.

Marie Curie's suggestion that atoms disintegrate contradicts Dalton's idea that atoms are indivisible, and requires an extension of Dalton's theory.

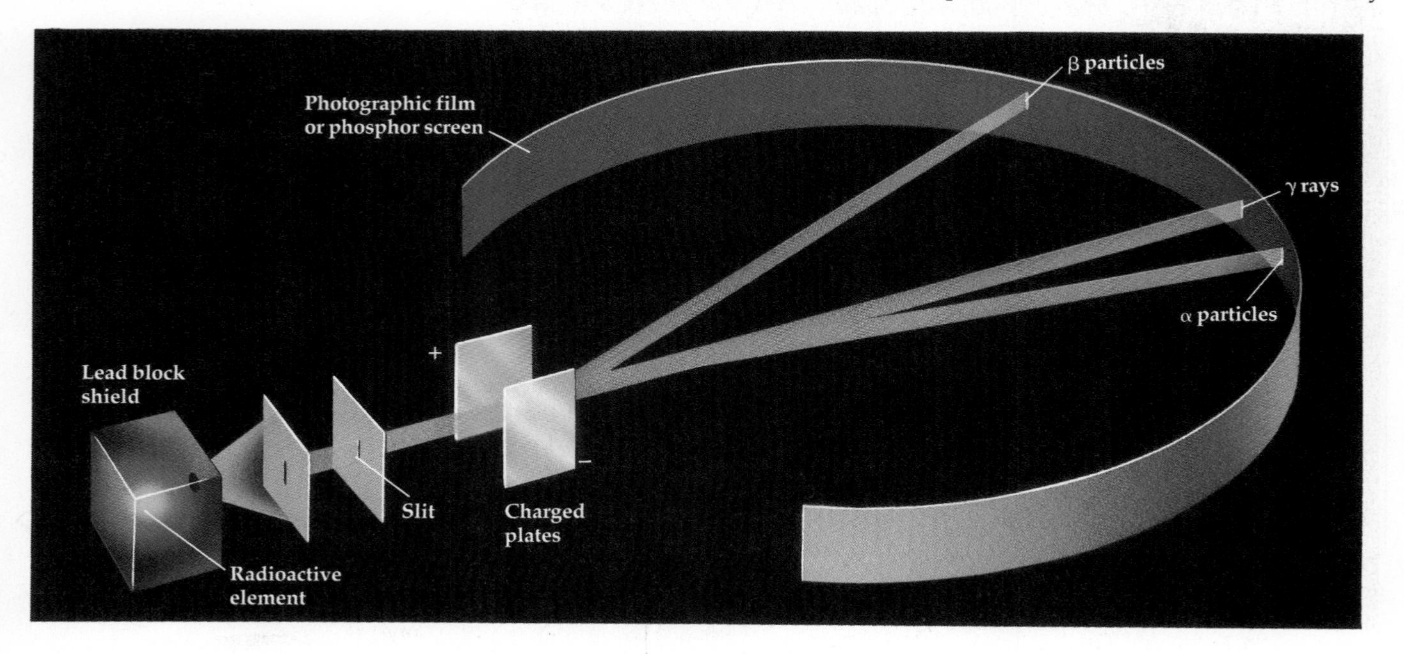

Figure 3.3
Separation of α, β, and γ rays from a radioactive element by an electrical field. Positively charged α particles are attracted to the negative plate, while the negative β particles are attracted to the positive plate. (Note that the heavier α particles are deflected less than the β particles.) Gamma rays have no electric charge and pass undeflected between the charged plates.

If atoms can break apart, there must be something smaller than an atom; that is, atomic structure must involve subatomic particles.

The Electron

Electrolysis is a common technique for separating metals from their compounds. An electric current causes a chemical reaction, such as plating of a metal from a compound onto a wire. In 1833 Michael Faraday (1791–1867) showed that the same current caused different quantities of different metals to deposit, and that those quantities were related to the relative masses of the atoms of those elements. Such experiments were interpreted to mean that, just as an atom is the fundamental particle of an element, there must be a fundamental particle of electricity. This "atom" of electricity was given the name **electron.**

Chlorine is made by the electrolysis of salt in water. See Figure 2.21.

Further evidence that atoms are composed of smaller particles came from experiments with glass tubes from which most of the air had been removed and which had a piece of metal called an electrode sealed into each end (Figure 3.4). When a sufficiently high voltage is applied to the electrodes, a beam of cathode rays flows from the negatively charged elec-

The deflection of cathode rays (an electron beam) by charged plates is used to paint the picture on a television picture tube or a computer's CRT (cathode-ray tube) screen. Neon and fluorescent lights also involve cathode rays.

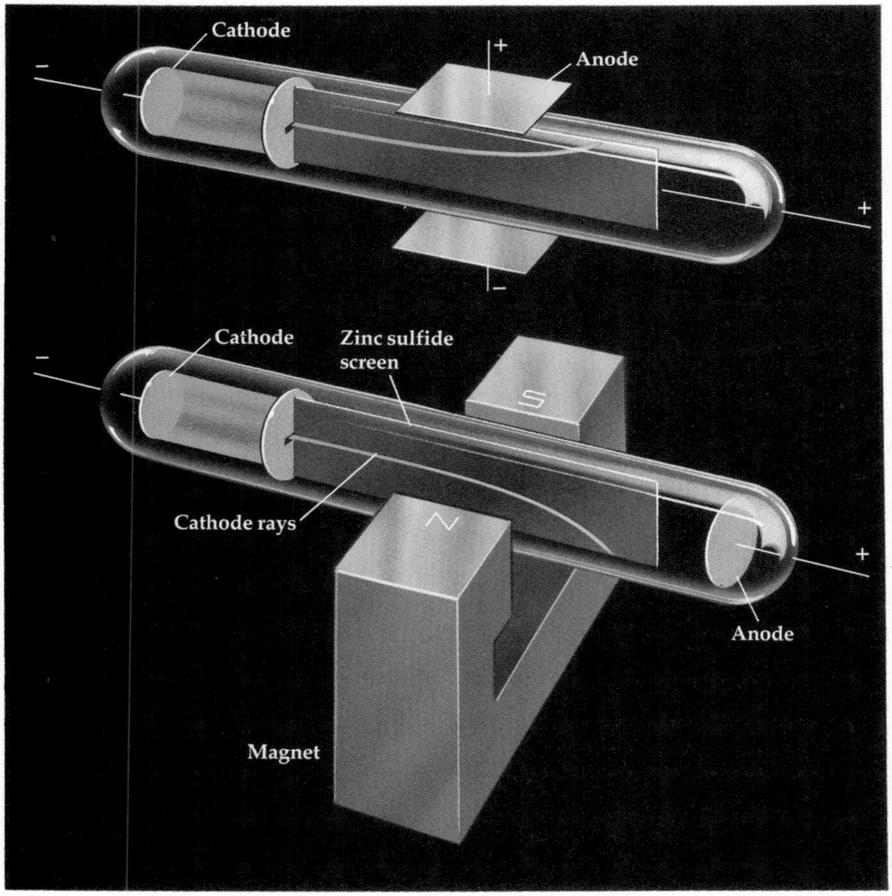

Figure 3.4

Deflection of a cathode ray by an electric field (*top*) and by a magnetic field (*bottom*). When an external electric field is applied, the cathode ray is deflected toward the positive pole. When a magnetic field is applied, the cathode ray is deflected from its normal straight path into a curved path. In both cases, the curvature is related to the mass of the particles of the cathode rays and the magnitude of the field.

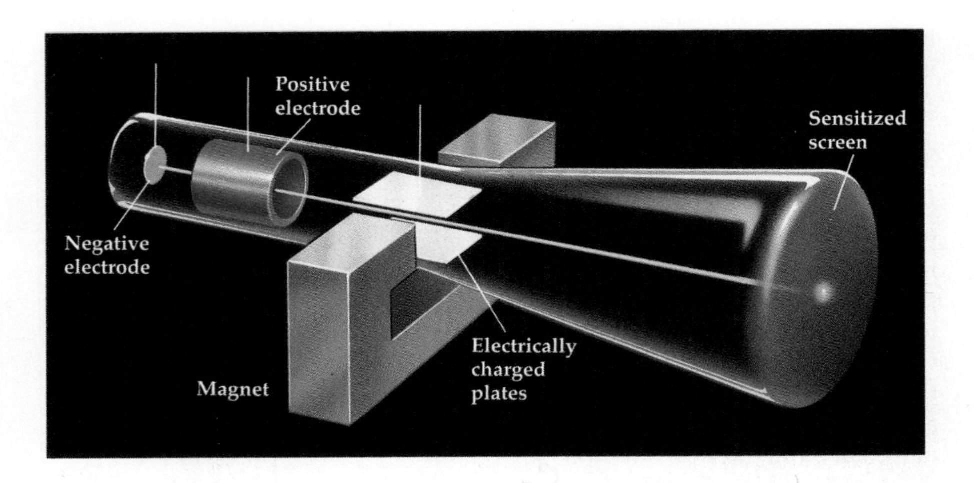

Figure 3.5
Thomson's experiment to measure the charge/mass ratio of the electron. A beam of electrons (cathode rays) passes through an electric field and a magnetic field. The experiment is arranged so that the electric field causes the beam to be deflected in one direction, while the magnetic field deflects the beam in the opposite direction. By balancing the effects of these fields, the charge/mass ratio of the electron can be determined.

trode (cathode) to the positive electrode (anode). Cathode rays travel in straight lines, cast sharp shadows, cause gases and fluorescent materials to glow, can heat metal objects red hot, can be deflected by a magnetic field, and are attracted toward positively charged plates. When cathode rays strike a fluorescent screen, light is given off in a series of tiny flashes. Thus a cathode ray appears to consist of a beam of negatively charged particles, each one of which produces a flash of light when it hits a fluorescent screen.

In 1897 Sir Joseph John Thomson (1856–1940) used a specially designed cathode-ray tube (Figure 3.5) to apply electric and magnetic fields simultaneously to a beam of cathode rays. By balancing the effect of the electric field against that of the magnetic field and using basic laws of electricity and magnetism, Thomson was able to calculate the ratio of charge to mass for the particles in the beam. However, he was not able to determine either charge or mass independently. On the basis of the fact that cathode rays are beams of negatively charged particles, Thomson suggested that they are the same as the electrons associated with Faraday's experiments. He obtained the same charge-to-mass ratio in experiments with 20 different metals and several different gases in the cathode-ray tube. These results suggested that electrons are present in all kinds of matter and that they presumably exist in atoms of all elements.

It remained for Robert Andrews Millikan (1868–1953) to measure the charge of an electron and thereby enable calculation of its mass. His apparatus is shown schematically in Figure 3.6. Tiny droplets of oil were sprayed into a chamber. As they settled slowly through the air, they were exposed to x rays, which caused air molecules to transfer electrons to them. Millikan used a small telescope to observe individual droplets, and adjusted the electric charge on plates above and below the droplets until the electrostatic attraction just balanced the force of gravity, causing a droplet to be suspended, motionless. From the equations describing these forces, Millikan calculated the charge on the droplet. Different droplets had different charges, but Millikan found that each was a whole-number multiple of the same smaller charge. That smaller charge was 1.60×10^{-19} C. Millikan assumed this to be the fundamental unit of charge, the charge on an electron. From this the mass of an electron could be calculated. The currently ac-

The coulomb, abbreviated C, is the SI unit of electric charge.

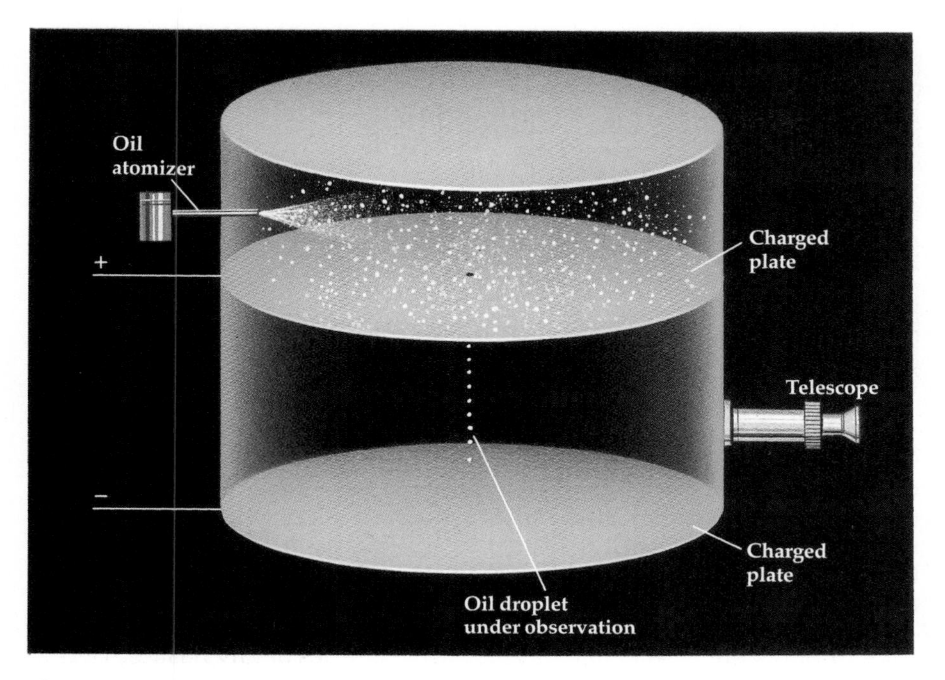

Figure 3.6
Millikan oil drop experiment. A fine mist of oil droplets is introduced into the chamber. The gas molecules in the chamber are ionized (split into electrons and positive ions) by a beam of x rays. The electrons adhere to the oil droplets, some droplets having one electron, some two electrons, and so forth. These negatively charged oil droplets fall under the force of gravity into the region between the electrically charged plates. By carefully adjusting the voltage on the plates, the force of gravity is exactly counterbalanced by the attraction of the negative oil drop to the upper, positively charged plate. Analysis of these forces leads to a value for the charge on the electron.

cepted value is 9.109389×10^{-28} g, and the currently accepted value of the electron's charge is $-1.60217733 \times 10^{-19}$ C. When referring to atoms and molecules, this much charge is represented as -1.

Additional experiments showed that cathode rays have the same properties as the beta particles emitted by radioactive elements, providing further evidence that the electron is a fundamental particle of matter.

Protons

The first experimental evidence of a fundamental positive particle came from the study of so-called canal rays (Figure 3.7), which were observed in a special cathode-ray tube with a perforated cathode. When high voltage is applied to the tube, cathode rays can be observed as in any cathode-ray tube. However, on the other side of the perforated cathode, a different kind of ray is observed. Since these rays are attracted toward a negatively charged plate, they must be composed of positively charged particles.

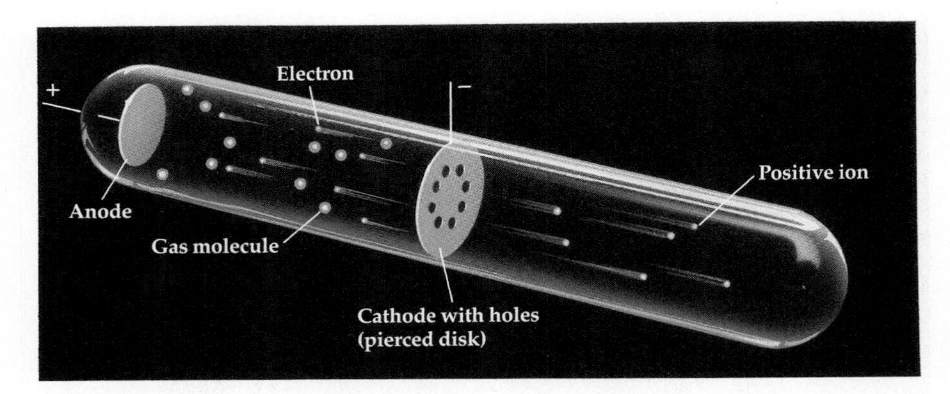

Figure 3.7
Cathode-ray tube with perforated cathode. Electrons collide with gas molecules and produce positive ions, which are attracted to the negative cathode. Some of the positive ions pass through the holes and form a positive ray. Like cathode rays, positive rays (or "canal rays") are deflected by electric and magnetic fields, but much less so for a given value of the field because positive ions are much heavier than electrons.

Each gas used in the tube gives a different charge-to-mass ratio for the positively charged particles (unlike the cathode rays, which are the same no matter what the gas is). When the tube contains hydrogen, the largest charge-to-mass ratio is obtained, suggesting that hydrogen provides positive particles with the smallest mass. The emissions from hydrogen were considered to be the fundamental positively charged particles of atomic structure and were called **protons** (from a Greek word meaning "the primary one").

The mass of a proton is known from experiment to be 1.672623×10^{-24} g, about 1800 times the mass of an electron. The charge on the proton ($+1.60217733 \times 10^{-19}$ C) is equal in size but opposite in sign to the charge on the electron; it is designated as +1.

Neutrons

Because atoms normally have no charge, there must be equal numbers of protons and electrons in an atom. Most atoms have masses greater than would be predicted from the sum of the masses of their protons and electrons, indicating that some other particles must be present in the atom. Since this third type of particle has no charge, the usual methods of detecting particles could not be used. Nonetheless, in 1932, many years after the discovery of the proton, James Chadwick (1891–1974) devised a clever experiment that produced these expected neutral particles and then detected them by having them knock hydrogen ions, which are detectable, out of paraffin. It is now known that the fundamental particle called a **neutron** has no electric charge and a mass of $1.6749286 \times 10^{-24}$ g, nearly the same as the mass of a proton.

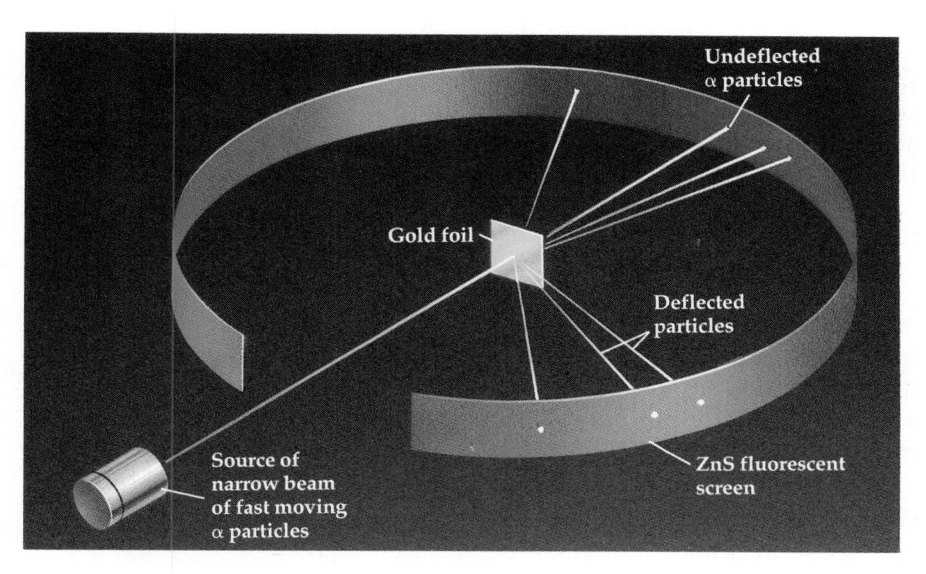

Figure 3.8
The experimental arrangement of the Rutherford experiment. A beam of positively charged α particles was directed at a very thin piece of gold foil. A luminescent screen coated with zinc sulfide (ZnS) was used to detect particles passing through or deflected by the foil. Most particles passed straight through. Some were deflected to some extent, and a few were even deflected backward. (Note that a circular luminescent screen is shown for simplicity; actually a smaller, movable screen was used.)

The Nucleus of the Atom

J. J. Thomson had supposed that an atom was a uniform sphere of positively charged matter within which thousands of electrons circulated in rings. But how many electrons were circulating within this sphere? To try to find an answer, Thomson and his students directed a beam of electrons at a very thin metal foil. As they passed through the foil, the electrons of the beam would encounter the very large number of electrons within the atoms, and the negative charges would repel one another. A tiny deflection of the beam from its straight path should be observed at each encounter, with the size of the total deflection related to the number of electrons in the atom. Thomson did indeed observe a deflection, but it was much smaller than he expected; he was forced to revise his estimate of the number of electrons, but not his model of the atom.

In about 1910 Ernest Rutherford (1871–1937) decided to test Thomson's model further. Rutherford had earlier discovered that alpha rays were positively charged particles having the same mass as helium atoms. He reasoned that, if Thomson's atomic model were correct, a beam of such relatively massive particles would be deflected very little as they passed through the atoms in a very thin sheet of gold foil. Rutherford's associate, Hans Geiger, and a young student, Ernst Marsden, set up the apparatus diagrammed in Figure 3.8 and observed what happened when alpha particles hit the gold

Alpha particles are four times heavier than the lightest atoms, which are hydrogen atoms.

foil. Most passed almost straight through, but Geiger and Marsden were amazed to find that a *very few* alpha particles were deflected through large angles, and some came almost straight back! Rutherford later described this unexpected result by saying, "It was about as credible as if you had fired a 15-inch [artillery] shell at a piece of paper and it came back and hit you."

The only way to account for this was to discard Thomson's model and to conclude that all of the positive charge and most of the mass of the atom is concentrated in a very small volume (Figure 3.9). Rutherford called this tiny core of the atom the **nucleus.** The electrons occupy the rest of the space in the atom. From their results Rutherford, Geiger, and Marsden calculated that the positive charge on the gold nucleus is in the range of 100 ± 20 and that the nucleus has a radius of about 10^{-14} m. The currently accepted values for these results are 79 for the charge and about 10^{-15} m for the radius, which makes the nucleus about 100,000 times smaller than the atom.

The experiments just described can be interpreted in terms of three primary constituents of atoms: protons, electrons, and neutrons. The nucleus or core of the atom contains most of the mass and all of the positive charge; its radius is about 100,000 times smaller than that of the atom itself. Protons and neutrons make up the nucleus. Negatively charged electrons occupy most of the volume of an atom but contribute very little mass. Since an atom has no net electric charge, the number of electrons outside the nucleus must equal the number of protons inside the nucleus. To chemists the electrons are the important part, because they are the first part of an atom that con-

Figure 3.9
Rutherford's interpretation of the results of the experiment done by Geiger and Marsden.

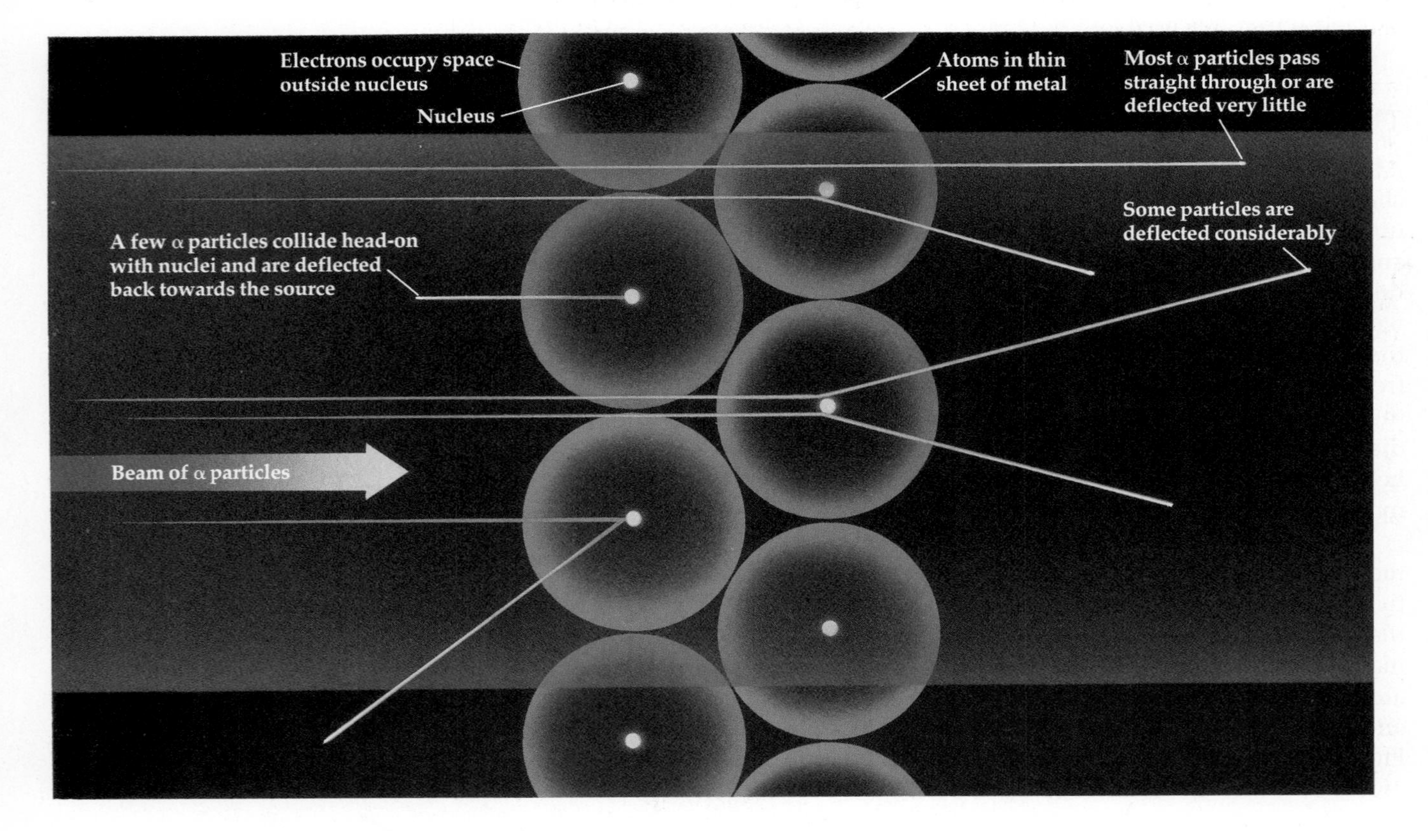

Portrait of a Scientist

Ernest Rutherford (1871–1937)
Lord Rutherford, one of the most interesting people in the history of science, was born in New Zealand in 1871 but went to Cambridge University in England to pursue his Ph.D. in physics in 1895. His original interest was in a phenomenon that we now call radio waves, and he apparently hoped to make his fortune in the field, largely so he could marry his fiancée back in New Zealand. However, his professor at Cambridge, J.J. Thomson, convinced him to work on the newly discovered phenomenon of radioactivity. Rutherford discovered α and β radiation while at Cambridge. In 1899 he moved to McGill University in Canada where he did further experiments to prove that alpha radiation is actually composed of helium nuclei and that beta radiation consists of electrons. For this work he received the Nobel Prize in Chemistry in 1908.

Rutherford was fortunate to have a talented student, Frederick Soddy, work with him at McGill. The two studied a radioactive gas coming from the radioactive element thorium. Their experiments showed that the gas was argon, which meant that they had made the first observation of the spontaneous disintegration of a radioactive element, one of the great discoveries of 20th century physics.

In 1903 Rutherford and his young wife visited Pierre and Marie Curie in Paris, on the very day that Madame Curie received her doctorate in physics (see Chapter 19). That evening during a party in the garden of the Curies' home, Pierre Curie brought out a tube coated with a phosphor and containing a large quantity of radioactive radium in solution. The phosphor glowed brilliantly from the radiation given off by the radium. Rutherford later said the light was so bright that he could clearly see Pierre Curie's hands were "in a very inflamed and painful state due to exposure to radium rays."

In 1907 Rutherford moved from Canada to Manchester University in England, and there he performed the experiments that gave us the modern view of the atom. In 1919 he moved back to Cambridge and assumed the position formerly held by J.J. Thomson. Not only was Rutherford responsible for very important work in physics and chemistry, but he also guided the work of no less than ten future recipients of the Nobel Prize.

Ernest Rutherford. (Oesper Collection in the History of Chemistry, University of Cincinnati)

---

tacts another atom when the two approach one another. It is the electrons that largely control the chemical combination of atoms.

It was obvious that the negative electrons occupied the space outside the nucleus, but their arrangement was completely unknown to Rutherford or other physicists of his time. This arrangement is now known and is the subject of Chapter 9.

EXERCISE 3.2 • *Describing Atoms*

If an atom were a macroscopic object with a radius of 100 meters, it would approximately fill a football stadium. What would be the radius of the nucleus of such an atom? Can you think of an object that is about that size?

3·3 ATOMIC COMPOSITION

The experiments done in the early part of this century clearly showed that the three primary constituents of atoms are **electrons, protons,** and **neutrons,** and that the **nucleus** or core of the atom is made up of protons with

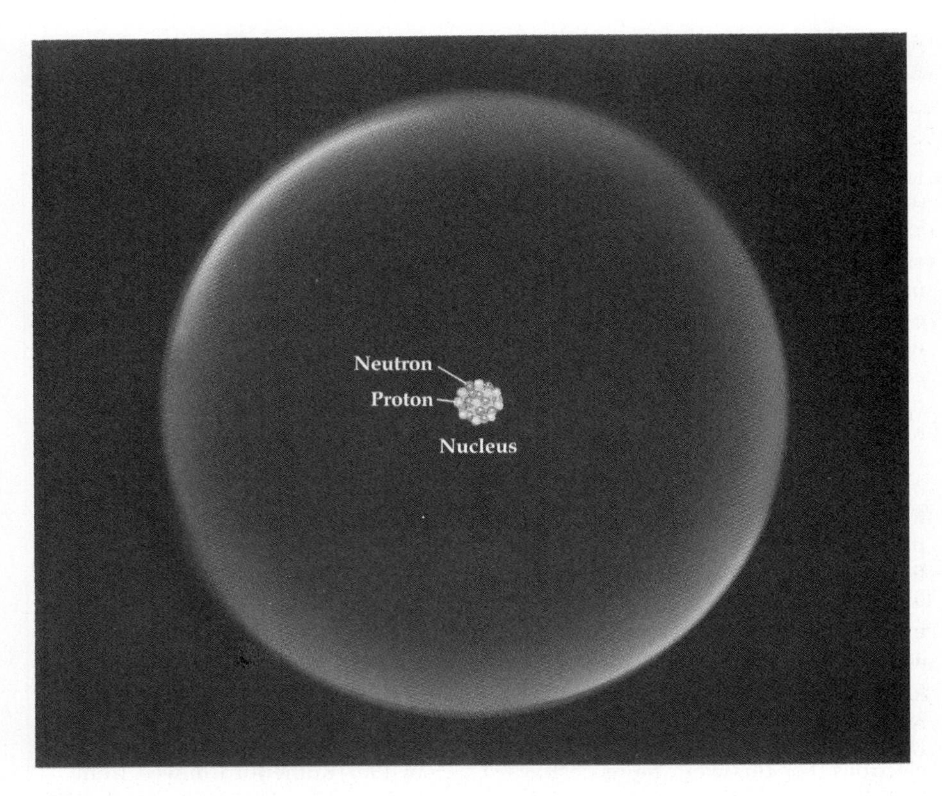

Figure 3.10
All atoms consist of one or more protons (positively charged) and usually at least as many neutrons (no charge) packed into an extremely small nucleus. Electrons (negatively charged) are arranged in space as a "cloud" about the nucleus. In an electrically neutral atom the number of electrons equals the number of protons.

positive electrical charge and neutrons with no charge. The electrons, with a negative electrical charge, are found in the space around the nucleus (Figure 3.10). For an atom, which has no net electrical charge, *the number of negatively charged electrons around the nucleus equals the number of positively charged protons in the nucleus.*

Atoms are extremely small; the radius of the typical atom is between 30 and 300 pm. To get a feeling for the incredible smallness of an atom, consider that

- One teaspoon of water (about 1 cm^3) contains about three times as many atoms as the Atlantic Ocean contains teaspoons of water.
- If, after the flood of about 3000 BC, Noah had started to string hydrogen atoms on a thread at the rate of one atom a second for 8 hours a day, the chain would be only about 1.6 meters (about 5 feet) long today!

All atoms of the same element have the same numbers of protons in the nucleus. This number is called the **atomic number,** and it is given the symbol **Z.** In the periodic table at the front of the book, the atomic number for each element is given above the element's symbol. Sodium, for example, has a nucleus containing 11 protons, so its atomic number is 11, while uranium has 92 nuclear protons and $Z = 92$.

Just as our clocks and longitude are set relative to the time and longitude at Greenwich, England, so a scale of atomic masses is established relative to a standard. This standard is the mass of a carbon atom that has six protons and six neutrons in its nucleus. Such an atom is defined to have a mass of exactly 12 **atomic mass units** (or 12 amu), and the mass of every other element is established relative to this mass. Thus, for example, experiment shows that an oxygen atom is, on the average, 1.33 times heavier than a carbon atom, so an oxygen atom has a mass of 1.33×12.0 amu or 16.0 amu.

1 amu = $\frac{1}{12}$ the mass of a carbon atom having 6 protons and 6 neutrons in the nucleus.

Masses of the basic atomic particles in amu have been determined experimentally (Table 3.1). Notice that the proton and neutron have masses very close to 1 amu, while the electron is nearly 2000 times lighter.

Once we have established a relative scale of atomic masses, we can estimate the mass of any atom for which the nuclear composition is known. The proton and neutron have masses so close to 1 amu that the difference can often be ignored. Electrons are so light that even a large number of them will not greatly affect the mass of the atom. Therefore, we need only to add up the number of protons and neutrons in an atom to estimate its mass. The result is called the **mass number** of that particular atom, a number given the symbol **A.** For example, a sodium atom has 11 protons and 12 neutrons in its nucleus; its mass number, A, is 23. The most common atom of uranium has 92 protons and 146 neutrons, so $A = 238$. With this information, an atom of known composition can be represented by the notation

$$\text{mass number} \longrightarrow \atop \text{atomic number} \longrightarrow \; {}^{A}_{Z}\mathrm{X} \longleftarrow \text{element symbol}$$

(where the subscript Z is optional because the element symbol tells you what the atomic number must be). For example, the sodium atom described above would have the symbol ${}^{23}_{11}\mathrm{Na}$ or just ${}^{23}\mathrm{Na}$. In words, we would say "sodium-23."

An obvious thing to do is to add up all the masses of the protons, neutrons, and electrons of an atom to see whether we obtain the atom's experimental mass. However, this sum is slightly greater than the actual mass. This is not a mistake. Rather, the difference (sometimes called the mass defect*) is related to the energy binding the particles of the nucleus together. More will be said about binding energy in Section 19.3.*

Table 3.1

Properties of Subatomic Particles

Particle	Mass		Charge	Symbol
	Grams	*amu*		
electron	9.109389×10^{-28}	0.0005485799	−1	${}^{0}_{-1}\mathrm{e}$
proton	1.672623×10^{-24}	1.007276	+1	${}^{1}_{1}\mathrm{p}$
neutron	1.674929×10^{-24}	1.008665	0	${}^{1}_{0}\mathrm{n}$

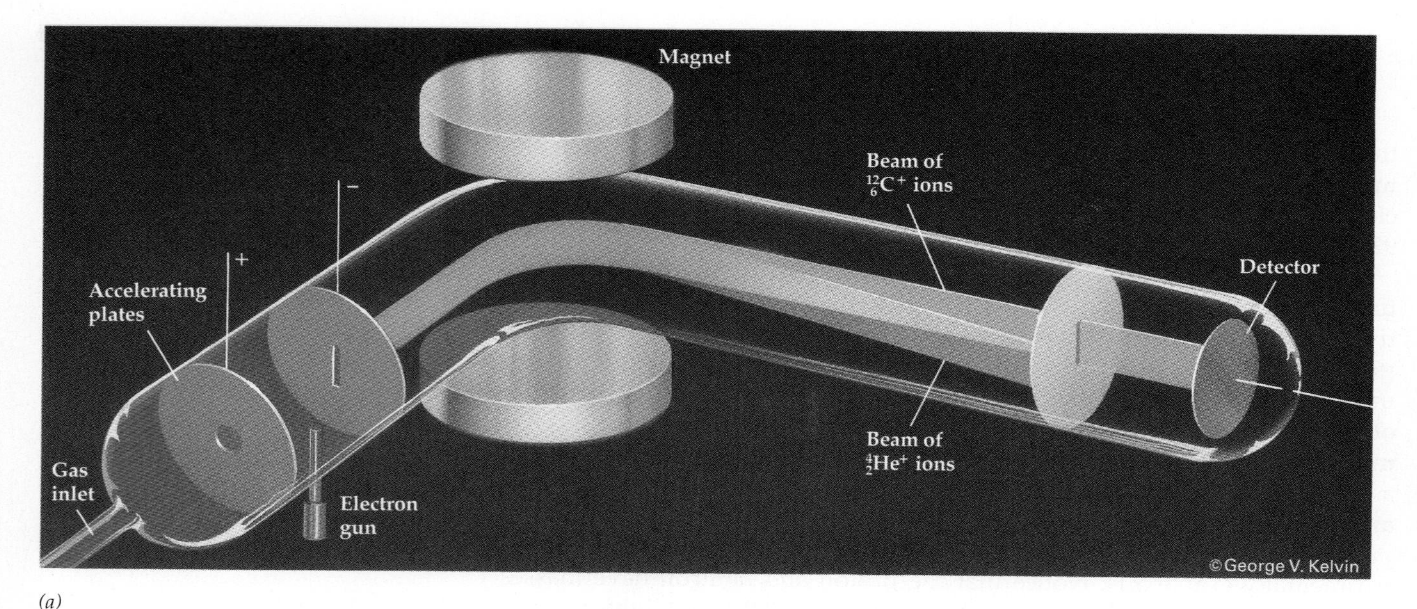

(a)

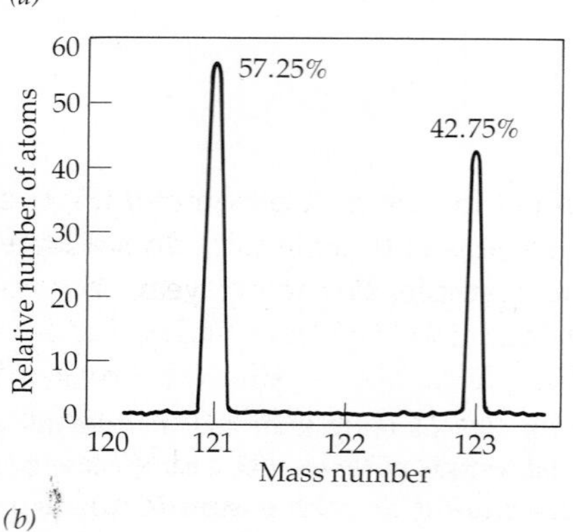

(b)

Figure 3.11
Mass spectrometer and spectrum. (a) A gas sample is injected into an evacuated tube. An electron beam ionizes the gas sample by knocking electrons from the neutral atoms or molecules. Charged plates are arranged to accelerate positive ions toward the first slit and into the rest of the apparatus. Positive ions that pass the first slit move into a magnetic field perpendicular to their path, where they follow a curved path determined by the charge-to-mass ratio of the ion. A detector detects charged particles passing through the second slit. (b) Result of separating the ions formed by the different isotopes of antimony in a mass spectrometer. The principal peak corresponds to the most abundant isotope, antimony-121. Percent relative abundance of the isotopes is shown.

The actual masses of atoms have been determined experimentally by using mass spectrometers (Figure 3.11). It is always observed that, while the actual mass approximately equals the mass number, the actual mass is not an integral number. For example, the actual mass of an iron atom with 32 neutrons is 57.933272 amu, slightly less than the mass number of 58.

EXAMPLE 3.1

Atomic Composition

How many neutrons are there in an atom of platinum with a mass number of 195?

SOLUTION Platinum has the symbol Pt, and its atomic number (shown in the periodic table inside the front cover) is 78. Therefore, the atom has 78 protons in the nucleus (and 78 electrons arranged outside the nucleus). The mass number of the atom is the sum of the numbers of protons and neutrons in the nucleus. Therefore,

$$\text{mass number} = 195 = \text{number of protons} + \text{number of neutrons}$$
$$= 78 + \text{number of neutrons}$$

$$\text{number of neutrons} = 195 - 78 = 117$$

EXERCISE 3.3 • *Atomic Composition*

a. What is the mass number of a copper atom with 34 neutrons?

b. How many protons, neutrons, and electrons are there in a $^{59}_{28}Ni$ atom?

Table 3.2

Masses of the Stable Isotopes of Some Elements

Element	Symbol	Atomic Weight	Mass Number	Isotopic Mass	Percent Isotopic Abundance
hydrogen	H	1.00794	1	1.007825	99.9855
	D		2	2.0141022	0.0145
boron	B	10.811	10	10.012939	19.91
			11	11.009305	80.09
magnesium	Mg	24.305	24	23.985042	78.99
			25	24.985370	10.00
			26	25.982593	11.01

3.4 ISOTOPES

If two atoms differ in the number of protons *they contain, they are different* elements. *If they differ only in the number of* neutrons, *they are* isotopes. *It is the number of protons in the nucleus and the number of electrons outside the nucleus that determine the chemistry of the atom.*

Frederick Soddy, Ernest Rutherford's assistant, coined the word "isotope" to describe the different forms of the same element.

If we examine a natural sample of an element in a mass spectrometer (Figure 3.11) we find, in most cases, that all of the atoms of that element do not have the same mass number. Take boron, for example. For many years boron-containing minerals such as borax have been mined in Death Valley, California. If we examine the boron atoms of these minerals, we find that while all boron atoms have five nuclear protons, some of them have five neutrons and others have six. That is, we would find a collection of $^{10}_{5}B$ and $^{11}_{5}B$ atoms, which we call *isotopes*. **Isotopes** are atoms having the same atomic number, *Z*, but different mass numbers, *A*. In other words, isotopes have different numbers of neutrons. Table 3.2 lists the masses of several common isotopes.

Most elements have at least two stable (nonradioactive) isotopes, but a few have only one isotope (aluminum, fluorine, and phosphorus, for example). Conversely, there are others with many isotopes (tin, for example, has ten stable isotopes). Generally, we refer to a particular isotope by giving its mass number (for example, uranium-238 or ^{238}U), but some isotopes are so important that they have special names and symbols. Hydrogen atoms all have one proton. When that is the only nuclear particle, the element is called simply "hydrogen." When one neutron is also present, the isotope $^{2}_{1}H$ is called deuterium or heavy hydrogen (symbol = D), and the presence of two neutrons gives radioactive hydrogen, $^{3}_{1}H$, or tritium (symbol = T).

EXAMPLE 3.2

Isotopes

Silver has two isotopes, one with 60 neutrons and the other with 62 neutrons. What are the mass numbers and symbols of these isotopes?

Water containing ordinary hydrogen (^1_1H) forms a solid that is less dense than the liquid and so floats in the liquid (*right*). In this form water is very unique, although this is not often recognized since it is so commonly observed. The solid phase of virtually all other substances sinks in the liquid phase of that substance. This is true of water made from deuterium or "heavy hydrogen." Solid D_2O sinks in liquid D_2O (*left*). Changing from one isotope to another of an element can have a significant effect. *(C.D. Winters)*

SOLUTION Silver has an atomic number of 47, so it has 47 protons in the nucleus. Therefore, the two isotopes have mass numbers of

Isotope 1: A = 47 protons + 60 neutrons = 107
Isotope 2: A = 47 protons + 62 neutrons = 109

The first isotope has the symbol $^{107}_{47}Ag$ and the second is $^{109}_{47}Ag$.

EXERCISE 3.4 • *Isotopes*

Silicon has three isotopes with 14, 15, and 16 neutrons, respectively. What are the mass numbers and symbols of these three isotopes?

3.5 ATOMIC WEIGHT

Boron atoms have two different isotopic masses, 10.0129 and 11.0093. This means that in a macroscopic collection of boron atoms the average mass of the atoms will be neither 10 nor 11 but somewhere in between, the actual value depending on the proportion of each isotope.

The percentage of atoms of a particular isotope in a natural sample of an element is called the **percent abundance** of that isotope, and can be found by using a mass spectrometer.

$$\text{percent abundance} = \frac{\text{number of atoms of a given isotope}}{\text{total number of atoms of all isotopes of that element}} \times 100\%$$

Chemists usually use the term atomic weight *of an element rather than "atomic mass." Although the quantity is more properly called a mass than a weight, the term "atomic weight" is so commonly used that it has become accepted.*

Once the percent abundance and isotopic mass of each isotope are known, they can be used to find the average mass of atoms of that element. The boron isotopes, ^{10}B and ^{11}B, have percent abundances of 19.91% and 80.09%, respectively. This means that, if you could count out 10,000 boron atoms from an "average" natural sample, 1991 of them would have a mass of 10.0129 amu and 8009 of them would have a mass of 11.0093 amu. The average mass of a representative sample of atoms, expressed in atomic mass units, is called the **atomic weight.** For boron, the atomic weight is 10.81 amu, as shown by the following calculation, in which the mass of each isotope is multiplied by its percent abundance (expressed as a decimal fraction):

$$\begin{aligned}\text{atomic weight} &= (19.91\%)(10.0129) + (80.09\%)(11.0093)\\ &= (0.1991)(10.0129) + (0.8009)(11.0093) = 10.81 \text{ amu}\end{aligned}$$

The periodic table entry for copper

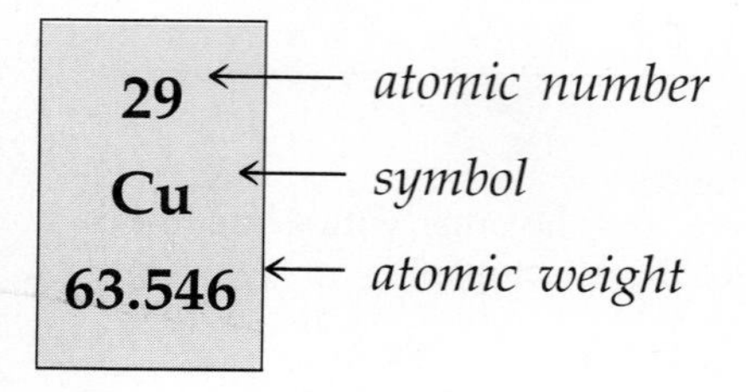

The atomic weight of each stable element has been determined—although not necessarily by the method just shown—and these are the masses that appear in the table in the front of the book. In the periodic table, each element's box contains the atomic number, the element symbol, and the atomic weight.

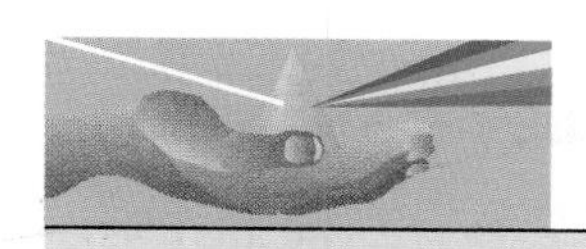

CHEMISTRY YOU CAN DO

Preparing a Pure Sample of an Element

You will need the following items to do this experiment.

- two glasses or plastic cups that will each hold about 250 mL of liquid
- about 100 mL of vinegar
- soap
- an iron nail, paper clip, or other similar-sized piece of iron
- something abrasive such as a piece of steel wool, Brillo, sandpaper, or nail file
- about 40–50 cm of thin string or thread
- some table salt
- a magnifying glass (optional)
- 15–20 dull pennies (shiny pennies will not work)

Wash the piece of iron with soap, dry it, and clean the surface further with steel wool or a nail file until the surface is shiny. Tie one end of the string around one end of the piece of iron.

Place the pennies in one cup (A) and pour in enough vinegar to cover them. Sprinkle on a little salt, swirl the liquid around so it contacts all the pennies, and observe what happens. When nothing more seems to be happening, pour the liquid into the second cup (B), leaving the pennies in the first cup (A) (that is, decant off the liquid). Suspend the piece of iron from the thread so that it is half submerged in the liquid in the second cup (B).

Observe the piece of iron over a period of 10 minutes or so, and then use the thread to pull it out of the liquid. Observe it carefully, using a magnifying glass if you have one. Compare the part that was submerged with the part that remained above the surface of the liquid.

1. What did you observe happening to the pennies?
2. How could you account for what happened to the pennies in terms of a microscopic model? Cite observations that support your conclusion.
3. What did you observe happening to the piece of iron?
4. Interpret the experiment in terms of a microscopic model, citing observations that support your conclusions.
5. Would this method be of use in purifying copper? If so, can you suggest ways that it could be used effectively to obtain copper from ores?

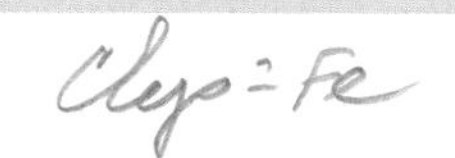

EXAMPLE 3.3

Calculating Average Atomic Weight from Isotopic Abundances

Bromine is used to make silver bromide, the important component of photographic film. The element has two naturally occurring isotopes. One has a mass of 78.918336 amu and a percent abundance of 50.69%. The other isotope, of mass 80.916289, has a percent abundance of 49.31%. Calculate the atomic weight.

SOLUTION The average value for a series of numbers can be found by adding the individual values and dividing by the number of values. For example, suppose you have two grades of 9 and three grades of 8 on five quizzes.

$$\text{average} = \frac{9 + 9 + 8 + 8 + 8}{5} = 8.4$$

This is the same thing as

$$\text{average} = (\tfrac{2}{5})(9) + (\tfrac{3}{5})(8) = (0.4)(9) + (0.6)(8) = 8.4$$

which says $\frac{2}{5}$ of the time you had a 9 and $\frac{3}{5}$ of the time you had an 8. The fraction

$\frac{2}{5}$ (or its decimal equivalent 0.4) is the "fractional abundance" of the grade of 9. Expressed as a percentage, 40% of the time you received a 9. Atomic weights are found in the same way—by multiplying the masses of the isotopes by their fractional abundances (which are the percent abundances divided by 100).

average atomic mass = (fractional abundance of isotope 1)(mass of isotope 1) + (fractional abundance of isotope 2)(mass of isotope 2) + . . .

For the bromine sample, the calculation would be as follows:

average atomic mass of bromine
= atomic weight
= (0.5069)(78.918336 amu) + (0.4931)(80.916289 amu)
= 79.90 amu

EXERCISE 3.5 • *Calculating Atomic Weight*

Verify that the atomic weight of chlorine is 35.45 amu, given the following information:

^{35}Cl, mass = 34.96885 amu, percent abundance = 75.77%
^{37}Cl, mass = 36.96590 amu, percent abundance = 24.23%

3.6 THE PERIODIC TABLE

The periodic table of elements in Figure 3.12 and in the front of the book is one of the most useful tools in chemistry. Not only does it contain a wealth of information, but it can be used to organize many of the ideas of chemistry. We shall refer to it often, since much of this book is devoted to examining the chemical and physical properties of the elements and their interrelationships as expressed in the periodic table. It is important that you be familiar with the history of its development and with its main features and terminology.

The History of the Periodic Table—Mendeleev

Until the discovery of element 104, elements were named for countries, states, cities, or people. Beginning with 104, the International Union of Pure and Applied Chemistry recommended three-letter symbols. These are based on the numerical roots nil (0), un (1), bi (2), tri (3), quad (4), pent (5), and so on. Thus, the symbol for 104 is Unq, and it is named unnilquadium.

On the evening of February 17, 1869, at the University of St. Petersburg in Russia, a 35-year-old professor of general chemistry, Dmitri Ivanovitch Mendeleev (1834–1907) was writing a chapter of his soon-to-be-famous textbook on chemistry. He had the properties of each element written on separate cards. While shuffling the cards trying to gather his thoughts before writing his manuscript, he realized that if the elements were arranged in order of increasing atomic weight, there were properties that repeated several times! That is, Mendeleev saw that there was a periodicity to the properties of elements (Figure 3.13), and he summarized this in a table, the most complete version of which was published in 1871 (Figure 3.14). He built the table by lining up the elements in vertical columns in order of increasing

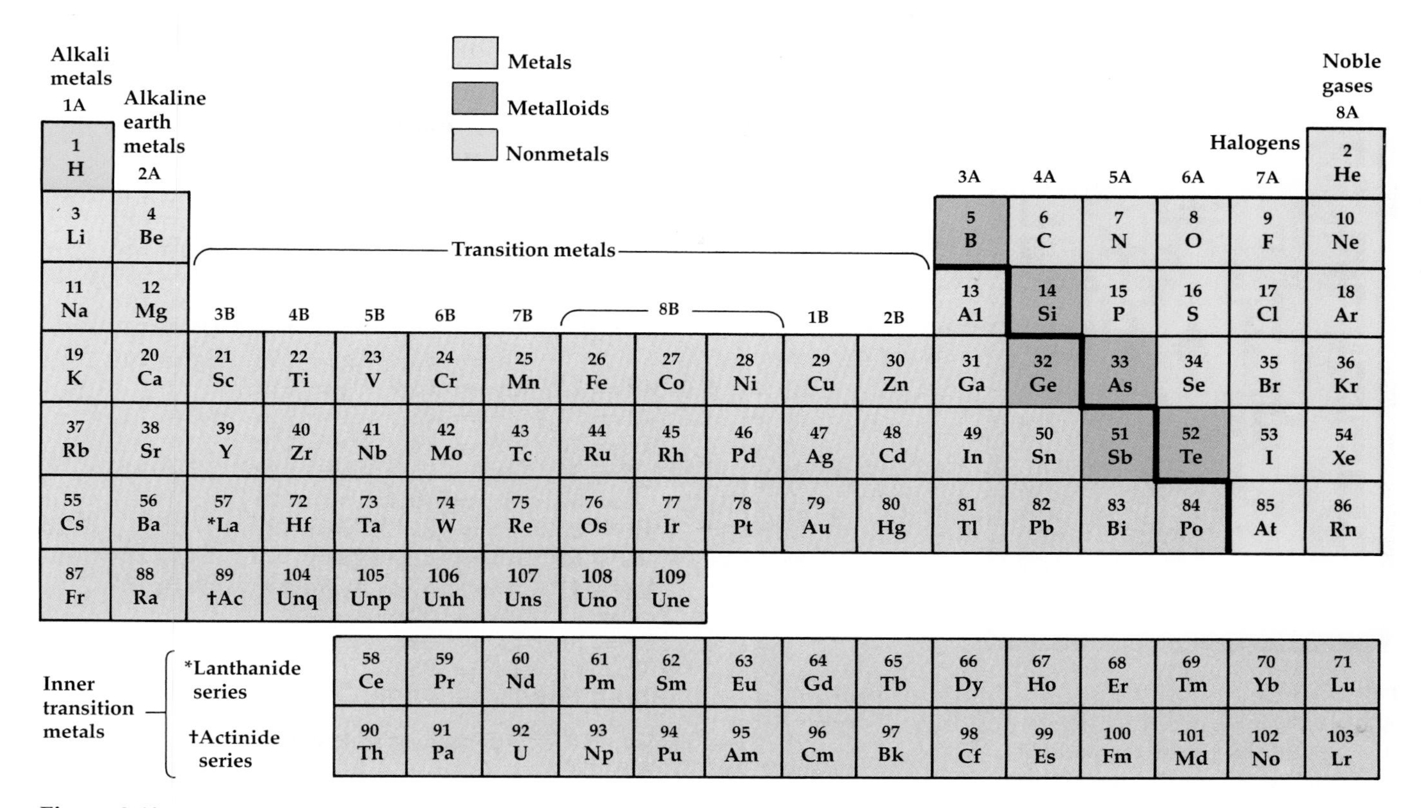

Figure 3.12

The periodic table of elements. Elements are listed in ascending order of atomic number. The following points are important: (a) Metals are shown in blue, the metalloids are green, and the nonmetals are yellow. (b) Periods are horizontal rows of elements and groups are vertical columns. (c) Groups are labeled by a number between 1 and 8 with a label of A (main group elements) or B (transition elements), the system used most commonly at the present in the United States. The new international system is to number the groups from 1 to 18. (d) Some groups have common names: Group 1A = alkali metals; Group 2A = alkaline earth metals; Group 7A = halogens; Group 8A = noble gases.

atomic weight. Every time he came to an element with properties similar to one already in the column, he started a new column. The rows, then, contained elements with similar properties.

The most important feature of Mendeleev's table—and a mark of his genius and daring—was that he left empty spaces for yet-to-be-discovered elements in the table. He left the empty spaces to retain the rationale of an ordered arrangement based on the periodic reoccurrence of similar chemical and physical properties. For example, in order of increasing atomic weight copper (Cu) is followed by zinc (Zn) and then arsenic (As). If arsenic had been placed next to zinc, arsenic would have fallen under aluminum (Al). But arsenic forms compounds similar to those formed by phosphorus (P) and antimony (Sb), not aluminum. Mendeleev reasoned, therefore, that arsenic belonged in a position under phosphorus, and that there were two un-

Figure 3.13
A rug woven by American Indians (the Navajo) illustrates the periodic recurrence of a design. This rug uses the "Two Gray Hills" pattern.
(C.D. Winters)

I.	II.	III.	IV.	V.	VI.*)
			Ti = 50	Zr = 90	? = 180
			V = 51	Nb = 94	Ta = 182
			Cr = 52	Mo = 96	W = 186
			Mn = 55	Rh = 104,4	Pt = 197,4
			Fe = 56	Ru = 104,4	Ir = 198
		Ni =	Co = 59	Pd = 106,6	Os = 199
H = 1			Cu = 63,4	Ag = 108	Hg = 200
	Be = 9,4	Mg = 24	Zn = 65,2	Cd = 112	
	B = 11	Al = 27,4	? = 68	Ur = 116	Au = 197 ?
	C = 12	Si = 28	? = 70	Sn = 118	
	N = 14	P = 31	As = 75	Sb = 122	Bi = 210
	O = 16	S = 32	Se = 79,4	Te = 128?	
	F = 19	Cl = 35,5	Br = 80	J = 127	
Li = 7	Na = 23	K = 39	Rb = 85,4	Cs = 133	Tl = 204
		Ca = 40	Sr = 87,6	Ba = 137	Pb = 207
		? = 45	Ce = 92		
		? Er = 56	La = 94		
		? Yt = 60	Di = 95		
		? In = 75,6	Th = 118 ?		

*) **Diese Zahlen sind der Uebersichtlichkeit halber vom Herausgeber beigesetzt worden.**

Figure 3.14
The 1871 periodic table of Dmitri Mendeleev. The spaces marked with question marks represent elements that Mendeleev recognized as unknown at the time, and places were left for them in the table. *(Oesper Collection in the History of Chemistry, University of Cincinnati)*

discovered elements whose atomic weights were between zinc and arsenic. The two missing elements were soon discovered: gallium (Ga) in 1875 and germanium (Ge) in 1886. In later years other gaps in Mendeleev's periodic table were filled as other predicted elements were discovered.

Mendeleev aided the discovery of the new elements by predicting their properties with remarkable accuracy, and he even suggested the geological regions in which minerals containing the elements could be found. The properties of a missing element were predicted by consideration of the properties of its neighboring elements in the table. An example of Mendeleev's prediction of the properties of an undiscovered element is shown in Table 3.3. The term *eka* comes from Sanskrit, and means "one"; thus, *ekasilicon* means "one place away from silicon." He also predicted the properties of ekaboron (scandium) and ekaaluminum (gallium).

Table 3.3

Some of Mendeleev's Predicted Properties of Ekasilicon and the Corresponding Observed Properties of Germanium

Property	Predicted properties of Ekasilicon	Observed properties of Germanium (Ge)
atomic weight	72	72.6
color of element	gray	gray
density of element (g/cm^3)	5.5	5.36
formula of oxide	EsO_2	GeO_2
density of oxide (g/cm^3)	4.7	4.228
formula of chloride	$EsCl_4$	$GeCl_4$
density of chloride	1.9	1.884
boiling point of chloride (°C)	under 100	84

The empty spaces in the table and Mendeleev's predictions of the properties of missing elements stimulated a flurry of prospecting for elements in the 1870s and 1880s. As a result, gallium (Ga) was discovered in 1875; scandium (Sc), samarium (Sm), holmium (Ho), and thulium (Tm) in 1879; gadolinium (Gd) in 1880; neodymium (Nd) and praseodymium (Pr) in 1885; and germanium (Ge) and dysprosium (Dy) in 1886. Many of these elements are not plentiful even today, yet they are important as ingredients in the compounds that produce the color in television screens.

Not only did Mendeleev believe he could predict the existence of then unknown elements, but he also thought that he could detect inaccurate atomic weights. Given the chemical methods of that day, such inaccuracies were not unexpected. Therefore, when confronted with the elements Te and I, chemical similarities meant that he had to place Te in the same group with sulfur (Group 6A) and I in the same group as chlorine (Group 7A), even though this inverted their atomic weight order. He simply assumed the atomic weight of Te must be incorrect. Time has proved him wrong, however, since we know now that tellurium atoms are indeed slightly heavier on average than iodine atoms.

Tellurium and iodine are not the only pair of elements for which the element with a smaller atomic weight follows the heavier one in the periodic table. Other reversed pairs in the modern periodic table include argon (Ar)–potassium (K) and cobalt (Co)–nickel (Ni).

The problem is that atomic weight is not the property that governs periodicity. This was discovered in 1913 by H.G.J. Moseley (1888–1915), a young scientist working with Ernest Rutherford. Moseley bombarded many different metals with electrons in a cathode-ray tube and observed the x-rays emitted by the metals. Most importantly, he found that the wavelengths of x-rays emitted by a particular element are related in a precise way to the *atomic number* of that element. He quickly realized that other atomic properties may be similarly related to atomic number and not, as Mendeleev had believed, to atomic weight. Indeed, if the elements are arranged in order of increasing atomic number, the defects in the Mendeleev table are corrected. That is, the **law of chemical periodicity** should be stated as "the properties of the elements are periodic functions of *atomic number*."

The wavelength of an x-ray, or of any other type of radiation, is the distance between two crests or two troughs of the wave. See Section 9.1.

Portrait of a Scientist

Dmitri Ivanovitch Mendeleev (1834–1907)

Mendeleev was born in Tobolsk, Siberia, but was educated in St. Petersburg where he lived virtually all his life. He taught at St. Petersburg University and while there wrote books and published his concept of chemical periodicity.

It is interesting that Mendeleev did little else with chemical periodicity after his initial articles. He went on to other interests, among them studying the natural resources of Russia and their commercial applications. In 1876 he visited the United States to study the fledgling oil industry and was much impressed with the industry but not with the country. He found Americans uninterested in science, and he felt the country carried on the worst features of European civilization.

By the end of the 19th century, political unrest was growing in Russia and Mendeleev lost his position at the university. He was appointed Chief of the Chamber of Weights and Measures for Russia, however, and established an inspection system for guaranteeing the honesty of weights and measures used in Russian commerce.

All pictures of Mendeleev show him with long hair. He made it a rule to cut his hair only once a year, in the spring, whether he had to appear at an important occasion or not.

Dmitri Ivanovitch Mendeleev. (Oesper Collection in the History of Chemistry, University of Cincinnati)

Features of the Periodic Table

There is a movement to adopt a new set of group designations as an international standard; in this table the groups are simply numbered 1 through 18 from left to right. We shall use the "A/B table" in this text.

The elements are arranged in the periodic table in such a way that *elements having similar chemical and physical properties lie in vertical columns* called **groups.** The table commonly used in the United States has groups numbered 1 through 8, with each number followed by a letter A or B. Using this system, chemists often designate the A groups as **main group elements** and the B groups as **transition elements.** The horizontal rows of the table are called **periods,** and they are numbered beginning with 1 for the period containing only H and He. For example, sodium, Na, is in Group 1A and is the first element in the third period. Mercury, Hg, is in Group 2B and in the seventh period.

The table can be divided into several regions according to the properties of the elements, which can be classified as **metals** (blue in Figure 3.12 and in the front of the book), **nonmetals** (yellow), and **metalloids** (green). Elements gradually become less metallic as one moves from left to right across a period, and in the metalloid region their properties are intermediate between those of metals and nonmetals. Photographs of a number of elements are collected in Figure 3.15.

Graphite, a form of the nonmetal carbon, conducts electricity because of its special structure.

You are probably familiar with many properties of metals from your everyday experience. Metals are solids (except for mercury, page 81), conduct electricity, are ductile (which means they can be drawn into wires), are malleable (which means they can be rolled into sheets), and can form alloys (solutions of one or more metals in another metal). Iron (Fe) and aluminum (Al) are used in automobile parts because of their ductility and malleability and their low cost relative to other metals. Copper (Cu) is used in electrical wiring because it conducts electricity better than most metals. Chromium (Cr) is plated onto automobile parts because its metallic luster makes them look better and protects them from reacting with oxygen in air.

It is more difficult to characterize nonmetals because they have a wider variety of properties. Some are solids, bromine is a liquid, and some, like nitrogen and oxygen in the air, are gases at room temperature. None of them conduct electricity, which is the main property that distinguishes them from metals. All nonmetals lie to the right of a zigzag line that passes between Al and Si, Ge and As, Sb and Te, and Po and At in Figure 3.12. Most of the elements next to this line have some properties that are typically metallic and other properties that are characteristic of nonmetals. These elements (B, Si, Ge, As, Sb, and Te) are called metalloids. They conduct electricity less well than metals, but are not insulators. Many are semiconductors and form the basis for the electronics revolution of the past several decades.

Several groups of elements with quite similar properties have distinctive names that are useful to know. The leftmost column, **Group 1A,** is known as the **alkali metals** (except for hydrogen, which is not a metal). "Alkali" comes from the Arabic language. Ancient Arabian chemists discovered that ashes of certain plants, which they called *al-qali,* gave water solutions that felt slippery and burned the skin. These ashes contain compounds of Group 1A elements that produce alkaline (basic) solutions. The elements themselves are very reactive, decomposing water to produce alkaline solutions (Figure 3.16). Because of their reactivity, the metals are found in nature only in compounds, never free. Sodium (Na) as sodium chloride has played an important role in history, for it is a fundamental part of the diet of humans and animals. Potassium (K) compounds are important plant nutrients. Besides having similar reactivities, all of these elements form compounds with oxygen that have formulas X_2O, where X represents the alkali metal: Li_2O, Na_2O, K_2O, Rb_2O, Cs_2O. Hydrogen also forms a compound having the same general formula: water, H_2O. This similarity of formulas was important to Mendeleev when he set up the periodic table.

The second group from the left, **Group 2A,** is also composed entirely of metals that occur naturally only in compounds. Except for beryllium (Be), these elements also react with water to produce alkaline solutions, and some of their compounds form alkaline solutions; hence they are known as the **alkaline earth elements.** Magnesium (Mg) and calcium (Ca) are the sixth and fifth most abundant elements in the earth's crust, respectively. Calcium is especially well known, since it is an important element in teeth and bones, and it occurs geologically in vast limestone deposits. Calcium carbonate ($CaCO_3$) is the chief constituent of limestone and of corals, sea shells, marble, and chalk (Figure 3.17). Radium (Ra), the heaviest alkaline earth element, is radioactive and is used in radiation treatment of some cancers. When alkaline earths combine with oxygen, the general formula is XO; for example, beryllium forms BeO and calcium gives CaO.

(text continues on p. 82)

Figure 3.15
Some common elements. *(C.D. Winters)*

Group 4A: *bottom,* carbon (C); *left middle,* silicon (Si); *right middle,* tin (Sn); *top,* lead (Pb).

Group 1A: Sodium (Na).

Group 2A: *left,* magnesium (Mg); *right,* calcium (Ca).

Group 6A: *left,* sulfur (S); *right,* selenium, (Se).

Fourth period transition metals: *left to right,* Ti, V, Cr, Mn, Fe, Co, Ni, Cu.

Group 5A: white phosphorus (P_4).

Group 2B: *left,* zinc (Zn); *right*, mercury (Hg).

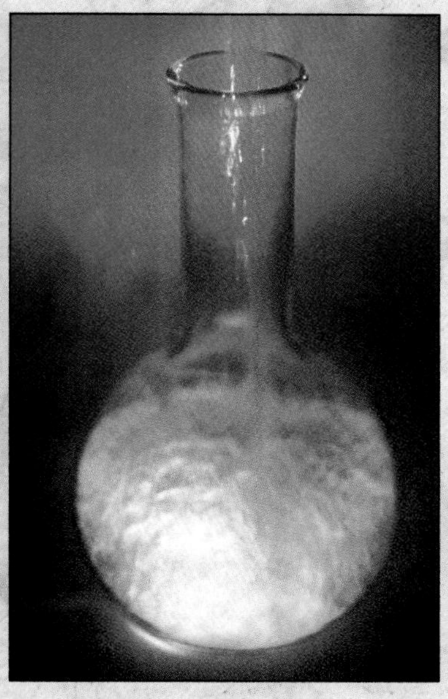

Group 5A: nitrogen (N_2).

Group 7A: *left,* bromine (Br_2); *right,* iodine (I_2).

Group 5A: *left,* arsenic (As); *right,* antimony (Sb); *top,* bismuth (Bi).

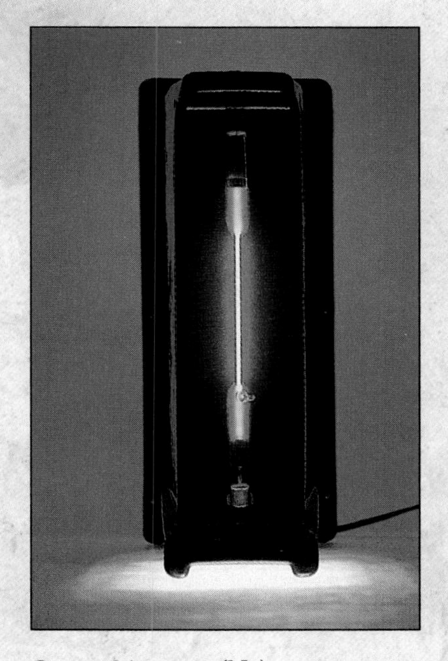

Group 8A: neon (Ne).

Group 3A: *top,* indium (In); *bottom,* aluminum (Al).

Figure 3.16
When water drops onto potassium, the reaction produces hydrogen, which burns in air. *(C.D. Winters)*

Figure 3.17
Various forms of calcium carbonate, $CaCO_3$: a clear crystal of calcite, a seashell, and a piece of limestone.
(C.D. Winters)

Following Group 2A is a series of **transition elements** that fill the fourth, fifth, and sixth periods in the center of the table. These are all metals. Some, such as iron (Fe), are abundant in nature and very important commercially. Others, including silver (Ag), gold (Au), and platinum (Pt), are much less abundant but also less reactive; they can be found in nature as the pure element and are coveted for their beauty. Two rows at the very bottom of the table encompass the **inner transition elements,** which are subdivided further into the **lanthanides** (beginning with the element lanthanum) and the **actinides** (beginning with actinium). These are much less abundant and less important commercially, though some lanthanide compounds are used in color television picture tubes.

The next column to the right is **Group 3A,** which has no special name. It contains a metalloid, boron (B), and four metals. Aluminum (Al) is the most abundant metal in the earth's crust at 8.3% by mass. It is exceeded only by the nonmetal oxygen (O; 45.5%) and the metalloid silicon (Si; 25.7%). These three elements are found combined in clays and other common minerals. Boron occurs in the mineral borax, which is mined in Death Valley, California, and in the late 19th century was hauled out of the valley in wagons drawn by 20 mules. Hence the name of a well known washing powder (Figure 3.18). Gallium, indium, and thallium are much less familiar because they have fewer important uses. All the Group 3A elements form oxygen compounds with the formula X_2O_3.

The word "plumbing" comes from the Latin word for lead, plumbum. This is also the origin of the symbol (Pb) for lead.

Group 4A contains one nonmetal, carbon (C); two metalloids, silicon (Si) and germanium (Ge); and two metals, tin (Sn) and lead (Pb). Because of the change from nonmetal to metals, there is more variation in properties of the elements from top to bottom in this group than in most; however, they all form oxygen compounds having the general formula XO_2. Carbon is responsible for the great variety of chemical compounds that make up living things. On earth it is found in carbonates such as limestone (Figure 3.17) and in coal, petroleum, and natural gas—the fossil fuels. Silicon is the basis of many minerals such as quartz and the beautiful gemstones of amethyst (Figure 3.19) Tin and lead have been known for centuries, because they are easily smelted from their ores. Tin can be alloyed with copper to make bronze, which was used for centuries in utensils and weapons. Lead has been used in pipes and paint, but the element is quite toxic to humans. Indeed, there is a theory that the Roman Empire fell because Romans suffered from lead poisoning induced by lead pipes they used for water supply.

Nitrogen (N), the first element in **Group 5A,** makes up about three fourths of the earth's atmosphere, but this is nearly all the nitrogen there is at the earth's surface. The element is unreactive, and so finding methods for fixing atmospheric nitrogen (forming compounds from the element) has played an essential role in the chemical industry. Phosphorus (P) glows in the dark, and this is the origin of the word "phosphorescent." Phosphorus is also essential to life as a constituent in bones and teeth. Bismuth (Bi) is the heaviest element that is not naturally radioactive; all elements whose atomic numbers are greater than 83 emit alpha, beta, or gamma rays. As in Group 4A, elements at the top of this group are nonmetals, but a metal, Bi, is at the bottom. Nevertheless, all these elements form oxygen-containing compounds with the general formula X_2O_5 as well as X_2O_3.

Figure 3.18
Boron is found in borax, the chief ingredient of a washing powder called "20 Mule Team Borax." The name comes from the fact that the borax was mined in Death Valley in California in the late 19th century, and teams of 20 mules each were used to haul out the heavy wagons loaded with borax. *(C.D. Winters)*

Group 6A begins with oxygen (O), which constitutes about 20% of the atmosphere and combines readily with most other elements. The energy that powers life on earth is derived from reactions in which oxygen combines with other substances. Sulfur (S) has been known in elemental form since ancient times as brimstone. Sulfur, selenium (Se), and tellurium (Te) are referred to collectively as **chalcogens** (from Greek, *khalkos,* copper) because copper ores contain them. Their compounds are foul smelling and poisonous; nevertheless sulfur and selenium are essential components of the human diet. Polonium was isolated in 1898 by Marie Curie, who separated it from tons of uranium ore and named it for her native country, Poland. General formulas of oxygen compounds of these elements are XO_3 and XO_2.

At the far right of the periodic table are two groups composed entirely of nonmetals. Within each group the elements are quite similar, but they are completely different from elements in the other group. **Group 7A** elements—fluorine, chlorine, bromine, and iodine—combine violently with alkali metals to form salts—table salt, NaCl, for example. The name for this group, **halogens,** comes from Greek words *hals,* salt, and *genes,* forming. The halogens

Be sure to notice the spelling of fluorine (u before o). A common misspelling is "flourine," which would be pronounced "flower-ene."

Figure 3.19
Amethyst, a form of silicon dioxide, SiO_2. Silicon dioxide is normally colorless (as in the quartz crystal at the right), but a trace of manganese oxide gives the solid a violet color. *(C.D. Winters)*

(a) (b)

Figure 3.20
Like all the alkali metals, sodium reacts vigorously with halogens. (a) A flask containing yellow chlorine gas (Cl_2). (b) A piece of sodium was placed in the flask, and the sodium reacted with the chlorine to give salt (sodium chloride, NaCl). The reaction evolves energy in the form of light and heat. *(C.D. Winters)*

react with many other metals to form salts, and they also combine with most nonmetals. They are among the most reactive of all elements (Figure 3.20).

By contrast, the **Group 8A** elements are the least reactive elements. Helium, neon, argon, krypton, xenon, and radon are all gases and are not very abundant on earth. Because of this, and their chemical unreactivity, they were not discovered until the end of the 19th century. Until 1962, when a compound of xenon was first prepared, it was thought that none of these elements would combine chemically. For this reason they are still called inert gases or **noble gases.** Because its atoms are very light, helium is used in lighter-than-air craft such as blimps; neon and argon are used in advertising signs. Radon is radioactive and the cause of indoor air pollution problems when it seeps out of the ground into a building.

EXERCISE 3.6 • *The Periodic Table*

How many elements are there in the third period of the periodic table? Give the name and symbol of each. Tell whether each element in the period is a metal, metalloid, or nonmetal.

Chemical Periodicity

The chemical basis of the periodic table is that properties of the elements in a group are similar to one another, but this does not mean they are identical. Some properties of elements in a group vary somewhat in a regular pattern.

Mercury Poses Benefits as Well as Risks

MERCURY, OR QUICKSILVER, IS about as abundant in the earth's crust as silver. Since the chief mercury-containing mineral—called cinnabar or vermilion—is a red solid that occurs in concentrated deposits, mercury is one of the oldest known elements. Prehistoric people used vermilion as a pigment to draw on the walls of caves and discovered that heating the red rock in a fire gave a shiny liquid metal. Samples of the metal have been found in graves of the 16th century BC, and Aristotle in the 4th century BC was the first to call it "liquid silver."

The ability of mercury to form **amalgams**—intimate mixtures with other minerals such as gold and silver—has been known for centuries, and this property has been used to extract metallic gold from natural sources. Mixing gold-bearing soil with mercury produces a liquid amalgam that is easily separated from dirt. Heating the amalgam then drives off the mercury and leaves gold behind. This is the reason the 16th century Spanish fleet carried casks of mercury to the New World. The practice of extracting gold with mercury continues today in the mines of Brazil, where 2 pounds of mercury are used for every pound of gold recovered. Unfortunately, it has also led to mercury poisoning of gold miners and severe pollution of the Amazon River basin.

There may be as many as 3000 uses for mercury in our modern society. Dentists use an amalgam with silver and tin to fill cavities; a drop of mercury goes into every fluorescent lamp; and the liquid metal is used in thermometers and barometers and in electrical equipment of many kinds. Compounds of mercury are used in calculator or camera batteries; others, such as mercurochrome, are antiseptics that inhibit the growth of bacteria and mildew.

But mercury and its salts have a darker side, and none has expressed it better than Alfred Stock (1876–1946), a German chemist. Stock said that "mercury is a strong poison, particularly dangerous because of its liquid form and noticeable volatility even at room temperature. . . . [He] found from personal experience . . . that protracted stay in an atmosphere charged with only 1/100 of the amount of mercury required for its saturation, [can] induce chronic mercury poisoning. [It] reveals itself [first] as an affection of the nerves, causing headaches, numbness, mental lassitude, depression, and the loss of memory."

Environmental scientists have estimated that volcanos and other natural sources emit 2000 to 20,000 tons of mercury every year, and industrial sources—such as mining, the manufacture of chemicals, the burning of coal and oil, and the incineration of wastes such as paint and batteries—contribute another 21,000 tons per year. This mercury finds its way into the soil, lakes, and oceans, where it accumulates in sediments and is eventually taken up by fish. If the fish are consumed by humans, mercury poisoning is possible. In the 1960s in Minamata, Japan, more than 100 people developed tremors, fell into comas, and died because their diet of local fish had a high mercury content owing to industrial pollution.

Cinnabar is a common mercury-containing mineral. It consists chiefly of deep red mercury sulfide. *(C.D. Winters)*

Mercury and its compounds are extraordinarily useful in our economy, but they can also be harmful. As a result, the 1972 Clean Air Act required industry to install equipment to prevent mercury from entering the environment. More recently, officials have called for a ban on mercury compounds in latex paints, in which they protect the paint from mildew and bacteria while it is stored.

For example, the melting points of the Group 1A elements from Li through Cs are (in °C) 179, 98, 64, 39, and 28. Lithium reacts slowly with water, sodium reacts faster, and potassium still faster (Figure 3.16); for the elements at the bottom of the group, just exposure to moist air produces a vigorous reaction or even explosion. All Group 1A elements are soft, silvery white, metallic solids except cesium, which is yellow and would be liquid on a warm day (Figure 3.15).

Other properties differ to some degree but not in a regular pattern. For example, the densities (in g/mL) of the solids Li through Cs are 0.53, 0.97, 0.86, 1.53, and 1.87. Lithium, sodium, and potassium have densities less than 1.00 g/mL (the approximate density of water at room temperature) and so float on water as they react vigorously.

Some properties are similar for every member of a group. For example, elements in a group generally react with other elements to form similar compounds. *This is the most useful and powerful inference that can be made from the periodic table.* Taking the alkali metals as an example, we find that they all can combine with halogens to give compounds whose formula is MX, that is, the combination of one atom of metal (M) and one atom of the halogen (X). Table salt, NaCl, and the compound in iodized salt, KI, are just two examples. Just this one observation from the periodic table gives us the ability to write formulas for compounds of cesium and rubidium, elements less familiar to us than the other members of the group. Hence, we can have confidence that the chlorides and bromides of cesium and rubidium must have the formulas CsCl, CsBr, RbCl, and RbBr.

Why do elements in the same group in the periodic table have similar chemical behavior? Why do metals and nonmetals have different properties? The answer is that all the elements in a group (particularly the main group A elements and the noble gases) have atoms with similar arrangements of electrons in atoms. In fact, the chemical view of matter is primarily concerned with what happens to the electrons in atoms during the course of a chemical reaction. Therefore, we shall return to the matter of chemical periodicity once you have been introduced to the more detailed structure of the atom in Chapter 9.

EXERCISE 3.7 • *Periodic Properties*

The element aluminum forms aluminum chloride, whose formula is $AlCl_3$. Predict a formula for a compound formed from indium and fluorine. Compare the formulas of sodium oxide and sodium chloride. Can you see a relationship? A general formula for alkaline earth oxides was given during discussion of Group 2A elements. Using this and the relationship between chloride and oxide formulas, predict the formula for a compound formed from magnesium and chlorine.

IN CLOSING

Having studied this chapter, you should be able to

- explain the historical development of the atomic theory and identify some of the scientists who made important contributions (Sections 3.1–3.2).
- describe electrons, protons, and neutrons, and the general structure of the atom (Section 3.3).
- define isotope and give the mass number and number of neutrons for a specific isotope (Section 3.4).
- calculate the atomic weight of an element from isotopic abundances (Section 3.4).

- explain the difference between the atomic number and atomic weight of an element and find this information for any element (Sections 3.3–3.5).
- identify the periodic table location of groups, periods, metals, metalloids, nonmetals, alkali metals, alkaline earth metals, halogens, noble gases, and the transition and inner transition elements (Section 3.6).
- use the periodic table to predict properties of elements and formulas of simple compounds (Section 3.6).

STUDY QUESTIONS

Review Questions

1. Which of John Dalton's ideas, which were first put forth early in the 19th century, are not true according to our present views of atomic theory?
2. What are the three fundamental particles from which most atoms are built? What are their electrical charges? Which of these particles constitute the nucleus of the atom? Which is the least massive particle of the three?
3. What is the definition of the atomic mass unit?
4. What is the difference between the mass number and the atomic number of an atom?
5. What did the discovery of radioactivity reveal about the structure of atoms?
6. What is the relationship between the work of J.J. Thomson and R. Millikan? How was Rutherford's work related to J.J. Thomson's work?
7. If the nucleus of an atom were the size of a medium-sized orange (let us say with a diameter of about 6 cm), what would be the diameter of the atom?
8. Element 25 is found in the form of oxide "nodules" at the bottom of the sea. What are the name and symbol of this element?

A manganese "nodule" from the floor of the Pacific Ocean. *(C.D. Winters)*

9. The volcanic eruption of Mt. St. Helens in Washington produced a considerable quantity of a radioactive element in the gaseous state. The element has atomic number 86. What are the symbol and name of this element?

The eruption of Mt. St. Helens in Washington in May 1980. *(Pat & Tom Leeson/Photo Researchers)*

10. Lithium has two stable isotopes, ^{6}Li and ^{7}Li. One of them has an abundance of 92.5% and the other has an abundance of 7.5%. Knowing that the atomic weight of lithium is 6.941, which is the more abundant isotope?
11. The beautifully written autobiography of Primo Levi, *The Periodic Table,* was mentioned in Chapter 1. In this book Levi says of zinc that "it is not an element which says much to the imagination, it is gray and its salts are colorless, it is not toxic, nor does it produce striking chromatic reactions; in short, it is a boring metal. It has been known to humanity for two or three centuries, so it is not a veteran covered with glory like copper, nor even one of these newly minted elements which are still surrounded with the glamour of their discovery." From this description, and from reading this chapter,

make a list of the properties of zinc. For example, include in your list the position of the element in the periodic table, and tell how many electrons and protons an atom of zinc has. What are its atomic number and atomic weight? Zinc is important in our economy. Check your dictionary (or a book such as *The Handbook of Chemistry and Physics*) and make a list of the uses of the element.

12. What was incorrect about Mendeleev's original concept of the periodic table? What is the modern "law of chemical periodicity" and how is this related to Mendeleev's ideas?
13. What is the difference between a group and a period in the periodic table?
14. Name and give symbols for (a) three elements that are metals; (b) four elements that are nonmetals; and (c) two elements that are metalloids. In each case also locate the element in the periodic table by giving the group and period in which the element is found.
15. Name and give symbols for three transition metals in the fourth period. Look up each of your choices in a dictionary and make a list of the properties that the dictionary entry gives. Also list the uses of the element as given by the dictionary.
16. Name three transition elements, two halogens, and one alkali metal.
17. Name an alkali metal, an alkaline earth metal, and a halogen.
18. Name two halogens. Look up each of your choices in a dictionary (or a book such as *The Handbook of Chemistry and Physics*) and make a list of the properties that the dictionary entry gives. Also list any uses of the element that are given by the dictionary.
19. Name an element discovered by Madame Curie. Give its name, symbol, and atomic number. Use a dictionary to find the origin of the name of this element.

The Composition of Atoms

20. Give the mass number of each of the following atoms: (a) beryllium with 5 neutrons, (b) titanium with 26 neutrons, and (c) gallium with 39 neutrons.

21. Give the mass number of (a) an iron atom with 30 neutrons, (b) an americium atom with 148 neutrons, and (c) a tungsten atom with 110 neutrons.

22. Give the complete symbol (${}^{A}_{Z}X$) for each of the following atoms: (a) sodium with 12 neutrons; (b) argon with 21 neutrons; and (c) gallium with 29 neutrons.

23. Give the complete symbol (${}^{A}_{Z}X$) for each of the following atoms: (a) nitrogen with 8 neutrons; (b) zinc with 34 neutrons; and (c) xenon with 75 neutrons.

24. How many electrons, protons, and neutrons are there in an atom of (a) calcium-40, ${}^{40}Ca$; (b) tin-119, ${}^{119}Sn$; and (c) plutonium-244, ${}^{244}Pu$?

25. How many electrons, protons, and neutrons are there in an atom of (a) carbon-13, ${}^{13}C$; (b) chromium-50, ${}^{50}Cr$; and bismuth-205, ${}^{205}Bi$?

26. Fill in the columns of blanks in the table (one column per element).

symbol	${}^{45}Sc$	${}^{33}S$	_____	_____
number of protons	_____	_____	8	_____
number of neutrons	_____	_____	9	31
number of electrons in the neutral atom	_____	_____	_____	25

27. Fill in the columns of blanks in the table (one column per element).

symbol	${}^{65}Cu$	${}^{37}Cl$	_____	_____
number of protons	_____	_____	34	_____
number of neutrons	_____	_____	46	46
number of electrons in the neutral atom	_____	_____	_____	36
name of element	_____	_____	_____	_____

Isotopes

28. Radioactive americium-241 is used in household smoke detectors and in bone mineral analysis. Give the number of electrons, protons, and neutrons in an atom of americium-241.

29. The artificial radioactive element technetium is used in many medical studies. Give the number of electrons, protons, and neutrons in an atom of technetium-99.

30. Which of the following are isotopes of element X, whose atomic number is 9: ${}^{10}_{9}X$, ${}^{13}_{9}X$, ${}^{12}_{10}X$, and ${}^{9}_{9}X$?

31. Cobalt has three radioactive isotopes used in medical studies. Atoms of these isotopes have 30, 31, and 33 neutrons, respectively. Give the symbol for each of these isotopes.

Atomic Weight

32. Verify that the atomic weight of lithium is 6.941 amu, given the following information:
^{6}Li, exact mass = 6.015121 amu, percent abundance = 7.500%
^{7}Li, exact mass = 7.016003 amu, percent abundance = 92.50%

33. Verify that the atomic weight of magnesium is 24.3050 amu, given the following information:
^{24}Mg, exact mass = 23.985042 amu, percent abundance = 78.99%
^{25}Mg, exact mass = 24.98537 amu, percent abundance = 10.00%
^{26}Mg, exact mass = 25.982593 amu, percent abundance = 11.01%

34. Gallium has two naturally occurring isotopes, ^{69}Ga and ^{71}Ga, with masses of 68.9257 amu and 70.9249 amu, respectively. Calculate the abundances of these isotopes of gallium.

35. Copper has two stable isotopes, ^{63}Cu and ^{65}Cu, with masses of 62.939598 amu and 64.927793 amu, respectively. Calculate the abundances of these isotopes of copper.

The Periodic Table

36. How many elements are there in Group 4A of the periodic table? Give the name and symbol of each of these elements. Tell whether each is a metal, nonmetal, or metalloid.

37. How many elements are there in the fourth period of the periodic table? Give the name and symbol of each of these elements. Tell whether each is a metal, metalloid, or nonmetal.

38. Which period in the periodic table is as yet incomplete? What is the name given to the majority of these elements and what well known property characterizes them?

39. How many periods of the periodic table have 8 elements, how many have 18 elements, and how many have 32 elements?

40. The chart at the right is a plot of the logarithm of the relative abundances of elements 1 through 36 in the solar system. [The abundances are given on a scale that gives silicon a relative abundance of 1.00×10^6 (the logarithm of which is 6).]
(a) What is the most abundant metal?
(b) What is the most abundant nonmetal?
(c) What is the most abundant metalloid?
(d) Which of the transition elements is most abundant?
(e) How many halogens are considered on this plot and which is the most abundant?

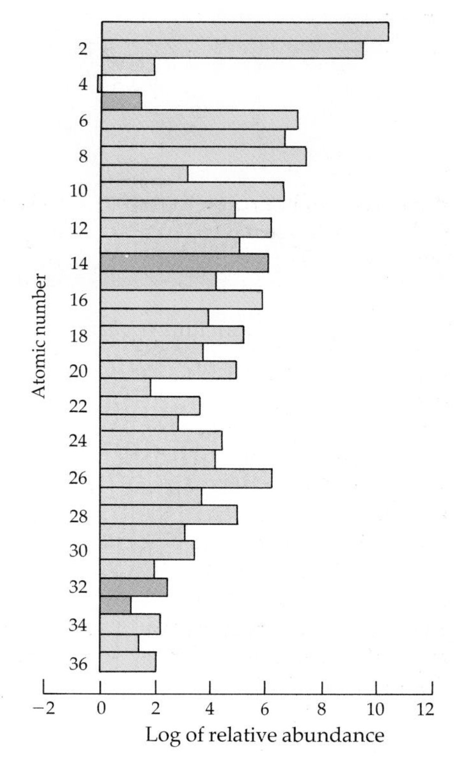

The relative abundances of elements 1–36 in the solar system.

41. Again consider the plot of relative abundance versus atomic number. Can you uncover any relation between abundance and atomic number? Is there any difference between elements of even atomic number and those of odd atomic number?

42. Calcium reacts with air to form a compound in which the ratio of calcium to oxygen atoms is 1:1. That is, the formula is CaO. Based on this, predict the formulas for all the compounds formed between oxygen and the Group 2A elements.

43. Compare the formulas of the compounds formed between Cl and Na, Mg, and Al. (See Exercise 3.7.) Can you see a pattern in these formulas? Describe this pattern. Based on this pattern, predict the formula of a compound formed between Si and Cl.

General Questions

44. Name three elements that you have encountered today. (Name only those that you have seen as elements, not those combined into compounds.) Give the location of each of these elements in the periodic table by specifying the group and period in which it is found.
45. Potassium has three stable (nonradioactive) isotopes, ^{39}K, ^{40}K, and ^{41}K, but ^{40}K has a very low natural abundance. Which of the other two is the more abundant?
46. Figure 3.16 shows how potassium reacts with water. The photo below shows the reactions of magnesium and calcium with water. Based on their relative reactivities, what might you expect to see when barium, another Group 2A element, is placed in water? Give the period in which each element (Mg, Ca, and Ba) is found. What correlation do you think you might find between the reactivity of these elements and their position in the periodic table?

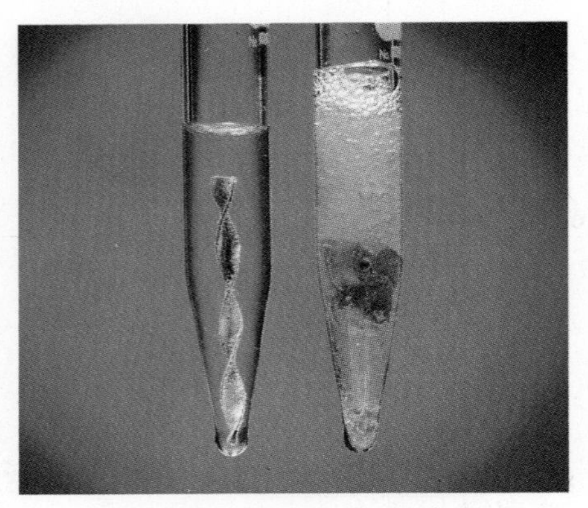

Magnesium (*left*) and calcium (*right*) in water. *(C.D. Winters)*

47. Figure 3.11 shows the mass spectrum of the isotopes of antimony. What are the symbols of the isotopes? Which is the more abundant isotope? How many protons, neutrons, and electrons does this isotope have? Without looking at a periodic table, give the approximate atomic weight of antimony.
48. The elements of Group 4A can combine to give compounds such as carbon tetrachloride, CCl_4. Give the formula for the compounds formed between Cl and two other Group 4A elements, silicon and germanium.
49. When an athlete tears ligaments and tendons, they can be surgically attached to bone to keep them in place until they reattach themselves. A problem with current techniques, though, is that the screws and washers used are often too big to be positioned accurately or properly. Therefore, a new titanium-containing device has been invented and is coming into use. (See figure below.)
    (a) What are the symbol, atomic number, and atomic weight of titanium?
    (b) In what group and period is it found? Name the other elements of its group.
    (c) What chemical properties do you suppose make titanium an excellent choice for this and other surgical applications?

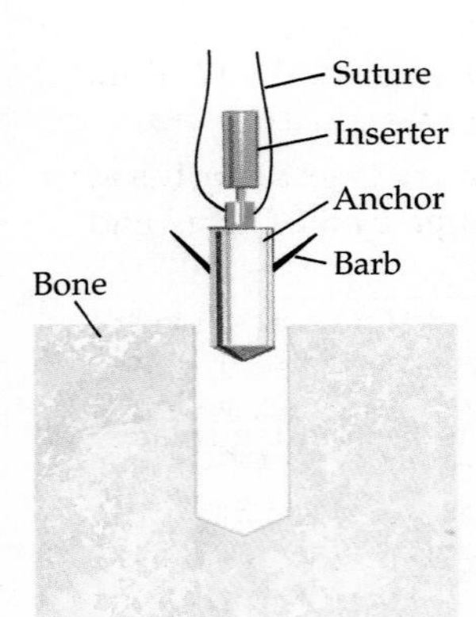

1. Sutures are attached to the anchor, which is then placed on the inserter.

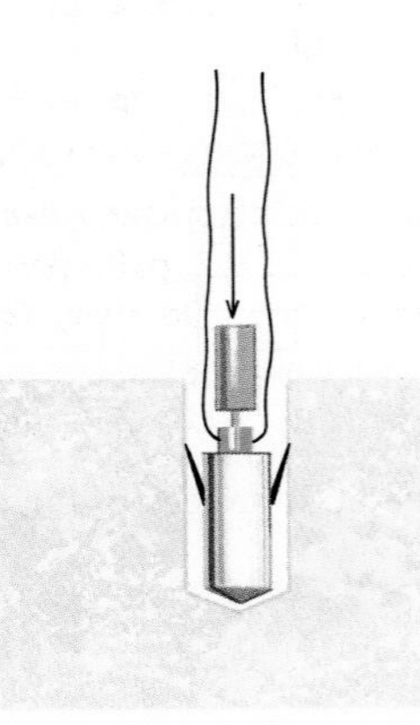

2. The anchor is pushed into the hole, causing the wire barbs to compress.

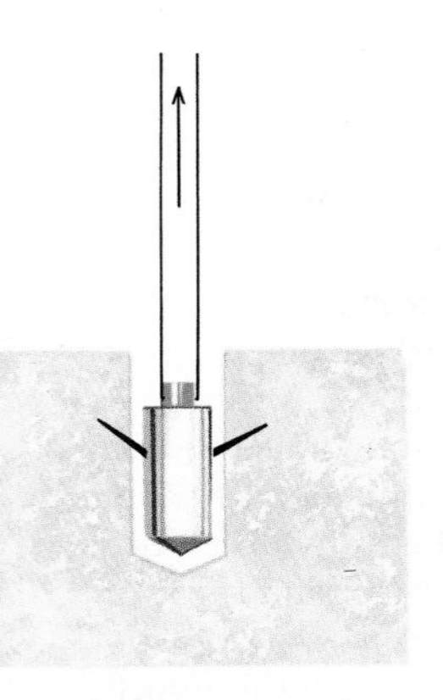

3. Tension is applied, causing the barbs to spread out, securing the anchor.

A new method of anchoring ligaments and tendons to bone. (*New York Times*, July 1, 1992, page D7.) (Redrawn with permission from the *New York Times*.)

(d) Use a dictionary (or a book such as *The Handbook of Chemistry and Physics*) to make a list of the properties of the element and its uses.

50. The following plot shows the variation in density with atomic number for the first 36 elements. Use this plot to answer the following questions:
    (a) What three elements in this series have the highest values of density? What is their approximate density? Are these elements metals or nonmetals?
    (b) Which element in the second period has the greatest density? Which element in the third period has the largest density? What do these two elements have in common?
    (c) Some elements have densities so low that they do not show up in the plot. What elements are these? What property do they have in common?

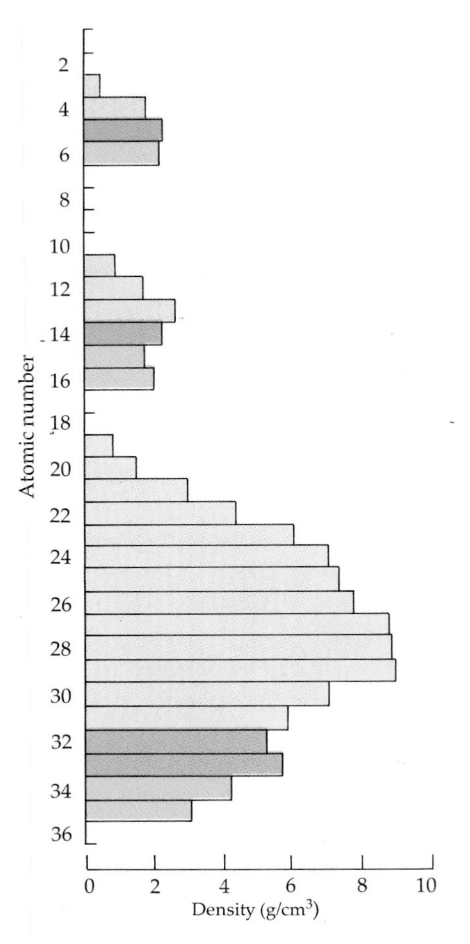

The densities of elements 1–36.

51. **Crossword Puzzle:** In the 2 × 2 crossword below, each letter must be correct four ways: horizontally, vertically, diagonally, and by itself. Instead of words, use symbols of elements. When the puzzle is complete, the four spaces will contain the overlapping symbols of 10 elements. There is only one correct solution.

1	**2**
3	**4**

Horizontal
1–2: Two-letter symbol for a metal used in ancient times
3–4: Two-letter symbol for a metal that burns in air and is found in Group 5A

Vertical
1–3: Two-letter symbol for a metalloid
2–4: Two-letter symbol for a metal used in U.S. coins

Single squares (all one-letter symbols)
1: A colorful nonmetal
2: A colorless gaseous nonmetal
3: An element that makes fireworks green
4: An element that has medicinal uses

Diagonal
1–4: Two-letter symbol for an element used in electronics
2–3: Two-letter symbol for a metal used with Zr to make wires for superconducting magnets

This puzzle appeared in *Chemical and Engineering News*, December 14, 1987, p. 86 (submitted by S.J. Cyvin) and in *Chem Matters*, October, 1988.

52. Draw a picture showing the approximate positions of all protons, electrons, and neutrons in an atom of helium-4. Make certain that your diagram indicates both the number and position of each type of particle.

53. Answer the following questions using the figures at the end of this question. (Each question may have more than one answer.)

____ Which represents submicroscopic particles in a sample of solid?
____ Which represents submicroscopic particles in a sample of liquid?

____ Which represents submicroscopic particles in a sample of gas?
____ Which represents submicroscopic particles in a sample of an element?
____ Which represents submicroscopic particles in a sample of a compound?
____ Which represents submicroscopic particles in a sample of a solution?
____ Which represents submicroscopic particles in a sample of a heterogeneous mixture?

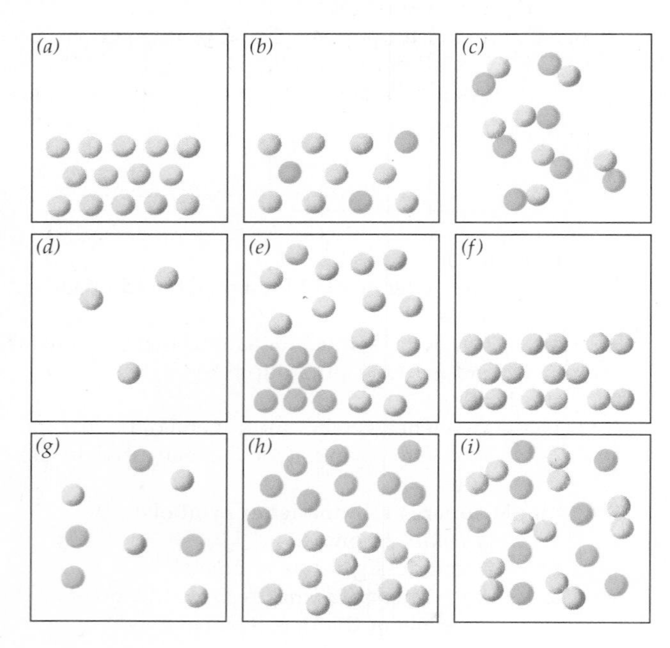

54. The spaceship of a group of astronauts accidentally encounters a spacewarp that traps them in an alternative universe where the chemical elements are quite different from the ones they are used to. The astronauts find these properties for the elements that they have discovered:

Symbol	*Atomic Weight*	*State*	*Color*	*Electrical Conductivity*	*Reactivity*
A	3.2	solid	silvery	high	medium
D	13.5	gas	colorless	very low	very high
E	5.31	solid	golden	very high	medium
G	15.43	solid	silvery	high	medium
J	27.89	solid	silvery	high	medium
L	21.57	liquid	colorless	very low	medium
M	11.23	gas	colorless	very low	very low
Q	8.97	liquid	colorless	very low	medium
R	1.02	gas	colorless	very low	very high
T	33.85	solid	colorless	very low	medium
X	23.68	gas	colorless	very low	very low
Z	36.2	gas	colorless	very low	medium
Ab	29.85	solid	golden	very high	medium

(a) Arrange these elements into a periodic table.
(b) If a new element, X, with atomic weight 25.84 is discovered, what would its properties be? Where would it fit in the periodic table you constructed?
(c) Are there any elements that have not yet been discovered? If so, what would their properties be?

Using Computer Programs and Videodiscs

55. Use the find function of *KC? Discoverer* to locate all the elements that have the following characteristics:
(a) Elements that are orange.
(b) Elements that were discovered between 1800 and 1850.
(c) Elements that cost less than $5 per 100 grams.
56. The data reported in *KC? Discoverer* have units. This question explores some of these units.
(a) What is the temperature, in kelvins, at which sodium melts? In degrees Celsius?
(b) Which has the lower melting temperature, sodium or potassium?
(c) What is the density of platinum and what are the units of this density? Which is the more dense, platinum or gold?
57. Use the *KC? Discoverer* program to find one or more elements that match each set of properties listed below. For each element list five other properties.
(a) The element reacts vigorously with water and is used in photocells.
(b) The element comes from the ore spodumene and is used in batteries.
(c) The element is sweet-tasting but very toxic. It is used in special alloys and in x-ray windows.
(d) This Group 2A element is used in signal flares and in fireworks.
(e) This element can be obtained from seawater. It is used to make PVC plastics and to bleach paper pulp.
58. Use the *KC? Discoverer* computer program to plot the abundance of the first 36 elements in the earth's crust and compare the plot with the abundance plot in Study Question 40.
(a) What similarities and differences do you see between these plots?
(b) What is the most abundant metal in the earth's crust? Most abundant nonmetal? Most abundant metalloid?

59. Use the sorting function of *KC? Discoverer* to find the 10 most expensive elements (judged on the basis of the cost of the pure element). Are these largely metals, nonmetals, or metalloids? Use *KC? Discoverer* to find the uses for the five most expensive elements.
60. Question 49 described titanium in a surgical application. Using *KC? Discoverer,* look up the properties of titanium.
    (a) What is the melting point of the element? Is this comparable to other transition metals of the fourth period? Is it higher or lower?
    (b) What is the density of titanium relative to the other metals in the transition elements of the fourth period?
    (c) Does titanium react readily with air or water? How is this related to its use in the human body?
61. Find the image of copper on the *Periodic Table Videodisc* (frame number 28534.) What qualitative observations can you make about this element? What are some uses of this element? (See frame numbers 36664 and 36671.)
62. The *Chemical Demonstrations* videodisc describes the properties of liquid oxygen (frame 8910 on Disc 2, Side 2). (See Demonstration 38 on the videotape version.) Oxygen boils at -183 °C, so if the temperature is lower than that, it can be converted readily to a liquid.
    (a) What observations can you make as you watch this demonstration? Are you seeing chemical or physical changes?
    (b) When the oxygen is liquefied, what is its color? Why don't you notice this color when the oxygen is in the form of a gas?
    (c) The demonstrator pours the liquid oxygen between the poles of a magnet. What do you see? What can you conclude about the magnetic properties of liquid oxygen?

"Circular Forms. Sun and Moon" painted in 1912–1913 by Robert Delauney (1885–1941).

(The Bettmann Archive)

Chemical Compounds

CHAPTER

4

We began Chapter 1 with the statement by Roald Hoffmann, a recipient of the 1981 Nobel Prize in Chemistry, who defined chemistry as "the science of molecules and their transformations." The present chapter begins the discussion of molecules and chemical compounds. Some molecules contain only one kind of atom and constitute elements, but most of them contain two or more kinds of atoms and make up chemical compounds. Some compounds do not consist of discrete molecules but rather are built up from ions—atoms that have lost or gained electrons. Nevertheless, all compounds have properties that can be understood better the more we know about the submicroscopic particles, molecules or ions, from which they are made. Of particular importance is the formula of a compound, which tells what kinds of atoms are combined and in what ratio. Names of compounds can be related to their formulas, as can the masses of elements that are found in a given mass of compound. Because atoms, molecules, and ions are so very small, a counting unit, the mole, is used to measure the numbers of particles that combine when compounds form.

The French abstract painter Robert Delauney (1885–1941) said that "color is form and subject," and you can see how well he used color in the painting that opens this chapter. But why is your eye drawn to the painting? Surely it is in part the brilliant colors—reds, yellows, oranges, blues, and greens. All the colors come from pigments, which are chemical compounds that, until the development of modern organic pigments, were almost exclusively made from minerals containing metals such as chromium, copper, iron, or lead. For centuries red pigments were oxides of iron or the sulfide of mercury, while yellow colors came from highly poisonous lead- and arsenic-based compounds. Blue has always been a valued color, and medieval artists used pigments based on the minerals azurite and lapis lazuli. Ultramarine blue was made from lapis lazuli, which was brought to Europe from Afghanistan and was as valuable as gold.

The recipe for one blue pigment used in medieval times called for the artist to mix strong vinegar, lime, and sal ammoniac in a copper pot—and then store the mixture under hot horse manure for fifteen days.

Our knowledge of ancient pigments has come from careful analysis of microscopic particles from old paintings by modern analytical methods. Meanwhile, chemists have created a new generation of chemical compounds that are the basis of the color photos in this book, the paint on your car, and the color of the ink in your pen. Much of the excitement of modern chemistry lies in the synthesis of new compounds thar are tailor-made for a specific application, so this chapter introduces you to the major types of chemical compounds.

4.1 ELEMENTS THAT EXIST AS MOLECULES

Before looking at chemical compounds, in which atoms of different elements are combined, it is important to realize that some elements exist naturally with their atoms combined in molecules. All but the heaviest of the elements have been isolated in large amounts in pure form. As you can see from Figures 3.12 and 3.15, most elements are metals and most of those are solids. In the solid state, a metal consists of atoms packed together as closely as possible. In general, metals have greater densities than nonmetals, which are often gases, liquids, or solids consisting of discrete atoms or even molecules. The elements of Group 8A, the noble gases, for example, are found as uncombined atoms in nature. In contrast, pure hydrogen (H_2), nitrogen (N_2), oxygen (O_2), and the Group 7A elements (the halogens, such as F_2, Cl_2, Br_2, and I_2) exist as two-atom or **diatomic molecules** under ordinary conditions of temperature and pressure (Figure 4.1).

You can smell the odor of ozone in an intense thunder storm. Lightning bursts provide the energy to convert O_2 in the atmosphere to ozone.

Oxygen, phosphorus, sulfur, and carbon are also interesting because each of these elements can exist in different forms called **allotropes.** The allotropes of oxygen are O_2, sometimes called dioxygen, and O_3, ozone. The latter is an unstable blue gas with a characteristic pungent odor. In fact, it was first detected by its odor, and its name comes from *ozein,* a Greek word meaning "to smell."

The most common allotrope of elemental phosphorus consists of four-atom (tetratomic) molecules (P_4) with the P atoms arranged at the corners of a tetrahedron (Figure 4.2); it is known as white phosphorus (Figure 3.15).

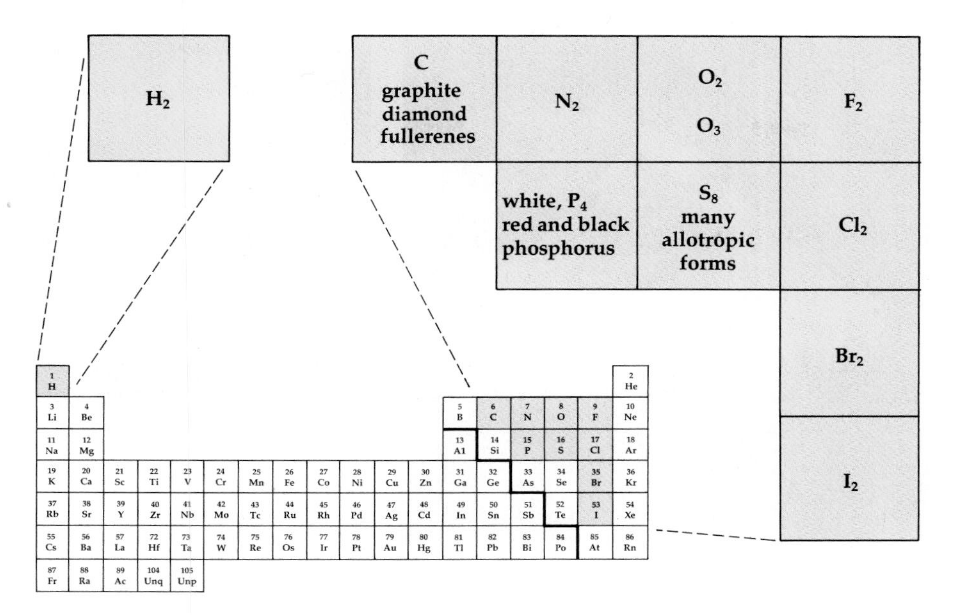

Figure 4.1
The nonmetallic elements that exist as polyatomic molecules. Notice that several of these elements have allotropic forms. Among them is sulfur, which has more allotropes than any other element.

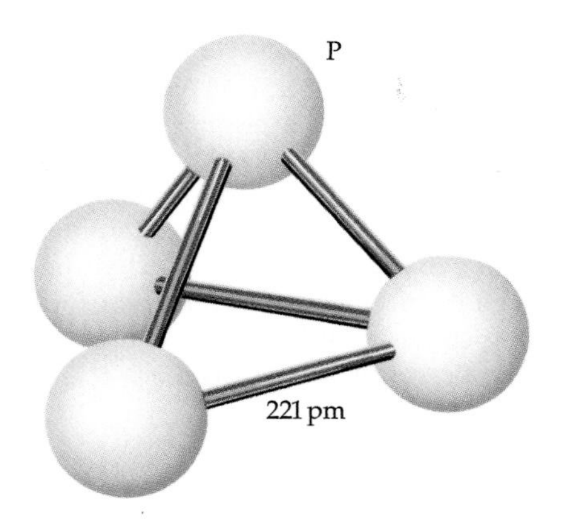

Figure 4.2
The structure of P_4, the white allotrope of phosphorus. See Figure 3.15 for a photograph of this form of the element.

Two other allotropes, red phosphorus (Figure 4.3) and the less common black form, consist of more complex networks of phosphorus atoms.

The common lemon-yellow form of sulfur consists of crown-shaped S_8 molecules (Figure 4.4). However, if this form is heated above 150 °C, the sulfur first melts and then the rings break open to give S_8 chains. These chains combine with one another to give even longer chains, and the tangled mass of chains results in a viscous or syrupy liquid. When heated to an even higher temperature, the sulfur chains break into shorter lengths to give a liquid that, when poured slowly into water, results in a flexible orange-red thread, a form sometimes called "plastic sulfur."

Figure 4.3
Red phosphorus, another allotrope of the element phosphorus. Its structure consists of P_4 units bonded to one another in a chain. *(C.D. Winters)*

(a)

(b)

Figure 4.4
Allotropes of sulfur. (a) At room temperature sulfur is a bright yellow solid (see Figure 2.16). At the submicroscopic level it consists of eight-membered, puckered rings of sulfur atoms. (b) When sulfur is heated the rings break open and eventually form long chains of sulfur atoms called "plastic sulfur." *(C.D. Winters)*

C_{60} is thought to be the only molecule made of a single kind of atom to form a spherical cage.

Finally, pure carbon is found as extended networks of C atoms in two well-known allotropic forms: graphite (Figure 2.8) and diamond (Section 14.4). Another allotrope consists of a newly discovered class of molecules, the fullerenes or "buckyballs" and related structures. Diamonds and graphite have been known for centuries, and it was generally believed that these were the only allotropic forms of carbon having well defined structures. Therefore, the scientific world was excited in the late 1980s when it learned that another form of carbon with a beautifully regular structure had been found and that it could be produced in quantities large enough to be studied. Black soot, the stuff that collects when carbon-containing materials are burned with very little oxygen, was found to contain small amounts of a new form of carbon composed of cages of 60 carbon atoms, molecules of C_{60} (Figure 4.5).

Figure 4.5
A representative of a third form of carbon, a new class of molecules called the fullerenes. One of the most important is C_{60}, buckminsterfullerene or "buckyball." The structure was proposed by R. Smalley, H. Kroto, and their co-workers in 1985 to explain an unusually stable material that they observed. In 1990, D. Huffman, W. Krätschmer, and their co-workers succeeded in isolating macroscopic quantities, and this has spurred extensive world-wide research efforts. This new form of carbon is very different from graphite and diamond, so research chemists and physicists are trying to understand and exploit such unique properties as superconductivity and the ability to trap other atoms.
(C.D. Winters)

Fullerene Researchers Make Tubular Discovery

RICHARD SMALLEY, ONE OF THE discoverers of buckyballs, recently said that "prior to 1985, as far as we were aware, no one had ever suggested that closed carbon cages would be formed spontaneously in a condensing carbon vapor. Now it turns out that there are readily obtainable conditions where this is effectively *all* that forms!"* The discovery of C_{60}, C_{70}, and other cage-like fullerenes has stimulated teams of scientists worldwide to look for other cages, to attempt to modify the ones already discovered, and to look for unusual properties. One such team is T. Ebbeson and P.M. Ajayan at the NEC laboratories. They discovered that by slightly modifying the conditions of the experiment used to make C_{60}, they could deposit a hard lump of material about 5 mm in diameter in their apparatus. When they broke open the lump, they were astonished to find a core of nearly pure *tubes*, yet another form of carbon. As the computer graphic shows, buckytubes resemble graphite sheets wrapped into a cylinder, one tube inside another! They are as narrow as two nanometers, and as long as a micron (1×10^{-4} cm). Although little is yet known of their properties, early indications are that they should be excellent conductors of electricity and that they should be incredibly strong. Just imagine what exciting new materials can be made of such a substance! And it is certain that continuing work on carbon cages and tubes will turn up fascinating new substances with useful and interesting properties.

Concentric tubules of carbon atoms called "buckytubes." They may be considered as "stretched" fullerenes. (BioSym Technologies, Inc./Photo Researchers)

*Smalley, R.E., *Accounts of Chemical Research, Vol. 25,* pp. 98–105, **1992.**

A C_{60} molecule resembles a hollow soccer ball; the surface is made up of five-membered rings linked to six-membered rings (like those seen in the other forms of carbon and in so many carbon-containing compounds). The shape also reminded its discoverers of a structure called a geodesic dome, which was invented years ago by the innovative American philosopher and engineer R. Buckminster Fuller. Therefore, the official name of the allotrope is buckminsterfullerene, but chemists often call it simply a "buckyball." We know now it is only one member of a larger family of carbon cages, generally called "fullerenes." Some fullerenes have fewer than 60 C atoms (called "buckybabies") and some have an even larger number—C_{70} and the giant fullerenes such as C_{240}, C_{540}, and C_{960}. The original buckyball, C_{60}, though, is the molecule found in greatest abundance in a pile of soot.

Professor Richard Smalley of Rice University, one of the discoverers of buckyballs, has said that "To a chemist [the discovery of buckyballs is] like Christmas" and that it is "a gift from the gods." The fullerenes have extraordinary properties, and dozens of uses have been proposed: microscopic ball bearings, lightweight batteries, new lubricants, new plastics, anti-tumor therapy for cancer patients (by enclosing a radioactive atom within the cage), and many others. Indeed, one chemist described C_{60} as "a Swiss army knife of a molecule."

The geodesic dome at Epcot Center in Florida. Such structures were designed originally by R. Buckminster Fuller, after whom the new class of carbon compounds is named. *(Dana Hyde/Photo Researchers)*

EXERCISE 4.1 • *Allotropes*

Graphite feels slippery if rubbed between your fingers, and the property has led to the use of graphite as a lubricant. How is the microscopic structure of graphite related to its macroscopic properties?

4.2 MOLECULAR COMPOUNDS

Compounds were defined in Chapter 2 as substances that can be decomposed into two or more different substances. You saw that sucrose, the sugar used in peanut brittle, is a compound with the **molecular formula** $C_{12}H_{22}O_{11}$. When sucrose is mixed with sulfuric acid, the sucrose decomposes in a chemical reaction to black carbon, an element, and transparent water, a compound (Figure 2.17).

$$1 \text{ molecule of } C_{12}H_{22}O_{11} \xrightarrow{\text{will give}} 12 \text{ atoms of C} + 11 \text{ molecules of } H_2O$$

The fact that sugar can be decomposed to carbon and water is the origin of the name of the class of compounds to which sugar belongs: carbohydrates. That is, it seems to be a "hydrate" of carbon.

Each molecule, the smallest unit of a molecular compound like sucrose, is changed into twelve *atoms* of carbon and eleven *molecules* of water. To describe this chemical change (or chemical reaction) on paper, the composition of each compound has been represented by its molecular formula, a shorthand way of expressing the number of atoms of each type in one molecule.

A striking feature of compounds is that, whether they are formed directly from the elements or from other compounds, the characteristics of the constituent elements are lost. Red phosphorus, shown in Figure 4.3, reacts violently with the element bromine, a foul-smelling, red-orange liquid (Fig-

(a)

(b)

Figure 4.6
The reaction of phosphorus with bromine. (a) Red-orange liquid bromine, Br_2, in a graduated cylinder and red phosphorus in an evaporating dish. (b) When bromine is poured over the phosphorus, they react to give phosphorus tribromide, PBr_3.
(C.D. Winters)

ure 4.6). The formula of the product, PBr_3, shows that there are four atoms per molecule: one atom of phosphorus and three atoms of bromine. The subscript to the right of the element's symbol indicates the number of atoms of that element in the molecule. If the subscript is omitted, it is understood to be one, as for P in PBr_3. These same principles apply to the formulas of some other, more common molecules.

Molecule	*Formula*
carbon dioxide	CO_2
ammonia	NH_3
sulfuric acid	H_2SO_4
ethyl alcohol	C_2H_6O
styrene	C_8H_8

Ethyl alcohol and styrene are examples of **organic compounds.** Such compounds are of great interest because they are the basis of the clothes you wear, the food you eat, and the living organisms in your environment. Organic compounds invariably contain carbon and hydrogen and may also have one or more atoms of oxygen, nitrogen, sulfur, or phosphorus.

Ethyl alcohol is made by the fermentation of sugar, starch, or cellulose. Alternatively, it can be synthesized from simpler molecules, a process that yields about 550 million pounds per year in the United States. The alcohol is found in many popular beverages. It is also commonly used to dissolve other chemicals, so you see it in toilet articles and pharmaceuticals. In addition, it is used as a germicide, and it may soon serve more widely in the United States as an automotive fuel.

There are often several ways to write the formulas of compounds. The formula written above for ethyl alcohol, C_2H_6O, is a simple molecular formula denoting that there are two C atoms, six H atoms, and one O atom per molecule.

C_2H_6O	CH_3CH_2OH
molecular formula of ethyl alcohol	condensed formula of ethyl alcohol

In the molecular formula of an organic compound the symbols of the elements are frequently written in alphabetical order, each with a subscript indicating the total number of atoms of that type in the molecule. On the right, the formula is written in a modified form to show how the atoms are grouped together in the molecule. Such formulas, called **condensed formulas,** emphasize chemically important groups of atoms in the molecule. Here the important group is an OH attached to a C atom, a grouping present in all alcohols. This is useful information, as this so-called **functional group** is often the point of attack on the molecule when it reacts with another atom or molecule.

Functional groups are responsible for the characteristic chemical behavior of organic compounds. For example, the chemistry of alcohols depends on the presence of an OH group attached to a C atom. Thus, ethyl alcohol (CH_3CH_2OH) differs little in its reactions from propyl alcohol ($CH_3CH_2CH_2OH$).

You will also see the formulas of molecules written to show how the atoms of the molecule are connected to one another. These are called **structural formulas.** You can determine the molecular formula of the compound from the structural formula or condensed formula simply by counting up the atoms.

Two lines connecting atomic symbols in a structural formula, such as C═C and C═O, show the presence of two bonds between those atoms (a double bond, *Section 10.2).*

C_2H_6O
ethyl alcohol

C_9H_8O
cinnamaldehyde, the source of the odor in cinnamon

$Pt(NH_3)_2Cl_2$
cisplatin, a cancer chemotherapy agent

The lines in structural formulas represent **covalent chemical bonds,** which are the forces that hold atoms together in molecules.

Remember that there is a distinction between a compound of two or more elements and a mixture of the same elements. The two gases hydrogen (H_2) and oxygen (O_2) can be mixed in all proportions without reaction. However, under certain conditions these two elements can and do react chemically to form the compound water. Not only does H_2O exhibit properties peculiar to itself and different from those of hydrogen and oxygen, but it also has a definite percentage composition by weight (88.8% oxygen and 11.2% hydrogen). Thus, there are two major differences between pure compounds and mixtures of elements: (a) compounds have distinctly different properties from their parent elements, and (b) compounds have a definite percentage composition by weight of their combining elements. We shall consider the latter difference in Section 4.7.

EXERCISE 4.2 • *Writing Molecular Formulas*

Write the molecular formula of each of the following compounds. (a) Oxalic acid, a compound found in many plants, has two atoms of carbon, two hydrogen atoms, and four oxygen atoms per molecule. (b) A molecule of the pesticide DDT has fourteen carbon atoms, nine hydrogen atoms, and five chlorine atoms.

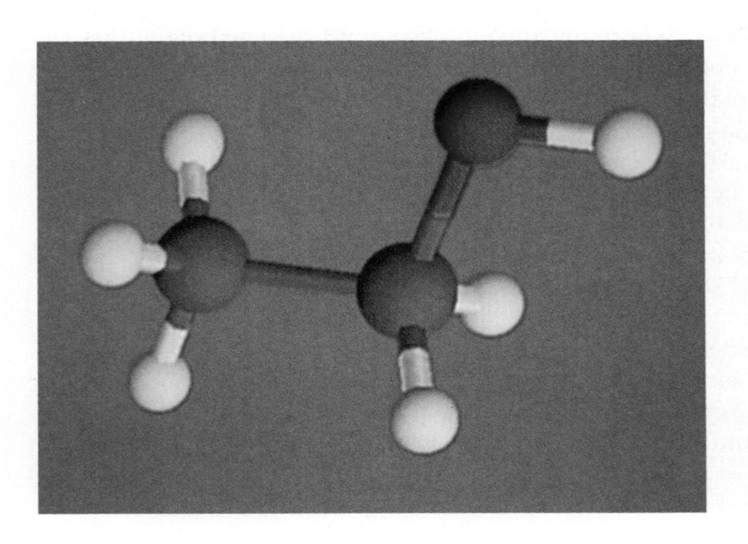

A computer-generated model of ethyl alcohol, CH_3CH_2OH. The C atoms are gray, the H atoms white, and the O atom is red. *(C.D. Winters)*

EXERCISE 4.3 • M lecular Formulas

Glycine is an important ami acid, a constituent of many living things. Its structural formula is

```
   H  O
   |  ||
H—C—C—O—H
   |
   N—H
   |
   H
```

Wh olecular formula? How many bonds can you count in the molecule? ite a reasonable condensed formula for glycine?

NS

Atoms of almost all the elements can gain or lose electrons in chemical reactions to form **ions,** which are atoms or groups of atoms bearing a net electrical charge. Many of the most important and familiar compounds are composed of ions and are thus known as **ionic compounds;** table salt (NaCl) and alum [$KAl(SO_4)_2$], which is used in making paper and pickles, are just two. Therefore, it is important to know the charges on commonly encountered ions, how to use those charges to predict formulas of ionic compounds, and how to name those compounds.

It is extremely important that you know the ions commonly formed by the elements shown in Figure 4.8 as well as the common polyatomic ions shown in Table 4.1.

A characteristic of metals is that *metal atoms lose electrons* in the course of their reactions *to form ions with a positive electrical charge,* ions commonly called **cations.** (The name is pronounced "cat'-ion.") Figure 4.7 shows how

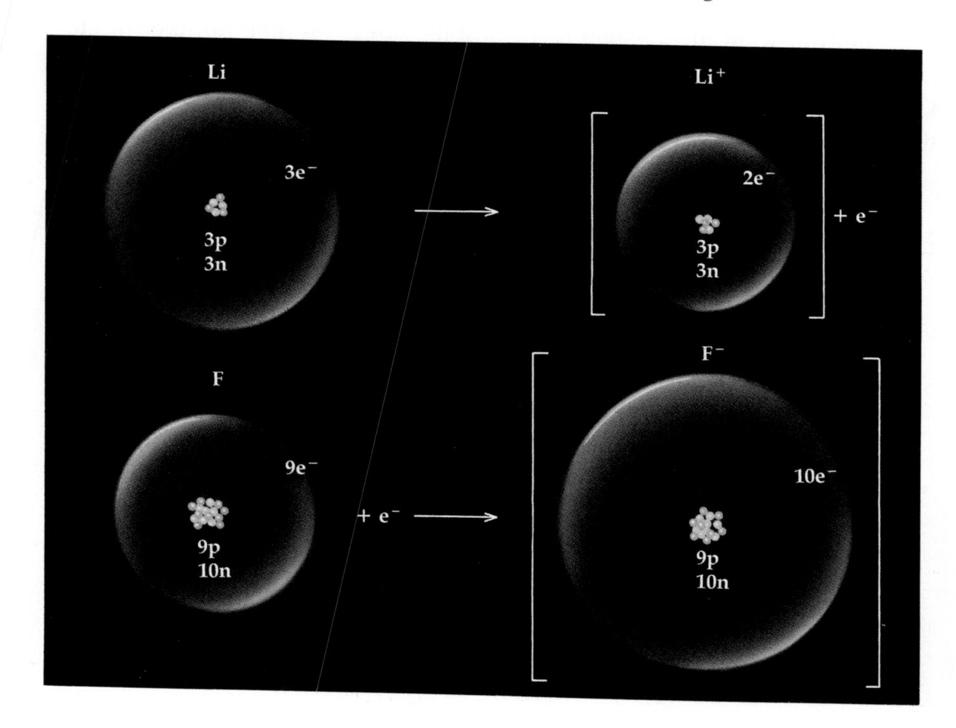

Figure 4.7
A lithium atom is electrically neutral because the number of positive charges (3 protons) and negative charges (3 electrons) is the same. When the atom loses one electron, it has one more positive charge than negative charge, so it has a net charge of 1+. We symbolize the resulting lithium cation as Li^+. A fluorine atom is also electrically neutral because there are nine protons and nine electrons. Because it is a nonmetal it gains electrons. As an F^- anion, it has one more electron than it has protons, so it has a net charge of 1−.

a lithium atom, which is electrically neutral because it has three protons and three electrons, can lose one of its electrons. Because it now has one more positive charge than negative charge (three protons the nucleus and only two electrons outside the nucleus), it has a net charge of 1+. We symbolize the resulting lithium cation as Li^+. When more than one electron has been lost, the number is written with the charge sign; for e... when the calcium ion is formed by the loss of two electrons from a ca... atom, the symbol is Ca^{2+}.

In contrast with metals, *nonmetals frequently gain electrons to g... a negative electrical charge* in the course of their reactions. Such ions a... **anions.** (The name is pronounced "ann′-ion.") Figure 4.7 shows how ... rine atom with nine protons and nine electrons gains an electron to give an ion having nine protons and ten electrons.

Charges on Monatomic Ions

Monatomic ions are single atoms that have lost or gained electrons. Typical charges on such ions are indicated in the periodic table in Figure 4.8. Notice that *metals of Groups 1A through 3A form positive ions whose charge is equal to the group number of the metal.*

Group	*Metal Atom*	*Electrons Lost*	*Metal Ion*
1A	Na (11 protons, 11 electrons)	1	Na^+ (11 protons, 10 electrons)
2A	Ca (20 protons, 20 electrons)	2	Ca^{2+} (20 protons, 18 electrons)
3A	Al (13 protons, 13 electrons)	3	Al^{3+} (13 protons, 10 electrons)

The transition metals also form cations, but a variable number of electrons may be lost. These elements may therefore form ions of different charges (Figure 4.8), but 2+ and 3+ ions are typical of many of them.

Figure 4.8
Charges on some common monatomic cations and anions. Note that metals usually form cations, where the charge is given by the group number in the case of the main group metals (*light blue*). For transition metals (*darker blue*), the positive charge is variable, and other ions in addition to those illustrated are possible. Nonmetals generally form anions with a charge equal to 8 minus the group number.

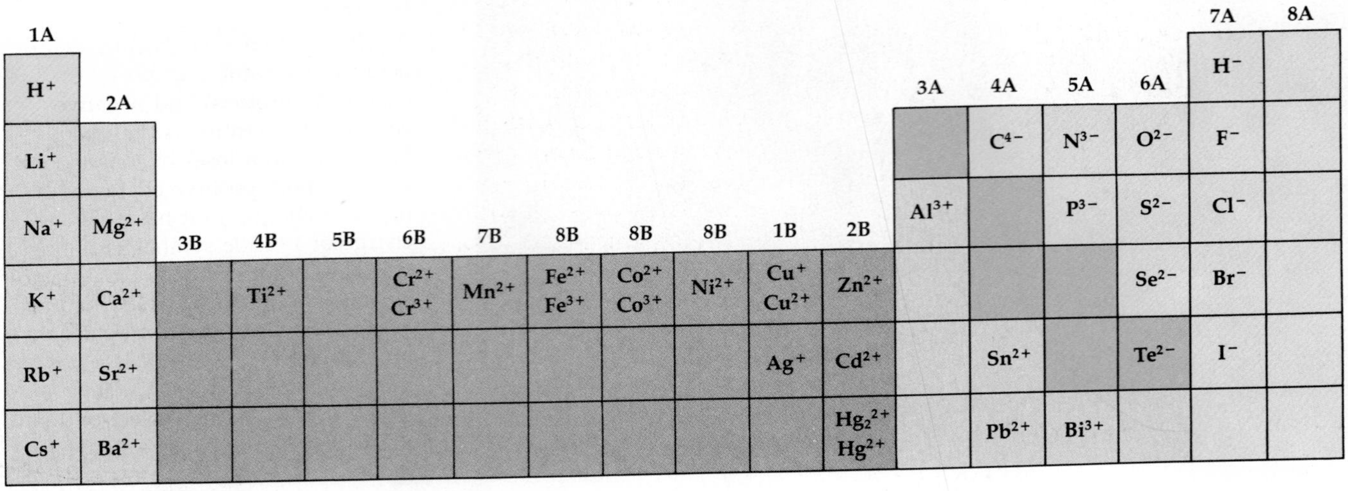

1A	2A	3B	4B	5B	6B	7B	8B	8B	8B	1B	2B	3A	4A	5A	6A	7A	8A
H^+																H^-	
Li^+													C^{4-}	N^{3-}	O^{2-}	F^-	
Na^+	Mg^{2+}											Al^{3+}		P^{3-}	S^{2-}	Cl^-	
K^+	Ca^{2+}		Ti^{2+}		Cr^{2+} Cr^{3+}	Mn^{2+}	Fe^{2+} Fe^{3+}	Co^{2+} Co^{3+}	Ni^{2+}	Cu^+ Cu^{2+}	Zn^{2+}				Se^{2-}	Br^-	
Rb^+	Sr^{2+}									Ag^+	Cd^{2+}		Sn^{2+}		Te^{2-}	I^-	
Cs^+	Ba^{2+}										Hg_2^{2+} Hg^{2+}		Pb^{2+}	Bi^{3+}			

Group	Metal Atom	Electrons Lost	Metal Ion
7B	Mn (25 protons, 25 electrons)	2	Mn^{2+} (25 protons, 23 electrons)
8B	Fe (26 protons, 26 electrons)	2	Fe^{2+} (26 protons, 24 electrons)
8B	Fe (26 protons, 26 electrons)	3	Fe^{3+} (26 protons, 23 electrons)

Nonmetals often form ions whose negative charge equals 8 minus the group number of the element. For example, N is in Group 5A and so can gain $8 - 5 = 3$ electrons to form a 3− ion.

Group	Nonmetal Atom	Electrons Gained	Nonmetal Ion
5A	N (7 protons, 7 electrons)	3 (=8 − 5)	N^{3-} (7 protons, 10 electrons)
6A	S (16 protons, 16 electrons)	2 (=8 − 6)	S^{2-} (16 protons, 18 electrons)
7A	Br (35 protons, 35 electrons)	1 (=8 − 7)	Br^{-} (35 protons, 36 electrons)

Also notice in Figure 4.8 that hydrogen appears at two locations in the table; it can either gain or lose an electron, depending on the other atoms it encounters.

$$\text{H (1 proton, 1 electron)} \rightarrow e^- + H^+ \text{ (1 proton, 0 electrons)}$$
$$\text{H (1 proton, 1 electron)} + e^- \rightarrow H^- \text{ (1 proton, 2 electrons)}$$

Finally, the noble gases lose or gain electrons in only a few reactions. These elements are extremely stable chemically and have no common ions listed in Figure 4.8.

Valence Electrons

Another important idea is illustrated by the charges on many of the ions listed in Figure 4.8. Consider that the metals of Groups 1A, 2A, and 3A form ions having 1+, 2+, and 3+ charges; that is, their atoms have lost one, two, or three electrons. In each case the number of electrons remaining on the ions is the same as the number of electrons in an atom of a noble gas. For example, a magnesium atom has 12 electrons. It loses two of them to give Mg^{2+}, which has ten electrons—the same number as in an atom of Ne.

Atoms of the nonmetals at the right of the periodic table would have to lose a great many electrons to achieve the same number as a noble gas atom, but if an atom of a nonmetal *gains* a few electrons, it can end up with the same number as a noble gas atom of higher atomic number. For example, an oxygen atom has eight electrons and would have to lose six to be left with the number present in He. But by gaining two electrons an oxygen atom can form O^{2-}, which has ten electrons, as does Ne. Since the noble gases are in periodic Group 8, eight minus the group number gives the number of electrons gained and hence the negative charge on the ion. It seems clear from the charges given in Figure 4.8 that ions having the same number of electrons as a noble gas atom are especially favored when compounds form.

While trying to help his introductory chemistry students to learn the typical charges on ions and other facts of chemical combination, the Ameri-

This idea of valence electrons occupying the outermost shell in an atom and determining the numbers of different kinds of atoms that can combine will be developed further in Chapter 9.

can chemist G. N. Lewis (1875–1946) came up with the idea that electrons in atoms might be arranged in shells, with the nucleus at the center. Each shell could hold a characteristic number of electrons, and only those electrons in the outermost shell were involved when one atom combined with another by forming ions or molecules. These outermost electrons came to be known as **valence electrons,** and they are the ones that determine how many atoms of one kind combine with atoms of another kind in a compound. For example, a magnesium atom has two valence electrons (two more electrons than a neon atom), and when these valence electrons are lost, a Mg^{2+} ion forms. An oxygen atom has six valence electrons (six more than a helium atom), and has room for two more (which would give it the same number as a neon atom); thus an O^{2-} ion is reasonable.

EXAMPLE 4.1

Predicting Ion Charges

Predict the charges on ions of gallium and of sulfur.

SOLUTION Gallium is a metal in Group 3A of the periodic table, so it is predicted to lose three electrons to give the Ga^{3+} cation.

$$Ga \rightarrow Ga^{3+} + 3e^-$$

Sulfur is a nonmetal in Group 6A, so it is predicted to gain electrons to give an anion. The number of electrons is 8 − 6 = 2. Therefore,

$$S + 2e^- \rightarrow S^{2-}$$

EXERCISE 4.4 • *Predicting Ion Charges*

Using the guidelines above, predict possible charges for ions formed from (a) K, (b) Se, (c) Be, (d) V, (e) Co, and (f) Cs.

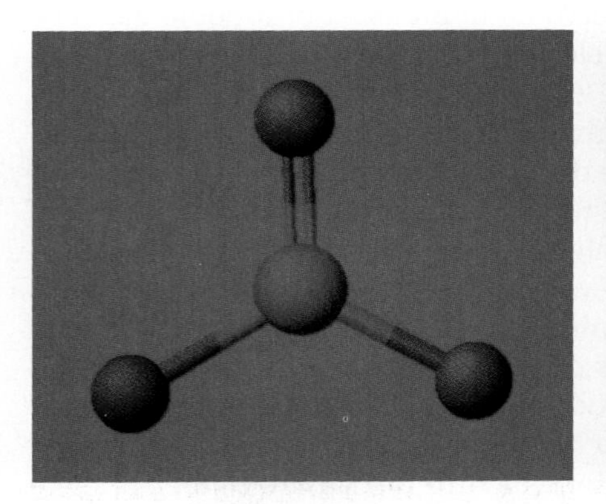

Computer-generated models of the carbonate ion, CO_3^{2-} (*left*), and ammonium ion, NH_4^+ (*right*). The colors are C = gray, O = red, N = blue, and H = white. *(C.D. Winters)*

Table 4.1

Names and Composition of Some Common Polyatomic Ions

Cation (Positive Ion)			
NH_4^+	ammonium ion		
Anions (Negative Ions)			
OH^-	hydroxide ion	NO_2^-	nitrite ion
CN^-	cyanide ion	NO_3^-	nitrate ion
$CH_3CO_2^-$	acetate ion		
CO_3^{2-}	carbonate ion	SO_3^{2-}	sulfite ion
HCO_3^-	hydrogen carbonate ion (or bicarbonate ion)	SO_4^{2-}	sulfate ion
		HSO_4^-	hydrogen sulfate ion (or bisulfate ion)
MnO_4^-	permanganate ion	PO_4^{3-}	phosphate ion
ClO_3^-	chlorate ion	HPO_4^{2-}	hydrogen phosphate ion
ClO_4^-	perchlorate ion	$H_2PO_4^-$	dihydrogen phosphate ion

Polyatomic Ions

A **polyatomic ion** contains two or more atoms, and the *group* bears a charge. For example, carbonate ion, CO_3^{2-}, is a common polyatomic anion. It consists of one C atom and three O atoms with two units of negative charge spread over the group of four atoms. One of the most common polyatomic cations is NH_4^+, the ammonium ion. In this case, four H atoms surround an N atom, and the group bears a 1+ charge. *It is extremely important to know the names, formulas, and charges of the common polyatomic ions listed in Table 4.1.*

4.4 IONIC COMPOUNDS

Ions of opposite charge (for example, a positive ion such as Al^{3+} and a negative ion such as O^{2-}) are attracted to one another by *electrostatic forces* (Figure 3.2). These forces are described by **Coulomb's law,**

$$\text{force of attraction} = k\frac{(n_+e)(n_-e)}{d^2}$$

where n_+ is the charge on the positive ion (*e.g.*, +3 for Al^{3+}), n_- is the charge on the negative ion (*e.g.*, −2 for O^{2-}), e is the charge on the electron (1.602×10^{-19} C), d is the distance between the ions, and k is a proportionality constant. A close look at the equation shows that the force of attraction increases as the charges on the ions become larger and as the distance between the ions becomes smaller. Smaller ions having larger charges are attracted more strongly to each other.

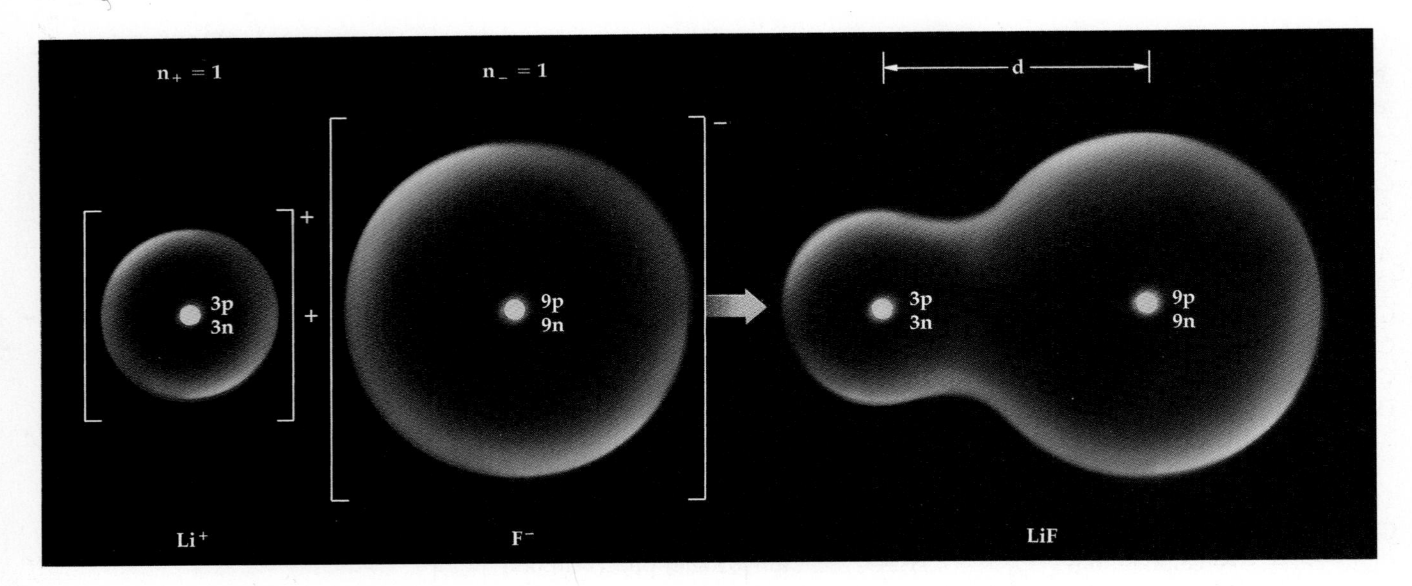

The formation of LiF from Li^+ and F^-. Here $n_+ = +1$ for Li^+ and $n_- = -1$ for F^-; d is the distance between the centers of the Li^+ and F^- ions.

The strength of electrostatic forces directly influences the properties of compounds formed from ions, as will be described below.

The first question to consider is how one can predict whether a compound will be ionic. As an extension of the guidelines on charges given in Figure 4.8, we can say the following:

a. Metals (the blue elements in Figure 4.8) almost always form positive ions and form ionic compounds.

b. Nonmetals (the yellow elements in Figure 4.8) give monatomic negative ions in ionic compounds *only* when combined with a metal.

c. It is difficult to predict when the metalloids (the green elements in Figure 4.8) form ions.

For example, you can be sure that Mg^{2+} with Br^-, K^+ with S^{2-}, and Ca^{2+} with O^{2-} are combinations that are suitable for ionic compound formations. On the other hand, although the nonmetal chlorine can combine with the metalloid boron to form a compound, it is not considered ionic.

When positive and negative ions combine to form an ionic compound, the outcome is an electrically neutral substance. This means you can predict the formulas of ionic compounds, because the ions must combine so that the number of positive charges equals the number of negative charges. Let us

Figure 4.9
Aluminum and oxygen form aluminum oxide, Al_2O_3, which in pure form is a stable white powder. (Indeed, all aluminum metal forms an invisible, microscopically thin coating of aluminum oxide on exposure to air. It is this coating that protects the metal from corrosion and allows aluminum to be widely used as a structural material for airplanes, furniture, and cooking utensils, among other things.) The red gem in the photo is a ruby and is a form of aluminum oxide, where some of the Al^{3+} ions have been replaced with Cr^{3+} ions. It is the transition metal ions that give rise to the color of rubies. *(C.D. Winters)*

take the compound formed when aluminum interacts with oxygen as an example (Figure 4.9). Aluminum is a metal in Group 3A, so it loses three electrons to form the Al^{3+} cation. Oxygen is a nonmetal in Group 6A and so gains two electrons to form a 2− anion. To have a compound with the same number of negative and positive charges, two Al^{3+} ions [total charge is $6+ = 2 \times (3+)$] combine with three O^{2-} ions [total charge is $6- = 3 \times (2-)$]. This means $2\ Al^{3+} + 3\ O^{2-} \xrightarrow{\text{will give}} Al_2O_3$.

Calcium ion has a 2+ charge because the metal is a member of Group 2A. It can combine with a number of anions to form ionic compounds. For example, the following combinations are possible.

Ion Combination	*Compound*	*Overall Charge on Compound*
$Ca^{2+} + 2\ Cl^-$	$CaCl_2$	$(2+) + 2 \times (1-) = 0$
$Ca^{2+} + CO_3^{2-}$	$CaCO_3$	$(2+) + (2-) = 0$
$3\ Ca^{2+} + 2\ PO_4^{3-}$	$Ca_3(PO_4)_2$	$3 \times (2+) + 2 \times (3-) = 0$

Notice that in writing all these formulas, *the symbol of the cation is always given first, followed by the symbol for the anion*. Also notice the use of parentheses when there is more than one of a polyatomic ion.

EXAMPLE 4.2

Ionic Compounds

For each of the following ionic compounds, write the symbols for the ions present and give the number of each: (a) $MgBr_2$; (b) Li_2CO_3, and (c) $Fe_2(SO_4)_3$.

SOLUTION (a) $MgBr_2$ is composed of one Mg^{2+} ion and two Br^- ions. When a halogen such as bromine is associated only with a metal, you can assume the halogen is an anion with a charge of 1−. Magnesium is a metal in Group 2A and *always* has a charge of 2+ in its compounds.

(b) Li_2CO_3 is composed of two lithium ions, Li^+, and one carbonate ion, CO_3^{2-}. You should learn to recognize the carbonate ion and recall its formula and charge. To help you remember that carbonate has a 2− charge, you can see that Li is a Group 1A ion and so always has a 1+ charge in its compounds. Because the two 1+ charges neutralize the negative charge of the carbonate ion, the latter must be 2−.

(c) $Fe_2(SO_4)_3$ comes from two Fe^{3+} ions and three sulfate ions, SO_4^{2-}. The only way to tell this is to recall that sulfate is 2−. Since you have three of them (for a total charge of 6−), the two iron cations must add up to 6+. This is possible only if each is 3+.

The moral of this tale: *It is imperative to learn the formulas* ***and*** *charges of the common polyatomic ions in Table 4.1.*

EXAMPLE 4.3

Writing Formulas for Ionic Compounds

Write formulas for ionic compounds composed of an aluminum cation and (a) a fluoride ion, (b) a sulfide ion, and (c) a nitrate ion.

SOLUTION First, the aluminum cation must have a charge of 3+ since Al is a metal in Group 3A.

(a) Fluorine is a Group 7A element and so is a nonmetal. Its charge is predicted to be 1− (from $8 - 7 = 1$). Therefore, we need three F^- ions (total charge for the three ions is 3−) to combine with Al^{3+}. The formula of the compound is AlF_3.

(b) Sulfur is a nonmetal in Group 6A and so forms a 2− anion. Thus, we need to combine two Al^{3+} ions [total charge is $6+ = 2 \times (3+)$] with three S^{2-} ions [total charge is $6- = 3 \times (2-)$]. The compound is Al_2S_3.

(c) The nitrate ion has the formula NO_3^- (Table 4.1). Therefore, the question is similar to the AlF_3 case, and the compound is $Al(NO_3)_3$. Here there are parentheses around the NO_3 portion of the formula to show that we have three of the NO_3^- "units" present.

EXERCISE 4.5 • *Ionic Compounds*

a. For each of the following, tell what ions are present and the number of each per formula unit: (a) NaF, (b) $Cu(NO_3)_2$, and (c) $NaCH_3CO_2$.

b. Iron is a transition metal and so can form ions with at least two different charges. Write the formulas of the compounds formed between iron and chlorine.

c. Write the formulas of all of the neutral ionic compounds that can be formed by combining the cations Na^+ and Ba^{2+} with the anions S^{2-} and PO_4^{3-}.

The Ionic Crystal Lattice

Ions having opposite charges attract one another, and ions having charges of the same sign repel one another. Coulomb's law tells us that the attraction or repulsion of one ion for another increases as the ions move closer and as their charges increase. For example, the large charges of 3+ and 2− on Al^{3+} and O^{2-} ions cause these ions to attract each other strongly if they are a few hundred picometers apart. By contrast, sodium and chloride ions (Na^+ and Cl^-) have smaller charges and attract less strongly at the same distance.

The attractive and repulsive forces are independent of the orientation of the ions; the attraction and repulsion will be the same whether one ion is above, below, to the left of, or in front of the other, as long as the charges and distance are the same. For this reason the most favorable arrangement of ions is one in which oppositely charged ions are brought as close together as possible, and ions of like charge remain somewhat farther apart on average. The most efficient way in which this can be done is for the ions to form an extended three-dimensional network like the one shown in Figure 4.10. This regular array of positive and negative ions is called a **crystal lattice.**

Careful examination of the crystal lattice in Figure 4.10 reveals that each sodium ion is surrounded by six chloride ions and each chloride ion is surrounded by six sodium ions. It is impossible to say that a certain cation is paired with a certain anion to form a single unit. That is, in an ionic compound such as NaCl there are no NaCl molecules, in the sense that there are H_2 molecules in a sample of hydrogen or H_2O molecules in water. Although they do not contain molecules, ionic compounds do have well-defined ratios of one kind of ions to another, and these are represented by their formulas. For example, NaCl represents the simplest ratio of sodium ions to chloride

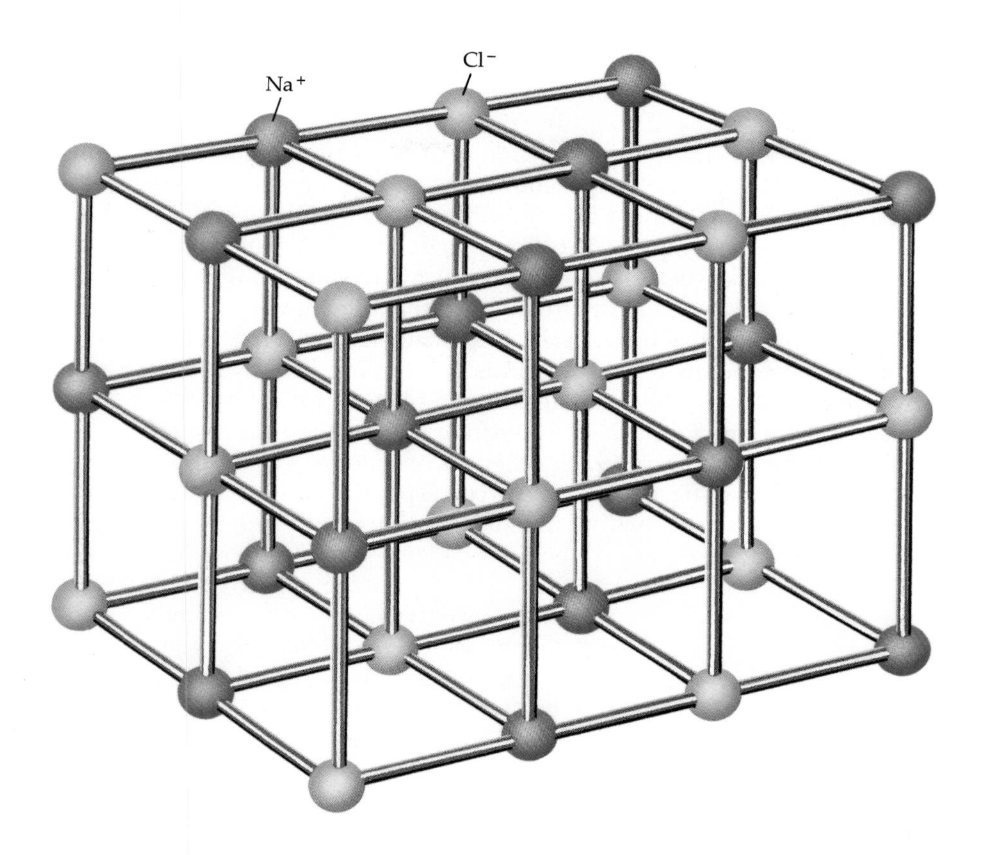

Figure 4.10
A model of a crystal of sodium chloride. The lines between the ions are not chemical bonds but are simply reference lines to show the relationship of Na^+ (*red*) and Cl^- (*green*) in space.

ions in the lattice shown in Figure 4.10, namely 1:1. The combination of one Na^+ ion and one Cl^- ion is referred to as a **formula unit** of sodium chloride.

Properties of Ionic Compounds

Compounds made up of ions have characteristic properties that can be understood in terms of the charges of the ions and their arrangement in a crystal lattice. Because each ion is surrounded by many oppositely charged nearest neighbors, each ion is held fairly tightly in its allotted location. At room temperature each ion can vibrate a little around its average location, but much more energy must be added before an ion can move fast enough and far enough to escape the confinement of its neighbors. Therefore, a fairly high temperature is required to melt an ionic compound, and most ionic compounds are solids at room temperature. Moreover, the melting point correlates with the charges on the ions. For example, NaCl, which consists of 1+ and 1− ions, melts at 801 °C; Al_2O_3, with 3+ and 2− ions, melts at a much higher temperature, 2072 °C. High melting point is a useful property. For example, white Al_2O_3 is used in fire bricks, ceramics, and other materials that must withstand high temperatures. In combination with another ionic oxide, ZrO_2, aluminum oxide can be used to make very fine fibers that can withstand temperatures up to 1400 °C and can be incorporated into molten metals to strengthen the metal after it cools back to a solid.

Crystals of ionic solids are transparent, hard, and brittle, have characteristic shapes, and can be cleaved along certain planes (Figure 4.11). Hardness is a consequence of the strong attractive forces between oppositely charged ions. It is very difficult to move one ion relative to another, and so the solid maintains its shape. The shape and cleavage of a crystal depend on the submicroscopic arrangement of ions within the crystal lattice. For example, Figure 4.10 shows 90° angles between planes of ions; crystals of NaCl also have 90° angles between flat faces. (If you made some salt crystals as described in Chapter 2, you have already seen that they have 90° angles; examination of some crystals from a salt shaker will also confirm this.) Cleavage can be understood in terms of repulsions among like-charged ions. As shown in Figure 4.11, only a small shift of a plane of ions is needed to bring positive ions in one layer close to positive ions in the next layer. The two portions of the lattice repel each other and separate. Macroscopically, the crystal cleaves, leaving two flat faces corresponding to the layers of ions.

In an ionic solid the ions are in fixed positions, but when the solid melts the ions can move. Most ionic solids do not conduct electricity, but molten ionic compounds do. This happens because an electric current is a movement of charged particles from one place to another. In a metal wire the charged particles are electrons, while in an ionic liquid they are ions, but the

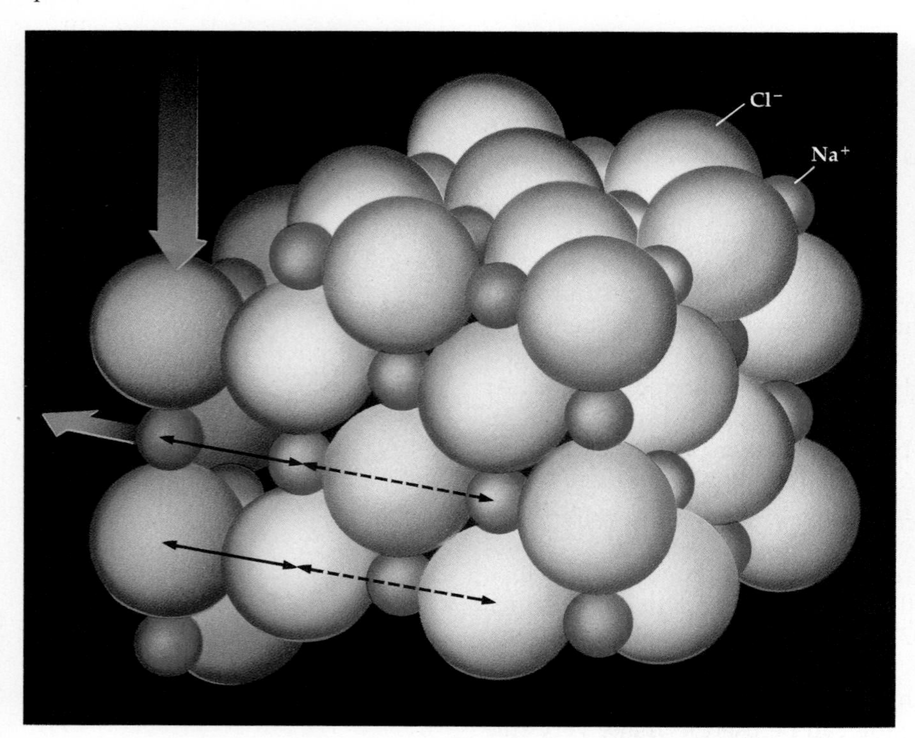

(a)

(b)

Figure 4.11
Microscopic view of cleavage of an ionic crystal. (a) When an external force (large gray arrow) causes one layer of ions to shift slightly with respect to another, positive ions are brought close to other positive ions, and negative ions become nearest neighbors to other negative ions. The strong repulsive forces produced by this arrangement of ions (*double-headed black arrows*) cause the two layers to split apart. (b) A sharp blow on a knife edge lying along a plane of a salt crystal causes the crystal to split. *(b, C.D. Winters)*

Portrait of a Scientist

Michael Faraday (1791–1867)

The terms "anion," "cation," "electrode," and "electrolyte" originated with Michael Faraday, one of the most influential men in the history of chemistry. Faraday was apprenticed to a bookbinder in London (England) when he was only 13. This suited him, however, as he enjoyed reading the books sent to the shop for binding. One of these chanced to be a small book on chemistry, and his appetite for science was whetted. He soon began performing experiments on electricity, and in 1812 a patron of the shop invited Faraday to accompany him to a lecture at the Royal Institution by one of the most famous chemists of his day, Sir Humphry Davy. Faraday was so intrigued by Davy's lecture that he wrote to ask Davy for a position as an assistant. Faraday was accepted and began his work in 1813. His work was very fruitful and Faraday was so talented that he was made the Director of the Laboratory of the Royal Institution about 12 years later.

It has been said that Faraday's contributions are so enormous that, had there been Nobel Prizes when he was alive, he would have received at least six. These could have been for contributions such as

- the explanation of electromagnetic induction, which led to the first transformer and electric motor
- the proposal of the laws of electrolysis (the effect of electric current on chemicals)
- the discovery of the magnetic properties of matter
- the discovery of benzene and other organic chemicals (which led to important chemical industries)
- the discovery of the "faraday effect" (the rotation of the plane of polarized light by a magnetic field)
- the introduction of the notion of electric and magnetic fields

Michael Faraday. (Oesper Collection in the History of Chemistry/University of Cincinnati)

In addition to making discoveries that had profound effects on science, Faraday was an educator. He wrote and spoke about his work in memorable ways, especially in lectures to the general public that helped to popularize science.

---

effect is the same. Figure 4.12 shows the result of placing two electric conductors called **electrodes** into a molten ionic compound. The bulb lights because an electric circuit is completed by movement of ions through the molten substance.

A very important property of ionic solids is that many of them dissolve in water. The water that you drink every day contains small concentrations of many ionic substances that have been dissolved from solid materials in our environment; some of these, such as calcium ions, are necessary nutrients. The oceans contain much higher concentrations of substances such as NaCl that have dissolved in the water. Dissolving an ionic solid involves separating each ion from the oppositely charged ions that surround it in a crystal lattice, and this is not energetically favorable. Water is especially good at doing this because each water molecule has a positively charged end and a negatively charged end. Therefore a water molecule can attract a positive ion to its negative end or it can attract a negative ion to its positive end. When an ionic compound dissolves, each negative ion becomes surrounded by water molecules with their positive ends pointing toward it,

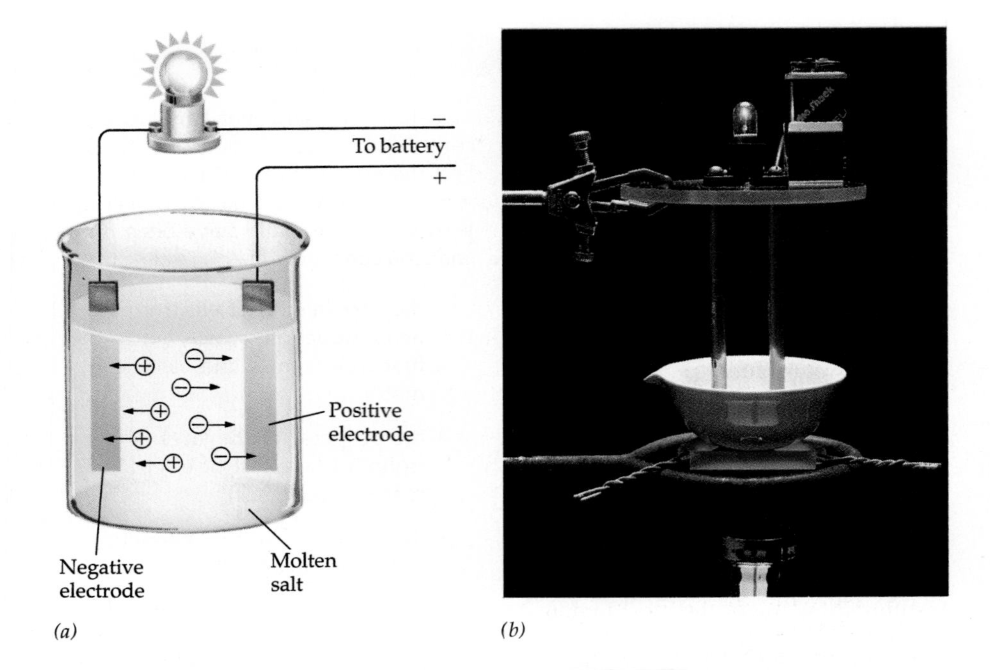

Figure 4.12
The conductivity of a molten salt. When an ionic compound is a solid, it will not conduct electricity. However, when the salt is melted, ions are free to move and migrate to the electrodes dipping into the melt (a). The lighted bulb shows that the electric circuit is complete (b).
(b, C.D. Winters)

and each positive ion becomes surrounded by the negative sides of many water molecules. Figure 4.13 shows how this works.

The water-encased ions coming from a dissolved ionic compound are free to move about in the solution. Under normal conditions, the movement of ions is random, and the Na^+ and Cl^- ions from dissolved NaCl are dispersed uniformly throughout the solution. However, if two electrodes are placed in the solution and connected to a battery, the cations migrate

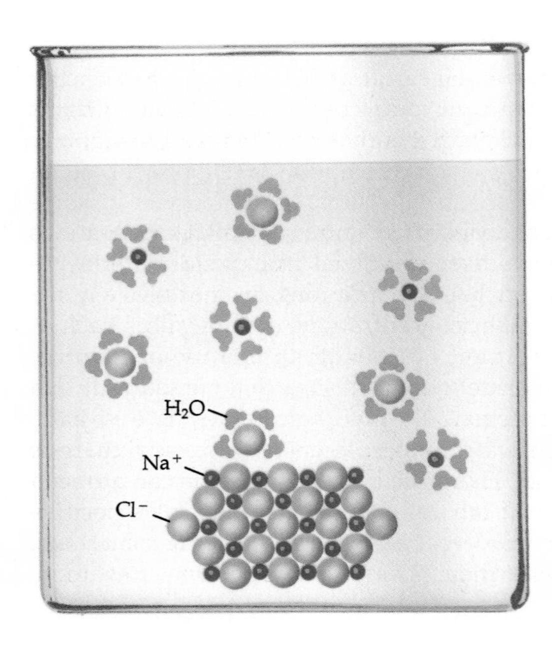

Figure 4.13
A model for the process of dissolving NaCl in water. The crystal dissociates to give Na^+ and Cl^- in aqueous (water) solution. These ions, which are cloaked in water molecules, are free to move about. Such solutions conduct electricity, so the substance dissolved is called an electrolyte.

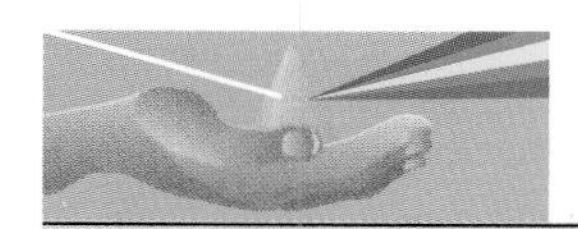

CHEMISTRY YOU CAN DO

Experiments with Electricity and Water

Here is an experiment you can do to illustrate how water is affected by electric charges, and how electrostatic attractions vary with distance. You will need the following:

- a nylon, rubber, or plastic comb; or a plastic knife, fork, or spoon like those from a fast-food restaurant
- a water faucet and sink
- a piece of fur or cloth

Adjust the faucet to produce a thin stream of water; it should be about 1 to 2 mm (1/16 in.) in diameter. Hold the comb horizontal and a few centimeters lower than the end of the faucet. Bring the end of the comb close to (but not into) the stream of water. Can you see anything happening? Now rub the comb with the fur or cloth, or run it through your hair. Then immediately bring the comb close to the stream of water. What do you observe now?

Move the comb closer to and farther from the stream of water. What difference does this make? Run the comb through your hair several more times and try the experiment again. Does the number of times you comb your hair make a difference? If so, what is different? Are your observations consistent with Coulomb's law?

What happens if you get the comb too close and the stream of water hits it? Do your observations change when the comb is wet? Once the comb has been wetted by the water, does combing your hair or rubbing the comb with fur or cloth still give the same effect?

Try some additional experiments with small items other than the comb. Try rubbing a plastic pen, wooden pencil, or other similar-sized article with cloth, fur, or through your hair and see if it has any effect on the stream of water.

If time permits, repeat this experiment on a damp, humid day and on a relatively dry day. What difference does humidity make with respect to what you observe?

Discuss your results with other students. Did they obtain the same results you did? If not, can you identify different conditions in their experiments from yours?

As a corollary experiment, see what happens when you bring a piece of paper close to the screen of a TV set that has been on for a few minutes. Can you hear something as well as see something? What do you observe when you bring the back of your hand close to the TV screen? A TV picture tube is just a glorified example of a cathode-ray tube (Figure 3.4). Given this information, can you propose an idea about the observations you made with the paper and TV screen? What do you think would happen to a thin stream of water falling past the front of a TV set? (It would not be a good idea to try this as an experiment.)

Interpret the observations you have made in terms of a submicroscopic model involving electrons and the structure of atoms and molecules.

through the solution to the negative plate and anions move to the positive plate. If a light bulb is inserted into the circuit as in Figure 4.12, the bulb will light, showing that the circuit is complete. Compounds that behave this way in water are called **electrolytes,** and all ionic compounds that are soluble in water are good electrolytes.

The most important chemical property of ionic substances is that *each ion has its own characteristics, and these are different from the characteristics of the atom from which the ion was derived.* For example, a sodium ion, Na^+, has quite different properties from a sodium atom, Na, and a chloride ion, Cl^-, is very different from either a chlorine atom, Cl, or a molecule of chlorine, Cl_2. You eat sodium ions and chloride ions every day in table salt, but you would not want to eat sodium, a metal that would react vigorously with the saliva in your mouth, or chlorine, a poisonous green gas. The sodium ions and chloride ions do not undergo such reactions. This difference between ions and atoms or molecules is related to the difference between the properties of ionic compounds and the properties of the elements that form them. We shall explore the chemical reactions of ions, particularly ions in solution, in the next chapter.

EXERCISE 4.6 • *Chemical Formulas*

Explain some differences among the species Br_2, K, and KBr.

4·5 NAMES OF COMPOUNDS

As you are beginning to see, chemistry tries to use precise language to describe the world. However, assigning clear, unambiguous names to compounds has always been a problem in chemistry, and it is a problem that continues today as new, ever more complicated molecules are discovered. Nonetheless, there are rules for naming the kinds of compounds you will read about in this book, and you should learn these rules thoroughly.

Naming Ionic Compounds

The names of ionic compounds are built from the names of the positive and negative ions in the compounds.

Naming Positive Ions With a few exceptions, the positive ions described in this book are metal ions. Positive ions are named by the following rules:

a. For a monatomic positive ion, that is, a metal cation, the name is simply the name of the metal plus the word "ion." For example, we have already referred to Al^{3+} as the aluminum ion.

b. There are cases, especially in the transition series, in which a metal can form more than one positive ion. A common practice today is to indicate the charge of the ion by a Roman numeral in parentheses immediately following the ion's name (the Stock system). For example, Co^{2+} is the cobalt(II) ion and Co^{3+} is the cobalt(III) ion. Another cation that you will see on occasion is Hg_2^{2+}, the name of which is mercury(I) ion. The reason for the Roman numeral (I) is that the ion is composed of two Hg^+ ions bonded together.

The Stock system came from Alfred Stock (1876–1946), a German chemist famous for his work on the hydrogen compounds of boron and silicon. Because he worked closely with mercury metal for many years, he contracted mercury poisoning, which leads to memory loss and severe physical disabilities.

Finally, you will encounter the nonmetal cation NH_4^+ or *ammonium* ion many times in this book, in the laboratory, and in your environment.

Naming Negative Ions Two types of negative ions must be considered: those having only one atom *(monatomic)* and those having several atoms *(polyatomic).*

a. A *monatomic negative ion* is named by adding *-ide* to the stem of the name of the nonmetal element from which the ion is derived (see Figure 4.14). As a group, the anions of the Group 7A elements, the halogens, are called the *halide ions.*

b. *Polyatomic negative ions* are quite common, especially those containing oxygen (called *oxoanions*). The names of some of the most common are given in Table 4.1. Most of these names must simply be memorized. However, there are some guidelines that can help. For example, consider the following pairs of ions.

NO_3^- is the nitr*ate* ion while NO_2^- is the nitr*ite* ion
SO_4^{2-} is the sulf*ate* ion while SO_3^{2-} is the sulf*ite* ion

The oxoanion with the *greater number of oxygen atoms* is given the suffix *-ate,* while the oxoanion with the *smaller number of oxygen atoms* has the suffix *-ite.* For a series of oxoanions with more than two members, the ion with the largest number of oxygen atoms has the prefix *per-* and the suffix *-ate*. The ion with the smallest number of oxygen atoms has the prefix *hypo-* and the suffix *-ite.* The oxoanions containing chlorine are good examples.

ClO_4^- is the *per*chlor*ate* ion
ClO_3^- is the chlor*ate* ion
ClO_2^- is the chlor*ite* ion
ClO^- is the *hypo*chlor*ite* ion

Oxoanions that contain hydrogen are named by adding the word "hydrogen" before the name of the oxoanion. If there are two hydrogens we say *di*hydrogen. Many of these hydrogen-containing oxoanions have common names that are used so often that you should know them. For example, the hydrogen carbonate ion, HCO_3^-, is often called bicarbonate ion.

Ion	*Systematic Name*	*Common Name*
HCO_3^-	hydrogen carbonate ion	bicarbonate ion
HSO_4^-	hydrogen sulfate ion	bisulfate ion
HSO_3^-	hydrogen sulfite ion	bisulfite ion
HPO_4^{2-}	hydrogen phosphate ion	
$H_2PO_4^-$	dihydrogen phosphate ion	

Figure 4.14
Names and charges of some monatomic ions of the nonmetals.

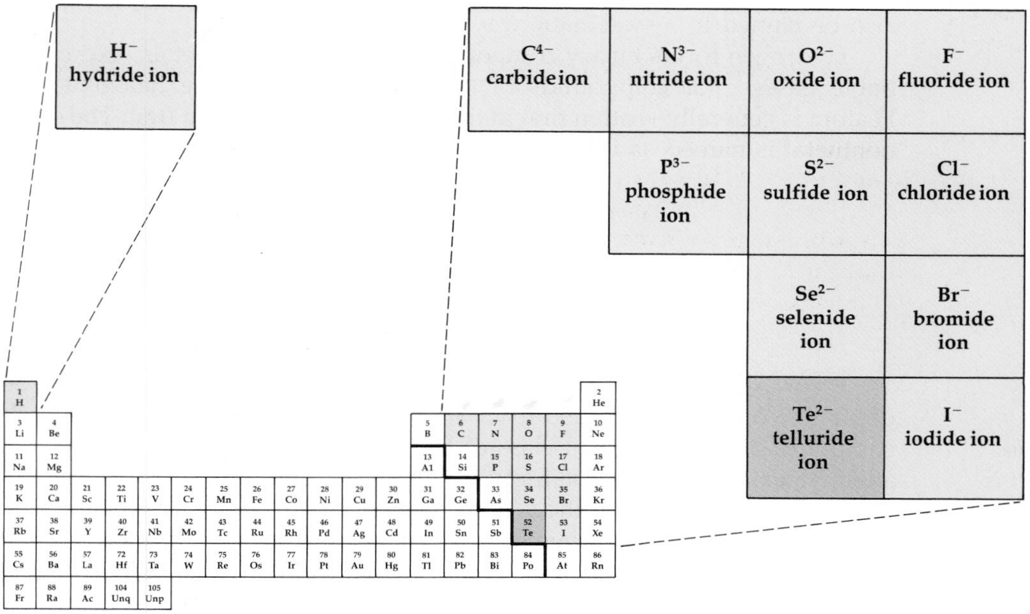

H^- hydride ion
C^{4-} carbide ion
N^{3-} nitride ion
O^{2-} oxide ion
F^- fluoride ion
P^{3-} phosphide ion
S^{2-} sulfide ion
Cl^- chloride ion
Se^{2-} selenide ion
Br^- bromide ion
Te^{2-} telluride ion
I^- iodide ion

Figure 4.15
Some common ionic compounds. Clockwise from the top are a box of salt (sodium chloride, $NaCl$), a clear crystal of calcite (calcium carbonate, $CaCO_3$), a pile of $CoCl_2 \cdot 6H_2O$ [cobalt(II) chloride hexahydrate], and an octahedral crystal of fluorite (calcium fluoride, CaF_2). *(C.D. Winters)*

Naming Ionic Compounds When naming ionic compounds, *the positive ion name is given first followed by the name of the negative ion.* Some examples are given below and others are shown in Figure 4.15.

Ionic Compound	*Ions Involved*	*Name*
$CaBr_2$	Ca^{2+} and 2 Br^-	calcium bromide
$NaHSO_4$	Na^+ and HSO_4^-	sodium hydrogen sulfate
$(NH_4)_2CO_3$	2 NH_4^+ and CO_3^{2-}	ammonium carbonate
$Mg(OH)_2$	Mg^{2+} and 2 OH^-	magnesium hydroxide
$TiCl_2$	Ti^{2+} and 2 Cl^-	titanium(II) chloride
Co_2O_3	2 Co^{3+} and 3 O^{2-}	cobalt(III) oxide

EXERCISE 4.7 • *Names and Formulas of Ionic Compounds*

1. Give the formula for each of the following ionic compounds:
   (a) ammonium nitrate (d) vanadium(III) oxide
   (b) cobalt(II) sulfate (e) barium oxide
   (c) nickel(II) cyanide (f) calcium hypochlorite
2. Name the following ionic compounds:
   (a) $MgBr_2$ (d) $KMnO_4$
   (b) Li_2CO_3 (e) $(NH_4)_2S$
   (c) $KHSO_3$ (f) $CuCl$ and $CuCl_2$

There are no ions in binary nonmetal compounds. The bonding in these compounds is said to be "covalent" and is described in Section 10.1.

Naming Binary Compounds of the Nonmetals

Thus far we have described ions and ionic compounds. Another kind of compound is formed by the combination of two nonmetals and is composed of molecules. These "two-element" or **binary compounds** of nonmetals can also be named in a systematic way.

Hydrogen forms binary compounds with all the nonmetals (except the noble gases). For compounds of oxygen, sulfur, and the halogens, the H atom is generally written first in the formula and is named first. The other nonmetal is named as if it were a negative ion.

Compound	*Name*
HF	hydrogen fluoride
HCl	hydrogen chloride
H_2S	hydrogen sulfide

Methane, the simplest alkane, is the major component of natural gas. It is also thought to be an important contributor to the "greenhouse effect." Ethane, propane, and butane are all important fuels.

Hydrocarbons, compounds composed only of C and H, are a major type of binary compounds formed from nonmetals. There are hundreds of such compounds. One important class of hydrocarbons consists of the **alkanes,** all of which have the general formula C_xH_{2x+2} (Table 4.2). Notice that all are traditionally written with the C atom first followed by the H atom, and all have *-ane* as the suffix in their name. When $x = 1$ to 4, the first part of the name is something of historical origin; these are common names that we just have to remember.

Table 4.2

*Selected Hydrocarbons of the Alkane Family, C_xH_{2x+2}**

Molecular Formula	Name		State at Room Temperature
CH_4	methane		
C_2H_6	ethane		gas
C_3H_8	propane		
C_4H_{10}	butane		
C_5H_{12}	pentane	(pent- =5)	
C_6H_{14}	hexane	(hex- =6)	
C_7H_{16}	heptane	(hept- =7)	liquid
C_8H_{18}	octane	(oct- =8)	
C_9H_{20}	nonane	(non- =9)	
$C_{10}H_{22}$	decane	(dec- =10)	
$C_{18}H_{38}$	octadecane	(octadec- =18)	solid
$C_{20}H_{42}$	eicosane	(eicos- =20)	

*This table lists only selected alkanes. Liquid compounds with 11 to 16 carbon atoms are well known, and other solid alkanes include $C_{17}H_{36}$ and $C_{19}H_{40}$.

When $x = 5$ or greater, the prefix tells how many carbon atoms are present. For example, the compound with six carbons is called *hex*ane.

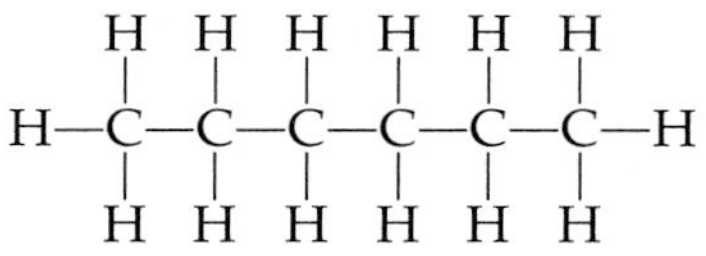

```
    H  H  H  H  H  H
    |  |  |  |  |  |
H—C—C—C—C—C—C—H
    |  |  |  |  |  |
    H  H  H  H  H  H
```

hexane = C_6H_{14}

Except for the hydrocarbons, virtually all the binary nonmetal compounds you will see contain at least one element from Groups 5A, 6A, or 7A. This element is always listed second in the formula and is named second. The number of atoms of a given type in the compound is designated with a prefix such as *di-*, *tri-*, *tetra-*, *penta-*, and so on.

Compound	*Name*
NF_3	nitrogen trifluoride
NO	nitrogen monoxide
NO_2	nitrogen dioxide
N_2O	dinitrogen oxide
N_2O_4	dinitrogen tetraoxide
PCl_3	phosphorus trichloride
PCl_5	phosphorus pentachloride
SF_6	sulfur hexafluoride
S_2F_{10}	disulfur decafluoride

Finally, many of the binary compounds of nonmetals were discovered years ago and have names so common they continue to be used. These names must simply be learned.

By tradition, the water formula is written with H before O and the hydrogen halides are HX. Other H-containing compounds are generally written like the hydrocarbon formulas, with the H atoms after the other atom.

Compound	*Name*
H_2O	water
NH_3	ammonia
N_2H_4	hydrazine
PH_3	phosphine
NO	nitric oxide
N_2O	nitrous oxide ("laughing gas")

The reasons that nonmetals combine in the various ways shown here will be described in Chapter 10. For now you should try to remember the formulas of common compounds such as water, ammonia, hydrogen fluoride, hydrogen chloride, and hydrogen sulfide and the C_1 to C_4 hydrocarbons.

EXERCISE 4.8 • *Names and Formulas of Compounds*

1. Give the formula for each of the following binary nonmetal compounds:
   (a) carbon dioxide
   (b) phosphorus triiodide
   (c) sulfur dichloride
   (d) xenon trioxide
   (e) boron trifluoride
   (f) dioxygen difluoride
   (g) nonane
2. Name the following binary nonmetal compounds:
   (a) N_2F_4
   (b) HBr
   (c) SF_4
   (d) ClF_3
   (e) BCl_3
   (f) P_4O_{10}
   (g) C_7H_{16}

EXERCISE 4.9 • *Alkanes*

1. Use the general formula for alkanes to write the molecular formulas for alkanes with 12 and 24 carbon atoms.
2. How many carbon and hydrogen atoms are there in hexadecane?

4.6 THE MOLE—THE MACRO/MICRO CONNECTION

Suppose we need to make some phosphorus tribromide by the reaction shown earlier in Figure 4.6.

1 molecule of P_4 + 6 molecules of Br_2 $\xrightarrow{\text{will give}}$ 4 molecules of PBr_3

How can we know that six molecules of Br_2 are needed for every one of P_4? We need some method of counting atoms and molecules, no matter how small they are. That is, there has to be a way to connect the macroscopic world, the world we can see, with the submicroscopic world of atoms and molecules. The solution to this problem is to define a convenient unit of matter that contains a known number of particles, and the chemical counting unit that has come into use is the **mole.**

Portrait of a Scientist

Amedeo Avogadro (1776–1856) and His Number

Lorenzo Romano Amedeo Carlo Avogadro, an Italian, was educated as a lawyer and practiced the profession for many years. However, in about 1800 he turned to science and was the first professor of mathematical physics in Italy. In 1811 he first suggested the hypothesis, which we now call a law, that "equal volumes of gases under the same conditions contain equal numbers of molecules." From this eventually came the concept of the mole. Avogadro did not see the acceptance of his ideas in his own lifetime. Scientists were not convinced until another Italian, Stanislao Cannizzaro, described experiments proving them at the great chemical conference in Karlsruhe, Germany, in 1860.

One of the difficulties presented by Avogadro's number is comprehending its size. It may help to write it out in full as

$$6.022 \times 10^{23} = 602{,}200{,}000{,}000{,}000{,}000{,}000{,}000$$

or as

$$602{,}200 \times 1 \text{ million} \times 1 \text{ million} \times 1 \text{ million}$$

But think of it this way: If you have Avogadro's number of unpopped popcorn kernels, and pour them over the continental United States, the country would be covered to a depth of 9 miles! Or, if you could divide one mole of pennies equally among every man, woman, and child in the United States, one person could pay off the national debt (currently \$2.9 trillion or 2.9×10^{12}) and still have 20 trillion dollars left over for an ice cream cone or two.

Amedeo Avogadro. (National Foundation for History of Chemistry)

The word "mole" was apparently introduced in about 1896 by Wilhelm Ostwald, who derived the term from the Latin word *moles* meaning "heap" or "pile." The mole, whose symbol is **mol,** is the SI base unit for measuring *amount of substance* (Table 2.2) and is defined as follows:

> A **mole** is the amount of substance that contains as many elementary entities (atoms, molecules, or other particles) as there are atoms in *exactly* 12 grams of the carbon-12 isotope.

The key to understanding the concept of the mole is that *one mole always contains the same number of particles,* no matter what the substance. But how many particles? Many, many experiments over the years have established that number as

$$1 \text{ mole} = 6.0221367 \times 10^{23} \text{ particles}$$

This value is commonly known as **Avogadro's number** in honor of Amedeo Avogadro, an Italian lawyer and physicist who conceived the basic idea (but never determined the number). There is nothing "special" about the value. It is determined by the definition of the mole as exactly 12 grams of carbon-12. If one mole were exactly 10 grams of carbon, then Avogadro's number would have a different value. It is interesting that the number was revised (from 6.022045×10^{23}) in the 1980s to the new value as a result of new and better measurements. Even the most fundamental values and ideas of science are under constant scrutiny.

The mole is the chemist's six-pack or dozen; it is just a counting unit. Many objects in our lives come in similar counting units. Shoes, socks, and gloves are sold by the pair, soft drinks by the six-pack, and eggs and donuts by the dozen. Atoms are counted by the mole.

The atomic weights of the elements are given in the periodic table in the front of the book.

As you saw in Section 3.3, the atomic mass scale is a relative scale with the ^{12}C atom chosen as the standard. The masses of all other atoms have been established by experiment and placed on this scale. For example, experiments show that a ^{16}O atom is 1.33 times heavier than a ^{12}C atom or that a ^{19}F atom is 1.58 times heavier than a ^{12}C atom. Now, since a mole of carbon-12 has a mass of exactly 12 grams and contains 6.0221367×10^{23} atoms, and since *a mole of one kind of atoms always contains the same number of particles as a mole of another kind of atoms,* this means that a mole of ^{16}O atoms has a mass in grams given by 1.33 times 12.0 or 16.0 g. Similarly, the mass of a mole of ^{19}F atoms is 1.58 times greater than 12.0 g or 19.0 g.

Moles of Elements and Atoms

The mass in grams of 1 mole of atoms of any element (or 6.0221367×10^{23} atoms of that element) is the **molar mass** of that element. Molar mass is abbreviated with a capital italicized ***M***, it has units of "g/mol," and for most elements it is *numerically equal to the atomic weight in atomic mass units.* Thus, based on experiments that take into account all isotopes of a particular element,

$$\begin{aligned}\text{molar mass of sodium (Na)} &= \text{mass of 1 mole of Na atoms}\\ &= \text{mass of } 6.02214 \times 10^{23} \text{ Na atoms}\\ &= 22.9898 \text{ g/mol}\end{aligned}$$

$$\begin{aligned}\text{molar mass of lead (Pb)} &= \text{mass of 1 mole of Pb atoms}\\ &= \text{mass of } 6.022 \times 10^{23} \text{ Pb atoms}\\ &= 207.2 \text{ g/mol}\end{aligned}$$

Although Avogadro's number is known to eight significant figures, we shall generally use it to four significant figures in calculations (6.022×10^{23}).

Figure 4.16 shows 1-mole quantities of some common elements. Although each of these "piles of atoms" has a different volume and different mass, each contains 6.022×10^{23} atoms.

The mole concept is the cornerstone of quantitative chemistry. It is *essential* to be able to make the conversions "moles ⇒ mass" and "mass ⇒ moles." Dimensional analysis, which is described in Appendix A, shows that this can be done in the following way:

Mass ⇔ Moles Conversions

moles ⇒ mass

$$\text{moles} \cdot \frac{\text{grams}}{\text{1 mole}} = \text{grams}$$

↑ molar mass

mass ⇒ moles

$$\text{grams} \cdot \frac{\text{1 mole}}{\text{grams}} = \text{moles}$$

↑ 1/molar mass

Figure 4.16
One-mole quantities of common elements. The 25-mL graduated cylinders contain *(left to right)* mercury (200.59 g), lead (207.2 g), and copper (63.546 g). The 125-mL Erlenmeyer flask at the left contains sulfur (32.066 g), and the one on the right contains magnesium (24.305 g). The watch glass holds chromium (51.996 g). All rest on sheets of aluminum foil (26.98 g).

For example, suppose you wish to have 0.35 mol of aluminum for a reaction. What mass, in grams, of aluminum must you weigh out? Using the molar mass of aluminum (27.0 g/mol).

$$0.35\ \cancel{\text{mol Al}} \cdot \frac{27.0\ \text{g Al}}{1\ \cancel{\text{mol Al}}} = 9.5\ \text{g of Al}$$

The molar masses of the elements are the masses in grams equal to the atomic weights, which are given in the periodic table in the front of the book.

you find that a mass of 9.5 g of aluminum is required.

Looking at the periodic table in the front of the book, you will notice that some molar masses are known to more significant figures and decimal places than others. When using molar masses in a calculation, the convention followed in this book is to *use one more significant figure in the molar mass than in any of the other data* (see Appendix A). For example, if you have weighed out 16.5 g of carbon, you would use 12.01 g/mol for the molar mass of C to find the number of moles of carbon present.

$$16.5\ \cancel{\text{g C}} \cdot \frac{1\ \text{mol C}}{12.01\ \cancel{\text{g C}}} = 1.37\ \text{mol C}$$

Note that 4 significant figures are used in the molar mass of the element while only 3 were in the sample mass.

By using one more significant figure we guarantee that the precision of the molar mass is greater than that of the other numbers and does not limit the precision of the result.

A piece of elemental silicon on top of a silicon wafer. The tiny rectangles on the wafer are electronic circuits etched into the surface of the wafer. *(C.D. Winters)*

EXAMPLE 4.4

Mass to Moles

How many moles are represented by 1.00 pound of silicon, an element used in semiconductors? (1.00 pound = 454 g)

SOLUTION The molar mass of silicon is 28.09 g/mol, so the number of moles of silicon is

$$454 \cancel{\text{g Si}} \cdot \frac{1 \text{ mol Si}}{28.09 \cancel{\text{g Si}}} = 16.2 \text{ mol Si}$$

EXAMPLE 4.5

Moles to Mass

How many grams are there in 2.50 mol of sulfur (S)?

SOLUTION For a conversion between mass and moles, you will always need the molar mass, which is 32.066 g/mol for S. Thus, the number of grams of sulfur in 2.50 mol is

$$2.50 \cancel{\text{mol S}} \cdot \frac{32.07 \text{ g S}}{1 \cancel{\text{mol S}}} = 80.2 \text{ g S}$$

The photograph below shows 80.2 g of S in a 250 mL beaker.

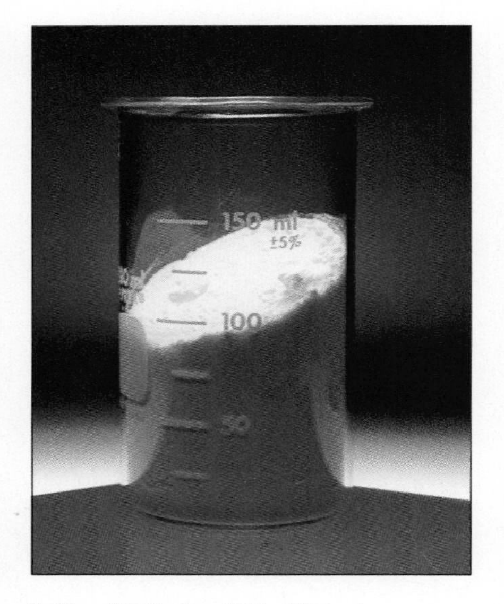

Sulfur (80.2 g). *(C.D. Winters)*

EXAMPLE 4.6

Moles and Density

The graduated cylinder in Figure 2.9 contains 25.4 mL of mercury. The density of mercury at 25 °C is 13.534 g/cm^3. How many moles of mercury are there in the cylinder?

SOLUTION The molar mass of mercury, 200.59 g/mol, is needed for conversions between mass and moles. Given the volume of mercury, you first have to convert mL of mercury into grams of mercury, and the conversion factor for this is density. Note that in dimensional analysis, the operation is

$$\text{given data} \cdot \frac{\text{desired units}}{\text{units of given data}} = \text{answer in desired units}$$

$$25.4 \cancel{\text{cm}^3 \text{ Hg}} \cdot \frac{13.534 \text{ g Hg}}{1 \cancel{\text{cm}^3 \text{ Hg}}} = 344 \text{ g Hg}$$

Only after you have the mass of mercury can this be converted to moles.

$$344 \cancel{\text{g Hg}} \cdot \frac{1 \text{ mol Hg}}{200.6 \cancel{\text{g Hg}}} = 1.71 \text{ mol Hg}$$

EXAMPLE 4.7

The Number of Atoms in a Sample

How many atoms of mercury are there in 25.4 mL of the liquid metal (Figure 2.9)?

SOLUTION In Example 4.6 we calculated that 25.4 mL of mercury is equivalent to 1.71 mol of the metal. Since we know the relation between atoms and moles (Avogadro's number), we can find the number of atoms in our sample.

$$1.71\ \text{mol Hg} \cdot \frac{6.022 \times 10^{23}\ \text{atoms Hg}}{1\ \text{mol Hg}} = 1.03 \times 10^{24}\ \text{atoms Hg}$$

EXERCISE 4.10 • ***Mass/Mole Conversions***

a. What is the mass, in grams, of 2.5 mol of aluminum?

b. How many moles are represented by 1.00 pound of lead?

Moles of Compounds and Molecules

The formula of a compound tells you the type of atoms or ions in the compound and the relative number of each. For example, in one molecule of phosphorus tribromide, PBr_3, there is one atom of P combined with three atoms of Br. But suppose you have Avogadro's number of P atoms (6.022×10^{23}) combined with Br atoms. It is clear that three times as many Br atoms are required (18.07×10^{23} Br atoms) and that there must be Avogadro's number of PBr_3 molecules. What masses of atoms are combined and what is the mass of this many PBr_3 molecules?

P	*combined with* *Br*	*in the compound* PBr_3
6.022×10^{23} P atoms	$3 \times 6.022 \times 10^{23}$ Br atoms	6.022×10^{23} PBr_3 molecules
or 1.000 mol of P	or 3.000 mol of Br atoms	or 1.000 mol of PBr_3 molecules
or 30.97 g of P atoms	or 239.7 g of Br atoms	or 270.7 g of PBr_3 molecules

Since we know the number of moles of P and Br atoms, we know the masses of phosphorus and bromine that combine to form PBr_3, so the mass of PBr_3 is the sum of these masses. That is, one mole of PBr_3 has a mass equivalent to the mass of one mole of P atoms plus 3 moles of Br atoms, or 270.7 g. This is the **molar mass, *M*,** of PBr_3.

For molecular substances the molar mass, in grams, is numerically equivalent to the **molecular weight,** which is the sum of the atomic weights of all the atoms appearing in the formula. The molecular weight is the average mass of a molecule of the substance, expressed in atomic mass units.

Molecular Weights and Molar Masses

Element or Compound	*Molecular Weight (amu)*	*Molar Mass (g/mol)*	*Mass of One Molecule* (grams)*
O_2	32.00	32.00	5.314×10^{-23}
P_4	123.9	123.9	2.057×10^{-22}
NH_3	17.03	17.03	2.828×10^{-23}
H_2O	18.02	18.02	2.992×10^{-23}
CH_2Cl_2	84.93	84.93	1.410×10^{-22}

*See text for the calculation of the mass of one molecule.

Ionic compounds such as NaCl do not exist as individual molecules. Thus, no *molecular* formula can be given; rather, one can only write the simplest formula, which shows only the *relative* number of each kind of atom in a sample. Nonetheless, we talk about the molecular weight or molar mass of such compounds and calculate it from the simplest formula. To differentiate substances such as NaCl that do not contain molecules, however, chemists do sometimes refer to their *formula weight* instead of their molecular weight.

Formula Weights and Molar Masses of Ionic Compounds

Compound	*Formula Weight (amu)*	*Molar Mass (g/mol)*
NaCl	58.44	58.44
$CaBr_2$	199.9	199.9
KNO_3	101.1	101.1

Figure 4.17
One-mole quantities of some ionic compounds. Clockwise from front right, they are NaCl, white (M = 58.44 g/mol); $CuSO_4 \cdot 5\,H_2O$, blue (M = 249.7 g/mol); $NiCl_2 \cdot 6\,H_2O$, green (M = 237.70 g/mol); $K_2Cr_2O_7$, orange (M = 294.2 g/mol); and $CoCl_2 \cdot 6H_2O$, red (M = 237.9 g/mol). *(C.D. Winters)*

Figure 4.17 is a photograph of 1-mole quantities of several common compounds. To find the molecular weight of any compound, and thus its molar mass, you need only to add up the atomic weights for all of the elements in one formula unit, each multiplied by the number of atoms of that element in the formula. As an example, let us find the molar mass of ammonium sulfate, $(NH_4)_2SO_4$, a compound used as fertilizer, in flameproofing fabrics, in tanning leather, and in many other applications. Each formula unit contains two ammonium ions, NH_4^+, and one sulfate ion, SO_4^{2-}. Therefore, in one formula unit there are two nitrogen atoms, eight hydrogen atoms, one sulfur atom, and four oxygen atoms. By combining masses in grams equivalent to the atomic weight, the molar mass of ammonium sulfate is found to be 132.15 g.

$$\text{2 mol of N per mole of } (NH_4)_2SO_4 = 2\ \cancel{\text{mol N}} \cdot \frac{14.01\text{ g N}}{1\ \cancel{\text{mol N}}} = 28.02\text{ g N}$$

$$\text{8 mol of H per mole of } (NH_4)_2SO_4 = 8\ \cancel{\text{mol H}} \cdot \frac{1.008\text{ g H}}{1\ \cancel{\text{mol H}}} = 8.064\text{ g H}$$

$$\text{1 mol of S per mole of } (NH_4)_2SO_4 = 1\ \cancel{\text{mol S}} \cdot \frac{32.07\text{ g S}}{1\ \cancel{\text{mol S}}} = 32.07\text{ g S}$$

$$\text{4 mol of O per mole of } (NH_4)_2SO_4 = 4\ \cancel{\text{mol O}} \cdot \frac{16.00\text{ g O}}{1\ \cancel{\text{mol O}}} = 64.00\text{ g O}$$

$$\text{molar mass of } (NH_4)_2SO_4 = 132.15\text{ g}$$

As in the case of elements, it is very important to be able to convert "mass of compound ⇒ moles of compound" or "moles ⇒ mass." For example, if you have one pound (454 g) of ammonium sulfate, how many moles of the compound do you have? The molar mass just calculated was 132.15 g, much less than one pound. In fact, it can be estimated before doing any detailed calculations that one pound of ammonium sulfate should be about three or four moles. The actual result is indeed between 3 and 4 moles.

$$454\text{ g } \cancel{(NH_4)_2SO_4} \cdot \frac{1\text{ mol } (NH_4)_2SO_4}{132.2\text{ g } \cancel{(NH_4)_2SO_4}} = 3.43\text{ mol } (NH_4)_2SO_4$$

The molar mass is the mass in grams of Avogadro's number of molecules (or formula units of an ionic compound). With this knowledge, it is possible to determine the number of molecules in any sample from its mass, or even to determine the mass of one molecule. For example, the mass of one buckyball, C_{60}, is

$$\frac{720.7\text{ g } C_{60}}{1\ \cancel{\text{mol } C_{60}}} \cdot \frac{1\ \cancel{\text{mol } C_{60}}}{6.022 \times 10^{23}\text{ molecules}} = 1.197 \times 10^{-21}\text{ g/molecule}$$

The masses of individual molecules in the table of molecular weights on the previous page were calculated in this way.

EXAMPLE 4.8

Molar Mass and Moles

You have 23.2 grams of ethyl alcohol, C_2H_6O.
(a) How many moles are represented by this mass of alcohol?
(b) How many molecules of ethyl alcohol are there in 23.2 g?
(c) How many atoms of carbon are there in 23.2 g of ethyl alcohol?
(d) What is the mass of one molecule of ethyl alcohol?

SOLUTION The first step in any problem involving the interconversion of mass and moles is to find the molar mass of the compound in question.

$$2 \text{ mol of C per mole of alcohol} = 2\ \cancel{\text{mol C}} \cdot \frac{12.01 \text{ g C}}{1\ \cancel{\text{mol C}}} = 24.02 \text{ g C}$$

$$6 \text{ mol of H per mole of alcohol} = 6\ \cancel{\text{mol H}} \cdot \frac{1.008 \text{ g H}}{1\ \cancel{\text{mol H}}} = 6.048 \text{ g H}$$

$$1 \text{ mol of O per mole of alcohol} = 1\ \cancel{\text{mol O}} \cdot \frac{16.00 \text{ g O}}{1\ \cancel{\text{mol O}}} = 16.00 \text{ g O}$$

$$\text{molar mass of } C_2H_6O = 46.07 \text{ g/mol}$$

(a) The molar mass with units of "g/mol" is the conversion factor in all mass ⇔ mole conversions.

$$23.2 \text{ g}\ \cancel{C_2H_6O} \cdot \frac{1 \text{ mol } C_2H_6O}{46.07 \text{ g}\ \cancel{C_2H_6O}} = 0.504 \text{ mol } C_2H_6O$$

(b) Since you know there are Avogadro's number of molecules in a mole, you can find the number of molecules in 0.504 mol of C_2H_6O.

$$0.504\ \cancel{\text{mol } C_2H_6O} \cdot \frac{6.022 \times 10^{23} \text{ molecules } C_2H_6O}{1\ \cancel{\text{mol } C_2H_6O}} = 3.04 \times 10^{23} \text{ molecules of } C_2H_6O$$

(c) Since you know that each molecule contains two carbon atoms, you can find the number of carbon atoms in 23.2 g of the alcohol.

$$3.04 \times 10^{23}\ \cancel{\text{molecules } C_2H_6O} \cdot \frac{2 \text{ atoms of C}}{1\ \cancel{\text{molecule } C_2H_6O}} = 6.08 \times 10^{23} \text{ atoms of C}$$

(d) The units of the desired answer are "grams/molecule," which indicates that you should divide the starting unit of molar mass (g/mol) by Avogadro's number (molecules/mol), so that the unit "mol" cancels.

$$\frac{46.07 \text{ g } C_2H_6O}{1\ \cancel{\text{mol}}} \cdot \frac{1\ \cancel{\text{mol}}}{6.022 \times 10^{23} \text{ molecules}} = \frac{7.650 \times 10^{-23} \text{ g}}{1\ C_2H_6O \text{ molecule}}$$

EXERCISE 4.11 • *Molar Mass and Moles ⇔ Mass Conversions*

1. Calculate the molar mass of (a) limestone, $CaCO_3$, and (b) caffeine, $C_8H_{10}N_4O_2$.
2. If you have 1.00 pound (454 g) of $CaCO_3$, how many moles does this represent?
3. In order to have 2.50×10^{-3} mol of caffeine, how many grams must you have?

4.7 DESCRIBING COMPOUND FORMULAS

Percent Composition

The law of constant composition states that *any sample of a pure compound always consists of the same elements combined in the same proportions by mass.* Thus, it would seem that there are at least two equivalent ways to express molecular composition: (a) in terms of the number of atoms of each type per molecule or (b) in terms of the mass of each element per mole of compound. Actually, there is at least one more way to express molecular composition, a way derived from (b). Composition can be given by the mass of each element in the compound relative to the total mass of the compound—that is, as the mass percent of each element or **percent composition** by mass. For ammonia, this is

Equivalent Ways of Expressing Molecular Composition:
(a) A formula giving the number of atoms of each element per molecule
(b) Mass of each element per mole of compound
(c) Mass of each element per 100 g of compound (percent composition)

$$\text{mass percent N in NH}_3 = \frac{\text{mass of N in 1 mol of NH}_3}{\text{mass of 1 mol of NH}_3} = \frac{14.01 \text{ g N}}{17.030 \text{ g NH}_3} \times 100\%$$

$$= 82.27\% \text{ (or 82.27 g N per 100.00 g NH}_3\text{)}$$

$$\text{mass percent H in NH}_3 = \frac{3.024 \text{ g H}}{17.030 \text{ g NH}_3} \times 100\%$$

$$= 17.76\% \text{ H (or 17.76 g H per 100.00 g NH}_3\text{)}$$

EXERCISE 4.12 • *Percent Composition*

Express the composition of each compound in terms of the mass of each element in 1.00 mole of compound and the mass percent of each element:

a. NaCl, sodium chloride

b. C_8H_{18}, octane

c. $(NH_4)_2SO_4$, ammonium sulfate

Empirical and Molecular Formulas

Now consider the *reverse* of the procedure we just discussed: using relative mass or percent composition data to find a molecular formula. Let us say you know the identity of the elements in a sample and have determined the mass of each element in a given mass of compound (the percent composition) by chemical analysis. You can then calculate the relative number of moles of each element in one mole of compound, and from this the relative number of atoms of each element in the compound.

Consider hydrazine, a close relative of ammonia. The mass percentages in a sample of hydrazine are 87.42% N and 12.58% H. Taking a 100.00-g sample of hydrazine, the percent composition data tell us that the sample contains 87.42 g of N and 12.58 g of H. Therefore, the number of moles of each element in the 100.0-g sample is

Hydrazine is used to remove dissolved oxygen in hot water heating systems and to remove metal ions from waste water. In the past it was also used as a rocket fuel.

$$87.42 \cancel{\text{g N}} \cdot \frac{1 \text{ mol N}}{14.007 \cancel{\text{g N}}} = 6.241 \text{ mol of N}$$

$$12.58 \cancel{\text{g H}} \cdot \frac{1 \text{ mol H}}{1.0079 \cancel{\text{g H}}} = 12.48 \text{ mol of H}$$

Notice that we divided the larger number by the smaller one. This is usually the easiest way to find the atom ratio.

What is important now is to use the number of moles of each element in 100.0 g of sample to find the number of moles of one element *relative* to the other. For hydrazine, this ratio is 2 moles of H to 1 of N,

$$\frac{12.48 \text{ mol H}}{6.241 \text{ mol N}} = \frac{2.000 \text{ mol H}}{1.000 \text{ mol N}}$$

showing that there are 2 moles of H atoms for every mole of N atoms in hydrazine. Thus, in one molecule there are 2 atoms of H for every atom of N. That is, the atom ratio is represented by the formula NH_2.

Empirical and molecular formulas can differ only for molecular compounds. The formula of any ionic compound is its empirical formula.

Percent composition data enable you to calculate the atom ratios in the compound. However, a molecular formula must convey *two* pieces of information: (a) the *relative number* of atoms of each element in a molecule (the atom ratios) and (b) the *total number* of atoms in the molecule. For hydrazine you know that there are twice as many H atoms as N atoms. This means the molecular formula could be NH_2. However, because *percent composition data give only the simplest possible ratio of atoms in a molecule,* the formulas N_2H_4, N_3H_6, N_4H_8, and so forth are also possible. A formula such as NH_2, in which the atom ratio is the simplest possible, is called the **empirical formula.** In contrast, the molecular formula shows the true number of atoms of each kind in a molecule; it can always be derived by multiplying the empirical formula by a whole number (which may be 1, 2, or 3, and so on).

To find the molecular formula of a compound after calculating the empirical formula, you must know the molar mass from an experiment.

To determine the molecular formula, the molar mass must be obtained *from experiment.* For example, experiments show that the molar mass of hydrazine is 32.0 g/mol, or twice the formula mass of NH_2, which is 16.0 g/mol. This must mean that the molecular formula of hydrazine is 2 times the empirical formula of NH_2, that is, N_2H_4.

As another example of the usefulness of percent composition data, let us say that you collect the following information for the compound naphthalene in the laboratory:

$$\% \text{ carbon} = 93.71$$
$$\% \text{ hydrogen} = 6.29$$
$$M = 128 \text{ g/mol}$$

and then calculate the empirical and molecular formulas for naphthalene, a substance commonly encountered in the form of moth balls. The percent composition data tell you that there are 93.71 g of C and 6.29 g of H in a 100.00-g sample. Therefore, you can find the number of moles of each element in the sample.

$$\frac{93.71 \text{ g C}}{100.00 \text{ g naphthalene}} \cdot \frac{1 \text{ mol C}}{12.011 \text{ g C}} = \frac{7.802 \text{ mol C}}{100.00 \text{ g naphthalene}}$$

$$\frac{6.29 \text{ g H}}{100.00 \text{ g naphthalene}} \cdot \frac{1 \text{ mol H}}{1.008 \text{ g H}} = \frac{6.24 \text{ mol H}}{100.00 \text{ g of naphthalene}}$$

This means that, in any sample of naphthalene, the ratio of moles of C to moles of H is

$$\text{mole ratio} = \frac{7.802 \text{ mol C}}{6.24 \text{ mol H}} = \frac{1.25 \text{ mol C}}{1.00 \text{ mol H}}$$

Now your task is to turn this decimal fraction into a whole-number ratio of

C to H. To do this, recognize that 1.25 is the same as $1 + \frac{1}{4} = \frac{4}{4} + \frac{1}{4} = \frac{5}{4}$. Therefore, the ratio of C to H is

$$\text{mole ratio} = \frac{\frac{5}{4}\ \text{mol C}}{1\ \text{mol H}} = \frac{5\ \text{mol C}}{4\ \text{mol H}}$$

and you know now that there are 5 C atoms for every 4 H atoms in naphthalene. Thus the simplest or *empirical formula* is C_5H_4. If C_5H_4 were the molecular formula, the molar mass would be 64 g/mol. However, experiments gave the actual molar mass as 128 g/mol, twice the value for the empirical formula.

$$\frac{128\ \text{g/mol of naphthalene}}{64.0\ \text{g/mol of}\ C_5H_4} = 2.00\ \text{mol}\ C_5H_4\ \text{per mol of naphthalene}$$

Therefore, the true molecular formula is $(C_5H_4)_2$ or $C_{10}H_8$.

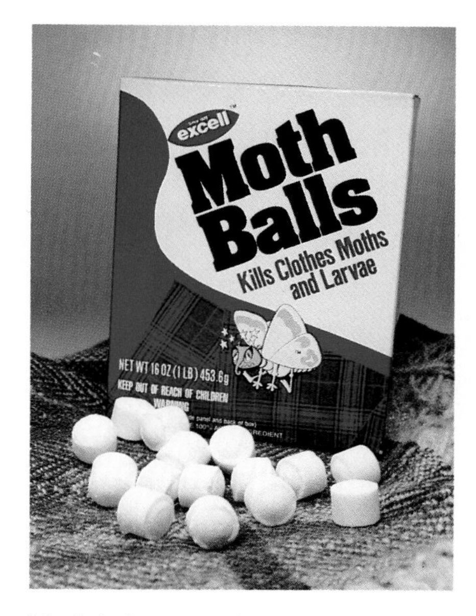

Naphthalene "moth balls." (See text.) *(C.D. Winters)*

EXAMPLE 4.9

Calculating a Formula from Percent Composition

Vanillin is a common flavoring agent. It has a molar mass of 152 g/mol and is 63.15% C and 5.30% H; the remainder is oxygen. What are the empirical and molecular formulas for vanillin?

SOLUTION We shall first find the number of moles of each element in a 100.00-g sample of vanillin.

$$63.15\ \text{g C} \cdot \frac{1\ \text{mol C}}{12.011\ \text{g C}} = 5.258\ \text{mol C}$$

$$5.30\ \text{g H} \cdot \frac{1\ \text{mol H}}{1.008\ \text{g H}} = 5.26\ \text{mol H}$$

But what about oxygen? Since the total mass is 100.00 g, this means that the mass of oxygen is

$$100.00\ \text{g} = 63.15\ \text{g C} + 5.30\ \text{g H} + \text{mass of O}$$
$$\text{mass of O} = 31.55\ \text{g O}$$

Now we can calculate moles of O as well.

$$31.55\ \text{g O} \cdot \frac{1\ \text{mol O}}{15.999\ \text{g O}} = 1.972\ \text{mol O}$$

To find the mole ratio, the easiest approach is to base the ratios on the element with the smallest number of moles present. In this case it is oxygen. Therefore,

$$\frac{\text{mole C}}{\text{mole O}} = \frac{5.258\ \text{mol C}}{1.972\ \text{mol O}} = \frac{2.666\ \text{mol C}}{1.000\ \text{mol O}} = \frac{2\frac{2}{3}\ \text{mol C}}{1\ \text{mol O}} = \frac{8\ \text{mol C}}{3\ \text{mol O}}$$

$$\frac{\text{mole H}}{\text{mole O}} = \frac{5.26\ \text{mol H}}{1.972\ \text{mol O}} = \frac{2.67\ \text{mol H}}{1.00\ \text{mol O}} = \frac{2\frac{2}{3}\ \text{mol H}}{1\ \text{mol O}} = \frac{8\ \text{mol H}}{3\ \text{mol O}}$$

Now we know that the empirical formula has the same number of atoms of C and H, and both are in an 8:3 ratio to O. Therefore, the empirical formula is $C_8H_8O_3$. This formula has a molar mass of 152.2 g/unit, which corresponds to the experimentally determined molar mass. Therefore, the molecular formula of vanillin, $C_8H_8O_3$, is the same as the empirical formula.

EXERCISE 4.13 • *Empirical and Molecular Formulas*

Boron hydrides, compounds containing only boron and hydrogen, form a large class of compounds. One boron hydride consists of 78.14% B and 21.86% H; its molar mass is 27.7 g/mol. What are the empirical and molecular formulas of this compound?

Determining Formulas

The empirical formula of a compound can be determined if the percent composition of the compound is known. Then, if the molar mass is determined by any of a variety of experimental methods, it is possible to convert the empirical formula to a molecular formula. But where do the necessary percent composition and molar mass data come from? Both come from methods of *chemical analysis,* methods that we will describe as we cover the necessary background in later chapters.

Elements often combine with oxygen, sulfur, and the halogens to produce simple binary compounds (see Figure 4.6). For example, tin metal and purple iodine will react to give an orange tin iodide (Figure 4.18).

$$\text{Sn metal} + I_2 \text{ — will give} \longrightarrow Sn_xI_y$$

Based on the law of conservation of matter, there are several ways to determine the masses of combining elements and thus the ratio of atoms in a molecule. For example, the formula of the compound containing tin and iodine can be determined by warming a mixture containing weighed quantities of Sn and I_2 in an organic solvent. The quantity of Sn present is far in

A solvent is a component of a solution, usually the major component. Solutions having water as the solvent are perhaps most familiar. In Figure 4.18 the solvent, ethyl acetate, is a liquid organic compound.

(a)

(b)

Figure 4.18
The reaction of tin (Sn) and iodine (I_2) to give SnI_4. (a) Weighed quantities of metallic tin (*right*) and the nonmetal iodine (*left*). The tin is present in excess of that required to consume all the iodine. The solvent for the reaction, ethyl acetate, is in the flask at the rear. (b) The dark mixture of reactants in ethyl acetate is heated in the Erlenmeyer flask. After the reaction mixture is cooled, it is filtered to remove excess tin. The orange product SnI_4 (*on the filter paper at the front*) precipitates and is recovered by filtration. *(C.D. Winters)*

excess of that needed to react with all of the I_2 present. Therefore, after the orange tin/iodine compound has been formed and dissolves in the organic solvent, the solid tin metal can be separated from it by filtration (Figure 4.18). Collecting the following experimental information in the laboratory will make it possible to find the values of x and y in Sn_xI_y:

mass of tin (Sn) in the beginning	1.056 g
mass of iodine (I_2) consumed in the reaction	1.947 g
mass of tin (Sn) recovered after reaction	0.601 g

The first step is to find the mass of Sn that combined with 1.947 g of I_2. These masses will then be converted to moles of Sn and I_2, and finally the ratio of moles of Sn and I_2 will lead to the empirical formula.

mass of Sn at the beginning	1.056 g
mass of Sn recovered after reaction	−0.601 g
mass of Sn consumed in the reaction	0.455 g

With the masses of Sn and I_2 that combined now known, the moles of Sn and I_2 that combined are found.

$$0.455 \cancel{\text{g Sn}} \cdot \frac{1 \text{ mol Sn}}{118.7 \cancel{\text{g Sn}}} = 3.83 \times 10^{-3} \text{ mol Sn}$$

$$1.947 \cancel{\text{g I}_2} \cdot \frac{1 \text{ mol I}_2}{253.81 \cancel{\text{g I}_2}} = 7.671 \times 10^{-3} \text{ mol I}_2$$

This leads to the ratio of moles of Sn and I_2 that combined.

$$\frac{7.671 \times 10^{-3} \text{ mol I}_2}{3.83 \times 10^{-3} \text{ mol Sn}} = \frac{2.00 \text{ mol I}_2}{1.00 \text{ mol Sn}} \rightarrow \frac{4.00 \text{ mol I}}{1.00 \text{ mol Sn}}$$

This ratio shows that there are twice as many moles of I_2 as moles of Sn combined in the tin iodide. But each I_2 molecule contains *two I atoms* to the combination. Therefore, the ratio of atoms of I to atoms of Sn is 4 to 1 and gives the empirical formula of SnI_4. More experimentally determined information would now be needed to find the molecular formula, and this would show that the molecular formula is the same as the empirical formula in this case.

This example and Example 4.10 illustrate only a few of the ways to determine formulas. The most important aspect of determining formulas is that you should *always focus on using the data to find the number of moles of elements combined in a given mass of compound and then find the ratio of those moles.*

EXAMPLE 4.10

Determining the Formula of a Binary Oxide

Analysis shows that 0.586 g of potassium can combine with 0.480 g of O_2 gas to give a white solid ionic compound with a formula of K_xO_y. What is the formula of the white solid?

SOLUTION Our problem is to find the values of x and y in K_xO_y. To do this, we simply need to find the number of moles of K and O and then find their ratio.
(a) Calculate moles of K:

$$0.586\ \text{g K} \cdot \frac{1\ \text{mol K}}{39.10\ \text{g K}} = 0.0150\ \text{mol K}$$

(b) Calculate moles of O:

$$0.480\ \text{g}\ O_2 \cdot \frac{1\ \text{mol}\ O_2}{32.00\ \text{g}\ O_2} \cdot \frac{2\ \text{mol O}}{1\ \text{mol}\ O_2} = 0.0300\ \text{mol O}$$

In this step we had to take into account the fact that 1 mol of O_2 molecules contains 2 mol of O atoms, and it is the moles of O atoms that we need to know.
(c) Ratio of moles:

$$\frac{0.0300\ \text{mol O}}{0.0150\ \text{mol K}} = \frac{2\ \text{mol of O atoms}}{1\ \text{mol of K atoms}}$$

(d) The empirical formula of the compound is KO_2. (This compound is called potassium superoxide, since the oxygen is in the form of the O_2^- ion, the name of which is "superoxide.")

EXERCISE 4.14 • *Empirical Formula of a Binary Compound*

Gaseous chlorine was combined with 0.532 g of titanium, and 2.108 g of Ti_xCl_y was collected. What is the empirical formula of the titanium chloride?

4.8 HYDRATED COMPOUNDS

When solid ionic compounds are prepared in water solution and then isolated, the crystals often include molecules of water. These are called **hydrated compounds,** compounds in which molecules of water are associated with the ions of the compounds. The beautiful green nickel(II) compound shown in Figure 4.17, for example, has the formula $NiCl_2 \cdot 6\ H_2O$. The dot between $NiCl_2$ and $6\ H_2O$ indicates that six moles of water are associated with every mole of $NiCl_2$, and the name of the compound, nickel(II) chloride *hexa*hydrate, reflects the presence of six moles of water. The molar mass of $NiCl_2 \cdot 6\ H_2O$ is 129.6 g/mol (for $NiCl_2$) plus 108.1 g/mol (for $6\ H_2O$), or 237.7 g/mol.

Hydrated compounds are quite common (Table 4.3). The walls of your home may be covered with wallboard or "plaster board." These sheets contain hydrated calcium sulfate or gypsum, $CaSO_4 \cdot 2\ H_2O$, as well as unhydrated $CaSO_4$, sandwiched between paper. Gypsum is a mineral that can be mined. However, it is now more commonly a byproduct in the manufacture of hydrofluoric acid or phosphoric acid, or it comes from cleaning sulfur dioxide from the exhaust gases of power plants.

If gypsum is heated to 120 to 180 °C the water is partly driven off to give calcium sulfate hemihydrate, $CaSO_4 \cdot \frac{1}{2}\ H_2O$, a compound commonly called

Table 4.3

Some Common Hydrated Ionic Compounds

Compound	Systematic Name	Common Name	Uses
$Na_2SO_4 \cdot 10\ H_2O$	sodium sulfate decahydrate	Glauber's salt	cathartic
$Na_2CO_3 \cdot 10\ H_2O$	sodium carbonate decahydrate	washing soda	water softener
$Na_2S_2O_3 \cdot 5\ H_2O$	sodium thiosulfate pentahydrate	hypo	photography
$MgSO_4 \cdot 7\ H_2O$	magnesium sulfate heptahydrate	Epsom salt	cathartic, dyeing and tanning
$CaSO_4 \cdot 2\ H_2O$	calcium sulfate dihydrate	gypsum	wallboard
$CaSO_4 \cdot \frac{1}{2}\ H_2O$	calcium sulfate hemihydrate	plaster of Paris	casts, molds
$CuSO_4 \cdot 5\ H_2O$	copper(II) sulfate pentahydrate	blue vitriol	insecticide

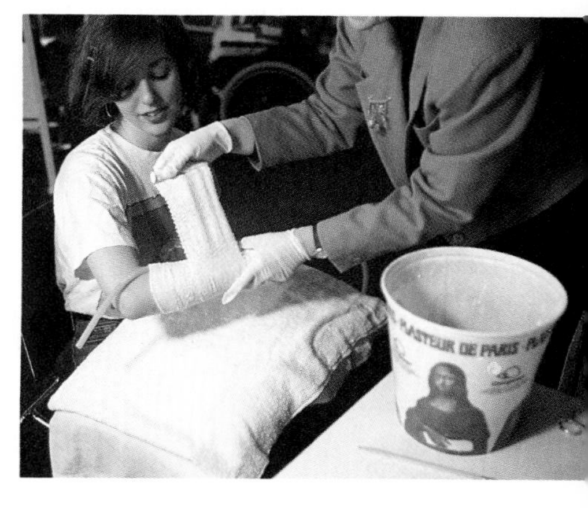

Figure 4.19
A plaster of Paris cast being applied to an arm. *(C.D. Winters)*

"plaster of Paris" (Figure 4.19). If you have ever broken an arm or leg and had to have a cast, the cast may have been made of this partly hydrated calcium sulfate. It is an effective casting material because, when added to water, it forms a thick slurry that can be poured into a mold or spread out over a part of a body. As it takes on more water, the material increases in volume and forms a hard, inflexible cast.

Hundreds of hydrated compounds are known. Since there is no simple way to predict how much water will be involved, it must be determined experimentally. Such an experiment may involve heating the hydrated material so that all the water is released from the solid and driven away (Figure 4.20). Only the **anhydrous** compound, a substance "without water," is left. According to the *law of the conservation of matter* (Section 3.1), the mass of the original hydrated compound must equal the sum of the mass of the water driven away and the mass of the anhydrous compound left behind. For example, heating 100.00 g of copper sulfate pentahydrate will give the following results:

$$\begin{array}{lll} \text{hydrated compound} & \rightarrow \text{water} & + \text{ anhydrous compound} \\ 100.00\text{ g } CuSO_4 \cdot 5\ H_2O & \rightarrow 36.08\text{ g } H_2O & + \ 63.92\text{ g } CuSO_4 \end{array}$$

The masses of the water evolved and the anhydrous compound can be determined by experiment. Converting these masses to moles, and finding their ratio, gives their ratio in the original hydrated material.

Hydrated cobalt(II) sulfate is the deep red solid shown in Figure 4.17. When heated, it turns purple and then deep blue as it loses water to form anhydrous $CoSO_4$. On exposure to moist air, anhydrous $CoSO_4$ takes up water and is converted back into the red hydrated compound. Because of this property, crystals of the blue compound are used as a humidity indica-

Figure 4.20
A crucible containing a hydrated compound [$CuSO_4 \cdot 5\ H_2O$] is heated to drive off all the water and leave anhydrous $CuSO_4$ (i.e., $CuSO_4$ without water of hydration. *(C.D. Winters)*

tor. (You may have seen them in a small bag packed with a piece of electronic equipment.) The compound also makes a good "invisible ink." A solution of cobalt(II) sulfate in water is red, but if you write on paper with the solution it cannot be seen. When the paper is warmed, however, the cobalt compound dehydrates to give the deep blue anhydrous compound, and you can read the writing.

Suppose that you do not know the value of x in the cobalt(II) sulfate hydrate, $CoSO_4 \cdot x\,H_2O$. To find how many water molecules there are for each unit of $CoSO_4$, you weigh out 1.023 g of the solid and then heat it in a porcelain crucible (Figure 4.20). After the water has been driven off completely, you are left with 0.603 g of blue anhydrous cobalt(II) sulfate, that is, with pure $CoSO_4$.

$$1.023\text{ g } CoSO_4 \cdot x\,H_2O + \text{heat} \rightarrow 0.603\text{ g } CoSO_4 + ?\text{ g } H_2O$$

Because mass is conserved, 0.420 g of water must have been driven off.

$$\text{total sample mass before heating} = 1.023\text{ g} = \text{mass of } CoSO_4 + \text{mass of water}$$
$$= 0.603\text{ g } CoSO_4 + 0.420\text{ g } H_2O$$

The masses of $CoSO_4$ and H_2O can now be converted to moles,

$$0.420\,\cancel{\text{g } H_2O} \cdot \frac{1\text{ mol } H_2O}{18.02\,\cancel{\text{g } H_2O}} = 0.0233\text{ mol } H_2O$$

$$0.603\,\cancel{\text{g } CoSO_4} \cdot \frac{1\text{ mol } CoSO_4}{155.0\,\cancel{\text{g } CoSO_4}} = 0.00389\text{ mol } CoSO_4$$

and the value of x determined from the molar ratio.

$$\frac{\text{moles } H_2O}{\text{moles } CoSO_4} = \frac{0.0233\text{ mol } H_2O}{0.00389\text{ mol } CoSO_4} = \frac{5.99\text{ mol } H_2O}{1.00\text{ mol } CoSO_4}$$

This tells us the water-to-$CoSO_4$ ratio is 6:1, so the formula of the hydrated compound is $CoSO_4 \cdot 6\,H_2O$, and its name is cobalt(II) sulfate hexahydrate.

EXERCISE 4.15 • *Determining the Formula of a Hydrated Compound*

Ruthenium is certainly one of the less familiar elements, but its chemistry is quite interesting. A good material for making other ruthenium compounds is $RuCl_3 \cdot x\,H_2O$. If you heat 1.056 g of this hydrated compound and find that only 0.838 g of $RuCl_3$ remains when all of the water has been driven off, what is the value of x?

IN CLOSING

Having studied this chapter, you should be able to

- interpret the meaning of molecular formulas, condensed formulas, and structural formulas (Section 4.1).
- define allotropes and give several examples (Section 4.2)
- list the elements that exist as diatomic molecules (Section 4.2).
- predict the charges on simple ions of metals and nonmetals (Section 4.3).

- give the names or formulas of polyatomic ions, knowing their formulas or names, respectively (Section 4.3).
- write the formulas for selected ionic compounds (Section 4.4).
- describe the properties of ionic compounds (Section 4.4).
- name ionic compounds and simple binary compounds of the nonmetals (Section 4.5).
- thoroughly explain the concept of the mole (Section 4.6).
- calculate the molar mass of a compound (Section 4.6).
- calculate the number of moles of an element or compound that are represented by a given mass, and vice versa (Section 4.6).
- express molecular composition in terms of percent composition (Section 4.7).
- use percent composition and molar mass to determine the empirical and molecular formulas of a compound (Section 4.7).
- determine the number of water molecules in a hydrated compound (Section 4.8).

STUDY QUESTIONS

Review Questions

1. A dictionary defines the word "compound" as a "combination of two or more parts." What are the "parts" of a chemical compound? Can you think of three pure (or nearly pure) compounds you have encountered today?
2. Roald Hoffmann has said that "Today chemistry is the science of molecules and their transformations." What does that statement mean in the context of your own experience?
3. Which of the following compounds might you properly call a hydrocarbon?
   (a) ethyl alcohol, C_2H_6O
   (b) ammonium sulfate, $(NH_4)_2SO_4$
   (c) toluene, $C_6H_5CH_3$
4. Three elements composed of diatomic molecules are among the top ten chemicals produced in the United States (nitrogen, oxygen, and chlorine). For example, 22.65 billion pounds of chlorine were made in 1991. By reading the newspapers, or asking local citizens, can you find out how this element might be used in your community?
5. Ozone, O_3, is an allotrope of oxygen. In the *Saunders Chemistry Update* newsletter for Spring 1992 it was reported that "Scientific data obtained in the last six months have provided additional disturbing news about the decline in stratospheric ozone, and the consequent increase in surface exposure to increased ultraviolet radiation." Scan recent newspapers and magazines to find out what is happening in this area of scientific investigation. Is stratospheric ozone (*i.e.*, ozone at very high altitudes) still on the decline? What are the United States and other countries doing to restore the ozone level?
6. The compound $(NH_4)_2CO_3$ consists of two kinds of polyatomic ions. What are the names and electrical charges of these ions?
7. Octane is a member of the alkane family of hydrocarbons (Table 4.2). For the sake of this problem, we shall assume that gasoline, a complex mixture of hydrocarbons, is represented by octane. If you fill the tank of your car with 18 gallons of gasoline, how many grams and how many pounds of liquid have you put into the car? Information you may need is: (a) the density of octane, 0.692 g/cm^3; (b) the volume of one gallon in milliliters (3790 mL).
8. The "mole" is simply a convenient unit for counting molecules and atoms. Name four "counting units" (such as a dozen for eggs and cookies) that you commonly encounter.
9. If you divide Avogadro's number of pennies among the 250 million men, women, and children in the United States, and if each person could count one penny each second every day of the year for 8 hours a day, how long would it take to count all of the pennies? How long would it take you to get bored doing this?
10. Why do you think it is more convenient to use some chemical counting unit when doing calculations—

chemists have adopted the unit of the mole, but it could have been something different—rather than using individual molecules?

11. Chemists often express the composition of compounds in terms of the percentage of a particular element that is present. Look for some food product that gives the composition in terms of percentages. What data are given? Is percent by weight of an element or of a compound listed?

12. What is the difference between an empirical formula and a molecular formula? Use the compound ethane, C_2H_6, to illustrate your answer.

13. Draw a diagram showing the crystal lattice of sodium chloride (NaCl). Show clearly why such a crystal can be cleaved easily by tapping on a knife blade properly aligned along the crystal. Describe in words why the cleavage occurs as it does.

Molecular Formulas

14. Which of the following molecules contains more O atoms, and which contains more atoms of all kinds: (a) sucrose, $C_{12}H_{22}O_{11}$, or (b) glutathione, $C_{10}H_{17}N_3O_6S$ (the major low-molecular-weight sulfur-containing compound in plant or animal cells)?

15. Write the molecular formula of each of the following compounds: (a) Anatase, a common titanium-containing mineral, has a titanium atom and two oxygen atoms per formula unit. (b) One of a series of compounds called the boron hydrides has four boron atoms and ten hydrogen atoms per molecule. (c) Aluminum trimethyl has two aluminum atoms, six carbon atoms, and eighteen hydrogen atoms per molecule.

16. Write the molecular formula of each of the following compounds: (a) Benzene, a liquid hydrocarbon, has six carbon atoms and six hydrogen atoms per molecule. (b) Vitamin C or ascorbic acid has six carbon atoms, eight hydrogen atoms, and six oxygen atoms per molecule. (c) The mineral barite has one barium atom, one sulfur atom, and four oxygen atoms per formula unit.

17. Write the formula for
   (a) a molecule of the organic compound octane, which has eight carbon atoms and eighteen hydrogen atoms.
   (b) a molecule of acrylonitrile, the basis of Orlon and Acrilan fibers, which has three carbon atoms, three hydrogen atoms, and one nitrogen atom.
   (c) a molecule of Fenclorac, an anti-inflammatory drug, which has fourteen carbon atoms, sixteen hydrogen atoms, two chlorine atoms, and two oxygen atoms.

18. Give the total number of atoms of each element in one formula unit of each of the following compounds:
(a) CaC_2O_4 (d) $Pt(NH_3)_2Cl_2$
(b) $C_6H_5CHCH_2$ (e) $K_4Fe(CN)_6$
(c) $(NH_4)_2SO_4$

19. Give the total number of atoms of each element in each of the molecules below.
(a) $Co_2(CO)_8$ (d) $C_{10}H_9NH_2Fe$
(b) $HOOCCH_2CH_2COOH$ (e) $C_6H_2CH_3(NO_2)_3$
(c) $CH_3NH_2CHCOOH$

20. Give a molecular formula for each of the following organic acids.

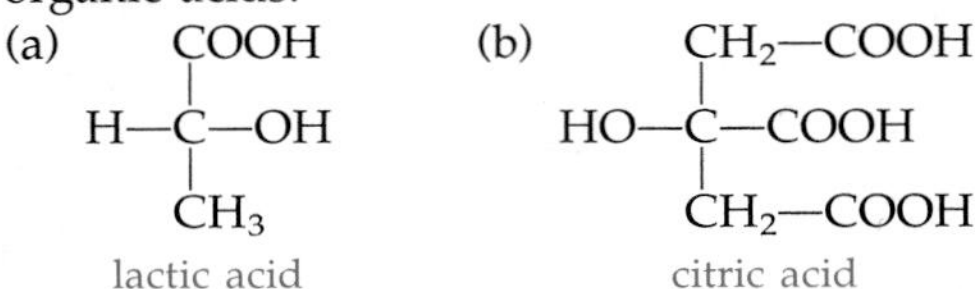

21. Give a molecular formula for each of the following molecules.

(a) benzene ring (HC, HC, C–H, CH, C–H) bearing CH=CH$_2$

styrene

(b) $H_3COC(=O)$—C$_6$H$_4$—$C(=O)OCH_3$

dimethyl terephthalate

The hydrocarbon styrene is the basis of the polymer polystyrene, and dimethyl terephthalate is one of the two types of molecules used to make the polymers Dacron and Mylar.

Predicting Ion Charges

22. You have atoms of aluminum and selenium. Predict the charges on the ions that they form.

23. Predict the charges of the ions in an ionic compound composed of barium and bromine.

24. Predict the charges for ions of the following elements:
(a) magnesium (c) iron
(b) zinc (d) gallium

25. Predict the charges for ions of the following elements:
(a) selenium (c) nickel
(b) fluorine (d) nitrogen

Ionic Compounds

26. For each of the following compounds, tell what ions are present and how many there are per formula unit:
(a) K_2S (b) $NiSO_4$ (c) $(NH_4)_3PO_4$

27. For each of the following compounds, tell what ions are present and how many there are per formula unit:
(a) $Ca(CH_3CO_2)_2$ (b) $Co_2(SO_4)_3$ (c) $Al(OH)_3$

28. Cobalt is a transition metal and so can form ions with at least two different charges. Write the formulas for the compounds formed between cobalt ions and oxide ion.

29. Although not a transition element, lead can also form two cations: Pb^{2+} and Pb^{4+}. Write the formulas for the compounds of these ions with the chloride ion.

30. Which of the following are the correct formulas of compounds? For those that are not, give the correct formula.
(a) AlCl (c) Ga_2O_3
(b) NaF_2 (d) MgS

31. Which of the following are the correct formulas of compounds? For those that are not, give the correct formula.
(a) Ca_2O (c) Fe_2O_5
(b) $SrCl_2$ (d) K_2O

32. Solid magnesium oxide melts at 2800 °C. This property, combined with the fact that it is not an electrical conductor, makes it an ideal heat insulator for electric wires in cooking ovens and toasters (photo at right). In contrast, solid NaCl melts at the relatively low temperature of 801 °C. What is the formula of magnesium oxide? Suggest a reason that it has a melting temperature so much higher than that of NaCl.

Magnesium oxide, MgO, is used as an electrical insulator in heating elements. *(C.D. Winters)*

33. Assume you have an unlabeled bottle containing a white crystalline powder. The powder melts at 310 °C. You are told that it could be NH_3, NO_2, or $NaNO_3$. What do you think it is and why?

Naming Compounds

34. Name each of the following ionic compounds (see Question 26):
(a) K_2S (b) $NiSO_4$ (c) $(NH_4)_3PO_4$

35. Name each of the following ionic compounds (see Question 27):
(a) $Ca(CH_3CO_2)_2$ (b) $Co_2(SO_4)_3$ (c) $Al(OH)_3$

36. Give the formula for each of the following ionic compounds:
(a) ammonium carbonate (c) copper(II) bromide
(b) calcium iodide (d) aluminum phosphate

37. Give the formula for each of the following ionic compounds:
(a) calcium hydrogen carbonate
(b) potassium permanganate
(c) magnesium perchlorate
(d) potassium hydrogen phosphate

38. Give the name for each of the following binary nonmetal compounds:
(a) NF_3 (c) BBr_3
(b) HI (d) C_6H_{14}

39. Give the name for each of the following binary nonmetal compounds:
(a) C_8H_{18} (c) OF_2
(b) P_2S_3 (d) XeF_4

40. Give the formula for each of the following nonmetal compounds:
(a) butane
(b) dinitrogen pentaoxide
(c) nonane
(d) silicon tetrachloride
(e) diboron trioxide (commonly called boric oxide)

41. Give the formula for each of the following nonmetal compounds:
(a) bromine trifluoride
(b) xenon difluoride
(c) diphosphorus tetrafluoride
(d) pentadecane
(e) hydrazine

The Mole

42. Calculate the number of grams in
(a) 2.5 mol of boron
(b) 0.015 mol of oxygen
(c) 1.25×10^{-3} mol of iron
(d) 653 mol of helium

43. Calculate the number of grams in
(a) 6.03 mol of gold
(b) 0.045 mol of uranium
(c) 15.6 mol of Ne
(d) 3.63×10^{-4} mol of plutonium

44. Calculate the number of moles represented by each of the following:
(a) 127.08 g of Cu
(b) 20.0 g of calcium
(c) 16.75 g of Al
(d) 0.012 g of potassium
(e) 5.0 mg of americium

45. Calculate the number of moles represented by each of the following:
(a) 16.0 g of Na
(b) 0.0034 g of platinum
(c) 1.54 g of P
(d) 0.876 g of arsenic
(e) 0.983 g of Xe

46. A piece of sodium metal, Na, if thrown into a bucket of water, produces a dangerously violent explosion from the reaction of sodium with water. If the piece contains 50.4 g of sodium, how many moles of sodium does that represent?

47. Krypton really does not give Superman his strength. If you have 0.00789 g of the gaseous element, how many moles does this represent?

48. Bromine is a liquid at room temperature (and atmospheric pressure). If you want to have 25 g of Br_2 for a reaction in the lab, how many milliliters of the liquid should you use? (The density of Br_2 is 3.12 g/cm^3.)

49. In an experiment, you need 0.125 mol of sodium metal. Sodium can be cut easily with a knife, so if you cut out a block of sodium, what should be the volume of the block in cubic centimeters? If you cut a perfect cube, what will be the length of the edge of the cube? (The density of sodium is 0.968 g/cm^3.) (Caution: Sodium is *very* reactive with water. The metal should be handled only by a knowledgeable chemist.)

50. If you have a 35.67-g piece of chromium metal on your car, how many atoms of chromium do you have?

51. If you have a ring that contains 1.94 g of gold, how many atoms of gold are in the ring?

52. What is the average mass of one copper atom?

53. What is the average mass of one atom of titanium?

54. Calculate the molar mass of each of the following compounds:
(a) Fe_2O_3, iron(III) oxide
(b) BF_3, boron trifluoride
(c) N_2O, dinitrogen oxide (laughing gas)
(d) $MnCl_2 \cdot 4\,H_2O$, manganese(II) chloride tetrahydrate
(e) $C_6H_8O_6$, ascorbic acid or vitamin C

55. Calculate the molar mass of each of the following compounds:
(a) $B_{10}H_{14}$, a boron hydride once considered as a rocket fuel
(b) $C_6H_2(CH_3)(NO_2)_3$, TNT, an explosive
(c) $PtCl_2(NH_3)_2$, a cancer chemotherapy agent called cisplatin
(d) $CH_3CH_2CH_2CH_2SH$, a compound that has a skunk-like odor
(e) $C_{20}H_{24}N_2O_2$, quinine, used as an antimalarial drug

56. How many moles are represented by 1.00 g of each of the following compounds?
(a) CH_3OH, methyl alcohol
(b) Cl_2CO, phosgene, a poisonous gas
(c) NH_4NO_3, ammonium nitrate
(d) $MgSO_4 \cdot 7\,H_2O$, magnesium sulfate heptahydrate (Epsom salt)
(e) $AgCH_3CO_2$, silver acetate

57. Assume you have 0.250 g of each of the following compounds. How many moles of each are represented?
(a) $C_7H_5NO_3S$, saccharin, an artificial sweetener
(b) $C_{13}H_{20}N_2O_2$, procaine, a "pain killer" used by dentists
(c) $C_{20}H_{14}O_4$, phenolphthalein, a dye

58. Vinyl chloride, C_2H_3Cl, is used to make polyvinylchloride (PVC), a plastic from which many useful items are made. If you have 2.50 kg of vinyl chloride, how many moles of the compound are present?

59. Acetone, $(CH_3)_2CO$, is an important industrial solvent. It was reported that 2130 million pounds of this organic compound were produced in 1991. How many moles does this represent? (1.00 lb = 454 g)

60. An Alka-Seltzer tablet contains 324 mg of aspirin ($C_9H_8O_4$), 1904 mg of $NaHCO_3$, and 1000. mg of citric acid ($C_6H_8O_7$). (The last two compounds react with each other to provide the "fizz," bubbles of CO_2, when the tablet is put into water.) (a) Calculate the number of moles of each substance in the tablet. (b) If you take one tablet, how many molecules of aspirin are you consuming?

61. Some types of chlorofluorocarbons (CFCs) have been used as the propellant in spray cans of paint, hair spray, and other consumer products. However, the use of CFCs is being curtailed, because there is strong suspicion that they cause environmental damage. If a spray can contains 250 g of one of these compounds, CCl_2F_2, how many molecules are you releasing to the air when you empty the can?

62. Sulfur trioxide, SO_3, is made in enormous quantities by combining oxygen and sulfur dioxide, SO_2. The trioxide is not usually isolated but is converted to sulfuric acid. If you have 1.00 pound (454 g) of sulfur trioxide, how many moles does this represent? How many molecules? How many sulfur atoms? How many oxygen atoms?

63. CFCs, or chlorofluorocarbons, are strongly suspected of causing environmental damage. A substitute may be CF_3CH_2F. If you have 25.5 g of this new compound, how many moles does this represent? How many atoms of fluorine are contained in 25.5 g of the compound?

Percent Composition

64. Calculate the molar mass of each of these compounds and the weight percent of each element.
(a) PbS, lead(II) sulfide, galena
(b) C_2H_6, ethane, a hydrocarbon fuel
(c) CH_3CO_2H, acetic acid, an important ingredient in vinegar
(d) NH_4NO_3, ammonium nitrate, a fertilizer

65. Calculate the molar mass of each of these compounds and the weight percent of each element.
(a) $MgCO_3$, magnesium carbonate
(b) C_6H_5OH, phenol, an organic compound used in some cleaners
(c) $C_2H_3O_5N$, peroxyacetyl nitrate, an objectionable compound in photochemical smog
(d) $C_4H_{10}O_3NPS$, acephate, an insecticide

66. Acrylonitrile, H_2CCHCN, is the basis of many important plastics and fibers. (a) Calculate the molar mass. (b) Calculate the mass percent of each element in the compound.

67. The copper-containing compound $Cu(NH_3)_4SO_4 \cdot H_2O$ is a beautiful blue solid. Calculate the molar mass of the compound and the mass percent of each element.

The blue copper compound $Cu(NH_3)_4SO_4 \cdot H_2O$. *(C.D. Winters)*

Empirical and Molecular Formulas

68. The empirical formula of maleic acid is CHO. Its molar mass is 116.1 g/mol. What is its molecular formula?

69. A well-known reagent in analytical chemistry, dimethylglyoxime, has the empirical formula C_2H_4NO. If its molar mass is 116.1 g/mol, what is the molecular formula of the compound?

70. Acetylene is a colorless gas that is used as a fuel in welding torches, among other things. It is 92.26% C and 7.74% H. Its molar mass is 26.02 g/mol. Calculate the empirical and molecular formulas.

71. There is a large family of boron-hydrogen compounds called boron hydrides. All have the formula B_xH_y and almost all react with air and burn or explode. One member of this family contains 88.5% B; the remainder is hydrogen. Which of the following is its empirical formula: BH_3, B_2H_5, B_5H_7, B_5H_{11}, BH_2?

72. Nitrogen and oxygen form an extensive series of at least seven oxides of general formula N_xO_y. One of them is a blue solid that comes apart or "dissociates," reversibly, in the gas phase. It contains 36.84% N. What is the empirical formula of this oxide?

73. Cumene is a hydrocarbon, a compound composed only of C and H. It is 89.94% carbon, and the molar mass is 120.2 g/mol. What are the empirical and molecular formulas of cumene?

74. Acetic acid is the important ingredient in vinegar. It is composed of carbon (40.0%), hydrogen (6.71%), and oxygen (53.29%). Its molar mass is 60.0 g/mol. Determine the empirical and molecular formulas of the acid.

75. An analysis of nicotine, a poisonous compound found in tobacco leaves, shows that it is 74.0% C, 8.65% H, and 17.35% N. Its molar mass is 162 g/mol. What are the empirical and molecular formulas of nicotine?

76. Cacodyl, a compound containing arsenic, was reported in 1842 by the German chemist Bunsen. It has an almost intolerable garlic-like odor. Its molar mass is 210 g/mol, and it is 22.88% C, 5.76% H, and 71.36% As. Determine its empirical and molecular formulas.

77. The action of bacteria on meat and fish produces a poisonous compound called cadaverine. As its name and origin imply, it stinks! It is 58.77% C, 13.81% H, and 27.42% N. Its molar mass is 102.2 g/mol. Determine the molecular formula of cadaverine.

78. If Epsom salt, $MgSO_4 \cdot x\ H_2O$, is heated to 250 °C, all the water of hydration is lost. After heating a 1.687-g sample of the hydrate, 0.824 g of $MgSO_4$ remains. How many molecules of water are there per formula unit of $MgSO_4$?

79. The alum used in cooking is potassium aluminum sulfate hydrate, $KAl(SO_4)_2 \cdot x\ H_2O$ (see photo at right). To find the value of x, you can heat a sample of the compound to drive off all of the water and leave only $KAl(SO_4)_2$. Assume that you heat 4.74 g of the hydrated compound and that it loses 2.16 g of water. What is the value of x?

Alum crystals. *(C.D. Winters)*

80. Elemental sulfur with a mass of 1.256 g is combined with fluorine, F_2, to give a compound with the formula SF_x, a very stable, colorless gas. If you isolate 5.722 g of SF_x, what is the value of x?

81. A new compound containing xenon and fluorine is isolated by shining sunlight on a mixture of Xe (0.526 g) and an excess of F_2 gas. If you isolate 0.678 g of the new compound, what is its empirical formula?

General Questions

82. Give the molecular formula for each of the following molecules.

(a)

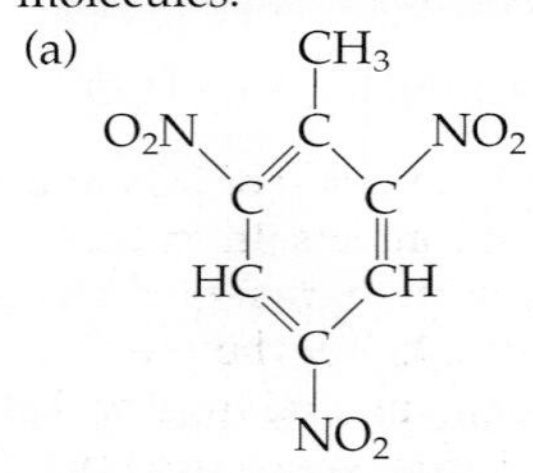

trinitrotoluene, TNT

(b)

$$HO{-}CH_2{-}\underset{NH_2}{\overset{H}{C}}{-}\overset{O}{\overset{\|}{C}}{-}OH$$

serine, an essential amino acid

83. Calculate the mass of one molecule of nitrogen. Now assume that someone decided to make Avogadro's number have a simpler value, say 1.000×10^{20}. Now what is the mass of a molecule of nitrogen?

84. Gems and precious stones are measured in carats, a weight unit equivalent to 200. mg. if you have a 2.3-carat diamond in a ring, how many moles of carbon do you have?

85. The international markets in precious metals operate in the weight unit "troy ounce" (where 1 troy ounce is equivalent to 31.1 g). Platinum sells for \$325 per troy ounce. (a) How many moles are there in one troy ounce? (b) If you have \$5000 to spend, how many grams and how many moles of platinum can be purchased?

86. Gold prices fluctuate, depending on the international situation. If gold currently sells for \$338.70 per troy ounce, how much must you spend to purchase 1.00 mol of gold? (1 troy ounce is equivalent to 31.1 g)

87. The Statue of Liberty in New York harbor is made of 2.00×10^5 pounds of copper sheets bolted to an iron framework. How many grams and how many moles of copper does this represent? (1 lb = 454 g)

88. A piece of copper wire is 25 feet long and has a diameter of 2.0 mm. Copper has a density of 8.92 g/cm^3. How many moles of copper and how many atoms of copper are there in the piece of wire?

89. Which of the following pairs of elements are likely to form ionic compounds? Write appropriate formulas for the compounds you expect to form and give the name of each.
(a) chlorine and bromine
(b) lithium and tellurium
(c) sodium and argon
(d) magnesium and fluorine
(e) nitrogen and bromine
(f) indium and sulfur
(g) selenium and bromine

90. Name each of the following compounds and tell which ones are best described as ionic:

(a) $ClBr_3$ (e) XeF_4 (h) Al_2S_3
(b) NCl_3 (f) OF_2 (i) PCl_5
(c) $CaSO_4$ (g) NaI (j) K_3PO_4
(d) C_7H_{16}

91. Write the formula for each of the following compounds and tell which ones are best described as ionic:
(a) sodium hypochlorite
(b) aluminum perchlorate
(c) potassium permanganate
(d) potassium dihydrogen phosphate
(e) chloride trifluoride
(f) boron tribromide
(g) calcium acetate
(h) sodium sulfite
(i) disulfur tetrachloride
(j) phosphorus trifluoride

92. Precious metals such as gold and platinum are sold in units of "troy ounces," where 1 troy ounce is equivalent to 31.1 grams. If you have a block of platinum with a mass of 15.0 troy ounces, how many moles of the metal do you have? What is the size of the block in cubic centimeters? (The density of platinum is 21.45 g/cm^3 at 20 °C.)

93. Dilithium is the fuel for the *Starship Enterprise.* However, because its density is quite low, you will need a large space to store a large mass. As an estimate for the volume required, we shall use the element lithium. If you want to have 256 mol for an interplanetary trip, what must the volume of a piece of lithium be? If the piece of lithium is a cube, what is the dimension of an edge of the cube? (The density of lithium is 0.534 g/cm^3 at 20 °C.)

94. Fluorocarbonyl hypofluorite was recently isolated, and analysis showed it to be 14.6% C, 39.0% O, and 46.3% F. If the molar mass of the compound is 82 g/mol, determine the empirical and molecular formulas of the compound.

95. Azulene, a beautiful blue hydrocarbon, is 93.71% C and has a molar mass of 128.16 g/mol. What are the empirical and molecular formulas of azulene?

96. A major oil company has used a gasoline additive called MMT to boost the octane rating of its gasoline. What is the empirical formula of MMT if it is 49.5% C, 3.2% H, 22.0% O, and 25.2% Mn?

97. Direct reaction of iodine (I_2) and chlorine (Cl_2) produces an iodine chloride, I_xCl_y, a bright yellow solid. If you completely used up 0.678 g of iodine, and produced 1.246 g of I_xCl_y, what is the empirical formula of the compound? A later experiment showed that the molar mass of I_xCl_y is 467 g/mol. What is the molecular formula of the compound?

98. Pepto-Bismol, which helps provide relief for an upset stomach, contains 300. mg of bismuth subsalicylate, $C_7H_5BiO_4$, per tablet. If you take two tablets for your stomach distress, how many moles of the "active ingredient" are you taking? How many grams of Bi are you consuming in two tablets?

99. Iron pyrite, often called "fool's gold," has the formula FeS_2 (see photo). If you could convert 15.8 kg of iron pyrite to iron metal and remove the sulfur, how many kilograms of the metal could you obtain?

Iron pyrite or "fool's gold." *(C.D. Winters)*

100. Ilmenite is a mineral that is an oxide of iron and titanium, $FeTiO_3$. If an ore that contains ilmenite is 6.75% titanium, what is the mass (in grams) of ilmenite in 1.00 metric ton (exactly 1000 kg) of the ore?

101. Stibnite, Sb_2S_3, is a dark gray mineral from which antimony metal is obtained. If you have one pound of an ore that contains 10.6% antimony, what mass of Sb_2S_3 (in grams) is there in the ore?

102. Draw diagrams in the following boxes to indicate the arrangement of submicroscopic particles of each substance. Consider each box to hold a very tiny portion of each substance. Each drawing should contain at least 16 particles, and it need not be three-dimensional.

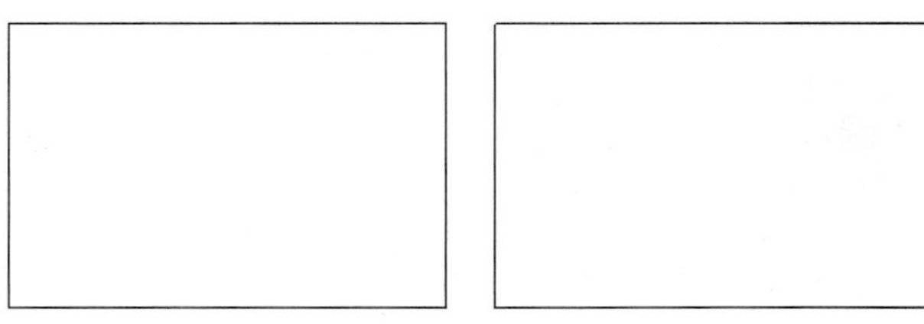

103. Draw submicroscopic diagrams of each situation listed below. Represent atoms or monatomic ions as circles, represent molecules or polyatomic ions by overlapping circles for the atoms that make up the molecule or ion, and distinguish among different kinds of atoms by labeling or shading the circles. In each case draw at least ten submicroscopic particles. Your diagrams can be two dimensional.
   (a) A crystal of solid sodium chloride.
   (b) The sodium chloride from (a) after it has been melted.
   (c) A sample of molten aluminum oxide, Al_2O_3.
   (d) A crystal of copper(II) sulfate pentahydrate.

104. Draw submicroscopic diagrams of each situation listed below. Represent atoms or monatomic ions as circles, represent molecules or polyatomic ions by overlapping circles for the atoms that make up the molecule or ion, and distinguish among different kinds of atoms by labeling or shading the circles. In each case draw at least ten submicroscopic particles. Your diagrams can be two dimensional.
   (a) A sample of solid lithium nitrate, $LiNO_3$.
   (b) A sample of molten lithium nitrate.
   (c) The same sample of lithium nitrate after electrodes have been placed into it and a direct current applied to the electrodes.
   (d) A sample of solid lithium nitrate in contact with a solution of lithium nitrate in water.

105. Assume that each of the ions listed below is in aqueous (water) solution. Draw a submicroscopic picture of the ion and the water molecules in its immediate vicinity.
   (a) Cl^- (c) Mg^{2+}
   (b) K^+ (d) Al^{3+}

106. Of the ions listed in Question 107, which will have the strongest attraction for water molecules in its vicinity? Why?

107. A sample of cobalt(II) chloride hexahydrate, a pink substance, was heated in a crucible and turned blue. Explain how you could use this observation to
   (a) keep track of changes in the weather.
   (b) test a series of compounds to see which ones were hydrates.
   (c) make a device that would indicate whether an electronic component had been exposed to moisture.
   (d) determine whether a substance that was burned in pure oxygen contained the element hydrogen.

108. Which of the following is impossible?
   (a) Silver foil that is 1.2×10^{-4} m thick.
   (b) A sample of potassium that contains 1.784×10^{24} atoms.
   (c) Liquid water heated to a temperature of 345 K.
   (d) A gold coin of mass 1.23×10^{-3} kg.
   (e) 3.43×10^{-27} mol of S_8.

Summary Questions

109. A piece of nickel foil, 0.550 mm thick and 1.25 cm square, was allowed to react with fluorine, F_2, to give a nickel fluoride. (a) How many moles of nickel foil were used? (b) If you isolate 1.261 g of the nickel fluoride, what is its formula? (c) What is its name? (The density of nickel is 8.908 g/cm^3.)

110. Uranium is used as a fuel, primarily in the form of uranium(IV) oxide, in nuclear power plants. This question considers some uranium chemistry.
   (a) A small sample of uranium metal (0.169 g) is heated to 800 to 900 °C in air to give 0.199 g of a dark-green oxide, U_xO_y. How many moles of uranium metal were used? What is the empirical formula of the oxide, U_xO_y? What is the name of the oxide? How many moles of U_xO_y must have been obtained?
   (b) The naturally occurring isotopes of uranium are ^{234}U, ^{235}U, and ^{238}U. Which is the most abundant?
   (c) The oxide U_xO_y is obtained if $UO_2(NO_3) \cdot n\,H_2O$ is heated to temperatures greater than 800 °C in the air. However, if you heat it gently, only the water of hydration is lost. If you have 0.865 g of $UO_2(NO_3) \cdot n\,H_2O$, and obtain 0.679 g of $UO_2(NO_3)$ on heating, how many molecules of water of hydration were there in each formula unit of the original compound?

Using Computer Programs and Videodiscs

111. Use the *KC? Discoverer* computer program to find the formula or formulas for
   (a) the oxides formed by the Group 2A metals Mg, Ca, Ba, and Sr.
   (b) the hydride formed by magnesium.
   (c) the halides of iron.

112. Use the *KC? Discoverer* computer program to find the formula or formulas for the oxides of the Group 4A elements, carbon through lead. Are all of these formulas the same? Assuming all oxides are formed from the oxide ion, O^{2-}, what is the apparent charge on the Group 4A element in each case?

113. Find the images on the *Periodic Table Videodisc* for the reactions of calcium with acids (frame number 20119). What qualitative observations can you make about the reactions of this element with hydrochloric acid (HCl) and nitric acid (HNO_3)? The reaction with HCl gives calcium chloride, and the reaction with nitric acid gives calcium nitrate. Write the formulas for these two compounds. One of the products of the reaction with nitric acid is the brown gas nitrogen dioxide. Write the formula of this gas.

114. Find the images on the *Periodic Table Videodisc* for the reactions of barium with hydrochloric acid (HCl) and nitric acid (HNO_3) (frame number 47293). Compare these reactions with those of calcium (20119, see Study Question 113). What qualitative observations can you make about the reactions of barium compared with those of calcium? Consider the formation of insoluble solids, obvious gas production, and speed of the reaction.

The reaction of aluminum and bromine
(C.D. Winters)

Chemical Reactions

CHAPTER

5

A chemical compound may be made of molecules or oppositely charged ions, and the compound's properties can be interpreted in terms of the behavior of those molecules or ions. The chemical properties of a substance consist of the transformations that the molecules or ions can undergo when the substance reacts. Usually such transformations can be described by a balanced chemical equation. A great many chemical reactions take place between substances that are dissolved in water, and these reactions are important because much of the human environment, as well as humans themselves, depends on aqueous solutions. The very large number of known chemical reactions can be assigned to a few general categories: combination, decomposition, precipitation, acid-base, and oxidation-reduction reactions. The ability to recognize when a given type of reaction may occur will enable you to predict the kinds of compounds that can be produced. You can use such information to design experiments in which compounds with desirable properties are prepared.

Chemists have studied hundreds of thousands of reactions, and there are many more waiting to be investigated. Hundreds of reactions are used by the chemical industry to manufacture the products we all use, and the mere act of reading this sentence involves an untold number of reactions in your body. Finally, thousands of reactions occur in our environment, to both our benefit and our harm. After looking at some basic principles of how to write chemical equations and the nature of compounds when they are dissolved in water, we classify reactions into a few types so that you can begin to predict what might happen in some reactions that you or even professional chemists have never seen before.

5.1 CHEMICAL EQUATIONS

The reaction illustrated in Figure 5.1 is dangerous! Under no circumstances should you perform this reaction yourself.

To take the photographs in Figure 5.1, we cut ordinary kitchen aluminum foil into small pieces, dropped them into a beaker of liquid bromine in a laboratory fume hood, and moved back! The aluminum pieces soon began to burn with a brilliant light, and glowing globs of molten aluminum appeared in the beaker. In fact, they were so hot that they melted through the bottom of the beaker. The **chemical reaction** that occurs is the combination of aluminum atoms from the solid metal with Br_2 molecules from the liquid element to give a white solid consisting of Al_2Br_6 molecules. We can depict

(a)

(b)

(c)

Figure 5.1
Bromine, Br_2, an orange-brown liquid, and aluminum metal (a) react so vigorously that the aluminum becomes molten and glows white hot (b). The vapor in (b) consists of vaporized Br_2 and some of the product, white Al_2Br_6. At the end of the reaction (c), the beaker is coated with aluminum bromide and the products of its reaction with atmospheric moisture. (NOTE: This reaction is dangerous! Under no circumstances should it be done except under properly supervised conditions.) *(C.D. Winters)*

Portrait of a Scientist

Antoine Laurent Lavoisier (1743–1794)

On Monday, August 7, 1774, the Englishman Joseph Priestley (1733–1804) became the first person to isolate oxygen. He heated solid mercury(II) oxide, HgO, causing the oxide to decompose to mercury and oxygen.

$$2\,HgO(s) \rightarrow 2\,Hg(\ell) + O_2(g)$$

Priestley did not immediately understand the significance of the discovery, but he mentioned it to the French chemist Lavoisier in October, 1774. One of Lavoisier's contributions to science was his recognition of the importance of exact scientific measurements and of carefully planned experiments, and he applied these methods to the study of oxygen. From this work he came to believe Priestley's gas was present in all acids and so he named it "oxygen," from the Greek words meaning "to form an acid." In addition, Lavoisier observed that the heat produced by a guinea pig in exhaling a given amount of carbon dioxide is similar to the quantity of heat produced by burning carbon to give the same amount of carbon dioxide. From this and other experiments he concluded that "Respiration is a combustion, slow it is true, but otherwise perfectly similar to that of charcoal." Although he did not understand the details of the process, this was an important step in the development of biochemistry.

Lavoisier was a prodigious scientist, and the principles of naming chemical substances that he introduced are still in use today. Further, he wrote a textbook in which he applied for the first time the principle of the conservation of matter to chemistry and used the idea to write early versions of chemical equations.

As Lavoisier was an aristocrat, he came under suspicion during the Reign of Terror of the French Revolution. He was an investor in the Ferme Générale, the infamous tax collecting organization in 18th century France. Tobacco was a monopoly product of the Ferme Générale, and it was common to cheat the purchaser by adding water to the tobacco, a practice that Lavoisier opposed. Nonetheless, because of his involvement with the Ferme, his career was cut short by the guillotine on May 8, 1794, on the charge of "adding water to the people's tobacco."

Lavoisier and his wife, painted in 1788 by Jacques-Louis David. (The Bettmann Archive)

For a fascinating account of Lavoisier's life and his friendship with Benjamin Franklin, see "The Passion of Antoine Lavoisier" in *Bully for Brontosaurus* by Stephen Jay Gould (Norton, 1991).

---

this by the following **balanced chemical equation** that shows the relative amounts of **reactants** (the substances combined) and **products** (the substances obtained).

$$\underset{\text{reactants}}{2\,Al(s) + 3\,Br_2(\ell)} \longrightarrow \underset{\text{product}}{Al_2Br_6(s)}$$

Al_2Br_6 has the empirical formula $AlBr_3$. However, the substance exists as a true molecule with the formula Al_2Br_6 in the gaseous or solid state.

The physical states of the reactants and products usually are also indicated in an equation. The symbol (s) indicates a solid, (g) a gas, and (ℓ) a liquid. What the equation does *not* show are the conditions of the experiment or whether any energy (in the form of heat or light) is involved. Lastly, a chemical equation does not tell you whether the reaction happens very quickly or whether it takes 100 years.

The speeds of chemical reactions will be discussed in Chapter 8.

In the 18th century, the great French scientist Antoine Lavoisier introduced the law of conservation of matter, which later became part of Dalton's atomic theory (Section 3.1). Lavoisier showed that *matter can neither be created nor destroyed.* This law means that if you use 10 g of reactants, then, if the reaction is complete, you must end up with 10 g of products. Combined with Dalton's atomic theory, this also means that if 1000 atoms of a particular element react, then those 1000 atoms must appear in the products in some fashion. When applied to the reaction of aluminum and bromine, the conservation of matter means that two atoms of aluminum and three diatomic molecules of Br_2 (or six atoms of Br) are required to produce one molecule of Al_2Br_6 (in which the two atoms of Al and six atoms of Br are combined). The numbers in front of the formulas in the balanced equation are put there to show how matter is conserved. The equality of the number of atoms of each kind in the reactants and in the products is what makes the equation "balanced."

The balanced equation for the reaction of aluminum with bromine also shows that 2000 atoms of Al, for example, will react with 3000 Br_2 molecules to give 1000 molecules of Al_2Br_6. To carry this argument even further, it is also true that $2 \times 6.022 \times 10^{23}$ atoms of Al (2 mol) react with $3 \times 6.022 \times 10^{23}$ molecules of Br_2 (3 mol) to provide 6.022×10^{23} molecules of Al_2Br_6 (1 mol). As demanded by the conservation of matter, the total number of atoms in the reactants is 4.818×10^{24}, and the total number of atoms in the product is 4.818×10^{24}.

Recall from Section 4.6 that there are 6.022×10^{23} atoms or 6.022×10^{23} molecules in a mole of any element or in a mole of any molecular compound.

	2 Al(s)	**+**	**3 $Br_2(\ell)$**	**⟶**	**$Al_2Br_6(s)$**
①	2 atoms Al	+	3 molecules Br_2	⟶	1 molecule Al_2Br_6
②	2 atoms Al	+	3×2 atoms Br	⟶	2 atoms Al + 6 atoms Br
③	2 mol Al	+	3 mol Br_2	⟶	1 mol Al_2Br_6
④	2 mol Al	+	3 mol Br_2 × 2 mol Br/mol Br_2		(2 mol Al and 6 mol Br)
	$\times 6.022 \times 10^{23}$ atoms/ mol $= 1.204 \times 10^{24}$ atoms		$\times 6.022 \times 10^{23}$ atoms/mol $= 3.612 \times 10^{24}$ atoms	⟶	$\times 6.022 \times 10^{23}$ atoms/mol $= 4.816 \times 10^{24}$ atoms
	4.816×10^{24} total atoms in reactants			⟶	**4.816×10^{24} total atoms in product**

The atomic weights show that 2.000 mol of Al is equivalent to 53.96 g of Al and that 3.000 mol of Br_2 is equivalent to 479.4 g of Br_2, so the total mass of reactants must be

$$\begin{array}{lrl} & 2.000\text{ mol Al} & = \ \ 53.96\text{ g} \\ + & 3.000\text{ mol Br}_2 & = 479.4\text{ g} \\ \hline \multicolumn{2}{r}{\text{total mass of reactants}} & = 533.4\text{ g} \end{array}$$

Conservation of matter demands that the same mass, 533.4 g of Al_2Br_6, must result from the reaction. Of course, the balanced equation shows that this is the case.

$$1.000\ \cancel{\text{mol Al}_2\text{Br}_6} \cdot \frac{533.4\text{ g Al}_2\text{Br}_6}{1\ \cancel{\text{mol Al}_2\text{Br}_6}} = 533.4\text{ g Al}_2\text{Br}_6$$

The ratios of the amounts of reactants and products given by the coefficients in the balanced equation apply to any quantity of reactants. For example, if 1 mol of aluminum was used, then only $\frac{1}{2}$ mol of Al_2Br_6 would be expected as the product. The relationship between the masses of chemical reactants and products is called **stoichiometry** (*stoy-key-AHM-uh-tree*), and the coefficients (or multiplying numbers) in a balanced equation are the **stoichiometric coefficients.** Balanced chemical equations are fundamental to the quantitative understanding of chemistry, so this chapter discusses the kinds of chemical reactions and methods for balancing their equations in order that the subject of stoichiometry can be explored in the next chapter.

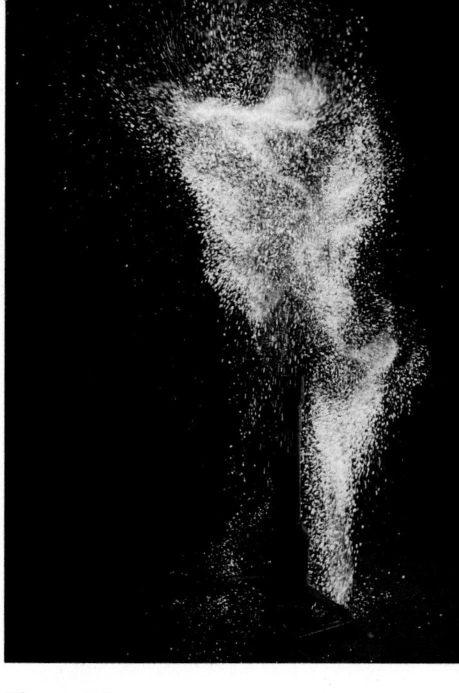

Figure 5.2
Powdered iron burns in air to form the iron oxides FeO and Fe_2O_3. The flame and the energy released in the reaction heat the particles to incandescence. *(C.D. Winters)*

EXERCISE 5.1 • *Chemical Reactions*

The reaction of iron with oxygen is shown in Figure 5.2. The equation for the reaction is

$$4\ Fe(s) + 3\ O_2(g) \longrightarrow 2\ Fe_2O_3(s)$$

a. What are the stoichiometric coefficients in this equation?

b. Assuming the iron and oxygen are present in the correct stoichiometric ratio, and starting with iron and oxygen with a total mass of 1.56 g, what is the maximum mass of Fe_2O_3 that can be formed?

c. If you were to use 8000 atoms of Fe, how many molecules of O_2 would be required to consume the iron completely?

5.2 BALANCING CHEMICAL EQUATIONS

A chemical equation must be balanced before useful quantitative information can be obtained about a chemical reaction. Balancing equations ensures that the same number of atoms of each element appear on both sides of the equation. Many chemical equations can be balanced by trial and error, although some will involve more trial than others.

One general class of chemical reactions is the reaction of metals or nonmetals with oxygen to give **oxides** with a general formula of M_xO_y (see Section 5.4). For example, iron reacts with oxygen to give iron(III) oxide (Figure 5.2),

$$4\ Fe(s) + 3\ O_2(g) \longrightarrow 2\ Fe_2O_3(s)$$

magnesium gives magnesium oxide (Figure 5.3),

$$2\ Mg(s) + O_2(g) \longrightarrow 2\ MgO(s)$$

and phosphorus, P_4, reacts vigorously with oxygen to give tetraphosphorus decaoxide, P_4O_{10} (Figure 5.4).

$$P_4(s) + O_2(g) \xrightarrow{\text{(unbalanced equation)}} P_4O_{10}(s)$$

The equations involving iron and magnesium are balanced as we have written them, since there are the same numbers of metal and oxygen atoms on each side. Similarly, the phosphorus atoms are balanced in the $P_4 + O_2$ equation. However, the oxygen is not. To balance it, we must place a 5 in

Figure 5.3
A piece of magnesium ribbon burns in air to give the white solid magnesium oxide, MgO. *(C.D. Winters)*

front of the O_2 on the left to balance the ten oxygen atoms appearing on the right side of the equation.

$$P_4(s) + 5\,O_2(g) \xrightarrow{\text{(balanced equation)}} P_4O_{10}(s)$$

Reactions in which an element or compound burns in O_2 are often referred to as **combustion reactions** (Section 5.4). The combustion of octane, a component of gasoline, is such a reaction.

$$2\,C_8H_{18}(g) + 25\,O_2(g) \longrightarrow 16\,CO_2(g) + 18\,H_2O(\ell)$$

In all combustion reactions, the elements in the reactants end up in compounds containing oxygen, that is, as oxides. For hydrocarbons (Table 4.2) and for compounds containing only C, H, and O, the products of complete combustion are always carbon dioxide and water.

As an example of equation balancing, let us write the balanced equation for the combustion reaction of propane, C_3H_8.

Step 1. As the first step in writing any equation, *write down the correct formulas of the reactants and products.*

$$C_3H_8(g) + O_2(g) \xrightarrow{\text{(unbalanced equation)}} CO_2(g) + H_2O(\ell)$$

Here propane and oxygen are reactants, and carbon dioxide and water are products.

Step 2. *Balance the number of C atoms.* In combustion reactions it usually works best to balance the carbon atoms first and leave the oxygen atoms to the end (because the oxygen atoms are not all found in one compound). In this case there are three carbon atoms in the reactants, so there must be three in the products. Therefore, three CO_2 molecules are required on the right side.

$$C_3H_8 + O_2 \longrightarrow 3\,CO_2 + H_2O$$

Step 3. *Balance the number of H atoms.* There are eight H atoms in the reactants. Each molecule of water has two hydrogen atoms, so four molecules of water will give the required eight hydrogen atoms on the right side.

$$C_3H_8 + O_2 \longrightarrow 3\,CO_2 + 4\,H_2O$$

Step 4. *Balance the number of O atoms.* There are ten oxygen atoms on the right side ($3 \times 2 = 6$ in CO_2 and $4 \times 1 = 4$ in water). Therefore, five O_2 molecules are needed to supply the required ten oxygen atoms.

$$C_3H_8 + 5\,O_2 \longrightarrow 3\,CO_2 + 4\,H_2O$$

Step 5. *Verify that the number of atoms of each element is balanced.* The equation shows three carbon atoms, eight hydrogen atoms, and ten oxygen atoms on each side.

Balancing equations by trial and error is usually simple enough if you are organized in your approach. Some may take a bit longer than others, but don't get discouraged. All equations that have the correct formulas for all reactants and products can be balanced. Finally, it is important to understand that *subscripts in the formulas of reactants and products cannot be changed*

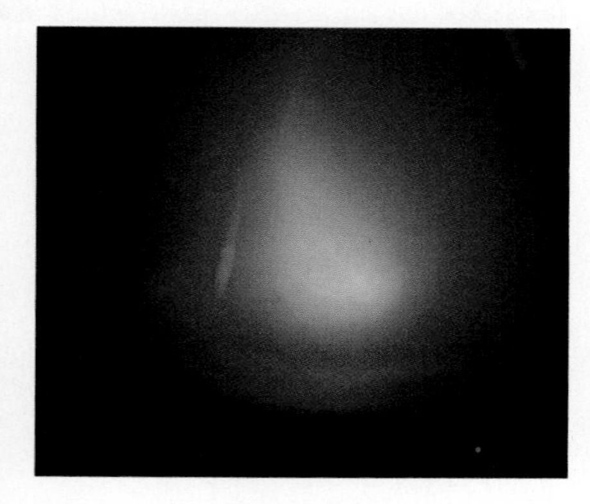

Figure 5.4
When white phosphorus is exposed to oxygen, it first glows (top; a process called phosphorescence) and then bursts into flame (bottom). To prevent this, white phosphorus is usually stored under water (Figure 3.15). *(C.D. Winters)*

to balance equations. The subscripts identify these substances, and changing the subscripts would change their identities. Carbon monoxide, CO, is a very different compound from carbon dioxide, CO_2, for example.

EXAMPLE 5.1

Balancing the Equation for a Combustion Reaction

Write the balanced equation for the combustion of butane, C_4H_{10}.

SOLUTION

Step 1. *Write the correct formulas for reactants and products.* As is always the case for compounds containing only C, H, and O, the products of a combustion will be CO_2 and H_2O if the reaction goes to completion. Therefore, the unbalanced equation is

$$C_4H_{10}(g) + O_2(g) \longrightarrow CO_2(g) + H_2O(\ell)$$

Step 2. *Balance the C atoms.* The four carbon atoms in butane require the production of four CO_2 molecules.

$$C_4H_{10}(g) + O_2(g) \longrightarrow 4\ CO_2(g) + H_2O(\ell)$$

Step 3. *Balance the H atoms.* There are ten hydrogen atoms on the left, so five molecules of H_2O, each having two hydrogen atoms, are required on the right.

$$C_4H_{10}(g) + O_2(g) \longrightarrow 4\ CO_2(g) + 5\ H_2O(\ell)$$

Step 4. *Balance the O atoms.* As the reaction stands after step 3, there are 13 oxygen atoms on the right ($4 \times 2 = 8$ in CO_2 plus $5 \times 1 = 5$ in H_2O) and two on the left. That is, there is an odd number on the right and an even number on the left. There are two equally valid ways to balance the oxygen.

Solution 1. To have 13 atoms of oxygen on the left side, use a stoichiometric coefficient of $\frac{13}{2}$; you can do this because

$$\frac{13}{2} \cdot \frac{2 \text{ oxygen atoms}}{1 \text{ molecule of } O_2} = 13 \text{ oxygen atoms}$$

Therefore, the balanced equation will be

$$C_4H_{10}(g) + \tfrac{13}{2}\ O_2(g) \longrightarrow 4\ CO_2(g) + 5\ H_2O(\ell)$$

Solution 2. Taking the equation as it stands after step 3, multiply each coefficient by 2 so that there is an even number of oxygen atoms on the right side (that is, 26).

$$2\ C_4H_{10}(g) + \underline{\quad\quad}\ O_2(g) \longrightarrow 8\ CO_2(g) + 10\ H_2O(\ell)$$

Now the O_2 on the left can be balanced by multiplying by a whole number instead of a fraction.

$$2\ C_4H_{10}(g) + 13\ O_2(g) \longrightarrow 8\ CO_2(g) + 10\ H_2O(\ell)$$

While you will find it convenient at times to use fractional coefficients, generally equations are balanced using whole-number coefficients.

Step 5. *As verification,* notice that there are 8 carbon atoms, 20 hydrogen atoms, and 26 oxygen atoms on each side of the equation from Solution 2.

EXERCISE 5.2 • *Balancing the Equation for a Combustion Reaction*

Pentane can burn completely in air to give carbon dioxide and water. Write a balanced equation for this combustion reaction. (If you don't recall the formula for pentane, refer to Table 4.2.)

5.3 PROPERTIES OF COMPOUNDS IN AQUEOUS SOLUTION

The properties of molecular compounds in solution are described in Chapters 15 and 16.

Many reactions occur between solids and gases, or between two gases. However, most of the reactions in your body occur among substances dissolved in water, that is, in an **aqueous solution.** This is also true of many reactions that you will see in your laboratory and that occur all around you in nature. Therefore, it is important to understand something about the behavior of compounds in water and how this affects their reactions. The focus here is on compounds that produce ions in aqueous solution.

If magnesium ribbon is put into an aqueous solution of the compound HCl, the mixture will bubble furiously as hydrogen gas is given off, according to the following equation (Figure 5.5).

$$Mg(s) + 2\,HCl(aq) \longrightarrow MgCl_2(aq) + H_2(g)$$

Figure 5.5
A ribbon of magnesium reacts with aqueous HCl to give H_2 gas and aqueous $MgCl_2$. *(C.D. Winters)*

In this balanced equation both HCl and $MgCl_2$ are followed by (aq), indicating that they are dissolved in water and are present as aqueous solutions. Also, these two substances exist as ions in the aqueous solution: $MgCl_2$ exists as $Mg^{2+}(aq)$ and $Cl^-(aq)$, and HCl exists as $H^+(aq)$ and $Cl^-(aq)$. But how do we know that these substances will dissolve and that ions are present? As we shall soon see, $MgCl_2$ is a water-soluble ionic compound, while HCl is completely dissociated into ions in aqueous solution.

Ions in Aqueous Solution: Electrolytes

One property of ionic compounds is that many dissolve in water to give aqueous solutions of ions and thus produce solutions that can conduct electricity. That is, ionic compounds that dissolve in water are **electrolytes** (Section 4.4). Electrolytes can be classified as *strong* or *weak.* When sodium chloride and many other ionic compounds dissolve in water, the ions separate or dissociate completely. For every mole of NaCl that dissolves, one mole of Na^+ and one mole of Cl^- ions are found in solution.

$$NaCl(aq) \equiv Na^+(aq) + Cl^-(aq)$$

100% dissociation = strong electrolyte

The idea that salts such as NaCl dissociate completely to give only ions in solution is a simplification. In fact, there can be a measurable concentration of species such as NaCl(aq), that is, species called "ion pairs."

Since there is a high concentration of ions in solution, the solution is a good conductor of electricity. Substances whose solutions conduct well are known as **strong electrolytes.**

Other substances produce only a few ions when they dissolve, and so are poor conductors of electricity; they are known as **weak electrolytes** (Fig-

(a) (b) (c)

Figure 5.6
When an electrolyte is dissolved in water in the beaker and provides ions that are free to move about, the electrical circuit is completed, and the light bulb included in the circuit glows. (See Figures 4.12 and 4.13.) (a) Pure water is a nonelectrolyte, and the bulb does not light. (b) In the case of a weak electrolyte, such as acetic acid, only a few of the dissolved molecules ionize. The bulb glows weakly, indicating only a small electrical current flow. (c) Dilute K_2CrO_4, potassium chromate, is a strong electrolyte. The bulb glows brightly, indicating that virtually every K_2CrO_4 unit has dissociated into its ions, K^+ and CrO_4^{2-}.
(C.D. Winters)

ure 5.6). For example, when acetic acid dissolves, less than 5% of the molecules ionize to produce a cation and anion.

$$CH_3CO_2H(aq) \rightleftharpoons H^+(aq) + CH_3CO_2^-(aq)$$

acetic acid, weak electrolyte; hydrogen ion; acetate ion

We use the term "dissociate" for ionic compounds that separate into their constituent ions in water. The term "ionize" is used for nonionic compounds, like acetic acid, whose constituent molecules form ions in water.

A characteristic of acetic acid and other weak electrolytes is that a dynamic *equilibrium* is established in solution. This means that CH_3CO_2H molecules undergo ionization while $H^+(aq)$ and $CH_3CO_2^-(aq)$ ions simultaneously recombine to give the un-ionized acid and water. In general, in an

equilibrium a forward reaction and its reverse take place simultaneously so that some of both the reactants and products are present. This situation is represented by the double arrows $\leftrightarrows$ in the ionization equation.

There are also substances that dissolve in water but do not ionize. These are called **nonelectrolytes** because their solutions do not conduct electricity (Figure 5.6). Common nonelectrolytes are sugar, starch, ethyl alcohol (CH_3CH_2OH), and antifreeze (ethylene glycol, HOC_2H_4OH).

When an ionic compound is water-soluble, you can assume it is a strong electrolyte. It will not always be obvious when you encounter a weak electrolyte or a nonelectrolyte, but it is useful to remember that acetic acid and ammonia (NH_3) are common weak electrolytes, and sugar, starch, alcohol, and antifreeze are typical nonelectrolytes.

Chemical equilibria are quite common and will be discussed in more detail in Chapters 8, 15, and 16.

The weak electrolyte acetic acid is the important ingredient in vinegar. See the experiment "Chemistry You Can Do" on page 159.

EXERCISE 5.3 • *Electrolytes*

Epsom salt, $MgSO_4 \cdot 7\,H_2O$, is sold in drugstores and used as a solution in water for various medical purposes. Methyl alcohol, CH_3OH, is dissolved in gasoline in the winter in colder climates to prevent the formation of ice in automobile fuel lines. Which of these compounds might be an electrolyte and which might be a nonelectrolyte?

Solubility of Ionic Compounds in Water

Table salt dissolves readily in water, but we cannot say from this that all ionic compounds will dissolve in water. There are many that do not, and still others that dissolve only to a small extent. Fortunately, we can make some general statements about which types of ions are present in water-soluble ionic compounds.

Figure 5.7 lists a set of guidelines that can help you predict whether a particular ionic compound will be soluble in water. For example, sodium nitrate, $NaNO_3$, contains both an alkali metal cation, Na^+, and the nitrate anion, NO_3^-. According to Figure 5.7, the presence of either of these ions ensures that the compound will be soluble in water. Further, since *ionic compounds that dissolve in water are strong electrolytes,* $NaNO_3$ in aqueous solution really consists of the separated ions $Na^+(aq)$ and $NO_3^-(aq)$.

$$NaNO_3(aq) \xrightarrow{\text{consists of}} Na^+(aq) + NO_3^-(aq)$$

For every mole of $NaNO_3$ that dissolves, a mole of Na^+ ions and a mole of NO_3^- ions are present in solution. On the other hand, CuS is only slightly soluble, as are all sulfides except those containing an alkali metal cation or the NH_4^+ ion. Only a tiny amount of CuS dissolves in water and produces Cu^{2+} cations and anions such as HS^- or S^{2-} ions. If solid CuS is placed in water, nearly all of it remains as the solid, and a heterogeneous mixture is the result.

Other examples of common ionic compounds are given in Figure 5.8. On the basis of the guidelines and the examples in Figure 5.8, be sure to notice the following:

> If a compound contains *at least one* of the ions leading to solubility (Figure 5.7), it is at least moderately soluble in water.

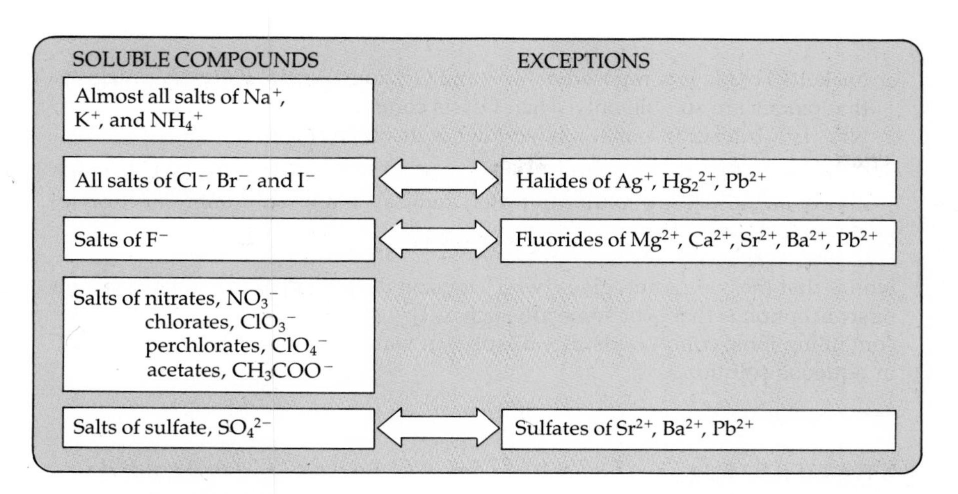

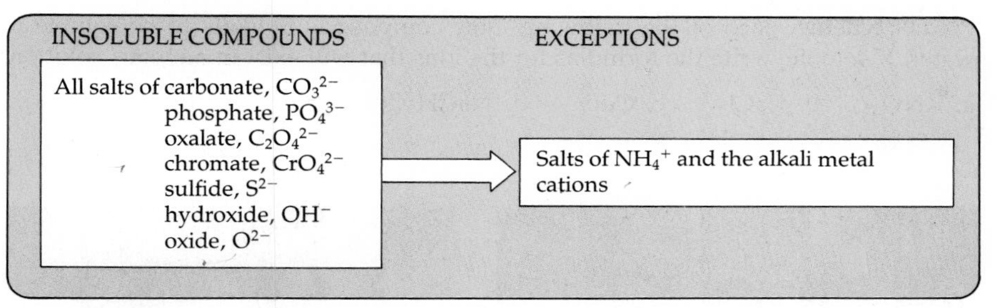

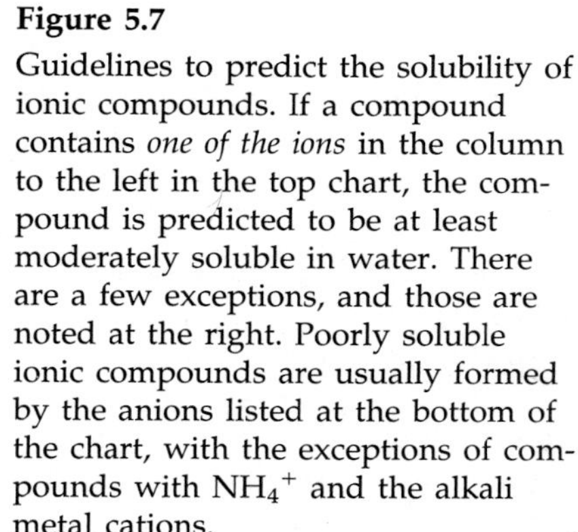

Figure 5.7

Guidelines to predict the solubility of ionic compounds. If a compound contains *one of the ions* in the column to the left in the top chart, the compound is predicted to be at least moderately soluble in water. There are a few exceptions, and those are noted at the right. Poorly soluble ionic compounds are usually formed by the anions listed at the bottom of the chart, with the exceptions of compounds with NH_4^+ and the alkali metal cations.

EXAMPLE 5.2

Solubility Guidelines

For each of the following ionic compounds, predict whether it is likely to be water-soluble. If it is soluble, tell what ions exist in solution.

a. KCl b. $MgCO_3$ c. NiO d. CaI_2

SOLUTION You must first recognize the cation and anion involved and then decide the probable water solubility. As long as one ion is listed in Figure 5.7 as leading to solubility, the compound is likely to be at least moderately soluble.

a. KCl is composed of K^+ and Cl^-. According to Figure 5.7, the presence of either of these ions means that the compound is likely to be soluble in water. Its actual solubility is about 35 g per 100 mL of water at 20 °C. Thus, a solution of KCl actually consists of K^+ and Cl^- ions, and KCl is an electrolyte.

$$\text{KCl in water} \longrightarrow K^+(aq) + Cl^-(aq)$$

b. Magnesium carbonate is composed of Mg^{2+} and CO_3^{2-} ions. Mg^{2+} is in the alkaline earth group, a group that often does not form water-soluble compounds. The carbonate ion usually gives insoluble compounds (Figure 5.7), unless combined with an ion like Na^+ or NH_4^+. Therefore, $MgCO_3$ is insoluble in water. The actual solubility of $MgCO_3 \cdot 3\,H_2O$ is less than 0.2 g per 100 mL of water.

c. Nickel(II) oxide is composed of Ni^{2+} and O^{2-} ions. Again, Figure 5.7 suggests that oxides are soluble only when O^{2-} is combined with an alkali metal ion; Ni^{2+} is a transition metal ion, so NiO is insoluble.

d. Calcium iodide is composed of Ca^{2+} and I^- ions. According to Figure 5.7 almost all iodides are soluble in water, and CaI_2 is a water-soluble electrolyte.

$$CaI_2 \text{ in water} \longrightarrow Ca^{2+}(aq) + 2\,I^-(aq)$$

Notice that the compound gives two I^- ions on dissolving in water. (A common misconception is that I_2 or some ion such as I_2^{2-} is found in solution. All halide-containing ionic compounds that dissolve in water produce F^-, Cl^-, Br^-, or I^- in aqueous solution.)

EXERCISE 5.4 • *Solubility of Ionic Compounds*

Predict whether each of the following ionic compounds is likely to be soluble in water. If soluble, write the formulas for the ions that will exist in aqueous solution.

a. KNO_3 b. $CaCl_2$ c. CuO d. $NaCH_3CO_2$

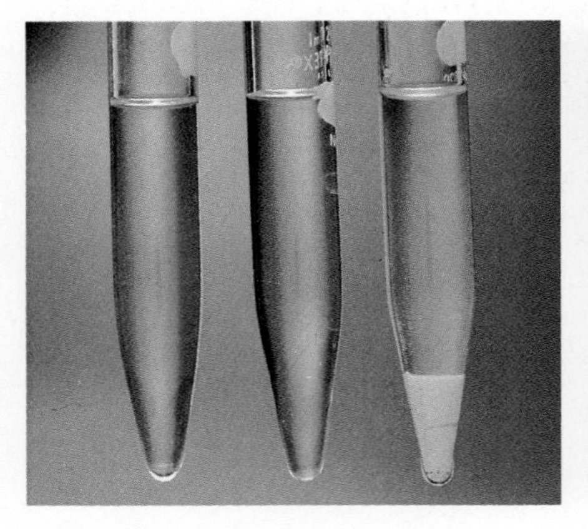

(a) $Ba(NO_3)_2$, $BaCl_2$, $BaSO_4$

(b) $Cu(NO_3)_2$, $CuSO_4$, $Cu(OH)_2$

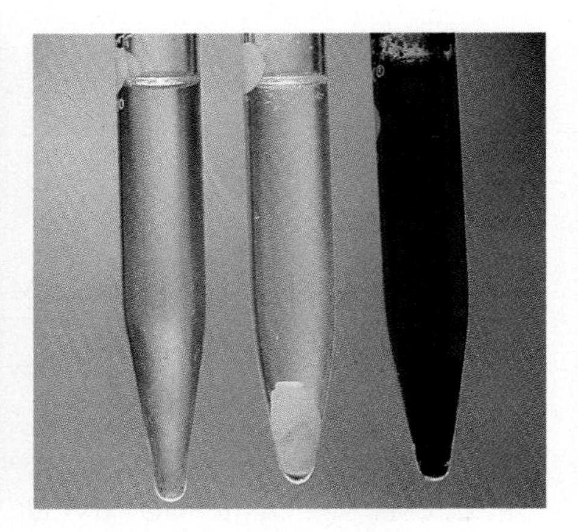

(c) $AgNO_3$, AgCl, AgOH

(d) $(NH_4)_2S$, CdS, Sb_2S_3, PbS

(e) NaOH, $Ca(OH)_2$, $Fe(OH)_3$, $Ni(OH)_2$

Figure 5.8
Illustration of the solubility guidelines of Figure 5.7. With certain exceptions, ionic compounds containing Cl^- and NO_3^- are water soluble. Although many sulfates are water-soluble, there are a few that are not. Finally, metal sulfides are almost invariably insoluble, except for $(NH_4)_2S$ and Na_2S, for example. *(C.D. Winters)*

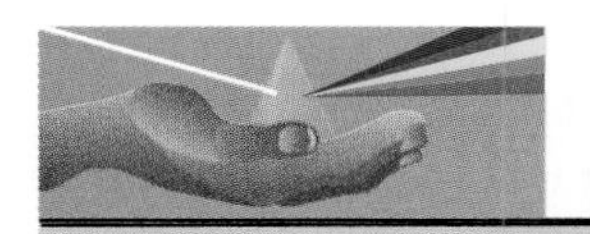

CHEMISTRY YOU CAN DO

Acids and Bases

Many everyday substances belong to two major classes of chemicals: acids and bases. One of the most useful and interesting properties of acids and bases is their ability to change the color of certain vegetable materials, one of them being red cabbage juice. To do this experiment you will need the following:

- Some red cabbage and about 125 mL (or about 1/2 cup) of vinegar. You can get some in a grocery store or at your campus food service.
- A strong plastic bag.
- A funnel and filter paper, or clean cloth.
- A bar of soap.
- A colorless soft drink (optional).
- Other household chemicals such as baking soda, ammonia, or bathroom cleaner (optional).

First make an extract of red cabbage. To do this, put about 500 mL (about 2 cups) of red cabbage into a strong plastic bag and mash the pieces into a pulp. Add about 250 mL of water (about 1 cup) and then mash the cabbage further. (A blender or food processor is a good alternative to the plastic bag.) Filter the mixture through a coffee filter or a piece of cloth, and save the juice.

Get a bottle of white vinegar from the grocery store or campus food service. If there is a label on the vinegar, does the label say it contains an acid? Does it say what acid? To see what effect this acid has on cabbage juice, pour about 125 mL (about 1/2 cup) of vinegar into a glass, add about 5 mL (about a teaspoonful) of the red cabbage juice, and stir. What is the color of the mixture? What can you conclude about the effect of an acid on red cabbage juice? (Save this mixture to compare with the result in the next part of the experiment.)

Next examine the effect of soap on the cabbage juice. Shave a bar of soap with a knife to get small pieces, dissolve them in warm water, and add about 5 mL of cabbage juice; stir thoroughly. Now what is the color of the solution? Does it differ from the color obtained with vinegar? Soap is a base, so the color you see with the cabbage juice "acid–base indicator" should be quite different from the color with an acid.

Now try some other household chemicals with your cabbage juice solution. Some things you might examine for their acid–base properties include soft drinks (colorless ones), soda water, white wine, a solution of baking soda, household ammonia, or various bathroom cleaners. Finally, can you think of any other vegetable materials that change color when in the presence of acids and bases? Try some common spices, colored paper, or tea.

Based on your observations, predict whether the following are most likely to be acidic, basic, or neutral (having no acid or base properties): (a) lemon juice, (b) a solution of dishwasher or laundry detergent, (c) tomato juice, (d) tap water, and (e) sugar water. Finally, predict what will happen if you pour some vinegar into a red cabbage solution that already contains some soap.

Acids and Bases

Acids and bases are two important classes of compounds. Members of each class have a number of properties in common, and some of these are related to properties of the other class. Solutions of acids change the colors of vegetable pigments in specific ways; for example, all acids change the color of litmus from blue to red and cause the dye phenolphthalein to be colorless. Similarly, bases affect pigments, but bases turn red litmus to blue and make phenolphthalein pink. If an acid has made litmus red, adding a base will eventually reverse the effect, making the litmus blue again. Thus, acids and bases seem to be opposites. A base can *neutralize* the effect of an acid, and an acid can neutralize the effect of a base.

Litmus is a dye derived from certain lichens. Phenolphthalein is a synthetic dye made from compounds related to naphtha.

Acids have other characteristic properties. They taste sour, they produce bubbles of gas when added to limestone, and they dissolve many metals while producing a flammable gas. Although tasting substances is never done in a chemistry laboratory, you have probably experienced the sour

(a)

(b)

(c)

Acids and bases. (a) Common foods and household products are acidic or basic. Citrus fruits contain citric acid, and household ammonia and oven cleaner are bases. (b) The acid in lemon juice turns blue litmus paper red, while household ammonia turns red litmus paper blue (c). *(C.D. Winters)*

taste of at least one acid—vinegar, which is a solution of acetic acid in water. Bases, in contrast, have a bitter taste. Rather than dissolving metals, bases often cause metal ions to form insoluble compounds that precipitate from solution. Such precipitates can be made to dissolve by adding an acid, another case in which an acid counteracts a property of a base.

The properties of acids can be explained by a common feature of acid molecules, and a different common feature can explain the properties of bases. An **acid** is any substance that, when dissolved in pure water, increases the concentration of hydrogen ions, H^+ (or, more accurately, **hydronium ions, H_3O^+**, the combination of a water molecule and an H^+ ion). In much the same way that water separates cations and anions when a salt dissolves, water molecules capture an H^+ ion from acid molecules and produce hydronium ions.

$$HCl(aq) + H_2O(\ell) \longrightarrow H_3O^+(aq) + Cl^-(aq)$$

hydrochloric acid (strong electrolyte = 100% ionized) → hydronium ion

The properties that acids have in common are the properties of $H_3O^+(aq)$.

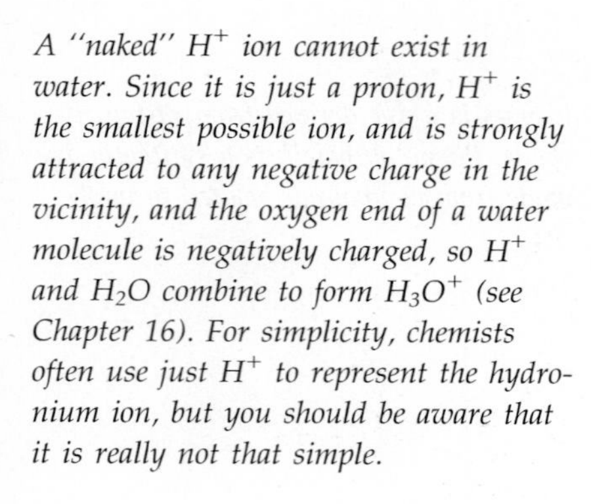

A "naked" H^+ ion cannot exist in water. Since it is just a proton, H^+ is the smallest possible ion, and is strongly attracted to any negative charge in the vicinity, and the oxygen end of a water molecule is negatively charged, so H^+ and H_2O combine to form H_3O^+ (see Chapter 16). For simplicity, chemists often use just H^+ to represent the hydronium ion, but you should be aware that it is really not that simple.

One of the most common acids is hydrochloric acid, which directly supplies an H^+ ion to form a hydronium ion and also releases a chloride ion (Cl^-) in aqueous solution. (See Table 5.1 for other common acids.) Because it is completely converted to ions in aqueous solution, HCl is classified as a strong electrolyte and as a **strong acid.**

Some common acids, such as sulfuric acid, can provide more than one mole of H^+ ions per mole of acid:

$$H_2SO_4(aq) + H_2O(\ell) \longrightarrow H_3O^+(aq) + HSO_4^-(aq)$$

sulfuric acid (100% ionized); hydronium ion; hydrogen sulfate ion

$$HSO_4^-(aq) + H_2O(\ell) \rightleftharpoons H_3O^+(aq) + SO_4^{2-}(aq)$$

hydrogen sulfate ion (<100% ionized); hydronium ion; sulfate ion

The first ionization reaction is essentially complete, so sulfuric acid is considered a strong electrolyte (and so a strong acid as well). However, the

Table 5.1

Common Acids and Bases

Strong acids (strong electrolytes)

HCl	hydrochloric acid
HNO_3	nitric acid
$HClO_4$	perchloric acid
H_2SO_4	sulfuric acid

Strong bases (strong electrolytes)

NaOH	sodium hydroxide
KOH	potassium hydroxide
$Ca(OH)_2$	calcium hydroxide

Weak acids (weak electrolytes)

H_3PO_4	phosphoric acid
CH_3CO_2H	acetic acid
H_2CO_3	carbonic acid

Weak base (weak electrolyte)

NH_3	ammonia

hydrogen sulfate ion, like acetic acid, is a weak electrolyte; that is, HSO_4^- is only partially ionized in aqueous solution. Both the hydrogen sulfate ion and acetic acid are therefore classified as **weak acids.**

A **base** is a substance that increases the concentration of the **hydroxide ion, OH^-,** when dissolved in pure water. The properties that bases have in common are the properties of $OH^-(aq)$. Compounds that contain hydroxide ions, such as sodium hydroxide or potassium hydroxide, are obvious bases. As ionic compounds they are strong electrolytes and **strong bases.**

$$NaOH(s) \xrightarrow{H_2O} Na^+(aq) + OH^-(aq)$$

strong electrolyte = 100% dissociated; hydroxide ion

Ammonia, NH_3, is another very common base. Although the compound does not have an OH^- ion as part of its formula, it produces the ion on reaction with water.

$$NH_3(aq) + H_2O(\ell) \rightleftharpoons NH_4^+(aq) + OH^-(aq)$$

ammonia, base weak electrolyte; ammonium ion; hydroxide ion

In the equilibrium between NH_3 and the NH_4^+ and OH^- ions, only a small concentration of the ions is present, so ammonia is a weak electrolyte (<5% ionized) and a **weak base.**

EXERCISE 5.5 • *Acids and Bases*

(a) What ions are produced when perchloric acid dissolves in water? (b) Calcium hydroxide is not very soluble in water. What little does dissolve, however, is dissociated. What ions are produced?

Acidic and Basic Oxides Each acid shown in Table 5.1 has one or more H atoms in the molecular formula that can be released to form H^+ ions in water. Less obvious examples of acids, though, are *oxides of nonmetals,* such

Figure 5.9
The white solid calcium oxide (lime) is a basic oxide because it reacts with water to produce the base $Ca(OH)_2$ (slaked lime). The test tube at the left also contains the dye phenolphthalein, which turns red in basic solution, while the tube at the right contains only CaO in water. *(C.D. Winters)*

The term "acid rain" was first used in 1872 by Robert Angus Smith, an English chemist and climatologist, to describe the acidic precipitation that fell on Manchester, England at the beginning of the Industrial Revolution.

as carbon dioxide and sulfur trioxide, which have no H atoms but react with water to produce H_3O^+ ions. Carbon dioxide, for example, dissolves in water to a small extent, and some of the dissolved molecules react with water to form a weak acid, carbonic acid. This acid then ionizes to a small extent to form the hydronium ion, H_3O^+, and the bicarbonate ion, HCO_3^-.

Dissolving CO_2	$CO_2(g) + H_2O(\ell) \rightleftharpoons \underset{\text{carbonic acid}}{H_2CO_3(aq)}$
H_2CO_3 ionization	$H_2CO_3(aq) + H_2O(\ell) \rightleftharpoons HCO_3^-(aq) + H_3O^+(aq)$
Overall	$CO_2(g) + 2\,H_2O(\ell) \rightleftharpoons HCO_3^-(aq) + H_3O^+(aq)$

This reaction is very important in our environment; since carbon dioxide is normally found in small amounts in the atmosphere, rainwater is always slightly acidic. Oxides such as CO_2 that can react with water to produce H^+ ions are known as **acidic oxides.**

Oxides of sulfur and nitrogen are present in significant amounts in polluted air, and both of these nonmetal oxides also react with water to produce acids. For example,

$$SO_3(g) + H_2O(\ell) \longrightarrow \underset{\text{sulfuric acid}}{H_2SO_4(aq)}$$

$$2\,NO_2(g) + H_2O(\ell) \longrightarrow \underset{\text{nitric acid}}{HNO_3(aq)} + \underset{\text{nitrous acid}}{HNO_2(aq)}$$

These reactions are the origin of acid rain in the United States, Canada, and other industrialized countries. As described in more detail in Chapter 20, the burning of fossil fuels such as coal and gasoline produces oxides of sulfur and nitrogen. These oxides mix with other chemicals in the atmosphere and with water and then come back to earth in the form of acidified rainfall. When the rain falls on areas that cannot easily tolerate greater than normal acidity, such as the northeastern parts of the United States and the eastern provinces of Canada, serious environmental problems may occur.

Oxides of metals can give basic solutions if they dissolve appreciably in water. Perhaps the best known example is calcium oxide, CaO, often called lime or quicklime. This metal oxide reacts with water to give calcium hydroxide, commonly called slaked lime. The latter compound, while not very soluble in water, dissolves sufficiently to give a strongly basic solution (Figure 5.9).

$$\underset{\text{lime}}{CaO(s)} + H_2O(\ell) \longrightarrow \underset{\text{slaked lime}}{Ca(OH)_2(s)} \rightleftharpoons Ca^{2+}(aq) + 2\,OH^-(aq)$$

Oxides such as CaO that react with water to produce OH^- ions are known as **basic oxides.** Almost 34 billion pounds of lime were produced in the United States in 1991 for use in the metals industry, in sewage and pollution control, in water treatment, in agriculture, and in the construction industry. Indeed, lime is one of the oldest construction materials. Lime plaster was used in Crete in 1500 BC, the Great Wall of China was largely laid with lime mortar, and many structures built with lime mortar by the Romans are still standing.

EXERCISE 5.6 • *Acidic and Basic Oxides*

For each of the following, indicate whether you expect an acidic or basic solution when the compound dissolves in water.

a. SeO_2 b. MgO c. P_4O_{10}

Balancing Equations for Reactions in Aqueous Solution: Net Ionic Equations

Let us again consider the reaction of magnesium with aqueous HCl (Figure 5.5).

$$Mg(s) + 2\,HCl(aq) \longrightarrow MgCl_2(aq) + H_2(g)$$

Hydrochloric acid and magnesium chloride both dissolve in water, and both are strong electrolytes. This means the solution really contains the ionization products of HCl

$$HCl(aq) \xrightarrow{\text{consists of}} H_3O^+(aq) + Cl^-(aq)$$

and the ions of $MgCl_2(aq)$.

$$MgCl_2(aq) \xrightarrow{\text{consists of}} Mg^{2+}(aq) + 2\,Cl^-(aq)$$

Therefore, to be more informative we should rewrite the balanced equation as

$$Mg(s) + \underbrace{2\,H_3O^+(aq) + 2\,Cl^-(aq)}_{\text{from 2 HCl(aq)}} \longrightarrow \underbrace{Mg^{2+}(aq) + 2\,Cl^-(aq)}_{\text{from }MgCl_2(aq)} + H_2(g) + 2\,H_2O(\ell)$$

Two Cl^- ions appear on both the reactant and product sides of the equation. Such ions are often called **spectator ions** because they are not involved in the net reaction process; they only look on from the sidelines. Therefore, little chemical or stoichiometric information is lost if the equation is written without them, and we can simplify the equation to

$$Mg(s) + 2\,H_3O^+(aq) \longrightarrow Mg^{2+}(aq) + H_2(g) + 2\,H_2O(\ell)$$

The balanced equation that results from leaving out the spectator ions is the **net ionic equation** *for the reaction.* Here we include only the solid element (Mg), the molecule (H_2), and the ions [$H_3O^+(aq)$, $Mg^{2+}(aq)$] that are involved in the *changes* that occur in the course of the reaction. That is, *the net ionic equation emphasizes the chemical changes* that occur.

Leaving out the spectator ions does not imply that Cl^- is totally unimportant in the Mg + HCl reaction. Indeed, H_3O^+ or Mg^{2+} ions cannot exist alone in solution; a negative ion of some kind *must* be present to balance the positive-ion charge. However, *any* anion will do, as long as it forms water-soluble compounds with H_3O^+ and Mg^{2+}. Thus, we could have used $HClO_4$ or HBr as the source of H_3O^+, and then ClO_4^- or Br^- would be the spectator ion.

As a final point concerning net ionic equations, there must always be a *conservation of charge* as well as mass in a balanced chemical equation. Thus, in the Mg + HCl net ionic equation there are two positive electrical charges on each side of the equation. On the left side there are two H_3O^+ ions, each with a charge of 1+, for a total charge of 2+. This is balanced by a Mg^{2+} ion on the right side.

EXAMPLE 5.3

Writing and Balancing Net Ionic Equations

Write a balanced net ionic equation for the reaction of $AgNO_3$ and $CaCl_2$ to give AgCl and $Ca(NO_3)_2$.

SOLUTION

Step 1. Write the complete, balanced equation using the correct formulas for reactants and products.

$$2\,AgNO_3 + CaCl_2 \longrightarrow 2\,AgCl + Ca(NO_3)_2$$

Step 2. Decide on the solubility of each compound from Figure 5.7. One general guideline is that nitrates are almost always soluble, so $AgNO_3$ and $Ca(NO_3)_2$ are water-soluble. Further, with a few exceptions (one of which is AgCl), chlorides are water-soluble. Therefore, we can write

$$2\,AgNO_3(aq) + CaCl_2(aq) \longrightarrow 2\,AgCl(s) + Ca(NO_3)_2(aq)$$

Step 3. Recognizing that all soluble ionic compounds dissociate to form ions in aqueous solutions, we have

$$AgNO_3(aq) \longrightarrow Ag^+(aq) + NO_3^-(aq)$$

$$CaCl_2(aq) \longrightarrow Ca^{2+}(aq) + 2\,Cl^-(aq)$$

$$Ca(NO_3)_2(aq) \longrightarrow Ca^{2+}(aq) + 2\,NO_3^-(aq)$$

This results in the following complete ionic equation:

$$2\,Ag^+(aq) + 2\,NO_3^-(aq) + Ca^{2+}(aq) + 2\,Cl^-(aq) \longrightarrow 2\,AgCl(s) + Ca^{2+}(aq) + 2\,NO_3^-(aq)$$

Step 4. There are two spectator ions in the complete ionic equation (Ca^{2+} and NO_3^-), so these are eliminated to give the net ionic equation.

$$2\,Ag^+(aq) + 2\,Cl^-(aq) \longrightarrow 2\,AgCl(s)$$

To complete the equation, realize that each species in the net equation is preceded by a coefficient of 2. Therefore, the equation can be simplified by dividing through by 2.

$$Ag^+(aq) + Cl^-(aq) \longrightarrow AgCl(s)$$

Step 5. Finally, notice that the sum of the ion charges is the same on both sides of the equation. On the left, 1+ and 1− give zero; on the right the electrical charge on AgCl is also zero.

EXERCISE 5.7 • *Net Ionic Equations*

Balance each of the following equations and write net ionic equations.

a. $BaCl_2(aq) + Na_2SO_4(aq) \longrightarrow BaSO_4(s) + NaCl(aq)$

b. Lead(II) nitrate reacts with potassium chloride to give lead(II) chloride and potassium nitrate.

5.4 COMBINATION REACTIONS

Combination reactions, in which two reactants combine to give a single product, form a large class of chemical reactions. For example, oxygen forms binary oxides with the general formula M_xO_y with almost all the elements (with the exception of He, Ne, Ar, and possibly Kr), although not always by *direct* combination. Halogens combine with all metals and most nonmetals to form halides. Thus, if one of the reactants is oxygen or a halogen it is reasonable to expect that a combination reaction will occur.

Recall that the halogens are F_2, Cl_2, Br_2, and I_2.

Combination of Metals and Oxygen

The reaction of a metal and oxygen produces an ionic compound, a metal oxide. As for any ionic compound, the formula of the metal oxide can be predicted from the fact that it must be electrically neutral. Since you can predict a reasonable positive charge for the metal ion by using the guidelines in Section 4.3, you can determine the formula of the metal oxide. For example, when aluminum (Al, Group 3A, 3+ charge) reacts with O_2 the product must be Al_2O_3 (2 Al^{3+} and 3 O^{2-} ions), a compound known as *alumina* or *corundum* (Figure 4.9).

$$4\,Al(s) + 3\,O_2(g) \longrightarrow \underset{\text{aluminum oxide}}{2\,Al_2O_3(s)}$$

Because transition metals form cations having different charges, their combinations with oxygen can produce more than one compound. Iron, for example, forms both Fe^{2+} and Fe^{3+} ions, so it can react with oxygen to form either FeO or Fe_2O_3 or both (Figure 5.2). The latter, in the hydrated form ($Fe_2O_3 \cdot H_2O$), is well known as rust.

$$2\,Fe(s) + O_2(g) \longrightarrow \underset{\text{iron(II) oxide}}{2\,FeO(s)}$$

$$4\,Fe(s) + 3\,O_2(g) \longrightarrow \underset{\text{iron(III) oxide}}{2\,Fe_2O_3(s)}$$

Combination of Nonmetals and Oxygen

When nonmetals or metalloids combine with O_2, the compounds formed are not ionic, but are composed of molecules. Oxides of carbon are good examples. Carbon dioxide forms from the combustion of hydrocarbons (Section 5.2) and whenever carbon burns in an excess of air.

$$C(s) + O_2(g) \longrightarrow \underset{\text{carbon dioxide}}{CO_2(g)}$$

The other oxide of carbon, carbon monoxide, is the product when carbon is burned in a limited supply of oxygen.

$$2\,C(s) + O_2(g) \longrightarrow \underset{\text{carbon monoxide}}{2\,CO(g)}$$

However, CO can then burn in more oxygen to form the dioxide.

$$2\,CO(g) + O_2(g) \longrightarrow 2\,CO_2(g)$$

There is concern that the concentration of CO_2 in the atmosphere is climbing beyond acceptable limits and that this will lead to a long-term warming of the earth. This so-called "greenhouse effect" is discussed in Section 20.7.

Carbon dioxide is the nontoxic gas dissolved in soft drinks, beer, and champagne. In contrast, carbon monoxide, CO, is toxic because it can combine with hemoglobin, the molecule in blood that picks up oxygen in the lungs and carries it to the cells throughout the body. Because CO interacts even more strongly with hemoglobin than does O_2, the carbon monoxide prevents hemoglobin from absorbing O_2, and the oxygen-carrying capacity of that hemoglobin is lost.

There are at least seven known nitrogen oxides: N_2O, NO, NO_2, N_2O_3, N_2O_4, N_2O_5, and NO_3. The first of these, N_2O, has the common name "nitrous oxide." It is often used as an anesthetic and is commonly called "laughing gas." It is not, however, produced by the direct combination of the elements.

The formation of two compounds of carbon on reaction with O_2 is typical of nonmetals. The common nonmetals C, N, P, and S all form several binary oxides. Elemental nitrogen, N_2, and O_2 are the major components of the atmosphere. Small amounts of these gases combine at high temperatures (in a hot automobile engine or when lightning passes through the air during a thunderstorm) to give NO, nitrogen monoxide.

$$N_2(g) + O_2(g) \longrightarrow 2\,NO(g)$$

nitrogen monoxide (colorless gas)

However, just as CO can be converted to CO_2, NO can react with excess O_2 to give NO_2 (Figure 5.10).

$$2\,NO(g) + O_2(g) \longrightarrow 2\,NO_2(g)$$

nitrogen dioxide (red-brown gas)

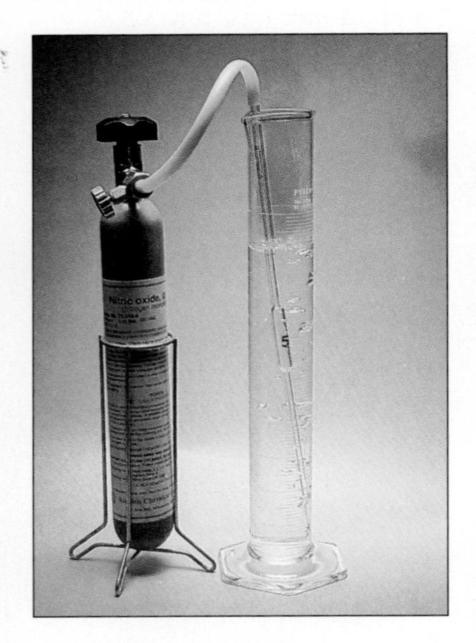

Figure 5.10
The gas NO, nitrogen monoxide, is stored in a tank. It is bubbled through water, where it is evident that the gas is colorless. However, as soon as the bubbles of NO enter the atmosphere, the NO is oxidized to brown NO_2, nitrogen dioxide. *(C.D. Winters)*

Both NO and NO_2 are important in a series of chemical reactions that produces smog from air, sunlight, and automobile exhaust.

Phosphorus in its most common form is a yellowish-white solid, P_4 (Figure 4.2). The slight amount of P_4 in the vapor above the solid at room temperature reacts with O_2 and produces a phosphorescent glow, a fact that led to the name of the element. At least six different products are possible when P_4 is burned in air (Figure 5.4), but the most common one is the white solid P_4O_{10}.

$$P_4(s) + 5\,O_2(g) \longrightarrow P_4O_{10}(s)$$

tetraphosphorus decaoxide (white solid)

Sulfur, the Group 6A neighbor of oxygen, combines with oxygen to form two oxides, SO_2 (Figure 5.11) and SO_3, in reactions of great environmental and industrial significance.

$$S_8(s) + 8\,O_2(g) \longrightarrow 8\,SO_2(g)$$

sulfur dioxide (colorless; choking odor)

$$2\,SO_2(g) + O_2(g) \longrightarrow 2\,SO_3(g)$$

sulfur trioxide (colorless)

Like NO and NO_2, most of the sulfur oxides in the atmosphere come from human activities such as burning coal. All coal contains sulfur, some in the form of the mineral pyrite, FeS_2 (Figure 2.6), and some combined with carbon. Sulfur is present as 1 to 4% by weight in many coal samples. The equation for the combustion of pyrite is

$$4\ FeS_2(s) + 11\ O_2(g) \longrightarrow 2\ Fe_2O_3(s) + 8\ SO_2(g)$$

Sulfur dioxide enters the atmosphere from natural sources as well as from human activities. For example, the eruption of Mt. St. Helens in Washington in May 1980 injected millions of tons of SO_2 into the stratosphere. This stratospheric SO_2 may facilitate the destruction of the earth's protective layer of ozone, O_3. Meanwhile, SO_2 in the lower atmosphere contributes to the problem of acid rain, and it can damage vegetation and produce respiratory problems in humans.

Although the sulfur oxides can cause environmental damage, they are also the starting material for the manufacture of the most important chemical made in industrialized countries, sulfuric acid.

$$H_2O(\ell) + SO_3(g) \longrightarrow H_2SO_4(aq)$$

Almost 86 billion pounds of sulfuric acid were made in the United States in 1991, placing it first in the list of chemicals produced in this country. About 60% of the production is used to manufacture the fertilizers that help to increase worldwide food crops. The remainder is used in the manufacture of inorganic chemicals (such as alum, $Al_2(SO_4)_3$, for the paper industry) and organic chemicals (such as dyes, explosives, soaps, and detergents).

"H_2SO_4 production is now taken as a reliable measure of a nation's industrial strength because it enters into so many industrial and manufacturing processes." N. N. Greenwood and A. Earnshaw, Chemistry of the Elements, *Pergamon Press, New York, 1984, page 842.*

Figure 5.11
Sulfur burns in pure oxygen with a bright blue flame to give sulfur dioxide, SO_2. *(C.D. Winters)*

The eruption of Mt. St. Helens (Washington) in May 1980 ejected tons of the acidic oxide SO_2 into the atmosphere. *(David Weintraub/Photo Researchers)*

Figure 5.12
The reaction of iron and chlorine. When hot steel wool was plunged into a flask of yellow chlorine gas, the steel wool glowed brightly and a cloud of brown iron(III) chloride formed. *(C.D. Winters)*

Combination of Elements with Halogens

Metals and most nonmetals react with halogens to give halides. In the case of metals, the reaction leads to metal halides, whose formulas you can generally predict from the guidelines of Section 4.4. The reaction of aluminum with bromine that was shown in Figure 5.1 is one example. Others are the reaction of iron metal with chlorine gas to give iron(III) chloride (Figure 5.12),

$$2\ Fe(s) + 3\ Cl_2(g) \longrightarrow 2\ FeCl_3(s)$$

and the reaction of zinc with iodine to give zinc iodide (Figure 5.13).

$$Zn(s) + I_2(s) \longrightarrow ZnI_2(s)$$

When metalloids or nonmetals react with halogens, as in the case of phosphorus and chlorine (Figure 5.14),

$$P_4(s) + 6\ Cl_2(g) \longrightarrow 4\ PCl_3(\ell)$$

the guidelines for predicting the product formula are more complex, and we shall leave that for a later chapter (Chapter 10).

Reactions of elements with halogens are used in industry for special purposes. For example, $AlCl_3$ for use in the petroleum industry is made by the reaction of scrap aluminum with chlorine, and thousands of tons of the colorless, reactive gas silicon tetrachloride are made each year by the reaction of silicon with chlorine.

$$Si(s) + 2\ Cl_2(g) \longrightarrow SiCl_4(g)$$

Hydrogen and chlorine react directly to give very pure HCl, which is used in the food industry and in making hydrochloric acid,

$$H_2(g) + Cl_2(g) \longrightarrow 2\ HCl(g)$$

and fluorine reacts with sulfur at 100 °C to give SF_6, an inert, nontoxic gas used in high voltage installations.

Figure 5.13
Gray, powdered zinc metal reacts with dark purple iodine in a vigorous reaction. The product is ZnI_2, but the heat of the reaction is great enough that excess iodine sublimes as a purple vapor. *(C.D. Winters)*

EXERCISE 5.8 • *Combination Reactions*

1. Write a balanced equation for the reaction of copper with oxygen in air to form copper(II) oxide.
2. Sulfur reacts with fluorine to give SF_6. Write a balanced equation for this reaction.
3. Write a balanced equation for the reaction of barium with bromine.

Figure 5.14
When a stream of chlorine gas is directed onto a piece of white phosphorus, the phosphorus burns. The product is phosphorus trichloride, PCl_3. *(C.D. Winters)*

5·5 DECOMPOSITION REACTIONS

Another large class of reactions consists of **decompositions,** in which compounds break down or decompose to form simpler compounds or elements. Decomposition can be caused by applying heat or electricity (Figure 5.15). For example, a few metal oxides can be decomposed by heating to give the metal and oxygen, in the reverse of combination reactions. One of the best known decomposition reactions is the reaction by which Joseph Priestley discovered oxygen in 1774.

$$2\,HgO(s) \longrightarrow 2\,Hg(\ell) + O_2(g)$$

Figure 5.15
The "dichromate volcano." Ammonium dichromate decomposes according to the equation $(NH_4)_2Cr_2O_7(s) \rightarrow N_2(g) + Cr_2O_3(s) + 4\,H_2O(\ell)$. In the first photo, the dichromate compound is placed in a beaker with an alcohol-soaked paper wick. After ignition, the heat evolved by the reaction allows it to continue. On completion, the beaker contains a pile of green chromium(III) oxide, and water droplets cling to the side of the beaker. (NOTE: Compounds containing the Cr(VI) ion, such as dichromates, are carcinogenic. These compounds should be handled only under properly supervised conditions.) *(C.D. Winters)*

Caribbean Crater Suggests Meteors Caused Dinosaur Extinction

DURING THE MESOZOIC geological period that stretched roughly from 180 million years ago to 65 million years ago, great mountain ranges were formed and dinosaurs roamed the earth. But then something happened. The dinosaurs disappeared, seemingly over a short period of time. The earth entered a new era, the Cenozoic, that stretches from 65 million years ago to the present. The boundary between these two eras, called the Cretaceous-Tertiary transition (abbreviated K-T), was a time of great change on the earth. According to Stephen Jay Gould, it "stands among the five great episodes of mass extinction that have punctuated the history of life. It removed the dinosaurs and their kin, along with some 50 percent of all marine species." What happened?

In 1980 Dr. Walter Alvarez and his colleagues discovered that there was an unusual accumulation of the element iridium (Ir) in sediments laid down at the end of the Cretaceous period. Since iridium is much more abundant in meteors than on earth, they proposed that the iridium came from the collision of a giant meteor with the earth. The meteor would have created an enormous cloud of dust and debris, rich in iridium, that soon spread around the earth. This cloud would have significantly cooled the atmosphere and would have altered the earth's climate over a period of years. This in turn could have led to a reduction in the food supplies for such large creatures as dinosaurs and to their eventual demise.

Many thought the Alvarez theory was plausible, but solid scientific proof was needed. Finally, in 1992 Alvarez and his colleagues at many universities found hard evidence, an enormous crater in the Caribbean sea floor off the coast of Mexico. Using radiochemical dating techniques (see Section 19.4), the date of this crater was found to be 64.98 ± 0.06 million years. Globules of tiny glassy beads that could have been thrown off by the meteor impact that made this crater, and that have been discovered in K-T sediments elsewhere in the Caribbean, were found to have an age of 65.06 ± 0.18 million years. This agreement in age between the crater and its supposed debris is too great to be ignored. It is clear that a major impact of a meteor occurred in the Caribbean basin 65 million years ago at the K-T boundary and that it spread debris over the planet.

Figure 5.16
A bed of limestone, primarily composed of calcium carbonate ($CaCO_3$), along the Verde River in Arizona. *(Image Enterprises/James Cowlin)*

Metal Carbonates

A very common, and important, type of decomposition reaction is illustrated by the chemistry of *metal carbonates* and calcium carbonate in particular. Calcium is the fifth most abundant element in the earth's crust and is the third most abundant metal (after Al and Fe). Most naturally occurring calcium is in the form of calcium carbonate, $CaCO_3$, from the fossilized remains of early marine life (Figure 5.16). Limestone, a form of calcium carbonate, is one of the basic raw materials of industry and has been compared to one leg of a six-legged stool upon which modern industry rests, the other legs being coal, oil, iron ore, sulfur, and salt. A characteristic reaction of limestone and most other metal carbonates is that, when heated, they decompose to form CO_2 gas and the corresponding solid metal oxide.

$$\underset{\text{limestone}}{CaCO_3(s)} \xrightarrow{800\text{–}1000\,°} \underset{\text{lime}}{CaO(s)} + CO_2(g)$$

In 1991, this reaction was used to obtain more than 33 billion pounds of lime, CaO, in the United States.

The dust and debris from the meteor impact could have altered earth's climate. Just as importantly, though, the meteor slammed into an area rich in limestone. The heat created by the impact not only melted rocks and created the glassy beads discovered elsewhere, but it also would have decomposed the calcium carbonate of the limestone to calcium oxide and carbon dioxide. The vast amount of CO_2 released would have led to a "greenhouse" effect that, after the dust of the impact had settled for several years, could have increased global temperatures by as much as 5 °C over a period of 10,000 to 100,000 years. The climate of the earth would have been altered, and mass extinctions of certain species could have resulted.

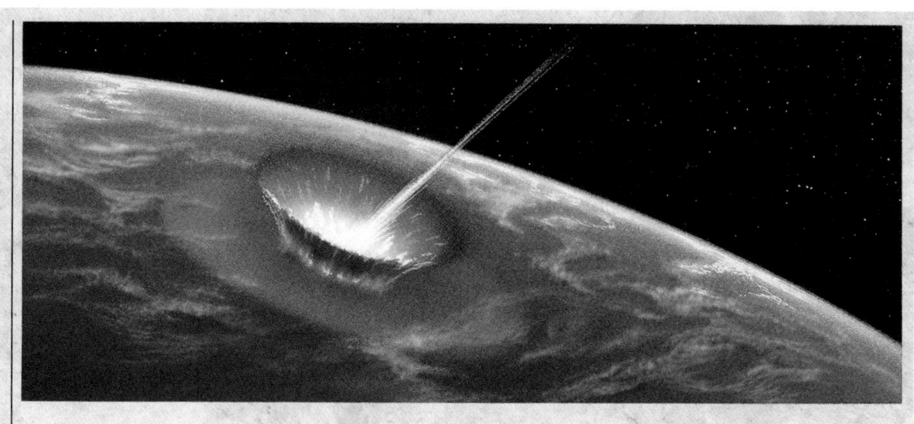

Artist's conception of the impact of a giant meteor with the earth off the coast of Mexico in the Caribbean Sea. (See Science, *Vol. 257, p. 878, 1992.)*

EXERCISE 5.9 • *Decomposition Reactions*

Cerussite, $PbCO_3$ (commonly called white lead ore), is an important lead-containing mineral. Write a balanced equation for the decomposition of the compound on heating.

5.6 EXCHANGE REACTIONS IN AQUEOUS SOLUTION

The general types of reactions described so far—the combination of elements with oxygen and halogens, combustion reactions of hydrocarbons, and decomposition reactions—are not carried out in aqueous solution. One major category of reaction that occurs *only* in aqueous solution is that of exchange reactions involving ions.

Exchange reactions, which are sometimes called *metathesis* or *double-displacement* reactions, result in the interchange of reactant partners as anions and cations in aqueous solution.

$$AB + XY \longrightarrow AY + XB$$

One important type of exchange reaction is **precipitation** (Figure 5.17), in which ions combine in solution to form an insoluble reaction product. For example, in the following reaction the insoluble solid lead(II) iodide forms.

Overall equation: $Pb(NO_3)_2(aq) + 2\,KI(aq) \longrightarrow 2\,KNO_3(aq) + PbI_2(s)$

Net ionic equation: $Pb^{2+}(aq) + 2\,I^-(aq) \longrightarrow PbI_2(s)$

Another important type of exchange reaction is an **acid-base reaction,** in which H^+ ions (or H_3O^+ ions) and OH^- ions combine to form water.

Overall equation: $HNO_3(aq) + KOH(aq) \longrightarrow KNO_3(aq) + H_2O(\ell)$

Net ionic equation: $H^+(aq) + OH^-(aq) \longrightarrow H_2O(\ell)$

or

$H_3O^+(aq) + OH^-(aq) \longrightarrow 2\,H_2O(\ell)$

Gas-forming reactions, chiefly between metal carbonates and acids (Figure 5.18), constitute a third type of exchange reaction.

One product from the reaction of a metal carbonate with an acid is carbonic acid, H_2CO_3, most of which decomposes to H_2O and CO_2. The latter is the gas in the bubbles you see during this reaction.

Overall equation: $NiCO_3(s) + 2\,HNO_3(aq) \longrightarrow Ni(NO_3)_2(aq) + H_2CO_3(aq)$

$H_2CO_3(aq) \longrightarrow CO_2(g) + H_2O(\ell)$

Net ionic equation: $NiCO_3(s) + 2\,H^+(aq) \longrightarrow Ni^{2+}(aq) + CO_2(g) + H_2O(\ell)$

Mixing two compounds together does not ensure that an exchange reaction will occur. For example, mixing aqueous solutions of NaCl and KNO_3 will not produce a net chemical change. The reactant solutions contain the ions $Na^+(aq)$, $Cl^-(aq)$, $K^+(aq)$, and $NO_3^-(aq)$, and these same ions remain long after the reactant solutions have been mixed.

$$NaCl(aq) + KNO_3(aq) \longrightarrow Na^+(aq) + Cl^-(aq) + K^+(aq) + NO_3^-(aq)$$

no net chemical change

Why then do exchange reactions occur and how can you predict when they occur? Two compounds will undergo an exchange reaction if

- a *water-insoluble solid product* is produced from two soluble reactants;
- a *stable molecule* that removes ions from solution, such as water, is formed; or
- a *gas* such as CO_2 is produced.

The first is the basis of precipitation reactions, the second explains why many acid–base reactions occur, and the third is the driving force behind the decomposition of metal carbonates by acids.

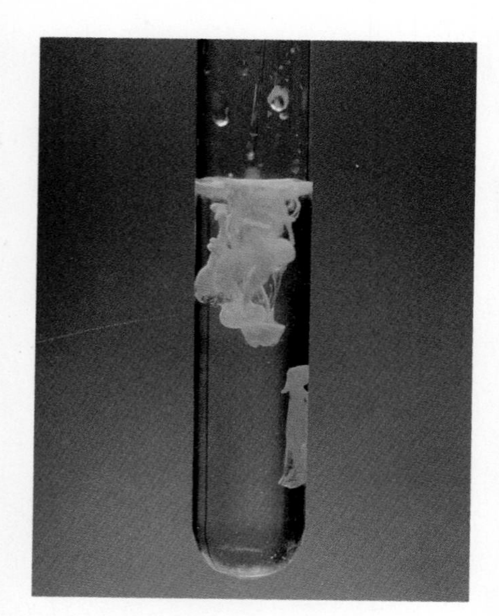

Figure 5.17
Adding a drop of aqueous potassium iodide to a solution of lead(II) nitrate leads to the formation of a yellow precipitate of lead(II) iodide and leaves water-soluble potassium nitrate in solution. *(C.D. Winters)*

Precipitation Reactions

A precipitation reaction produces an insoluble product, a **precipitate,** *from soluble reactants.* There are many positive/negative ion combinations that give insoluble substances (Figure 5.7), and so many precipitation reactions are possi-

Figure 5.18
A piece of coral that is largely calcium carbonate, $CaCO_3$, reacts readily with hydrochloric acid to give CO_2 gas and aqueous calcium chloride. *(C.D. Winters)*

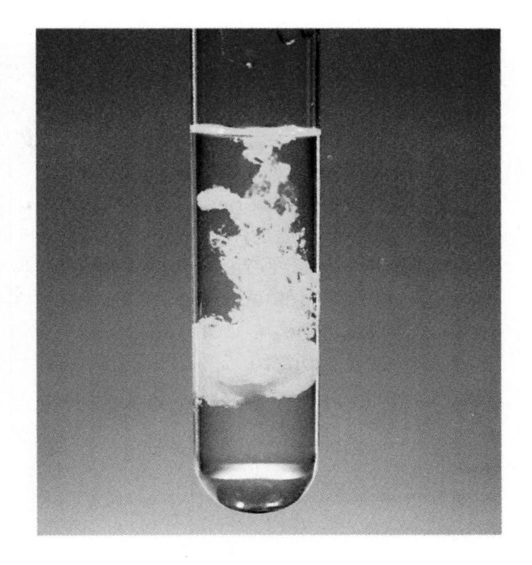

Figure 5.19
Adding a drop of aqueous potassium chromate to a solution of lead(II) nitrate leads to the formation of a yellow precipitate of lead(II) chromate and leaves water-soluble potassium nitrate in solution. *(C.D. Winters)*

ble. For example, since most chromates are insoluble, lead(II) chromate is easily precipitated by the reaction of a water-soluble lead(II) compound with a water-soluble chromate compound (Figure 5.19).

Overall equation: $Pb(NO_3)_2(aq) + K_2CrO_4(aq) \longrightarrow PbCrO_4(s) + 2\,KNO_3(aq)$

Net ionic equation: $Pb^{2+}(aq) + CrO_4^{2-}(aq) \longrightarrow PbCrO_4(s)$

Almost all metal sulfides are insoluble in water. Therefore, if a soluble metal compound in nature comes in contact with a source of sulfide ions (say from a volcano, a natural gas pocket in the earth, or a "black smoker" in the ocean), the metal sulfide precipitates. In fact, this is how many metal sulfide-containing minerals are believed to have been formed.

Many reactions you see in nature and in the laboratory are precipitation reactions. Let us say you pour an aqueous solution of silver nitrate into one containing potassium bromide.

$$AgNO_3(aq) + KBr(aq) \longrightarrow ?$$

What will you see and how do you express the result in a balanced chemical equation? From the information in Figure 5.7 you know that silver nitrate is water-soluble because it contains the NO_3^- ion. Similarly, potassium bromide is water-soluble because it contains K^+ and Br^- ions. This means these ions are released into solution when the compounds are dissolved, and an exchange reaction may occur.

$$AgNO_3(aq) + KBr(aq) \longrightarrow AgBr + KNO_3$$

Black Smokers

THE SUN PROVIDES THE energy to drive some of the chemical reactions that are fundamental to life on our planet, but some living things find other sources of energy. There is abundant heat energy available within the earth, and it fuels some very strange forms of life on the floor of the oceans.

In 1977 scientists were exploring the tectonic plates that form the earth's surface on the ocean floor hundreds of feet down in the equatorial Pacific. Along the ridge at the junction of two plates they found direct evidence for the great energy within the earth: thermal springs gushing a hot, black soup of minerals through chimneys in the ocean's floor. Water seeping into cracks in the thin surface along the ridge is superheated to 300 to 400 °C by the magma from the earth's mantle that comes close to the surface. This superhot water dissolves minerals in the crust and provides conditions for the conversion of sulfates in seawater to hydrogen sulfide, H_2S. The hot water, now laden with dissolved minerals and rich in sulfides, gushes through the surface and cools, causing metal sulfides such as those of copper, manganese, iron, zinc, and nickel to precipitate.

$$H_2S(aq) + Cu^{2+}(aq) + 2\,H_2O(\ell) \longrightarrow CuS(s) + 2\,H_3O^+(aq)$$

Since many metal sulfides are black, smoke appears to be coming from the earth, and the vents have been called "black smokers." The "smoke" settles around the edges of

A black-smoker chimney and plume. These result when superheated, mineral-laden water emerges from a crack in the earth's surface. The black plume, and the chimney beneath it, result from the precipitation of metal sulfides that had been dissolved in the water. (V. Tunnicliffe)

Will precipitation occur? Again referring to Figure 5.7, you see that salts containing the Br^- ion are generally soluble *except* for those involving Ag^+. Therefore, silver bromide, AgBr, is *not* water-soluble. Potassium nitrate, KNO_3, however, is water-soluble because it contains both the K^+ and NO_3^- ions that generally lead to water-soluble compounds. Based on this information, it is possible now to write the overall and net ionic equations.

Overall equation: $AgNO_3(aq) + KBr(aq) \longrightarrow AgBr(s) + KNO_3(aq)$

Net ionic equation: $Ag^+(aq) + Br^-(aq) \longrightarrow AgBr(s)$

This means that, on mixing the two water-soluble reactants, you should observe the immediate formation of a white cloud of silver bromide precipitate (Figure 5.20).

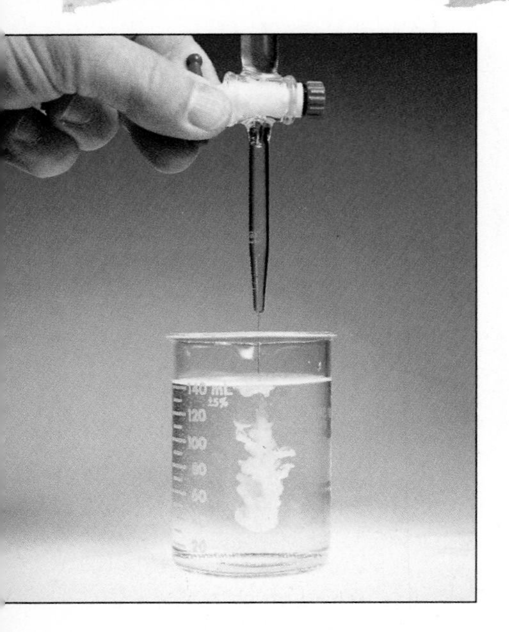

Figure 5.20
Adding aqueous potassium bromide, KBr, to aqueous silver nitrate, $AgNO_3$, leads to a precipitate of white silver bromide, AgBr, and an aqueous solution of potassium nitrate, KNO_3. *(C.D. Winters)*

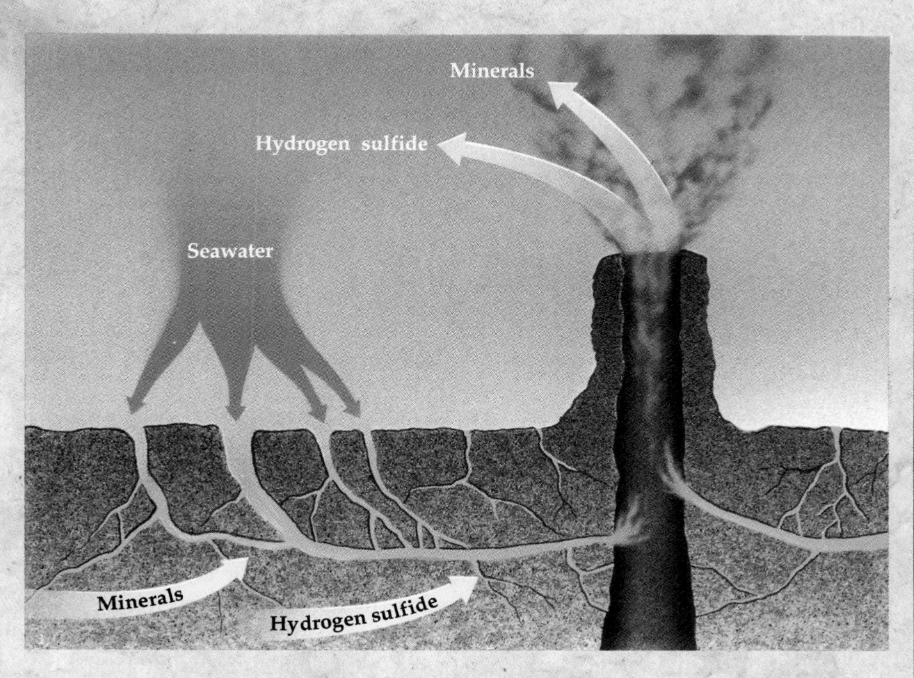

Hydrothermal vents form in the valleys created in the earth's surface where tectonic plates separate. The crust is thin in those regions, and water seeping through the crust becomes heated as hot as 400 °C. The hot water leaches minerals from the rocks. When the plume of hot water emerges from the sea floor the ambient water temperature is only about 4 °C, so the minerals in the plume precipitate to form chimneys and "smoke."

the vent and eventually forms a chimney of precipitated minerals.

The people studying the "black smokers" were amazed to discover that the chimneys were surrounded by dense fields of peculiar, primitive animals living in the hot, sulfide-rich environment. The "smokers" are under hundreds of feet of water, and sunlight does not penetrate to these depths, so the animals have developed a way to live without the energy from sunlight. It is currently believed that they derive the energy needed to make organic compounds from the reaction of oxygen with hydrogen sulfide or the hydrogen sulfide ion, HS^-.

$$HS^-(aq) + 2\,O_2(g) \longrightarrow HSO_4^-(aq) + \text{energy}$$

See V. Tunnicliffe, *American Scientist*, Vol. 80, pp. 336–349, 1992, for more information.

Silver halides in general are sensitive to light. If you let the AgBr made in the preceding reaction stand in room light, the product gradually becomes more and more gray. Indeed, this light-driven reaction is the fundamental process used in photographic film, and the AgBr is made in industry by a precipitation reaction such as that between $AgNO_3$ and KBr.

Aqueous silver nitrate reacts with aqueous potassium chromate.

EXERCISE 5.10 • *Precipitation Reactions*

Write balanced overall and net ionic equations for the precipitation reaction that occurs when aqueous silver nitrate is mixed with an aqueous solution of potassium chromate, K_2CrO_4.

Acid–Base Reactions

An exchange reaction between an acid and a base produces a **salt** *and water.* For example,

$$\underset{\text{sodium hydroxide}}{NaOH(aq)} + \underset{\text{hydrochloric acid}}{HCl(aq)} \longrightarrow \underset{\text{sodium chloride}}{NaCl(aq)} + \underset{\text{water}}{H_2O(\ell)}$$

Here sodium hydroxide and hydrochloric acid give common table salt and water. Because of this, the word "salt" has come into our language to describe any *ionic compound whose cation comes from a base* (here Na^+ from NaOH) *and whose anion comes from an acid* (here Cl^- from HCl). The reaction of any of the acids listed in Table 5.1 with any of the bases listed there produces a salt and water.

Because hydrochloric acid is a strong electrolyte in water (Table 5.1), the complete ionic equation for the reaction of HCl(aq) and NaOH(aq) should be written as

$$\underbrace{Na^+(aq) + OH^-(aq)}_{\text{from NaOH(aq)}} + \underbrace{H_3O^+(aq) + Cl^-(aq)}_{\text{from HCl(aq)}} \longrightarrow \underbrace{Na^+(aq) + Cl^-(aq)}_{\text{salt}} + \underbrace{2\,H_2O(\ell)}_{\text{water}}$$

Since Na^+ and Cl^- ions appear on both sides of the equation, the net ionic equation is simply the combination of the ions H_3O^+ and OH^- to give water.

$$H_3O^+(aq) + OH^-(aq) \longrightarrow 2\,H_2O(\ell)$$

Indeed, *this will always be the net ionic equation for the reaction between any* ***strong*** *acid and any* ***strong*** *base.* Reactions between strong acids and strong bases are sometimes called **neutralization reactions** because the result of the complete reaction is a neutral solution, neither acidic nor basic. The other ions (the cation of the base and the anion of the acid) remain unchanged. If we evaporate the water, however, the cation and anion form a solid salt. In the preceding example, NaCl could be obtained; similarly, NaOH and nitric acid, HNO_3, would give the salt sodium nitrate, $NaNO_3$.

$$\text{Overall equation:}\quad NaOH(aq) + HNO_3(aq) \longrightarrow NaNO_3(aq) + H_2O(\ell)$$

It was mentioned earlier that the basic oxide of calcium, lime, is used in waste and pollution control. One of its major uses is in "scrubbing" sulfur oxides from the exhaust gases of power plants fueled by coal and oil. The oxides of sulfur dissolve in water to produce acids, and these acids can react with a base. Lime is inexpensive and produces the base calcium hydroxide when suspended in water. This suspension is sprayed into the exhaust stack

Figure 5.21
The white mineral gypsum is calcium sulfate dihydrate, $CaSO_4 \cdot 2\,H_2O$. *(C.D. Winters)*

Figure 5.22
Coal is removed from the strip mine in the photograph, and burned in a power plant. Sulfur oxides in the waste gas from the plant are converted to white calcium sulfate, $CaSO_4 \cdot 2\,H_2O$, which is then put back into a part of the mine from which the coal has been removed (at the center of the photograph).

of the plant, where it reacts with the sulfur-containing acids. One such reaction is

$$Ca(OH)_2(s) + H_2SO_4(aq) \longrightarrow \underset{\text{gypsum}}{CaSO_4 \cdot 2\,H_2O(s)}$$

$Ca(OH)_2$ is known as hydrated or slaked lime.

The compound $CaSO_4 \cdot 2\,H_2O$, hydrated calcium sulfate, is found in the earth as the mineral gypsum (Table 4.3 and Figure 5.21). Assuming the gypsum from a coal-burning power plant is not contaminated with heavy metals or other pollutants, it is environmentally acceptable to put it into the earth (Figure 5.22).

EXAMPLE 5.4

Acid–Base Reactions

Write the balanced equation for the reaction of aqueous ammonia and nitric acid.

SOLUTION Ammonia is a common and very important chemical, and you should understand something of its chemistry. It is a base because it reacts with water, to a small extent, to produce the hydroxide ion.

$$NH_3(aq) + H_2O(\ell) \rightleftharpoons NH_4^+(aq) + OH^-(aq)$$

Nitric acid produces hydronium and nitrate ions.

$$HNO_3(aq) + H_2O(\ell) \longrightarrow H_3O^+(aq) + NO_3^-(aq)$$

Since the OH^- produced by NH_3 will react with the H_3O^+ produced by HNO_3 to form water, ammonia and nitric acid must react according to the equation.

$$NH_3(aq) + HNO_3(aq) \longrightarrow NH_4^+(aq) + NO_3^-(aq)$$

To see how we got here, simply *add* the two ionization equations together. This gives

Ammonia is an exception to the rule that acid + base → salt + water. When acid reacts with NH_3, the acid transfers H^+ to NH_3 to give NH_4^+ and the anion of the acid.

from the left side of the two equations

$$NH_3(aq) + 2\,H_2O(\ell) + HNO_3(aq) \longrightarrow NH_4^+(aq) + OH^-(aq) + H_3O^+(aq) + NO_3^-(aq)$$

from the right side of the two equations

Finally, the two H_2O molecules on the left side are canceled by the two H_2O molecules [from $OH^-(aq) + H_3O^+(aq)$] on the right. In essence, the acid transfers its proton, H^+, to the base NH_3.

EXERCISE 5.11 • *Acid–Base Reactions*

Write the balanced overall equation and the net ionic equation for the reaction of magnesium hydroxide with hydrochloric acid.

Gas–Forming Reactions

The final type of exchange reaction is one in which a gas is formed; it is loss of this gas that drives the reaction from reactants to products. As mentioned at the beginning of this section, an excellent example is the reaction of metal carbonates with acids (Figure 5.18).

$$CaCO_3(s) + 2\,HCl(aq) \longrightarrow CaCl_2(aq) + H_2CO_3(aq)$$

$$\downarrow$$

$$H_2O(\ell) + CO_2(g)$$

A salt and H_2CO_3, carbonic acid, are *always* the products from an acid and a metal carbonate, and their formation shows the exchange reaction pattern. Carbonic acid is unstable, however, and much of it is rapidly converted to water and CO_2 gas. If the reaction is done in an open beaker, most of the gas bubbles out of the solution.

EXERCISE 5.12 • *A Gas-Forming Reaction*

As mentioned in Exercise 5.9, cerussite, $PbCO_3$, is an important lead-containing mineral. Write a balanced equation that shows what happens when cerussite is treated with nitric acid. Give the name of each of the reaction products.

EXERCISE 5.13 • *Classifying Reactions*

Classify each of the following exchange reactions; that is, tell whether the reaction is an acid–base reaction, a precipitation, or a gas-forming reaction. Predict the products of the reaction, and then balance the completed equation.

a. $CuCO_3(s) + H_2SO_4(aq) \longrightarrow$

b. $Ba(OH)_2(s) + HNO_3(aq) \longrightarrow$

c. $ZnCl_2(aq) + (NH_4)_2S(aq) \longrightarrow$

Preparation of Compounds by Exchange Reactions

The main occupation of many chemists is the preparation of new materials. Exchange reactions are one way to do this. However, there may be several exchange reactions that give a desired product, and you may have to decide which to use. There is also the problem of isolating the product from the solution. For example, in the case of the reaction of NaOH and HCl in aqueous solution, the final solution contains only NaCl in water. To obtain the NaCl as a solid, you could simply evaporate the water from the solution, leaving the salt as a white, crystalline solid.

Alternatively, the desired salt may be obtained by a precipitation reaction. The desired compound could be *either* the insoluble salt *or* the soluble species left in aqueous solution. For example, you could prepare insoluble barium chromate, $BaCrO_4$, by the precipitation reaction illustrated in Figure 5.23. Aqueous barium chloride is added to a solution of water-soluble potassium chromate to give insoluble $BaCrO_4$ and soluble potassium chloride.

$$BaCl_2(aq) + K_2CrO_4(aq) \longrightarrow BaCrO_4(s) + 2\ KCl(aq)$$

Filtration is then used to separate the insoluble and soluble species. Insoluble barium chromate is trapped on the paper filter, and the solution containing potassium chloride passes through.

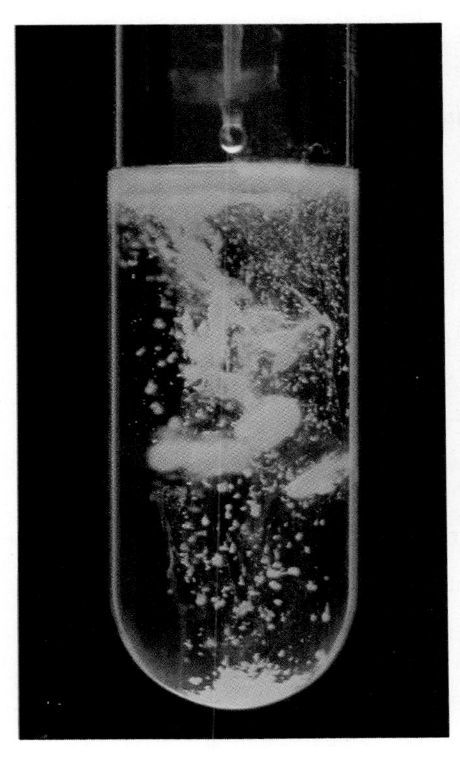

(a)

(b)

Figure 5.23
Preparation of barium chromate. (a) Aqueous barium chloride is added from a dropper to a solution of potassium chromate, K_2CrO_4. Barium chromate, $BaCrO_4$, precipitates and is separated from the solution by collecting it on a filter paper (b).

EXAMPLE 5.5

Preparing a Compound by a Precipitation Reaction

Write a balanced overall equation and a net ionic equation for the preparation of calcium carbonate, $CaCO_3$, by a precipitation reaction.

SOLUTION

Step 1. Is the compound soluble or insoluble in water? According to Figure 5.7, carbonates are generally insoluble, except when the cation is from Group 1A (*e.g.*, Na^+ or K^+). Therefore, a precipitation reaction is appropriate.

Step 2. Choose water-soluble salts as reactants, one containing Ca^{2+} and the other containing CO_3^{2-}. Nitrates are generally water-soluble, so calcium nitrate, $Ca(NO_3)_2$, is a good choice. As a source of carbonate ion, sodium carbonate, Na_2CO_3, is reasonable, since sodium salts are generally soluble.

Step 3. Write the balanced equation.

Complete equation: $Ca(NO_3)_2(aq) + Na_2CO_3(aq) \longrightarrow CaCO_3(s) + 2\ NaNO_3(aq)$

Net ionic equation: $Ca^{2+}(aq) + CO_3^{2-}(aq) \longrightarrow CaCO_3(s)$

The insoluble product could be isolated by filtration, $CaCO_3$ being held by the filter and $NaNO_3$ remaining in the water.

EXAMPLE 5.6

Preparing a Compound by an Acid-Base Reaction

Write a balanced equation for the preparation of potassium acetate, KCH_3CO_2, by an acid-base reaction.

SOLUTION

Step 1. Is the compound soluble or insoluble in water? According to Figure 5.7, acetates are generally soluble. Therefore, it is appropriate to use an acid–base reaction in which one product is the desired salt and the other is water.

Step 2. Acetic acid (Table 5.1) is the source of acetate ion, $CH_3CO_2^-$, and potassium hydroxide can supply potassium ion, K^+.

Step 3. Write the balanced equation

$$CH_3CO_2H(aq) + KOH(aq) \longrightarrow KCH_3CO_2(aq) + H_2O(\ell)$$

A precipitation reaction could be used as an alternative, but the product other than KCH_3CO_2 must then be insoluble. For example, mixing the water-soluble salts barium acetate and potassium sulfate gives the desired soluble salt, potassium acetate, plus insoluble barium sulfate.

$$Ba(CH_3CO_2)_2(aq) + K_2SO_4(aq) \longrightarrow 2\ KCH_3CO_2(aq) + BaSO_4(s)$$

Finally, a gas-forming reaction is a good alternative. Here the K^+ ion must be part of a carbonate, and the acetate ion must come from an acid, acetic acid.

$$2\ CH_3CO_2H(aq) + K_2CO_3(aq) \longrightarrow 2\ KCH_3CO_2(aq) + H_2O(\ell) + CO_2(g)$$

If the water is evaporated, the desired potassium acetate will be left as a white solid.

EXERCISE 5.14 • *Preparation of Compounds*

Using a balanced equation, show how to prepare barium sulfate by (a) an acid-base reaction and (b) a gas-forming reaction.

5.7 OXIDATION-REDUCTION REACTIONS

The terms "oxidation" and "reduction" come from reactions that have been known for centuries. Ancient civilizations learned how to change metal oxides and sulfides to the metal, that is, how to *reduce* ore to the metal. For example, cassiterite or tin(IV) oxide, SnO_2, is a tin ore discovered in Britain centuries ago, and it is very easily reduced to tin by heating with carbon.

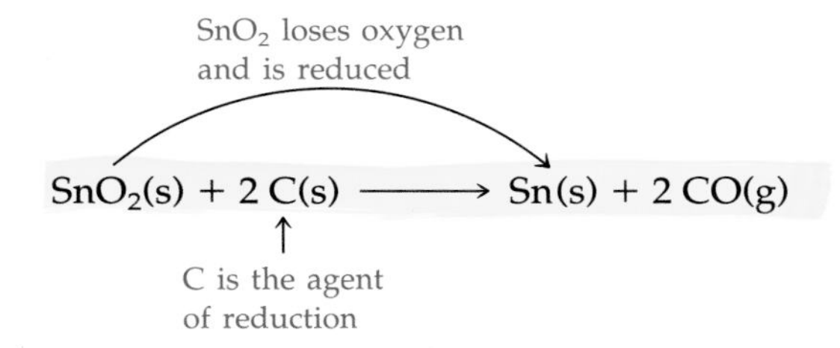

In this reaction carbon is the agent that brings about the reduction of tin ore to tin metal, so carbon is called the **reducing agent.**

When SnO_2 is reduced by carbon, oxygen is removed from the tin and added to the carbon, which is "oxidized" by the addition of oxygen. In fact, *any process in which oxygen is added to another substance is an oxidation.* This too is a process known for centuries, and Figures 5.2 through 5.4 show excellent examples. In the reaction with magnesium, oxygen is the **oxidizing agent** because it is the agent that is responsible for the oxidation.

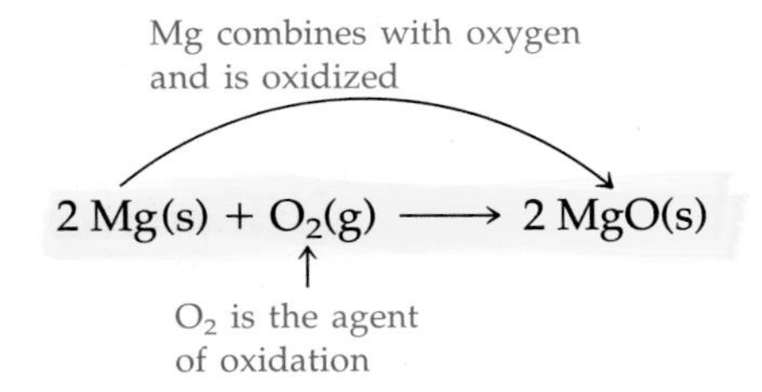

The experimental observations we have just outlined point to several fundamental conclusions: (a) If one substance is oxidized, another substance in the same reaction must be reduced. For this reason, we refer to such reactions as oxidation-reduction reactions, or **redox reactions** for short. (b) The reducing agent is itself oxidized, and the oxidizing agent is reduced. (c) Oxidation is the reverse of reduction. For example, the reactions we have described show that addition of oxygen is oxidation and removal of oxygen is reduction.

Redox Reactions and Electron Transfer

Many oxidation and reduction reactions involve transfer of electrons. When a substance *accepts electrons,* it is said to be **reduced.** The language is descriptive because there is a reduction in the real or apparent electric charge on an atom of the substance when it takes on electrons. In the following net ionic equation, Ag^+ is reduced to uncharged Ag(s) by accepting electrons from copper metal. Since copper metal supplies the electrons and causes the Ag^+ ion to be reduced, Cu is called the **reducing agent** (Figure 5.24).

Ag^+ accepts electrons and is reduced to Ag;
Ag^+ is the oxidizing agent

$+2e^-$ from Cu

$$2\,Ag^+(aq) + Cu(s) \longrightarrow 2\,Ag(s) + Cu^{2+}(aq)$$

$-2e^-$ to Ag

Cu donates electrons and is oxidized to Cu^{2+};
Cu is the reducing agent

When a substance *releases electrons,* it is said to be **oxidized.** In **oxidation,** the real or apparent electric charge on an atom of the substance increases when it gives up electrons. In our example, copper metal releases electrons on going to Cu^{2+}; since its electric charge has increased, it is said to have been oxidized. In order for this to happen, there must be something available to take the electrons offered by the copper. In this case, Ag^+ is the electron acceptor and its charge is reduced (to zero in the element). Therefore, Ag^+ is the "agent" that causes Cu metal to be oxidized, so Ag^+ is the oxidizing agent. In every oxidation-reduction reaction, something is re-

X loses one or more electrons, is oxidized, and is the reducing agent. **Oxidation** *is the loss of electrons.*

$$X \longrightarrow X^+ + e^-$$

Y gains one or more electrons, is reduced, and is the oxidizing agent. **Reduction** *is the gain of electrons.*

$$Y + e^- \longrightarrow Y^-$$

Figure 5.24
The oxidation of copper metal by silver ion. A clean piece of copper screen is placed in a solution of silver nitrate, $AgNO_3$. With time, the copper reduces Ag^+ ions to silver metal crystals, and the copper metal is oxidized to Cu^{2+} ions. The blue color of the solution is due to the presence of aqueous copper(II) ion.

duced (and therefore is the oxidizing agent) and something is oxidized (and therefore is the reducing agent).

In the reaction of magnesium and oxygen (Figure 5.3) oxygen is the oxidizing agent because it gains electrons on going to the oxide ion.

Mg releases $2e^-$ per atom;
Mg is oxidized and is the reducing agent

$$2\,Mg(s) + O_2(g) \longrightarrow 2\,[Mg^{2+}, O^{2-}]$$

O_2 gains $4e^-$ per molecule;
O_2 is reduced and is the oxidizing agent

In the same reaction, magnesium is the reducing agent because it loses two electrons per atom on forming the Mg^{2+} ion. All redox reactions can be analyzed in a similar manner.

Common Oxidizing and Reducing Agents

Like oxygen, the halogens (F_2, Cl_2, Br_2, and I_2) are always oxidizing agents in their reactions with metals and most nonmetals. For example, Figure 3.20 illustrates the reaction of sodium metal with chlorine.

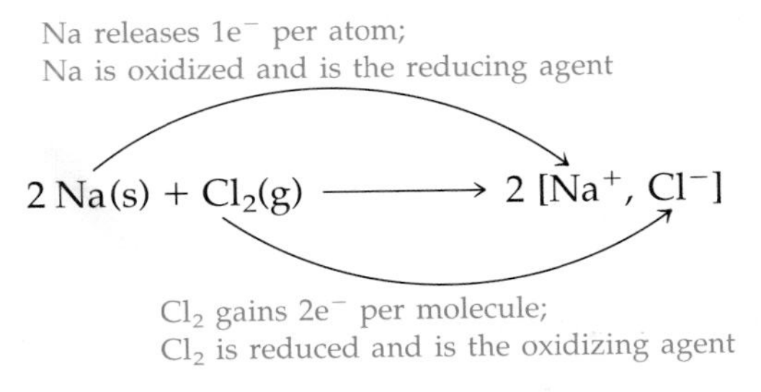

Here sodium begins as the element, but it ends up as the Na^+ ion after combining with chlorine. Thus, sodium is oxidized and is the reducing agent. Chlorine ends up as Cl^-; having acquired two electrons per Cl_2 molecule, it has been reduced and therefore is the oxidizing agent.

When the halogens are combined with metals they always end up as halide ions, X^- (that is, as F^-, Cl^-, Br^-, or I^-) in the product (page 162). Therefore, like oxygen, the halogens are oxidizing agents in their reactions with metals.

$$\underset{\text{oxidizing agent}}{X_2} + 2e^- \xrightarrow{\text{reduction reaction}} 2\,X^-$$

That is, a halogen will always oxidize a metal to give a metal halide (or, conversely, a metal will always reduce a halogen to a halide), and the formula of the product can be predicted from the charge on the metal ion and the charge of the halide.

Chlorine is widely used as an oxidizing agent in water and sewage treatment. A common contaminant of water is hydrogen sulfide, H_2S, which

Table 5.2

Recognizing Oxidation-Reduction Reactions

	Oxidation	Reduction
In terms of oxygen:	gain of oxygen	loss of oxygen
In terms of halogen:	gain of halogen	loss of halogen
In terms of electrons:	loss of electrons	gain of electrons

may come from the decay of organic matter or from underground mineral deposits. Hydrogen sulfide gives a thoroughly unpleasant rotten-egg odor to the water, but chlorine oxidizes H_2S to insoluble, elemental sulfur, which is easily removed.

$$8\ Cl_2(g) + 8\ H_2S(aq) \longrightarrow S_8(s) + 16\ HCl(aq)$$

Oxidation and reduction occur readily when a strong oxidizing agent comes into contact with a strong reducing agent. Knowing that there are easily recognized oxidizing and reducing agents enables you to predict whether a reaction should be classified as an oxidation-reduction, and in some cases to predict what the products might be. Tables 5.2 and 5.3, and the following text, contain some guidelines.

- If something has combined with oxygen, it has been oxidized. In the process the oxygen, O_2, is changed to the oxide ion, O^{2-} (as in a metal oxide) by adding electrons or is combined in a molecule such CO_2 or

Table 5.3

Common Oxidizing and Reducing Agents

Oxidizing Agent	Reaction Product	Reducing Agent	Reaction Product
O_2 (oxygen)	O^{2-}(oxide ion)	H_2 (hydrogen)	H^+ (hydrogen ion) or H combined in H_2O
F_2, Cl_2, Br_2, or I_2 (halogens)	F^-, Cl^-, Br^-, or I^- (halide ion)	M, metals such as Na, K, Fe, or Al	M^{n+}, metal ions such as Na^+, K^+, Fe^{2+} or Fe^{3+}, or Al^{3+}
HNO_3 (nitric acid)	nitrogen oxides such as NO and NO_2	C (carbon), used to reduce metal oxides	CO and CO_2
$Cr_2O_7^{2-}$ (dichromate ion)	Cr^{3+} (chromium(III) ion), in acid solution		
MnO_4^- (permanganate ion)	Mn^{2+} (manganese(II) ion), in acid solution		

H_2O (as occurs in the combustion reaction of a hydrocarbon). Therefore, the oxygen has been reduced. Since it must have taken on electrons, oxygen is the oxidizing agent, and it is a fairly strong one.

- If something has combined with a halogen, it has been oxidized. In the process the halogen, X_2, is changed to halide ions, X^-, by adding electrons, or it is combined in a molecule such as HCl. Therefore, the halogen has been reduced, and it is the oxidizing agent. Among the halogens, fluorine and chlorine are particularly strong oxidizing agents.
- If a metal combines with something, it has been oxidized. In the process, the metal has lost electrons, usually to form a positive ion (as in metal oxides or halides, for example).

$$\underset{\text{reducing agent}}{M} \xrightarrow{\text{oxidation reaction}} M^{n+} + ne^-$$

Therefore, the metal (an electron donor) has been oxidized and has functioned as a reducing agent. Most metals are reasonably good reducing agents, and metals such as sodium, magnesium, and aluminum from Groups 1A, 2A, and 3A are particularly good ones.

There are exceptions to the guideline that metals are always positively charged in compounds. However, you will not encounter them in introductory chemistry.

- Other common oxidizing and reducing agents are listed in Table 5.3, and some are described below. When one of these agents takes part in a reaction, it is reasonably certain that it is a redox reaction. (Nitric acid can be an exception. In addition to being a good oxidizing agent, it is also an acid and functions only in this role in reactions such as the decomposition of a metal carbonate.)

Figure 5.25 illustrates one of the best oxidizing agents, nitric acid, HNO_3. Here the acid oxidizes copper metal to give copper(II) nitrate, and

(a)

(b)

Figure 5.25
Copper reacts vigorously with nitric acid to give brown NO_2 gas (a). When the solution is evaporated, blue crystals of copper(II) nitrate, $Cu(NO_3)_2$, remain (b).

Figure 5.26
Potassium metal reacts very vigorously with water to give hydrogen gas and to leave potassium hydroxide in solution. *(C.D. Winters)*

the acid is reduced to the brown gas NO_2. The net ionic equation for the reaction is

$$\underset{\text{reducing agent}}{Cu(s)} + \underset{\text{oxidizing agent}}{2\,NO_3^-(aq)} + 4\,H_3O^+(aq) \longrightarrow Cu^{2+}(aq) + 2\,NO_2(g) + 6\,H_2O(\ell)$$

The metal is clearly the reducing agent, since it is the substance oxidized. The most common reducing agents in the laboratory, in fact, are metals. The alkali and alkaline earth metals are quite strong reducing agents in most of their reactions. For example, potassium is capable of reducing water to H_2 gas (Figure 5.26).

$$\underset{\text{reducing agent}}{2\,K(s)} + \underset{\text{oxidizing agent}}{2\,H_2O(\ell)} \longrightarrow 2\,KOH(aq) + H_2(g)$$

Some metal ions such as Fe^{2+} can also be reducing agents because they can be oxidized to ions of higher charge. Aqueous Fe^{2+} ion reacts readily with the strong oxidizing agent MnO_4^-, the permanganate ion. The Fe^{2+} ion is oxidized to Fe^{3+} and the MnO_4^- ion is reduced to the Mn^{2+} ion.

$$5\,Fe^{2+}(aq) + MnO_4^-(aq) + 8\,H_3O^+(aq) \longrightarrow 5\,Fe^{3+}(aq) + Mn^{2+}(aq) + 12\,H_2O(\ell)$$

Just as carbon was found centuries ago to reduce tin(IV) oxide to tin metal, it can reduce many other metal oxides to the metal. Thus, it is widely used in the metals industry as a reducing agent. For example, titanium is produced by treating a mineral containing titanium(IV) oxide with carbon and chlorine.

$$TiO_2(s) + C(s) + 2\,Cl_2(g) \longrightarrow TiCl_4(\ell) + CO_2(g)$$

In effect, the carbon reduces the oxide to titanium metal, and the chlorine then oxidizes it to titanium(IV) chloride. Since $TiCl_4$ is easily converted to a gas, it can be removed from the reaction mixture and captured. The $TiCl_4$ is then reduced with another metal, such as magnesium, to give titanium metal.

$$TiCl_4(\ell) + 2\,Mg(s) \longrightarrow Ti(s) + 2\,MgCl_2(s)$$

Finally, H_2 gas is a common reducing agent, widely used in the laboratory and in industry. For example, it readily reduces copper(II) oxide to copper metal (Figure 5.27).

$$\underset{\text{reducing agent}}{H_2(g)} + \underset{\text{oxidizing agent}}{CuO(s)} \longrightarrow Cu(s) + H_2O(g)$$

Balancing equations for the reactions of metals with oxygen and halogens, or those between metal oxides and hydrogen, is usually straightforward, and you should be able to predict the outcome of such reactions and balance the equations. However, it is somewhat more difficult to balance reactions involving species such as HNO_3, $Cr_2O_7^{2-}$, and MnO_4^-, and you will not be asked to do so at this time.

There are hundreds of compounds that are good oxidizing and reducing agents and that, when mixed, will lead to a reaction. However, you should also be aware that it is not a good idea to mix a strong oxidizing agent with a strong reducing agent; a violent reaction, even an explosion, may take place. Chemicals are no longer stored on laboratory shelves in alphabetical order, because such an ordering may place a strong oxidizing agent next to a strong reducing agent.

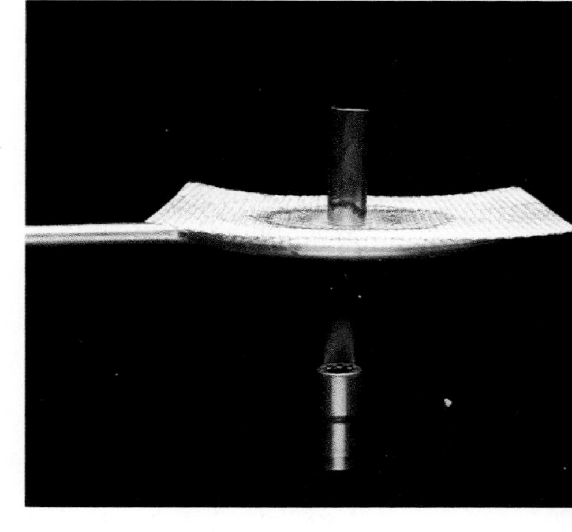

(a)

EXERCISE 5.15 • *Oxidation-Reduction Reactions*

Decide which of the following reactions are oxidation-reduction reactions. In each case explain your choice and identify the oxidizing and reducing agents.

a. $NaOH(aq) + HNO_3(aq) \longrightarrow NaNO_3(aq) + H_2O(\ell)$

b. $4\,Cr(s) + 3\,O_2(g) \longrightarrow 2\,Cr_2O_3(s)$

c. $NiCO_3(s) + 2HCl(aq) \longrightarrow NiCl_2(aq) + H_2O(\ell) + CO_2(g)$

d. $Cu(s) + Cl_2(g) \longrightarrow CuCl_2(s)$

Oxidation-Reduction Reactions in the Environment

Now that you know something about oxidation-reduction reactions, let us apply these ideas to a few of the reactions that take place in our environment. Many of them occur because the earth's atmosphere is a reasonably strong oxidizing environment, containing about 20% oxygen. As you saw in Figure 5.11, sulfur burns in air to give sulfur dioxide, SO_2.

$$S_8(s) + 8\,O_2(g) \longrightarrow 8\,SO_2(g)$$

(b)

Figure 5.27
Reduction of copper oxide with hydrogen. (a) A piece of copper has been heated in air to form a film of black copper(II) oxide on the surface. (b) When the hot copper metal, with its film of CuO, is placed in a stream of hydrogen gas (from the yellow tank at the rear), the oxide is reduced to copper metal, and water forms as the by-product.

Since the reaction involves the combination of an element with oxygen, oxidation-reduction has occurred. Furthermore, you can be certain that O_2 is the oxidizing agent and that it is sulfur that has been oxidized. And finally, to complete the analysis, sulfur must be the reducing agent, the substance that has caused the O_2 to be reduced.

As described earlier, sulfur dioxide is a notorious air pollutant. It is 1300 times more soluble in water than is O_2, and after further oxidation it can form sulfuric acid solutions, which contribute to acid rain. These solutions greatly accelerate the deterioration of buildings and art objects in a polluted environment. Furthermore, oxidation-reduction reactions such as

$$Fe(s) + SO_2(g) + O_2(g) \longrightarrow FeSO_4(s)$$

and

$$4\,FeSO_4(s) + O_2(g) + 6\,H_2O(\ell) \longrightarrow 2\,Fe_2O_3 \cdot H_2O(s) + 4\,H_2SO_4(aq)$$

are believed to be responsible for the corrosion of iron objects when SO_2 is present. Once these reactions have begun, the sulfuric acid that forms is difficult to remove. In fact, even if an iron object is carefully cleaned, corrosion will continue as long as sulfates are present.

Corrosion, which is described in more detail in Section 17.8, is a major problem in our economy and occurs by oxidation-reduction reactions. In some areas salt is spread on roads to melt ice and snow. Although the salt helps to prevent accidents, it also accelerates the formation of rust on car bodies, bridges, and other steel structures. Near the ocean, the salt suspended in the air assists in corroding metal objects. In each of these cases the oxidation of the metal (usually iron or aluminum) to its oxide is central to the corrosion process. For example, in the presence of water, iron can be oxidized readily to iron(II) hydroxide.

$$2\,Fe(s) + O_2(g) + 2\,H_2O(\ell) \longrightarrow 2\,Fe(OH)_2(s)$$

If there is a good supply of air, this hydroxide is further oxidized to red-brown rust.

$$4\,Fe(OH)_2(s) + O_2(g) \longrightarrow 2\,Fe_2O_3 \cdot H_2O(s) + 2\,H_2O(\ell)$$

EXERCISE 5.16 • *Oxidation-Reduction Reactions in the Environment*

Terpinene, $C_{10}H_{16}$, is an organic oil that can be isolated from cardamon seeds. Write a balanced equation to show what happens when the oil is burned in oxygen. What are the oxidizing and reducing agents?

IN CLOSING

Having studied this chapter, you should be able to

- interpret the information conveyed by a balanced chemical equation (Section 5.1).
- balance simple chemical equations (Section 5.2).
- explain how aqueous solutions are formed, and summarize the differences between electrolytes and nonelectrolytes (Section 5.3).
- predict the solubility of ionic compounds in water (Section 5.3).
- recognize common acids and bases and understand their behavior in aqueous solution (Section 5.3).
- write net ionic equations and show how to arrive at such an equation for a given reaction (Section 5.3).
- predict products of common types of chemical reactions: combination, decomposition, exchange, precipitation, acid-base, and gas-forming (Sections 5.4–5.6).
- give examples of the preparation of compounds by common types of reactions, particularly exchange reactions (Section 5.6).
- recognize oxidation-reduction reactions and common oxidizing and reducing agents (Section 5.7).

STUDY QUESTIONS

Review Questions

1. What information is provided by a balanced chemical equation?
2. In the reaction in Figure 5.1, how many molecules of Br_2 would you need if you have 2000 atoms of Al? How many molecules of Al_2Br_6 would you obtain from the reaction?
3. Find in the chapter one example of each of the following reaction types and write the balanced equation for the reaction: (a) combustion; (b) reaction of O_2 with a metal; (c) reaction of O_2 with a nonmetal; (d) a decomposition reaction; and (e) an oxidation-reduction reaction. Name the products of each reaction.

4. Find two examples in the chapter of the reaction of a metal with a halogen, write a balanced equation for each example, and name the product.
5. What is an electrolyte? How can we differentiate between a weak and a strong electrolyte? Give an example of each.
6. Tell how to predict that $Ni(NO_3)_2$ is soluble in water while $NiCO_3$ is not soluble in water.
7. Name two acids that are strong electrolytes and one that is a weak electrolyte. Name two bases that are strong electrolytes and one that is a weak electrolyte.
8. Name the spectator ions in the reaction of nitric acid and magnesium hydroxide, and write the net ionic equation from the following overall equation:

$$2\,H_3O^+(aq) + 2\,NO_3^-(aq) + Mg(OH)_2(s) \longrightarrow 4\,H_2O(\ell) + Mg^{2+}(aq) + 2\,NO_3^-(aq)$$

What type of reaction is this?
9. Name the water-insoluble product in each reaction.
(a) $CuCl_2(aq) + H_2S(aq) \longrightarrow CuS + 2\,HCl$
(b) $CaCl_2(aq) + K_2CO_3(aq) \longrightarrow 2\,KCl + CaCO_3$
(c) $AgNO_3(aq) + NaI(aq) \longrightarrow AgI + NaNO_3$
10. Find two examples of acid-base reactions in the chapter. Write balanced equations for these reactions and name the reactants and products.
11. Find two examples of precipitation reactions in the chapter. Write balanced equations for these reactions and name the reactants and products.
12. Find an example of a gas-forming reaction in this chapter. Write a balanced equation for the reaction and name the reactants and products.
13. Explain how you could prepare barium sulfate by (a) an acid-base reaction, (b) a precipitation reaction, and (c) a gas-forming reaction. The materials you have to start with are $BaCO_3$, $Ba(OH)_2$, Na_2SO_4, and H_2SO_4.
14. Explain the difference between oxidation and reduction. Give an example of each.
15. Explain the difference between an oxidizing agent and a reducing agent. Give an example of each.

Balancing Equations

16. Balance the following equations.
(a) $Al(s) + O_2(g) \longrightarrow Al_2O_3(s)$
(b) $N_2(g) + H_2(g) \longrightarrow NH_3(g)$
(c) $C_6H_6(\ell) + O_2(g) \longrightarrow H_2O(\ell) + CO_2(g)$
17. Balance the following equations.
(a) $Fe(s) + Cl_2(g) \longrightarrow FeCl_3(s)$
(b) $SiO_2(s) + C(s) \longrightarrow Si(s) + CO(g)$
(c) $Fe(s) + H_2O(g) \longrightarrow Fe_3O_4(s) + H_2(g)$
18. Balance the following equations.
(a) $UO_2(s) + HF(\ell) \longrightarrow UF_4(s) + H_2O(\ell)$
(b) $B_2O_3(s) + HF(\ell) \longrightarrow BF_3(g) + H_2O(\ell)$
(c) $BF_3(g) + H_2O(\ell) \longrightarrow HF(\ell) + H_3BO_3(s)$
19. Balance the following equations.
(a) $MgO(s) + Fe(s) \longrightarrow Fe_2O_3(s) + Mg(s)$
(b) $H_3BO_3(s) \longrightarrow B_2O_3(s) + H_2O(\ell)$
(c) $NaNO_3(s) + H_2SO_4(\ell) \longrightarrow Na_2SO_4(s) + HNO_3(g)$
20. Balance the following equations.
(a) Reaction to produce hydrazine, N_2H_4, a good industrial reducing agent.

$$H_2NCl(aq) + NH_3(g) \longrightarrow NH_4Cl(aq) + N_2H_4(aq)$$

(b) Reaction of the fuel (dimethylhydrazine and dinitrogen tetraoxide) used in the moon lander and space shuttle.

$$(CH_3)_2N_2H_2(\ell) + N_2O_4(\ell) \longrightarrow N_2(g) + H_2O(g) + CO_2(g)$$

(c) Reaction of calcium carbide to produce acetylene, C_2H_2.

$$CaC_2(s) + H_2O(\ell) \longrightarrow Ca(OH)_2(s) + C_2H_2(g)$$

21. Balance the following equations.
(a) Reaction of calcium cyanamide to produce ammonia.

$$CaNCN(s) + H_2O(\ell) \longrightarrow CaCO_3(s) + NH_3(g)$$

(b) Reaction to produce diborane, B_2H_6.

$$NaBH_4(s) + H_2SO_4(aq) \longrightarrow B_2H_6(g) + H_2(g) + Na_2SO_4(aq)$$

(c) Reaction to rid water of hydrogen sulfide, H_2S, a foul-smelling compound.

$$H_2S(aq) + Cl_2(aq) \longrightarrow S_8(s) + HCl(aq)$$

Properties of Aqueous Solutions

22. Which compound or compounds in each of the following groups is (are) expected to be soluble in water?
(a) CuO, $CuCl_2$, and $CuCO_3$
(b) $AgCl$, Ag_3PO_4, and $AgNO_3$
(c) KCl, K_2CO_3, and $KMnO_4$
23. Which compound or compounds in each of the following groups is (are) expected to be soluble in water?
(a) $PbSO_4$, $Pb(NO_3)_2$, and $PbCO_3$
(b) Na_2SO_4, $NaClO_4$, $NaCH_3CO_2$
(c) $AgBr$, KBr, $AlBr_3$

24. Give the formula for
(a) a soluble compound containing the acetate ion
(b) an insoluble sulfide
(c) a soluble hydroxide
(d) an insoluble chloride

25. Give the formula for
(a) a soluble compound containing the chloride ion
(b) an insoluble hydroxide
(c) an insoluble carbonate

26. Each of the following compounds is water-soluble. What ions are present in water?
(a) NaI (c) $KHSO_4$
(b) K_2SO_4 (d) NaCN

27. Each of the following compounds is water-soluble. What ions are present in water?
(a) KOH (c) $NaNO_3$
(b) Na_3PO_4 (d) $(NH_4)_2SO_4$

28. Tell whether each of the following is water-soluble or not. If soluble, tell what ions are present.
(a) $BaCl_2$ (c) $Pb(NO_3)_2$
(b) $Cr(NO_3)_2$ (d) $BaSO_4$

29. Tell whether each of the following is water-soluble or not. If soluble, tell what ions are present.
(a) Na_2CO_3 (c) CdS
(b) $CuSO_4$ (d) $CaBr_2$

30. If you wanted to make a water-soluble compound containing Cu^{2+}, name two anions you could choose to combine with the copper ion. Conversely, if you wanted an insoluble Cu^{2+} compound, what two anions could you choose?

31. If you wanted to make a water-soluble compound containing Ca^{2+}, name two anions you could use. Conversely, if you wanted an insoluble Ca^{2+} compound, what two anions could you choose?

Chemical Reactions

32. Write a balanced equation for the ionization of nitric acid in water.

33. Write a balanced equation for the interaction of the basic oxide magnesium oxide with water.

34. Balance each of the following equations, and then write the net ionic equation.
(a) $Zn(s) + HCl(aq) \longrightarrow H_2(g) + ZnCl_2(aq)$
(b) $Mg(OH)_2(s) + HCl(aq) \longrightarrow MgCl_2(aq) + H_2O(\ell)$
(c) $HNO_3(aq) + CaCO_3(s) \longrightarrow Ca(NO_3)_2(aq) + H_2O(\ell) + CO_2(g)$
(d) $HCl(aq) + MnO_2(s) \longrightarrow MnCl_2(aq) + Cl_2(g) + H_2O(\ell)$

35. Balance each of the following equations, and then write the net ionic equation.
(a) $(NH_4)_2CO_3(aq) + Cu(NO_3)_2(aq) \longrightarrow CuCO_3(s) + NH_4NO_3(aq)$
(b) $Pb(NO_3)_2(aq) + HCl(aq) \longrightarrow PbCl_2(s) + HNO_3(aq)$
(c) $BaCO_3(s) + HCl(aq) \longrightarrow BaCl_2(aq) + H_2O(\ell) + CO_2(g)$

36. Balance each of the following equations, and then write the net ionic equation. Refer to Table 5.1 and Figure 5.7 for information on acids and bases and on solubility. Show phases for all reactants and products.
(a) $Ba(OH)_2 + HNO_3 \longrightarrow Ba(NO_3)_2 + H_2O$
(b) $BaCl_2 + Na_2CO_3 \longrightarrow BaCO_3 + NaCl$
(c) $Na_3PO_4 + Ni(NO_3)_2 \longrightarrow Ni_3(PO_4)_2 + NaNO_3$

37. Balance each of the following equations, and then write the net ionic equation. Refer to Table 5.1 and Figure 5.7 for information on acids and bases and on solubility. Show phases for all reactants and products.
(a) $ZnCl_2 + KOH \longrightarrow KCl + Zn(OH)_2$
(b) $AgNO_3 + KI \longrightarrow AgI + KNO_3$
(c) $NaOH + FeCl_2 \longrightarrow Fe(OH)_2 + NaCl$

38. Complete and balance the following equations involving oxygen reacting with an element. Name the product in each case.
(a) $Mg(s) + O_2(g) \longrightarrow$
(b) $Ca(s) + O_2(g) \longrightarrow$
(c) $In(s) + O_2(g) \longrightarrow$

39. Complete and balance the following equations involving oxygen reacting with an element.
(a) $Ti(s) + O_2(g) \longrightarrow$ titanium(IV) oxide
(b) $S_8(s) + O_2(g) \longrightarrow$ sulfur dioxide
(c) $Se(s) + O_2(g) \longrightarrow$ selenium dioxide

40. Complete and balance the following equations involving the reaction of a halogen with a metal. Name the product in each case.
(a) $K(s) + Cl_2(g) \longrightarrow$
(b) $Mg(s) + Br_2(\ell) \longrightarrow$
(c) $Al(s) + F_2(g) \longrightarrow$

41. Complete and balance the following equations involving the reaction of a halogen with a metal.
(a) $Cr(s) + Cl_2(g) \longrightarrow$ chromium(III) chloride
(b) $Cu(s) + Br_2(\ell) \longrightarrow$ copper(II) bromide
(c) $Pt(s) + F_2(g) \longrightarrow$ platinum(IV) fluoride

42. Write a balanced equation for the formation of each of the following compounds from the elements.
(a) carbon monoxide
(b) nickel(II) oxide
(c) chromium(III) oxide

43. Write a balanced equation for the formation of each of the following compounds from the elements.
(a) copper(I) oxide
(b) arsenic(III) oxide
(c) zinc oxide

44. Write a balanced equation for each of the following decomposition reactions. Name each product.

(a) $BeCO_3(s)$ + heat ⟶
(b) $NiCO_3(s)$ + heat ⟶
(c) $Al_2(CO_3)_3(s)$ + heat ⟶

45. Write a balanced equation for each of the following decomposition reactions. Name each product.
(a) $ZnCO_3(s)$ + heat ⟶
(b) $MnCO_3(s)$ + heat ⟶
(c) $PbCO_3(s)$ + heat ⟶

46. Barium hydroxide is used in lubricating oils and greases. Write a balanced equation for the reaction of this hydroxide with nitric acid to give barium nitrate, a compound used in pyrotechnics such as green flares.

47. Aluminum is obtained from bauxite, which is not a specific mineral but a name applied to a mixture of minerals. One of those minerals, which can dissolve in acids, is gibbsite, $Al(OH)_3$. Write a balanced equation for the reaction of gibbsite with sulfuric acid.

48. Balance the equation for the following precipitation reaction and then write the net ionic equation. Indicate the state of each species (s, ℓ, aq, or g).

$$CdCl_2 + NaOH \longrightarrow Cd(OH)_2 + NaCl$$

49. Balance the equation for the following precipitation reaction and then write the net ionic equation. Indicate the state of each species (s, ℓ, aq, or g).

$$Ni(NO_3)_2 + Na_2CO_3(aq) \longrightarrow NiCO_3 + NaNO_3$$

50. Write an overall, balanced equation for the precipitation reaction that occurs when aqueous lead(II) nitrate is mixed with an aqueous solution of potassium chloride. Name each reactant and product.

51. Write an overall, balanced equation for the precipitation reaction that occurs when aqueous copper(II) nitrate is mixed with an aqueous solution of sodium carbonate. Name each reactant and product.

52. The beautiful mineral rhodochrosite (*above, right*) is manganese(II) carbonate. Write an overall, balanced equation for the reaction of the mineral with hydrochloric acid. Name each reactant and product.

53. Many minerals are metal carbonates, and siderite (*right*) is a mineral that consists largely of iron(II) carbonate (see Question 52). Write an overall, balanced equation for the reaction of the mineral with nitric acid and name each reactant and product.

54. Classify each of the following exchange reactions; that is, tell whether the reaction is an acid-base reaction, a precipitation, or a gas-forming reaction. Predict the products of the reaction, and then balance the completed equation.
(a) $MnCl_2(aq) + Na_2S(aq)$ ⟶
(b) $Na_2CO_3(aq) + ZnCl_2(aq)$ ⟶
(c) $K_2CO_3(aq) + HClO_4(aq)$ ⟶

Rhodochrosite. *(Brian Parker/Tom Stack & Associates)*

55. Classify each of the following exchange reactions; that is, tell whether the reaction is an acid-base reaction, a precipitation, or a gas-forming reaction. Predict the products of the reaction, and then balance the completed equation.
(a) $Fe(OH)_3(s) + HNO_3(aq)$ ⟶
(b) $FeCO_3(s) + H_2SO_4(aq)$ ⟶
(c) $FeCl_2(aq) + (NH_4)_2S(aq)$ ⟶
(d) $Fe(NO_3)_2(aq) + Na_2CO_3(aq)$ ⟶

Siderite. *(C.D. Winters)*

Preparing Compounds by Exchange Reactions

56. Write a balanced equation showing how each of the following salts could be prepared by an acid-base reaction.
(a) $NaNO_3$ (c) $Ca_3(PO_4)_2$
(b) KCl (d) Cs_2SO_4

57. Write an overall, balanced equation showing how each of the following salts could be prepared by an acid-base reaction.
(a) $NaCH_3CO_2$ (c) $MgSO_4$
(b) NH_4NO_3 (d) Na_3PO_4

58. Write an overall, balanced equation showing how each of the following insoluble salts could be prepared by a precipitation reaction.
(a) $Ni(OH)_2$ (c) NiS
(b) $SrCO_3$ (d) $BaSO_4$

59. Write an overall, balanced equation showing how each of the following salts could be prepared by a precipitation reaction.
(a) $FeCO_3$ (c) ZnS
(b) Ag_3PO_4 (d) KNO_3

60. Write an overall, balanced equation showing how each of the following salts could be prepared by a gas-forming reaction.
(a) KNO_3 (b) $CaCl_2$ (c) $Fe(NO_3)_2$

61. Write a balanced equation showing how each of the following salts could be prepared by a gas-forming reaction.
(a) $AgNO_3$ (b) Na_3PO_4 (c) $MnSO_4$

62. Write a balanced equation showing how each of the following compounds could be prepared by an exchange reaction.
(a) $NaNO_3$ (b) $SrSO_4$ (c) ZnS

63. Write a balanced equation showing at least one way to prepare each of the following compounds by an exchange reaction.
(a) Na_2SO_4 (b) PbI_2 (c) CdS

Oxidation-Reduction Reactions

64. Tell which of the following reactions are oxidation-reduction reactions. Explain your answer briefly. Classify the remaining reactions.
(a) $CdCl_2(aq) + Na_2S(aq) \longrightarrow CdS(s) + 2\,NaCl(aq)$
(b) $2\,Ca(s) + O_2(g) \longrightarrow 2\,CaO(s)$
(c) $Ca(OH)_2(s) + 2\,HCl(aq) \longrightarrow CaCl_2(aq) + 2\,H_2O(\ell)$

65. Tell which of the following reactions are oxidation-reduction reactions. Explain your answer in each case. Classify the remaining reactions.
(a) $Zn(s) + 2\,NO_3^-(aq) + 4\,H_3O^+(aq) \longrightarrow Zn^{2+}(aq) + 2\,NO_2(g) + 6\,H_2O(\ell)$
(b) $Zn(OH)_2(s) + H_2SO_4(aq) \longrightarrow ZnSO_4(aq) + 2\,H_2O(\ell)$
(c) $Ca(s) + 2\,H_2O(\ell) \longrightarrow Ca(OH)_2(s) + H_2(g)$

66. Tell which of the following substances are common oxidizing agents.
(a) Zn (d) MnO_4^-
(b) O_2 (e) H_2
(c) HNO_3 (f) H^+

67. Tell which of the following substances are reducing agents.
(a) Ca (d) Al
(b) Ca^{2+} (e) Br_2
(c) $Cr_2O_7^{2-}$ (f) H_2

General Questions

68. Name the spectator ions in the reaction of calcium carbonate and hydrochloric acid and write the net ionic equation.

$$CaCO_3(s) + 2\,H_3O^+(aq) + 2\,Cl^-(aq) \longrightarrow CO_2(g) + Ca^{2+}(aq) + 2\,Cl^-(aq) + 3\,H_2O(\ell)$$

What type of reaction does this appear to be?

69. Magnesium metal reacts readily with HNO_3, and the following compounds are all involved.

$$Mg(s) + HNO_3(aq) \longrightarrow Mg(NO_3)_2(aq) + NO_2(g) + H_2O(\ell)$$

(a) Balance the equation for the reaction.
(b) Name each compound.
(c) Write the net ionic equation for the reaction.
(d) What type of reaction does this appear to be?

70. The compound $(NH_4)_2S$ reacts with $Hg(NO_3)_2$ to give HgS and NH_4NO_3.
(a) Write the overall balanced equation for the reaction. Indicate the phase (s or aq) for each compound.
(b) Name each compound.
(c) Write the net ionic equation for the reaction.
(d) What type of reaction does this appear to be?

71. Write an overall, balanced equation showing how you could prepare each of the following compounds by an exchange reaction. In each case, tell what kind of exchange reaction you have used.
(a) $BaCl_2$ (c) $MgCl_2$
(b) $(NH_4)_2SO_4$ (d) $NaClO_4$

72. Azurite is a copper-containing mineral that often forms beautiful crystals (see photo). Its formula is $Cu_3(CO_3)_2(OH)_2$. Write a balanced equation for the reaction of this mineral with hydrochloric acid.

73. Explain how you could prepare zinc chloride by (a) an acid-base reaction, (b) a combination reaction, and (c) a

gas-forming reaction. The materials you have to start with are $ZnCO_3$, HCl, Cl_2, HNO_3, $Zn(OH)_2$, NaCl, $Zn(NO_3)_2$, and Zn.

74. Draw a submicroscopic diagram of each situation below. Indicate atoms or monatomic ions as circles, indicate molecules or polyatomic ions by overlapping circles for the atoms that make up a molecule, and distinguish among different kinds of atoms by labeling or shading the circles. In each case draw at least ten submicroscopic particles other than water molecules, and make certain that the numbers of different kinds of particles are appropriate. Your diagrams can be two-dimensional.
    (a) A concentrated solution of sodium chloride, NaCl, in water.
    (b) A concentrated solution of magnesium nitrate, $Mg(NO_3)_2$, in water.
    (c) A concentrated solution of ammonia, NH_3, in water.
    (d) A concentrated solution of hydrochloric acid, HCl, in water.
75. What species (atoms, molecules, ions) are present in an aqueous solution of each of the following compounds?
    (a) NH_3
    (b) CH_3CO_2H
    (c) NaOH
    (d) HBr

Summary Question

76. Much has been written recently about chlorofluorocarbons and their impact on our environment. Their manufacture begins with the preparation of HF from the mineral fluorspar according to the following *unbalanced* equation.

$$CaF_2(s) + H_2SO_4(aq) \longrightarrow HF(g) + CaSO_4(s)$$

The HF is combined with, for example, CCl_4 in the presence of $SbCl_5$ to make CCl_2F_2, called dichlorodifluoromethane or CFC-12, and other chlorofluorocarbons.

$$2\,HF(g) + CCl_4(\ell) \longrightarrow CCl_2F_2(g) + 2\,HCl(g)$$

(a) Balance the first equation above and name each substance.
(b) Is the first reaction best classified as an acid-base reaction, an oxidation-reduction, or a precipitation reaction?
(c) Give the names of the compounds CCl_4, $SbCl_5$, and HCl.
(d) Another chlorofluorocarbon produced in the reaction is composed of 8.74% C, 77.43% Cl, and 13.83% F. What is the empirical formula of the compound?

Using Computer Programs and Videodiscs

77. Use *The Periodic Table Videodisc* to explore the reactions of magnesium.
    (a) Describe the reaction of magnesium with oxygen. Write a balanced equation for the reaction and name the reaction product.
    (b) Describe the reactions of magnesium with aqueous HCl and HNO_3, noting similarities and differences. Would you describe these reactions as acid-base reactions or oxidation-reduction reactions?
78. Use *The Periodic Table Videodisc* and *KC? Discoverer* to explore the reactions of the Group 2A elements with water.
    (a) Is there an observable trend in the reactivity of the Group 2A elements with water as one goes from Be to Ba? That is, does the ease of reaction of the elements increase or decrease on going down the group?
    (b) Write a balanced equation for the reaction of barium with water, and name the reaction products.
    (c) What is the white solid in the beaker when the reactions of calcium and barium with water are complete?
79. Figure 5.25 illustrates the oxidation-reduction reaction of copper with nitric acid. Such reactions are typical of many, but not all, transition metals. Use *The Periodic Table Videodisc* to examine the oxidation-reduction reactions of the metals from chromium to copper with nitric acid and hydrochloric acid. (The necessary frame numbers on the disc are: chromium, 22055; manganese, 23281; iron, 24981; cobalt, 26677; nickel, 28046; and copper, 29501.)
    (a) Compare the similarities and differences in these reactions; among other things you should observe relative reaction speed and the color of reaction products. What metals react with both acids or only with HCl or only with HNO_3?
    (b) When a metal reacts with nitric acid, is the brown gas NO_2 generally one of the products?
    (c) Where a metal reacts with aqueous HCl, write a balanced equation to describe that reaction. You can use the reaction of zinc with hydrochloric acid (see Example 6.6) as a model.

$$Zn(s) + 2\,HCl(aq) \longrightarrow ZnCl_2(aq) + H_2(g)$$

Sodium bicarbonate and vinegar (acetic acid) react to produce carbon dioxide gas, sodium acetate, and water. *(C.D. Winters)*

Stoichiometry

CHAPTER

6

Doing chemistry requires observing chemical reactions and physical changes. When making peanut brittle, for example, you see the color turn brown and water vapor evolve when the sugar is heated. These are qualitative observations. No measurements and numbers are involved, but a chemical reaction obviously happened.

To further understand the caramelization reaction in making peanut brittle or the mineral-forming reactions in the earth, or any other reaction, chemists first apply fundamental principles. Then, with this understanding, they try to predict what will happen in some related change or process. At the same time, they usually make quantitative measurements. For example, if two compounds react with one another when mixed, are they both consumed completely or will one of them be left over when the reaction has run its course? How much product will form? The objective of this chapter is to describe the tools necessary to answer such questions. That is, we will describe how to deal with chemical reactions in a quantitative way by using the principles of chemical stoichiometry.

Figure 6.1
Phosphorus burns in chlorine gas with a bright flame to produce phosphorus trichloride, PCl_3.
(C.D. Winters)

The study of chemical reactions is the essence of chemistry, and one such reaction is that between an acidic oxide, SO_3, a base, $Ca(OH)_2$, and water.

$SO_3(g)$	$+ Ca(OH)_2(s) + H_2O(\ell)$	$\longrightarrow$	$CaSO_4 \cdot 2\,H_2O(s)$
acidic oxide	base		salt
sulfur trioxide	calcium hydroxide		calcium sulfate dihydrate
	(slaked lime)		(gypsum)

This is a useful reaction because it is one way to remove sulfur trioxide from the flue gas of coal- or oil-fired power generating plants. As discussed in Section 5.6, this prevents the sulfur trioxide from being released into the atmosphere, where it would contribute to acid rain. Consider the quantitative aspects of the reaction. If you know how much sulfur trioxide is in the flue gas, how many tons of slaked lime are needed to remove the sulfur trioxide? How many tons of calcium sulfate are produced? And where will all of the calcium sulfate be put after it has been made? The principles used to answer at least the first two of these questions apply to all chemical reactions and are the subject of this chapter.

6.1 WEIGHT RELATIONS IN CHEMICAL REACTIONS: STOICHIOMETRY

One reason to study chemical reactions is the hope of finding some new material that can be useful to society. If a product turns out to have potential, perhaps for curing a disease, improving a structural plastic, or treating polluted water or air, the efficiency of the reaction is of interest. Can the product be made in sufficient quantity and purity that it can be sold for a reasonable price? Questions such as this mean that *reactions must be studied quantitatively*.

A chemical equation shows the quantitative relationships between reactants and products (Section 5.1). These relationships, which are founded in the law of conservation of matter, are applied in the study of **stoichiometry.** The **stoichiometric coefficients** in a balanced equation give the relative numbers of atoms, molecules, or ions, or the number of moles, and therefore can be used to also calculate the masses of reactants and products.

The reaction of phosphorus with chlorine is a combination reaction involving a nonmetal and a halogen (see Section 5.4).

A piece of phosphorus will burn in chlorine to give phosphorus trichloride, PCl_3 (Figure 6.1). Suppose you use 1.00 mol of phosphorus (P_4, 124 g/mol) in this reaction. The following balanced equation shows that 6.00 mol or 425 g of Cl_2 must be used and that 4.00 mol or 549 g of PCl_3 can be produced.

Reactants			**Product**
$P_4(s)$	$+ 6\,Cl_2(g)$	$\longrightarrow$	$4\,PCl_3(\ell)$
1 molecule	6 molecules	$\longrightarrow$	4 molecules
1.00 mol	6.00 mol	$\longrightarrow$	4.00 mol
124 g	425 g	$\longrightarrow$	549 g (= 124 g + 425 g)

The balanced equation applies no matter how much P_4 is used. If 0.0100 mol of P_4 (1.24 g) is used, then 0.0600 mol of Cl_2 (4.25 g) will be re-

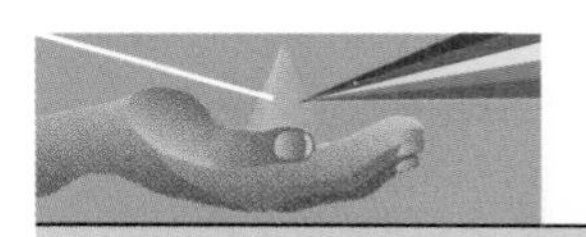

CHEMISTRY YOU CAN DO

Vinegar and Baking Soda: A *Stoichiometry* Experiment

This experiment focuses on the reactions of metal carbonates with acid. For example, limestone reacts with hydrochloric acid to give calcium chloride, carbon dioxide, and water.

$$CaCO_3(s) + 2\,HCl(aq) \longrightarrow CaCl_2(aq) + CO_2(g) + H_2O(\ell)$$

In a similar way, baking soda (sodium bicarbonate) and vinegar (aqueous acetic acid) react to give sodium acetate, carbon dioxide, and water.

$$NaHCO_3(s) + CH_3CO_2H(aq) \longrightarrow NaCH_3CO_2(aq) + CO_2(g) + H_2O(\ell)$$

In this experiment we want to explore the relation between the quantity of acid or carbonate used and the quantity of carbon dioxide evolved. To do the experiment you need some baking soda, vinegar, a small balloon, and a bottle with a narrow neck and a volume of 50 to 100 mL. The balloon should fit tightly but easily over the top of the bottle. (It may slip on more easily if the bottle and balloon are wet.)

Place one level teaspoon of baking soda in the balloon. (You can make a funnel out of a rolled-up piece of paper to help get the baking soda into the balloon.) Add three teaspoons of vinegar to the bottle, and then slip the balloon over the neck of the bottle. Turn the balloon over so that the baking soda runs into the bottle, and then shake the bottle to make sure that the vinegar and baking soda are well mixed. What do you see? Does the balloon inflate? If so, why?

The set-up for studying the reaction of baking soda with acetic acid. (C.D. Winters)

Now repeat the experiment, this time with the same amount of baking soda but with more or less vinegar. Does the balloon blow up to a greater or lesser extent?

Repeat the experiment with the same quantity of baking soda but with differing quantities of vinegar. Is there a connection between the quantity of vinegar or baking soda used and the extent to which the balloon inflates? If so, how can you explain this connection?

quired, and 0.0400 mol of PCl_3 should form. You could confirm by experiment that if 1.24 g of P_4 and 4.25 g of Cl_2 are consumed, then a maximum of 5.49 g (= 1.24 g + 4.25 g) of PCl_3 is produced.

Following this line of reasoning further, suppose you have a piece of phosphorus with a mass of 1.45 g. What mass of Cl_2 is required if all the phosphorus is to react? The following procedure leads to the solution.

Step 1. *Write the balanced equation* (using correct formulas for reactants and products). This is always the first thing to do when dealing with chemical reactions.

$$P_4(s) + 6\,Cl_2(g) \longrightarrow 4\,PCl_3(\ell)$$

Step 2. *Calculate moles from masses.* From the mass of P_4, calculate the number of moles of P_4 available. This must be done because the balanced equation shows mole relationships, not mass relationships.

$$1.45\ \cancel{g\ P_4} \cdot \frac{1\ \text{mol}\ P_4}{123.9\ \cancel{g\ P_4}} = 0.0117\ \text{mol}\ P_4\ \text{available}$$

Step 3. *Use a stoichiometric factor.* Now the available number of moles of one reactant (P_4) must be related to the required number of moles of the other reactant (Cl_2).

$$0.0117\ \cancel{\text{mol}\ P_4} \cdot \frac{6\ \text{mol}\ Cl_2\ \text{required}}{1\ \cancel{\text{mol}\ P_4}\ \text{available}} = 0.0702\ \text{mol}\ Cl_2\ \text{required}$$

a stoichiometric factor (from the balanced equation)

To do this, the number of moles of available reactant has been multiplied by a **stoichiometric factor,** *a mole-ratio factor relating moles of the required reactant to moles of the available reactant.* The stoichiometric factor comes *directly* from the coefficients in the balanced chemical equation; this is why you must balance the chemical equation before proceeding with calculations. Here the calculation shows that 0.0702 mol of Cl_2 is required to use up all the available phosphorus.

Step 4. *Calculate mass from moles.* From the number of moles of Cl_2 calculated in step 3, you can now calculate the mass of Cl_2 required.

$$0.0702\ \cancel{\text{mol}\ Cl_2} \cdot \frac{70.91\ g\ Cl_2}{1\ \cancel{\text{mol}\ Cl_2}} = 4.98\ g\ Cl_2$$

Since the object of this example was to find the mass of Cl_2 required, the problem is solved.

By saying a reaction "goes to completion" we mean that all of at least one reactant is completely converted to products.

You may also want to know the mass of PCl_3 produced in the reaction of 1.45 g of phosphorus with the required mass of chlorine (4.98 g) when the reaction goes to completion. Because matter is conserved, this can be answered by adding the masses of P_4 and Cl_2 used (giving 1.45 g + 4.98 g = 6.43 g of PCl_3 produced). Alternatively, steps 3 and 4 could be repeated, but with the appropriate stoichiometric factor and molar mass.

Step 3′. *Use a stoichiometric factor.* Relate the number of moles of P_4 to the number of moles of PCl_3 produced.

$$0.0117\ \cancel{\text{mol}\ P_4} \cdot \frac{4\ \text{mol}\ PCl_3\ \text{produced}}{1\ \cancel{\text{mol}\ P_4}\ \text{available}} = 0.0468\ \text{mol}\ PCl_3\ \text{produced}$$

another stoichiometric factor (from the balanced equation)

Step 4′. *Calculate mass from moles.* Convert moles of PCl_3 produced to a mass in grams.

$$0.0468\ \cancel{\text{mol}\ PCl_3}\ \text{produced} \cdot \frac{137.3\ g\ PCl_3}{1\ \cancel{\text{mol}\ PCl_3}} = 6.43\ g\ PCl_3$$

Using this alternative may seem a bit silly here, but adding the masses of reactants to obtain the total product mass is useful only if the reaction gives a *single* product. When more than one product results from a reaction, it is usually necessary to apply steps 3 and 4 for each product.

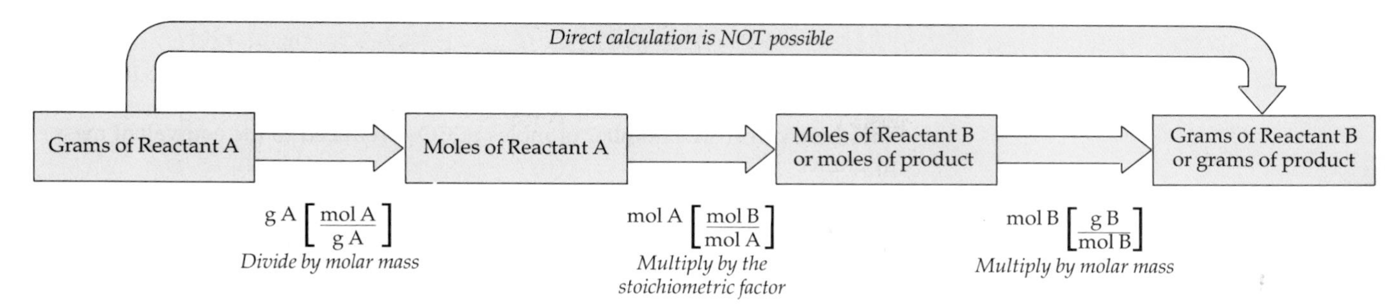

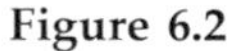

Figure 6.2
A scheme outlining the relation between the mass of a reactant (A) and the mass of another reactant or product (B). The stoichiometric factor is always (moles of B/moles of A) and is given by the coefficients of A and B in the balanced equation.

The preceding example illustrates the general approach to follow in a stoichiometric calculation. To see this more clearly, consider the scheme in Figure 6.2 and work through Example 6.1 and Exercise 6.1. All stoichiometry problems fit the model we have outlined. There will be some differences from one situation to another, but that should not obscure the basic outline in Figure 6.2.

EXAMPLE 6.1

Weight Relations in Reactions

Propane, C_3H_8, can be used as a fuel in your home, car, or barbecue grill because it is easily liquefied and transported. If 1.00 pound or 454 g of propane is burned, what mass of oxygen (in grams) is required for complete combustion and what masses of carbon dioxide and water (in grams) are formed?

SOLUTION Remember that the first step must always be to write a balanced equation.

$$C_3H_8(g) + 5\,O_2(g) \longrightarrow 3\,CO_2(g) + 4\,H_2O(g)$$

Having balanced the equation, proceed to the stoichiometric calculation. First, convert the mass of propane to moles.

$$454\,\cancel{g\,C_3H_8} \cdot \frac{1\text{ mol } C_3H_8}{44.10\,\cancel{g\,C_3H_8}} = 10.3\text{ mol } C_3H_8$$

Now the number of moles of propane available is related to the number of moles of O_2 required by using a stoichiometric factor that is based on the coefficients in the balanced chemical equation.

$$10.3\,\cancel{\text{mol } C_3H_8} \cdot \frac{5\text{ mol } O_2\text{ required}}{1\,\cancel{\text{mol } C_3H_8}\text{ available}} = 51.5\text{ mol } O_2\text{ required}$$

Next, convert the number of moles of O_2 required to the equivalent mass in grams.

$$51.5\,\cancel{\text{mol } O_2}\text{ required} \cdot \frac{32.00\text{ g } O_2}{1\,\cancel{\text{mol } O_2}} = 1650\text{ g } O_2\text{ required}$$

To find the mass of CO_2 produced, for example, simply repeat the last two steps. First, relate the number of moles of C_3H_8 available to the number of moles of CO_2 produced by using a stoichiometric factor,

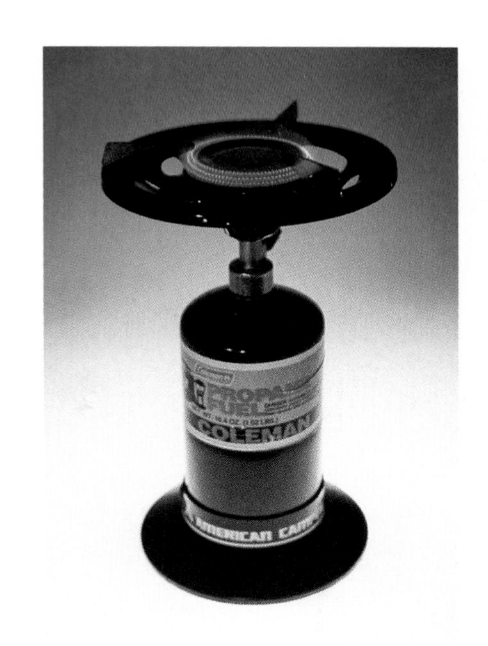

Propane is easily liquefied and can be safely stored in low pressure tanks that can then be attached to a home heating system, a barbecue grill, or a camping stove. *(C.D. Winters)*

$$10.3\ \cancel{\text{mol } C_3H_8}\text{ available} \cdot \frac{3\text{ mol } CO_2\text{ produced}}{1\ \cancel{\text{mol } C_3H_8}\text{ available}} = 30.9\text{ mol } CO_2\text{ produced}$$

and then convert the number of moles of CO_2 produced to the equivalent mass in grams.

$$30.9\ \cancel{\text{mol } CO_2} \cdot \frac{44.01\text{ g } CO_2}{1\ \cancel{\text{mol } CO_2}} = 1360\text{ g } CO_2$$

Now, how can we find the mass of H_2O produced? Go through steps 3′ and 4′ again? Of course! But it would be easier to recognize that the total mass of the reactants

$$454\text{ g } C_3H_8 + 1650\text{ g } O_2 = 2100\text{ g of reactants}$$

must be the same as the total mass of products. Therefore, the mass of water produced is found from

$$\text{total mass of products} = 2100\text{ g} = 1360\text{ g } CO_2\text{ produced} + ?\text{ g } H_2O$$
$$\text{mass of } H_2O = 2100\text{ g} - 1360\text{ g} = 740\text{ g}$$

EXERCISE 6.1 • *Weight Relations in Reactions*

What mass of carbon, in grams, can be consumed by 454 g of O_2 in a combustion to give carbon monoxide? What mass of CO will be produced?

$$2\text{ C(s)} + O_2\text{(g)} \longrightarrow 2\text{ CO(g)}$$

6.2 REACTIONS WITH ONE REACTANT IN LIMITED SUPPLY

The objective of the example in the previous section was to find the quantity of material required to consume completely a given quantity of some reactant or to find how much product would be formed from a given quantity of a reactant. However, in carrying out a reaction, a chemist—or nature for that matter—rarely supplies precise, stoichiometric amounts of reactants. As an example, consider the preparation of cisplatin, $Pt(NH_3)_2Cl_2$, a compound used to treat cancer (see Chapter 1).

$$(NH_4)_2PtCl_4\text{(s)} + \underset{\text{ammonia}}{2\ NH_3\text{(aq)}} \longrightarrow 2\ NH_4Cl\text{(aq)} + \underset{\text{cisplatin}}{Pt(NH_3)_2Cl_2\text{(s)}}$$

The goal in carrying out a reaction is usually to produce the largest possible quantity of a useful compound from a given quantity of starting material. Often a large excess of one reactant is supplied to ensure that the more expensive reactant is consumed completely. In the case of the synthesis of cisplatin, it makes sense to combine the more expensive substance $(NH_4)_2PtCl_4$ (roughly $40 per gram) with a much greater amount of the much less expensive NH_3 (only pennies per gram) than is called for by the balanced equation. Thus, on completion of the reaction, all the $(NH_4)_2PtCl_4$ will have been converted to product, and some NH_3 will remain. How much

$Pt(NH_3)_2Cl_2$ will be formed? It depends on the amount of $(NH_4)_2PtCl_4$ present at the start, not on the amount of NH_3, since more of the latter was present than required by stoichiometry. A compound such as $(NH_4)_2PtCl_4$ in this example is called the **limiting reactant** because its amount determines or limits the amount of product formed.

As an analogy to a chemical "limiting reactant" situation, consider what happens when you make some cheeseburgers. Suppose you have enough meat to make ten hamburger patties, and you also have two dozen buns and 30 slices of cheese. If each cheeseburger requires one bun, two slices of cheese, and one hamburger patty, you can make only ten sandwiches. There will be 14 buns and 10 cheese slices left over. The cheese and buns are "excess reactants" and the hamburger patties are the "limiting reactant," since the amount of hamburger meat has limited or determined the number of sandwiches that can be made. Furthermore, although meat is bought by the pound, buns by the dozen, and cheese by the slice, the combining units are the number of buns or cheese slices required by each hamburger patty, and your calculation had to be done in those units, just as chemical stoichiometry problems are done in units of moles.

An example of a reaction that involves a limiting reactant is the manufacture of the pure silicon that is used in computer chips and solar cells (Figure 6.3). The starting place for the production of silicon is quartz or sand, SiO_2. The quartz is first reduced with high purity coke (carbon) in an electric furnace at temperatures above 2000 °C.

$$SiO_2(s) + C(s) \longrightarrow Si(s) + 2\,CO(g)$$

The solid obtained from this process contains only 96 to 99% silicon and is not pure enough for electronic use. The silicon is removed from the solid by reaction with an excess of chlorine, which produces liquid silicon tetrachloride ($SiCl_4$).

$$Si(s) + 2\,Cl_2(g) \longrightarrow SiCl_4(\ell)$$

After the $SiCl_4$ is purified, it is then reduced back to pure elemental silicon with an excess of very pure magnesium.

$$SiCl_4(\ell) + 2\,Mg(s) \longrightarrow Si(s) + 2\,MgCl_2(s)$$

While the chemistry of this process is relatively simple, great care must be exercised to get the silicon pure enough. As a result, it is not surprising that solar cells and computer chips are relatively expensive.

Now, suppose that 225 g of $SiCl_4$ is mixed with 225 g of Mg in the final step. Are these reactants mixed in the correct stoichiometric ratio or is one of them in short supply? That is, will one of them limit the quantity of silicon that can be produced? If so, how much silicon can be formed if the reaction goes to completion? And how much of the excess reactant is left when the maximum amount of silicon is formed?

Since the quantities of both starting materials are given, the first step in answering our questions involves finding the number of moles of each.

$$225\,\cancel{g\,SiCl_4} \cdot \frac{1\text{ mol }SiCl_4}{169.9\,\cancel{g\,SiCl_4}} = 1.32\text{ mol }SiCl_4\text{ available}$$

$$225\,\cancel{g\,Mg} \cdot \frac{1\text{ mol Mg}}{24.31\,\cancel{g\,Mg}} = 9.26\text{ mol Mg available}$$

(a)

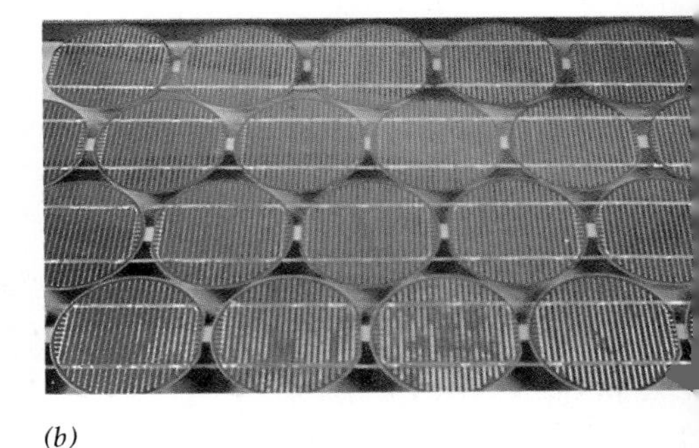

(b)

Figure 6.3
(a) Pure silicon is used in solar cells on the wings of the solar-powered aircraft "Solair I." (b) A closer view of the solar cells. *(a, b courtesy of Wacker Silicones Corporation)*

Are these reactants present in the correct stoichiometric ratio as given by the balanced equation?

$$\text{mole ratio of reactants required by the balanced equation} = \frac{2 \text{ mol Mg}}{1 \text{ mol SiCl}_4}$$

$$\text{mole ratio of reactants actually available} = \frac{9.26 \text{ mol Mg available}}{1.32 \text{ mol SiCl}_4 \text{ available}}$$

$$= \frac{7.02 \text{ mol Mg}}{1 \text{ mol SiCl}_4}$$

Dividing the number of moles of Mg available by the number of moles of $SiCl_4$ available shows that the ratio is much larger than the (2 mol Mg/1 mol $SiCl_4$) ratio required according to the balanced equation. Therefore, there is more magnesium available than is needed to react with all the available $SiCl_4$. It also follows that $SiCl_4$ is the limiting reactant, as we can prove by the calculations that follow.

Let us determine what quantity of elemental silicon can be formed if we begin with 1.32 mol of $SiCl_4$ and unlimited magnesium, or if we begin with 9.26 mol of Mg and unlimited $SiCl_4$.

Quantity of Si produced from 1.32 mol of $SiCl_4$ and unlimited Mg

$$1.32 \cancel{\text{mol SiCl}_4} \cdot \frac{1 \cancel{\text{mol Si}}}{1 \cancel{\text{mol SiCl}_4}} \cdot \frac{28.09 \text{ g Si}}{1 \cancel{\text{mol Si}}} = 37.1 \text{ g Si}$$

Quantity of Si produced from 9.26 mol of Mg and unlimited $SiCl_4$

$$9.26 \cancel{\text{mol Mg}} \cdot \frac{1 \cancel{\text{mol Si}}}{2 \cancel{\text{mol Mg}}} \cdot \frac{28.09 \text{ g Si}}{1 \cancel{\text{mol Si}}} = 130. \text{ g Si}$$

Comparison of the quantities of silicon produced shows that 225 g of $SiCl_4$ is capable of producing less silicon (37.1 g) than 225 g of Mg can produce (130 g). Therefore, *silicon tetrachloride, $SiCl_4$, is the limiting reactant.*

In this case, $SiCl_4$ is the equivalent of the hamburger patties in the cheeseburger analogy for limiting reactants.

Our calculation shows that there is more than enough magnesium available to consume the available silicon tetrachloride, $SiCl_4$. Magnesium is the "excess reactant," and it is possible to calculate the quantity of the metal that remains after all the $SiCl_4$ has been used. To do this, we first need to know how much magnesium is required to consume all the limiting reactant, $SiCl_4$.

$$1.32 \cancel{\text{mol SiCl}_4 \text{ available}} \cdot \frac{2 \text{ mol Mg required}}{1 \cancel{\text{mol SiCl}_4 \text{ available}}} = 2.64 \text{ mol Mg required}$$

Since there is 9.26 mol of magnesium available, the number of moles of excess magnesium can be calculated,

$$\text{excess Mg} = 9.26 \text{ mol Mg available} - 2.64 \text{ mol Mg consumed}$$
$$= 6.62 \text{ mol Mg remaining}$$

and then converted to a mass.

$$6.62 \cancel{\text{mol Mg}} \cdot \frac{24.31 \text{ g Mg}}{1 \cancel{\text{mol Mg}}} = 161 \text{ g Mg in excess of that required}$$

Finally, since 161 g of magnesium is left over, this means that 225 g − 161 g = 64 g of magnesium has been consumed.

Titanium tetrachloride, $TiCl_4$, is a liquid. When exposed to air it forms a dense fog of titanium(IV) oxide, TiO_2. *(C.D. Winters)*

EXAMPLE 6.2

Limiting Reactants

Titanium tetrachloride is an important industrial chemical. For example, from it one can make TiO_2, the substance used as a white pigment in paper and paints. The tetrachloride can be made by combining titanium-containing ore (which is often impure TiO_2) with carbon and chlorine.

$$TiO_2(s) + 2\,Cl_2(g) + C(s) \longrightarrow TiCl_4(\ell) + CO_2(g)$$

If you begin with 125 g each of Cl_2 and C, but plenty of TiO_2-containing ore, which is the limiting reactant in this reaction? What quantity of $TiCl_4$, in grams, can be produced?

SOLUTION As is always the case, when you have a mass of a pure element or compound, you should find the equivalent number of moles.

$$\text{Calculate moles } Cl_2\text{: } 125\,\cancel{\text{g } Cl_2} \cdot \frac{1 \text{ mol } Cl_2}{70.91\,\cancel{\text{g } Cl_2}} = 1.76 \text{ mol } Cl_2$$

$$\text{Calculate moles C: } 125\,\cancel{\text{g C}} \cdot \frac{1 \text{ mol C}}{12.01\,\cancel{\text{g C}}} = 10.4 \text{ mol C}$$

Are these reactants, in the quantities available, present in a perfect stoichiometric ratio?

$$\frac{\text{moles C available}}{\text{moles } Cl_2 \text{ available}} = \frac{10.4 \text{ mol C}}{1.76 \text{ mol } Cl_2} = \frac{5.91 \text{ mol C}}{1 \text{ mol } Cl_2}$$

According to the balanced equation, the required mole ratio is 1 mol of C to 2 mol of Cl_2, or 0.5 mol C/1 mol Cl_2. Clearly there is much more carbon than is required. Conversely, there is not enough Cl_2 present to use up all of the carbon, so Cl_2 is the limiting reactant. Let us prove that by the following calculations.

1. Calculate the mass of product, $TiCl_4$, that can be formed from 1.76 mol of Cl_2 and unlimited C and TiO_2.

$$1.76\,\cancel{\text{mol } Cl_2} \cdot \frac{1\,\cancel{\text{mol } TiCl_4}}{2\,\cancel{\text{mol } Cl_2}} \cdot \frac{189.7 \text{ g } TiCl_4}{1\,\cancel{\text{mol } TiCl_4}} = 167 \text{ g } TiCl_4$$

2. Calculate the mass of product, $TiCl_4$, that can be formed from 10.4 mol of C and unlimited Cl_2 and TiO_2.

$$10.4\,\cancel{\text{mol C}} \cdot \frac{1\,\cancel{\text{mol } TiCl_4}}{1\,\cancel{\text{mol C}}} \cdot \frac{189.7 \text{ g } TiCl_4}{1\,\cancel{\text{mol } TiCl_4}} = 1970 \text{ g } TiCl_4$$

Clearly the quantity of $TiCl_4$ that can be formed using the given quantities of C and Cl_2 is controlled or limited by the quantity of Cl_2 available.

EXERCISE 6.2 • *Limiting Reactants*

You have 20.0 g of elemental sulfur, S_8, and 160. g of O_2. Which is the limiting reactant in the combustion of S_8 in oxygen? What amount of which reactant (in moles) is left after complete reaction? What mass of SO_2, in grams, is formed in the complete reaction? The *unbalanced* equation is

$$S_8(s) + O_2(g) \longrightarrow SO_2(g)$$

6.3
PERCENT YIELD

The theoretical yield is calculated by the methods described in Section 6.2.

The maximum quantity of product that can be obtained from a chemical reaction of a given quantity of reactants is the **theoretical yield.** However, the desired reaction may not be the only one that occurs, and some reactants may be converted to products other than the one you want. Also, there is invariably some waste in the isolation and purification of products; you will "lose" small quantities of material. For these reasons, the **actual yield** of a compound—the quantity of material you actually obtain in the laboratory or chemical plant—may be less than the theoretical yield. The efficiency of a chemical reaction and the techniques used to obtain the desired compound in pure form can be evaluated by calculating the ratio of the actual yield to the theoretical yield. We call the result the **percent yield** (Figure 6.4).

Figure 6.4
We began with 20 pieces of popcorn and found that only 16 of them popped. The percent yield of popcorn from our "reaction" was (16/20) · 100% or 80%. *(C.D. Winters)*

$$\text{percent yield} = \frac{\text{actual yield}}{\text{theoretical yield}} \cdot 100\%$$

Suppose you made aspirin in the laboratory by the following reaction (Figure 6.5):

$$\underset{\text{salicylic acid}}{2\ C_7H_6O_3(s)} + \underset{\text{acetic anhydride}}{C_4H_6O_3(\ell)} \longrightarrow \underset{\text{aspirin}}{2\ C_9H_8O_4(s)} + H_2O(\ell)$$

and that you began with 14.4 g of salicylic acid. Further assume that the experiment was done with excess acetic anhydride; this means that the acid reacted completely but some acetic anhydride remained in the reaction vessel after the reaction was complete. Therefore, the acid was the limiting reactant, and the mass of aspirin depended on the mass of the salicylic acid. If you obtained 6.26 g of aspirin, what was the percent yield of this product?

First calculate the number of moles of limiting reactant, salicylic acid in this case.

$$14.4\ \cancel{\text{g}\ C_7H_6O_3} \cdot \frac{1\ \text{mol}\ C_7H_6O_3}{138.1\ \cancel{\text{g}\ C_7H_6O_3}} = 0.104\ \text{mol}\ C_7H_6O_3$$

Next, use the stoichiometric factor from the balanced equation to find the number of moles of aspirin expected based on the limiting reactant, $C_7H_6O_3$.

$$0.104\ \cancel{\text{mol}\ C_7H_6O_3} \cdot \frac{2\ \text{mol aspirin}}{2\ \cancel{\text{mol}\ C_7H_6O_3}} = 0.104\ \text{mol aspirin expected}$$

The maximum amount of aspirin that could be produced is 0.104 mol; that is, this is the theoretical yield in moles. Since the quantity that you measure directly is the mass of the product, it is useful to express the theoretical yield as a mass in grams.

$$0.104\ \cancel{\text{mol aspirin}} \cdot \frac{180.2\ \text{g aspirin}}{1\ \cancel{\text{mol aspirin}}} = 18.8\ \text{g aspirin}$$

Finally, the actual yield is only 6.26 g, so we calculate the percent yield of aspirin as

$$\text{percent yield} = \frac{6.26 \text{ g aspirin actually isolated}}{18.8 \text{ g aspirin expected}} \cdot 100\% = 33.3\% \text{ yield}$$

When a chemist makes a new compound or carries out a reaction, the percent yield is usually reported. It is useful information because it gives other chemists some idea of the quantity of product that can be reasonably expected from the reaction. But be aware that it is difficult to obtain yields of 90% or better. It is impossible to keep track of every drop of liquid or crumb of solid, and reactions other than the desired one may occur. Further, many reactions do not go completely to products (as you may know from popping a cup of popcorn kernels). Thus, reaction yields are usually less than 90% and often much lower than that.

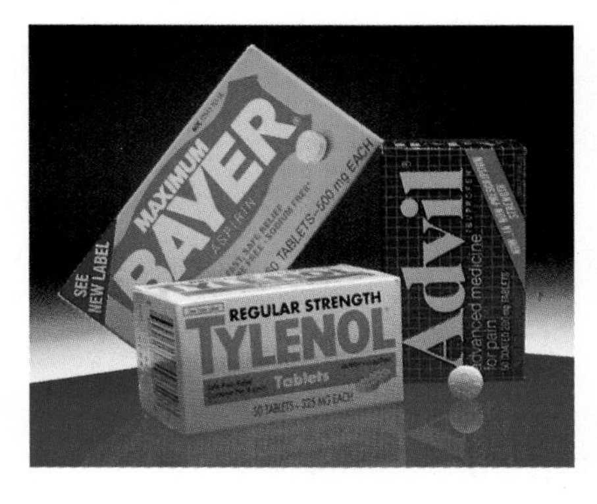

Figure 6.5
Aspirin, a white solid ($C_9H_8O_4$), is made by the reaction of salicylic acid ($C_7H_6O_3$) and acetic anhydride ($C_4H_6O_3$). Other pain-relief medications contain other active compounds such as acetaminophen ($C_8H_9NO_2$) and ibuprofen ($C_{13}H_{18}O_2$), which are quite different from aspirin. *(C.D. Winters)*

EXERCISE 6.3 • *Percent Yield*

Professor H. C. Brown of Purdue University received the Nobel Prize in Chemistry in 1979 for his work on the chemistry of diborane, B_2H_6, and its use in preparing new organic compounds. This gas can be prepared by the following reaction (which is carried out in a nonaqueous solvent, that is, a solvent other than water):

$$3\ NaBH_4 + 4\ BF_3 \longrightarrow 3\ NaBF_4 + 2\ B_2H_6$$

If you begin with 18.9 g of $NaBH_4$ (and excess BF_3), and you isolate 7.50 g of B_2H_6, what is the percent yield of B_2H_6?

6.4 CHEMICAL EQUATIONS AND CHEMICAL ANALYSIS

With an increased awareness of environmental problems in recent years, we need to know just what chemicals are in our environment and in what quantities. This need has made analytical chemistry an increasingly important field. Chemists in this field use their creativity to identify pure substances as well as to measure the quantities of components of mixtures. Analytical chemistry is often done now by instrumental methods (Figure 6.6), but classical chemical reactions and stoichiometry still play a central role.

Figure 6.6
A student using a modern analytical instrument. This instrument measures the extent to which a sample absorbs visible and ultraviolet light. *(C.D. Winters)*

Analysis of a Mixture

The analysis of mixtures is often challenging. It can take a great deal of imagination to figure out how to use chemistry to determine what, and how much, is there. We can illustrate how analytical chemistry problems are solved, though, with a reasonably straightforward example.

Suppose you have a white powder that you know is a mixture of magnesium oxide (MgO) and magnesium carbonate ($MgCO_3$), and you are asked to find out what percent of the mixture is $MgCO_3$. In Section 5.5 you learned that metal carbonates decompose on heating to give metal oxides and carbon dioxide. For magnesium carbonate the reaction is

$$MgCO_3(s) \longrightarrow MgO(s) + CO_2(g)$$

Therefore, if you heat the powder strongly the magnesium carbonate in the mixture decomposes to MgO, and CO_2 will be evolved as a gas. The solid left after heating the mixture will consist only of MgO, and its mass will be the sum of the mass of the MgO that was originally in the mixture *plus* the MgO remaining from the $MgCO_3$. The difference in mass of the solid before and after heating therefore gives the mass of CO_2 that left the solid and, by stoichiometry, the mass of $MgCO_3$ in the original mixture.

Assume you have 1.598 g of a mixture of MgO and $MgCO_3$ and that heating evolves CO_2 and leaves 1.294 g of white MgO. What was the weight percent of $MgCO_3$ in the original mixture? Knowing the mass of the mixture before and after heating, you can find the mass difference, which equals the mass of CO_2 evolved when the mixture was heated.

$$\begin{array}{lr} \text{mass of mixture } (\text{MgO} + \text{MgCO}_3) = & 1.598 \text{ g} \\ -\text{mass after heating (pure MgO)} = & -1.294 \text{ g} \\ \hline \text{mass of CO}_2 = & 0.304 \text{ g} \end{array}$$

Now the mass of CO_2 can be converted to the number of moles of CO_2.

$$0.304 \text{ g CO}_2 \cdot \frac{1 \text{ mol CO}_2}{44.01 \text{ g CO}_2} = 0.00691 \text{ mol CO}_2$$

The balanced equation for the decomposition of $MgCO_3$ shows that every mole of CO_2 given off on heating means that there was a mole of $MgCO_3$ in the mixture. Therefore, there must have been 0.00691 mol of $MgCO_3$ in the mixture, and the mass of $MgCO_3$ in the mixture was

$$0.00691 \text{ mol CO}_2 \cdot \frac{1 \text{ mol MgCO}_3}{1 \text{ mol CO}_2} \cdot \frac{84.31 \text{ g MgCO}_3}{1 \text{ mol MgCO}_3} = 0.582 \text{ g MgCO}_3$$

This means the weight percent of magnesium carbonate in the mixture was

$$\frac{\text{mass of MgCO}_3}{\text{sample mass}} \cdot 100\% = \frac{0.582 \text{ g MgCO}_3}{1.598 \text{ g sample}} \cdot 100\% = 36.4\% \text{ MgCO}_3$$

EXAMPLE 6.3

Analysis of a Mixture

Butyl lithium, LiC_4H_9, is a very reactive compound used by chemists to make new compounds. One way to determine the quantity of LiC_4H_9 in a mixture is to add aqueous hydrochloric acid. The following reaction occurs:

$$\text{LiC}_4\text{H}_9 + \text{HCl(aq)} \longrightarrow \text{LiCl(aq)} + \text{C}_4\text{H}_{10}\text{(g)}$$

Suppose you have 5.606 g of a solution of LiC_4H_9 dissolved in benzene. On adding HCl(aq), a mass of 0.636 g of C_4H_{10} is evolved. What is the weight percent of LiC_4H_9 in the sample?

SOLUTION Let us first calculate the number of moles of C_4H_{10} evolved.

$$0.636 \text{ g C}_4\text{H}_{10} \cdot \frac{1 \text{ mol C}_4\text{H}_{10}}{58.12 \text{ g C}_4\text{H}_{10}} = 0.0109 \text{ mol C}_4\text{H}_{10}$$

Next, calculate the number of moles and then the mass of LiC_4H_9 in the sample.

$$0.0109 \text{ mol } C_4H_{10} \cdot \frac{1 \text{ mol } LiC_4H_9}{1 \text{ mol } C_4H_{10}} \cdot \frac{64.06 \text{ g } LiC_4H_9}{1 \text{ mol } LiC_4H_9} = 0.701 \text{ g } LiC_4H_9$$

Finally, calculate the weight percent of LiC_4H_9,

$$\frac{0.701 \text{ g } LiC_4H_9}{5.606 \text{ g sample}} \cdot 100\% = 12.5\%$$

EXERCISE 6.4 • *Chemical Analysis*

You have 2.357 g of a mixture of $BaCl_2$ and $BaCl_2 \cdot 2\,H_2O$. If experiment shows that the mixture has a mass of only 2.108 g after heating to drive off all the water of hydration in $BaCl_2 \cdot 2\,H_2O$, what is the weight percent of $BaCl_2 \cdot 2\,H_2O$ in the original mixture?

Determining the Empirical Formula of a Compound

The empirical formula of a compound can be determined if the percent composition of the compound is known (Section 4.7). But where do the percent composition data come from? Various methods are used, and many depend on reactions that decompose the unknown but pure compound into known products. The reaction products can be isolated and their masses determined. For example, suppose you have isolated a new compound containing only boron and hydrogen. Such compounds can be decomposed by reaction with water to produce boric acid [$B(OH)_3$] and hydrogen gas. One such reaction is

$$B_2H_6(g) + 6\,H_2O(\ell) \longrightarrow 2\,B(OH)_3(aq) + 6\,H_2(g)$$

and in general,

$$B_xH_y + y\,H_2O(\ell) \longrightarrow x\,B(OH)_3(aq) + y\,H_2(g)$$

$$1 \text{ mol B} \longrightarrow 1 \text{ mol } B(OH)_3$$

$$1 \text{ mol H} \longrightarrow 1 \text{ mol } H_2$$

The important observation is that each mole of H in the original compound combines with one mole of H from water to produce one mole of H_2 (and each mole of B leads to one mole of boric acid). Suppose that 0.125 g of your new compound produced 0.036 g of H_2. What is the empirical formula of the boron-hydrogen compound? Since you know the mass and formula of H_2, the first step is to calculate the number of moles of H_2 isolated,

$$0.036 \cancel{\text{g } H_2} \cdot \frac{1 \text{ mol } H_2}{2.02 \cancel{\text{g } H_2}} = 0.018 \text{ mol } H_2$$

The balanced general equation shows that for each mole of H_2 isolated, there must have been one mole of H in the compound B_xH_y.

$$0.018 \cancel{\text{mol } H_2} \cdot \frac{1 \text{ mol H in } B_xH_y}{1 \cancel{\text{mol } H_2}} = 0.018 \text{ mol H in } B_xH_y$$

Therefore, the original 0.125-g sample of compound contained 0.018 mol of H, or 0.018 g of H, since

$$0.018\ \text{mol H in B}_x\text{H}_y \cdot \frac{1.01\ \text{g H}}{1\ \text{mol H}} = 0.018\ \text{g H}$$

The remainder of the original mass was boron, so the original sample must have contained 0.018 g of H and 0.107 g of B (= 0.125 g sample − 0.018 g H).

Now that you know the mass of boron in the original 0.125-g sample, you can calculate the number of moles of boron,

$$0.107\ \text{g B} \cdot \frac{1\ \text{mol B}}{10.81\ \text{g B}} = 0.00990\ \text{mol B}$$

and compare the numbers of moles of boron and hydrogen in the sample:

$$\frac{0.018\ \text{mol H}}{0.00990\ \text{mol B}} = \frac{1.8\ \text{mol H}}{1\ \text{mol B}} = \frac{(1 + 4/5)\ \text{mol H}}{1\ \text{mol B}} = \frac{9/5\ \text{mol H}}{1\ \text{mol B}} = \frac{9\ \text{mol H}}{5\ \text{mol B}}$$

Therefore, the empirical formula of the new compound is B_5H_9.

Many different approaches to determining empirical formulas have been developed by chemists over the years. Another common method is used for combustible compounds, as illustrated in the following example.

EXAMPLE 6.4

Determining an Empirical Formula by Combustion Analysis

A method that works well for determining empirical formulas of compounds that burn in oxygen is *analysis by combustion.* Each element (except oxygen) in the compound combines with oxygen to produce the appropriate oxide. For example,

$$\begin{array}{lcl} CH_4(g) + 2\,O_2(g) & \longrightarrow & CO_2(g) + 2\,H_2O(g) \\ \text{1 mole C} & \longrightarrow & \text{1 mole } CO_2 \\ \text{4 moles H} & \longrightarrow & \qquad\qquad \text{2 moles } H_2O \end{array}$$

If a compound contains carbon, each mole of carbon in the original compound is converted to one mole of CO_2. For those containing hydrogen, each mole of hydrogen in the original compound gives *half* a mole of H_2O. Finally, every other element is converted to its oxide. Because carbon dioxide and water from the combustion are gases, they can be separated and their masses determined as illustrated in Figure 6.7. Since these masses can be converted to the masses of C and H in CO_2 and H_2O, respectively, the masses of carbon and hydrogen in a given mass of the original compound can be found, and this leads to the empirical formula.

Suppose you have isolated an acid from clover leaves and do not know its identity. However, you do know that it contains only the elements C, H, and O. Burning 0.513 g of the acid in oxygen produces 0.501 g of CO_2 and 0.103 g of H_2O. What is the empirical formula of the acid? Given that another experiment has shown that the molar mass of the acid is 90.04 g/mol, what is its molecular formula?

SOLUTION Because every atom of C in the molecule must be converted to one molecule of CO_2, and every two atoms of H in the original compound must be converted to one molecule of H_2O, the first step is to convert the masses of CO_2

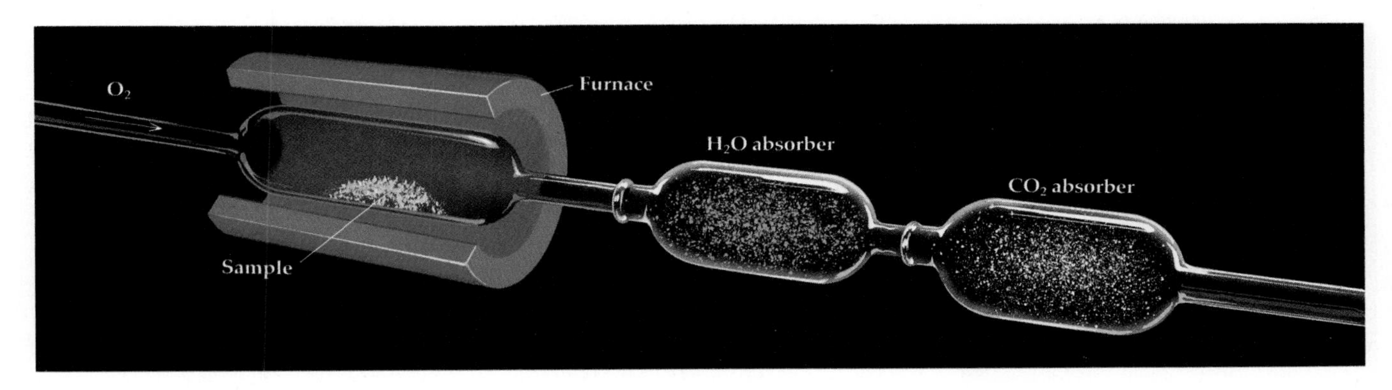

Figure 6.7
If a compound containing C and H is burned in oxygen, CO_2 and H_2O are formed, and the mass of each can be determined. The H_2O is absorbed by magnesium perchlorate, and the CO_2 is absorbed by finely divided NaOH supported on asbestos. The mass of each absorbent before and after combustion will give the masses of CO_2 and H_2O. Only a few milligrams of a combustible compound are needed for analysis.

and H_2O to moles in order to find the numbers of moles of C and H, respectively, in the original compound.

$$0.501\ \cancel{\text{g CO}_2} \cdot \frac{1\ \text{mol CO}_2}{44.01\ \cancel{\text{g CO}_2}} = 0.0114\ \text{mol CO}_2$$

$$0.103\ \cancel{\text{g H}_2\text{O}} \cdot \frac{1\ \text{mol H}_2\text{O}}{18.02\ \cancel{\text{g H}_2\text{O}}} = 0.00572\ \text{mol H}_2\text{O}$$

The numbers of moles of CO_2 and H_2O can now be converted to the masses of C and H that were in the original compound.

$$0.0114\ \cancel{\text{mol CO}_2} \cdot \frac{1\ \cancel{\text{mol C}}}{1\ \cancel{\text{mol CO}_2}} \cdot \frac{12.01\ \text{g C}}{1\ \cancel{\text{mol C}}} = 0.137\ \text{g C in CO}_2\ \textit{and}\ \text{formerly in the acid sample}$$

$$0.00572\ \cancel{\text{mol H}_2\text{O}} \cdot \frac{2\ \cancel{\text{mol H}}}{1\ \cancel{\text{mol H}_2\text{O}}} \cdot \frac{1.008\ \text{g}}{1\ \cancel{\text{mol H}}} = 0.0115\ \text{g H in H}_2\text{O}\ \textit{and}\ \text{formerly in the acid sample}$$

↑ *Be sure to notice this stoichiometric factor*

These calculations reveal that the 0.513-g sample of acid contains 0.137 g of C and 0.0115 g of H; the remaining mass, 0.365 g, must be oxygen.

$$0.137\ \text{g C} + 0.0115\ \text{g H} + 0.365\ \text{g O} = 0.513\ \text{g acid sample}$$

To find the empirical formula of the acid, you need only to find the number of moles of each element in the acid sample (see Section 4.7).

$$0.137\ \cancel{\text{g C}} \cdot \frac{1\ \text{mol C}}{12.01\ \cancel{\text{g C}}} = 0.0114\ \text{mol C}$$

$$0.0115\ \cancel{\text{g H}} \cdot \frac{1\ \text{mol H}}{1.008\ \cancel{\text{g H}}} = 0.0114\ \text{mol H}$$

$$0.365\ \cancel{\text{g O}} \cdot \frac{1\ \text{mol O}}{16.00\ \cancel{\text{g O}}} = 0.0228\ \text{mol O}$$

Then, to find the mole ratios of the elements, divide the number of moles of each element by the *smallest* number of moles.

$$\frac{0.0114\ \text{mol H}}{0.0114\ \text{mol C}} = \frac{1.00\ \text{mol H}}{1.00\ \text{mol C}} \qquad \frac{0.0228\ \text{mol O}}{0.0114\ \text{mol C}} = \frac{2.00\ \text{mol O}}{1.00\ \text{mol C}}$$

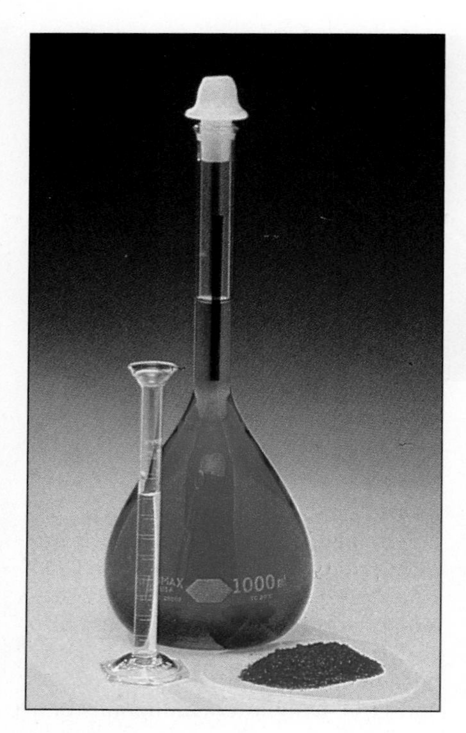

Figure 6.8
To make a 0.100-M solution of $CuSO_4$, 25.0 g or 0.100 mol of $CuSO_4 \cdot 5\ H_2O$ (the blue crystalline solid) was placed in a 1.00-L volumetric flask. In this photo, exactly 1.00 L of water was measured out and slowly added to the volumetric flask. When enough water had been added so that the *solution volume* was exactly 1.00 L, approximately 8 mL of water (the quantity in the small graduated cylinder) was left over. This unused solvent emphasizes that molar concentrations are defined as moles per liter of solution and not per liter of water or other solvent. *(C.D. Winters)*

The mole ratios show that for every C atom in the molecule, there are one H atom and two O atoms. Therefore, the *empirical formula* of the acid is CHO_2, and the molar mass of the empirical formula unit is 45.02 g/mol.

Finally, to determine the molecular formula, compare the experimental molar mass of the compound and the molar mass of one empirical formula unit.

$$\frac{90.04\ \text{g/mol unknown acid}}{45.02\ \text{g/mol of}\ CHO_2} = \frac{2.000\ \text{mol}\ CHO_2}{1.000\ \text{mol unknown acid}}$$

Thus, the *molecular formula* of the acid is twice the empirical formula, that is, $C_2H_2O_4$.

EXERCISE 6.5 • *Formula Determination by Combustion Analysis*

A molecule of vitamin C contains only C, H, and O. Determine the empirical formula of vitamin C from the following data: burning 0.400 g of solid vitamin C in pure oxygen gives 0.600 g of CO_2 and 0.163 g of H_2O.

6.5 WORKING WITH SOLUTIONS

Many of the chemicals in your body or in a plant are dissolved in water, that is, they are in an aqueous solution. Just as a living system uses chemistry in solution, so too do chemists, and we need to do our work quantitatively. To accomplish this, we continue to use balanced equations and moles, but we measure volumes of solution rather than masses of solids, liquids, or gases. Solution concentration expressed as *molarity* relates the volume of solution in liters or milliliters to the amount of substance in moles.

Solution Concentration: Molarity

The concept of concentration is useful in many contexts. For example, there are about 4,900,000 people in Wisconsin and the state has a land area of roughly 56,000 square miles; therefore, the average concentration of people is about 88 per square mile. In chemistry the amount of solute dissolved in a given volume of solution can be found in the same way and is known as the **concentration** of the solution. Solution concentration is usually reported as *moles of solute per liter of solution;* this is called the **molarity** of the solution.

The substance that is present in the same phase as the solution, and generally in greater amount, is called the ***solvent.*** *The other substance (or substances) is called the* ***solute.***

$$\text{molarity} = \frac{\text{moles of solute}}{\text{liters of solution}}$$

For example, if 58.4 g, or 1.00 mol, of NaCl is dissolved in enough water to give a total solution volume of 1.00 liter, the concentration is 1.00 mole per liter or 1.00 *molar.* This is often abbreviated as 1.00 M, where the capital M stands for "moles per liter." Another common notation is the use of square brackets to indicate molarity.

$$[NaCl] = 1.00 \text{ M}$$

Placing the formula of the compound in square brackets implies that the quantity specified is the concentration of the solute in moles of compound per liter of solution.

It is important to notice that molarity refers to moles of solute per liter of *solution* (and not to liters of solvent). If one liter of water is added to one mole of a solid compound, the final volume probably will not be exactly one liter and the final concentration will not be exactly 1 molar (Figure 6.8). Therefore, when making solutions of a given molarity, always *add the solvent to the solute until the desired solution volume is reached.*

The terms "moles per liter" and "molar" are used interchangeably.

A capital M *stands for "moles/liter," whereas an italicized M indicates "molar mass."*

The strong oxidizing agent potassium permanganate, $KMnO_4$, which was used at one time as a germicide in the treatment of burns, is a common laboratory chemical. It is a shiny, purple-black solid that dissolves readily in water to give deep purple solutions. Suppose 0.435 g of $KMnO_4$ has been dissolved in enough water to give 250. mL of solution (Figure 6.9). What is

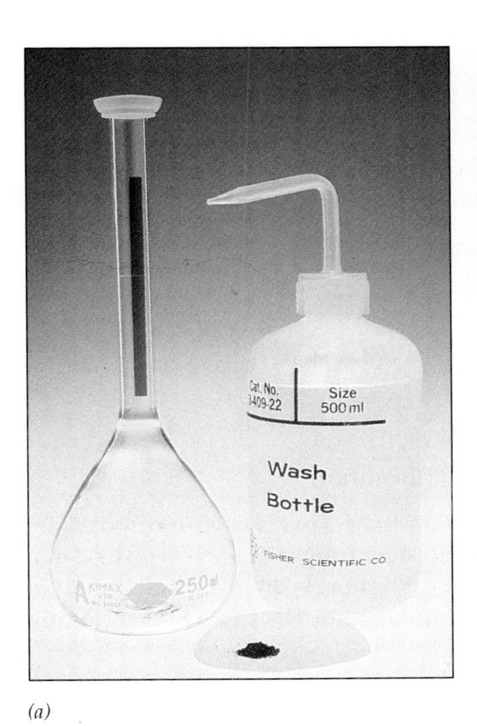

(a)

(b)

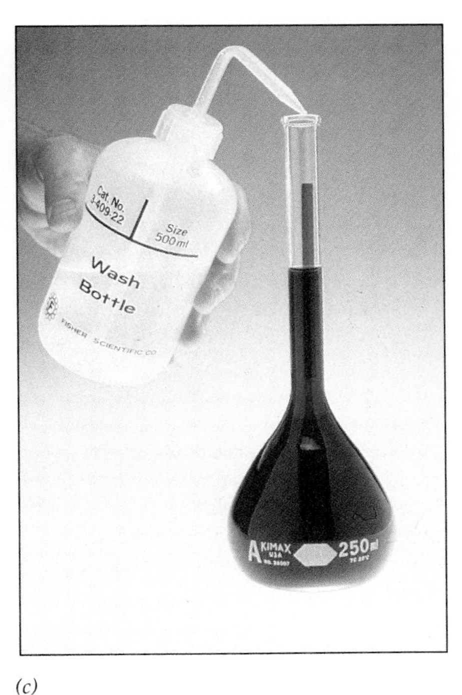

(c)

Figure 6.9

A 0.0110-M solution of $KMnO_4$ is made by adding enough water to 0.435 g of $KMnO_4$ to make 0.250 L of solution (a). To ensure the correct solution volume, the $KMnO_4$ is placed in a volumetric flask and dissolved in a small amount of water (b). After the solid is completely dissolved, sufficient water is added to fill the flask to the mark on the neck. The flask now contains 0.250 L of solution (c). *(C.D. Winters)*

Figure 6.10
Ion concentrations on dissolving ionic compounds. (a) When $KMnO_4$ dissolves in water, one mole of K^+ ions and one mole of MnO_4^- ions form for every mole of $KMnO_4$ dissolved. (b) Dissolving Na_2CO_3 produces two Na^+ ions and one CO_3^{2-} ions per Na_2CO_3. Therefore, three moles of Na_2CO_3 produce nine moles of ions, whereas three moles of $KMnO_4$ produce only six moles of ions.

the molar concentration of $KMnO_4$? As is almost always the case, the first step is to convert the mass of material to number of moles.

$$0.435 \text{ g KMnO}_4 \cdot \frac{1 \text{ mol KMnO}_4}{158.0 \text{ g KMnO}_4} = 0.00275 \text{ mol KMnO}_4$$

Now that the number of moles of substance is known, this can be combined with the volume of solution—*which must be in liters*—to give the molarity. Since 250. mL is equivalent to 0.250 L,

$$\text{molarity of KMnO}_4 = \frac{0.00275 \text{ mol KMnO}_4}{0.250 \text{ L solution}} = 0.0110 \text{ M}$$

The $KMnO_4$ concentration, $[KMnO_4]$, is 0.0110 M. This is useful information, but it is often equally useful to know the concentration of each type of ion in a solution. In Section 5.3 you learned that $KMnO_4$ is a strong electrolyte in water; that is, it dissociates completely into its ions, K^+ and MnO_4^- (Figure 6.10).

$$\underset{\text{100\% dissociation}}{KMnO_4(aq) \longrightarrow} K^+(aq) + MnO_4^-(aq)$$

The coefficients in the dissociation reaction show that 1 mol of $KMnO_4$ breaks up into 1 mol of K^+ and 1 mol of MnO_4^-. Accordingly, 0.0110 M $KMnO_4$ gives a concentration of K^+ in the solution of 0.0110 M; similarly, the concentration of MnO_4^- is 0.0110 M.

Another example of ion concentrations is provided by the dissociation of sodium carbonate, Na_2CO_3 (Figure 6.10).

$$Na_2CO_3(aq) \longrightarrow 2\ Na^+(aq) + CO_3^{2-}(aq)$$

100% dissociation

If one mole of Na_2CO_3 is dissolved in enough water to make one liter of solution, the concentration of the sodium ion will be $[Na^+] = 2$ M, since the dissociation of the compound releases two Na^+ ions for each mole of Na_2CO_3 added. Similarly, the concentration of the carbonate ion will be $[CO_3^{2-}] = 1$ M, and the total concentration of ions will be 3 M.

Chemists often use the expression "compound XY is 6 M in solution." However, if XY is an ionic compound, it will form X and Y ions on dissolving; XY does not exist as such in solution. Thus, the expression "compound XY is 6 M" is a way of saying that 6 moles of XY units are dissolved in enough water to make one liter of solution.

EXERCISE 6.6 • *Solution Molarity*

Sodium bicarbonate, $NaHCO_3$, is used in baking powder formulations, in fire extinguishers, and in the manufacture of plastics and ceramics, among other things. If you have 26.3 g of the compound and dissolve it in enough water to make 200. mL of solution, what is the molar concentration of $NaHCO_3$?

EXERCISE 6.7 • *Ion Concentrations in Solution*

Both HCl and Na_2SO_4 are strong electrolytes when dissolved in water. For each compound, write a balanced equation for its dissociation. Assume enough of each has been dissolved so that [HCl] = 1 M and $[Na_2SO_4] = 0.5$ M, and then state the molar concentration of each ion and the total concentration of all ions in each solution.

Preparing Solutions of Known Concentration

A situation a chemist often faces is the opposite of the preceding example. Instead of finding the concentration when the solution volume and solute mass are known, a given volume of solution of known concentration must be prepared. The problem is to find out what mass of solute to use.

Suppose you wish to prepare 2.00 L of a 1.5 M solution of Na_2CO_3. You are given a bottle of solid Na_2CO_3, some distilled water, and a 2.00-L volumetric flask. To make the solution, you must weigh the necessary quantity of Na_2CO_3 as accurately as possible, carefully place all the solid in the volumetric flask, and then add some water to dissolve the solid. After the solid has completely dissolved, you can add more water to bring the solution volume to 2.00 L. The solution will then have the desired concentration and the volume specified.

A volumetric flask is a special flask with a line marked on its neck (Figures 6.8 and 6.9). If it is filled with solution to this line (at 20 °C), the flask contains exactly the volume of solution specified.

But what mass of Na_2CO_3 is required to make the 2.00 L of 1.50 M Na_2CO_3? As usual, the number of moles of substance required must first be calculated.

$$2.00\ \cancel{L} \cdot \frac{1.50\ \text{mol}\ Na_2CO_3}{1.00\ \cancel{\text{L solution}}} = 3.00\ \text{mol}\ Na_2CO_3\ \text{required}$$

Now that you know the number of moles of Na_2CO_3 required, this can be converted to a mass in grams.

$$3.00\ \cancel{\text{mol}\ Na_2CO_3} \cdot \frac{106\ \text{g}\ Na_2CO_3}{1\ \cancel{\text{mol}\ Na_2CO_3}} = 318\ \text{g}\ Na_2CO_3$$

EXERCISE 6.8 • *Making Solutions of Known Concentration*

An experiment in your laboratory requires 500. mL of a 0.0200 M solution of $KMnO_4$. You are given a bottle of solid $KMnO_4$, some distilled water, and a 500. mL volumetric flask. Describe how you would go about making up the required solution.

Making a sodium carbonate solution as just described illustrates the most common way to make a solution of known concentration. However, another method is to *begin with a concentrated solution and add water to make it more dilute until the desired concentration is reached.* Many of the solutions prepared for your laboratory course are probably made by this *dilution* method. It is often most efficient to store a few liters of a concentrated solution and then add water to make it into many liters of a dilute solution.

As an example of the dilution method, suppose you need 1.00 L of 0.0025 M potassium dichromate, $K_2Cr_2O_7$, for use in analyzing another solution for its iron content. You have available a few liters of 0.100 M $K_2Cr_2O_7$ and some distilled water and glassware. How can you make the required 0.0025 M solution?

Just as water is added to the antifreeze in a car radiator to make the concentration of the antifreeze solution lower, the same can be done with the $K_2Cr_2O_7$ solution. The approach will be to take some of the more concentrated $K_2Cr_2O_7$ solution, put it in a flask, and then add water until the $K_2Cr_2O_7$ is contained in a larger volume of water, that is, until it is less concentrated (or more dilute).

The problem is to make a specified quantity of solution of known concentration. If the volume and concentration of the final solution are known,

Figure 6.11
Making a solution by dilution. (a) A 100.-mL volumetric flask is filled to the mark with 0.100 M $K_2Cr_2O_7$. (b) This is transferred to a 1.00-L volumetric flask. (c) The 1.00-L flask is then filled with distilled water to the mark on the neck, and the concentration of the now diluted solution is 0.0100 M. *(C.D. Winters)*

(a)

(b)

(c)

then the number of moles of solute it contains is also known. Therefore, the number of moles of $K_2Cr_2O_7$ that must be in the final dilute solution is

$$\text{amount of } K_2Cr_2O_7 \text{ in final solution} = 1.00 \text{ L} \cdot 0.0025 \text{ mol/L} = 0.0025 \text{ mol } K_2Cr_2O_7$$

A more concentrated solution containing this number of moles of $K_2Cr_2O_7$ must be placed in the flask and then diluted to a total volume of 1.00 L. But what volume of 0.100 M $K_2Cr_2O_7$ contains the required number of moles?

$$\text{volume required} = 0.0025 \cancel{\text{mol } K_2Cr_2O_7} \cdot \frac{1.00 \text{ L}}{0.100 \cancel{\text{mol}}} = 0.025 \text{ L or } 25 \text{ mL}$$

The answer is 25 mL of the more concentrated $K_2Cr_2O_7$ solution, which can easily be measured precisely as pictured in Figure 6.11.

A second look at the preceding argument suggests a simple way to remember how to do these calculations. The central idea is that the number of moles of $K_2Cr_2O_7$ in the final dilute solution has to be equal to the number of moles of $K_2Cr_2O_7$ taken from the more concentrated solution. If c is the concentration (molarity) and V is the volume, we can use subscripts to identify the dilute (d) and concentrated (c) solutions. Then the number of moles of solute in either solution can be calculated:

$$\text{amount of } K_2Cr_2O_7 \text{ in the final dilute solution} = c_dV_d = 0.0025 \text{ mol}$$

$$\text{amount of } K_2Cr_2O_7 \text{ taken from the more concentrated solution} = c_cV_c = 0.0025 \text{ mol}$$

Since both cV products are equal to the same number of moles, we can use the following equation.

$$c_cV_c = c_dV_d$$

moles of solute in concentrated solution = moles of solute in dilute solution

This equation is valid for all cases in which a more concentrated solution is used to make a more dilute one. It can be used, for example, to find the molarity of the dilute solution, c_d, from values of c_c, V_c, and V_d.

The equation $c_cV_c = c_dV_d$ can be applied to all dilution problems. Although you may be tempted to use it for the calculations involved in acid-base titrations, it is true only *in certain cases.*

EXAMPLE 6.5

Preparing a Solution by Dilution

Suppose you are doing an experiment to find out the quantity of iron in a vitamin pill. The method of analysis calls for the preparation of a standard solution of iron(III) by placing 1.00 mL of 0.236 M iron(III) nitrate in a volumetric flask and diluting it to exactly 100.0 mL. What is the concentration of the diluted iron(III) solution?

SOLUTION From the volume and molarity of the original iron solution, you can find the number of moles of iron(III) nitrate in the first solution.

$$1.00 \times 10^{-3} \text{ L} \cdot 0.236 \text{ mol/L} = 2.36 \times 10^{-4} \text{ mol of iron(III) nitrate}$$

This number of moles is also found in the 100.0 mL of dilute solution. Therefore, the new concentration is found to be

$$\frac{2.36 \times 10^{-4} \text{ mol of iron(III) nitrate}}{0.1000 \text{ L}} = 2.36 \times 10^{-3} \text{ M}$$

The solution was diluted by a factor of 100, so the new concentration is 100 times smaller than the original concentration.

EXERCISE 6.9 • *Preparing a Solution by Dilution*

An experiment calls for you to use 250. mL of 1.00 M NaOH, but you are given a large bottle of 2.00 M NaOH. Tell how you would make up the 1.00 M NaOH in the desired volume.

EXERCISE 6.10 • *Preparing a Solution by Dilution*

In one of your laboratory experiments you are given a solution of $CuSO_4$ that has a concentration of 0.15 M. If you mix 6.0 mL of this solution with enough water to have a total volume of 10.0 mL, what is the concentration of $CuSO_4$ in this new solution?

Figure 6.12
A commercial remedy for stomach acid, which contains a metal carbonate, reacts with aqueous hydrochloric acid, the acid present in the digestive system. The most obvious product is CO_2 gas. *(C.D. Winters)*

6.6 STOICHIOMETRY OF REACTIONS IN AQUEOUS SOLUTION

General Solution Stoichiometry

A common type of reaction is that between a metal carbonate and an aqueous acid to give a salt and CO_2 gas, the type of reaction that occurs when you take a popular remedy for an upset stomach (Figure 6.12).

$$\underset{\text{metal carbonate}}{CaCO_3(s)} + \underset{\text{acid}}{2\,HCl(aq)} \longrightarrow \underset{\text{salt}}{CaCl_2(aq)} + \underset{\text{carbon dioxide}}{CO_2(g)} + \underset{\text{water}}{H_2O(\ell)}$$

In such a case, we might want to know what mass in grams of $CaCO_3$ is required to react completely with 25 mL of 0.75 M HCl. This can be solved in the same way as all the stoichiometry problems you have seen so far, except that the amount of one reactant is given in volume and concentration units. It is therefore necessary first to find the number of moles of HCl,

$$0.025\ \cancel{\text{L HCl}} \cdot \frac{0.75 \text{ mol HCl}}{1\ \cancel{\text{L HCl}}} = 0.019 \text{ mol HCl}$$

and then relate this to the number of moles of $CaCO_3$ required.

$$0.019\ \cancel{\text{mol HCl}} \cdot \frac{1 \text{ mol } CaCO_3}{2\ \cancel{\text{mol HCl}}} = 0.0094 \text{ mol } CaCO_3$$

Finally, the number of moles of $CaCO_3$ is converted to a mass in grams.

$$0.0094 \text{ mol CaCO}_3 \cdot \frac{100. \text{ g CaCO}_3}{1 \text{ mol CaCO}_3} = 0.94 \text{ g CaCO}_3$$

Chemists do such calculations many times in the course of their work in research and product development. If you follow the general scheme outlined in Figure 6.2, and pay attention to the units of the numbers, you can successfully carry out any kind of stoichiometry calculations involving concentrations.

Zinc reacts readily with aqueous hydrochloric acid to give hydrogen gas and aqueous zinc chloride. *(C.D. Winters)*

EXAMPLE 6.6

Stoichiometry of a Reaction in Solution

Metallic zinc reacts with aqueous solutions of acids such as HCl, as do many other metals.

$$\text{Zn(s)} + 2 \text{ HCl(aq)} \longrightarrow \text{ZnCl}_2\text{(aq)} + \text{H}_2\text{(g)}$$

Such reactions are often used to produce hydrogen gas for laboratory uses. If you have 10.0 g of zinc, what volume of 2.50 M HCl (in milliliters) would you need for complete reaction?

SOLUTION The balanced equation for the reaction is given and shows that two moles of HCl are required for each one of zinc. Therefore, once you know the number of moles of zinc you have, then you can find the number of moles of HCl required and from that the volume of solution needed.

You can begin by calculating the number of moles of zinc available.

$$10.0 \text{ g Zn} \cdot \frac{1.00 \text{ mol Zn}}{65.39 \text{ g Zn}} = 0.153 \text{ mol Zn}$$

Knowing the number of moles of zinc, then you can use the stoichiometric factor "2 mol HCl/1 mol Zn" to find the number of moles of HCl required.

$$0.157 \text{ mol Zn} \cdot \frac{2 \text{ mol HCl}}{1 \text{ mol Zn}} = 0.306 \text{ mol HCl required}$$

Finally, knowing the number of moles of HCl required and the concentration of the acid, you can calculate the volume of acid required.

$$0.306 \text{ mol HCl} \cdot \frac{1.00 \text{ L solution}}{2.50 \text{ mol HCl}} = 0.122 \text{ L of HCl solution or 122 mL}$$

Since the answer is needed in milliliters, you can convert liters to milliliters as a final step.

EXERCISE 6.11 • *Solution Stoichiometry*

As described above, a common type of reaction is that between a metal carbonate and an aqueous acid to give a salt and CO_2 gas.

$$\text{Na}_2\text{CO}_3\text{(aq)} + 2 \text{ HCl(aq)} \longrightarrow 2 \text{ NaCl(aq)} + \text{CO}_2\text{(g)} + \text{H}_2\text{O}(\ell)$$

If you combine 50.0 mL of 0.450 M HCl and an excess of Na_2CO_3, what mass in grams of NaCl is produced?

Solving a Problem in Industrial Chemistry

A transportable acid recovery pilot-plant for transforming metal-bearing spent acids into reusable acid and a reclaimable metal salt. (The WADR pilot-plant developed for the Department of Energy by Pacific Northwest Laboratory courtesy of Viatec/Recovery Systems, Inc., Richland, Washington)

MANY OF THE REACTIONS YOU have seen in Chapter 5 and in this chapter involve acids. For instance, in Example 6.6 you saw how zinc reacts with hydrochloric acid. One reason for the emphasis on acid chemistry is that they are some of the most important chemicals in our economy. Five acids are listed among the top 50 chemicals produced in the United States in 1992.

Acid	*Billions of pounds*	*Rank*
Sulfuric acid	88.80	1
Phosphoric acid	25.36	7
Nitric acid	16.08	13
Hydrochloric acid	5.75	26
Acetic acid	3.60	35

Over 15,000 companies use these acids to make other chemicals, to clean and refinish metals, to plate metals onto other metals or onto plastics, and in many other applications. A problem faced by all of these industries is what to do with acid-containing waste. For example, when acids are used to wash a metal surface, the washings contain unused acid along with ions of metals such as copper(II), vanadium(II), silver(I), nickel(II), and lead(II). It is estimated that over 8 billion pounds of acid-containing wastes are generated annually, and they cannot simply be flushed into the nearest lake or river. Not only will the acid damage aquatic life, but heavy metals are toxic to plants and animals.

Because of the enormous quantity of acid wastes generated every year, many laboratories have been involved in finding a way to recover unused acids and to remove heavy metals and other dissolved substances. A process developed at the Department of Energy's Pacific Northwest Laboratory holds promise for significantly reducing the volume and toxicity of acid waste.

Typically the waste from a chemical or metallurgical operation may contain sulfuric, phosphoric, and nitric acids, along with some heavy metal ions dissolved in the aqueous acid. When the mixture is heated, the acids vaporize, and the heavy metal ions remain in the liquid phase. The vapor is purified to yield a clean acid, in some cases cleaner than industrial-grade acids. The solution containing heavy metals is collected in a tank, and the metal ions are removed by adding salts of anions that form precipitates with the heavy metal ions. The metals can be reclaimed from these precipitates and can be sold or reused.

It is claimed that this new approach to recovering and detoxifying waste acids can lead to a significant cost saving for chemical industries. Furthermore, it is a technology that can be adapted to both large and small producers of waste. A large steel company needs a system to process thousands of gallons of waste a week, while a small company plating metals or plastic may need to clean only a few hundred gallons.

Titrations

Your study of stoichiometry so far should have convinced you that if you know (1) the balanced equation for a reaction and (2) the quantity of one of the reactants, then you can calculate the quantity of any other substance consumed or produced in the reaction. This is the essence of any technique of **quantitative chemical analysis,** the determination of the *quantity* of a given constituent in a mixture.

Another requirement for a reaction to be useful as an analytical method is that it must go to completion in a reasonable length of time.

__Qualitative analysis__ is the determination of the identity of the constituents of a mixture. __Quantitative analysis__ is the determination of the quantity of a constituent in a mixture.

An acid isolated from clover leaves was shown in Example 6.4 to have the molecular formula $H_2C_2O_4$. This compound, called oxalic acid, is commercially important in manufacturing paint and textiles, in metal treatment, and in photography. Its main use is to remove calcium ions from aqueous solutions by forming insoluble calcium oxalate.

$$Ca^{2+}(aq) + H_2C_2O_4(aq) + 2\,H_2O(\ell) \longrightarrow CaC_2O_4(s) + 2\,H_3O^+(aq)$$

Unfortunately, this reaction can also occur in your body and lead to kidney stones.

Suppose a textile company has received a shipment of impure oxalic acid powder. As an analytical chemist working for the company, your job is to analyze the powder to determine its purity. Because the compound is an acid, it reacts with the base sodium hydroxide in aqueous solution according to the balanced equation

$$H_2C_2O_4(aq) + 2\,NaOH(aq) \longrightarrow Na_2C_2O_4(aq) + 2\,H_2O(\ell)$$

Therefore, you can tell how much oxalic acid is present in a given mass of powder if all of the following conditions are met in the reaction with NaOH:

1. You can tell when the amount of sodium hydroxide added is *just enough* to react with *all* of the oxalic acid present in solution.
2. You know the volume of the sodium hydroxide solution added at the point of complete reaction.
3. You know the concentration of the sodium hydroxide solution.

These conditions are fulfilled in a **titration,** a procedure illustrated in the series of photographs in Figure 6.13. The solution containing oxalic acid is placed in a flask along with an acid-base indicator. An **indicator** is a dye that changes color when the reaction used for analysis is complete. In this case, the dye is colorless in acid solution but pink in basic solution. Aqueous sodium hydroxide of accurately known concentration is placed in a *buret*, a measuring cylinder that most commonly has a volume of 50.0 mL and is calibrated in 0.1-mL divisions. As the sodium hydroxide in the buret is added slowly to the acid solution in the flask, the acid reacts with the base according to the net ionic equation

The "Chemistry You Can Do" experiment in Chapter 5 involved making an acid-base indicator from red cabbage juice.

$$H_2C_2O_4(aq) + 2\,OH^-(aq) \longrightarrow C_2O_4^{2-}(aq) + 2\,H_2O(\ell)$$

As long as some acid is present in the solution, all the base supplied from the buret is consumed, and the indicator remains colorless. However, at some point the number of moles of OH^- added will exactly equal the number of moles of H^+ that can be supplied by the acid. This is called the **equivalence point.** To indicate when this point has been reached, an **acid-**

Oxalic acid is a relatively weak acid; that is, it ionizes to a relatively small extent.

(a)

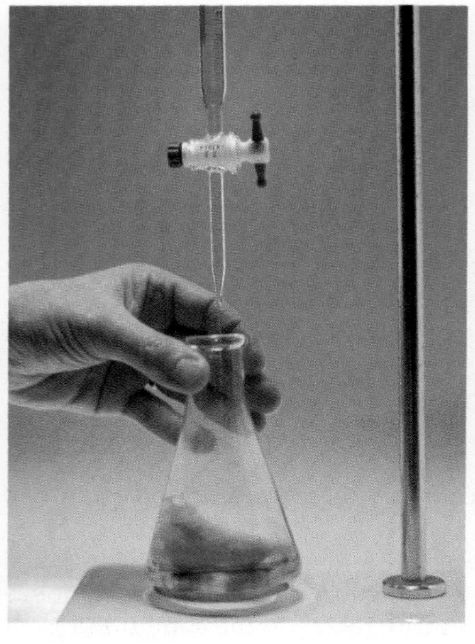
(b)

(c)

Figure 6.13
Titration of an acid in aqueous solution with a base. (a) A buret, a volumetric measuring device calibrated in divisions of 0.1 mL, is filled with an aqueous solution of a base of known concentration. (b) Base is added slowly from the buret to the solution. (c) A change in color of an indicator signals the equivalence point. *(C.D. Winters)*

In a titration, before the equivalence point is reached, the substance being added from the buret is the limiting reactant. After the equivalence point, the substance added from the buret is in excess.

base indicator, the dye already mentioned, was added to the solution prior to titration. As soon as an excess of base has been added, the solution becomes basic, and the dye changes color (Figure 6.14).

When the equivalence point has been reached in a titration, the volume of base added since the beginning of the titration can be determined by reading the calibrated buret (Figure 6.13). From this volume and the concentration of the base, the number of moles of base used can be found.

$$\text{amount of base added} = \text{molarity of base (mol/L)} \cdot \text{volume of base (L)}$$

Then, using the stoichiometric factor from the balanced equation, the number of moles of base added is related to the number of moles of acid present in the original sample.

*There is a subtle difference between the $[H^+]$ at which the color changes—called the **endpoint** of the titration—and the equivalence point. For our purposes, the difference is negligible.*

Now, let us illustrate this procedure by analyzing the impure oxalic acid received by your textile company. Suppose you dissolve a 1.034-g sample in some water, add an acid-base indicator, and titrate it with 0.485 M NaOH. The sample requires addition of 34.47 mL of the NaOH solution to reach the equivalence point. What is the mass of oxalic acid and what is its weight percent in the sample?

To find the mass of oxalic acid, or its concentration, requires knowing how many moles of acid were present. This is certainly not given directly here, but the number of moles of NaOH consumed in the reaction can be found from the volume and molarity of the NaOH solution. Therefore, re-

Figure 6.14
The juice of a red cabbage turns color when the acidity of a solution changes. When the solution is highly acidic, the juice gives the solution a red color. As the solution becomes less acidic (more basic), the color changes from red to violet to yellow. *(C.D. Winters)*

membering that the volume of NaOH must be in liters, we use it to find the number of moles of NaOH.

$$0.03447\,\cancel{\text{L}} \cdot \frac{0.485\text{ mol NaOH}}{1\,\cancel{\text{L}}} = 0.0167\text{ mol NaOH}$$

Now, using the appropriate stoichiometric factor, the number of moles of oxalic acid in the sample can be found,

$$0.0167\,\cancel{\text{mol NaOH}} \cdot \frac{1\text{ mol } H_2C_2O_4}{2\,\cancel{\text{mol NaOH}}} = 0.00836\text{ mol } H_2C_2O_4$$

and this leads to the mass of the acid.

$$0.00836\,\cancel{\text{mol } H_2C_2O_4} \cdot \frac{90.04\text{ g } H_2C_2O_4}{1\,\cancel{\text{mol } H_2C_2O_4}} = 0.753\text{ g } H_2C_2O_4$$

Finally, this mass of oxalic acid represents 72.8% of the total sample mass:

$$\frac{0.753\text{ g } H_2C_2O_4}{1.034\text{ g sample}} \cdot 100\% = 72.8\%\ H_2C_2O_4$$

In this example the concentration of the base, NaOH, was given. In real life this usually has to be determined by a prior measurement. The procedure by which the concentration of an analytical reagent is determined is called **standardization,** and there are two general approaches.

One approach is to weigh accurately a sample of a pure, solid acid or base (known as a **primary standard**) and then titrate this sample with a solution of the base or acid to be standardized. For example, solutions of acids such as HCl can be standardized by using them to titrate a base such as Na_2CO_3, a solid that can be obtained in pure form, that can be weighed accurately, and that reacts completely with an acid. Suppose that 0.263 g of

A substance used as the "reference standard" in the standardization procedure is called a "primary standard." It is a substance that can be obtained in very pure form and whose chemistry is well understood.

Na_2CO_3 requires 28.35 mL of aqueous HCl for titration to the equivalence point. What is the concentration of the HCl? Since the goal is to find the number of moles of HCl per liter, what is really needed is the number of moles of HCl in 28.35 mL of the HCl solution. Then that number of moles can be divided by 0.02835 L to find the concentration in moles/L.

As usual, the balanced equation for the reaction is written first.

$$Na_2CO_3(aq) + 2\ HCl(aq) \longrightarrow 2\ NaCl(aq) + CO_2(g) + H_2O(\ell)$$

Next, the mass of Na_2CO_3 is converted to moles.

$$0.263\ \cancel{g\ Na_2CO_3} \cdot \frac{1.00\ mol\ Na_2CO_3}{106.0\ \cancel{g\ Na_2CO_3}} = 0.00248\ mol\ Na_2CO_3$$

Knowing the number of moles of Na_2CO_3, we can now find the number of moles of HCl in the solution by using the appropriate stoichiometric factor.

$$0.00248\ \cancel{mol\ Na_2CO_3} \cdot \frac{2\ mol\ HCl\ required}{1\ \cancel{mol\ Na_2CO_3}\ available} = 0.00496\ mol\ HCl$$

The 28.35-mL (or 0.02835-L) sample of aqueous HCl therefore contains 0.00496 mol of HCl, so the concentration of the HCl solution is 0.175 M.

$$[HCl] = \frac{0.00496\ mol\ HCl}{0.02835\ L} = 0.175\ M$$

You might think that solid NaOH would make a good primary standard. However, solid NaOH is difficult to weigh accurately because it is deliquescent, *that is, it absorbs water from humid air. Furthermore, its solutions readily take up CO_2 from the air, thus changing the concentration of OH^-:*
$2\ NaOH(aq) + CO_2(g) \longrightarrow Na_2CO_3(aq) + H_2O(\ell)$

Another approach to standardizing a solution is to titrate it with another solution that is already standardized. This is often done with standard solutions purchased from chemical supply companies, and an exercise at the end of this section illustrates that approach.

Acid-base titrations are extremely useful for quantitative chemical analysis, and some questions at the end of this chapter illustrate their scope. You should not get the impression, however, that a titration can be done only with an acid reacting with a base. Oxidation-reduction reactions, another class of reactions described in Chapter 5, lend themselves very well to chemical analysis by titration because many of these reactions go rapidly to completion in aqueous solution and there are ways of finding their equivalence points. As an example, consider the analysis of an ore for its iron content. In this case, the iron in the ore can be converted completely to the iron(II) ion, Fe^{2+}, in aqueous solution, and this solution can then be titrated with potassium permanganate, $KMnO_4$. The balanced net ionic equation for the reaction that occurs in the course of the titration is

$$\underset{\substack{\text{purple} \\ \text{oxidizing agent}}}{MnO_4^-(aq)} + \underset{\substack{\text{colorless} \\ \text{reducing agent}}}{5\ Fe^{2+}(aq)} + \underset{\text{colorless}}{8\ H_3O^+(aq)} \longrightarrow \underset{\text{colorless}}{Mn^{2+}(aq)} + \underset{\text{pale yellow}}{5\ Fe^{3+}(aq)} + 12\ H_2O(\ell)$$

This is a useful analytical reaction, because it is easy to detect when all the iron(II) has reacted. The MnO_4^- ion is deep purple, but when it reacts with Fe^{2+} the color disappears because the reaction product, the Mn^{2+} ion, is colorless. Thus, as $KMnO_4$ is added from a buret, the purple color disappears as the solutions mix. When all the Fe^{2+} has been converted to Fe^{3+},

The Chemical World

The Mole

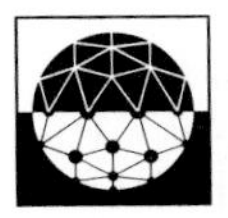

The mole concept is the basis for quantitative chemistry. It also provides the foundation for some highly significant industrial processes, like the preparation of intravenous (IV) solutions so important to health.

The Baxter-Travenol Corporation produces very large amounts of IV solutions for hospital use. This company as well as the Food and Drug Administration recognize that the molecular concentrations of sodium chloride and glucose (a simple sugar used as a nutrient) are critical because improper concentrations can result in cell destruction.

The solutions are initially prepared in approximately correct concentrations in very large quantities by mixing materials of known weight with known volumes of distilled water. Once mixed, a key process is then the certification of the concentrations. Here the mole concept enters the picture.

An exact amount of solution is withdrawn from the batch and a titration method for determining chloride is used. A standard solution of silver nitrate is carefully withdrawn from a buret until all the chloride from the sodium chloride is precipitated as silver chloride.

$$AgNO_3(aq) + NaCl(aq) \longrightarrow AgCl(s) + NaNO_3(aq)$$

At the end of the reaction, a special indicator that reacts with the first drop of excess silver nitrate produces a salmon-pink color. From the volume of silver nitrate the chemist knows how many moles of silver nitrate were consumed because one mole of silver nitrate reacts with one mole of sodium chloride in the reaction. Therefore, the chemist can immediately determine the number of moles of sodium chloride and thus its concentration in the IV solution. (*The World of Chemistry*, Program 11, "The Mole.")

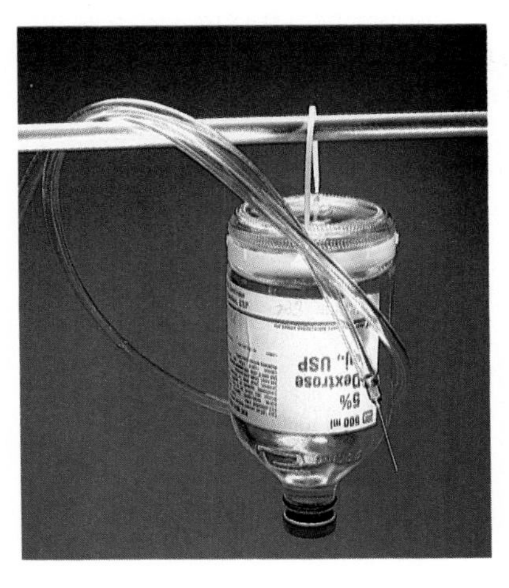

Intravenous, or IV, solutions must be prepared very carefully and exactly to maintain the stability of blood and other body fluids. (Charles Steele)

---

any additional $KMnO_4$ will give the solution a permanent purple color. Therefore, $KMnO_4$ solution is added from the buret until the initially colorless, Fe^{2+}-containing solution just turns a faint purple color, the signal that the equivalence point has been reached.

Let us assume that a 1.026-g sample of iron-containing ore requires 24.34 mL of 0.0195 M $KMnO_4$ to reach the equivalence point. What is the weight percent of iron in the ore? Since the volume and concentration of the $KMnO_4$ solution are known, the number of moles of $KMnO_4$ used in the titration can be calculated. Remembering first to change the volume to liters, we have

$$0.02434\,\cancel{L} \cdot \frac{0.0195 \text{ mol}}{1\,\cancel{L}} = 0.000475 \text{ mol KMnO}_4$$

Based on the balanced chemical equation for the reaction of $KMnO_4$ with Fe^{2+}, the number of moles of iron that were present in the solution (and in the ore sample) can be calculated.

$$0.000475\,\cancel{\text{mol KMnO}_4} \cdot \frac{5 \text{ mol Fe}^{2+}}{1\,\cancel{\text{mol KMnO}_4}} = 0.00237 \text{ mol Fe}^{2+}$$

The mass of iron is calculated from this,

$$0.00237 \text{ mol Fe} \cdot \frac{55.85 \text{ g Fe}}{1 \text{ mol Fe}} = 0.133 \text{ g iron}$$

and, finally, the weight percent of iron in the ore can be calculated.

$$\frac{0.133 \text{ g iron}}{1.026 \text{ g sample}} \cdot 100\% = 12.9\% \text{ iron}$$

Oxidation-reduction reactions, as well as acid-base reactions, are clearly useful in chemical analysis. However, no matter what type of reaction is used in analysis, you must remember that *before you use any reaction as a quantitative analytical method, you must know the balanced chemical equation* (or at least the stoichiometric relation between the key reactants and products) *in order to know the necessary stoichiometric factors.*

EXAMPLE 6.7

Acid-Base Titration

Acid-base titrations can be used in the analysis of mixtures in which one component is an acid. Tartaric acid, $C_2H_4O_2(CO_2H)_2$, occurs naturally in many fruits. Suppose you are given a sample of a mixture containing tartaric acid and other (chemically inert) solids. A 0.873-g sample of the mixture requires 25.78 mL of 0.423 M NaOH for titration to the equivalence point according to the following reaction.

$$C_2H_4O_2(CO_2H)_2(aq) + 2\,NaOH(aq) \longrightarrow Na_2C_2H_4O_2(CO_2)_2(aq) + 2\,H_2O(\ell)$$

What is the weight percent of tartaric acid in the mixture?

SOLUTION The volume and concentration of the NaOH solution used in the titration are known. From this, you can calculate the number of moles of NaOH needed to react with all the tartaric acid. Remembering to change milliliters of NaOH solution into liters, you have

$$0.02578 \text{ L NaOH} \cdot \frac{0.423 \text{ mol NaOH}}{1 \text{ L}} = 0.0109 \text{ mol NaOH}$$

Now the number of moles of tartaric acid in the unknown sample can be calculated from the number of moles of NaOH consumed in the titration.

$$0.0109 \text{ mol NaOH} \cdot \frac{1 \text{ mol } C_2H_4O_2(CO_2H)_2}{2 \text{ mol NaOH}} = 0.00545 \text{ mol } C_2H_4O_2(CO_2H)_2$$

Finally, the mass of tartaric acid can be calculated from moles of the substance,

$$0.00545 \text{ mol } C_2H_4O_2(CO_2H)_2 \cdot \frac{150.1 \text{ g } C_2H_4O_2(CO_2H)_2}{1 \text{ mol } C_2H_4O_2(CO_2H)_2} = 0.818 \text{ g } C_2H_4O_2(CO_2H)_2$$

and then the weight percent of the acid in the sample can be determined.

$$\frac{0.818 \text{ g } C_2H_4O_2(CO_2H)_2}{0.873 \text{ g sample}} \cdot 100\% = 93.7\% \; C_2H_4O_2(CO_2H)_2$$

EXERCISE 6.12 • *Acid-Base Titration*

A 25.0-mL sample of vinegar requires 28.33 mL of a 0.953 M solution of NaOH for titration to the equivalence point. What mass in grams of acetic acid is in the vinegar sample and what is the concentration of acetic acid in the vinegar?

$$\underset{\text{acetic acid}}{CH_3CO_2H(aq)} + NaOH(aq) \longrightarrow \underset{\text{sodium acetate}}{NaCH_3CO_2(aq)} + H_2O(\ell)$$

EXERCISE 6.13 • *Standardization of a Base*

Hydrochloric acid, HCl, can be purchased from chemical supply houses in a solution with a concentration of 0.100 M, so such a solution can be used to standardize the solution of a base. If titrating 25.00 mL of a sodium hydroxide solution to the equivalence point requires 29.67 mL of 0.100 M HCl, what is the concentration of the base?

IN CLOSING

Having studied this chapter, you should be able to

- calculate the mass of one reactant or product from the mass of another reactant or product by using the balanced chemical equation (Section 6.1).
- determine which of two reactants is the limiting reactant (Section 6.2).
- explain the differences among actual yield, theoretical yield, and percent yield, and calculate theoretical and percent yields (Section 6.3).
- use stoichiometry principles in the chemical analysis of a mixture or to find the empirical formula of an unknown compound (Section 6.4).
- define concentration (molarity) and calculate concentrations (Section 6.5).
- describe how to prepare a solution of a given molarity from the solute and water or by dilution from a more concentrated solution (Section 6.5).
- solve stoichiometry problems using solution concentrations (Section 6.6).
- explain how a titration is carried out, explain standardization, and calculate concentrations or amounts of reactants from titration data (Section 6.6).

STUDY QUESTIONS

General Stoichiometry

1. If a 10.0-g mass of carbon is combined with an exact stoichiometric amount of oxygen (26.6 g) to get carbon dioxide, what mass in grams of CO_2 can be isolated? That is, what is the theoretical yield of CO_2?
2. Assume 16.04 g of methane, CH_4, is burned in pure oxygen.
   (a) What are the products of the reaction?
   (b) What is the balanced equation for this reaction?
   (c) What mass of O_2, in grams, is required for complete combustion?
   (d) What total mass of products, in grams, is expected?
3. If you want to synthesize 1.45 g of the semiconducting material GaAs, what masses of Ga and of As, in grams, are required?
4. Nitrogen monoxide is oxidized in air to give brown nitrogen dioxide.

$$2\ NO(g) + O_2(g) \longrightarrow 2\ NO_2(g)$$

Starting with 2.2 mol of NO, how many moles and how many grams of O_2 are required for complete reaction? What mass of NO_2, in grams, is produced?

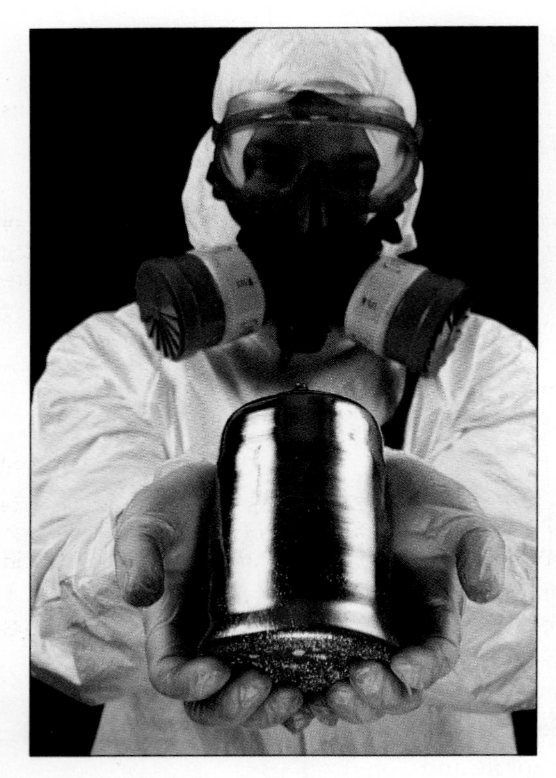

A freshly prepared ingot of the semiconductor gallium arsenide. This material is a much better electrical conductor than silicon. *(H. Morgan/Science Source/Photo Researchers)*

5. Aluminum reacts with oxygen to give aluminum oxide.

$$4\ Al(s) + 3\ O_2(g) \longrightarrow 2\ Al_2O_3(s)$$

If you have 6.0 mol of Al, how many moles of O_2 are needed for complete reaction? What mass of Al_2O_3, in grams, is produced?

6. Many metals react with halogens to give metal halides. For example, iron gives iron(II) chloride, $FeCl_2$.

$$Fe(s) + Cl_2(g) \longrightarrow FeCl_2(s)$$

Beginning with 10.0 g of iron, what mass of Cl_2, in grams, is required for complete reaction? What quantity of $FeCl_2$, in moles and in grams, is expected?

7. Like many metals, aluminum reacts with a halogen to give a metal halide (see Figure 5.1).

$$2\ Al(s) + 3\ Br_2(\ell) \longrightarrow Al_2Br_6(s)$$

If you begin with 2.56 g of Al, what mass in grams of Br_2 is required for complete reaction? What mass in grams of the white solid Al_2Br_6 is expected?

8. The final step in the manufacture of platinum metal (for use in automotive catalytic converters and other products) is the reaction

$$3\ (NH_4)_2PtCl_6(s) \longrightarrow 3\ Pt(s) + 2\ NH_4Cl(s) + 2\ N_2(g) + 16\ HCl(g)$$

If 12.35 g of $(NH_4)_2PtCl_6$ is heated, what mass of platinum metal, in grams, is expected? What mass of HCl would be obtained as well?

9. The equation for one of the reactions in the process of turning iron ore into the metal is

$$Fe_2O_3(s) + 3\ CO(g) \longrightarrow 2\ Fe(s) + 3\ CO_2(g)$$

(a) What is the maximum mass of iron, in grams, that can be obtained from 1.00 pound (454 g) of iron(III) oxide? (b) What mass of CO is required to reduce the iron(III) oxide to iron metal?

10. Many metal halides react with water to produce the metal oxide (or hydroxide) and the appropriate hydrogen halide. For example,

$$TiCl_4(\ell) + 2\ H_2O(g) \longrightarrow TiO_2(s) + 4\ HCl(g)$$

(See the photograph of this reaction in Example 6.2.) (a) Name the four compounds involved in this reaction. (b) If you begin with 14.0 g of $TiCl_4$, what mass of water, in grams, is required for complete reaction? (c) What mass of each product is expected?

11. Gaseous sulfur dioxide, SO_2, can be removed from smokestacks by treatment with limestone and oxygen (see photo on page 177).

$$2\ SO_2(g) + 2\ CaCO_3(s) + O_2(g) \longrightarrow 2\ CaSO_4(s) + 2\ CO_2(g)$$

(a) Name the compounds involved in this reaction. (b) What mass of $CaCO_3$ is required to remove 150. g of SO_2? (c) What mass of $CaSO_4$ is formed when 150. g of SO_2 is consumed completely?

12. Iron reacts with oxygen to give iron(III) oxide, Fe_2O_3. (a) Write a balanced equation for this reaction. (b) If an ordinary iron nail (assumed to be pure iron) has a mass of 5.58 g, what mass in grams of Fe_2O_3 would be produced if the nail is converted completely to this oxide? (c) What mass of O_2 (in grams) is required for the reaction?

13. Careful decomposition of ammonium nitrate, NH_4NO_3, gives laughing gas (dinitrogen monoxide, N_2O) and water. (a) Write a balanced equation for this reaction. (b) Beginning with 10.0 g of NH_4NO_3, what masses of N_2O and water are expected?

14. Cisplatin, $Pt(NH_3)_2Cl_2$, can be made by the reaction of K_2PtCl_4 with ammonia, NH_3. Besides cisplatin, the other product is KCl. (a) Write a balanced equation for this reaction. (b) In order to obtain 2.50 g of cisplatin, what masses in grams of K_2PtCl_4 and ammonia do you need?

Limiting Reactant

15. Aluminum chloride, $AlCl_3$, is an inexpensive reagent used in many industrial processes. It is made by treating scrap aluminum with chlorine according to the following balanced equation.

$$2\,Al(s) + 3\,Cl_2(g) \longrightarrow 2\,AlCl_3(s)$$

(a) Which reactant is limiting if 2.70 g of Al and 4.05 g of Cl_2 are mixed? (b) What mass of $AlCl_3$ can be produced? (c) What mass of the excess reactant will remain when the reaction is completed?

16. The reaction of methane and water is one way to prepare hydrogen.

$$CH_4(g) + 2\,H_2O(g) \longrightarrow CO_2(g) + 4\,H_2(g)$$

If you begin with 995 g of CH_4 and 2510 g of water, what is the maximum possible yield of H_2?

17. Methyl alcohol, CH_3OH, is a clean-burning, easily handled fuel. It can be made by the direct reaction of CO and H_2 (obtained from heating coal with steam).

$$CO(g) + 2\,H_2(g) \longrightarrow CH_3OH(\ell)$$

Starting with a mixture of 12.0 g of H_2 and 74.5 g of CO, which is the limiting reactant? What mass of the excess reactant, in grams, is left after reaction is complete? What mass of methyl alcohol can be obtained in theory?

18. Disulfur dichloride, S_2Cl_2, is used to vulcanize rubber. It can be made by treating molten sulfur with gaseous chlorine.

$$S_8(\ell) + 4\,Cl_2(g) \longrightarrow 4\,S_2Cl_2(g)$$

Starting with a mixture of 32.0 g of sulfur and 71.0 g of Cl_2, which is the limiting reactant? What mass of S_2Cl_2, in grams, can be produced? What mass of which starting material remains after the maximum amount of S_2Cl_2 has been formed?

19. Ammonia gas can be prepared by the following reaction:

$$CaO(s) + 2\,NH_4Cl(s) \longrightarrow 2\,NH_3(g) + H_2O(g) + CaCl_2(s)$$

If 112 g of CaO and 224 g of NH_4Cl are mixed, what is the maximum possible yield of NH_3?

20. Aspirin is produced by the reaction of salicylic acid and acetic anhydride.

$$\underset{\text{salicylic acid}}{2\,C_7H_6O_3(s)} + \underset{\text{acetic anhydride}}{C_4H_6O_3(\ell)} \longrightarrow \underset{\text{aspirin}}{2\,C_9H_8O_4(s)} + H_2O(\ell)$$

If you mix 100. g of each of the reactants, what is the maximum mass of aspirin that can be obtained?

Percent Yield

21. Ammonia gas can be prepared by the reaction of a basic oxide such as calcium oxide with ammonium chloride, an acidic salt.

$$CaO(s) + 2\,NH_4Cl(s) \longrightarrow 2\,NH_3(g) + H_2O(g) + CaCl_2(s)$$

If exactly 100. g of ammonia is isolated, but the theoretical yield is 136 g, what is the percent yield of this gas?

22. Diborane, B_2H_6, is valuable for the synthesis of new organic compounds. The boron compound can be made by the reaction

$$2\,NaBH_4(s) + I_2(s) \longrightarrow B_2H_6(g) + 2\,NaI(s) + H_2(g)$$

Suppose you use 1.203 g of $NaBH_4$ and excess iodine, and you isolate 0.295 g of B_2H_6. What is the percent yield of B_2H_6?

23. The reaction of zinc and chlorine has been used as the basis of a car battery.

$$Zn(s) + Cl_2(g) \longrightarrow ZnCl_2(s)$$

What is the theoretical yield of $ZnCl_2$ if 35.5 g of zinc is allowed to react with excess chlorine? If only 65.2 g of zinc chloride is isolated, what is the percent yield of the compound?

24. Disulfur dichloride, which has a revolting smell, can be prepared by directly combining S_8 and Cl_2, but it can also be made by the reaction

$$3\,SCl_2(\ell) + 4\,NaF(s) \longrightarrow SF_4(g) + S_2Cl_2(\ell) + 4\,NaCl(s)$$

Assume you begin with 5.23 g of SCl_2 and excess NaF. What is the theoretical yield of S_2Cl_2? If only 1.19 g of S_2Cl_2 is isolated, what is the percent yield of the compound?

Chemical Analysis

25. A mixture of $CuSO_4$ and $CuSO_4 \cdot 5\,H_2O$ has a mass of 1.245 g, but, after heating to drive off all the water, the mass is only 0.832 g. What is the weight percent of $CuSO_4 \cdot 5\,H_2O$ in the mixture?

Dehydrating hydrated copper sulfate. *(C.D. Winters)*

26. A sample of limestone and other soil materials is heated, and the limestone decomposes to give calcium oxide and carbon dioxide. A 1.506-g sample of limestone-containing material gives 0.711 g of CaO, in addition to gaseous CO_2, after being strongly heated. What was the weight percent of $CaCO_3$ in the original sample?

27. A 1.25-g sample contains some of the very reactive compound $Al(C_6H_5)_3$. On treating the compound with HCl, 0.951 g of C_6H_6 is isolated.

$$Al(C_6H_5)_3(s) + 3\,HCl(aq) \longrightarrow AlCl_3(aq) + 3\,C_6H_6(\ell)$$

What was the weight percent of $Al(C_6H_5)_3$ in the original 1.25-g sample?

28. Bromine trifluoride reacts with metal oxides to evolve oxygen quantitatively. For example,

$$3\,TiO_2(s) + 4\,BrF_3(\ell) \longrightarrow 3\,TiF_4(s) + 2\,Br_2(\ell) + 3\,O_2(g)$$

Suppose you wish to use this reaction to determine the weight percent of TiO_2 in a sample of ore. To do this, the O_2 gas from the reaction is collected. If 2.367 g of the TiO_2-containing ore evolves 0.143 g of O_2, what is the weight percent of TiO_2 in the sample?

Determination of Empirical Formulas

29. Styrene, the building block of polystyrene, is a hydrocarbon (a compound consisting only of C and H). If 0.438 g of the compound is burned and produces 1.481 g of CO_2 and 0.303 g of H_2O, what is the empirical formula of the compound?

30. Mesitylene is a liquid hydrocarbon. If 0.115 g of the compound is burned in pure O_2 to give 0.379 g of CO_2 and 0.1035 g of H_2O, what is the empirical formula of the compound?

31. Propionic acid, an organic acid, contains only C, H, and O. If 0.236 g of the acid burns completely in O_2 and gives 0.421 g of CO_2 and 0.172 g of H_2O, what is the empirical formula of the acid?

32. Quinone, which is used in the dye industry and in photography, is an organic compound containing only C, H, and O. What is the empirical formula of the compound if 0.105 g of the compound gives 0.257 g of CO_2 and 0.0350 g of H_2O when burned completely?

Solution Concentration

33. Assume 6.73 g of Na_2CO_3 is dissolved in enough water to make 250. mL of solution. What is the molarity of the sodium carbonate? What are the concentrations of the Na^+ and CO_3^{2-} ions?

34. Some $K_2Cr_2O_7$, with a mass of 2.335 g, is dissolved in enough water to make 500. mL of solution. What is the molarity of the potassium dichromate? What are the concentrations of the K^+ and $Cr_2O_7^{2-}$ ions?

35. What is the mass, in grams, of solute in 250. mL of a 0.0125 M solution of $KMnO_4$?

36. What is the mass, in grams, of solute in 100. mL of a 1.023×10^{-3} M solution of Na_3PO_4? What are the concentrations of the Na^+ and PO_4^{3-} ions?

37. What volume of 0.123 M NaOH, in milliliters, contains 25.0 g of NaOH?

38. What volume of 2.06 M $KMnO_4$, in liters, contains 322 g of solute?

39. If 6.00 mL of 0.0250 M $CuSO_4$ is diluted to 10.0 mL with pure water, what is the concentration of copper(II) sulfate in the diluted solution?

40. If you dilute 25.0 mL of 1.50 M HCl to 500. mL, what is the molar concentration of the diluted HCl?

41. If you need 1.00 L of 0.125 M H_2SO_4, which method would you use to prepare this solution?

(a) Dilute 36.0 mL of 1.25 M H_2SO_4 to a volume of 1.00 L

(b) Dilute 20.8 mL of 6.00 M H_2SO_4 to a volume of 1.00 L
(c) Add 950. mL of water to 50.0 mL of 3.00 M H_2SO_4
(d) Add 500. mL of water to 500. mL of 0.500 M H_2SO_4

42. If you need 300. mL of 0.500 M K_2CrO_7, which method would you use to prepare this solution?
(a) Dilute 250. mL of 0.600 M K_2CrO_7 to 300. mL
(b) Add 50.0 mL of water to 250. mL of 0.250 M K_2CrO_7
(c) Dilute 125 mL of 1.00 M K_2CrO_7 to 300. mL
(d) Add 30.0 mL of 1.50 M K_2CrO_7 to 270. mL of water

43. You have a 0.12 M solution of $BaCl_2$. What ions exist in solution and what are their concentrations?

44. A solution is 0.25 M in $(NH_4)_2SO_4$. What ions exist in solution and what are their concentrations?

Stoichiometry in Solution

45. What mass in grams of Na_2CO_3 is required for complete reaction with 25.0 mL of 0.155 M HNO_3?

$$Na_2CO_3(aq) + 2\ HNO_3(aq) \longrightarrow 2\ NaNO_3(aq) + CO_2(g) + H_2O(\ell)$$

46. What volume in milliliters of 0.125 M HNO_3 is required to react completely with 1.30 g of $Ba(OH)_2$?

$$2\ HNO_3(aq) + Ba(OH)_2(s) \longrightarrow Ba(NO_3)_2(aq) + 2\ H_2O(\ell)$$

47. One of the most important industrial processes in our economy is the electrolysis of brine solutions (aqueous solutions of NaCl). That is, when an electric current is passed through an aqueous solution of salt, the NaCl and water produce $H_2(g)$, $Cl_2(g)$, and NaOH.

$$2\ NaCl(aq) + 2\ H_2O(\ell) \longrightarrow H_2(g) + Cl_2(g) + 2\ NaOH(aq)$$

What mass of NaOH can be formed beginning with 10.0 L of 0.15 M NaCl?

48. Diborane, B_2H_6, can be produced by the following reaction.

$$2\ NaBH_4(aq) + H_2SO_4(aq) \longrightarrow 2\ H_2(g) + Na_2SO_4(aq) + B_2H_6(g)$$

What volume in milliliters of 0.0875 M H_2SO_4 should be used to completely consume 1.35 g of $NaBH_4$?

49. In the photographic process silver bromide is dissolved by adding sodium thiosulfate.

$$AgBr(s) + 2\ Na_2S_2O_3(aq) \longrightarrow Na_3Ag(S_2O_3)_2(aq) + NaBr(aq)$$

If you want to dissolve 0.250 g of AgBr, what volume in milliliters of 0.0138 M $Na_2S_2O_3$ should you add?

50. Hydrazine, N_2H_4, a base like ammonia, can react with an acid such as sulfuric acid.

$$2\ N_2H_4(aq) + H_2SO_4(aq) \longrightarrow 2\ N_2H_5^+(aq) + SO_4^{2-}(aq)$$

What mass of hydrazine can be consumed by 250. mL of 0.225 M H_2SO_4?

(a)

(b)

(a) A precipitate of AgBr formed by adding $AgNO_3(aq)$ to $KBr(aq)$. (b) On adding $Na_2S_2O_3(aq)$, sodium thiosulfate, the solid AgBr dissolves. *(C.D. Winters)*

51. What volume in milliliters of 0.812 M HCl would be required to titrate 1.33 g of NaOH to the equivalence point?

$$NaOH(aq) + HCl(aq) \longrightarrow NaCl(aq) + H_2O(\ell)$$

52. If a volume of 32.45 mL of HCl is used to titrate 2.050 g of Na_2CO_3 according to the following equation,

$$Na_2CO_3(aq) + 2\,HCl(aq) \longrightarrow 2\,NaCl(aq) + CO_2(g) + H_2O(\ell)$$

what is the molarity of the HCl?

53. What volume in milliliters of 0.955 M HCl would be needed to titrate 2.152 g of Na_2CO_3 to the equivalence point?

$$Na_2CO_3(aq) + 2\,HCl(aq) \longrightarrow 2\,NaCl(aq) + CO_2(g) + H_2O(\ell)$$

54. Potassium acid phthalate, $KHC_8H_4O_4$ (molar mass = 204.22 g/mol), is used to standardize solutions of bases. The acidic anion reacts with bases according to the following net ionic equation:

$$HC_8H_4O_4^-(aq) + OH^-(aq) \longrightarrow H_2O(\ell) + C_8H_4O_4^{2-}(aq)$$

If a 0.902-g sample of potassium acid phthalate is dissolved in water and titrated to the equivalence point with 26.45 mL of NaOH, what is the molarity of the NaOH?

55. A soft drink contains an unknown amount of citric acid, $C_3H_5O(CO_2H)_3$. If a volume of 100. mL of the soft drink requires 33.51 mL of 0.0102 M NaOH to neutralize the citric acid completely, what mass in grams of citric acid does the soft drink contain per 100. mL? The reaction of citric acid and NaOH is

$$C_3H_5O(CO_2H)_3(aq) + 3\,NaOH(aq) \longrightarrow Na_3C_3H_5O(CO_2)_3(aq) + 3\,H_2O(\ell)$$

56. A sample of a mixture of oxalic acid, $H_2C_2O_4$, and sodium chloride has a mass of 4.554 g. If a volume of 29.58 mL of 0.550 M NaOH is required to titrate the sample to the equivalence point, what is the weight percent of oxalic acid in the mixture? Oxalic acid and NaOH react according to the equation

$$H_2C_2O_4(aq) + 2\,NaOH(aq) \longrightarrow Na_2C_2O_4(aq) + 2\,H_2O(\ell)$$

57. You are given an acid and told only that it could be citric acid (molar mass = 192.1 g/mol) or tartaric acid (molar mass = 150.1 g/mol). To determine which acid you have, you titrate it with NaOH. The appropriate reactions are:

citric acid: $C_6H_8O_7 + 3\,NaOH \longrightarrow Na_3C_6H_5O_7 + 3\,H_2O$

tartaric acid: $C_4H_6O_6 + 2\,NaOH \longrightarrow Na_2C_4H_4O_6 + 2\,H_2O$

You find that a 0.956-g sample requires 29.1 mL of 0.513 M NaOH for titration to the equivalence point. What is the unknown acid?

58. You are given 0.954 g of an unknown acid, H_2A, which reacts with NaOH according to the balanced equation

$$H_2A(aq) + 2\,NaOH(aq) \longrightarrow Na_2A(aq) + 2\,H_2O(\ell)$$

If a volume of 36.04 mL of 0.509 M NaOH is required to titrate the acid to the equivalence point, what is the molar mass of the acid?

59. Vitamin C is the simple compound $C_6H_8O_6$. Besides being an acid, it is also a reducing agent. Therefore, one method for determining the amount of vitamin C in a sample is to titrate it with a solution of bromine, Br_2, a good oxidizing agent.

$$C_6H_8O_6(aq) + Br_2(aq) \longrightarrow 2\,HBr(aq) + C_6H_6O_6(aq)$$

Suppose a 1.00-g "chewable vitamin C tablet" requires 27.85 mL of 0.102 M Br_2 for titration to the equivalence point. What mass in grams of vitamin C is in the tablet?

60. To analyze an iron-containing compound, you convert all the iron to Fe^{2+} in aqueous solution and then titrate the solution with $KMnO_4$ according to the following balanced net ionic equation.

$$MnO_4^-(aq) + 5\,Fe^{2+}(aq) + 8\,H^+(aq) \longrightarrow Mn^{2+}(aq) + 5\,Fe^{3+}(aq) + 4\,H_2O(\ell)$$

If a 0.598-g sample of the compound requires 22.25 mL of 0.0123 M $KMnO_4$ for titration to the equivalence point, what is the weight percent of iron in the compound?

61. What volume in milliliters of 0.750 M $Pb(NO_3)_2$ solution is required to react completely with 1.00 L of 2.25 M NaCl solution? The balanced equation is

$$Pb(NO_3)_2(aq) + 2\,NaCl(aq) \longrightarrow PbCl_2(s) + 2\,NaNO_3(aq)$$

62. What volume in milliliters of 0.512 M NaOH is required to react completely with 25.0 mL of 0.234 M H_2SO_4?

63. What is the maximum mass, in grams, of AgCl that can be precipitated by mixing 50.0 mL of 0.025 M $AgNO_3$ solution with 100.0 mL of 0.025 M NaCl solution? Which reactant is in excess? What is the concentration of the excess reactant remaining in solution after the maximum mass of AgCl has been precipitated?

64. Suppose you mix 25.0 mL of 0.234 M $FeCl_3$ solution with 42.5 mL of 0.453 M NaOH. What is the maximum mass, in grams, of $Fe(OH)_3$ that will precipitate? Which reactant is in excess? What is the concentration of the excess reactant remaining in solution after the maximum mass of $Fe(OH)_3$ has precipitated?

General Questions

65. Nitrogen gas can be prepared in the laboratory by the reaction of ammonia with copper(II) oxide according to the following *unbalanced* equation.

$$NH_3(g) + CuO(s) \longrightarrow N_2(g) + Cu(s) + H_2O(g)$$

If 26.3 g of gaseous NH_3 is passed over a bed of solid CuO (in stoichiometric excess), what mass in grams of N_2 can be isolated?

66. In an experiment, 1.056 g of a metal carbonate containing an unknown metal M was heated to give the metal oxide and 0.376 g CO_2.

$$MCO_3(s) + \text{heat} \longrightarrow MO(s) + CO_2(g)$$

What is the identity of the metal M?
(a) M = Ni (c) M = Zn
(b) M = Cu (d) M = Ba

67. Aluminum bromide is a valuable laboratory chemical. What is the maximum theoretical yield in grams of Al_2Br_6 if 25.0 mL of liquid bromine (density = 3.1023 g/mL) and excess aluminum metal are used?

$$2\,Al(s) + 3\,Br_2(\ell) \longrightarrow Al_2Br_6(s)$$

68. The cancer chemotherapy agent cisplatin is made by the following reaction:

$$(NH_4)_2PtCl_4(s) + 2\,NH_3(aq) \longrightarrow 2\,NH_4Cl(aq) + Pt(NH_3)_2Cl_2(s)$$

Assume that 15.5 g of $(NH_4)_2PtCl_4$ is combined with 2.55 g of NH_3 to make cisplatin. (a) Which reactant is in excess and which is the limiting reactant? (b) What mass in grams of cisplatin can be formed? (c) After all the limiting reactant has been consumed, and the maximum quantity of cisplatin has been formed, what mass in grams of the other reactant remains?

69. Uranium(VI) oxide reacts with bromine trifluoride to give uranium(IV) fluoride, an important step in the purification of uranium ore.

$$6\,UO_3(s) + 8\,BrF_3(\ell) \longrightarrow 6\,UF_4(s) + 4\,Br_2(\ell) + 9\,O_2(g)$$

If you begin with 365 g of each of UO_3 and BrF_3, what is the maximum yield, in grams, for UF_4?

70. Silicon and hydrogen form a series of interesting compounds, Si_xH_y. To find the formula of one of them, a 6.22-g sample of the compound is burned in oxygen. On doing so, all of the Si is converted to 11.64 g of SiO_2 and all of the H to 6.980 g of H_2O. What is the empirical formula of the silicon compound?

71. To find the formula of a compound composed of iron and carbon monoxide, $Fe_x(CO)_y$, the compound is burned in pure oxygen, a reaction that proceeds according to the following *unbalanced* equation.

$$Fe_x(CO)_y(s) + O_2(g) \longrightarrow Fe_2O_3(s) + CO_2(g)$$

If you burn 1.959 g of $Fe_x(CO)_y$ and find 0.799 g of Fe_2O_3 and 2.200 g of CO_2, what is the empirical formula of $Fe_x(CO)_y$?
(a) $Fe(CO)_4$ (c) $Fe(CO)_5$
(b) $Fe_2(CO)_9$ (d) $Fe(CO)_6$

72. Boron forms an extensive series of compounds with hydrogen, all with the general formula B_xH_y. To analyze one of these compounds, you burn it in air and isolate the boron in the form of B_2O_3 and the hydrogen in the form of water according to the following *unbalanced* equation.

$$B_xH_y(s) + \text{excess } O_2(g) \longrightarrow B_2O_3(s) + H_2O(g)$$

If 0.148 g of B_xH_y gives 0.422 g of B_2O_3 when burned in excess O_2, what is the empirical formula of B_xH_y?
(a) BH_3 (c) B_3H_5
(b) B_2H_5 (d) B_5H_7

73. What are the concentrations of ions in a solution made by diluting 10.0 mL of 2.56 M HCl to 250. mL?

74. Half a liter (500. mL) of 2.50 M HCl is mixed with 250. mL of 3.75 M HCl. Assuming the total solution volume after mixing is 760. mL, what is the concentration of hydrochloric acid in the resulting solution?

75. Diborane, B_2H_6, can be produced by the following reaction.

$$2\,NaBH_4(aq) + H_2SO_4(aq) \longrightarrow 2\,H_2(g) + Na_2SO_4(aq) + B_2H_6(g)$$

What is the maximum yield, in grams, of B_2H_6 that can be prepared starting with 250. mL of 0.0875 M H_2SO_4 and 1.55 g of $NaBH_4$?

76. Sodium thiosulfate, $Na_2S_2O_3$, is used as a "fixer" in black and white photography. Assume you have a bottle of sodium thiosulfate and want to determine its purity. The thiosulfate ion can be oxidized with I_2 according to the equation

$$I_2(aq) + 2\,S_2O_3^{2-}(aq) \longrightarrow 2\,I^-(aq) + S_4O_6^{2-}(aq)$$

This reaction occurs rapidly and quantitatively, and there is a simple method of observing when the reaction has reached the equivalence point. Therefore, it can be used as a method of analysis by titration. If you use 40.21 mL of 0.246 M I_2 in the titration, what is the weight percent of $Na_2S_2O_3$ in a 3.232-g sample of impure $Na_2S_2O_3$?

77. The lead content of a sample can be estimated by converting the lead to PbO_2

$$\text{Pb in sample} + \text{oxidizing agent} \longrightarrow PbO_2(s)$$

and then dissolving the PbO_2 in an acid solution of KI. This liberates I_2 according to the equation

$$PbO_2(s) + 4\,H^+(aq) + 2\,I^-(aq) \longrightarrow Pb^{2+}(aq) + I_2(aq) + 2\,H_2O(\ell)$$

The liberated I_2 is then titrated with $Na_2S_2O_3$ (see Study Question 76).

$$I_2(aq) + 2\,S_2O_3^{2-}(aq) \longrightarrow 2\,I^-(aq) + S_4O_6^{2-}(aq)$$

The amount of titrated I_2 is related to the amount of lead in the sample. If 0.576 g of lead-containing mineral requires 35.23 mL of 0.0500 M $Na_2S_2O_3$ for titration of the liberated I_2, calculate the weight percent of lead in the mineral.

78. Cobalt(III) ion forms many compounds with ammonia. To find the formula of one of these compounds, you titrate the NH_3 with standardized acid.

$$Co(NH_3)_xCl_3(aq) + x\,HCl(aq) \longrightarrow x\,NH_4^+(aq) + Co^{3+}(aq) + (x+3)Cl^-(aq)$$

Assume that 23.63 mL of 1.500 M HCl is used to titrate 1.580 g of $Co(NH_3)_xCl_3$. What is the value of x?
(a) 2 (b) 3 (c) 4 (d) 6

79. You wish to determine the weight percent of copper in a copper-containing alloy. After dissolving a sample in acid, you add an excess of KI, and the Cu^{2+} and I^- undergo the reaction

$$2\,Cu^{2+}(aq) + 5\,I^-(aq) \longrightarrow 2\,CuI(s) + I_3^-(aq)$$

The liberated I_3^- is titrated with sodium thiosulfate according to the equation

$$I_3^-(aq) + 2\,S_2O_3^{2-}(aq) \longrightarrow S_4O_6^{2-}(aq) + 3\,I^-(aq)$$

If a volume of 26.32 mL of 0.101 M $Na_2S_2O_3$ is required for titration to the equivalence point, what is the weight percent of Cu in 0.251 g of the alloy?

80. A weighed sample of a metal is added to liquid bromine and allowed to react completely. The product substance is then separated from any leftover reactants and weighed. This experiment is repeated with several masses of the metal but with the same volume of bromine. The following graph indicates the results. Explain why the graph has the shape that it does.

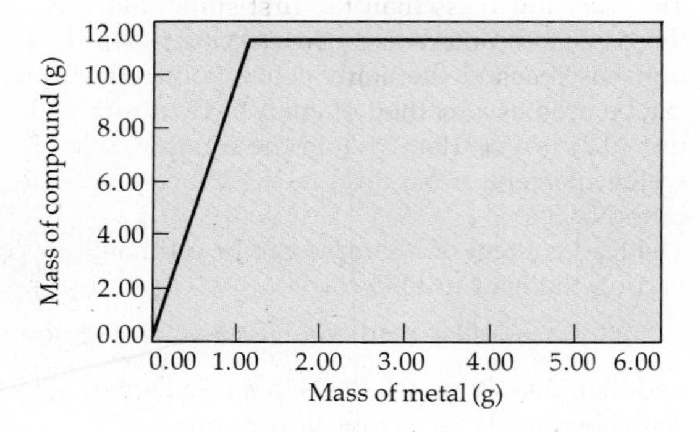

81. A series of experimental measurements like the ones described in Study Question 80 is carried out for iron reacting with bromine. The following graph is obtained. What is the empirical formula of the compound formed between iron and bromine? Write a balanced equation for the reaction between iron and bromine and name the product.

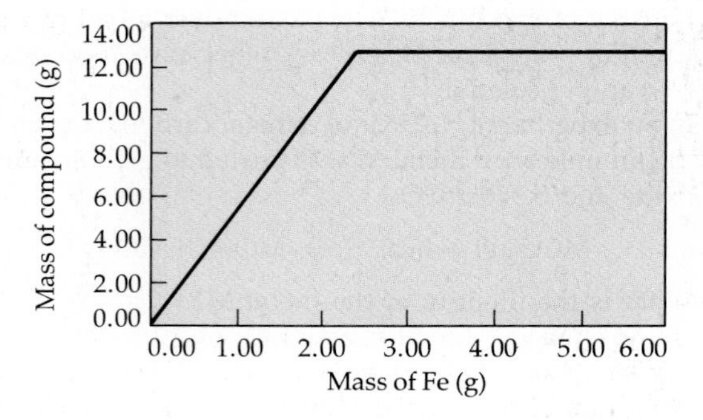

82. Four groups of students from an introductory chemistry laboratory are studying the reactions of solutions of alkali metal halides with aqueous silver nitrate, $AgNO_3$. They use the following salts:
Group A: NaCl Group C: NaBr
Group B: KCl Group D: KBr
Each of the four groups dissolves 0.004 mol of their salt in some water. Each then adds various masses of silver nitrate, $AgNO_3$, to their solutions. After each group collects the precipitated silver halide, the mass of this product is plotted versus the mass of $AgNO_3$ added. The results are given on the following graph.
(a) Write the balanced net ionic equation for the reaction observed by each group.
(b) Explain why the data for groups A and B lie on the same line, while those for groups C and D lie on a different line.
(c) Explain the shape of the curve observed by each group. Why do they level off at about 0.6 g of added $AgNO_3$ (for groups A and B) or at about 0.8 g of added $AgNO_3$ (for groups C and D)?

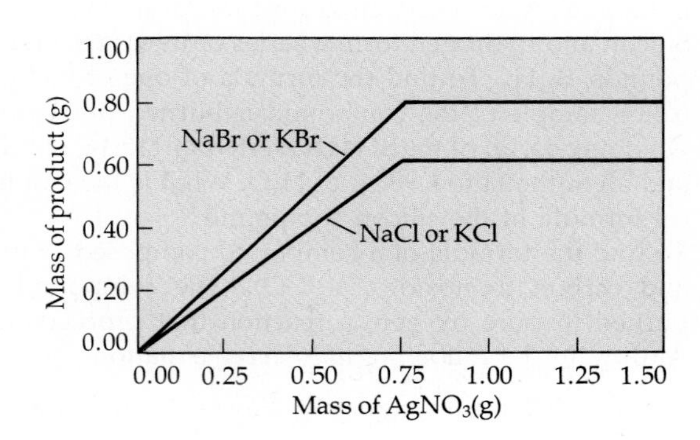

83. Ammonia is formed in a direct reaction of nitrogen and hydrogen.

$$N_2(g) + 3\,H_2(g) \longrightarrow 2\,NH_3(g)$$

The starting mixture is represented by the following diagram, where the blue circles represent N and the white circles represent H.

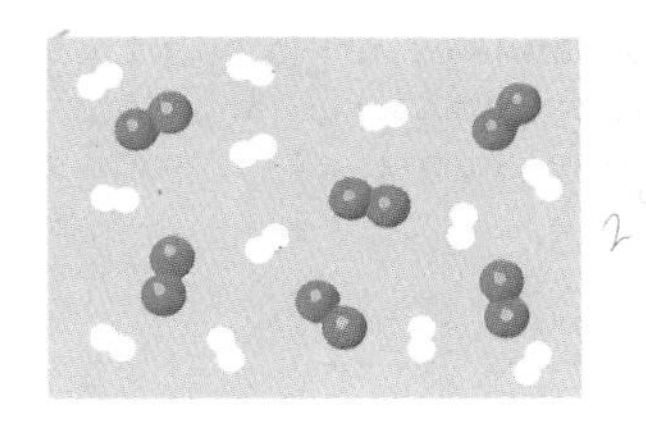

Which of the following represents the product mixture?

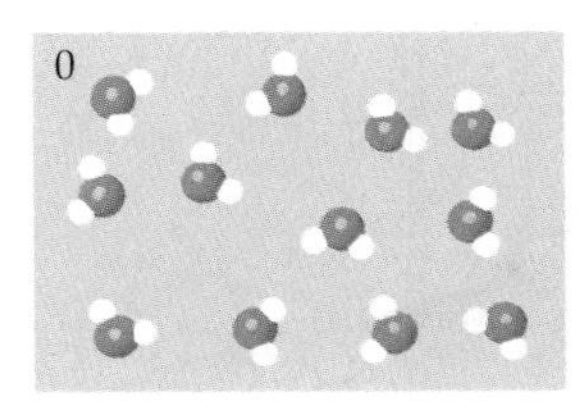

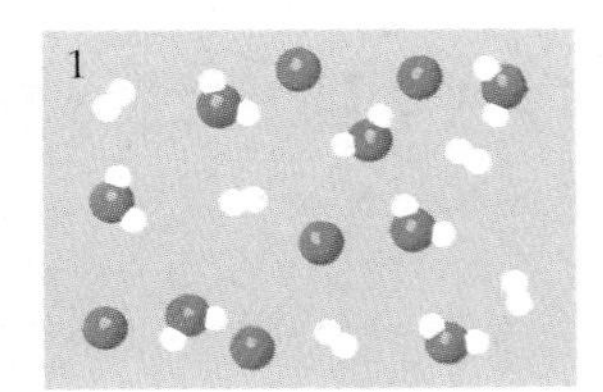

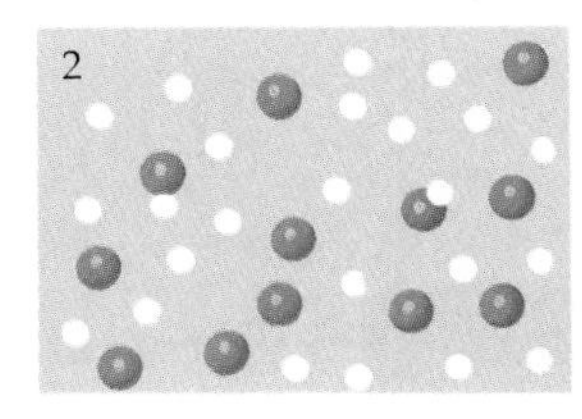

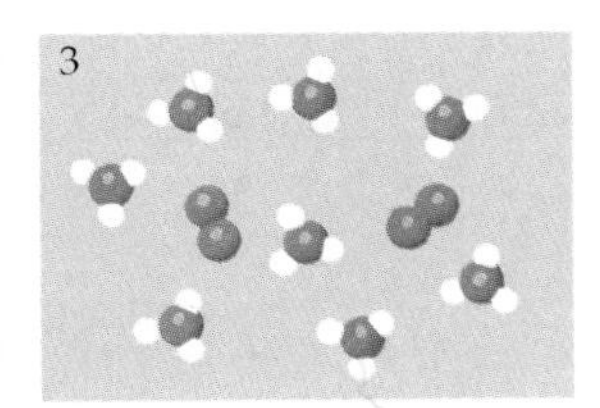

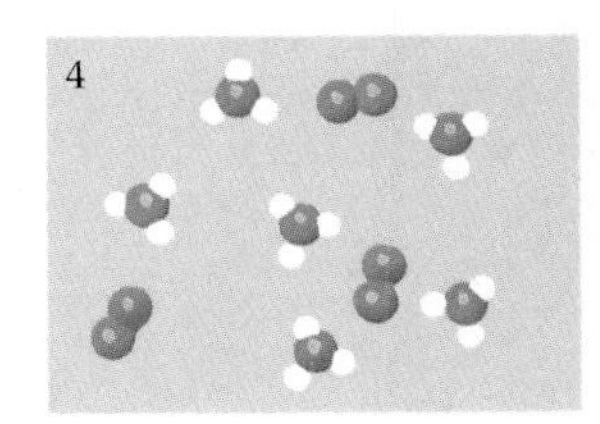

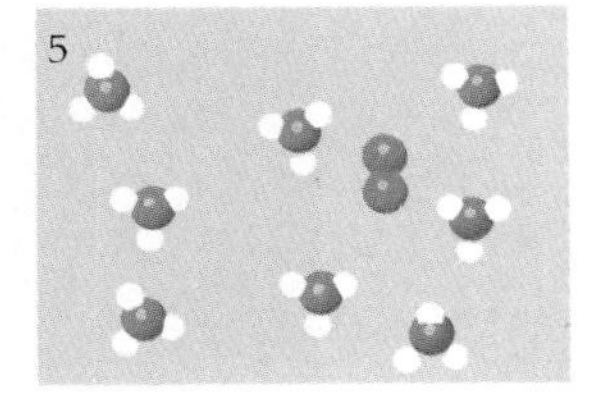

For the reaction of the given sample, which of the following is true?
(a) N_2 is the limiting reactant.
(b) H_2 is the limiting reactant.
(c) NH_3 is the limiting reactant.
(d) No reactant is limiting; they are present in the correct stoichiometric ratio.

84. Ten milliliters of a solution of an acid is mixed with 10 mL of a solution of a base. When the mixture was tested with litmus paper (see page 160), the blue litmus turned red, and the red litmus remained red. Which of the following interpretations is (are) correct?
(a) The mixture contains more hydronium ions than hydroxide ions.
(b) The mixture contains more hydroxide ions than hydronium ions.
(c) When an acid and base react, water is formed, and so the mixture cannot be acidic or basic.
(d) If the acid was HCl and the base was NaOH, the concentration of HCl in the initial acidic solution must have been greater than the concentration of NaOH in the initial basic solution.
(e) If the acid was H_2SO_4 and the base was NaOH, the concentration of H_2SO_4 in the initial acidic solution must have been greater than the concentration of NaOH in the initial basic solution.

85. In a reaction 1.2 grams of Element A reacts with exactly 3.2 grams of oxygen to form an oxide, AO_x; 2.4 g of Element A reacts with exactly 3.2 g of oxygen to form a second oxide AO_y.
(a) What is the ratio x/y?
(b) If $x = 2$, what is the identity of element A?

86. Two students titrate different samples of the same solution of HCl using 0.100 M NaOH solution and phenolphthalein indicator (Figure 6.13). The first student pipets 20.0 mL of acid into a flask, adds 20 mL of distilled water and a few drops of phenolphthalein solution, and titrates until a lasting pink color appears. The second student pipets 20.0 mL of acid into a flask, adds 60 mL of distilled water, and a few drops of phenolphthalein solution, and titrates to the first lasting pink color. Each student correctly calculates the molarity of a HCl solution. The second student's result will be
(a) four times less than the first student's.
(b) four times more than the first student's.
(c) two times more than the first student's.
(d) two times less than the first student's.
(e) the same as the first student's.

87. Dilute sulfuric acid is added from a buret to dilute, aqueous calcium hydroxide in a beaker. Assuming that the concentrations of acid and base are equal, make a graph of electrical conductivity of the solution in the beaker on the vertical axis versus volume of sulfuric acid added on the horizontal axis. Before starting your graph, write balanced net ionic equations for all reactions that occur in the beaker.

Summary Problem

88. Various masses of the three Group 2A elements magnesium, calcium, and strontium were allowed to react with liquid bromine, Br_2. After the reaction was complete the reaction product was freed of excess reactant(s) and weighed. In each case the mass of product was plotted against the mass of metal used in the reaction.
    (a) Based on your knowledge of the reactions of metals with halogens, what product is predicted for each reaction? What are the name and formula for the reaction product in each case?
    (b) Write a balanced equation for the reaction occurring in each case.
    (c) What kind of reaction occurs between the metals and bromine? That is, is the reaction a gas-forming reaction, a precipitation reaction, or an oxidation-reduction?
    (d) Each plot shows that the mass of product increases with increasing mass of metal used, but the plot levels out at some point. Use these plots to verify your prediction of the formula of each product and explain why the plots become level at different masses of metal and different masses of product.

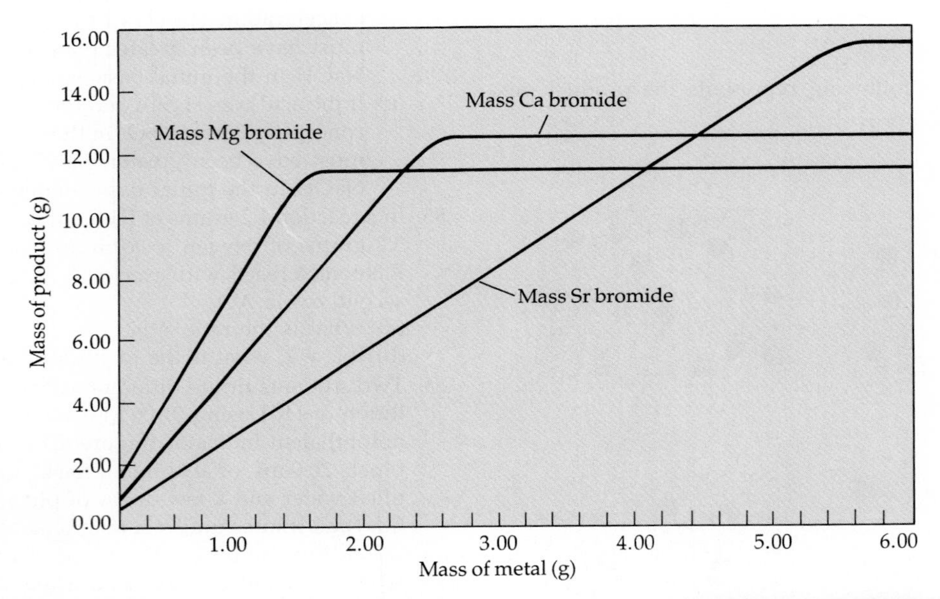

89. Gold can be dissolved from gold-bearing rock by treating the rock with sodium cyanide in the presence of the oxygen in air.

$$4\ Au(s) + 8\ NaCN(aq) + O_2(g) + 2\ H_2O(\ell) \longrightarrow 4\ NaAu(CN)_2(aq) + 4\ NaOH(aq)$$

Once the gold is in solution in the form of the $[Au(CN)_2]^-$ ion, it can be precipitated as the metal according to the following *unbalanced* equation

$$[Au(CN)_2]^-(aq) + Zn(s) \longrightarrow Zn^{2+}(aq) + Au(s) + CN^-(aq)$$

(a) Are the two reactions above acid-base or oxidation-reduction reactions? Briefly describe your reasoning.
(b) How many liters of 0.075 M NaCN will you need to use to extract the gold from 1000 kg of rock if the rock is 0.019% gold?

One face of an open pit gold mine in Lead, South Dakota.

(c) How many kilograms of metallic zinc will you need to use to recover the gold from the $[Au(CN)_2]^-$ obtained from the gold in the rock?
(d) If the gold is recovered completely from the rock, and the metal is made into a cylindrical rod 15.0 cm long, what is the diameter of the rod? (The density of gold is 19.3 g/cm^3).

Using Computer Programs and Videodiscs

90. Use *The Periodic Table Videodisc* to explore the reaction of copper with acids (frame 29501).
    (a) Describe the reactions of copper with aqueous HCl and HNO_3, noting similarities and differences.
    (b) Write the balanced net ionic equation for the reaction of copper with nitric acid. (See Section 5.7 and Figure 5.25.)
    (c) Give the name and formula for the salt of copper(II) that gives rise to the blue-green color observed in the solution in this reaction.
    (d) If you began with 1.25 g of copper metal, what mass of NO_2 gas would be produced? How many grams of copper-containing product could be isolated?

The exothermic reaction of magnesium with solid carbon dioxide (to give magnesium oxide and carbon).
(C.D. Winters)

Principles of Reactivity: Thermodynamics

CHAPTER

7

Many of the benefits that we receive from chemistry result from transformation of inexpensive, readily available substances into new substances with more desirable properties. Important examples are manufacture of rubber from petroleum and the manufacture of medicines from simple chemicals. Chemists want to be able to predict what will happen when potential reactants are mixed. Will most of the reactants be converted to products? Will some be converted? Or virtually none? If we know the properties of elements and compounds, and are familiar with classes of chemical reactions such as combination, exchange, and oxidation-reduction, we can often make good predictions. However, there are more general rules that can be applied to *any* reaction, not just reactions in a particular class. One of these rules states that if heat is given off when a reaction occurs, and if the products of the reaction are more dispersed or mixed up than the reactants, then the reaction will produce appreciable quantities of products. If heat must be supplied and the products are more ordered, then almost all of the reactants will remain unchanged, and few or no products will be produced. But how do we decide whether heat will be given off and whether products are more disordered? Answering these questions requires knowledge of thermodynamics, which is the subject of this chapter.

Figure 7.1

Some reactions begin as soon as the reactants come into contact, and continue until one or more reactants have been consumed. Reactions such as these are said to be "product-favored," and one example is the reaction of the alkali metals with water. Here potassium reacts explosively to give hydrogen gas and potassium hydroxide. *(C.D. Winters)*

$$2\,K(s) + 2\,H_2O(\ell) \longrightarrow 2\,KOH(aq) + H_2(g)$$

"A theory is the more impressive the greater the simplicity of its premises is, the more different kinds of things it relates, and the more extended is its area of applicability. Therefore, the deep impression which classical thermodynamics made upon me. It is the only physical theory of universal content concerning which I am convinced that, within the framework of the applicability of its basic concepts, it will never be overthrown." Albert Einstein. *Autobiographical Notes, page 33 in The Library of Living Philosophers, Vol. VII,* Albert Einstein: Philosopher-Scientist, *edited by P.A. Schilpp, Evanston, Illinois, 1949.*

Some chemical reactions begin as soon as the reactants come into contact and continue until at least one reactant (the limiting reactant) is completely consumed. Drop a clean piece of potassium into water and it reacts so violently that sparks fly, and you see a purple flame characteristic of very hot potassium salts (Figure 7.1). Or combine aluminum and bromine, and they react rapidly with evolution of a great deal of energy to give aluminum bromide (Figure 5.1). Other reactions happen much more slowly, but reactants are still converted almost completely to products. An example is rusting of iron at room temperature; given many years and enough flaking of iron(III) oxide from its surface, a piece of iron exposed to air will be completely converted—it will rust away. Still other reactions occur so slowly at room temperature that even a lifetime is not enough to observe a measurable change. An example is combustion of gasoline, which burns rapidly at high temperatures but can be stored for long periods in direct contact with air so long as the temperature is not raised by a spark or flame. Nevertheless chemists are convinced that if only one could wait long enough, gasoline would be converted to CO_2 and H_2O.

Two questions about chemical reactions are important to chemists: will a reaction occur and how fast will it go. We shall put off discussing the factors that determine the speed of a reaction until Chapter 8. In Chapter 7 we wish to investigate the first question. Here *we will categorize reactions that produce appreciable quantities of products as* ***product-favored systems.*** Examples are decomposition of water by potassium, rusting of iron, and combustion of gasoline. If a system is product-favored, most of the reactants will eventually be converted to products without outside intervention, although "eventually" may mean a very, very long time.

There are other reactions that have virtually no tendency to occur by themselves. For example, nitrogen and oxygen have coexisted in the earth's atmosphere for at least a billion years without any sign of their combining to form nitrogen oxides such as NO or NO_2. Similarly, deposits of salt, NaCl, have existed on earth for millions of years without forming the elements Na and Cl_2. The same is true of aluminum bromide, which is not transformed into aluminum and bromine at room temperature no matter how long we wait. We shall label such combinations as **reactant-favored systems.** A reactant-favored system involves a transformation that is exactly the opposite of a product-favored system. For example, $Na(s) + Cl_2(g)$ react to form NaCl(s), but NaCl(s) does not react in the opposite direction to form $Na(s) + Cl_2(g)$. Without some continuous outside intervention, reactants in a reactant-favored system are not transformed into appreciable quantities of products.

What do we mean by outside intervention? Usually it is some flow of energy. For example, at very high temperatures, small but significant quantities of NO can be formed from air. Such high temperatures are found in power plants and automobile engines, and the large number of such sources can produce enough NO and other nitrogen oxides to cause significant air pollution problems. Salt can be decomposed to its elements by providing heat to melt it and electrical energy to separate the ions and form the elements:

$$2NaCl(\ell) \xrightarrow{\text{electricity}} 2Na(\ell) + Cl_2(g)$$

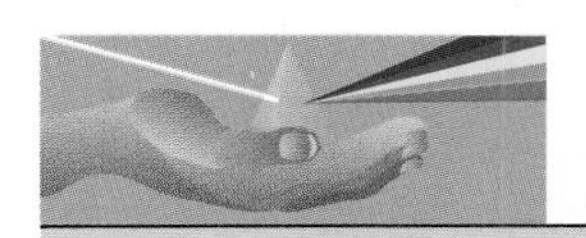

CHEMISTRY YOU CAN DO

Rusting and Heat

Chemical reactions can evolve heat, and a simple experiment demonstrates this very well. To do it you will need a steel wool pad (without soap), 1/4 cup of vinegar, a cooking or outdoor thermometer, and a large jar with a lid. (The thermometer must fit inside the jar.)

Soak the steel wool pad in vinegar for several minutes. While doing so, place the thermometer in the jar, close the lid, and let it stand for several minutes. Read the temperature.

Squeeze the excess vinegar out of the steel wool pad, wrap the pad around the bulb of the thermometer, and place both in the jar. Close the lid. After about five minutes again read the temperature. What has happened?

Repeat the experiment with another steel pad, but wash it with water instead of vinegar. Try a third pad that is not washed at all. Allow each pad to stand in air for a few hours or for a day and observe the pad carefully. Do you see any change in the metal? Suggest an explanation for your observations of temperature changes and appearance of the steel wool.

(C.D. Winters)

In each case energy can cause a reactant-favored system to produce products.

Clearly it is useful to know something about interactions between energy and matter. The most common of these is transfer of energy as heat when chemical reactions occur. Transfer of energy as heat is a main theme of **thermodynamics,** the science of heat and work.

7.1 ENERGY: ITS FORMS AND UNITS

Just what is energy and where does the energy we use come from? **Energy** is defined as the capacity to do work, and work is something you experience all the time. If you climb a mountain, you do some work against the force of gravity as you carry yourself and your equipment upward. You can do this work because you have the energy or capacity to do so, the energy having been provided by the food you have eaten. Food energy is chemical energy—energy stored in chemical compounds and released when the compounds undergo the chemical reactions of metabolism.

Energy can be assigned to one of two classes: kinetic or potential. **Kinetic energy** is energy that something has because it is moving. Examples are

Thermal energy is often referred to as heat. However, it is useful to distinguish it from heat, a process in which energy is transferred as a result of a temperature difference, from energy itself.

- mechanical energy of a macroscopic object such as a moving baseball or automobile
- thermal energy of atoms, molecules, or ions in motion at the submicroscopic level
- electric energy of electrons moving through a conductor
- sound, which corresponds to compression and expansion of the spaces between molecules

All matter has thermal energy because, according to the kinetic-molecular theory (Section 2.1), its submicroscopic particles are in constant motion.

Potential energy is energy that something has as a result of its position. Examples are

- gravitational energy such as that of a ball held well above the floor or that of water at the top of a waterfall (Figure 7.2)
- electrostatic energy such as that of positive and negative ions a small distance apart
- chemical potential energy resulting from attractions among electrons and atomic nuclei in molecules

Potential energy can be converted to kinetic energy. An example is water falling over a waterfall: as the droplets fall, the potential energy they had at the top is converted to kinetic energy, and they move faster and faster. Similarly, kinetic energy can be converted to potential energy. The kinetic energy of falling water can drive a water wheel to pump water to an elevated reservoir where its potential energy is higher.

Figure 7.2
Water on the brink of a waterfall has stored or potential energy, energy that can be used to generate electricity, for example. *(Image Enterprises, James Cowlin)*

What resources are available to supply energy for humans to use? Historically the first was *biomass,* material produced by living organisms that contains significant quantities of chemical potential energy. Biomass, mostly wood, still provides one third of the energy resources for some developing countries. However, *fossil fuels*—petroleum, coal, and natural gas—made the Industrial Revolution possible and provide most of the energy resources used in the industrialized world today. For 200 years coal was the principal fuel for industrializing nations; it was largely replaced by petroleum and natural gas around the middle of this century (Figure 7.3). Petroleum and natural gas are easier to handle and cleaner to use. The world's appetite for energy has greatly inflated the prices of petroleum and natural gas because their supply is quite limited. Greater quantities of coal are available, and so attention is shifting back toward coal for energy and as a new source for many of the chemicals currently obtained from petroleum and natural gas. Relatively small but still significant quantities of energy are now supplied by hydroelectric sources and nuclear energy, and still smaller quantities are obtained in the form of direct solar energy, geothermal energy, and wind and ocean currents. These **primary energy resources** may be converted to electricity, a **secondary resource** and a form of energy we find particularly useful.

Our most important resource is the sun, whose energy is available to us through

- biomass as a result of photosynthesis

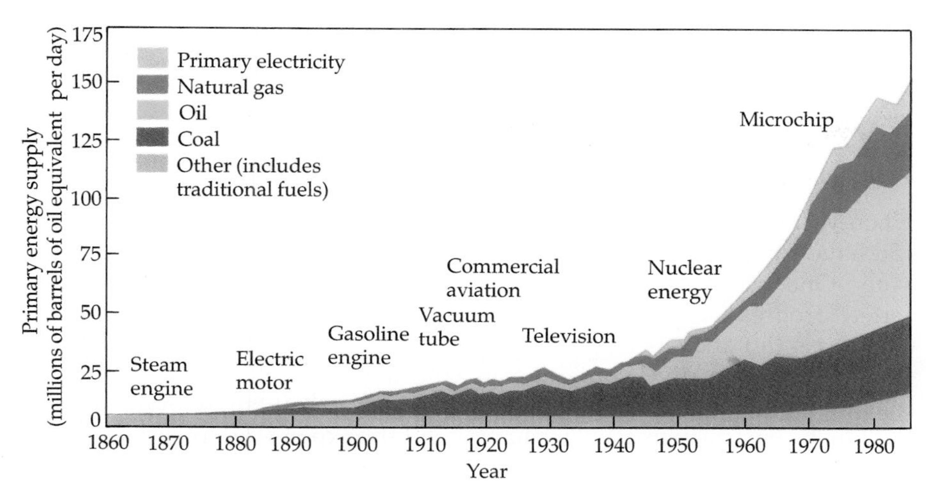

Figure 7.3
The rate of use of primary energy from 1860 to the present. (Primary energy exists in a crude form such as fossil fuels.) At the beginning of this period wood was the predominant resource, but coal became more important with the advent of the steam engine and the electric motor. With the invention of the automobile at the end of the 19th century, oil began to contribute. Coal continued to be the most common fuel until the 1920s, when it contributed more than 70% of the world's energy supply. Today coal meets only about 26% of global energy needs, while oil is the major resource. Natural gas and electricity generated by hydro and nuclear power plants have contributed a steadily rising portion since the 1940s. For more information see G. R. Davis, *Scientific American,* September 1990, pp. 55–62.

- fossil fuels that are believed to have been formed from biomass as a result of changes that required millions of years
- wind and water that store kinetic and potential energy
- direct absorption of solar energy by our bodies and the materials around us
- recently, photovoltaic cells capable of transforming solar energy into electrical energy

Renewable energy resources such as biomass or hydroelectric energy can figure in long-range energy planning, in contrast to fossil fuels which, once used, are gone.

Photovoltaic cells are illustrated in Figure 6.3.

In our industrialized, high-tech, appliance-oriented society, the average use of energy per individual is almost at its highest point in history. The United States, with only 5% of the world's population, accounts for 30% of the world's consumption of energy resources. Only Canada uses more energy per capita. Since 1958, the United States has consumed more energy resources than it has produced. However, our voracious appetite has been curbed somewhat in recent years as a result of increased energy costs and the prospect of actual shortages. Emphasis has been placed on more efficient use of energy rather than curtailment of our standard of living. Since the oil shortage of 1973, the energy required to produce $1 of the U.S. gross national product has fallen by 28%.

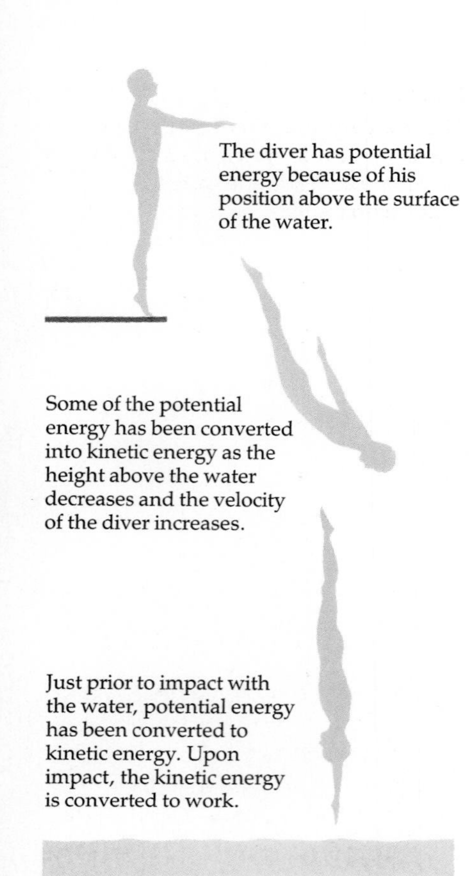

Figure 7.4
Energy conservation and the interconversion of potential energy, kinetic energy, and work.

Temperature, Heat, and the Conservation of Energy

When you stand on a diving board, poised to dive into a swimming pool, you have considerable potential energy because of your position above the water. Once you jump off the board, some of that potential energy is converted progressively into kinetic energy (Figure 7.4), which depends on your mass (m) and velocity (v). Kinetic energy is given by the expression $\frac{1}{2}mv^2$. During the dive your mass is constant, while the force of gravity accelerates your body to move faster and faster as you fall, so your velocity and kinetic energy increase. This happens at the expense of potential energy. At the moment you hit the water, your velocity is abruptly reduced, and much of your kinetic energy is converted to mechanical energy of the water, which splashes as your body moves it aside by doing work on it. Eventually you float on the surface and the water becomes still again. However, if you could see them, you would find that the water molecules were moving a little faster in the vicinity of your dive; that is, the temperature of the water is a little higher.

This series of energy conversions, from potential to kinetic, and from kinetic to thermal, illustrates the **law of conservation of energy,** which states that energy can neither be created nor destroyed—the total energy of the universe is constant. This law summarizes a great many experiments in which heat, work, and other forms of energy transfer have been measured and the total energy found to be the same before and after an event. The law of conservation of energy is the reason we have been careful not to say that when a lump of coal is burned its energy has been used up. What has been used up is an *energy resource:* the coal's capacity to transfer energy by heating to its surroundings when it is burned. If the coal is burned in a power plant, its chemical energy can be changed to an equal quantity of energy in other forms. These are mainly electricity, which can be very useful, and thermal energy in the gases going up the smokestack and in the immediate surroundings of the plant. It is not the coal's energy, but rather its ability to store energy and release it in a form which people can use that has been used up.

Notice also that in our diving example the potential energy you originally had on the diving board ended up heating the water. Thermal energy, or heat, is associated with temperature, but heat is not the same as temperature. Transferring energy by heating an object increases the object's temperature, and the temperature increase can be measured with a thermometer. For example, Figure 7.5 shows a thermometer containing mercury. When the thermometer is placed into hot water, heat is transferred from the water to the thermometer. The increased energy of the mercury atoms means that they are moving about more rapidly, which slightly increases the volume of the spaces between the atoms. Consequently the mercury expands (as most substances do upon heating), and the upper end of the column of mercury rises higher in the thermometer tube.

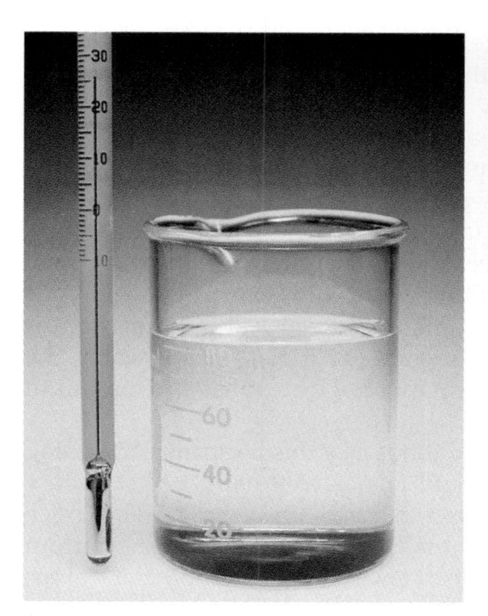

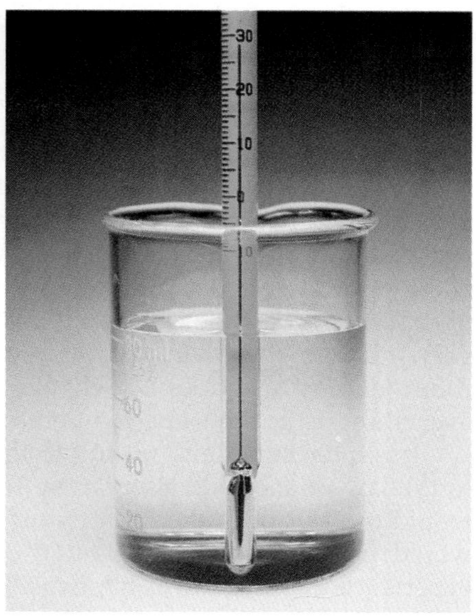

Figure 7.5
The mercury in a thermometer expands as heat is transferred to the liquid metal from hot water.
(C.D. Winters)

Energy transfer, as heat, happens when two objects at different temperatures are brought into contact. Energy is always transferred from the hotter to the cooler object. For example, a piece of metal being heated in a bunsen burner flame and a beaker of cold water (Figure 7.6, *left*) are two objects with different temperatures or hotness. When the hot metal is plunged into the cold water (Figure 7.6, *right*), heat is transferred from the metal to the water until the two objects reach the same temperature. The quantity of energy transferred may be sufficient to raise the temperature of the water above 100 °C immediately around the metal, and this causes some water to boil.

Transferring energy by heating is a process, but it is common to talk about that process as if heat were a form of energy. We will use such language in the remainder of this chapter, saying that one object transfers heat to another when we mean that one object transfers energy by heating the other.

There are two aspects of thermal energy to be understood: (1) the faster an atom or molecule moves, the more energy it has; and (2) the total thermal energy in an object is the sum of the individual energies of all the atoms or molecules in that object. The average speed of motion of atoms and molecules is related to the temperature. *The total thermal energy of a substance depends on temperature, the type of atoms or molecules, and how many of them there are in a sample.* For a given substance, thermal energy depends on tem-

Figure 7.6
Water in a beaker is warmed when a hotter object (a brass bar) is plunged into the water. Heat is transferred from the hotter metal bar to the cooler water (and enough heat is transferred that the water immediately around the metal boils). Eventually the bar and the water will reach the same temperature (at which point they will have achieved thermal equilibrium). *(C.D. Winters)*

perature and the amount of substance. Thus a cup of steaming coffee may contain less thermal energy than a bathtub full of warm water, even though the coffee is at a higher temperature.

Under most circumstances, most objects in a given space, such as your room, are at about the same temperature. If an object such as a cup of coffee is very much hotter than this, it transfers energy by heating to the rest of the objects in your room until it cools off (and they warm up a bit). If an object, say a glass of ice water, is much cooler than its surroundings in your room, energy is transferred to it from everything else until it warms up (and your room cools off a little). Since the total amount of material in your room is very much greater than that in a cup of coffee or a glass of ice water, the room temperature does not change very much. Heat transfer occurs until everything is at the same temperature.

It is interesting (and useful) to think about *why* this happens. According to the law of energy conservation, it could just as well happen that energy would transfer from the rest of your room to a hot cup of coffee. The coffee would then get hotter and hotter, eventually boiling. But you know from experience that this never happens. It also never happens that a glass of ice water freezes, warming the rest of the room around it. There is *directionality* in heat transfer: energy always transfers from hotter to colder, never the reverse. This directionality corresponds to *spreading out of energy* over the greatest possible number of atoms or molecules. Whether a relatively small number of molecules in a hot cup of coffee cools by transferring energy to a large number of atoms and molecules surrounding the cup, or the large number of particles in the surroundings heats a glass of ice water by transferring some of their energy to relatively few molecules in the glass, the end result is that the total thermal energy is spread evenly over the maximum number of molecules. Concentrating energy in only a few particles at the expense of many, or even concentrating energy over a large number of particles at the expense of a few, is highly improbable and has never been observed on a macroscopic scale. Energy transfers always occur in the way that spreads energy over as many atoms and molecules as possible. This same idea can be used to help us predict directionality in the conversion of chemical reactants to products, that is, to say whether a mixture of substances will be product-favored or reactant-favored.

Figure 7.7
A packet of artificial sweetener from Australia. The sweetener in the packet supplies 16 kJ of nutritional energy. It is equivalent to two level metric teaspoons of sugar, a quantity that supplies 140 kJ of nutritional energy. *(C.D. Winters)*

EXERCISE 7.1 • *Energy*

You place a raw egg in a frying pan and fry it, over easy. As in jumping off a diving board, several kinds of energy are involved. Describe them and the changes they cause.

Energy Units

Since all forms of energy can be converted into thermal energy, some of the units in which energy is measured were originally designed to measure heat. A **calorie** was originally defined as the quantity of energy (transferred by heating) that was required to raise the temperature of 1.00 g of pure liquid water by 1.00 degree Celsius, from 14.5 °C to 15.5 °C. This is a very small quantity of heat, so the **kilocalorie** is often used. The kilocalorie (abbreviated kcal) is equal to 1000 calories.

1 kilocalorie = 1 kcal = 1000 calories

Most of us tend to think of heat as measured in calories, probably because we hear about dieting or read breakfast cereal boxes. However, the "calorie" used to describe food is Calorie with a capital C, a unit equivalent to the kilocalorie. Thus, a breakfast cereal that gives you 100 Calories of nutritional energy really provides 100 kcal or 100×10^3 calories (with a small c).

In this book we shall use the **joule (J),** the SI unit of energy. One calorie is now defined as 4.184 J. The joule is preferred because it is derived directly from the units used in the calculation of mechanical energy (*i.e.*, potential and kinetic energy). If a 2.0-kg object (about 4 pounds) is moving with a velocity of 1.0 meter per second (roughly 2 mph), the kinetic energy is

$$\text{kinetic energy} = \tfrac{1}{2}mv^2 = \tfrac{1}{2}(2.0\text{ kg})(1.0\text{ m/s})^2 = 1.0\text{ kg}\cdot\text{m}^2/\text{s}^2 = 1.0\text{ J}$$

To give you some feeling for joules, if a six-pack of soft drink cans is dropped on your foot, the kinetic energy at the moment of impact is about a calorie or two, that is, about 4 to 10 J. In many countries that have standardized on SI units, food energy is also measured in joules. For example, the packet of non-sugar sweetener shown in Figure 7.7 would provide 16 kJ of nutritional energy, instead of the 140 kJ in two teaspoons of sugar.

1 calorie = 4.184 J

The joule is named for James P. Joule (1818–1889). Joule was a student of John Dalton (Chapter 3) and the son of a brewer in Manchester, England.

EXAMPLE 7.1

Using Energy Units

A baseball with a mass of 114 g, thrown at a speed of 100. miles per hour, has a kinetic energy of 27.2 cal. What is this energy in units of joules?

SOLUTION To find the energy in joules, we only need to use the relation 1.000 cal = 4.184 J.

$$27.2\text{ cal}\cdot\frac{4.184\text{ J}}{1.000\text{ cal}} = 114\text{ J}$$

James Joule *(Oesper Collection in the History of Chemistry, University of Cincinnati)*

Like the calorie, the joule is a small unit for many purposes in chemistry, and so we shall often use the kilojoule (kJ), which is equal to 1000 joules. Consider, for example, the quantity of energy that enters the earth's atmosphere (about 8×10^{15} kJ/minute or about 8 J/cm^2 · minute). This is an enormous amount of energy, even though it is only three ten-millionths (0.0000003) of the total energy emitted by the sun. About half of this energy (4 J/cm^2 · minute) reaches the earth's surface; the rest is reradiated from the atmosphere into space or is absorbed and scattered by the lower portion of the atmosphere. The quantity that reaches the surface depends on location, the season, and weather conditions. However, even 4 J/cm^2 · minute is a large quantity of energy. For example, at this rate the roof of an average-sized house receives about 100,000 kJ/day, equivalent to the heat energy derived from the burning of about 32 pounds of coal a day or to 120 kW-h (kilowatt-hours) of electrical energy a day—more than enough to heat an average American home in the winter.

1 kilojoule = 1 kJ = 1000 joules.

EXERCISE 7.2 • *Energy Units*

a. If you eat a hot dog, it will provide 160 Calories of energy. What is this energy in joules?

b. The energy used by a light bulb of *W* watts for *S* seconds is *WS* joules. If you turn on a 75-watt bulb for 3.0 hours, how many joules of energy will it use?

c. The non-sugar sweetener in Figure 7.7 provides 16 kJ of nutritional energy. What is this energy in kilocalories?

7.2 HEAT CAPACITY AND SPECIFIC HEAT

In the 1770s Joseph Black of the University of Glasgow (Scotland) studied the nature of heat and was the first to distinguish clearly between the "hotness" or temperature of an object and its heat capacity. The **heat capacity** of an object, which is defined as *the quantity of energy required to change the temperature of that object by one degree,* depends on the size of the object, the substance of which it is made, and the temperature scale.

Recall that 1 degree has the same magnitude on both the Celsius and Kelvin scales.

To make useful comparisons between objects of different sizes, the *specific heat capacity,* which is often just called **specific heat,** is used instead. This is defined as the quantity of energy needed to change the temperature of 1 g of a substance by 1 degree Celsius (or 1 kelvin). For water, the specific heat is 1.000 cal/g · K or 4.184 J/g · K, whereas it is only about 0.8 J/g · K for common glass. That is, it takes about five times as much heat to raise the temperature of a gram of water by 1 K as it does for a gram of glass. Thus, specific heat is a property that can be used to distinguish one substance from another, just as density can.

Joseph Black (1728–1799) was a professor in Glasgow and Edinborough, Scotland. Among his important studies was that of heat, which later led his student James Watt to the invention of the steam engine. (Oesper Collection in The History of Chemistry, University of Cincinnati)

The specific heat of a substance can be determined experimentally by measuring the quantity of heat transferred by a known mass of the substance as its temperature rises or falls.

$$\text{specific heat} = \frac{\text{quantity of heat transferred}}{(\text{mass of material})(\text{temperature change})}$$

As an example, it has been found that 60.5 J is required to change the temperature of 25.0 g of ethylene glycol (a compound used as antifreeze in automobile engines) by 1.00 °C (or 1.00 K). This means the specific heat of the compound is

$$\text{specific heat of ethylene glycol} = \frac{60.5\ \text{J}}{(25.0\ \text{g})(1.00\ \text{K})} = 2.42\ \text{J/g} \cdot \text{K}$$

The specific heats of many substances have been determined, and a few values are listed in Table 7.1. Notice that water has a high specific heat, one of the highest known. This is important to all of us, since a high specific heat means that many joules must be absorbed by a large body of water to raise its temperature by just a degree or so. Conversely, a lot of energy must be lost before the temperature of the water drops by more than a degree. Thus, a lake can store an enormous quantity of energy, and so bodies of water have a profound influence on local weather.

Table 7.1

Specific Heat Values for Some Elements, Compounds, and Common Solids

Substance	Name	Specific Heat (J/g · K)
Elements		
Al	aluminum	0.902
C	graphite	0.720
Fe	iron	0.451
Cu	copper	0.385
Au	gold	0.128
Compounds		
$NH_3(\ell)$	ammonia	4.70
$H_2O(\ell)$	water (liquid)	4.184
$C_2H_5OH(\ell)$	ethyl alcohol	2.46
$(CH_2OH)_2(\ell)$	ethylene glycol (antifreeze)	2.42
$H_2O(s)$	water (ice)	2.06
$CCl_4(\ell)$	carbon tetrachloride	0.861
CCl_2F_2	a chlorofluorocarbon (CFC)	0.598
Common Solids		
wood		1.76
concrete		0.88
glass		0.84
granite		0.79

Knowing the specific heat of a substance, you can calculate the heat transferred to or from that substance by measuring its mass and the size of its temperature change. Conversely, you can calculate the temperature change that should occur when a given quantity of heat is transferred to or from a sample of known mass. The equation that defines specific heat allows you to do this, but it may be more convenient to rewrite it in the following form:

$$\text{heat transferred} = (\text{specific heat})(\text{mass})(\text{temperature change of substance})$$
$$q = (\text{specific heat in J/g} \cdot \text{K})(\text{mass in g})(\Delta T \text{ in kelvins})$$

where the *quantity of heat transferred by heating* is symbolized by ***q***.

The symbol q is used to designate the quantity of heat transferred between objects. The symbol Δ (the Greek letter delta) means "change in."

To this point, we have emphasized only the *quantity* of heat transferred. However, Equation (1) allows a determination not only of the quantity of heat but also the *direction* in which it is transferred. When ΔT is calculated as

$$\Delta T = \text{final temperature} - \text{initial temperature}$$

it will have an algebraic sign: positive (+) for an increase and negative (−) for a decrease. If the temperature of the substance increases, ΔT has a positive (+) sign *and so does q*. This means that heat was transferred *to* the substance. The opposite case, a decrease in the temperature of the substance,

A Deeper Look

Sign Conventions in Energy Calculations

Whenever you take the difference between two quantities in chemistry, you should *always subtract the initial quantity from the final quantity.* A natural consequence of this convention is that the algebraic sign of the calculated result indicates an increase (+) or a decrease (−) in the quantity for the substance being studied. This is an important point, as you will see in other chapters of this book.

Thus far, we have described temperature changes and the direction of heat transfer. The table below summarizes the conventions used.

Be sure to understand that the sign of q is just a "signal" to tell us the direction of heat transfer. Heat itself cannot be negative; it is simply a quantity of energy. As an example, consider your bank account. Assume you have \$26 in your account (A_{initial}), and after a withdrawal you have \$20 (A_{final}). The cash flow is thus

$$\text{cash flow} = A_{\text{final}} - A_{\text{initial}} = \$20 - \$26 = -\$6$$

The negative sign on the \$6 indicates that a withdrawal has been made; the cash itself is not a negative quantity. Thus, when we talk about heat we shall use an unsigned number. However, when we want to indicate the direction of transfer in a process, we shall attach a negative sign (heat transferred *from* the substance) or a positive sign (heat transferred *into* the substance) to the value of q.

Change in T *of Object*	*Sign of* ΔT	*Sign of* q	*Direction of Heat Transfer*
increase	+	+	heat transfer into substance
decrease	−	−	heat transfer out of substance

means that ΔT has a negative (−) sign and so does q; heat was transferred *from* the substance.

As an example of the use of specific heat, you can find the number of joules of heat energy transferred from a cup of coffee to your body and the surrounding air when the temperature of the coffee drops from 60.0 °C to 37.0 °C (normal body temperature). Assume the cup holds 250. mL of coffee with a density of 1.00 g/mL, so the coffee has a mass of 250. g; also assume the specific heat of coffee is the same as that of water. The quantity of heat transferred can be obtained from Equation (1).

$$\text{heat transferred} = q = (4.184 \text{ J/g}\cdot\text{K})(250.\text{ g})(\underset{\text{final temp.}}{37.0\text{ °C}} - \underset{\text{initial temp.}}{60.0\text{ °C}})$$

$$= -24.1 \times 10^3 \text{ J} = -24.1 \text{ kJ}$$

Notice that the final answer has a *negative* value. In this case heat energy is transferred from the coffee to the surroundings, and the temperature of the coffee decreases.

EXAMPLE 7.2

Heat Transfer Between Substances

Suppose a 55.0-g piece of iron is heated in the flame of a bunsen burner to 425 °C and then plunged into a beaker of water (Figure 7.6). The beaker holds 600. mL of water (density = 1.00 g/mL), and its temperature before the iron is dropped in is 25 °C. What is the *final* temperature of both the water and the piece of iron?

(Assume that no heat transfers through the walls of the beaker or to the atmosphere.)

SOLUTION The most important aspects of this problem are that (1) the water and the iron bar will end up at the same temperature (T_{final} is the same for both), and (2) as expected from the law of energy conservation, the amount of heat taken on by the water and the amount given up by the iron are equal in quantity but *not* in algebraic sign. Thus, q_{iron} has a negative value because its temperature dropped as heat was transferred *out* of the iron. Conversely, q_{water} has a positive value because its temperature increased as heat was transferred *into* the water. Thus, $q_{iron} = -q_{water}$. Using the specific heats of iron and water from Table 7.1, we have

$$q_{iron} = -q_{water}$$

$$(55.0\ g)(0.451\ J/g \cdot K)(T_{final} - 425\ °C) = -(600.\ g)(4.184\ J/g \cdot K)(T_{final} - 25\ °C)$$

where $T_{initial}$ for the iron is 425 °C and $T_{initial}$ for the water is 25 °C. Solving, we find T_{final} = 29 °C. The iron has indeed cooled down (ΔT_{iron} = −396 °C) and the water has warmed up (ΔT_{water} = 4 °C) to the same final temperature (T_{final}), a temperature between the two initial values.

A final comment: Don't be confused by the fact that the transfer of one quantity of heat has caused two different values of ΔT; this is because the specific heats and quantities of iron and water are so different.

$$\Delta T_{iron} = 29\ °C - 425\ °C = -396\ °C$$

$$\Delta T_{water} = 29\ °C - 25\ °C = 4\ °C$$

EXERCISE 7.3 • *Using Specific Heat*

If 24.1 kJ of heat is used to warm a piece of aluminum with a mass of 250. g from an initial temperature of 5.0 °C, what would be the final temperature of the aluminum? The specific heat of aluminum is 0.902 J/g · K.

EXERCISE 7.4 • *Specific Heat and Temperature Changes*

Suppose you heat a small piece of rock (granite) and a glass of water in a bunsen burner flame for 3 minutes. If the rock and the water have the same mass, and both are heated at the same rate, which has the higher temperature when heating is stopped?

EXERCISE 7.5 • *Heat Transfer Between Substances*

A piece of iron (400. g) is heated in a flame and then dropped into a beaker containing 1000. g of water. The original temperature of the water was 20.0 °C, but it is 32.8 °C after the iron bar has been dropped in and both have come to the same temperature. What was the original temperature of the hot iron bar?

7·3 ENERGY AND ENTHALPY

So far we have described transfers of energy that occur between objects as a result of temperature differences. But energy transfers also occur when matter is transformed from one form to another; that is, when physical or chemi-

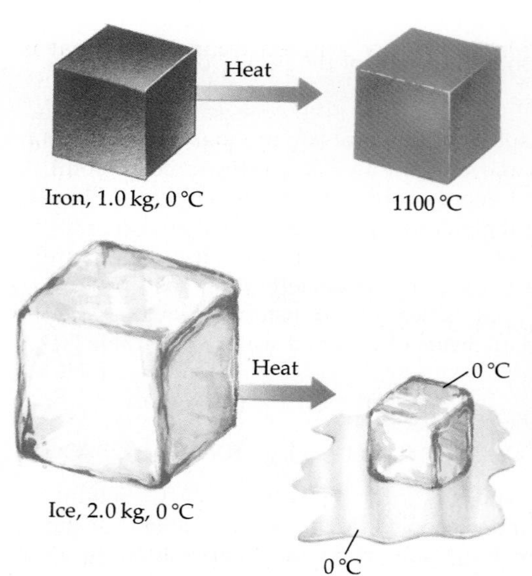

Figure 7.8
Heat transfer to an object can cause a temperature change, a phase change, or both. Here the same quantity of heat (500 kJ) has been transferred to a 1-kg block of iron at 0 °C as has been transferred to a 2-kg block of ice at 0 °C. The temperature of the iron block increases to 1100 °C, making the block red-hot. The same quantity of heat leads only to the melting of 1.5 kg of ice, and both the remaining ice and the melted water are still at 0 °C.

cal changes take place. The simplest of these processes are changes of state, and so we shall describe them before considering energy transfers that accompany chemical reactions.

Energy Transfers and Changes of State

Changes of state (from solid to liquid, liquid to gas, or solid to gas) take place without a change in temperature.

When a solid melts, its atoms, molecules, or ions move about vigorously enough to break partially free from their neighbors in the crystal lattice. When a liquid boils, particles move much farther apart from each other. In both cases attractive forces among the particles must be overcome, and this requires an input of energy. Joseph Black, whom we mentioned a few pages ago, was the first to recognize that heat is associated not only with temperature changes but also with changes of state (Figure 7.8), which always take place at constant temperature.

Fusion means melting.

The quantity of heat required to melt ice at 0 °C, which is 333 J/g, is called the **heat of fusion** of the ice. You see in Figure 7.8 that the same quantity of heat (500. kJ) required to raise the temperature of a 1.00-kg block of iron from 0 °C to 1100 °C, where it glows red-hot, will only lead to the melting of 1.50 kg of ice, where the ice and the melted water both have a temperature of 0 °C.

$$500.\ \cancel{kJ} \cdot \frac{1000\ \cancel{J}}{\cancel{kJ}} \cdot \frac{1.00\text{ g ice melted at 0 °C}}{333\ \cancel{J}} = 1.50 \times 10^3 \text{ g ice melted}$$

The quantity of heat required to convert liquid water to vapor, called the **heat of vaporization,** is 2260 J/g at 100 °C. What quantity of water can be vaporized if you transfer 500. kJ of energy to water at 100 °C? Only 221 grams!

$$500.\ \cancel{kJ} \cdot \frac{1000\ \cancel{J}}{\cancel{kJ}} \cdot \frac{1.00\text{ g water vaporized}}{2260\ \cancel{J}} = 221 \text{ g liquid water vaporized}$$

It is interesting that the mass of ice melted by a given quantity of heat is much greater than the mass of liquid water vaporized by the same quantity of heat. The difference is explained by the much greater separation of water molecules in the gas phase, which requires that forces between the molecules be almost completely overcome.

EXERCISE 7.6 • *Changes of State*

Assume you have 1 cup of ice (237 mL or 237 g) at 0.0 °C. How much heat is required to melt the ice and then warm the resulting water to 25.0 °C?

Figure 7.9
A system absorbing heat at constant volume. When heat is absorbed by solid CO_2 in a closed container, the pressure in the container increases as the solid is converted to gas. However, because there is no mechanical connection between the system and its surroundings, no work is done on the surroundings by the system. Therefore, q_v = heat absorbed at constant volume = ΔE. *(C.D. Winters)*

Thermodynamics and Changes of State

Carbon dioxide undergoes a change of state from solid to gas, a process called *sublimation*, at −78 °C.

$$CO_2(\text{solid}, -78\ °C) + \text{heat} \longrightarrow CO_2(\text{gas}, -78\ °C)$$

To describe the change from solid to gas in thermodynamic terms, we first need to extend the earlier discussion of heat as energy transferred between objects. In thermodynamics one of the "objects" is usually of primary concern. It may be a substance involved in a change of state, or one or more substances undergoing a reaction. It is called the **system.** The other "object" is called the **surroundings,** and it includes everything outside the system that can exchange energy with the system. In Figures 7.9 and 7.10 a CO_2 sample (both solid and vapor) is the system, and it interacts with its surroundings—the flask or plastic bag, the table on which they rest, and the air in the room surrounding the sample (and eventually the rest of the universe). In general, a system may be contained within an actual physical boundary, such as a flask or a cell in your body. Alternatively, the boundary may be purely imaginary. For example, you could study the solar system within its surroundings, the rest of the galaxy. In any case, the system can always be defined precisely.

Suppose that our system consists of solid and gaseous CO_2 and that both system and surroundings are at −78 °C. The CO_2 molecules are much more widely separated in the gas phase. Just as the potential energy of a ball is greater when it is farther above the earth, which attracts it, the potential energy of a CO_2 molecule is greater when it is farther from another CO_2 molecule, again because there is an attractive force to overcome as the two are separated. Thus, the energy of a mole of $CO_2(g)$ is greater than the energy of a mole of $CO_2(s)$. Where does this energy come from? That depends on the circumstances.

Suppose that the $CO_2(s)$ and $CO_2(g)$ are inside a *rigid container* so that the total volume cannot change (Figure 7.9). The energy needed for one molecule to escape from solid to gas can be supplied by neighboring molecules, which consequently vibrate a little less about their positions in the crystal lattice. As more and more molecules escape, the motion of the re-

Figure 7.10
(a) Pieces of dry ice and an empty plastic bag. (b) Crushed dry ice has been placed inside the plastic bag. (c) The bag has been closed. Some dry ice has sublimed, converting CO_2(s) at −78 °C to CO_2(g) at −78 °C. The gas has filled the bag and has done work by raising a book that has been placed on the bag. (d) A similar experiment but without the book. The expanding CO_2 in this case has pushed aside the air that formerly occupied the space taken up by the inflated bag. Pushing aside the atmosphere requires work just as pushing the book does to raise it. Work must always be done to push aside the atmosphere by any reaction whose products occupy greater volume than the reactants. The heat absorbed by the system under these constant pressure conditions ($= q_p$) is identified as the enthalpy change, ΔH.
(C.D. Winters)

maining solid-state molecules is more and more diminished, which corresponds to a lowering of the temperature. However, atoms and molecules in the surroundings (the rigid container) are in contact with CO_2 molecules in the solid, and energy exchange can occur. As soon as the temperature of solid CO_2 drops slightly below the temperature of the surroundings, more energy transfers into the CO_2 than transfers out, and the temperature of CO_2(s) is restored to −78 °C. In order for the temperature not to drop below −78 °C, the heat transfer from the surroundings must equal exactly the energy required to overcome attractive forces as solid CO_2 sublimes.

What has just been described is transfer of energy as heat from surroundings to system while the process $CO_2(s) \rightarrow CO_2(g)$ takes place. In such a case the temperature of the system and the immediate surroundings at first goes down, but then thermal energy is transferred from the surroundings and constant temperature is maintained. When heat must be transferred into a system to maintain constant temperature as a process occurs, the process is said to be **endothermic.** In contrast, the reverse process, converting CO_2 gas to solid CO_2, is an **exothermic** process; heat is transferred

It is assumed that the surroundings contain a great deal more matter than the system and hence have a much greater heat capacity; if so, the change in temperature of the surroundings would be so small that it cannot be measured.

out of the sample of CO_2 to the surroundings when some of the $CO_2(g)$ condenses to a solid. In summary,

Phase Change	*Direction of Heat Transfer*	*Sign of q_{system}*	*Type of Change*
solid $CO_2 \longrightarrow CO_2$ gas	surroundings $\longrightarrow$ system	positive	*endo*thermic
CO_2 gas $\longrightarrow$ solid CO_2	system $\longrightarrow$ surroundings	negative	*exo*thermic

When the endothermic process $CO_2(s) \rightarrow CO_2(g)$ occurs inside a rigid container, the only exchange of energy is the heat transferred into the system. Thus, by conservation of energy, the *heat absorbed at constant volume*, q_v, must equal the change in energy of the system, ΔE.

If the process does not occur in a rigid container the situation is more complicated, because there are two energy transfers, not just one. As shown in Figure 7.10c, $CO_2(s)$ vaporizing into a flexible container can raise a book above a table top. Raising a book is an example of doing work. Not only is there a transfer of energy into the solid CO_2 as heat, but there is also a transfer of energy out the CO_2 gas as work. Work is done whenever something is caused to move against an opposing force. In this case the gas has to work against the weight of the book to raise it.

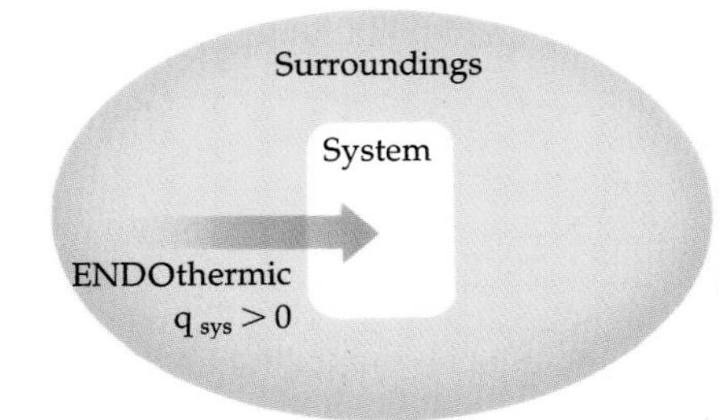

ENDOthermic: energy transferred from surroundings to system

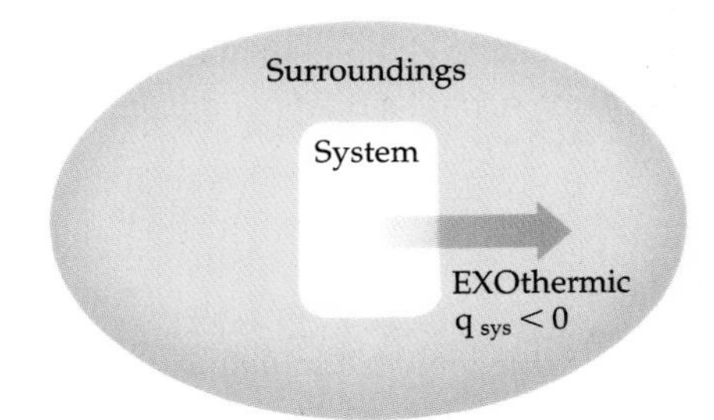

EXOthermic: energy transferred from system to surroundings

Since energy is now transferred in two ways between system and surroundings, it is no longer true that $\Delta E = q$. Energy transferred into the system by work, symbolized by w, must also be taken into account. The change in energy of the system is now

$$\Delta E = q + w$$

that is, the energy of the system is changed by the quantity of energy transferred as heat and by the quantity of energy transferred by work. This is a statement of the **first law of thermodynamics,** also called the law of energy conservation. Another way of saying the same thing is that *the total quantity of energy in the universe is constant.* Energy may be transferred as work or heat, but no energy can be lost, nor can heat or work be obtained from nothing. All the energy transferred between a system and its surroundings must be accounted for as heat and work. [This is true as long as no other energy transfer, such as the radiant energy of a glowing light stick (Figure 7.11) or the electric energy produced by a battery (Section 17.6) is involved.] The first law is important in all aspects of our lives; it plays a central role in our models of weather, in designing a power plant, or in understanding why diet and exercise together lead to weight loss.

Notice that w represents energy transferred, so it has a direction and a sign, just like q. The work w has a negative value if the system expends energy as work, and w has a positive value if the system receives energy in the form of work. For the process $CO_2(s) \rightarrow CO_2(g)$, w is negative because work is done on the surroundings when the atmosphere is pushed back.

Even if the book had not been on top of the plastic bag in Figure 7.10c, work would have been done by the expanding gas. This is because whenever a gas expands into the atmosphere it has to push back the atmosphere itself. Instead of raising a book the expanding gas raises a part of the atmosphere. For the process shown in Figure 7.10d, the energy transferred from the surroundings has two effects: (1) it overcomes the forces holding the molecules together in the solid state at −78 °C, and (2) it does work on the atmosphere as the gas expands. The energy that allows the system to perform work on its surroundings has to come from somewhere, and that

Figure 7.11
"Light sticks" give off light because a chemical reaction produces energy that is transferred to the surroundings in the form of light.
(C.D. Winters)

"somewhere" was energy, q, transferred as heat to the system from its surroundings. The system simply converted part of that heat into work. Therefore, the first law of thermodynamics tells us that the energy change, ΔE, for the CO_2 system must be less than q, since heat is transferred into the system from its surroundings (q has a positive sign) and work is done by the system on the surroundings (w has a negative sign). That is,

Systems that convert heat into work are called heat engines: an example is the engine in an automobile, which converts heat resulting from the combustion of fuel into work to move the car forward.

$$\Delta E = q + w = (q, \text{ positive number}) + (w, \text{ negative number})$$

and so

$$\Delta E < q$$

In plants and animals, as well as in the laboratory, reactions usually occur in contact with the atmosphere, that is, at constant atmospheric pressure. *The heat transfer into (or out of) a system at constant pressure,* q_p, *equals a quantity called the* **enthalpy change,** symbolized by $\mathbf{\Delta H}$ ("delta H"). Thus,

$$\Delta H = q_p = \text{heat transferred to or from a system at constant pressure}$$

Since $\Delta E = q + w$, then at constant pressure $\Delta E = \Delta H + w$, showing that ΔH accounts for all the energy transferred except the quantity that does the work of pushing back the atmosphere, which is usually small. Even if pressure is not constant,

$$\Delta H = (\text{enthalpy of the system at the end of a process}) - (\text{enthalpy of the system at the start of the process})$$

$$\Delta H = H_{\text{final}} - H_{\text{initial}}$$

If the enthalpy of the final system is greater than that of the initial system (as when solid CO_2 changes to CO_2 vapor at −78 °C), the enthalpy has in-

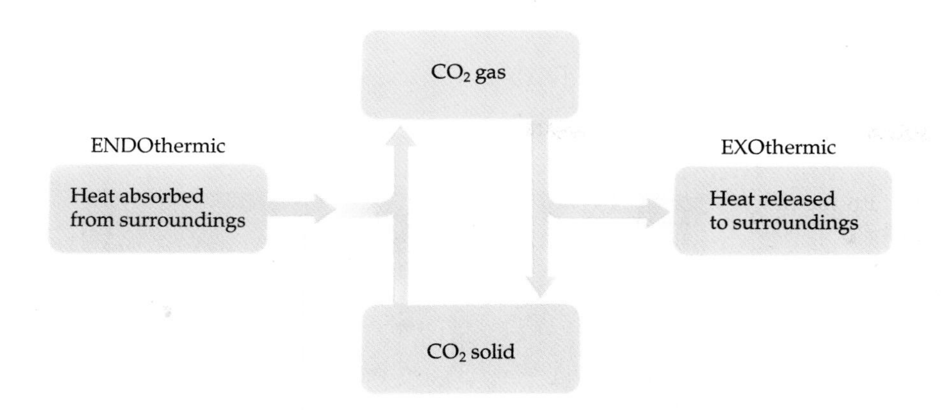

Figure 7.12
Energy diagram for the interconversion of solid CO_2 and CO_2 gas at constant pressure.

creased (left side of Figure 7.12), and the process has a positive ΔH; q_p must also be positive and the process is *endo*thermic. Conversely, if the enthalpy of the final system is less than that of the initial system, heat has transferred out of the system, and ΔH is negative; the process is *exo*thermic (right side of Figure 7.12).

Another endothermic process is the evaporation of water to vapor at 25 °C.

$$H_2O(\ell) + 44.0\text{ kJ} \longrightarrow H_2O(g) \qquad \Delta H = +44.0\text{ kJ}$$

Change is endothermic.
44.0 kJ of heat energy transferred from surroundings to the system (liquid H_2O).

Positive enthalpy change: ΔH has a positive value (ΔH > 0)
System ←heat transfer— Surroundings
***Endothermic** process with respect to system*

We have included heat as a "product" or "reactant" in some chemical equations as a signal to you that a process is exo- or endothermic, respectively. Be sure to understand that heat is not *a substance.*

But what about water vapor condensing to form liquid again? If 44.0 kJ of heat energy is required to break the attractions between H_2O molecules in one mole of the liquid so they can move into the gas phase, the same quantity of energy (44.0 kJ per mole) will be regained when the molecules in the vapor condense to form the liquid. Condensation of water is exothermic; 44.0 kJ is transferred from the water to the surroundings when a mole of H_2O molecules condenses to liquid.

$$H_2O(g) \longrightarrow H_2O(\ell) + 44.0\text{ kJ} \qquad \Delta H = -44.0\text{ kJ}$$

Change is exothermic.
44.0 kJ of heat energy transferred to surroundings from the system (H_2O vapor).

Negative enthalpy change: ΔH has a negative value (ΔH < 0)
System —heat transfer→ Surroundings
***Exothermic** process with respect to system*

Several very useful ideas that apply to all of thermodynamics may be noted here.

- When heat transfer occurs (at constant pressure) from a system to its surroundings, as when a vapor condenses to a liquid or solid, ΔH has a negative value. Conversely, when heat is absorbed from the surroundings, for example, when a solid or liquid changes to a vapor, ΔH has a positive value.
- For changes that are the reverse of each other, the ΔH values are numerically the same, but their signs are opposite. Thus, for evaporation of water $\Delta H = +44.0$ kJ/mol, while for the condensation of water $\Delta H = -44.0$ kJ/mol.

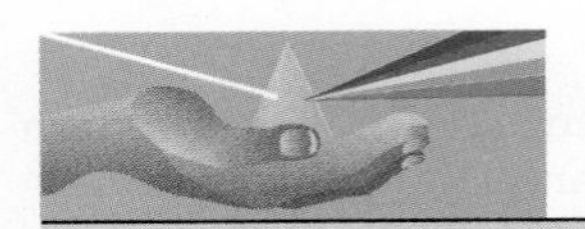

CHEMISTRY YOU CAN DO

Work and Heat

Here is an experiment that you can analyze with the first law of thermodynamics. All you need is a wide rubber band and a stiff upper lip. The rubber band is the thermodynamic system and your lip is part of the surroundings. Hold the rubber band against your lip and quickly stretch it to its limit. Your lip is a good detector of small temperature changes. What did you feel? Was the process of stretching the rubber band exothermic or endothermic? What was the sign of q and ΔE for this process?

When you stretched the rubber band you had to work to do it. What is the sign of w for the stretching process? Can you determine the sign of ΔE for the rubber band? If so, what is it? If not, why not? What kinds of experiments might you devise to make possible a determination of both the sign and the value of ΔE?

(C.D. Winters)

- The change in energy or enthalpy is directly proportional to the quantity of material undergoing a change. If two moles of water were evaporated, twice as much heat energy or 88.0 kJ would be required.
- The value of ΔH is always associated with a balanced equation for which the coefficients are read as moles, so that the equation shows the macroscopic amount of material to which the value of ΔH applies. Thus, for the evaporation of 2 mol of water

$$2\,H_2O(\ell) \longrightarrow 2\,H_2O(g) \qquad \Delta H = +88.0 \text{ kJ}$$

ΔH and Phase Changes
liquid ⟶ *solid + heat*
vapor ⟶ *liquid + heat*
ΔH = *negative number; change is* **exo***thermic*
liquid + heat ⟶ *vapor*
solid + heat ⟶ *liquid*
ΔH = *positive number; change is* **endo***thermic*

As an interesting footnote to the preceding discussion, we can calculate the energy transferred to the surroundings when water vapor in the air condenses to give rain in a thunderstorm. Suppose that one inch of rain falls over one square mile of ground, so that 6.6×10^{10} cm^3 or 6.6×10^{10} g of water (assuming a density of 1.0 g/cm^3) has fallen. The heat of vaporization of water at 25 °C is 44.0 kJ/mol. Therefore, the quantity of heat transferred to the surroundings from water vapor condensation is

$$6.6 \times 10^{10} \text{ g water} \cdot \frac{1 \text{ mol}}{18.0 \text{ g}} \cdot \frac{44.0 \text{ kJ}}{1 \text{ mol}} = 1.6 \times 10^{11} \text{ kJ}$$

Like many other examples in this chapter, this one assumes that temperature remains constant, so that all the energy associated with the phase change transfers to the surroundings.

Since the explosion of 1000 tons of dynamite is equivalent to 4.2×10^9 kJ, the energy transferred by our hypothetical thunderstorm is about the same as that released when 38,000 tons of dynamite explodes! This huge number tells you how much energy is "stored" in water vapor and why storms are such great forces of energy in nature.

EXERCISE 7.7 • *Changes of State and ΔH*

The enthalpy change for the vaporization of solid iodine is 62.4 kJ/mol.

$$I_2(s) \longrightarrow I_2(g) \qquad \Delta H = +62.4 \text{ kJ}$$

a. What quantity of heat energy must be used to vaporize 10.0 g of solid iodine?
b. If 3.45 g of iodine vapor condenses to solid iodine, what quantity of energy is involved? Is the process exo- or endothermic?

7.4 ENTHALPY CHANGES FOR CHEMICAL REACTIONS

The enthalpy change can be determined for any physical or chemical change. For a chemical reaction the products represent the "final system" and the reactants the "initial system." Therefore,

$$\Delta H = H_{products} - H_{reactants}$$

Like changes of state, chemical reactions can be exothermic or endothermic. Many of the reactions you have seen so far in photographs in this book [such as the Al + Br_2 reaction (Figure 5.1) and the reactions of elements with air (Figures 5.2 to 5.4)] are exothermic. To learn more about enthalpy changes that accompany chemical reactions, consider the decomposition of water vapor to its elements.

$$H_2O(g) \longrightarrow H_2(g) + \tfrac{1}{2}O_2(g) \qquad \Delta H = +241.8 \text{ kJ}$$

Change is endothermic; ΔH is positive. Decomposition of 1 mol of water vapor requires 241.8 kJ of energy to be transferred in from the surroundings.

(Note that in order to write an equation for the decomposition of one mole of H_2O it is necessary to use a fractional coefficient for O_2. This is acceptable with thermochemistry because coefficients are always taken to mean moles and not molecules.) The left side of Figure 7.13 shows that the enthalpy of the products of this reaction is greater than that of the reactants. Water vapor must *absorb* 241.8 kJ per mole of energy (at constant pressure) from the surroundings as it decomposes to its elements. That is, the enthalpy change for the *endo*thermic decomposition of water is ΔH = +241.8 kJ/mol.

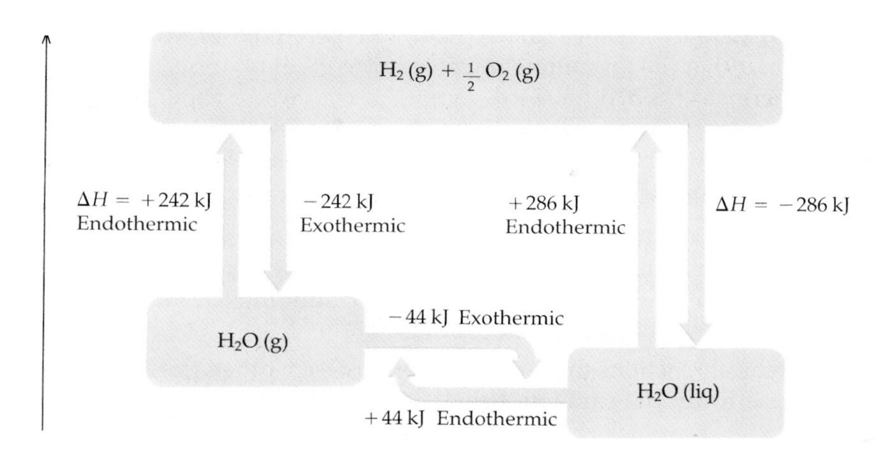

Figure 7.13
Enthalpy diagram for the interconversion of water vapor, $H_2O(g)$, liquid water, $H_2O(\ell)$, and the elements.

MREs *Are Heated in the* FRH

IF YOU HAVE BEEN IN THE army or Marine Corps you know about MREs—meals ready to eat—and FRHs—Flameless Ration Heaters. Soldiers need to eat when in the field, and so, until recently, they carried the canned C-rations made famous in World War II. However, beginning with Operation Desert Storm, the 1990–1991 war to push Iraq out of Kuwait, and in the recent effort to feed starving people in Somalia, soldiers carried MREs with an FRH.

The main course of a MRE is food such as chicken stew or spaghetti and meatballs in a pouch made of plastic and aluminum foil. The food can be heated by dropping the pouch into a pot of boiling water or leaning it on the exhaust manifold of an engine. But there may not be time to boil water or wait for dinner to heat on an engine. The alternative is the FRH.

To heat an MRE a soldier drops the food pouch into a bag-like sleeve, slides in the FRH, which is enclosed in a thin plastic sleeve of its own, and then adds a small amount of water to the sleeve. The FRH contains magnesium metal, which combines with water to form magnesium hydroxide in a very exothermic oxidation-reduction reaction.

$$Mg(s) + 2\,H_2O(\ell) \longrightarrow Mg(OH)_2(s) + H_2(g) + \text{heat}$$

Indeed, the reaction transfers enough energy to heat the meal without flame or smoke.

(U.S. Army Natick Research Development and Engineering Center)

The decomposition of water can be reversed; hydrogen and oxygen can combine to form water. The quantity of heat energy involved in the combination reaction (Figure 7.13)

$$H_2(g) + \tfrac{1}{2}\,O_2(g) \longrightarrow H_2O(g) \qquad \Delta H = -241.8\text{ kJ}$$

Change is exothermic; ΔH is negative.
1 mol of water vapor transfers 241.8 kJ of energy to the surroundings.

Because so much energy is provided per gram, H_2 is used as a fuel for the space shuttle and is being considered as fuel for cars and homes in the future. See R. Keating, Omni, *June, 1989, page 66.*

is the same as for the decomposition reaction except that the combination reaction is *exothermic.* That is, $\Delta H = -241.8$ kJ per mole of water vapor formed. The fact that the formation of water vapor gives off exactly the same amount of energy as needed for the decomposition of water vapor is another example of the law of the conservation of energy.

As in the case of phase changes, there are some useful ideas about enthalpy changes for reactions:

- When heat is evolved or transferred (at constant pressure) to the surroundings by an exothermic reaction, ΔH has a negative value. Conversely, when heat is absorbed from the surroundings by an endothermic reaction, ΔH has a positive value.
- For chemical reactions that are the reverse of each other, the ΔH values are numerically the same, but their signs are opposite.

The chemistry of the FRH is quite interesting. In addition to the magnesium, the inventors of the FRH found that the package had to contain some iron powder and salt. The function of the iron is not entirely clear, but the chloride ions of the salt promote the water-magnesium reaction. If the magnesium in the FRH has been exposed to moist air at any time, its surface will be coated with a thin layer of insoluble magnesium hydroxide. It is thought that the chloride ion of the NaCl in the FRH reacts with the surface $Mg(OH)_2$ to produce MgCl(OH), a more soluble species than $Mg(OH)_2$. This removes the $Mg(OH)_2$ film from the metal's surface, and the way is open for the pure magnesium beneath the coating to react with water.

If you have never seen an MRE with its FRH, you might find in a camping store a hand warmer that works on the same principle. One type of hand warmer contains iron powder and other chemicals. It also works by an oxidation-reduction reaction, but in this case it is the combination of iron with the O_2 of air.

$$4\ Fe(s) + 3\ O_2(g) \longrightarrow 2\ Fe_2O_3(s) + \text{heat}$$

This reaction is also very exothermic, and the hand warmer can maintain a temperature of 57 to 69 °C for several hours if the oxygen flow is somewhat restricted, as by keeping the warmer in one of your gloves. (See Scott, D., and Meadows, R., "Hot Meals," *Chem Matters*, February 1992, p. 12.)

(C.D. Winters)

- The change in energy or enthalpy is directly proportional to the quantity of material undergoing a change. If two moles of water decomposed, twice as much heat energy or 483.6 kJ would be required.
- In thermodynamics, the value of ΔH is always associated with a balanced equation for which the coefficients are read as moles, so that the equation shows the macroscopic amount of material to which the value of ΔH applies. Thus, for the decomposition of 2 mol of water,

$$2\ H_2O(g) \longrightarrow 2\ H_2(g) + O_2(g) \qquad \Delta H = +483.6\ \text{kJ}$$

Since energy is transferred when a substance undergoes a change of state, the quantity of energy associated with a chemical reaction must depend on the physical state (solid, liquid, or gas) of the reactants and products. For example, the decomposition of *liquid* water to H_2 and O_2

$$H_2O(\ell) \longrightarrow H_2(g) + \tfrac{1}{2}\ O_2(g) \qquad \Delta H = +285.8\ \text{kJ}$$

requires more energy than does decomposition of water vapor, as shown in Figure 7.13. The heat of vaporization, 44 kJ/mol, must be added to the enthalpy change for decomposition of $H_2O(g)$ to give the value for the decomposition of liquid water.

The Relationship of Reaction Heat and Enthalpy Change

reactant ⟶ product + heat
ΔH = *negative number; reaction is* ***exo****thermic*

reactant + heat ⟶ product
ΔH = *positive number; reaction is* ***endo****thermic*

Enthalpy changes for reactions have many practical applications. For instance, when enthalpies of combustion are known, the quantity of heat transferred by the combustion of a given mass of a fuel such as propane,

C_3H_8, can be calculated. Suppose you are designing a heating system, and you want to know how much heat can be provided by burning 454 g (1 pound) of propane gas in a furnace. The *exothermic* reaction that occurs is

$$C_3H_8(g) + 5\,O_2(g) \longrightarrow 3\,CO_2(g) + 4\,H_2O(\ell) \quad \Delta H = -2220 \text{ kJ}$$

and the enthalpy change is $\Delta H = -2220$ kJ *per mole of propane burned.* How much heat is transferred to the surroundings by burning 454 g of C_3H_8 gas? The first step is to find the number of moles of propane present in the sample.

$$454 \cancel{\text{g}} \cdot \frac{1 \text{ mol propane}}{44.10 \cancel{\text{g}}} = 10.3 \text{ mol propane}$$

Then you can multiply by the amount of heat transferred per mole of gas.

$$10.3 \cancel{\text{mol propane}} \cdot \frac{2220 \text{ kJ evolved}}{1.00 \cancel{\text{mol propane}}} = 22{,}900 \text{ kJ}$$

This is a substantial quantity of energy. When your body completely "burns" 454 g of milk, which is largely water, only about 1400 kJ is evolved.

EXERCISE 7.8 • *Heat Energy Calculation*

How much heat energy would be required to decompose 12.6 g of liquid water to the elements?

$$H_2O(\ell) \longrightarrow H_2(g) + \tfrac{1}{2}\,O_2(g) \qquad \Delta H = +285.8 \text{ kJ}$$

7.5 HESS'S LAW

Knowing how much heat is transferred as a chemical process occurs is very important. First of all, the direction of heat transfer is an important clue that helps predict the direction in which a chemical reaction will go: at room temperature most exothermic reactions are product-favored. Second, we can use ΔH to calculate the heat obtainable when a fuel is burned, as was done in the preceding section. Third, when reactions are carried out on a larger scale, as in a chemical plant that produces sulfuric acid or ethylene as a raw material for plastics, the surroundings must have enough cooling capacity to prevent an exothermic reaction from overheating and possibly damaging the plant. Therefore, we would like to know ΔH values for as many reactions as possible. For many reactions direct experimental measurements can be made by using a device called a calorimeter, but for many other reactions this is not a simple task. Besides, it would be very time-consuming to measure values for every conceivable reaction, and it would take a great deal of space to tabulate so many values. Fortunately there is a better way. It is based on the fact that mass *and* energy are conserved in chemical reactions.

Energy conservation is the basis of **Hess's Law,** which states that, *if a reaction is the sum of two or more other reactions, then ΔH for the overall process must be the sum of the ΔH values of the constituent reactions.* For example, as illustrated in the preceding section (Figure 7.13) for the decomposition of *liquid* water into its elements $H_2(g)$ and $O_2(g)$ (with all substances at 25 °C), the two successive changes are (1) the vaporization of liquid water and (2) the decomposition of water vapor to the elements (with all substances at 25 °C). The equation and the ΔH value for the overall process can be found by adding the equations and the ΔH values for the two steps:

$$
\begin{array}{llll}
(1) & H_2O(\ell) \longrightarrow H_2O(g) & & \Delta H = +\ 44.0\ kJ \\
(2) & H_2O(g) \longrightarrow H_2(g) + \frac{1}{2}\,O_2(g) & & \Delta H = +241.8\ kJ \\
\hline
(1) + (2) & H_2O(\ell) \longrightarrow H_2(g) + \frac{1}{2}\,O_2(g) & & \Delta H = +285.8\ kJ
\end{array}
$$

Here, $H_2O(g)$ is a product of the first reaction and a reactant in the second. Thus, as in adding two algebraic equations where the same quantity or term appears on both sides of the equation, $H_2O(g)$ can be canceled out. The net result is an equation for the overall reaction and its associated enthalpy change.

EXAMPLE 7.3

Using Hess's Law

Suppose you wish to know the enthalpy change for the removal of H_2 from C_2H_6, ethane, to give C_2H_4, ethylene, a key step in the production of the plastic polyethylene.

$$\underset{\text{ethane}}{C_2H_6(g)} \longrightarrow \underset{\text{ethylene}}{C_2H_4(g)} + H_2(g) \qquad \Delta H = ?$$

From experiments you know the following ΔH values:

(a) $2\,C_2H_6(g) + 7\,O_2(g) \longrightarrow 4\,CO_2(g) + 6\,H_2O(\ell)$ $\qquad \Delta H_a = -3119.4$ kJ

(b) $C_2H_4(g) + 3\,O_2(g) \longrightarrow 2\,CO_2(g) + 2\,H_2O(\ell)$ $\qquad \Delta H_b = -1410.9$ kJ

(c) $H_2(g) + \frac{1}{2}\,O_2(g) \longrightarrow H_2O(\ell)$ $\qquad \Delta H_c = -285.8$ kJ

Use this information to find the value of ΔH for the formation of ethylene from ethane.

SOLUTION The three reactions, (a), (b), and (c), as they are now written, cannot simply be added together to obtain the equation for the desired reaction. Equation (a) is the enthalpy change for the combustion of two moles of ethane when only one mole is required in the desired equation. Furthermore, C_2H_4 should be a product, but it is a reactant in equation (b), and H_2 is a reactant in equation (c), while it is a product in the desired equation. Therefore, we must change the three equations in some way so that they give the desired equation when added together.

First, since only one mole of ethane is required as a reactant in the desired equation, we multiply equation (a) by one half. Notice that the heat evolved must also be halved.

(a)′ = $\frac{1}{2}$ (a) $C_2H_6(g) + \frac{7}{2}\,O_2(g) \longrightarrow 2\,CO_2(g) + 3\,H_2O(\ell)$ $\qquad \Delta H_{a'} = -1559.7$ kJ

Next, we reverse equation (b) so that C_2H_4 is a reaction product. Recognize, though, that if the combustion of C_2H_4 is exothermic, the reverse reaction must be endothermic. This means the sign of the sign of the enthalpy change for equation (b), ΔH_b, must be reversed.

$$(b)' = -(b)\ \ 2\,CO_2(g) + 2\,H_2O(\ell) \longrightarrow C_2H_4(g) + 3\,O_2(g)$$
$$\Delta H_{b'} = -\Delta H_b = +1410.9\text{ kJ}$$

In equation (c) H_2 is a reactant, whereas it is a product in the desired equation. Therefore, we shall also reverse equation (c) and the sign of its enthalpy change.

$$(c)' = -(c)\ \ H_2O(\ell) \longrightarrow H_2(g) + \tfrac{1}{2}\,O_2(g) \qquad \Delta H_{c'} = -\Delta H_c = +285.8\text{ kJ}$$

Now it is possible to add equations (a)′, (b)′, and (c)′ to give the desired equation.

(a)′	$C_2H_6(g) + \tfrac{7}{2}\,O_2(g) \longrightarrow 2\,CO_2(g) + 3\,H_2O(\ell)$	$\Delta H_{a'} = -1559.7$ kJ
(b)′	$2\,CO_2(g) + 2\,H_2O(\ell) \longrightarrow C_2H_4(g) + 3\,O_2(g)$	$\Delta H_{b'} = +1410.9$ kJ
(c)′	$H_2O(\ell) \longrightarrow H_2(g) + \tfrac{1}{2}\,O_2(g)$	$\Delta H_{c'} = +285.8$ kJ
Net	$C_2H_6(g) \longrightarrow C_2H_4(g) + H_2(g)$	$\Delta H_{net} = +137.0$ kJ

When the equations are added you see that there are $3\tfrac{1}{2}$ moles of O_2 on opposite sides, so these cancel out. Similarly, there are three moles of H_2O on opposite sides, and these cancel out. We are left with the equation for the conversion of ethane to ethylene and hydrogen. Notice that the net result is an endothermic reaction, as we might expect when a molecule is broken down into simpler molecules.

EXERCISE 7.9 • *Using Hess's Law*

Lead has been known and used for centuries. To obtain the metal, lead sulfide (in the form of a common mineral called galena) is first roasted in air to form PbO.

$$PbS(s) + \tfrac{3}{2}\,O_2(g) \longrightarrow PbO(s) + SO_2(g) \qquad \Delta H = -413.7\text{ kJ}$$

Then lead(II) oxide is reduced with carbon to the metal.

$$PbO(s) + C(\text{graphite}) \longrightarrow Pb(s) + CO(g) \qquad \Delta H = +106.8\text{ kJ}$$

What is the enthalpy change for the following reaction?

$$PbS(s) + C(\text{graphite}) + \tfrac{3}{2}\,O_2(g) \longrightarrow Pb(s) + CO(g) + SO_2(g)$$

Is the reaction exothermic or endothermic? How much energy, in joules, is required (or evolved) when 454 g (1.00 pound) of PbS is converted to lead?

Galena. *(C.D. Winters)*

7.6 STANDARD ENTHALPIES OF FORMATION

Hess's law makes it possible to tabulate ΔH values for a few reactions and derive ΔH values for a great many other reactions by adding appropriate ones as just shown. Since ΔH values depend on temperature and pressure, it is necessary to specify both when a table of data is made. Usually a pressure of 1 bar and a temperature of 25 °C are specified, although other conditions of temperature are often used. The **standard state** of an element or compound is the most stable form of the substance in the physical state in which it exists at 1 bar and the specified temperature. Thus at 25 °C the standard state is $H_2(g)$ and for sodium chloride it is NaCl(s). For an element such as carbon that can exist in several different solid forms at 1 bar and 25 °C, the most stable form, graphite, is selected as the standard.

In 1982 the International Union of Pure and Applied Chemistry chose a pressure of 1 bar *for the standard state. This pressure is very close to the average atmospheric pressure observed near sea level (1 bar = 0.98692 atmosphere).*

When a reaction occurs with all the reactants and products in their standard states under standard conditions, the observed or calculated enthalpy change is known as the **standard enthalpy change of reaction, $\Delta H°$**, where the superscript ° indicates standard conditions. All the reactions we have discussed to this point have followed this convention, so all the ΔH values should have ° attached.

It is common to use the term "heat of reaction" interchangeably with "enthalpy of reaction." Understand that it is only the heat of reaction at ***constant pressure, q_p,*** *that is equivalent to the enthalpy change.*

The standard enthalpy change for *the formation of one mole of a compound from its elements* is called the **standard molar enthalpy of formation, ΔH_f°**, where the subscript $_f$ indicates that one mole of the compound in question has been *formed* in its standard state from its elements, also in their standard states. Some of the reactions already discussed define standard molar enthalpies of formation.

$$H_2(g) + \tfrac{1}{2}O_2(g) \longrightarrow H_2O(\ell) \qquad \Delta H_f^\circ = -285.8 \text{ kJ}$$
$$2\,C(s) + 2\,H_2(g) \longrightarrow C_2H_4(g) \qquad \Delta H_f^\circ = 52.26 \text{ kJ}$$

As another example, the following equation shows that 277.7 kJ would be *evolved* if graphite, the standard state form for carbon, were to combine with gaseous hydrogen and oxygen to form one mole of ethyl alcohol at 298 K.

$$2\,C(\text{graphite}) + 3\,H_2(g) + \tfrac{1}{2}O_2(g) \longrightarrow C_2H_5OH(\ell) \qquad \Delta H_f^\circ = -277.7 \text{ kJ}$$

Finally, it is important to understand that a standard molar enthalpy of formation, ΔH_f°, is just a special case of $\Delta H°$, the case where *one mole* of a compound is formed *from its elements*, as in the three preceding equations. In contrast, the enthalpy change for the exothermic reaction

To distinguish between enthalpies of formation, and enthalpy changes for other kinds of reactions, we will often use the symbols ΔH_f° and ΔH_{rxn}°, respectively.

$$CaO(s) + CO_2(g) \longrightarrow CaCO_3(s) \qquad \Delta H_{rxn}^\circ = -178.3 \text{ kJ}$$

is not an enthalpy of formation, since calcium carbonate has been formed from other compounds, not from its elements. And neither is $\Delta H°$ for the following reaction a standard enthalpy of formation.

$$P_4(s) + 6\,Cl_2(g) \longrightarrow 4\,PCl_3(\ell) \qquad \Delta H_{rxn}^\circ = -1278.8 \text{ kJ}$$

Here more than one mole of product is formed, even though it has formed from the elements, and $\Delta H_{rxn}^\circ = 4\,\Delta H_f^\circ\,[PCl_3(\ell)]$.

EXAMPLE 7.4

Writing Equations to Define Enthalpies of Formation

The standard enthalpy of formation of gaseous ammonia is −46.11 kJ/mol. Write the balanced equation for the formation reaction.

SOLUTION The equation to be written must show the formation of 1 mol of $NH_3(g)$ from the elements in their standard states; both N_2 and H_2 are gases at 25 °C and 1 bar. Therefore, the correct equation is

$$\frac{1}{2} N_2(g) + \frac{3}{2} H_2(g) \longrightarrow NH_3(g) \qquad \Delta H^\circ_{rxn} = \Delta H^\circ_f = -46.11 \text{ kJ}$$

where the subscript "rxn" is an abbreviation for "reaction").

Table 7.2 and Appendix J list values of ΔH°_f, obtained from the National Institute for Standards and Technology (NIST), for many other compounds. Notice that there are no values listed in these tables for elements such as C(graphite) or $O_2(g)$. *Standard enthalpies of formation for the elements in their standard states are zero,* because forming an element in its standard state from the same element in its standard state involves no chemical or physical change.

You can find the standard enthalpy change of any reaction if there is a set of reactions whose enthalpies are known and whose equations, when added together, will give the equation for the desired reaction. For example, suppose you are a chemical engineer and want to know how much heat is required to decompose limestone (calcium carbonate) to lime (calcium oxide) and carbon dioxide, with all substances at standard conditions.

$$CaCO_3(s) \longrightarrow CaO(s) + CO_2(g) \qquad \Delta H^\circ = ?$$

To do this, you find the following enthalpies of formation in a table such as Table 7.2 or Appendix J.

(a) enthalpy of formation of $CaCO_3$: $\Delta H^\circ_f[CaCO_3(s)] = -1206.9$ kJ/mol
(b) enthalpy of formation of CaO: $\Delta H^\circ_f[CaO(s)] = -635.1$ kJ/mol
(c) enthalpy of formation of CO_2: $\Delta H^\circ_f[CO_2(g)] = -393.5$ kJ/mol

Recall that giving the enthalpy of formation data in this format is just a convenient way of writing the following equations:

(a) $Ca(s) + C(\text{graphite}) + \frac{3}{2} O_2(g) \longrightarrow CaCO_3(s) \qquad \Delta H^\circ_f = -1206.9 \text{ kJ}$
(b) $Ca(s) + \frac{1}{2} O_2(g) \longrightarrow CaO(s) \qquad \Delta H^\circ_f = -635.1 \text{ kJ}$
(c) $C(\text{graphite}) + O_2(g) \longrightarrow CO_2(g) \qquad \Delta H^\circ_f = -393.5 \text{ kJ}$

Now add the three equations in such a way that the resulting equation is the one for the decomposition of limestone, that is,

$$CaCO_3(s) \longrightarrow CaO(s) + CO_2(g)$$

In equation (a), $CaCO_3(s)$ is a product, but it must appear in the final equation as a reactant. Therefore, equation (a) must be reversed *and* the sign of ΔH° must also be reversed. On the other hand, CaO(s) and $CO_2(g)$ must appear as products in the final equation, so equations (b) and (c) can be added with the same direction and sign of ΔH° as they have in the ΔH°_f equations:

$$\begin{array}{llr} -(a) & CaCO_3(s) \longrightarrow Ca(s) + C(\text{graphite}) + \frac{3}{2} O_2(g) & \Delta H^\circ = +1206.9 \text{ kJ} \\ +(b) & Ca(s) + \frac{1}{2} O_2(g) \longrightarrow CaO(s) & \Delta H^\circ_f = -635.1 \text{ kJ} \\ +(c) & C(\text{graphite}) + O_2(g) \longrightarrow CO_2(g) & \Delta H^\circ_f = -393.5 \text{ kJ} \\ \hline & CaCO_3(s) \longrightarrow CaO(s) + CO_2(g) & \Delta H^\circ = +178.3 \text{ kJ} \end{array}$$

Table 7.2

Selected Standard Molar Enthalpies of Formation at 298.15 K

Substance	Name	Standard Molar Enthalpy of Formation (kJ/mol)
$Al_2O_3(s)$	Aluminum oxide	−1675.7
$BaCO_3(s)$	Barium carbonate	−1216.3
$CaCO_3(s)$	Calcium carbonate	−1206.92
$CaO(s)$	Calcium oxide	−635.09
$CCl_4(\ell)$	Carbon tetrachloride	−135.44
$CH_4(g)$	Methane	−74.81
$CH_3OH(\ell)$	Methyl alcohol	−238.66
$C_2H_5OH(\ell)$	Ethyl alcohol	−277.69
$CO(g)$	Carbon monoxide	−110.525
$CO_2(g)$	Carbon dioxide	−393.509
$C_2H_2(g)$	Acetylene	+226.73
$C_2H_4(g)$	Ethylene	+52.26
$C_2H_6(g)$	Ethane	−84.68
$C_3H_8(g)$	Propane	−103.8
$n\text{-}C_4H_{10}(g)$	Butane	−888.0
$CuSO_4(s)$	Copper(II) sulfate	−771.36
$H_2O(g)$	Water vapor	−241.818
$H_2O(\ell)$	Liquid water	−285.830
$HF(g)$	Hydrogen fluoride	−271.1
$HCl(g)$	Hydrogen chloride	−92.307
$HBr(g)$	Hydrogen bromide	−36.40
$HI(g)$	Hydrogen iodide	+26.48
$KF(s)$	Potassium fluoride	−567.27
$KCl(s)$	Potassium chloride	−436.747
$KBr(s)$	Potassium bromide	−393.8
$MgO(s)$	Magnesium oxide	−601.70
$MgSO_4(s)$	Magnesium sulfate	−1284.9
$Mg(OH)_2(s)$	Magnesium hydroxide	−924.54
$NaF(s)$	Sodium fluoride	−573.647
$NaCl(s)$	Sodium chloride	−411.153
$NaBr(s)$	Sodium bromide	−361.062
$NaI(s)$	Sodium iodide	−287.78
$NH_3(g)$	Ammonia	−46.11
$NO(g)$	Nitrogen monoxide	+90.25
$NO_2(g)$	Nitrogen dioxide	+33.18
$PCl_3(\ell)$	Phosphorus trichloride	−319.7
$PCl_5(s)$	Phosphorus pentachloride	−443.5
$SiO_2(s)$	Silicon dioxide (quartz)	−910.94
$SnCl_2(s)$	Tin(II) chloride	−325.1
$SnCl_4(\ell)$	Tin(IV) chloride	−511.3
$SO_2(g)$	Sulfur dioxide	−296.830
$SO_3(g)$	Sulfur trioxide	−395.72

Taken from "The NBS Tables of Chemical Thermodynamic Properties," 1982.

When the equations are added in this fashion, one mole each of C(graphite) and Ca(s) and three-halves moles of O_2 appear on opposite sides and so are canceled out. Thus, the sum of these equations is the desired one for the decomposition of calcium carbonate, and the sum of the enthalpy changes of the three equations gives that for the desired equation.

Another very useful conclusion can be drawn from this example. The mathematics of the problem can be summarized by the expression

$$\text{enthalpy change for a reaction} = \Delta H^\circ_{rxn} = \Sigma\,[a\,\Delta H^\circ_f(\text{products})] - \Sigma\,[b\,\Delta H^\circ_f(\text{reactants})]$$

where Σ (the Greek letter sigma) means to "take the sum" and *a* and *b* represent the coefficients in the balanced equation. This expression shows that adding up the molar enthalpies of formation of the products (multiplying each by its stoichiometric coefficient in the balanced equation) and then subtracting the sum of the molar enthalpies of formation of the reactants (each having been multiplied by its coefficient) gives the enthalpy change of the reaction. *When the enthalpies of formation of* ***all*** *the compounds in a reaction are known, the enthalpy change of the reaction,* ΔH°_{rxn}, *can be calculated from the preceding equation.* This is a useful shortcut to writing the equations for the several formation reactions involved. Applying this to the decomposition of limestone, we have

$$\begin{aligned}\Delta H^\circ_{rxn} &= \Delta H^\circ_f[\text{CaO(s)}] + \Delta H^\circ_f[\text{CO}_2\text{(g)}] - \Delta H^\circ_f[\text{CaCO}_3\text{(s)}]\\ &= 1\text{ mol }(-635.1\text{ kJ/mol}) + 1\text{ mol }(-393.5\text{ kJ/mol}) - 1\text{ mol }(-1206.9\text{ kJ/mol})\\ &= 178.3\text{ kJ}\end{aligned}$$

EXAMPLE 7.5

Using Enthalpies of Formation

Nitroglycerin is a powerful explosive. Four different gases can form when it decomposes.

$$2\,C_3H_5(NO_3)_3(\ell) \longrightarrow 3\,N_2(g) + \tfrac{1}{2}\,O_2(g) + 6\,CO_2(g) + 5\,H_2O(g)$$

Given that the enthalpy of formation of nitroglycerin, ΔH°_f, is −364 kJ/mol, and using the other enthalpies found in Table 7.2, calculate the energy (heat at constant pressure) liberated when 10.0 g of nitroglycerin is detonated.

SOLUTION To solve this problem, we require the standard molar enthalpies of the products. Two are elements in their standard states (N_2 and O_2), so their values of ΔH°_f are zero. However, from Table 7.2, we have

$$\Delta H^\circ_f[CO_2(g)] = -393.5\text{ kJ/mol}$$
$$\Delta H^\circ_f[H_2O(g)] = -241.8\text{ kJ/mol}$$

The enthalpy change for the reaction can now be found:

$$\begin{aligned}\Delta H^\circ_{rxn} &= 6\text{ mol}\cdot\Delta H^\circ_f[CO_2(g)] + 5\text{ mol}\cdot\Delta H^\circ_f[H_2O(g)] - 2\text{ mol}\cdot\Delta H^\circ_f[C_3H_5(NO_3)_3(\ell)]\\ &= 6\text{ mol }(-393.5\text{ kJ/mol}) + 5\text{ mol }(-241.8\text{ kJ/mol}) - 2\text{ mol }(-364\text{ kJ/mol})\\ &= -2842\text{ kJ}\end{aligned}$$

This is the enthalpy change for the explosion of two moles of nitroglycerin. From this value, we can calculate the heat liberated by this exothermic reaction when only 10.0 g of nitroglycerin is used.

$$10.0 \text{ g} \cdot \frac{1 \text{ mol}}{227.1 \text{ g}} = 0.0440 \text{ mol nitroglycerin}$$

$$0.0440 \text{ mol nitro} \cdot \frac{2842 \text{ kJ}}{2 \text{ mol nitro}} = 62.6 \text{ kJ}$$

EXERCISE 7.10 • *Using Enthalpies of Formation*

Benzene, C_6H_6, is an important hydrocarbon. Calculate its enthalpy of combustion; that is, find the value of $\Delta H°$ for the reaction

$$C_6H_6(\ell) + \tfrac{15}{2}\ O_2(g) \longrightarrow 6\ CO_2(g) + 3\ H_2O(\ell)$$

The enthalpy of formation of benzene, $\Delta H_f^\circ[C_6H_6(\ell)]$, is +49.0 kJ/mol. Use Table 7.2 for any other values you may need.

7·7 MEASURING HEATS OF REACTION: CALORIMETRY

The heat evolved by a chemical reaction can be determined by a technique called **calorimetry.** Often, in finding heats of combustion or the caloric value of foods, the measurement is done in a *bomb calorimeter* (Figure 7.14). A

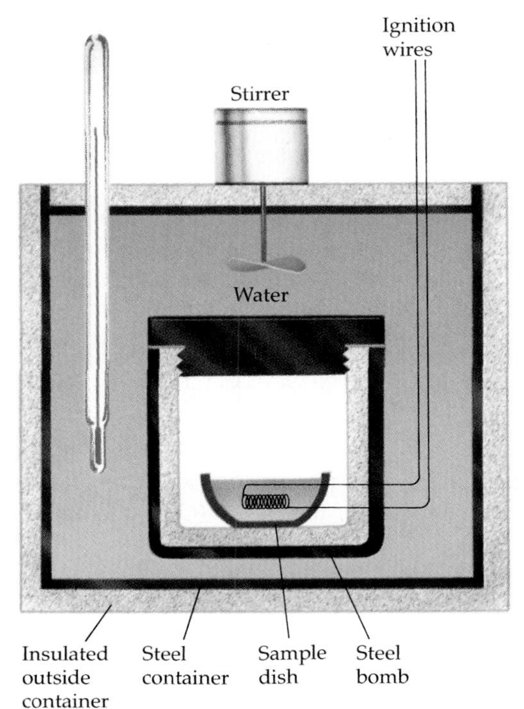

Figure 7.14
A combustion calorimeter. A combustible sample is burned in pure oxygen in a steel "bomb." The heat generated by the reaction flows into the bomb and into the water surrounding it, and warms both to the same temperature. By measuring the temperature increase, you can determine the heat evolved by the reaction.

weighed sample of a solid or liquid that is combustible is placed in a dish that is encased in a "bomb," a cylinder about the size of a large fruit juice can with heavy steel walls and ends. The bomb is then placed in a water-filled container with well-insulated walls. The bomb is filled with pure O_2, and the mixture of O_2 and sample is ignited, usually by an electrical spark. The heat generated when the sample burns warms up the bomb and the water around it, with both coming to the same temperature. In this configuration, the oxygen and the compound represent the *system* and the bomb and water around it are the *surroundings*. From the law of the conservation of energy, we can say that

heat transferred from the system = heat transferred into the surroundings

heat evolved by the reaction = heat absorbed by bomb and water

$$q_{reaction} = -(q_{bomb} + q_{water})$$

(where $q_{reaction}$ has a negative value because the combustion reaction is exothermic). The temperature change of the water, which is also equal to the change for the bomb, is measured. Then the total quantity of heat absorbed $(q_{bomb} + q_{water})$ can be calculated from the heat capacities of the bomb and the water. According to the equation above, this total gives the heat evolved by combustion of the compound.

Since the bomb is rigid, the heat transfer is measured at *constant volume* and is therefore equivalent to ΔE, the change in energy. As explained earlier (Section 7.3), $\Delta E = q_v$, while the change in enthalpy is the heat evolved or required at constant pressure, that is, $\Delta H = q_p$. However, because ΔE and ΔH are related in a relatively simple way, ΔH values can be calculated from ΔE values found in bomb calorimeter experiments.

To see how chemists can determine an enthalpy of combustion, consider octane, C_8H_{18}, a primary constituent of gasoline.

$$C_8H_{18}(\ell) + \tfrac{25}{2}\ O_2(g) \longrightarrow 8\ CO_2(g) + 9\ H_2O(\ell)$$

The heat capacity of the bomb, C_{bomb}, is usually found in a separate experiment by measuring the change in T produced by burning a measured mass of a compound of known ΔH.

Suppose that a 1.00-g sample of octane is burned in a calorimeter that contains 1.20 kg of water, and the temperature of the water and the bomb rises from 25.00 °C to 33.20 °C. If the heat capacity of the bomb, C_{bomb}, is known to be 837 J/K, calculate the heat transferred in the combustion of the 1.00-g sample of C_8H_{18}.

The heat that appears as a rise in temperature of the water surrounding the bomb is calculated as described in Section 7.2.

$$\begin{aligned} q_{water} &= (\text{specific heat of water})(m_{water})(\Delta T) \\ &= (4.184\ \text{J/g} \cdot \text{K})(1.20 \times 10^3\ \text{g})(33.20\ °\text{C} - 25.00\ °\text{C}) \\ &= +41.2 \times 10^3\ \text{J} \end{aligned}$$

The heat released by the reaction appears as a rise in temperature of the bomb and is calculated from the heat capacity of the bomb (C_{bomb}, units of J/K) and the temperature change, ΔT.

$$\begin{aligned} q_{bomb} &= C_{bomb} \cdot \Delta T \\ &= (837\ \text{J/K})(33.20\ °\text{C} - 25.00\ °\text{C}) \\ &= +6.86 \times 10^3\ \text{J} \end{aligned}$$

The total heat transferred by the reaction to the surroundings is equal to the negative of the sum of q_{bomb} and q_{water}. Thus,

$$\text{total heat transferred by 1.00 g of octane} = -(41.2 \times 10^3\ \text{J} + 6.86 \times 10^3\ \text{J})$$
$$= -(48.1 \times 10^3\ \text{J}) \text{ or } -48.1\ \text{kJ}$$

Any substance such as octane that burns exothermically in air and is available in large quantities at the surface of the earth constitutes an **energy resource.** Some examples are given in Table 7.3. In many cases we are interested in how much heat can be released per unit mass of an energy resource, and it is the energy released per gram burned that is tabulated. Octane, which is a constituent of gasoline, tops the list; it has the further advantage of being a liquid at room temperature and is therefore conveniently storable. One gram of a gaseous fuel would take up a lot more room unless it were highly compressed in a strong and heavy container. Even though other factors are also involved, it is primarily the amount of heat released by a reaction per unit mass of substance that determines its importance as an energy resource. That heat release can be calculated from standard enthalpies of formation for a wide variety of potential fuels.

EXERCISE 7.11 • *Determining the Heat Released by a Reaction*

A 1.00-g sample of ordinary table sugar (sucrose, $C_{12}H_{22}O_{11}$) is burned in a combustion calorimeter. The temperature of the 1.50×10^3 g of water in the calorimeter rises from 25.00 °C to 27.32 °C. If the heat capacity of the bomb is 837 J/K and the specific heat of the water is 4.184 J/g · K, calculate (a) the amount heat evolved per gram of sucrose and (b) the amount of heat evolved per mole of sucrose.

Table 7.3

Energy Released by Combustion of Some Substances

Substance	Energy Released (kJ/g)
gasoline	48
crude petroleum	43
typical animal fat	38
coal	29.3
charcoal	29
paper	20
dry biomass	16
air-dried wood or dung	15
bread	12
milk	3.0
beer	1.8

From J. Harte, *Consider a Spherical Cow,* University Science Books, Mill Valley, CA, 1988.

7.8 ENERGY IN INDUSTRY

For further reading on the use of energy in our society, see the special issue of Scientific American, *September 1990. See especially M. H. Ross and D. Steinmeyer, "Energy for Industry," page 89.*

Thermodynamic principles can be used to investigate energy use in our industrial society and how it relates to chemistry. Much of the chemicals industry is involved in the production of basic materials from air, minerals, and petroleum, and in doing so it uses about 15% of the energy resources consumed by all industries (Figure 7.15). Any energy saving is therefore significant.

There are three ways industries can save energy: (a) lower the energy costs of existing processes with energy-saving devices, (b) introduce process refinements, and (c) develop new, lower-energy methods of manufacture. The second and third of these are areas in which chemistry can most directly help.

Almost 34 billion pounds of ammonia were produced in the United States in 1990 (Figure 7.16). It is used directly as a fertilizer or converted to a plethora of other products such as nitric acid, urea (used as a fertilizer and in making plastics), and hydrazine (a widely used reducing agent). As described in Section 8.3, ammonia is made by the direct reaction of N_2 and H_2.

$$N_2(g) + 3\,H_2(g) \longrightarrow 2\,NH_3(g)$$

A recent refinement in the design of the reaction vessel raised the production rate of ammonia by 6% and reduced energy consumption by 5%. Of course, an even bigger saving would come about if biotechnologists could make new varieties of plants that could capture nitrogen directly from the air. Then we would not have to use ammonia as a fertilizer and would save the energy now used to make it. The manufacture of fertilizer consumes about 2% of all the energy resources used in industry, and fertilizers contribute to "greenhouse gases" that may cause global warming.

Almost 38 billion pounds of ethylene were produced in the United States in 1990 from the dehydrogenation of ethane.

$$C_2H_6(g) \longrightarrow H_2(g) + C_2H_4(g) \qquad \Delta H^\circ = +136.94\text{ kJ}$$

Much of the product is used to make polyethylene, the plastic used in many consumer items. However, a significant amount is used to manufacture other organic chemicals. Since ethylene production is the largest single consumer of energy resources in the chemical industry, there has been great

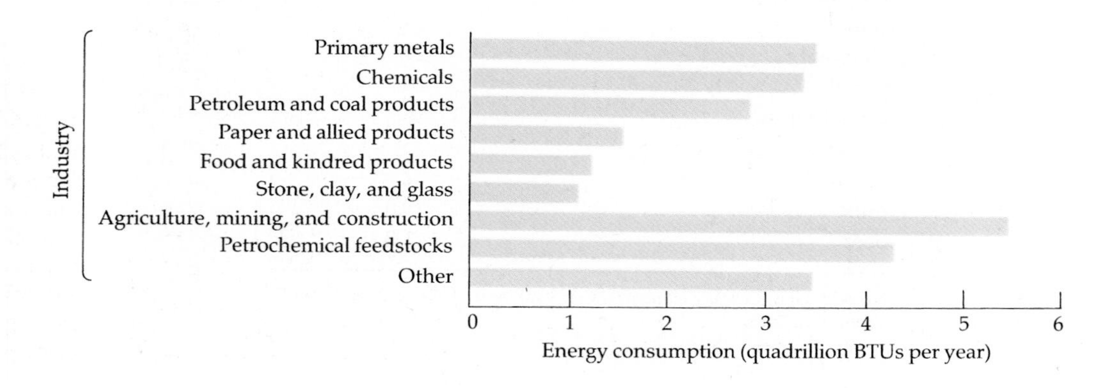

Figure 7.15
Industrial processes account for roughly 37% of the energy resources used in the United States, and the largest consumers are industries that convert raw materials. This figure shows the various uses of the energy consumed by industry. (Units are British thermal units, Btu, where 1 J = 9.484 × 10^{-4} Btu.)

Figure 7.16
A chemical plant in Houston, Texas, for the production of polymers. *(Courtesy Phillips Petroleum Co.)*

interest in improving the process to save energy and money. Many small improvements in ethane-to-ethylene conversion have led to a 60% decline in the energy requirement per pound of ethylene produced since 1960. Even so, the energy resources used to make ethylene from ethane are four times the minimum required by the first law of thermodynamics (that is, four times 136.94 kJ/mol of ethylene), largely due to inefficiencies in energy transfer.

Modern buildings use acres of glass, so the energy cost of glass manufacturing is an important consideration. Making glass requires large amounts of energy, first to melt the sand that makes up the glass and then to finish the glass. Here a development in manufacture led to significant savings. Beginning in the 1960s, molten glass was poured onto baths of molten tin (Figure 7.17). Since the melting point of tin (232 °C) is lower than that of

Figure 7.17
Plate glass is made by pouring molten glass onto a bath of molten tin. Since the surface of the liquid tin is smooth, as the glass cools the surface contacting the tin is completely smooth. The photo shows cooled plate glass emerging from the float-glass process. *(Courtesy PPG Industries)*

glass, and since glass has a lower density than molten tin, the glass floats on top of the liquid tin and solidifies. This process makes the glass smooth on both sides and eliminates the energy cost of grinding and polishing the glass, which was the practice several decades ago.

Recycling is also an important way to decrease energy resource consumption and is another application of thermodynamic principles. This topic is described in Section 13.8.

7·9 DIRECTIONALITY OF REACTIONS: ENTROPY

At room temperature most exothermic reactions are product-favored and therefore would be useful for carrying out chemical transformations from which we would like to have products. The explanation for this is very similar to our explanation of the one-way transfer of energy from a hotter to a colder object. When an exothermic reaction takes place, chemical potential energy that had been stored in relatively few atoms and molecules (the reactants) spreads over many more atoms and molecules as the products and especially the surroundings are heated. Since there are a great many more atoms and molecules in the surroundings than in the reactants and products, it is always true that after an exothermic reaction, energy will be distributed more randomly—dispersed over a much larger number of atoms and molecules—than it was before.

But why is dispersal of energy favored? The answer lies in probability. It is much more probable that energy will be dispersed than that it will be concentrated. To better understand energy dispersal and probability, consider the hypothetical case of a very small sample of matter consisting of two atoms, A and B, and suppose that this sample contains two units of energy. There are three ways the energy can be distributed over the two atoms: atom A could have both units of energy; atom A and atom B could each have one; or atom B could have both. Designate these three situations as A^2, AB, and B^2.

Now suppose that atoms A and B come into contact with two other atoms, C and D, that have no energy. Consider the possibilities for distributing the two units of energy over all four atoms. There are ten: A^2, AB, AC, AD, B^2, BC, BD, C^2, CD, and D^2. Only three of these (A^2, AB, and B^2) have all the energy in atoms A and B, which was the initial situation. When all four atoms are in contact, there are seven chances out of ten that some energy will have transferred from A and B to C and D. Thus, it is probable that the energy will become spread out over more than just the two atoms A and B. When this dispersal of energy occurs, we say that the atoms A, B, C, and D have gone from a situation where the energy was highly ordered to one where it has become more disordered.

Recall that for the same amount of substance or number of particles, higher thermal energy corresponds to higher temperatures. Therefore, a substance at a higher temperature has greater energy per particle on average. Dispersal of energy over a larger number of particles corresponds to transfer of energy from a substance at a higher temperature to another at a lower temperature.

The probability that energy will become disordered becomes overwhelming when large numbers of atoms or molecules are involved. For example, suppose that atoms A and B had been brought into contact with a mole of other atoms. There would still be only three arrangements in which

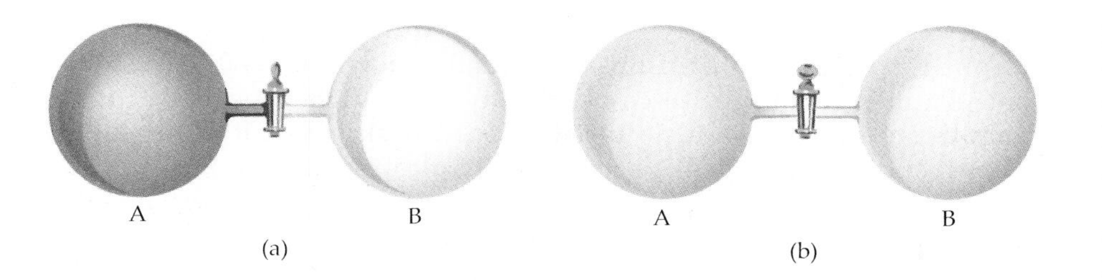

Figure 7.18
The expansion of a gas is a highly probable process. (a) A gas is confined in one flask, A. There are no atoms or molecules of gas in B. (b) When the valve between the flasks is opened, the gas atoms or molecules rush into flask B and eventually distribute themselves evenly between A and B.

all the energy was associated with atoms A and B, but there would be many, many more arrangements (more than 10^{47}) in which all the energy had been transferred to other atoms. In such a case it is so close to certain that energy would be transferred that it is not worth considering the alternative. In other words, if energy *can* be dispersed over a very much larger number of atoms or molecules, *it will be.*

Just as there is a tendency for highly concentrated energy to disperse, highly concentrated matter also tends to disperse, unless there is something to prevent it. For example, in a substance that is a gas at room temperature there are only weak forces *between* molecules. Suppose a gas is confined within a single flask connected through a stopcock to a second flask of equal size from which all gas molecules have been removed (Figure 7.18). What will happen if the stopcock is opened? Whoosh! The confined gas will expand to fill the entire volume available to it. Dispersal of the gas can be analyzed in the same way as dispersal of energy. Suppose that a single gas molecule occupies the two-flask system with the stopcock open. The molecule will move around within one flask until it hits the opening connecting to the other flask; then it will occupy the second flask for a while before returning to the first. Since the volumes are equal and the molecule's motion is random, it will spend half its time in each flask, on average. The probability is 1/2 that it can be found in flask A and 1/2 that it is in flask B. Now consider two molecules within the same system. There are four equally probable arrangements, as shown in Figure 7.19, but only one of them has both molecules in flask A. The probability that both molecules are in flask A is thus 1/4 or $(1/2)^2$. By making a similar diagram you can verify that for three molecules there are eight arrangements; only one of these corresponds to all three molecules in flask A, giving a probability of 1/8 or $(1/2)^3$. In general (with the stopcock open) the probability that all molecules will be in the same flask is $(1/2)^n$, where n is the number of molecules.

The low probability that a lot of energy will be associated with only a few atoms or molecules makes a substance with a lot of chemical potential energy valuable. We call substances like coal, oil, and natural gas "energy resources" and sometimes fight wars over them for this reason.

Atoms or molecules of a material like glass that is a solid at room temperature do not disperse, because there are strong attractive forces between them. Their tendency to disperse is only revealed if the temperature is raised a great deal, which causes melting and vaporization.

As the number of molecules increases, the probability of their all being in the same flask (a more highly ordered state) gets very, very small. For example, suppose that 0.100 mol of gas were in flask A. The number of

Figure 7.19
When two molecules are placed in two flasks of equal volume, four different arrangements are possible.

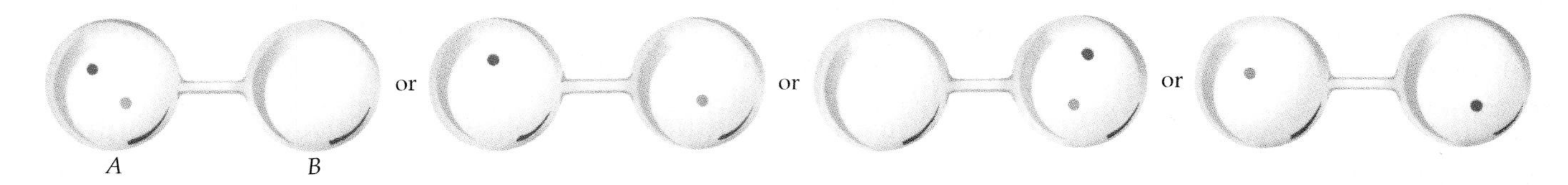

molecules in 0.100 mol is 0.602×10^{23}, and so the probability would be $(1/2)^{0.602 \times 10^{23}}$. This is an incredibly small number; if written as 0.000. . . there would be more than 10^{22} zeros before any other digit appeared. It is so highly improbable for all molecules to stay in flask A that it is absolutely certain that many molecules will move from flask A to flask B. This two-flask system is a very simple example of a product-favored system—the final arrangement with gas in both flasks is very much more probable than the initial one, and so the process occurs of its own accord. On the other hand, if we wanted to reverse the process by concentrating all the molecules into flask A, an outside influence such as a pump would be required—the pump could do work on the gas to force it into a highly improbable arrangement. The work done by the pump would be stored in the gas and could later be used for some other purpose.

To summarize, there are two ways that the final state of a system can be more probable than the initial one: having energy dispersed over a greater number of atoms and molecules, and having the atoms and molecules themselves more disordered. If both of these happen, then a reaction will definitely be product-favored, since the products and the distribution of energy will both be more probable. If only one of them happens, then quantitative information is needed to decide which effect is greater. (At room temperature, however, disordering of energy is more important than dispersal of matter, and *most exothermic reactions are product-favored.* At high temperatures the opposite is true—dispersal of matter becomes more important.) If neither matter nor energy is more spread out after a process occurs, then that process will be reactant-favored—the initial substances will remain no matter how long we wait.

Measuring Dispersal or Disorder: Entropy

The dispersal or disorder in a sample of matter can be measured with a calorimeter, the same instrument needed to measure the enthalpy change when a reaction occurs. The result is a thermodynamic function called **entropy** and symbolized by **S**. Measurement of entropy depends on the assumption that in a perfect crystal at the absolute zero of temperature (0 K or −273.15 °C) there is minimum disorder; this sets the zero of the entropy scale. When energy is transferred to matter in very small increments, so that the temperature change is very small, the entropy change can be calculated as $\Delta S = q/T$, the heat absorbed divided by the absolute temperature. By starting as close as possible to absolute zero and repeatedly introducing small quantities of energy, an entropy change can be determined for each small increase in temperature. These entropy changes can then be added to give the total (or absolute) entropy of a substance at any desired temperature.

Though it is impossible to cool anything all the way to absolute zero, it is possible to get very close, and there are ways of estimating the disorder that is already in a substance near 0 K; thus, accurate entropy values can be obtained for many substances.

The results of such measurements for several substances at 298 K are given in Table 7.4. These are standard molar entropy values, so they apply to one mole of each substance at the standard pressure of 1 bar. The units are joules per kelvin per mole.

Some interesting generalizations can be drawn from the examples given in Table 7.4.

- *Entropies of gases are generally much larger than those of liquids, which in turn are generally larger than those of solids.*

  In a solid the particles can only vibrate around their lattice positions. When a solid melts, its particles can move around more freely, and there is an increase in molar entropy. When a liquid vaporizes, the position restrictions due to forces between the particles nearly disappear, and there is another large entropy increase. For example, the entropies (in J/K · mol) of the halogens I_2(s), Br_2(ℓ), and Cl_2(g) are 116.1, 152.2, and 223.0, respectively. Similarly, the entropies of C(s, graphite) and C(g) are 5.7 and 158.1.

- *Entropies of more complex molecules are larger than those of simpler molecules, especially in a series of closely related compounds.*

  In a more complicated molecule there are more ways for the atoms to move about in three-dimensional space and hence greater entropy. For example, the entropies (in J/K · mol) of methane (CH_4), ethane (CH_3CH_3), and propane ($CH_3CH_2CH_3$) are 186.264, 229.60, and 269.9,

Table 7.4

Some Standard Molar Entropy Values at 298.15 K

Compound or Element	Entropy, $S°$ (J/K · mol)
C(graphite)	5.740
C(g)	158.096
CH_4(g)	186.264
C_2H_6(g)	229.60
C_3H_8(g)	269.9
CH_3OH(ℓ)	126.8
CO(g)	197.674
CO_2(g)	213.74
Ca(s)	41.42
Ar(g)	154.7
H_2(g)	130.684
O_2(g)	205.138
N_2(g)	191.61
H_2O(g)	188.825
H_2O(ℓ)	69.91
NH_3(g)	192.45
HCl(g)	186.908
F_2(g)	202.78
Cl_2(g)	223.036
Br_2(ℓ)	152.2
I_2(s)	116.135
NaF(s)	51.5
MgO(s)	26.94

Taken from "The NBS Tables of Chemical Thermodynamic Properties," 1982.

Portrait of a Scientist

Ludwig Boltzmann (1844–1906)

Ludwig Boltzmann was an Austrian mathematician who gave us a useful interpretation of entropy (and who also did much of the work on the kinetic theory of gases). Engraved on his tombstone in Vienna is his equation relating entropy and "chaos,"

$$S = k \log W$$

(where k is a fundamental constant of nature, now called Boltzmann's constant). Boltzmann said the symbol W was related to the number of ways that atoms or molecules can be arranged in a given state, always keeping their total energy fixed. Therefore, his equation tells us that if there are only a few ways to arrange the atoms of a substance—that is, if there are only a few places in which we can put our atoms or molecules—then the entropy is low. On the other hand, the entropy is high if there are many possible arrangements ($W \gg 1$), that is, if the level of chaos is high.

The tombstone of Boltzmann with the equation $S = k \log W$ engraved on it. (Oesper Collection in the History of Chemistry, University of Cincinnati)

respectively. For atoms or molecules of similar molar mass, we have Ar, CO_2, and $CH_3CH_2CH_3$ with entropies of 154.7, 213.74, and 269.9.

- *Entropies of ionic solids are larger the weaker the attractions among the ions.* The weaker such forces, the easier it is for ions to vibrate about their lattice positions. Examples are NaF(s) and MgO(s) with entropies of 51.5 and 26.8 J/K · mol; the 2+ and 2− charges on the magnesium and oxide ions result in greater attractive forces and hence lower entropy.
- *Entropy usually increases when a pure liquid or solid dissolves in a solvent.* Since Table 7.4 refers only to pure substances, no example values are available, but matter usually becomes more dispersed or disordered when a substance dissolves and different kinds of molecules mix together (Figure 7.20).
- *Entropy increases when a dissolved gas escapes from a solution.* Although gas molecules are dispersed among solvent molecules in solution, the very large entropy increase that occurs upon changing from the liquid phase to the gas phase results in a higher entropy for the separated gas and solvent than for the mixture.

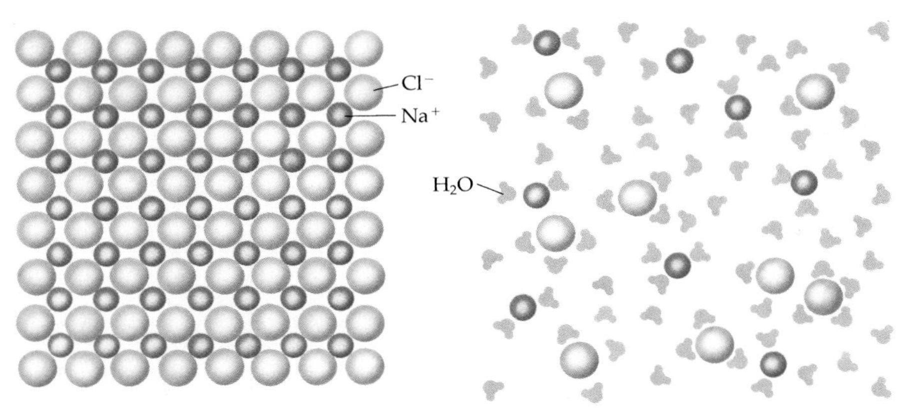

Figure 7.20
There is a great increase in entropy when a highly ordered solid dissolves in a solvent.

The entropies of reactant and product substances can be used to help predict the outcomes of physical and chemical processes. To do so, we assume that each reactant is present in the amount required by its stoichiometric coefficient and that each product is also present in the amount required by its coefficient. All substances are assumed to be at the standard pressure of 1 bar and at the temperature specified, so that the values in Table 7.4 apply. Then we can just add all the entropies of the products and subtract the entropies of the reactants to see whether there is an increase or decrease in entropy.

It is also convenient to use the generalizations given above to estimate whether there will be an increase in disorder of the substances involved when reactants are converted to products. Such predictions are much easier to make for entropy changes than for enthalpy changes. For the processes $H_2O(s) \longrightarrow H_2O(\ell)$ and $H_2O(\ell) \longrightarrow H_2O(g)$, we expect an entropy increase in each case, since water molecules in the solid are more ordered than in the liquid and much more ordered than in the gas (Figure 7.21). This is confirmed by entropy measurements. At 273.15 K, for example, ice can be converted to liquid water very slowly, and there is no temperature change. The quantity of energy transferred as heat is 6.02 kJ/mol, and the transfer of energy is into the water, so $q = 6020$ J/mol. This gives

$$\Delta S \text{ for } H_2O(s) \longrightarrow H_2O(\ell) = \frac{q}{T} = \frac{6020 \text{ J/mol}}{273.15 \text{ K}} = 22.0 \text{ J/K} \cdot \text{mol}$$

Similarly, in order to boil water we must transfer energy into the water, q must be positive, and so must ΔS; the value in this case is 109 J/K · mol.

$$\Delta S \text{ for } H_2O(\ell) \longrightarrow H_2O(g) = \frac{q}{T} = \frac{40{,}700 \text{ J/mol}}{373.15 \text{ K}} = 109 \text{ J/K} \cdot \text{mol}$$

For the decomposition of iron(III) oxide to its elements,

$$2\, Fe_2O_3(s) \longrightarrow 4\, Fe(s) + 3\, O_2(g) \qquad \Delta S° = +551.7 \text{ J/K}$$

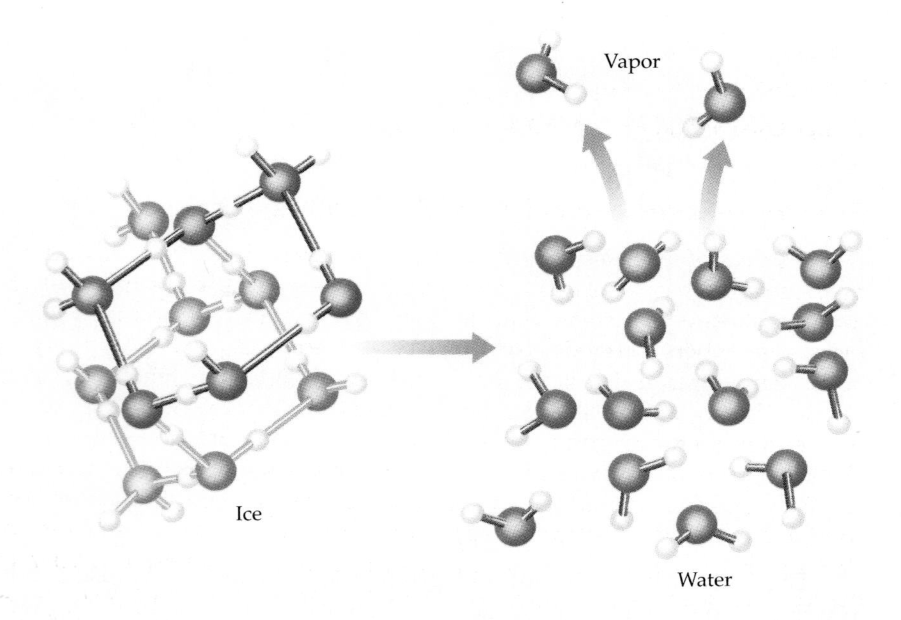

Figure 7.21
Increasing disorder in the melting of ice to liquid water and the evaporation of the liquid to vapor.

we would also predict an increase in entropy, because three moles of gaseous oxygen are present in the products, and the reactant is a solid. Since gases have much higher entropy than solids or liquids, gaseous substances are most important in determining entropy changes. An example where a decrease in entropy would be predicted is

$$2\ CO(g) + O_2(g) \longrightarrow 2\ CO_2(g) \qquad \Delta S^\circ = -173.0\ J/K$$

Here there are three moles of gaseous substances (two of CO and one of O_2) at the beginning but only two moles of gaseous substance at the end of the reaction. Two moles of gas almost always contain less entropy than three moles of gas, and so ΔS is negative. Another example of decreased entropy is the process

$$Ag^+(aq) + Cl^-(aq) \longrightarrow AgCl(s)$$

Here the reactant ions are free to move about among water molecules in aqueous solution, but those same ions are held in a crystal lattice in the solid, a situation with much greater constraint.

EXERCISE 7.12 • *Calculating ΔS Values for Phase Changes*

The enthalpy of vaporization of benzene (C_6H_6) is 30.8 kJ/mol at the boiling point of 80.1 °C. Calculate the entropy change for benzene going from (a) liquid to vapor and (b) vapor to liquid at 80.1 °C.

EXERCISE 7.13 • *Entropy*

For each of the following processes, tell whether you would expect entropy to be greater for the products than for the reactants; explain how you arrived at your prediction.

a. $CO_2(g) \longrightarrow CO_2(s)$

b. $NaCl(s) \longrightarrow NaCl(aq)$

c. $MgCO_3(s) + heat \longrightarrow MgO(s) + CO_2(g)$

Entropy and the Second Law of Thermodynamics

A great deal of experience with a great many chemical reactions and other processes in which energy is transferred has led to the **second law of thermodynamics,** which states that *the total entropy of the universe is continually increasing.* When a chemical or physical process occurs, matter, energy, or both become more dispersed or disordered. Evaluating whether this will happen during a proposed chemical reaction allows us to predict whether or not reactants will form appreciable quantities of products.

Such an evaluation can be made relatively easily if we consider only a carefully specified situation—a set of standard-state conditions defined in Section 7.6, namely 1 bar and a specified temperature. For reactions in solution a standard-state concentration of 1 mole/liter can be used. Once a prediction has been made, it can be adjusted to take into account differences from the standard-state conditions. Predicting whether a reaction is product-favored involves two steps: calculating how much entropy is created by dispersal of energy, and calculating how much entropy is created by dispersal of matter. Both calculations are carried out by assuming that reactants under standard conditions are converted into products under standard conditions. For example, if we want to evaluate the reaction

$$CO(g) + 2\,H_2(g) \longrightarrow \underset{\text{methanol}}{CH_3OH(\ell)}$$

as a possible way to manufacture liquid methanol for use in automobile fuel, we would base our prediction upon having as reactants 1 mol $CO(g)$ and 2 mol $H_2(g)$, each at a pressure of 1 atmosphere. We would further assume that the product would be 1 mol of methanol, also at a pressure of 1 atmosphere. If the total entropy is predicted to be higher after the product has been produced, then the reaction is product-favored under these conditions and might be useful. If not, perhaps some other conditions could be used, or perhaps we should consider some other reaction altogether.

Dispersal of energy by a chemical reaction can be evaluated by calculating $\Delta H°$ from tables such as Table 7.2 and assuming that this quantity of energy is transferred to or from the surroundings. (Recall that Table 7.2 contains values for the standard condition of 1 bar; the values also apply to a temperature of 298.15 K.) If the energy transfer is slow and occurs at a constant temperature, the entropy change for the surroundings can be calculated as:

We shall consider nonstandard conditions and the effect of temperature in the next chapter; rules will be developed to suggest how to change the conditions under which a reaction is carried out so as to produce more products than could be obtained under standard conditions.

$$\Delta S_{\text{surroundings}} = \frac{q_{\text{surroundings}}}{T} = \frac{-\Delta H_{\text{system}}}{T}$$

The equation says that for an exothermic reaction (negative ΔH_{system}) there will be an increase in entropy of the surroundings, a fact that we have already mentioned. For the proposed methanol-producing reaction ΔH_{system} is -128.14 kJ, and so the entropy change is

The minus sign in this equation comes from the fact that $-\Delta H_{system}$ is the energy transferred out of the system, and hence equals the energy transferred into the surroundings, $q_{surroundings}$.

$$\Delta S_{\text{surroundings}} = -\frac{(-128.1 \text{ kJ})(1000 \text{ J/kJ})}{298 \text{ K}} = 430. \text{ J/K}$$

To calculate the entropy change due to dispersal of matter, we use information in Appendix J and Table 7.4, subtracting the entropies of reactants from the entropy of the products. For example,

$$\begin{aligned}\Delta S_{\text{system}} &= \Sigma\, a\, S°(\text{products}) - \Sigma\, b\, S°(\text{reactants}) \\ &= S°\,[CH_3OH(\ell)] - \{S°\,[CO(g)] + 2\, S°\,[H_2(g)]\} \\ &= 1 \text{ mol } (126.8 \text{ J/K}\cdot\text{mol}) - [1 \text{ mol } (197.7 \text{ J/K}\cdot\text{mol}) \\ &\quad +2 \text{ mol } (130.7 \text{ J/K}\cdot\text{mol})] \\ &= -332.3 \text{ J/K}\end{aligned}$$

Notice that the equation for calculating ΔS *has the same form as that for calculating* ΔH *for a reaction (Section 7.6).*

Notice that this calculation gives the entropy change for the *system,* since it involves dispersal of the atoms and molecules that make up the system. In this case there is a large decrease in entropy because three moles of gaseous material are converted to one mole of a more complicated, but liquid-phase, product.

The *total* entropy change for a process, referred to as $\Delta S_{\text{universe}}$ (the entropy change for the universe), is the sum of the entropy change for the system and the entropy change for the surroundings. (We assume that nothing else but our reaction happens, and so there are no other entropy changes.) This total entropy change is

$$\Delta S_{\text{universe}} = \Delta S_{\text{system}} + \Delta S_{\text{surroundings}} = (-332.3 + 430.) \text{ J/K} = 98 \text{ J/K}$$

The process we are evaluating is accompanied by an increase in entropy of the universe. This means that if we had CO(g), $H_2(g)$, and $CH_3OH(\ell)$, each at 1 atmosphere pressure and all in contact with each other, some of the CO(g) and some of the $H_2(g)$ would react to form $CH_3OH(\ell)$. The process is product-favored and might be useful for manufacturing methanol.

Predictions of the sort we have just made by calculating $\Delta S_{\text{universe}}$ can also be made qualitatively, without calculating, if we know whether or not a reaction is exothermic and if we can predict whether there is dispersal of matter when the reaction takes place. *A reaction is sure to be product-favored if it is exothermic and proceeds from a state of lesser to greater disorder.* Also, *a reaction is certainly* ***not*** *product-favored if it is endothermic and there is a decrease in entropy for the system.* There are two other possible cases, as indicated in Table 7.5, but they are more difficult to predict without quantitative information.

Table 7.5

Predicting Whether a Reaction is Product-Favored

ΔH_{system}	ΔS_{system}	Product-Favored
−, reaction exothermic	+	yes
−, reaction exothermic	−	yes at low T; no at high T
+, reaction endothermic	+	no at low T; yes at high T
+, reaction endothermic	−	no

As examples of such predictions, consider gas-producing reactions of solids. They are product-favored because they are exothermic and produce highly disordered gases or liquids.

$$CaCO_3(s) + 2\,HCl(aq) \longrightarrow CaCl_2(aq) + H_2O(\ell) + CO_2(g) + \text{heat}$$

Similarly, combustion reactions are product-favored because they are exothermic and produce a larger number of product molecules from a few reactant molecules.

$$\underset{\text{butane}}{2\,C_4H_{10}(g)} + 13\,O_2(g) \longrightarrow 8\,CO_2(g) + 10\,H_2O(g) + \text{heat}$$

But what about a reaction such as the production of ethylene, C_2H_4, from ethane, C_2H_6? The reaction is very endothermic,

$$C_2H_6(g) \longrightarrow H_2(g) + C_2H_4(g) \quad \Delta H^\circ_{rxn} = +136.94\text{ kJ}$$

although the entropy is predicted to increase. (One gas-phase molecule forms two.) So, the enthalpy change would not lead to the reaction being product-favored, while the predicted entropy change suggests the opposite. Which is the more important? Calculating $\Delta S_{surroundings}$ involves dividing the enthalpy change by temperature, and ΔH° does not change much as temperature increases. Therefore, the greater the temperature the less important the ΔH° term is. At room temperature ΔH° is usually the deciding factor, so the ethylene-producing reaction is expected to be reactant-favored. This is indeed the case at 25 °C. To make this reaction work in industry, chemical engineers have designed a special process at about 1000 °C. At this higher temperature the ΔS° term is more important, and more products can be produced.

The actual value of ΔS° for "cracking" ethane to ethylene and H_2 is +120.6 J.

The goal of chemists is often to take small molecules and assemble them into larger molecules that can be sold for much more than the cost of the reactants. An example is the process described earlier, assembling CO and H_2 into methyl alcohol, CH_3OH,

$$CO(g) + 2\,H_2(g) \longrightarrow CH_3OH(\ell)$$

$$\Delta H^\circ = -128.14\text{ kJ} \qquad \Delta S_{system} = -332.3\text{ J/K} \qquad \Delta S_{universe} = 98\text{ J/K}$$

and then turning the methyl alcohol into gasoline (that is, into molecules such as octane, C_8H_{18}). The problem with this is that we are fighting a losing battle with entropy, at least locally. The way to get this to work is to increase the entropy somewhere else in the universe. Indeed, Roald Hoffmann, who shared the Nobel Prize in Chemistry in 1981, has said that "One amusing way to describe synthetic chemistry, the making of molecules that is at the intellectual and economic center of chemistry, is that it is the local defeat of entropy" [*American Scientist*, Nov.–Dec. 1987, pages 619–621].

EXERCISE 7.14 • *Predicting the Direction of Reactions*

Classify the following reactions into the four types of reactions summarized in Table 7.5.

Reaction	ΔH° (298 K) kJ	ΔS_{system} (298 K) J/K
(a) $CH_4(g) + 2\,O_2(g) \longrightarrow 2\,H_2O(\ell) + CO_2(g)$	−890.3	−243.0
(b) $2\,Fe_2O_3(s) + 3\,C(graphite) \longrightarrow 4\,Fe(s) + 3\,CO_2(g)$	+467.9	+560.3
(c) $C(graphite) + O_2(g) \longrightarrow CO_2(g)$	−393.5	2.9
(d) $2\,Ag(s) + 3\,N_2(g) \longrightarrow 2\,AgN_3(s)$	+617.6	−451.5

EXERCISE 7.15 • *Product- or Reactant-Favored?*

Is the direct reaction of hydrogen and chlorine to give hydrogen chloride gas predicted to be product-favored or reactant-favored?

$$H_2(g) + Cl_2(g) \longrightarrow 2\,HCl(g)$$

Answer the question by calculating the value for $\Delta S_{\text{universe}}$.

7.10 GIBBS FREE ENERGY

Calculations of the sort done in the previous section would be simpler if we did not have to separately evaluate the entropy change of the surroundings from a table of ΔH_f° values and the entropy change of the system from a table of S° values. To make this possible, a new thermodynamic function was defined by J. Willard Gibbs (1838–1903), a professor at Yale University. In Gibb's honor this function is now called the **Gibbs free energy** and given the symbol G. In the previous section we showed that the total entropy change accompanying a chemical reaction carried out slowly at constant temperature and pressure is

$$\Delta S_{\text{universe}} = \Delta S_{\text{surroundings}} + \Delta S_{\text{system}}$$

$$= \frac{-\Delta H_{\text{system}}}{T} + \Delta S_{\text{system}}$$

J. Willard Gibbs. *(Burndy Library/courtesy AIP Emilio Segre Visual Archives)*

Gibbs defined his free energy function so that under these same conditions $\Delta G_{\text{system}} = -T\Delta S_{\text{universe}}$. Notice that because of the minus sign, if the entropy of the universe increases, the free energy of the system must decrease. Then the equation for total entropy change can be algebraically transformed to one of the most useful equations in all of science:

$$\Delta G_{\text{system}} = -T\Delta S_{\text{universe}} = \Delta H_{\text{system}} - T\Delta S_{\text{system}}$$

or, under standard state conditions

$$\Delta G^\circ_{\text{system}} = \Delta H^\circ_{\text{system}} - T\Delta S^\circ_{system}$$

The Gibbs free energy change provides a way of predicting whether a reaction will be product-favored that depends only on the system—the chemical substances undergoing reaction. Therefore, we can tabulate values of ΔG_f° for a variety of substances and from them calculate ΔG° values for a great many reactions, just as we used ΔH_f° values from Table 7.2 to calculate ΔH° for a reaction. This equation confirms the conclusions we drew earlier and summarized in Table 7.5 for reactions under standard conditions. If the reaction is exothermic (negative $\Delta H°$) and if the entropy of the system increases (positive $\Delta S°$) then $\Delta G°$ must be negative and the reaction will be product-favored. If $\Delta H°$ and $\Delta S°$ have the same sign, then the magnitude of T determines whether $\Delta G°$ will be negative and whether the reaction will be product-favored. If $\Delta H°$ is positive and $\Delta S°$ is negative, then $\Delta G°$ must be positive and the reaction cannot be product-favored under standard conditions and at the temperature for which the data were tabulated.

EXERCISE 7.16 • *Gibbs Free Energy*

In the text we concluded that the reaction to produce methanol from CO and H_2 is product-favored because the entropy of the universe increases.

$$CO(g) + 2\,H_2(g) \longrightarrow CH_3OH(\ell)$$

Verify this result by calculating $\Delta G°$ from $\Delta H°$ and $\Delta S°$ for the system. (The required values of $\Delta H_f°$ and $S°$ are in Appendix J.) Is the sign of $\Delta G°$ positive or negative? What does this mean in terms of the product favorability of the reaction?

Gibbs Free Energy and Energy Resources

Suppose that we have predicted that a reaction we would like to carry out is reactant-favored. Does this mean that we can never obtain the products? No. It means that this reaction cannot occur *unless there is some continuous outside intervention*, which at the beginning of this chapter we identified as a transfer of energy. For example, a dead car battery will not charge itself, but it can be charged if it is connected to a battery charger that is in turn powered by electricity generated in a power plant. Probably the power plant generates electricity by burning coal, which is mainly carbon and burns in air according to the equation

$$C(s) + O_2(g) \longrightarrow CO_2(g) \qquad \Delta G° = -394.4\ \text{kJ}$$

The large negative Gibbs free energy change for combustion of coal more than offsets the positive Gibbs free energy change of the battery-charging reactions, and so the total process results in a decrease in Gibbs free energy, even though the part we are interested in does not.

This is an example of coupling of a product-favored reaction with a reactant-favored process to cause the latter to take place. There are a great many similar situations. Other examples are obtaining aluminum, iron, or other metals from their ores; synthesizing large, complicated molecules from simple reactants to make medicines, plastics, and other useful materials; and maintaining a comfortable temperature in a house on a day when the outside temperature is below zero. All of these processes involve decreasing entropy (increasing order) in the area of our interest, but all can be made to occur provided there is a larger increase in entropy at a power plant or somewhere else.

The Gibbs free energy change indicates a chemical reaction's capacity to drive a reactant-favored system so that it produces products. The word "free" in the name indicates not "zero cost," but rather "available." Gibbs free energy is available to do useful tasks that would not happen on their

A Deeper Look

Using Gibbs Free Energy

The Gibbs equation ($\Delta G° = \Delta H° - T\Delta S°$) can be used to calculate the free energy change for a chemical reaction if the enthalpy and entropy changes for the reaction are known or can be calculated. However, just as values of the enthalpy of formation have been determined for compounds, so too have values of the free energy of formation, ΔG_f°. For example, the free energy of formation of water vapor is −228.572 kJ/mol.

$$H_2(g) + \frac{1}{2} O_2(g) \longrightarrow H_2O(g) \qquad \Delta G_f^\circ = -228.572 \text{ kJ/mol}$$

Values for many other compounds are listed in Appendix J.

If the enthalpies of formation of all the compounds involved in a reaction are known, we can calculate the enthalpy change for the reaction by summing the values of ΔH_f° for the products and subtracting the sum for the reactants (page 266). *If **all** the values of ΔG_f° are known* the same may be done to find the free energy change for a reaction. As an example, let us calculate the free energy change for the combustion of methane. The values of ΔG_f° are found in Appendix J. (Notice that elements in their standard states have $\Delta G_f^\circ = 0$, just as they have $\Delta H_f^\circ = 0$.)

The free energy change for this combustion reaction, ΔG°_{rxn}, is a large negative number, clearly indicating the reaction is product-favored under standard conditions.

$$\begin{array}{lcccc} & CH_4(g) + & 2\,O_2(g) \longrightarrow & 2\,H_2O(g) + & CO_2(g) \\ \Delta G_f^\circ \text{ (kJ/mol):} & -50.7 & 0 & -228.6 & -394.4 \end{array}$$

$$\begin{aligned} \Delta G^\circ_{rxn} &= 2\,\Delta G_f^\circ[H_2O(g)] + \Delta G_f^\circ[CO_2(g)] - \{\Delta G_f^\circ[CH_4(g)] + 2\,\Delta G_f^\circ[O_2(g)]\} \\ &= 2 \text{ mol } (-228.6 \text{ kJ/mol} + 1 \text{ mol } (-394.4 \text{ kJ/mol}) \\ &\qquad - [1 \text{ mol } (-50.7 \text{ kJ/mol}) + 2 \text{ mol } (0 \text{ kJ/mol})] \\ &= -800.9 \text{ kJ} \end{aligned}$$

---

own. Another way of saying this is that Gibbs free energy is a measure of the *quality* of the energy contained in a chemical system. If it contains a lot of Gibbs free energy, a chemical system can do a lot of useful work for us; the energy is of high quality—potentially useful to humankind. When the system's reactants are transformed into products, that available free energy can do useful work, but only if the system is coupled to some other, reactant-favored process we want to carry out. If systems are not coupled, then the free energy will be wasted.

Energy conservation does not really consist of conserving energy—nature takes care of that automatically, as recognized in the law of conservation of energy. Nature does not, however, automatically conserve Gibbs free energy. Systems with high Gibbs free energies are energy resources, and it is their free energy that we must take pains to conserve. Once a product-favored reaction has taken place, its free energy cannot be restored, except at the expense of another product-favored reaction coupled to it. Analysis of chemical systems in terms of Gibbs free energy can lead to important insights into how energy resources can be conserved effectively.

EXERCISE 7.17 • *Using Gibbs Free Energy*

One way to produce iron metal is to reduce iron(III) oxide with aluminum. You can think about the reaction as occurring in two steps. The first is the loss of oxygen from iron(III) oxide,

(i) $Fe_2O_3(s) \longrightarrow 2\,Fe(s) + \frac{3}{2} O_2(g)$

and the second is the combination of aluminum with the oxygen.

(ii) $2\,Al(s) + \frac{3}{2} O_2(g) \longrightarrow Al_2O_3(s)$

a. Calculate the enthalpy, entropy, and free energy changes for each step. Decide if each step is product- or reactant-favored. Comment on the signs of $\Delta H°$, $\Delta S°$, and $\Delta G°$ for each step.

b. What is the net reaction that occurs when aluminum is combined with iron(III) oxide? What are the enthalpy, entropy, and free energy changes for the net reaction? Is it product- or reactant-favored? Comment on the signs of $\Delta H°$, $\Delta S°$, and $\Delta G°$ for the net reaction.

c. Discuss briefly the effect of coupling reaction (i) with reaction (ii) [that is, adding equations (i) and (ii)] on the ability to change iron(III) oxide to the metal.

IN CLOSING

Having studied this chapter, you should be able to

- describe the various forms of energy and the nature of heat and heat transfer (Section 7.1).
- use specific heat and the sign conventions for heat transfer (Section 7.2).
- recognize and use the language of thermodynamics: the system and its surroundings; exothermic and endothermic reactions; and the first law of thermodynamics (the law of energy conservation) (Section 7.3).
- use the fact that the enthalpy change for a reaction, ΔH, is proportional to the quantity of material present (Sections 7.3–7.5).
- apply Hess's law to find the enthalpy change for a reaction (Sections 7.5 and 7.6).
- describe how one can measure the quantity of heat energy transferred in a reaction by using calorimetry (Section 7.7).
- explain the concept of entropy and know how to use it to determine the direction of the entropy change for a process (Section 7.9).
- use entropy and enthalpy changes to predict whether a reaction is product-favored (Section 7.9).
- describe the connection between enthalpy and entropy changes for a reaction and the Gibbs free energy change (Section 7.10).

STUDY QUESTIONS

Review Questions

1. Name two laws stated in this chapter and explain each in your own words.
2. Based on your experience, when ice melts to liquid water is the process exothermic or endothermic (with respect to the ice)? When liquid water freezes to ice at 0 °C is this exothermic or endothermic (with respect to the liquid water)?
3. You pick up a six-pack of soft drinks from the floor, but it slips from your hand and smashes onto your foot. Comment on the work and energy involved in this sequence. What forms of energy are involved and at what stages of the process?
4. For each of the following, define a system and its surroundings and give the direction of heat transfer:
   (a) Propane is burning in a bunsen burner in the laboratory.
   (b) Water drops, sitting on your skin after a dip in the ocean, evaporate.
   (c) Water, originally at 25 °C, is placed in the freezing compartment of a refrigerator.

(d) Two chemicals are mixed in a flask sitting on a laboratory bench. A reaction occurs and heat is evolved.

5. What is the value of the standard enthalpy of formation for any element under standard conditions?
6. In the *Chemistry You Can Do* experiment with the rubber band, you found that the relaxation of the rubber is endothermic. The molecules that make up the rubber are long chains of carbon atoms, and when the band is stretched the chains are straightened out. When the rubber relaxes, the chains return to a tangled, curled state. What happens to the entropy of the rubber band when a stretched band relaxes?
7. In the other *Chemistry You Can Do* experiment in this chapter you explored the heat from the rusting of iron to form iron oxide. Look at the enthalpies of formation of other metal oxides in Table 7.2 or Appendix J and comment on your observations. Are oxidations of metals generally endothermic or exothermic?
8. Explain how the entropy of the universe is increased when an aluminum metal can is made from aluminum ore. The first step is to extract the ore, which is primarily a form of Al_2O_3, from the ground. After it is purified by freeing it from oxides of silicon and iron, aluminum oxide is changed to the metal by an input of electrical energy.

$$2\ Al_2O_3(s) + \text{electrical energy} \longrightarrow 4\ Al(s) + 3\ O_2(g)$$

9. A house is made of wood and glass. Assuming an equal amount of sunshine falls on a wooden wall and a piece of glass (of equal mass), which will warm more? Explain briefly.
10. Explain why the entropy of the system increases when solid NaCl dissolves in water.
11. Criticize the following statements:
   (a) An enthalpy of formation refers to a reaction in which 1 mole of one or more reactants produces some quantity of product.
   (b) The standard enthalpy of formation of O_2 as a gas at 25 °C and a pressure of 1 atmosphere is 15.0 kJ/mol.
   (c) The entropy change when water vapor is condensed to liquid water is positive.
12. Is the following reaction predicted to favor the reactants or products? Explain your answer briefly.

$$Mg(s) + \tfrac{1}{2}\ O_2(g) \longrightarrow MgO(s) \quad \Delta H_f^\circ = -601.70\text{ kJ}$$

13. Explain briefly why each of the following reactions is product-favored.
   (a) The combustion of propane.

$$C_3H_8(g) + 5\ O_2(g) \longrightarrow 3\ CO_2(g) + 4\ H_2O(g)$$

   (b) The reaction of a metal carbonate with an acid.

$$CuCO_3(s) + H_2SO_4(aq) \longrightarrow CuSO_4(aq) + CO_2(g) + H_2O(\ell)$$

14. Add a column for the Gibbs free energy to Table 7.5.
   (a) For the first and last lines in the table, tell whether ΔG is positive or negative.
   (b) When ΔH_{system} and ΔS_{system} are both negative, is ΔG positive or negative or does it depend on temperature? If it is temperature dependent, what is that dependence?
15. Discuss the following statements from the text in Section 7.10. In each case tell how the statement leads to the conclusion cited.
   (a) If the reaction is exothermic (negative ΔH°) and if the entropy of the system increases (positive ΔS°) then ΔG° must be negative and the reaction will be product-favored.
   (b) If ΔH° and ΔS° have the same sign, then the magnitude of T determines whether ΔG° will be negative and whether the reaction will be product-favored.

Energy Units

16. A two-inch piece of two-layer chocolate cake with frosting provides 1670 kilojoules of energy. What is this in Calories?

17. If you are on a diet that calls for eating no more than 1200 Calories per day, how many joules would this be?

18. Melting lead requires a heat input of 5.91 calories per gram. How many joules are required to melt 1.00 pound (454 g) of lead?

19. Sulfur dioxide, SO_2, is found in wines and in polluted air. If a 32.1-g sample of sulfur is burned in the air to get 64.1 g of SO_2, 297 kJ of energy is released. Express this energy in (a) joules, (b) calories, and (c) kilocalories.

Specific Heat, Heat Capacity, and Changes of State

20. Which requires more energy: (a) warming 15.0 g of water from 25 °C to 37 °C or (b) warming 60.0 g of aluminum from 25 °C to 37 °C?

21. You hold a gram of copper in one hand and a gram of aluminum in the other. Each metal was originally at 0 °C. (Both metals are in the shape of a little ball that

fits into your hand.) If they both take up heat at the same rate, which will come to your body temperature first?

22. How much heat energy in kilojoules is required to heat all the aluminum in a roll of aluminum foil (500. g) from room temperature (25 °C) to the temperature of a hot oven (250 °C)?

23. One way to cool a cup of coffee is to plunge an ice-cold piece of aluminum into it. Suppose a 20.0-g piece of aluminum is stored in the refrigerator at 32 °F (0.0 °C) and then dropped into a cup of coffee. The coffee's temperature drops from 90.0 °C to 75.0 °C. How much heat energy (in kilojoules) did the aluminum block absorb?

24. The heat energy required to melt 1.00 g of ice at 0 °C is 333 J. If one ice cube has a mass of 62.0 g, and a tray contains 20 ice cubes, how much energy is required to melt a tray of ice cubes at 0 °C?

25. Ethylene glycol, $(CH_2OH)_2$, is often used as an antifreeze in cars. (a) Which requires more heat energy to warm from 25.0 °C to 100.0 °C, pure water or an equal mass of pure ethylene glycol? (b) If the cooling system in an automobile has a capacity of 5.00 quarts of liquid, compare water and ethylene glycol as to the quantity of heat energy (in joules) the liquid in the system absorbs on raising the temperature from 25.0 °C to 100.0 °C. (Assume the densities of water and ethylene glycol are both 1.00 g/cm^3.) (1 quart = 0.946 L)

26. How much heat energy (in joules) would be required to raise the temperature of 1.00 pound of lead (1.00 pound = 454 g) from room temperature (25 °C) to its melting point, 327 °C, and then melt the lead at 327 °C? The specific heat of lead is 0.159 J/g · K, and the metal requires 24.7 J/g to convert the solid to a liquid. (See the figure below.)

27. The hydrocarbon benzene, C_6H_6, boils at 80.1 °C. How much heat energy (in joules) is required to heat 1.00 kg of this liquid from 20.0 °C to the boiling point and then change the liquid completely to a vapor at that temperature? (The specific heat capacity of liquid C_6H_6 is 1.74 J/g · K and the enthalpy of vaporization is 395 J/g.)

28. Calculate the quantity of heat required to convert the water in four ice cubes (60.1 g) from H_2O(s) at 0 °C to H_2O(g) at 100 °C. The heat of fusion of ice at 0 °C is 333 J/g and the heat of vaporization of liquid water at 100 °C is 2260 J/g.

29. Mercury, with a freezing point of −39 °C, is the only metal that is liquid at room temperature. How much heat must be released by mercury if 1.00 mL of the metal is cooled from room temperature (23.0 °C) to −39 °C and then frozen to a solid? (The density of mercury is 13.6 g/cm^3. Its specific heat is 0.138 J/g · K and its heat of fusion is 11 J/g.)

30. A piece of iron (400. g) is heated in a flame and then plunged into a beaker containing 1.00 kg of water. The original temperature of the water was 20.0 °C, but it is 32.8 °C after the iron bar is dropped in. What was the original temperature of the hot iron bar?

31. A 192-g piece of copper was heated to 100.0 °C in a boiling water bath and then was dropped into a beaker containing 750. mL of water (density = 1.00 g/cm^3) at 4.0 °C. What is the final temperature of the copper and water after they come to thermal equilibrium? (The specific heat of copper is 0.385 J/g · K.)

32. When a mass of 182 g of gold is added to 22.1 g of water at a temperature of 25.0 °C, the final temperature of the resulting mixture is 27.5 °C. If the heat capacity of gold is 0.129 J/g · K, what was the initial temperature of the gold sample?

33. When 108 g of water at a temperature of 22.5 °C is mixed with 65.1 g of water at an unknown temperature, the final temperature of the resulting mixture is 47.9 °C. What was the temperature of the second sample of water?

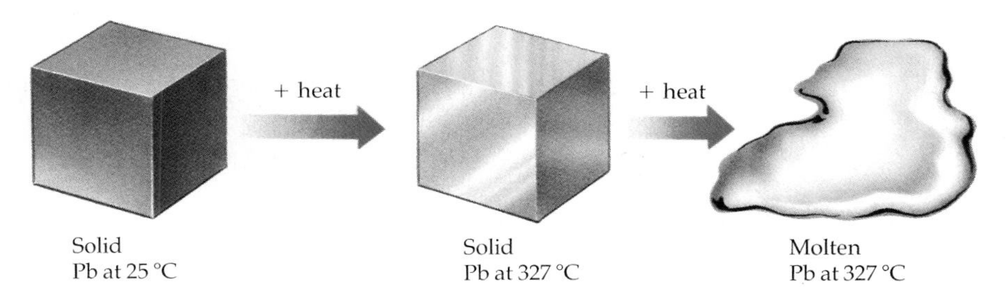

Enthalpy

34. Energy is stored in the body in the form of adenosine triphosphate, ATP. It is formed by the reaction between adenosine diphosphate, ADP, and phosphoric acid.

$$ADP(aq) + H_3PO_4(aq) + 38\ kJ \longrightarrow ATP(aq) + H_2O(\ell)$$

Is the reaction endothermic or exothermic?

35. Calcium carbide, CaC_2, is manufactured by reducing lime with carbon at high temperature. (The carbide is used in turn to make acetylene, an industrially important organic chemical.)

$$CaO(s) + 3\ C(s) \longrightarrow CaC_2(s) + CO(g) \qquad \Delta H^\circ_{rxn} = +464.8\ kJ$$

Is the reaction endothermic or exothermic?

36. "Gasohol," a mixture of gasoline and ethyl alcohol, C_2H_5OH, is a possible automobile fuel. The alcohol releases energy in a combustion reaction with O_2.

$$C_2H_5OH(\ell) + 3\ O_2(g) \longrightarrow 2\ CO_2(g) + 3\ H_2O(\ell)$$

If 0.115 g of alcohol evolves 3.62 kJ when burned at constant pressure, what is the molar enthalpy (or heat) of combustion for ethyl alcohol?

37. White phosphorus, P_4, ignites in air to produce heat, light, and P_4O_{10} (Figure 5.4).

$$P_4(s) + 5\ O_2(g) \longrightarrow P_4O_{10}(s)$$

If 3.56 g of P_4 is burned, 37.4 kJ of heat is evolved at constant pressure. What is the molar enthalpy of combustion of P_4?

38. A laboratory "volcano" can be made from ammonium dichromate (Figure 5.15). When ignited, the compound decomposes in a fiery display.

$$(NH_4)_2Cr_2O_7(s) \longrightarrow N_2(g) + 4\ H_2O(g) + Cr_2O_3(s)$$

If the decomposition produces 315 kJ per mole of ammonium dichromate at constant pressure, how much heat energy would be produced by 28.4 g (1.00 ounce) of the solid?

39. The thermite reaction, the reaction between aluminum and iron(III) oxide,

$$2\ Al(s) + Fe_2O_3(s) \longrightarrow Al_2O_3(s) + 2\ Fe(s) \qquad \Delta H^\circ_{rxn} = -851.5\ kJ$$

produces a tremendous amount of heat (see Figure 2.18). If you begin with 10.0 g of Al and excess Fe_2O_3, how much heat (in kilojoules) is evolved at constant pressure?

Hess's Law

40. Calculate the enthalpy change, ΔH, for the formation of 1 mol of strontium carbonate (the material that gives the red color in fireworks) from its elements.

$$Sr(s) + C(graphite) + \tfrac{3}{2}\ O_2(g) \longrightarrow SrCO_3(s)$$

The information available is:

$$Sr(s) + \tfrac{1}{2}\ O_2(g) \longrightarrow SrO(s) \qquad \Delta H^\circ_{rxn} = -592\ kJ$$
$$SrO(s) + CO_2(g) \longrightarrow SrCO_3(s) \qquad \Delta H^\circ_{rxn} = -234\ kJ$$
$$C(graphite) + O_2(g) \longrightarrow CO_2(g) \qquad \Delta H^\circ_{rxn} = -394\ kJ$$

41. What is the enthalpy change for the reaction of lead(II) chloride with chlorine to give lead(IV) chloride?

$$PbCl_2(s) + Cl_2(g) \longrightarrow PbCl_4(\ell) \qquad \Delta H^\circ_{rxn} = ?$$

It is known that $PbCl_2(s)$ can be formed from the metal and $Cl_2(g)$,

$$Pb(s) + Cl_2(g) \longrightarrow PbCl_2(s) \qquad \Delta H^\circ_{rxn} = -359.4\ kJ$$

and that $PbCl_4(\ell)$ can be formed directly from the elements.

$$Pb(s) + 2\ Cl_2(g) \longrightarrow PbCl_4(\ell) \qquad \Delta H^\circ_{rxn} = -329.3\ kJ$$

42. Using the following reactions, find the enthalpy change for the formation of 1 mole of PbO(s) from lead metal and oxygen gas.

$$PbO(s) + C(graphite) \longrightarrow Pb(s) + CO(g) \qquad \Delta H^\circ_{rxn} = +106.8\ kJ$$
$$2\ C(graphite) + O_2 \longrightarrow 2\ CO(g) \qquad \Delta H^\circ_{rxn} = -221.0\ kJ$$

If 250 g of lead reacts with oxygen to form lead(II) oxide, what quantity of heat (in kJ) is absorbed or evolved?

43. Three reactions very important to the semiconductor industry are

(a) the reduction of silicon dioxide to crude silicon,

$$SiO_2(s) + 2\ C(s) \longrightarrow Si(s) + 2\ CO(g) \qquad \Delta H^\circ_{rxn} = +689.9\ kJ$$

(b) the formation of silicon tetrachloride from the crude silicon,

$$Si(s) + 2\ Cl_2(g) \longrightarrow SiCl_4(g) \qquad \Delta H^\circ_f = -657.0\ kJ$$

(c) and the reduction of silicon tetrachloride to pure silicon with magnesium.

$$SiCl_4(g) + 2\ Mg(s) \longrightarrow 2\ MgCl_2(s) + Si(s) \qquad \Delta H^\circ_{rxn} = -625.6\ kJ$$

What is the overall enthalpy change for changing 1.00 mol of sand (SiO_2) into very pure silicon?

Standard Enthalpies of Formation

44. The standard molar enthalpy of formation of AgCl(s) is −127.1 kJ/mol. Write the balanced equation for which the enthalpy of reaction is −127.1 kJ.

45. The standard molar enthalpy of formation of methyl alcohol, $CH_3OH(\ell)$, is −238.7 kJ/mol. Write the balanced equation for which the enthalpy of reaction is −238.7 kJ.

46. For each compound below, write a balanced equation

depicting the formation of 1 mol of the compound. Look up the standard molar enthalpy of formation for each compound in Table 7.2 or Appendix J.
(a) $Al_2O_3(s)$ (b) $TiCl_4(\ell)$ (c) $NH_4NO_3(s)$

47. The standard molar enthalpy of formation of glucose, $C_6H_{12}O_6(s)$, is -1260 kJ/mol. (a) Is the formation of glucose from its elements exo- or endothermic? (b) Write a balanced equation depicting the formation of glucose from its elements and for which the enthalpy of reaction is -1260 kJ.

48. An important reaction in the production of sulfuric acid is

$$SO_2(g) + \tfrac{1}{2} O_2(g) \longrightarrow SO_3(g)$$

It is also a key reaction in the formation of acid rain, beginning with the air pollutant SO_2. Using the data in Table 7.2, calculate the enthalpy change for the reaction.

49. In photosynthesis, the sun's energy brings about the combination of CO_2 and H_2O to form O_2 and a carbon-containing compound such as a sugar. In its simplest form, the reaction would be

$$CO_2(g) + 2\,H_2O(\ell) \longrightarrow 2\,O_2(g) + CH_4(g)$$

Using the enthalpies of formation in Table 7.2, (a) calculate the enthalpy of reaction and (b) decide whether the reaction is exo- or endothermic.

50. The first step in the production of nitric acid from ammonia involves the oxidation of NH_3.

$$4\,NH_3(g) + 5\,O_2(g) \longrightarrow 4\,NO(g) + 6\,H_2O(g)$$

Use the information in Table 7.2 or Appendix J to find the enthalpy change for this reaction. Is the reaction exo- or endothermic?

51. The Romans used CaO as mortar in stone structures. The CaO was mixed with water to give $Ca(OH)_2$, and this slowly reacted with CO_2 in the air to give limestone.

$$Ca(OH)_2(s) + CO_2(g) \longrightarrow CaCO_3(s) + H_2O(g)$$

Calculate the enthalpy change for this reaction.

52. A key reaction in the processing of uranium for use as fuel in nuclear power plants is the following:

$$UO_2(s) + 4\,HF(g) \longrightarrow UF_4(s) + 2\,H_2O(g)$$

Calculate the enthalpy change, $\Delta H°$, for the reaction using the data in Table 7.2, Appendix J and the following: $\Delta H_f°$ for $UO_2(s) = -1085$ kJ/mol; $\Delta H_f°$ for $UF_4(s) = -1914$ kJ/mol.

53. Oxygen difluoride, OF_2, is a colorless, very poisonous gas that reacts rapidly with water vapor to produce O_2, HF, and heat.

$$OF_2(g) + H_2O(g) \longrightarrow 2\,HF(g) + O_2(g) \qquad \Delta H°_{rxn} = -318 \text{ kJ}$$

Using this information, and Table 7.2 or Appendix J, calculate the molar enthalpy of formation of $OF_2(g)$.

54. Iron can react with oxygen to give iron(III) oxide. If 5.58 g of Fe is heated in pure O_2 to give $Fe_2O_3(s)$, how much heat is liberated (at constant pressure)?

55. The formation of aluminum oxide from its elements is highly exothermic. If 2.70 g Al metal is burned in pure O_2 to give Al_2O_3, how much heat energy (in kilojoules) is evolved in the process (at constant pressure)?

Calorimetry

56. How much heat energy (in kilojoules) is evolved by a reaction in a bomb calorimeter (Figure 7.11) in which the temperature of the bomb and water increases from 19.50 °C to 22.83 °C? The bomb has a heat capacity of 650. J/K; the calorimeter contains 320. g of water.

57. Sulfur (2.56 g) was burned in a bomb calorimeter with excess $O_2(g)$. The temperature increased from 21.25 °C to 26.72 °C. The bomb had a heat capacity of 923 J/K and the calorimeter contained 815 g of water. Calculate the heat evolved, per mole of SO_2 formed, in the course of the reaction $S(s) + O_2(g) \rightarrow SO_2(g)$.

58. You can find the amount of heat evolved in the combustion of carbon by carrying out the reaction in a combustion calorimeter. Suppose you burn 0.300 g of C(graphite) in an excess of $O_2(g)$ to give $CO_2(g)$.

$$C(\text{graphite}) + O_2(g) \longrightarrow CO_2(g)$$

The temperature of the calorimeter, which contains 775 g of water, increases from 25.00 °C to 27.38 °C. The heat capacity of the bomb is 893 J/K. What quantity of heat is evolved per mole of C?

59. Benzoic acid, $C_7H_6O_2$, occurs naturally in many berries. Suppose you burn 1.500 g of the compound in a combustion calorimeter and find that the temperature of the calorimeter increases from 22.50 °C to 31.69 °C. (Water is produced in the combustion as water vapor.) The calorimeter contains 775 g of water, and the bomb has a heat capacity of 893 J/K. How much heat is evolved per mole of benzoic acid?

Entropy

60. For each of the following pairs, tell which has the higher entropy.
(a) A sample of solid CO_2 at -78 °C or CO_2 vapor at 0 °C.
(b) Sugar, as a solid or dissolved in a cup of tea.
(c) Two 100-ml beakers, one containing pure water and the other containing pure alcohol, or a beaker containing a mixture of the water and alcohol.

61. Tell which substance has the higher entropy in each of the following pairs.
   (a) A sample of pure silicon (to be used in a computer chip) or a piece of silicon containing a trace of some other atoms such as B or P.
   (b) An ice cube or liquid water, both at 0 °C.
   (c) A sample of pure solid I_2 or iodine vapor, both at room temperature.

62. Comparing the formulas or states for each pair of compounds, tell which you would expect to have the higher entropy at the same temperature.
   (a) $KCl(s)$ or $Al_2O_3(s)$
   (b) $CH_3I(\ell)$ or $CH_3CH_2I(\ell)$
   (c) $NH_4Cl(s)$ or $NH_4Cl(aq)$

63. Comparing the formulas or states for each pair of compounds, tell which you would expect to have the higher entropy at the same temperature.
   (a) $NaCl(s)$ or $MgCl_2(s)$
   (b) $Cl_2(g)$ or $P_4(g)$
   (c) $CH_3NH_2(g)$ or $(CH_3)_2NH$
   (d) $Au(s)$ or $Hg(\ell)$

64. Calculate the entropy change, ΔS, for the vaporization of ethyl alcohol, C_2H_5OH, at the boiling point of 78.0 °C. The heat of vaporization of the alcohol is 39.3 kJ/mol.

$$C_2H_5OH(\ell) \longrightarrow C_2H_5OH(g) \qquad \Delta S = ?$$

65. Not too many years ago diethyl ether, $(C_2H_5)_2O$, was used as an anesthetic. What is the entropy change, ΔS, for the vaporization of the ether if its heat of vaporization is 26.0 kJ/mol at the boiling point of 35.0 °C?

66. What are the signs of the enthalpy and entropy changes for the splitting of water to give gaseous hydrogen and oxygen, a process that requires considerable energy? Is this reaction likely to be product-favored or not? Explain your answer briefly.

67. Octane is the product of adding hydrogen to 1-octene.

$$\underset{\text{octene}}{C_8H_{16}(g)} + H_2(g) \longrightarrow \underset{\text{octane}}{C_8H_{18}(g)}$$

The enthalpies of formation are

$\Delta H_f^\circ[C_8H_{16}(g)] = -82.93$ kJ/mol
$\Delta H_f^\circ[C_8H_{18}(g)] = -208.45$ kJ/mol

Is this reaction likely to be product-favored or reactant-favored? Explain your reasoning briefly.

68. Classify each of the reactions according to one of the four reaction types summarized in Table 7.5.
   (a) $Fe_2O_3(s) + 2\,Al(s) \longrightarrow 2\,Fe(s) + Al_2O_3(s)$
   $\Delta H^\circ_{rxn} = -851.5$ kJ $\quad \Delta S^\circ = -37.5$ J/K
   (b) $N_2(g) + 2\,O_2(g) \longrightarrow 2\,NO_2(g)$
   $\Delta H^\circ_{rxn} = 66.4$ kJ $\quad \Delta S^\circ = -122$ J/K

69. Classify each of the reactions according to one of the four reaction types summarized in Table 7.5.
   (a) $C_6H_{12}O_6(s) + 6\,O_2(g) \longrightarrow 6\,CO_2(g) + 6\,H_2O(\ell)$
   $\Delta H^\circ_{rxn} = -673$ kJ $\quad \Delta S^\circ = 60.4$ J/K
   (b) $MgO(s) + C(graphite) \longrightarrow Mg(s) + CO(g)$
   $\Delta H^\circ_{rxn} = 491.18$ kJ $\quad \Delta S^\circ = 197.67$ J/K

70. Is the combustion of ethane, C_2H_6, likely to be a product-favored reaction?

$$C_2H_6(g) + \tfrac{7}{2}\,O_2(g) \longrightarrow 2\,CO_2(g) + 3\,H_2O(\ell)$$

Answer the question by calculating the value of $\Delta S_{universe}$. Required values of ΔH_f° and S° are in Tables 7.2 and 7.4, respectively. Does your calculated answer agree with your preconceived idea of this reaction?

71. In the discussion of "meals ready to eat" (MREs) in the text, it was noted that the reaction of magnesium with water provides the heat.

$$Mg(s) + 2\,H_2O(\ell) \longrightarrow Mg(OH)_2(s) + H_2(g)$$

Is this reaction in fact predicted to be product-favored? Answer the question by calculating the value of $\Delta S_{universe}$. In addition to the values of ΔH_f° and S° in Tables 7.2 and 7.4, respectively, you need to know the following:

Compound/Element	*ΔH_f° (kJ/mol)*	*S° (J/K · mol)*
Mg	0	32.68
$Mg(OH)_2(s)$	−924.54	63.18

Gibbs Free Energy

72. Predict whether the reaction below is product- or reactant-favored by calculating ΔG° from the entropy and enthalpy changes for the reaction at 25 °C.

$$H_2(g) + CO_2(g) \longrightarrow H_2O(g) + CO(g)$$
$$\Delta H^\circ_{rxn} = 41.17 \text{ kJ} \qquad \Delta S^\circ = 42.08 \text{ J/K}$$

73. One way to produce iron metal is to reduce iron(III) oxide with carbon.

$$2\,Fe_2O_3(s) + 3\,C(graphite) \longrightarrow 4\,Fe(s) + 3\,CO_2(g)$$
$$\Delta H^\circ_{rxn} = +467.9 \text{ kJ} \qquad \Delta S^\circ = 560.3 \text{ J/K}$$

Is this reaction product- or reactant-favored at 25 °C? Does the reaction become more product-favored or more reactant-favored as the temperature increases?

74. Calculate ΔG° for the reactions of sand (SiO_2) with hydrogen fluoride and hydrogen chloride. Explain why hydrogen fluoride attacks glass while hydrogen chloride does not.

$$SiO_2(s) + 4\,HF(g) \longrightarrow SiF_4(g) + 2\,H_2O(g)$$
$$SiO_2(s) + 4\,HCl(g) \longrightarrow SiCl_4(g) + 2\,H_2O(g)$$

(Hint: Decide whether each reaction is product- or reactant-favored by calculating $\Delta G°$ from the enthalpy and entropy changes for the reaction at 25 °C. (See Appendix J for values of $\Delta H_f°$ and $S°$.)

75. Many metal carbonates can be decomposed to the metal oxide and carbon dioxide by heating.

$$MgCO_3(s) \longrightarrow MgO(s) + CO_2(g)$$

What are the enthalpy, entropy, and free energy changes for this reaction? Is it product- or reactant-favored? Comment on the signs of $\Delta H°$, $\Delta S°$, and $\Delta G°$ for the reaction.

General Questions

76. The specific heat of copper metal is 0.385 J/g · K, while it is 0.128 J/g · K for gold. Assume you place 100. g of each metal, originally at 25 °C, in a boiling water bath at 100 °C. If each metal takes up heat at the same rate (the number of joules of heat absorbed per minute is the same), which piece of metal reaches 100 °C first?

77. Calculate the molar heat capacity, in J/mol · K, for the four metals in Table 7.1. What observation can you make about these values? Are they widely different or very similar? Using this information, can you calculate the heat capacity in the units J/g · K for silver? (The correct value for silver is 0.23 J/g · K.)

78. Suppose you add 100.0 g of water at 60.0 °C to 100.0 g of ice at 0.00 °C. Some of the ice melts and cools the warm water to 0.00 °C. When the ice/water mixture has come to a uniform temperature of 0.00 °C, how much ice has melted?

79. The combustion of diborane, B_2H_6, proceeds according to the equation

$$B_2H_6(g) + 3\,O_2(g) \longrightarrow B_2O_3(s) + 3\,H_2O(\ell)$$

and 2166 kJ is liberated per mole of $B_2H_6(g)$ (at constant pressure). Calculate the molar enthalpy of formation of $B_2H_6(g)$ using this information, the data in Table 7.2, and the fact that $\Delta H_f°$ for $B_2O_3(s)$ is -1273 kJ/mol.

80. In principle, copper could be used to generate valuable hydrogen gas from water.

$$Cu(s) + H_2O(g) \longrightarrow CuO(s) + H_2(g)$$

(a) Is the reaction exo- or endothermic?

(b) If 2.00 g of copper metal reacts with excess water vapor, how much heat (at constant pressure) is involved (either absorbed or evolved) in the reaction?

81. As explained in Study Question 37, P_4 ignites in air to give P_4O_{10} and a large quantity of heat. This is an important reaction, since the phosphorus oxide can then be treated with water to give phosphoric acid for use in making detergents, toothpaste, soft drinks, and other consumer products. About 500×10^3 tons of elemental phosphorus is made annually in the United States. If you oxidize just one ton of P_4 (9.08×10^5 g) to the oxide, how much heat (in kJ) is evolved at constant pressure?

82. The enthalpies of the following two reactions are available.

$$2\,C(\text{graphite}) + 2\,H_2(g) \longrightarrow C_2H_4(g) \qquad \Delta H° = +52.3 \text{ kJ}$$

$$C_2H_4Cl_2(\ell) \longrightarrow Cl_2(g) + C_2H_4(g) \qquad \Delta H° = +217.5 \text{ kJ}$$

Calculate the molar enthalpy of formation of $C_2H_4Cl_2(\ell)$.

83. Given the following information and the data in Table 7.2, calculate the molar enthalpy of formation for liquid hydrazine, N_2H_4.

$$N_2H_4(\ell) + O_2(g) \longrightarrow N_2(g) + 2\,H_2O(g) \qquad \Delta H° = -534 \text{ kJ}$$

84. The combination of coke and steam produces a mixture called coal gas, which can be used as a fuel or as a starting material for other reactions. If we assume coke can be represented by graphite, the equation for the production of coal gas is

$$2\,C(s) + 2\,H_2O(g) \longrightarrow CH_4(g) + CO_2(g)$$

Determine the standard enthalpy change for this reaction from the following standard enthalpies of reaction:

$$C(s) + H_2O(g) \longrightarrow CO(g) + H_2(g) \qquad \Delta H°_{rxn} = +131.3 \text{ kJ}$$

$$CO(g) + H_2O(g) \longrightarrow CO_2(g) + H_2(g) \qquad \Delta H°_{rxn} = -41.2 \text{ kJ}$$

$$CH_4(g) + H_2O(g) \longrightarrow 3\,H_2(g) + CO(g) \qquad \Delta H°_{rxn} = +206.1 \text{ kJ}$$

85. Some years ago Texas City, Texas, was devastated by the explosion of a shipload of ammonium nitrate, a compound intended to be used as a fertilizer. When heated, however, ammonium nitrate can decompose exothermically to N_2O and water.

$$NH_4NO_3(s) \longrightarrow N_2O(g) + 2\,H_2O(g)$$

If the heat from this exothermic reaction is contained, higher temperatures are generated, at which point ammonium nitrate can decompose explosively to N_2, H_2O, and O_2.

$$2\,NH_4NO_3(s) \longrightarrow 2\,N_2(g) + 4\,H_2O(g) + O_2(g)$$

If oxidizable materials are present, fires can break out, as was the case at Texas City. Using the information in Appendix J, answer the following questions.

(a) If the standard enthalpy of formation of $N_2O(g)$ is 82.1 kJ/mol, how much heat is evolved (at constant pressure and under standard conditions) by the first reaction?

(b) If 8.00 kg of ammonium nitrate explodes (the second reaction), how much heat is evolved (at constant pressure and under standard conditions)?

86. Uranium-235 is used as a fuel in nuclear power plants. Since natural uranium contains only a small amount of this isotope, the uranium must be enriched in uranium-235 before it can be used. To do this, uranium(IV) oxide is first converted to a gaseous compound, UF_6, and the isotopes are separated by a gaseous diffusion technique (Chapter 12). Some key reactions are

$$UO_2(s) + 4\,HF(g) \longrightarrow UF_4(s) + 2\,H_2O(g)$$
$$UF_4(s) + F_2(g) \longrightarrow UF_6(g)$$

How much heat energy (at constant pressure) would be involved in producing 225 tons of $UF_6(g)$ from UO_2? (1 ton = 9.08×10^5 g) Some necessary standard enthalpies of formation are: $\Delta H_f^\circ[UO_2(s)] = -1081.2$ kJ/mol; $\Delta H_f^\circ[UF_4(s)] = -1912.7$ kJ/mol; $\Delta H_f^\circ[UF_6(g)] = -2141.3$ kJ/mol. See also Table 7.2 and Appendix J.

87. One method of producing H_2 on a large scale is the following chemical cycle.

Step 1: $SO_2(g) + 2\,H_2O(g) + Br_2(g) \longrightarrow H_2SO_4(\ell) + 2\,HBr(g)$

Step 2: $H_2SO_4(\ell) \longrightarrow H_2O(g) + SO_2(g) + \frac{1}{2}\,O_2(g)$

Step 3: $2\,HBr(g) \longrightarrow H_2(g) + Br_2(g)$

Using the table of standard enthalpies of formation in Appendix J, calculate ΔH° for each step. What is the equation for the *overall* process and what is its enthalpy change? Is the overall process exo- or endothermic?

88. One reaction involved in the conversion of iron ore to the metal is

$$FeO(s) + CO(g) \longrightarrow Fe(s) + CO_2(g)$$

Calculate the standard enthalpy change for this reaction from the following reactions of iron oxides with CO:

$$3\,Fe_2O_3(s) + CO(g) \longrightarrow 2\,Fe_3O_4(s) + CO_2(g) \quad \Delta H^\circ = -47\text{ kJ}$$

$$Fe_2O_3(s) + 3\,CO(g) \longrightarrow 2\,Fe(s) + 3\,CO_2(g) \quad \Delta H^\circ = -25\text{ kJ}$$

$$Fe_3O_4(s) + CO(g) \longrightarrow 3\,FeO(s) + CO_2(g) \quad \Delta H^\circ = +19\text{ kJ}$$

89. If you want to convert 56.0 g of ice (at 0 °C) to water at 75.0 °C, how much propane (C_3H_8) would you have to burn in order to supply the heat (at constant pressure) to melt the ice and then warm it to the final temperature?

90. Suppose you want to heat your house with natural gas (CH_4). Assume your house has 1800 ft^2 of floor area and that the ceilings are 8.0 ft from the floors. The air in the house has a molar heat capacity of 29.1 J/mol · K. (The number of moles of air in the house can be found by assuming that the average molar mass of air is 28.9 g/mol and that the density of air at these temperatures is about 1.22 g/L.) How much methane do you have to burn to heat the air from 15.0 °C to 22.0 °C?

91. Companies around the world are constantly researching compounds that can be used as a substitute for gasoline in automobiles. Perhaps the most promising of these is methyl alcohol, CH_3OH, a compound that can be made relatively inexpensively from coal. The alcohol has a smaller energy content than gasoline, but, with its higher octane rating, it burns more efficiently than gasoline in combustion engines. (It also has the added advantage of contributing to a lesser degree to some air pollutants.) Compare the amount of heat produced per gram of CH_3OH and C_8H_{18} (octane), the latter being representative of the compounds in gasoline. (See Section 7.7 for thermochemical information on octane.)

92. Hydrazine and 1,1 dimethylhydrazine both react spontaneously with O_2 and can be used as rocket fuels.

$$N_2H_4(\ell) + O_2(g) \longrightarrow N_2(g) + 2\,H_2O(g)$$
hydrazine

$$N_2H_2(CH_3)_2(\ell) + 4\,O_2(g) \longrightarrow 2\,CO_2(g) + 4\,H_2O(g) + N_2(g)$$
dimethylhydrazine

The molar enthalpy of formation of liquid hydrazine is +50.6 kJ/mol and that of liquid dimethylhydrazine is +49.2 kJ/mol. By doing appropriate calculations, decide whether the reaction of hydrazine or dimethylhydrazine with oxygen gives more heat *per gram* (at constant pressure). (Other enthalpy of formation data can be obtained from Table 7.2.)

93. Some metal oxides can be decomposed to the metal and oxygen under reasonable conditions. Is the decomposition of silver(I) oxide product-favored at 25 °C?

$$Ag_2O(s) \longrightarrow 2\,Ag(s) + \tfrac{1}{2}\,O_2(g)$$

If not, can it become so if the temperature is raised? At what temperature is the reaction product-favored? (Hint: Calculate the temperature at which $\Delta G^\circ = 0$.)

Summary Question

94. Sulfur dioxide, SO_2, is a major pollutant in our industrial society, and it is often found in wine.
    (a) In wine-making, SO_2 is commonly added to kill microorganisms in the grape juice when it is put into vats before fermentation. Further, it is used to neutralize byproducts of the fermentation process, enhance wine flavor, and prevent oxidation. Wine usually contains 80 to 150 ppm SO_2 (1 ppm = 1 part per million = 1 g of SO_2 per 10^6 g of wine). The United States produced 440 million gallons of wine in 1987. Assuming the density of wine is 1.00 g/cm^3, and that the average bottle of wine contains 100. ppm of SO_2, how many grams and how many moles of SO_2 are contained in this wine?
    (b) When SO_2 is given off by an oil- or coal-burning power plant, it can be trapped by reaction with MgO in air to form $MgSO_4$.

    $$MgO(s) + SO_2(g) + \tfrac{1}{2}O_2(g) \longrightarrow MgSO_4(s)$$

    If 20. million tons of SO_2 are given off by coal burning power plants each year, how much MgO would you have to supply to remove all of this SO_2? How much $MgSO_4$ would be produced?
    (c) If ΔH_f° for $MgSO_4(s)$ is -1284.9 kJ/mol, how much heat (at constant pressure) is evolved or absorbed per mole of $MgSO_4$ by the reaction in part (b)?
    (d) Sulfuric acid comes from the oxidation of sulfur, first to SO_2 and then to SO_3. The SO_3 is then absorbed by water to make H_2SO_4.

    $$S(s) + O_2(g) \longrightarrow SO_2(g) \qquad \Delta H^\circ = -296.8\ kJ$$

    $$SO_2(g) + \tfrac{1}{2}O_2(g) \longrightarrow SO_3(g) \qquad \Delta H^\circ = -98.9\ kJ$$

    $$SO_3(g) + H_2O(\text{in } 98\%\ H_2SO_4) \longrightarrow H_2SO_4(\ell) \qquad \Delta H^\circ = -132.5\ kJ$$

    The typical plant produces 750. tons of H_2SO_4 per day (1 ton = 9.08×10^5 g). Calculate the amount of heat produced by these reactions per day.

Using Computer Programs and Videodiscs

95. Use the *KC? Discoverer* program to find the heat of fusion, ΔH_{fusion}, and the melting point, T_{fusion}, for Li, B, La, Ir, and Hg.
    (a) Calculate the heat required to melt 10.0 g of each element.
    (b) Calculate the entropy change when 1 mol of each element is melted.
    (c) Is there an obvious relationship among your results in part (a)? In part (b)? If so, suggest a reason that such a relationship might exist.
96. Use the *KC? Discoverer* program to find the heat of vaporization, ΔH_{vap}, and the boiling point, T_b, for Al, Si, Cu, Ag, La, Ce, and Ge.
    (a) Calculate the heat required to vaporize 20.0 g of each element.
    (b) Calculate the entropy change when 1 mol of each element is boiled.
    (c) Is there an obvious relationship among your results in part (a)? In part (b)? If so, suggest a reason that such a relationship might exist.
97. Use *KC? Discoverer* to graph the ratio $\Delta H_{fusion}/T_{fusion}$ versus atomic number for all elements. Look carefully at the graph and analyze it with respect to how the ratio varies with atomic number. Is the ratio similar from one element to another? Are there some elements that seem to fall way out of line from the rest? Suggest explanations or reasons for the shape of the graph.
98. Use the *Shakhashiri Chemical Demonstrations* videodiscs (frame 45915 on Side 2 of Disc 1) or videotapes to observe the reaction of zinc with sulfur. Three reactions occur:
    (i) zinc with sulfur to produce zinc sulfide.
    (ii) zinc with oxygen to give zinc oxide.
    (iii) sulfur with oxygen to give sulfur dioxide.
    (a) What observations do you make when the reactions occur?
    (b) Write balanced equations for the three possible reactions.
    (c) Which is the most exothermic of these reactions?
    (d) If you have 1.50 g of powdered zinc, and the only reaction that occurs is the production of ZnS, what quantity of heat is evolved?

Manganese dioxide catalyzes the decomposition of hydrogen peroxide to water vapor and oxygen.
(C.D. Winters)

Principles of Reactivity: Kinetics and Equilibrium

CHAPTER

8

Successful product-favored reactions are usually exothermic, although endothermic reactions can occur if the products have greater entropy than the reactants. But this criterion of chemical reactivity is insufficient in a practical sense: a reaction may be exothermic and involve an increase in entropy but still require such a long time to occur that it would take several lifetimes to produce appreciable quantities of products.

Chemical kinetics is concerned with how rapidly reactions occur and what happens to the atoms and molecules as they change from reactants to products. The first part of this chapter describes how to measure the speed of a chemical reaction and determine how reactant concentrations, changes in temperature, and the presence of catalysts affect that speed. In addition, the pathways taken by atoms and molecules as a reaction occurs—reaction mechanisms—are described.

How can we predict how far a reaction can proceed? Are reactants converted completely to products or is there a mix of both when all reaction has apparently stopped? Is a reaction product- or reactant-favored? If reaction conditions change, will the reaction become more or less product-favored? The answers to these important and useful questions come from a study of chemical equilibria.

The three main subjects of this chapter and the previous one, thermodynamics, kinetics, and equilibria, are essential to understanding chemical reactions of all types.

The theme of the last chapter and this one is **chemical reactivity**—whether a reaction occurs at all and, if so, whether the reaction produces a useful amount of product in a reasonable length of time. If you are a pharmaceutical chemist and have synthesized a new drug, you would certainly want to know whether the compound will react as intended and how long it will take to do so. If you buy hydrogen peroxide, H_2O_2, to use as an antiseptic, will it decompose to H_2O and O_2; if so, how long will it be until most of it is gone?

$$2\,H_2O_2(aq) \longrightarrow 2\,H_2O(\ell) + O_2(g)$$

If you are a chemical engineer, you might be trying to find a new way to make hydrazine, N_2H_4. This is a valuable reducing agent and can be used as a component of rocket fuels, for example. What about making it from the very inexpensive, easily made compound urea?

$$\underset{\text{urea}}{H_2N\!-\!\overset{\overset{\displaystyle O}{\|}}{C}\!-\!NH_2} \longrightarrow \underset{\text{hydrazine}}{H_2N\!-\!NH_2} + CO$$

Is this reaction product-favored? If so, is it rapid enough to be commercially useful? What are the reaction conditions that will produce the most product quickly enough at the least cost?

The answers to such questions, and many others of practical importance, can come from a discussion of chemical reactivity. In this chapter we shall combine the conclusions of Chapter 7 with additional ideas to help you develop some feeling for when a reaction will be successful, how it can be made to happen in a reasonable amount of time, and what conditions will produce the greatest conversion of reactants to products.

8.1 RATES AND MECHANISMS OF CHEMICAL REACTIONS

Turn on the valve of a bunsen burner in your laboratory, bring up a lighted match, and a rapid combustion reaction begins with a whoosh or a bang!

$$\underset{\text{methane}}{CH_4(g)} + 2\,O_2(g) \longrightarrow CO_2(g) + 2\,H_2O(g) \qquad \Delta H^\circ_{rxn} = -802.34\text{ kJ}$$

But what would happen if you didn't put a lighted match in the methane/air stream? Nothing! If methane and oxygen were mixed in a closed flask at room temperature, the mixture would be stable for centuries.

Three moles of gaseous reactants produce three moles of gaseous products in this methane combustion reaction, so the entropy of the system changes only a little. There is, however, a large negative enthalpy change, and the heat transferred out of the system causes a significant increase in the entropy of the surroundings. Thus, thermodynamics leads to the conclusion that the reaction should occur without a continuous input of energy—but it reveals nothing at all about the rate or speed of the reaction or the need for a match to get the reaction started. Experiments show that, although the reaction of methane with oxygen is *very* slow at room temperature, it is much faster at high temperatures. Once the lighted match gets things started, heat released by the reaction keeps it going. These observations lie within the realm of **chemical kinetics**—the study of the rates of reactions and the pathways or **reaction mechanisms** by which atoms and molecules are transformed from reactants to products.

In order for a chemical reaction to occur, reactant molecules must come together so that atoms can be exchanged or rearranged. Atoms and molecules are mobile in the gas phase or in solution, so reactions are often carried out in a mixture of gases or among solutes in a solution. Under these circumstances, three factors affect the speed of a reaction.

- The *concentrations* and physical state of reactants and products
- The *temperature* of the reaction
- The presence or absence of a *catalyst*

We will describe each of these factors and explore their effects in the submicroscopic world of atoms and molecules, but first we need to define what is meant by the rate or speed of a reaction.

A catalyst (described below) speeds up a reaction but undergoes no net chemical change.

Reaction Rate

The speed of a reaction or any other process is expressed as its "rate," which is the change in some measurable quantity per unit of time. A car's rate of travel, for example, is found by measuring the change in its position, Δp, during a given time interval, Δt. For instance, if you travel 110 miles in 2 hours, you are traveling at an average rate of $\Delta p/\Delta t = 55$ miles per hour.

Recall that the Greek letter Δ (delta) means that a change *in some quantity has been measured.*

When describing chemical reactions, the **reaction rate** is expressed as the change in concentration of a reactant or product per second (or per hour, or per day). To illustrate this, consider what can happen to the cancer chemotherapy agent cisplatin, $Pt(NH_3)_2Cl_2$, in the presence of water.

It is obviously important for us to know something about the speed with which Cl^- is replaced by water in cisplatin. Once the compound is placed in the aqueous environment of the human body, some other molecule capable of binding to Pt^{2+} can replace the Cl^- ion.

$$H_2O + Cl{-}\underset{\displaystyle NH_3}{\overset{\displaystyle Cl}{Pt}}{-}NH_3 \longrightarrow \left[H_2O{-}\underset{\displaystyle NH_3}{\overset{\displaystyle Cl}{Pt}}{-}NH_3\right]^+ + Cl^-$$

One of the Cl^- ions bound to the central Pt^{2+} ion can be replaced by a water molecule. The rate at which this can occur is found by measuring, over a given time interval, the amount of $Pt(NH_3)_2Cl_2$ consumed *or* the amount of Cl^- released per unit volume of the reaction mixture. That is, reaction rates are measured as changes in concentrations with time, thus making the rate independent of the total volume of reaction mixture. For example, suppose we focus on the cisplatin starting material. If the concentration at some time, t_1, is measured in moles per liter, and the measurement is repeated at time t_2, then the rate of reaction could be given as the number of moles of cisplatin per liter consumed per minute (or per second).

$$\text{rate of change of } [Pt(NH_3)_2Cl_2] = \frac{\text{amount of } Pt(NH_3)_2Cl_2 \text{ reacting per unit volume (mol/L)}}{\text{elapsed time } (t)}$$

$$\frac{\Delta[Pt(NH_3)_2Cl_2]}{\Delta t} = \frac{[Pt(NH_3)_2Cl_2] \text{ at } t_2 - [Pt(NH_3)_2Cl_2] \text{ at } t_1}{t_2 - t_1}$$

Recall that square brackets around a compound's molecular formula indicate the compound's concentration in moles per liter.

The change in concentration of $Pt(NH_3)_2Cl_2$ and Cl^- with time are shown in Figure 8.1a and b. Each point on these curves gives the concentration of that compound or ion at a particular time. In Figure 8.1a we illustrate the way to find the average rate during some period of the reaction. For example, after 20 minutes of reaction, the concentration of $Pt(NH_3)_2Cl_2$ has declined from 0.0100 M to 0.00970 M. Therefore, the rate of change of the cisplatin concentration is

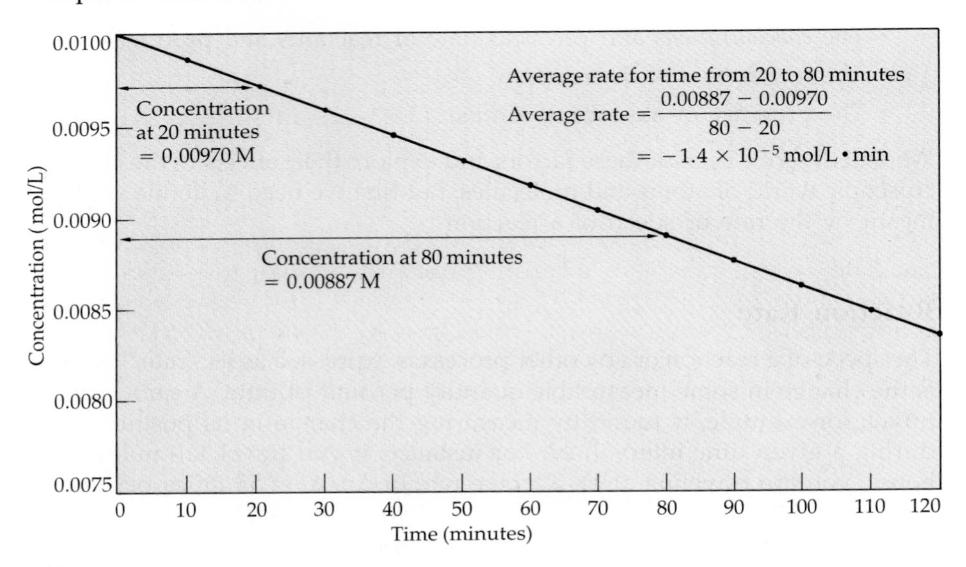

(a)

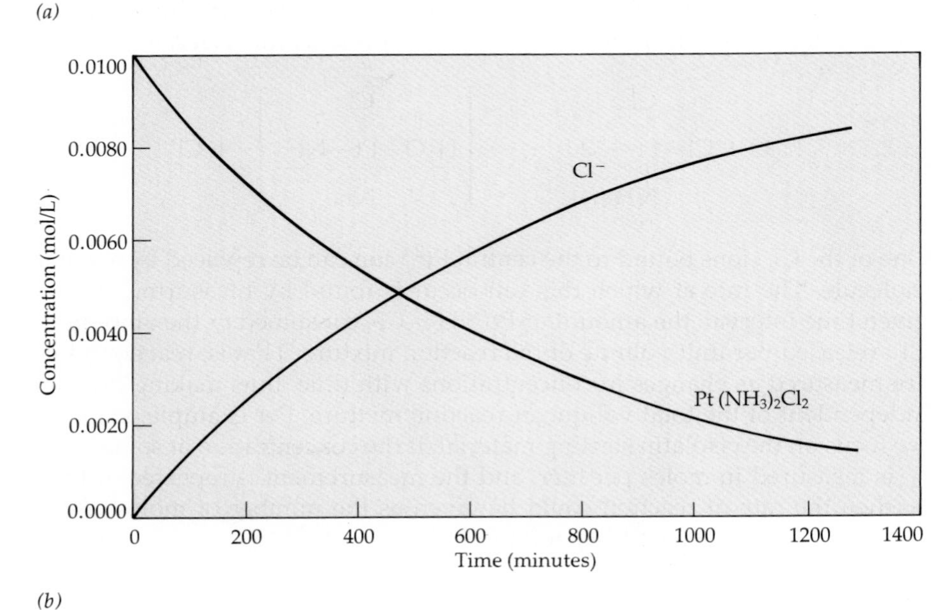

(b)

Figure 8.1
The concentrations of $Pt(NH_3)_2Cl_2$ and Cl^- as functions of time as Cl^- is replaced in the compound by H_2O. (a) Change in concentration of $Pt(NH_3)_2Cl_2$ over the first 120 minutes. (b) Changes in concentration of $Pt(NH_3)_2Cl_2$ and Cl^- over a period of 1300 minutes. The shaded areas show that the decrease in concentration of $Pt(NH_3)_2Cl_2$ is equal to the increase in concentration of Cl^- over the same period.

Therefore, the rate of change of the cisplatin concentration is

$$\text{rate of change of } [Pt(NH_3)_2Cl_2] = \frac{\Delta[Pt(NH_3)_2Cl_2]}{\Delta t}$$
$$= \frac{(0.00887 - 0.00970)\text{ mol/L}}{(80. - 20.)\text{ min}}$$
$$= \frac{-0.00083\text{ mol/L}}{60.\text{ min}}$$
$$= -1.4 \times 10^{-5}\text{ mol/L}\cdot\text{min}$$

where the negative sign indicates that the concentration of $Pt(NH_3)_2Cl_2$ is *de*creasing.

Figure 8.1b extends the time over a longer period and includes the concentration of the product ion Cl^-. Notice that as the concentration of $Pt(NH_3)_2Cl_2$ decreases, the concentration of Cl^- increases. Because one Cl^- ion appears in solution for every molecule of $Pt(NH_3)_2Cl_2$ that disappears, the curve for the *appearance* of Cl^- is exactly "upside down" from the curve for the *disappearance* of $Pt(NH_3)_2Cl_2$. That is, the rate of change in the concentration of $Pt(NH_3)_2Cl_2$ is exactly the same as that for Cl^-, but the value of $\Delta[Pt(NH_3)_2Cl_2]/\Delta t$ is negative while that for $\Delta[Cl^-]/\Delta t$ is positive. As long as we take the stoichiometry and mathematical sign into account, we can calculate the reaction rate by measuring *either* the decrease in $[Pt(NH_3)_2Cl_2]$ *or* the increase in $[Cl^-]$. The reaction rate is defined as the rate of decrease of concentration of a reactant or the rate of increase of concentration of a product, and is always given as a positive number. This means that when a rate is expressed in terms of $\Delta[\text{reactant}]/\Delta t$, which is a negative quantity, a minus sign is always used to make the rate positive.

EXERCISE 8.1 • *Rate of Reaction*

Sucrose decomposes to fructose and glucose in acid solution. A plot of the concentration of sucrose as a function of time is given below. What is the rate of change of the sucrose concentration over the first two hours? What is the rate of change over the last two hours?

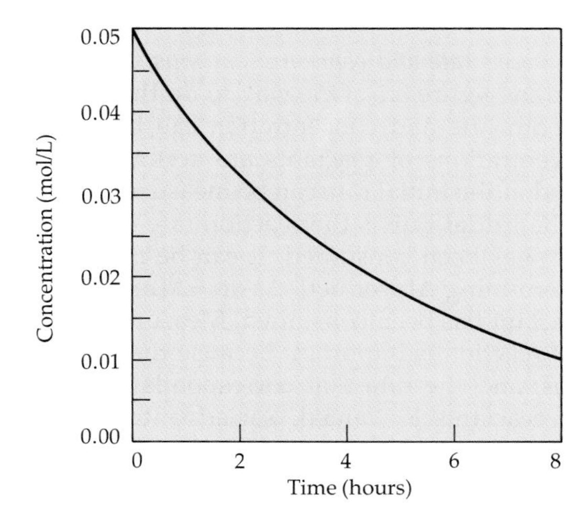

Figure 8.2
The speed of the reaction of aluminum with aqueous NaOH depends on the concentration of the base. With dilute NaOH the reaction is slow (*left*), but it is more rapid in more concentrated NaOH (*right*).

Effect of Concentration on Reaction Rate

It is often possible to change the rate of a reaction by changing the reactant concentrations, and Figure 8.2 shows you an example. The reaction of aluminum metal with sodium hydroxide produces hydrogen gas

$$2\,Al(s) + 6\,H_2O(\ell) + 2\,NaOH(aq) \longrightarrow 2\,NaAl(OH)_4(aq) + 3\,H_2(g)$$

but it is clearly more rapid in 6 M NaOH than in 1 M NaOH. One goal in studying kinetics is to find out whether a reaction speeds up when the concentration of one reactant is increased and, if so, by how much. The effect of concentration on rate can be determined by measuring the rate of a reaction in several experiments in which there are different concentrations of each reactant (and temperature is held constant). When such experiments were carried out for cisplatin, it was discovered that if the concentration of cisplatin is doubled, the reaction rate is likewise doubled. If the concentration of cisplatin is halved, then the reaction rate is halved. This relation between concentration and time gives a plot of $[Pt(NH_3)_2Cl_2]$ versus time with the shape shown in Figure 8.1, which shows that the reaction rate is directly proportional to the reactant concentration.

$$\text{rate of reaction} \propto [Pt(NH_3)_2Cl_2] \qquad \text{(where the symbol } \propto \text{ means ``proportional to'')}$$

A direct or 1:1 relation between reaction rate and a reactant concentration is only one kind of behavior that is observed. Some reactions are known in which the rate increases much faster than the increase in concentration, while there are others in which it decreases as the concentration increases. Often, though, to speed up a reaction one simply increases the concentrations of the reactants.

The relationship between reactant concentration and reaction rate is expressed by an equation called a **rate equation** or **rate law.** For the cisplatin reaction the rate equation is

$$\text{rate of reaction} = k[Pt(NH_3)_2Cl_2]$$

where the proportionality constant, k, is called the **rate constant.** The rate constant is independent of concentration, but it generally increases with temperature.

The relation between rate and concentration must be determined experimentally (as illustrated by Figure 8.1). One way to do this is to carry out experiments in which both rate and concentration can be measured simultaneously. Often concentrations of reactants are accurately known before the reactants are mixed; if the initial rate can be measured, the rate equation can be determined. The **initial rate** is the reaction rate during the period when very little reactant has been consumed. It can be measured by mixing the reactants and determining $\Delta[\text{product}]/\Delta t$ or $-\Delta[\text{reactant}]/\Delta t$ after 1 to 2% of the limiting reactant has been consumed. Measuring the rate during only a small percent of reaction is important because the rate changes with time for most reactions, and the rate that corresponds to the measured initial concentrations of reactants is wanted. An advantage of measuring initial rates is that as a reaction proceeds, more and more products are formed. In some cases products can alter the rate; comparing initial rates with rates when products are present can reveal such a complication.

As an example of the determination of rates, consider the reaction of a base with methyl acetate, an organic compound and a widely used industrial solvent. The reaction produces acetate ion and methyl alcohol.

$$\underset{\text{methyl acetate}}{CH_3\overset{\overset{O}{\|}}{C}{-}O{-}CH_3} + OH^- \longrightarrow \underset{\text{acetate ion}}{CH_3\overset{\overset{O}{\|}}{C}{-}O^-} + \underset{\text{methyl alcohol}}{CH_3OH}$$

Since reaction rates change with temperature, several experiments were done at the same temperature, and the following data were collected:

	Initial Concentrations		
Experiment	*[CH_3COOCH_3]*	*[OH^-]*	*Initial Reaction Rate (mol/L·s)*
1	0.050 M	0.050 M	0.00034
	↓ no change	↓ × 2	↓ × 2
2	0.050 M	0.10 M	0.00069
	↓ × 2	↓ no change	↓ × 2
3	0.10 M	0.10 M	0.00137

When the initial concentration of either CH_3COOCH_3 or OH^- was doubled, and the concentration of the other reactant was held constant, the reaction rate doubled. This rate doubling shows that the rate for the reaction is directly proportional to the concentrations of *both* CH_3COOCH_3 and OH^-, that is, to the *product* of the two concentrations. Therefore, the rate equation that reflects these experimental observations is

$$\text{reaction rate} = k[CH_3COOCH_3][OH^-]$$

From this equation we can also conclude that doubling both concentrations at the same time causes the rate to go up by a factor of four. And what would happen if one concentration is doubled and the other halved? The rate equation tells us the rate will not change, and that is what happens!

For the methyl acetate–hydroxide ion reaction, or any other, the value for k, the rate constant, can be found by substituting rate and concentration data for any one experiment into the rate equation. To find k for the methyl acetate–hydroxide ion reaction, data from the first experiment could be substituted, for example,

$$\text{reaction rate} = 0.00034 \text{ mol/L·s} = k(0.050 \text{ mol/L})(0.050 \text{ mol/L})$$

$$k = \frac{0.00034 \text{ mol/L·s}}{(0.050 \text{ mol/L})(0.050 \text{ mol/L})} = 0.14 \text{ L/mol·s}$$

and we would calculate that k is 0.14 L/mol·s at 25 °C.

EXAMPLE 8.1

Determining a Rate Equation for the NO + O_2 Reaction

The rate of the reaction of nitrogen oxide and oxygen

$$2\,NO(g) + O_2(g) \longrightarrow 2\,NO_2(g)$$

was measured at 25 °C starting with various concentrations of NO and O_2, and the following data were collected:

Experiment	*Initial Concentrations (mol/L)* [NO]	$[O_2]$	*Initial Rate (mol/L·s)*
1	0.020	0.010	0.028
2	0.020	0.020	0.057
3	0.020	0.040	0.114
4	0.040	0.020	0.227
5	0.010	0.020	0.014

Based on these data, what is the rate equation? What is the value of the rate constant *k*?

SOLUTION In the first three experiments the concentration of NO is constant while the O_2 concentration increases from 0.010 to 0.020 to 0.040 mol/L. Each time the O_2 concentration is doubled, the initial rate also doubles. For example, when $[O_2]$ is doubled from 0.020 to 0.040 mol/L, the initial rate increases by a factor of 2 from 0.057 to 0.114 mol/L·s. This means that the initial rate is directly proportional to $[O_2]$.

In Experiments 2, 4, and 5 the O_2 concentration is constant, while [NO] varies. From Experiment 2 to 4, [NO] is doubled, while the initial rate increases by a factor of 4 or 2^2.

$$\frac{\text{Experiment 4 rate}}{\text{Experiment 2 rate}} = \frac{0.227 \text{ mol/L·s}}{0.057 \text{ mol/L·s}} = \frac{4}{1}$$

This same result is found on comparing Experiments 4 and 5, and it means that the initial rate is proportional to the *square* of [NO]. Therefore, the rate equation is

$$\text{initial rate} = k[O_2][NO]^2$$

The rate constant *k* can be found by inserting data for one of the experiments into the rate equation. For Experiment 1, for example,

$$\text{initial rate} = 0.028 \text{ mol/L·s} = k(0.010 \text{ mol/L})(0.020 \text{ mol/L})^2$$

$$k = \frac{0.028 \text{ mol/L·s}}{(0.010 \text{ mol/L})(0.020 \text{ mol/L})^2} = 7.0 \times 10^3 \text{ L}^2\text{/mol}^2\text{·s}$$

A better value of *k* would be obtained by calculating *k* for each experiment and then averaging the values. Once this value is available, it can be used to calculate the initial rate for any set of NO and O_2 concentrations.

For the reaction of methyl acetate with hydroxide ion, or for many other reactions, the rate equation has the general form

$$\text{reaction rate} = k[\text{A}]^m[\text{B}]^n$$

where [A] and [B] represent the concentrations of the reactants, which are raised to the powers m and n, respectively. These exponents, which were determined by experiment, are both equal to 1 for the methyl acetate–hydroxide ion reaction. By coincidence they are also the same as the stoichiometric coefficients. But be careful! This is not always true. For example, in dilute sodium hydroxide, hydrogen peroxide decomposes to water and oxygen

$$2\,H_2O_2(aq) \longrightarrow 2\,H_2O(\ell) + O_2(g)$$

$$\text{reaction rate} = k[H_2O_2] \qquad \text{where } k = 1.77 \times 10^{-5}\text{/s at 20 °C}$$

and experiments clearly show that the exponent of $[H_2O_2]$ is 1 in the rate equation, even though the stoichiometric coefficient for H_2O_2 is 2 in the balanced equation.

The important point to remember is that *the effects of changing concentrations on reaction rate, which are summarized by the rate equation, must be determined experimentally. Concentration effects, and therefore the rate equation, cannot be predicted from the balanced chemical equation.* The rate equation often has the form

$$\text{reaction rate} = k[\text{A}]^m[\text{B}]^n \ldots$$

where [A] and [B] are the concentrations of reactants, products, or a catalyst at a given time. The exponents m and n are usually positive whole numbers, but they can be negative numbers or even fractions. These exponents define the **order** of the reaction. If m is 1, for example, the reaction is first order with respect to A; if n is 2, then the reaction is second order with respect to B. The exponents m and n define the **order** of the reaction. If m is 1, for example, the reaction is first order with respect to A; if n is 2, then the reaction is second order with respect to B. The sum of m and n, and the exponents on any other concentration terms, gives the **overall reaction order.** For example, the overall order would be 3 if $m = 1$ and $n = 2$. We would say it is a third order reaction.

EXERCISE 8.2 • *Interpreting a Rate Law*

The rate equation for the reduction of NO to N_2 with hydrogen is

$$2\,NO(g) + 2\,H_2(g) \longrightarrow N_2(g) + 2\,H_2O(g)$$

$$\text{reaction rate} = k[NO]^2[H_2]$$

a. What is the order of the reaction with respect to the NO? With respect to H_2?

b. If the concentration of NO is doubled, what happens to the reaction rate?

c. If the concentration of H_2 is halved, what happens to the reaction rate?

EXERCISE 8.3 • *Using Rate Laws*

The rate constant k is 0.090/hr for the reaction

$$Pt(NH_3)_2Cl_2 + H_2O \longrightarrow [Pt(NH_3)_2(H_2O)Cl]^+ + Cl^-$$

and the rate equation is reaction rate $= k[Pt(NH_3)_2Cl_2]$. Calculate the initial rate of reaction when the concentration of $Pt(NH_3)_2Cl_2$ is 0.020 M. What is the rate of change in the concentration of Cl^- under these conditions?

trans-2-butene (*top*) and *cis*-2-butene (bottom). *(C.D. Winters)*

A straight line between two atomic symbols in a structural formula represents the covalent bond that holds two atoms together in the molecule. A pair of straight lines represents a double bond, which is a pair of covalent bonds. Chemical bonds are discussed in Chapter 10.

A Microscopic View of Reactions: Mechanisms

You know that reactant concentrations can affect reaction rates. This is a macroscopic observation. We can also describe reaction rates in terms of what happens in the microscopic world of atoms and molecules by using the *kinetic theory of matter*, which was first introduced in Section 2.1.

According to kinetic theory, molecules in a gas or liquid move rapidly and sometimes bump into one another. Atoms within molecules move as well; molecules constantly flex or vibrate along or around the bonds that hold the atoms together. If a molecule has enough energy, the arrangement of atoms can be changed, resulting in a different molecule. An example is the conversion of *cis*-2-butene to *trans*-2-butene

$$(H_3C)(H)C{=}C(CH_3)(H)\ (g) \longrightarrow (H)(H_3C)C{=}C(CH_3)(H)\ (g)$$

cis-2-butene → *trans*-2-butene

The experimentally determined rate equation is

$$\text{reaction rate} = k[\textit{cis}\text{-2-butene}]$$

which shows that the reaction is first order in the reactant, *cis*-2-butene. The models in the photograph show that the *cis* form (*bottom*) can become the *trans* form (*top*) if one half of the *cis* molecule is twisted relative to the other. Thus, it is a reasonable hypothesis that the molecular pathway or reaction mechanism that converts *cis*-2-butene to *trans*-2-butene reaction involves twisting the molecule.

We know from experiment, though, that carbon-carbon double bonds are like springs. Springs can be stretched, twisted, and bent, although it takes the input of energy to do so. Consequently, some kinetic energy must be converted to potential energy when one end of the *cis*-2-butene molecule twists relative to the other, just as it would if a spring were stretched or bent. At room temperature most of the molecules do not have enough kinetic energy to twist far enough so that *cis*-2-butene can be changed to *trans*-2-butene, and so *cis*-2-butene can be kept in a sealed flask at room temperature for a long time without any appreciable quantity of *trans*-2-butene being formed.

Figure 8.3 shows a plot of potential energy versus the angle of twist in *cis*- and *trans*-2-butene. The potential energy of a *cis*-2-butene molecule is 435×10^{-21} J higher when one end is twisted by 90° from the initial, flat molecule. This is similar to the increased potential energy that an object like a car has at the top of a hill compared with its energy at the bottom. Just as a car cannot reach the top of a hill unless it has enough energy, a molecule cannot reach the top of the "hill" for a reaction unless it has enough energy. Notice that the top of the hill can be approached from either side, and from the top a twisted molecule can go downhill energetically to either the *cis* or the *trans* form.

Since molecules are very small, the energy required to twist one cis-2-butene *molecule is also very small. However, if we wanted to twist a mole of molecules, all at once, it would take a lot of energy. The energy required to reach the top of the "hill" is often reported per mole of molecules, that is, as* (435×10^{-21} *J/molecule*)$\cdot$(6.022×10^{23} *molecules/mol*) *or 262 kJ/mol.*

Rotation around a double bond, as in the interconversion of *cis*- and *trans*-2-butene, occurs in the reactions that allow you to see. A yellow-orange compound, β-carotene, the natural coloring agent in carrots, breaks

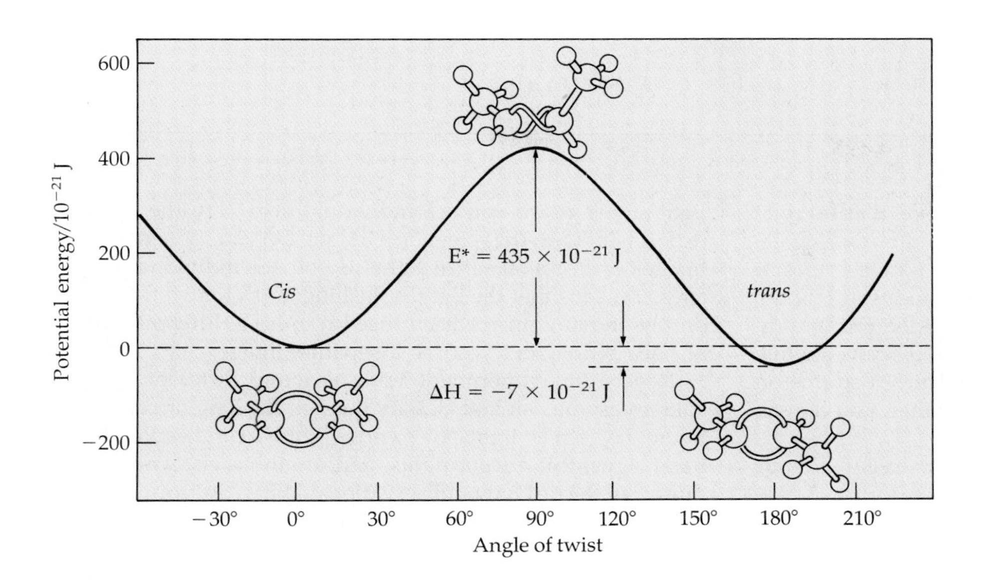

Figure 8.3

Energy profile for the conversion of *cis*-2-butene to *trans*-2-butene. The double bond between the two central C atoms is stiff and resists twisting. However, when enough energy is added, one half of the molecule can twist with respect to the other half. When the angle between the ends of the molecule is 90°, the potential energy has risen to 435×10^{-21} J/molecule. A molecule of *cis*-2-butene must have at least this amount of energy before it can convert to *trans*-2-butene.

down in your body to produce vitamin A, and this compound is converted in the liver to a compound called 11-*cis*-retinal. In the retina of your eye 11-*cis*-retinal combines with the protein opsin to form a light-sensitive substance called rhodopsin. When light strikes the retina, enough energy is transferred to a rhodopsin molecule to allow rotation around a carbon-carbon double bond, transforming rhodopsin into metarhodopsin II, a molecule whose shape is quite different, as you can see from the structural formulas below. This change in molecular shape causes a nerve impulse to be sent to your brain, and you see the light.

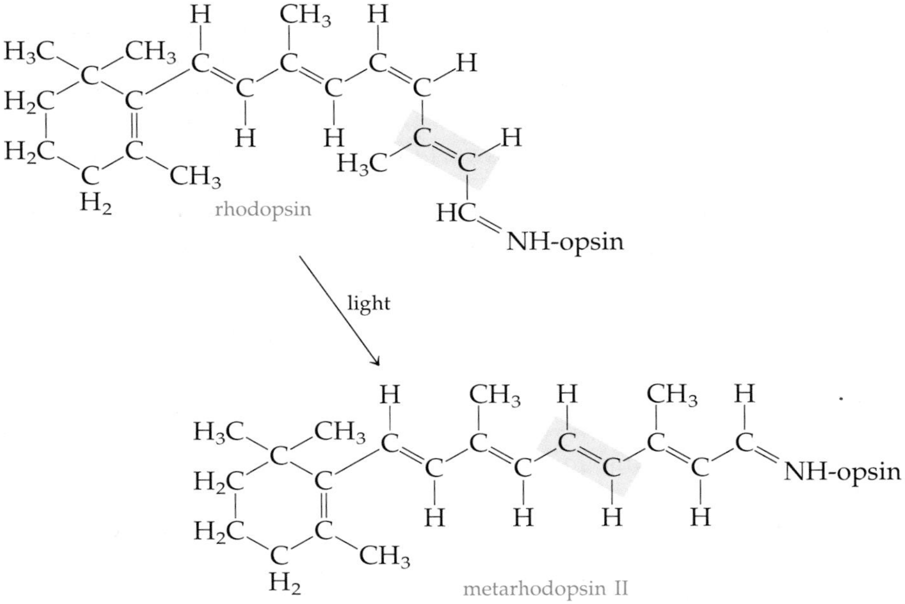

rhodopsin

light

metarhodopsin II

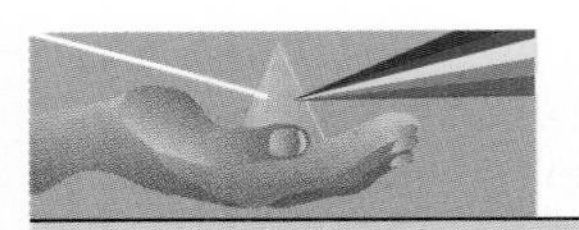

CHEMISTRY YOU CAN DO

Kinetics and Vision

Vision is not an instantaneous process. It takes a little while after a bright light goes out before you stop seeing its image. An example is a flash lamp, which blinds you for a short time even though it is on only for an instant. Some sources of light flash on and off very rapidly, but you do not notice that their images continue to be perceived while they are off. However, if you can focus such a source on different parts of your retina at different times, you can see whether it is flashing. Here's how.

Find a small mirror that you can hold easily in your hand. Now use the mirror to reflect the image of an incandescent light bulb onto your eye. You should be far enough away from the light so that its image is small. Now move the mirror quickly back and forth so that the image of the light bulb moves quickly across your eyeball. Does the light smear or do you see individual dots? Try the same experiment with the screen of a TV set. Do you see separate images or just a smear of light? If you see individual images, it means that the light is flashing. Each time it flashes on, the moving mirror has caused it to hit a different part of your retina, and you see a separate image.

Repeat this experiment with as many different light sources as you can and classify them as flashing or continuous. Try street lights of various kinds, car headlights, neon signs, fluorescent lights, and anything else you can think of. Record your observations. What do you think would happen if you photographed these light sources and moved the camera quickly while the shutter was open? What would happen if you moved the camera more slowly or more quickly?

Eventually the metarhodopsin II reacts chemically to produce a different form of retinal, which is then converted back to vitamin A, and the cycle of chemical changes can begin again. However, decomposition of metarhodopsin II is not as rapid as its formation, and an image formed on the retina persists for a tenth of a second or so. This persistence of vision allows you to perceive videos as continuously moving images, even though they actually consist of separate pictures, each painted on a screen for a thirtieth of a second.

Returning to the simple case of *cis-* and *trans*-2-butene, there is another interesting relationship shown in Figure 8.3 that connects kinetics and thermodynamics. The energy of the product, a molecule of *trans*-2-butene, is 7×10^{-21} J *lower* than that of reactant, a molecule of *cis*-2-butene. This means that the *cis* → *trans* reaction is *exothermic* by 7×10^{-21} J/molecule, which translates to 4 kJ/mol. Conversely, *cis*-2-butene is *higher* in energy by 7×10^{-21} J/molecule, and so the reverse reaction requires that 4 kJ/mol be absorbed from the surroundings; it is *endothermic*. The energy hill that has to be climbed when the reverse reaction occurs is $(435 + 7) \times 10^{-21}$ J or 442×10^{-21} J high (= 266 kJ/mol).

Every chemical reaction has an energy barrier that must be surmounted if molecules are to react. The heights of such barriers vary greatly—from almost zero to hundreds of kilojoules per mole. *For similar reactions at a given temperature, the higher the energy barrier the slower the reaction.* The minimum energy required to surmount the barrier is called the **activation energy, E*,** for the reaction. For the *cis*-2-butene → *trans*-2-butene reaction the activation energy is 442×10^{-21} J/molecule or 266 kJ/mol.

The molecular pathway or reaction mechanism followed by the *cis* → *trans* reaction involves the twisting of molecules. This is a very simple mechanism, which involves a single reactant and a reaction that takes place in a single step. The mechanism can be written in the form of a chemical equation as

$$cis\text{-2-butene} \longrightarrow trans\text{-2-butene}$$

Because only one reactant molecule is involved, the reaction is said to be **unimolecular.** Because it takes place in a single step, the energy barrier has only a single hump and the reaction is said to be an **elementary process** or **elementary reaction.** (Just as compounds can be built up from chemical elements, more complicated reaction mechanisms can be built as a series of elementary reactions.) But how is this microscopic view related to the experimental rate equation (a macroscopic observation)?

Suppose a flask contains 0.005 mol/L of *cis*-2-butene vapor at room temperature. The molecules have a wide range of energies, but only a few of them have enough energy at this temperature to get over the activation energy barrier. Thus, during a given period of time at this temperature only a few molecules twist sufficiently to become *trans*-2-butene.

Now suppose that another flask contains 0.010 mol/L of *cis*-2-butene, that is, the concentration of molecules is twice that of the first flask. If both flasks are at the same temperature, both have the same fraction of molecules with enough energy to cross over the barrier. However, in the flask containing twice as many molecules, there must be twice as many molecules crossing the barrier in any given time. Therefore, the rate of the *cis* → *trans* reaction is twice as great. That is, reaction rate is proportional to the concentration of *cis*-2-butene, which is exactly a statement of the rate equation, rate = k[*cis*-2-butene]. For a unimolecular reaction the microscopic mechanism predicts a first-order rate equation.

There are many reactions in which two molecules of different kinds must collide with one another. These are called **bimolecular** reactions. One example is the reaction of nitrogen monoxide and ozone

$$NO(g) + O_3(g) \longrightarrow NO_2(g) + O_2(g)$$

for which the experimental rate equation is

$$\text{reaction rate} = k[NO][O_3]$$

Here the elementary reaction involves the collision of an NO molecule and an O_3 molecule, but there is still only a single step. Since the molecules must collide to exchange atoms, however, the rate depends on the number of collisions per unit time. Figure 8.4a represents a flask containing one NO molecule (the red ball) and several O_3 molecules (the blue balls). In a given time, the NO molecule collides with, let us say, five O_3 molecules. If the concentration of NO molecules is doubled to two in the flask (Figure 8.4b), each NO collides with five different O_3 molecules, and so the total number of collisions with O_3 molecules is now ten. Doubling the concentration of NO has doubled the rate. The same thing would happen if the O_3 concentration were doubled. This description of the NO + O_3 reaction applies to all processes involving the collision of two molecules A and B. Such reactions have the general rate equation "reaction rate = k[A][B]" where the reaction is first order in each of the two reactants.

Like the *cis* → *trans*-2-butene reaction, the NO + O_3 reaction is exothermic (by about 200 kJ/mol), and this is reflected in the activation energy

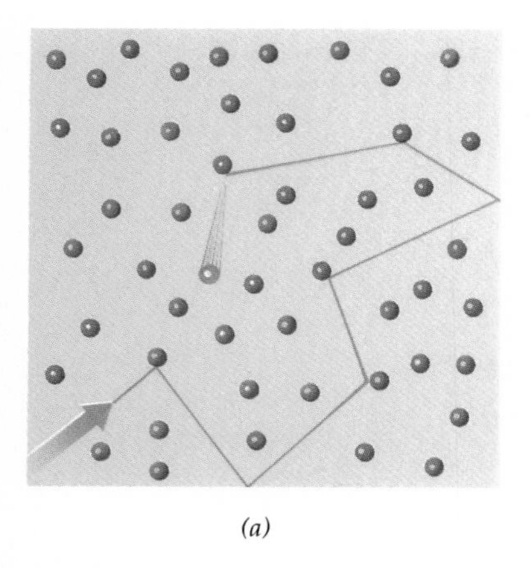

(a)

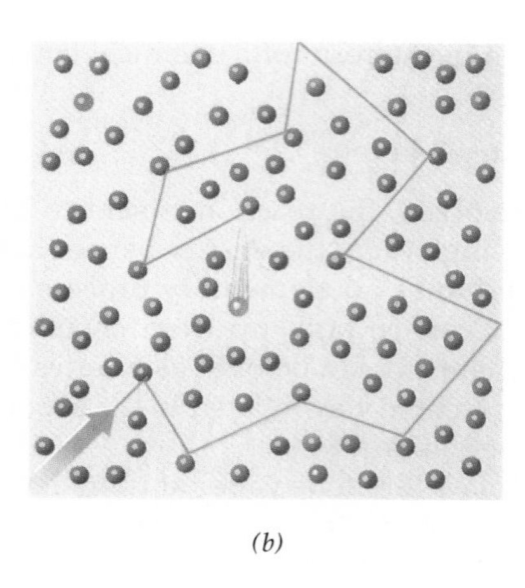

(b)

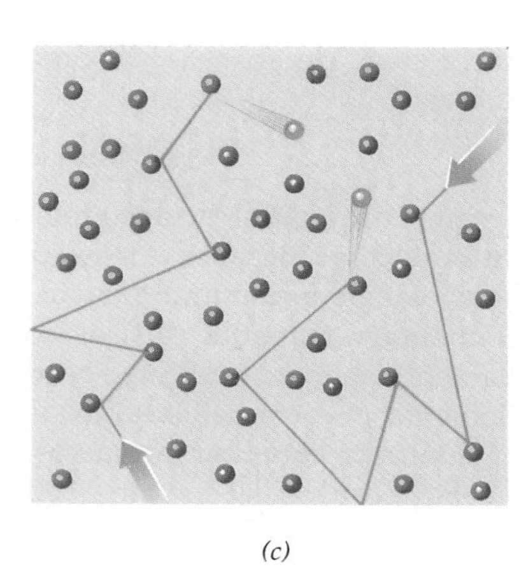

(c)

Figure 8.4

The effect of concentration on the frequency of molecular collisions. (a) A single red molecule moves among 50 blue molecules and collides with five of them per second. (b) If the number of blue molecules is doubled to 100, the frequency of red-blue collisions is also doubled, to 10 per second. (c) Two red molecules now move among 50 blue molecules, and there are 10 red-blue collisions per second. The number of collisions is thus seen to be proportional to *both* the concentration of red molecules *and* the concentration of blue molecules.

diagram for the reaction (Figure 8.5). At the top of the barrier, 10 kJ/mol higher than the energy of the separated molecules, NO and O_3 have collided and the O atom is just beginning to transfer from O_3 to NO.

The NO + O_3 reaction is interesting and important for another reason: molecules must collide in the proper orientation for the reaction to be effective. Having a sufficiently high energy is necessary, but it is not all that is required to ensure that reactants will form products. In the case of the NO + O_3 reaction, the N of NO must come together with one of the O atoms at the end of the O_3 molecule. This so-called "steric factor" is an important factor in determining how fast a reaction can be, and it is reflected in the value of the rate constant, *k*. The more difficult it is to achieve the proper alignment, the slower the process.

An excellent example of the importance of the correct geometry for the collision of reactants is the reaction of methyl iodide, CH_3I, with hydroxide ion, OH^-.

$$OH^- + CH_3I \longrightarrow CH_3OH + I^-$$

methyl iodide — methyl alcohol

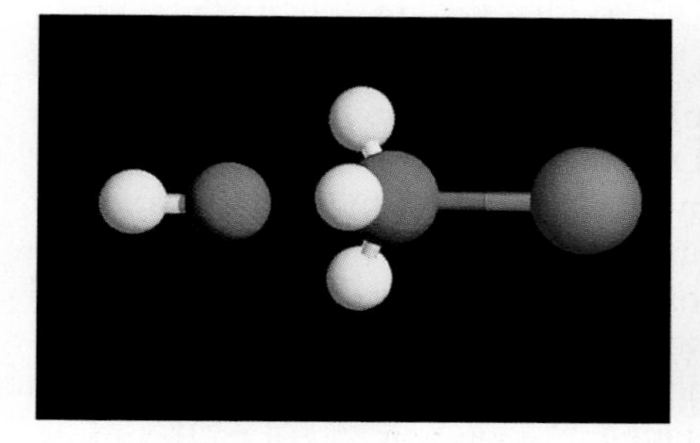

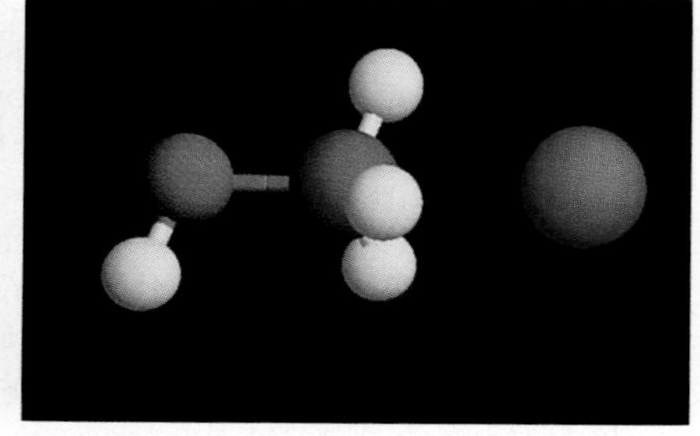

Hydroxide ion–methyl iodide reaction. *(C.D. Winters)*

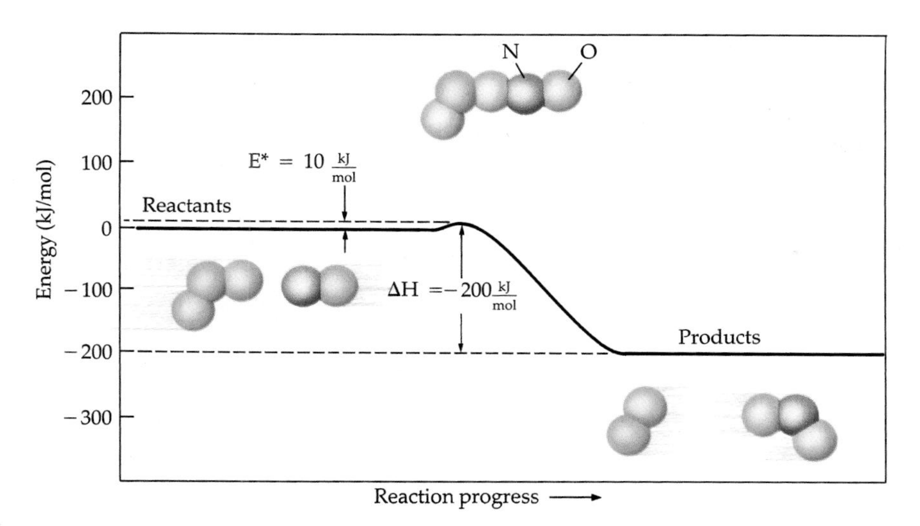

Figure 8.5
Energy profile for the $NO + O_3$ reaction. As an NO molecule and an O_3 molecule approach one another, energy is needed to squeeze them together, until the N atom of NO and one of the O atoms of O_3 are close enough to form a bond. If the two colliding molecules do not have a total energy of at least 10 kJ/mol, then an O atom of O_3 cannot be forced close enough to transfer to the NO molecule. Also notice that the NO and O_3 molecules must come together in the proper orientation: one of the terminal O atoms of O_3 must approach the N atom of NO. (Reaction is not possible if the O_3 approaches the O atom of NO, or if the central O atom of O_3 is the one that makes contact.) This "steric factor" is important in determining how large the reaction rate can be; the more difficult it is to achieve the proper orientation, the slower the reaction is.

The methyl iodide molecule has a tetrahedral shape, and numerous experiments suggest that the reaction occurs most rapidly in solution when the OH^- ion attacks the methyl iodide on the side of the tetrahedron opposite from the iodine atom. That is, an attack on only one of the four sides of CH_3I, or a maximum of $\frac{1}{4}$ of all of the collisions between reactants, can be effective. Furthermore, even when OH^- does attack on the correct face of the tetrahedron, it must attack with the O end and not the H end. This further reduces the effectiveness of collisions to only $\frac{1}{8}$ of all collisions. You can see that for complicated molecules, these geometry restraints mean only a very small fraction of the total collisions can lead to reaction. No wonder some chemical reactions are so slow. Conversely, it is amazing that some are so fast!

EXERCISE 8.4 • *Activation Energy*

For the hypothetical reaction $A \rightarrow B$, the activation energy is 24 kJ/mol. For the reverse reaction, $B \rightarrow A$, the activation energy is 36 kJ/mol. Draw a diagram similar to Figures 8.3 and 8.5 to illustrate this reaction. Is the reaction $A \rightarrow B$ exothermic or endothermic?

Temperature and Reaction Rate

As a rough rule of thumb, the reaction rate usually increases by a factor of 2 to 4 for each 10 °C rise in temperature.

The most common way to speed up a reaction is to increase the temperature (Figure 8.6). We mentioned earlier that a mixture of methane and air is stable for centuries at room temperature; the reaction to give CO_2 and H_2O is extraordinarily slow. However, if a lighted match is brought up to the mixture, its temperature is raised, the reaction rate increases, and the gas ignites. Thereafter the heat evolved by the combustion reaction maintains a high temperature and the reaction continues at a rapid rate.

For the same reason reactions speed up when the temperature is raised, they slow down when the temperature is lowered. Food is often stored in refrigerators or freezers because the rate of spoilage is slowed by lowering the temperature.

The increase in reaction rate with temperature occurs because higher temperature means a greater fraction of reactant molecules with enough energy to cross the activation energy barrier. Consider again the conversion of *cis*- to *trans*-2-butene (Figure 8.3). Although the molecules are constantly in motion, at any temperature they have a wide distribution of speeds and energies. At room temperature relatively few *cis*-2-butene molecules are sufficiently energetic to cross the energy barrier. However, as the temperature goes up more and more molecules have enough energy, and the reaction rate increases.

Increasing the rate of a reaction by increasing the temperature is illustrated in Figure 8.7, which shows a reaction of a cobalt(III) compound in which a Co—Cl bond is replaced with a bond between Co and the O atom of a water molecule. The original compound is green, while the final product is red. Here you see that after 20 minutes no reaction has occurred in the flask held at room temperature. However, heating the reaction to about 80 °C in a hot water bath causes the reaction to be nearly complete in 10 minutes or less.

Figure 8.6
Magnesium will react only slowly with water at room temperature. However, warming the water causes a vigorous oxidation-reduction reaction that produces hydrogen gas and magnesium hydroxide. *(C.D. Winters)*

$$Mg(s) + 2H_2O(\ell) \longrightarrow H_2(g) + Mg(OH)_2(s)$$

Multi-Step Mechanisms

Both the *cis* → *trans*-2-butene reaction and the NO + O_3 reaction take place in a single elementary process. Most chemical reactions, however, involve a sequence of elementary reactions—a multi-step mechanism. For example, iodide ion can be oxidized by hydrogen peroxide in acidic solution to form iodine and water according to this overall equation:

$$2\,I^-(aq) + H_2O_2(aq) + 2\,H_3O^+(aq) \longrightarrow I_2(aq) + 4\,H_2O(\ell)$$

When the acid concentration is between 10^{-3} M and 10^{-5} M, experiments show that the rate equation is

$$\text{reaction rate} = k[I^-][H_2O_2]$$

The reaction is first order in the concentrations of I^- and H_2O_2, and second order overall. The concentration of the hydronium ion does not affect the reaction rate.

Looking at the balanced equation for the oxidation of iodide by hydrogen peroxide, you might get the idea that reaction occurs when two iodide ions, one hydrogen peroxide molecule, and two hydronium ions come together. However, it is highly unlikely that these five ions or molecules would all be at the same place at the same time. Instead, those who have studied this reaction propose that first a H_2O_2 molecule and an I^- ion come together:

Step 1 $$\text{HOOH} + \text{I}^- \xrightarrow{\text{slow}} \text{HOI} + \text{OH}^-$$

This step forms hypoiodous acid, HOI, and hydroxide ion, both known substances. The HOI then reacts with another I^- to form the product I_2:

Step 2 $$\text{HOI} + \text{I}^- \xrightarrow{\text{fast}} \text{I}_2 + \text{OH}^-$$

In each of steps 1 and 2, a hydroxide ion was produced. Since the solution is acidic, these OH^- ions will react immediately with H_3O^+ ions to form water:

In the reaction mechanism, H_2O_2 has been written as HOOH to emphasize that the two oxygen atoms are connected to each other in the molecule.

Step 3 $$2\,\text{OH}^- + 2\,\text{H}_3\text{O}^+ \xrightarrow{\text{fast}} 4\,\text{H}_2\text{O}$$

Each of the three steps in this mechanism is an elementary reaction and has its own activation energy, E^*, and its own rate constant, k. When the three steps are summed by putting all the reactants on the left, putting all the products on the right, and eliminating formulas that appear as both reactants and products, the overall stoichiometric equation is obtained. This is a requirement for any valid mechanism.

$$
\begin{array}{ll}
\textit{Step 1} & \text{HOOH} + \text{I}^- \xrightarrow{\text{slow}} \cancel{\text{HOI}} + \cancel{\text{OH}^-} \\
\textit{Step 2} & \cancel{\text{HOI}} + \text{I}^- \xrightarrow{\text{fast}} \text{I}_2 + \cancel{\text{OH}^-} \\
\textit{Step 3} & 2\,\cancel{\text{OH}^-} + 2\,\text{H}_3\text{O}^+ \xrightarrow{\text{fast}} 4\,\text{H}_2\text{O} \\
\hline
\textit{Overall} & 2\,\text{I}^- + \text{HOOH} + 2\,\text{H}_3\text{O}^+ \longrightarrow \text{I}_2 + 4\,\text{H}_2\text{O}
\end{array}
$$

Step 1 of the mechanism is labeled as slow, while steps 2 and 3 are fast. Step 1 is called the **rate-limiting step;** because it is the slowest in the sequence, it limits the rate at which I_2 and H_2O can be produced. Steps 2 and 3 are rapid and therefore not rate-limiting. As soon as some HOI and OH^- are produced by step 1, they are transformed into I_2 and H_2O by steps 2 and 3. *The rate of the overall reaction is limited by, and equal to, the rate of the slowest step in the mechanism.*

You are familiar with the concept of the rate-limiting or rate-determining step. No matter how quickly you shop in the supermarket, it seems that the time to get out of the store depends on the rate at which you move through the check-out line.

Figure 8.7
The reaction shown in this photo is the combination of a green cobalt(II) compound with water (called a *hydrolysis* reaction). The compound has two Co—Cl bonds, one of which is replaced by a bond between Co^{2+} and H_2O to give a red compound. The reaction is slow at room temperature, so much of the original compound is still present after 20 minutes, as indicated by the green color. When warm water is used, however, the solution turns red quickly, because the rate of reaction increases with increasing temperature. *(C.D. Winters)*

Step 1 is a bimolecular process, and its rate is first order in HOOH and first order in I^-. Therefore the mechanism predicts that the rate equation should be

$$\text{reaction rate} = k[\text{HOOH}][\text{I}^-]$$

which agrees with the experimentally observed rate equation. This is a second requirement for a valid mechanism—it must correctly predict the observed rate equation.

The species HOI and OH^-, which are produced in step 1 and used up in steps 2 or 3, are called **reaction intermediates.** Very small concentrations of HOI and OH^- will be produced while the reaction is going on, but once the HOOH, the I_2, or both are used up, these intermediates will be consumed by steps 2 and 3. HOI and OH^- are crucial to the overall reaction, but neither of them appears in the net stoichiometric equation. Reaction intermediates usually have a very fleeting existence, but some have long enough lifetimes to be observed. If one is proficient enough to demonstrate experimentally that a particular intermediate was present, this provides additional evidence that a mechanism involving that intermediate is the correct one. In fact, that is an exciting aspect of chemistry—devising an experiment to track down an elusive reaction intermediate, thereby opening up an understanding of a family of related reactions.

Having described some aspects of rate equations and reaction mechanisms, it is useful to summarize some important points.

- The rate equation for a reaction can be determined *only* by experiment. You observe the change in the concentration of reactants or products at a constant temperature and then derive the rate equation from this experimental information.
- After determining the rate equation, you can try to devise a reaction mechanism that accounts for the experimental results at the submicroscopic or molecular level. Be sure to understand that a reaction mechanism is just an educated guess—a hypothesis—about the way the reaction occurs. Sometimes several mechanisms can agree with the same set of experiments, so one postulated mechanism can be quite wrong and provoke disputes among scientists. This is the situation today in some areas of environmental chemistry. Nonetheless, trying to account for experimental observations in terms of molecular reaction mechanisms is one of the most interesting and rewarding areas of chemistry.

EXERCISE 8.5 • *Reaction Mechanisms*

The Raschig reaction produces the industrially important reducing agent hydrazine, N_2H_4, from NH_3 and OCl^- in basic aqueous solution. A proposed mechanism is

Step 1 $NH_3(aq) + OCl^-(aq) \xrightarrow{\text{fast}} NH_2Cl(aq) + OH^-(aq)$

Step 2 $NH_2Cl(aq) + NH_3(aq) \xrightarrow{\text{slow}} N_2H_5^+(aq) + Cl^-(aq)$

Step 3 $N_2H_5^+(aq) + OH^-(aq) \xrightarrow{\text{fast}} N_2H_4(aq) + H_2O(\ell)$

What is the overall stoichiometric equation? Which step is rate-limiting? What reaction intermediates are involved?

Figure 8.8
Decomposition of hydrogen peroxide can be accelerated by the heterogeneous catalyst MnO_2. Here a 30% aqueous solution of H_2O_2 is dropped onto the black solid MnO_2 and rapidly decomposes to O_2 and H_2O. The water is given off as steam because of the high heat of reaction.
(C.D. Winters)

Catalysts and Reaction Rate

Raising the temperature increases a reaction rate because it increases the fraction of molecules that are energetic enough to cross a potential energy barrier. Increasing reactant concentrations can also increase the rate, because the absolute number of molecules capable of reacting has increased; the fraction with sufficient energy remains the same. There is yet another way to increase reaction rates: using a catalyst. A **catalyst** is a substance that increases the rate of a reaction but is not consumed in the overall reaction. For example, you have already learned that hydrogen peroxide, H_2O_2, can decompose to water and oxygen.

$$2\,H_2O_2(\ell) \longrightarrow O_2(g) + 2\,H_2O(\ell)$$

If the peroxide is stored in a cool place in a clean plastic container, it is reasonably stable for months; the rate of the decomposition reaction is exceedingly slow. However, in the presence of a manganese salt, an iodide-containing salt, or a biological substance called an enzyme, the reaction occurs with explosive speed (Figure 8.8). In fact, an insect called a bombardier beetle uses a very similar reaction as its defense mechanism (Figure 8.9). By combining the organic compound hydroquinone with the peroxide in the presence of an enzyme, it produces superheated steam and an irritating chemical to spray its enemies.

$$\text{hydroquinone } (C_6H_4(OH)_2) + H_2O_2 \xrightarrow{\text{enzyme, a catalyst}} \text{quinone, an irritant } (C_6H_4O_2) + 2\,H_2O + \text{heat}$$

Figure 8.9
A bombardier beetle uses the catalyzed decomposition of hydrogen peroxide as a defense mechanism. The heat of reaction lets the insect eject steam and other irritating chemicals with explosive force.
(Thomas Eisner with Daniel Aneshansley)

Let us again consider the conversion of *cis*- to *trans*-butene, this time as an example of the operation of a catalyst.

$$H_3C(H)C{=}C(H)CH_3\ (g) \longrightarrow H_3C(H)C{=}C(CH_3)H\ (g)$$

cis-2-butene → *trans*-2-butene

As described earlier, the rate equation for the reaction is reaction rate = k[*cis*-2-butene].

If a trace of gaseous molecular iodine, I_2, is added to *cis*-2-butene, the iodine acts as a catalyst, accelerating the change to *trans*-2-butene. The iodine is neither consumed nor produced in the overall reaction and so does not appear in the overall balanced equation. However, since the reaction rate depends on the concentration of I_2, this is a term in the rate equation:

$$\text{reaction rate} = k[cis\text{-2-butene}][I_2]^{1/2}$$

An exponent of 1/2 for the concentration of I_2 in the rate equation indicates the square root of the concentration.

The rate of the conversion of *cis*- to *trans*-2-butene changes because the presence of I_2 somehow changes the reaction mechanism. The best hypothesis is that iodine molecules first dissociate to form iodine atoms.

Step 1: I_2 dissociation

$$\tfrac{1}{2}\,I_2(g) \rightleftharpoons I(g)$$

(This equation has a coefficient of $\frac{1}{2}$ for I_2 to emphasize that only one of the two I atoms from the I_2 molecule is needed in subsequent steps of the mechanism.) An iodine atom then attaches to the *cis*-2-butene molecule, breaking one of the bonds between the carbon atoms and allowing the ends of the molecule to twist freely relative to each other.

Step 2: Attachment of I atom to cis-2-butene

$$I(g) + H_3C(H)C{=}C(H)CH_3\ (g) \longrightarrow I{-}C(CH_3)(H){-}C(CH_3)(H)\ (g)$$

cis-2-butene

Step 3: Rotation around the C—C bond

$$I{-}C(CH_3)(H){-}C(CH_3)(H)\ (g) \longrightarrow I{-}C(CH_3)(H){-}C(H)(CH_3)\ (g)$$

Step 4: Loss of an I atom and re-formation of the carbon-carbon double bond

$$I{-}C(CH_3)(H){-}C(H)(CH_3)\ (g) \longrightarrow H_3C(H)C{=}C(H)CH_3\ (g) + I(g)$$

trans-2-butene

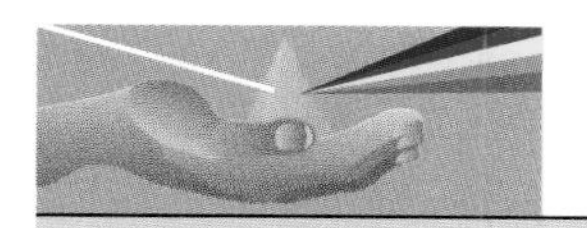

CHEMISTRY YOU CAN DO

Enzymes, Biological Catalysts

Raw potatoes contain an enzyme called *catalase,* which converts hydrogen peroxide to water and oxygen. You can demonstrate this by the following experiment:

Purchase a small bottle of hydrogen peroxide at a pharmacy. The peroxide is usually sold as a 3% solution in water. Pour about 50 mL of the peroxide solution into a clear glass or plastic cup. Add a small slice of a fresh potato to the cup. (Since potato is less dense than water, the potato will float.)

Almost immediately you will see bubbles of oxygen gas on the potato slice. Does the rate of evolution of oxygen change with time? If so, how does it change? If you cool the hydrogen peroxide solution in a refrigerator and then do the experiment, is there a perceptible change in the initial rate of O_2 evolution? Is there a difference between the time at which O_2 evolution begins for warm and for cold hydrogen peroxide?

After the double bond re-forms to give *trans*-2-butene, and the iodine atom falls away, two iodine atoms come together to re-form molecular iodine.

Step 5: I_2 formation

$$I(g) \rightleftharpoons \tfrac{1}{2} I_2(g)$$

There are four important points concerning this mechanism.

- Notice that I_2 dissociates to atoms and then re-forms. To an "outside" observer the concentration of I_2 is unchanged; it is not involved in the balanced, stoichiometric equation even though it has appeared in the rate equation. *This is generally true of catalysts.*
- Figure 8.10 shows that the activation energy barrier to reaction is changed (because the mechanism changed), and it is significantly lower. Thus, the reaction rate has gone up. In fact, dropping the activation energy from 262 kJ/mol for the uncatalyzed reaction to 115 kJ/mol for the catalyzed process makes the catalyzed reaction 10^{15} times faster!

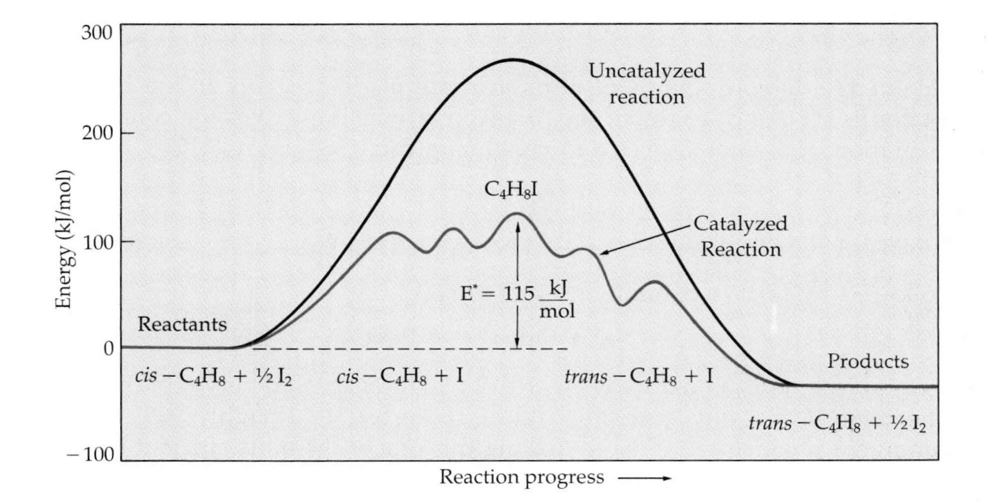

Figure 8.10
A catalyst accelerates a reaction by altering the mechanism so that the activation energy is reduced. With a smaller barrier to overcome, there are more reactant molecules with enough energy to cross the barrier, and reaction occurs more readily. The energy profile for the uncatalyzed conversion of *cis*-2-butene to *trans*-2-butene is shown by the black curve, and that for the iodine-catalyzed reaction is represented by the red curve. Notice that the shape of the barrier has changed because the mechanism has changed. See the text for a description of the steps involved.

- The catalyzed mechanism has five reaction steps, and the energy versus reaction progress diagram (Figure 8.10) has five energy barriers (five humps appear in the curve).
- The catalyst I_2 and the reactant *cis*-2-butene are both in the gas phase during the reaction.

When a catalyst is present in the same phase as the reacting substance, it is called a **homogeneous catalyst.**

Catalysis in Industry and the Environment

An expert in the field of industrial chemistry has said that "Every year more than a trillion dollars' worth of goods is manufactured with the aid of man-made catalysts. Without them, fertilizers, pharmaceuticals, fuels, synthetic fibers, solvents, and surfactants would be in short supply. Indeed, 90 percent of all manufactured items use catalysts at some stage of production." The major areas of catalyst use are in petroleum refining, industrial production of chemicals, and environmental controls. We shall look at only a few examples here, since more will be described later in the book.

*A recent article by J. M. Thomas (*Scientific American*, April 1992, pp. 112–118) describes some industrial uses of catalysts. See also Ann M. Taylor,* Chemical and Engineering News, *March 9, 1992, pp. 27–49.*

Virtually every industrial reaction uses a **heterogeneous catalyst,** one that is present in a different phase than the reactants being catalyzed. An example is the manganese compound in Figure 8.8. Heterogeneous catalysts are used in industry because they are more easily separated from the products and leftover reactants than are homogeneous catalysts. (However, enzymes, the substances that catalyze reactions in plants and animals, are often homogeneous catalysts.)

Catalysts for chemical processing are generally metal-based and often contain precious metals such as platinum and palladium. In the United States more than $600 million worth of such catalysts are used annually by the chemical processing industry, almost half of them in the preparation of polymers. For example, ethylene can be made into polyethylene, a long chain of CH_2 groups made by stringing together ethylene molecules into what is known as a polymer (Figure 8.11).

More than 18 billion pounds of polyethylene, which you know in the form of sandwich bags and milk bottles, are produced annually in the United States. The catalysts used to make these and similar polymers (see Chapter 13) are often called Ziegler-Natta catalysts. These are named after Karl Ziegler, a German chemist, and Guilio Natta, an Italian chemist. They shared the 1963 Nobel Prize in Chemistry for this work.

$$\underset{\text{ethylene}}{n\,C_2H_4(g)} \xrightarrow{\text{catalyst}} \underset{\text{polyethylene}}{\left(CH_2{-}CH_2{-}CH_2{-}CH_2{-}CH_2\right)_n CH_2{-}}$$

About 15.5 billion pounds of nitric acid are made annually in the United States in the *Ostwald process,* the first step of which involves the controlled oxidation of ammonia over a Pt-containing catalyst (Figure 8.12)

$$4\,NH_3(g) + 5\,O_2(g) \xrightarrow[\text{catalyst}]{\text{Pt-containing}} 4\,NO(g) + 6\,H_2O(g) \quad \Delta H^\circ_{rxn} = -905.5\text{ kJ}$$

followed by further oxidation of NO to NO_2.

$$2\,NO(g) + O_2(g) \longrightarrow 2\,NO_2(g) \quad \Delta H^\circ_{rxn} = -114.1\text{ kJ}$$

In a typical plant, a mixture of air with 10% NH_3 is passed very rapidly over the catalyst at a high pressure and at about 850 °C. Roughly 96% of the ammonia is converted to NO_2, making this one of the most efficient industrial catalytic reactions. The final step is to absorb the NO_2 into water to give the acid and NO, the latter being recycled into the process.

$$3\,NO_2(g) + H_2O(\ell) \longrightarrow 2\,HNO_3(aq) + NO(g) \quad \Delta H^\circ_{rxn} = -138.2\text{ kJ}$$

Nitric acid is produced as a concentrated aqueous solution, but careful procedures can convert this to the anhydrous acid. At room temperature, HNO_3 is a colorless liquid with a pungent, choking odor. By far the largest amount of the acid is turned into ammonium nitrate by neutralization of nitric acid with ammonia.

$$HNO_3(aq) + NH_3(g) \longrightarrow NH_4NO_3(aq)$$

Most of the ammonium salt is consumed as a fertilizer, but another use depends on the fact that it decomposes in a product-favored reaction and is potentially explosive.

$$2\,NH_4NO_3(s) \xrightarrow{> 300\text{ °C}} 2\,N_2(g) + O_2(g) + 4\,H_2O(g)$$

$$NH_4NO_3(s) \xrightarrow{200\text{–}260\text{ °C}} N_2O(g) + 2\,H_2O(g)$$

A mixture of ammonium nitrate and fuel oil has largely replaced dynamite as an industrial explosive. Ammonium nitrate is thermodynamically unstable, but quite safe to handle at normal temperatures when it is pure and present in small amounts.

Acetic acid, CH_3CO_2H, has a place in the organic chemicals industry comparable to that of sulfuric acid in the inorganic chemicals industry; more than 3.6 billion pounds of acetic acid were made in the United States in 1991. Acetic acid is used widely in industry to make plastics and synthetic fibers, as a fungicide, and as the starting material for preparing many dietary sup-

Figure 8.11
A model of polyethylene (the black balls represent C atoms and the white balls are H atoms) and a laboratory bottle made of polyethylene. *(C.D. Winters)*

Figure 8.12
The platinum-rhodium gauze catalyst used for the oxidation of ammonia in manufacturing nitric acid. *(Johnson Matthey)*

New Life for Natural Gas

YOUR HOME MAY BE HEATED with natural gas, which consists largely of methane, CH_4. Although widely used, it is also widely wasted! The world's known and projected reserves of this, the simplest hydrocarbon, are 250,000 trillion liters with an energy equivalent to 1.5 trillion barrels of oil. Unfortunately, much of it is in areas such as Southeast Asia or the northern reaches of Canada that are far removed from the centers of fuel consumption. Methane can be carried as a gas to consumers in pipelines, but this is expensive if the distances are great. Alternatively, it can be liquefied and carried by ship, but this risks a fiery catastrophe. The drawbacks to transportation mean that methane is often simply burned or "flared off" when it comes out of the ground with a substance that is currently more useful, oil.

One solution to making methane useful is to convert it, where it is found, to a more readily transportable liquid such as methanol, CH_3OH. The methanol can then be used directly as a fuel, added to gasoline (as is currently done in some areas of the United States), or used to make other chemicals.

It has been known for some time that methane can be converted to carbon monoxide and hydrogen

$$CH_4(g) + \tfrac{1}{2}O_2(g) \longrightarrow CO(g) + 2\,H_2(g)$$

and this mixture of gases can readily be turned into methanol in another step.

$$CO(g) + 2\,H_2(g) \longrightarrow CH_3OH(\ell)$$

Unfortunately, the first step is a high-temperature (> 900 °C), energy-intensive process. Catalytica, Inc. of California announced in 1993, however, that methane can be converted to methanol with a 43% yield by using a mercury(II) salt as a catalyst in the presence of sulfuric acid and water.

$$CH_4(g) + \tfrac{1}{2}O_2(g) \xrightarrow[180\ °C]{Hg(II)\ salt,\ H_2SO_4,\ H_2O} CH_3OH(\ell)$$

Although there are potential problems in a process that uses toxic mercury compounds and highly corrosive sulfuric acid, the discovery is important to consumers of energy and chemicals worldwide.

Just as exciting is another discovery regarding methane. Chemical engineers at the University of Minnesota have found that methane can in fact be converted to CO and H_2 under very mild conditions of temperature. They simply found the right catalyst! The photograph below shows what happens when a room-temperature mixture of methane and oxygen flowed through a heated, sponge-like ceramic disk coated with platinum or rhodium. Rather than oxidizing the methane all the way to water and carbon dioxide, the process produces a hot mixture of CO and H_2, which can be converted in good yield to methanol.

For more information on these discoveries see *Science News*, January 16, 1993, page 36, and *Science*, January 15, 1993, pp. 340–346.

Methane flowing through a catalyst.
(Schmidt/University of Minnesota)

plements. One way of synthesizing the acid is an excellent example of homogeneous catalysis. A rhodium-containing compound is used to speed up the combination of carbon monoxide and methyl alcohol, both inexpensive chemicals, to form acetic acid.

$$\underset{\text{methyl alcohol}}{CH_3OH} + CO \xrightarrow{\text{Rh-containing catalyst}} \underset{\text{acetic acid}}{CH_3\overset{\overset{\displaystyle O}{\|}}{C}{-}OH}$$

The role of the rhodium-containing catalyst in this reaction is to bring the reactants together and allow them to rearrange to the products. The first step in the process is the reaction of the alcohol with hydrogen iodide to give CH_3I

$$CH_3OH + HI \longrightarrow CH_3I + H_2O$$

which then reacts with the catalyst, a molecule containing a rhodium(I) ion and CO. This gives a new molecule with CH_3, I, and CO attached to the metal center. Acetic acid is the product after these fragments rearrange and the intermediate reacts with water.

$$\{Rh(CH_3)(CO)I\} + H_2O \longrightarrow \text{Rh catalyst} + HI + CH_3CO_2H$$

Ammonium nitrate explosion. *(C.D. Winters)*

Besides producing acetic acid, this final step regenerates the Rh-containing catalyst and produces HI, which is then available to react with more CH_3OH to begin a new catalytic cycle.

The largest growth in catalyst use is predicted to be in **emissions control**, for both automobiles and power plants. This market consumes very large quantities of platinum group metals: platinum, palladium, rhodium, and iridium. Some 9200 kg of platinum, and 1319 kg each of palladium and rhodium, were sold in the United States in the first half of 1991 for automotive uses. In contrast, chemical processing used only 1369 kg of all three metals, and the petroleum industry used 1782 kg of platinum and rhodium.

The purpose of the catalysts in the exhaust system of an automobile is to ensure that the combustion of carbon monoxide and hydrocarbons is complete (Figure 8.13)

$$2\,CO(g) + O_2(g) \xrightarrow{\text{Pt-NiO catalyst}} 2\,CO_2(g)$$

$$\underset{\substack{\text{isooctane}\\\text{a component of gasoline}}}{2\,C_8H_{18}(g)} + 25\,O_2(g) \xrightarrow{\text{Pt-NiO catalyst}} 16\,CO_2(g) + 18\,H_2O(g)$$

Figure 8.13
An automobile's catalytic converter, which has been cut apart to show the pellets that contain the platinum-palladium-rhodium catalyst. *(General Motors)*

and to convert nitrogen oxides to molecules less harmful to the environment. At the high temperature of combustion, some N_2 from air reacts with O_2 to give NO, a serious air pollutant. Thermodynamics informs us that nitrogen oxide is unstable and should revert to N_2 and O_2. But remember that thermodynamics says nothing about rate, and rate of reversion of NO to N_2 and O_2 is slow. Fortunately, catalysts have been developed that greatly speed this reaction.

$$2\,NO(g) \xrightarrow{\text{catalyst}} N_2(g) + O_2(g)$$

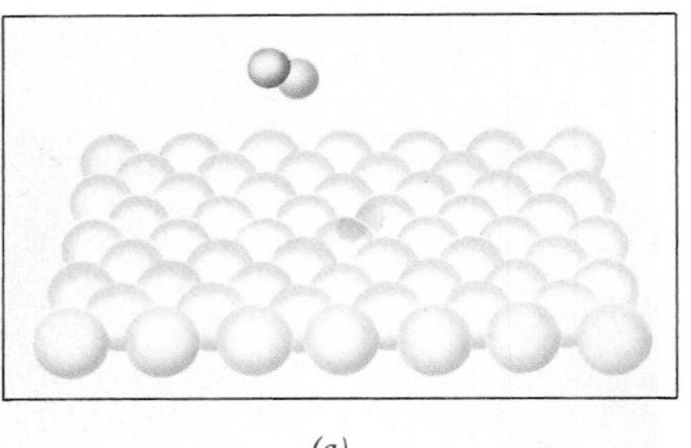
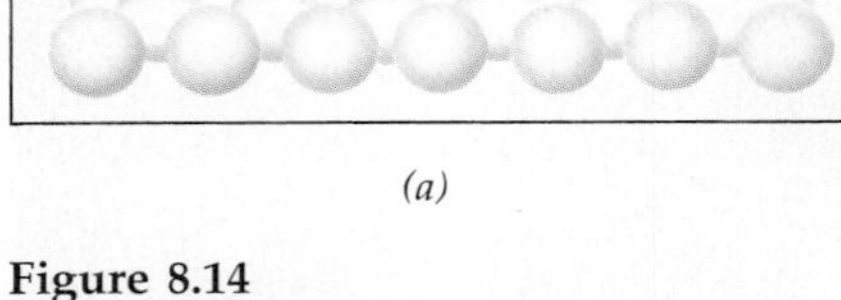

(a)

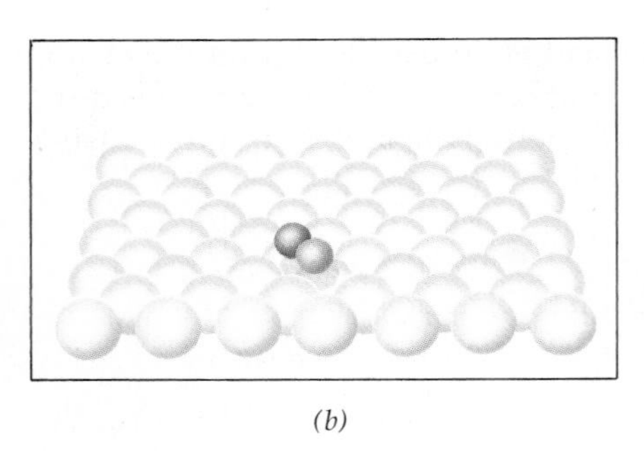

(b)

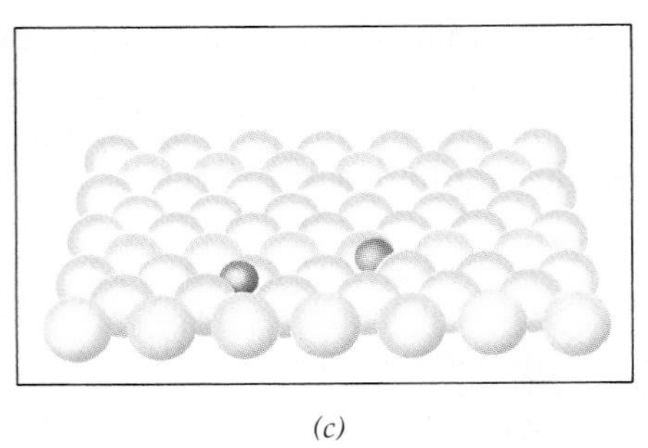

(c)

Figure 8.14
The dissociation of NO into N (red) and O (blue) atoms on a platinum surface. In (a) the NO approaches the surface. After 1.5×10^{-12} seconds it interacts with the surface (b), and then (c) it dissociates 1.7×10^{-12} seconds after approaching the surface.

The role of the heterogeneous catalyst in the preceding reactions is probably to weaken the bonds of the reactants and to assist in product formation. For example, in Figure 8.14 you see how NO molecules can dissociate into N and O atoms on the surface of a platinum metal catalyst.

EXERCISE 8.6 • *Catalysis*

Which of the following statements is (are) true? If any are false, change the wording to make them true.

a. The concentration of a homogeneous catalyst may appear in the rate equation.

b. A catalyst is always consumed in the overall reaction.

c. A catalyst must always be in the same phase as the reactants.

Thermodynamic and Kinetic Stability

Chemists are often heard to remark "Molecule A is stable" in the same way and for the same reason you might say that the aluminum foil you used to wrap a piece of cake is "stable." But what do you mean by that? Probably you mean that the foil shows no sign of decomposing even though you use it to wrap food or even heat food in it. Chemists mean something similar when using the word.

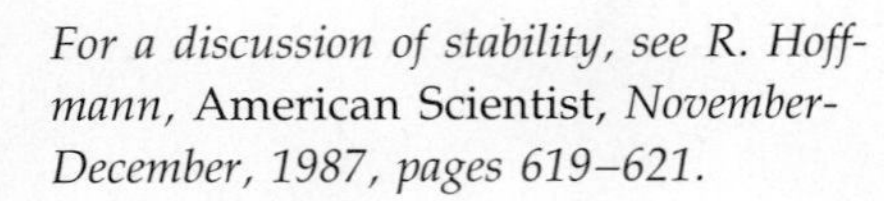

For a discussion of stability, see R. Hoffmann, American Scientist, *November-December, 1987, pages 619–621.*

There are actually two kinds of stability. A *thermodynamically stable* substance does not undergo product-favored reactions that release energy and increase disorder. A *kinetically stable* substance may have the potential to undergo product-favored reactions, but it does this so slowly that it remains essentially unchanged for a long time. Roald Hoffmann, who received the Nobel Prize in chemistry in 1981, has said that 90% of the time chemists use the word "stable" to mean something is *kinetically* stable.

The aluminum in foil is *thermodynamically* unstable since it should literally burn in air with the evolution of an enormous amount of heat.

$$4\,Al(s) + 3\,O_2(g) \longrightarrow 2\,Al_2O_3(s)$$

$$\Delta H^\circ = -3351.4\ kJ \qquad \Delta S^\circ = -626.8\ J/K \qquad \Delta G^\circ = -3164.6\ kJ$$

This would indeed happen if we ground up the aluminum into a fine powder and threw it into a flame; the powder would burn, and the evolved heat would lead to an entropy increase in the little piece of the universe around the burning metal. Although the metal is *thermodynamically unstable,* it is *kinetically stable* in the form of kitchen foil since the foil is covered with a

The Chemical World

Discovery of a Catalyst for Acrylonitrile Production

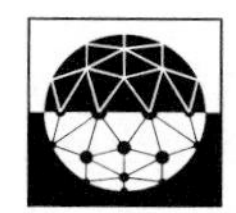

Oil companies invest considerable sums in equipment and human effort to develop new catalysts to make fuels and chemicals from petroleum. Research and development in this area is a multimillion-dollar gamble. There is no guarantee the money spent will produce anything useful. But if it does, the payoff can be enormous. Just one catalyst breakthrough made more than half a billion dollars for Standard Oil of Ohio, now part of BP America. In the late 1950s, SOHIO researchers came up with a new catalytic process for the manufacture of acrylonitrile, a monomer used to make polyacrylonitrile, the polymer used to make textiles, tires, and car bumpers. Oddly enough, SOHIO researchers weren't even trying to produce acrylonitrile at first. They simply wanted to make a metal oxide catalyst to convert waste propane gas from petroleum refining into something more valuable. As Dr. Jeanette Grasselli said:

"The theory, the hypothesis at the time, was that we could take the oxygen from the catalyst and insert it into the propane, a relatively unreactive molecule. So this was a tough technical objective. And, in turn, we wanted to generate or take the catalyst back to its original oxidized form by using oxygen from the air."

But the theory didn't hold up. Propane was too stable to react, and the catalyst particles broke down in service. Management gave the research team three more months to show results, so they made some changes. They replaced propane with a more reactive refinery gas, propylene. They made their catalyst out of different metal oxides, and they added ammonia to promote, or speed up, the reaction.

To their surprise, ammonia reacted. Rather than just encouraging the reaction to go faster, as a promoter, ammonia reacted and became part of the reaction sequence, and acrylonitrile was made in one step. The researchers had struck paydirt. Their new catalyst, combined with ammonia, had made a valuable product out of a cheap gas. Management quickly saw the value of the new process and wasted no time building a plant to use it. As Dr. Grasselli recalls, "In 1960, when our plant came on stream, acrylonitrile was selling for 28¢/lb. We were making it for 14¢/lb. And we shut down every other commercial process for making acrylonitrile. Today 90% of the world's acrylonitrile is manufactured by the SOHIO process."

The World of Chemistry (Program 20), "On the Surface."

Jeanette Grasselli

microscopically thin coating of Al_2O_3 that slows or prevents a thermodynamically favored reaction.

8.2 CHEMICAL EQUILIBRIUM

Now we are ready to consider another quantitative question: how far will a reaction proceed before it has apparently ceased? That is, how much product can we expect to get by the time the rate of increase of product concen-

Figure 8.15
Calcium carbonate stalactites hang from the ceiling of a cave, and stalagmites grow from the floor. The process that produces these formations is an excellent illustration of chemical equilibria. *(Tom Stack & Associates)*

tration has gone to zero? To answer this question, and to extend our understanding of chemical reactivity, we take up a subject closely related to thermodynamics and kinetics: chemical equilibrium.

The concept of equilibrium is fundamental to chemistry, but it is not peculiar to chemistry. You participate in social situations and live in an economy that represent equilibria of competing forces. You and your family or roommate have arrived at some arrangements that keep your personal relations as smooth as possible. That is, you have achieved an equilibrium. Of course, that equilibrium is upset easily when another child is added to the family or another roommate moves in. In international affairs, countries achieve an equilibrium of competing interests, but that equilibrium can be upset, for example, when a third country intervenes. Whether in personal or international relations, a new equilibrium is often achieved, however temporarily, when the old one has been upset.

Chemical systems are analogous to our social and political arrangements and can also be forced to move to a new equilibrium position by outside influences. Our goal here is to explore briefly the facts that *most chemical reactions are reversible* and that *a state of equilibrium can eventually be achieved between reactants and products.* A major result of this exploration will be an ability to describe chemical reactivity in more quantitative terms. As you will see, *the concentrations of reactants and products at equilibrium are a measure of the tendency of chemical reactions to proceed from reactants to products,* that is, *a measure of the tendency of* a reaction to be product-favored or reactant-favored.

The Nature of the Equilibrium State

As an example of chemical equilibrium, consider the reaction that leads to the formation of limestone stalactites and stalagmites in some caves (Figure 8.15). These structures originate from the reaction of limestone, a form of calcium carbonate, with CO_2 and water.

$$CaCO_3(s) + CO_2(aq) + H_2O(\ell) \rightleftharpoons Ca^{2+}(aq) + 2\,HCO_3^-(aq)$$

The $CaCO_3$ is present in underground deposits, a leftover of ancient oceans. If water seeping through the ground contains dissolved CO_2, the preceding reaction can give aqueous Ca^{2+} and HCO_3^- ions. However, *most chemical reactions are reversible,* and this reaction is no exception. To prove the reversibility of the reaction, you could do an experiment with some soluble salts containing the Ca^{2+} and HCO_3^- ions (say $CaCl_2$ and $NaHCO_3$). If you put them in an open beaker of water in the laboratory, you will eventually see bubbles of CO_2 gas and solid $CaCO_3$ (Figure 8.16). Because the CO_2 is swept away into the air, thus upsetting the equilibrium, all the dissolved Ca^{2+} and HCO_3^- ions will eventually disappear from the solution, having been converted completely to solid $CaCO_3$.

The set of double arrows, $\rightleftharpoons$, in an equation indicates that the reaction is reversible and, in general chemistry courses, is often a signal to you that the reaction will be studied by using the concepts of chemical equilibria.

$$Ca^{2+}(aq) + 2\,HCO_3^-(aq) \rightleftharpoons CaCO_3(s) + CO_2(aq) + H_2O(\ell)$$

This is the reverse of the reaction that dissolves limestone and is the process that redeposits the $CaCO_3$ in caves as beautiful limestone formations.

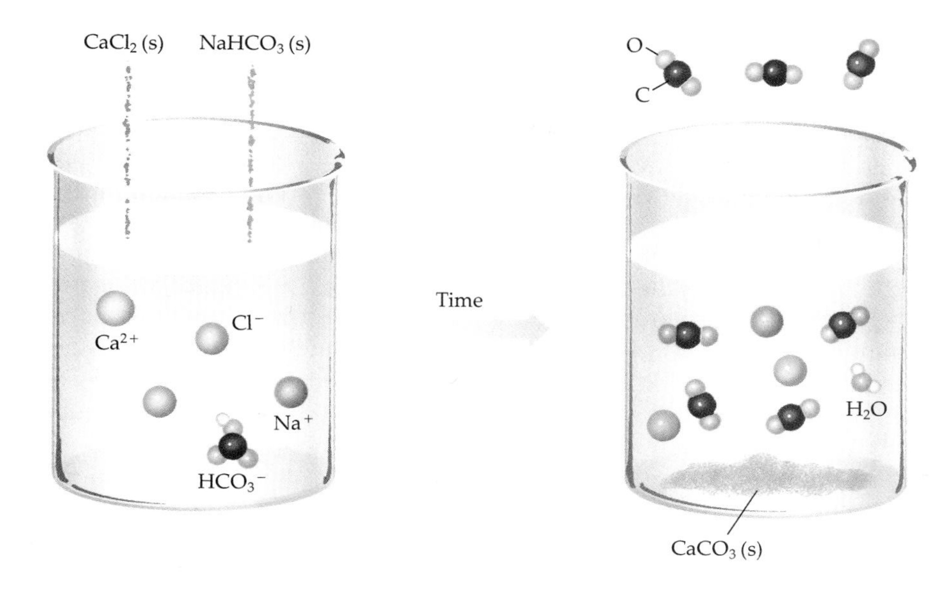

Figure 8.16
Some $CaCl_2$ and $NaHCO_3$ are added to a beaker of water. With time, the reaction $Ca(HCO_3)_2(aq) \rightleftharpoons CaCO_3(s) + CO_2(g) + H_2O(\ell)$ occurs; CO_2 is evolved, $CaCO_3$ precipitates, and NaCl is left in solution.

If the solutions containing Ca^{2+} and HCO_3^- ions are mixed in a closed container, after a while no further observable change will occur. It looks as though the reaction has stopped, but this is not the case. Instead, the reaction has reached equilibrium. *Chemical equilibria are* **dynamic.** *When a system is at equilibrium, the forward and reverse reactions continue, and they take place at equal rates.* The concentrations of reactants and products do not change any longer.

A common introductory chemistry experiment is the study of the equilibrium that exists in the reaction of aqueous iron(III) ion, which we designate for simplicity as $Fe(H_2O)^{3+}$, and aqueous thiocyanate ion, SCN^-.

$$\underset{\text{nearly colorless}}{Fe(H_2O)^{3+}(aq)} + \underset{\text{colorless}}{SCN^-(aq)} \rightleftharpoons \underset{\text{red-orange}}{Fe(SCN)^{2+}(aq)} + H_2O(\ell)$$

When colorless solutions of these two ions are mixed (Figure 8.17), the SCN^- ion rapidly replaces water in the $Fe(H_2O)^{3+}$ ion to give a red-orange ion in which SCN^- is bonded to Fe^{3+}. As the concentration of $Fe(SCN)^{2+}$ begins to build up, this ion reacts with H_2O to release SCN^- and revert to aqueous $Fe(H_2O)^{3+}$ at a greater and greater rate. Eventually, the rate at which SCN^- replaces H_2O on Fe^{3+} to form $Fe(SCN)^{2+}$ (the "forward" reaction) becomes equal to the rate at which $Fe(SCN)^{2+}$ sheds SCN^- ion to go back to the simple ions in solution (the "reverse" reaction). At this point—the point where the rates of the forward and reverse reactions become equal—equilibrium has been achieved.

When equilibrium has been achieved in the $Fe(H_2O)^{3+} + SCN^-$ reaction, the forward and reverse reactions do not stop. To prove this, a drop of an aqueous solution of radioactive SCN^- ion can be added to the solution. (The ion is made radioactive by using radioactive ^{14}C in place of normal, nonradioactive ^{12}C.) Sampling the solution shortly after adding the radioactive SCN^- ion would show that the radioactive ion has been incorporated into $Fe(SCN)^{2+}$.

Figure 8.17
The reaction of aqueous iron(III) ion and thiocyanate ion, SCN^-. The colorless solutions are mixed to give the red-orange ion $Fe(SCN)^{2+}$, which is in equilibrium with the reactants.
(C.D. Winters)

$$Fe(SCN)^{2+}(aq) + H_2O(\ell) \rightleftharpoons Fe(H_2O)^{3+}(aq) + SCN^-(aq)$$
$$Fe(H_2O)^{3+}(aq) + S^*CN^-(aq) \rightleftharpoons Fe(S^*CN)^{2+}(aq) + H_2O(\ell)$$

(The symbol means S^*CN^- contains radioactive ^{14}C.)

The only way for the radioactive SCN^- ion to be incorporated into the red-orange $Fe(SCN)^{2+}$ ion is if the reaction of $Fe(SCN)^{2+}$ with water is dynamic and reversible, and continues even at equilibrium.

The nature of the equilibrium state is the same no matter what the direction of approach only if the number of atoms of each element used is the same in the forward and reverse directions.

Not only are equilibrium processes dynamic and reversible, but, for a specific reaction, *the nature of the equilibrium state will be the same, no matter what the direction of approach.* Let us say you measure the concentrations at equilibrium of acetic acid and the ions that come from its ionization in aqueous solution.

$$\underset{\text{acetic acid}}{CH_3CO_2H(aq)} + H_2O(\ell) \rightleftharpoons \underset{\text{acetate ion}}{CH_3CO_2^-(aq)} + \underset{\text{hydronium ion}}{H_3O^+(aq)}$$

Since acetic acid is a weak acid (see Section 5.6), the concentrations of the acetate and hydronium ions will be small. Now mix sodium acetate and hydrochloric acid,

$$NaCH_3CO_2(aq) + HCl(aq) \longrightarrow CH_3CO_2H(aq) + NaCl(aq)$$

a reaction that has the following net ionic equation

$$\underset{\text{acetate ion}}{CH_3CO_2^-(aq)} + \underset{\text{hydronium ion}}{H_3O^+(aq)} \longrightarrow \underset{\text{acetic acid}}{CH_3CO_2H(aq)} + H_2O(\ell)$$

and measure the equilibrium concentrations. As long as you had begun with, say, one mole of acetic acid in the first experiment and one mole each of sodium acetate and HCl in the second experiment (all in the same volume), the concentrations of acetic acid, acetate ion, and hydronium ion at equilibrium would be identical.

The Equilibrium Constant

One way to describe the equilibrium position of a chemical reaction is to give equilibrium concentrations of the reactants and products. The equilibrium constant expression *relates equilibrium concentrations of reactants and products at a given temperature to a numerical constant.* This expression is of great importance in chemistry because it allows us to predict the amounts of the reactants and products that will be present when equilibrium has been achieved. Since many reactions that you will see in the laboratory move quickly to equilibrium, this means you often have a good idea of the composition of a solution, for example. What follows is a brief introduction to this useful mathematical relationship.

The notion that the equilibrium concentrations of reactants and products are related in a simple manner is easy to prove by experiments such as those for the $H_2 + I_2$ gas phase reaction.

$$H_2(g) + I_2(g) \rightleftharpoons 2\,HI(g)$$

A very large number of experiments have shown that at equilibrium the ratio

$$\frac{[HI]^2}{[H_2][I_2]} = 55.64$$

is always the same within experimental error for all experiments done at 425 °C! For example, assume enough H_2 and I_2 has been placed in a flask so that the concentration of each at 425 °C is 0.0175 mol/L. With time, the concentrations of H_2 and I_2 will decline and that of HI will increase; an equilibrium state will be reached eventually. If the gases in the flask were then analyzed, the result would be that $[H_2] = [I_2] = 0.0037$ mol/L while $[HI] = 0.0276$ mol/L.

	Initial and Equilibrium Concentrations (mol/L)		
	$H_2(g)$	$I_2(g)$	$HI(g)$
Initial concentration	0.0175	0.0175	0
Change in concentration as reaction proceeds to equilibrium	−0.0138	−0.0138	+0.0276
Equilibrium concentration	0.0037	0.0037	0.0276

Putting these equilibrium concentration values into the preceding expression

$$\frac{[HI]^2}{[H_2][I_2]} = \frac{(0.0276)^2}{(0.0037)(0.0037)} = 56$$

gives a ratio of about 56 (or 55.64 if the experimental information contains more significant figures). This ratio is always the same for all experiments at 425 °C, no matter from which direction the reaction is approached (mixing H_2 and I_2 or allowing HI to decompose) and no matter what the initial concentrations! If you start with 2.0 mol of H_2 and 2.0 mol of I_2, equilibrium will still be reached with concentrations of reactants and products that satisfy the equilibrium constant expression ($[H_2] = [I_2] = 0.42$ mol/L; $[HI] = 3.16$ mol/L).

The number 55.64 is the **equilibrium constant** for the equilibrium of H_2, I_2, and HI at 425 °C, and hundreds of experiments have proved that it and other equilibrium constants can be calculated from the following general expression for an equilibrium constant. For the general reaction

$$a\text{A} + b\text{B} + \cdots \rightleftharpoons c\text{C} + d\text{D} + \cdots$$

experiments show that equilibrium concentrations of reactants and products are always related by the **equilibrium constant expression**

$$\text{equilibrium constant} = K = \frac{\overbrace{[\text{C}]^c[\text{D}]^d}^{\text{product concentrations}}}{\underbrace{[\text{A}]^a[\text{B}]^b}_{\text{reactant concentrations}}}$$

In equilibrium constant expressions the product concentrations appear in the numerator and reactant concentrations in the denominator. Each concentration is raised to the power of its stoichiometric coefficient in the balanced equation. The value of the constant K depends on the particular reaction and on the temperature. (Units are traditionally not given with *K*.)

The value of the equilibrium constant for a given reaction provides a great deal of information about the extent of reaction when equilibrium has been achieved. After learning a few characteristics of the equilibrium expression, we will turn to that topic.

Writing Equilibrium Constant Expressions

There are a few rules that you should know about writing equilibrium constant expressions for chemical reactions. One of the most important is that *the concentrations of pure solids, of pure liquids, and of solvents in dilute solutions* ***never*** *appear in equilibrium constant expressions.* For example, the oxidation of solid yellow sulfur produces colorless sulfur dioxide gas (Figure 5.11).

$$\tfrac{1}{8}\,S_8(s) + O_2(g) \rightleftharpoons SO_2(g)$$

Following the general principle that products appear in the numerator and reactants in the denominator, you would write

$$K' = \frac{\overset{\text{product}}{[SO_2(g)]}}{\underset{\text{reactants}}{[S_8(s)]^{\frac{1}{8}}[O_2(g)]}}$$

However, since sulfur is a molecular solid and since the concentration of molecules within any solid is fixed, the sulfur concentration is not changed either by reaction or by addition or removal of some solid. Further, it is an experimental fact that the equilibrium concentrations of O_2 and SO_2 are not changed by the amount of sulfur, as long as there is some solid sulfur present at equilibrium. Therefore, it is conventional not to include the concentrations of any solid reactants and products in the equilibrium constant expression, and you should write the equilibrium constant expression as

$$K = \frac{[SO_2(g)]}{[O_2(g)]} \quad \text{where } K = 4.2 \times 10^{52} \quad (\text{at } 25\ ^\circ\text{C})$$

A reminder: It is very *important to remember that pure solids and liquids do not appear in the equilibrium constant expression.*

There are special considerations for reactions occurring in aqueous solution. Consider ammonia, which is a weak base in water owing to its interaction with the water.

$$NH_3(aq) + H_2O(\ell) \rightleftharpoons NH_4^+(aq) + OH^-(aq)$$

Since the molar concentration of water is effectively constant for reactions involving dilute solutions, the concentration of water (like that of solids) is not included in the equilibrium constant expression (particularly for reactions of weak acids and bases; Table 8.1). Thus, we write

For ionization reactions of dilute aqueous solutions of weak acids and bases the K we write is actually the product of the ionization constant for the substance and the molar concentration of water (55 M).

$$K = \frac{[NH_4^+][OH^-]}{[NH_3]} \quad \text{where } K = 1.8 \times 10^{-5} \quad (\text{at } 25\ ^\circ\text{C})$$

EXAMPLE 8.2

Writing and Manipulating Equilibrium Constant Expressions

A mixture of nitrogen, hydrogen, and ammonia can be brought to equilibrium. When the equation is written using whole-number coefficients, as follows, the value of K is 3.2×10^8 at 25 °C.

Equation 1 $N_2(g) + 3\,H_2(g) \rightleftharpoons 2\,NH_3(g)$ $K_1 = 3.5 \times 10^8$ at 25 °C

However, the equation can also be written as

Equation 2 $\frac{1}{2}\,N_2(g) + \frac{3}{2}\,H_2(g) \rightleftharpoons NH_3(g)$ $K_2 = ?$

in order to make the stoichiometric coefficient of NH_3 equal to 1. What is the value of K_2, the equilibrium constant for Equation 2? What is the value of K_3, the equilibrium constant for the reverse of Equation 1, that is, the decomposition of ammonia to the elements?

Equation 3 $2\,NH_3(g) \rightleftharpoons N_2(g) + 3\,H_2(g)$ $K_3 = ?$

SOLUTION To see the relation between K_1 and K_2, let us first write the equilibrium expressions for these two balanced equations.

$$K_1 = \frac{[NH_3]^2}{[N_2][H_2]^3} = 3.5 \times 10^8$$

and

$$K_2 = \frac{[NH_3]}{[N_2]^{1/2}[H_2]^{3/2}}$$

Writing the expressions makes it clear that K_1 is the square of K_2; that is, $K_1 = K_2^2$. Therefore, the answer to our question is

$$K_2 = \sqrt{K_1} = (3.5 \times 10^8)^{1/2} = 1.9 \times 10^4$$

In general, *when the stoichiometric coefficients of a balanced equation are multiplied by some factor, the equilibrium constant of the new equation (K_{new}) is the old equilibrium constant (K_{old}) raised to the power of the multiplication factor.*

Equation 3 is the reverse of Equation 1, and its equilibrium expression is

$$K_3 = \frac{[N_2][H_2]^3}{[NH_3]^2}$$

In this case K_3 is the reciprocal of K_1. That is, $K_3 = 1/K_1$.

$$K_3 = 1/K_1 = 1/(3.5 \times 10^8) = 2.9 \times 10^{-9}$$

In general, *the equilibrium constants for a reaction and its reverse are the reciprocals of one another.*

As a final comment, notice that the production of ammonia from the elements has a large equilibrium constant. As expected, the reverse reaction, the decomposition of ammonia to its elements, has a small equilibrium constant.

EXERCISE 8.7 • *Writing Equilibrium Expressions*

Write equilibrium expressions for the following chemical equations.

a. $H_2(g) + \frac{1}{8}\,S_8(s) \rightleftharpoons H_2S(g)$

b. $HCl(g) + LiH(s) \rightleftharpoons H_2(g) + LiCl(s)$

c. $AgCl(s) \rightleftharpoons Ag^+(aq) + Cl^-(aq)$

d. $[Cu(NH_3)_4]^{2+}(aq) \rightleftharpoons Cu^{2+}(aq) + 4\,NH_3(aq)$

EXERCISE 8.8 • *Manipulating Equilibrium Constants*

The conversion of 1.5 mol of oxygen to 1.0 mol of ozone has a very small value of K.

$$\tfrac{3}{2}\,O_2(g) \rightleftharpoons O_3(g) \qquad K = 2.5 \times 10^{-29}$$

a. What is the value of K when the conversion is written using whole-number stoichiometric coefficients?

$$3\,O_2(g) \rightleftharpoons 2\,O_3(g)$$

b. What is the value of K for the conversion of 2 mol of ozone to 3 mol of oxygen?

$$2\,O_3(g) \rightleftharpoons 3\,O_2(g)$$

The Meaning of the Equilibrium Constant

The value of the equilibrium constant indicates whether a reaction is product- or reactant-favored. In addition, it can be used to calculate how much product will be formed at equilibrium, which is very useful information to chemists and chemical engineers.

A large value of K means that reactants are converted largely to products when equilibrium has been achieved. That is, the products are strongly favored over the reactants at equilibrium. An example is the $NO + O_3$ reaction for which we described the energy barrier earlier.

$K \gg 1$: Reaction is product-favored; equilibrium concentrations of products are greater than equilibrium concentrations of reactants.

$$NO(g) + O_3(g) \rightleftharpoons NO_2(g) + O_2(g) \qquad K = 6 \times 10^{34} \text{ at } 25\ ^\circ C$$

$$K = \frac{[NO_2][O_2]}{[NO][O_3]} \text{ means } [NO_2][O_2] \gg [NO][O_3] \text{ since } K \gg 1$$

The very large value of K tells us that, if stoichiometric amounts of NO and O_3 are mixed in a flask and allowed to come to equilibrium, virtually none of the reactants will be found; essentially only NO_2 and O_2 will be found in the flask. A chemist would say "the reaction has gone to completion."

Conversely, a small K (as for the formation of ozone from oxygen) means that very little of the reactants have formed products when equilibrium has been achieved. In other words, the reactants are favored over the products at equilibrium.

SUMMARY: K and the Extent of Reaction at Equilibrium
Large value of K ($\gg 1$) means most reactants converted to products. The reaction is product-favored.
Small value of K ($\ll 1$) means very little of the reactants converted to products. The reaction is reactant-favored.

$K \ll 1$: Reaction is reactant-favored; equilibrium concentrations of reactants are greater than equilibrium concentrations of products.

$$\tfrac{3}{2}\,O_2(g) \rightleftharpoons O_3(g) \qquad K = 2.5 \times 10^{-29} \text{ at } 25\ ^\circ C$$

$$K = \frac{[O_3]}{[O_2]^{3/2}} \text{ means } [O_3] \ll [O_3]^{3/2} \text{ since } K \ll 1$$

Table 8.1

Selected Equilibrium Constants

Reaction	Equilibrium Constant, K (at 25 °C)
Nonmetal Reactions	
$\frac{1}{8} S_8(s) + O_2(g) \rightleftharpoons SO_2(g)$	4.2×10^{52}
$2 H_2(g) + O_2(g) \rightleftharpoons 2 H_2O(g)$	3.2×10^{81}
$N_2(g) + 3 H_2(g) \rightleftharpoons 2 NH_3(g)$	3.5×10^{8}
$N_2(g) + O_2(g) \rightleftharpoons 2 NO(g)$	1.7×10^{-3} (at 2300 K)
Weak Acids and Bases	
$HCO_2H(aq) + H_2O(\ell) \rightleftharpoons HCO_2^-(aq) + H_3O^+(aq)$ formic acid	1.8×10^{-4}
$CH_3CO_2H(aq) + H_2O(\ell) \rightleftharpoons CH_3CO_2^-(aq) + H_3O^+(aq)$ acetic acid	1.8×10^{-5}
$H_2CO_3(aq) + H_2O(\ell) \rightleftharpoons HCO_3^-(aq) + H_3O^+(aq)$ carbonic acid	4.2×10^{-7}
$NH_3(aq) + H_2O(\ell) \rightleftharpoons NH_4^+(aq) + OH^-(aq)$ ammonia (weak base)	1.8×10^{-5}
"Insoluble" Solids	
$CaCO_3(s) \rightleftharpoons Ca^{2+}(aq) + CO_3^{2-}(aq)$	3.8×10^{-9}
$AgCl(s) \rightleftharpoons Ag^+(aq) + Cl^-(aq)$	1.8×10^{-10}

The very small value of K indicates that, if O_2 is placed in a flask, *very* little O_2 will have been converted to O_3 when equilibrium is achieved.

Equilibrium constants for a few reactions are given in Table 8.1. These reactions occur to widely differing extents, as shown by the wide range of values of K. Let us use some of the reactions to explore further the relation of K and the extent of reaction.

Calculating Equilibrium Concentrations

Most metals and nonmetals react spontaneously with oxygen to give metal or nonmetal oxides, respectively. The favorable nature of such reactions is revealed by the very large value of the equilibrium constant for the reaction of sulfur and oxygen to give SO_2.

$$\frac{1}{8} S_8(s) + O_2(g) \rightleftharpoons SO_2(g) \qquad K = 4.2 \times 10^{52} \text{ at } 25\ ^\circ\text{C}$$

Let us say we began with 4.0 mol/L of O_2 and a large excess of sulfur. If the reaction is done in a closed flask, so that the system can come to equilibrium, we can calculate the quantity of O_2 left and the quantity of SO_2 formed at equilibrium. The best way to do this is to organize the information about the reaction into a small table.

	[O_2]	[SO_2]
Initial concentration (mol/L)	4.0	0.0
Change in concentration on reaction (mol/L)	$-x$	$+x$
Concentration at equilibrium (mol/L)	$4.0 - x$	x

The concentrations of reactants and products before the reaction are known. However, we do not know how many moles per liter of O_2 are used in the reaction, so this is designated as x. Since the stoichiometric factor is [1 mol SO_2/1 mol O_2], we know that x mol/L of SO_2 is formed when x mol/L of O_2 is consumed. This means that, when equilibrium has been reached, the concentration of O_2 must be what was present initially (= 4.0 mol/L) minus what was consumed in the reaction (= x), while the equilibrium concentration of SO_2 is x. Putting these values into the equilibrium expression, we have

$$K = 4.2 \times 10^{52} \text{ at } 25\ °\text{C} = \frac{[SO_2]}{[O_2]} = \frac{x}{4.0 - x}$$

Solving algebraically for x (and following the usual rules for significant figures),

$$4.2 \times 10^{52}(4.0 - x) = x$$
$$17 \times 10^{52} - 4.2 \times 10^{52}x = x$$
$$17 \times 10^{52} = 4.2 \times 10^{52}x$$
$$x = 4.0 \text{ mol/L}$$

we find its value to be 4.0 mol/L. This means the equilibrium concentration of SO_2 is 4.0 mol/L and that of O_2 is $4.0 - x$, or 0 mol/L. That is, essentially all the O_2 has been converted to SO_2! A very large value of K does indeed imply that most of the reactants have been converted to products. The reaction is product-favored.

To see quantitatively that a *small* value of K means that a reaction is reactant-favored, let us look at the reduction of carbon dioxide by hydrogen at 420 °C to give water vapor and carbon monoxide.

$$H_2(g) + CO_2(g) \rightleftharpoons H_2O(g) + CO(g) \quad K = 0.10 \text{ at } 420\ °\text{C}$$

Suppose you place enough H_2 and CO_2 in a flask so that their concentrations are both 0.050 mol/L. You supply energy to induce a reaction and then wait for equilibrium to be achieved. What are the concentrations of reactants and products at equilibrium? To solve the problem, let us again set up a table of information.

	[H_2]	[CO_2]	[H_2O]	[CO]
Initial concentration (mol/L)	0.050	0.050	0.00	0.00
Change in concentration on reaction (mol/L)	$-x$	$-x$	$+x$	$+x$
Concentration at equilibrium (mol/L)	$0.050 - x$	$0.050 - x$	x	x

Stoichiometry tells us that equal numbers of moles of H_2 and CO_2 are consumed on approaching equilibrium. Since the volume of the flask remains the same, this means that equal numbers of moles per liter (equal concentrations) must also be consumed, and we designate each of these as x. Therefore, the equilibrium concentrations of H_2 and CO_2 must be the same, and they must be equal to the initial concentration minus the change (x) as the reactants are consumed. If the concentration of H_2 decreases by x moles per liter, stoichiometry tells us that the concentration of H_2O (and the concentration of CO) must increase by the same quantity x. Substituting these values into the expression for K, we have

$$K = 0.10 = \frac{[H_2O][CO]}{[H_2][CO_2]} = \frac{(x)(x)}{(0.050 - x)(0.050 - x)} = \frac{(x)^2}{(0.050 - x)^2}$$

Solving this equation for the value of x is much more straightforward than it would seem at first glance. Taking the square root of both sides, we have

$$\sqrt{K} = 0.32 = \frac{x}{(0.050 - x)}$$

Now we solve for x.

$$0.32(0.050 - x) = x$$
$$0.016 - 0.32x = x$$
$$0.016 = 1.32x$$
$$x = 0.012$$

This means that the concentrations of the products, H_2O and CO, are both 0.012 mol/L, while the concentrations of the reactants are both $(0.050 - x)$ or 0.038 mol/L. The reactant concentrations are greater than the product concentrations, so the reaction is indeed reactant-favored, as suggested by the fact that K is less than 1.

If a reaction has a large tendency to occur in one direction, it makes sense that the reverse reaction has little tendency to occur. We have already seen the effect of this fact on heats of reaction—if ΔH is negative in one direction, it is positive and of equal magnitude for the reverse direction. For example, you can see in Table 8.1 that the oxidation of hydrogen to form water has an enormous equilibrium constant, and in Section 7.4 you learned that the reaction has a large negative value for $\Delta H°$ (= −483.64 kJ). The reaction is product-favored, largely owing to its very exothermic nature. Conversely, the reverse reaction, the decomposition of water to its elements,

$$2\,H_2O(g) \rightleftharpoons 2\,H_2(g) + O_2(g) \qquad K = \frac{[H_2]^2[O_2]}{[H_2O]^2} = 3.1 \times 10^{-82} \text{ at } 25\ °C$$

is strongly reactant-favored, as indicated by the *very* small value of K and a large positive value for $\Delta H°$ (= +483.64 kJ).

Finally, look briefly at the ionization reactions of the acids and bases listed in Table 8.1. As an example consider acetic acid, the main ingredient in vinegar.

$$CH_3CO_2H(aq) + H_2O(\ell) \rightleftharpoons CH_3CO_2^-(aq) + H_3O^+(aq)$$

$$K = \frac{[CH_3CO_2^-][H_3O^+]}{[CH_3CO_2H]} = 1.8 \times 10^{-5}$$

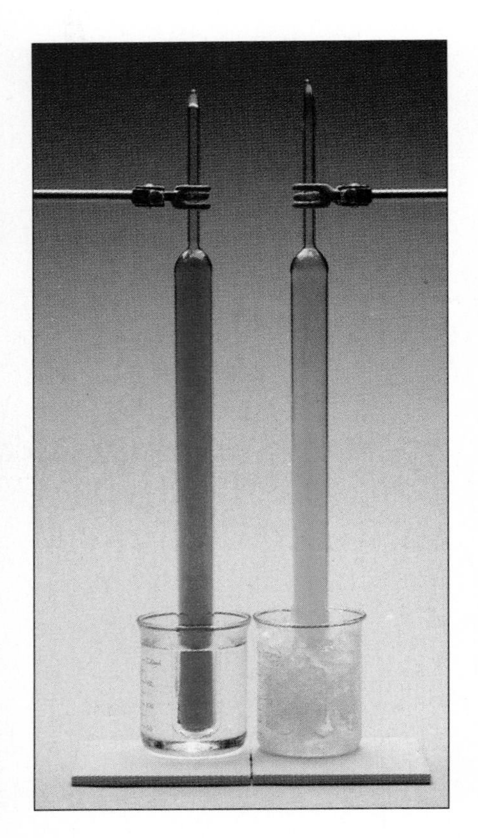

Figure 8.18
The effect of temperature on the N_2O_4/NO_2 equilibrium. Both tubes in the photo contain a mixture of the colorless gas N_2O_4 and the brown gas NO_2. As predicted by Le Chatelier's principle, at lower temperatures the equilibrium shifts toward N_2O_4. This is seen in the tube at the right: the gas in the ice bath at 0 °C is only slightly brown because the concentration of NO_2 is small. At 50 °C (in the tube at the left) the equilibrium is shifted toward brown NO_2.
(C.D. Winters)

The values of K for acids such as formic acid and acetic acid are small, so the concentrations of products relative to the concentrations of reactants at equilibrium are very small. No wonder that chemists speak of "weak acids," a large family of acids that you will learn more about in Chapter 16.

EXAMPLE 8.3

Calculating Equilibrium Concentrations

Nitrogen monoxide (NO), an air pollutant, can react with O_2 and other oxidants in the atmosphere to give NO_2, the source of some of the air pollution in urban areas. Nitrogen dioxide, NO_2, is a brown gas that can join with another molecule of the same kind to give colorless N_2O_4.

$$2\,NO_2(g) \rightleftharpoons N_2O_4(g) \qquad K = 170 \text{ at } 298\text{ K}$$

Find the concentration of N_2O_4 in a flask such as that in Figure 8.18, if the equilibrium concentration of NO_2 is 0.041 mol/L.

SOLUTION As the first step in solving this problem, write the equilibrium constant expression.

$$K = \frac{[N_2O_4]}{[NO_2]^2} = 170 \text{ at } 298\text{ K}$$

You are given a value for $[NO_2]$, the equilibrium concentration of NO_2, so substitute this value into the expression and calculate the equilibrium concentration of N_2O_4, that is, calculate $[N_2O_4]$.

$$K = 170 = \frac{[N_2O_4]}{(0.041)^2}$$

$$[N_2O_4] = 170(0.041)^2 = 0.29 \text{ mol/L}$$

Solving, we find that $[N_2O_4]$ is 0.29 mol/L, a concentration greater than that of NO_2. This result was anticipated, since K is larger than 1, indicating a product-favored reaction.

EXERCISE 8.9 • *Equilibrium Constants*

If solid AgCl and AgI are placed in 1.0 L of water in separate beakers, in which beaker would the silver ion concentration, $[Ag^+]$, be larger?

$$AgCl(s) \rightleftharpoons Ag^+(aq) + Cl^-(aq) \qquad K = 1.8 \times 10^{-10}$$

$$AgI(s) \rightleftharpoons Ag^+(aq) + I^-(aq) \qquad K = 1.5 \times 10^{-16}$$

EXERCISE 8.10 • *Manipulating Equilibrium Constants*

The equilibrium constant for the ionization of acetic acid

$$CH_3CO_2H(aq) + H_2O(\ell) \rightleftharpoons CH_3CO_2^-(aq) + H_3O^+(aq)$$

is 1.8×10^{-5} at 25 °C. What is the equilibrium constant for the reverse reaction, the union of hydronium ion and acetate ion to give acetic acid? Is it very large or very small? What does this tell you about the extent to which reaction can occur between the acetate ion and the hydronium ion? Is the reaction product- or reactant-favored?

$$CH_3CO_2^-(aq) + H_3O^+(aq) \rightleftharpoons CH_3CO_2H(aq) + H_2O(\ell)$$

acetate ion

EXERCISE 8.11 • *Calculating Equilibrium Concentrations*

Two molecules of nitrogen dioxide, NO_2, a brown gas, can join to form colorless N_2O_4.

$$2\ NO_2(g) \rightleftharpoons N_2O_4(g)$$

Enough NO_2 is placed in a flask to give a concentration of 0.90 mol/L. After the reaction forms some N_2O_4 and achieves equilibrium, the NO_2 concentration is only 0.26 mol/L. Is the reaction product- or reactant-favored at this temperature (85 °C)? What is the value of K?

Henri Le Chatelier. *(Oesper Collection in the History of Chemistry/University of Cincinnati)*

Henri Le Chatelier (1850–1936) was a French chemist who developed his concept of effects on equilibrium in connection with the chemistry of cement.

Disturbing a Chemical Equilibrium: Le Chatelier's Principle

Suppose you are an environmental engineer, biologist, or geologist and that you have just measured the concentration of hydronium ion, H_3O^+, in a lake. You know that the H_3O^+ ions are involved in many different equilibrium reactions in the lake. How can you predict the influence of changing conditions? For example, what happens if there is a large increase in rainfall that has an acidity different from that of the lake? Or what happens if lime (calcium oxide), a strong base, is added to the lake? These questions and many others like them can be answered qualitatively by applying a useful guideline known as **Le Chatelier's principle:** *a change in any of the factors that determine the equilibrium conditions of a system will cause the system to change so that the effect of the change is reduced or counteracted.*

There are at least two ways to disturb a chemical equilibrium: change the concentrations of reactants or products, change the temperature, or both. If a reaction is at equilibrium and the concentrations of reactants or products are changed, the system can no longer be at equilibrium. This effect leads to stalactites and stalagmites in caves or to the crust of limestone in your teapot if you have hard water. Recall that the reaction responsible for these deposits is the dissolving of otherwise insoluble limestone, calcium carbonate, in the presence of CO_2 and water.

Hard water contains dissolved CO_2 and metal ions such as Ca^{2+} and Mg^{2+}.

$$CaCO_3(s) + CO_2(aq) + H_2O(\ell) \rightleftharpoons Ca^{2+}(aq) + 2\ HCO_3^-(aq)$$

If ground water that is saturated with CO_2 encounters a bed of limestone deep inside the earth, the reaction can occur until equilibrium is reached. Now let us say the stream of water containing dissolved CO_2, Ca^{2+}, and HCO_3^- comes into a cave (or into your teapot). Carbon dioxide bubbles out of the solution, just as it does when you pop the top of a soda can (and this is especially vigorous if the soda is warm). The concentration of CO_2 decreases on the reactant side, so the equilibrium shifts to compensate for the CO_2 loss, that is, to raise the CO_2 concentration again. The only way this can happen is if some of the calcium and bicarbonate ions in solution are consumed, and a necessary byproduct is $CaCO_3$, which precipitates in the cave or in your teapot.

In summary, if you add a reactant or product to a reaction at equilibrium at a given temperature, some of the added substance will always be consumed as the reaction compensates for the disturbance. If excess reactant is added, for example, more product will be formed. Similarly, if some product is removed from solution as a gas or a precipitate, the reaction will compensate by producing more product and further consuming reactants.

You can make a qualitative prediction about the effect of a temperature change on the equilibrium position in a chemical reaction if you know whether the reaction is exo- or endothermic. As an example, consider the endothermic reaction of N_2 with O_2 to give nitrogen monoxide, NO.

$$N_2(g) + O_2(g) \rightleftharpoons 2\,NO(g) \qquad \Delta H^\circ_{rxn} = +180.5 \text{ kJ}$$

$$K = \frac{[NO]^2}{[N_2][O_2]} \qquad \begin{aligned} K &= 4.5 \times 10^{-31} \text{ at } 298 \text{ K} \\ K &= 6.7 \times 10^{-10} \text{ at } 900 \text{ K} \\ K &= 1.7 \times 10^{-3} \text{ at } 2300 \text{ K} \end{aligned}$$

Le Chatelier's principle tells us that putting in energy (as heat) to raise the temperature will cause the equilibrium to shift in the direction that reduces or counteracts this input. The way to counteract the energy input here is for the endothermic reaction to consume some of the added heat. N_2 and O_2 should react to produce more NO. Therefore, an increase in temperature should result in a new equilibrium situation in which the concentration of NO is higher and the concentrations of N_2 and O_2 are lower. Since this raises the value of the numerator in the K expression and lowers the denominator, K must be larger at the higher temperature, and this is confirmed by the experimental values of K.

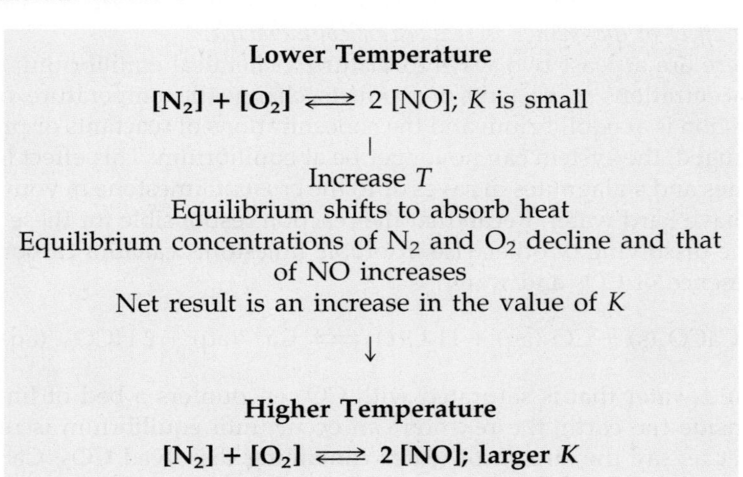

SUMMARY: The Effect of Temperature on K Values. The value of the equilibrium constant is constant as long as the temperature is held constant.
For an endothermic reaction, K *increases with increasing* T.
For an exothermic reaction, K *decreases with increasing* T.

The effect on K of raising the temperature of the $N_2 + O_2$ reaction leads us to a general conclusion. *For an endothermic reaction, an increase in tempera-*

ture always means an increase in K; the reaction will become more product-favored. That is, when equilibrium is achieved at the higher temperature, the concentration of products is greater and that of the reactants is smaller. By the same analysis, and as illustrated by Exercise 8.12, the opposite is true for an exothermic reaction.

EXAMPLE 8.4

Le Chatelier's Principle

Consider the equilibrium established in a mixture of nitrogen, hydrogen, and ammonia.

$$N_2(g) + 3\,H_2(g) \rightleftharpoons 2\,NH_3(g) \qquad \Delta H° = -92.2 \text{ kJ}$$

If more H_2 is added (at constant temperature) to the equilibrium mixture, will the concentration of ammonia increase or decrease? If the temperature is increased, will the value of K increase or decrease? That is, is the reaction more product-favored or more reactant-favored at higher temperatures?

SOLUTION The equilibrium expression for the reaction is

$$K = \frac{[NH_3]^2}{[N_2][H_2]^3} = 3.5 \times 10^8 \text{ at } 25\ °\text{C}$$

When more H_2 is added at constant temperature, the value of K cannot change. However, adding H_2 makes the value of the denominator larger. Because K is a constant, the value of the numerator must also increase. This can happen because the excess H_2 can react with N_2 to make more NH_3. That is, adding excess reactant molecules to a reaction at equilibrium will be countered by a reaction to produce more product and consume reactants.

The effect of increasing the temperature depends on whether the reaction is endothermic or exothermic. Here it is exothermic. Therefore, on increasing the temperature, the system will reduce the evolution of heat by the reaction by shifting the equilibrium to the left (toward the reactants.) This leads to a decrease in the NH_3 concentration, an increase in the concentrations of H_2 and N_2, and a decrease in the value of K.

EXERCISE 8.12 • *Temperature Effects on an Equilibrium*

Two moles of the brown gas NO_2 can combine to form N_2O_4, and equilibrium is achieved in a closed system.

$$2\,NO_2(g) \rightleftharpoons N_2O_4(g) \qquad \Delta H° = -57.2 \text{ kJ}$$

$$K = \frac{[N_2O_4]}{[NO_2]^2} \qquad \begin{array}{l} K = 1300 \text{ at } 273 \text{ K} \\ K = 170 \text{ at } 298 \text{ K} \end{array}$$

Describe what happens as the temperature of this reaction, at equilibrium, is lowered. Verify your answer by looking at Figure 8.18.

8.3
CHEMICAL REACTIVITY

We can summarize our findings in the preceding sections by saying that *a product-favored reaction is one that has an equilibrium constant larger than 1.* Further, reactions that are exothermic and for which the entropy increases are product-favored. Since entropy changes usually play a smaller role at room temperature than do enthalpy changes, it is reasonable to use the following rule of thumb: *product-favored reactions are those that are strongly exothermic and that increase the disorder of the system or leave it essentially unchanged.* Thus the following reaction is not likely to produce appreciable quantities of products

$$N_2(g) + O_2(g) \rightleftharpoons 2\,NO(g) \qquad \Delta H^\circ = +180.5\text{ kJ}$$
$$K = 4.5 \times 10^{-31} \text{ at } 25\text{ °C}$$

while the formation of ammonia from its elements is an exothermic reaction and is more likely to be product-favored.

$$N_2(g) + 3\,H_2(g) \rightleftharpoons 2\,NH_3(g) \qquad \Delta H^\circ_{rxn} = -92.22\text{ kJ}$$
$$\Delta S^\circ_{rxn} = -198.8\text{ J/K}$$
$$K = 3.5 \times 10^{8} \text{ at } 25\text{ °C}$$

in spite of the obvious decline in entropy in the course of the reaction.

So, it is often possible to predict whether a reaction can produce products. But don't forget that this is a thermodynamic concept; it gives us no clue about the rates of reactions. Here are some useful generalizations about reaction rates.

1. Reactions occur more rapidly in the gas phase or in solution, where molecules of one reactant are completely mixed with molecules of another.
2. Reactions occur more rapidly at higher temperatures.
3. Reactions are faster when the reactant concentrations are high. This means, for example, that reactions between gases occur most rapidly at higher pressures. This is one reason that the mixture of air and gasoline vapor is compressed in the cylinder of an automobile engine before it is ignited.
4. Reactions between a solid and a gas, or a solid and something dissolved in solution, are usually much faster when the solid particles are as small as possible. This effect is obvious in Figure 8.19. In Figure 8.19a, we have tried to burn a combustible solid in a dish, but combustion is slow. When the solid is blown into the air as a dust, though, combustion is much more rapid, since the surface area open to interaction with oxygen is now much greater (Figure 8.19b). For this reason coal is ground to a powder before it is burned for generating electricity.

Many of the predictions chemists make about reaction rates are based on experience, but even these are not always correct. Nonetheless, here are

(a)

(b)

Figure 8.19

The combustion of lycopodium powder. (a) The spores of this common moss burn only with difficulty when piled in a dish. (b) If the surface area is increased by spraying the finely divided powder into the flame, combustion is rapid. *(C.D. Winters)*

some things that may increase the rate of a reaction that may otherwise be too slow to be useful:

- increase reactant concentrations
- try to reduce the size of the particles that must react (to increase the surface area open to reaction) or dissolve the reactants in a solvent
- increase the temperature
- find an appropriate catalyst

The last of the possibilities—finding a catalyst—is much easier said than done. It is not easy to predict what type of material will catalyze a reaction, which is why so many chemists study catalysts. Biological catalysts called enzymes work very well in plants and animals, so chemists have tried to apply these lessons to the laboratory and chemical plant.

One of the best examples of the application of the principles of chemical reactivity is the **Haber process** for the synthesis of ammonia from its elements (Figure 8.20). Even though the earth is bathed in tons of N_2 gas, nitrogen cannot be used by plants until it is "fixed," that is, converted into biologically useful forms. Although nitrogen fixation is done naturally by organisms such as blue-green algae and some field crops such as alfalfa and soybeans, most plants cannot fix N_2. Fixed nitrogen must be supplied by an external source, and this is especially important for the new varieties of wheat, corn, and rice that were developed during the past several decades and that have led to much improved food production, particularly in Third World countries.

A commercially feasible process for manufacturing ammonia was devised by the German chemist Fritz Haber. Although it should be possible to fix nitrogen in some usable form from simple molecules by using any of several possible reactions, Haber chose the *direct synthesis of ammonia from its elements*.

$$N_2(g) + 3\,H_2(g) \rightleftharpoons 2\,NH_3(g)$$

Temperature	*K*	$\Delta H°$
25 °C	3.5×10^8 (calculated)	−92.2 kJ
450 °C	0.16 (experimental)	−111.3 kJ

At first this process seems a poor choice for several reasons. Hydrogen is available naturally only in combined form, for example in water or hydrocarbons, meaning that some H-containing compounds must be destroyed first at the cost of considerable free energy. Not only that, but this energy expense is completely wasted because the hydrogen of ammonia is converted to water and the nitrogen to nitrate ion by soil bacteria *before* the nitrogen can be used by plants. The plant must then expend more energy, derived from photosynthesis, to reduce the nitrate ion back to ammonia. Nonetheless, the Haber process has been so well developed that ammonia is very inexpensive (about $150 per ton). For this reason it is widely used as a fertilizer and so is often in the "top 5" chemicals produced in the United States. In 1992 about 34 billion pounds of NH_3 were produced by the Haber process.

The direct synthesis of ammonia is an equilibrium process that has been carefully fine-tuned by industry.

- The reaction is strongly exothermic and so is predicted to be product-favored (K is greater than 1 at 25 °C), in spite of the decline in entropy. However, the reaction is quite slow, so the reaction is carried out at a higher temperature to increase the rate.

Figure 8.20

The Haber process for ammonia synthesis. (*left*) A block diagram of the process. (*right*) A schematic diagram. The mixed gases are pumped over the catalytic surface to produce ammonia. Note that the ammonia is collected as a liquid, and the uncombined nitrogen and hydrogen are recycled to the catalytic chamber.

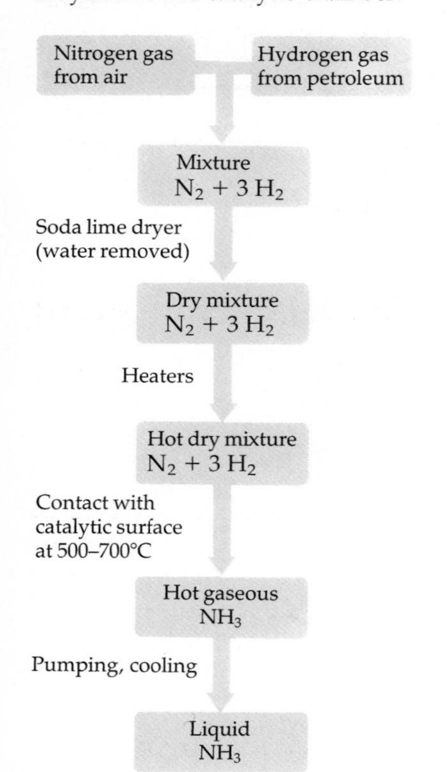

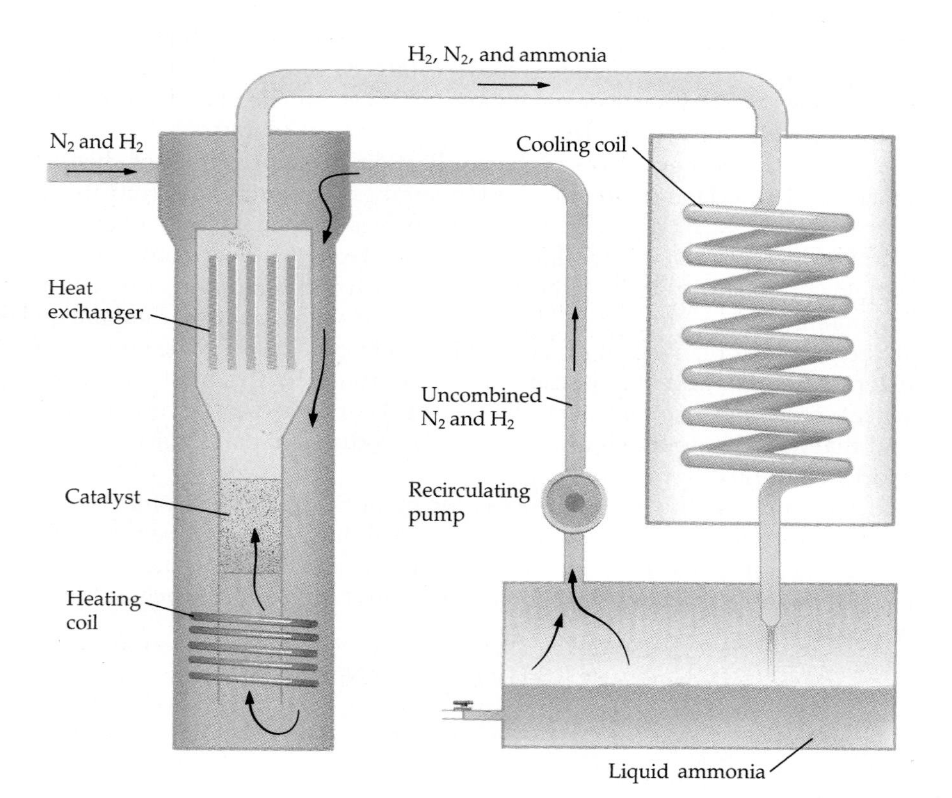

A Deeper Look

The Relation Between Free Energy Changes and Equilibrium Constants for Reactions

The change in Gibbs free energy when a reaction occurs under standard conditions indicates whether the reaction is product-favored (negative $\Delta G°$) or reactant-favored (positive $\Delta G°$). The equilibrium constant provides similar information: a reaction is product-favored when $K > 1$, and a reaction is reactant-favored when $K < 1$. This suggests that there must be a connection between $\Delta G°$ and the equilibrium constant, and indeed there is:

$$\Delta G° = -RT \ln K_{th}$$

In this equation $\Delta G°$ and T are already familiar, and R is a constant whose value is 8.314 J/K·mol. K_{th} is called the thermodynamic equilibrium constant, and its value depends on the choice of standard state for the substances involved in a reaction. If the reaction occurs in solution, for example, K_{th} has the same form as the equilibrium constants you are familiar with, and the concentrations are expressed in moles per liter because the standard state involves each substance at a concentration of 1 mol/L. For gases, where the standard state involves pressure, K_{th} must be expressed in terms of pressures of the various gases involved, not their concentrations. Therefore, K_{th} values for gas-phase reactions are different from the K values that appear in Table 8.1, which are for ratios of concentrations.

Regardless of the choice of standard state, the equation indicates that the Gibbs free energy change for a reaction is minus a constant times the temperature times the natural logarithm of the equilibrium constant. If K_{th} is larger than 1, then $\ln K_{th}$ is positive, and $\Delta G°$ will be negative because of the minus sign. Both of these—a negative $\Delta G°$ and $K > 1$—indicate the reaction is product-favored under standard-state conditions. Conversely, if K_{th} is less than 1, then $\ln K_{th}$ is negative, and $\Delta G°$ must be positive, indicating a reactant-favored process.

ΔG is the difference in Gibbs free energy between products in their standard states and reactants in their standard states. For example, the reaction might be as expressed below by Equation 1.

When 1 mole of HCOOH(aq) at a concentration of 1 mol/L is converted into 1 mole of $HCOO^-$(aq) and 1 mole of H^+(aq), each at a concentration of 1 mol/L, the change in free energy is 21.4 kJ. Since there is a positive free energy change, and so we predict that the process will be reactant-favored.

Now consider the same system reacting to reach equilibrium (Equation 2).

Since K is much smaller than 1, there will be smaller concentrations products and a larger concentration of reactant when equilibrium is reached. In order to achieve such concentrations, the reverse reaction must occur. This is what we would expect of a reactant-favored system: products are converted to reactants if both are present at standard conditions. The *size* of $\Delta G°$ and the *size* of K indicate quantitatively *how far* from standard-state conditions the reaction will go before equilibrium is achieved.

A similar argument could have been made if the equilibrium constant had been greater than one. In that case $\Delta G°$ would be negative, and we would expect the reaction to proceed from left to right to produce products. The size of $\Delta G°$—how negative it was—would tell us how far from left to right the reaction would proceed. Remember also that $\Delta G°$ can be calculated from the Gibbs equation

$$\Delta G° = \Delta H° - T\Delta S°$$

If we know or can estimate changes in enthalpy and entropy for a reaction, then we can calculate or estimate the Gibbs free energy change and hence the equilibrium constant. All of these thermodynamic quantities are connected to the concepts of chemical equilibria.

	HCOOH(aq) $\rightleftharpoons$	$HCOO^-$(aq) +	H^+(aq)	
Eq. 1	1 mol/L	1 mol/L	1 mol/L	$\Delta G° = 21.4$ kJ
Eq. 2	HCOOH(aq) $\rightleftharpoons$	$HCOO^-$(aq) +	H^+(aq)	$K = 1.8 \times 10^{-4}$
Initial conditions	1 mol/L	1 mol/L	1 mol/L	
At equilibrium	> 1 mol/L	< 1 mol/L	< 1 mol/L	

- Although the reaction rate increases with increasing temperature, the equilibrium constant declines, as predicted by Le Chatelier's principle for an exothermic process. This means that, for a given concentration of starting material, the equilibrium concentration of NH_3 is smaller at a higher temperature.

Portrait of a Scientist

Fritz Haber (1868–1934)

In 1898 William Ramsay, the discoverer of the rare gases, pointed out the depletion of fixed nitrogen in the world, and he predicted world disaster due to a "fixed nitrogen shortage" by mid-20th century. Such a shortage would have prevented food production from keeping pace with population, resulting in widespread famine. That this has not occurred is due to the work of Fritz Haber. Haber studied the reaction

$$N_2(g) + 3\,H_2(g) \longrightarrow 2\,NH_3(g)$$

in the early 1900s and concluded that direct ammonia synthesis should be possible. However, it was not until 1914 that the engineering problems were solved by Carl Bosch, and ammonia production began just in time for the start of World War I. Ammonia is the starting material to make nitric acid, a vital material in the manufacture of the explosives TNT and nitroglycerin. Therefore, ammonia is thought to be the first synthetic chemical used on a large scale for purposes of warfare.

Haber's contract with the manufacturer of ammonia called for him to receive 1 pfennig per kilogram of ammonia, and he soon became not only famous but rich! Unfortunately, he joined the German Chemical Warfare Service at the start of World War I and became its director in 1916. The primary mission of the service was to develop gas warfare, and in 1915 he supervised the first use of Cl_2 at the battle of Ypres. It was a tragedy not only of modern warfare, but to Haber personally as well. His wife pleaded with him to stop his work in this area and when he refused, she committed suicide. In 1918 he was awarded the Nobel prize for the ammonia synthesis, but the choice was criticized because of his role in chemical warfare.

After World War I Haber did some of his best work, continuing to study thermodynamics. However, because of his Jewish background, Haber left Germany in 1933, worked for a time in England, and died in Switzerland in 1934.

Fritz Haber. *(Oesper Collection in the History of Chemistry/University of Cincinnati)*

- To increase the equilibrium concentration of NH_3, the reaction is carried out at higher pressure. This does not change the value of K, but an increase in pressure can be compensated by converting N_2 and H_2 to NH_3; two moles of $NH_3(g)$ exert less pressure than a total of four moles of gaseous reactants $[N_2(g) + 3\,H_2(g)]$ in the same size container.
- Since the temperature cannot be raised too much in an attempt to increase the rate, a rate increase can be achieved with a catalyst. An effective catalyst for the Haber process is Fe_3O_4 mixed with KOH, SiO_2, and Al_2O_3. Since the catalyst is not effective below about 400 °C, the optimum temperature, considering all the factors controlling the reaction, is about 450 °C.

Making predictions about chemical reactivity is part of the challenge, the adventure, and the art of chemistry. Many chemists enjoy trying to make useful new materials, and this usually means choosing to make them by reactions that we believe will be product-favored and reasonably rapid. Such predictions are based on the ideas outlined in this chapter.

IN CLOSING

Having studied this chapter, you should be able to

- define reaction rate and explain the terms in a rate equation or rate law (Section 8.1).
- describe the effect on reaction rate of changes in temperature and reactant concentrations (Section 8.1).
- show by using an energy plot what happens as two reactant molecules interact to form product molecules and how a catalyst affects a reaction rate (Section 8.1).
- use rate data to write a rate equation, and use a rate equation to predict how a reaction rate varies with changing concentrations (Section 8.1).
- define reaction mechanism and identify rate-determining steps, catalysts, and intermediates (Section 8.1).
- write equilibrium constant expressions, given balanced chemical equations (Section 8.2).
- make qualitative predictions about the extent of reaction based upon equilibrium constant values; that is, be able to predict whether a reaction is product- or reactant-favored (Section 8.2).
- calculate a value of K using equilibrium concentrations or calculate an equilibrium concentration if K is known (Section 8.2).
- show by using Le Chatelier's principle how changes in concentrations and temperature affect chemical equilibria (Section 8.2).
- list the factors affecting chemical reactivity (Section 8.3).

STUDY QUESTIONS

Review Questions

1. The rate expression for a chemical reaction can be determined by which of the following?
   (a) Theoretical calculations.
   (b) Measuring the rate of the reaction as a function of the concentration of the reacting species.
   (c) Measuring the rate of the reaction as a function of temperature.
2. Name at least three factors that affect the rate of a chemical reaction.
3. Refer to Figure 8.1 and explain why the rate of change of the concentration of $Pt(NH_3)_2Cl_2$ *de*creases with time but the concentration of Cl^- *in*creases with time.
4. Using the rate expression "rate = $k[A]^2[B]$," define the order of the reaction with respect to A and B and the overall reaction order.
5. If a reaction has the experimental rate expression "rate = $k[A]^2$," explain what happens to the rate when the concentration of A is tripled, and when the concentration of A is halved.
6. A reaction has the experimental rate expression "rate = $k[A]^2[B]$." If the concentration of A is doubled and the concentration of B is halved, what happens to the reaction rate?
7. Draw a reaction energy diagram for an exothermic process. Mark the activation energies of the forward and reverse processes and explain how the net energy change of the reaction can be calculated.
8. Indicate whether each of the following statements is true or false. Change the wording of each false statement to make it true.
   (a) It is possible to change the rate constant for a reaction by changing the temperature.
   (b) As a first order reaction proceeds at a constant temperature, the rate remains constant.
   (c) The rate constant for a reaction is independent of reactant concentrations.
   (d) As a first order reaction proceeds at a constant temperature, the rate constant changes.
9. What is a catalyst? What is its effect on the energy barrier for a reaction?

10. Explain the difference between a homogeneous and a heterogeneous catalyst. Give an example of each.
11. Name three important features of the equilibrium condition.
12. Which of the following is true about a chemical system at equilibrium?
    (a) No reactions take place.
    (b) Temperature increases will no longer increase reaction rates.
    (c) The rates of forward and reverse reactions are equal.
    (d) All reaction products will be solids.
13. Which is true about the equilibrium constant expression?
    (a) It determines the activation energy needed to perform a reaction.
    (b) It relates reactant and product concentrations.
    (c) It relates concentrations to rates.
    (d) It tells which reactant is highest in concentration.
14. Tell whether each of the following statements is true or false. If false, change the wording to make it true.
    (a) The magnitude of the equilibrium constant is always independent of temperature.
    (b) The equilibrium constant, K, for a reaction is equal to the equilibrium constant for the reverse reaction.
    (c) Only the concentration of CO_2 appears in the equilibrium constant expression for the reaction $CaCO_3(s) \rightleftarrows CaO(s) + CO_2(g)$.
15. Neither $PbCl_2$ nor PbF_2 is appreciably soluble in water. If solid $PbCl_2$ and solid PbF_2 are placed in equal amounts of water in separate beakers, in which beaker is the concentration of Pb^{2+} greater? Equilibrium constants for these solids dissolving in water are

$$PbCl_2(s) \rightleftharpoons Pb^{2+}(aq) + 2\,Cl^-(aq) \qquad K = 1.7 \times 10^{-5}$$

$$PbF_2(s) \rightleftharpoons Pb^{2+}(aq) + 2\,F^-(aq) \qquad K = 3.7 \times 10^{-8}$$

16. Using Le Chatelier's principle, explain how increasing the temperature would affect the equilibrium $CaCO_3(s) + \text{heat} \rightleftarrows CaO(s) + CO_2(g)$. If more $CaCO_3$ is added to a flask in which this equilibrium exists, how is the equilibrium affected? What if some additional CO_2 is placed in the flask?
17. Chlorine atoms are thought to lead to the destruction of the earth's ozone layer by the following sequence of reactions:

$$Cl(g) + O_3(g) \longrightarrow ClO(g) + O_2(g)$$

$$ClO(g) + O(g) \longrightarrow Cl(g) + O_2(g)$$

where the O atoms in the second step come from the decomposition of ozone by sunlight.

$$O_3(g) \rightleftharpoons O(g) + O_2(g)$$

What is the net equation on summing these three equations? Why does this lead to ozone loss in the stratosphere? What is the role played by Cl in this sequence of reactions? What name is given to the species ClO?

Reaction Rates and the Effect of Concentration

18. Experimental data are listed below for the hypothetical reaction $A \rightarrow 2\,B$.

Time (s)	*[A] (mol/L)*
0.00	1.000
10.0	0.833
20.0	0.714
30.0	0.625
40.0	0.555

(a) Plot these data, connect the points with a smooth line, and calculate the rate of change of [A] for each 10-second interval from 0 to 40 seconds. Why might the rate of change decrease from one time interval to the next?
(b) How is the rate of change of [B] related to the rate of change of [A] in the same time interval? Calculate the rate of change of [B] for the time interval from 10 to 20 seconds.

19. A compound called phenylacetate reacts with water according to the equation

$$\underset{\text{phenylacetate}}{CH_3\overset{O}{\overset{\|}{C}}{-}O{-}C_6H_5} + H_2O(\ell) \longrightarrow \underset{\text{acetic acid}}{CH_3\overset{O}{\overset{\|}{C}}{-}O{-}H(aq)} + \underset{\text{phenol}}{C_6H_5{-}OH(aq)}$$

The following data were collected at 5 °C.

Time (min)	*[Phenyl acetate] (mol/L)*
0	0.55
0.25	0.42
0.50	0.31
0.75	0.23
1.00	0.17
1.25	0.12
1.50	0.085

(a) Plot these data, and describe the shape of the curve. Compare it with Figure 8.1.

(b) Calculate the rate of change of the concentration of phenyl acetate during the period from 0.20 min to 0.40 min and then during the period from 1.2 min to 1.4 min. Compare the values and tell why one is smaller than the other.

(c) What is the rate of change of the phenol concentration during the period from 1.00 min to 1.25 min?

20. The reaction of $CO(g) + NO_2(g)$ is second order in NO_2 and zero order in CO at temperatures less than 500 K. (a) Write the rate expression for the reaction. (b) How will the reaction rate change if the NO_2 concentration is halved? (c) How will the reaction rate change if the concentration of CO is doubled?

21. Nitrosyl bromide, NOBr, is formed from NO and Br_2.

$$2\,NO(g) + Br_2(g) \longrightarrow 2\,NOBr(g)$$

Experiment shows that the reaction is first order in Br_2 and second order in NO.

(a) Write the rate expression for the reaction.

(b) If the concentration of Br_2 is tripled, how will the reaction rate change?

(c) What happens to the reaction rate when the concentration of NO is doubled?

22. For each of the following expressions, state the reaction order with respect to each reagent.

(a) rate = $k[A][B]^2$ (c) rate = $k[A]$

(b) rate = $k[A][B]$ (d) rate = $k[A]^3[B]$

23. A reaction between molecules A and B (A + B → products) is found to be second order in B. Which rate equation *cannot* be correct?

(a) rate = $k[A][B]$

(b) rate = $k[A][B]^2$

(c) rate = $k[B]^2$

24. For the reaction of $Pt(NH_3)_2Cl_2$ with water (Section 8.1)

$$Pt(NH_3)_2Cl_2 + H_2O \longrightarrow Pt(NH_3)_2(H_2O)Cl^+ + Cl^-$$

the rate expression was given as "rate = $k[Pt(NH_3)_2Cl_2]$" with k = 0.090/hr. Calculate the initial rate of reaction when the concentration of $Pt(NH_3)_2Cl_2$ is (a) 0.010 M, (b) 0.020 M, and (c) 0.040 M. How does the rate of disappearance of $Pt(NH_3)_2Cl_2$ change with its initial concentration? How is this related to the rate law? How does the initial concentration of $Pt(NH_3)_2Cl_2$ affect the rate of appearance of Cl^- in the solution?

25. Methyl acetate, CH_3COOCH_3, reacts with base to break one of the C—O bonds.

$$CH_3\overset{\overset{\displaystyle O}{\|}}{C}{-}O{-}CH_3 + OH^-(aq) \longrightarrow CH_3\overset{\overset{\displaystyle O}{\|}}{C}{-}O^-(aq) + HO{-}CH_3(aq)$$

The rate expression is "rate = $k[CH_3COOCH_3][OH^-]$" where k = 0.14 L/mol·s at 25 °C. What is the initial rate at which the methyl acetate disappears when both reactants, CH_3COOCH_3 and OH^-, have a concentration of 0.025 M? How rapidly does the methyl alcohol, CH_3OH, appear in the solution?

26. The transfer of an oxygen atom from NO_2 to CO has been studied at 540 K:

$$CO(g) + NO_2(g) \longrightarrow CO_2(g) + NO(g)$$

The following data were collected. Use them to (a) write the rate expression, (b) determine the reaction order with respect to each reactant, and (c) calculate the rate constant, giving the correct units for k.

Initial Concentration (mol/L)		
[CO]	*[NO₂]*	*Initial Rate (mol/L·hr)*
5.1×10^{-4}	0.35×10^{-4}	3.4×10^{-8}
5.1×10^{-4}	0.70×10^{-4}	6.8×10^{-8}
5.1×10^{-4}	0.18×10^{-4}	1.7×10^{-8}
1.0×10^{-3}	0.35×10^{-4}	6.8×10^{-8}
1.5×10^{-3}	0.35×10^{-4}	10.2×10^{-8}

27. A study of the reaction 2A + B → C + D gave the following results:

	Initial Concentration (mol/L)		
Experiment	*[A]*	*[B]*	*Initial Rate (mol/L·s)*
1	0.10	0.05	6.0×10^{-3}
2	0.20	0.05	1.2×10^{-2}
3	0.30	0.05	1.8×10^{-2}
4	0.20	0.15	1.1×10^{-1}

(a) What are the rate expression and order with respect to A and B for this reaction?

(b) Calculate the rate constant for the reaction, giving the correct units for k.

28. The bromination of acetone is catalyzed by acid.

$$CH_3COCH_3 + Br_2 + H_2O(\ell) \xrightarrow{\text{acid catalyst}} CH_3COCH_2Br + H_3O^+ + Br^-$$

The rate of disappearance of bromine was measured for several different initial concentrations (all in mol/L) of acetone, bromine, and hydronium ion.

Initial Concentration (mol/L)			*Initial Rate of Change of Br_2 (mol/L·s)*
$[CH_3COCH_3]$	*$[Br_2]$*	*$[H_3O^+]$*	
0.30	0.05	0.05	5.7×10^{-5}
0.30	0.10	0.05	5.7×10^{-5}
0.30	0.05	0.10	12.0×10^{-5}
0.40	0.05	0.20	31.0×10^{-5}
0.40	0.05	0.05	7.6×10^{-5}

(a) Deduce the rate expression for the reaction and give the order with respect to each reactant.
(b) What is the numerical value of k, the rate constant?
(c) If $[H_3O^+]$ is maintained at 0.050 M, whereas both $[CH_3COCH_3]$ and $[Br_2]$ are 0.10 M, what is the rate of the reaction?

29. One of the major eye irritants in smog is formaldehyde, CH_2O, formed by reaction of ozone with ethylene.

$$C_2H_4(g) + O_3(g) \longrightarrow 2\,CH_2O(g) + \tfrac{1}{2}\,O_2(g)$$

(a) Determine the rate expression for the reaction using the data in the following table. What is the reaction order with respect to O_3? What is the order with respect to C_2H_4?
(b) Calculate the rate constant, k.
(c) What is the rate of reaction when $[C_2H_4]$ and $[O_3]$ are both 2.0×10^{-7} M?

Initial Concentration (mol/L)		*Initial Rate of Formation of CH_2O (mol/L·s)*
$[O_3]$	*$[C_2H_4]$*	
0.50×10^{-7}	1.0×10^{-8}	1.0×10^{-12}
1.5×10^{-7}	1.0×10^{-8}	3.0×10^{-12}
1.0×10^{-7}	2.0×10^{-8}	4.0×10^{-12}

Reaction Mechanisms

30. For the hypothetical reaction $A + B \rightarrow C + D$, the activation energy is 32 kJ/mol. For the reverse reaction $(C + D \rightarrow A + B)$, the activation energy is 58 kJ/mol. Is the reaction $A + B \rightarrow C + D$ exothermic or endothermic?

31. Use the diagram below to answer the following questions.
(a) Is the reaction exothermic or endothermic?
(b) What is the approximate value of ΔE for the forward reaction?
(c) What is the activation energy in each direction?
(d) A catalyst is found that lowers the activation energy of the reaction by about 10 kJ/mol. How will this catalyst affect the rate of the reverse reaction?

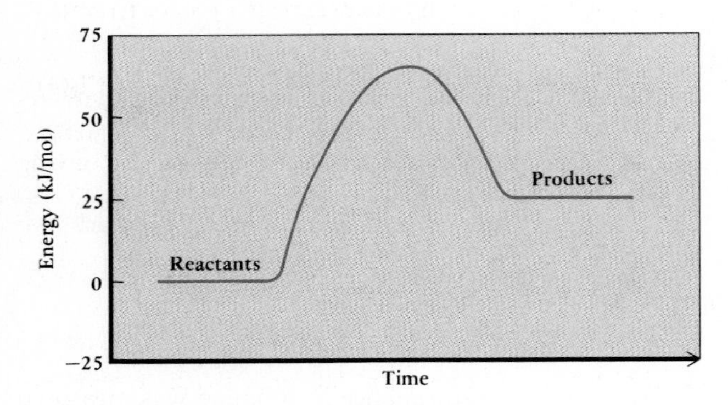

32. Experiments show that the reaction of nitrogen dioxide with fluorine

Overall reaction: $2\,NO_2(g) + F_2(g) \longrightarrow 2\,FNO_2(g)$

has the following rate expression:

$$\text{initial reaction rate} = k[NO_2][F_2]$$

and the reaction is thought to occur in two steps, the first being very slow and the second rapid.

Step 1 (Slow): $NO_2(g) + F_2(g) \longrightarrow FNO_2(g) + F(g)$

Step 2 (Fast): $NO_2(g) + F(g) \longrightarrow FNO_2(g)$

That is, NO_2 and F_2 first produce one molecule of the product (FNO_2) plus an F atom, and the F atom then reacts with additional NO_2 to give one more molecule of product. Show that the sum of this sequence of reactions gives the balanced equation for the overall reaction. Which step is rate determining?

33. Nitrogen oxide is reduced by hydrogen to give water and nitrogen

$$2\,H_2(g) + 2\,NO(g) \longrightarrow N_2(g) + 2\,H_2O(g)$$

and one possible mechanism for this reaction is a sequence of three elementary steps.

$$2\,NO(g) \rightleftharpoons N_2O_2(g)$$
$$N_2O_2(g) + H_2(g) \longrightarrow N_2O(g) + H_2O(g)$$
$$N_2O(g) + H_2(g) \longrightarrow N_2(g) + H_2O(g)$$

Show that the sum of these steps gives the net reaction.

Catalysis

34. Which of the following statements is (are) true?
(a) The concentration of a homogeneous catalyst may appear in the rate expression.
(b) A catalyst is always consumed in the reaction.
(c) A catalyst must always be in the same phase as the reactants.
(d) A catalyst can change the course of a reaction and allow different products to be produced.

35. Hydrogenation reactions, processes in which H_2 is added to a molecule, are usually catalyzed. An excellent catalyst is a very finely divided metal suspended in the reaction solvent. Tell why finely divided rhodium, for example, is a much more efficient catalyst than a small block of the metal.

36. Which of the following reactions *appear* to involve a catalyst? In those cases where a catalyst is present, tell whether it is homogeneous or heterogeneous.
(a) $CH_3CO_2CH_3(aq) + H_2O(\ell) + H_3O^+(aq) \longrightarrow CH_3CO_2H(aq) + CH_3OH(aq) + H_3O^+(aq)$
(b) $2\ H_2(g) + O_2(g) \longrightarrow 2\ H_2O(g)$
(c) $2\ H_2(g) + O_2(g) + Pt(s) \longrightarrow 2\ H_2O(g) + Pt(s)$
(d) $NH_3(aq) + CH_3Cl(aq) + H_2O(\ell) \longrightarrow Cl^-(aq) + NH_4^+(aq) + CH_3OH(aq)$

37. In acid solution, methyl formate forms methyl alcohol and formic acid.

$$\underset{\text{methyl formate}}{HCO_2CH_3(aq)} + H_2O(\ell) \longrightarrow \underset{\text{formic acid}}{HCO_2H(aq)} + \underset{\text{methyl alcohol}}{CH_3OH(aq)}$$

The rate expression is as follows: "rate = $k[HCOOCH_3][H_3O^+]$." Why does H_3O^+ appear in the rate expression but not in the overall equation for the reaction?

Writing Equilibrium Constant Expressions

38. Write equilibrium constant expressions, in terms of reactant and product concentrations, for the following reactions:
(a) $2\ H_2O_2(g) \rightleftharpoons 2\ H_2O(g) + O_2(g)$
(b) $PCl_3(g) + Cl_2(g) \rightleftharpoons PCl_5(g)$
(c) $SiO_2(s) + 3\ C(s) \rightleftharpoons SiC(s) + 2\ CO(g)$
(d) $H_2(g) + \frac{1}{8}\ S_8(s) \rightleftharpoons H_2S(g)$

39. Write equilibrium constant expressions, in terms of reactant and product concentrations, for the following reactions:
(a) $3\ O_2(g) \rightleftharpoons 2\ O_3(g)$
(b) $SiH_4(g) + 2\ O_2(g) \rightleftharpoons SiO_2(s) + 2\ H_2O(g)$
(c) $MgO(s) + SO_2(g) + \frac{1}{2}\ O_2(g) \rightleftharpoons MgSO_4(s)$
(d) $2\ PbS(s) + 3\ O_2(g) \rightleftharpoons 2\ PbO(s) + 2\ SO_2(g)$

40. Write equilibrium constant expressions, in terms of concentrations, for each of the following reactions:
(a) $TlCl_3(s) \rightleftharpoons TlCl(s) + Cl_2(g)$
(b) $CuCl_4^{2-}(aq) \rightleftharpoons Cu^{2+}(aq) + 4\ Cl^-(aq)$
(c) $CO(g) + H_2O(g) \rightleftharpoons CO_2(g) + H_2(g)$
(d) $4\ H_3O^+(aq) + 2\ Cl^-(aq) + MnO_2(s) \rightleftharpoons Mn^{2+}(aq) + 6\ H_2O(\ell) + Cl_2(g)$

41. Write equilibrium constant expressions for each of the following reactions:
(a) The oxidation of ammonia with ClF_3 in a rocket motor.

$$NH_3(g) + ClF_3(g) \rightleftharpoons 3\ HF(g) + \tfrac{1}{2}\ N_2(g) + \tfrac{1}{2}\ Cl_2(g)$$

(b) The simultaneous oxidation and reduction of a chlorine-containing ion.

$$3\ ClO_2^-(aq) \rightleftharpoons 2\ ClO_3^-(aq) + Cl^-(aq)$$

(c) $IO_3^-(aq) + 6\ OH^-(aq) + Cl_2(g) \rightleftharpoons IO_6^{5-}(aq) + 2\ Cl^-(aq) + 3\ H_2O(\ell)$

42. Consider the following two equilibria involving $SO_2(g)$ and their corresponding equilibrium constants.

$$SO_2(g) + \tfrac{1}{2}\ O_2(g) \rightleftharpoons SO_3(g) \qquad K_1$$
$$2\ SO_3(g) \rightleftharpoons 2\ SO_2(g) + O_2(g) \qquad K_2$$

Which of the following expressions correctly relates K_1 to K_2?
(a) $K_2 = K_1^2$ (b) $K_2^2 = K_1$ (c) $K_2 = 1/K_1$ (d) $K_2 = K_1$ (e) $K_2 = 1/K_1^2$

43. The reaction of hydrazine (N_2H_4) with chlorine trifluoride (ClF_3) was used in experimental rocket motors at one time.

$$N_2H_4(g) + \tfrac{4}{3}\ ClF_3(g) \rightleftharpoons 4\ HF(g) + N_2(g) + \tfrac{2}{3}\ Cl_2(g)$$

How is the equilibrium constant, K, for this reaction related to K' for the reaction written in the following way?

$$3\ N_2H_4(g) + 4\ ClF_3(g) \rightleftharpoons 12\ HF(g) + 3\ N_2(g) + 2\ Cl_2(g)$$

(a) $K = K'$ (b) $K = 1/K'$ (c) $K^3 = K'$ (d) $K = (K')^3$ (e) $3\ K = K'$

44. Hydrogen can react with elemental sulfur to give the smelly, toxic gas H_2S according to the reaction

$$H_2(g) + \tfrac{1}{8}\ S_8(solid) \rightleftharpoons H_2S(g)$$

If the equilibrium constant for this reaction is 7.6×10^5 at 25 °C, determine the value of the equilibrium constant for the reaction written as

$$8\,H_2(g) + S_8(s) \rightleftharpoons 8\,H_2S(g)$$

45. At 450 °C, the equilibrium constant for the Haber synthesis of ammonia is 0.16 for the reaction written as

$$3\,H_2(g) + N_2(g) \rightleftharpoons 2\,NH_3(g)$$

Calculate the value of K for the same reaction written as

$$\tfrac{3}{2}\,H_2(g) + \tfrac{1}{2}\,N_2(g) \rightleftharpoons NH_3(g)$$

Using Equilibrium Constants

46. Consider the transformation of butane to isobutane.

$$\underset{\text{butane}}{CH_3{-}CH_2{-}CH_2{-}CH_3} \xrightarrow{\text{catalyst}} \underset{\text{isobutane}}{H{-}C(CH_3)_2{-}CH_3}$$

$$K = \frac{[\text{isobutane}]}{[\text{butane}]} = 2.5$$

If the concentration of butane is 1.0 mol/L at equilibrium, what is the concentration of isobutane at equilibrium?

47. A mixture of the butanes in Study Question 46 has [butane] = 2.5 mol/L and [isobutane] = 3.5 mol/L. Is the system at equilibrium? If the equilibrium concentration of butane is 2.5 mol/L, what must [isobutane] be at equilibrium?

48. Consider the following equilibrium: $2\,A(aq) \rightleftharpoons B(aq)$. At equilibrium, [A] = 0.056 M and [B] = 0.21 M. Calculate the equilibrium constant for the reaction as written.

49. The following reaction was examined at 250 °C:

$$PCl_5(g) \rightleftharpoons PCl_3(g) + Cl_2(g)$$

At equilibrium, $[PCl_5] = 4.2 \times 10^{-5}$ M, $[PCl_3] = 1.3 \times 10^{-2}$ M, and $[Cl_2] = 3.9 \times 10^{-3}$ M. Calculate the equilibrium constant for the reaction.

50. At high temperature, hydrogen and carbon dioxide react to give water and carbon monoxide.

$$H_2(g) + CO_2(g) \rightleftharpoons H_2O(g) + CO(g)$$

(a) Laboratory measurements at 986 °C show that there are 0.11 mol of each of CO and water vapor and 0.087 mol of each of H_2 and CO_2 at equilibrium in a 1.0-L container. Calculate the equilibrium constant for the reaction at 986 °C.

(b) If there were 0.050 mol of each of H_2 and CO_2 in a 2.0-L container at equilibrium at 986 °C, what amounts of CO(g) and H_2O(g), in moles, would be present?

51. Carbon dioxide reacts with carbon to give carbon monoxide according to the equation $C(s) + CO_2(g) \rightleftharpoons 2\,CO(g)$. At 700 °C, a 2.0-L flask is found to contain at equilibrium 0.10 mol of CO, 0.20 mol of CO_2, and 0.40 mol of C. Calculate the equilibrium constant for this reaction at the specified temperature.

Le Chatelier's Principle

52. Hydrogen, bromine, and HBr are in equilibrium in the gas phase.

$$H_2(g) + Br_2(g) \rightleftharpoons 2\,HBr(g) \qquad \Delta H^\circ_{rxn} = -103.7$$

How will each of the following changes affect the indicated quantities? Write *increase, decrease,* or *no change.*

Change	*[Br₂]*	*[HBr]*	K
Some H_2 is added to the container.	____	____	____
The temperature of the gases in the container is increased.	____	____	____
The pressure of HBr is increased.	____	____	____

53. The equilibrium constant for the following reaction is 0.16 at 25 °C, and the enthalpy change at standard conditions is +16.1 kJ.

$$2\,NOBr(g) \rightleftharpoons 2\,NO(g) + Br_2(\ell)$$

Predict the effect of each of the following changes on the position of the equilibrium; that is, state which way the equilibrium will shift (left, right, or no change) when each of the following changes is made: (a) adding more Br_2; (b) removing some NOBr(g); (c) decreasing the temperature.

54. The formation of hydrogen sulfide from the elements is exothermic.

$$H_2(g) + \tfrac{1}{8}\,S_8(s) \rightleftharpoons H_2S(g) \qquad \Delta H^\circ_{rxn} = -20.6 \text{ kJ}$$

Predict the effect of each of the following changes on the position of the equilibrium; that is, state which way the equilibrium will shift (left, right, or no change) when each of the following changes is made: (a) adding more sulfur; (b) adding more $H_2(g)$; (c) raising the temperature.

55. The oxidation of NO to NO_2 [$2\ NO(g) + O_2(g) \rightleftharpoons 2\ NO_2(g)$] is exothermic. Predict the effect of each of the following changes on the position of the equilibrium; that is, state which way the equilibrium will shift (left, right, or no change) when each of the following changes is made: (a) adding more $O_2(g)$; (b) adding more $NO_2(g)$; (c) lowering the temperature.

56. Consider the following equilibrium:

$$PbCl_2(s) \rightleftharpoons Pb^{2+}(aq) + 2\ Cl^-(aq)$$

What will happen to the equilibrium concentration of lead(II) ion if some solid NaCl is added to the flask? (a) It will increase. (b) It will decrease. (c) It will not change. (d) You cannot tell with the information provided.

57. Phosphorus pentachloride is in equilibrium with phosphorus trichloride and chlorine.

$$PCl_5(g) \rightleftharpoons PCl_3(g) + Cl_2(g)$$

What will happen to the concentration of Cl_2 if additional $PCl_5(g)$ is added to the flask? (a) It will increase. (b) It will decrease. (c) It will not change. (d) You cannot tell with the information provided.

General Questions

58. Nitrogen monoxide can be reduced with hydrogen.

$$2\ H_2(g) + 2\ NO(g) \longrightarrow 2\ H_2O(g) + N_2(g)$$

Experiment shows that when the concentration of H_2 is halved, the reaction rate is halved. Furthermore, raising the concentration of NO by a factor of three raises the rate by a factor of nine. Write the rate equation for this reaction.

59. One reaction that may occur in air polluted with nitrogen monoxide is

$$2\ NO(g) + O_2(g) \longrightarrow 2\ NO_2(g)$$

Using the data in the table, answer the questions that follow.

Experiment	*Initial Concentration (M)* [NO]	*Initial Concentration (M)* [O_2]	*Initial Rate of Formation of NO_2 (mol/L·s)*
1	0.001	0.001	7×10^{-6}
2	0.001	0.002	14×10^{-6}
3	0.001	0.003	21×10^{-6}
4	0.002	0.003	84×10^{-6}
5	0.003	0.003	189×10^{-6}

(a) What is the order of reaction with respect to each reactant?
(b) Write the rate expression for the reaction.
(c) Calculate the rate of formation of NO_2 when $[NO] = [O_2] = 0.005$ mol/L.

60. Solid barium sulfate is in equilibrium with barium ions and sulfate ions in solution.

$$BaSO_4(s) \rightleftharpoons Ba^{2+}(aq) + SO_4^{2-}(aq)$$

What will happen to the barium ion concentration if more solid $BaSO_4$ is added to the flask? (a) It will increase. (b) It will decrease. (c) It will not change. (d) You cannot tell with the information provided.

61. Consider the following equilibrium:

$$N_2O_4(g) \rightleftharpoons 2\ NO_2(g) \qquad \Delta H^\circ_{rxn} = +57.2\ kJ$$

What will happen to the concentration of N_2O_4 if the temperature is increased? (a) It will increase. (b) It will decrease. (c) It will not change. (d) You cannot tell with the information provided.

62. Assume you place 0.010 mol of $N_2O_4(g)$ in a 2.0-L flask at 50 °C. After the system reaches equilibrium, $[N_2O_4] = 0.00090$ M. What is the value of K for the following reaction?

$$N_2O_4(g) \rightleftharpoons 2\ NO_2(g)$$

63. Cyclohexane, C_6H_{12}, a hydrocarbon, can isomerize or change into methylcyclopentane, a compound with the same formula but with a different molecular structure.

$$\underset{\text{cyclohexane}}{C_6H_{12}(g)} \rightleftharpoons \underset{\text{methylcyclopentane}}{C_5H_9CH_3(g)}$$

The equilibrium constant has been estimated to be 0.12 at 25 °C. If you had originally placed 3.79 g of cyclohexane in a 2.80-L flask, how much cyclohexane (in grams) is present when equilibrium is established?

64. The hydrocarbon C_4H_{10} can exist in two forms, butane and isobutane. The value of K for the interconversion of the two forms is 2.5 at 25 °C.

$$\underset{\text{butane}}{CH_3{-}CH_2{-}CH_2{-}CH_3} \rightleftharpoons \underset{\text{isobutane}}{H{-}C(CH_3)_2{-}CH_3}$$

If you place 0.017 mol of butane in a 0.50-L flask at 25 °C, what will be the equilibrium concentrations of the two forms of butane?

65. Nitrosyl chloride, NOCl, decomposes to NO and Cl_2 at high temperatures.

$$2\,NOCl(g) \rightleftharpoons 2\,NO(g) + Cl_2(g)$$

Suppose you place 2.00 mol of NOCl in a 1.00-L flask and raise the temperature to 462 °C. When equilibrium has been established, 0.66 mol of NO is present. Calculate the equilibrium constant for the decomposition reaction from these data.

66. Consider the transformation of butane into isobutane, a reaction with an equilibrium constant of $K = 2.5$ at 25 °C (Study Question 64). The system is originally at equilibrium with [butane] = 1.0 M and [isobutane] = 2.5 M.
   (a) Some isobutane (0.50 mol/L) is suddenly added to the equilibrium system, and the system shifts to a new equilibrium position. What is the new equilibrium concentration of each gas?
   (b) If 0.50 mol/L of butane is added to the original mixture, and the system shifts to a new equilibrium position, what is the new equilibrium concentration of each gas?

67. The deep blue compound CrO_5 can be made from the chromate ion by using hydrogen peroxide in an acidic solution.

$$HCrO_4^-(aq) + 2\,H_2O_2(aq) + H_3O^+(aq) \longrightarrow CrO_5(aq) + 4\,H_2O(\ell)$$

The kinetics of this reaction have been studied, and the rate equation is found to be

$$\text{rate of disappearance of } HCrO_4^- = k[HCrO_4^-][H_2O_2][H_3O^+]$$

One of the mechanisms suggested for the reaction is

$$HCrO_4^- + H_3O^+ \rightleftharpoons H_2CrO_4 + H_2O$$
$$H_2CrO_4 + H_2O_2 \longrightarrow H_2CrO_5 + H_2O$$
$$H_2CrO_5 + H_2O_2 \longrightarrow CrO_5 + 2\,H_2O$$

   (a) Give the order of the reaction with respect to each reactant.
   (b) Show that the steps of the mechanism agree with the overall equation for the reaction.

Summary Problem

68. The chemistry of compounds composed of a transition metal and carbon monoxide has been an interesting area of research for the past 30 years. For example, the replacement of CO by another molecule in $Ni(CO)_4$ (in the nonaqueous solvents toluene and hexane) was studied some years ago to understand the general principles that govern the chemistry of such compounds. (See J. P. Day, F. Basolo, and R. G. Pearson, Journal of the American Chemical Society, vol. 90, p. 6927, 1968.)

$$Ni(CO)_4 + L \longrightarrow Ni(CO)_3L + CO$$

In this case the molecule L is a compound such as $P(CH_3)_3$. A detailed study of the kinetics of the reaction led to the following mechanism:

Slow $\quad Ni(CO)_4 \rightleftharpoons Ni(CO)_3 + CO$

Fast $\quad Ni(CO)_3 + L \longrightarrow Ni(CO)_3L$

   (a) Tell whether each step in the mechanism is uni- or bimolecular.
   (b) When the steps of the mechanism are added together, show that the result is the balanced equation for the observed reaction.
   (c) Is there an intermediate in this reaction? If so, what is its identity?
   (d) It was found that doubling the concentration of $Ni(CO)_4$ led to an increase in reaction rate by a factor of 2. Doubling the concentration of L had no effect on the reaction rate. Based on this information, write the rate expression for the reaction.
   (e) $Ni(CO)_4$ is formed by the reaction of nickel metal with carbon monoxide. If you have 2.05 g of CO, and you combine it with 0.125 g of nickel metal, how much $Ni(CO)_4$ (in grams) can be formed?
   (f) An excellent way to make pure nickel metal is to decompose $Ni(CO)_4$ in a vacuum at a temperature slightly higher than room temperature. What is the enthalpy change for the decomposition reaction

$$Ni(CO)_4(g) \longrightarrow Ni(s) + 4\,CO(g)$$

   if the molar enthalpy of formation of $Ni(CO)_4$ gas is −602.91 kJ/mol?

Using Computer Programs and Videodiscs

69. Oscillating reactions are among the most interesting kinds of chemical changes. The videodisc of *Chemical Demonstrations* contains several such demonstrations on Side 2 of Disc 2. See The "Briggs-Rauscher Reaction" beginning at frame 40239. The basic chemistry of the reaction is the decomposition of hydrogen peroxide to give water and oxygen (as described in Section 8.1). Hydrogen peroxide can function both as an oxi-

dizing agent and a reducing agent. One reaction involves reduction of iodate ion to iodine (with accompanying oxidation of the peroxide to O_2 gas). The yellow color you observe is from iodine, and the black color comes from a complex of iodine, iodide ion, and starch. (Iodine is formed in the first reaction below, starch is added to the mixture, and I^- is formed in a reaction other than those written below.)

$$5\,H_2O_2(aq) + 2\,IO_3^-(aq) + 2\,H_3O^+(aq) \longrightarrow I_2(aq) + 5\,O_2(g) + 8\,H_2O(\ell)$$

Another reaction is the oxidation of iodine to iodate ion by hydrogen peroxide.

$$5\,H_2O_2(aq) + I_2(aq) \longrightarrow 2\,IO_3^-(aq) + 2\,H_3O^+(aq) + 2\,H_2O(\ell)$$

The net result is the decomposition of hydrogen peroxide to oxygen and water.

$$2\,H_2O_2(aq) \longrightarrow O_2(g) + 2\,H_2O(\ell)$$

(a) Since the iodate ion is not shown in the net reaction, what is its role in this oscillating chemical system?

(b) The system is complicated, and reactions other than those above occur. Nonetheless, can you use the equations above to explain why the system color oscillates?

STM image of copper clusters on a silica surface.

(X. Xu, S.M. Vesecky, D.W. Goodman, Science, *Vol. 58, p. 788, 1992)*

Electron Configurations, Periodicity, and Properties of Elements

CHAPTER

9

The periodic table (Section 3.6) was created by Mendeleev to summarize experimental observations. He had no submicroscopic theory or model to explain why all alkaline earths, for example, combined with oxygen in a 1:1 atom ratio—they just did. In the early years of this century, however, it became evident that atoms contain electrons, and explanations of periodic trends in physical and chemical properties began to be based on the arrangement of electrons within atoms—on what we now call electron configurations. Experiments involving interaction of light with atoms and molecules revealed that electrons in atoms are arranged roughly in concentric shells, like the layers in an onion. Electrons in the outermost shell are called valence electrons; their number and location are the chief factors that determine chemical reactivity. Atomic shell structure and valence electrons are closely related to the periodic table and can be derived from it. In this chapter the relationship of electron configurations of atoms to their properties will be described. Special emphasis is placed on the use of the periodic table to derive representations for valence electrons, since these are the electrons that take part in chemical reactions and form chemical bonds.

Elements that exhibit similar properties are found in the same column of the periodic table. But what is the fundamental reason for these similarities? The discovery of the electron, the proton, and the nuclear atom (Section 3.2) prompted scientists to look for relationships between atomic structure and chemical behavior. As early as 1902, Gilbert N. Lewis (1875–1946) hit upon the idea that electrons in atoms might be arranged in shells, starting close to the nucleus and building outward. Lewis explained the similarity of chemical properties for elements in a given group by assuming that all elements in that group have the same number of electrons in their outer shell. These are the **valence electrons,** introduced in Section 4.3. But where are the electrons in an atom? Do electrons in an atom have different energies? These questions were the basis for many experimental and theoretical studies, beginning around 1900 and continuing to this day. Some of the answers that have been found are discussed in this chapter.

9.1 ELECTROMAGNETIC RADIATION

Theories about the energy and arrangement of electrons in atoms are based on experimental studies of the interaction of electromagnetic radiation with matter. Since our eyes see the spectrum of colors that make up visible light, we are most familiar with what happens when visible light, a small portion of the electromagnetic spectrum, interacts with matter. For example, the red glow of a neon sign comes from neon atoms excited by electricity, and fire-

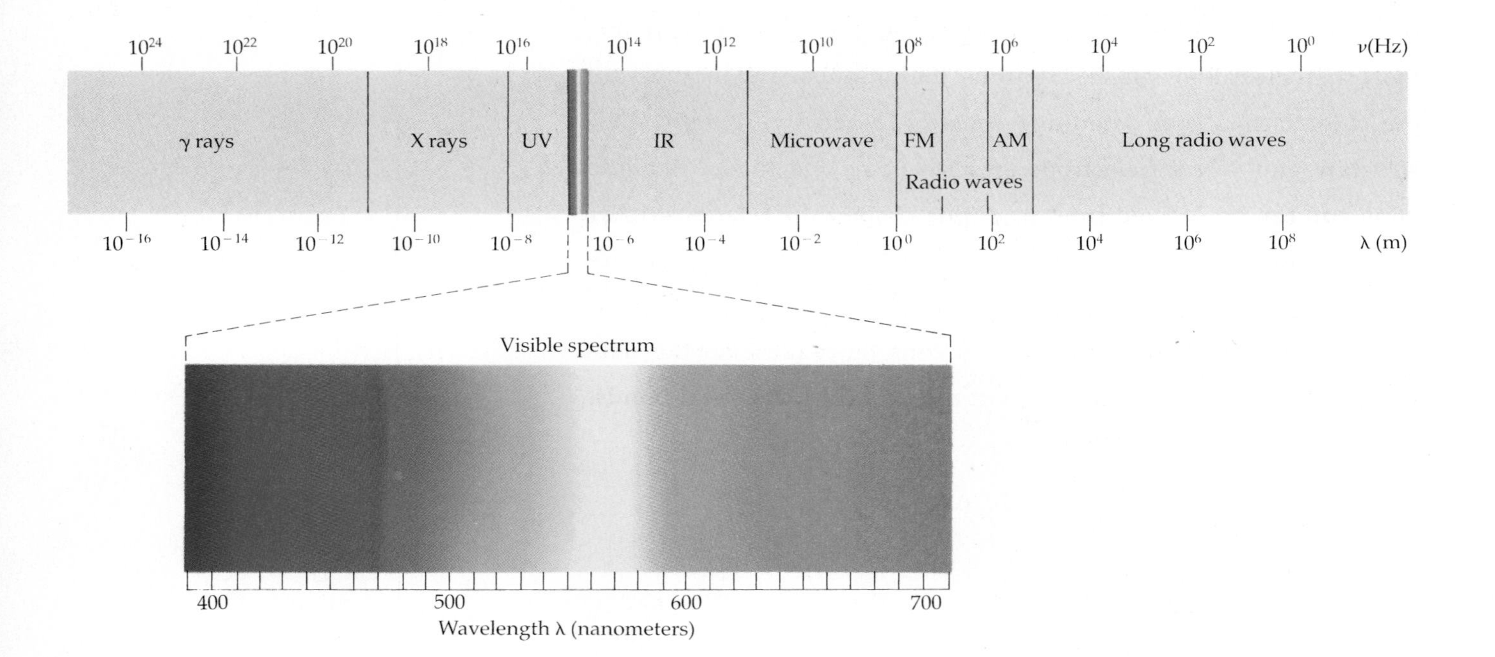

Figure 9.1
The electromagnetic spectrum. Visible light (enlarged section) is but a small part of the entire spectrum. The radiation's energy increases from the radio wave end of the spectrum (low frequency, ν, and long wavelength, λ) to the gamma ray end (high frequency and short wavelength).

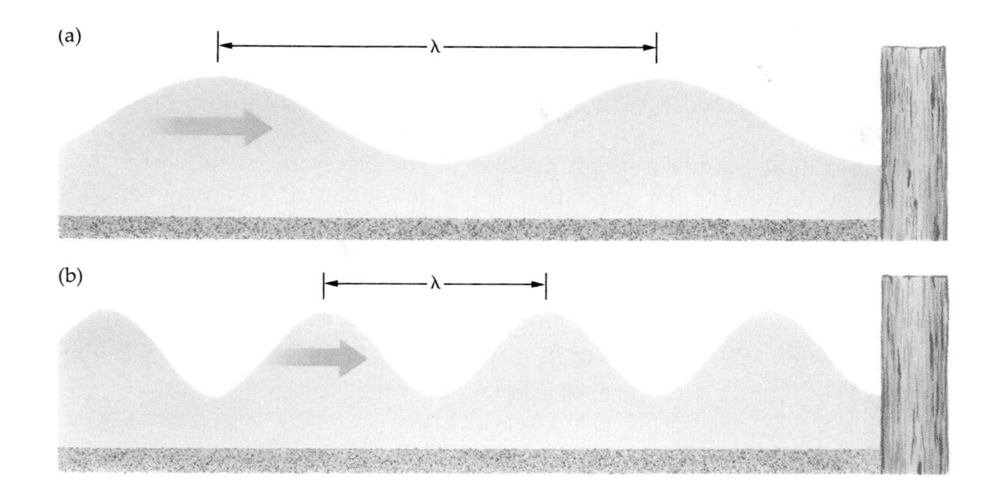

Figure 9.2
Illustrations of wavelength and frequency with water waves. The waves are moving toward the post. (a) The wave has a long wavelength (large λ) and low frequency (the number of times per second its peak hits the post). (b) The wave has a shorter wavelength and a higher frequency (it hits the post more often per unit of time).

works displays are the result of excited electrons of metal ions. Have you ever wondered how these colors are produced?

When the atoms of certain elements are "excited"—when we add energy to these atoms in any of a variety of ways—the electrons of the atoms absorb some of this energy and then return the absorbed energy to us in the form of electromagnetic radiation, some of which is in the visible light region. Electromagnetic radiation and its applications are familiar to all of us; sunlight, headlights on automobiles, dentist's x rays, microwave ovens, and radio and TV waves that we use for communications are a few examples (Figure 9.1). These kinds of radiation seem very different, but they are actually very similar in some properties. All **electromagnetic radiation** consists of oscillating electric and magnetic fields and travels through space at the same rate (the "speed of light": 186,000 miles/second or 2.998×10^8 m/s in a vacuum). Any of the various kinds of electromagnetic radiation can be described in terms of frequency (ν) and wavelength (λ). As illustrated in Figure 9.2, the **wavelength** is the distance between crests (or troughs) in a wave, and the frequency is the number of complete waves passing a point in a given amount of time (cycles per second). The **frequency** of electromagnetic radiation is related to its wavelength by

The velocity of light through a substance (air, glass, water, etc.) depends on the chemical constitution of the substance and the wavelength of the light. This is the basis for using a glass prism to disperse light and is the explanation for rainbows. The velocity of sound is also dependent on the material through which it passes.

$$\nu\lambda = c$$

where c is the speed of light. Figure 9.1 gives wavelength and frequency values for several regions of the electromagnetic spectrum.

A sample calculation will illustrate the wavelength-frequency relationship. If orange light has a wavelength of 625 nm, what is its frequency? Since the speed of light is in meters/second, the wavelength in nanometers must be converted to meters before substituting into the equation $\nu\lambda = c = 2.998 \times 10^8$ m/s.

$$625 \text{ nm} \cdot (1 \times 10^{-9} \frac{\text{m}}{\text{nm}}) = 6.25 \times 10^{-7} \text{ m}$$

$$\nu = \frac{c}{\lambda} = \frac{2.998 \times 10^8 \text{ m/s}}{6.25 \times 10^{-7} \text{ m}} = 4.80 \times 10^{14}/\text{s or } 4.80 \times 10^{14} \text{ hertz}$$

The hertz was named in honor of Heinrich Hertz (1857–1894), a German physicist.

Reciprocal units such as 1/s are often represented in the negative exponent form, s^{-1}. A frequency of $4.80 \times 10^{14}\ s^{-1}$ means that 4.80×10^{14} waves pass a fixed point every second. The unit s^{-1} is usually given the name **hertz (Hz)**.

Visible light is only a small portion of the electromagnetic spectrum. Ultraviolet radiation, the type that leads to sunburn, has wavelengths shorter than those of visible light; x rays and γ rays (the latter emitted in the process of radioactive disintegration of some atoms) have even shorter wavelengths. At longer wavelengths than visible light, we first encounter infrared radiation, the type that is sensed as heat from a fire. Longer still is the wavelength of the radiation in a microwave oven or in television and radio transmissions.

EXAMPLE 9.1

Wavelength-Frequency Conversions

Compact disc players use lasers that emit light at a wavelength of 785 nm. What is the frequency of this light in hertz?

SOLUTION

$$\nu = \frac{c}{\lambda} = \frac{2.998 \times 10^{8}\ \text{m/s}}{7.85 \times 10^{-7}\ \text{m}} \cdot \frac{1\ \text{Hz}}{1\ \text{s}^{-1}} = 3.82 \times 10^{14}\ \text{Hz}$$

EXERCISE 9.1 • ***Wavelength-Frequency Conversions***

If your favorite FM radio station broadcasts at a frequency of 104.5 MHz (where 1 MHz = 1 megahertz = $10^{6}\ s^{-1}$), what is the wavelength (in meters) of the radiation emitted by this station?

EXERCISE 9.2 • ***Frequency and Wavelength***

Does the frequency of light increase or decrease as its wavelength increases?

Planck's Quantum Theory

Have you ever sat near an electric resistance heater as it warms up? Of course, you cannot see the metal atoms in the heater's wire, but as electric energy flows through the wire, the atoms gain some of this energy and then emit it as radiation. First the wire emits a slight amount of heat that you can feel (infrared radiation). As the wire heats more, it begins to glow (emit visible light); first red light, and then orange, is emitted. If the wire gets very hot, it appears almost white.

At the close of the 19th century, scientists were trying to explain the nature of these emissions from hot objects. They assumed that vibrating atoms in a hot wire caused electromagnetic vibrations (light waves) to be emitted, and that those light waves could have any frequency along a continuously varying range. The classical wave theory predicted that as the object got hotter and acquired more energy, its color should shift to the blue

Portrait of a Scientist

Max Planck (1858–1947)

Max Karl Ernst Ludwig Planck was raised in Munich, Germany, where his father was a distinguished professor at the University. When still in his teens Planck decided to become a physicist, in spite of the advice of the head of the physics department at Munich that "The important discoveries (in physics) have been made. It is hardly worth entering physics anymore." Fortunately, Planck did not take this advice, and he worked for a while at the University of Munich before going to Berlin to study thermodynamics. His deep interest in thermodynamics led him eventually to his revolutionary hypothesis. The discovery was announced two weeks before Christmas in 1900, and he was awarded the Nobel Prize in 1918. Einstein later said it was a longing to find harmony and order in nature, a "hunger in his soul," that spurred Planck on.

Max Planck. (Lande Collection/AIP Emilio Segre Visual Archives)

and finally all the way to the violet and beyond, but no object was ever observed to do this.

In 1900 Max Planck (1858–1947) offered an explanation for the spectrum of a heated body that contained the seeds of a revolution in scientific thought. (In this context, "spectrum" means a plot of the intensity of emitted radiation as a function of its wavelength.) He made what was at that time an incredible assumption: when a hot object emits radiation, there is a minimum quantity of energy that can be emitted at any given time. That is, there must be a small packet of energy such that no smaller quantity can be emitted, just as an atom is the smallest packet of an element. Planck called his "atom" of energy a **quantum.** He further asserted that the energy of a quantum is related to the frequency of the radiation by the equation

$$E_{\text{quantum}} = h\nu_{\text{radiation}}$$

The proportionality constant h is called **Planck's constant** in his honor; it relates the frequency of radiation to its energy and has the value 6.626×10^{-34} J·s.

Earlier in this section we calculated the frequency of orange light to be $4.8 \times 10^{14}\ \text{s}^{-1}$. The energy of one quantum of orange light is therefore

$$E = h\nu = (6.626 \times 10^{-34}\ \text{J·s})(4.80 \times 10^{14}\ \text{s}^{-1}) = 3.18 \times 10^{-19}\ \text{J}$$

The theory based on Planck's work is called the **quantum theory.** Using his quantum theory, Planck was able to calculate the number of quanta of each frequency that would be emitted by a hot object. The number of quanta per second gives the intensity of the radiation, and since the frequency is related to the wavelength, Planck was able to calculate the spectrum of a hot object. His results agreed very well with the experimentally measured spectrum.

When a theory can accurately predict experimental results, the theory is usually regarded as useful. At first, however, Planck's quantum theory was not widely accepted because of its radical assumption that energy is quantized. But after Planck's quanta were used by Albert Einstein to explain another phenomenon called the **photoelectric effect,** the quantum theory of electromagnetic energy was firmly accepted.

In the early 1900s it was known that certain metals exhibit a photoelectric effect, that is, they emit electrons when illuminated by light of certain wavelengths. For each metal there is a minimum frequency. For example, the metal cesium (Cs) will emit electrons when illuminated by red light, whereas some metals require yellow light and others require ultraviolet light. Figure 9.3 shows how an electric current suddenly increases when light of a frequency above a certain value shines on a photosensitive metal. Einstein explained these observations by assuming that Planck's quanta were *massless* "particles" of light, now called **photons.** That is, light could be described as a stream of photons that had particle-like properties as well as wave-like properties. Removing one electron from a metal surface requires a certain quantity of energy. Since $E = h\nu$, only photons whose frequency is high enough will have enough energy to knock an electron loose. Photons with lower frequencies (left side of Figure 9.3b) do not have enough energy, and no photoelectrons appear. The current produced by a photocell depends on how many electrons are ejected, which in turn depends on the number of photons of sufficient energy striking the surface. The more in-

Cesium is often used in "electric eye" devices such as automatic door openers because all visible wavelengths of light cause it to emit electrons.

Example of an automatic door operated by an electric eye device. *(C.D. Winters)*

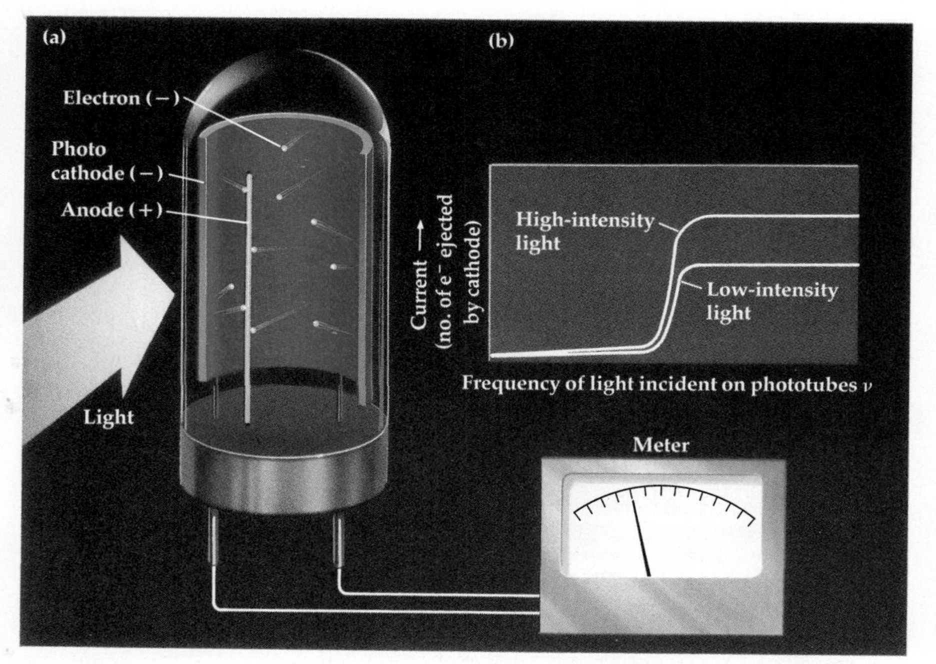

Figure 9.3

The photoelectric effect. (a) A photocell's metallic cathode is struck by photons. Light with any frequency above a certain threshold causes electrons to be ejected. These are attracted to the anode, causing an electric current to flow through the cell. (b) As the frequency of light striking the metal cathode is increased, some frequency is reached at which electrons begin to be ejected. Once current is flowing, the number of electrons depends on the light's intensity, not its frequency.

A Deeper Look

Wave-Particle Dual Nature of Light

Such developments as Einstein's explanation of the photoelectric effect led eventually to acceptance of what is referred to as the dual nature of light. Depending on the experimental circumstances, radiation appears to have either "wave" or "particle" (photon) behavior. For example, classical wave theory fails to explain the photoelectric effect but it does explain the diffraction of light by a prism or diffraction grating. However, it is important to realize that this "dual nature" description arises because of our attempts to explain observations by inadequate models. Light is not changing back and forth from being a wave to being a particle but has a single consistent nature that can be described by modern quantum theory. The dual nature description arises when we try to explain our observations by using our familiarity with classical models for "wave" or "particle" behavior.

Although this view of the nature of light may be disconcerting, it did lead to a revolutionary idea—namely, if light can be viewed in terms of both "wave" and "particle" properties, why can't matter? This is exactly the question posed by Louis de Broglie that led to his hypothesis about the wave properties of matter (see Section 9.2).

---

tense the light, the greater the number of photons and the greater the photoelectron current. Albert Einstein received a Nobel Prize in 1921 for his explanation of the photoelectric effect.

EXAMPLE 9.2

Calculating Photon Energies

What is the energy associated with one photon of the laser light described in Example 9.1, which has a frequency of $3.82 \times 10^{14}\ s^{-1}$?

SOLUTION According to Planck's quantum theory, the energy and frequency of radiation are related by $E = h\nu$.

$$E = h\nu = (6.626 \times 10^{-34}\ J{\cdot}s)(3.82 \times 10^{14}\ s^{-1}) = 2.53 \times 10^{-19}\ J$$

EXERCISE 9.3 • *Energy of Electromagnetic Radiation*

Which has more energy, (a) one photon of microwave radiation or one photon of ultraviolet radiation, (b) one photon of blue light or one photon of green light?

EXERCISE 9.4 • *Calculating Photon Energies*

Compare the energy of one photon of orange light (3.18×10^{-19} J) with the energy of one photon of x radiation having a wavelength of 2.36 nm.

Line Spectra

The final piece of information that played a major role in the modern view of atomic structure is the observation of the properties of light emitted by atoms after they absorb extra energy. The spectrum of white light, such as that from the sun or an incandescent light bulb, is the rainbow display of

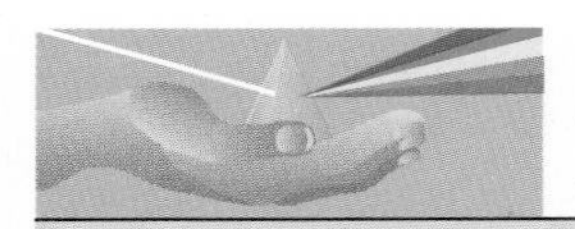

CHEMISTRY YOU CAN DO

CD As a Diffraction Grating

Seeing the visible spectrum by using a prism to refract visible light is a familiar experiment (Figure 9.4). Less familiar is the use of a diffraction grating for the same purpose. A diffraction grating consists of many equally spaced parallel lines—thousands of lines per centimeter. A grating that transmits diffracted light is made by cutting thousands of grooves on a piece of glass or clear plastic; a grating that reflects diffracted light is made by cutting the grooves on a piece of metal or opaque plastic. You can get an idea of how diffraction gratings work by using a compact disc as a diffraction grating. Compare the spectra of light from two different sources—a mercury vapor lamp and a white incandescent light bulb. Stand about 20 to 60 meters away from a mercury street lamp and hold the CD about waist high with the print side down. Tilt the CD down until you see the reflected image of the street lamp. Close one eye, and view along the line from the light source to your body as you slowly tilt the CD up toward you. What colors do you see? Use the same procedure with an incandescent light bulb. Do you see the same colors? Explain your results. (Source: R.C. Mebane and T.R. Rybolt, *Journal of Chemical Education,* Vol. 69, p. 401, 1992.)

separated colors shown in Figure 9.4. This rainbow spectrum, containing light of all wavelengths, is called a **continuous spectrum.**

If a high voltage is applied to an element in the gas phase at low pressure, the atoms absorb energy and are said to be "excited." The excited atoms emit light (Figure 9.5). An example of this is a neon advertising sign, in which excited neon atoms emit orange-red light. When light from such a source passes through a prism onto a white surface, only a few colored lines are seen. (Neon has an orange-red line, for example.) This is called a **line**

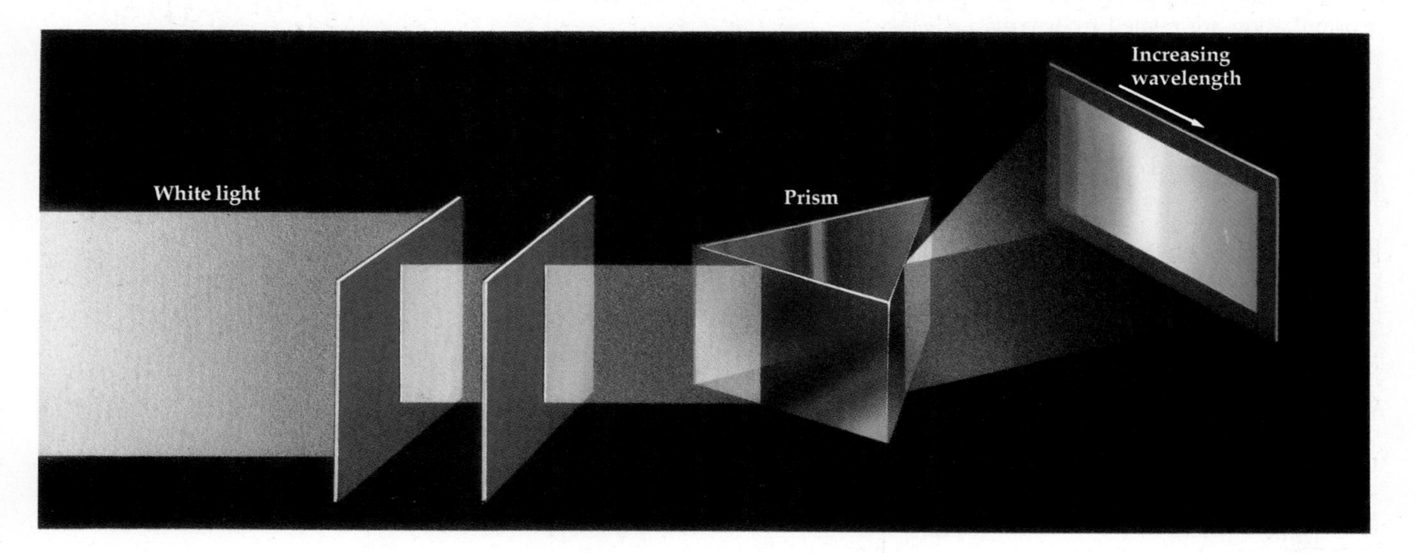

Figure 9.4

A spectrum from white light, produced by refraction in a glass prism. The various colors blend smoothly into one another.

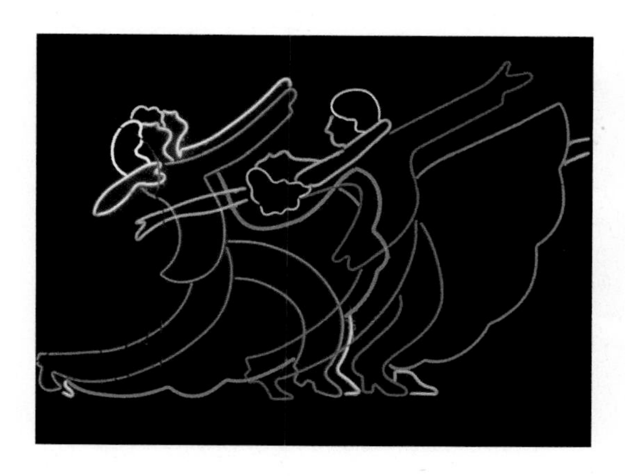

Figure 9.5
Any gas-discharge sign of this type is referred to as a neon sign even though many of them don't contain neon. Colors are emitted by excited atoms of rare gases—neon, reddish-orange; argon, blue; helium, yellowish-white. A helium-argon mixture emits an orange light while a neon-argon mixture gives a dark lavender light. Mercury vapor, used in fluorescent lights, is also used in gas mixtures to obtain a wide range of colors. By using these various gas mixtures and colored glass tubes, most of the colors in the visible spectrum can be produced. *(Steve Drexler/The Image Bank)*

emission spectrum (Figure 9.6). The line spectra of the visible light emitted by excited atoms of hydrogen, mercury, and neon are shown in Figure 9.7.

Every element has a unique line emission spectrum. The characteristic lines in the emission spectrum of an element can be used in chemical analysis, especially in metallurgy, both to identify the element and to determine how much of it is present.

Figure 9.6
A line emission spectrum of excited H atoms is measured by passing the emitted light through a series of slits to isolate a narrow beam of light, and this beam is then passed through a prism to separate the light into its component wavelengths. A photographic plate or other instrument detects the separate wavelengths as individual lines. Hence the name "line spectrum" for the light emitted by excited atoms or molecules.

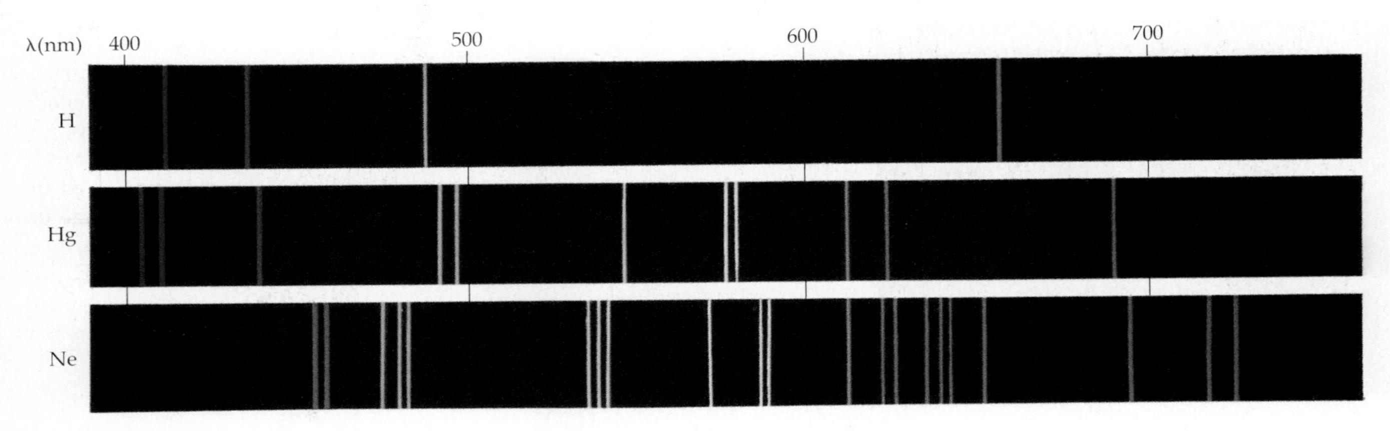

Figure 9.7
Line emission spectra of hydrogen, mercury, and neon. Excited gaseous elements produce characteristic spectra that can be used to identify the element as well as to determine how much of the element is present in a sample.

9.2 MODELS OF THE ATOM

Bohr Model of the Atom

In 1913 Niels Bohr provided the connection between the line emission spectrum of hydrogen and the quantum ideas of Planck and Einstein. Bohr introduced the notion that *the single electron of the hydrogen atom could occupy only certain energy levels.* He referred to these energy levels as orbits, and represented the energy difference between any two adjacent orbits as a single **quantum** of energy. The energy of the electron in an atom is said to be "quantized."

In Bohr's model, each allowed orbit is assigned an integer, n, known as the **principal quantum number.** The values of n for the orbits range from 1 to infinity. The radii of the circular orbits increase as n increases. The orbit of lowest energy, with $n = 1$, is closest to the nucleus, and the electron of the hydrogen atom is normally in this energy level. Any atom with its electrons in their normal, lowest energy levels is said to be in the **ground state.** Energy must be supplied to move the electron farther away from the nucleus because the positive nucleus and the negative electron attract each other. When the electron of a hydrogen atom occupies an orbit with n greater than 1, the atom has more energy than in its ground state and is said to be in an **excited state.** The excited state is an unstable state, and the extra energy is emitted when the electron returns to the ground state. Bohr introduced the assumption that the energy, $h\nu$, of the photon that is emitted corresponds to the *difference* between two energy levels of the atom. Think of the Bohr orbit model as a set of stairs in which the higher stair steps are closer together. Each step represents a quantized energy level; as you climb the stairs, you can stop on any step but not between steps.

Each step represents a quantized energy level.

According to Bohr, the light forming the lines in the emission spectrum of hydrogen (Figure 9.7) comes from an electron in the hydrogen atom moving from higher orbits to lower orbits closer to the nucleus (and eventually to the $n = 1$ orbit) after having first been excited to orbits with $n = 2, 3, 4,$ and higher that are farther from the nucleus (Figure 9.8). The emission spec-

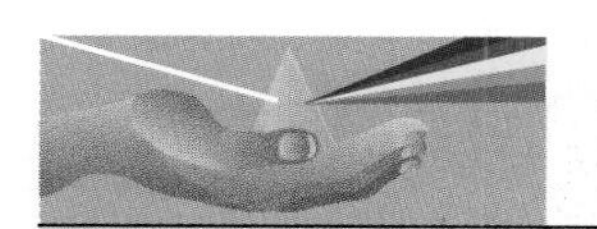

CHEMISTRY YOU CAN DO

Characteristic Flame Colors

Fireworks displays are the result of excited electrons of metal ions. If you have a fireplace, you can observe the different colors that metal ions give to flames. Take two small pieces of kindling or long *used* wooden matches. Soak one in a solution of table salt (NaCl) for about five minutes. Soak the other piece in a solution of calcium chloride ($CaCl_2$) for about five minutes. (Calcium chloride is available at farm supply stores and some hardware stores. It is used to salt sidewalks to remove ice.) Observe the color given off when each piece of wood is placed on the fire in a fireplace (or ignited with a match). What colors do you see? Explain your observations.

trum of hydrogen has several regions because the energy level differences vary from ultraviolet energy (electrons moving from energy levels with n greater than 1 to $n = 1$) to visible (from levels with n greater than 2 to $n = 2$) to infrared energy (from levels with n greater than 3 or 4 to $n = 3$ or 4).

From his model, Bohr was able to calculate the wavelengths of the lines in the hydrogen spectrum, some of which are shown in Table 9.1. Note the close agreement between the measured values and the values predicted by the calculations of the Bohr theory. Niels Bohr had tied the unseen (the interior of the atom) to the seen (the observable lines in the hydrogen spectrum)—a fantastic achievement!

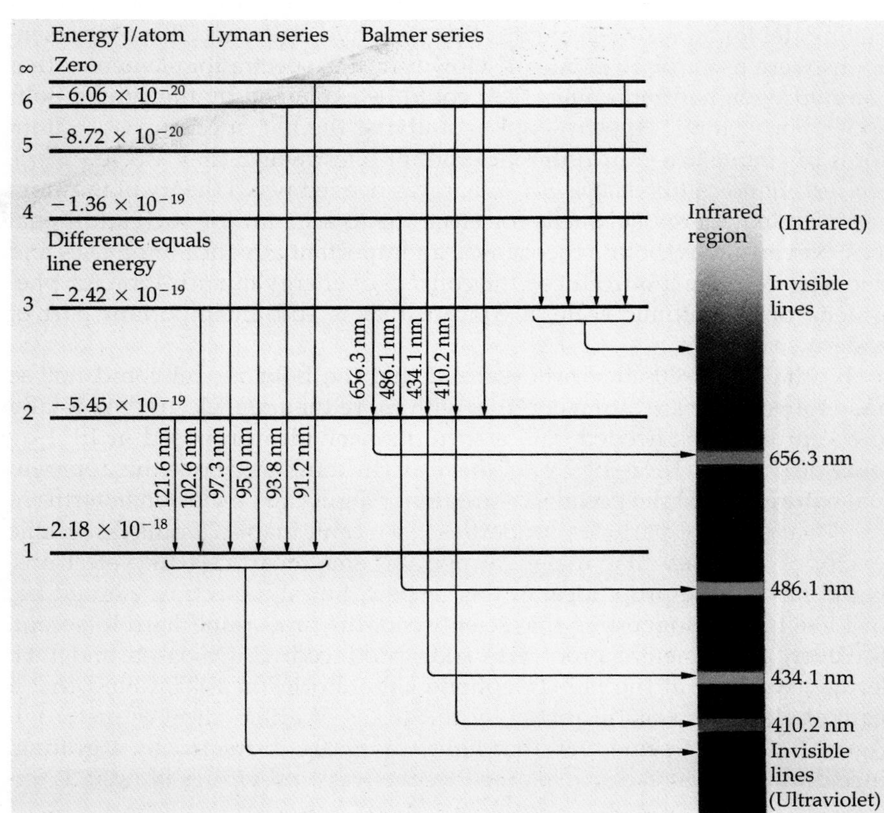

Figure 9.8
Some of the electronic transitions that can occur in an excited H atom. The lines in the ultraviolet region result from all transitions to the $n = 1$ level. Transitions from levels with values of n greater than 2 to the $n = 2$ level occur in the visible region. Lines in the infrared region result from transitions from levels with values of n greater than 3 or 4 to the $n = 3$ or 4 levels (only the series for transitions to the $n = 3$ level is shown).

Table 9.1

Agreement Between Bohr's Theory and the Lines of the Hydrogen Spectrum*

Changes in Energy Levels	Wavelength Predicted by Bohr's Theory (nm)	Wavelength Determined from Laboratory Measurement (nm)	Spectral Region
$2 \rightarrow 1$	121.6	121.7	ultraviolet
$3 \rightarrow 1$	102.6	102.6	ultraviolet
$4 \rightarrow 1$	97.28	97.32	ultraviolet
$3 \rightarrow 2$	656.6	656.3	visible red
$4 \rightarrow 2$	486.5	486.1	visible blue-green
$5 \rightarrow 2$	434.3	434.1	visible blue
$4 \rightarrow 3$	1876	1876	infrared

*These lines are typical; other lines could be cited as well, with equally good agreement between theory and experiment. The unit of wavelength is the nanometer (nm), 10^{-9} m.

The Bohr theory was accepted almost immediately after its presentation. Bohr's success with the hydrogen atom soon led to attempts both by him and by others to extend the same model to other atoms. For example, Lewis's model of shells for electrons gave the same representation as Bohr's orbit model for atoms of elements other than hydrogen, and these representations were used for all elements. However, line spectra for elements other than hydrogen had more lines than could be explained by the simple Bohr model. For example, spectroscopists studying the line spectrum of sodium atoms distinguished four different types of lines, which they labeled *sharp, principal, diffuse,* and *fundamental.* What was needed was a theory of the atom that included energy subshells for electrons to account for such additional lines. Nevertheless, Bohr's theory was an important advance in physics and chemistry because it introduced the concept of energy quantization for phenomena on the atomic scale, a concept that is still an important part of modern science.

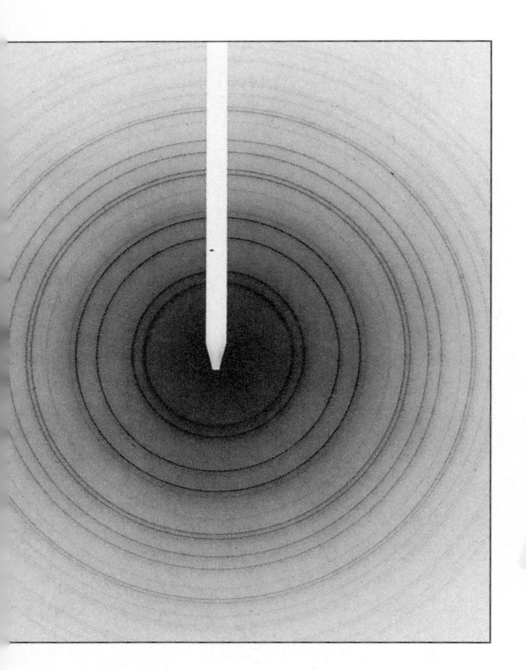

Figure 9.9
Electron diffraction pattern obtained for aluminum foil. *(Donald Potter, Department of Metallurgy, University of Connecticut)*

By the mid-1920s theorists realized that the Bohr model could not be made to work for any atoms or ions with more than one electron. A totally new approach was needed, and a revolutionary idea proposed in 1924 by Louis de Broglie (1892–1987) led the way. In thinking about the apparent dual nature of light, he posed the question: If light can be viewed in terms of both "wave" and "particle" properties, why can't matter? Louis de Broglie developed his idea that matter, especially small particles like electrons, should have wave properties. In this respect, he said, electrons should behave like light, a suggestion that scientists of the time found hard to accept. However, experimental proof was soon produced. C. Davisson and L.H. Germer, working at the Bell Telephone Laboratories in 1927, found that a beam of electrons was diffracted by the atoms of a thin sheet of metal foil (Figure 9.9) in the same way that light waves are diffracted by a grating. Since diffraction is readily explained by the wave properties of light, it fol-

Portrait of a Scientist

Niels Bohr (1885–1962)

Niels Bohr was born in Copenhagen, Denmark. He earned a Ph.D. in physics in Copenhagen in 1911 and then went to work first with J.J. Thomson in Cambridge, England, and later with Ernest Rutherford in Manchester, England.

It was in England that he began to develop the ideas that a few years later led to the publication of his theory of atomic structure and his explanation of atomic spectra. He received the Nobel Prize in 1922 for this work.

After working with Rutherford for a very short time, Bohr returned to Copenhagen, where he eventually became the director of the Institute of Theoretical Physics.

Many young physicists carried on their work in this Institute, and seven of them later received Nobel Prizes for their studies in chemistry and physics. Among them were such well-known scientists as Werner Heisenberg, Wolfgang Pauli, and Linus Pauling.

Bohr was a major figure in science in this century, and his life and work are described in the book *The Making of the Atomic Bomb* by R. Rhodes (Simon and Schuster, 1986).

Niels Bohr. (American Institute of Physics)

lowed that electrons also can be described by the equations of waves under some circumstances.

A few years after de Broglie's hypothesis about the wave nature of the electron, Werner Heisenberg (1901–1976) proposed the **uncertainty principle,** which states that *it is impossible to know simultaneously both the exact position and the exact momentum of an electron.* The uncertainty principle arises from an experimental limitation on observations of subatomic particles. This limitation is not a problem for a macroscopic object because the energy of photons used to locate such an object does not cause a measurable change in the position or momentum of that object. However, the very act of measurement would affect the position and momentum of the electron because of its very small size and mass. High energy photons would be required to locate the small electron, and when the photons collide with the electron, the momentum of the electron would be changed. If lower energy photons were used to avoid affecting the momentum, little information would be obtained about the location of the electron. Consider an analogy in photography. If you are taking a picture of a car race at a high shutter speed setting, you get a clear picture of the cars but you can't tell how fast they are going or even whether they are moving. With a slow shutter speed, you can tell from the blur of the car images something about the speed and direction, but you have less information about where the cars are.

The Heisenberg uncertainty principle illustrated another inadequacy in the Bohr model—its representation of the electron in the hydrogen atom in terms of well-defined orbits about the nucleus. In practical terms the best we can do is to represent the **probability** of finding an electron (of a given energy and momentum) within a given space.

A Deeper Look

Electron Microscopes

Louis de Broglie proposed that the characteristic wavelength of an electron (or any other particle) depends on its mass, m, and velocity, v:

$$\lambda = \frac{h}{mv}$$

(where h is Planck's constant)

Louis de Broglie (1892–1987). (Oesper Collection in the History of Chemistry/University of Cincinnati)

Since Planck's constant is very small (6.626×10^{-34} J·s), the product of m and v must be very small in order for the wavelength to be large enough to measure. For example, a 114-g baseball traveling at 110 mph has a large mv product (5.6 kg·m/s) and therefore the incredibly small wavelength of 1.2×10^{-34} km. Such a tiny value cannot be measured by any instrument now available, so the wave property of a moving baseball cannot be observed experimentally. However, a fast-moving electron gives a measurable wavelength. For example, the mass of an electron is 9.109×10^{-31} kg. If the electron has a velocity of 6.0×10^6 m/s, which is about 2% the speed of light, the wavelength of the electron is 0.12 nm, which is similar to the wavelength of x rays.

The experiments of C. Davisson and L.H. Germer demonstrated that a beam of electrons was diffracted like light waves by the atoms of a thin sheet of metal foil (Figure 9.9). This fact led the way for development of the electron microscope. An electron microscope uses the wave behavior of a stream of electrons in the same way that a conventional microscope uses the wave behavior of a beam of light. The ordinary microscope is limited in its magnification by the wavelength of light. The lower limit is about 500 nm for an optical microscope with a visible light source. However, objects as small as 0.1 nm can be observed if a beam of electrons is used because the wavelength of electrons traveling at high speed is much shorter than that of visible light.

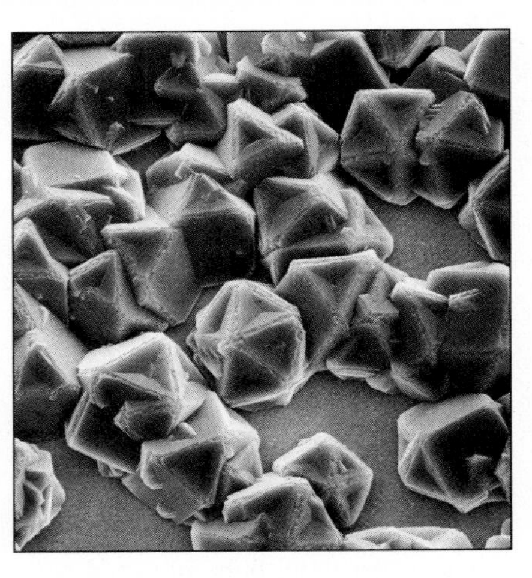

Electron microscope image of diamond thin film. (General Electric)

EXERCISE 9.5 • *Bohr Model*

How does the Bohr model explain the many lines in the line spectrum of hydrogen even though the hydrogen atom contains only one electron?

Quantum Mechanical Model of the Atom

In 1926 Erwin Schrödinger (1887–1961) combined de Broglie's hypothesis with classical equations for wave motion. From these and other ideas he derived a new equation called the **wave equation** to describe the behavior of an electron in the hydrogen atom, and this theoretical approach has come to be called **quantum mechanics.** Solutions to the wave equation predict the allowed energy states of the electron and the probability of finding that electron in a given region of space. These solutions are called **wave functions.**

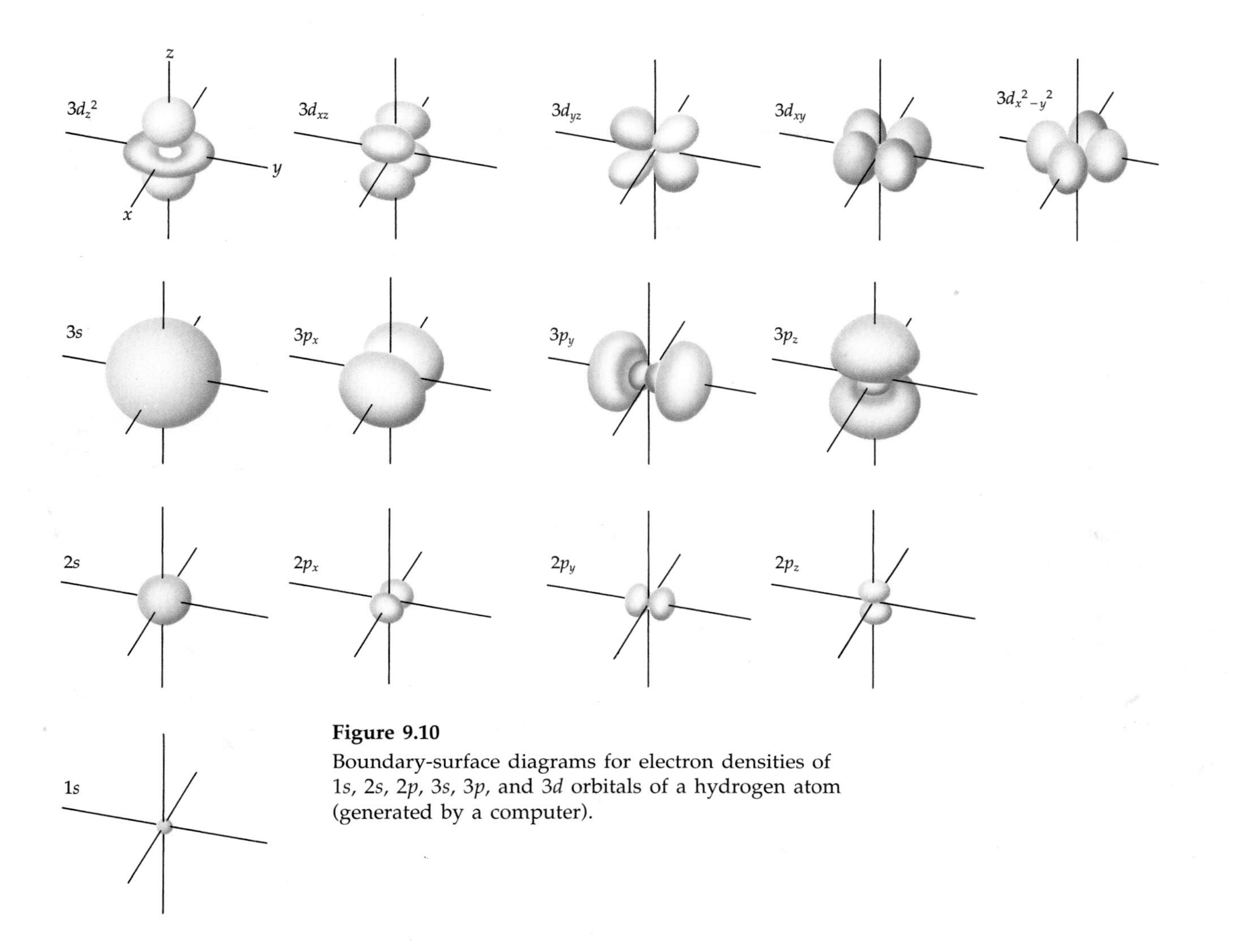

Figure 9.10
Boundary-surface diagrams for electron densities of 1*s*, 2*s*, 2*p*, 3*s*, 3*p*, and 3*d* orbitals of a hydrogen atom (generated by a computer).

Although the wave functions themselves are complex mathematical equations, it is possible to represent the square of the wave function in graphic form—as a picture of the region where an electron is most likely to be found. One way to make such a picture is to draw a surface within which there is a 90% probability that the electron will be found. That is, nine times out of ten an electron will be somewhere inside such a **boundary surface;** there is one chance in ten that the electron is outside.

A 100% probability isn't chosen because such a surface would have an indefinite boundary. Imagine a dart board without any bullseye. Think how hard it would be to draw the circle for the bullseye if you had to include all the holes left by the darts rather than 90% of them.

A series of such boundary-surface diagrams for the hydrogen atom is shown in Figure 9.10. These three-dimensional boundary surfaces are called **orbitals.** Note that an *orbital* (quantum mechanical model) is not the same as an *orbit* (Bohr model), but the allowed energy levels for the hydrogen atom are the same as those predicted by the Bohr model. In the quantum mechanical model, the principal quantum number, n, is a measure of the radius of the orbital or the most probable distance of the electron from the nucleus.

A collection of orbitals with the same value of the principal quantum number, n, is called an electron **shell.** For example, all the orbitals with

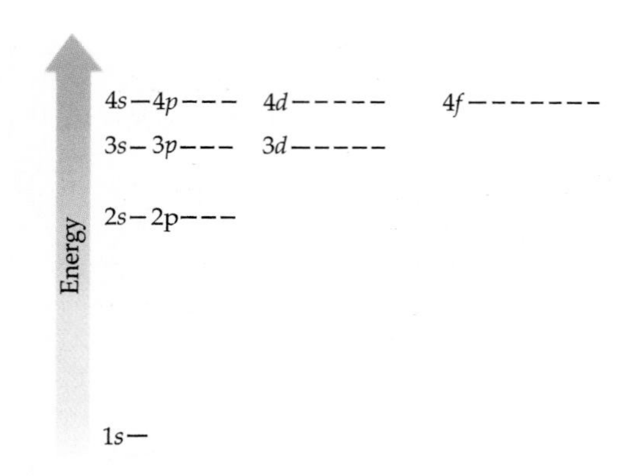

Figure 9.11

A representation of orbital energy levels in the hydrogen atom. Each short horizontal line represents one orbital. Orbitals with the same principal quantum number, n, have the same energy.

$n = 2$ are in the second shell. Each shell is divided into **subshells,** and the number of subshells is equal to the value of n for that shell; for example, the second shell has two subshells. Each subshell is designated by a number (the value of n) and one of the letters s, p, d, f. For example, the $n = 3$ shell has three subshells: $3s$, $3p$, and $3d$. By predicting energies of electron subshells, and thus the differences in energy that correspond to spectral lines, the Schrödinger wave equation provides a direct connection between quantum mechanical theory and experimental atomic spectra.

The letters s, p, d, f *were derived from the terms in spectroscopy (sharp, principal, diffuse, and fundamental) for the extra lines in emission spectra of elements other than hydrogen. They emphasize that atomic theory was developed in an attempt to understand and describe experimental atomic spectra.*

The subscripts on the labels of orbitals in Figure 9.10 distinguish between orbitals that have the same shape and are in the same subshell, but have different orientations in space. Any s subshell has one spherical s orbital; a p subshell has three p orbitals, one along each of the three axes, labelled p_x, p_y, and p_z. A d subshell contains five d orbitals, while an f subshell has seven f orbitals. A summary of the wave equation representations of energy levels for the hydrogen atom is given in Figure 9.11. Each line represents an orbital; orbitals of the same subshell, such as the $2p$, are grouped together. Notice that in a hydrogen atom in its ground state, all subshells with the same value of n have the same energy.

Now that it appears that we have an appropriate model for representing electron energies in atoms, we encounter another stumbling block. The

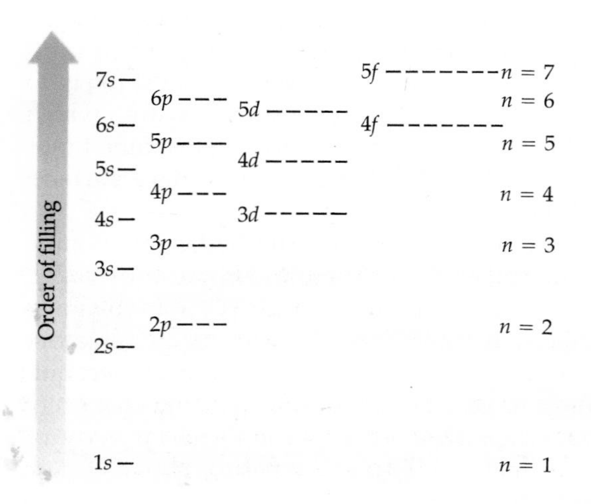

Figure 9.12

A representation of the order of filling of subshells for many-electron atoms.

Schrödinger equation cannot be solved exactly for any atom containing more than one electron! Fortunately, methods of approximation have been developed. With these methods and the Schrödinger equation, a computer can calculate atomic (or molecular) properties to a high degree of accuracy, and these results have led chemists and physicists to rely on quantum mechanical theory for understanding atomic structure.

In a many-electron atom, the electron-electron repulsions cause different subshells to have different energies, as shown in Figure 9.12. Before we use this set of energy levels for many-electron atoms, we need to consider one additional property of the electron—its spin.

Erwin Schrödinger was born in Vienna, Austria, and studied at the university there. Following service in World War I as an artillery officer, he became a professor of physics at several universities; in 1928 he succeeded Max Planck as professor of theoretical physics at the University of Berlin. He shared the Nobel Prize in physics (with Paul Dirac) in 1933. *(Oesper Collection in the History of Chemistry/University of Cincinnati)*

EXAMPLE 9.3

Drawing Atomic Orbitals

Draw pictures of the 90% boundary surfaces of the $2p_y$ and $3d_{xz}$ orbitals.

SOLUTION

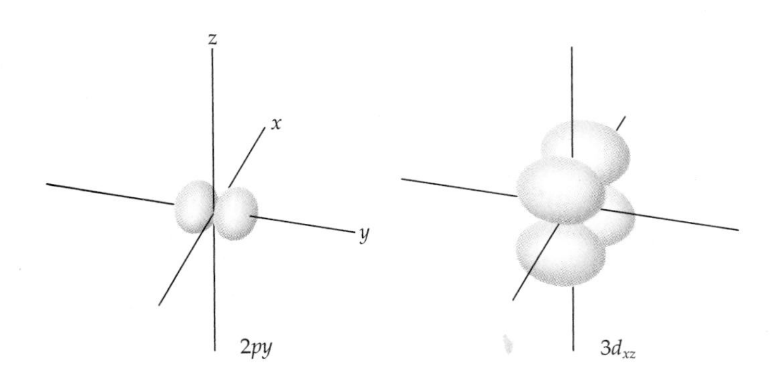

EXERCISE 9.6 • *Atomic Orbitals*

What is the maximum number of *s* orbitals that may be found in a given electron shell? The maximum number of *p* orbitals? Of *d* orbitals? Of *f* orbitals?

9.3 ELECTRON SPIN

When spectroscopists studied the emission spectra of hydrogen and sodium atoms in greater detail, they discovered that lines originally thought to be single were actually closely spaced pairs of lines. In 1925 the Dutch physicists George Uhlenbeck and Samuel Goudsmit proposed that the splitting could be explained by assuming that the energy of the spinning electron is quantized in a magnetic field. The magnetic properties of the electron can be understood if we think of it as a charged sphere rotating about an axis

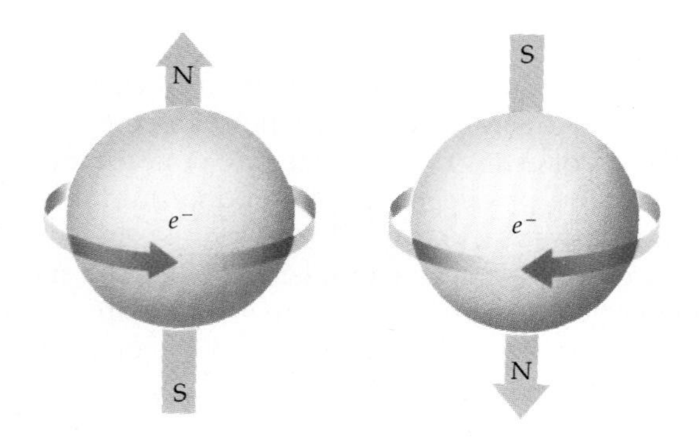

Figure 9.13
The electron behaves as though it were a charged sphere spinning about an axis through its center. The electron can have only two directions of spin; any other position is forbidden. Therefore, the spin of the electron is said to be quantized.

through its center (Figure 9.13). Since a spinning charge produces a magnetic field, the electron behaves like a tiny bar magnet with north and south magnetic poles. Uhlenbeck and Goudsmit proposed that the interaction of the electron's magnetic field with the external magnetic field causes the electron's energy to be quantized. Only two orientations of the spin are allowed with respect to the direction of the external field; these orientations are called clockwise and counterclockwise. The two opposite directions of spin produce oppositely directed magnetic fields, as shown in Figure 9.13. The interactions of the oppositely directed fields with the external magnetic field result in two slightly different energies, which in turn lead to the splitting of the spectral lines into closely spaced pairs.

The Pauli Exclusion Principle

Born in Vienna, Wolfgang Pauli received his Ph.D. in 1921 from the University of Munich. Following further study with Niels Bohr, he became a professor at the University of Zurich. Pauli received the Nobel Prize in physics in 1945. *(Oesper Collection in the History of Chemistry/University of Cincinnati)*

To make the quantum theory consistent with experiment, Wolfgang Pauli stated in 1925 what is now known as the **Pauli exclusion principle:** *at most two electrons can be assigned to the same orbital in the same atom, and these two electrons must have opposite spins.* When electrons of equal energy have opposite spin directions (which we represent by arrows pointing in opposite directions, ↓ ↑), they are said to be **paired** and their magnetic fields cancel. Two electrons spinning in the same direction (↑ ↑) are said to have **parallel** spins. According to the Pauli exclusion principle, two electrons with parallel spins must be in different orbitals. Also, since their magnetic fields do not cancel, an atom (or molecule) that contains one or more unpaired electrons will be attracted into a magnetic field.

The limitation that only two electrons can occupy a single orbital determines the maximum number of electrons for each principal quantum number, as summarized in Table 9.2. The $n = 1$ energy level has only one s orbital and therefore can accommodate only two electrons. The $n = 2$ energy level has one s orbital and three p orbitals, each of which can accommodate a pair of electrons; therefore, this level can accommodate a total of eight electrons (two in the s orbital and six in the p orbitals). For each principal energy level, the maximum number of electrons is $2n^2$.

EXERCISE 9.7 • *Pauli Exclusion Principle*

Calculate the total number of electrons in the $n = 3$ level and identify the orbital of each electron.

Table **9.2**

Number of Electrons Accommodated in Electron Shells and Subshells

Electron Shell (n)	Subshells Available ($=n$)	Orbitals Available	Number of Electrons Possible in Subshell	Maximum Electrons Possible for nth Shell ($2n^2$)
1	*s*	1	2	2
2	*s*	1	2	8
	p	3	6	
3	*s*	1	2	18
	p	3	6	
	d	5	10	
4	*s*	1	2	32
	p	3	6	
	d	5	10	
	f	7	14	
5	*s*	1	2	50
	p	3	6	
	d	5	10	
	f	7	14	
	*g**	9	18	
6	*s*	1	2	72
	p	3	6	
	d	5	10	
	*f**	7	14	
	*g**	9	18	
	*h**	11	22	
7	*s*	1	2	

*These orbitals are not used in the ground state of any known element.

The results expressed in this table were predicted by the Schrödinger theory and have been confirmed by experiment.

9.4 ATOM ELECTRON CONFIGURATIONS

The complete description of the orbitals occupied by all the electrons in an atom or ion is called its **electron configuration.** Now it is possible to make some sense of the electron configurations of atoms of the elements, and we shall largely use the periodic table as our guide to these configurations. As suspected by Lewis, the similarities of elements in the same periodic table groups are explained by their similar electron configurations.

The following points should be kept in mind as electron configurations are written:

1. Orbitals with the same value of the principal quantum number, n, occupy a principal energy level called a **shell.** There are n^2 orbitals in a shell.
2. Each shell is divided into a number of **subshells** equal to the principal quantum number, n, for that shell. For example, the $n = 3$ shell has three subshells, $3s$, $3p$, and $3d$.
3. Each subshell is divided into orbitals (s subshell, 1 orbital; p subshell, 3 orbitals; d subshell, 5 orbitals; f subshell, 7 orbitals). For example, the $n = 2$ subshell has one $2s$ orbital and three $2p$ orbitals, for a total of 4 as given by n^2.
4. Because each orbital is limited to occupancy by two electrons of opposite spin, the maximum number of electrons in a shell is $2n^2$.
5. Electrons pair up only after each orbital in a subshell is occupied by a single electron (this is Hund's rule, explained in the following subsection).

Electron Configurations of Main Group Elements

The main group elements are those in the A groups in the periodic table (see inside front cover).

The atomic numbers of the elements increase in numerical order across the periodic table. As a result, atoms of each element contain one more electron than atoms of the element to the left in the table. How do we know which shell and orbital each new electron occupies? In addition to the guidelines summarized above, there is one important additional principle: *For atoms in their ground states, electrons are found in the lowest energy shells, subshells, and orbitals available to them.* In other words, electrons fill orbitals in pairs starting with the 1*s* orbital at the bottom of Figure 9.12 and working upward.

To understand what this means, consider the experimentally determined electron configurations of the first ten elements, which are written in three different ways in Table 9.3. Since electrons assigned to the $n = 1$ shell are closest to the nucleus and therefore lowest in energy, electrons are assigned to it first (H and He). At the left in Table 9.3, the occupied orbitals and the number of electrons in each orbital are represented in the following notation:

H $1s^1$ ← one electron in s subshell

principal quantum number, n → 1; subshell (s) → s

At the right in the table, each occupied orbital is represented by a box in which electrons are shown as arrows that point in opposite directions for paired electrons.

Although chemists often say that electrons "occupy an orbital" or are "placed in an orbital," orbitals are not literally things or boxes in which electrons are placed. An orbital is an electron matter-wave. It is more correct to say an electron is "assigned to an orbital."

In helium the two electrons are paired in the 1*s* orbital so that the lowest energy shell is filled. After the $n = 1$ shell is filled, electrons are found in the next highest energy level, the $n = 2$ shell, beginning with lithium. This second shell can hold eight electrons, and one by one its orbitals are filled in the eight elements from lithium to neon. Notice in the periodic table inside the front cover that these are the elements of the second period.

Table 9.3

Electron Configurations of the First Ten Elements

	Electron Configurations (Condensed)	(Expanded)	Orbital Box Diagrams 1s	2s	$2p_x$	$2p_y$	$2p_z$
H	$1s^1$		↑				
He	$1s^2$		↑↓				
Li	$1s^22s^1$		↑↓	↑			
Be	$1s^22s^2$		↑↓	↑↓			
B	$1s^22s^22p^1$		↑↓	↑↓	↑		
C	$1s^22s^22p^2$	$1s^22s^22p_x{}^12p_y{}^1$	↑↓	↑↓	↑	↑	
N	$1s^22s^22p^3$	$1s^22s^22p_x{}^12p_y{}^12p_z{}^1$	↑↓	↑↓	↑	↑	↑
O	$1s^22s^22p^4$	$1s^22s^22p_x{}^22p_y{}^12p_z{}^1$	↑↓	↑↓	↑↓	↑	↑
F	$1s^22s^22p^5$	$1s^22s^22p_x{}^22p_y{}^22p_z{}^1$	↑↓	↑↓	↑↓	↑↓	↑
Ne	$1s^22s^22p^6$	$1s^22s^22p_x{}^22p_y{}^22p_z{}^2$	↑↓	↑↓	↑↓	↑↓	↑↓

As happens in each principal energy level (and each period), the first two electrons fill the *s* orbital, giving Li $1s^22s^1$ and Be $1s^22s^2$. Since the *p* subshell has three *p* orbitals, we need to examine where electrons are found as these orbitals fill up. Carbon, with an electron configuration of $1s^22s^22p^2$, is the first element for which there is a choice for placement of the second electron in a *p* subshell. Does this electron pair with the existing electron in a *p* orbital or does it occupy another *p* orbital? It has been shown experimentally that both *p* electrons have the same spin. Hence, they occupy different *p* orbitals (by the Pauli exclusion principle). The expanded electron configurations in the middle of Table 9.3 show the locations of electrons in the three *p* orbitals (p_x, p_y, p_z). Since electrons are negatively charged particles, this arrangement minimizes electron-electron repulsions, making the total energy of the set of electrons as low as possible. **Hund's rule** summarizes how subshells are filled: the most stable arrangement of electrons in the same subshell is that with the maximum number of unpaired electrons, all with the same spin direction. The general result of Hund's rule is that in *p*, *d*, or *f* orbitals each successive electron enters a different orbital of the subshell until the subshell is half-full, after which electrons pair in the orbitals one by one.

Suppose you need to know the electron configuration of phosphorus, which means identifying the locations of all its electrons. Checking the periodic table inside the front cover, you find that phosphorus has atomic num-

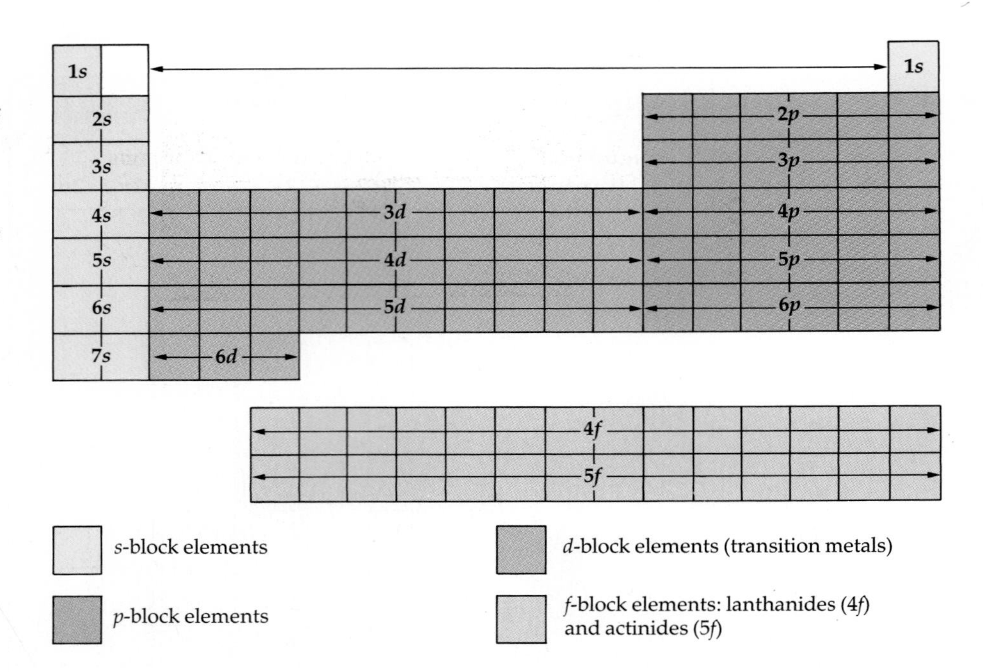

Figure 9.14
Electron configurations and the periodic table. The outermost electrons of atoms of the main group elements or the next-to-outermost electrons of atoms of the transition elements are assigned to the indicated orbitals. See Table 9.4.

ber 15 and therefore has 15 electrons. Figure 9.14 illustrates how the periodic table can be used to determine the orbital locations of the electrons. Write an electron configuration by starting with H and putting electrons into subshells until you reach the element in question. Phosphorus is in the third period. According to Figure 9.14, the $n = 1$ and $n = 2$ shells are filled in the first two periods, giving the first ten electrons in phosphorus the configuration $1s^22s^22p^6$ (the same as neon). Figure 9.14 shows that the next two electrons are assigned to the 3*s* orbital, giving $1s^22s^22p^63s^2$ so far. The final three electrons have to be assigned to the 3*p* orbitals, so the electron configuration for phosphorus is

$$\mathrm{P} \qquad 1s^22s^22p^63s^23p^3$$

According to Hund's rule, the three electrons in *p* orbitals must be unpaired. To show this, you can write the expanded electron configuration or the orbital box diagram:

$$\mathrm{P} \qquad 1s^22s^22p_x^{\;2}2p_y^{\;2}2p_z^{\;2}3s^23p_x^{\;1}3p_y^{\;1}3p_z^{\;1}$$

	1s	2s	$2p_x$	$2p_y$	$2p_z$	3s	$3p_x$	$3p_y$	$3p_z$
P	↑↓	↑↓	↑↓	↑↓	↑↓	↑↓	↑	↑	↑

Thus, all the electrons are paired except for the three electrons in 3*p* orbitals, and these occupy different orbitals with parallel spins.

Notice that Table 9.4 uses an abbreviated representation of electron configurations in which the symbol of the preceding noble gas represents filled inner shells. This is called the **noble gas notation.** Thus the noble gas notation for Ca, atomic number 20, is $[\mathrm{Ar}]4s^2$ where the symbol [Ar] represents the filled shells $1s^22s^22p^63s^23p^6$.

Table 9.4

Electron Configurations of Atoms in the Ground State

Z	Element	Configuration	Z	Element	Configuration	Z	Element	Configuration
1	H	$1s^1$	38	Sr	$[Kr]5s^2$	75	Re	$[Xe]4f^{14}5d^5 6s^2$
2	He	$1s^2$	39	Y	$[Kr]4d^1 5s^2$	76	Os	$[Xe]4f^{14}5d^6 6s^2$
3	Li	$[He]2s^1$	40	Zr	$[Kr]4d^2 5s^2$	77	Ir	$[Xe]4f^{14}5d^7 6s^2$
4	Be	$[He]2s^2$	41	Nb	$[Kr]4d^4 5s^1$	78	Pt	$[Xe]4f^{14}5d^9 6s^1$
5	B	$[He]2s^2 2p^1$	42	Mo	$[Kr]4d^5 5s^1$	79	Au	$[Xe]4f^{14}5d^{10}6s^1$
6	C	$[He]2s^2 2p^2$	43	Tc	$[Kr]4d^5 5s^2$	80	Hg	$[Xe]4f^{14}5d^{10}6s^2$
7	N	$[He]2s^2 2p^3$	44	Ru	$[Kr]4d^7 5s^1$	81	Tl	$[Xe]4f^{14}5d^{10}6s^2 6p^1$
8	O	$[He]2s^2 2p^4$	45	Rh	$[Kr]4d^8 5s^1$	82	Pb	$[Xe]4f^{14}5d^{10}6s^2 6p^2$
9	F	$[He]2s^2 2p^5$	46	Pd	$[Kr]4d^{10}$	83	Bi	$[Xe]4f^{14}5d^{10}6s^2 6p^3$
10	Ne	$[He]2s^2 2p^6$	47	Ag	$[Kr]4d^{10}5s^1$	84	Po	$[Xe]4f^{14}5d^{10}6s^2 6p^4$
11	Na	$[Ne]3s^1$	48	Cd	$[Kr]4d^{10}5s^2$	85	At	$[Xe]4f^{14}5d^{10}6s^2 6p^5$
12	Mg	$[Ne]3s^2$	49	In	$[Kr]4d^{10}5s^2 5p^1$	86	Rn	$[Xe]4f^{14}5d^{10}6s^2 6p^6$
13	Al	$[Ne]3s^2 3p^1$	50	Sn	$[Kr]4d^{10}5s^2 5p^2$	87	Fr	$[Rn]7s^1$
14	Si	$[Ne]3s^2 3p^2$	51	Sb	$[Kr]4d^{10}5s^2 5p^3$	88	Ra	$[Rn]7s^2$
15	P	$[Ne]3s^2 3p^3$	52	Te	$[Kr]4d^{10}5s^2 5p^4$	89	Ac	$[Rn]6d^1 7s^2$
16	S	$[Ne]3s^2 3p^4$	53	I	$[Kr]4d^{10}5s^2 5p^5$	90	Th	$[Rn]6d^2 7s^2$
17	Cl	$[Ne]3s^2 3p^5$	54	Xe	$[Kr]4d^{10}5s^2 5p^6$	91	Pa	$[Rn]5f^2 6d^1 7s^2$
18	Ar	$[Ne]3s^2 3p^6$	55	Cs	$[Xe]6s^1$	92	U	$[Rn]5f^3 6d^1 7s^2$
19	K	$[Ar]4s^1$	56	Ba	$[Xe]6s^2$	93	Np	$[Rn]5f^4 6d^1 7s^2$
20	Ca	$[Ar]4s^2$	57	La	$[Xe]5d^1 6s^2$	94	Pu	$[Rn]5f^6 7s^2$
21	Sc	$[Ar]3d^1 4s^2$	58	Ce	$[Xe]4f^1 5d^1 6s^2$	95	Am	$[Rn]5f^7 7s^2$
22	Ti	$[Ar]3d^2 4s^2$	59	Pr	$[Xe]4f^3 6s^2$	96	Cm	$[Rn]5f^7 6d^1 7s^2$
23	V	$[Ar]3d^3 4s^2$	60	Nd	$[Xe]4f^4 6s^2$	97	Bk	$[Rn]5f^9 7s^2$
24	Cr	$[Ar]3d^5 4s^1$	61	Pm	$[Xe]4f^5 6s^2$	98	Cf	$[Rn]5f^{10}7s^2$
25	Mn	$[Ar]3d^5 4s^2$	62	Sm	$[Xe]4f^6 6s^2$	99	Es	$[Rn]5f^{11}7s^2$
26	Fe	$[Ar]3d^6 4s^2$	63	Eu	$[Xe]4f^7 6s^2$	100	Fm	$[Rn]5f^{12}7s^2$
27	Co	$[Ar]3d^7 4s^2$	64	Gd	$[Xe]4f^7 5d^1 6s^2$	101	Md	$[Rn]5f^{13}7s^2$
28	Ni	$[Ar]3d^8 4s^2$	65	Tb	$[Xe]4f^9 6s^2$	102	No	$[Rn]5f^{14}7s^2$
29	Cu	$[Ar]3d^{10}4s^1$	66	Dy	$[Xe]4f^{10}6s^2$	103	Lr	$[Rn]5f^{14}6d^1 7s^2$
30	Zn	$[Ar]3d^{10}4s^2$	67	Ho	$[Xe]4f^{11}6s^2$	104	Unq*	$[Rn]5f^{14}6d^2 7s^2$
31	Ga	$[Ar]3d^{10}4s^2 4p^1$	68	Er	$[Xe]4f^{12}6s^2$	105	Unp*	$[Rn]5f^{14}6d^3 7s^2$
32	Ge	$[Ar]3d^{10}4s^2 4p^2$	69	Tm	$[Xe]4f^{13}6s^2$	106	Unn*	$[Rn]5f^{14}6d^4 7s^2$
33	As	$[Ar]3d^{10}4s^2 4p^3$	70	Yb	$[Xe]4f^{14}6s^2$	107	Uns*	$[Rn]5f^{14}6d^5 7s^2$
34	Se	$[Ar]3d^{10}4s^2 4p^4$	71	Lu	$[Xe]4f^{14}5d^1 6s^2$	108	Uno*	$[Rn]5f^{14}6d^6 7s^2$
35	Br	$[Ar]3d^{10}4s^2 4p^5$	72	Hf	$[Xe]4f^{14}5d^2 6s^2$	109	Une*	$[Rn]5f^{14}6d^7 7s^2$
36	Kr	$[Ar]3d^{10}4s^2 4p^6$	73	Ta	$[Xe]4f^{14}5d^3 6s^2$			
37	Rb	$[Kr]5s^1$	74	W	$[Xe]4f^{14}5d^4 6s^2$			

*In 1992 the following names were proposed for elements 107 to 109: 107, nielsbohrium (Ns); 108, hassium (Hs); and 109, meiterium (Mt). These names will not become official until they are approved by the International Union of Pure and Applied Chemistry (IUPAC). There is still disagreement about element 104. Americans proposed rutherfordium after Ernest Rutherford. Russians want element 104 to be named kurchatovium, after Igor Kurchatov, who led the development of Soviet atomic and hydrogen bombs. Previously, the name hahnium had been proposed for element 105. No name has been proposed for element 106.

At this point, you should be able to write electron configurations for main group elements through Ca, atomic number 20, using the periodic table as a guide. Pick one of these elements and write its electron configuration, and do Exercise 9.8. Then check your electron configurations against the ones given in Table 9.4 for all the elements.

EXAMPLE 9.4

Electron Configuration of Si

Give the electron configuration of silicon, Si.

SOLUTION The periodic table shows silicon is in the third period with atomic number 14 and therefore has 14 electrons. The first ten electrons are represented by $1s^22s^22p^6$, the electron arrangement for Ne. The last four electrons placed in the atom have the configuration $3s^23p^2$. Therefore, the electron configuration of Si is

$$\text{Si} \quad 1s^22s^22p^63s^23p^2 \quad \text{or} \quad [\text{Ne}]3s^23p^2$$

EXERCISE 9.8 • *Electron Configuration*

For sulfur, write the electron configuration in noble gas notation. Determine how many unpaired electrons sulfur atoms have by drawing the orbital box diagram.

EXERCISE 9.9 • *Electron Configuration*

Identify the number of electrons in the highest occupied energy level (highest n) in chlorine and in sulfur.

Valence Electrons

In our earlier discussion of chemical compounds (Section 4.3), we identified valence electrons as those that participate in the formation of ions and molecular compounds. Now we can answer a fundamental question: How are electron configurations related to the periodic table and the concept of valence electrons?

Since the electrons in the filled orbitals of noble gases and the filled d subshells of the Group 3A through 7A elements are not greatly affected by reactions with other atoms, we can focus on the behavior of the s and p electrons in the highest occupied n levels (and d electrons in unfilled subshells of the transition metals)—these outermost electrons are the valence electrons. A few examples of valence electron locations are:

Element	*Core Electrons*	*Valence Electrons*	*Total Electron Configuration*	*Periodic Group*
Na	$1s^22s^22p^6$	$3s^1$	$[\text{Ne}]3s^1$	1A
Si	$1s^22s^22p^6$	$3s^23p^2$	$[\text{Ne}]3s^23p^2$	4A
As	$1s^22s^22p^63s^23p^63d^{10}$	$4s^24p^3$	$[\text{Ar}]3d^{10}4s^24p^3$	5A

"Core electrons" can be represented by the noble gas notation for atoms of elements in the first three periods, but Group 3A to 7A elements in later

Table 9.5

Lewis Dot Symbols for Atoms

1A ns^1	2A ns^2	3A ns^2np^1	4A ns^2np^2	5A ns^2np^3	6A ns^2np^4	7A ns^2np^5	8A ns^2np^6
Li·	·Be·	·B·	·C·	·N·	:O·	:F·	:Ne:
Na·	·Mg·	·Al·	·Si·	·P·	:S·	:Cl·	:Ar:

periods also have a filled *d* subshell as part of the inner filled shells. For example, the core electrons of As are represented by $[Ar]3d^{10}$. Notice that for all of these examples of main group elements *the number of valence electrons of each element is equal to the group number.*

The concept of valence electrons was first proposed by G.N. Lewis more than 90 years ago. He assumed that each noble gas atom had a completely filled outermost shell, which he regarded as a stable configuration because of the lack of reactivity of noble gases. Since all noble gases (except He) have eight valence electrons, this observation is called the **octet rule.** Lewis used the element's symbol to represent the atomic nucleus together with all but the outermost shell of electrons; he called this the kernel of the atom. The valence electrons, represented by dots, are placed around the symbol one at a time until they are used up or until all four sides are occupied; any remaining electrons are paired with the ones already there. **Lewis dot symbols** for atoms of elements in periods 2 and 3 are shown in Table 9.5. Main-group elements in Groups 1A through 7A are referred to as ***s*- and *p*-block** elements because they have electron configurations of ns^1, ns^2, ns^2np^1, ns^2np^2, ns^2np^3, ns^2np^4, ns^2np^5, or ns^2np^6 for valence electrons, where *n* is the principal quantum number.

Although Lewis didn't publish his ideas about valence electrons until 1916, he had written a memorandum in 1902 that outlined his ideas, which developed from his attempt to explain periodic trends to students in his introductory chemistry course.

The Lewis symbol emphasizes the ns^2np^6 octet, the electron configuration of the noble gases, as an especially stable arrangement. When ions form from atoms of elements in *s*- and *p*-block groups, electrons are removed or added so that a noble-gas configuration is achieved. Atoms from Group 1A, 2A, and 3A lose 1, 2, or 3 electrons to form 1+, 2+, or 3+ ions. Atoms from Groups 7A, 6A, and some in 5A gain 1, 2, or 3 electrons to form 1−, 2−, or 3− ions. *Metals form cations with a charge equal to the group number, and nonmetals form anions with a charge equal to the group number minus eight.* Ions with identical electron configurations are said to be **isoelectronic.** Table 9.6 lists some isoelectronic ions and the noble gas that has the same electron configuration. Note from the table that metal ions are isoelectronic with the *preceding* noble gas, while nonmetal ions have the electron configuration of the *next* noble gas.

Table 9.6

Noble Gas Atoms and Isoelectronic Ions

He, Li^+, Be^{2+}, H^-
Ne, Na^+, Mg^{2+}, Al^{3+}, F^-, O^{2-}
Ar, K^+, Ca^{2+}, Ga^{3+}, Cl^-, S^{2-}
Kr, Rb^+, Sr^{2+}, Br^-, Se^{2-}
Xe, Cs^+, Ba^{2+}, I^-, Te^{2-}

EXAMPLE 9.5

Valence Electrons

Draw the Lewis dot symbols showing valence electrons for Na, C, and I.

SOLUTION Look at the periodic table and find the group for the element. All of these elements are in A groups, so the group number is the number of valence electrons. Na is in Group 1A, C is in Group 4A, and I is in Group 7A.

Na· ·Ċ· :Ï·

EXERCISE 9.10 • *Electron Configurations*

Using the noble gas notation, write the electron configurations for Se and Te. What do these configurations illustrate about elements in the same main group? Also draw the Lewis structures for these two atoms.

EXERCISE 9.11 • *Valence Electrons*

Give Lewis dot symbols for Rb, Si, Br, and Ba.

EXAMPLE 9.6

Main Group Ions

Give the valence electron configuration for Se, and predict the charge on an ion formed from an Se atom. What noble gas has the same electron configuration as the ion?

SOLUTION Se is a Group 6A element in the fourth period, so its valence electron configuration is $4s^2 4p^4$. An Se atom needs two electrons to achieve the configuration of the nearest noble gas, Kr. As a result, the stable ion expected for Se is Se^{2-}.

$$Se + 2e^- \longrightarrow Se^{2-}$$

EXERCISE 9.12 • *Main Group Ions*

(a) Use the periodic table to write the general electron configuration for Group 3A atoms. (b) Predict the charge of the ion formed from a Ga atom. (c) What is the electron configuration for the Ga ion you predicted in (b)?

Electron Configurations of Transition Elements

The elements of the fourth through the sixth periods in the middle of the periodic table are those elements in which a *d* or *f* subshell is being filled (Figure 9.14). Elements filling *d* subshells are often collectively called the **transition elements.** Those for which *f* subshells are filling are sometimes called the inner transition elements or, more usually, **lanthanides** (filling the 4*f* subshell) and **actinides** (filling the 5*f* subshell).

In each period, the transition elements are immediately preceded by two *s*-block elements. As shown in Figure 9.12, once the 4*s* subshell is filled, the subshell with the next higher energy is 3*d*. The use of *d* orbitals begins with the first transition metal, scandium, which has the configuration $[Ar]3d^1 4s^2$. After scandium come titanium with $[Ar]3d^2 4s^2$ and vanadium with $[Ar]3d^3 4s^2$. Results from magnetic and spectroscopic measurements indicate that the electron configuration for chromium is $[Ar]3d^5 4s^1$ rather than $[Ar]3d^4 4s^2$. This illustrates one of several "anomalies" in predicting electron configurations from transition and inner transition elements. When half-filled *d* or *f* shells are possible, they will generally be favored. As a result, elements that are close to the middle and the end of filling a *d* or *f* subshell will have a stable configuration that leaves the *s* orbital half-filled and the *d* or *f* orbital half-filled or filled. For example, copper has the electron configuration $[Ar]3d^{10} 4s^1$ instead of the expected $[Ar]3d^9 4s^2$.

The ns *and* (n − 1)d *orbital energies are not very different for the transition elements. For Cr and Cu (and some others in the 5th and 6th periods) they are nearly the same, so the electron configurations often vary slightly from those predicted.*

The lanthanide series starts with lanthanum (La), which has the electron configuration $[Xe]5d^1 6s^2$. The next element, cerium (Ce), is set out in a separate row at the bottom of the periodic table, and it is with these elements that *f* orbitals come into play (Figure 9.14 and Table 9.4). The electron configuration of Ce is $[Xe]4f^1 5d^1 6s^2$. Moving across the lanthanide series, the pattern continues with some variation, until the seven 4*f* orbitals are filled by 14 electrons in lutetium atoms, Lu ($[Xe]4f^{14} 5d^1 6s^2$). Note that both the $n = 5$ and $n = 6$ levels are partially filled before the 4*f* starts to be occupied.

The number of unpaired electrons for atoms of most transition and inner transition elements can be predicted by placing valence electrons in orbital diagrams according to Hund's rule. For example, the electron configuration of Co ($Z = 27$) is $[Ar]3d^7 4s^2$ and the number of unpaired electrons is three, as seen from the following orbital box diagram.

	3*d*					4*s*
Co [Ar]	↑↓	↑↓	↑	↑	↑	↑↓

The importance of the periodic table as a guide to electron configurations cannot be overemphasized. Select some elements from different groups of the periodic table and practice writing their electron configurations and orbital diagrams. Use Figure 9.14 as a guide. For example, Se, atomic number 34, is in Group 6A. With this information you immediately know that it has *six* valence electrons with a configuration of $ns^2 np^4$. Since Se is in the fourth period, $n = 4$, the complete electron configuration is given by starting with the electron configuration of argon $[1s^2 2s^2 2p^6 3s^2 3p^6]$, the noble gas at the end of the preceding $n = 3$ period. Then add the filled $3d^{10}$ shell and the six valence electrons of Se ($4s^2 4p^4$) to give $1s^2 2s^2 2p^6 3s^2 3p^6 3d^{10} 4s^2 4p^4$, or $[Ar]3d^{10} 4s^2 4p^4$. If you want to predict the number of unpaired electrons, you only need to look at the valence electron configuration, since the inner shells are completely filled and all electrons in the inner shells are paired. For Se, $[Ar]3d^{10} 4s^2 4p_x^2 4p_y^1 4p_z^1$ indicates two *p* orbitals with unpaired electrons for a total of two unpaired electrons.

Paramagnetism and Unpaired Electrons

The magnetic properties of a spinning electron were described in Section 9.3. Atoms and ions that have filled shells are said to be **diamagnetic** because all the electrons are paired, and the magnetic fields of the electrons effectively cancel each other. Diamagnetic substances are slightly repelled by a magnetic field. Atoms or ions with unpaired electrons are **paramagnetic,** and these substances are attracted to a magnetic field; the more unpaired electrons, the greater is the attraction.

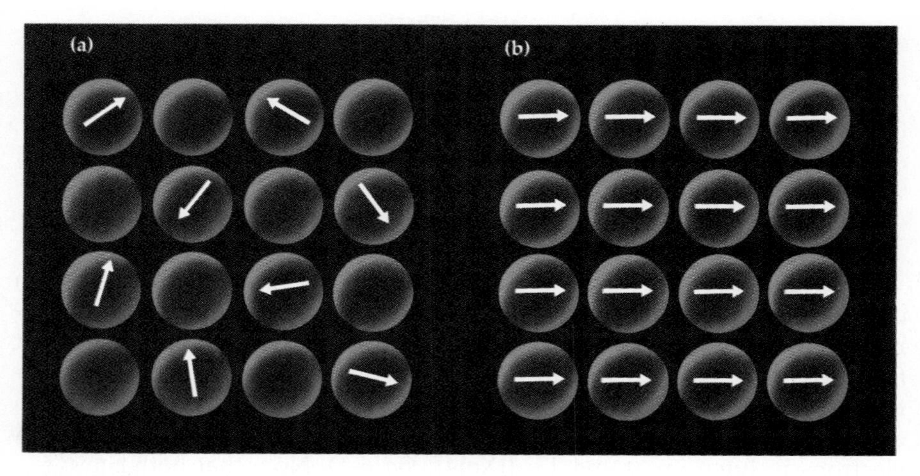

Figure 9.15

Types of magnetic behavior. (a) Paramagnetism: the centers (atoms or ions) with magnetic moments are not aligned unless the substance is in a magnetic field. (b) Ferromagnetism: the spins of unpaired electrons in a cluster of atoms or ions align in the same direction.

The difference between paramagnetic and ferromagnetic materials is that domains of aligned magnets do not form in paramagnetic substances. Ferromagnets are thus "superparamagnets."

Alnico magnets contain Al, Co, and Ni, as well as iron and copper.

Substances that retain their magnetism (are permanent magnets) are **ferromagnetic.** The magnetic effect in ferromagnetic materials is much larger than for paramagnetic materials. Ferromagnetism occurs when the spins of unpaired electrons in a cluster of atoms (called a domain) in the solid align themselves in the same direction (Figure 9.15). Only the metals of the iron, cobalt, and nickel subgroups in the periodic table exhibit this property. They are also unique in that, once the domains are aligned in a magnetic field, the metal is permanently magnetized. In such a case the magnetism can be eliminated only by heating or vibrating the metal to rearrange the electron spin domains. Many alloys exhibit greater ferromagnetism than do the pure metals themselves. Some metal oxides (for example, CrO_2 and Fe_3O_4) are also ferromagnetic and are used in magnetic recording tape (Figure 9.16).

Transition Metal Ions

It is particularly important to know something about the electron configurations of transition metal ions. Metal ions are the center of such important molecules as hemoglobin, myoglobin, vitamin B_{12}, and many biological catalysts called enzymes. The configuration of the ion plays a role in determining the chemistry of these systems.

The closeness in energy of the 4*s* and 3*d* subshells was discussed earlier in connection with the electron configurations of transition metal atoms. The 5*s* and 4*d* subshells, the 6*s*, 4*f*, and 5*d* subshells, and the 7*s*, 5*f*, and 6*d* subshells are also close to each other. The *ns* and $(n - 1)d$ orbitals are so close in energy that once *d* orbital electrons are added, the *d* orbitals become slightly lower in energy than the *s* orbitals. As a result, the *ns* electrons are at higher energy and will be removed first when transition and inner transition metals form cations.

For example, Fe^{2+} is formed from Fe by the loss of two 4*s* electrons,

$$\text{Fe [Ar]}3d^6 4s^2 \longrightarrow \text{Fe}^{2+}\ \text{[Ar]}3d^6 + 2e^-$$

and Fe^{3+} is formed by loss of a 3*d* electron.

$$Fe^{2+}\ [Ar]3d^6 \longrightarrow Fe^{3+}\ [Ar]3d^5 + e^-$$

There are five unpaired electrons in the Fe^{3+} ion, as shown by using orbital box diagrams for the *d* electrons:

	3*d*				
Fe^{3+} [Ar]	↑	↑	↑	↑	↑

Mn^{2+} is another example of a transition metal ion that has five unpaired electrons. Atoms or ions of inner transition elements can have as many as seven unpaired electrons in the *f* subshell, as occurs in the Eu^{2+} and Gd^{3+} ions.

Figure 9.16
Several products that use magnetic recording tape. *(C.D. Winters)*

EXAMPLE 9.7

Electron Configurations for Transition Elements

(a) Write the electron configuration for the Co atom, using the noble gas notation. Then draw the orbital box diagram for the electrons beyond the preceding noble gas. (b) Cobalt commonly exists as 2+ and 3+ ions. How does the orbital box diagram given in (a) have to be changed to represent the outer electrons of Co^{2+} and Co^{3+}? (c) How many unpaired electrons do Co, Co^{2+}, and Co^{3+} have?

SOLUTION

(a) Co is in the fourth period with an atomic number of 27. It has nine more electrons than Ar, with two of the nine in the 4*s* subshell and seven in the 3*d* subshell, so its electron configuration is $[Ar]3d^74s^2$. For the orbital box diagram, all *d* subshells get one electron before pairing (Hund's rule).

3*d*					4*s*
↑↓	↑↓	↑	↑	↑	↑↓

(b) To form Co^{2+}, two electrons are removed from the 4*s* subshell. To form Co^{3+}, one of the paired electrons is removed from a *d* subshell.

	3*d*				
Co^{2+}	↑↓	↑↓	↑	↑	↑
Co^{3+}	↑↓	↑	↑	↑	↑

(c) The numbers of unpaired electrons are three for Co, three for Co^{2+}, and four for Co^{3+}.

EXERCISE 9.13 • *Electron Configurations for Transition Elements*

(a) Write the electron configuration for the nickel atom, using the noble gas notation. (b) Draw an orbital box diagram for the outer electrons. How many unpaired electrons does nickel have? (c) Draw an orbital box diagram for the Ni^{2+} ion. How many unpaired electrons does Ni^{2+} have?

9.5 PERIODIC TRENDS IN ATOMIC PROPERTIES

Atomic Radii

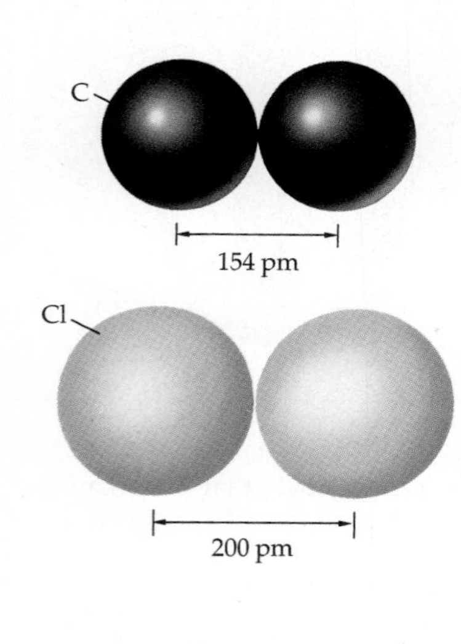

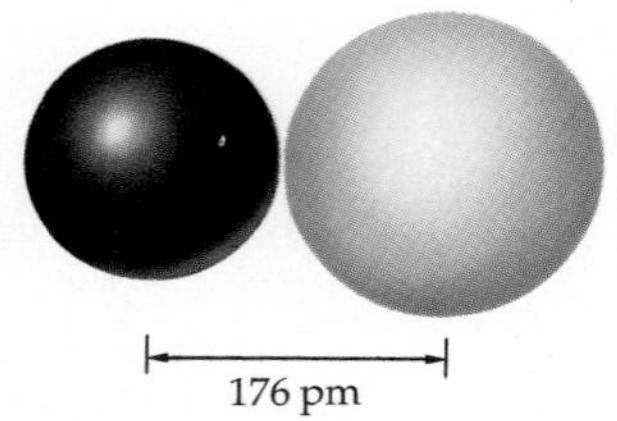

Orbitals represent the probable locations of atoms and have no sharp boundaries. However, throughout this section atoms and ions are drawn as well-defined spheres to make trends more easily apparent. As you look at these, keep in mind the "fuzzy" nature of the outer regions of atoms and ions.

For atoms that form simple diatomic molecules, such as Cl_2, the **atomic radius** can be defined experimentally by first finding the distance between the centers of the two atoms. One half of this distance is assumed to be a good estimate of the atom's radius. In the Cl_2 molecule, the atom-atom distance (the distance from the center of one atom to the center of the other) is 200 pm. Dividing by 2 shows that the Cl radius is 100 pm. Similarly, the C—C distance in diamond is 154 pm, and so the radius of the carbon atom is 77 pm. To test these estimates, we can add them together to estimate the distance between Cl and C in CCl_4. The estimated distance of 177 pm is in good agreement with the experimentally measured C—Cl distance of 176 pm.

This approach can then be extended to other atomic radii. For example, the radii of O, C, and S can be estimated by measuring the O—H, C—Cl, and H—S distances in H_2O, CCl_4, and H_2S and then subtracting the H and Cl radii found from H_2 and Cl_2. By this and other techniques, a reasonable set of atomic radii for main-group elements has been assembled (Figure 9.17).

For the main group elements, atomic radii increase going down a group in the periodic table and decrease going across a period. These trends reflect two important effects: (1) In going from the top to the bottom of a group in the periodic table, electrons are assigned to orbitals that are successively farther from the nucleus, and as a result the atomic radii increase. (2) For a given period, the principal quantum number n of the outermost orbitals stays the same. This means the radius of the orbitals to which the electrons are assigned would be expected to remain approximately constant. However, with the addition of each successive electron the nuclear charge has also increased. The result is that attraction between the nucleus and electrons increases, and, because this attraction is somewhat stronger than the increasing repulsion between electrons, the atomic radius decreases. Note the large increase in atomic radius in going from any noble gas atom to the following Group 1A atom, where the outermost electron is assigned to the next higher energy level.

The periodic trend in the atomic radii of transition metal atoms is illustrated in Figure 9.18. One feature to notice is that the sizes of the transition metal atoms change very little across a series, especially beginning at Group 5B (V, Nb or Ta), because they are all determined by the radius of an ns orbital (n = 4, 5, or 6), occupied by at least one electron. The variation in the number of electrons occurs instead in the $(n - 1)d$ orbitals. As the number of electrons in these $(n - 1)d$ orbitals increases, they increasingly repel the ns electrons, partly compensating for the increased nuclear charge across the periods. Consequently, the ns electrons experience only a slightly increasing nuclear attraction, and the radii remain nearly constant until the slight rise at Groups 1B and 2B due to the continually increasing electron-electron repulsions as the d subshell is filled.

The similar radii of the transition metals and their ions have an important effect on their chemistry—they tend to be more alike in their properties than other elements in the same groups. For example, the nearly identical

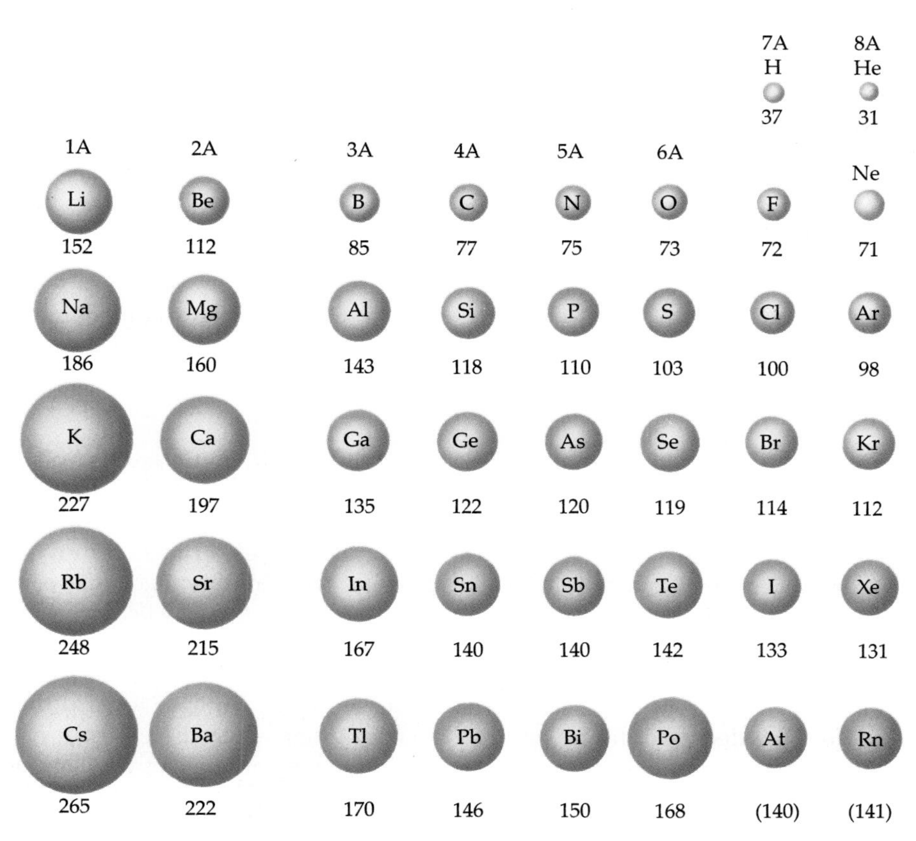

Figure 9.17
Atomic radii in picometers (1 pm = 10^{-12} m) for the main group elements.

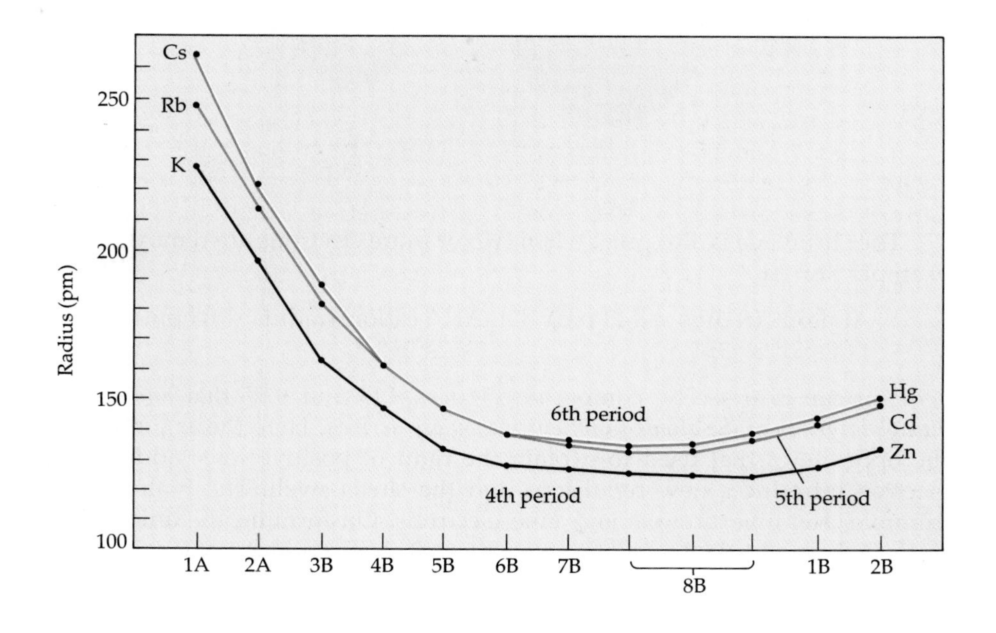

Figure 9.18
Atomic radii of the transition metals (and the *s*-block metals of the same periods) as a function of periodic group.

radii of fifth and sixth period transition elements lead to difficult problems of metal recovery. The metals Ru, Os, Rh, Ir, Pd, and Pt are called the "platinum group metals" because they occur together in nature. Apparently their radii and chemistry are so similar that their minerals are similar and are found in the same geologic zones.

EXERCISE 9.14 • *Periodic Trends in Atomic Radii*

Place the three elements Al, C, and Si in order of increasing atomic radius.

Ionic Radii

Figure 9.19 shows clearly that the periodic trends in the **ionic radii** are the same as the trends in radii for neutral atoms: *positive or negative ions of elements in the same group increase in size down the group.* But pause for a moment and compare Figure 9.19 with Figure 9.17. When an electron is removed from an atom to form a cation, the size shrinks considerably; *the radius of a cation is always smaller than that of the atom from which it was derived.* For example, the radius of Li is 152 pm, whereas that for Li^+ is only 90 pm. This is understandable, because when an electron is removed, the other electrons no longer feel its repulsion, so the remaining electrons contract toward the nucleus. The decrease in ion size is especially great when the electron removed comes from a higher energy level than the new outer electron. This is the case for Li, for which the "old" outer electron was from a 2*s* orbital and the "new" outer electron is in a 1*s* orbital.

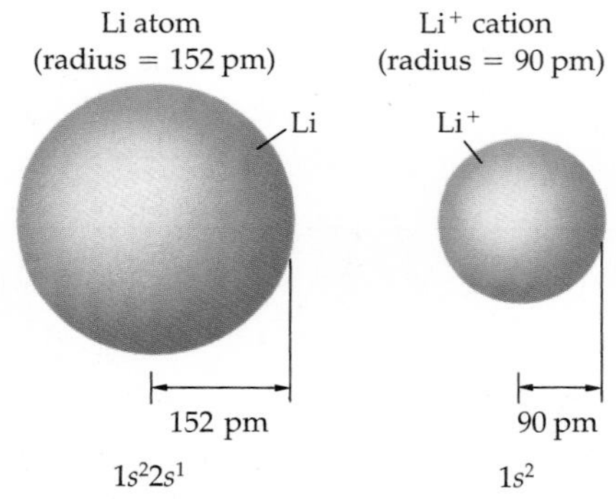

The shrinkage is also great when two or more electrons are removed; for example, for Al^{3+}:

Al atom (radius = 143 pm)	Al^{3+} cation (radius = 68 pm)
$1s^22s^22p^63s^23p^1$	$1s^22s^22p^6$

You can also see by comparing Figures 9.19 and 9.17 that *anions are always larger than the atoms from which they are derived.* Here the argument is the opposite of that used to explain the radii of positive ions: adding an electron introduces new repulsions and the shells swell. The F atom, for example, has nine protons and nine electrons. On forming the anion, the nuclear charge is still 9+, but there are now 10 electrons in the anion. The F^-

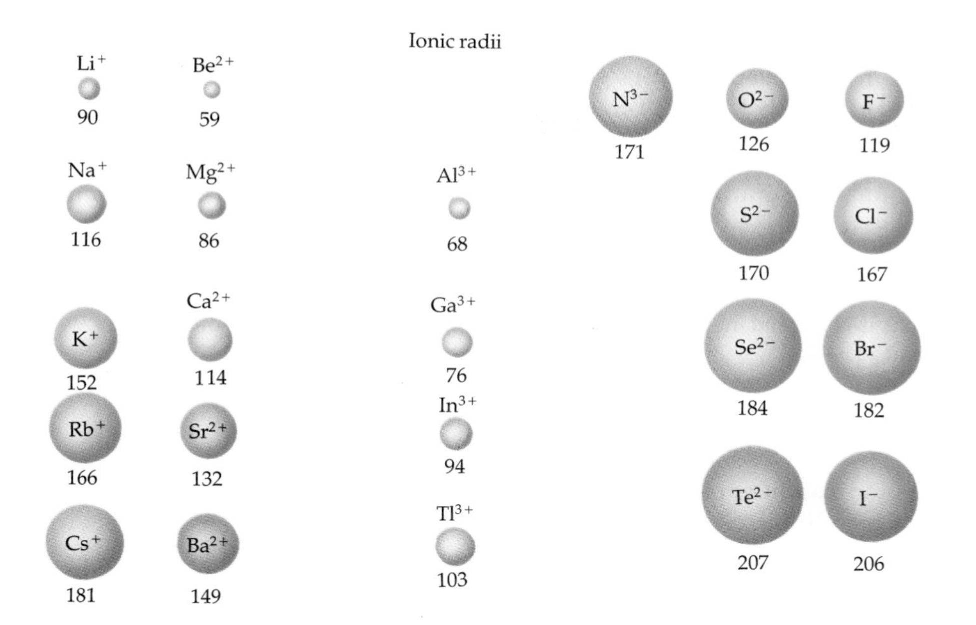

Figure 9.19
Relative sizes of some common ions. Radii are given in picometers ($1\ pm = 10^{-12}\ m$).

ion is much larger than the F atom because of increased electron-electron repulsions.

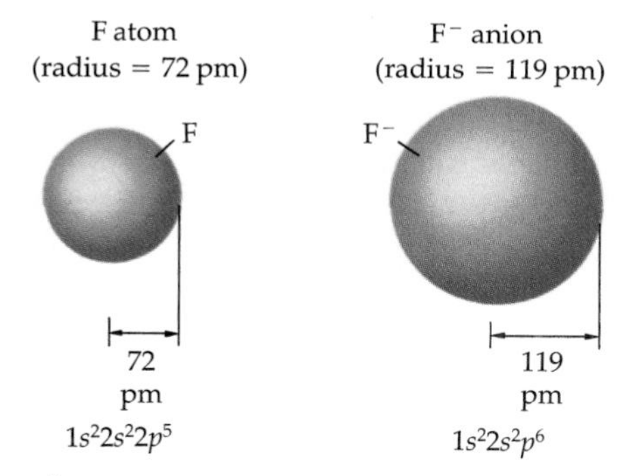

The oxide ion, O^{2-}, is isoelectronic with F^-, that is, they both have the same electron configuration (the neon configuration). However, the oxide ion is larger than the fluoride ion because the oxide ion has only eight protons available to attract ten electrons, whereas F^- has more protons (nine) to attract the same number of electrons. It is useful to compare the sizes of isoelectronic ions across the periodic table. For example, consider O^{2-}, F^-, Na^+, and Mg^{2+}.

Ion	O^{2-}	F^-	Na^+	Mg^{2+}
Ionic radius (pm)	126	119	116	86
Number of protons	8	9	11	12
Number of electrons	10	10	10	10

Each ion contains ten electrons. However, the O^{2-} ion has only eight protons in its nucleus to attract these electrons, while F^- has nine, Na^+ has eleven, and Mg^{2+} has twelve. As the proton/electron ratio increases in a series of isoelectronic ions, the balance in electron-proton attraction and electron-electron repulsion shifts in favor of attraction, and the ion shrinks. As you can see in Figure 9.19, this is true for all isoelectronic series of ions.

EXERCISE 9.15 • *Ion Sizes*

What is the trend in sizes of the ions N^{3-}, O^{2-}, and F^-? Briefly explain why this trend exists.

Ionization Energies

The **ionization energy** of an atom is the energy needed to remove an electron from the atom in the gas phase.

$$A(g) + \text{energy} \longrightarrow A^+(g) + e^- \qquad \text{where A is an atom of an element}$$

Energy is always required to remove an electron, so the process is endothermic and the sign of the ionization energy is always positive. For *s*- and *p*-block elements, *first ionization energies (the energy needed to remove one electron from the neutral atom) generally increase across a period and decrease down a group* (Figure 9.20 and Table 9.7). The decrease down a group reflects the increasing radii of the atoms—it is easier to remove an electron from a larger atom. Similarly, the increase across a period occurs for the same reason that radii decrease across a period. The trend across a given period is not smooth, however, particularly in the second period. A Group 3A element (ns^2np^1) has a smaller ionization energy than the preceding Group 2A element (ns^2). This difference indicates that the single np electron of the Group 3A element is more easily removed than one of the ns electrons of the pre-

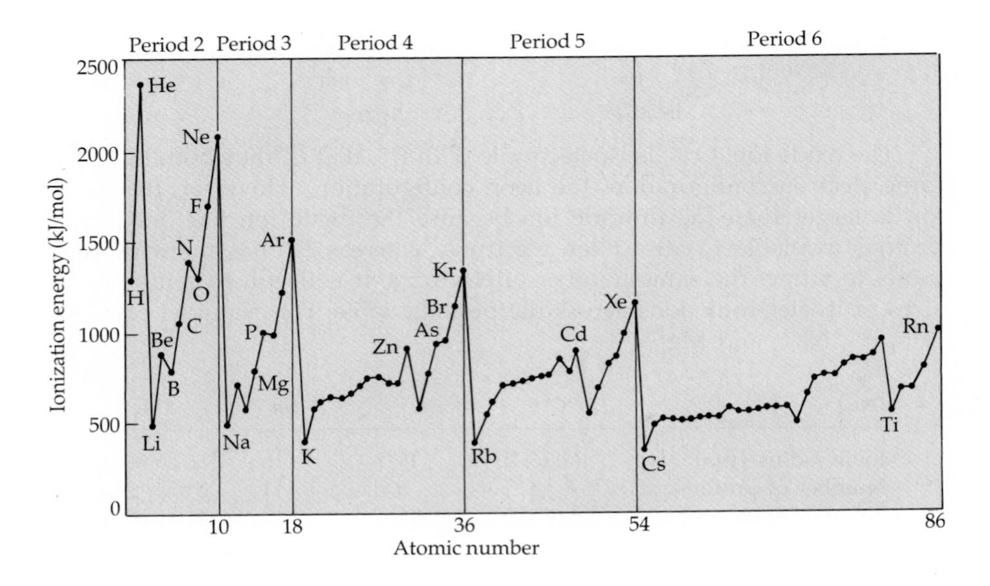

Figure 9.20
Values of first ionization energies for the elements in the first five periods, plotted against atomic number.

Table 9.7

First Ionization Energies of the Elements (kJ/mol)

1A (1)	2A (2)	3B (3)	4B (4)	5B (5)	6B (6)	7B (7)	8B (8, 9, 10)			1B (11)	2B (12)	3A (13)	4A (14)	5A (15)	6A (16)	7A (17)	8A (18)
H 1312																	He 2371
Li 520	Be 899											B 801	C 1086	N 1402	O 1314	F 1681	Ne 2081
Na 496	Mg 738											Al 578	Si 786	P 1012	S 1000	Cl 1251	Ar 1521
K 419	Ca 599	Sc 631	Ti 658	V 650	Cr 652	Mn 717	Fe 759	Co 758	Ni 757	Cu 745	Zn 906	Ga 579	Ge 762	As 947	Se 941	Br 1140	Kr 1351
Rb 403	Sr 550	Y 617	Zr 661	Nb 664	Mo 685	Tc 702	Ru 711	Rh 720	Pd 804	Ag 731	Cd 868	In 558	Sn 709	Sb 834	Te 869	I 1008	Xe 1170
Cs 377	Ba 503	La 538	Hf 681	Ta 761	W 770	Re 760	Os 840	Ir 880	Pt 870	Au 890	Hg 1007	Tl 589	Pb 715	Bi 703	Po 812	At 890	Rn 1037

The data in this table are taken from *KC? Discoverer*, a computer database of information about the elements. The program is available for several types of computers through *Journal of Chemical Education: Software*. (See the preface for details.)

ceding Group 2A element. Another deviation occurs for Group 6A elements (ns^2np^4), which have smaller ionization energies than the Group 5A elements that precede them. Beginning in Group 6A, two electrons are assigned to the same p orbital. Thus, greater electron repulsion is experienced by the fourth p electron and this makes it easier to remove.

For the transition and inner transition elements, ionization energies increase much more gradually across the period than for the main group elements (Table 9.7). Since the atomic radii of B group elements change very little across a period, the ionization energy for the removal of an ns electron from B group elements also shows small changes across the period. As a result, there is less difference between adjacent elements in the B groups of the periodic table.

Each atom has a series of ionization energies, since more than one electron can always be removed (except for H). For example, the first three ionization energies of Mg(g) are

$$\underset{1s^2 2s^2 2p^6 3s^2}{Mg(g)} + \text{energy} \longrightarrow \underset{1s^2 2s^2 2p^6 3s^1}{Mg^+(g)} + e^- \qquad IE(1) = 738 \text{ kJ/mol}$$

$$\underset{1s^2 2s^2 2p^6 3s^1}{Mg^+(g)} + \text{energy} \longrightarrow \underset{1s^2 2s^2 2p^6}{Mg^{2+}(g)} + e^- \qquad IE(2) = 1450 \text{ kJ/mol}$$

$$\underset{1s^2 2s^2 2p^6}{Mg^{2+}(g)} + \text{energy} \longrightarrow \underset{1s^2 2s^2 2p^5}{Mg^{3+}(g)} + e^- \qquad IE(3) = 7734 \text{ kJ/mol}$$

Notice that removing each subsequent electron requires more energy, and the jump from the second [IE(2)] to the third [IE(3)] ionization energy of Mg is particularly great. The first electron removed from the magnesium atom comes from the 3s orbital. The second ionization energy corresponds to removing a 3*s* electron from the positive ion Mg^+. As expected, the second ionization energy is higher than the first ionization energy because the electron is being removed from a positive ion, which strongly attracts the electron. The third ionization energy corresponds to removing a 2*p* electron from Mg^{2+}, a stable ion with a noble gas configuration. The very great difference between the second and third ionization energies for Mg is excellent experimental evidence for the existence of electron shells in atoms.

EXERCISE 9.16 • *Ionization Energy*

Arrange the following atoms in the order of increasing ionization energy: F, Al, P, Mg, and Ba.

EXERCISE 9.17 • *Ionization Energy*

Between which two ionization energies (second, third, fourth, etc., . . .) would you expect the largest increase for a phosphorus atom?

9.6 PERIODIC TRENDS IN PROPERTIES OF THE ELEMENTS

The periodic trends in atom size, ion size, and ionization energy provide a basis for understanding the chemical and physical properties of elements. For example, metals have low ionization energies and nonmetals have high ionization energies, which helps to explain the usual formation of positive ions by metals and negative ions by nonmetals. Since ionization energy decreases down a group and generally increases across a period, the elements toward the bottom left-hand corner of the periodic table are the most active metals, meaning that they react easily with many other elements by loss of electrons. Correspondingly, the elements at the top right of the periodic table are the most active nonmetals. Thus the trends in ionization energies are a good indication of the trends in metallic or nonmetallic character of elements.

Although the elements within a given group have similar properties, there are differences in reactivity and extent of metallic or nonmetallic character. In some groups the differences are more dramatic than in other groups. For example, Group 1A elements are all active metals, with Rb and Cs being the most active, while in Group 4A the first element carbon is a nonmetal; silicon and germanium are metalloids; and tin and lead are metals. The following sections use periodic trends to organize some of the physical and chemical properties of elements.

Hydrogen and the *s*-Block Elements

Hydrogen was the first chemical element formed in the moment of creation, and it is now the most abundant element in the universe. All other elements were formed from it, and virtually every element, with the exception of the noble gases, combines chemically with hydrogen. Indeed, hydrogen-containing compounds are of such importance that you have already seen many of them in this book and will be introduced to many more.

Except for hydrogen, the elements of Groups 1A and 2A of the periodic table are all metals and have valence electron configurations of ns^1 and ns^2, respectively. The Na^+, K^+, Mg^{2+}, and Ca^{2+} ions constitute 99% of the positive ions in the human body. Seven of the top 50 chemicals produced in the United States are based on metal ions of Groups 1A and 2A.

Hydrogen ($1s^1$) Hydrogen has three isotopes, two of them stable (^1_1H and deuterium, ^2_1H or D) and one radioactive (tritium, ^3_1H). Of the three isotopes, only H and D are found in measurable quantities in nature (their abundances are H = 99.985% and D = 0.015%). In any sample of natural water there is always a small concentration of D_2O. Since D has twice the atomic mass of H, deuterium compounds react slightly more slowly than similar H-containing compounds, and a method for producing D_2O or "heavy water" is based on this difference. Electrolysis of water gives H_2 and O_2.

$$2\,H_2O(\ell) + \text{electrical energy} \longrightarrow 2\,H_2(g) + O_2(g)$$

When water is electrolyzed, H_2 is liberated about six times more rapidly than D_2. Thus, as the electrolysis proceeds, the liquid remaining is increasingly D_2O, and nearly pure D_2O can be obtained eventually. Heavy water is very valuable for use in slowing down neutrons in some nuclear power reactors (see Section 19.6).

Although hydrogen-containing compounds seem to come in a bewildering variety, there are only three different types of binary compounds containing hydrogen.

a. *Anionic hydrides,* which form when H_2 interacts with active metals such as Na, Li, and Ca.

$$2\,Na(s) + H_2(g) \longrightarrow 2\,NaH(s)$$
$$Ca(s) + H_2(g) \longrightarrow CaH_2(s)$$

Electrolysis of water. Notice that the volume ratio is 2 (H_2, left) to 1 (O_2, right), as expected from the 2:1 ratio of H to O in H_2O. *(C.D. Winters)*

Hydrogen: Fuel *of the* Future?

WHAT WILL WE USE FOR energy sources when fossil fuels are depleted? Hydrogen has long been touted as "the fuel of the future," but there are currently two major barriers to extensive use of hydrogen as an alternative to fossil fuels. The first is the need for an inexpensive method of making H_2 that avoids the use of fossil fuels, which are now the major starting materials for the production of hydrogen. Electrolysis of water is an obvious source of hydrogen, but the electrical energy needed is too expensive to yield hydrogen at a reasonable cost. However, solar cells may soon be cheap enough to provide the electrical energy needed for this purpose. For example, the HYSOLAR project, carried out jointly by Saudi Arabia and Germany, uses solar energy to power an electrolysis plant that produces hydrogen. (See I. Dostrovsky, *Scientic American*, December 1991, pp. 102–107.)

The second barrier to using hydrogen as a fuel is a means of convenient storage. The space program has demonstrated that hydrogen can be stored relatively easily and safely as a liquid even though cold temperatures and high pressures are required. This might be appropriate for large-scale industrial applications where hydrogen could be piped to the plant and burned to heat water to steam, which in turn could generate electricity. However, use of hydrogen as a fuel for vehicles requires a more convenient form of storage, such as an interstitial hydride. A prototype car run by hydrogen combustion in a slightly modified combustion engine is shown in the photo. The hydrogen is stored in a metal hydride from which it is released by heat from the exhaust.

The hydrogen-oxygen reaction may also see further use in fuel cells like those that provide electricity and drinking water in the U.S. space shuttle.

H_2 is used as the fuel in the space shuttle. The cigar-shaped tank contains 385,265 gallons of liquid H_2 and 143,351 gallons of liquid O_2 at liftoff. (NASA)

b. *Molecular hydrides,* which form with nonmetals such as N_2, O_2, and F_2, and include hydrocarbons.

$$N_2(g) + 3\,H_2(g) \longrightarrow 2\,NH_3(g)$$
$$F_2(g) + H_2(g) \longrightarrow 2\,HF(g)$$

c. *Interstitial metallic hydrides,* in which neutral hydrogen atoms combine with metals. Hydrogen (H_2) is a very small molecule, and so it can be absorbed, apparently as H atoms, into the holes or interstices of the crystal lattice of a metal. For example, when a piece of palladium metal is used as an electrode for the electrolysis of water, the metal can soak up a thousand times its volume of H_2 at 1 atmosphere pressure and 0 °C. However, the H_2 can be driven out of the metal by heating. Interstitial hydrides have been considered for storage of H_2 for later use, just as a sponge stores water.

EXERCISE 9.18 • *Reactions of Hydrogen*

(a) Write a balanced chemical equation for the reaction of magnesium with hydrogen gas. (b) Use the periodic table to predict which metal besides palladium might form an interstitial hydride with hydrogen.

A prototype car run by combustion of hydrogen in a slightly modified engine. The hydrogen is stored in the form of an iron-titanium hydride, and heat from the engine exhaust is used to release H_2 from the hydride. *(Mercedes-Benz)*

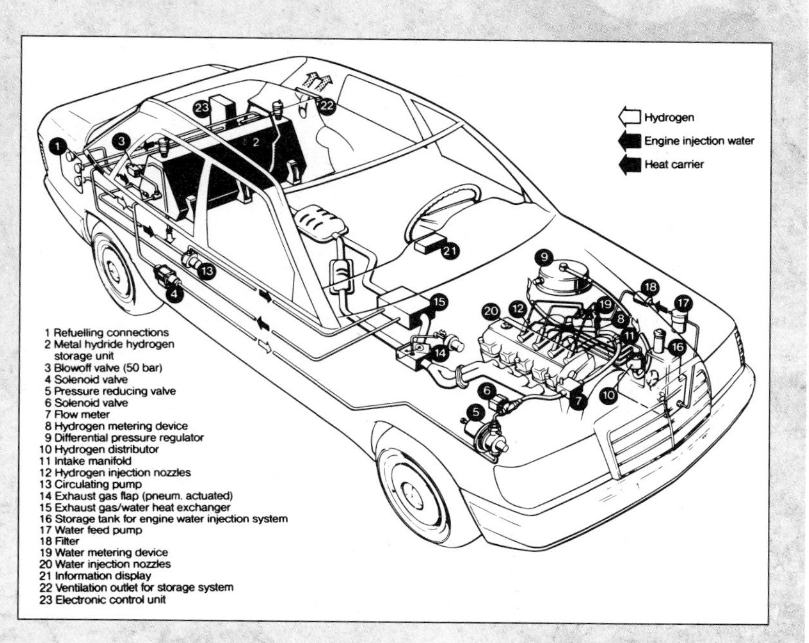

A hydrogen-oxygen fuel cell like the kind used on spacecraft. *(United Technologies)*

A 4.5-megawatt fuel cell power plant in Tokyo. Unlike the fuel cells used in space vehicles, which have an alkaline (KOH) electrolyte, larger plants such as the one in this photo use phosphoric acid as the electrolyte. *(Johnson-Matthey)*

Relative reactivities of Li and K with water. Notice that K is the most reactive and Li the least reactive. *(C.D. Winters)*

The Alkali Metals: Group 1A (ns^1) All the alkali metals are soft enough to be cut with a knife. None are found in nature as free elements, since all combine rapidly and completely with virtually all the nonmetals. Alkali metals readily lose the ns^1 valence electron to form 1+ ions.

Figure 1.3 illustrates the vigorous reaction of sodium with water.

One of the best known reactions of the alkali metals is that with water to form hydroxides, historically known as alkalies. When a small piece of sodium is placed in water, it reacts vigorously, and the heat of reaction sometimes ignites the hydrogen that is produced.

$$2\,Na(s) + 2\,H_2O(\ell) \longrightarrow 2\,NaOH(aq) + H_2(g)$$

The more general reaction $2\,M(s) + 2\,H_2O(\ell) \rightarrow 2\,MOH(aq) + H_2(g)$ can be written, where M = Li, Na, K, Rb, and Cs. Lithium reacts relatively slowly, Na reacts much more rapidly, and K reacts explosively, illustrating the increase in reactivity on going down the group.

1A
1 H hydrogen
3 Li lithium
11 Na sodium
19 K potassium
37 Rb rubidium
55 Cs cesium
87 Fr francium

All of the alkali metals are so reactive that they readily combine with the O_2 and H_2O vapor in the atmosphere and must therefore be stored under an inert liquid, such as mineral oil. Francium, the last member of Group 1A, is found only in trace amounts because all its isotopes are radioactive.

The reactions of alkali metals with oxygen are not as predictable as those with water. Lithium reacts with O_2 to give the expected Li_2O.

$$4\,Li(s) + O_2(g) \longrightarrow 2\,Li_2O(s)$$

Sodium instead gives mainly sodium peroxide, a yellowish-white solid containing the peroxide ion, O_2^{2-}.

$$2\,Na(s) + O_2(g) \longrightarrow Na_2O_2(s)$$

Because of the great oxidizing power of sodium peroxide, it is widely used in the paper and textile industries as a bleaching agent.

Potassium, rubidium, and cesium react with oxygen to give superoxides, MO_2, which are ionic compounds of M^+ and O_2^-.

$$K(s) + O_2(g) \longrightarrow KO_2(s)$$

The most important commercial use of KO_2 is as a source of oxygen in an emergency breathing apparatus. The so-called "oxygen masks" are designed so that the CO_2 and water vapor in the exhaled breath of the wearer react with the superoxide to provide oxygen.

$$4\,KO_2(s) + 4\,CO_2(g) + 2\,H_2O(g) \longrightarrow 4\,KHCO_3(s) + 3\,O_2(g)$$

The reaction of alkali metals with oxygen is an example of one series where the periodic table prediction—namely that all alkali metals would react with oxygen to give M_2O—is not quite correct. However, only the major products are given in the preceding equations. For example, Na also forms some Na_2O, and K forms some K_2O and K_2O_2.

2A
4 Be beryllium
12 Mg magnesium
20 Ca calcium
38 Sr strontium
56 Ba barium
88 Ra radium

EXERCISE 9.19 • *Alkali Metals*

Write a balanced equation for the reaction of Rb with water and compare its reactivity with that of K.

The Alkaline Earth Metals: Group 2A (ns^2) The formulas of compounds of alkaline earth metals with nonmetals are easily predicted because these elements usually are present as 2+ ions, as in CaO, MgF_2, $SrCl_2$, and Ba_3N_2. Compared with the alkali metals, the alkaline earth metals are harder, are more dense, and melt at higher temperatures. The first ionization energies of the alkaline earth elements are low, but not as low as those of the alkali metals. Consequently, the alkaline earths are less reactive than their alkali metal neighbors. The trend of increasing reactivity within the family is illustrated by the behavior of the elements toward water. Beryllium does not react with water or steam, even when heated. Although magnesium does not react with liquid water, it does react with steam to form magnesium oxide and hydrogen.

$$Mg(s) + H_2O(g) \longrightarrow MgO(s) + H_2(g)$$

Calcium and the elements below it react readily with water at room temperature.

$$Ca(s) + 2\,H_2O(\ell) \longrightarrow Ca(OH)_2(aq) + H_2(g)$$

In the presence of O_2, magnesium metal is protected from many chemicals by a thin surface coating of water-insoluble MgO. As a result, Mg can be incorporated into structural alloys even though it is a very reactive metal. Several hundred thousand tons of magnesium are produced annually, and most of it is used in lightweight alloys for aircraft and automotive parts. Magnesium is generally isolated from ocean water (Section 18.1).

The heavier alkaline earth metals (Ca, Sr, and Ba) are even more reactive toward nonmetals than magnesium and must be protected from oxidation by O_2 and H_2O. Magnesium reacts with acids to give hydrogen gas.

$$Mg(s) + 2\,H_3O^+(aq) \longrightarrow Mg^{2+}(aq) + H_2(g) + 2\,H_2O(\ell)$$

Calcium, strontium, and barium also react with acids to produce hydrogen gas but, because they also react with water, two different reactions occur simultaneously.

Calcium reacts with water to give H_2 and $Ca(OH)_2$. *(C.D. Winters)*

Gypsum wall board. *(C.D. Winters)*

The principal source of commercial calcium compounds is limestone ($CaCO_3$). Gypsum, or hydrated calcium sulfate, is also extensively mined. Some of it is used in portland cement (Section 14.4), but most is heated in large kilns to form "plaster of Paris," a process called *calcining.*

$$CaSO_4 \cdot 2\,H_2O(s) \longrightarrow CaSO_4 \cdot \tfrac{1}{2}\,H_2O(s) + \tfrac{3}{2}\,H_2O(g)$$

If enough water is added to plaster of Paris to make a paste, it quickly hardens as it reverts to gypsum. The mixture also expands as it hardens, so it forms a sharp impression of anything molded in it (Figure 4.19). Most calcined gypsum is used to make wallboard (sheet rock or plaster board), and the rest is used for industrial and building plasters. In fact, this use of gypsum is very old; there is evidence that the interiors of some of the great pyramids of Egypt were coated with gypsum plaster.

EXERCISE 9.20 • *Alkaline Earth Metals*

Use your knowledge of valence electrons and the periodic table to predict formulas for the following compounds. (a) Sr and N, (b) Mg and Br, (c) Ca and P, (d) Ba and O.

EXERCISE 9.21 • *Alkali and Alkaline Earth Metals*

Explain why the heavier alkali metals and alkaline earth metals are more reactive than the lighter elements in these groups.

The *p*-Block Elements

3A
5 B boron
13 Al aluminum
31 Ga gallium
49 In indium
81 Tl thallium

The elements of Groups 3A and 4A form a bridge between the metals of Groups 1A and 2A and the largely nonmetallic elements of Groups 5A through 8A. Boron is a metalloid but the other elements of Group 3A are metals. Group 4A begins with carbon, a nonmetal, followed by the metalloids silicon and germanium, and the metals tin and lead.

Group 3A (ns^2np^1) Although boron is very low in relative abundance, its common minerals are found in concentrated deposits, especially in California. Large deposits of *borax,* $Na_2B_4O_7 \cdot 10\,H_2O$, are currently mined in the Mojave Desert near the town of Boron (Figure 9.21). Borax has been used for centuries as a low-melting flux in metallurgy because of the ability of molten borax to dissolve other metal oxides, cleaning the surfaces to be joined, and permitting good metal-to-metal contact. After refinement, borax can be treated with sulfuric acid and converted to boric acid, $B(OH)_3$.

A flux *is added to a metal or mineral to lower its melting point and protect it from forming oxides.*

$$Na_2B_4O_7 \cdot 10\,H_2O(s) + H_2SO_4(aq) \longrightarrow 4\,B(OH)_3(aq) + Na_2SO_4(aq) + 5\,H_2O(\ell)$$

Boric acid is dehydrated to boric oxide when heated strongly.

$$2\,B(OH)_3(s) \longrightarrow B_2O_3(s) + 3\,H_2O(g)$$

The largest use of borax and of boric oxide is in the manufacture of borosilicate glass. This type of glass is composed of about 76% SiO_2, 13% B_2O_3, and much smaller amounts of Al_2O_3 and Na_2O. The presence of boric oxide

Figure 9.21
An aerial view of the open pit borax mine of the U.S. Borax and Chemical Corporation in the Mojave Desert near Boron, California. *(Rick McIntyre, Tom Stack & Associates)*

gives the glass a higher softening temperature and a better resistance to attack by acids, and makes it expand less on heating.

One common brand of borosilicate glass is Pyrex, the trademark used by Corning Glass Company.

The production of aluminum has grown rapidly because aluminum has thousands of uses as a structural material and in packaging. However, pure aluminum is rarely used, because it is soft and weak, and it loses strength rapidly above 300 °C. To strengthen the metal and improve its properties, it is alloyed with small amounts of other metals. A typical alloy, for example, may contain about 4% copper with smaller amounts of silicon, magnesium, and manganese, and a large passenger plane today may use more than 50 tons of this alloy. However, to make a softer, more corrosion-resistant alloy for window frames, furniture, highway signs, and cooking utensils, a combination of aluminum and manganese is used.

Much of the usefulness of aluminum comes from its corrosion resistance, which is due to the formation of a thin, tough, and transparent skin of oxide, Al_2O_3. If you scratch a piece of aluminum, it forms a new layer of oxide that covers the damaged area.

Pyrex glass is a borosilicate glass. *(C.D. Winters)*

Alloys of aluminum and magnesium are used in the manufacture of airplanes. *(Kolvoord/The Image Works)*

Figure 9.22
Copper reacts vigorously with nitric acid (right) to give $Cu(NO_3)_2(aq)$ and $NO_2(g)$, but aluminum (left) is unreactive in nitric acid. *(C.D. Winters)*

Figure 9.23
Metallic Ga melts at 30 °C, which is less than body temperature (37 °C). *(C.D. Winters)*

A gallium arsenide (GaAs) semiconductor.

Aluminum will dissolve in acids such as HCl, but not in nitric acid (Figure 9.22).

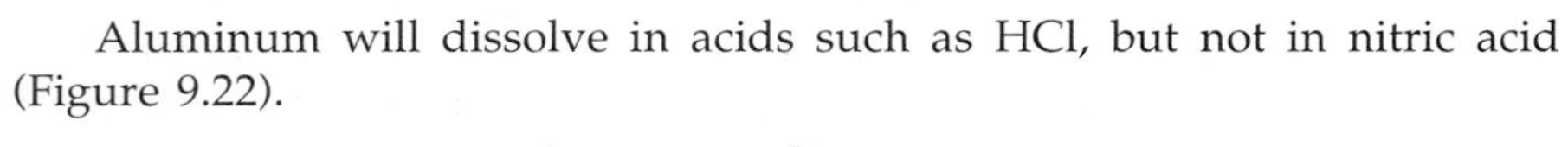

$$2\ Al(s) + 6\ H_3O^+ \longrightarrow 2\ Al^{3+}(aq) + 3\ H_2(g) + 6\ H_2O(\ell)$$

Because HNO_3 is a powerful oxidizing agent, it oxidizes the surface of the aluminum rapidly, and Al_2O_3 protects the metal from further attack. In fact, nitric acid is often shipped in aluminum tanker trucks.

Gallium was one of the elements that was not known at the time Mendeleev developed his periodic table (Section 3.6), but his prediction of its existence and properties helped greatly in its discovery just a few years later. It is truly a remarkable element. It has the greatest liquid range of all known elements; it can melt in your hand (mp = 30 °C) (Figure 9.23), but it does not boil until the temperature reaches 2403 °C. Finally, like water, gallium is one of the few known materials that expands upon freezing. The greatest use for gallium, and one that may continue to grow, is in the semiconductor gallium arsenide, GaAs. Integrated circuits based on GaAs have achieved operating speeds up to five times that of the fastest silicon chips currently available, and they will operate over a wider temperature range than silicon circuits.

4A
6 C carbon
14 Si silicon
32 Ge germanium
50 Sn tin
82 Pb lead

Group 4A (ns^2np^2) Carbon is found in fossil fuels and in all living matter. More than ten million of the twelve million known compounds are compounds of carbon, and a separate branch of chemistry, **organic chemistry,** is devoted to the study of these compounds (see Chapter 13). The largest natural sources of carbon, however, are the carbonate minerals limestone and dolomite. The allotropic forms of carbon were discussed in Section 4.1. Carbon also occurs as CO_2 in the atmosphere. Carbon burns in the presence of excess oxygen to give CO_2, while CO is the primary product when insufficient oxygen is available.

More than 30 million tons of CO_2 are produced annually in the United States, and roughly half of it is used as a refrigerant in the solid form. Solid CO_2, known as *Dry Ice,* is a useful refrigerant because it sublimes at −78 °C at atmospheric pressure (Section 7.3). Although once used mostly in the form of the solid, CO_2 is increasingly being used as a liquid refrigerant and as the propellant gas in aerosol cans. Approximately 25% of the CO_2 manufactured is used to carbonate beverages. More than 400 bottles of carbonated beverages are produced per person in the United States each year.

Carbon monoxide is a toxic gas. It reacts with the iron of the hemoglobin in the blood and displaces O_2. Exposure to even small concentrations (120 parts per million, or ppm) can impair your abilities, and concentrations as high as 100 ppm are not unusual in tunnels, garages, and even the streets of large cities (Section 20.4). One of the major uses of carbon monoxide is as a fuel, primarily in a mixture with hydrogen known as **synthesis gas.** This mixture is prepared by passing steam over a bed of glowing coke.

$$C(s) + H_2O(g) \longrightarrow CO(g) + H_2(g)$$
$$\text{coal + steam} \longrightarrow \text{synthesis gas}$$

Synthesis gas is also used to make methanol and other chemicals (Section 13.4).

Pewter is an alloy that contains 85% Sn, 7% Cu, and small amounts of Bi and Sb. *(C.D. Winters)*

Silicon, second only to oxygen as the most abundant element in the earth's crust, occurs in silica (SiO_2) and silicates (Section 14.4). Reasonably pure silicon is made in large quantities by heating pure silica sand with purified coke to approximately 3000 °C in an electric furnace.

$$SiO_2(s) + 2\,C(s) \longrightarrow Si\ (\text{liquid at 3000 °C}) + 2\,CO(g)$$

The molten silicon is drawn from the bottom of the furnace and allowed to cool to a shiny blue-gray solid.

Tin is obtained from *cassiterite*, SnO_2, an oxide that can be reduced easily to the metal with glowing charcoal.

The solder used in electrical circuits averages 33% tin.

$$SnO_2(s) + C(s) \longrightarrow Sn(s) + CO_2(g)$$

The 210,000 tons of cassiterite mined today come from such countries as Malaysia (25%) and Bolivia (14%), and the United States is the largest consumer. Tin is an expensive and fairly rare metal. It is very resistant to corrosion, so almost 40% of the metal produced is used to plate the surfaces of other metals, as in the tin-coated iron cans that you have always called "tin cans." The coating, typically 0.0004 to 0.025 mm thick, is applied by dipping the object in molten tin or by electroplating.

Lead is obtained from galena ore, PbS, by roasting the ore with oxygen to produce PbO, then reducing the PbO with charcoal to give Pb. Approximately one million tons of lead are produced in the United States annually, and most of this (60%) is used in an alloy for storage battery plates.

EXERCISE 9.22 • *Group 4A*

Complete and balance the following equations.

a. $GeO_2(s) + C(s) \longrightarrow$

b. $PbS(s) + O_2(g) \longrightarrow$

A lead storage battery. *(C.D. Winters)*

Group 5A (ns^2np^3) Nitrogen and phosphorus are nonmetals, arsenic and antimony are metalloids, and bismuth is a metal. Ammonia (Section 8.3),

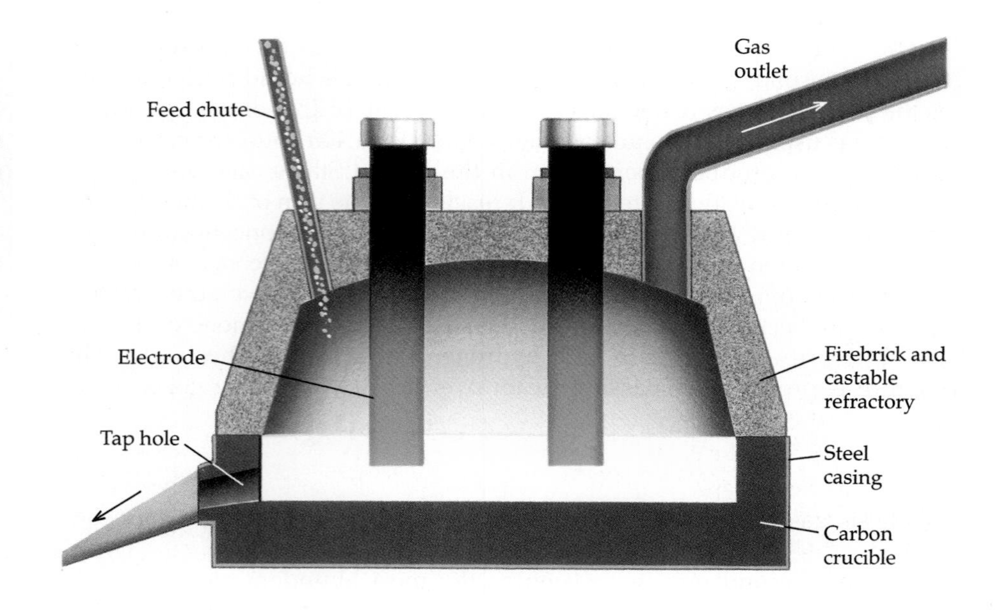

Figure 9.24
An electric furnace for production of phosphorus. The "feed" is a mixture of $Ca_3(PO_4)_2$, SiO_2, and C. A mixture of P_4 gas and CO is driven off at the top of the furnace, and molten slag containing calcium silicate and other substances is drawn off at the bottom.

nitric acid, and phosphoric acid are among the most important compounds in our economy.

Nitrogen forms seven binary oxides, all of them thermodynamically unstable with respect to decomposition to N_2 and O_2. The most important oxides include N_2O, NO, and NO_2. Dinitrogen monoxide, N_2O, can be made by the careful decomposition of ammonium nitrate at 250 °C.

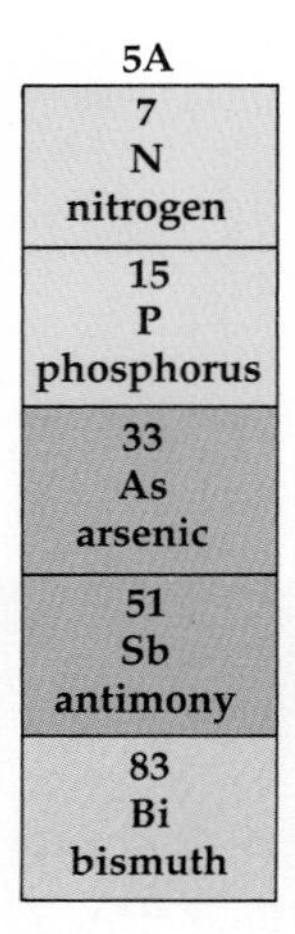

$$NH_4NO_3(s) \longrightarrow N_2O(g) + 2\,H_2O(g)$$

Nitrogen monoxide, NO, is an intermediate in the synthesis of nitric acid by the oxidation of ammonia, and is also produced in low yields from the direct reaction of N_2 and O_2 at high temperatures.

Matches. The head of a "strike anywhere" match contains, among other things, P_4S_3 and an oxidizing agent, potassium chlorate ($KClO_3$). Safety matches have sulfur (3 to 5%) and $KClO_3$ (45 to 55%) in the head and red phosphorus in the striking strip. *(C.D. Winters)*

Much of the chemistry of phosphorus compounds can be organized around the reactions of the P_4 tetrahedron of elemental phosphorus (Sections 4.1 and 11.3). Most phosphorus is obtained from the calcium phosphate that occurs naturally in apatite minerals with the general formula $Ca_5X(PO_4)_3$, where X may be F^-, Cl^-, or OH^-. An electric furnace is used to heat a mixture of the phosphate ore, silica (SiO_2), and coke (Figure 9.24). At 1400–1500 °C, the following reaction produces elemental phosphorus vapor:

$$2\,Ca_3(PO_4)_2(\ell) + 6\,SiO_2(\ell) + 10\,C(s) \longrightarrow 6\,CaSiO_3(\ell) + 10\,CO(g) + P_4(g)$$

The phosphorus is then condensed from the gaseous state and purified. The low melting point of white phosphorus (44.1 °C) means that it can be stored and handled as a liquid, although it must be protected from contact with air because it ignites spontaneously. Red phosphorus is stable in dry air.

About 90% of the solid P_4 produced (Figure 9.25) is oxidized with air to give P_4O_{10}, which reacts with water to produce phosphoric acid, H_3PO_4. (Most of the remaining 10% of P_4 production is used to make phosphorus sulfides for matches.)

$$P_4(s) + 5\,O_2(g) \longrightarrow P_4O_{10}(s)$$

$$P_4O_{10}(s) + 6\,H_2O(\ell) \xrightarrow{H_2O} 4\,H_3PO_4(aq)$$

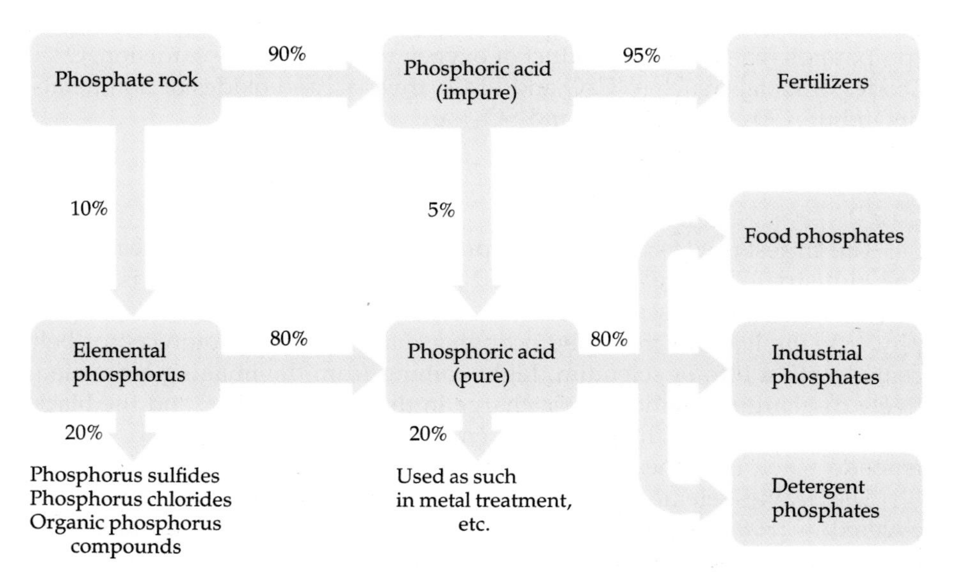

Figure 9.25 Uses of phosphate-containing rock.

Some common products that contain phosphate. *(C.D. Winters)*

When arsenic is present in the original phosphorus ore, it is carried through the process and must be removed from the product if the phosphoric acid is to be used in food products.

Phosphoric acid and its salts are used to manufacture many materials we encounter daily, including baking powder, carbonated beverages, detergents, fertilizers, and fire-resistant textiles.

Arsenic, antimony, and bismuth are among the oldest known elements, in spite of their low abundances. Although all three form stable oxides, they resemble the transition metals in forming very stable sulfides, and they are commonly found in this form. Lemon-yellow *orpiment*, As_2S_3, was used by physicians and assassins for hundreds of years, and black *stibnite*, Sb_2S_3, was used as a cosmetic hundreds of years ago. Like most metals, these elements can be obtained by reducing a mineral source. Scrap iron, for example, is used to reduce stibnite to elemental antimony.

$$3\ Fe(s) + Sb_2S_3(s) \longrightarrow 3\ FeS(s) + 2\ Sb(s)$$

A principal use of arsenic and antimony is in automobile batteries. The battery plates are made of lead, but their performance is greatly improved by adding 1–3% Sb and a trace of As. The consumption of antimony in the United States is about 12,000 tons per year.

The more metallic nature of bismuth is clear from the chemistry of the oxides of Group 5A elements. Nitrogen and phosphorus oxides are decidedly acidic; arsenic and antimony oxides behave both as acids and bases; and Bi_2O_3 is basic. This means that, while As_4O_6 and Sb_4O_6 are soluble in both acid and base, Bi_2O_3 is soluble only in acid.

6A
8 O oxygen
16 S sulfur
34 Se selenium
52 Te tellurium
84 Po polonium

Group 6A (ns^2np^4) Oxygen, sulfur, and selenium are nonmetals; tellurium is a metalloid; polonium is a metal. The major industrial source of oxygen is from air, and sulfur is mined as the free element.

Oxygen reacts with virtually all the elements. Some elements such as W, Pt, and Au do not combine *directly* with O_2, but all the elements—except the noble gases He, Ne, Ar, and possibly Kr—can be made to form oxygen compounds. The ultimate product of oxygen reduction is the oxide ion, O^{2-}. Oxides of metals (such as CaO and Li_2O) are *basic*, and oxides of nonmetals (including CO_2, NO_2, P_4O_{10}, and SO_3) are *acidic*.

$$CaO(s) + H_2O(\ell) \longrightarrow Ca(OH)_2(aq)$$
$$SO_3(g) + H_2O(\ell) \longrightarrow H_2SO_4(aq)$$

The largest use of sulfur is in the production of sulfuric acid, the number one industrial chemical in the United States. Although selenium is rare, it has a wide range of uses. The use you are most familiar with is in "xerography." At the heart of most photocopying machines is a photoreceptor belt coated with a film of selenium. Light coming from the imaging lens selectively discharges a static electric charge in the selenium film, and the black "toner" sticks only to the areas that remained charged (Figure 9.26). A copy is made when the toner is transferred to a sheet of plain paper.

The heaviest element in Group 6A, polonium, is radioactive. It was discovered in 1898 by Marie Sklodowska Curie (1867–1934) and her husband, Pierre Curie (1869–1906) (Section 19.1). The Curies painstakingly separated the elements in a large quantity of uranium-containing ore, *pitchblende*, and

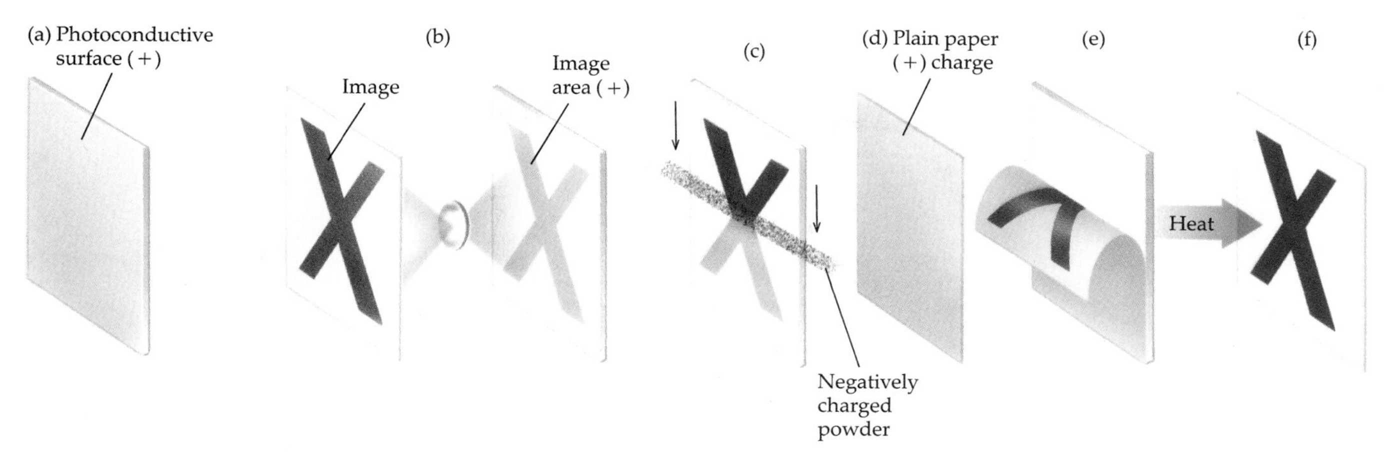

Figure 9.26

Basic xerography. (a) A photoconductive surface is given a positive electric charge (+). (b) The image of the document is exposed on the surface. The light energy causes the charge to drain away from the surface in all but the image area, which remains unexposed and charged. (c) Negatively charged carbon powder is cascaded over the surface. It adheres electrostatically to the positively charged image area, making a visible image. (d) A piece of plain paper is placed over the surface and given a positive charge. (e) The negatively charged powder image on the surface is electrostatically attracted to the positively charged paper. (f) The powder image is fused to the paper by heat.

found the new elements radium and polonium. The latter was named in honor of Poland, Marie Curie's home country.

7A
9 **F** **fluorine**
17 **Cl** **chlorine**
35 **Br** **bromine**
53 **I** **iodine**
85 **At** **astatine**

Group 7A (ns^2np^5) The Group 7A elements, the halogens, are nonmetals. Fluorine is the most reactive of all the elements, forming compounds with all the other elements with the exception of He, Ne, and Ar. Halogens form ionic salts with metals (NaF, $BaCl_2$) and molecular compounds with other nonmetals (CCl_4, PBr_3). Almost 55% of the fluorine produced is used for processing and reprocessing uranium for fueling nuclear power plants (Section 12.3).

Other major uses of fluorine and its compounds are in the polymer Teflon, the nonstick surface on cooking utensils (Section 13.6), and in NaF, used in fluoridation of water and toothpaste.

The estimated worldwide production of hydrogen fluoride is about one million tons annually, virtually all by the action of concentrated sulfuric acid on *fluorspar,* the chief mineral containing fluorine.

Fluorspar. *(C.D. Winters)*

$$CaF_2(s) + H_2SO_4(\ell) \longrightarrow CaSO_4(s) + 2\,HF(g)$$

A large fraction of the HF produced is used to make *cryolite* (Na_3AlF_6), which is necessary for making aluminum (Section 18.1) but is found in only small amounts in nature.

$$6\,HF(aq) + Al(OH)_3(s) + 3\,NaOH(aq) \longrightarrow Na_3AlF_6(s) + 6\,H_2O(\ell)$$

Hydrogen chloride ranks 26th among industrial chemicals, with the annual worldwide production being about three million tons. Although it can

Figure 9.27
Hydrogen reacts with chlorine to form hydrogen chloride, a colorless gas. Hydrogen chloride combines with moisture in the air to form the fog visible in the photo. *(C.D. Winters)*

be made by burning hydrogen in chlorine (Figure 9.27), over 90% is

$$H_2(g) + Cl_2(g) \longrightarrow 2\,HCl(g)$$

obtained as a by-product of the organic chemicals industry.

Hydrochloric acid is an aqueous solution of hydrogen chloride, which ionizes when it is mixed with water.

Group 8A (ns^2np^6) All the elements in Group 8A exist as monatomic gases. Helium, the lightest element of the group, is the second most abundant element in the universe after hydrogen. The noble gases together make up about 1% of the earth's atmosphere. Historically, they became known as noble gases because of their lack of reactivity. However, the discovery of xenon compounds in 1962 proved that the heavier members of the group have some reactivity.

Casino neon signs on Fremont Street in Las Vegas, Nevada. *(Dale E. Boyer/Photo Researchers)*

The discovery of the first noble gas compound provides us with a lesson: the most interesting discoveries are often the result of lucky accidents, and we must always be prepared to interpret an observation in a new way. While studying the chemistry of PtF_6, Neil Bartlett noticed quite accidentally that exposing it to air led to a new compound that he showed to be $[O_2^+][PtF_6^-]$. Just as importantly, Bartlett recognized that the ionization energy of O_2 to form O_2^+ (1175 kJ/mol) is comparable to that of Xe (1170 kJ/mol). He quickly proceeded to treat PtF_6 with Xe and isolated a red crystalline solid, which was assigned the formula $[Xe^+][PtF_6^-]$. (Later work indicated that the reaction at 25 °C gives a mixture of $[XeF^+][PtF_6^-]$ and PtF_5 which combines when heated to 60 °C to give $[XeF^+][Pt_2F_{11}^-]$.)

Discovery by chance is sometimes attributed to serendipity but Louis Pasteur observed that in science "chance favors the prepared mind."

Very soon after Bartlett's work, other xenon fluorides were discovered, and the field has been an active one ever since. Isolable compounds have been obtained only for Xe and Kr, but were it not for the intense radioactivity of Rn, compounds of this element would surely be isolated. Virtually all known noble gas compounds contain bonds to F or O.

Among the best known noble gas compounds are the first discovered, the simple xenon fluorides XeF_2, XeF_4, and XeF_6 (Figure 9.28). All are colorless, volatile solids that can be prepared by combining the elements under carefully controlled conditions. It is ironic, in view of the historical belief in the inertness of these elements, that XeF_2 can be obtained simply by exposing a mixture of Xe and F_2 to sunlight.

$$Xe(g) + F_2(g) \longrightarrow XeF_2(s)$$

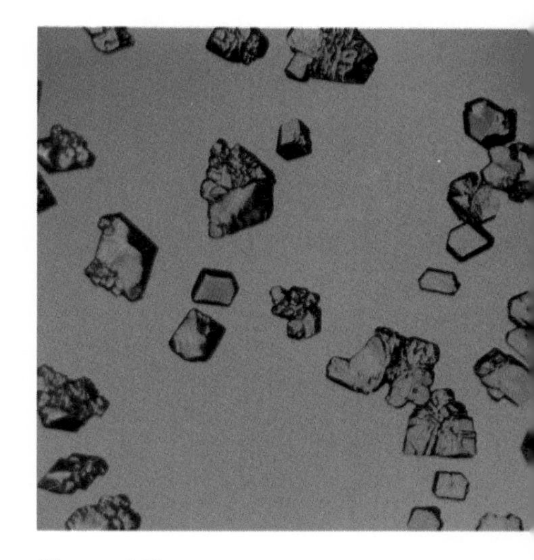

Figure 9.28
Crystals of xenon tetrafluoride, XeF_4. *(Argonne National Laboratory)*

The *d*- and *f*-Block Elements

In main group elements you've seen that chemical similarities occur mainly within a group, and the chemistry changes markedly across a given period as the number of valence electrons changes. In contrast, the transition and inner transition metals have similar chemistry within a given period as well as within a group. This behavior can be explained by the similarities in atomic radii and ionization energies for these elements (Section 9.5). Many transition elements form ionic compounds, and the charges on the ions are

The series of compounds $MnCl_2$, MnF_3, MnO_2, and $KMnO_4$ illustrates the variety of compounds formed by manganese, a transition metal.

Some common minerals: *left*, iron pyrite or "fool's gold"; *center*, calcite; and *right*, fluorite. *(C.D. Winters)*

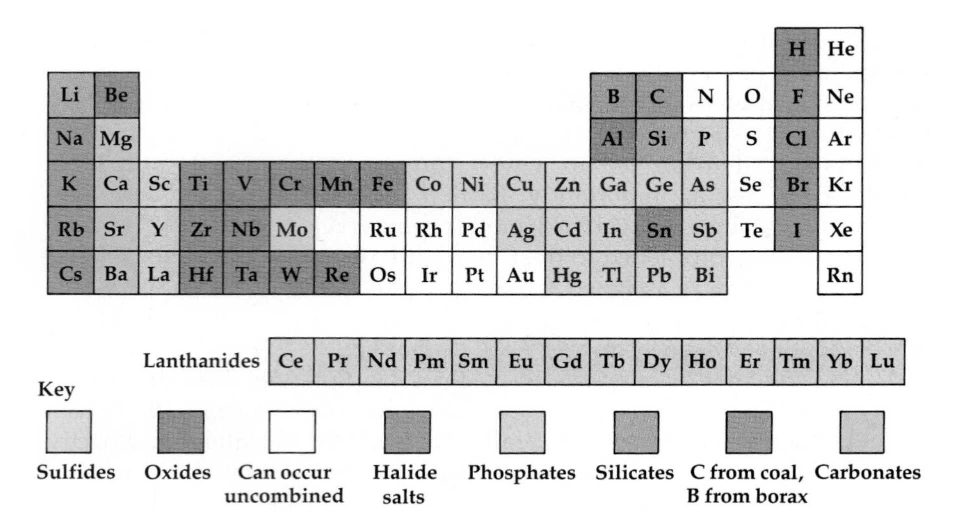

Figure 9.29
Sources of the elements. The *d*-block elements are found uncombined or as oxides, sulfides, or halides. The lanthanides occur predominantly as phosphate salts. The blank spaces in the table are for Tc (below Mn), Po (below Te), and At (below I), which are not found in nature.

often determined by loss of electrons from the *s* orbitals in the highest occupied energy levels. In some cases, however, the transition metals also form ions in which electrons have been lost from their incompletely filled *d* orbitals. Such participation in bonding by electrons from *d* orbitals accounts for the variety of compounds formed by transition metals and the difficulty of predicting the charges on their ions.

Most metals are found in nature as oxides, sulfides, halides, carbonates, or other salts (Figure 9.29). Many metal-containing mineral deposits are of little value, either because the deposit is impure or because the metal is too difficult to separate from impurities. The relatively few minerals from which elements can be obtained profitably are called **ores,** and these are the ones listed in Figure 9.29. Very few ores are chemically pure substances. They are usually mixtures of the desired mineral and large quantities of impurities such as sand and clay minerals, called **gangue** (pronounced "gang"). Therefore, a major step in obtaining the desired metal is to separate its mineral from the gangue. This procedure is discussed for iron and copper ores in Section 18.2.

The *d*-block elements are important in living organisms, in certain industrial processes, and in alloys and structural materials. For example, cobalt is the crucial element in vitamin B_{12}, where it acts as a catalyst. Iron is the key element in biochemical oxidation-reduction processes using hemoglobin or myoglobin. Molybdenum and iron, together with sulfur, form the reactive portion of nitrogenase, a biological catalyst used by plants to convert atmospheric nitrogen into ammonia, and copper and zinc are important in other biological catalysts. Iron is a catalyst in the Haber process for ammonia, and rhodium and platinum are used in automobile catalytic converters (Section 8.3).

The most apparent uses of transition elements are as the metals themselves (Figure 9.30). They are used in coins in many countries around the world and are the primary structural materials in cars, appliances, and many large buildings. Approximately 700 million tons of raw steel, 8 million tons of copper, and 750,000 tons of nickel are produced annually worldwide to meet these needs.

Lanthanides are used in the fluorescent screens of color television sets.
(C.D. Winters)

(a) *(b)*

Figure 9.30
Some uses of transition metals. (a) Steel is used in skyscraper construction. (b) Titanium is used in construction of space and high-speed aircraft.
(*a*, J. Sohm/The Image Works; *b*, R. Isear/Photo Researchers)

Lanthanides have fewer applications than transition metals. Mixtures of lanthanides are used in the fluorescent screens of color television sets. Praseodymium and neodymium are used in sunglasses and glassblowers' goggles to filter out ultraviolet radiation.

In 1982 the U.S. penny was changed from 95% copper and 5% zinc to 97.6% zinc and 2.4% copper (added as a thin coating to a zinc slug). The change was made because the value of the copper in the pre-1982 penny was greater than one cent.

IN CLOSING

Having studied this chapter, you should be able to

- explain the relationship between Planck's quantum theory and the energy absorbed or emitted when electrons in atoms change energy levels (Sections 9.1 and 9.2).
- describe the limitations of the Bohr model of the atom (Section 9.2).
- explain the use of the quantum mechanical model of the atom to represent the energy and probable location of electrons (Section 9.2).
- define and explain the relationships between shells, subshells, and orbitals (Section 9.2).
- use the periodic table to write the electron configurations of atoms and ions of main group and transition elements (Section 9.4).
- give the orbital locations and numbers of valence electrons for main group elements (Section 9.4).
- explain variations in electron configurations, ion formation, and paramagnetism of transition metals (Section 9.4).
- use the periodic table to predict trends in the radii of atoms and ions, ionization energies, and periodic properties (Sections 9.5 and 9.6).
- list some properties of elements and their compounds from each region of the periodic table (Section 9.6).

STUDY QUESTIONS

Review Questions

1. How is the frequency of electromagnetic radiation related to its wavelength?
2. What is a photon? How are the energies of photons calculated?
3. Bohr pictured the electrons of the atom as being located in definite orbits about the nucleus, just as planets orbit the sun. Criticize this model in view of the quantum mechanical model.
4. Light is given off by a sodium- or mercury-containing streetlight when the atoms are excited in some way. The light you see arises for which of the following reasons?
   (a) electrons moving from a given quantum level to one of higher n.
   (b) electrons being removed from the atom, thereby creating a metal cation.
   (c) electrons moving from a given quantum level to one of lower n.
   (d) electrons whizzing about the nucleus in an absolute frenzy.
5. What is the Pauli exclusion principle?
6. What is Hund's rule? Give an example of the use of this rule.
7. Explain what it means when an electron occupies the $3p_x$ orbital.
8. How many electrons can be accommodated in the n = 4 shell?
9. Tell what happens to atomic size and ionization energy when proceeding across a period and down a group.
10. Why is the radius of Li^+ so much smaller than the radius of Li? Why is the radius of F^- so much larger than the radius of F?
11. Write electron configurations to show the first two ionization processes for potassium. Explain why the second ionization energy is much higher than the first.
12. Explain how the sizes of atoms change, and why they change, when proceeding across a period of the periodic table.
13. What is meant by the term the noble gas notation? Write an electron configuration using this notation.
14. Name an element of Group 3A. What does the group designation tell you about the electron configuration of the element?
15. Name an element of Group 6B. What does the group designation tell you about the electron configuration of the element?
16. Which ions in the following list are likely to be formed: K^{2+}, Cs^+, Al^{4+}, F^{2-}, and Se^{2-}? Which, if any, of these ions have a noble gas configuration?
17. Look at a periodic table and write the electron configurations for the valence electrons in Groups 1A through 8A.

Electromagnetic Radiation

18. The regions of the electromagnetic spectrum are shown in Figure 9.1. Answer the following questions on the basis of this figure.
    (a) Which type of radiation involves less energy: radar or infrared light?
    (b) Which radiation has the higher frequency: radar or microwaves?
19. The colors of the visible spectrum, and the wavelengths corresponding to the colors, are given in Figure 9.1.
    (a) What colors of light involve less energy than yellow light?
    (b) Which color of visible light has photons of greater energy: green or violet?
    (c) Which color of light has the greater frequency: blue or green?
20. Assume a microwave oven operates at a frequency of $1.00 \times 10^{11}\ s^{-1}$. What is the wavelength of this radiation in meters? What is the energy in joules per photon? What is the energy per mole of photons?
21. The U.S. Navy has a system for communicating with submerged submarines. The system uses radio waves with a frequency of $76\ s^{-1}$. What is the wavelength of this radiation in meters? In miles? (1 mile = 1.61 km)
22. Place the following types of radiation in order of increasing energy per photon:
    (a) green light from a mercury lamp
    (b) x rays from an instrument in a dentist's office
    (c) microwaves in a microwave oven
    (d) an FM music station at 96.3 MHz
23. Place the following types of radiation in order of increasing energy per photon:
    (a) radar signals
    (b) radiation from a microwave oven
    (c) gamma rays from a nuclear reaction
    (d) red light from a neon sign
    (e) ultraviolet radiation from a sun lamp

Atomic Spectra and the Bohr Atom

24. The energy emitted when an electron moves from a higher energy level to one of lower energy in any atom can be observed as electromagnetic radiation. Which involves the emission of less energy in the H atom, an electron moving from $n = 4$ to $n = 3$ or an electron moving from $n = 3$ to $n = 1$? (See Figure 9.8.)

25. If energy is absorbed by a hydrogen atom in its ground state, the atom is excited to a higher energy state. For example, the excitation of an electron from the energy level with $n = 1$ to a level with $n = 4$ requires radiation with a wavelength of 97.3 nm. Which of the following transitions would require radiation of *longer wavelength* than this? (See Figure 9.8.)
(a) $n = 2$ to $n = 4$ (c) $n = 1$ to $n = 5$
(b) $n = 1$ to $n = 3$ (d) $n = 3$ to $n = 5$

Quantum Mechanics

26. From memory, sketch the shape of the boundary surface for each of the following atomic orbitals:
(a) $2p_z$ (b) $3d_{yz}$

27. How many subshells are there in the electron shell with the principal quantum number $n = 4$?

28. How many subshells are there in the electron shell with the principal quantum number $n = 5$?

Electron Configurations of Main Group Elements

29. Write electron configurations for Mg and Cl atoms.

30. Write electron configurations for Al and S atoms.

31. Write electron configurations for atoms of the following:
(a) Strontium, Sr, named for a town in Scotland.
(b) Tin, Sn, a metal used in the ancient world. Alloys of tin (solder, bronze, and pewter) are important.

32. Germanium had not been discovered when Mendeleev formulated his ideas of chemical periodicity. He predicted its existence, however, and it was found in 1886 by Winkler. Write its electron configuration.

Valence Electrons

33. Locate the following elements in the periodic table, and draw a Lewis dot symbol that represents the number of valence electrons for an atom of each element.
(a) F (b) In (c) Te (d) Cs

34. Locate the following elements in the periodic table, and draw a Lewis dot symbol that represents the number of valence electrons for an atom of each element.
(a) Sr (b) Br (c) Ga (d) Sb

35. Give the electron configurations of the following ions, and indicate which ions are isoelectronic: (a) Na^+, (b) Al^{3+}, and (c) Cl^-.

36. Give the electron configurations of the following ions, and indicate which ions are isoelectronic: (a) Mg^{2+}, (b) K^+, and (c) O^{2-}.

Electron Configurations of Transition Elements

37. Give the electron configurations of Mn, Mn^{2+}, and Mn^{3+}. Use orbital box diagrams to determine the number of unpaired electrons for each species.

38. Write the electron configurations of chromium, Cr, Cr^{2+}, and Cr^{3+}. Use orbital box diagrams to determine the number of unpaired electrons for each species.

39. Give the electron configuration of vanadium, V. (The name of the element was derived from Vanadis, a Scandinavian goddess.)

40. Write electron configurations for the following:
(a) Zirconium, Zr. This metal is exceptionally resistant to corrosion and so has important industrial applications. Moon rocks show a surprisingly high zirconium content compared with rocks on earth.
(b) Rhodium, Rh, used in jewelry and in industrial catalysts.

41. The lanthanides, or rare earths, are now only "medium rare." All can be purchased for a reasonable price. Give electron configurations for atoms of the following elements.
(a) Europium, Eu. It is the most expensive of the rare earth elements; one gram can be purchased for $50 to $100.
(b) Ytterbium, Yb. It is less expensive than Eu, as Yb costs only about $15 per gram. It was named for the village of Ytterby in Sweden, where a mineral source of the element was found.

Periodic Trends

42. Arrange the following atoms in the order of increasing ionization energy: F, Al, P, and Mg.
43. Arrange the following atoms in the order of increasing ionization energy: Li, K, C, and N.
44. Which of the following groups of elements is arranged correctly in order of increasing ionization energy?
(a) C < Si < Li < Ne (c) Li < Si < C < Ne
(b) Ne < Si < C < Li (d) Ne < C < Si < Li
45. Rank the following ionization energies in order from the smallest value to the largest value. Briefly explain your answer.
(a) first IE of Be (d) second IE of Na
(b) first IE of Li (e) first IE of K
(c) second IE of Be
46. Predict which of the following elements would have the greatest difference between the first and second ionization energies: Si, Na, P, and Mg. Briefly explain your answer.
47. Arrange the following elements in order of increasing size: Al, B, C, K, and Na. (Try doing it without looking at Figure 9.17 and then check yourself by looking up the necessary atomic radii.)
48. Arrange the following elements in order of increasing size: Ca, Rb, P, Ge, and Sr. (Try doing it without looking at Figure 9.17 and then check yourself by looking up the necessary atomic radii.)
49. Select the atom or ion in each pair that has the larger radius.
(a) Cl or Cl^-
(b) Al or N
(c) In or Sn
50. Select the atom or ion in each pair that has the larger radius.
(a) Cs or Rb (b) O^{2-} or O (c) Br or As
51. Compare the elements Li, K, C, and N.
(a) Which has the largest atomic radius?
(b) Place the elements in order of increasing ionization energy.
52. Compare the elements B, Al, C, and Si.
(a) Which has the most metallic character?
(b) Which has the largest atomic radius?
(c) Place the three elements B, Al, and C in order of increasing first ionization energy.
53. (a) Place the following elements in order of increasing ionization energy: F, O, and S.
(b) Which has the largest ionization energy: O, S, or Se?
54. (a) Rank the following in order of increasing atomic radius: O, S, and F.
(b) Which has the largest ionization energy: P, Si, S, or Se?
(c) Place the following in order of increasing radius: Ne, O^{2-}, N^{3-}, and F^-.
(d) Place the following in order of increasing ionization energy: Cs, Sr, Ba.
55. When magnesium burns in air, it forms both an oxide and a nitride. Write a balanced equation for the formation of the nitride.
56. Write balanced equations for the following:
(a) the reaction of potassium with hydrogen
(b) the reaction of chlorine with hydrogen
(c) the reaction of sulfur with hydrogen
(d) the reaction of potassium with water
57. Complete and balance the following equations:
(a) $SiO_2(s) + C(s) \longrightarrow$
(b) $Si(s) + Cl_2(g) \longrightarrow$
(c) $PbS(s) + O_2(g) \longrightarrow$
(d) $PbO(s) + CO(g) \longrightarrow$
(e) $Ga(s) + O_2(g) \longrightarrow$
(f) $In(s) + Br_2(\ell) \longrightarrow$
58. Aluminum sulfate, with a worldwide production of about 3 million tons per year, is the most important aluminum compound after aluminum oxide and aluminum hydroxide. Write a balanced equation for the reaction of aluminum oxide with sulfuric acid to give aluminum sulfate. If you want to manufacture 1.00 kg of aluminum sulfate, what masses (in kilograms) of aluminum oxide and sulfuric acid must you use?
59. Complete and balance the following equations:
(a) $HNO_3(aq) + KOH(aq) \longrightarrow$
(b) $NH_4NO_3(s) + \text{heat} \longrightarrow$
(c) $K(s) + O_2(g) \longrightarrow$
(d) $UO_2(s) + 4\ HF(aq) \longrightarrow$
60. The annual U.S. and Canadian fluorine production is 5.00×10^3 tons (1 ton is 2.00×10^3 pounds), and 55% of this is used to manufacture UF_6. What mass in tons of uranium(VI) fluoride is manufactured every year (assume 100% efficiency in using the F_2)?

General Questions

61. A neutral atom has two electrons with $n = 1$, eight electrons with $n = 2$, eight electrons with $n = 3$, and one electron with $n = 4$. Assuming this element is in its ground state, supply the following information: (a) atomic number and name; (b) total number of *s* electrons; (c) total number of *p* electrons; and (d) total number of *d* electrons.
62. How many *p* orbital electron pairs are there in an atom of selenium, Se, in its ground state?
63. For which element is the last electron placed in the third ($n = 3$) quantum shell of an atom?
64. Answer the following questions about the elements A and B, which have the electron configurations shown.

$A = \ldots 4p^6 5s^1 \quad B = \ldots 3p^6 3d^{10} 4s^2 4p^4$

(a) Is element A a metal, a nonmetal, or a metalloid?
(b) Which element would have the greater ionization energy?
(c) Which element has a larger atomic radius?

65. Name the element corresponding to each of the following characteristics:
(a) The element whose atoms have the electron configuration $1s^2 2s^2 2p^6 3s^2 3p^4$.
(b) The element in the alkaline earth group that has the largest atomic radius.
(c) The element in Group 5A whose atoms have the largest ionization energy.
(d) The element whose +2 ion has the configuration $[Kr]4d^6$.
(e) The element whose atoms have the electron configuration $[Ar]3d^{10}4s^1$.

66. The ionization energies for the removal of the first electron from atoms of Si, P, S, and Cl are listed in the following table. Briefly rationalize this trend.

Element	*First Ionization Energy (kJ/mol)*
Si	780
P	1060
S	1005
Cl	1255

67. Answer the following questions about the elements with electron configurations shown.

$A = \ldots 3p^6 4s^2 \quad B = \ldots 3p^6 3d^{10} 4s^2 4p^5$

(a) Is element A a metal, a metalloid, or a nonmetal?
(b) Is element B a metal, a metalloid, or a nonmetal?
(c) An atom of which element is expected to have the larger ionization energy?
(d) An atom of which element would be the smaller of the two?

68. Place the following elements and ions in order of decreasing size: Ar, K^+, Cl^-, S^{2-}, and Ca^{2+}.

69. Which of the following ions are unlikely, and why: Cs^+, In^{4+}, Fe^{6+}, Te^{2-}, Sn^{5+}, and I^-?

70. Rank the following in order of increasing ionization energy: Zn, Ca, Ca^{2+}, and Cl^-. Briefly explain your answer.

71. Worldwide production of silicon carbide, SiC, a widely used abrasive, is several hundred thousand tons annually. In order to produce 100,000 tons of the carbide, what mass in tons of silica sand (SiO_2) will you have to use if 70% of the sand is converted to SiC?

72. Suppose a new element, extraterrestium, tentatively given the symbol Et, has just been discovered. Its atomic number is 113.
(a) Depict the electron configuration of the element.
(b) Name another element you would expect to find in the same group as Et.
(c) Give the formulas for the compounds of Et with O and Cl.

Summary Question

73. When sulfur dioxide reacts with chlorine, the products are thionyl chloride, $SOCl_2$, and dichlorine monoxide, Cl_2O.

$$SO_2(g) + 2\,Cl_2(g) \longrightarrow SOCl_2(g) + Cl_2O(g)$$

(a) In what period of the periodic table is S located?
(b) Give the complete electron configuration of S. Do *not* use the noble gas notation.
(c) An atom of which element involved in this reaction (O, S, or Cl) should have the smallest ionization energy? The smallest radius?
(d) If you want to make 675 g of $SOCl_2$, what mass in grams of Cl_2 is required?
(e) If you use 10.0 g of SO_2 and 20.0 g of Cl_2, what is the theoretical yield of $SOCl_2$?

Using Computer Programs and Videodiscs

74. Give the name of an element that reacts vigorously with water and is the softest metal. List three other properties of this element. From what ore is it obtained (what is the source of the element)?

75. Describe the reactions (if any) of aluminum with air, water, acids (HCl, HNO_3), and base (NaOH). Using the data base as a guide, write balanced equations for all reactions. Were you surprised by what happened to aluminum in contact with HNO_3? If so, why? Provide an explanation of what happened to the aluminum when it was in contact with nitric acid.

76. Tungsten (W) is used in electric light bulbs as the filament because of its high melting point. Suggest two other elements that might work well as a filament. Think about what properties (in addition to high melting point) are needed for a light-bulb filament. List the names of these properties. Check whether the two elements you selected have appropriate properties other than high melting points. Based on its properties, what are the advantages and disadvantages of each element you selected?

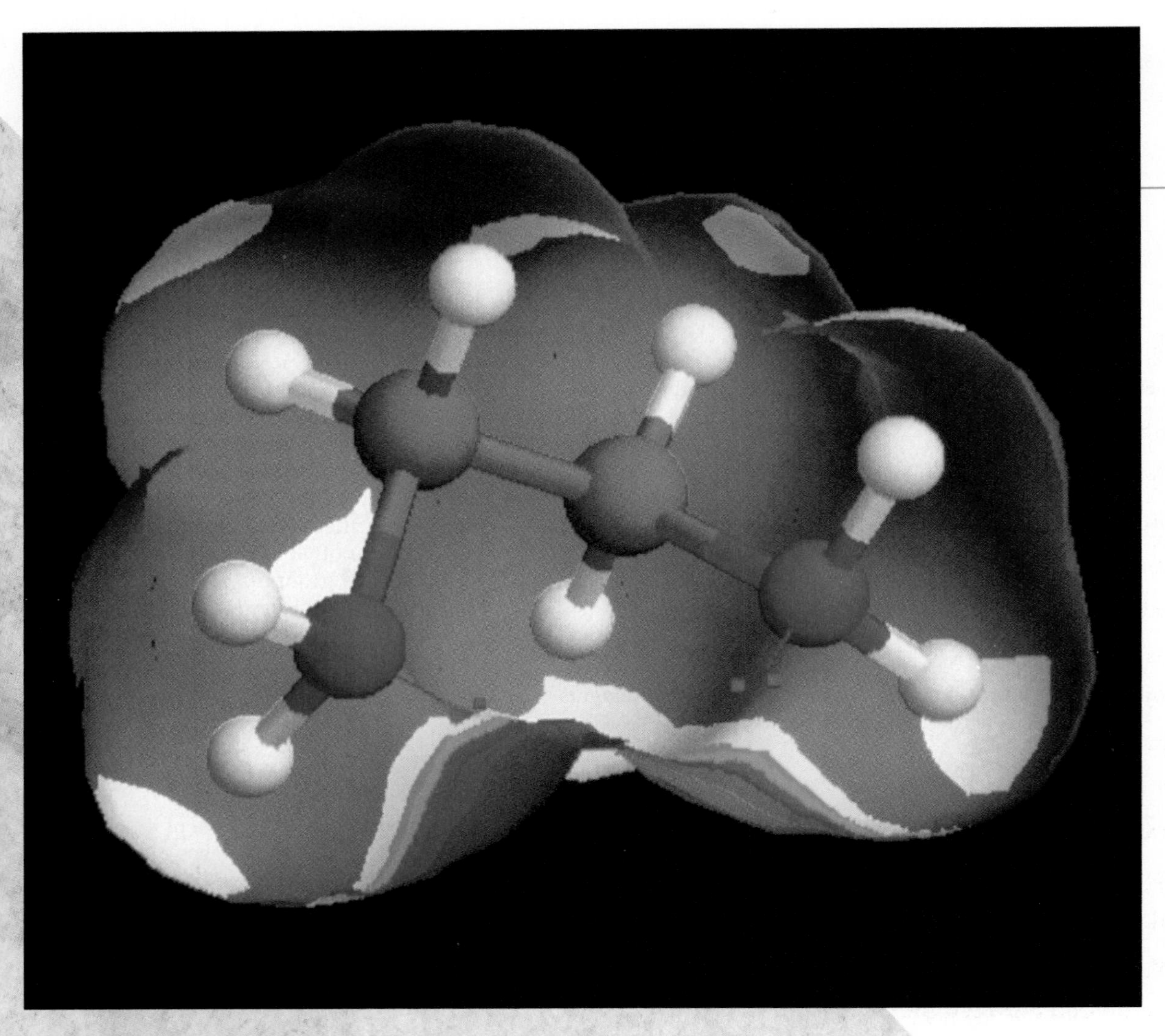

The molecular structure of ethylenediamine, $H_2NCH_2CH_2NH_2$, inside a surface representing the electron density of the molecule. The colors of the surface represent sites of reactivity.

(C.D. Winters and CAChe Scientific, Inc.)

Covalent Bonding

CHAPTER

10

Atoms of elements are only rarely found by themselves in nature. Most nonmetallic elements consist of molecules, and in a solid metal each atom is closely surrounded by eight or twelve neighbors. Only the noble gases consist of individual atoms. What makes atoms stick to one another? Valence electrons form the glue, but how? In ionic compounds, the transfer of one or more valence electrons from an atom of a metal to an atom of a nonmetal produces ions whose opposite charges attract ions into a crystal lattice. But many compounds do not conduct electricity strongly when in the liquid state or when dissolved in water, and apparently do not consist of ions. Examples are carbon monoxide (CO), methane (CH_4), water (H_2O), quartz (SiO_2), ammonia (NH_3), and about ten million organic compounds. What holds atoms together in molecules that make up these compounds? G. N. Lewis suggested that the bonds that hold atoms in molecules consist of one or more pairs of electrons shared between the bonded atoms. The attraction of positively charged nuclei for electrons between them pulls the nuclei together. This simple idea can account for the covalent bonding in more than ten million compounds, allowing us to correlate their molecular structures with physical and chemical properties. This chapter describes the use of the Lewis model to explain the bonding found in various types of molecules ranging from simple diatomic gases to coordination compounds.

Portrait of a Scientist

Gilbert Newton Lewis (1875–1946)

In a paper published in the *Journal of the American Chemical Society* in 1916, Gilbert N. Lewis introduced the theory of the shared electron pair chemical bond and revolutionized chemistry. It is to honor this contribution that we often refer to "electron dot" structures as Lewis structures. However, he also made major contributions to other fields such as thermodynamics, isotope studies, and the interaction of light with substances. Of particular interest in this text is the extension of his theory of bonding to a generalized theory of acids and bases (Section 16.9).

G. N. Lewis was born in Massachusetts but raised in Nebraska. After earning his B.A. and Ph.D. degrees at Harvard University, he began his academic career. In 1912 he was appointed Chairman of the Chemistry Department at the University of California, Berkeley, and he remained there for the rest of his life. Lewis felt that a chemistry department should both teach and advance fundamental chemistry, and he was not only a productive researcher but also a teacher who profoundly affected his students. Among his ideas was the use of large problem sets in teaching, an idea that is much in use today.

G.N. Lewis. (Oesper Collection in the History of Chemistry/University of Cincinnati)

Hydrogen gas (H_2) and lithium hydride (LiH), with molecular weights of 2 and 8 amu, respectively, have the two smallest molecular weights of any known substances. Both H and Li are in Group 1A of the periodic table, and atoms of each element have one valence electron. But H_2 is a gas at room temperature, is practically insoluble in water, is an electric insulator, and burns easily in air. Lithium hydride is a solid, reacts with water to form H_2, conducts electricity when in the molten state, and bursts into flame when exposed to moist air at high temperatures. Why is there such a difference in properties for two substances that, on an atomic scale, might seem similar?

A lithium atom can lose its one valence electron (the 2*s* electron) relatively easily to form Li^+, which has the electron configuration of He, the nearest noble gas. A hydrogen atom has only one electron and that electron is in the innermost shell, quite close to the nucleus. Because of the first shell's small size, electrons in it are tightly held and H has a much higher ionization energy than Li (Section 9.5). However, a H atom can add an electron to form H^-, which also has the noble-gas electron configuration of He. Hence, lithium hydride consists of Li^+ cations and H^- anions, each of which has the same electron configuration as a He atom. Lithium hydride has properties characteristic of an ionic compound—it is a crystalline solid at room temperature and has a high melting point.

Hydrogen gas, on the other hand, has none of the properties of an ionic compound and is a collection of H_2 molecules. Each H_2 molecule is held together by a **covalent bond**—an attractive force between two atoms that results from sharing one or more pairs of electrons. Covalent bonding generally occurs when atoms of nonmetals react with each other to form molecular compounds.

G. N. Lewis suggested that valence electrons rearrange to give noble-gas electron configurations when chemical bonds form. His idea can be applied to molecular compounds as well as ionic compounds. Lewis proposed that when a pair of electrons is shared between two atoms, that pair occupies the same shell as the valence electrons of each atom and contributes to a noble-gas configuration on each atom. By counting the valence electrons of an atom, you can predict how many bonds that atom can form. The number of bonds is the number of electrons that must be shared to achieve a noble-gas configuration.

But why does sharing electrons provide an attractive force between atoms? If an electron pair is located between two atomic nuclei, the electrons will attract the nuclei together (Figure 10.1). The region between the nuclei is called the bonding region. When two atoms share a pair of electrons, those electrons have a higher probability of being in the bonding region than outside it, so the attraction of the positively charged nuclei for the electrons between them pulls the nuclei together.

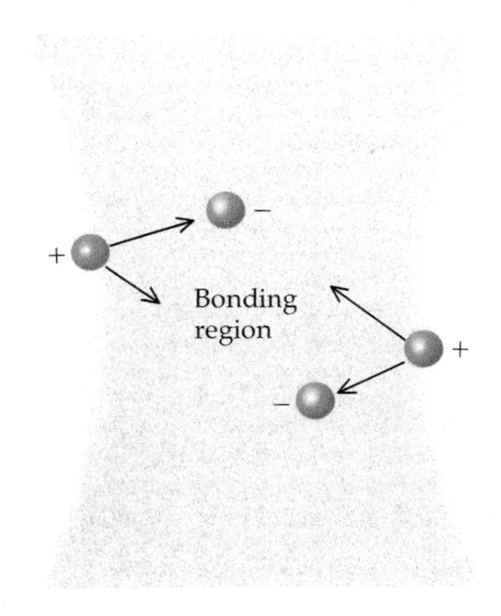

Figure 10.1
An electron in the bonding region attracts two nuclei together.

10.1
SINGLE COVALENT BONDS

A **single covalent bond** is formed when two atoms share a pair of electrons. The simplest examples are diatomic (two-atom) molecules such as H_2, F_2, and Cl_2. **Lewis structures** for many molecules can be drawn by starting with Lewis dot symbols for atoms (Table 9.5) and arranging the valence electrons in the molecule until each atom has a noble-gas configuration. For example, the Lewis structure for H_2 shows two electrons (two dots) shared between two hydrogen nuclei (two H·). The two bonding electrons are often represented by a line instead of a pair of dots.

$$\mathrm{H\!:\!H} \qquad \text{or} \qquad \mathrm{H{-}H}$$

To obtain the Lewis structure for F_2 we start with the Lewis dot symbol for a fluorine atom, $:\ddot{\underset{\cdot\cdot}{\mathrm{F}}}\cdot$ (fluorine is in Group 7A and has seven valence electrons). The dot symbol indicates that each fluorine atom has one unpaired electron, so the octet rule will be followed if each F atom in F_2 has contributed an electron to form a single covalent bond.

$$2\ :\ddot{\underset{\cdot\cdot}{\mathrm{F}}}\cdot \longrightarrow\ :\ddot{\underset{\cdot\cdot}{\mathrm{F}}}:\ddot{\underset{\cdot\cdot}{\mathrm{F}}}: \qquad \text{or} \qquad :\ddot{\underset{\cdot\cdot}{\mathrm{F}}}{-}\ddot{\underset{\cdot\cdot}{\mathrm{F}}}:$$

Shared electrons are counted with both atoms.

(Recall from Section 9.4 that achieving the noble gas configuration is also referred to as obeying the octet rule since all noble gases except He have eight valence electrons.)

Only the shared pair of electrons represented between the two symbols (the two F's) are **bonding electrons,** and these are referred to as a **bonding pair** or **bond pair** of electrons. The other six pairs of electrons are called **nonbonding** or **lone pair** valence electrons, and the pairs repel each other. In drawing Lewis structures, the bonding pairs of electrons are usually indi-

Magnetic measurements (Section 9.3) support the concept that a "pair of electrons," regardless of whether they are bonding or nonbonding, have opposite spins.

The term "lone pairs" will be used in this text to refer to nonbonding electron pairs.

cated by lines connecting the atoms they hold together and lone pairs are usually represented by dots.

What about Lewis structures for molecules such as H_2O or NH_3? Oxygen (Group 6A) has the Lewis dot symbol $\cdot\ddot{\underset{\cdot\cdot}{\mathrm{O}}}\cdot$ and must share two electrons to satisfy the octet rule. This can be accomplished by forming covalent bonds with two hydrogen atoms.

$$2\,\mathrm{H}\cdot + \cdot\ddot{\underset{\cdot\cdot}{\mathrm{O}}}\cdot \longrightarrow \mathrm{H}:\ddot{\underset{\cdot\cdot}{\mathrm{O}}}:\mathrm{H} \qquad \text{or} \qquad \mathrm{H}{-}\ddot{\underset{\cdot\cdot}{\mathrm{O}}}{-}\mathrm{H}$$

Nitrogen ($\cdot\ddot{\underset{\cdot}{\mathrm{N}}}\cdot$, Group 5A) in NH_3 must share three electrons to achieve a noble gas configuration, which can be done by forming covalent bonds with three hydrogen atoms.

$$3\,\mathrm{H}\cdot + \cdot\ddot{\underset{\cdot}{\mathrm{N}}}\cdot \longrightarrow \mathrm{H}:\ddot{\underset{\underset{\mathrm{H}}{\cdot\cdot}}{\mathrm{N}}}:\mathrm{H} \qquad \text{or} \qquad \mathrm{H}{-}\ddot{\underset{\underset{\mathrm{H}}{|}}{\mathrm{N}}}{-}\mathrm{H}$$

Although Lewis structures are useful for predicting the number of covalent bonds an atom will form, they do not give an accurate representation of where electrons are located in a molecule. Bonding electrons do not stay in fixed positions between nuclei, as Lewis's dots might imply. Instead, quantum mechanics tells us that there is a high probability of finding the bonding electrons between the nuclei, but they could be found elsewhere as well (Figure 10.1). Also, since bond pairs occupy the same orbital, their spins must be opposite; the same applies to lone pairs. Finally, Lewis structures are not meant to convey the shapes of molecules. The angle between O—H bonds in a water molecule is not 180°, as the Lewis diagram above seems to imply. However, Lewis structures are used to predict geometries by a method based on the repulsions between valence-shell electron pairs (Section 11.1).

Although H is given first in the formulas of H_2O and H_2O_2, for example, it is not the central atom. H is never the central atom in a molecule or ion.

Suppose we want to draw the Lewis structure for PCl_3. From the periodic table we see that P is in Group 5A and Cl is in Group 7A. The Lewis dot symbols are $\cdot\ddot{\underset{\cdot}{\mathrm{P}}}\cdot$ and $:\ddot{\underset{\cdot\cdot}{\mathrm{Cl}}}\cdot$, showing that P can form three bonds and Cl one bond. Usually the atom that can form more bonds comes first in the molecular formula, has the smallest subscript, and is the central atom in the molecule. First we draw a skeleton structure with P as the central atom and the others arranged around it, sharing pairs of electrons.

$$\mathrm{Cl}:\underset{\underset{\mathrm{Cl}}{\cdot\cdot}}{\mathrm{P}}:\mathrm{Cl} \qquad \text{or} \qquad \mathrm{Cl}{-}\underset{\underset{\mathrm{Cl}}{|}}{\mathrm{P}}{-}\mathrm{Cl}$$

skeleton structure

Then we add lone pairs to complete the octets, first for each chlorine atom and then for the central phosphorus atom.

Phosphorus trichloride (PCl_3) is a colorless liquid that fumes as it reacts with water in moist air.

$$:\ddot{\underset{\cdot\cdot}{\mathrm{Cl}}}:\ddot{\underset{\underset{:\ddot{\underset{\cdot\cdot}{\mathrm{Cl}}}:}{\cdot\cdot}}{\mathrm{P}}}:\ddot{\underset{\cdot\cdot}{\mathrm{Cl}}}: \qquad \text{or} \qquad :\ddot{\underset{\cdot\cdot}{\mathrm{Cl}}}{-}\ddot{\underset{\underset{:\ddot{\underset{\cdot\cdot}{\mathrm{Cl}}}:}{|}}{\mathrm{P}}}{-}\ddot{\underset{\cdot\cdot}{\mathrm{Cl}}}:$$

To check the structure, verify that the total number of dots or dots + (lines × 2) equals the total number of valence electrons. The number of valence electrons is 5 for P and 7 for each Cl, which gives $5 + (3 \times 7) = 26$. Counting dots and lines in the Lewis structure to the right also gives 26.

EXAMPLE 10.1

Lewis Structures

Draw the Lewis structures of (a) Cl_2O and (b) H_2O_2.

SOLUTION (a) The Lewis dot symbols ·Ö· (Group 6A) and :C̤̈l· (Group 7A) show that oxygen can form two bonds and chlorine can form one bond. The total number of valence electrons is 20 (14 from the two Cl atoms and 6 from the O atom). The central atom is O, and the skeleton structure is

Cl—O—Cl

Placing three lone pairs on each Cl and two lone pairs on the O atom uses all the electrons and satisfies the octet rule for all atoms.

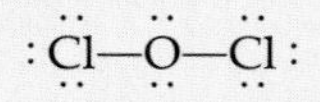

(b) The atoms in H_2O_2 are H· , which can never form more than one bond, and ·Ö̤· , which can form two bonds, meaning that the two O atoms must be bonded to each other. The total number of valence electrons is 14, including 2 from the two H atoms and 12 from the two O atoms. The skeleton structure must be

H—O—O—H

which uses 6 electrons. Placing two lone pairs on each O atom satisfies the octet rule and uses all valence electrons.

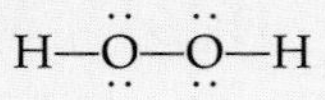

EXERCISE 10.1 • *Drawing Lewis Structures*

Draw the Lewis structures for (a) NF_3, (b) H_2S, and (c) N_2H_4.

Single Bonds in Hydrocarbons

Carbon is unique among the elements of the periodic table because of the ability of its atoms to form strong bonds with one another as well as with atoms of hydrogen, oxygen, nitrogen, sulfur, and the halogens. The strength of the carbon-carbon bond permits long chains to form:

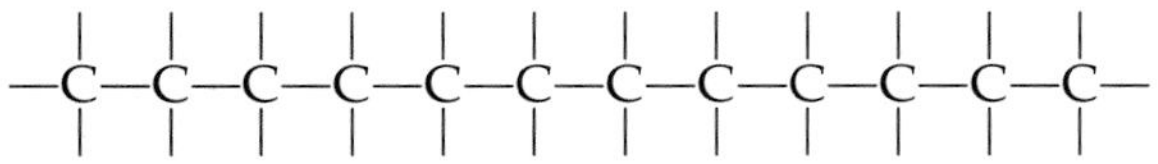

This behavior is called **catenation.** Such chains contain numerous sites to which other atoms (including more carbon atoms) can bond, leading to a great variety of carbon compounds. In fact, all of the ten million or so organic compounds reported in the chemical literature are carbon compounds, and several hundred thousand new carbon compounds are synthesized each year. The largest class of organic compounds consists of the **hydrocarbons,** compounds containing only carbon and hydrogen. **Alkanes** and **cycloalkanes** are hydrocarbons that contain only C—C and C—H single

Review the discussion on alkanes in Section 4.5. See Table 4.2 for a list of selected alkanes.

bonds. Alkanes are often referred to as **saturated hydrocarbons** because they have the highest possible ratio of hydrogen to carbon atoms bonded in a molecule.

Rules for naming organic compounds are given in Appendix E.

The two simplest members of the alkane family are **methane** (CH_4), the principal component of natural gas, and **ethane** (C_2H_6). As demonstrated by these two compounds, the general formula for alkanes is C_nH_{2n+2} where $n = 1, 2, 3$, etc. The Lewis structures of organic molecules are often called **structural formulas;** the structural formulas of methane and ethane are

```
     H            H  H
     |            |  |
  H—C—H       H—C—C—H
     |            |  |
     H            H  H
  methane       ethane
```

EXERCISE 10.2 • *Alkanes*

Draw the Lewis structural formula of propane, C_3H_8, the principal constituent of liquefied petroleum gas (LPG).

Cycloalkanes are saturated hydrocarbons consisting of rings of —CH_2— units. The simplest cycloalkane is cyclopropane:

```
    H     H
     \   /
       C
 H    / \    H
  \  /   \  /
   C —— C          or   △
  /        \
 H          H
```

C_3H_6

The cycloalkanes are commonly represented by polygons in which each corner represents a carbon atom and two hydrogen atoms, and each line represents a C—C bond. The C—H bonds usually are not shown, but are understood to be present. Other common cycloalkanes include cyclobutane, cyclopentane, and cyclohexane. These are represented as:

□	⬠	⬡
C_4H_8	C_5H_{10}	C_6H_{12}
cyclobutane	cyclopentane	cyclohexane

EXERCISE 10.3 • *Cyclic Hydrocarbons*

Draw the polygon that represents cyclooctane and write the molecular formula for this compound.

10.2 MULTIPLE COVALENT BONDS

An atom with fewer than seven valence electrons can form covalent bonds in two ways. The atom may share a single electron with another atom, which can also contribute a single electron. This leads to a single covalent bond. But the atom can also share two (or three) pairs of electrons with another atom. In this case there will be two (or three) bonds between two atoms. When two shared pairs of electrons join the same pair of atoms, the bond is called a **double bond,** and when three shared pairs are involved, the bond is called a **triple bond.**

How can you predict when a molecule has multiple bonds rather than single bonds? Let's draw the Lewis structure of the CO_2 molecule. The Lewis dot symbols for the atoms are $\cdot\dot{\underset{\cdot}{C}}\cdot$ and $\cdot\ddot{O}:$ and so C should form four bonds and O two bonds. Because C forms more bonds, it is likely to be the central atom. There is a total of 16 electrons available (4 from $\cdot\dot{\underset{\cdot}{C}}\cdot$, 12 from two $\cdot\ddot{\underset{\cdot\cdot}{O}}\cdot$). The skeleton structure uses 4 electrons:

$$O—C—O$$

Adding lone pairs to give each O an octet of electrons uses up the remaining 12 electrons, but leaves C needing four more valence electrons to complete an octet.

$$:\ddot{\underset{\cdot\cdot}{O}}—\not{C}—\ddot{\underset{\cdot\cdot}{O}}:$$

With no more valence electrons available, the only way that carbon can have four more valence electrons is to have one pair of electrons on each oxygen form covalent bonds to carbon.

$$:\underset{\cdot\cdot}{O}=C=\underset{\cdot\cdot}{O}:$$

As a check, we see that 16 valence electrons are accounted for and each atom has an electron octet.

As another example, let's draw the Lewis structure for molecular nitrogen, N_2. The total number of valence electrons available is 10 (5 from each $\cdot\ddot{\underset{\cdot}{N}}\cdot$). A skeleton structure with a single bond and an octet on each N atom would require 14 electrons, more than we have.

$$:\ddot{\underset{\cdot\cdot}{N}}\not{—}\ddot{\underset{\cdot\cdot}{N}}:$$

If two nonbonding pairs of electrons (one pair from each N) become bonding pairs to give a triple bond, the octet rule is satisfied. This is the correct Lewis structure of N_2.

$$:N\equiv N:$$

The formation of double and triple bonds is not as widespread among the atoms of the periodic table as one might expect. At least one of the atoms involved in a multiple bond is almost always C, N, or O, and in many cases both atoms are members of this trio. Other elements complete their octets by forming additional single bonds rather than multiple bonds. An example of this is the binary compound of Si with O. Earlier we saw that C and O form the molecule carbon dioxide with the Lewis structure

$$:\ddot{O}=C=\ddot{O}:$$

On this basis we might predict that Si and O would give a stable molecule SiO_2 with the Lewis structure

$$:\ddot{O}=Si=\ddot{O}:$$

However, silicon does not readily form double bonds. Each silicon atom can form single bonds to four oxygen atoms, but as the Lewis structure shows, each O still needs one electron to have an octet.

$$\begin{matrix} & :\dot{O}: & \\ \cdot\ddot{O} : & Si & :\ddot{O}\cdot \\ & :\dot{O}: & \end{matrix} \qquad \text{or} \qquad \begin{matrix} & :\dot{O}: & \\ & | & \\ \cdot\ddot{O}- & Si & -\ddot{O}\cdot \\ & | & \\ & :\dot{O}: & \end{matrix}$$

If each of the oxygen atoms links to another silicon atom, the octet rule will be satisfied, but then the added silicon atoms will not have an octet of electrons.

$$\begin{matrix} & & \cdot\dot{Si}\cdot & & \\ & & :\ddot{O}: & & \\ \cdot\dot{Si} : & \ddot{O} : & Si & : \ddot{O} : & \dot{Si}\cdot \\ & & :\ddot{O}: & & \\ & & \cdot\dot{Si}\cdot & & \end{matrix} \qquad \text{or} \qquad \begin{matrix} & & \cdot\dot{Si}\cdot & & \\ & & | & & \\ & & :\ddot{O}: & & \\ & & | & & \\ \cdot\dot{Si}- & \ddot{O}- & Si & -\ddot{O}- & \dot{Si}\cdot \\ & & | & & \\ & & :\ddot{O}: & & \\ & & | & & \\ & & \cdot\dot{Si}\cdot & & \end{matrix}$$

The process of adding oxygen or silicon atoms can continue indefinitely, producing a giant lattice of covalently bonded atoms (Figure 10.2). In this giant molecule, each silicon is bonded to four oxygen atoms and each oxygen is bonded to two silicon atoms. The molecular formula can be written $(SiO_2)_n$, where n is a very large number. For simplicity it is usually written as SiO_2.

This example illustrates the need to consider whether the proposed Lewis structure matches the known properties of a substance. For example, carbon dioxide is a gas at room temperature, so the proposed Lewis structure that depicts it as individual CO_2 molecules is reasonable. However, silicon dioxide (also called silica, of which sand is a common form) is a solid at room temperature, and a three-dimensional, giant molecular structure is more in keeping with this property.

The procedure used for drawing correct Lewis structures in this section is summarized in the following guidelines.

Guidelines for Drawing Lewis Structures

1. *Count the total number of valence electrons.* Use the group number in the periodic table as a guide. For a neutral molecule, the total equals the sum of the valence electrons of the atoms in the molecule. For an ion, add electrons equal to the ion charge for a negative ion or subtract electrons equal to the ion charge for a positive ion. For example, add one electron for OH^-; subtract one electron for NH_4^+. Draw the Lewis dot symbol for each atom in the molecule or ion.

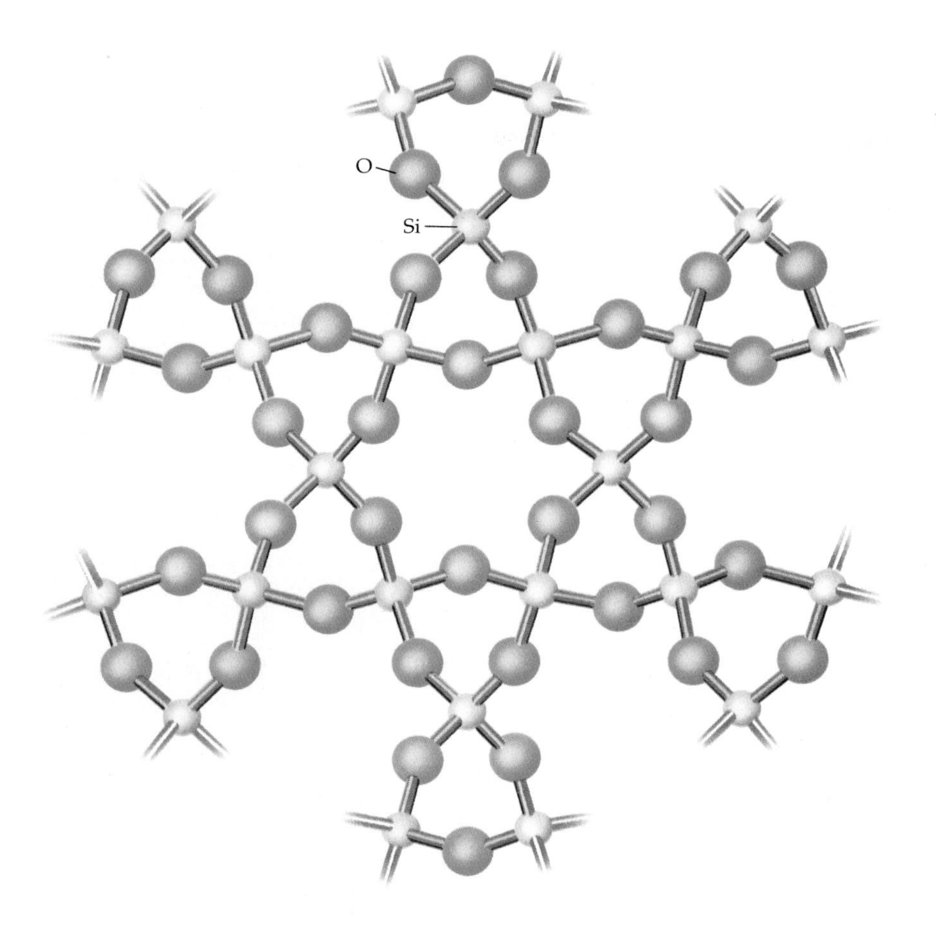

Figure 10.2

A portion of the giant covalent molecule $(SiO_2)_n$. The lattice shown would extend indefinitely in all directions in a macroscopic crystal. Each silicon atom (gold) is covalently bonded to four oxygen atoms (red). Each oxygen atom bonds to two silicon atoms. The ratio of silicon to oxygen is 2:4 or 1:2, in accord with the formula.

2. *Use the symbols for the atoms to draw a skeleton structure.* Usually the central atom is the one with the lowest subscript in the molecular formula and the one that can form the most bonds.
3. *Place a bonding pair of electrons between the central atom and each terminal atom.*
4. *Place lone pairs of electrons around each terminal atom (except H) to satisfy the octet rule.* Remember that shared electrons are counted as ''belonging'' to each of the atoms bonded by the shared pair.
5. *Place any leftover electrons on the central atom, even if it will give the central atom more than an octet.* If the atom is from the third or higher period, it can accommodate more than an octet of electrons. See Section 10.4 for examples.
6. *If the number of electrons around the central atom is less than eight, change single bonds to multiple bonds.* Use one or more lone pairs of electrons on the terminal atoms to form double (two shared pairs) or triple (three shared pairs) bonds until the central atom and all terminal atoms have octets.

Exceptions to the octet rule are described in Section 10.4.

Silica Aerogels Put to Use as Cosmic Dust Catchers

What looks like fog, is only four times as dense as air, but can support 1600 times its own weight? A silica aerogel! In the summer of 1992, silica aerogels were sent into earth orbit to serve as cosmic dust catchers. After almost a year of orbiting 300 miles above the earth's surface, the aerogel-carrying satellite will be retrieved. Scientists are hoping to find micrometeroids 1/30th the width of a human hair trapped in the fine lattice of silicon dioxide that makes up the aerogel.

A chemist at Lawrence Livermore National Laboratory (LLNL) applies a propane torch to the top of an inch-thick silica aerogel brick, while his hand below remains totally unaffected by the heat. The silica aerogels are 25 times lighter than the previous airiest aerogel and rank as the best insulators ever tested. (Lawrence Livermore National Laboratory)

Aerogels are prepared from an ordinary gel, a substance much like a gelatin dessert, that contains huge silicon dioxide molecules dispersed in water. Gentle removal of the water leaves behind a network of silicon dioxide, or silica, the same substance in beach sand. The lightest silica aerogels are 99.9% air surrounding tiny clusters of silicon dioxide with diameters of about 2 nm; they are visible only as a blue haze, although they are solid to the touch. These remarkable materials are such excellent heat insulators that a 1-inch thick piece protects the hand of the chemist in the photo from the heat of a propane torch.

One of many possible uses of silica aerogels is in refrigerator insulation. They would replace the current foams that contain chlorofluorocarbons (CFCs), which are being phased out because of the damaging effect of the CFCs on the ozone layer (Section 20.6). The insulating ability and transparency of aerogels may also be valuable in greatly improved insulating window glass, and they show promise as electrical and sound insulators. Perhaps, if the cosmic dust experiment is successful, silica aerogels will also contribute to finding out whether there is life in space—scientists plan to analyze the retrieved micrometeroids for carbon compounds.

A piece of silica aerogel looks like a blue haze floating on beaten egg whites. (Lawrence Livermore National Laboratory)

Science, Vol. 247, p. 807, 1990; *Chemecology,* February 1992, p. 9.

EXAMPLE 10.2

Lewis Structures

Draw Lewis structures for (a) formaldehyde, H_2CO, and (b) the nitronium ion, NO_2^+.

SOLUTION (a) There are 12 valence electrons in the molecule (2 from two H; 4 from C; 6 from O). Carbon can form four bonds, O can form two bonds, and H can form one bond. Carbon must be the central atom. The skeleton drawing with a single bond between each pair of atoms requires six electrons.

```
H—C—H
  |
  O
```

If the six remaining electrons are placed around the O atom, the octet rule is not satisfied for C.

```
H—C—H   (C crossed out)
  |
 :O:
  ..
```

To remedy this, change one of the lone pairs on O to a bonding pair between C and O. Now both C and O have a share in four pairs of electrons and the octet rule is obeyed.

```
H—C—H
  ||
 :O
  ..
```

(b) The total number of valence electrons is 16 (12 from two O; 5 from N; −1 for the positive charge on the ion). Nitrogen can form the most bonds and should be the central atom. The skeleton drawing with single bonds between N and O and lone pairs around each O uses all 16 electrons.

$$[:\ddot{\underset{..}{O}}—N—\ddot{\underset{..}{O}}:]^+$$

However, the nitrogen atom is four electrons short of an octet, so a lone pair from each O atom is converted to a bonding pair. Now each atom in the ion satisfies the octet rule.

$$[\ddot{\underset{..}{O}}=N=\ddot{\underset{..}{O}}]^+$$

Formaldehyde, a gas at room temperature, is the simplest compound with an aldehyde functional group, —C(=O)—H. (Appendix E includes a list of functional groups.)

EXERCISE 10.4 • *Lewis Structures*

Which of the following are correct Lewis structures and which are not?

a. $:\ddot{\underset{..}{O}}—\ddot{\underset{..}{C}}—\ddot{\underset{..}{O}}:$ b. H—C(H)(H)—$\ddot{\underset{..}{O}}$—H c. :C≡O: d. $:\ddot{\underset{..}{F}}=\ddot{N}—\ddot{\underset{..}{Cl}}:$ with $:\underset{..}{Cl}:$ bonded to N

EXERCISE 10.5 • *Lewis Structures*

Draw Lewis structures for the following: (a) NO^+, (b) HCN.

Multiple Bonds in Hydrocarbons

Molecules of **alkenes** have one or more carbon-carbon double bonds (C=C). The general formula for alkenes with one double bond is C_nH_{2n} where n = 2, 3, 4, etc. The first two members of the alkene series are **ethene** (C_2H_4) and **propene** (C_3H_6), commonly called ethylene and propylene, particularly when referring to the polymers polyethylene and polypropylene.

$$\begin{array}{ccccccc} & & H & & H & & \\ & & | & & | & & \\ H & - & C & = & C & - & H \end{array} \qquad \begin{array}{ccccccccc} & & H & & H & & H & & \\ & & | & & | & & | & & \\ H & - & C & - & C & = & C & - & H \\ & & | & & & & & & \\ & & H & & & & & & \end{array}$$

ethene (ethylene) propene (propylene)

The structural formulas illustrate why alkenes are said to be **unsaturated hydrocarbons;** they contain fewer hydrogen atoms than the corresponding alkanes. Alkenes are named by using the name of the corresponding alkane to indicate the number of carbons and the suffix *-ene* to indicate one or more double bonds. The first member, ethene or ethylene, is the most important raw material used in the organic chemical industry. It ranks fourth in the top 25 chemicals (see inside back cover) and is the number-one organic chemical. More than 40 billion pounds were produced in 1992 for use in making polyethylene, antifreeze (ethylene glycol), ethyl alcohol, and other chemicals.

Alkenes are more reactive than alkanes because the double bonds provide sites where atoms of halogens or hydrogen, or even functional groups can become bonded to carbon atoms. Such reactions, in which one or more atoms are added to a molecule, are called **addition reactions.** For example, alkenes react with chlorine in the dark at room temperature to give substituted alkanes.

$$\begin{array}{ccccccc} & & H & & H & & \\ & & | & & | & & \\ H & - & C & = & C & - & H \end{array} + Cl{-}Cl \xrightarrow[25\,^{\circ}C]{dark} \begin{array}{ccccccc} & & H & & H & & \\ & & | & & | & & \\ H & - & C & - & C & - & H \\ & & | & & | & & \\ & & Cl & & Cl & & \end{array}$$

ethylene 1,2-dichloroethane

In contrast, alkanes undergo **substitution reactions** with chlorine, but only in the presence of heat or ultraviolet light.

$$\begin{array}{ccccc} & & H & & \\ & & | & & \\ H & - & C & - & H \\ & & | & & \\ & & H & & \end{array} + Cl{-}Cl \xrightarrow[\text{or } 120\,^{\circ}C]{uv\ light} \begin{array}{ccccc} & & H & & \\ & & | & & \\ H & - & C & - & H \\ & & | & & \\ & & Cl & & \end{array} + H{-}Cl$$

methane chloromethane

In a substitution reaction, one or more atoms in a molecule is replaced by another type of atom.

Hydrogenation, reaction with hydrogen, is another important addition reaction of alkenes (Figure 10.3).

$$\begin{array}{ccccccc} & & H & & H & & \\ & & | & & | & & \\ H & - & C & = & C & - & H \end{array} + H_2 \xrightarrow[]{Pt\ catalyst} \begin{array}{ccccccc} & & H & & H & & \\ & & | & & | & & \\ H & - & C & - & C & - & H \\ & & | & & | & & \\ & & H & & H & & \end{array}$$

ethylene (unsaturated) ethane (saturated)

Cycloalkenes also exist but are not as common as cycloalkanes. One example is cyclohexene, which is used as a stabilizer in high-octane gasoline.

Figure 10.3
Parr shaker-type hydrogenation apparatus. Materials to be hydrogenated and a catalyst are sealed in the reaction bottle which is connected to a hydrogen reservoir. (The reaction bottle is located inside the circular steel mesh shield in the photo.) Air is removed either by evacuating the bottle or by flushing with hydrogen. Pressure is then applied from the hydrogen reservoir and the bottle is shaken vigorously to initiate the reaction. After the reaction is completed, the shaker is stopped, the bottle vented, and the product and catalyst are recovered. *(Parr Instrument Co., Moline, Ill.)*

Hydrocarbons with one or more triple bonds (—C≡C—) per molecule are called **alkynes.** The general formula for simple alkynes is C_nH_{2n-2} where $n = 2, 3, 4$, etc. The simplest one is **ethyne,** commonly called acetylene (C_2H_2).

$$\underset{\text{ethyne (acetylene)}}{\mathrm{H{-}C{\equiv}C{-}H}}$$

A mixture of acetylene and oxygen burns with a flame hot enough (3000 °C) to cut steel (Figure 10.4).

Figure 10.4
Cutting steel with an oxyacetylene torch. *(Joseph Nettis/Photo Researchers, Inc.)*

Like alkenes, alkynes undergo addition reactions. An important difference between alkenes and alkynes, however, is that two molecules of reagent can add across a triple bond of an alkyne because the addition of the first molecule gives an alkene. This is illustrated for chlorination of acetylene.

$$\underset{\text{acetylene}}{\mathrm{H{-}C{\equiv}C{-}H}} + \mathrm{Cl{-}Cl} \xrightarrow[25\,^\circ\mathrm{C}]{\text{dark}} \underset{\text{1,2-dichloroethene}}{\mathrm{H{-}\overset{\overset{\displaystyle Cl}{|}}{C}{=}\overset{\overset{\displaystyle Cl}{|}}{C}{-}H}}$$

$$\underset{\text{1,2-dichloroethene}}{\mathrm{H{-}\overset{\overset{\displaystyle Cl}{|}}{C}{=}\overset{\overset{\displaystyle Cl}{|}}{C}{-}H}} + \mathrm{Cl{-}Cl} \longrightarrow \underset{\text{1,1,2,2-tetrachloroethane}}{\mathrm{H{-}\overset{\overset{\displaystyle Cl}{|}}{\underset{\underset{\displaystyle Cl}{|}}{C}}{-}\overset{\overset{\displaystyle Cl}{|}}{\underset{\underset{\displaystyle Cl}{|}}{C}}{-}H}}$$

10.3 COORDINATE COVALENT BONDS

In the Lewis structures discussed so far, each bonded atom contributes one electron for each bond formed. However, atoms of some elements, such as nitrogen, phosphorus, and sulfur, tend to contribute both electrons to the pair shared between both atoms. This type of covalent bond is called a **coordinate covalent bond.** As an example, consider the formation of the ammonium ion:

```
                                          coordinate covalent bond
                      H                     ┌  \   H    ┐+
                      |                     │   \  |    │
         H+       + :N—H  ──→               │ H—N—H     │
                      |                     │      |    │
                      H                     └      H    ┘
hydrogen ion       ammonia                    ammonium ion
(proton)
```

Once the coordinate covalent bond is formed, it is impossible to distinguish between the equivalent N—H bonds.

Coordinate covalent bonding theory is very useful for explaining the bonding in metal coordination compounds (Section 10.10).

10.4 EXCEPTIONS TO THE OCTET RULE

Experiments that determine the stoichiometry of molecules show that the formulas and bond properties for many molecules are not consistent with the octet rule. However, we can use the periodic table to organize these exceptions in categories that require Lewis structures with either fewer or more than eight electrons.

Fewer than Eight Valence Electrons

Structural data indicate that the boron trifluoride molecule, BF_3, has single bonds between B and F. A Lewis structure in agreement with the structural data has only six electrons around the B atom.

```
  ..
 :F:
  |   ..
  B—F:
  |   ..
 :F:
  ..
```

BF_3 is very reactive and readily combines with NH_3 to form a compound with the formula BF_3NH_3. This can be explained by using the lone pair of electrons in NH_3 to form a coordinate covalent bond with B in BF_3. Both N and B now have an octet of electrons.

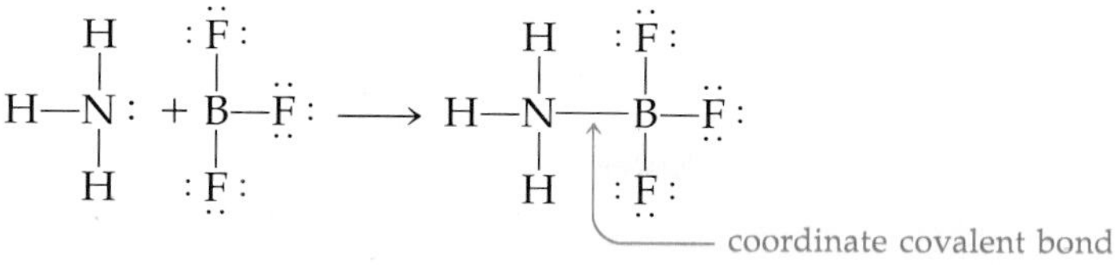

$$H_3N: + BF_3 \longrightarrow H_3N{-}BF_3$$

coordinate covalent bond

More than Eight Valence Electrons

Atoms of the third or higher periods can be surrounded by more than four electron pairs in certain compounds because they have empty *d* orbitals low enough in energy to accommodate the extra electrons (see Figure 9.12). For example, phosphorus and sulfur commonly form stable molecules in which they are surrounded by more than eight electrons (Table 10.1). In using the periodic table for making comparisons of elements in a given group, it is important to recognize this difference between elements in the second period and elements in later periods. The octet rule is reliable for predicting stable molecules with C, N, O, or F as central atoms, but later members of these groups such as Sn, P, S, and Br can also form stable molecules or polyatomic ions that have more than eight electrons around the central atom.

Table 10.1

Lewis Structures for Some Ions and Molecules with More than Eight Electrons Around the Central Atom

	Group 4A	Group 5A	Group 6A	Group 7A	Group 8A
Central atoms with five valence pairs		PF_5	SF_4	ClF_3	XeF_2
bond pairs		5	4	3	2
lone pairs		0	1	2	3
Central atoms with six valence pairs	$SnCl_6^{2-}$	PF_6^-	SF_6	BrF_5	XeF_4
bond pairs	6	6	6	5	4
lone pairs	0	0	0	1	2

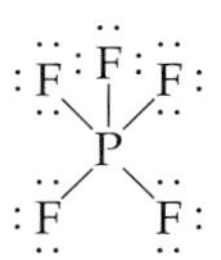

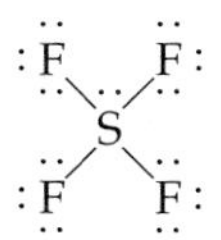

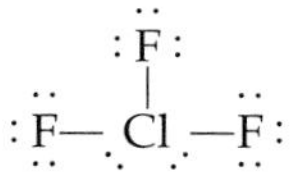

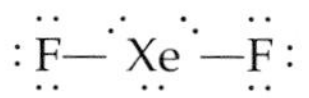

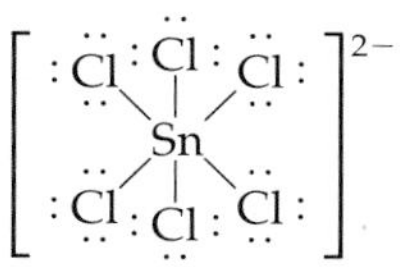

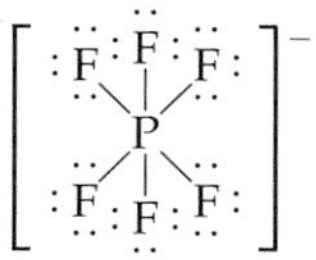

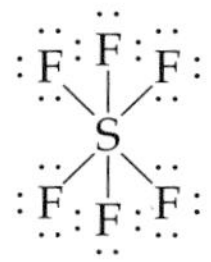

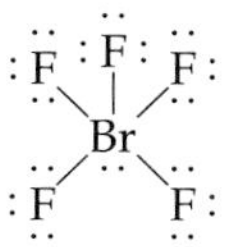

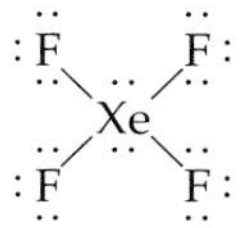

In each case, the numbers of bond pairs and lone pairs about the central atom are given.

Figure 10.5
Liquid oxygen is suspended between the poles of a magnet because O_2 is paramagnetic. Paramagnetic substances are attracted into the magnetic field. *(S. Ruren Smith)*

Molecules with an Odd Number of Valence Electrons

All the molecules we have discussed up to this point have contained only *pairs* of valence electrons. However, there are a few stable molecules that have an odd number of valence electrons. For example, NO has 11 valence electrons, and NO_2 has 17 valence electrons. The most plausible Lewis structures of these molecules are

$$:\dot{N}=\ddot{O}: \qquad \text{(bent } \dot{N} \text{ with single bond to } \vdots\ddot{O}\text{ and double bond to } \ddot{O}\text{)}$$

Molecules such as NO and NO_2 are often called **free radicals** because of the presence of the unpaired electron. Atoms that contain unpaired electrons are also free radicals. How do unpaired electrons affect reactivity? Simple free radicals such as atoms of H· and Cl· are very reactive and readily combine with other atoms to give molecules such as H_2, Cl_2, and HCl. Therefore, we would expect free radical molecules to be more reactive than molecules that have all paired electrons, and they are. A free radical either combines with another free radical to form a more stable molecule in which the electrons are paired, or it reacts with other molecules to produce new free radicals. These kinds of reactions are central to the formation of addition polymers (Section 13.6) and air pollutants (Section 20.4). For example, when small amounts of NO and NO_2 are released from vehicle exhausts, the NO_2 decomposes in the presence of sunlight to give NO and O.

$$NO_2 + \text{sunlight} \longrightarrow :\dot{N}=\ddot{O}: + :\ddot{O}\cdot$$

The free O atom reacts with O_2 in the air to give ozone, O_3, an air pollutant that affects the respiratory system. Free radicals also have a tendency to combine with themselves to form **dimers.** For example, when NO_2 gas is cooled it dimerizes to N_2O_4.

As expected, NO and NO_2 are paramagnetic because of the odd number of electrons. Experimental evidence indicates that O_2 is also paramagnetic (Figure 10.5) with two unpaired electrons and a double bond. The predicted Lewis structure for O_2 shows a double bond but all the electrons would be paired. It is impossible to write a conventional Lewis structure of O_2 that is in agreement with the experimental results.

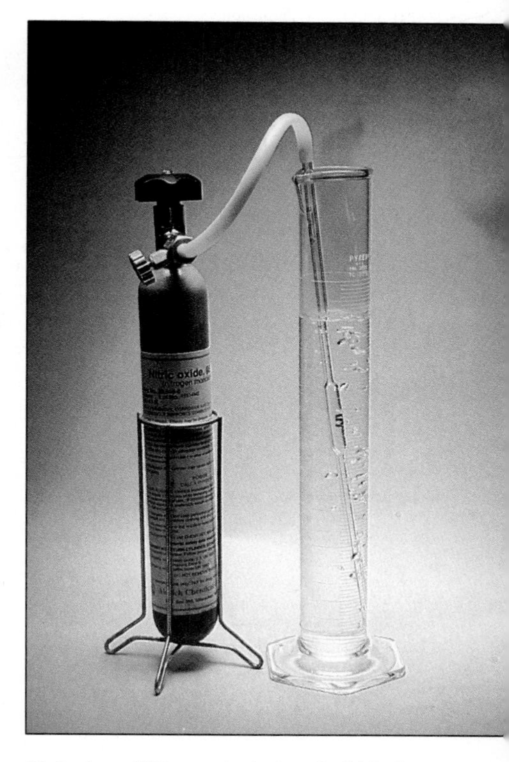

Colorless NO gas is being bubbled into water. As it escapes into the air above the water, it reacts with O_2 to give brown NO_2 gas. *(C.D. Winters)*

EXAMPLE 10.3

Exceptions to the Octet Rule

Draw the Lewis structure for ICl_2^-.

SOLUTION There are a total of 22 valence electrons, 14 from two Cl atoms and 7 from I and 1 for the −1 charge on the ion. Forming two I—Cl single bonds and then distributing six of the remaining nine electron pairs as lone pairs on Cl atoms to satisfy the octet rule uses a total of 8 electron pairs.

$$[:\underset{..}{\ddot{Cl}}—I—\underset{..}{\ddot{Cl}}:]^-$$

The remaining three electron pairs are placed on iodine, which can accommodate more than eight electrons because it is from the fifth period.

$$[:\underset{..}{\ddot{Cl}}—\underset{..}{\overset{..}{\dot{I}}}—\underset{..}{\ddot{Cl}}:]^-$$

EXERCISE 10.6 • ***Lewis Structures of Molecules that are Exceptions to the Octet Rule***

Draw the Lewis structure for each of the following molecules: (a) BeF_2, (b) ClO_2, and (c) PCl_5.

10.5 BOND PROPERTIES

Bond Length

The most important factor determining **bond length,** the distance between nuclei of two bonded atoms, is the sizes of the atoms themselves (Figure 9.17). When you compare bonds of the same type (single, double, or triple), the bond length will be greater for the larger atoms. Thus, single bonds with carbon would increase in length along the following series:

$$\underrightarrow{\text{C—N} < \text{C—C} < \text{C—P}}$$
Increase in bond length

Table 10.2

Some Average Single and Multiple Bond Lengths (in picometers, pm)*

Single Bond Lengths

	Group										
	1A	*4A*	*5A*	*6A*	*7A*	*4A*	*5A*	*6A*	*7A*	*7A*	*7A*
	H	C	N	O	F	Si	P	S	Cl	Br	I
H	74	110	98	94	92	145	138	132	127	142	161
C		154	147	143	141	194	187	181	176	191	210
N			140	136	134	187	180	174	169	184	203
O				132	130	183	176	170	165	180	199
F					128	181	174	168	163	178	197
Si						234	227	221	216	231	250
P							220	214	209	224	243
S								208	203	218	237
Cl									200	213	232
Br										228	247
I											266

Multiple Bond Lengths

C=C	134	C≡C	121
N=N	120	N≡N	110
C=N	127	C≡N	115
C=O	122	C≡O	113
N=O	115	N≡O	108
O=O	112		

*1 pm = 10^{-12} m.

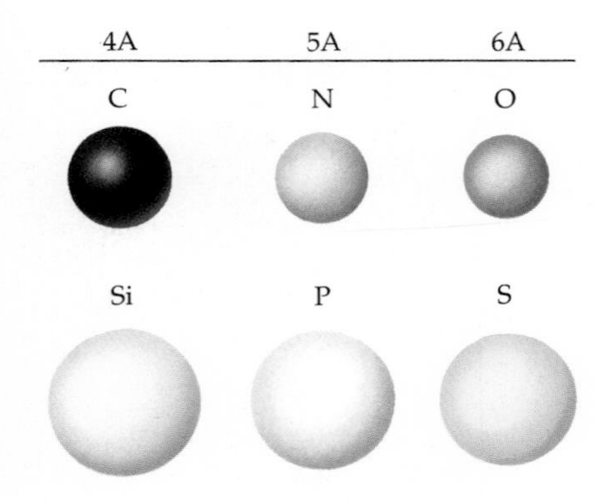

Relative atom sizes for Groups 4A, 5A, and 6A.

Similarly, a C=O bond will be shorter than a C=S bond, and a C≡N bond will be shorter than a C≡C bond. Each of these trends can be predicted from the relative sizes shown in the margin, and some common bond lengths given in Table 10.2.

The effect of bond type is evident when bonds between the same two atoms are compared. For example, structural data show that the bonds become shorter in the series C—O, C=O, and C≡O.

bond	C—O	C=O	C≡O
bond length (pm)	143	122	113

The bond lengths decrease because the atoms are pulled together more closely as the electron density between the atoms increases.

Bond lengths between different atoms in Table 10.2 are average values, because variations in neighboring parts of a molecule can affect the length of a particular bond. For example, the C—H bond has a length of 105.9 pm in acetylene, H—C≡C—H, but a length of 109.3 pm in methane, CH_4. Although there can be a variation of as much as 10% from the average values listed in Table 10.2, the average bond lengths are useful for estimating bond lengths and building models of molecules (Chapter 11).

Bond distances are given in picometers (pm) in Table 10.2, but many scientists use nanometers (nm) or the older unit of angstroms (Å). 1 Å equals 100 pm. 1nm = 1000 pm.

EXAMPLE 10.4

Bond Lengths

In each pair of bonds, predict which will be shorter.

(a) Si—O or P—O (b) C=C or C—C (c) C=C or C=O

SOLUTION (a) P—O is shorter than Si—O because the P atom is smaller than the Si atom.

(b) C=C is shorter than C—C because the more electrons are shared by atoms, the more closely the atoms are pulled together.

(c) C=O is shorter than C=C because the O atom is smaller than the C atom.

EXERCISE 10.7 • *Bond Lengths*

Explain the increasing order of bond lengths in the following series.

a. C—S is shorter than C—Si b. C—Cl is shorter than C—Br c. N≡O is shorter than N=O

Bond Energy

The bond energy, ***D***, is the enthalpy charge, ΔH, required to break a particular bond in a mole of gaseous molecules. Bond energies are always positive numbers because *the process of breaking bonds in a molecule is always endothermic.*

You've seen in the previous section that the greater the number of bonding electrons between a pair of atoms, the shorter the bond. It is therefore reasonable to expect that multiple bonds are stronger than single bonds, and the greater the multiplicity, the stronger is the bond. For example, the bond energy of C=O in CO_2 is 803 kJ/mol and that of C≡O is 1075 kJ/mol. In fact, the C≡O triple bond in carbon monoxide is the strongest known covalent bond.

Represented as chemical reactions, the energies for breaking the carbon-carbon bonds in ethane ($H_3C—CH_3$), ethylene ($H_2C=CH_2$), and acetylene (HC≡CH) are:

$$\text{molecule} \underset{\text{energy released}}{\overset{\text{energy supplied}}{\rightleftharpoons}} \text{molecular fragments}$$

$$H_3C{-}CH_3(g) \longrightarrow H_3C(g) + CH_3(g) \qquad \Delta H = +347 \text{ kJ/mol}$$

$$H_2C{=}CH_2(g) \longrightarrow H_2C(g) + CH_2(g) \qquad \Delta H = +611 \text{ kJ/mol}$$

$$HC{\equiv}CH(g) \longrightarrow HC(g) + CH(g) \qquad \Delta H = +837 \text{ kJ/mol}$$

Table 10.3

Some Average Single and Multiple Bond Energies (in kJ/mol)

Single Bonds

	H	C	N	O	F	Si	P	S	Cl	Br	I
H	436	414	389	463	569	293	318	339	431	368	297
C		347	293	351	439	289	264	259	330	276	238
N			159	201	272	—	209	—	201	243?	—
O				138	184	368	351	—	205	—	201
F					159	540	490	285	255	197?	—
Si						176	213	226	360	289	213
P							213	230	331	272	213
S								213	251	213	—
Cl									243	218	209
Br										192	180
I											151

Multiple Bonds

Bond	Energy	Bond	Energy
N=N	418	C=C	611
N≡N	946	C≡C	837
C=N	615	C=O (in O=C=O)	803
C≡N	891	C=O (as in H_2C=O)	745
O=O (in O_2)	498	C≡O	1075

The quantity of energy supplied to break the carbon-carbon bonds in these molecules must be the same as the amount of energy released when the same bonds form. *The formation of bonds from atoms in the gas phase is always exothermic.* For example, ΔH for the formation of $H_3C—CH_3$ from two $CH_3(g)$ fragments is -347 kJ/mol.

Some experimental bond energies are tabulated in Table 10.3, and you should be aware of several important points: (1) The energies listed are all positive; they are energies required to break the bond in question. (2) The energies of Table 10.3 are *average* bond energies. The energy of a C—H bond, for example, is given as 414 kJ/mol. However, this energy may vary by as much as 30 to 40 kJ/mol from molecule to molecule, for the same reason that bond lengths vary from one molecule to another. (3) The energies of Table 10.3 are for breaking bonds in molecules in the gaseous state. If a reactant or product is in the solid or liquid state, energy must first be provided to convert it to a gas before using these values. (4) Finally, notice once again the connection between bond energy and bond type. For example, the bond energy for C—C is much smaller (347 kJ/mol) than the bond energy for C=C (611 kJ/mol).

EXERCISE 10.8 • *Bond Lengths and Bond Energies*

Arrange C═N, C≡N, and C—N in order of decreasing bond distance. Is the order for decreasing bond energy the same or the reverse order? Explain.

Estimating Heats of Reaction from Bond Energies

In any reaction between molecules, bonds in the reactants are broken and new bonds are formed in the products. *If the total energy released when new bonds are formed exceeds the energy required to break the original bonds, the overall reaction is exothermic. If the opposite is true, then the overall reaction is endothermic.* Usually an exothermic reaction corresponds to the breaking of weak bonds (with small bond energies) and the making of strong bonds (with large energies).

Bond energies can be used in the following equation to estimate heats of reaction for reactions in the gaseous state:

$$\Delta H^\circ_{rxn} = \Sigma D(\text{bonds broken}) - \Sigma D(\text{bonds formed})$$

The best ΔH°_{rxn} values are obtained from ΔH°_f values (Section 7.6).

Multiply the bond energy, D, for each bond broken by the number of bonds of that type. Adding the results of these multiplications gives a positive number because bond breaking is endothermic. Then, because the sign of energies for bonds made is negative, subtract the sum of the energies of bonds formed.

Suppose you want to estimate the heat of combustion of methane by using the average bond energies in Table 10.3.

$$CH_4(g) + 2\,O_2(g) \longrightarrow CO_2(g) + 2\,H_2O(g)$$

It is best to sketch the molecules and their bonds in order to be sure that none are missed.

$$\begin{array}{c}\text{H}\\|\\\text{H—C—H}\\|\\\text{H}\end{array} + \begin{array}{c}\text{O═O}\\\text{O═O}\end{array} \longrightarrow \text{O═C═O} + \begin{array}{c}\text{H—O—H}\\\text{H—O—H}\end{array}$$

Then,

$$\begin{aligned}\Delta H^\circ_{rxn} &= \Sigma D(\text{bonds broken}) - \Sigma D(\text{bonds formed})\\ &= 4\,D_{C-H} + 2\,D_{O=O} - [2\,D_{C=O} + 4\,D_{O-H}]\\ &= [(4 \times 414) + (2 \times 498)]\text{ kJ} - [(2 \times 803) + (4 \times 463)]\text{ kJ}\\ &= -806\text{ kJ}\end{aligned}$$

The experimental value for the heat of combustion of methane is −802 kJ/mol so use of average bond energies gives a good estimate. This is an example of the formation of stronger bonds at the expense of weaker ones. The bond energy of the O—H bond is not much different in magnitude from those of the C—H and O═O bonds it replaces; all lie between 400 and 500 kJ/mol (Table 10.3). The determining factor making this reaction exothermic is the

Burning natural gas, which is primarily methane with small amounts of ethane, propane, and butane. *(C.D. Winters)*

exceedingly large bond energy of the C=O bond, which at 803 kJ/mol is almost twice as great as for the other bonds involved in the reaction. Not only this reaction but virtually all reactions in which CO_2 is produced are exothermic.

When the bonds being broken and formed are of equal strength, making more moles of bonds than are broken often results in the release of energy. An example of this is the highly exothermic [$\Delta H° = -484$ kJ/mol] reaction between hydrogen and oxygen to form water that was discussed in Section 7.4:

$$2\,H_2(g) + O_2(g) \longrightarrow 2\,H_2O(g)$$

If reactants and products are not all gases, additional thermodynamic data for phase charges are needed (Section 7.3).

All three types of bonds involved have comparable bond energies:

$$D_{H-H} = 436\text{ kJ/mol} \qquad D_{O=O} = 498\text{ kJ/mol} \qquad D_{O-H} = 463\text{ kJ/mol}$$

but the reason the reaction is exothermic becomes obvious if we rewrite it to make the bonds visible:

H—H / H—H + O=O ⟶ H—O—H / H—O—H

2mol (436 kJ/mol) 1 mol (498 kJ/mol) → 1370 kJ Energy required to break bonds

4 mol (463 kJ/mol) → 1852 kJ Energy released in bond formation

While three moles of bonds must be broken (two moles of H—H bonds and one mole of O=O double bonds), a total of four moles of O—H bonds are made. Since all the bonds are similar in strength, making more moles of bonds than are broken means a release of energy. In mathematical terms, the difference in energy of the bonds broken and made is -482 kJ.

$$\begin{aligned}\Delta H^\circ_{rxn} &= 2\,D_{H-H} + D_{O=O} - 4\,D_{O-H} \\ &= [(2 \times 436) + (1 \times 498)]\text{ kJ} - 4 \times 463\text{ kJ} \\ &= -482\text{ kJ}\end{aligned}$$

As described in the discussion of reactivity in Chapters 7 and 8, exothermic reactions are often product-favored.

In most cases the bond energies of the reactants and products are different, so the relative strengths of the bonds that are being broken and formed is the most important consideration. In summary, an exothermic reaction always corresponds to the formation of a stronger set of bonds.

EXAMPLE 10.5

Estimating Heats of Reaction from Bond Energies

Estimate the heat of combustion of 1 mole of heptane vapor, C_7H_{16}.

SOLUTION

$$\mathrm{H{-}\overset{\overset{H}{|}}{\underset{\underset{H}{|}}{C}}{-}\overset{\overset{H}{|}}{\underset{\underset{H}{|}}{C}}{-}\overset{\overset{H}{|}}{\underset{\underset{H}{|}}{C}}{-}\overset{\overset{H}{|}}{\underset{\underset{H}{|}}{C}}{-}\overset{\overset{H}{|}}{\underset{\underset{H}{|}}{C}}{-}\overset{\overset{H}{|}}{\underset{\underset{H}{|}}{C}}{-}\overset{\overset{H}{|}}{\underset{\underset{H}{|}}{C}}{-}H(g) + 11\,O{=}O(g) \longrightarrow 7\,O{=}C{=}O(g) + 8\,H{-}O{-}H(g)}$$

Bonds Broken		*Bonds Formed*	
6 C—C	6 × 347 = 2082 kJ	14 C=O	14 × 803 = 11,242 kJ
16 C—H	16 × 414 = 6624 kJ	16 O—H	16 × 463 = 7408 kJ
11 O=O	11 × 498 = 5478 kJ		
Total	14,184 kJ	Total	18,650 kJ

Thus the heat of combustion of heptane is

$$\Delta H^\circ_{rxn} = 14{,}184\text{ kJ} - 18{,}650\text{ kJ}$$
$$= -4466\text{ kJ/mol heptane}$$

EXERCISE 10.9 • *Estimating Heats of Reaction from Bond Energies*

Use the average bond energies in Table 10.3 to estimate the heat of combustion for pentane vapor, C_5H_{12}.

10.6 RESONANCE

Ozone, O_3, is both beneficial and harmful. The ozone layer in the upper stratosphere protects the earth and its inhabitants from intense ultraviolet radiation from the sun, but ozone pollution in the lower atmosphere causes respiratory problems (see Chapter 20). Ozone is an unstable, blue, diamagnetic gas with a pungent odor.

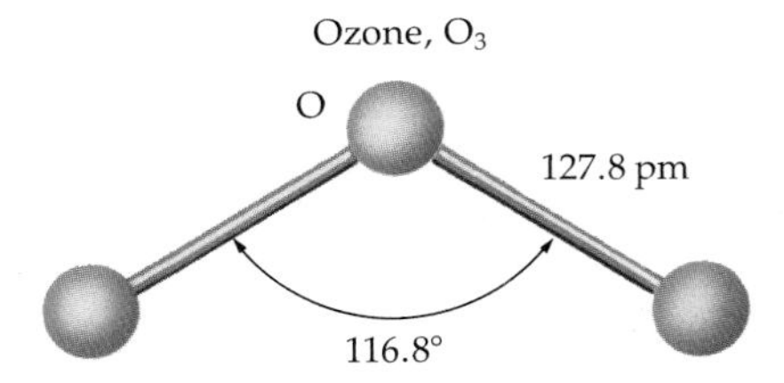

As you have seen, the number of bonding electron pairs between two atoms is important in determining bond length and strength. Ozone has equal bond lengths of 128 pm, implying that both bonds contain the same number of bond pairs. However, using the guidelines for drawing Lewis structures, you might come to a different conclusion. Two possible Lewis structures are

$$\mathrm{:\ddot{O}{=}\ddot{O}{-}\ddot{\underset{..}{O}}:} \quad \text{or} \quad \mathrm{:\ddot{\underset{..}{O}}{-}\ddot{O}{=}\ddot{O}:}$$

Each structure shows a double bond on one side of the central O atom and a single bond on the other side. If either one were the electron structure of O_3, one bond should be shorter (O=O) than the other (O—O), but this is not the case. One way to reconcile the experimental observation of equal O—O bond lengths with the Lewis structures is to propose that the true structure of O_3 is a **resonance hybrid** of the two Lewis structures shown above. It is conventional to connect Lewis structures that contribute to the resonance hybrid structure with a double-headed arrow, ↔, to emphasize that the actual bonding is a composite of the contributing Lewis structures. The contributing Lewis structures are often referred to as **resonance structures.**

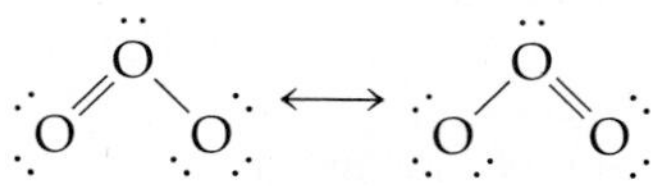

resonance structures

The resonance hybrid is often written as a composite picture in which a dotted line represents two electrons spread over both bonding areas.

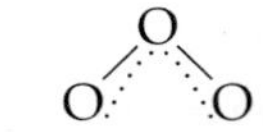

The resonance concept is useful whenever there is a choice about which of two or three atoms contribute lone pairs to achieve an octet of electrons about a central atom by multiple bond formation.

Writing the Lewis structures of oxygen-containing anions often requires resonance.

When applying the concept of **resonance,** there are several important things to keep in mind. First, *Lewis structures contributing to the resonance hybrid structure differ only in the assignment of electron pair positions, never atom positions.* Second, *contributing Lewis structures differ in the number of bond pairs between a given pair of atoms.* Third, *the resonance hybrid structure represents a single intermediate structure and not different structures which are continually changing back and forth.*

To illustrate the use of resonance, consider what happens in writing the Lewis structure of the carbonate ion, CO_3^{2-}, which has 24 valence electrons (4 from C, 18 from three O atoms, and 2 for the 2− charge). Writing the skeleton structure and putting in lone pairs so that each O has an octet uses 24 electrons, but leaves carbon without an octet:

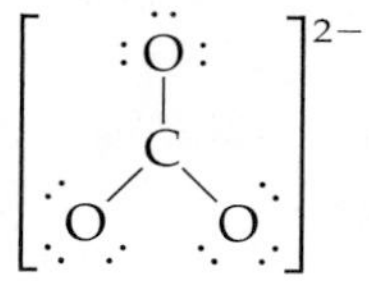

To give carbon an octet requires changing a single bond to a double bond and this can be done in three equivalent ways:

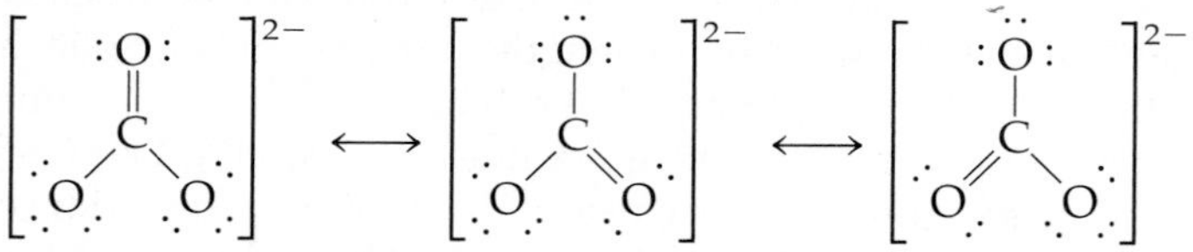

These three Lewis structures are resonance structures that contribute to the resonance hybrid structure, which is drawn below with dotted lines representing the two electrons spread over the three C—O bonds.

This representation is in agreement with experimental results: all three C—O bond distances are 129 pm, a distance intermediate between the C—O single bond (143 pm) and the C═O double bond (122 pm) distance.

Benzene is the simplest aromatic hydrocarbon.

Resonance is also useful for explaining the structure of the benzene molecule, C_6H_6. Structural data for benzene indicate a planar, symmetrical structure in which all carbon-carbon bonds are equivalent. Each carbon-carbon bond is 139 pm long, intermediate between the length of a C—C single bond (154 pm) and a C═C double bond (134 pm). This is represented by the two resonance structures on the left and the resonance hybrid structure on the right.

When hydrogen and carbon atoms are not shown, benzene is represented by a circle in a hexagon. Each corner in the hexagon represents one carbon atom and one hydrogen atom, each line represents a single C—C bond, and the circle represents the even distribution of six bonding electrons spread evenly (delocalized) over all of the carbon atoms.

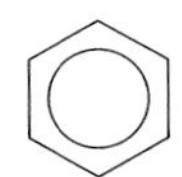

EXAMPLE 10.6

Drawing Resonance Structures

Draw the resonance structures for NO_3^-.

SOLUTION The NO_3^- ion contains 24 valence electrons (5 from the N atom, 18 from three O atoms, and 1 for the 1− charge). Placing single bonds and lone pairs to give each O atom an octet produces a structure without an octet on the central N atom. A lone pair can be converted to a double bond in three equivalent ways, to give three resonance structures.

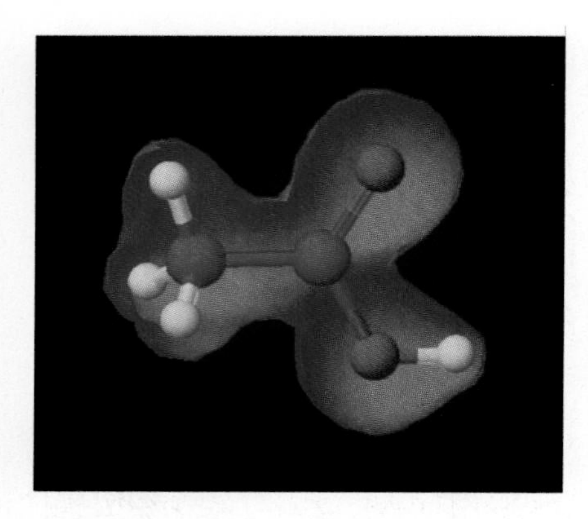

Computer-drawn model of the polar acetic acid molecule. The molecule is enclosed in a surface (purple) that represents the boundary surface of the electron density of the molecule. The boundary surface is larger around the electronegative O atom where there are lone pairs and smaller around the —OH hydrogen atom and the carbon atom, which have less electron density. *(C.D. Winters/CAChe Scientific)*

EXERCISE 10.10 • *Resonance*

The N—O bond lengths in NO_2^- are both 124 pm. Compare this with the bond distances given in Table 10.2 for N—O and N=O bond lengths. Account for any difference.

10.7 BOND POLARITY AND ELECTRONEGATIVITY

In a molecule such as H_2 or F_2, where both atoms are the same kind, there is equal sharing of the bonding electrons and the bond is **nonpolar.** When two different atoms are bonded, however, the sharing of the bonding electrons is unequal and results in a displacement of the bonding electrons toward one of the atoms. If the displacement is complete, the bond is ionic because electrons have been transferred. If the displacement is less than complete, the bond is said to be a **polar covalent bond.** As you will see in Chapters 12 and 14, properties of molecules are dramatically affected by bond polarity. The most dramatic example of this is the fact that water is a liquid and not a gas at room temperature.

A measure of the ability of an atom in a covalent bond to attract shared electrons to itself is called the **electronegativity** of the atom. In 1932, Linus Pauling first proposed the concept of electronegativity based on an analysis of bond energies, and the currently accepted values for electronegativities are shown in Figure 10.6.

Although electronegativities show a periodic trend (Figure 10.7), the pattern is not as regular as that for ionization energies (Figure 9.20). As expected, metals are the least electronegative elements and nonmetals are

Figure 10.6
Electronegativity values in a periodic table arrangement.

<1.0 | 1.0–1.4 | 1.5–1.9 | 2.0–2.4 | 2.5–2.9 | 3.0–4.0

1 H 2.1																
3 Li 1.0	4 Be 1.5											5 B 2.0	6 C 2.5	7 N 3.0	8 O 3.5	9 F 4.0
11 Na 1.0	12 Mg 1.2											13 Al 1.5	14 Si 1.8	15 P 2.1	16 S 2.5	17 Cl 3.0
19 K 0.9	20 Ca 1.0	21 Sc 1.3	22 Ti 1.4	23 V 1.5	24 Cr 1.6	25 Mn 1.6	26 Fe 1.7	27 Co 1.7	28 Ni 1.8	29 Cu 1.8	30 Zn 1.6	31 Ga 1.7	32 Ge 1.9	33 As 2.1	34 Se 2.4	35 Br 2.8
37 Rb 0.9	38 Sr 1.0	39 Y 1.2	40 Zr 1.3	41 Nb 1.5	42 Mo 1.6	43 Tc 1.7	44 Ru 1.8	45 Rh 1.8	46 Pd 1.8	47 Ag 1.6	48 Cd 1.6	49 In 1.6	50 Sn 1.8	51 Sb 1.9	52 Te 2.1	53 I 2.5
55 Cs 0.8	56 Ba 1.0	57 La 1.1	72 Hf 1.3	73 Ta 1.4	74 W 1.5	75 Re 1.7	76 Os 1.9	77 Ir 1.9	78 Pt 1.8	79 Au 1.9	80 Hg 1.7	81 Tl 1.6	82 Pb 1.7	83 Bi 1.8	84 Po 1.9	85 At 2.1
87 Fr 0.8	88 Ra 1.0	89 Ac 1.1														

Portrait of a Scientist

Linus Pauling (1901–)

Linus Pauling was born in Portland, Oregon, in 1901, where he grew up as the son of a druggist. He earned a B.Sc. degree in chemical engineering from Oregon State College in 1922 and completed his Ph.D. in chemistry at the California Institute of Technology in 1925. Before joining Cal Tech as a faculty member, he traveled to Europe where he worked briefly with Erwin Schrödinger and Niels Bohr (see Chapter 9).

In chemistry Pauling is best known for his work on chemical bonding. Pauling, along with R.B. Corey, proposed the helical and sheetlike secondary structures for proteins. For his bonding theories and his work with proteins, Pauling was awarded the Nobel Prize in Chemistry in 1954. His book *The Nature of the Chemical Bond* has influenced several generations of scientists. Shortly after World War II, Pauling and his wife began a crusade to limit nuclear weapons, a crusade that came to fruition in the limited test ban treaty of 1963. For this effort, Pauling was awarded the 1963 Nobel Prize for Peace. Never before had any person received two unshared Nobel Prizes.

Linus Pauling. (Oesper Collection in the History of Chemistry/University of Cincinnati)

the most electronegative. *In general, electronegativity increases diagonally upward and to the right of the periodic table.*

Pauling's electronegativity values are relative numbers with an arbitrary value of 4.0 for the most electronegative element, fluorine. The nonmetal with the next highest electronegativity is oxygen with a value of 3.5, followed by chlorine and nitrogen, which have the same value of 3.0. Elements with electronegativities of 2.5 or more are all nonmetals in the top right-hand corner of the periodic table. By contrast, elements with electronegativities of 1.3 or less are all metals in the lower left of the periodic table. These elements are often referred to as the most **electropositive** elements,

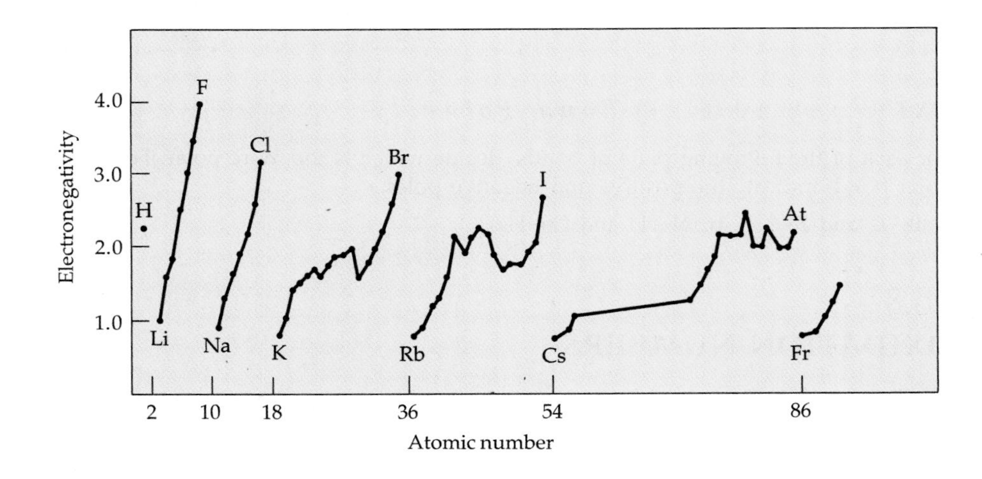

Figure 10.7
Periodic nature of electronegativities when plotted against atomic number.

and they are the metals that invariably form ionic compounds. Between these two extremes are most of the remaining metals (largely transition metals) with electronegativities between 1.4 and 1.9, the metalloids with electronegativities between 1.8 and 2.1, and some nonmetals with electronegativities between 2.1 and 2.4.

Electronegativity values are approximate and are primarily used to predict the polarity of covalent bonds. The bond polarity is indicated by writing $\delta+$ by the less electronegative atom and $\delta-$ by the more electronegative atom, where δ stands for partial charge. For example, the polar HF bond in hydrogen fluoride can be represented as

$$\overset{\delta+}{\text{H}}\text{—}\overset{\delta-}{\text{F}}$$

All bonds have some degree of polarity (except those between identical atoms) and the difference in electronegativity values is a qualitative measure of the degree of polarity. The change from nonpolar covalent bonds to slightly polar covalent bonds to very polar bonds to ionic bonds can be regarded as a continuum. Examples are H_2 (nonpolar), HI (slightly polar), HF (very polar), Na^+Cl^- (ionic).

EXAMPLE 10.7

Bond Polarity

For each of the following bond pairs, tell which is the more polar and indicate the positive and negative poles.

(a) Cl—F and Cl—Br (b) Si—Br and C—Br

SOLUTION (a) The Cl—F bond is more polar because the difference in electronegativity is greater between Cl and F than between Cl and Br. The F atom is the negative end in ClF, but the Cl atom is the negative end in ClBr since Cl is more electronegative than Br.

$$\overset{\delta+}{\text{Cl}}\text{—}\overset{\delta-}{\text{F}} \text{ more polar than } \overset{\delta-}{\text{Cl}}\text{—}\overset{\delta+}{\text{Br}}$$

(b) Si—Br is more polar than C—Br because Si is less electronegative than C, so the electronegativity difference is greater between Si and Br.

$$\overset{\delta+}{\text{Si}}\text{—}\overset{\delta-}{\text{Br}} \text{ more polar than } \overset{\delta+}{\text{C}}\text{—}\overset{\delta-}{\text{Br}}$$

EXERCISE 10.11 • ***Bond Polarity***

For each of the following pairs of bonds, decide which is the more polar. For each polar bond, indicate the positive and negative poles.

a. B—C and B—Cl b. N—H and O—H

10.8 OXIDATION NUMBERS

The concept of electronegativity has a number of practical applications besides predicting polarity of bonds. For example, it can be useful in assigning oxidation numbers to atoms within a covalent molecule.

The **oxidation number** of an atom in a molecule is the charge that atom would have if each bonding pair of electrons belonged completely to the more electronegative atom. For example, since carbon is more electronegative than hydrogen, the pair of electrons in each C—H bond in CH_4 is assigned to the C atom. Therefore, each H has an oxidation number of +1, the charge it would have if it had actually lost the electron. The C atom has an oxidation number of −4, the charge of the ion that would result if it had gained four electrons.

$$\overset{-4}{C}\overset{+1}{H_4}$$

The use of oxidation numbers is simply a "bookkeeping" device—invented by chemists, not by molecules—that indicates the distribution of electrons in molecules and polyatomic ions and helps keep track of electrons in oxidation-reduction reactions. The following set of rules is used to determine oxidation numbers.

1. *The oxidation number of the atoms in a pure element is zero.* For example, the oxidation number of oxygen in O_2 or of nitrogen in N_2 or of C in graphite is zero.
2. *The oxidation number of a monatomic ion is the same as its charge.* For example, the oxidation number of calcium in Ca^{2+} is +2 and that of bromine in Br^- is −1.
3. *Some elements have the same oxidation number in nearly all their compounds.*
   a. Hydrogen has an oxidation number of +1 unless it is combined with metals, in which case its oxidation number is −1.
   b. Fluorine has an oxidation number of −1 in all its compounds.
   c. Oxygen usually is assigned an oxidation number of −2; exceptions include H_2O_2, in which O has an oxidation number of −1 (because H has an oxidation number of +1) and OF_2 where O has an oxidation number of +2 (because F has an oxidation number of −1).
4. *In binary compounds of two different elements, the element with greater electronegativity is assigned a negative oxidation number equal to its charge in simple ionic compounds of the element.* For example, in PCl_3, Cl is assigned an oxidation number of −1 (as in the chloride ion, Cl^-), so P has an oxidation number of +3.
5. *The sum of the oxidation numbers in a neutral species is 0; in a polyatomic ion, it equals the charge on that ion.* For example, in NO_2, the oxidation number of O is −2, so that of N is +4; the sum of $+4 + (2 \times -2)$ is zero. In NO_3^- the oxidation number of O is −2 and the oxidation number of N is +5; the sum of $+5 + (3 \times -2)$ is −1.

Suppose we wanted to determine the oxidation numbers of all atoms in H_2SO_4. The oxidation number of H is +1 and the oxidation number of O is −2. The total for these two atoms is +2 for two H plus −8 for four O atoms to give −6. Since the sum of the oxidation numbers for the molecule of H_2SO_4 must be zero, S must have an oxidation number of +6.

Oxidation numbers are used in the same way that ion charges are used in naming compounds. You've seen, for example, that iron forms two ionic chlorides known as iron(II) chloride ($FeCl_2$) and iron(III) chloride ($FeCl_3$).

Arsenic forms two fluorides that are named by the same system—arsenic(III) fluoride (AsF_3) and arsenic(V) fluoride (AsF_5)—although neither one is an ionic compound.

In Section 5.7, oxidation-reduction (redox) reactions were defined as reactions that involve the transfer of electrons. Many oxidation-reduction reactions involve polyatomic ions or neutral molecules, for which it is difficult to determine what is oxidized and what is reduced. Oxidation numbers are a useful bookkeeping device for determining the oxidized and reduced species. In all oxidation-reduction reactions, the oxidation numbers increase for the species that are oxidized, and they decrease for the species that are reduced. Suppose you want to determine what is oxidized and what is reduced in the following reaction.

$$2\,Cr(s) + 3\,ClO_4^-(aq) + 3\,H_2O(\ell) \longrightarrow 2\,Cr(OH)_3(s) + 3\,ClO_3^-(aq)$$

Oxidation: Oxidation number increases.
Reduction: Oxidation number decreases.

Reactants: The chromium metal has an oxidation number of 0. The oxidation number of Cl in ClO_4^- is +7. (To find the oxidation number of Cl, reason as follows: The oxidation number of O is −2. The total for four O atoms is −8. Since the ion has a −1 charge, the oxidation number of Cl must be +7, so that $-8 + 7 = -1$.)

Products: The oxidation number of Cr in $Cr(OH)_3$ is +3 (because the charge of the three OH^- ions totals −3). The oxidation number of Cl in ClO_3^- is +5. (Reason as before: The total for three O atoms is −6; since the ion has a −1 charge, the oxidation number of Cl must be +5.)

In the reaction, the oxidation number of Cr increases from 0 to +3, so Cr is *oxidized* to $Cr(OH)_3$. The oxidation number of Cl decreases from +7 to +5, so ClO_4^- is *reduced* to ClO_3^-.

EXAMPLE 10.8

Redox Reactions and Oxidation Numbers

Which of the following equations represent redox reactions?

a. $Sn(s) + 2\,HCl(aq) \longrightarrow SnCl_2(aq) + H_2(g)$

b. $KOH(aq) + HCl(aq) \longrightarrow KCl(aq) + H_2O(\ell)$

SOLUTION The question can be answered by assigning oxidation numbers to the atoms in each reactant and product.

(a)

$$\overset{0}{Sn}(s) + 2\,\overset{+1\,-1}{HCl}(aq) \longrightarrow \overset{+2\,-1}{SnCl_2}(aq) + \overset{0}{H_2}(g)$$

The Sn has increased in oxidation number and H has decreased in oxidation number, so this is a redox reaction.

(b)

$$\overset{+1\,-2\,+1}{KOH}(aq) + \overset{+1\,-1}{HCl}(aq) \longrightarrow \overset{+1\,-1}{KCl}(aq) + \overset{+1\,-2}{H_2O}(\ell)$$

None of the atoms has undergone a change in oxidation number, so this is not a redox reaction. (It is an exchange reaction between an acid and a base.)

EXERCISE 10.12 • *Oxidation Numbers*

Assign the oxidation number to the underlined atom in each of the following:

a. $\underline{S}F_4$ b. $\underline{C}O_3^{2-}$ c. $\underline{Mn}F_3$ d. $\underline{Mn}O_4^-$ e. $H_2\underline{S}O_3$ f. $H_3\underline{P}O_4$

10.9 FORMAL CHARGES ON ATOMS

The method used to assign oxidation numbers is especially unrealistic for estimating the electron distribution on atoms in covalent bonds. However, by assuming that each bond pair is *shared equally* by two atoms, we can derive a more reasonable estimate of the atomic charges in molecules. The values found in this way are called **atom formal charges** and are given by the expression

$$\text{atom formal charge} = (\text{no. of valence e}^-) - (\text{no. of nonbonding e}^-) - \tfrac{1}{2}(\text{no. of bonding electrons})$$

Here we count the nonbonding electrons on an atom plus half of the bonding electrons that it shares. To illustrate the difference between oxidation numbers and formal charges, let's look at ClO_4^-. Begin with the Lewis structure.

```
[        ..        ]-
[       :O:        ]
[  ..    |    ..   ]
[ :O — Cl — O:     ]
[  ..    |    ..   ]
[       :O:        ]
[        ..        ]
```

The oxidation number of O is −2; four oxygen atoms give a total of −8. Since the ion has a −1 charge, the oxidation number of Cl is +7. However, if you consider how much ionization energy would be required to remove seven electrons from a chlorine atom, it is obvious that this cannot be the actual charge on the Cl atom.

The formal charges in ClO_4^- are calculated as follows:

$$\text{atom formal charge} = (\text{no. of valence e}^-) - (\text{no. of nonbonding e}^-) - -\tfrac{1}{2}(\text{no. of bonding e}^-)$$

$$\text{formal charge for O} = 6 - 6 - \tfrac{1}{2}(2)$$
$$= -1 \text{ compared to } -2 \text{ for the oxidation number}$$

$$\text{formal charge for Cl} = 7 - 0 - \tfrac{1}{2}(8)$$
$$= +3 \text{ compared to } +7 \text{ for the oxidation number}$$

The signs of the formal charges and oxidation numbers are the same (as is often the case), but the magnitudes of the formal charges are a more realistic estimate of the relative positive and negative character of the atoms.

Formal charges, like oxidation numbers, sum to give the charge on the ion.

In the resonance structures for O_3 and C_6H_6 drawn in Section 10.6, all the possible electron configurations are equally likely. Therefore, the actual structure of the molecule or ion is a symmetrical distribution of electrons over all the atoms involved—that is, an equal "mixture" of the resonance structures. This is not the case for all molecules that can be represented by resonance structures, however. In many such molecules, the actual structure is more like one of the resonance structures than like the others. We can use formal charges together with the following two rules to decide which resonance structure is most stable and therefore the most important.

1. Atoms in molecules (or ions) should have formal charges as small as possible.
2. A molecule (or ion) is most stable when any negative formal charge resides on the most electronegative atom.

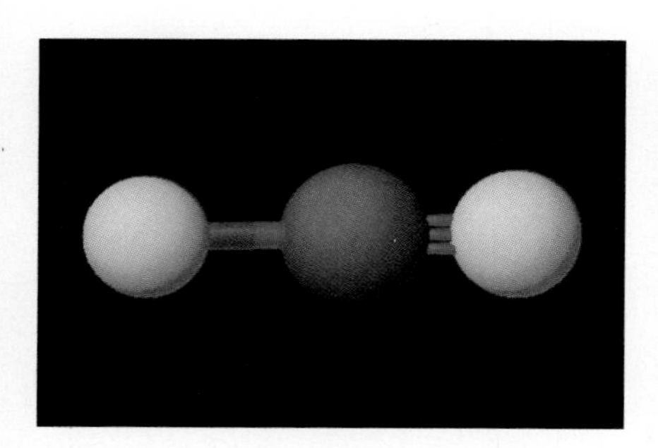

Computer-drawn model of a resonance structure of OCN^-. The green spheres indicate the atom (O and N) is negatively charged, and the purple sphere indicates a positive charge (C atom). *(C.D. Winters/CAChe Scientific)*

For example, there are three possible resonance structures for the cyanate ion, OCN^-. Using atom formal charges, we can determine which is the most important of these structures.

formal charges:	+1 0 −2	0 0 −1	−1 0 0
	$[:O{\equiv}C{-}\ddot{\underset{\cdot\cdot}{N}}:]^-$	$[:\ddot{O}{=}C{=}\ddot{N}:]^-$	$[:\ddot{\underset{\cdot\cdot}{O}}{-}C{\equiv}N:]^-$
	A	B	C

As an example of formal charge, consider structure A.

	Valence e⁻	−	*Nonbonding e⁻*	− $\frac{1}{2}$*(Bonding e⁻)*		
formal charge for O =	6	−	2	$-\frac{1}{2}$	(6)	= +1
formal charge for C =	4	−	0	$-\frac{1}{2}$	(8)	= 0
formal charge for N =	5	−	6	$-\frac{1}{2}$	(2)	= −2
sum of formal charges = charge on ion						= −1

In structure A, the O atom has a formal charge of +1, a very unfavorable situation for this electronegative atom. Therefore, A contributes little to the overall electronic structure of the cyanate ion. Structure B places a −1 charge on nitrogen and 0 on oxygen, not an unfavorable situation. However, structure C is favored slightly because the negative atom in the structure is the most electronegative one in the ion.

In the actual structure, we would expect to find an average electron density between the C and N atoms that is greater than in a double bond but less than in a triple bond. Similarly, the electron density in the C—O bond region should be between those of single and double bonds. These expectations are confirmed by measurements of bond lengths.

EXAMPLE 10.9

Formal Charges

Draw the two possible resonance structures for the HOCN molecule and use formal charges to determine which is the more important structure.

SOLUTION The two resonance structures are

$H{-}\ddot{\underset{\cdot\cdot}{O}}{-}C{\equiv}N:$	$H{-}\ddot{O}{=}C{=}\ddot{N}:$
A	B

The formal charge of each atom is given by

$$\text{formal charge} = (\text{no. of valence electrons}) - (\text{no. of nonbonding electrons}) - \tfrac{1}{2}(\text{no. of bonding electrons})$$

Structure A	*Structure B*
$H = 1 - 0 - \frac{1}{2}(2) = 0$	$H = 1 - 0 - \frac{1}{2}(2) = 0$
$O = 6 - 4 - \frac{1}{2}(4) = 0$	$O = 6 - 2 - \frac{1}{2}(6) = +1$
$C = 4 - 0 - \frac{1}{2}(8) = 0$	$C = 4 - 0 - \frac{1}{2}(8) = 0$
$N = 5 - 2 - \frac{1}{2}(6) = 0$	$N = 5 - 4 - \frac{1}{2}(4) = -1$

Structure A gives all atoms a zero formal charge, while structure B has nonzero formal charges on two atoms. According to the rule that the most important structure is the one with the smallest possible formal charges, structure A makes the largest contribution to the resonance hybrid for HOCN.

EXERCISE 10.13 • *Atom Formal Charges*

The Lewis structure of H_2BF is given in A below. To satisfy the octet rule for the boron atom, one could write resonance structure B. Calculate the formal charge on each atom in structure B and comment on the relative importance of the two structures.

:F: — B(H)(H)   :F═B(H)(H)

A   B

10.10 COORDINATION COMPOUNDS

A **metal complex** or **coordination compound** is a compound in which a metal ion or atom is bonded to one or more neutral molecules or anions in an integral structural unit. Much of the chemistry of *d*-transition metals is related to their ability to form a variety of coordination compounds. The molecules or ions bonded to the central metal ion are called **ligands,** from the Latin verb *ligare,* "to bind." Each ligand has one or more atoms with lone pairs, and these atoms are bonded to the metal atom or ion by *coordinate covalent bonds.*

Electron pair donors (ligands in this case) and electron pair acceptors (metal ions in this case) are examples of Lewis bases and Lewis acids, respectively (see Section 16.9).

Metal complexes can occur as cations, anions, or neutral molecules. The group of ligands bonded to the metal is said to constitute the metal's **coordination sphere;** square brackets usually enclose the coordination sphere. For example, the hydrate $NiCl_2 \cdot 6\,H_2O$ (Figure 4.19) is a coordination compound whose formula is more properly written as $[Ni(H_2O)_6]Cl_2$ to emphasize that the six water molecules are bound to Ni^{2+} in a **complex ion,** $[Ni(H_2O)_6]^{2+}$, and this 2+ ion is associated with two Cl^- ions.

Positive or negative ions that contain a metal ion bound to ligands are called complex ions. *For example,* $[Cu(NH_3)_4]^{2+}$ *is a complex ion with 4* NH_3 *molecules bound to* Cu^{2+}. *An example of a coordination compound containing this complex ion is* $[Cu(NH_3)_4]SO_4$. *An example of a negative complex ion is* $[NiCl_4]^{2-}$, *which has 4* Cl^- *bound to* Ni^{2+} *to give a net charge of* −2. *An example of a coordination compound containing this complex ion is* $K_2[NiCl_4]$.

coordinate covalent bond

$$\left[Ni(OH_2)_6\right]^{2+} \quad 2\ Cl^-$$

The charge of a complex ion is determined by the charge of the metal ion and any negative ions bonded to it. In the hydrated nickel compound, the water ligands are neutral so the charge of the complex ion is the same as that of the nickel ion, Ni^{2+}. A neutral ionic compound is formed by the complex ion and two Cl^- ions.

The compound $K_3[Fe(CN)_6]$ contains the anion $[Fe(CN)_6]^{3-}$ and three K^+ cations, and the anticancer drug $[Pt(NH_3)_2Cl_2]$ is a neutral molecule that contains the ligands NH_3 and Cl^- coordinated to the Pt^{2+} ion. The number of coordinate covalent bonds between the ligands and the central metal ion is its **coordination number** and is usually 2, 4, or 6.

Ligands such as H_2O, NH_3, and Cl^- that form only one coordinate covalent bond to the metal are termed **monodentate.** The word "dentate" stems from the Latin word *dentis* for tooth, so NH_3 is a "one-toothed" ligand. Common monodentate ligands are shown in Figure 10.8. Some molecules can form two or more coordinate covalent bonds to the same metal ion because they have two or more atoms with lone pairs separated by several intervening atoms. A good example is the **bidentate** molecule 1,2-diaminoethane ($H_2NCH_2CH_2NH_2$), commonly called ethylenediamine and abbreviated *en*. When lone pairs of electrons from both nitrogen atoms coordinate to a metal ion, a stable five-membered ring is formed. The word "*chelating*," derived from the Greek *chele*, "claw," describes the pincerlike way in which a ligand can grab a metal ion. Some common **chelating ligands** are shown in Figure 10.9.

Rules for naming coordination compounds are given in Appendix E.

A chelating agent that forms several bonds to a metal without greatly straining its own structure is usually able to replace a simpler ligand. For example, although both ligands form coordinate covalent bonds via the N: atoms, ethylenediamine can readily replace ammonia from most complexes:

$$\left[Ni(NH_3)_6\right]^{2+} + 3\ H_2\ddot{N}CH_2CH_2\ddot{N}H_2 \longrightarrow \left[Ni(NH_2CH_2CH_2NH_2)_3\right]^{2+} + 6\ NH_3$$

Monodentate ligands

Ligand	Structure	Ligand	Structure
Water molecule	$:\ddot{O}-H$ with H below O	Chloride ion	$:\ddot{Cl}:^-$
		Cyanide ion	$:C\equiv N:^-$
Ammonia molecule	$:N-H$ with two more H on N	Carbon monoxide molecule	$:C\equiv O:$
		Hydroxide ion	$:\ddot{O}-H^-$

Figure 10.8
Some common monodentate ligands.

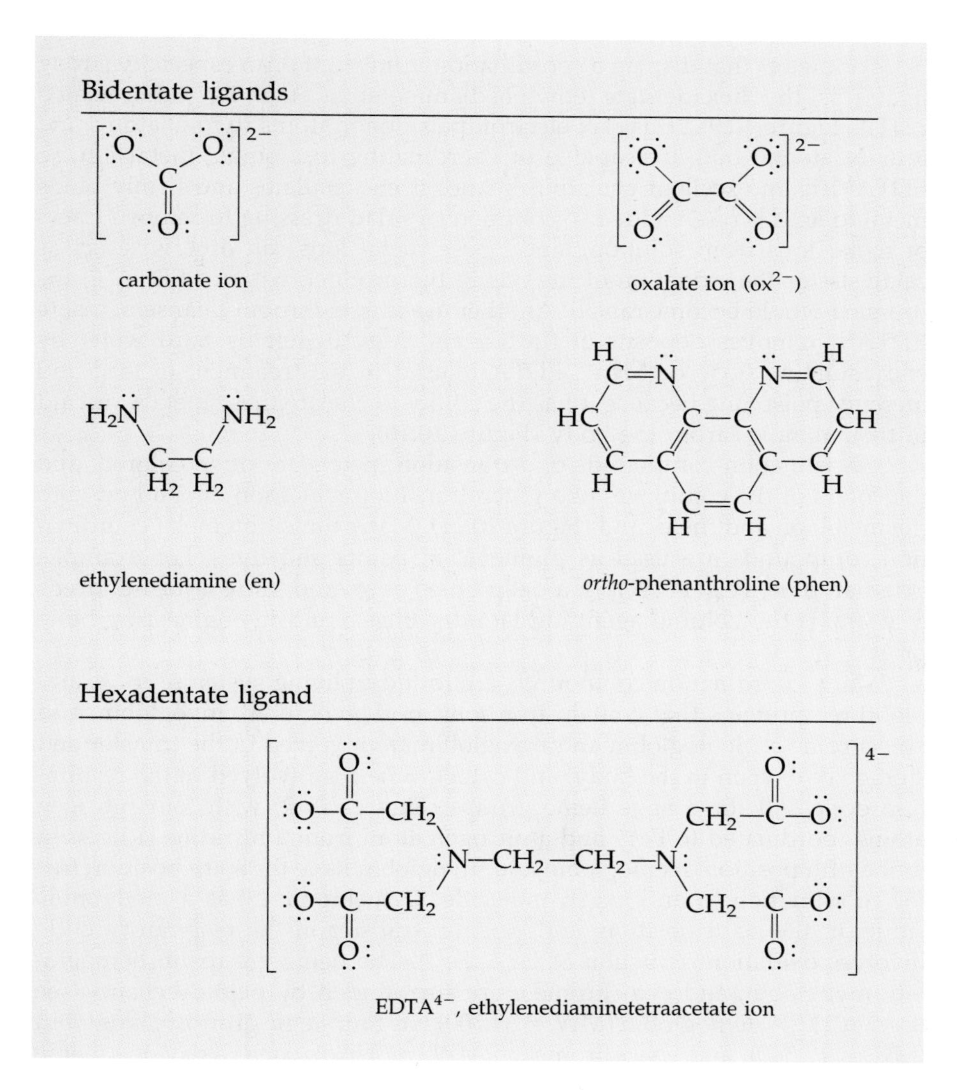

Figure 10.9
Some chelating ligands.

Some household products that contain EDTA. *(C.D. Winters)*

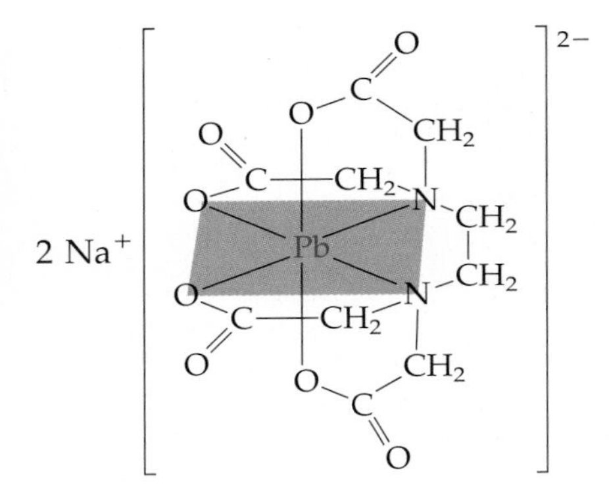

Figure 10.10
The structure of the chelate formed when the EDTA anion forms a complex with Pb^{2+}.

For metals that display a coordination number of 6, an especially strong ligand is the **hexadentate** ethylenediaminetetraacetate ion (abbreviated EDTA, Figure 10.9). It has six electron-pair donor atoms (two O atoms and four N atoms) that are capable of coordinating to a single metal ion, so $EDTA^{4-}$ is an excellent chelating ligand; it encapsulates and firmly binds metal ions. It is often added to commercial salad dressing to remove traces of metal ions from solution, since these metal ions can otherwise act as catalysts for the oxidation of the oils in the product; without $EDTA^{4-}$, the dressing would become rancid. Another use is in bathroom cleansers, where $EDTA^{4-}$ removes deposits of $CaCO_3$ and $MgCO_3$ left by hard water by encapsulating Ca^{2+} or Mg^{2+}. EDTA is used in the treatment of lead and mercury poisoning because it has the ability to chelate these metals and aid in their removal from the body (Figure 10.10).

Coordination compounds of *d*-transition metals are often colored, and the colors of the complexes of a given transition metal ion depend on both the metal ion and the ligand (Figure 10.11). Many transition metal coordination compounds are used as pigments in paints and dyes. For example, Prussian blue, $Fe_4[Fe(CN)_6]_3$, a deep blue compound known for hundreds of years, is the "bluing agent" in laundry bleach and in engineering blueprints.

Prussian blue, $Fe_4[Fe(CN)_6]_3$, is a deep blue compound that has been known for hundreds of years. Here it was formed by the reaction of $Fe^{3+}(aq)$ with $[Fe(CN)_6]^{4-}(aq)$. However, the earliest recipe used ox blood in an iron kettle. *(C.D. Winters)*

Many coordination compounds are found in living systems. For example, three proteins that contain iron ions are hemoglobin, myoglobin, and cytochrome *c*. Hemoglobin and myoglobin are involved in the transfer and storage of oxygen in the body, and cytochrome *c* is involved in the respiration cycle. All three have heme groups (Figure 10.12) with four nitrogen atoms coordinated to Fe^{2+} and a nitrogen atom from a histidine side chain in the fifth position. Hemoglobin and myoglobin have the sixth position free for coordination to an oxygen molecule. Cytochrome *c* has a methionine sulfur in the sixth position, and electron transfer in the respiratory cycle involves oxidation-reduction of Fe^{2+}/Fe^{3+}. The heme groups in hemoglobin, myoglobin, and cytochrome *c* are surrounded by peptide chains (see Section 21.2). Hemoglobin (MW = 64,500) has four heme groups, myoglobin (MW = 17,500) and cytochrome *c* (MW = 12,500) each have one heme group.

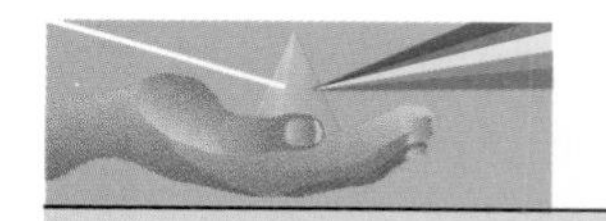

CHEMISTRY YOU CAN DO

Do All Detergents Contain Water Softeners?

Certain minerals make water hard, which is less able to make soap suds and clean things. Obtain some distilled water, three plastic soft-drink bottles with screw caps, about a teaspoon each of table salt and of epsom salts ($MgSO_4 \cdot 7\,H_2O$), and a little liquid dishwashing detergent (not the kind for automatic dishwashers). Rinse any remaining soft drink out of the bottles. Pour about 250 mL of distilled water into each bottle. Add a teaspoonful of table salt to one bottle. Add a teaspoonful of epsom salts to a second bottle. Swirl the bottles until the solid salts dissolve. Keep the third bottle with just distilled water as a control.

Now add a few drops of the liquid dishwashing detergent to each bottle. Screw on the bottle caps, and shake the bottles. What do you observe? Do all salts make water hard? Try the same experiment with other soluble solids that you can obtain. Do any of them have the same effect as table salt? As epsom salts? Repeat the experiment, but this time use some automatic dishwasher detergent. What do you observe? If you get different results than with the first detergent, compare the list of ingredients and explain how metal complexes might account for your results.

EXERCISE 10.14 • *Coordination Compounds*

For the compound $[Co(NH_3)_4(CO_3)]Cl$, identify (a) the complex ion, (b) the ligands, (c) the metal ion and its charge.

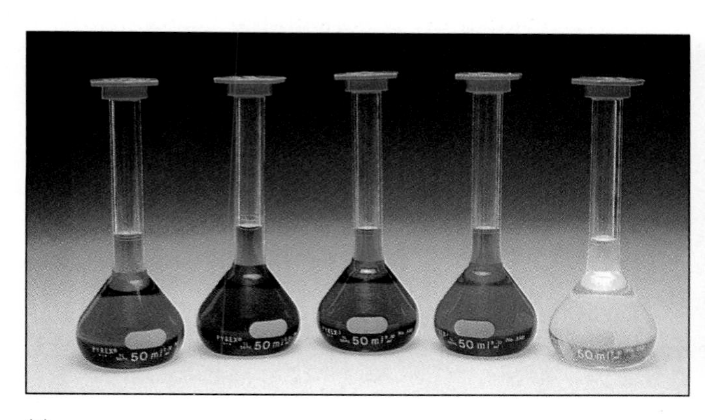

(a)

(b)

Figure 10.11

(a) Compounds of the transition of elements are often colored. Pictured are aqueous solutions of the nitrate salts of (left to right) Fe^{3+}, Co^{2+}, Ni^{2+}, Cu^{2+}, and Zn^{2+}. (b) The colors of the complexes of a given transition metal ion depend on the ligand. All of the complexes pictured here contain the Ni^{2+} ion. The green solid is $[Ni(H_2O)_6](NO_3)_2$; the purple solid is $[Ni(NH_3)_6]Cl_2$; the red solid is $Ni(dimethylglyoximate)_2$. The dimethylgloximate anion

$$\begin{array}{c} CH_3C{-}CCH_3 \\ \| \quad \| \\ HO{-}\ddot{N} \quad \ddot{N}{-}O^- \end{array}$$

is a bidentate ligand that forms coordinate covalent bonds to metal ions through the two nitrogen atoms. *(C.D. Winters)*

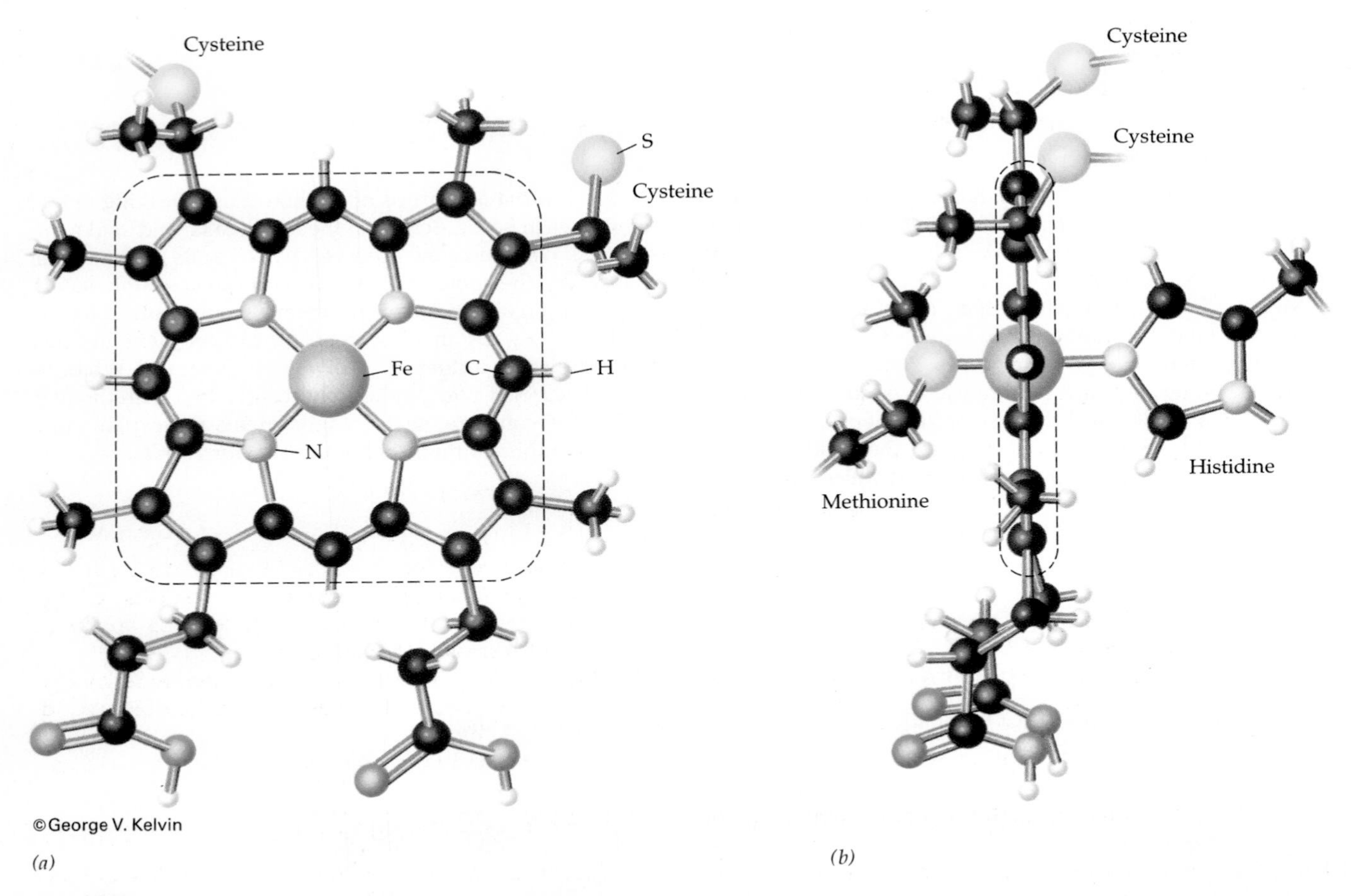

Figure 10.12
(a) Heme group of cytochrome *c*. The portion within the dotted lines is the same for hemoglobin and myoglobin, which have —CH=CH_2 groups in place of the cysteine groups. (b) Edge view of Part a. The four heme groups in hemoglobin and the one heme group in myoglobin do not have methionine in the sixth coordination position, which is open for bonding to an oxygen molecule.

IN CLOSING

Having studied this chapter, you should be able to

- recognize the different types of covalent bonding (Sections 10.1, 10.2, 10.3, 10.10).
- use Lewis structures to represent the different types of covalent bonds in molecules and polyatomic ions (Sections 10.1, 10.2).
- explain why there are exceptions to the octet rule (Section 10.4).
- predict bond lengths from periodic trends in atomic radii (Section 10.5).
- use bond energies to estimate enthalpies of reaction (Section 10.5).
- use resonance structures to model multiple bonding in molecules and polyatomic ions such as O_3, C_6H_6, and CO_3^{2-} (Section 10.6).
- predict bond polarity from electronegativity trends (Section 10.7).
- calculate oxidation numbers for atoms in molecules or polyatomic ions (Section 10.8).

- use oxidation numbers to identify what is oxidized and reduced in redox reactions (Section 10.8)
- calculate formal charges for atoms in molecules or polyatomic ions and use formal charges to choose the most likely resonance structures (Section 10.9).
- give examples of coordination compounds and their uses (Section 10.10).

STUDY QUESTIONS

Review Questions

1. Explain the difference between an ionic bond and a covalent bond.
2. What characteristics must atoms A and B have if they are able to form a covalent bond A—B with each other?
3. Boron compounds often do not obey the octet rule. Illustrate this with BCl_3. Show how the molecule can obey the octet rule by forming a coordinate covalent bond with NH_3.
4. Indicate the difference between alkanes, alkenes, and alkynes by drawing the structural formula of a compound in each class that contains three carbon atoms.
5. Refer to Table 10.1 and answer the following questions:
   (a) Do any molecules with more than eight electrons have a second period element as the central atom?
   (b) What is the maximum number of bond pairs and lone pairs that surround the central atom in any of these molecules?
6. While sulfur forms the compounds SF_4 and SF_6, no equivalent compounds of oxygen, OF_4 and OF_6, are known. Explain.
7. Which of the following molecules have an odd number of valence electrons? NO_2, SCl_2, NH_3, and NO_3.
8. Draw resonance structures for NO_2^-. Predict a value for the N—O bond length based on bond lengths given in Table 10.2, and explain your answer.
9. Consider the following structures for the formate ion, HCO_2^-. Designate which two are resonance structures and which is equivalent to one of the resonance structures.

   (a) $[:\ddot{\underset{..}{O}}-C(H)=\ddot{O}:]^-$ (b) $[:\ddot{O}=C(H)-\ddot{\underset{..}{O}}:]^-$

   (c) $[:\ddot{\underset{..}{O}}-C(H)=\underset{..}{O}:]^-$

10. Consider a series of molecules in which the C atom is bonded to atoms of second period elements: C—O, C—F, C—N, C—C, and C—B. Place these bonds in order of increasing bond length.
11. What are the trends in bond length and bond energy for a series of related bonds, say single, double, and triple C—C bonds?
12. If you wished to calculate the enthalpy change of the reaction

$$2\ C_2H_6(g) + 7\ O_2(g) \longrightarrow 6\ H_2O(g) + 4\ CO_2(g)$$

    what bond energies would you need? Outline the calculation, being careful to show correct algebraic signs.
13. Define and give an example of a polar covalent bond. Give an example of a nonpolar bond.
14. Define electronegativity and describe the trends in electronegativity in the periodic table.
15. Describe the difference between the oxidation number of an atom and its formal charge. Which is often the more realistic description of the charge on an atom in a molecule?
16. Define the following words or phrases and give an example for each: (a) coordination compound, (b) complex ion, (c) ligand, (d) chelate, (e) bidentate ligand.

Lewis Structures

17. Draw Lewis structures for the following molecules or ions:
   (a) $SiCl_4$ (b) ClO_3^- (c) HOCl (d) SO_3^{2-}
18. Draw Lewis structures for the following molecules or ions:
   (a) ClF (b) H_2Se (c) BF_4^- (d) PO_4^{3-}
19. Draw Lewis structures for the following molecules:
   (a) $CHClF_2$, one of the many chlorofluorocarbons that have been used in refrigeration
   (b) methyl alcohol, CH_3OH
   (c) methyl amine, CH_3NH_2
20. Draw Lewis structures for the following molecules or ions:
   (a) CH_3Cl (b) SiO_4^{4-} (c) PH_4^+ (d) C_2H_6

21. Draw Lewis structures for the following molecules:
(a) formic acid, HCO_2H, in which the atomic arrangement is

$$\begin{array}{c} \text{O} \\ | \\ \text{H—C—O—H} \end{array}$$

(b) acetonitrile, CH_3CN
(c) vinyl chloride, CH_2CHCl, the molecule from which PVC plastics are made

22. Draw Lewis structures for the following molecules:
(a) tetrafluoroethylene, C_2F_4, the molecule from which Teflon is made
(b) acrylonitrile, CH_2CHCN, the molecule from which Orlon is made

Exceptions to the Octet Rule

23. Draw the Lewis structure for each of the following molecules or ions.
(a) BrF_3 (b) I_3^- (c) XeF_4

24. Draw the Lewis structure for each of the following molecules or ions.
(a) BrF_5 (b) SeF_6 (c) IBr_2^-

25. Which of the following elements can form compounds with five or six pairs of valence electrons surrounding their atoms?
(a) C (d) F (g) Se
(b) P (e) Cl (h) Sn
(c) O (f) B

Bond Properties

26. In each pair of bonds, predict which will be the shorter. If possible, check your prediction in Table 10.2.
(a) Si—N or P—O
(b) Si—O or C—O
(c) C—F or C—Br
(d) The C=C or the C≡N bond in acrylonitrile, $H_2C{=}CH{-}C{\equiv}N$

27. In each pair of bonds, predict which will be the shorter. If possible, check your prediction in Table 10.2.
(a) B—Cl or Ga—Cl
(b) C—O or Sn—O
(c) P—S or P—O
(d) The C=C or the C=O bond in acrolein,

$$\begin{array}{l} \text{H}_2\text{C=CH—C=O} \\ \qquad\qquad\quad | \\ \qquad\qquad\quad \text{H} \end{array}$$

28. Compare the nitrogen-nitrogen bonds in hydrazine, N_2H_4, and in "laughing gas," N_2O. In which molecule is the nitrogen-nitrogen bond shorter? In which should the nitrogen-nitrogen bond be stronger?

29. Consider the carbon-oxygen bonds in formaldehyde (H_2CO) and in carbon monoxide (CO). In which molecule is the CO bond shorter?

30. Which bond will require more energy to break, the CO bond in formaldehyde (H_2CO) or the CO bond in carbon monoxide (CO)?

Estimating Reaction Energies

31. The compound oxygen difluoride reacts with water to give oxygen and HF.

$$OF_2(g) + H_2O(g) \longrightarrow O{=}O(g) + 2\,HF(g) \qquad \Delta H^\circ_{rxn} = -325.1 \text{ kJ}$$

Using bond energies, calculate the average O—F bond energy in OF_2.

32. Using bond energies in Table 10.3, calculate the enthalpy change for the following:
(a) $H_2(g) + Br_2(g) \longrightarrow 2\,HBr(g)$
(b) $C_2H_2(g) + H_2(g) \longrightarrow C_2H_4(g)$
(c) $CO(g) + H_2O(g) \longrightarrow CO_2(g) + H_2(g)$

33. Phosgene, $COCl_2$, is a highly toxic gas that was used as a war gas in World War I. Using the bond energies in Table 10.3, estimate the enthalpy change for the reaction of carbon monoxide and chlorine to produce phosgene.

$$CO(g) + Cl_2(g) \longrightarrow COCl_2(g)$$

34. The equation for the combustion of methyl alcohol is

$$2\,CH_3OH(g) + 3\,O_2(g) \longrightarrow 2\,CO_2(g) + 4\,H_2O(g)$$

Using the bond energies in Table 10.3, estimate the heat of combustion of methyl alcohol in kJ/mol.

35. Hydrogenation reactions, the addition of H_2 to compounds, are widely used in industry. For example, propene is converted to propane by addition of H_2.

$$\underset{\text{propene}}{H_3C{-}\overset{\text{H}}{\overset{|}{C}}{=}\overset{\text{H}}{\overset{|}{C}}{-}H(g)} + H_2(g) \longrightarrow \underset{\text{propane}}{H_3C{-}\underset{\text{H}}{\underset{|}{\overset{\text{H}}{\overset{|}{C}}}}{-}\underset{\text{H}}{\underset{|}{\overset{\text{H}}{\overset{|}{C}}}}{-}H(g)}$$

Use the bond energies in Table 10.3 to estimate the enthalpy change for this reaction.

Resonance

36. The following molecules have two or more resonance structures. Show all the resonance structures for each molecule.
    (a) nitric acid, H—O—N(=O)O (N bonded to two O atoms)
    (b) nitrous oxide (laughing gas), N—N—O
37. The following molecules or ions have two or more resonance structures. Show all the resonance structures for each molecule or ion.
    (a) SO_3 (b) SCN^-
38. Compare the carbon-oxygen bond lengths in the formate ion, HCO_2^-, and in the carbonate ion, CO_3^{2-}. In which ion is the bond longer? Explain briefly.
39. Compare the nitrogen-oxygen bond lengths in NO_2^+ and in NO_3^-. In which ion are the bonds longer? Explain briefly.

Electronegativity and Bond Polarity

40. Given the bonds C—N, C—H, C—Br, and S—O,
    (a) tell which atom in each is the more electronegative.
    (b) which of these bonds is the most polar?
41. In each pair of bonds, indicate the more polar bond, and use $\delta+$ and $\delta-$ to show the direction of polarity in each bond.
    (a) C—O and C—N
    (b) B—O and P—S
    (c) P—H and P—N
    (d) B—H and B—I
42. The molecule below is urea, a compound used in plastics and fertilizers.

    H₂N̈—C(=Ö:)—N̈H₂

    (a) Which bonds in the molecule are polar and which are nonpolar?
    (b) Which is the most polar bond in the molecule? Which atom is the negative end of this bond?
43. The molecule below is acrolein, the starting material for certain plastics.

    H—C(H)=C(H)—C(H)=Ö:

    (a) Which bonds in the molecule are polar and which are nonpolar?
    (b) Which is the most polar bond in the molecule? Which atom is the negative end of this bond?

Oxidation Numbers and Atom Formal Charges

44. Determine the oxidation number of each atom in the following molecules or ions.
    (a) ClF_3 (d) HCN
    (b) XeF_2 (e) HNO_2
    (c) OF_2 (f) HCO_3^-
45. Determine the oxidation number of each atom in the following molecules or ions.
    (a) H_2O (e) ClO^-
    (b) H_2O_2 (f) HNO_3
    (c) SO_2 (g) BiOCl
    (d) N_2O
46. For each of the following reactions, find the oxidation number for each atom in the reactants and products, and then decide if the equation represents a redox reaction.
    (a) $SnO_2(s) + 2\,C(s) \longrightarrow Sn(s) + 2\,CO(g)$
    (b) $AgNO_3(aq) + NaCl(aq) \longrightarrow AgCl(s) + NaNO_3(aq)$
47. For each of the following reactions, find the oxidation number for each atom in the reactants and products, and then decide whether the equation represents a redox reaction.
    (a) $2\,NaHCO_3(s) \longrightarrow Na_2CO_3(s) + H_2O(\ell) + CO_2(g)$
    (b) $Ti(s) + 2\,Cl_2(g) \longrightarrow TiCl_4(\ell)$
    (c) $2\,H_2O(\ell) \longrightarrow 2\,H_2(g) + O_2(g)$
48. Calculate the formal charge on each atom in each of the following molecules or ions, and compare the formal charge with the oxidation number.
    (a) ICl_3 (b) NH_3 (c) ClO^- (d) SF_4
49. Calculate the formal charge on each atom in each of the following molecules or ions, and compare the formal charge with the oxidation number.
    (a) H_2O (b) CH_4 (c) NO_2^+ (d) HOF
50. Two resonance structures are possible for NO_2^-. Draw these structures and then determine the formal charge on each atom in the resonance structures.
51. Three resonance structures are possible for dinitrogen oxide, N_2O.
    (a) Draw the three resonance structures.
    (b) Calculate the formal charge on each atom in each resonance structure.
    (c) On the basis of the formal charges, which is the most reasonable resonance structure?

52. The cyanate ion, OCN^-, has the least electronegative atom, C, in the center. The very unstable fulminate ion, CNO^-, has the same formula, but the N atom is in the center.
    (a) Draw the three possible resonance structures of CNO^-.
    (b) On the basis of formal charges, decide on the resonance structure with the most reasonable distribution of charge.
    (c) Mercury fulminate is so unstable it is used in blasting caps. Can you offer an explanation for this instability? (Hint: Are the formal charges in any resonance structure reasonable in view of the relative electronegativities of the atoms?)

General Questions

53. Suggest a reason why $SiCl_4$ reacts with chloride ions to form $SiCl_6^{2-}$, but CCl_4 does not react with chloride ions.
54. Is it a good generalization that elements that are close together in the periodic table form covalent bonds while elements that are far apart form ionic bonds? Why or why not?
55. What is the oxidation number of carbon in each compound below?
    (a) CH_4 (d) $H_2C{=}O$
    (b) CH_3OH (e) HCO_2H
    (c) $H_2C{=}CH_2$ (f) CO_2
56. What is the oxidation number of S in thiosulfate ion, $S_2O_3^{2-}$? Use the rules given in Section 10.8 to calculate the oxidation number. Then draw a Lewis structure and determine the formal charge of each S atom (Section 10.9). The skeleton structure of the thiosulfate ion is

$$\left[\begin{array}{ccc} & S & \\ & | & \\ O- & S & -O \\ & | & \\ & O & \end{array}\right]^{2-}$$

57. The C—Br bond length in CBr_4 is 191 pm; the Br—Br distance in Br_2 is 228.4 pm. Estimate the radius of a C atom in CBr_4. Use this value to estimate the C—C distance in ethane, $H_3C{-}CH_3$. How does your calculated bond length agree with the measured value of 154 pm? Are radii of atoms exactly the same in every molecule?
58. Answer this question without looking up bond energies or standard enthalpies of formation or standard entropies. Predict which of the following reactions are exothermic. Predict the sign of $\Delta S°$ for each reaction; that is, predict which reactions are product favored.
    (a) $CH_4(g) + 2\,F_2(g) \longrightarrow CH_2F_2(g) + 2\,HF(g)$
    (b) $2\,CO_2(g) + 4\,H_2O(g) \longrightarrow 2\,CH_3OH(g) + 3\,O_2(g)$
    (c) $N_2H_2(g) + 2\,H_2(g) \longrightarrow 2\,NH_3(g)$
59. Check your predictions in Question 58 by using bond energies to estimate $\Delta H°$ for each reaction.
60. Urea is widely used as a fertilizer because of its high nitrogen content, so better methods for its production are always being sought. Using Table 10.3, estimate the enthalpy change of the following reaction used to make urea.

$$2\,NH_3(g) + CO(g) \longrightarrow H_2N{-}\overset{\overset{\displaystyle O}{\|}}{C}{-}NH_2(g) + H_2(g)$$

61. The molecule pictured below is acrylonitrile, the building block of the synthetic fiber Orlon.

$$H{-}\overset{\overset{\displaystyle H}{|}}{C}{=}\overset{\overset{\displaystyle H}{|}}{C}{-}C{\equiv}N{:}$$

    (a) Which is the shorter carbon-carbon bond?
    (b) Which is the stronger carbon-carbon bond?
    (c) Which is the most polar bond and what is the negative end of the bond?
62. In nitryl chloride, NO_2Cl, there is no oxygen–oxygen bond. Draw a Lewis structure for the molecule. Are there resonance structures for this molecule?
63. The bunsen burners in your laboratory are fueled either by natural gas, which is primarily methane (CH_4), or by propane (C_3H_8). Using the bond energies in Table 10.3, calculate the heat of combustion of each of these substances in the gas phase. Which provides the greater heat *per gram?*

Summary Problem

64. Chlorine trifluoride, ClF_3, is one of the most reactive compounds known. It reacts violently with many substances generally thought to be inert, and was used in incendiary bombs in World War II. It can be made by heating Cl_2 and F_2 in a closed container.
    (a) Write a balanced equation to depict the reaction of Cl_2 and F_2 to give ClF_3.
    (b) If you mix 0.71 g of Cl_2 with 1.00 g of F_2, what mass in grams of ClF_3 is expected?
    (c) Draw the Lewis structure of ClF_3.
    (d) Calculate the standard enthalpy of formation of ClF_3, using bond energies.

Using Computer Programs and Videodiscs

65. Complexes of nickel(II) are prepared in a series of reactions shown on the *Chemical Demonstrations* videodisc (Side 1 of Disc 1, frame 43914).
    (a) What is the color of aqueous nickel(II) sulfate? In what form is the Ni^{2+} ion found in aqueous solution?
    (b) Name the ligands added to the aqueous nickel(II) solution. Give the colors of the resulting complex ions.
    (c) Show how Ni^{2+} combines with NH_3 and ethylenediamine.
66. When copper metal reacts with nitric acid (*Periodic Table Videodisc,* frame 30261) the following reaction occurs.

$$Cu(s) + 4\,H_3O^+(aq) + 2\,NO_3^-(aq) \longrightarrow Cu^{2+}(aq) + 2\,NO_2(g) + 6\,H_2O(\ell)$$

    (a) Prove that this reaction is an oxidation-reduction reaction.
    (b) Based on this observation, what types of reactions do you believe occur when cobalt metal is dropped into acid? (See the *Periodic Table Videodisc,* frame 27398).

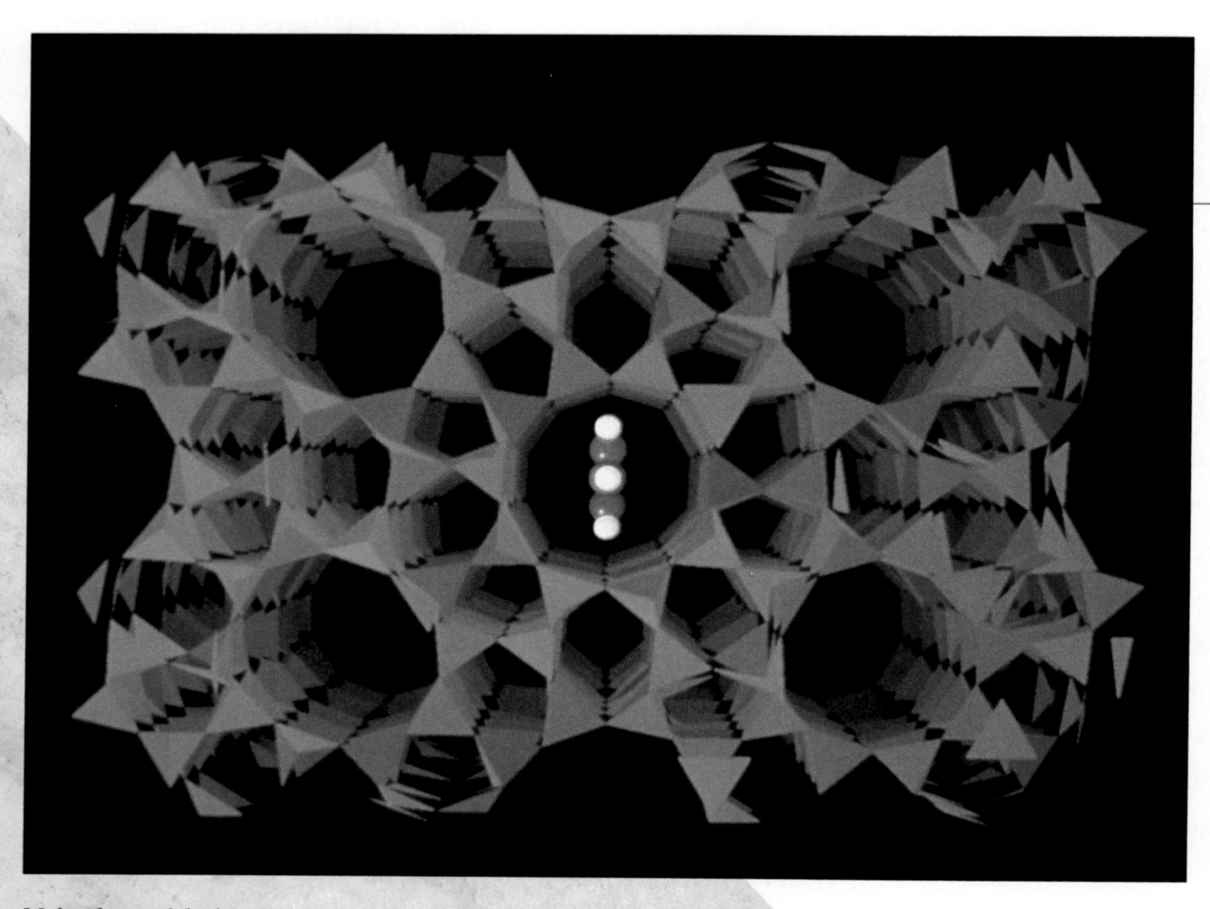

Molecular model of zeolite catalyst used in the catalytic cracking process in the production of gasoline. A model of the benzene molecule is shown in the center of the catalyst.

(Mobil Corporation)

Molecular Geometry and Isomerism

C H A P T E R

11

The composition, empirical formula, and molecular formula of a substance provide important information, but they are not sufficient to predict or explain properties of most molecular compounds. Drawing a Lewis structure tells us which atoms are connected to which, but even this is not enough. The arrangement of the atoms in three-dimensional space—the shape of a molecule—is also very important. Molecules can have the same numbers of the same kinds of atoms and yet be different; the substances made up of those molecules will have different properties. Graphite and diamond are both pure carbon, but the carbon atoms are bonded differently, and graphite has quite different properties from diamond. Only a small variation in molecular shape makes corn starch vastly different from the cellulose in cotton, and that small variation makes it possible for humans to digest starch but not cellulose. The ideas about molecular shape presented in this chapter are crucial to understanding the behavior of molecules in living organisms, the design of molecules that serve effectively as drugs, and many other aspects of modern chemistry.

Our ability to understand three-dimensional structures of molecules is helped by the use of models. Probably the best example of the impact a model can have on the advancement of science is the model of the double helix of DNA built by James Watson and Francis Crick (Figure 11.1), which revolutionized the understanding of human heredity and genetic disease (Section 21.7). They used x-ray structural data to build a model of DNA with two long strands coiled together and held in place by intermolecular forces acting between the strands. Encoded in this structure is the genetic information that is passed from one generation to the next.

Watson and Crick's model shown in Figure 11.1 is a ball-and-stick model of the DNA helix.

Two types of models are generally used—the "ball-and-stick" model and the "space-filling" model. Both are shown in Figure 11.2 to represent the methane molecule. The ball-and-stick model uses balls to represent the atoms and short pieces of wood or plastic to represent bonds. For example, the ball-and-stick model for methane (Figure 11.2a) has a black ball representing carbon, with holes at the correct angles connected by sticks to four white balls representing hydrogen atoms. The space-filling model (Figure 11.2b) is a more realistic representation because it depicts both the relative sizes of the atoms and their spatial orientation in the molecule. This is done by scaling the pieces in the model according to the experimental values for atom sizes. The pieces are held together by links that are not visible when the model is assembled.

Ball-and-stick model kits are often available in college book stores. They are easy to assemble and will help you visualize the molecular geometries described in this chapter. They are relatively inexpensive in comparison with the space-filling models.

To convey a sense of three-dimensionality for a molecule drawn on a piece of flat paper, we can make a perspective drawing that uses solid wedges (▬) to represent bonds extending above the page, and dashed lines (---) to represent bonds below the page. Bonds that lie in the plane of the page are indicated by a line (—). Figure 11.2c illustrates this type of sketch for the tetrahedral methane molecule.

Figure 11.1
Francis H.C. Crick (b. 1916) (*right*) and James D. Watson (b. 1928) (*left*), working in the Cavendish Laboratory at Cambridge, built scale models of the double helical structure of DNA based on the x-ray data of Rosalind Franklin (1920–1958) and Maurice H.F. Wilkins (b. 1916). Knowing distances and angles between atoms, they compared the task to the working of a three-dimensional jig-saw puzzle. Watson, Crick, and Wilkins received the Nobel Prize in 1962 for their work relating to the structure of DNA.

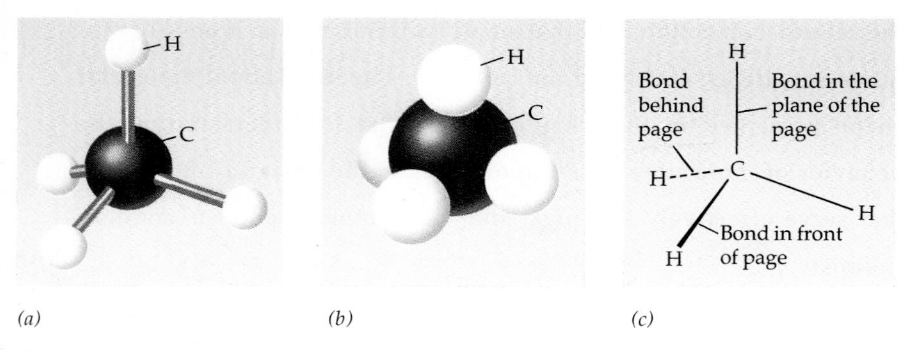

Figure 11.2
The structure of methane, CH_4. The figure shows the various ways in which molecular structures are represented. (a) Ball-and-stick model. (b) Space-filling model. (c) Perspective drawing: the dashed line represents a bond extending behind the page, the solid lines represent bonds in the plane of the page, and the wedge represents a bond extending in front of the page.

Advances in computer graphics have made it possible to draw convincing, scientifically accurate pictures of molecules. Figure 11.3 is a computer-generated model of DNA. Computer graphics are also used to study interactions of molecules. For example, the spatial "fit" of molecules in the grooves of DNA is important to the study of compounds that cause cancer as well as compounds that cure cancer. Scientists can use computer graphics to bring molecules close to DNA on the computer screen to see how the molecules fit together and then use this information to explain the properties of cancer-causing substances as well as to design better drugs for treating cancer.

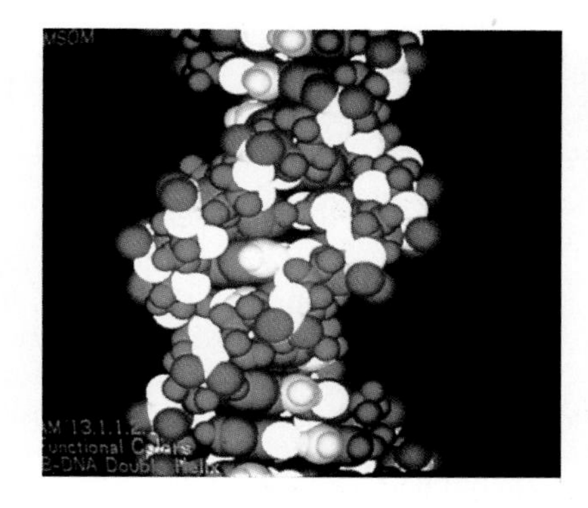

Figure 11.3
Computer-generated space-filling model of DNA double helix.
(Photo Researchers)

11.1
PREDICTING MOLECULAR SHAPES

A simple, reliable method for predicting the shapes of covalent molecules and polyatomic ions is the **Valence Shell Electron Pair Repulsion (VSEPR) model,** devised by Ronald J. Gillespie (1924–) and Ronald S. Nyholm (1917–1971). The VSEPR model is based on the idea that repulsions among the pairs of bonding or nonbonding electrons of an atom control the angles between bonds from that atom to other atoms surrounding it. The central atom and its core electrons are represented by the atom's symbol. This atomic core is surrounded by pairs of valence electrons, the number of pairs corresponding to the number of pairs of dots in the Lewis structure. The geometric arrangement of the electron pairs is predicted on the basis of their repulsions, and the geometry of the molecule or polyatomic ion depends on the numbers of lone pairs and bonding pairs.

Gillespie was born in England and Nyholm in Australia, but both received the Ph.D. in chemistry at University College, London. Nyholm made many contributions to chemistry as Professor of Chemistry at University College, London, until his untimely death. Gillespie has been a Professor of Chemistry at McMaster University (Canada) since 1960.

How do repulsions among electron pairs result in different shapes? Imagine that a balloon represents each electron pair. Each balloon's volume represents a repulsive force that prevents other balloons from occupying the same space. When two, three, four, five, or six balloons are tied together at a central point (the central point represents the nucleus and core electrons of a central atom), the balloons form the shapes shown in Figure 11.4. These geometric arrangements minimize interactions among the balloons (electron-pair repulsions).

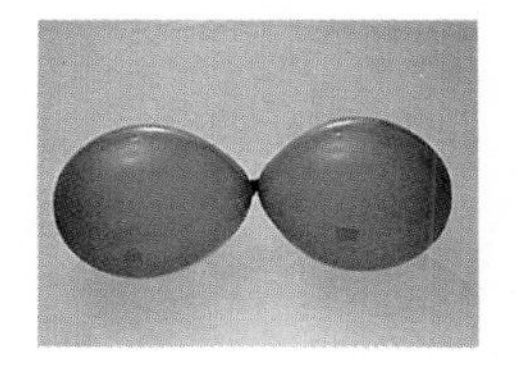

Figure 11.4
Balloon models of the geometries predicted by the VSEPR theory.

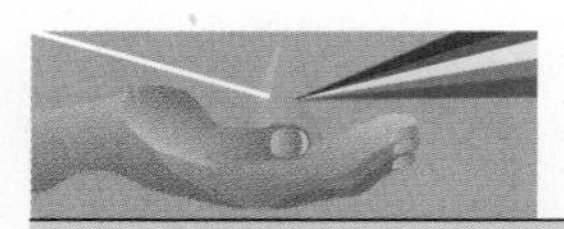

CHEMISTRY YOU CAN DO

Using Balloons as Models

Blow up balloons and tie the ends together as shown in Figure 11.4 for combinations of two, three, four, five, and six balloons. These make excellent models for visualizing the linear, triangular planar, tetrahedral, triangular bipyramidal, and octahedral geometries assumed by two, three, four, five, and six electron pairs, respectively. After the balloon models are assembled, write answers to the following questions.

1. Identify the apexes for each shape (the farthest point of each balloon from the central point). Sketch the figure you would get if you connected the apexes with lines.
2. Identify the faces (the flat surfaces bounded by the lines between apexes) in each sketch, and count how many apexes and faces there are for each shape.
3. Does the name octahedron indicate the number of bonds or the number of faces?

Central Atoms with Only Bond Pairs

The simplest application of VSEPR is to molecules in which all the electron pairs around the central atom are in single covalent bonds. Figure 11.5 illustrates the geometries predicted by the VSEPR model for molecules of the types AX_2 to AX_6 that contain only single covalent bonds, where A is the central atom.

The **linear** geometry for two bond pairs and the **triangular planar** geometry for three bond pairs contain a central atom that does not have an octet of electrons (Section 10.4). The central atom in a **tetrahedral** molecule obeys the octet rule with four bond pairs. The central atoms in **triangular bipyramidal** and **octahedral** molecules do not obey the octet rule since they have five and six bonding pairs, respectively. Hence, triangular bipyramidal and octahedral geometries would be expected only when the central atom is an element in Period 3 or higher (Section 10.4). The geometries illustrated in Figure 11.5 are by far the most common in molecules and ions, and you should be thoroughly familiar with them. The **bond angles** predicted for the

A bond angle is the angle between two atoms bonded to a third atom.

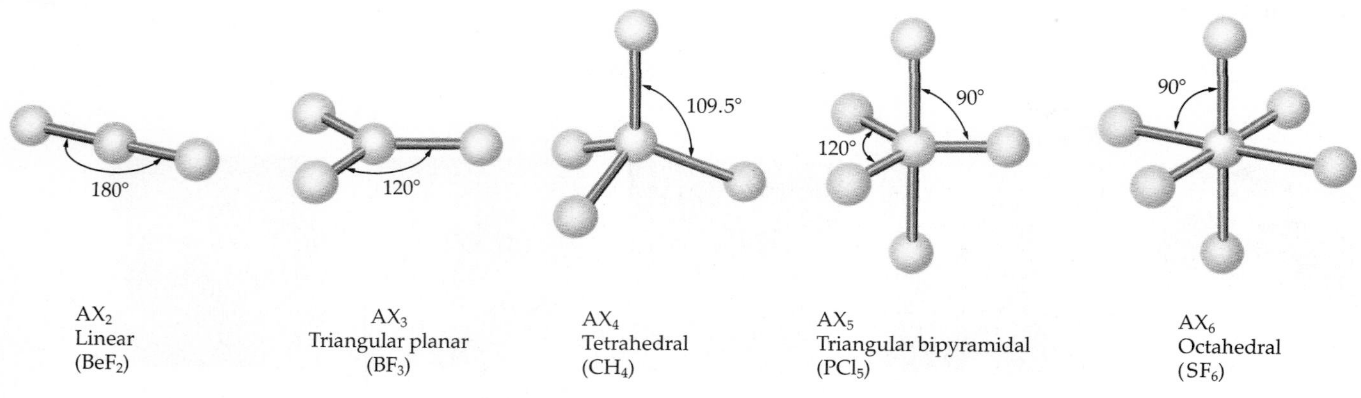

Figure 11.5

Geometries predicted by the VSEPR model for molecules of the types AX_2 to AX_6 that contain only single covalent bonds.

examples given are in agreement with experimental values obtained from structural studies.

Suppose you want to predict the shape of $SiCl_4$. First, draw the Lewis structure as shown in the margin. Since there are four bond pairs forming four single covalent bonds to Si, you would predict a tetrahedral structure for the $SiCl_4$ molecule, and this is in agreement with structural results for this molecule.

:Cl:
:Cl—Si—Cl:
:Cl:

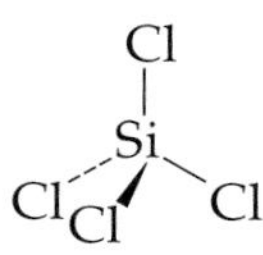

Central Atoms with Bond Pairs and Lone Pairs

How does the presence of lone pairs on the central atom affect the geometry of the molecule or polyatomic ion? The easiest way to visualize this situation is to return to the balloon model and notice that we did not say that the electron pairs had to be bonding pairs. We can predict the geometry of a molecule by applying the VSEPR model to the *total* number of electron pairs around the central atom. However, the shape that this method predicts is the electron-pair geometry rather than the molecular geometry, and we must then decide which positions are occupied by bond pairs and which by lone pairs. The **electron-pair geometry** around a central atom includes the spatial positions of all bond pairs and lone pairs of electrons, whereas the **molecular geometry** of a molecule or ion is the arrangement in space of its atoms. The distinction is necessary because only the atoms are located by structural techniques such as x-ray crystallography (Section 14.2), and positions occupied by lone pairs are not specified in the shapes of molecules. The success of the VSEPR model in predicting molecular shapes indicates that it is correct to account for the effects of lone pairs in this way. In other words, lone pairs of electrons around the central atom occupy spatial positions even though they are not included in the description of the shape of the molecule or ion.

Let's examine the steps in using the VSEPR model to predict the molecular geometry and bond angles in a molecule that includes lone pairs on the central atom, the NH_3 molecule. First, draw the Lewis structure and count the total number of electron pairs around the central N atom.

H—N̈—H
H

Since there are three bond pairs and one lone pair for a total of four pairs of electrons, we predict that the *electron-pair geometry* is tetrahedral. Draw a tetrahedron with N as the central atom and the three bond pairs represented by lines since they are single covalent bonds. The lone pair is drawn as follows to indicate its spatial position in the tetrahedron:

N
H H H

The *molecular geometry* is described as a triangular pyramid because the nitrogen atom is at the apex of the pyramid, while the three hydrogen atoms form its triangular base. (This can be seen by covering up the lone pair of electrons and looking at the molecular geometry.)

What is the predicted value for the H—N—H bond angle? Since the electron-pair geometry is tetrahedral, we would expect the H—N—H bond angle to be 109.5°. However, the experimentally determined bond angles in NH_3 are 107.5°. Does this different value indicate a flaw in the model, or could there be a difference between the spatial requirements of lone pairs and bond pairs? Actually, the latter is the case. Bond pairs are attracted into the bond region between atoms by the strong attractive forces of protons in two nuclei and are, therefore, relatively compact; they are "skinny." For a lone pair, however, there is only one nucleus attracting the electron pair, and this nuclear charge is not so effective in overcoming the normal repulsive forces between two negative electrons; as a result, lone pairs are "fat." Their increased volume spreads the lone pairs farther apart and squeezes the bond pairs closer together. Hence, the relative strength of repulsions is

lone pair–lone pair > lone pair–bond pair > bond pair–bond pair

Using the balloon analogy, a lone pair is like a fatter balloon that takes up more room and squeezes the thinner balloons closer together.

Gillespie and Nyholm recognized the importance of the different spatial requirements of lone pairs and bond pairs and included this as part of their VSEPR model. For example, they used the VSEPR model to predict variations in the bond angles in the series of molecules CH_4, NH_3, and H_2O. Notice in Figure 11.6 that the bond angles decrease in the series CH_4, NH_3,

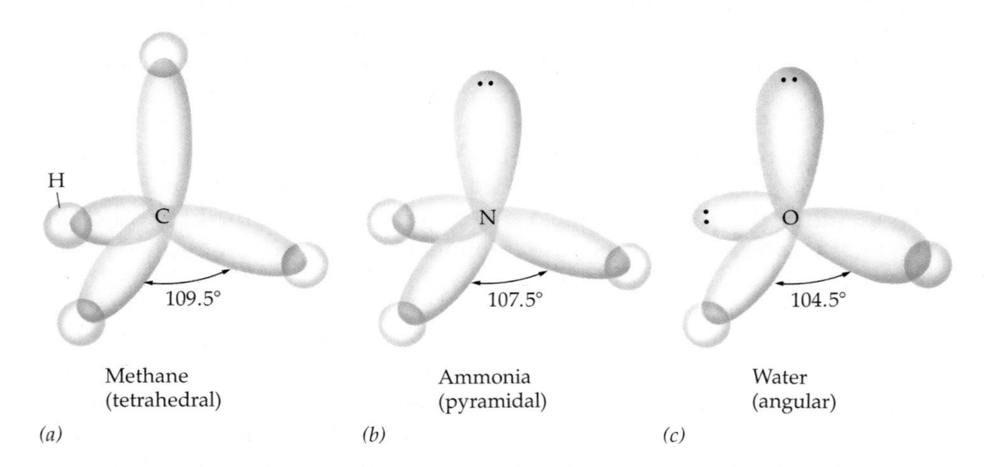

Figure 11.6
In molecules of methane, ammonia, and water there are four electron pairs around the central atom, giving a tetrahedral electron-pair geometry. (a) Methane has four bond pairs, so it has a tetrahedral molecular shape. (b) Ammonia has three bond pairs and one lone pair, so it has a triangular pyramidal molecular shape. (c) Water has two bond pairs and two lone pairs, so it has an angular molecular shape. The decrease in bond angles in the series can be explained by the larger spatial requirements of lone pairs, which squeeze the bond pairs closer together.

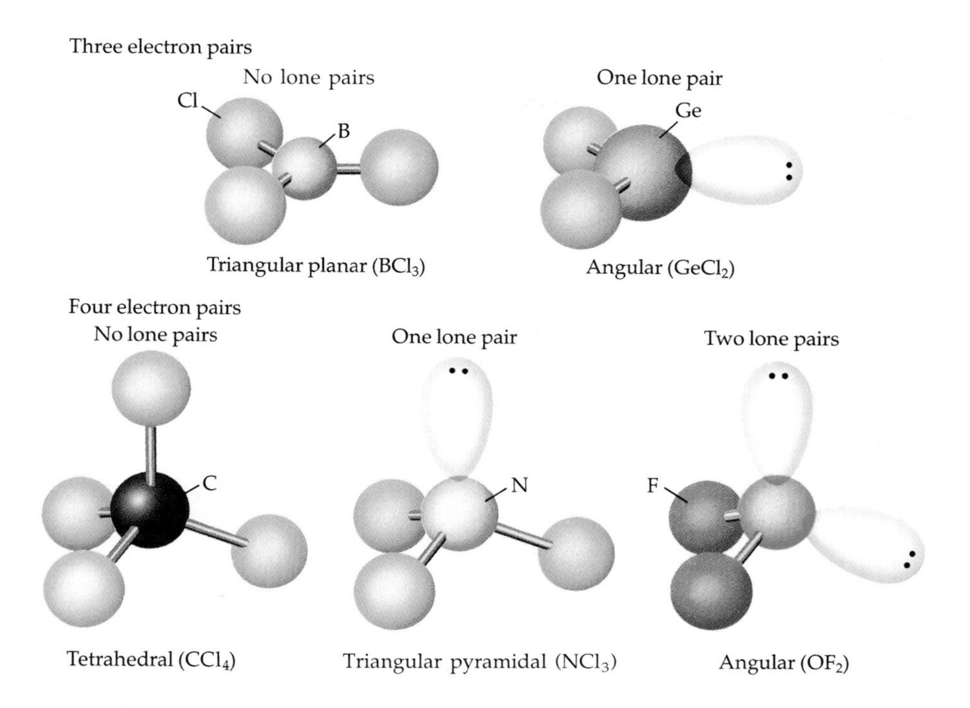

Figure 11.7
Additional examples of electron-pair geometries and molecular shapes for molecules and ions with three and four electron pairs around the central atom.

and H_2O as the number of lone pairs increases. Figure 11.7 gives additional examples of electron-pair geometries and molecular geometries for molecules and ions with three and four electron pairs around the central atom. The experimentally determined bond angles are given for the examples. To check your understanding of the VSEPR model, try to explain the molecular geometry and bond angles in each case.

The situation becomes more complicated if the central atom has five or six electron pairs, some of which are lone pairs. Let's look first at the entries in Figure 11.8 for the case of five electron pairs. The three angles in the triangular plane are all 120°. The angles between any of the pairs in this plane and an upper or lower pair are only 90°. Thus, the triangular bipyramidal structure has two sets of positions that are not equivalent. Because the positions in the triangular plane lie in the equator of an imaginary sphere around the central atom, they are called the equatorial positions. The north and south poles are called the axial positions. Each equatorial position is closely flanked by only two other positions (the axial ones), while an axial position is closely flanked by three positions (the equatorial ones). This means that any lone pairs, which we assume to be fatter than bonding pairs, will occupy equatorial positions rather than axial positions. For example, consider the ClF_3 molecule, which has three bond pairs and two lone pairs. The two lone pairs in ClF_3 are equatorial; two bond pairs are axial and the third is allowed into the equatorial plane, so the molecular geometry is T-shaped (Figure 11.8).

All of the angles in the octahedron are 90°. Unlike the triangular bipyramid, the octahedron has no distinct axial and equatorial positions; all posi-

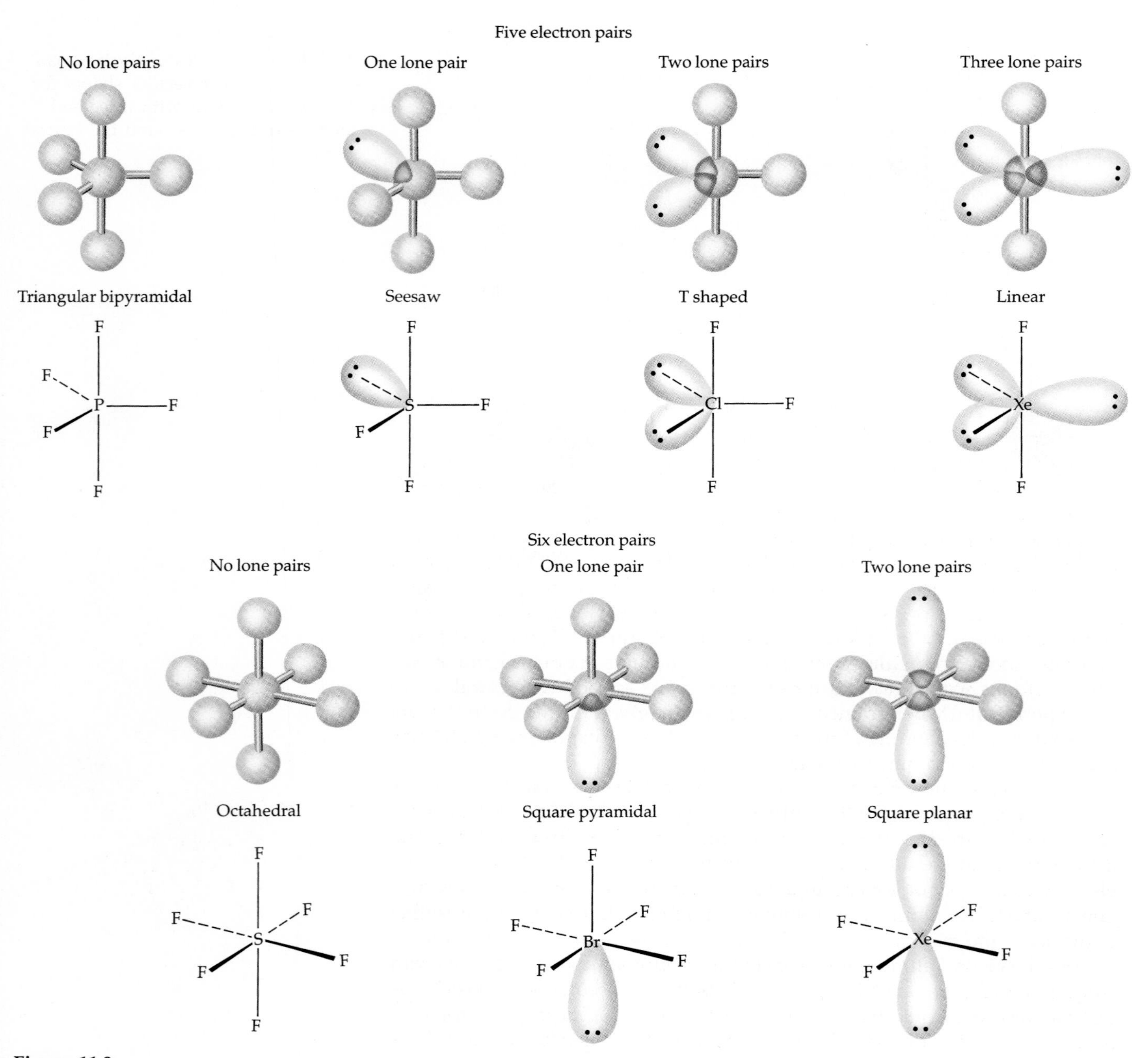

Figure 11.8

Electron-pair geometries and molecular shapes for molecules and ions with five and six electron pairs around the central atom.

tions are the same. The diagram at the top of the next page shows two views of an octahedral molecule.

The drawing on the left emphasizes that the regular octahedron is a figure with six corners and eight faces, each of which is an equilateral triangle. The drawing on the right shows that the six atoms bonded to the central atom are located along three axes at right angles to one another (the x, y, and z axes) and are *equidistant* from the central atom. Therefore, if the molecule

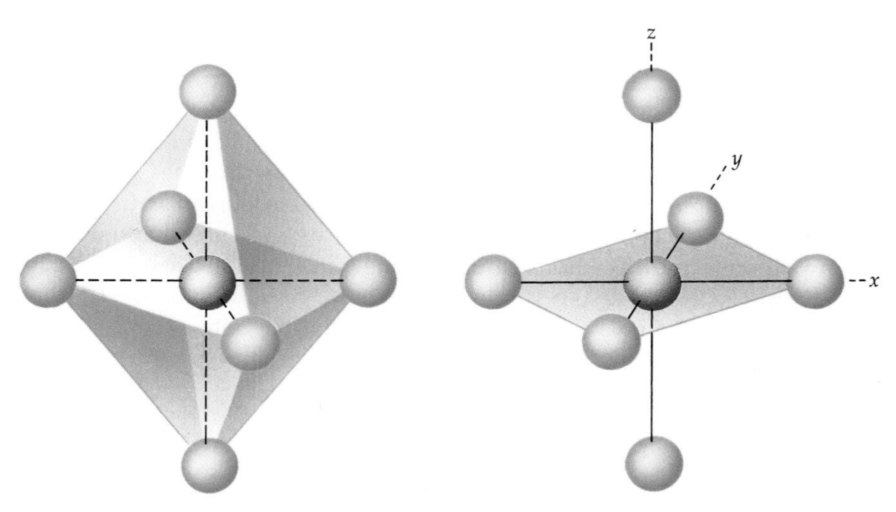

has one lone pair, as in BrF_5, it makes no difference which apex it occupies. Usually the lone pair is drawn in the top or bottom position to make it easier to visualize the molecular geometry, which in this case is square pyramidal (Figure 11.8).

The drawing on the right also shows that we can consider an octahedron as a *square plane* containing the central atom and four of the atoms bonded to it, with two other atoms or lone pairs above and below the plane. For example, if the molecule or ion has two lone pairs, as in ICl_4^-, each of the lone pairs needs as much room as possible. This is best achieved by placing the lone pairs above and below the square plane that contains the I atom and the four Cl atoms, so the molecular geometry of ICl_4^- is square planar.

An example of the power of the VSEPR model is the correct prediction of the shape of XeF_4. At one time the noble gases were not expected to form compounds, since they have a stable octet of valence electrons (Section 9.6). The synthesis of XeF_4 (Figure 9.28) created a challenge for theorists because it could not be explained with existing bonding theories. Let's see how you can use the VSEPR model to predict the correct geometry. The molecule has 36 valence electrons (8 from Xe and 28 from four F atoms). There are eight electrons in four bond pairs around Xe, and a total of 24 electrons in the lone pairs on the four F atoms. That leaves four electrons in two lone pairs on the Xe atom.

:F:
:F—Xe—F:
:F:

Since Xe is in Period 5, it can accommodate more than an octet of electrons. The total of six electron pairs on Xe leads to a prediction of an octahedral electron-pair geometry. Where do you put the lone pairs? As explained above for the ICl_4^- ion, the lone pairs are placed at opposite corners of the octahedron to provide them as much room as possible. The result is a square planar molecular geometry for the XeF_4 molecule (cover the lone pairs in the XeF_4 drawing in Figure 11.8 to see this), and that shape agrees with experimental structural results.

A number of other xenon compounds have been prepared, and the VSEPR model has been useful in predicting their geometries as well. For example, the XeF_2 molecule is predicted to have a linear molecular geometry (Figure 11.8) because the three lone pairs occupy the equatorial positions of a triangular bipyramid while the two bond pairs are in the axial positions.

Multiple Bonds and Molecular Geometry

Although double bonds and triple bonds are shorter and stronger than single bonds (Section 10.5), they do not affect predictions of molecular shape. Why? Electron pairs involved in a multiple bond are all shared between the same two nuclei and therefore occupy the same region. Because they must remain in that region, two electron pairs in a double bond (or three in a triple bond) are like a single balloon, rather than two or three balloons. All electron pairs in a multiple bond count as one bond, and contribute to molecular geometry the same as a single bond would. For example, the C atom in CO_2 has no lone pairs and participates in two double bonds. Each double bond counts as one for the purpose of predicting geometry, so the structure of CO_2 is linear.

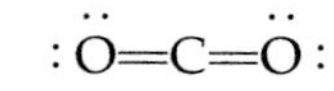

$$:\ddot{O}=C=\ddot{O}:$$

When resonance structures are possible, the geometry can be predicted from any of the Lewis resonance structures or from the resonance hybrid structure. For example, the geometry of the CO_3^{2-} ion is predicted to be triangular planar because the carbon atom has three sets of bonds and no lone pairs. This can be predicted from either of the representations below.

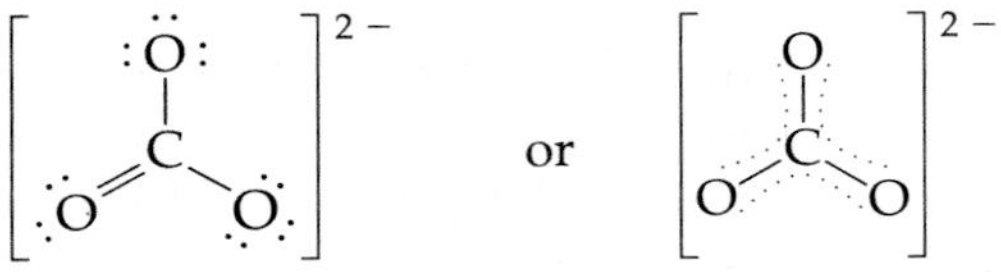

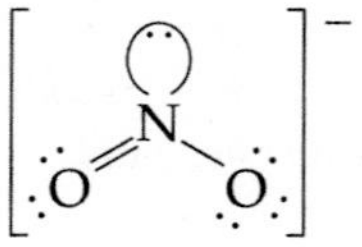

or

The NO_2^- ion is described as angular because it has a lone pair in one position and two bonds in the other two positions of a triangular planar electron-pair geometry.

The structure of sulfur dioxide is a good illustration of the fact that chemistry is a dynamic, constantly developing field of science. Following the guidelines given in Section 10.6 for drawing resonance structures, two resonance structures (*a* and *b* below) would be drawn and considered to be of equal importance. However, it has recently been argued that a third resonance structure, *c*, is actually the most important one.*

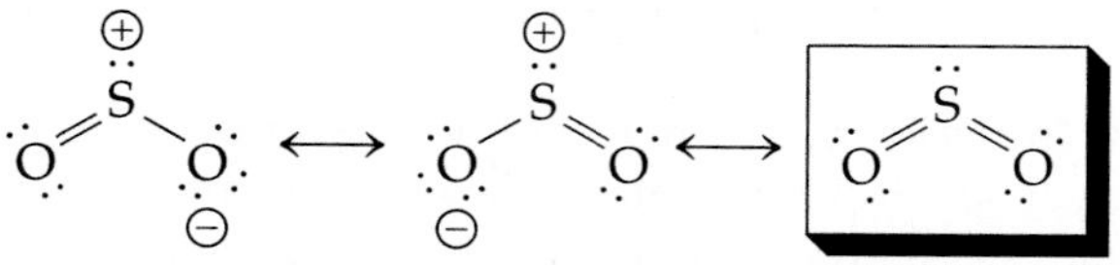

(*a*) (*b*) (*c*)

Although structure *c* violates the octet rule for the central S atom, elements from the third and higher periods may have more than eight valence electrons (Section 10.4). Since all the atoms in structure *c* have 0 formal charge, it would be predicted from rules for formal charge (Section 10.9) to be the most important structure. The merits of the bonding argument aside, the

*G.H. Purser, *J. Chem. Educ.* **1989**, 66, 710–713.

SO_2 molecule would be predicted to be angular by using any of the resonance structures above since they all show the S atom with a lone pair of electrons in one position and two bonds in the other two positions of a triangular planar electron pair geometry.

To summarize, you can describe the shape of virtually any molecule or polyatomic ion by thinking through the following steps:

1. Draw the Lewis structure.
2. Determine the total number of single bond pairs (including multiple bonds counted as single pairs).
3. Pick the appropriate electron-pair geometry, and then choose the molecular shape that matches the total number of single bond pairs and lone pairs.
4. Predict the bond angles, remembering that lone pairs occupy more volume than do bonding pairs.

Table 11.1 (on page 465) gives additional examples of molecules and ions whose shapes can be predicted by using the VSEPR model.

EXAMPLE 11.1

Electron-Pair Geometry

Predict the electron-pair geometries of (a) SO_3, (b) H_2CO, and (c) $XeOF_4$.

SOLUTION First draw the Lewis structure for each molecule.

The resonance structure

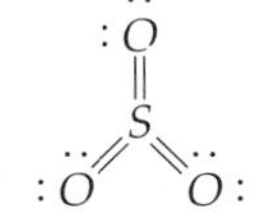

could also be drawn *(see earlier discussion of SO_2 on page 462).*

(a) Sulfur trioxide, SO_3: Three bond pairs surround the central S atom. Therefore, the electron-pair geometry around S is triangular planar. (Since three resonance structures can be drawn, the best representation uses a dotted line to indicate that the bonding involves all the atoms in the molecule.)

(b) Formaldehyde, H_2CO: The C atom is surrounded by three bond pairs. (As in SO_3 the two bond pairs in the double bond are counted as one bond pair for determining the geometry of the molecule.) Therefore, the electron-pair geometry around C is triangular planar.

(c) Xenon oxytetrafluoride, $XeOF_4$: The central Xe atom is surrounded by five bond pairs and one lone pair. Because there are six pairs around the Xe atom, the electron-pair geometry is octahedral. The O and F atoms could be placed so that either the O atom or one of the F atoms is across the molecule from the lone pair. The O atom occupies the axial position since it is larger than the F atom.

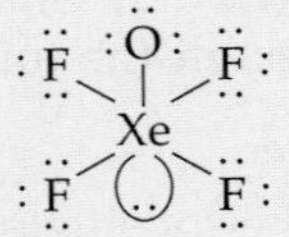

EXAMPLE 11.2

VSEPR and Molecular Shape

What are the shapes of (a) PH_4^+ and (b) NO_3^-?

SOLUTION (a) The Lewis structure of $[PH_4]^+$ reveals that the central P atom has four electron pairs, so the electron-pair geometry is tetrahedral. Since all four electron pairs are used to bond terminal atoms, the ion has a tetrahedral shape.

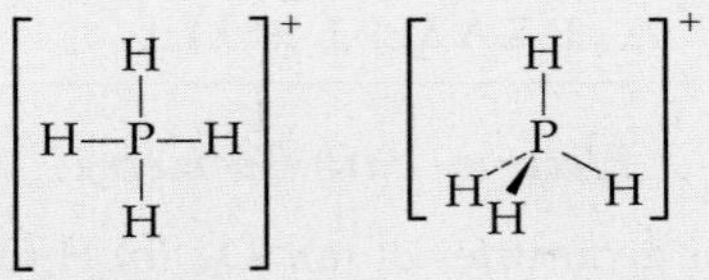

(b) The NO_3^- ion has the same number of valence electrons as the CO_3^{2-} ion described earlier in this section. Thus, like the carbonate ion, the nitrate ion is triangular planar.

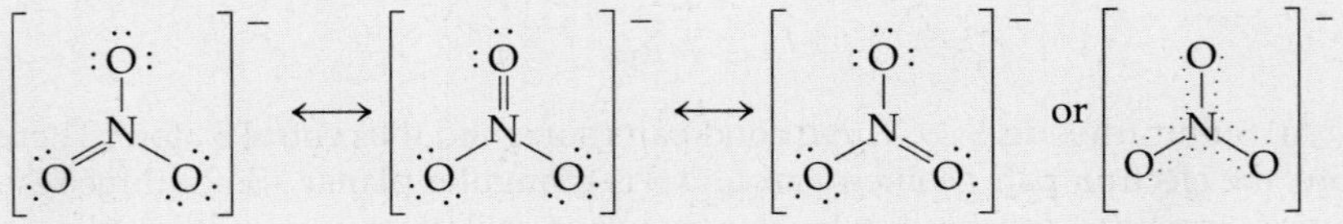

or

EXERCISE 11.1 • *Electron-Pair Geometry and Molecular Geometry*

Use Lewis structures and the VSEPR model to determine the electron-pair and molecular geometries for (a) Cl_2O, (b) SO_3^{2-}, (c) SiO_4^{4-}.

EXERCISE 11.2 • *Molecular Shapes*

Draw the Lewis structure for ICl_2^- and then decide the electron-pair and molecular geometries of the ion.

EXERCISE 11.3 • *Bond Angles*

Predict the bond angles in SCl_2 and PCl_3.

Table 11.1

Examples of Molecular Geometries Predicted by VSEPR Model

Formula (X = electron pairs)	Number of Lone Pairs on Central Atom	Example	Geometry of Molecule or Ion*
AX_2	no lone pairs	CO_2, $BeCl_2$	linear
AX_3	no lone pairs	BCl_3, CO_3^{2-}, SO_3	triangular planar
	one lone pair	NO_2^-, O_3, $SnCl_2$	angular
AX_4	no lone pairs	SiH_4, BF_4^-, NH_4^+	tetrahedral
	one lone pair	PF_3, ClO_3^-, NH_3	triangular pyramidal
	two lone pairs	H_2F^+, H_2O, NH_2^-	angular
AX_5	no lone pairs	PF_5, $AsCl_5$	triangular bipyramidal
	one lone pair	SF_4	seesaw
	two lone pairs	ClF_3, BrF_3	T-shaped
	three lone pairs	ICl_2^-, XeF_2	linear
AX_6	no lone pairs	SF_6, PF_6^-	octahedral
	one lone pair	BrF_5, ClF_5	square pyramidal
	two lone pairs	XeF_4, ICl_4^-	square planar

*"Triangular" and "angular" are often referred to as "trigonal" and "bent."

11.2 MOLECULAR POLARITY

Recall from Section 10.7 that the polarity of a covalent bond can be predicted from the difference in electronegativity of the two atoms joined by the bond. However, *a molecule that has polar bonds may or may not be polar.* Depending on the three-dimensional shape of the molecule, the contributions of two or more polar bonds may cancel each other, leading to a nonpolar molecule. In a polar molecule, there is an accumulation of electron density toward one end of the molecule, giving that end a slight negative charge, $\delta-$, and leaving the other end with a slight positive charge of equal value, $\delta+$ (Figure 11.9).

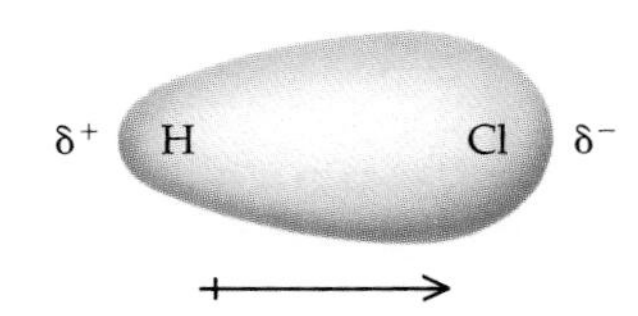

Figure 11.9
In a polar molecule the valence electron density has shifted slightly to one side of the molecule. An arrow (+⟶) is used to show the direction of molecular polarity, with the arrow head pointing toward the negative end of the molecule and the plus sign at the positive end of the molecule.

Before examining the factors that determine whether a molecule is polar, let's look at the experimental measurement of the polarity of molecules. Polar molecules experience a force in an electric field that tends to align them with the field (Figure 11.10). When the electric field is created by a pair of oppositely charged plates, the positive end of each molecule is attracted toward the negative plate and the negative end is attracted toward the positive plate. The extent to which the molecules line up with the field depends on their **dipole moment,** which is defined as the product of the magnitude of the partial charges (δ) times the distance of separation be-

Table 11.2

Dipole Moments

Molecule	Dipole Moment (D)
H_2	0
H_2O	1.86
NH_3	1.46
CH_4	0
CH_3Cl	1.86

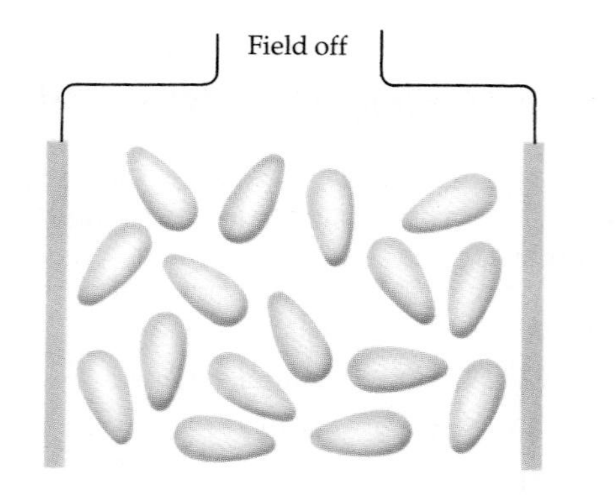

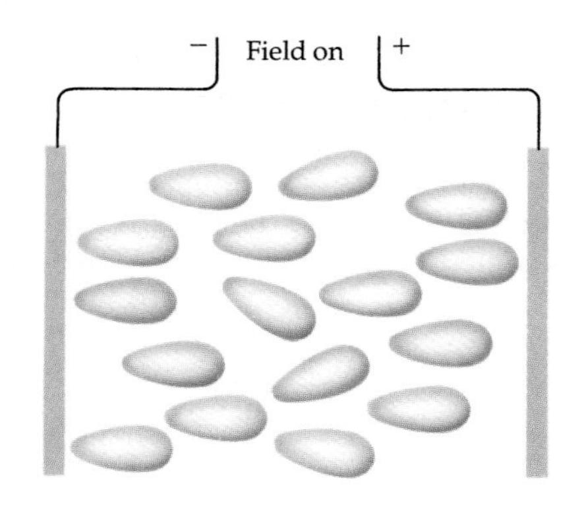

Figure 11.10

Polar molecules in an electric field experience a force that tends to align them so that oppositely charged ends of adjacent molecules are closer to each other.

tween the $\delta+$ and $\delta-$ charges. The SI unit of the dipole moment is the coulomb-meter; a more convenient, derived unit is the debye (D), defined as $1\ D = 3.34 \times 10^{-30}\ C \cdot m$. Dipole moments can be determined experimentally, and some typical values are listed in Table 11.2.

The force of attraction between the negative end of one polar molecule and the positive end of another (called a dipole-dipole force, and discussed further in Section 12.5) has an extraordinarily important effect on the properties of water and other polar substances. For example, molecular polarity is important in determining the temperatures at which a liquid freezes or boils, whether a liquid will dissolve certain gases or solids or mix with other liquids, whether it will adhere to glass or other solids, and how it may react with other substances.

Peter Debye (1884–1966) was born and educated in Europe, and became Professor of Chemistry at Cornell University in 1940. He was noted for his work on x-ray diffraction, electrolyte solutions, and the properties of polar molecules. He received the Nobel Prize in Chemistry in 1936. *(Oesper Collection in the History of Chemistry, University of Cincinnati)*

To predict whether a molecule will be polar, we need to consider whether the molecule has polar bonds and how these bonds are positioned relative to one another. Carbon dioxide, CO_2, is a linear triatomic molecule. Each C=O bond is polar because O is more electronegative than C, so O is the negative end of the bond dipole. If we represent the dipole moment contribution from each bond (the bond dipole) by the symbol $+\!\!\longrightarrow$ in which the plus sign indicates the positive partial charge, we can use the arrows to help estimate whether a molecule is polar.

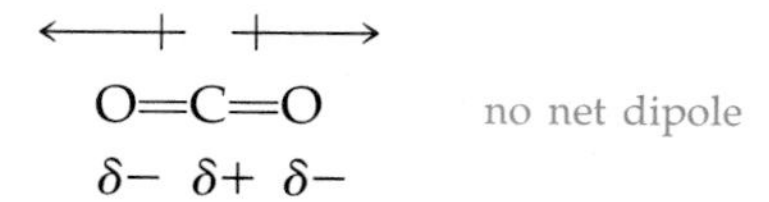

The O atoms are at the same distance from the C atom, they both have the same $\delta-$ charge, and they are on opposite sides of C. Therefore, their bond dipoles cancel each other and CO_2 is a nonpolar molecule. It has a zero dipole moment, even though each bond is polar.

The situation is different for water, an angular triatomic molecule. Here both O—H bonds are polar, with the H atoms having the same δ^+ charge.

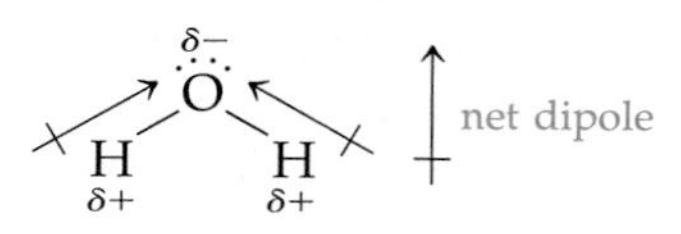

Note that the two bond dipoles do not point directly toward or away from each other, but add together to give a molecular dipole moment. Thus, water is a polar molecule.

Using the same approach, can you explain why CCl_4, a tetrahedral molecule, is nonpolar? First draw the molecule with the bond dipoles represented:

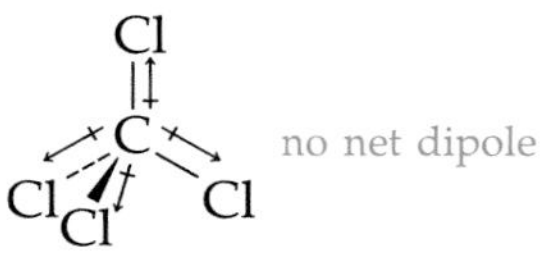

no net dipole

Remember that all four positions at the corners of the tetrahedron are geometrically equivalent. Therefore, the bond dipoles of the four C—Cl atoms will cancel to give a net molecular dipole moment of zero. However, changing one of the Cl atoms to H, as in $CHCl_3$ (chloroform), gives a polar molecule because the H atom is not as electronegative as the Cl atoms and the C—H distance is different from the C—Cl distances. Since the electronegativities of the atoms in $CHCl_3$ are H (2.1) $<$ C (2.5) $<$ Cl (3.0), the Cl atoms draw the electron density toward their side of the molecule. This means that the positive end of the molecular dipole is at the H atom and the negative end is on the CCl_3 side of the molecule.

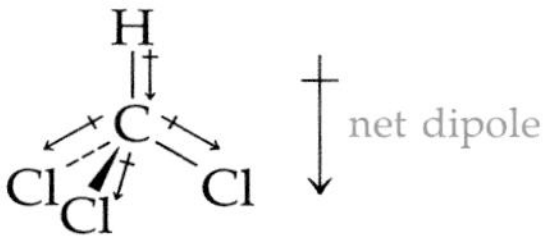

net dipole

EXAMPLE 11.3

Molecular Polarity

Are dichloromethane (CH_2Cl_2) and boron trifluoride (BF_3) polar or nonpolar? If polar, indicate the direction of polarity.

SOLUTION In CH_2Cl_2, a tetrahedral molecule, Cl is more electronegative than C, while H is less electronegative. Therefore, negative charge is drawn away from H toward Cl, and CH_2Cl_2 has a net dipole with the negative end at the Cl atoms.

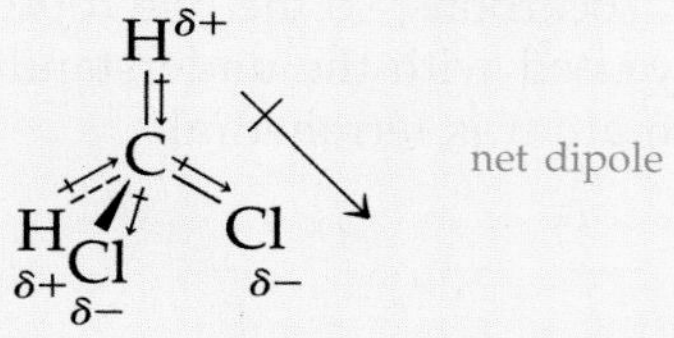

$H^{\delta+}$, $H\ \delta+$, $Cl\ \delta-$, $Cl\ \delta-$

net dipole

As predicted for a molecule with three electron pairs around the central atom, BF_3 is triangular planar.

no net dipole

Since F is more electronegative than B, the B—F bonds are polar with F being the negative end. The molecule is nonpolar, though, since the three terminal atoms are identical; the dipole of each B—F is canceled by the dipoles of the opposite two B—F bonds.

EXERCISE 11.4 • *Molecular Polarity*

For each of the following molecules, decide whether the molecule is polar, and if so, which side is positive and which is negative: (a) $BFCl_2$, (b) NH_2Cl, (c) SCl_2.

11.3 THE TETRAHEDRON

The VSEPR model focuses on several common shapes of molecules—linear, triangular planar, pyramidal, tetrahedral, triangular bipyramidal, and octahedral. Of these, the tetrahedral shape is probably the most important because of its predominance in the chemistry of carbon, silicon, and phosphorus compounds and ions.

Tetrahedral Carbon

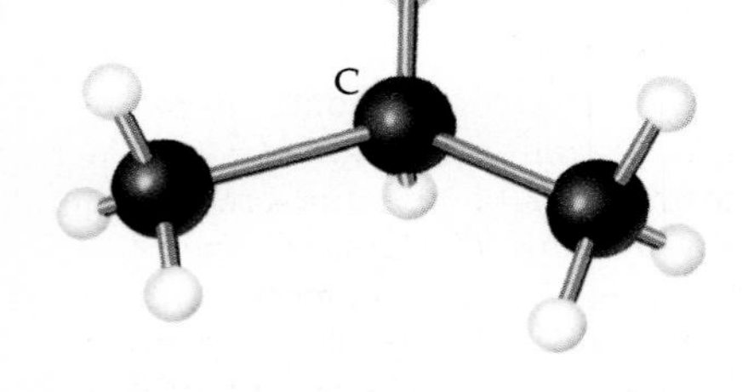

Figure 11.11
Ball-and-stick model of propane, C_3H_8.

Carbon compounds hold the key to life on earth, and about ten million of the more than twelve million known compounds are carbon-containing compounds. The saturated hydrocarbons (alkanes) discussed in Section 10.1 are an example of a large class of organic compounds in which every carbon atom has a tetrahedral environment. The tetrahedral shape of CH_4 has been discussed, but it is important to recognize that every carbon in an alkane has a tetrahedral environment. For example, notice that the three carbon atoms in the ball-and-stick model of propane in Figure 11.11 do not lie in a straight line because of the tetrahedral geometry about each carbon atom. Thus these line drawings of hydrocarbons are not accurate representations of the tetrahedral H—C—H and C—C—C bond angles. Perspective drawings such as those for methane (Figure 11.2c) and ethane (Figure 11.12a) take time to do, especially for larger hydrocarbons, so the line representation shown in Figure 11.12b is commonly used with the understanding that the actual environment of each carbon atom is tetrahedral.

Names of alkanes were given in Section 4.5. Nomenclature rules are given in Appendix E.

```
H       H          H  H
 \     /           |  |
  C—C            H—C—C—H
H/ |  | \H         |  |
  H    H           H  H

   (a)              (b)
```

Figure 11.12
(a) Perspective drawing of ethane.
(b) Line representation of ethane.

Tetrahedral Silicon

The tetrahedron is also important to the chemistry of silicon and because silicon-oxygen compounds make up 75% of the crust of the earth; the bonding between these two elements in clays and rocks literally holds together the earth's skin (see Chapter 14).

Often the tetrahedron is the building block of three-dimensional structures. For example, carbon in the form of diamond is a three-dimensional

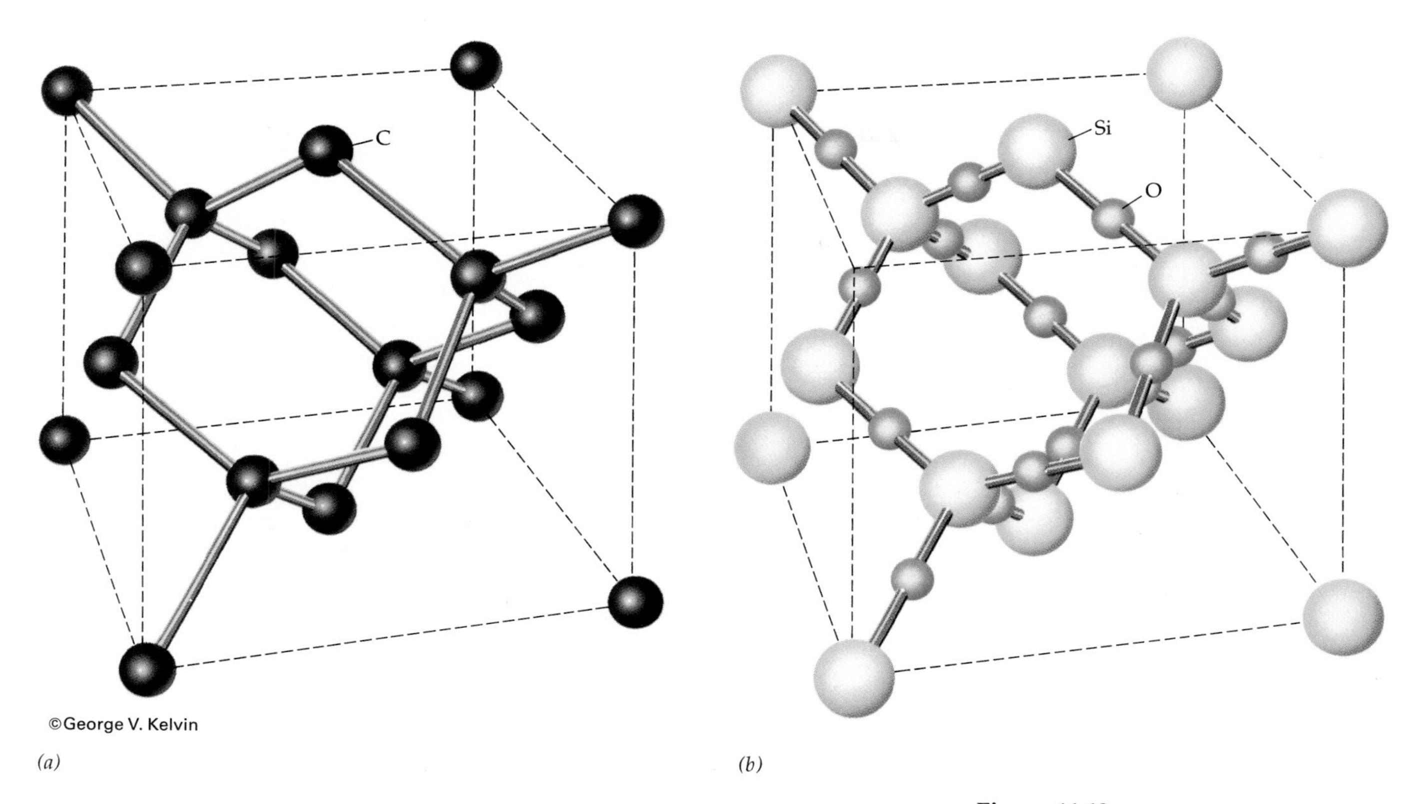

Figure 11.13
(a) Structure of diamond. (b) Structure of cristobalite, one of the solid forms of SiO_2. Note the similarity to the structure of diamond. Each C—C—C bond in diamond is replaced by a Si—O—Si linkage in cristobalite. The Si—O—Si linkage is bent as shown in Figure 11.14.

structure of tetrahedral covalently bound carbon (Figure 11.13a). Elemental silicon has the diamond structure, and one of the solid forms of SiO_2, cristobalite, can be derived from the diamond structure by replacing the C—C—C bonds with Si—O—Si bonds (Figure 11.13b). The structures of all the silicate minerals can be classified on the basis of whether SiO_4^{4-} tetrahedra are sharing one or more corners (oxygen atoms) to give chains, rings, sheets, or three-dimensional crystals.

Zeolites

The shapes of molecules are not only of academic interest, since they have many practical everyday applications. If you've done your laundry today, a zeolite may have helped produce the clean clothes. Zeolites are silicates in which some silicon atoms have been replaced by aluminum atoms.

The three-dimensional framework of zeolites consists essentially of linked SiO_4 tetrahedra with different numbers of Al^{3+} ions substituted for Si^{4+} (Figure 11.14). Within the framework are channels of just the right diameter to hold small molecules such as the water molecule. The negative charge of the aluminosilicate anion is balanced by cations from Group 1A (Na^+ or K^+) or Group 2A (Ca^{2+} or Mg^{2+}), which lie close to the negative aluminosilicate anions in the channels of the structure. Since the cations are not part of the framework, they can be exchanged for other cations. This **ion-exchange** property led to one of the first commercial uses of zeolites—water softeners. If hard water is passed through zeolites that have Na^+ ions

Zeolites make up as much as 25% of many detergents. More than 300 million pounds of zeolites are used in the United States annually, and about 75% of that is in detergents.

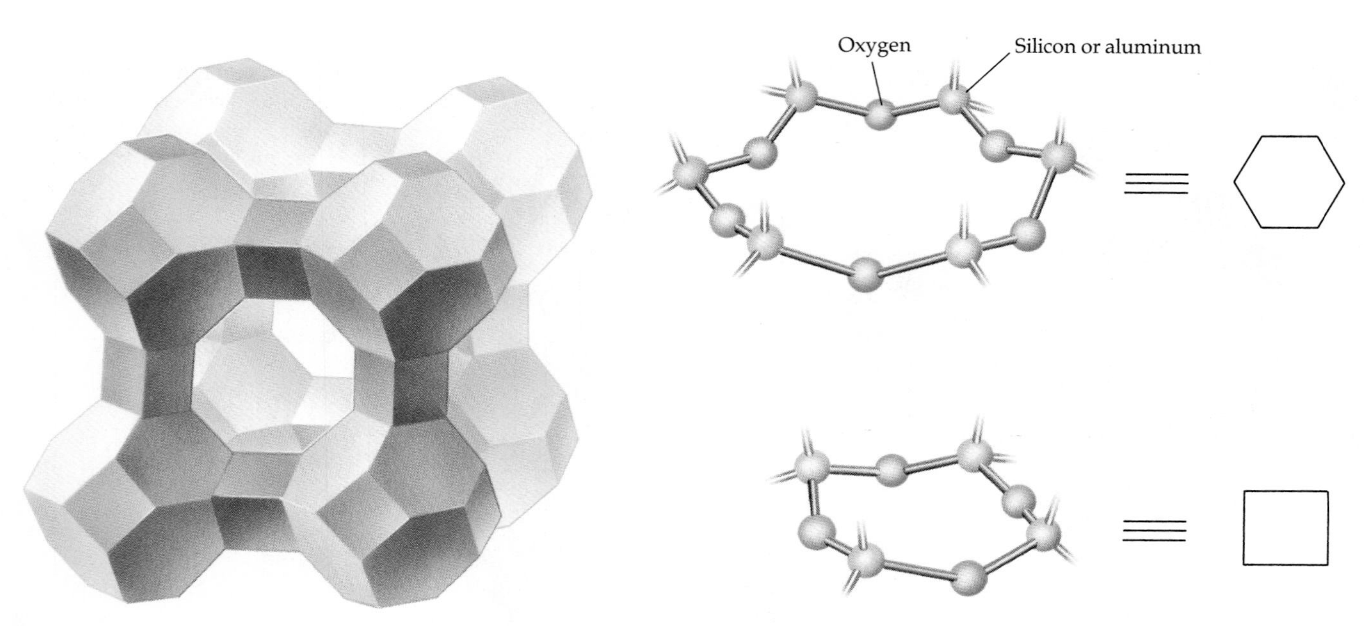

Figure 11.14
The general structure for natural and synthetic zeolites is a network of aluminum, silicon, and oxygen atoms (the aluminosilicate framework). The negative charge of the aluminosilicate framework is balanced by cations of Group 1A or 2A. The structure shown here is that of Linde A, a synthetic zeolite.

in their channels, the Ca^{2+} and Mg^{2+} ions that cause water to be hard displace the Na^{+} ions from the zeolite. In this way, the Ca^{2+} and Mg^{2+} ions are removed, and the water is softened. Water softeners are needed in homes in hard-water areas, because the calcium and magnesium ions cause the soap scum and build-up in water pipes that come with hard water. The cleaning power of detergents is also enhanced by the addition of zeolites to soften the water.

Because of their uniformly sized channels, zeolites are excellent drying agents and can be used to remove water from air or an organic solvent. The zeolites can then be regenerated by heating to remove the solvent. In another application, small amounts of a zeolite are sealed into multipane windows to prevent condensation of moisture on the inner surfaces.

One current theory of the origin of life is that zeolites and clay minerals provided an organizing influence so that simple molecules could form the precursors to the biologically important molecules we know today.

Many zeolites are found in nature, but they can also be synthesized under conditions that favor cavities of uniform size and shape (Figure 11.14). The size of the cavity can be varied by changing the Si/Al ratio and the reaction conditions. Most natural zeolites have Si/Al ratios from 3 to 5, but synthetic zeolites have been prepared with Si/Al ratios as low as 1 to 1.5 and as high as 100. Synthetic zeolites are used as catalysts in the production of gasoline (Figure 11.15).

Tetrahedral Phosphorus

A molecule of white phosphorus, P_4, the most common allotrope of phosphorus, is a simple tetrahedron of atoms (Section 9.6). All compounds of phosphorus with oxygen can be thought of as structurally derived from the P_4 tetrahedron of elemental phosphorus. For example, if P_4 is carefully oxidized, tetraphosphorus hexoxide, P_4O_6, is formed; an oxygen atom has been

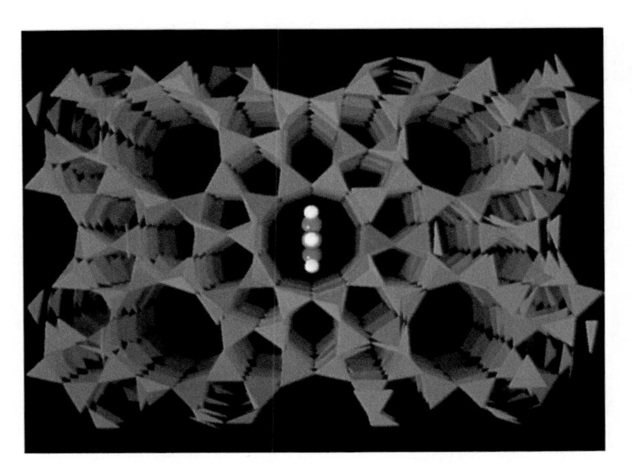

Figure 11.15
The framework of a zeolite catalyst used in the catalytic cracking process in the production of gasoline. The dimensions of the zeolite channels are similar to the sizes of the hydrocarbon molecules in gasoline. During the cracking process, the larger molecules of high-boiling petroleum fractions are converted to small hydrocarbons in the gasoline fraction (Section 13.1). A model of the benzene molecule is shown in the center of the catalyst. *(Mobil Corporation)*

inserted into each of the P—P bonds of the tetrahedron. The most common and important oxide of phosphorus is tetraphosphorus decaoxide P_4O_{10}, in which each P atom is surrounded tetrahedrally by O atoms. Addition of water to P_4O_{10} gives phosphoric acid, which also has a tetrahedral structure around phosphorus. The structures of P_4, P_4O_6, and P_4O_{10} are shown below.

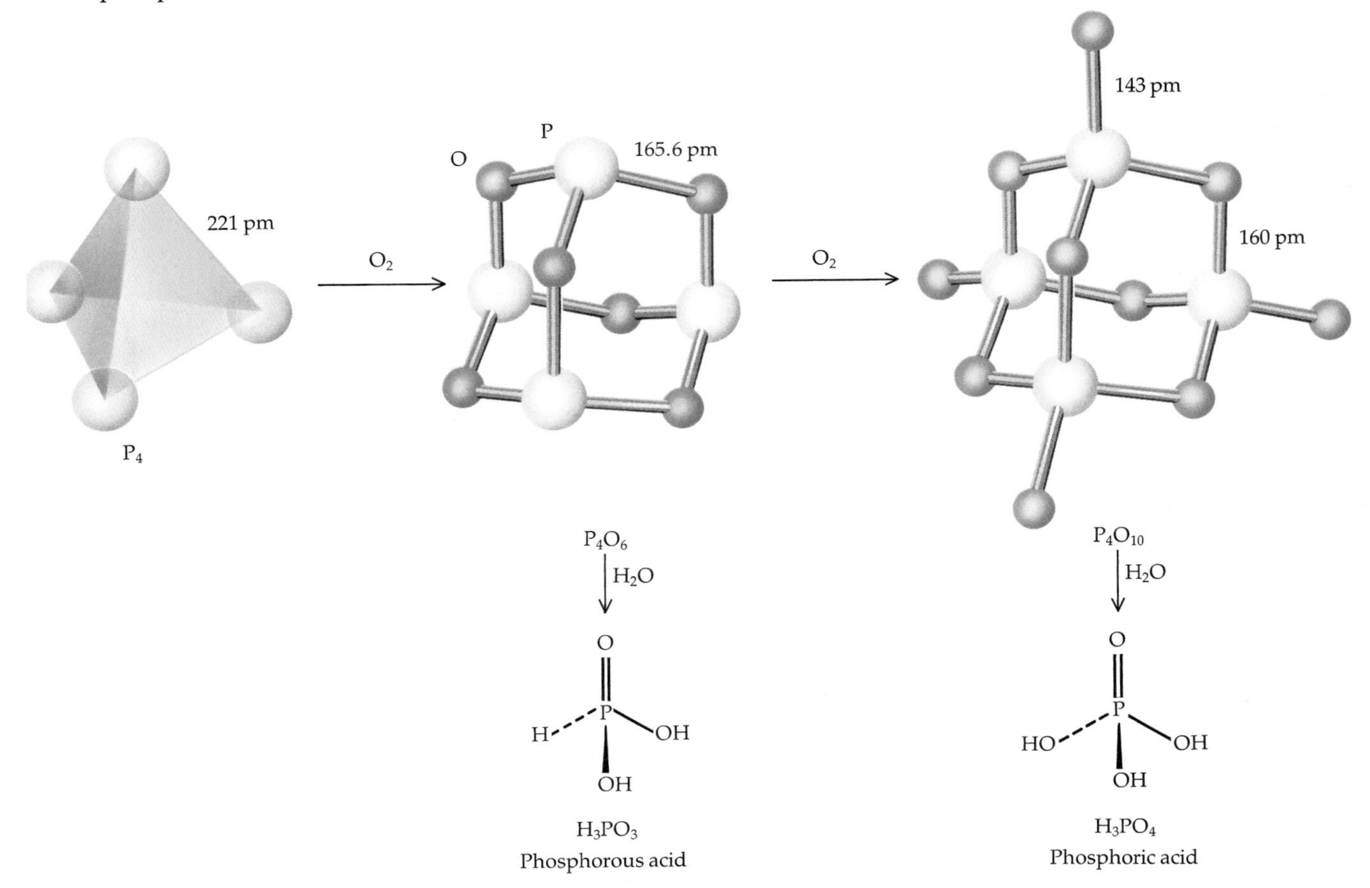

Tetraphosphorus trisulfide, P_4S_3, one of the substances used in "strike-anywhere" matches, has the structure

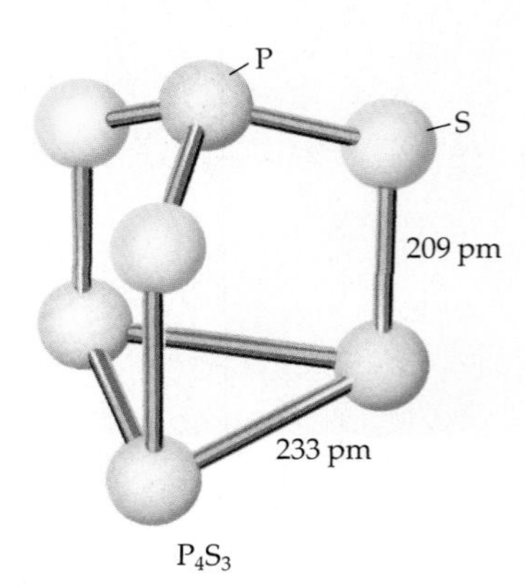

P_4S_3

The other common P—S compound is P_4S_{10}, structurally analogous to P_4O_{10}. Since P_4S_{10} is used as a starting material for compounds that contain P—S bonds, such as organophosphorus pesticides, world production of P_4S_{10} exceeds 250,000 tons.

Polymers of PO_4 tetrahedra can form by P—O—P bonding, analogous to the linking of SiO_4 tetrahedra in silicates.

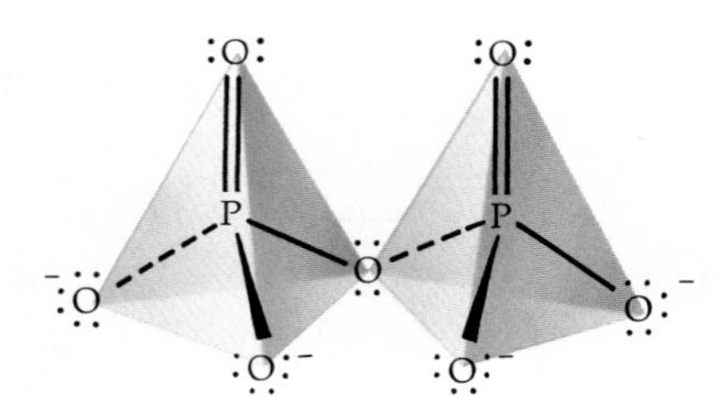

The pyrophosphate anion, $P_2O_7^{4-}$

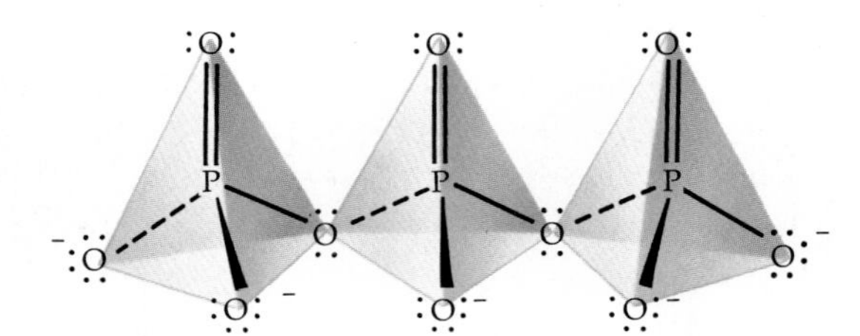

The tripolyphosphate ion, $P_3O_{10}^{5-}$

Figure 11.16
Structure of adenosine triphosphate (ATP), a principal source of energy for biological reactions. Energy is released when ATP reacts with water to give ADP and HPO_4^{2-}. Energy is absorbed when the reverse reaction occurs.

Sodium tripolyphosphate is the "phosphate builder" that is added to many detergents. It has many functions, one of which is to remove Mg^{2+} and Ca^{2+} from hard water by forming complex ions (Section 10.10).

Did you identify phosphates as the complexing agent in the Chemistry You Can Do experiment in Section 10.10?

Phosphate groups are part of many important molecules found in nature. For example, ATP, the chief energy source of all biochemical reactions, contains the triphosphate group (Figure 11.16), and the polymeric backbone of nucleic acids is a phosphate-sugar linkage (Figure 11.1).

There is considerable controversy in the United States over the use of phosphates. Since inorganic phosphate is an essential nutrient for plant growth, introducing large quantities into natural waters can lead to eutrophication or overfertilization of water. It can promote a massive growth of algae that in turn depletes the oxygen in water and kills off the fish life.

11.4 CONSTITUTIONAL ISOMERS

Two or more compounds with the same molecular formula but different arrangements of atoms are called **isomers.** Isomers differ in one or more physical or chemical properties such as boiling point, color, solubility, reactivity, and density. Several different kinds of isomerism are possible, particularly in organic compounds and coordination compounds. **Constitutional isomers** (also called structural isomers) are compounds with the same formula that differ in the order in which their atoms are bonded together. For example, consider molecules with the molecular formula C_2H_6O. Possible arrangements are

$$\begin{array}{ccccccc} & H & H & & \\ & | & | & & \\ H- & C- & C- & O- & H \\ & | & | & & \\ & H & H & & \end{array} \qquad \begin{array}{ccccccc} & H & & H & \\ & | & & | & \\ H- & C- & O- & C- & H \\ & | & & | & \\ & H & & H & \end{array}$$

ethyl alcohol dimethyl ether

The molecular structure on the left represents a molecule of ethyl alcohol, found in alcoholic beverages; the molecular structure on the right represents a molecule of dimethyl ether, a colorless gas that can be used as a refrigerant. Obviously, these two completely different compounds have different properties.

	CH_3CH_2OH *Ethyl Alcohol*	*CH_3OCH_3* *Dimethyl Ether*
boiling point (1 atm)	78.5 °C	−24.8 °C
melting point	−115 °C	−141 °C
reaction with Na	reacts to give H_2	no reaction

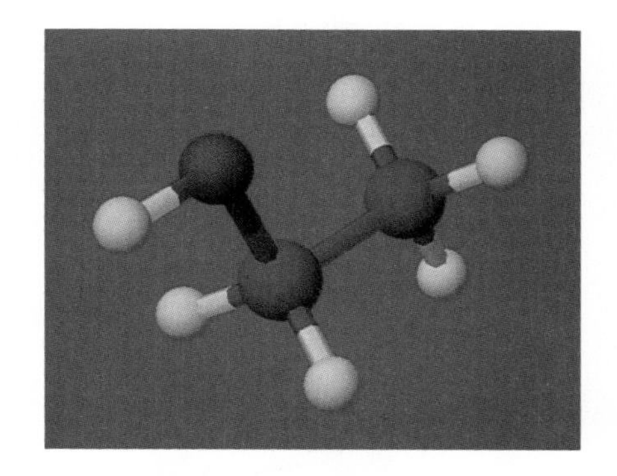

Models of ethyl alcohol and dimethyl ether. *(C.D. Winters)*

The different types of constitutional isomers of alkanes, alkenes, and aromatic compounds are explained in the following sections.

Straight- and Branched-Chain Isomers of Alkanes

The first three alkanes, CH_4, C_2H_6, and C_3H_8, were introduced in Section 10.1. Each of these molecules has only one possible structural arrangement. However, constitutional isomers are possible for C_4H_{10} and higher members of the series because of the possibility of having **"branched-chain" isomers.** Two structural arrangements are possible for C_4H_{10}. These drawings are an example of the two ways of writing structural formulas—in expanded form or in condensed form (Section 4.5). The expanded form is the same as the Lewis structure.

STRUCTURAL FORMULAS:

$$\begin{array}{ccccccccc} & & H & & H & & H & & H & & \\ & & | & & | & & | & & | & & \\ H & — & C & — & C & — & C & — & C & — & H \\ & & | & & | & & | & & | & & \\ & & H & & H & & H & & H & & \end{array} \qquad \begin{array}{ccccccc} & & & & H & & & & \\ & & & & | & & & & \\ & & H & H— & C & —H & H & & \\ & & | & & | & & | & & \\ H & — & C & —— & C & —— & C & — & H \\ & & | & & | & & | & & \\ & & H & & H & & H & & \end{array}$$

CONDENSED FORMULAS:

	$CH_3CH_2CH_2CH_3$ *butane*	CH_3CHCH_3 (with CH_3 on the central C) *methylpropane (isobutane)*
melting point	−138.3 °C	−160 °C
boiling point (1 atm)	0.5 °C	−12 °C
density (at 20 °C)	0.579 g/mL	0.557 g/mL

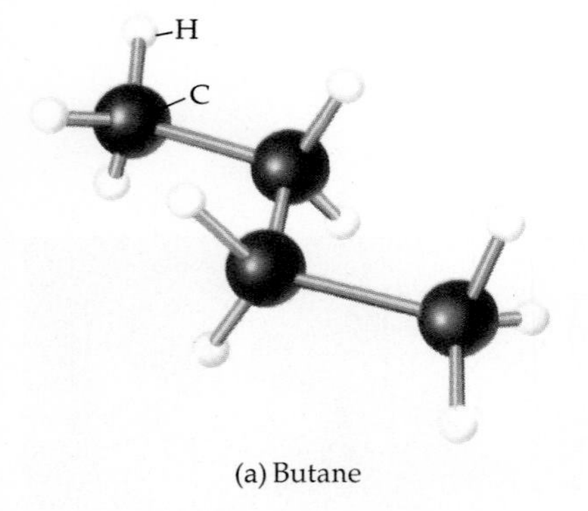

(a) Butane

(b) Methylpropane (isobutane)

Figure 11.17
Ball-and-stick models of butane and methylpropane, the isomers of C_4H_{10}.

These are different compounds with different properties, but because they have the same formula, they are examples of constitutional isomers. Ball-and-stick models of these two isomers are shown in Figure 11.17. Constitutional isomerism can be compared to the results you might expect from a child building many different structures with the same collection of building blocks and using all of the blocks in each structure.

The branched-chain isomer of C_4H_{10}, methylpropane, has a "methyl" group (—CH_3) attached to the central carbon atom. This is the simplest example of the fragments of alkanes known as **alkyl groups.** In this case, removal of an H atom from methane gives the **methyl** group.

$$\begin{array}{ccc} & H & \\ & | & \\ H— & C & —H \\ & | & \\ & H & \end{array} \xrightarrow{-H} \begin{array}{cc} & H \\ & | \\ H— & C— \\ & | \\ & H \end{array} \quad \text{or} \quad —CH_3 \left(\begin{array}{l}\text{also written}\\ \text{as } CH_3—\end{array}\right)$$

Removal of an H atom from ethane gives an **ethyl** group.

$$\begin{array}{cccc} & H & H & \\ & | & | & \\ H— & C— & C & —H \\ & | & | & \\ & H & H & \end{array} \xrightarrow{-H} \begin{array}{ccc} & H & H \\ & | & | \\ H— & C— & C— \\ & | & | \\ & H & H \end{array} \quad \text{or} \quad —C_2H_5 \left(\begin{array}{l}\text{also written}\\ \text{as } CH_3CH_2—\end{array}\right)$$

Notice that more than one alkyl group is possible when an H atom is removed from C_3H_8.

```
          H  H  H
          |  |  |
        H—C—C—C—H
          |  |  |
          H  H  H
 – H from /          \ – H from
 either end           middle carbon
 carbon  ↙              ↘

  H  H  H                              H  H  H
  |  |  |                              |  |  |
H—C—C—C— or CH3CH2CH2—             H—C—C—C—H or (CH3)2CH—
  |  |  |                              |     |
  H  H  H                              H     H

       propyl                              isopropyl
```

Historically, straight-chain hydrocarbons were referred to as normal *hydrocarbons, and n- was used as a prefix in their names. The current practice is not to use n-. If a hydrocarbon's name is given without indicating that the compound is branched-chain, assume it is a straight-chain hydrocarbon.*

Alkyl groups are named by dropping "-ane" from the parent alkane and adding "-yl." Theoretically, any alkane can be converted to an alkyl group. Some of the more common examples of alkyl groups are given in Table 11.3.

EXERCISE 11.5 • *Straight-Chain and Branched-Chain Isomers*

Three isomers are possible for the isomeric pentanes (five-carbon hydrocarbons). Draw both expanded and condensed formulas for these isomers.

Table 11.3

Some Common Alkyl Groups

Name	Condensed Structural Representation
methyl	CH_3—
ethyl	CH_3CH_2—
propyl	$CH_3CH_2CH_2$—
isopropyl (branching on second carbon)	CH_3CH— or $(CH_3)_2CH$— $\quad$ | $\quad CH_3$
butyl	$CH_3CH_2CH_2CH_2$—
sec-butyl*	CH_3CHCH_2— or $(CH_3)_2CHCH_2$— $\quad$ | $\quad CH_3$
t-butyl**	$\quad CH_3$ $\quad$ | CH_3C— or $(CH_3)_3C$— (subject to steric hindrance) $\quad$ | $\quad CH_3$

**sec* stands for *secondary*, which means that the central C atom is bonded to two other C atoms and an H atom.
***t* stands for *tertiary*, sometimes abbreviated *tert*, which means that the central C atom is bonded to three other C atoms.

Table 11.4

Number of Predicted Constitutional Isomers for Some Alkanes*

Formula	Isomers Predicted	Found
C_6H_{14}	5	5
C_7H_{16}	9	9
C_8H_{18}	18	18
$C_{15}H_{32}$	4,347	
$C_{20}H_{42}$	366,319	
$C_{30}H_{62}$	4,111,846,763	
$C_{40}H_{82}$	62,491,178,805,831	

*R. E. Davies and P. J. Freyd, "$C_{167}H_{336}$ Is the Smallest Alkane with More Realizable Isomers than the Observed Universe Has 'Particles'," *J. Chem. Educ.*, Vol. 66, p. 278, 1989.

Table 11.4 gives the number of constitutional isomers predicted for some larger alkane molecules, starting with C_6H_{14}. Every predicted isomer, *and no more*, has been isolated and identified for the C_6, C_7, and C_8 alkanes. Although not all of the C_{15} molecules have been isolated, there is reason to believe that they could be made with enough time and effort, so constitutional isomerism certainly helps to explain the vast number of carbon compounds. However, for compounds with more than 17 carbon atoms, the number of isomers that are likely to be isolated is much smaller than predicted. This difference arises because the simple calculation of the number of ways to construct isomers does not consider the space requirements of atoms within the molecules. Many of the isomers predicted for these larger molecules would require such overcrowding of the atoms that the molecules would not be stable. (See the reference listed in Table 11.4 for further discussion of this point.)

Naming Branched-Chain Alkanes

The rules for systematic names were formulated by the International Union of Pure and Applied Chemistry and are described in Appendix E.

Many alkanes and other organic compounds have both common names and systematic names. Usually the common name was assigned first and is widely known. Many consumer products are labeled with their common names; when only a few isomers are possible, the common name adequately identifies the product for the consumer. However, as the examples in this section will illustrate, a system of common names quickly fails when several constitutional isomers are possible.

For example, 2,2,4-trimethylpentane is the systematic name of the following branched-chain isomer of octane (C_8H_{18}). The "pentane" part, which means a straight five-carbon chain, identifies the longest chain in the molecule. The numbers "2,2,4-" indicate the locations of the three ("tri-") methyl groups, according to the numbering of the pentane chain from one end to the other. This particular isomer of octane is used as a standard in assigning the "octane ratings" of various types of gasoline.

```
        CH3        CH3
        |          |
CH3—C—CH2—CH—CH3
1      2|  3       4      5
        CH3
```

EXERCISE 11.6 • *Branched-Chain Alkanes*

Draw the condensed structural formula for each of the following compounds: (a) 2-methylpentane, (b) 3-methylpentane, (c) 2,2-dimethylbutane, (d) 2,3-dimethylbutane.

Constitutional Isomers of Alkenes

In the alkene series, the possibility of locating the double bond between different pairs of carbon atoms adds additional constitutional isomers. In ethene and propene there is only one possible location of the double bond.

```
H       H          H       H
  \   /              \   /
   C=C                C=C
  /   \              /   \
H       H        H3C       H
 ethene             propene
(ethylene)        (propylene)
```

However, the next alkene in the series, butene, has two possible locations for the double bond.

```
   H  H  H  H              H  H  H  H
   |  |  |  |              |  |  |  |
H—C=C—C—C—H          H—C—C=C—C—H
   1  2  3|  4|             1|  2  3  4|
         H  H              H        H
   1-butene                2-butene
```

When groups such as methyl or ethyl are attached to carbon atoms in an alkene, the longest hydrocarbon chain is numbered from the end that will give the double bond the lowest number, and then numbers are assigned to the attached groups. For example, in the following compound the longest chain has 7 carbons (heptene); the double bond is between C2 and C3 (2-heptene); and the three (tri-) methyl groups are on the third, fourth, and sixth carbons (3,4,6-trimethyl-).

```
   H  H  CH3  CH3  H  CH3  H
   |  |  |    |    |  |    |
H—C—C=C——C——C—C——C—H
   1| 2  3    4|   5| 6|   7|
   H          H    H  H    H
```

Hence the name is 3,4,6-trimethyl-2-heptene.

EXERCISE 11.7 • *Drawing Constitutional Isomers of Alkenes*

Draw the structure of 2,3-dimethyl-2-pentene.

Constitutional Isomers of Aromatic Compounds

Hydrocarbons containing one or more benzene rings (Figure 11.18) are called **aromatic compounds.** The word *"aromatic"* was derived from *"aroma,"* which describes the rather strong and often pleasant odor of these compounds. Since the benzene molecule has a planar structure (Section 10.6), constitutional isomers are possible when two or more groups are substituted for hydrogen atoms on the benzene ring.

Three isomers are possible if two groups are substituted for two hydrogen atoms on the benzene ring. Either the prefixes *ortho-*, *meta-*, and *para-* or numbers are used to distinguish between them. When the prefixes are used, usually only the first letter is given. For example, when the two groups are methyl groups, the three isomers are commonly known as *o*-xylene, *m*-xylene, and *p*-xylene (Figure 11.18). If more than two groups are attached to the benzene ring, numbers must be used to identify the positions. Consider the following compounds:

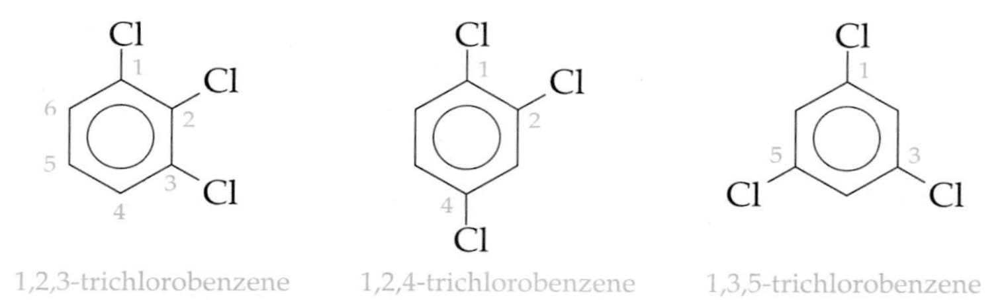

1,2,3-trichlorobenzene 1,2,4-trichlorobenzene 1,3,5-trichlorobenzene

There is no other way to attach three atoms of chlorine to a benzene ring, and only three trichlorobenzenes have been isolated in the laboratory.

Benzene and many of its derivatives, such as toluene and ethylbenzene, are on the list of the top 25 chemicals produced in United States (see inside back cover) because of their use in manufacturing plastics, detergents, pesticides, drugs, and other organic chemicals. There are other aromatic compounds, such as naphthalene, anthracene, and benzopyrene, that contain two or more benzene rings sharing ring edges (Figure 11.18).

Carcinogens are cancer-causing agents. The type of cancer caused may vary from one carcinogen to another. Benzene may cause a form of leukemia, and benzopyrene may cause skin cancer and lung cancer.

Benzene and several of the aromatic compounds are known **carcinogens.** One of these, benzopyrene, is found in smoke from burning fossil fuels (Section 20.4), tobacco, and meat grilling on charcoal.

EXERCISE 11.8 • *Constitutional Isomers of Aromatic Compounds*

Draw the structure of 1,2,4-trimethylbenzene.

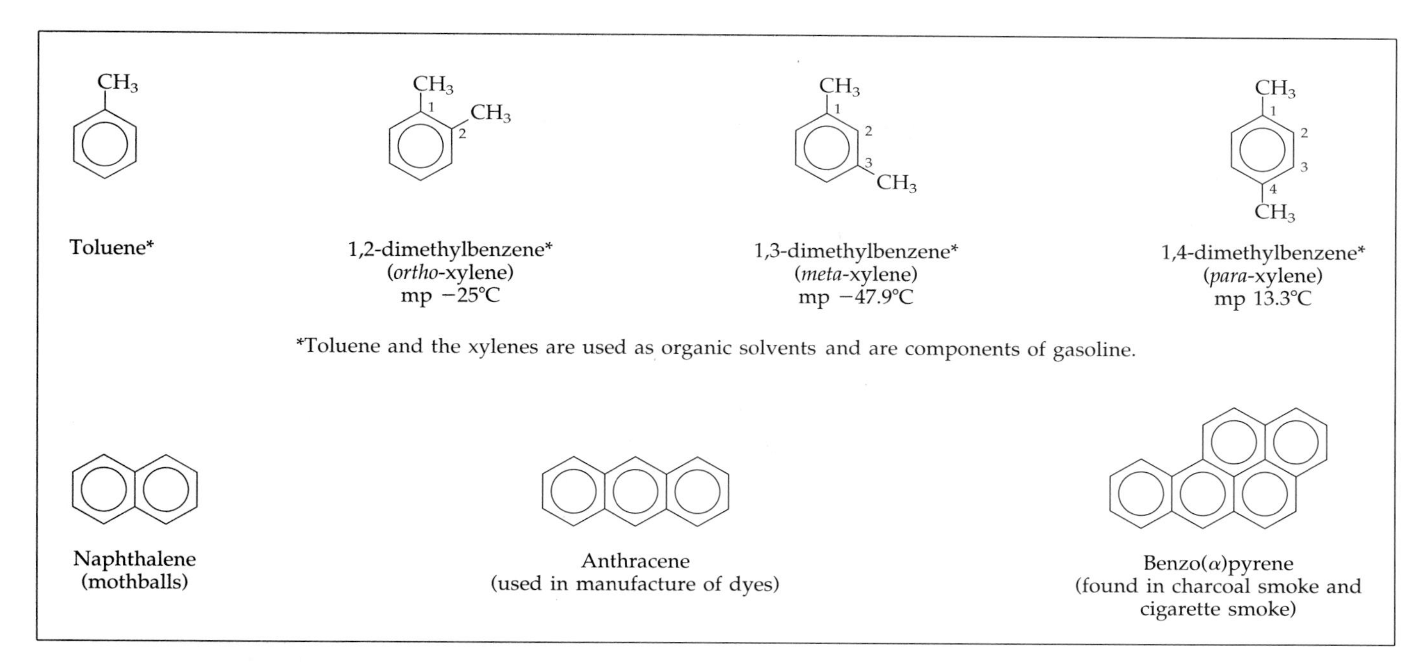

Figure 11.18
Aromatic hydrocarbons.

11.5 STEREOISOMERISM

The second major type of isomerism is **stereoisomerism.** Here the isomers have the same molecular formulas and the same atom-to-atom bonding sequences, *but the atoms differ in their arrangement in space.* There are two forms of stereoisomerism, ***cis-trans*** **isomerism** and **optical isomerism.**

Cis and *Trans* Isomers in Alkenes

An important difference between alkanes and alkenes is the degree of flexibility of the carbon-carbon bonds in the molecules. Rotation around single carbon-carbon bonds in alkanes occurs readily at room temperature, but the carbon-carbon double bond in alkenes is strong enough to prevent free rotation about the bond (Section 8.1, Figure 8.3). Consider the compound ethene, C_2H_4. Its six atoms lie in the same plane, with bond angles of approximately 120 °.

The difficulty of rotation about a carbon-carbon double bond was illustrated in Section 8.1 by the kinetics of the interconversion of cis- *and* trans-2-*butene.*

```
H       H
 \     /
  C═C
 /     \
H       H
```

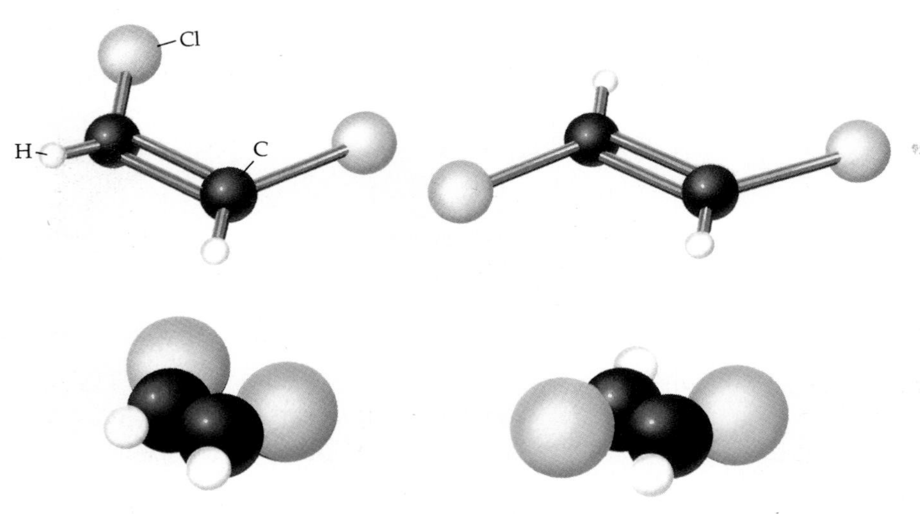

Figure 11.19
Cis and *trans* isomers of 1,2-dichloroethene.

If free rotation occurred around a carbon-carbon double bond, these two compounds would be the same.

If two chlorine atoms replace two hydrogen atoms, one on each carbon atom of ethene, the result is ClHC=CHCl. Experimental evidence confirms the existence of two compounds with the same set of bonds. The difference between the two compounds is the location in space of the two chlorine atoms: the ***cis* isomer** has two chlorine atoms on the same side in the plane of the double bond, and the ***trans* isomer** has two chlorine atoms on opposite sides of the double bond (Figure 11.19). Both compounds are called 1,2-dichloroethene (the 1 and 2 indicate that the two chlorine atoms are attached to different carbon atoms). Note that the two isomeric compounds have different properties.

	Cl Cl C=C H H *cis-1,2-dichloroethene*	H Cl C=C Cl H *trans-1,2-dichloroethene*
melting point	−80.5 °C	−50 °C
boiling point (1 atm)	60.3 °C	47.5 °C
density (at 20 °C)	1.284 g/mL	1.265 g/mL
dipole moment	1.90 D	0 D

The third possible isomer, 1,1-dichloroethene (a constitutional isomer of the *cis* and *trans* isomers), does not have *cis* and *trans* structures since each carbon atom is attached to two identical atoms. *Cis-trans* isomerism in alkenes is possible only when both of the double-bond carbon atoms have two different groups attached. (For the sake of simplicity, the word "groups" refers to both atoms and groups.)

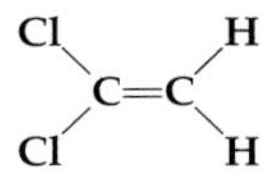

Cl H
C=C
Cl H

1,1-dichloroethene

	1,1-dichloroethene
melting point	−122.1 °C
boiling point (1 atm)	37 °C
density (at 20 °C)	1.218 g/mL
dipole moment	1.34 D

When there are four or more carbon atoms in an alkene, the possibility exists for *cis* and *trans* isomers even when only carbon and hydrogen atoms are present. For example, 2-butene has both *cis* and *trans* isomers.

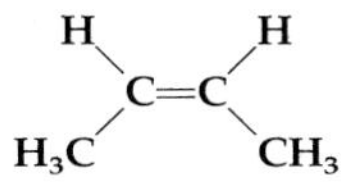

H H
C=C
H_3C CH_3

cis-2-butene

H CH_3
C=C
H_3C H

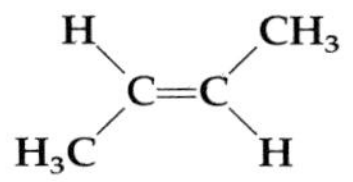

trans-2-butene

	cis-2-butene	*trans*-2-butene
melting point	−138.9 °C	−105.5 °C
boiling point (1 atm)	3.7 °C	0.9 °C
density (at 20 °C)	0.621 g/mL	0.604 g/mL

EXAMPLE 11.4

Cis *and* Trans *Isomers*

Which of the following molecules can have *cis* and *trans* isomers? For those that do, draw the two isomers and label them *cis* and *trans*.

a. $(CH_3)_2C{=}CCl_2$ b. $CH_3ClC{=}CClCH_3$ c. $CH_3BrC{=}CClCH_3$

SOLUTION (b) and (c). (a) cannot have *cis* and *trans* isomers because both the groups on each carbon are the same.

b.

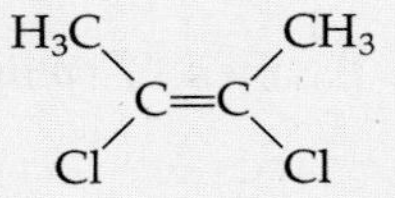

H_3C CH_3
C=C
Cl Cl

cis

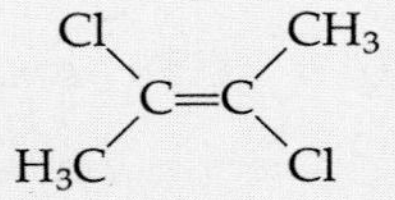

Cl CH_3
C=C
H_3C Cl

trans

c.

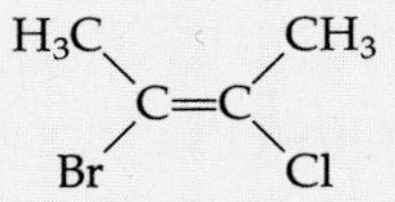

H_3C CH_3
C=C
Br Cl

cis

Br CH_3
C=C
H_3C Cl

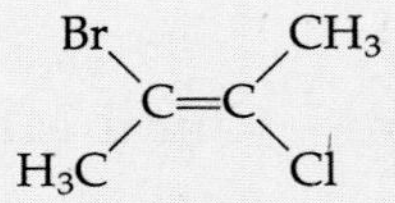

trans

Cis *and* Trans *Fatty Acids and Your Health*

THE DIRECT LINK BETWEEN DIETS high in saturated fats and heart disease is well known. Saturated fatty acids are usually found in solid or semisolid fats, while unsaturated fatty acids are usually found in oils. As a result, nutritionists recommend the use of liquid vegetable oils for cooking, and people are generally aware of this. However, less well known is the process by which margarine and semisolid cooking fats are made. Hydrogen is added to the C=C double bonds in unsaturated fats (oils) to convert them to a solid or semisolid fat that has better consistency and less chance for spoilage. This hydrogenation process decreases the number of double bonds but also forms synthetic *trans*-fatty acids from natural *cis*-fatty acids. The *trans*-fatty acids are not metabolized in the human system, but they can be stored for the life of the individual because the *trans*-fatty acids are "straight" molecular structures and pack together like the saturated fatty acids. By contrast, the *cis*-fatty acids are bent and do not pack well.

The health risks of *trans*-fatty acids are reduced by not eating processed vegetable fats. How do you know which products were made with processed vegetable fats? Read the label. For example, a box of cookies or crackers may say on the label "made with 100% pure vegetable shortening . . . (partially hydrogenated soybean oil with hydrogenated cottonseed oil)." Although soybean oil and cottonseed oil are low in saturated fat, the hydrogenation process converts cottonseed oil to a saturated fat, and the partially hydrogenated soybean oil contains *trans*-fatty acid.

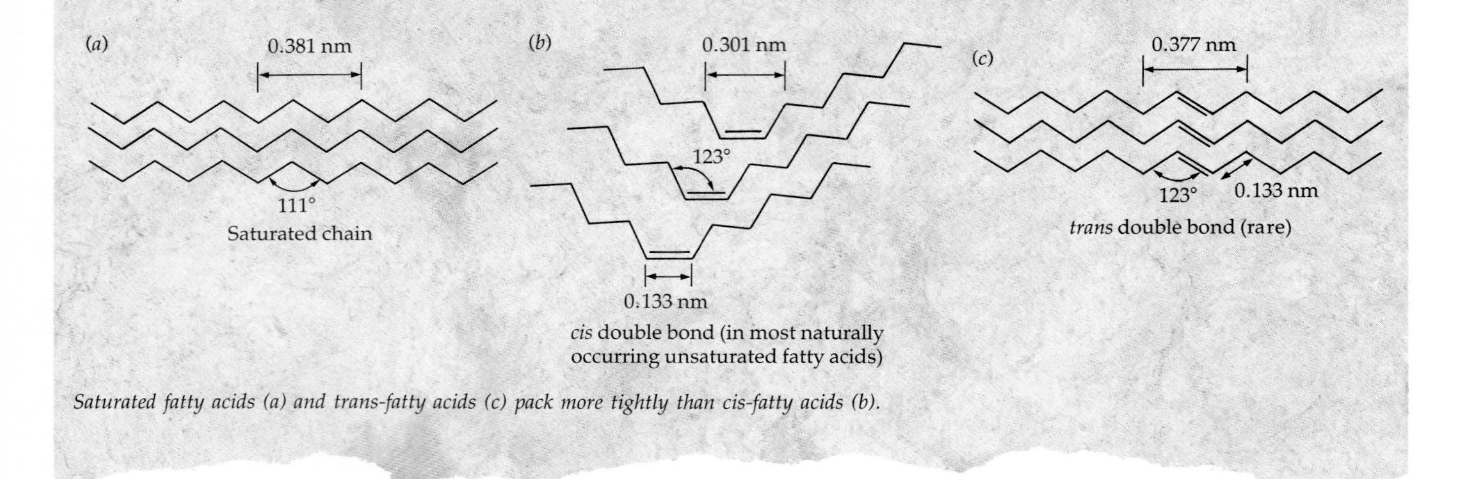

Saturated fatty acids (a) and trans-fatty acids (c) pack more tightly than cis-fatty acids (b).

EXERCISE 11.9 • Cis *and* Trans *Isomers*

Which of the following molecules can have *cis* and *trans* isomers? For those that do, draw the two isomers and label them *cis* and *trans*.

a. 2-methyl-2-butene b. 1-butene c. 1-bromo-2-chloro-2-butene

Optical Isomerism—Chiral Molecules

Chiral is pronounced "ki-ral" and is derived from the Greek cheir, *meaning hand.*

Are you right-handed or left-handed? Regardless of our preference, we learn at a very early age that a right-handed glove doesn't fit the left hand and vice versa. Our hands are mirror images of one another and are not superimposable (Figure 11.20). *An object that cannot be superimposed on its mirror image is called* ***chiral.*** Objects that are superimposable on their mirror images are **achiral.** Stop and think about the extent to which chirality is a part of our everyday life. We've already discussed the chirality of hands (and feet). Helical sea shells are chiral, and most spiral to the right like a right-handed screw. Many creeping vines show a chirality when they wind around a tree or post.

What is not as well known is that a large number of the molecules in plants and animals are chiral, and usually only one form (left-handed or right-handed) of the chiral molecule is found in nature. For example, all but one of the 20 naturally occurring amino acids are chiral, and only the left-handed amino acids are found in nature! Most of the natural sugars are right-handed, including the sugar found in DNA.

A chiral molecule and its nonsuperimposable mirror image are called **enantiomers.** Enantiomers are two different molecules, just as your left hand and right hand are different. Enantiomers are possible when a molecular structure is **asymmetrical** (without symmetry). The simplest case is a tetrahedral carbon atom bonded to four *different* atoms or groups of atoms. Such a carbon atom is said to be chiral (without symmetry), and a molecule that contains a chiral atom is a chiral molecule.

Some compounds are found in nature in both enantiomeric forms. For example, both enantiomers of lactic acid are found in nature. During the contraction of muscles the body produces only one enantiomer of lactic acid. However, the other enantiomer is produced when milk sours. Let's look at the structure of lactic acid to see why two enantiomers might be possible. The central carbon atom of lactic acid has four different groups bonded to it: $—CH_3$, —OH, —H, and —COOH.

H
|
C
HO CH_3 COOH

lactic acid

As a result of the tetrahedral geometry around the central carbon atom, it is possible to have two different arrangements of the four groups. If a lactic

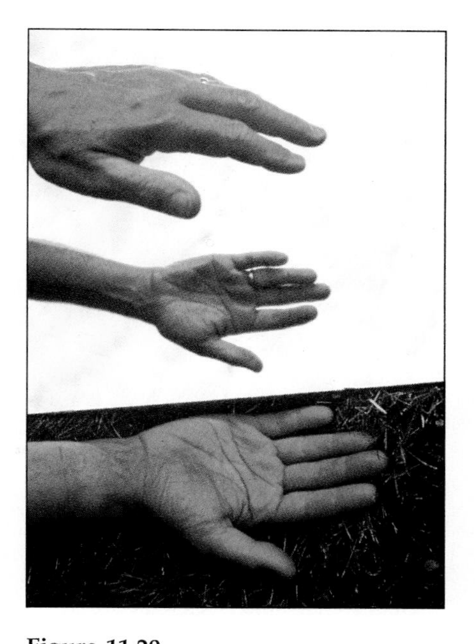

Figure 11.20
Mirror images. Your left hand is a nonsuperimposable mirror image of your right hand. For example, the mirror image of your right hand looks like your left hand. However, if you place one hand directly over the other, they are not identical; hence, they are nonsuperimposable mirror images.

Sea shells have a handedness—that is, they are chiral—and virtually all of them are right-handed. If you cup a shell in your right hand (with your thumb extended), your fingers will follow the curl of the shell as it proceeds from the outside toward the center. Your thumb will point along the axis of the shell from the narrow end toward the wider end. *(C.D. Winters)*

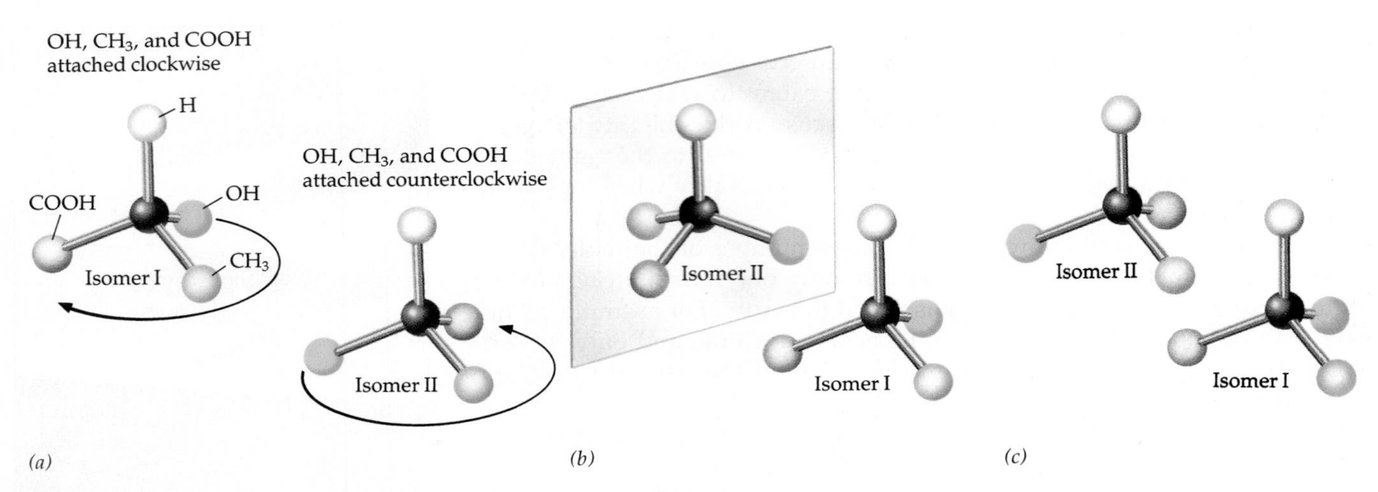

Figure 11.21
The enantiomers of lactic acid, $CH(CH_3)(OH)(COOH)$. (a) Isomer I: $-OH$, $-CH_3$, and $-COOH$ are attached in a clockwise manner. Isomer II: $-OH$, $-CH_3$, and $-COOH$ are attached in a counterclockwise manner. (b) Isomer I is placed in front of a mirror, and its mirror image is Isomer II. (c) The isomers are nonsuperimposable.

acid molecule is placed so that the C—H bond is vertical, as illustrated in Figure 11.21, one possible arrangement of the remaining groups would be that in which $-OH$, $-CH_3$, and $-COOH$ are attached in a clockwise sequence (isomer I). Alternatively, these groups can be attached in a counterclockwise sequence (isomer II). To see further that the arrangements are different, we place isomer I in front of a mirror (Figure 11.21b). Now you see that isomer II is the mirror image of isomer I. What is important, however, is that these mirror-image molecules *cannot be superimposed* on one another. *These two nonsuperimposable, mirror-image chiral molecules are enantiomers.*

The "handedness" of enantiomers is sometimes represented by D for right-handed (D stands for "dextro" from the Latin *dexter* meaning "right") and L for left-handed (L stands for "levo" from the Latin *laevus* meaning "left"). In the case of lactic acid, the D-form is found in souring milk and the L-form is found in muscle tissue, where it accumulates during vigorous exercise and can cause cramps.

Plane polarized light consists of electromagnetic waves vibrating in one direction.

Most of the properties of enantiomers of a chiral compound are identical—they have the same melting point, the same boiling point, the same density, and many other identical physical and chemical properties. However, they always differ with respect to one physical property: they rotate a beam of **plane polarized light** in opposite directions (Figure 11.22). For this reason, chiral molecules are sometimes referred to as **optical isomers** and are said to be **optically active.** A **polarimeter,** an instrument based on the design shown in Figure 11.22, is used to identify enantiomers because the direction and angle of rotation are unique properties of each enantiomer.

In 1811 Jean Baptiste Biot (1774–1862) discovered optical activity when he observed that a quartz crystal rotates plane polarized light. Three years later he found that solutions of sugar show the same behavior. However, Louis Pasteur was the first person to relate optical activity to the spatial arrangement of atoms in molecules. In 1848, Pasteur observed that sodium ammonium tartrate forms two different kinds of crystals, which are mirror images of each other. Pasteur used a hand lens and tweezers to separate the two kinds of crystals into two piles; he prepared a solution of each pile, and found that the solutions rotated plane polarized light in opposite directions.

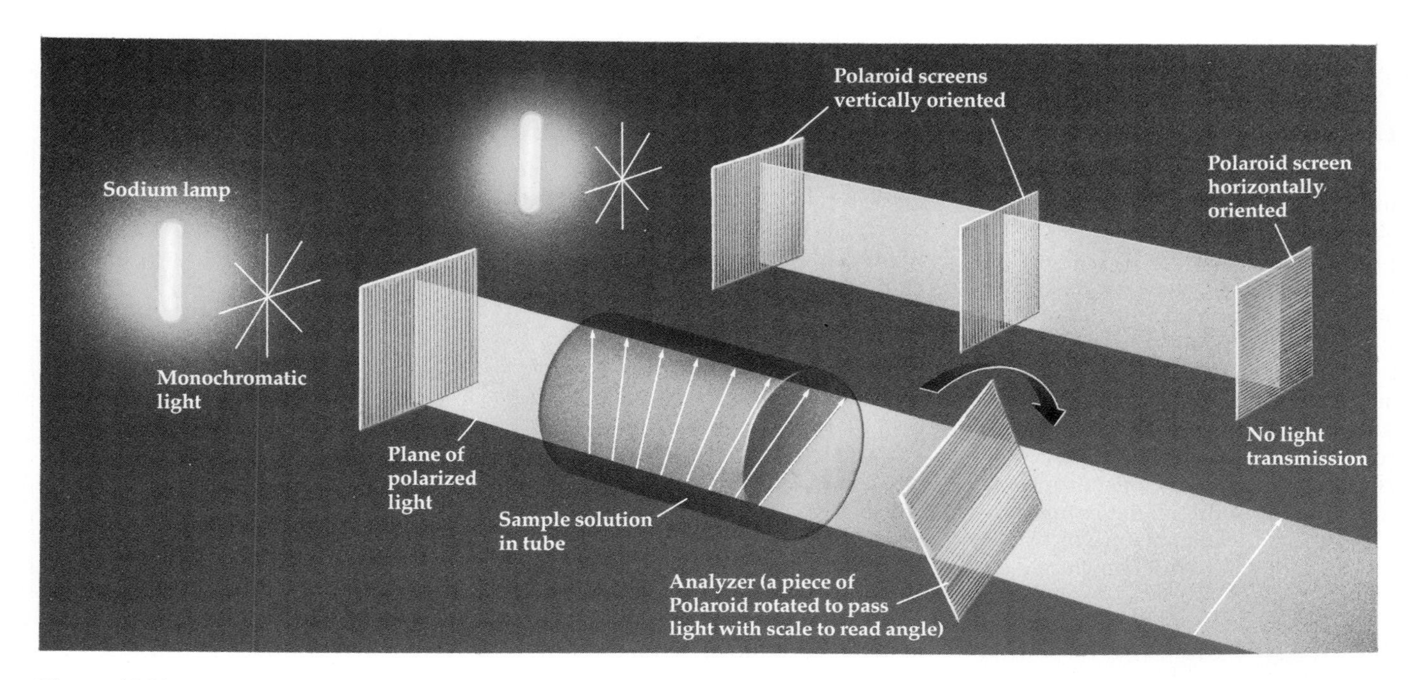

Figure 11.22

Schematic drawing of the rotation of plane polarized light by an optical isomer. (Top) Monochromatic light (light of only one wavelength) is provided by a sodium lamp. After it passes through a Polaroid filter, the light is vibrating in only one direction—it is polarized. Polarized light will pass through a second Polaroid filter if the two filters are parallel, but not if the second filter is perpendicular to the first. (Bottom) If a solution of an optical isomer is placed between the first and second Polaroid filters, the plane of polarization of the light is rotated. The angle of rotation can be determined by rotating the second Polaroid filter until maximum transmission of light occurs. The size and the direction (clockwise or counterclockwise) of the angle of rotation are unique physical properties of the isomer being tested.

Pasteur explained this by suggesting that optical activity is caused by an asymmetric arrangement of atoms in the individual molecule, and that the molecules that rotate the plane of polarized light to the right and those that rotate it to the left are mirror images. This was a remarkable deduction since the tetrahedral nature of carbon was not known at that time and so Pasteur was not able to explain the actual spatial arrangement of atoms. It was not until 1874 that Joseph Achille LeBel (1847–1930) and Jacobus Henricus van't Hoff (1852–1911) independently demonstrated that an arrangement of four different atoms or groups at the corners of a regular tetrahedron would produce two structures, one of which is the mirror image of the other.

Enantiomers also differ with respect to their biological properties. Enantiomers react at different rates in a chiral environment, and many of the molecules in plants and animals are chiral. To understand why this difference in activity might exist, think of the hand-in-glove analogy. Although you can put a right-handed glove on your left hand, it takes longer to put it

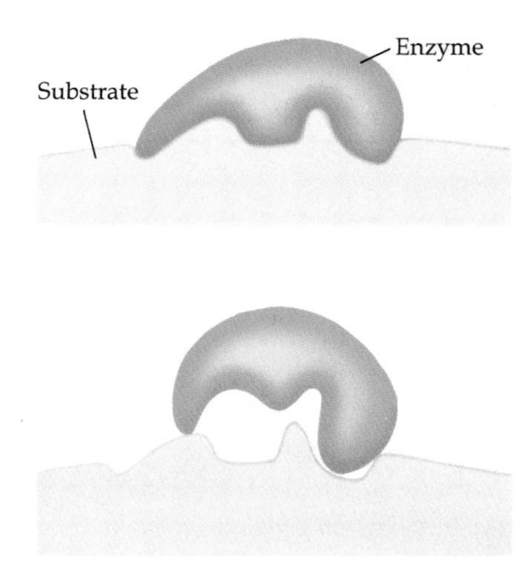

Schematic drawing illustrating the importance of shape for enzyme activity.

Portrait of a Scientist

Louis Pasteur (1822–1895)

Although Pasteur is best known for his accomplishments in bacteriology and medicine, his doctorate was in physics and chemistry. In 1847 his research work for the doctor's degree at the École Normale Supérieure in Paris was in crystallography. To gain some experience, he was repeating another chemist's earlier work on salts of tartaric acid when he saw something that no one had noticed before—a batch of crystals of optically inactive sodium ammonium tartrate actually was a mixture of two different kinds of crystals, which were mirror images of each other. He mechanically separated the two types of crystals and observed that their solutions rotated plane polarized light in opposite directions. His proposal that the molecules making up the crystals were mirror images of each other earned him wide recognition and laid the groundwork for the field of stereochemistry.

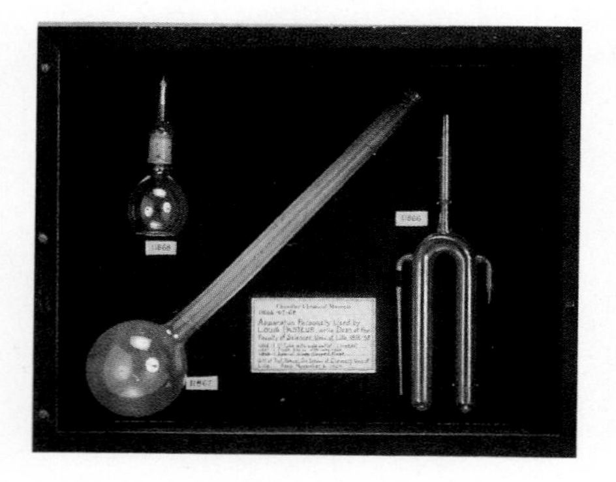

Glassware used by Louis Pasteur, which is on display at Columbia University. (C.D. Winters)

Pasteur was appointed Professor of Chemistry at the University of Strasbourg in 1848, where he continued his work in crystallography. In 1854 at the age of 32 he accepted the chair of chemistry at the newly organized University of Lille, and in 1857 he returned to École Normale Supérieure in Paris as assistant director in charge of scientific studies. Here he continued his studies of microorganisms begun while he was at the University of Lille, and in the same year discovered a biological method of separation of optical isomers when he observed that a penicillin mold selectively destroyed the right-handed tartaric acid but would not touch the left-handed tartaric acid. This was the first evidence of the handedness of most organic molecules associated with living organisms.

Louis Pasteur. (Oesper Collection in the History of Chemistry/University of Cincinnati)

Pasteur's continuing studies of microorganisms led to his development of sterilization methods (pasteurization), and his study of infectious diseases led to development of vaccines against chicken cholera, anthrax, and rabies.

An excellent biography of Pasteur is "Pasteur and Modern Science" by René Dubos, Science Tech Publishers, Madison, Wisconsin.

---

on and it doesn't fit very well. Nature has a preference for L-isomers of amino acids. Enzymes, the catalysts for biochemical reactions, are proteins made from these L-amino acids linked together in chains, so enzymes are chiral molecules. The catalytic activity of enzymes is a consequence of their three-dimensional structure, which in turn depends on their L-amino acid sequence. As a result, an enzyme has a binding preference for one of the enantiomers of the reactants.

Since enantiomers rotate the plane of polarized light an equal angle but in opposite directions, a solution of a racemic mixture will not rotate the plane of polarized light.

Although biological systems have a preference for one enantiomer, laboratory synthesis of a chiral compound gives a mixture of equal amounts of the enantiomers, which is called a **racemic mixture.** The separation and purification of enantiomers is difficult because of the similarity of their physical properties. Although the first separation of optical isomers by Pas-

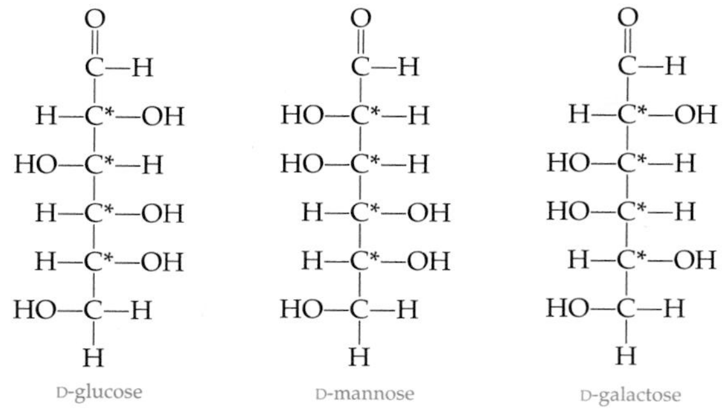

D-glucose	D-mannose	D-galactose
O=C—H	O=C—H	O=C—H
H—C*—OH	HO—C*—H	H—C*—OH
HO—C*—H	HO—C*—H	HO—C*—H
H—C*—OH	H—C*—OH	HO—C*—H
H—C*—OH	H—C*—OH	H—C*—OH
HO—C—H	HO—C—H	HO—C—H
H	H	H

(C* = Asymmetric Carbon Atom)

Figure 11.23
Three of the 16 possible isomers for simple sugars with the formula $C_6H_{12}O_6$. (a) D-glucose is the principal source of chemical energy in human metabolism. (b) D-mannose is found in plants. (c) D-galactose is found in milk.

teur was based on his observation of different kinds of crystals in a racemic mixture, this method is rarely an option because chiral compounds seldom form crystals recognizable as mirror images. In fact, even sodium ammonium tartrate does not give two different types of crystals unless it crystallizes at a temperature below 28 °C. Thus, a contributing factor to Pasteur's discovery was the cool Parisian climate!

The usual method of separating enantiomers is to treat them with optically active reagents that have a greater affinity for one enantiomer than for the other. Pasteur's discovery of a mold that would selectively destroy one enantiomer of tartaric acid was the first example of this approach. Chiral reagents, chiral solvents, and chiral catalysts are currently used to separate enantiomers.

Nature's preference for one form of optically active amino acids has provoked much discussion and speculation among scientists since Pasteur's discovery of optical activity in 1848. However, there is still no widely accepted explanation of this preference.

Large organic molecules may have a number of chiral carbon atoms within the same molecule. At each such carbon atom there are two possible arrangements of the molecule. The total number of possible molecules, then, increases exponentially with the number of different chiral centers. With two different asymmetric carbon atoms there are 2^2, or four, possible structures. It should be emphasized that each of the four isomers can be made from the same set of atoms with the same set of chemical bonds. Glucose, a simple blood sugar also known as dextrose, contains four different asymmetric carbon atoms per molecule. However, of the 16 possible isomers, only three are biologically important (Figure 11.23). These are D-glucose, D-mannose, and D-galactose. Of these, D-glucose is by far the most important.

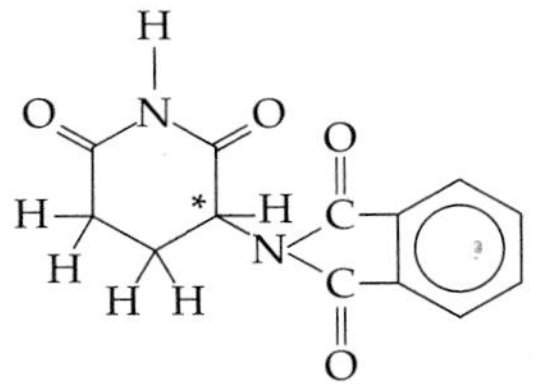

The chirality of carbohydrates, proteins, and DNA makes the human body highly sensitive to enantiomers. For example, one enantiomer of a drug is usually more active (or more toxic) than the other enantiomer. A tragic example that called attention to the need to test both enantiomers occurred in 1963 when horrible birth defects were induced by thalidomide.

Thalidomide

*chiral carbon

Chiral Drugs

ONE ENANTIOMER OF A CHIRAL drug is usually more active than the other, but 80% of chiral drugs are still sold as racemic mixtures. Only in those cases where one enantiomer is toxic or has harmful side effects is the single-enantiomer drug used. The primary reason for this is the large increase in cost required to isolate an enantiomer from a racemic mixture.

Before long, though, many more drugs may be brought to market as pure optical isomers. You may soon be seeing advertisements for a drug that is *optically pure* or *twice as effective as . . . [the racemic mixture, half of which is inactive]* or a *new and improved* version of a familiar over-the-counter medication. In recent years, techniques for separating en-

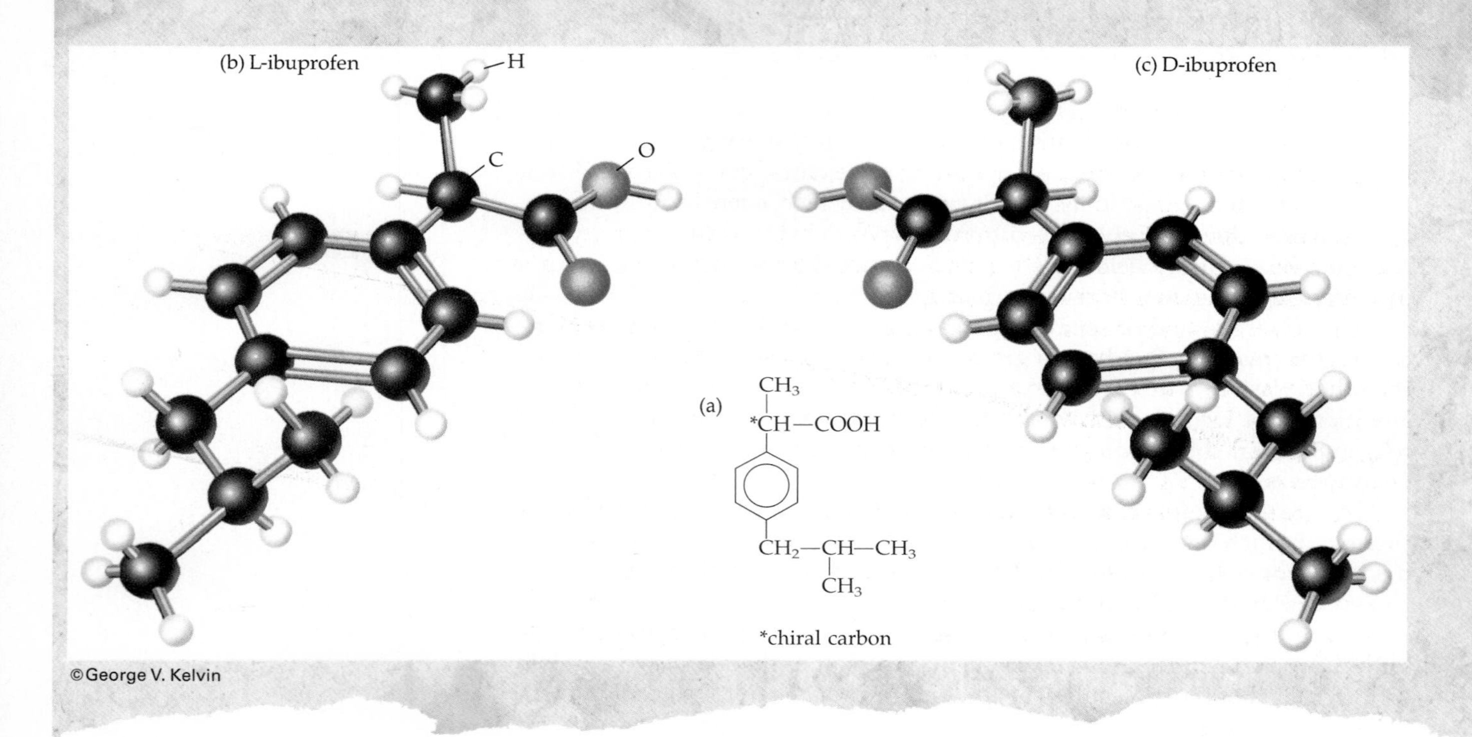

After this was discovered, it was determined that one enantiomer cured morning sickness and the other enantiomer caused birth defects.

Often, the difference is one of activity or effectiveness with no difference in toxicity. Aspartame (NutraSweet®), widely used as an artificial sweetener, has two enantiomers. However, one enantiomer has a sweet taste while the other enantiomer is bitter. This indicates that the receptor sites on our taste buds must be chiral, since they respond differently to the "handedness" of aspartame enantiomers! This becomes more clear when looking at the properties of the simple sugars. D-glucose is sweet and nutritious while L-glucose is tasteless and cannot be metabolized by the body.

antiomers have improved greatly. Also, the U.S. Food and Drug Administration (FDA) in 1992 released long-awaited guidelines on marketing chiral drugs.

The decision on whether to sell a chiral drug in the racemic mixture or enantiomerically pure form has been left to the drug's manufacturer (although the decision is subject to FDA approval). With the regulations finally in place, drug companies will be looking for situations where production of a single enantiomer can give them a competitive edge. The racemic mixture, however, may remain the best choice in most cases. For example, ibuprofen, the pain reliever contained in Advil, Motrin, and Nuprin, is now sold as a racemic mixture. The left-handed enantiomer of ibuprofen (b) is the active pain reliever and the right-handed isomer is inactive (c). But D-ibuprofen is converted to L-ibuprofen in the body, so there is probably no therapeutic advantage to the patient to switch from the racemic mixture to the more costly L-ibuprofen.

Naproxen is an example of a prescription chiral drug that is sold as an enantiomer rather than the racemic mixture. In this case, one enantiomer is a pain reliever while the other enantiomer causes liver damage.

The sale of enantiomeric drugs is already big business, with world sales of $18 billion in 1990. The top ten are listed below with their uses.

Naproxen

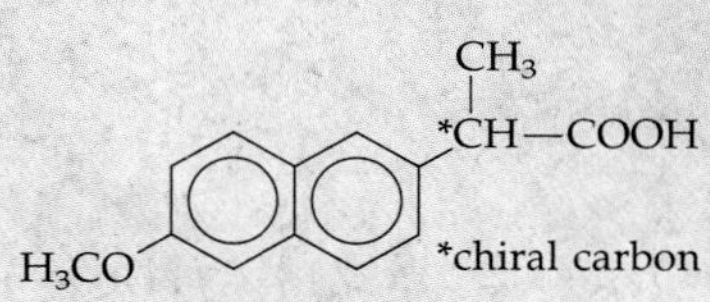

CH_3
$^{*}CH{-}COOH$
H_3CO
*chiral carbon

Annual Sales of Chiral Drugs

Drug	*Use*	*1990 sales*
Chiral antibiotics	antibiotic	$6,360,000
Captopril	antihypertensive	1,501,000
Enalapril	diuretic, antihypertensive	1,482,000
Lovastatin	cholesterol-lowering agent	751,000
Diltiazem	heart disease (calcium blocker)	746,000
Naproxen	anti-inflammatory agent	686,000
Cefadroxil	antibiotic	690,000
Lisinopril	antihypertensive	430,000
Timolol	heart disease (beta blocker)	297,000
Deprenyl	Parkinson's disease	141,000

Chem. and Eng. News, September 28, 1992, page 46.

EXAMPLE 11.5

Chiral Molecules

For each of the following molecules, decide whether the underlined carbon atom is or is not a chiral center.

a. $\underline{C}H_2Cl_2$ b. $H_2N{-}\underline{C}H(CH_3){-}COOH$ c. $Cl{-}\underline{C}H(OH){-}CH_2Cl$

d. $H{-}\underline{C}(=O){-}OH$

SOLUTION Molecules (b) and (c) are chiral. The carbon atoms in (a) and (d) do not have four different groups bonded to them, but the underlined carbon atoms in (b) and (c) do.

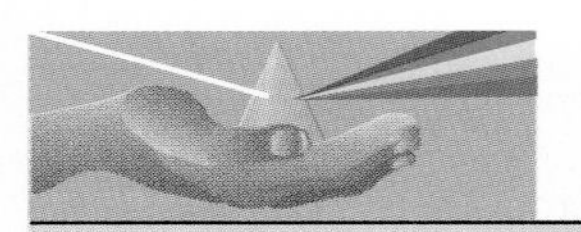

CHEMISTRY YOU CAN DO

Homemade Polarimeter

Optical isomers rotate the plane of polarized light in opposite directions. A device that measures this rotation is called a polarimeter. A polarimeter for demonstration purposes can be made from a flat-sided jar (a "Taster's Choice" coffee jar will work). Obtain two pieces of Polaroid sheet that have diameters at least as large as the width of the jar. These can be lenses from old Polaroid sunglasses. Tape one piece of Polaroid to the side of the jar. Darken the room, and either mount a flashlight or have someone hold it against the taped piece of Polaroid. Rotate the other piece of Polaroid on the opposite face of the jar until a minimum amount of light comes through. Then tape a flat wooden stick such as a tongue depressor to the edge of the untaped Polaroid piece so that the stick points vertically upward.

Position the flashlight so that its beam of light will shine through the taped piece of Polaroid, and turn on the flashlight. Rotate the other piece of Polaroid on the opposite face until a minimum of light gets through. The stick should be pointing upward. Now add water to a level above the taped piece of Polaroid. Rotate the untaped piece until a minimum of light gets through. The stick should be pointing upward again because water molecules do not have "handedness" or optical isomers.

Now make enough sucrose solution to fill the jar to the same level as before. The solution needs to be fairly concentrated; 20 g of sucrose per 100 mL of water will work. Empty the water out of the jar and replace it with the sucrose solution. (Clear Karo corn syrup can also be used.) Now rotate the untaped piece until a minimum of light gets through. What is the position of the wooden stick? Is it to the right or left? The angle of the rotation can be measured with a clear plastic protractor. The amount of the rotation and the direction of rotation are specific physical properties of optical isomers.

References: E. F. Silversmith, *J. Chem. Educ.* Vol. 65, p. 70, 1988; G. F. Hambly, *J. Chem. Educ.*, Vol. 65, p. 623, 1988.

EXERCISE 11.10 • *Chiral Molecules and Enantiomers*

Which of the following molecules is chiral? Draw the enantiomers for any chiral molecule.

a. C with substituents OH, Cl, Cl, CH_3 b. C with substituents OH, H, Cl, CH_3 c. C with substituents H, H_2N, H, COOH

11.6 GEOMETRY AND ISOMERISM OF COORDINATION COMPOUNDS

Common Geometries

The common coordination numbers for a metal or metal ion, represented by M, are two, four, and six. If we consider for a moment only complexes of M with the monodentate ligand L, common coordination complexes would have stoichiometries of ML_2, ML_4, and ML_6. Complexes of ML_2 are always linear, and the silver-containing ion that comes from dissolving AgCl in aqueous ammonia is a good example.

$$AgCl(s) + 2\,NH_3(aq) \longrightarrow \underbrace{[H_3N{-}Ag{-}NH_3]^+(aq)}_{\text{linear } ML_2 \text{ complex}} + Cl^-(aq)$$

If we consider only M—L bond pairs in coordination complexes, the VSEPR theory leads us to expect tetrahedral structures for ML_4 complexes. Indeed, this is observed for many complexes, such as $[CoCl_4]^{2-}$, $Ni(CO)_4$, and $[Zn(H_2O)_4]^{2+}$.

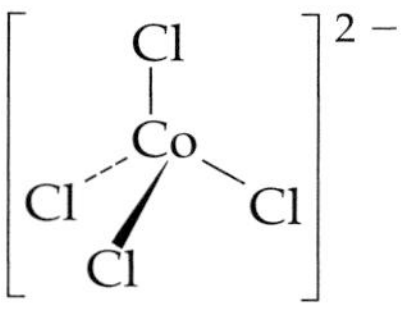

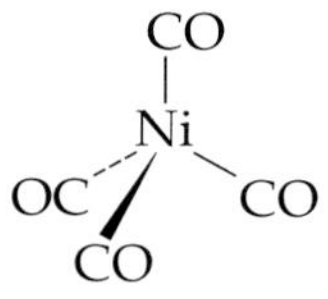

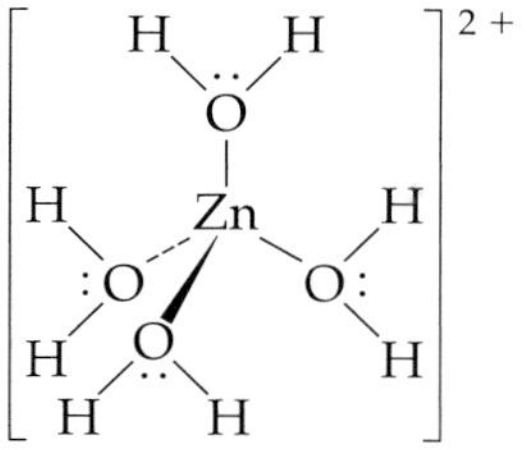

However, the VSEPR theory is rarely reliable for transition metal complexes; for ML_4 complexes, *square planar* geometry is observed more often than tetrahedral geometry. This is particularly true for ML_4 complexes of Pt^{2+} and Pd^{2+}, and sometimes Ni^{2+} and Cu^{2+}.

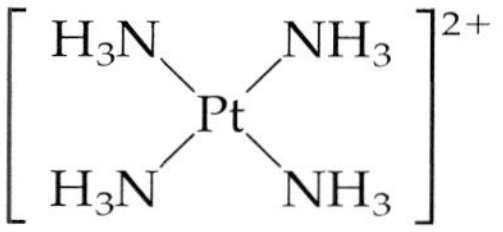

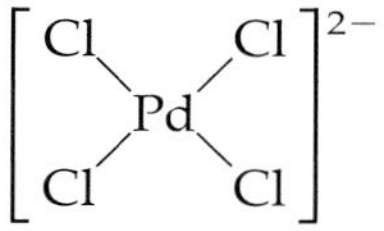

With *very* rare exceptions, ML_6 complexes have the six ligands arranged at the corners of an octahedron, as in the $[Ni(NH_3)_6]^{2+}$ ion.

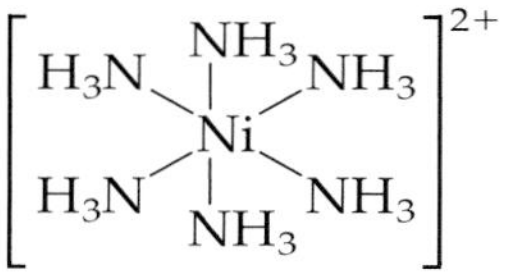

Cis and *Trans* Isomers in Coordination Compounds

To write the formula of a compound that contains a complex ion, add ions of opposite charge equal to the charge on the complex ion. Examples in this section are $K_2[CoCl_4]$, $[Zn(H_2O)_4](NO_3)_2$, $[Ni(NH_3)_6]Cl_2$.

Cis and *trans* isomers are commonly found in coordination compounds with square planar or octahedral geometry. The simplest example is in a square planar complex such as $[Pt(NH_3)_2Cl_2]$. This complex is formed from Pt^{2+}, two NH_3 molecules, and two Cl^- ions. The two Cl^- ions can be adjacent to one another (*cis*) or located on opposite sides of the molecule (*trans*). The *cis* isomer is effective in the treatment of testicular, ovarian, bladder, and osteogenic sarcoma cancers, while the *trans* isomer has no effect (*below, left and center*).

The discovery by Dr. Barnett Rosenberg that $Pt(NH_3)_2Cl_2$, cisplatin, can be used in cancer chemotherapy is described in Chapter 1. It is interesting that only the cis *isomer is physiologically active.*

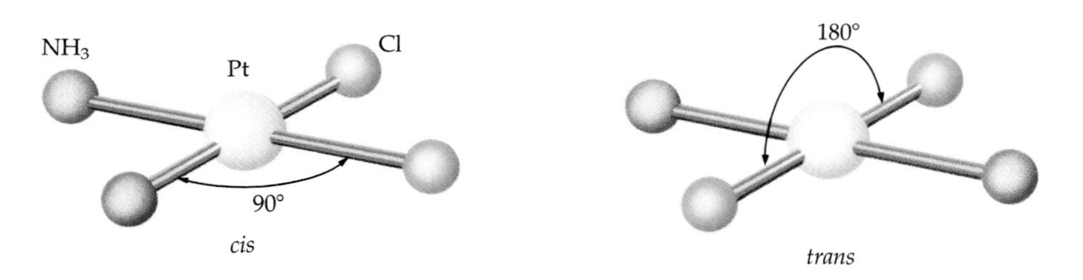

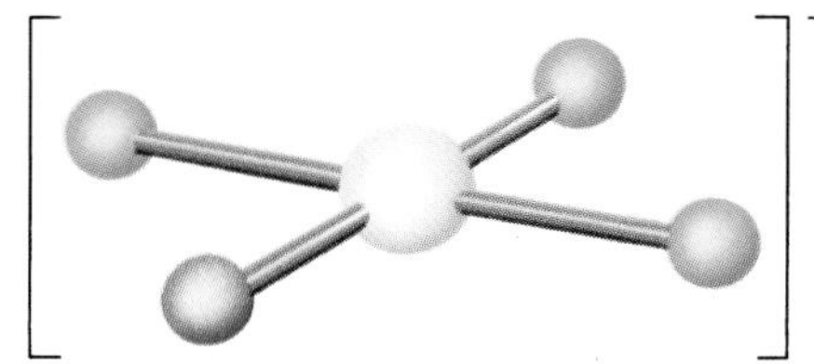

Now consider square planar $[Pt(NH_3)Cl_3]^-$, a complex in which three of the four ligands are the same (*right*).

Cis-trans isomers are not possible here. No matter how you draw the molecule, there are always two Cl^- ions 180° apart and one Cl ion 180° from NH_3. On the other hand, *cis* and *trans* isomers can be constructed for $[Pt(NH_3)_2(Cl)(NO_2)]$.

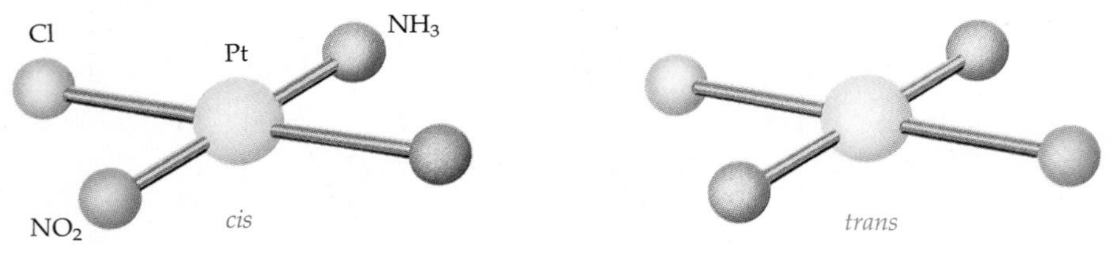

Four-coordinate metal complexes can also have tetrahedral geometry. However, *cis-trans isomerism is not possible for tetrahedral complexes* because it is not possible for two ligands to be "across" the molecule from one another. All possible angles between tetrahedral bonds are 109.5°, so all ligands are effectively adjacent.

Cis-trans isomerism is widely observed for octahedral complexes. As an example consider $[Co(H_2NCH_2CH_2NH_2)_2Cl_2]^+$, an octahedral complex ion with two different kinds of ligands. Here the two Cl^- ions can occupy adjacent or opposite positions of the octahedron. However, the N atoms of ethylenediamine (en) can coordinate only to adjacent *(cis)* positions in the octahedron. The resulting isomers of $[Co(en)_2Cl_2]^+$ are those illustrated in Figure 11.24. The different ways of connecting the ligands lead to different colors; the *cis* isomer is purple, but the *trans* isomer is green.

Enantiomers of Coordination Compounds

The lactic acid molecule can exist in two chiral forms. In coordination chemistry, however, there are no examples of stable complexes with a metal bonded tetrahedrally to four *different types* of ligands, and so enantiomers of

Figure 11.24
The *cis* (left) and *trans* (right) isomers of $[Co(NH_2CH_2CH_2NH_2)_2Cl_2]^+$.

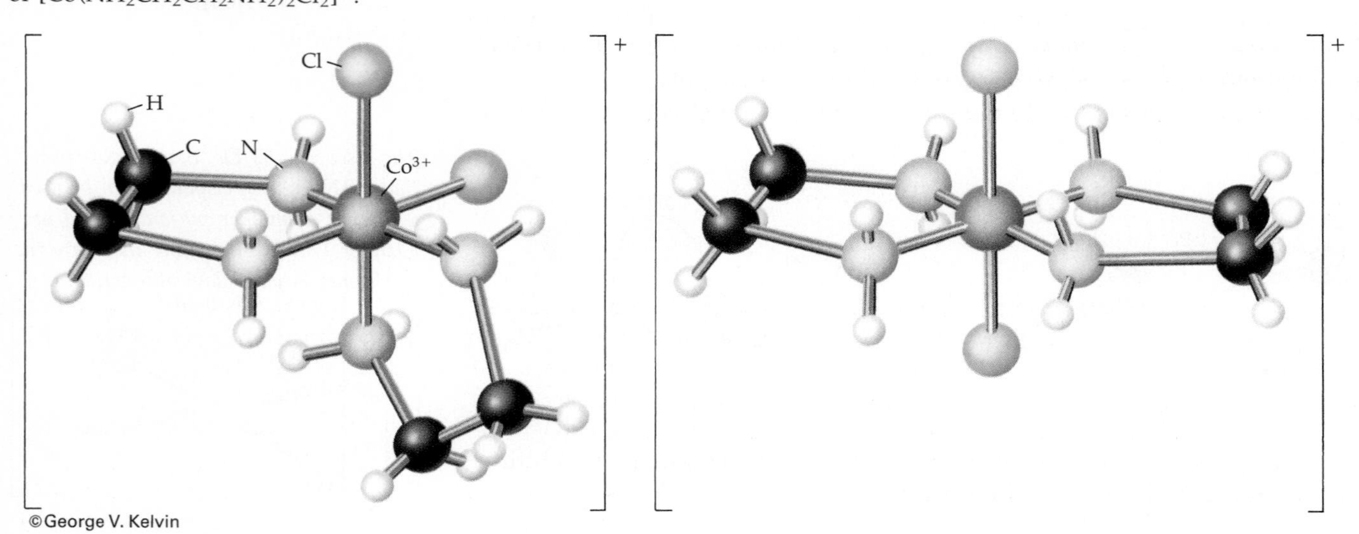

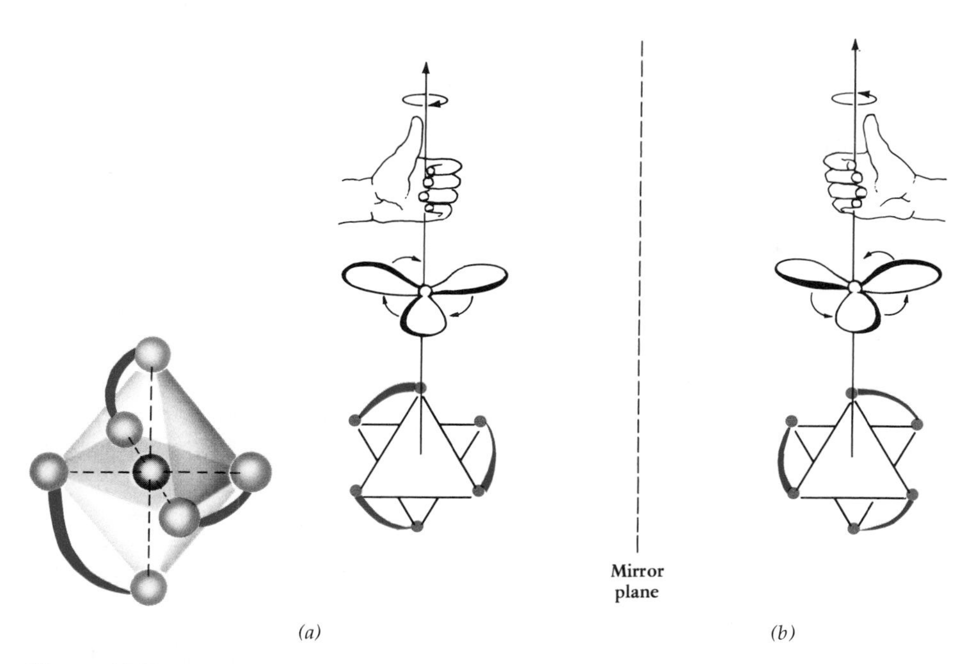

Figure 11.25

Optical isomerism in complexes of the type M(bidentate)$_3$. The three chelate rings are arranged so that the complex looks like a three-bladed propeller. One of the propellers twists clockwise (a) and the other twists counterclockwise (b). The mirror images cannot be superimposed.

tetrahedral coordination complexes need not be considered further. The situation is quite different for octahedral complexes, though, where optical isomerism can arise in a variety of ways. Only one of these ways will be described here. When three bidentate ligands, such as ethylenediamine, bind to a metal ion or atom, three metal-containing rings are formed, as represented by the drawing in Figure 11.25. Two enantiomers result from the two ways of arranging these rings in space. Think of the complex as a three-bladed propeller. The blades can be arranged so that the propeller twists clockwise or counterclockwise. One arrangement is the mirror image of the other, and neither is superimposable on the other.

11.7 MOLECULAR STRUCTURE DETERMINATION BY SPECTROSCOPY

How do we determine the structure of molecules? Many of the methods rely on the interaction of electromagnetic radiation with matter. Probing matter with electromagnetic radiation is called **spectroscopy,** and each area of the electromagnetic spectrum (see Figure 9.1) can be used as the basis for a particular spectroscopic method. Recall from Section 9.1 that electromagnetic radiation is emitted or absorbed in quantized packets of energy called **photons,** and that the energy of the photon is represented by $E = h\nu$ where ν is the frequency of the light. Molecules may absorb several different electro-

X-ray diffraction is used to determine the crystal structure of solids (Section 14.2).

Table **11.5**

Spectroscopic Experiments

Spectral Region	Frequency (s^{-1})	Energy Levels Involved	Information Obtained
Radio waves	10^7–10^9	Nuclear spin states	Electronic structure near the nucleus
Microwave, far infrared	10^9–10^{12}	Rotational	Bond lengths and bond angles
Near infrared	10^{12}–10^{14}	Vibrational	Stiffness of bonds
Visible, ultraviolet	10^{14}–10^{17}	Valence electrons	Electron configuration
X-ray	10^{17}–10^{19}	Core electrons	Core-electron energies

Imagine the matching of frequencies as being similar to jumping rope. To jump a rotating rope, the person's frequency of jumping must match that of the rope's rotation.

magnetic radiation frequencies depending on the energy differences between their allowed energy levels. Each frequency absorbed must provide the exact package of energy needed to lift a molecule from one energy level to the next. For example, ultraviolet and visible spectroscopy are often referred to as **electronic spectroscopy** because the energy of photons in the ultraviolet or visible region matches the energy needed to promote electrons from a lower energy level to a higher one in a molecule.

Infrared Spectroscopy

Infrared spectroscopy uses the interaction of infrared radiation with matter to study molecular structure. It is particularly useful for learning about molecules because the energy of the internal motions of molecules is similar to that of photons whose frequency is in the infrared region. Imagine that covalent bonds between atoms in a molecule are like springs that can only bend or stretch in specified amounts (Figure 11.26). Bending or stretching of the bonds of the water molecule occurs at specific energy levels. The strength of the covalent bonds determines what frequency of infrared light is necessary for changing from one stretching or bending energy level to another. In order to get "signals from within" the molecule, the molecule must be excited from one of these allowed energy states to another by an exact amount of energy.

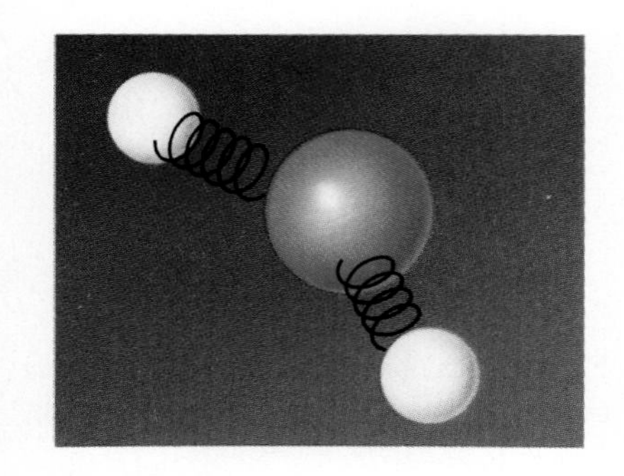

Figure 11.26
Model of the water molecule, with springs for bonds.

For example, a hydrogen chloride molecule (Figure 11.27) vibrates at a specific energy level. Photons with too low or too high an energy do not cause vibration at the next higher energy level, and the radiation passes through the molecule without being absorbed.

Since the covalent bonds in molecules differ in strength and number, the molecular motions and the number of vibrational energy levels vary; hence, the infrared radiation that is absorbed by different molecules differs. As a

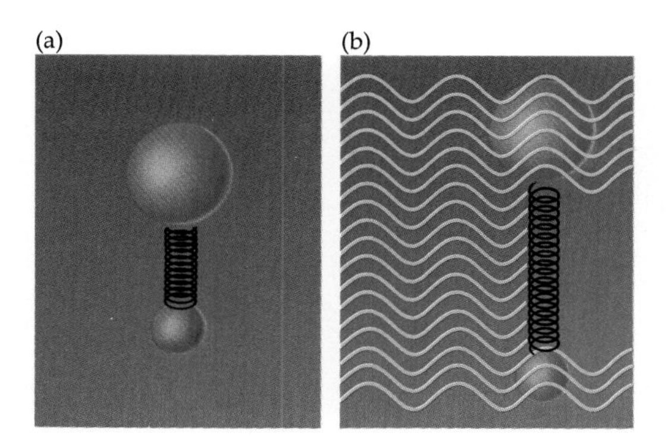

Figure 11.27
Model for vibration of the HCl molecule. (a) The vibration of the HCl molecule is quantized. (b) Electromagnetic radiation at the quantized frequency will be absorbed by the molecule.

result, infrared spectroscopy can be used to learn about the structure of molecules and even to analyze an unknown material by matching its infrared spectrum with that of a known compound. In fact, the infrared frequencies absorbed by a molecule are so characteristic that the infrared spectrum of a molecule is regarded as its *fingerprint*. An example of the use of infrared spectroscopy in the identification of vitamin C is shown in Figure 11.28.

Understanding how molecules absorb infrared light is also important to understanding the "greenhouse effect" described in Section 20.7. Although the primary concern has been with the increasing concentration of CO_2 in the atmosphere from the combustion of fossil fuels, water vapor and trace gases—chlorofluorocarbons (CFCs), N_2O, CH_4, and O_3—also contribute to the effect. The gases of the atmosphere allow the sun's ultraviolet and visible radiation to pass through because they do not absorb energy of these wavelengths (Figure 11.29). Although about half of the visible light striking the earth is reflected back into space, the remainder reaches the earth's surface and causes warming. These warmed surfaces then re-radiate this energy into the atmosphere as heat, that is, as energy in the infrared region of the spectrum. The H_2O and CO_2 molecules in the atmosphere are capable of absorbing infrared radiation, so the atmosphere is warmed. Some of this heat energy is returned to earth and some energizes other molecules. The net effect is an increase in the temperature of the lower atmosphere, similar to the rise in temperature in a greenhouse due to its glass enclosure. Although this may seem entirely detrimental to the people of the earth, it is

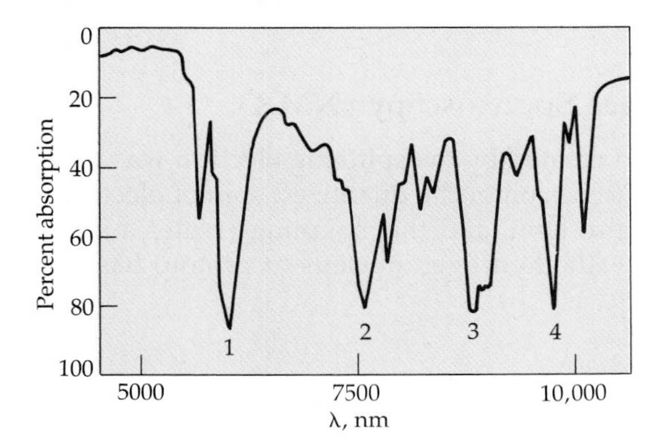

Figure 11.28
Infrared spectrum of vitamin C (ascorbic acid). Strong absorptions at **1** (6100 nm), **2** (7600 nm), **3** (8800 nm), and **4** (9600 nm) can be used to identify vitamin C.

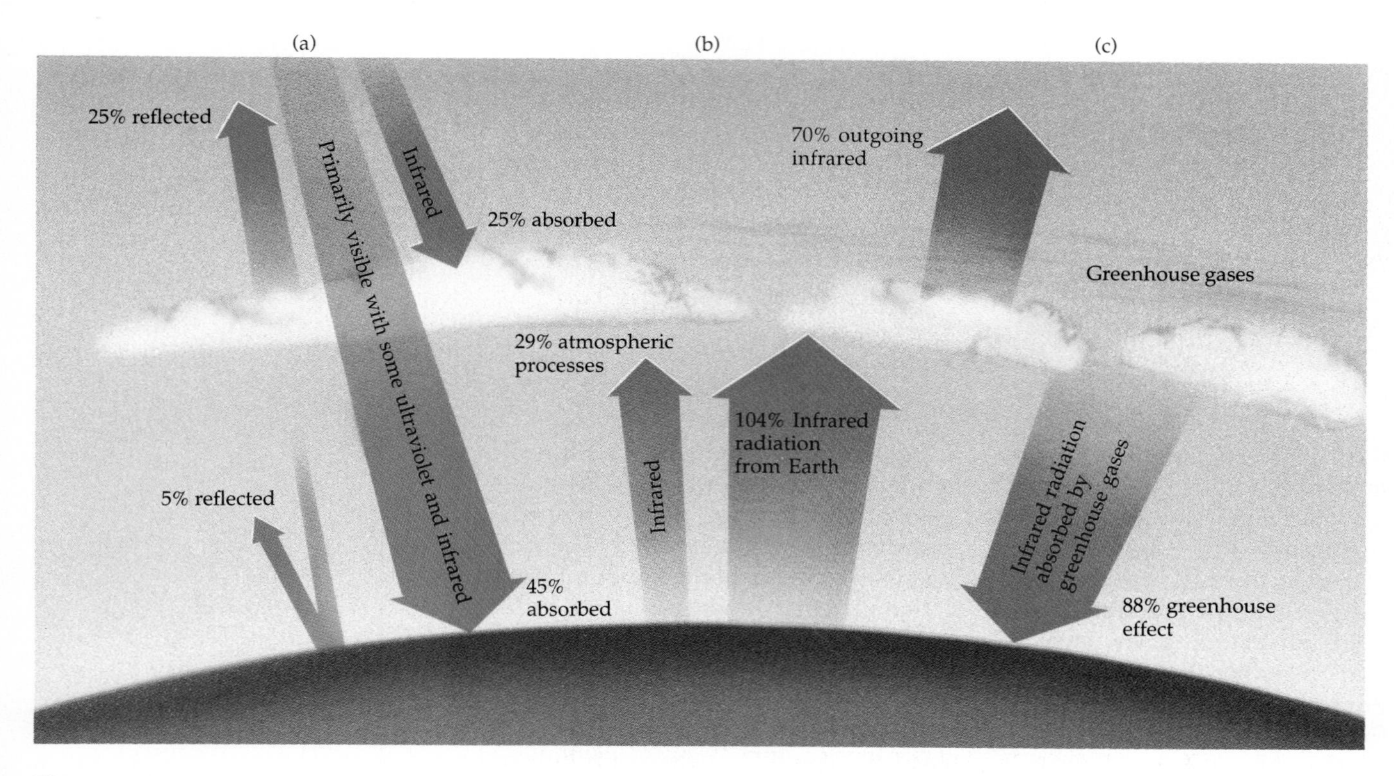

Figure 11.29

(a) About 30% of the incoming solar energy is reflected, either from clouds and particles in the atmosphere or from the earth's surface. Of the 70% that is absorbed, 45% consists of visible light with some ultraviolet radiation, and the other 25% is infrared radiation. (b) The visible and ultraviolet radiation absorbed by the atmosphere and by the surface is re-emitted as infrared (heat) radiation. (c) Gases such as CO_2, H_2O vapor, CFCs, CH_4, and N_2O absorb the infrared radiation and warm the atmosphere. The 70% outgoing infrared radiation in (b) is the sum of the 25% absorbed infrared radiation (a), the 29% infrared radiation from atmospheric processes (b), and the 16% difference between the 104% emitted infrared radiation (b) and the 88% infrared radiation absorbed by greenhouse gases (c).

important not to lose sight of the fact that the insulating effect of water vapor and carbon dioxide is also essential to the existence of life on this planet. Without it, the daily temperature changes would be much more extreme.

Nuclear Magnetic Resonance Spectroscopy (NMR)

In Section 9.3, the magnetic field created by the spinning electron was mentioned in connection with the discussion of the quantized spin of electrons. The nuclei of certain isotopes also spin, and this spinning creates a small magnetic field. For example, ^1H (the hydrogen nucleus or proton) has two

MAGNETIC FIELD DIRECTION ↑

——— $-\frac{1}{2}$	$\xrightarrow{+h\nu}$	—↓— $-\frac{1}{2}$	$\xrightarrow{-h\nu}$	——— $-\frac{1}{2}$
—↑— $+\frac{1}{2}$		——— $+\frac{1}{2}$		—↑— $+\frac{1}{2}$

Figure 11.30

Spinning ^1H nuclei have a slight preference to spin in one direction in the presence of a magnetic field. Energy in the radio frequency range ($h\nu$) causes nuclei to change spin direction. When the spinning ^1H nuclei return to the more stable spin state, $h\nu$ is emitted.

The Chemical World

Signals from Within

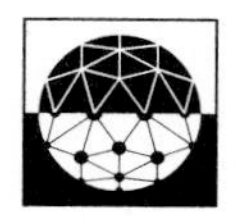

The applications of electromagnetic spectroscopy in our everyday lives and in the work of research chemists are so numerous that one can fill volumes describing them. However, some particularly exciting examples are used in the laboratories of the Drug Enforcement Agency (DEA), and others in the research of the Department of Agriculture devoted to developing pheromones, the sex lures of insects.

In the laboratories of the DEA widespread use is made of infrared spectroscopy. These techniques make it possible not only to identify existing drugs rapidly, but actually to develop a library of important signature spectra, or fingerprints of new drugs, so necessary for future identification. Thus, infrared spectroscopy becomes one of the most powerful weapons in the arsenal for fighting the scourge of drugs.

In the laboratories of the Department of Agriculture infrared spectroscopy has made it possible to identify readily the composition of the pheromones, or sex attractants, upon which insects depend for survival as a species. Once a pheromone is identified, it becomes possible to synthesize it and to use it to keep harmful insects from proliferating.

The World of Chemistry (Program 10) "Signals from Within."

---

quantized spin states (Figure 11.30). In the absence of a magnetic field, the two spin states have the same energy. When an external magnetic field is applied, there is a slight preference for the hydrogen nucleus to spin so that the magnetic field of the nucleus is aligned with the external field. The energy difference, ΔE, between the two nuclear spin states is small enough that radiation in the radio frequency region of the electromagnetic spectrum (Figure 9.1) can change the direction of spin.

Picture the aligned hydrogen nuclei absorbing a particular radio frequency that changes their spin to the less stable direction. When the nuclei return to the more stable spin direction, the same radio frequency is emitted and can be measured with a radio receiver. A schematic diagram of the important components of an NMR spectrometer is shown in Figure 11.31.

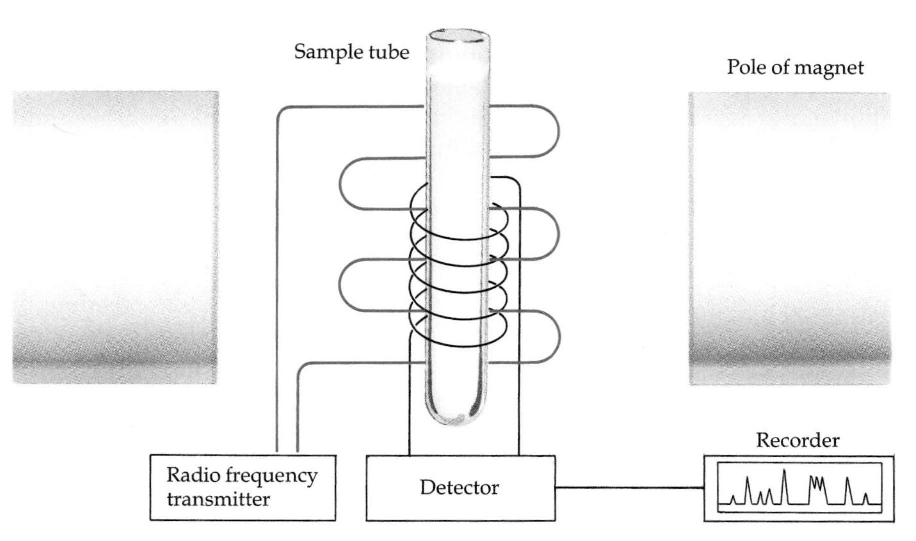

(a)

(b)

Figure 11.31
(a) Schematic diagram of an NMR spectrometer. (b) A modern NMR spectrometer is a highly automated instrument that is controlled by a built-in computer. *(b,* General Electric)

Edward Purcell. *(AIP Meggers Gallery of Nobel Laureates)*

Felix Bloch. *(AIP Emilio Segre Visual Archives)*

Felix Bloch and Edward Purcell were awarded the Nobel prize in 1952 for their discovery of nuclear magnetic resonance.

NMR is used extensively by chemists because the radio frequency absorbed and then emitted depends on the chemical environment of the hydrogen atoms in the sample. Chemists can deduce the kinds of atoms bonded to hydrogen as well as the number of hydrogen atoms present. The study of hydrogen atoms with NMR (known as proton nuclear magnetic resonance, ^{1}H NMR) has quickly developed into an indispensable structural and analytical tool, particularly for the organic chemists.

Magnetic Resonance Imaging (MRI)

The use of NMR in medical imaging is now referred to as Magnetic Resonance Imaging or MRI.

Since our bodies are 70% water, and the protons in water give a strong NMR signal, ^{1}H is the most logical candidate for the application of NMR to medical imaging. The first use of NMR for this purpose was not reported until 1973. The delay was related to the technical difficulties associated with getting a uniform magnetic field with a diameter large enough to enclose a patient. In addition, advances in computer technology for analysis of data and construction of an image from these data were needed.

NMR imaging in medicine is based on the time it takes for the protons (hydrogen nuclei) in the unstable high-energy nuclear spin position to "relax" or return to the low-energy nuclear spin position. These times are different for protons in fat, muscle, blood, and bone, and these differences provide the contrast in the magnetic resonance image.

In **magnetic resonance imaging** (MRI), the patient is placed in an opening of a large magnet (Figures 11.32 and 11.33). The magnetic field aligns the magnetic spin of the protons (as well as those of other magnetic nuclei). A radio frequency transmitter coil is placed in position near the region of the body to be examined. The radio frequency energy absorbed by the spinning

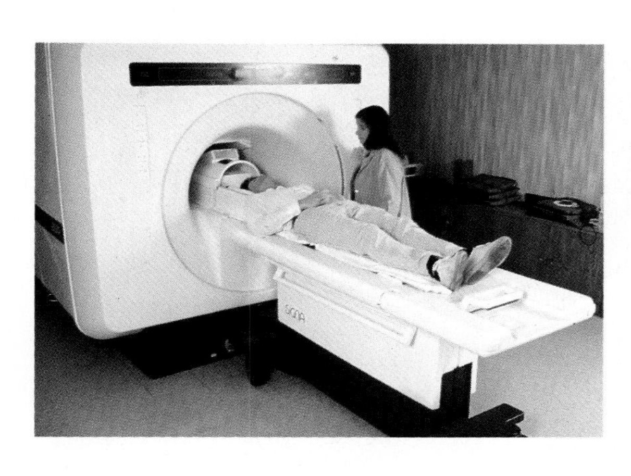

Figure 11.32
A magnetic resonance imaging (MRI) machine. The patient is placed on the platform and rolled forward into the magnet opening. *(Peter Arnold)*

hydrogen nuclei causes the aligned nuclei to flip to the less stable, high-frequency spin direction. When the nuclei flip back to the more stable spin state, a radio frequency signal is emitted. The intensity of the emitted signal is related to the density of hydrogen nuclei in the region being examined, and the time it takes for the signal to be emitted (relaxation time) is related to the type of tissue. The emitted radio frequency signal is received by a

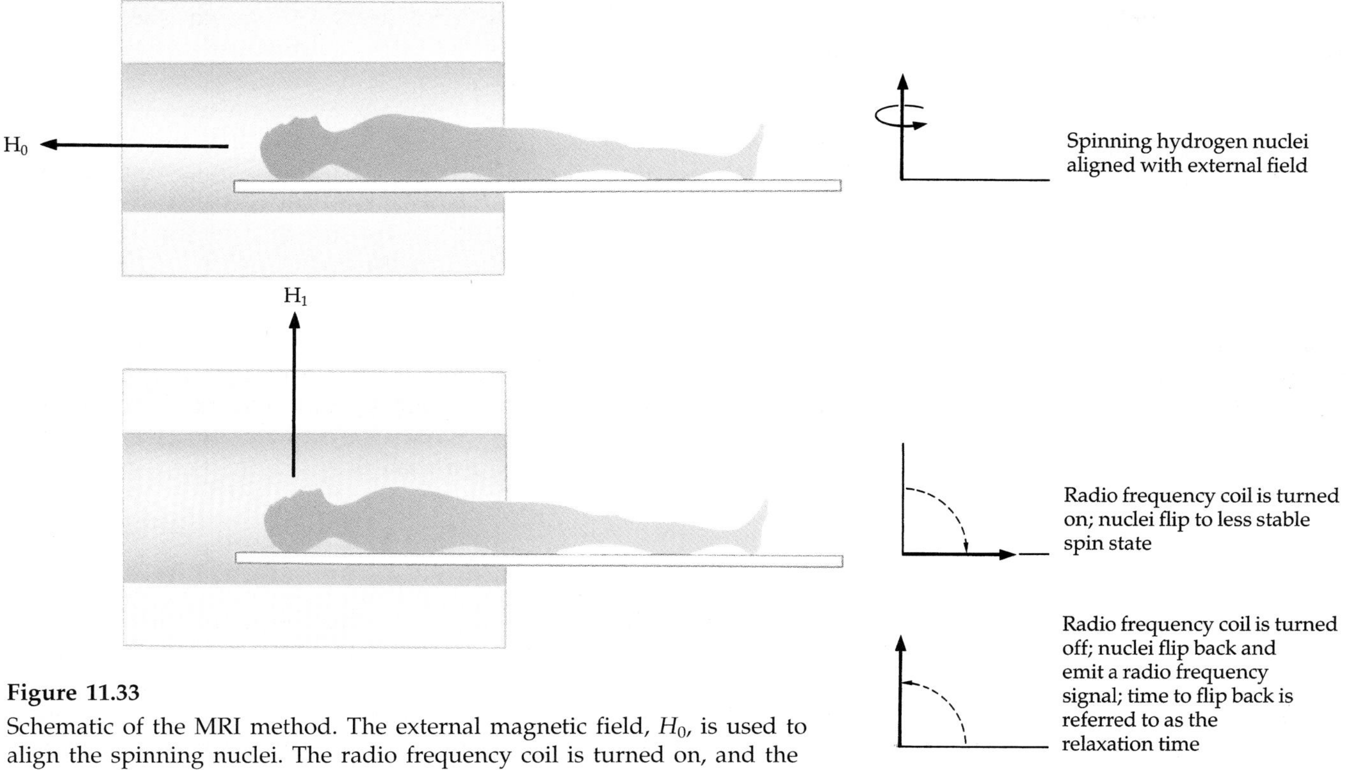

Figure 11.33
Schematic of the MRI method. The external magnetic field, H_0, is used to align the spinning nuclei. The radio frequency coil is turned on, and the resulting radio frequency (H_1) flips the nuclei to the less stable spin state. A radio frequency signal is generated when the nuclei flip back to the more stable spin state.

Potential New Drug Isolated from Frog Skin

SOME SCIENTISTS BECOME TRAVelers in unknown lands in their search for new chemicals of medical value. Consider the adventures of Dr. John W. Daly of the National Institutes of Health (NIH). Together with a zoologist from the American Museum of Natural History in New York, he has traveled to tropical rain forests throughout South and Central America collecting dozens of new species of frogs. Frogs? Yes, Daly and his co-workers have identified more than 225 chemical compounds in extracts from the skins of frogs. Their studies have yielded information about the ancestry of frogs as well as about local anesthetics, since several of the compounds act in this manner.

In 1978 Daly traveled to Ecuador with the goal of collecting a particular species of frog. He needed enough frogs to provide a bigger sample of a compound he had discovered the year before. In mice, the compound acted like morphine, but did so by a clearly different mechanism.

Armed with 60 mg of a mixture containing frog skin extract, he set out with the patience and determination typical of many scientific investigators. First, he separated the active compound from others, using a technique based on differences in how compounds in a mixture are adsorbed from a solution onto a solid (liquid chromatography). Then, using a mass spectrometer (Figure 3.11), he found that his compound had a molecular weight of either 208 or 210. Next, by breaking down the large molecule he was able to identify one fragment that contained six C atoms, ten H atoms, and one N atom, and a second fragment that contained five C atoms, one N atom, three H atoms, and one Cl atom.

By then, his supply of the compound was down to 500 μg. Afraid of destroying what was left, he stored the tiny remaining quantity and turned his attention elsewhere—for nine years. Meanwhile, advances were made in instruments used for chemical analysis. Finally, Daly dared to take up the abandoned sample again, encouraged by improvements in the sensitivity of infrared and NMR spectrometers. A sensitive infrared spectrometer allowed identification of one fragment—a six-membered ring containing one N atom and five C atoms, with one Cl atom bonded to one of the C atoms in the ring. At last, by making a minor chemical modification of the other fragment, it too was identified by using ^{1}H NMR. The compound is named epibatidine, after the Ecuadoran poison frog, *Epipedobates tricolor*, that was the source of the skin extract.

Red-eyed tree frog. (Courtesy National Aquarium in Baltimore/George Grall)

Cl N NH

epibatidine

The challenge of the analysis has now been replaced with a new challenge—how to synthesize epibatidine in the lab to provide material for further study of its remarkable properties as a painkiller.

References: J. W. Daly et al., *J. Amer. Chem. Soc.*, Vol. 114, p. 3475, 1992; E. Pennisi, *Science News*, July 18, 1992, Vol. 142, p. 40.

radio receiver coil, which then sends it to a computer for mathematical construction of the image.

MRI can readily distinguish brain tumors from normal brain tissue, not only on the basis of altered anatomy but also on the basis of high water content, which may be caused by hypervascularity (abnormal growth of blood vessels) or reactive edema (excessive absorption of water by the cells). Figure 11.34 shows an MRI scan of a normal human brain.

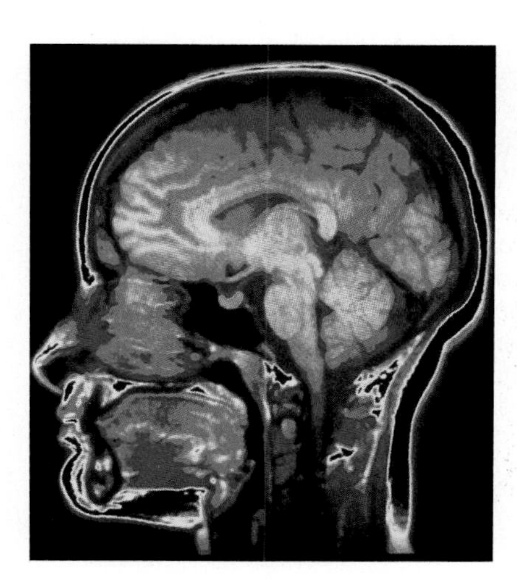

Figure 11.34
Computer-enhanced MRI scan of a normal human brain with the pituitary gland highlighted. *(Scott Camazine/Photo Researchers, Inc.)*

IN CLOSING

Having studied this chapter, you should be able to

- predict shapes of molecules and polyatomic ions by using the VSEPR model (Section 11.1).
- predict the polarities of molecules (Section 11.2).
- explain the importance of the tetrahedral shape in the chemistry of carbon, silicon, and phosphorus (Section 11.3).
- define and identify examples of the different types of isomers (Sections 11.4, 11.5, and 11.6).
- list the common geometries of coordination compounds (Section 11.6).
- describe the importance of spectroscopic methods in probing the structures of molecules (Section 11.7).
- give some examples of the importance of molecular shape in determining the properties of substances.

STUDY QUESTIONS

1. What is the VSEPR model? What is the physical basis of the model?
2. What is the difference between the electron-pair geometry and the molecular geometry of a molecule? Use the water molecule as an example in your discussion.
3. Designate the electron-pair geometry for each case of from two to six electron pairs around a central atom.
4. What are the molecular geometries for each of the following?

   $$\mathrm{H{-}\underset{\cdot\cdot}{\ddot{X}}\!:} \qquad \mathrm{H{-}\underset{\cdot\cdot}{\ddot{X}}{-}H} \qquad \mathrm{H{-}\underset{|\atop H}{\ddot{X}}{-}H} \qquad \mathrm{H{-}\underset{|\atop H}{\overset{H\atop |}{X}}{-}H}$$

   Give the H—X—H bond angle for each of the last three.
5. If you have three electron pairs around a central atom, how can you have a triangular planar molecule? An angular molecule? What bond angles are predicted in each case?
6. Draw a triangular bipyramid of electron pairs. Designate the axial and equatorial pairs. Are there axial and equatorial pairs in an octahedron?
7. Use VSEPR to explain why ethylene is a planar molecule.
8. How can a molecule with polar bonds be nonpolar? Give an example.

9. Why are *trans*-fatty acids a health concern?
10. Give examples that illustrate the importance of the tetrahedral shape to a better understanding of chemistry.
11. Why is *cis-trans* isomerism not possible for alkynes?
12. Give an example of a coordination compound that has (a) a linear shape, (b) a tetrahedral shape, (c) a square planar shape, and (d) an octahedral shape.
13. Why are zeolites important commercially?
14. In the Chemistry You Can Do experiment on optical isomers, why does a sugar solution rotate a plane of polarized light?
15. Explain the current interest in chiral drugs.
16. Use butene alkenes to explain the difference between constitutional isomers and *cis-trans* isomers.
17. What is a racemic mixture?
18. How do CO_2, CFCs, and other trace gases cause global warming?
19. Explain why the infrared spectrum of a molecule is referred to as its "fingerprint."
20. For infrared energy to be absorbed by a molecule, what frequency of motion must the molecule have?
21. How does MRI work?
22. One of the three isomers of dichlorobenzene, $C_6H_4Cl_2$, has a dipole moment of zero. Draw the structure of the isomer and explain your choice.

Molecular Shape

23. Draw the Lewis structure for each of the following molecules or ions. Describe the electron-pair geometry and the molecular geometry.
(a) NH_2Cl (b) OCl_2 (c) SCN^- (d) HOF

24. Determine the electron-pair geometry and molecular geometry for each of the following.
(a) ClF_2^+ (b) $SnCl_3^-$ (c) PO_4^{3-} (d) CS_2

25. In each of the following molecules or ions, two oxygen atoms are attached to a central atom. Draw the Lewis structure for each one and then describe the electron-pair geometry and the molecular geometry. Comment on similarities and differences in the series.
(a) CO_2 (b) NO_2^- (c) SO_2
(d) O_3 (e) ClO_2^-

26. In each of the following molecules or ions, three oxygen atoms are attached to a central atom. Draw the Lewis structure for each one and then describe the electron-pair geometry and the molecular geometry. Comment on similarities and differences in the series.
(a) BO_3^{3-} (b) CO_3^{2-} (c) SO_3^{2-} (d) ClO_3^-

27. The following are examples of molecules and ions that do not obey the octet rule. After drawing the Lewis structure, describe the electron-pair geometry and the molecular geometry.
(a) ClF_2^- (b) ClF_3 (c) ClF_4^- (d) ClF_5

28. The following are examples of molecules and ions that do not obey the octet rule. After drawing the Lewis structure, describe the electron-pair geometry and the molecular geometry.
(a) SiF_6^{2-} (b) SF_4 (c) PF_5 (d) XeF_4

29. Give the approximate values for the indicated bond angles.
(a) O—S—O angle in SO_2
(b) F—B—F angle in BF_3
(c) H—O—N(=O)—O (angles 1 and 2 indicated)
(d) H—C(H)=C(H)—C≡N: (angles 1 and 2 indicated)

30. Give approximate values for the indicated bond angles.
(a) Cl—S—Cl angle in SCl_2
(b) N—N—O angle in N_2O
(c) H_2N—C(=O)—NH_2 (angles 1 and 2 indicated)
(d) H—C(H)(H)—O—H (angles 1 and 2 indicated)

31. Give approximate values for the indicated bond angles.
(a) F—Se—F angles in SeF_4
(b) O—S—F angles in SOF_4 (the O atom is in an equatorial position)
(c) F—Br—F angles in BrF_5

32. Give approximate values for the indicated bond angles.
(a) F—S—F angles in SF_6
(b) F—Xe—F angle in XeF_2
(c) F—Cl—F angle in ClF_2^-

33. Which would have the greater O—N—O bond angle, NO_2^- or NO_2^+? Explain your answer.

34. Compare the F—Cl—F angles in ClF_2^+ and ClF_2^-. From Lewis structures, determine the approximate bond angle in each ion. Explain which ion has the greater angle and why.

Molecular Polarity

35. Consider the following molecules:
H_2O NH_3 CO_2 ClF CCl_4
(a) Which compound has the most polar bonds?
(b) Which compounds in the list are *not* polar?
(c) Which atom in ClF is more negatively charged?

36. Consider the following molecules:
CH_4 NCl_3 BF_3 CS_2
(a) Which compound has the most polar bonds?
(b) Which compounds in the list are *not* polar?

37. Which of the following molecules is (are) polar? For each polar molecule, what is the direction of polarity, that is, which is the negative and which is the positive end of the molecule?
(a) CO_2 (b) HBF_2 (c) CH_3Cl (d) SO_3

38. Which of the following molecules is (are) *not* polar? Which molecule has the most polar bonds?
(a) CO (b) PCl_3 (c) BCl_3
(d) GeH_4 (e) CF_4

Constitutional Isomers of Organic Compounds

39. Draw condensed structural formulas for all five constitutional isomers of C_6H_{14}.

40. Draw condensed structural formulas for all constitutional isomers of C_4H_9Cl.

41. Draw the condensed structural formulas for the following hydrocarbons and indicate which one is not a constitutional isomer of the others.
(a) 2-pentene (c) 2-methyl-2-butene
(b) 2-methyl-1-butene (d) 2-methyl-1-pentene

42. Draw the condensed structural formulas for the following hydrocarbons and indicate which one is not a constitutional isomer of the others.
(a) 1-pentene (c) 2-pentene
(b) 2,2-dimethylpropane (d) 3-methyl-1-butene

43. Draw structural formulas for the following:
(a) 1,3,5-trimethylbenzene (c) 1,2,4-tribomobenzene
(b) 2,3-dichlorotoluene (d) 1,4-dimethylbenzene

44. Draw structural formulas for the following:
(a) 2,4-dichlorotoluene (c) 1,2,5-trichlorobenzene
(b) 3,5-dichlorotoluene (d) 1,2-dimethylbenzene

Cis-Trans *Isomers of Organic Compounds*

45. Draw the *cis* and *trans* isomers of 2-pentene.

46. In each case tell whether *cis* and *trans* isomers exist. If they do, draw the two and label them *cis* and *trans*.
(a) 1,1-dibromoethene (c) 2-butene
(b) 3-hexene (d) 1-butene

Optical Isomers of Organic Compounds

47. Circle the chiral carbon atoms, if any, in the following molecules.

(a)
```
       H  H
       |  |
HO—C—C—C—C—H
   ‖  |  |  ‖
   O  OH OH O
```

(b)
```
CH3—C—C—OH
    ‖  ‖
    O  O
```

(c)
```
            H
            |
CH3—CH2—C—COOH
            |
           NH2
```

48. Underline the chiral carbon atoms, if any, in the following molecules.

(a)
```
       H  H
       |  |
CH3—C—C—H
       |  |
       OH OH
```

(b)
```
H—C=C—CH2—OH
   |  |
   H  H
```

(c)
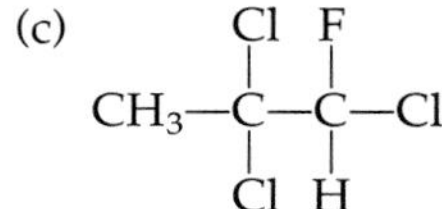

49. Draw the structures of the following molecules. If they contain chiral carbon atoms, underline those atoms.
(a) 2-chloro-2-methylpropane
(b) 1-chloro-2-methylpropane
(c) 2-bromopentane
(d) 3-bromopentane

50. Draw the structures of the following molecules. If they contain chiral carbon atoms, underline those atoms.
(a) 2-bromo-2-chlorobutane
(b) 2-methylpropane
(c) 4-bromo-2-pentene

Geometry and Isomerism of Coordination Compounds

51. Draw the *cis* and *trans* isomers of $[Pd(NH_3)_4Cl_2]$.

52. Which of the following complexes can have *cis-trans* isomers? If isomers are possible, draw the structures of the isomers and label them as *cis* or *trans.*
(a) $[Co(H_2O)_4Cl_2]^+$ (b) $[Pt(NH_3)Br_3]^-$
(c) $[Co(en)_2(NH_3)Cl]^{2+}$

53. Draw all of the isomers (*cis-trans* and optical) that are possible for $[Co(en)_2Cl_2]^+$.

54. Draw all the isomers (*cis-trans* and optical) for the complex ion $[Co(en)(NH_3)_2(H_2O)Cl]^{2+}$.

General Questions

55. The formula for nitryl chloride is NO_2Cl. Draw the Lewis structure for the molecule, including all resonance structures. Describe the electron-pair and molecular geometries, and give values for all bond angles.

56. Vanillin is the flavoring agent in vanilla extract and in vanilla ice cream. Its structure is

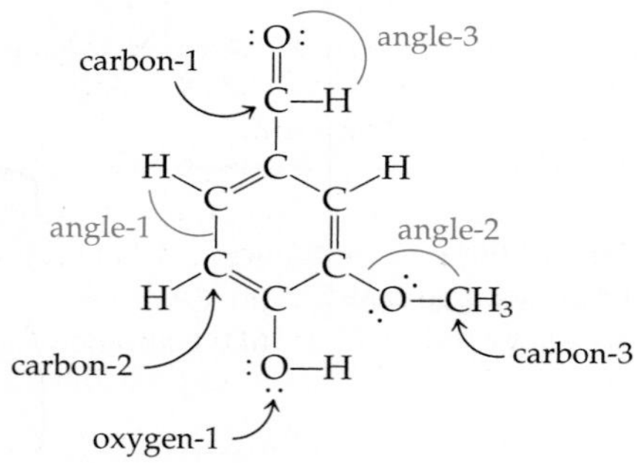

(a) Give values for the three bond angles indicated.
(b) Indicate the most polar bond in the molecule.
(c) Circle the shortest carbon-oxygen bond.

57. Draw the *cis* and *trans* isomers of 1,2-dibromoethene. Does either of these have a dipole moment? If so, give the direction of the dipole moment.

58. Each of the following molecules or ions has fluorine atoms attached to an atom from Groups 1A or 3A to 6A. Draw the Lewis structure for each one and then describe the electron-pair geometry and the molecular geometry. Comment on similarities and differences in the series.
(a) BF_3 (b) CF_4 (c) NF_3 (d) OF_2
(e) HF

59. In 1962 Watson and Crick received the Nobel Prize for their simple but elegant model of the "heredity molecule" DNA. The key to their structure (the "double helix") was an understanding of the geometry and bonding capabilities of nitrogen-containing bases such as the thymine molecule below. (a) Give approximate values for the indicated bond angles. (b) Which are the most polar bonds in the molecule?

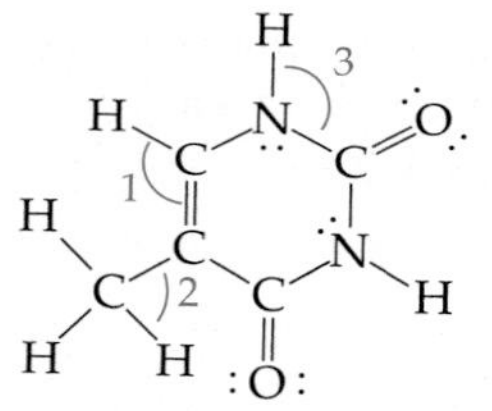

60. The dipole moment of the HCl molecule is 3.43×10^{-30} C m and the bond length is 127.4 pm; the dipole moment of HF is 6.37×10^{-30} C m, with bond length of 91.68 pm. Use the definition of dipole moment as a product of partial charge on each atom times distance of separation (see Section 11.2) to calculate the quantity of charge in coulombs that is separated by the bond length in each dipolar molecule. Use your result to show that fluorine is more electronegative than chlorine.

61. In the gas phase positive and negative ions form ion pairs that are like molecules. An example is KF, which is found to have a dipole moment of 28.7×10^{-30} C m and a distance of separation between the two ions of 217.2 pm. Use this information and the definition of dipole moment to calculate the partial charge on each atom. Compare your result with the expected charge, which is the charge on an electron, -1.602×10^{-19} C. Based on your result, is KF really completely ionic?

62. Sketch the geometry of a molecule or ion containing carbon in which the angle between two atoms bonded to carbon is
(a) exactly 109.5°
(b) slightly different from 109.5°
(c) exactly 120°
(d) exactly 180°

63. Draw structural formulas for two alkenes with formula C_5H_{10} and a straight chain of five carbon atoms. Are these the only two structures that meet these specifications?

64. Judging from the number of carbon and hydrogen atoms in their formulas, which of these formulas represent alkanes? Which are probably aromatic? Which fall into neither category? (It may help to try to draw structural formulas.)
(a) C_8H_{10} (b) $C_{10}H_8$ (c) C_6H_{12}
(d) C_6H_{14} (e) C_8H_{18} (f) C_6H_{10}

65. Benzene is an aromatic hydrocarbon that consists of a ring of six carbon atoms and has the formula C_6H_6. Cyclohexane is a hydrocarbon that consists of a ring of six carbon atoms and has the formula C_6H_{12}. Draw a structural formula for each molecule and contrast the shapes of the molecules. Describe each shape in words. Why do the two molecules have different shapes?

Summary Problem

66. The following compound is commonly called acetylacetone. As shown it exists in two forms, one called the *enol* form and the other called the *keto* form.

$$H_3C-C(OH)=C(H)-C(=O)-CH_3 \rightleftharpoons H_3C-C(=O)-C(H)(H)-C(=O)-CH_3$$

enol form *keto* form

While in the *enol* form, the molecule can lose H^+ from the —OH group to form an anion. One of the most interesting aspects of this anion (sometimes called the *acac* ion) is that one or more of them can react with a transition metal cation to give very stable, highly colored compounds (see photo).

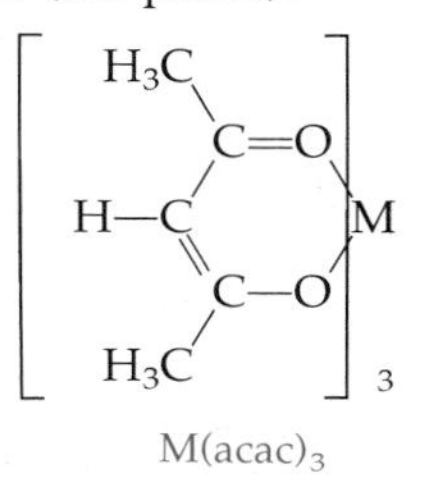

$M(acac)_3$

Three complexes of the type $M(acac)_3$ where M is (*left to right*) Co, Cr, and Fe. *(C.D. Winters)*

(a) Using bond energies, calculate the enthalpy change for the *enol* → *keto* change. Is the reaction predicted to be exo- or endothermic?
(b) What are the electron pair and "molecular" geometries around each C atom in the *keto* and *enol* forms? What changes (if any) occur when the *keto* form changes to the *enol* form?
(c) If you wanted to prepare 15.0 g of deep red $Cr(acac)_3$ using the following reaction,

$$CrCl_3 + 3\,H_3C-C(OH)=CH-C(O)-CH_3 + 3\,NaOH \longrightarrow Cr(acac)_3 + 3\,H_2O + 3\,NaCl$$

how many grams of each of the reactants would you need?

Using Computers and Videodiscs

67. Use *KC? Discoverer* and the *Periodic Table Videodisc* to examine the reaction of red phosphorus with oxygen in air (frame 12319).
(a) Write a balanced equation for the reaction of P and O_2 to give P_4O_{10}.
(b) Write a balanced equation for the reaction of P_4O_{10} with H_2O to give H_3PO_4.
(c) Examine the reactions of the other elements of Group 5A with oxygen in air and comment on similarities and differences in their reactivity.

68. Use the *Chemical Demonstrations* videodisc (Side 1 of Disc 1, frame 26230) to view the explosive reaction of carbon disulfide and nitrogen monoxide (commonly called nitric oxide).
(a) Draw Lewis dot structures for carbon disulfide and nitrogen monoxide. What is unique about nitrogen monoxide?
(b) What is the molecular geometry of CS_2?
(c) The reaction produces carbon dioxide and monoxide, nitrogen, sulfur, and sulfur dioxide. Which of these products are gases? Which are solids?

Liquid nitrogen
(C.D. Winters)

Gases, Intermolecular Forces, and Liquids

CHAPTER

12

A great many chemical reactions take place in the gas phase, including combustion of gasoline in an automobile engine, burning of natural gas in a home furnace, and formation of air pollutants above a city. When gases react, the volumes consumed or produced are small whole number multiples of each other, and volumes of different gases respond in nearly the same way to changes in temperature or pressure. The kinetic molecular theory explains these general properties by saying that gas molecules are relatively far apart, move rapidly and randomly, and collide with the walls of a container, exerting pressure. Molecules that are far apart have very little influence on each other, and so it matters little what kind of molecules make up a gas—the gas behavior will be similar.

At high pressures, however, gas molecules are much closer together, and at low temperatures the average energies of gas molecules become less than the energy required to overcome attractive forces among molecules. If the temperature is lowered (or the pressure raised) enough, most gases will liquefy. Liquid phases of different substances have much less in common than the gaseous forms of the same substances. In a liquid molecules are nearly touching as they roll and tumble past one another, and so forces between molecules are very important. Understanding how intermolecular forces are related to molecular structure allows properties of liquids to be predicted.

Early chemists studied gases and their chemistry extensively. They carried out reactions that generated gases, bubbled the gases through water into glass containers, transferred the gases to animal bladders for storage, and then mixed gases together to see whether they would react and, if so, how much of each gas would be consumed or formed. They discovered that gases have a great many properties in common—much more so than liquids or solids. All gases are transparent, although some are colored. (F_2 is light yellow, Cl_2 is greenish yellow, Br_2 and NO_2 are reddish brown, and I_2 is violet.) Gases are also mobile—they expand to fill any space available, they mix together in any proportions, and they can escape through very tiny holes in their containers. The volume of the same sample of gas can easily be changed by changing temperature or pressure or both—that is, gas densities are quite variable, depending on the conditions of measurement. At first this might seem to make the study of gases more complicated, but it really does not. The way that gas volume depends on temperature and pressure is the same for all gases, and so it is very easy to take it into account. Before dealing with reactions among gases, then, it is useful to consider how gases respond to changes in pressure and temperature.

See pages 14 and 15 for photos of some of the colored gases.

Because gases can easily pass through tiny holes or an incompletely closed valve, leaks from gas pipes are common. Therefore, a substance that provides a tell-tale odor is added to natural gas before it is pumped into mains.

12.1 GASES AND THEIR PROPERTIES

Gases Exert Pressure

The tightness of a balloon filled with air indicates that the air inside exerts pressure. If too much gas is forced into a balloon, that pressure bursts the balloon and the gas rapidly escapes (Figure 12.1). The balloon's tightness is caused by gas molecules striking its inner surface. Each collision of a gas molecule with the balloon surface exerts a force on the surface. The force per

Pressure units
1 atm = 760 mm Hg (exact)
= 760 torr (after Torricelli)
= 101.3 kPa
= 1.013 bar
= 14.7 lb/sq inch (psi)

Figure 12.1
Gas under pressure escaping. This famous photo shows what happens the instant a bullet pierces a balloon filled with a gas under pressure. Suddenly the gas molecules escape. *(© The Harold E. Edgerton 1992 Trust)*

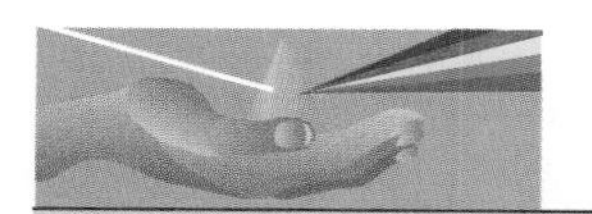

CHEMISTRY YOU CAN DO

Cartesian Diver

Here is a simple device that is easy to make, fun to play with, and illustrates some of the properties of gases and liquids. Obtain the following items:

- An empty 1- or 2-L transparent plastic soft drink bottle, with cap
- A plastic soda straw, preferably transparent
- A large water glass or a bowl
- Several bobby pins or paper clips that can clip tightly to the end of the straw

Work at a sink and have some towels available—you will almost certainly spill some water! Rinse out the bottle and remove as much of the label as you can; you need to be able to see what happens inside the bottle. Fill both the bottle and the glass with hot water to about an inch from the top. Fold the soda straw in half and fasten the two open ends together with a bobby pin or paper clip. Immerse the straw in the water in the glass with the open ends up; shake or tap the straw until some water flows in. Attach two more bobby pins to the open ends of the straw. Now turn the straw so that the open ends point down and see if there is enough air left in it to float it. You want to make it just barely float on the water in the glass. If the straw sinks, raise it just above the water, let a drop or two of water out, and try again. If it floats more than a quarter of an inch above the water, invert it, let more water flow in, and try again.

Once you have the straw barely floating, you have to get it into the bottle without losing any water from the straw. While it is still under water, turn the straw so that the open ends are up; then remove it from the glass. Quickly and smoothly, so as not to shake any water out, invert the straw into the soda bottle. If you lose a little water and the straw floats more than a quarter of an inch out of the water, squeeze the sides of the bottle until the straw sticks out the top. (Some water may come out, too; don't squeeze too hard!) Remove the straw and adjust its buoyancy again in the glass. Repeat until you have the straw barely floating in the bottle.

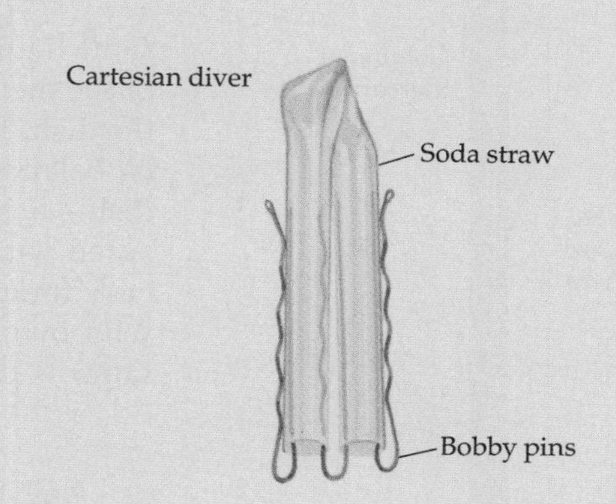

Now for the fun part. Screw the cap on the bottle tightly. Squeeze the sides of the bottle, and watch what happens inside. If the straw does not dive to the bottom of the bottle, squeeze really hard and it probably will. (You can make a strength tester out of this if you want.) If the straw still does not dive, take it out of the bottle, adjust it so it floats even lower in the water, and try again. If the straw sinks, you will need to upend the bottle and squeeze some water (and the straw) out; then adjust buoyancy and try again.

You now have a Cartesian Diver (named after the French mathematician/philosopher René Descartes) to experiment with. Try to figure out how it works. Refer to Section 2.1, where the properties of gases and liquids were described, and see if that helps you come up with ideas. Here are some things to try. Loosen the bottle cap, squeeze gently until water just runs out the top, and tighten the cap; is it easier or harder to make the diver dive? Place the bottle and diver in a refrigerator for an hour or so; what happens? While the bottle is cold, loosen and retighten the cap. Now let the bottle warm to room temperature; what happens? Loosen the bottle cap and retighten; what happens? Think up other experiments you can do with your diver. Then write down a theory about how it works that explains as many as possible of your observations.

unit area is the **pressure** of the gas. A gas exerts pressure on every surface it contacts, no matter what the direction of contact.

Our atmosphere also exerts pressure on everything it contacts. Atmospheric pressure can be measured with a **barometer,** which can be made by filling a tube with a liquid and inverting the tube in a dish containing the

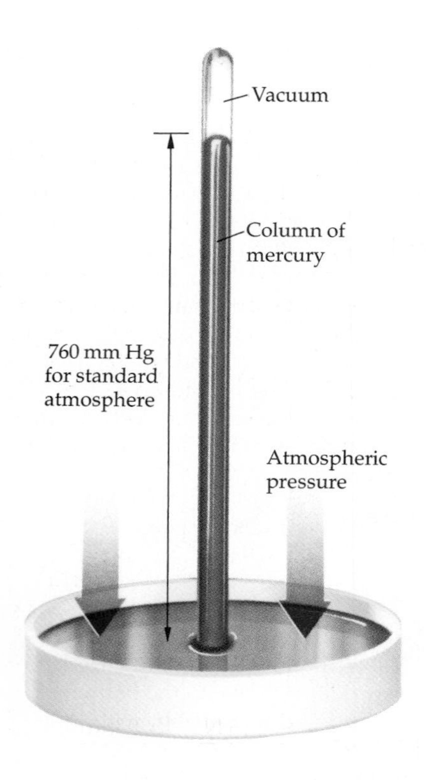

Figure 12.2
A Torricellian barometer. The pressure at the bottom of the mercury in the tube is balanced by atmospheric pressure on mercury in the dish.

same liquid. Figure 12.2 shows a mercury barometer. At sea level the height of the mercury column is about 760 mm above the surface of the mercury in the dish. The pressure at the bottom of a column of mercury 760 mm tall can be balanced by the pressure at the bottom of the column of air above the dish—a column that extends to the top of the atmosphere. Pressure measured with a mercury barometer is usually reported in **millimeters of mercury (mm Hg),** a unit that is also called the **torr** after Evangelista Torricelli, who invented the mercury barometer in 1643. The **standard atmosphere (atm)** is defined as

$$1 \text{ standard atmosphere} = 1 \text{ atm} = 760 \text{ mm Hg (exactly)}$$

Any liquid can be used in a barometer, but the height of the column depends on the density of the liquid. A water barometer would be about 34 feet tall, far too big to be practical. However, many water wells (especially in regions of the world without readily available electricity) bring up water from rock strata less than 33 feet below the surface. A simple hand pump can create a vacuum at the top of the well casing that allows atmospheric pressure on the water at the bottom of the well to cause it to rise upward. When the water column reaches the pump, it flows out. Before the invention of submersible electric pumps, well diggers knew from experience that it was fruitless to try to dig deeper than 33 feet for water, because it could not be pumped out even if it were found.

Water wells without submersible pumps can be dug deeper than 33 feet, but some means of getting a container down to the water must be provided. These shafts can pose dangers to animals or small children who might fall in.

EXERCISE 12.1 • *Pressure Calculations*

In weather reports, barometric pressure is often given in inches of mercury. Convert a barometric pressure reading of 29.10 inches Hg to (a) mm Hg and (b) atm.

Water wells deeper than 33 feet are common and have been used since ancient times as sources for water. Note that this well, in Sinai, Egypt, has a wide opening. It was dug long before the invention of electric pumps. *(Photo Researchers, Inc.)*

A Deeper Look

Force and Pressure

Pressure is defined as the force exerted on an object divided by the area over which the force is exerted:

$$\text{pressure} = \frac{\text{force}}{\text{area}}$$

A force can make an object accelerate, and the force equals the mass of the object times its acceleration.

$$\text{force} = \text{mass} \times \text{acceleration}$$

The SI units for mass and acceleration are kilograms (kg) and meters per second squared (m/s^2), so force has the units $kg{\cdot}m/s^2$; this is called a **newton (N)**.

To illustrate the meaning of pressure units, we can calculate the pressure exerted on the bottom of a glass by 500 mL of water. Suppose the water has a mass of 0.50 kg, and the bottom of the glass is a circle whose radius is 3.0 cm. The force exerted on the glass by the water is

$$\begin{aligned}\text{force} &= \text{mass} \\ &\quad \cdot \text{acceleration due to gravity} \\ &= 0.50 \text{ kg} \cdot 9.81 \text{ m/s}^2 \\ &= 4.9 \text{ kg}{\cdot}\text{m/s}^2\end{aligned}$$

To find the pressure, we need to know the area over which the force is exerted. The cross-sectional area of the glass at its base is

$$\begin{aligned}\pi \cdot (\text{radius})^2 &\cong 3.14 \cdot (0.030 \text{ m})^2 \\ &= 2.8 \times 10^{-3} \text{ m}^2\end{aligned}$$

Therefore the pressure is

$$\begin{aligned}\text{pressure} &= \frac{\text{force}}{\text{area}} = \frac{4.9 \text{ kg}{\cdot}\text{m/s}^2}{2.8 \times 10^{-3} \text{ m}^2} \\ &= 1.8 \times 10^3 \text{ N/m}^2\end{aligned}$$

The unit of pressure N/m^2 is given a special name, a **pascal** (Pa). The pressure of the atmosphere is about 100,000 Pa, and 760 mm Hg = 1 atm = 101.3 kPa. A related unit, a **bar,** which equals 100,000 Pa, is sometimes reported in connection with atmospheric pressure and weather. For a gaseous substance the standard thermodynamic properties given in Chapter 7 are for a gas pressure of 1 bar; see Section 7.6.

To show that 760.0 mm Hg equals 101.3 kPa, a calculation similar to the preceding one can be done. Suppose that a column of mercury 760.0 mm high is in a tube whose cross-sectional area is 1.000 mm^2. Since the volume of the cylinder is the area of the base (1.000 mm^2) times the height (760.0 mm), the volume of the mercury is 760.0 mm^3. The density of mercury is 13.596 g/cm^3, which can be used to calculate the mass of the mercury.

$$\begin{aligned}&\text{mass of Hg} \\ &= 760.0 \text{ mm}^3 \cdot \left(\frac{1 \text{ cm}}{10 \text{ mm}}\right)^3 \cdot \frac{13.596 \text{ g}}{1 \text{ cm}^3} \\ &= 10.33 \text{ g} = 0.01033 \text{ kg}\end{aligned}$$

$$\begin{aligned}\text{force} &= 0.01033 \text{ kg} \cdot 9.807 \text{ m/s}^2 \\ &= 0.1013 \text{ N}\end{aligned}$$

$$\begin{aligned}\text{pressure} &= \frac{0.1013 \text{ N}}{1 \text{ mm}^2} \cdot \frac{1000 \text{ mm}}{1 \text{ m}}^2 \\ &= 101.3 \times 10^3 \text{ N/m}^2 = 101.3 \text{ kPa}\end{aligned}$$

Notice that the pressure at the bottom of a glass of water, 1.7 kPa, is less than 2% as large as atmospheric pressure, 101.3 kPa.

Gases Are Compressible

When you pump up a tire with a bicycle pump, the gas is squeezed into a smaller volume by application of pressure. This property is called **compressibility.** In contrast to gases, liquids and solids are almost completely noncompressible.

Robert Boyle studied the compressibility of gases in 1661 by pouring mercury into an inverted J-shaped tube containing a sample of trapped gas. Each time he added more mercury, the volume of the trapped gas decreased (Figure 12.3). The mercury additions increased the pressure on the gas and changed the gas volume in a predictable fashion. Boyle was the first to observe that *the volume of a fixed amount of gas at a given temperature varies inversely with the applied pressure.* This is known as **Boyle's law.**

Boyle's law can be demonstrated by the experiment shown in Figure 12.4. A hypodermic syringe is partially filled with air and sealed. When pressure is applied to the movable plunger of the syringe, the volume of air

Early chemists routinely used mercury for all kinds of experiments. The toxic properties of mercury are now well known; it is a central nervous system poison and causes numerous chronic toxic effects. Mercury should not be used without adequate ventilation and should never be allowed to contact the skin.

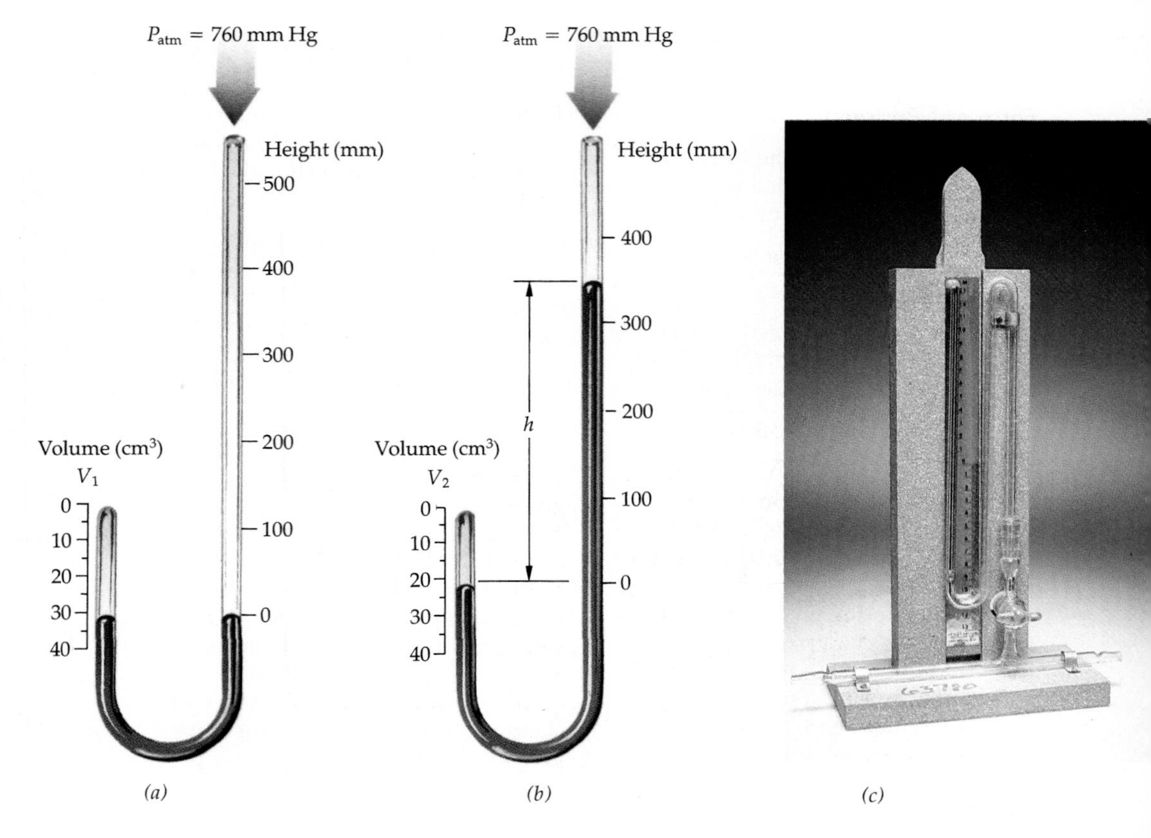

Figure 12.3
Boyle's law. Mercury confines a gas in a J-shaped tube. (a) When the mercury levels are the same on both sides of the J, the gas pressure equals atmospheric pressure. (b) When more mercury has been added, atmospheric pressure is augmented by the pressure of a mercury column of height h, where h = 340 mm Hg. At this higher pressure the gas volume is smaller, as predicted by Boyle's law. The temperature remains constant. (c) A manometer used for measuring gas pressures in the laboratory.
(C.D. Winters)

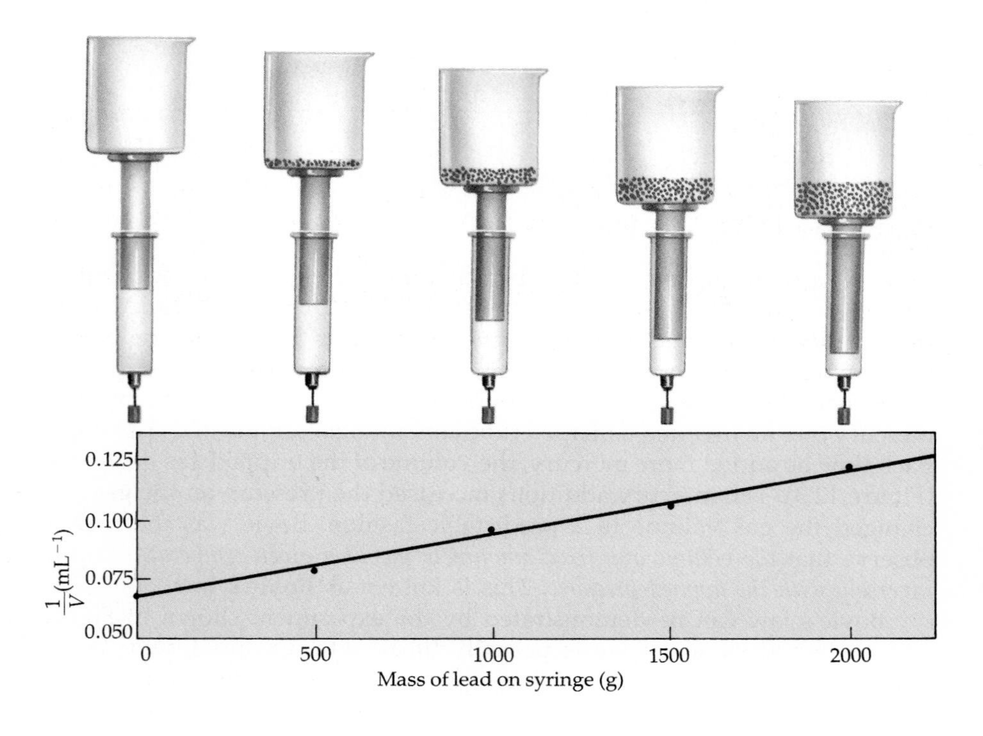

Figure 12.4
A simple apparatus demonstrating Boyle's law. A sample of a gas is compressed by placing a beaker on the plunger of a syringe and successively adding lead shot. This increases the pressure on the gas inside the syringe. The volume of the gas sample decreases according to Boyle's law. The plot of $\frac{1}{V}$ vs. mass of lead is a straight line. *(See D. Davenport,* Journal of Chemical Education, ***1962****, 39, 252)*

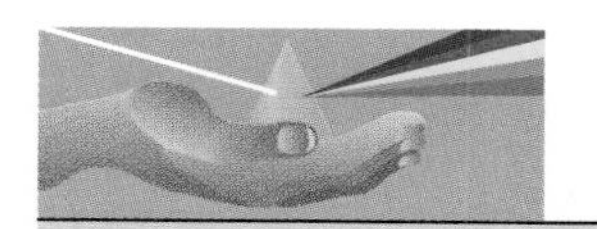

CHEMISTRY YOU CAN DO

Atmospheric Pressure

Here is a simple way to illustrate that atmospheric pressure is much greater than the pressure at the bottom of a 500-mL glass of water. Obtain a glass or other container for water, and a piece of stiff paper or cardboard that will completely cover the open top of the container. Do this experiment over a sink. Fill the container to the very top with water. Then carefully slide the paper or cardboard over the top of the container so that no air bubbles remain inside. Now, holding the cardboard so that it will not slip off, invert the container. Let go of the cardboard. What happens? What happens if there is a little air inside the container after you slide the cardboard on? Can you explain the difference in terms of force and pressure?

in the sealed apparatus decreases. When the reciprocal of the volume, $1/V$, is plotted as a function of pressure, a straight line that passes through the origin is observed. This type of plot demonstrates that the pressure and volume of a gas change in opposite directions; they are inversely proportional. Pressure is proportional to the reciprocal of the volume, or volume is proportional to the reciprocal of pressure.

$$P \propto \frac{1}{V} \quad \text{or equivalently} \quad V \propto \frac{1}{P}$$

When two quantities are proportional, they can be equated if a proportionality constant is used. Thus

$$P = \text{a constant} \times \frac{1}{V} \quad \text{or} \quad PV = \text{a constant}$$

The last equation, PV = a constant, is a mathematically useful statement of Boyle's law: *For a given amount of gas at a given temperature, the product of pressure and volume is a constant.* This means that if the pres‿ure-volume product is known for one set of conditions (P_1 and V_1), it is known for all other conditions of pressure and volume (P_2 and V_2), (P_3 and V_3), and so on. Therefore, because P_1V_1 and P_2V_2 both equal the same constant, we can write

When any quantity A is inversely proportional to another quantity B, the product A × B always has the same value; A × B = a constant.

$$P_1V_1 = P_2V_2 \quad \text{(at constant amount and temperature)}$$

and use it whenever we want to know, for example, what happens to the volume of a given amount of gas when the pressure changes (at a constant temperature).

Suppose you have a sample of a gas in a 100-mL container (V_1) at a pressure of 3 atm (P_1) and you need to know the volume (V_2) if the pressure is decreased to 1 atm (P_2) at the same temperature. The inverse relationship between P and V tells us that V_2 will be larger than V_1 because P_2 is smaller than P_1.

The value of V_2 can be found by rearranging $P_1V_1 = P_2V_2$ to solve for V_2.

$$V_2 = \frac{P_1V_1}{P_2}$$

Next, we substitute the values we know, and calculate V_2.

$$V_2 = \frac{3 \text{ atm} \cdot 100 \text{ mL}}{1 \text{ atm}} = 300 \text{ mL}$$

EXERCISE 12.2 • *Boyle's Law*

At a pressure of 1.00 atm and some temperature, a gas sample occupies 400. mL. What will be the volume of the gas at the same temperature if the pressure is decreased to 0.750 atm?

Gas Volume Changes with Temperature

Temperature as well as pressure can affect the volume of a sample of gas. In 1787, Jacques Charles discovered that the volume of a fixed quantity of a gas at constant pressure increased with increasing temperature. Figure 12.5 shows how the volumes of two different samples of gas change with the temperature (pressure remains constant). When the plots of volume vs. temperature for different gases are extended toward lower temperatures, they all reach zero volume at a common temperature, −273.15 °C. The volume of any gas would appear to be zero at −273.15 °C, but this does not actually happen—all gases liquefy before reaching this temperature.

The Kelvin temperature scale is also known as the absolute temperature scale or the thermodynamic temperature scale.

In 1848, William Thompson, also known as Lord Kelvin, proposed that it would be convenient to have a temperature scale in which the zero point was −273.15 °C. This temperature scale has been named for Lord Kelvin and the units of the scale are known as kelvins. As noted in Section 2.7, the kelvin is the same size as the Celsius degree and has been adopted as the SI

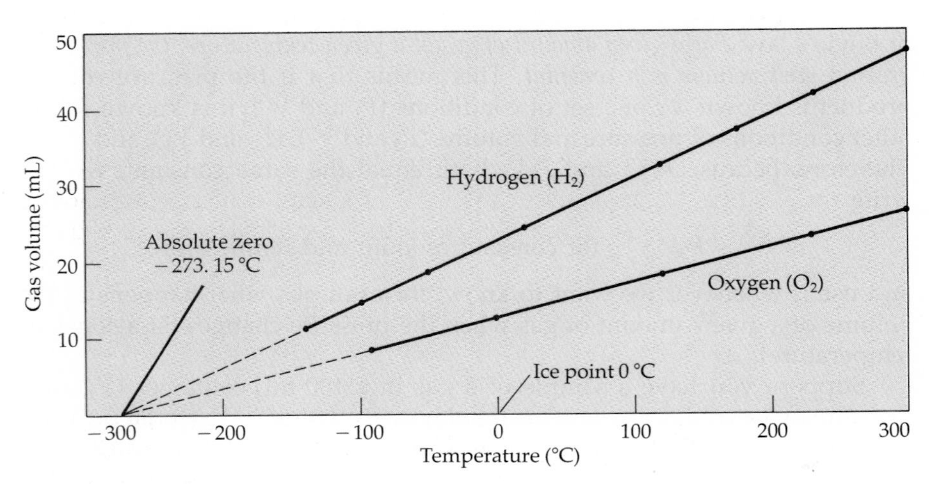

Figure 12.5
Charles's law. The volumes of two different samples of gases decrease with decreasing temperature (at constant pressure). These graphs (as would those of all gases) intersect the temperature axis at about −273 °C.

Portrait of a Scientist

Robert Boyle, the "Sceptical Chymist" (1627–1691)

Robert Boyle was born in Ireland, in a home that still stands, as the 14th and last child of the first Earl of Cork. He published his earliest studies of gases in 1660, and his book, *The Sceptical Chymist,* in 1680. Although he was the first to define elements in modern terms, he retained medieval views about what the elements were. For example, Boyle thought that gold was not an element, but rather a metal that could be formed from other metals. Boyle was also a physiologist—he was the first to show that the healthy human body has a constant temperature. Not everyone applauded all aspects of Boyle's work. Isaac Newton, a young man when Boyle was at the peak of his career, wrote concerning Boyle, "I question . . . what he speaks about."

Robert Boyle. (Oesper Collection in the History of Chemistry/University of Cincinnati)

---

unit for temperature measurement. When the Kelvin temperature scale is used, the volume-temperature relationship, now known as **Charles's law,** becomes: *the volume of a fixed amount of a gas at a constant pressure is directly proportional to the absolute temperature.*

Written as a proportion, Charles's law for any gas at constant amount and pressure becomes

$$V \propto T$$

or using a proportionality constant,

$$V = \text{a constant} \times T$$

or

$$\frac{V}{T} = \text{a constant}$$

where T is the temperature in kelvins.

If the volume, V_1, and temperature, T_1, of a sample of gas are known, then the volume, V_2, at some other temperature, T_2, at the same pressure is given by

$$\frac{V_1}{T_1} = \frac{V_2}{T_2} \quad \text{or} \quad V_1T_2 = V_2T_1 \quad \text{(at constant amount and pressure)}$$

or

$$V_2 = \frac{V_1T_2}{T_1}$$

When using the gas-law relationships, temperatures *must* be expressed on the absolute scale, using kelvins. For example, suppose you want to use Charles's law to calculate the new volume when 450.0 mL of a gas is cooled from 60.0 °C to 20.0 °C, at constant pressure. First, convert the temperatures to kelvins by adding 273.15 to the Celsius values: 60.0 °C becomes 333.2 K

Portrait of a Scientist

Jacques Alexandre Cesar Charles (1746–1823)

The French chemist Charles was most famous in his lifetime for his experiments in ballooning. The first such flights were made by the Montgolfier brothers in June 1783, using a large spherical balloon made of linen and paper and filled with hot air. In August 1783, however, a different group, supervised by Jacques Charles, tried a different approach. Exploiting his recent discoveries in the study of gases, Charles decided to inflate the balloon with hydrogen. Since hydrogen would easily escape a paper bag (see Section 12.3), Charles made a bag of silk coated with a rubber solution. Inflating the bag to its final diameter took several days and required nearly 500 pounds of acid and 1000 pounds of iron to generate the hydrogen gas. A huge crowd watched the ascent on August 27, 1783. The balloon stayed aloft for almost 45 minutes and traveled about 15 miles, but, when it landed in a village, the people were so terrified that they tore it to shreds.

Jacques Charles. (The Bettmann Archive)

and 20.0 °C becomes 293.2 K. Use the third form of Charles's law for the two sets of conditions at constant pressure,

$$V_2 = \frac{V_1 T_2}{T_1} = \frac{450.0 \text{ mL} \cdot 293.2 \text{ K}}{333.2 \text{ K}} = 396.0 \text{ mL}$$

EXERCISE 12.3 • *Charles's Law*

Because of the proportionality between temperature and volume, a gas sample at constant pressure can serve as a thermometer. For example, a certain gas sample occupies 100. mL at 25 °C. If the pressure is held constant and the temperature is increased, the sample occupies 175 mL. What is the final temperature in kelvins and Celsius degrees? Hint: first convert the starting temperature to kelvins.

It is also important to remember that neither Boyle's nor Charles's law depends on the identity of the gas being studied. These laws reflect properties of all gases and therefore must depend on the behavior of *any* gaseous atoms or molecules, regardless of their exact identities.

Equal Volumes of Gases Contain Equal Numbers of Molecules (at Constant T and P)

Neither Boyle's law nor Charles's law mentions anything about numbers of gaseous atoms or molecules, and yet it is easy to imagine that packing more of them into the same container at the same temperature would increase the pressure. In fact, we now know that if the pressure, volume, and temperature of a gas sample are specified, the number of gaseous atoms or molecules it contains can be calculated. The development of this idea goes back to the early 1800s.

In 1809 Joseph Gay-Lussac conducted some experiments in which he measured the volumes of gases reacting with one another to form gaseous products. He found that, at constant temperature and pressure, the gas volumes were always in the *ratios of small whole numbers.* For example, to produce 2 L of water vapor requires the reaction of 2 L of H_2 with 1 L of O_2. Similarly, to produce 1 L of water vapor requires reaction of 1 L of H_2 with 0.5 L of O_2.

	$2\,H_2(g)$ +	$O_2(g)$ $\longrightarrow$	$2\,H_2O(g)$
Experiment 1:	2 L	1 L	2 L
Experiment 2:	1 L	0.5 L	1 L

In 1811 Amedeo Avogadro suggested that Gay-Lussac's observations actually showed that equal volumes of all gases under the same temperature and pressure conditions contain the same number of molecules (Figure 12.6). Avogadro further proposed something that until then had not been considered—that hydrogen and oxygen consisted of the diatomic molecules H_2 and O_2.

Avogadro's idea was controversial because it was inconsistent with one of John Dalton's major ideas. Dalton, who had developed the concept of the atom, thought that the correct *molecular* formulas for the elements hydrogen and oxygen were H and O and that the formula for water was HO. However, Gay-Lussac's experiments showed that two volumes of hydrogen combined with one volume of oxygen to form two volumes of water vapor. Clearly, the only way Dalton's ideas could agree with Avogadro's interpretation of Gay-Lussac's experiments was for an oxygen atom to be split in two on reacting with hydrogen to form water vapor! Although it is now known that an oxygen *molecule* (not an atom) does split in two on reacting with hydrogen molecules to form water, Dalton strongly opposed Avogadro's ideas, and never did accept them. It took about 50 years—long after Avogadro and Dalton had died—for Avogadro's explanation of Gay-Lussac's experiments to be generally accepted.

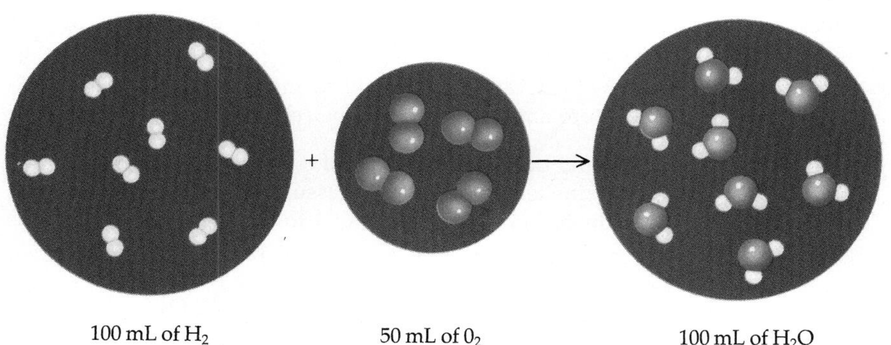

Figure 12.6
Gay-Lussac's law. When gases at the same temperature and pressure combine with one another their volumes are in the ratio of small whole numbers. Here, one volume of O_2 gas, say 50 mL at 100 °C and 1 atm combines with two volumes of H_2 gas (100 mL) to give two volumes (100 mL) of H_2O vapor also at 100 °C and 1 atm.

What is now known as **Avogadro's law** states that *the volume of a sample of a gas, at a given temperature and pressure, is directly proportional to the amount of the gas.* With V representing the volume and n representing the number of moles of the gas, the mathematical relationships are

$$V \propto n \qquad \text{or} \qquad V = \text{a constant} \times n$$

Avogadro's law means, for example, that at constant temperature and pressure, if the number of moles of gas doubles, the volume occupied doubles. It also means that at the same temperature and pressure, the volumes of two different amounts of gases are related as follows:

$$\frac{V_1}{V_2} = \frac{n_1}{n_2} \text{ (at constant temperature and pressure)}$$

Knowing any three variables in this expression allows calculation of the fourth. Also, this expression means that in balanced chemical equations, the coefficients for gaseous reactants or products represent relative *volumes* as well as numbers of moles.

Suppose you are going to carry out the reaction between nitrogen and hydrogen to produce ammonia.

$$N_2(g) + 3\,H_2(g) \longrightarrow 2\,NH_3(g)$$

If 1 L of nitrogen was consumed by this reaction, what volume of hydrogen would be consumed and what volume of ammonia would be produced, assuming the temperature and pressure remained constant? The 1 L of nitrogen represents some number of nitrogen molecules by Avogadro's law. Since the balanced equation tells you that three hydrogen molecules are required to react with one nitrogen molecule, there must be three times the volume of hydrogen to contain three times as many molecules. The same reasoning can apply to the ammonia produced—one volume of nitrogen produces two volumes of ammonia. The answers are therefore 3 L of hydrogen and 2 L of ammonia.

EXAMPLE 12.1

Using Avogadro's Law

Carbon monoxide burns in oxygen to form carbon dioxide.

$$2\,CO(g) + O_2(g) \longrightarrow 2\,CO_2(g)$$

If the volume occupied by the CO is 400 mL at 40 °C and 1 atm, what volume of O_2, at the same temperature and pressure, will be required in the reaction?

SOLUTION We want to find the volume of O_2 and the given information is the volume of the CO at the same temperature and pressure. Therefore the problem can be solved by using the reaction coefficients of 1 for O_2 and 2 for CO as the volume ratio.

$$\text{volume of } O_2 = 400 \text{ mL CO} \cdot \frac{1 \text{ mL } O_2}{2 \text{ mL CO}} = 200 \text{ mL } O_2$$

EXERCISE 12.4 • *Avogadro's Law*

The gas burner in a stove or furnace admits enough air so that methane gas can react completely with oxygen according to the equation

$$CH_4(g) + 2\,O_2(g) \longrightarrow CO_2(g) + 2\,H_2O(g)$$

Air is 1/5 oxygen by volume. Both air and gas are supplied to the flame by passing through a small tube. Compared to the tube for the gas, how much bigger does the tube for the air need to be? Assume that both gases are at the same T and P.

12.2 THE IDEAL GAS LAW

The properties of gases described by Boyle's, Charles's, and Avogadro's laws make it possible to write a single gas law that uses all four variables—pressure, volume, temperature, and amount—to describe any gas. The relationships to be combined are

$V \propto \frac{1}{P}$ (Boyle's law. Temperature and amount are constant.)

$V \propto T$ (Charles's law. Pressure and amount are constant.)

$V \propto n$ (Avogadro's law. Temperature and pressure are constant.)

The gas law that combines all four variables is

$$V \propto \frac{nT}{P}$$

This law states that the volume of a gas is directly proportional to the amount of gas present in the sample, n; directly proportional to the absolute temperature, T; and inversely proportional to the pressure, P. To make this proportionality into an equation, a proportionality constant, R, named the **ideal gas constant,** is used. The equation becomes

$$V = R\frac{nT}{P}$$

and on rearranging,

$$PV = nRT$$

This equation is called the **ideal gas law.** It correctly predicts the amount, pressure, volume, and temperature for samples of most gases at pressures of a few atmospheres or less and at temperatures near or above room temperature. The constant R can be calculated from the experimental fact that at 0 °C and 1 atm the volume of one mole of gas is 22.414 L. (This temperature and pressure are called **standard temperature and pressure,** STP, and the volume is called the **standard molar volume.**) Solving the ideal gas law for R, and substituting, we have

$$R = \frac{PV}{nT} = \frac{22.414\ \text{L} \cdot 1\ \text{atm}}{1\ \text{mol} \cdot 273.15\ \text{K}}$$

$$= 0.082057\ \text{L·atm/mol·K}$$

which is usually rounded to 0.0821 L·atm/mol·K.

The ideal gas law can be used to find P, V, n, or T whenever three of the four variables are known, provided the conditions of temperature and pressure are not extreme. For example, what volume will 0.20 g of oxygen occupy at 1.0 atm pressure and 20. °C? To use $PV = nRT$ it is convenient to have all the variables in the same units as the gas constant, $R =$ 0.0821 L·atm/mol·K. Temperatures should be in kelvins, pressure in atmospheres, volume in liters, and the amount of gas in moles.

The gas constant R *appeared in the discussion of thermodynamics and equilibrium on page 339, but in different units. If* P *is measured in SI units of pascals* ($kg/m \cdot s^2$) *and* V *is measured in* m^3, *then* R *has the value*

$$R = \frac{(1.01325 \times 10^5\ \text{kg/m·s}^2)(22.414 \times 10^{-3}\ \text{m}^3)}{(1\ \text{mol})(273.15\ \text{K})}$$

$$= 8.3145\ \text{kg} \cdot \text{m}^2/\text{s}^2\text{·mol·K}$$

Because 1 kg $\cdot$ m^2/s^2 *is 1 joule, the gas constant is also*

$$R = 8.3145\ \text{J/mol·K}$$

Begin by converting the mass of oxygen to moles.

$$n_{\text{oxygen}} = 0.20\ \text{g}\ O_2 \cdot \frac{1\ \text{mol}\ O_2}{32.0\ \text{g}\ O_2} = 0.0063\ \text{mol}\ O_2$$

Next, convert the temperature to kelvins:

$$T = (20. + 273.15)\ \text{K} = 293\ \text{K}$$

Now solve for V in the ideal gas law equation. The result will be in liters.

$$V = \frac{nRT}{P} = \frac{0.0063\ \cancel{\text{mol}} \cdot 0.0821\ \text{L}\cdot\cancel{\text{atm}}/\cancel{\text{mol}}\cdot\cancel{\text{K}} \cdot 293\ \cancel{\text{K}}}{1.0\ \cancel{\text{atm}}}$$

$$= 0.15\ \text{L} = 150\ \text{mL}$$

Notice which units cancel on the right side of the equation.

EXAMPLE 12.2

The Ideal Gas Law and the General Gas Equation

Helium-filled balloons are used to carry scientific instruments high into the atmosphere. Suppose that such a balloon is launched on a summer day when the temperature is 22.5 °C and the barometer reads 754 mm Hg. If the balloon's volume is 1.00×10^6 L at launch, what will it be at a height of 20 miles, where the pressure is 76.0 mm Hg and the temperature is 240. K?

SOLUTION Assume that no gas escapes from the balloon. Then only T, P, and V change. The ideal gas law can be used to relate these three variables.

$$\frac{P_1V_1}{P_2V_2} = \frac{n_1RT_1}{n_2RT_2}$$

Here a subscript 1 indicates initial conditions (at launch) and a subscript 2 indicates final conditions (high in the atmosphere). Because R is a constant, it has no subscript. Since $n_1 = n_2$,

$$\frac{P_1V_1}{P_2V_2} = \frac{T_1}{T_2} \quad \text{or} \quad \frac{P_1V_1}{T_1} = \frac{P_2V_2}{T_2}$$

The resulting equation is useful and is known as the **general gas equation.** Initial conditions are $P_1 = 754$ mm Hg, $V_1 = 1.00 \times 10^6$ L, and $T_1 = (22.5 +$

273.15)K. Final conditions are $P_2 = 76.0$ mm Hg, and $T_2 = 240.$ K. Solving for V_2 gives

$$V_2 = \frac{P_1V_1T_2}{P_2T_1} = \frac{754 \text{ mm Hg} \cdot 1.00 \times 10^6 \text{ L} \cdot 240. \text{ K}}{76.0 \text{ mm Hg} \cdot 295.7 \text{ K}} = 8.05 \times 10^6 \text{ L}$$

Thus the volume would be about eight times larger. The volume has increased because the pressure has dropped. For this reason weather balloons are never completely filled at launch—a great deal of room has to be left so that the helium can expand at high altitudes.

EXERCISE 12.5 • *Gas Law Equations*

Show how the equations for Boyle's law and Charles's law for a fixed amount of a gas can be derived from the ideal gas law equation.

Gas Law Stoichiometry

The ideal gas law makes it possible to use volumes as well as masses or molar amounts in calculations based on reaction stoichiometry. For example, some commercial drain cleaners contain sodium hydroxide and small pieces of aluminum (Figure 12.7). When the mixture is poured into a clogged drain, the reaction that occurs is

$$2 \text{ Al(s)} + 2 \text{ NaOH(aq)} + 6 \text{ H}_2\text{O}(\ell) \longrightarrow 2 \text{ NaAl(OH)}_4\text{(aq)} + 3 \text{ H}_2\text{(g)}$$

The heat generated by the reaction helps the sodium hydroxide break up the grease, and the hydrogen gas being generated stirs up the mixture and speeds up the unclogging of the drain. If 6.5 g of Al and an excess of NaOH are used, what volume of gaseous H_2 (measured at 742 mm Hg and 22.0 °C) is produced?

The first step in solving this problem is to calculate the number of moles of Al available and then find the number of moles of H_2 generated. Once the number of moles of H_2 is known, the volume is obtained from $PV = nRT$.

$$\text{Amount of Al available} = 6.5 \text{ g Al} \cdot \frac{1.0 \text{ mol Al}}{27.0 \text{ g Al}} = 0.24 \text{ mol Al}$$

The amount of H_2 produced by the aluminum is next calculated from the mole ratio in the balanced equation.

$$\text{Amount of H}_2 \text{ expected} = n = 0.24 \text{ mol Al} \cdot \frac{3 \text{ mol H}_2}{2 \text{ mol Al}} = 0.36 \text{ mol H}_2$$

Next, solve $PV = nRT$ for V and substitute values for T, P, and n.

When substituting into the rearranged gas equation, you must make sure the units of P, V, and T are compatible with the units of R. This means P must be converted to atmospheres.

$$V = \frac{nRT}{P} = \frac{0.36 \text{ mol} \cdot 0.0821 \text{ L·atm/mol·K} \cdot 295.2 \text{ K}}{0.976 \text{ atm}} = 8.9 \text{ L}$$

$$P = 742 \text{ mm} \cdot \frac{1 \text{ atm}}{760. \text{ mm}} = 0.976 \text{ atm}$$

Figure 12.7
A drain cleaner containing sodium hydroxide and pieces of aluminum. When water is added the hydroxide ion attacks the aluminum, and hydrogen gas is produced. *(C.D. Winters)*

EXAMPLE 12.3

Gas Law Stoichiometry

Octane, C_8H_{18}, is one of the hydrocarbons in gasoline. In an automobile engine, octane burns to produce CO_2 and H_2O. What volume in liters of oxygen, entering the engine at 0.950 atm and 20. °C, is required to burn 1.00 g of octane? What volume would be required on a cold winter day, when the temperature is −20. °C (assuming the pressure is still 0.950 atm)?

SOLUTION First write the balanced equation for the combustion reaction. The molar mass of octane is 114.2 g/mol.

$$2\,C_8H_{18}(\ell) + 25\,O_2(g) \longrightarrow 16\,CO_2(g) + 18\,H_2O(\ell)$$

Next, calculate the number of moles of O_2 required, using the stoichiometric factor of 25 mol O_2 to 2 mol C_8H_{18}.

$$n_{O_2} = 1.00\,\cancel{g\,C_8H_{18}} \cdot \frac{1\,\text{mol}\,\cancel{C_8H_{18}}}{114.2\,\cancel{g\,C_8H_{18}}} \cdot \frac{25\,\text{mol}\,O_2}{2\,\text{mol}\,\cancel{C_8H_{18}}} = 0.109\,\text{mol}\,O_2$$

Then, using the ideal gas law, calculate the volume of oxygen at 20. °C and 0.950 atm.

$$V_{O_2} = \frac{nRT}{P} = \frac{0.109\,\text{mol}\,O_2 \cdot 0.0821\,\text{L·atm/mol·K} \cdot 293\,\text{K}}{0.950\,\text{atm}} = 2.76\,\text{L}$$

At −20. °C or 253 K, the same type of calculation gives

$$V_{O_2} = 2.38\,\text{L}$$

Because the volume of oxygen (and hence of air) required is significantly less in cold weather, automobile engines have a choke—a valve that chokes off part of the air supply until the engine warms up.

EXERCISE 12.6 • *Gas Law Stoichiometry*

If you carried out the Chemistry You Can Do experiment at the beginning of Chapter 6, you observed the reaction of vinegar (acetic acid solution) with baking soda (sodium hydrogen carbonate) to generate carbon dioxide gas and inflate a balloon. The reaction is

$$H_3O^+(aq) + HCO_3^-(aq) \longrightarrow CO_2(g) + 2\,H_2O(\ell)$$

If there is plenty of vinegar, what mass of $NaHCO_3$ is required to inflate a balloon to a diameter of 20. cm at room temperature 20. °C? Assume the balloon is a sphere ($V = \frac{4}{3}\pi r^3$), and, because the rubber is stretched, the pressure inside the balloon is twice normal atmospheric pressure.

Gas Densities

Density is mass per unit volume. Because the volume of a gas sample (but not its mass) varies with temperature and pressure, gas densities are extremely variable. However, once T and P are specified, the density of a gas can be calculated from the ideal gas law. Also, because equal volumes of gas at the same T and P contain equal numbers of molecules, the densities of different gases are proportional to their molar masses. In other words, measured gas densities can be used to determine molecular weights.

Consider the densities of the three gases: He, O_2, and SF_6 (used as an insulator in high voltage transmission lines). Taking 1 mol of each of these gases at 25 °C and 0.750 atm, the densities are

Gas	*Density (25°C)*
He	0.123 g/L
O_2	0.981 g/L
SF_6	4.48 g/L

The ideal gas equation can be used to calculate the density for helium at 25 °C and 0.750 atm by substituting m/M (mass over molar mass) for n,

$$PV = nRT$$

$$PV = \frac{m}{M}RT$$

On rearranging, we get an expression for density:

$$\frac{m}{V} = d = \frac{PM}{RT}$$

$$d = \frac{0.750 \text{ atm} \cdot 4.003 \text{ g/mol}}{0.0821 \text{ L} \cdot \text{atm/mol K} \cdot 298 \text{ K}}$$

$$= 0.123 \text{ g/L}$$

As can be seen in the equation above, the density of a gas is proportional to its molar mass. Also note that helium is about 8000 times less dense than liquid water at the same temperature.

EXAMPLE 12.4

Gas Density and Molar Mass

A 0.100-g sample of a gaseous compound with the empirical formula CH_2F_2 occupies 0.0470 L at 298 K and 755 mm Hg. What is the molar mass of the compound?

SOLUTION This problem is typical of the laboratory measurement of the molar mass of a gas. Begin by organizing the data.

$$V = 0.0470 \text{ L} \qquad P = 755 \text{ mm Hg} \qquad T = 298 \text{ K} \qquad n = ?$$

$$P = 755 \text{ mm Hg} \cdot \frac{1 \text{ atm}}{760 \text{ mm Hg}} = 0.993 \text{ atm}$$

Using the ideal gas law equation to find n, the number of moles of gas,

$$n = \frac{PV}{RT}$$

Therefore,

$$n = \frac{0.993 \text{ atm} \cdot 0.0470 \text{ L}}{0.0821 \text{ L·atm/mol·K} \cdot 298 \text{ K}} = 0.00191 \text{ mol}$$

The mass of the sample is already known to be 0.100 g, and so

$$\text{molar mass} = \frac{m}{n} = \frac{0.100 \text{ g}}{0.00191 \text{ mol}} = 52.4 \text{ g/mol}$$

EXERCISE 12.7 • *Gas Densities*

A glass flask contains 1.00 L of a gas at 0.850 atm and 20. °C. The density of the gas is 1.13 g/L. What is the molar mass of the gas? What is its identity?

Dalton's Law of Partial Pressures

Our atmosphere is a mixture of nitrogen, oxygen, argon, carbon dioxide, water vapor, and small amounts of several other gases (Table 12.1). Atmospheric pressure is the sum of the pressures exerted by all of these individual gases. The same is true of every gas mixture. Consider the mixture of nitrogen and oxygen illustrated in Figure 12.8. The pressure exerted by the mixture is equal to the sum of the pressures that the nitrogen alone and the oxygen alone would exert in the same volume at the same temperature and

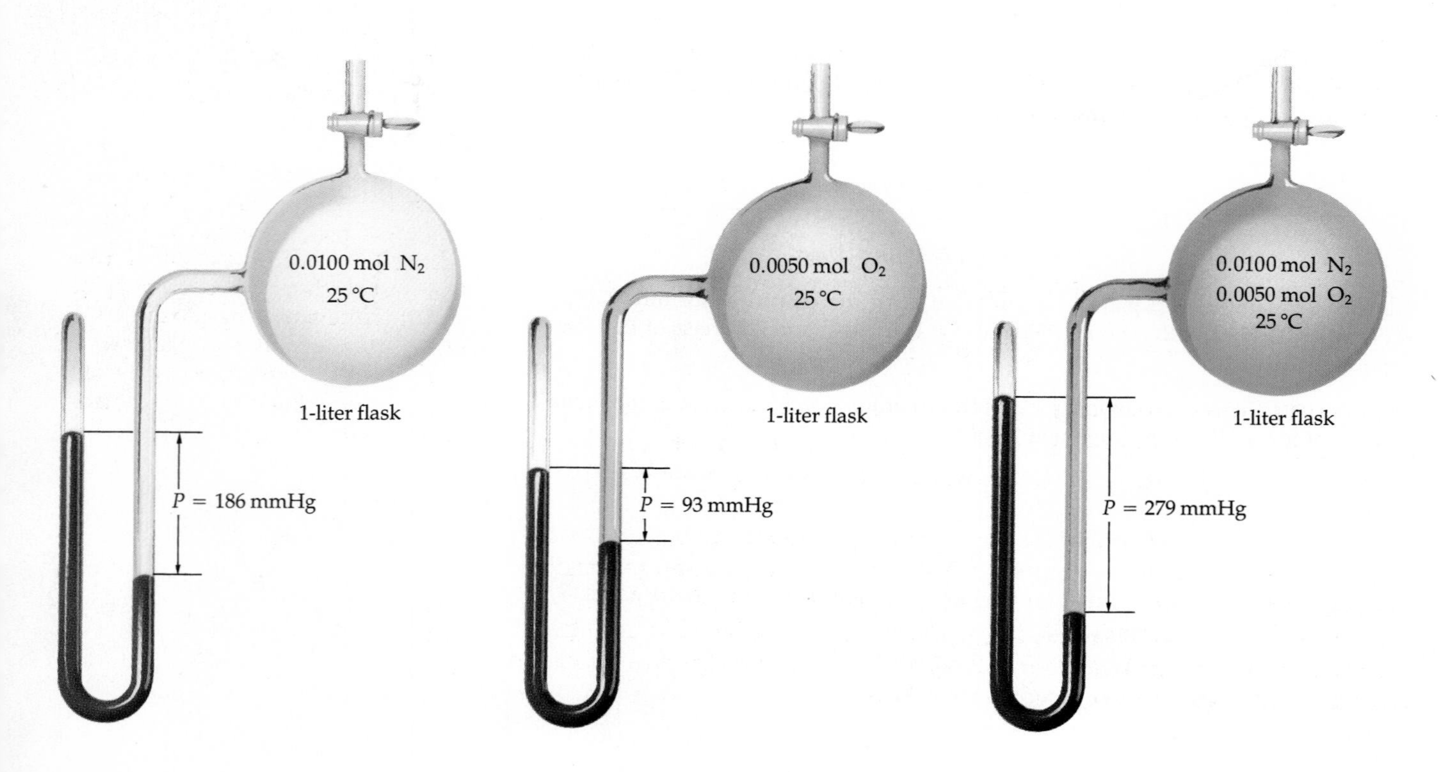

Figure 12.8
Dalton's law. (*Left*) 0.0100 mol of N_2 in a 1.00-L flask of 25 °C exerts a pressure of 186 mm Hg. (*Middle*) 0.0050 mol of O_2 in a 1.00-L flask at 25 °C exerts a pressure of 93 mm Hg. (*Right*) The N_2 and O_2 samples are mixed in the same 1.00-L flask at 25 °C. The total pressure, 279 mm Hg, is the sum of the pressures the individual gases would have if each were alone in the flask.

Table 12.1

The Composition of Dry Air

Component	Mole Percent*	Partial Pressure at STP (atm)
nitrogen, N_2	78.084	0.78084
oxygen, O_2	20.946	0.20946
argon, Ar	0.934	0.00934
carbon dioxide, CO_2	0.033	0.00033
neon, Ne	0.002	0.00002
helium, He	0.001	0.00001

*Mole percent is the number of moles of a component divided by the total number of moles of gas and multiplied by 100 percent.

pressure. The pressure of one gas in a mixture of gases is called the **partial pressure** of that gas.

John Dalton was the first to observe that *the total pressure exerted by a mixture of gases is the sum of the partial pressures of the individual gases in the mixture.* This statement is known as **Dalton's law of partial pressures.** Dalton's law is a consequence of the fact that the ideal gas law applies to all gases regardless of composition. For the first three components of our atmosphere, the total number of moles is

$$n_{\text{total}} = n_{N_2} + n_{O_2} + n_{\text{Ar}}$$

If we replace n_{total} in the ideal gas law with the summation of the individual numbers of moles of gases, the equation becomes

$$P_{\text{total}}V = n_{\text{total}}RT$$

$$P_{\text{total}} = \frac{n_{\text{total}}RT}{V} = \frac{(n_{N_2} + n_{O_2} + n_{\text{Ar}})RT}{V}$$

Expanding the right side of this equation and rearranging,

$$P_{\text{total}} = \frac{n_{N_2}RT}{V} + \frac{n_{O_2}RT}{V} + \frac{n_{\text{Ar}}RT}{V} = P_{N_2} + P_{O_2} + P_{\text{Ar}}$$

This means that the pressure exerted by the atmosphere is the sum of the pressures due to nitrogen, oxygen, argon, and the other components. The quantities P_{N_2}, P_{O_2}, and P_{Ar} are the partial pressures of the three major components.

Figure 12.9
A gas-mixing manifold used by an anesthesiologist to prepare a gas mixture to keep a patient unconscious during an operation. By proper mixing, the anesthetic gas can slowly be added to the breathing mixture. Near the end of the operation, the anesthetic gas can be replaced by air of normal composition or pure oxygen. *(C.D. Winters)*

To see how partial pressures work, let's consider a gas mixture containing cyclopropane (C_3H_6) and oxygen. Such mixtures were used as anesthetics before the discovery of more modern anesthetic gases (Fig. 12.9). Suppose 63.0 g of cyclopropane and 42.0 g of oxygen are placed in a tank and the total pressure inside the tank is 1.40 atm. What are the partial pressures due to cyclopropane and oxygen in the tank?

cyclopropane

To calculate the partial pressures, you will need the numbers of moles of the two gases. The pressure due to cyclopropane is proportional to the number of moles of cyclopropane, and the pressure due to oxygen is proportional to the number of moles of oxygen. The total pressure (1.40 atm) is the sum of the two partial pressures.

$$P_{\text{total}} = P_{\text{cyclopropane}} + P_{\text{oxygen}} = 1.40 \text{ atm}$$

The number of moles of each gas can be found from the masses given.

$$\text{amount of cyclopropane} = 63.0 \text{ g } C_3H_6 \cdot \frac{1.0 \text{ mol } C_3H_6}{42.08 \text{ g } C_3H_6} = 1.50 \text{ mol } C_3H_6$$

$$\text{amount of oxygen} = 42.0 \text{ g } O_2 \cdot \frac{1.0 \text{ mol } O_2}{32.00 \text{ g } O_2} = 1.31 \text{ mol } O_2$$

$$\text{total amount} = 1.50 \text{ mol} + 1.31 \text{ mol} = 2.81 \text{ mol}$$

According to Dalton's law of partial pressures,

$$P_{\text{cyclopropane}} = n_{\text{cyclopropane}}RT/V$$

and

$$P_{\text{total}} = n_{\text{total}}RT/V$$

Dividing the first equation by the second gives an expression that can be solved for the pressure due to cyclopropane in this mixture.

$$\frac{P_{\text{cyclopropane}}}{P_{\text{total}}} = \frac{n_{\text{cyclopropane}}RT/V}{n_{\text{total}}RT/V} = \frac{n_{\text{cyclopropane}}}{n_{\text{total}}}$$

The ratio $n_{\text{cyclopropane}}/n_{\text{total}}$ is the amount of cyclopropane in the mixture divided by the total amount of gas. Another name for this ratio is the **mole fraction (X)** of cyclopropane in the mixture.

By rearranging, we can see that the pressure in the mixture due to cyclopropane, $P_{\text{cyclopropane}}$, is

$$P_{\text{cyclopropane}} = P_{\text{total}} \cdot \frac{n_{\text{cyclopropane}}}{n_{\text{total}}} = 1.40 \text{ atm} \cdot \frac{1.50 \text{ mol}}{2.81 \text{ mol}} = 0.747 \text{ atm}$$

The pressure due to oxygen must be 1.40 atm − 0.747 atm = 0.65 atm.

In general, for a gas A in a mixture, its partial pressure, P_A, is equal to the total pressure of the mixture, P_{total}, times the mole fraction of that gas, X_A.

$$P_A = P_{\text{total}}X_A$$

EXERCISE 12.8 • *Dalton's Law of Partial Pressures*

A mixture of 7.0 g of N_2 and 6.0 g of H_2 is confined in a 5.0-L reaction vessel at 500. °C. Assume that no reaction has occurred and calculate the total pressure. Then calculate the mole fraction and partial pressure of each gas.

12.3 KINETIC-MOLECULAR THEORY

The fact that all gases behave in very similar ways can be interpreted by means of the **kinetic-molecular theory.** A qualitative introduction to this theory was given in Section 2.1, where it was mentioned that the theory applies to properties of liquids and solids as well as gases. According to the kinetic-molecular theory, a gas consists of very tiny molecules in constant motion. The pressure of a gas results from continual bombardment by rapidly moving molecules upon any surface the gas contacts (Figure 12.10).

In the kinetic-molecular theory the word "molecule" includes atoms of the monoatomic rare gases He, Ne, Ar, Kr, and Xe.

Five fundamental concepts make up the kinetic-molecular theory. Each of them is consistent with the results of various experiments with gases.

1. *A gas is composed of molecules whose size is much smaller than the distances between them.* This concept accounts for the ease with which gases can be compressed and for the fact that gases at ordinary temperature and pressure mix completely with other gases. These facts imply that there must be plenty of room for additional molecules in a sample of gas.
2. *Gas molecules move randomly—at various speeds and in every possible direction.* This concept is consistent with the fact that gases quickly and completely fill any container in which they are placed.
3. *Except when molecules collide, forces of attraction and repulsion between them are negligible.* This concept is consistent with the fact that a gas will remain a gas indefinitely at a fixed pressure and temperature in spite of the countless collisions that take place.
4. *When collisions occur, they are elastic;* the speeds of colliding molecules may change but the total kinetic energy of two colliding molecules is the same after a collision as before. This concept is consistent with the fact that a gas sample at constant temperature never "runs down," with all molecules falling to the bottom of the container.
5. *The average kinetic energy of gas molecules is proportional to the absolute temperature.* This concept is consistent with the fact that the rate of escape of gas molecules from a tiny hole is faster the higher the temperature is, and with the fact that rates of chemical reactions are faster at higher temperatures.

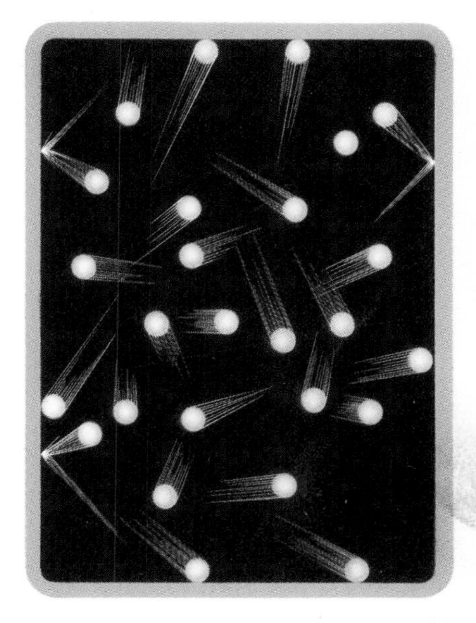

Figure 12.10
Gas pressure is caused by bombardment of the container walls by gas molecules.

Like any moving object, a gas molecule has kinetic energy. An object's kinetic energy, E_K, depends on its mass, m, and its speed, v, according to the equation

$$E_K = \tfrac{1}{2}(\text{mass})(\text{speed})^2 = \tfrac{1}{2}mv^2$$

All of the molecules in a gas are moving, but they do not all move at the same speed, and so they do not all have the same kinetic energy. At a given instant a few molecules are going very fast, most are moving at close to the average speed, and a few others may be in the process of colliding with a surface, in which case their speed is zero. If we could watch an individual molecule, we would find that its speed would continually change as it collides with and exchanges energy with other molecules.

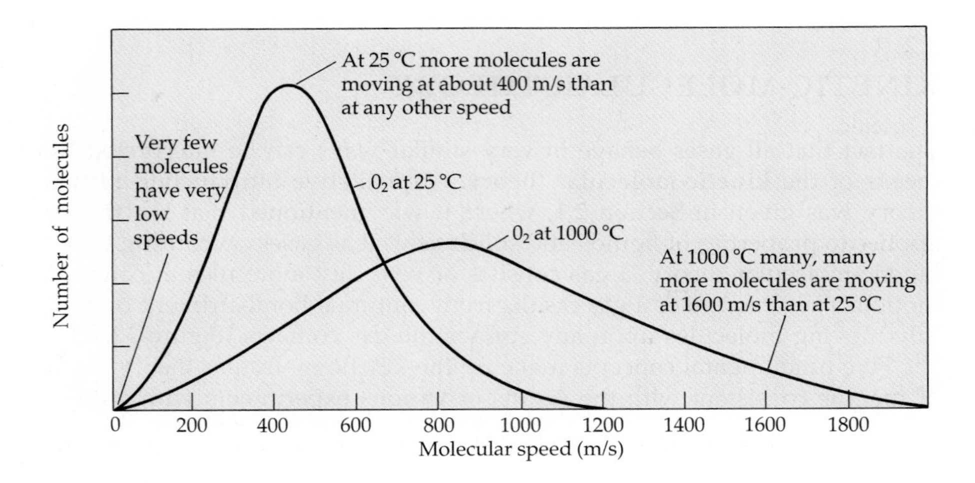

Figure 12.11
Distribution of molecular speeds. A plot of the relative number of gas molecules with a given speed versus that speed (in meters per second). The curve for O_2 at 1000 °C shows the effect of a temperature increase on the distribution of speeds.

The relative number of molecules that have a given speed can be measured experimentally. Figure 12.11 shows a graph of number of molecules vs. speed. The higher the curve, the greater the number of molecules going at that speed. For oxygen gas at 25 °C, for example, the maximum in the curve comes at a speed of 400 m/s, and most of the molecules are within the range from 200 m/s to 700 m/s. When the temperature is increased, the most common speed goes up, and the number of molecules travelling very fast goes up a lot. The kinetic-molecular theory states that the average kinetic energy of the molecules of any gas will be the same at a given temperature. Since $E_k = \frac{1}{2}mv^2$, this means that the larger m is the smaller v must be. That is, the heavier the molecules, the slower their average speed must be, and vice versa. Figure 12.12 illustrates this: the peak in the curve for the heaviest molecule, O_2, occurs at a much lower speed than for the lightest, He. Because the curves are not symmetric around their maxima, the average speed for each type of molecule is a little faster than the most common speed (which is at the top of the peak). You can also see from the graph that average speeds range from a few hundred to a few thousand meters per second.

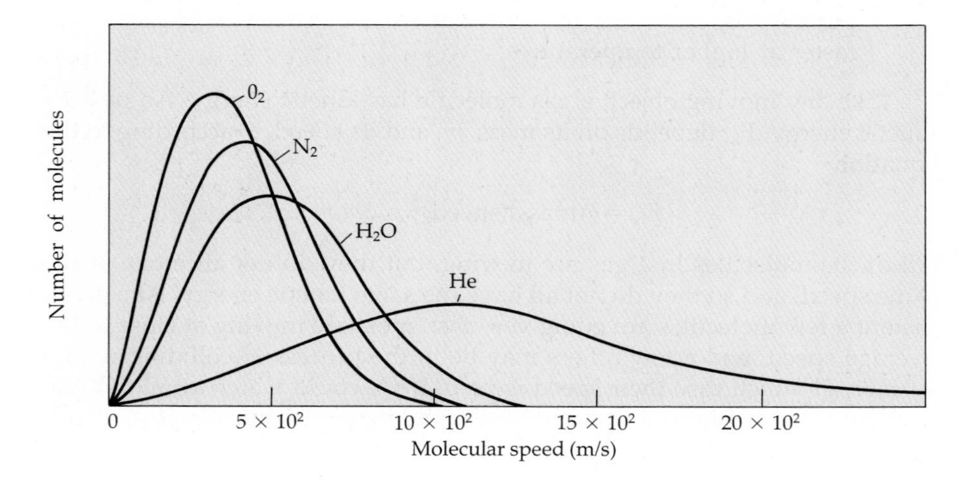

Figure 12.12
The effect of molecular mass on the distribution of molecular speeds at a given temperature. Notice the similarity of the effect of increasing mass to the effect of decreasing temperature shown in Figure 12.11.

EXERCISE 12.9 • *The Kinetic-Molecular Theory*

Use the kinetic-molecular theory to explain why the pressure goes up when more gas molecules are added to a sample of gas in a fixed-volume container at constant temperature.

Diffusion and Effusion

When a person wearing a strong perfume enters a room, the smell of the perfume is immediately noticed by those close by, and eventually by everyone in the room. The perfume contains easily vaporized **(volatile)** compounds that gradually mix with the oxygen, nitrogen, argon, carbon dioxide, water vapor, and other gases in the room's atmosphere. Even if there were no movement of the air in the room caused by fans or by people walking around, the smell of perfume would eventually reach everywhere in the room. This mixing of molecules of two or more gases due to their molecular motions is called **gaseous diffusion;** it results from the random molecular motion present in all gases. Given time, the molecules of one component in a gas mixture will thoroughly and completely mix with all of the other components of the mixture. Closely related to diffusion is **effusion,** which is the rate at which a gas can escape through a tiny hole into a container in which the pressure is low (Figure 12.13).

In 1830, Thomas Graham discovered the mathematical relation between the effusion rates of gases and their molar masses. If the effusion rate of the first gas is r_1, and the effusion rate of the second is r_2, then

$$\frac{r_1}{r_2} = \sqrt{\frac{M_2}{M_1}}$$

Graham's law can be stated as: *At constant pressure and temperature, the rate of effusion of a gas is inversely proportional to the square root of its molar mass.*

Graham's law explains why a helium-filled balloon loses pressure faster than an identical balloon filled with air or nitrogen. This is why helium balloons now are usually made of metallized plastic—the metallized balloon walls offer fewer holes for the helium to effuse through, so the balloon stays inflated longer. Besides that, metallized balloons are prettier than ordinary rubber balloons.

An interesting historical note is that when Gay-Lussac first experimented with balloons in 1783, he used a paper balloon filled with hot air. When he tried making a hydrogen-filled balloon, paper would not work; he had to use silk coated with rubber. This is understandable, since light hydrogen molecules have a higher average speed than the nitrogen and oxygen molecules in air; hydrogen molecules are also smaller. Not only did hydro-

The driving force for mixing of gases by diffusion is entropy, which was discussed in Section 7.9. A mixture of gases has considerably more entropy than its pure components, and so mixing is a product-favored process.

Gasoline is a common mixture that contains volatile components. These help you get your car started by supplying sufficient vapors to ignite with air in the combustion chambers of the engine, even on a cold day. When you pump gasoline you will usually notice the odors of the more volatile components of gasoline.

Although it is a more complex phenomenon, diffusion is also described by Graham's law.

N_2

H_2

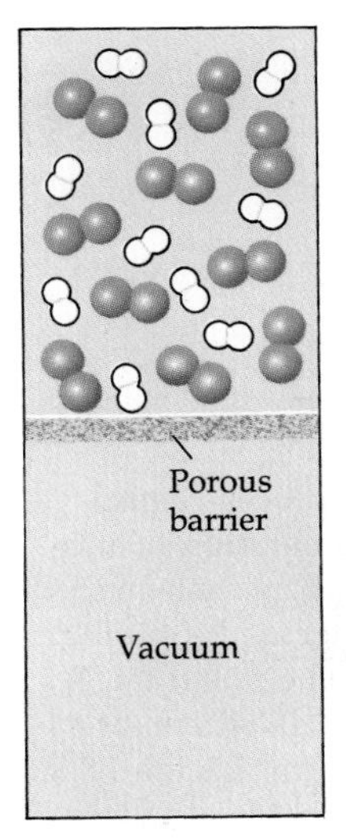

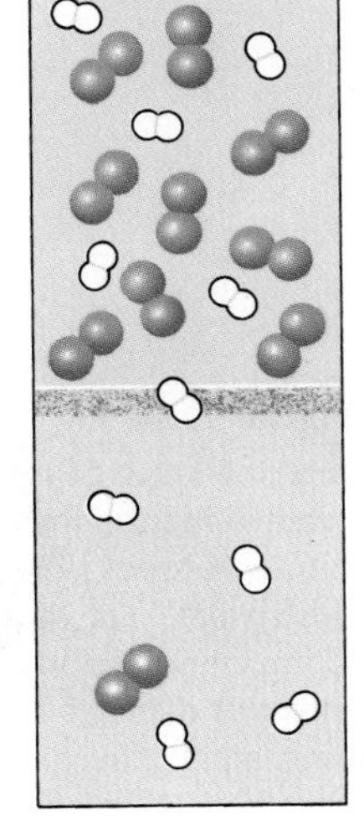

Before effusion

During effusion

Figure 12.13
An illustration of the effusion of gas molecules through the pores of a membrane or other porous barrier. Lighter molecules with greater average speeds strike the membrane more often and pass more rapidly through it than the heavier, slower molecules at the same temperature.

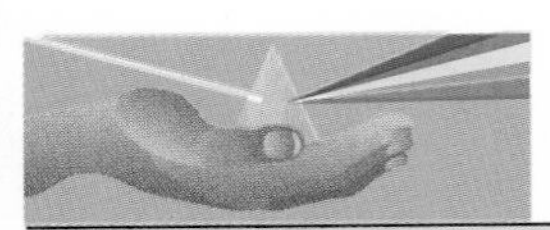

CHEMISTRY YOU CAN DO

Gas Diffusion

Using a medicine dropper, or using a soda straw as a pipet, carefully place a few drops of vanilla extract in a balloon. (If vanilla extract is not available, try perfume or after-shave lotion.) Inflate the balloon, tie it off, and take it into another room. In a few minutes you will notice the odor of vanilla, a moderately volatile organic compound that is released from the extract. Explain what is happening in terms of the kinetic-molecular theory. Suggest one or two other substances you could use in this experiment and try one of them.

gen molecules collide with the balloon walls more often, but being smaller, they were more likely to pass through tiny openings in the wall.

One of the most important industrial applications of Graham's law was in the separation of two isotopes of uranium. Natural uranium consists of ^{238}U (99.27%) and ^{235}U (0.73%). The uranium-235 isotope is the one that undergoes nuclear fission (Section 19.6) and therefore provides the energy when uranium is used as a fuel in nuclear reactors or in nuclear bombs. In order for a sample of uranium to be used in a nuclear reactor or in a nuclear bomb, the ^{235}U content must somehow be increased. This process is known as *enrichment*. A uranium fuel rod has been enriched to about 3% ^{235}U; bomb-grade uranium needs to be enriched much more.

The molar masses of $^{235}UF_6$ and $^{238}UF_6$ can be calculated using the isotopic masses: $^{235}U = 235.044$ g/mol and $^{238}U = 238.051$ g/mol.

In the late 1930s it was discovered that uranium could be enriched by first treating it with fluorine to produce uranium hexafluoride, UF_6, a white solid that readily vaporizes. When UF_6 vapor is allowed to diffuse through a series of porous membranes, a process known as the **gaseous diffusion**

Figure 12.14
Interior view of the process equipment used for the separation of uranium isotopes. The large tank-shaped "diffusers" are connected by piping. Compressors pump the volatile UF_6 into the diffuser vessels where, at each stage, the uranium is slightly enriched in the fissionable ^{235}U isotope. *(Oak Ridge National Laboratory)*

process, the lighter molecules of $^{235}UF_6$ diffuse more rapidly than the heavier $^{238}UF_6$ molecules. With Graham's law it is possible to calculate this difference in rates.

$$\frac{\text{rate of diffusion of } ^{235}UF_6}{\text{rate of diffusion of } ^{238}UF_6} = \sqrt{\frac{352.041 \text{ g/mol}}{349.034 \text{ g/mol}}} = 1.00430$$

Thus the lighter $^{235}UF_6$ molecules diffuse about 0.4% faster. When this process is repeated many thousands of times through successive porous membranes, the uranium becomes enriched in ^{235}U. The first bomb-grade uranium was prepared in this way at the Oak Ridge, Tennessee gaseous diffusion plant during World War II (Figure 12.14).

Gaseous diffusion is still being used to enrich uranium, but other processes have also been developed.

EXAMPLE 12.5

Graham's Law

When an unknown gas at 2.0 atm and 20 °C is allowed to diffuse through a porous plug, the measured rate is 7.20×10^{-3} mol/s. When oxygen at the same temperature and pressure diffuses through the same barrier, the rate is 5.09×10^{-3} mol/s. What is the molar mass of the unknown gas?

SOLUTION According to Graham's law the rates of diffusion should be in inverse proportion to the square root of the ratio of molar masses. Let subscript 1 identify the unknown, and subscript 2 oxygen.

$$\frac{\text{rate of diffusion}_1}{\text{rate of diffusion}_2} = \sqrt{\frac{M_2}{M_1}}$$

Substituting the measured rates of effusion and the molar mass of O_2, we have

$$\frac{7.20 \text{ mmol/s}}{5.09 \text{ mmol/s}} = \sqrt{\frac{32.00 \text{ g/mol}}{M_1}}$$

Solving algebraically for M_1 gives

$$M_1 = 32.00 \text{ g/mol} \cdot \left(\frac{5.09 \text{ mmol/s}}{7.20 \text{ mmol/s}}\right)^2 = 16.0 \text{ g/mol}$$

Can you guess what this gas might be?

EXERCISE 12.10 • ***Calculating the Diffusion Ratio of Two Gases***

Use Graham's law to calculate the ratio of rates of diffusion of nitrogen (N_2) to hydrogen (H_2).

12.4 NONIDEAL BEHAVIOR: REAL GASES

For pressures of approximately 1 atm or less and temperatures near or above room temperature, the ideal gas law predicts the pressures, volumes, temperatures, and amounts of air and many other gases quite accurately. At much higher pressures or much lower temperatures, however, the ideal gas

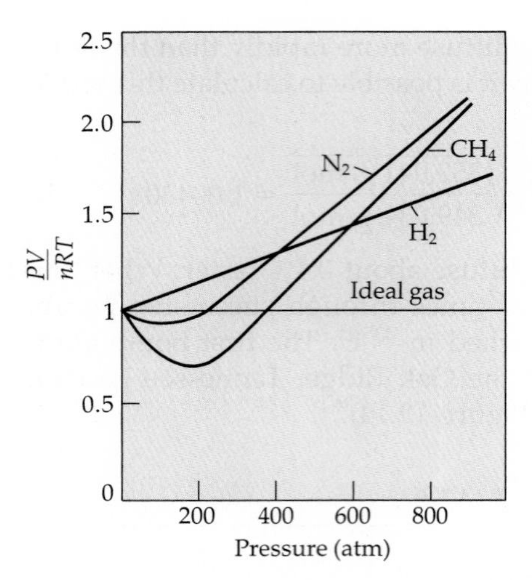

Figure 12.15
The nonideal behavior of real gases compared to that of an ideal gas. For all pressures the quotient PV/nRT for an ideal gas is 1.

law doesn't work nearly so well. As shown in Figure 12.15, for methane, CH_4, the quotient PV/nRT (which should equal 1) at first dips below 1 and then rises well above 1 with increasing pressure. The figure does not show it, but the dip below 1 becomes larger at lower temperatures. This failure of the ideal gas law happens for all gases at temperatures just above their boiling points and for very high pressures.

At standard temperature and pressure (STP), the volume occupied by a single molecule is very small relative to its share of the total gas volume. Recall that there are 6.02×10^{23} molecules in a mole and that 1 mol of a gas occupies 22.4 L (22.4×10^{-3} m^3) at STP. The volume, V, that each molecule has to move around in is given by

$$V = \frac{22.4 \times 10^{-3}\ \text{m}^3}{6.02 \times 10^{23}\ \text{molecules}} = 3.72 \times 10^{-26}\ \text{m}^3/\text{molecule}$$

The volume of a sphere is given by $\frac{4}{3}\pi r^3$. *Solving for* r:

$$r = \left(\frac{3V}{4\pi}\right)^{1/3} \cong \left(\frac{3(3.72 \times 10^{-26}\ \text{m}^3)}{4(3.14)}\right)^{1/3}$$

$$= 2.11 \times 10^{-9}\ \text{m} = 2110\ \text{pm}$$

If this volume is assumed to be a sphere, then the radius, r, of the sphere is about 2000 pm. The radius of the smallest gas molecule, the helium atom, is 31 pm (see Figure 9.17), so a helium atom has a space to move around in that is similar to the room a pea has inside a basketball. Now suppose the pressure is increased significantly, to 1000 atm. The volume available to each molecule is now a sphere only about 200 pm in radius, which means the situation is now like that of a pea inside a sphere a bit larger than a ping-pong ball. The volume occupied by the gas molecules themselves is no longer negligible, which violates the first concept of the kinetic-molecular theory. The kinetic-molecular theory and the ideal gas law deal with the volume available for the molecules to move around in, not the volume of the molecules themselves, but the measured volume of the gas must include both. Therefore, at very high pressures, the measured volume will be larger than predicted by $PV = nRT$.

Attractions between molecules cause PV/nRT to drop below the ideal value of 1 at low temperatures and medium pressures. Consider a gas molecule that is about to hit the wall of the container, as shown in Figure 12.16. The kinetic-molecular theory assumes that the other molecules exert no forces on the molecule, but in fact such forces do exist. This means that when a molecule is about to hit the wall, most other molecules are farther from the wall and therefore pull it *away* from the wall. This causes the molecule to hit the wall with less impact—the collision is softer than if there were no attraction among the molecules. Since all collisions with the walls are softer, the pressure is less than that predicted by the ideal gas law and PV/nRT is less than 1.

The lower the temperature, the greater is the deviation from ideal behavior. At lower temperatures the molecules are moving slower and on average their kinetic energies are smaller. The potential energy resulting from attractive forces between molecules becomes comparable to the average kinetic energy of the molecules at low temperatures; this causes a significant reduction of the observed pressure.

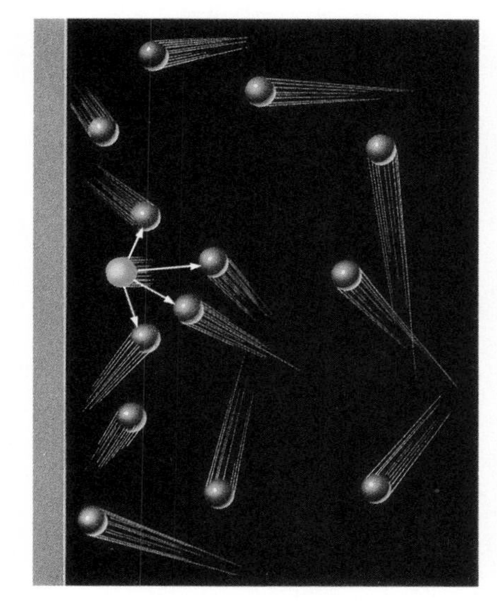

Figure 12.16
A gas molecule strikes the walls of a container with less force due to the attractive forces between it and its neighbors.

12.5 INTERMOLECULAR FORCES

If there were a universe somewhere with no gravity and without any attractions among molecules, it would be nothing but a vast collection of isolated particles, each moving about as if it were alone in space. That is, everything would be in the gaseous state. Of course, this kind of universe has never been observed. Atoms and molecules in our universe are attracted to one another. In a sense, they are "sticky."

Atoms within molecules are held together by strong chemical bonds. The table of bond energies in Section 10.5 indicates bond strengths ranging from 150 to 1000 kJ/mol. The attractive forces between two separate molecules, however, are much weaker. For example, only 1.23 kJ/mol is required to pull methane molecules away from one another. The forces between molecules are called **intermolecular forces** or **intermolecular attractions.** The difference between chemical bond forces within a molecule and intermolecular forces between molecules is illustrated in Figure 12.17.

Besides being responsible for the nonideal behavior of gases, intermolecular attractions are also responsible for the fact that gases condense to form liquids and solids. When the pressure is high, the molecules of a gas are close together and the effect of attractive forces is appreciable. When the temperature of a gas is lowered, the average kinetic energy of the molecules

$$\begin{array}{c}\mathrm{H}\\ |\\ \mathrm{H{-}C{-}H}\\ |\\ \mathrm{H}\end{array}\cdots\begin{array}{c}\mathrm{H}\\ |\\ \mathrm{H{-}C{-}H}\\ |\\ \mathrm{H}\end{array}\xrightarrow[\substack{\text{force}\\ 1.23\ \text{kJ/mol}}]{\substack{\text{overcome}\\ \text{intermolecular}}}\begin{array}{c}\mathrm{H}\\ |\\ \mathrm{H{-}C{-}H}\\ |\\ \mathrm{H}\end{array}\quad\begin{array}{c}\mathrm{H}\\ |\\ \mathrm{H{-}C{-}H}\\ |\\ \mathrm{H}\end{array}$$

$$\begin{array}{c}\mathrm{H}\\ |\\ \mathrm{H{-}C{-}H}\\ |\\ \mathrm{H}\end{array}\xrightarrow[\substack{\text{bond}\\ \text{forces}\\ 4\times 413\ \text{kJ/mol}\\ =1652\ \text{kJ/mol}}]{\substack{\text{overcome}\\ \text{chemical}}}\mathrm{C + 4\,H}$$

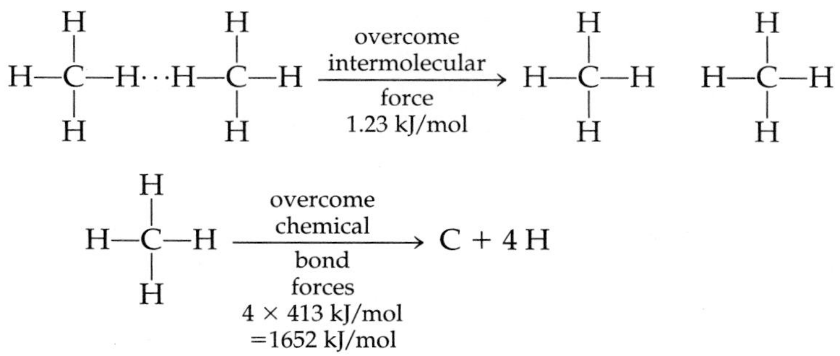

Figure 12.17
Intermolecular forces must be overcome to separate one molecule from another molecule. Chemical bond forces must be overcome to separate atoms that were originally in the same molecule.

is lowered until the molecules are not energetic enough to overcome one another's attractions. At this temperature enough gas molecules stick together to form small droplets of liquid. This phenomenon is called **condensation.** As the sample cools further, the liquid eventually solidifies as intermolecular forces of attraction confine the molecules to specific positions. But even then they still possess a vibratory motion.

The boiling point of a liquid depends upon intermolecular forces. The molecules in a liquid are very close to one another; in order for the liquid to boil, they must have enough energy to overcome the attractive forces among them. If more energy is required to overcome the intermolecular attractions between molecules of A than the intermolecular attractions between molecules of B, then the boiling point of A will be higher than that of B. Conversely, lower intermolecular attractions result in lower boiling points. In general, melting points of solids depend on intermolecular attractions in a similar way.

At the molecular level, particles are attracted together when one has a region of positive charge and another has a region of negative charge. The charges might be on ions, polar molecules, or nonpolar molecules in which there is a momentary redistribution of electron density.

Dispersion Forces

The most common intermolecular forces, found in all molecular substances, are known as **dispersion forces.** They are electrostatic in nature and result from an attraction between *induced dipoles* in adjacent molecules. An **induced dipole** is caused in a molecule when the electrons of a neighboring molecule are momentarily unequally distributed. The neighbor is said to have an instantaneous dipole, and it can then induce a dipole in the first molecule. Figure 12.18 illustrates how one molecule with a momentary unevenness in its electrical distribution can induce a dipole in a neighbor. Even nonpolar noble gas atoms, molecules of diatomic gases such as oxygen, nitrogen, and chlorine, and nonpolar hydrocarbon molecules such as CH_4 can have instantaneous dipoles that induce dipoles in their neighbors. This process of inducing a dipole is called **polarization.** In general, the more electrons there are in a molecule, the more easily they can be polarized because they are further from the restraining forces of atomic nuclei.

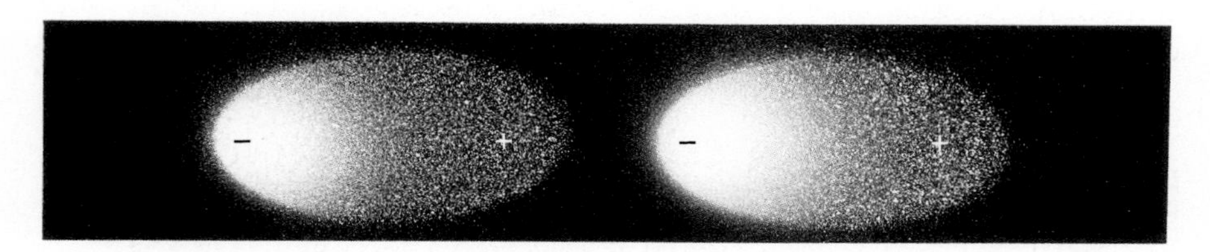

Figure 12.18

Origin of dispersion forces. If electrons are momentarily distributed unevenly in the molecule on the left, a positive charge is close to the molecule on the right. This positive charge attracts electrons toward it, creating an unbalanced electron distribution in the right-hand molecule.

Table 12.2

Effect of Molar Mass on Boiling Points of Nonpolar Molecular Substances

Noble Gases			Halogens			Hydrocarbons		
	M (g/mol)	b.p. (°C)		M (g/mol)	b.p. (°C)		M (g/mol)	b.p. (°C)
He	4	−269	F_2	38	−188	CH_4	16	−161
Ne	20	−246	Cl_2	71	−34	C_2H_6	30	−88
Ar	40	−186	Br_2	160	59	C_3H_8	44	−42
Kr	84	−152	I_2	254	184	C_4H_{10}*	58	0

*Butane

When we look at the boiling points of several groups of nonpolar molecules, the effect of number of electrons becomes readily apparent (Table 12.2). (This effect also correlates with molar mass—the heavier an atom, the more electrons it has.) Interestingly, molecular shape can also play a role in dispersion forces. Two of the isomers of pentane, straight-chain pentane and 2,2-dimethylpropane, both with the molecular formula C_5H_{12}, differ in boiling point by 27 °C. The linear shape of the pentane molecule allows close contact with adjacent molecules over its entire length, while the more spherical 2,2-dimethylpropane molecule does not allow this much close contact.

$H_3C—CH_2—CH_2—CH_2—CH_3$
pentane
b.p. 36 °C

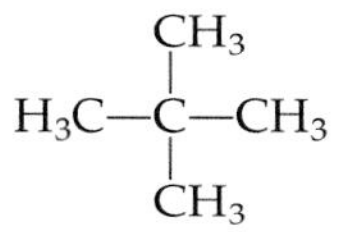

CH_3
$H_3C—C—CH_3$
CH_3
2,2-dimethylpropane
b.p. 9 °C

Dipole-Dipole Attractions

Another type of intermolecular force is the **dipole-dipole attraction** between polar molecules. Section 11.2 explained how molecules containing permanent dipoles are formed when atoms of different electronegativity bond together in an unsymmetrical manner. Molecules that are dipoles attract each other when the positive end of one is close to the negative end of another (Figure 12.19). The boiling points of several nonpolar molecules and polar molecules of comparable molar masses reflect this attraction (Table 12.3). In general, the boiling point of compound *A* will be higher than that of compound *B* (both having about the same molar mass) if the molecules of compound *A* contain permanent dipoles while those of *B* are nonpolar.

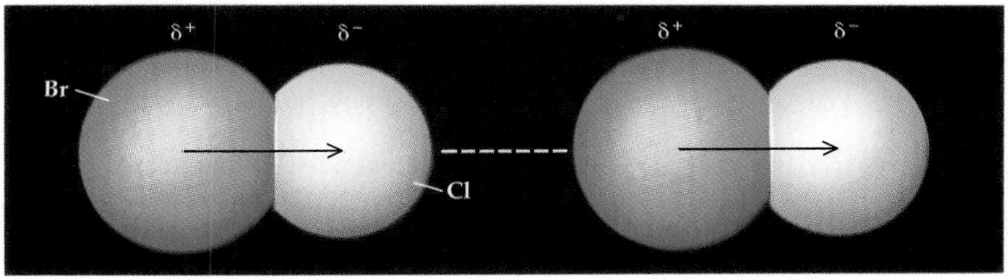

δ+ δ− δ+ δ−
Br
Cl

Figure 12.19
Dipole–dipole attractions between two molecules of BrCl. The Cl atom is more electronegative and attracts electrons away from the Br atom, which becomes slightly positive. The negative end of one molecule is attracted to the positive end of the other. The dipole–dipole force is indicated by a dotted line.

Table 12.3

Molar Masses and Boiling Points of Nonpolar and Polar Substances

Nonpolar			Polar		
	M (g/mol)	b.p. (°C)		M (g/mol)	b.p. (°C)
N_2	28	−196	CO	28	−192
SiH_4	32	−112	PH_3	34	−88
GeH_4	77	−90	AsH_3	78	−62
Br_2	160	59	ICl	162	97

Hydrogen Bonds

An especially strong intermolecular force called **hydrogen bonding** can occur when a molecule contains a hydrogen atom that is covalently bonded to a highly electronegative atom with lone-pair electrons. Usually the highly electronegative atom is fluorine, oxygen, or nitrogen. The hydrogen bond is the attraction between a slightly positive hydrogen atom on one molecule (positive because it is attached to a very electronegative atom such as F, O, or N) and a small, very electronegative atom (F, O, or N) of another molecule. The shifting of electron density toward F, O, or N causes this atom to take on a *partial negative charge.* As a result, a hydrogen atom bonded to the

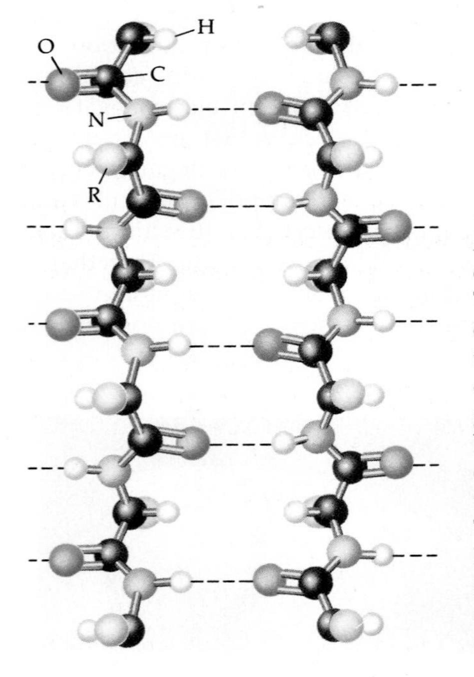

A number of natural materials consist of chains of amino acids (see Chapter 21). These chains interact with a neighboring chain by hydrogen bonding. Here are two chains that have a "backbone" consisting of C and N atoms, with R groups (such as CH_3) on every third backbone atom, an H atom attached to each N atom, and an O atom attached to the C atoms not having an R group. The polar N—H groups in one chain are hydrogen bonded to the polar C—O groups in a neighboring chain. The result is a protein found in silk and some other insect fibers.

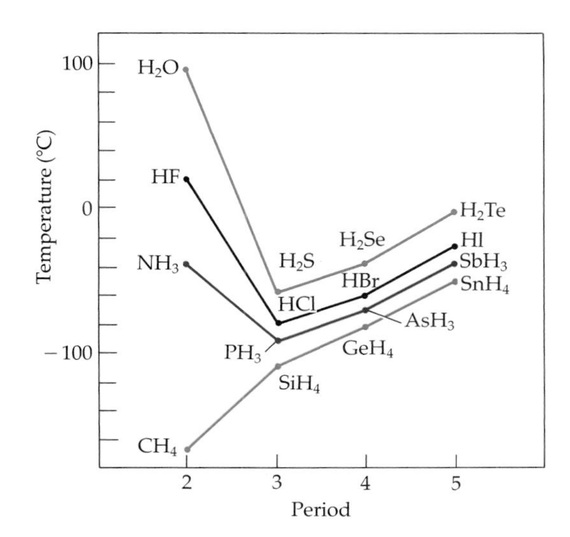

Figure 12.20
The boiling points of some simple hydrogen-containing compounds. Lines connect molecules containing atoms from the same periodic group. The effect of hydrogen bonding is evident in the high boiling points of H_2O, HF, and NH_3.

nitrogen, oxygen, or fluorine atom acquires *partial positive charge.* The greater the electronegativity of the atom connected to H, the greater the partial positive charge and hence, the stronger the hydrogen bond that H atom can take part in. Hydrogen bonds are typically shown as dotted lines (···) between the atoms.

The hydrogen bond, then, is a "bridge" between two highly electronegative atoms. A hydrogen atom is bonded covalently to one of them, and electrostatically (positive-to-negative attraction) to a lone pair on the other. Because the H atom is very small, it can come very close to the lone pair, and there is an abnormally strong dipole–dipole force. Hydrogen bonds have a strength of about one-tenth to one-fifteenth that of an average single covalent bond. However, a great many hydrogen bonds often occur in a sample of matter, and the overall effect can be very dramatic.

Hydrogen bonds can form between molecules or within a molecule. These are known as intermolecular and intramolecular hydrogen bonds, respectively.

The hydrogen halides illustrate the significant effects of hydrogen bonding. The general relationship between boiling points and molecular weight holds for hydrogen chloride (HCl), hydrogen bromide (HBr), and hydrogen iodide (HI), as shown in Figure 12.20. However, the boiling point of hydrogen fluoride (HF), the lightest compound, is much higher than expected; this is attributed to hydrogen bonding. The association between HF molecules is illustrated in Figure 12.21. The strong hydrogen bonding between HF molecules, compared with the ordinary dipole attractions between HCl molecules, explains the unusually high boiling point.

Water provides another good example of hydrogen bonding. Hydrogen compounds of oxygen's neighbors and family members in the periodic table are gases at room temperature: CH_4, NH_3, H_2S, H_2Se, H_2Te, PH_3, and HCl.

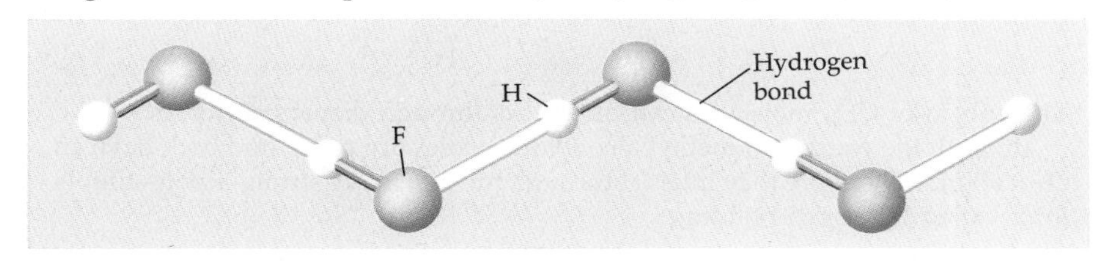

Figure 12.21
Hydrogen bonding between HF molecules. A nonbonding electron pair on an F atom in one molecule attracts an H atom in an adjacent molecule. This repeats over and over, forming a chain of HF molecules.

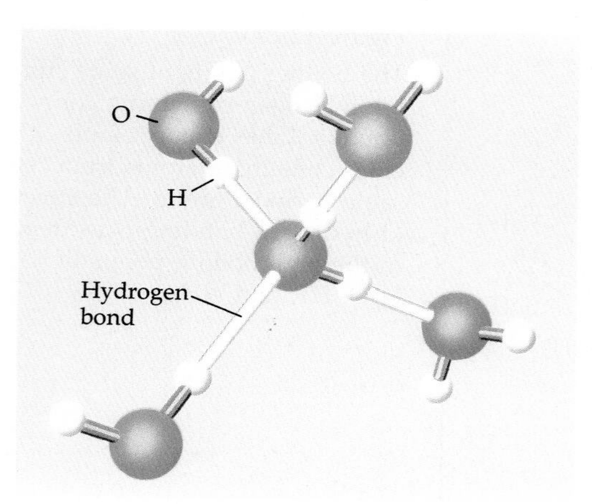

Figure 12.22
Hydrogen bonds between one water molecule and its neighbors. Each water molecule can participate in four hydrogen bonds—two through its two hydrogens and two through the two lone pairs of oxygen.

But H_2O is a liquid at room temperature, indicating a strong degree of intermolecular attraction. Figure 12.20 also illustrates that the boiling point of H_2O is about 200 °C higher than would be predicted if hydrogen bonding were not present.

The structure of solid water is described in Section 14.1.

In liquid and solid water, where the molecules are close enough to interact, the hydrogen atom on one water molecule is attracted to the lone pair of electrons on the oxygen atom of an adjacent water molecule. Since each hydrogen atom can form a hydrogen bond to an oxygen atom in another water molecule, and since each oxygen atom has two lone pairs, each water molecule can form a maximum of four hydrogen bonds to four other water molecules (Figure 12.22). The result is a tetrahedral cluster of water molecules around the central water molecule.

Relative strengths of various intermolecular forces are summarized in Table 12.4.

EXAMPLE 12.6

Intermolecular Forces

Decide which type of intermolecular force is represented by the dotted line in each of the following cases, and rank the interactions in order of increasing strength: (a) $CH_4 \cdots CH_4$, (b) $CH_3OH \cdots H_2O$, (c) $HCl \cdots HCl$.

SOLUTION (a) In the tetrahedral structure of CH_4, all of the H atoms are equivalent. Therefore, even though each C—H bond has some polarity, these polarities cancel and the CH_4 molecule is nonpolar.

```
   H          H
   |          |
H—C—H        C
   |        / \
   H       H   H H
```

The only way CH_4 molecules can interact is through dispersion forces.

(b) Both the water and methyl alcohol molecules are polar, and both have an O—H bond. Therefore they interact through the unusually strong dipole–dipole force called hydrogen bonding.

$$\overset{\delta+}{} \begin{matrix} H \\ \end{matrix} \overset{\delta-}{O}\cdots\overset{\delta+}{H}-\overset{\delta-}{O}-CH_3 \quad \text{or} \quad H-\overset{\delta-}{O}-H\cdots\overset{\delta-}{O}(CH_3)-H^{\delta+}$$

(c) The HCl molecule is polar because the Cl atom is more electronegative than the H atom. The molecules that are adjacent can interact through dipole–dipole forces in the following ways:

$$H{-}Cl\cdots H{-}Cl \text{ or } \begin{matrix} H{-\!\!-}Cl \\ \vdots \quad \vdots \\ Cl{-\!\!-}H \end{matrix}$$

In order of increasing strength, the interactions are

$$CH_4\cdots CH_4 < HCl\cdots HCl < CH_3OH\cdots H_2O$$

EXERCISE 12.11 • *Strengths of Hydrogen Bonds*

Of the three hydrogen bonds, F—H···F, O—H···O, and N—H···N, which is the strongest? Explain why.

EXERCISE 12.12 • *Intermolecular Forces*

Decide what type of intermolecular force is involved in (a) $N_2\cdots N_2$, (b) $CO_2\cdots H_2O$, (c) $CH_3OH\cdots NH_3$.

Table **12.4**

Summary of Intermolecular Forces

Type of Interaction	Principal Factors Responsible for Interaction Energy	Approximate Magnitude (kJ/mol)
Ion–Dipole	Ion charge; dipole moment	40–600
Dipole–Dipole (including H-bonding)	Dipole moment	5–25
Dipole–Induced Dipole	Dipole moment; polarizability	2–10
Induced Dipole–Induced Dipole	Polarizability	0.05–40

Increasing strength of interaction ↑

12.6 THE LIQUID STATE

At low enough temperatures most gases will condense to liquids. Condensation occurs when most molecules no longer have enough kinetic energy to overcome their intermolecular attractions. Substances that we commonly think of as liquids, such as water, alcohol, or gasoline, all condense at temperatures well above room temperature.

In a liquid the molecules are close enough together that, unlike a gas, a liquid is only slightly compressible. However, the molecules remain mobile enough that the liquid flows. Liquids have a regular structure only in very small regions, and most of the molecules continue to move about randomly. Because they are difficult to compress and their molecules are moving in all directions, confined liquids can transmit applied pressure equally in all directions. This property is used in the hydraulic fluids that operate automotive brakes and airplane wing surfaces, tail flaps, and rudders.

Unlike gases, liquids have *surface properties.* Molecules beneath the surface of the liquid are completely surrounded by other molecules and experience forces in all directions due to intermolecular attractions. By contrast, molecules at the surface are attracted only by molecules below or beside them (Figure 12.23). This unevenness of attractive forces at the surface of the liquid causes the surface to contract, making it act like a skin. The energy required to overcome the "toughness" of this liquid skin is called the **surface tension,** and it is higher for liquids that have strong intermolecular attractions. For example, water's surface tension is high compared to those of most other liquids (Table 12.5) because of the extensive hydrogen bonding that holds water molecules together. The very high surface tension of mercury is due to the strong metallic bonding that holds Hg atoms together in the liquid, while the very low surface tension of neon reflects the weakness of the dispersion forces holding the small Ne atoms together. Note that neon's surface tension must be measured at a very low temperature at which neon is a liquid.

A class of chemicals called surfactants *(soaps and detergents are examples) can dissolve in water and dramatically lower its surface tension. When this happens, water becomes "wetter," and does a better job of cleansing. You can see the effect of a surfactant by comparing how water alone beads on a surface such as the hood of a car, and how a soap solution fails to bead on the same surface. Surfactants are discussed in Section 15.6.*

Even though their densities are greater than water's density, water bugs can walk on the surface and small metal objects can be "floated." Surface tension prevents the objects from breaking through and sinking. Surface tension accounts for the nearly spherical shape of water droplets that bead

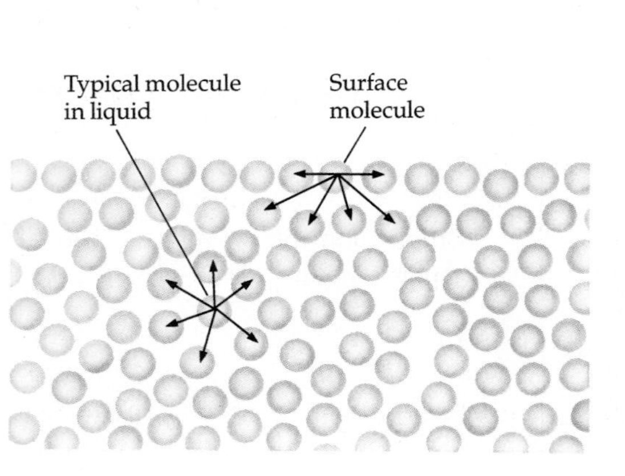

Figure 12.23
The difference between the forces acting on a molecule within the liquid phase and those acting on a molecule at the surface of the liquid.

Table 12.5

Surface Tensions of Some Liquids

Compound	Formula	Surface Tension (J/m^{2*} at 20 °C)
benzene	C_6H_6	2.85×10^{-2}
chloroform	$CHCl_3$	2.68×10^{-2}
ethyl alcohol	CH_3CH_2OH	2.23×10^{-2}
octane	C_8H_{18}	2.16×10^{-2}
mercury	Hg	46×10^{-2}
neon	Ne	0.54×10^{-2} (at −245 °C)
water	H_2O	7.29×10^{-2}

*The units of surface tension are related to the work required (in joules) to expand a unit of surface area (m^2).

The surface tension of water allows insects to walk on its surface easily. *(Hans Pfletschinger/Peter Arnold, Inc.)*

up on waxed surfaces. A sphere is the geometrical shape that has least surface area per unit volume; in a spherical droplet fewer molecules experience uneven attractive forces at the surface. Water's surface tension also causes capillary action, which is important to living organisms and to the movement of water in soils.

EXERCISE 12.13 • *Surface Tension of Liquids*

Chloroform ($CHCl_3$) has a surface tension at 20 °C of 2.68×10^{-2} J/m^2, while bromoform ($CHBr_3$) has a higher surface tension of 4.11×10^{-2} J/m^2 at the same temperature. Explain this observation in terms of intermolecular forces.

Vaporization and Condensation

Like the molecules in a gas, the molecules in a liquid are in constant motion and have a range of kinetic energies like that shown in Figure 12.11. Some molecules have more kinetic energy than the potential energy of intermolecular attractive forces holding the liquid molecules together. If such a molecule is at the surface of the liquid and moving in the right direction, it will leave the liquid phase and enter the gaseous phase (Figure 12.24). This process is **vaporization** or **evaporation.** Since high-energy molecules leave the liquid and take some of their energy with them, the vaporization process can continue only if additional energy is supplied to the liquid to produce more molecules with enough energy to vaporize. Therefore, the process of vaporization is *endothermic;* the heat required is called the **heat of vaporization.**

Capillary action is observed when the molecules of a liquid are attracted to a solid, causing the liquid level to rise up the surface of the solid until the force of gravity and the attractive forces are at equilibrium. A small glass capillary tube will allow water to rise in it due to attractions between O atoms on the surface of the glass and water molecules.

A molecule in the gas phase will eventually transfer some of its kinetic energy by colliding with slower gaseous molecules and solid objects. If it should happen to come in contact with the liquid's surface again, it can reenter the liquid phase in a process of **condensation.** The overall effect of

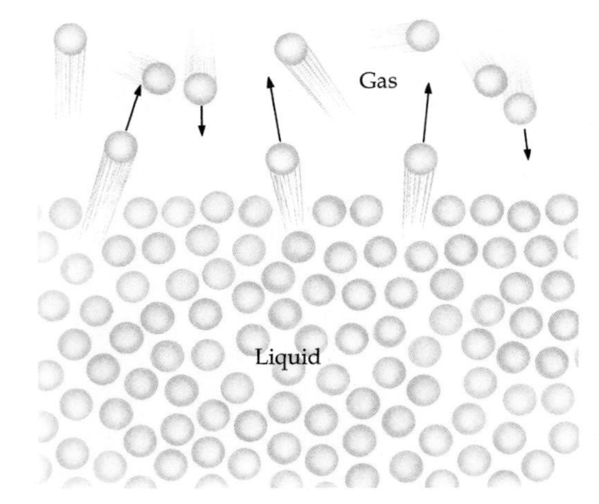

Figure 12.24
Molecules in the liquid and gas phases. All the molecules in both phases are moving, although the distances traveled in the liquid before collision with another molecule are small. Some of the liquid-phase molecules are moving with a kinetic energy large enough to overcome the intermolecular forces in the liquid and escape to the gas phase. At the same time, some molecules of the gas re-enter the liquid surface.

molecules reentering the liquid phase is the release of the heat of vaporization. This is an exothermic process. The heat evolved is equal but opposite in sign to the heat of vaporization.

The heat absorbed during evaporation is sometimes called the latent heat of vaporization. *It depends on the temperature somewhat: at 100 °C, ΔH_{vap} (H_2O) = 40.7 kJ/mol, but at 25 °C the value is 44 kJ/mol. This latter value would be obtained from Table 7.2, for example.*

The heat associated with a change in state (Section 7.3), like the heat of reaction, is represented by an enthalpy change, ΔH. For example, the quantity of heat required to vaporize completely one mole of water once it has reached the boiling point of 100 °C, and the quantity of heat released when one mole water at 100 °C condenses to liquid water at 100 °C, are

$$H_2O(\ell) \longrightarrow H_2O(g) \qquad \Delta H_{vap} = +40.7 \text{ kJ/mol}$$
$$H_2O(g) \longrightarrow H_2O(\ell) \qquad \Delta H_{cond} = -40.7 \text{ kJ/mol}$$

Table 12.6 illustrates the influence of intermolecular forces on heats of vaporization and boiling points. In the series of nonpolar hydrocarbons and rare gases, the increasing dispersion forces with increasing molecular weights are shown by the increasing ΔH_{vap} and boiling point values. Comparison of HF and H_2O with the similar molecules listed in the table shows the effect of hydrogen bonding.

EXAMPLE 12.7

Enthalpy of Vaporization

You put 1.00 L of water (about 4 cupsful) in a pan at 100 °C, and it evaporates. How much heat must have been absorbed (at constant pressure) by the water for it all to vaporize?

SOLUTION There are three pieces of information you need to solve this problem:

(a) ΔH_{vap} for water is 40.7 kJ/mol at 100 °C.

(b) The density of water at 100 °C is 0.958 g/cm^3. (This is needed because ΔH_{vap} has units of kJ/mol, so you first find the mass of water and then the number of moles.)

(c) The molar mass of water is 18.02 g/mol.

Given the density of water, a volume of 1.00 L (or 1.00×10^3 cm^3) is equivalent to 958 g, and this mass in turn is equivalent to 53.2 mol of water. Therefore,

$$\text{heat required for vaporization} = 53.2 \text{ mol } H_2O \cdot \frac{40.7 \text{ kJ}}{\text{mol}} = 2.17 \times 10^3 \text{ kJ}$$

This enthalpy change of 2170 kJ is equivalent to about one quarter of the energy in the daily food intake of an average person in the U.S.

Dew drops forming on a spider's web on a cool morning. During a warm day, water evaporates to form water vapor in the air, and a rough equilibrium is established between bodies of water and water in the air. During the night, however, the temperature drops, disturbing the equilibrium. As the temperature falls, the equilibrium vapor pressure of the water decreases. Condensation occurs if the partial pressure of water vapor in the air exceeds the equilibrium vapor pressure of water. *(C.D. Winters)*

EXERCISE 12.14 • *Enthalpy of Vaporization*

Using the data from Table 12.6, estimate the ΔH_{vap} and boiling point for Kr. Do the same for NO_2. Hint: pick a molecule in the table with a similar molar mass.

Table 12.6

Molar Enthalpies of Vaporization and Boiling Points for Some Common Substances

Substance	Molar Mass (g/mol)	ΔH_{vap} (kJ/mol)*	Boiling Point (° C) (Vapor Pressure = 760 mm Hg)
HF	19.0	25.2	19.7
HCl	36.5	17.5	−84.8
HBr	80.9	19.3	−66.5
HI	127.9	21.2	−35.1
CH_4 (methane)	16.0	8.9	−161.5
C_2H_6 (ethane)	30.1	15.7	−88.6
C_3H_8 (propane)	44.1	19.0	−42.1
C_4H_{10} (butane)	58.1	24.3	−0.5
NH_3	17.0	25.1	−33.4
H_2O	18.0	40.7	100.0
SO_2	64.1	26.8	−10.0
He	4.0	0.08	−269.0
Ne	20.2	1.8	−246.0
Ar	39.9	6.5	−185.9
Xe	131.3	12.6	−107.1
H_2	2.0	0.90	−252.8
O_2	32.0	6.8	−183.0
F_2	38.0	6.54	−188.1
Cl_2	70.9	20.39	−34.6
Br_2	159.8	29.54	59.6

*ΔH_{vap} is given at the normal boiling point of the liquid.

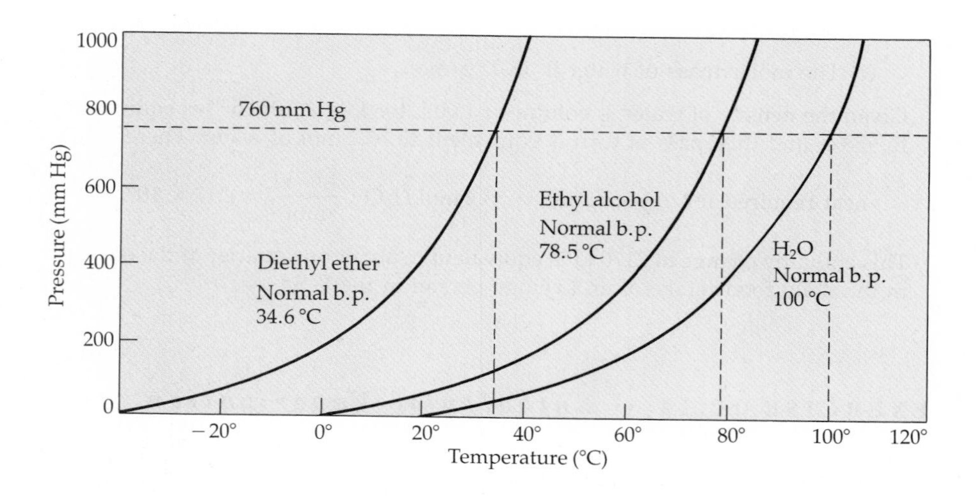

Figure 12.25
Vapor pressure curves for diethyl ether ($C_2H_5OC_2H_5$), ethyl alcohol (C_2H_5OH), and water. Each curve represents the conditions of T and P where the two phases (pure liquid and its vapor) are in equilibrium. The compound exists as a liquid under conditions of T and P defined by points to the left of the curve, and it exists as a vapor for all temperatures and pressures to the right of the curve.

Vapor Pressure

The tendency of a liquid to vaporize is called the **volatility** of the substance. The higher the temperature, the greater the volatility because a larger fraction of the molecules have sufficient energy to overcome the attractive forces at the liquid's surface. Our everyday experiences such as heating water on a stove or spilling a liquid on a hot pavement in the summer tell us that raising the temperature of the liquid makes evaporation take place more readily. Conversely, at lower temperatures, the volatility of a liquid will be lower.

A liquid in an open container will eventually evaporate completely, because air currents and diffusion take away most of the gas-phase molecules before they can reenter the liquid phase. In a closed container, however, no molecules can escape. If a liquid is injected into a container that contains a perfect vacuum, so that no other substance is present, the rate of vaporization (number of molecules vaporizing per unit time) will at first far exceed the rate of condensation. The pressure of gas above the liquid will increase as the number of gas-phase molecules increases. Eventually the system will attain a state of *dynamic equilibrium* of the sort described in Section 8.2. Once equilibrium is achieved it will appear that no further vaporization is occurring, but actually both vaporization and condensation continue at equal rates. At this point the pressure of the gas will no longer increase, and this pressure is known as the **equilibrium vapor pressure** (or just the **vapor pressure**) of the liquid. As you would expect, the vapor pressure of a liquid increases with increasing temperature (Figure 12.25).

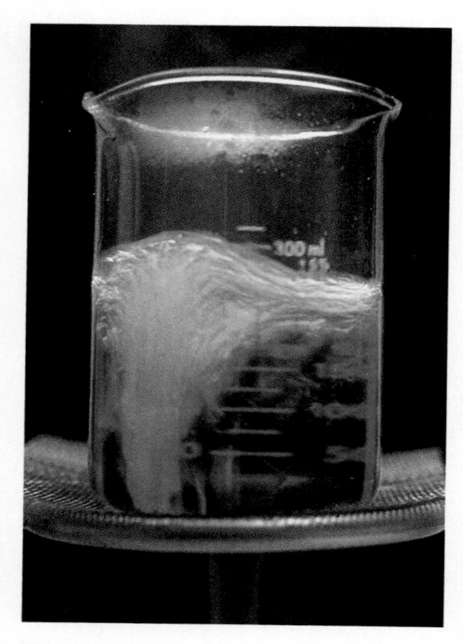

Figure 12.26
A liquid boils when its equilibrium vapor pressure equals the atmospheric pressure. *(C.D. Winters)*

If a liquid is placed in an open container and heated, a temperature eventually is reached at which the vapor pressure of the liquid is equal to the atmospheric pressure. Below this temperature, only molecules at the surface of the liquid can go into the gas phase. But when the vapor pressure equals the atmospheric pressure, the liquid begins vaporizing throughout. Bubbles of vapor form and immediately rise to the surface due to their lower density. The liquid is said to be **boiling** (Figure 12.26). The temperature at which the equilibrium vapor pressure equals the atmospheric pressure is the **boiling point** of the liquid. If the atmospheric pressure is 1 atm, the temper-

VOCs *and the* Ethanol-in-Gasoline Controversy

THE CLEAN AIR ACT OF 1990 contains provisions for reducing emissions of volatile organic compounds (VOCs). Most hydrocarbons in automotive fuels are VOCs; they enter the atmosphere when gas tanks are filled and when combustion of fuel is incomplete. To reduce VOC emission, gasoline can be changed so that it contains a smaller fraction of components with relatively high vapor pressures. (In general, this means using fewer compounds with low molecular weights.)

Hydrocarbon emissions from automobiles are a major factor in the formation of ozone in the air of cities where there are large concentrations of cars (see Section 20.3). Nine large metropolitan areas where ozone problems are the greatest, including New York City, Los Angeles, Chicago, and Philadelphia, must now use reformulated gasolines all year. Other areas of the United States may choose to require reformulated gasolines, especially in the summer months when the higher temperatures cause greater loss of VOCs.

While the regulations were being written, producers of ethanol (C_2H_5OH) succeeded in introducing a provision that 10% ethanol in gasoline (gasohol) would, by statute, meet the reformulation guidelines. There is one major problem with this provision—the presence of ethanol in gasoline actually *raises* its overall vapor pressure, and hence the VOC emissions. Ethanol does add oxygen to the fuel blend, thus reducing carbon monoxide emissions in the engine exhaust, which is good; but the increase in VOCs that it causes has generated a major controversy. Environmentalists and oil companies oppose the ethanol-in-gasoline provision, while farmers who grow corn (from which ethanol is made) favor it. One of the farmers' arguments is that ethanol use decreases the United States' reliance on foreign oil.

Gasoline can contain up to 10% by volume ethanol in many parts of the country. Federal regulations require that the pumps be clearly marked. (J. Wood)

ature is designated the **normal boiling point.** The boiling points shown in Table 12.6 are normal boiling points.

The lower the atmospheric pressure, the smaller is the vapor pressure required for boiling and hence the lower is the boiling temperature. It takes longer to hard-boil an egg high in the mountains (where the atmospheric pressure is lower) than it does at sea level, because the water at the higher elevation is boiling at a lower temperature. In Salt Lake City, Utah, where the average barometric pressure is about 650 mm Hg, water boils at about 95 °C. Refer to Figure 12.25.

To shorten cooking times, one can use a pressure cooker. This is a sealed pot (with a relief valve for safety) that allows water vapor to build up pressures slightly greater than the external atmospheric pressure. At the higher pressure, the boiling point of water is higher and foods cook faster.

EXERCISE 12.15 • *Using Vapor Pressure vs. Temperature Curves*

Use Figure 12.25 to estimate the boiling point of water at 400 mm Hg. Also estimate the boiling point of water at 1000 mm Hg.

IN CLOSING

Having studied this chapter, you should be able to

- explain the properties of gases and the relations expressed by the gas laws (Sections 12.1, 12.2).
- solve mathematical problems using the appropriate gas laws (Sections 12.1, 12.2).
- apply the ideal gas law to finding gas densities and partial pressures (Section 12.2).
- state the fundamental concepts of the kinetic-molecular theory and use them to explain gas behavior, including diffusion and effusion (Section 12.3).
- describe the differences between real and ideal gases (Section 12.4).
- define dispersion forces, dipole-dipole attractions, and hydrogen bonding and recognize when they occur (Section 12.5).
- explain the liquid properties of surface tension, vaporization and condensation, vapor pressure, and boiling point, and describe how these properties are influenced by intermolecular forces (Section 12.6).

STUDY QUESTIONS

Review Questions

1. Name the three gas laws and explain how they interrelate P, V, and T. Explain the relationships in words and with equations.
2. What are the conditions represented by STP?
3. What is the volume occupied by one mole of an ideal gas at STP?
4. What is the definition of pressure?
5. State Avogadro's law. Explain why two volumes of hydrogen react with one volume of oxygen to form two volumes of steam.
6. State Dalton's law of partial pressures. If the air we breathe is 78% N_2 and 22% O_2 on a mole basis, what is the mole fraction of O_2? What is the partial pressure of O_2 if the total pressure is 720 mm Hg?
7. List the five basic concepts of the kinetic-molecular theory. Which assumption is incorrect at high pressures? Which one is incorrect at low temperatures? Which assumption is probably most nearly correct?
8. Explain Boyle's law on the basis of the kinetic-molecular theory.
9. State the two concepts of the kinetic-molecular theory that assume certain general characteristics of the particles making up a gas. What are these assumptions?
10. State Graham's law of diffusion. Name an industrial application of this law.
11. State Graham's law of diffusion in terms of gas densities.
12. List the concepts of the kinetic-molecular theory that apply to liquids.
13. Explain on the molecular scale the process of condensation and vaporization.
14. What causes surface tension in liquids? Name a compound that has a very high surface tension. What kinds of intermolecular forces account for the high value?
15. What is the heat of vaporization of a liquid? How is it related to the heat of condensation of that liquid?
16. Explain how the equilibrium vapor pressure of a liquid might be measured.
17. Define boiling point and normal boiling point.
18. Hydrogen bonding is most important when H is bonded to certain electronegative atoms. What are those atoms?
19. Both HF (b.p. 19.7 °C) and H_2O (b.p. 100 °C) are extensively hydrogen bonded in the liquid state. Explain why H_2O has a much higher boiling point than HF.
20. Tell which member of each of the following pairs of compounds has the higher boiling point.
    (a) O_2 and N_2
    (b) SO_2 and CO_2
    (c) HF and HBr
    (d) SiH_4 and GeH_4
21. Which of the following molecules would you expect to form hydrogen bonds with water?
    (a) $CH_3—O—CH_3$
    (b) CH_4

(c) HF

(d) I_2

(e) $H{-}\overset{\overset{\displaystyle O}{\|}}{C}{-}OH$

(f) CH_3OH

22. The surface tension of a liquid decreases with increasing temperature. Using the idea of intermolecular attractions, explain why this is so.

Pressure

23. Gas pressures can be expressed in units of mm Hg, atm, torr, and kPa. Do the following conversions.
(a) 720 mm Hg to atm
(b) 1.25 atm to mm Hg
(c) 542 mm Hg to torr
(d) 740 mm Hg to kPa
(e) 700 kPa to atm

24. Convert the following pressure measurements (1 atm = 14.7 psi).
(a) 120 mm Hg to atm
(b) 2 atm to mm Hg
(c) 100 kPa to mm Hg
(d) 200 kPa to atm
(e) 36 kPa to atm
(f) 600 kPa to mm Hg

25. Mercury has a density of 13.956 g/cm^3. A barometer is constructed using an oil with a density of 0.75 g/cm^3. If the atmospheric pressure is 1.0 atm, what will be the height in meters of the oil column in the barometer?

26. A vacuum pump is connected to the top of an upright tube whose lower end is immersed in a pool of mercury. How high will the mercury rise in the tube when the pump is turned on?

The Gas Laws

27. A sample of a gas has a pressure of 100. mm Hg in a 125-mL flask. If this gas sample is transferred to another flask with a volume of 200. mL, what will be the new pressure? Assume that the temperature remains constant.

28. A sample of a gas is placed in a 256-mL flask, where it exerts a pressure of 75.0 mm Hg. What is the pressure of this gas if it is transferred to a 125-mL flask? (The temperature stays constant.)

29. A sample of a gas has a pressure of 62 mm Hg in a 100-mL flask. This sample of gas is transferred to another flask, where its pressure is 29 mm Hg. What is the volume of the new flask? (The temperature does not change.)

30. Some butane, the fuel used in backyard grills, is placed in a 3.50-L container at 25 °C; its pressure is 735 mm Hg. If you transfer the gas to a 15.0-L container, also at 25 °C, what is the pressure of the gas in the larger container?

31. A sample of gas at 30 °C has a pressure of 2 atm in a 1-L container. What pressure will it exert in a 4-L container? The temperature does not change.

32. Suppose you have a sample of CO_2 in a gas-tight syringe with a movable piston. The gas volume is 25.0 mL at a room temperature of 20. °C. What is the final volume of the gas if you hold the syringe in your hand to raise its temperature to 37 °C?

33. A balloon is inflated with helium to a volume of 4.5 L at 23 °C. If you take the balloon outside on a cold day (−10 °C), what will be the new volume of the balloon?

34. A sample of gas has a volume of 2.50 L at 670 mm Hg pressure and a temperature of 80 °C. If the pressure remains constant but the temperature is decreased, the gas occupies 1.25 L. What is this new temperature, in degrees C?

35. A sample of 9.0 L of CO_2 at 20 °C and 1 atm pressure is cooled so that it occupies 1.0 L at some new temperature. The pressure remains constant. What is the new temperature, in kelvins?

36. A bicycle tire is inflated to a pressure of 3.74 atm at 15 °C. If the tire is heated to 35 °C, what is the pressure in the tire? Assume the tire volume doesn't change.

37. An automobile tire is inflated to a pressure of 3.05 atm on a rather warm day when the temperature is 40 °C. The car is then driven to the mountains and parked overnight. The morning temperature is −5 °C. What will be the pressure of the gas in the tire? Assume the volume of the tire doesn't change.

38. A sample of gas occupies 754 mL at 22 °C and a pressure of 165 mm Hg. What is its volume if the temperature is raised to 42 °C and the pressure is raised to 265 mm Hg? Note that the number of moles does not change.

39. A balloon is filled with helium to a volume of 1.05×10^3 L on the ground, where the pressure is 745 mm Hg and the temperature is 20 °C. When the balloon ascends to a height of 2 miles where the pressure is only 600 mm Hg and the temperature is −33 °C, what is the volume of the helium in the balloon?

40. What is the pressure exerted by 1.55 g of Xe gas at 20 °C in a 560-mL flask?

41. A 1.00-g sample of water is allowed to vaporize completely inside a 10.0-L container. What is the pressure of the water vapor at a temperature of 150. °C?

42. A sample of SiH_4 weighing 4.25 g is placed in a 580-mL container. The resulting pressure is 1.2 atm. What is the temperature in ° C? (SiH_4 is a gas.)

43. To find the volume of a flask, it is first evacuated so it contains no gas at all. Next 4.4 g of CO_2 is introduced into the flask. On warming to 27 °C, the gas exerts a pressure of 730 mm Hg. What is the volume of the flask in milliliters?

44. What mass of helium in grams is required to fill a 5.0-L balloon to a pressure of 1.1 atm at 25 °C?

45. A hydrocarbon with the general formula C_xH_y is 92.26% carbon. Experiment shows that 0.293 g of the hydrocarbon fills a 185-mL flask at 23 °C with a pressure of 374 mm Hg. What is the molecular formula for this compound?

46. Forty miles above the earth's surface the temperature is −23 °C and the pressure is only 0.20 mm Hg. What is the density of air (M = 29.0 g/mol) at this altitude?

47. A newly discovered gas has a density of 2.39 g/L at 23.0 °C and 715 mm Hg. What is the molar mass of the gas?

Gas Law Stoichiometry

48. Water can be made by combining gaseous O_2 and H_2. If you begin with 1.5 L of $H_2(g)$ at 360 mm Hg and 23 °C, what volume in liters of $O_2(g)$ would you need for complete reaction if the O_2 gas is also measured at 360 mm Hg and 23 °C?

49. Gaseous silane, SiH_4, ignites spontaneously in air according to the equation

$$SiH_4(g) + 2\,O_2(g) \longrightarrow SiO_2(s) + 2\,H_2O(g)$$

If 5.2 L of SiH_4 is treated with O_2, what volume in liters of O_2 is required for complete reaction? What volume of H_2O vapor is produced? Assume all gases are measured at the same temperature and pressure.

50. Hydrogen can be made in the "water gas reaction."

$$C(s) + H_2O(g) \longrightarrow H_2(g) + CO(g)$$

If you begin with 250 L of gaseous water at 120 °C and 2.0 atm pressure, what mass in grams of H_2 can be prepared?

51. If the boron hydride B_4H_{10} is treated with pure oxygen, it burns to give B_2O_3 and H_2O.

$$2\,B_4H_{10}(s) + 11\,O_2(g) \longrightarrow 4\,B_2O_3(s) + 10\,H_2O(g)$$

If a 0.050-g sample of the boron hydride burns completely in O_2, what will be the pressure of the gaseous water in a 4.25 L-flask at 30. °C?

52. If 1.0×10^3 g of uranium metal is converted to gaseous UF_6, what pressure of UF_6 would be observed at 62 °C in a chamber that has a volume of 3.0×10^2 L?

53. Metal carbonates decompose to the metal oxide and CO_2 on heating according to this general equation.

$$M_x(CO_3)_y(s) \longrightarrow M_xO_y(s) + y\,CO_2(g)$$

You heat 0.158 g of a white, solid carbonate of a Group 2A metal and find that the evolved CO_2 has a pressure of 69.8 mm Hg in a 285-mL flask at 25 °C. What is the molar mass, M, of the metal carbonate?

54. Nickel carbonyl, $Ni(CO)_4$, can be made by the room-temperature reaction of finely divided nickel metal with gaseous CO. This is the basis for purifying nickel on an industrial scale. If you have CO in a 1.50-L flask at a pressure of 418 mm Hg at 25.0 °C, what is the maximum mass in grams of $Ni(CO)_4$ that can be made?

Kinetic-Molecular Theory and Graham's Law

55. You are given two flasks of equal volume. Flask A contains H_2 at 0 °C and 1 atm pressure. Flask B contains CO_2 gas at 0 °C and 2 atm pressure. Compare these two samples with respect to each of the following:
(a) average kinetic energy per molecule
(b) average molecular velocity
(c) number of molecules

56. Place the following gases in order of increasing average molecular speed at 25 °C:
(a) Kr (b) CH_4 (c) N_2 (d) CH_2Cl_2

57. The reaction of SO_2 with Cl_2 to give dichlorine oxide is

$$SO_2(g) + 2\,Cl_2(g) \longrightarrow SOCl_2(g) + Cl_2O(g)$$

All of the molecules involved in the reaction are gases. Place them in order of increasing average molecular speed.

58. Rank the following gases in order of increasing rate of effusion:
(a) He (b) Xe (c) CO (d) C_2H_6

59. A gas whose molar mass you wish to know effuses through an opening at a rate only one third as great as the effusion rate of helium. What is the molar mass of the unknown gas?

60. Which graph (see top of next page) would best represent the distribution of molecular speeds for the gases

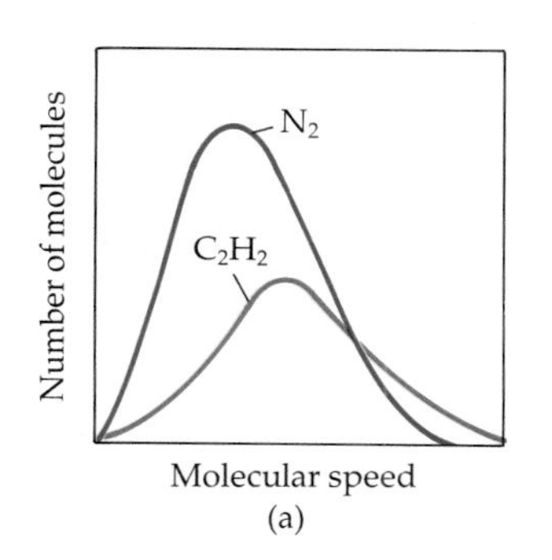

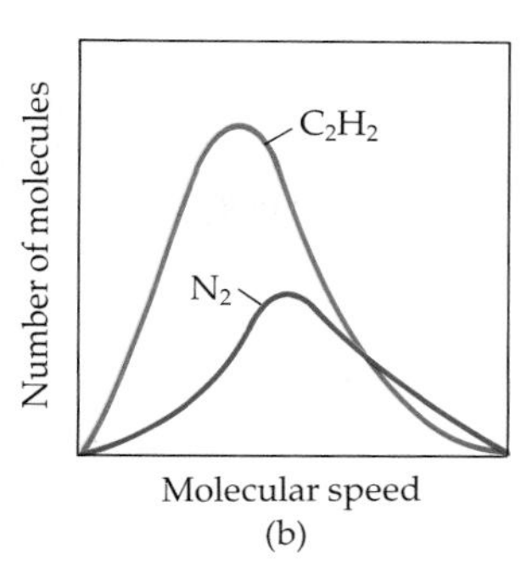

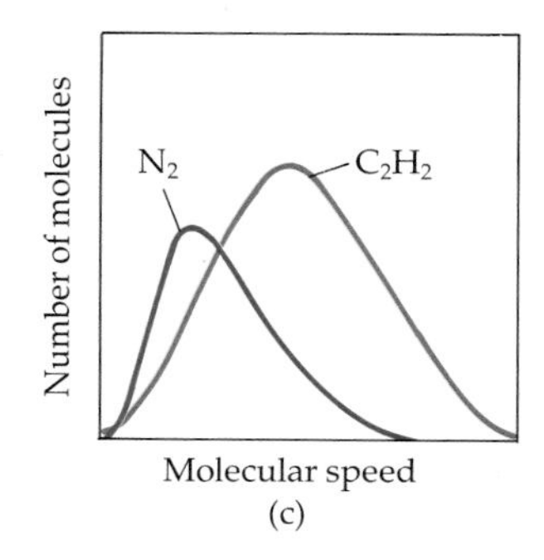

acetylene (C_2H_2) and N_2? Both gases are in the same flask with a total pressure of 750 mm Hg. The partial pressure of N_2 is 500 mm Hg.

61. Assume you have a glass tube 50. cm long (see figure).

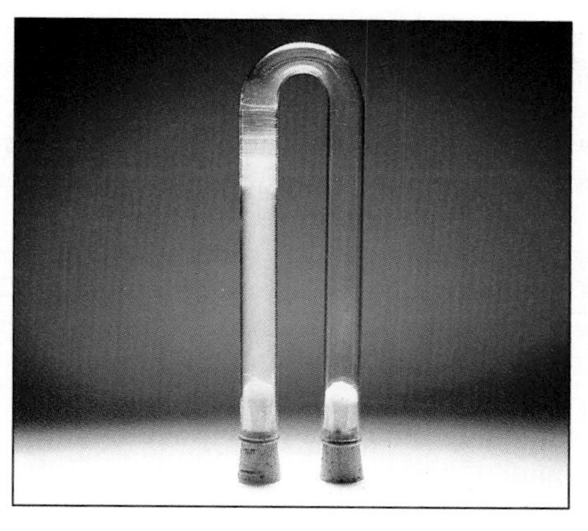

Ammonia and HCl are injected into the bent tube through the caps at opposite ends of the tube. The gases diffuse along the tube and eventually meet, where they form the white ring of solid NH_4Cl (just below the bend on the left-hand side of the tube). *(C.D. Winters)*

You allow some NH_3 gas to diffuse along the tube from one end and some HCl gas to diffuse along the tube from the opposite end. The gases are introduced into the tube at the same time. At which end of the tube (left or right) was the HCl gas introduced?

General Questions

62. Which of the following gas samples contains the largest number of molecules and which contains the smallest?
    (a) 1.0 L of H_2 at STP
    (b) 1.0 L of N_2 at STP
    (c) 1.0 L of H_2 at 27 °C and 760 mm Hg
    (d) 1.0 L of CO_2 at 0 °C and 800 mm Hg
63. The air pollutant sulfur dioxide, SO_2, is known to increase mortality in people exposed to it for 24 hours at a concentration of 0.175 ppm. (One part per million [1 ppm] is the equivalent of 1 L of SO_2 dispersed in a million liters of air.)
    (a) What is the partial pressure of SO_2 when its concentration is 0.175 ppm?
    (b) What is the mole fraction of SO_2 at the same concentration?
    (c) Assuming the air is at STP, what mass in micrograms of SO_2 is present in one cubic meter?
64. Acetone, $(CH_3)_2C{=}O$, is a common laboratory solvent. However, it is often contaminated with water. Why does acetone absorb water so readily? Draw molecular structures showing how water and acetone molecules interact. What intermolecular force(s) is (are) involved in the interaction?
65. Rationalize the observation that 1-propanol, $CH_3CH_2CH_2OH$, has a boiling point of 97.2 °C, whereas a compound with the same empirical formula, ethyl methyl ether, $CH_3CH_2OCH_3$, boils at 7.4 °C.
66. Briefly explain the variations in the following boiling points. In your discussion be sure to mention the types of intermolecular forces involved.

Compound	*Boiling Point (°C)*
NH_3	−33.4
PH_3	−87.5
AsH_3	−62.4
SbH_3	−18.4

67. Gaseous CO exerts a pressure of 45.6 mm Hg in a 56.0-L tank at 22.0 °C. If this gas is released into a room with a volume of 2.70×10^4 L, what is the partial pressure of CO (in mm Hg) in the room at 22.0 °C?

68. How much heat is required to vaporize 1.0 metric ton of ammonia? (1 metric ton = 10^3 kg) ΔH_{vap} for ammonia is 25.1 kJ/mol.
69. The chlorofluorocarbon CCl_3F has a heat of vaporization of 24.8 kJ/mol. To vaporize 1.00 kg of the compound, how much heat is required?
70. Methyl alcohol (CH_3OH) has a normal boiling point of 64.7 °C and a vapor pressure of 100 mm Hg at 21.2 °C. Another compound of the same elements, formaldehyde ($H_2C{=}O$), has a normal boiling point of −19.5 °C and a vapor pressure of 100 mm Hg at −57.3 °C. Explain why these two compounds have different boiling points and require different temperatures to achieve the same vapor pressure.
71. The molar heat of vaporization of methyl alcohol is 38.0 kJ/mol at 25 °C. How much heat is required to convert 250. mL of the alcohol from liquid to vapor? The density of CH_3OH is 0.787 g/mL at 25 °C.
72. The density of air at 20.0 km above the earth's surface is 92 g/m^3. The pressure is 42 mm Hg and the temperature is −63 °C. Assuming the atmosphere contains only O_2 and N_2, calculate (a) the average molar mass of the atmosphere, and (b) the mole fraction of each gas.
73. The atmosphere is a mixture of gases with a total pressure equal to the barometric pressure. Assume for this problem that this pressure is 740. mm Hg. Further assume that the gases of the air have the following partial pressures: $P(N_2)$ = 557 mm Hg; $P(CO_2)$ = 23 mm Hg; $P(H_2O)$ = 40 mm Hg.
(a) What is the partial pressure of O_2?
(b) List the gases in order of increasing average molecular speed.

Summary Problems

74. Acetylene can be made by allowing calcium carbide to react with water.

$$CaC_2(s) + 2\,H_2O(\ell) \longrightarrow C_2H_2(g) + Ca(OH)_2(s)$$

Assume that you place 2.65 g of CaC_2 in excess water, and collect the acetylene over water as shown in the figure. The volume of the acetylene and water vapor is 795 mL at 25.0 °C and a pressure of 735.2 mm Hg. Correct for the partial pressure of water vapor in the gas sample and calculate the percent yield of acetylene. The vapor pressure of water at 25 °C is 23.8 mm Hg.

75. Hydrogen chloride, HCl, can be made by the direct reaction of H_2 and Cl_2 in the presence of light. Assume that 3.0 g of H_2 and 140 g of Cl_2 are mixed in a 10-L flask at 28 °C.
Before reaction:

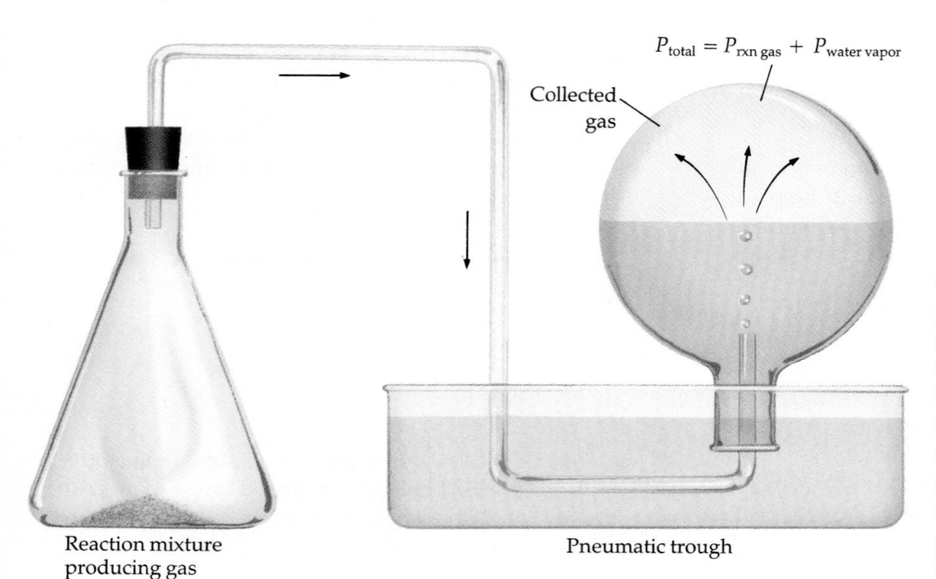

Collecting a gas over water. The pressure of the generated gas forces some of the water from the collection flask. Thus, when the pressure of the collected gas is measured (P_{total}), it is the sum of the pressure of the gas from the reaction (P_{gas}) plus the pressure of the water vapor from the evaporation of water in the collection flask ($P_{water\ vapor}$).

(a) what are the partial pressures of the two reactants?
(b) what is the total pressure due to the gases in the flask?

After reaction:

(c) what is the total pressure in the flask?
(d) what reactant remains in the flask? How many moles of it remain?
(e) what are the partial pressures of the gases in the flask?

What will be the pressure inside the flask if the temperature is increased to 40 °C?

Using Computer Programs and Videodiscs

76. Use the "KC? Discoverer" program to make a graph of heat of vaporization on the vertical axis versus boiling point on the horizontal axis for all elements.
    (a) Do most of the points lie close to a straight line?
    (b) If so, can you think of a reason why? Consider what the slope of the line is in terms of ΔH_{vap} and T_b (the boiling temperature). A straight line means constant slope.
    (c) Reread the discussion of measuring entropy in Section 7.9 and try to connect it with parts (a) and (b).
77. If "KC? Discoverer" is not available, use a spreadsheet program or the "Notebook" program (*J. Chem. Educ.: Software*, Vol. IVB, No. 1, April 1991) to graph ΔH_{vap} vs. boiling point, using data from Table 12.6. Answer the same questions as in Study Question 76.
78. Use the Shakhashiri **Chemical Demonstrations** videodisc (or videotape) to view the experiment on "Volume Changes on Mixing of Liquids." When two liquids are mixed, the total volume may be less than, or even greater than, the sum of the volumes. Account for these observations based on what you have learned about intermolecular forces.
79. Locate the experiment titled "Diffusion of Bromine Vapor" on the videotape or videodisc version of *Chemical Demonstrations* by Shakhashiri. What are the main observations you make regarding the rate of diffusion of bromine vapor in air and in a vacuum? Use the kinetic molecular theory to explain your observations.
80. Locate the experiment titled "The Collapsing Can" on the videotape or videodisc version of *Chemical Demonstrations* by Shakhashiri. In one portion of this demonstration, the narrator places a soda can, which contains a small amount of boiling water, upside down in a large beaker of cool water. What is the consequence of this action? Explain your observations in terms of the properties of liquids and gases.

Organic chemicals packaged in containers made of organic polymers.
(C.D. Winters)

Energy, Organic Chemicals, and Polymers

CHAPTER

13

The combustion of coal, natural gas, and petroleum provides 88% of all the energy used in the world. At current usage rates, world reserves of these three fuels are estimated to last 1500, 120, and 60 years, respectively. Fossil fuels are also the major source of hydrocarbons, and about 6% of the petroleum refined today is the source of most of the organic chemicals used to make consumer products. For this reason the organic chemical industry is often referred to as the petrochemical industry.

The millions of organic compounds are either hydrocarbons or derivatives of hydrocarbons. Hydrocarbons are therefore the starting materials for the synthesis of organic chemicals used in making plastics, synthetic rubber, synthetic fibers, fertilizers, and thousands of other consumer products. A few of the major classes of organic compounds and some of their reactions, especially those used in making polymers, are discussed in this chapter with the goal of introducing you to a major segment of chemistry and the chemical industry.

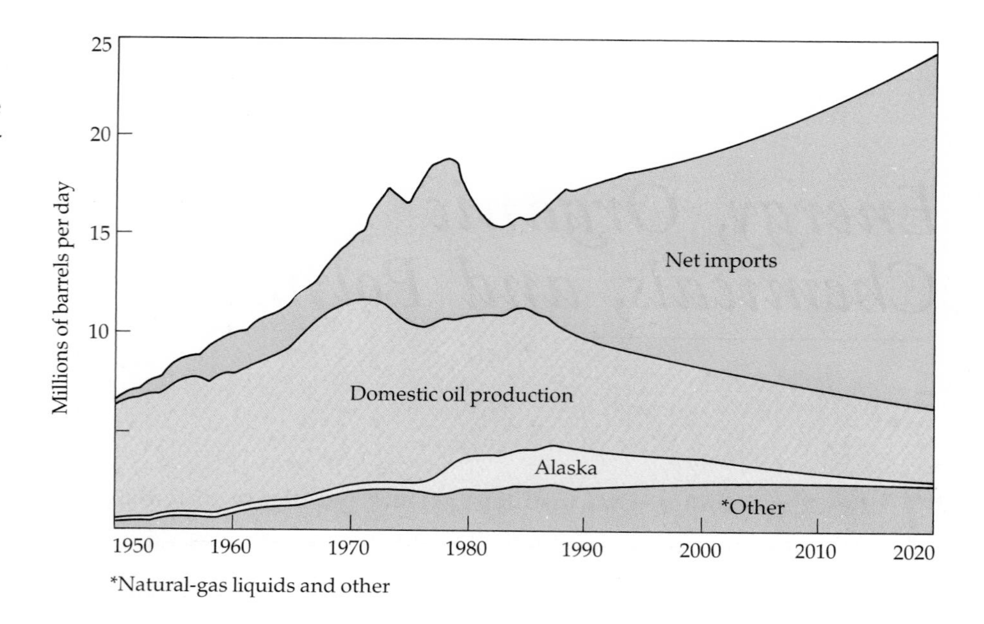

Figure 13.1
United States oil production and oil imports. At present, about 50% of the total oil used is imported, and projections show oil imports will continue to increase.

The economic importance of the organic chemical industry can be seen in the list of the chemicals produced in the largest quantities in the United States. Of the top 25 listed inside the back cover, 13 are organic chemicals. At present, petroleum is the major source of the hydrocarbons used as raw materials for the organic chemical industry. How long will petroleum be viable as a source for energy and starting materials for consumer products? At the current rate of use, all of the known petroleum reserves will be consumed by the year 2050. Oil production and oil imports by the United States over the last several years are shown in Figure 13.1. With only 4% of global reserves in 1990, the United States produced 12% of the world output of oil. Current oil use worldwide averages approximately 4.5 barrels of oil a year for each person, ranging from 24 barrels per person in the United States to less than 1 in sub-Saharan Africa. The world use rate is expected to fall to 1.5 barrels per person per year by 2030, which will require extensive changes in the global energy economy and a reduction in the use of petroleum as the major source of organic chemicals. However, both natural gas and coal are important sources of organic chemicals, and coal will likely become more important as petroleum reserves are depleted.

There are 42 gallons of oil per barrel.

13.1 PETROLEUM AND NATURAL GAS

The different classes of hydrocarbons (compounds of hydrogen and carbon) were discussed in Section 11.4. Alkanes have C—C bonds; alkenes have one or more C=C bonds; aromatics include benzene and benzene derivatives.

Petroleum is a complex mixture of alkanes, cycloalkanes, alkenes, and aromatic hydrocarbons formed from the remains of plants and animals from millions of years ago. Thousands of compounds are present in crude petroleum, and its composition varies with the location in which it is found. For example, Pennsylvania crude oils are primarily straight-chain hydrocar-

Table 13.1

Hydrocarbon Fractions from Petroleum

Fraction	Size Range of Molecules	Boiling Point Range (°C)	Uses
gas	C_1–C_4	0 to 30	gas fuels
straight-run gasoline	C_5–C_{12}	30 to 200	motor fuel
kerosene	C_{12}–C_{16}	180 to 300	jet fuel, diesel oil
gas-oil	C_{16}–C_{18}	over 300	diesel fuel, cracking stock
lubricating stock	C_{18}–C_{20}	over 350	lubricating oil, cracking stock
paraffin wax	C_{20}–C_{40}	low-melting solids	candles, wax paper
asphalt	above C_{40}	gummy residues	road asphalt, roofing tar

A petroleum refinery tower. *(Ashland Oil, Inc.)*

bons, whereas California crude oil is composed of a larger portion of aromatic hydrocarbons. The components of petroleum can be classified according to their boiling point ranges (fractions) (Table 13.1). The mixture of compounds in each of these ranges has important uses.

Petroleum Refining

The refining of petroleum begins with the separation of groups of compounds with distinct boiling point ranges by a process called **fractional distillation.** The difference between simple distillation and fractional distillation is the degree of separation achieved for the mixture being distilled. For example, water that contains dissolved solids or other liquids can be purified by distillation. The impure solution is heated to boiling; the pure water vapor is condensed and collected in a separate container (Figure 13.2, p. 556). Since petroleum contains thousands of hydrocarbons, separation of the pure compounds is neither feasible nor necessary. The products obtained from the fractional distillation of petroleum are still mixtures of hundreds of hydrocarbons, so they are called **petroleum fractions.**

Petroleum fractions are mixtures of hundreds of hydrocarbons with boiling points within a certain range.

Figure 13.3 (p. 557) is a schematic drawing of a fractional distillation tower used in the petroleum refining process. The crude oil is first heated to about 400 °C to produce a hot vapor and liquid mixture that enters the fractionating tower. The vapor rises and condenses at various points along the tower. The lower boiling petroleum fractions (those that are more volatile) will remain in the vapor stage longer than the less volatile higher boiling fractions. These differences in boiling point ranges allow the separation of fractions. Some of the gases do not condense and are drawn off the top of the tower, while the unvaporized residual oil is collected at the bottom of the tower. Typical products of the fractional distillation of petroleum are listed in Table 13.1.

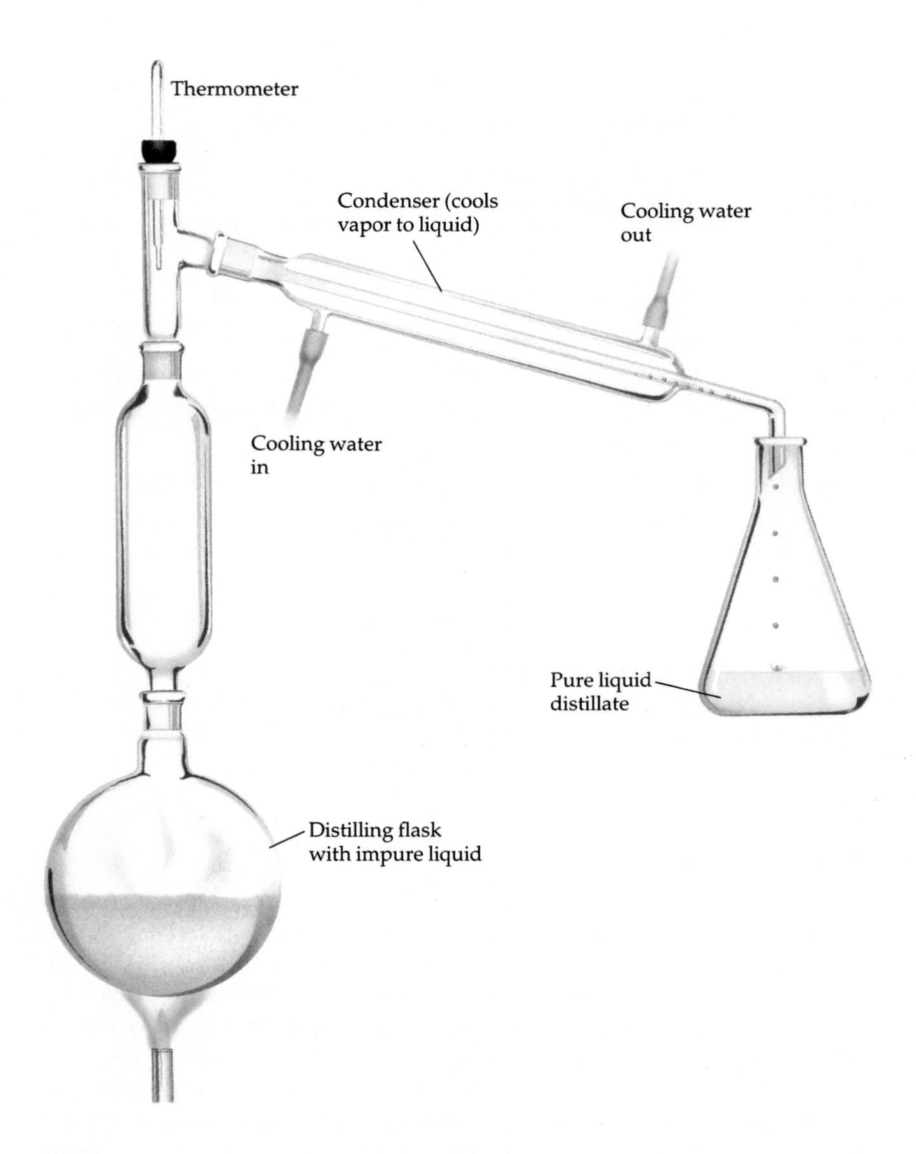

Figure 13.2
Water that contains dissolved solids or other liquids usually can be purified by distillation. When the solution is heated above the boiling point, water vaporizes and passes into the cool condenser where it liquefies and runs into the collection flask. (Open flames should not be used when heating flammable liquids.)

Octane Number

The octane number of a gasoline is a measure of its ability to prevent engine knocking. Premature ignition causes a "knocking" or "pinging" in the engine that reduces engine power and may damage the engine. Straight-chain alkanes burn less smoothly than branched-chain alkanes. For example, the "straight-run" gasoline fraction obtained from the fractional distillation of petroleum is only 55 octane and needs additional refinement because it contains primarily straight-chain hydrocarbons that burn too rapidly to be suitable for use as a fuel in internal combustion engines.

The octane number of a gasoline is determined by comparing its knocking characteristics in a one-cylinder test engine with those obtained for mixtures of heptane and 2,2,4-trimethylpentane (often called isooctane). Heptane knocks considerably and is assigned an octane number of 0 while

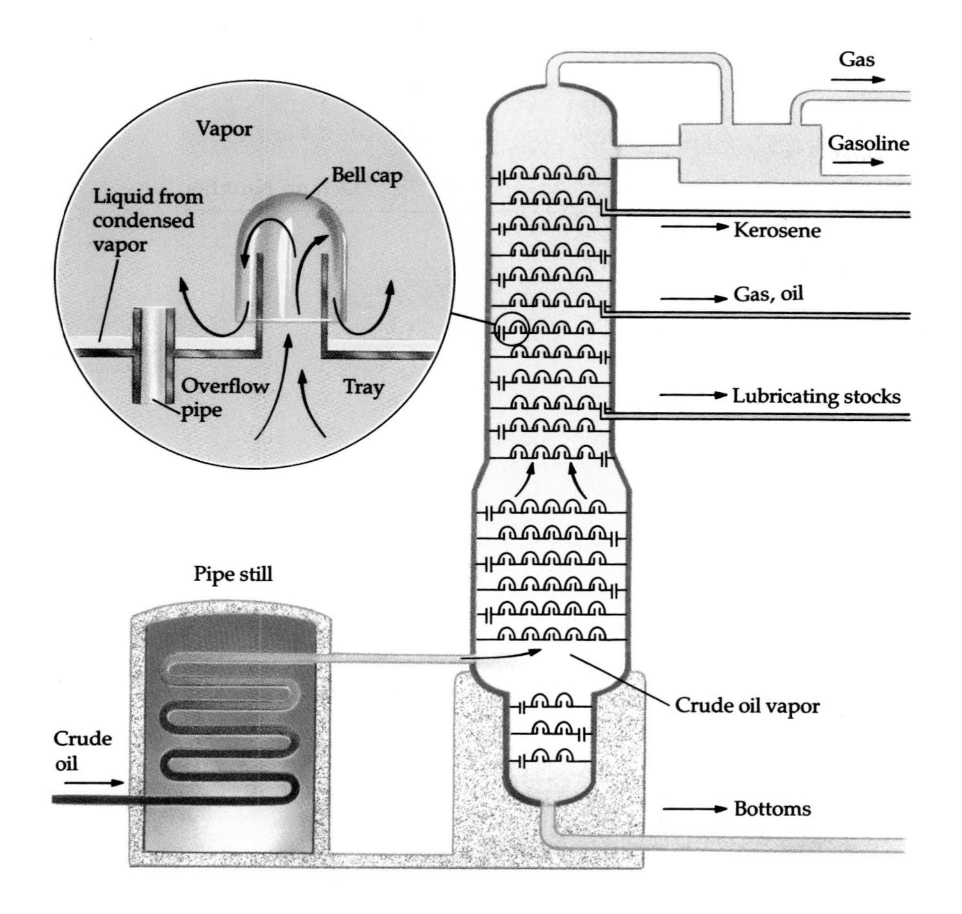

Figure 13.3
Schematic diagram of petroleum fractionation. Crude oil is first heated to 400 °C in the pipe still. The vapors then enter the fractionation tower. As vapors rise in the tower, they cool down and condense so that different fractions can be drawn at different heights.

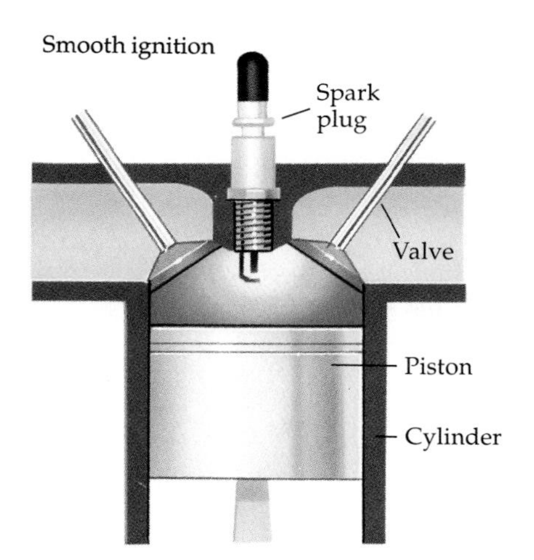

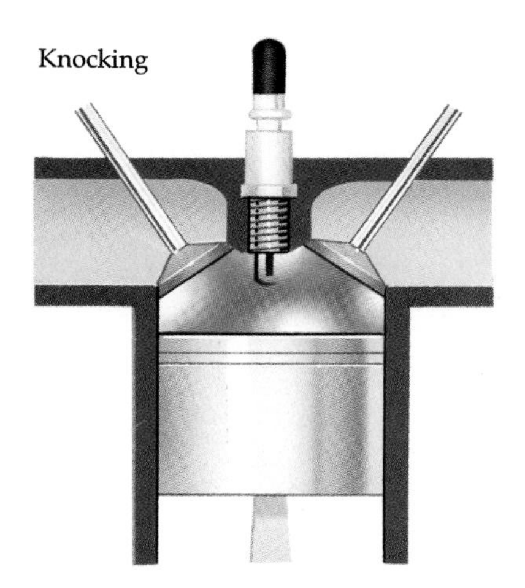

Smooth ignition and knocking.

2,2,4-trimethylpentane burns smoothly and is assigned an octane number of 100. Thus, if a gasoline has the same knocking characteristics as a mixture of 13% heptane and 87% 2,2,4-trimethylpentane, it is assigned an octane number of 87. This corresponds to the octane number of regular unleaded gasoline. Other higher grades of gasoline available at gas stations have octane numbers of 89 (regular plus) and 92 (premium).

The octane number of a gasoline can be increased either by increasing the percentage of branched chain and aromatic hydrocarbon fractions or by adding octane enhancers (or a combination of both). Since the method for determining octane numbers was established, fuels superior to 2,2,4-trimethylpentane have been developed that have octane numbers above 100. Table 13.2 lists octane numbers for some hydrocarbons and octane enhancers.

Catalytic Reforming

Increasing the octane number of straight-run gasoline by converting straight-chain hydrocarbons to branched-chain hydrocarbons and aromatics is accomplished by the **catalytic reforming process.** Under the influence of

Typical octane ratings available at gas stations. *(C.D. Winters)*

Table 13.2

Octane Numbers of Some Hydrocarbons and Gasoline Additives

Name	Octane Number
octane	−20
heptane	0
hexane	25
pentane	62
1-pentene	91
2,2,4-trimethylpentane (isooctane)	100
benzene	106
methanol	107
ethanol	108
tertiary-butyl alcohol	113
methyl *tertiary*-butyl ether	116
toluene	118

The straight-run gasoline fraction is also called the naphtha fraction.

certain catalysts, such as finely divided platinum on a support of Al_2O_3, straight-chain hydrocarbons with low octane numbers can be re-formed into their branched-chain isomers, which have higher octane numbers.

$$\underset{\substack{\text{pentane} \\ \text{62 octane}}}{CH_3CH_2CH_2CH_2CH_3} \longrightarrow \underset{\substack{\text{2-methylbutane} \\ \text{94 octane}}}{CH_3\overset{\overset{\displaystyle CH_3}{|}}{C}HCH_2CH_3}$$

Catalytic reforming is also used to produce aromatic hydrocarbons such as benzene, toluene, and xylenes by using different catalysts and petroleum mixtures. For example, when the vapors of straight-run gasoline, kerosene, and light oil fractions are passed over a copper catalyst at 650 °C, a high percentage of the original material is converted into a mixture of aromatic hydrocarbons from which benzene, toluene, xylenes, and similar compounds may be separated by fractional distillation. This process can be represented by the equation for converting hexane into benzene.

$$\underset{\substack{\text{hexane} \\ \text{25 octane}}}{CH_3CH_2CH_2CH_2CH_2CH_3} \longrightarrow \underset{\substack{\text{benzene} \\ \text{106 octane}}}{C_6H_6} + 4\,H_2$$

Catalytic reforming is one example of the use of expensive noble metals (Pt, Pd, Rh, Ir, Au, Ag) as catalysts in many industrial processes. The surfaces of these metals are often used to catalyze gas-phase reactions. For example, the platinum catalyst in the catalytic reforming process adsorbs low-octane alkanes such as hexane and converts them to compounds with higher octane numbers, such as benzene and branched or cyclic alkanes. The

To adsorb *molecules is to bind them to a surface.*

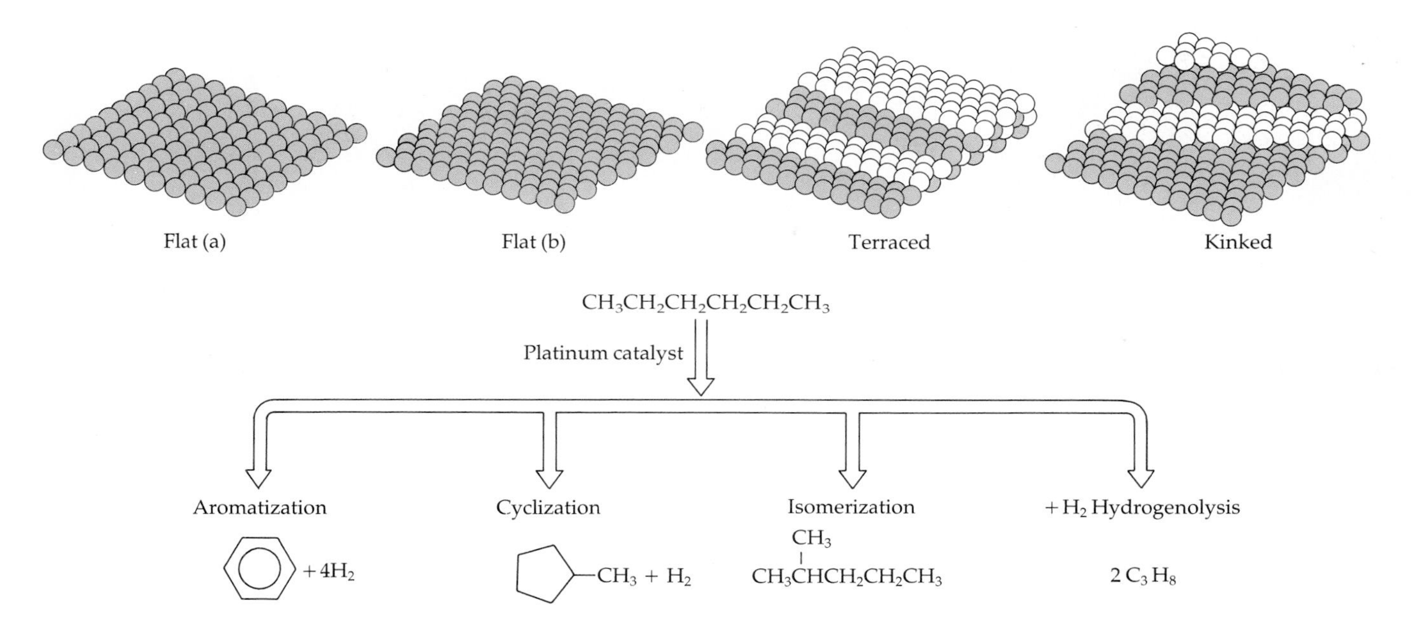

Figure 13.4
Chemistry on a platinum surface depends on which surface is exposed. (Top) Types of platinum surfaces. (Bottom) Different reactions occur on different surfaces. *(Redrawn from George C. Pimentel, "Surfaces and Condensed Phases," Chem Tech,* September 1986, p. 537.)

product depends on the pattern of platinum atoms at the exposed crystal surface where adsorption takes place, as illustrated in Figure 13.4. Note in Figure 13.4 that undesirable straight-chain alkanes are produced at the "kinked" surface. The catalytic converter found in American automobiles produced since 1975 also provides a good example of a catalytic process that uses platinum and other platinum group metals as a surface catalyst.

See Section 8.1 for a discussion of catalytic converters.

Octane Enhancers

The octane number of a given blend of gasoline can also be increased by adding "anti-knock" agents or octane enhancers. Prior to 1975, the most widely used antiknock agent was tetraethyllead, $(C_2H_5)_4Pb$. The addition of 3 g of $(C_2H_5)_4Pb$ per gallon increases the octane number by 10 to 15, and before the Environmental Protection Agency (EPA) required reductions in lead content, both regular and premium gasoline contained an average of 3 g of $(C_2H_5)_4Pb$ or $(CH_3)_4Pb$ per gallon. However, in the Clean Air Act of 1970, Congress required that 1975 model cars emit no more than 10% of the carbon monoxide and hydrocarbons emitted by 1970 models. The platinum-based catalytic converter chosen to reduce emissions of carbon monoxide and hydrocarbons required lead-free gasolines, since lead deactivates the platinum catalyst by coating its surface. For this reason, new automobiles manufactured since 1975 have been required to use lead-free gasoline to protect the catalytic converter.

As little as two tanks of leaded gasoline can destroy the activity of a catalytic converter.

Since tetraethyllead can no longer be used, other octane enhancers are added to gasoline to increase the octane number. These include toluene,

Completely automated unit in a Kentucky refinery for the production of MTBE, which is blended into gasoline to improve its octane rating, decrease its rate of evaporation, and reduce tailpipe emissions of pollutants. *(Ashland Oil, Inc.)*

Laboratory research on reformulated gasoline for automobile engines. *(Phillips Petroleum Co.)*

2-methyl-2-propanol (also called *tertiary*-butyl alcohol), methyl-*tertiary*-butyl ether (MTBE), methanol, and ethanol. The most popular octane enhancer is MTBE, which joined the top 50 chemical list for the first time in 1984 and was number 20 in 1992.

$$\begin{array}{c} \qquad\qquad CH_3 \\ \qquad\qquad | \\ CH_3{-}O{-}C{-}CH_3 \\ \qquad\qquad | \\ \qquad\qquad CH_3 \end{array}$$

MTBE

Oxygenated and Reformulated Gasolines

The 1990 amendments to the Clean Air Act require cities with excessive carbon monoxide pollution to use oxygenated gasolines during the winter season. Oxygenated gasolines are blends of gasoline with organic compounds that contain oxygen, such as MTBE, methanol, ethanol, and *tertiary*-butyl alcohol. Oxygenated gasolines burn more completely and are expected to reduce carbon monoxide emissions in urban areas by 17%. The 1990 regulations require oxygenated gasolines to contain 2.7% oxygen by weight, and this is accomplished either by adding MTBE to the gasoline at the refinery or by adding ethanol or methanol to the gasoline at distribution terminals. The EPA requires 41 cities to start using oxygenated gasolines in 1995, and several cities, including Phoenix and Denver, are already using them.

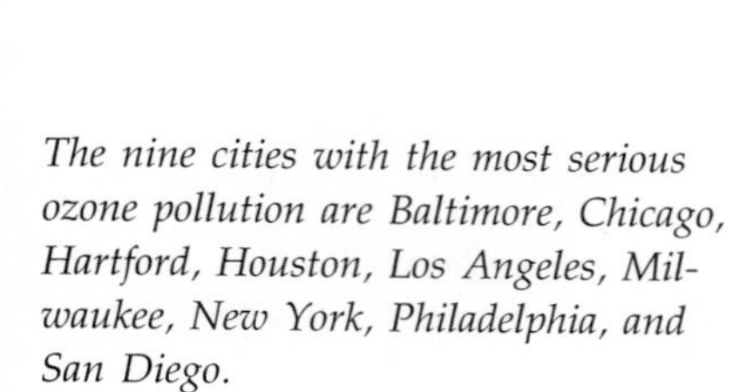

The nine cities with the most serious ozone pollution are Baltimore, Chicago, Hartford, Houston, Los Angeles, Milwaukee, New York, Philadelphia, and San Diego.

Reformulated gasolines are oxygenated gasolines that contain a lower percentage of aromatic hydrocarbons and have a lower rate of evaporation. Nine cities with the most serious ozone pollution are required by the 1990 regulations to use reformulated gasolines starting in 1995, and another 87 cities that are not meeting the ozone air quality standards can choose to use them.

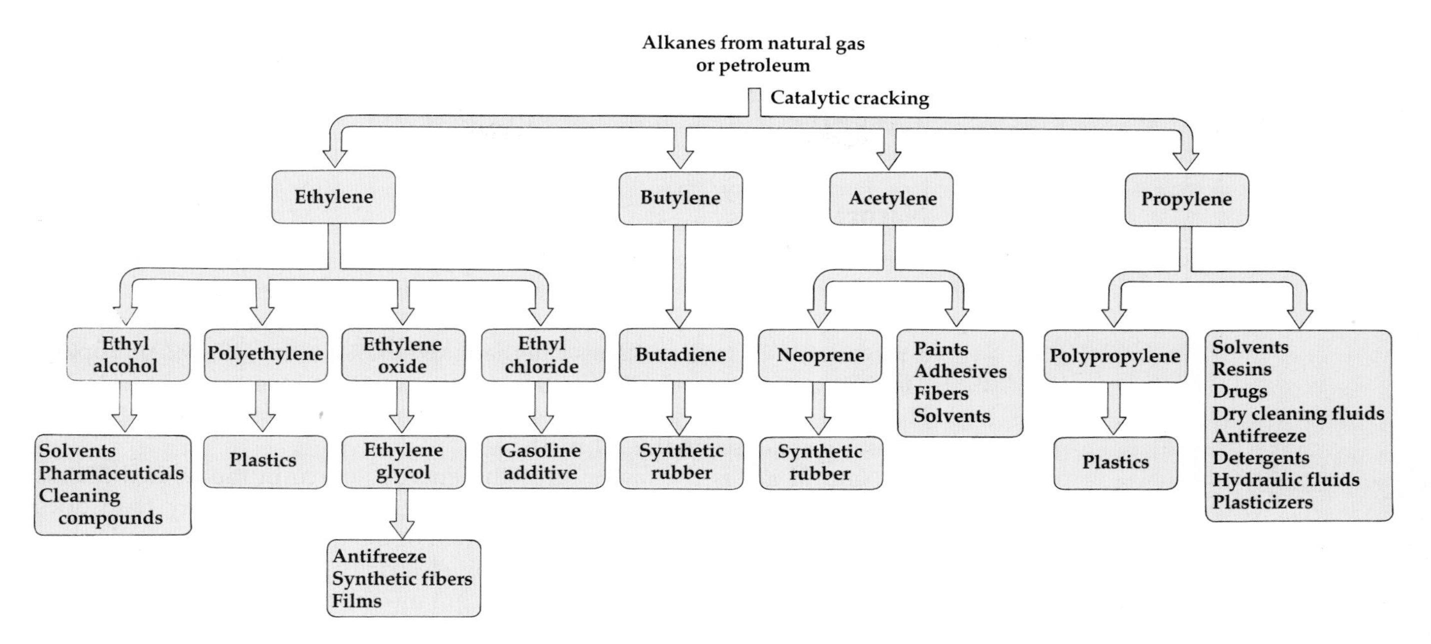

Figure 13.5
Some of the organic chemicals obtained from petroleum and natural gas and their uses as raw materials.

Catalytic Cracking

Part of the petroleum refining process involves adjusting the percentage of each fraction to match the market demand. For example, there is more demand for gasoline than for kerosene. Therefore, refiners use chemical reactions to convert the larger kerosene fraction molecules into smaller molecules in the gasoline range, in a process called "cracking." The **catalytic cracking process** uses a zeolite catalyst (Figure 11.15) and involves heating saturated hydrocarbons under pressure in the absence of air. The hydrocarbons break into shorter-chain hydrocarbons including both alkanes and alkenes, some of which will be in the gasoline range. Since alkenes have higher octane numbers than alkanes, the catalytic cracking process also increases the octane number of the mixture. Different refiners of petroleum use their own special methods, which offer different advantages in cost and in type of crude oil handled. The hydrocarbon molecules produced are much the same regardless of the methods used. Catalytic cracking is also important for the production of alkenes used as starting materials in the organic chemical industry (Figure 13.5).

Cracking breaks larger molecules into smaller ones.

$$C_{16}H_{34} \xrightarrow[\text{heat}]{\text{pressure}} C_8H_{18} + C_8H_{16}$$

an alkane — *an alkane* — *an alkene*
in the gasoline range

EXAMPLE 13.1

Octane Number

Place the following organic compounds in order of decreasing octane number: heptane, 1-pentene, benzene, 2,2,4-trimethylpentane, methanol.

SOLUTION Aromatics and alcohols have the highest octane numbers, followed by branched-chain alkanes and alkenes; straight-chain alkanes have the lowest octane numbers (see Table 13.2). The decreasing order is benzene, methanol, 2,2,4-trimethylpentane, 1-pentene, heptane.

EXERCISE 13.1 • *Catalytic Reforming*

Write a balanced equation for the catalytic reforming of heptane (C_7H_{16}) to toluene ($C_6H_5CH_3$).

Natural Gas

Natural gas is a mixture of gases trapped with petroleum in the earth's crust. It can be recovered from oil wells or from gas wells where the gases have migrated through the rock. The natural gas found in North America is a mixture of C_1 to C_4 alkanes [methane (60–90%), ethane (5–9%), propane (3–18%), and butane (1–2%)] with a number of other gases, such as CO_2, N_2, H_2S, and the noble gases, in varying amounts. In Europe and Japan the natural gas is essentially all methane.

Natural gas is the fastest growing energy source in the United States, and U.S. production of natural gas supplies 17% more energy than does U.S.-produced oil (Figure 13.6). About half of the homes in the United States are heated by natural gas, followed by electricity (18.5%), fuel oil (14.9%), wood (4.8%), and liquefied gas such as butane and propane (4.6%). Coal and kerosene come in at a low 0.5%, and solar heating of homes is even lower. However, the United States has only about 5% of the known world reserves of natural gas; at the present rate of use, this is enough to last until 2050.

Natural gas is now also used as a vehicle fuel; worldwide, there are about 700,000 vehicles powered by natural gas. Although the number of natural gas vehicles in the United States (30,000) is much smaller than in countries like Italy (300,000) and New Zealand (100,000), California and several other states are encouraging the use of natural gas vehicles to help meet new air quality regulations. Vehicles powered by natural gas emit minimal amounts of carbon monoxide, hydrocarbons, and particulates; and the price of natural gas is about one-third that of gasoline. The main disadvantages of natural gas vehicles include the need for a pressurized gas tank and the lack of service stations that sell compressed gas.

Although most natural gas is used as an energy source, it is also an important source of raw materials for the organic chemical industry (Figure 13.5).

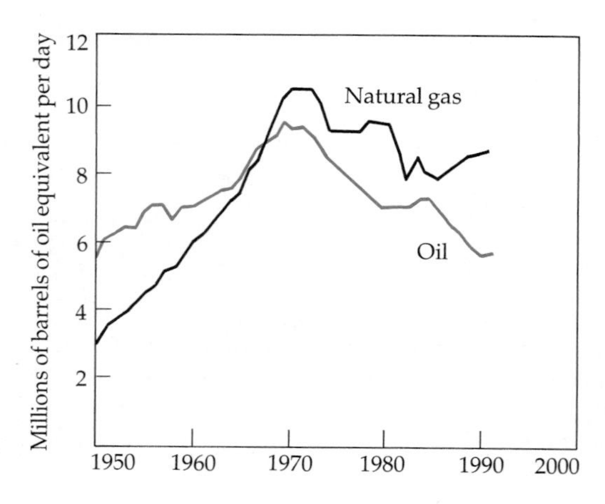

Figure 13.6
Natural gas and oil production in the continental United States, 1950–1991.

13.2
COAL

About 75% of our annual coal production is burned to produce electricity. Only 1% is used for residential and commercial heating. Coal declined as a heating fuel because it is a relatively dirty fuel, bulky to handle, and a major cause of air pollution (because of its sulfur content). The dangers of deep coal mining and the environmental disruption caused by strip mining contributed to the decline in the use of coal. Like petroleum, coal also supplies raw materials to the chemical industry.

Coal tar, a black viscous liquid, is an important source of aromatic hydrocarbons.

Most of the useful compounds obtained from coal are in the aromatic class of hydrocarbons. Heating coal at high temperatures in the absence of air, a process called pyrolysis, produces a mixture of coke, coal tar, and coal gas. One ton of bituminous (soft) coal yields about 1500 pounds of coke, 8 gallons of coal tar, and 10,000 cubic feet of coal gas. **Coal gas** is a mixture of H_2, CH_4, CO, C_2H_6, NH_3, CO_2, H_2S, and other gases, and at one time it was used as a fuel. Coal tar can be distilled to yield the aromatic fractions listed in Table 13.3. Some uses of these compounds as starting materials in the preparation of organic chemicals of commercial importance are shown in Figure 13.7.

Coal reserves are vast compared to the other fossil fuels. The known world reserves are estimated to be about 1024 billion tons, of which about 29% is in the United States. How much coal has been used and how long coal is expected to last are summarized in Figure 13.8 (p. 564).

Coal can be converted into a combustible gas (coal gasification) or a liquid fuel (coal liquefaction). In each case environmental problems can be averted, but at additional costs per energy units obtained from these fuels.

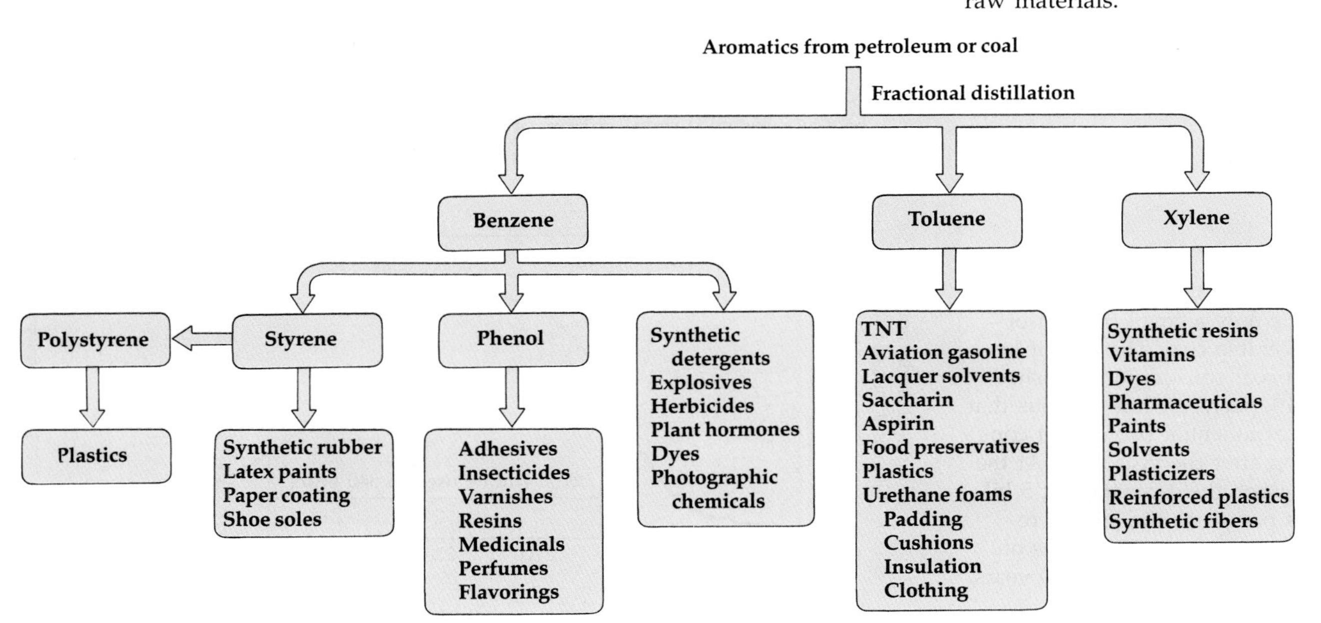

Figure 13.7
Aromatic compounds obtained from petroleum or coal and their uses as raw materials.

Coal liquefaction plant in Daggett, California. *(H.R. Bramaz/Peter Arnold, Inc.)*

Table 13.3

Fractions from Distillation of Coal Tar

Boiling Point Range (°C)	Name	Mass %	Primary Constituents
below 200	light oil	5	benzene, toluene, xylenes
200–250	middle oil (carbolic oil)	17	naphthalene, phenol, pyridine
250–300	heavy oil (creosote oil)	7	naphthalenes and methylnaphthalenes, cresols
300–350	green oil	9	anthracene, phenanthrene
residue		62	pitch or tar

Coal Gasification

When coal is pulverized and treated with superheated steam, a mixture of CO and H_2, **synthesis gas,** is obtained (Figure 13.9).

$$\underset{\text{coal}}{C(s)} + \underset{\text{steam}}{H_2O(g)} \longrightarrow CO(g) + H_2(g) \qquad \Delta H = 131 \text{ kJ/mol C}$$

Synthesis gas is used both as a fuel and as a starting material for the production of organic chemicals. The heats of combustion of its component gases are

$$2\,CO(g) + O_2(g) \longrightarrow 2\,CO_2(g) \qquad \Delta H = -566 \text{ kJ } (-283 \text{ kJ/mol CO})$$

$$2\,H_2(g) + O_2(g) \longrightarrow 2\,H_2O(g) \qquad \Delta H = -484 \text{ kJ } (-242 \text{ kJ/mol } H_2)$$

The sum of these two reactions gives the heat of combustion of synthesis gas, $(-283) + (-242) = -545$ kJ/mol. This is less heat per mole than methane ($\Delta H = -802$ kJ/mol).

In a newer coal gasification process, methane is the end product. In the process, crushed coal is mixed with a catalyst; the mixture is dried and CO

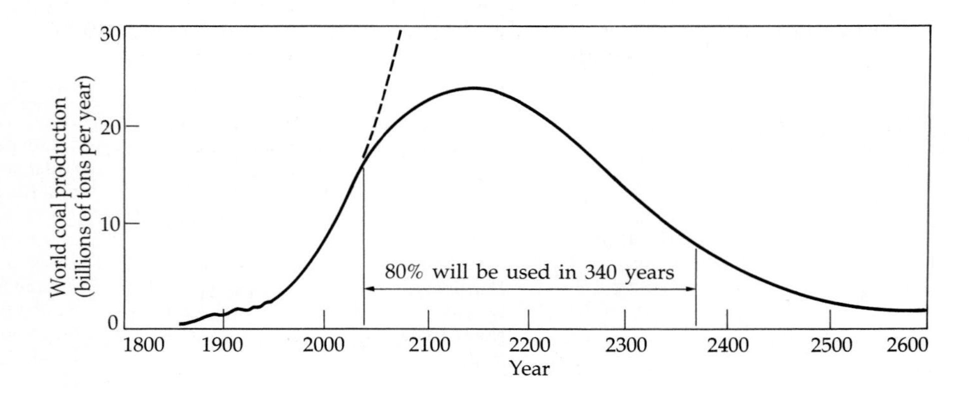

Figure 13.8
The coal mined to date (shaded area) represents only a small fraction of the recoverable coal. The rate of increase in coal consumption (dashed line) is 4% per year. It is obvious that such an exponential rise cannot continue long after the year 2000. At the present usage (held constant at 5 billion tons per year), the known reserves of about one trillion tons of coal would last for another 400 years.

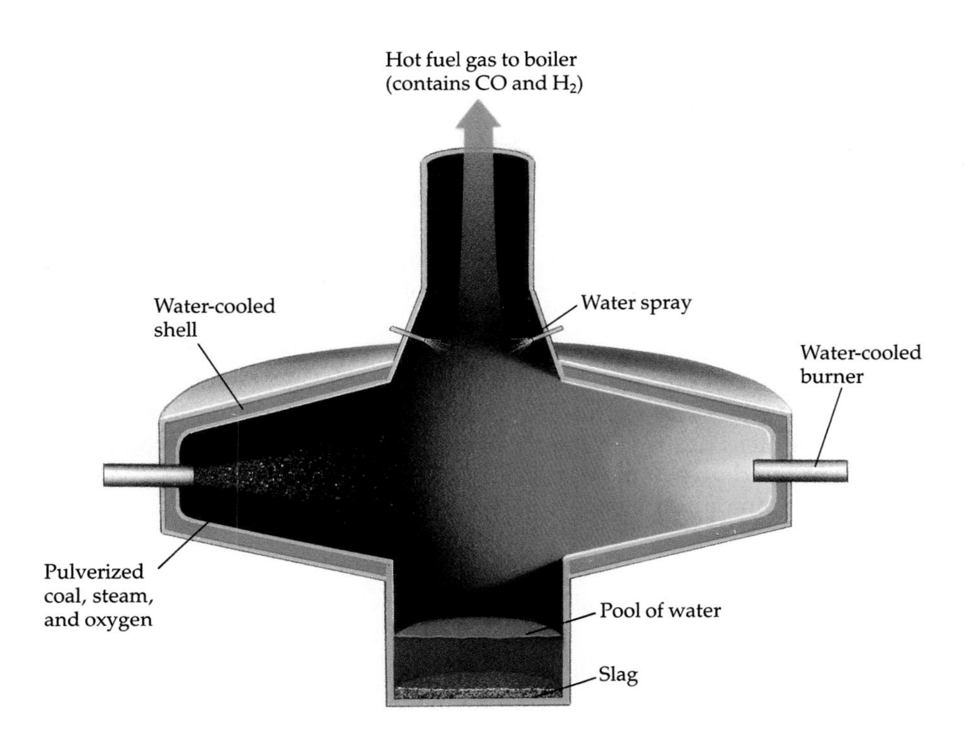

Figure 13.9
Schematic drawing of coal gasifier. A relatively cool combustion of powdered coal in a limited supply of oxygen produces a mixture of carbon monoxide and hydrogen along with other gases. The minerals in the coal collect in the slag.

and H_2 are added. The resulting mixture is then heated to 700 °C to produce methane and carbon dioxide. The overall reaction is

$$2\,C(s) + 2\,H_2O(g) \longrightarrow CH_4(g) + CO_2(g) \qquad \Delta H = 15.3\ \text{kJ/mol}\ CH_4$$

Although the overall reaction is endothermic, the process is an energy-efficient way to obtain methane, an environmentally clean fuel that releases 802 kJ/mol when burned.

Coal Liquefaction

Liquid fuels are made from coal by treating the coal with H_2 under high pressure in the presence of catalysts (hydrogenating the coal). The process produces hydrocarbons like those in petroleum. The product is similar to crude oil and can be fractionally distilled as described for petroleum in Section 13.1 to give fuel, gasoline, and hydrocarbons used in the manufacture of plastics, medicines, and other commodities. About 230 gallons of liquid are produced for each ton of coal. Today the cost of a barrel of liquid from coal liquefaction is about double that of a barrel of crude oil. However, as petroleum supplies diminish and the cost of crude oil increases, coal liquefaction should become economically feasible.

EXERCISE 13.2 • *Synthesis Gas*

Explain why the production of synthesis gas is economically feasible even though the production of synthesis gas from coal is endothermic.

13.3
ORGANIC CHEMICALS

Organic chemicals were once obtained only from plants, animals, and fossil fuels. Living organisms are still direct sources for hydrocarbons and many other important chemicals, such as sucrose from sugar cane or ethanol from fermented grain mash. However, the development of organic chemistry has led to cheaper methods for the synthesis of both naturally occurring substances and new substances.

Prior to 1828, it was widely believed that chemical compounds found in living matter could not be made without living matter—a "vital force" was thought to be necessary for the synthesis. In 1828, a young German chemist, Friedrich Wöhler, destroyed the vital force myth when he prepared the organic compound urea, a major product in urine, by evaporating an aqueous solution of ammonium cyanate, an inorganic compound.

Friedrich Wöhler (1800–1882) was professor of chemistry at the University of Berlin and later at Göttingen. His preparation of the organic compound urea from the inorganic compound ammonium cyanate did much to overturn the theory that organic compounds must be prepared in living organisms. He was also one of the first to study the properties of aluminum and the first to isolate the element beryllium, among many other outstanding contributions to chemistry. *(The Oesper Collection in the History of Chemistry/University of Cincinnati)*

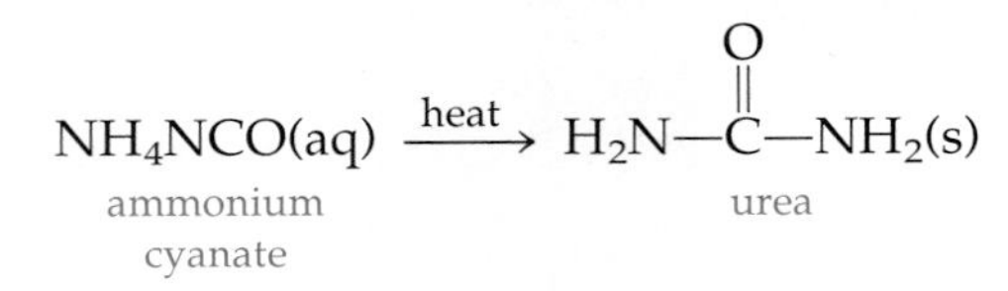

The notion of a mysterious vital force declined as other chemists began to prepare more and more organic chemicals without the aid of a living system, and millions of organic compounds have been prepared in the laboratories of the world since Wöhler's discovery.

Organic molecules can all be viewed as derived from hydrocarbons, and many of them are prepared in this manner. Saturated hydrocarbons (alkanes), however, do not react with concentrated strong acids, strong bases, or strong oxidizing agents. The only reaction that alkanes undergo readily is combustion, which produces carbon dioxide and water when the reaction is complete.

In the presence of heat and a catalyst, alkanes also undergo **substitution reactions,** in which the hydrogen atoms are replaced by other atoms. The reaction of methane with chlorine in the presence of ultraviolet light is an example of substitution:

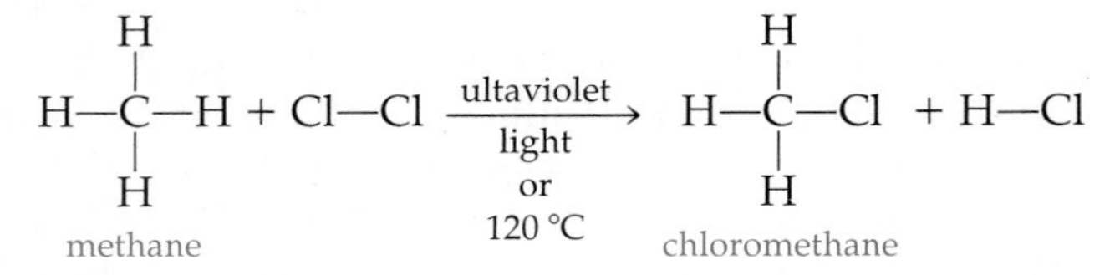

Further reaction can occur to give CH_2Cl_2 (dichloromethane), $CHCl_3$ (chloroform), or CCl_4 (carbon tetrachloride).

If alkanes are not very reactive, how can alkanes from petroleum be the starting materials for making organic chemicals with functional groups, such as the alcohols? The answer is that alkanes must first be subjected to catalytic cracking (Section 13.1) to give alkenes and alkynes. Because of the presence of double and triple bonds to which other atoms can be added (Section 10.2), alkenes and alkynes are more reactive than alkanes and can be used as starting materials for the preparation of other organic chemicals

(Figure 13.5). For example, they undergo **addition reactions,** which occur more quickly than the substitution reactions of alkanes. Compared to substitutions, which require heat and catalysts, the reactions of alkenes and alkynes with chlorine can take place at room temperature in the dark.

$$H_2C{=}CH_2 + Cl{-}Cl \xrightarrow[25\ ^\circ C]{dark} H{-}CHCl{-}CHCl{-}H$$

ethene (ethylene) — 1,2-dichloroethane

Other addition reactions include hydrogenation (Section 11.2) and formation of an alcohol by addition of the H and OH of water to the double bond in the presence of an acid.

A table of functional groups and further information on how organic compounds are named is given in Appendix E.

$$H_2C{=}C(CH_3)_2 + H_2O \xrightarrow{H_3O^+} H{-}CH_2{-}C(CH_3)(OH){-}CH_3$$

2-methylpropene — 2-methyl-2-propanol (*tertiary*-butyl alcohol)

EXERCISE 13.3 • *Addition and Substitution Reactions*

What is the difference between addition and substitution reactions of hydrocarbons? Illustrate by writing equations for each type of reaction.

13.4 ALCOHOLS AND THEIR OXIDATION PRODUCTS

Alcohols, which contain one or more —OH groups, are a major class of organic compounds. Some examples of commercially important alcohols and their uses are listed in Table 13.4.

Alcohols are classified according to the number of carbon atoms bonded to the —C—OH carbon as primary (one other C atom), secondary (two other C atoms), and tertiary (three other C atoms).

Primary	*Secondary*	*Tertiary*
$R{-}CH_2{-}OH$	$R'{-}CH(R){-}OH$	$R'{-}C(R)(R''){-}OH$

The use of R, R′, R″ indicates all R groups can be different.

Ethanol and 1-propanol are primary alcohols. 2-Propanol is a secondary alcohol and is one of the two constitutional isomers of an alcohol with three carbon atoms (Section 11.4). For an alcohol with four carbon atoms, there are

The nomenclature of alcohols is described in Appendix E.

Table 13.4

Some Important Alcohols

Condensed Formula	B.p. (°C)	Systematic Name	Common Name	Use
CH_3OH	65.0	methanol	methyl alcohol	fuel, gasoline additive, making formaldehyde
CH_3CH_2OH	78.5	ethanol	ethyl alcohol	beverages, gasoline additive, solvent
$CH_3CH_2CH_2OH$	97.4	1-propanol	propyl alcohol	industrial solvent
$CH_3\underset{\displaystyle OH}{\underset{\vert}{C}}HCH_3$	82.4	2-propanol	isopropyl alcohol	rubbing alcohol
$\underset{\displaystyle OH}{\underset{\vert}{CH_2}}\underset{\displaystyle OH}{\underset{\vert}{CH_2}}$	198	1,2-ethanediol	ethylene glycol	antifreeze
$\underset{\displaystyle OH}{\underset{\vert}{CH_2}}\underset{\displaystyle OH}{\underset{\vert}{CH}}\underset{\displaystyle OH}{\underset{\vert}{CH_2}}$	290	1,2,3-propanetriol	glycerol (glycerin)	moisturizer in foods

four constitutional isomers, including 2-methyl-2-propanol, which is commonly known as *tertiary*-butyl alcohol and is used as a gasoline additive:

$$CH_3CH_2CH_2CH_2OH \qquad CH_3\overset{\displaystyle CH_3}{\overset{\vert}{C}}HCH_2OH$$

1-butanol, primary alcohol — 2-methyl-1-propanol, primary alcohol

$$CH_3\overset{\displaystyle OH}{\overset{\vert}{C}}HCH_2CH_3 \qquad (CH_3)_3C—OH$$

2-butanol, secondary alcohol — 2-methyl-2-propanol, tertiary alcohol

EXAMPLE 13.2

Alcohols

Write the condensed structural formula for 2-methyl-2-butanol. Is this a primary, secondary, or tertiary alcohol?

SOLUTION

$$CH_3CH_2—\overset{\displaystyle CH_3}{\overset{\vert}{\underset{\displaystyle OH}{\underset{\vert}{C}}}}—CH_3$$

This is a **tertiary** alcohol.

Stepwise oxidation of primary alcohols produces aldehydes and carboxylic acids. For example, the stepwise oxidation of ethanol with aqueous potassium permanganate or aqueous sodium dichromate produces acetaldehyde, a member of the **aldehyde** functional group class, followed by acetic acid, a member of the **carboxylic acid** functional group class.

Oxidation of organic compounds is usually the addition of oxygen or the removal of hydrogen, and reduction is usually the removal of oxygen or the addition of hydrogen.

$$\underset{\text{ethanol (a primary alcohol)}}{CH_3CH_2OH} \xrightarrow{KMnO_4(aq)} \underset{\text{acetaldehyde}}{CH_3-\overset{\overset{\displaystyle O}{\|}}{C}-H} \xrightarrow{KMnO_4(aq)} \underset{\text{acetic acid}}{CH_3\overset{\overset{\displaystyle O}{\|}}{C}-OH}$$

Members of the **ketone** class of functional groups are prepared by the oxidation of secondary alcohols. Acetone, commercially the most important ketone, can be prepared by the oxidation of 2-propanol. This method is also an important commercial source of hydrogen peroxide.

Aldehydes contain the $-\overset{\overset{\displaystyle H}{|}}{C}=O$ (or —CHO) functional group. Carboxylic acids contain the $-\overset{\overset{\displaystyle O}{\|}}{C}-OH$ (or —COOH) functional group.

$$\underset{\text{2-propanol (a secondary alcohol)}}{H-\overset{\overset{\displaystyle CH_3}{|}}{\underset{\underset{\displaystyle CH_3}{|}}{C}}-OH} + O_2 \xrightarrow[90\text{–}140\ ^\circ C]{15\text{–}20\text{ atm}} \underset{\text{acetone (a ketone)}}{\overset{\overset{\displaystyle CH_3}{|}}{\underset{\underset{\displaystyle CH_3}{|}}{C}}=O} + \underset{\text{hydrogen peroxide}}{H_2O_2}$$

Ketones contain a carbonyl group, $-\overset{\overset{\displaystyle O}{\|}}{C}-$, bonded to two carbon atoms: $-\overset{|}{\underset{|}{C}}-\overset{\overset{\displaystyle O}{\|}}{C}-\overset{|}{\underset{|}{C}}-$

In analogy with water, alcohols react with alkali metals to produce hydrogen. The other product, instead of hydroxide ion, is the alkoxide ion, OR^-, which is also a strong base.

$$2\ HOH(\ell) + 2\ Na(s) \longrightarrow H_2(g) + 2\ Na^+(aq) + 2\ OH^-(aq)$$
$$2\ ROH(\ell) + 2\ Na(s) \longrightarrow H_2(g) + 2\ Na^+(\text{in alcohol}) + 2\ OR^-\ (\text{in alcohol})$$

The physical properties of alcohols offer an example of the effects of hydrogen bonding between molecules in liquids. In Table 13.4 the boiling points of alcohols are listed. Hydrogen bonding in methanol (32 g/mol) explains why it is a liquid, whereas propane, which has a larger molar mass (44 g/mol), has no hydrogen bonding and so is a gas at the same temperature. The boiling point of methanol is lower than that of water because methanol has only one —OH hydrogen atom available for hydrogen bonding. The higher boiling point of ethylene glycol can be attributed to the presence of two —OH groups per molecule. Glycerol, with three —OH groups, has an even higher boiling point.

The alcohols listed in Table 13.4 are very soluble in water because of hydrogen bonding between water molecules and the —OH group in alcohol molecules. However, as the length of the hydrocarbon chain increases, the alcohols become less soluble because the nonpolar hydrocarbon portion has greater influence on the solubility than the hydrogen bonding by the —OH groups.

Reaction of sodium metal with ethanol. *(C.D. Winters)*

EXAMPLE 13.3

Oxidation of Alcohols

Write the condensed structural formula of the alcohol that can be oxidized to make the following:

a. $CH_3CH_2CH_2\overset{O}{\overset{\|}{C}}—H$ b. $CH_3CH_2\overset{O}{\overset{\|}{C}}CH_3$ c. $CH_3CH_2CH_2CH_2\overset{O}{\overset{\|}{C}}—OH$

SOLUTION

a. The oxidation of the four-carbon primary alcohol, 1-butanol, will produce this aldehyde. The condensed structural formula of 1-butanol is $CH_3CH_2CH_2CH_2OH$.

b. To produce this ketone, choose the secondary alcohol, 2-butanol, which has two carbons to the left and one carbon to the right of the C—OH group. The condensed structural formula of 2-butanol is

$$CH_3CH_2\overset{OH}{\overset{|}{C}}HCH_3.$$

c. The oxidation of the five-carbon primary alcohol, 1-pentanol, will produce this carboxylic acid. The condensed structural formula of 1-pentanol is $CH_3CH_2CH_2CH_2CH_2OH$.

EXERCISE 13.4 • *Oxidation of Alcohols*

Draw the structure of the expected oxidation products of

a. $CH_3CH_2CH_2OH$ b. $CH_3CH(OH)CH_3$

Methanol

Methanol, CH_3OH, is the simplest of all alcohols and is highly toxic. Drinking as little as 30 mL can cause death, and smaller amounts (10 to 15 mL) cause blindness. More than 8 billion pounds of methanol are produced annually in the United States because it is so useful. About 50% is used in the production of formaldehyde (used in plastics, embalming fluid, germicides, and fungicides); 30% is used in the production of other chemicals; and the remaining 20% is used for jet fuels, antifreeze mixtures, and solvents, as a gasoline additive, and as a denaturant (a poison added to make ethanol unfit for beverages).

Methanol currently is prepared from synthesis gas. High pressure, high temperature, and a mixture of catalysts are used to increase the yield.

$$\underset{\text{coal}}{C(s)} + \underset{\text{steam}}{H_2O(g)} \longrightarrow \underset{\text{synthesis gas}}{CO(g) + H_2(g)}$$

$$CO(g) + 2\,H_2(g) \xrightarrow[300\,°C]{ZnO,\ Cr_2O_3} CH_3OH(g)$$

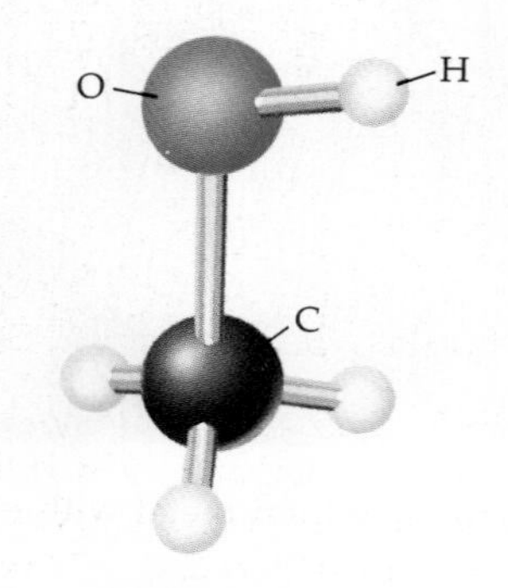

A model of methanol.

An old method of producing methanol involved heating a hardwood such as beech, hickory, maple, or birch in the absence of air. For this reason methanol is sometimes called *wood alcohol.*

Methanol will likely continue to increase in importance as petroleum and natural gas become too expensive as sources of both energy and chemicals. Although most of the world's methanol currently comes from synthesis gas made from natural gas, coal gasification will become a more important source of methanol as natural gas reserves are used up. Since methanol is relatively cheap, its potential as a fuel and as a starting material for the synthesis of other chemicals is receiving more attention.

Many race cars use methanol fuel. *(Bernard Asset/Photo Researchers, Inc.)*

Methanol as a Fuel

Methanol is being considered as a replacement for gasoline, especially in urban areas that have extremely high levels of air pollution caused by motor vehicles. For example, Southern California has been testing methanol-powered cars since 1981. About half the test cars use 100% methanol **(M100)** while the other half are flexible-fueled vehicles (FFVs) that use either **M85,** a blend of 85% methanol and 15% gasoline, or gasoline.

What are the advantages and disadvantages of switching to methanol-powered vehicles? Methanol burns more cleanly than gasoline, and levels of troublesome pollutants such as carbon monoxide, unreacted hydrocarbons, nitrogen oxides, and ozone are reduced. However, there is concern about the greater exhaust emissions of carcinogenic formaldehyde from methanol-powered vehicles. Since the number of methanol-powered vehicles is limited, it is still difficult to assess the extent to which these formaldehyde emissions will contribute to the total aldehyde levels from other sources.

Cars at the Indianapolis 500 are powered by methanol.

The technology for methanol-powered vehicles has existed for many years, particularly for racing cars that burn methanol because of its high octane number of 107. However, the same volume of methanol has only about half the energy content of gasoline, which would require fuel tanks to be twice as large to give the same distance per tankful. This is partially compensated by the fact that methanol costs about half as much as gasoline, so the price per mile would be competitive. Since methanol burns with a colorless flame, something needs to be added (a small amount of gasoline, for example) to methanol so it can be seen when it burns. Another disadvantage is the tendency for methanol to corrode regular steel. Therefore, it will be necessary to use stainless steel for the fuel system or have a methanol-resistant coating. Until sufficient numbers of methanol-powered vehicles are on the road, cars equipped to run on both methanol and gasoline will be necessary because of the lack of service stations selling methanol. As the problems of distribution and storage are solved, better engineered methanol-fueled engines will be designed and produced, which will lead to more efficient utilization of methanol as a fuel.

Methanol-to-Gasoline

Another option is to use methanol to make gasoline. Mobil Oil Company has developed a methanol-to-gasoline process that is currently not competitive with refined gasoline prices in the United States, but is competitive in those regions of the world, such as New Zealand, where the price of gasoline is much higher.

$$2\,CH_3OH \xrightarrow[\text{catalyst}]{\text{ZSM-5}} \underset{\text{dimethyl ether}}{(CH_3)_2O} + H_2O$$

$$2\,(CH_3)_2O \xrightarrow[\text{catalyst}]{\text{ZSM-5}} \underset{\text{ethylene}}{2\,C_2H_4} + 2\,H_2O$$

Plant in New Zealand that converts natural gas to methanol, which then reacts in the presence of a zeolite catalyst to produce gasoline. *(Mobil Corporation)*

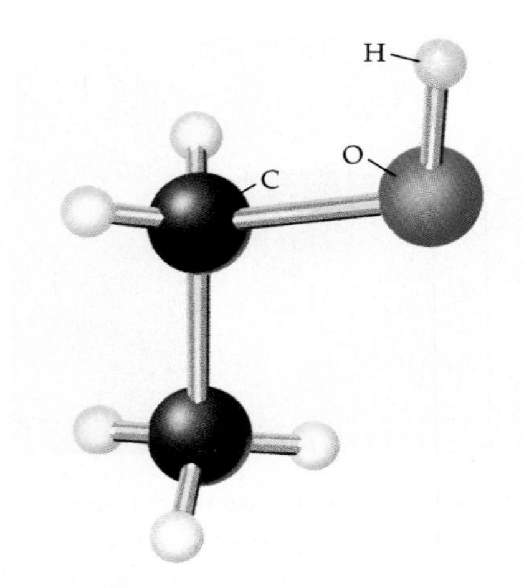

A model of ethanol.

$$C_2H_4 \xrightarrow[\text{catalyst}]{\text{ZSM-5}} \underset{\text{gasoline}}{\text{hydrocarbon mixture in the } C_5\text{–}C_{12} \text{ range}}$$

The ZSM-5 catalyst is a zeolite catalyst similar to the one shown in Figure 11.15. The New Zealand plant is currently producing 14,000 barrels per day of gasoline with an octane number of 92 to 94, which is about one-third the amount of gasoline used in New Zealand.

Ethanol

Ethanol, also called ethyl alcohol or grain alcohol, is the "alcohol" of alcoholic beverages and is prepared for this purpose by fermentation of carbohydrates (starch, sugars) from a wide variety of plant sources. For example, glucose is converted into ethanol and carbon dioxide by the action of yeast in the absence of oxygen.

$$\underset{\text{glucose}}{C_6H_{12}O_6} \xrightarrow{\text{yeast}} 2\ \underset{\text{ethanol}}{C_2H_5OH} + 2\ CO_2$$

Some of the most commonly encountered alcoholic beverages and their characteristics are presented in Table 13.5. The growth of yeast is inhibited at alcohol concentrations greater than 12%, and fermentation comes to a stop. Beverages with a higher alcohol concentration are prepared either by distillation or by fortification with alcohol that has been obtained by the distillation of another fermentation product. The maximum concentration of ethanol that can be obtained by distillation of alcohol/water mixtures is 95% ethanol. The "proof" of an alcoholic beverage is twice the volume percent of ethanol; 80 proof vodka, for example, contains 40% ethanol by volume.

Although ethanol is not as toxic as methanol, one pint of pure ethanol, rapidly ingested, would kill most people. Ethanol is a depressant, and the effects of various blood levels of alcohol are shown in Table 13.6. Rapid consumption of two 1-oz "shots" of 90-proof whiskey or of two 12-oz beers

A 5% ethanol solution. *(C.D. Winters)*

Table 13.5

Common Alcoholic Beverages

Name	Source of Fermented Carbohydrate	Amount of Ethyl Alcohol	Proof
beer	barley, wheat	5%	10
wine	grapes or other fruit	12% maximum, unless fortified	20–24
brandy	distilled wine	40–45%	80–90
whiskey	barley, rye, corn, etc.	45–55%	90–110
rum	molasses	~45%	90
vodka	potatoes	40–50%	80–100

can cause one's blood alcohol level to reach 0.05%. Ethanol is quickly absorbed by the blood and metabolized by enzymes produced in the liver. The rate of detoxification is about 1 oz of pure alcohol per hour. Ethanol is oxidized to acetaldehyde, which is further oxidized to acetic acid; eventually CO_2 and H_2O are produced and eliminated through the lungs and kidneys.

$$\underset{\text{ethanol}}{\mathrm{H{-}\overset{H}{\underset{H}{\overset{|}{\underset{|}{C}}}}{-}\overset{H}{\underset{H}{\overset{|}{\underset{|}{C}}}}{-}OH}} \xrightarrow[\text{enzymes}]{\text{liver}} \underset{\text{acetaldehyde}}{\mathrm{H{-}\overset{H}{\underset{H}{\overset{|}{\underset{|}{C}}}}{-}\overset{O}{\overset{\|}{C}}{-}H}}$$

Table 13.6

Alcohol Levels in Blood and Their Effects

% by Volume	Effect
0.05–0.15	lack of coordination
0.15–0.20	intoxication
0.30–0.40	unconsciousness
0.50	possible death

The breathalyzer test used to detect drunken drivers is based on the color change that occurs when ethanol is oxidized to acetic acid by dichromate anion ($Cr_2O_7^{2-}$) in acidic solution.

$$16\,H^+ + \underset{\text{yellow-orange}}{2\,Cr_2O_7^{2-}} + 3\,CH_3CH_2OH \longrightarrow 3\,CH_3COOH + \underset{\text{green}}{4\,Cr^{3+}} + 11\,H_2O$$

Breathalyzer for measuring level of ethanol in exhaled breath. *(C.D. Winters)*

In the United States, the federal tax on alcoholic beverages is about $20 per gallon. Since the cost of producing ethanol is only about $1 per gallon, ethanol intended for industrial use must be *denatured* to avoid the beverage tax. **Denatured alcohol** contains small amounts of a toxic substance, such as methanol or gasoline, that cannot be removed easily by chemical or physical means.

Apart from being used in the alcoholic beverage industry, ethanol is used widely in solvents, in the preparation of many other organic compounds, and as a gasoline additive. For many years industrial ethanol was also made by fermentation. However, in the last several decades it became cheaper to make the ethanol from petroleum byproducts, specifically by the catalyzed addition of water to ethylene. More than 1 billion pounds of ethanol are produced each year by this process.

Ethanol is receiving increased attention as an alternative fuel. At present, most of it is used in a blend of 90% gasoline and 10% ethanol (known as **gasohol** when introduced in the 1970s). Although the methanol fuels M85 and M100 have received more attention than the corresponding ethanol fuels E85 and E100, vehicles are being built to test the use of both M85 and E85.

The major component of antifreeze is ethylene glycol. The other products in the photo contain glycerol. *(C.D. Winters)*

Ethylene Glycol and Glycerol

Ethylene glycol is a di-alcohol (Table 13.4) used in permanent antifreeze and in synthesis of polymers (Section 13.6). Glycerol, a tri-alcohol (Table 13.4), is a byproduct from the manufacture of soaps. Because of its moisture-holding properties, glycerol has many uses in foods and tobacco as a digestible and nontoxic humectant (an ingredient that gathers and holds moisture), and in the manufacture of drugs and cosmetics. It is also used in the manufacture of nitroglycerin and numerous other chemicals. Perhaps the most important derivatives of glycerol are its natural esters (fats and oils), which are discussed in Section 21.4.

EXERCISE 13.5 • *Uses of Alcohols*

Give the names of the alcohols that are (a) used as a rubbing alcohol, (b) used as an antifreeze, (c) found in alcoholic beverages, (d) found in fats, (e) oxidized to formaldehyde.

13.5 CARBOXYLIC ACIDS AND ESTERS

Carboxylic Acids

Three ways of representing the carboxylic acid group are —COOH, —CO_2H, and

$$-\overset{\overset{\displaystyle O}{\|}}{C}-OH.$$

Carboxylic acids, which contain the —COOH functional group, are prepared by the oxidation of alcohols or aldehydes. These reactions occur quite readily, as evidenced by the souring of wine, which is the oxidation of ethanol to acetic acid in the presence of oxygen from the air.

Carboxylic acids are polar and readily form hydrogen bonds. This hydrogen bonding results in relatively high boiling points for the acids, even higher than those of alcohols of comparable molar mass. For example, formic acid (46 g/mol) has a boiling point of 101 °C, while ethanol (46 g/mol) has a boiling point of only 78.5 °C.

All carboxylic acids are weak acids (see Sections 5.6 and 16.4) and react with bases to form salts. For example,

$$\underset{\text{acetic acid}}{CH_3\overset{\overset{\displaystyle O}{\|}}{C}-OH(aq)} + NaOH(aq) \longrightarrow \underset{\text{sodium acetate}}{CH_3\overset{\overset{\displaystyle O}{\|}}{C}O^-Na^+(aq)} + H_2O(\ell)$$

A number of carboxylic acids are found in nature and have been known for many years. As a result, some of the familiar carboxylic acids are almost

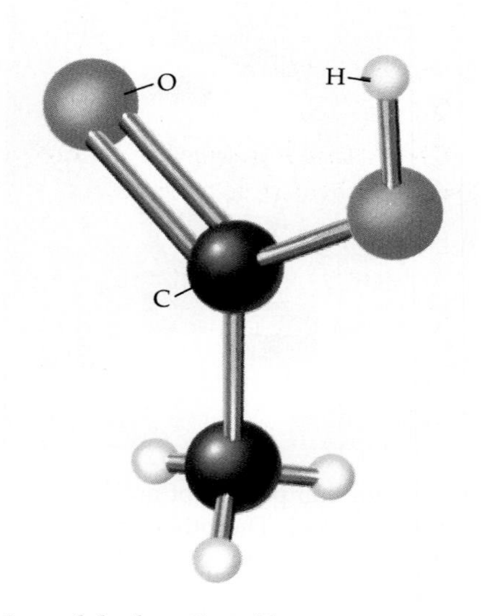

A model of acetic acid.

Table 13.7

Some Simple Carboxylic Acids

Structure	Common Name	Systematic Name	B.p. (°C)
$H\overset{\overset{O}{\|}}{C}OH$	formic acid	methanoic acid	101
$CH_3\overset{\overset{O}{\|}}{C}OH$	acetic acid	ethanoic acid	118
$CH_3CH_2\overset{\overset{O}{\|}}{C}OH$	propionic acid	propanoic acid	141
$CH_3(CH_2)_2\overset{\overset{O}{\|}}{C}OH$	butyric acid	butanoic acid	163
$CH_3(CH_2)_3\overset{\overset{O}{\|}}{C}OH$	valeric acid	pentanoic acid	187

always referred to by their common names (Table 13.7). The systematic names of carboxylic acids are easily derived: "-e" is dropped from the name of the corresponding alkane and "-oic" is added followed by the word "acid." Both common and systematic names are given in Table 13.7.

The simplest common acid is formic acid, the substance responsible for the sting of an ant bite. Therefore, the name of the acid comes from the Latin word (*formica*) for ant. Acetic acid gives the sour taste to vinegar, and the name comes from the Latin word for this substance (*acetum*). Butyric acid gives rancid butter its unpleasant odor, and the name is related to the Latin word for butter, *butyrum*. The names for caproic (C_6), caprilic (C_8), and capric (C_{10}) acids are derived from the Latin word for goat, *caper*, since these acids combine to give goats their characteristic odor. Most of the simple carboxylic acids have unpleasant odors.

Acetic acid is produced in large quantities (number 35 on the top 50 chemicals list) for manufacturing cellulose acetate, a polymer used to make photographic film base, synthetic fibers, plastics, and other products. The catalytic production of acetic acid was described in Chapter 8.

The only other carboxylic acids produced in large quantity are several with two acid groups known as dicarboxylic acids:

$$HO\overset{\overset{\displaystyle O}{\|}}{C}(CH_2)_4\overset{\overset{\displaystyle O}{\|}}{C}OH \qquad HO-\overset{\overset{\displaystyle O}{\|}}{C}-C_6H_4-\overset{\overset{\displaystyle O}{\|}}{C}-OH \qquad C_6H_4\left(\overset{\overset{\displaystyle O}{\|}}{C}-OH\right)_2$$

adipic acid — terephthalic acid — phthalic acid

These three acids are used to make polymers (Section 13.6). Other carboxylic acids whose names may be familiar to you, because they occur in nature, are listed in Table 13.8.

Vinegar is 5% acetic acid. *(C.D. Winters)*

Esters

Carboxylic acids react with alcohols in the presence of strong mineral acids to produce **esters,** which contain the —COOR functional group. In an ester the —OH of the carboxylic acid is replaced by the OR group from the alcohol. For example, when ethanol is mixed with acetic acid in the presence of sulfuric acid, an ester called ethyl acetate is formed. This reaction is a dehydration in which sulfuric acid acts as a catalyst and dehydrator.

Esters contain the $-\overset{\overset{\displaystyle O}{\|}}{C}-OR$ *(also represented as* $-CO_2R$ *and —COOR) functional group.*

$$\underset{\text{acetic acid}}{CH_3\overset{\overset{\displaystyle O}{\|}}{C}-OH} + \underset{\text{ethanol}}{H-OCH_2CH_3} \xrightarrow{H^+} \underset{\text{ethyl acetate}}{CH_3\overset{\overset{\displaystyle O}{\|}}{C}-OCH_2CH_3} + H_2O$$

Ethyl acetate is a common solvent for lacquers and plastics and is often used as fingernail polish remover.

Unlike the acids from which esters are derived, esters often have pleasant odors (Table 13.9). The characteristic odors and flavors of many flowers and fruits are caused by the presence of natural esters. For example, the odor and flavor of bananas is primarily caused by the ester 3-methylbutyl acetate

Sources of some naturally occurring carboxylic acids. *(C.D. Winters)*

Table 13.8

Some Other Naturally Occurring Carboxylic Acids

Name	Structure	Natural Source
citric acid	$HOOC-CH_2-\overset{\displaystyle OH}{\underset{\displaystyle COOH}{C}}-CH_2-COOH$	citrus fruits
lactic acid	$CH_3-\underset{\displaystyle OH}{CH}-COOH$	sour milk
malic acid	$HOOC-CH_2-\underset{\displaystyle OH}{CH}-COOH$	apples
oleic acid	$CH_3(CH_2)_7-CH=CH-(CH_2)_7-COOH$	vegetable oils
oxalic acid	$HOOC-COOH$	rhubarb, spinach, cabbage, tomatoes
stearic acid	$CH_3(CH_2)_{16}-COOH$	animal fats
tartaric acid	$HOOC-\underset{\displaystyle OH}{CH}-\underset{\displaystyle OH}{CH}-COOH$	grape juice, wine

(also known as isoamyl acetate). Although the odor and flavor of a fruit or flower may be due to a single compound, usually they are due to a complex mixture in which a single ester predominates.

Food and beverage manufacturers often use mixtures of esters as food additives. The ingredient label of a brand of imitation banana extract reads "water, alcohol 40%, isoamyl acetate and other esters, orange oil and other essential oils, and FD&C Yellow #5." Except for the water, these are all organic compounds.

As a class, esters are not very reactive. Their most important reaction is splitting (hydrolysis) in the presence of a strong base to give the constituents

Fruits containing esters. *(C.D. Winters)*

Table 13.9

Some Acids, Alcohols, and Their Esters

Acid	Alcohol	Ester	Odor of Ester
CH_3COOH acetic acid	$CH_3CH(CH_3)CH_2CH_2OH$ 3-methyl-1-butanol	$CH_3C(=O)OCH_2CH_2CH(CH_3)CH_3$ 3-methylbutyl acetate	banana
$CH_3CH_2CH_2CH_2COOH$ pentanoic acid	$CH_3CH(CH_3)CH_2CH_2OH$ 3-methyl-1-butanol	$CH_3CH_2CH_2CH_2C(=O)OCH_2CH_2CH(CH_3)CH_3$ 3-methylbutyl pentanoate	apple
$CH_3CH_2CH_2COOH$ butanoic acid	$CH_3CH_2CH_2CH_2OH$ 1-butanol	$CH_3CH_2CH_2C(=O)OCH_2CH_2CH_2CH_3$ butyl butanoate	pineapple
$CH_3CH_2CH_2COOH$ butanoic acid	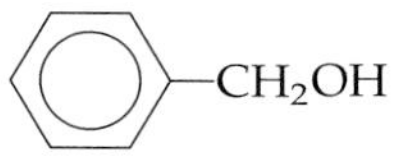 C_6H_5—CH_2OH benzyl alcohol	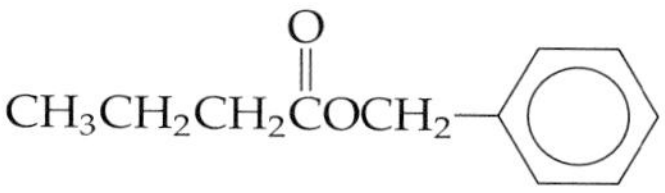 $CH_3CH_2CH_2C(=O)OCH_2$—C_6H_5 benzyl butanoate	rose

of the ester: the alcohol and a salt of the acid from which the ester was formed.

$$\underset{\text{ester}}{RC(=O)—O—R'} + NaOH(aq) \xrightarrow{\text{heat}} \underset{\text{carboxylate salt}}{RC(=O)—O^-Na^+(aq)} + \underset{\text{alcohol}}{R'OH}$$

$$\underset{\text{ethyl acetate}}{CH_3C(=O)—O—CH_2CH_3} + NaOH(aq) \xrightarrow{\text{heat}} \underset{\text{sodium acetate}}{CH_3C(=O)—O^-Na^+(aq)} + \underset{\text{ethanol}}{CH_3CH_2OH}$$

EXAMPLE 13.4

Carboxylic Acids and Esters

Write the equation for the formation of butyl acetate.

SOLUTION The name of the ester indicates what alcohol and acid are needed to make the ester. The first part, butyl, identifies the alcohol, 1-butanol, and the last part, acetate, identifies the acid, acetic acid. 1-Butanol and acetic acid react in the presence of sulfuric acid catalyst to give butyl acetate.

$$\underset{\text{acetic acid}}{CH_3COOH} + \underset{\text{1-butanol}}{CH_3CH_2CH_2CH_2OH} \longrightarrow \underset{\text{butyl acetate}}{CH_3C(=O)OCH_2CH_2CH_2CH_3}$$

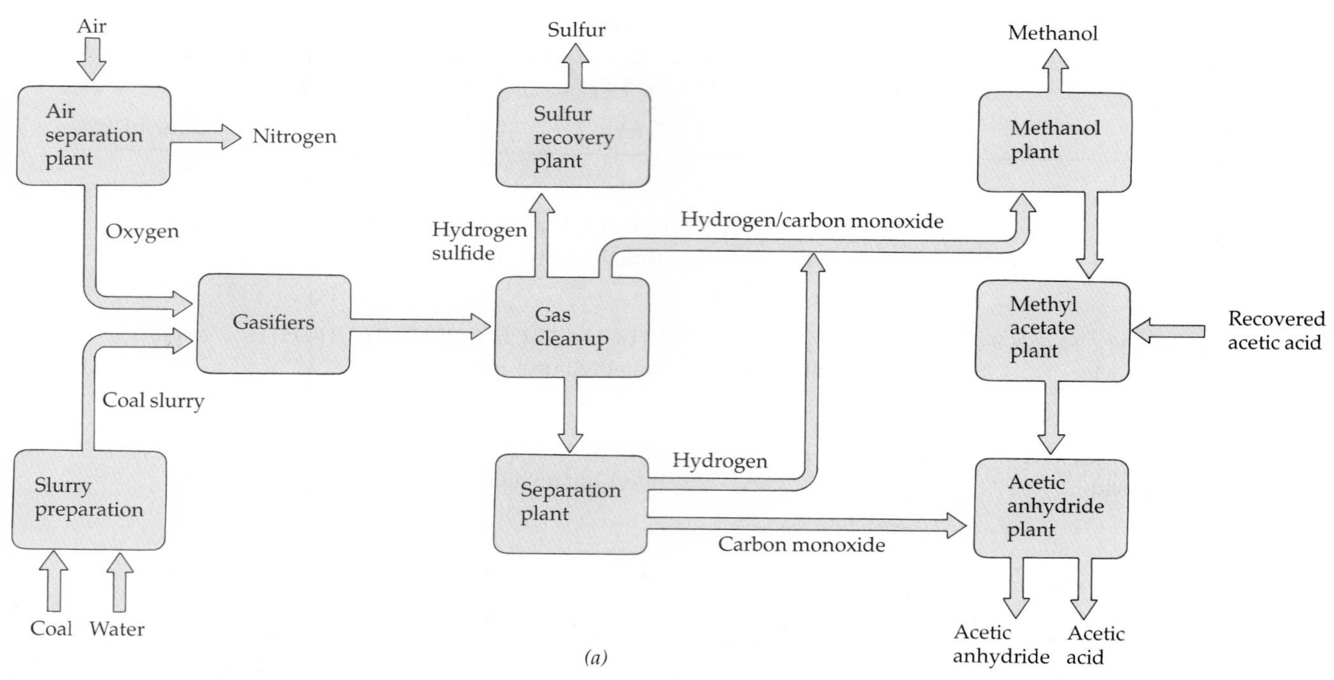

Figure 13.10
(a) Schematic drawing showing the production of methanol, methyl acetate, and acetic anhydride from coal. (b) Eastman Kodak's chemicals-from-coal facility in Kingsport, Tennessee. Numbers in the photograph represent different parts of the plant: 1, coal unloading; 2, coal silos; 3, steam plant; 4, slurry preparation; 5, coal gasification plant; 6, gas cleanup and separation; 7, sulfur recovery plant; 8, gas flare stack; 9, chemical storage; 10, methanol plant; 11, methyl acetate plant; 12, acetic anhydride plant. *(Courtesy of Tennessee Eastman Kodak.)*

EXERCISE 13.6 • *Ester Formation*

Draw the structural formula of the ester formed in the reaction of (a) acetic acid and methanol, (b) excess propionic acid and ethylene glycol.

Organic Chemicals from Coal

An example of the use of coal for the industrial synthesis of organic chemicals is an Eastman Kodak process that produces acetic anhydride used to make cellulose acetate. The first complete "chemicals from coal" plant, built by Eastman Kodak in Kingsport, Tennessee, started production in 1983. Figure 13.10a is a schematic drawing of the various components of the plant, which is pictured in Figure 13.10b.

Within the complex pictured in Figure 13.10b are nine separate plants, four related to the gasification of coal, two for synthesis gas preparation, and three for the synthesis of methanol, methyl acetate, and acetic anhydride. The main chemical reactions used in the process are

Acetic anhydride reacts with water to give acetic acid.

1. $\underset{\text{coal}}{C} + \underset{\text{steam}}{H_2O} \longrightarrow \underset{\text{synthesis gas}}{CO + H_2}$
2. $CO + 2\,H_2 \longrightarrow \underset{\text{methanol}}{CH_3OH}$
3. $CH_3OH + \underset{\text{acetic acid}}{CH_3\overset{O}{\overset{\|}{C}}OH} \longrightarrow \underset{\text{methyl acetate}}{CH_3\overset{O}{\overset{\|}{C}}OCH_3} + H_2O$
4. $CH_3\overset{O}{\overset{\|}{C}}OCH_3 + CO \longrightarrow \underset{\text{acetic anhydride}}{CH_3{-}\overset{O}{\overset{\|}{C}}{-}O{-}\overset{O}{\overset{\|}{C}}{-}CH_3}$

An acid anhydride has the general formula

$R{-}\overset{O}{\overset{\|}{C}}{-}O{-}\overset{O}{\overset{\|}{C}}{-}R.$

About 900 tons per day of high-sulfur coal from nearby Appalachian coal mines are ground in water to form a slurry of 55% to 65% by weight of coal in water. The slurry is fed into two gasifiers to make synthesis gas. To produce the same amount of acetic anhydride from petroleum would require the equivalent of two million barrels per year.

The plant design uses the latest environmental control technologies to protect the environment. For example, the sulfur recovery unit converts the hydrogen sulfide gas that was removed during the gasification of coal into elemental sulfur. This process removes more than 99% of the sulfur from the coal, and the sulfur is sold to chemical companies.

13.6 SYNTHETIC ORGANIC POLYMERS

It is impossible for us to get through a day without using a dozen or more synthetic organic **polymers.** The word "polymer" means "many parts" (Greek, *poly* and *meros*). Polymers are giant molecules with molecular weights ranging from thousands to millions. Our clothes are made with polymers; our food is packaged in polymers; our appliances and cars contain a number of polymer components.

Approximately 80% of the organic chemical industry is devoted to the production of synthetic polymers; about half of the top 50 chemicals produced in the United States are used in making plastics, fibers, and rubbers. The average production of synthetic polymers in the United States exceeds 200 pounds per person annually.

Many synthetic organic polymers are **plastics** of one sort or another (Figure 13.11). There are two broad categories of plastics. One type, when heated repeatedly, softens and flows; when it is cooled, it hardens again. Materials that undergo such reversible changes when heated and cooled are called **thermoplastics;** polyethylene and polystyrene are thermoplastics.

"People don't realize what polymer chemistry does for the world. If you were to stand in the middle of your living room and have somebody come and take out everything in the room that had any polymer material in it, the room would be denuded. The paint would be gone, the carpet would be gone, the upholstery would be gone. If you took away the manmade polymer materials you'd be utterly amazed how little you'd have left." Dr. Mary Good, Senior Vice President for Technology, Allied-Signal, Inc., Morristown, New Jersey.

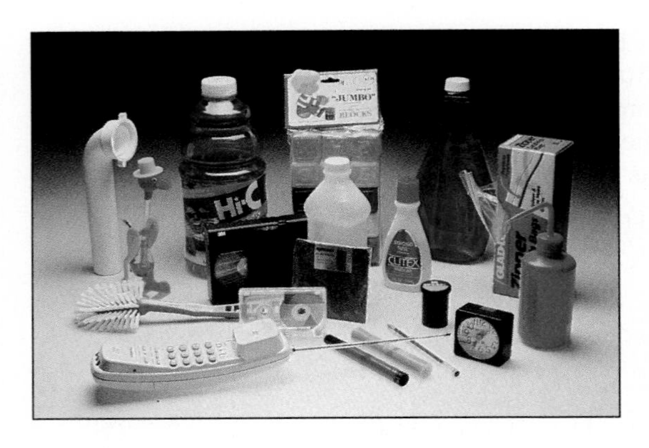

Figure 13.11
Assorted plastics. *(C.D. Winters)*

The other type is plastic when first heated, but when heated further it forms a highly cross-linked structure. When reheated, it cannot be softened and reformed without extensive degradation. These materials are called **thermosetting plastics;** the Formica used for kitchen counter tops is an example of a thermosetting plastic.

Some of our most useful polymer chemistry has resulted from copying giant molecules in nature. Rayon is remanufactured cellulose (discussed in Chapter 21); synthetic rubber is copied from natural latex rubber. As useful as these polymers may be, however, polymer chemistry is not restricted to nature's models. Polystyrene, nylon, and Dacron are a few examples of synthetic molecules that do not have analogues in nature. We have learned from nature and extended our knowledge to produce synthetic polymers that are more useful than natural polymers.

Polymers are made by chemically joining together many small molecules into one giant molecule or macromolecule. The small molecules used to synthesize polymers are called **monomers.** Synthetic polymers can be classified as **addition polymers,** made by monomer units directly joined together, or **condensation polymers,** made by monomer units combining so that a small molecule, usually water, is split out between them.

Addition Polymers

Polyethylene

The monomer for addition polymers normally contains one or more double bonds. The simplest monomer of this group is ethylene, C_2H_4. When ethylene is heated to 100 to 250 °C at a pressure of 1000 to 3000 atm in the presence of a catalyst, polymers with molecular weights of up to several million may be formed. A reaction of ethylene usually begins with breaking of one of the bonds in the carbon-carbon double bond, so that an unpaired electron that is a reactive site remains at each end of the molecule. This step, the **initiation** of the polymerization, can be accomplished with chemicals such as organic peroxides that are unstable and easily break apart into **free radicals,** which have unpaired electrons. The free radicals react readily with

An organic peroxide, RO—OR, produces free radicals, RO·, each with an unpaired electron. See Section 10.4 for a discussion of free radicals.

molecules containing carbon-carbon double bonds to produce new free radicals.

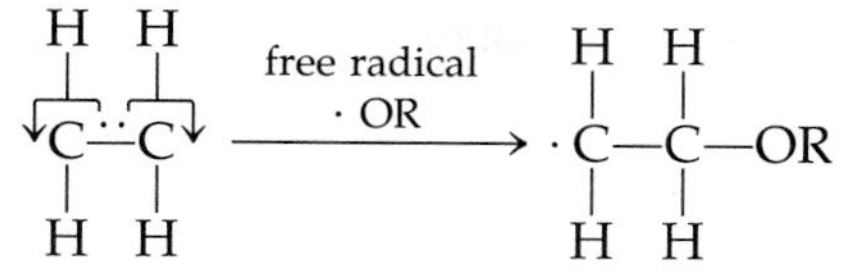

$$\mathrm{H_2C{=}CH_2} \xrightarrow[\cdot\,\mathrm{OR}]{\text{free radical}} \cdot\mathrm{CH_2{-}CH_2{-}OR}$$

The growth of the polyethylene chain then begins, as the unpaired electron bonds to a double-bond electron in another ethylene molecule. This leaves another unpaired electron to bond with yet another ethylene molecule. For example,

$$\cdot\mathrm{CH_2{-}CH_2}\cdot + \cdot\mathrm{CH_2{-}CH_2{-}OR} \longrightarrow \cdot\mathrm{CH_2{-}CH_2{-}CH_2{-}CH_2{-}OR} \xrightarrow{n\,\mathrm{CH_2{=}CH_2}} \left(\mathrm{CH_2{-}CH_2}\right)_n$$

polyethylene
n ranges from 1000 to 50,000

In the process, the unsaturated hydrocarbon monomer, ethylene, is changed to a saturated hydrocarbon polymer, polyethylene.

Polyethylenes formed under various pressures and catalytic conditions have different molecular structures and hence different physical properties. For example, chromium oxide as a catalyst yields almost exclusively the linear polyethylene shown below—a polymer with no branches on the carbon chain. The zig-zag structure represents the shape of the chain more closely than does a linear drawing, because of the tetrahedral arrangement of bonds around each carbon in the saturated polyethylene chain (Figure 13.12).

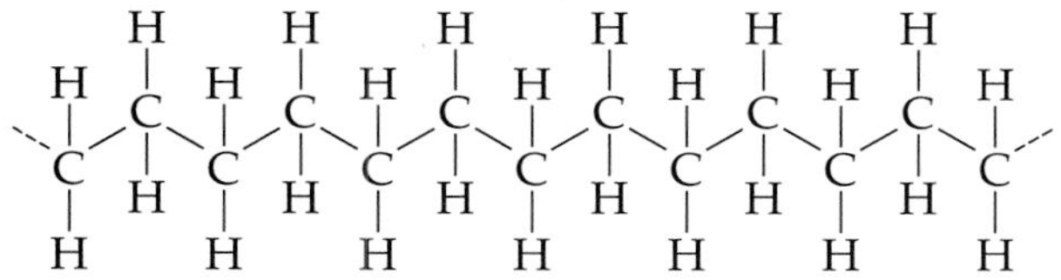

$$\cdots\mathrm{CH_2{-}CH_2{-}CH_2{-}CH_2{-}CH_2{-}CH_2{-}CH_2{-}CH_2{-}CH_2{-}CH_2{-}CH_2{-}CH_2{-}CH_2}\cdots$$

a portion of a polyethylene molecule

or

(zig-zag line) each apex corner represents a CH_2 group

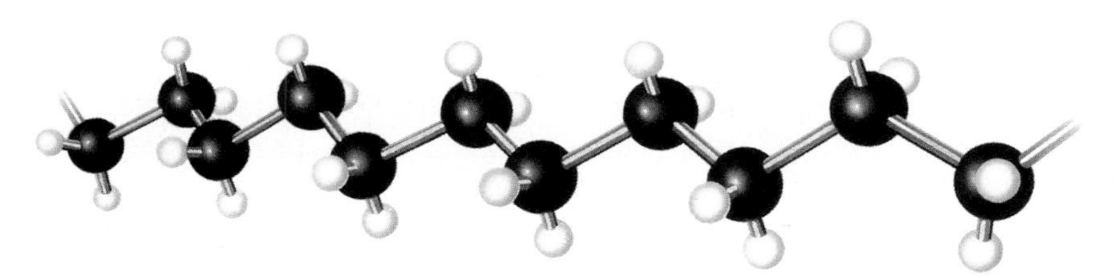

Figure 13.12
Model of linear polyethylene.

If ethylene is heated to 230 °C at a pressure of 200 atm, free radicals attack the chain at random positions, causing irregular branching (Figure 13.13). Other conditions can lead to cross-linked polyethylene, in which branches connect long chains to each other.

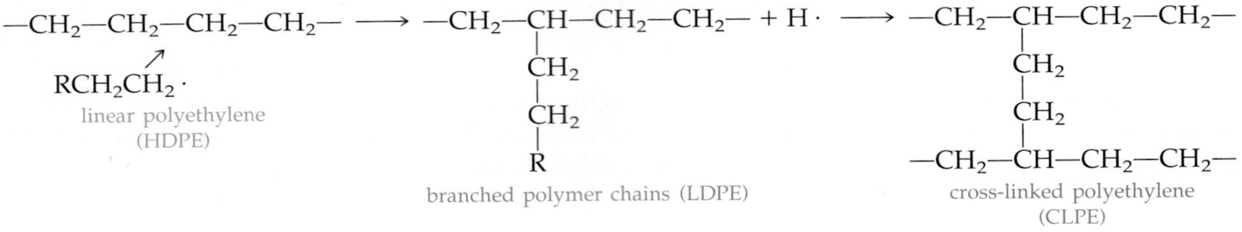

$$—CH_2—CH_2—CH_2—CH_2— \longrightarrow —CH_2—CH(CH_2CH_2R)—CH_2—CH_2— + H\cdot \longrightarrow —CH_2—CH—CH_2—CH_2— \;(\text{linked via } CH_2—CH_2 \text{ to}) —CH_2—CH—CH_2—CH_2—$$

$RCH_2CH_2\cdot$ (attacking the chain)

linear polyethylene (HDPE)

branched polymer chains (LDPE)

cross-linked polyethylene (CLPE)

Polyethylene is the world's most widely used polymer. The wide range of properties of polyethylene leads to many uses (Figure 13.14). Long, linear chains of polyethylene can pack closely together (Figure 13.15a) and give a material with high density (0.97 g/mL) and high molecular weight, referred to as high-density polyethylene (HDPE). This material is hard, tough, and rigid. The plastic milk bottle is a typical application of HDPE.

Branched chains of polyethylene cannot pack closely together (Figure 13.15b), so the resulting material has a lower density (0.92 g/mL) and is called low-density polyethylene (LDPE). This material is soft and flexible. Sandwich bags are made from LDPE. If the linear chains of polyethylene are treated in a way that causes crosslinks to form between chains, the result is cross-linked polyethylene (CLPE), a very tough material (Figure 13.13b). The plastic caps on soft drink bottles are made from CLPE.

Some Other Addition Polymers

Many different kinds of addition polymers are made from monomers in which one or more of the hydrogen atoms in ethylene have been replaced with either halogen atoms or a variety of organic groups. If the formation of polyethylene is represented as

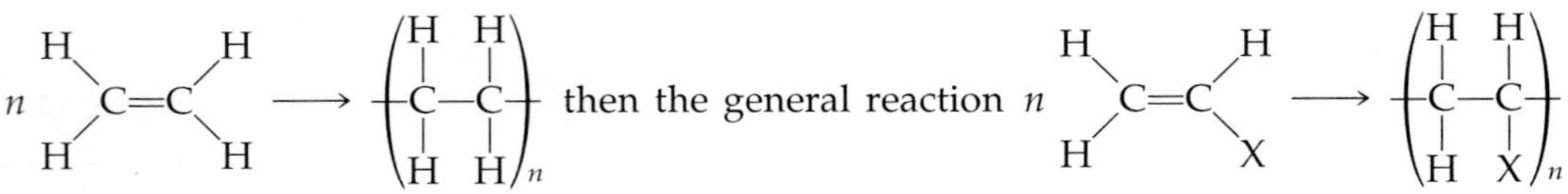

$$n\, H_2C{=}CH_2 \longrightarrow \left(\begin{matrix} H & H \\ | & | \\ C & — C \\ | & | \\ H & H \end{matrix} \right)_n$$

then the general reaction

$$n\, H_2C{=}CHX \longrightarrow \left(\begin{matrix} H & H \\ | & | \\ C & — C \\ | & | \\ H & X \end{matrix} \right)_n$$

can be used to represent a number of other important addition polymers in which X is Cl, F, or an organic group (Table 13.10).

For example, the monomer for making polystyrene is styrene, and n = 5700.

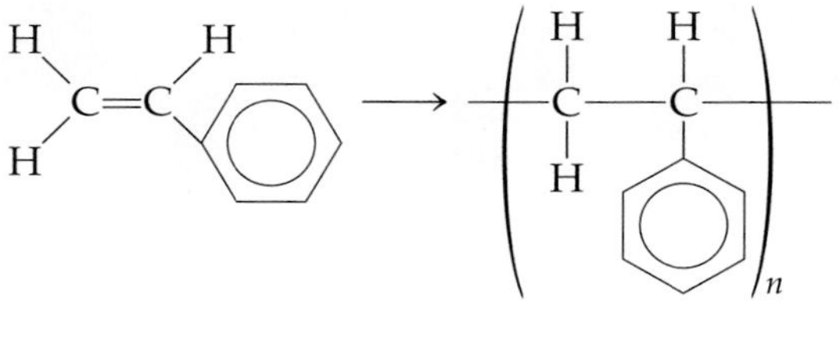

$$H_2C{=}CH—C_6H_5 \longrightarrow \left(—CH_2—CH(C_6H_5)— \right)_n$$

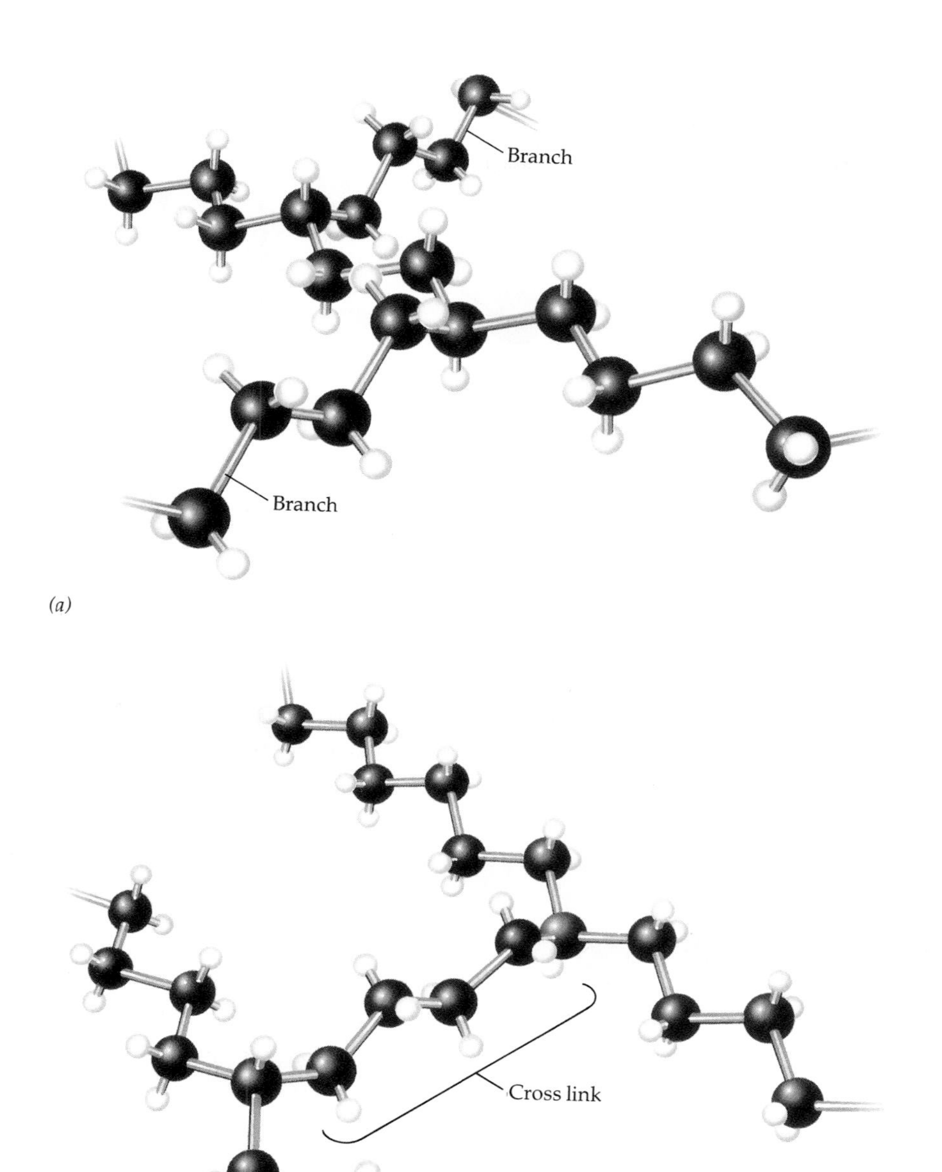

Figure 13.13
Models of (a) branched and (b) cross-linked polyethylene.

(a)

(b)

(c)

Figure 13.14
(a) Production of polyethylene. (b), (c) The wide range of properties of different structural types of polyethylene leads to a variety of applications. *(b,c The World of Chemistry* (Program 22), "The Age of Polymers.") (*a* Gary Gladstone/The Image Bank).

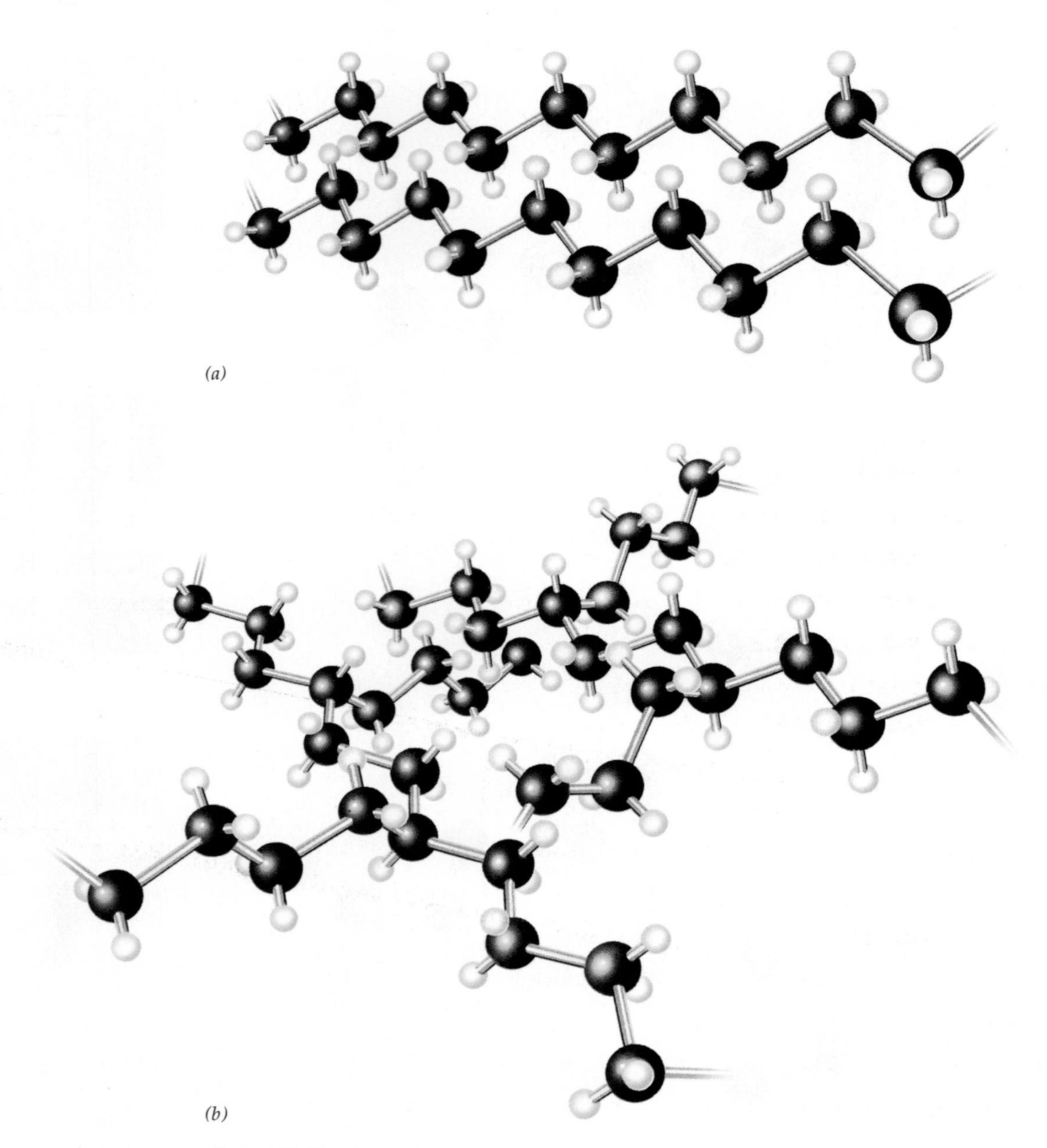

Figure 13.15
(a) Model of high-density polyethylene. (b) Model of low-density polyethylene.

Table 13.10

Ethylene Derivatives That Undergo Addition Polymerization

Formula	Monomer Common Name	Polymer Name (Trade Names)	Uses	U.S. Polymer Production (Tons/Yr)
$H_2C{=}CH_2$	ethylene	polyethylene (Polythene)	squeeze bottles, bags, films, toys and molded objects, electrical insulation	8 million
$H_2C{=}CH(CH_3)$	propylene	polypropylene (Vectra, Herculon)	bottles, films, indoor–outdoor carpets	2.7 million
$H_2C{=}CHCl$	vinyl chloride	poly(vinyl chloride) (PVC)	floor tile, raincoats, pipe	3.5 million
$H_2C{=}CH(CN)$	acrylonitrile	polyacrylonitrile (Orlon, Acrilan)	rugs, fabrics	1 million
$H_2C{=}CH(C_6H_5)$	styrene	polystyrene (Styrene, Styrofoam, Styron)	food and drink coolers, building material insulation	2 million
$H_2C{=}CH{-}O{-}C({=}O){-}CH_3$	vinyl acetate	poly(vinyl acetate) (PVA)	latex paint, adhesives, textile coatings	500,000
$H_2C{=}C(CH_3){-}C({=}O){-}O{-}CH_3$	methyl methacrylate	poly(methyl methacrylate) (Plexiglas, Lucite)	high-quality transparent objects, latex paints, contact lenses	450,000
$F_2C{=}CF_2$	tetrafluoroethylene	polytetrafluoroethylene (Teflon)	gaskets, insulation, bearings, pan coatings	7000

Polystyrene is a clear, hard, colorless solid at room temperature, but it can be molded easily at 250 °C. More than 5 billion pounds of polystyrene are produced in the United States each year to make food containers, toys, electrical parts, insulating panels, appliance and furniture components, and many other items. The variation in properties shown by polystyrene products is

Styrene is 21st on the list of top 50 chemicals, primarily because of its use in making polystyrene.

Figure 13.16
(a) Polystyrene coffee cup is soft. (b) Clear polystyrene "glass" is brittle. *(C.D. Winters)*

typical of synthetic polymers. For example, a clear polystyrene drinking glass that is brittle and breaks into sharp pieces somewhat like glass is quite different from a polystyrene coffee cup that is soft and pliable (Figure 13.16).

A major use of polystyrene is in the production of Styrofoam® by "expansion molding." In this process, polystyrene beads are placed in a mold and heated with steam or hot air. The beads, 0.25 to 1.5 mm in diameter, contain 4% to 7% by weight of a low-boiling liquid such as pentane. The steam causes the low-boiling liquid to vaporize and expand the beads; as the foamed particles expand, they are molded in the shape of the mold cavity. Styrofoam is used for egg cartons, meat trays, coffee cups, and packing material (Figure 13.17).

Poly(vinyl chloride) (PVC), used in floor tile, garden hose, plumbing pipes, and trash bags, is made from vinyl chloride, which has a chlorine atom substituted for one of the hydrogen atoms in ethylene.

$$n\ \mathrm{H_2C{=}CHCl} \longrightarrow \left(\begin{array}{cc} \mathrm{H} & \mathrm{H} \\ | & | \\ -\mathrm{C} & -\mathrm{C}- \\ | & | \\ \mathrm{H} & \mathrm{Cl} \end{array}\right)_n$$

vinyl chloride → poly(vinyl chloride)

Although the representation

$$\left(\begin{array}{cc} \mathrm{H} & \mathrm{H} \\ | & | \\ -\mathrm{C} & -\mathrm{C}- \\ | & | \\ \mathrm{H} & \mathrm{X} \end{array}\right)_n$$

saves space, keep in mind how large the polymer molecules are. Generally n is 500 to 50,000 and this gives molecules with molecular weights ranging from 10,000 to several million. The molecules that make up a given polymer sample are of different lengths and thus are not all of the same molecular weight. As a result, only the average molecular weight can be determined.

In summary, the numerous variations in substituents, length, branching, and crosslinking make it possible to produce a variety of properties for each type of addition polymer. Chemists and chemical engineers can fine-tune the properties of the polymer to match the desired properties by appropriate selection of monomer and reaction conditions, thus accounting for the widespread and growing use of polymers.

Figure 13.17
A Styrofoam ® cooler. *(C.D. Winters)*

EXAMPLE 13.5

Addition Polymers

Draw the structural formula of the repeating unit for each of the following addition polymers: (a) polypropylene, (b) poly(vinyl acetate), (c) poly(vinyl alcohol)

SOLUTION The names show that the monomers for these polymers are propylene ($CH_2{=}CHCH_3$), vinyl acetate ($CH_2{=}CHOOCCH_3$), and vinyl alcohol ($CH_2{=}CHOH$). The repeating units in the polymers therefore have the same structures, but without the double bonds.

a. $\left(\begin{array}{cc} H & H \\ | & | \\ -C & -C- \\ | & | \\ H & CH_3 \end{array}\right)_n$ b. $\left(\begin{array}{cc} H & H \\ | & | \\ -C & -C- \\ | & | \\ H & O \\ & | \\ & O{=}C \\ & | \\ & CH_3 \end{array}\right)_n$ c. $\left(\begin{array}{cc} H & H \\ | & | \\ -C & -C- \\ | & | \\ H & OH \end{array}\right)_n$

Compact disks are made of poly(vinyl chloride). *(C.D. Winters)*

EXERCISE 13.7 • *Addition Polymers*

Draw the structural formulas of the monomers used to prepare the following polymers:

a. polyethylene b. poly(vinyl chloride) c. polystyrene

Natural and Synthetic Rubbers

Natural rubber, a product of the *Hevea brasiliensis* tree, is a hydrocarbon with the empirical formula C_5H_8. When rubber is decomposed in the absence of oxygen, the monomer isoprene is obtained:

$$\begin{array}{c} CH_3 \\ | \\ CH_2{=}C{-}CH{=}CH_2 \end{array} \quad \text{or} \quad \begin{array}{l} H \qquad\quad CH_3 \\ \quad \diagdown \;\; \diagup \\ \quad\; C{=}C \qquad\quad H \\ \quad \diagup \qquad \diagdown \quad\;\; \diagup \\ H \qquad\quad C{=}C \\ \qquad\quad \diagup \qquad \diagdown \\ \qquad H \qquad\quad\; H \end{array}$$

isoprene or 2-methyl-1,3-butadiene

Natural rubber occurs as *latex* (an emulsion of rubber particles in water) that oozes from rubber trees when they are cut. Precipitation of the rubber particles yields a gummy mass that is not only elastic and water-repellent but also very sticky, especially when warm. In 1839, after five years' work on natural rubber, Charles Goodyear (1800–1860) discovered that heating gum rubber with sulfur produces a material that is no longer sticky but is still elastic, water-repellent, and resilient.

Vulcanized rubber, as the type of rubber Goodyear discovered is now known, contains short chains of sulfur atoms that bond together the polymer chains of the natural rubber and reduce its unsaturation. The sulfur chains help to align the polymer chains, so the material does not undergo a permanent change when stretched but springs back to its original shape and size when the stress is removed (Figure 13.18). Substances that behave this way are called **elastomers.**

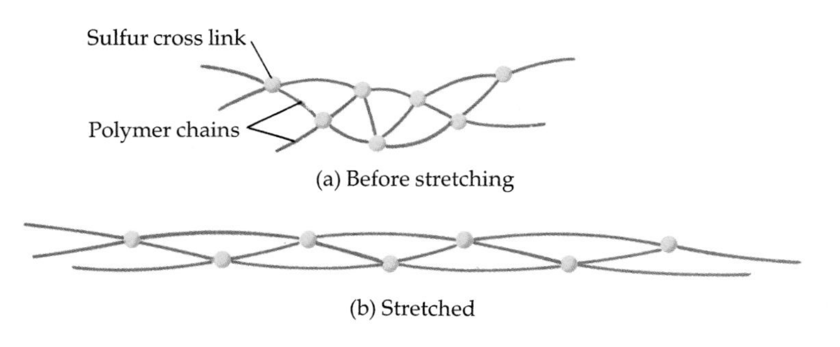

Figure 13.18
Stretched and unstretched vulcanized rubber.

In later years chemists searched for ways to make a synthetic rubber so we would not be completely dependent on imported natural rubber during emergencies, such as during the first years of World War II. In the mid-1920s, German chemists polymerized butadiene (obtained from petroleum and structurally similar to isoprene, but without the methyl group side chain).

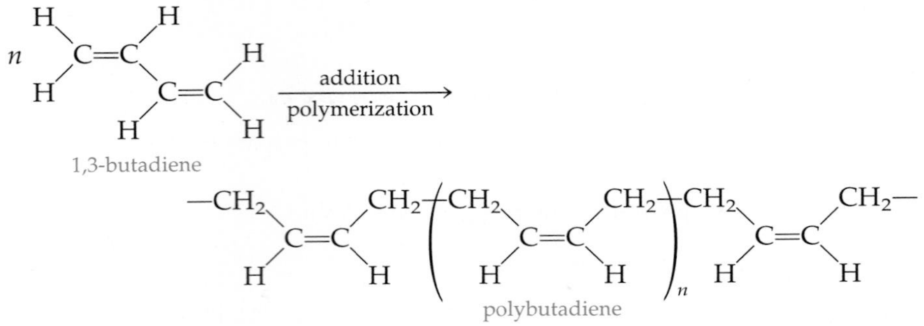

$$n\ H_2C{=}CH{-}CH{=}CH_2 \xrightarrow[\text{polymerization}]{\text{addition}} {-}CH_2{-}CH{=}CH{-}CH_2{-}\left(CH_2{-}CH{=}CH{-}CH_2\right)_n{-}CH_2{-}CH{=}CH{-}CH_2{-}$$

1,3-butadiene

polybutadiene

Polybutadiene is used in the production of tires, hoses, and belts.

The behavior of natural rubber (polyisoprene), it was learned later, is due to the specific molecular geometry within the polymer chain. We can write the formula for polyisoprene with the CH_2 groups on opposite sides of the double bond (the *trans* arrangement):

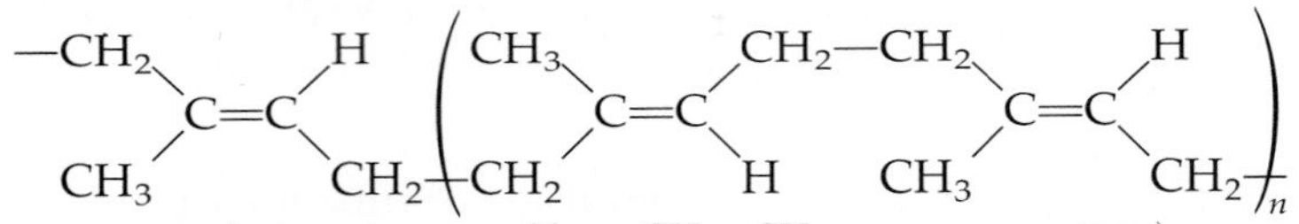

$${-}CH_2{-}C(CH_3){=}CH{-}CH_2{-}\left(CH_2{-}C(CH_3){=}CH{-}CH_2{-}CH_2{-}C(CH_3){=}CH{-}CH_2\right)_n$$

poly-*trans*-isoprene (the $-CH_2-CH_2-$ groups are *trans*)

or with the CH_2 groups on the same side of the double bond (the *cis* arrangement).

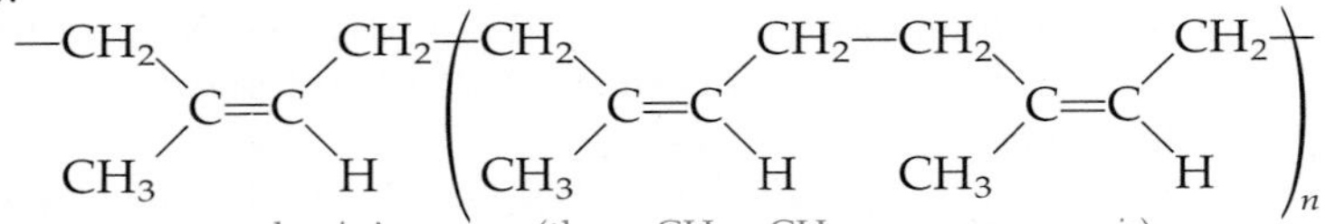

$${-}CH_2{-}C(CH_3){=}CH{-}CH_2{-}\left(CH_2{-}C(CH_3){=}CH{-}CH_2{-}CH_2{-}C(CH_3){=}CH{-}CH_2\right)_n$$

poly-*cis*-isoprene (the $-CH_2-CH_2-$ groups are *cis*)

Natural rubber is poly-*cis*-isoprene. However, the *trans* material also occurs in nature in the leaves and bark of the sapotacea tree and is known as *gutta-percha.* It is brittle and hard and is used for golf ball covers, electrical insulation, and other such applications. Without an appropriate catalyst, polymerization of isoprene yields a solid that is like neither rubber nor gutta-percha because it is a random mixture of the *cis* and *trans* geometries. Neither the *trans* polymer nor the randomly arranged material is as good as natural rubber *(cis)* for making automobile tires.

Stereoregulation catalysts catalyze reactions that favor the formation of one geometric isomer.

In 1955, chemists at the Goodyear and Firestone companies almost simultaneously discovered how to use **stereoregulation** catalysts to prepare synthetic poly-*cis*-isoprene. This material is structurally identical to natural rubber. Today, synthetic poly-*cis*-isoprene can be manufactured cheaply and is used almost equally well (there is still an increased cost) when natural rubber is in short supply. More than 2.4 million tons of synthetic rubber are produced in the United States every year.

Neoprene

One of the first synthetic rubbers produced in the United States was neoprene, an addition polymer of the monomer 2-chlorobutadiene, which has a chlorine atom substituted for the methyl group in isoprene. Neoprene is used in the production of gaskets, garden hoses, and adhesives.

$$n\ \mathrm{H_2C{=}C(Cl){-}CH{=}CH_2} \xrightarrow[\text{polymerization}]{\text{addition}}$$

2-chlorobutadiene

$$\mathrm{{-}CH_2{-}C(Cl){=}CH{-}CH_2}\left(\mathrm{{-}CH_2{-}C(Cl){=}CH{-}CH_2{-}}\right)_n\mathrm{CH_2{-}C(Cl){=}CH{-}CH_2{-}}$$

neoprene

Tires are made from synthetic rubber polymers. *(The World of Chemistry* (Program 22), "The Age of Polymers.")

Copolymers

Many commercially important addition polymers are **copolymers,** polymers obtained by polymerizing a mixture of two or more monomers. A copolymer of styrene with butadiene is the most important synthetic rubber produced in the United States. More than 1.4 million tons of styrene-butadiene rubber (SBR) are produced each year in the United States for making tires. A 3:1 mole ratio of butadiene to styrene is used to make SBR.

$$3n\ \mathrm{H_2C{=}CH{-}CH{=}CH_2} + n\ \mathrm{H_2C{=}CH(C_6H_5)} \xrightarrow[\text{polymerization}]{\text{addition}}$$

1,3-butadiene   styrene

$$\left(\mathrm{{-}CH_2{-}CH{=}CH{-}CH_2{-}CH_2{-}CH{=}CH{-}CH_2{-}CH_2{-}CH(C_6H_5){-}CH_2{-}CH{=}CH{-}CH_2{-}}\right)_n$$

styrene-butadiene rubber (SBR)

Other important copolymers are made by polymerizing mixtures of ethylene and propylene or acrylonitrile, butadiene, and styrene. Saran Wrap is an example of a copolymer of vinyl chloride with 1,1-dichloroethylene.

Condensation Polymers

A chemical reaction in which two molecules react by splitting out or eliminating a small molecule is called a **condensation reaction.** The reactions of alcohols with carboxylic acids to give esters (Section 13.5) are examples of condensation. This important type of chemical reaction does not depend on the presence of a double bond in the reacting molecules. Rather, it requires the presence of two different kinds of functional groups on two different molecules. If each reacting molecule has two functional groups, both of which can react, it is possible for condensation reactions to produce long-chain polymers.

Polyesters

A molecule with two carboxylic acid groups, such as terephthalic acid, and another molecule with two alcohol groups, such as ethylene glycol, can react with each other at both ends.

$$\underset{\text{terephthalic acid}}{2\,HO{-}\overset{\overset{\large O}{\|}}{C}{-}C_6H_4{-}\overset{\overset{\large O}{\|}}{C}{-}OH} + \underset{\text{ethylene glycol}}{2\,HO{-}CH_2{-}CH_2{-}OH} \longrightarrow$$

$$HO{-}\overset{\overset{\large O}{\|}}{C}{-}C_6H_4{-}\overset{\overset{\large O}{\|}}{C}{-}O{-}CH_2{-}CH_2{-}O{-}\overset{\overset{\large O}{\|}}{C}{-}C_6H_4{-}\overset{\overset{\large O}{\|}}{C}{-}O{-}CH_2{-}CH_2{-}OH + 2\,H_2O$$

If n molecules of acid and alcohol react in this manner, the process will continue until a large polymer molecule, known as a **polyester,** is produced.

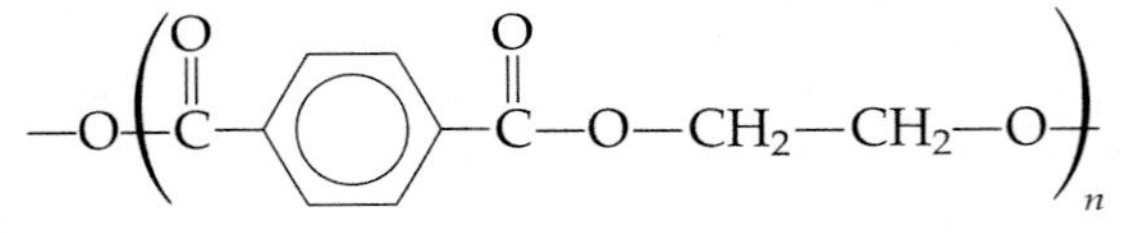

Figure 13.19
The model is wearing a Dacron dress. *(C.D. Winters)*

More than 2 million tons of poly(ethylene terephthalate), commonly referred to as PET, are produced in the United States each year for use in making beverage bottles, apparel, tire cord, film for photography and magnetic recording, food packaging, coatings for microwave and conventional ovens, and home furnishings. A variety of trade names are associated with the various applications. Polyester textile fibers are marketed under such names as Dacron and Terylene (Figure 13.19). Films of the same polyester, when magnetically coated, are used to make audio and TV tapes. This film, Mylar, has unusual strength and can be rolled into sheets one-thirtieth the thickness of a human hair.

The inert, nontoxic, noninflammatory, and non-blood-clotting characteristics of Dacron polymers make Dacron tubing an excellent substitute for human blood vessels in heart bypass operations (Figure 13.20), and Dacron sheets are used as skin substitute for burn victims.

Amines are classified as primary, secondary, or tertiary according to how many of the H atoms in the $-NH_2$ group are replaced by alkyl groups: RNH_2 (primary), R_2NH (secondary), and R_3N (tertiary).

Polyamides (Nylons)

Another useful and important type of condensation reaction is that between a carboxylic acid and a primary **amine,** which is an organic compound containing an $-NH_2$ functional group. Amines can be considered derivatives of ammonia (NH_3) and most of them are weak bases, similar in strength to

The Chemical World

Inventor of the Poly(ethylene terephthalate) Bottle

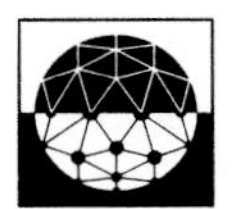

The inventor of the poly(ethylene terephthalate) soft-drink bottle is Nathaniel Wyeth, who comes from the internationally famous family of artists. His brother, Andrew Wyeth, expresses his creativity on canvas, but Nat Wyeth expresses his through chemical engineering. He has an intriguing story:

"I got to thinking about the work that Wallace Carothers did for Du Pont way back in the days when nylon was born, where he found that, if you took a thread of nylon when it was cold, that is, way below the melt point, and stretched it, it would orient itself. That is, the molecules of the polymer would align themselves. This is what you're doing to the molecules when you orient them, you're lining them up, so they can give you the most strength. They're all pulling in the direction you want them to pull in."

Wyeth tried this approach to make a poly(ethylene terephthalate) bottle, but the bottles kept splitting. Wyeth estimates that he made 10,000 tries and 10,000 failures before he made a simple observation.

"Well, then I realized what we've got to do now is to align these molecules in the sidewall of the bottle; not only in one direction, but in two directions. So I thought I'd play a trick on this mold, on this problem. I took two pieces of poly(ethylene terephthalate) and turned one of them ninety degrees with the other. So then I had one that split in this direction, and one that would split in that direction. Well, one piece reinforced the other. As soon as I did that, I could blow bottles. That seems almost dirt simple. But as I've often said, quoting Einstein, the biggest part of a problem and the easiest way to solving a problem is to understand it, have the problem in a form you can understand what's going on. And what I was doing here was learning about what was going on. Once I knew it, it was simple to solve."

Nathaniel Wyeth.

The World of Chemistry (Program 22), "The Age of Polymers."

---

ammonia. An amine reacts with a carboxylic acid at high temperature to split out a water molecule and form an **amide:**

$$\underset{\text{carboxylic acid}}{\mathrm{R{-}\overset{\overset{\displaystyle O}{\|}}{C}{-}OH}} + \underset{\text{amine}}{\mathrm{H{-}\underset{\underset{\displaystyle H}{|}}{N}{-}R}} \xrightarrow{\text{heat}} \underset{\text{amide}}{\mathrm{R{-}\overset{\overset{\displaystyle O}{\|}}{C}{-}\underset{\underset{\displaystyle H}{|}}{N}{-}R}} + H_2O$$

Polymers are produced when diamines (compounds containing two $—NH_2$ groups) react with dicarboxylic acids (compounds containing two —COOH groups). Reactions of this type yield a group of polymers that perhaps have had a greater impact on society than any other type. These are the **polyamides,** or nylons.

In 1928, the Du Pont Company embarked on a program of basic research headed by Dr. Wallace Carothers (1896–1937), who came to Du Pont from the Harvard University faculty. His research interests were high-molecular-weight compounds, such as rubber, proteins, and resins, and the reaction

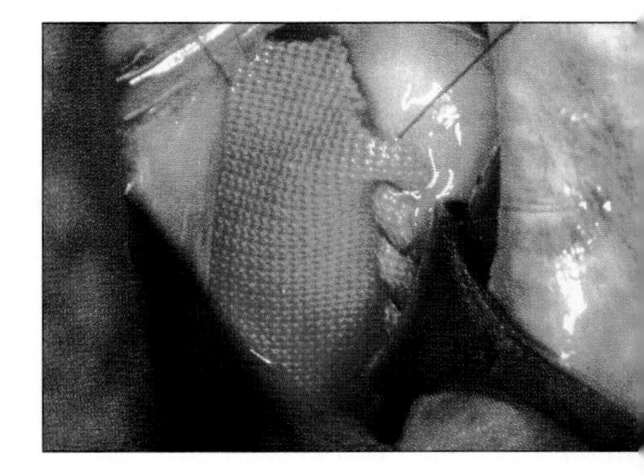

Figure 13.20
A Dacron patch is used to close an atrial septal defect in a heart patient. *(Courtesy of Drs. James L. Monro and Gerald Shore and the Wolfe Medical Publications, London, England.)*

Figure 13.21
Nylon-66. Hexamethylenediamine is dissolved in water (*bottom layer*), and a derivative of adipic acid (adipoyl chloride) is dissolved in hexane (*top layer*). The two compounds mix at the interface between the two layers to form nylon, which is being wound onto a stirring rod. *(C.D. Winters)*

Amides contain the $-\overset{O}{\overset{\|}{C}}-NH_2$ *or* $-\overset{O}{\overset{\|}{C}}-NH-$, *or* $-\overset{O}{\overset{\|}{C}}-\underset{|}{N}-$ *functional group.*

mechanisms that produced these compounds. In February 1935, his research yielded a product known as nylon-66 (Figure 13.21), prepared from adipic acid (a diacid) and hexamethylenediamine (a diamine):

$$n\,HO-\overset{O}{\overset{\|}{C}}-(CH_2)_4-\overset{O}{\overset{\|}{C}}-OH + n\,H_2N-(CH_2)_6-NH_2 \longrightarrow$$

adipic acid   hexamethylenediamine

$$-\overset{O}{\overset{\|}{C}}-(CH_2)_4-\overset{O}{\overset{\|}{C}}\left(-\underset{H}{\underset{|}{N}}-(CH_2)_6-\underset{H}{\underset{|}{N}}-\overset{O}{\overset{\|}{C}}-(CH_2)_4-\overset{O}{\overset{\|}{C}}-\right)_n\underset{H}{\underset{|}{N}}-(CH_2)_6-\overset{H}{\overset{|}{N}}- + n\,H_2O$$

nylon-66
(The amide groups are outlined for emphasis.)

The name of nylon-66 is based on the number of carbon atoms in the diamine and diacid, respectively, that are used to make the polymer. Since both hexamethylenediamine and adipic acid have six carbon atoms, the product is nylon-66.

This material could easily be extruded into fibers that were stronger than natural fibers and chemically more inert. The discovery of nylon jolted the American textile industry at almost precisely the right time. Natural fibers were not meeting the needs of 20th-century Americans. Silk was not durable and was very expensive, wool was scratchy, linen crushed easily, and cotton did not have a high fashion image. All four had to be pressed after cleaning. As women's hemlines rose in the mid-1930s, silk stockings were in great demand, but they were very expensive and short-lived. Nylon changed all that almost overnight. It could be knitted into the sheer hosiery women wanted, and it was much more durable than silk. The first public sale of nylon hose took place in Wilmington, Delaware (the hometown of Du Pont's main office), on October 24, 1939. World War II caused all commercial use of nylon to be abandoned until 1945, as the industry turned to making parachutes and other war materials. Not until 1952 was the nylon industry able

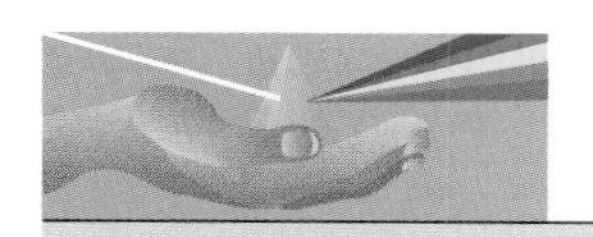

CHEMISTRY YOU CAN DO

Making "Gluep"

White school glue, such as Elmer's glue, contains poly(vinyl acetate), water, and other ingredients. A "gluep" similar to Silly Putty can be made by mixing 1/2 cup of glue with 1/2 cup of water and then adding 1/2 cup of liquid starch and stirring. Roll it together in your hands until it has a putty consistency. Roll it into a ball and let it sit on a flat surface undisturbed. What do you observe? Roll a piece into a ball and drop it on a hard surface. Does it bounce? The gluep can be stored in a sealed plastic bag for several weeks. Although gluep does not readily stick to clothes, walls, desks, or carpets, it does leave a water mark on wooden furniture, so be careful where you set it down. Mold will form on the gluep after a few weeks, but the addition of a few drops of Lysol to the gluep will retard formation of mold.

to meet the demands of the hosiery industry and to release nylon for other uses as a fiber and as a thermoplastic.

Figure 13.22 illustrates another facet of the structure of nylon—hydrogen bonding—which explains why nylons make such good fibers. To have good tensile strength, the chains of atoms in a polymer should be able to attract one another, but not so strongly that the plastic cannot be initially extended to form the fibers. Ordinary covalent chemical bonds linking the chains together would be too strong. Hydrogen bonds, with a strength of about one tenth that of an ordinary covalent bond, link the chains in the desired manner.

The amide linkage in nylon is the same bonding found in proteins, where it is called the peptide linkage.

Hair, wool, and silk are examples of nature's version of nylon. However, these natural polymers have only one carbon between each pair of $-\overset{O}{\overset{\|}{C}}-NH-$ *units instead of the half dozen or so found in synthetic nylons.*

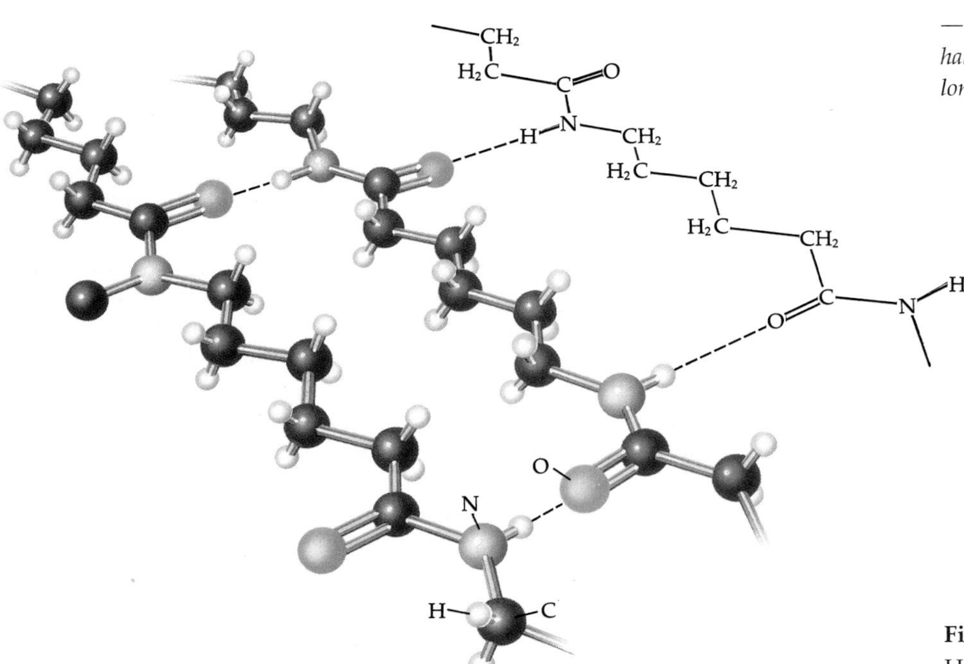

Figure 13.22
Hydrogen bonding in nylon-66.

p-phenylenediamine

Kevlar, another polyamide, is used to make bulletproof vests (Figure 13.23) and fireproof garments. Kevlar is made from *p*-phenylenediamine and terephthalic acid.

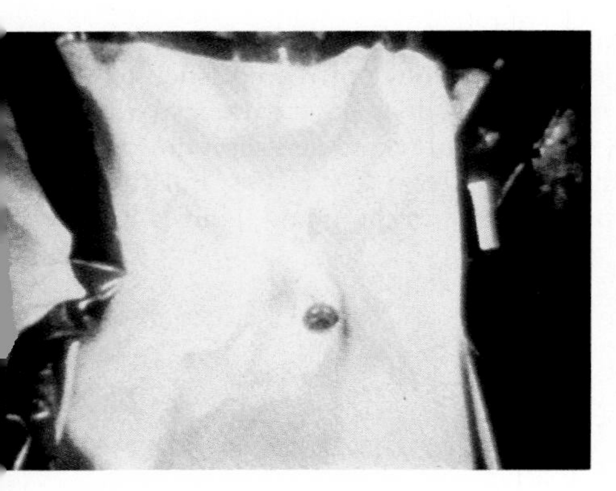

Figure 13.23
This vest, made of Kevlar fiber, has deflected a bullet. A recent TV commercial shows a large group of police officers who were saved by Kevlar vests. *(E.I. Du Pont de Nemours)*

EXAMPLE 13.6

Condensation Polymers

Write the repeating unit of the condensation polymer obtained by combining $HOOCCH_2CH_2COOH$ and $H_2NCH_2CH_2NH_2$.

SOLUTION A condensation polymer composed of a diacid and a diamine forms by loss of water between monomers to give an amide bond. The repeating unit is therefore

$$\left(-\overset{O}{\overset{\|}{C}}CH_2CH_2\overset{O}{\overset{\|}{C}}\underset{H}{\underset{|}{N}}CH_2CH_2\underset{H}{\underset{|}{N}}- \right)_n$$

EXERCISE 13.8 • ***Condensation Polymers***

Draw the structure of the repeating unit of the condensation polymer obtained from the reaction of terephthalic acid with ethylene glycol.

13.7 NEW POLYMER MATERIALS

Few plastics produced today find end uses without some kind of modification. For example, body panels for the GM Saturn and Corvette are made of **reinforced plastics,** which contain fibers embedded in a matrix of a polymer. These are often referred to as **composites.** The strongest geometry for a solid is a wire or a fiber, and the use of a polymer matrix prevents the fiber from bending or buckling. As a result, reinforced plastics are stronger than steel. In addition, the composites have a low density—from 1.5 to 2.25 g/cm^3, compared to 2.7 g/cm^3 for aluminum, 7.9 g/cm^3 for steel, and 2.5 g/cm^3 for concrete. The only structural material with a lower density is wood, which has an average value of 0.5 g/cm^3. In addition, polymers do not corrode. The low density, high strength, and high chemical resistance of composites are the basis for their increased use in the automobile, airplane, construction, and sporting goods industries.

Bumpers on new cars are made from synthetic rubber polymers. *(C.D. Winters)*

Glass fibers currently account for more than 90% of the fibrous material used in reinforced plastics because glass is inexpensive and glass fibers possess high strength, low density, good chemical resistance, and good insulating properties. In principle, any polymer can be used for the matrix material. Polyesters are the number one polymer matrix at the present time so glass-reinforced polyester composites have been used in structural applications such as boat hulls, airplanes, missile casings, and automobile body panels.

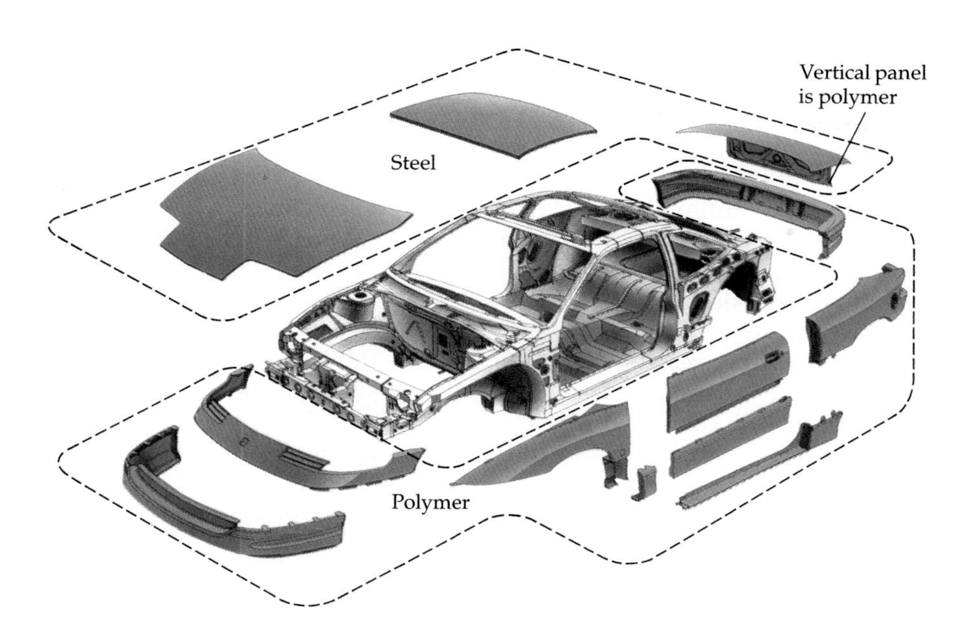

Expanded view of Saturn sports coupe showing the extensive use of polymers for body parts. *(Saturn Corporation)*

Other fibers and polymers have been used, and the trend is toward increased utilization of composites in automobiles and aircraft. For example, a composite of graphite fibers in an epoxy matrix is used in the construction of the Lear jet. Graphite/epoxy composites are used in a number of sporting goods such as golf-club shafts, tennis racquets, fishing rods, and skis. The F-16 military aircraft was the first to contain graphite/epoxy composite material, and the technology has advanced to the point where many aircraft, such as the F-18, use graphite composites for up to 26% of the aircraft's structural weight. This percentage is expected to increase to 40 to 50% in future aircraft.

Although few automobiles have exterior body panels made of plastics, most contain a number of components that are plastic. Examples include bumpers, trim, light lenses, grilles, dashboards, seat covers, steering wheels—enough plastics to account for an average of 250 pounds per car. The increased emphasis on improving fuel efficiency will lead to greater amounts of plastics in the construction of automobiles, in both interior components and exterior body panels. General Motors predicts that during the 1990s it will manufacture one million plastic-body automobiles per year.

A graphite composite reinforces the skin of the B-2 advanced-technology aircraft. *(U.S. Air Force/Northrup Corporation)*

13.8 RECYCLING PLASTICS

Disposal of plastics has been the subject of considerable debate in recent years as municipalities face increasing problems in locating sufficient landfill space. The number one waste is paper products, which make up about 40% of the volume in landfills. (Newspaper alone accounts for 16% of the volume). Next are plastics, which make up about 20% of the volume in landfills.

Trailer used for curbside recycling. *(C.D. Winters)*

At the present time, 1% of plastics waste is being recycled as compared to recycling of 60% of aluminum cans, 20% of paper, and 10% of glass. Recycling aluminum cans has been a successful money-making venture for individuals and nonprofit organizations for about 20 years, but comprehensive, community-wide recycling plans are a fairly recent development. The early success of aluminum recycling is based on economics, with scrap cans selling for 25–30 cents or more per pound. The demand for other recycled items, such as paper, has been much lower. For example, recycled newspapers often have no market.

Four phases are needed for a successful recycling of any waste material: collection, sorting, reclamation, and end-use. Public enthusiasm for recycling and state laws requiring recycling have resulted in a dramatic increase in the collection of recycled items. Between 1989 and 1991, the number of U.S. households with curbside collection of recyclables increased from 9 million to 16 million, which is estimated to be 29% of U.S. households. Codes are stamped on plastic containers to help consumers identify and sort their recyclable plastics (Figure 13.24).

Milk jug made of HDPE. *(C.D. Winters)*

Code	Material	Percent of total bottles
1 PETE	Polyethylene terephthalate (PET)*	20–30
2 HDPE	High-density polyethylene	50–60
3 V	Vinyl polyvinyl chloride (PVC)*	5–10
4 LDPE	Low-density polyethylene	5–10
5 PP	Polypropylene	5–10
6 PS	Polystyrene	5–10
7 OTHER	All other resins and layered multi-material	5–10

*Bottle codes are different from standard industrial identification to avoid confusion with registered trademarks.

Figure 13.24
Plastic container codes.

Elephants, Piano Keys, and Polymers

UNTIL RECENTLY, NO SYNthetic material has matched ivory in responding to the delicate touch of a concert pianist. Piano keys covered with plastic veneers have been rejected as too slippery and too cool. Eventually, however, a replacement for ivory must be found. In an effort to halt the slaughter of elephants, a global ban on trading in ivory was initiated in 1990 and is reported to be very effective.*

A team of scientists from Rensselaer Polytechnic Institute has patented a new material that may solve the problem. The essential step in their work was analysis of the surface of natural ivory at the microscopic level by a tribologist, an engineer who studies friction and materials that slide against each other. An ivory surface, they found, is covered with ridges, valleys, and tiny pores. When a sweaty finger slides over such a surface, it alternately sticks and slips, creating the feeling that pianists need for better control. The pores make an important contribution by absorbing sweat and oil from the finger.

To duplicate the ridges and valleys, the scientists worked with a finely made cast of a natural ivory surface. Duplicating the pores was challenging. Ultimately, they developed a synthetic ivory made from a mixture of a liquid polyester, a white titanium pigment, and finely powdered poly(ethylene glycol), a water-soluble polymer. By soaking the material in hot water after it has hardened, the poly(ethylene glycol) is dissolved away, leaving behind pores like those in natural ivory. Their new material has met the ultimate test. Concert pianists have failed to detect a difference between Steinway pianos with keys covered by the new material, known as RPIvory, and natural ivory.

**New York Times, May 25, 1993, p C3.*

Polyethylene terephthalate (PET), widely used as soft-drink bottles, is the most commonly recycled plastic. The used bottles are available from retailers in states requiring refundable deposits or from curbside pickups. Over 327 million pounds of PET were recycled in the United States in 1991 (36% of annual production). Major end uses for recycled PET include fiberfill for ski jackets and sleeping bags, carpet fibers, and tennis balls. Coca Cola is using two-liter bottles made of 25% recycled PET.

High-density polyethylene (HDPE) is the second most widely recycled resin, with 281 million pounds processed in 1991 (6.3% of annual production). Milk, juice, and water jugs are the principal source of recycled HDPE. Some products made from recycled HDPE are trash containers, drainage pipe, garbage bags, and fencing.

Although recycling of plastics has shown a dramatic increase in recent years, recycling companies will not be able to increase the percentage of recycled plastics to the 50% goal by the year 2000 without a significant increase in the demand for products made from recycled materials.

"Plastic lumber" is a composite material made from recycled plastics. *(Greg Vaughn/Tom Stack & Associates)*

IN CLOSING

Having studied this chapter, you should be able to

- describe petroleum refining and methods used to improve the gasoline fraction. (Section 13.1)
- identify processes used to obtain organic chemicals from coal and name some of their products. (Section 13.2)

- define and give examples of addition reactions of alkenes and substitution reactions of alkanes. (Section 13.3)
- name and draw the structures of three functional groups produced by the oxidation of alcohols. (Section 13.4)
- name and give examples of the uses of some important alcohols. (Section 13.4)
- list some properties of carboxylic acids and write equations for the formation of esters from carboxylic acids and alcohols. (Section 13.5)
- explain formation of polymers by addition or condensation; give examples of polymers formed by each type of reaction. (Section 13.6)
- draw the structures of the repeating units in some common types of polymers. (Section 13.6)
- identify or write the structures of the functional groups in alcohols, aldehydes, ketones, carboxylic acids, esters, amines, and amides. (Sections 13.4 to 13.6)
- identify the types of plastics most successfully being recycled and discuss some of the issues that influence successful recycling. (Section 13.8)

STUDY QUESTIONS

Review Questions

1. What is the difference between *catalytic cracking* and *catalytic reforming*?
2. Explain how an octane number of a gasoline is determined.
3. Why is coal receiving increased attention as a source of organic compounds?
4. What is synthesis gas? How can it be used to produce chemicals?
5. Methanol is now 22nd in the list of top chemicals produced in the United States. What factors are likely to lead to an increased demand for methanol in the next decade?
6. What is the difference between *oxygenated gasoline* and *reformulated gasoline*? Why are they being produced?
7. Explain why world use of natural gas is estimated to double between 1990 and 2010.
8. Table 13.2 lists several compounds with octane numbers above 100 and one compound with an octane number below zero. Explain why this is possible.
9. Explain why methanol (molecular weight = 32) has a lower boiling point (65.0 °C) than water (molecular weight of 18, boiling point = 100.0 °C).
10. Describe the Mobil process for converting methanol to gasoline.
11. Outline the steps necessary to obtain 89 octane gasoline, starting with a barrel of crude oil.
12. What is the major difference between crude oil and coal as a source of hydrocarbons?
13. Describe the structural difference between an aldehyde, a ketone, and a carboxylic acid.
14. Write the structural formula of a representative compound for each of the following classes of organic compounds: alcohols, aldehydes, ketones, carboxylic acids, esters, and amines.
15. Explain why esters have lower boiling points than carboxylic acids of the same molecular weight.
16. What structural features must a molecule have in order to undergo addition polymerization?
17. What feature do all condensation polymerization reactions have in common?
18. Give examples of (a) a synthetic addition polymer, (b) a synthetic condensation polymer, (c) a natural addition polymer.
19. How does *cis-trans* isomerism affect the properties of rubber?
20. Discuss what plastics are currently being recycled, and give examples of some products being made from these recycled plastics.
21. What are the four conditions that must be met for successful recycling of any solid waste?

Functional Groups

22. Draw the structural formula for
(a) an amine
(b) a carboxylic acid
(c) an alcohol
(d) a ketone
(e) an aldehyde
(f) an ester

23. Classify each compound according to its functional group.
(a) $CH_3CH_2NH_2$
(b) $CH_3CH_2CH_2\overset{O}{\overset{\|}{C}}H$
(c) $CH_3CH_2\overset{O}{\overset{\|}{C}}CH_3$
(d) $CH_3CH_2\underset{OH}{\underset{|}{C}}HCH_3$
(e) $CH_3CH_2\overset{O}{\overset{\|}{C}}OCH_2CH_3$
(f) $CH_3CH_2\overset{O}{\overset{\|}{C}}OH$

24. Many organic compounds have more than one functional group. Identify the functional groups in the following structures.
(a) $CH_3\underset{OH}{\underset{|}{C}}H\overset{O}{\overset{\|}{C}}OH$
(b) $C_6H_4(\overset{O}{\overset{\|}{C}}OH)(O\overset{O}{\overset{\|}{C}}CH_3)$ (benzene ring bearing $\overset{O}{\overset{\|}{C}}OH$ and, on the adjacent carbon, $-O\overset{O}{\overset{\|}{C}}CH_3$)
(c) $CH_3\overset{O}{\overset{\|}{C}}CH_2OH$
(d) $H_2NCH_2\overset{O}{\overset{\|}{C}}OH$
(e) $CH_3O\overset{O}{\overset{\|}{C}}CH_2\overset{O}{\overset{\|}{C}}OH$
(f) $HOCH_2\overset{NH_2}{\overset{|}{C}}HCH_2\overset{O}{\overset{\|}{C}}H$

25. Identify the functional groups in neosynephrine, a decongestant.

$HO-C_6H_4-\overset{OH}{\overset{|}{C}}HCH_2-\overset{CH_3}{\overset{|}{N}}H$

neosynephrine

26. Draw the structure of methyl *tertiary*-butyl ether, one of the gasoline additives.

Alcohols

27. Write a balanced equation for the reaction of ethanol with potassium metal.

28. Write a balanced equation for the reaction of methanol with sodium metal.

29. Classify each of the following alcohols as primary, secondary, or tertiary.
(a) $CH_3CH_2CH_2CH_2OH$
(b) $CH_3\underset{OH}{\underset{|}{C}}HCH_2CH_3$
(c) $CH_3\underset{OH}{\underset{|}{C}}HCH_2CH_2OH$
(d) $CH_3\overset{CH_3}{\overset{|}{\underset{OH}{\underset{|}{C}}}}CH_2CH_3$
(e) $CH_3\overset{CH_3}{\overset{|}{\underset{OH}{\underset{|}{C}}}}CH_3$

30. Draw the structures of the first two oxidation products of each of the following alcohols.
(a) CH_3CH_2OH (b) $CH_3CH_2CH_2CH_2OH$

31. Write the condensed structural formula for each of the following:
(a) 2-methyl-2-pentanol
(b) 2,3-dimethyl-1-butanol
(c) 4-methyl-2-pentanol
(d) 2-methyl-3-pentanol
(e) *tertiary*-butyl alcohol
(f) isopropyl alcohol

32. What is the percentage of ethanol in 90 proof vodka?

33. Explain the common names of *wood alcohol* for methanol and *grain alcohol* for ethanol.

Aldehydes and Ketones

34. Draw the structures of the aldehyde and ketone having the fewest number of carbon atoms.

35. Write the condensed formula of the aldehyde produced when ethanol is oxidized in the liver.

36. Write the structural formulas for (a) the aldehyde and (b) the ketone with the formula C_3H_6O.

Carboxylic Acids and Esters

37. Write the structural formula of the ester that can be formed from
(a) $CH_3COOH + CH_3CH_2OH$
(b) $CH_3CH_2COOH + CH_3CH_2CH_2OH$
(c) $CH_3CH_2COOH + CH_3OH$

38. Write the structural formula of the ester that can be produced in the following reactions:
(a) formic acid + methanol
(b) butyric acid + ethanol
(c) acetic acid + 1-butanol
(d) propionic acid + 2-propanol

39. Write the condensed formula of the alcohol and acid that will react to form each of the following esters.
(a) $CH_3CH_2\overset{\overset{O}{\|}}{C}OCH_3$ (b) $H\overset{\overset{O}{\|}}{C}OCH_2CH_3$
(c) $CH_3\overset{\overset{O}{\|}}{C}OCH_2CH_3$

40. Explain why carboxylic acids are more soluble in water than are esters with the same molecular weight.

Organic Polymers

41. Draw the structure of the repeating unit in a polymer in which the monomer is
(a) 1-butene (b) 1,1-dichloroethylene
(c) vinyl acetate

42. Methyl methacrylate has the structural formula shown in Table 13.10. When polymerized it is very transparent, and it is sold in the United States under the trade names Lucite and Plexiglas. Draw the repeating unit for the poly(methyl methacrylate) chain.

43. What is the principal structural difference between low-density and high-density polyethylene?

44. What monomers are used to prepare the following polymers?
(a) $-CH_2CH_2CH_2CH_2CH_2CH_2CH_2CH_2CH_2-$
(b) $-\overset{\overset{CH_3}{|}}{C}HCH_2\overset{\overset{CH_3}{|}}{C}HCH_2\overset{\overset{CH_3}{|}}{C}HCH_2-$
(c) $-CH_2-\overset{\overset{H}{|}}{\underset{\underset{C_6H_5}{|}}{C}}CH_2-\overset{\overset{H}{|}}{\underset{\underset{C_6H_5}{|}}{C}}CH_2-\overset{\overset{H}{|}}{\underset{\underset{C_6H_5}{|}}{C}}CH_2-\overset{\overset{H}{|}}{\underset{\underset{C_6H_5}{|}}{C}}-$
(d) $-CH_2-\overset{\overset{CH_3}{|}}{\underset{\underset{Cl}{|}}{C}}-CH_2-\overset{\overset{CH_3}{|}}{\underset{\underset{Cl}{|}}{C}}-CH_2-\overset{\overset{CH_3}{|}}{\underset{\underset{Cl}{|}}{C}}-$
(e) $-CH_2\underset{\underset{\underset{OC_2H_5}{|}}{C=O}}{\underset{|}{C}}HCH_2\underset{\underset{\underset{OC_2H_5}{|}}{C=O}}{\underset{|}{C}}HCH_2\underset{\underset{\underset{OC_2H_5}{|}}{C=O}}{\underset{|}{C}}HCH_2\underset{\underset{\underset{OC_2H_5}{|}}{C=O}}{\underset{|}{C}}H-$

45. Draw structures of monomers that could form each of the following condensation polymers.
(a) $-\underset{\underset{O}{\|}}{C}-(CH_2)_8\underset{\underset{O}{\|}}{C}-NH(CH_2)_6NH-$
(b) $-\underset{\underset{O}{\|}}{C}-C_6H_4-\underset{\underset{O}{\|}}{C}-OCH_2-C_6H_4-CH_2O-$

46. Orlon has a polymeric chain structure of
$$-CH_2-\underset{\underset{CN}{|}}{C}H-CH_2-\underset{\underset{CN}{|}}{C}H-CH_2-\underset{\underset{CN}{|}}{C}H-$$
What is the monomer from which this structure can be made?

47. How many ethylene units are in a polyethylene molecule that has a molecular weight of approximately 42,000?

48. Kevlar, a bulletproof plastic, is made from *p*-phenylenediamine, $H_2N-C_6H_4-NH_2$ and terephthalic acid, $HOOC-C_6H_4-COOH$. Draw the repeating unit of Kevlar.

49. Write the chemical structures for the repeating units of
(a) natural rubber (b) neoprene
(c) polybutadiene

General Questions

50. Compounds A and B both have the molecular formula C_2H_6O. Compound A reacts with sodium metal to give hydrogen gas. Compound B does not react with sodium metal. The boiling points of compounds A and B are 78.5 °C and −23.7 °C, respectively. Write the structural formulas and names of the two compounds.

51. Explain why ethanol, CH_3CH_2OH, is soluble in water in all proportions, but decanol, $CH_3(CH_2)_9OH$, is almost insoluble in water.
52. Nitrile rubber (Buna N) is a copolymer of two parts of 1,3-butadiene to one part of acrylonitrile. Draw the repeating unit of this polymer.
53. How are rubber molecules modified by vulcanization?

Summary Questions

54. Polytetrafluoroethylene is made by first treating HF with chloroform followed by cracking the resultant difluorochloromethane.

$$CHCl_3 + 2\,HF \longrightarrow CHClF_2 + 2\,HCl$$

$$CHClF_2 + \text{heat} \longrightarrow F_2C{=}CF_2 + 2\,HCl$$

$$F_2C{=}CF_2 + \text{peroxide catalyst} \longrightarrow \text{Teflon}$$

If you wished to make 1.0 kilogram of Teflon, what mass of chloroform and HF must you use to make the starting material, $CHClF_2$? (Although it is not realistic, assume each reaction step proceeds in 100% yield.)

55. Acetic anhydride is an important intermediate in making acetic acid.
(a) How many chemical bonds are there in a molecule of acetic anhydride?
(b) Give the approximate value of the indicated bond angles.

```
 1          2
  ⌒H   O   ⌒O   H
   |   ‖    ‖   |
 H—C—C—O—C—C—H
   |     ‿      |
   H     3      H
```

Computer and Videodisc Questions

56. Nylon 6,10 is made from sebacoyl chloride and 1.6-diaminohexane. (See the *Chemical Demonstrations* Videodisc 1, Side 2 , frame 14086). As you view this demonstration, what evidence is there that a reaction has occurred?
57. When you place acetone in a cup made of polystyrene foam, the cup collapses. Packing "peanuts" made of polystyrene foam also collapse when they are put into the beaker of acetone. (See the *Chemical Demonstrations* Videodisc 1, Side 2, frame 18188.) The reason for this is that the polystyrene absorbs acetone, and the polymer structure collapses. Why does the volume of the "peanuts" decrease when put into acetone?

A model of the structure of diamond, a network solid.

(C.D. Winters)

Solids

CHAPTER 14

A sample of solid matter has a rigid shape and its volume varies only a little with changes in temperature and pressure. The kinetic-molecular theory interprets these observations by saying that in a crystalline solid the atoms, molecules, or ions are packed closely together in a regular arrangement; they can vibrate about their average positions, but seldom squeeze past their neighbors and move to a new position. The regular structure of a crystalline solid can be built up by repeating many times, in all three dimensions, the structure of a very small group of particles called a unit cell. This atomic-scale regularity of structure gives rise to many beautiful macroscopic shapes, such as those of snowflakes or gemstones. Metals are composed of like atoms, each bonded to its neighbors by valence electrons that are delocalized—free to move throughout the metallic crystal. Many properties of metals, especially electrical conductivity, depend on the behavior of these valence electrons. Many nonmetallic crystals are built up of networks of atoms (or groups of atoms) bonded together by localized covalent bonds; such solids have properties vastly different from those of metals. The properties of semiconductors and superconductors also depend on their solid-state structures. Finally, some solids are not crystalline at all. They have very irregular structures, somewhat like liquids whose randomly moving molecules have somehow become immobilized. These solids are referred to as amorphous or glassy.

Why are solids interesting? One reason is that their rigidity makes them very useful. We live on solid land; we build with wood, stone, glass, metal, and plastic; we travel on concrete highways and metal rails in cars, buses, ships, and planes made of metal; we clothe ourselves with solid fibers; and we wear metal jewelry that often includes beautiful solid minerals such as diamonds and other gemstones.

Solids are fascinating because their regular, repeating arrangement of atoms, ions, or molecules makes it possible for us to study and understand how their atomic-level structure influences their macroscopic properties. Who hasn't seen a Sci-Fi movie in which some alien culture has a lightweight material of such fantastic strength that it cannot be cut or penetrated using puny Earth tools? Well, materials scientists actually look for ways to make solids with those kinds of properties. They also work to develop new solids that can be used in medical applications, in transistors and other semiconductor devices for communications and computers, in lasers for printing words and pictures on paper and reading music from CDs, and in low-cost devices that can capture sunlight and convert it to electricity.

14.1 SOLIDS COMPARED TO GASES AND LIQUIDS

The kinetic-molecular theory (see Sections 2.1 and 12.3) postulates that molecules are in continual motion. In gases the molecules are far apart and moving in random directions, while in liquids they are nearly touching but still moving randomly. In solids, molecular movement is restricted to vibration and sometimes rotation around an average position. This leads to an orderly array of molecules, atoms, or ions, and to properties very different from those of liquids or gases. Because the particles that make up a solid are very close together, solids (like liquids) are very difficult to compress. However, unlike liquids, solids are rigid—they cannot transmit pressure in all directions. Solids have definite shapes and varying degrees of hardness. Hardness depends on the kinds of bonds that hold the particles of the solid together. Talc or soapstone (Figure 14.1) is one of the softest solids known;

Figure 14.1
Although they do not appear to be soft in this photo, these talc crystals are so soft that they can be crushed between one's fingers.
(C.D. Winters)

diamond is one of the hardest. Talc is used as a lubricant and in talcum powder. At the atomic level it consists of layered sheets that contain silicon, magnesium, and oxygen atoms. Attractive forces between these sheets are very weak, and so one sheet can slide along another and be removed easily from the rest. In diamond each carbon atom is strongly bonded to four neighbors, each of those neighbors is strongly bonded to three more carbon atoms, and so on throughout the solid. Because of the number and strength of the bonds holding each carbon atom to its neighbors, diamond is so hard that it can scratch or cut almost any other solid. For this reason diamonds are used in cutting tools and abrasives, which are more important commercially than diamonds as gemstones.

Snowflakes. It has long been said that no two are exactly alike, yet they all have hexagonal symmetry.
(John Shaw/Tom Stack & Associates)

The Beauty of Molecular Architecture: Structure of Ice

When atoms, molecules, or ions are packed closely together in a solid, the result is often strikingly beautiful because of its symmetry. The structure of ice is a good example (Figure 14.2). To achieve maximum hydrogen bonding, each water molecule in a perfect ice crystal is surrounded by four others; each of those is hydrogen bonded to three more (in addition to the first one), and so on. Only water molecules at the surfaces of the crystal are without four neighbors. Look closely at Figure 14.2; water molecules in the ice structure are arranged at the corners of puckered, six-sided rings or hexagons. Each edge of a six-sided ring consists of a normal O—H covalent

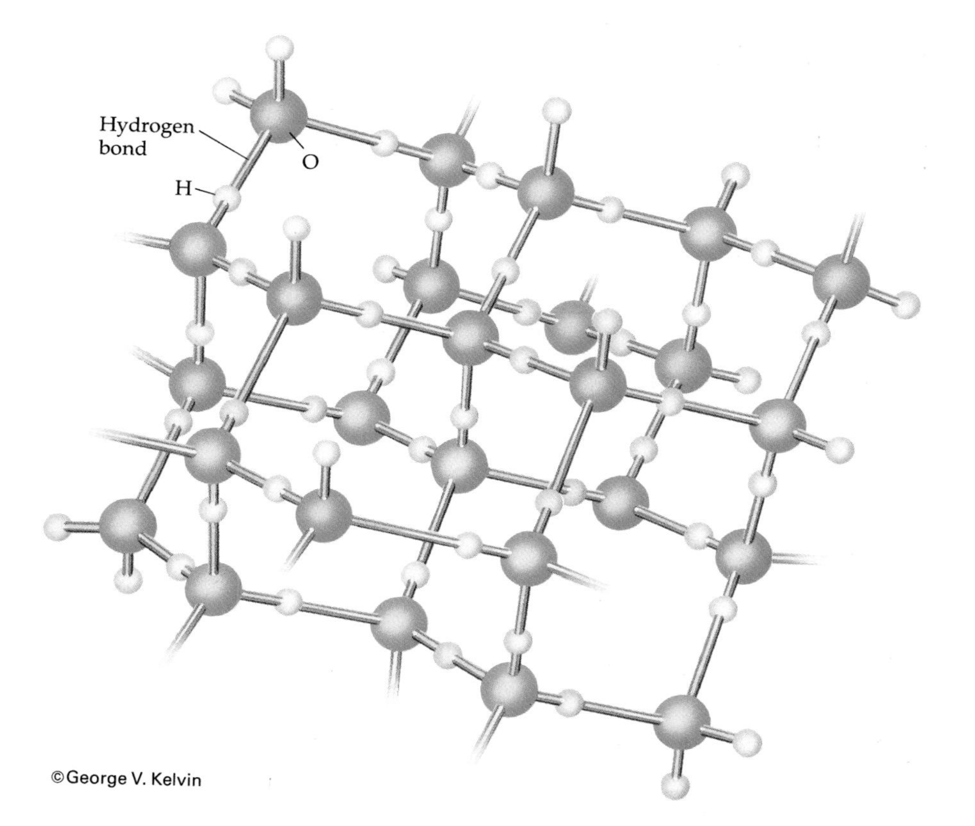

Figure 14.2
The structure of ice. Each oxygen atom is covalently bound to two hydrogen atoms and hydrogen bonded to two other hydrogen atoms. The hydrogen bonds are longer than the covalent H—O bonds. Due to the hydrogen bonding, ice has a more open structure than water. Because the water molecules are farther apart, on average, in ice than in liquid water, the density of ice is less than that of liquid water.

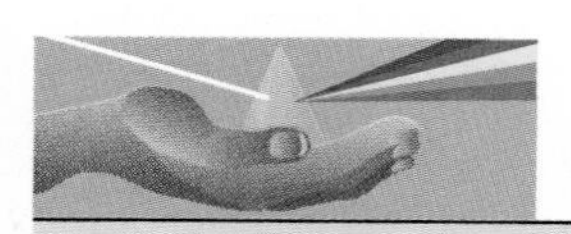

CHEMISTRY YOU CAN DO

Melting Ice with Pressure

Do this experiment in a sink. Obtain a piece of thin, strong, single-strand wire about 50 cm long, two weights of about 1 kg each, and a piece of ice about 25 cm by 3 cm by 3 cm. The ice can be either a cylinder or a bar, and can be made by pouring some water into a mold made of several thicknesses of aluminum foil and placing it into a freezer. A piece of plastic pipe, not tightly sealed, will also work fine as a mold. You can use metal weights or make weights by filling two one-quart milk jugs with water. (One quart is about one liter, which is 1 kg of water.)

Fasten one weight to each end of the wire. Support each end of the bar of ice so that you can hang the wire over it, with one weight on each side without breaking the bar. Observe the bar of ice and the wire every minute or so and record what happens. Suggest an explanation for what you have observed; the effect of pressure on melting of ice and the heat of fusion of ice are good ideas to consider when thinking about your explanation.

bond and a hydrogen bond. Snowflakes always have six-sided geometry, in part a result of the hexagonal arrangement of molecules in the solid-state structure.

Another important observation from Figure 14.2 is that the structure of ice includes a good deal of empty space. Water molecules are farther apart, on average, in ice than they are in liquid water near the freezing point. The same number of molecules in a larger volume means less mass per unit volume, and so ice is less dense than water. This means that ice will float on water. It also means that when water freezes its volume increases. It is for this reason that water pipes exposed to sub-freezing temperatures must be drained—water remaining inside them would expand and burst the pipe when it froze. The same effect causes weathering of rocks; liquid water that

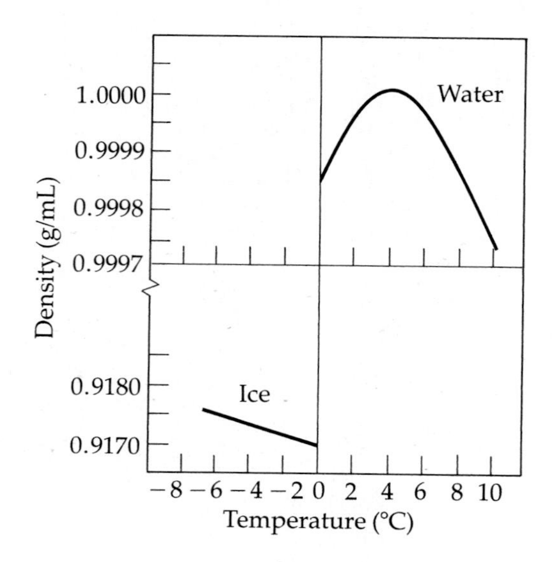

The temperature dependence of the densities of ice and liquid water. Not only are the densities of ice and water different, but the density of water *increases* as temperature increases from 0 °C to 4 °C. This happens because many hydrogen-bonded structures remain in the cold liquid, holding water molecules apart so that empty spaces remain between them. As temperature increases, increased movement of water molecules breaks up some of these structures, allowing water molecules to fill in the empty spaces so that density increases. Above 4 °C the density decreases with increasing temperature, which is what happens with most liquids.

seeps into cracks can freeze and break the rock apart, helping to convert it into fine gravel that can support plant growth.

Another important consequence of ice's structure is that increased pressure can cause ice to melt. Consider the equilibrium

$$H_2O(\text{solid, larger volume}) \rightleftharpoons H_2O(\text{liquid, smaller volume})$$

If increased pressure is applied to the equilibrium system, LeChatelier's principle predicts that the equilibrium will shift to minimize the increase. This means a shift so that volume decreases, which for water is a shift toward liquid and away from solid. Higher pressure occurs, for example, under the thin blade of an ice skate. Even though the rest of the ice remains frozen, ice melts under the blade where the pressure is highest. Liquid between the skate blade and the ice lubricates the blade, making it slide along easily with very little friction, and making ice skating fun.

American speed skater Bonnie Blair at Albertville, France, on her way to winning the women's Olympic 1000 meter event. *(Reuters/Bettmann)*

Phase Changes: Fusion and Sublimation

When a solid is heated to a temperature at which molecular motions are violent enough to partially overcome intermolecular forces, the orderliness of the solid's structure collapses and the solid melts (Figure 14.3). The temperature is the **melting point.** Melting requires transfer of energy (the heat of fusion, see Section 7.3) from the surroundings into the system to overcome intermolecular attractions.

Solids with high heats of fusion usually melt at high temperatures, and solids with low heats of fusion usually melt at low temperatures. The reverse of melting is **solidification** or **crystallization,** which is always an exothermic process. The heat evolved on crystallization is equal in magnitude to the heat of fusion.

(a)

(b)

(c)

Figure 14.3
The melting of naphthalene, $C_{10}H_8$, at 80.22 °C. (a) The crystals of naphthalene are at a temperature just below the melting temperature. (b) Melting commences at 80.22 °C. (c) The entire sample is molten. Note that some of the naphthalene has sublimed.
(C.D. Winters)

$$\text{Solid} \underset{-\text{ heat of crystallization}}{\overset{+\text{ heat of fusion}}{\rightleftharpoons}} \text{Liquid}$$

Some solids do not have measureable melting points and some liquids do not have measureable boiling points because increasing temperature causes them to decompose chemically before they melt or boil. Making peanut brittle (described at the beginning of Chapter 2) provides an example: melted sucrose chars before it boils, but it produces a great-tasting candy.

Table 14.1 lists melting points and heats of fusion for examples of three classes of compounds: (a) nonpolar molecular solids, (b) polar molecular solids, some capable of hydrogen bonding, and (c) ionic solids. Solids composed of low-molecular-weight nonpolar molecules have the lowest melting temperatures, because intermolecular attractions are weakest. These molecules are held together by dispersion forces only (Section 12.5), and they form solids with melting points so low that we seldom encounter them in the solid state. Melting points and heats of fusion of nonpolar molecular solids increase with increasing number of electrons (which corresponds to increasing molecular weight) as the dispersion forces become stronger. The ionic compounds in Table 14.1 have the highest melting points and the highest heats of fusion owing to the very strong ionic bonding that holds the ions together.

Molecules can escape directly from the solid to the gas phase, a process known as **sublimation.** The heat required is the **heat of sublimation.**

$$\text{solid} + \text{heat of sublimation} \longrightarrow \text{gas}$$

Figure 14.4
Dry Ice. In this photo the cold vapors of CO_2 are causing moisture, which is seen as wispy white clouds, to condense. Being more dense than air at room temperature, the vapors glide slowly toward the table top or the floor. *(C.D. Winters)*

Table 14.1

Enthalpies of Fusion and Melting Points of Some Solids

Solid	Melting Point (°C)	Enthalpy of Fusion (kJ/mol)	Type of Intermolecular Forces
Molecular Solids: Nonpolar Molecules			
O_2	−248	0.445	These molecules have only dispersion forces (which increase with number of electrons).
F_2	−220	1.020	
Cl_2	−103	6.406	
Br_2	−7.2	10.794	
Molecular Solids: Polar Molecules			
HCl	−114	1.990	All of these molecules have dispersion forces enhanced by dipole-dipole forces. H_2O also has significant hydrogen bonding.
HBr	−87	2.406	
HI	−51	2.870	
H_2O	0	6.020	
H_2S	−86	2.395	
Ionic Solids			
NaCl	800	30.21	All ionic solids have strong attractions between oppositely charged ions.
NaBr	747	25.69	
NaI	662	21.95	

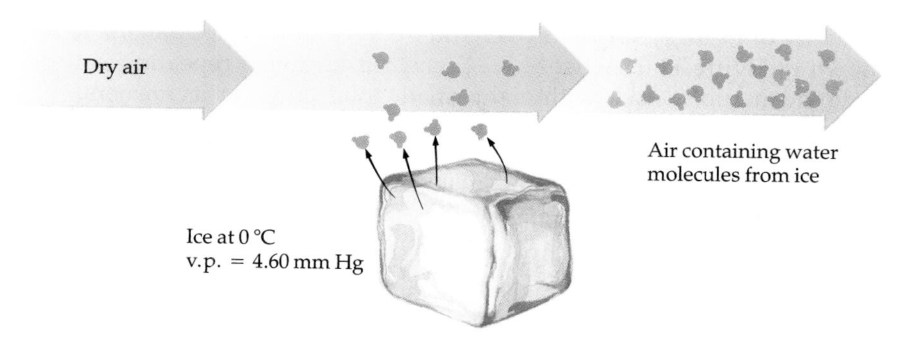

Figure 14.5
Ice subliming. As dry air passes over a sample of ice, the water molecules leaving the surface are carried away.

A common substance that sublimes at normal atmospheric pressure is solid carbon dioxide, whose vapor pressure at −78 °C is 1 atm (Figure 14.4). Because there is no temperature at which liquid CO_2 has a vapor pressure less than 1 atm, $CO_2(\ell)$ cannot exist except under a pressure well above atmospheric pressure. Hence it sublimes rather than melting, and solid CO_2 is known by the trade name Dry Ice. For water the heat of sublimation is 51 kJ/mol and the vapor pressure of ice at 0 °C is 4.60 mm Hg. Have you ever noticed that ice and snow slowly disappear even if the temperature never gets above freezing? The reason is that ice sublimes readily in dry air (Figure 14.5). Given enough air passing over it, a sample of ice will sublime away, leaving no trace behind. This is what happens in a frost-free refrigerator. A current of dry air periodically blows across any ice formed in the freezer compartment, taking away water vapor (and hence the ice) without warming the freezer.

EXERCISE 14.1 • *Heat of Fusion*

Calculate the heat required to melt 100.0 g of NaCl at its melting point.

14.2 CRYSTALLINE SOLIDS

The beautiful regularity of ice crystals, crystalline salts, or gemstones suggests that there must also be some *internal* regularity. Toward the end of the eighteenth century, scientists found that shapes of crystals can be used to identify minerals. The angles at which crystal faces meet are characteristic of a crystal's composition, but do not depend on its size and shape. The shapes of all crystalline solids, whether ionic, molecular, or metallic, reflect the shape of a **crystal lattice**—an orderly, repeating arrangement of ions, molecules, or atoms. One example of a crystal lattice is the structure of ice shown in Figure 14.2. Each water molecule occupies a specific position, and the

Figure 14.6
"Cubic Space Division," by M.C. Escher. *(© 1953 M.C. Escher Foundation, Baarn, Holland)*

structure of hexagons repeats over and over. Another example of a lattice is shown in Figure 14.6. Artist M. C. Escher has drawn a repeating pattern of large, open cubes, each with eight small, solid cubes at its corners.

Unit Cells

As a convenient way to describe and classify this kind of repeating pattern, a small segment of a crystal lattice is defined as a **unit cell**—the smallest part of the lattice that, when repeated along the directions defined by its edges, reproduces the entire crystal structure. To help understand the idea of the unit cell, look at the simple two-dimensional array of circles shown in Figure 14.7. The same size circle is repeated over and over, but a circle is not a good unit cell because it gives no indication of its relationship to all the other circles. A better choice is to recognize that the centers of four adjacent circles lie at the corners of a square, and to draw four lines connecting those centers. A square unit cell results. In Figure 14.7, four unit cells are drawn. As you look at the unit cell drawn in dark purple, notice that each of four circles contributes one quarter of itself to the unit cell, so a net of one circle is located within the unit cell. When this unit cell is repeated by moving the square parallel to its edges (that is, when unit cells are placed side by side), the two-dimensional lattice results. Notice that the corners of a unit cell are equivalent to each other, and collectively they define the crystal lattice.

The three-dimensional unit cells from which all known crystal lattices can be constructed fall into only seven categories. These seven types of unit cells have sides of different relative lengths and edges that meet at different angles (Figure 14.8). Only cubic unit cells composed of atoms or monatomic ions are described here. These are quite common in nature and also are simpler and easier to visualize. The principles illustrated, however, apply to all unit cells and all crystal structures, including those composed of polyatomic ions and complicated molecules.

Cubic unit cells have equal-length edges that meet at 90° angles. There are three types: simple cubic (sc), body-centered cubic (bcc), and face-

Figure 14.7
Unit cells for a two-dimensional solid made from flat, circular objects. Here the unit cell is a square. Each circle at the corner of the square contributes $\frac{1}{4}$ of its area to the area inside the square. Thus, there is a *net* of one circle per unit cell.

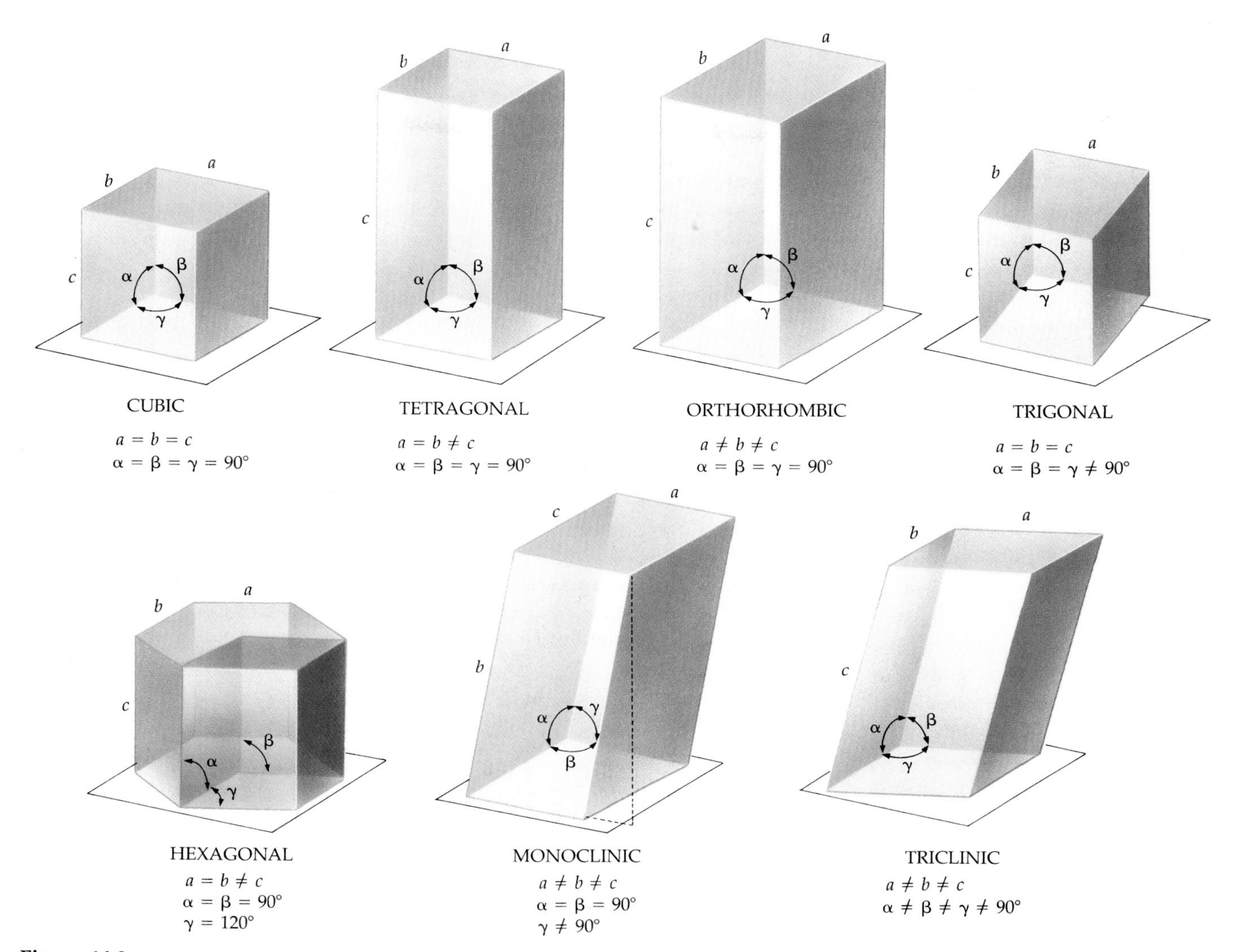

Figure 14.8
Shapes of the unit cells of the seven crystal systems.

centered cubic (fcc) (Figure 14.9, p. 612). All three cubic unit cells have identical atoms or ions centered at each corner of the cube. When cubes pack into three-dimensional space, an atom or ion at a corner is shared among eight cubes (Figure 14.10a, p. 613); this means that only one-eighth of each corner atom or ion is actually within the unit cell. Since a cube has eight corners and one-eighth of the atom at each corner belongs to the unit cell, the net result is $8 \times \frac{1}{8} = 1$ atom within the unit cell. In the bcc unit cell there is an additional atom at the exact center of the cube; it lies entirely within the unit cell. This combined with the net of one atom from the corners gives a total of two atoms per unit cell. In the fcc unit cell there are six atoms or ions that lie in the centers of the faces of the cube. One half of each of these belongs to the

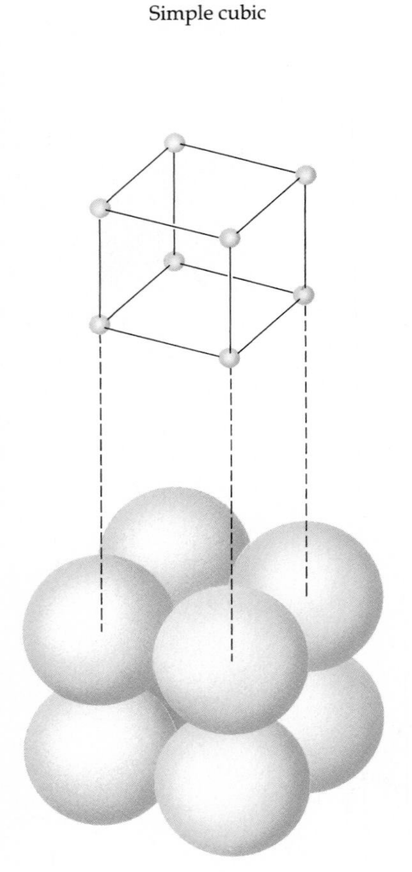

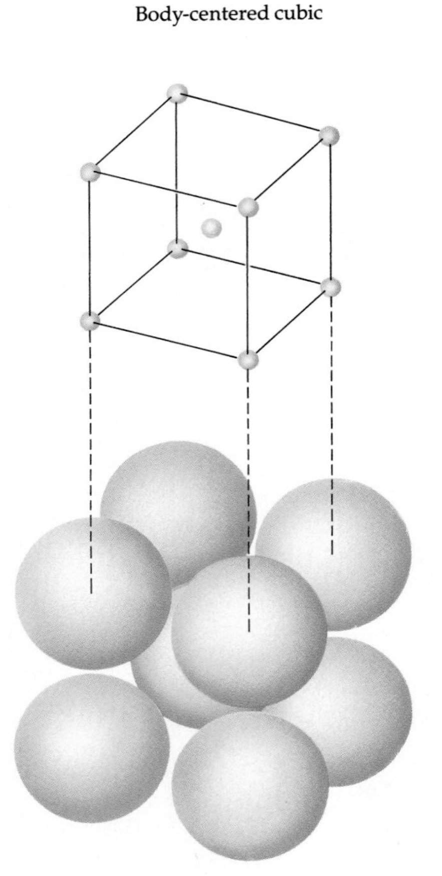

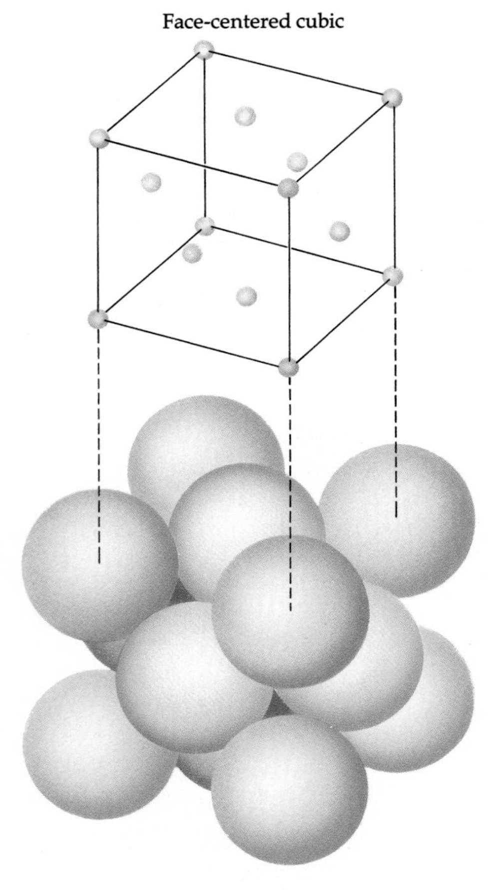

Figure 14.9
The three different types of cubic unit cells. The top row shows the lattice points of the three cells. In the bottom row the points are replaced with space-filling spheres representing the atoms or ions of the lattice. All spheres—no matter what their color—represent identical atoms or ions centered on the lattice points. Notice that the spheres at the corners of the body-centered and face-centered cubes do not touch each other. Rather, each corner atom in the body-centered cell touches the center atom, and each corner atom in the face-centered cell touches spheres in the three adjoining faces.

unit cell (Figure 14.10b); in this case there is a net result of $6 \times \frac{1}{2} = 3$ atoms within the unit cell, in addition to the net of one atom contributed by the corners, for a total of four atoms per unit cell.

EXERCISE 14.2 • *Cubic Structures*

Crystalline polonium has a simple cubic unit cell, lithium has a body-centered cubic unit cell, and calcium has a face-centered cubic unit cell. How many Po atoms belong to one unit cell? How many Li atoms belong to one unit cell? How many Ca atoms belong to one unit cell? Draw each unit cell. Indicate on your drawing what fraction of each atom lies within the unit cell.

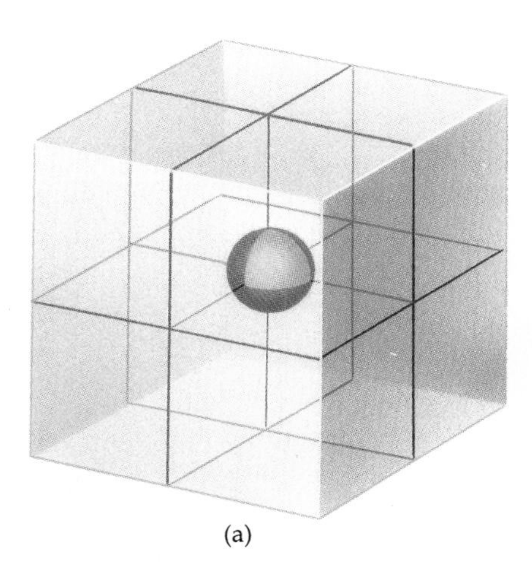

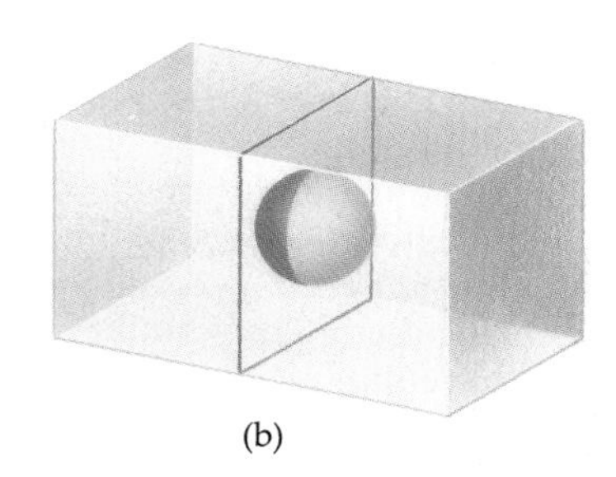

Figure 14.10
Atom sharing at cube corners and faces. (a) In any cubic lattice each corner atom (or ion) is shared equally among 8 cubes; $\frac{1}{8}$ of each atom (or ion) is within a particular cube. (b) In a face-centered lattice, each atom (or ion) in a cube face is shared equally between 2 cubes. Each atom (or ion) of this type contributes $\frac{1}{2}$ of itself to a given cube.

Unit Cells and Density

What has been described about unit cells so far allows us to check whether a proposed unit cell is reasonable. Aluminum has a density of 2.699 g/cm^3 at 20 °C, making it one of the lowest density metals with good structural properties (magnesium is another). An aluminum atom has a radius of 143 pm. Are these data consistent with a fcc unit cell for aluminum? If they are, we should be able to calculate the volume of a unit cell and then, using the density of aluminum, calculate the mass of one unit cell. Since a fcc unit cell contains four atoms (8 corners × $\frac{1}{8}$ atom per corner + 6 faces × $\frac{1}{2}$ atom per face), the mass of four aluminum atoms can be compared with the mass of the unit cell. If they are equal, a fcc unit cell is reasonable.

Since the volume of a cube is the cube of its edge, we need the length of the edge of the unit cell. One face of a fcc unit cell is shown in Figure 14.11. Notice that the face-centered atom touches each of the corner atoms, but the corner atoms do not touch each other. The diagonal of the face is equal to four times the radius of an aluminum atom (143 pm). To figure out the length of an edge we use the Pythagorean theorem from geometry: the sum of the squares of the two edges equals the square of the hypotenuse. The hypotenuse is the diagonal distance;

$$(\text{diagonal distance})^2 = \text{edge}^2 + \text{edge}^2$$

or

$$(\text{diagonal distance})^2 = 2\ (\text{edge})^2$$

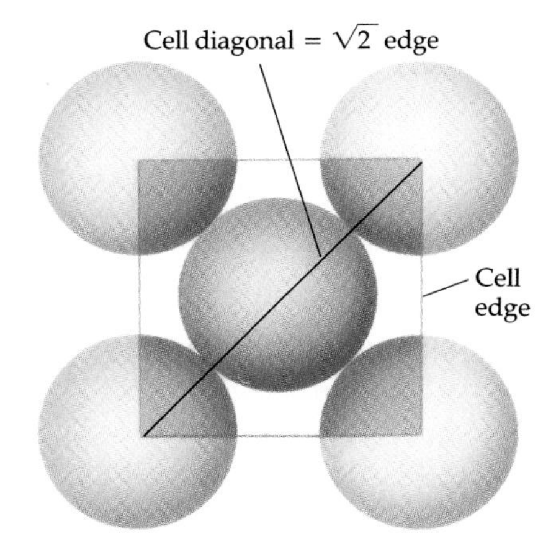

Figure 14.11
One face of a face-centered cubic crystal.

Memory Metal

ONE OF THE MOST INTRIGUING properties of metals ever discovered is shape memory. One metal with this property is called **Nitinol** because it contains equal numbers of **ni**ckel and **ti**tanium atoms and was discovered at the **N**aval **O**rdnance **L**aboratory. Nitinol can be formed into one shape at a high temperature (above 500 °C), then cooled to room temperature and formed into another shape. When the Nitinol is warmed to 50 or 60 °C, it will *revert* to its high-temperature shape.

Nitinol crystallizes with each Ni atom at the center of a cube surrounded by eight Ti atoms, one at each corner. When the cubic unit cell is repeated in three dimensions, each Ti atom is found to be surrounded by eight Ni atoms as well. At room temperature each cubic unit cell is distorted slightly; there are 24 different ways this can happen, and any of these can be changed to any other. This makes the alloy soft and easily formed into different shapes. The original shape is "remembered" because when the low-temperature form is warmed it reverts to the undistorted cubic unit cell. There is only one pathway from the low-temperature, distorted crystalline structure back to the original structure that retains the arrangement of eight Ti atoms around each Ni and vice versa. This pathway exactly reverses the distortions introduced when the low-temperature shape was formed.

The shape memory phenomenon is exhibited by several alloys and has found some very interesting uses. One medical application is in hip-joint replacements. Up to now metal hip joints have been sealed into the femur (thigh bone) by driving them into an undersized hole in the bone and applying glue. The impact of driving the joint into the

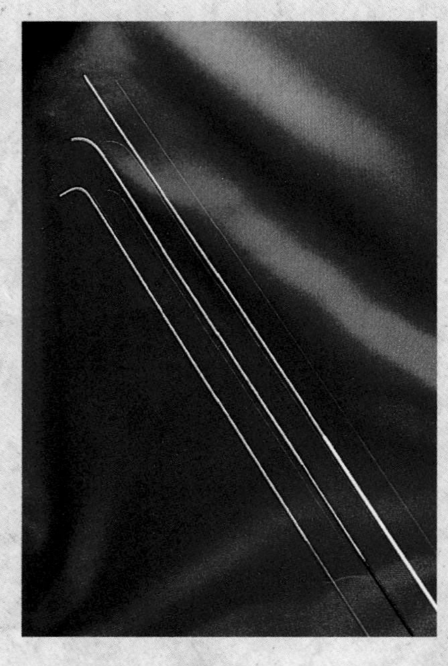

Memory metal used in medical applications. (FlexMedics Corporation)

By taking the square root of both sides of the equation, we get

$$\text{diagonal distance} = \sqrt{2}\,(\text{edge})$$

The diagonal distance = 4 × 143 pm = 572 pm, and so the length of the edge is (diagonal/$\sqrt{2}$) = 404 pm. In centimeters,

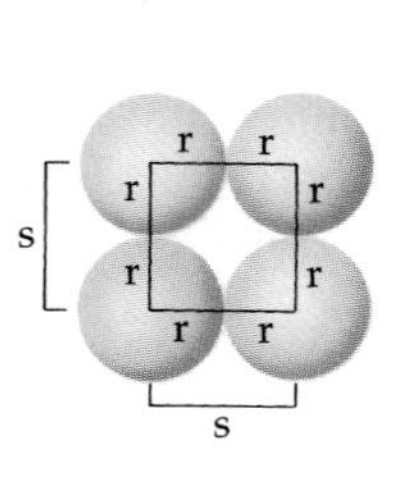

Simple cubic

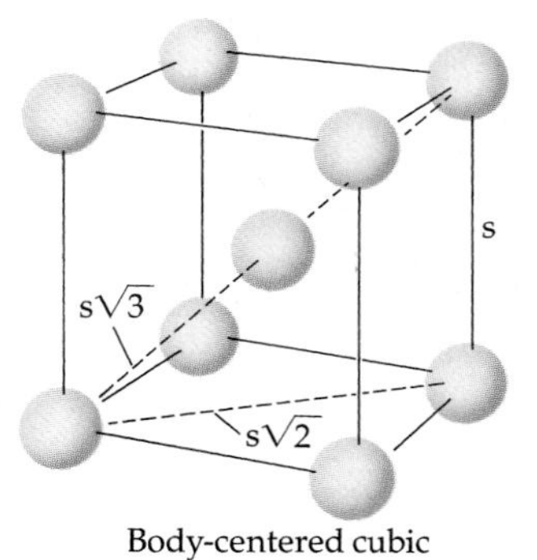

Body-centered cubic

Dimensions of some other single cubic unit cells.

bone often causes hairline fractures that delay healing and cause pain and discomfort for the patient. Instead, a memory metal tube is shaped in the form of a C at a temperature above body temperature, cooled and squeezed together, and then inserted into the hole in the bone. As the patient's body temperature warms the metal, it opens to its C shape while tightly pressing against the inside of the bone. The metal joint is then attached to the tube.

(a)

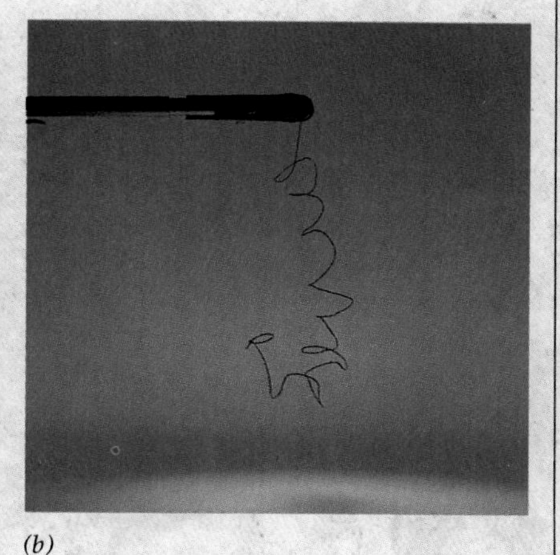

(b)

(c)

A memory metal. This piece of Nitinol wire (a) was bent and shaped to spell the letters "ICE" (for Institute of Chemical Education) at a high temperature. When it is cooled (b) it can be bent into a random shape. When it is reheated (c), it reverts to the shape spelling the word "ICE." (C.D. Winters)

$$\text{cube edge} = 404\ \text{pm} \cdot \frac{1\ \text{m}}{1 \times 10^{12}\ \text{pm}} \cdot \frac{100\ \text{cm}}{1\ \text{m}} = 4.04 \times 10^{-8}\ \text{cm}$$

So, the volume of the cubic unit cell is

$$\text{unit cell volume} = (\text{edge})^3 = (4.04 \times 10^{-8}\ \text{cm})^3 = 6.59 \times 10^{-23}\ \text{cm}^3$$

Now the mass of the unit cell is

$$\text{mass of unit cell} = \text{density} \times \text{volume} = \frac{2.699\ \text{g}}{\text{cm}^3} \cdot \frac{6.59 \times 10^{-23}\ \text{cm}^3}{\text{unit cell}}$$

$$= 1.78 \times 10^{-22}\ \text{g/unit cell}$$

How does this mass compare with the mass of one Al atom?

$$\text{mass of one Al atom} = \frac{26.98\ \text{g}}{\text{mol}} \cdot \frac{1\ \text{mol}}{6.022 \times 10^{23}\ \text{atoms}}$$

$$= 4.480 \times 10^{-23}\ \text{g/atom}$$

So, the number of atoms per unit cell is

Notice that in this calculation we used Avogadro's number to figure out the mass of one Al atom. Since it is an experimental fact that aluminum has a fcc crystal structure that has four atoms per unit cell, we could have reversed the last part of the calculation and calculated Avogadro's number. For some years the best way of determining Avogadro's number was a calculation like this.

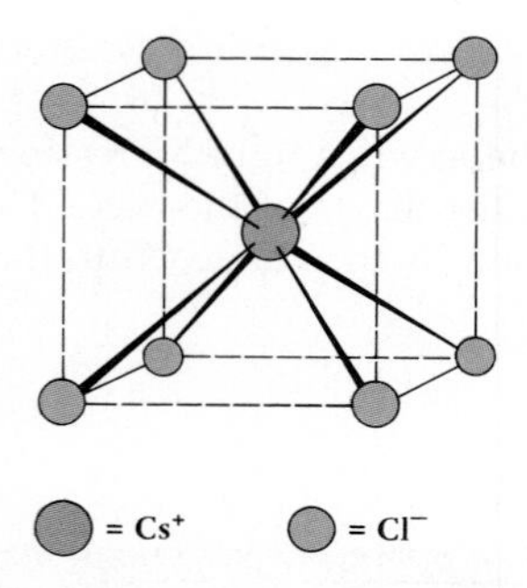

Figure 14.12
Cesium chloride (CsCl) crystal structure.

$$\text{number of atoms} = \frac{1.78 \times 10^{-22}\ \text{g}}{\text{unit cell}} \cdot \frac{1\ \text{atom}}{4.480 \times 10^{-23}\ \text{g}}$$

$$= 3.97\ \text{atoms/unit cell}$$

Within experimental error, our answer, 3.97 atoms per unit cell, agrees with the expected 4 atoms per unit cell for the fcc unit cell.

EXERCISE 14.3 • *Calculating The Density of a Metal*

Gold crystals have a bcc structure. The radius of a gold atom is 144 pm, and the atomic weight of gold is 196.97 amu. Calculate the density of gold.

Unit Cells and Simple Ionic Compounds

Remember that negative ions are usually larger than positive ions. Therefore, building an ionic crystal is a lot like placing marbles (positive ions) in the spaces between ping-pong balls (negative ions).

Crystal structures of many ionic compounds can be described as simple cubic or face-centered cubic lattices of spherical negative ions, with positive ions occupying spaces (called holes) between the negative ions. The number and locations of the occupied holes are the keys to understanding the relation between the lattice structure and the formula of a salt. Perhaps the simplest example is the hole in a simple cubic unit cell (Figure 14.9). The ionic compound cesium chloride, CsCl, has such a structure. In it, each simple cube of Cl^- ions has a Cs^+ ion at its center (Figure 14.12). The spaces occupied by the Cs^+ ions are called cubic holes, and each Cs^+ has eight nearest-neighbor Cl^- ions.

The structure of sodium chloride, NaCl, is one of the most common ionic crystal lattices. It consists of a fcc lattice of the larger Cl^- ions (Figure

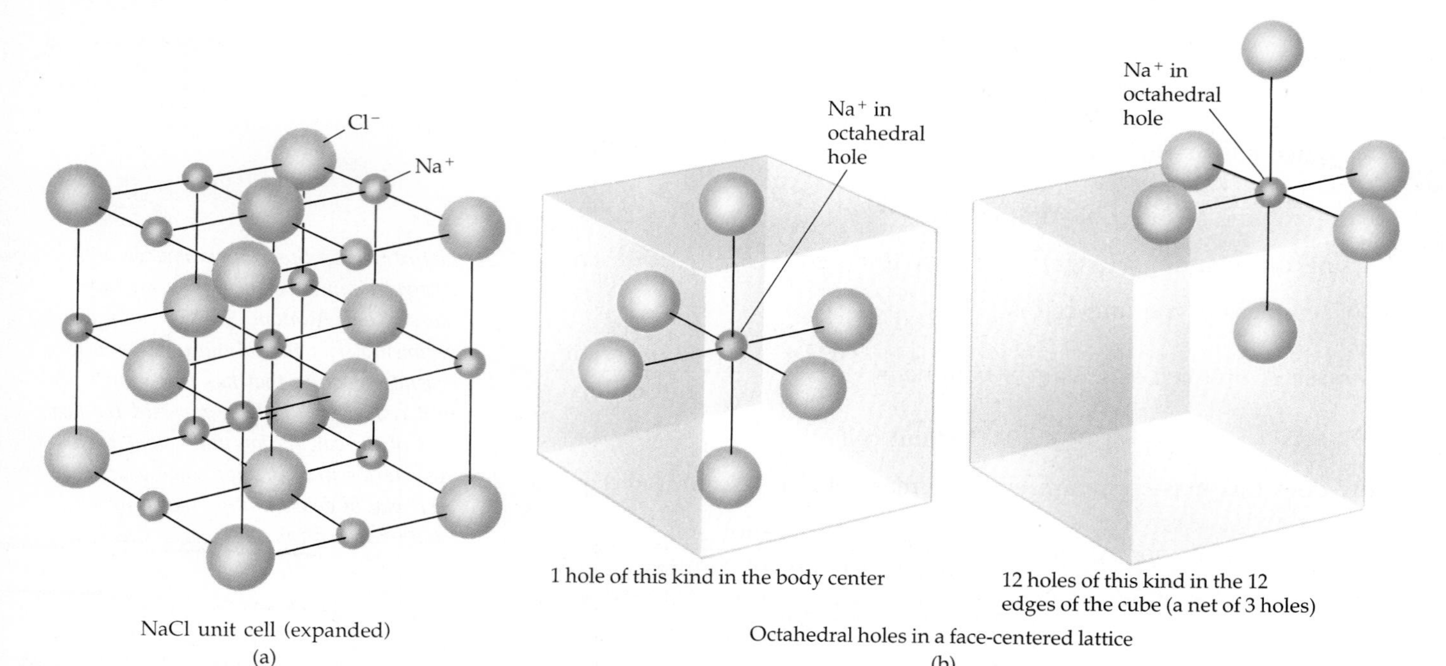

NaCl unit cell (expanded)
(a)

Octahedral holes in a face-centered lattice
(b)

14.13a), into which Na^+ ions have been placed in what are called *octahedral* holes—octahedral because each Na^+ ion is surrounded by six Cl^- ions at the corners of an octahedron (Figure 14.13b). It is easy to see this for the central Na^+ ion in Figure 14.13b, but the unit cell would have to be repeated in several directions to see six Cl^- ions around each of the other Na^+ ions. Figure 14.13c shows a space-filling model of the NaCl lattice, where each ion is drawn to scale based on its ionic radius.

It is not appropriate to describe the CsCl structure as bcc; the names simple cubic, body-centered cubic, and face-centered cubic refer to the arrangement of only one type of ion in the lattice, not both.

If you look carefully at Figure 14.13a, it is possible to determine the number of Na^+ and Cl^- ions in the NaCl unit cell. There is $\frac{1}{8}$ of a Cl^- ion at each corner of the unit cell, and $\frac{1}{2}$ of a Cl^- in the middle of each face. The total number of Cl^- ions within the unit cell is

Since the edge of a cube in a cubic lattice is surrounded by four cubes, one fourth of a spherical ion at the midpoint of an edge is within any one of the cubes.

$$\tfrac{1}{8}\ Cl^- \text{ per corner} \times 8 \text{ corners} = 1\ Cl^-$$
$$\tfrac{1}{2}\ Cl^- \text{ per face} \times 6 \text{ faces} = 3\ Cl^-$$

There is $\frac{1}{4}$ of a Na^+ at the midpoint of each edge, and a whole Na^+ in the center of the unit cell.
For Na^+ ions, the total is

$$\tfrac{1}{4}\ Na^+ \text{ per edge} \times 12 \text{ edges} = 3\ Na^+$$
$$1\ Na^+ \text{ per center} \times 1 \text{ center} = 1\ Na^+$$

Thus, the unit cell contains 4 Na^+ and 4 Cl^- ions. This result agrees with the formula of NaCl for sodium chloride.

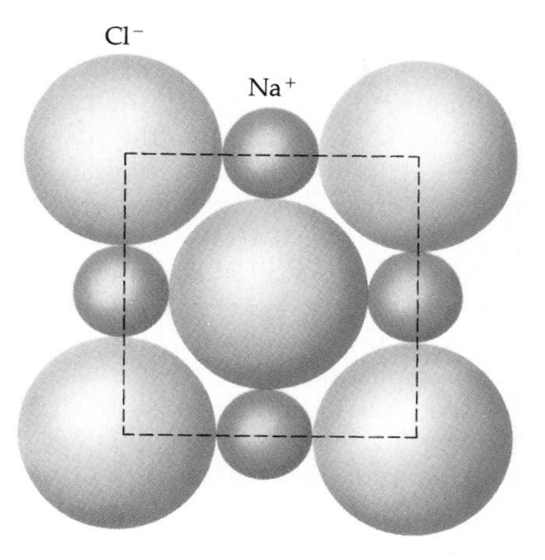

One face of a face-centered cubic unit cell.

EXERCISE 14.4 • *Ionic Structure and Formula*

Cesium chloride has a simple cubic unit cell as seen in Figure 14.12. Show that the formula for the salt must be CsCl.

Based on what we know about the crystal lattice of NaCl, let's calculate the volume of its unit cell. The radius of a Na^+ ion is 116 pm, and the radius

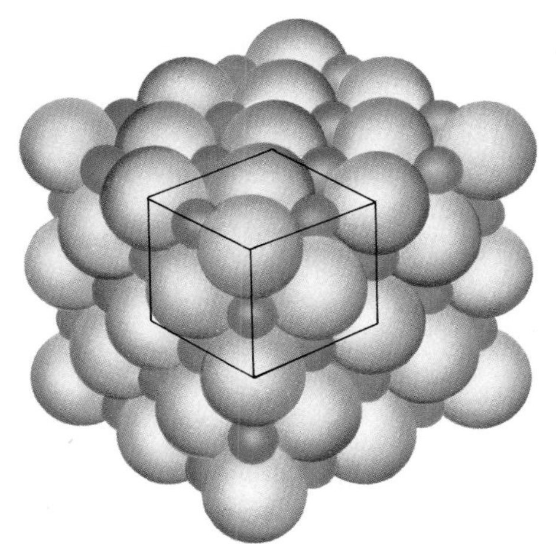

Figure 14.13
The NaCl unit cell. (a) An expanded view of one unit cell. The lines represent only the connections between lattice points. (b) A close-up view of the octahedral holes within the lattice. (c) A more extended view showing the ions as spheres, the smaller Na^+ ions packed into a face-centered cubic lattice of larger Cl^- ions.

A Deeper Look

Using X Rays to Probe Solid Structures

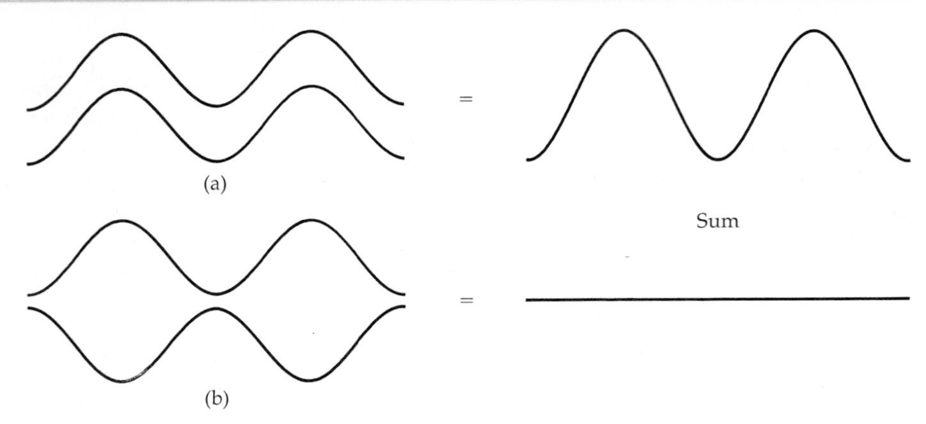

Interference of waves. (a) Constructive interference occurs when two in-phase waves combine to produce a wave of greater amplitude. (b) Destructive interference results from the combination of two waves of equal magnitude that are exactly out of phase. The result is zero amplitude.

Shortly after x rays were discovered by W.C. Röentgen in 1895, scientists began arguing about whether x rays were particles or waves (see Section 9.2 for a similar argument regarding the electron). Max von Laue and his co-workers guessed that x rays were light of very short wavelength, and that the distances between the repetitive arrays of atoms in a crystal would be about the same as the wavelengths of x rays. Therefore, a crystal should act as a grating, diffracting x rays in the same way that finely divided scratches on the surface of a CD will diffract visible light into a rainbow of colors. In an important experiment x rays were focused on a crystal of blue vitriol, copper(II) sulfate pentahydrate, and diffracted rays were detected on a photographic plate.

Upon hearing of this experiment, William and Lawrence Bragg, father and son, proposed that crystals diffract x rays because the regularly spaced atoms form semitransparent mirror planes. A small part of an incoming x-ray beam is reflected by each mirror plane, and certain arrangements of these mirror planes cause the reflections to interfere with one another. When the interference is *constructive,* the waves combine to produce diffracted x rays that can be observed because they expose a spot on a photographic film. When the interference is *destructive,* the waves partly or entirely cancel each other and no x-ray diffraction is observed. Whether interference is constructive or destructive depends on the wavelength of the x rays and the distance between the planes of atoms. When the wavelength is known, the distance can be calculated from the observed x-ray diffraction.

After the Braggs' pioneering work, others showed that the intensity of the scattering of the x-ray beam depends on the electron density of the atoms in the crystal. This means that a hydrogen atom is least effective at causing the x rays to scatter, while atoms such as lead or mercury are quite effective. **X-ray crystallography,** the science of determining atomic-scale crystal structure, is based on these ideas. Today crystallographers use computers and photocells to accurately measure the angles of reflection and the intensities of x-ray beams diffracted by crystals. While the early work by the Braggs and others was with minerals, the crystal structures of organic molecules in solids are now routinely determined with x rays, even though the lighter carbon, oxygen, nitrogen, and hydrogen atoms reflect the x rays with relatively low intensities.

X-ray crystallography has been increasingly used to discover important molecular structures. In 1944 Dorothy Crowfoot Hodgkin determined the structure of penicillin, and in 1953 James Watson and Francis Crick used x-ray crystallography results obtained by Rosalind Franklin to discover the double-helix structure of deoxyribonucleic acid (DNA), the molecule that carries genetic information (Section 21.6 and Figure 11.1).

---

of a Cl^- ion is 167 pm. The ions are touching along the edge of the unit cell, and so the edge length is equal to two Cl^- radii plus two Na^+ radii.

$$\text{edge} = 167\text{ pm} + 2 \cdot 116\text{ pm} + 167\text{ pm} = 566\text{ pm}$$

The volume of the cubic unit cell is the cube of the edge length.

$$\text{volume of unit cell} = (\text{edge})^3 = (566 \text{ pm})^3 = 1.81 \times 10^8 \text{ pm}^3$$

Converting this to cm^3,

$$\text{volume of unit cell} = 1.81 \times 10^8 \text{ pm}^3 \cdot \left(\frac{10^{-10} \text{ cm}}{\text{pm}}\right)^3 = 1.81 \times 10^{-22} \text{ cm}^3$$

Next we can calculate the mass of a unit cell and divide it by the volume to get the density. With four NaCl formula units per unit cell,

mass of NaCl in unit cell

$$= 4 \text{ NaCl formula units} \cdot \frac{58.44 \text{ g}}{\text{mol NaCl}} \cdot \frac{1 \text{ mol}}{6.022 \times 10^{23} \text{ formula units}}$$

$$= 3.88 \times 10^{-22} \text{ g}$$

This means the density of NaCl is

$$\text{density of NaCl} = \frac{3.88 \times 10^{-22} \text{ g}}{1.81 \times 10^{-22} \text{ cm}^3} = 2.14 \text{ g/cm}^3$$

The experimental density is 2.164 g/cm^3.

The general relationship among the unit cell type, ion or atom size, and solid density for most metallic or ionic solids is

mass of 1 formula unit (e.g. NaCl)

↓ *x number of formula units per unit cell (e.g., 4 for NaCl)*

mass of unit cell

↓ ÷ *unit cell volume (= edge^3 for NaCl)*

density

EXERCISE 14.5 • *Unit Cell Dimensions*

KCl has the same crystal structure as NaCl. Calculate the volume of the unit cell for KCl, given that the ionic radii are K^+ = 152 pm and Cl^- = 167 pm. Which has the larger unit cell, NaCl or KCl? Now compute the density of KCl.

14.3 METALS AND SEMICONDUCTORS

All the metals are solids at room temperature, except for mercury (m.p. −38.8 °C). All metals exhibit common properties that we call metallic:

- **High electrical conductivity**—metal wires carry electricity from power plants to homes and offices, because electrons in metals are highly mobile.
- **High thermal conductivity**—we learn early not to touch any part of a metal pot on a heated stove, because it will transfer heat rapidly and painfully.
- **Ductility and malleability**—most metals are easily drawn into wire (ductile), or hammered into thin sheets (malleable); some (gold, for example) are more easily formed into shapes than others.
- **Luster**—polished metal surfaces reflect light; most metals have a silvery white color because they reflect all wavelengths equally well.
- **Insolubility in water and other common solvents**—no metal dissolves in water, but a few (mainly from Groups 1A and 2A) react with water to form hydrogen and solutions of metal hydroxides.

Enthalpies of fusion and melting points of metals vary greatly. Low melting points correlate with low enthalpies of fusion, which implies

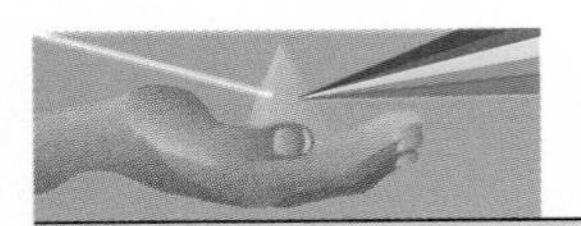

CHEMISTRY YOU CAN DO

Closest Packing of Spheres

In a metal the crystal lattice is occupied by identical atoms, which can be pictured as identical spheres. Obtain a bag of marbles, some expanded polystyrene foam spheres, some ping-pong balls, or some other set of reasonably small, identical spheres. Use them to construct models of each situation described here.

Consider how identical spheres can be arranged in two dimensions. Pack your spheres into a layer so that each sphere touches four other spheres; this should resemble the arrangement in part a of the figure. (You will find this somewhat hard to do; it may be necessary to use four pieces of wood to enclose the layer and keep the spheres in line. Making a square array is hard because the spheres are rather inefficiently packed, and they readily adopt a more efficient packing in which each sphere touches six other spheres.)

To make a three-dimensional lattice requires stacking layers of spheres. Start with the layer of spheres in a square array, and stack another, identical layer directly on top of the first; then stack another on that one, and so on. (This is really hard and will require vertical walls to hold the spheres as you stack the layers. If you are not able to do it, use sugar cubes, dice, or some other kind of small cubes instead of spheres.) If you are successful in making this 3-D lattice, it will consist of simple cubic unit cells; in such a lattice 52.4% of the available space is occupied by the spheres.

Next, make the more efficiently packed hexagonal layer in which each sphere is surrounded by six others (part b of the figure). Notice that in this layer, as in the square one, there are little holes left between the spheres. This time, however, the holes are smaller than they were in the square array; in fact, they are as small as possible, and for this reason the layer you have made is called a **closest-packed** layer. Now add a second layer, and let the spheres nestle into the holes in the first layer. (Trying to prevent this from happening was probably the hardest task when you made your simple cubic array.) Now add a third layer. If you look carefully, you will see that there are two ways to put down the third layer. One of these places spheres directly above *holes* in the first layer. This is called **cubic closest-packing,** and the unit cell is face-centered cubic (fcc). The other arrangement, also found for many metals, places third-layer spheres directly above *spheres* in the first layer. This is **hexagonal closest-packing,** and its unit cell is *not* cubic. Cubic and hexagonal closest packing are the most efficient ways known for filling space with spheres; 74% of the available space is occupied. The atoms in most metals are arranged in cubic closest-packing, in hexagonal closest-packing, or in lattices composed of body-centered cubic unit cells. Efficient packing of the atoms or ions in the crystal lattice allows stronger bonding, which gives greater stability to the crystal.

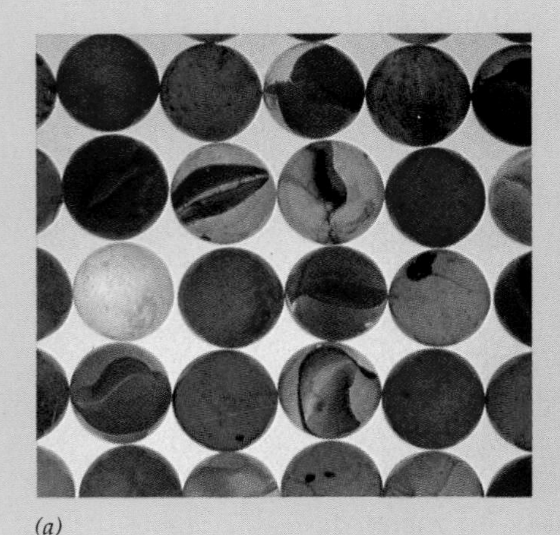

(a)

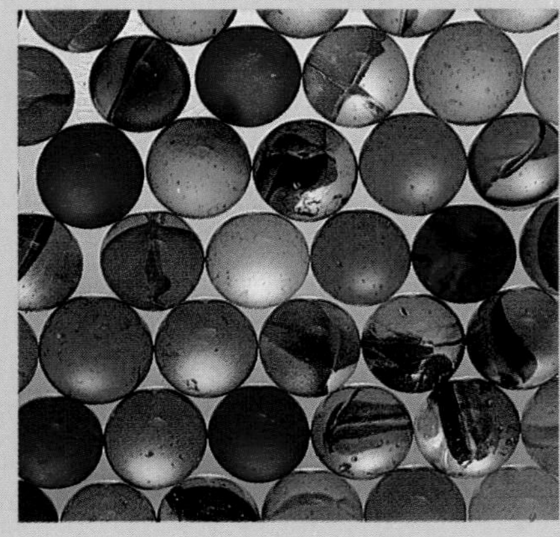

(b)

The packing of spheres in one layer. (a) Marbles are arranged in a square pattern where each marble contacts four others. (b) Each marble contacts six others at the corners of a hexagon. The hexagonal arrangement is a more efficient packing arrangement than the square pattern. *(C.D. Winters)*

Table 14.2

Enthalpies of Fusion and Melting Points of Some Metals

Metal	ΔH_{fusion} (kJ/mol)	Melting Point (°C)
Hg	2.3	−38.8
Li	3.0	180.5
Na	2.59	97.9
Ga	7.5	29.78*
Al	10.7	660.4
U	12.6	1132.1
Ti	20.9	1660.1
W	35.2	3410.1

*This means that gallium metal would melt in the palm of your hand from the warmth of your body. In addition, gallium is a liquid over the largest range of temperatures of any metal. Its boiling point is approximately 2250 °C.

weaker attractive forces holding the metal atoms together. Mercury, a liquid at room temperature, has an enthalpy of fusion of only 2.3 kJ/mol. Compare this with the values in Table 14.1 for nonmetals. The alkali metals and gallium also have very low enthalpies of fusion and notably low melting points (Table 14.2).

Figure 14.14 (p. 622) shows the enthalpies of fusion of most of the elements related to their positions in the periodic table. The metals are shown in blue. The transition metals, especially those of the third transition series, have very high melting points and extraordinarily high enthalpies of fusion. Tungsten (W) has the highest melting point of all the metals and is second only to carbon (graphite), whose melting point is 3550 °C. Pure tungsten is used in light bulbs as the filament—the wire that glows white hot. No other material has been found to be better since the invention of light bulbs in 1908 by Thomas Edison and his co-workers.

EXERCISE 14.6 • *Heats of Fusion*

Use data from Table 14.2 to calculate the heat required to melt 1.00 cm^3 of mercury and the heat required to melt 1.00 cm^3 of aluminum. (Use a handbook or computer database to obtain densities.)

Metallic Bonding

The properties of solids can be explained by the type of bonding that holds their constituent particles together. In molecular solids, molecules are held together by intermolecular forces; in ionic solids, ions are held together by

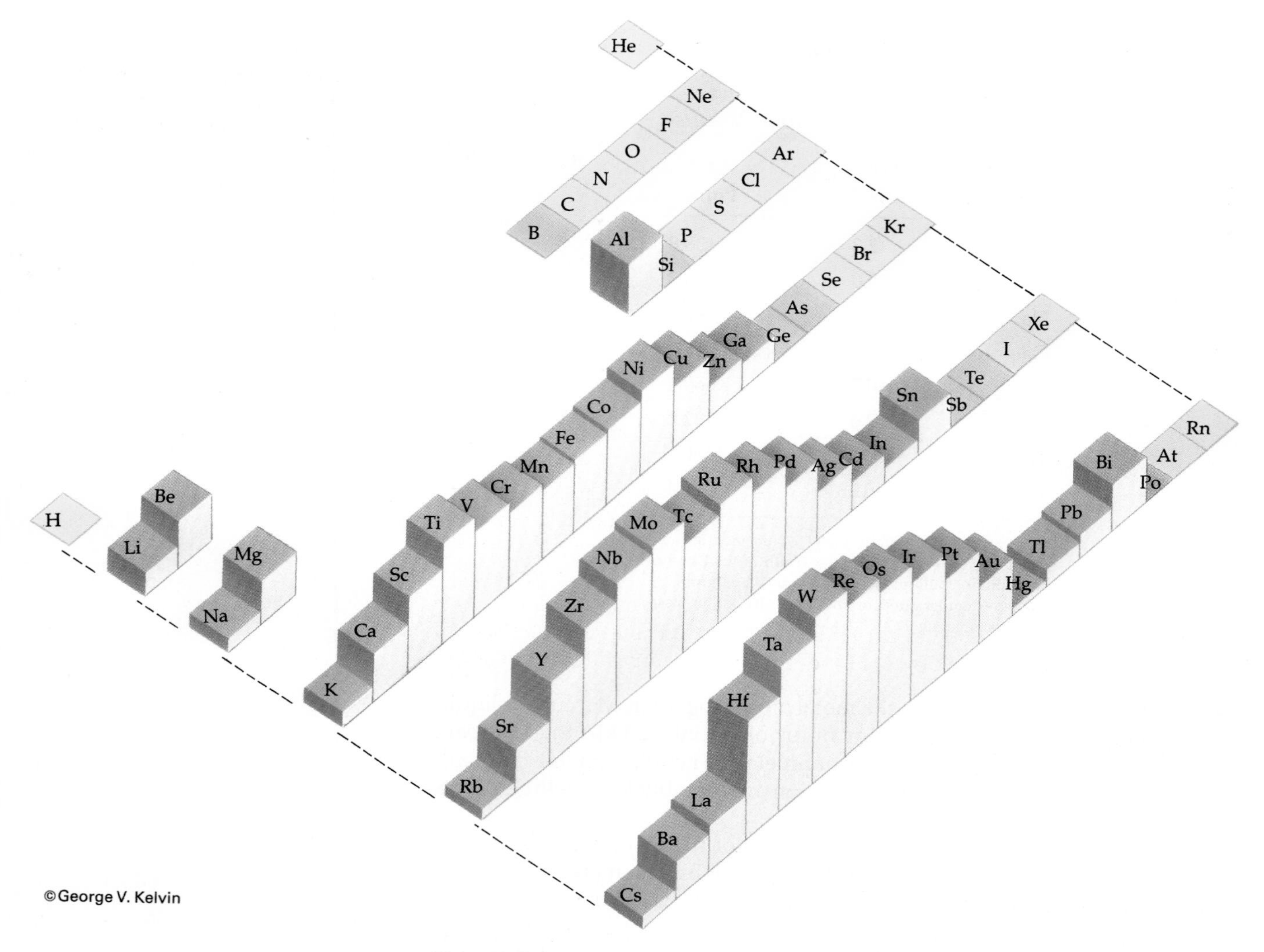

Figure 14.14
Relative enthalpies of fusion for the metals in the periodic table. To provide a scale, refer to Table 14.2. (Symbols of nonmetals are included only to show their positions in the periodic table.)

electrostatic attractions between positive and negative ions. Metals behave as though metal ions exist in a "sea" of mobile electrons. **Metallic bonding** is the nondirectional attraction between positive metal ions and the surrounding sea of negative charge. Each metal ion has a large number of near neighbors. The valence electrons move throughout the array of spheres, holding things together. When an electric field is applied to a metal, these valence electrons move toward the positively charged end, and the metal conducts electricity. Because of the uniform charge distribution provided by the mobile valence electrons in a metal lattice, the positions of the positive ions can be changed without destroying the attractions among them. Thus most metals can be bent and drawn into wire; they are malleable and duc-

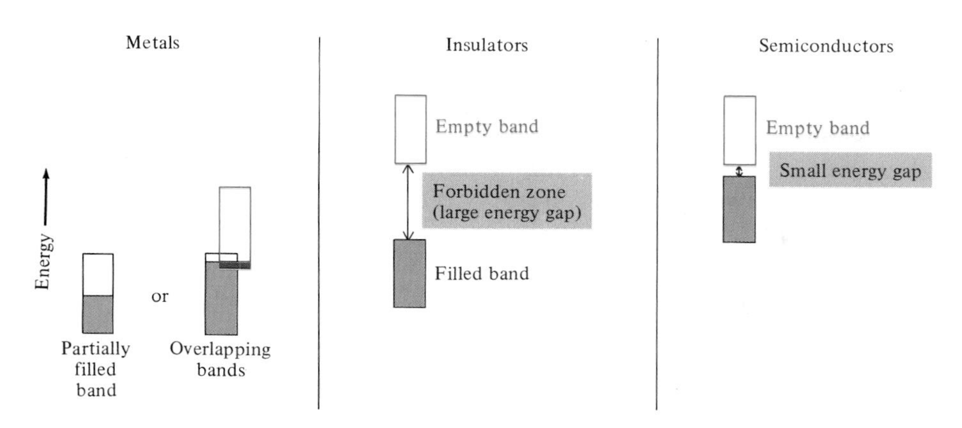

Figure 14.15
Distinction among metals, insulators, and semiconductors. In each case, an unshaded area represents a conduction band.

tile. When we try to deform an ionic solid, the crystal usually shatters because the balance of positive ions surrounded by negative ions and vice versa is disrupted (see Section 4.4).

To visualize how bonding electrons behave in a metal, first consider the arrangement of electrons in an individual atom far enough away from any neighbor so that there is no bonding. In such an atom the electrons occupy orbitals that have definite energy levels. In a large number of separated, identical atoms, all of the energy levels are identical. If the atoms are brought closer together, they begin to influence one another and the identical energy levels shift up or down, broadening to become bands. An **energy band** is a large group of orbitals whose energies are closely spaced and whose average energy is the same as the energy of the corresponding orbital in an individual atom. In some cases energy bands for different types of electrons (s, p, d, etc.) overlap; in other cases there is a gap between different energy bands.

Within each band, electrons fill the lowest-energy orbitals much as electrons fill orbitals in atoms. The number of electrons a given energy band can hold depends on the number of metal atoms in the crystal, as does the number of electrons that occupy the bands. As in the case of atoms, it is usually necessary to consider only valence electrons, since other electrons all occupy completely filled bands. In a band that is completely filled, two electrons occupy every orbital. No electron can move from one orbital to another, because there is no empty spot for it to go to. In a band that is partially filled, only a little added energy can cause an electron to be excited to a slightly higher-energy orbital. Such a small increment of energy can be provided by applying an electric field, for example. The presence of low-energy, empty orbitals that electrons can occupy allows the electrons to be mobile, and allows electric current to be conducted. In metals the energy bands overlap, and so there are many empty orbitals that electrons can easily occupy. This accounts for the high electrical conductivity of metals (Figure 14.15).

Electrons, orbitals, and energy levels were discussed in Sections 9.2 to 9.4.

This band theory also explains why some solids are **insulators** and do not conduct electricity. In an insulator the energy bands do not overlap; rather, there is a large gap between them (Figure 14.15). There are very few electrons that have enough energy to jump across the large gap from a lower-energy filled energy band to an empty higher-energy band, and so no current flows when an external electric field is applied.

In **semiconductors** there are very *narrow* energy gaps between fully occupied energy bands and empty energy bands (Figure 14.15). At quite low temperatures, electrons remain in the lower, filled energy band and semiconductors are not good conductors. At higher temperatures, or when an electric field is applied, however, some electrons have enough energy to jump across the band gap into an empty energy band. This allows an electric current to flow.

Silicon polymers contain

$$-O-\underset{R}{\overset{R}{|}}{Si}-O-\underset{R}{\overset{R}{|}}{Si}-$$

structures and find uses as putties and rubber-like materials.

Silicon and the Chip

Silicon is known as an "intrinsic" semiconductor because the element itself is a semiconductor. (Many semiconductors typically found in electronic devices are mixtures of elements or compounds.) Silicon of about 98% purity can be obtained by heating silica (purified sand) and coke (an impure form of carbon) at 3000 °C in an electric arc furnace.

$$SiO_2(s) + 2\,C(s) \xrightarrow{\text{heat}} Si(s) + 2\,CO(g)$$

Silicon of this purity can be alloyed with aluminum and magnesium to increase the hardness and durability of the metals and is used for making silicone polymers. For use in electronic devices, however, a much higher degree of purification is needed. High-purity silicon can be prepared by reducing $SiCl_4$ with magnesium.

$$SiCl_4(\ell) + 2\,Mg(s) \longrightarrow Si(s) + 2\,MgCl_2(s)$$

The magnesium chloride, which is water soluble, is then washed from the silicon. The final purification of the silicon takes place by a melting process called **zone refining** (Figure 14.16), which produces silicon containing less than one part per billion of impurities such as boron, aluminum, and arsenic (Figure 14.17).

Figure 14.16
Zone refining. The melted region (zone) of a crystal is slowly moved upward in a crystal. This molten zone carries in it any impurities, leaving behind a highly purified crystal.

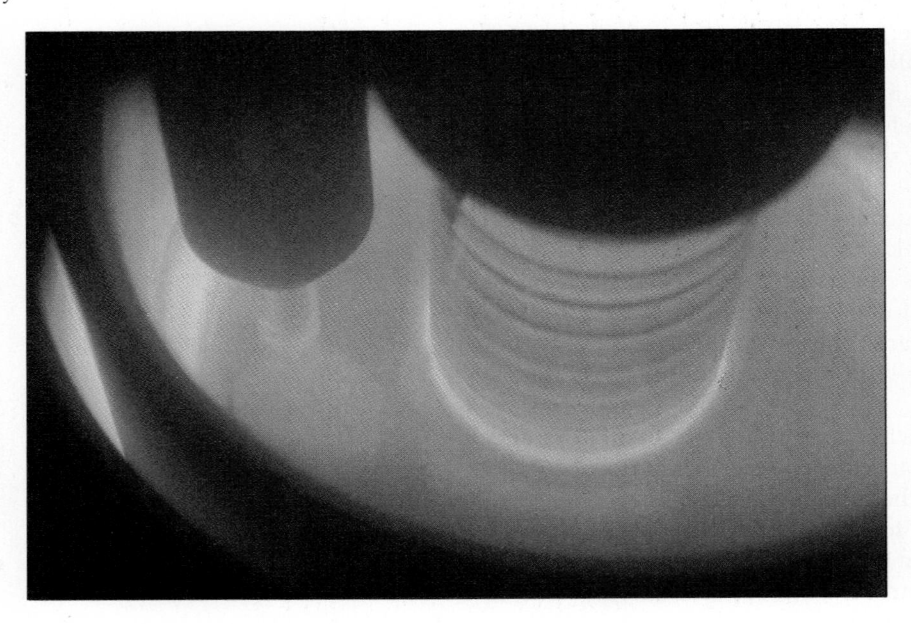

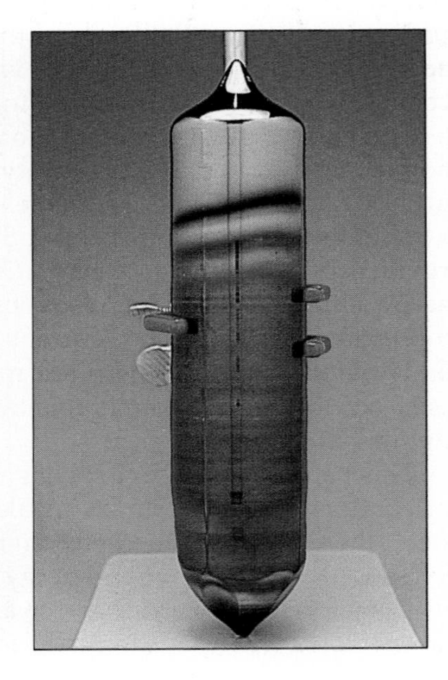

Like all semiconductors, high-purity silicon fails to conduct an electric current until a certain electrical voltage is applied, but at higher voltages it conducts moderately well. Silicon's semiconducting properties can be improved by a process known as doping. **Doping** involves adding a tiny amount of some other element (a *dopant*) to the silicon. For example, suppose a small number of boron atoms (or atoms of some other Group 3A element) replace silicon atoms (Group 4A) in solid silicon. Boron has only three valence electrons, whereas silicon has four. This leaves a deficiency of one electron around the B atom, creating what is called a positive hole for every B atom added. Hence silicon doped in this manner is referred to as positive-type or *p*-type silicon (Figure 14.18). When a few atoms of a Group 5A element such as arsenic are added to silicon, only four of the five valence electrons of As are used for bonding with four Si atoms, leaving one electron relatively free to move. This type of silicon is referred to as negative-type or *n*-type silicon, because it has extra (negative) valence electrons.

Figure 14.17
A wafer of highly purified silicon.

In 1947 an electronic device called the **transistor** was invented. The simplest device used layers of *n-p-n-* or *p-n-p*-doped silicon. Germanium, a Group 4A element just below silicon in the periodic table, was also used in place of silicon. The most revolutionary application of silicon's semiconductor properties has been the design of integrated electrical circuits (ICs), computer memories, and even whole computers called microprocessors on tiny chips of silicon scarcely larger than a millimeter in diameter. These devices permeate our whole society. You will find them in calculators, cameras, watches, toys, coin changers, cardiac pacemakers, and many other products. Truly, silicon is both the world we walk on (sand and silicate rock) and at the same time our constant companion in communications and electronic controls (the chip).

A positive hole in an energy band of a semiconductor is a place where there is one less electron than normal and hence extra positive charge.

Suppose that a layer of *n*-type doped silicon is next to a *p*-type layer. There is a strong tendency for the extra electrons in the *n*-type layer to pair with the unpaired electrons in the holes of the *p*-type layer. If the two layers are connected by an external electrical circuit, light that strikes the silicon provides enough energy to cause an electrical current to flow. When a photon is absorbed, it excites an electron to a higher-energy orbital, allowing the

Figure 14.18
Schematic drawing of semiconductor crystals derived from silicon.

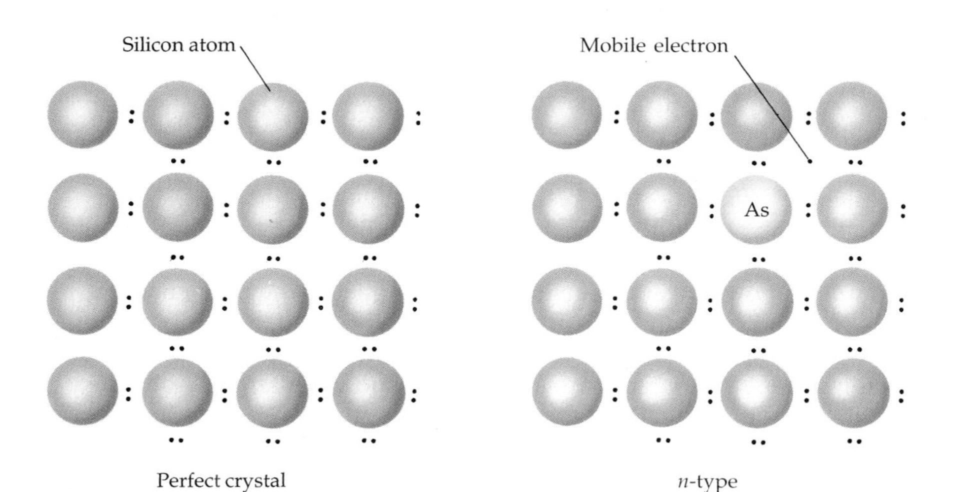

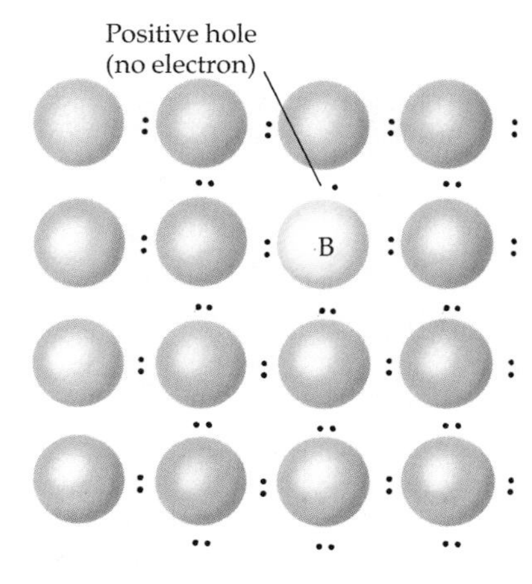

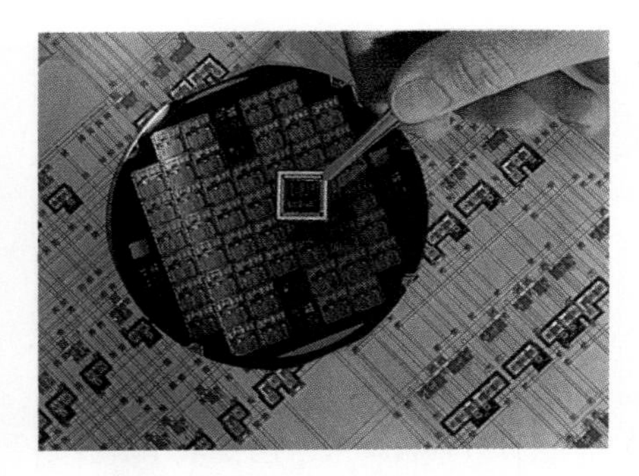

A tiny microcomputer can be fabricated from a single piece of highly purified silicon. *(Courtesy of AT&T Bell Laboratories.)*

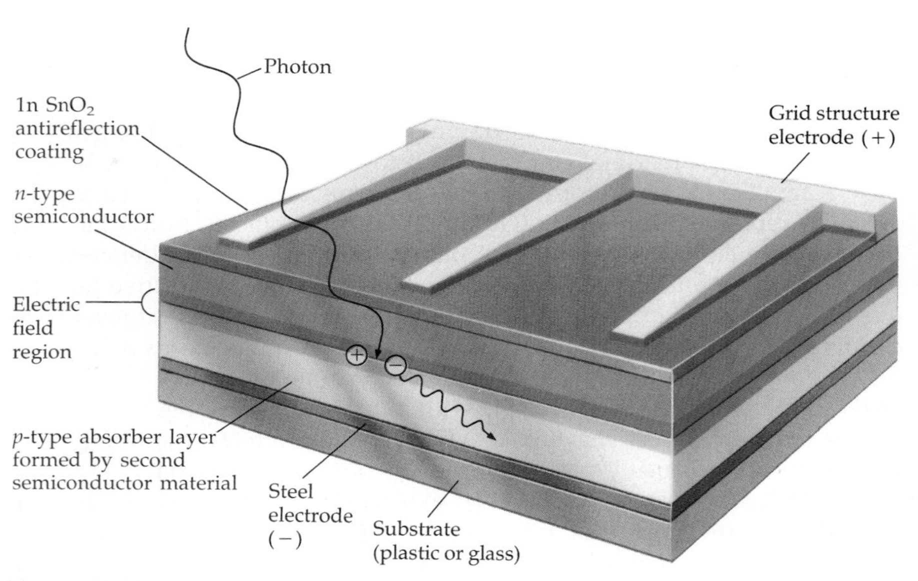

Figure 14.19

Typical photovoltaic cell using layers of doped silicon. *(Adapted from* Scientific American.*)*

The GM Hughes Electronics Sun Raycer. This experimental solar powered car has logged many miles showing how solar energy can be used for transportation. *(G.M. Hughes Electronics)*

Photovoltaic cells undergoing efficiency testing at the Pacific Gas and Electric facility in San Ramon, California. *(Hank Morgan/Science Library/Photo Researchers, Inc.)*

electron to leave the *n*-type silicon layer and flow through the external circuit to the *p*-type layer. As the *p*-type layer becomes more negative (because of added electrons), the extra electrons are repelled *internally* back into the *n*-type layer (which has become positive because of loss of electrons via the circuit). This process can continue as long as the silicon layers are exposed to sunlight and the circuit remains closed. This is how a **solar cell** works.

A typical solar cell is constructed on a substrate (support) layer of plastic or glass (Figure 14.19). Next to the substrate is a thin sheet of metal that acts as an electrode, transferring electrons to the *p*-type semiconductor layer. The topmost semiconductor layer is *n*-type; it receives the sun's rays and is nearly transparent. The solar cell is covered with a thin film of indium tin oxide ($InSnO_2$), which acts as an antireflection coating. A metallic grid structure on top of the cell functions as the second electrode, allowing as much light as possible to strike the *n*-type layer. The efficiency of a solar cell depends on how well it can absorb photons and convert the light to electrical energy. Photons that are reflected, pass through, or produce only heat, decrease the efficiency of the cell.

Solar cells are on the threshold of being the next great technological breakthrough, perhaps comparable to the computer chip. Although experimental solar-powered automobiles are now available and many novel applications of solar cells already exist (such as powering a fan to remove hot air from a luxury automobile parked in the sun), the real breakthrough will be general use of banks of solar cells as a utility power plant to produce huge quantities of electricity. One plant already operating in California uses solar cells to produce 20 MW of power—enough to supply the electricity needs of a city the size of Tampa, Florida.

14.4 NETWORK SOLIDS

A number of solids are composed of atoms connected by a network of covalent bonds. These solids are nonconductors. Such **network solids** really consist of one huge molecule—a super molecule—in which all the atoms are connected to all the others via a network of bonds. Separate small molecules do not exist in a network solid.

Graphite and Diamond

Allotropes of carbon were described in Section 4.1.

Graphite and diamond are allotropes of carbon. Graphite's name comes from the Greek *graphein* meaning "to write" because one of its earliest uses was for writing on parchment. Artists today still draw with charcoal, an impure form of graphite, and we write with graphite pencil leads. Graphite is an example of a **planar network solid** (Figure 14.20). The planes consist of six-membered rings of carbon atoms (like those in benzene, Section 11.6). Each hexagon shares all six of its sides with other hexagons around it, form-

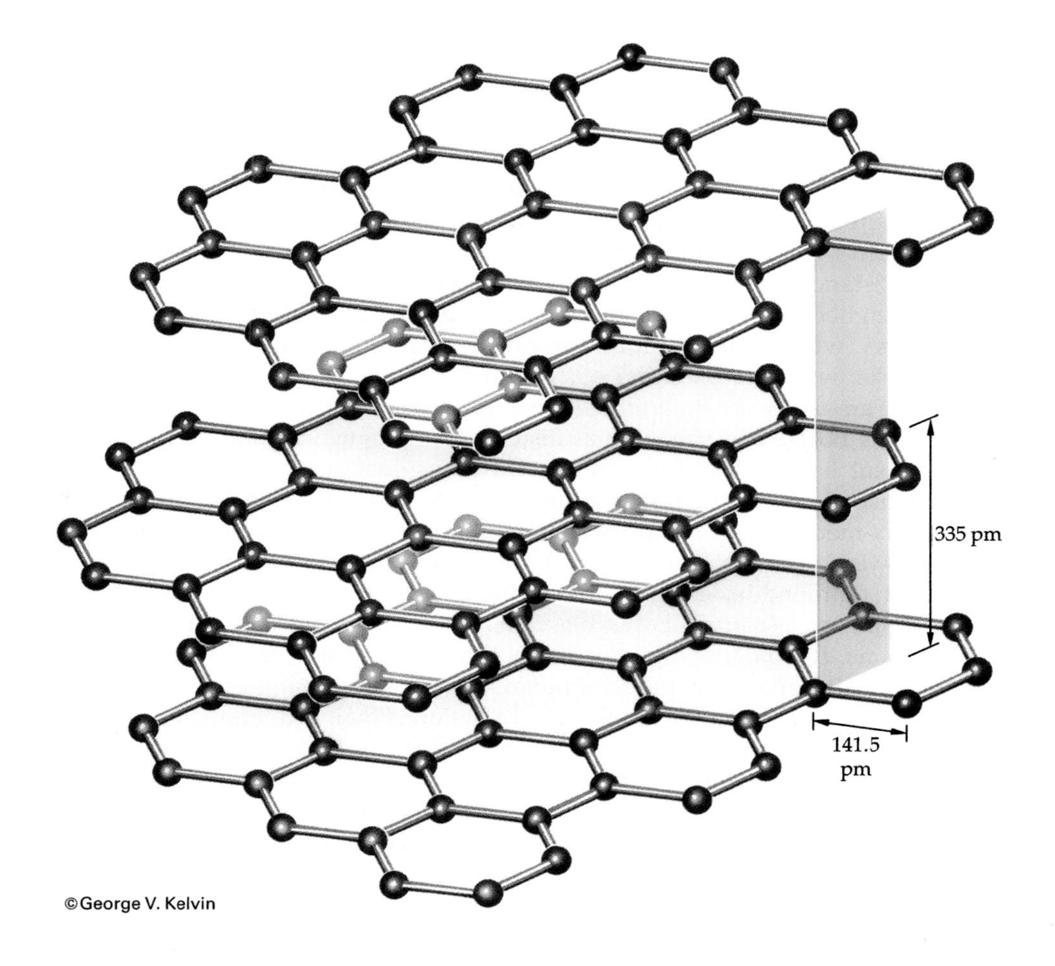

Figure 14.20
Graphite.

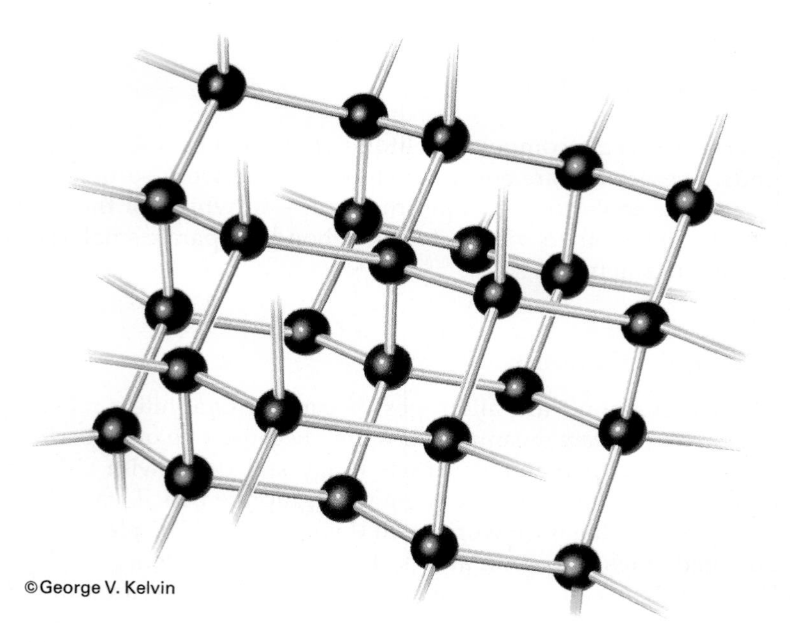
©George V. Kelvin

Figure 14.21
Diamond.

ing a network. Some of the bonding electrons are able to move freely around this network, and so graphite is a conductor of electricity. There are strong covalent bonds between carbon atoms in the same plane, but attractions between the planes are caused by dispersion forces and hence are weaker. The distance between planes is more than twice the distance between nearest neighbors within a plane. Also, the planes can easily slip across one another, which makes graphite an excellent solid lubricant for uses, such as in locks, where greases and oils are undesirable.

Diamonds are also built of six-membered carbon rings, but each carbon atom is bonded to *four* others (instead of three, as in graphite) by single covalent bonds. This forms a **three-dimensional network** (Figure 14.21). Because of the tetrahedral arrangement of bonds around each carbon atom, the six-membered rings in the diamond structure are puckered. In graphite the layers are much farther apart than normal C—C bond distances; as a result, diamond is denser than graphite (3.51 g/cm^3 and 2.22 g/cm^3, respectively). Also, because all of its electrons are localized between carbon atoms, diamond does not conduct electricity. Diamond is one of the hardest materials and also one of the best conductors of heat known. It is also transparent to visible, infrared, and ultraviolet radiation. Finally, diamonds have low density and high rigidity, making them ideal for producing sounds with frequencies as high as 60,000 Hz, far higher than the range of human hearing (recall the specifications on audio equipment—about 80 to 15,000 Hz). What more could a scientist want in a material—except a cheap, practical way to make it!

EXERCISE 14.7 • *Making Diamonds*

Given the properties of graphite and diamond, predict what conditions of temperature and pressure might be used if you wanted to convert inexpensive graphite into valuable diamonds within a reasonable period of time. (Hint: Apply LeChatelier's principle.)

In the 1950s, scientists at General Electric in Schenectady, New York, achieved something alchemists had attempted for centuries—the synthesis of diamonds. The GE scientists even used such diverse carbon-containing materials as wood and peanut butter to prove their prowess at diamond synthesis! Their technique, still in use today, was to heat graphite to a temperature of 1500 °C in the presence of a metal, such as nickel or iron, and under a pressure of 50,000 to 65,000 atmospheres. Under these conditions, the carbon dissolves in the metal and recrystallizes in its higher-density form, slowly becoming diamond. The worldwide market for diamonds made this way is now worth about $500 million; many of these diamonds are used for abrasives and diamond-coated cutting tools for drilling.

Figure 14.22
A pure quartz crystal (pure SiO_2). Quartz is one of the most common minerals on earth. Pure, colorless quartz was used as an ornamental material as early as the Stone Age, and by the Roman times it was known that a wedge of quartz could be used to disperse the sun's rays. Quartz crystals today are used in the oscillators in watches, radios, VCRs, and computers. *(C.D. Winters)*

Crystalline SiO_2 and Silicates

There are more than 22 forms of pure silica, SiO_2, and the most common of them is α-quartz. It is a major component of many rocks such as granite and sandstone, and it occurs alone or as pure rock crystal or in a variety of less pure forms (Figure 14.22). The main crystalline modifications of SiO_2 consist of infinite arrays of SiO_4 tetrahedra sharing corners. When α-quartz is heated to almost 1500 °C, it transforms into another crystalline form, cristobalite, which has a structure derived from that of elemental silicon. (Elemental silicon has the diamond structure, Figure 14.21.) If an oxygen atom is placed between each pair of silicon atoms in the crystal lattice, the structure of cristobalite is generated (see Figure 11.13).

The simplest network silicates are the *pyroxenes,* which contain extended chains of linked SiO_4 tetrahedra (Figure 14.23). All are based on the metasili-

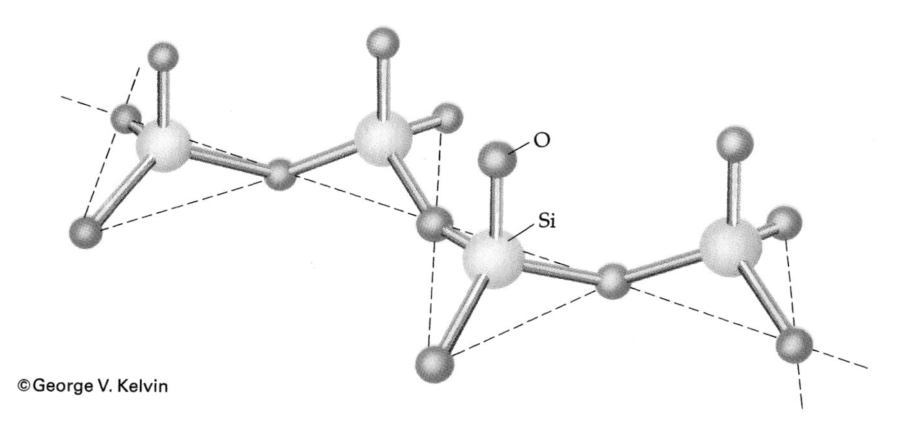

Figure 14.23
Pyroxene. The SiO_4 units share a common O atom.

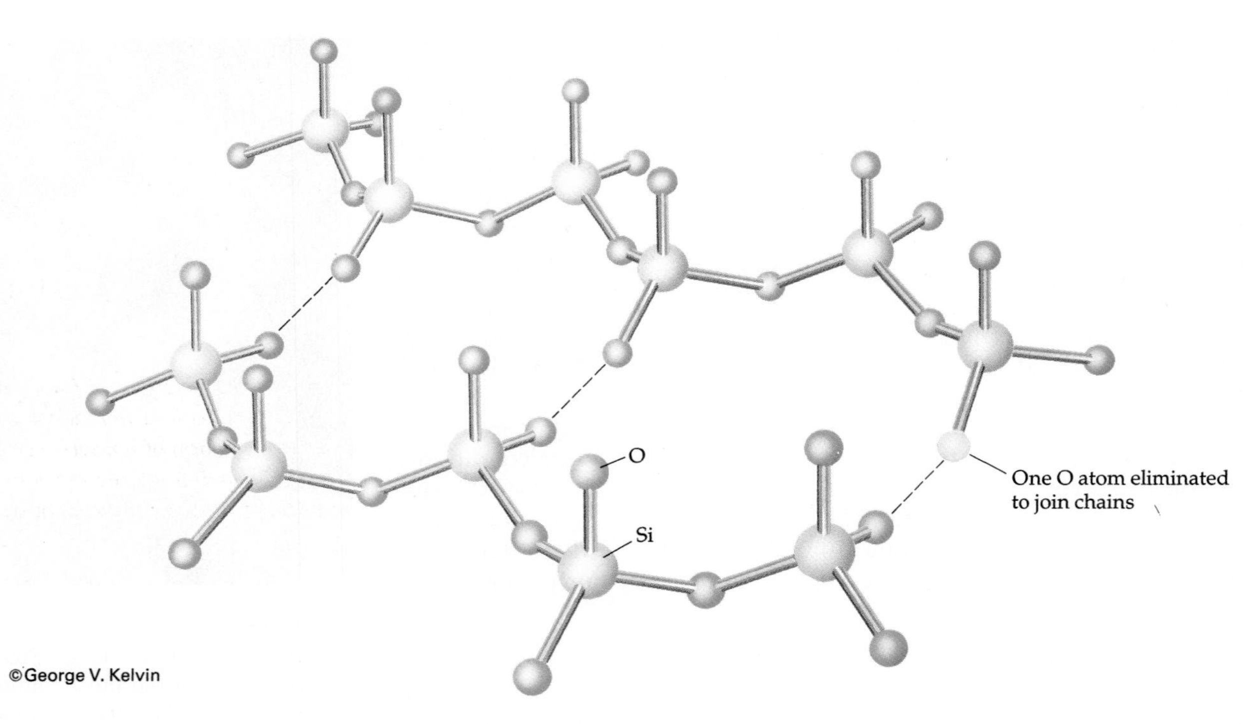

Figure 14.24
Amphibole structure. Chains of SiO_4 units share an O atom between them.

cate ion, SiO_3^{2-}. A typical formula might be $Mg_2Si_2O_6$. If two such chains are laid side by side, they may link together by sharing oxygen atoms in adjoining chains (Figure 14.24). The result is an *amphibole,* of which the *asbestos* minerals are excellent examples. Because of their double-stranded chain structure, asbestos minerals are fibrous. Prior to the discovery of its tendency to cause lung cancer, asbestos found wide use as an insulating and fireproofing material because of its low thermal conductivity. Older buildings often contain large quantities of asbestos insulation, and if left undisturbed, this asbestos is unlikely to cause harm.

The material we call "asbestos" is not a single substance, however. Rather, the name asbestos applies broadly to a family of naturally occurring hydrated silicates that crystallize in a fibrous manner. These minerals are generally subdivided into two forms, serpentine and amphibole fibers. Approximately 5 million tons of the serpentine form of asbestos, chrysotile, are mined each year, chiefly in Canada and the former Soviet Union; this is essentially the only form used commercially in the United States. Another form, the amphibole crocidolite, is mined in small quantities, mainly in South Africa. The two minerals differ greatly in composition, color, shape, solubility, and persistence in human tissue. Crocidolite is blue, relatively insoluble, and persists in tissue. Its fibers are long, thin, and straight and can penetrate narrow lung passages. In contrast, chrysotile is white, and it tends to be soluble and disappear in tissue. Its fibers are curly; they ball up like yarn and are more easily rejected by the body. Long-term occupational exposures to certain asbestos minerals can lead to lung cancer. Although there is some disagreement in the medical and scientific communities, evidence

A sample of chrysotile, one of the minerals in the asbestos family.

strongly suggests that the amphiboles such as crocidolite are much more potent cancer-causing agents than the serpentines such as chrysotile. Since most asbestos in public buildings is of the chrysotile type, the drive to remove asbestos insulation may be misguided overaction in many cases. Nevertheless, most asbestos-containing materials have been removed from the market and strict standards now exist for the handling and use of asbestos.

If the linking of silicate chains continues in two dimensions, sheets of SiO_4 tetrahedral units result (Figure 14.25). Various clay minerals and mica have this sheet-like silicate structure. Mica, for example, is used to prepare "metallic" looking paint on new automobiles.

Clays are essential components of soils that come from the weathering of igneous rocks. They have been used since the beginning of human history for pottery, bricks, tiles, and writing materials. Clays are actually *aluminosilicates,* in which some Si^{4+} ions are replaced by Al^{3+} ions. **Feldspar,** a component of many rocks, is weathered in the following reaction to form the clay *kaolinite.*

Igneous rocks were formed from molten material such as lava from a volcano. Feldspar is a form of igneous rock containing crystalline silicate minerals. Feldspars make up 60% of the earth's crust.

$$\underset{\text{feldspar}}{2\ KAlSi_3O_8(s)} + CO_2(g) + 2\ H_2O(\ell) \longrightarrow \underset{\text{kaolinite}}{Al_2(OH)_4Si_2O_5(s)} + 4\ SiO_2(s) + K_2CO_3(aq)$$

The structure of kaolinite consists of layers of SiO_4 tetrahedra linked in sheets, as in mica, but these sheets are interleaved with layers of Al^{3+} ions, each of which is surrounded by six oxygen atoms of the silicon-oxygen

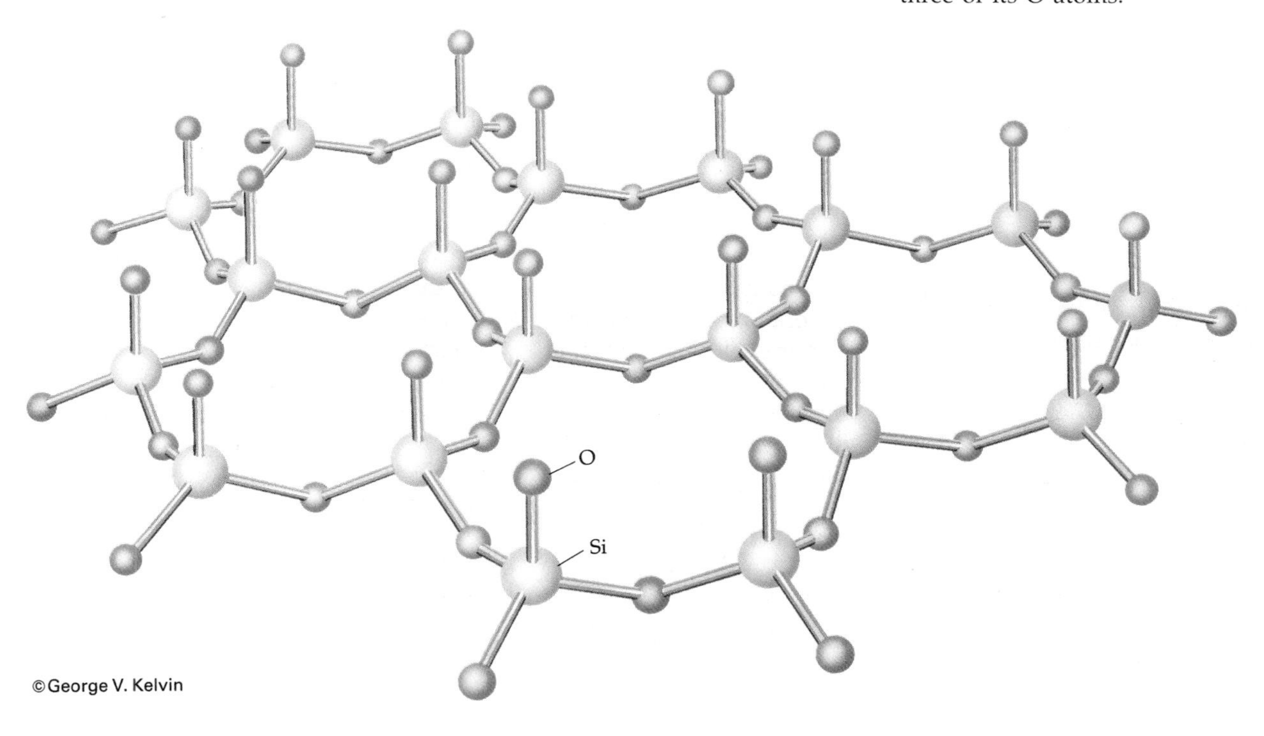

Figure 14.25
Mica structure. Each SiO_4 unit shares three of its O atoms.

Ancient writing on a clay tablet. *(The University Museum, University of Pennsylvania)*

Red clay in a dried river bottom in Arizona. The red color is due to Fe^{3+} ions. *(Shirley Richards/Photo Researchers, Inc.)*

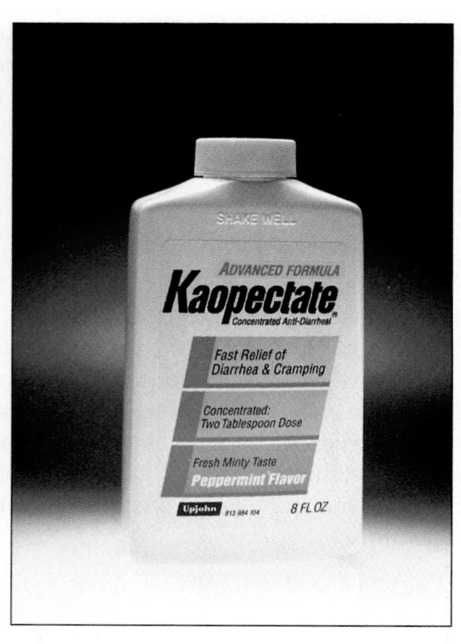

A commercial stomach remedy containing purified clay. *(C.D. Winters)*

sheets as well as by OH^- ions (Figure 14.26). When Al^{3+} ions are replaced by other 3+ metal ions, the clay becomes colored. For example, a red clay contains Fe^{3+} ions in place of some Al^{3+} ions.

Interestingly, in some societies clays are eaten for medicinal purposes. Several pharmaceuticals sold in the United States for relief of stomach upset contain highly purified clay, which absorbs excess stomach acid as well as possibly harmful bacteria and their toxins.

The fascinating colors of some gemstones come from impurities in natural silicate crystals that, if they were pure, would be colorless. For example, pure **corundum,** *a crystalline form of aluminum oxide (Al_2O_3), is a colorless crystalline solid in which each Al^{3+} ion is enclosed in a cage of six O^{2-} ions. If about 1% of the aluminum ions are replaced by Cr^{3+} ions, the crystal is a* **ruby.** *A ruby is red because chromium ions absorb light in the blue end of the visible spectrum while the ruby-red light passes through the crystal.*

Figure 14.26
A ball and stick model of kaolinite, an aluminosilicate. Each silicon (black) is surrounded tetrahedrally by oxygen atoms (red) to give rings consisting of six O and six Si atoms. The layer of aluminum ions (light blue) is attached through O atoms to the Si—O rings. Hydroxide ions (OH^-, light green) act as bridges between Al ions. The net result is a layered structure that gives clays their slipperiness and workability, especially when wet. *(C.D. Winters)*

EXERCISE 14.8 • *Drawing Network Solids*

Use SiO_4 units, shown as 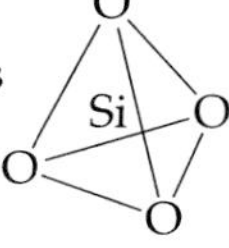

, to draw a linear chain structure, a flat sheet structure, and a three-dimensional structure.

Synthetic ruby crystals *(Kurt Nassau).*

Cement and Ceramics

You probably know that "cement" is something used between bricks to hold them together, or something used to make sidewalks. What about "ceramics"? Perhaps you associate the term with pottery vases at a craft show, or bathroom tile, or even components of electrical equipment. **Ceramics** form an extremely large and diverse class of materials. What they have in common is that all ceramics are made by baking ("firing") a nonmetallic mineral or other kind of nonmetallic compound to make a solid containing myriad tiny particles bound together. **Cement** consists of microscopic particles containing compounds of calcium, iron, aluminum, silicon, and oxygen in varying proportions. Cement reacts in the presence of water to form a hydrated colloid with a large surface area, which subsequently undergoes recrystallization and reaction to bond to itself and to bricks, stone, or other silicate materials.

Cement is made by roasting a powdered mixture of calcium carbonate (limestone or chalk), silica (sand), aluminosilicate mineral (kaolin, clay, or shale), and iron oxide at a temperature of up to 870 °C in a rotating kiln. As the materials pass through the kiln, they lose water and carbon dioxide and ultimately form a "clinker," in which the materials are partially fused. A small amount of calcium sulfate (gypsum) is added and the cooled clinker is then ground to a very fine powder. A typical composition of cement can be expressed as follows: 60% to 67% CaO, 17% to 25% SiO_2, 3% to 8% Al_2O_3, up to 6% Fe_2O_3, and small amounts of magnesium oxide, magnesium sulfate, and potassium and sodium oxides. The oxides are not isolated into molecules or ionic crystals, and the submicroscopic structure—which tends to be composed of very large molecular species—is quite complex.

Cement is usually mixed with other substances. **Mortar** is a mixture of cement, sand, water, and lime. **Concrete** is a mixture of cement, sand, and aggregate (crushed stone or pebbles) in proportions that vary according to the application and the strength required.

Many different reactions occur during the setting of cement. Various constituents react with water and subsequently at the surface with carbon dioxide in the air. The initial reaction of cement with water is the hydrolysis of the calcium silicates, which forms a gel that sticks to itself and to the other particles (sand, crushed stone, or gravel). The gel has a very large surface area and ultimately is responsible for the great strength of concrete once it has "set." The setting process also involves the formation of small, densely interlocked crystals after the initial solidification of the wet mass. This interlocking crystal formation continues for a long time after the initial setting, and it increases the compressive strength of the cement. Water is required for the setting process because the reactions involve hydration. For this reason, freshly poured concrete is kept moist for several days. More than 800 million tons of cement are manufactured each year, most of which is used to

Compressive strength refers to a solid's resistance to shattering under compression.

Tensile strength refers to a substance's resistance to undergo stretching.

make concrete. Concrete, like many other materials containing Si—O bonds, is highly noncompressible but lacks tensile strength. If concrete is to be used where it is subject to tension, it must be reinforced with steel.

Ceramic materials have been made since well before the dawn of recorded history. They are generally fashioned from clay or other natural earths at room temperature and then permanently hardened by heat. **Silicate ceramics** include objects made from clays, such as pottery, bricks, and table china. The techniques developed with natural clay have been applied to a wide range of other inorganic materials in recent years. The result has been a considerable increase in the kinds of ceramic materials available.

China clay, or kaolin, is primarily kaolinite that is practically free of iron. As a result it is almost colorless and particularly valuable in making fine pottery. The first pieces of fine Oriental clayware, which was named "china," arrived in Europe in the Middle Ages; European potters envied and admired the obviously superior product. Clays, mixed with water, form a moldable paste consisting of tiny silicate sheets that can easily slide past one another (recall the lubricant action of graphite, caused by sheets sliding over one another). When this paste is mixed to just the right consistency, it can be molded into almost any form. Then it is heated, the water is driven off, and new Si—O—Si bonds form so that the mass becomes permanently rigid. If silica (SiO_2) and feldspar are added in the right proportions, the mixture will not crack after heating. (Cracking occurs as a result of shrinkage when the new bonds form between the silicate sheets.)

Oxide ceramics are metal oxides that are capable of forming ceramic materials. Examples are alumina (Al_2O_3) and magnesia (MgO). Because it has a high electrical resistivity, alumina is used in spark-plug insulators. High-density alumina has very high mechanical strength, and it is used in armor plating and in high-speed cutting tools for machining metals. Magnesia is an insulator with a high melting point (2800 °C), so it is often used as insulation in electric heaters and electric stoves.

A third class of ceramics includes the **nonoxide ceramics** such as silicon nitride (Si_3N_4), silicon carbide (SiC), and boron nitride (BN). All of these compounds form ceramics that are hard and strong, but brittle. Two structural modifications of boron nitride are the graphite structure and the diamond structure (boron nitride has the same average number of electrons per atom as elemental carbon). The form of BN with the diamond structure is comparable in hardness to diamond and is more resistant to oxidation. One application of these properties is the use of boron nitride cups and tubes to contain molten metals that are being evaporated.

Will some researcher discover a "buckyball" form of BN?

Silicon carbide has the trade name Carborundum, and can be regarded as diamond with half of the C atoms replaced by Si atoms. Widely used as an abrasive material, SiC is receiving more attention recently as a ceramic material, particularly for high-temperature engines.

A market of $10 billion per year for advanced ceramics, industrial need for new materials, and accelerating fundamental research are causing an explosion of new ceramic materials. These include new forms of the ceramics described earlier, as well as **ceramic composites,** mixtures of ceramic materials that have improved properties. The one severely limiting problem in using ceramics is their brittleness. Ceramics deform very little before they fail catastrophically, the failure resulting from a weak point in the bonding within the ceramic matrix. However, such weak points are not consistent from object to object so that the failure is not very predictable.

Since stress failure of ceramic materials is due to molecular irregularities resulting from impurities or disorder in the atomic arrangements, much attention is now being given to using purer starting materials and controlling the processing steps. The *solution-sol-gel* technique for making ceramics, for example, allows for intimate mixing of the ingredients down to the level of 0.5 nm. The oxides are dissolved, usually in an aqueous medium, so that solution species are homogeneously mixed. The solution concentrations are gradually changed, usually by changing pH, so that some of the dissolved species begin to precipitate. After the sol stage (colloidal solid particles dispersed in a liquid) the mixture reaches a gel stage, in which the solid particles link into a network that traps the liquid—just like making a gelatin dessert. The gel can be formed in the desired shape, dried, and then fired to produce the ceramic object.

The possibilities for placing chemically active structures or catalytically active surfaces on ceramic materials appear unlimited in what has become known as ultrastructure processing. For example, imagine a ceramic structure put together like a stack of corrugated pieces of cardboard (Figure 14.27). At 1000 °C, pass a fuel such as gasoline and air through adjacent channels. If the ceramic material between the air and fuel is chemically modified to pass ions but not electrons, if the walls of the fuel channels are catalytically active to facilitate oxidation (loss of electrons from fuel molecules), if the air channels are active to reduce the oxygen to oxide ions, and if separate strips of electron-conducting ceramics connect together all the fuel-channel walls on one side and also all the air-channel walls on the other side, we have a fuel cell (Section 17.7). Fuel cells currently used in space travel are too heavy to power automobiles efficiently. But researchers at Argonne National Laboratory are working on a ceramic fuel cell that will weigh 600 pounds and will develop 231 horsepower of electrical energy. Test models

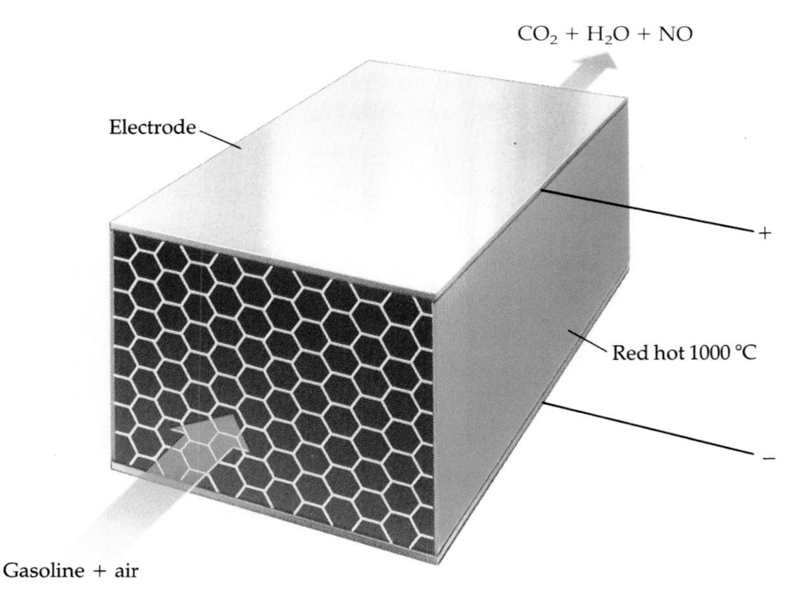

Figure 14.27
A ceramic fuel cell operating at 1000 °C burning gasoline vapors in air. The products are CO_2, H_2O, some NO, and electric energy.

show that fuel efficiency—which is about 30% for the internal combustion engine—can be as much as 50% in the fuel-cell electric car. The speed of the electric motor is controlled by current flow, which in turn is controlled by fuel flow. The exhaust products are carbon dioxide and water, without the polluting wastes of internal combustion.

14.5 SUPERCONDUCTORS

One interesting property of metals is that their electrical conductivity *decreases* with increasing temperature. In Figure 14.28b a metal in a burner flame shows higher resistance (lower conductivity) than the same metal at room temperature (Figure 14.28a). Lower conductivity at higher temperature is understandable if valence electrons in the lattice of metal ions are thought of as waves. As an electron wave moves through the metal crystal under the influence of an electrical voltage, it encounters lattice positions where the metal ions are close enough together to scatter it. This scattering is analogous to the scattering of x rays caused by atoms in crystals (see *A Deeper Look: Using X Rays* in Section 14.2). The scattered electron wave moves off in another direction, only to be scattered again when it encounters some other occupied lattice position. All this scattering *lowers* the conductivity of the metal. At higher temperatures, the metal ions vibrate more, and the distances between lattice positions change more from their average values. This causes more scattering of electron waves as they move through the crystal, because there are now more possibilities of unfavorable lattice spacings. Hence there is lower electrical conductivity.

Recall from Section 9.2 that electrons have wave-like properties as well as particle-like properties.

From this picture of electrical conductivity, it might be expected that the conductivity of a metal crystal at absolute zero (0 K) might be very large. In fact, the conductivity of a pure metal crystal approaches infinity as absolute

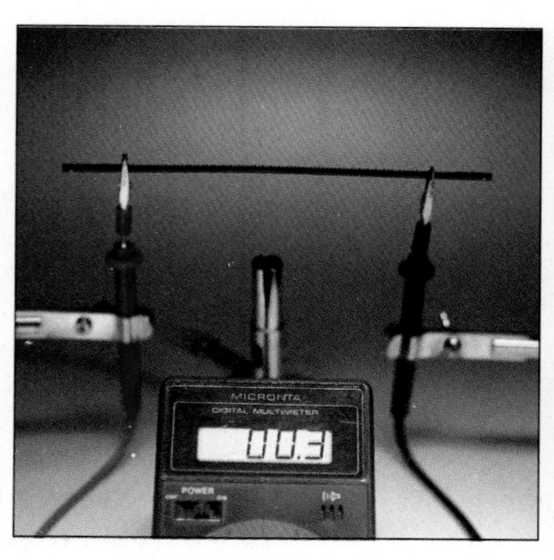

(a)

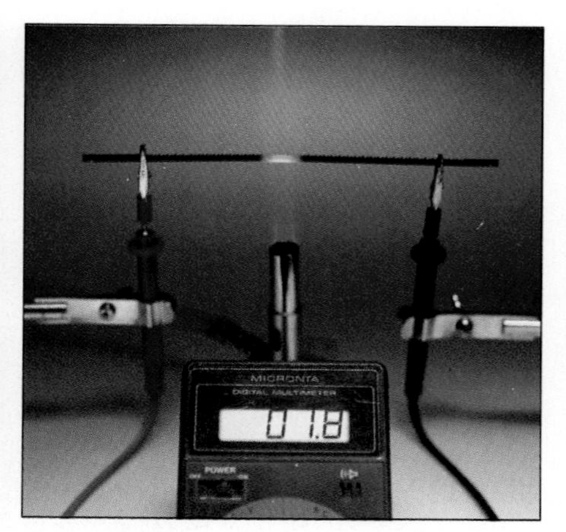

(b)

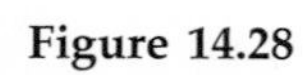

Figure 14.28
(a) A piece of metal at room temperature exhibits a small resistance (high electrical conductivity). (b) While being heated, this same piece of metal exhibits a higher resistance value, indicating a lower conductivity. *(C.D. Winters)*

Table 14.3

*Superconducting Transition Temperatures of Some Metals**

Metal	Superconducting Transition Temperature (K)
aluminum	1.183
gallium	1.087
lanthanum	4.8
lead	7.23
niobium	9.17

*Not all metals have superconducting properties. Those that can become superconductors at atmospheric pressure are Al, Ti, V, Zn, Ga, Zr, Nb, Mo, Tc, Ru, Cd, In, Sn, La, Hf, Ta, W, Re, Os, Ir, Hg, Tl, Pb, Th, Pa, and U.

zero is approached. But in many metals, a more interesting thing happens. At a critical temperature that is low but finite, the conductivity abruptly increases to infinity, which means that the resistance drops to zero. The metal becomes a **superconductor** of electricity. A superconductor offers no resistance whatever to electric current. No clear theory explaining superconductivity has emerged, but it appears that the scattering of electron waves by vibrating atoms is replaced by some cooperative action that allows the electrons to move through the crystal unhindered.

If a material could be made superconducting at a high enough temperature, it would find uses in the transmission of electrical energy, in high efficiency motors, and in computers and other devices. Table 14.3 lists some metals that have superconducting transition temperatures. While some useful devices can be fabricated from these metals, the low temperatures at which they become superconductors make them impractical for most applications. Shortly after superconductivity of metals was discovered, alloys were prepared that had higher transition temperatures. Niobium alloys were the best, but still required cooling to below 23 K (−250 °C) to exhibit superconductivity. To maintain such a low temperature required liquid helium, which costs $4 per liter—an expensive proposition.

All that abruptly changed in January 1986, when K. Alex Müller and J. Georg Bednorz, IBM scientists in Switzerland, discovered that a barium-lanthanum-copper oxide became superconducting at 35 K. This type of mixed metal oxide is a ceramic with the same structure as the mineral perovskite ($CaTiO_3$), and therefore would be expected to have insulating properties. The announcement of the superconducting properties of $LaBa_2Cu_3O_x$ rocked the world of science and provoked a flurry of activity to prepare related compounds in the hope that even higher transition temperatures could be found. Within four months a material that became superconducting at 90 K, $YBa_2Cu_3O_x$, was announced! This was a major breakthrough because 90 K exceeds the boiling point of liquid nitrogen (77 K). At less than 6¢ per liter, liquid nitrogen is a much cheaper refrigerant than liquid helium.

The structure of $YBa_2Cu_3O_x$, called a "1-2-3 superconductor" because of the subscripts of Y, Ba, and Cu in the formula, is shown in Figure 14.29. There are four sheets of Cu and O atoms in the structure. When $x = 7$, the O:Cu ratio in the top and bottom sheets is 1:1, while it is 2:1 in the two inner sheets. Barium ions and oxide ion bridges connect the top and bottom sheets to the inner sheets, and yttrium(III) ions are located between the inner sheets. The copper ions in the top and bottom sheets have an oxidation state of +3, unusual for copper, while the inner sheets have copper ions with an oxidation number of +2. By carefully varying the number of oxide ions that act as bridges between the O-Cu sheets, some of the Cu ions in the top and bottom sheets can be reduced to the +1 oxidation state, thereby giving a sheet of mixed-valence copper ions. If half the oxide ions are removed, the Cu^{3+} sheet will contain equal numbers of +1 and +3 copper ions. However, if less than half are removed, there are more Cu^{3+} than Cu^+, and the com-

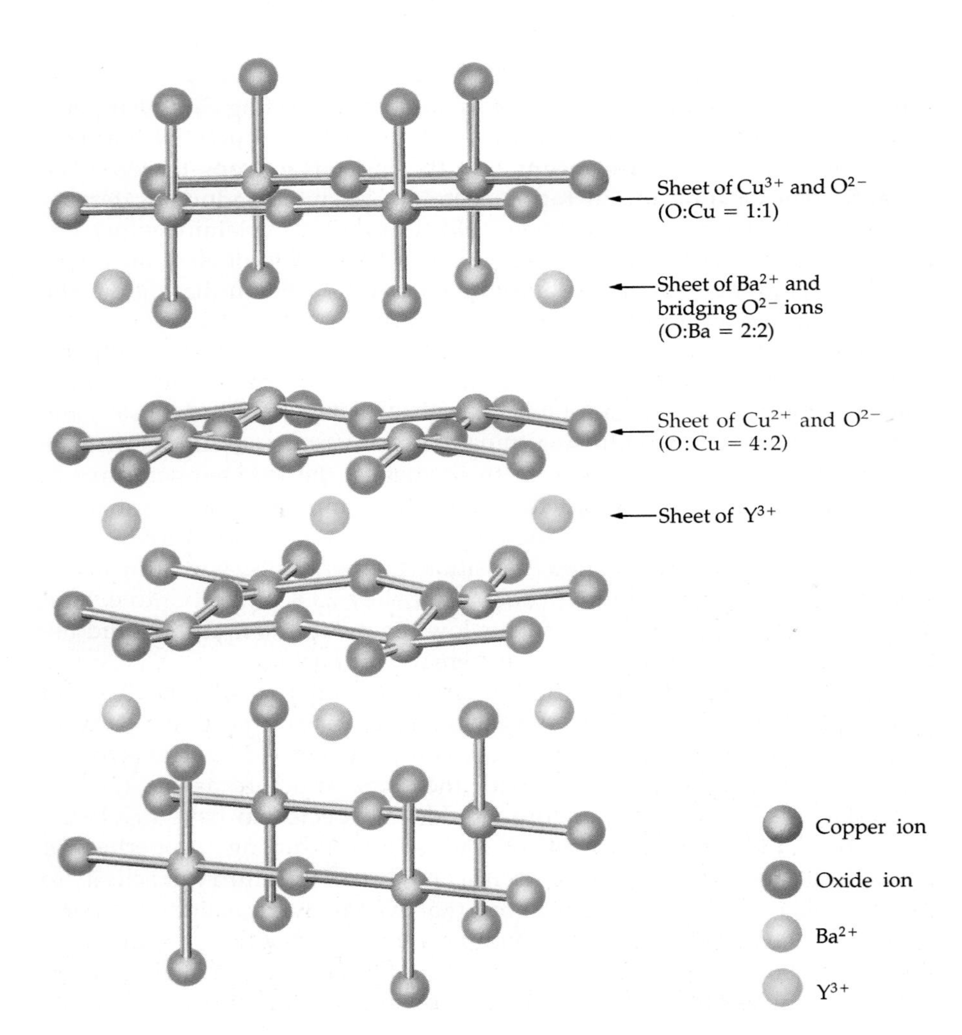

Figure 14.29

A representation of the unit cell of $YBa_2Cu_3O_7$. The structure consists of sheets of copper and oxygen ions held together by bridging oxygen ions, together with Ba^{2+} and Y^{3+} ions. (Note: The oxygens extending above and below the unit cell [four on the top and four on the bottom], and to the front and rear from the inner CuO_2 sheets, are in the neighboring unit cells.) When the O^{2-} ions are lost (as O_2 from the inter-sheet bridges), the solid becomes superconducting.

pound becomes superconducting. It appears that, somehow, the extra two electrons (compared to Cu^{3+}) in each Cu^{+} ion are free to roam over a sheet of copper and oxide ions. So far, hundreds of variations of the "1-2-3" structure have been tested by substituting other lanthanides for Y and other alkaline-earth metals for Ba. Here is another example of using the periodic table to predict the properties of materials.

Müller and Bednorz received the 1987 Nobel Prize in physics for their discovery of the superconducting properties of $LaBa_2Cu_3O_x$.

Why the great excitement over the potential of superconductivity? Superconducting materials will allow the building of more powerful electromagnets such as those used in nuclear particle accelerators and in magnetic resonance imaging (MRI) machines for medical diagnosis. The electromagnets will allow higher energies to be maintained for longer periods of time (and at lower cost) in the case of the particle accelerators, and allow better imaging of problem areas in a patient's body. The main barrier to wider application is the cost of cooling the magnets used in these devices. Many scientists say that this discovery is more important than the discovery of the transistor because of its potential effect on electrical and electronic technology. For example, the use of superconducting materials for transmission of electric power could save as much as 30% of the energy now lost because of the resistance of the wire. Superchips for computers could be up to 1000 times faster than existing silicon chips. Electromagnets could be both more powerful and smaller, which could hasten the day of a practical nuclear fusion reactor. Since a superconductor repels magnetic materials, it is conceivable that cars and trains could be levitated above a track and move with no friction other than air resistance to slow them down.

A demonstration of magnetic levitation of a sample of the superconductor $YBa_2Cu_3O_x$. The vapor is from the liquid nitrogen which maintains the sample within its superconducting temperature range. *(David Parker/IMI/ University of Birmingham TC Consortium/ Photo Researchers, Inc.)*

Although the discovery of superconductivity is a significant one, translating the research into practical applications such as those described here will take time. After all, these superconductors have many of the properties expected of ceramics—they are brittle and fragile. The technology is just beginning to be developed, but recently there has been some progress toward making ribbons and wire filaments from superconducting materials (Figure 14.30). In addition, the maximum superconducting transition temperature has risen to 125 K, well above the boiling point of liquid nitrogen.

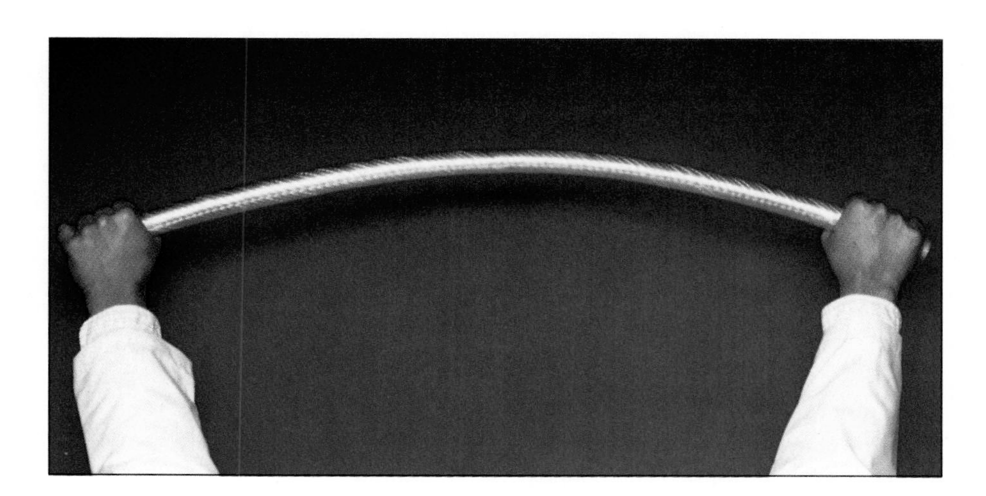

Figure 14.30
A piece of flexible high temperature superconducting cable of the type that may soon find use in underground electrical power cable and possible large electric motors. *(Courtesy of American Superconductor Corporation, Westboro, MA—Garrett Killen.)*

14.6
AMORPHOUS SOLIDS

Amorphous solids are somewhat like liquids in that they exhibit very little regular structure, and yet they are like solids in being hard and having a definite shape. Ordinary glass is an amorphous solid, as are organic polymers such as polyethylene and polystyrene.

Glasses

Glasses occur naturally or can be prepared synthetically. The manufacture of glass goes back to at least 5000 BC, when Phoenician sailors used blocks of sodium carbonate, Na_2CO_3, and sand, SiO_2, to insulate fires from the wooden planks of their ships. The metal carbonate and sand melted in the heat of the fire and formed a material that resembled *obsidian,* a natural glassy material that has been valued since antiquity.

One of the more common glasses today is soda-lime glass. It is clear and colorless if the purity of the ingredients is carefully controlled. If, for example, too much iron oxide is present, the glass will have a green color. Of course, color may also be a desirable property. By adding certain metal oxides to the basic ingredients of a glass, many beautiful colors can be obtained (see Table 14.4).

"Lead glass," as the name implies, contains lead as PbO and is highly prized for its massive feel, acoustic properties, and high index of refraction. Recent regulations concerning toxic water pollutants in California (Proposition 65) have caused manufacturers of lead glass to withdraw some of their products from the marketplace.

The main glass-forming oxides are SiO_2, B_2O_3, GeO_2, and P_4O_{10}, all of which involve elements close to one another in the periodic table. Several other metal oxides, including Al_2O_3 and Na_2O, are also important in forming commercial glasses. The simplest glass is probably amorphous silica, SiO_2 (known as *vitreous silica*), which is built up of corner-sharing SiO_4 tetrahedra linked into a three-dimensional network that lacks symmetry or long range order (Figure 14.31). Vitreous silica can be prepared by melting and quickly cooling either quartz or cristobalite.

Table **14.4**

Substances Used to Color Glass

Substance	Color
copper(I) oxide	red, green, blue
tin(IV) oxide	opaque
calcium fluoride	milky white
manganese(IV) oxide	violet
cobalt(II) oxide	blue
finely divided gold	red, purple, blue
uranium compounds	yellow, green
iron(II) compounds	green
iron(III) compounds	yellow

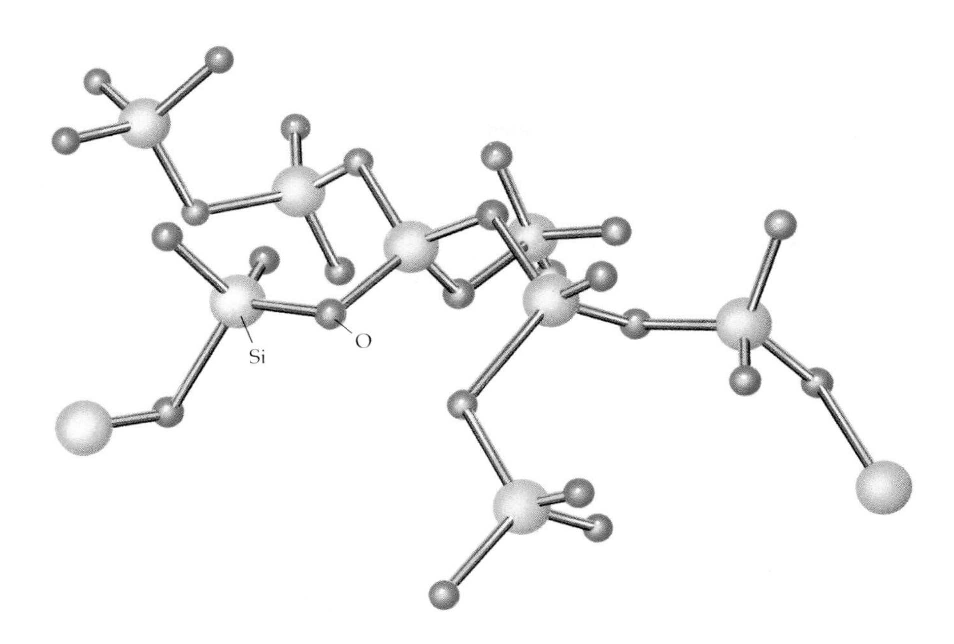

Figure 14.31
Vitreous silica. There is a lack of organization in the arrangement of the SiO_4 units.

If another oxide is added to SiO_2, the melting point of the mixture is lowered considerably (from 1800 °C for quartz to about 800 °C if about 25 mole percent Na_2O is added). The resulting melt cools to form a glass that is somewhat water soluble and is definitely soluble in strongly basic solutions. It is also prone to convert back to a crystalline solid. If other metal oxides such as CaO, MgO, or Al_2O_3 are added, the mixture still melts at a fairly low temperature, but it becomes resistant to chemical attack. Common glass like that used for windows, bottles, and lamps contains these metal oxides in addition to SiO_2.

Mole percent is just another way of expressing concentration. If 0.75 mol of SiO_2 and 0.25 mol of Na_2O are mixed, there is 1.00 mol of matter present. The mole percent of SiO_2 is 75%, and the mole percent of Na_2O is 25%.

It is important that glass be *annealed* properly during the manufacturing process. Annealing means cooling the glass slowly as it passes from a vis-

Obsidian. A natural glass formed from molten volcanic rock similar in composition to granite. *(C.D. Winters)*

Table 14.5

Structures and Properties of Various Types of Solid Substances

Type	Examples	Structural Units
ionic	$NaCl$, K_2SO_4, $CaCl_2$, $(NH_4)_3PO_4$	positive and negative ions; no discrete molecules
metallic	iron, silver, copper, other metals and alloys	metal atoms (or positive metal ions surrounded by an electron sea)
molecular	H_2, O_2, I_2, H_2O, CO_2, CH_4, CH_3OH, CH_3CO_2H	molecules held together by covalent bonds
network	graphite, diamond, quartz, feldspars, mica	atoms held in an infinite one-, two-, or three-dimensional network
amorphous (glassy)	glass, polyethylene, nylon	covalently bonded networks with no long-range regularity

cous, liquid state to a solid at room temperature. If a glass is cooled too quickly, bonding forces become uneven because small regions of crystallinity develop. Poorly annealed glass may crack or shatter when subjected to mechanical shocks or sudden temperature changes. High quality glass, such as that used in optics, must be annealed very carefully. The 200-inch mirror for the telescope at Mt. Palomar, California, was annealed from 500 °C to 300 °C over a period of nine months!

Recently, a new type of *glass ceramic* with unusual but very valuable properties has been commercialized. Ordinary glass breaks because once a

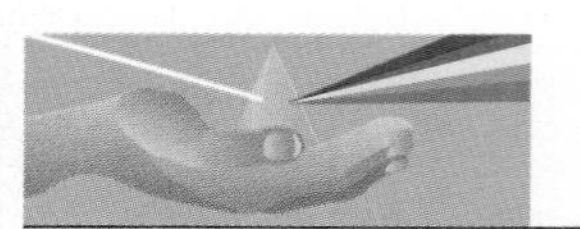

CHEMISTRY YOU CAN DO

The Importance of Annealing

Glasses are not the only solids whose properties are affected by annealing. Obtain a glass of water, at least four bobby pins, and some kind of tongs or tweezers that you can use to hold the bobby pins in a flame without burning your fingers. Scrape off any plastic coating from the ends of the bobby pins. Hold one bobby pin by its open end and heat the bend red hot in the flame of a gas stove or other gas burner. Remove the bobby pin from the flame and place it on a nonflammable surface to cool. After it has cooled, try bending the bobby pin by pulling both ends apart; what do you observe?

Take a second bobby pin and also heat its bend until it is red hot; when you remove it from the flame, plunge it into the glass of water so that it cools very rapidly. Try to bend this bobby pin; what happens?

Heat a third bobby pin red hot and plunge it into the water, but then gently heat it again until an iridescent blue coating forms on the surface; now slowly cool the bobby pin. What happens this time when you try to bend it?

When you test each bobby pin, compare its response to bending with that of the fourth bobby pin, which serves as a control. What can you conclude about the influence of heat treatment and speed of cooling of a metal on its properties?

Table 14.5
(continued)

Forces Holding Units Together	Typical Properties
ionic; attractions among charges on positive and negative ions	hard; brittle; high melting point; poor electrical conductivity as solid, good as liquid; often water-soluble
metallic; electrostatic attraction among metal ions and electrons	malleable; ductile; good electrical conductivity in solid and liquid; good heat conductivity; wide range of hardnesses and melting points
dispersion forces, dipole-dipole forces, hydrogen bonds	low to moderate melting points and boiling points; soft; poor electrical conductivity in solid and liquid
covalent; directional electron-pair bonds	wide range of hardnesses and melting points (3-dimensional bonding $>$ 2-dimensional bonding $>$ 1-dimensional bonding); poor electrical conductivity, with some exceptions
covalent; directional electron-pair bonds	noncrystalline; wide temperature range for melting; poor electrical conductivity, with some exceptions

crack starts, there is nothing to stop the crack from spreading. It was discovered that if glass is treated by heating until many tiny crystals have developed throughout the sample, the resulting material, when cooled, is much more resistant to breaking than normal glass. In molecular terms, the randomness of the glass structure has been partially replaced by the order in a crystalline silicate. The process must be controlled carefully to obtain the desired properties. Materials produced in this way are generally opaque and are used for cooking utensils and kitchen ware, as in products marketed under the name *Pyroceram*. The initial manufacturing process is similar to that of other glass objects, but once the materials have been formed into their final shapes, they are heat treated to develop their special properties.

Dinnerware made of Pyroceram. *(C.D. Winters)*

14.7 SOLID-STATE STRUCTURE AND PROPERTIES OF MATERIALS

Throughout this and earlier chapters, we have emphasized that submicroscopic structure is closely related to macroscopic properties. This relationship is used every day by practicing chemists and other scientists as they try to create new substances that will be beneficial to humankind. Nowhere is the relationship of structure and function more evident than in solids. Table 14.5 summarizes the submicroscopic structure and the macroscopic properties of all the types of substances we have discussed so far. By categorizing materials into the classes listed there, you will be able to form a reasonably good idea of what properties to expect, even in a new substance that you have never encountered before.

IN CLOSING

Having studied this chapter, you should be able to

- explain the differences between gases, liquids, and solids in terms of distance between particles, motion of particles, and structure (Section 14.1)
- explain the heats of fusion and crystallization and their relative magnitudes for different types of solids (Section 14.1)
- explain sublimation of a solid in terms of the transition that takes place (Section 14.1)
- define the unit cell in a crystalline lattice and do simple calculations of unit cell mass and substance density for cubic unit cells (Section 14.2)
- explain the observable properties of metals in terms of metallic bonding (Section 14.3)
- describe why some substances conduct electricity, some are insulators, and some are semiconductors (Section 14.3)
- contrast the properties of graphite and diamond (Section 14.4)
- describe how the SiO_4 unit is used in different silica and silicate structures (Section 14.4)
- define ceramics and give some examples (Section 14.4)
- describe the property of superconduction (Section 14.5)
- explain some of the properties of glasses in terms of their structure (Section 14.6)
- for each type of solid substance, describe the structural units, interparticle forces, and general properties, and give some examples (Section 14.7).

STUDY QUESTIONS

Review Questions

1. Name three properties of solids that are different from those of liquids. Explain the differences for each.
2. What type of solid exhibits each of these sets of properties? (a) Melts below 100 °C and is insoluble in water. (b) Conducts electricity only when melted. (c) Is insoluble in water and conducts electricity. (d) Noncrystalline and melts over a wide temperature range.
3. Explain why (a) the density of ice decreases as the temperature changes from −8 °C to 0 °C, and (b) the density of water at 0 °C is greater than that of ice at 0 °C.
4. What does a low heat of fusion for a solid tell you about the solid (its bonding or type)?
5. What does a high melting point and a high heat of fusion tell you about a solid?
6. Which would you expect to have the higher heat of fusion, N_2 or I_2? Explain your choice.
7. What is sublimation?
8. Describe how each of the following would behave as they were deformed by a hammer strike. Explain why they behave as they do.
   (a) a metal, such as gold
   (b) a nonmetal, such as sulfur
   (c) an ionic compound, such as NaCl
9. What is the heat of crystallization of a substance and how is it related to the substance's heat of fusion?
10. Why is solid CO_2 called Dry Ice?
11. What is the unit cell of a crystal?
12. Name and draw the three cubic unit cells. Describe their similarities and differences.
13. Assuming the same substance could form crystals with its atoms or ions in *either* simple cubic packing *or* hexagonal closest packing, which form would have the higher density?
14. Explain how the volume of a simple cubic unit cell is related to the radius of the atoms in the cell.
15. Name three properties of metals and explain them by using a theory of metallic bonding.
16. In terms of band theory, what is the difference between a conductor and a nonconductor? Between a conductor and a semiconductor?

17. What is the process of doping, as applied to semiconductors?
18. Using what you know about the bonding present in graphite and diamond, explain why diamond is denser than graphite.
19. Explain why the conductivity of metals decreases with increasing temperature while the conductivity of semiconductors increases with increasing temperature.
20. Define a superconductor.
21. What makes a glass different from a solid such as NaCl? Under what conditions could NaCl become glass-like?
22. Define the term "amorphous."
23. How would you convert a sample of liquid to vapor without changing the temperature?
24. How does conductivity vary with temperature for (a) a conductor, (b) a nonconductor, (c) a semiconductor, and (d) a superconductor? In your answer, begin at high temperature and come down to low temperatures.

Unit Cells

25. Solid xenon forms crystals with a face-centered unit cell that has an edge of 0.620 nm. Calculate the atomic radius of xenon.

26. Gold (atomic radius = 144 pm) crystallizes in a face-centered unit cell. What is the length of a side of the cell?

27. Consider the CsCl unit cell shown in Figure 14.14. How many Cs^+ ions are there per unit cell? How many Cl^- ions?

28. Using the NaCl structure shown in Figure 14.15, how many unit cells share each of the Na^+ ions in the front face of the unit cell? How many unit cells share each of the Cl^- ions in this face?

29. The ionic radii of Cs^+ and Cl^- are 0.169 and 0.181 nm, respectively. What is the length of the body diagonal in the CsCl unit cell? What is the length of the side of this unit cell? (See Figure 14.14.)

30. Thallium chloride, TlCl, crystallizes in either a simple cubic lattice or a face-centered cubic lattice of Cl^- ions with Tl^+ ions in the holes. If the density of the solid is 7.00 g/cm^3 and the edge of the unit cell is 3.85×10^{-8} cm, what is the unit cell geometry?

31. Could $CaCl_2$ possibly have the NaCl structure? Explain your answer briefly.

32. Each diagram below represents an array of like atoms that would extend indefinitely in two dimensions. Draw a two-dimensional unit cell for each array. How many atoms are there in each unit cell?

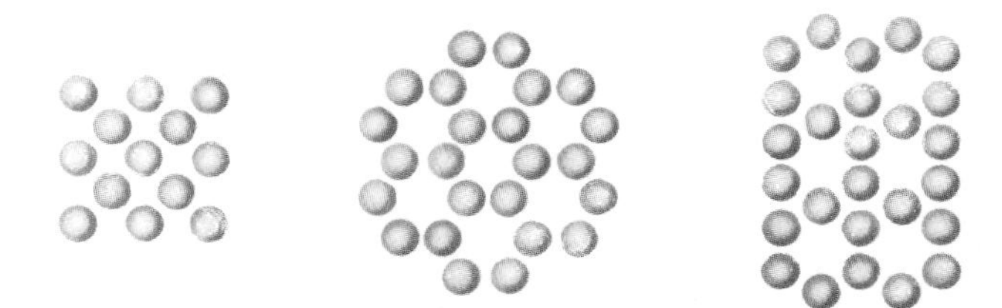

Phase Changes

33. The heat of fusion for H_2O is about 2.5 times larger than the heat of fusion for H_2S. What does this say about the relative strengths of the forces holding the molecules together in their respective solids? Explain.

34. Which requires more heating? (a) melting 1.0 mol of ice starting at −10 °C and ending at 2 °C, or (b) melting 1.0 mol of ice at −2 °C and ending at 10 °C? (Specific heat capacities are given in Table 7.1.)

35. What is the total quantity of heat required to change 0.50 mol of ice at −5 °C to 0.50 mol of steam at 100 °C?

36. Benzene is an organic liquid that freezes at 5.5 °C to beautiful, featherlike crystals. How much heat is evolved when 15.5 g of benzene freezes at 5.5 °C? The heat of fusion of benzene is 127 J/g. If the 15.5 g sample is remelted, again at 5.5 °C, what quantity of heat is required to convert it to a liquid?

37. Mercury is a highly toxic metal. Although it is a liquid at room temperature, it has a high vapor pressure and a low heat of vaporization (294 J/g). What quantity of heat is required to vaporize 0.500 mL of mercury at 357 °C, its normal boiling point? (The density of $Hg(\ell)$ is 13.6 g/mL.)

38. Your air conditioner probably contains the chlorofluorocarbon CCl_2F_2 as the heat transfer fluid. (Because of environmental damage caused by these compounds, they will be phased out soon; see Chapter 20.) Its normal boiling point is −30 °C and the heat of vaporization is 165 J/g. The gas and the liquid have specific heat capacities of 0.61 J/g·K and 0.97 J/g·K, respectively. How much heat is evolved when 10.0 g of CCl_2F_2 is cooled from +40 °C to −40 °C?

39. Liquid ammonia, $NH_3(\ell)$, was used as a refrigerant fluid before the discovery of the chlorofluorocarbons, and is still widely used today, even though it is a somewhat toxic gas and a strong irritant when it is released into the air. Its normal boiling point is −33.4 °C and the heat of vaporization is 23.5 kJ/mol. The gas and liquid have specific heat capacities of 2.2 J/g·K and 4.7 J/g·K, respectively. How much heat must be supplied to 10. kg of liquid ammonia to raise its temperature from −50.0 °C to −33.4 °C, and then to 0.0 °C?

General Questions

40. The ions of NaF and MgO all have the same number of electrons, and the internuclear distances are about the same (235 pm and 212 pm). Why then are the melting points of NaF and MgO so different (992 °C and 2642 °C)?

41. For the pair of compounds LiF and CsI, tell which compound is expected to have the higher melting point and briefly explain.

42. During thunderstorms, very large hailstones can fall from the sky. (Some are the size of golfballs!) To preserve some of these stones, we want to place them in the freezer compartment of our frost-free refrigerator. Our friend, who is a chemistry student, tells us to use an older, non-frost-free model. Why?

43. Some camping stoves contain liquid butane (C_4H_{10}). They work only when the outside temperature is warm enough to allow the butane to have a reasonable vapor pressure (and so are not very good for camping in temperatures below about 0 °C). Assume the enthalpy of vaporization of butane is 24.3 kJ/mol. If the camp stove fuel tank contains 190. g of liquid C_4H_{10}, how much heat is required to vaporize all of the butane?

44. A simple cubic unit cell is formed so that the spherical atoms or ions just touch one another along the edge. Calculate the percentage of empty space within the unit cell. (Recall that the volume of a sphere is $\frac{4}{3}\pi r^3$, where r is the radius of the sphere.)

45. Classify each of the following solids as ionic, metallic, molecular, network, or amorphous.
(a) KF (c) SiO_2
(b) I_2 (d) polypropylene

46. Classify each of the following solids as ionic, metallic, molecular, network, or amorphous.
(a) tetraphosphorus decoxide
(b) graphite
(c) brass
(d) ammonium phosphate

47. Which substance has the highest melting point? The lowest melting point? Explain your choice briefly.
(a) LiBr (c) CO
(b) CaO (d) CH_3OH

48. Which substance has the highest melting point? The lowest melting point? Explain your choice briefly.
(a) SiC (c) Rb
(b) I_2 (d) $CH_3CH_2CH_2CH_3$

49. Which substance has the greatest electrical conductivity? The smallest electrical conductivity? Explain your choice briefly.
(a) Si (c) Ag
(b) Ge (d) P_4

50. Which substance has the greatest electrical conductivity? The smallest electrical conductivity? Explain your choice briefly.
(a) RbCl(ℓ) (c) Rb
(b) NaBr(s) (d) diamond

51. On the basis of the description given, classify each of the following solids as molecular, metallic, ionic, network, or amorphous. Explain your reasoning.
(a) A brittle, yellow solid that melts at 113 °C; neither solid nor liquid conducts electricity.
(b) A soft, silvery solid that melts at 40 °C; both solid and liquid conduct electricity.
(c) A hard, colorless, crystalline solid that melts at 1713 °C; neither solid nor liquid conducts electricity.
(d) A soft, slippery solid that melts at 63 °C; neither solid nor liquid conducts electricity.

52. On the basis of the description given, classify each of the following solids as molecular, metallic, ionic, network, or amorphous. Explain your reasoning.
(a) A soft, slippery solid that has no definite melting point but decomposes at temperatures above 250 °C; the solid does not conduct electricity.
(b) Violet crystals that melt at 114 °C and whose vapor irritates the nose; neither solid nor liquid conducts electricity.
(c) Hard, colorless crystals that melt at 2800 °C; the liquid conducts electricity, but the solid does not.
(d) A hard solid that melts at 3410 °C; both solid and liquid conduct electricity.

53. Metallic lithium has a body-centered cubic structure, and its unit cell is 351 pm on a side. Lithium iodide has the same crystal lattice structure as sodium chloride. The cubic unit cell is 600 pm on a side. (a) Assume that the metal atoms in lithium touch along the body diagonal of the cubic unit cell and estimate the radius of a lithium atom. (b) Assume that in lithium iodide the I^- ions touch along the face diagonal of the cubic unit cell, and that the Li^+ and I^- ions touch along the edge of the cube; calculate the radius of an I^- ion and of a Li^+ ion. (c) Compare your results in parts (a) and (b) for the radius of a lithium atom and a lithium ion. Are your results reasonable? If not, how could you account for the unexpected result? Could any of the assumptions that were made be in error?

54. Potassium chloride and rubidium chloride both have the sodium chloride structure. X-ray diffraction experiments indicate that their cubic unit cell dimensions are 629 pm and 658 pm, respectively. (1) One mole of KCl

and one mole of RbCl are ground together in a mortar and pestle to a very fine powder, and the x-ray diffraction pattern of the pulverized solid is measured. Two patterns are observed, each corresponding to a cubic unit cell—one with an edge length of 629 pm and one with an edge length of 658 pm. Call this Sample 1. (2) One mole of KCl and one mole of RbCl are heated until the entire mixture is molten and then cooled to room temperature. In this case there is a single x-ray diffraction pattern that indicates a cubic unit cell with an edge length of roughly 640 pm. Call this Sample 2.

(a) Suppose that Samples 1 and 2 were analyzed for their chlorine content. What fraction of each sample is chlorine? Could the samples be distinguished by means of chemical analysis?

(b) Interpret the two x-ray diffraction results in terms of the structures of the crystal lattices of Samples 1 and 2.

(c) What chemical formula would you write for Sample 1? For Sample 2?

(d) Suppose that you dissolved 1.00 g of Sample 1 in 100 mL of water in a beaker and did the same with 1.00 g of Sample 2. Which sample would conduct electricity better, or would both be the same? What ions would be present in each solution at what concentrations?

Summary Questions

55. Sulfur dioxide, SO_2, is found in polluted air. It is formed during the combustion of fossil fuels containing small percentages of sulfur, and from industrial plants that convert certain metal-containing ores to metals or metal oxides.
    (a) Draw the electron dot structure for SO_2. From this describe first the O-S-O angle, and second the structural geometry and molecular geometry.
    (b) What type of forces are responsible for binding SO_2 molecules to one another in the solid or liquid phase?
    (c) Using the information below, place the compounds listed in order of increasing intermolecular forces.

Compound	*Normal Boiling Point (°C)*
SO_2	−10
NH_3	−33.4
CH_4	−161.5
H_2O	100

56. Copper is an important metal in our economy, most of it being mined in the form of the mineral chalcopyrite, $CuFeS_2$.
    (a) To obtain one metric ton (1000 kilograms) of copper metal, how many metric tons of chalcopyrite would you have to mine?
    (b) If the sulfur in chalcopyrite is converted to SO_2, how many metric tons of the gas would you get from one metric ton of chalcopyrite?
    (c) Copper crystallizes as a face-centered cube. Knowing that the density of copper is 8.95 g/cm^3, calculate the radius of the copper atom.
57. Extremely high purity silicon is required to manufacture semiconductors such as memory chips found in calculators and computers. If a silicon wafer is 99.99999999% pure, approximately how many silicon atoms per gram have been replaced by impurity atoms of some other element?

Using Computer Programs and Videodiscs

58. For each of the following elements, use *KC? Discoverer* to obtain the density and atomic radius. Use these data to decide whether a body-centered cubic or a closest-packed structure is likely for the metal.
    (a) Fe (d) Au
    (b) Na (e) V
    (c) Po (f) Ni
59. Use *KC? Discoverer* to explore whether enthalpy of fusion is correlated with melting point for all elements. This might be done by graphing one property on each axis or by sorting on one property and listing the other at the same time. If there is a relationship for many elements, suggest reasons for the exceptions that you find.
60. Use *KC? Discoverer* to explore whether thermal conductivity and electrical conductivity are correlated. This might be done by graphing one property on each axis or by sorting on one property and listing the other at the same time. If there is a relationship for many elements, suggest reasons for the exceptions that you find. Do all or most of the elements for which there is a correlation fall into some category, that is, is there some other relationship between them as well as the thermal and electrical conductivity relationship?
61. Use *KC? Discoverer* to make a graph of enthalpy of fusion versus atomic number for each period in the periodic table. Do the enthalpies of fusion for metals vary regularly across a period? Use the electron-sea theory of metallic bonding to explain the trend you see.
62. Use *KC? Discoverer* to make a graph of melting point versus atomic number for each period in the periodic table. Do the melting points of metals vary regularly across a period? Use the electron-sea theory of metallic bonding to explain the trend you see.

The oceans (here surrounding the island of Weilangilala in the Figis) are an aqueous solution of many salts.
(J. Kotz)

Solution Chemistry

CHAPTER

15

Solutions are important because we all encounter them every day, and because a great many different kinds of chemical reactions occur in them. In a solution the atoms, molecules, or ions are thoroughly mixed and intermingled. Such homogeneous mixtures can be solids (metal alloys), liquids (tap water), or gases (air). Forces between particles can often account for solubility; substances whose intermolecular forces are of the same type as those of a solvent usually dissolve readily. The composition of a solution—how much solute there is for a given quantity of solvent—can be expressed in several ways. Some of these are more useful for very small traces of pollutants; others apply to typical laboratory situations.

Certain solution properties are proportional to the concentration of solute particles, regardless of the type of particle. These are called colligative properties and have several interesting applications. When a substance dissolves only partially in a solvent, an equilibrium is set up between pure solute and solution. The solubility of the substance is related to the equilibrium constant for the dissolution reaction, and tables of such constants can be used to estimate solubilities. When a mixture consists of particles larger than atoms, molecules, or ions, but still too small to separate rapidly, we have a colloid; examples are homogenized milk and smog. Water that contains either too much or too little of one or more solutes or colloids is said to be polluted. Dealing with water pollution requires considerable knowledge of solutions and their properties.

A **solution** is a homogeneous mixture of two or more substances. It exists in a single phase, which may be solid, liquid, or gas. The component present in the largest amount is usually called the **solvent;** the other components are called **solutes.** For example, the ocean is an aqueous solution; water is the solvent, and sodium chloride, many other salts, a little oxygen, traces of oil and other pollutants, and many other substances are solutes. Other commonly encountered solutions are surface waters (rivers and lakes), groundwater (water beneath the earth's surface), steel (a solution of one or more elements in iron) and other alloys, and air (a solution of oxygen, argon, water vapor, carbon dioxide, and other gases in nitrogen). Groundwater and surface waters are our sources for drinking water; too high a concentration of certain solutes can be harmful. Similarly, above-normal concentrations in air of sulfur dioxide or ozone, for example, are categorized as air pollution (see Chapter 20).

Solutions are important not only because they are so commonplace, but because so many chemical reactions take place in them. Particles in a solution (atoms, molecules, or ions) are thoroughly mixed, and so they can come into contact and react. In gas-phase or liquid-phase solutions the particles move about and collide, increasing the opportunities for them to react with each other. Liquid solutions have the added advantage over gaseous solutions that the particles are close together and therefore collide more often. Liquid solutions are therefore the media for many reactions used to produce polymers, medicines, and other commercial products. They are also the media within our bodies and those of other living organisms in which biochemical reactions occur (Chapter 21). Because of the importance of liquid solutions, especially aqueous solutions, much of this chapter will be devoted to them.

15.1 THE SOLUTION PROCESS

Some solutes dissolve to a much greater extent than others. For example, consider silver nitrate ($AgNO_3$) and silver chloride ($AgCl$) dissolving in water. About 330 g of $AgNO_3$ will dissolve in 100 mL of water at 25 °C, and we say this is a (very) soluble salt. In contrast, only 0.00035 g of AgCl will dissolve in 100 mL of water at 25 °C, and we say AgCl is insoluble, even though a little bit does dissolve. A substance's **solubility** is defined as its concentration in a solution that is in equilibrium with pure solute at a given temperature. If almost none of a solute dissolves, then the solute is said to be **insoluble** in that solvent.

When more solute has been added to a liquid than can dissolve at a given temperature, there is a *dynamic equilibrium* between undissolved and dissolved solute. Some solute molecules or ions are going into solution, while others are separating from solvent molecules and entering the pure solute phase. Both processes are going on all the time, at identical rates. If we could observe all these changes, they would appear very frenzied indeed.

When the concentration of a solute equals its solubility, the solution is said to be **saturated.** A solution is **unsaturated** if the solute concentration is

(a) Unsaturated: crystal dissolves when added to solution

(b) Supersaturated: when a crystal is added to the solution it grows bigger until the solution becomes a saturated solution

(c) Saturated: quantity of solid does not change with time

Figure 15.1
The difference between unsaturated, saturated, and supersaturated solutions. In (a), some solute is added to a solution, and the solute dissolves. The solution was unsaturated with respect to solute. In (b), the solution is supersaturated. Adding extra solute leads to the precipitation of some dissolved solute. The solution in (c) is saturated, since no more solute will dissolve; the concentration of solute does not change with time.

less than the equilibrium concentration. For some substances it is possible to prepare solutions that contain *more* than the equilibrium concentration of solute; a solution like this is **supersaturated** (see Figure 15.1). Separation of solute from a supersaturated solution is a product-favored process, but it often occurs very slowly; some solutions can remain supersaturated for days or months. The formation of a crystal lattice requires that several ions or molecules be arranged very near to appropriate lattice positions, and it can take a long time for such an alignment to occur. However, precipitation of a solid from a supersaturated solution occurs rapidly if a tiny crystal of the solute is added to the solution. The crystal's lattice provides a template onto which more ions or molecules can be added. Sometimes other actions, such as stirring a supersaturated solution or scratching the walls of its container, will cause solute to precipitate rapidly.

When a solution forms, molecules of one kind become mixed together with molecules of a different kind. Because of this mixing, there is usually a positive entropy change when one substance dissolves in another. As described in Section 7.9, this entropy increase favors solution formation. For this reason all gases dissolve in each other in all proportions at normal pressures. However, when a solid or liquid dissolves in a liquid, the atoms, molecules, or ions are much closer together, and intermolecular forces become important. Intermolecular attractions among solvent molecules are disrupted as room is made for solute particles. Attractions among solute particles must also be overcome as they separate and mix among solvent molecules. New attractions between solvent and solute particles come into play. If the new solvent-solute forces are not strong enough to overcome the solute-solute or solvent-solvent forces, dissolution may not occur.

Mixing of gases depends almost entirely on entropy, because gas particles are relatively far apart and intermolecular forces are very small.

Dissolving Liquids in Liquids

Mixing of molecules in the liquid phase is favored by entropy, just as mixing of molecules in the gas phase is. However, the intermolecular attractions are much stronger because the molecules are closer together. Many liquids will dissolve in another liquid in any proportion. When this is the case, the liq-

(a)

(b)

Figure 15.2
Miscibility. (a) The colorless, more dense bottom layer is nonpolar carbon tetrachloride, CCl_4. The middle layer is a solution of $CuSO_4$ in water, and the colorless, less dense top layer is nonpolar octane (H_3C-CH_2-CH_2-CH_2-CH_2-CH_2-CH_2-CH_3). Neither the weak intermolecular attractions between carbon tetrachloride and water nor those between octane and water can overcome the very strong forces attracting water molecules to each other, and so carbon tetrachloride and octane are not miscible with water. (b) After stirring the mixture, the two nonpolar liquids form a homogeneous mixture, since they are miscible with one another. This layer of mixed liquids sits on top of the water layer because the water layer is the more dense of the two. *(C.D. Winters)*

uids are said to be completely **miscible.** The solvent-solvent and solute-solute intermolecular attractions must be very similar to the solvent-solute intermolecular attractions in order for liquids to be miscible.

To see how this works, consider the situation shown in Figure 15.2. Octane (C_8H_{18}, a component of gasoline) and carbon tetrachloride (CCl_4) are completely miscible, but water is insoluble in octane, insoluble in carbon tetrachloride, and insoluble in a solution of the two. Octane and carbon tetrachloride are nonpolar, and the only intermolecular forces in either liquid are dispersion forces. The strengths of such forces depend on the number of electrons in a molecule and are nearly independent of other features of molecular structure. Therefore, it does not matter whether a carbon tetrachloride molecule is next to an octane molecule or another carbon tetrachloride molecule—the intermolecular attractions will be about the same. There is little net change in intermolecular attractions when the molecules mix, and the entropy increase causes mixing to be a product-favored process.

Now think about what would happen if water molecules mixed with octane molecules. The water molecules are polar and can form hydrogen bonds to each other, producing strong intermolecular attractions within liquid water that are not possible within liquid octane. If each water molecule became surrounded by octane molecules, there would be no hydrogen bonding or dipole-dipole forces. The energy required to overcome these forces when the water molecules were separated would not be provided by formation of new intermolecular attractions, and there would have to be a net input of energy. Similarly, since a water molecule has fewer electrons than an octane molecule, the dispersion forces between water and octane would be smaller than between octane molecules. Therefore, energy effects favor water and octane remaining separated. If the entropy increase is not large enough to counteract this energy effect, there will be little mixing, and the two liquids will be immiscible. The same reasoning applies to mixing water and carbon tetrachloride.

An old adage says that "oil and water don't mix." Chemists use a similar saying about solubility: "like dissolves like," where "like" refers to similar intermolecular forces. It should be fairly clear from the solubilities of alcohols in water shown in Table 15.1 that this is the case. In a low-molecular-weight alcohol such as ethanol (Figure 15.3), hydrogen bonding with the solvent water molecules is strong compared to the dispersion forces between hydrocarbon parts of the molecules. As the hydrocarbon part of the alcohol becomes larger, the dispersion forces between the alcohol molecules

Table 15.1

Solubilities of Some Alcohols in Water

Name	Formula	Solubility in Water (g/100 g H_2O at 20 °C)
methanol	CH_3OH	miscible
ethanol	C_2H_5OH	miscible
1-propanol	C_3H_7OH	miscible
1-butanol	C_4H_9OH	7.9
1-pentanol	$C_5H_{11}OH$	2.7
1-hexanol	$C_6H_{13}OH$	0.6
1-heptanol	$C_7H_{15}OH$	0.09

become greater and the hydrogen bonding between the alcohol and solvent water becomes relatively less important in determining solubility. The polar —OH part of an alcohol molecule is commonly called **hydrophilic,** which means "water loving." Any polar part of a solute molecule will be hydrophilic. The hydrocarbon part of an alcohol molecule is called **hydrophobic,** which means "water fearing." Any nonpolar part of a solute molecule will be hydrophobic. As the alcohol molecule gets bigger, it becomes more hydrophobic than hydrophilic, and its water solubility becomes very small.

Gasoline, a mixture of nonpolar hydrocarbons attracted to each other by dispersion forces, readily dissolves greases and oils, which are also nonpolar hydrocarbons. Gasoline can dissolve grease because the intermolecular forces between the molecules in gasoline are similar to those between the molecules in a grease. If water is added to gasoline, or gasoline to water, the hydrogen bonding between water molecules is so strong that very little intermingling of the hydrocarbon molecules with water molecules occurs. In fact, water, with a density of about 1 g/mL, will sink to the bottom of a

Although gasoline dissolves grease, it is not a good idea to use gasoline as a solvent. It is flammable, and it contains numerous harmful hydrocarbons. One of these, benzene, can cause cancer in humans. Others, such as hexane and heptane, can damage the central nervous system and cause unconsciousness upon prolonged exposure.

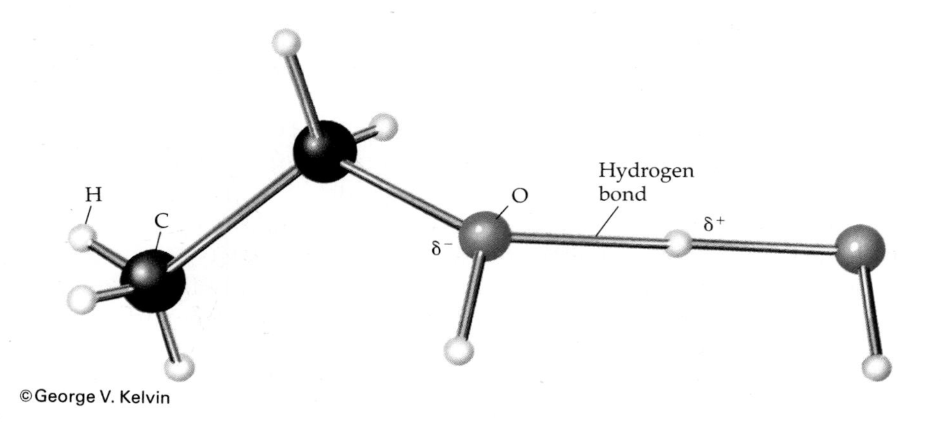

Figure 15.3
Because of the hydrogen bonding between water molecules and ethanol molecules, the two liquids dissolve completely in one another in all proportions.

(a)

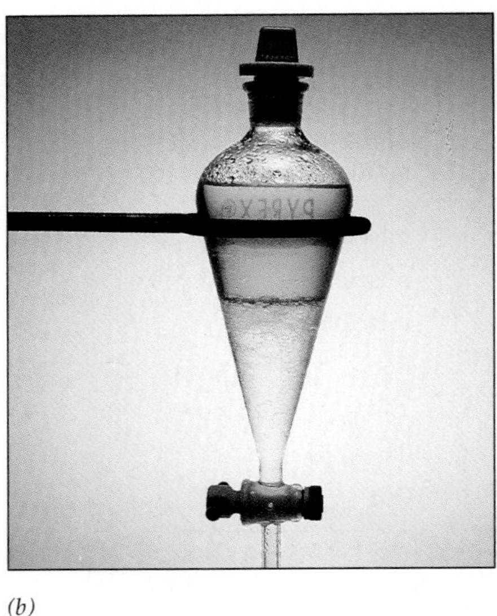

(b)

(c)

Figure 15.4
Gasoline and water will not mix, at least not for long. (a) In a separatory funnel, a device used for separating immiscible liquids, a sample of gasoline and water remains as two distinct layers. (b) After vigorous mixing, some gasoline droplets are seen mixed with water droplets at the interface between the two layers.
(c) But given time, the two layers re-form. *(C.D. Winters)*

sample of gasoline, which has a density of about 0.7 g/mL (Figure 15.4). Gasoline spills on water are always a serious problem because the hydrocarbon floats on top of the water and evaporates, producing flammable vapors that can be ignited by a flame or spark.

EXERCISE 15.1 • *Liquid Solubilities*

Which would be more soluble in water, octanol or methanol? Which would be more soluble in gasoline? Explain your choice in terms of intermolecular attractions.

Dissolving Solids in Liquids

Iodine, I_2, is an example of a solid in which nonpolar molecules are attracted to one another by dispersion forces. For I_2 to dissolve in a liquid, these forces must be overcome and some of the intermolecular attractions among solvent molecules must also be disrupted. Iodine dissolves only slightly in H_2O—just enough to form a colored solution (Figure 15.5). The hydrogen bonds attracting water molecules to each other are much stronger than forces between I_2 and H_2O molecules. Carbon tetrachloride, CCl_4, a nonpolar liquid, dissolves I_2 readily—so readily that it can dissolve I_2 that has already been dissolved in water, a process known as extraction.

Structural formula for sucrose.

Sucrose (cane sugar), $C_{12}H_{22}O_{11}$, is also a molecular solid. Each sucrose molecule has eight —OH groups, and sucrose molecules in the solid state are hydrogen bonded to each other. When sugar dissolves in water, the new hydrogen bonds formed between sucrose molecules and water molecules are strong enough to overcome the sucrose-sucrose and water-water interactions. On the other hand, sucrose is insoluble in CCl_4 (and other nonpolar liquids) because the sucrose-CCl_4 intermolecular attractions are fairly weak.

Sodium chloride is an ionic compound. Its crystal lattice consists of Na^+ and Cl^- ions in a cubic array (Sections 4.4 and 14.2). The strong attractions

(a)

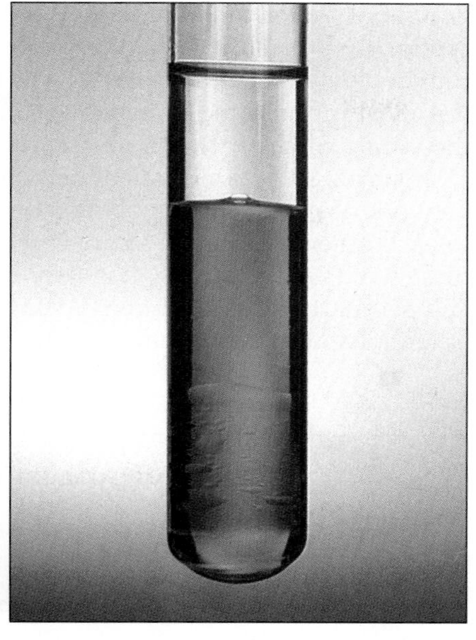

(b)

Figure 15.5

Water, carbon tetrachloride (CCl_4), and iodine. (a) Water (polar molecules) and CCl_4 (nonpolar molecules) are not miscible, and the less dense water layer is found on top of the more dense CCl_4 layer. A small amount of iodine is dissolved in water to give a brown solution (top). However, the nonpolar molecule I_2 is more soluble in nonpolar CCl_4, as indicated by the fact that I_2 dissolves preferentially in CCl_4 to give a purple solution after the mixture is shaken (b). *(C.D. Winters)*

between oppositely charged ions hold them tightly in the lattice, accounting for sodium chloride's high melting point of 800 °C. If you try to dissolve NaCl (or any other ionic compound) in carbon tetrachloride or hexane, you will have little success. Nonpolar molecules have very little attraction for ions, and so the strong interionic attractions in the crystal lattice predominate; table salt does not dissolve in gasoline. On the other hand, water molecules are polar. The negative (oxygen) end of a water molecule is attracted to a positive Na^+ ion, helping pull the ion from its lattice position. Similarly, the positive side (hydrogen atoms) of a water molecule is attracted to a negative Cl^- ion, and again this attraction helps overcome interionic forces in the lattice. When they dissolve, ions become surrounded by many water molecules, as shown in Figure 15.6. When solute particles are surrounded by

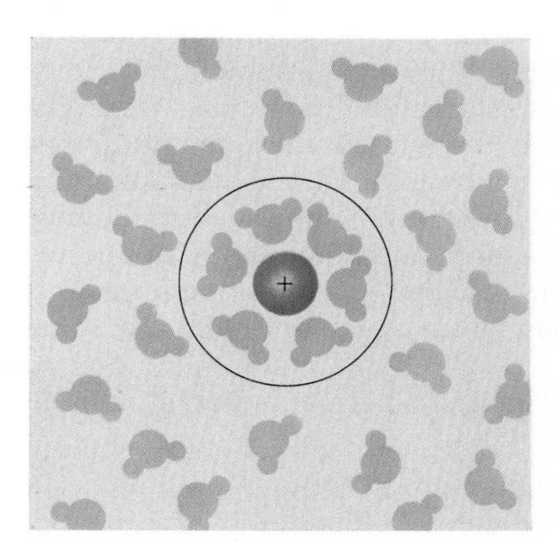

(a)

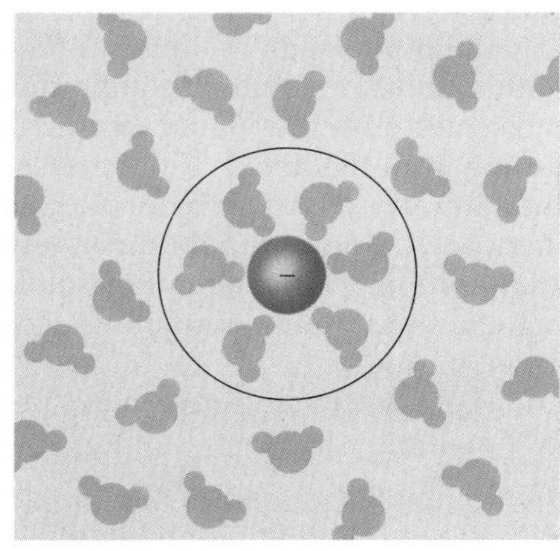

(b)

Figure 15.6

The hydration of (a) a positive ion; (b) a negative ion. When ions are dissolved in water, they attract and hold several water dipoles around them as shown in the circular area in the center of each part of the diagram.

groups of water molecules, they are said to be **hydrated.** Hydration is a special case of **solvation,** the process in which solute particles are surrounded by molecules of any solvent. Solvation occurs in every solution, but it is most pronounced in aqueous solutions of ionic compounds, because of the strong forces between ions and dipolar water molecules.

Quartz, SiO_2, is a covalent network solid (Section 14.4). Its structure is a network of strong covalent bonds that connect all silicon and oxygen atoms to each other, making an entire crystal of SiO_2 one huge molecule. Because these bonds are so strong, it is very difficult to separate the Si and O atoms, and so quartz (and sand derived from quartz) is insoluble in water or any other solvent at room temperature.

EXAMPLE 15.1

Solubility and Interparticle Attractions

Predict the solubility of iron in water and in hexane.

SOLUTION If the iron is to dissolve, iron atoms must be separated and dispersed among the solvent molecules. In metallic iron, the atoms are held tightly together by a sea of valence electrons (metallic bonding, Section 14.3). These strong metallic bonds would have to be broken in order for iron to dissolve. In solution, iron atoms would be nonpolar and there would be only dispersion forces between them and solvent molecules. Since water molecules are polar and hydrogen bond with each other, the intermolecular forces in the pure substances will be much stronger than in a solution, and iron will be insoluble. Although there are only weak attractions among hexane molecules, the strong metallic bonding in iron is the most important factor. The relatively weak dispersion forces that would exist between iron atoms and hexane molecules in a solution could not overcome the metallic bonding, and so iron is predicted not to dissolve in hexane. Both predictions are confirmed by observation; water and gasoline can be stored in iron or steel containers.

Dissolving Gases in Liquids: Henry's Law

The solubility of a gas in a liquid increases as the pressure of the gas increases (Figure 15.7). (In contrast, pressure does not measurably affect the solubilities of solids or liquids in liquids.) When a gas sample is in contact with a liquid, a dynamic equilibrium is established. The rate at which gas molecules enter the liquid phase equals the rate at which gas molecules escape from the liquid. If the pressure is increased, solute molecules strike the surface of the liquid more often, and consequently the rate of dissolution increases. A new equilibrium is established when the rate of escape increases to match the rate of dissolution. A higher rate of escape requires a higher concentration of solute molecules, which corresponds to higher solubility.

The relationship between gas pressure and solubility is known as **Henry's law:**

$$S_g = k_H P_g$$

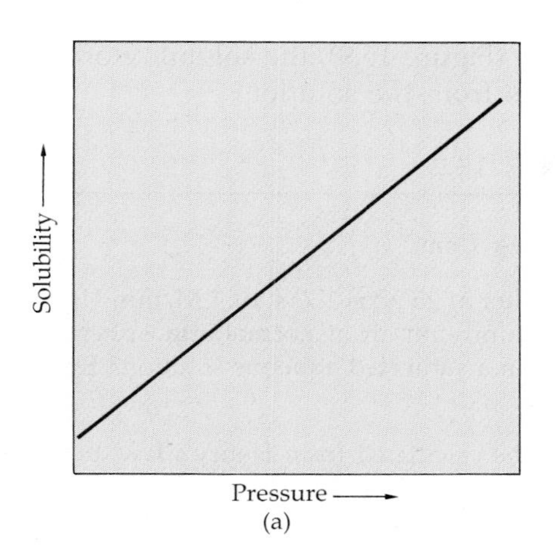

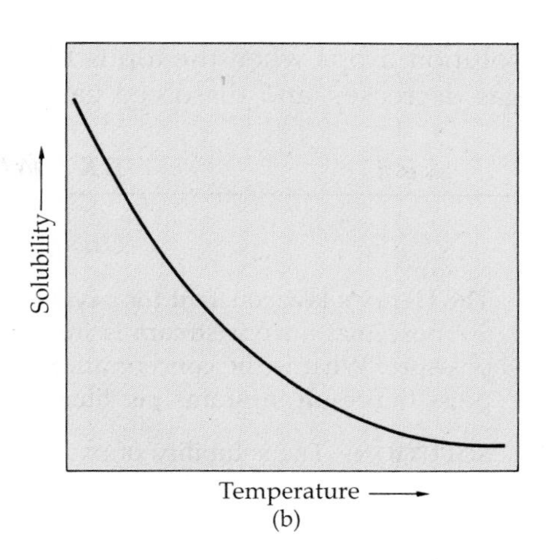

Figure 15.7
Solubility of a gas in a liquid. (a) Pressure dependency. (b) Temperature dependency.

where S_g is the solubility of the gas in the solution, P_g is the pressure of the gas above the solution (or the partial pressure of the gas if the solution is in contact with a mixture of gases), and k_H is a constant, known as the Henry's law constant, whose value depends on the identities of both the solute and the solvent and on the temperature. Figure 15.8 illustrates how gas solubility depends on pressure. The behavior of a carbonated drink when the cap is removed is a common illustration of the solubility of gases in liquids under pressure. The fizz is observed because the partial pressure of CO_2 over the

The partial pressure of a gas in a mixture of gases is the pressure that a pure sample of the gas would exert if it occupied the same volume as the mixture. Partial pressure is proportional to the mole fraction of the gas, Section 12.2.

Henry's law holds quantitatively only for gases that do not interact chemically with the solvent. It does not work perfectly for NH_3, for example, which gives small concentrations of NH_4^+ and OH^- in water, or for CO_2, which reacts with water to form carbonic acid.

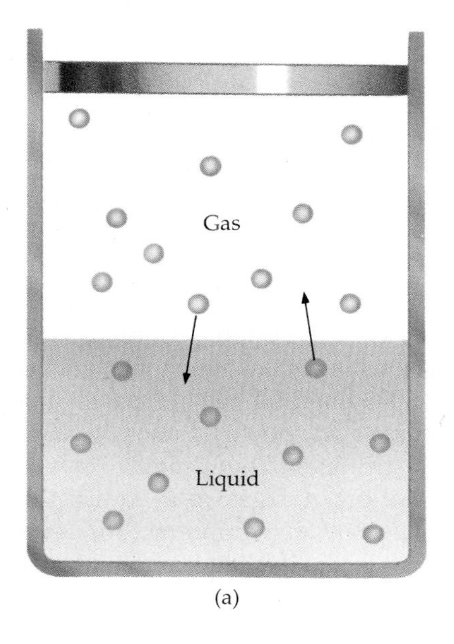

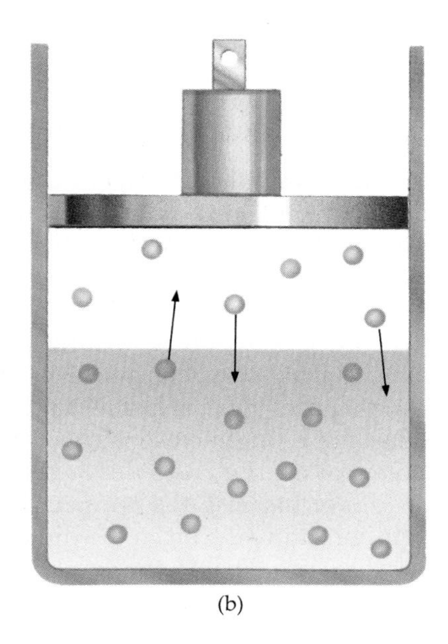

Figure 15.8
Gas solubility in a liquid. At constant temperature, a pressure increase causes the gas molecules to have a smaller volume to occupy. There are more collisions of gas molecules with the liquid surface, and so more molecules dissolve in the liquid.

solution drops when the top is removed (Figure 15.9), the solubility of the gas decreases, and dissolved gas escapes from the solution.

EXAMPLE 15.2

Using Henry's Law

The Henry's law constant for oxygen in water at 25 °C is 1.7×10^{-6} M/mm Hg. Suppose that a trout stream is in equilibrium with air at normal atmospheric pressure. What is the concentration of O_2 in a saturated aqueous solution? Express the result in grams per liter (g/L).

SOLUTION The solubility of oxygen can be calculated from Henry's law, but first we must calculate the partial pressure of oxygen in air. From Section 12.2 you know that air is 21 mole % oxygen, which means that the mole fraction of O_2 is 0.21. If the total pressure is 1.0 atm,

$$\text{pressure of } O_2 = 1.0 \text{ atm} \cdot \frac{760 \text{ mm Hg}}{1 \text{ atm}} \cdot 0.21 = 160 \text{ mm Hg}$$

$$\text{solubility of } O_2 = 1.7 \times 10^{-6} \frac{\text{M}}{\text{mm Hg}} \cdot 160 \text{ mm Hg} = 2.7 \times 10^{-4} \text{ M}$$

This concentration can be converted to the desired units by using molar mass.

$$\text{solubility of } O_2 = \frac{2.7 \times 10^{-4} \text{ mol}}{\text{L}} \cdot \frac{32.0 \text{ g}}{\text{mol}} = 0.0086 \text{ g/L or } 8.6 \text{ mg/L}$$

Using Henry's law, you see that the oxygen concentration is very low, and yet this is sufficient to provide the oxygen required by aquatic life.

Figure 15.9
Illustration of Henry's law. The greater the partial pressure of CO_2 over the soft drink in the can, the greater the amount of CO_2 dissolved. When the partial pressure of CO_2 is lowered by opening the can, CO_2 begins to bubble out of the solution. *(C.D. Winters)*

EXERCISE 15.2 • *Using Henry's Law*

The Henry's law constant for N_2 in water at 25 °C is 8.4×10^{-7} M/mm Hg. What is the solubility of N_2 in g/L if its partial pressure is 1520 mm Hg? What is the solubility when the N_2 partial pressure is 20. mm Hg?

Solubility and Temperature

To understand how temperature affects solubility, we can apply Le Chatelier's principle. Consider the equilibrium between a pure solute gas and a saturated solution.

$$\text{gas} + \text{solvent} \rightleftharpoons \text{saturated solution} + \text{heat}$$

When a gas dissolves to form a saturated liquid solution, the process is almost always exothermic. For similar reasons, condensation of a gas to form a pure liquid is exothermic. Gas molecules that had been relatively far apart are brought close to other molecules that attract them, lowering their potential energy and releasing some energy to the surroundings.

If we increase the temperature of a solution of a gas in a liquid, the equilibrium will shift in the direction that will partially counteract the temperature rise. That is, the equilibrium will shift to the left, since energy must be transferred from the surroundings when gas molecules and pure solvent

Figure 15.10
A warm glass rod is placed in a glass of ginger ale. The heat from the rod warms a cold solution of CO_2 in water and causes the CO_2 to be less soluble. The Henry's law constant for CO_2 in water is 4.45×10^{-5} M/mm Hg at 25 °C, whereas it is 2.5×10^{-5} M/mm Hg at 50 °C. *(C.D. Winters)*

Table **15.2**

Solubility of Oxygen in Water at Various Temperatures*

Temperature (°C)	Solubility of O_2 (g/L)
0	0.0141
10	0.0109
20	0.0092
25	0.0083
30	0.0077
35	0.0070
40	0.0065

*These data are for water in contact with air at 760 mm Hg pressure.

are formed from the solution, and this would reduce the temperature of the surroundings. Thus a dissolved gas becomes less soluble with increasing temperature (Figure 15.10). Conversely, cooling a solution that is at equilibrium with undissolved gas will cause the equilibrium to shift in the direction that liberates heat (to the right), and so more of the gas dissolves. Table 15.2 shows the solubility of oxygen in water at several temperatures.

Cooler water in contact with the atmosphere can contain more dissolved oxygen than the same water at a higher temperature. For this reason fish seek out cooler (usually deeper) waters in the summer; their gills have an easier time obtaining oxygen when the concentration in the water is higher. The decrease in gas solubility as temperature increases makes *thermal pollution* a problem for aquatic life in rivers and streams. Natural heating of water by sunlight and by warmer air can usually be accommodated, but excess heat from such sources as industrial facilities and electrical power plants can reduce the concentration of dissolved oxygen to the point where some species of fish die.

As a glass of cold water slowly warms to room temperature, tiny bubbles can be seen on the inner wall of the glass (Figure 15.11). These bubbles are formed by dissolved air as it comes out of solution. A carbonated drink will lose carbonation as it warms to room temperature at atmospheric pressure. Bubbles of CO_2 form so rapidly that they can be seen rising to the surface. If enough CO_2 is lost from the drink, it tastes "flat."

Figure 15.11
The effect of temperature on gas solubility is illustrated by the formation of air bubbles on the inside walls of a container filled with cool tap water as the water warms. *(C.D. Winters)*

Figure 15.12 shows the temperature dependence of water solubility for several solids. Most are much more soluble at higher temperatures. When a solid is at equilibrium with a saturated solution, the process of dissolving is usually endothermic.

$$\text{solid} + \text{solvent} + \text{heat} \rightleftharpoons \text{saturated solution}$$

When discussing solubility as a function of temperature, the relevant ΔH is for dissolving the last little bit of solute to make a saturated solution. It is not the ΔH that would be calculated for the process solid → solution from data in Appendix J. These are for a solution at a standard-state concentration of solute, which may be quite different from the concentration of a saturated solution.

Le Chatelier's principle predicts that if the temperature is increased, this equilibrium will shift so that heat is absorbed, which implies increased solubility as observed.

One compound in Figure 15.12, lithium sulfate, Li_2SO_4, is more soluble at low temperatures than at higher temperatures. In the case of Li_2SO_4, heat is given off when a small quantity of $Li_2SO_4 \cdot H_2O$ dissolves in a nearly saturated solution. Thus Le Chatelier's principle predicts decreasing solubility as temperature increases. ($Li_2SO_4 \cdot H_2O$, rather than Li_2SO_4, is involved because the anhydrous salt is converted to the hydrate when it comes into contact with water.)

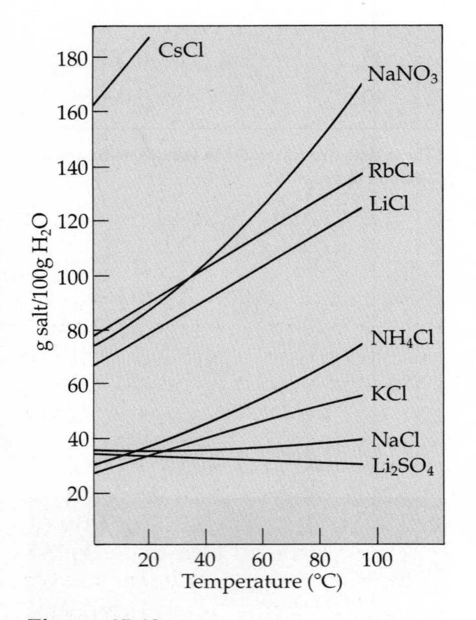

Figure 15.12
The temperature dependence of the solubility of some ionic compounds in water.

EXERCISE 15.3 • *Solubility and Heat of Solution*

The solubility of $NaNO_3$ in water at 5 °C is about 78 g/100 g H_2O. At 60 °C, its solubility is about 120 g/100 g H_2O. Is the process of dissolving $NaNO_3$ in a nearly saturated solution exothermic or endothermic?

15.2 SOLUTION COMPOSITION

The composition of a solution can be expressed as the quantity of solute (or solutes) in a given quantity of solvent, or of solution. A solution that has a relatively large quantity of solute dissolved in a given quantity of solvent is said to be *concentrated,* while a solution with only a relatively small quantity of solute in the same quantity of solvent is said to be *dilute.* These qualitative terms are not sufficient for most applications, however, and you will need to become familiar with several quantitative methods of describing solution composition.

Molarity

As defined in Section 6.5, the **molarity** of a solution (abbreviated *c*) is the amount (moles) of solute dissolved per unit volume (liter) of solution. Molarity is a useful measure of solution concentration because it is easy to measure volumes of solutions and to transport and deliver known volumes of solution with pipets or burets.

$$\text{molarity of compound A} = c_A = \frac{\text{amount (moles) of A}}{\text{volume (liters) of solution}}$$

A capital M signifies the units mol/L; an italic c signifies the quantity molarity.

Consider the solution shown on the right of Figure 15.13. It is a 0.100 M aqueous solution of potassium chromate that was made by dissolving 0.100 mol (19.4 g) of K_2CrO_4 in water, and adding enough water to make 1.000 L of solution. The quantity of water (the solvent) was not measured, only the final (total) volume of the solution.

Molality

Another quantity, molality, is useful if we are interested in how much solute is present for a given quantity of *solvent*. **Molality** (abbreviated *m*) is defined as the amount (moles) of solute per kilogram of solvent.

$$\text{molality of compound A} = m_A = \frac{\text{amount (moles) of A}}{\text{mass (kg) of solvent}}$$

The solution on the left in Figure 15.13 has a molality of

$$\text{molality of } K_2CrO_4 = m_{K_2CrO_4} = \frac{19.4 \cdot 1 \text{ mol}/194.1 \text{ g}}{1.00 \text{ kg water}} = 0.100 \text{ m}$$

Notice the different quantities of water used to make the 0.100 M and the 0.100 m solutions in Figure 15.13. You can see that the molarity and molality of a given solution are not the same (although the difference becomes negligibly small for dilute solutions, say less than 0.01 M). It is interesting to note that the molarity of a solution depends on the temperature because the solution's volume varies with temperature. Strictly speaking, you should report the temperature when giving a solution's molarity. On the other hand, because the amount of solute and the mass of solvent do not change with temperature, the molality of a solution is independent of temperature.

Figure 15.13
Molarity and molality. The photo shows a 0.10-molal solution (0.10 m) of potassium chromate (*flask at left*) and a 0.10-molar solution (0.10 M) solution of potassium chromate (*flask at right*). Each solution contains 0.10 mol (19.4 g) of yellow K_2CrO_4 shown in the dish at the front. The 0.10-molar (0.10 M) solution on the right was made by placing the solid in the flask and adding enough water to make 1.0 L of solution. The 0.10-molal (0.10 m) solution on the left was made by placing the solid in the flask and adding 1000 g (1 kg) of water. Adding 1 kg of water leads to a solution clearly having a volume greater than 1 L. *(C.D. Winters)*

Mass Fraction

The **mass fraction** of a solute in a solution is the fraction of the total mass that is contributed by that solute. That is, we add up the total mass of all solutes plus solvent and divide it into the mass of a single solute. Mass fraction is commonly expressed as a percentage and called **weight percent,** which is the mass fraction multiplied by 100%. This is the same as the number of grams of solute per 100 grams of solution. For example, the mass fraction of A in a solution containing solutes A and B dissolved in solvent C is

$$\text{mass fraction of A} = \frac{\text{mass of solute A}}{\text{mass of solute A} + \text{mass of solute B} + \text{mass of solvent C}}$$

$$\text{weight percent of A} = \text{mass fraction of A} \times 100\%$$

$$= \frac{\text{number of grams of solute A}}{100 \text{ g solution}}$$

EXAMPLE 15.3

Mass Fraction/Weight Percent

What is the weight percent of NaCl in a sterile saline (salt) solution made by dissolving 4.6 g of NaCl in 500. g of pure water, boiled to kill any bacteria? This is a typical intravenous saline solution used in medicine (see page 223).

SOLUTION Using the definitions of mass fraction and weight percent we have

$$\text{mass fraction NaCl} = \frac{4.6\text{ g NaCl}}{4.6\text{ g NaCl} + 500.\text{ g H}_2\text{O}} = 0.0091$$

$$\text{weight percent NaCl} = \text{mass fraction NaCl} \times 100\% = 0.0091 \times 100\% = 0.91\%$$

Mass fractions of very dilute solutions are often expressed in **parts per million** (abbreviated **ppm**). To express the mass fraction as parts per million, multiply it by 1,000,000 ppm. This makes a very small number bigger and easier to handle. One part per million is equivalent to one gram of solute per million grams of solution or one milligram of solute per thousand grams of solution (1 mg/kg).

For even smaller mass fractions, **parts per billion (ppb)** and **parts per trillion (ppt)** are often used. As the names imply, we can express a mass fraction in parts per billion by multiplying by 1,000,000,000 ppb (10^9 ppb), and in parts per trillion by multiplying by 1,000,000,000,000 ppt (10^{12} ppt).

Often, particularly in discussing water pollutants or trace elements in the diet, reference is made to certain elements by name; "lead," "iron," "zinc," "chromium," etc. Usually the metal ions rather than the metals themselves are meant.

Parts per million, billion, or trillion are often used to report the concentration of **water pollutants** of various kinds. A pollutant is any unwanted or harmful matter in air, water, or soil. For example, the U.S. Environmental Protection Agency (EPA) has set limits for pollutants in drinking water. The limit for lead, as Pb^{2+} ion, is 0.05 ppm, or 50 ppb. That means for every million grams of water, about 1000 liters, the allowed quantity of dissolved lead is 0.05 g or less.

EXERCISE 15.4 • *Mass Fraction*

A 1.0-kg sample of water from Lake Michigan is found to contain 8.7 ng (8.7×10^{-9} g) of mercury. What is the mass fraction of mercury in the water in ppb? In ppt?

Use of Concentration: Lead in the Environment

Lead is a widely encountered metal, and it is a poison. Dealing with the problem of unwanted lead in our environment requires knowledge of the variety of units used to express the concentrations of dilute solutions. Consider a sample of water that is found to contain 0.010 ppm (10. ppb) lead. What is the mass of lead per liter of this solution? Since the solution is almost entirely water, its density will be the same as that of water, namely 1.0 g/mL, and so 1 L of solution has a mass of 1.0×10^3 g.

Recall that 1 μg (microgram) = 10^{-6} g; 1 mg (milligram) = 10^{-3} g = 1000 μg

$$\text{mass of Pb} = \frac{0.010\text{ g Pb}}{1 \times 10^6\text{ g solution}} \cdot \frac{1.0 \times 10^3\text{ g solution}}{\text{L solution}} = 1.0 \times 10^{-5}\text{ g Pb/L solution}$$

This concentration can also be expressed in micrograms per liter as

$$\text{mass of Pb} = \frac{1.0 \times 10^{-5}\text{ g}}{\text{L}} \cdot \frac{1\ \mu\text{g}}{1 \times 10^{-6}\text{ g}} = 10.\ \mu\text{g/L}$$

That is, a mass fraction of 10. ppb corresponds to 10. μg per liter of solution.

Lead concentrations in the environment are in the range of parts per billion. For liquid solutions they are usually reported in micrograms per liter (which corresponds to ppb as illustrated above); for solids 1 μg/kg corresponds to 1 ppb. Lead ions are present in some foods (up to 100 to 300 μg/kg), beverages (up to 20 to 30 μg/L), public water supplies (up to 100 μg/L, from lead-sealed pipes), and even air (up to 2.5 $\mu g/m^3$ from lead compounds in auto exhausts). With this many sources and contacts per day (Figure 15.14), it is obvious that the body must be able to rid itself of lead—otherwise everyone would have died long ago of lead poisoning! The aver-

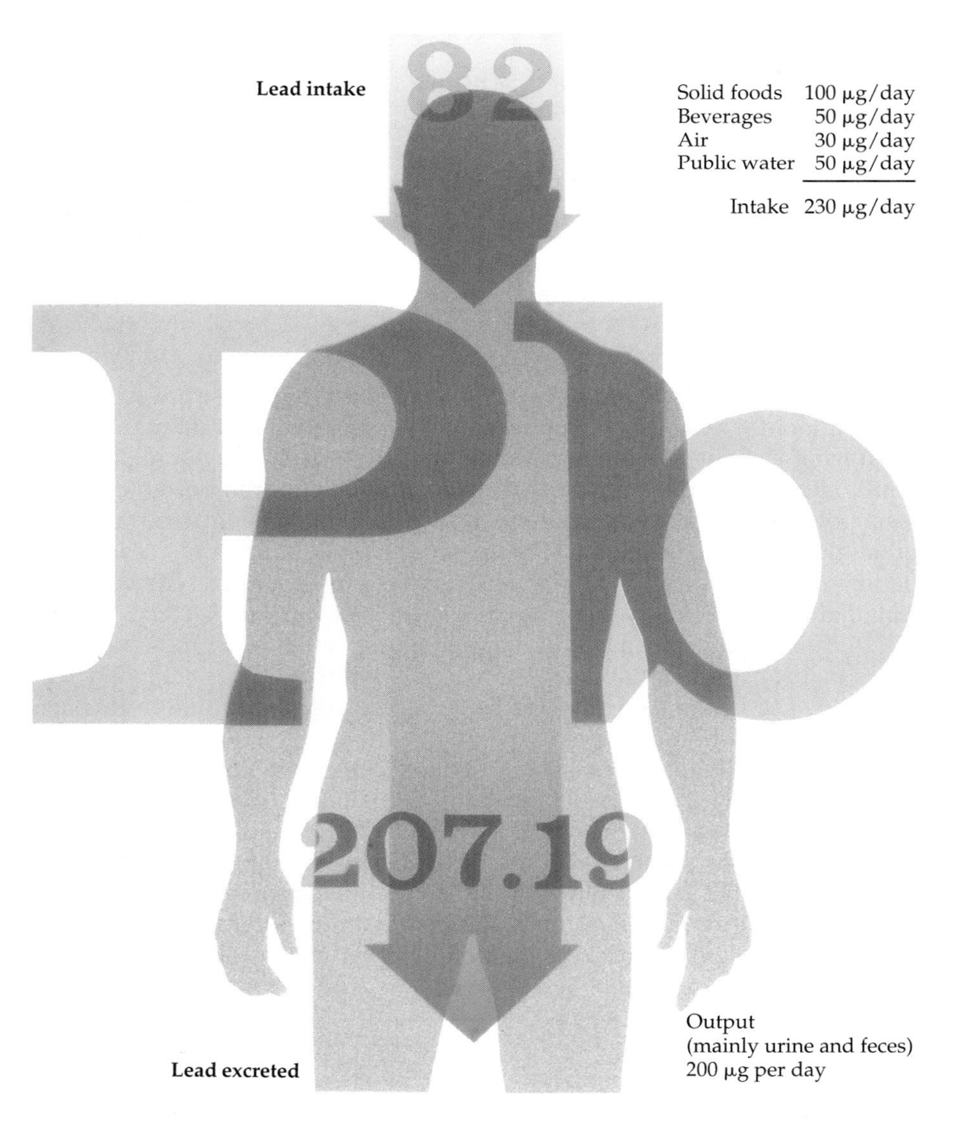

Figure 15.14
Lead equilibrium in humans. Figures chosen for intake are probable upper limits.

age person can excrete about 230 μg of lead a day through the kidneys and intestinal tract. The daily intake is normally less than this; if it is not, accumulation and storage result. In the body lead can accumulate in bone cells, where it acts on the bone marrow; in tissues it behaves like other heavy-metal poisons, such as mercury and arsenic, by reacting with *sulfhydryl groups* (—SH groups) found in proteins containing the amino acid cysteine. Enzymes, the catalysts for reactions in the body, are proteins; if an enzyme containing a —SH group becomes bound to a metal ion such as Pb^{2+}, the enzyme will likely cease to function. Lead, like mercury and arsenic, can also affect the central nervous system, binding to various active sites in nerves.

At the height of the Roman empire, lead production worldwide was about 80,000 tons per year. Today it is about 3 million tons per year.

Unless they are very insoluble, lead salts are always toxic. Metallic lead can even be absorbed through the skin because it reacts with weak acids in perspiration and dissolves. Cases of lead poisoning have resulted from repeated handling of lead foil, bullets, and other lead objects.

EXERCISE 15.5 • *Lead in Drinking Water*

Assume that you eat, on average, about 2 kg of food, drink about 3 L of water and other beverages, and breathe in about 11,500 L of air per day. Use the figures for Pb content given in the text to verify that the quantity of lead consumed in a day would be roughly the same as that shown in Figure 15.14. Compare this with the figure of 230 μg/day for elimination of Pb from the body. (Remember that the lead-content figures are upper limits—the quantity of lead in the foods, beverages, and air you consume is probably lower than this.)

The exact number of children poisoned by lead is shadowed by controversy. Whatever the true number, it is nevertheless very high.

One of the truly tragic aspects of lead poisoning is that even though lead-pigmented paints have not been used for interior painting in this country during the past 30 years, children are still poisoned by lead from old paint. Health experts estimate that up to 225,000 children become ill from lead poisoning each year, with many experiencing mental retardation or other neurological problems. The reason for this is twofold. Lead-based paints still cover the walls of many older dwellings. Coupled with this is the fact that many children in poverty-stricken areas are ill fed and anemic. These children develop a peculiar appetite trait called **pica,** and among the items that satisfy their cravings are pieces of flaking paint, which may contain lead. Lead salts have a sweet taste, which may contribute to this consumption of lead-based paint.

Lead compounds, such as yellow lead chromate, $PbCrO_4$, and white lead oxide, PbO_2, have long been used in paints as pigments. If a child consumes as little as 200 mg (200,000 μg) of an older, lead-based paint, he or she may ingest as much as 2600 μg of lead, of which up to 550 μg may be absorbed into the blood. Since 1977, lead in all types of paints has been limited to 600 ppm. Let's calculate the maximum mass of lead that can, by law, be in a gallon of paint. If a gallon of paint weighs about 8 lb,

$$\text{mass of lead} = 8 \text{ lb paint} \cdot \frac{600 \text{ parts Pb}}{10^6 \text{ parts paint}} = 0.005 \text{ lb Pb}$$

That is, there should be no more than about 0.005 lb of lead in a gallon of paint. What is this mass in μg? Since there are 454 grams in 1 lb,

$$\text{mass of Pb} = 0.005\text{ lb} \cdot \frac{454\text{ g}}{\text{lb}} \cdot \frac{10^6\ \mu\text{g}}{\text{g}} = 2{,}000{,}000\ \mu\text{g}$$

Now if this paint is spread uniformly over 400 ft^2 or 3.72×10^5 cm^2, a paint chip 1 cm square will contain less than 6 μg of lead, about the same mass one might find in a loaf of bread or a liter of soft drink. However, it should be noted that some older paints contained up to 50,000 ppm lead (or about 500 μg of lead/cm^2 of paint). Eating a couple of chips would exceed the body's capacity to get rid of the lead.

Lead is a special problem for children because they retain a larger fraction of absorbed lead than do adults, and children do not immediately tie up absorbed lead in their bones as adults do. This inability to absorb lead quickly into their bones means the lead stays in a child's blood longer, where it can exert its toxic effects on various organs. Table 15.3 shows the effects of lead in children's blood.

Studies in the mid-1980s by the U.S. Department of Health showed that about 1.5 million children in the United States have blood levels of lead above 15 μg/dL, and about 900,000 children have blood levels above 20 μg/dL. In October 1991 the Centers for Disease Control decreased the intervention level for lead in childrens' blood to 10 μg/dL from 25 μg/dL. For adults in a workplace where lead exposure would be expected, the corresponding acceptable blood-lead level is 40 μg/dL. In May 1993 the U.S. En-

EDTA, a chelating agent discussed in Section 10.10, can be used to treat lead poisoning.

Table 15.3

Effects of Lead in Children's Blood

Pb Levels		
μg/dL*	***ppb***	**Effect**
~5	~50	Elevated blood pressure
~10	~100	Lowered intelligence
15–25	150–250	Decreased heme synthesis, decreased vitamin-D and calcium metabolism
25–40	250–400	Impaired central nervous system functions, delayed cognitive development, reduced IQ scores, impaired hearing, reduced hemoglobin formation
40–80	400–800	Peripheral neuropathy, anemia
>80	>800	Coma, convulsions, irreversible mental retardation, possible death

* μg/dL = micrograms per deciliter of blood.
Source: Agency for Toxic Substance and Disease Registry, U.S. Department of Health and Human Services, Atlanta, 1988.

dL means deciliter (1/10 L, or 100 mL). It is used here because it is a common unit of measure for health-related studies. 1 mg/dL = 10 mg/L = 10 ppb.

vironmental Protection Agency released a list of public water supplies that exceeded its acceptable level of 15 ppb lead. Hundreds of cities and towns were on the list, and lead concentrations ranged up to 484 ppb. The EPA estimates that one in six American children under six years of age has a blood-lead level above the 10 μg/dL acceptable level.

EXERCISE 15.6 • *Using ppm*

If the concentration of Pb^{2+} ions in tap water is found to be 0.025 ppm, what volume of this water (in liters) will contain 100.0 μg of Pb?

15.3 COLLIGATIVE PROPERTIES

In liquid solutions, particles are close together and the solute molecules or ions disrupt intermolecular forces between the solvent molecules, causing changes in those properties of the solvent that depend on intermolecular attractions. For example, the freezing point of a solution is lower than that of the pure solvent and the boiling point is higher. **Colligative properties** of solutions are those that *depend on the concentration of solute particles* in the solution, regardless of what kinds of particles are present. The greater the concentration of *any* solute, the lower the freezing point and the higher the boiling point of a solution.

Boiling-Point Elevation

The discussion in this paragraph assumes that both systems, the solution and the pure solvent, are at the same temperature.

Compare a small portion of the liquid/gas boundary for pure water with that for seawater (an aqueous solution of mainly sodium chloride) as shown at the molecular scale in Figure 15.15. Recall from Section 12.6 that in a closed container there is a dynamic equilibrium between a pure liquid and its vapor—the rate at which molecules escape the liquid phase equals the rate at which vapor-phase molecules return to the liquid. In the solution there are sodium ions and chloride ions as well as water molecules at the liquid surface. The rate at which water molecules can escape to the gas phase is therefore less than for pure solvent. Because sodium chloride is a nonvolatile solute, there are essentially no sodium ions or chloride ions in the vapor phase. Thus the rate at which vapor-phase water molecules return to the solution is not affected. Because the rate of escape is slower than the rate of return, some of the water vapor will condense into the solution until a new equilibrium is achieved. At this new equilibrium, the pressure of water vapor in equilibrium with the solution will be less than the vapor pressure of pure water. Therefore, at 100 °C the vapor pressure of the solution will be less than 1 atm, and the aqueous solution will have to be heated above 100 °C for it to boil.

The difference between the normal boiling point of water and the higher boiling point of an aqueous solution of a nonvolatile solute is known as the **boiling point elevation,** ΔT_b. It is proportional to the concentration of the

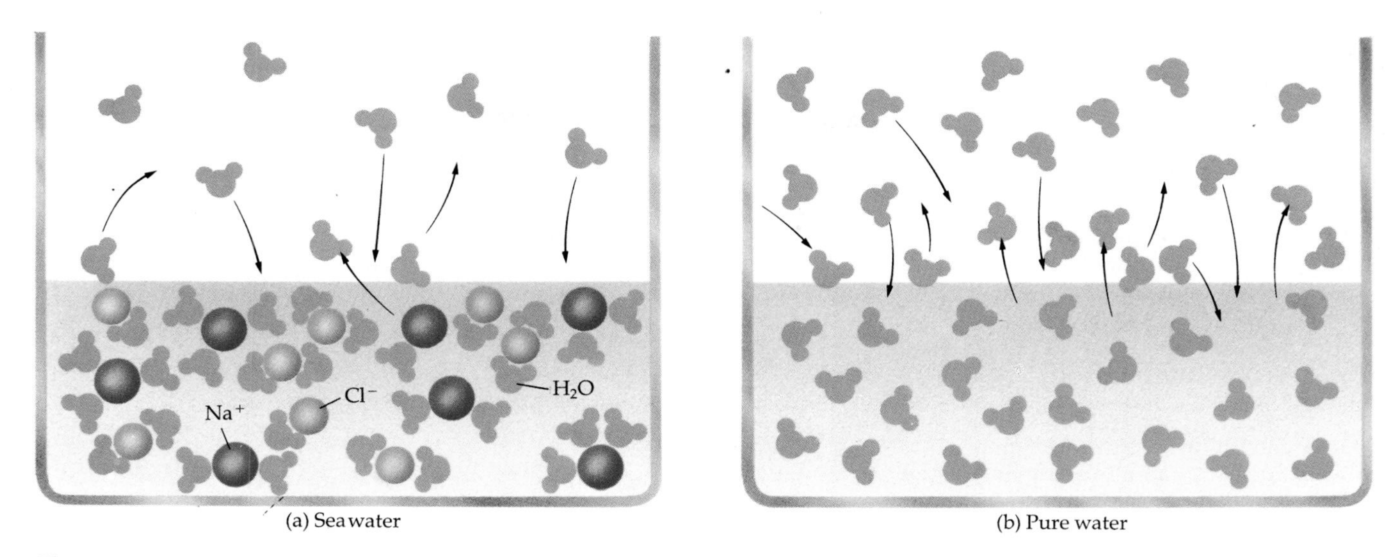

Figure 15.15
Seawater is an aqueous solution of sodium chloride and many other salts. The vapor pressure of water over an aqueous solution is not as large as the vapor pressure of water over pure water at the same temperature.

solute, which is usually expressed as molality, m_{solute}, and to the number of particles per formula unit of solute, i_{solute}.

$$\Delta T_b = K_b \, m_{\text{solute}} \, i_{\text{solute}}$$

(For sodium chloride $i_{\text{solute}} = 2$, because there is one Na^+ ion and one Cl^- ion in solution per formula unit of NaCl.) The proportionality constant, K_b, depends only on the identity of the solvent and not the kind of solute. This is because it makes almost no difference whether the particles among water molecules at the surface of a liquid are Na^+ and Cl^-, as in Figure 15.15, or some other kind of molecules or ions. In any case the concentration of water molecules is less. For example, K_b for water is 0.512 °C/m. If $CaCl_2$ is dissolved in water at a concentration of 0.20 m, the boiling point of the solution can be obtained by first calculating the boiling point elevation, and then adding that value to the boiling point for water. Since there are two Cl^- ions and one Ca^{2+} ion per formula unit of $CaCl_2$, i_{solute} is 3, and so

$$\Delta T_b = (0.512\ ^\circ\text{C/m}) \cdot 0.20\ \text{m} \cdot 3 = +0.31\ ^\circ\text{C}$$

$$\text{boiling point of solution} = 100.00\ ^\circ\text{C} + 0.31\ ^\circ\text{C} = 100.31\ ^\circ\text{C}$$

Boiling point elevation is a colligative property because only the number (moles) of solute particles is important, not their identity.

The units for the molal boiling point elevation constant are °C/m, which is equivalent to (°C · kg solvent)/(mol solute).

EXERCISE 15.7 • *Boiling Point Elevation*

The boiling point elevation constant for benzene is 2.53 °C/m. If a solute's concentration in benzene is 0.10 m, what will be the boiling point of the solution? The boiling point of pure benzene is 80.10 °C. Assume that there is a single solute particle per formula unit.

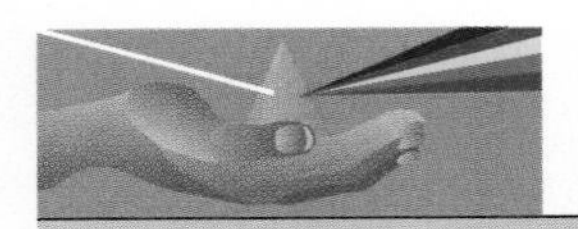

CHEMISTRY YOU CAN DO

Picking Up an Ice Cube

Is it possible to use a string to pick up an ice cube from a plate without touching the ice cube with your hands or tying a knot in the string? Here's one way to do it. Obtain a six-inch piece of string or thread, some table salt (preferably in a shaker), a shallow plate, saucer, or plastic cup, a little water, and an ice cube. Put the ice cube on the plate. Moisten a small area of the string, and lay the wet part over the top of the ice cube. Sprinkle salt on the ice, especially where the string is. After a minute or so, gently raise the ends of the string and see if you can lift the ice cube. If you cannot, add some more salt and try again a minute or so later.

Once the string has become frozen to the ice cube and you can lift it, think about why this has happened. Suggest an explanation in terms of the heat of fusion of water and the freezing-point lowering of a salt solution.

Freezing-Point Lowering

A liquid begins to freeze when the temperature is lowered to the substance's freezing point and the first few molecules cluster together into a crystal lattice to form a tiny quantity of solid. As long as both solid and liquid phases are present at the freezing point, the rate of crystallization equals the rate of melting and there is a dynamic equilibrium. When a *solution* freezes, a few molecules of solvent cluster together to form pure solid *solvent* (Figure 15.16), and a dynamic equilibrium is set up between solution and solid solvent.

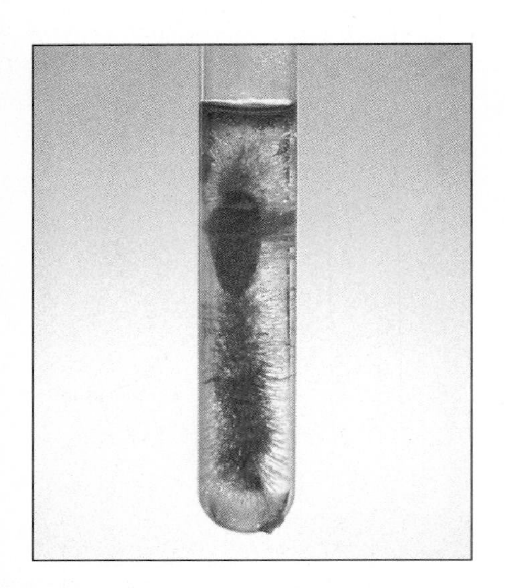

Figure 15.16

A purple dye was dissolved in water, and the solution was frozen slowly. When a solution freezes, the solvent solidifies as the pure substance. Thus, pure ice forms along the walls of the tube, and the dye stays in solution. As more and more solvent is frozen out, the solution becomes more concentrated and the resulting solution has a lower and lower freezing point. When equilibrium is reached, we see pure, colorless ice along the walls of the tube with concentrated solution in the center of the tube.
(C.D. Winters)

In the case of a solution, the molecules in the liquid in contact with the solid solvent are not all solvent molecules. The rate at which molecules move from solution to solid is therefore smaller than in the pure liquid. To achieve dynamic equilibrium there must be a corresponding smaller rate of escape of molecules from the solid crystal lattice. This slower rate occurs at a lower temperature, and so the freezing point of the solution is lower than that of the pure liquid solvent.

The change in the freezing point, ΔT_f, is proportional to the concentration of the solute in the same way as the boiling point elevation.

$$\Delta T_f = K_f \; m_{\text{solute}} \; i_{\text{solute}}$$

Here also, the proportionality constant, K_f, depends on the solvent and not the kind of solute, and i_{solute} represents the number of particles per formula unit of solute. For water, the freezing point constant is −1.86 °C/m. Suppose an aqueous solution is 0.50 m in antifreeze, ethylene glycol, which is a molecular substance. Since there is one molecule per formula unit, the **freezing point** lowering is

$$\Delta T_f = (-1.86\ ^\circ\text{C/m})(0.50\ \text{m})(1) = -0.93\ ^\circ\text{C}$$

The freezing point of this solution would be

$$\text{freezing point of solution} = 0.00\ ^\circ\text{C} - 0.93\ ^\circ\text{C} = -0.93\ ^\circ\text{C}.$$

The use of ethylene glycol ($HOCH_2CH_2OH$), a relatively nonvolatile alcohol, in automobile cooling systems (Figure 15.17) is a practical applica-

Table 15.4

Salts That Could Be Used for Road De-icing

Name	Formula	No. Ions	ΔH_{sol} (kJ/mol)	Environmental Effects
sodium chloride	$NaCl$	2	3.9	Cl^- harmful to plants
calcium chloride	$CaCl_2$	3	90.8	Cl^- harmful to plants
sodium acetate	$NaC_2H_3O_2 \cdot 3H_2O$	2	19.7	acetate ion is biodegradable
sodium nitrate	$NaNO_3$	2	20.5	causes algal blooms in ponds and slow-moving streams*

*Algae growing rapidly ("blooming") remove oxygen from water and thereby harm aquatic life.

Figure 15.17
Adding an ethylene glycol–based antifreeze to an automobile's cooling system. *(C.D. Winters)*

tion of boiling-point elevation and freezing-point lowering. In the summer, when the air temperature is high, ethylene glycol raises the boiling temperature of the coolant to a level that prevents "boil over." Ethylene glycol also lowers the freezing point of the coolant and protects the solution from freezing in the winter.

EXERCISE 15.8 • *Freezing-Point Lowering*

Suppose that you are closing a cabin in the north woods for the winter and you do not want the pipes to freeze. You know that the temperature might get as low as −30. °C, and you want to protect about 4.0 L of water in a toilet tank from freezing. What volume of ethylene glycol, density = 1.113 g/mL, should be added to the 4.0 L of water?

Ethylene glycol is toxic, so do not allow it to contact drinking-water supplies.

Another practical application of freezing point lowering can be seen in areas where winters produce lots of frozen precipitation. To remove snow and particularly ice, roads and walkways are often salted (Figure 15.18). Although sodium chloride is usually used, calcium chloride is particularly good for this purpose. $CaCl_2$ has three ions per formula unit, and it dissolves exothermically; not only is the freezing point of water lowered, but the heat of solution helps melt ice. Table 15.4 lists some salts that are used or could be used for de-icing. Factors to consider in choosing a salt include cost, availability, number of ions per formula unit, and environmental effects.

Figure 15.18
Salt truck applying road salt. *(C.D. Winters)*

Osmotic Pressure

A *membrane* is a thin sheet of material that will allow molecules or ions to pass through it. A **semipermeable membrane** allows only certain kinds of molecules or ions to pass through while excluding others (Figure 15.19). Examples of semipermeable membranes are animal bladders, cell walls in

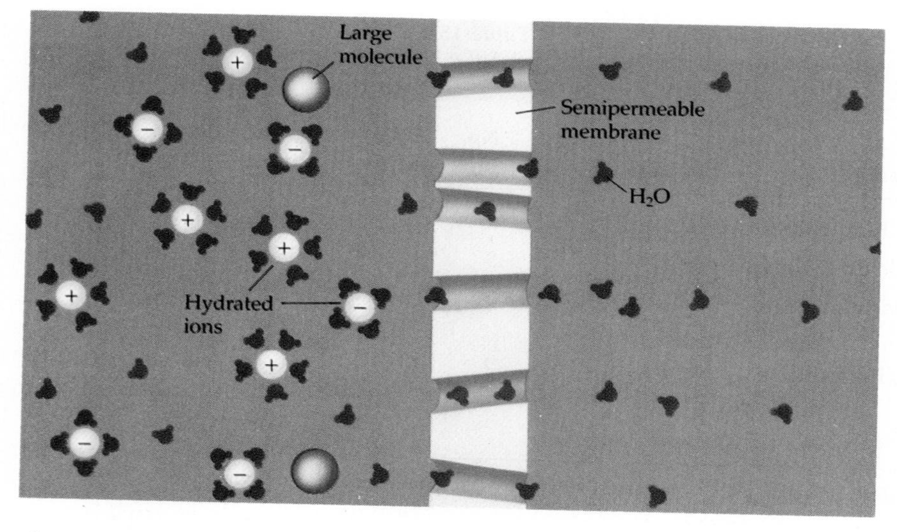

Figure 15.19
Osmotic flow of a solvent through a semipermeable membrane to a solution. The semipermeable membrane is shown acting as a sieve. Many membranes operate in a different way, but the ultimate effect is the same.

plants and animals, and cellophane (which is a polymer derived from cellulose). When two solutions containing the same solvent are separated by a membrane that is permeable only to solvent molecules, osmosis will occur. **Osmosis** is *movement of a solvent through a semipermeable membrane from a region of lower solute concentration to a region of higher solute concentration*. The **osmotic pressure** of a solution is *the pressure that must be applied to the solution to stop osmosis from a sample of pure solvent*.

Consider the example shown in Figure 15.20. A dilute aqueous solution of glucose has been placed in a cellophane bag attached to a glass tube. When the bag is submerged in pure water, some water flows into it and raises the liquid level in the tube. The flow of water stops when the pressure at the bottom of the tube is equal to the osmotic pressure. When the bag is first submerged, there are more collisions of solvent molecules per unit area of the membrane on the pure solvent side than there are on the solution side, and hence solvent passes through the membrane into the solution. As water rises in the tube and pressure builds up, the number of collisions of water molecules on the solution side increases, and when the osmotic pressure is reached, a dynamic equilibrium is achieved where the rate of passage of water molecules is the same in both directions.

Like boiling-point elevation and freezing-point lowering, osmotic pressure results from the unequal rates at which solvent molecules pass through an interface or boundary. In the case of boiling it is the solution/vapor interface; for freezing it is the solution/solid interface, and for osmosis it is the semipermeable membrane. All of these colligative properties can be understood in terms of differences in entropy between a pure solvent and a solution. This is perhaps most easily seen in the case of osmosis. When solvent and solute molecules mix, there is usually an increase in entropy. If pure solvent is added to a solution, a higher entropy state will be achieved when the solvent and solute molecules have diffused among one another to form a more dilute solution. Unless there are strong intermolecular forces, there will be a negligible enthalpy change, and so the increase in entropy makes mixing of solvent and solution a product-favored process. A semipermeable membrane prevents solute molecules from passing into pure solvent. The

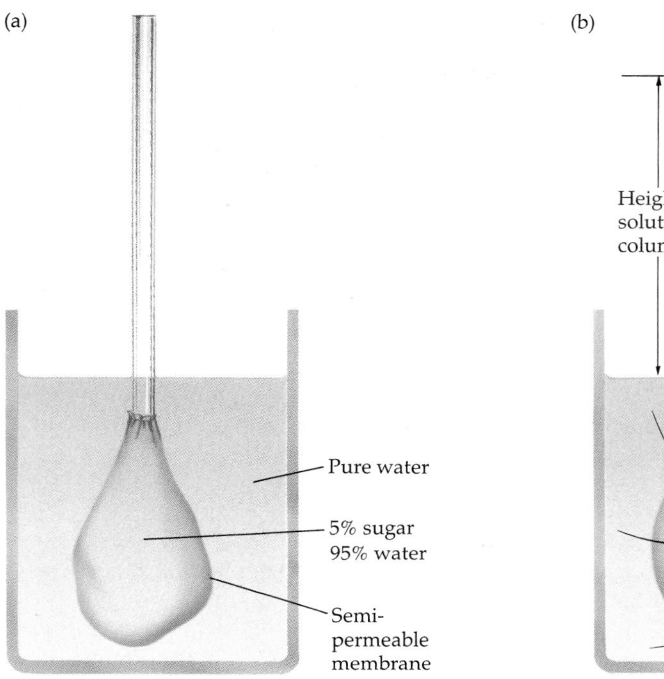

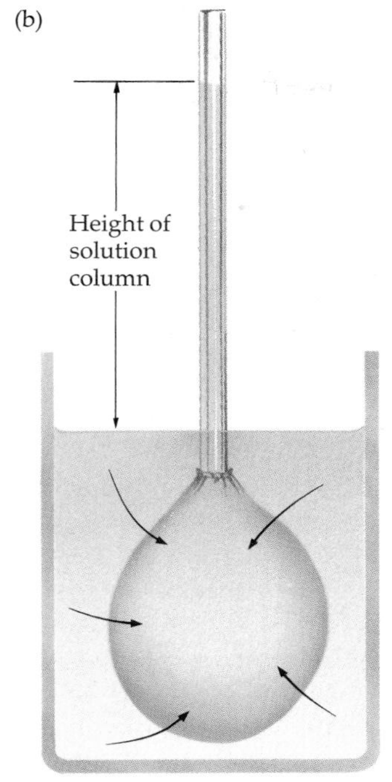

Figure 15.20

Osmosis. (a) The bag attached to the glass tube contains a solution that is 5% sugar and 95% water, and there is pure water in the beaker. The material from which the bag is made is selectively permeable (semipermeable) in that it will allow water molecules to pass through but not sugar molecules. (b) With time, water flows through the walls of the bag from a region of low solute concentration (pure water) to one of higher solute concentration (the sugar solution). Flow will continue until the pressure exerted by the column of solution in the tube above the water level in the beaker is large enough to result in equal rates of passage of water molecules in both directions. The height of the column of solution (b) is a measure of the osmotic pressure, Π.

only way mixing can occur (and entropy can increase) is for solvent to flow into the solution, and so it does.

The more concentrated the solution, the more product-favored the mixing and the greater the pressure required to prevent it. Osmotic pressure is proportional to the molarity of the solution, c:

The osmotic pressure equation also depends on the number of particles of solute and can be written as $\Pi = cRTi$.

$$\Pi = cRT$$

where R is the gas constant and T is the absolute temperature.

Osmotic pressure can be quite large, even though the solution concentration is small; for example, the osmotic pressure of a 0.020 M solution at 25 °C is

$$\Pi = cRT = \frac{0.020 \text{ mol}}{\text{L}} \cdot \frac{0.0821 \text{ L} \cdot \text{atm}}{\text{mol} \cdot \text{K}} \cdot 298 \text{ K}$$

$$= 0.49 \text{ atm}$$

The osmotic pressure equation $\Pi = cRT$ *is similar to the ideal gas equation,* $PV = nRT$, *which can be rearranged to* $P = (n/V)RT = cRT$, *where* n/V *is the molar concentration of the gas.*

This pressure would support a water column more than 15 feet high. Since one way to determine osmotic pressure involves measuring the height of a column of solution in a tube (Figure 15.20), and heights of a few centimeters can be measured accurately, quite small concentrations can be determined by osmotic pressure experiments. If the mass of solute dissolved in a measured volume of solution is known, it is possible to calculate the molar mass of the solute by using the definition of molar concentration, $c = n/V =$ amount (mol)/volume (L). Osmotic pressure is especially useful in studying large molecules whose molar mass is difficult to determine by other means.

EXAMPLE 15.4

Molar Mass from Osmotic Pressure

The osmotic pressure at 25 °C is 1.79 atm for a sugar solution that contains 2.50 g sucrose, empirical formula $C_{12}H_{22}O_{11}$, dissolved in enough water to give a solution volume of 100 mL. What is the molecular formula for sucrose?

SOLUTION We can determine the molecular formula from the empirical formula if we know the molar mass. To find the molar mass, calculate how many moles and how many grams are in the same volume of solution, say 1 L.

First, rearrange the osmotic pressure equation to calculate molar concentration. This tells how many *moles* of sucrose there are per liter of solution.

$$c = \frac{\Pi}{RT} = \frac{1.79 \text{ atm}}{(0.0821 \text{ L atm/mol K})(298 \text{ K})} = 7.32 \times 10^{-2} \text{ mol/L}$$

This concentration was achieved by dissolving 2.50 g sucrose in 100. mL of water, and so the *mass* of sucrose per liter is

$$\text{mass of sucrose per liter} = \frac{2.50 \text{ g}}{100. \text{ mL}} \cdot \frac{1000 \text{ mL}}{\text{L}} = 25.0 \text{ g/L}$$

We now know both the mass and the amount of sucrose per liter of solution, so we can divide to get molar mass:

$$\text{molar mass of sucrose} = M_{\text{sucrose}} = \frac{25.0 \text{ g/L}}{7.32 \times 10^{-2} \text{ mol/L}} = 342 \text{ g/mol}$$

The molar mass that corresponds to the empirical formula, $C_{12}H_{22}O_{11}$, is 342 g/mol; therefore the molecular formula must be the same as the empirical formula.

Freezing point–lowering measurements can be used to find molar mass in the same manner as shown in Example 15.4 for osmotic pressure measurements.

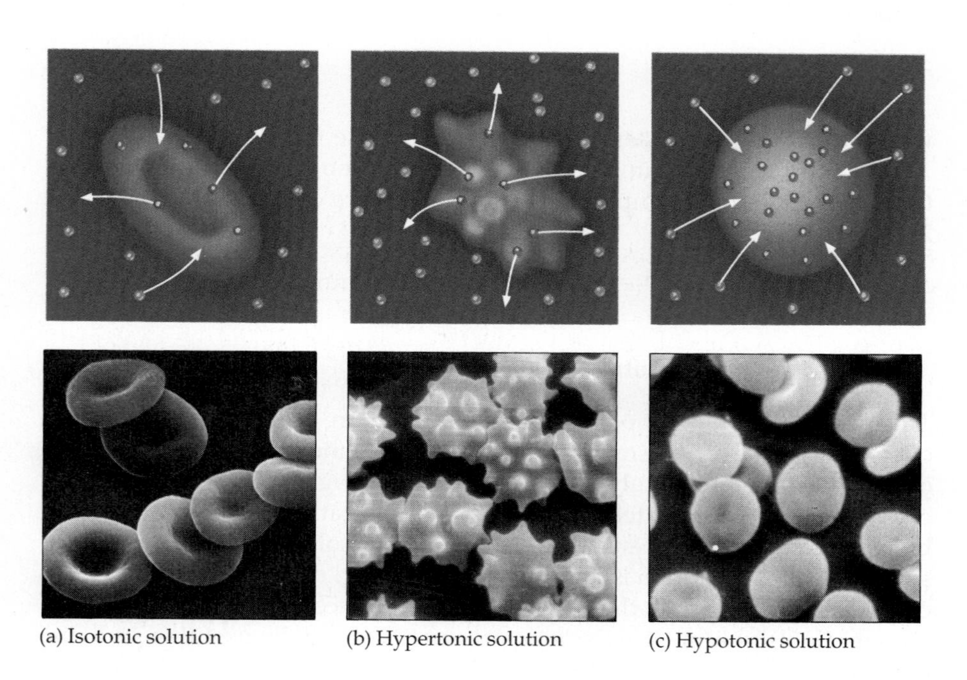

Figure 15.21
Osmosis and the living cell. (a) A cell placed in an *iso*tonic solution. The net movement of water in and out of the cell is zero because the concentration of solutes inside and outside the cell is the same. (b) In a *hyper*tonic solution, the concentration of solutes outside the cell is greater than inside. There is a net flow of water out of the cell, causing the cell to dehydrate, shrink, and perhaps die. (c) In a *hypo*tonic solution, the concentration of solutes outside the cell is less than inside. There is a net flow of water into the cell, causing the cell to swell and perhaps to burst.

Blood and other fluids inside living cells contain many different solutes, and the osmotic pressures of these solutions play an important role in the distribution of solutes within the body. Patients who have become dehydrated are often given water and nutrients intravenously. However, one cannot simply drip pure water into a patient's veins, because the water would flow into the red blood cells by osmosis, causing them to burst (Figure 15.21c). A solution that causes this condition is called a *hypotonic* solution. To prevent cells from bursting, an intravenous solution must have the same total concentration of solutes (and therefore the same osmotic pressure) as the patient's blood. Such a solution is called iso-osmotic (or *isotonic*, according to the language of the medical profession; see Figure 15.21a). A solution of 0.9% sodium chloride is isotonic with fluids inside cells in the body.

If an intravenous solution more concentrated than the solution inside a blood cell were added to blood, the cell would lose water and shrivel up. A solution that causes this condition is a *hypertonic* solution (Figure 15.21b). Cell shriveling by osmosis happens when vegetables or meats are cured in *brine*, a concentrated solution of NaCl. If you put a fresh cucumber or carrot into brine, water will flow out of its cells and into the brine, leaving behind a shriveled vegetable (Figure 15.22). With the proper spices added to the brine, a cucumber will become a tasty pickle.

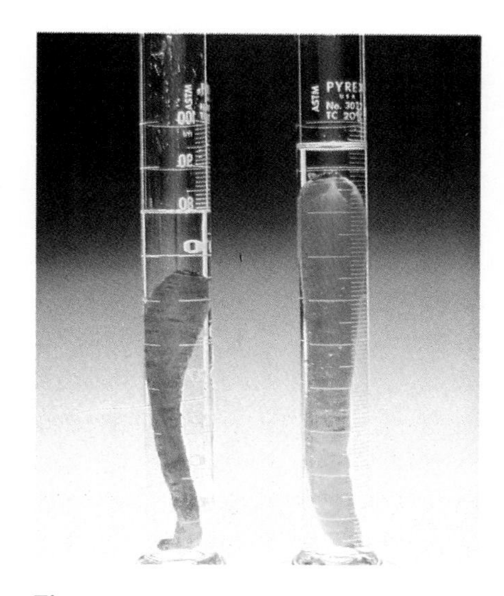

Figure 15.22
Illustration of osmosis. When a carrot is soaked in a concentrated salt solution, water flows out of the plant cells by osmosis. A carrot soaked overnight in salt solution (*left*) has lost much water and become limp. A carrot soaked overnight in pure water (*right*) is little affected.
(C.D. Winters)

EXERCISE 15.9 • *Osmotic Pressure*

The osmotic pressure found for a solution of 5.0 g of horse hemoglobin in 1.0 L of water is 1.8×10^{-3} atm at 25 °C. What is the molar mass of the hemoglobin?

Reverse Osmosis

Application of pressure greater than the osmotic pressure will cause the solvent to flow through a semipermeable membrane from a concentrated solution to a solution of lower concentration. In effect the semipermeable membrane serves as a filter with very tiny pores through which only the solvent can pass. This makes possible removal of particles as small as molecules or ions, purifying the water. This process is called **reverse osmosis**. Sea water contains a high concentration of dissolved salts; its osmotic pressure is 24.8 atm. If a pressure greater than 24.8 atm is applied to a chamber containing sea water, water molecules can be forced to flow through a semipermeable membrane to a region containing purer water (Figure 15.23). Pressures up to 100 atm are used to provide reasonable rates of purification. The largest reverse osmosis plant in operation today is in Arizona, where more than 100 million gal/day of water is purified. This plant was built to reduce the salt concentration of irrigation wastewater discharged to the Colorado River from 3200 ppm to just under 300 ppm. Sea water, which contains upwards of 35,000 ppm of dissolved salts (Table 15.5), can be purified to between 400 and 500 ppm (which is well within the World Health Organization's limits for drinking water) by reverse osmosis.

Table 15.5

Ions Present in Sea Water at 1 ppm or More

Ion	Mass Fraction *g/kg*	*ppm*
Cl^-	19.35	19,350
Na^+	10.76	10,760
SO_4^{2-}	2.710	2710
Mg^{2+}	1.290	1290
Ca^{2+}	0.410	410
K^+	0.400	400
HCO_3^-, CO_3^{2-}	0.106	106
Br^-	0.067	67
$H_2BO_3^-$	0.027	27
Sr^{2+}	0.008	8
F^-	0.001	1
total	35.129	35,129

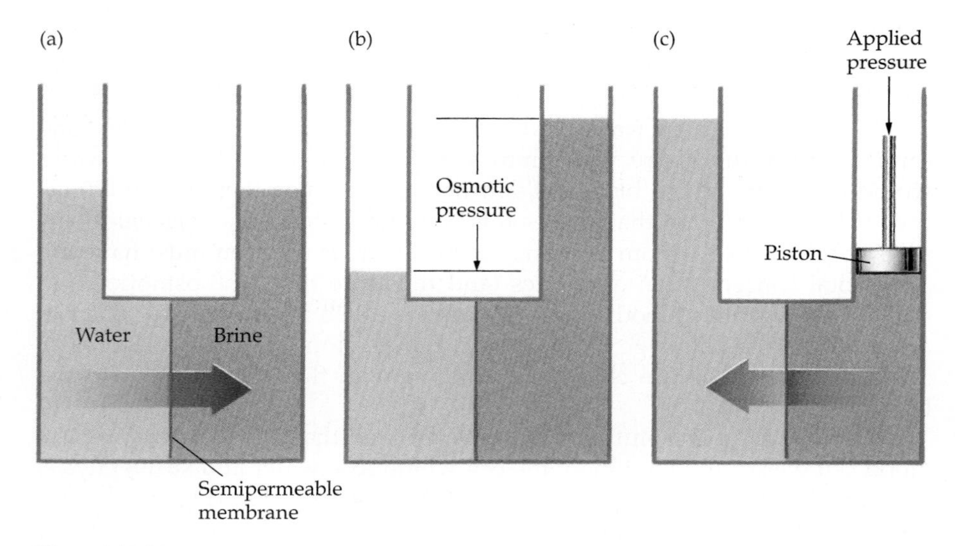

Figure 15.23

Normal osmosis is represented by (a) and (b). Water molecules pass through the semipermeable membrane to dilute the brine solution until the height of the solution creates sufficient pressure to stop the net flow. Reverse osmosis, represented in (c), is the application of an external pressure in excess of osmotic pressure to force water molecules to the pure water side.

15.4 AQUEOUS SOLUTIONS

If it could exist at all, life on earth would be very different without water and aqueous solutions. Water plays important roles in most of the chemical reactions in our environment, and in ourselves and other organisms. Water can serve as a reactant or product, as a ligand in coordination compounds, and as a solvent. Certainly there are other media in which interesting reactions occur, but on earth the chemistry of aqueous solutions and the chemistry of water itself predominate.

The Unique Solvent Properties of Water

Water is a very good solvent for many substances. When a substance dissolves in water, solute molecules or ions become hydrated by groups of water molecules. Polar water molecules strongly interact with polar portions of organic molecules, hydrating them to form molecular solutions. Aqueous sugar solutions (such as soft drinks, syrups, and fruit juices) and aqueous alcohol solutions (such as wine, vodka, beer, and whiskey) are examples. Inorganic substances that consist of polar molecules (such as ammonia, NH_3; hydrogen chloride, HCl; and sulfur dioxide, SO_2) are water-soluble; in addition to becoming hydrated, many of them react with water as they dissolve. Salts consist of ions in a crystal lattice. These ions, being permanently charged, interact even more strongly with the polar water molecules than do neutral solute molecules. Therefore, aqueous solutions of ionic compounds contain hydrated positive and negative ions.

Most of the solutions you deal with on a daily basis are aqueous, the chief exceptions being petroleum products such as gasoline. Natural waters (such as raindrops, lakes, and rivers) contain dissolved cations and anions, nitrogen, oxygen, and carbon dioxide as well as other gases and solids the water may have contacted. These solutes include a wide variety of pollutants. Surface waters such as rivers dissolve air, minerals, natural organic substances such as tannic acid from decaying leaves, and pollutants such as metal salts and solvents. Groundwater often contains many of the same kinds of solutes as surface waters. Because groundwaters come in contact with so many mineral deposits, they may have much higher concentrations of salts than do surface waters. For this reason, well water is often "hard"—its mineral content interferes with the ability of soaps to remove soil from people or clothing.

While many substances dissolve in water, many do not. Except for those that react with water (such as sodium, potassium, and calcium), metals do not dissolve in water to any appreciable extent. Diamond, graphite, and many other network solids (Section 14.4) such as polymers fail to dissolve in water. Nonpolar substances (such as hydrocarbons) have low solubilities in water. There are even some ionic compounds whose water solubilities are low, as you will see in the next few pages.

Aqueous Solutions of Ionic Solids

Solutions of ionic compounds (salts) were discussed in Section 5.3. Such solutions conduct electricity, and substances that produce ionic solutions are called electrolytes. In an electrolyte solution each ion exhibits its own, individual properties. Figure 5.7 (solubility rules) indicated which ions confer solubility on compounds containing them, and which are usually found in insoluble substances. The concentration of an electrolyte solution can be expressed in terms of the concentration of the ionic compound itself, or in terms of individual ions. For example, the concentration of dissolved sodium chloride could be expressed as 0.10 M NaCl. This means that there is 0.10 mol of NaCl formula units per liter of solution. The concentration of the same solution could also be expressed as 0.10 M sodium ion or as 0.10 M chloride ion, since the concentrations of Na^+ and Cl^- would each be 0.10 M. This distinction becomes important when several solutes are dissolved in the same solution. Table 15.5, for example, indicates the concentrations of *ions* (not compounds like NaCl) in seawater, because a great many different compounds are dissolved, and several contribute Na^+ ions or Cl^- ions. Since each ion has its own properties, the concentrations of the ions, not the compounds from which they came, are important, and the concentrations are expressed accordingly.

When salts dissolve in water there is often a noticeable temperature change. Some dissolve exothermically, and the solution becomes warmer. Other salts dissolve endothermically, and the solution becomes colder. The quantity of energy transferred as heat when a substance dissolves to form a solution of standard concentration is called the **enthalpy of solution.** Enthalpy of solution is the basis for instant hot and cold packs used for treatment of sprains and other injuries.

Ammonium nitrate, NH_4NO_3, is found in cold packs (Figure 15.24). It dissolves endothermically, cooling its surroundings:

$$NH_4NO_3(s) + \text{heat} \xrightarrow{H_2O} NH_4^+(aq) + NO_3^-(aq) \qquad \Delta H_{soln} = 25.69 \text{ kJ}$$

Calcium chloride, $CaCl_2$, is found in hot packs. It dissolves exothermically, warming its surroundings:

$$CaCl_2(s) \xrightarrow{H_2O} Ca^{2+}(aq) + 2\ Cl^-(aq) + \text{heat} \qquad \Delta H_{soln} = -82.88 \text{ kJ}$$

Whether ΔH_{soln} is positive or negative depends on two main factors: the more strongly the ions are hydrated in solution, the more negative the enthalpy of solution; and the stronger the ionic attractions in the crystal lattice, the more positive the enthalpy of solution.

As we indicated earlier in this chapter, a positive entropy change usually accompanies dissolving and favors solubility. In aqueous solutions of a few salts, those that contain +2 or +3 ions, the charges on the ions are so large that they align water molecules extremely strongly around them. Since the water molecules are locked into place by strong hydration, the entropy of solution may actually be negative, which does not favor solubility.

The Solubility Product

The solubility rules in Figure 5.7 provide no quantitative information about solubility. They indicate which ions form compounds that are soluble enough to make solutions about 0.1 M, but they say nothing about how much more of a soluble compound would dissolve, nor do they indicate the maximum possible concentration of an insoluble salt. There is a way to tabulate such information. As we indicated in the first section of this chapter, solubility is an equilibrium process. In a saturated aqueous solution, solid and solution are at dynamic equilibrium. The equilibrium constant for this process, then, can tell us something about solubility.

To see how this works, consider the synthesis of silver bromide, AgBr, which is used in photographic film as one of the light-sensitive ingredients. According to Figure 5.7, silver bromide is insoluble. Therefore it can be made by adding a water-soluble silver salt, such as $AgNO_3$, to an aqueous solution of a bromide-containing salt, such as KBr. The net ionic equation for the reaction that occurs is

$$Ag^+(aq) + Br^-(aq) \longrightarrow AgBr(s)$$

Although we say AgBr is insoluble in water, if some of this precipitated AgBr is placed in pure water, a tiny bit will dissolve and an equilibrium will be established between solid AgBr and the Ag^+ and Br^- ions in solution.

$$AgBr(s) \rightleftharpoons Ag^+(aq, 5.7 \times 10^{-7}\text{ M}) + Br^-(aq, 5.7 \times 10^{-7}\text{ M})$$

When equilibrium has been established, the solution is saturated. Experiments show that at 25 °C the concentrations of both Ag^+ and Br^- ions are 5.7×10^{-7} M. The equilibrium expression is given by

$$K = [Ag^+(aq)][Br^-(aq)]$$

This constant is the product of the equilibrium concentrations of the ions of the slightly soluble salt, and so it is called the **solubility product constant.** It is designated K_{sp}.

Figure 15.24
A cold pack used for athletic injuries. This one contains ammonium nitrate in a separate inner container, which is broken when the desired cooling effect is needed. *(C.D. Winters)*

$$K_{sp}(AgBr) = [Ag^+(aq)][Br^-(aq)]$$

Since the concentrations of both the Ag^+ and Br^- ions are 5.7×10^{-7} M when silver bromide is in equilibrium with its ions at 25 °C, this means that K_{sp} at 25 °C is

$$K_{sp} = [5.7 \times 10^{-7}\ M][5.7 \times 10^{-7}\ M] = 3.2 \times 10^{-13}\ M^2$$

The ion concentrations in the K_{sp} expression have units, and so K_{sp} also has units. However, you can always figure out what those units would be from the equation for dissolution of an isoluble salt. For example, in the case of AgBr two ions are formed and the units are M^2. Because the units can always be figured out, they are usually not included when K_{sp} values are reported, and we will not include them from now on.

In general, a slightly soluble salt could have the formula A_xB_y. The equilibrium expression for dissolving the salt would be

$$A_xB_y(s) \rightleftharpoons x\,A^{n+}(aq) + y\,B^{m-}(aq)$$

This results in the general K_{sp} expression

$$K_{sp} = [A^{n+}]^x[B^{m-}]^y$$

As is true for all pure solids, the concentration of solid salt is not included in the solubility product expression. Here are two examples of K_{sp} expressions for slightly soluble salts:

$$CaF_2(s) \rightleftharpoons Ca^{2+}(aq) + 2\,F^-(aq) \qquad K_{sp} = [Ca^{2+}][F^-]^2 = 3.9 \times 10^{-11}$$

$$Ag_2SO_4(s) \rightleftharpoons 2\,Ag^+(aq) + SO_4^{2-}(aq) \qquad K_{sp} = [Ag^+]^2[SO_4^{2-}] = 1.7 \times 10^{-5}$$

Additional K_{sp} values are listed in Table 15.6 and in Appendix H.

If the K_{sp} of a slightly soluble salt is known, it is possible to estimate the solubility of the salt. For example, the K_{sp} of $BaSO_4$ is 1.1×10^{-10} at 25 °C. The equation for dissolving $BaSO_4$ is

$$BaSO_4(s) \rightleftharpoons Ba^{2+}(aq) + SO_4^{2-}(aq)$$

K_{sp} provides only an estimate of the solubility because sulfate ions react with water to a small extent in a process called hydrolysis. Therefore the concentration of sulfate ions is slightly less than would be predicted from the K_{sp} of $BaSO_4$ and the concentration of barium ions is slightly higher. Hydrolysis of heavy-metal compounds often causes higher than expected concentrations of toxic metals in the environment. Hydrolysis is discussed in Section 16.5.

Suppose that you start with one liter of pure water and then add solid $BaSO_4$, allowing it to dissolve until equilibrium is reached. According to the equation, one mole of Ba^{2+} ions and one mole of SO_4^{2-} ions are produced for every mole of $BaSO_4$ that dissolves. Therefore, the concentration of either Ba^{2+} or SO_4^{2-} ions indicates how many moles of $BaSO_4$ dissolve per liter. If S represents the solubility of $BaSO_4$, both $[Ba^{2+}]$ and $[SO_4^{2-}]$ must also be equal to S at equilibrium. Writing this in a table,

	$[Ba^{2+}]$	$[SO_4^{2-}]$
initial concentrations (before $BaSO_4$ has been added)	0	0
change in concentration on reaction	$+S$	$+S$
concentration at equilibrium	S	S

The equilibrium-constant expression derived from the equation for dissolving $BaSO_4$ is

$$K_{sp} = [Ba^{2+}][SO_4^{2-}]$$

Where the ions in a K_{sp} expression are not in a 1:1 ratio, S must be multiplied by the appropriate coefficient; for example, for CaF_2, $K_{sp} = [Ca^{2+}]\,[F^-]^2 = (S)(2S)^2$.

Substituting from the table into the K_{sp} expression and then solving for S, we get

$$K_{sp} = [Ba^{2+}][SO_4^{2-}] = (S)(S) = S^2$$

$$S = \sqrt{K_{sp}} = \sqrt{1.1 \times 10^{-10}} = 1.0 \times 10^{-5}\ M$$

K_{sp} values cannot be used to compare the relative solubilities of salts with different ion ratios (such as AgCl and $PbCl_2$) unless calculations such as the one in Exercise 15.10 are done.

The solubility of $BaSO_4$ in pure water at 25 °C is estimated to be 1.0×10^{-5} mol/L. To find the solubility in g/L, we need only to multiply the molar solubility by the molar mass of $BaSO_4$.

$$\text{solubility} = \frac{1.0 \times 10^{-5} \text{ mol BaSO}_4}{\text{L}} \cdot \frac{233 \text{ g BaSO}_4}{\text{mol BaSO}_4} = 0.0023 \text{ g/L}$$

EXERCISE 15.10 • *Using The Solubility Product*

The K_{sp} of AgCl is 1.8×10^{-10}. Calculate the solubility of AgCl; express your result in mol/L and in g/L.

Table **15.6**

K_{sp} *Values for Some Slightly Soluble Salts*

Compound	K_{sp} at 25 °C
AgBr	3.3×10^{-13}
AuBr	5.0×10^{-17}
$AuBr_3$	4.0×10^{-36}
CuBr	5.3×10^{-9}
Hg_2Br_2*	1.3×10^{-22}
$PbBr_2$	6.3×10^{-6}
AgCl	1.8×10^{-10}
AuCl	2.0×10^{-23}
$AuCl_3$	3.2×10^{-25}
CuCl	1.9×10^{-7}
Hg_2Cl_2*	1.1×10^{-18}
$PbCl_2$	1.7×10^{-5}
AgI	1.5×10^{-16}
AuI	1.6×10^{-23}
AuI_3	1.0×10^{-46}
CuI	5.1×10^{-12}
Hg_2I_2*	4.5×10^{-29}
HgI_2	4.0×10^{-29}
PbI_2	8.7×10^{-9}
Ag_2SO_4	1.7×10^{-5}
$BaSO_4$	1.1×10^{-10}
$PbSO_4$	1.8×10^{-8}
Hg_2SO_4*	6.8×10^{-7}
$SrSO_4$	2.8×10^{-7}

*These compounds contain the diatomic ion Hg_2^{2+}.

For salts having the same ion ratios, the bigger K_{sp} the greater is the solubility. For example, AgCl, AgBr, and AgI are all 1:1 salts. Their K_{sp} values are

$$K_{sp}(\text{AgCl}) = 1.8 \times 10^{-10}$$
$$K_{sp}(\text{AgBr}) = 3.3 \times 10^{-13}$$
$$K_{sp}(\text{AgI}) = 1.5 \times 10^{-16}$$

and so the solubility of AgCl is greater than the solubility of AgBr, which in turn is greater than the solubility of AgI. Relative solubilities can be predicted this way because in each case the relation between solubility, S, and K_{sp} is the same, namely $S = \sqrt{K_{sp}}$.

EXERCISE 15.11 • *Relative Solubility*

Which substance in Table 15.6 with general formula MX_2 is least soluble?

The Effect of a Common Ion on Solubility

It is often desirable to remove an ion from solution by forming a precipitate of one of its insoluble compounds. For example, barium is quite effective in making the intestinal tract visible when x-ray photographs are taken, but barium ions are poisonous and must not be allowed to dissolve in body fluids. The insoluble compound barium sulfate can be used as an x-ray contrast medium, but both physician and patient want to be certain that no barium ions will be in solution. Earlier in this section we estimated the solubility of $BaSO_4$ in water at 25 °C to be 1.0×10^{-5} mol $BaSO_4$/L, which means that the concentration of Ba^{2+} ions would be 1.0×10^{-5} M as well. But the concentration of Ba^{2+} ions can be reduced still further by adding a soluble sulfate salt, such as Na_2SO_4. The solubility of $BaSO_4$ becomes smaller because of the increased concentration of the SO_4^{2-} ion, which is present in both $BaSO_4$ and Na_2SO_4. Sulfate is called a "common ion" because it is common to both substances dissolved in the solution.

This displacement of an equilibrium by having more than one source of a reactant or product ion is called the **common ion effect.** It can be interpreted by using Le Chatelier's principle. Consider the solubility equilibrium

$$BaSO_4(s) \rightleftharpoons Ba^{2+}(aq) + SO_4^{2-}(aq)$$

Suppose that some Na_2SO_4 solution is added to a saturated solution of $BaSO_4$. This increases the concentration of SO_4^{2-} ions, which in turn causes the equilibrium to shift to offset the effect of the change. The equilibrium shifts left to use up some of the added sulfate ions, and at the same time Ba^{2+} ions are consumed. The outcome is that the salt solubility is lower in the presence of the common ion.

EXAMPLE 15.5

The Common Ion Effect

The solubility of AgCl in pure water is 1.3×10^{-5} M (0.0019 g/L). If you put some AgCl in a solution that is 0.55 M in NaCl, what mass of AgCl will dissolve per liter of this solution?

SOLUTION The solubility of silver chloride can be expressed by the silver ion concentration, $[Ag^+]$. As usual, we assign the variable S to this concentration.

$$\text{Solubility of AgCl defined by } [Ag^+] = S$$

In *pure* water, S is equal to both $[Ag^+]$ and $[Cl^-]$. However, in salt water containing the common ion Cl^-, S is equal only to $[Ag^+]$, and it must have a smaller value than in pure water owing to the fact that excess Cl^- ion causes the equilibrium to shift to the left, thereby lowering the concentration of Ag^+.

$$AgCl(s) \rightleftharpoons Ag^+(aq) + Cl^-(aq)$$

←excess Cl^- shifts equilibrium left—

The following table shows the concentrations of Ag^+ and Cl^- when equilibrium is attained in the presence of extra Cl^-.

	$[Ag^+]$	$[Cl^-]$
Initial concentration (M)	0	0.55
Change on proceeding to equilibrium	$+S$	$+S$
Equilibrium concentration (M)	S	$S + 0.55$

Some AgCl dissolves in the presence of chloride ion and produces Ag^+ and Cl^- ion concentrations of S mol/L. However, there was already some chloride ion present, so *the total chloride ion concentration* is the amount coming from AgCl (equals S) *plus* what was already there (0.55 M).

Now the K_{sp} expression can be written, using the equilibrium concentrations from the table.

$$K_{sp} = 1.8 \times 10^{-10} = [Ag^+][Cl^-] = (S)(S + 0.55)$$

and rearranged to

$$S^2 + 0.55\,S - K_{sp} = 0$$

This is a quadratic equation and can be solved by the methods in Appendix B.4. The easiest approach is to make the approximation that S is *very* small with respect to 0.55; that is, it will make a negligible difference to the answer if we assume that $(S + 0.55) \approx 0.55$. This is a very reasonable assumption since we know that the solubility equals 1.3×10^{-5} M *without* the common ion Cl^-, and it will be even smaller in the presence of added Cl^-. Therefore,

$$K_{sp} = 1.8 \times 10^{-10} \approx (S)(0.55)$$
$$S = [Ag^+] \approx 3.3 \times 10^{-10}$$

(If we use the method of successive approximations to solve the more "accurate" expression, the answer is the same as the "approximate" answer to two significant figures.)

As predicted by Le Chatelier's principle, the solubility of AgCl in the presence of added Cl^-, an ion common to the equilibrium, is clearly less (3.3×10^{-10} M) than in pure water (1.3×10^{-5} M).

As a final step, let us check the approximation we made. To do this, we substitute the approximate value of S into the exact expression $K_{sp} = (S)(S + 0.55)$. Then, if the product $(S)(S + 0.55)$ is the same as the given value of K_{sp}, the approximation is valid. (The validity is also confirmed by the fact that the method of successive approximations gave the same answer as the approximate expression.)

Example 15.5 could also be solved by the method of successive approximations (Appendix B.4). Try it to see if you get the same answer.

$$K_{sp} = (S)(S + 0.55) = (3.3 \times 10^{-10})(3.3 \times 10^{-10} + 0.55) \approx 1.8 \times 10^{-10}$$

Lastly, you should be sure to understand the problem solving strategy used here. It is identical to that used to solve common ion problems in acid–base chemistry (see Example 18.3). In the present example we assume that Cl^- is already present in solution from the added NaCl. Then we add AgCl and additional Cl^- enters the solution as the silver salt dissolves. Thus, $[Cl^-]$ at equilibrium is the *sum* of the two concentrations of the ion from the two sources.

EXERCISE 15.12 • *Common Ion Effect*

Calculate the solubility of $PbCl_2$ at 25 °C in a solution that is 0.50 M in NaCl. Compare your result with the solubility you calculated in Exercise 15.10.

15.5 COLLOIDS

Around 1860, Thomas Graham found that substances such as starch, gelatin, glue, and albumin from eggs diffuse in water only very slowly, compared with sugar or salt. In addition, the former substances differ significantly from the latter ones in their ability to diffuse through a thin membrane: sugar or salt will diffuse through many types of membranes, but glue, starch, albumin, and gelatin will not. Graham also found that he could not crystallize these substances, while he could crystallize sugar and salt from their solutions. Therefore, Graham coined the word "colloid" (from the Greek meaning "glue") to describe a class of substances distinctly different from sugar and salt and similar materials.

Because of the sudden loss of visibility caused by fog on the morning of December 11, 1990, 99 vehicles in the northbound and southbound lanes of I-75 north of Chattanooga, Tennessee, crashed, killing 12 people and injuring 42 others. *(Nick Arroyo)*

Colloids are now understood to be mixtures in which relatively large particles, the **dispersed phase,** are distributed uniformly throughout a solvent-like medium called the **continuous phase** or the dispersing medium. Like true solutions, colloids are found in the gas, liquid, and solid states. Although both true solutions and colloids appear homogeneous to the naked eye, at the microscopic level colloids are not homogeneous. Colloids can often be distinguished from solutions by their lack of long-term stability—they will often "settle out." The large colloid particles coalesce into even larger particles that eventually separate from the continuous phase.

In colloids, the dispersed-phase particles might be as large as 10 to 1000 times the size of a single small molecule. Colloidal particles are so large, about 1000 nm in diameter, that they can readily scatter visible light that passes through the continuous medium, a phenomenon known as the **Tyndall effect** (Figure 15.25).

True solutions also scatter light, but only weakly. The blue sky is the result of light scattering by molecules in the air. Because blue light is more strongly scattered by molecule-sized particles than is red light, the sky appears a diffuse blue.

Figure 15.26 shows a beam of light passing through two glass bottles. The bottle on the left contains a colloidal mixture of a gelatin in water, while the bottle on the right holds a solution of NaCl. Colloidal particles of dust and smoke in the air of a room can easily be observed in a beam of sunlight because they scatter the light; you have probably seen such a well-defined sunbeam many times. A common colloid, *fog*, consists of water droplets (the dispersed phase) in air (the continuous phase, and itself a solution).

Figure 15.25
The Tyndall effect. Shafts of light are visible coming through the trees in the forest along the Oregon coast.
(Bob Pool/Tom Stack and Associates)

Types of Colloids

Colloids are classified according to the state of the dispersed phase (solid, liquid, or gas) and the state of the continuous phase. Table 15.7 lists several types of colloids and some examples of each. Liquid-liquid colloids form only in the presence of an emulsifier—a third substance that coats and stabilizes the particles of the dispersed phase. Such colloidal dispersions are called **emulsions.** In mayonnaise, for example, egg yolk contains a protein that stabilizes the tiny drops of oil that are dispersed in the aqueous continuous phase. As you can see from Table 15.7, colloids are very common in everyday life.

Colloids with water as the continuous phase can be classified as hydrophilic or hydrophobic. In a **hydrophilic colloid** there is a *strong attraction*

Table 15.7

Types of Colloids

Continuous Phase	Dispersed Phase	Type	Examples
gas	liquid	aerosol	fog, clouds, aerosol sprays
gas	solid	aerosol	smoke, airborne viruses, automobile exhaust
liquid	gas	foam	shaving cream, whipped cream
liquid	liquid	emulsion	mayonnaise, milk, face cream
liquid	solid	sol	gold in water, milk of magnesia, mud
solid	gas	foam	foam rubber, sponge, pumice
solid	liquid	gel	jelly, cheese, butter
solid	solid	solid sol	milk glass, many alloys such as steel, some colored gemstones

Figure 15.26
A narrow beam of light passing through an NaCl solution on the left and then a colloidal mixture illustrates the light scattering ability of the colloidal sized particles.
(C.D. Winters)

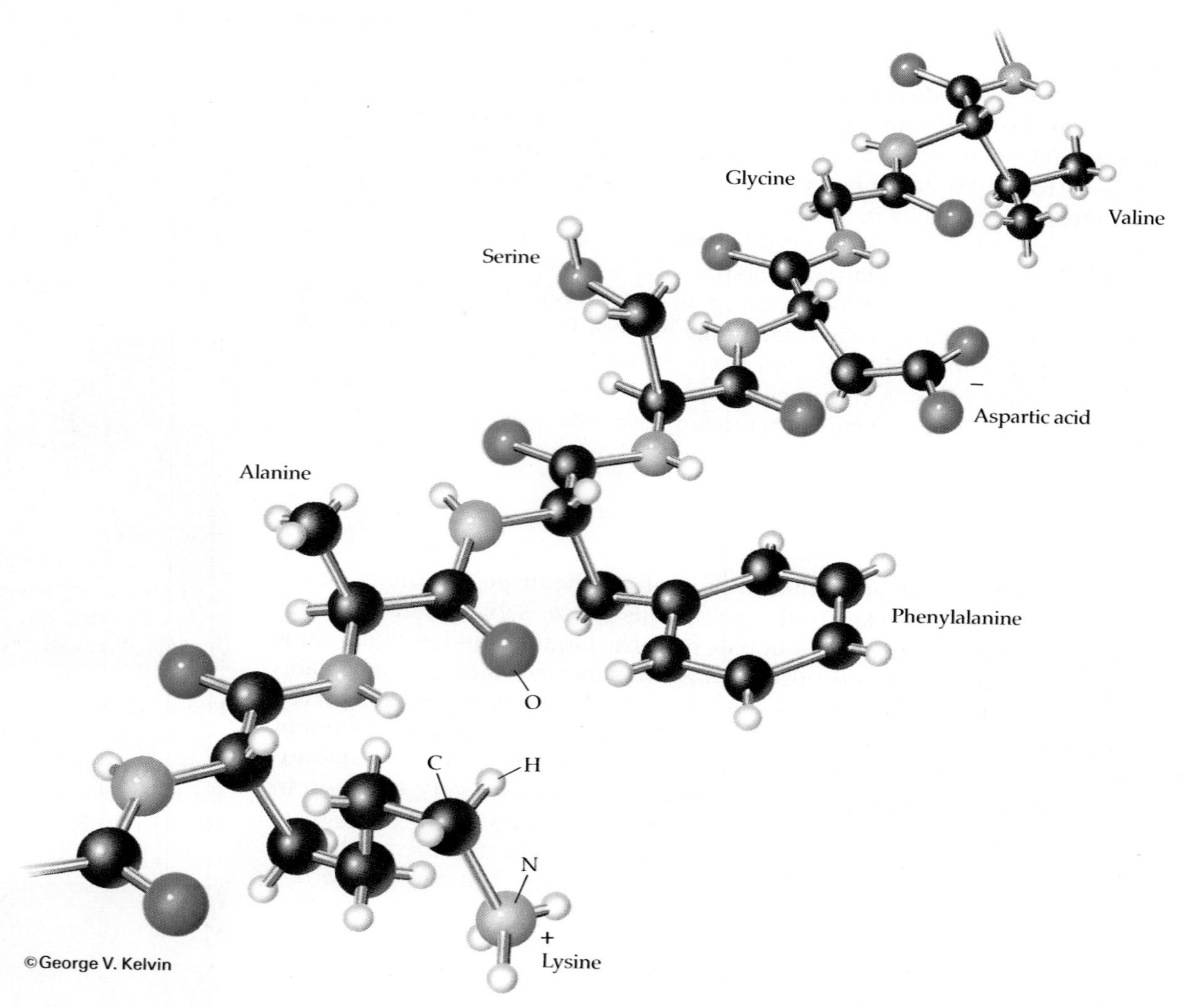

Figure 15.27
Proteins in aqueous solution are hydrophilic owing to the sites on the molecule that can hydrogen bond with water molecules.

between the dispersed phase and the continuous (aqueous) phase. Hydrophilic colloids are formed when the molecules of the dispersed phase have multiple sites that interact with water through hydrogen bonding and dipole-dipole attraction. Proteins in aqueous media are hydrophilic colloids (Figure 15.27).

In a **hydrophobic colloid** there is *lack of attraction between the dispersed phase and the continuous phase.* Although you might assume that such colloids would tend to separate quickly, hydrophobic colloids can be quite stable once they are formed. A colloidal sol of gold particles prepared in 1857 is still preserved in the British Museum. In hydrophobic colloids the surfaces of the colloidal particles apparently become covered with one kind of charge by some process that is not completely understood. Oppositely charged ions in solution are then attracted to the surfaces, forming a second layer. These ions then stabilize the sol and prevent the particle from sticking to other particles of the same kind to form larger particles.

A stable hydrophobic colloid can be made to **coagulate** by introducing ions into the dispersed phase. Milk is a colloidal suspension of hydrophobic particles. When milk ferments, lactose (milk sugar) is converted to lactic acid, which forms lactate ions and hydronium ions. The protective charge layer on the surfaces of the colloidal particles is overcome, and the milk coagulates—the milk solids separate in clumps called "curds." The coagulated milk may be used to make buttermilk or various kinds of cheese, depending on the species of bacteria that caused the fermentation. Similarly, the soil particles carried in rivers are hydrophobic sols. When river water containing large amounts of colloidally suspended soil particles meets sea water with a high ionic concentration, the particles coagulate to form silt. The deltas of the Mississippi and Nile rivers are formed in this way (Figure 15.28).

Figure 15.28
Silt forms at a river delta as colloidal particles in the contact of fresh water with salt water. The high concentration of salt causes the colloidal particles to coagulate. *(NASA/Peter Arnold, Inc.)*

Surfactants

There are a large number of natural and synthetic molecules that have both a hydrophobic part and a hydrophilic part. Such molecules are called **surfactants** because they tend to act at the surface of a substance that is in contact with the solution that contains them. The classic surfactant, **soap,** dates back to the Sumerians in 2500 BC. Soaps are salts of fatty acids and have always been made by the reaction of a fat with an alkali, a process known as *saponification.* The Greek physician Galen referred to this recipe and stated further that soap removed dirt from the body as well as serving as a treatment for wounds.

$$\begin{array}{l} CH_3(CH_2)_{16}COO\text{—}CH_2 \\ \qquad\qquad\qquad\quad | \\ CH_3(CH_2)_{16}COO\text{—}CH \\ \qquad\qquad\qquad\quad | \\ CH_3(CH_2)_{16}COO\text{—}CH_2 \end{array} + 3\,NaOH \longrightarrow 3\,CH_3(CH_2)_{16}COO^-Na^+ + \begin{array}{l} HO\text{—}CH_2 \\ \qquad\ | \\ HO\text{—}CH \\ \qquad\ | \\ HO\text{—}CH_2 \end{array}$$

tristearin (glyceryl tristearate) — sodium stearate (a soap) — glycerol

Sodium stearate is a typical soap. The long-chain hydrocarbon part of the molecule is hydrophobic, while the polar carboxylate group (COO^-) is hydrophilic.

$$CH_3CH_2CH_2CH_2CH_2CH_2CH_2CH_2CH_2CH_2CH_2CH_2CH_2CH_2CH_2CH_2C\begin{matrix} //O \\ \\ \backslash O^- \end{matrix}\quad Na^+$$

hydrophobic end — hydrophilic end

sodium stearate

In addition to soaps, which are made from naturally occurring fats and oils, many synthetic surfactants are made from refined petroleum or coal products. These **detergents** have molecular structures somewhat like those of soaps; that is, they have a long hydrocarbon end that is hydrophobic, and

Hand soaps are generally pure soap to which dyes and perfumes are added.

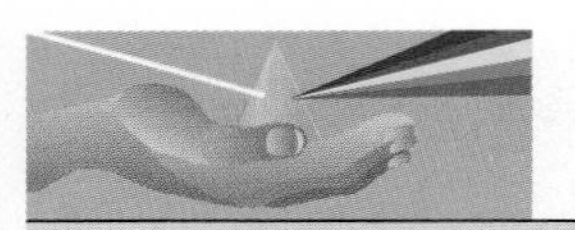

CHEMISTRY YOU CAN DO

Curdled Colloids

Regular (whole) milk contains about 4% fat. Skim milk contains considerably less. In addition, milk contains protein. Both the fat and the proteins are in the form of colloids.

Add about 2 tablespoons of vinegar or lemon juice to about 100 mL of whole milk (do the same with skim milk, and 1% or 2% butterfat milk, if you have it), stir, and watch what happens. Let it stand overnight at room temperature. Observe and record the results. Write an explanation for what you observed.

An additional experiment you can try is adding salt to similar samples of milk and recording your observations. Does salt have the same effect on milk as the acid does?

You may discard this milk down the drain. You should **not** drink this milk because it has been unrefrigerated and might contain harmful bacteria.

a polar end that is hydrophilic. One common synthetic surfactant is sodium lauryl sulfate, which is used in many shampoos.

$$CH_3CH_2CH_2CH_2CH_2CH_2CH_2CH_2CH_2CH_2CH_2CH_2OSO_3^-\ Na^+$$

sodium lauryl sulfate

In water solutions, surfactants tend to aggregate (stick together) to form hollow, colloid-sized particles called **micelles** (Figure 15.29) that can transport various materials within them. The hydrophobic ends of the molecules point inward to the center of the micelle, and the hydrophilic ends point outward so that they interact with water molecules. Ordinary soap is a surfactant and, in water, forms micelles.

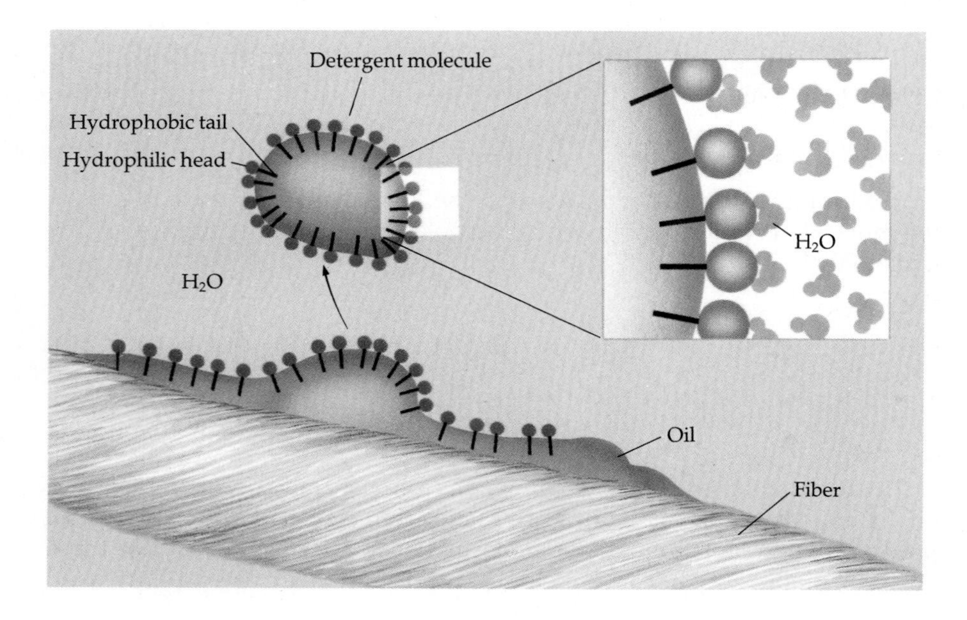

Figure 15.29
The cleansing action of soap and detergents. Soap molecules interact strongly with water through charge interaction at the hydrophilic salt ends of the molecule. The hydrocarbon ends, having more attraction for the other hydrocarbons than for water, are "pulled" into solution by the water–salt attraction. The soap molecules become oriented at the oil–water interface so that the hydrophobic hydrocarbon ends are dissolved in the oil. When greasy dirt is broken up by soapy water, a process that is aided by mechanical agitation, the oily particles are surrounded and insulated from each other by the soap molecules.

Molecular Harpoons

CHEMISTS HAVE LONG SOUGHT molecules that would kill disease-causing microorganisms without harming the patient. The current search for a cure for AIDS is based on this principle, which originated with the work of Paul Ehrlich in 1907. He discovered that a dye called trypan red, if injected in proper doses into the veins of a person suffering from African sleeping sickness, would kill the protozoan *Trypanosoma* without harming the patient. Looking for an even better cure, Ehrlich tried a compound of arsenic, arsphenamine, and found that it killed the microbe that causes syphilis instead. In Ehrlich's eyes he had found "magic bullets" for the microorganisms that caused both of those dreaded diseases.

In a continuation of the search for "magic bullets," in 1992, Steven L. Regen and his coworkers at Lehigh University in Bethlehem, Pennsylvania discovered a class of surfactants they call "molecular harpoons." They gave the molecules this name because they attack a class of unwanted bacteria by injecting an arrow-shaped part of the molecule into a stretched portion of the outer covering of the bacterium, just as a barbed harpoon digs into an animal and then holds on.

Regen's compounds are surfactants because they have a hydrophilic end and a hydrophobic end. The hydrophilic end enables them to be mobile in aqueous solutions such as blood. The hydrophobic end can harpoon a bacterium's cell walls, and it does this when the bacterium is under osmotic stress. This stress happens because the solution around the bacterium is less concentrated than its own cell contents, which causes the cell wall to stretch. The remarkable property of Regen's molecular harpoons is that they can be very specific to osmotic pressure differences. This may enable researchers to design them for specific microorganisms. Regen and his coworkers are looking at molecular harpoons as a means of attacking organisms with highly curved outer membranes, such as the HIV virus that causes AIDS.

Soaps cleanse because oil and grease (which are also hydrophobic) become associated with the hydrophobic centers of the soap micelles and are washed away with the rinse water (Figure 15.29). One problem with soaps is that they tend to form insoluble salts that appear as a scum when ordinary soap is used with "hard water," which contains Ca^{2+}, Mg^{2+}, and Fe^{2+} ions. Synthetic surfactants do not form such insoluble salts and for that reason are widely used for washing, especially in those regions of the country where water hardness is common.

A surfactant, water, and oil together form an emulsion, with the surfactant acting as the emulsifying agent.

15.6 WATER POLLUTION

Water is the most abundant substance on the earth's surface. Oceans (with an average depth of 2.5 miles) cover over 72% of the earth and are a reservoir for 97.2% of the earth's water. The rest consists of 2.16% in glaciers, 0.0197% in fresh water in lakes and rivers, 0.61% in groundwater, 0.01% in brine wells and brackish waters, and 0.001% in atmospheric water. Water is also a major component of all living things. For example, the water content of human adults is 70%—the same proportion as the earth's surface (Table 15.8).

Brackish water contains dissolved salts but at a lower concentration than sea water.

Table 15.8

Water Content of Some Common Organic Materials

marine invertebrates	97%
human fetus (1 month)	93%
adult human	70%
body fluids	95%
nerve tissue	84%
muscle	77%
skin	71%
connective tissue	60%
vegetables	89%
milk	88%
fish	82%
fruit	80%
lean meat	76%
potatoes	75%
cheese	35%

Water pollutants are a source of great concern because water is so important to life on this planet. Water that is judged unsuitable for drinking, washing, irrigation, or industrial use is **polluted water.** Natural bodies of water such as lakes, rivers, and salt water bays that no longer can support their usual population of microorganisms, fish, or birds are also described as polluted. The pollutants may be heat, radioisotopes, toxic metal ions or organic molecules, acids, alkalies, colloidal particles such as silt, or plant nutrients that can cause excessive growth of aquatic microorganisms. Water suitable for some uses may be considered polluted and therefore unsuitable for other uses. Human activities such as industrialization and land development are the cause for much of the water pollution, but natural leaching of metal ions from soils, organic substances from decaying animal and vegetable matter, animal wastes, and soil erosion can also pollute otherwise clean water.

As human activities have continued to pollute water, various federal, state, and local governments have passed laws designed to cause us to keep our water clean and unpolluted. The U.S. Public Health Service now classifies water pollutants into eight categories (Table 15.9), and the U.S. Environmental Protection Agency sets limits for contaminants in drinking water and requires that water supplies be monitored continually.

Biochemical Oxygen Demand

Industrial processes as well as organisms that live in and near bodies of water constantly discharge organic substances into surface waters. These substances eventually find their way into groundwater and the oceans. Bacteria and other microorganisms change the organic substances (which consist mostly of carbon, hydrogen, oxygen, and nitrogen atoms) into simpler compounds. To do so, however, the microorganisms need oxygen to metabolize the organic compounds that constitute their food. The quantity of oxygen required to oxidize a given quantity of organic material is called the **biochemical oxygen demand (BOD).** Given enough oxygen, a moderate temperature, and enough time, the microorganisms will convert huge quantities of organic matter into CO_2, H_2O, and NO_3^- or N_2.

Highly polluted water often has a high concentration of organic material, so it has a large biochemical oxygen demand. When organic matter is introduced into a river from a source such as a city's sewage treatment plant, the BOD suddenly jumps and the concentration of dissolved oxygen goes down, as shown in Figure 15.30. In extreme cases, more oxygen is required than can be replenished by air dissolving in the water, and fish and other freshwater organisms can no longer survive. The aerobic bacteria (those that require oxygen for the decomposition process) also die. The death of these organisms produces even more lifeless organic matter, and the BOD soars. However, nature has a backup system for such conditions. A whole new group of microorganisms, the anaerobic bacteria, take oxygen from oxygen-containing compounds to convert organic matter to CO_2 and water, while converting organic nitrogen and nitrates to elemental nitrogen.

Anaerobic bacteria cause organic matter to decay in the absence of oxygen; they produce foul-smelling gases such as H_2S in a process called putrefaction.

BOD values can be greatly reduced by treating industrial wastes and sewage with oxygen or ozone, or both. Numerous commercial cleanup pro-

Table 15.9

Classes of Water Pollutants, with Some Examples

oxygen-demanding wastes	plant and animal material
infectious agents	bacteria, viruses
plant nutrients	fertilizers, such as nitrates and phosphates
organic chemicals	pesticides such as DDT*, detergent molecules
minerals and other chemicals	acids from coal mine drainage, inorganic chemicals such as iron from steel plants
sediment from land erosion	clay silt deposited on stream bed, which may reduce or even destroy life forms living at the solid-liquid boundary
radioactive substances	waste products from mining and processing of radioactive material, radioactive isotopes
heat from industry	cooling water used in steam generation of electricity

*Banned in the United States, but still produced and exported.

cesses, both in use and under development, employ this type of "burning" of organic wastes. Another benefit of treating waste water with oxygen is that some of the **nonbiodegradable** material becomes **biodegradable** as a result of partial oxidation.

A substance is biodegradable if it can be broken down by microorganisms. Some organic compounds are nonbiodegradable, presumably because their structures are such that microorganisms cannot use them for food, or because they are so toxic that they kill the microorganisms. Interestingly, some microorganisms can "eat" certain highly toxic substances.

One way to determine the quantity of organic pollution in a sample of water is to measure how much oxygen the bacteria require for complete

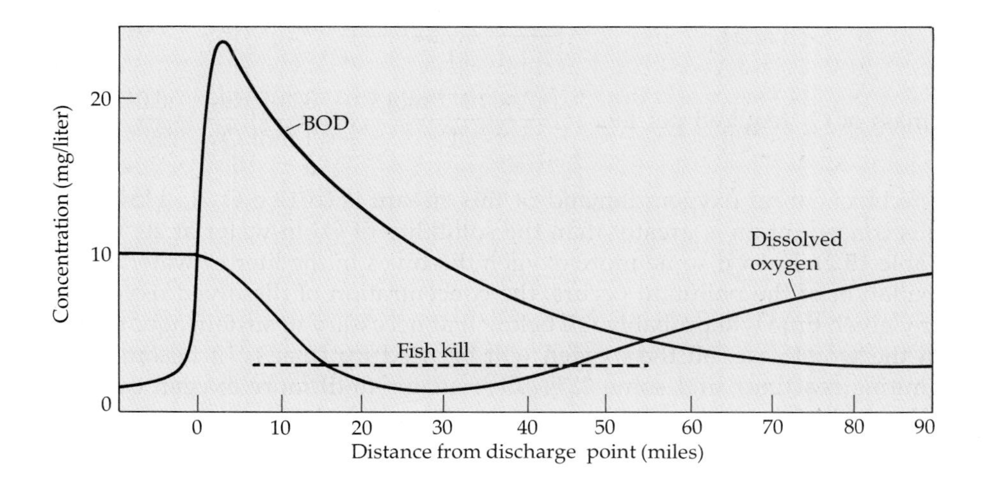

Figure 15.30
Graph showing oxygen content and oxidizable nutrients (BOD) as a result of sewage introduced into a river by a city. The results are approximated on the basis of a river flow of 750 gal/s. Note that it takes 70 miles for the stream to recover from a BOD of 0.023 g oxygen per liter.

oxidation of the compounds it contains. Usually, a known volume of the polluted water is diluted with a known volume of standardized sodium chloride solution that has a known oxygen content. After some bacteria are added, the mixture is held in a closed bottle at 20 °C for 15 days. (Closing the bottle prevents additional oxygen from dissolving in the solution as O_2 is used in the reaction.) At the end of this time the quantity of oxygen that has been consumed is taken to be the biochemical oxygen demand.

For example, suppose that a stream (at 20 °C) contains 10. ppm by weight (just 0.0010%) of an organic pollutant, the formula of which is represented by $C_6H_{10}O_5$. What is the biochemical oxygen demand caused by this pollutant?

The mass of $C_6H_{10}O_5$ present in 1 liter of this water is

$$\text{mass of } C_6H_{10}O_5 = \frac{1000 \text{ g } H_2O}{1 \text{ L}} \cdot \frac{10. \text{ g } C_6H_{10}O_5}{10^6 \text{ g } H_2O} = 0.010 \text{ g}$$

To transform this pollutant to CO_2 and H_2O, the bacteria present would use oxygen according to the equation

$$\underset{\text{pollutant}}{C_6H_{10}O_5(aq)} + 6\,O_2(g) \longrightarrow 6\,CO_2(g) + 5\,H_2O(\ell)$$

The concentration of $C_6H_{10}O_5$ in moles per liter is

$$\text{concentration of } C_6H_{10}O_5 = \frac{0.010 \text{ g } C_6H_{10}O_5}{\text{L}} \cdot \frac{1 \text{ mol } C_6H_{10}O_5}{162 \text{ g } C_6H_{10}O_5}$$

$$= \frac{6.2 \times 10^{-5} \text{ mol } C_6H_{10}O_5}{\text{L}}$$

Using the stoichiometric factor from the balanced equation (recall Section 6.1) gives the concentration of oxygen required and the corresponding mass.

$$\text{concentration of } O_2 \text{ required} = \frac{6.2 \times 10^{-5} \text{ mol } C_6H_{10}O_5}{\text{L}} \cdot \frac{6 \text{ mol } O_2}{1 \text{ mol } C_6H_{10}O_5}$$

$$= \frac{3.7 \times 10^{-4} \text{ mol } O_2}{\text{L}}$$

$$\text{mass of } O_2 \text{ required per liter} = \frac{3.7 \times 10^{-4} \text{ mol } O_2}{\text{L}} \cdot \frac{32.0 \text{ g } O_2}{1 \text{ mol } O_2} = \frac{0.012 \text{ g } O_2}{\text{L}}$$

The biochemical oxygen demand of this stream is 0.012 g O_2/L. However, this concentration is greater than the solubility of O_2 in water at 20 °C (see Table 15.2). Even if some more oxygen dissolves in the liter of water as the oxidation of the pollutant occurs, the concentration of dissolved oxygen at any given time will probably fall below that necessary to sustain aquatic life. In the worst case, all the oxygen will be used up (that is, it becomes the limiting reactant), and some $C_6H_{10}O_5$ remains until more oxygen can dissolve in the water.

EXERCISE 15.13 • *Calculating BOD*

Calculate the BOD for 1000. mL of water at 10 °C containing 2.5 ppm of a pollutant with the formula C_2H_6O. Express the answer in grams of O_2. Is this value greater or less than the solubility of O_2 in water at this temperature?

The Impact of Industrial and Household Waste on Water Quality

Industry produces a wide variety of wastes along with its intended products. It was once considered good engineering practice to put all wastes into landfills, but many of the waste compounds were partially dissolved by rainfall and leached into the groundwater, causing serious pollution of water supplies. Today, hazardous industrial wastes must be placed in secure landfills, incinerated, or treated in some way to render them nonhazardous. Secure landfills usually have plastic linings to prevent leaching (Figure 15.31), are built on a thick layer of clay, and have carefully spaced monitor wells for detecting any leaks.

We often don't think about what we throw away, or how our household wastes can affect groundwater, lakes, rivers, and coastlines. Table 15.10 lists some common household products and the kinds of chemicals they contain. The bulk of this waste still goes into landfills. As consumers of industrial products we often put into our groundwater the very same chemicals that industry is required to clean up. Households, however, have a greater problem disposing of hazardous chemicals than does industry. Although many cities have active recycling programs for glass, paper, metals, and plastics, most municipalities have no provision for picking up chemicals separately from ordinary trash, most of which goes to landfills. (Some cities burn part of their garbage, but then hazardous chemicals can contribute to air pollution; see Chapter 20.)

The EPA estimates that each year 350 million gallons of waste oil are poured on the ground or flushed down the drain. That's 35 times more oil than the Exxon Valdez spilled in Alaska.

Recycling of materials such as glass, paper, metals, and plastics helps conserve energy resources and raw materials. Recycling also conserves valuable landfill space and keeps some otherwise harmful chemicals out of groundwater.

(a)

(b)

Figure 15.31
Plastic linings are required for all landfills. This one, under construction (a), will have a capacity of millions of cubic yards of waste when completed. The thick plastic is sealed (b) so it will not allow leakage of contaminants into the groundwater.

Table 15.10

Some Common Household Hazardous Wastes and Recommended Disposal

Type of Product	Harmful Ingredients	Disposal*
bug sprays	pesticides, organic solvents	special
oven cleaner	caustics	drain
bathroom cleaners	acids or caustics	drain
furniture polish	organic solvents	special
aerosol cans (empty)	solvents, propellants	trash
nail-polish remover	organic solvents	special
nail polish	solvents	trash
antifreeze	organic solvents, metals	special
insecticides	pesticides, solvents	special
auto battery	sulfuric acid, lead	special
medicine (expired)	organic compounds	drain
paint (latex)	organic polymers	drain
gasoline	organic solvents	special
motor oil	organic compounds, metals	special
drain cleaners	caustics	drain
shoe polish	waxes, solvents	trash
paints (oil-based)	organic solvents	special
mercury batteries	mercury	special
moth balls	chlorinated organic compound	special
batteries	heavy metals such as Hg	special

*Special: Professional disposal as a hazardous waste. Drain: disposal down the kitchen or bathroom drain. Trash: Treat as normal trash—no harm to the groundwater. In most households, the items marked special are disposed of as normal trash, which results in groundwater pollution.
Source: "Household Hazardous Waste: What You Should and Shouldn't Do," Water Pollution Control Federation, 1986.

Ordinary garbage costs about $27 per ton for disposal, but proper disposal of hazardous waste costs about $1000 per ton.

How can we dispose of hazardous household wastes without increasing the risks to our water supply? We can ask our local waste authorities to provide disposal sites for such wastes. In some cities in the United States and Europe (such as in the Netherlands), special trucks pick up paint, oil, batteries, and other products for disposal. Increasingly, municipalities sponsor hazardous waste disposal days on which citizens can bring materials to a central location. Another approach is to consider the pollution potential of products when deciding what to buy. For example, alkaline batteries often work just as well as mercury batteries (see Section 17.6), but their ingredients are less toxic. When you change the oil in your automobile, buy from a merchant who will accept the used oil and dispose of it properly.

IN CLOSING

Having studied this chapter, you should be able to

- explain the factors that influence the solubility of one substance in another (Section 15.1).
- predict whether one substance will dissolve in another (Section 15.1).
- predict the influence of pressure and temperature on gas solubility (Section 15.1).
- define and convert among molarity, molality, and mass fraction (weight percent, parts per million, parts per billion, and parts per trillion) (Section 15.2).
- identify the colligative properties, explain the cause of each, and explain the variations expected for solutions of nonvolatile solutes (Section 15.3).
- predict the boiling point and freezing point of a solution of a nonvolatile solute (Section 15.3).
- find a molar mass from the results of an osmotic pressure experiment (Section 15.3).
- write the K_{sp} expressions for slightly soluble ionic compounds and use them to estimate solubilities (Section 15.4).
- explain the influence of a common ion on the solubility of a salt, and calculate salt solubility in the presence of a common ion (Section 15.4).
- describe the difference between colloids and solutions, and name some common colloids (Section 15.5).
- define *surfactant*, *hydrophobic*, and *hydrophilic*, and explain how soap works to cleanse (Section 15.5).
- give examples of various types of water pollutants and their sources (Section 15.6).
- define and explain biochemical oxygen demand and indicate why it is important to aquatic life (Section 15.6).

STUDY QUESTIONS

Review Questions

1. Define the following terms:
   (a) solvent
   (b) solute
   (c) aqueous
   (d) groundwater
   (e) molarity
   (f) molality
   (g) part per million
   (h) solubility
   (i) saturated
   (j) miscible
   (k) enthalpy of solution
   (l) colligative property
2. Define the following terms:
   (a) solubility product
   (b) common ion
   (c) semipermeable membrane
   (d) osmosis
   (e) reverse osmosis
   (f) colloid
   (g) dispersed phase
   (h) continuous phase
   (i) surfactant
   (j) polluted water
   (k) biochemical oxygen demand
3. Describe the difference between the terms *concentrated* and *dilute*, and between *saturated* and *unsaturated*.
4. Under what conditions would the molarity and the molality of a solution be approximately the same?
5. When you read about lead or chromium in drinking water, in what forms are those elements found?

6. How is lead harmful to humans? What group seems to be most susceptible to the harmfulness of lead?
7. Which is the highest concentration, 50 ppm, 500 ppb, or 0.05% by weight?
8. Estimate your concentration on campus in parts per million and parts per thousand.
9. If 5 g of solvent, 0.2 g of some solute A, and 0.3 g of solute B are mixed to form a solution, what is the weight percent concentration of A?
10. In general, how does the water solubility of most ionic compounds change as the temperature is increased? Explain in general terms why this is so.
11. How does the solubility of gases in liquids change with increased temperature? Explain why this is so.
12. If a solution of a certain salt in water is *saturated* at some temperature and a small amount of the salt in the form of tiny crystals is added to the solution, what do you expect to happen? What happens if the same quantity of salt crystals is added to an *unsaturated* solution of the salt? What would you expect to happen if the temperature of this second salt solution is slowly lowered?
13. Explain why some liquids are miscible in one another while other liquids are immiscible. Using only three liquids, give an example of a miscible pair and an immiscible pair.
14. Why would the same solid readily dissolve in one liquid and be almost insoluble in another liquid? Give an example of such behavior.
15. Describe what happens when an ionic solid dissolves in water. Draw a picture that includes at least one positive ion, one negative ion, and a dozen or so water molecules in the vicinity of the ions.
16. Write a general solubility product constant expression for a salt of the form A_xB_y.
17. Which is more soluble in water at 25 °C, CuS ($K_{sp} = 8.7 \times 10^{-29}$) or CuCl ($K_{sp} = 1.9 \times 10^{-7}$)?
18. Would you expect $CaCO_3$ ($K_{sp} = 3.8 \times 10^{-9}$ at 25 °C) to be more or less soluble in a solution that was 0.01 M in the soluble salt $CaCl_2$? Explain your answer using Le Chatelier's principle.
19. State Henry's law. Name three factors that govern the solubility of a gas in a liquid.
20. Explain why the vapor pressure of a solvent would be lowered by the presence of a nonvolatile solute.
21. Why is a higher temperature required for boiling a solution containing a nonvolatile solute than for boiling the pure solvent?
22. What happens on the molecular level when a liquid freezes? What effect does a nonvolatile solute have on this process? Comment on the purity of water obtained by melting an iceberg.
23. Describe benefits of using ethylene glycol with water in the cooling system of an automobile.
24. Which would have the lowest freezing point? (a) a 1.0 m NaCl solution, (b) a 1.0 m $CaCl_2$ solution, (c) a 1.0 m methanol solution. Explain your choice.
25. Write the osmotic pressure equation and explain all terms.
26. How can the presence of a strong electrolyte cause a colloid to coagulate?
27. List three classes of industrial wastes and illustrate how each can cause water pollution.
28. What can happen to aquatic life if the BOD of organic pollutants in a stream exceeds the solubility of O_2 at the stream's temperature? Would this problem be helped or made worse if the stream is slow-moving?

Concentration Units

29. Calculate the molarity of a solution prepared with 0.125 g of $K_2Cr_2O_7$ and sufficient water to make 500. mL of solution.

30. Calculate the molarity of a solution containing 0.65 g of $KMnO_4$ and enough water to just reach the mark on a 100-mL volumetric flask.

31. Convert 2.5 ppm to weight percent. Assume that water is the solvent.

32. What mass (in grams) of sucrose is in 1.0 kg of a 0.25% sucrose solution?

33. A sample of lead-based paint is found to contain 60.5 ppm lead. The density of the paint is 8.0 lb/gal. What mass of lead (in grams) would be present in 50. gal of this paint?

34. A paint contains 200. ppm lead. Approximately what mass of lead (in grams) will be in 1.0 cm^2 of this paint (density = 8.0 lb/gal) when 1 gal is uniformly applied to 500. ft^2 of a wall?

35. Assume you dissolve 45.0 g of ethylene glycol [$C_2H_4(OH)_2$] in half a liter of water (500. g). Calculate the molality and weight percent of ethylene glycol in the solution.

36. Dimethylglyoxime (DMG) reacts with nickel(II) ion in aqueous solution to form a bright red coordination compound. However, DMG is insoluble in water. In order to get it into aqueous solution where it can encounter Ni^{2+} ions, it must first be dissolved in a suitable solvent such as ethanol. Suppose you dissolve 45.0 g of DMG ($C_4H_8N_2O_2$, molar mass = 116.1 g/mol) in 500. mL of ethanol (C_2H_5OH, molar mass = 46.07 g/mol, density = 0.7893 g/mL). What are the molality and weight percent of DMG in this solution?

37. You need an aqueous solution of methanol (CH_3OH) with a concentration of 0.050 molal. What mass of methanol would you need to dissolve in 500. g of water to make this solution?
38. You want to prepare a 1.0 m solution of ethylene glycol [$C_2H_4(OH)_2$] in water. What mass of ethylene glycol (in grams) do you need to mix with 950. g of water?
39. Hydrochloric acid is sold as a concentrated aqueous solution. The concentration of commercial HCl is 12.0 M and its density is 1.18 g/cm^3. Calculate (a) the molality of the solution and (b) the weight percent of HCl in the solution.
40. Concentrated sulfuric acid has a density of 1.84 g/cm^3 and is 95.0% by weight H_2SO_4. What is the molality of this acid? What is its molarity?
41. A 10.7 m aqueous solution of NaOH has a density of 1.33 g/cm^3 at 20 °C. Calculate the weight percent of NaOH and the molarity of the solution.
42. Concentrated aqueous ammonia is 14.8 M and has a density of 0.90 g/cm^3. What is the molality of the solution? Calculate the weight percent of NH_3 in the solution.

Henry's Law

43. The partial pressure of O_2 in your lungs varies from 25 mm Hg to 40 mm Hg. What concentration of O_2 (in grams per liter) can dissolve in water at 37 °C when the O_2 partial pressure is 40. mm Hg? The Henry's law constant for O_2 at 37 °C is 1.5×10^{-6} M/mm Hg.
44. The Henry's law constant for nitrogen in blood serum is approximately 8×10^{-7} M/mm Hg. What is the N_2 concentration in a diver's blood at a depth where the total pressure is 2.5 atm? The air the diver is breathing is 78% N_2 by volume.

Colligative Properties

45. What is the boiling point of a solution containing 0.200 mol of a nonvolatile solute in 100. g of benzene? The normal boiling point of benzene is 80.10 °C, and K_b = 2.53 °C/molal.
46. What is the boiling point of a solution composed of 15.0 g of urea, $(NH_2)_2CO$, in 0.500 kg of water?
47. Place the following aqueous solutions in order of increasing boiling point: (a) 0.10 m NaCl, (b) 0.10 m sugar, (c) 0.080 m $CaCl_2$.
48. Arrange the following aqueous solutions in order of increasing boiling point: (a) 0.20 m ethylene glycol, (b) 0.12 m Na_2SO_4, (c) 0.10 m $CaCl_2$, (d) 0.12 m KBr.
49. The solubility of NaCl in water at 100 °C is 39.1 g/100. g of water. Calculate the boiling point of a saturated solution of NaCl.
50. The organic salt $(C_4H_9)_4NClO_4$ consists of the ions $(C_4H_9)_4N^+$ and ClO_4^-; it will dissolve in chloroform. What mass (in grams) of the salt must have been dissolved if the boiling point of a solution of the salt in 25.0 g of chloroform is 63.20 °C? The normal boiling point of chloroform is 61.70 °C and K_b = 3.63 °C/molal. Assume that the salt dissociates completely into its ions in solution.
51. You add 0.255 g of an orange crystalline compound with an empirical formula of $C_{10}H_8Fe$ to 11.12 g of benzene. The boiling point of the solution is 80.26 °C. The normal boiling point of benzene is 80.10 °C, and K_b = 2.53 °C/molal. What are the molar mass and molecular formula of the compound?
52. Anthracene is a hydrocarbon obtained from coal, and it has an empirical formula of C_7H_5. To find its molecular formula you dissolve 0.500 g of anthracene in 30.0 g of benzene. The boiling point of the solution is 80.34 °C. The normal boiling point of benzene is 80.10 °C, and K_b = 2.53 °C/molal. What is the molecular formula of anthracene?
53. List the following aqueous solutions in order of decreasing freezing point: (a) 0.1 m sugar, (b) 0.1 m NaCl, (c) 0.08 m $CaCl_2$, (d) 0.04 m Na_2SO_4. (Assume all of the salts dissociate completely into their ions in solution.)
54. Arrange the following aqueous solutions in order of decreasing freezing point: (a) 0.20 m ethylene glycol, (b) 0.12 m K_2SO_4, (c) 0.10 m NaCl, (d) 0.12 m KBr.
55. If you use only water and pure ethylene glycol, $C_2H_4(OH)_2$, in your car's cooling system, what mass (in grams) of the glycol must you add to each quart of water to give freezing protection down to −31.0 °C? (One quart of water has a mass of 946 g.)
56. Some ethylene glycol, $C_2H_4(OH)_2$, was added to your car's cooling system along with 5.0 kg of water. (a) If the freezing point of the solution is −15.0 °C, what mass (in grams) of the glycol must have been added? (b) What is the boiling point of the coolant mixture?
57. Calculate the concentration of solute particles in human blood if the osmotic pressure is 7.53 atm at 37 °C, the temperature of the body.
58. Cold-blooded animals and fish have blood that is isotonic with sea water. If sea water freezes at −23 °C, what is the osmotic pressure of the blood of these animals at 20.0 °C?

Solubility Product

59. Calculate the water solubility of AuCl in mol/L.

60. Calculate the water solubility of $PbSO_4$ in mol/L.

61. Calculate the water solubility of $PbBr_2$ in mol/L.

62. Calculate the water solubility of Ag_2SO_4 in mol/L.

63. What is the concentration of gold(III) ion (in milligrams per liter of solution) in a saturated aqueous solution of gold(III) iodide? ($K_{sp} = 1.0 \times 10^{-46}$)

$$AuI_3(s) \rightleftharpoons Au^{3+}(aq) + 3\ I^-(aq)$$

64. Estimate the solubility of lead(II) iodide in (a) moles per liter and (b) grams per liter of pure water. ($K_{sp} = 8.7 \times 10^{-9}$)

$$PbI_2(s) \rightleftharpoons Pb^{2+}(aq) + 2\ I^-(aq)$$

65. In a saturated aqueous solution of silver sulfate, Ag_2SO_4, what is the concentration of Ag^+ ion in milligrams per 100. mL of solution at 25 °C? ($K_{sp} = 1.7 \times 10^{-5}$)

66. The solubility of $PbCl_2$ in water is 1.62×10^{-2} M. Calculate K_{sp} for $PbCl_2$.

The Common Ion Effect

67. Calcium phosphate [$Ca_3(PO_4)_2$, $K_{sp} = 1 \times 10^{-33}$] is a water-insoluble mineral used as a starting material for making phosphate-containing fertilizers. Calculate
(a) the solubility of $Ca_3(PO_4)_2$.
(b) the solubility of $Ca_3(PO_4)_2$ in a solution that is 0.010 M in Ca^{2+} ion. (Since the solubility of $Ca_3(PO_4)_2$ is so small, approximate the Ca^{2+} ion concentration as 0.010 M.)
(c) the solubility of $Ca_3(PO_4)_2$ in a solution that is 0.0050 M in PO_4^{3-} ion.

68. What is the Cl^- concentration (in mol/L) in a solution that is 0.05 M in $AgNO_3$ and contains some undissolved AgCl ($K_{sp} = 1.8 \times 10^{-10}$)?

69. Calculate the molar solubility in moles per liter of silver thiocyanate (AgSCN, $K_{sp} = 1.0 \times 10^{-12}$) in pure water and in water containing 0.01 M NaSCN.

70. Calculate the molar solubility (in moles per liter) of lead(II) chloride, $PbCl_2$, in pure water. Compare this value to the molar solubility of $PbCl_2$ in 255 mL of water to which 25.0 mL of 6.0 M HCl has been added. (Assume that the volumes are additive.)

General Questions

71. A solution is prepared by dissolving 9.41 g of $NaHSO_3$ in 1.00 kg of water. It freezes at −0.33 °C. From these data, decide which of the following equations is the correct expression for the ionization of the salt.
(a) $NaHSO_3(aq) \longrightarrow Na^+(aq) + HSO_3^-(aq)$
(b) $NaHSO_3(aq) \longrightarrow Na^+(aq) + H^+(aq) + SO_3^{2-}(aq)$

72. You have just isolated a new compound with the empirical formula $(C_2H_5)_2AlF$. You believe from your knowledge of bonding theory, however, that the molecular form of the compound consists of either two units combined to give $[(C_2H_5)_2AlF]_2$ or four of them combined to give $[(C_2H_5)_2AlF]_4$. To decide which is correct, you dissolve 0.115 g of the compound in 15.65 g of benzene. The freezing point of the solution is lowered by 0.090 °C. [K_f(benzene) = −5.12 °C/m.] Which molecular formula is correct?

73. You are given a white solid that is suspected of being pure cocaine (molar mass = 303.4 g/mol, a molecular solid). When 1.22 g of the solid is dissolved in 15.60 g of benzene, the freezing point is lowered by 1.32 °C.
(a) Calculate the molar mass of the sample. [K_f(benzene) = −5.12 °C/m.]
(b) Assume that the mass of the solid was measured to ±0.01 g; the mass of benzene was measured to ±0.01 g; the freezing point lowering was measured to ±0.04 °C; and K_f is known to ±0.01 °C/m. Given these uncertainties in the experimental data, can you be sure that the substance is not codeine (molar mass = 299.4 g/mol, a molecular solid)? Support your answer with a calculation.

74. Consider the process of vaporization of a solution containing a nonvolatile solute. Which has greater entropy, the solution or the solvent vapor in contact with it? Which has greater entropy, a sample of pure solvent or the same quantity of solvent with a solute dissolved in it? In which case is there a greater difference in entropy between liquid and vapor phases, the pure solvent or the solution? Use the answers to these questions and the equation for calculating entropy change upon a phase transition (page 277) to show why the boiling point of a solution is higher than the boiling point of pure solvent.

75. The Ca^{2+} ion in hard water is often precipitated as $CaCO_3$ by adding soda ash, Na_2CO_3. If the calcium ion concentration in hard water is 0.010 M, and if the Na_2CO_3 is added until the carbonate ion concentration

is 0.050 M, what percentage of the calcium ion has been removed from the water? (You may neglect hydrolysis of the carbonate ion.)

76. In chemical research we often send newly synthesized compounds to commercial laboratories for analysis. These laboratories determine the weight percent of C and H by burning the compound and collecting the evolved CO_2 and H_2O. They determine the molar mass by measuring the osmotic pressure of a solution of the compound. Calculate the empirical and molecular formulas of a compound, C_xH_yCr, given the following information: (a) The compound contains 73.94% C and 8.27% H; the remainder is chromium. (b) At 25 °C, the osmotic pressure of 5.00 mg of the unknown dissolved in 100. mL of chloroform solution is 3.17 mmHg.

Summary Question

77. Aluminum chloride reacts with phosphoric acid to give aluminum phosphate, $AlPO_4$. The system $Al^{3+}-PO_4^{3-}-H_2O$ exists in many of the same crystal forms as SiO_2 and is used industrially as the basis of adhesives, binders, and cements.
    (a) Write a balanced equation for the reaction of aluminum chloride and phosphoric acid.
    (b) If you begin with 152 g of aluminum chloride and 3.00 L of 0.750 M phosphoric acid, what mass of $AlPO_4$ can be isolated?
    (c) If you place 25.0 g of $AlPO_4$ in enough pure water to have a volume of exactly one liter, what are the concentrations of Al^{3+} and PO_4^{3-} at equilibrium?
    (d) If you mix 1.50 L of 0.0025 M Al^{3+} (in the form of $AlCl_3$) with 2.50 L of 0.035 M Na_3PO_4, will a precipitate of $AlPO_4$ form? If so, what mass of $AlPO_4$ precipitates?

Using Computer Programs and Videodiscs

78. Use *KC? Discoverer* to find all elements that are used in alloys. What else are these elements used for? Do you think that other elements are used in alloys as well as the ones the program lists? (The "Uses" category of *KC? Discoverer* lists the most important uses of the elements, but not all the uses of each element.)
79. Use *KC? Discoverer* to find all elements that are obtained from solutions. (Air and water are the most obvious possibilities.) Which elements are obtained from seawater?
80. Use *KC? Discoverer* to find all elements that are poisonous. Based on what you have read in this chapter, is there an element that is obviously missing from the list the program provides? What is it? Look up the definition and source of data for the toxicity property. Can you think of a reason that the element in question might be missing from the list?
81. A few crystals of a salt, sodium acetate, are placed on a board, and a clear, colorless solution of sodium acetate is poured onto the crystals. (See the *Chemical Demonstrations* videodisc, Disc 2, Side 2, frame 29510 or the videotape version, Experiment 41.)
    (a) Describe your observations.
    (b) Was the sodium acetate solution unsaturated, saturated, or supersaturated? Explain your answer briefly.
    (c) The crystallization process that you see is exothermic. What would you do to get the crystals to dissolve once again?

Many of these fruits contain organic acids such as citric acid.

(C.D. Winters)

The Importance of Acids and Bases

CHAPTER

16

In Chapters 5 and 6 we defined acids and bases and discussed their reactions. Acids and bases were classified as strong or weak, but we did not say *how* strong or weak. In this chapter we provide a more quantitative approach to acid-base strength, using the ideas developed in Chapter 8 about equilibrium constants.

It is difficult to overstate the importance of acids and bases. Aqueous solutions almost always are acidic or basic to some degree, and aqueous solutions abound in our environment and in ourselves. Carbon dioxide, CO_2, is the most important acid-producing compound in nature. Rainwater is slightly acidic because of dissolved CO_2. Acid rain results from further acidification by the pollutants SO_2 and NO_2. The oceans are slightly basic, as are many ground and surface waters, but natural water can also be acidic; the more acidic the water, the more easily metals such as lead can be dissolved from pipes or solder joints. Acids and bases are important catalysts in natural and industrial processes. Photosynthesis and respiration, the two most important biological processes on earth, depend on acid-base reactions. In addition, the relative amounts of acid and base in various body fluids such as blood often make the difference between sickness and health.

Figure 16.1
Acids and bases react with organic dyes like litmus. The color change indicates whether an acid or base has reacted. An acid turns litmus from blue to pink. A base turns litmus from pink to blue. *(C.D. Winters)*

You should never taste anything in a laboratory, but taste-testing used to be common. Robert Boyle, who studied acids as well as gases, thought that acids tasted sour because they consisted of sharp-edged particles.

Knowledge of acids, bases, and the salts they produce goes back to ancient times. The word "acid" comes from the Latin word *acidus,* meaning "sour," which is how acids taste. Other properties most acids have in common were discussed in Section 5.3. Acids change the colors of vegetable pigments such as litmus (Figure 16.1), they produce bubbles of $CO_2(g)$ when added to limestone, they dissolve many metals and produce $H_2(g)$, and they react with a class of compounds called *bases* or *alkalis,* which neutralize their properties. The word alkali comes from the Arabic word *al-qali,* meaning "plant ashes." Potassium carbonate, commonly known as potash, is found in ashes from wood fires. It dissolves in water to yield a solution that feels slippery, tastes bitter, and reacts with acids. Such properties are characteristic of **alkalis.**

Figure 16.2
Svanté Arrhenius. *(The Oesper Collection in the History of Chemistry/University of Cincinnati)*

In 1648 the German Johann R. Glauber described the contrasting properties of acids and alkalis and was the first to state that a **salt** is formed when an acid reacts with an alkali. About 100 years later, Glauber's salt-formation concept was used to define a base as a substance that reacts with an acid to form a salt. In 1777 Antoine Lavoisier proposed that oxygen was the element that made an acid acidic; he even derived the name *oxygen* from Greek words meaning "acid former." But in 1808 it was discovered that the gaseous compound HCl, which dissolves in water to give hydrochloric acid, contains only hydrogen and chlorine, and by 1830 it became clear that hydrogen, not oxygen, was common to all acids in aqueous solution. Also, it was shown that aqueous solutions of both acids and bases conduct electrical current, implying the presence of ions. In 1887 the Swedish chemist Svanté Arrhenius (Figure 16.2) proposed that **acids** *ionize in aqueous solution to produce hydrogen ions and anions;* **bases** *ionize to produce hydroxide ions and cations.*

A hydrogen ion, H^+, is just a proton, which means that it is very small compared to an atom and has a very high concentration of positive charge. In water, therefore, a hydrogen ion always attracts electrons strongly, and the electrons most readily at hand are lone pairs on water molecules. Consequently a hydrogen ion always associates with at least one water molecule, forming a **hydronium ion,** H_3O^+.

$$H^+ + H_2\ddot{O}: \longrightarrow [H_3\ddot{O}]^+$$

hydrogen ion + water molecule ⟶ hydronium ion

Water molecules themselves are extensively hydrogen bonded in liquid water, and so more than one water molecule can be associated with the same H^+ ion. Therefore more complicated formulas could also be written for aqueous hydrogen ion. Nevertheless, we shall use $H_3O^+(aq)$, which is sufficient to remind us that the proton is associated with at least one water molecule.

The hydroxide ion, OH^-, also hydrogen bonds to water molecules to form more complicated species, but we shall simply use $OH^-(aq)$ to describe it.

16.1 THE BRØNSTED CONCEPT OF ACIDS AND BASES

A major problem with Arrhenius' acid-base concept is that certain substances, such as ammonia, NH_3, produce basic solutions and react with acids, yet contain no hydroxide ions. In 1923 J. N. Brønsted in Denmark and T. M. Lowry in England independently proposed a new way of defining acids and bases. Because Brønsted and his students developed these ideas to a much greater extent than Lowry, this has become known as the **Brønsted concept of acids and bases.** According to Brønsted and Lowry an **acid** can *donate a proton to another substance,* while a **base** can *accept a proton from another substance.* For an acid-base reaction to occur, an acid must donate a proton (that is, a H^+ ion) and a base must accept the proton.

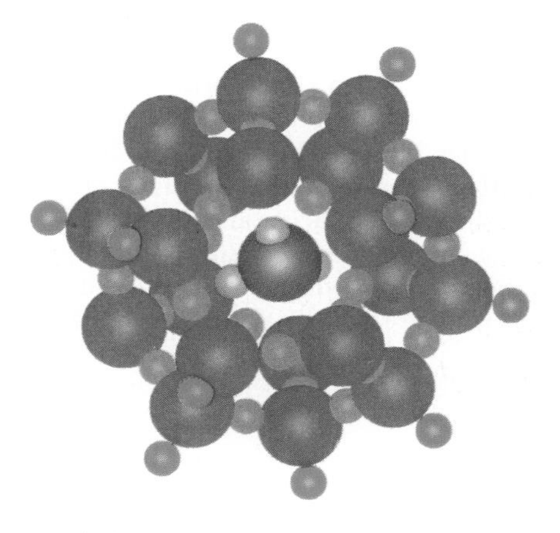

A clathrate cage comprising 20 maximally hydrogen-bonded water molecules with a hydronium ion, H_3O^+, in the center of the interior cavity. *(A. Welford Castleman, Jr. Pennsylvania State University)*

Proton Transfers in Acid-Base Reactions

To see how Brønsted's idea works, let's look at a typical acid and a typical base in aqueous solution. Nitric acid (HNO_3) is a strong electrolyte and a strong acid; it ionizes essentially completely to hydronium ions and nitrate ions when it dissolves in water.

$$HNO_3(aq) + H_2O(\ell) \longrightarrow H_3O^+(aq) + NO_3^-(aq)$$

nitric acid
a strong acid
100% ionized

A typical base is gaseous ammonia, NH_3 (Figure 16.3). Ammonia is a weak electrolyte and therefore a weak base. When it dissolves, it establishes an equilibrium with ammonium ions, $NH_4^+(aq)$, and hydroxide ions, $OH^-(aq)$. This is a reactant-favored process, and, unless the solution is very dilute, less than 5% of the ammonia ionizes. A solution of ammonia in water is often called an *ammonium hydroxide* solution, but this is not a good name, because most of what is dissolved is actually ammonia molecules.

Consideration of the equation for the reaction of NH_3 with H_2O and Le Chatelier's principle shows why a larger percentage of NH_3 will be ionized in a very dilute solution.

$$NH_3(g) + H_2O(\ell) \rightleftharpoons NH_4^+(aq) + OH^-(aq)$$

ammonia
a weak base
<5% ionized

The most important point about these two reactions is that *each involves a water molecule as a reactant.* In the first reaction a water molecule *accepts a*

Portrait of a Scientist

Johannes Nicolaus Brønsted (1879–1947)

The son of a civil engineer, Brønsted planned a career in engineering, but became interested in chemistry and switched his major in college. He earned his doctorate in 1908 and was selected for a new professorship of chemistry at the University of Copenhagen that same year. Brønsted published many important papers and books on solubility, the interaction of ions in solution, and thermodynamics. From 1921 to about 1935 his laboratories in Copenhagen were filled with numerous foreign students. His studies of how acids and bases catalyze reactions caused him to attempt to clarify just what acids and bases were. The fact that acids neutralize bases and bases neutralize acids led Brønsted to define acids and bases in relation to one another. Thus he suggested that if an acid molecule *donates* hydrogen ions in solution, then a base ought to be the opposite—a molecule that is able to *take up* hydrogen ions. This was quite different from Arrhenius' concept that a base would donate hydroxide ions to the solution, which implied a set of acids and a set of bases, each independent of the other. According to Brønsted each acid was related to a base (called its conjugate base) and each base to an acid. This simplified acid-base chemistry, because the properties of bases could be related to the properties of their conjugate acids and vice versa.

During World War II Brønsted distinguished himself by firmly opposing the Nazi takeover of Denmark. In 1947 he was elected to a seat in the Danish Folketing (parliament), but he died in December of that year, before he could take his new office.

J.N. Brønsted (The Oesper Collection in the History of Chemistry/University of Cincinnati)

proton from nitric acid, while in the second reaction, a water molecule *donates a proton* to the ammonia molecule. According to the Brønsted definition water serves as a base in the first reaction and as an acid in the second. Since all water molecules are the same, water must display both acid and base properties, depending on whether it reacts with a base or an acid. Water is said to be **amphiprotic**—*it can donate or accept protons,* depending on the circumstances.

Another example of water's ability to serve as either an acid or a base is provided by the experimental fact that, even when carefully purified, water conducts a very tiny electrical current. This indicates that pure water contains a very small concentration of ions. These ions are formed when two water molecules react to produce a hydronium ion and a hydroxide ion in a process called **auto-ionization.**

$$\underset{\text{base}}{H_2O(\ell)} + \underset{\text{acid}}{H_2O(\ell)} \rightleftharpoons \underset{\text{acid}}{H_3O^+(aq)} + \underset{\text{base}}{OH^-(aq)}$$

Here one water molecule serves as a proton acceptor (base) while the other is a proton donor (acid). The equilibrium between the water molecules and the hydronium and hydroxide ions is very reactant-favored, and so the concentrations of the ions in pure water are very low (1.0×10^{-7} M at 25 °C). Nevertheless, auto-ionization of water is very important to understanding how acids and bases function in aqueous solutions.

Figure 16.3
Ammonia being placed underground. *(Grant Heilman/Grant Heilman Photography, Inc.)*

Table 5.1 (p. 161) lists examples of common strong and weak acids and bases; remembering this information is useful in deciding whether acid-base reactions will occur. However, the Brønsted concept provides a way to relate acid-base behavior to molecular structure. A Brønsted acid must contain a proton that can be donated. Often hydrogen atoms that are bonded to rather electronegative atoms such as oxygen or a halogen are acidic. The highly electronegative atom attracts electrons in the bond away from the hydrogen atom, making it more like an unbonded proton and making it easier for the proton to be transferred to a base. The strong acids listed in Table 5.1 fit this rule nicely. In HCl the hydrogen is bonded to a halogen. In HNO_3, $HClO_4$, and H_2SO_4, the hydrogen is bonded to an oxygen that is connected to another electronegative atom.

```
         O             O                 O
        /              |                 |
H—O—N          H—O—Cl—O        H—O—S—O—H
        \              |                 |
         O             O                 O
  nitric acid    perchloric acid    sulfuric acid
```

These last three acids are called oxoacids, since they have hydrogen bonded to oxygen. For an inorganic oxoacid to be a strong acid, there must be at least two more oxygen atoms than hydrogen atoms. If there are fewer, then the oxoacid will be a weak acid.

Most of the strong acids that were listed in Table 5.1 are fairly common. Hydrochloric, nitric, and sulfuric acids are used in many industrial processes, and you might see a tank truck filled with one of these strong acids on an interstate highway (Figure 16.4). Many chemical plants have large storage containers for these acids (Figure 16.5). Perchloric acid is used mainly in the laboratory for smaller scale chemical reactions and chemical analysis.

Figure 16.4
Tank trucks like this one are common on highways. This one carries fuming sulfuric acid, as indicated by the number 1831 on the corrosive placard on the side of the vehicle. *(Courtesy of Leaseway Transportation)*

Figure 16.5
Acids are used in large quantities in chemical manufacturing plants. This tank holds several thousand gallons of an acid. *(Courtesy of Du Pont Company)*

Many weak acids are organic acids (carboxylic acids, Section 13.5). Although carboxylic acid molecules contain many hydrogen atoms, only the ones bound to oxygen atoms of carboxyl groups ionize in aqueous solution. The C—H bonds are relatively nonpolar and strong as well, and so most of the hydrogen atoms are not acidic. In the carboxyl group the second oxygen atom on carbon helps pull electrons away from hydrogen, making the H—O bond polar and making the H acidic.

```
    H      O
    |     //           acidic hydrogen
H—C—C                 ↙
    |     \
    H      O—H
   acetic acid
    CH3COOH
```

To accept a proton and serve as a Brønsted base, a molecule or ion must have an unshared pair of electrons. It also helps if the ion has a negative charge, since that can attract and hold a proton. Examples of strong Brønsted bases are hydroxide ion, OH^-, sulfide ion, S^{2-}, oxide ion, O^{2-}, amide ion, NH_2^-, and hydride ion, H^-. All have lone pairs and negative charges. Notice that this broadens the number of strong-base species considerably over what appears in Table 5.1. In addition, there are a great many weak bases that are negative ions with unshared pairs of electrons. Exam-

Acetic acid is a very common weak acid. It is found in vinegar and many foods. Acetic acid and acetate ion play an important role in metabolism.

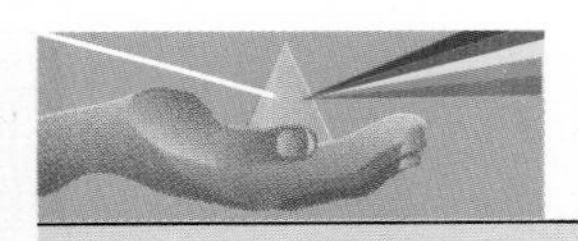

CHEMISTRY YOU CAN DO

Acid-Base Reactions

When acids react with bases, there is often little indication that anything is going on. In the Chemistry You Can Do in Section 5.3, you used red-cabbage indicator to show whether a substance was acidic or basic. But what if you want to see an acid-base reaction itself? Some bases react with acids to form $CO_2(g)$. One of these is baking soda, sodium bicarbonate, $NaHCO_3$. It is used around the lab to react with any spilled acid, and around the home for cooking and cleaning. It can also be used to test for acids.

$$NaHCO_3(s) \xrightarrow{\text{acid}} H_2CO_3(aq) \longrightarrow CO_2(g) + H_2O(\ell)$$

Some common acids that will react with baking soda include soft drinks and vinegar. Obtain some sodium bicarbonate from the grocery store and test it with a list of potential acids. (A good start would be the list you generated in the Chemistry You Can Do in Section 5.3.) You can place dry sodium bicarbonate in a small container such as a jar cap, add some of the acid being tested, and (if the acid is not a liquid) some water. Make a list of all the acids you can find that generate $CO_2(g)$ from baking soda.

ples are Cl^-, F^-, and CH_3COO^-. Neutral molecules that have lone pairs, such as NH_3 and H_2O, have already been seen to serve as weak Brønsted bases.

EXERCISE 16.1 • *Brønsted Acids and Bases*

Identify each molecule or ion as a Brønsted acid or base. Which acids are strong acids? (a) HBr, (b) Br^-, (c) HNO_2, (d) PH_3, (e) H_2SO_3, (f) $HClO_4$, (g) CN^-

Conjugate Acid-Base Pairs

Whenever an acid donates a proton to a base, a new acid and a new base are formed. This can be understood by looking at the reaction between acetic acid, CH_3COOH, and water. The new acid is H_3O^+ and the new base is CH_3COO^-.

$$\underset{\text{acid}}{CH_3COOH(aq)} + \underset{\text{base}}{H_2O(\ell)} \rightleftharpoons \underset{\text{acid}}{H_3O^+(aq)} + \underset{\text{base}}{CH_3COO^-(aq)}$$

The CH_3COOH and CH_3COO^- are related to one another by the transfer of a single proton, just as the H_2O and H_3O^+ are.

A pair of molecules or ions *related to one another by the gain or loss of a single proton* is called a **conjugate acid-base pair.** Every Brønsted acid has a conjugate base and every Brønsted base has a conjugate acid. The conjugate base can be derived from the formula of an acid by removing a H^+ ion; that is, removing H and making the charge of the acid one unit more negative. The conjugate acid can be derived from the formula of a base by adding a H^+ ion; that is, adding H and making the charge of the base one unit more positive.

EXERCISE 16.2 • *Conjugate Acid-Base Pairs*

(a) What is the conjugate base of H_2CO_3? Of HNO_3? (b) What is the conjugate acid of NH_3? Of CN^-?

Polyprotic Acids and Bases

So far we have discussed mainly Brønsted acids that can donate a single proton, such as hydrogen fluoride (HF), hydrogen chloride (HCl), nitric acid (HNO_3), and hydrocyanic acid (HCN); these are called **monoprotic acids.** Similarly, a base that can accept only one proton, such as NH_3, is called a **monoprotic base.**

There are many acids that can donate more than one proton. These are called **polyprotic acids** and include sulfuric acid (H_2SO_4), carbonic acid (H_2CO_3), phosphoric acid (H_3PO_4), oxalic acid ($H_2C_2O_4$ or $HO_2C—CO_2H$), and other organic acids with two or more carboxyl ($—CO_2H$) groups (Table 13.8). In aqueous solution, a polyprotic acid such as H_2SO_4 donates its protons to water molecules in a stepwise manner. In the first step, hydrogen sulfate ion is formed. Sulfuric acid is a strong acid, and so this first ionization is complete.

Many chemical reactions take place in steps that can be represented by individual equations. Sometimes only the overall reaction is seen and only the overall equation is written.

$$\underset{\text{acid}}{H_2SO_4(aq)} + \underset{\text{base}}{H_2O(\ell)} \longrightarrow \underset{\text{base}}{HSO_4^-(aq)} + \underset{\text{acid}}{H_3O^+(aq)}$$

Hydrogen sulfate ion, HSO_4^-, which is amphiprotic and can serve as a monoprotic base or a monoprotic acid, then donates a proton to another water molecule. In this case an equilibrium is established—hydrogen sulfate ion is a weak acid.

$$\underset{\text{acid}}{HSO_4^-(aq)} + \underset{\text{base}}{H_2O(\ell)} \rightleftharpoons \underset{\text{base}}{SO_4^{2-}(aq)} + \underset{\text{acid}}{H_3O^+(aq)}$$

There are also **polyprotic bases,** substances that can accept more than one proton, although they do it one proton at a time. The most common polyprotic bases are the anions of the polyprotic acids such as PO_4^{3-} (from phosphoric acid), CO_3^{2-} (from carbonic acid), and $C_2O_4^{2-}$ (from oxalic acid). Another example of a polyprotic base is sulfide ion, S^{2-}, which can accept two protons. The first reaction produces hydrogen sulfide ion.

$$\underset{\text{base}}{S^{2-}(aq)} + \underset{\text{acid}}{H_2O(\ell)} \longrightarrow \underset{\text{acid}}{HS^-(aq)} + \underset{\text{base}}{OH^-(aq)}$$

Then the second reaction produces hydrosulfuric acid, H_2S.

$$\underset{\text{base}}{HS^-(aq)} + \underset{\text{acid}}{H_2O(\ell)} \rightleftharpoons \underset{\text{acid}}{H_2S(aq)} + \underset{\text{base}}{OH^-(aq)}$$

Again an amphiprotic species, HS^-, is formed in the first step and reacts in the second.

For an amphiprotic species like HS^- a conjugate acid-base pair is defined by whether a proton is gained or lost. When HS^- acts as an acid, the conjugate pair is HS^-/S^{2-}; when it acts as a base, the conjugate pair is H_2S/HS^-. The relations among polyprotic acids and bases and their amphiprotic anions are shown in Table 16.1.

EXERCISE 16.3 • *Polyprotic Acids and Bases*

(a) Write equations for three stepwise reactions that occur when PO_4^{3-} ion acts as a polyprotic base in aqueous solution. (b) Write equations for two stepwise reactions that occur when $H_2C_2O_4$ acts as a polyprotic acid in aqueous solution.

Table **16.1**

Polyprotic Acids and Bases

Acid Form	Amphiprotic Form	Base Form
H_2S (hydrosulfuric acid or hydrogen sulfide)	HS^- (hydrogen sulfide ion)	S^{2-} (sulfide ion)
H_3PO_4 (phosphoric acid)	$H_2PO_4^-$ (dihydrogen phosphate ion)	HPO_4^{2-} (hydrogen phosphate ion)
$H_2PO_4^-$ (dihydrogen phosphate ion)	HPO_4^{2-} (hydrogen phosphate ion)*	PO_4^{3-} (phosphate ion)
H_2CO_3 (carbonic acid)	HCO_3^{2-} (hydrogen carbonate or bicarbonate ion)*	CO_3^{2-} (carbonate ion)
$H_2C_2O_4$ (oxalic acid)	$HC_2O_4^-$ (hydrogen oxalate ion)	$C_2O_4^{2-}$ (oxalate ion)

*The amphiprotic nature of HPO_4^{2-} and HCO_3^- is especially important in the chemical reactions that occur in living organism. (See Section 16.7.)

16.2 THE pH SCALE

Recall from Section 8.2 that an equilibrium-constant expression includes concentrations of solutes, but not the concentration of the solvent, which in this case is water.

Autoionization of water is an equilibrium reaction that lies far to the left; that is, at equilibrium, only a tiny fraction of all water molecules have ionized. As in the case of any equilibrium reaction, an equilibrium-constant expression can be written for auto-ionization of water:

$$2\,H_2O(\ell) \rightleftharpoons H_3O^+(aq) + OH^-(aq) \qquad K_w = [H_3O^+][OH^-]$$

This equilibrium constant is given a special symbol, K_w, and is known as the **ionization constant for water.** From electrical conductivity measurements of pure water, we know that $[H_3O^+] = [OH^-] = 1.0 \times 10^{-7}$ M at 25 °C, and so

Although a product of concentrations has units (in this case M^2), once an equilibrium constant has been defined, its units are usually not included, a practice that we shall follow from now on.

$$\begin{aligned} K_w &= [H_3O^+][OH^-] \\ &= (1.0 \times 10^{-7}\text{ M})(1.0 \times 10^{-7}\text{ M}) = 1.0 \times 10^{-14}\text{ M}^2 \text{ (at 25 °C)} \end{aligned}$$

According to this mathematical expression, the product of the hydronium-ion concentration and the hydroxide-ion concentration will always remain the same. If hydronium-ion concentration increases (because an acid was added to the water, for example), then the hydroxide-ion concentration must decrease, and vice versa. The equation also tells us that if we know one concentration, the other can be calculated, which means that it is unnecessary to try to measure it.

The equation $K_w = [H_3O^+][OH^-]$ is valid in pure water and in any aqueous solution. However, K_w is temperature-dependent; the auto-ionization reaction is endothermic, and so K_w increases with increasing temperature.

When the concentrations of $[H_3O^+]$ and $[OH^-]$ are equal, a solution is said to be **neutral.** If either an acid or a base is added to a neutral solution, the auto-ionization equilibrium between H_3O^+ and OH^- will be disturbed. Recall that according to Le Chatelier's principle (Section 8.2), an equilibrium shifts in such a way as to oppose the effect of any disturbance. When an acid is added, the concentration of H_3O^+ ions increases; to oppose this increase a small fraction of the added H_3O^+ ions react with OH^- ions to form H_2O, thereby reducing the $[OH^-]$ until once again $[H_3O^+][OH^-] = 1.0 \times 10^{-14}$ at 25°C. Similarly, if a base is added to water, a few of the added OH^- ions react with H_3O^+ ions to form H_2O, thereby decreasing the $[H_3O^+]$. When a new equilibrium is achieved, the product $[H_3O^+][OH^-]$ again equals 1.0×10^{-14}. For aqueous solutions at 25 °C we can write:

T (°C)	K_w
10	0.29×10^{-14}
15	0.45×10^{-14}
20	0.68×10^{-14}
25	1.01×10^{-14}
30	1.47×10^{-14}
50	5.48×10^{-14}

- **Neutral solution:** $[H_3O^+]$ equals $[OH^-]$
  both are equal to 1.0×10^{-7} M
- **Acidic solution:** $[H_3O^+]$ greater than 1.0×10^{-7} M
  $[OH^-]$ less than 1.0×10^{-7} M
- **Basic solution:** $[OH^-]$ greater than 1.0×10^{-7} M
  $[H_3O^+]$ less than 1.0×10^{-7} M

The $[H_3O^+]$ and $[OH^-]$ in an aqueous solution can vary widely depending on the acid or base present and its concentration. A 6.0 M nitric acid solution, which is a fairly concentrated solution, has a $[H_3O^+]$ of 6.0 M, so

$$[H_3O^+][OH^-] = 1.0 \times 10^{-14}$$

$$(6.0)[OH^-] = 1.0 \times 10^{-14}$$

$$[OH^-] = \frac{1.0 \times 10^{-14}}{6.0} = 1.7 \times 10^{-15}\ \text{M}$$

On the other hand, consider the strong base NaOH. A 6.0 M solution of NaOH would have a $[OH^-]$ of 6.0 M, and a similar calculation gives $[H_3O^+] = 1.7 \times 10^{-15}$ M. The $[H_3O^+]$ in aqueous solutions can range from about 10 down to about 10^{-15}.

In terms of the Brønsted definition of a base, it is OH^- rather than NaOH that is the base.

Because the H_3O^+ and OH^- concentrations vary over such a great range in aqueous solutions, it is more convenient to express them in terms of logarithms (exponents). Because the H_3O^+ and OH^- concentrations are usually small and involve negative powers of ten, it is convenient to take the negative of the logarithm. The **pH** of a solution is defined as *the negative of the base-10 logarithm (log) of the hydronium ion concentration.*

$$\text{pH} = -\log[H_3O^+]$$

Thus, the pH of pure water at 25 °C is given by

$$\text{pH} = -\log[1.0 \times 10^{-7}] = 7.00$$

Notice that, as in the values of equilibrium constants, the concentration units of mol/L are ignored when the logarithm is taken. It is impossible to take the logarithm of a unit.

We can define a pOH in a manner similar to the definition of pH:

$$\text{pOH} = -\log[OH^-]$$

The pOH of pure water at 25 °C is also 7.00. Because the values of $[H_3O^+]$ and $[OH^-]$ are related by the K_w expression, for all aqueous solutions at 25 °C,

$$\begin{aligned} K_w = \quad [H_3O^+] \quad &\times \quad [OH^-] = 1.0 \times 10^{-14} \\ -\log[H_3O^+] + (-&\log[OH^-]) = -\log(1.0 \times 10^{-14}) \\ \text{pH} \quad &+ \quad \text{pOH} = 14.00 \end{aligned}$$

A Deeper Look

Sulfuric Acid

It may surprise you to know that a diprotic acid, H_2SO_4, is the number one chemical in commerce. More than 80 billion pounds of it are produced and sold each year in the United States alone. Sulfuric acid is used to manufacture fertilizers, in the petroleum industry, in the production of steel, in automobile batteries, and in the manufacture of organic dyes, plastics, drugs, and many other products. Since sulfuric acid costs less to manufacture than any other acid, it is the first to be considered when an acid is needed. Its cost, about 1 cent per pound, has not changed much in 300 years, a tribute to improving technology for its production from natural sources. It is now also produced from wastes that otherwise would pollute the environment.

Sulfur, in very pure form, is found in underground deposits in the United States along the coast of the Gulf of Mexico. It is recovered by the **Frasch process,** which melts the sulfur with superheated steam. The molten sulfur is raised to the surface of the earth by means of compressed air, and is then allowed to cool in large vats. Large quantities of sulfur are also produced from petroleum (which often contains sulfur compounds), and sulfuric acid can be produced directly from copper or lead smelter wastes that consist of sulfur dioxide.

Sulfur is converted to sulfuric acid in four steps, collectively called the **contact process.** In the first step the sulfur is burned in air to give mostly sulfur dioxide:

$$S_8(s) + 8\,O_2(g) \longrightarrow 8\,SO_2(g)$$

The gaseous SO_2 is then converted to SO_3 by passing it over a hot, catalytically active surface, such as platinum or vanadium pentoxide:

$$2\,SO_2(g) + O_2(g) \xrightarrow{\text{catalyst}} 2\,SO_3(g)$$

Although SO_3 could be converted directly into H_2SO_4 by passing the SO_3 into water, the enormous heat release from this reaction would cause the formation of a stable fog of $H_2SO_4(\ell)$. This is avoided by passing the SO_3 into H_2SO_4 to form pyrosulfuric acid:

$$\underset{\text{}}{SO_3(g)} + \underset{\text{sulfuric acid}}{H_2SO_4(\ell)} \longrightarrow \underset{\text{pyrosulfuric acid}}{H_2S_2O_7(\ell)}$$

and then diluting the $H_2S_2O_7$ with water:

$$H_2S_2O_7(\ell) + H_2O(\ell) \longrightarrow 2\,H_2SO_4(aq)$$

The Frasch process for mining sulfur. Superheated steam is injected along with compressed air into a sulfur-bearing stratum underground. A molten sulfur froth pours out of the inner tube. Most of the sulfur mined is converted to sulfuric acid. (See the endpapers of this book.)

If you know pH, then pOH is just 14.00 − pH; if you know pOH, then pH = 14.00 − pOH.

The relation between pH and pOH can be used to find one value when the other is known. A 0.0010 M solution of the strong base NaOH, for example, has an OH^- concentration of 0.0010 M and a pOH given by

$$\text{pOH} = -\log[1.0 \times 10^{-3}] = 3.00$$

and therefore a pH given by

$$\text{pH} = 14.00 - \text{pOH} = 14.00 - 3.00 = 11.00$$

For solutions in which $[H_3O^+]$ or $[OH^-]$ has a value other than an exact power of ten (1, 0.1, 0.01, . . .) a calculator is convenient for finding the pH (see Appendix B.3). For example, the pH of a solution that contains 0.0040 mol of the strong acid HNO_3 per liter is 2.40.

$$\text{pH} = -\log[4.0 \times 10^{-3}] = 2.40$$

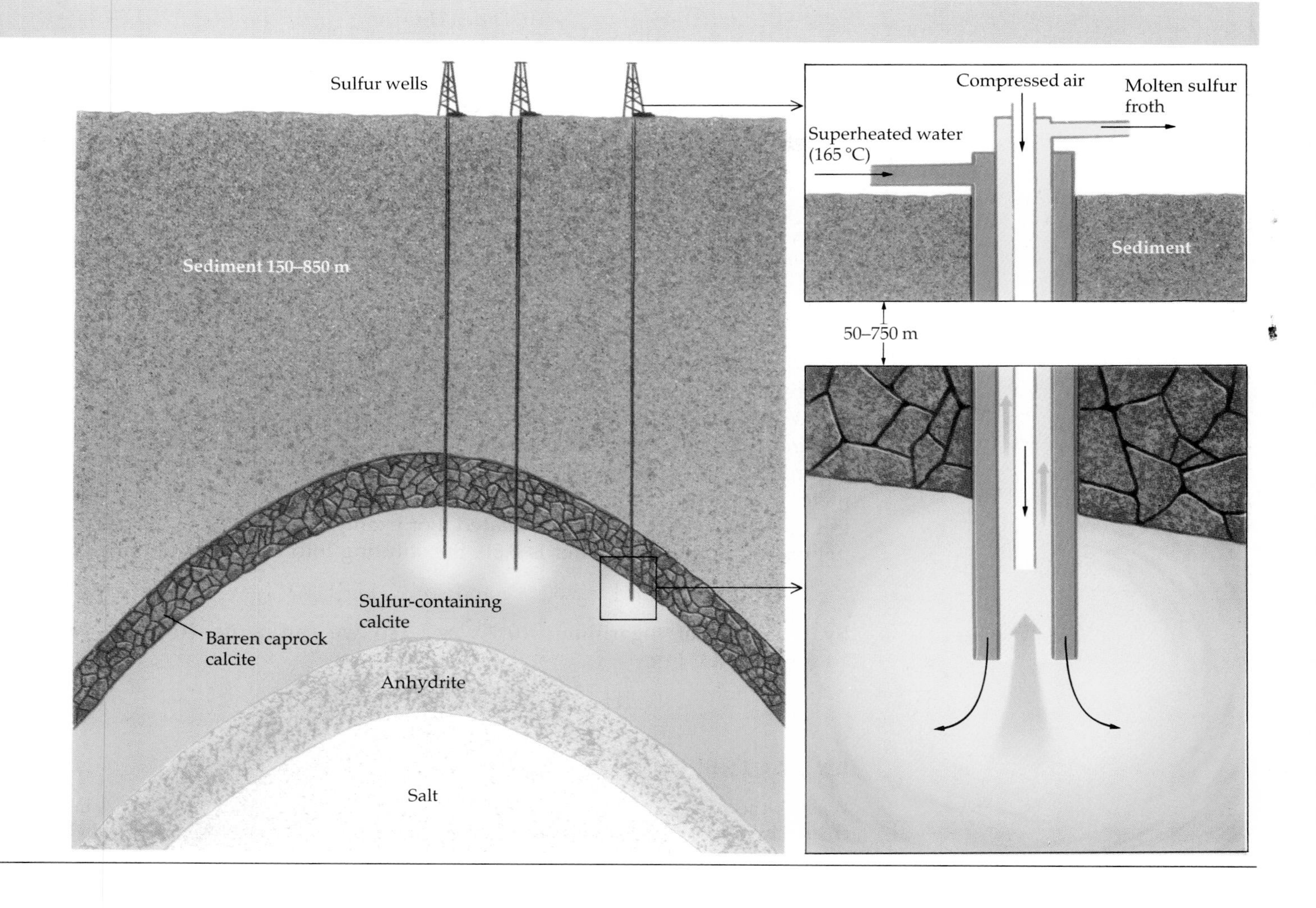

EXAMPLE 16.1

Calculating pH from $[H_3O^+]$

If an aqueous solution of HCl has a volume of 500. mL and contains 1.25 g of HCl, what is its pH?

SOLUTION Hydrochloric acid is a strong acid, and so every mole of HCl that dissolves produces a mole of H_3O^+ and a mole of Cl^-. First, determine the amount (moles) of HCl.

$$\text{amount of HCl} = 1.25 \text{ g HCl} \cdot \frac{1 \text{ mol HCl}}{36.46 \text{ g HCl}} = 0.0343 \text{ mol HCl}$$

Next, calculate the H_3O^+ concentration.

The digits to the left of the decimal point in a pH represent a power of ten; only the digits to the right of the decimal point are significant. In Example 16.1, where pH = $-\log(6.86 \times 10^{-2})$ = $-\log(6.86) + (-\log(10^{-2})) = -0.836 + 2.000 = 1.164$, there are only three significant figures in the result, because there were only three significant figures in -0.836, which is log(6.86).

$$[H_3O^+] = \frac{0.0343 \text{ mol } H_3O^+}{0.500 \text{ L}} = 0.0686 \text{ M}$$

Then express this concentration as pH.

$$pH = -\log(6.86 \times 10^{-2}) = 1.164$$

EXERCISE 16.4 • *pH from [OH⁻]*

Calculate the pH of a 0.040 M NaOH solution.

Figure 16.6 shows the pH values of some common solutions. When studying this figure, it is important to keep in mind that a change of one pH unit represents a 10-fold change in H_3O^+ concentration, two pH units represent a 100-fold change, and so on. Thus, according to Figure 16.6, the $[H_3O^+]$ in lemon juice is more than 100 times greater than in tomato juice. Values in Figure 16.6 were probably obtained by using an instrument called a pH meter to measure the pH of each solution. Once pH has been measured, the H_3O^+ concentration can be calculated. For example, suppose the measured pH of a sample of sea water is 8.30. Substituting into the definition of pH,

$$-\log[H_3O^+] = 8.30 \quad \text{and so} \quad \log[H_3O^+] = -8.30$$

By the rules of logarithms, $10^{\log(x)} = x$, so we can write $10^{\log[H_3O^+]} = 10^{-pH} = [H_3O^+]$ or

$$[H_3O^+] = 10^{-8.30} = 5.0 \times 10^{-9} \text{ M}$$

EXERCISE 16.5 • *Working with pH*

What is the pH of a solution that has $[H_3O^+] = 1.5 \times 10^{-10}$ M? If the $[H_3O^+]$ increases by a factor of 3000, what will the new pH be? Is this new solution acidic?

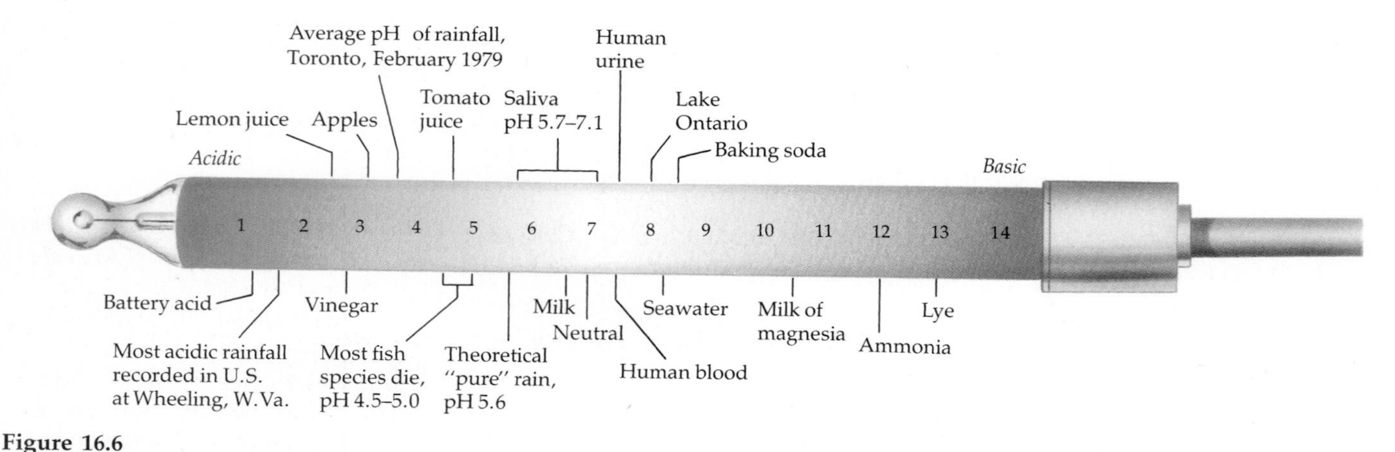

Figure 16.6

pH values of some common aqueous solutions. The scale is superimposed on the drawing of a pH electrode from an instrument used to measure pH.

16.3
RELATIVE STRENGTHS OF ACIDS AND BASES

The acidity of a solution of a single acid depends on two factors: the concentration of the acid and the strength of the acid. *For a given acid, the more concentrated the solution, the lower the pH; for a given concentration, the stronger the acid, the lower the pH.* How concentration of a strong acid affects pH has been shown in Example 16.1. Now it is time to consider the effect of acid strength. It is useful to define acid strength in terms of a numeric quantity, so that bigger numbers correspond to stronger acids. In this section we will show that equilibrium constants for acid ionization reactions provide this information.

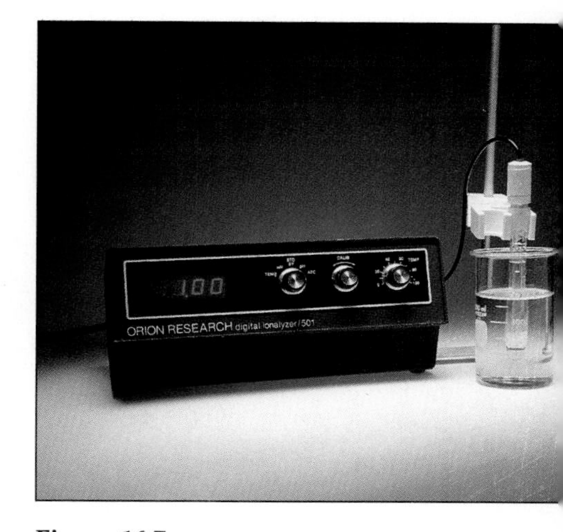

Figure 16.7
A pH meter electrode in a dilute solution of 0.1 M nitric acid.

As an example of a strong acid, consider a 0.10 M solution of nitric acid (Figure 16.7). The pH of this solution is 1.00, which indicates that the concentration of H_3O^+ must be 0.10 M. This also means that the solution contains almost no undissociated HNO_3; by stoichiometry, 0.10 mol/L of HNO_3 must have been consumed if 0.10 mol/L of H_3O^+ was produced in the solution. Clearly the ionization reaction of HNO_3 with water is product-favored, and the equilibrium constant must therefore be greater than one. In fact, the equilibrium constant has a value of about 20.

$$HNO_3(aq) + H_2O(\ell) \longrightarrow H_3O^+(aq) + NO_3^-(aq) \qquad K \approx 20$$

Nitric acid is a liquid at room temperature, and it is rather difficult to handle safely because it is highly corrosive and a very strong oxidizing agent. However, suppose that you could take 0.10 mol of $HNO_3(\ell)$ and dissolve it in enough water to make 1.00 L of solution. As soon as the HNO_3 dissolves in the water, HNO_3 molecules begin to react with water molecules and ionize according to the preceding equation. This reaction will proceed from left to right until almost every HNO_3 molecule has been consumed, at which point equilibrium will be achieved. Remember that this equilibrium is dynamic. HNO_3 molecules continue to ionize, and at the same time H_3O^+ and NO_3^- ions recombine to form HNO_3 and water molecules. The large value of the equilibrium constant means that the equilibrium lies quite far to the right.

What happens when nitric acid dissolves and ionizes in water can be described quantitatively in a table of concentrations that includes all solutes in the solution.

	$HNO_3(aq)$ + $H_2O(\ell)$	$\longrightarrow$ $H_3O^+(aq)$	+ $NO_3^-(aq)$
initial concentration (mol/L)	0.10	1.0×10^{-7}	0
change in concentration on reaction (mol/L)	−0.10	+0.10	+0.10
concentration at equilibrium (mol/L)	$\approx 5 \times 10^{-4}$	0.10	0.10

We usually do not write the ionization reaction of a strong acid as an equilibrium, with double arrows, even though it is. A single arrow is used instead; this indicates that the equilibrium lies quite far to the right.

The initial concentration of $HNO_3(aq)$ is the concentration before any ionization takes place. This concentration decreases very rapidly as soon as the nitric acid contacts the water; when equilibrium is reached it is very small (about 5×10^{-4} M). Since you started with pure water, the initial concentra-

Auto-ionization of water results in $[H_3O^+] = 1.0 \times 10^{-7}$ mol/L initially, but the presence of even a little weak acid adds H_3O^+ to the solution and causes the auto-ionization equilibrium

$$2\ H_2O(\ell) \rightleftharpoons H_3O^+(aq) + OH^-(aq)$$

to shift to the left, making water's contribution of H_3O^+ negligible at equilibrium.

tion of $H_3O^+(aq)$ is 1.0×10^{-7} M, which is negligibly small compared to the concentration of $H_3O^+(aq)$ produced by the nitric acid. The initial concentration of $NO_3^-(aq)$ is obviously zero. These latter two concentrations build up very rapidly once the nitric acid contacts the water and eventually reach 0.10 M.

Although there are almost no HNO_3 molecules remaining at equilibrium, this solution is still referred to as 0.10 M HNO_3. The 0.10 M concentration is referred to as the analytic concentration of this HNO_3 solution, because it is the concentration that would be determined by a titration in which all the acid reacted with base (Section 6.6). It is important to distinguish the actual concentrations of ions and molecules in a solution from the analytic concentration of the solution.

In contrast to strong acids, weak acids such as acetic acid ionize only to a small extent; they establish equilibria in which significant concentrations of weak-acid molecules are still present. The ionization of a weak acid is reactant-favored. If a pH measurement is made on a solution of 0.10 M acetic acid, the pH is found to be 2.88. This means that the concentration of H_3O^+ is only 1.3×10^{-3} M. According to the following ionization equation, one mole of acetate ion is produced for every mole of hydronium ion produced, and so the concentration of CH_3COO^- must be the same as the concentration of H_3O^+. These concentrations are only a little more than 1% of the initial concentration of acetic acid. Therefore, almost 99% of the acetic acid remains in the nonionized form.

	$CH_3COOH(aq)$ +	$H_2O(\ell)$ $\rightleftharpoons$	$H_3O^+(aq)$ +	$CH_3COO^-(aq)$
	acid	base	acid	base
initial concentration (mol/L)	0.100		1.0×10^{-7}	0
change in concentration on reaction (mol/L)	−0.0013		+0.0013	+0.0013
concentration at equilibrium (mol/L)	0.099		0.0013	0.0013

Another way of looking at this acetic acid solution is to recall that according to the Brønsted concept there are really two bases and two acids present. You can think of the two bases as competing for protons that can be donated from the two acids. Since the equilibrium favors nonionized acetic acid molecules, the acetate ion must be a more effective (stronger) base than the water molecule. In other words, the acetate ion accepts H^+ more strongly than does the water molecule, and so acetic acid is formed. It can also be said that H_3O^+ must be a stronger acid than CH_3COOH, because H_3O^+ donates its proton more effectively, causing the equilibrium to lie toward the reactants.

Ionization Constants of Acids and Bases

In Section 8.2 you learned that the greater the equilibrium constant for a reaction, the more product-favored that reaction is. Thus the equilibrium constant for ionization of an acid indicates whether the reaction is product-favored. The more product-favored the reaction, the stronger the acid and the base on the left-hand side of the ionization equation. Consequently, equilibrium constants can be used to express the relative strengths of weak acids and bases.

For any acid, represented by the general formula HA, an ionization reaction can be written as

$$HA(aq) + H_2O(\ell) \rightleftharpoons H_3O^+(aq) + A^-(aq)$$

and the corresponding equilibrium-constant expression is

$$K_a = \frac{[H_3O^+][A^-]}{[HA]}$$

The equilibrium constant, K_a, is called the **acid ionization constant** for the acid. The larger the acid ionization constant, the stronger is the acid. For a strong acid the ionization constant is greater than one; for a weak acid the ionization constant is less than one and its size indicates the relative strength of the acid.

A similar general equation can be written for ionization of a base, B.

$$B(aq) + H_2O(\ell) \rightleftharpoons BH^+(aq) + OH^-(aq)$$

(For example, if the base B were NH_3, then BH^+ would be NH_4^+.) The corresponding equilibrium-constant expression is

$$K_b = \frac{[BH^+][OH^-]}{[B]}$$

When the base is an anion, A^- (such as the anion of a weak acid, HA), the ionization reaction can be written as

$$A^-(aq) + H_2O(\ell) \rightleftharpoons HA(aq) + OH^-(aq)$$

and the corresponding equilibrium-constant expression is

$$K_b = \frac{[HA][OH^-]}{[A^-]}$$

The equilibrium constant, K_b, is called the **base ionization constant** for the base. The larger the base ionization constant, the more product-favored the ionization reaction and the stronger the base. For a strong base the ionization constant is greater than one; for a weak base the ionization constant is less than one.

EXERCISE 16.6 • *Writing Ionization Constant Expressions*

Write the ionization equations and ionization constant expressions for these acids: H_2SO_3, HF, HSO_3^-, and HCN; and for these bases: NO_2^-, HS^-, CN^-, and PO_4^{3-}.

There are several experimental methods for determining acid or base ionization constants. The simplest (but not the most accurate) is based on the ideas we have just been discussing. The stronger an acid, the more its ionization equation lies toward the right-hand side; that is, the greater the concentration of H_3O^+ ions in the solution. Therefore, when we measure the pH of an acid solution of known concentration, we find that the lower the pH, the stronger the acid. In addition, it is possible to use the pH and the acid concentration to calculate the value of K_a for the acid.

EXAMPLE 16.2

Determining K_a *of a Weak Acid from pH Measurement*

The pH of a 0.10 M solution of propanoic acid, CH_3CH_2COOH, a weak organic acid, is measured after equilibrium has been established and it is found to be 2.93. What is K_a for propanoic acid?

SOLUTION The acid ionizes according to the balanced equation

$$CH_3CH_2COOH(aq) + H_2O(\ell) \rightleftharpoons H_3O^+(aq) + CH_3CH_2COO^-(aq)$$

The pH gives us the equilibrium concentration of H_3O^+. Using the definition of pH,

$$[H_3O^+] = 10^{-pH} = 10^{-2.93} = 1.2 \times 10^{-3}$$

The equilibrium concentrations of the other species can be determined by using the following table.

	$[CH_3CH_2COOH]$	$[H_3O^+]$	$[CH_3CH_2COO^-]$
initial concentration (mol/L)	0.10	1.0×10^{-7}	0
change in concentration on reaction (mol/L)	$-x$	$+x$	$+x$
concentration at equilibrium (mol/L)	$0.10 - x$	x	x

The quantity x represents the concentration of both H_3O^+ and $CH_3CH_2COO^-$ at equilibrium. As we have already calculated, $[H_3O^+] = 1.2 \times 10^{-3}$ M, and so $x = [CH_3CH_2COO^-] = 1.2 \times 10^{-3}$ M. By stoichiometry, x also represents the amount of acid that has ionized per liter of solution at equilibrium. Using these values, we can now derive K_a for propanoic acid.

$$K_a = \frac{[H_3O^+][CH_3CH_2COO^-]}{[CH_3CH_2COOH]} = \frac{(x)(x)}{0.10 - x} = \frac{(0.0012)(0.0012)}{0.10 - 0.0012} = 1.5 \times 10^{-5}$$

This K_a is small, indicating that propanoic acid is a weak acid. It is similar to acetic acid in strength, and a 0.10 M solution has a pH nearly the same as that of 0.10 M acetic acid.

When a molecular formula is used to represent an organic acid, the acidic hydrogen or hydrogens are often written first. Thus, $HC_3H_5O_3$ shows that lactic acid has one acidic hydrogen atom.

EXERCISE 16.7 • K_a *from pH Measurement*

Lactic acid is a monoprotic acid that occurs naturally in sour milk and arises from metabolism in the human body. A 0.10 M aqueous solution of lactic acid, $HC_3H_5O_3$, has a pH of 2.43. What is the value of K_a for lactic acid? Is lactic acid stronger or weaker than propanoic acid?

The results of a great many measurements of the type just discussed (as well as measurements of several other types) are summarized in Table 16.2, which gives the ionization constants for a number of weak acids and their conjugate bases.

The ionization constants for strong acids (those above H_3O^+ in Table 16.2) and strong bases (those below OH^- in Table 16.2) are too large to be measured easily. Fortunately, since the ionization reactions are virtually complete, these K_a and K_b values are hardly ever needed. For weak acids K_a values show relative strengths quantitatively, and for weak bases K_b values do the same. Consider acetic acid and boric acid. Since boric acid is below acetic acid in Table 16.2, boric acid must be weaker than acetic acid; the K_a values tell us how much weaker. The K_a for boric acid is 7.3×10^{-10}, 2.5×10^4 times smaller than K_a for acetic acid (1.8×10^{-5}), and so boric acid is more than 10^4 times weaker. In fact, boric acid is such a weak acid that it can be used safely as an eyewash. Don't try that with acetic acid!

EXERCISE 16.8 • K_b *and Base Strength*

Which is the stronger base, acetate ion or hydrogen sulfide ion? By what numeric factor is it stronger?

Relationship between K_a and K_b Values

The right-hand side of Table 16.2 gives K_b for the conjugate base of each acid. Try an experiment with these data: multiply each K_a value by the K_b value for the conjugate base. What do you find? Within a very small error you ought to find that $K_a \times K_b = 1.0 \times 10^{-14} = K_w$, the autoionization constant for water! Why should this be? To see why, multiply the equilibrium constant expressions for K_a and K_b and see what you get:

$$K_a \cdot K_b = \frac{[H_3O^+][\cancel{A^-}]}{[\cancel{HA}]} \cdot \frac{[\cancel{HA}][OH^-]}{[\cancel{A^-}]} = [H_3O^+][OH^-] = K_w$$

This relation tells you that, if you know K_a for an acid, you can find K_b for its conjugate base by using K_w. Furthermore, the larger the K_a, the smaller the K_b, and vice versa (because they always have to multiply to give the same product, namely K_w). For example, K_a for HCN is 4.0×10^{-10}. The value of K_b for the conjugate base, CN^-, is

$$K_b \text{ (for } CN^-) = \frac{K_w}{K_a \text{ (for HCN)}} = \frac{1.0 \times 10^{-14}}{4.0 \times 10^{-10}} = 2.5 \times 10^{-5}$$

HCN has a relatively small K_a and lies fairly far down in Table 16.2, which means it is a relatively weak acid. However, CN^- is a fairly strong base; its K_b of 2.5×10^{-5} is nearly the same as the K_a (1.8×10^{-5}) for acetic acid, making CN^- a little stronger as a base than acetic acid is as an acid.

The cyanide ion is a deadly poison, attacking the enzyme cytochrome oxidase found in the mitochondria of almost every cell in living organisms.

EXERCISE 16.9 • K_b *from* K_a

Phenol, or carbolic acid, C_6H_5OH, is a weak acid, $K_a = 1.3 \times 10^{-10}$. Calculate K_b for the phenolate ion, $C_6H_5O^-$. Which base in Table 16.2 is closest in strength to the phenolate ion? How did you make your choice?

Table 16.2

Ionization Constants for Some Acids and Their Conjugate Bases

Increasing Acid Strength ↑ — Increasing Base Strength ↓

Acid Name	Acid	K_a	Base	K_b	Base Name
perchloric acid	$HClO_4$	large	ClO_4^-	very small	perchlorate ion
sulfuric acid	H_2SO_4	large	HSO_4^-	very small	hydrogen sulfate ion
hydrochloric acid	HCl	large	Cl^-	very small	chloride ion
nitric acid	HNO_3	≈ 20	NO_3^-	$\approx 5 \times 10^{-16}$	nitrate ion
hydronium ion	H_3O^+	1.0	H_2O	1.0×10^{-14}	water
sulfurous acid	H_2SO_3	1.2×10^{-2}	HSO_3^-	8.3×10^{-13}	hydrogen sulfite ion
hydrogen sulfate ion	HSO_4^-	1.2×10^{-2}	SO_4^{2-}	8.3×10^{-13}	sulfate ion
phosphoric acid	H_3PO_4	7.5×10^{-3}	$H_2PO_4^-$	1.3×10^{-12}	dihydrogen phosphate ion
hexaaquairon(III) ion	$Fe(H_2O)_6^{3+}$	6.3×10^{-3}	$Fe(H_2O)_5OH^{2+}$	1.6×10^{-12}	pentaaquahydroxoiron(III) ion
hydrofluoric acid	HF	7.2×10^{-4}	F^-	1.4×10^{-11}	fluoride ion
nitrous acid	HNO_2	4.5×10^{-4}	NO_2^-	2.2×10^{-11}	nitrite ion
formic acid	HCO_2H	1.8×10^{-4}	HCO_2^-	5.6×10^{-11}	formate ion
benzoic acid	$C_6H_5CO_2H$	6.3×10^{-5}	$C_6H_5CO_2^-$	1.6×10^{-10}	benzoate ion
acetic acid	CH_3CO_2H	1.8×10^{-5}	$CH_3CO_2^-$	5.6×10^{-10}	acetate ion
propanoic acid	$CH_3CH_2CO_2H$	1.4×10^{-5}	$CH_3CH_2CO_2^-$	7.1×10^{-10}	propanoate ion
hexaaquaaluminum ion	$Al(H_2O)_6^{3+}$	7.9×10^{-6}	$Al(H_2O)_5OH^{2+}$	1.3×10^{-9}	pentaaquahydroxoaluminum ion
carbonic acid	H_2CO_3	4.2×10^{-7}	HCO_3^-	2.4×10^{-8}	hydrogen carbonate ion
hexaaquacopper(II) ion	$Cu(H_2O)_6^{2+}$	1.6×10^{-7}	$Cu(H_2O)_5OH^+$	6.25×10^{-8}	pentaaquahydroxocopper(II) ion
hydrogen sulfide	H_2S	1×10^{-7}	HS^-	1×10^{-7}	hydrogen sulfide ion
dihydrogen phosphate ion	$H_2PO_4^-$	6.2×10^{-8}	HPO_4^{2-}	1.6×10^{-7}	hydrogen phosphate ion
hydrogen sulfite ion	HSO_3^-	6.2×10^{-8}	SO_3^{2-}	1.6×10^{-7}	sulfite ion
hypochlorous acid	$HClO$	3.5×10^{-8}	ClO^-	2.9×10^{-7}	hypochlorite ion
hexaaqualead(II) ion	$Pb(H_2O)_6^{2+}$	1.5×10^{-8}	$Pb(H_2O)_5OH^+$	6.7×10^{-7}	pentaaquahydroxolead(II) ion
hexaaquacobalt(II) ion	$Co(H_2O)_6^{2+}$	1.3×10^{-9}	$Co(H_2O)_5OH^+$	7.7×10^{-6}	pentaaquahydroxocobalt(II) ion
boric acid	$B(OH)_3(H_2O)$	7.3×10^{-10}	$B(OH)_4^-$	1.4×10^{-5}	tetrahydroxoborate ion
ammonium ion	NH_4^+	5.6×10^{-10}	NH_3	1.8×10^{-5}	ammonia
hydrocyanic acid	HCN	4.0×10^{-10}	CN^-	2.5×10^{-5}	cyanide ion
hexaaquairon(II) ion	$Fe(H_2O)_6^{2+}$	3.2×10^{-10}	$Fe(H_2O)_5OH^+$	3.1×10^{-5}	pentaaquahydroxoiron(II) ion
hydrogen carbonate ion	HCO_3^-	4.8×10^{-11}	CO_3^{2-}	2.1×10^{-4}	carbonate ion
hexaaquanickel(II) ion	$Ni(H_2O)_6^{2+}$	2.5×10^{-11}	$Ni(H_2O)_5OH^+$	4.0×10^{-4}	pentaaquahydroxonickel(II) ion
hydrogen phosphate ion	HPO_4^{2-}	3.6×10^{-13}	PO_4^{3-}	2.8×10^{-2}	phosphate ion
water	H_2O	1.0×10^{-14}	OH^-	1.0	hydroxide ion
hydrogen sulfide ion	HS^-	$\approx 8 \times 10^{-18}$	S^{2-}	1×10^3	sulfide ion
ethanol	C_2H_5OH	very small	$C_2H_5O^-$	large	ethoxide ion
ammonia	NH_3	very small	NH_2^-	large	amide ion
hydrogen	H_2	very small	H^-	large	hydride ion
methane	CH_4	very small	CH_3^-	large	methide ion

Polyprotic Acids

The successive K_a values for weak polyprotic acids indicate that each ionization step occurs to a lesser extent than the one before it. The weak acid H_3PO_4, for example, has three protons to donate, and hence three ionization reactions.

Phosphoric acid, H_3PO_4, along with carbonic acid, H_2CO_3, imparts taste to cola drinks.

First ionization: $H_3PO_4(aq) + H_2O(\ell) \rightleftharpoons H_3O^+(aq) + H_2PO_4^-(aq)$ $\quad K_a = 7.5 \times 10^{-3}$

Second ionization: $H_2PO_4^-(aq) + H_2O(\ell) \rightleftharpoons H_3O^+(aq) + HPO_4^{2-}(aq)$ $\quad K_a = 6.2 \times 10^{-8}$

Third ionization: $HPO_4^{2-}(aq) + H_2O(\ell) \rightleftharpoons H_3O^+(aq) + PO_4^{3-}(aq)$ $\quad K_a = 3.6 \times 10^{-13}$

K_a for the second ionization is about 10^5 times smaller than K_a for the first, which shows that it is more difficult to remove H^+ from a negatively charged $H_2PO_4^-$ ion than from a neutral H_3PO_4 molecule. The much smaller K_a value for the third ionization shows that it is even more difficult to remove H^+ from a doubly negative HPO_4^{2-} ion.

16.4 ACID AND BASE STRENGTH APPLIED TO CHEMICAL REACTIVITY

The fact that K_a times K_b equals K_w indicates that *the stronger an acid, the weaker its conjugate base,* and conversely, *the stronger a base, the weaker its conjugate acid.* Table 16.2 shows this relationship clearly. The very strongest acids are at the top of the table. They have such weak conjugate bases that in aqueous solution the acids ionize completely to give H_3O^+ ions and their conjugate base anions. An example is HCl, which reacts with water according to the equation

$$HCl(aq) + H_2O(\ell) \longrightarrow H_3O^+(aq) + Cl^-(aq)$$

(The single arrow indicates that this equilibrium lies almost completely to the right.)

On the opposite end of the acid-strength scale (at the bottom of the table) are species such as H_2 and CH_4. They are such weak acids that almost no base exists that is strong enough to take away a proton. Their conjugate bases H^- and CH_3^- are extremely strong; they can take protons from almost any molecule that can act as an acid. In aqueous solution these strong bases will react completely with water molecules to form OH^- and their conjugate-acid molecules.

A compound containing H^- ion is lithium hydride, LiH. What do you think this compound does on contact with water? If you said react with water to form hydrogen, you got it right! The reaction (and a vigorous one at that) is

$$\underset{\text{strong base}}{H^-(aq)} + \underset{\text{acid}}{H_2O(\ell)} \longrightarrow \underset{\text{base}}{OH^-(aq)} + \underset{\text{acid}}{H_2(g)}$$

The basic properties of the hydride ion, H^-. Calcium hydride, CaH_2, is a source of the H^- ion. This ion is such a powerful proton acceptor that it reacts vigorously with water, a proton donor, to give H_2 gas.

Oxide ion, O^{2-}, is another example of a very strong Brønsted base. In aqueous solution it reacts with water to form hydroxide ions.

$$\underset{\text{strong base}}{O^{2-}(aq)} + \underset{\text{acid}}{H_2O(\ell)} \longrightarrow \underset{\text{acid}}{OH^-(aq)} + \underset{\text{base}}{OH^-(aq)}$$

In this equation one OH^- ion is the conjugate acid of the oxide ion, while the other OH^- ion is the conjugate base of the water molecule.

Carbonate ion, CO_3^{2-}, is a weak Brønsted base. It reacts with water to form bicarbonate ion and OH^-. Soluble carbonates such as Na_2CO_3 (washing soda) produce concentrated solutions that are basic enough to burn human skin or to strip old paint from furniture. However, K_b for carbonate ion is less than one (Table 16.2), and not all of the carbonate has reacted to form OH^- when equilibrium is reached. Hence its ionization equation is written with a double arrow.

$$\underset{\text{base}}{CO_3^{2-}(aq)} + \underset{\text{acid}}{H_2O(\ell)} \rightleftharpoons \underset{\text{acid}}{HCO_3^-(aq)} + \underset{\text{base}}{OH^-(aq)}$$

Weaker Brønsted bases such as ammonia, NH_3, dissolve in water to produce solutions containing much smaller concentrations of OH^- ions.

$$\underset{\text{base}}{NH_3(aq)} + \underset{\text{acid}}{H_2O(\ell)} \rightleftharpoons \underset{\text{acid}}{NH_4^+(aq)} + \underset{\text{base}}{OH^-(aq)}$$

EXERCISE 16.10 • *Acid Strength, Base Strength, and Reactivity*

Examine each of the acid-base reactions given in this section so far. Use Table 16.2 to identify the stronger acid and the stronger base in each equation. Can you make a generalization regarding acid-base strength and whether a reaction is product favored?

What you probably discovered from Exercise 16.10 is that *the stronger acid and the stronger base will always react to form a weaker conjugate base and a weaker conjugate acid*. If the stronger acid and base are on the reactant side of an equation, the process will be product favored. If not, it will be reactant favored.

Another important fact is that no matter how strong an acid is, the hydronium ion will always be the strongest acid species in an aqueous solution. Any acid stronger than H_3O^+ will donate a proton to H_2O and be converted to its conjugate base. Conversely, no matter how strong a base is, the hydroxide ion is the strongest base that can exist in aqueous solutions. A base such as S^{2-} that is stronger than OH^- will accept a proton from H_2O and be converted to its conjugate acid. This effect is called the **leveling effect,** because acids above H_3O^+ and bases below OH^- in Table 16.2 cannot exist in aqueous solution. Their strength is leveled (reduced) to the strength of H_3O^+ or OH^-.

It is possible to use Table 16.2 to predict the direction of the equilibrium for many acid/base reactions. For example, what happens when acetic acid, CH_3COOH, and sodium cyanide, NaCN, are mixed in water solution? Will

HCN, a poisonous gas, be produced to any significant extent? Acetic acid is a weak acid and its conjugate base is the acetate ion, CH_3COO^-. When dissolved in water, NaCN dissociates to form Na^+ ions and CN^- ions. The CN^- ion is a fairly strong base and has a weak conjugate acid, HCN. We can neglect the Na^+ ion because it is neither a Brønsted acid nor base in water. Looking in Table 16.2, we can see that HCN is a weaker acid than CH_3COOH, and CH_3COO^- is a weaker base than CN^-. Applying the rule that the stronger acid and base react to form the weaker conjugate base and acid, the reaction

$$CH_3COOH(aq) + CN^-(aq) \longrightarrow CH_3COO^-(aq) + HCN(aq)$$

is predicted to occur. Mixing acetic acid with NaCN is dangerous because HCN is not very soluble in water, so HCN(g) would be formed. The average fatal dose of HCN(g) is 50 to 60 mg or about 0.002 mol.

EXERCISE 16.11 • *Acid-Base Reactions*

Which of these reactions will occur as written? What will the reaction products be? Write complete equations, indicate by a longer arrow in which direction the reaction will be favored, and identify the acids and bases on each side of each equilibrium.

a. $S^{2-} + H_2O \longrightarrow$ b. $HPO_4^{2-} + OH^- \longrightarrow$
c. $CO_3^{2-} + HSO_4^- \longrightarrow$ d. $H_2PO_4^- + HCO_3^- \longrightarrow$

Calculating pH from an Ionization Constant

Once a table of acid-base ionization constants (such as Table 16.2) has been set up, you can use it to calculate the approximate pH of a solution of a weak acid from its concentration. For example, suppose you want to estimate the pH of a 0.050 M solution of benzoic acid, $C_6H_5CO_2H$ ($K_a = 6.3 \times 10^{-5}$). First write the equilibrium equation and equilibrium-constant expression:

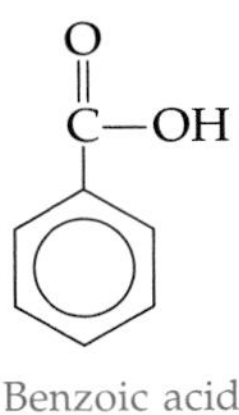

Benzoic acid

$$C_6H_5CO_2H(aq) + H_2O(\ell) \rightleftharpoons C_6H_5CO_2^-(aq) + H_3O^+(aq)$$

$$K_a = \frac{[C_6H_5CO_2^-][H_3O^+]}{[C_6H_5CO_2H]}$$

Next, define equilibrium concentrations and organize the known information in a small table based on the balanced chemical equation.

These small tables, introduced in Section 8.2, are helpful in solving many kinds of equilibrium problems.

	$[C_6H_5CO_2H]$	$[C_6H_5CO_2^-]$	$[H_3O^+]$
Initial concentration (mol/L)	0.050	0	1.0×10^{-7}
Change in concentration on reaction (mol/L)	$-x$	$+x$	$+x$
Concentration at equilibrium (mol/L)	$0.050 - x$	x	x

Since all equilibrium concentrations are defined in terms of the single unknown, x, the equilibrium-constant expression can be rewritten as

$$K_a = \frac{[C_6H_5CO_2^-][H_3O^+]}{[C_6H_5CO_2H]} = \frac{(x)(x)}{(0.050 - x)} = 6.3 \times 10^{-5}$$

There are at least two ways to obtain a value for x that will satisfy this equation. One is to rearrange the equation into the form $ax^2 + bx + c = 0$ and use the *quadratic formula*. But this much effort probably isn't needed. Instead you can reason that, because K_a is very small, not very much product will form and the concentrations of $C_6H_5CO_2^-$ and H_3O^+ will be very small when equilibrium is reached. Therefore, x must be quite small. When it is subtracted from 0.050, the result will still be almost exactly 0.050, and we can write

When 0.0018 is subtracted from 0.050, the answer is 0.048 because of the significant figure rules, and so the simplifying assumption is a reasonable one. When simplifying assumptions are not reasonable, it is still possible to solve an equation like the one shown here by using the quadratic formula.

$$\frac{x^2}{0.050} \approx 6.3 \times 10^{-5}$$

and solve for x, which gives

$$x = \sqrt{(0.050)(6.3 \times 10^{-5})} = \sqrt{3.2 \times 10^{-6}} = 1.8 \times 10^{-3}$$

Since $x = [H_3O^+]$, then $[H_3O^+] = 1.8 \times 10^{-3}$, and so $pH = -\log[H_3O^+] = -\log(1.8 \times 10^{-3}) = 2.74$, which is acidic as expected.

EXERCISE 16.12 • *pH from* K_a

Boric acid is a weak acid often used as an eyewash. K_a for boric acid is 7.3×10^{-10}. Estimate the pH of a 0.10 M solution of boric acid.

16.5 ACID-BASE REACTIONS, SALTS, AND HYDROLYSIS

In Section 5.6 we indicated that an exchange reaction between an acid and a base produces a salt plus a water. A salt is any ionic compound that could have been formed by the reaction of an acid with a base; a salt's positive ion comes from a base and its negative ion comes from an acid. Now that you know more about the Brønsted acid-base concept and the strengths of acids and bases, it is useful to consider acid-base reactions and salt formation in more detail.

Salts of Strong Bases and Strong Acids

The strong acid HCl reacts with the strong base NaOH to form the salt NaCl. If the amounts of HCl and NaOH are in the stoichiometric ratio (1 mol HCl per 1 mol NaOH), this reaction results in complete neutralization of the acidic properties of HCl and the basic properties of NaOH. The reaction can be described by any of three equations, each of which contains useful information.

$$HCl(aq) + NaOH(aq) \longrightarrow NaCl(aq) + H_2O(\ell)$$

$$H_3O^+(aq) + Cl^-(aq) + Na^+(aq) + OH^-(aq) \longrightarrow Na^+(aq) + Cl^-(aq) + 2\,H_2O(\ell)$$

$$\underset{\text{acid}}{H_3O^+(aq)} + \underset{\text{base}}{OH^-(aq)} \longrightarrow \underset{\text{base}}{H_2O(\ell)} + \underset{\text{acid}}{H_2O(\ell)}$$

Table 16.3

Some Salts Formed by Neutralization of Strong Acids with Strong Bases

Acid \ Base	NaOH	KOH	$Ba(OH)_2$
HCl	NaCl	KCl	$BaCl_2$
HNO_3	$NaNO_3$	KNO_3	$Ba(NO_3)_2$
H_2SO_4	Na_2SO_4	K_2SO_4	$BaSO_4$
$HClO_4$	$NaClO_4$	$KClO_4$	$Ba(ClO_4)_2$

The first equation shows the substances that were dissolved or that could be recovered at the end of the reaction. The second equation indicates all of the ions that are present before and after reaction. The third equation is the net ionic equation, which emphasizes that a Brønsted acid (H_3O^+) is reacting with a Brønsted base (OH^-). The reaction goes to completion because H_3O^+ is a strong acid, OH^- is a strong base, and water is a very weak acid and a very weak base.

If the water were evaporated from the NaCl salt solution formed by the reaction of HCl and NaOH, the resulting solid NaCl would be identical to NaCl that could be prepared by the reaction of metallic sodium with gaseous chlorine.

The resulting solution contains only sodium ions and chloride ions, with a few more water molecules than before. Its properties are the same as if it had been prepared by simply dissolving some NaCl(s) in water. It has a neutral pH because it contains no acids or bases. The Cl^- ion is the conjugate base of a strong acid and hence is such a weak base that it does not react with water. The Na^+ ion also does not react with water as either an acid or a base.

A salt such as a metal chloride, MCl_x, can be thought of as the reaction product of HCl and the metal hydroxide, $M(OH)_x$. Similarly, metal bromides, iodides, nitrates, sulfates, and perchlorates can be considered as the products of neutralization reactions between the strong acids and metal hydroxides. Examples of such salts are given in Table 16.3.

The monatomic ions of Groups IA, IIA (except Be^{2+}), and VIIA (except F^-) do not undergo any appreciable acid/base reactions with water.

Salts of Strong Bases and Weak Acids

Suppose, for example, that 0.010 mol of NaOH is added to 0.010 mol of acetic acid in 1 L of solution. The reaction is

$$NaOH(aq) + CH_3COOH(aq) \longrightarrow NaCH_3COO(aq) + H_2O(\ell)$$

$$Na^+(aq) + OH^-(aq) + CH_3COOH(aq) \longrightarrow Na^+(aq) + CH_3COO^-(aq) + H_2O(\ell)$$

$$\underset{\text{strong base}}{OH^-(aq)} + \underset{\text{weak acid}}{CH_3COOH(aq)} \longrightarrow \underset{\text{base}}{CH_3COO^-(aq)} + \underset{\text{acid}}{H_2O(\ell)}$$

In this case a weak base, acetate ion, remains in the solution when the reaction is complete. This means that the solution is slightly basic; its pH is greater than 7, even though exactly the stoichiometric amount of acetic acid was added to the sodium hydroxide. The reaction that makes the solution basic is just the reaction of acetate ion as a weak Brønsted base:

$$CH_3COO^-(aq) + H_2O(\ell) \rightleftharpoons CH_3COOH(aq) + OH^-(aq)$$

This process splits a water molecule into a proton (which becomes attached to the acetate ion) and a hydroxide ion. A reaction in which a water molecule is split is called a **hydrolysis** reaction (derived from *hydro,* water, and *lysis,* to break apart). The extent of hydrolysis is determined by the value of K_b for acetate ion.

The hydrolysis of salts is one of the reasons that solubilities calculated using K_{sp} expressions for slightly soluble salts are only estimates. If an ion reacts with water, the ion's concentration is less than expected and solubility of a precipitate containing that ion is greater than expected.

All of the weak bases in Table 16.2, except for the very weak bases above water, undergo hydrolysis reactions in aqueous solution. The larger their K_b values, the more basic the solutions they produce. The pH of a solution of a salt of a strong base and a weak acid can be estimated from K_b as shown in Example 16.3.

EXAMPLE 16.3

pH of a Salt Solution

Sodium hypochlorite, NaClO, is used as a source of chlorine in some laundry bleaches, swimming-pool disinfectants, and water treatment plants. Estimate the pH of a 0.010 M solution of NaClO. (Use Table 16.2 to obtain K_b.)

SOLUTION Sodium hypochlorite consists of sodium ions and hypochlorite ions. Na^+ does not react with water, but ClO^- is the conjugate base of a weak acid and reacts with water to produce a basic solution.

$$ClO^-(aq) + H_2O(\ell) \rightleftharpoons HClO(aq) + OH^-(aq)$$

$$K_b = 2.9 \times 10^{-7} = \frac{[HClO][OH^-]}{[ClO^-]}$$

The concentrations of hypochlorite ion, hypochlorous acid, and hydroxide ion initially and at equilibrium are

	$[ClO^-]$	$[HClO]$	$[OH^-]$
Initial concentration (mol/L)	0.010	0	1.0×10^{-7}
Change in concentration on reaction (mol/L)	$-x$	$+x$	$+x$
Concentration at equilibrium (mol/L)	$0.010 - x$	x	x

Hypochlorite ion has a very small K_b and is a very weak base. It is safe to assume that x will be negligibly small compared to 0.010, and so

$$K_b = 2.9 \times 10^{-7} \approx \frac{x^2}{0.010}$$

Solving for x gives $x = 5.4 \times 10^{-5}$. Since $0.010 - 5.4 \times 10^{-5} = 0.010$, our assumption that x is negligible is justified. Therefore, at equilibrium

$$[OH^-] = [HClO] = 5.4 \times 10^{-5}\text{ M, and } [ClO^-] = 0.010\text{ M}$$

Finally, the pH of the solution is found as follows:

$$K_w = [H_3O^+][OH^-] = [H_3O^+](5.4 \times 10^{-5})$$
$$[H_3O^+] = 1.9 \times 10^{-10}$$
$$pH = 9.73$$

Products that contain an aqueous solution of hypochlorite, NaClO. The hypochlorite ion is a weak base. See Example 16.3. *(C.D. Winters)*

EXERCISE 16.13 • *Calculating pH of a Solution of a Weak Base*

Sodium carbonate is an environmentally safe paint stripper. It is water soluble, and carbonate ion is a strong enough base to make a solution with pH high enough that it can attack paint so it may be scraped off. What is the pH of a 1.0 M solution of Na_2CO_3? The hydrolysis reaction is $H_2O + CO_3^{2-} \rightleftarrows HCO_3^- + OH^-$.

Salts of Weak Bases and Strong Acids

When a weak base reacts with a strong acid, the conjugate acid of the weak base determines the pH of the resulting salt solution. For example, suppose equal volumes of 0.10 M ammonia and 0.10 M HCl are mixed. The reaction is

$$NH_3(aq) + HCl(aq) \longrightarrow NH_4Cl(aq)$$
$$NH_3(aq) + H_3O^+(aq) + Cl^-(aq) \longrightarrow NH_4^+(aq) + Cl^-(aq) + H_2O(\ell)$$
$$\underset{\text{weak base}}{NH_3(aq)} + \underset{\text{strong acid}}{H_3O^+(aq)} \longrightarrow \underset{\text{acid}}{NH_4^+(aq)} + \underset{\text{base}}{H_2O(\ell)}$$

As soon as it is formed, the weak acid NH_4^+ reacts with water and establishes an equilibrium. The resulting solution is slightly acidic, with a pH below 7.00, because of the reaction

$$NH_4^+(aq) + H_2O(\ell) \rightleftharpoons NH_3(aq) + H_3O^+(aq)$$

Some of the weak acids listed in Table 16.2 are hydrated metal ions. When a salt containing a metal ion dissolves in water, the metal ion becomes hydrated. That is, a complex ion forms with water molecules as the ligands. This complex ion usually has a coordination number of six, $[M(H_2O)_6]^{n+}$, where M represents a metal ion whose charge is $n+$. Metal ions other than those in Groups IA and IIA have great enough charges and small enough sizes to attract electrons away from the ligands. This makes the O—H bonds in the water-molecule ligands more polar and the protons more acidic than in a water molecule that is not bonded to a metal ion (Figure 16.8). Thus the metal ion causes water molecules to donate protons, and the solution becomes acidic.

The ionization reaction and ionization constant for a hydrated metal ion such as $Fe^{3+}(aq)$ can be written as

$$Fe(H_2O)_6^{3+}(aq) + H_2O(\ell) \rightleftharpoons Fe(H_2O)_5(OH)^{2+}(aq) + H_3O^+(aq)$$

$$K_a = \frac{[Fe(H_2O)_5(OH)^{2+}][H_3O^+]}{[Fe(H_2O)_6^{3+}]} = 6.3 \times 10^{-3}$$

Thus, a solution of $FeCl_3$ will have about the same pH as a solution of phosphoric acid ($K_a = 7.5 \times 10^{-3}$) of equal concentration. Many metal ions are weak acids, and this property is important in the chemistry of such ions in the environment as well as in numerous biochemical reactions.

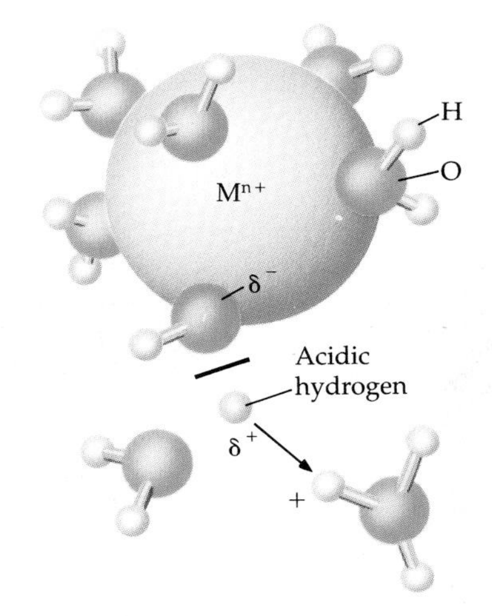

Figure 16.8
The attraction between water molecules bound to a positively charged metal ion is strong enough that one water molecule may become acidic, losing its proton to a base. The resulting ion is $[M^{n+}(H_2O)_x(OH)]^{(n-1)^+}$.

EXERCISE 16.14 • *Estimating the pH of a Metal-Ion Solution*

Estimate the pH of a 0.20 M solution of aluminum chloride, $AlCl_3$. Use Table 16.2 to find K_a.

Salts of Weak Bases and Weak Acids

What is the pH of a solution of a salt containing an acidic cation and a basic anion, such as NH_4F or $Ni(CH_3COO)_2$? There are two possible reactions that can determine the pH of the solution: formation of H_3O^+ by the cation, and formation of OH^- by the anion. In the case of NH_4F,

$$NH_4^+(aq) + H_2O(\ell) \rightleftharpoons H_3O^+(aq) + NH_3(aq) \qquad K_a(NH_4^+) = 5.6 \times 10^{-10}$$

$$F^-(aq) + H_2O(\ell) \rightleftharpoons HF(aq) + OH^-(aq) \qquad K_b(F^-) = 1.4 \times 10^{-11}$$

Since $K_a(NH_4^+) > K_b(F^-)$, the ammonium ion is a stronger acid than the fluoride ion is a base. Therefore, the resulting solution is slightly acidic. In the case of $Ni(CH_3COO)_2$,

$$Ni(H_2O)_6^{2+}(aq) + H_2O(\ell) \rightleftharpoons Ni(H_2O)_5(OH)^+ + H_3O^+(aq)$$
$$K_a(Ni(H_2O)_6^{2+}) = 2.5 \times 10^{-11}$$

$$CH_3COO^-(aq) + H_2O(\ell) \rightleftharpoons CH_3COOH(aq) + OH^-(aq)$$
$$K_b(CH_3COO^-) = 5.6 \times 10^{-10}$$

$K_b(CH_3COO^-) > K_a(Ni(H_2O)_6^{2+})$, and so the resulting solution is slightly basic. Notice that when the weak acid is stronger than the weak base, an acidic solution results; when the weak base is stronger than the weak acid, the solution is basic.

In summary, *bases listed above water on the right-hand side of Table 16.2 are extremely weak*—they do not affect the pH of aqueous solutions. *Acids listed below water on the left-hand side of Table 16.2 are extremely weak* and do not make aqueous solutions acidic. (Some of them are amphiprotic, however, and do make a solution basic.) Although all of the necessary information is

Table 16.4

Acid-Base Properties of Typical Ions in Aqueous Solution

	Neutral		Basic			Acidic
Anions	Cl^-	NO_3^-	CH_3COO^-	CN^-	SO_4^{2-}	HSO_4^-
	Br^-	ClO_4^-	$HCOO^-$	PO_4^{3-}	HPO_4^{2-}	$H_2PO_4^-$
	I^-		CO_3^{2-}	HCO_3^-	SO_3^{2-}	HSO_3^-
			S^{2-}	HS^-	ClO^-	
			F^-	NO_2^-		
Cations	Li^+	Mg^{2+}	None			Al^{3+}
	Na^+	Ca^{2+}				NH_4^+
	K^+	Ba^{2+}				Transition metal ions

in Table 16.2, it is convenient to use Table 16.4 as a concise statement of acid-base behavior (hydrolysis) of anions and cations found in salts. We can also make the following generalizations about acid-base reactions:

Solution of strong acid + Solution of strong base $\longrightarrow$
Solution of salt with pH = 7

Solution of strong acid + Solution of weak base $\longrightarrow$
Solution of salt with pH < 7 (acidic)

Solution of weak acid + Solution of strong base $\longrightarrow$
Solution of salt with pH > 7 (basic)

Solution of weak acid + Solution of weak base $\longrightarrow$
Solution of salt with pH determined by relative strengths of conjugate base and conjugate acid formed

EXERCISE 16.15 • *Acidity and Basicity of Salt Solutions*

Which of these salts would give a neutral solution? An acidic solution? A basic solution?

NH_4NO_3 $\quad$ $NaBr$ $\quad$ K_3PO_4 $\quad$ $Fe(NO_3)_2$ $\quad$ NH_4HCO_3

The Solubility of Salts in Water and Acids

An insoluble salt can be made to dissolve if one or both of its ions can be removed from solution somehow. Consider calcium carbonate, $CaCO_3$, which is found in minerals such as limestone and marble. $CaCO_3$ is not very soluble in pure water.

(a) $$CaCO_3(s) \rightleftharpoons Ca^{2+}(aq) + CO_3^{2-}(aq) \quad K_{sp} = 3.8 \times 10^{-9}$$

Since K_{sp} is small, the equilibrium concentrations of Ca^{2+} and CO_3^{2-} must also be small. However, if acid is added, the calcium carbonate will dissolve and CO_2 will foam out of the solution. (This is one of the typical properties of acids mentioned at the beginning of this chapter.) Adding acid adds hydronium ions, which react with carbonate ions:

(b) $$CO_3^{2-}(aq) + H_3O^+(aq) \rightleftharpoons HCO_3^-(aq) + H_2O(\ell)$$

(c) $$HCO_3^-(aq) + H_3O^+(aq) \rightleftharpoons H_2CO_3(aq) + H_2O(\ell)$$

Reaction (b) involves a fairly strong base and a strong acid on the left-hand side, and so nearly all of the carbonate is converted to hydrogen carbonate ion. Reaction (c) produces a product, carbonic acid, that is unstable. It breaks down to $CO_2(g)$ and water in a very product-favored reaction:

(d) $$H_2CO_3(aq) \longrightarrow CO_2(g) + H_2O(\ell) \qquad K \approx 10^5$$

As CO_2 gas escapes from the solution, the H_2CO_3 concentration decreases. This shifts reaction (c) to the right, which decreases the concentration of HCO_3^-. This in turn shifts reaction (b) to the right, decreasing the concentration of CO_3^{2-} to an even lower value. To oppose this decrease in carbonate-ion concentration, reaction (a) shifts to the right, and more $CaCO_3(s)$ dissolves. Because the acidity of the solution determines the positions of equilibria (b) and (c), small changes in pH can cause limestone and marble to dissolve or precipitate. Acid rain can dissolve a marble statue, and layers of sedimentary rock (limestone) have been precipitated on the ocean floor because of a slight increase in pH of seawater.

In general, *insoluble salts containing anions that are Brønsted bases dissolve in solutions of low pH.* This rule includes carbonates, sulfides (which produce $H_2S(g)$), hydroxides, phosphates, and other anions listed as bases in Table 16.2. The principal exceptions to this rule are a few sulfides, such as HgS, CuS, and CdS, that have extremely small K_{sp} values and therefore do not dissolve even when the pH is extremely low.

In contrast, an insoluble salt such as AgCl, which contains the conjugate base of a strong acid, is not soluble in strongly acidic solution, because Cl^- is a very weak base and so does not react with H_3O^+.

$$AgCl(s) \rightleftharpoons Ag^+(aq) + Cl^-(aq) \qquad K_{sp} = 1.8 \times 10^{-10}$$

$$H_3O^+(aq) + Cl^-(aq) \rightleftharpoons HCl(aq) + H_2O(\ell) \qquad K \text{ very small}$$

EXERCISE 16.16 • *Predicting Solubilities of Salts in Acid*

Predict which of the following insoluble salts would be soluble in 1 M HCl solution.

a. $ZnCO_3$ b. FeS c. AgBr d. $BaCO_3$ e. $Ca(OH)_2$

16.7 BUFFER SOLUTIONS

Often, changing the pH of a solution can cause significant problems. Many aquatic organisms can survive only within a narrow pH range. If acid rain lowers the pH of a lake or stream, fish such as trout may die. In addition, if such organisms are to be studied in the laboratory, the solutions they live in must have a fairly constant pH. The blood of mammals is another example. The normal pH of human blood is 7.40 ± 0.05. If pH decreases below 7.35, a condition known as *acidosis* occurs; increasing pH above 7.45 leads to *alkalosis*. Both of these conditions can be life-threatening. Acidosis, for example, causes hemoglobin to stop transporting oxygen and also depresses the central nervous system, leading in extreme cases to coma and death. In addition acidosis can cause weakening and irregularity of cardiac contractions—symptoms of heart failure. To prevent such problems your body must keep the pH of your blood nearly constant. A solution, such as human blood, that resists pH change upon addition of a small amount of acid or base is called a **buffer solution.**

Adding a small amount of acid or base can radically affect the pH of pure water. Consider what happens if 0.01 mol of HCl is added to 1.0 L of water. The pH changes from 7 to 2 because $[H_3O^+]$ changes from 10^{-7} M to 10^{-2} M. This pH change represents a *100,000-fold increase* in $[H_3O^+]$. Similarly, if 0.01 mol of NaOH is added to 1 L of pure water, the pH goes from 7 to 12, a *100,000-fold decrease* in $[H_3O^+]$. If those same amounts of strong acid and strong base were added to two 1-L samples of human blood, the pH would decrease or increase by only 0.1 pH unit.

How does a buffer solution maintain its pH at a nearly constant value? A buffer must contain an *acid that can react with any added base,* and at the same time it must contain a *base that can react with any added acid.* It is also necessary that the acid and base components of a buffer solution not react with each other. Buffers usually consist of approximately equal quantities of a weak acid and its conjugate base, or a weak base and its conjugate acid. In a conjugate pair, if the acid and base react with each other they just produce conjugate base and conjugate acid—no observable change occurs. For example, when acetic acid reacts with acetate ion, acetate ion and acetic acid are the products.

$$\underset{\text{acid}}{CH_3COOH(aq)} + \underset{\text{base}}{CH_3COO^-(aq)} \rightleftharpoons \underset{\text{base}}{CH_3COO^-(aq)} + \underset{\text{acid}}{CH_3COOH(aq)}$$

To see how a buffer works, consider human blood. Carbon dioxide is the most important compound (but not the only one) affecting the pH in blood. In solution, CO_2 reacts with water to form H_2CO_3, which can then dissociate to produce H_3O^+ and HCO_3^- ions. The reactions are

$$CO_2(aq) + H_2O(\ell) \rightleftharpoons H_2CO_3(aq)$$
$$H_2CO_3(aq) \rightleftharpoons H_3O^+(aq) + HCO_3^-(aq)$$

An enzyme, *carbonic anhydrase,* found in blood and numerous other body fluids, causes the equilibria between CO_2 and H_3O^+ to be established quickly. The normal concentrations of H_2CO_3 and HCO_3^- in blood are 0.0025 M and 0.025 M. Since H_2CO_3 is a weak acid and HCO_3^- is a weak base, they constitute a buffer. As long as the ratio of H_2CO_3 to HCO_3^- concentrations remains about 1 to 10, the pH of the blood remains near 7.4.

If a strong base such as NaOH is added to this buffer, carbonic acid will react with the OH^-. Since OH^- is the strongest base that can exist in water solution, this reaction is essentially complete.

$$H_2CO_3(aq) + OH^-(aq) \longrightarrow HCO_3^-(aq) + H_2O(\ell)$$
$$K = 1/K_b(HCO_3^-) = 4.2 \times 10^7$$

Here, the equilibrium constant is $1/K_b$ for hydrogen carbonate ion, because the reaction is the reverse of the hydrolysis of the hydrogen carbonate ion. If a strong acid such as HCl is added to this buffer, HCO_3^- will react with the hydronium ions from the acid. Since the H_3O^+ ion is such a strong acid, the reaction between HCO_3^- and H_3O^+ is essentially complete.

$$HCO_3^-(aq) + H_3O^+(aq) \longrightarrow H_2CO_3(aq) + H_2O(\ell)$$
$$K' = 1/K_a(H_2CO_3) = 5.6 \times 10^4$$

In this case, the equilibrium constant is $1/K_a$ for carbonic acid, because the reaction is the reverse of the ionization of carbonic acid.

It is often desirable in the laboratory to make a buffer with a specified pH. The pH of a buffer solution can be estimated if the concentrations of the conjugate acid and the conjugate base are known. For example, if blood contains 0.0025 M carbonic acid and 0.025 M hydrogen carbonate ion, these concentrations can be substituted into the ionization-constant expression for carbonic acid. When we do this, we find that the only concentration remaining in the expression is $[H_3O^+]$, and pH is easily calculated.

$$K_a = \frac{[H_3O^+][HCO_3^-]}{[H_2CO_3]} = \frac{[H_3O^+](0.025)}{(0.0025)} = 4.2 \times 10^{-7}$$

$$[H_3O^+] = 4.2 \times 10^{-7} \cdot \frac{0.0025}{0.025} = 4.2 \times 10^{-8}$$

$$\text{pH} = -\log(4.2 \times 10^{-8}) = 7.38$$

Notice that in this calculation the hydronium ion concentration is one-tenth the K_a for carbonic acid, because the ratio $[H_2CO_3]/[HCO_3^-]$ equals 1/10. If we are to have appreciable quantities of both acid and conjugate base in the buffer solution, this ratio cannot get much smaller than one-tenth (or much bigger than ten). Consequently we can use this same acid/base pair to prepare buffer solutions over a range of $[H_3O^+]$ from about one-tenth to ten times the K_a. The range of pH is limited to about one pH unit above or below $-\log(K_a)$. Other acid/base pairs could be used to prepare buffers with much different pH ranges, as determined by the K_a value of the acid.

EXERCISE 16.17 • *Selecting an Acid/Base Pair for a Buffer Solution*

Use data from Table 16.2 to select an acid/base conjugate pair you could use to make a buffer solution having each of these hydrogen-ion concentrations:

a. 3.2×10^{-4} M b. 5.0×10^{-5} M c. 7.0×10^{-8} M d. 6.0×10^{-11} M

The pH of a buffer solution, or the ratio of acid to base concentrations needed to achieve a given pH, is conveniently calculated by using an equation called the **Henderson-Hasselbalch equation.** This equation is obtained by writing the equilibrium constant expression for a weak acid, and solving for $[H_3O^+]$.

$$HA(aq) + H_2O(\ell) \rightleftharpoons H_3O^+(aq) + A^-(aq)$$

$$K_a = \frac{[H_3O^+][A^-]}{[HA]}$$

$$[H_3O^+] = K_a \cdot \frac{[HA]}{[A^-]}$$

Taking the base-10 logarithm of each side of this equation gives

$$\log[H_3O^+] = \log K_a + \log\frac{[HA]}{[A^-]}$$

Multiplying both sides by -1 and using the relation $-\log(x) = \log(1/x)$, we get

$$-\log[H_3O^+] = -\log K_a + \log\frac{[A^-]}{[HA]}$$

Table 16.5

Buffer Systems That Are Useful at Various pH Values*

Desired pH	Buffer System: *Weak Acid*	Buffer System: *Weak Base*	K_a(Weak Acid)	pK_a
4	lactic acid ($CH_3CHOHCOOH$)	lactate ion ($CH_3CHOHCOO^-$)	1.4×10^{-4}	3.85
5	acetic acid (CH_3COOH)	acetate ion (CH_3COO^-)	1.8×10^{-5}	4.74
6	carbonic acid (H_2CO_3)	hydrogen carbonate ion (HCO_3^-)	4.2×10^{-7}	6.38
7	dihydrogen phosphate ion ($H_2PO_4^-$)	hydrogen phosphate ion (HPO_4^{2-})	6.2×10^{-8}	7.21
8	hypochlorous acid ($HClO$)	hypochlorite ion (ClO^-)	3.5×10^{-8}	7.46
9	ammonium ion (NH_4^+)	ammonia (NH_3)	5.6×10^{-10}	9.25
10	hydrogen carbonate ion (HCO_3^-)	carbonate ion (CO_3^{2-})	4.8×10^{-11}	10.32

*Adapted from W. L. Masterton and C. N. Hurley: *Chemistry—Principles and Reactions,* 2nd ed. Philadelphia, Saunders College Publishing, 1993.

Using the definition of pH, and defining $-\log K_a$ as pK_a, the equation becomes

$$pH = pK_a + \log\frac{[A^-]}{[HA]}$$ the Henderson-Hasselbalch equation

As the equation shows, if $[A^-] = [HA]$, the pH $= pK_a$ because $\log(1) = 0$. A buffer's pH equals pK_a of the weak acid when the concentrations of the acid and its conjugate base are equal. A buffer for maintaining a desired pH can easily be chosen by examining pK_a values, which are often tabulated along with K_a values. Table 16.5 lists pK_a values of several common acids that could be used to prepare buffers over the pH range from 4 to 10.

As we noted earlier, carbonic acid/hydrogen carbonate is not the only buffer system in human blood. Another is dihydrogen phosphate/hydrogen phosphate. Let's calculate the ratio of HPO_4^{2-} to $H_2PO_4^-$ in blood with a pH of 7.40. The equilibrium equation is

$$H_2PO_4^-(aq) + H_2O(\ell) \rightleftharpoons H_3O^+ + HPO_4^{2-}(aq)$$

From Table 16.2 we find $K_a = 6.2 \times 10^{-8}$, and so $pK_a = 7.21$. Using the Henderson-Hasselbalch equation,

$$7.40 = 7.21 + \log\frac{[HPO_4^{2-}]}{[H_2PO_4^-]}$$

which leads to

$$\log\frac{[HPO_4^{2-}]}{[H_2PO_4^-]} = 7.40 - 7.21 = 0.19$$

Finally,

$$\frac{[HPO_4^{2-}]}{[H_2PO_4^-]} = 10^{0.19} = 1.55$$

Thus the blood buffer contains a little more than one and a half times as much hydrogen phosphate ion as dihydrogen phosphate ion.

While the pH of a buffer varies with pK_a and the ratio of acid/base concentrations, the capacity of the buffer to absorb acid and base without a pH change depends on the *amounts* of acid and base present. When nearly all of the acid in a buffer has reacted with base, adding a little more base can change pH significantly, because there is almost no acid left to consume the base. Similarly, if enough acid is added to a buffer to make the buffer's weak base the limiting reactant, the pH will go down a lot. For example, 1 L of a buffer solution that is 0.25 M in CH_3COOH and 0.25 M in CH_3COO^- contains 0.25 mol CH_3COOH and 0.25 mol CH_3COO^-; it will not be able to handle the addition of 0.30 mol of strong acid or 0.30 mol of strong base, because such additions would use up all of the weak base or all of the weak acid.

EXERCISE 16.18 • *Buffers and pH*

Estimate the pH of each of these buffers: H_2CO_3(0.10 M)/HCO_3^-(0.25 M); $H_2PO_4^-$(0.10 M)/HPO_4^{2-}(0.25 M); NH_4^+(0.10 M)/NH_3(0.25 M)

16.7 ACID-BASE TITRATIONS

In Section 6.6 we discussed titration as a method for determining an unknown concentration of a solution. In a titration a measured volume of one solution is placed in a flask, and another solution that reacts with the first one is added from a buret until the equivalence point is reached. At the equivalence point the ratio of amount of substance added from the buret to amount of substance in the flask is exactly the stoichiometric ratio derived from the balanced equation for whatever reaction is occurring. Often an indicator is added so that a color change occurs at the equivalence point. But how does a titration work? Why is it possible for an indicator to undergo a sudden color change even though the reactant in the buret is being added at constant rate (or perhaps at a very slow rate near the end point)? You now have enough background about acid-base reactions to be able to examine what happens during a titration in more detail and answer such questions.

First let's look at a titration involving a strong acid and a strong base. A typical example is titration of 50.0 mL of a 0.100 M solution of HCl with 0.100 M NaOH. The reaction that occurs is

$$HCl(aq) + NaOH(aq) \longrightarrow NaCl(aq) + H_2O(\ell)$$

and so the stoichiometric ratio is (1 mol HCl)/(1 mol NaOH). Before any base has been added, $[H_3O^+] = 0.100$ M and the pH is 1.000. The amount of HCl present initially is $0.0500 \text{ L} \times 0.100 \text{ mol/L} = 0.00500$ mol. Suppose you add 10.0 mL of 0.100 M NaOH. This increases the solution volume to 60.0 mL. Also, the 10.0 mL of NaOH contains $0.0100 \text{ L} \times 0.100 \text{ mol/L} = 0.00100$ mol of NaOH. This will react with 0.00100 mol of HCl, decreasing

Several differences are obvious from the figures. For acetic acid (1) the initial pH is higher; (2) pH rises rapidly at first, then rises more slowly, (3) pH again rises very rapidly at the equivalence point. The curves for both acids are very similar after the equivalence point has been reached. These differences can be understood in terms of situations discussed earlier.

(1) The pH at the beginning of the acetic-acid titration can be calculated from K_a as described in Section 16.4. It is higher than for HCl because acetic acid is a weak acid and only partly ionized, providing a smaller concentration of hydronium ions than HCl.

(2) Adding some NaOH uses up some of the acetic acid, converting it into acetate ions. The pH rises quickly at first because the added NaOH produces a weak base, CH_3COO^-, which remains in solution. (In the HCl titration all of the NaOH was converted into water and Na^+ ions, neither of which is very basic, and so the pH did not rise as fast.) As soon as some acetate ion has been produced, the solution contains a conjugate acid/base pair (CH_3COOH/CH_3COO^-), which makes it a buffer. Therefore adding more base causes only a slow rise in pH. In this "buffer region" of the titration curve, the Henderson-Hasselbalch equation can be used to calculate the pH. When half as much NaOH has been added as is needed to reach the equivalence point, half of the acetic acid has been converted to sodium acetate. According to the Henderson-Hasselbalch equation, at this point pH = pK_a for acetic acid. The buffer region for another weak acid would lie above or below that for acetic acid, depending on whether the other acid was weaker or stronger than acetic acid.

(3) The rapid increase in pH near the equivalence point occurs when the buffer capacity of the acetic acid/acetate ion buffer is exceeded. When nearly all of the acetic acid has been converted to acetate ion, there is no acid left to react with added OH^-, and pH increases rapidly. At the equivalence point the solution is identical to one prepared by dissolving sodium acetate in water. Since acetate ion is a weak base, the pH is above 7; it could be calculated from the concentration of $NaCH_3COO$ and K_b for acetate ion, as was done in Example 16.3. The weaker the acid being titrated, the stronger its conjugate base, and the higher the pH at the equivalence point.

Beyond the equivalence point the pH is governed by the concentration of added strong base, NaOH, that has not been consumed by reaction with acetic acid. Since only the NaOH concentration is important here, the HCl and CH_3COOH curves look the same.

EXERCISE 16.19 • *Titration Curves*

Without doing any calculations, sketch the curve of pH versus volume of titrant added for each titration. Make certain that the equivalence point is at the appropriate volume of titrant and that the buffer region is at approximately the correct pH.

a. 50.0 mL of 0.100 M HClO (in flask) with 0.100 M NaOH (in buret)

b. 50.0 mL of 0.100 M NaOH (in flask) with 0.100 M HNO_3 (in buret)

c. 25.0 mL of 0.100 M HNO_2 (in flask) with 0.200 M NaOH (in buret)

Acid-Base Indicators

The goal of a titration is to add exactly enough solution from a buret to react with all of the substance in a flask. The equivalence point is defined to occur at exactly the stoichiometric ratio of reactants, but when this equivalence point is reached it must be detected in some way. If one of the reactants were colored, the intensity of its color might be an indicator of the amount remaining, so that when the color disappeared the reactant would be all used up, but most acids and bases are colorless. There are, however, some organic dyes that change color depending on pH. Common litmus paper contains a plant juice that is red at pH < 5 and blue at pH > 8.2. Phenolphthalein is colorless at pH < 8.2 and red at pH > 9.8. This color change occurs because phenolphthalein is itself a weak, diprotic acid, and has the rather complicated molecular structure shown below. When it is in the protonated form, H_2In, it is colorless. As the pH rises from 8.2 to 9.8, two protons are lost and the conjugate base, In^{2-}, forms. This form has a beautiful red color. The reaction can be written as

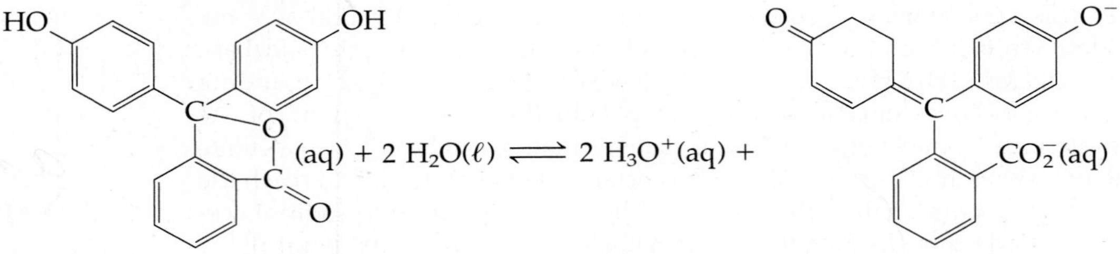

$$H_2In(aq) + 2\,H_2O(\ell) \rightleftharpoons 2\,H_3O^+(aq) + In^{2-}(aq)$$

Colorless phenolphthalein H_2In

Red conjugate base of phenolphthalein In^{2-} (with CO_2^-(aq))

As you can see from Figure 16.11, there are a great many indicators available. Each has an ionization equation and K_a of general form

$$HIn(aq) + H_2O(\ell) \rightleftharpoons H_3O^+(aq) + In^-(aq) \qquad K_a = \frac{[H_3O^+][In^-]}{[HIn]}$$

The acid form of the indicator (HIn) has one color, while the conjugate base (In^-) has another. In Figure 16.11, the color of the acid form is shown on the left, or acid side, of the color band for each indicator, while the color for the basic form is shown on the right. According to Le Chatelier's principle, addition of H_3O^+ or OH^- to an indicator solution will cause one color or the other to appear, depending on whether HIn or In^- predominates. The best indicator for a specific acid-base titration has a color change that is easy to see so that only a negligible amount of the indicator needs to be added. (Remember that the indicator is an acid and will be titrated along with the acid being analyzed.) Most important of all is that the point at which the indicator color changes, the **end point** of the titration, must be as close to the equivalence point as possible.

If the K_a expression for the indicator is rearranged

$$[H_3O^+] = K_a\left(\frac{[HIn]}{[In^-]}\right) \qquad \text{or} \qquad pH = pK_a + \log\frac{[In^-]}{[HIn]}$$

we see that the ratio $[HIn]/[In^-]$ is controlled by changes in $[H_3O^+]$. (The second equation is just the Henderson-Hasselbalch equation applied to the indicator.) If pH is one less than pK_a, then the ratio $[In^-]/[HIn] = 0.1$ and most of the indicator is in its acid form (colorless for phenolphthalein). If pH

Some acid-base indicators

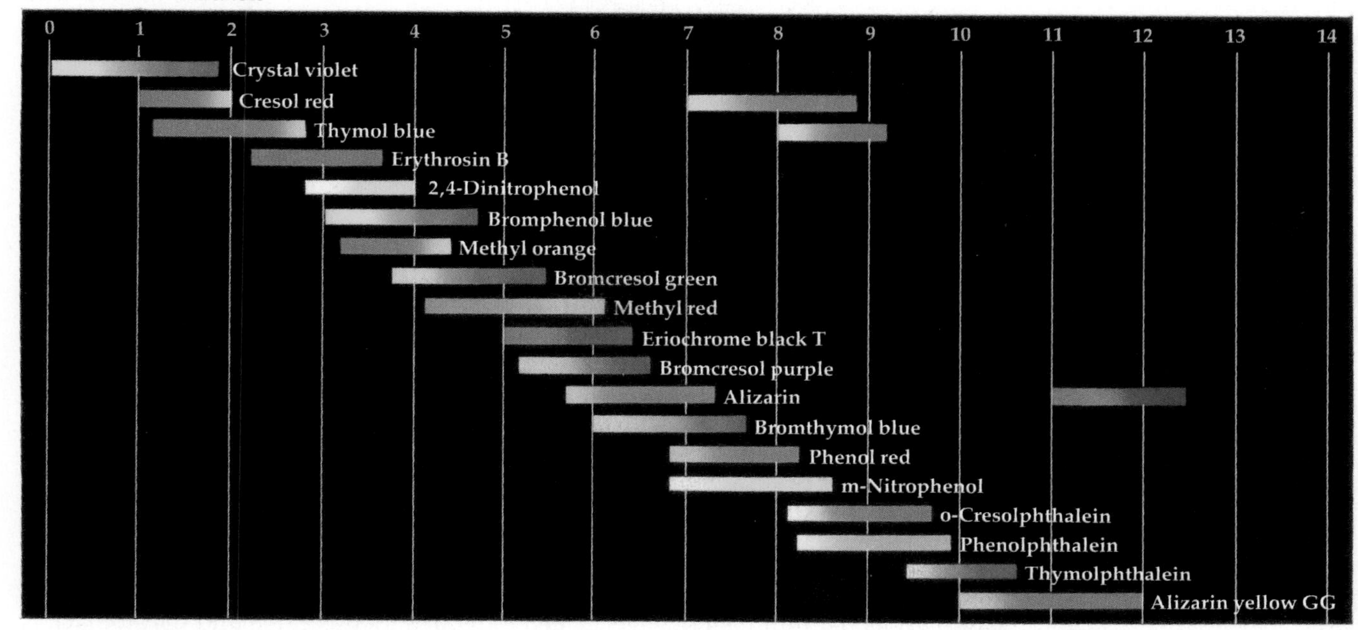

Figure 16.11
Some acid-base indicators. Color changes occur over a range of pH values. Some, like Cresol red, undergo two distinct color changes in two different pH ranges.
(Hach Company)

is one more than pK_a, then $[In^-]/[HIn] = 10$, and nearly all of the indicator is in its base form (red for phenolphthalein). Thus, if pK_a for an indicator corresponds to a pH in the vertical part of a titration curve (such as those in Figures 16.9 and 16.10), where $[H_3O^+]$ and pH are changing rapidly near the equivalence point, then the ratio $[HIn]/[In^-]$ also changes rapidly. The indicator's acid-form color will suddenly change to its base-form color, signaling the end point of the titration. If the indicator's pK_a corresponds to the buffer region of a titration curve, where the pH is changing slowly, the color change will also be slow and the indicator will not correctly signal the equivalence point. Thus an indicator such as methyl orange, whose pK_a is about 3.8, would not be appropriate for the titration in Figure 16.10. Its color change would begin to be seen almost as soon as some NaOH was added to the acetic acid solution, and would be complete by the time 30 mL of NaOH had been added—well before the equivalence point. Methyl orange would be suitable for the strong-acid/strong-base titration in Figure 16.9, however, because its pK_a corresponds to the vertical part of the curve.

EXERCISE 16.20 • *Indicators*

Pick two indicators from Figure 16.11 that would accurately signal the equivalence point of the titration shown in Figure 16.10.

Stomach Acidity

The Cl^- ions secreted by the cells in our stomach walls come mostly from table salt (NaCl) we eat, and salt-containing foods like vegetables and fish.

The pH of human stomach fluids is approximately 1. This very acidic pH is caused by HCl, which is secreted by thousands of cells in the wall of the stomach that specialize in transporting $H_3O^+(aq)$ and $Cl^-(aq)$ from the

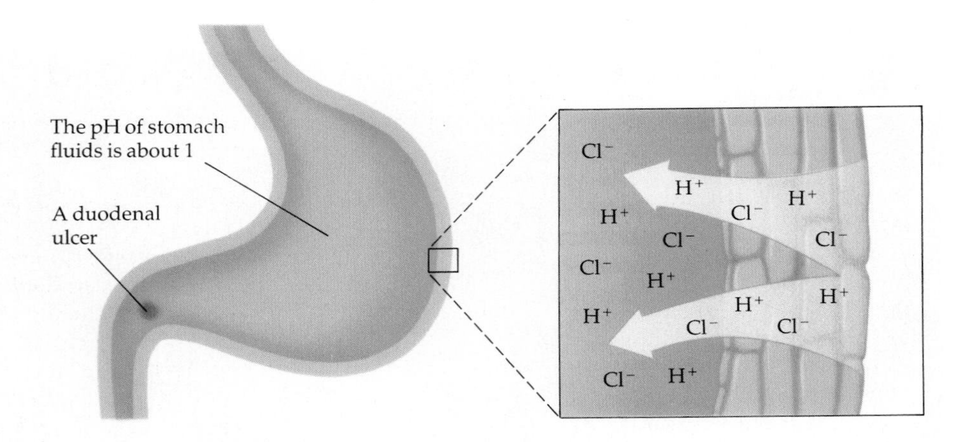

Figure 16.12
The lining of the stomach contains cells that secrete a solution of hydrochloric acid. The pH of the solution is about 1. Some people's stomachs produce far more acid than is needed for the primary digestion of food. When this happens, an ulcer can form.

blood (Figure 16.12). The main purpose of this acid is to suppress the growth of bacteria and to aid in the digestion of certain foods. Normally, the stomach is not harmed by the presence of hydrochloric acid because the mucosa, the inner lining of the stomach, is replaced at the rate of about half a million cells per minute. However, when too much food is eaten and the stomach is stretched, or when the stomach is irritated by very spicy food, it responds with an outpouring of so much acid that the pH is lowered to a point where discomfort is felt.

An antacid is a base that is used to decrease the amount of hydrochloric acid in the stomach. Several antacids and their acid/base reactions are shown in Table 16.6. People who need to restrict the quantity of sodium (Na^+) in their diets should avoid antacids such as sodium bicarbonate.

Table **16.6**

The Acid-Base Chemistry of Some Antacids

Compound	Reaction in Stomach	Examples of Commercial Products
milk of magnesia: $Mg(OH)_2$ in water	$Mg(OH)_2(s) + 2\,H_3O^+(aq) \longrightarrow Mg^{2+}(aq) + 4\,H_2O(\ell)$	Phillips Milk of Magnesia
calcium carbonate: $CaCO_3$	$CaCO_3(s) + 2\,H_3O^+(aq) \longrightarrow Ca^{2+}(aq) + 3\,H_2O(\ell) + CO_2(g)$	Tums, Di-Gel
sodium bicarbonate: $NaHCO_3$	$NaHCO_3(s) + H_3O^+(aq) \longrightarrow Na^+(aq) + H_2O(\ell) + CO_2(g)$	baking soda, Alka-Seltzer
aluminum hydroxide: $Al(OH)_3$	$Al(OH)_3(s) + 3\,H_3O^+(aq) \longrightarrow Al^{3+}(aq) + 6\,H_2O(\ell)$	Amphojel
dihydroxyaluminum sodium carbonate: $NaAl(OH)_2CO_3$	$NaAl(OH)_2CO_3(s) + 4\,H_3O^+(aq) \longrightarrow Na^+(aq) + Al^{3+}(aq) + 7\,H_2O(\ell) + CO_2(g)$	Rolaids

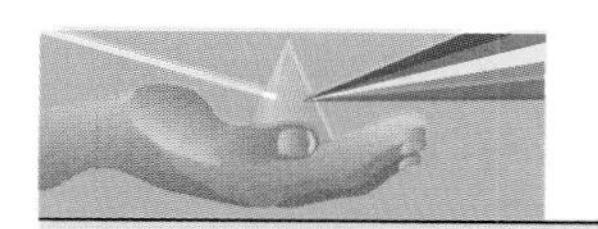

CHEMISTRY YOU CAN DO

Aspirin and Digestion

Aspirin is a potent drug capable of relieving pain, fever, and inflammation. Recent studies indicate it may also have an effect on blood clotting and heart disease. It is made from salicylic acid, which is naturally found in a variety of plants. The effect of pure salicylic acid on your stomach, however, makes it quite unpleasant as a pain remedy. Commercial aspirin is a derivative of salicylic acid, called acetylsalicylic acid, which has all the benefits of salicylic acid without the discomfort.

Aspirin is still somewhat acidic, however, and can sometimes cause discomfort or worse in people susceptible to stomach irritation. As a result, there are a couple of different forms of aspirin on the market today. The first and most common is plain aspirin. For people with stomach problems, there is also the option of taking buffered aspirin, which includes buffer in the aspirin tablet to lessen the effect of aspirin's acidity. A more recent development is enteric aspirin, which is plain aspirin in a coated tablet which prevents the aspirin from dissolving in stomach acid, but does allow it to dissolve in the small intestine, which is basic.

For this experiment, obtain at least three tablets of each kind of aspirin. Examples of regular aspirin are Bayer and Anacin. Buffered aspirin can be found as Bufferin or as Bayer Plus. Enteric aspirin is most commonly seen as Bayer Enteric or as Ecotrin. Fill three transparent cups or glasses with water. Drop one intact tablet of regular aspirin into the first glass, one tablet of buffered aspirin into the second glass, and one tablet of enteric aspirin into the third glass. Note the changes in the tablets at one-minute intervals until no further changes occur. Now repeat this experiment using vinegar instead of water to dissolve the tablets. Observe each of the tablets in the vinegar at 30-second intervals until no further change is seen. For the final experiment, fill the glasses with water and add 2 teaspoons of baking soda and stir until dissolved. Once you have made your baking soda solution, add one of each type of aspirin tablet and observe what happens.

How did each of the tablets react to each of the different solutions? The acidity of the vinegar should have mimicked the acidity of your stomach. The basic nature of the baking soda should have mimicked the environment of your intestines, which follow your stomach in the digestive system. How do you suppose each type of aspirin works according to its ability to dissolve in your experiments? Does this lead you to think about the kind of aspirin you take?

EXAMPLE 16.4

Neutralizing Stomach Acid

How many moles (and what mass) of HCl could be neutralized by 0.750 g of the antacid $CaCO_3$?

SOLUTION The balanced equation for this reaction is found in Table 16.6. From the equation we see that 1 mol of the antacid reacts with 2 mol of HCl, molar mass = 36.46 g.

$$\text{amount of } CaCO_3 = 0.750 \text{ g} \cdot \frac{1 \text{ mol } CaCO_3}{100.1 \text{ g } CaCO_3} = 7.49 \times 10^{-3} \text{ mol } CaCO_3$$

Using the stoichiometric ratio,

$$\text{amount of HCl} = 7.49 \times 10^{-3} \text{ mol } CaCO_3 \cdot \frac{2 \text{ mol HCl}}{1 \text{ mol } CaCO_3}$$

$$= 1.50 \times 10^{-2} \text{ mol HCl}$$

$$\text{mass of HCl} = 1.50 \times 10^{-2} \text{ mol HCl} \cdot \frac{36.46 \text{ g HCl}}{1 \text{ mol HCl}} = 0.547 \text{ g HCl}$$

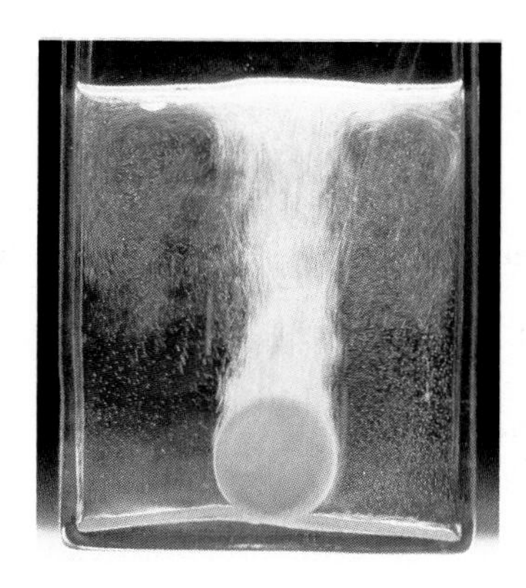

A commercial remedy for excess stomach acid. The bubbles are carbon dioxide from the reaction between the Brønsted acid citric acid and the Brønsted base bicarbonate ion, from sodium bicarbonate.

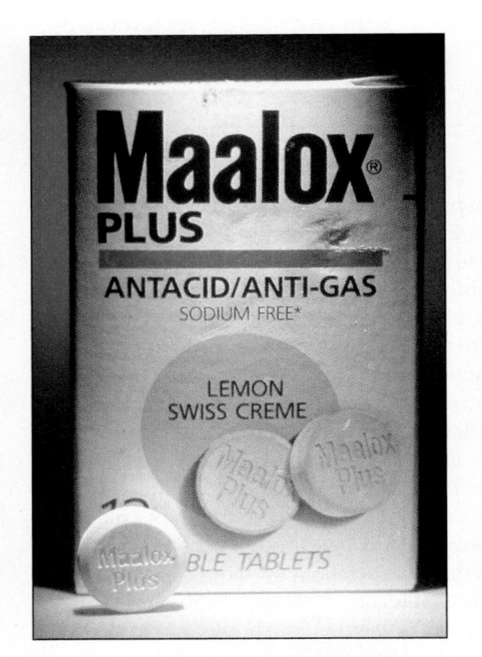

Maalox is composed of dried *ma*gnesium hydroxide and *al*uminum hydr*ox*ide.

EXERCISE 16.21 • *Neutralizing Stomach Acidity*

Using the reactions in Table 16.6, determine which antacid, on a per gram basis, can neutralize the most stomach acid (assume 1.0 M HCl).

16.8 LEWIS ACIDS AND BASES

In 1923 when Brønsted and Lowry independently proposed their acid-base concept, Gilbert N. Lewis also was developing a new theory of acids and bases. By the early 1930s Lewis had proposed definitions of acids and bases that are more general than those of Brønsted because they are based on sharing of electron pairs rather than on proton transfers. A **Lewis acid** is *a substance that can accept a pair of electrons to form a new bond,* and a **Lewis base** is *a substance that can donate a pair of electrons to form a new bond.* This means that an acid-base reaction in the Lewis sense can occur when there is a molecule (or ion) with a pair of electrons that can be donated and a molecule (or ion) that can accept an electron pair.

$$\underset{\text{acid}}{A} + \underset{\text{base}}{B:} \longrightarrow \underset{\text{new bond}}{B:A}$$

In Section 10.3 we defined this type of bond, which is present in many neutral molecular compounds and also in complex ions, as a *coordinate covalent bond.*

A simple example of a Lewis acid-base reaction is formation of a hydronium ion from H^+ and water. The H^+ ion has no electrons, while the water molecule has two unshared pairs of electrons on the oxygen atom. One of the electron pairs can be shared between H^+ and water, thus forming an O—H bond. A similar interaction occurs between H^+ and the base ammonia to form the ammonium ion.

$$\underset{\text{acid}}{H^+} + \underset{\text{base}}{:NH_3} \longrightarrow [NH_4]^+$$

Such reactions are very common. In general, they involve Lewis acids that are cations or neutral molecules with an available, empty orbital, and Lewis bases that are anions or neutral molecules with a lone pair of electrons.

Cationic Lewis Acids

All metal cations are potential Lewis acids. Not only do they attract electrons due to their positive charge, but all have at least one empty orbital. This empty orbital can accommodate an electron pair donated by a base and thereby form a two-electron chemical bond. Consequently metal ions read-

(a)

(b)

(c)

Figure 16.13
The amphoteric nature of $Al(OH)_3$. (a) Adding aqueous ammonia to a solution of Al^{3+} causes a precipitate of $Al(OH)_3$. (b) Adding a strong base (NaOH) to the $Al(OH)_3$ dissolves the precipitate. Here the aluminum hydroxide acts like a Lewis acid toward the Lewis base OH^- and forms a soluble salt of the complex ion $Al(OH)_4^-$. (c) If we begin again with freshly precipitated $Al(OH)_3$, it is dissolved as strong acid (HCl) is added. In this case $Al(OH)_3$ acts as a Brønsted base and forms a soluble aluminum salt and water.

ily form coordination complexes (Section 10.10) and also are hydrated in aqueous solution. One of the lone pairs on the oxygen atom in each of several water molecules forms a coordinate covalent bond to a metal ion when the ion becomes hydrated, and so the ion is a Lewis acid and water is a Lewis base.

The hydroxide ion ($:\ddot{\underset{..}{O}}—H^-$) is an excellent Lewis base and so binds readily to metal cations to give metal hydroxides. An important feature of the chemistry of many metal hydroxides is that they are *amphoteric,* meaning that they can react as both a base and an acid. The amphoteric aluminum hydroxide, for example, is behaving as a Lewis acid when it dissolves in a basic solution to form a complex ion containing one additional OH^- ion.

$$Al(OH)_3(s) + OH^-(aq) \rightleftharpoons [Al(OH)_4]^-(aq)$$

This reaction is shown in Figure 16.13. The same compound behaves as a Brønsted base when it reacts with a Brønsted acid (Table 16.7).

$$Al(OH)_3(s) + 3\,H_3O^+(aq) \rightleftharpoons Al^{3+}(aq) + 6\,H_2O(\ell)$$

Table 16.7

Some Common Amphoteric Metal Hydroxides

Hydroxide	Reaction as a Base	Reaction as an Acid
$Al(OH)_3$	$Al(OH)_3(s) + 3\,H_3O^+(aq) \longrightarrow Al^{3+}(aq) + 6\,H_2O(\ell)$	$Al(OH)_3(s) + OH^-(aq) \longrightarrow [Al(OH)_4]^-(aq)$
$Zn(OH)_2$	$Zn(OH)_2(s) + 2\,H_3O^+(aq) \longrightarrow Zn^{2+}(aq) + 4\,H_2O(\ell)$	$Zn(OH)_2(s) + 2\,OH^-(aq) \longrightarrow [Zn(OH)_4]^{2-}(aq)$
$Sn(OH)_4$	$Sn(OH)_4(s) + 4\,H_3O^+(aq) \longrightarrow Sn^{4+}(aq) + 8\,H_2O(\ell)$	$Sn(OH)_4(s) + 2\,OH^-(aq) \longrightarrow [Sn(OH)_6]^{2-}(aq)$
$Cr(OH)_3$	$Cr(OH)_3(s) + 3\,H_3O^+(aq) \longrightarrow Cr^{3+}(aq) + 6\,H_2O(\ell)$	$Cr(OH)_3(s) + OH^-(aq) \longrightarrow [Cr(OH)_4]^-(aq)$

Metal ions also form many complex ions with the Lewis base ammonia, $:NH_3$. For example, silver ion readily forms a water-soluble, colorless complex ion in liquid ammonia or in aqueous ammonia. Indeed, this complex is so stable that the very insoluble compound AgCl can be dissolved in aqueous ammonia:

$$AgCl(s) + 2\,NH_3(aq) \longrightarrow [H_3N:Ag:NH_3]^+(aq) + Cl^-(aq)$$

Neutral Molecules as Lewis Acids

Lewis's ideas about acids and bases account nicely for the fact that oxides of nonmetals behave as acids. Two important examples are carbon dioxide and sulfur dioxide, whose Lewis structures are

$$:\ddot{O}=C=\ddot{O}: \qquad :\underset{..}{O}=\ddot{S}\diagdown\!\!:\underset{..}{O}$$

In each case, there is a double bond; an "extra" pair of electrons is being shared between an oxygen atom and the central atom. Because oxygen is highly electronegative, electrons in these bonds are attracted away from the central atom, which becomes slightly positively charged. This makes the central atom a likely site to attract a pair of electrons. A Lewis base such as OH^- can bond to the carbon atom in CO_2 to give bicarbonate ion, HCO_3^-. This displaces one double-bond pair of electrons back onto an oxygen atom.

$$:\underset{..}{O}=C=\underset{..}{O}: + :\underset{..}{\ddot{O}}-H^- \longrightarrow :\underset{..}{O}=C\begin{matrix}\diagup \ddot{O}-H \\ \diagdown :\underset{..}{O}:^-\end{matrix}$$

As seen in Figure 16.14, carbon dioxide from the air can react to form sodium carbonate around the mouth of a bottle of sodium hydroxide. SO_2 can react similarly with hydroxide ion, and both CO_2 and SO_2 react with water (a Lewis base) to give acidic solutions.

Figure 16.14
Carbon dioxide in the air reacts with spilled base such as NaOH, forming Na_2CO_3. If the mouth of a glass-stoppered bottle such as the one shown here is not routinely cleaned, the sodium carbonate formed can virtually cement the top of the bottle to the neck, making it difficult to open the bottle. *(C.D. Winters)*

A Deeper Look

Toxicity of Carbon Monoxide, a Lewis Base

The interference of carbon monoxide with oxygen transport through the bloodstream is one of the best understood processes of metabolic poisoning. As early as 1895, it was noted that carbon monoxide deprives body cells of oxygen (asphyxiation), but it was not found until much later that carbon monoxide, like oxygen, combines with hemoglobin:

$$O_2 + \text{hemoglobin} \rightleftharpoons \text{oxyhemoglobin}$$

$$CO + \text{hemoglobin} \rightleftharpoons \text{carboxyhemoglobin}$$

Experimental measurements show that when carbon monoxide reacts as a Lewis base with the Fe^{2+} ion in a hemoglobin molecule, it forms a compound (carboxyhemoglobin) that is 140 times more stable than the compound of hemoglobin and oxygen (oxyhemoglobin). Since hemoglobin is so effectively tied up by carbon monoxide, it cannot perform its vital function of transporting oxygen.

When any organic material undergoes incomplete combustion, it liberates carbon monoxide. Sources include auto exhausts, smoldering leaves, lighted cigars or cigarettes, and charcoal burners. In the United States alone, combustion sources of all types release about 200 million tons of carbon monoxide per year into the atmosphere.

While the best estimates of the maximum global background level of carbon monoxide are of the order of 0.1 ppm, the background concentration in cities is higher. In heavy traffic, sustained levels of 100 ppm or more are common; for offstreet sites in large cities, about 7 ppm is typical. A concentration of 30 ppm for 8 h is sufficient to cause headache and nausea. Breathing an atmosphere that is 0.1% (1000 ppm) carbon monoxide for 4 h converts approximately 60% of the hemoglobin of an average adult to carboxyhemoglobin, and death is likely to result.

Since both the carbon monoxide and oxygen reactions with hemoglobin can be reversed, the concentrations, as well as relative strengths of bonds, affect the equilibrium. In air that contains 0.1% CO, oxygen molecules outnumber CO molecules 200 to 1. The larger concentration of oxygen helps to counteract the greater combining power of CO with hemoglobin by shifting the oxygen reaction equilibrium to the right. Consequently, if a carbon monoxide victim is exposed to fresh air or, still better, pure oxygen (provided he or she is still breathing), the carboxyhemoglobin (HbCO) is gradually decomposed, owing to the greater concentration of oxygen:

$$HbCO + O_2 \longrightarrow HbO_2 + CO$$

equilibrium shifted to right because of greater concentration of oxygen (Le Chatelier's principle)

Although carbon monoxide is not a cumulative poison, permanent damage can occur if certain vital cells (e.g., brain cells) are deprived of oxygen for more than a few minutes.

Individuals differ in their tolerance of carbon monoxide, but generally those with anemia or an otherwise low reserve of hemoglobin (including children) are more susceptible. No one is helped by carbon monoxide, and smokers suffer chronically from its effects. There is also a strong relationship between low infant birth weight and mothers who smoke. Finally, it should be noted that carbon monoxide is a subtle poison, since it is colorless, odorless, and tasteless.

EXERCISE 16.22 • *Lewis Acids and Bases*

Predict whether each of the following is a Lewis acid or a Lewis base. Drawing a Lewis structure for a molecule or ion is often helpful in making such a prediction.

a. PH_3 b. BCl_3 c. H_2S d. NO_2 e. Ni^{2+} f. CO

IN CLOSING

Having studied this chapter, you should be able to

- identify Brønsted acids and bases and tell which ones might be amphiprotic. (Section 16.1)
- identify conjugate bases of weak acids and conjugate acids of weak bases. (Section 16.1)

- calculate the pH of a solution, given either hydrogen ion concentration or hydroxide ion concentration. (Section 16.2)
- determine relative strengths of acids and bases by using a table of K_a and K_b values; from such values, predict the direction of the favored reaction in an equilibrium involving acids and bases. (Sections 16.3 and 16.4)
- calculate the K_a or K_b of a weak acid (or base), given the pH of its solution. (Section 16.3)
- estimate the pH of a solution of weak acid or base, given the concentration of the acid or base and its K_a or K_b value. (Section 16.4)
- predict the effect of changing pH on solubility of salts containing anions that are weak bases or cations that are weak acids. (Section 16.5)
- write balanced equations for simple hydrolysis reactions. (Section 16.5)
- select an acid/base pair from which a buffer of desired pH can be made; calculate the pH of a such a buffer, given information about the ratio of the concentrations of the conjugate acid/base pair. (Section 16.6)
- sketch the shapes of titration curves for titrations involving strong acid/strong base, weak acid/strong base, and strong acid/weak base, approximating the pH at the equivalence point and halfway to the equivalence point. (Section 16.7)
- select an appropriate indicator for a titration, given information such as that in Figure 16.11. (Section 16.7)
- identify Lewis acids and Lewis bases from their formulas or Lewis diagrams. (Section 16.8)

STUDY QUESTIONS

Review Questions

1. Trace the history of ideas about acids and bases; describe how acids were defined by Arrhenius, Brønsted, and Lewis.
2. Define the terms strong electrolyte, weak electrolyte, and nonelectrolyte. Give two examples of each type of substance. Give one example of a strong electrolyte that is an acid and one that is a base. Do the same for weak electrolytes.
3. Write the names and formulas of three strong Brønsted acids and three strong Brønsted bases.
4. Write a chemical equation showing a proton reacting with a water molecule to form a hydronium ion.
5. Write the chemical equation for the auto-ionization of water. Write the equilibrium-constant expression for this reaction. What is the value of the equilibrium constant at 25 °C? What is this constant called?
6. Write balanced chemical equations that show phosphoric acid, H_3PO_4, ionizing stepwise as a polyprotic acid.
7. Show algebraically by carrying out calculations for acids and conjugate bases in Table 16.2 that the ionization constant for a weak acid times the ionization constant for its conjugate base equals the auto-ionization constant for water.
8. Write ionization equations for a weak acid and its conjugate base. Show that adding these two equations gives the auto-ionization equation for water. How does this result relate to Study Question 7?
9. Designate the acid and the base on the left side of the following equations, and designate the conjugate partner of each on the right side.
   (a) $HNO_3(aq) + H_2O(\ell) \longrightarrow H_3O^+(aq) + NO_3^-(aq)$
   (b) $NH_4^+(aq) + CN^-(aq) \longrightarrow NH_3(aq) + HCN(aq)$
10. Dissolving ammonium bromide in water gives an acidic solution. Write a balanced equation showing how that can occur.
11. Write balanced equations showing how HPO_4^{2-} can be both a Brønsted acid and a Brønsted base.
12. State whether the pH is equal to 7, less than 7, or greater than 7 when equal molar amounts of
   (a) a weak base and a strong acid react,
   (b) a strong base and a strong acid, react, and
   (c) a strong base and a weak acid react.
13. Briefly describe how a buffer solution can control the

pH of a solution when strong acid is added, and when strong base is added. Use NH_3/NH_4Cl as an example buffer and HCl and NaOH as the strong acid and strong base.

14. Contrast the main ideas of the Brønsted and Lewis acid-base concepts. Name and write the formula for a substance that behaves as a Lewis acid, but not as a Brønsted acid.
15. Write an equation to show how water can act as a Brønsted base; write another equation to show how water can act as a Lewis base.
16. Define the term "amphoteric." Give an example of a metal hydroxide that is amphoteric.
17. Assuming that Al^{3+} ion exists in aqueous solution as the complex ion $[Al(H_2O)_6]^{3+}$, write a balanced equation to show how hydrolysis leads to an acidic solution.

Brønsted Acids and Bases

18. Write an equation to describe the proton transfer that occurs when each of the following acids is added to water.
(a) HBr (c) HSO_4^-
(b) CF_3COOH (d) HNO_2

19. Write an equation to describe the proton transfer that occurs when each of the following bases is added to water.
(a) H^- (c) NO_2^-
(b) HCO_3^- (d) PO_4^{3-}

20. Write the formula for and name the conjugate partner for each acid or base.
(a) CN^- (d) S^{2-}
(b) SO_4^{2-} (e) HSO_4^-
(c) HS^- (f) HCOOH (formic acid)

21. Write the formula for and name the conjugate partner for each acid or base.
(a) HI (d) H_2CO_3
(b) NO_3^- (e) HSO_4^-
(c) CO_3^{2-} (f) SO_3^{2-}

22. Identify the acid and the base that are reactants in each equation; identify the conjugate base and conjugate acid on the product side of each equation.
(a) $HI(aq) + H_2O(\ell) \rightleftharpoons H_3O^+(aq) + I^-(aq)$
(b) $OH^-(aq) + NH_4^+(aq) \rightleftharpoons H_2O(\ell) + NH_3(aq)$
(c) $NH_3(aq) + H_2CO_3(aq) \rightleftharpoons NH_4^+(aq) + HCO_3^-(aq)$

23. Identify the acid and the base that are reactants in each equation; identify the conjugate base and conjugate acid on the product side of the equation.
(a) $HS^-(aq) + H_2O(\ell) \rightleftharpoons H_2S(aq) + OH^-(aq)$
(b) $S^{2-}(aq) + NH_4^+(aq) \rightleftharpoons NH_3(aq) + HS^-(aq)$
(c) $HCO_3^-(aq) + HSO_4^-(aq) \rightleftharpoons H_2CO_3(aq) + SO_4^{2-}(aq)$

24. Write stepwise equations for protonation or deprotonation of each of these polyprotic acids and bases.
(a) H_2SO_3 (c) $NH_3CH_2COOH^+$ (glycinium ion, a diprotic acid)
(b) S^{2-}

25. Write stepwise equations for protonation or deprotonation of each of these polyprotic acids and bases.
(a) CO_3^{2-} (c) $NH_2CH_2COO^-$ (glycinate ion, a diprotic base)
(b) H_3AsO_4

pH Calculations

26. A popular soft drink has a pH of 3.30. What is the hydronium ion concentration? Is the drink acidic or basic?

27. Milk of magnesia, $Mg(OH)_2$, has a pH of 10.5. What is the hydronium ion concentration of the solution? Is this solution acidic or basic?

28. What is the pH of a 0.0013 M solution of HNO_3? What is the pOH of this solution?

29. What is the pH of a solution that is 0.025 M in NaOH? What is the pOH of this solution?

30. The pH of a $Ba(OH)_2$ solution is 10.66 at 25 °C. What is the hydroxide-ion concentration of this solution? If the solution volume is 250. mL, what mass (in grams) of $Ba(OH)_2$ must have been used to make this solution?

31. A 1000. mL solution of hydrochloric acid has a pH of 1.3. What mass (in grams) of HCl is dissolved in the solution?

32. Make the following interconversions. In each case tell whether the solution is acidic or basic.

	pH	*[H_3O^+] (M)*	*[OH^-] (M)*
(a)	1.00	______	______
(b)	10.5	______	______
(c)	______	1.8×10^{-4}	______
(d)	______	5.6×10^{-10}	______
(e)	______	______	2.3×10^{-5}

33. Make the following interconversions. In each case tell whether the solution is acidic or basic.

	pH	*[H_3O^+] (M)*	*[OH^-] (M)*
(a)	______	6.1×10^{-7}	______
(b)	______	2.2×10^{-9}	______
(c)	4.67	______	______
(d)	______	2.5×10^{-2}	______
(e)	9.12	______	______

Acid-Base Strengths

34. Classify each of the following as a strong acid, weak acid, strong base, weak base, amphiprotic substance, or neither acid nor base.
(a) HCl (d) CH_3COO^-
(b) NH_4^+ (e) CH_4
(c) H_2O (f) CO_3^{2-}

35. Classify each of the following as a strong acid, weak acid, strong base, weak base, amphiprotic substance, or neither acid nor base.
(a) CH_3COOH (d) NH_3
(b) Na_2O (e) $Ba(OH)_2$
(c) H_2SO_4

36. Which acid in each pair is stronger?
(a) HNO_2 or HNO_3 (c) $HClO_4$ or $HClO$
(b) HCl or HF (d) H_2SO_4 or H_2SO_3

37. Which base in each pair is stronger?
(a) Cl^- or CH_3COO^- (c) ClO^- or ClO_4^-
(b) NO_2^- or NO_3^- (d) S^{2-} or CH_3COO^-

38. Place the following acids (a) in order of increasing strength, and (b) in order of increasing pH, assuming you have a 0.10 M solution of each acid.
(a) valeric acid, $K_a = 1.5 \times 10^{-5}$
(b) glutaric acid, $K_a = 3.4 \times 10^{-4}$
(c) hypobromous acid, $K_a = 1.3 \times 10^{-3}$

39. Write the chemical equation for proton transfer to water and the equilibrium-constant expression for K_a for each acid. If an acid is polyprotic, write a separate equation and equilibrium-constant expression for each stepwise loss of a proton.
(a) CH_3COOH (d) H_3PO_4
(b) HCN (e) NH_4^+
(c) H_2CO_3 (f) H_2SO_4

40. Write the chemical equation for proton transfer to water and the equilibrium-constant expression for K_b for each base. If a base is polyprotic, write a separate equation and equilibrium-constant expression for each stepwise protonation.
(a) SO_3^{2-} (d) NH_3
(b) PO_4^{3-} (e) CH_3COO^-
(c) F^- (f) S^{2-}

41. Barbituric acid, $C_4H_4N_2O_3$, has a K_a of 9.9×10^{-5}, while for nicotinic acid, $C_6H_5NO_2$, K_a is 1.4×10^{-5}. (a) Which is the stronger acid? (b) For a 0.010 M solution of each of these monoprotic acids, which will have the higher pH?

42. Which solution will be more acidic?
(a) 0.10 M H_2CO_3 or 0.10 M NH_4Cl
(b) 0.10 M HF or 0.10 M $KHSO_4$

43. Which solution will be more basic?
(a) 0.10 M NH_3 or 0.10 M NaF
(b) 0.10 M K_2S or 0.10 M K_3PO_4

44. Without doing any calculations, assign each of these 0.10 M aqueous solutions to one of these pH ranges: pH < 2; pH between 2 and 6; pH between 6 and 8; pH between 8 and 12; pH > 12
(a) HNO_2 (e) BaO
(b) NH_4Cl (f) $KHSO_4$
(c) NaF (g) $NaHCO_3$
(d) $Mg(CH_3COO)_2$ (h) $BaCl_2$

45. Calculate the pH of each solution in Study Question 44 to verify your prediction.

46. A 0.015 M solution of cyanic acid has a pH of 2.67. (a) What is the hydronium ion concentration in the solution? (b) What is the ionization constant, K_a, of the acid?

47. What are the equilibrium concentrations of H_3O^+, acetate ion, and acetic acid in a 0.20 M aqueous solution of acetic acid (CH_3COOH)?

48. The ionization constant of a very weak acid HA is 4.0×10^{-9}. Calculate the equilibrium concentrations of H_3O^+, A^-, and HA in a 0.040 M solution of the acid.

49. For each of these basic solutions, a pH measured at 25 °C is given. Calculate K_b and pK_b for each base.
(a) 0.113 M NH_3, pH = 11.15
(b) 0.074 M N_2H_4, pH = 10.42
(c) 0.175 M NaCN, pH = 11.28
(d) 0.437 M $NaCH_3COO$, pH = 9.19

50. Calculate pK_a and K_a for the conjugate acid of each base in Study Question 49.

51. The weak base methylamine, CH_3NH_2, has $K_b = 5.0 \times 10^{-4}$. It reacts with water according to the equation

$$CH_3NH_2(aq) + H_2O(\ell) \rightleftharpoons CH_3NH_3^+(aq) + OH^-(aq)$$

Calculate the equilibrium hydroxide concentration in a 0.23 M solution of the base. What is the pH of the solution?

52. Calculate the pH of a 0.12 M aqueous solution of the base aniline, $C_6H_5NH_2$ ($K_b = 4.2 \times 10^{-10}$).

53. By now you may be wishing you had an aspirin. Aspirin is a weak acid with $K_a = 3.27 \times 10^{-4}$ for the reaction

$$HC_9H_7O_4(aq) + H_2O(\ell) \rightleftharpoons C_9H_7O_4^-(aq) + H_3O^+(aq)$$

Two aspirin tablets, each containing 0.325 g of aspirin (along with a nonreactive "binder" to hold the tablet together), are dissolved in 200.0 mL of water. What is the pH of this solution?

54. Lactic acid, $C_3H_6O_3$, occurs in sour milk as a result of the metabolism of certain bacteria. What is the pH of a solution of 56 mg of lactic acid in 250 mL of water? K_a for lactic acid is 1.4×10^{-4}.

55. For each aqueous solution, predict what ions and

molecules will be present. Without doing any calculations, list the ions and molecules in order of decreasing concentration.
(a) HCl (c) HNO_2
(b) $NaClO_4$ (d) NaClO

56. For each aqueous solution, predict what ions and molecules will be present. Without doing any calculations, list the ions and molecules in order of decreasing concentration.
(a) NaOH (c) HCN
(b) $MgCl_2$ (d) $NaCH_3CH_2COO$

Acid-Base Reactions

57. Predict whether each reaction is product favored, and explain your reasoning.
(a) $NH_4^+(aq) + Br^-(aq) \rightleftharpoons NH_3(aq) + HBr(aq)$
(b) $CH_3COOH(aq) + CN^-(aq) \rightleftharpoons CH_3COO^-(aq) + HCN(aq)$
(c) $NH_2^-(aq) + H_2O(\ell) \rightleftharpoons NH_3(aq) + OH^-(aq)$

58. Complete each of these reactions by filling in the blanks. Predict whether each reaction is product favored or reactant favored, and explain your reasoning.
(a) ______(aq) + $HSO_4^-(aq) \rightleftharpoons HCN(aq) + SO_4^{2-}(aq)$
(b) $H_2S(aq) + H_2O(\ell) \rightleftharpoons H_3O^+(aq) +$ ______(aq)
(c) $H^-(aq) + H_2O(\ell) \rightleftharpoons OH^-(aq) +$ ______(aq)

59. Several acids and their respective equilibrium constants are:

$$HF(aq) + H_2O(\ell) \rightleftharpoons H_3O^+(aq) + F^-(aq) \quad K_a = 7.2 \times 10^{-4}$$

$$HS^-(aq) + H_2O(\ell) \rightleftharpoons H_3O^+(aq) + S^{2-}(aq) \quad K_a = 8 \times 10^{-18}$$

$$CH_3COOH(aq) + H_2O(\ell) \rightleftharpoons H_3O^+ + CH_3COO^-(aq) \quad K_a = 1.8 \times 10^{-5}$$

(a) Which is the strongest acid? Which is the weakest acid?
(b) Which acid has the weakest conjugate base?
(c) Which acid has the strongest conjugate base?

60. Predict which of the following acid-base reactions are product favored and which are reactant favored. In each case write a balanced equation for any reaction that might occur, even if the reaction is reactant favored. Consult Table 16.2 if necessary.
(a) $H_2O(\ell) + HNO_3(aq)$ (c) $CN^-(aq) + HCl(aq)$
(b) $H_3PO_4(aq) + H_2O(\ell)$ (d) $NH_4^+(aq) + F^-(aq)$

61. Predict which of the following acid-base reactions are product favored and which are reactant favored. In each case write a balanced equation for any reaction that might occur, even if the reaction is reactant favored. Consult Table 16.2 if necessary.
(a) $NH_4^+(aq) + HPO_4^{2-}(aq)$
(b) $CH_3COOH(aq) + OH^-(aq)$
(c) $HSO_4^-(aq) + H_2PO_4^-(aq)$
(d) $CH_3COOH(aq) + F^-(aq)$

62. For each salt, predict whether an aqueous solution will have pH less than, equal to, or greater than 7.
(a) $NaHSO_4$ (c) $KClO_4$
(b) NH_4Br (d) NaH_2PO_4

63. For each salt, predict whether an aqueous solution will have pH less than, equal to, or greater than 7.
(a) $AlCl_3$ (c) $NaNO_3$
(b) Na_2S (d) Na_2HPO_4

64. For each salt, predict whether an aqueous solution will have pH less than, equal to, or greater than 7.
(a) NH_4NO_3 (c) $SrCl_2$
(b) KCH_3COO (d) $(NH_4)_2S$

65. Explain why $BaCO_3$ is soluble in aqueous HCl, but $BaSO_4$, which is used for making the intestines visible in x-ray photographs, remains sufficiently insoluble in the HCl in a human stomach so that poisonous barium ions do not get into the bloodstream.

66. For which of the following substances would solubility be greater at pH = 2 than at pH = 7? Write a balanced net ionic equation for the reaction that increases solubility at the lower pH in each case.
(a) $Cu(OH)_2$ (b) $CuSO_4$ (c) $CuCO_3$ (d) CuS
(e) $Cu_3(PO_4)_2$

67. How could you prepare pure $CaSO_4$ from limestone, $CaCO_3$? Write balanced equations for reactions that you would carry out, and explain the conditions and amounts of substances you would use for each reaction.

68. Beginning with magnesium carbonate, describe how you would prepare (a) magnesium sulfate, (b) magnesium bromide, (c) magnesium perchlorate, and (d) magnesium acetate.

Buffer Solutions

69. A buffer solution can be made from benzoic acid (C_6H_5COOH) and sodium benzoate (NaC_6H_5COO). What mass (in grams) of the acid would you have to mix with 14.4 g of the sodium salt in order to have a liter of a solution with a pH of 3.88?

70. If a buffer solution is prepared from 5.15 g of

NH_4NO_3 and 0.10 L of 0.15 M NH_3, what is the pH of the solution? What is the new pH if the solution is diluted with pure water to a volume of 500. mL?

71. Many natural processes can be studied in the laboratory but only in an environment of controlled pH. Which of the following combinations would be the best choice to buffer the pH at approximately 7?
(a) H_3PO_4/NaH_2PO_4
(b) NaH_2PO_4/Na_2HPO_4
(c) Na_2HPO_4/Na_3PO_4

72. Which of the following combinations would be the best to buffer the pH at approximately 9?
(a) $CH_3COOH/NaCH_3COO$
(b) $HCl/NaCl$
(c) NH_3/NH_4Cl

73. Select from Table 16.2 an acid/base conjugate pair that would be suitable for preparing a buffer solution whose concentration of hydronium ions is
(a) 4.5×10^{-3} (c) 8.3×10^{-6}
(b) 5.2×10^{-8} (d) 9.7×10^{-11}

74. Select from Table 16.2 an acid/base conjugate pair that would be suitable for preparing a buffer solution with pH equal to
(a) 3.45 (c) 8.32
(b) 5.48 (d) 10.15

75. In order to buffer a solution at a pH of 4.57, what mass of sodium acetate, $NaCH_3COO$, should you add to 500. mL of a 0.150 M solution of acetic acid, CH_3COOH? (Assume that there is negligible change in volume.)

76. You dissolve 0.425 g of NaOH in 2.00 L of a solution that originally had $[H_2PO_4^-] = [HPO_4^{2-}] = 0.132$ M. Calculate the resulting pH.

77. A buffer solution is prepared by adding 0.125 mol of ammonium chloride to 500. mL of 0.500 M aqueous ammonia. What is the pH of the buffer? If 0.0100 mol of HCl gas is bubbled into 500. mL of the buffer, what is the new pH of the solution?

78. What mass (in grams) of ammonium chloride, NH_4Cl, would have to be added to 500. mL of 0.10 M NH_3 solution to have a pH of 9.00?

Titration Curves

79. What volume (in mL) of 0.0997 M NaOH is required to titrate each of the following solutions to an end point indicated by phenolphthalein?
(a) 50.0 mL of 0.1023 M HNO_3
(b) 22.4 mL of 0.1549 M CH_3COOH
(c) 25.0 mL of 0.0983 M H_2SO_4
(d) 0.347 g of ammonium chloride dissolved in 20. mL of pure water

80. Without doing calculations, sketch the curve for the titration of 25.0 mL of 0.10 M NaOH with 0.10 M HCl. Indicate the pH at the beginning of the titration, at the equivalence point, and after 30.0 mL of HCl has been added.

81. Without doing calculations, sketch the titration curve for the titration of 10.0 mL of 0.10 M $NaCH_3COO$ with 0.10-M HCl.

82. Without doing calculations, sketch the titration curve for the titration of 25.0 mL of 0.10 M aqueous aluminum ion ($pK_a = 5.14$) with 0.10 M NaOH.

83. Four titration curves, A, B, C, and D, are shown in the figure. These are for titrations of 50.0-mL samples of four weak acids, each with 0.10 M NaOH.
A (a) Which curve corresponds to the weakest acid?
D (b) Which curve corresponds to the strongest acid?
(c) What is the concentration of each weak acid?

All are .1

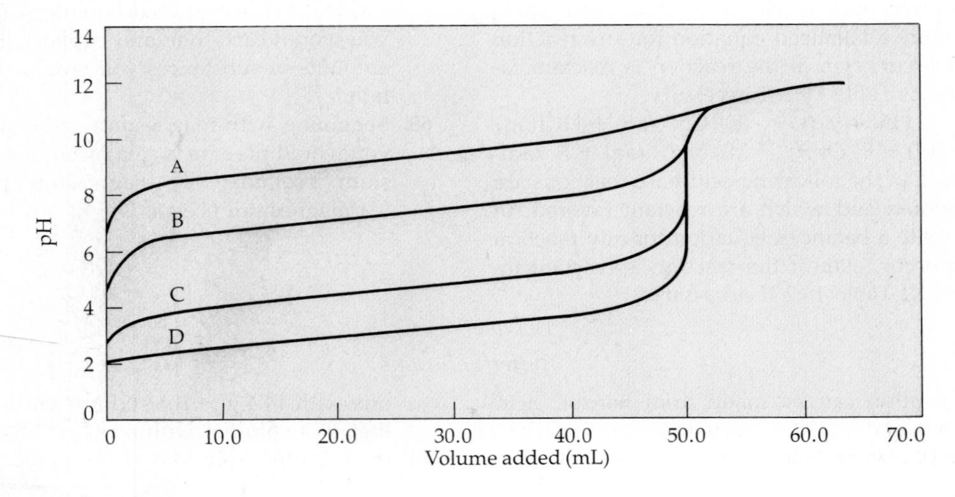

(d) For which acid or acids would phenolphthalein be a suitable end-point indicator?
(e) For which acid or acids would methyl red be a suitable end-point indicator?
(f) What are the approximate pK_a values for acids A, B, C, and D?
(g) If one of the curves is for acetic acid, which is it?
(h) Use Table 16.2 to try to identify acid A based upon its titration curve.

84. Three titration curves, A, B, and C, are shown in the figure. These are for titrations of 50.0-mL samples of three weak bases, each with 0.10 M HCl.
(a) Which curve corresponds to the weakest base?
(b) Which curve corresponds to the strongest base?
(c) What is the concentration of each weak base?
(d) For which base or bases would methyl orange be a suitable end-point indicator?
(e) For which base or bases would phenol red be a suitable end-point indicator?
(f) What are the approximate pK_a values for the conjugate acids of bases A, B, and C?

85. Calculate the pH corresponding to a point on the titration curve after 30.00 mL of 0.10 M HCl is added to 25.00 mL of 0.10 M NaOH.

86. Calculate the pH corresponding to a point on the titration curve after 15.00 mL of 0.10 M NaOH is added to 10.00 mL of 0.10 M acetic acid.

87. Assume you titrate 20.0 mL of 0.10 M NH_3 with 0.10 M HCl.
(a) What is the pH of the NH_3 solution before the titration begins?
(b) What is the pH at the equivalence point?
(c) What is the pH halfway to the equivalence point?
(d) Which indicator in Figure 16.11 would be best to detect the equivalence point?

88. At best the human eye can distinguish about a range of 100 in the intensity of color. When one color is changing into another, the eye will notice changes from about 90% color A and 10% color B to about 10% color A and 90% color B. Based on this information, use the Henderson-Hasselbalch equation to explain why the color change of most indicators occurs within a pH range of about two.

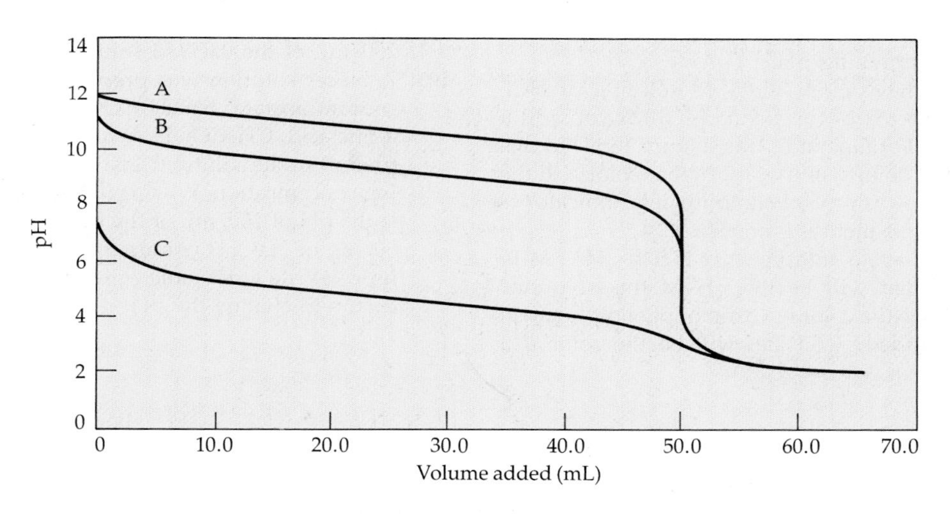

Lewis Acids and Bases

89. Which of these is a Lewis acid? A Lewis base?
(a) NH_3 (d) Al^{3+}
(b) $BeCl_2$ (e) H_2O
(c) BCl_3

90. Which of these is a Lewis acid? A Lewis base?
(a) O^{2-} (d) Cr^{3+}
(b) CO_2 (e) SO_3
(c) H^-

91. Identify the Lewis acid and the Lewis base in each reaction.
(a) $H_2O(\ell) + SO_2(aq) \longrightarrow H_2SO_3(aq)$
(b) $H_3BO_3(aq) + OH^-(aq) \longrightarrow B(OH)_4^-(aq)$
(c) $Cu^{2+}(aq) + 4\,NH_3(aq) \longrightarrow [Cu(NH_3)_4]^{2+}(aq)$
(d) $2\,Cl^-(aq) + SnCl_2(aq) \longrightarrow SnCl_4^{2-}(aq)$

92. Trimethylamine, $(CH_3)_3N:$, interacts readily with diborane, B_2H_6. The diborane dissociates to two BH_3 fragments, each of which can react with trimethylamine to form a complex, $(CH_3)_3N:BH_3$. Write an equation for this reaction and interpret it in terms of Lewis's acid-base theory.

93. Draw a Lewis structure for ICl_3. Predict the shape of this molecule. Does it function as a Lewis acid or base when it reacts with chloride ion to form ICl_4^-? What is the structure of this ion?

General Questions

94. When direct electric current at the same voltage is passed through the same cell containing an aqueous solution of each substance below and the current is measured, the data in the table are obtained. Based on these results, decide which ions or molecules are present in each solution. List these ions or molecules in order of decreasing concentration.

Substance	*Current (mA)*
NaCl	1.06
$KHCO_3$	1.31
$KHSO_4$	4.82

95. Sulfurous acid, H_2SO_3, is a weak diprotic acid ($K_{a1} = 1.2 \times 10^{-2}$, $K_{a2} = 6.2 \times 10^{-8}$). What is the pH of a 0.45 M solution of H_2SO_3? (Assume that only the first ionization is important in determining pH.)

96. Ascorbic acid (vitamin C, $C_6H_8O_6$) is a diprotic acid ($K_{a1} = 7.9 \times 10^{-5}$, $K_{a2} = 1.6 \times 10^{-12}$). What is the pH of a solution that contains 5.0 mg of the acid per mL of water? (Assume that only the first ionization is important in determining pH.)

97. Suppose you have a 5.0×10^{-4} M solution of HCl. (a) What is the pH of this solution? (b) If you add a drop of methyl orange indicator (Figure 16.11) to this solution, what color will it be? (c) What color will the solution be if a drop of phenolphthalein indicator is added instead of methyl orange?

98. Suppose you have a solution that is 5.0×10^{-3} M in NaOH. (a) What will be the pH of this solution? (b) What will be the color if a drop of phenolphthalein indicator is added? (c) What will be the color if a drop of alizarin indicator is added instead of phenolphthalein?

99. Does the pH of the solution increase, decrease, or stay the same when you
 (a) add solid ammonium chloride to 100 mL of 0.10 M NH_3?
 (b) add solid sodium acetate to 50.0 mL of 0.015 M acetic acid?
 (c) add solid NaCl to 25.0 mL of 0.10 M NaOH?

100. Does the pH of the solution increase, decrease, or stay the same when you
 (a) add solid sodium oxalate, $Na_2C_2O_4$, to 50.0 mL of 0.015 M oxalic acid?
 (b) add solid ammonium chloride to 100 mL of 0.016 M HCl?
 (c) add 20.0 g of NaCl to 1.0 L of 0.012 M sodium acetate, $NaCH_3COO$?

101. What is the pH of a 0.15 M acetic acid solution? If you add 83 g of sodium acetate to 1.50 L of the 0.15 M acetic acid solution, what is the new pH of the solution?

102. Calculate the pH of a 0.050 M solution of HF. What is the pH of the solution if you add 1.58 g of NaF to 250. mL of the 0.050 M solution?

103. A buffer solution was prepared by adding 4.95 g of sodium acetate, $NaCH_3COO$, to 250. mL of 0.150 M acetic acid, CH_3COOH. What ions and molecules are present in the solution? List them in order of decreasing concentration. What is the pH of the buffer? What is the pH of 100. mL of the buffer solution if you add 80. mg of NaOH? (Assume negligible change in volume.) Write a net ionic equation for the reaction that occurs to change the pH.

Summary Questions

104. Strongly acidic and strongly basic solutions are classified as hazardous wastes in many locales. Solutions whose pH is near 7 are not classified as hazardous on the basis of their acidity or basicity, though they may be hazardous for other reasons. Suppose that your company generates large volumes of acidic and basic solutions in a variety of locations, and you want to have a test that can be carried out by untrained personnel to determine whether it is safe (and legal) to dispose of these solutions. Devise a simple, inexpensive procedure (that does not require any electronic instruments) that a nonchemist could use to test a solution, decide whether it was too acidic or basic, and adjust the acidity until its pH was between 6 and 8.

105. Aniline, $C_6H_5NH_2$, is a weak base used in the dye industry. Its conjugate acid forms a salt, aniline hydrochloride, $[C_6H_5NH_3]Cl$, that can be titrated with a strong base such as NaOH. Assume you titrate 25.0 mL of 0.100 M aniline hydrochloride with 0.115 M NaOH. (K_a for aniline hydrochloride is 2.4×10^{-5}).
 (a) What is the pH of the aniline hydrochloride solution before the titration begins?
 (b) What is the pH at the equivalence point?
 (c) What is the pH halfway to the equivalence point?
 (d) Which indicator in Figure 16.11 would be best to detect the equivalence point?

106. A common situation in humans is hyperventilation. A person in a hysterical state breathes more rapidly than normal, and more carbon dioxide is excreted than normal. When a person hyperventilates, the pH of his or her blood rises and alkalosis can occur. Alkalosis causes overexcitation of the nervous system, making the problem worse. Write chemical equations for the reactions by which rapid breathing can raise the pH of blood. Use your equations and LeChatelier's principle to explain why breathing into a paper bag is a standard treatment for hyperventilation.

Using Computer Programs and Videodiscs

107. Use the *KC? Discoverer* program to find all elements that react with acids. Do most of these elements have something else in common besides their reaction with acids? If so, what?
108. Use the *KC? Discoverer* program to find all elements that react with base. Identify at least one metal that reacts with base. What is a metal such as this called?
109. Use the *Notebook* program (*J. Chem. Educ.: Software,* 4B(1), 1991) or a spreadsheet program to calculate and plot pH as a function of volume of base added in the titration of 100.0 mL of 0.050 M HCl(aq) with 0.050 M NaOH(aq). Set up the columns of the spreadsheet as shown.

volume NaOH	*mol NaOH*	*mol HCl left*	*total volume*	*conc. HCl left*	*pH*
0.00	0.00	0.0050	100.0	0.050	1.30
5.00	0.00025	0.0048	105.0	0.045	1.34
etc.					

Once the equivalence point is reached, HCl becomes the limiting reactant, and so the third and fifth columns and the calculation of pH need to be modified.

110. The *Chemical Demonstrations* videodisc (and videotape) describes the use of an acid-base indicator made from the juice of a red cabbage.
   (a) Describe the general effect of an acid on the cabbage juice. What is the effect of a base?
   (b) List the household products tested and tell if they have a pH less than 7, equal to 7, or greater than 7.

Galvanized objects. A thin coating of zinc helps prevent the oxidation of iron.
(C.D. Winters)

Electrochemistry

CHAPTER

17

In Section 5.7 you learned that an important class of chemical reactions, known as oxidation-reduction (redox) reactions, involves transfer of electrons from one atom, molecule, or ion to another. Electrochemistry is the study of the relationship between electricity and chemical reactions, and it deals with redox reactions because they involve electron transfer. Electrochemistry is involved in many practical devices and processes. In an electrochemical cell (a battery), electrons transferred as the result of a product-favored oxidation-reduction reaction are directed through a wire so that an electrical current flows. The voltage of such a cell depends on the strengths of the oxidizing agent and reducing agent, and voltage measurements can be used to construct a table of numerical values of oxidizing and reducing strength (similar to Table 16.2 for acid-base strengths). Knowledge of oxidizing and reducing strengths help us understand and prevent corrosion of metals, because corrosion is oxidation of the metal. Electrolysis and electroplating also involve redox reactions. In an electrolysis cell, an external source of power causes electrical current to flow through a solution of molten salt, and the flow of electrons forces a reactant-favored process to produce products. Electrolysis is important in the manufacture of items as diverse as aluminum foil, vinyl plastics, chlorine for disinfecting water supplies, and compact audio discs.

If you carried out Chemistry You Can Do: Preparing a Pure Sample of an Element in Chapter 3, you generated pure copper on the surface of an iron nail by means of an oxidation-reduction reaction. (If you did not do this, you might want to try it now; all you need are some pennies, vinegar, a nail, and salt.) Figure 17.1 shows a similar reaction. A piece of zinc metal is immersed in a solution containing copper ions (from a soluble compound such as $CuSO_4$). As time passes, the results of a chemical change can be seen. The blue color of the solution fades and a coating of orange metallic copper forms on the zinc. This is clearly a product-favored reaction. The reaction comes to a halt when all of the zinc has dissolved or the copper-ion concentration becomes very small, depending on which is the limiting reactant.

These observations can be interpreted by saying that Cu^{2+} ions in solution were transformed into Cu atoms on the surface of the zinc. Since the zinc dissolved, Zn atoms were apparently changed into Zn^{2+} ions in the solution. There has been a change in the number of electrons associated with each kind of atom. Each copper ion has gained two electrons to become a copper atom. Each zinc atom has lost two electrons to become a zinc ion. In Section 5.7 we defined a process in which an atom, ion, or molecule *gains* electrons as a **reduction.** A process in which an atom, ion, or molecule *loses* electrons was defined as an **oxidation.** The overall process is called a "redox" reaction, combining the first letters of the words *red*uction and *ox*idation.

Electrons must be transferred from one kind of atom, molecule, or ion to another in a redox reaction. Since transfer of electrons through a wire constitutes an electrical current, it is reasonable to ask whether the electrons that are transferred from the zinc atoms to the copper ions could be made to pass through a circuit. If so, we would have a way of generating an electrical current from a chemical reaction. This can be done in a variety of ways, and it is the basis for all kinds of batteries, such as those that power flashlights,

Figure 17.1

(a) A piece of zinc metal is immersed in a solution containing Cu^{2+} ions. (b) After a few minutes, the solution's blue color diminishes and metallic copper builds up on the zinc. (c) In an hour or so, the solution contains a very low concentration of Cu^{2+} ions as indicated by its colorlessness, and the amount of copper built up on the remaining zinc is considerable. The Zn^{2+} ions formed are colorless. *(C.D. Winters)*

(a)

(b)

(c)

toys, portable radios, and many other consumer products. This chapter will help you understand how batteries work.

Extending the idea of electron transfer further, what do you expect would happen if electrons were forced into a chemical system from a source of electrical current such as a battery? If you said that electrical energy passing through a chemical system could force reactant-favored systems to produce products, you were exactly right. A process in which this occurs is called an **electrolysis.** Electrolytic processes are even more important in our economy than the redox reactions that power batteries. They are used in the production of many metals, including copper and aluminum, and in electroplating processes that produce a thin coating of metal on many different kinds of items.

17.1 REDOX REACTIONS

Since electrochemistry deals with redox reactions, it is useful to be able to recognize them and to be familiar with oxidation, reduction, oxidizing agents, and reducing agents. In Section 5.7 (Table 5.3) common oxidizing and reducing agents were listed, and you could predict that a redox process would occur when a strong oxidizing agent and a strong reducing agent were involved as reactants. Table 5.2 defined oxidation as combination with oxygen or a halogen, or a loss of electrons, and reduction as loss of oxygen or of halogen, or as gain of electrons. Oxidation can also be defined as an increase in oxidation number and reduction as a decrease in oxidation number. (Section 10.8 showed how to use oxidation numbers to keep track of electrons gained or lost during chemical reactions.) Redox reactions can be recognized because they involve changes in oxidation numbers of two or more species as reactants are converted to products.

You may want to review the definitions of oxidation and reduction in Section 5.7 and the rules for assigning oxidation numbers in Section 10.8.

Oxidation occurs when an atom, molecule, or ion loses electrons; the oxidation number of at least one atom increases.

$$\overset{0}{Mg}(s) + 2\,H_3O^+(aq) \longrightarrow \overset{+2}{Mg^{2+}}(aq) + H_2(g) + 2\,H_2O(\ell)$$

An Mg atom in the solid metal is oxidized: its oxidation number has increased from 0 to +2.

$$\overset{0}{H_2}(g) + Cl_2(g) \longrightarrow 2\,\overset{+1}{H}Cl(g)$$

Two H atoms in an H_2 molecule are oxidized: the oxidation number of each H atom has increased from 0 to +1. We say that the H_2 molecule has been oxidized.

Reduction occurs when an atom, molecule, or ion gains electrons; the oxidation number of at least one atom decreases. Let's look at the same two reactions.

$$Mg(s) + 2\,\overset{+1}{H_3}O^+(aq) \longrightarrow Mg^{2+}(aq) + \overset{0}{H_2}(g) + 2\,H_2O(\ell)$$

Two H_3O^+ ions are reduced: the oxidation number of one H from each H_3O^+ ion has decreased from +1 to 0.

$$H_2(g) + \overset{0}{Cl_2}(g) \longrightarrow 2\,H\overset{-1}{Cl}(g)$$

Two Cl atoms in a Cl_2 molecule are reduced: the oxidation number of each Cl atom has decreased from 0 to −1. The Cl_2 molecule is said to have been reduced.

A **reducing agent** is an atom, molecule, or ion that causes *something else* to be reduced by providing electrons; because it loses electrons, it is itself oxidized.

$$\overset{0}{Mg}(s) + 2\,\overset{+1}{H_3}O^+(aq) \longrightarrow \overset{+2}{Mg^{2+}}(aq) + \overset{0}{H_2}(g) + 2\,H_2O(\ell)$$

Mg causes H_3O^+ to be reduced; it is the reducing agent and is itself oxidized.

$$\overset{0}{H_2}(g) + \overset{0}{Cl_2}(g) \longrightarrow 2\,\overset{+1}{H}\overset{-1}{Cl}(g)$$

H_2 causes Cl_2 to be reduced; it is the reducing agent and gets oxidized.

An **oxidizing agent** is an atom, molecule, or ion that causes *something else* to be oxidized by taking away electrons; because it accepts electrons, it is itself reduced.

$$\overset{0}{Mg}(s) + 2\,\overset{+1}{H_3}O^+(aq) \longrightarrow \overset{+2}{Mg^{2+}}(aq) + \overset{0}{H_2}(g) + 2\,H_2O(\ell)$$

H_3O^+ causes Mg to be oxidized; it is the oxidizing agent and gets reduced.

$$\overset{0}{H_2}(g) + \overset{0}{Cl_2}(g) \longrightarrow 2\,\overset{+1}{H}\overset{-1}{Cl}(g)$$

Cl_2 causes H_2 to be oxidized; it is the oxidizing agent and gets reduced.

EXAMPLE 17.1

Determining the Oxidation Number

What is the oxidation number of Mn in the permanganate ion, MnO_4^-?

SOLUTION Using the rules you learned in Section 10.8, the oxidation number of oxygen is usually −2. Also, the sum of the oxidation numbers of all the atoms in an ion is equal to the charge on the ion.

$$\text{charge on } MnO_4^- = -1 = (\text{oxidation number Mn}) + 4(\text{oxidation number O})$$
$$= (\text{oxidation number Mn}) + 4(-2)$$
$$\text{oxidation number Mn} = (+8) + (-1) = +7$$

The permanganate ion can be written as $\overset{+7}{Mn}O_4^-$ to show the oxidation number of Mn.

EXERCISE 17.1 • *Identifying Oxidizing and Reducing Agents*

Give the oxidation number for each atom and identify the oxidizing and reducing agents in the following balanced chemical equations.

a. $2\,Fe(s) + 3\,Cl_2(g) \longrightarrow 2\,FeCl_3(s)$

b. $2\,H_2(g) + O_2(g) \longrightarrow 2\,H_2O(\ell)$

c. $Cu(s) + 2\,NO_3^-(aq) + 4\,H_3O^+(aq) \longrightarrow Cu^{2+}(aq) + 2\,NO_2(g) + 6\,H_2O(\ell)$

d. $C(s) + O_2(g) \longrightarrow CO_2(g)$

e. $6\,Fe^{2+}(aq) + Cr_2O_7^{2-}(aq) + 14\,H_3O^+(aq) \longrightarrow 6\,Fe^{3+}(aq) + 2\,Cr^{3+}(aq) + 21\,H_2O(\ell)$

Electrons in Redox Reactions

Now apply what you know about redox reactions to the reaction between zinc metal and copper(II) ions pictured in Figure 17.1. The products were found to be zinc ions and copper metal, and so the equation must be

$$Zn(s) + Cu^{2+}(aq) \longrightarrow Zn^{2+}(aq) + Cu(s)$$

In order to see more clearly how electrons are transferred, this overall reaction can be thought of as resulting from two simultaneous "half-reactions": one for the oxidation of Zn, and one for the reduction of Cu^{2+} ions. The oxidation half-reaction is

$$Zn(s) \longrightarrow Zn^{2+}(aq) + 2\,e^-$$

This shows that each atom of the reducing agent, Zn, loses two electrons when it is oxidized to a Zn^{2+} ion. These two electrons are accepted by a Cu^{2+} ion in the reduction half-reaction,

$$Cu^{2+}(aq) + 2\,e^- \longrightarrow Cu(s)$$

As Cu^{2+} ions are converted to Cu(s) in this half-reaction, the blue color of the solution becomes less intense, and metallic copper forms on the surface of the zinc.

The net reaction is the sum of the oxidation and the reduction half-reactions.

$Zn(s) \longrightarrow Zn^{2+}(aq) + 2e^-$		(oxidation half-reaction)
$Cu^{2+}(aq) + 2\,e^- \longrightarrow Cu(s)$		(reduction half-reaction)
$Cu^{2+}(aq) + Zn(s) \longrightarrow Zn^{2+}(aq) + Cu(s)$		(net reaction)

Notice that no electrons appear in this net reaction; the number of electrons donated in the oxidation half-reaction exactly equals the number of electrons gained in the reduction half-reaction. This must always be true in a net reaction; otherwise electrons would be created from nothing or destroyed, violating the law of conservation of mass.

Consider another example, as shown in Figure 17.2. A piece of copper screen is immersed in a solution of silver nitrate. As the reaction proceeds, the solution gradually turns blue, and fine hair-like crystals form on the copper screen. We can conclude that copper(II) ions and silver metal are

Figure 17.2
Copper metal screen in a solution of $AgNO_3$. The blue color intensifies as more copper is oxidized to Cu^{2+} ion. *(C.D. Winters)*

being formed, and write two half-reactions. One of them must involve oxidation of Cu to Cu^{2+} and the other must involve reduction of Ag^+ to Ag.

$$\text{Cu(s)} \longrightarrow \text{Cu}^{2+}\text{(aq)} + 2\,\text{e}^- \qquad \text{(oxidation half-reaction)}$$
$$\text{Ag}^+\text{(aq)} + \text{e}^- \longrightarrow \text{Ag(s)} \qquad \text{(reduction half-reaction)}$$

In this case two electrons are produced in the oxidation half-reaction, but only one is needed for the reduction half-reaction. *One* atom of copper provides enough electrons to reduce *two* Ag^+ ions, and so the reduction half-reaction must occur twice every time the oxidation half-reaction occurs once. To indicate this, we multiply the reduction half-reaction by two:

$$2\,\text{Ag}^+\text{(aq)} + 2\,\text{e}^- \longrightarrow 2\,\text{Ag(s)} \qquad \text{(reduction half-reaction} \times 2)$$

Adding this to the oxidation half-reaction gives the net equation

$$\text{Cu(s)} + 2\,\text{Ag}^+\text{(aq)} \longrightarrow \text{Cu}^{2+}\text{(aq)} + 2\,\text{Ag(s)}$$

The method shown here is a general one. A net equation can always be generated by writing oxidation and reduction half-reactions and then adjusting the half-reaction equations so that the number of electrons produced by the oxidation equals the number used up by the reduction.

EXERCISE 17.2 • *Using Half-Reactions*

Write two half-reactions for the following redox equation. Show that their sum is the net reaction.

$$\text{Zn(s)} + 2\,\text{H}_3\text{O}^+\text{(aq)} \longrightarrow \text{Zn}^{2+}\text{(aq)} + \text{H}_2\text{(g)} + 2\,\text{H}_2\text{O}(\ell)$$

Balancing Redox Equations

Redox reactions in aqueous solution are often difficult to balance by the methods described in Section 5.3, because they often involve water, hydronium ions, and hydroxide ions as reactants or products. It is difficult to tell by observing the reaction which (if any) of H_2O, H_3O^+, and OH^- are involved, but there is a way to figure this out.

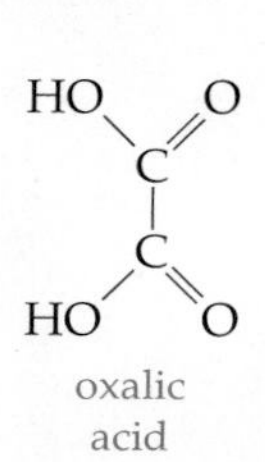

oxalic acid

As an example, consider the reaction of permanganate ion with oxalic acid in acidic solution. The products are manganese(II) ion and carbon dioxide, so the unbalanced equation is

$$\text{MnO}_4^-\text{(aq)} + \text{H}_2\text{C}_2\text{O}_4\text{(aq)} \longrightarrow \text{Mn}^{2+}\text{(aq)} + \text{CO}_2\text{(g)}$$

If you try to balance this equation by trial and error, you will almost certainly have a hard time with hydrogen and oxygen, because it turns out that water and hydronium ions are involved. Here is a series of steps that will show you how they are involved and also produce a balanced equation.

Step 1. Recognize the reaction as an oxidation-reduction process. Then determine what is reduced and what is oxidized. The oxidation number of Mn changes from +7 in MnO_4^- to +2 in Mn^{2+}, and the oxidation number of C changes from +3 in $H_2C_2O_4$ to +4 in CO_2.

Step 2. Break the overall reaction into half-reactions.

$$H_2C_2O_4(aq) \longrightarrow CO_2(g) \qquad \text{(oxidation half-reaction)}$$
$$MnO_4^-(aq) \longrightarrow Mn^{2+}(aq) \qquad \text{(reduction half-reaction)}$$

Step 3. Balance the atoms in each half-reaction. Begin by balancing all atoms except for O and H.

Oxalic acid half-reaction: First, balance the carbon atoms in the half-reaction.

$$H_2C_2O_4(aq) \longrightarrow 2\,CO_2(g)$$

This balanced the O atoms as well, so only H atoms remain. Because the product side is deficient by two H atoms, we put 2 H^+ there.

$$H_2C_2O_4(aq) \longrightarrow 2\,CO_2(g) + 2\,H^+(aq)$$

Permanganate half-reaction: The Mn atoms are already balanced, but more oxygen atoms are needed on the right side in order to balance the oxygens in MnO_4^-.

$$MnO_4^-(aq) \longrightarrow Mn^{2+}(aq) + \text{(4 oxygen atoms)}$$

In acidic aqueous solutions either H_3O^+ or H_2O could also be a reactant or product in this half-reaction. In this case oxygen atoms are needed on the right; therefore H_2O must be a product.

$$MnO_4^-(aq) \longrightarrow Mn^{2+}(aq) + 4\,H_2O(\ell)$$

Now there are eight H atoms on the right and none on the left. The fact that H atoms do not balance tells us that H^+ must be involved in the reaction. To balance hydrogen atoms, eight H^+ are placed on the left side of the half-reaction.

$$8\,H^+(aq) + MnO_4^-(aq) \longrightarrow Mn^{2+}(aq) + 4\,H_2O(\ell)$$

(Strictly speaking we ought to use H_3O^+ instead of H^+, but this would result in adding eight more water molecules to each side of the equation, which is rather cumbersome; it is clearer to add just H^+ now, and add the water molecules at the end.)

Step 4. Balance the half-reactions for charge. The oxalic acid half-reaction has a net charge of 0 on the left side and 2+ on the right. The reactants have lost two electrons. To show this, 2 e^- should appear on the positive (in this case, right) side.

$$H_2C_2O_4(aq) \longrightarrow 2\,CO_2(g) + 2\,H^+(aq) + 2\,e^-$$

This means that $H_2C_2O_4$ is the reducing agent (it loses electrons and gets oxidized). The loss of two electrons is also in keeping with the observation in Step 1 that the oxidation number of each of two C atoms increases by one, from +3 to +4.

The MnO_4^- half-reaction has a charge of 7+ on the left and 2+ on the right. Therefore, to achieve a net 2+ charge on each side, 5 e^- must appear on the left. Because it gains electrons, MnO_4^- is the oxidizing agent.

$$5\,e^- + 8\,H^+(aq) + MnO_4^-(aq) \longrightarrow Mn^{2+}(aq) + 4\,H_2O(\ell)$$

Step 5. Multiply the half-reactions by appropriate factors so that the reducing agent donates as many electrons as the oxidizing agent accepts. The oxalic acid reaction should be multiplied by 5, and the MnO_4^- reaction by 2, so that each half-reaction involves 10 electrons.

$$5[H_2C_2O_4(aq) \longrightarrow 2\,CO_2(g) + 2\,H^+(aq) + 2\,e^-]$$
$$2[5\,e^- + 8\,H^+(aq) + MnO_4^-(aq) \longrightarrow Mn^{2+}(aq) + 4\,H_2O(\ell)]$$

Step 6. Add the half-reactions to give the overall reaction.

$$5\,H_2C_2O_4(aq) \longrightarrow 10\,CO_2(g) + 10\,H^+(aq) + 10\,e^-$$
$$10\,e^- + 16\,H^+(aq) + 2\,MnO_4^-(aq) \longrightarrow 2\,Mn^{2+}(aq) + 8\,H_2O(\ell)$$
$$5\,H_2C_2O_4(aq) + 16\,H^+(aq) + 2\,MnO_4^-(aq) \longrightarrow 10\,CO_2(g) + 10\,H^+(aq) + 2\,Mn^{2+}(aq) + 8\,H_2O(\ell)$$

Step 7. Cancel common reactants and products. Since 16 H^+ appear on the left and 10 H^+ appear on the right, 10 H^+ are cancelled, leaving 6 H^+ on the left.

$$5\,H_2C_2O_4(aq) + 6\,H^+(aq) + 2\,MnO_4^-(aq) \longrightarrow 10\,CO_2(g) + 2\,Mn^{2+}(aq) + 8\,H_2O(\ell)$$

Step 8. Check the final results to make sure both atoms and charge are balanced.

Atom balance: Both sides of the equation have 2 Mn, 28 O, 10 C, and 16 H atoms.

Charge balance: Each side has a net charge of 4+.

To balance a redox equation for a reaction in basic solution, first balance the equation as though it were in acidic solution as described here (Steps 1 to 8), then add to both sides of the equation the number of OH^- ions needed to convert the H^+ ions to water.

Step 9. Add enough water molecules to both sides of the equation to convert all H^+ to H_3O^+. In this case six water molecules are needed.

$$5\,H_2C_2O_4(aq) + 6\,H_3O^+(aq) + 2\,MnO_4^-(aq) \longrightarrow 5\,CO_2(g) + 2\,Mn^{2+}(aq) + 14\,H_2O(\ell)$$

EXERCISE 17.3 • *Balancing Oxidation-Reduction Equations*

Balance this equation for the reaction of Zn with $Cr_2O_7^{2-}$ in acidic aqueous solution. Use the nine steps outlined in the text.

$$Zn(s) + Cr_2O_7^{2-}(aq) \longrightarrow Cr^{3+}(aq) + Zn^{2+}(aq)$$

17.2 ELECTROCHEMICAL CELLS

It is easy to see by the color changes in the two redox reactions shown in Figures 17.1 and 17.2 that these reactions favor the formation of products—as soon as the reactants are mixed, changes take place. All product-favored reactions release Gibbs free energy, and that energy can do useful work if the reactants—the oxidizing agent and the reducing agent—are separated in such a way that electrons cannot be transferred directly from one to the other. When electrons are forced to flow outside the reaction system and back again, they form an electric current that can operate a motor, cause a reactant-favored system to produce products, heat a cup of coffee, or do other useful things.

An arrangement of an oxidizing agent and a reducing agent in such a way that they can react only if electrons flow through an outside conductor is called an **electrochemical cell,** a **voltaic cell,** or in everyday terms, a **battery.** Figure 17.3 diagrams how this can be done for the Zn/Cu^{2+} reaction that was shown in Figure 17.1. The two half-reactions are allowed to occur in separate beakers, each of which is called a **half-cell.** When Zn atoms are oxidized, the electrons that are given up pass from the piece of zinc through a wire and a voltmeter (this could also be a lamp or a small motor) to the piece of copper, where they are made available to reduce Cu^{2+} ions from the solution. The strips of zinc and copper are called electrodes. An **electrode** conducts electrical current (electrons) into or out of something—in this case,

Strictly speaking, many devices we call batteries consist of several voltaic cells connected together, but the term battery has taken on the same meaning as a voltaic cell.

An electrode is most often a metal plate or wire, but it may also be a piece of graphite or something else that conducts electricity.

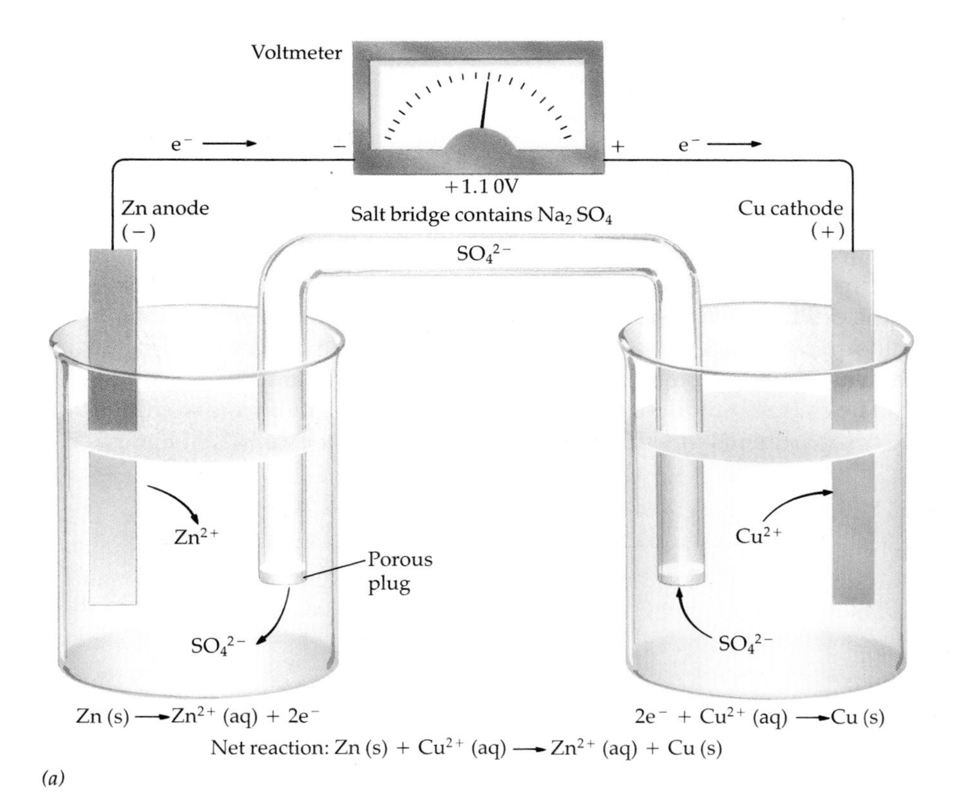

(a)

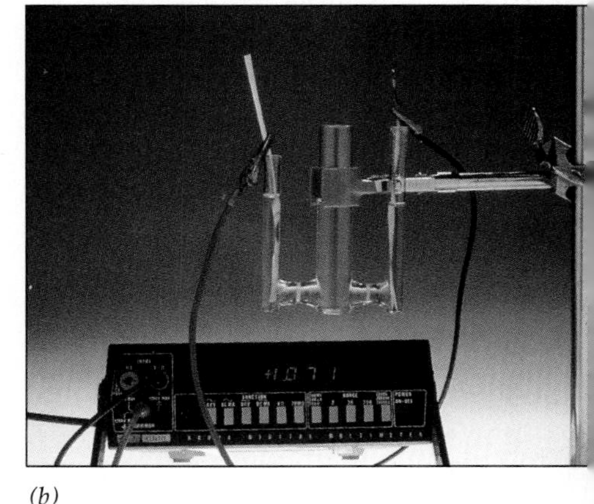

(b)

Figure 17.3
A voltaic cell using Cu^{2+} (aq)/Cu(s) and Zn^{2+}(aq)/Zn(s) half-cells. (a) A potential of 1.10 V is generated if the cell is set up under the conditions shown. Electrons flow through the external wire from the Zn electrode (anode) to the Cu electrode (cathode). A salt bridge provides a connection between the half-cells for ion flow; thus SO_4^{2-} ions flow from the copper to the zinc compartment. (b) An actual cell operating under nearly standard conditions. The negative zinc electrode is at the left, and the positive copper electrode is at the right. The compartments are separated from one another by porous glass disks. The center compartment is the salt bridge that contains Na_2SO_4. (b, C.D. Winters)

To identify the anode and cathode, remember that oxidation takes place at the anode (both words begin with vowels), and reduction takes place at the cathode (both words begin with consonants.)

a solution. The electrode where oxidation occurs is named the **anode** and the electrode where reduction takes place is called the **cathode.** On a flashlight battery the anode is marked "−" because oxidation produces electrons that make the anode negative; conversely, the cathode is marked "+" because reduction consumes electrons, leaving the metal electrode positive.

The voltaic cell is named after the Italian scientist Alessandro Volta, who, in about 1800, constructed the first series of electrochemical cells—a stack of alternating disks of zinc and silver separated by pieces of paper soaked in salt water (an electrolyte). Later, Volta showed that anv two different metals and an electrolyte could be used to make a battery. Figure 17.4 shows a cell constructed by sticking a strip of zinc and a strip of copper into a grapefruit, for example.

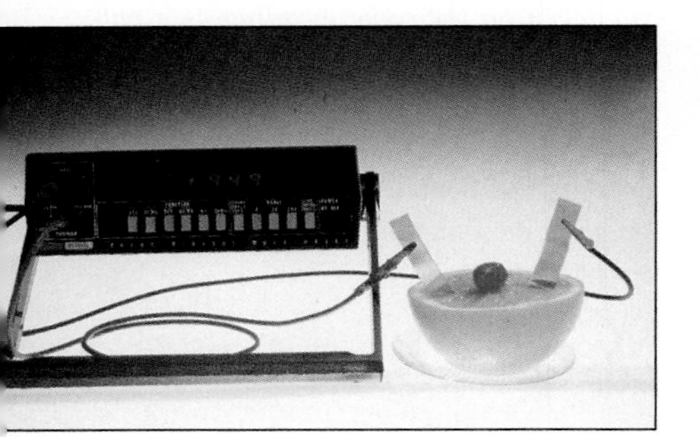

Figure 17.4
A voltaic cell made by inserting zinc and copper electrodes into a grapefruit. A potential of 0.9 V is obtained. (The water and citric acid of the fruit allow for ion conduction between electrodes.) *(C.D. Winters)*

Suppose that the voltmeter connected to the battery diagrammed in Figure 17.4 is replaced by a light bulb. Electrons are transferred to the anode by the half-reaction

$$Zn(s) \longrightarrow Zn^{2+}(aq) + 2\,e^-$$

They then flow from the anode through the bulb, causing it to light up, and eventually travel to the cathode, where they react with copper(II) ions in the half-reaction

$$Cu^{2+}(aq) + 2\,e^- \longrightarrow Cu(s)$$

If nothing else but the flow of electrons took place, the concentration of Zn^{2+} ions in the anode compartment would increase, building up positive charge in the solution, and the concentration of Cu^{2+} ions in the cathode compartment would decrease, making that solution less positive. Because of this buildup of charge, the flow of electrons would very quickly stop. In order for the cell to work, there has to be a way for the positive charge that would build up in the anode half-cell to be balanced by addition of negative ions or removal of positive ions, and vice versa for the cathode half-cell.

The charge buildup can be avoided by using a salt bridge to connect the two compartments. A **salt bridge** is a solution of salt (Na_2SO_4 in Figure 17.3)

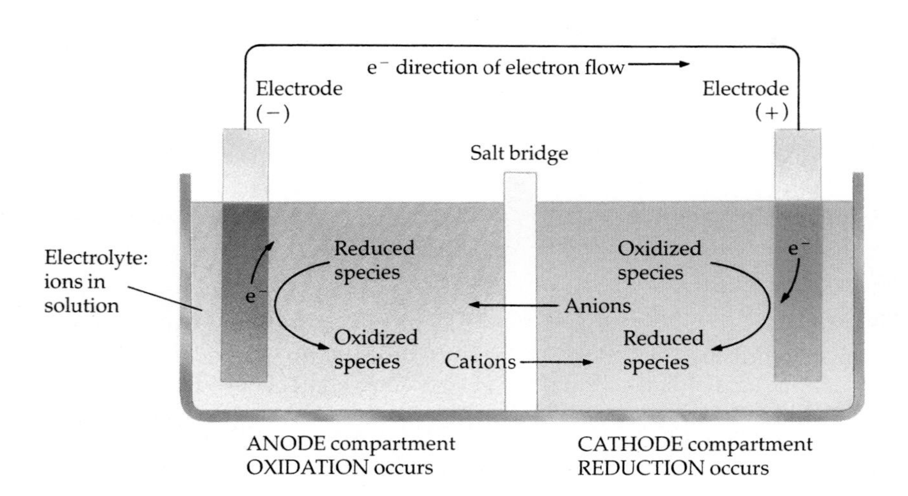

Figure 17.5
The essential components of a battery.

arranged so that the bulk of the solution cannot flow into the battery solutions, but salt ions (K^+ and Cl^-) are allowed to pass freely. As electrons flow through the wire from the zinc electrode to the copper electrode, negative ions (SO_4^{2-}) move through the salt bridge from the solution containing the copper electrode toward the solution containing the zinc electrode, and positive ions (Na^+) move in the opposite direction. This flow of ions completes the electrical circuit, allowing current to flow. If the salt bridge is removed from this battery, the flow of electrons will stop. No voltage will register on the meter.

In commercial batteries, the salt bridge is often a porous membrane.

All voltaic cells and batteries operate in a similar fashion. The oxidation-reduction reaction must favor the formation of products. There must be an external circuit through which electrons flow, and there must be a salt bridge to allow ions to flow between the electrode compartments. Figure 17.5 diagrams these essential components of a battery.

EXAMPLE 17.2

Electrochemical Cells

A simple voltaic cell is assembled with Ni(s) and $Ni(NO_3)_2(aq)$ in one compartment and Cd(s) and $Cd(NO_3)_2(aq)$ in the other. An external wire connects the two electrodes, and a salt bridge containing KNO_3 connects the two solutions. The net reaction is

$$Ni^{2+}(aq) + Cd(s) \longrightarrow Ni(s) + Cd^{2+}(aq)$$

What half-reaction occurs at each electrode? Which is the anode, and which is the cathode? What are the directions of electron flow in the external wire and of the ion flow in the salt bridge?

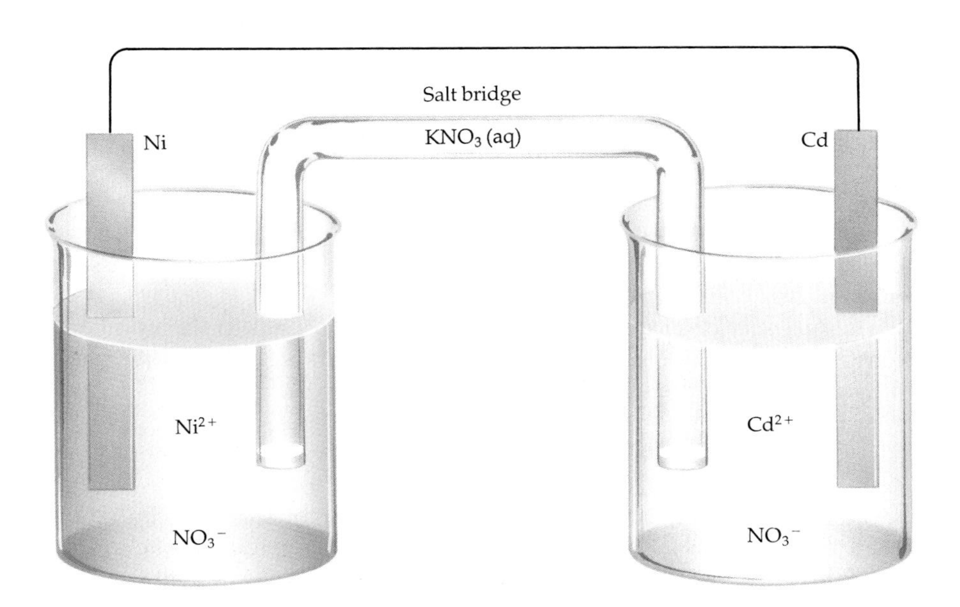

SOLUTION The net equation shows that Cd(s) is the reducing agent and $Ni^{2+}(aq)$ is the oxidizing agent. Thus the half-reactions are

$$\text{oxidation at the anode:} \quad Cd(s) \longrightarrow Cd^{2+}(aq) + 2\,e^-$$

$$\text{reduction at the cathode:} \quad Ni^{2+}(aq) + 2\,e^- \longrightarrow Ni(s)$$

Electrons flow from their source (the oxidation of cadmium at the Cd electrode or anode) through the wire to the electrode where they are used (the Ni electrode or cathode).

Since Cd^{2+} ions are being formed in the anode compartment, anions must move into that compartment from the salt bridge or cations must move out. The Ni^{2+} concentration in the cathode compartment is being depleted, and fewer anions are needed to balance the charge on the Ni^{2+}. Anions can move out of the cathode compartment, or cations can move in. The "circle" of flow of negative charge is now complete: electrons flow from Cd to Ni and anions move from Ni to Cd (or cations move from Cd to Ni).

EXERCISE 17.4 • *Battery Design*

A voltaic cell is assembled to use the following net reaction.

$$Ni(s) + 2\,Ag^+(aq) \longrightarrow Ni^{2+}(aq) + 2\,Ag(s)$$

Write half-reactions for this cell, and indicate which is the oxidation and which is the reduction. Name the electrodes in which these reactions take place. What is the direction of flow of electrons in an external wire connected between the electrodes? If a salt bridge connecting the two electrode compartments contains KNO_3, what is the direction of flow of the nitrate ions? Of the K^+ ions?

17.3 ELECTROCHEMICAL CELLS AND VOLTAGE

Because electrons flow from the anode to the cathode in an electrochemical cell, they can be thought to be "driven" or "pushed" by an **electromotive force** or **emf.** The emf is produced by the difference in electrical potential energy between the two electrodes. Just as a ball rolls downhill in response to a difference in gravitational potential energy, so an electron moves from an electrode of higher electrical potential energy to another of lower potential energy. The moving ball can do work, and so can moving electrons—for example, they could make a motor run.

The quantity of electrical work done is proportional to the number of electrons (quantity of electrical charge) that go from higher to lower potential energy, and to the size of the potential-energy difference.

$$\text{electrical work} = \text{charge} \times \text{potential-energy difference}$$

Charge is measured in coulombs. A **coulomb** is the quantity of charge that passes a fixed point in an electrical circuit when a current of one ampere flows for one second. The charge on a single electron is very small (1.6022×10^{-19} C), so it takes 6.24×10^{18} electrons to product just one coulomb of charge. Electrical potential-energy difference is measured in volts. The **volt** is defined so that one joule of work is performed when 1 coulomb of charge moves through a potential difference of one volt:

$$1 \text{ volt} = \frac{1 \text{ joule}}{1 \text{ coulomb}} \quad \text{or} \quad 1 \text{ joule} = 1 \text{ volt} \times 1 \text{ coulomb}$$

Figure 17.6
Two 1.5-V dry cell batteries. The one on the left is capable of more work since it contains more oxidizing and reducing agents. *(C.D. Winters)*

The cell emf or voltage of an electrochemical cell, therefore, shows how much work a cell can produce for each coulomb of charge that the chemical reaction produces.

The emf of an electrochemical cell depends on the substances that make up the cell, and, if they are gases or solutes in solution, on their concentrations. The quantity of charge depends on how much of each substance reacts. Look at Figure 17.6. Here are two 1.5-V batteries. Both have the same voltage because they rely on electrodes with the same potential difference between them, yet one battery is capable of far more work than the other, because it contains a larger quantity of reactants. In this section and the next section we consider how cell emf depends on the materials from which a cell is made; in Section 17.5 we shall return to the question of how much electrical work a cell can do.

A cell's voltage or emf is readily measured by inserting a voltmeter into the circuit as was shown in Figure 17.4. Since the emf depends on concentrations, **standard conditions** are defined for voltage measurements. These are the same as those used for $\Delta H°$ in Section 7.6: all reactants and products must be present as pure solids or liquids, gases at 1 atm pressure, or solutes at concentrations of 1 M. Voltages measured under these conditions are **standard emfs,** symbolized by $E°$. Unless specified otherwise, all values of $E°$ are given at 25 °C (298 K). By definition, cell voltages for product-favored electrochemical reactions are *positive.* The standard cell potential for the Zn/Cu^{2+} cell discussed earlier, for example, is +1.10 V at 25 °C.

Since any redox reaction can be thought of as the sum of two half-reactions, it is convenient to assign an emf to every possible half-reaction. Then, the cell potential for any reaction can be obtained by adding the emfs of the half-reactions. If $E°$ is positive, the reaction is product-favored; if not, it is reactant-favored. However, because only *differences* in potential energy can be measured, it is not possible to measure the emf for a single half-reaction. Instead, we choose one half-reaction as a standard, and then compare all others to it. The half-reaction chosen is the one that occurs at the **standard hydrogen electrode,** which consists of a platinum electrode immersed in 1 M aqueous acid over which H_2 gas is bubbled at a pressure of 1 atm.

The convention of assigning emfs to half-reactions is similar to the convention of tabulating standard enthalpies of formation; in both cases a relatively small table of data can provide information about a large number of different reactions.

$$2\,H_3O^+(aq, 1\text{ M}) + 2\,e^- \longrightarrow H_2(g, 1\text{ atm}) + 2\,H_2O(\ell)$$

A potential of exactly zero volts has been arbitrarily *assigned* to this half-reaction. When a cell is constructed that combines another half-reaction with the standard hydrogen electrode, the cell voltage is the difference between the two electrodes. Since the potential of the hydrogen electrode is assigned to be zero, the cell voltage gives the voltage of the other electrode. When the standard hydrogen electrode is paired with another half-cell that contains a better reducing agent than H_2, then $H_3O^+(aq)$ is reduced to H_2.

H_3O^+ reduced: $2\,H_3O^+(aq,\,1\,M) + 2\,e^- \longrightarrow$
$$H_2(g,\,1\,atm) + 2\,H_2O(\ell) \qquad E^\circ_{red} = 0\,V$$

If the other half-cell contains a better oxidizing agent than H_3O^+, then H_2 is oxidized.

H_2 oxidized: $H_2(g,\,1\,atm) + 2\,H_2O(\ell) \longrightarrow$
$$2\,H_3O^+(aq,\,1\,M) + 2\,e^- \qquad E^\circ_{ox} = 0\,V$$

In either direction, the standard H_2/H^+ half-cell has a potential of exactly zero volts. The potential of any electrochemical cell that includes the standard hydrogen electrode is then *assigned* as the potential of the other half-cell.

Figure 17.7 diagrams a cell in which one compartment contains the H_2/H_3O^+ reaction mixture, and the other half-cell contains a zinc electrode dipping into a 1 M solution of Zn^{2+}. A voltmeter is connected between the two electrodes to measure the difference in electrical potential energy. A voltmeter gives a positive reading when its positive terminal is connected to the positive electrode of the cell and its negative terminal is connected to the negative electrode of the cell. With the terminals connected in reverse the voltmeter gives a negative reading, and so it not only measures the voltage but also tells which electrode is positive and which negative. For the cell diagrammed in Figure 17.7, it is found that the H_2/H_3O^+ electrode is positive, the zinc electrode is negative, and the measured potential is 0.76 V. The

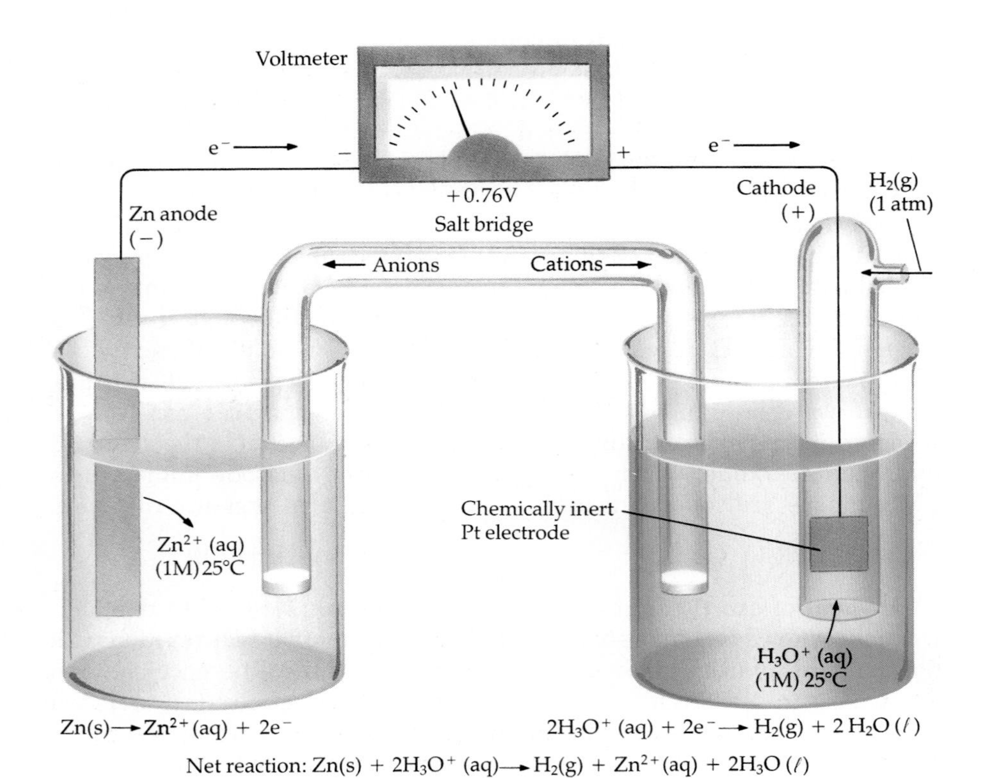

Figure 17.7
A voltaic cell consisting of a zinc electrode at standard conditions and a standard hydrogen electrode, connected by a salt bridge. The potential for this cell, under standard conditions, would be +0.76 V.

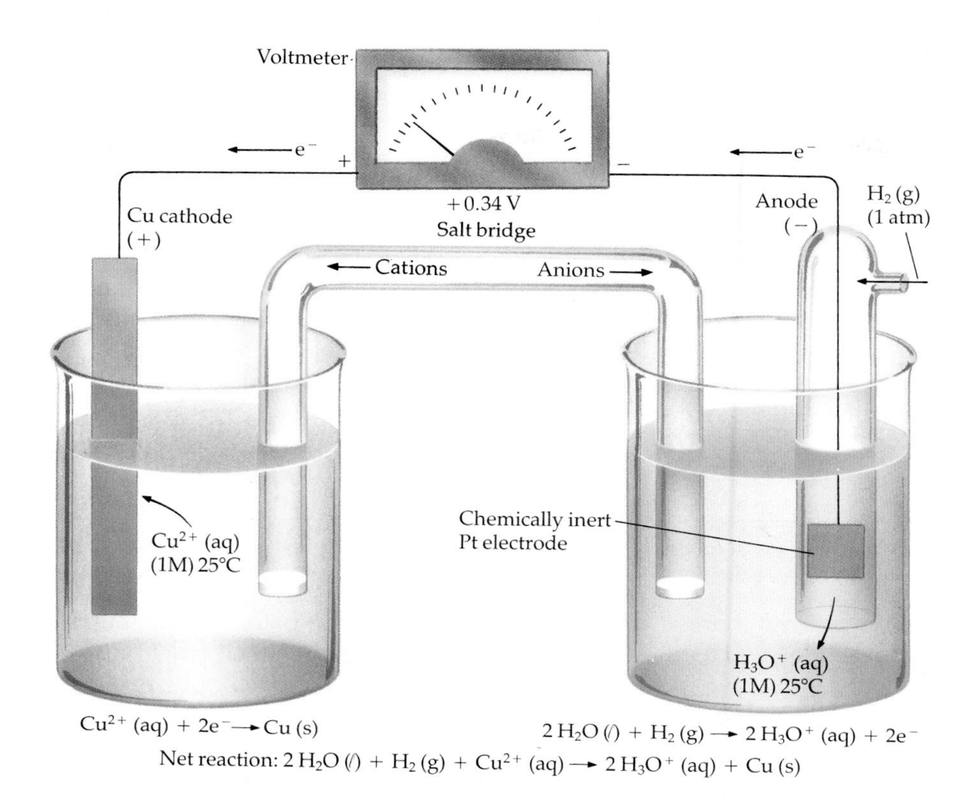

Figure 17.8
A voltaic cell consisting of a copper electrode at standard conditions and a standard hydrogen electrode, connected by a salt bridge. The potential for this cell, under standard conditions, would be +0.34 V.

fact that the Zn electrode is negative means that electrons must be given off by the half-reaction in that cell, making it an oxidation; therefore the Zn electrode must be the anode, and the hydrogen electrode must be the cathode. The cell reaction is therefore the sum of the half-cell reactions:

anode, oxidation: $Zn(s) \longrightarrow Zn^{2+}(aq, 1\ M) + 2\ e^-$ $\quad E^\circ_{ox} = ?\ V$

cathode, reduction: $2\ H_3O^+(aq, 1\ M) + 2\ e^- \longrightarrow H_2(g, 1\ atm) + 2\ H_2O(\ell)$ $\quad E^\circ_{red} = 0\ V$

net cell reaction: $Zn(s) + 2\ H_3O^+(aq, 1\ M) \longrightarrow Zn^{2+}(aq, 1\ M) + H_2(g, 1\ atm) + 2\ H_2O(\ell)$ $\quad E^\circ_{net} = +0.76\ V$

The voltmeter tells us that the potential at the Zn electrode is 0.76 V higher than at the H_2/H_3O^+ electrode. Since the half-cell potential for H_2/H_3O^+ is assigned to be 0 V, the half-cell potential for oxidation of zinc must be +0.76 V:

Quantities defined as exact, such as the voltage of the hydrogen electrode, do not limit the number of significant digits in the answer when they are used in calculations.

$$Zn(s) \longrightarrow Zn^{2+}(aq, 1\ M) + 2\ e^- \qquad E^\circ_{ox} = +0.76\ V$$

The half-cell potentials of many different half-reactions can be measured by comparing them with the H_2/H_3O^+ half-cell. Figure 17.8 shows a Cu/Cu^{2+} half-cell connected to a standard hydrogen electrode. In this cell the copper electrode is positive, the H_2/H_3O^+ electrode is negative, and the voltmeter reads +0.34 V. This means that the reactions are

anode:	$H_2(g, 1\ atm) + 2\ H_2O(\ell) \longrightarrow 2\ H_3O^+(aq, 1\ M) + 2\ e^-$	$E^\circ_{ox} = 0\ V$
cathode:	$Cu^{2+}(aq, 1\ M) + 2\ e^- \longrightarrow Cu(s)$	$E^\circ_{red} = ?\ V$
net cell reaction:	$H_2(g, 1\ atm) + Cu^{2+}(aq, 1\ M) + 2\ H_2O(\ell) \longrightarrow 2\ H_3O^+(aq, 1\ M) + Cu(s)$	$E^\circ_{net} = +0.34\ V$

The half-cell potential for $Cu^{2+}(aq, 1\ M) + 2\ e^- \longrightarrow Cu(s)$ must be +0.34 V. Note that, in this cell, the standard hydrogen electrode is the anode, not the cathode, as it was in the combination with a Zn/Zn^{2+} half cell.

We can now return to the first electrochemical cell we looked at, in which Zn(s) reduces Cu^{2+} ions to Cu. Since we now have the potentials for the half-reactions, we can write

anode:	$Zn(s) \longrightarrow Zn^{2+}(aq, 1\ M) + 2\ e^-$	$E^\circ_{ox} = +0.76\ V$
cathode:	$Cu^{2+}(aq, 1\ M) + 2\ e^- \longrightarrow Cu(s)$	$E^\circ_{red} = +0.34\ V$
net cell reaction:	$Zn(s) + Cu^{2+}(aq, 1\ M) \longrightarrow Zn^{2+}(aq, 1\ M) + Cu(s)$	$E^\circ_{net} = +1.10\ V$

This is an important result because the sum of the potentials of the two half-reactions equals the measured potential for the net cell reaction.

EXAMPLE 17.3

Determining a Half-Reaction Potential

The cell illustrated in the following drawing generates a potential of $E^\circ = 0.51\ V$ under standard conditions at 25 °C. The net cell reaction is

$$Zn(s) + Ni^{2+}(aq, 1\ M) \longrightarrow Zn^{2+}(aq, 1\ M) + Ni(s)$$

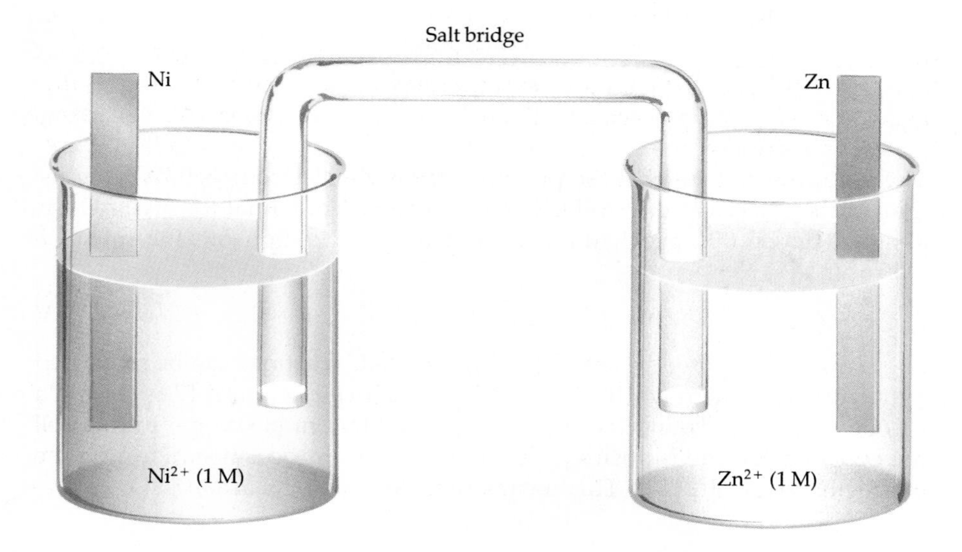

Determine which of the electrodes is the anode and which is the cathode, give the signs of the electrodes, and calculate the half-cell potential for $Ni^{2+}(aq) + 2\,e^- \longrightarrow Ni(s)$.

SOLUTION The electrode where oxidation occurs is the anode (and, since it is the source of electrons, it is negative). Because Zn(s) is oxidized to $Zn^{2+}(aq)$, the Zn electrode is the anode. Nickel(II) ions are reduced at the Ni electrode, so Ni metal is the positive cathode.

Since the overall cell potential is known, and the potential for the $Zn(s)/Zn^{2+}(aq, 1\,M)$ half-cell is known, the value of $E°$ for $Ni^{2+}(aq, 1\,M) + 2\,e^- \rightarrow Ni(s)$ can be calculated.

anode:	$Zn(s) \longrightarrow Zn^{2+}(aq) + 2\,e^-$	$E°_{ox} = 0.76$ V
cathode:	$Ni^{2+}(aq) + 2\,e^- \longrightarrow Ni(s)$	$E°_{red} = ?$ V
net cell reaction:	$Zn(s) + Ni^{2+}(aq) \longrightarrow Zn^{2+}(aq) + Ni(s)$	$E°_{net} = 0.51$ V

At 25 °C, the value of $E°$ for the $Ni^{2+}(aq, 1\,M) + 2\,e^- \rightarrow Ni(s)$ half-reaction is $0.51\text{ V} - 0.76\text{ V} = -0.25$ V.

EXERCISE 17.5 • *Determining a Half-Reaction Potential*

Given that the reaction of aqueous copper(II) ions with iron metal has an $E°$ value of 0.78 V, what is the value of $E°$ for the half-cell $Fe(s) \rightarrow Fe^{2+}(aq) + 2\,e^-$?

$$Fe(s) + Cu^{2+}(aq, 1\,M) \longrightarrow Fe^{2+}(aq, 1\,M) + Cu(s) \quad E°_{net} = +0.78\text{ V}$$

17.4 USING STANDARD CELL POTENTIALS

The results of a great many measurements such as the ones just described are summarized in Table 17.1. The values reported are called **standard reduction potentials** because they are the potentials that would be measured for a cell in which a half-reaction *occurred as a reduction* and was paired with the standard hydrogen electrode. If a half-reaction would occur as an oxidation when paired with the standard hydrogen electrode, this is indicated by giving its voltage a negative sign. For example, we saw earlier that for the half-reaction

$$Zn(s) \longrightarrow Zn^{2+}(aq) + 2\,e^- \qquad E°_{ox} = +0.76\text{ V}$$

But this is an oxidation reaction; the reaction involving Zn and Zn^{2+} that appears in Table 17.1 is the reduction

$$Zn^{2+}(aq) + 2\,e^- \longrightarrow Zn(s) \qquad E°_{red} = -0.76\text{ V}$$

and its standard potential is equal in magnitude but opposite in sign to that for the oxidation. It is always true that if a half-reaction is written in the reverse direction, the sign of the corresponding $E°$ must be changed. That is,

Table 17.1

Standard Reduction Potentials in Aqueous Solution at 25 °C*

Reduction Half-Reaction			$E°$ (V)
$F_2(g) + 2\,e^-$	$\longrightarrow$	$2\,F^-(aq)$	+2.87
$H_2O_2(aq) + 2\,H_3O^+(aq) + 2\,e^-$	$\longrightarrow$	$4\,H_2O(\ell)$	+1.77
$PbO_2(s) + SO_4^{2-}(aq) + 4\,H_3O^+(aq) + 2\,e^-$	$\longrightarrow$	$PbSO_4(s) + 6\,H_2O(\ell)$	+1.685
$MnO_4^-(aq) + 8\,H_3O^+(aq) + 5\,e^-$	$\longrightarrow$	$Mn^{2+}(aq) + 12\,H_2O(\ell)$	+1.52
$Au^{3+}(aq) + 3\,e^-$	$\longrightarrow$	$Au(s)$	+1.50
$Cl_2(g) + 2\,e^-$	$\longrightarrow$	$2\,Cl^-(aq)$	+1.360
$Cr_2O_7^{2-}(aq) + 14\,H_3O^+(aq) + 6\,e^-$	$\longrightarrow$	$2\,Cr^{3+}(aq) + 21\,H_2O(\ell)$	+1.33
$O_2(g) + 4\,H_3O^+(aq) + 4\,e^-$	$\longrightarrow$	$6\,H_2O(\ell)$	+1.229
$Br_2(\ell) + 2\,e^-$	$\longrightarrow$	$2\,Br^-(aq)$	+1.08
$NO_3^-(aq) + 4\,H_3O^+ + 3\,e^-$	$\longrightarrow$	$NO(g) + 6\,H_2O$	+0.96
$OCl^-(aq) + H_2O(\ell) + 2\,e^-$	$\longrightarrow$	$Cl^-(aq) + 2\,OH^-(aq)$	+0.89
$Hg^{2+}(aq) + 2\,e^-$	$\longrightarrow$	$Hg(\ell)$	+0.855
$Ag^+(aq) + e^-$	$\longrightarrow$	$Ag(s)$	+0.80
$Hg_2^{2+}(aq) + 2\,e^-$	$\longrightarrow$	$2\,Hg(\ell)$	+0.789
$Fe^{3+}(aq) + e^-$	$\longrightarrow$	$Fe^{2+}(aq)$	+0.771
$I_2(s) + 2\,e^-$	$\longrightarrow$	$2\,I^-(aq)$	+0.535
$O_2(g) + 2\,H_2O(\ell) + 4\,e^-$	$\longrightarrow$	$4\,OH^-(aq)$	+0.40
$Cu^{2+}(aq) + 2\,e^-$	$\longrightarrow$	$Cu(s)$	+0.337
$Sn^{4+}(aq) + 2\,e^-$	$\longrightarrow$	$Sn^{2+}(aq)$	+0.15
$2\,H_3O^+(aq) + 2\,e^-$	$\longrightarrow$	$H_2(g) + 2\,H_2O(\ell)$	0.00
$Sn^{2+}(aq) + 2\,e^-$	$\longrightarrow$	$Sn(s)$	−0.14
$Ni^{2+}(aq) + 2\,e^-$	$\longrightarrow$	$Ni(s)$	−0.25
$PbSO_4(s) + 2\,e^-$	$\longrightarrow$	$Pb(s) + SO_4^{2-}(aq)$	−0.356
$Cd^{2+}(aq) + 2\,e^-$	$\longrightarrow$	$Cd(s)$	−0.40
$Fe^{2+}(aq) + 2\,e^-$	$\longrightarrow$	$Fe(s)$	−0.44
$Zn^{2+}(aq) + 2\,e^-$	$\longrightarrow$	$Zn(s)$	−0.763
$2\,H_2O(\ell) + 2\,e^-$	$\longrightarrow$	$H_2(g) + 2\,OH^-(aq)$	−0.8277
$Al^{3+}(aq) + 3\,e^-$	$\longrightarrow$	$Al(s)$	−1.66
$Mg^{2+}(aq) + 2\,e^-$	$\longrightarrow$	$Mg(s)$	−2.37
$Na^+(aq) + e^-$	$\longrightarrow$	$Na(s)$	−2.714
$K^+(aq) + e^-$	$\longrightarrow$	$K(s)$	−2.925
$Li^+(aq) + e^-$	$\longrightarrow$	$Li(s)$	−3.045

Increasing strength of oxidizing agents ↑ (left column); Increasing strength of reducing agents ↓ (right column)

*In volts (V) versus the standard hydrogen electrode.

if a half-reaction is written as an oxidation, the sign of the $E°$ in Table 17.1 must be reversed. Here are some other important points to notice about Table 17.1

1. All of the half-reactions are written as reductions. This means that the species on the left-hand side of each half-reaction is an oxidizing agent and the species on the right-hand side is a reducing agent.
2. All of the half-reactions listed in the table can occur in either direction. A given substance can react at the anode or the cathode, de-

pending on the conditions. We have already seen examples where H_2 is oxidized to H_3O^+ and where H_3O^+ is reduced to H_2 by different reactants.

3. The more positive the value of the reduction potential, $E°$, the more easily the substance on the left side of a half-reaction can be reduced. When a substance is easy to reduce, it is a strong oxidizing agent. (Remember that an oxidizing agent must be reduced when it oxidizes something else.) Thus, $F_2(g)$ is the best oxidizing agent in the table and Li^+ is the poorest oxidizing agent in the table. Other strong oxidizing agents are at the top left of the table: $H_2O_2(aq)$, $PbO_2(s)$, $MnO_4^-(aq)$, $Au^{3+}(aq)$, $Cl_2(g)$, $Cr_2O_7^{2-}(aq)$, and $O_2(g)$.

4. The more negative the value of the reduction potential, $E°$, the less likely the reaction will occur as a reduction, and the more likely the reverse reaction (an oxidation) will occur. That is, the farther down we go in the table, the better the reducing ability of the atom, ion, or molecule on the right. Thus Li(s) is the strongest reducing agent in the table and F^- is the weakest reducing agent in the table. Other strong reducing agents are alkali and alkaline earth metals and hydrogen at the lower right of the table.

5. Under standard conditions, any species on the left of a half-reaction will oxidize any species on the right that is farther down in the table. For example, we can predict that $Fe^{3+}(aq)$ will oxidize Al(s); $Br_2(\ell)$ will oxidize Mg(s); and even $Na^+(aq)$ will oxidize Li(s). The net reaction and the cell emf are obtained by adding the half-reactions and their voltages; for example,

$$Br_2(\ell) + 2\,e^- \longrightarrow 2\,Br^-(aq) \qquad E^\circ_{red} = 1.08\ V$$
$$Mg(s) \longrightarrow Mg^{2+}(aq) + 2\,e^- \qquad E^\circ_{ox} = 2.37\ V$$
$$Br_2(\ell) + Mg(s) \longrightarrow Mg^{2+}(aq) + 2\,Br^-(aq) \qquad E^\circ_{net} = 3.45\ V$$

6. Electrode potentials depend on the nature and concentration of reactants and products, but not on the quantity of each that reacts. This means that changing the stoichiometric coefficients for a half-reaction does not change the value of $E°$. For example, the reduction of Fe^{3+} has an $E°$ of +0.771 V whether the reaction is written as

Recall that a half-cell emf is energy per unit charge (1 volt = 1 joule/1 coulomb). Multiplying a half-reaction by some number causes both the energy and the charge to be multiplied by that number. Thus the ratio (emf) does not change.

$$Fe^{3+}(aq,\ 1\ M) + e^- \longrightarrow Fe^{2+}(aq,\ 1\ M) \qquad E^\circ_{red} = +0.771\ V$$

or as

$$2\,Fe^{3+}(aq,\ 1\ M) + 2\,e^- \longrightarrow 2\,Fe^{2+}(aq,\ 1\ M) \qquad E^\circ_{red} = +0.771\ V$$

Using the preceding guidelines and the table of standard reduction potentials, let's make some *predictions* about whether reactions will occur and then check our results by calculating $E°$. Will aluminum dissolve in a solution of tin(IV) ion? $Sn^{4+}(aq)$ is about half-way down the table on the left, while Al(s) is fifth up from the bottom on the right. Since $Sn^{4+}(aq)$ is above Al(s), we predict that it can oxidize aluminum, causing the metal to dissolve. Adding the half-cell reactions to give the balanced equation and adding the half-cell potentials gives

anode:	$2[Al(s) \longrightarrow Al^{3+}(aq, 1\ M) + 3\ e^-]$	$E^\circ_{ox} = +1.66$ V
cathode:	$3[Sn^{4+}(aq, 1\ M) + 2\ e^- \longrightarrow Sn^{2+}(aq, 1\ M)]$	$E^\circ_{red} = +0.15$ V
net cell reaction:	$2\ Al(s) + 3\ Sn^{4+}(aq, 1\ M) \longrightarrow 2\ Al^{3+}(aq, 1\ M) + 3\ Sn^{2+}(aq, 1\ M)$	$E^\circ_{net} = +1.81$ V

Since E° is positive, this reaction is product favored, as we predicted.

What about Cd^{2+} oxidizing Cu? Do you predict that this reaction will occur?

$$Cd^{2+}(aq) + Cu(s) \longrightarrow Cd(s) + Cu^{2+}(aq)$$

$Cd^{2+}(aq)$ is about three-quarters of the way down the table on the left, but Cu(s) is a little less than half way down on the right. Therefore, $Cd^{2+}(aq)$ is not strong enough as an oxidizing agent to oxidize Cu(s), and we predict no reaction. When the half-cell potentials are added, we get

anode:	$Cu(s) \longrightarrow Cu^{2+}(aq) + 2\ e^-$	$E^\circ_{ox} = -0.34$ V
cathode:	$Cd^{2+} + 2\ e^- \longrightarrow Cd(s)$	$E^\circ_{red} = -0.40$ V
net cell reaction:	$Cd^{2+}(aq) + Cu(s) \longrightarrow Cd(s) + Cu^{2+}(aq)$	$E^\circ_{net} = -0.74$ V

The negative E° value shows that this process is reactant favored and does not form appreciable quantities of products under standard conditions.

Standard reduction potentials can be used to explain an annoying experience many of us have had. Have you ever experienced pain in your teeth when you accidentally touched a filling with a stainless steel fork or a piece of aluminum? A common filling material for tooth cavities is dental amalgam—tin and silver dissolved in mercury to form solid solutions having compositions approximating Ag_2Hg_3, Ag_3Sn, and Sn_xHg (where x ranges from 7 to 9). All of these may undergo electrochemical reactions; for example,

$3\ Hg_2{}^{2+}(aq) + 4\ Ag(s) + 6\ e^- \longrightarrow 2\ Ag_2Hg_3(s)$	$E^\circ_{red} = +0.85$ V
$Sn^{2+}(aq) + 3\ Ag(s) + 2\ e^- \longrightarrow Ag_3Sn(s)$	$E^\circ_{red} = -0.05$ V

The E° values in Table 17.1 indicate that both iron and aluminum have much more negative reduction potentials and therefore are much better reducing agents than any of the solid solutions. If a piece of iron or aluminum comes in contact with a dental filling, the saliva and gum tissue can act as a salt bridge, and an electrochemical cell results. The iron or aluminum donates electrons, producing a tiny electrical current that results in a complaint from your tooth nerves.

17.5 E° AND GIBBS FREE ENERGY

For the remainder of this chapter E° will be used for the overall reaction in place of E°_{net}.

The sign of E° indicates whether a redox reaction will occur. If E° is positive, the reaction is product-favored; if E° is negative, it is reactant-favored. Section 7.10 described another way to decide whether a reaction is product favored: the change in standard Gibbs free energy, ΔG°, must be negative. Since both E° and ΔG° tell something about whether a reaction will occur, there may be some relationship between them. Here's how to show what it is.

The "free" in Gibbs free energy indicates that it is energy available to do work. The electrical work a cell can do can be calculated by multiplying the quantity of electrical charge transferred times the cell voltage, $E°$. The quantity of charge is given by the number of moles of electrons transferred in the overall reaction, n, multiplied by the number of coulombs per mole of electrons.

quantity of charge = moles of electrons × coulombs per mole of electrons

The charge on one mole of electrons can be calculated from the charge on one electron and Avogadro's number.

$$\text{charge on 1 mol of electrons} = (1.6022 \times 10^{-19}\ \text{C/electron})(6.022046 \times 10^{23}\ \text{electrons/mol}) = 9.6485 \times 10^{4}\ \text{C/mol}$$

The quantity 9.6485×10^4 C/mol of electrons is commonly rounded to 96,500 C/mole of electrons and is known as the **Faraday constant** *(F)* in honor of Michael Faraday, who first explored the quantative aspects of electrochemistry.

The electrical work that can be done by a cell is equal to the Faraday constant multiplied by the number of moles of electrons transferred and by the cell voltage.

$$\text{electrical work} = nFE°$$

Notice that, unlike the cell voltage, the electrical work a cell can do *does* depend on the quantity of reactants in the cell reaction. More reactants mean more moles of electrons transferred and hence more work. Equating the electrical work of a cell at standard conditions with $\Delta G°$, we get

$$\Delta G° = -nFE°$$

The negative sign on the right side of the equation accounts for the fact that, for a product-favored process, $\Delta G°$ is always negative, but $E°$ is always positive. Thus their signs must be opposite.

Using this equation we can calculate $\Delta G°$, which represents the maximum work that the cell can do, for the Cu^{2+}/Zn cell. The reaction is

$$Cu^{2+}(aq) + Zn(s) \longrightarrow Cu(s) + Zn^{2+}(aq) \qquad E° = +1.10\ \text{V}$$

so two moles of electrons are transferred per mole of copper reduced. The Gibbs free energy change when this quantity of reactants is consumed is

$$\Delta G° = -2\ \text{mol electrons transferred} \cdot \frac{9.65 \times 10^4\ \text{J}}{\text{V} \cdot \text{mol e}^-} \cdot 1.10\ \text{V} \cdot \frac{1\ \text{kJ}}{10^3\ \text{J}} = -212\ \text{kJ}$$

EXERCISE 17.6 • *The Relation Between $E°$ and $\Delta G°$*

Use half-cell potentials to calculate $E°$ and then $\Delta G°$ for the following reaction. Is the reaction product-favored as written?

$$Zn^{2+}(aq) + H_2(g) + 2\ H_2O(\ell) \longrightarrow Zn(s) + 2\ H_3O^+(aq)$$

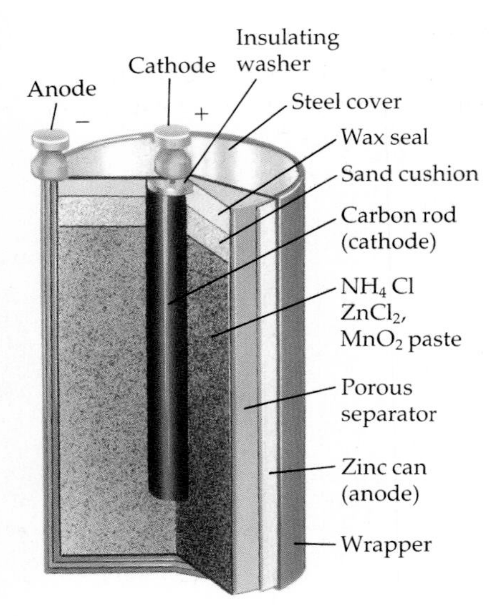

Figure 17.9
A diagrammatic representation of the Leclanché dry cell. It consists of a zinc anode (the battery container), a graphite cathode, and an electrolyte consisting of a moist paste of MnO_2 and NH_4Cl.

17.6 COMMON BATTERIES

Voltaic cells include the convenient, portable sources of energy that we call **batteries.** Some batteries consist of a single cell, while others contain multiple cells. Batteries may be classified as primary or secondary. In a **primary battery** the electrochemical reactions cannot be easily reversed, so when the reactants are used up the battery is "dead" and must be discarded. A **secondary battery** (sometimes called a **storage battery** or a **rechargeable battery**), in contrast, uses an electrochemical reaction that can be reversed, so such a battery can be recharged.

Primary Batteries

For a long time the "dry cell" battery, invented by Georges Leclanché in 1866, was the major source of energy for flashlights and toys. The container of the dry cell is made of zinc, which acts as the anode. The zinc is separated from the other chemicals by a liner of porous paper (Figure 17.9), which functions as the salt bridge. In the center of the dry cell is a graphite cathode, inserted into a moist mixture of ammonium chloride (NH_4Cl), zinc chloride ($ZnCl_2$), and manganese dioxide (MnO_2). As electrons flow from the cell through a flashlight bulb, for example, the zinc is oxidized

$$\text{anode, oxidation:}\quad Zn(s) \longrightarrow Zn^{2+}(aq) + 2\,e^-$$

and the ammonium ions are reduced.

$$\text{cathode, reduction:}\quad 2\,NH_4^+(aq) + 2\,e^- \longrightarrow 2\,NH_3(g) + H_2(g)$$

The ammonia that is formed reacts with zinc ions to form a zinc-ammonia complex ion; this reaction prevents a buildup of gaseous ammonia.

$$Zn^{2+}(aq) + 2\,NH_3(g) \longrightarrow [Zn(NH_3)_2]^{2+}(aq)$$

The hydrogen that is produced is oxidized by the MnO_2 in the cell. In this way, hydrogen gas does not accumulate.

$$H_2(g) + 2\,MnO_2(s) \longrightarrow Mn_2O_3(s) + H_2O(\ell)$$

All these reactions lead to the following net process, which produces 1.5 V.

$$2\,MnO_2(s) + 2\,NH_4^+(aq) + Zn(s) \longrightarrow Mn_2O_3(s) + H_2O(\ell) + [Zn(NH_3)_2]^{2+}(aq)$$

This battery has two major disadvantages. First, if current is withdrawn rapidly, the gases NH_3 and H_2 produced in the reduction reaction cannot react rapidly enough with the Zn^{2+} and MnO_2. As a result, the cell voltage drops, although it can be restored by letting the battery sit undisturbed for a while. Second, there is a slow direct reaction between the zinc electrode and the ammonium ions even when current is not being drawn, so stored dry cells run down and tend to have a poor "shelf life." Since the rates of almost all chemical reactions decrease with decreasing temperature, the shelf life of a dry cell can be doubled or tripled by storing it at about 4 °C.

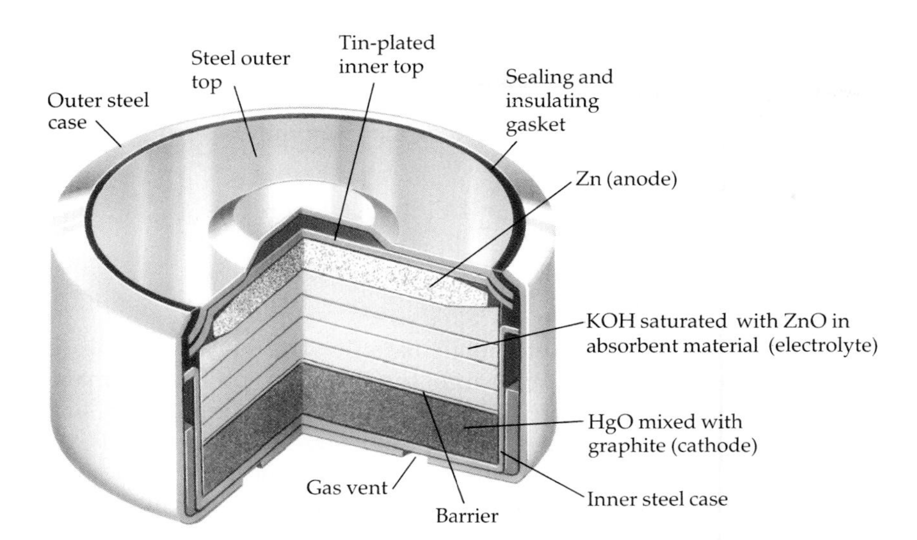

Figure 17.10
The mercury battery. The reducing agent is zinc and the oxidizing agent is mercury(II)oxide.

Some of the problems of the dry cell are overcome by the more expensive "alkaline" battery. An **alkaline battery,** which produces 1.54 V, also uses the oxidation of zinc as the anode reaction, but under alkaline (pH > 7) conditions.

$$\text{anode, oxidation:}\quad Zn(s) + 2\,OH^-(aq) \longrightarrow ZnO(aq) + H_2O(\ell) + 2\,e^-$$

The electrons that pass through the external circuit are consumed by reduction of manganese dioxide at the cathode.

$$\text{cathode, reduction:}\quad 2\,MnO_2(s) + H_2O(\ell) + 2\,e^- \longrightarrow Mn_2O_3(s) + 2\,OH^-(aq)$$

In contrast to the Lechlanché dry cell, no gases are formed in the alkaline battery, and there is no decline in voltage under high current loads.

In the **mercury battery** (Figure 17.10) the oxidation of zinc is again the anode reaction. The cathode reaction, however, is the reduction of mercury(II) oxide.

$$HgO(s) + H_2O(\ell) + 2\,e^- \longrightarrow Hg(\ell) + 2\,OH^-(aq)$$

The HgO, mixed with graphite, is in a tightly compacted powder separated from a KOH electrolyte and the zinc by a moist paper or polyvinyl chloride barrier that serves as the salt bridge. The voltage of this battery is about 1.35 V. Mercury batteries are used in calculators, watches, hearing aids, cameras, and other devices where small size is an advantage. However, mercury and its compounds are poisonous, and careless disposal of mercury batteries can lead to environmental problems such as contamination of groundwater or even mercury vapor in the atmosphere.

The **lithium cell** (Figure 17.11) is another popular battery, primarily because of its light weight. Instead of zinc (density = 7.14 g/cm^3) as the anode material, lithium (density = 0.534 g/cm^3) is used. Because lithium is such a strong reducing agent compared to zinc (see Table 17.1), this battery

Figure 17.11
The lithium cell finds many uses where a high energy density is desired. *(C.D. Winters)*

Mercury batteries are hermetically sealed (to prevent leakage of mercury) and should never be heated. Heating increases the pressure of vapors within the battery, ultimately causing it to explode.

Lithium has the lowest density of any nongaseous element.

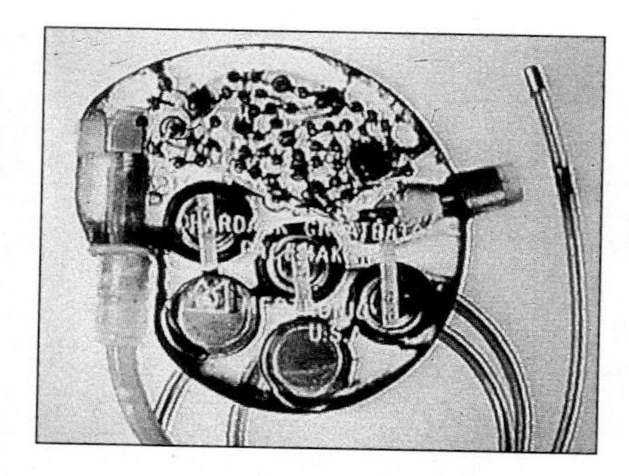

A cardiac pacemaker. *(Martin M. Rotker)*

has a very large voltage (3.4 V). Because lithium has such a low density, a lot of stored energy can be packed into a very lightweight package.

Some lithium batteries use MnO_2 as the oxidizer, and others, such as some cardiac pacemaker batteries, use exotic compounds such as sulfuryl chloride ($SOCl_2$), whose cathode reaction is

$$2\,SOCl_2(\ell) + 4\,e^- \longrightarrow 4\,Cl^-(aq) + S(s) + SO_2(g)$$

Secondary Batteries

Secondary batteries, such as automobile batteries and ni-cad batteries, are rechargeable because, as they discharge, the oxidation products remain at the anode and the reduction products remain at the cathode. As a result, if the direction of electron flow is reversed, the anode and cathode reactions can be reversed, and the reactants are regenerated. Under favorable conditions, secondary batteries may be discharged and recharged hundreds or thousands of times.

The familiar automobile battery, also called the **lead storage battery,** is a secondary battery containing porous lead electrodes and lead(IV) oxide electrodes immersed in aqueous sulfuric acid (Figure 17.12). As this battery is discharged, metallic lead is oxidized to lead sulfate at the anode, and lead(IV) oxide is reduced to lead sulfate at the cathode.

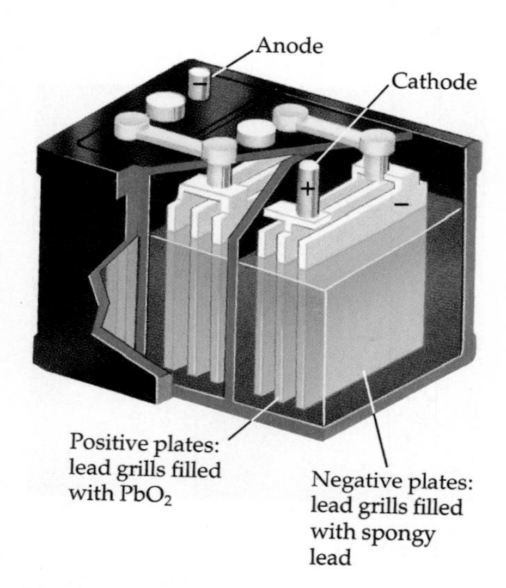

Figure 17.12

A schematic diagram of the lead storage battery.

anode reaction: $Pb(s) + HSO_4^-(aq) + H_2O(\ell) \longrightarrow PbSO_4(s) + H_3O^+(aq) + 2\,e^-$

cathode reaction: $PbO_2(s) + 3\,H_3O^+(aq) + HSO_4^-(aq) + 2\,e^- \longrightarrow PbSO_4(s) + 5\,H_2O(\ell)$

The lead sulfate formed at both electrodes is an insoluble compound that *stays on the electrode surface,* which keeps it available and makes it possible to reverse the electrode reactions when the battery is recharged. Sulfuric acid is consumed in both the anode and the cathode reactions, causing the concentration of the sulfuric acid electrolyte to decrease as the battery discharges. Before the introduction of modern sealed automotive batteries, the measured density of this battery acid was used to indicate the state of charge of the battery. The lower the density, the lower the charge. It is now almost impossible to measure the density of battery acid, because the cells are tightly sealed by the manufacturer.

To recharge a secondary battery, a source of direct electrical current is connected so that electrons are forced to flow in the direction opposite from when the battery was discharging. This causes the overall battery reaction to be reversed and regenerates the reactants that originally produced the battery's voltage and current. For the lead storage battery the overall reaction is

$$Pb(s) + PbO_2(s) + 2\,HSO_4^-(aq) + 2\,H_3O^+(aq) \underset{\text{charge}}{\overset{\text{discharge}}{\rightleftharpoons}} 2\,PbSO_4(s) + 4\,H_2O(\ell)$$

The lead-acid battery was first presented to the French Academy of Sciences in 1860 by Gaston Planté.

Normal charging of an automobile lead storage battery occurs during driving. In addition to reversing the overall battery reaction, charging reduces a little water at the cathode and oxidizes a little water at the anode, reduction of water, $4\,H_2O(\ell) + 4\,e^- \longrightarrow 2\,H_2(g) + 4\,OH^-(aq)$

oxidation of water, $6\,H_2O(\ell) \longrightarrow O_2 + 4\,H_3O^+(aq) + 4\,e^-$

These reactions produce a mixture of hydrogen and oxygen in the compartment at the top of the battery. If this mixture is accidentally sparked, it can explode, and so no sparks or open flames should be brought near a lead storage battery that is not sealed.

Rapid recharging, whether during normal driving or using a recharger, often causes elongated crystals to grow on the electrode surfaces as the lead and lead oxide are redeposited; these crystals can grow between the electrodes and cause internal short circuits. Usually, when this happens, the battery is "dead," will not accept a recharge, and must be replaced. If electrolyte fluid runs low, the electrode surfaces dry, which prevents the electrode from recharging properly; the result is again a dead battery.

The lead storage battery is relatively inexpensive, reliable, and simple, and it has an adequate life. Its high weight is its major fault. A typical automobile battery contains about 15 to 20 kg of lead, which is required to provide the large number of electrons needed to crank an automobile engine, especially on a cold morning. (Recall that the number of electrons that a battery can move from the anode to the cathode is proportional to the amount of reactants involved.) Another problem with lead batteries is that the lead they contain can contaminate air and groundwater, possibly causing lead poisoning (Section 15.2). Auto batteries should be recycled by companies equipped with the proper safeguards to protect the environment.

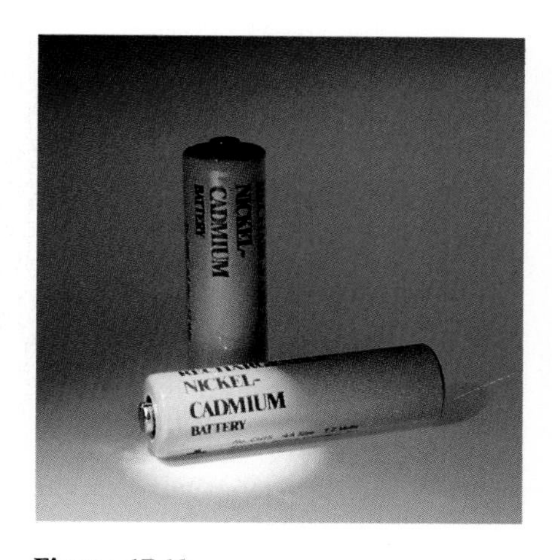

Figure 17.13
Ni-cad batteries come in a variety of sizes and shapes. *(C.D. Winters)*

Nickel-cadmium ("ni-cad") batteries are another popular type of secondary battery. Ni-cad batteries are lightweight, can be quite small, and produce a constant voltage until completely discharged, making them useful in cordless appliances, video camcorders, portable radios, and other applications (Figure 17.13). They suffer somewhat from discharge "memory"; that is, if the battery is repeatedly used for only a short time and then recharged, it develops a tendency to need recharging after only a short use time. This obviously causes problems if the full charge on the battery is needed.

Users of ni-cad batteries should carefully follow the manufacturer's charging recommendations for maximum battery life.

Ni-cad batteries can be recharged because the reaction products are insoluble hydroxides that remain at the electrode surfaces. The anode reaction during discharge is the oxidation of cadmium, and the cathode reaction is the reduction of the nickel compound NiO(OH).

anode reaction: $Cd(s) + 2\,OH^-(aq) \longrightarrow Cd(OH)_2(s) + 2\,e^-$

cathode reaction: $NiO(OH)(s) + H_2O(\ell) + e^- \longrightarrow Ni(OH)_2(s) + OH^-(aq)$

net cell reaction: $Cd(s) + 2\,NiO(OH)(s) + 2\,H_2O(\ell) \longrightarrow Cd(OH)_2(s) + 2\,Ni(OH)_2(s)$

Like mercury batteries, ni-cad batteries should be disposed of properly because of the toxicity of cadmium and its compounds. It is likely that such batteries will soon be replaced by others that have similar characteristics but are less harmful to the environment.

Batteries for Electric Cars

A LIGHTWEIGHT STORAGE BATtery that could be recharged quickly and that would store enough energy to drive a car 100 miles or more would make electric-powered vehicles much more attractive and at the same time spare our environment much air pollution. A battery that is expected to have these characteristics is under development by Energy Conversion Devices of Troy, Michigan under an $18.5 million grant from the U.S. Advanced Battery Consortium, which was set up by the U.S. Energy Department, the big three automakers, and the Electric Power Research Institute. Called the Ovonic battery, it has a negative electrode made of nickel alloyed with several other metals—vanadium, titanium, zirconium, and chromium—instead of the usual pure-metal electrode.

The half-cell reactions for this battery are

$$MH(s) + OH^-(aq) \underset{\text{charge}}{\overset{\text{discharge}}{\rightleftharpoons}} M(s) + H_2O(\ell) + e^-$$

$$NiOOH(s) + H_2O(\ell) + e^- \underset{\text{charge}}{\overset{\text{discharge}}{\rightleftharpoons}} Ni(OH)_2(s) + OH^-(aq)$$

MH(s) represents a metal hydride compound similar to those used to store hydrogen in other applications (Section 9.6). The amount of hydrogen that can be absorbed into this electrode determines the number of electrons that the battery can deliver as it discharges, and hence the energy storage capacity of the battery. The metal-hydride electrode has the advantage of being metallic and hence electrically conducting; in many other batteries metal oxides, which do not conduct electricity, are formed on the electrode surface and reduce the number of times the battery can be recharged.

Each component of the metal alloy that makes up the negative electrode has a role to play in the battery's excellent performance. Vanadium, titanium, and zirconium are in the alloy because they readily absorb hydrogen. For an effective metal-hydride battery the strength of the bonding between the metal atoms and hydrogen atoms must be just right—from 25 to 50 kJ/mol. If it is too low, recharging will release hydrogen as $H_2(g)$, instead of incorporating hydrogen atoms into holes in the metal crystal lattice. If hydrogen-to-metal bonding is too strong, the electrode metal will be oxidized instead of the hydrogen atoms, and the battery will not be able to discharge properly. Alloying the other metals with nickel allows the bond strength to hydrogen to be carefully adjusted for maximum efficiency. Chromium limits corrosion of vanadium in the alloy, and both zirconium and chromium affect the structure of the alloy, leading to high surface area that promotes rapid cell reactions and hence high power output.

The metal-hydride electrode consists of amorphous metal. Its atoms are in an irregular, disordered structure (Section 14.6) instead of the orderly, crystalline arrangement of most metals. This increases hydrogen storage capacity and speeds up electrode reactions. According to Energy Conversion Devices spokesperson Stanford R. Ovshinsky, the use of an amorphous material is "a fundamentally different approach" to battery design. The company claims that the battery can be charged in as little as 15 minutes and can undergo more than 1,000 charge-discharge cycles, which translates to a lifetime of 10 years and more than 100,000 miles of travel in an automobile.

For further information, see *Wall Street Journal,* April 9, 1993, page B5; *Science,* Vol. 260, April 9, 1993, page 176.

EXERCISE 17.7 • ***Recharging a Ni-cad Battery***

Write the electrode reactions that take place when a ni-cad battery is recharged; identify the anode and cathode reactions.

17.7 FUEL CELLS

A **fuel cell** is an electrochemical cell, but, in contrast to a battery, its reactants are continually supplied from an external reservoir. The best known fuel cell is the hydrogen-oxygen cell (Figure 17.14) used in the Gemini,

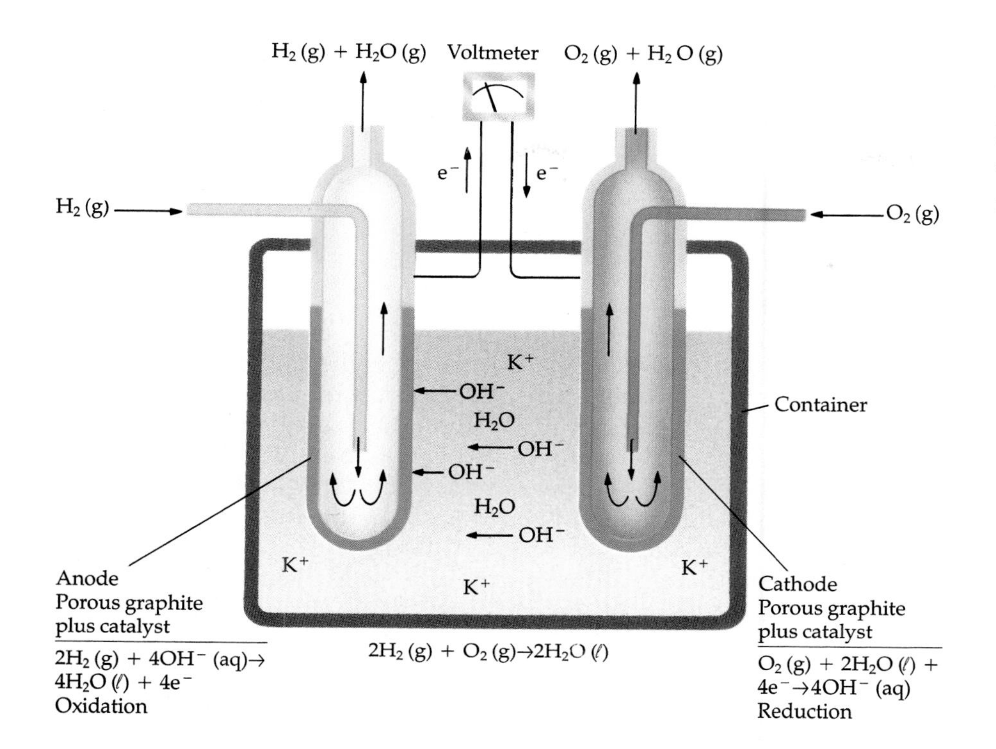

Figure 17.14
Schematic diagram of an H_2/O_2 fuel cell. The anode chamber oxidizes H_2. The cathode chamber reduces O_2. The water produced is often purified for drinking purposes.

Apollo, and Space Shuttle programs. The net cell reaction is simply the oxidation of hydrogen to give water. If a mixture of hydrogen and oxygen is sparked, energy is released suddenly in a violent explosion. On a platinum gauze that acts as a catalyst, these gases will react at room temperature, slowly heating the catalytic surface to incandescence. In a fuel cell, hydrogen and oxygen are made to react in such a way that the energy is produced in the form of an electrical current. A stream of H_2 gas is pumped into the anode compartment of the cell, and pure O_2 gas is directed onto the cathode. The cell contains concentrated KOH, so the reactions are

$$\begin{array}{ll} \text{anode, oxidation:} & 2\,H_2(g) + 4\,OH^-(aq) \longrightarrow 4\,H_2O(\ell) + 4\,e^- \\ \text{cathode, reduction:} & O_2(g) + 2\,H_2O(\ell) + 4\,e^- \longrightarrow 4\,OH^-(aq) \\ \hline \text{net cell reaction:} & 2\,H_2(g) + O_2(g) \longrightarrow 2\,H_2O(\ell) \\ & (E = 0.9\text{ V at 70 to 140 °C}) \end{array}$$

A hydrogen-oxygen fuel cell. Three of these units provide the power for the Space Shuttle and its astronauts. *(Courtesy of International Fuel Cells Corporation)*

The electrons lost by the hydrogen molecules at the cathode flow out of the fuel cell, through a circuit, and then back into the cell at the cathode, where oxygen is reduced. This electron flow powers the electrical needs of the spacecraft, or whatever else is connected to the fuel cell. The water produced in the fuel cell can be purified for drinking purposes.

Fuel cells are about 60% efficient in converting chemical energy into electricity.

Because of their light weight and their high efficiency compared to batteries, fuel cells have proved valuable in the space program. Beginning with Gemini 5, alkaline fuel cells have logged more than 10,000 hours of operation in space. The fuel cells used aboard the Space Shuttle deliver the same power that batteries weighing ten times as much would provide. On a typical seven-day mission, the Shuttle fuel cells consume 1500 lb of hydrogen and generate 190 gal of potable water (suitable for drinking).

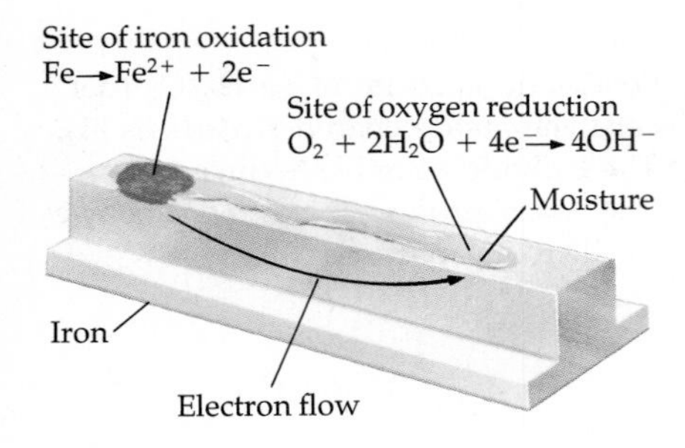

Figure 17.15
The site of iron oxidation may be different from the point of oxygen reduction because of the ability of the electrons to flow through the iron.

Other types of fuel cells that have been developed use air as the oxidizer and hydrogen or carbon monoxide as the fuel. Considerable research is currently aimed at developing fuel cells capable of direct air oxidation of cheap gaseous fuels such as natural gas.

17.8 CORROSION

Corrosion is oxidation of metal that is exposed to the environment; usually it results in loss of structural strength. Corrosion reactions are invariably product favored. Rusting of iron, for example, takes place quite readily and is difficult to prevent; it results in red-brown rust, which is hydrated iron(III) oxide [$Fe_2O_3 \cdot xH_2O$, where x varies from 2 to 4]. About 25% of the annual steel production in the United States is destined for replacement of material lost to corrosion.

For corrosion to occur at the surface of a metal, there must be anodic areas where the oxidation of the metal can occur. The general reaction is

$$\text{anode reaction:}\quad M(s) \longrightarrow M^{n+} + ne^-$$

There must also be cathodic areas where electrons are consumed by any or all of several possible half-reactions, such as

$$\text{cathode reactions:}\quad 2\,H_2O(\ell) + 2\,e^- \longrightarrow 2\,OH^-(aq) + H_2(g)$$
$$O_2(g) + 2\,H_2O(\ell) + 4\,e^- \longrightarrow 4\,OH^-(aq)$$

Anodic areas occur at cracks in the oxide coating that protects many metals, and they may also occur around impurities. Cathodic areas occur at the metal oxide coating, at less reactive metallic impurity sites, or around other metal compounds trapped at the surface, such as sulfides or carbides.

The other requirements for corrosion are an electrical connection between the anode and cathode and an electrolyte with which both anode and cathode are in contact. Both requirements are easily fulfilled—the metal itself is the conductor, and ions dissolved in moisture from the environment provide the electrolyte (Figures 17.15 and 17.16).

In the corrosion of iron, the anodic reaction is clearly the oxidation of iron. If both water and O_2 gas are present, the cathode reaction is the reduction of oxygen, giving the net reaction

$$\text{anode:}\quad 2[Fe(s) \longrightarrow Fe^{2+}(aq) + 2\,e^-]$$
$$\text{cathode:}\quad O_2(g) + 2\,H_2O(\ell) + 4\,e^- \longrightarrow 4\,OH^-(aq)$$
$$\text{net cell reaction:}\quad 2\,Fe(s) + O_2(g) + 2\,H_2O(\ell) \longrightarrow \underset{\text{iron(II) hydroxide}}{2\,Fe(OH)_2(s)}$$

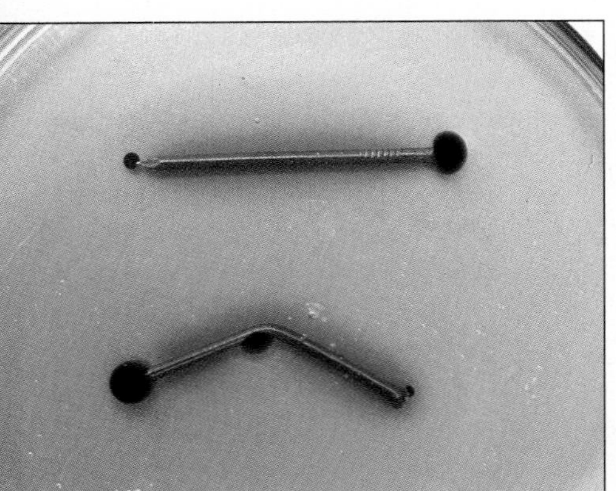

Figure 17.16
Corroding iron nails. Two nails were placed in an agar gel, which also contained the indicator phenolphthalein and $[Fe(CN)_6]^{3-}$. The nails began to corrode and gave Fe^{2+} ions at the tip and where the nail is bent. (These are points of stress and corrode more quickly.) These points are the anode as indicated by the formation of the blue colored compound called Prussian blue ($Fe_3[Fe(CN)_6]_2$). The remainder of the nail is the cathode, since oxygen is reduced in water to give OH^-. The presence of OH^- ions causes phenolphthalein to turn pink in color. *(C.D. Winters)*

In the presence of an ample supply of oxygen and water, as in the open air or in flowing water, the iron(II) hydroxide is oxidized to the red-brown iron(III) oxide (Figure 17.17).

$$4\,Fe(OH)_2(s) + O_2(g) \longrightarrow 2\,H_2O(\ell) + \underset{\text{red brown}}{2\,Fe_2O_3 \cdot H_2O(s)}$$

Figure 17.17
The formation of rust destroys the structural integrity of objects made of iron and steel. Given time, this old car will completely rust away. The process may take several decades but in the end there will remain only a pile of rust, glass, rubber tires, plastic, and possibly the copper radiator. *(C.D. Winters)*

This hydrated iron oxide is the familiar rust you see on cars and buildings, and the substance that colors the water red in some mountain streams or in your home water supply at times. Rust does not adhere strongly to iron or steel and for that reason fails to form a protective coating. It is easily removed by mechanical shaking, rubbing, or even the action of rain or freeze-thaw cycles, thus exposing more iron at the surface and allowing iron objects to eventually deteriorate completely.

If oxygen is not freely available to the corroding iron, further oxidation of the iron(II) hydroxide is limited to the formation of magnetite (Fe_3O_4), which can be thought of as a mixed oxide of Fe_2O_3 and FeO in a 1:1 ratio.

$$6\,Fe(OH)_2(s) + O_2(g) \longrightarrow \underset{\text{hydrated magnetite}}{2\,Fe_3O_4 \cdot H_2O(s)} + 4\,H_2O(\ell)$$

On the loss of water, the hydrated form of magnetite (which is green) forms black magnetite, also known as *lodestone.*

Lodestone, natural magnetite, was the first known magnetic material. By 1200 AD it was used by European sailors in compasses that would point to the magnetic north pole. Highly purified magnetite is used in making recording tape.

$$Fe_3O_4 \cdot H_2O(s) \longrightarrow \underset{\text{black magnetite}}{Fe_3O_4(s)} + H_2O(\ell)$$

Other substances in air and water can hasten corrosion. Chlorides, from sea air or from salt spread on roadways in the winter, function as a salt bridge between anodic and cathodic regions, thus speeding up corrosion reactions. Sulfur dioxide is a major air pollutant formed in the combustion of oil and coal. It forms sulfuric acid on iron surfaces and is responsible for a great deal of rusting.

Corrosion Protection

How can you stop a metal object from corroding? The general approaches are (a) to inhibit the anodic process, (b) to inhibit the cathodic process, or (c) to do both. The most common method is **anodic inhibition,** attempting directly to limit or prevent the oxidation half-reaction by painting the metal surface, coating it with grease or oil, or allowing a thin film of metal oxide to form. More recently developed methods of anodic protection are illustrated by the following reaction, in which the surface is treated with a solution of sodium chromate.

$$2\,Fe(s) + 2\,Na_2CrO_4(aq) + 2\,H_2O(\ell) \longrightarrow Fe_2O_3(s) + Cr_2O_3(s) + 4\,NaOH(aq)$$

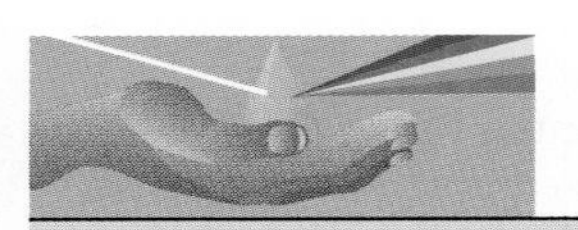

CHEMISTRY YOU CAN DO

Corrosion of Common Nails

Corrosion is a surprisingly fast process. Go to a hardware store and purchase several common nails (10d or 12d size will do). Also purchase several galvanized nails of the same size. Common nails are made of mild steel, with no corrosion protection (except a thin coating of shellac in some cases). Wash several of each kind of nail with warm soapy warm to get rid of any oil or grease that might prevent corrosion, and place them, undried, in a container. After several days look at the nails. Which ones have started to corrode? Now repeat the experiment, this time coating some clean, uncorroded nails with motor oil, fingernail polish, or some similar coating. Did the nails corrode after you protected them? What electrochemical process did you inhibit by coating the nails?

The surface iron is oxidized by the chromate salt to give Fe(III) and Cr(III) oxides. These together form a coating that is impervious to O_2 and water, and further atmospheric oxidation is inhibited.

Cathodic protection is accomplished by forcing the metal to become the cathode instead of the anode. Usually, this is achieved by attaching another, more readily oxidized metal to the metal being protected. The best example of this is **galvanized** iron, iron that has been coated with a thin film of zinc (Figure 17.18). E°_{ox} for zinc oxidation is considerably more positive than E°_{ox} for iron oxidation (look at Table 17.1, but remember to change the sign of E°, since the reactions in the table are written as reductions). Therefore, the zinc metal film is oxidized before any of the iron and the zinc coating forms what is called a *sacrificial anode.* In addition, when the zinc is corroded, $Zn(OH)_2$ forms an insoluble film on the surface (K_{sp} of $Zn(OH)_2 = 4.5 \times 10^{-17}$) that further slows corrosion.

EXERCISE 17.8 • *Corrosion Rates*

Rank the following environments for their relative rates of corrosion of iron. Place the fastest first. Explain your answers. (a) Moist clay, (b) sand by the sea shore, (c) the surface of the moon, (d) desert sand in Arizona.

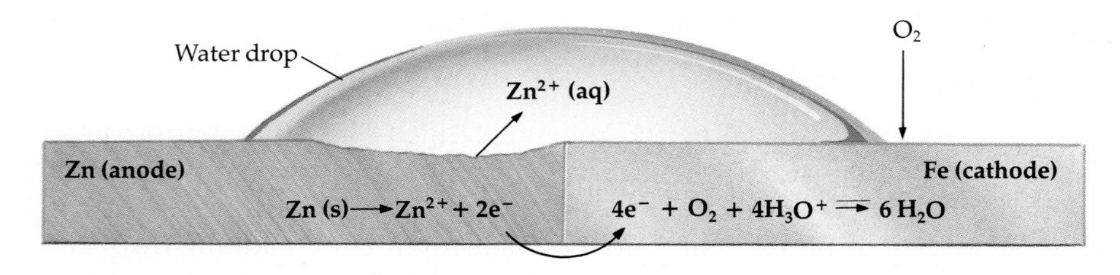

Figure 17.18

Cathodic protection of an iron-containing object. The iron is coated with a film of zinc, a metal more easily oxidized than iron. Therefore, the zinc acts as an anode and forces iron to become the cathode, thereby preventing the corrosion of the iron.

17.9 ELECTROLYSIS

Lysis means "splitting," so electrolysis means "splitting with electricity." Electrolysis reactions are chemical reactions caused by the flow of electricity.

Electrolysis is the use of an electrical current to produce a chemical change. It provides a way to carry out reactions that will not take place by themselves. Electrolysis is used to produce common metals such as aluminum and magnesium from their ores (Chapter 18), as well as some less common ones such as sodium and beryllium. Electrolysis also enables us to deposit (plate) such metals as copper, silver, or chromium on various metals—or even on plastics—in order to beautify and protect these materials.

Like voltaic cells, electrolysis cells have electrodes in contact with a conducting medium and an external circuit. By contrast with voltaic cells, however, the external circuit connected to an electrolysis cell must contain a *source* of electrons. The conducting medium in contact with the electrodes is often the same for both electrodes, and it can be a molten salt or an aqueous solution. Finally, the electrodes in electrolysis cells are often inert, and they only furnish a path for electrons to enter and leave the cell.

The same definition is used for electrodes in voltaic cells and in electrolysis cells: the electrode where reduction takes place is called the cathode, and the electrode where oxidation takes place is the anode.

The decomposition of molten sodium chloride, NaCl, is a simple example of a reaction that must be done by electrolysis. A pair of electrodes dip into pure sodium chloride that has been heated above its melting temperature (Figure 17.19). In the liquid sodium chloride, the Na^+ and Cl^- ions are free to move about. A battery can be used as a source of electrical current when an electrolysis is carried out on a small scale. The battery forces electrons into one of the electrodes (which becomes negative), and removes electrons from the other electrode (which becomes positive). In the molten sodium chloride, Cl^- ions are attracted to the positive electrode and Na^+ ions are attracted to the negative electrode. Reduction of Na^+ ions to Na atoms occurs at the negative electrode; this electrode is therefore the cathode. Oxidation of Cl^- ions occurs at the positive electrode (the anode).

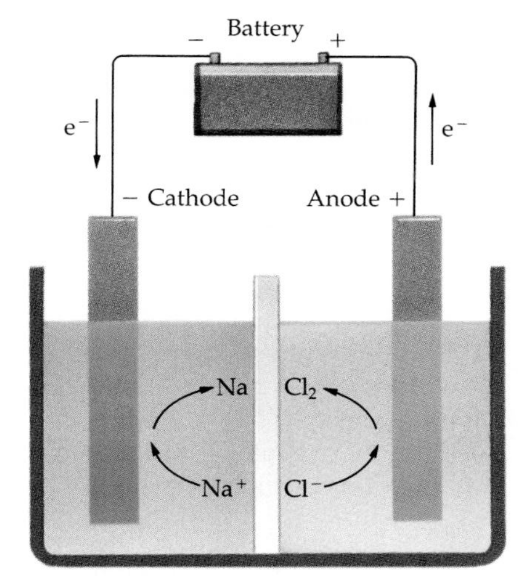

Figure 17.19
Electrolysis of molten sodium chloride.

anode reaction (oxidation):	$2\,Cl^- \longrightarrow Cl_2(g) + 2\,e^-$
cathode reaction (reduction):	$Na^+ + e^- \longrightarrow Na(\ell)$
net cell reaction:	$2\,Cl^- + 2\,Na^+ \longrightarrow 2\,Na(\ell) + Cl_2(g)$

What happens if we pass electricity through an *aqueous solution* of some salt, such as potassium iodide, KI? To predict the outcome of the electrolysis we must first decide what is in the solution that can be oxidized and reduced. For KI(aq), the solution contains K^+ ions, I^- ions, and H_2O molecules. In K^+ potassium is already in its highest possible oxidation state, so the possible anode half-reaction oxidations are

$$2\,I^-(aq) \longrightarrow I_2(s) + 2\,e^- \qquad E^\circ_{ox} = -0.535\text{ V}$$

$$6\,H_2O(\ell) \longrightarrow O_2(g) + 4\,H_3O^+(aq) + 4\,e^- \qquad E^\circ_{ox} = -1.229\text{ V}$$

Whenever two or more reactions are possible at a single electrode, the one with the more positive E° will occur under standard-state conditions. Judging by the values of E°_{ox} (which we obtained from Table 17.1), the iodide ion will be oxidized more readily than will water.

Since I^- is in its lowest possible oxidation state, there are two species that could be reduced at the cathode: K^+ ion and water.

$$2\,H_2O(\ell) + 2\,e^- \longrightarrow H_2(g) + 2\,OH^-(aq) \qquad E^\circ_{red} = -0.8277\text{ V}$$
$$K^+(aq) + e^- \longrightarrow K(s) \qquad E^\circ_{red} = -2.925\text{ V}$$

In this case we predict that $H_2O(\ell)$ will be reduced.

An experiment in which electrons are passed through aqueous KI (Figure 17.20) shows that this prediction is correct. At the anode, on the right, the I^- ion is oxidized to I_2, which produces a yellow-brown color in the solution. At the cathode, water is reduced and hydroxide ions are formed, as shown by the pink color of the acid-base indicator phenolphthalein that has been added to the solution. A close look at Figure 17.20b reveals that a gas, presumed to be hydrogen, is also being produced at the surface of the inert platinum electrode.

When several reactions are possible in an electrolysis, the one that will occur is the one that is least reactant favored.

When an electrolysis is carried out by passing electrical current through an aqueous solution, the electrode reactions most likely to take place are those that require the least voltage, that is, the half-reactions that combine to give the least negative overall cell emf. This means that in aqueous solution the following conditions apply:

1. A metal ion or other species can be reduced if it has a reduction potential more positive than −0.8 V, the potential for reduction of water. Table 17.1 shows that most metal ions are in this category. If a species has a reduction potential more negative than −0.8 V, then water will be reduced to $H_2(g)$ preferentially. Metal ions in this latter category include Na^+, K^+, Mg^{2+}, and Al^{3+}. To produce these metals

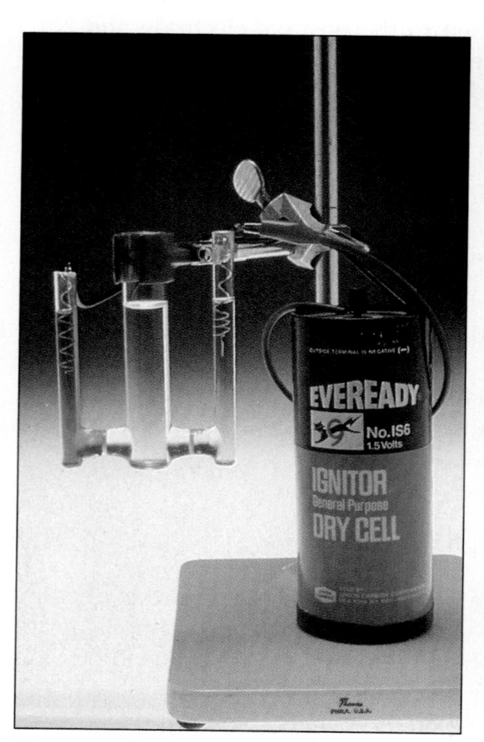

(a)

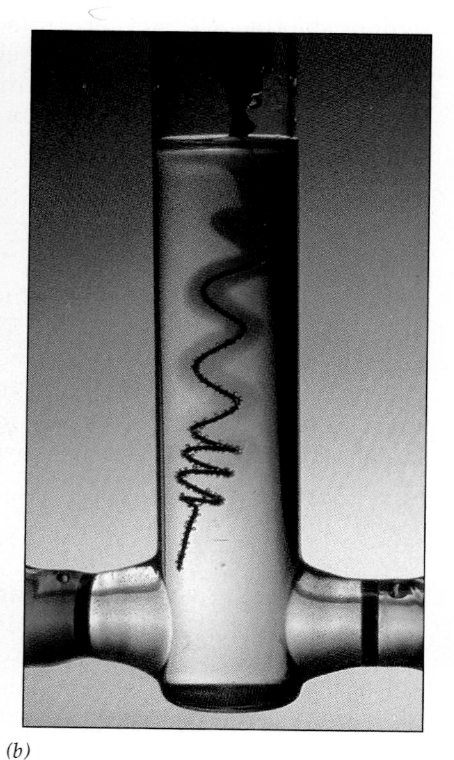
(b)

Figure 17.20
The electrolysis of aqueous potassium iodide. (a) Aqueous KI is contained in all three compartments of the cell, and both electrodes are platinum. At the positive electrode or anode (right), the I^- ion is oxidized to iodine, which gives the solution a yellow-brown color.

$$2\,I^-(aq) \longrightarrow I_2(aq) + 2\,e^-$$

At the negative electrode or cathode (left), water is reduced, and the presence of OH^- ion is indicated by the red color of the acid-base indicator phenolphthalein.

$$2\,H_2O(\ell) + 2\,e^- \longrightarrow H_2(g) + 2\,OH^-(aq)$$

(b) In a close-up of the cathode, bubbles of H_2 and evidence of OH^- being generated at the electrode are clearly seen.
(C.D. Winters)

from their ions requires electrolysis of a molten salt in which no water is present.

2. A species can be oxidized in aqueous solution if it has an oxidation potential more positive than −1.2 V, the potential for oxidation of water to $O_2(g)$. Most species on the right-hand sides of the half-equations in Table 17.1 are in this category. If a species has an oxidation potential more negative than −1.2 V (that is, if its half-equation is above the water-oxygen half-equation in Table 17.1), water will be oxidized preferentially. Thus, for example, $F^-(aq)$ cannot be oxidized electrolytically to $F_2(g)$, because water will be oxidized to $O_2(g)$ instead.

EXAMPLE 17.4

Electrolysis of Aqueous NaOH

Predict the result of passing an electrical current through an aqueous solution of NaOH.

SOLUTION First, list all the species in the solution. In this case they are Na^+, OH^-, and H_2O. Next, use Table 17.1 to decide which of these species can be oxidized and which can be reduced, and note the potential of each possible reaction.

Reductions:

$$Na^+(aq) + e^- \longrightarrow Na(s) \qquad E^\circ_{red} = -2.71\ V$$

$$2\ H_2O(\ell) + 2\ e^- \longrightarrow H_2(g) + 2\ OH^-(aq) \qquad E^\circ_{red} = -0.83\ V$$

Oxidations:

$$4\ OH^-(aq) \longrightarrow O_2(g) + 2\ H_2O(\ell) + 4\ e^- \qquad E^\circ_{ox} = -0.40\ V$$

$$6\ H_2O(\ell) \longrightarrow O_2(g) + 4\ H_3O^+(aq) + 4\ e^- \qquad E^\circ_{ox} = -1.229\ V$$

It is evident that water will be reduced to H_2 at the cathode and that OH^- will be oxidized at the anode. The net cell reaction is $2\ H_2O(\ell) \rightarrow 2\ H_2(g) + O_2(g)$, and the potential under standard conditions is (−0.83 V) + (−0.40 V) = −1.23 V.

EXERCISE 17.9 • ***Prediction of Electrolysis Products***

Predict the results of passing an electrical current through (a) molten NaBr, (b) aqueous NaBr, and (c) aqueous $SnCl_2$.

17.10 COUNTING ELECTRONS

Metallic silver is produced at the cathode during the electrolysis of aqueous $AgNO_3$.

$$Ag^+(aq) + e^- \longrightarrow Ag(s)$$

The balanced equation tells you that one mole of electrons is required to produce one mole of silver from one mole of silver ions. In contrast, two moles of electrons are required to produce one mole of metallic copper from one mole of copper ions.

$$Cu^{2+}(aq) + 2\,e^- \longrightarrow Cu(s)$$

If you could measure the number of moles of electrons flowing through the electrolysis cell, you would know the number of moles of silver or copper produced. Conversely, if you knew the amount of silver or copper produced, you could calculate the number of moles of electrons that had passed through the circuit.

The number of moles of electrons transferred during a redox reaction is usually determined by measuring the current flowing in the external electrical circuit during a given time. The charge (in units of coulombs, C) equals the current (measured in amperes, A) multiplied by the time interval (in seconds, s):

$$\text{charge} = \text{current} \times \text{time}$$
$$1 \text{ coulomb} = 1 \text{ ampere} \times 1 \text{ second}$$

The Faraday constant can then be used to find the number of coulombs of charge from a known number of moles of electrons, or to find the number of moles of electrons from a known number of coulombs of charge.

EXAMPLE 17.5

Using the Faraday Constant

What mass of nickel will be deposited at the cathode of an electrolysis cell if a current of 20. mA passes through an aqueous solution containing Ni^{2+} ions for 1 hr?

SOLUTION The reaction at the cathode is

$$Ni^{2+}(aq) + 2\,e^- \longrightarrow Ni(s)$$

The charge that passes through the cell is

$$\text{charge} = 20. \times 10^{-3}\text{ A} \times 3600\text{ s}$$
$$= 72\text{ C}$$

Now, use the Faraday constant, the coefficients of the balanced cathode half-reaction, and the molar mass of nickel as conversion factors to find the mass of nickel deposited:

$$\text{mass of Ni} = 72\text{ C} \cdot \frac{1\text{ mol e}^-}{9.65 \times 10^4\text{ C}} \cdot \frac{1\text{ mol Ni}}{2\text{ mol e}^-} \cdot \frac{58.7\text{ g Ni}}{1\text{ mol Ni}}$$
$$= 0.022\text{ g Ni}$$

EXERCISE 17.10 • *Using the Faraday Constant*

In the commercial production of sodium by electrolysis, the cell operates at 7.0 V and a current of 25×10^3 A. What mass of sodium can be produced in 1 hr?

Hydrogen holds great promise as a fuel in our economy because it is a gas and can be easily transported through pipelines, it burns without producing pollutants, and it could be used in fuel cells to generate electricity on demand. Hydrogen can be produced by the electrolysis of a solution of sulfuric acid. Thė minimum voltage required for this reaction is 1.24 V. Let's consider how much electrical energy would be required to produce 1.00 kg of gaseous H_2 (about 11,200 L at STP). We will first calculate the required charge in coulombs by using Faraday's constant, and then use the definition 1 joule = 1 volt × 1 coulomb to get energy units.

The reduction half-reaction shows that 2 mol of electrons is required to produce 1 mol (2.02 g) of $H_2(g)$.

$$2\,H_3O^+(aq) + 2\,e^- \longrightarrow H_2(g) + 2\,H_2O(\ell)$$

The amount (number of moles) of electrons required to produce 1.00 kg of H_2 is found as follows:

$$\text{amount of } e^- = 1.00 \text{ kg } H_2 \cdot \frac{1 \times 10^3 \text{ g}}{1 \text{ kg}} \cdot \frac{1 \text{ mol } H_2}{2.016 \text{ g } H_2} \cdot \frac{2 \text{ mol } e^-}{1 \text{ mol } H_2}$$
$$= 9.92 \times 10^2 \text{ mol } e^-$$

Now we can calculate the charge by using Faraday's constant.

$$\text{charge} = 9.92 \times 10^2 \text{ mol } e^- \cdot \frac{9.65 \times 10^4 \text{ C}}{1 \text{ mol } e^-} = 9.57 \times 10^7 \text{ C}$$

The energy (in joules) can be calculated from the charge and the cell voltage.

$$\text{energy} = \text{charge} \times \text{voltage}$$
$$= 9.57 \times 10^7 \text{ C} \cdot 1.24 \text{ V} = 1.19 \times 10^8 \text{ J}$$

Now we convert joules to kilowatt-hours, which is the unit we see when we pay the electric bill. The conversion factor is 1 kwh = 3.60×10^6 J.

$$\text{energy} = 1.19 \times 10^8 \text{ J} \cdot \frac{1 \text{ kwh}}{3.60 \times 10^6 \text{ J}} = 33.1 \text{ kwh}$$

At a rate of 10 cents per kilowatt-hour, the production of 1.00 kg of hydrogen costs $3.31.

The kilowatt-hour is a unit of energy.

EXERCISE 17.11 • *Calculations Based On Electrolysis*

In the production of aluminum metal, Al^{3+} is reduced to Al. The currents are generally about 50,000 A. A low voltage of about 4.0 V is used. How much energy (in kilowatt-hours) would be required to produce 2000. tons of aluminum metal?

17.11 ELECTROPLATING

If a metal or other electrical conductor is made the cathode in an electrolysis cell, it can be plated with another metal to protect it against corrosion, decorate it, or purify the deposited metal. Consider, for example, an electrolysis

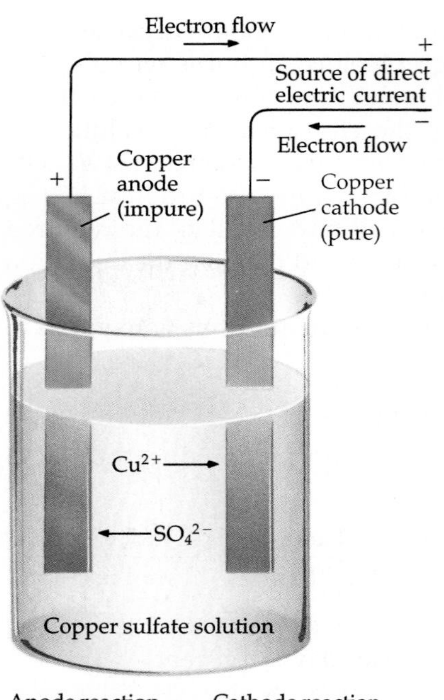

Figure 17.21
Electroplating from a copper sulfate solution. Copper from the impure copper anode is plated on the cathode, which consists of highly purified copper.

cell like the one shown on Figure 17.21. Here, the cathode is pure copper and the anode is a bar of less pure copper. The electrolysis solution contains a soluble copper salt such as $CuSO_4$. The cell half-reactions are

anode:	$Cu(s) \longrightarrow Cu^{2+}(aq) + 2\,e^-$	$E^\circ_{ox} = -0.34$ V
cathode:	$Cu^{2+}(aq) + 2\,e^- \longrightarrow Cu(s)$	$E^\circ_{red} = +0.34$ V
net cell reaction:	*nothing (all reactants and products cancel)*	

The use of a net cell reaction here is misleading. Certainly something is taking place in the cell, but it is merely transport of copper from the impure anode to the pure cathode. We could write the net cell reaction as

$$Cu(s)_{impure} \xrightarrow{\text{electrical current}} Cu(s)_{pure}$$

The pure copper has less entropy than the mixture of copper and impurities; hence energy must be supplied to overcome the entropy decrease.

To plate an object with copper, we have only to render the surface conducting and make the object the cathode in a cell containing a solution of a soluble copper salt. The object will become coated with copper, and the copper coating will grow thicker as the electrolysis is continued and more electrons reduce Cu^{2+} ions to Cu atoms. If the object is a metal, it will conduct electricity by itself. If the object is a nonmetal, its surface can be lightly dusted with graphite powder to render it conducting (Figure 17.22).

Precious metals such as gold are often plated onto cheaper metals such as copper to make jewelry. If the current and time of the plating reaction are known, it is possible to calculate the mass of gold that plates out. For exam-

A Deeper Look

Making Compact Audio Discs

Compact audio discs (CDs), like vinyl records before them, depend on electroplating to create a "stamper" master. The polycarbonate CDs are mass-produced by pressing the melted plastic between two metal plates, one of which is the stamper. The laser beam in the CD player derives information from "pits and lands" placed in the plastic disc by the stamping process. Making a CD requires several steps involving electrochemistry and the electroplating of metals.

After the audio tracks for the CD have been assembled, a high-powered laser beam cuts a digital audio signal consisting of small pits into a special plastic or glass substrate. This "master" is then cleaned and sprayed with a dilute solution of silver diammine complex $[Ag(NH_3)_2]^+$ followed by a reducing agent such as formaldehyde. A silver mirror results, similar to the thin layer of metal on the back of an ordinary mirror. The reaction is

$$2\,Ag^+(aq) + HCHO(aq) + 3\,H_2O(\ell) \longrightarrow 2\,Ag(s) + HCOOH(aq) + 2\,H_3O^+(aq)$$

This thin silver coating forms the surface on which Ni atoms are plated to make a "mother" disc, which is a reversal of the original cut made by the recording laser. Another coating of nickel is electrodeposited on the mother, and when peeled away from the mother this copy becomes the stamping master. The stamping master is a reversal of the mother, so it is identical to the original cut made by the recording laser. The nickel deposition reaction is

$$Ni^{2+}(aq) + 2\,e^- \longrightarrow Ni(s)$$

The stamping master is then used to make thousands of copies of the original, called replicates. If the stamper wears out, or breaks, another stamper is electrodeposited on the mother. If a really big hit is being pressed, one mother might be used to make scores of stamping masters.

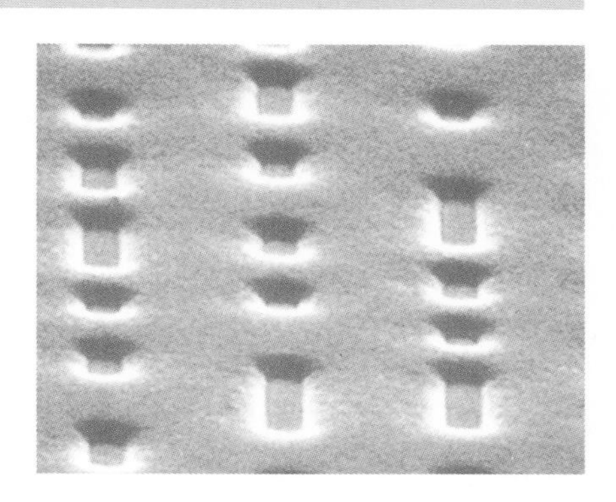

Information is physically stored on a compact disc in the form of pits and "lands," the flat portion of the disc between the pits. The digital data are defined by the length of the pits and the distance between them. The pits may vary in length from 0.833 to 3.56 micrometers (10^{-6} m), are 0.5 micrometer wide, and are 0.11 micrometer deep. About 7×10^9 total bits of information can be placed on the surface of a standard compact disc. *(Courtesy of DADC)*

---

ple, suppose the object is immersed in a solution of $AuCl_3$ and is made a cathode by connecting it to the negative pole of a battery. The circuit is completed by immersing an inert anode in the solution, and gold is plated out on the cathode for 60. min at a current of 0.25 A. The mass of gold that plates out is

$$\text{mass Au} = 0.25\text{ A} \cdot 60.\text{ min} \cdot \frac{60\text{ s}}{1\text{ min}} \cdot \frac{1\text{ C}}{1\text{ A}\cdot\text{s}} \cdot \frac{1\text{ mol e}^-}{9.65 \times 10^4\text{ C}} \cdot \frac{1\text{ mol Au}}{3\text{ mol e}^-} \cdot \frac{197\text{ g Au}}{1\text{ mol Au}}$$

$$= 0.61\text{ g Au}$$

That's about $7.83 worth of gold if gold is selling for $400 per ounce.

EXERCISE 17.12 • *Electroplating*

Calculate the mass of gold that could be plated from solution with a current of 0.50 A for 20. min. The cathode reaction is $Au^{3+}(aq) + 3\,e^- \rightarrow Au(s)$.

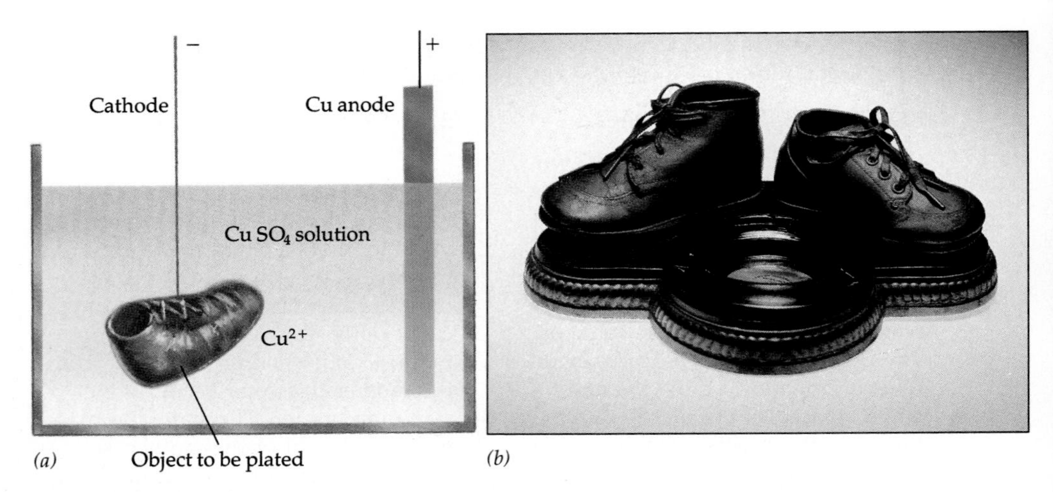

Figure 17.22
(a) The plating of baby shoes with copper. The surface of the shoe is made conducting with a fine powdering with graphite or powdered metal. (b) The finished product can last many years. *(b, C.D. Winters)*

IN CLOSING

Having studied this chapter, you should be able to

- Identify the oxidizing and reducing agents in a redox reaction, write equations for the oxidation and reduction half-reactions, and use them to balance the net equation (Section 17.1).
- Identify and describe the functions of the parts of an electrochemical cell; describe the direction of electron flow outside the cell and the direction of the ion flow inside the cell (Section 17.2).
- Describe how standard reduction potentials are defined and use them to predict whether a reaction will be product-favored as written (Sections 17.3 and 17.4).
- Calculate $\Delta G°$ from the value of $E°$ for a redox reaction (Section 17.5).
- Explain how product-favored electrochemical reactions can be used to do useful work, and list the requirements for using such reactions in rechargeable batteries (Section 17.6).
- Describe the chemistry of the dry cell, the mercury battery, and the lead storage battery (Section 17.6).
- Describe how a fuel cell works and indicate how it is different from a battery (Section 17.7).
- Explain how iron corrodes and explain some ways in which iron corrosion can be controlled (Section 17.8).
- Use standard reduction potentials to predict the products of electrolysis of an aqueous salt solution (Section 17.9).
- Calculate the quantity of product formed at an electrode during an elec-

trolysis reaction, given the current passing through the cell and the time during which the current flows (Section 17.10).

- Explain how electroplating works (Section 17.11).

STUDY QUESTIONS

1. In each of the following reactions, tell which substance is oxidized and which is reduced. Tell which is the oxidizing agent and which is the reducing agent.
   (a) $2\ Al(s) + 3\ Cl_2(g) \longrightarrow 2\ AlCl_3(s)$
   (b) $8\ H_3O^+(aq) + MnO_4^-(aq) + 5\ Fe^{2+}(aq) \longrightarrow 5\ Fe^{3+}(aq) + Mn^{2+}(aq) + 12\ H_2O(\ell)$
   (c) $FeS(s) + 3\ NO_3^-(aq) + 4\ H_3O^+(aq) \longrightarrow 3\ NO(g) + SO_4^{2-}(aq) + Fe^{3+}(aq) + 6\ H_2O(\ell)$
2. Explain the function of a salt bridge in an electrochemical cell.
3. Tell whether each of the following statements is true or false. If false, rewrite it to make it a correct statement.
   (a) Oxidation always occurs at the anode of an electrochemical cell.
   (b) The anode of a battery is the site of reduction and is negative.
   (c) Standard conditions for electrochemical cells are a concentration of 1.0 M for dissolved species and a pressure of 1 atm for gases.
   (d) The potential of a cell does not change with temperature.
   (e) All product-favored oxidation-reduction reactions have an $E°$ with a negative sign.
4. Tell whether each of the following statements is true or false. If false, rewrite it to make it a correct statement.
   (a) The value of an electrode potential changes when the half-reaction is multiplied by a factor. That is, $E°$ for $Li^+ + e^- \rightarrow Li$ is different from that for $2\ Li^+ + 2\ e^- \rightarrow 2\ Li$.
   (b) Al is the strongest reducing agent listed in Table 17.1.
5. Are standard half-cell reactions always written as (a) oxidation reactions or (b) reduction reactions?
6. Write half-reactions for the following:
   (a) oxidation of zinc to Zn^{2+} ion
   (b) reduction of H_3O^+ ion to hydrogen gas
   (c) oxidation of hydrogen gas to H_3O^+ ion
   (d) reduction of chlorine to Cl^- ion
   (e) reduction of MnO_4^- ion to Mn^{2+} ion in acid solution
   (f) reduction of $Cr_2O_7^{2-}$ ion to Cr^{3+} ion in acid solution
   (g) reduction of Sn^{4+} ion to Sn^{2+} ion
7. What are the advantages and disadvantages of lead storage batteries?
8. How does a fuel cell differ from a battery?
9. What are the products of the electrolysis of a concentrated aqueous solution of NaCl? What species are present in the solution? What is formed at the cathode? What is formed at the anode?
10. What is cathodic protection? Give a common example.
11. What three chemical substances are necessary for the corrosion of iron? Which of these species does a coating of oil or grease eliminate from the reaction?
12. Describe the principal parts of an electrochemical cell. What is the sign of the charge on the anode? On the cathode? Draw a hypothetical cell, indicating the direction of electron flow outside the cell and the direction of ion flow within the cell.
13. Describe the principal parts of a H_2/O_2 fuel cell. What is the reaction at the cathode? At the anode? What is the product of the fuel cell reaction?
14. Using the reduction potentials in Table 17.1, place the following elements in order of increasing ability to function as reducing agents: (a) Cl_2, (b) Fe, (c) Ag, (d) Na.
15. What is the strongest oxidizing agent in Table 17.1? What is the strongest reducing agent? What is the weakest oxidizing agent? What is the weakest reducing agent?
16. Four metals, A, B, C, and D, exhibit the following properties:
   (a) Only A and C react with 1.0 M HCl to give $H_2(g)$.
   (b) When C is added to solutions of ions of the other metals, metallic B, D, and A are formed.
   (c) Metal D reduces B^{n+} ions to give metallic B and D^{n+} ions.

   On the basis of this information, arrange the four metals in order of increasing ability to act as reducing agents.

Cells and Cell Potentials

17. Copper can reduce silver(I) to metallic silver, a reaction that could in principle be used in a battery.

$$Cu(s) + 2\ Ag^+(aq) \longrightarrow Cu^{2+}(aq) + 2\ Ag(s)$$

   (a) Write equations for the half-reactions involved.
   (b) Which half-reaction is an oxidation and which is a reduction? Which half-reaction occurs in the anode

compartment and which in the cathode compartment?

18. Chlorine gas can oxidize zinc metal in a reaction that has been suggested as the basis of a battery.
    (a) Write the half-reactions involved. Label which is the oxidation and which is the reduction reaction.
    (b) According to data from Table 17.1, what is $E°$ for this reaction? What is $\Delta G°$?
19. One of the most energetic redox reactions is that between F_2 gas and lithium metal (it is *the* most energetic using the reactants in Table 17.1).
    (a) Write the half-reactions involved. Label which is the oxidation and which is the reduction reaction.
    (b) According to data from Table 17.1, what is $E°$ for this reaction? What is $\Delta G°$?
20. Calculate the value of $E°$ for each of the following reactions. Decide whether each is product-favored.
    (a) $I_2(s) + Mg(s) \longrightarrow Mg^{2+}(aq) + 2\,I^-(aq)$
    (b) $Ag(s) + Fe^{3+}(aq) \longrightarrow Ag^+(aq) + Fe^{2+}(aq)$
    (c) $Sn^{2+}(aq) + 2\,Ag^+(aq) \longrightarrow Sn^{4+}(aq) + 2\,Ag(s)$
    (d) $2\,Zn(s) + O_2(g) + 2\,H_2O(\ell) \longrightarrow 2\,Zn^{2+}(aq) + 4\,OH^-(aq)$
21. Consider the following half-reactions:

Half-Reaction	*E° (V)*
$Cl_2(g) + 2\,e^- \longrightarrow 2\,Cl^-(aq)$	+1.36
$I_2(s) + 2\,e^- \longrightarrow 2\,I^-(aq)$	+0.535
$Pb^{2+}(aq) + 2\,e^- \longrightarrow Pb(s)$	−0.126
$V^{2+}(aq) + 2\,e^- \longrightarrow V(s)$	−1.18

    (a) Which is the weakest oxidizing agent in the list?
    (b) Which is the strongest oxidizing agent in the list?
    (c) Which is the strongest reducing agent?
    (d) Which is the weakest reducing agent?
    (e) Will Pb(s) reduce $V^{2+}(aq)$ to V(s)?
    (f) Will $I_2(g)$ oxidize $Cl^-(aq)$ to $Cl_2(g)$?
    (g) Name the elements or ions that can be reduced by Pb(s).
22. Consider the following half-reactions:

Half-Reaction	*E° (V)*
$Ce^{4+}(aq) + e^- \longrightarrow Ce^{3+}(aq)$	+1.61
$Ag^+(aq) + e^- \longrightarrow Ag(s)$	+0.80
$Hg_2^{2+}(aq) + 2\,e^- \longrightarrow 2\,Hg(\ell)$	+0.79
$Sn^{2+}(aq) + 2\,e^- \longrightarrow Sn(s)$	−0.14
$Ni^{2+}(aq) + 2\,e^- \longrightarrow Ni(s)$	−0.25
$Al^{3+}(aq) + 3\,e^- \longrightarrow Al(s)$	−1.66

    (a) Which is the weakest oxidizing agent in the list?
    (b) Which is the strongest oxidizing agent in the list?
    (c) Which is the strongest reducing agent?
    (d) Which is the weakest reducing agent?
    (e) Will Sn(s) reduce $Ag^+(aq)$ to Ag(s)?
    (f) Will $Hg(\ell)$ reduce $Sn^{2+}(aq)$ to Sn(s)?
    (g) Name the ions that can be reduced by Sn(s).
    (h) What metals can be oxidized by $Ag^+(aq)$?
23. Ni-cad batteries are rechargeable and are commonly used in cordless appliances. Although such batteries actually function under basic conditions, imagine an electrochemical cell using the following setup.

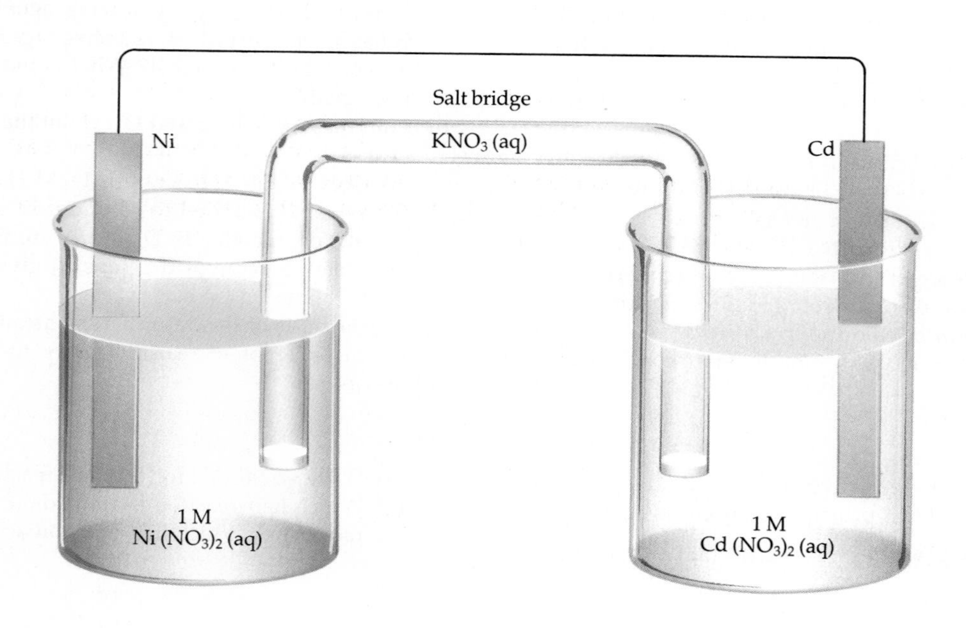

(a) Write a balanced net ionic equation depicting the reaction occurring in the cell.
(b) What is oxidized? What is reduced? What is the reducing agent and what is the oxidizing agent?
(c) Which is the anode and which is the cathode? What is the polarity of the Cd electrode?
(d) What is $E°$ for the cell?
(e) What is the direction of electron flow in the external wire?
(f) If the salt bridge contains KNO_3, toward which compartment will the NO_3^- ions migrate?

24. Hydrazine, N_2H_4, can be used as the reducing agent in a fuel cell.

$$N_2H_4(aq) + O_2(g) \longrightarrow N_2(g) + 2\ H_2O(\ell)$$

If $\Delta G°$ for the reaction is -607 kJ, calculate the value of $E°$ expected for the reaction.

25. The standard cell potential for the oxidation of Mg by Br_2 is 3.45 V.

$$Br_2(\ell) + Mg(s) \longrightarrow Mg^{2+}(aq) + 2\ Br^-(aq)$$

Calculate $\Delta G°$ for this reaction.

26. The standard cell potential, $E°$, for the reaction of Zn(s) and $Cl_2(g)$ is +2.12 V. What is the standard free energy change, $\Delta G°$, for the reaction?

27. In principle, a battery could be made from aluminum metal and chlorine gas.
(a) Write a balanced equation for the reaction that would occur in a battery using $Al^{3+}(aq)/Al(s)$ and $Cl_2(g)/Cl^-(aq)$ half-reactions.
(b) Tell which half-reaction occurs at the anode and which at the cathode. What are the polarities of these electrodes?
(c) Calculate the standard potential, $E°$, for the battery.

Electrolysis and Electrical Energy

28. For each of the following solutions, tell what reactions take place at the anode and at the cathode during electrolysis.
(a) $NiBr_2(aq)$
(b) NaI(aq)
(c) $CdCl_2(aq)$

29. A current of 0.015 A is passed through a solution of $AgNO_3$ for 155 min. What mass of silver is deposited at the cathode?

30. Current is passed through a solution containing $Ag^+(aq)$. How much silver was in the solution if all the silver was removed as Ag metal by electrolysis for 14.5 min at a current of 1.0 mA?

31. A current of 2.50 A is passed through a solution of $Cu(NO_3)_2$ for 2.00 hr. What mass of copper is deposited at the cathode?

32. A current of 0.0125 A is passed through a solution of $CuCl_2$ for 2.00 hr. What mass of copper is deposited at the cathode and what volume of Cl_2 gas (in mL at STP) is produced at the anode?

33. The major reduction half-reaction occurring in the cell in which Al_2O_3 and aluminum salts are electrolyzed is $Al^{3+} + 3\ e^- \rightarrow Al(s)$. If the cell operates at 5.0 V and 1×10^5 A, what mass (in grams) of aluminum metal can be produced in 8.0 hr?

34. The vanadium(II) ion can be produced by electrolysis of a vanadium(III) salt in solution. How long must you carry out an electrolysis if you wish to convert completely 0.125 L of 0.015 M $V^{3+}(aq)$ to $V^{2+}(aq)$ using a current of 0.268 A?

35. The reactions occurring in a lead storage battery are given in Section 17.6. A typical battery might be rated at "50 ampere-hours." This means that it has the capacity to deliver 50. amperes for 1.0 hour or 1.0 ampere for 50. hours. If it does deliver 1.0 ampere for 50. hours, what mass of lead would be consumed to accomplish this?

36. It has been demonstrated that an effective battery can be built using the reaction between Al metal and O_2 from the air. If the Al anode of this battery consists of a 3-ounce piece of aluminum (84 g), for how many hours can the battery produce 1.0 A of electricity?

37. A dry-cell battery is used to supply a current of 250 mA for 20 min. What mass of Zn is consumed?

38. If the same current as in Study Question 37 were supplied by a mercury battery, what mass of Hg would be produced at the cathode?

39. Assuming that the anode reaction for the lithium battery is

$$Li(s) \longrightarrow Li^+(aq) + e^-$$

and the anode reaction for the lead storage battery is

$$Pb(s) + HSO_4^-(aq) \longrightarrow PbSO_4(s) + 2\ e^- + H^+(aq)$$

compare the masses of metals consumed when each of these batteries supplies a current of 1 A for 10 minutes.

40. A hydrogen/oxygen fuel cell operates on the simple reaction

$$2\ H_2(g) + O_2(g) \longrightarrow 2\ H_2O(\ell)$$

If the cell is designed to produce 1.5 A of current, how long can it operate if there is an excess of oxygen and only sufficient hydrogen to fill a 1.0 L tank at 200. atm pressure at 25 °C?

General Questions

41. Consider the ni-cad cell in Study Question 23.
(a) If the concentration of Cd^{2+} is reduced to 0.010 M, and $[Ni^{2+}]$ = 1.0 M, will the cell emf be smaller or larger than when the concentration of $Cd^{2+}(aq)$ was 1.0 M? Explain your answer in terms of Le Chatelier's principle.
(b) Begin with 1.0 L of each of the solutions, both initially 1.0 M in dissolved species. Each electrode weighs 50.0 g in the beginning. If 0.050 A is drawn from the battery, how long can it last?

42. Hydrazine, N_2H_4, has been proposed as the fuel in a fuel cell in which oxygen is the oxidizing agent. The reactions are

$$N_2H_4(aq) + 4\,OH^-(aq) \longrightarrow N_2 + 4\,H_2O(\ell) + 4\,e^-$$
$$O_2(g) + 2\,H_2O(\ell) + 4\,e^- \longrightarrow 4\,OH^-(aq)$$

(a) Which reaction occurs at the anode and which at the cathode?
(b) What is the net cell reaction?
(c) If the cell is to produce 0.50 A of current for 50.0 hr, what mass in grams of hydrazine must be present?
(d) What mass in grams of O_2 must be available to react with the mass of N_2H_4 determined in part (c)?

43. Fluorine, F_2, is made by the electrolysis of anhydrous HF.

$$2\,HF(\ell) \longrightarrow H_2(g) + F_2(g)$$

Typical electrolysis cells operate at 4000 to 6000 A and 8 to 12 V. A large-scale plant can produce about 9 tons of F_2 gas per day. (a) What mass in grams of HF is consumed? (b) Using the conversion factor of 3.60×10^6 J/kwh, how much energy in kilowatt-hours is consumed by a cell operating at 6.0×10^3 A at 12 V for 24 hours?

Summary Questions

44. A 12-V automobile battery consists of six cells of the type described in Section 17.6. The cells are connected in series so that the same current flows through all of them. Calculate the theoretical minimum electrical potential difference needed to recharge an automobile battery. (Assume standard-state concentrations.) How does this compare with the maximum voltage that could be delivered by the battery? Assuming that the lead plates in an automobile battery weigh 10.0 kg, and that there is sufficient PbO_2 available, what is the maximum possible work that could be obtained from the battery?

45. Three electrolytic cells are connected in series, so that the same current flows through all of them for 20. min. In Cell A, 0.0234 g Ag plates out from a solution of $AgNO_3(aq)$, Cell B contains $Cu(NO_3)_2(aq)$, and Cell C contains $Al(NO_3)_3(aq)$. What mass of Cu will plate out in Cell B? What mass of Al will plate out in Cell C?

46. Fluorinated organic compounds are important commercially, as they are used as herbicides, flame retardants, and fire extinguishing agents, among other things. A reaction such as

$$CH_3SO_2F + 3\,HF \longrightarrow CF_3SO_2F + 3\,H_2$$

is actually carried out electrochemically in liquid HF as the solvent.
(a) Draw a dot structure for CH_3SO_2F. (S is the "central" atom and is bonded to two O atoms, an F atom, and a CH_3 group by a S—C bond.) What is the geometry around the S atom? What are the O—S—O and O—S—F bond angles?
(b) If you electrolyze 150 grams of CH_3SO_2F, how many grams of HF are required and how many grams of each product can be isolated?
(c) Is H_2 produced at the anode or the cathode of the electrolysis cell?
(d) A typical electrolysis cell operates at 8.0 V and a low current such as 250 amps. How many kilowatt-hours of energy does one such cell consume in 24 hours?

Using Computer Programs and Videodiscs

47. Use *KC? Discoverer* and/or the Periodic Table Videodisc to view reactions of the first-row transition metals with acids. Compare what you observe with what would be predicted under standard-state conditions from Table 17.1. Do any of the observations surprise you? Make a list of the cases where your predictions do not agree with what is observed. Use *KC? Discoverer* to try to find information about why what you expected did not happen.
48. Use *KC? Discoverer* to find all elements that react with water. For each element decide from the information given whether the reaction is a redox reaction. See whether each element that reacts with water is listed in Table 17.1. If so, is the reaction with water predicted by data in the table?
49. Recall from Chapter 9 the definition of ionization energy. How does the process corresponding to ionization energy differ from the process of oxidation of a metal in an electrochemical cell? How are the two processes similar? Use *KC? Discoverer* to sort the elements in order of increasing ionization energy. Does this list of elements correlate with Table 17.1? What are the similarities and differences?

The Earth.
(G. Kelvin)

Elements from the Land, Sea, and Air

CHAPTER

18

The long view from space has dramatized what we already knew: the crust of the earth is a very unusual environment, uniquely suited—at least in this solar system—for the production and support of life forms. Our environment is also quite heterogeneous in nature. Mixtures abound; everywhere we look the elements and compounds are almost lost in the complicated array of mixtures that have resulted from natural forces acting over very long periods of time.

Throughout most of our history, we had not developed the ability to alter our environment significantly. Most of the materials used, such as the stone hammer or the wooden plow, were only physically changed from the natural materials. Then came the chemical reduction of copper from its ores, followed by iron, and now a flood of new chemicals are produced each year. We have now developed, beyond question, the power to change the chemical mixtures that are found around us.

In this chapter we shall look at the chemistry of some major industrial operations for recovery of elements from the land, sea, and air.

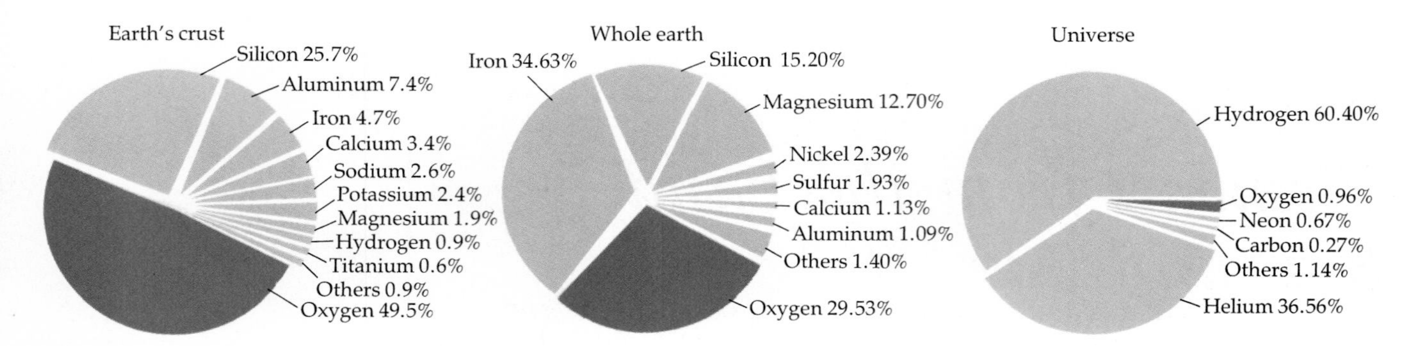

Figure 18.1
Relative abundances (by mass) of the most common elements in the earth's crust, the whole earth, and the universe.

The chemical composition of the earth's crust differs dramatically from that of the universe or even the composition of the earth as a whole (Figure 18.1). The element oxygen dominates at about 50%, since it is found in the air, water, and rocks. Silicon, at about 25%, is a major part of silicate rocks, clays, and sand. Then come the major metals—iron, calcium, sodium, potassium, and magnesium—found mostly in mineral deposits and in the sea. Hydrogen, at less than 1% by mass, is ninth in abundance, and carbon, the central element in all life forms, is present in little more than trace amounts.

18.1 COMMERCIAL PRODUCTION OF ELEMENTS BY ELECTROCHEMICAL METHODS

This chapter extends the discussion of electrochemistry begun in Chapter 17 to nonspontaneous chemical reactions brought about by the flow of electricity, processes known as electrolysis.

Much of the chlor-alkali industry is centered near Niagara Falls, where low-cost electric power is available from hydroelectric plants.
(International Stock Photo)

Sodium by Electrolysis

The first commercial production of sodium metal was adapted from the method used by Humphrey Davy in 1807, when he discovered the metal. Davy passed electricity through molten NaOH, which has a melting point of 318 °C, fairly low for an ionic compound. The half-reactions are

anode, oxidation: $4\,OH^-(\ell) \longrightarrow O_2(g) + 2\,H_2O(g) + 4\,e^-$

cathode, reduction: $4\,[Na^+(\ell) + e^- \longrightarrow Na(\ell)]$

net cell reaction: $4\,Na^+(\ell) + 4\,OH^-(\ell) \longrightarrow 4\,Na(\ell) + O_2(g) + 2\,H_2O(g)$

Sodium was a laboratory curiosity until 1824, when it was found to reduce aluminum chloride to aluminum. The net reaction is

$$3\,Na(s) + AlCl_3(s) \longrightarrow Al(s) + 3\,NaCl(s)$$

The possibility of producing commercially useful aluminum led to considerable interest in manufacturing sodium.

When an electrolytic method for the production of aluminum was discovered in 1886 (discussed later in this section), interest in sodium waned.

Before long, however, other uses were found for the metal, which led to the development in 1921 of an efficient electrolytic method called the Downs process. One of the first major uses for the low-cost sodium from the Downs process was in the manufacturing of tetraethyllead [$Pb(C_2H_5)_4$], the octane-improving compound in "leaded gasoline," which is no longer used in the United States (Section 13.1). The capacity for sodium manufacture in the United States is now about 76,000 tons per year, and much of that is centered near Niagara Falls, New York, where relatively low-cost power is available from hydroelectric plants.

The Downs cell for the electrolysis of molten NaCl (Figure 18.2) operates at 7 to 8 volts and currents of 25,000 to 40,000 amps. The cell is filled with a mixture of dry NaCl and $CaCl_2$ in a 1:3 ratio. Pure NaCl is not used because its melting point is 800 °C. As you learned in Section 15.3, however, the melting point of a solvent can be lowered by adding a solute; in this case addition of $CaCl_2$ to NaCl gives a mixture melting at approximately 600 °C.

The sodium in the Downs cell is produced at a copper or iron cathode that surrounds a circular graphite anode. Immediately over the cathode there is an inverted trough in which the low-density molten sodium collects (melting point of pure sodium = 97.8 °C), and the byproduct, gaseous Cl_2, is collected in an inverted cone that extends through the molten salt mixture almost to the level of the anode.

In sodium-containing lamps, electrically excited sodium atoms emit a bright yellow light that is useful for street lighting in foggy areas.

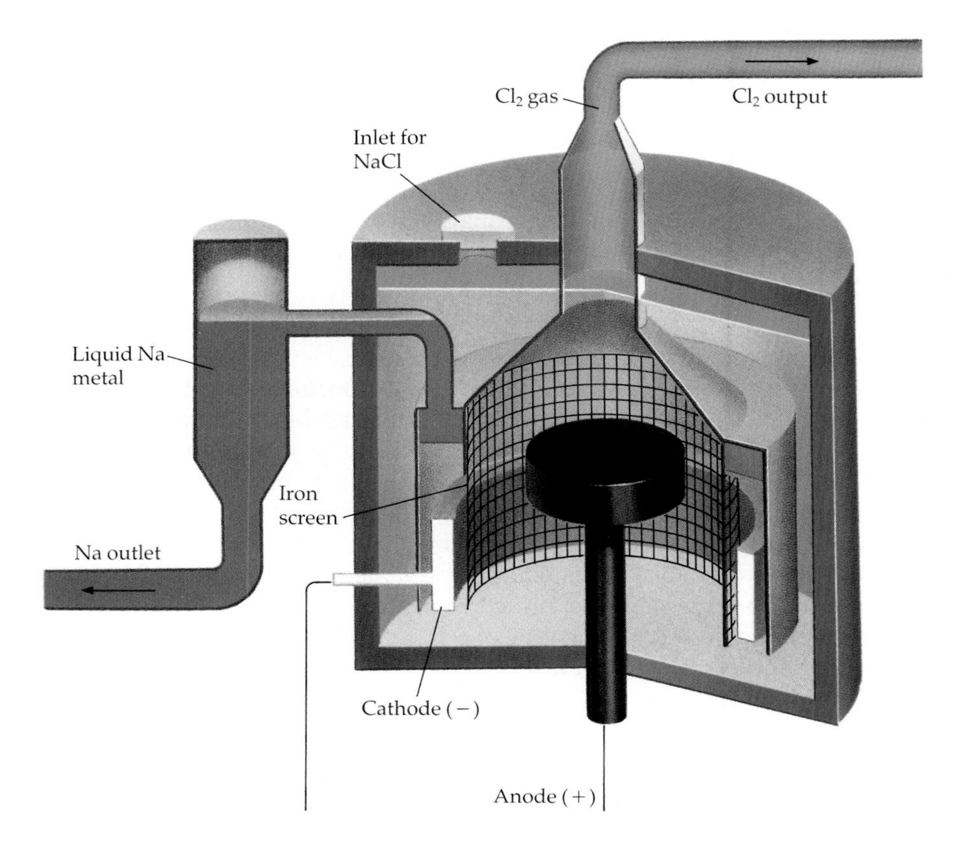

Figure 18.2
The Downs cell for the electrolysis of molten NaCl. The circular iron cathode is separated from the graphite anode by an iron screen. Since the cell operates at about 600 °C, the sodium is produced in the molten state. The liquid metal has a low density, so it floats to the top and can be drawn off. The chlorine gas bubbles out of the anode compartment and is collected.

Since the use of sodium in making tetraethyllead has been curtailed, its primary use, as it was before, is to reduce valuable metal halides to their metals. For instance, titanium has important uses in the aircraft industry, and it can be prepared from its chloride by reduction with sodium.

$$TiCl_4(s) + 4\,Na(s) \longrightarrow 4\,NaCl(s) + Ti(s)$$

How many tons of titanium could be prepared annually from this reaction if half of the sodium production capacity were used? Since the annual capacity for sodium is about 76,000 tons, half would be 38,000 tons. Using the balanced equation for the reduction of titanium chloride by sodium, we see that four moles of sodium are required to produce one mole of titanium. This is our stoichiometric factor (see Section 6.1). Calculating the number of moles of sodium and applying the stoichiometric factor,

$$\text{moles of Na} = 38{,}000 \text{ tons} \cdot \frac{2000 \text{ lb}}{1 \text{ ton}} \cdot \frac{454 \text{ g}}{1 \text{ lb}} \cdot \frac{1 \text{ mol Na}}{23.0 \text{ g}}$$

$$= 1.5 \times 10^9 \text{ mol Na}$$

$$\text{moles of Ti} = 1.5 \times 10^9 \text{ mol Na} \cdot \frac{1 \text{ mol Ti}}{4 \text{ mol Na}}$$

$$= 3.8 \times 10^8 \text{ mol Ti}$$

Finally,

$$\text{mass of Ti} = 3.8 \times 10^8 \text{ mol Ti} \cdot \frac{47.9 \text{ g}}{1 \text{ mol Ti}} \cdot \frac{1 \text{ lb}}{454 \text{ g}} \cdot \frac{1 \text{ ton}}{2000 \text{ lb}}$$

$$= 2.0 \times 10^4 \text{ tons Ti}$$

Useful information for working electrolysis problems in this chapter:
coulombs = amperes × seconds
charge on 1 mol e^- = 9.65×10^4 coulombs = 1 Faraday
1 watt = 1 joule/second
energy (joules) = coulombs × volts

EXERCISE 18.1 • *The Downs Cell*

How many tons of sodium can be prepared from a Downs cell operating at 2.0×10^4 A for 24 hr? How many tons of Cl_2 will be produced?

Chlorine and Sodium Hydroxide

Chlorine is used to treat water and sewage, to produce organic chemicals such as pesticides and vinyl chloride [the building block of plastics called PVCs, poly(vinyl chlorides) (Section 13.6)], and to bleach pulp used to make white paper. In 1992 chlorine was tenth on the list of chemicals produced in the United States. Almost all Cl_2 is made by electrolysis, with 95% coming from the electrolysis of NaCl brine (aqueous NaCl). The other product coming from these cells, NaOH, is equally valuable, and more than 24 billion pounds were produced in the United States in 1992.

A simplified version of one type of cell used for **chlor-alkali** production, the mercury cell, is illustrated in Figure 18.3. Chloride ion, Cl^-, is oxidized at the graphite anode to give Cl_2. Sodium can form a liquid solution with mercury called an *amalgam*.

$$Na^+ + e^- + Hg \longrightarrow \underset{\text{sodium amalgam}}{Na(Hg)}$$

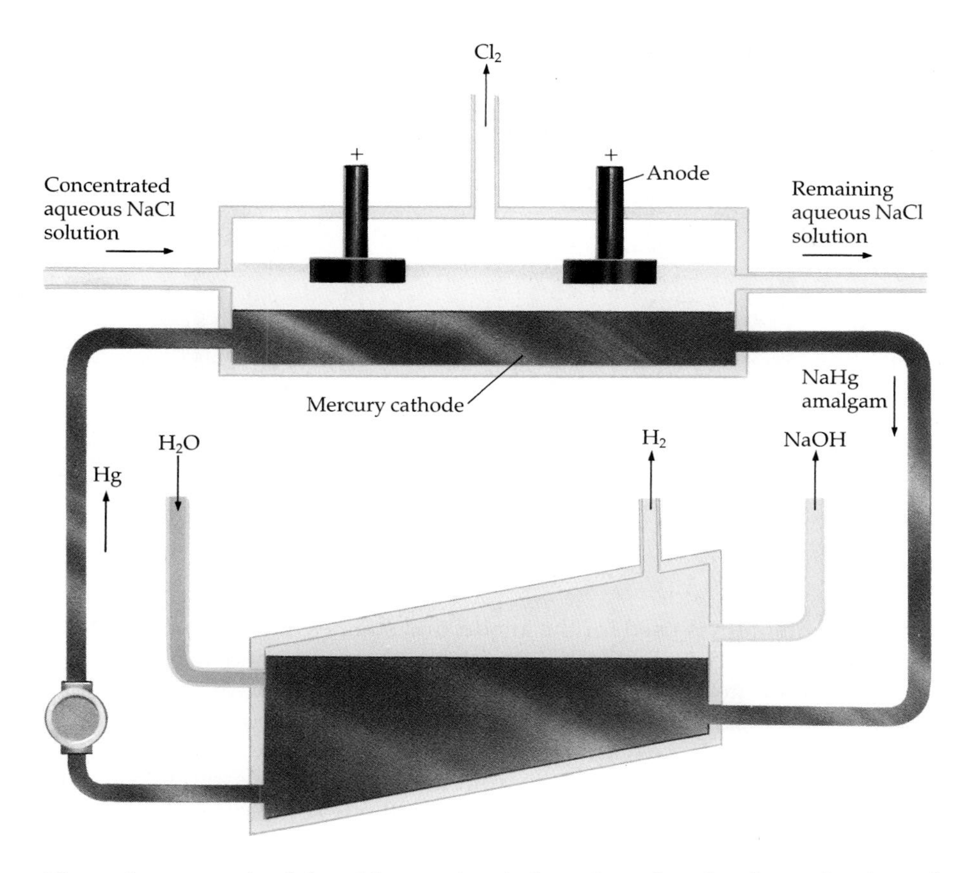

Figure 18.3
In a mercury cell, the dense liquid mercury at the bottom of the electrolytic chamber acts as the cathode. The mercury amalgam of sodium is continually run off into a separate chamber, where the reaction of water with Na(Hg) produces NaOH(aq) and H_2(g). The mercury is then cycled back to the electrolytic cell.

The voltage required for this reaction is less than that for the reduction of water to H_2. Thus, Na^+ is reduced to Na(Hg), and this liquid solution flows out of the cell. When the Na(Hg) solution is vigorously stirred with water, the sodium reacts to give NaOH and H_2.

$$2\,Na(Hg)(\ell) + 2\,H_2O(\ell) \longrightarrow 2\,NaOH(aq) + H_2(g) + 2\,Hg(\ell)$$

In this manner, electrolysis of brine can yield two pure, commercially useful gases as well as pure NaOH.

One major problem with the mercury cell is the environmental damage caused by mercury spills. Many years ago, mercury was allowed to run into nearby rivers and bays during routine cleaning of the cells. In response to environmental problems, the mercury cell has been replaced by the chlor-alkali membrane cell (Figure 18.4, next page). In this cell the anode and cathode compartments are separated by a special plastic membrane that allows only cations to pass through it. Sodium chloride solution is added to the anode compartment, where chlorine gas is produced. Sodium ions pass through the membrane to the cathode compartment, where H_2 and OH^- are produced, and sodium hydroxide solution is removed at the bottom. The reactions are

$$\begin{array}{ll} \text{anode reaction:} & 2\,Cl^-(aq) \longrightarrow Cl_2(g) + 2\,e^- \\ \text{cathode reaction:} & 2\,H_2O(\ell) + 2\,e^- \longrightarrow H_2(g) + 2\,OH^-(aq) \\ \hline \text{net cell reaction:} & 2\,Cl^-(aq) + 2\,H_2O(\ell) \longrightarrow H_2(g) + Cl_2(g) + 2\,OH^-(aq) \end{array}$$

A chlor-alkali plant. Each membrane cell shown contains many anodes and cathodes arranged in series. *(Oxychem)*

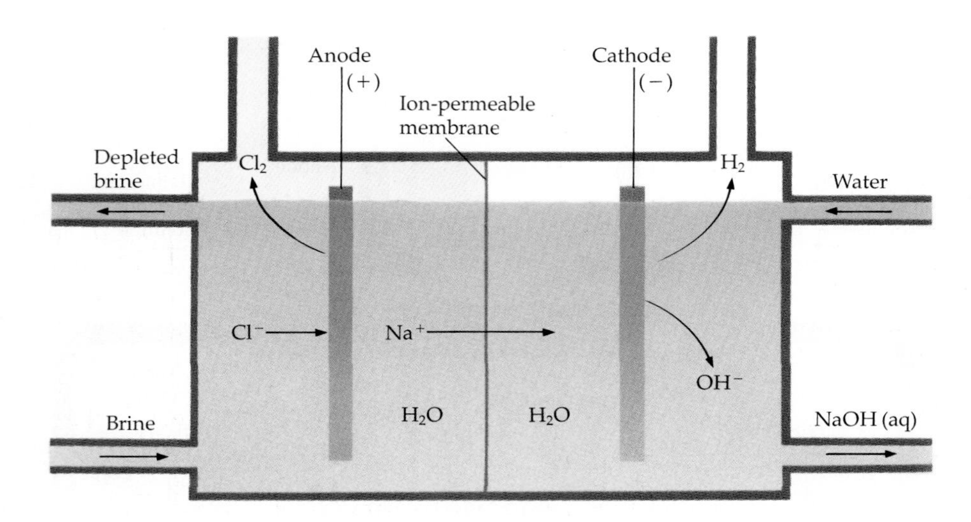

Figure 18.4
A membrane cell for the production of NaOH(aq) and Cl_2(g) from brine solution. Here the anode and cathode compartments are separated by a water-impermeable but ion-conducting membrane. A widely used membrane is made of Nafion, a fluorine-containing polymer that is a relative of Teflon. Brine is fed into the anode compartment and dilute sodium hydroxide or water into the cathode compartment.

Activated titanium is used for the anode, and stainless steel or nickel is used for the cathode. The membrane separating the anode and cathode is not permeable to water, but it does allow ions to pass through; that is, the membrane is just a salt bridge between the anode and cathode. Therefore, to maintain charge balance within the cell, as Cl^- ions are reduced in the anode compartment, Na^+ ions must pass from the anode to the cathode compartment. Since OH^- ions are produced in the cathode compartment, the product there is aqueous NaOH with a concentration of 20 to 35% by weight. The average electrical consumption of these cells is from 2000 to 2500 kwh/ton of NaOH produced. A large percentage of the NaOH produced is used in the Bayer process to purify aluminum ore.

EXERCISE 18.2 • *The Membrane Cell for NaOH*

How many tons of NaOH will be prepared in a membrane cell operating at a current of 2.00×10^4 A for 100. hours?

Fluorine by Electrolysis

The first nonelectrolytic synthesis of F_2 (from the reaction of K_2MnF_6(s) with SbF_5(g) at 150 °C) was reported in 1986.

Henri Moissan won the 1906 Nobel Prize in chemistry for isolating elemental fluorine.

Fluorine (F_2), one of the strongest oxidizing agents, reacts violently with all elements except the noble gases. As a result, fluorine was not isolated until 1886 even though hydrogen fluoride was first prepared in 1771 by Carl Wilhelm Scheele. Finally, in 1886, the same year electrolytic methods for aluminum were developed, Henri Moissan prepared F_2 by electrolysis of KF in anhydrous HF at −50 °C, which is still the only practical method for preparing F_2. In the Moissan method, KF (a good conductor) is dissolved in liquid HF (a weak acid and a poor conductor of electricity) to give a mixture with the approximate composition of KF · 2 HF, which can be regarded as a solution of $K^+HF_2^-$ in liquid HF. The electrode processes are

$$\text{anode reaction:} \quad HF_2^-(\ell) \longrightarrow F_2(g) + H^+(\ell) + 2\,e^-$$
$$\text{cathode reaction:} \quad 2\,H^+(\ell) + 2\,e^- \longrightarrow H_2(g)$$
$$\text{net cell reaction:} \quad H^+(\ell) + FHF^-(\ell) \longrightarrow H_2(g) + F_2(g)$$

Since HF is consumed during the electrolysis, it must be replenished.

The gaseous products of this electrolysis will spontaneously react with each other, so great care must be taken in the design of the cell to keep the products separated. Fluorine is extremely reactive toward most metals, and especially so at about 70 °C. Moissan's early cell was constructed of platinum. He no doubt reasoned that since platinum was so inert to oxidation by oxygen it would also be inert toward fluorine. Not so. In his early experiments, the mass of platinum lost exceeded the mass of fluorine produced. Presumably this was due to the reaction producing the compound PtF_4 (melting point 56.7 °C).

$$Pt(s) + 2\,F_2(g) \xrightarrow{70\,°C} PtF_4(\ell)$$

These problems were later solved by using other electrode materials such as nickel and carbon. From 1939 to 1945 elemental fluorine was in great demand to prepare volatile UF_6, which was used to make enriched uranium for the atomic bomb (Section 12.3).

The electrolytic cell used to prepare fluorine today contains a carefully constructed baffle arrangement to keep the H_2 and F_2 gases separated (Figure 18.5). Because of the volatility of HF (boiling point 19.5 °C), both the H_2 and F_2 produced by the cell are contaminated with HF. The process exploits the stability of the HF_2^- ion to remove the HF impurity; the gases pass through a container of sodium fluoride, and $NaHF_2$ is formed.

$$NaF(s) + HF(g) \longrightarrow NaHF_2(s)$$

Heating the $NaHF_2$ adduct regenerates the NaF and releases HF.

$$NaHF_2(s) \xrightarrow{heat} NaF(s) + HF(g)$$

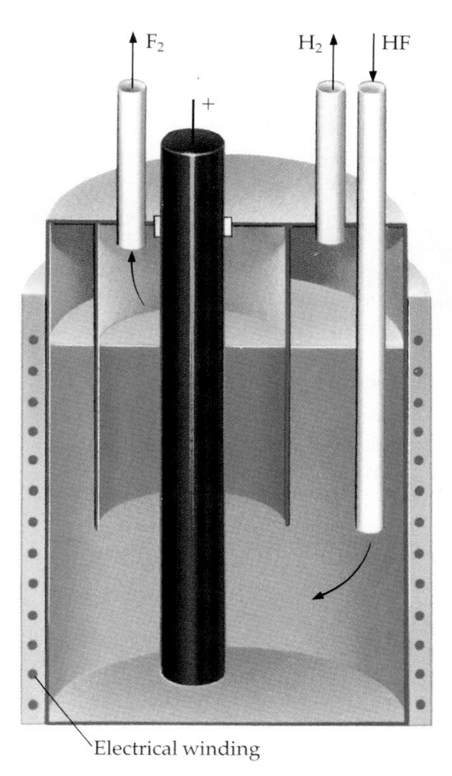

Figure 18.5
Electrolytic cell for the production of fluorine. Gaseous F_2 is produced at the carbon anode and $H_2(g)$ is evolved at the steel cathode.

EXERCISE 18.3 • *Fluorine Production*

If 10. Faradays of electricity are used in a fluorine cell, what mass in grams of F_2 will be produced? What mass of H_2? What mass of HF will be required?

Aluminum Production

Aluminum, in the form of Al^{3+} ions, is the third most abundant element in the earth's crust (7.4%) and has many important uses in our economy (Figure 18.6). You probably know it best as a food wrap, a use demonstrating its excellent formability, or as a stepladder, which illustrates its low density (2.70 g/cm^3) and high strength. Just as importantly, aluminum has excellent corrosion resistance because a transparent, chemically inert film of aluminum oxide (Al_2O_3) clings tightly to the metal's surface.

$$2\,Al(s) + 3\,O_2(g) \longrightarrow 2\,Al_2O_3(s)$$

Figure 18.6
Some consumer products made from aluminum. *(C.D. Winters)*

Figure 18.7
The medieval village of Les Baux in the south of France. The aluminum-containing mineral bauxite gets its name from this town, near which the mineral is still found in commercial quantities.

The top of the Washington Monument is aluminum, made in 1884 by the sodium-reduction method.

However, Al was considered a precious metal until the late 1800's because of the difficulty of reducing Al^{3+} to Al.

From the time of its discovery in 1825 by Hans Christian Oersted, aluminum was made by reducing $AlCl_3$ with a more active metal. Oersted used potassium; soon others substituted sodium, but only at a very high cost. Even though aluminum was commercially produced near Paris by 1854, it was considered a precious metal, like gold or platinum, and one of its early uses was for jewelry. In the 1855 Exposition in Paris, some of the first aluminum metal produced was exhibited along with the crown jewels of France. Napoleon II saw its possibility for military use, however, and commissioned studies on improving its production. The French had a ready source of aluminum-containing ore, bauxite (Figure 18.7), and in 1886 a 23-year-old Frenchman, Paul Héroult, conceived the electrochemical method that is still in use today. In an interesting coincidence, an American, Charles Hall, who was 22 at the time, announced his invention of the identical process in the same year. Hence, the commercial process is now known as the Hall-Héroult process.

In the Hall-Héroult process, metallic aluminum is obtained from Al_2O_3 by using molten cryolite, Na_3AlF_6 (melting point 1000 °C), as a solvent for the oxide (Figure 18.8). Considerable amounts of aluminum oxide dissolve in cryolite, which gives a cryolite solution with a lower melting point. This mixture of cryolite and aluminum oxide is electrolyzed in a cell with carbon anodes and a carbon cell lining that serves as the cathode on which aluminum is deposited. As the cell operates, molten aluminum sinks to the bottom of the cell. From time to time the cell is tapped and the molten aluminum is allowed to run into molds. The cells operate at the very low voltage of 4.0 to 5.5 V, but at a current of 50,000 to 150,000 A.

Each kilogram of aluminum requires about 13 to 16 kwh of energy, excluding that required to heat the molten solution. This is why there is so much interest in recycling soft drink cans and other aluminum objects: the

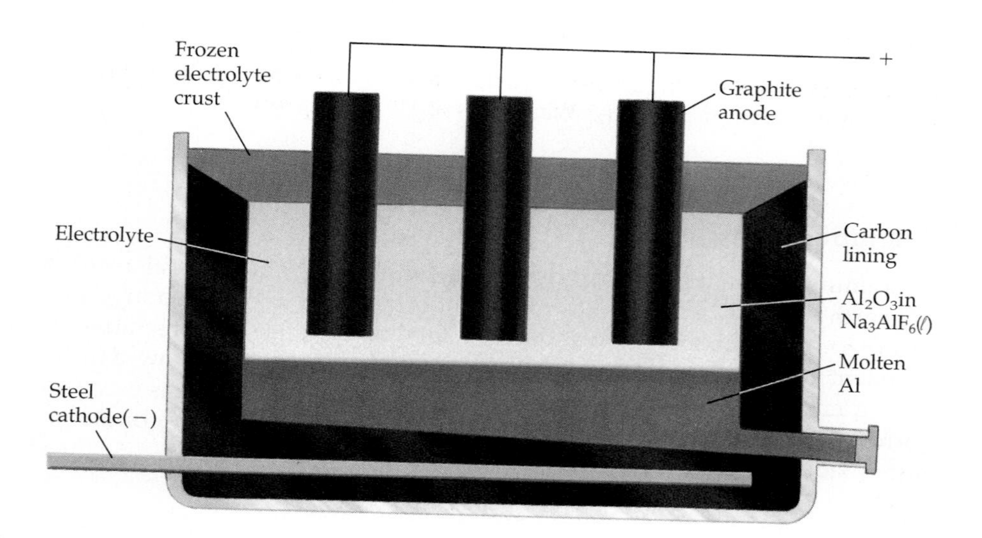

Figure 18.8
An electrolytic cell used in the Hall-Héroult process for making aluminum.

The Chemical World

A Better Aluminum Foil

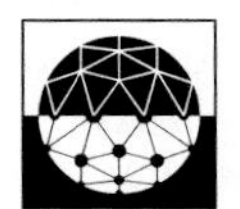

The specialty of the materials scientist is to try to produce, by means of changes in chemistry and processing, a material that possesses the unique properties required for a particular application. It is relatively easy to have ideas and dreams for new materials for new technological applications, but we can achieve these dreams only if we can transform the raw materials into the new substances required.

One example is the story of a new kind of aluminum foil developed by researchers at Allied Signal Corporation. Chemically and physically, it is very different from household aluminum foil: it is an alloy containing iron, silicon, and other elements, and it is stronger and more resistant to high temperatures than household aluminum foil. In ordinary aluminum casting, the metal cools slowly, so many added elements can separate partially and form different kinds of crystals within the foil. The interfaces between the different crystalline forms are weak points along which cracks could easily develop. Researchers at Allied Signal Corporation found that if they cooled the molten aluminum alloy quickly, up to a hundred degrees in a fraction of a second, the aluminum solidified before the different crystals could form. This quick cooling of the metal casting created a new aluminum alloy. Its enhanced strength comes from the iron and its heat resistance comes from the silicon.

The World of Chemistry (Program 19), "Metals."

---

recycled metal can be purified and made into new objects at a small fraction of the cost of making aluminum from the ore.

EXAMPLE 18.1

Aluminum Production

How much electricity, in kilowatt-hours, is required to produce 1.00 ton of aluminum in a typical Hall-Héroult cell operating at 5.00 V and 1.00×10^5 A?

SOLUTION First calculate the number of moles of Al produced.

$$\text{number of moles of Al} = 1.00 \text{ ton Al} \cdot \frac{2000 \text{ lb}}{\text{ton}} \cdot \frac{454 \text{ g}}{\text{lb}} \cdot \frac{1 \text{ mol Al}}{26.98 \text{ g}}$$

$$= 3.37 \times 10^4 \text{ mol Al}$$

The reduction half-reaction is $Al^{3+} + 3\,e^- \rightarrow Al$, so 1 mol of Al requires 3 mol of electrons. Thus,

$$\text{number of moles of } e^- = 3.37 \times 10^4 \text{ mol Al} \cdot \frac{3 \text{ mol } e^-}{1 \text{ mol Al}} = 1.01 \times 10^5 \text{ mol } e^-$$

$$\text{charge} = 1.01 \times 10^5 \text{ mol } e^- \cdot \frac{9.65 \times 10^4 \text{ C}}{1 \text{ mol } e^-} = 9.74 \times 10^9 \text{ C}$$

The energy required is

$$\text{energy} = 9.74 \times 10^9 \text{ C} \times 5.00 \text{ V} = 4.87 \times 10^{10} \text{ J}$$

$$= 4.87 \times 10^{10} \text{ J} \cdot \frac{1 \text{ kwh}}{3.60 \times 10^6 \text{ J}} = 1.35 \times 10^4 \text{ kwh}$$

Note that the current was not involved in calculating the energy consumption.

Charles Martin Hall. As a chemistry student at Oberlin College, Hall was intrigued by the potential uses of aluminum and the difficulties involved in reducing this chemically active metal from its oxide. In electrolysis experiments in his family woodshed, Hall used batteries and a blacksmith's fire in 1886 to reduce Al_2O_3 dissolved in cryolite to metallic aluminum. Hall was 22 years old when he made this great discovery. Later he founded the Aluminum Corporation of America (ALCOA), and died a multimillionaire in 1914. Paul Héroult, a Frenchman, independently made the same discovery at approximately the same time. *(Oesper Collection in the History of Chemistry/University of Cincinnati)*

Magnesium from the Sea

Magnesium, with a density of 1.74 g/cm^3, is the lightest structural metal in common use. For this reason magnesium is most often used in alloys designed for light weight and great strength (Section 9.6). Magnesium is a reactive metal because it loses electrons easily (see Table 17.1).

$$Mg(s) \longrightarrow Mg^{2+}(aq) + 2\,e^- \quad E^\circ_{ox} = 2.37\text{ V}$$

Since magnesium is so abundant in the oceans (one liter of sea water contains 10.5 g Na^+, 19 g Cl^-, and 1.35 g Mg^{2+}) this has been one of the major sources of the metal. In fact, sea water can be considered an "ore" for magnesium. Because there are 6 million tons of magnesium present as Mg^{2+} in every cubic mile of sea water, the sea can furnish an almost limitless amount of this element.

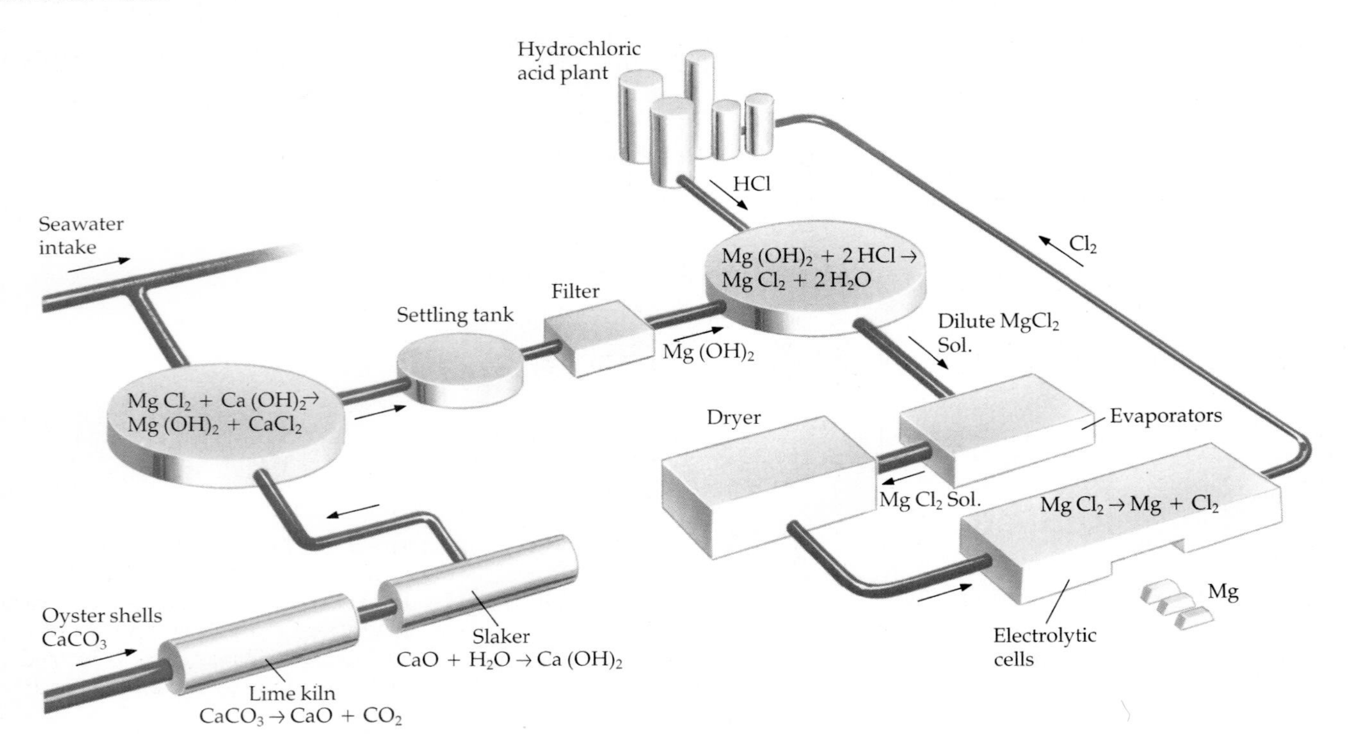

Figure 18.9
Diagram of an industrial plant showing steps for recovering magnesium from sea water.

The recovery of magnesium from sea water (Figure 18.9) begins with the precipitation of the insoluble magnesium hydroxide [$K_{sp} = 1.5 \times 10^{-11}$ for $Mg(OH)_2$]. The only thing needed to precipitate the hydroxide is a ready supply of an inexpensive base, a need fulfilled nicely by sea shells, which contain calcium carbonate. As you learned in Section 5.5, heating a metal carbonate generally leads to loss of CO_2 and formation of the metal oxide. Thus, calcium carbonate is converted to lime and then to calcium hydroxide, the source of hydroxide ion used to precipitate $Mg(OH)_2$.

$$\underset{\text{sea shells}}{CaCO_3(s)} \longrightarrow \underset{\text{lime}}{CaO(s)} + CO_2(g)$$

$$CaO(s) + H_2O(\ell) \longrightarrow Ca(OH)_2(s)$$

$$Mg^{2+}(aq) + Ca(OH)_2(s) \longrightarrow Mg(OH)_2(s) + Ca^{2+}(aq)$$

The magnesium hydroxide is isolated by filtration and then neutralized by another inexpensive chemical, hydrochloric acid.

$$Mg(OH)_2(s) + 2\,HCl(aq) \longrightarrow MgCl_2(aq) + 2\,H_2O(\ell)$$

When the water is evaporated, solid hydrated magnesium chloride is left. After drying, it is melted at 708 °C and then electrolyzed in a huge steel pot (Figure 18.10) that serves as the cathode. Graphite bars serve as the anode. The electrode reactions are

$$\begin{array}{ll} \text{anode reaction:} & 2\,Cl^-(\ell) \longrightarrow Cl_2(g) + 2\,e^- \\ \text{cathode reaction:} & Mg^{2+}(\ell) + 2\,e^- \longrightarrow Mg(\ell) \\ \hline \text{net cell reaction:} & Mg^{2+}(\ell) + 2\,Cl^-(\ell) \longrightarrow Mg(\ell) + Cl_2(g) \end{array}$$

As the magnesium melts, it floats to the surface of the molten $MgCl_2$ where it is removed. The chlorine produced at the anode is recovered by

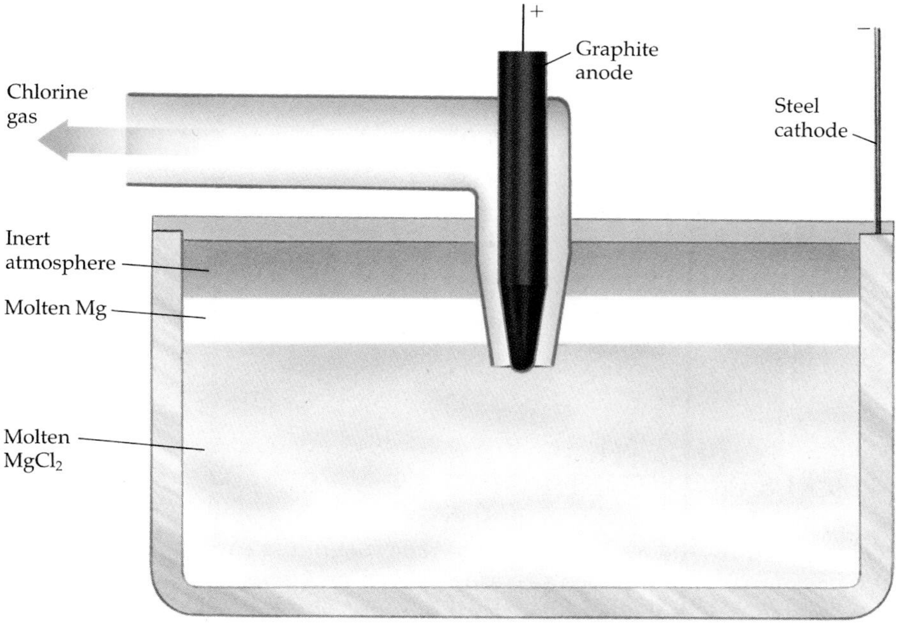

Beryllium, a rare and highly toxic metal in the same group as magnesium, is also produced by the electrolysis of the metal chloride.

$$BeCl_2(s) \xrightarrow{\text{electricity}} Be(s) + Cl_2(g)$$

Some beryllium is also produced by treating BeF_2 with magnesium.

$$BeF_2(s) + Mg(s) \longrightarrow Be(s) + MgF_2(s)$$

Figure 18.10
A cell for electrolyzing molten $MgCl_2$. Liquid magnesium metal is formed on the steel cathode and rises to the top, where it is dipped out periodically. Chlorine gas is formed on the graphite anode and is piped off.

mixing it with natural gas (methane, CH_4) and burning the mixture to form HCl.

$$4\,Cl_2(g) + 2\,CH_4(g) + O_2(g) \longrightarrow 2\,CO(g) + 8\,HCl(g)$$

Although it is potentially available in much larger quantities, the total world production of magnesium is only about 250,000 tons per year. Much of this production is used to make an alloy of aluminum, since most aluminum has about 5% magnesium metal added to it to improve its mechanical properties and to make it more resistant to corrosion under basic conditions. Alloys are also made with the reverse formulation, that is, more magnesium than aluminum. These alloys are used where a high strength-to-weight ratio is needed and where corrosion resistance is important.

Alloy: a solution consisting of two or more metals.

Because of magnesium's reactivity, it is used in *sacrificial anodes* to protect the hulls of ships and underground pipelines against corrosion (see Section 17.8).

EXAMPLE 18.2

Magnesium Production

How long must a cell operate at a current of 4.00×10^3 A to produce 500. kg of Mg?

SOLUTION The reduction of one mole of Mg requires two moles of e^-. Thus, the number of moles of electrons needed is

$$\text{moles of } e^- = 500.\text{ kg Mg} \cdot \frac{1000\text{ g}}{\text{kg}} \cdot \frac{1\text{ mol Mg}}{24.31\text{ g}} \cdot \frac{2\text{ mol } e^-}{1\text{ mol Mg}}$$

$$= 4.1 \times 10^4 \text{ mol } e^-$$

This amount of electrons is provided by a charge of

$$\text{charge} = 4.11 \times 10^4 \text{ mol } e^- \cdot \frac{9.65 \times 10^4\text{ C}}{\text{mol } e^-} = 3.97 \times 10^9\text{ C}$$

Since charge = current × time, the cell must operate for

$$\text{time} = \frac{3.97 \times 10^9\text{ C}}{4.00 \times 10^3\text{ A}} = 9.93 \times 10^5\text{ s or } 276\text{ hr}$$

EXERCISE 18.4 • *Magnesium Production*

How long (in hours) must a cell operate at a current of 5.00×10^3 A to produce 2.00 tons of Mg?

18.2 METALS AND THEIR ORES

Metals occur mostly as compounds in the crust of the earth, although some of the less active metals such as copper, silver, and gold can be found also as free elements. Fortunately, the distribution of elements in the crust is not

Table 18.1

Some Common Metals and Their Minerals

Metal	Chemical Formula	Mineral
aluminum	$Al_2O_3 \cdot x\ H_2O$	bauxite
calcium	$CaCO_3$	limestone
chromium	$FeO \cdot Cr_2O_3$	chromite
copper	Cu_2S	chalcocite
iron	Fe_2O_3	hematite
	Fe_3O_4	magnetite
lead	PbS	galena
manganese	MnO_2	pyrolusite
tin	SnO_2	cassiterite
zinc	ZnS	sphalerite
	$ZnCO_3$	smithsonite

uniform. Some elements that are not particularly abundant are familiar to us because they tend to occur in very concentrated, localized deposits, called ores, from which they can be extracted economically. Examples of these are lead, copper, and tin, none of which is among the most abundant elements in the crust of the earth (Figure 18.1). Other elements that actually form a much larger percentage of the crust are almost unknown to us because concentrated deposits of their ores are less commonly found or because the metal is difficult to extract from its ore. An example is titanium, the tenth most abundant element in the crust of the earth. Although the ores of titanium, rutile (mostly TiO_2) and ilmenite ($FeTiO_3$), are common, the metal is rare because it is difficult to reclaim it from the ores.

The preparation of many metals from their ores, called **metallurgy,** involves chemical reduction. Some useful metals and related common ores are listed in Table 18.1.

Figure 18.11
Open-pit mining of iron ore.
(Bethlehem Steel Corporation)

Iron

Iron is the fourth most abundant element in the earth's crust and the second most abundant metal (Figure 18.1). Our economy depends on iron and its alloys, particularly steel. The sources of most of the world's iron are large deposits of iron oxides in Minnesota, Sweden, France, Venezuela, Russia, Australia, and England (Figure 18.11). In nature these oxides are frequently mixed with impurities, so the production of iron usually incorporates steps to purify the ores. Iron ores are then reduced to the metal by using carbon, in the form of coke, as the reducing agent.

Iron ores are mixtures containing iron compounds. To get iron from the ores, the iron in the compounds must be reduced.

Iron ore is reduced in a blast furnace (Figure 18.12). The solid material fed into the top of the blast furnace consists of a mixture of an oxide of iron (Fe_2O_3), coke (C), and limestone ($CaCO_3$). A blast of heated air is forced into

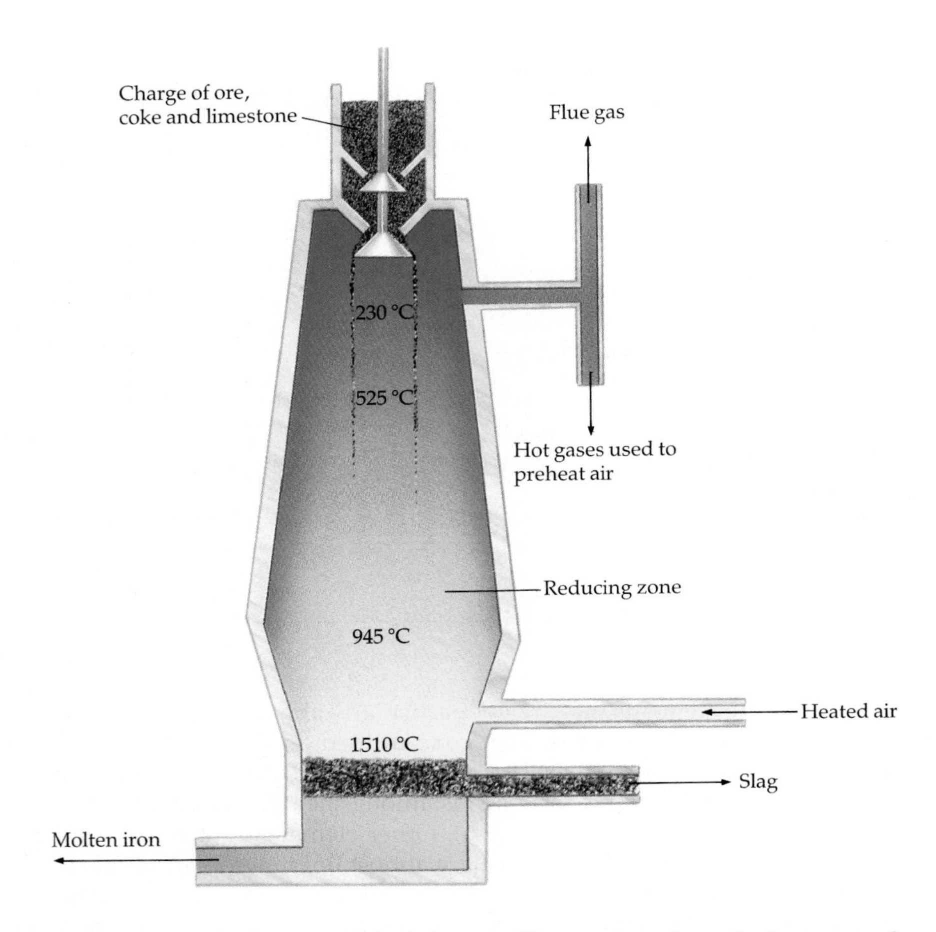

Figure 18.12
Diagram of a blast furnace used for the reduction of iron from iron ore.

the furnace near the bottom. Much heat is liberated as the coke burns, and the heat speeds up the reaction, which is important in making the process economical. The reactions that occur within the blast furnace are

$$2\,C(s) + O_2(g) \longrightarrow 2\,CO(g) + \text{heat}$$
$$Fe_2O_3(s) + 3\,CO(g) \longrightarrow 2\,Fe(s) + 3\,CO_2(g) + \text{heat}$$

Limestone (calcium carbonate) is added to remove the silica (SiO_2) impurity.

$$CaCO_3(s) \xrightarrow{\text{heat}} CaO(s) + CO_2(g)$$
$$CaO(s) + SiO_2(s) \longrightarrow CaSiO_3(s)$$

The calcium silicate, or **slag,** exists as a liquid in the furnace. Consequently, as the blast furnace operates, two molten layers collect in the bottom. The lower, denser layer is mostly liquid iron that contains a fair amount of dissolved carbon and often smaller amounts of other impurities. The upper, lighter layer is primarily molten calcium silicate with some impurities. From time to time the furnace is tapped at the bottom, and the molten iron is drawn off. Another outlet somewhat higher in the blast furnace can be opened to remove the liquid slag.

Iron ore samples. *(Bethlehem Steel Corporation)*

The iron that comes from the blast furnace contains many impurities and is called *pig iron.* Pig iron typically contains up to 4.5% carbon, 1.7%

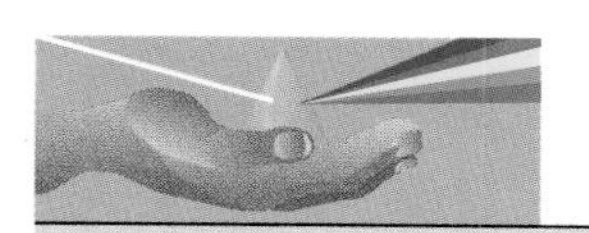

CHEMISTRY YOU CAN DO

Remove Tarnish the Easy Way

Silverware tarnishes when exposed to air because the silver reacts with the hydrogen sulfide gas in the air to form a thin coating of black silver sulfide, Ag_2S. You can chemically remove the tarnish from silverware and other silver utensils by using a solution of baking soda and some aluminum foil. The chemical cleaning of silver is an electrochemical process in which electrons move from aluminum atoms to silver ions in the tarnish, reducing the silver ions to silver atoms while aluminum atoms are oxidized to aluminum ions. The sodium bicarbonate provides a conductive ionic solution for the flow of electrons and also helps to remove the aluminum oxide coating from the surface of the aluminum foil.

Get a large pan. Put 1 to 2 liters of water in the pan. Add 7 to 8 tablespoons of baking soda. Heat the solution, but do not boil it. Place some aluminum foil in the bottom of the pan, and put the tarnished silverware on the aluminum foil. Make sure the silverware is covered with water. Heat the water almost to boiling. After a few minutes remove the silverware and rinse it in running water.

This method of cleaning silverware is better than using polish, because polish removes the silver sulfide, including the silver it contains; instead, the process described here restores the silver from the tarnish to the surface. If you have aluminum pie pans or aluminum cooking pans, you can use them as both the container and the aluminum source.

manganese, 0.3% phosphorus, 0.04% sulfur, and 1% silicon. Iron reacts with the carbon impurity at the temperatures of the blast furnace to form **cementite,** an iron carbide (Fe_3C), which causes pig iron to be brittle.

$$3\ Fe(s) + C(s) \longrightarrow Fe_3C(s) \qquad \Delta H^\circ_{rxn} = 25\ kJ$$

When molten pig iron is poured into molds of a desired shape (for engine blocks, brake drums, transmission housings, and the like), it is called *cast iron.* However, pig iron and cast iron contain too much carbon and other impurities for most uses. A structurally stronger material, known as **steel,** is obtained by removing the phosphorus, sulfur, and silicon impurities and reducing the carbon content to about 1.3%.

Steels

Many alloys of iron with different structural properties are collectively known as *steels.* One of the most common is *carbon steel,* an alloy of iron with about 1.3% carbon. In order to convert pig iron into carbon steel, the excess carbon is burned out with oxygen. There are several techniques for burning the excess carbon; one of the most common is the basic oxygen process (Figure 18.13, next page). In this process pure oxygen is blown into molten iron through a refractory tube (oxygen gun), which is pushed below the surface of the iron. At elevated temperatures, the dissolved carbon reacts very rapidly with the oxygen to give gaseous carbon monoxide and carbon dioxide, which then escape.

A refractory tube is a ceramic tube that can withstand high temperatures without melting.

During the processing of steel, silicon or transition metals (such as chromium, manganese, and nickel) can be added to make alloys having specific physical, chemical, and mechanical properties. Table 18.2 lists the composi-

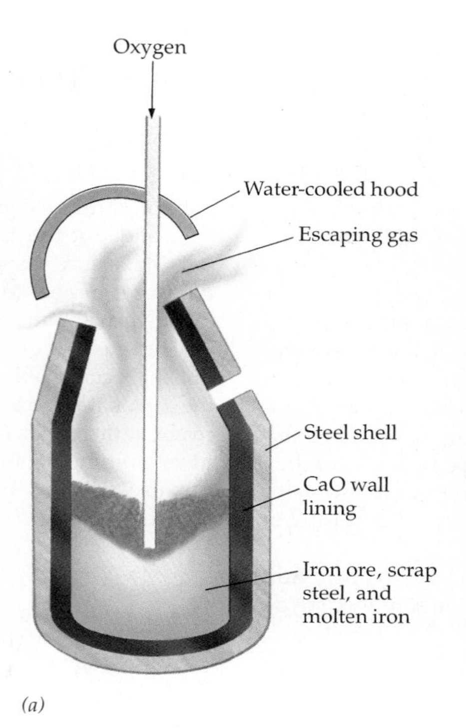

Figure 18.13
(a) The basic oxygen process furnace. Much of the steel manufactured today is produced by blowing oxygen through a furnace charged with scrap and molten iron from a blast furnace. Measured amounts of alloying elements determine the particular steel produced. (b) Molten steel being poured from a basic oxygen furnace. *(b, Bethlehem Steel Corporation)*

tions of some common steel alloys and their uses. Magnetic alloys are solutions of iron and other elements that are permanently magnetic (and are used in loudspeaker magnets and the like) or that can be temporarily magnetized (and so are used in electric motors, generators, and transformers). Alnico V, as its name implies, contains five elements: 8% Al, 14% Ni, and 24% Co, as well as 51% Fe and 3% Cu (Figure 18.14).

The properties of steel can also be affected by the processing temperature and the rate of cooling, as well as by hammering, rolling, or extrusion. If

Figure 18.14
Alnico magnet. *(C.D. Winters)*

Table 18.2

Some Steels and Their Uses

Name	Composition	Properties	Uses
carbon steel	1.3% C, 98.7% Fe	hard	sheet steel, tools
manganese steel	10–18% Mn, 90–82% Fe, 0.5% C	hard, resistant to wear	railroad rails, safes, armor plate
stainless steel	14–18% Cr, 7–9% Ni, 79–73% Fe, 0.2% C	resistant to corrosion	cutlery, instruments
nickel steel	2–4% Ni, 98–96% Fe, 0.5% C	hard, elastic, resistant to corrosion	drive shafts, gears, cables
Invar steel	36% Ni, 64% Fe, 0.5% C	low coefficient of expansion	meter scales, measuring tapes
silicon steel	1–5% Si, 99–95% Fe, 0.5% C	hard, strong, highly magnetic	magnets
Duriron	12–15% Si, 88–85% Fe, 0.85% C	resistant to corrosion, acids	pipes
high-speed steel	14–20% W, 86–80% Fe, 0.5% C	retains temper at high speeds	high-speed cutting tools

the steel is cooled rapidly by quenching the hot steel in water or oil, the carbon in the steel will remain in the form of cementite, Fe_3C, and the steel will be hard, brittle, and light-colored. Since the reaction of iron with carbon to form cementite is endothermic, slow cooling favors the formation of crystals of carbon (graphite) rather than cementite. The resulting steel is more ductile. The properties of the steel can be varied further by rapid cooling followed by controlled reheating to adjust the ratio of cementite to graphite.

All of the processes in steelmaking—from the blast furnace to the final heat treatment—use tremendous quantities of energy, most in the form of heat. In the production of a ton of steel, approximately one ton of coal or its energy equivalent is consumed.

EXAMPLE 18.3

Iron Production

Carbon monoxide is used as a reducing agent to obtain iron from its ores. (a) How much carbon monoxide is required to form 1.00 kg of iron from hematite (Fe_2O_3) and from magnetite (Fe_3O_4)? (b) How much carbon in the form of coke is needed to prepare the total amount of carbon monoxide used in part (a)?

Native copper and two minerals of copper. Azurite [$2CuCO_3 \cdot Cu(OH)_2$] is blue and malachite [$CuCO_3 \cdot Cu(OH)_2$] is green. Based on the colors you may have observed, which mineral is likely to be found on copper statues? *(C.D. Winters)*

SOLUTION (a) The reduction of hematite is represented by the equation

$$Fe_2O_3(s) + 3\,CO(g) \longrightarrow 2\,Fe(\ell) + 3\,CO_2(g)$$

so the mass of CO required to form 1.00 kg of Fe from hematite is

$$\text{mass of CO} = 1.00 \text{ kg Fe} \cdot \frac{1000 \text{ g}}{1 \text{ kg}} \cdot \frac{1 \text{ mol Fe}}{55.85 \text{ g Fe}} \cdot \frac{3 \text{ mol CO}}{2 \text{ mol Fe}} \cdot \frac{28.00 \text{ g CO}}{1 \text{ mol CO}}$$

$$= 752 \text{ g}$$

The reduction of magnetite is represented by the equation

$$Fe_3O_4(s) + 4\,CO(g) \longrightarrow 3\,Fe(\ell) + 4\,CO_2(g)$$

so the mass of CO required to form 1.00 kg of Fe from magnetite is

$$\text{mass of CO} = 1.00 \text{ kg Fe} \cdot \frac{1000 \text{ g}}{1 \text{ kg}} \cdot \frac{1 \text{ mol Fe}}{55.85 \text{ g Fe}} \cdot \frac{4 \text{ mol CO}}{3 \text{ mol Fe}} \cdot \frac{28.00 \text{ g CO}}{1 \text{ mol CO}}$$

$$= 668 \text{ g}$$

(b) The carbon monoxide is prepared by the reaction

$$2\,C(s) + O_2(g) \longrightarrow 2\,CO(g)$$

The total mass of CO needed is 752 g + 668 g = 1420 g. To obtain this mass of CO we need

$$\text{mass of C} = 1420 \text{ g CO} \cdot \frac{1 \text{ mol CO}}{28.00 \text{ g CO}} \cdot \frac{2 \text{ mol C}}{2 \text{ mol CO}} \cdot \frac{12.01 \text{ g C}}{1 \text{ mol C}} = 609 \text{ g C}$$

Copper

Gangue is the unwanted substances mixed with the desired mineral.

Although copper metal occurs in the free state in some parts of the world, the supply available from such sources is quite insufficient for the world's needs. The majority of the copper used today is obtained from various copper sulfide ores, such as $CuFeS_2$ (chalcopyrite), Cu_2S (chalcocite), and CuS (covellite). Because the copper content of these ores is about 1% to 2%, the powdered ore is first concentrated by the flotation process.

In the flotation process, the powdered ore is mixed with water and a frothing agent such as pine oil. A stream of air is blown through the mixture to produce froth (Figure 18.15). The gangue in the ore, which is composed of sand, rock, and clay, is easily wetted by the water and sinks to the bottom of the container. In contrast, a copper sulfide particle is hydrophobic—it is not wetted by the water. The copper sulfide particle becomes coated with oil and is carried to the top of the container in the froth. The froth is removed continuously, and the floating copper sulfide minerals are recovered from it.

Roasting *is a common step in recovery of metals from their ores. It consists of heating the ore, often a metal sulfide, in air to convert the sulfide to the oxide. If the metal oxide product from roasting is less stable than SO_2, the free metal is obtained by sulfide roasting. This is the case for Cu_2S, where further heating of the mixture of Cu_2S and Cu_2O produces Cu and SO_2. Unfortunately, if the SO_2 is vented to the atmosphere, it contributes to air pollution and is one source of "acid rain" (Section 20.5).*

The preparation of copper metal from copper sulfide ore involves **roasting** the ore in air to convert some of the copper sulfide and any iron sulfide present to the oxides:

$$2\,Cu_2S(s) + 3\,O_2(g) \longrightarrow 2\,Cu_2O(s) + 2\,SO_2(g)$$

$$2\,FeS(s) + 3\,O_2(g) \longrightarrow 2\,FeO(s) + 2\,SO_2(g)$$

Swords of Damascus Leave Legacy of Superplastic Steel

ABOUT 3000 YEARS AGO, A TYPE of steel called wootz steel was produced in India by heating a mixture of pure iron ore and wood in a sealed pot or crucible. Some of the carbon reduced the iron ore to metallic iron, which then absorbed some of the remaining carbon to form an excellent steel. The wootz steel later became famous as Damascus steel, used for making swords that retained their sharpness and strength after countless battles. The blacksmiths who forged this steel into swords used a process that they carefully kept secret—one that produced a steel that was much more pliable at high temperatures than normal steel. The knowledge of how to make this special steel was lost sometime in the nineteenth century.

Professor Oleg Sherby of Stanford University and specialists at the Lawrence Livermore National Laboratory have developed types of steels with properties similar to those of Damascus steel. The new steels have a higher percentage of carbon (about 1.8%) than normal carbon steels and are referred to as ultrahigh-carbon steel alloys or superplastic steels.

Unlike most steels, which fail after being stretched to 2 times their original size, superplastic steel at elevated temperatures can be stretched to 11 times its size without cracking or pulling apart. As a result, heated superplastic steel can be formed into complex shapes. It can even flow like molasses and be poured into a mold. This property eliminates the need for machining, which typically results in about 50% scrap. Superplastic steel is also similar to stainless steel in its resistance to corrosion, but is made with less scarce and expensive materials than the nickel and chromium in stainless steel.

The superior properties of superplastic steel are attributed to a much finer grain structure than that in ordinary carbon steels, which are quite brittle. A group of industrial firms has banded together to develop the potential of using superplastic steel in manufactured objects. The first exploitation of the lower manufacturing cost and greater toughness of this material is expected to be in bulldozers, gears, and other complex parts subjected to high stress.

Stanford University Professor Oleg D. Sherby holds a 300-year-old sword of Damascus steel. (For further information, see *Chemecology*, p. 6, March 1992.) (Sam Forencich)

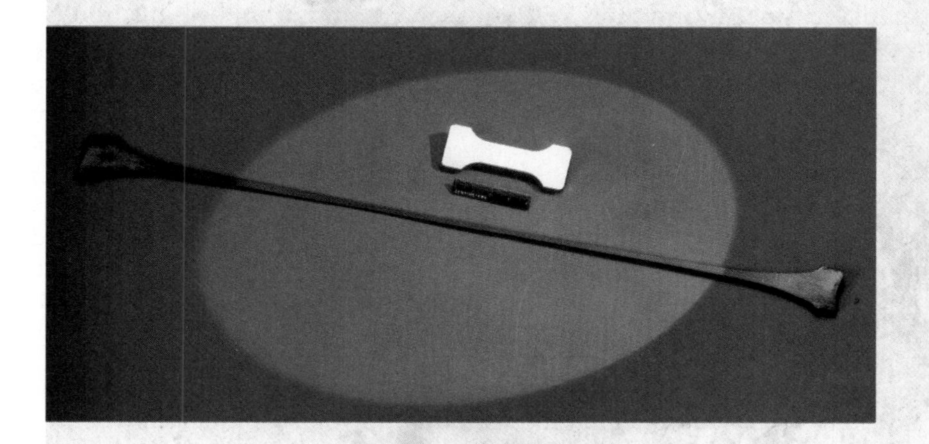

Superplastic, ultrahigh-carbon steel can be stretched to 11 times its size at 900 °C. The original size of the bottom steel piece was 1 inch, but it was pulled to a length of 14 inches. Conventional steels collapse when they are extended to two times their size. (James Stoots/ Livermore Laboratory)

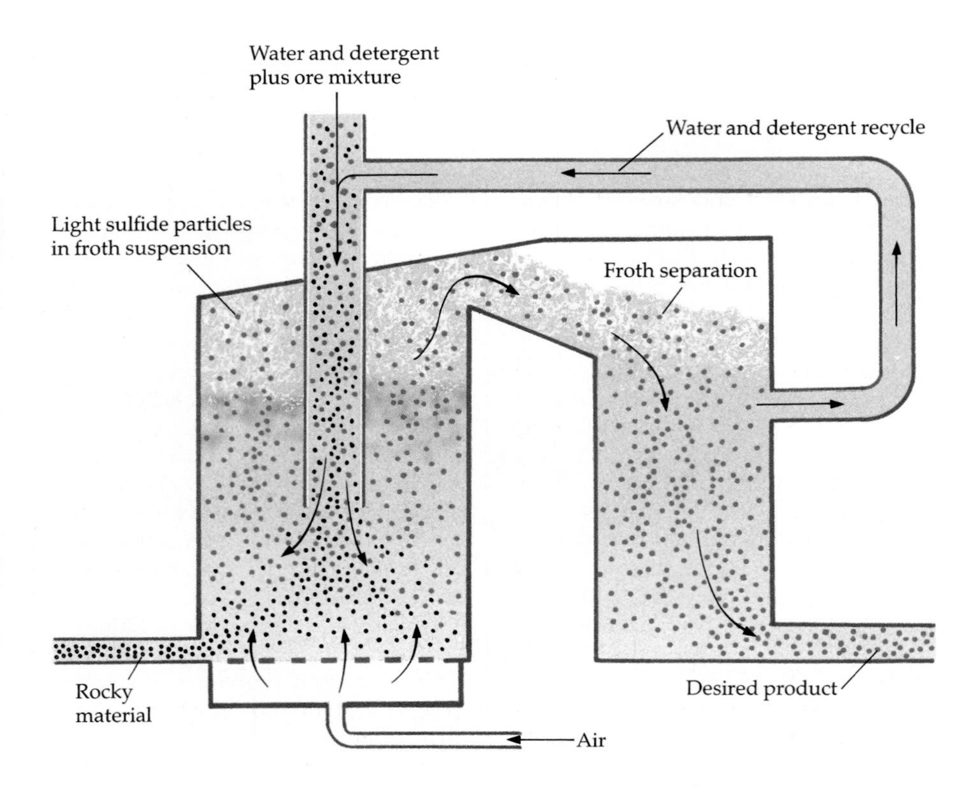

Figure 18.15
The flotation process. The lighter copper sulfide particles are trapped in the bubbles of the foam and are passed over the spillway, while the heavier waste is removed from the bottom of the mixing container. (See Figure 2.20 on p. 35 for a photograph of the flotation process.)

Subsequently the mixture is heated to a higher temperature, and some copper is produced by the reaction

$$Cu_2S(s) + 2\,Cu_2O(s) \xrightarrow{\text{heat}} 6\,Cu(s) + SO_2(g)$$

The product of this operation is a mixture of copper metal and sulfides of copper, iron, other ore constituents, and slag. The molten mixture is heated in a converter with silica materials. When air is blown through the molten material in the converter, two reaction sequences occur. In one, the iron is converted to a slag:

$$2\,FeS(s) + 3\,O_2(g) \longrightarrow 2\,FeO(s) + 2\,SO_2(g)$$

$$FeO(s) + SiO_2 \longrightarrow \underset{\text{molten slag}}{FeSiO_3(s)}$$

In the other, the remaining copper sulfide is converted to copper metal by the two reactions shown above for Cu_2S. The copper produced in this manner is crude or "blister" copper, the blistered surface resulting from the escaping gas. The blister copper is later purified electrolytically.

Open-pit copper mining near Bagdad, Arizona. *(James Cowlin/Image Enterprises)*

In the electrolytic purification of copper, the anodes are crude copper and the cathodes are made of pure copper (Figure 18.16). As electrolysis proceeds, copper is oxidized at the anode, moves through the solution as Cu^{2+} ions, and is deposited on the cathode. The voltage of the cell is regulated so that more active impurities (such as iron) are left in the solution, and less active ones are not oxidized at all. The less active impurities include

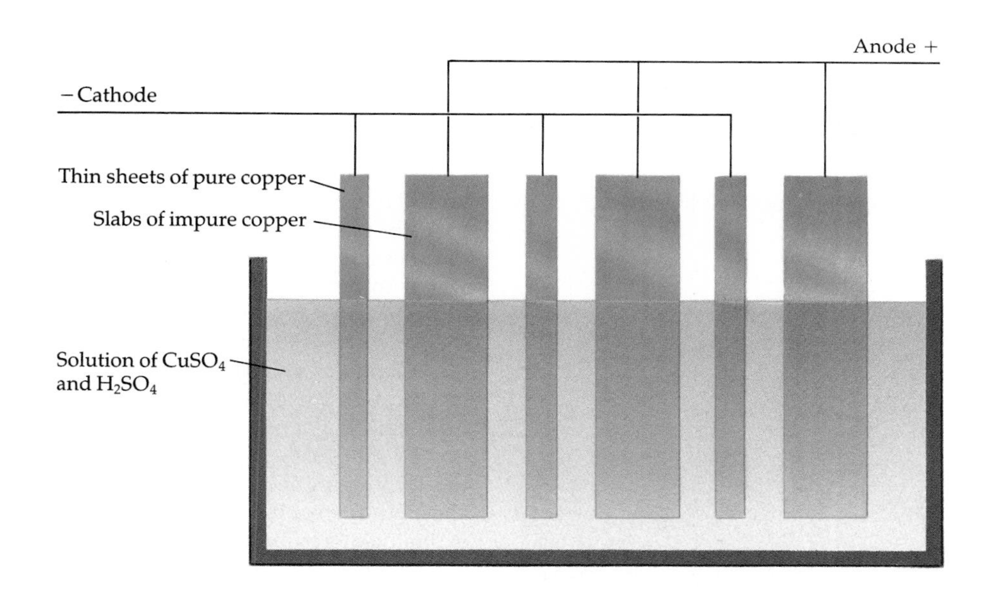

Figure 18.16
Copper from a smelter is purified in an electrolysis cell. The impure copper ore is oxidized at the anode and passes into solution. The oxidized copper is reduced from the solution onto the cathode as 99.95% pure copper.

gold and silver, which collect as "anode slime," an insoluble residue beneath the anode. The anode slime is subsequently treated to recover the valuable metals.

The copper produced by the electrolytic cell is 99.95% pure and is suitable for use as an electrical conductor. Copper for this purpose must be pure because very small amounts of impurities, such as arsenic, considerably reduce the electrical conductivity of copper.

Rack of copper-plated sheets being removed from an electroplating bath during the first stage in the preparation of flexible electronic circuits. Based on a polyester laminate, flexible circuits are typically used in circumstances where the use of the normal rigid printed circuit board is impossible—in the nose-cone of a missile, for instance. *(Simon Fraser/Northumbria Circuits/Photo Researchers, Inc.)*

EXAMPLE 18.4

Copper Production

What mass of pure copper is deposited in an electrolytic cell operating at 300. A for 10.0 hours?

SOLUTION

$$\text{charge} = 300.\ \text{A} \cdot 10.0\ \text{hr} \cdot \frac{3600\ \text{s}}{\text{hr}} = 1.08 \times 10^7\ \text{C}$$

$$\text{moles of e}^- = 1.08 \times 10^7\ \text{C} \cdot \frac{1\ \text{mol e}^-}{9.65 \times 10^4\ \text{C}} = 112\ \text{mol e}^-$$

$$\text{mass of Cu} = 112\ \text{mol e}^- \cdot \frac{1\ \text{mol Cu}}{2\ \text{mol e}^-} \cdot \frac{63.55\ \text{g Cu}}{1\ \text{mol Cu}} = 3.56 \times 10^3\ \text{g Cu}$$

EXERCISE 18.5 • *Copper Production*

If an electrolytic cell for the refinement of copper operates at 200. A for 24 hours a day at 90.0% efficiency, what mass in kilograms of pure copper is produced in a year?

The Chemical World

Materials from the Earth

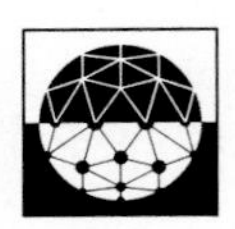

In the South African bush veldt, an area the size of New England, interesting and valuable natural chemical separations have been made. Just beneath the surface lie some of the world's richest deposits of platinum, rhodium, and chromium. A mineral is a naturally occurring substance with a characteristic chemical composition. When minerals are concentrated and have economic value, they are called ores. Some places are rich with valuable ores, like South Africa, while others are not, and the fortunes of whole countries can rise and fall based on the wealth found in the natural treasuries.

Since prehistoric times, we have recovered and used a variety of minerals. How did our ancestors obtain the minerals and elements they needed from the earth? One of the techniques used in an ancient tin mine in Cornwall, England, was to build large fires at the base of the rock cliffs. The intense heat cracked the rocks, exposing the tin ore, which was then removed. Today we use dynamite and ammonium nitrate to blast rock away from ore-rich veins. But although we have greatly increased our power to remove useful ores from the earth, we are still limited by effective and economical mining techniques and, to an ever increasing degree, the availability of exploitable ore deposits.

The World of Chemistry (Program 18), "The Chemistry of Earth."

Ancient tin mines at Cornwall, England. (*The World of Chemistry*, Program 18, "The Chemistry of the Earth.")

18.3 THE FRACTIONATION OF AIR

The composition of dry air at sea level is given in Table 20.1.

The atmosphere of the earth is a fantastically large source of the elements nitrogen and oxygen and much smaller amounts of certain of the noble gases, including argon, neon, and xenon.

Before pure oxygen and nitrogen can be obtained from the air, water vapor and carbon dioxide must be removed. This is usually done by precooling the air by refrigeration or by using silica gel to absorb water and lime to absorb carbon dioxide. Afterward, the air is compressed to a pressure exceeding 100 times normal atmospheric pressure, cooled to room temperature, and allowed to expand into a chamber. This expansion produces a cooling effect (the **Joule-Thompson effect**) due to disruption of the intermolecular forces between the molecules. Recall that overcoming intermolecular forces requires energy, so the expanding gas absorbs kinetic energy from the motion of its own molecules, which cools the gas. If this expansion is repeated and controlled properly, the expanding air cools to the point of liquefaction (Figure 18.17). The temperature of the liquid air is usually well below the normal boiling points of nitrogen (−195.8 °C), oxygen (−183 °C), and argon (−189 °C). This liquid air is then allowed to vaporize partially again, and since N_2 is more volatile than O_2 or Ar (N_2 has a lower boiling point), the liquid becomes more concentrated in O_2 and Ar. This process, known as the **Linde process,** produces high-purity nitrogen (99.5+ %) and

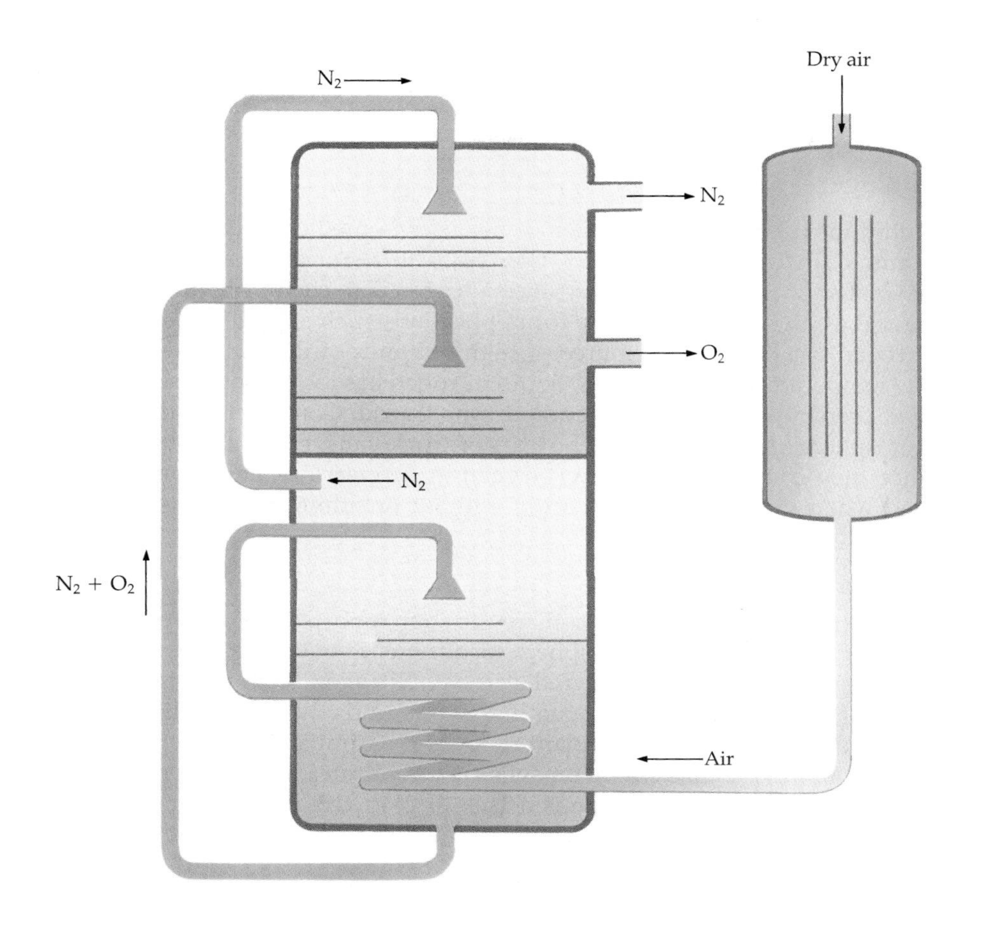

Figure 18.17
Diagram of a column for the fractional distillation of liquid air.

oxygen with a purity of 99.5%. Further processing produces pure Ar and Ne (b.p. −246 °C).

A 1983 U.S. postage stamp commemorating the 250th anniversary of the birth of Priestley, (1733–1804), who discovered elemental oxygen in 1774. He used sunlight and a magnifying glass to heat an oxide of mercury that released oxygen gas. Priestley left his native England for religious reasons to spend the last ten years of his life in Northumberland, Pennsylvania. *[Stamp from the private collection of Professor C. M. Lang, photography by Gary J. Shulfer, University of Wisconsin–Stevens Point. "United States #2038 (1983)"; Scott Standard Postage Stamp Catalogue, Scott Pub. Co., Sidney, Ohio.]*

Oxygen

Most oxygen produced by the fractionation of liquid air is used in steelmaking, although some is used in rocket propulsion (to oxidize hydrogen) and in controlled oxidation reactions of other types. Liquid oxygen (LOX) can be shipped and stored at its boiling temperature of −183 °C under atmospheric pressure. Substances this cold are called **cryogens** (from Greek *kryos*, meaning "icy cold"). Cryogens represent special hazards since contact produces instantaneous frostbite, and structural materials such as plastics, rubber gaskets, and some metals become brittle and fracture easily at these low temperatures. Liquid oxygen can accelerate oxidation reactions to the point of explosion because of the high oxygen concentration. For this reason, contact between liquid oxygen and substances that will ignite and burn in air must be prevented.

Special cryogenic containers holding liquid oxygen incorporate huge vacuum-walled bottles much like those used to carry hot soup or hot coffee.

Liquid oxygen tank. *(C.D. Winters)*

These containers can be seen outside hospitals or industrial complexes, on highways and railroads, and even aboard ocean-going vessels.

Nitrogen

Liquid nitrogen is also a cryogen. It has uses in medicine (cryosurgery), for example, in cooling an area of skin prior to removal of a wart or other unwanted or pathogenic tissue. Since nitrogen is so chemically unreactive, it is used as an inert atmosphere for applications such as welding, and liquid nitrogen is a convenient source of high volumes of the gas. Because of its low temperature and inertness, liquid nitrogen has found wide use in frozen food preparation and preservation during transit. Containers with nitrogen atmospheres, such as railroad boxcars or truck vans, present health hazards since they contain little (if any) oxygen to support life, and workers have died when they entered such areas without breathing apparatus.

Nitrogen Fixation

Some nitrogen species ranked in terms of decreasing oxidation number: NO_3^-, NO_2, NO_2^-, NO, N_2O, N_2, NH_4^+.

The **primary nutrients** of the soil are nitrogen, phosphorus, and potassium. Although bathed in an atmosphere of nitrogen, most plants are unable to use the air as a supply of this vital element. **Nitrogen fixation** is the process of changing atmospheric nitrogen into compounds that can be dissolved in water, absorbed through the plant roots, and assimilated by the plant (Figure 18.18). Most plants thrive on soils rich in nitrates, but many plants that grow in swamps, where there is a lack of oxidized materials, can use reduced forms of nitrogen such as the ammonium ion. The nitrate ion is the

Liquid nitrogen. *(C.D. Winters)*

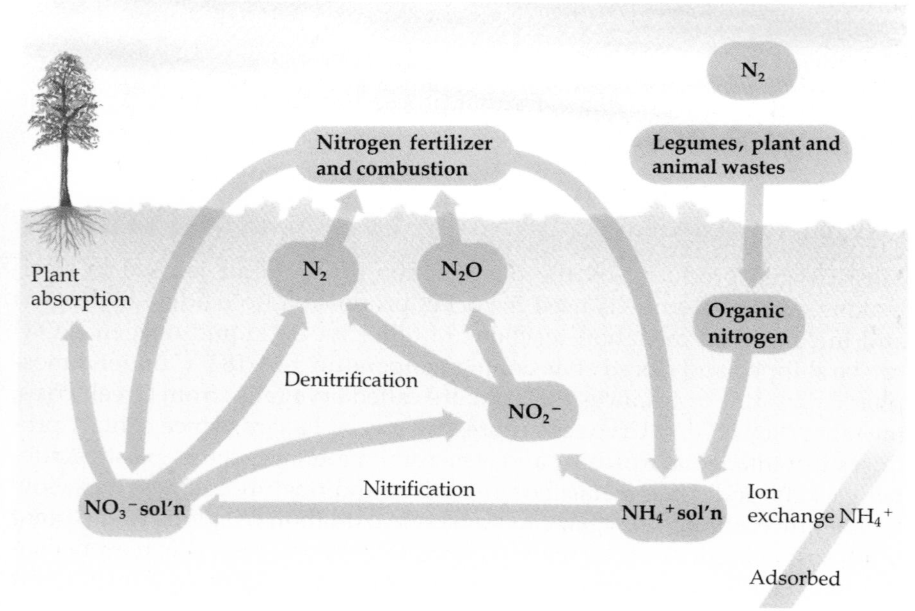

Figure 18.18
Nitrogen pathways through the soil.

The Chemical World

Agriculture

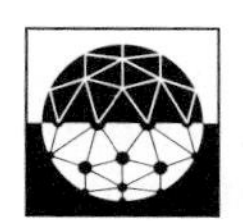

When any covalent bond forms, energy is released. In turn the same amount of energy is needed in order to break a bond.

Consider nitrogen. There is nitrogen in all living things; muscles, hair, and DNA all contain nitrogen bonded to other elements. But 80% of the atmosphere consists of nitrogen molecules held together by strong triple bonds. How do living things get the form of nitrogen they need? Lightning helps. The electrical flash in the sky has enough energy to break apart nitrogen molecules, which then react with oxygen in the air, eventually forming nitric acid. The natural acid dissolves in rain and falls to earth as a dilute solution. There it is absorbed and metabolized by plants.

Some plants, though, convert molecular nitrogen in a different way. Soybeans, peas, peanuts, and other legumes host a unique bacterium in their roots. It is this bacterium that converts the nitrogen molecule into a nitrogen compound, ammonia, which the plant can then use to make amino acids.

Exactly how the bacterium works is the subject of vigorous research. Don Keister, of the U.S. Department of Agriculture, claims, "This is one of the very unique enzymes in all of nature, because it is the only solution that nature has evolved for biologically reducing nitrogen."

The soybean and the bacterium have a symbiotic relationship. The plant houses and feeds the bacterium and, in turn, it receives the nitrogen it needs. But not all plants can host these nitrogen fixers. They have to rely on rain, natural fertilizers, or expensive manufactured fertilizers, such as ammonium nitrate.

Keister goes on to point out, "We are currently using something like 300 million barrels of oil per year in the United States alone to produce nitrogen fertilizers. We forget sometimes that we're going to need to double the food supply over the next 20 years." He further asks the unanswered questions, "Where is that energy going to come from? Where is the fertilizer going to come from?"

For feeding the world, there are two basic options: We can either produce more fertilizer at greater cost and some risk to the environment, or we can create new varieties of nitrogen-fixing plants. Both options are being pursued worldwide.

The World of Chemistry (Program 8), "Chemical Bonds."

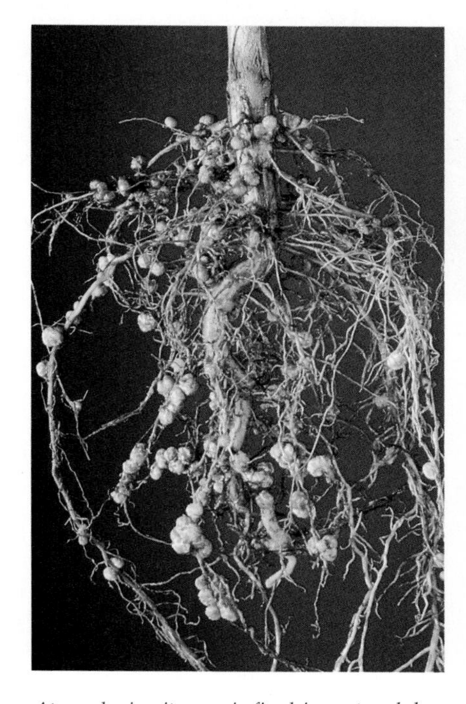

Atmospheric nitrogen is fixed in root nodules on leguminous plants such as the soybean pictured here. Rather than requiring fertilization for nitrogen, such plants can be grown to enrich the soil with nitrogen compounds and produce a valuable crop at the same time. (David M. Dennis/Tom Stack & Associates)

most highly oxidized form of combined nitrogen, and the ammonium ion is the most reduced form of nitrogen.

Nature fixes nitrogen on a massive scale in two ways. In the first method, nitrogen is oxidized under highly energetic conditions, such as in the discharge of lightning (or, to a lesser extent, in a fire). The initial reaction is the reaction of nitrogen and oxygen to form nitrogen monoxide, NO:

$$N_2(g) + O_2(g) \longrightarrow 2\,NO(g)$$

Nitrogen monoxide is easily oxidized in air to nitrogen dioxide, NO_2, which dissolves in water to form nitrous acid, HNO_2, and nitric acid, HNO_3:

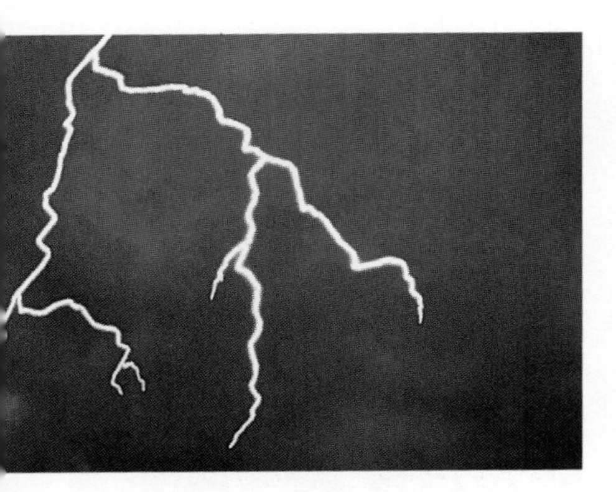

A vast amount of atmospheric nitrogen is fixed as a result of natural electric discharges in the atmosphere. The energy in a bolt of lightning is sufficient to break the very stable triple bond in a nitrogen molecule. Oxidation of nitrogen results. *(Patrick Eden/The Image Bank)*

$$2\,NO(g) + O_2(g) \longrightarrow 2\,NO_2(g)$$

$$H_2O(\ell) + 2\,NO_2(g) \longrightarrow HNO_2(aq) + HNO_3(aq)$$

Nitric acid is readily soluble in rain, clouds, or ground moisture and thus increases nitrate concentration in soil.

In the second method, bacteria that live in the roots of plants called *legumes* convert atmospheric nitrogen into ammonia. This complex series of reactions depends on enzyme catalysis. Under ideal conditions, legume fixation can add more than 100 pounds of nitrogen per acre of soil in one growing season.

Noble Gases

Approximately 250,000 tons of argon, the most abundant noble gas in the air, are isolated each year in the United States. Most of the argon is used to provide inert atmospheres for high-temperature metallurgical processes. If air is not excluded from these processes, unwanted oxides of the hot metals would form. Argon is also used as a filler gas in incandescent light bulbs in order to prolong the life of the hot filament.

Relatively small amounts of neon, krypton, and xenon are recovered from air for commercial purposes. While neon is used in the greatest amounts, all three are used in "neon" signs. Xenon, which is readily soluble in blood, acts as an inhalation anesthetic in much the same way as does laughing gas, dinitrogen monoxide (N_2O).

Helium is obtained from natural gas wells where it is present in concentrations up to 7% by volume of the natural gas. Helium is used for inert atmospheres, especially in welding. Liquid helium, boiling point −271 °C, is used as a refrigerant when extremely cold temperatures are required. Helium/oxygen mixtures are used in place of air for deep-sea divers since helium is less soluble in blood than nitrogen. If air were used under high pressure for breathing by deep-sea divers, the nitrogen would dissolve in their blood and, on return to lower pressure, bubble out of the blood and cause the fatal condition called "the bends."

Tank of liquid helium. *(C.D. Winters)*

18.4 SEABED MINERALS—A VAST POTENTIAL RESOURCE

Minerals and ores result from the natural separation processes occurring in the crust of the earth. The separation processes under the ocean waters, where hot and cold solids interact with hot and cold liquids, are very different from those on the continents where the air and weathering processes greatly modify the materials thrust up from the magma under the crust. Since the world is mostly covered with water, the amounts of chemicals under the seas are relatively great even though the earth's crust is thinner under the oceans than under the continents.

In the deep sea, extensive deposits of high quality minerals have been found around the vents in the developing undersea ridges and mountains.

The next island that will surface in Hawaii, known as Loihi, is now within 980 meters of the surface. Low-temperature vents (30 °C), at depths of 980 meters to 1360 meters, are surrounded with volcanic cones of iron oxides. The vents are often called "black smokers" because the hot, mineral-laden water coming from the vents produces a dark convection steam. Solid material carried along in the stream tends to settle around the vent, as do solids from the crystallization of dissolved minerals as the hot water cools. A second class of vents along the faults and fissures of the steep slopes of the mountain have developed iron oxide deposits as thick as 30 meters. A third class of vents inside the developing craters produce rich deposits of iron, manganese, cobalt, copper, chromium, zinc, nickel, and gold minerals.

See "Black Smoker" box, Section 5.6

Large hydrothermal fields occur in the deep sea where the tectonic plates of the crust are separating and allowing seawater to contact the underlying volcanic rocks. Water temperatures as high as 500 °C have been measured in hydrothermal fields containing mineral deposits of copper, iron, lead, and zinc. Theoretical computer models of the separation of these minerals show conduits of superheated water (at about 350 °C) as large as 200 meters in diameter moving up from just above the liquid magma to the ocean floor. Such models can account for as much as 50% of the zinc and 90% of the copper being extracted from the crust material.

A tectonic plate in the surface of the earth is a solid mass moving relative to an adjacent mass. Interfaces between plates form the earthquake fault lines.

Sedimentary deposits of minerals do not appear to be extensive over the deep sea floor, probably because of the fluid nature of the medium and the great ability of water to dissolve most minerals. However, nodules of minerals are found on some areas of the ocean floor; an example is the extensive field of manganese nodules in the Pacific Ocean.

In the continental coastal shelf regions, there are often mineral deposits that are extensions of the deposits in the nearby land areas. For example, as the phosphate ores in North Carolina and Florida are exhausted in the near future, further consideration of the mining of the nearby underwater phosphate beds may be necessary.

Except for oil and gas, it is still an open question whether there are enough ores of sufficient quality and accessibility to justify the development of the seabed as a source of needed materials until the land sources are much more depleted than they are now.

IN CLOSING

Having studied this chapter, you should be able to

- describe electrochemical methods for the production of sodium, chlorine, sodium hydroxide, fluorine, aluminum, and magnesium. (Section 18.1)
- calculate the quantities of electrical energy used and products produced in electrolysis reactions. (Section 18.1)
- explain the process and chemistry of iron production. (Section 18.2)
- outline the steps in the production of copper. (Section 18.2)
- apply stoichometry to industrial-scale processes. (Sections 18.1 and 18.2)
- identify the products of fractionation of air and some of their uses. (Section 18.3)

STUDY QUESTIONS

Review Questions

1. Why is CaO necessary for the production of iron in a blast furnace?
2. What is the primary reducing agent in the production of iron from its ore? Write a balanced equation for the reduction process.
3. What chemical is obtained from oyster shells in the production of magnesium from sea water? What is the role of this chemical in the process?
4. Write balanced equations for the recovery of magnesium from sea water, beginning with the precipitation of magnesium hydroxide by addition of lime to sea water.
5. Explain the difference between *pig iron* and *carbon steel.*
6. Write equations for the roasting and reduction of Cu_2S ore.
7. Why is it important to purify industrial quantities of copper electrolytically to a level about 99.95% pure?
8. What is the Joule-Thompson effect and why is it important in the commercial production of nitrogen and oxygen?
9. Why are nitrogen and oxygen important commercial chemicals?
10. How does nature fix nitrogen in the atmosphere? In the soil? Why is nitrogen fixation necessary?
11. Discuss the relative advantages and disadvantages of membrane and mercury cells for chlor-alkali electrolysis.
12. Briefly explain why different products are obtained from the electrolysis of molten NaCl and the electrolysis of brine solution.
13. Explain why each of the following materials is needed in the blast furnace reduction of iron ore to iron: (a) air, (b) limestone, (c) coke. Include balanced chemical equations to illustrate your answers.
14. Explain how the properties of steel are affected by the rate of cooling of the molten steel.

Electrolysis

15. The electrolysis of aqueous NaCl gives NaOH, Cl_2, and H_2.
   (a) Write a balanced equation for the process.
   (b) In 1991 in the United States, 24.4 billion pounds of NaOH and 22.6 billion pounds of Cl_2 were produced. Does the ratio of masses of NaOH and Cl_2 produced agree with the ratio of masses expected from the balanced equation? If not, what does this tell you about the ways in which NaOH and Cl_2 are actually produced? Is the electrolysis of aqueous NaCl the only source of these chemicals?

16. (a) When 1000. kg of molten $MgCl_2$ is electrolyzed to produce magnesium what mass in kilograms of metal is produced at the cathode?
   (b) What is produced at the anode? What mass of the other product is produced?
   (c) What is the total number of faradays of electricity used in the process?
   (d) One industrial process has an energy consumption of 8.4 kwh per pound of Mg. How much energy in joules is therefore required per mole?

17. An electrolysis cell for aluminum production operates at 5.0 V and 1.0×10^5 A. How long will it take to produce 1.0 ton of aluminum?

18. Electrolysis of molten NaCl was carried out in a Downs cell operating at 7.0 V and 4.0×10^4 A.
   (a) How much Na(s) and Cl_2(g) can be produced in 24 hours in such a cell?
   (b) What is the energy consumption in kilowatt-hours? (Assume 100% efficiency.)

19. Typical electrolysis cells for the production of F_2 operate at 4.0×10^3 to 6.0×10^3 A and 8.0 to 12. V; a large-scale plant can produce about 9 tons of liquefied fluorine per day.
   (a) How much energy in kilowatt-hours is consumed by a cell operating at 6.0×10^3 A and 12 V for 24 hours?
   (b) What mass in grams of HF is consumed, and what masses of H_2 and F_2 are produced, in a cell under these conditions?

20. In the first step of recovery of copper, an ore such as chalcocite, Cu_2S, is roasted in air to give Cu_2O. If you begin with 1.0 ton of the ore, what mass of SO_2 is produced?

21. What current in amperes is required to produce 1.00×10^3 kg of Cl_2 per day in a chlor-alkali cell operating at 90.0% efficiency?

22. How much energy in kilowatt-hours of electricity is required to prepare a ton of sodium in a typical Downs cell operating at 25,000 A and 7.0 V?

23. How much aluminum can be produced when 6.0×10^4 A is passed through a series of 100 Hall-Héroult electrolytic cells operating at 85% efficiency for 24 hours?

24. What mass in grams of aluminum can be produced

from the electrolysis of molten $AlCl_3$ in an electrolytic cell operating at 100. A for 2.00 hours?

25. (a) Write the electrode reactions for the electrolysis of molten $CaCl_2$.
(b) What mass in grams of calcium metal can be produced by an electrolytic cell operating at 50. A for 45 minutes?

26. How much current is needed in a Hall-Héroult electrolytic cell to produce 1.0 ton of aluminum in 24 hours?

General Questions

27. Electrolysis of molten NaCl is carried out in a Downs cell operating at 1.00×10^4 A for 24 hours.
(a) How much sodium is produced?
(b) What volume of Cl_2 in liters will be produced? Assume the Cl_2 is collected from the outlet tube at 20 °C and 15 atm pressure.

28. Bauxite, the principal source of aluminum oxide, contains about 55% Al_2O_3. How much bauxite is required to produce the 5.0 million tons of aluminum metal produced each year by electrolysis?

29. How long will it take an electric current of 1.50 A to deposit all the copper from 1.25 L of 0.225 M $CuSO_4(aq)$?

30. $Ca(OH)_2$ has a K_{sp} of 7.9×10^{-6}, whereas that for $Mg(OH)_2$ is 1.5×10^{-11}. Calculate the equilibrium constant for the reaction

$$Ca(OH)_2(s) + Mg^{2+}(aq) \longrightarrow Ca^{2+}(aq) + Mg(OH)_2(s)$$

and explain why this reaction can be used in the commercial isolation of magnesium from sea water.

Using Computer Programs and Videodiscs

31. Using *KC? Discoverer* and the *Periodic Table* videodisc, list the uses of the elements described in this chapter: sodium, chlorine, fluorine, aluminum, magnesium, iron, copper, oxygen, and nitrogen. Do any of these elements appear in the "top 25" list of chemicals inside the back cover of this book?

32. Use *KC? Discoverer* to list the elements described in this chapter (see the Study Question above) in order of increasing cost. What is the most expensive of these elements?

33. The "thermite reaction" is illustrated on the *Chemical Demonstrations* videodisc (Disc 2, frame 10). Write a balanced equation to describe the reaction. What observations can you make about this reaction? That is, what evidence is there that a reaction has occurred? Is the reaction product- or reactant-favored? Can you prove this by appropriate calculations? Explain why this reaction is not a practical way to make iron in large quantities.

34. Copper metal can be produced by the reaction of hydrogen gas with copper(II) oxide (see the *Chemical Demonstrations* videodisc, Disc 1, frame 31918).
(a) What observations can you make about the reaction that occurs when hot copper metal, coated with copper(II) oxide, is exposed to hydrogen gas? That is, what evidence is there that a reaction has occurred?
(b) Write a balanced chemical equation for the reaction between copper(II) oxide and hydrogen gas. Identify the oxidizing and reducing agents and the substance oxidized and the substance reduced.
(c) What evidence for reaction do you see when pure copper metal is heated in air? Write a balanced equation for this reaction and identify the oxidizing and reducing agents.

A home smoke alarm uses radioactive americium in the detection circuit. **See page 825.**

(C.D. Winters)

Nuclear Chemistry

CHAPTER

19

A recent report issued by the National Research Council stated that "The future vigor and prosperity of American medicine, science, technology, and national defense clearly depend on continued use and development of nuclear techniques and use of radioactive isotopes." Nuclear chemistry, a subject that bridges chemistry and physics, has a significant impact on our society. Radioactive isotopes are now widely used in medicine, and some of the latest diagnostic techniques such as PET scans depend on radioactivity. Similarly, your home may be protected with a smoke detector that contains a radioactive element, and research in all fields of science uses radioactive elements and their compounds. The national security of the United States since World War II has depended on nuclear weapons, and a number of nations around the world depend on nuclear reactors as a source of electricity. No matter what your reason for taking a college course in chemistry—to prepare for a career in one of the sciences or simply to gain knowledge as a concerned citizen—you should know something about nuclear chemistry. Therefore, this chapter considers changes in the atomic nucleus and their effects, the fissioning and fusion of nuclei and the energy that can be derived from such changes, the units used to measure radioactivity, and the uses of radioactive isotopes.

On August 2, 1939, as the world was on the brink of World War II, Albert Einstein sent a letter to President Franklin D. Roosevelt. In this letter, which profoundly changed the course of history, Einstein called attention to work being done on the physics of the atomic nucleus. He said he and others believed this work suggested the possibility that "uranium may be turned into a new and important source of energy . . . and [that it was] conceivable . . . that extremely powerful bombs of a new type may thus be constructed . . . ."

The Making of the Atomic Bomb *by Richard Rhodes (Simon and Schuster, 1986) is a comprehensive history of the exploration of atomic physics in this century and of the events leading up to the development of atomic weapons. This very readable book is highly recommended.*

Powerful indeed! Einstein's letter was the beginning of the Manhattan Project, the project that led to the detonation of the first atomic bomb at 5:30 AM on July 16, 1945, in the desert of New Mexico. The rest of the world would learn the truth of the power locked in the atomic nucleus a few weeks later, on August 6 and August 9, when the United States used atomic weapons against Japan. J. Robert Oppenheimer, the director of the atomic bomb project, is said to have recalled the following words from the sacred Hindu epic, Bhagavad-Gita, at the moment of the explosion of the first atomic bomb.

"If the radiance of a thousand suns
Were to burst at once upon the sky,
That would be like the splendor of the Mighty One . . .
I am become Death,
The shatterer of worlds."

In the almost 50 years since the first—and thankfully only—use of an atomic weapon in war, more powerful weapons have been developed and stockpiled by a number of nations. With the end of the Cold War, fears of a nuclear holocaust are fading, but they are replaced to some extent by the concern that Third World nations have developed nuclear weapons. The respected magazine *The Bulletin of Atomic Scientists* has used for many years the symbol of a clock with its hands near midnight, illustrating the danger faced by the world from atomic weapons. Even with the end of the Cold War, the hands have moved back only a little.

Although nuclear reactions have been associated in the public consciousness with weapons for more than 40 years, radioactivity has been continuously developed for the generation of electric power, for the diagnosis and treatment of disease, and for food preservation.

19.1 THE NATURE OF RADIOACTIVITY

Becquerel, Marie and Pierre Curie, and Rutherford

Many minerals, called phosphors, glow for some time after being stimulated by exposure to sunlight or ultraviolet light. (You may have some on the hands of a wristwatch or clock.) In 1896, French physicist Henri Becquerel was studying this phenomenon, called *phosphorescence,* when he accidentally discovered radioactivity.

For more on the experiments done by Becquerel and the Curies, see H.F. Walton, Journal of Chemical Education, *Vol. 69, p. 10, 1992.*

The German physicist Wilhelm Röntgen discovered x rays in November, 1895, and Becquerel was intrigued by a possible connection between x

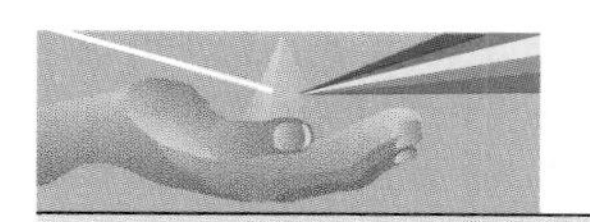

CHEMISTRY YOU CAN DO

Detecting Radioactive Isotopes

Some home smoke alarms use the disintegration of a radioactive element as a way to detect smoke particles in the air. As sketched below, a weak radioactive source in the smoke alarm ionizes the air, thus setting up a small current in an electrical circuit. If smoke is present, the ions become attached to the smoke particles. The slower movement of the heavier, charged smoke particles reduces the current in the circuit and sets off the alarm.

Go to a hardware store or other place that sells household appliances, find a smoke alarm, and get some information about the device. Find out the identity of the radioactive element used in the alarm.

There is some concern in certain areas of the United States that the radioactive gas radon can collect in the basements of homes (see Section 19.8). While you are in the hardware store, see if you can find a home radon detector. Answer the following questions with information that is on or in the box.

- What is the source of radon?
- How does it enter your home?
- What are the health hazards of radon?
- Is radon a health hazard outdoors?

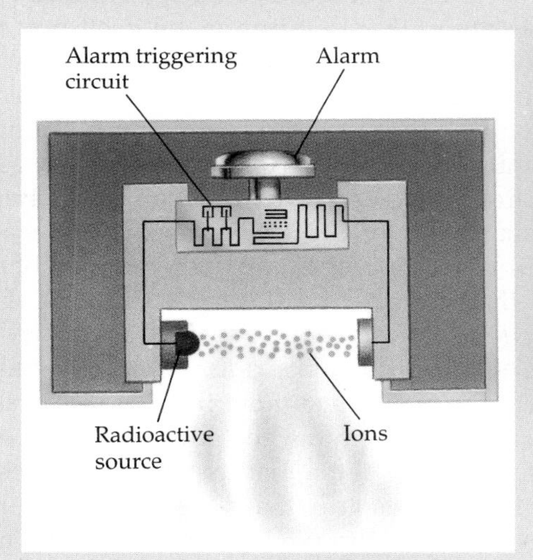

rays and phosphorescence. To test this idea he decided to use uranium salts, which he knew were phosphorescent. He tightly wrapped a photographic plate in paper, placed a uranium salt on the paper, and exposed the salt to sunlight. Becquerel knew that x rays will darken a photographic plate, so he believed that, if the phosphorescence of the salt after it was exposed to sunlight included x rays, he should see an image of the uranium salt after the plate had been developed. The experiment was a success! But it misled him. He wanted to duplicate his first experiment, but Paris was very cloudy in February 1896, so he put the wrapped plate away in a drawer until the sun reappeared. Much to his amazement, the image of the uranium salt appeared on the plate that had been in the dark drawer. He quickly realized that he had observed penetrating radiation from matter that had not been stimulated by light.

Figure 19.1
Marie and Pierre Curie and their daughter Irene, in the garden of their home near Paris. The Curies and H.A. Becquerel shared the Nobel Prize in physics in 1903 for their work on radioactivity. Marie Curie received a second Nobel Prize, this in chemistry in 1911 for the discovery of radium and polonium. The latter element was named in honor of her homeland, Poland. Pierre Curie also did important work on magnetic phenomena. He discovered the piezoelectric effect, an effect that is used to generate the sound in a digital watch, for example. *(Mutter Museum, Philadelphia College of Physicians)*

Becquerel performed many more experiments on the radiation from uranium and its salts, completing them in 1897. The results showed that the emissions were undiminished after many months in the dark, and so the radiation could not have been stimulated by sunlight. Further, pure uranium metal gave the same emissions, but even more strongly. This would be expected if the radiation was the property of the metal and not dependent on its form of chemical combination. But no pure metal had been observed to phosphoresce before. Becquerel was completely baffled. Where did all of this energy come from? Becquerel gave up his work for several years, but it was taken up by Marie Curie and her husband Pierre, who later named the phenomenon **radioactivity** (Figure 19.1).

Marie Curie was a chemist at the École Supérieure de Physique et de Chimie de Paris. As a chemist she was interested not so much in the origin of the energy of the radiation but in whether other elements could be radioactive. After Pierre Curie had developed a highly sensitive way to detect the radiation from a radioactive source, Marie Curie tested every substance she could find in order to answer this question. One of her first findings was to confirm Becquerel's observation that uranium metal itself was radioactive, and that the degree to which a uranium-bearing sample was radioactive depended on the percentage of uranium present. Thus, she was astonished when she tested pitchblende, a common ore containing uranium and other metals (such as lead, bismuth, and copper), and found that it was even more radioactive than pure uranium. There was only one explanation: pitchblende contained an element (or elements) more radioactive than uranium.

In 1898 the Curies published their work on the separation of pitchblende into the various metals that it contained. Using techniques of qualitative analysis that are still in use today in the introductory chemistry laboratory, she separated a sample of bismuth hydroxide, $Bi(OH)_3$, from the pitchblende. This was converted to bismuth sulfide, Bi_2S_3, which was dried and heated in a vacuum. A vapor rose from the heated sulfide and condensed as a black film in a cooler portion of the apparatus. The radioactivity of the sample was there, and the Curies stated that "We therefore think that the substance that we have extracted from pitchblende contains a metal previously unknown, a neighbor of bismuth in its analytical properties. If the existence of the new metal is confirmed, we propose to call it *polonium*, after the native land of one of us." They had discovered the next element after bismuth in the periodic table, and further investigation of pitchblende by the Curies uncovered another new, highly radioactive element, radium.

To study radioactivity further, the Curies needed to isolate larger samples of radium, but they needed much larger amounts of pitchblende. The Austrian government soon sent them a ton of the mineral and, after months of back-breaking labor, they isolated 100 milligrams of pure radium!

Encouraged by the Curies, Becquerel returned to the study of radiation. He found that the radiation from uranium was affected by magnetic fields and consisted of two kinds of particles, which we know to be α and β particles.

In England at about the same time, Sir J.J. Thomson and his student Ernest Rutherford were studying the radiation from uranium and thorium. Rutherford found that "There are present at least two distinct types of radiation—one that is readily absorbed, which will be termed for convenience α [alpha] radiation, and the other of a more penetrative character, which will be termed β [beta] radiation." **Alpha radiation,** he discovered, was composed of particles that, when passed through an electric field, were attracted to the negative side of the field (Figure 3.3); indeed, his later studies showed

Table **19.1**

Characteristics of α, β, and γ Emissions

Name	Symbol	Charge	Mass (g/particle)
alpha	${}^{4}_{2}He^{2+}$, ${}^{4}_{2}\alpha$	+2	6.65×10^{-24}
beta	${}^{0}_{-1}e$, ${}^{0}_{-1}\beta$	−1	9.11×10^{-28}
gamma	${}^{0}_{0}\gamma$, γ	0	0

these particles to be helium nuclei, ${}^{4}_{2}He^{2+}$, which were ejected at high speeds from a radioactive element (Table 19.1). As might be expected for such massive particles, they have limited penetrating power and can be stopped by several sheets of ordinary paper or clothing (Figure 19.2).

In the same experiment, Rutherford also found that β radiation must be composed of negatively charged particles, since the beam of radiation was attracted to the electrically positive plate. Later work by Becquerel showed that these particles have an electric charge and mass equal to those of an electron. Thus, **beta particles** are electrons ejected at high speeds from some radioactive nuclei. They are more penetrating than alpha particles, since at least a 1/8-inch piece of aluminum is necessary to stop beta particles, and they will penetrate several millimeters of living bone or tissue.

See "Portrait of a Scientist: Ernest Rutherford *(1871–1937)," Chapter 3.*

Rutherford hedged his bets when he said there were *at least* two types of radiation. Indeed, a third type was later discovered by P. Villard, a Frenchman, who named it **γ (gamma) radiation,** using the third letter in the Greek alphabet in keeping with Rutherford's scheme. Unlike alpha and beta radiation, which are particulate in nature, γ radiation is a form of electromagnetic radiation like x radiation, although γ rays are even more energetic than x rays (Figure 9.1). Furthermore, γ rays have no electric charge and so are not affected by an electric field (Figure 3.3). Finally, gamma radiation is the most penetrating, since it can pass completely through the human body. Thick layers of lead or concrete are required to minimize penetration.

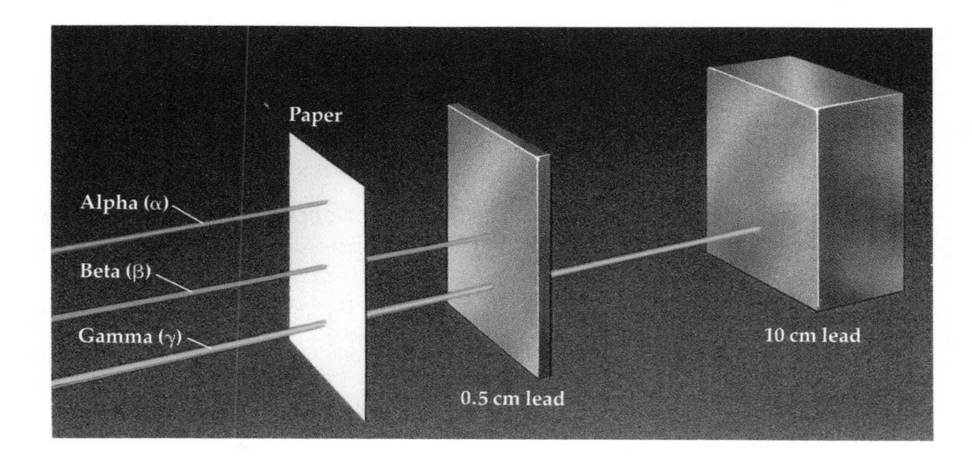

Figure 19.2
The relative penetrating abilities of the three major types of nuclear radiation. Heavy, highly charged alpha particles interact with matter most strongly and so are stopped by a piece of paper or a layer of skin. Beta particles and positrons are lighter and have a smaller charge, so they interact less strongly with matter; they are stopped by about one half centimeter of lead. Gamma rays are uncharged, massless particles and are the most penetrating.

19.2 NUCLEAR REACTIONS

Equations for Nuclear Reactions

Ernest Rutherford found that radium not only emits alpha particles but that it also produces the radioactive gas radon in the process. Such observations led Rutherford and Frederick Soddy, in 1902, to propose the revolutionary theory that *radioactivity is the result of a natural change of the isotope of one element into the isotope of a* different *element.* In such changes, called **nuclear reactions** or *transmutations,* an unstable nucleus emits radiation and is converted into a more stable nucleus of a different element. Thus, a nuclear reaction results in a change in atomic number and often a change in mass number as well. For example, the reaction studied by Rutherford can be written as

$$^{226}_{88}Ra \longrightarrow {}^{4}_{2}He + {}^{222}_{86}Rn$$

In this balanced equation the subscripts are the atomic numbers and the superscripts are the mass numbers.

Be sure to notice that, when a radioactive atom decays, the emission of a charged particle leaves a charged atom. Thus, when Ra-226 decays it gives a helium-4 cation (He^{2+}) and a Rn-222 anion (Rn^{2-}). By convention, the ion charges are not shown in balanced equations for nuclear reactions.

The atoms in molecules and ions are rearranged in a chemical change; they are not created or destroyed. The number of atoms remains the same. Similarly, in nuclear reactions the total number of nuclear particles, or **nucleons** (protons plus neutrons), remains the same. The essence of nuclear reactions, however, is that one nucleon can change into a different nucleon. A proton can change to a neutron or a neutron can change to a proton, but the total number of nucleons remains the same. Therefore, *the sum of the mass numbers of reacting nuclei must equal the sum of the mass numbers of the nuclei produced.* Furthermore, to maintain charge balance, *the sum of the atomic numbers of the products must equal the sum of the reactants.* These principles may be verified for the preceding nuclear equation.

	$^{226}_{88}Ra$ radium-226	$\longrightarrow$	$^{4}_{2}He$ alpha particle	+	$^{222}_{86}Rn$ radon-222
mass number: (protons + neutrons)	226	$\longrightarrow$	4	+	222
atomic number: (protons)	88	$\longrightarrow$	2	+	86

Reactions Involving Alpha and Beta Particles

One way a radioactive isotope can disintegrate or decay is to eject an alpha particle from the nucleus. This is illustrated by the conversion of radium to radon and by the following reaction.

	$^{234}_{92}U$ uranium-234	$\longrightarrow$	$^{4}_{2}He$ alpha particle	+	$^{230}_{90}Th$ thorium-230
mass number:	234	$\longrightarrow$	4	+	230
atomic number:	92	$\longrightarrow$	2	+	90

You will notice that in alpha emission *the atomic number decreases by two units and the mass number decreases by four units for each alpha particle emitted.*

Emission of a beta particle is another way for an isotope to decay. For example, loss of a beta particle by uranium-235 is represented by

	$^{235}_{92}U$	$\longrightarrow$	$^{0}_{-1}\beta$	+	$^{235}_{93}Np$
	uranium-235		beta particle		neptunium-235
mass number:	235	$\longrightarrow$	0	+	235
atomic number:	92	$\longrightarrow$	−1	+	93

Since a beta particle has a charge of −1, electrical balance makes the atomic number of the product *greater* by one than that of the reacting nucleus. However, the mass number does not change. The mass number of 0 for the electron is due to the small mass of the particle (only $\frac{1}{1836}$ the mass of a proton).

How does a nucleus, composed only of protons and neutrons, eject an electron? It is generally accepted that a series of reactions is involved, but the net process is

$$^{1}_{0}n \longrightarrow {}^{0}_{-1}\beta + {}^{1}_{1}p$$

neutron electron proton

where we use the symbol p for a proton. *The ejection of a beta particle always means that a new element is formed with an atomic number one unit greater than the decaying nucleus.*

In many cases, the emission of an alpha or beta particle results in the formation of an isotope that is also unstable and therefore radioactive. The new radioactive isotope may therefore undergo a number of successive transformations until a stable, nonradioactive isotope is finally produced. Such a series of reactions is called a **radioactive series.** One such series begins with uranium-238 and ends with lead-206, as illustrated in Figure 19.3. The first step in the series is

A nucleus formed as a result of an alpha or beta emission is generally in an excited state and so also emits a γ ray.

$$^{238}_{92}U \longrightarrow {}^{4}_{2}He + {}^{234}_{90}Th$$

and the equation for the final step, the conversion of polonium-210 to lead-206, is

$$^{210}_{84}Po \longrightarrow {}^{4}_{2}He + {}^{206}_{82}Pb$$

EXAMPLE 19.1

Radioactive Series

The second, third, and fourth steps in the uranium-238 series in Figure 19.3 involve emission of first a β particle, then another β particle, and finally an α particle. Write equations to show the products of these steps.

SOLUTION The product of the first step, thorium-234, is our starting point. Figure 19.3 shows that the mass remains the same during the second step, but that the atomic number increases by 1 to 91, a result shown by the balanced equation

$$\,^{234}_{90}Th \longrightarrow \,^{0}_{-1}\beta + \,^{234}_{91}Pa$$
thorium-234 protactinium-234

In the third step Figure 19.3 shows that the mass again stays constant, and the atomic number increases once again by 1.

$$\,^{234}_{91}Pa \longrightarrow \,^{0}_{-1}\beta + \,^{234}_{92}U$$
protactinium-234 uranium-234

Finally, the fourth step involves alpha particle emission, so both the mass number and atomic number decline. This is again confirmed in Figure 19.3.

$$\,^{234}_{92}U \longrightarrow \,^{4}_{2}He + \,^{230}_{90}Th$$
uranium-234 thorium-230

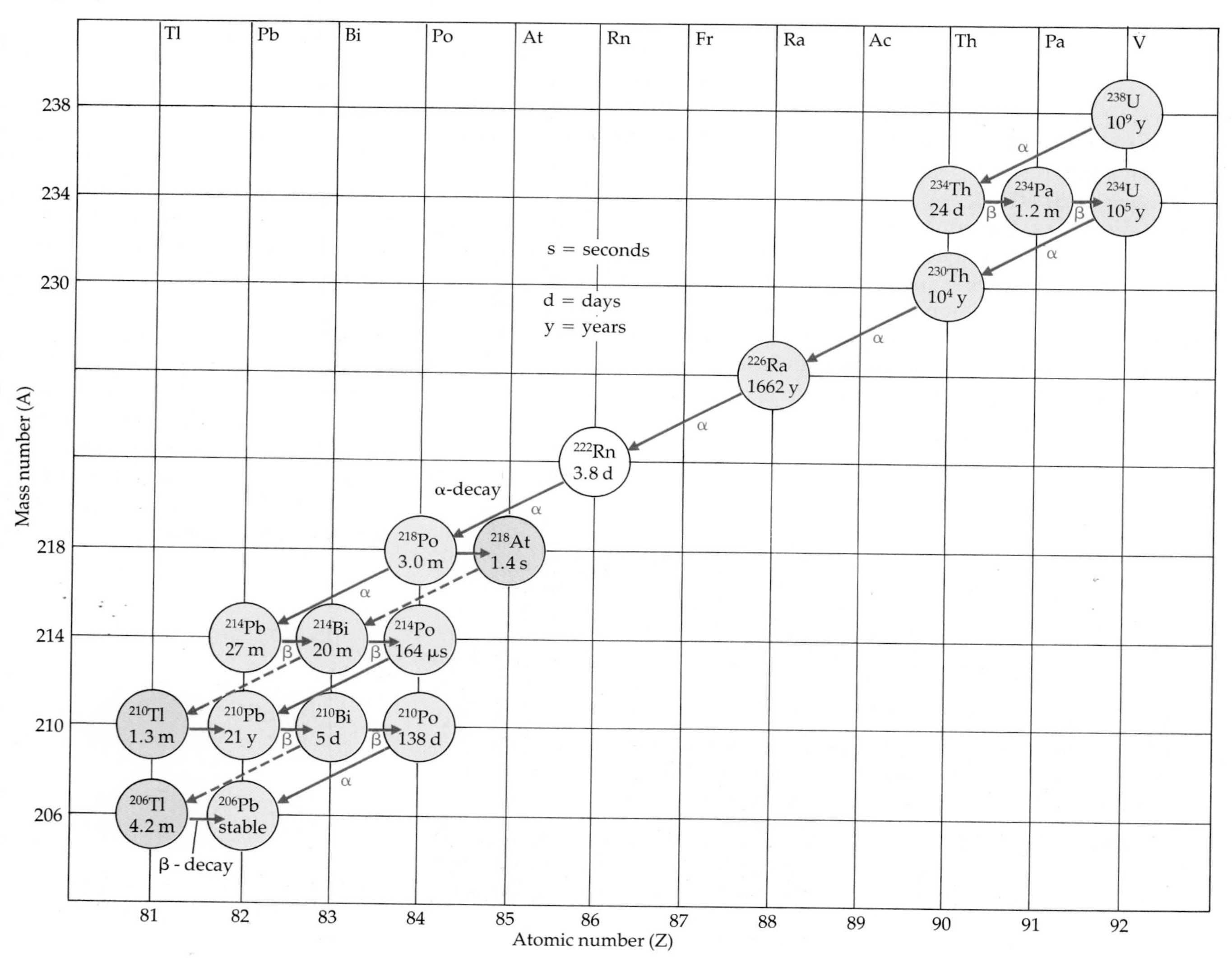

Figure 19.3
A radioactive series beginning with uranium-238 and ending with lead-206. In the first step, for example, ^{238}U emits an α particle to give thorium-234. This radioactive isotope then emits a β particle to give protactinium-234. The $^{234}_{91}Pa$ then emits another β particle to continue the series, which finally ends at $^{206}_{82}Pb$. The half-life is given for each isotope (see Section 19.4). The time units are s = seconds, m = minutes, d = days, and y = years.

EXERCISE 19.1 • *Nuclear Reactions: Alpha and Beta Emission*

(a) Write an equation showing the emission of an alpha particle by an isotope of neptunium, ${}^{237}_{93}Np$, to produce an isotope of protactinium. (b) Write an equation showing the emission of a beta particle by an isotope of sulfur-35, ${}^{35}_{16}S$, to produce an isotope of chlorine.

EXERCISE 19.2 • *Radioactive Series*

The actinium series begins with uranium-235, ${}^{235}_{92}U$, and ends with lead-207, ${}^{207}_{82}Pb$. The first five steps involve the emission of α, β, α, α, and β particles, respectively. Identify the radioactive isotope produced in each of the steps beginning with uranium-235.

Other Types of Radioactive Decay

In addition to radioactive decay by emission of alpha, beta, or gamma radiation, other decay processes are observed. Some nuclei decay, for example, by emission of a **positron,** ${}^{0}_{+1}\beta$, which is effectively a positively charged electron. Positron emission by polonium-207 leads to the formation of bismuth-207, for example.

	${}^{207}_{84}Po$	$\longrightarrow$	${}^{0}_{+1}\beta$	+	${}^{207}_{83}Bi$
	polonium-207		beta particle		bismuth-207
mass number:	207	$\longrightarrow$	0	+	207
atomic number:	84	$\longrightarrow$	+1	+	83

Notice that this is the opposite of beta decay, because positron decay leads to a *decrease* in the atomic number.

The atomic number is also reduced by one when **electron capture** occurs. In this process an inner shell electron is captured by the nucleus.

The positron was discovered by Carl Anderson in 1932. It is sometimes called an "antielectron," one of a group of particles that have become known as "antimatter." Contact between an electron and a positron leads to mutual annihilation of both particles with production of two high-energy photons.

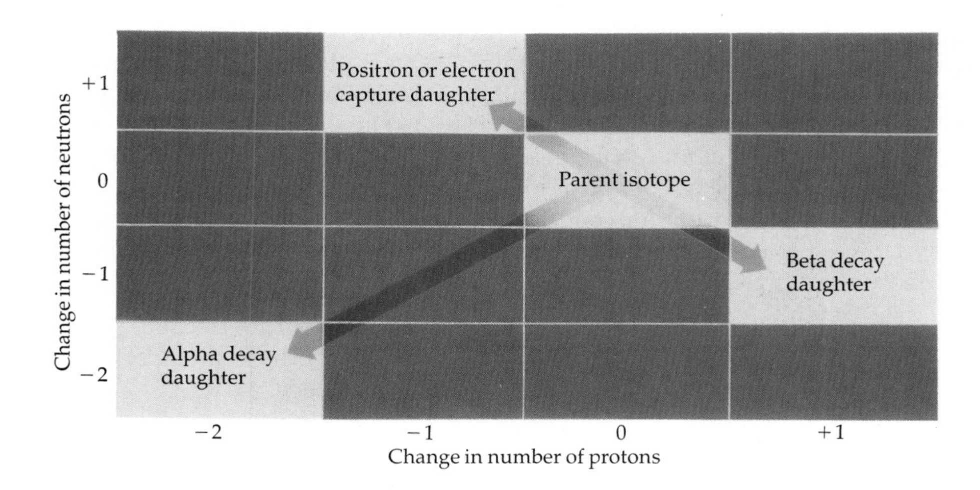

A memory aid for predicting the product of alpha, beta, or positron emission.

	$^{7}_{4}Be$	+	$^{0}_{-1}e$	$\longrightarrow$	$^{7}_{3}Li$
	beryllium-7		electron		lithium-7
mass number:	7	+	0	$\longrightarrow$	7
atomic number:	4	+	−1	$\longrightarrow$	3

In the old nomenclature of atomic physics the innermost shell was called the K-shell, so the electron capture decay mechanism is sometimes called *K-capture.*

In summary, there are four common ways that a radioactive nucleus can decay, as summarized in the figure on the previous page. In nuclear chemistry, the radioactive isotope that begins a process is called the "parent" and the product is called a "daughter" isotope.

EXERCISE 19.3 • *Nuclear Reactions*

Balance the following nuclear reactions. Indicate the symbol, the mass number, and the atomic number of "?".

a. $^{13}_{7}N \longrightarrow {}^{13}_{6}C + ?$
b. $^{41}_{20}Ca + {}^{0}_{-1}e \longrightarrow ?$
c. $^{90}_{38}Sr \longrightarrow {}^{90}_{39}Y + ?$
d. $^{11}_{6}C \longrightarrow {}^{11}_{5}B + ?$
e. $^{43}_{21}Sc \longrightarrow ? + {}^{1}_{1}H$

19.3 STABILITY OF ATOMIC NUCLEI

The fact that some nuclei are unstable (radioactive), while others are stable (nonradioactive), leads us to consider the reasons for stability. Figure 19.4 shows the naturally occurring isotopes of the elements from hydrogen to bismuth. It is quite astonishing that there are so few. Why not hundreds more?

In its simplest and most abundant form, hydrogen has only one nuclear particle, the proton. In addition, the element has two other well known isotopes: nonradioactive deuterium, with one proton and one neutron ($^{2}_{1}H$ = D), and radioactive tritium, with one proton and two neutrons ($^{3}_{1}H$ = T). Helium, the next element, has two protons and two neutrons in its most stable isotope. At the end of the actinide series is element 103, lawrencium, one isotope of which has a mass number of 257 and 154 neutrons. From hydrogen to lawrencium, except for $^{1}_{1}H$ and $^{3}_{2}He$, *the mass numbers of stable isotopes are always at least twice as large as the atomic number.* In other words, except for $^{1}_{1}H$ and $^{3}_{2}He$, every isotope of every element has a nucleus containing *at least* one neutron for every proton. Apparently the tremendous *repulsive* forces between the positively charged protons in the nucleus are moderated by the presence of neutrons with no electrical charge.

As illustrated by Figure 19.4, there are very few stable combinations of protons and neutrons. Examining those combinations can give us some insight into what factors affect nuclear stability.

1. For light elements up to Ca (Z = 20), the stable isotopes usually have equal numbers of protons and neutrons, or perhaps one more neutron than protons. Examples include $^{7}_{3}Li$, $^{12}_{6}C$, $^{16}_{8}O$, and $^{32}_{16}S$.
2. Beyond calcium the neutron/proton ratio becomes increasingly greater than 1. The band of stable isotopes deviates more and more

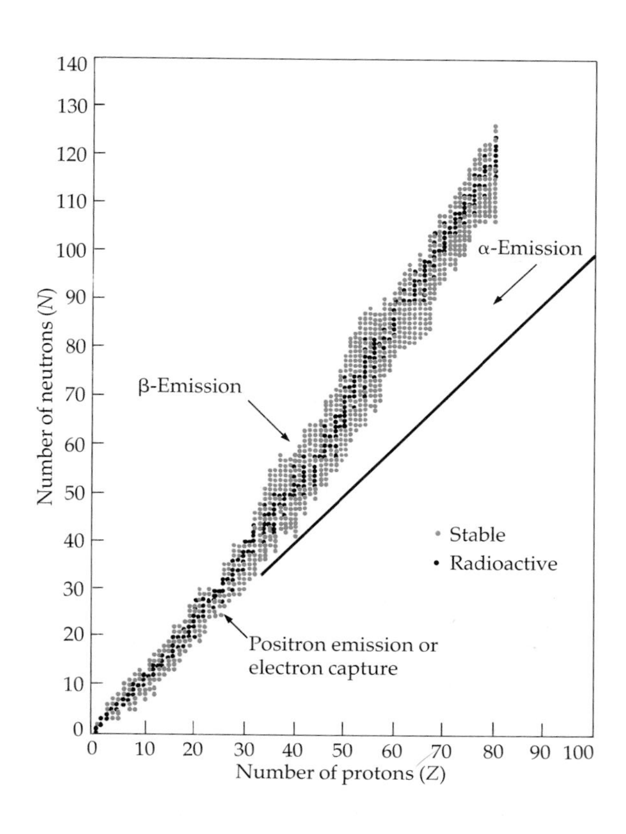

Figure 19.4
A plot of the number of neutrons (N) versus the number of protons (Z) for stable and radioactive isotopes from hydrogen ($Z = 1$) through bismuth ($Z = 83$). The effects of α, β, and positron emission or electron capture are indicated by arrows. For example, radioactive isotopes (indicated by red dots) that lie above the band of stable isotopes (indicated by black dots) decay by β emission. The arrow indicates that this raises the value of Z (by one unit per beta particle) and lowers the value of N by one unit. *(Redrawn from Oxtoby, Nachtrieb, and Freeman:* Chemistry: Science of Change*)*

from the line $N = Z$. It is evident that more neutrons are needed for nuclear stability in the heavier elements. For example, whereas one stable isotope of Fe has 26 protons and 30 neutrons, one of the stable isotopes of platinum has 78 protons and 117 neutrons.

3. Above bismuth (83 protons and 126 neutrons) all isotopes are unstable and radioactive. Beyond this point there is apparently no nuclear "super glue" strong enough to hold heavy nuclei together. Furthermore, the rate of disintegration becomes greater the heavier the nucleus. For example, half of a sample of $^{238}_{92}U$ disintegrates in a billion years, whereas half of a sample of $^{257}_{103}Lr$ is gone in only 8 seconds.
4. A very careful look at Figure 19.4 shows even more interesting features. First, elements of even atomic number have more stable isotopes than do those of odd atomic number. Second, stable isotopes generally have an *even* number of neutrons. For elements of odd atomic number, the most stable isotope has an even number of neutrons. To emphasize these points, of the more than 300 stable isotopes represented in Figure 19.4, roughly 200 have an even number of neutrons *and* an even number of protons. Only about 120 have an odd number of either protons *or* neutrons. Only four isotopes ($^{2}_{1}H$, $^{6}_{3}Li$, $^{10}_{5}B$, and $^{14}_{7}N$) have odd numbers of *both* protons and neutrons.

The Band of Stability and Type of Radioactive Decay

The narrow "band" of stable isotopes in Figure 19.4 (the black dots) is sometimes called the *peninsula of stability* in a "sea of instability." Any isotope not on this peninsula (the red dots) will decay in such a way that it can come ashore on the peninsula, and the chart can help us predict what type of decay will be observed.

All elements beyond Bi ($Z = 83$) are unstable—that is, radioactive—and most decay by ejecting an alpha particle. For example, americium, the radioactive element used in smoke alarms, decays in this manner.

$$^{243}_{95}\text{Am} \longrightarrow {}^{4}_{2}\text{He} + {}^{239}_{93}\text{Np}$$

Beta emission occurs in isotopes that have too many neutrons to be stable, that is, isotopes *above* the peninsula of stability in Figure 19.4. When beta decay converts a neutron to a proton and an electron, which is then ejected, the mass number remains constant, but the number of neutrons drops.

$$^{60}_{27}\text{Co} \longrightarrow {}^{0}_{-1}\beta + {}^{60}_{28}\text{Ni}$$

Conversely, lighter isotopes that have too few neutrons—isotopes *below* the peninsula of stability—attain stability by positron emission (because this converts a proton to a neutron in one step) or by electron capture.

$$^{13}_{7}\text{N} \longrightarrow {}^{0}_{+1}\beta + {}^{13}_{6}\text{C}$$

$$^{41}_{20}\text{Ca} + {}^{0}_{-1}\text{e} \longrightarrow {}^{41}_{19}\text{K}$$

Decay by these routes is most commonly observed for elements with $Z < 40$, that is, with atomic numbers less than that of calcium.

EXERCISE 19.4 • *Nuclear Stability*

For each of the following unstable isotopes, write an equation for its probable mode of decay.

a. silicon-32, $^{32}_{14}\text{Si}$ b. titanium-45, $^{45}_{22}\text{Ti}$ c. plutonium-239, $^{239}_{94}\text{Pu}$

Binding Energy

As proved by Ernest Rutherford's experiments (Chapter 3), the nucleus of the atom is extremely small. Yet the nucleus can contain up to 83 protons before becoming unstable. This is evidence that there must be a very strong short-range binding force that can overcome the electrostatic repulsive force of a number of protons packed into such a tiny volume. A measure of the force holding the nucleus together is the nuclear **binding energy.** This energy (E_b) is defined as the negative of the energy change (ΔE) that would occur if a nucleus were formed directly from its component protons and neutrons. For example, if a mole of protons and a mole of neutrons directly formed a mole of deuterium nuclei, the energy change would be more than 200 million kilojoules!

$$^{1}_{1}\text{H} + {}^{1}_{0}\text{n} \longrightarrow {}^{2}_{1}\text{H} \qquad \Delta E = -2.15 \times 10^{8}\text{ kJ}$$

$$\text{binding energy} = -\Delta E = E_b = +2.15 \times 10^{8}\text{ kJ}$$

This nuclear synthesis reaction is highly exothermic (and so E_b is very positive), an indication of the strong attractive forces holding the nucleus together. The deuterium nucleus is more stable than an isolated proton and an isolated neutron, just as the H_2 molecule is more stable than two isolated H atoms. However, recall that the energy released when a mole of H—H covalent bonds form is only 436 kJ, a tiny fraction of the energies released when protons and neutrons coalesce to form a nucleus.

To understand the enormous energy release during the formation of an atomic nucleus, we turn to an experimental observation and a theory. The experimental observation is that the mass of a nucleus is always less than the sum of the masses of its constituent protons and neutrons.

$$\underset{1.007825\ \text{g/mol}}{{}^{1}_{1}\text{H}} + \underset{1.008665\ \text{g/mol}}{{}^{1}_{0}\text{n}} \longrightarrow \underset{2.01410\ \text{g/mol}}{{}^{2}_{1}\text{H}}$$

$$\begin{aligned}\text{Change in mass} = \Delta m &= \text{mass of product} - \text{sum of masses of reactants}\\ &= 2.01410\ \text{g/mol} - 2.016490\ \text{g/mol}\\ &= -0.00239\ \text{g/mol}\end{aligned}$$

The theory is that the "missing mass," Δm, has been converted to energy, and it is this energy that we described as the binding energy.

The quantity Δm *is sometimes called the* mass defect.

The relation between mass and energy is contained in Albert Einstein's 1905 theory of special relativity, which holds that mass and energy are simply different manifestations of the same quantity. Einstein stated that the energy of a body is equivalent to its mass times the square of the speed of light, $E = mc^2$. So, to calculate the energy change in a process where the mass has changed, the equation becomes

$$\Delta E = (\Delta m)c^2$$

We can calculate ΔE in joules if the change in mass is given in kilograms and the velocity of light is in meters per second (because $1\ \text{J} = 1\ \text{kg} \cdot \text{m}^2/\text{s}^2$). For the formation of deuterium nuclei from protons and neutrons, we have

$$\Delta E = (-2.39 \times 10^{-6}\ \text{kg})(3.00 \times 10^{8}\ \text{m/s})^2 = -2.15 \times 10^{11}\ \text{J}$$
$$\Delta E = -2.15 \times 10^{8}\ \text{kJ}$$

This is the value of ΔE given at the beginning of this section for the change in energy when a mole of protons and a mole of neutrons form a mole of deuterium nuclei.

A helium nucleus is composed of two protons and two neutrons. As expected, the binding energy, E_b, is very large, even larger than for deuterium.

$$2\ {}^{1}_{1}\text{H} + 2\ {}^{1}_{0}\text{n} \longrightarrow {}^{4}_{2}\text{He} \qquad E_b = +2.73 \times 10^{9}\ \text{kJ/mol of helium nuclei}$$

To compare nuclear stabilities more directly, however, nuclear scientists generally calculate the **binding energy per nucleon.** For helium-4 this is

$$E_b \text{ per mol nucleons} = \frac{2.73 \times 10^{9}\ \text{kJ}}{4\ \text{mol nucleons}} = 6.83 \times 10^{8}\ \text{kJ/mol nucleons}$$

The greater the binding energy per nucleon, the greater is the stability of the nucleus. Scientists have calculated the binding energies of a great number of nuclei and have plotted them as a function of mass number (Figure 19.5). It is very interesting—and important—that the point of maximum stability occurs in the vicinity of iron-56, ${}^{56}_{26}\text{Fe}$. This means that *all elements are thermodynamically unstable with respect to iron.* That is, very heavy nuclei may split or **fission,** with the release of enormous quantities of energy, to give more

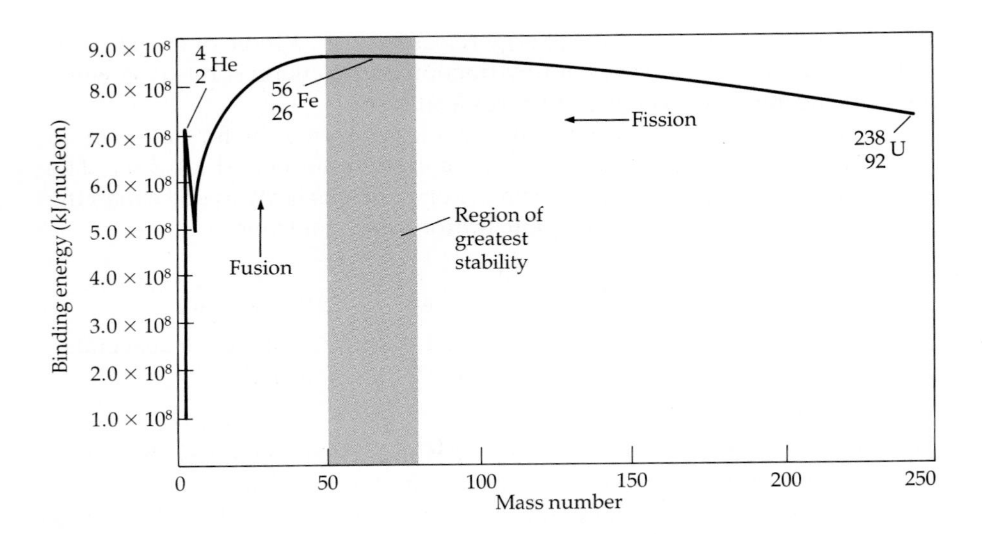

Figure 19.5
The relative stability of nuclei. This "curve of binding energy" was derived by calculating the binding energy per nucleon (in kJ/mol) for the most abundant isotope of each element from hydrogen to uranium.

stable nuclei with atomic numbers nearer iron. In contrast, two very light nuclei may come together and undergo **fusion** exothermically to form heavier nuclei. Finally, *this is the reason that iron is the most abundant of the heavier elements in the universe.*

EXERCISE 19.5 • *Binding Energy*

Calculate the binding energy, in kJ/mol, for the formation of lithium-6.

$$3\,{}^{1}_{1}\mathrm{H} + 3\,{}^{1}_{0}\mathrm{n} \longrightarrow {}^{6}_{3}\mathrm{Li}$$

The necessary masses are ${}^{1}_{1}\mathrm{H}$ = 1.00783 g/mol, ${}^{1}_{0}\mathrm{n}$ = 1.00867 g/mol, and ${}^{6}_{3}\mathrm{Li}$ = 6.015125 g/mol. Is the binding energy greater than or less than that for helium-4? Finally, compare the binding energy per nucleon of ${}^{6}_{3}\mathrm{Li}$ and helium-4. Which nucleus is the more stable?

19.4 RATES OF DISINTEGRATION REACTIONS

Cobalt-60 is used as a source of β particles and γ rays to treat malignancies in the human body. Although the isotope is radioactive, it is nonetheless reasonably stable, since only half of a sample of cobalt-60 will decay in a little over 5 years. On the other hand, copper-64, which is used in the form of copper acetate to detect brain tumors, decays much more rapidly; half of the radioactive copper decays in slightly less than 13 hours. These two radioactive isotopes are clearly different in their stabilities; one of them decays much faster, that is, at a greater rate, than the other.

Half-Life

The relative stabilities of radioactive isotopes are often expressed just as we have done: in terms of the time required for half of the sample to decay. This is called the **half-life, $t_{1/2}$,** of a radioactive isotope. As illustrated by Table

Table 19.2

Half-Lives of Some Common Radioactive Isotopes

Isotope	Decay Process	Half-Life
$^{238}_{92}U$	$^{238}_{92}U \longrightarrow ^{234}_{90}Th + ^{4}_{2}He$	4.51×10^{9} years
$^{3}_{1}H$ (tritium)	$^{3}_{1}H \longrightarrow ^{3}_{2}He + ^{0}_{-1}\beta$	12.26 years
$^{14}_{6}C$ (carbon-14)	$^{14}_{6}C \longrightarrow ^{14}_{7}N + ^{0}_{-1}\beta$	5730 years
$^{131}_{53}I$	$^{131}_{53}I \longrightarrow ^{131}_{54}Xe + ^{0}_{-1}\beta$	8.05 days

19.2, isotopes have widely varying half-lives; some take years for half of the sample to decay, while others decay to half the original number of atoms in fractions of seconds.

As an example of the concept of half-life, consider the decay of oxygen-15, $^{15}_{8}O$, by positron emission.

$$^{15}_{8}O \longrightarrow ^{15}_{7}N + ^{0}_{+1}\beta$$

The half-life of oxygen-15 is 2.0 minutes. This means that half of the quantity of $^{15}_{8}O$ present at any given time will disintegrate every 2.0 minutes. Thus, if we begin with 20 mg of $^{15}_{8}O$, 10 mg of the isotope will remain after 2.0 minutes. After 4.0 minutes (two half-lives), only half of the remainder, 5.0 mg, will still be there. After 6.0 minutes (three half-lives), only half of the 5.0 mg will still be present, or 2.5 mg. The amounts of $^{15}_{8}O$ present at various times are illustrated in Figure 19.6.

For nuclear disintegration reactions, the half-life is a constant, independent of temperature and of the number of radioactive nuclei present.

EXAMPLE 19.2

Half-Life

Tritium ($^{3}_{1}H$), a radioactive isotope of hydrogen, has a half-life of 12.3 years.

$$^{3}_{1}H \longrightarrow ^{0}_{-1}\beta + ^{3}_{2}He$$

If you begin with 1.5 mg of the isotope, how many milligrams remain after 49.2 years?

SOLUTION First, we find the number of half-lives in the given time period of 49.2 years. Since the half-life is 12.3 years, the number of half-lives is

$$49.2 \text{ years} \cdot \frac{1 \text{ half-life}}{12.3 \text{ years}} = 4.00 \text{ half-lives}$$

This means that the initial quantity of 1.5 mg is reduced four times by 1/2.

$$1.5 \text{ mg} \times \tfrac{1}{2} \times \tfrac{1}{2} \times \tfrac{1}{2} \times \tfrac{1}{2} = 1.5 \times (\tfrac{1}{2})^4 = 1.5 \text{ mg} \times \tfrac{1}{16} = 0.094 \text{ mg}$$

After 49.2 years, only 0.094 mg of the original 1.5 mg remains.

EXERCISE 19.6 • *Radioactivity and Half Life*

Strontium-90, $^{90}_{38}Sr$, is a radioisotope ($t_{1/2}$ = 28 years) produced in atomic bomb explosions. Its long life and tendency to concentrate in bone marrow make it particularly dangerous to people and animals.

a. The isotope decays with loss of a β particle; write a balanced equation showing the other product of decay.

b. A sample of the isotope emits 2000 β particles per minute. How many half-lives and how many years are necessary to reduce the emission to 125 β particles per minute?

Rate of Radioactive Decay

The ideas used in describing the rate of radioactive decay are those of kinetics (see Chapter 8).

To determine the half-life of a radioactive element, the *rate of decay* must be measured. That is, we must measure the number of atoms that disintegrate per second or per hour or per year.

The rate of nuclear decay is often described in terms of the **activity** (A) of the sample, the number of disintegrations observed per unit time. The activity is *proportional* to the number of radioactive atoms present (N).

$$\text{rate of radioactive decay} \equiv \text{activity } (A) \propto \text{number of radioactive atoms present } (N)$$

This proportionality can also be expressed in the form

$$A = kN \tag{19.1}$$

$$\frac{\text{disintegrations}}{\text{time}} = \frac{\text{disintegrations}}{(\text{number of atoms})(\text{time})} \cdot \text{number of atoms}$$

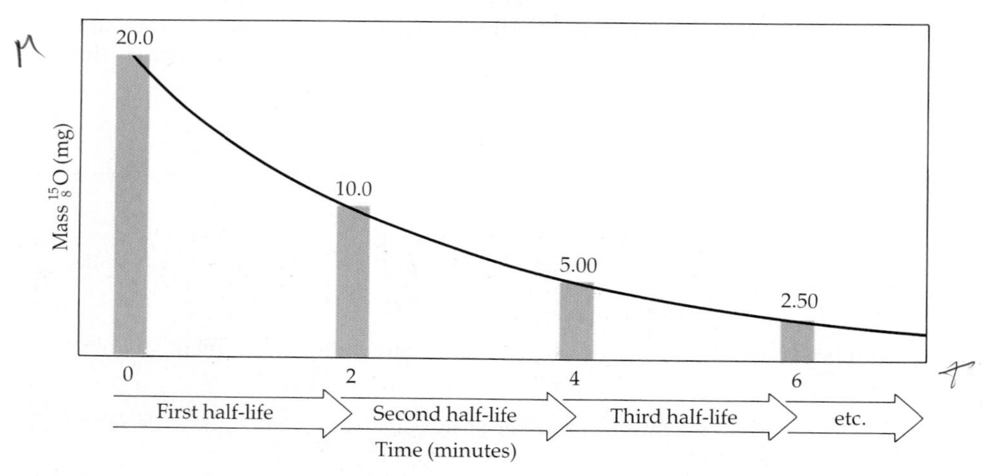

Figure 19.6

Decay of 20 mg of oxygen-15. After each half-life period of 2.0 minutes, the quantity present at the beginning of the period is reduced by half. The plot is based on the following data:

Number of Half-Lives	*Fraction of Initial Quantity Remaining*	*Quantity Remaining (mg)*
0	1	20.0 (initial)
1	1/2	10.0
2	1/4	5.00
3	1/8	2.50
4	1/16	1.25
5	1/32	0.625

where k is the proportionality constant or *decay constant*. In the language of kinetics, Equation 19.1 is simply a rate law that is first order in the number of atoms in the sample, and k is the rate constant.

The activity of a sample can be measured with a device such as a Geiger counter (Figure 19.7). Let us say the activity is measured at some time t_0 and then measured again after a few minutes, hours, or days. If the initial activity is A_0 at t_0, then a second measurement will give a smaller activity A at a later time t. From Equation 19.1, you can see that the ratio of the activity A at some time t to the activity at the beginning of the experiment (A_0) must be equal to the ratio of the number of radioactive atoms N that are present at time t to the number present at the beginning of the experiment (N_0).

$$\frac{A}{A_0} = \frac{kN}{kN_0}$$

or

$$\frac{A}{A_0} = \frac{N}{N_0} = \text{fraction of radioactive atoms still present in a sample after some time has elapsed}$$

This means that experimental information is related directly to the fraction of radioactive atoms remaining in a sample after some time has passed.

An extraordinarily useful equation relates the time period over which a sample is observed (t) to the fraction of radioactive atoms present after that amount of time has passed.

$$\ln \frac{N}{N_0} = -kt \qquad (19.2)$$

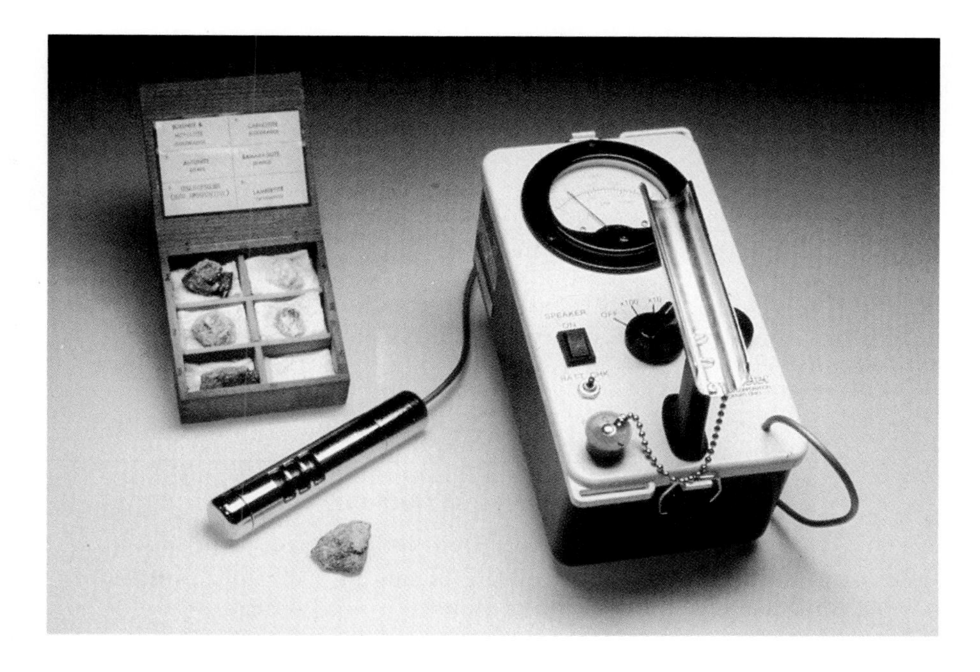

Figure 19.7

A Geiger counter with a sample of carnotite, a mineral containing uranium oxide. The Geiger counter was invented by Hans Geiger and Ernest Rutherford in 1908. A charged particle (such as an α or β particle), when entering a gas-filled tube, ionizes the gas. These gaseous ions are attracted to electrically charged plates and cause a "pulse" or momentary flow of electric current. The current is amplified and used to operate a counter. *(C.D. Winters)*

Equation 19.2 can be derived from Equation 19.1 by calculus.

In words, the equation says

$$\text{natural logarithm}\frac{\text{number of radioactive atoms at some time } t}{\text{number of radioactive atoms at start of experiment}}$$

$= \ln$ (fraction of radioactive atoms remaining from original sample)

$= -$(decay constant)(time)

Notice the negative sign in the equation. The ratio N/N_0 is less than 1 because N is always less than N_0. This means that the logarithm of N/N_0 is negative, and so the other side of the equation must also bear a negative sign. Equation 19.2 is useful in three ways:

- If A/A_0 (and thus N/N_0) is measured in the laboratory over some time period t, then k can be calculated. The decay constant k can then be used to determine the half-life of the sample, as illustrated in Example 19.3.
- If k is known, the fraction of a radioactive sample still present after some time t has elapsed can be calculated.
- If k is known for a particular radioactive isotope, you can calculate the time required for that isotope to decay to a certain activity.

Now we are in a position to see how the half-life of a radioactive isotope, $t_{1/2}$, is determined. The half-life is the time needed for half of the material present at the beginning of the experiment (N_0) to disappear. Thus, when time $= t_{1/2}$, then $N = \frac{1}{2}N_0$. This means that

$$\ln \frac{\frac{1}{2}N_0}{N_0} = -kt_{1/2}$$

or

$$\ln \tfrac{1}{2} = -kt_{1/2}$$

$$-0.693 = -kt_{1/2}$$

and we arrive at a simple equation that connects the half-life and decay constant.

$$t_{1/2} = \frac{0.693}{k} \qquad (19.3)$$

The half-life, $t_{1/2}$, is found by calculating k from Equation 19.2, where N and N_0 in turn come from laboratory measurements over the time period t.

EXAMPLE 19.3

Determination of Half-Life

A sample of radon initially undergoes 7.0×10^4 alpha particle disintegrations per second (dps). After 6.6 days, it undergoes only 2.1×10^4 alpha particle dps. What is the half-life of this isotope of radon?

SOLUTION Experiment has provided us with both A and A_0.

$$A = 2.1 \times 10^4 \text{ dps} \qquad A_0 = 7.0 \times 10^4 \text{ dps}$$

and the time (t = 6.6 days). Therefore, we can find the value of k. Since $N/N_0 = A/A_0$,

$$\ln\left(\frac{2.1 \times 10^4}{7.0 \times 10^4}\right) = -k(6.6 \text{ d})$$

$$\ln(0.30) = -k(6.6 \text{ d})$$

$$k = -\frac{\ln(0.30)}{6.6 \text{ d}} = -\frac{(-1.20)}{6.6 \text{ d}} = 0.18 \text{ d}^{-1}$$

and from k we can obtain $t_{1/2}$.

$$t_{1/2} = \frac{0.693}{k} = \frac{0.693}{0.18 \text{ d}^{-1}} = 3.8 \text{ d}$$

EXAMPLE 19.4

Time and Radioactivity

Some high-level radioactive waste with a half-life, $t_{1/2}$, of 200. years is stored in underground tanks. What time is required to reduce an activity of 6.50×10^{12} disintegrations per minute (dpm) to a fairly harmless activity of 3.00×10^{-3} dpm?

SOLUTION The data give you the initial activity ($A_0 = 6.50 \times 10^{12}$ dpm) and the activity after some elapsed time ($A = 3.00 \times 10^{-3}$ dpm). In order to find the elapsed time t, you must first find k from the half-life.

$$k = \frac{0.693}{t_{1/2}} = \frac{0.693}{200. \text{ y}} = 0.00347 \text{ y}^{-1}$$

With k known, the time t can be calculated.

$$\ln\left(\frac{3.00 \times 10^{-3}}{6.50 \times 10^{12}}\right) = -[0.00347 \text{ y}^{-1}]t$$

$$-35.312 = -[0.00347 \text{ y}^{-1}]t$$

$$t = \frac{-35.312}{-(0.00347) \text{ y}^{-1}}$$

$$t = 1.02 \times 10^4 \text{ y}$$

EXERCISE 19.7 • *Rate of Radioactive Decay*

Gallium citrate, containing the radioactive isotope gallium-67, is used medically as a tumor-seeking agent. It has a half-life of 77.9 hours. How much time is needed for a sample of gallium citrate to decay to 10% of its original activity?

Radiochemical Dating

Scientists have used radiochemical dating to determine the ages of rocks, fossils, and artifacts that date back many years. For example, radiochemical methods were used recently to show that the Shroud of Turin was created somewhere around 1300 AD, and not at the time of Christ as has been alleged for many centuries (Figure 19.8).

For excellent articles on the Shroud of Turin and carbon-14 dating, see Chem Matters *magazine, February, 1989, pages 8–15.*

In 1946 Willard Libby developed the technique of age determination using radioactive carbon-14 ($^{14}_{6}C$). Carbon is an important building block of all living systems, and so all organisms contain the three isotopes of carbon: ^{12}C, ^{13}C, and ^{14}C. The first two are stable and have been around since the universe was created. Carbon-14, however, is radioactive and decays to nitrogen-14 by β emission.

$$^{14}_{6}C \longrightarrow {}^{0}_{-1}\beta + {}^{14}_{7}N$$

Since the half-life of ^{14}C is known to be 5.73×10^3 years, the amount of the isotope present (N) can be measured from the activity of a sample. *If* the amount of ^{14}C originally in the sample (N_0) is known, then the age of the sample can be found from Equation 19.2.

This method of age determination clearly depends on knowing how much ^{14}C was originally in the sample. The answer to this question comes from work by physicist Serge Korff who discovered, in 1929, that ^{14}C is continually generated in the upper atmosphere. High-energy cosmic rays smash into gases in the upper atmosphere and force them to eject neutrons. These free neutrons collide with nitrogen atoms in the atmosphere and produce carbon-14.

$$^{14}_{7}N + {}^{1}_{0}n \longrightarrow {}^{14}_{6}C + {}^{1}_{1}H$$

Throughout the *entire* atmosphere, only about 7.5 kg of ^{14}C is produced per year. However, this tiny amount of radioactive carbon is incorporated into

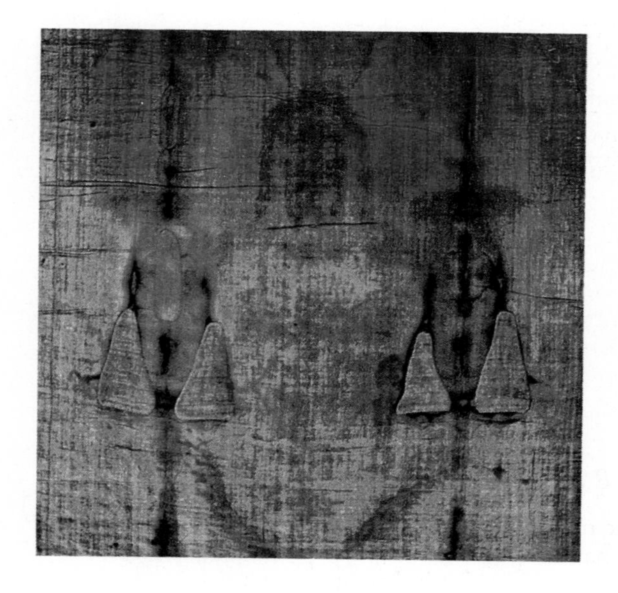

Figure 19.8
The Shroud of Turin is a linen cloth more than 4 m long. It bears a faint, straw-colored image of an adult male of average build who had apparently been crucified. Reliable records of the shroud date to about 1350, and for the past 600 years it has been alleged to be the burial shroud of Jesus Christ. Numerous chemical and other tests have been done on tiny fragments of the shroud in recent years. The general conclusion has been that the image was not painted on the cloth by any traditional method, but no one could say exactly how the image had been created. Recent advances in radiochemical dating methods, however, led to a new effort in 1987–1988 to estimate the age of the cloth. Using radioactive ^{14}C, these methods show that the flax from which the linen was made was grown between 1260 and 1390 AD. There is no chance that the cloth was made at the time of Christ. *(Jean Lorre, Donald Lynne, and the Shroud of Turin Project)*

CO_2 and then becomes part of the carbon cycle and is distributed worldwide. The continual formation of ^{14}C; exchange of the isotope within the oceans, atmosphere, and biosphere; and decay of living matter keep the supply of ^{14}C constant.

Plants absorb carbon dioxide from the atmosphere, convert it into food, and so incorporate the carbon-14 into living tissue. It has been established that the beta activity of carbon-14 in *living* plants and in the air is constant at about 14 disintegrations per minute per gram of carbon. However, when the plant dies or is ingested by an animal, carbon-14 disintegration continues *without the* ^{14}C *being replaced;* consequently, the activity decreases with passage of time. The smaller the activity of carbon-14, the longer the period between the death of the plant and the present time. Assuming that ^{14}C activity was about the same hundreds of years ago as it is now, measurement of the ^{14}C beta activity of an artifact can be used to date the article.

Willard Libby and his apparatus for carbon-14 dating. *(Oesper Collection in the History of Chemistry/University of Cincinnati)*

EXAMPLE 19.5

Radiochemical Dating

The so-called Dead Sea Scrolls, Hebrew manuscripts of the books of the Old Testament, were found in 1947. The activity of carbon-14 in the linen wrappings of the book of Isaiah is about 11 disintegrations per minute per gram (d/min · g). Calculate the approximate age of the linen.

SOLUTION We will use Equation 19.2

$$\ln\left(\frac{N}{N_0}\right) = -kt$$

where N is proportional to the activity at the present time (11 d/min · g) and N_0 is proportional to the activity of carbon-14 in the living material (14 d/min · g). In order to calculate the time elapsed since the linen wrappings were part of a living plant, we first need k, the rate constant. From the text you know that $t_{1/2}$ is 5.73×10^3 years, so

$$k = \frac{0.693}{t_{1/2}} = \frac{0.693}{5.73 \times 10^3\ \text{y}} = 1.21 \times 10^{-4}\ \text{y}^{-1}$$

Now everything is in place to calculate t.

$$\ln\left(\frac{11\ \text{d/min} \cdot \text{g}}{14\ \text{d/min} \cdot \text{g}}\right) = -[1.21 \times 10^{-4}\ \text{y}^{-1}]t$$

$$t = \frac{\ln 0.79}{-[1.21 \times 10^{-4}\ \text{y}^{-1}]}$$

$$= \frac{-0.24}{-[1.21 \times 10^{-4}\ \text{y}^{-1}]}$$

$$= 2.0 \times 10^3\ \text{y}$$

Therefore, the linen is about 2000 years old.

EXERCISE 19.8 • *Radiochemical Dating*

A wooden Japanese temple guardian statue of the Kamakura period (AD 1185–1334) had a carbon-14 activity of 12.9 d/min · g in 1990. What is the age of the statue? In what year was the statue made? The initial activity of carbon-14 was 14 d/min · g, and $t_{1/2} = 5.73 \times 10^3$ y.

19.5 ARTIFICIAL TRANSMUTATIONS

In the course of his experiments, Rutherford found in 1919 that alpha particles ionize atomic hydrogen, knocking off an electron from each atom. If atomic nitrogen was used instead, he found that bombardment with alpha particles *also produced protons.* Quite correctly he concluded that the alpha particles had knocked a proton out of the nitrogen nucleus and that an isotope of another element had been produced. Nitrogen had undergone a *transmutation* to oxygen.

$$^{4}_{2}He + ^{14}_{7}N \longrightarrow ^{17}_{8}O + ^{1}_{1}H$$

Rutherford had proposed that protons and neutrons are the fundamental building blocks of nuclei. Although Rutherford's search for the neutron was not successful, it was found by James Chadwick in 1932 as a product of the alpha-particle bombardment of beryllium.

$$^{9}_{4}Be + ^{4}_{2}He \longrightarrow ^{12}_{6}C + ^{1}_{0}n$$

Changing one element into another by alpha-particle bombardment has its limitations. Before a positively charged particle (such as the alpha particle) can be captured by a positively charged nucleus, the particle must have sufficient kinetic energy to overcome the repulsive forces developed as the particle approaches the nucleus. But the neutron is electrically neutral, so Enrico Fermi (1934) reasoned that a nucleus would not oppose its entry. By this approach, practically all elements have since been transmuted, and a number of *transuranium elements* (elements beyond uranium) have been prepared. For example, uranium-238 forms neptunium-239 on neutron bombardment,

Enrico Fermi (1901–1954), the Italian physicist who first experimentally observed nuclear fission and who demonstrated a nuclear chain reaction. *(AIP/Niels Bohr Library)*

$$^{238}_{92}U + ^{1}_{0}n \longrightarrow ^{239}_{92}U \longrightarrow ^{239}_{93}Np + ^{0}_{-1}\beta$$

and the latter decays to plutonium-239

$$^{239}_{93}Np \longrightarrow ^{0}_{-1}\beta + ^{239}_{94}Pu$$

Of the 109 elements known at present, only elements up to uranium exist in nature (except for Tc, Pm, At, and Fr). The transuranium elements are all synthetic. Up to element 101, mendelevium, all of the elements can be made by bombarding the nucleus of a lighter element with small particles such as $^{4}_{2}He$ or $^{1}_{0}n$. Beyond 101, though, special techniques using heavier particles are required and are still being developed. For example, lawrencium is made by bombarding californium-252 with boron nuclei,

$$^{252}_{98}Cf + ^{10}_{5}B \longrightarrow ^{257}_{103}Lr + 5\,^{1}_{0}n$$

The Chemical World

A Revision to the Periodic Table

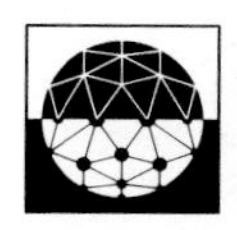

Among the most significant contributions to the modern periodic chart is that made by Nobel Laureate Glenn Seaborg (born 1912). Among other things he demonstrated the importance of maintaining the courage of one's convictions. Thanks to his insights, it is now very well established that the transuranium elements (atomic numbers greater than 92), a number of which he either discovered or helped to discover during the Manhattan Project, are members of the actinide series. Actinides are the elements following actinium and belonging in a grouping off the main periodic chart.

Until Seaborg offered his version of the periodic table, chemists were convinced that Th, Pa, and U belonged in the main body of the table, Th under Hf, Pa under Ta, and U under W. When Seaborg proposed that Th was the beginning of the actinides and that the transuranium elements belonged as a group under the rare earths, some prominent and famous inorganic chemists, many of them Seaborg's friends, tried to discourage his publication of this proposal in the open literature. One very prominent inorganic chemist felt that Seaborg would ruin his scientific reputation. Nevertheless Seaborg, strongly convinced, persisted. As a result, with Seaborg's expansion of the periodic table it was possible to predict accurately the properties of many of the as-yet-undiscovered transuranium elements. Subsequent preparation of these elements in atomic accelerators proved him right, and he was awarded the Nobel Prize in 1951 for his work.

The World of Chemistry (Program 7) "The Periodic Table"

Glenn Theodore Seaborg (1912–) began his college education as a literature major, but changed to science in his junior year at the University of California. For his preparation and discovery of several transuranium elements, he shared the 1951 Nobel prize in chemistry with E. M. McMillan (1907–), who started Seaborg in this area of research. (Lawrence Berkeley Laboratory)

---

and the latest element to be discovered, 109, was made by firing iron atoms at bismuth atoms.

EXERCISE 19.9 • *Nuclear Transmutations*

Balance the following nuclear reactions, indicating the symbol, the mass number, and the atomic number of the remaining product.

a. $^{13}_{6}C + ^{1}_{0}n \longrightarrow ^{4}_{2}He + ?$

b. $^{14}_{7}N + ^{4}_{2}He \longrightarrow ^{1}_{0}n + ?$

c. $^{253}_{99}Es + ^{4}_{2}He \longrightarrow ^{1}_{0}n + ?$

19.6 NUCLEAR FISSION

In 1938 the radiochemists Otto Hahn and Fritz Strassman found some barium in a sample of uranium that had been bombarded with neutrons. Further work by Lise Meitner, Otto Frisch, Niels Bohr, and Leo Sziland confirmed that a uranium-235 nucleus had captured a neutron to form

Nuclear power plants generate large amounts of electricity worldwide, but they also generate highly radioactive wastes. Some of these "high level" wastes have very long half-lives, up to tens of thousands of years. In addition, they can be very dangerous in high concentrations. Therefore, the storage of these wastes is a formidable problem. They are now stored in large double-walled tanks buried in the ground, but better long-term solutions must be found. *(U.S. Department of Energy)*

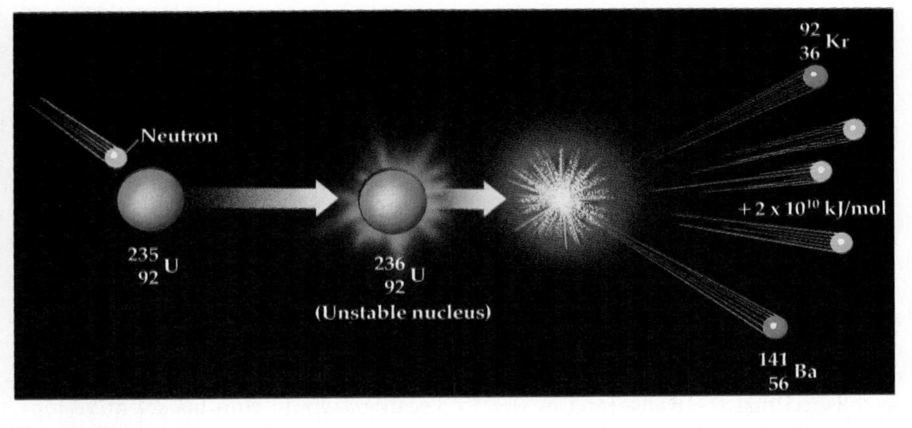

Figure 19.9
The fission of a $^{236}_{92}U$ nucleus that arises from the bombardment of $^{235}_{92}U$ with a neutron. The electrical repulsion between protons rips the nucleus apart.

uranium-236 and then this heavier isotope had undergone **nuclear fission;** that is, the nucleus had split in two (Figure 19.9)

$$^{235}_{92}U + ^{1}_{0}n \longrightarrow ^{236}_{92}U \longrightarrow ^{141}_{56}Ba + ^{92}_{36}Kr + 3\,^{1}_{0}n \qquad \Delta E = -2 \times 10^{10}\ kJ/mol$$

The fact that the fission reaction produces more neutrons than are required to begin the process is important. In the above nuclear reaction bombardment with a single neutron produces three neutrons capable of inducing three more fission reactions, which release nine neutrons to induce nine more fissions, from which 27 neutrons are obtained, and so on. Since the fission of uranium-236 is extremely rapid, this sequence of reactions can be an explosive chain reaction as illustrated in Figure 19.10. If the amount of uranium-235 is small, so few neutrons are captured by ^{235}U nuclei that the chain reaction cannot be sustained. In an atomic bomb, two small pieces of uranium, neither capable of sustaining a chain reaction, are brought together to form one piece capable of supporting a chain reaction, and an explosion results.

Rather than allow a fission reaction to run away explosively, engineers can slow it by limiting the number of neutrons available, and energy can be derived safely and used as a heat source in a power plant (Figure 19.11). In a **nuclear** or **atomic reactor,** the rate of fission is controlled by inserting cadmium rods or other "neutron absorbers" into the reactor. The rods absorb the neutrons that cause fission reactions; by withdrawing or inserting the rods, the rate of the fission reaction can be increased or decreased.

If the uranium-235 content of a sample is over 90%, it is considered of weapons quality.

Not all nuclei can be made to fission on colliding with a neutron, but ^{235}U and ^{239}Pu are two isotopes for which fission is possible. Natural uranium contains an average of only 0.72% of the fissionable 235 isotope; more than 99% of the natural element is nonfissionable uranium-238. Since the percentage of natural ^{235}U is too small to sustain a chain reaction, uranium

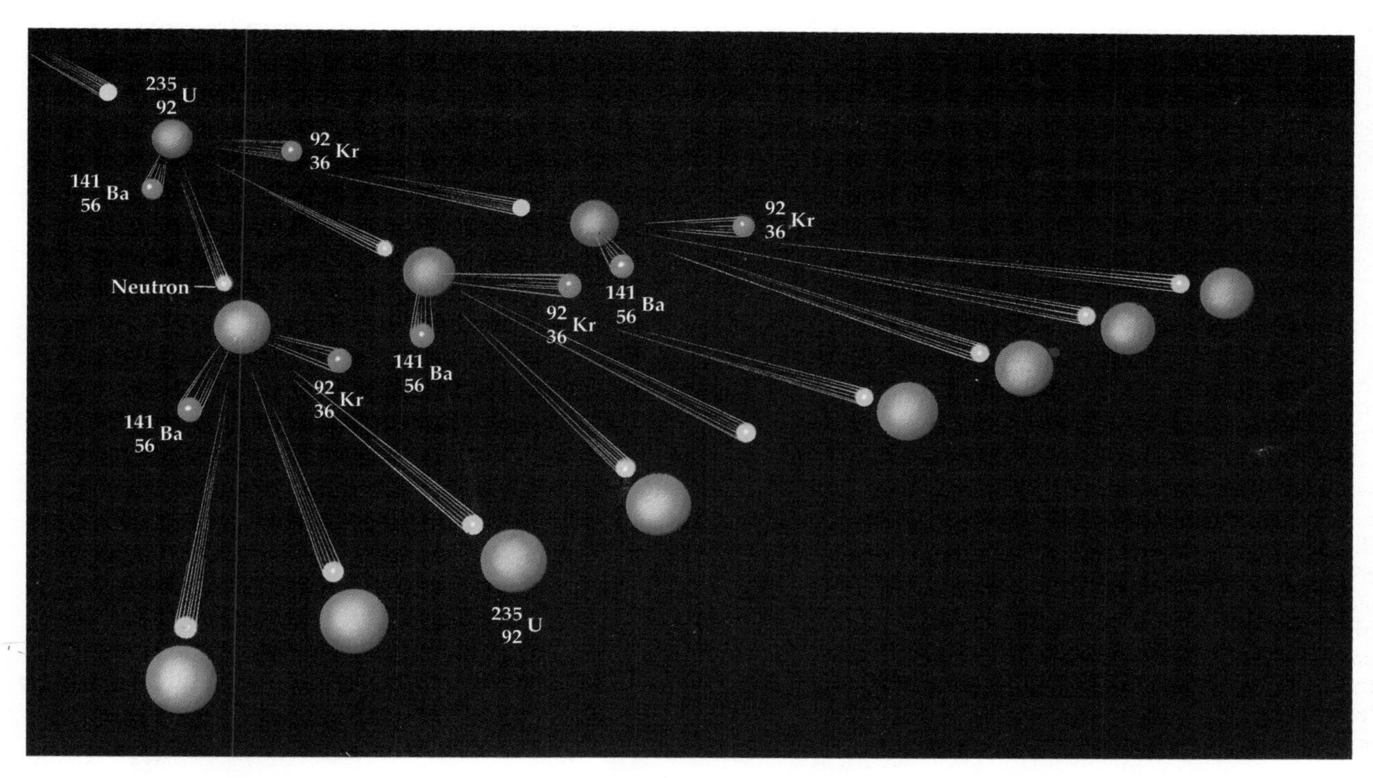

Figure 19.10
Illustration of a chain reaction initiated by capture of a stray neutron. (Many pairs of different isotopes are produced, but only one kind of pair is shown.)

for nuclear power fuel must be enriched. To accomplish this, some of the ^{238}U isotope in a sample is effectively discarded, thereby raising the concentration of ^{235}U, and one way to do this is by gaseous diffusion (as described in Section 12.2).

There is of course some controversy surrounding the use of nuclear power plants, particularly in the United States. Their proponents regard nuclear power to be an essential part of an advancing, technologically dependent society. The health of our economy and our standard of living are dependent on inexpensive, reliable, and safe sources of energy. Just within the past few years the demand for electric power has once again begun to exceed the supply, so many believe nuclear power plants should be built to meet the demand. Nuclear power plants are capable of supplying these demands, and they can be the source of "clean" energy in that they do not pollute the atmosphere with ash, smoke, or oxides of sulfur, nitrogen, or carbon. In addition, they help to ensure that our supplies of fossil fuels will not be depleted in the near future, and they free us of dependence on such fuels from other countries. There are currently more than 100 operating plants in the United States, and more than 350 worldwide. The nuclear

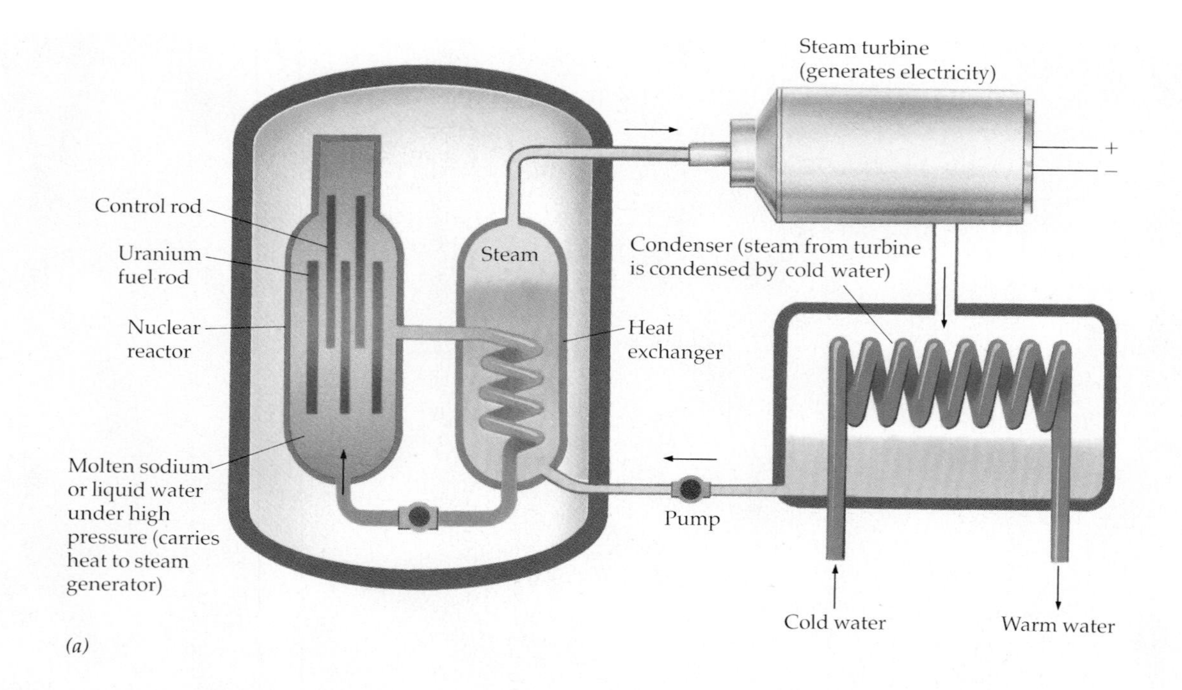

(a)

(b)

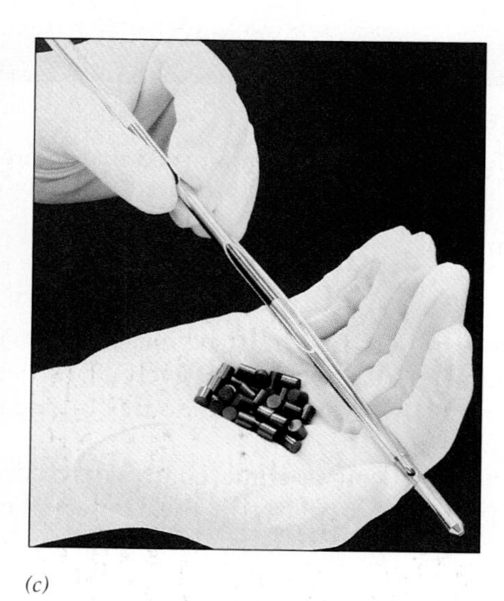

(c)

Figure 19.11

Nuclear power plants. (a) Liquid water (or liquid sodium) is circulated through the reactor, where the liquid is heated to about 325 °C. When this hot liquid is circulated through a steam generator, water in the generator is turned to steam, which in turn drives a steam turbine. After it passes through the turbine, the steam is converted back to liquid water and is recirculated through the steam generator. Enormous quantities of outside cooling water from rivers or lakes are necessary to condense the steam. (This basic system is the same as in any power plant, except that the water or circulating liquid may be heated initially by coal, gas, or oil-fired burners.) (b) A nuclear power plant at Indian Pointe, New York. (c) Uranium pellets used in the reactor fuel rods. *(b, Joe Azzara/The Image Bank; c, D.O.E./Science Source/Photo Researchers, Inc.)*

plants in the United States supply about 20% of the nation's electric energy; only coal-fired plants contribute a greater share (57%) (Figure 19.12).

There are *no* new nuclear power plants now under construction in the United States because these plants do have disadvantages. One problem is presented by the reactor fission products. Although some are put to various uses (Section 19.9), many are not suitable as a fuel or for other purposes. Since these products are often highly radioactive, their disposal poses an enormous problem. Perhaps the most reasonable suggestion is that radioactive wastes can be converted to a glassy material having a volume of about 2 m^3 per reactor per year; this relatively small volume of material can then be stored underground in geological formations, such as salt deposits, that are known to be stable for hundreds of millions of years.

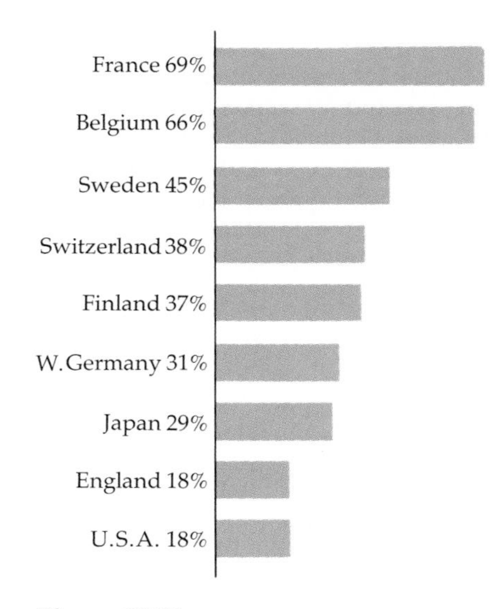

Figure 19.12
The approximate share of electricity generated by nuclear power in various countries.

19.7 NUCLEAR FUSION

Tremendous amounts of energy are generated when comparatively light nuclei combine to form heavier nuclei. Such a reaction is called **nuclear fusion,** and one of the best examples is the fusion of hydrogen nuclei (protons) to give helium nuclei.

$$4\,{}^{1}_{1}\mathrm{H} \longrightarrow {}^{4}_{2}\mathrm{He} + 2\,{}^{0}_{+1}\beta \qquad \Delta E = -2.5 \times 10^{9}\ \mathrm{kJ}$$

This reaction is the source of the energy from our sun and other stars, and it is the beginning of the synthesis of the elements in the universe. Temperatures of 10^6 to 10^7 K, found in the core and radiative zone of the sun, are required to bring the positively charged nuclei together with enough kinetic energy to overcome nuclear repulsions.

Deuterium—heavy hydrogen—can also be fused to give helium-3,

$${}^{2}_{1}\mathrm{H} + {}^{2}_{1}\mathrm{H} \longrightarrow {}^{3}_{2}\mathrm{He} + {}^{1}_{0}\mathrm{n} \qquad \Delta E = -3.2 \times 10^{8}\ \mathrm{kJ}$$

or deuterium can be fused with tritium, a radioactive isotope of hydrogen, to give helium-4.

$${}^{2}_{1}\mathrm{H} + {}^{3}_{1}\mathrm{H} \longrightarrow {}^{4}_{2}\mathrm{He} + {}^{1}_{0}\mathrm{n} \qquad \Delta E = -1.7 \times 10^{9}\ \mathrm{kJ}$$

Both of these reactions evolve an enormous quantity of energy, so it has been the dream of nuclear physicists to try to harness them to provide power for the nations of the world.

At the very high temperatures that allow fusion reactions to occur rapidly, atoms do not exist as such; instead, there is a **plasma** consisting of unbound nuclei and electrons. In order to achieve the high temperatures required for the fusion reaction of the hydrogen bomb, a fission bomb (atomic bomb) is first set off. One type of hydrogen bomb depends on the production of tritium (${}^{3}_{1}\mathrm{H}$) in the bomb. In this type, lithium-6 deuteride (LiD, a solid salt) is placed around an ordinary ${}^{235}_{92}\mathrm{U}$ or ${}^{239}_{94}\mathrm{Pu}$ fission bomb, and the fission is set off in the usual way. A ${}^{6}_{3}\mathrm{Li}$ nucleus absorbs one of the neutrons produced and splits into tritium and helium.

$${}^{6}_{3}\mathrm{Li} + {}^{1}_{0}\mathrm{n} \longrightarrow {}^{3}_{1}\mathrm{H} + {}^{4}_{2}\mathrm{He}$$

The temperature reached by the fission of uranium or plutonium is high enough to bring about the fusion of tritium and deuterium and the release of

Figure 19.13

Schematic diagram of an apparatus for laser-induced fusion. Tiny glass pellets (microballoons about 0.1 mm in diameter) filled with frozen deuterium and tritium are subjected to a powerful laser beam, and the contents undergo nuclear fusion.

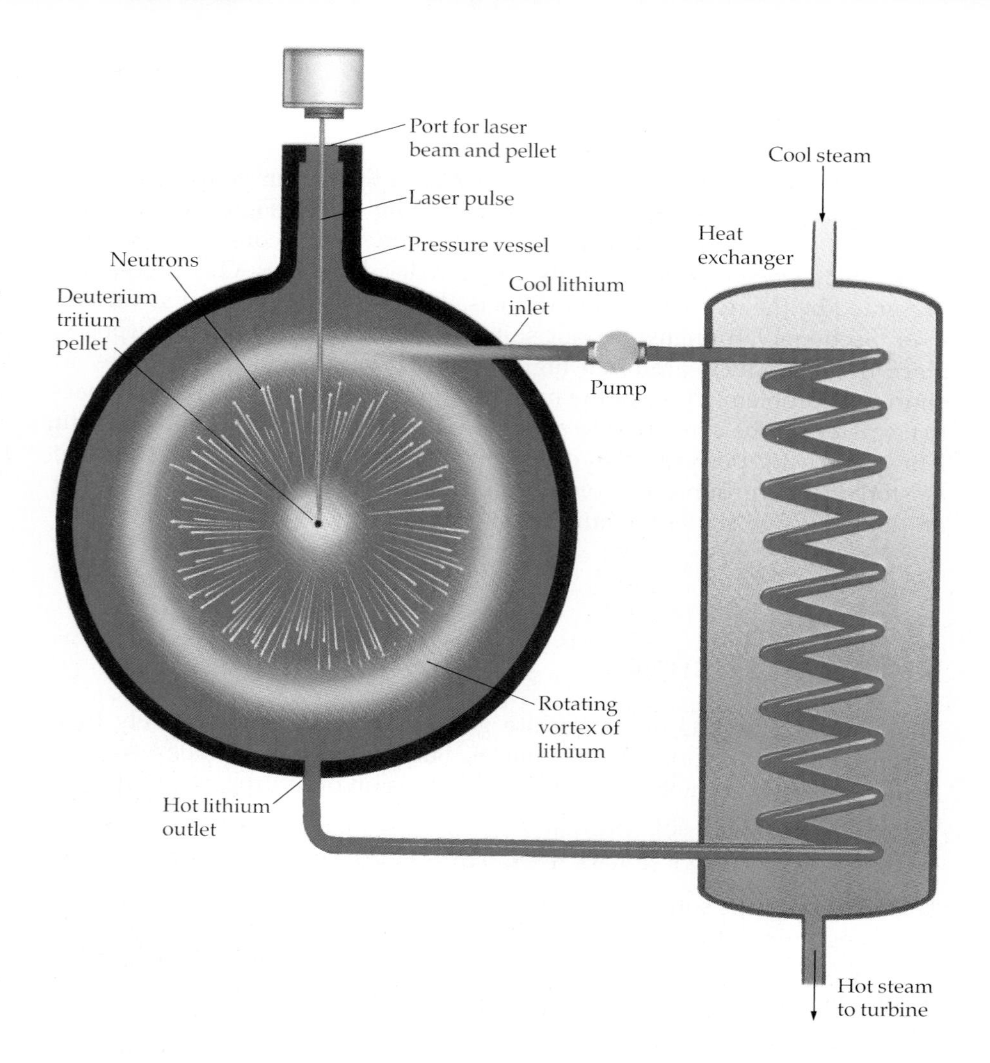

1.7×10^9 kJ per mole of ^3He. A 20-megaton bomb usually contains about 300 lb of lithium deuteride, as well as a considerable amount of plutonium and uranium.

Controlling a nuclear fusion reaction, in order to harness it for peaceful uses, has been extraordinarily difficult and has not yet been achieved. Three critical requirements must be met for controlled fusion. First, the temperature must be high enough for fusion to occur. The fusion of deuterium and tritium, for example, requires a temperature of 100 million degrees or more. Second, the plasma must be confined long enough to release a net output of energy. Third, the energy must be recovered in some usable form.

One experimental approach to controlling fusion is based on a laser system. Many high-energy laser beams simultaneously strike a tiny hollow glass sphere called a **microballoon,** which encloses a fuel consisting of equal parts of deuterium and tritium gas at high pressure. Figure 19.13 shows how the experiment can be set up and how energy is derived from the fusion reaction by hot, circulating lithium vapor.

Containment is one of the biggest problems in developing controlled fusion.

There are a number of attractive features that encourage research in controlled nuclear fusion. For example, the hydrogen fuel (in water) is cheap and abundant. Furthermore, most radioisotopes produced by fusion have short half-lives and so are a serious radiation hazard for only a short time.

Unfortunately, fusion reactions have not yet been "controlled." No physical container can contain the plasma without cooling it below the critical fusion temperature. Magnetic "bottles" (enclosures in space bounded by magnetic fields) have confined the plasma, but not for long enough periods.

The possibility of "cold fusion," the room-temperature fusion of deuterium atoms to provide energy, was announced in early 1989. Although this "discovery" is now largely discredited, work continues. The reaction to the announcement by the international scientific community illustrates well how the process of scientific investigation works. See J. Chem. Educ., *Vol. 66, p. 449, 1989.*

19.8 RADIATION EFFECTS AND UNITS OF RADIATION

All three types of radiation (alpha, beta, and gamma) disrupt normal cell processes in living organisms, and the potential for serious radiation damage to humans is well known. The biological effects of the atomic bombs exploded at Hiroshima and Nagasaki, Japan, at the close of World War II in 1945 have been well documented. However, controlled exposure can be beneficial in destroying unwanted tissue, as in the radiation therapy used in treating some types of cancer.

To quantify radiation and its effects, particularly on humans, several units have been developed. For example, the **röntgen** (R) is used to give the dosage of x rays and γ rays, where one röntgen corresponds to the *deposition* of 93.3×10^{-7} J per gram of tissue. The **rad** is similar to the röntgen, but it measures the amount of radiation *absorbed;* 1 rad represents a dose of 1.00×10^{-5} J absorbed per gram of material.

To quantify the biological effects of radiation in general, the **rem** (standing for *r*öntgen *e*quivalent *m*an) is used. One rem is a dose of any radiation that has the effect of 1 R. Since one rem is a large amount of radiation, the millirem or mrem is commonly used, where 1 mrem = 10^{-3} rem.

Another unit applied to radioactive substances is the bequerel (Bq), where 1 Bq = 1 disintegration per second.

Finally, the **curie** (Ci) is commonly used as a unit of activity. One curie represents the quantity of any radioactive isotope that undergoes 3.7×10^{10} disintegrations per second.

Humans are constantly exposed to natural and artificial **background radiation,** estimated to be about 200 mrem per year (Table 19.3). More than half of this is from natural background radiation sources: cosmic radiation and radioactive elements and minerals found naturally in the earth and air.

Cosmic radiation, emitted by the sun and other stars, continually bombards the earth and accounts for about 40% of background radiation. The remainder comes from elements such as ^{40}K. Since potassium (which is present to the extent of about 0.3 g/kg of soil) is essential to all living organisms, we all carry some radioactive potassium. Other radioactive elements found in some abundance on the earth are thorium-232, uranium-238, and radium-226. Thorium, for example, is found to the extent of 12 g/1000 kg of soil. Its oxide, ThO_2, glows very brightly when heated, so it was used until very recently in the mantles of lanterns that are used by campers.

Roughly 17% of our annual exposure comes from medical procedures such as diagnostic x rays and the use of radioactive compounds to trace the body's functions. Finally, another 17% comes from such sources as the radioactive products from testing nuclear explosives in the atmosphere, x-ray generators, televisions, nuclear power plants and their wastes, nuclear weapons manufacture, and nuclear fuel processing.

Burning fossil fuels (coal and oil) releases naturally occurring radioactive isotopes into the atmosphere. This has added significantly to the background radiation in recent years.

Table **19.3**

Radiation Exposure for One Year from Natural and Artificial Sources*

	Millirem/Year	Percentage
Natural Sources		
cosmic radiation	50.0	25.8
the earth	47.0	24.2
building materials	3.0	1.5
inhaled from the air	5.0	2.6
elements found naturally in human tissues	21.0	10.8
Subtotal	126.0	64.9
Medical Sources		
diagnostic x rays	50.0	25.8
radiotherapy	10.0	5.2
internal diagnosis	1.0	0.5
Subtotal	61.0	31.5
Other Artificial Sources		
nuclear power industry	0.85	0.4
luminous watch dials, TV tubes, industrial wastes	2.0	1.0
fallout from nuclear testing	4.0	2.1
Subtotal	6.9	3.5
Total	193.9	99.9

*From J. R. Amend, B. P. Mundy, and M. T. Armold: *General, Organic, and Biochemistry,* 2nd ed., p. 356. Philadelphia, Saunders College Publishing, 1993.

Radon

Radon is a chemically inert gas, in the same periodic group as helium, neon, argon, and krypton. The trouble with radon is that it is radioactive. As Figure 19.3 shows, $^{222}_{86}\text{Rn}$ is part of the chain of events beginning with the decay of uranium-238. (Other isotopes of Rn are products of other decay series.) In February 1989, *Chemical and Engineering News* said that "In three short years, the radioactive gas radon has progressed from relative obscurity to a cause of high anxiety as an indoor air pollutant."*

Radon occurs naturally in our environment. Since it comes from natural uranium deposits, the amount depends on local geology, but it is believed to account for less than half of normal background radioactivity. Furthermore, since the gas is chemically inert, and has a relatively long half-life (3.82 days), it is not trapped by chemical processes in the soil or water, and it is free to seep up from the ground and into underground mines or into homes through pores in block walls, cracks in the basement floor or walls, or around pipes. When breathed by humans occupying that space, the radon-222 isotope can decay inside the lungs to give polonium, a radioactive element that is not a gas and is not chemically inert.

$$^{222}_{86}\text{Rn} \longrightarrow {}^{4}_{2}\text{He} + {}^{218}_{84}\text{Po} \qquad t_{1/2} = 3.82\ \text{d}$$
$$^{218}_{84}\text{Po} \longrightarrow {}^{4}_{2}\text{He} + {}^{214}_{82}\text{Pb} \qquad t_{1/2} = 3.05\ \text{m}$$

*D.J. Hanson, *Chemical and Engineering News,* February 6, 1989, page 7.

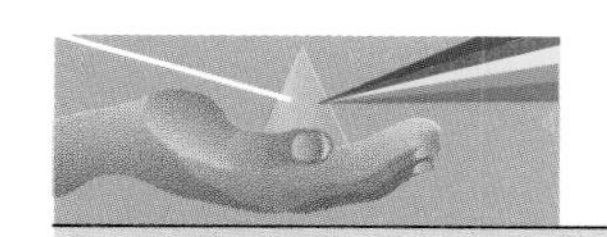

CHEMISTRY YOU CAN DO

The Committee on Biological Effects of Ionizing Radiation of the National Academy of Sciences issued a report in 1980 that contained a survey for an individual to evaluate his or her exposure to ionizing radiation. The following table is adapted from this report. By adding up your exposure, you can compare your annual dose to the United States annual average of 180 to 200 mrem.

Adapted from A.R. Hinrichs, *Energy*, pp. 335–336. Philadelphia, Saunders College Publishing, 1992.

	Common Sources of Radiation	**Your Annual Dose (mrem)**
Where You Live	**Location:** Cosmic radiation at sea level	26
	For your elevation (in feet), add this number of mrem	______
	Elevation 1000: 2 mrem; 2000: 5; 3000: 9; 4000: 15; 5000: 21; 6000: 29; 7000: 40; 8000: 53; 9000: 70	
	Ground: U.S. average	26
	House construction: For stone, concrete, or masonry building, add 7	______
What You Eat, Drink, and Breathe	**Food, water, air:** U.S. Average	24
	Weapons test fallout	4
How You Live	**X-ray and radiopharmaceutical diagnosis**	
	Number of chest x-rays ______ × 10	______
	Number of lower gastrointestinal tract x-rays ______ × 500	______
	Number of radiopharmaceutical examinations ______ × 300	______
	(Average dose to total U.S. population = 92 mrem)	
	Jet plane travel: For each 2500 miles add 1 mrem	______
	TV viewing: Number of hours per day ______ × 0.15	______
How Close You Live to a Nuclear Plant	**At site boundary:** average number of hours per day ______ × 0.2	______
	One mile away: average number of hours per day ______ × 0.02	______
	Five miles away: average number of hours per day ______ × 0.002	______
	Over 5 miles away: .... none	______
	Note: Maximum allowable dose determined by "as low as reasonably achievable" (ALARA) criteria established by the U.S. Nuclear Regulatory Commission. Experience shows that your actual dose is substantially less than these limits.	
	Your total annual dose in mrem	______

The elevation sub-table from the "Where You Live" section:

Elevation	mrem	Elevation	mrem	Elevation	mrem
1000	2	4000	15	7000	40
2000	5	5000	21	8000	53
3000	9	6000	29	9000	70

Compare your annual dose to the U.S. annual average of 180 mrem.
One mrem per year is equal to increasing your diet by 4%, or taking a 5-day vacation in the Sierra Nevada (CA) mountains.

*Based on the "BEIR Report III"—National Academy of Sciences, Committee on Biological Effects of Ionizing Radiation, "The Effects on Populations of Exposure to Low Levels of Ionizing Radiation," National Academy of Sciences, Washington, DC, 1980.

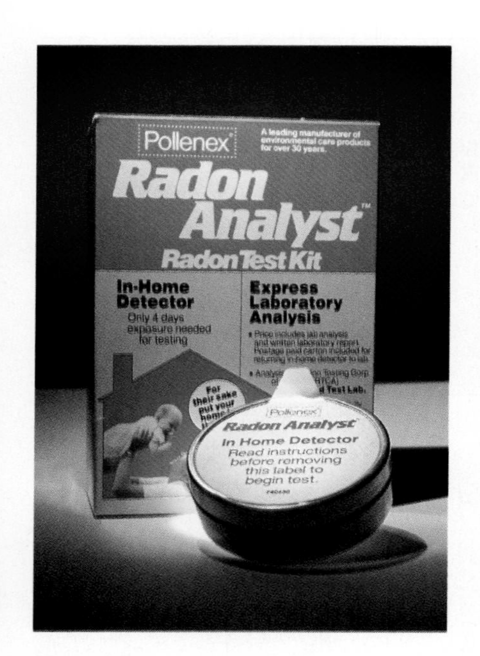

Figure 19.14
A commercially available kit for testing for radon gas in the home.
(C.D. Winters)

Therefore, polonium-218 can lodge in body tissues where it undergoes alpha decay to give lead-214, itself a radioactive isotope. The range of an alpha particle is quite small, perhaps 0.7 mm (about the thickness of a sheet of paper). However, this is approximately the thickness of the epithelial cells of the lungs, so the radiation can damage these tissues and induce lung cancer.

Virtually every home in the United States is believed to have some level of radon gas. To test for the presence of the gas, you can purchase testing kits of various kinds (Figure 19.14). There is currently a great deal of controversy over the level of radon that is considered "safe." The U.S. Environmental Protection Agency has set a standard of 4 picocuries per liter of air as an "action level." There are some who believe 1.5 picocuries is close to the average level, and that only about 2% of the homes will contain over 8 picocuries per liter. If your home shows higher levels of radon gas than this, you should probably have it tested further and perhaps take corrective actions such as sealing cracks around the foundation and in the basement. But keep in mind the relative risks involved (see Chapter 1). A 1.5 picocurie/liter level of radon leads to a lung cancer risk about the same as the risk of your dying in an accident in your home.

19.9 APPLICATIONS OF RADIOACTIVITY

Food Irradiation

Although uncontrolled radioactivity may be harmful, the radiation from radioisotopes can be put to beneficial use. For example, consider the importance of killing pests that would destroy food during storage. In some parts of the world stored-food spoilage may claim up to 50% of the food crop. In our society, refrigeration, canning, and chemical additives lower this figure considerably. Still, there are problems with food spoilage, and food protection costs amount to a sizable fraction of the final cost of food. Food irradiation with gamma rays from sources such as ^{60}Co and ^{137}Cs is commonly used in European countries, Canada, and Mexico. Some irradiated foods are sold in the United States as well. Foods may be pasteurized by irradiation to retard the growth of organisms such as bacteria, molds, and yeasts. This irradiation prolongs shelf life under refrigeration in much the same way that heat pasteurization protects milk. Chicken normally has a three-day refrigerated shelf life; after irradiation, it may have a three-week refrigerated shelf life.

The FDA may soon permit irradiation up to 100 kilorads for the pasteurization of foods. Radiation levels in the 1- to 5-megarad range sterilize; that is, every living organism is killed. Foods irradiated at these levels will keep indefinitely when sealed in plastic or aluminum-foil packages. However, the FDA is unlikely to approve irradiation sterilization of foods in the near future because of potential problems caused by as yet undiscovered, but possible, "unique radiolytic products." For example, irradiation sterilization might produce a chemical substance that is capable of causing genetic damage. To prove or disprove the presence of these substances, animal feeding studies using irradiated foods are now being conducted.

More than 40 classes of foods are already irradiated in 24 countries. In the United States, only a small number of foods may be irradiated (Figure 19.15 and Table 19.4).

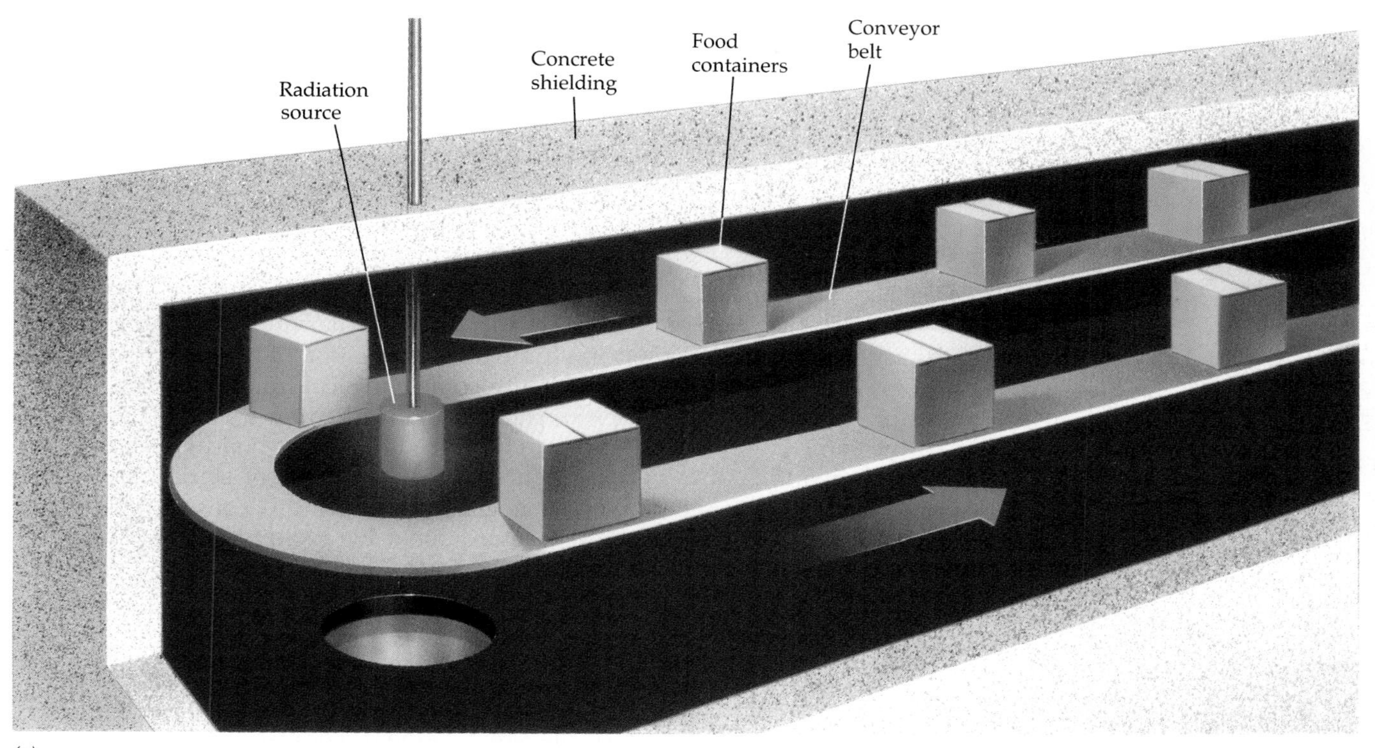

(a)

(b)

Figure 19.15

A typical commercial food irradiator. (a) Boxes of food are conveyed into the shielded chamber and around the radiation source (center). When not in use, the source can be lowered into a pool of water below. (b) Irradiated strawberries. *(b, Nordion International)*

Table 19.4

Examples of Irradiated Foodstuffs

Food	Purpose	Status
potatoes	retardation of sprouts	FDA approved
wheat	insect disinfection	FDA approved
wheat flour	insect disinfection	FDA approved
spices	retardation of microbe growth	FDA approved
grapefruit	mold control	approved for export
strawberries	mold control	approved for export
fish	microbe control	approved for export
shrimp	microbe control	approved for export

Recent studies indicate that there may be harmful health effects from several common agricultural fumigants; irradiation of fruits and vegetables could be an effective alternative to some chemical fumigants. The agricultural products may be picked, packed, and readied for shipment. After that, the entire shipping container can be passed through a building containing a strong source of radiation (Figure 19.15). This type of sterilization offers greater worker safety because it lessens chances of exposure to harmful chemicals, and it protects the environment by avoiding contamination of water supplies with these toxic chemicals.

Radioactive Tracers

The chemical behavior of a radioisotope is almost identical to that of the nonradioactive isotopes of the same element, because the energies of the valence electrons are nearly the same in both atoms. Therefore, chemists can use radioactive isotopes as **tracers** in chemical reactions and biological processes. To use a tracer, a chemist prepares a reactant compound in which one of the elements consists of both radioactive and stable isotopes, and introduces it into the reaction (or feeds it to an organism). After the reaction, the chemist measures the radioactivity of the products (or determines which parts of the organism contain the radioisotope) by using a Geiger counter or similar instrument. Several radioisotopes commonly used as tracers are listed in Table 19.5.

For example, plants are known to take up phosphorus-containing compounds from the soil through their roots. The use of the radioactive phosphorus isotope ^{32}P, a beta-emitter, presents a way not only of detecting the uptake of phosphorus by a plant but also of measuring the speed of uptake under various conditions. Plant biologists can grow hybrid strains of plants that can absorb phosphorus quickly, and then they can test this ability with the radioactive phosphorus tracer. This type of research leads to faster maturing crops, better yields per acre, and more food or fiber at less expense.

Important characteristics of a pesticide can be measured by tagging the pesticide with radioisotopes that have short half-lives and then applying it to a test field. Following the tagged pesticide can provide information on its

Table 19.5

Radioisotopes Used as Tracers

Isotope	Half-Life	Use
^{14}C	5730 yr	CO_2 for photosynthesis research
^{3}H	12.26 yr	tag hydrocarbons
^{35}S	87.9 d	tag pesticides, measure air flow
^{32}P	14.3 d	measure phosphorus uptake by plants

tendency to accumulate in the soil, to be taken up by the plant, and to accumulate in run-off surface water. This is done with a high degree of accuracy by counting the disintegrations of the radioactive tracer. After these tests are completed, the radioactive isotopes in the tagged pesticides decay to a harmless level in a few days or a few weeks because of the short half-lives of the species used. This type of research leads to safer, more effective pesticides.

Medical Imaging

Radioactive isotopes are also used in **nuclear medicine** in two different ways, diagnosis and therapy. In the diagnosis of internal disorders such as tumors, physicians need information on the locations of abnormal tissue. This is done by **imaging,** a technique in which the radioisotope, either alone or combined with some other chemical, accumulates at the site of the disorder. There, acting like a homing device, the radioisotope disintegrates and emits its characteristic radiation, which is detected. Modern medical diagnostic instruments not only determine where the radioisotope is located in the patient's body but also construct an image of the area within the body where the radioisotope is concentrated.

Four of the most common diagnostic radioisotopes are given in Table 19.6. All are made in a particle accelerator in which heavy, charged nuclear

Table 19.6

Diagnostic Radioisotopes

Radioisotope	Name	Half-Life (Hours)	Uses
^{99m}Tc*	technetium-99m	6.0	as TcO_4^- to the thyroid, brain, kidneys
^{201}Tl	thallium-201	74	to the heart
^{123}I	iodine-123	13.3	to the thyroid
^{67}Ga	gallium-67	77.9	to various tumors and abscesses

*The technetium-99m isotope is the one most commonly used for diagnostic purposes. The *m* stands for "metastable," a term explained in the text.

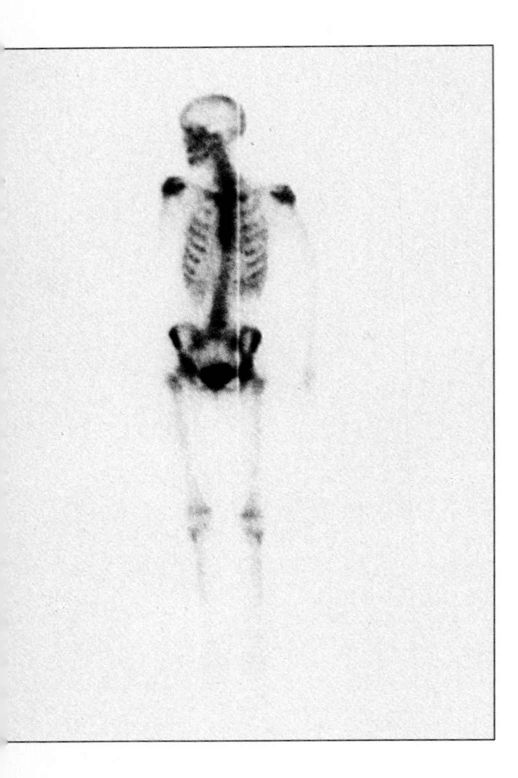

Figure 19.16
A whole body scan. Phosphate with technetium-99m was injected into the blood and then absorbed by the bones and kidneys. This picture was taken three hours after injection. *(SUNY Upstate Medical Center)*

particles are made to react with other atoms. Each of these radioisotopes produces gamma radiation, which in low doses is less harmful to the tissue than ionizing radiations such as beta or alpha particles. By the use of special carrier compounds, these radioisotopes can be made to accumulate in specific areas of the body. For example, the pyrophosphate ion, $P_4O_7^{4-}$, can bond to the technetium-99m radioisotope; together they accumulate in the skeletal structure where abnormal bone metabolism is occurring (Figure 19.16). The technetium-99m radioisotope is metastable, as denoted by the letter *m*; this term means that the nucleus loses energy by disintegrating to a more stable version of the same isotope,

$$^{99m}Tc \longrightarrow {}^{99}Tc + \gamma$$

and the gamma rays are detected. Such investigations often pinpoint bone tumors.

Positron emission tomography (PET) is a form of nuclear imaging that uses **positron emitters,** such as carbon-11, fluorine-18, nitrogen-13, or oxygen-15. All these radioisotopes are neutron deficient, have short half-lives, and therefore must be prepared in a cyclotron immediately before use. When these radioisotopes decay, a proton is converted into a neutron, a positron, and a neutrino (ν),

$${}^1_1p \longrightarrow {}^1_0n + {}^0_1\beta + \nu$$

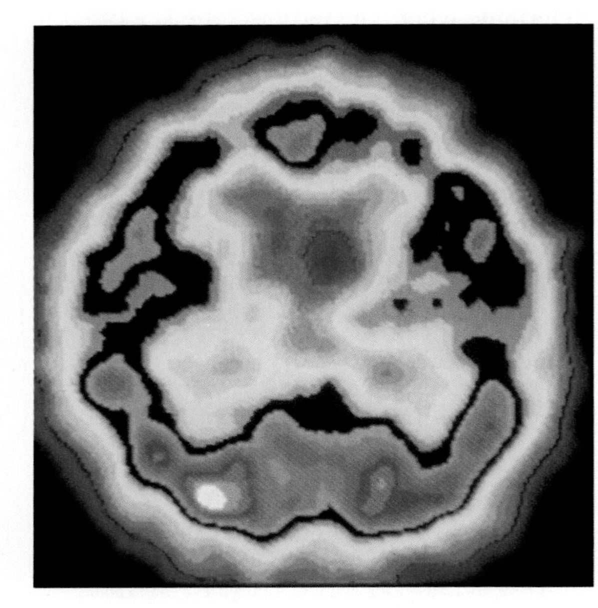

Figure 19.17
PET scan (positron emission tomography) of an axial section through a normal human brain. PET scans are obtained by injecting a tracer labelled with a short-lived radioactive isotope into the bloodstream, which concentrates in brain tissue and emits positrons that are recorded by a circular detector when the scan is performed. Here, radioactive methionine (an amino acid) has been used to show the level of activity of protein synthesis in the brain. *(CEA-ORSAY/CNRI/Science Photo Library/Photo Researchers, Inc.)*

Since matter is virtually transparent to neutrinos, they escape undetected, but the positron travels less than a few millimeters before it encounters an electron and undergoes antimatter-matter annihilation.

$$^{0}_{1}\beta + {}^{0}_{-1}\beta \longrightarrow 2\gamma$$

The annihilation event produces two gamma rays that radiate in opposite directions and are detected by two scintillation detectors located 180° apart in the PET scanner. By detecting several million annihilation gamma rays within a circular slice around the subject over approximately 10 minutes, the region of tissue containing the radioisotope can be imaged with computer signal-averaging techniques (Figure 19.17).

IN CLOSING

Having studied this chapter, you should be able to

- characterize the three major types of radiation observed in radioactive decay: alpha, beta, and gamma (Section 19.1).
- write a balanced equation for a nuclear reaction or transmutation (Section 19.2).
- decide whether a particular radioactive isotope will decay by alpha, beta, or positron emission or by electron capture (Sections 19.2 and 19.3).
- calculate the binding energy for a particular isotope and understand what this energy means in terms of nuclear stability (Section 19.3).
- use the equation $\ln(N/N_0) = -kt$ (Equation 19.2), which relates (through the decay constant k) the time period over which a sample is observed (t) to the number of radioactive atoms present at the beginning (N_0) and end (N) of the time period (Section 19.4).
- calculate the half-life of a radioactive isotope ($t_{1/2}$) from the activity of a sample, or use the half-life to find the time required for an isotope to decay to a particular activity (Section 19.4).
- describe nuclear chain reactions, nuclear fission, and nuclear fusion (Sections 19.6 and 19.7).
- describe some sources of background radiation and the units used to measure radiation (Section 19.8).
- relate some uses of radioisotopes (Section 19.9).

STUDY QUESTIONS

Review Questions

1. Name three people who made important contributions to nuclear chemistry and indicate their contributions.
2. Tell the similarities and differences in α particles, β particles, positrons, and gamma rays.
3. What is the binding energy of a nucleus?
4. If the mass number of an isotope is *much greater than* twice the atomic number, what type of radioactive decay might you expect?
5. If the number of neutrons in an isotope is *much less than* the number of protons, what type of radioactive decay might you expect?
6. What is the difference between fission and fusion? Illustrate your answer with an example of each.
7. Is there a nuclear reactor producing electric power in your state? If so, where is it located?
8. Part of the "trigger" for the hydrogen bomb is a reaction between tritium and deuterium. What type of reaction is this, fusion or fission?

$$^{3}_{1}H + ^{2}_{1}H \longrightarrow ^{1}_{0}n + ^{4}_{2}He$$

9. What are some of the advantages and disadvantages of nuclear power plants?
10. Name at least two uses of radioactive isotopes (outside of their use in power reactors and weapons).

Nuclear Reactions

11. Balance the following nuclear reactions. Write the mass number and atomic number for the remaining particle, as well as the element symbol where possible.
(a) $^{54}_{26}Fe + ^{4}_{2}He \longrightarrow 2\,^{1}_{1}H + ?$
(b) $^{27}_{13}Al + ^{4}_{2}He \longrightarrow ^{30}_{15}P + ?$
(c) $^{32}_{16}S + ^{1}_{0}n \longrightarrow ^{1}_{1}H + ?$
(d) $^{96}_{42}Mo + ^{2}_{1}H \longrightarrow ^{1}_{0}n + ?$
(e) $^{98}_{42}Mo + ^{1}_{0}n \longrightarrow ^{99}_{43}Tc + ?$

12. Balance the following nuclear reactions. Write the mass number and atomic number for the remaining particle, as well as the element symbol where possible.
(a) $^{9}_{4}Be + ? \longrightarrow ^{6}_{3}Li + ^{4}_{2}He$
(b) $? + ^{1}_{0}n \longrightarrow ^{24}_{11}Na + ^{4}_{2}He$
(c) $^{40}_{20}Ca + ? \longrightarrow ^{40}_{19}K + ^{1}_{1}H$
(d) $^{241}_{95}Am + ^{4}_{2}He \longrightarrow ^{243}_{97}Bk + ?$
(e) $^{246}_{96}Cm + ^{12}_{6}C \longrightarrow 4\,^{1}_{0}n + ?$
(f) $^{238}_{92}U + ? \longrightarrow ^{249}_{100}Fm + 5\,^{1}_{0}n$

13. Balance the following nuclear reactions. Write the mass number and atomic number for the remaining particle, as well as the element symbol where possible.
(a) $^{104}_{47}Ag \longrightarrow ^{104}_{48}Cd + ?$
(b) $^{87}_{36}Kr \longrightarrow ^{0}_{-1}\beta + ?$
(c) $^{231}_{91}Pa \longrightarrow ^{227}_{89}Ac + ?$
(d) $^{230}_{90}Th \longrightarrow ^{4}_{2}He + ?$
(e) $^{82}_{35}Br \longrightarrow ^{82}_{36}Kr + ?$
(f) $? \longrightarrow ^{24}_{12}Mg + ^{0}_{-1}\beta$

14. Balance the following nuclear reactions. Write the mass number and atomic number for the remaining particle, as well as the element symbol where possible.
(a) $^{19}_{10}Ne \longrightarrow ^{0}_{+1}\beta + ?$
(b) $^{59}_{26}Fe \longrightarrow ^{0}_{-1}\beta + ?$
(c) $^{40}_{19}K \longrightarrow ^{0}_{-1}\beta + ?$
(d) $^{37}_{18}Ar + ^{0}_{-1}e$ (electron capture) $\longrightarrow ?$
(e) $^{55}_{26}Fe + ^{0}_{-1}e$ (electron capture) $\longrightarrow ?$
(f) $^{26}_{13}Al \longrightarrow ^{25}_{12}Mg + ?$

15. One radioactive series that begins with uranium-235 and ends with lead-207 undergoes the sequence of emission reactions: $\alpha, \beta, \alpha, \beta, \alpha, \alpha, \alpha, \alpha, \beta, \beta, \alpha$. Identify the radioisotope produced in each of the *first five steps.*

16. One radioactive series that begins with uranium-235 and ends with lead-207 undergoes the sequence of emission reactions: $\alpha, \beta, \alpha, \beta, \alpha, \alpha, \alpha, \alpha, \beta, \beta, \alpha$. Identify the radioisotope produced in each of the *last six steps.* (See Study Question 15.)

Nuclear Stability

17. Boron has two stable isotopes, ^{10}B (abundance = 19.78%) and ^{11}B (abundance = 80.22%). Calculate the binding energies of these two nuclei per nucleon and compare their stabilities.

$$5\,^{1}_{1}H + 5\,^{1}_{0}n \longrightarrow ^{10}_{5}B$$

$$5\,^{1}_{1}H + 6\,^{1}_{0}n \longrightarrow ^{11}_{5}B$$

The required masses (in g/mol) are: $^{1}_{1}H = 1.00783$; $^{1}_{0}n = 1.00867$; $^{10}_{5}B = 10.01294$; and $^{11}_{5}B = 11.00931$.

18. Calculate the binding energy in kJ per mole of P for the formation of $^{30}_{15}P$

$$15\,^{1}_{1}H + 15\,^{1}_{0}n \longrightarrow ^{30}_{15}P$$

and for the formation of $^{31}_{15}P$.

$$15\,^{1}_{1}H + 16\,^{1}_{0}n \longrightarrow ^{31}_{15}P$$

Which is the more stable isotope? The required masses (in g/mol) are $^{1}_{1}H = 1.00783$; $^{1}_{0}n = 1.00867$; $^{30}_{15}P = 29.97832$; and $^{31}_{15}P = 30.97376$.

Rates of Disintegration Reactions

19. Copper-64 is used in the form of copper acetate to study brain tumors. It has a half-life of 12.8 hours. If you begin with 15.0 mg of copper acetate, what mass in milligrams remains after 2 days and 16 hours?

20. Gold-198 is used as the metal in the diagnosis of liver problems. The half-life of ^{198}Au is 2.7 days. If you begin with 5.6 mg of this gold isotope, what mass remains after 10.8 days?

21. Iodine-131 is used in the form of sodium iodide to treat cancer of the thyroid. (a) The isotope decays by ejecting a β particle. Write a balanced equation to show this process. (b) The isotope has a half-life of 8.05 days. If you begin with 25.0 mg of radioactive $Na^{131}I$, what mass remains after 32.2 days (about a month)?

22. Phosphorus-32 is used in the form of Na_2HPO_4 in the treatment of chronic myeloid leukemia, among other things. (a) The isotope decays by emitting a β particle. Write a balanced equation to show this process. (b) The half-life of ^{32}P is 14.3 days. If you begin with 9.6 mg of radioactive Na_2HPO_4, what mass remains after 28.6 days (about one month)?

23. In 1984, cobalt-60 was involved in the worst accident with radioactive isotopes in North America (*Science 84,* December 1984, page 28). The half-life of the isotope is 5.3 years. Starting with 10.0 mg of ^{60}Co, how much will remain after 21.2 years? Approximately how much will remain after a century?

24. Gallium-67 ($t_{1/2}$ = 77.9 hours) is used in the medical diagnosis of certain kinds of tumors. If you ingest a compound containing 0.15 mg of this isotope, what mass in milligrams will remain in your body after 13 days?

25. Radioisotopes of iodine are widely used in medicine. For example, iodine-131 ($t_{1/2}$ = 8.05 days) is used to treat thyroid cancer. If you ingest a sample of NaI containing ^{131}I, how much time is required for the isotope to fall to 5.0% of its original activity?

26. The rare gas radon has been the focus of much attention recently because it can be found in homes. Radon-222 emits α particles and has a half-life of 3.82 days. (a) Write a balanced equation to show this process. (b) How long does it take for a sample of radon to decrease to 10.0% of its original activity?

27. A sample of wood from a Thracian chariot found in an excavation in Bulgaria has a ^{14}C activity of 11.2 disintegrations per minute per gram. Estimate the age of the chariot and the year it was made. ($t_{1/2}$ for ^{14}C is 5.73×10^3 years and the activity of ^{14}C in living material is 14.0 disintegrations per minute per gram.)

28. A piece of charred bone found in the ruins of an American Indian village has a ^{14}C to ^{12}C ratio of 0.72 times that found in living organisms. Calculate the age of the bone fragment. (See Study Question 27 for required data on carbon-14.)

Nuclear Transmutations

29. There are two isotopes of americium, both with half-lives sufficiently long to allow the handling of massive quantities. Americium-241, for example, has a half-life of 248 years as an α emitter, and it is used in gauging the thickness of materials and in smoke detectors. The isotope is formed from ^{239}Pu by absorption of two neutrons followed by emission of a β particle. Write a balanced equation for this process.

30. Americium-240 is made by bombarding a plutonium-239 atom with an α particle. In addition to ^{240}Am, the products are a proton and two neutrons. Write a balanced equation for this process.

31. To synthesize the heavier transuranium elements, one must bombard a lighter nucleus with a relatively large particle. If you know the products are californium-246 and 4 neutrons, with what particle would you bombard uranium-238 atoms?

32. The element with the highest known atomic number is 109. It is thought that still heavier elements are possible, especially with $Z = 114$ and $N = 184$. To this end, serious attempts have been made to force calcium-40 and curium-248 to merge. What would be the atomic number of the element formed?

Nuclear Fission and Power

33. The average energy output of a good grade of coal is 2.6×10^7 kJ/ton. Fission of one mole of ^{235}U releases 2.1×10^{10} kJ. Find the number of tons of coal needed to produce the same energy as one pound of ^{235}U. (See Appendix C for conversion factors.)

34. A concern in the nuclear power industry is that, if nuclear power becomes more widely used, there may be serious shortages in worldwide supplies of fissionable uranium. One solution is to build "breeder" reactors

that manufacture more fuel than they consume. One such cycle works as follows:

(a) A "fertile" ^{238}U nucleus collides with a neutron to produce ^{239}U.

(b) ^{239}U decays by β emission ($t_{1/2}$ = 24 min) to give an isotope of neptunium.

(c) This neptunium isotope decays by β emission to give a plutonium isotope.

(d) The plutonium isotope is fissionable. On collision of one of these plutonium isotopes with a neutron, fission occurs with energy, at least two neutrons, and other nuclei as products.

Write an equation for each of the steps, and explain how this process can be used to breed more fuel than the reactor originally contained and still produce energy.

Uses of Radioisotopes

35. In order to measure the volume of the blood system of an animal, the following experiment was done. A 1.0-mL sample of an aqueous solution containing tritium with an activity of 2.0×10^6 disintegrations per second (dps) was injected into the bloodstream. After time was allowed for complete circulatory mixing, a 1.0-mL blood sample was withdrawn and found to have an activity of 1.5×10^4 dps. What was the volume of the circulatory system? (The half-life of tritium is 12.3 years, so this experiment assumes that only a negligible amount of tritium has decayed in the time of the experiment.)

36. Radioactive isotopes are often used as "tracers" to follow an atom through a chemical reaction, and the following is an example. Acetic acid reacts with methyl alcohol, CH_3OH, by eliminating a molecule of H_2O to form methyl acetate, $CH_3CO_2CH_3$. Explain how you would use the radioactive isotope ^{18}O to show whether the oxygen atom in the water product comes from the —OH of the acid or the —OH of the alcohol.

$$H_3C{-}\overset{\overset{\large O}{\|}}{C}{-}OH + HOCH_3 \longrightarrow H_3C{-}\overset{\overset{\large O}{\|}}{C}{-}O{-}CH_3 + H_2O$$

acetic acid — methyl alcohol — methyl acetate

General Questions

37. Balance the following nuclear reactions. Write the mass number and atomic number for the remaining particle, as well as the element symbol where possible.

(a) $^{13}_{6}C + ? \longrightarrow {}^{14}_{6}C$

(b) $^{40}_{18}Ar + ? \longrightarrow {}^{43}_{19}K + {}^{1}_{1}H$

(c) $^{250}_{98}Cf + {}^{11}_{5}B \longrightarrow 4\,{}^{1}_{0}n + ?$

(d) $^{53}_{24}Cr + {}^{4}_{2}He \longrightarrow ? + {}^{56}_{26}Fe$

(e) $^{212}_{84}Po \longrightarrow {}^{208}_{82}Pb + ?$

(f) $^{122}_{53}I \longrightarrow {}^{0}_{+1}\beta + ?$

(g) $? \longrightarrow {}^{23}_{11}Na + {}^{0}_{-1}\beta$

(h) $^{137}_{53}I \longrightarrow {}^{1}_{0}n + ?$

38. The following reaction sequence involves four unknown elements, Q, Δ, Σ, and Π. Based on the reactions in the sequence, identify the unknown elements.

$$Pb + Cr \longrightarrow Q$$
$$Cf + O \longrightarrow Q$$
$$Q \longrightarrow \alpha + \Delta$$
$$\Delta \longrightarrow \alpha + \Sigma$$
$$\Sigma \longrightarrow \alpha + \Pi$$

39. The oldest known fossil cells form a biological cluster found in South Africa. The fossil has been dated by the reaction

$$^{87}Rb \longrightarrow {}^{87}Sr + {}^{0}_{-1}\beta \qquad t_{1/2} = 4.9 \times 10^{10}\ y$$

If the ratio of the present quantity of ^{87}Rb to the original quantity is 0.951, calculate the age of the fossil cells.

40. Balance the following reactions used for the synthesis of transuranium elements.

(a) $^{238}_{92}U + {}^{14}_{7}N \longrightarrow ? + 5\,{}^{1}_{0}n$

(b) $^{238}_{92}U + ? \longrightarrow {}^{249}_{100}Fm + 5\,{}^{1}_{0}n$

(c) $^{253}_{99}Es + ? \longrightarrow {}^{256}_{101}Md + {}^{1}_{0}n$

(d) $^{246}_{96}Cm + ? \longrightarrow {}^{254}_{102}No + 4\,{}^{1}_{0}n$

(e) $^{252}_{98}Cf + ? \longrightarrow {}^{257}_{103}Lr + 5\,{}^{1}_{0}n$

41. On December 2, 1942, the first man-made self-sustaining nuclear fission chain reactor was operated by Enrico Fermi and others under the University of Chicago Stadium. In June, 1972, *natural* fission reactors, which operated billions of years ago, were discovered in Oklo, Gabon. At present, natural uranium contains 0.72% ^{235}U. How many years ago did natural uranium contain 3.0% ^{235}U, sufficient to sustain a natural reactor? ($t_{1/2}$ for ^{235}U is 7.04×10^8 y.)

Using Computer Programs and Videodiscs

42. Use *KC? Discoverer* to find all the elements that are radioactive. How many are there?
43. Using *KC? Discoverer* and the *Periodic Table Videodisc,* find uses for the actinide elements actinium, thorium, uranium, plutonium, and americium.
44. A nuclear chain reaction is simulated on the *Chemical Demonstrations* videodisc and videotapes (Demonstration 49; Disc 2, side 2).
    (a) What observations do you make in the course of this demonstration?
    (b) How does the rate of "reaction" change as the demonstration proceeds? Why does it change?
    (c) What change in "reaction" rate would you observe if only one rubber stopper was used on each mouse trap instead of two?

Clear air in the early morning over Grand Teton National Park, Wyoming.
(J. Kotz)

Chemistry of the Atmosphere

CHAPTER

20

You could survive for a few weeks without food, and for a few days without water, but you need a supply of fresh air every few minutes. Earth's atmosphere is a thin, protective blanket of gases that nurtures life and protects it from the hostile environment of outer space. Air contains oxygen needed for animal respiration and carbon dioxide needed for plant photosynthesis, it carries water vapor from the oceans to the land masses where it is needed to sustain life, and it receives, dilutes, and destroys by chemical reactions many pollutants produced both naturally and by human activities.

With so many substances mixed together in the atmosphere, it should come as no surprise that there are many chemical reactions going on. Sunlight plays an important role in atmospheric chemistry, because photons of visible and ultraviolet light contain sufficient energy to break chemical bonds or raise molecules to higher-energy, more reactive states. Absorption of photons by ozone molecules in the stratosphere protects us from harmful ultraviolet radiation from the sun, but absorption of photons by nitrogen dioxide molecules in urban atmospheres results in photochemical smog.

The higher the density of population and industry, the more varied the air pollutants and the higher their concentrations. It has become apparent that global atmospheric concentrations of some substances can be increased to the point where they may cause major changes in the natural environment. This chapter deals with the most important substances and reactions in earth's atmosphere.

More than 99% of the total mass of the atmosphere is found within 30 km (20 miles) of the earth's surface. Compared with the earth's diameter of 7918 miles, the atmosphere is like two or three outer layers of an onion compared to the whole onion.

A metric ton is 1000 kg.

Planet Earth is enveloped by a few vertical miles of chemicals that compose the gaseous medium in which we exist—the atmosphere. The total mass of the atmosphere is approximately 5.1×10^{15} metric tons, a huge figure, but still only about one millionth of the earth's total mass. Close to the earth's surface the atmosphere is mostly nitrogen and life-sustaining oxygen, but a fraction of a percent of other chemicals can make a difference in the quality of life. Extra water in the atmosphere can mean a rain forest; a little less water produces a balanced rainfall; and practically no water results in a desert.

Urbanization often creates an unhealthy, unpleasant medium for the existence of human life. With its vast numbers of vehicles and increased industrialization, urbanization has produced an unwanted (and for a while ignored) increase in some of the naturally occurring minor chemicals in the atmosphere (nitrogen oxides, sulfur dioxide, carbon monoxide, carbon dioxide, and ozone). In this chapter we will look at the composition of the atmosphere right down to the air in your house, classroom, or office. By understanding what air is, perhaps we can be better equipped to deal with problems related to achieving and maintaining clean, healthful air.

20.1 THE ATMOSPHERE

Natural Composition of Air at Sea Level

The two major chemicals in our atmosphere are nitrogen, a rather unreactive gas, and oxygen, a highly reactive one. The composition of dry air at sea level shows that nitrogen is the most common atmospheric gas, followed by oxygen, and then 13 other gases, each at less than 1% by volume (Table 20.1). If we round off the number for oxygen to 21%, this means that for every 100 volume units of air, 21 units are oxygen. When it is pure, oxygen supports combustion at an explosive rate (Figure 20.1), but when diluted with nitrogen, oxygen's oxidizing capability is tamed somewhat. Except for helium, which occurs in some deposits of natural gas, the atmosphere is our only source for the noble gases—argon, neon, helium, krypton, and xenon. Think

(a)

(b)

Figure 20.1
(a) Liquid oxygen is poured into a beaker. (b) When a glowing splint is inserted into the mouth of the beaker, the high oxygen concentration causes the splint to burn at an explosive rate. Other oxidizing agents would react in the same way. The nitrogen in our atmosphere acts as a diluting gas to help control the rate of burning of most substances in air. *(C.D. Winters)*

Table 20.1

The Composition of Dry Air at Sea Level

Gas	Percentage by Volume
nitrogen	78.084
oxygen	20.948
argon	0.934
carbon dioxide	0.033*
neon	0.00182
hydrogen	0.0010
helium	0.00052
methane	0.0002*
krypton	0.0001
carbon monoxide	0.00001*
xenon	0.000008
ozone	0.000002*
ammonia	0.000001
nitrogen dioxide	0.0000001*
sulfur dioxide	0.00000002*

*Trace gases of environmental importance discussed in this chapter.

Figure 20.2
Neon lights at work. Times Square, New York City. *(Rafael Macia/Photo Researchers, Inc.)*

what New York's Times Square would look like at night without "neon" lights (Figure 20.2).

Besides the percentages by volume used in Table 20.1, parts per million (ppm) and parts per billion (ppb) by volume are also used in describing the concentrations of components of the atmosphere. Since volume is proportional to the number of molecules, these units also give a ratio of molecules of one kind to another kind. For example, "10 ppm SO_2" means that for every 1 million air molecules, 10 of them are SO_2 molecules. This may not sound like much until you consider that in just 1 cm^3 of air there are about 2.7×10^{13} million molecules. If this air contains 10 ppm SO_2, then there are 2.7×10^{8} SO_2 molecules. That's a lot of SO_2 molecules, and we're only talking about 1 cm^3 of air!

To convert percent to ppm, multiply by 10,000. Divide by 10,000 to convert ppm to percent.

The Troposphere and the Stratosphere

The earth's atmosphere can be roughly divided into layers, as shown in Figure 20.3. Near the earth, in the region named the **troposphere,** the temperature of the atmosphere decreases with increasing altitude. In this region the most violent mixing of air and the biggest variations in moisture content and temperature occur. Winds, clouds, storms, and precipitation are the result: the phenomena we know as **weather.** The troposphere is where we

The troposphere was named by the British meteorologist Sir Napier Shaw from the Greek word tropos, *meaning "turning."*

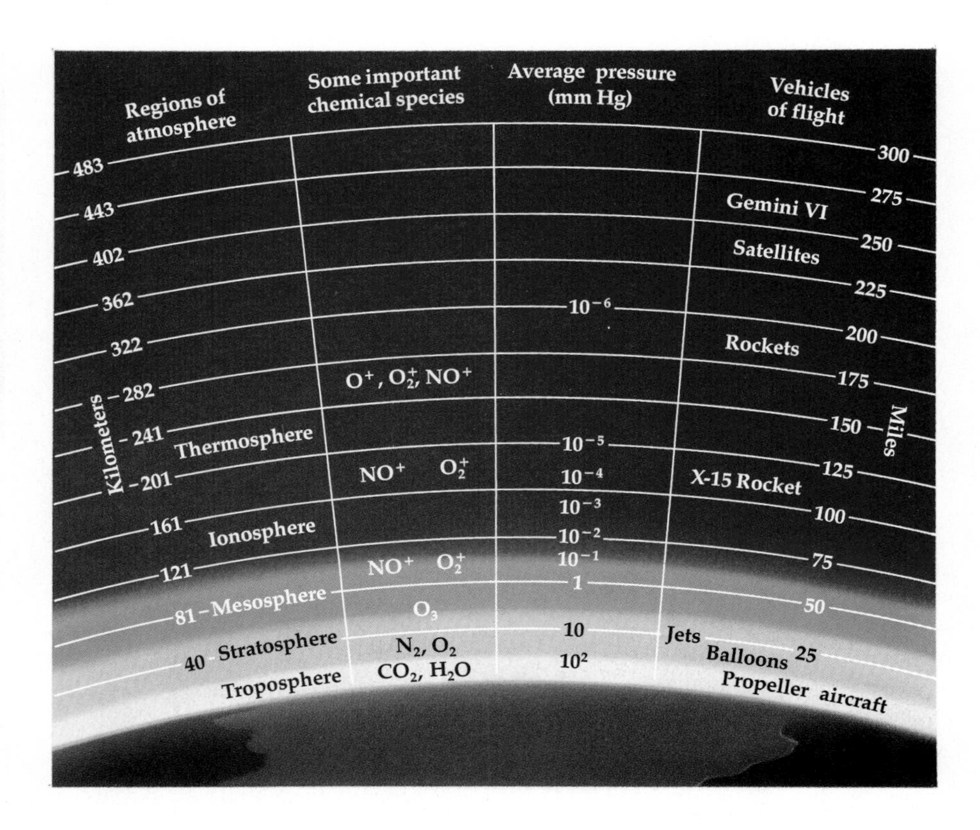

Figure 20.3
Some facts about our Earth's atmosphere.

live. Even in high-flying commercial jet airplanes, we are still in the troposphere, although near its upper limits. The composition of the troposphere is roughly that of dry air near sea level (Table 20.1), except that the concentration of water vapor varies considerably; on average, it is about 10 ppm.

The stratosphere was named by the French meteorologist Leon Phillipe Treisserenc de Bort, who believed this region consisted of orderly layers with no turbulence or mixing. Stratum *is a Latin word meaning "layer."*

Just above the troposphere, from about 12 to 50 km above the Earth's surface, lies the region called the **stratosphere.** If you take a ride in a Concorde supersonic aircraft, you will cruise in the stratosphere at about 20 km. In the stratosphere, the average kinetic energy of the air molecules increases with altitude because some of the solar energy absorbed by ozone molecules is imparted to other air molecules as kinetic energy. The pressures in the stratosphere are extremely low, and there is little mixing between the stratosphere and the troposphere. The lower limit of the stratosphere varies from night to day over the globe, and at the polar regions the top of the troposphere may be as low as 8 to 9 km.

The names of the oxides of nitrogen. The common names given in parentheses are often used in discussions of air pollutants.

N_2O	*dinitrogen monoxide (nitrous oxide)*
NO	*nitrogen monoxide (nitric oxide)*
NO_2	*nitrogen dioxide*
N_2O_4	*dinitrogen tetraoxide*
N_2O_5	*dinitrogen pentaoxide*

Defining Air Pollutants

An **air pollutant** is a substance that degrades air quality. Nature pollutes the air on a massive scale with volcanic ash, mercury vapor, hydrogen chloride, hydrogen fluoride, and hydrogen sulfide from volcanos and with reactive, odorous organic compounds from coniferous plants such as pine trees. Decaying vegetation, ruminant animals, and even termites add methane gas to the atmosphere, and decaying animal carcasses and other protein materials

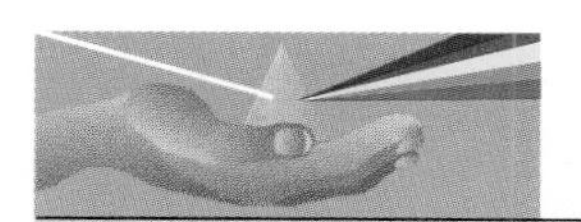

CHEMISTRY YOU CAN DO

Particle Size and Visibility

A common feature of all aerosols is that they decrease visibility. This can be observed in a city or along a busy highway, for example. Here is a way to simulate the effect of air pollutants on visibility. You will need a flashlight, a little milk, a transparent container (if possible, with flat, parallel sides) full of water, and something to stir the water with.

Turn off the lights and shine the beam of the flashlight through the container perpendicular to the flat sides. What do you observe? Can you see the beam? Now add a couple of drops of milk to the container and stir. Can you see the flashlight beam now? What color is it? What color is the light that passes through the milky water? Keep adding milk dropwise, stirring and observing until the beam of the flashlight is no longer visible from the far side of the container. Based on your observations of the milky water, devise an explanation of the fact that at midday on a sunny day the sun appears to be white or yellow, while at sunset it appears orange or red.

add dinitrogen monoxide (N_2O). But automobiles, electric power plants, smelting and other metallurgical processes, and petroleum refining also add significant quantities of chemicals to the atmosphere, especially in heavily populated areas. Atmospheric pollutants cause burning eyes, coughing, decreased lung capacity, harm to vegetation, and even the destruction of ancient monuments.

Particle Size of Pollutants

Pollutant particles range in size from fly ash particles, which are big enough to see, down to individual molecules, ions, or atoms. Many pollutants are attracted into water droplets and form **aerosols,** which are colloids (Section 15.6) consisting of liquid droplets or finely divided solids dispersed in a gas. Fogs and smoke are common examples of aerosols. Larger solid particles in the atmosphere are called **particulates.** The solids in an aerosol or particulate may be metal oxides, soil particles, sea salt, fly ash from electrical generating plants and incinerators, elemental carbon, or even small metal particles. Aerosols range in diameter from about 1 nm to about 10,000 nm and may contain about a trillion (10^{12}) atoms, ions, or small molecules. Particles in the 2000 nm range are largely responsible for the deterioration of visibility often observed in highly populated urban centers such as Los Angeles and New York.

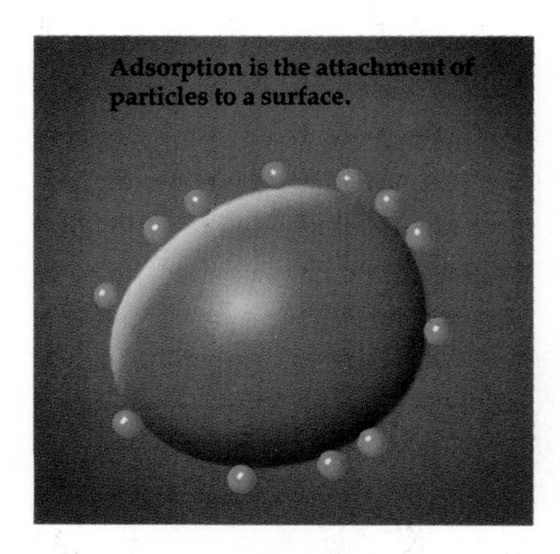

Adsorption is the attachment of particles to a surface.

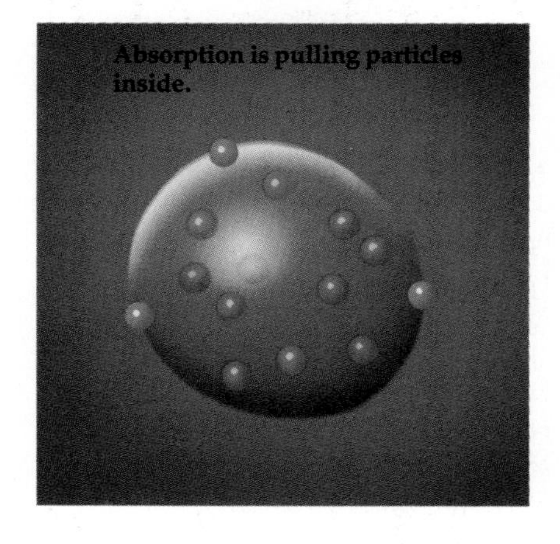

Absorption is pulling particles inside.

Aerosols are small enough to remain suspended in the atmosphere for long periods. Such small particles are easily breathable and can cause lung disease. They may also contain **mutagenic** or **carcinogenic** compounds. Because of their relatively large surface area, aerosol particles have great capacities to **adsorb** and concentrate chemicals on their surfaces. Liquid aerosols or particles covered with a thin coating of water may also **absorb** air pollutants, thereby concentrating them and providing a medium in which reactions may occur. A typical urban aerosol is shown schematically in Figure 20.4.

Millions of tons of soot, dust, and smoke particles are emitted into the atmosphere every year (Figure 20.5). The average suspended particulate

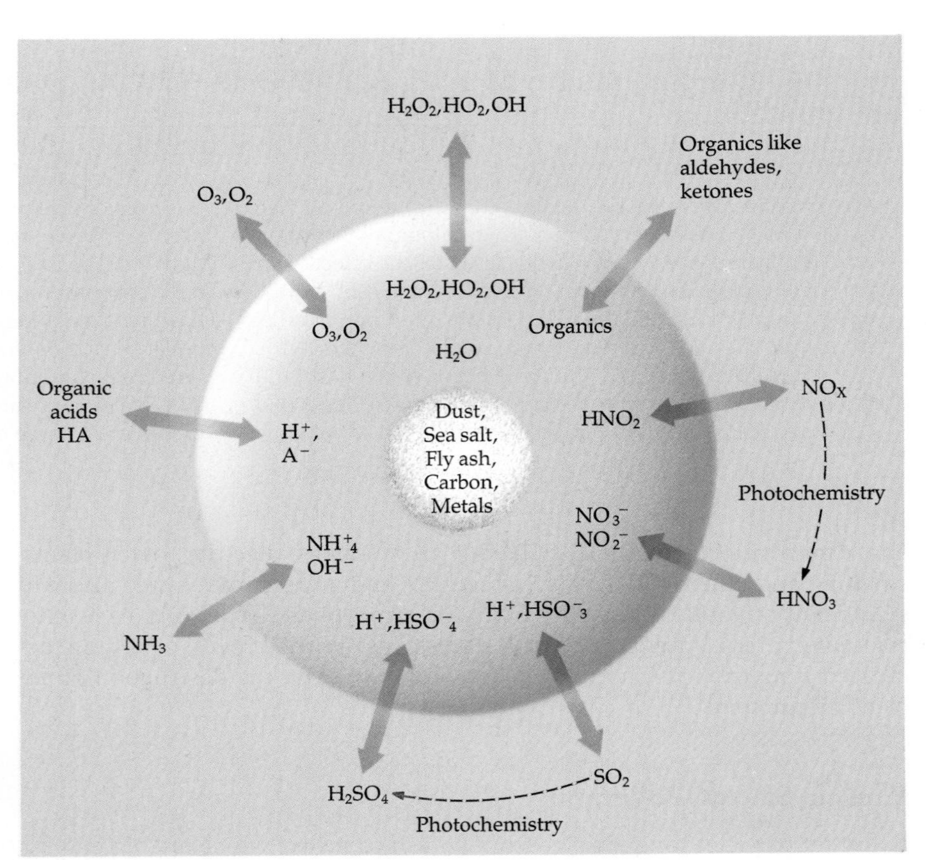

Figure 20.4
Schematic of an aerosol particle and some of its chemical reactions involving urban air pollutants.

Figure 20.5
Dark smoke comes from industrial stacks. In spite of air pollution regulations in the U.S. and elsewhere, various operating permits make such emissions possible. *(Gary Milburn/Tom Stack & Associates)*

concentrations in the United States vary from about 0.00001 g/m^3 of air in rural areas to about six times that amount in urban locations. In heavily polluted areas, concentrations of particulates may increase to 0.02 g/m^3. Particulates can be prevented from entering the atmosphere by using physical methods of separation such as filtration, centrifugation, spraying, and electrostatic separation (Figure 20.6). Once in the atmosphere, particulates will settle out naturally by gravitation or wash out in rain and snow.

EXERCISE 20.1 • *Using ppm Units*

Convert each of the entries in Table 20.1 to ppm.

20.2 CHEMICAL REACTIONS IN THE ATMOSPHERE

The atmosphere is a gaseous solution, bathed in a constant flux of light during the daylight hours and containing suspended aerosols and particulates. Because gas molecules are in constant motion, collisions are taking place all the time, and many of these collisions can lead to chemical reactions. The absorption of light by molecules in the atmosphere can cause reactions, called **photochemical reactions,** which would not otherwise occur at normal atmospheric temperatures. These photochemical reactions play an important role in determining the nature and fate of many chemical species found in the atmosphere. Aerosols and particulates also play an important role in atmospheric reactions, as you will see.

Nitrogen dioxide (NO_2) is one of the most photochemically active species found in the atmosphere. When a NO_2 molecule absorbs a photon of light, $h\nu$, the molecule is raised to a higher energy level; it becomes an **electronically excited molecule,** designated by an asterisk (*).

$$NO_2(g) + h\nu \longrightarrow NO_2^*(g)$$

This excited molecule may quickly re-emit a photon of light, or it may break apart to form two other species: a molecule of nitrogen monoxide (NO) and an oxygen atom (O).

$$NO_2^*(g) \longrightarrow NO(g) + O(g)$$

As described in Section 10.4, the NO_2 molecule, the NO molecule, and the oxygen atom, are examples of highly reactive species containing unpaired electrons, called **free radicals.** Some free radicals, such as the oxygen atom, are so reactive that they are very short-lived. Others, such as the NO_2 molecule, are not quite so reactive and are stable enough to exist for a long time. In general, if a radical is produced from another more stable molecule or radical, it can be highly reactive and short-lived. A molecule of acetaldehyde, CH_3CHO, for example, will absorb a photon and split in two, forming two radicals. This is called a **photodissociation** reaction.

$$CH_3{-}C(=O)H + h\nu \longrightarrow \underset{\text{methyl radical}}{\cdot CH_3} + \underset{\text{formyl radical}}{\cdot C(=O)H}$$

The methyl radical might react with another methyl radical to form ethane,

$$H_3C\cdot + \cdot CH_3 \longrightarrow H_3C{-}CH_3$$

and the formyl radical might react with an oxygen molecule to form a hydroperoxyl radical, $HOO\cdot$, and a CO molecule.

$$\cdot C(=O)H(g) + O_2(g) \longrightarrow HOO\cdot(g) + CO(g)$$

Many other reactions are possible for each of these radicals.

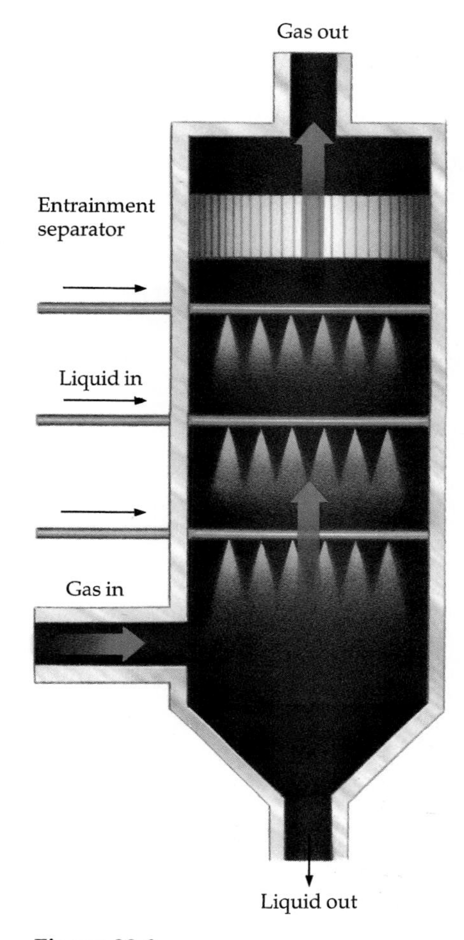

Figure 20.6
Removing particulates by scrubbing. The fine mist of water droplets trap many of the particulates entering with the gas stream, which are referred to as being "entrained" in the gas.

O=N–N=O structure (two O on each N)

structure of N_2O_4

Recall from your study of polymerization reactions in Section 13.5 that the formation of free radicals results in chain reactions that proceed until no new radicals are formed.

A radical usually reacts in either of two ways:

1. It combines with another radical. Each of the radicals contributes an electron to the formation of a bond, as when two NO_2 molecules combine to form N_2O_4 (dinitrogen tetraoxide.)

$$NO_2(g) + NO_2(g) \longrightarrow N_2O_4(g)$$

2. It reacts with a molecule to form one or more new radicals, or a new molecule, as illustrated by three important atmospheric reactions. The first is the formation of ozone, O_3, from O_2 and O.

$$\cdot O(g) + O_2(g) \longrightarrow O_3(g)$$

The second is the formation of the hydroxyl radical ($HO\cdot$) by the reaction of O with H_2O,

$$\cdot O(g) + H_2O(g) \longrightarrow 2\,HO\cdot(g)$$

and the third is the formation of a new molecule by a radical reacting with a molecule, as in the case of O reacting with SO_2 to form SO_3, a precursor to acid rain.

$$\cdot O(g) + SO_2(g) \longrightarrow SO_3(g)$$

EXERCISE 20.2 • *Reactions Involving Radicals*

Predict the reaction products for

a. the photolysis of water, $H_2O + h\nu \longrightarrow$

b. a methane molecule reacting with a hydroxyl radical, $CH_4 + HO\cdot \longrightarrow$

c. a hydrogen atom reacting with oxygen, $H\cdot + O_2 \longrightarrow$

20.3 KINDS OF SMOG

Now that you have seen some of the reactions that can take place in the atmosphere, let's look at air pollution in more detail. We will begin with a form of pollution familiar to most who live in or near cities, smog. The poisonous mixture of smoke (particulate matter), fog (an aerosol), air, and other chemicals was first called **smog** in 1911 by Dr. Harold de Voeux in his report on a London air pollution disaster that caused the deaths of 1150 people. Two general kinds of smog have been identified. One is the *chemically reducing type* that is derived largely from the combustion of coal and oil and contains sulfur dioxide (a strong reducing agent) mixed with soot, fly ash, smoke, and partially oxidized organic compounds. This is the **industrial** or **London type,** which is becoming less common as less coal is burned and more pollution controls are installed. The other is the *chemically oxidizing type* that contains strong oxidizing agents such as ozone and oxides of nitrogen, NO_x. It is called **photochemical smog** because light—in this in-

The common oxides of nitrogen found in air, NO and NO_2, are collectively called NO_x.

stance sunlight—is important in initiating the reactions that cause it. Another name commonly given this type of smog is **urban smog,** because of its formation around urban as opposed to industrial areas.

London Smog

London smog, which was first described in London, England and which forms around some industrial and power plants, is thought to be caused by sulfur dioxide. At a concentration of 5 ppm for 1 hour, SO_2 can cause constriction of bronchial tubes. A level of 10 ppm for 1 hour can cause severe distress. In the 1962 London smog, readings as high as 1.98 ppm of SO_2 were recorded. Laboratory experiments have shown that sulfur dioxide increases aerosol formation, particularly in the presence of mixtures of hydrocarbons, nitrogen oxides, and air energized by sunlight. Within an aerosol, sulfur dioxide is converted to sulfuric acid. For example, mixtures of 3 ppm unsaturated hydrocarbons, 1 ppm NO_2, and 0.5 ppm SO_2 at 50% relative humidity form aerosols in which sulfuric acid is a major product. Even with 10 to 20% relative humidity, sulfuric acid is readily formed. Sulfuric acid is very harmful to people suffering from respiratory diseases such as asthma or emphysema. The sulfur dioxide and sulfuric acid are thought to be the primary causes of deaths in the London smogs.

Relative humidity is a ratio of the concentration of water vapor in air (the partial pressure of water at the air temperature) compared with the maximum concentration it can contain (the vapor pressure of water at that temperature).

Photochemical Smog

Photochemical smog is typical in Los Angeles and other cities where sunshine is abundant and internal combustion engines exhaust large quantities of pollutants to the atmosphere. This type of smog is practically free of sulfur dioxide but contains substantial amounts of nitrogen oxides, ozone, ozonated hydrocarbons, and organic peroxides, together with hydrocarbons of varying complexity. The concentrations of these substances vary during the day, building up in the morning hours and dropping off at night (Figure 20.7).

Organic peroxides have the R—O—O—R′ structure and are produced by ozone reacting with organic molecules. Hydrogen peroxide is H—O—O—H.

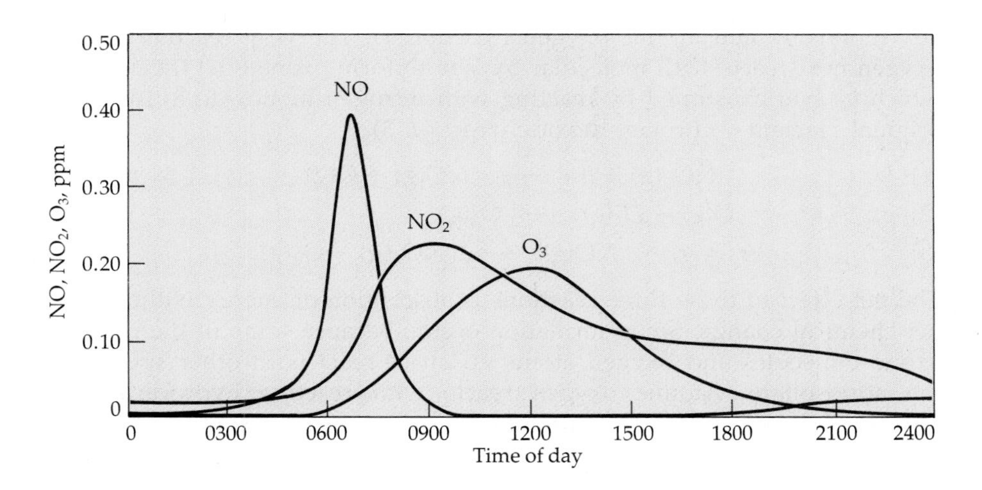

Figure 20.7
The average concentrations of pollutants NO, NO_2, and O_3 on a smoggy day in Los Angeles, California. The NO concentration builds up during the morning rush hour. Later in the day the concentrations of NO_2 and O_3 build up.

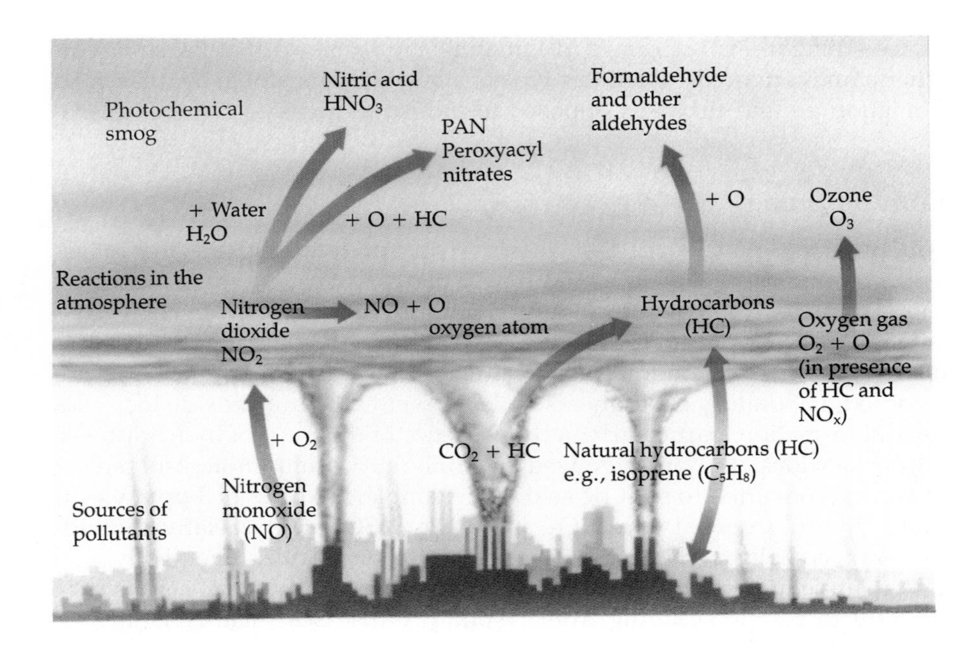

Figure 20.8
The processes involved in photochemical smog formation. Sunlight is the energy source. When the sources of the pollutants are highly concentrated, as in highly populated urban or industrial areas, photochemical smog can be very concentrated at times. Many of the components of photochemical smog shown in this figure are discussed in this chapter.

A city's atmosphere is an enormous mixing bowl of frenzied chemical reactions. Ferreting out the exact chemical reactions that produce photochemical smog has been a tedious job, but in 1951, insight into the formation process was gained when smog was first duplicated in the laboratory. Detailed studies have subsequently revealed that the chemical reactions involved in the smog-making process are photochemical and that aerosols serve to keep the **primary pollutants,** the pollutants emitted directly into the air, together long enough to form other pollutants, called **secondary pollutants,** by chemical reaction. Ultraviolet radiation from the sun is the energy source for the formation of photochemical smog.

The exact reaction scheme by which primary pollutants are converted into the secondary pollutants found in smog is still not completely understood (Figure 20.8). The process is thought to begin with the absorption of a photon of light by nitrogen dioxide, which causes its breakdown into nitrogen monoxide and atomic oxygen (reaction 1). The very reactive atomic oxygen next reacts with molecular oxygen to form ozone (O_3) (reaction 2), which is then consumed by reacting with nitrogen monoxide to form the original reactant—nitrogen dioxide (reaction 3).

$$(1) \qquad \cdot NO_2(g) + h\nu \longrightarrow \cdot NO(g) + O(g)$$

$$(2) \qquad O(g) + O_2(g) \longrightarrow O_3(g)$$

$$(3) \qquad O_3(g) + \cdot NO(g) \longrightarrow \cdot NO_2(g) + O_2(g)$$

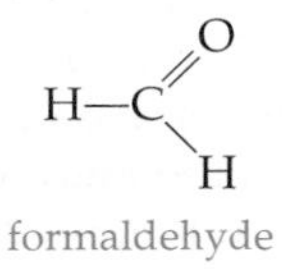

The net effect of these three reactions is absorption of energy without any net chemical change. Smog formation occurs because some of the reactive ozone molecules and oxygen atoms go on to react with other species in the atmosphere. Atomic oxygen reacts with reactive hydrocarbons—unsaturated compounds and aromatics—to form other free radicals. These radicals, in turn, react to form yet other radicals and secondary pollutants such as aldehydes (e.g., formaldehyde). About 0.2 ppm of nitrogen oxides and 1 ppm of reactive hydrocarbons are sufficient to initiate these reactions. The hydrocarbons involved come mostly from unburned petroleum products such as gasoline.

EXERCISE 20.3 • *Smog Ingredients*

Write the formulas and give sources for three ingredients of industrial smog. Do the same for three ingredients of photochemical smog.

20.4 MAJOR ATMOSPHERIC POLLUTANTS AND THEIR REACTIONS

In this section we will look at the major ingredients of photochemical smog—the primary pollutants, which are nitrogen oxides and hydrocarbons, and a secondary pollutant, ozone. We will describe how these substances interact with oxygen in the atmosphere to produce urban pollution. We will also look at sulfur dioxide, the primary pollutant that plays a major role in London smog and acid rain, and carbon monoxide, a poisonous component of air always found in abundance wherever there is a large number of automobiles.

Nitrogen Oxides: Primary Pollutants

Most of the nitrogen oxides found in the atmosphere originate from nitrogen monoxide, NO, a colorless, reactive gas. Whenever air is heated (as in a fire, in the combustion chamber of an internal combustion engine, or in a lightning bolt), some of the atmospheric nitrogen reacts with oxygen to produce NO, which is a reactive radical.

$$N_2(g) + O_2(g) + \text{heat} \longrightarrow 2\,NO(g)$$

Since this is an endothermic reaction, NO production is favored by higher temperatures. This temperature dependence is illustrated by Figure 20.9, which shows the equilibrium concentration of NO in a mixture of 3% O_2 and 75% N_2, typical of the composition inside the combustion chamber of an automobile engine just prior to exhausting to the atmosphere. At room temperature (25 °C) the equilibrium concentration of NO is only 1.1×10^{-10} ppm, but it is much higher at higher temperatures.

The speed with which this reaction takes places also increases with increasing temperature, so high temperatures favor both a higher equilibrium concentration and a faster rate of formation of NO. After combustion in the engine, the exhaust gases rapidly expand and cool, effectively "freezing" the NO that has formed at a relatively high concentration.

In the atmosphere NO reacts rapidly with atmospheric oxygen to produce NO_2.

$$2\,NO(g) + O_2(g) \longrightarrow \underset{\text{nitrogen dioxide}}{2\,NO_2(g)}$$

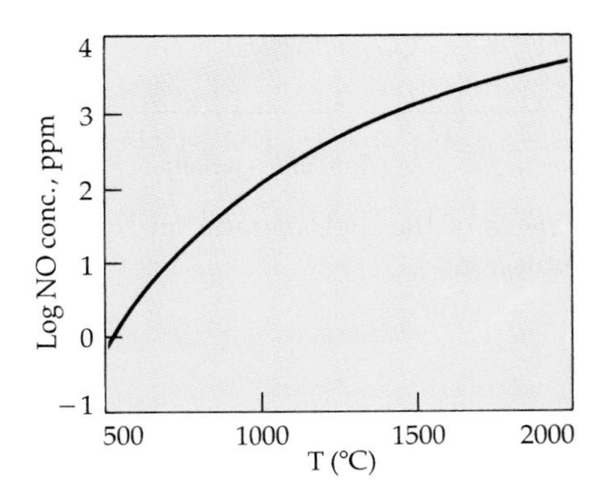

Figure 20.9

The log of the equilibrium concentration of NO as a function of the temperature in a mixture containing 75% N_2 and 3% O_2 and varying amounts of oxides of carbon and unburned hydrocarbons. While lowering the combustion temperature would result in lower NO concentrations and hence, emissions, this is not always feasible.

Table **20.2**

Emissions of NO_x

Source	Emissions (millions of tons)	
	United States	*Global*
fossil fuel combustion	66	231
biomass burning	1.1	132
lightning	3.3	88
microbial activity in soil	3.3	88
input from the stratosphere	0.3	5.5
total (uncertainty in estimates)	74 (±1)	554.5 (±275)*

*The large uncertainty for global emissions is due to incomplete data for much of the world.
Source: Stanford Research Institute, 1983.

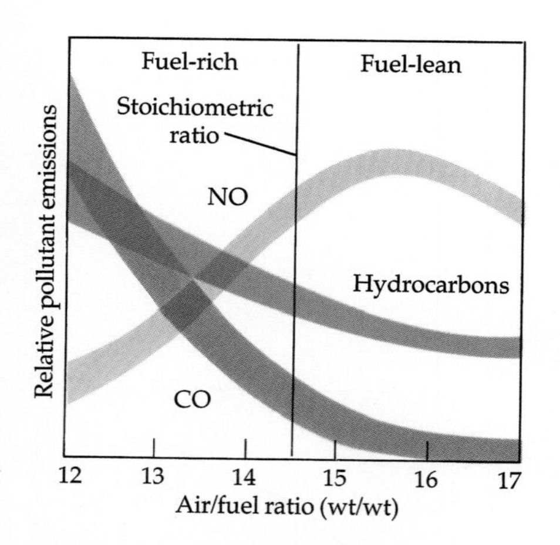

Effects of the fuel/air ratio on NO emissions.

Vast quantities of nitrogen oxides are formed each year throughout the world (Table 20.2). Normally the atmospheric concentration of NO_2 is a few parts per billion or less. In the United States, most oxides of nitrogen are produced from fossil fuel combustion such as that of gasoline in automobile engines, with significantly less coming from natural sources such as lightning. Elsewhere in the world, large amounts are also produced by burning trees and other biomass. Most NO_2 produced, either from human activities or natural causes, eventually washes out of the atmosphere in precipitation. This is one way green plants obtain the nitrogen necessary for growth. In the troposphere, however, especially around urban centers, excessive NO_2 causes problems.

In laboratory studies, nitrogen dioxide in concentrations of 25 to 250 ppm inhibits plant growth and causes defoliation. The growth of tomato and bean seedlings is inhibited by 0.3 to 0.5 ppm of NO_2 applied continuously for 10 to 20 days. Breathing NO_2 at a concentration of 3 ppm for 1 hour causes bronchial constriction in humans, and short exposures at high levels (150–220 ppm) cause changes in the lungs that produce fatal results. A seemingly harmless exposure at this level one day can cause death a few days later.

During the day, as you saw in the previous section, nitrogen dioxide photodissociates to form nitrogen monoxide and free oxygen atoms. The oxygen atoms can then react to form ozone and regenerate NO_2. At night, another oxide of nitrogen, the nitrogen trioxide radical, NO_3, is produced by the reaction

$$\cdot NO_2(g) + O_3(g) \longrightarrow \cdot NO_3(g) + O_2(g)$$

Nitric acid is present in acid rain.

In daylight, the nitrogen trioxide radical would quickly photodissociate, but in the absence of light it accumulates and reacts with nitrogen dioxide to form N_2O_5, which in turn reacts with water to form nitric acid.

$$\cdot NO_3(g) + \cdot NO_2(g) \longrightarrow N_2O_5(g)$$
$$N_2O_5(g) + H_2O(g) \longrightarrow 2\,HNO_3(g)$$

Nitrogen dioxide can also react with water, for example in aerosol particles, to form nitric acid and nitrous acid.

$$2\,\dot{N}O_2(g) + H_2O(g) \longrightarrow \underset{\text{nitric acid}}{HNO_3(g)} + \underset{\text{nitrous acid}}{HNO_2(g)}$$

In addition, nitrogen dioxide, water, and oxygen yield nitric acid:

$$4\,NO_2(g) + 2\,H_2O(g) + O_2(g) \longrightarrow 4\,HNO_3(g)$$

These acids in turn can react with ammonia or metallic particles in the atmosphere to produce nitrate salts. For example, ammonia reacts with nitric acid to form ammonium nitrate.

Ammonia is released into the atmosphere by the decay of organic matter.

$$\underset{\text{ammonia}}{NH_3(g)} + HNO_3(g) \longrightarrow \underset{\text{ammonium nitrate (a salt)}}{NH_4NO_3(s)}$$

The acids and salts form aerosols that eventually settle from the air or dissolve in raindrops. Nitrogen dioxide, then, is a primary cause of haze in urban or industrial atmospheres because of its participation in aerosol formation. Normally nitrogen dioxide has a lifetime of about three days in the atmosphere.

EXERCISE 20.4 • *Calculations Involving Pollutants*

If 400 metric tons of N_2 are converted to NO, and then to HNO_3, what mass (in metric tons) of HNO_3 is produced?

Ozone: A Secondary Pollutant

Ozone (O_3) has a pungent odor that can be detected at concentrations as low as 0.02 ppm. We often smell the ozone produced by sparking electric appliances, or after a thunderstorm when lightning-caused ozone washes out with the rainfall. Ozone can be either beneficial or harmful, depending on where it is found. In the troposphere (the air we breathe), ozone is harmful because it is a component of photochemical smog and because it can damage human health and decomposes material such as rubber. In the stratosphere, ozone is beneficial because it protects us from damaging ultraviolet radiation.

Ozone is an allotropic form of oxygen.

The only significant chemical reaction producing ozone in the atmosphere is the combination of molecular oxygen and atomic oxygen. In the lower atmosphere the major source of oxygen atoms is the photodissociation of NO_2 molecules, which are especially plentiful whenever there is an abundance of NO from automobile exhaust or other high-temperature combustion reactions. At high altitudes, oxygen atoms are produced by the photodissociation of oxygen molecules, which is caused by ultraviolet photons.

Ozone photodissociates to give an oxygen atom and an oxygen molecule whenever it is struck by photons in the near-ultraviolet range (200–300 nm), at any altitude.

Figure 20.10
Automobile emissions testing is mandated in many communities that have failed to meet EPA's ozone standards. Cars that fail the emissions standards for hydrocarbons and carbon monoxide are required to be repaired. Often the local government will not reissue operating licenses until satisfactory emissions scores are achieved. *(J. Wood)*

$$O_3(g) + h\nu \longrightarrow O_2(g) + O(g)$$

Oxygen atoms can then react with water to produce hydroxyl radicals (OH).

$$O(g) + H_2O(g) \longrightarrow 2 \cdot OH(g)$$

In the daytime, when they are produced in large numbers, hydroxyl radicals can react with nitrogen dioxide to produce nitric acid.

$$\cdot NO_2(g) + \cdot OH(g) \longrightarrow HNO_3(g)$$

Any reaction between two molecules or radicals in the atmosphere is an elementary reaction (Section 8.1). The elementary reactions given above constitute a reaction mechanism for the production of nitric acid from ozone, nitrogen dioxide, and water. Summing the steps in the mechanism gives the net reaction.

$$\begin{array}{l} O_3 + h\nu \longrightarrow O_2 + O \\ O + H_2O \longrightarrow 2 \cdot OH \\ 2\,(\cdot NO_2 + OH \cdot \longrightarrow HNO_3) \\ \hline O_3 + h\nu + 2\,NO_2 + H_2O \longrightarrow 2\,HNO_3 + O_2 \end{array}$$

In this mechanism, as is usually the case, the reactive free radicals are intermediates that do not appear in the net reaction. Given the diversity of photodissociation reactions that form free radicals, there are a great many possible mechanisms for reactions in the atmosphere.

Ozone is a secondary air pollutant, and is the most difficult pollutant to control because its formation depends on ever-present sunlight and on hydrocarbons, which almost every automobile emits to some degree. According to the EPA, the standard limit for ozone of 0.12 ppm was exceeded in 76 urban areas of the United States between 1983 to 1985. These high ozone concentrations were primarily related to emissions of nitrogen oxides from automobiles, buses, and trucks. Most major urban areas have vehicle inspection centers for passenger automobiles in an effort to control NO_x emissions as well as those of carbon monoxide and unburned hydrocarbons (Figure 20.10).

Children at play in Hoboken, New Jersey, with New York City in the background. The high ozone concentrations in and near cities present dangers for respiratory damage. *(Jeff Isaac Greenberg/Photo Researchers, Inc.)*

As difficult as it is to attain, the ozone standard may not be low enough for good health. Exposure to concentrations of ozone at or near 0.12 ppm lowers the volume of air a person breathes out in 1 s (the forced expiratory volume, or FEV_1). Studies of children who were exposed to ozone concentrations slightly below the EPA standard showed a 16% decrease in the FEV_1. Some scientists have been urging the EPA to lower the standard to 0.08 ppm. No matter what the standard becomes, present ozone concentrations in many urban areas represent health hazards to children at play, joggers, others doing outdoor exercise, and older persons who may have diminished respiratory capabilities.

EXERCISE 20.5 • *Ozone Production*

Write two photodissociation reactions that produce atomic oxygen. Then write a reaction in which ozone is formed.

Table 20.3

The Ten Most Abundant Hydrocarbons Found in the Air of Cities

	Concentration (ppb)	
Compound	*Median*	*Maximum*
2-methylbutane	45	3393
butane	40	5448
toluene	34	1299
propane	23	399
ethane	23	475
pentane	22	1450
ethylene	21	1001
m-xylene, *p*-xylene	18	338
2-methylpentane	15	647
2-methylpropane	15	1433

Results from 800 air samples taken in 39 cities.
Source: Air Pollution Control Association, 1988.

$$CH_2{=}C(CH_3){-}CH{=}CH_2$$
isoprene

CH_3, CH_3, CH_3
α-pinene

Hydrocarbons: Primary Pollutants

Hydrocarbons enter the atmosphere from both natural sources and human activities. Isoprene and α-pinene are produced in large quantities by both coniferous and deciduous trees. Methane gas is produced by such diverse sources as ruminant animals, termites, ants, and decay-causing bacteria acting on dead plants and animals. Human activities such as the use of industrial solvents, petroleum refining and distribution, and incomplete burning of gasoline and diesel fuel components account for a large amount of hydrocarbons in the atmosphere. The principal hydrocarbons found in urban air are listed in Table 20.3.

The hydroxyl radical ($\cdot OH$), produced indirectly from ozone, helps in the oxidation of hydrocarbons in the atmosphere. Like the reactions we have seen earlier, this oxidation involves several steps. Using RH as the general formula for a hydrocarbon, we can write the first two steps as

(1) $$RH(g) + HO\cdot(g) \longrightarrow R\cdot(g) + H_2O(g)$$

(2) $$R\cdot(g) + O_2(g) \longrightarrow ROO\cdot(g)$$

where atmospheric oxygen is the oxidizing agent. The $ROO\cdot$ represents organic peroxy radicals with a variety of R groups. The peroxy radicals are also oxidizing agents that oxidize NO to NO_2, and produce various aldehydes (RCHO) and ketones (R_2CO) and the hydroperoxyl radical ($HO_2\cdot$).

(3) $$ROO\cdot(g) + NO(g) \longrightarrow \text{aldehydes} + \text{ketones} + NO_2(g) + HO_2\cdot(g)$$

William Chameides, of Georgia Tech in Atlanta, published a report in Science *magazine in September, 1988, in which he stated that trees may account for more hydrocarbons in the atmosphere in some cities than are produced by human activities. The EPA has since found this hypothesis to be true, causing rethinking about how to control urban air pollution.*

Trees in urban environments may emit as many reactive hydrocarbons as automobiles. *(J. Wood)*

The NO_2 radical can then undergo the familiar photodissociation, which is followed quickly by the production of ozone.

$$\text{(4)} \qquad \cdot NO_2(g) + h\nu \longrightarrow \cdot NO(g) + O(g)$$

$$\text{(5)} \qquad O(g) + O_2(g) \longrightarrow O_3(g)$$

In the general manner illustrated in equations 1 through 5, oxidation of various hydrocarbons by hydroxyl radicals contributes to ozone formation.

The hydroperoxyl radical can also react with NO to produce NO_2 and more hydroxyl radicals.

$$HO_2 \cdot (g) + NO(g) \longrightarrow NO_2(g) + \cdot OH(g)$$

Hydroxyl radicals are also involved in the oxidation of aldehydes to give compounds that can react with NO_2. Acetaldehyde (CH_3CHO), for example, reacts with the hydroxyl radical and oxygen in the following way:

$$CH_3CHO(g) + \cdot OH(g) + O_2(g) \longrightarrow \underset{\text{peroxyacetyl radical}}{CH_3C(=O)—O—O \cdot (g)} + H_2O(g)$$

The peroxyacetyl radical can then react with NO_2 to produce peroxyacetyl-nitrate (PAN):

$$CH_3C(=O)—O—O \cdot (g) + \cdot NO_2(g) \longrightarrow \underset{\text{PAN}}{CH_3C(=O)—O—O—NO_2(g)}$$

PAN and related compounds are powerful eye irritants, helping to account for the discomfort we feel during episodes of photochemical smog. PAN formation stabilizes NO_2 so that it can be carried over great distances by prevailing winds. Eventually PAN decomposes and releases NO_2. In this way urban pollution in the form of NO_2 may be carried to outlying areas, where it may do additional damage to vegetation, human tissue, and fabrics—a direct result of an unfortunate combination of events involving hydrocarbons emitted into the atmosphere.

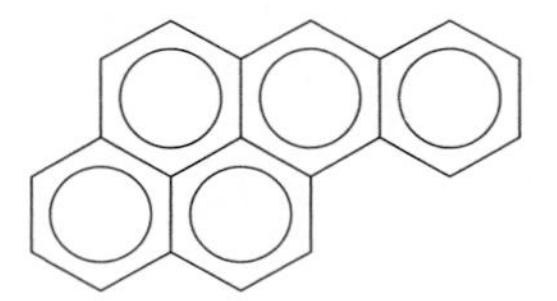

Benzo(α)pyrene, a carcinogenic polynuclear aromatic hydrocarbon found in smoke

Polynuclear hydrocarbons consist of multiple aromatic rings, fused together by sharing sides (see Figure 11.18). They are formed whenever complex organic substances such as heavy petroleum or wood products are incompletely burned. Many of these compounds are known or suspected carcinogens.

In addition to simpler hydrocarbons such as alkanes, alkenes, and alkynes, a large number of **polynuclear aromatic hydrocarbons** (PAH) are released into the atmosphere from coal and wood burning and from motor vehicle exhaust. These chemicals can react with hydroxyl radicals and oxygen much as simpler hydrocarbons do, but their greatest danger is their toxic properties. One PAH, benzo(α)pyrene (BAP), is a known carcinogen. Concentrations of BAP as high as 60 $\mu g/m^3$ have been found in urban air. Coal smoke contains about 300 $\mu g/m^3$ benzo(α)pyrene. Measurements have shown that for every million tons of coal burned, about 750 tons of benzo(α)pyrene are produced. British researchers reported that a typical London resident in the 1950s inhaled about 200 mg of BAP a year. Heavy smokers (those who smoke about two packs a day without filters) receive an additional 150 mg a year. This is about 40,000 times the amount necessary to produce cancer in mice. Extracts of urban air taken at various times during the past decade have in fact produced cancer in mice, but not all of these cancers were caused by PAHs like benzo(α)pyrene; other carcinogenic organic chemicals were present as well.

EXERCISE 20.6 • *Calculating Benzo(α)pyrene Pollution*

A 1000 megawatt coal-burning power plant burns about 700. metric tons of coal per hour. Calculate the mass of benzo(α)pyrene formed by the coal plant in 24 hours.

Sulfur Dioxide—A Primary Pollutant

Sulfur dioxide (SO_2), a pollutant that is a major contributor to London smog and acid rain, is produced when sulfur or sulfur-containing compounds are burned in air.

$$S(s) + O_2(g) \longrightarrow SO_2(g)$$

Most of the coal burned in the United States contains sulfur in the form of the mineral pyrite (FeS_2). The weight percent of sulfur in this coal ranges from 1 to 4%. The pyrite is oxidized as the coal is burned.

$$4\ FeS_2(s) + 11\ O_2(g) \longrightarrow 2\ Fe_2O_3(s) + 8\ SO_2(g)$$

Large quantities of coal are burned in this country to generate electricity. If the 1000 MW coal-fired generating plant described in Exercise 20.6 burns coal that contains 4% sulfur, it releases 56 tons of SO_2 per hour, or 490,000 tons of SO_2 every year. Oil-burning electrical generation plants produce comparable quantities of SO_2, since fuel oils can contain up to 4% sulfur. Table 20.4 shows the number of coal-fired power plants and the amounts of SO_2 emitted in those eight states that have the highest SO_2 emissions. Operators of these facilities are making efforts to eliminate most of the SO_2 before it leaves the smoke stack.

The sulfur in oil is in the form of mercapto compounds, in which sulfur atoms are bound to carbon and hydrogen atoms (the —SH functional group).

Most low-sulfur coal deposits are mined far from the major metropolitan areas where the coal is needed for power generation. Removing sulfur from the closer, high-sulfur coal is costly and not completely effective. One

Table 20.4

Characteristics of Coal-Fired Power Plants in Eight States

State	Plants	SO_2 Emissions (thousand tons/yr)	Capacity (gigawatts)
Ohio	99	2221	22.31
Indiana	66	1588	14.58
Pennsylvania	70	1427	17.93
Missouri	41	1214	9.97
Illinois	59	1136	15.75
West Virginia	33	966	14.46
Tennessee	37	950	9.41
Kentucky	54	947	11.82

Some of these plants have emissions controls installed; others do not. A gigawatt is 10^9 watts.
Source: U.S. Environmental Protection Agency, 1988.

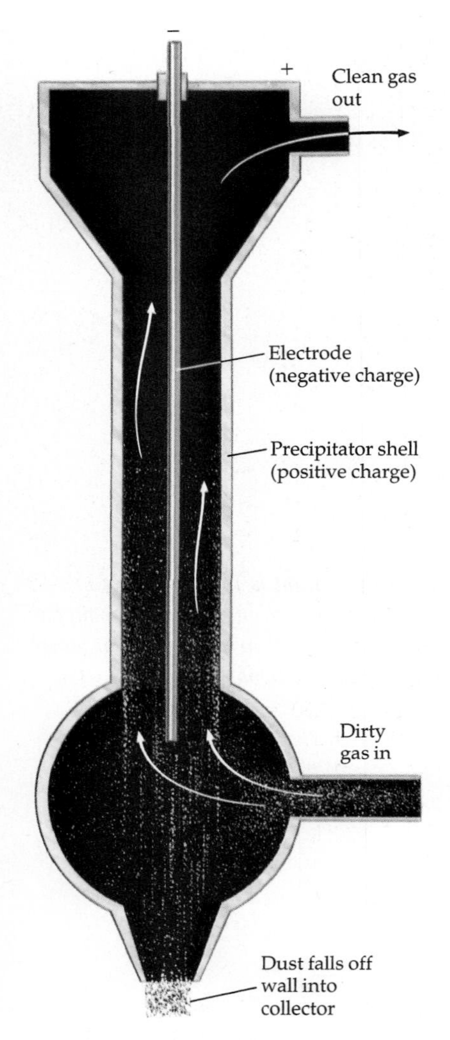

An electrostatic precipitator. The central electrode is negatively charged and imparts a negative charge to particles in the gas that contact it. These charged particles, in turn, are attracted to the positively charged walls.

method is to pulverize the coal to the consistency of a fine powder and remove the pyrite (FeS_2) by magnetic separation. Technology is available to decrease the sulfur content of fuel oil to 0.5%, but this process, too, is costly. It involves the formation of hydrogen sulfide (H_2S) by bubbling hydrogen through the oil in the presence of a metal catalyst.

Several efficient methods are available to remove SO_2 from the gases in smoke stacks. In one method, limestone is first heated to produce lime. The lime is then allowed to react with SO_2 to form calcium sulfite, a solid particulate that can be removed from an exhaust stack by an electrostatic precipitator.

$$\underset{\text{limestone}}{CaCO_3(s)} \xrightarrow{\text{heat}} \underset{\text{lime}}{CaO(s)} + CO_2(g)$$

$$CaO(s) + SO_2(g) \longrightarrow \underset{\text{calcium sulfite}}{CaSO_3(s)}$$

In another recovery method, the SO_2 is passed through molten sodium carbonate to form sodium sulfite.

$$SO_2(g) + \underset{\text{sodium carbonate}}{Na_2CO_3(\ell)} \xrightarrow{800\ ^\circ C} \underset{\text{sodium sulfite}}{Na_2SO_3(\ell)} + CO_2(g)$$

A less desirable method of dissipating SO_2 is to send it out through tall stacks that release the SO_2 far above the source and its surroundings. This allows the SO_2 to be diluted on the way down. The SO_2 will eventually come down, but the longer it stays in the upper atmosphere, the greater chance it has to become sulfuric acid and be precipitated some distance from the source. A 10-year study in Great Britain showed that although SO_2 emissions from power plants increased by 35%, the construction of tall stacks decreased the ground level concentrations of SO_2 by as much as 30%. The question is, who got the SO_2? In this case, Britain's solution was others' pollution. In the United States, the Environmental Protection Agency may have unwittingly added to a pollution problem with Clean Air Act rules in 1970 that caused plants to increase the height of smokestacks and caused pollutants to be carried longer distances by winds. There are about 179 stacks in the United States that are 500 ft high or taller, and 20 stacks are 1000 ft high or taller (Figure 20.11).

Most of the SO_2 in the atmosphere reacts with oxygen to form sulfur trioxide (SO_3). Several reactions are possible. SO_2 may react with atomic oxygen:

$$SO_2(g) + O(g) \longrightarrow SO_3(g)$$

It may react with molecular oxygen:

$$2\,SO_2(g) + O_2(g) \longrightarrow 2\,SO_3(g)$$

Or, it may react with hydroxyl radicals:

$$SO_2(g) + 2\cdot OH(g) \longrightarrow SO_3(g) + H_2O(g)$$

The SO_3 has a strong affinity for water and will dissolve in aqueous aerosol droplets to form sulfuric acid, a strong acid.

$$SO_3(g) + H_2O(\ell) \longrightarrow H_2SO_4(aq)$$

Sulfur dioxide can corrode metals and decay building stones, in particular marble and limestone. Both marble and limestone are forms of calcium carbonate ($CaCO_3$), which reacts readily with acid (H_3O^+) and with SO_2 and water. As the water-soluble calcium sulfite is formed it washes away, exposing the surface to further corrosion.

$$CaCO_3(s) + 2\,H_3O^+(aq) \longrightarrow Ca^{2+}(aq) + 2\,H_2O(\ell) + CO_2(g)$$

$$CaCO_3(s) + SO_2(g) + 2\,H_2O(\ell) \longrightarrow \underset{\text{calcium sulfite (soluble)}}{CaSO_3 \cdot 2\,H_2O(s)} + CO_2(g)$$

An alarming example is the disintegration of marble statues and buildings on the Acropolis in Athens, Greece. Coating the marble has failed to protect it adequately, and so the prized objects must be brought into air-conditioned museums to be protected from SO_2 and other corroding chemicals.

Sulfur dioxide is physiologically harmful to both plants and animals. Table 20.5 summarizes its effects. Most healthy adults can tolerate fairly high levels of SO_2 without apparent lasting ill effects. Individuals with chronic respiratory difficulties such as bronchitis or asthma tend to be much more sensitive to SO_2, accounting for many of the deaths during episodes of industrial smog.

Figure 20.11
One of the world's tallest chimneys, standing 1250 feet high in the Sudbury district of Ontario, Canada.
(Courtesy of Inco Limited)

EXERCISE 20.7 • *Getting Rid of SO_2*

If the SO_2 produced by a 1000. megawatt coal-burning power plant (see Exercise 20.6) is completely converted to calcium sulfite as described in the text, calculate the mass (in kilograms) and the weight (in pounds) of calcium sulfite produced per day. Assume a density of 2.3 g/cm^3 for calcium sulfite. What volume, in cubic meters and in cubic feet, does this mass of calcium sulfite occupy?

Table 20.5

Physiological and Corrosive Effects of SO_2

SO_2 Exposure (ppm)	Duration	Effect
0.03–0.12	annual average	corrosion, especially in moist temperate climates
0.3	8 hour	vegetation damage (bleached spots, suppression of growth, leaf drop, and low yield)
0.47	<1 hour	odor threshold (50% of subjects detect) varies with individuals
0.2	daily average	respiratory symptoms when community exposure exceeds 0.2 ppm more than 3% of the time
>0.05	long-term average	respiratory symptoms, including impairment of lung function in children with particulates >100 μg/m^3

Heavy traffic conditions contribute to high carbon monoxide levels. *(Marc Solomon/The Image Bank)*

Carbon Monoxide: An Urban Problem

Carbon monoxide (CO) is the most abundant and most widely distributed air pollutant. Like ozone, carbon monoxide is difficult to control. Cities such as Los Angeles with high densities of automobiles tend to have high concentrations of carbon monoxide in the air, and for this reason they are repeatedly cited by regulatory agencies for having unhealthful air quality.

Carbon monoxide is always produced when carbon or carbon-containing compounds are oxidized by insufficient oxygen.

$$2\ C(s) + O_2(g) \longrightarrow 2\ CO(g)$$

For every 1000 gal of gasoline burned, 2300 lb of carbon monoxide is formed in the combustion gases. Modern catalytic converters on car mufflers convert much of this carbon monoxide to carbon dioxide, but near heavily traveled streets at peak traffic times, concentrations as high as 50 ppm are common. In the countryside, carbon monoxide levels are closer to the global average of 0.1 ppm.

At least ten times more carbon monoxide enters the atmosphere from natural sources than from all industrial and automotive sources combined. Of the 3.8 billion tons of carbon monoxide emitted every year, about 3 billion tons are emitted by the oxidation of decaying organic matter in the topsoil.

About 10^{14} g of CO is released each year in the United States; that's about 1000 lb for every person.

It is generally agreed that CO is removed from the atmosphere by reaction with hydroxyl radicals, making it a problem mainly in cities where it is produced faster than it can be removed.

$$CO(g) + HO\cdot(g) \longrightarrow CO_2(g) + \cdot H(g)$$

20.5 ACID RAIN

The term **acid rain** was first used in 1872 by Robert Angus Smith, an English chemist and climatologist. He used the term to describe acidic precipitation that fell on Manchester early in the Industrial Revolution.

Pure water has a pH of 7, but rainwater becomes naturally acidified from dissolved carbon dioxide, a normal component of the atmosphere.

$$2\ H_2O(\ell) + CO_2(g) \longrightarrow H_3O^+(aq) + HCO_3^-(aq)$$

The pH of a solution at equilibrium with atmospheric CO_2 is 5.6 (Figure 20.12). Any rainfall or other precipitation with a pH below 5.6 is considered excessively acidic.

A more adequate term for acid rain might be acid precipitation, because snow and other forms of precipitation can also be excessively acidic. Some scientists use acid deposition.

As you have seen earlier in this chapter, NO_2 and SO_2 can undergo reactions in water droplets or aqueous aerosol particles to produce acids; this acidic moisture can precipitate as rain or snow. The NO_2 yields nitric acid (HNO_3) and nitrous acid (HNO_2); SO_2 is oxidized to SO_3, which yields sulfuric acid (H_2SO_4). Ice core samples taken in Greenland and dating back to 1900 contain sulfate (SO_4^{2-}) and nitrate (NO_3^-) ions, indicating that at least from 1900 onward there has been acid rain, and it has been deposited far from where the oxides of nitrogen and sulfur were formed.

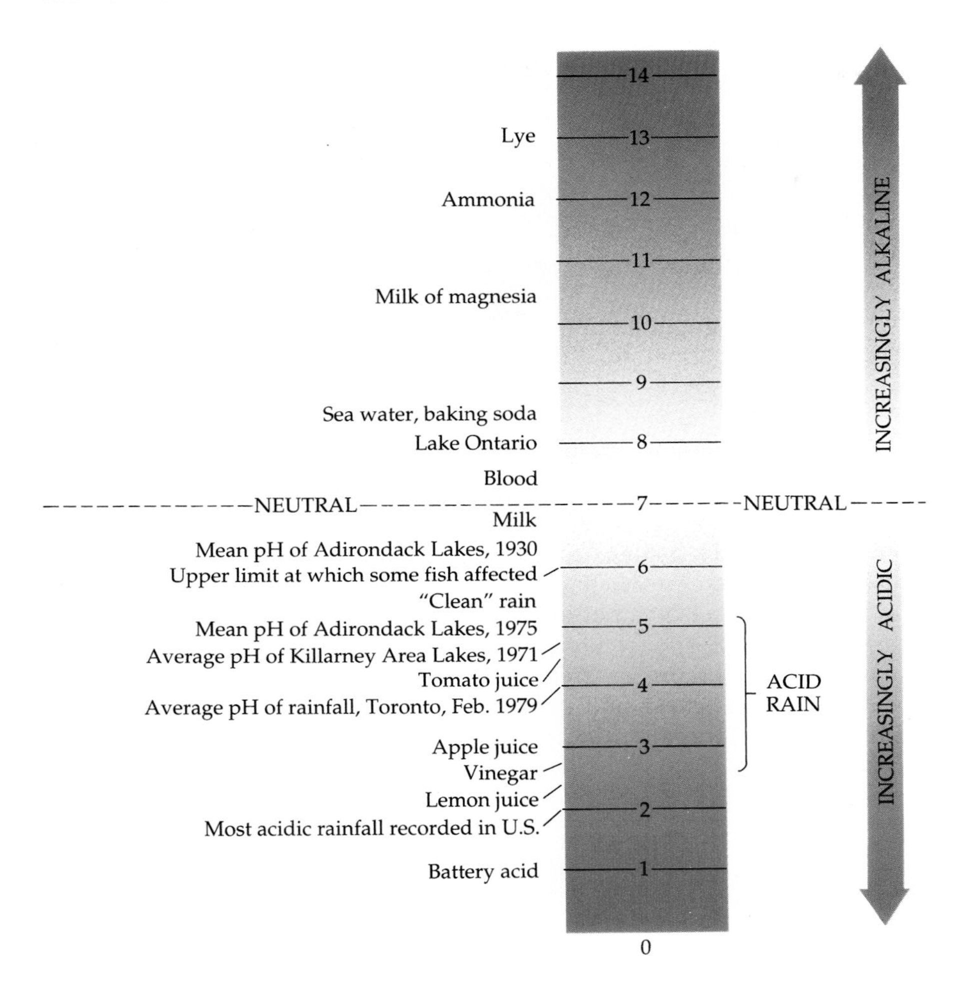

Figure 20.12
The pH of acid rain as compared with the pH of other mixtures.

Large quantities of acidic oxides are produced each year by human activities and emitted into the atmosphere (Figure 20.13). When the resulting acidic precipitation falls on areas that cannot easily tolerate such acidity, serious environmental problems occur. The annual average pH of precipitation falling on much of the northeastern United States and northeastern Europe is between 4 and 4.5. Some areas of West Virginia have had rainfall with pH values as low as 1.5. Furthermore, acid rain is an international problem—rain and snow don't observe borders. Many Canadian residents are angry with the government of the United States because some of the acidic oxides produced in the United States cause the rain falling on Canadian cities and forests to be acidified (Figure 20.14).

The extent of the problem with acid rain can be seen in dead, fishless ponds and lakes, dying or dead forests, and crumbling buildings. Because of wind patterns, Norway and Sweden have received the brunt of western Europe's emission of sulfur oxides and nitrogen oxides in acid rain. As a

Figure 20.13
Major sources and components of acid precipitation.

result, 4000 of the 100,000 lakes in Sweden have become fishless, and 14,000 other lakes have been acidified to some degree. In the United States, 6% of all ponds and lakes in the Adirondack Mountains of New York are now fishless, and 200 lakes in Michigan are dead. For the most part, these "dead" lakes are still picturesque, but no fish can live in the acidified water (Figure 20.15).

Acid rain also damages trees in several ways. It disturbs the stomata (openings) in tree leaves and causes increased transpiration, allowing too much water to evaporate from the tree. Acid rain acidifies the soil, damag-

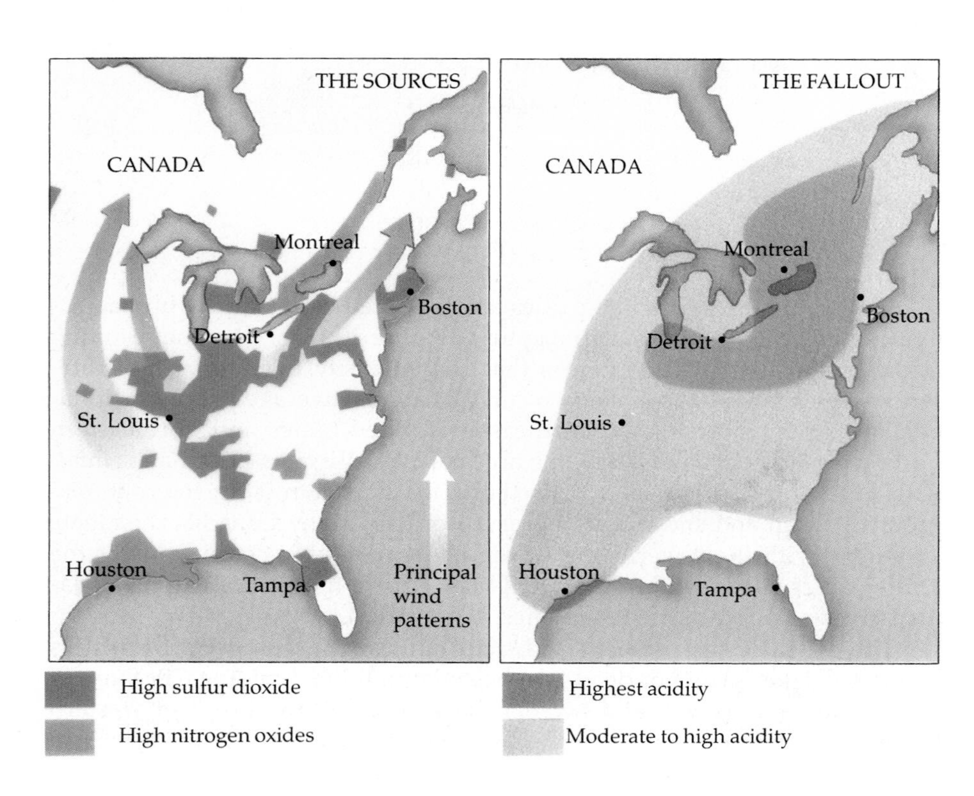

Figure 20.14
Most of the oxides of sulfur responsible for acid rain comes from the Midwestern states. Prevailing winds carry the acid droplets over the Northeast and into Canada. Oxides of nitrogen also contribute to acid rain formation.

Figure 20.15

Aerial photo of Walker Lake in Adirondack Park, New York. The pH of this lake is 4.7, caused by acid rain. The blue color and beauty of this lake is no indication that it is essentially lifeless because of its low pH. *(Mike Storey/ Adirondack Park Agency)*

ing fine root hairs and thus diminishing nutrient and water uptake. The surface structures of the bark and the leaves can also be destroyed by acid in the rain. The effects of acid rain on some forests have already been so severe that experts predict they will be lost, possibly forever. This is the case with Germany's famous Black Forest.

Acid rainfall has had a devastating impact on many forests. These dying trees are atop Clingman's Dome in The Great Smoky Mountains National Park. *(Kenneth Murray/Photo Researchers, Inc.)*

In addition to destroying vegetation directly, acid rain damages forests by leaching (dissolving) minerals from the soil. Leaching may have two effects. First, slightly soluble minerals that are necessary for the health of trees and other vegetation may be washed out of the soil and carried away by streams and rivers. Second, metal ions that are toxic to plant life may be made soluble enough at low pH to be absorbed by roots. These toxic minerals might otherwise remain in the soil indefinitely, because they are insoluble in water at neutral pH. For example, acid rain can dissolve aluminum hydroxide in the soil, releasing aluminum ions (Al^{3+}) to be taken up by roots of plants, where they have toxic effects.

$$Al(OH)_3(s) + 3\,H_3O^+(aq) \longrightarrow Al^{3+}(aq) + 6\,H_2O(\ell)$$

The effects of acid rain and other pollution on stone and metal structures are more subtle, but they are especially devastating because of their irreversibility. By damaging stone buildings in Europe, acid rain is slowly but surely dissolving the continent's historical heritage. The bas-reliefs on the Cologne (Germany) cathedral are barely recognizable. The Tower of London, St. Paul's Cathedral, and the Lincoln Cathedral in London (Figure 20.16) have suffered the same fate. Other beautifully carved statues and

Figure 20.16

The photo on the left was taken at the Lincoln Cathedral in London in 1910. The photo on the right was taken in 1984. *(Dean and Chapter of London)*

bas-reliefs on buildings throughout Europe and the eastern part of the United States and Canada are slowly passing into oblivion by the action of pollutants, and acid rain in particular.

Lime, CaO, also exists as a hydrated form called "slaked" lime. Its formula is $Ca(OH)_2$, which is sometimes written as $CaO \cdot H_2O$.

What can be done about acid rain? Some stopgap measures are being taken, such as spraying hydrated lime, $Ca(OH)_2$, into acidified lakes to neutralize at least some of the acid and raise the pH toward 7 (Figure 20.17).

$$Ca(OH)_2(s) + 2\,H_3O^+(aq) \longrightarrow Ca^{2+}(aq) + 4\,H_2O(\ell)$$

Sweden is spending $40 million a year to neutralize the acid in some of its lakes. Some lakes in the problem areas have their own safeguard against acid rain: they have limestone-lined bottoms, which supply calcium carbonate ($CaCO_3$) for neutralizing the acid (just as an antacid tablet relieves indigestion). Statues and bas-reliefs have been coated with a variety of plastics and other materials. None of these materials appear to be long-term protectors, however.

Figure 20.17
Aerial spraying of lime to raise the pH of an acidified lake.

The ultimate answers to acid rain problems may lie with governments, who will use regulations to control acid-producing emissions. Twenty-one European countries agreed in 1985 to reduce their SO_2 emissions by 30% or more over a 10-year period. By 1989, more than half of those countries had already reached that goal. In 1988, the Canadian government set a goal of lowering its SO_2 emissions by half by 1994. In the United States, progress has not been as rapid.

EXERCISE 20.8 • *Neutralizing Acid Rain*

What weight of CaO will be required to neutralize a lake containing 2.0×10^6 L with a pH of 3.0 if the acid is all HNO_3?

20.6 CHLOROFLUOROCARBONS AND THE OZONE LAYER

Chlorofluorocarbons (CFCs) are halogen-substituted alkanes such as $CFCl_3$ and CF_2Cl_2.

Most of the processes that remove pollutants from the atmosphere take place in the troposphere. Adsorption onto a particle or absorption by a water droplet can lead to removal by precipitation. Chemical reactions in the gas phase, on a particle, or in solution can convert a pollutant to a less harmful substance. But one class of industrial pollutants, the halogenated hydrocarbons collectively called **chlorofluorocarbons** or **CFCs,** are relatively unreactive. They decompose only when exposed to ultraviolet radiation and they are not eliminated in the troposphere. Instead, many of them eventually mix with air in the stratosphere, where they reside for many years as they are slowly photodissociated. It is a series of reactions of CFCs, their decomposition products, and ozone that cause so much concern about the presence of CFCs in our atmosphere.

CFCs are used as refrigeration fluids and degreasing solvents in the manufacture of such diverse products as electronic parts and machined metallic objects. They are even used as fire extinguishers because of their highly unreactive nature. The lower molecular weight CFCs were at one time used as propellants for aerosol products including hair sprays, deodorants, medicines, and foods, but they were banned from this use in the United States in 1978. CFCs are, however, still used worldwide in both developed and developing countries.

In 1974 M.J. Molina and F.S. Rowland of the University of California, Irvine, published a scientific paper in which they predicted that continued use of CFCs would lead to a serious depletion of the earth's stratospheric ozone layer. Why is this a matter of serious concern? Recall that ozone molecules photodissociate when struck by photons in the 200 to 300 nm (near-ultraviolet) range to produce oxygen atoms and oxygen molecules. In the process the ultraviolet radiation is converted to other kinds of energy.

The O atoms from the photodissociation of O_3 react with O_2 to regenerate O_3 so that under normal conditions there is no net O_3 loss.

$$O_3(g) + h\nu \longrightarrow O_2(g) + O(g)$$

This reaction is essential for living things on this planet. Ozone is so naturally abundant in the stratosphere (about 10 ppm) that it normally prevents 95 to 99% of the sun's near-ultraviolet radiation from reaching the earth's surface. Radiation in the near-ultraviolet range is of such high energy that it is damaging to living organisms. It can cause skin cancer in humans, is damaging to plants, and possibly has other harmful effects that we do not even know about yet. For every 1% decrease in the stratospheric ozone, an additional 2% of this most damaging radiation reaches the earth's surface. Ozone depletion therefore has the potential for drastically damaging our environment.

Destruction of the ozone layer by CFCs begins with photodissociation of the carbon-chlorine bond in a CFC molecule. This produces an active chlorine atom, as shown here for one of the most common CFCs, CFC-11.

$$CFCl_3(g) + h\nu \longrightarrow CFCl_2\cdot(g) + Cl\cdot(g)$$

The chlorine atom then combines with an ozone molecule, producing a chlorine oxide (ClO) radical and an oxygen molecule.

$$Cl\cdot(g) + O_3(g) \longrightarrow ClO\cdot(g) + O_2(g)$$

Thus, an ozone molecule has been destroyed. If this were the only reaction that particular CFC molecule caused, there would be little danger to the ozone layer. However, the $ClO\cdot$ radical can react with an oxygen atom to produce atomic chlorine again.

$$ClO\cdot(g) + O(g) \longrightarrow O_2(g) + Cl\cdot(g)$$

The net reaction obtained by adding these two steps is destruction of an ozone molecule:

$$O_3(g) + O(g) \longrightarrow 2\,O_2(g)$$

These two reaction steps can repeat over and over. The chlorine atom that reacts in the first step is regenerated in the second and serves as a catalyst for destruction of ozone. It has been estimated that a single Cl atom can destroy as many as 100,000 molecules of O_3 before it is inactivated or returned to the troposphere (probably as HCl).

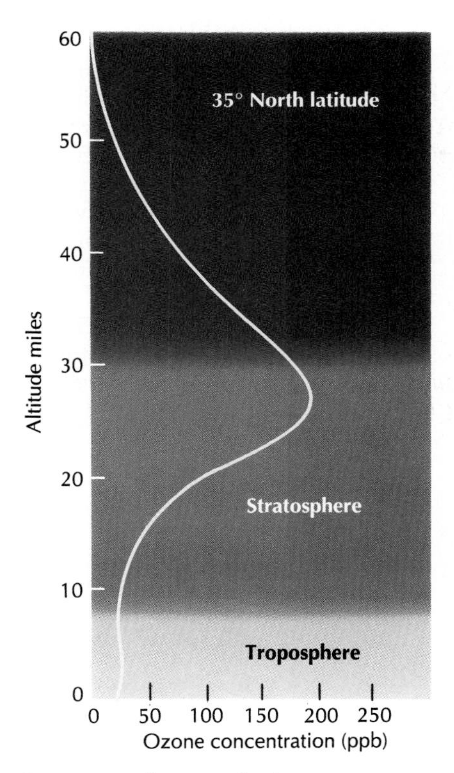

Variation of atmospheric ozone concentration with altitude.

The Space Shuttle and Gas Reactions in the Upper Atmosphere

THE ATMOSPHERE ABOVE THE earth where the Space Shuttles operate (about 190 miles up) is very thin, yet there are enough gas molecules to react with one another when the conditions are right. Astronauts have long known about an eerie red-orange glow seen on parts of the Shuttle facing the direction of motion. Sometimes the glow is so intense that it hampers the experiments being conducted by the Shuttle astronauts and astronomers back on earth. Experiments to duplicate this glow in the laboratory have been only partly successful.

In 1991, Shuttle mission STS 39 carried out an experiment (see the figure) designed to determine just what gases were causing the glow. The astronauts released samples of carbon dioxide, xenon, argon, and nitrogen monoxide (NO) into space from the shuttle's cargo bay and observed what happened. Unintentionally, some of the NO migrated to the far end of the spacecraft, where the tail section lit up quite noticeably. This glow is observable in the figure around the tail and shuttle main engine pods, both right and left.

Scientists think that the NO molecules collided with oxygen atoms (recall that ultraviolet radiation abounds at this altitude and that O_2 molecules readily photodissociate to O atoms) to produce an energetic form of nitrogen dioxide (NO_2) that radiates visible light. If that was so, then where did the NO come from in previous missions? One crew member, Edmond Murad, thinks nitrogen from the thruster rocket motors reacts with oxygen atoms to form NO, which then reacts with other oxygen atoms. In 1994, another set of experiments will release some pure nitrogen to see just what the effects will be.

One of the four compressed gas cylinders on the forward port side of Discovery's cargo bay releases NO gas during flight day 4 of the STS-39 mission. (NASA)

Molina and Rowland's warning in 1974 regarding CFCs has been confirmed by satellite and ground-based measurements, which show an average of about 2.5% decrease in the ozone layer worldwide during the decade from 1978 to 1988. Local decreases in ozone are even larger. In a latitude band that includes Dublin, Moscow, and Anchorage, ozone had decreased by 8% from January 1969 to January 1986. The reaction catalyzed by chlorine atoms from CFCs is thought to account for about 80% of the observed loss.

Near the North and South Poles ozone losses have been between 1% and 2.5% *per year*. Recently, huge losses, termed "holes," have been observed at the poles (Figure 20.18). Many scientists believe that similar holes may appear at the mid-latitudes in the future (Figure 20.19).

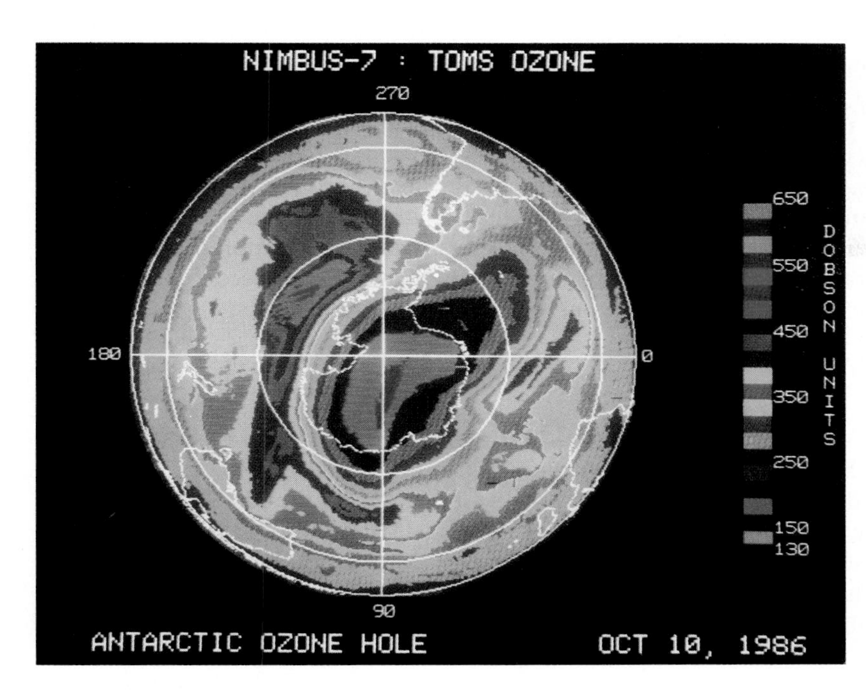

Figure 20.18
The hole in the ozone layer over the Antarctic continent. The purple region has the lowest ozone concentration. The Dobson unit is a measure of the thickness of the ozone layer. 300 Dobson units equal a 3 mm thick layer of ozone at 1 atmosphere pressure. *(Photo courtesy of NASA)*

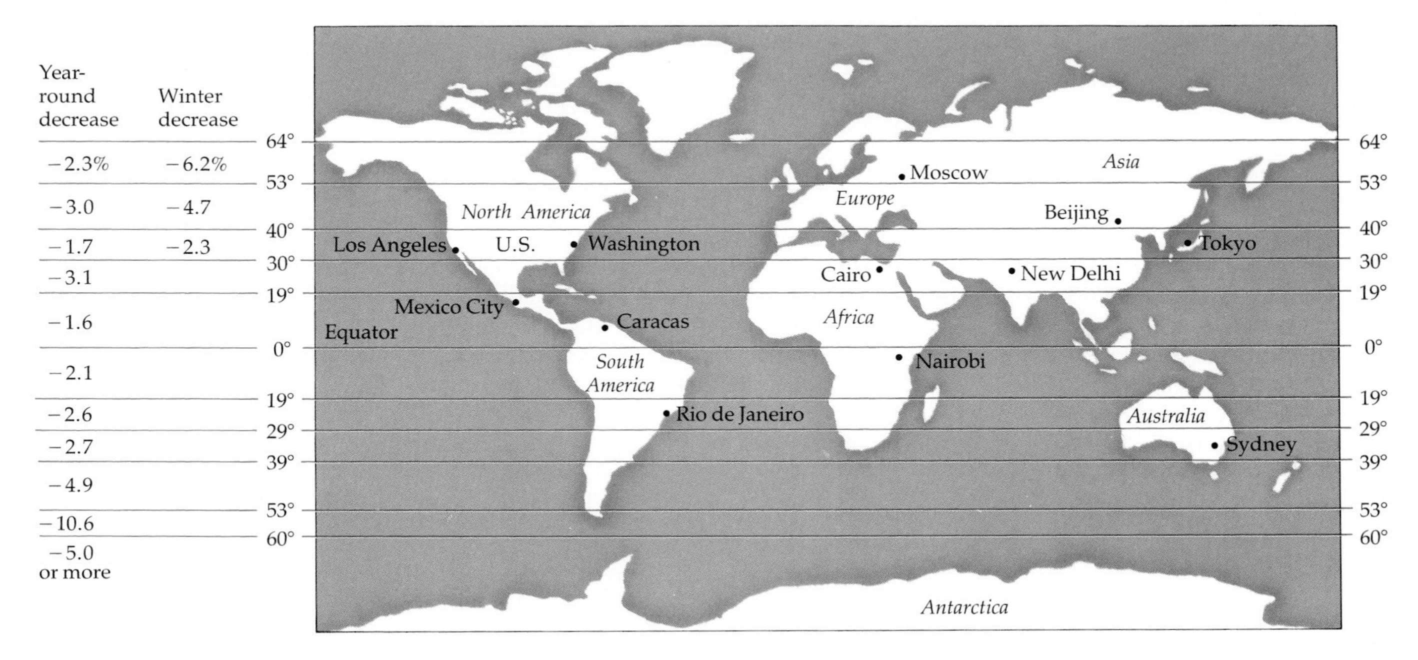

Note: Data for the area 30 to 64 degrees north of the equator is based on information gathered from satellites and ground stations from 1969–1986. Data for the area from 60 degrees south to the South Pole is based on information gathered from satellites and ground stations since 1979. All other information was complete after November 1978 from satellite data alone.

Figure 20.19
Atmospheric ozone levels show a global decline. *(Ozone Trends Panel)*

Figure 20.20
CFCs such as those used in automobile air conditioners must be recycled during repairs. Formerly, CFCs were simply vented into the atmosphere.
(Courtesy Robinair Corporation)

It will definitely cost you more to air condition your car. Production of CFC-12, the most popular auto refrigerant, will cease at the end of 1995. Although a lot of this refrigerant, along with others, will be recycled, service shops will extract a premium price. Older air-conditioning systems will probably not be compatible with substitutes, either. Some of the more recent substitutes cause seals to swell and break, and the lower molecular weight substitutes escape through tiny pores in hoses. Beware of "miracle" ozone-friendly refrigerants that are said to replace CFC-12 with no modifications necessary. They will probably ruin your car's air-conditioning system.

Intensive studies during the winters of 1987 and 1988 showed that other factors reduce the ozone concentration at the poles, in addition to the gas-phase reactions we have already described. In the dark Antarctic winter, a vortex of intensely cold air containing ice crystals builds up. On the surfaces of these crystals additional reactions produce hydrogen chloride and chlorine nitrate ($ClONO_2$), which can react with each other to form chlorine molecules. When sunlight returns in the spring, the Cl_2 molecules are readily photodissociated into chlorine atoms, which can then become involved in the ozone destruction reactions.

$$HCl(g) + ClONO_2(g) \longrightarrow Cl_2(g) + HNO_3(g)$$
$$Cl_2(g) + h\nu \longrightarrow 2\,Cl\cdot(g)$$

In an effort to reduce the harm done by CFCs to the stratospheric ozone layer, state legislatures began passing laws restricting CFC use. Vermont banned CFC refrigerants in 1990, and Connecticut banned CFCs from automobile air conditioners in 1993. In 1990 the U.S. Congress passed the Clean Air Act Amendments, which in effect ban the use of CFCs and similar compounds nationwide. For example, in November of 1992, under Clean Air Act regulations the EPA banned retail store sales of small containers (less than 20 pounds) of CFCs for motor vehicle air-conditioning systems. That same year the EPA prohibited service stations from emitting CFCs and required them to begin a recycling program for these compounds (Figure 20.20). Other governments are also acting. In January 1989, the Montreal Protocol on Substances That Deplete the Ozone Layer went into effect. Signed by 24 nations, the protocol initially called for reducing production and consumption of several of the long-lived CFCs. Later ratifications of this treaty now call for complete phaseouts of all chemicals that can harm the ozone layer (Figure 20.21). CFC manufacturers began seeking alternatives to these compounds almost immediately after it became clear that they would eventually be phased out.

One class of possible CFC substitutes includes molecules that contain a C—H bond. These compounds, called HCFCs, decompose in the troposphere due to the reactivity of the C—H bond and therefore do not reach the stratosphere to deplete the ozone there. Two such CFC substitutes are HCFC-22 and HCFC-141b.

```
     F             H  F
     |             |  |
  H—C—F        H—C—C—F
     |             |  |
     Cl            H  Cl
  HCFC-22       HCFC-141b
```

Even these compounds are somewhat harmful, however, so the latest phase-out programs call for their disappearance by the year 2030 (see Figure 20.21).

EXERCISE 20.9 • *Ozone Depletion*

Assume that 100 metric tons of the CFC with the formula CF_3Cl reaches the stratosphere. If one molecule of CF_3Cl can destroy 100,000 O_3 molecules, how many O_3 molecules can 100 tons destroy?

Portrait of a Scientist

Susan Soloman and Our View of Earth

In 1985, a British team at Halley Bay Station, Antarctica, discovered the existence of a hole in the ozone layer above that continent. This totally unexpected phenomenon needed an explanation, and it was Susan Soloman, a young NOAA (National Oceanic and Atmospheric Administration) scientist, who first proposed a good theory for it. While attending a lecture on polar stratospheric clouds, she realized that ice crystals in the clouds might do more than just scatter light over the Antarctic. Her chemist's intuition told her that the ice crystals could provide a surface on which chemical reactions of CFC compounds could take place.

In 1986, NASA chose Soloman (then 30 years old) to lead a team to Antarctica to sort out the right explanation for the ozone hole. Experiments during that visit to Antarctica showed that her cloud theory was correct, and a second expedition that year added further evidence of its validity. Soloman's team and their experiments led to the first solid proof that there is a connection between CFCs and ozone depletion.

At age 37, Susan Soloman is now the youngest member of the National Academy of Sciences. She decided to become a scientist at age 10, having been influenced by watching Jacques Cousteau on TV. At age 16 she won first place in the Chicago Science Fair for a project called "Using Light To Determine Percentage of Oxygen," and went on to place third in the national science fair that year. She said that her winters as a young girl in Chicago prepared her for her visits to Antarctica.

Susan Soloman in the Dry Valleys. (Courtesy of S. Soloman)

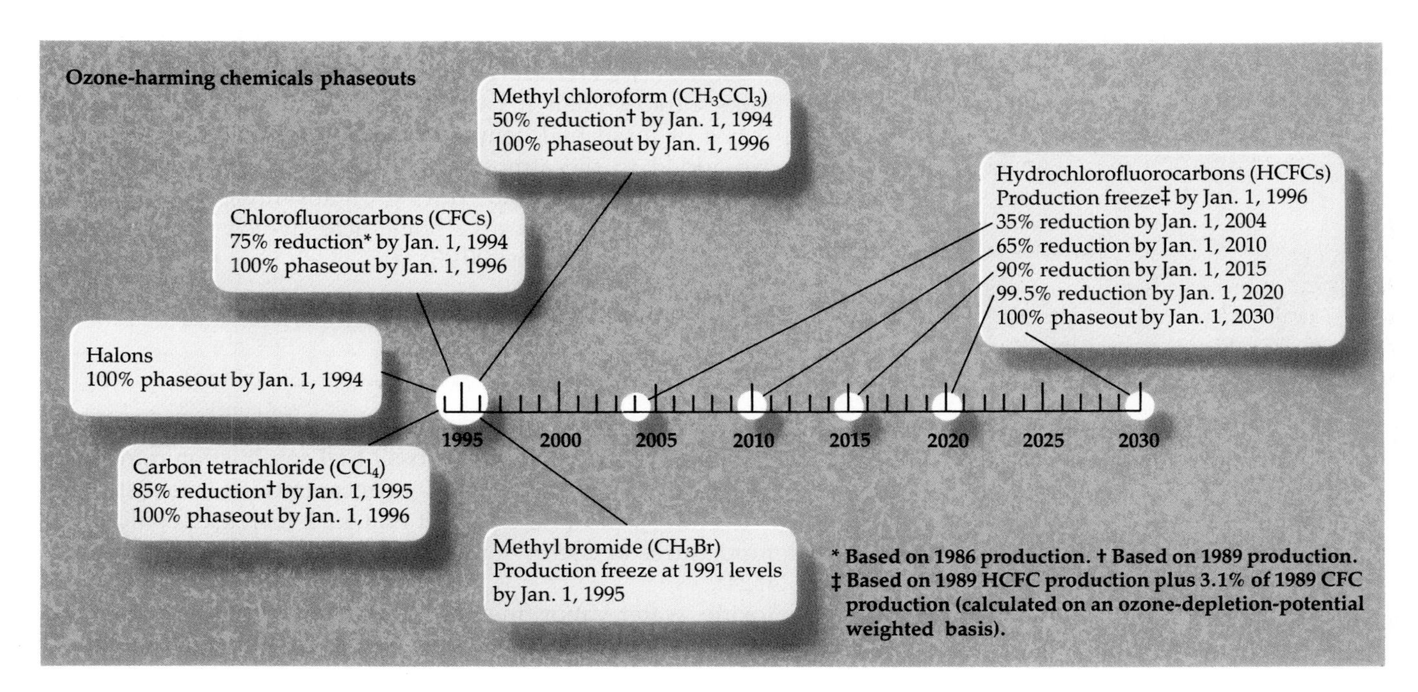

Figure 20.21
Phaseout schedule for ozone depleting chemicals according to the Montreal Protocol.

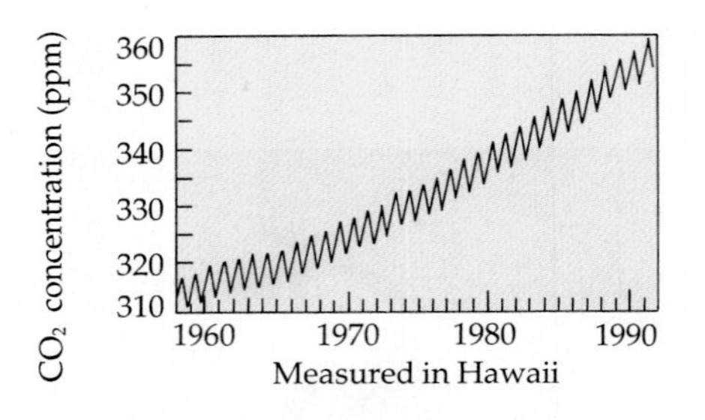

Figure 20.22
Atmospheric carbon dioxide concentration has been steadily rising.

20.7 CARBON DIOXIDE AND THE GREENHOUSE EFFECT

Without human influences, the flow of carbon dioxide between the air, plants, animals, and the oceans would be roughly balanced. However, between 1900 and 1970, the global concentration of CO_2 increased from 296 ppm to 318 ppm, an increase of 7.4%. By 1990 the concentration was about 343 ppm, and expectations are that the CO_2 concentration will continue to increase (Figure 20.22). For example, from the end of World War II, the growth in world energy use was about 5.3% per year, until the OPEC oil embargo in the mid-1970s. Rates of energy use have actually decreased since then.

To see how easily our everyday activities affect the quantity of CO_2 being put into the atmosphere, consider a round-trip flight from New York to Los Angeles. Each passenger pays for about 200 gal of jet fuel, which weighs 1400 lb. When burned, each pound of jet fuel produces about 3.14 lb of carbon dioxide. So 4400 lb or 2 metric tons of carbon dioxide are produced per passenger during that trip.

Population pressure is also contributing heavily to increased CO_2 concentrations. In the Amazon region of Brazil, for example, forests are being cut and burned to create cropland. This activity places a tremendous burden on the natural carbon dioxide cycle, since CO_2 is added to the atmosphere during burning at the same time that there are fewer trees to use the CO_2 in photosynthesis.

It took millions of years to form fossil fuels from organisms that ultimately obtained their carbon from atmospheric CO_2. By burning these fuels over just a few decades, we are returning that carbon dioxide to the atmosphere much more rapidly than it can be used up in natural processes. Counting all forms of fossil fuel combustion worldwide, about 50 billion tons of CO_2 are added to the atmosphere each year. About half of this amount is consumed—some by plants during photosynthesis, and the rest by dissolving in the oceans to form carbonic acid, which can then form bicarbonates and carbonates (Figure 20.23).

$$CO_2(g) + 2\,H_2O(\ell) \rightleftharpoons H_3O^+(aq) + \underbrace{HCO_3^-(aq)}_{\text{bicarbonate}}$$

$$HCO_3^-(aq) + H_2O(\ell) \rightleftharpoons H_3O^+(aq) + \underbrace{CO_3^{2-}(aq)}_{\text{carbonate}}$$

The oceans dissolve a huge quantity of carbon dioxide, storing it in the form of bicarbonates and carbonates. *(J. Kotz)*

The other half of the carbon dioxide from fossil fuel combustion remains in the atmosphere, increasing the global CO_2 concentration. Scientists who study the physics of the atmosphere predict that such increases could cause worldwide temperatures to rise by a mechanism called the **greenhouse effect** (Figure 20.24).

Carbon dioxide, water vapor, methane, and ozone all absorb radiation in various portions of the infrared region (see Section 11.7). Thus, all four are "greenhouse gases"; together they constitute an absorbing blanket that reduces loss of energy by radiation into space, which keeps the earth's atmo-

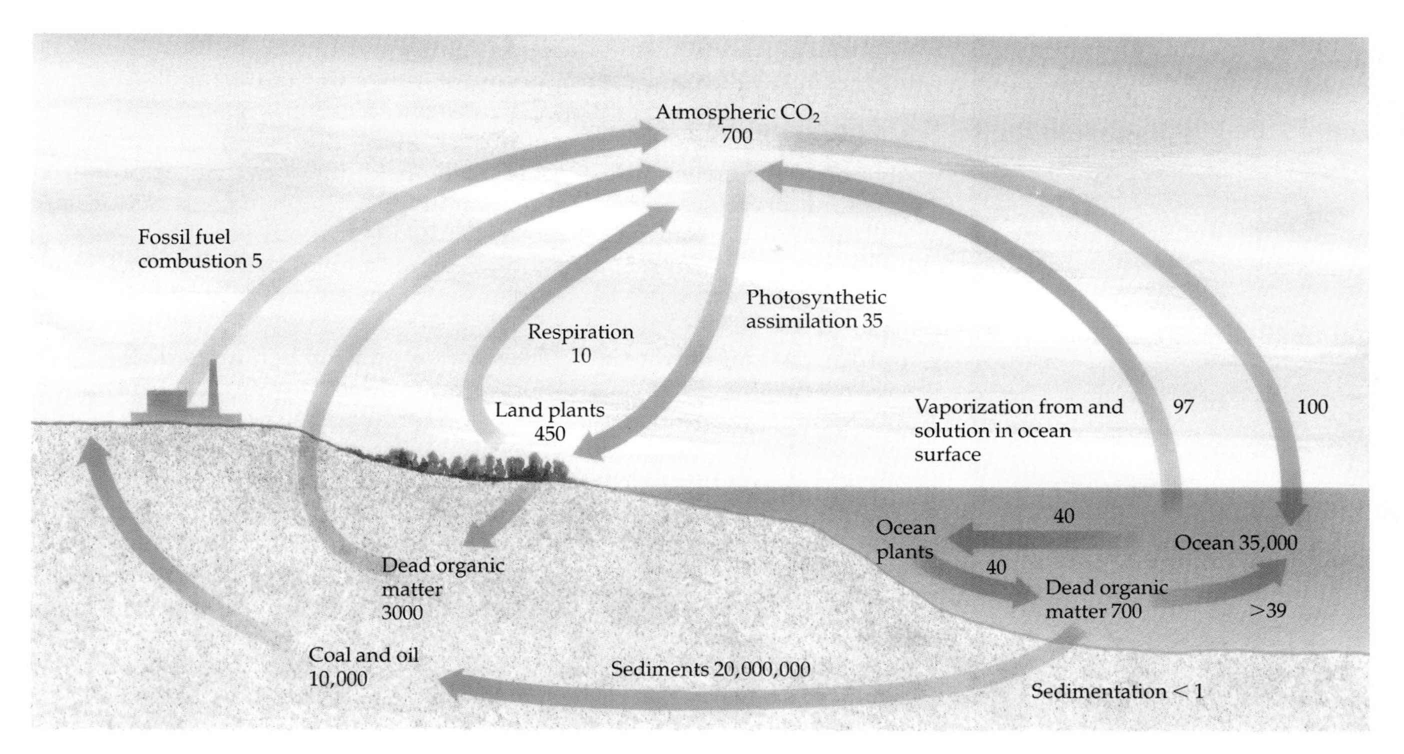

Figure 20.23
The carbon cycle in the biosphere. The numbers represent quantities of carbon in billions of metric tons in a particular reservoir or flowing from one reservoir to another. For example, plants on land contain about 450 billion metric tons of carbon and assimilate about 35 billion metric tons of carbon in the form of CO_2 from the atmosphere. *(adapted from J. Moore and E. Moore,* Environmental Chemistry, *Academic Press)*

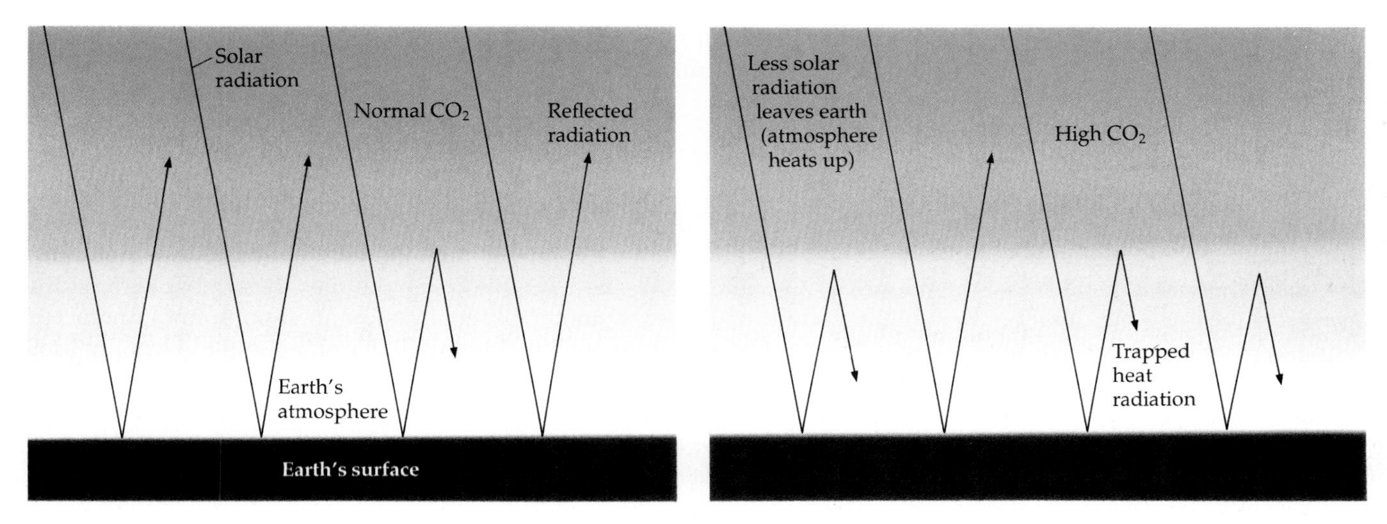

Figure 20.24
The greenhouse effect. Owing to the balance of incoming and outgoing energy in the earth's atmosphere, the mean temperature of the earth's surface is about 287 K (58 °F). Carbon dioxide permits the passage of visible radiation from the Sun to the earth, but traps some of the heat radiation attempting to leave the earth.

Venus. Its atmosphere is 96% carbon dioxide and the atmospheric pressure is 90 times that of earth. These conditions produce an extreme "greenhouse effect." The temperature at Venus's surface is about 480 °C, high enough to melt lead. *(NASA)*

spheric temperature comfortable (although not in all places at the same time!). There is such a vast reservoir of water in the oceans that human activity has a negligible influence on the concentration of water vapor in the atmosphere, and ozone is present in such small concentrations that its contribution to the greenhouse effect is small. In addition, methane is produced by natural processes in such large quantities that human contributions are negligible. Of the four gases, most attention is focused on CO_2.

Recently, Russian scientists took ice core samples that dated back as far as 160,000 years. In these ice samples were tiny pockets of air that could be analyzed for their CO_2 content. The scientists found a direct correlation between the carbon dioxide concentration and temperatures in the same period (which were known by other means): as the CO_2 level increased, global temperatures increased, and vice versa. Some scientists believe that today's rising carbon dioxide concentrations will lead to increasing global temperatures and to corresponding changes in climates.

If predictions by the U.S. National Academy of Science are correct, when (and if) the global concentration of CO_2 reaches 600 ppm, the average global temperature will have risen by 1.5 °C to 4.5 °C (or 2.7 to 8.1 °F). Warming by as little as 1.5 °C would produce the warmest climate seen on earth in the last 6000 years, and an increase of 4.5 °C would produce world temperatures higher than any since the Mesozoic era, the time of the dinosaurs. Some scientists also worry that rising temperatures will cause more of the polar ice caps to melt, raising the sea level and flooding coastal cities, and that atmospheric currents will change and produce significant changes in climate and agricultural productivity.

Clearly, global warming is potentially a major worldwide problem, and it is under continuous study. Computer models are used to predict future global temperature changes, although it is very difficult to include all possible influences on temperature in the models. The most obvious preventive measure is to control CO_2 emissions throughout the world. Given our dependence on fossil fuels, however, this will undoubtedly be more difficult than controlling emissions of CFCs or the precursors to acid rain.

20.8 INDUSTRIAL AIR POLLUTION

These regulations have been called "Community Right-to-Know" regulations because they inform communities about releases of harmful chemicals in their areas. They are a direct outgrowth of a tragic accidental chemical release that occurred in 1984 in Bhopal, India, where more than 2000 people were killed and tens of thousands were injured.

Besides CO_2, CFCs, NO_x, and SO_2, industry pollutes the atmosphere with a wide variety of solvents, metal particulates, acid vapors, and unreacted monomers. The extent to which this takes place became evident in 1989 when the first summary of annual releases was published from data received by the EPA. These release-reporting regulations were placed on manufacturers who use any of about 320 classes of substances representing special health hazards. The reporting was divided into releases to air, water, and land.

Since 1988, each industrial facility in the United States has been asked to count all releases to the atmosphere, regardless of type. This meant that leaky valves and fittings, accidental spills, vapor losses while filling tank

Aerial photo of a large chemical manufacturing plant. *(Courtesy of Du Pont Company)*

trucks and rail tank cars, emissions at stacks, and so forth, were all added together. As expected, heavily industrialized states and states with a lot of chemical industry had high releases (Table 20.6), but the quantities of some chemicals released were also surprisingly large (Table 20.7). The EPA has made these data available through a publicly accessible computerized database. Interested persons may call the EPA at 1-800-535-0202 for more information on releases. (When you call this number, you may be asked to use a computer connection to obtain the data you want.)

Table 20.6

The Top Ten States by Toxic Chemicals Released to Air in 1988

State	Emissions (millions of lb)
1. Texas	229
2. Louisiana	134
3. Tennessee	132
4. Virginia	131
5. Ohio	122
6. Michigan	106
7. Indiana	103
8. Illinois	103
9. Georgia	94
10. North Carolina	92

Note: The total emissions for all 50 states in 1988 was 2,396,915,248 lb.
Source: U.S. EPA, 1989.

Table 20.7

Top Ten Chemicals Released into Air in 1988

Chemical	Emissions (millions of lb)	Uses
toluene	235	gasoline, solvent
ammonia	233	refrigerant, reactant*
acetone	186	solvent in paints
methanol	182	solvent, reactant
carbon disulfide	137	solvent, reactant
1,1,1-trichloroethane	130	degreasing operation
methyl ethyl ketone	124	solvent in paints
xylene (mixed isomers)	120	gasoline, solvents
dichloromethane	112	solvent, reactant†
chlorine	103	reactant, bleach‡

*Ammonia's use as a fertilizer not reported.
†A known carcinogen.
‡Chlorine's use to disinfect water and wastewater not reported.
Source: U.S. EPA, 1989.

Testing a pipe joint to make sure no leaks are present. *(Courtesy of Du Pont Company)*

EXERCISE 20.10 • *Releases of Toxic Chemicals to the Atmosphere*

Add the annual emissions in Table 20.6, convert to metric tons, and express as a percentage of the mass of the earth's atmosphere over the continental United States. The mass of the atmosphere is about 5.1×10^{15} metric tons, and the United States covers 1.8% of the earth's surface area.

20.9 INDOOR AIR POLLUTION

We shouldn't be surprised that air in our homes is contaminated by industrial chemicals—after all, we bring industry's products into our homes.

As if the data about pollutants in the outside air were not enough to concern us, the air inside our homes and workplaces is also contaminated, usually by the same chemicals emitted by industry. Some scientists have concluded that air in our homes may be more harmful than the air outdoors, even in heavily industrialized areas. A study by the EPA indicated that indoor pollution levels in rural homes were about the same as for homes in industrialized areas. One cause for this is the emphasis on tighter, more energy-efficient homes, which tend to trap air inside for long periods.

What are the sources of home air pollution (see Figure 20.25)? Tobacco smoke, if present, is an obvious source. Benzene, a known carcinogen found in tobacco smoke, occurs at 30 to 50% higher levels in homes of smokers

Table 20.8

Some Common Household Products and the Chemicals They Contribute to Indoor Air Pollution

Product	Major Organic Chemicals
silicone caulk	methyl ethyl ketone, butyl propionate, 2-butoxyethanol, butanol, benzene, toluene
floor adhesive	nonane, decane, undecane, xylene
particleboard	formaldehyde, acetone, hexanal, propanal, butanone, benzaldehyde, benzene
moth crystals	p-dichlorobenzene, naphthalene
floor wax	nonane, decane, dimethyloctane, ethylmethylbenzene
wood stain	nonane, decane, methyloctane, trimethylbenzene
latex paint	2-propanol, butanone, ethylbenzene, toluene
furniture polish	trimethylpentane, dimethylhexane, ethylbenzene, limonene
room freshener	nonane, limonene, ethylheptane, various substituted aromatics (as fragrances)

Source: EPA, 1988.

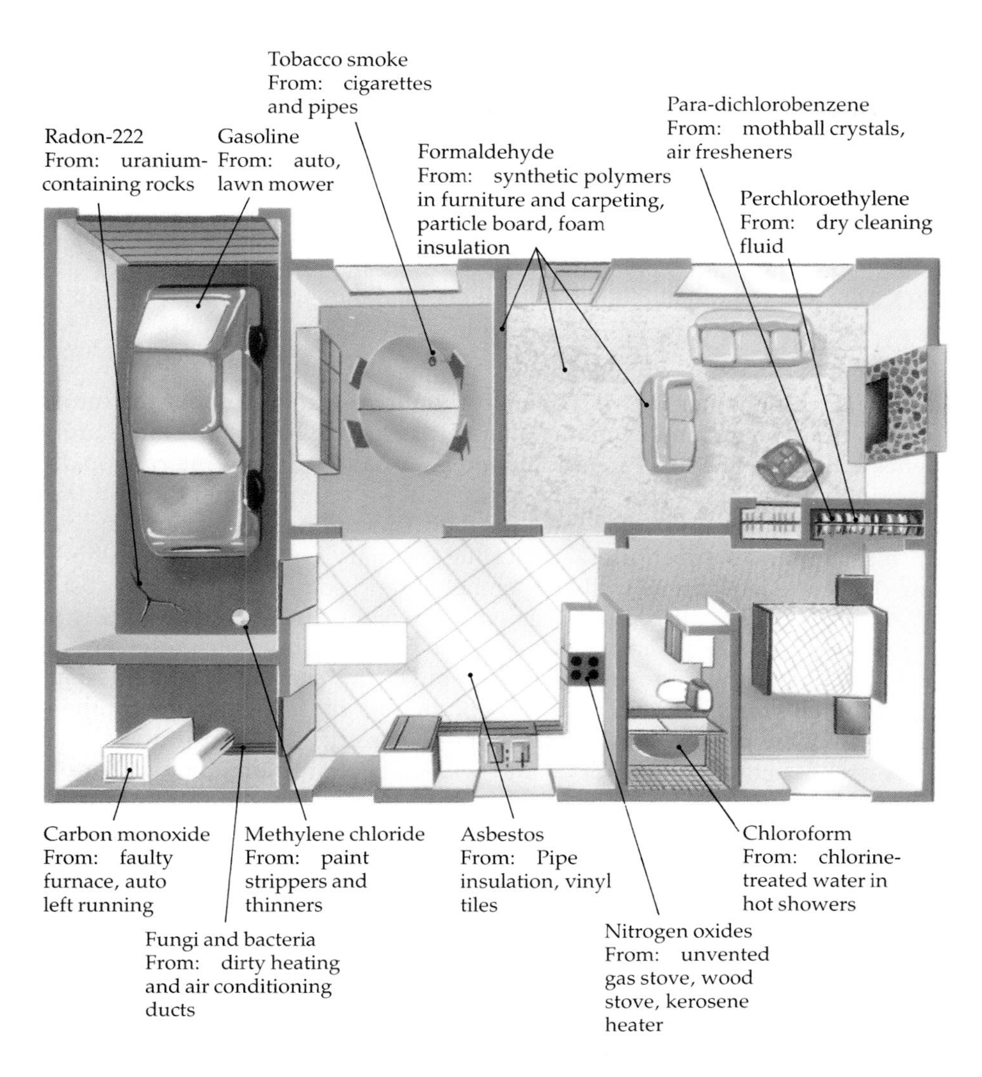

Figure 20.25
Some indoor air pollutants and their sources.

than in homes of nonsmokers. Building materials and other consumer products are also sources of pollutants (Table 20.8). Entire buildings can acquire a ''sick building syndrome'' when a particular chemical or group of chemicals is found in sufficiently high concentration to cause headaches, nausea, stinging eyes, itching nose, or some combination of these symptoms. Usually, the best cure for all forms of indoor air pollution is to limit the introduction of the offending chemicals and to have better interchange of inside and outside air.

Finally, it should be noted that radon gas is an indoor air pollutant (see Section 19.8). If a ^{222}Rn atom happens to decay inside the lungs, the β particles produced can likely cause a disruption of DNA, or trigger some other event that causes a cancer. There is also a link between radon exposure, smoking, and lung cancer. Clearly, limiting our exposure to radon *and* tobacco smoke seem to be the most effective steps to take to avoid lung cancer.

EXERCISE 20.11 • *Household Products and Air Pollution*

Find six commercial products in your home and try to list the volatile chemicals in each.

20.10 LOOKING TO THE FUTURE

Studying this chapter may have caused you to feel uneasy about the future of our atmosphere, and even our planet. It's true we have some large problems, but massive efforts are under way both nationally and internationally to bring about improvements. Users of CFCs, for example, first thought that they would never find substitutes for their favorite metal cleaners, and yet in most instances substitutes—including ordinary water and detergents—are doing the jobs just as well or even better (and sometimes at lower cost!). We will definitely all be inconvenienced as substitutes are phased in for home and automotive air-conditioning refrigerants, but substitutes are already here. If you go into an auto showroom today, chances are the salesperson will tell you proudly that the air conditioner in that car is environmentally safe. One possible down-side to the CFC refrigerant substitutes is that they are less efficient than CFCs and will probably require about 3% more electricity, thus causing more fossil fuels to be consumed and introducing even more CO_2 into the atmosphere. Getting rid of old refrigerators and the CFCs they contain will also be difficult. You can be sure the CFC phaseout will have an impact on your life in some way during the next 10 to 20 years.

On a very positive note, the quality of our air is improving, in spite of the difficult nature of the problem. In the 19th annual *National Air Quality and Emissions Trends Report,* published by the U.S. EPA in October, 1992, continuing reductions in pollutants were observed between 1982 and 1991. According to the report:

- Smog levels fell 8 percent
- Lead levels fell 89 percent
- Sulfur dioxide levels dropped 20 percent
- Carbon monoxide levels fell 30 percent
- Particulate levels declined 10 percent
- Nitrogen dioxide levels decreased 6 percent

The EPA report further stated that 41 of the 97 metropolitan areas designated as "non-attainment areas" for ozone under the Clean Air Act are now in compliance, and 13 of the 42 carbon monoxide non-attainment areas now meet the standard.

Work still needs to be done, however. In 1991, more than 86 million American lived in areas with unhealthful air, including 69 million in areas with high smog, and 21 million in areas with high quantities of particulates.

One of the greatest challenges to young scientists of the future will be to deal successfully with the problems of air pollution and its effects on the global environment.

IN CLOSING

Having studied this chapter, you should be able to

- Describe the atmosphere and its composition (Section 20.1).
- Describe and give examples of the kinds of reactions that take place in the atmosphere and indicate the role of sunlight in these reactions (Section 20.2).
- Distinguish between the two kinds of smog, the chemicals found in them, and the conditions that produce them (Section 20.3).
- Name the major atmospheric pollutants, and illustrate how they are formed and what reactions they take part in (Section 20.4).
- Explain how acid rain is formed, how it can be reduced, and the effects it has on the environment (Section 20.5).
- Discuss how CFCs and similar compounds can deplete ozone in the stratosphere, the timetable of the discovery of the ozone layer depletion problem, and steps governments have taken to correct it (Section 20.6).
- Describe the greenhouse effect caused by carbon dioxide (Section 20.7).
- Identify some of the chemicals emitted by industry and their applications (Section 20.8).
- Name some of the sources of indoor air pollution (Section 20.9).

STUDY QUESTIONS

General Questions

1. Explain the major roles played by nitrogen in the atmosphere. Do the same for oxygen.
2. Name two sources of the gaseous element helium.
3. Beginning nearest the earth, name the two most important layers or regions of the atmosphere. Describe, in general, the kinds of chemical reactions that occur in each layer.
4. Define air pollution in terms of the kinds of pollutants, their sources, and the ways they are harmful.
5. Describe the difference between primary and secondary air pollutants and give two examples of each.
6. Explain how particulates can contribute to air pollution.
7. What is adsorption? What is absorption?
8. Give an example of a photochemical reaction. Do all photons of light have sufficient energy to cause photochemical reactions? Explain with examples.
9. What is a free radical? Give an example of a chemical reaction that occurs in the atmosphere that produces a free radical.
10. What product is formed when two methyl radicals react with each other?
11. Complete the following equations (all reactants and products are gases):
    (a) $O + O_2 \longrightarrow$
    (b) $O_3 + h\nu \longrightarrow$
    (c) $O + SO_2 \longrightarrow$
    (d) $H_2O + h\nu \longrightarrow$
    (e) $NO_2 + NO_2 \longrightarrow$
12. What two reactions account for the production of O free radicals in the troposphere?
13. The reducing nature of industrial (London) smog is due to what oxide? The burning of what two fuels produces this oxide? Write a reaction showing this oxide being further oxidized.
14. Photochemical smog contains quantities of what two

oxidizing gases? What is the energy source for photochemical smog?

15. Describe what is called a thermal inversion. Why is this phenomenon a problem?
16. What reaction favors the formation of nitrogen monoxide, NO?
17. Explain how the formation of NO in a combustion chamber is similar to the formation of NH_3 in a reactor designed to manufacture ammonia.
18. What two acidic gases are primarily responsible for acid rain? Write reactions for the formation of the respective acids these gases can form.
19. Give an example of ozone in the atmosphere that is beneficial and an example of ozone that is harmful. Explain how ozone is beneficial and how it is harmful.
20. Complete the following reactions (all reactants and products are gases):
    (a) $O + H_2O \longrightarrow$
    (b) $NO_2 + HO \longrightarrow$
    (c) $RH + HO \longrightarrow$
    (d) $CO + HO \longrightarrow$
21. What is the principal danger of polynuclear aromatic hydrocarbons? Name one and describe one way that it can be formed and released into the atmosphere.
22. Describe and write an equation for a chemical reaction that:
    (a) uses up CO in the atmosphere
    (b) uses up CO_2 in the atmosphere
23. Rainfall is naturally acidic. What is its approximate pH? What gives rise to this acidity? Is this acid a strong acid or a weak acid?
24. Explain (a) the adverse effect acid rain has on trees and (b) the effect acid rain has on marble statuary.
25. Write the products for the following reactions that take place in the stratosphere.
    (a) $\mathrm{F{-}\underset{\underset{F}{|}}{\overset{\overset{F}{|}}{C}}{-}Cl} + h\nu \longrightarrow$
    (b) $Cl + O_3 \longrightarrow$
    (c) $ClO + O \longrightarrow$
26. The molecule $\mathrm{F{-}\underset{\underset{F}{|}}{\overset{\overset{F}{|}}{C}}{-}F}$ is not implicated as having ozone depletion potential. Can you explain why?
27. The molecule $\mathrm{H{-}\underset{\underset{H}{|}}{\overset{\overset{H}{|}}{C}}{-}Cl}$ has much less ozone depletion potential than the corresponding molecule in which the H atom is substituted by a F atom. Can you explain why?
28. In what regions of the globe are the effects of ozone depletion most dramatic? Why is this?
29. What is chlorine nitrate? How is it formed and when does it, along with HCl, get involved in Cl atom production?
30. Explain the greenhouse effect. What gas is primarily responsible for this effect? What three gases can contribute to the greenhouse effect?
31. What trends have been observed globally for the gas you named in Study Question 30?
32. Name a favorable effect of the global increase of CO_2 in the atmosphere.

Pollution Stoichiometry

33. Assume that limestone ($CaCO_3$) is used to remove 90% of the sulfur from 4 metric tons of coal containing 2% S. The product is $CaSO_4$ ($CaO + SO_3 \longrightarrow CaSO_4$). Calculate the mass of limestone required. Express your answer in metric tons.

34. Approximately 65 million metric tons of SO_2 enter the atmosphere every year from the burning of coal. If coal, on average, contains 2% S, how many metric tons of coal were burned to produce this much SO_2? A 1000 MW power plant burns about 700 metric tons of coal per hour. Calculate the number of hours the quantity of coal will burn in one of these power plants.

35. What mass in metric tons of 5.00% S coal would be needed to yield the H_2SO_4 required to produce a 3.00 cm rainfall of pH 2.00 over a 100. km^2 area?

36. If the pH of a 1.00 in. rainfall over 200. mi^2 is 2.45, what mass of HNO_3 is present? Assume this is the only acid contributing to the pH. Express your answer in kg of HNO_3.

Gas Law Problems

37. What amount (number of moles) of CO is found in 1.0 L of air at STP that contains 950 ppm CO?

38. What mass of gasoline must be burned according to the reaction

$$C_8H_{18}(\ell) + \tfrac{17}{2}\, O_2(g) \longrightarrow 8\,CO(g) + 9\,H_2O(g)$$

to cause a garage with dimensions of $7 \times 3 \times 3$ m to contain a CO concentration of 1000. ppm? (Assume STP conditions.)

Greenhouse Gases

39. Assume that a car burns pure octane, C_8H_{18} (d = 0.692 g/cm^3).
(a) Write the balanced equation for burning octane in air, forming CO_2 and H_2O.
(b) If the car has a fuel efficiency of 32 miles per gallon of octane, what volume of CO_2 at 25 °C and 1.0 atm is generated when the car goes on a 10. mile trip?

40. Follow the directions in Study Question 39, but use methanol, CH_3OH (d = 0.791 g/cm^2) as the fuel. Assume the fuel efficiency is 20. miles per gallon.

Summary Problems

41. One of the major sources of SO_2 in the atmosphere is from the oxidation of H_2S, produced by the decay of organic matter. Worldwide, about 100 million metric tons of H_2S is produced from sources that include the oceans, bogs, swamps, and tidal flats. The reaction in which H_2S molecules are oxidized to SO_2 involves O_3. Write an equation showing that one molecule of each reactant combines to form two product molecules, one of them being SO_2. Then calculate the annual production in tons of H_2SO_4, assuming all of this SO_2 is converted to sulfuric acid.

42. Ozone molecules attack rubber and cause cracks to appear. If enough cracks occur in a rubber tire, for example, it will be weakened and it may burst from the pressure of the air inside. As little as 0.02 ppm O_3 will cause cracks to appear in rubber in about 1 hour. Assume that a 1.0 cm^3 sample of air containing 0.020 ppm O_3 is brought in contact with a sample of rubber that is 1.0 cm^2 in area. Calculate the number of O_3 molecules that are available to collide with the rubber surface. The temperature of the air sample is 25 °C and the pressure is 0.95 atm.

43. Benzene is a known carcinogen (causing leukemia and other cancers in both laboratory animals and humans). It has acute effects as well. For example, it causes mucous membrane irritation at a concentration of 100 ppm, and fatal narcosis at 20,000 ppm. Calculate the partial pressures in atmospheres at STP corresponding to these concentrations.

Using Computer Programs and Videodiscs

44. Use the *KC? Discoverer* program to find all elements that react with air. Are any of the products of these reactions air pollutants? If so, what problems do they cause?

45. Use the Notebook program (*J. Chem. Educ.: Software* **1991** *4B*(1).) or a spreadsheet program to estimate the equilibrium constant for the reaction

$$N_2(g) + O_2(g) \rightleftharpoons 2\,NO(g)$$

as a function of temperature from 100 °C to 10,000 °C in 100-°C increments. Use data from Tables 7.2 and 7.4, and assume that $\Delta H°$ and $\Delta S°$ do not vary with temperature.

46. Predictions of the effects of atmospheric reactions often depend on computer modeling. Given initial concentrations of the substances involved, the computer calculates subsequent concentrations based on the rates of the reactions that may occur. B. J. Huebert, *J. Chem. Educ.* **1974** *51*(10), 644–645, describes a simple example of such a computer model. Use a spreadsheet program to carry out calculations of the sort that Huebert describes. (Note: this is a very time-consuming project unless you are quite fluent with spreadsheet calculations and spreadsheet macro programming.)

Biochemistry is the study of the chemistry of living things.

(C.D. Winters)

Chemistry of Life

CHAPTER

21

The chemistry of life is referred to as biochemistry, and the organic chemicals found in living things are called biochemicals. Biochemicals common to living systems of various kinds are lipids, carbohydrates, proteins, enzymes, vitamins, hormones, nucleic acids, and compounds for the storage and exchange of energy, such as adenosine triphosphate (ATP). In addition to these biochemicals, certain minerals are required for proper functioning of living organisms.

Many biochemicals are polymers. Starches are condensation polymers of simple sugars; proteins are condensation polymers of amino acids; and nucleic acids are condensation polymers of simple sugars, nitrogenous bases, and phosphoric acid species. As you read about these biochemicals, you will see their similarities to the synthetic polymers discussed earlier.

21.1
AMINO ACIDS

D- and L- are used to identify the two enantiomers of α-amino acids. These symbols represent the spatial orientation of the enantiomer (see Figure 21.1) and do not necessarily indicate the direction of rotation of plane-polarized light.

All proteins are condensation polymers of **amino acids.** A large number of proteins exist in nature. For example, the human body contains an estimated 100,000 different proteins. What is amazing is that all of these proteins are derived from only 20 different naturally occurring amino acids (Table 21.1). Even more amazing is nature's exclusive preference for only the L-enantiomer of each amino acid (Figure 21.1).

All but one of the 20 amino acids found in nature have the general formula

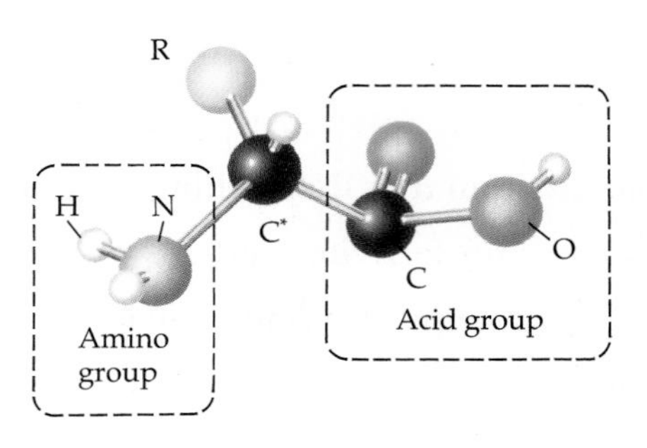

and are described as α-amino acids with an amino ($—NH_2$) group attached to the **alpha carbon,** the first carbon next to the —COOH. R is a characteristic group of each amino acid (see Table 21.1), and the asterisk identifies the chiral carbon. R is a hydrogen atom in glycine, the simplest amino acid, so glycine is not chiral and is the only naturally occurring amino acid that does not have enantiomers. The polarities of the R groups in amino acids affect the structure and function of proteins. The amino acids are grouped in Table 21.1 according to whether the R group is nonpolar, polar, acidic, or basic.

Proline has a secondary amine (>NH) group instead of a primary amine group because the N atom is part of a ring.

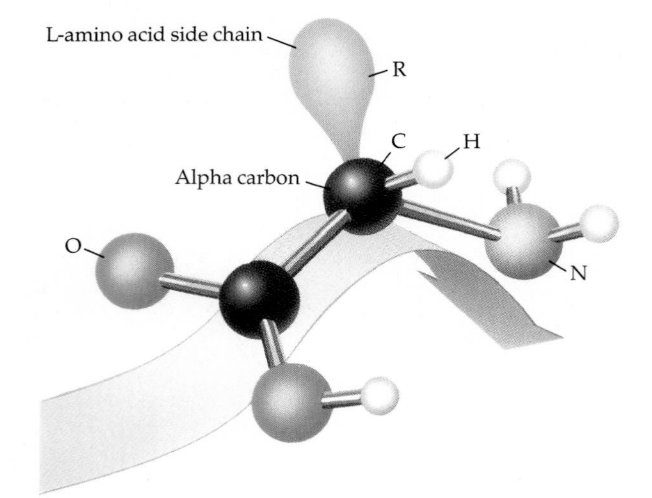

Figure 21.1
Spatial orientation of the L-form of amino acids. When walking along the C—C—N bridge from the C—O end to the NH_2 end, the acid is in the L-form if the R group is on the left, and if the acid is in the D-form the R group is on the right. Only L-isomers of amino acids are found in nature.

Table 21.1

Common L-Amino Acids Found in Proteins (The R Group in Each Amino Acid is Highlighted)

Amino Acid	Abbreviation	Structure
Nonpolar R Groups		
glycine	Gly	$H-CH(NH_2)-COOH$
alanine	Ala	$CH_3-CH(NH_2)-COOH$
*valine	Val	$CH_3-CH(CH_3)-CH(NH_2)-COOH$
*leucine	Leu	$CH_3-CH(CH_3)-CH_2-CH(NH_2)-COOH$
*isoleucine	Ile	$CH_3-CH_2-CH(CH_3)-CH(NH_2)-COOH$
proline	Pro	$H_2C—CH_2$ / H_2C, $CHCOOH$ joined through $N-H$ (ring)
*phenylalanine	Phe	$C_6H_5-CH_2-CH(NH_2)-COOH$ (benzene ring)
*methionine	Met	$CH_3-S-CH_2CH_2-CH(NH_2)-COOH$
Polar but Neutral R Groups		
serine	Ser	$HO-CH_2-CH(NH_2)-COOH$
*threonine	Thr	$CH_3-CH(OH)-CH(NH_2)-COOH$
cysteine	Cys	$HS-CH_2-CH(NH_2)-COOH$
tyrosine	Tyr	$HO-C_6H_4-CH_2-CH(NH_2)-COOH$ (para-hydroxyphenyl ring)
asparagine	Asn	$H_2N-C(=O)-CH_2-CH(NH_2)-COOH$
glutamine	Gln	$H_2N-C(=O)-CH_2CH_2-CH(NH_2)-COOH$
*tryptophan	Trp	indole ring (N—H)$-CH_2-CH(NH_2)-COOH$

(table continues on next page)

Table 21.1

Common L-Amino Acids Found in Proteins (The R Group in Each Amino Acid is Highlighted) (Continued)

Acidic R Groups			Basic R Groups		
glutamic acid	Glu	$HO-C(=O)-CH_2CH_2-CH(NH_2)-COOH$	*lysine	Lys	$H_2N-CH_2CH_2CH_2CH_2-CH(NH_2)-COOH$
aspartic acid	Asp	$HO-C(=O)-CH_2-CH(NH_2)-COOH$	†arginine	Arg	$H_2N-C(=NH)-NH-CH_2CH_2CH_2-CH(NH_2)-COOH$
			histidine	His	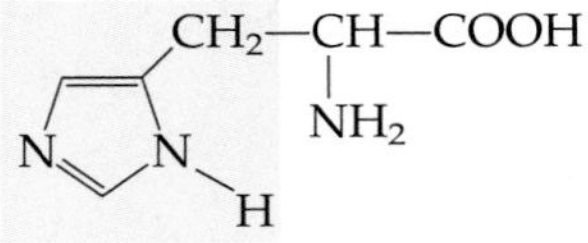 (imidazole ring with N and N–H)$-CH_2-CH(NH_2)-COOH$

*Essential amino acids that must be part of the human diet. The other amino acids can be synthesized by the body.
†Growing children also require arginine in their diet.

Most amino acids dissolve reasonably well in water. For example, 25 g of glycine dissolves per 100 g of water at 25 °C, and 3 g of phenylalanine dissolves under the same conditions. In every case, the following equilibrium exists in aqueous solution.

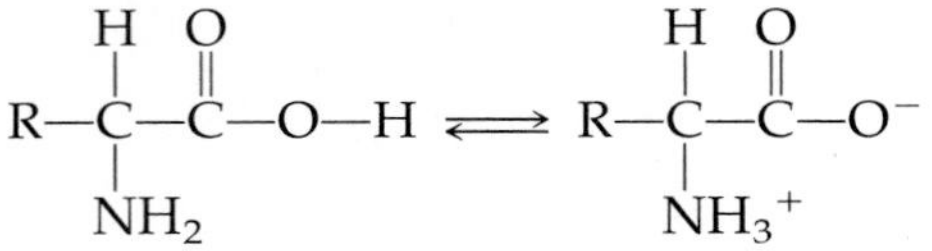

$$R-\overset{H}{\underset{NH_2}{C}}-\overset{O}{\overset{\|}{C}}-O-H \rightleftharpoons R-\overset{H}{\underset{NH_3^+}{C}}-\overset{O}{\overset{\|}{C}}-O^-$$

zwitterion

Since the amino group is more basic than the carboxylic acid group, the proton of the —COOH group is donated to the $-NH_2$ group to form an internal salt known as a **zwitterion,** a term taken from the German word *zwitter* meaning "double." In the pure solid state and in aqueous solution near pH 6, amino acids exist almost completely as zwitterions. The presence of the zwitterion structure explains why amino acids have properties typical of many ionic compounds—relatively high melting points and water solubility.

EXAMPLE 21.1

Amino Acids

Draw the zwitterion structure of alanine.

SOLUTION The R group in alanine is CH_3- (Table 21.1), and in the zwitterion form a proton has been transferred to give $-NH_3^+$ and $-COO^-$ ions.

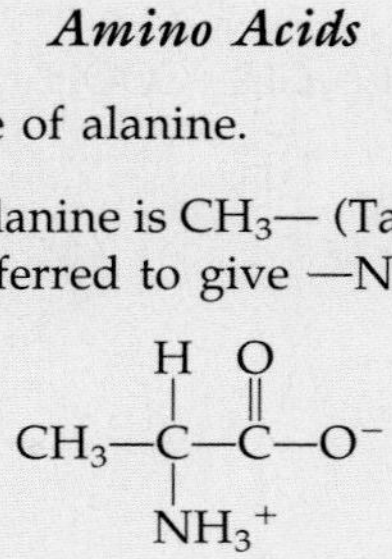

$$CH_3-\overset{H}{\underset{NH_3^+}{C}}-\overset{O}{\overset{\|}{C}}-O^-$$

EXERCISE 21.1 • *Amino Acids*

Why do amino acids have the zwitterion structure? Draw the zwitterion structure of valine.

21.2
PEPTIDES AND PROTEINS

How do amino acids polymerize to give proteins? The polymerization reaction is the formation of an amide from an amine and a carboxylic acid, which was described in Section 13.6.

$$\mathrm{R{-}\overset{\displaystyle O}{\overset{\|}{C}}{-}OH + H_2NR' \longrightarrow R{-}\overset{\displaystyle O}{\overset{\|}{C}}{-}\overset{\displaystyle H}{\overset{|}{N}}{-}R' + H_2O}$$

Since each amino acid has both an amine group and a carboxylic acid group, the —COOH of one amino acid can combine with the $-NH_2$ of a second amino acid.

$$\mathrm{H_2N{-}\underset{\displaystyle H}{\underset{|}{\overset{\displaystyle R}{\overset{|}{C}}}}{-}\overset{\displaystyle O}{\overset{\|}{C}}{-}OH + H{-}\underset{\displaystyle H}{\underset{|}{N}}{-}\underset{\displaystyle H}{\underset{|}{\overset{\displaystyle R}{\overset{|}{C}}}}{-}\overset{\displaystyle O}{\overset{\|}{C}}{-}OH}$$

$$\Big\downarrow -H_2O$$

$$\mathrm{H_2N{-}\underset{\displaystyle H}{\underset{|}{\overset{\displaystyle R}{\overset{|}{C}}}}{-}\overset{\displaystyle O}{\overset{\|}{C}}{-}\underset{\displaystyle H}{\underset{|}{N}}{-}\underset{\displaystyle H}{\underset{|}{\overset{\displaystyle R}{\overset{|}{C}}}}{-}\overset{\displaystyle O}{\overset{\|}{C}}{-}OH}$$

a peptide bond

In the condensation reaction, one molecule of water is eliminated between the carboxylic acid of one amino acid and the amine group of another. The result is a **peptide** bond (called an amide group in simpler molecules), and the new molecule is a **dipeptide.**

Names of peptides are written from left to right starting with the N-terminal end. The -ine *ending of all amino acid residues except the C-terminal end is changed to* -yl. *For example, Gly-Ala-Se is the tripeptide glycylalanylserine.*

When two different amino acids form a dipeptide, two different combinations are possible, depending on which amine reacts with which acid group. For example, when glycine and alanine react, both glycylalanine and alanylglycine can be formed. Either end of the dipeptide can react with a third amino acid to give a dipeptide.

peptide bonds

N-terminal end → $\mathrm{H_2N{-}\underset{\displaystyle H}{\underset{|}{\overset{\displaystyle H}{\overset{|}{C}}}}{-}\overset{\displaystyle O}{\overset{\|}{C}}{-}\underset{\displaystyle H}{\underset{|}{N}}{-}\underset{\displaystyle CH_3}{\underset{|}{\overset{\displaystyle H}{\overset{|}{C}}}}{-}COOH}$ $\mathrm{H_2N{-}\underset{\displaystyle CH_3}{\underset{|}{\overset{\displaystyle H}{\overset{|}{C}}}}{-}\overset{\displaystyle O}{\overset{\|}{C}}{-}\underset{\displaystyle H}{\underset{|}{N}}{-}\underset{\displaystyle H}{\underset{|}{\overset{\displaystyle H}{\overset{|}{C}}}}{-}COOH}$ ← C-terminal end

glycylalanine (Gly-Ala) alanylglycine (Ala-Gly)

The amino acids that have combined to form the peptide molecule are called **amino acid residues.** Since each dipeptide has a —COOH and an $—NH_2$ group, a tripeptide can be formed from each dipeptide, and the polymerization process can continue until a large **polypeptide** chain is formed. **Proteins** are polypeptides containing hundreds to thousands of amino acid residues. When is a peptide a protein, or in other words, when is a peptide considered a polypeptide? The number of amino acid residues required for a molecule to be classed as a protein is not well defined, but usually a protein is assumed to have a minimum of 50 amino acid residues.

Peptides

The order of the amino acid residues in a peptide molecule is called the **amino acid sequence** of that molecule. As the length of the chain increases, the number of variations in the sequence of amino acids quickly increases. Six tripeptides are possible if three different amino acids (for example, glycine, Gly; alanine, Ala; and serine, Ser) are linked in combinations that contain all three amino acids. They are

Gly-Ala-Ser	Ser-Ala-Gly	Ala-Ser-Gly
Gly-Ser-Ala	Ser-Gly-Ala	Ala-Gly-Ser

If n amino acids are all different, the number of arrangements is $n!$ (n factorial). For four different amino acids, the number of different arrangements is $4!$ or $4 \times 3 \times 2 \times 1 = 24$. For five different amino acids, the number of different arrangements is $5!$ or 120. If all 20 different naturally occurring amino acids were bonded in one peptide, the sequences would make 2.43×10^{18} (2.43 quintillion) unique 20-monomer molecules! *Since proteins can also include more than one molecule of a given amino acid, the possible combinations are effectively infinite.* However, of the many different proteins that could be made from a set of amino acids, a living cell will make only the relatively small number it needs.

Many short-chain peptides are important biochemicals. For example, enkephalins and endorphins are referred to as "natural opiates" because they moderate pain. For many years scientists speculated about the action of opiates in the brain and the possible relationship to the human response to pain. Solomon Snyder and co-workers at Johns Hopkins University discovered in 1973 that the brain and spinal cord contain specific bonding or receptor sites that the opiate molecules fit as a key fits into a lock. This enhanced the search for opiate-like neurotransmitters. In 1975 John Hughes and Hans Kosterlitz, of the University of Aberdeen, Scotland, isolated two peptides with opiate activity from pig brains. They decided to call these peptides **enkephalins** (from the Greek *en* and *kephale,* meaning "within the head"); specifically, the two pentapeptides they isolated are known as methionine-enkephalin and leucine-enkephalin (Figure 21.2).

A year later, Roger Guillemin and co-workers at the Salk Institute isolated a longer peptide, called **β-endorphin,** from extracts of the pig hypothalamus. β-Endorphin is 50 times more potent than morphine. Since this early work, other enkephalins and endorphins have been isolated. In addition, the relationship of these natural opiates to pain has been studied.

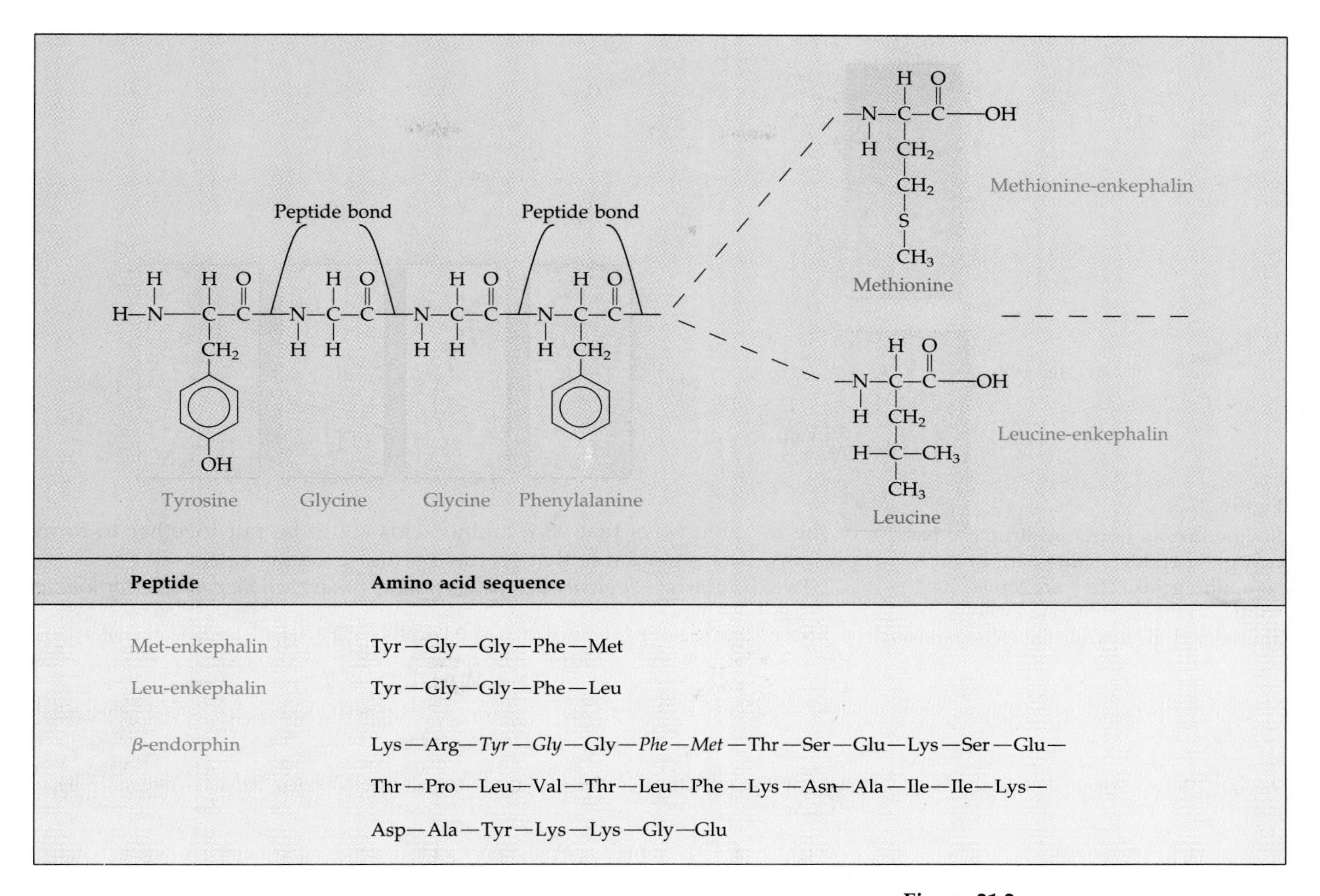

Peptide	Amino acid sequence
Met-enkephalin	Tyr—Gly—Gly—Phe—Met
Leu-enkephalin	Tyr—Gly—Gly—Phe—Leu
β-endorphin	Lys—Arg—*Tyr*—*Gly*—Gly—*Phe*—*Met*—Thr—Ser—Glu—Lys—Ser—Glu—Thr—Pro—Leu—Val—Thr—Leu—Phe—Lys—Asn—Ala—Ile—Ile—Lys—Asp—Ala—Tyr—Lys—Lys—Gly—Glu

Figure 21.2
Some enkephalins and endorphins, polypeptides that are natural opiates.

Our bodies synthesize enkephalins and endorphins to modulate pain, and our pain threshold is related to levels of these **neuropeptides** in our central nervous system. Individuals with a high tolerance for pain produce more neuropeptides and consequently tie up more receptor sites than normal; hence, they feel less pain. A dose of heroin temporarily bonds to a high percentage of the sites, resulting in feeling little or no pain. Continued use of heroin causes the body to reduce or cease its production of enkephalins and endorphins. If use of the narcotic is stopped, the receptor sites become empty and withdrawal symptoms occur.

Proteins occur in a variety of sizes. The common protein insulin has only 51 amino acid units in two linked chains (Figure 21.3). In contrast, human hemoglobin contains four protein chains, two identical ones having 141 amino acids and the other two, again identical, having 146. Considering all

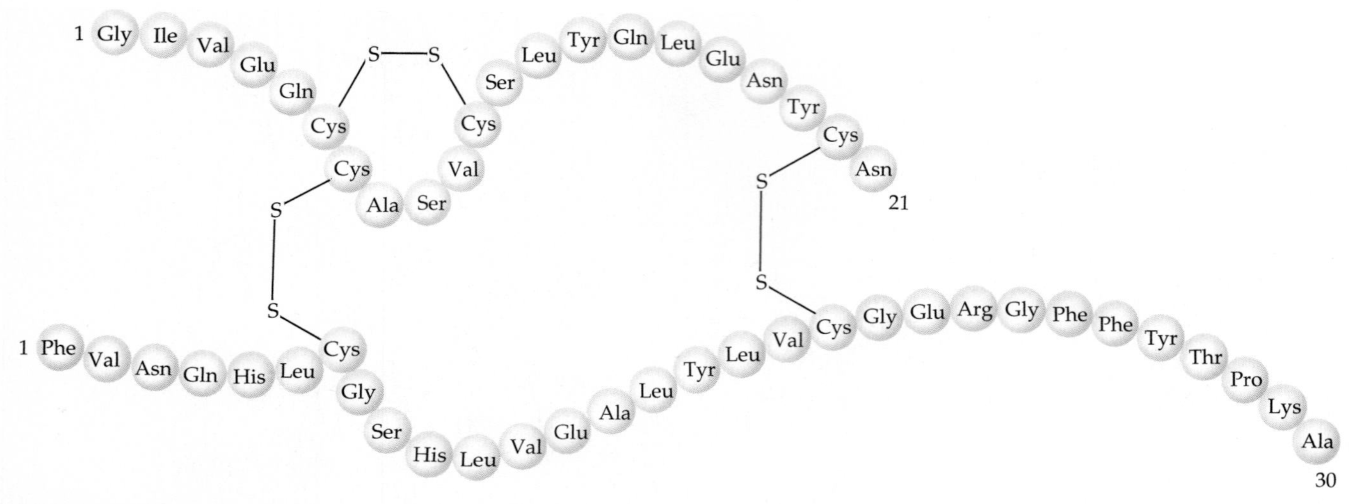

Figure 21.3

Bovine insulin hormone structure has two polypeptide chains with 21 and 30 amino acids. They are joined by sulfur bridges connecting cysteine amino-acid groups on the two chains.

of the possible ways that 20 α-amino acids could be put together to form proteins, it is remarkable that so few natural proteins exist.

To summarize, *proteins are polypeptides, condensation polymers of amino acids* (Figure 21.4).

EXAMPLE 21.2

Peptides

Draw the structure of the tripeptide represented by Ala-Ser-Gly, and give its name.

SOLUTION The amino acid sequence in the abbreviated name shows that alanine should be written at the left with a free H_2N— group, glycine should be written at the right with a free —COOH group, and both should be connected to serine by peptide bonds.

```
          H  O        H     O        H  O
          |  ‖        |     ‖        |  ‖
    H2N—C—C—N—C——C—N—C—C—OH
          |       |   |        |  |
         CH3      H  CH2OH     H  H
```

The name is alanylserylglycine.

EXERCISE 21.2 • *Peptides*

Draw the structure of the tetrapeptide Cys-Phe-Ser-Ala.

Proteins

The close relationship between proteins and living organisms was first noted in 1835 by the Dutch chemist G.T. Mulder. He named proteins from the Greek *proteios* ("first"), thinking that proteins are the starting point for a

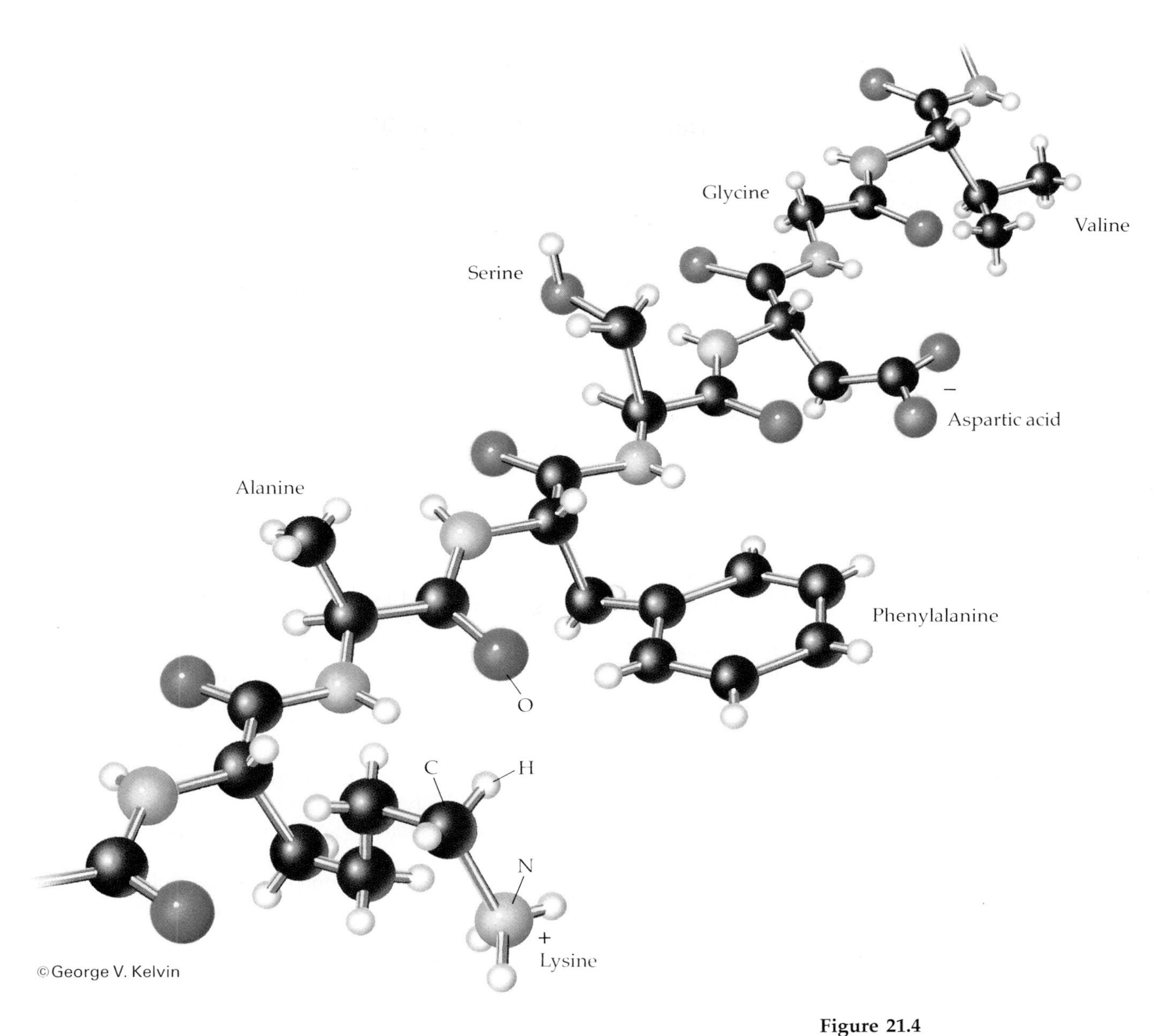

Figure 21.4
A polypeptide chain. This figure also shows (a) the planarity of the peptide link and (b) the fact that the chain bends only around the alpha carbon.

chemical understanding of life. Proteins are important in a variety of ways. As *enzymes* they serve as catalysts in biological synthesis and degradation reactions. As *hormones* they serve a regulatory role, and as *antibodies* they protect us against disease. The protein *hemoglobin* carries oxygen from the lungs to various parts of the body. Finally, proteins are the major constituents of cellular and intracellular membranes, skin, hair, muscle, and tendons. Each kind of protein is composed of specific amino acids arranged in a definite molecular structure. In a few proteins the major fraction is only one kind of amino acid; the protein in silk, for example, is 44% glycine.

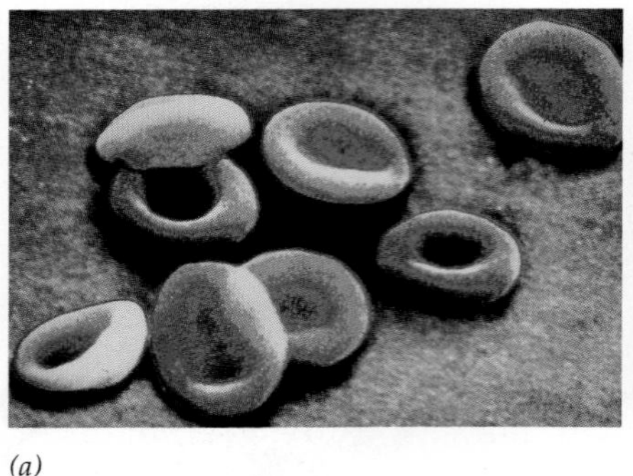

(a)

(b)

Figure 21.5
(a) A normal red blood cell. (b) A sickled red blood cell. Sickle cells are caused by the substitution of the nonpolar amino acid valine for the negatively charged amino acid glutamate in the protein structure of hemoglobin. This substitution produces a crucial alteration in the tertiary structure, which causes the sickling.
(Bill Longcore/Photo Researchers, Inc.)

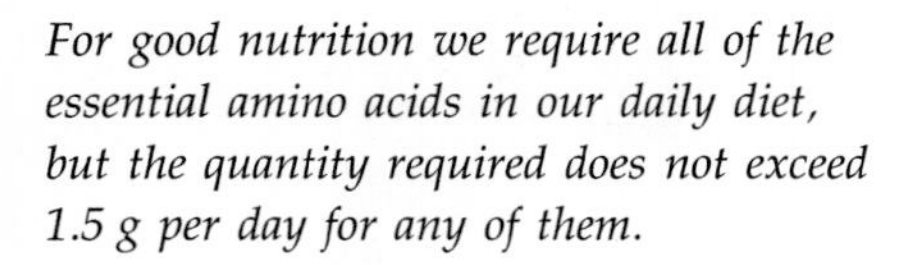

For good nutrition we require all of the essential amino acids in our daily diet, but the quantity required does not exceed 1.5 g per day for any of them.

The eight **essential amino acids** (indicated by asterisks in Table 21.1) must be taken in from food, because the human body cannot synthesize them. A diet that includes meat, milk, eggs, or cheese provides all the essential amino acids. Vegetarians who obtain their protein from grains and vegetables have to check their diet to make sure all the essential amino acids are included. For example, rice and wheat are deficient in lysine and corn is deficient in lysine and tryptophan. Peas are deficient in methionine, so a combination of peas and rice would provide the essential amino acids. The other amino acids can be synthesized by the human body.

Protein Structure The sequence of amino acids bonded to one another in a protein by peptide bonds is the protein's **primary structure.** To fully understand the functioning of the protein, we must know its primary structure. Changing the sequence alters the properties of a protein, and just one change may produce a new protein unable to function as well as the original one. For example, *sickle cell anemia,* a reduction in the ability of hemoglobin to transfer oxygen, is caused by the replacement of only one specific amino acid in two of the four protein chains that make up the hemoglobin molecule (Figure 21.5).

In some proteins, the shape of the backbone of the molecule (the chain containing peptide bonds) has a regular, repetitive pattern that is referred to as its **secondary structure.** The two most common secondary structures are the **α-helix** and the **β-pleated sheet.** What kind of secondary structure the protein adopts depends on the positions of the C=O···H—N hydrogen bonds. For example, the helical structure seen in some proteins is caused by *intramolecular* hydrogen bonding within the protein chain that holds the coils in place. In this case, an N—H group of one amino acid forms a hydrogen bond with the oxygen atom in the third amino acid down the chain (Figure 21.6). If the hydrogen bonding is *intermolecular* (between protein chains), β-pleated sheets form.

The α-helix is the basic structural unit of fibrous proteins known as α-keratins, which are found in wool, hair, skin, beaks, nails, and claws. Because of the helical protein chains, fibers such as those in human hair are elastic to some extent. Stretching the fibers involves breaking some or all of the relatively weak hydrogen bonds, but not the strong covalent bonds. If a hair is not stretched to the point where the covalent bonds begin to break, the fiber will snap back to its original length. The fiber is elastic because the hydrogen bonds can be re-formed.

Silk has the β-pleated sheet structure (Figure 21.7), in which several chains of amino acids are joined side-to-side by hydrogen bonds. The resulting structure is not elastic, because stretching the fibers would involve breaking either covalent bonds or the many hydrogen bonds holding the individual protein strands in the sheet. However, just as you can bend the stack of pages in this book, so can the stack of protein sheets be bent.

Tertiary structure refers to how a protein molecule is folded. The nature of the R groups of the amino acids in the primary structure determines the tertiary structure (Figure 21.8). Hydrogen bonding, disulfide bridge bonds (—S—S—), and ionic bonds are three types of interactions that affect the folding of protein molecules. One kind of tertiary structure is found in collagen, a fibrous protein: three amino acid chains are twisted into left-handed helices, which in turn are twisted into a right-handed superhelix to form an extremely strong fibril (Figure 21.8a). Bundles of fibrils make up the tough collagen. A second kind of tertiary structure is found in globular proteins, in which the polypeptide chain is folded and twisted in a complicated manner. Although these folds may seem random, they form a definite geometric pattern that is the same in all molecules of the same protein (Figure 21.8b). Many enzymes are globular proteins.

Quaternary structure is the shape assumed by the entire group of chains in a protein composed of two or more chains. In human hemoglobin, a globular protein with a molecular weight of 68,000, the four separate chains

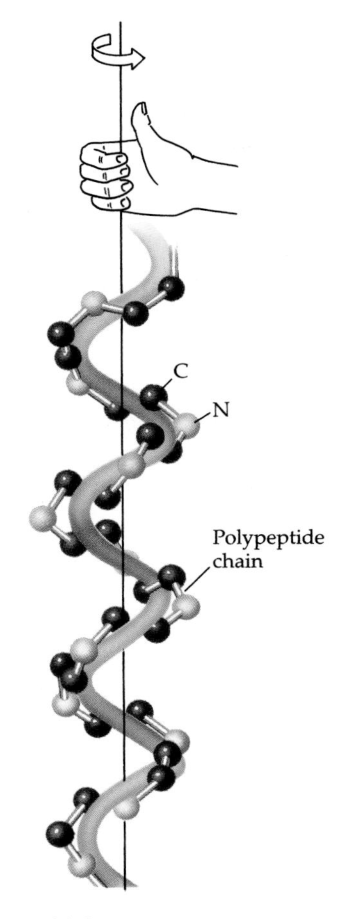

Figure 21.6
The helical structure of proteins. Hydrogen bonding within a polypeptide chain leads to a spiral arrangement called an α-helix. In proteins, the helix always has a right-handed direction, so that if the thumb of your right hand points along the axis of the helix, your fingers curl in the same direction as the curl of the spiral.

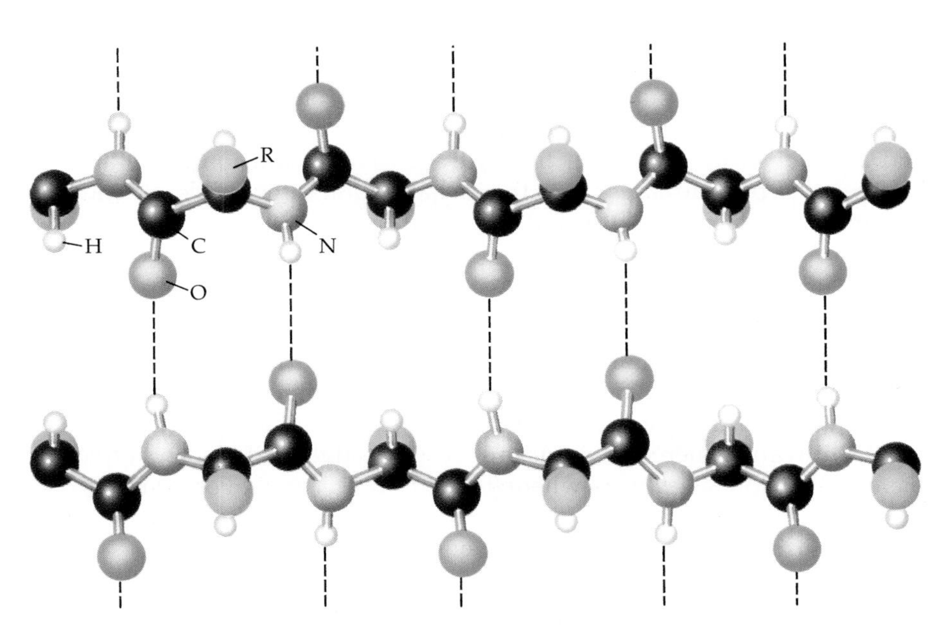

Figure 21.7
The β-pleated sheet structure. In silk and other insect fibers, the polypeptide chains are bound to one another by hydrogen bonds.

Figure 21.8
Tertiary structure of proteins. (a) The twisted structure of collagen. (b) The folded structure of the helix in a globular protein.

of amino acids must be positioned properly in order to form an active oxygen-carrying complex.

All of the structural features—primary, secondary, tertiary, and quaternary—are critical to the proper functioning of a protein. Any physical or chemical process that changes the protein structure and makes it incapable of performing its usual function is called a **denaturation** process. For example, heating an aqueous solution of a protein breaks hydrogen bonds in the secondary and tertiary structures and causes the protein molecule to unfold. Denaturing chemicals include reducing agents that break disulfide linkages, and acids or bases that affect the hydrogen bonds and ionic interactions between polypeptide chains. Whether denaturation is reversible depends on the protein and the conditions of denaturation.

Cooking eggs and permanent waving of hair are examples of protein denaturation.

Enzymes

Enzymes function as catalysts for chemical reactions in living systems. As mentioned earlier, a major part of the structure of an enzyme is globular protein. Like all catalysts, enzymes increase the rate of a reaction by lowering the activation energy. Enzymes are very effective catalysts, and typically increase reaction rates by anywhere from 10^9 to 10^{20} times. For example, one molecule of the enzyme carbonic anhydrase can catalyze the decomposition of about 36 million molecules of carbonic acid in one minute.

$$H_2CO_3(aq) \rightleftharpoons CO_2(g) + H_2O(\ell)$$

Most enzymes are very specific. For example, carbonic anhydrase catalyzes only the one reversible reaction shown above. Many biomolecules are broken down during digestion by **hydrolysis** reactions, which are essentially the reverse of condensation reactions. In hydrolysis, a larger molecule

The general equation for hydrolysis of an ester is

$$RCOOR + H_2O \longrightarrow RCOOH + ROH$$

The Chemical World

Unraveling the Protein Structure

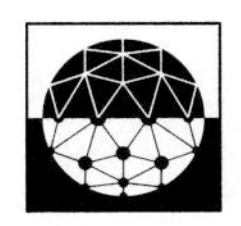

One of the key steps in unraveling the mystery of hydrogen bonds in protein structure involved Linus Pauling, a cold, and a Nobel Prize. Pauling now lives in the Big Sur region of California. His living room is his office. There he spends a large part of each day at a simple desk, working on a new research interest, metals. Earlier in his career Pauling had another interest, the structure of protein molecules. At that time there were several conflicting theories. Pauling and his colleagues thought that the first level of protein structure was a polypeptide chain. Then they asked themselves a fundamental question.

"We asked: How is the polypeptide chain folded? We couldn't answer the question, but we said it's probably held together by hydrogen bonds. The conclusion we reached was that there are . . . polypeptide chains in the protein, which . . . are coiled back and forth; and that they are coiled into a very well defined structure, configuration, with the different parts of the chain held together by hydrogen bonds. In 1937, I spent a good bit of the summer with models for [polypeptides]. I assumed that I knew what a polypeptide chain looks like except for the way in which it's folded. And I wanted to fold it to form the hydrogen bonds. I didn't succeed. The fact is I thought that there was something about proteins that perhaps I didn't know."

Pauling continued to work on this problem, but the solution eluded him. Then one day he had a crucial insight in a completely different and unexpected setting.

"I had a cold. I was lying in bed for two or three days, and I read detective stories, light reading, for awhile, and then I got sort of bored with that. So I said to my wife, "Bring me a sheet of paper, and . . . I think I'll work on that problem of how polypeptide chains are folded in proteins. So she brought me a sheet of paper and the slide rule and pencil, and I started working."

Using the knowledge gained from his years of model building, he drew the backbone of a polypeptide chain on a piece of paper. Then it occurred to him to try to fold the paper to see how hydrogen bonds could form along the polypeptide chain. The result was a structure that twisted around like a spring.

"Well, I succeeded. It only took a couple of hours of work that day, March of 1948, for me to find the structure, called the alpha helix."

The World of Chemistry (Program 23), "Proteins: Structure and Function"

Linus Pauling and his wife at a press conference in 1963 after it was announced that he had been awarded the 1962 Nobel Peace Prize. (Topham/The Image Works)

---

is split into smaller molecules with the addition of the H— and —OH of water where a bond was broken. The enzyme maltase catalyzes the hydrolysis of the sugar maltose into two molecules of D-glucose. This is the only function of maltase, and no other enzyme can substitute for it. Sucrase, another enzyme, hydrolyzes only sucrose. Some enzymes are less specific. The digestive enzyme trypsin, for example, primarily hydrolyzes peptide bonds in proteins. However, the structure and polarity of trypsin are such that it can also catalyze the hydrolysis of some esters.

The action of an enzyme in a hydrolysis is shown in Figure 21.9. The reactant molecule is called the **substrate.** Enzymes and substrates have electrically polar regions, partially charged groupings, or ionic sections that attract and guide the enzyme and substrate together; these regions of chemical activity are the **active sites.** Substrates sit down on active sites on enzymes in assembly-line fashion at a remarkably fast rate.

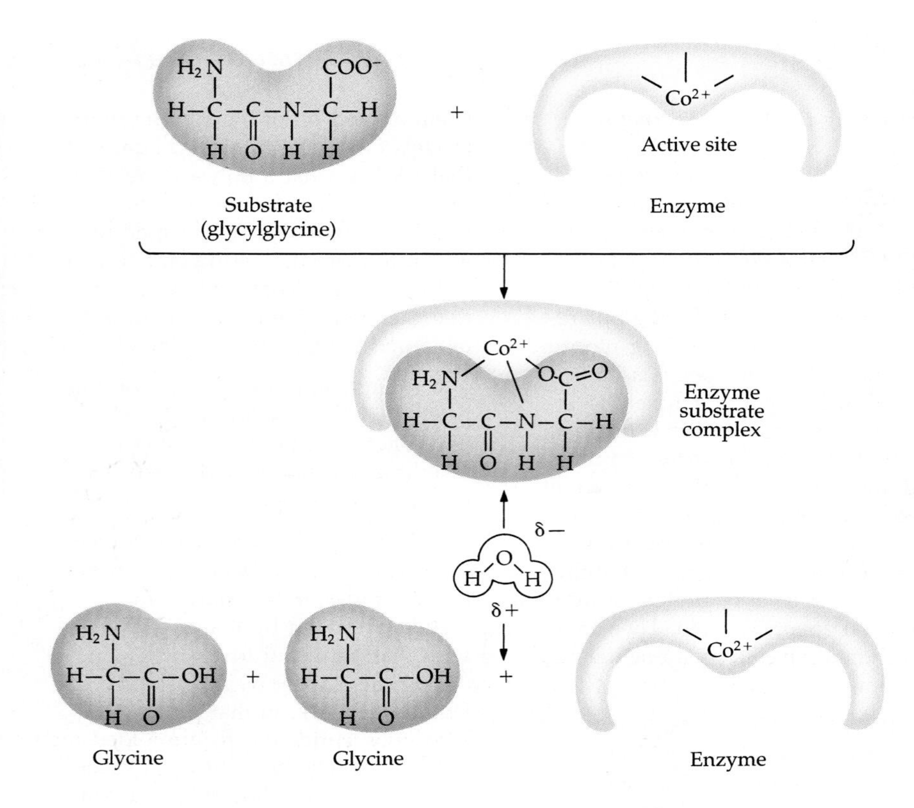

Figure 21.9

Action of an enzyme. The substrate molecule is chemically bonded to the enzyme (glycylglycine dipeptidase). The negative carboxylate group and the amine group of glycylglycine bond to the Co^{2+} in the enzyme. The bonding of the substrate makes it more susceptible to attack by water. Hydrolysis occurs and the glycine molecules are released by the enzyme, which is then ready to play its catalytic role again.

Since the structure of the active site of an enzyme is important, the same factors that cause denaturation will destroy the activity of the enzyme. For example, most enzymes are effective only over a narrow temperature range and a narrow pH range. Enzymes are denatured irreversibly at high temperatures or pH values outside their effective range.

EXAMPLE 21.3

Enzymes

Measurement of the activities of more than 40 enzymes in blood serum can provide information useful in the diagnosis of diseases. For example, high levels of amylase and lipase in blood serum are associated with pancreatic disorders. Why should the clinical tests be carried out at constant pH and constant temperature?

SOLUTION Changes in pH and temperature affect the secondary and tertiary structures of enzymes and hence their activity.

EXERCISE 21.3 • *Enzymes*

Explain why an enzyme that binds the dipeptide L-alanyl-L-lysine does not bind D-alanyl-D-lysine.

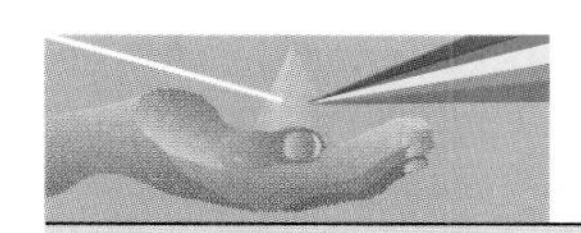

CHEMISTRY YOU CAN DO

Action of an Enzyme

Casein is the protein substance in milk that forms the curds used to make cheese. The casein also separates from milk when milk sours. Rennin is an enzyme obtained from the stomach lining of calves that catalyzes the coagulation of casein. Rennin is available in grocery stores under the tradename Junket and is used in making ice cream, jelly, and cheese.

For this experiment you will need four half-tablets of rennin, about 150 mL (5/8 cup) of milk (whole or 2%), a thermometer that reads to at least 80 °C (176 °F), a sauce pan or beaker, vinegar, a clock or watch, five clear plastic cups, paper towels, and ice water.

1. Warm about 60 mL (1/4 cup) of milk in a sauce pan or beaker to between 40 °C and 50 °C. Do not boil. Pour about 30 mL of the warmed milk into a clear plastic cup. Add half of a rennin tablet to the cup, stir for several minutes, and observe the milk form curds. Use your hand to gather the curds into a ball. Place the curdled mixture on a paper towel, squeeze the remaining liquid from the solid curds, and observe their properties.
2. Heat the milk remaining in the sauce pan to 80 °C and pour the hot milk into a clear plastic cup. Stir in half of a rennin tablet. Observe for about 10 minutes. Did the milk form curds? Allow the milk to cool to 40 °C, and stir for a few minutes. Did the milk form curds? Now add an additional half of a rennin tablet and stir for a few minutes. Did the milk form curds?
3. Pour about 30 mL of unheated milk into a clear plastic cup and cool with ice water to below 10 °C. Add half of a rennin tablet, stir for several minutes, and observe. Did the milk form curds?
4. Pour about 30 mL of unheated milk into a clear plastic cup and add 1 tablespoon of vinegar. Stir. Separate the curds from the milk as in part 1 and compare their properties to those of the curds formed in part 1.

Explain your observations.

21.3 CARBOHYDRATES

The word "carbohydrate" literally means "hydrate of carbon." Carbohydrates have the general formula $C_x(H_2O)_y$ in which x and y are integers. However, even though the reaction of the carbohydrate sucrose with sulfuric acid produces carbon, this does not mean that sucrose is a simple combination of carbon and water. The carbon, hydrogen, and oxygen in carbohydrates are arranged primarily in three organic functional groups:

$-O-H$	$-\overset{\overset{\displaystyle O}{\|\|}}{C}-H$	$-\overset{\overset{\displaystyle O}{\|\|}}{C}-$
alcohol group	aldehyde group	ketone group

Carbohydrates are simple polyhydroxy aldehydes or ketones, or larger molecules formed from them by condensation reactions.

Carbohydrates are divided into three groups, depending on how many monomers are combined by condensation polymerization: **monosaccharides** (from the Latin *saccharum*, "sugar"), **disaccharides,** and **polysaccharides.** Monosaccharides are simple sugars that cannot be broken down into smaller carbohydrate units by acid hydrolysis. In contrast, hydrolysis of a disaccharide yields two simple sugar molecules or monosaccharides (of either the same or different sugars), while complete hydrolysis of a polysaccharide produces many monosaccharides (sometimes thousands of them).

Carbohydrates, which make up about half of the average human diet, form an essential part of the energy cycle of living things. Plants make carbohydrates by photosynthesis from carbon dioxide, water, and energy from the sun.

$$x\,CO_2(g) + y\,H_2O(\ell) + \text{energy} \longrightarrow C_x(H_2O)_y(aq) + x\,O_2(g)$$

In animals, simple carbohydrates are oxidized to CO_2 and H_2O with the release of energy (in an overall reaction which is the reverse of the preceding reaction). It is this energy that drives the metabolic processes of the organism. Large amounts of energy are stored in polysaccharides such as starch. Living cells can use this energy only after the polysaccharides have been hydrolyzed into monosaccharides.

Besides playing a central role in energy storage, carbohydrates serve many biological purposes. Cellulose is the main structural component of plants; together with other carbohydrates it forms wood, cotton, paper, and other important natural materials. The nucleic acids (Section 21.6) incorporate carbohydrate units in their repeating structures.

Monosaccharides

Approximately 70 monosaccharides are known; 20 occur naturally. The most common simple sugar is D-glucose or dextrose, which is found in fruit, blood, and living cells. As discussed in Section 11.5, glucose is a chiral molecule with four asymmetric carbon atoms (Figure 11.23). However, the straight-chain structure shown in Figures 11.23 and 21.10a is an oversim-

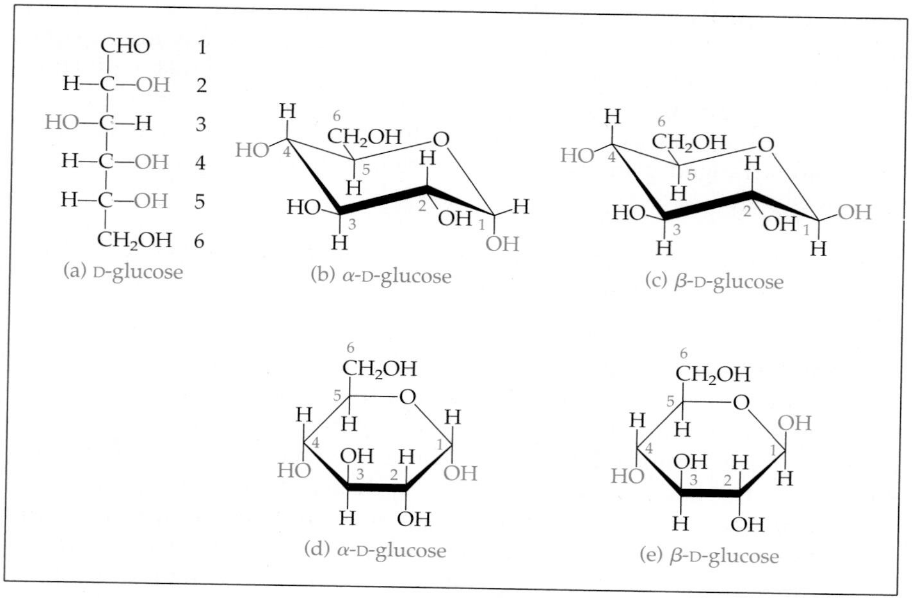

(a) D-glucose (b) α-D-glucose (c) β-D-glucose (d) α-D-glucose (e) β-D-glucose

Figure 21.10
The structures of D-glucose; (d) and (e) are two-dimensional representations of (b) and (c), respectively. Note the difference between the positions of the —OH groups (color) in the α and β forms of glucose: the —OH groups on carbons 1 and 4 are *trans* when the structure is β, and they are *cis* when the structure is α. In both α and β glucose, the —OH groups on carbon 4 must be in the same position.

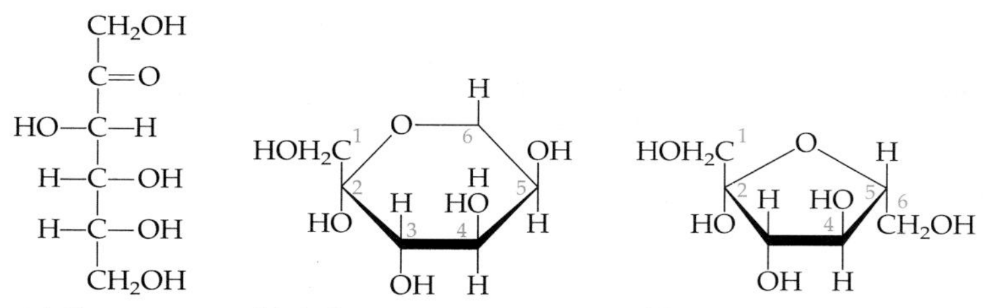

(a) Ketone structure

(b) β-ring structure (Pyranose structure: 6-membered ring with an oxygen atom in the ring)

(c) β-ring structure (Furanose structure: 5-membered ring with an oxygen atom in the ring)

Figure 21.11
The structures of D-fructose, which forms both five-membered and six-membered rings. The α-ring structure (not shown) differs from the β-ring structure in that the —CH_2OH and —OH groups are in reversed positions on carbon 2.

plification of the actual structure of D-glucose. An aqueous solution of D-glucose contains all three structures shown in Figure 21.10 in dynamic equilibria involving mostly the two ring forms, with less than one percent in the straight-chain form. The aldehyde group in the straight-chain structure of D-glucose qualifies this sugar as an **aldose** monosaccharide.

Because D-fructose, a monosaccharide found in many fruits, has a ketone group in its straight-chain form (Figure 21.11), it is classified as a **ketose** monosaccharide. Glucose and fructose are hexose monosaccharides because they have six carbons in a row in the straight-chain form. Two pentoses (five-carbon sugars), ribose and 2-deoxyribose, are part of RNA and DNA, respectively, and are discussed later in this chapter.

The names of saccharides end in -ose. *The general name for a saccharide with five carbon atoms is a pentose, while one with six carbon atoms is a hexose.*

There is one more structural feature of glucose that is important in understanding the structural differences among the polysaccharides made from it. Look at the different positions of the —OH groups in the ring forms of glucose in Figure 21.10. When the —OH groups on carbons 1 and 4 are on opposite sides of the ring *(trans)*, the structure is labelled β-D-glucose; when the —OH groups on carbons 1 and 4 are on the same side of the ring *(cis)*, the structure is labelled α-D-glucose.

Cis *and* trans *geometric isomers were discussed in Section 11.5.*

Disaccharides

Monosaccharide units can condense, with the elimination of water, to give disaccharides. The three most important disaccharides are

- **sucrose** (from sugar cane or sugar beets), a dimer of a D-glucose monomer and a D-fructose monomer.
- **maltose** (from starch), a dimer of two D-glucose monomers.
- **lactose** (from milk), a dimer of a D-glucose monomer and a D-galactose monomer.

The formula for these disaccharides ($C_{12}H_{22}O_{11}$) is not simply the sum of two monosaccharides, $C_6H_{12}O_6 + C_6H_{12}O_6$. A water molecule is eliminated as two monosaccharides are united to form the disaccharide. The

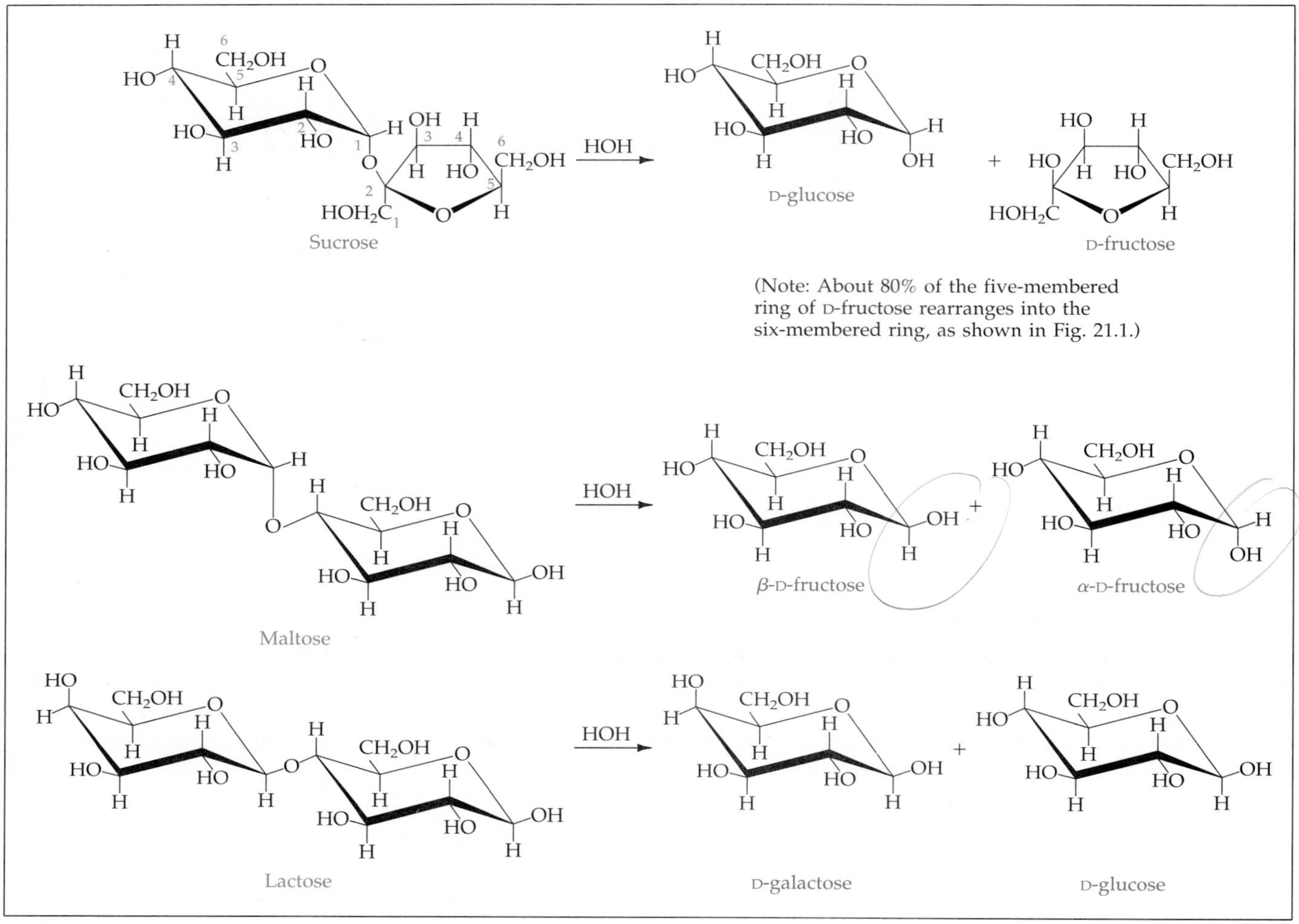

Figure 21.12
Hydrolysis of disaccharides.

structures of sucrose, maltose, and lactose, along with their hydrolysis reactions, are shown in Figure 21.12.

Sucrose is produced in a high state of purity on an enormous scale—more than 80 million tons per year. About 40% of the world sucrose production comes from sugar beets and 60% from sugar cane. Table 21.2 compares the sweetness of common sugars and artificial sweeteners relative to sucrose. Honey, a mixture of the monosaccharides glucose and fructose, has been used for centuries as a natural sweetener for foods. In contrast, sucrose, derived from sugar cane or sugar beets, is a disaccharide. Honey is popular because it is sweeter than cane sugar (Table 21.2). To convert cane sugar into glucose and fructose requires treatment with acid or with a natural enzyme called "invertase."

Table 21.2

Sweetness of Common Sugars and Artificial Sweeteners Relative to Sucrose

Substance	Sweetness Relative to Sucrose as 1.00
lactose	0.16
galactose	0.32
maltose	0.33
glucose	0.74
sucrose	1.00
fructose	1.74
aspartame*	180
saccharin*	300
sucralose*	650
alitame*	2000

* Artificial sweeteners.

Artificial Sweeteners

Saccharin was the first common artificial sweetener. Saccharin passes through the body undigested and consequently has no caloric value. It has a somewhat bitter aftertaste that is offset in commercial products by the addition of small amounts of naturally occurring sweeteners. Such products do have a small caloric value because of the natural sweeteners added.

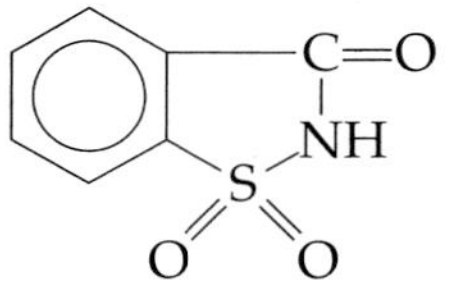

C=O
NH
S
O O

saccharin

High doses of saccharin have been shown to cause cancer in mice, so commercial products containing saccharin are required by law to have a warning label: "This product contains saccharin which has been determined to cause cancer in laboratory animals."

Aspartame (NutraSweet), which has replaced saccharin as the principal artificial sweetener, is used in more than 3000 products and accounts for 75% of the one billion dollar worldwide artificial sweetener market.

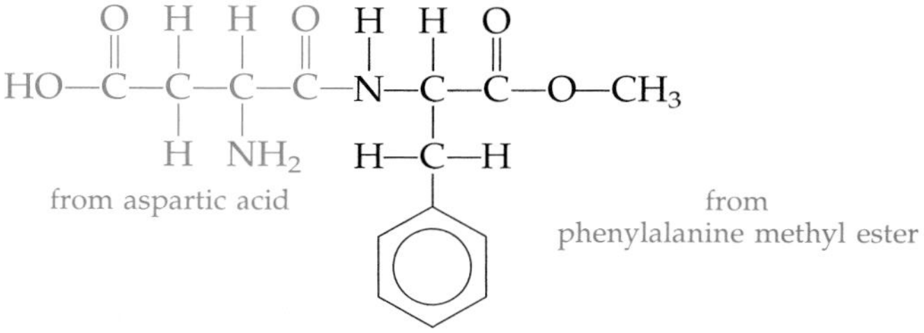

O H H O H H O

HO—C—C—C—C—N—C—C—O—CH_3

H NH_2 H—C—H

from aspartic acid

from phenylalanine methyl ester

aspartame

Examples of some monosaccharides, a disaccharide, and an artificial sweetener. *(C.D. Winters)*

It is a dipeptide derivative made from aspartic acid and the methyl ester of phenylalanine. Aspartame can be digested, and its caloric value is approximately equal to that of proteins. However, since much smaller amounts of aspartame than of table sugar are needed for sweetness, many fewer calories are consumed in the sweetened food. Aspartame is unstable at cooking temperatures, limiting its use as a sugar substitute to cold foods and soft drinks.

The sweetness of saccharin and aspartame was discovered accidentally after these compounds had been prepared. The terms "accident" or "serendipity" have been associated with the discovery and commercialization of the sweet taste of these compounds. In recent years, theoretical models of the interaction of molecules of sweet compounds with the biological receptors on the tongue have proved useful in predicting new compounds that should be sweet. Earlier it was noted that if chlorine atoms replaced hydroxyl atoms on the sucrose molecule, the derivative was sweeter than sucrose. In fact, a product that is predicted to become a great commercial success is **sucralose,** in which three hydroxyl groups are replaced by chlorine atoms. As shown in Table 21.2, sucralose is 650 times sweeter than sucrose.

sucralose

Alitame, a dipeptide sweetener, was first prepared in 1979 and has a sweetness more than 2000 times that of table sugar! It has a sucrose-like sweetness with no aftertaste. Also, alitame is more stable than aspartame during cooking. One potential problem with alitame and other high-potency sweeteners is the difficulty in controlling sweetness in food when such small amounts of the compound are required.

alitame

Control of Blood Sugar

Serious health problems can be encountered when the concentration of glucose in the blood is either too low or too high. After about ten hours without food, the normal person has from 80 to 120 mg of glucose per 100 mL of blood. People with low blood sugar, a condition known as **hypoglycemia,** must carefully control the amounts of carbohydrates in their diets in order to maintain the normal blood sugar levels. A low glucose concentration may lead to sluggish or dizzy feelings, and possibly fainting. Eating meals with a very high concentration of carbohydrates can also cause problems for people with hypoglycemia, as the body overcorrects by secreting an excess of the hormone insulin, which quickly drives the glucose concentration below the normal level.

Hyperglycemia is the condition of elevated blood sugar concentrations. Above a concentration of 160 mg of glucose per 100 mL of blood, the kidneys begin to excrete glucose in the urine. If such high blood sugar concentrations are chronic without medical or dietary control, the individual has **diabetes mellitus** and, without treatment, is likely to have symptoms of thirst, frequent urination, weakness, low resistance to infection, slowness to heal, and in later stages, blindness and coma. About 5% of Americans have diabetes.

There are two types of diabetes, which have been termed Type I (insulin-dependent, formerly known as juvenile-onset diabetes) and Type II (non-insulin dependent, formerly known as maturity-onset diabetes). Type-I diabetics, about 10% of those who have diabetes, do not produce enough insulin in the *islets of Langerhans* in the pancreas. Insulin, which is necessary for the glucose to move from the blood to the cells, has to be obtained by daily injections; it cannot be taken orally because it is destroyed in the digestive tract. Type-II diabetes is usually found in older, obese people. These people generally produce plenty of insulin, but the insulin receptors on their cells do not respond properly and consequently do not move glucose from the blood into the cells. Type-II diabetics can generally control the disease by diet and oral medication.

Polysaccharides

Nature's most abundant polysaccharides are the **starches, glycogen,** and **cellulose.** Some polysaccharides are known to combine more than 5000 monosaccharide monomers into molecules with molecular weights of over 1 million. The monosaccharide most commonly used to build polysaccharides is D-glucose.

Starches and Glycogen

Plant starch is found in protein-covered granules. If these granules are ruptured by heat, they yield a starch that is soluble in hot water, *amylose,* and an insoluble starch, *amylopectin.* Amylose constitutes about 25% of most natural starches. When tested with iodine solution, amylose turns blue-black, whereas amylopectin turns red.

Structurally, amylose is a straight-chain condensation polymer with an average of about 200 α-D-glucose monomers per molecule. Each monomer is bonded to the next with the loss of a water molecule, just as the two units are bonded in maltose (Figure 21.12). A representative portion of the structure of amylose is shown in Figure 21.13.

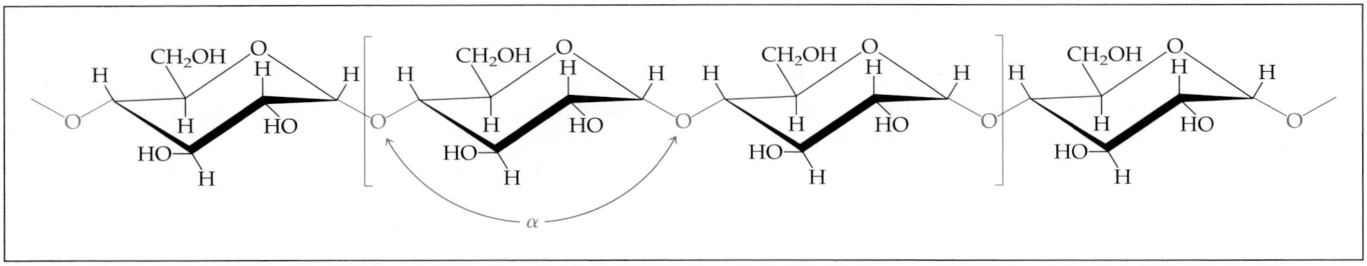

Figure 21.13
The structure of amylose. From 60 to 300 α-D-glucose units are bonded together by α linkages to form amylose molecules. In α linkages, only the α structure of glucose is used. The bonding between monomers is at carbons 1 and 4 (see Figure 21.10). The —OH groups on carbon atoms 1 and 4 are *cis* in α-glucose, so all bonds between α-glucose monomers (O) point in the same direction.

A typical amylopectin molecule has about 1000 α-D-glucose monomers arranged into branched chains (Figure 21.14). Complete hydrolysis yields D-glucose; partial hydrolysis produces mixtures called **dextrins.** Dextrins are used as food additives and in mucilage, paste, and finishes for paper and fabrics.

Glycogen is an energy reservoir in animals, just as starch is in plants. The α-glucose chains in glycogen are more highly branched than the chains in amylopectin. Glycogen is stored in the liver and muscle tissues and is used for "instant" energy until the process of fat metabolism can take over and serve as the energy source.

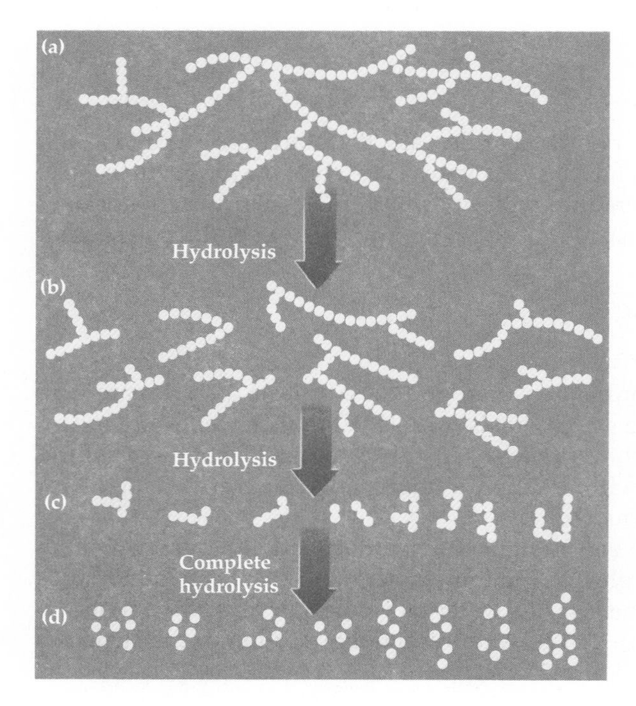

Figure 21.14
(a) Partial schematic of amylopectin structure. (b) Dextrins from incomplete hydrolysis of the structure in (a). (c) Small polysaccharides from hydrolysis of dextrins. (d) Final hydrolysis product: D-glucose. Each circle represents a glucose unit.

Figure 21.15
Cellulose structure. About 280 β-D-glucose units are bonded together by β linkages to form an unbranched cellulose structure. Cellulose contains only the β form of glucose. The —OH groups on carbon atoms 1 and 4 (see Figure 21.10) are *trans* in β-glucose, so the bonds between β-glucose monomers (⟍O⟋ and ⟋O⟍) alternate in direction. Compare cellulose with the amylose structure in Figure 21.13. Note in cellulose that every other β-glucose monomer is turned over, whereas in amylose all the α-glucose monomers are in the same position.

Cellulose

Cellulose is the most abundant organic compound on earth, and its purest natural form is cotton. This polysaccharide is also found as the woody part of trees and the supporting material in plants and leaves. Like amylose, it is composed of D-glucose units. The difference between the structures of cellulose and amylose lies in the bonding between the D-glucose units; in cellulose all of the glucose units are in the β-ring form (Figure 21.15), whereas in amylose they are in the α-ring form. (Review the ring forms in Figure 21.10 and compare the structures in Figures 21.13 and 21.15.) This subtle structural difference between starch and cellulose causes their differences in digestibility. Since cellulose is so abundant, it would be advantageous if humans could use it for food. Unfortunately, we cannot digest it, because we lack the necessary enzyme to break the β-1,4 bonds. However, termites, a few species of cockroaches, and ruminant mammals such as cows, sheep, goats, and camels, do have the proper internal chemistry laboratory for this purpose.

D-Glucose can be obtained from cellulose by heating a suspension of the polysaccharide in the presence of a strong acid. Unfortunately, wood cannot presently be hydrolyzed into food (D-glucose) economically enough to satisfy the world's growing need for an adequate food supply.

Cellulose is the most abundant organic compound on earth and is found in the woody part of trees and the supporting material of plants and leaves. *(Biophoto Associates/Photo Researchers, Inc.)*

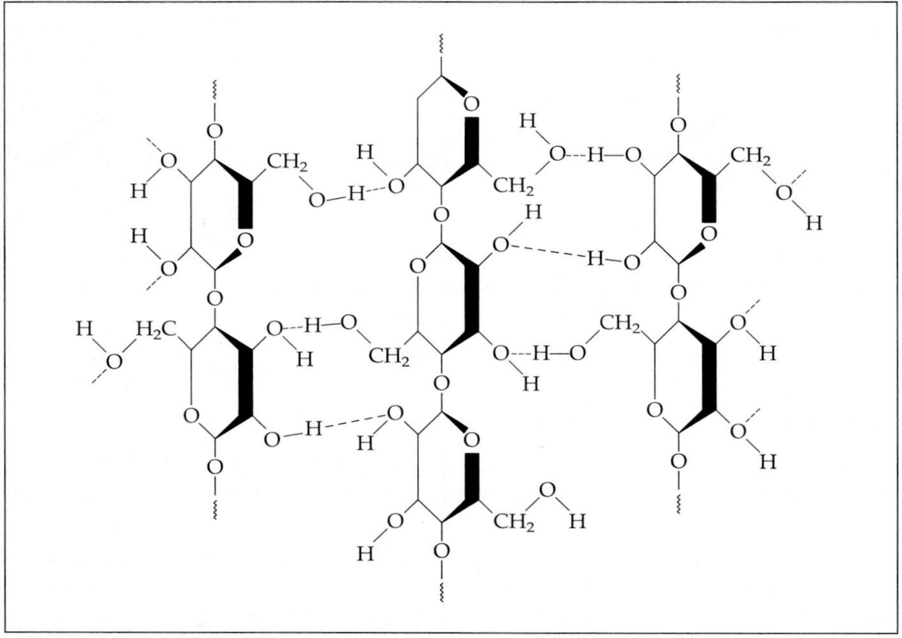

Figure 21.16
The properties of cotton, which is about 98% cellulose, can be explained in terms of this submicroscopic structure. A small group of cellulose molecules, each with 2000 to 9000 units of D-glucose, are held together in an approximately parallel fashion by hydrogen bonding (---). When several these chain bundles cling together in a relatively vast network of hydrogen bonds, a microfibril results; the microfibril is the smallest unit that can be seen under the microscope. The macroscopic fibril is a collection of many microfibrils. Cotton is absorbent because of the numerous capillaries, in which water molecules are held by hydrogen bonds.

Paper, rayon, cellophane, and cotton are principally cellulose. A representative portion of the structure of cotton is shown in Figure 21.16. Note the hydrogen bonding between cellulose chains.

EXAMPLE 21.4

Carbohydrates

Why is table sugar soluble in water?

SOLUTION The polar —OH groups in sucrose form hydrogen bonds with water molecules.

EXERCISE 21.4 • ***Carbohydrates***

Explain why humans cannot digest cellulose. Consult a reference and explain why ruminant animals can digest cellulose.

21.4 LIPIDS

A **lipid** is an organic substance found in living systems that is insoluble in water but soluble in nonpolar organic solvents. Because their classification is based on insolubility in water rather than a structural feature such as a functional group, lipids vary widely in their structure and, unlike proteins and polysaccharides, are not polymers. The predominant lipids are **fats** and **oils,** which make up 95% of the lipids in our diet. The other 5% include **steroids** and several other lipids that are important to cell function.

Fats and Oils

Fats and oils are esters of glycerol (glycerin) and fatty acids, and are collectively known as **triglycerides.** Fatty acids are rarely found in the free form in nature; instead, they occur in the combined ester form in fats and oils. The general equation for the formation of a triester of glycerol is

Lipids vary in their structures and, unlike proteins and polysaccharides, are not polymers.

$$\begin{array}{lclcll}
H_2C{-}OH & & HO{-}\overset{\overset{O}{\|}}{C}{-}R & & H_2C{-}O{-}\overset{\overset{O}{\|}}{C}{-}R & \\
\quad | & & & & \quad | & \\
HC{-}OH & + & HO{-}\overset{\overset{O}{\|}}{C}{-}R' & \rightleftharpoons & HC{-}O{-}\overset{\overset{O}{\|}}{C}{-}R' & +\ 3\,H_2O \\
\quad | & & & & \quad | & \\
H_2C{-}OH & & HO{-}\overset{\overset{O}{\|}}{C}{-}R'' & & H_2C{-}O{-}\overset{\overset{O}{\|}}{C}{-}R'' & \\
\text{glycerol} & & \text{fatty acid} & & \text{fat or oil} & \text{water} \\
\text{(one molecule)} & & \text{(three molecules that may} & & \text{(one molecule)} & \text{(three molecules)} \\
 & & \text{or may not be the same)} & & &
\end{array}$$

The three R groups can be the same or different within the same fat or oil, and they can be saturated or unsaturated. The acid portions of fats almost always have an even number of carbon atoms (C_{16} or C_{18}). The most common fatty acids found in fats and oils are listed in Table 21.3.

Fatty acids that contain only one double bond, such as oleic acid, are referred to as monounsaturated acids. One of the unsaturated acids, **linoleic acid,** is referred to as an **essential fatty acid.** The human body cannot syn-

Table 21.3

Common Fatty Acids

Saturated Acids	
lauric	$CH_3(CH_2)_{10}COOH$
myristic	$CH_3(CH_2)_{12}COOH$
palmitic	$CH_3(CH_2)_{14}COOH$
stearic	$CH_3(CH_2)_{16}COOH$
Unsaturated Acids	
oleic	$CH_3(CH_2)_7CH{=}CH(CH_2)_7COOH$
linoleic	$CH_3(CH_2)_4CH{=}CHCH_2CH{=}CH(CH_2)_7COOH$
linolenic	$CH_3CH_2CH{=}CHCH_2CH{=}CHCH_2CH{=}CH(CH_2)_7COOH$

thesize this acid, but it is required for the synthesis of an important group of compounds known as the prostaglandins.

Prostaglandins form a group of more than a dozen related compounds with potent effects on physiological activities such as blood pressure, relaxation and contraction of smooth muscle, gastric acid secretion, body temperature, food intake, and blood platelet aggregation. Their potential use as drugs is currently under widespread investigation. For example, one of the prostaglandins, PGE_2, lowers blood pressure but a closely related prostaglandin, $PGF_{2\alpha}$, raises blood pressure. The structures of these two prostaglandins differ only in one functional group (highlighted below).

O, HO, OH, COOH, CH_3

PGE_2

HO, HO, OH, COOH, CH_3

$PGF_{2\alpha}$

Many prostaglandins cause inflammation and fever. The fever-reducing effect of aspirin results from the inhibition of cyclooxygenase, the enzyme that catalyzes the synthesis of prostaglandins.

Fats and oils are hydrolyzed by strong bases to give glycerol and salts of the fatty acids. The acid salt is a *soap*, which is why the hydrolysis of fats and oils is given the name **saponification.**

$$\begin{array}{l} H_2C{-}O{-}\overset{O}{\overset{\|}{C}}{-}(CH_2)_{16}CH_3 \\ HC{-}O{-}\overset{O}{\overset{\|}{C}}{-}(CH_2)_{16}CH_3 \\ H_2C{-}O{-}\overset{O}{\overset{\|}{C}}{-}(CH_2)_{16}CH_3 \end{array} + 3\ NaOH \xrightarrow{\Delta} \begin{array}{l} H_2C{-}OH \\ HC{-}OH \\ H_2C{-}OH \end{array} + 3\ NaO{-}\overset{O}{\overset{\|}{C}}{-}(CH_2)_{16}CH_3$$

glycerol tristearate, a fat — glycerol — sodium stearate, a soap

Saturated fatty acid chains

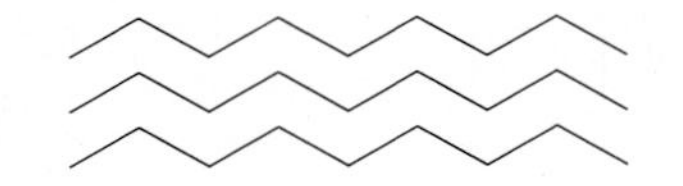

Unsaturated fatty acid chains

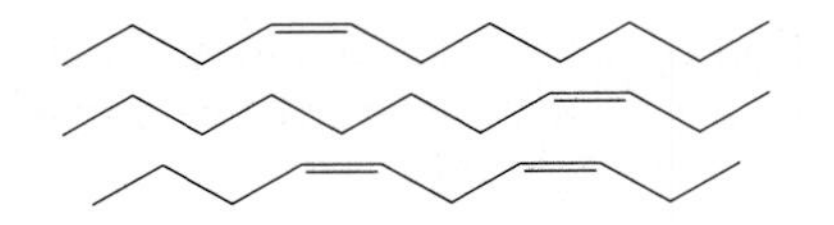

The term *fat* is usually reserved for solid triglycerides (such as butter, lard, or tallow), and *oil* is used for liquid triglycerides (castor, olive, linseed, tung, etc.). The R groups in the fatty acid portions of fats are generally saturated, with only C—C single bonds. The resulting regular zigzag shape of the fatty acid chains allows the molecules to pack close together and form a solid. The R groups in oils are usually either monounsaturated (one C=C double bond) or polyunsaturated (two or more C=C double bonds). Since the C=C bonds interrupt the zigzag pattern of tetrahedral angles with 120° angles, the molecules are irregular in shape and do not pack together efficiently enough to form a solid.

EXAMPLE 21.5

Saponification

Write the condensed formulas of the products of the following saponification reaction.

$$\begin{array}{l} \quad\quad\quad\quad\;\; O \\ \quad\quad\quad\quad\;\; \| \\ H_2C-O-C-(CH_2)_7CH{=}CH-(CH_2)_7CH_3 \\ \;| \\ \;|\quad\quad\quad\quad O \\ \;|\quad\quad\quad\quad \| \\ HC-O-C-(CH_2)_{16}CH_3 \quad\quad\quad\quad + 3\,NaOH \xrightarrow{\Delta} ? \\ \;| \\ \;|\quad\quad\quad\quad O \\ \;|\quad\quad\quad\quad \| \\ H_2C-O-C-(CH_2)_7CH{=}CH-(CH_2)_7CH_3 \end{array}$$

SOLUTION The saponification of esters with NaOH produces alcohols and the sodium salts of carboxylic acids. In all triglycerides, the alcohol is glycerol and there are three fatty acids, which may be the same or different. In this case the products are

$$HOCH_2CH_2OHCH_2OH$$
$$2\,CH_3(CH_2)_7CH{=}CH(CH_2)_7COO^-Na^+$$
$$CH_3(CH_2)_{16}COO^-Na^+$$

Fats in the Human Body

Fats are the most concentrated source of energy in our diets. They furnish 9000 cal/g when oxidized, compared with sugars, which provide about 3800 cal/g. Stored fat is a potential energy source for the body. Fats also insulate thermally, pad the body, and are packing material for various organs. Fatty tissue is composed mainly of specialized cells, each featuring a relatively large globule of triglycerides.

One food calorie, usually written Calorie (C), is 1000 calories or 1 kilocalorie.

Fat makes up about 40% of the average American diet, although nutritionists think it should be no higher than about 30%. The fat in today's diet is about 40% saturated, 40% monounsaturated, and 20% polyunsaturated, but nutritionists recommend lowering the saturated and monounsaturated and raising the polyunsaturated to make each category about one third of intake. What is the basis for these recommendations? Heart disease is the number one cause of death in the United States, and **atherosclerosis,** the buildup of fatty deposits called plaque on the inner walls of arteries, reduces the flow of blood to the heart. If a coronary artery is blocked by plaque, a heart attack occurs as a result of the reduced blood flow carrying oxygen to the heart. About 98% of all heart attack victims have atherosclerosis, and the major components of atherosclerotic plaque are saturated fatty acids and cholesterol. The percentages of saturated, monounsaturated, and polyunsaturated fat in common fats and oils are given in Table 21.4.

Hydrogenation of the double bonds in vegetable oils converts the liquid oil into a solid fat.

Hydrogen can be catalytically added to the double bonds of an oil to convert it into a semisolid fat. For example, liquid soybean and other vegetable oils are **hydrogenated** to produce cooking fats and margarine.

$$\begin{array}{l} H_2C{-}O{-}\underset{\underset{O}{\|}}{C}{-}(CH_2)_7CH{=}CH(CH_2)_7CH_3 \\ HC{-}O{-}\underset{\underset{O}{\|}}{C}{-}(CH_2)_7CH{=}CH(CH_2)_7CH_3 \\ H_2C{-}O{-}\underset{\underset{O}{\|}}{C}{-}(CH_2)_7CH{=}CH(CH_2)_7CH_3 \end{array} \xrightarrow[200^\circ\ C]{H_2,\ Ni} \begin{array}{l} H_2C{-}O{-}\underset{\underset{O}{\|}}{C}{-}(CH_2)_7CH_2CH_2(CH_2)_7CH_3 \\ HC{-}O{-}\underset{\underset{O}{\|}}{C}{-}(CH_2)_7CH_2CH_2(CH_2)_7CH_3 \\ H_2C{-}O{-}\underset{\underset{O}{\|}}{C}{-}(CH_2)_7CH_2CH_2(CH_2)_7CH_3 \end{array}$$

triolein (a liquid oil) — tristearin (a solid fat)

If it is better to consume unsaturated fats instead of saturated ones, why do food companies hydrogenate oils to reduce their unsaturation? There are several answers to this question. First, the double bonds in the fatty acid are reactive functional groups and oxygen can attack the fat at these bonds.

Table 21.4

Amounts of Saturated and Unsaturated Fatty Acids in Fats and Oils

Dietary oil/fat	Saturated fat	Polyunsaturated fat	Monounsaturated fat
canola oil	6%	36%	58%
safflower oil	9%	78%	13%
sunflower oil	11%	69%	20%
corn oil	13%	62%	25%
olive oil	14%	9%	77%
soybean oil	15%	61%	24%
peanut oil	18%	34%	48%
cottonseed oil	27%	54%	19%
lard	41%	12%	47%
palm oil	51%	10%	39%
beef tallow	52%	4%	44%
butterfat	66%	4%	30%
coconut oil	92%	2%	6%

Fake Fats Offer Diet Alternative

ANOTHER SOLUTION TO PROBLEMS with fats is to use "fake fats," substances that give the fat or oil taste and consistency but are not fats or oils. Some of the fake fats on the market are Simplesse, Olestra, emulsified starch, and emulsified protein. Simplesse, developed by G.D. Searle Co., is a butter substitute with only 15% of the real fat calories in butter. An ounce of cheese made from Simplesse instead of butterfat drops from 82 Cal to 36 Cal. Simplesse is made from egg white or milk proteins, and it feels creamy on the tongue. However, Simplesse is not suitable for cooking because heat makes it tough. Olestra, developed by Procter and Gamble Co., is a sucrose polyester made from sugar and fatty acids. Olestra contributes no calories because it is indigestible, and it can be used in cooking. Emulsified starch is used in Hellmann's light mayonnaise and salad dressing. The starch is not used for cooking but can be used in ice cream and yogurt. Emulsified protein (Unilever) is an emulsion of gelatin and water that cuts margarine calories in half. It can be used for baking and light frying.

An assortment of fat-free and reduced fat (light) food products. (C.D. Winters)

When the oil is oxidized, unpleasant odors and flavors develop. Hydrogenating an oil reduces the likelihood that the food will oxidize and become rancid. Second, hydrogenating an oil makes it less liquid. There are many times when a food manufacturer needs a solid fat in a food to improve the food. (For example, if liquid vegetable oil were used in a cake icing, the icing would slide off the cake.) Rather than use animal fat, which also contains cholesterol, the manufacturer turns to a hydrogenated or partially hydrogenated oil.

The hydrogenation process also forms unnatural trans-*fatty acids from natural* cis-*fatty acids. The news feature box in Section 11.5 describes the health problems associated with* trans-*fatty acids.*

Steroids

Steroids are found in all plants and animals and are derived from the following four-ring structure:

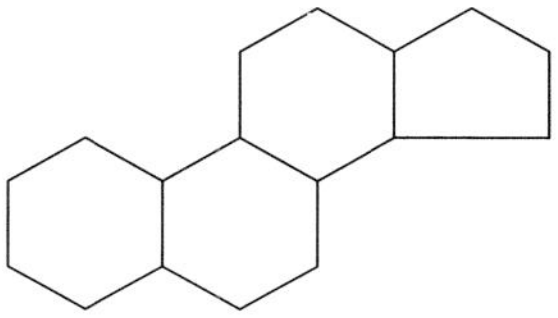

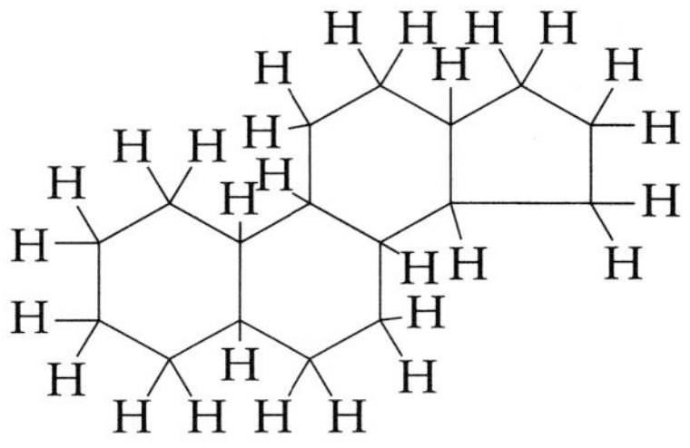

The skeletal four-ring structure drawing on the left is chemical shorthand similar to that described for cyclic hydrocarbons in Section 10.1. There is a carbon atom at each corner, and the lines represent C—C bonds. Since every carbon atom forms four bonds, additional bonds between carbon atoms and hydrogen atoms are understood to be present whenever the skeletal structure shows fewer than four bonds. The structure on the right shows the hydrogen atoms understood to be present in the structure on the left. Although all the rings in the skeletal drawing are shown as saturated rings, steroids often have one ring that is unsaturated or aromatic. For example, cholesterol has one double bond in the second ring. Note that its structure also includes alkyl groups and an alcohol group. These are substituted for hydrogen atoms in the skeletal representation shown above.

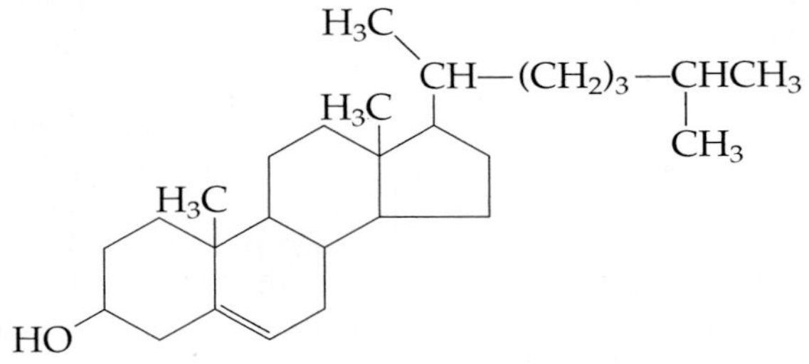

H_3C, $CH—(CH_2)_3—CHCH_3$, CH_3, H_3C, H_3C, HO

cholesterol

Cholesterol is the most abundant animal steroid. The human body synthesizes cholesterol and readily absorbs dietary cholesterol through the intestinal wall. An adult human contains about 250 g of cholesterol. Cholesterol receives a lot of attention because high blood cholesterol levels are associated with heart disease. However, it is important to realize that proper amounts of cholesterol are essential to health, because cholesterol undergoes biochemical alteration to give milligram quantities of many important hormones such as vitamin D, cortisone, and the sex hormones. Cholesterol combines with proteins to form **lipoproteins,** which transport cholesterol in the bloodstream. About 65% of the cholesterol in the blood is carried by low-density lipoproteins **(LDLs)** whereas 25% of the cholesterol in the blood is carried by high-density lipoproteins **(HDLs).** LDLs are "bad" cholesterol and HDLs are "good" cholesterol in discussions of problems concerning atherosclerotic plaque and heart disease. LDLs transport cholesterol away from the liver and throughout the body; they are therefore "bad" because they distribute cholesterol to arteries where it can form the deposits of atherosclerosis. HDLs are "good" because they transport excess cholesterol from body tissues to the liver, where it can be broken down and excreted.

Cholesterol is the starting material for the synthesis of steroid sex hormones. One female sex hormone, **progesterone,** differs only slightly in structure from an important male hormone, **testosterone.**

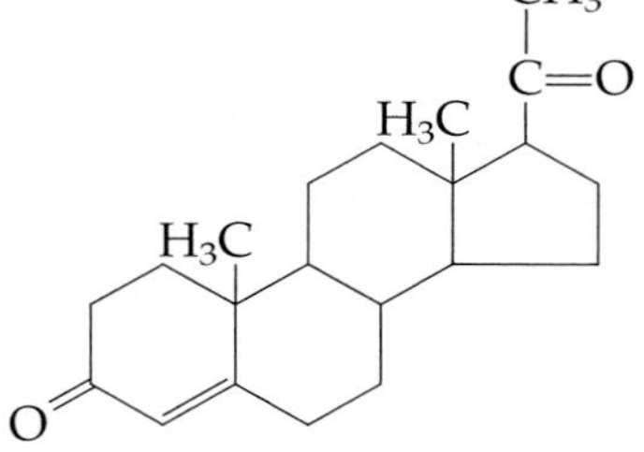

CH_3, $C=O$, H_3C, H_3C, O

progesterone

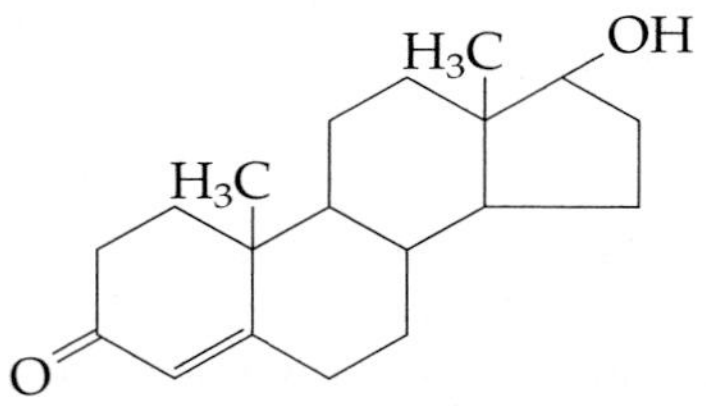

H_3C, OH, H_3C, O

testosterone

Other female hormones are estradiol and estrone, together called **estrogens.** The estrogens differ from the steroids shown above in that they contain an aromatic ring.

The estrogens and progesterone are produced by the ovaries. Estrogens are important to the development of the egg in the ovary, whereas progesterone causes changes in the wall of the uterus and after pregnancy prevents release of a new egg from the ovary (ovulation). Birth control drugs use derivatives of estrogens and progesterone to simulate the hormonal process resulting from pregnancy and thereby prevent ovulation.

EXERCISE 21.5 • *Steroids*

Our body needs cholesterol, but doctors and nutritionists recommend low-cholesterol diets. Explain.

21.5 ENERGY AND BIOCHEMICAL SYSTEMS

Energy for life's processes comes from the sun. During photosynthesis, green plants absorb energy from the sun to make glucose and oxygen from carbon dioxide and water. Glucose is a major energy source for all living organisms. The energy stored in glucose is eventually transferred to the bonds in molecules such as adenosine triphosphate (ATP, Figure 21.17). When living organisms need energy, phosphate bonds in the ATP molecules are broken to give adenosine diphosphate (ADP, Figure 21.18) and energy for other biochemical reactions.

In the complex process of photosynthesis, carbon dioxide is reduced to make sugar,

$$6\,CO_2 + 24\,H_3O^+ + 24\,e^- \longrightarrow C_6H_{12}O_6 + 30\,H_2O$$

and water is oxidized to oxygen:

$$36\,H_2O \longrightarrow 6\,O_2 + 24\,H_3O^+ + 24\,e^-$$

The oxidation and reduction reactions added together give the overall reaction:

$$\underset{\text{carbon dioxide}}{6\,CO_2} + \underset{\text{water}}{6\,H_2O} + \underset{\text{energy (sunlight)}}{2880\text{ kJ}} \longrightarrow \underset{\text{glucose}}{C_6H_{12}O_6} + \underset{\text{oxygen}}{6\,O_2}$$

The oxygen produced in photosynthesis is the source of all of the oxygen in our atmosphere. Only this life-giving gas, given off by trees, grass,

Adenine
Triphosphate
Ribose

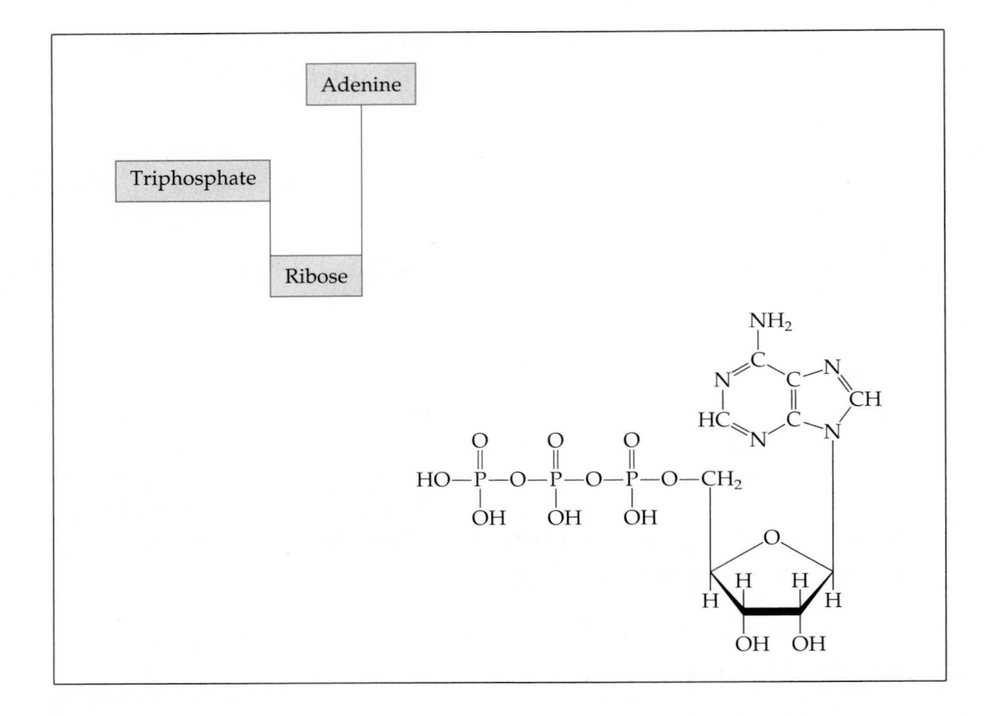

Figure 21.17
Structure of adenosine triphosphate (ATP).

greenery, and even by algae in the sea, makes possible human life and most animal life on Earth. We are dependent on the plant life of our planet, and we must live in balance with the oxygen output of that plant life, as well as with the food output of the same plant life. Photosynthesis is thus absolutely vital to life on Earth.

Photosynthesis is generally broken down into a series of **light reactions,** which occur only in the presence of light energy, and a series of **dark reactions,** which can occur in the dark because they do not depend on light energy. The dark reactions feed instead on high-energy compounds (such as ATP) produced by the light reactions.

The so-called "dark reactions" also occur during the day.

Photosynthesis is initiated by light energy. Green plants contain certain pigments that readily absorb light in the visible region of the spectrum. The

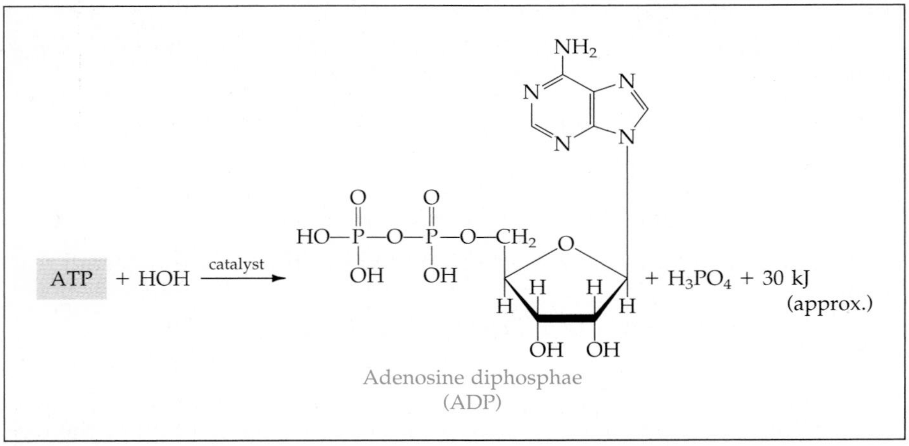

ATP
+ HOH
catalyst
+ H3PO4 + 30 kJ
(approx.)
Adenosine diphosphae
(ADP)

Figure 21.18
Hydrolysis of ATP to ADP.

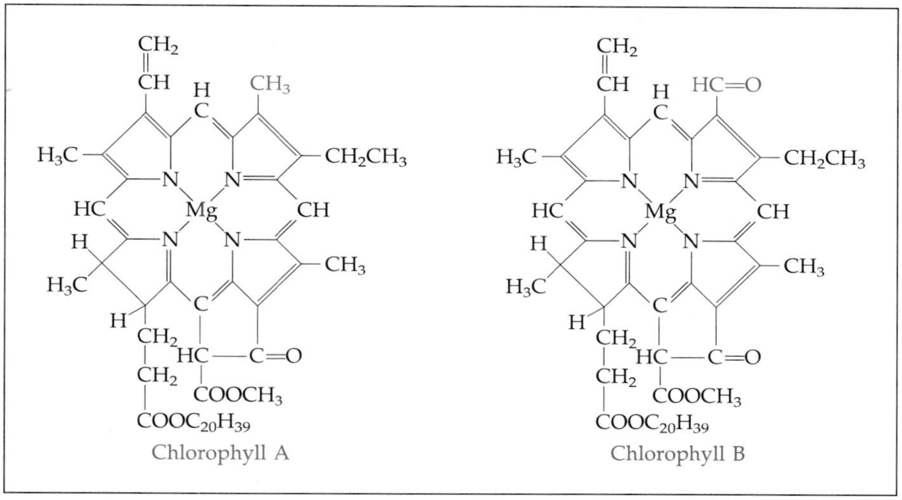

Chlorophyll A

Chlorophyll B

Figure 21.19
Structures of chlorophyll a and chlorophyll b.

most important of these are the chlorophylls, **chlorophyll a** and **chlorophyll b** (Figure 21.19). Note that both chlorophylls are coordination compounds of Mg^{2+} bonded to a complex ring structure similar to the one surrounding Fe^{2+} in cytochrome c and hemoglobin (Figure 10.12). Chlorophyll is green because it absorbs light in the violet region (about 400 nm) and the red region (about 650 nm) and allows the green light between those wavelengths to be reflected or transmitted.

When chlorophyll a and chlorophyll b absorb photons of light, electrons are raised to higher energy levels. As these electrons return to the ground state, very efficient subcellular components of the plant cell known as chloroplasts absorb the energy that is released. Through a series of reactions, water is oxidized to oxygen, and energy is stored in the bonds of energy-storage compounds such as ATP.

The function of the dark reactions is to convert CO_2 and hydrogen from water into glucose and other carbohydrates, and these reactions are driven by the release of energy stored in ATP by the light reactions. In the presence of a suitable catalyst, ATP releases energy by undergoing a three-step hydrolysis of the P—O bonds of its phosphate groups. In the first step ATP is hydrolyzed to adenosine diphosphate (ADP) and releases about 30 kJ/mol (Figure 21.18). The second hydrolysis step, from ADP to adenosine monophosphate (AMP), also produces about 30 kJ/mol. The last hydrolysis step, from AMP to adenosine, releases only about 14 kJ/mol.

The living plant may convert the glucose from photosynthesis to disaccharides, polysaccharides, starches, cellulose, proteins, or oils. The end product depends on the type of plant involved and the complexity of its biochemistry. The various compounds in the plants are then available to be eaten by a plant-eating animal, digested, transported to the cells of the body, and metabolized to provide energy and chemical compounds.

Plants, then, are the primary source of energy for animals and humans. This can be represented by the oxidation of glucose,

$$C_6H_{12}O_6 + 6\,O_2 \longrightarrow 6\,CO_2 + 6\,H_2O + 2880\text{ kJ}$$

which is the reverse of the reaction taking place in photosynthesis. Some of this energy is used immediately, and some is stored through a series of coupled reactions involving the conversion of ADP to ATP (the reverse of the reaction illustrated in Figure 21.18). This series of coupled reactions can be represented by the overall reaction below.

$$C_6H_{12}O_6 + 36\text{ ADP} + 36\,H_3PO_4 + 6\,O_2 \longrightarrow 6\,CO_2 + 36\text{ ATP} + 42\,H_2O$$

21.6 NUCLEIC ACIDS

The genetic information that makes each organism's offspring look and behave like its parents is encoded in molecules called **nucleic acids.** Together with a set of specialized enzymes that catalyze their synthesis and decomposition, the nucleic acids constitute a remarkable system that accurately copies millions of pieces of data with very few mistakes.

A heterocyclic organic compound has a ring that contains both carbon and another element.

Like polysaccharides and polypeptides, nucleic acids are condensation polymers. Each monomer in these polymers includes one of two simple sugars, one or more phosphoric acid groups, and one of a family of heterocyclic nitrogen compounds that behave chemically as bases. A particular nucleic acid is a **deoxyribonucleic acid (DNA)** if it contains the sugar α-2-deoxy-D-ribose, and it is a **ribonucleic acid (RNA)** if it contains the sugar α-D-ribose. The structures of these two sugars are shown in Figure 21.20.

The five organic bases that are the heart of the information encoding system are shown in Figure 21.21. **Adenine** (A) and **guanine** (G) are derivatives of the organic base purine, whereas **thymine** (T), **cytosine** (C), and **uracil** (U) are derivatives of pyrimidine. These bases are mentioned so often in any discussion of nucleic acid chemistry that, in order to save space, they are usually referred to only by the first letters of their names.

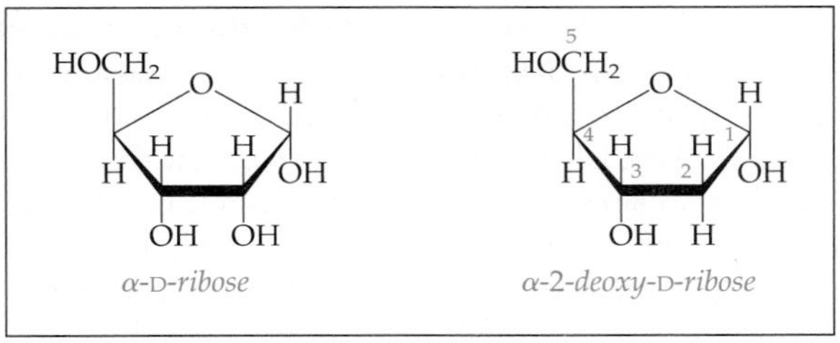

α-D-ribose *α-2-deoxy-D-ribose*

Figure 21.20
The structures of the pentose sugars α-D-ribose and α-2-deoxy-D-ribose. In the names of these sugars, α indicates one of the two possible ring forms, D distinguishes between the isomers that rotate plane-polarized light in opposite directions, and 2 indicates the carbon atom to which no oxygen is attached in the second sugar.

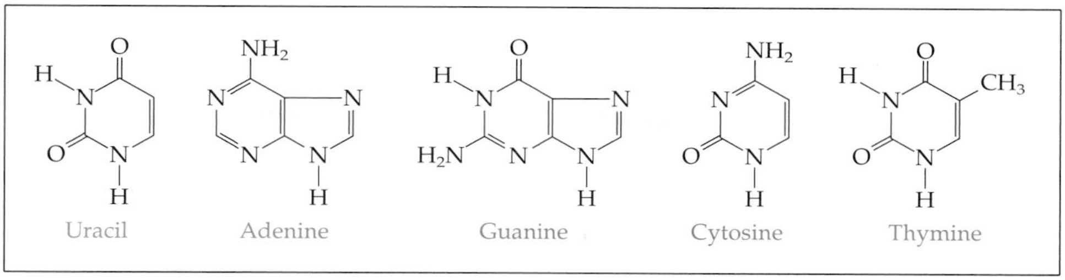

Figure 21.21

Purine and pyrimidine bases. Adenine and guanine are purine bases. Uracil, cytosine, and thymine are pyrimidine bases.

Nucleic acids are found in all living cells, with the exception of the red blood cells of mammals. DNA occurs primarily in the nucleus of the cell, and RNA is found mainly in the cytoplasm, outside the nucleus.

Three major types of RNA have been identified: messenger RNA (mRNA), transfer RNA (tRNA), and ribosomal RNA (rRNA). Each has a characteristic molecular weight and base composition. Messenger RNAs are generally the largest, with molecular weights between 25,000 and 1,000,000. They contain from 75 to 3000 monomer units. Transfer RNAs, the smallest type, have molecular weights between 23,000 and 30,000 and contain from 75 to 90 monomer units. Ribosomal RNAs, which have molecular weights between those of the other two types, make up as much as 80% of the total cell RNA. Besides having different molecular weights, the three types of RNAs differ in function, as described later in the discussion of natural protein synthesis.

The monomers that polymerize to make both DNA and RNA are known as **nucleotides.** They have structures like the one shown in Figure 21.22a. The nucleotides of DNA and RNA have two structural differences: (1) they contain different sugars, as described earlier (Figure 21.20), and (2) the base uracil occurs only in RNA, whereas the base thymine is found only in DNA. The other bases—adenine, guanine, and cytosine—are found in both DNA and RNA.

When the phosphate group is removed from a nucleotide, the remaining fragment (a pentose bonded to a nitrogenous base) is called a **nucleoside.**

Polynucleotides have molecular weights ranging from about 25,000 for tRNA molecules to billions for human DNA. The sequence of nucleotides in the polymer chain (shown by the base sequence) is its primary structure. Polynucleotides are formed by the polymerization of nucleotides to make esters. As an example, Figure 21.22b shows three monomers condensed to a trinucleotide.

In 1953, James D. Watson and Francis H.C. Crick (Figure 11.1) proposed a secondary structure of DNA that revolutionized our understanding of heredity and genetic diseases. Figure 21.23 illustrates a small portion of the structure, in which two polynucleotides are arranged in a double helix that is stabilized by hydrogen bonding between the base groups lying opposite each other in the two chains. The critical point of the Watson-Crick model is that hydrogen bonding can best occur between specific bases. A purine base

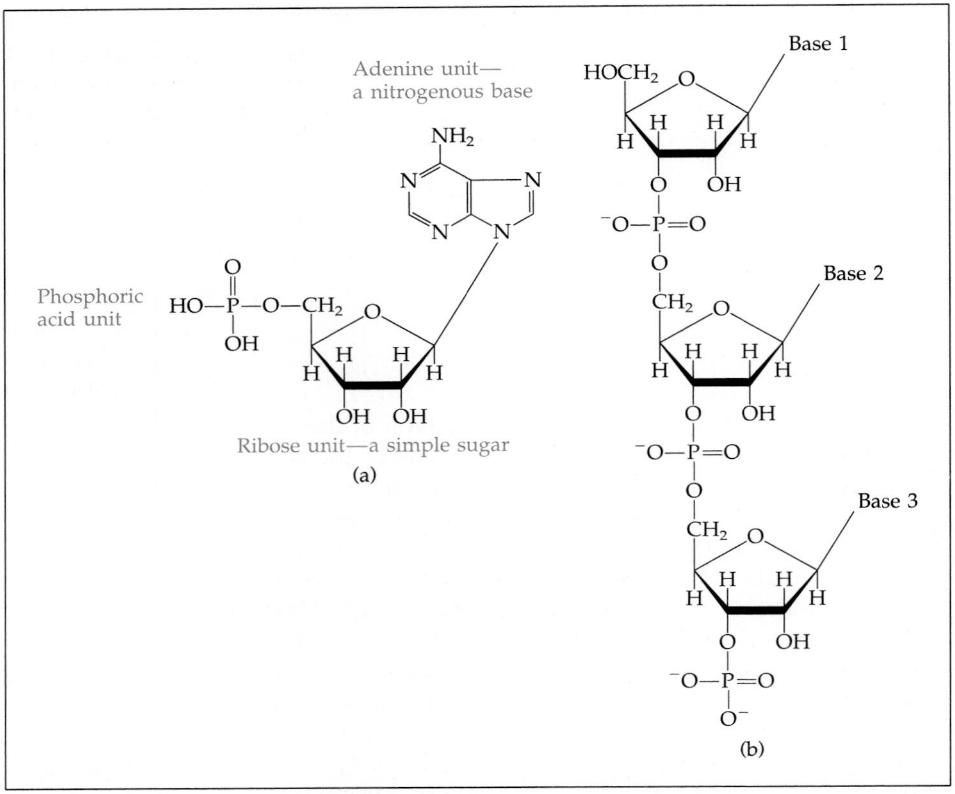

Figure 21.22
(a) A nucleotide. If other bases are substituted for adenine, several nucleotides are possible for each of the two sugars shown in Figure 21.20. (b) Bonding structure of a trinucleotide. Bases 1, 2, and 3 represent any of the bases that occur in DNA and RNA (Figure 21.21). The primary structures of both DNA and RNA are extensions of this structure.

The inherited traits of an organism are controlled by DNA molecules.

of one strand generally pairs with a pyrimidine base of the other strand. Adenine-thymine and guanine-cytosine pairs occur almost exclusively because they are very tightly hydrogen-bonded. The hydrogen bonding between these specific bases is called **complementary hydrogen bonding.** RNA is generally a single strand of helical polynucleotide.

The function of polynucleotides is to transcribe cellular and organism information so that like begets like. The almost infinite variety of primary structures of polynucleotides allows an almost infinite variety of information to be recorded in the molecular structures of the strands of nucleic acids. The different arrangements of just a few different bases give the large variety of structures. In a somewhat similar fashion, the multiple arrangements of just a few language symbols convey the many ideas in this book. The coded information in the polynucleotide is believed to control the inherited characteristics of the next generation as well as most of the continuous life processes of the organism.

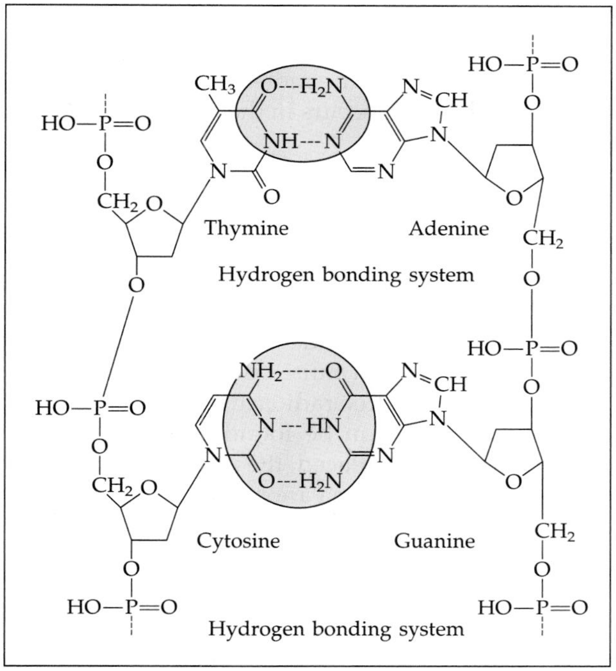

Figure 21.23

The double helix structure proposed for DNA by Watson and Crick. Hydrogen bonds in the thymine–adenine (T–A) and cytosine–guanine (C–G) pairs stabilize the double helix. Adenine also pairs with uracil (A–U) in mRNA, which contains no thymine.

Double-stranded DNA forms the 46 human chromosomes, which have special heredity areas called genes. Genes are segments of DNA that have as few as 1000 or as many as 100,000 base pairs such as those shown in Figure 21.23. Human DNA (the human **genome**) is estimated to have up to 100,000 genes and about 3 billion pairs of bases. However, genes are estimated to make up only 3% of DNA, with each gene sandwiched between "junk" or noncoding DNA sequences. There are also short segments that act as switches to signal where the coding sequence begins.

The total sequence of base pairs of a cell is called the genome.

The transfer of coded information begins with the replication of DNA and continues with natural protein synthesis as well as with the synthesis of body tissues. In the following sections we shall see how DNA replicates and how protein is synthesized naturally.

Replication of DNA: Heredity

Almost all nuclei in an organism's cells contain the same chromosomal composition. This composition remains constant regardless of whether the cell is starving or has an ample supply of food materials. Each organism begins life as a single cell with this same chromosomal composition; in sexual reproduction half of a chromosome comes from each parent. These well-known biological facts, along with recent discoveries concerning polynucleotide structures, have led scientists to the conclusion that the DNA structure is faithfully copied during normal cell division (mitosis—both strands) and that only half is copied in the cell division that produces reproductive cells (meiosis—one strand).

In **replication** the double helix of the DNA structure unwinds and each half of the structure serves as a template, or pattern, from which the other complementary half can be reproduced from the molecules in the cell environment (Figure 21.24). Replication of DNA occurs in the nucleus of the cell.

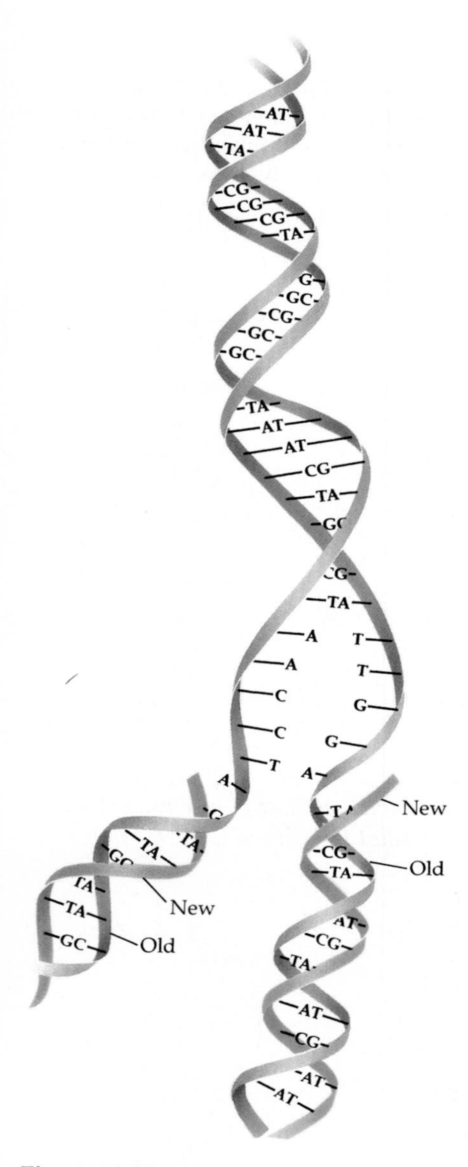

Figure 21.24
Replication of DNA. When the double helix of DNA (blue) unwinds, each half serves as a template on which to assemble subunits (pink) from the cell environment.

Natural Protein Synthesis

The proteins of the body are continually being replaced and resynthesized from the amino acids available to the body. The use of isotopically labeled amino acids has made possible studies of the average lifetimes of amino acids as constituents in proteins—that is, the time it takes the body to replace a protein in a tissue. For a process that must be extremely complex, replacement is very rapid. Only minutes after radioactive amino acids are injected into animals, radioactive protein can be found. Although all the proteins in the body are continually being replaced, the rates of replacement vary. Half of the proteins in the liver and plasma are replaced in 6 days; the time needed for replacement of muscle proteins is about 180 days, and replacement of protein in other tissues, such as bone collagen, takes even longer.

Recall that each organism has its own kinds of proteins. The number of possible unique arrangements of 20 amino acid units is 2.43×10^{18}, yet proteins characteristic of a given organism can be synthesized by the organism in a matter of a few minutes!

The DNA in the cell nucleus holds the code for protein synthesis, which is carried out in a series of steps summarized in Figure 21.25. First, messenger RNA, like all forms of RNA, is synthesized in the cell nucleus. The sequence of bases in one strand of the chromosomal DNA serves as the template from which a single strand of a messenger ribonucleotide (mRNA) is made in a process known as **transcription.** The bases of the mRNA strand complement those of the DNA strand. A pair of bases are complementary when each one fits the other and forms one or more hydrogen bonds. Messenger RNA contains only the four bases adenine (A), guanine (G), cytosine (C), and uracil (U). DNA contains principally the four bases adenine (A), guanine (G), cytosine (C), and thymine (T). The base pairs are as follows:

DNA	*mRNA*
A	U
G	C
C	G
T	A

This means that, provided the necessary enzymes and energy are present, wherever a DNA has an adenine base (A), the mRNA will contain a uracil base (U).

After transcription, mRNA passes from the nucleus of the cell to a ribosome, where mRNA serves as the template for the sequential ordering of amino acids during protein synthesis. As the name implies, messenger RNA contains the sequence message, in the form of a three-base code (called the

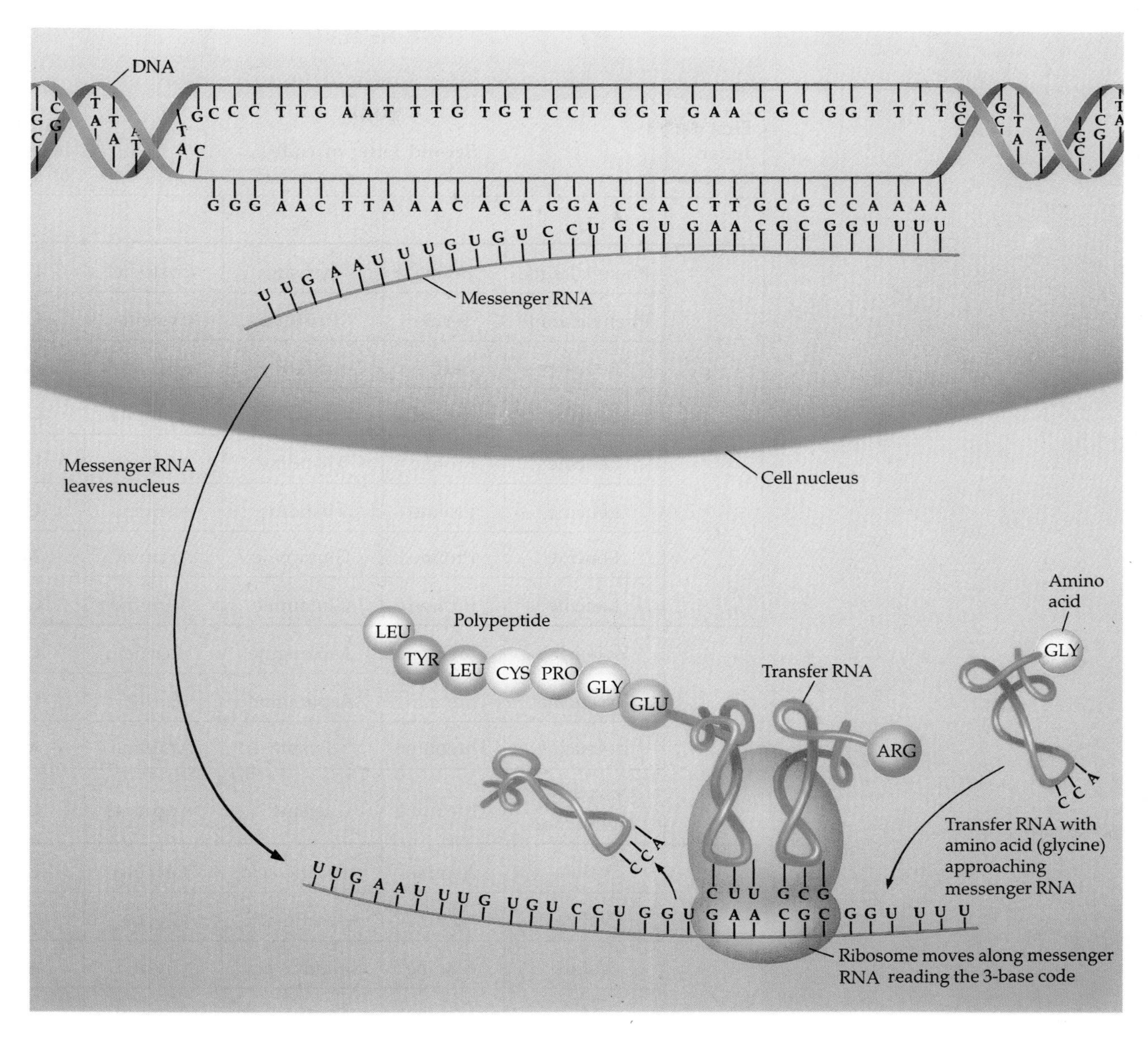

Figure 21.25

A schematic illustration of the roles of DNA and RNA in protein synthesis. A, G, C, T, and U are bases characteristic of the individual nucleotides. See Figure 21.21 for the structures of the bases and Table 21.1 for abbreviations of the amino acids.

codon), for ordering amino acids into proteins (Table 21.5). Each of the thousands of different proteins synthesized by cells is coded by a specific mRNA or segment of an mRNA molecule.

Transfer RNAs carry the amino acids to the mRNA one by one. Each of the 20 amino acids found in proteins has at least one corresponding tRNA, and some have multiple tRNAs (Table 21.5). For example, there are five distinctly different tRNA molecules specifically for the transfer of the amino acid leucine in cells of the bacterium *Escherichia coli.* Table 21.5 lists the RNA codes and shows in the first line that, for example, UUU codes for phenylalanine and UCU codes for serine.

Table 21.5

Messenger RNA Codes for Amino Acids*

First letter of code	Second letter of code				Third letter of code
	U	C	A	G	
U	Phenylalanine	Serine	Tyrosine	Cysteine	U
	Phenylalanine	Serine	Tyrosine	Cysteine	C
	Leucine	Serine	STOP	STOP	A
	Leucine	Serine	STOP	Tryptophan	G
C	Leucine	Proline	Histidine	Arginine	U
	Leucine	Proline	Histidine	Arginine	C
	Leucine	Proline	Glutamine	Arginine	A
	Leucine	Proline	Glutamine	Arginine	G
A	Isoleucine	Threonine	Asparagine	Serine	U
	Isoleucine	Threonine	Asparagine	Serine	C
	Isoleucine	Threonine	Lysine	Arginine	A
	START or methionine	Threonine	Lysine	Arginine	G
G	Valine	Alanine	Aspartic acid	Glycine	U
	Valine	Alanine	Aspartic acid	Glycine	C
	Valine	Alanine	Glutamic acid	Glycine	A
	Valine	Alanine	Glutamic acid	Glycine	G

*In groups of three (called codons), bases of mRNA code the order of amino acids in a polypeptide chain. A, C, G, and U represent adenine, cytosine, guanine, and uracil, respectively. Some amino acids have more than one codon, and hence more than one tRNA can bring the amino acid to mRNA.

At one end of a tRNA molecule is a trinucleotide base sequence (the **anticodon**) that fits a trinucleotide base sequence on mRNA (the codon). At the other end of a tRNA molecule is a specific base sequence of three terminal nucleotides —CCA— with a hydroxyl group on the sugar exposed on the terminal adenine nucleotide group. With the aid of enzymes, this hydroxyl group reacts with a specific amino acid by an esterification reaction.

$$\underset{\text{tRNA}}{(\text{mononucleotides})_{75-90}\text{CCA—OH}} + \underset{\text{amino acid}}{\text{HO}\overset{\text{O}}{\overset{\|}{\text{C}}}\text{CH(NH}_2\text{)R}} \longrightarrow$$

$$\underset{\text{tRNA-amino acid}}{(\text{mononucleotides})_{75-90}\text{CCA—O}\overset{\text{O}}{\overset{\|}{\text{C}}}\text{CH(NH}_2\text{)R}} + H_2O$$

The tRNA and its amino acid migrate to the ribosome, where the amino acid is used in the synthesis of a protein. The tRNA is then free to migrate back to the cell cytoplasm and repeat the process.

Messenger RNA is used at most only a few times before being depolymerized. Although this may seem to be a terrible waste, it enables the cell to produce different proteins on very short notice. As conditions change, different types of mRNA come from the nucleus, different proteins are made, and the cell responds adequately to a changing environment.

In the schematic illustration in Figure 21.25, which summarizes DNA → RNA → protein, pick a three-base sequence on the bottom DNA strand of the two DNA strands at the top of the figure and follow it through to the bottom, where a tRNA attaches to a three-base codon on mRNA. Assume the DNA strand you have selected serves as the template for the synthesis of the single strand of mRNA shown in the figure. Do you agree with the changes that occur in the letters representing the three-base sequence?

Use of Nucleoside Drugs to Treat AIDS

The most promising drugs for treatment of AIDS patients are nucleoside derivatives. The first to be approved for use in the United States is AZT or azidothymidine, a derivative of deoxythymidine.

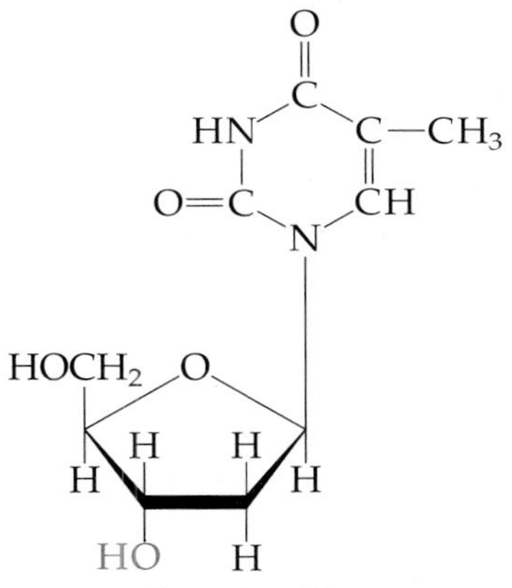

deoxythymidine
(a nucleoside)

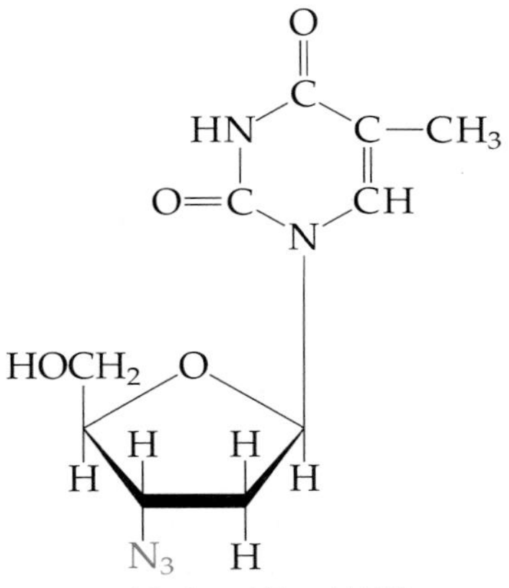

azidothymidine (AZT)

AIDS is caused by a **retrovirus,** a virus with an outer double layer of lipid material that acts as an envelope for several types of proteins, an en-

The prefix retro- *means "reverse."*

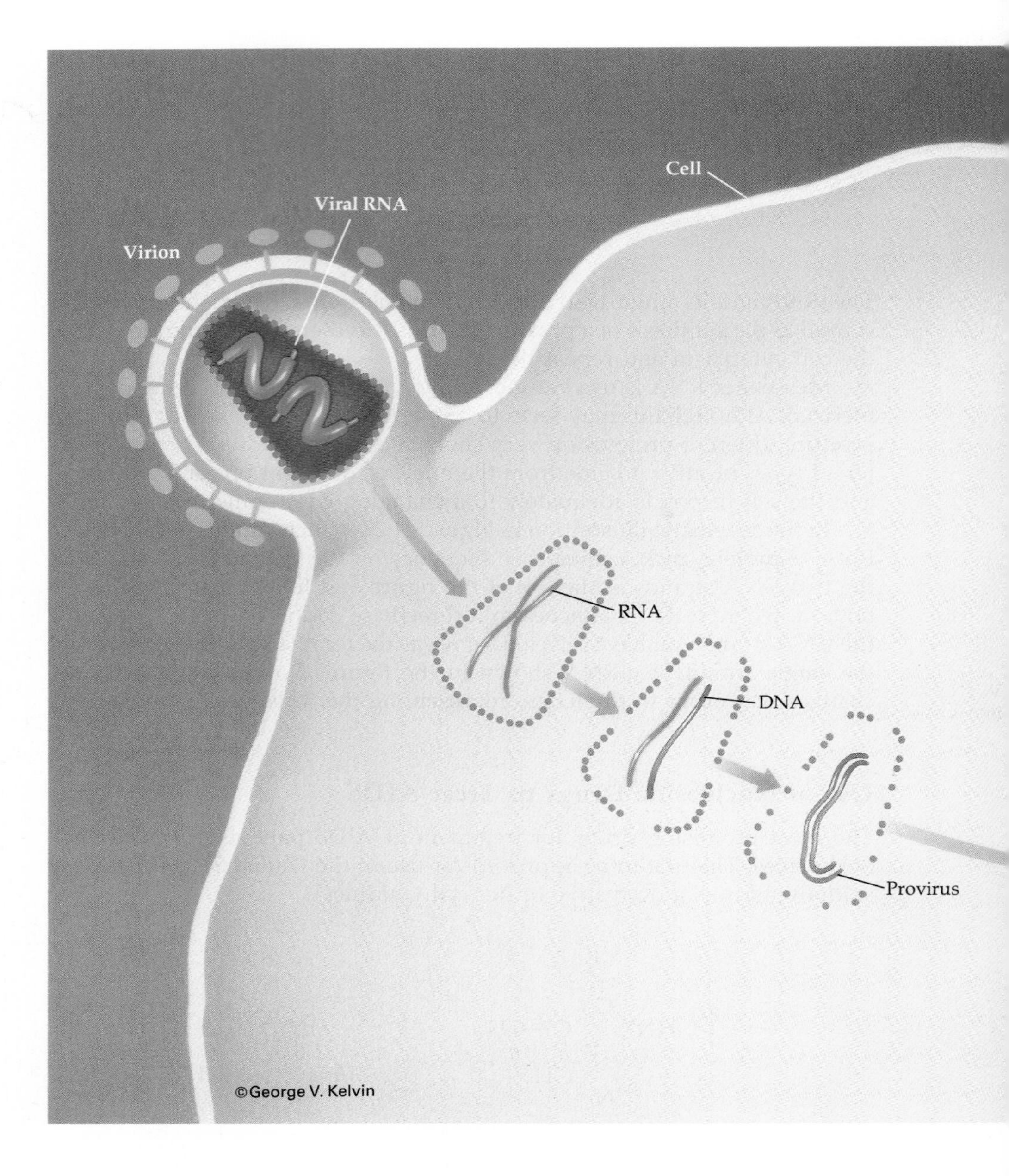

Figure 21.26
A schematic diagram of the HIV retrovirus attacking a T lymphocyte cell. HIV infection begins when a virus particle (virion shown in upper left of figure) attaches to the surface of a T cell and injects two strands of viral RNA, viral protein, and reverse transcriptase enzyme into the cell. The reverse transcriptase enzyme carries out RNA-directed synthesis of a single strand of DNA, which is duplicated to form the double-strand DNA "provirus." The provirus migrates to the nucleus where the viral DNA is inserted into the host DNA. The provirus can remain latent or it can direct the T cell to synthesize viral RNA and viral protein which are then assembled into new virions that bud from the T cell and are released when the T cell ruptures.

zyme called reverse transcriptase, and RNA. The term retrovirus is used because the virus enzyme carries out RNA-directed synthesis of DNA rather than the usual DNA-directed synthesis of RNA.

The AIDS retrovirus penetrates the T cell, a key cell in the immune system of the body. Once the retrovirus is inside the T cell, the reverse transcriptase of the AIDS virus translates the RNA code of the virus into the T cell's double-stranded DNA, directing the T cell to synthesize more AIDS viruses. Eventually the T cell swells and dies, releasing more AIDS viruses to attack other T cells (Figure 21.26).

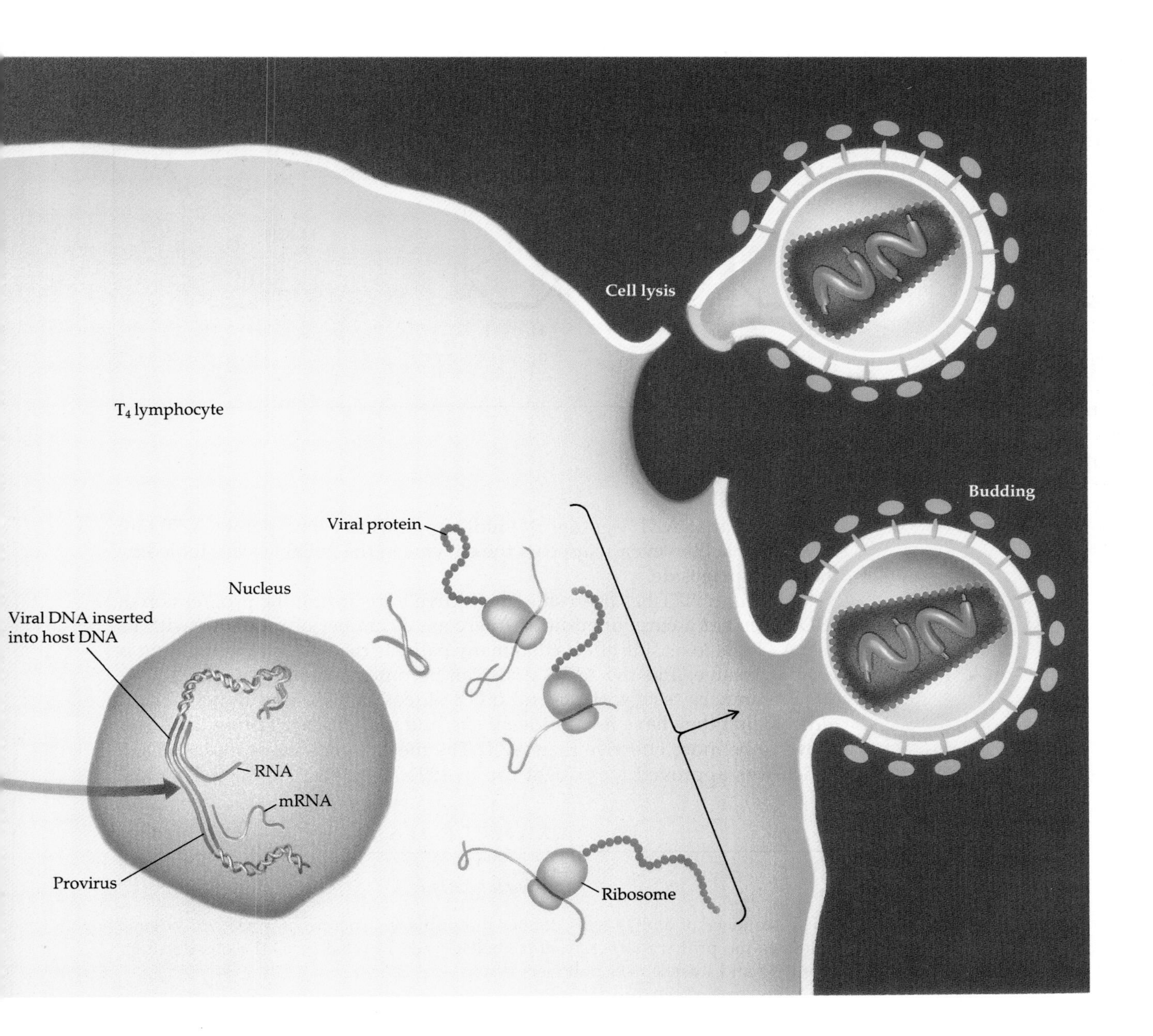

AZT apparently works because it is accepted by reverse transcriptase in place of thymidine. After AZT has become a part of the DNA chain, its structure prevents additional nucleosides from being added onto the DNA chain, and production of more AIDS virus ceases.

Azidothymidine was first synthesized in the 1960s by Jerome Horwitz, who was looking for a compound that would stop cancer cells from multiplying. He reasoned that the incorporation of a "fake nucleoside" into a DNA chain would prevent additional nucleosides from being added and thus cause cell division to stop. The idea didn't work because tumor cells

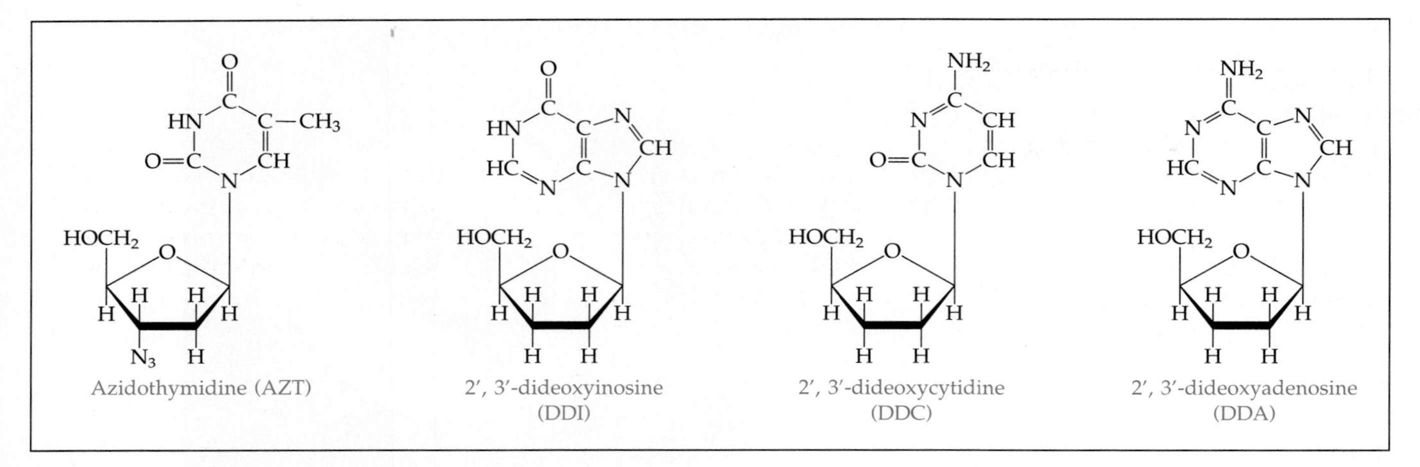

Figure 21.27
Nucleosides that are being used in the treatment of AIDS.

recognized that AZT was not thymidine and didn't incorporate AZT into DNA chains. However, it appears the enzyme in the AIDS virus is fooled by the fake nucleoside.

Although AZT has been shown effective in retarding the progression of AIDS, it is not a cure. In addition, there are a number of problems with its use. AZT has toxic side effects that many patients cannot tolerate; the drug is very expensive; and the AIDS virus can become resistant to AZT.

Preliminary tests with several other dideoxynucleosides (Figure 21.27) indicate that they have fewer side effects, and with further testing they may prove to be more effective than AZT. The most promising of these is DDI, which was approved for patient use in 1991.

EXAMPLE 21.6

Nucleic Acids

The sequence of amino acids defines the primary structure of a protein. What defines the primary structure of a nucleic acid? In what general ways do the protein and nucleic acid polymers differ?

SOLUTION The sequence of bases defines the primary structure of a nucleic acid. However, although proteins can have 20 different amino acids, nucleic acids can have only four different bases. The bases in DNA are adenine, guanine, cytosine, and thymine. Those in RNA are adenine, guanine, cytosine, and uracil. In addition, the molecular weight of nucleic acids is generally much greater than that of proteins, reaching billions.

EXERCISE 21.6 • *Nucleic Acids*

If the sequence of bases along an mRNA strand is . . . UCCGAU . . . , what was the sequence along the DNA template?

21.7
BIOGENETIC ENGINEERING

The field of biogenetic engineering started after the first successful gene-splicing and gene-cloning experiments produced **recombinant DNA** in the early 1970s. The basic idea is to use the rapidly dividing property of common bacteria, such as *E. coli,* as a microbe factory for producing recombinant DNA molecules that contain the genetic information for the desired product. Bacteria have been produced that can synthesize specific proteins, human growth hormone, and human insulin. The method of producing bacteria for a particular function involves removing a gene from the bacterium, splicing in part of a gene from a human or other organism (the part that produces human insulin, for example), placing the spliced gene back into the bacterium, and letting the bacterium make millions of other insulin-producing bacteria. (See Figure 21.28 for a schematic diagram of this process.) The process of splicing and recombining genes is referred to as **recombinant DNA technology** or **biogenetic engineering.**

Recombinant *means capable of genetic recombination, and* recombinant DNA *refers to human control over splitting, splicing, and recombining DNA structures.*

One of the earliest benefits of recombinant DNA technology was the biosynthesis of human insulin in 1978. Millions of diabetics depend on the availability of insulin, but many are allergic to animal insulin, which was the only previous source. Biosynthesized human insulin is now being marketed. Biotechnology firms are also producing human growth hormone, which is used in treating youth dwarfism.

A number of "transgenic animals" have been produced including goats, rabbits, and mice; these animals are used as drug "pharms" since their new genes cause them to produce marketable quantities of desirable pharmaceuticals. For example, tissue-plasminogen-activator (TPA), which is used to dissolve blood clots in emergency treatment of heart attack victims, was

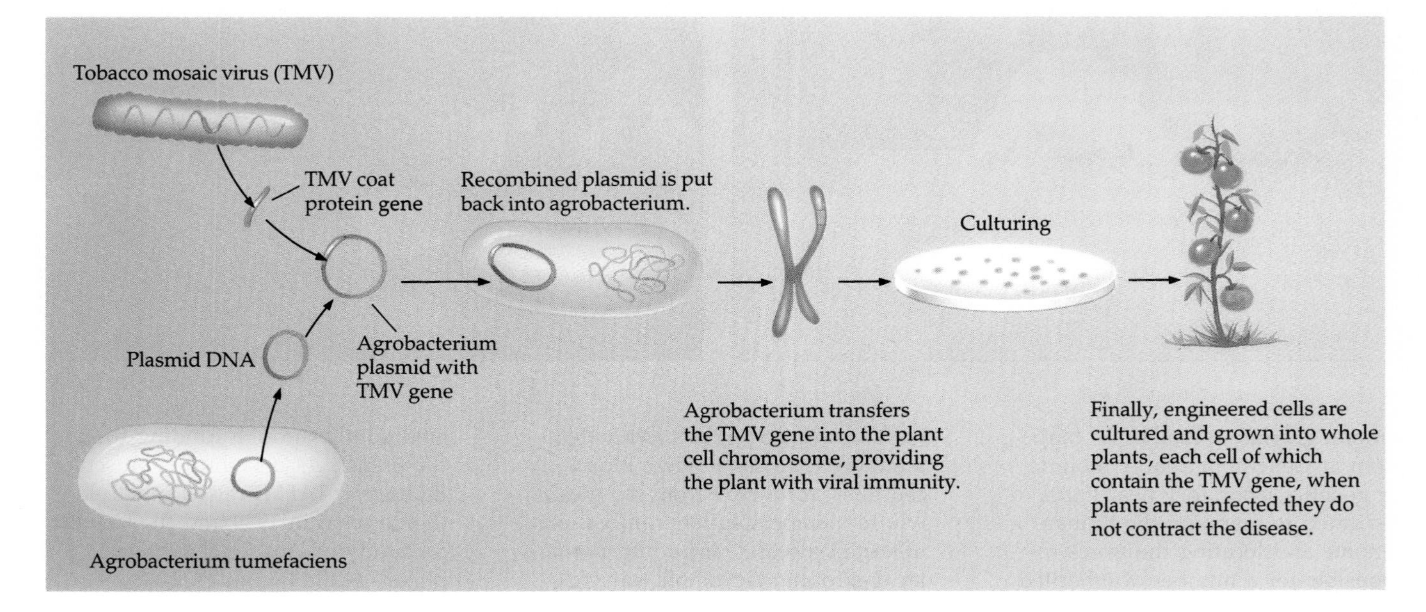

Figure 21.28
Steps necessary to insert a gene from tobacco mosaic virus into a tomato plant in order to make the plant resistant to a particular viral disease.

A field of biogenetically engineered plants.

originally obtained from a bioreactor of *E. coli,* then from the milk of transgenic mice, and now from the milk of transgenic goats.

Sometimes, transgenic animals are less than ideal. Transgenic pigs do produce leaner meat but they have arthritis, lethargy, and a low sex drive. A group of transgenic beef calves did better in general health, but it is as yet unclear if the quality of beef will be improved.

Biogenetic engineering has also led to developments in the following areas: growing plants that produce their own insecticide, making plants resistant to viral and bacterial infections (Figure 21.28), and matching the chemistry of a plant with a protecting herbicide.

21.8 HUMAN GENOME PROJECT

Until 1986, the experimental determination of the sequence of base pairs in a DNA strand was laboriously slow, taking place at a speed of about 200,000 base pairs per year. With the invention of an automatic DNA sequencer by a team headed by Professor Leroy Hood at California Institute of Technology in Pasadena, the number of base pairs that can be determined per day increased dramatically (to more than 10,000 base pairs per day). This made complete sequencing of genomes a possibility. In January, 1989, the National Institutes of Health (NIH) announced plans to determine the complete sequence of the human genome, estimated to contain 3 billion base pairs. The

(a)

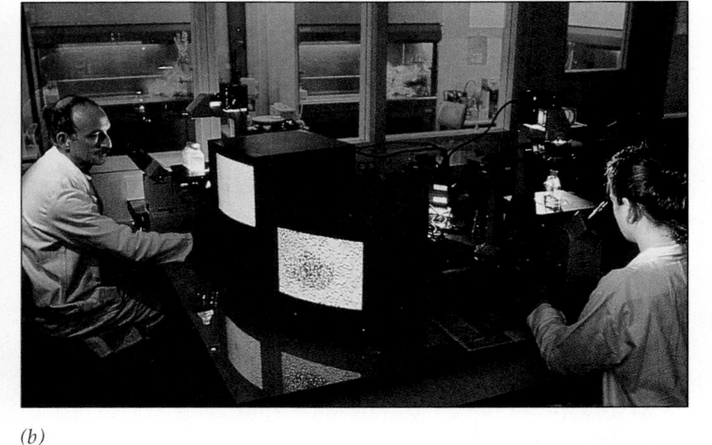

(b)

Human genome research. (a) Map room at Genethon, a large molecular genetics laboratory near Paris, dedicated to mapping the human genome and locating the genes responsible for a number of inherited diseases. (b) Scientists at Genethon work with microscopes and video monitors to analyze white blood cell lines, recovered from families whose members suffer from various inherited diseases, including muscular dystrophy. Genethon was created jointly through funding from the French Association for Muscular Dystrophy (AFM) raised in a Telethon appeal in 1989 and the Center for Study of Human Polymorphisms (CEPH).

Table 21.6

Advances in Human Gene Mapping*

Chromosome	1	2	3	4	5	6	7	8	9	10	11	12	13	14	15	16	17	18	19	20	21	22	X	Y
genes mapped	247	138	85	99	92	131	132	64	77	79	154	136	31	66	65	83	135	28	132	45	39	71	225	18
disease-related	55	24	24	25	23	27	25	22	26	15	46	23	13	19	17	17	26	9	24	14	8	17	111	1
kilobases of sequence	760	867	204	379	302	778	598	229	277	303	815	648	113	686	183	429	860	137	487	175	95	402	553	24

*Genome Maps III, *Science*, Vol. 258, October 2, 1992.

Human Genome Project is estimated to take 15 years. The first five-year phase of the project involves the study of some model genome systems, and research groups are working on sequencing the base pairs in *E. coli*, estimated to contain 4.5 million base pairs; a yeast genome estimated to contain 12.5 million base pairs; and a nematode genome with 100 million base pairs. It is hoped that the techniques and knowledge gained from the study of these model systems, along with continued increases in the number of base pairs that can be sequenced per day, will make it possible to sequence the base pairs in human DNA in 15 years. A complete mapping of the human genome will improve the knowledge of the estimated 4000 hereditary diseases and lead to better diagnosis and treatment of these diseases.

Advances in mapping the human genome are summarized in Table 21.6. The sequences mapped as of July 15, 1992 total more than 10.3 million bases (including overlapping sequences).

IN CLOSING

Having studied this chapter, you should be able to

- draw structures of common amino acids (Section 21.1).
- illustrate the formation of small peptides from amino acids (Section 21.2).
- explain the differences among primary, secondary, tertiary, and quaternary structures of proteins (Section 21.2).
- explain how enzymes function as catalysts (Section 21.2).
- explain the difference between the structures of α-D-glucose and β-D-glucose, and show how this difference affects the properties of starch and cellulose (Section 21.3).
- write equations for the formation of triglycerides from glycerol and fatty acids (Section 21.4).
- illustrate the difference between saturated and unsaturated fats and oils (Section 21.4).
- identify the structures of steroids (Section 21.4).
- describe the structure of ATP and its function (Section 21.5).
- describe and illustrate the structures of nucleotides and the difference between the structures of DNA and RNA (Section 21.6.)

- explain the functions of the different types of RNA (Section 21.6).
- use the DNA → RNA → protein pathway to explain the transfer of genetic information for use in protein synthesis (Section 21.6).
- describe the process for producing recombinant DNA molecules (Section 21.7).
- describe the Human Genome Project (Section 21.8).

STUDY QUESTIONS

Review Questions

1. What is an essential amino acid?
2. What functional groups are always present in each molecule of an amino acid?
3. What are the meanings of the terms *primary, secondary, and tertiary structures of proteins?*
4. Which of the following biochemicals are polymers: starch, cellulose, glucose, fats, glycylalanylcysteine, proteins, DNA, RNA?
5. How do amylopectin and glycogen differ? How are they similar?
6. What three molecular units are found in nucleotides?
7. What are the differences in structure between DNA and RNA?
8. What is recombinant DNA?
9. What is the difference between a nucleoside and a nucleotide?
10. Why are enzymes specific for only one enantiomer?
11. Name the three types of RNA and describe their structures and functions.
12. The human body can metabolize D-glucose but not L-glucose. Explain.
13. What is atherosclerotic plaque? What are its two major ingredients?
14. What do HDL and LDL stand for? Which one should be at high levels and which one should be at low levels to reduce the danger of heart attack? Explain.
15. What is a retrovirus? How does the AIDS retrovirus attack the immune system of the body?
16. Why is AZT effective in the treatment of AIDS?
17. What are transgenic animals? Give some examples.
18. What is the purpose of ATP?

Amino Acids and Peptides

19. Glutathione is an important tripeptide found in all living tissues. It is also named glutamylcystylglycine. Draw the structure of glutathione. Which enantiomeric forms of the amino acids would you predict are used to synthesize this peptide?

20. (a) How many tetrapeptides are possible if four amino acids are linked in different combinations that contain all four amino acids? (b) Write these combinations for tetrapeptides made from glycine, alanine, serine, and cystine. Use three-letter abbreviations for the amino acids. For example, one combination is Gly-Ala-Ser-Cys.

21. Use three-letter abbreviations to write the possible tripeptides that can be formed from phenylalanine, serine, and valine if each tripeptide contains all three amino acids.

22. Name the following tripeptide.

```
      H  O      H  O      H  O
      |  ||     |  ||     |  ||
H2N—C—C—N—C—C—N—C—C—OH
      |      |  |      |  |
     CH      H  CH2    H  CH2
    /  \        |         |
 CH3   CH3      SH     (C6H4)
                          |
                          OH
```

23. Draw the structure of alanylglycylphenylalanine.

24. Draw the structure of leucylmethionylalanylserine.

Proteins

25. What happens when a protein is denatured?

26. What is meant by the *chain conformation* of a protein?

27. Nylon is a polyamide. If it were a natural polymer, what would it be called? Explain.

28. Why are most enzymes effective only over a narrow temperature range?

Carbohydrates

29. What polysaccharide yields only α-D-glucose upon complete hydrolysis?

30. What is the chief function of glycogen in animal tissue?

31. How do amylose and amylopectin differ? How are they similar?

32. Which of the following structures represents α-D-glucose and which represents β-D-glucose?

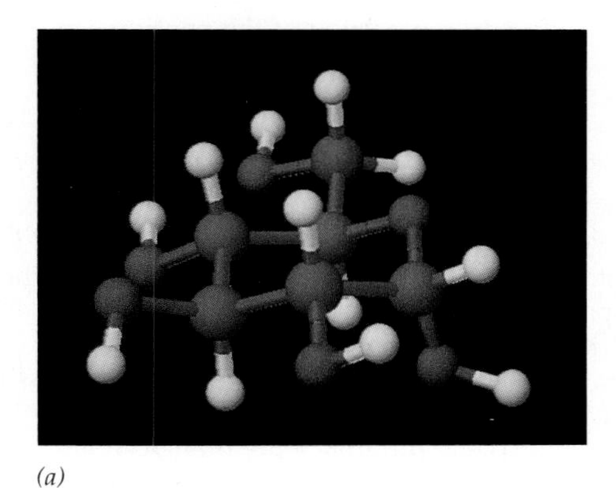

(a)

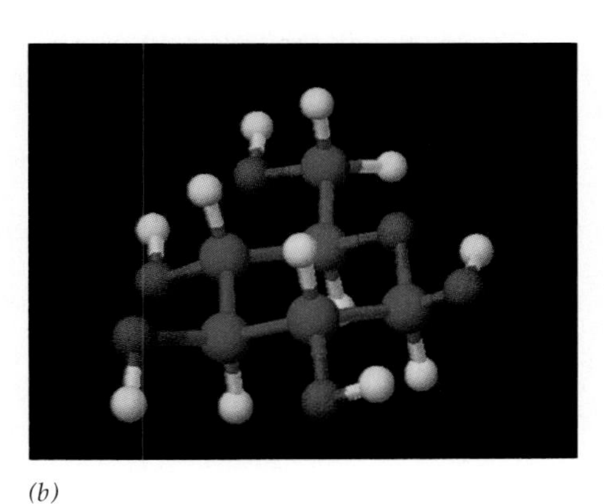

(b)

33. Which of the following structures is ribose and which is deoxyribose?

(a)

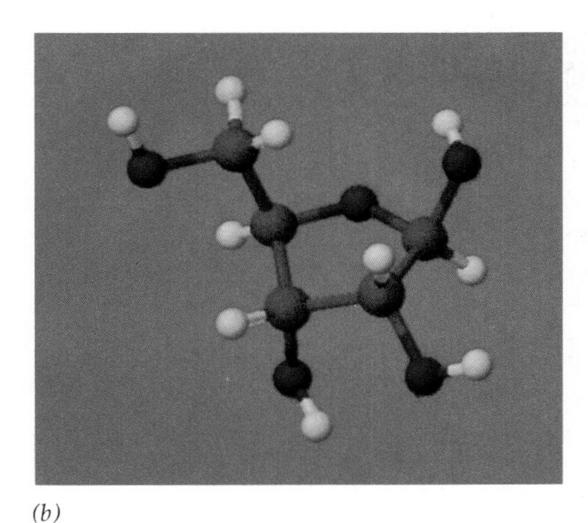

(b)

Lipids

34. Why are fats and oils known as triglycerides?

35. Look at Table 21.4 and explain why canola oil is highly recommended for use in cooking.

36. The following triglyceride is to be completely hydrogenated. Draw the structure of the hydrogenation product.

$$H_2C{-}O{-}\overset{\overset{\displaystyle O}{\|}}{C}{-}(CH_2)_7CH{=}CHCH_2CH{=}CH(CH_2)_4CH_3$$

$$HC{-}O{-}\overset{\overset{\displaystyle O}{\|}}{C}{-}(CH_2)_7CH{=}CH(CH_2)_7CH_3$$

$$H_2C{-}O{-}\overset{\overset{\displaystyle O}{\|}}{C}{-}(CH_2)_7CH{=}CHCH_2CH{=}CHCH_2CH{=}CHCH_2CH_3$$

Nucleic Acids

37. Explain the terms *codon* and *anticodon*.

38. Draw the structure for a three-nucleotide segment of RNA.

39. What is meant by the term *complementary bases*?

40. A segment of a DNA strand has the base sequence . . . GCTGTAACCGAT . . .
   (a) What is the base sequence in the complementary mRNA?
   (b) What is the anticodon order in tRNA?
   (c) Consult Table 21.5 and give the amino acid sequence in the portion of the peptide being synthesized.

41. A segment of a DNA strand has the base sequence . . . TGTCAGTGGGCCGCT . . .
   (a) What is the base sequence in the complementary mRNA?
   (b) What is the anticodon order in tRNA?
   (c) Consult Table 21.5 and give the amino acid sequence in the portion of the peptide being synthesized.

General Questions

42. Write equations for the hydrolysis of (a) sucrose and (b) starch.

43. How does an enzyme increase the rate of a biological reaction?

44. Monosodium-L-glutamate (MSG) is a popular flavor enhancer. Draw its structure and circle the chiral carbon atom.

45. Give a specific example of (a) a polysaccharide, (b) a sugar present in DNA, (c) a disaccharide, and (d) a sugar present in human blood.

46. The artificial sweetener aspartame is a dipeptide derivative made from aspartic acid and the methyl ester of phenylalanine. Write equations for the synthesis of aspartame.

47. Write the overall equation for photosynthesis.

48. Draw the structure of the amino acid in which R = $-CH_2-SH$. What is its name?

49. (a) What is the difference between ATP and ADP?
   (b) What is the important result of conversion of ATP to ADP?

Summary Problems

50. Write equations to show how alanine in its zwitterionic form can function as a buffer.

51. The complete metabolism of glucose, $C_6H_{12}O_6$, is represented by the equation

$$C_6H_{12}O_6 + 6\,O_2 \rightarrow 6\,CO_2 + 6\,H_2O + 2880\text{ kJ}$$

If 100 g of glucose are metabolized, calculate
   (a) the volume of oxygen required at one atmosphere pressure and 37 °C,
   (b) the volume of CO_2 formed at one atmosphere pressure and 37 °C,
   (c) the number of food calories released (one food Calorie = 1kcal).

Appendices

APPENDIX

A

Problem Solving

In Chapter 1 we described the problem of finding an adequate supply of the cancer-treatment drug taxol, and several approaches to solving it. Problem solving of this sort is a major aspect of the work of scientists, engineers, and technologists, and so it is an important aspect of any science course. Throughout this book we provide examples of, and guidance toward solving, a wide variety of chemistry-related problems. However, problem solving is not a simple skill that can be mastered in a few hours of study or practice. Because there are many different kinds of problems, and many different kinds of people who try to solve them, there are no hard and fast rules that will always lead you to a solution. This appendix provides some general guidelines for approaching problems, and some techniques that will help you check that your solution to a problem is reasonable.

A.1 GENERAL STRATEGIES

A very important problem-solving strategy is to imagine several possible approaches or ways of dealing with a problem, and then to try each one. In the taxol example, some scientists worked on the possibility of putting together simple, readily available substances to make taxol directly. Others looked for a closely related compound that could be extracted from yew-tree twigs and needles and then converted to taxol. The latter approach is working better now, but the former may eventually turn out to be the best. Another important technique is to try to break a problem into smaller pieces, each of which can be solved, and then put the pieces together as a complete solution. In the case of the total synthesis of taxol, this approach might involve several different groups of chemists, each working on part of an overall chemical pathway from inexpensive, readily available substances to taxol or a close relative. When a problem can be divided into simpler problems, it often helps to write down a plan that indicates what the simpler problems are and in what order they need to be put together to arrive at an overall solution.

A third strategy is to try to relate a new problem to one that has been solved before; if you can recognize that the new problem belongs to a class of problems you know how to solve, you can use the same method that worked before. Another good approach is to try to get a clear picture of the physical or chemical situation, and to apply what you know about the situation. You might ask, for example, what kind of molecule taxol is made of, what kinds of reactions can be used to create that kind of molecule, and what reactants might be needed to carry out each of those reactions. If all of the reactants for one of the reactions are readily available and the reaction works, the problem is solved; if not, the same strategy can be applied to the precursor compounds until a reasonable reaction is discovered.

In getting a clear picture of a problem and asking appropriate questions regarding the problem, you need to keep in mind all the principles of chemistry and other subjects that you think may apply. In many real-life problems there is not enough information available for you to arrive at an unambiguous solution; in such cases, try to look up or estimate what is needed and then go ahead, noting assumptions you have made. Often the hardest part is deciding which principle or idea is most likely to help solve the problem and what information is needed. To some degree this can be a matter of luck or chance. Nevertheless, in the words of Louis Pasteur, "In the field of observation, chance only favors those minds which have been prepared." The more practice you have had, and the more principles and facts you can keep in mind, the more likely you are to be able to solve the problems that you face.

In this book we have attempted to provide many illustrations of problem solving and many practice problems. Some of these are mathematical and calculational; others are descriptive and require application of chemical facts and principles; still others are based on use of atomic and molecular models that can predict macroscopic behavior of chemical substances. Like the scales and simple chords that a student pianist plays to develop skills that will later be used in sonatas and concertos, many of the problems in this book involve oversimplified situations that are not as difficult as those encountered in the real world. Nevertheless, we intend that they will help you to develop the sorts of skills that later can be applied to difficult, important problems whose solutions will benefit society.

A.2
NUMBERS, UNITS, AND QUANTITIES

Many scientific problems require you to use mathematics to calculate a result or draw a conclusion. Therefore, knowledge of mathematics and its application to problem solving is important. However, one aspect of scientific calculations is often absent from pure mathematical work: science deals with *measurements* in which an unknown quantity is compared with a standard or unit of measure. For example, using a balance to determine the mass of an object involves comparing the object's mass with standard masses, usually in multiples or fractions of one gram; the result is reported as some number of grams, say 4.357 g. *Both the number and the unit are important.* If the result had been 123.5 g, this would clearly be different, but a result of 4.357 oz (ounces) would also be different, because the unit "ounce" is different from the unit "gram." *A result that describes the magnitude of a property, such as 4.357 g, is called a* **quantity,** and chemical problem solving requires calculating with quantities. Notice that whether a quantity is large or small depends on the units as well as the number; the two quantities 123.5 g and 4.357 oz represent the *same* mass.

A quantity is always treated as though the number and the units are multiplied together; that is, 4.357 g can be handled mathematically as $4.357 \times g$. Keeping in mind this simple rule, calculations involving quantities follow the normal rules of algebra and arithmetic: $5\ g + 7\ g = (5 + 7) \times g = 12\ g$; or $6\ g \div 2\ g = (6\ g)/(2\ g) = 3$. (Notice that in the second calculation the unit g appears in numerator and denominator and cancels out, leaving a pure number, 3.) Treating units as algebraic entities has the advantage that *if a calculation is set up correctly, the units will cancel out or multiply together so that the final result has appropriate units.* For example, if you measured the size of a sheet of paper and found it to be 8.5 inches by 11 inches, the area A of the sheet could be calculated as $A = l \times w = 11\ in \times 8.5\ in = 94\ in^2$, or 94 square inches. If a calculation is set up incorrectly, the units of the result will be inappropriate. Using units to check whether a calculation has been properly set up is called **dimensional analysis.**

This idea of using algebra on units as well as numbers is useful in all kinds of situations. For example, suppose you are having a party for some friends who like pizza. A large pizza consists of 12 slices and costs \$10.75. You expect to need 36 slices of pizza and want to know how much you will have to spend. A strategy for solving the problem is first to figure out how many pizzas you need and then to figure the cost in dollars. This solution could be diagrammed as:

$$\text{slices} \xrightarrow[\text{Step 1}]{\text{slices per pizza}} \text{pizzas} \xrightarrow[\text{Step 2}]{\text{dollars per pizza}} \text{dollars}$$

Step 1. Find the number of pizzas required by dividing the number of slices per pizza into the number of slices, thus converting "units" of slices to "units" of pizzas:

$$\text{number of pizzas} = 36\ \cancel{\text{slices}} \left(\frac{1\ \text{pizza}}{12\ \cancel{\text{slices}}}\right) = 3\ \text{pizzas}$$

Strictly speaking, slices and pizzas are not units in the same sense that a gram is a unit; however, labeling things this way will often help you keep in mind what a number refers to—pizzas, slices, or dollars in this case.

Notice that if you had multiplied the number of slices times the number of slices per pizza, the result would have been labeled pizza × slices^2, which does not make sense. In other words, the labels indicate whether multiplication or division is appropriate.

Step 2 Find the total cost by multiplying the cost per pizza times the number of pizzas needed, thus converting "units" of pizzas to "units" of dollars:

$$\text{total price} = 3\ \cancel{\text{pizzas}} \left(\frac{\$10.75}{1\ \cancel{\text{pizza}}}\right) = \$32.25$$

Notice that in each step you have multiplied by a factor that allowed the initial units to cancel algebraically, giving the answer in the desired unit. A factor such as (1 pizza/12 slices) or (\$10.75/pizza) is referred to as a **propor-**

tionality factor. This name indicates that it comes from a proportion. For instance, in the pizza problem you could set up the proportion:

$$\frac{x \text{ pizzas}}{36 \text{ slices}} = \frac{1 \text{ pizza}}{12 \text{ slices}} \quad \text{or} \quad x \text{ pizzas} = 36 \text{ slices}\left(\frac{1 \text{ pizza}}{12 \text{ slices}}\right) = 3 \text{ pizzas}$$

A proportionality factor such as (1 pizza/12 slices) is also called a **conversion factor,** which indicates that it converts one kind of unit or label to another; in this case the label "slices" is converted to the label "pizzas."

Many everyday scientific problems involve proportionality. For example, the bigger the volume of a solid or liquid substance, the bigger its mass. When the volume is zero, the mass is zero also. These facts indicate that mass, m, is directly proportional to volume, V, or, symbolically,

$$m \propto V$$

where the symbol $\propto$ means "is proportional to." Whenever a proportion is expressed this way, it can also be expressed as an equality by using a proportionality constant; for example,

$$m = d \times V$$

In this case the proportionality constant, d, is called the density of the substance. This equation embodies the definition of density as mass per unit volume, since it can be rearranged algebraically to

$$d = \frac{m}{V}$$

As with any algebraic equation involving three variables, it is possible to calculate any one of the three quantities m, V, or d, provided the other two are known. If density is wanted, simply use the definition of mass per unit volume; if mass or volume is to be calculated, the density can be used as a proportionality factor.

Suppose that you are going to buy a ton of gravel and want to know how big a bin you will need to store it. You know the mass of gravel and want to find the volume of the bin; this implies that density will be useful. If the gravel is primarily limestone, you can assume that its density is about the same as for limestone and look it up. Limestone has the chemical formula $CaCO_3$ and its density is 2.7 kg/L. However, these mass units are different from the units for mass of gravel, namely tons. Therefore you need to recall or look up the mass of 1 ton (2000 pounds, lb) and the fact that there are 2.20 lb per kg. This provides enough information to calculate the volume needed.

Step 1 Figure out how many kilograms of gravel are in a ton.

$$m_{\text{gravel}} = 1 \text{ ton} = 2000 \text{ lb} = 2000 \cancel{\text{lb}}\left(\frac{1 \text{ kg}}{2.20 \cancel{\text{lb}}}\right) = 910 \text{ kg}$$

The fact that there are 2.20 lb per kg implies two proportionality factors: (2.20 lb/1 kg) and (1 kg/2.20 lb). The latter was used because it results in appropriate cancellation of units.

Step 2 Use the density to calculate the volume of 909 kg gravel.

$$V_{gravel} = \frac{m_{gravel}}{d_{gravel}} = \frac{909\ \text{kg}}{2.7\ \text{kg/L}} = 910\ \cancel{\text{kg}}\left(\frac{1\ \text{L}}{2.7\ \cancel{\text{kg}}}\right) = 340\ \text{L}$$

In this step we used the definition of density, solved algebraically for mass, substituted the two known quantities into the equation, and calculated the result. However, it is quicker simply to remember that mass and volume are related by a proportionality factor called density, and to use the units of the quantities to decide whether to multiply or divide by that factor. In this case we divided mass by density because the units kilograms canceled, leaving a result in liters, which is a unit of volume.

The liter is not the most convenient volume unit for this problem, however, because it does not relate well to what we want to find out—how big a bin to make. A liter is about the same volume as a quart, but whether you are familiar with liters or quarts or both, 300 of them is not easy to visualize. Let's convert these units to something we can understand better. A liter is a volume equal to a cube one-tenth of a meter (1 dm) on a side; that is, a liter is 1 dm^3. Consequently,

$$340\ \text{L} = 340\ \cancel{\text{L}}\left(\frac{1\ \text{dm}^3}{1\ \cancel{\text{L}}}\right)\left(\frac{1\ \text{m}}{10\ \text{dm}}\right)^3 = 340\ \cancel{\text{dm}^3}\left(\frac{1\ \text{m}^3}{1000\ \cancel{\text{dm}^3}}\right) = 0.34\ \text{m}^3$$

Thus the bin would need to have a volume of about one-third of a cubic meter; that is, it could be a meter wide, a meter long, and about a third of a meter high and it would hold the ton of gravel.

One more thing should be noted about this example. We don't need to know the volume of the bin very precisely, because being off a bit will make very little difference; it might mean getting a little too much wood to build the bin, or not making the bin quite big enough and having a little gravel spill out, but this isn't a big deal. In other cases, such as calculating the quantity of fuel needed to get a space shuttle into orbit, being off by a few percent could be a life-or-death matter. Because it is important to know how precise data are, and to be able to evaluate how important precision is, scientific results usually indicate precision. The simplest way to do this is by means of significant figures.

A.3 SIGNIFICANT FIGURES

The **precision** of a measurement indicates how well several determinations of the same quantity agree. Precision is illustrated by the results of throwing darts at a bullseye (Figure A.1.). In part (a) the darts are scattered all over the board; the dart thrower was apparently not very skillful (or threw the darts from a long distance away from the board), and the precision of their placement on the board is low. In Figure A.1b the darts are all clustered together, indicating much better reproducibility on the part of the thrower, that is, greater precision. In addition, every dart has come very close to the bull's

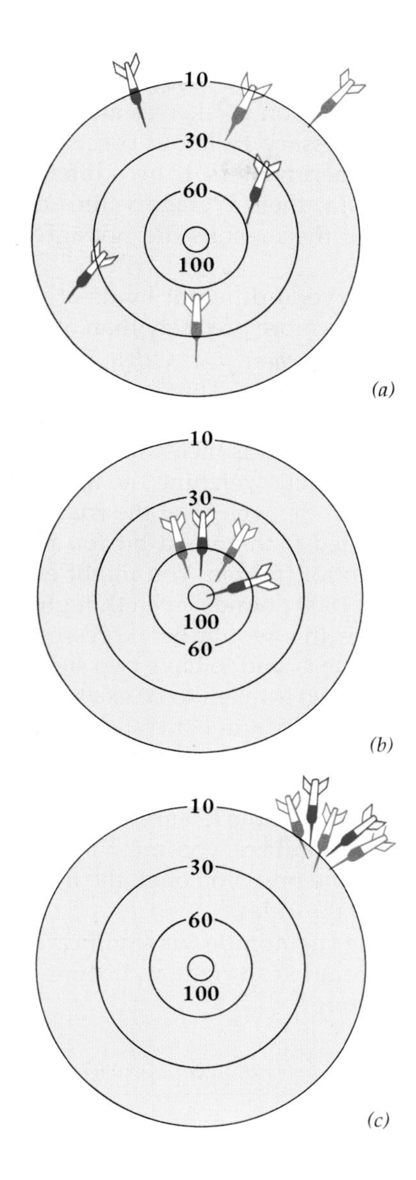

Figure A.1
Precision and accuracy: (a) poor precision and poor accuracy; (b) good precision and good accuracy; (c) good precision and poor accuracy.

eye; this is described by saying that the thrower has been quite **accurate**—the average of all throws is very close to the accepted position, namely the bull's eye. Figure A.1c illustrates that it is possible to be precise without being accurate—the dart thrower has consistently missed the bull's eye, although all darts are clustered very precisely around the wrong point on the board. This third case is like an experiment with some flaw (either in its design or in a measuring device) that causes all results to differ from the correct value by the same amount.

In the laboratory we attempt to set up experiments so that the greatest possible accuracy can be obtained. As a further check on accuracy, results are usually compared among different laboratories so that any flaw in experimental design or measurement can be detected. For each individual experiment, several measurements are usually made and their precision determined. Usually better precision is taken as an indication of better experimental work, and it is necessary to know precision in order to compare results among different experimenters. If two different experimenters both had results like Figure A.1a, their average values could differ quite a lot before they would say that their results did not agree within experimental error.

In most experiments several different kinds of measurements must be made, and some can be done more precisely than others. It is common sense that *a calculated result can be no more precise than the least precise piece of information that went into the calculation.* This is where the rules for *significant figures* come in. In the preceding example the quantity of gravel was described as "a ton." Usually gravel is measured by weighing a truck empty, putting some gravel in the truck, weighing the truck again, and subtracting the weight of the truck from the weight of the truck plus gravel. The quantity of gravel is not adjusted if there is a bit too much or a bit too little, because this would be a lot of trouble. You might end up with as much as 2200 pounds or as little as 1800 pounds, even though you asked for a ton. In terms of significant figures this would be expressed as 2.0×10^3 lb.

The quantity 2.0×10^3 lb is said to have two significant figures; it designates a quantity where the two is taken to be exactly right but the zero is not known precisely. (In this case the number could be as large as 2.2 or as small as 1.8, and so the zero obviously is not exactly right.) In general, in a number that represents a scientific measurement, the last digit on the right is taken to be inexact, but all digits farther to the left are assumed to be exact. When you do calculations using such numbers, you must follow some simple rules so that the results will reflect the precision of all the measurements that go into the calculations. Here are the rules:

Rule 1 To determine the number of significant figures in a measurement, read the number from left to right and count all digits, starting with the first digit that is *not* zero.

For a number written in scientific notation, all digits are significant.

Example	*Number of Significant Figures*
1.23 g	3
0.00123 g	3; the zeros to the *left* of the 1 simply locate the decimal point. To avoid confusion, write numbers of this type in scientific notation; thus, $0.00123 = 1.23 \times 10^{-3}$.
2.0 g and 0.020 g	2; both have two significant digits. When a number is greater than 1, *all zeros to the right of the decimal point are significant.* For a number less than 1, only zeros to the right of the first significant digit are significant.
100 g	1; in numbers that do not contain a decimal point, "trailing" zeros may or may not be significant. To eliminate possible confusion, the practice followed in this book is to include a decimal point if the zeros are significant. Thus, 100. has three significant digits, while 100 has only one. Alternatively, we write it in scientific notation as 1.00×10^2 (three significant

Example	Number of Significant Figures
	digits) or as 1×10^2 (one significant digit). For a number written in scientific notation, all digits are significant.
100 cm/m	Infinite number of significant figures, because this is a defined quantity.
$\pi = 3.1415926\ldots$	The value of π is known to a greater number of significant figures than any data you will ever use in a calculation.

The number π is now known to 1,011,196,691 digits. It is doubtful that you will need this accuracy in this course—or ever.

Rule 2 When adding or subtracting, the number of decimal places in the answer should be equal to the number of decimal places in the number with the *fewest* places.

0.12	2 significant figures	2 decimal places
1.6	2 significant figures	1 decimal place
10.976	5 significant figures	3 decimal places
12.696		

This answer should be reported as 12.7, a number with one decimal place, because 1.6 has only one decimal place.

Rule 3 In multiplication or division, the number of significant figures in the answer should be the same as that in the quantity with the *fewest* significant figures.

$$\frac{0.01208}{0.0236} = 0.512 \text{ or, in scientific notation, } 5.12 \times 10^{-1}$$

Since 0.0236 has only three significant figures, while 0.01208 has four, the answer is limited to three significant figures.

Rule 4 When a number is rounded off (the number of significant figures is reduced), the last digit retained is increased by 1 only if the following digit is 5 or greater.*

Full Number	*Number Rounded to Three Significant Figures*
12.696	12.7
16.249	16.2
18.35	18.4
18.351	18.4

One last word regarding significant figures and calculations. In working problems on a pocket calculator, you should do the calculation using all the digits allowed by the calculator and round off only at the end of the prob-

*A modification of this rule is sometimes used to reduce the accumulation of roundoff errors. If the digit following the last permitted significant figure is *exactly* 5 (with no following digits or with all following digits being zeros), then (a) increase the last significant figure by 1 if it is *odd* or (b) leave the last significant figure unchanged if it is *even*. Thus, both 18.35 and 18.45 are rounded to 18.4.

The ppm unit stands for "parts per million." If a substance is present with a concentration of 1 ppm, there is 1 gram of the substance in 1 million grams of sample.

lem. Rounding off in the middle can introduce errors. If your answers do not quite agree with those in the back of the book, this may be the source of the disagreement.

Now let us consider a problem that is of practical importance and that makes use of all the rules. Suppose you discover that young children are eating chips of paint that flake off a wall in an old house. The paint contains 200. ppm lead (200. mg Pb per kg paint). Suppose that a child eats five such chips. How much lead has the child gotten from the paint?

As stated this problem does not include enough information for a solution to be obtained; however, some reasonable assumptions can be made and they can lead to experiments that could be used to obtain the necessary information. The statement does not say how big the paint chips are. Let's assume that they are 1.0 cm by 1.0 cm so that the area is 1.0 cm^2. Then eating five chips means eating 5.0 cm^2 of paint. (This assumption could be improved by measuring similar chips from the same place.) Since the concentration of lead is reported in units of mass of lead per mass of paint, we need to know the mass of 5.0 cm^2 of paint. This could be determined by measuring the areas of several paint chips and determining the mass of each. Suppose that the results of such measurements were

Mass of Chip (mg)	*Area of Chip (cm^2)*	*Mass per Unit Area (mg/cm^2)*
29.6	2.34	12.65
21.9	1.73	12.66
23.6	1.86	12.69

$$\text{average mass per unit area} = \frac{(12.65 + 12.66 + 12.69)\ \text{mg/cm}^2}{3}$$

$$= 12.67\ \text{mg/cm}^2 = 12.7\ \text{mg/cm}^2$$

The average has been rounded to three significant figures because each experimental number has three significant figures. (Notice that more than three significant figures were kept in the intermediate calculations so as not to lose precision.) Now we can use this information to calculate how much lead the child has consumed.

$$m_{\text{paint}} = 5.0\ \cancel{\text{cm}^2}\ \text{paint}\left(\frac{12.67\ \cancel{\text{mg}}\ \text{paint}}{\cancel{\text{cm}^2}}\right)\left(\frac{1\ \cancel{\text{g}}}{1000\ \cancel{\text{mg}}}\right)\left(\frac{1\ \text{kg}}{1000\ \cancel{\text{g}}}\right)$$

$$= 6.335 \times 10^{-5}\ \text{kg paint}$$

$$m_{\text{Pb}} = 6.335 \times 10^{-5}\ \cancel{\text{kg paint}}\left(\frac{200.\ \text{mg Pb}}{1\ \cancel{\text{kg paint}}}\right) = 1.267 \times 10^{-2}\ \text{mg Pb}$$

$$= 0.013\ \text{mg Pb}$$

Notice that the final result was rounded to two significant figures because there were only two significant figures in the initial area of the paint chip. This is quite adequate precision, however, for you to determine whether this quantity of lead is likely to be harmful to the child.

The methods of problem solving presented here have been developed over time and represent a good way of keeping track of the precision of results, the units in which those results were obtained, and the correctness of calculations. These methods are not the only way that such goals can be achieved, but they do work well. We recommend that you include units in all calculations and check that they cancel appropriately. Also it is important not to overstate the precision of results by keeping too many significant figures. By solving a great many problems, you should be able to develop your problem-solving skills so that they become second nature and you can do them without thinking about the mechanics. This will allow you to devote all your thought to the logic of a problem solution.

APPENDIX

B

Some Mathematical Operations

The mathematical skills required in this introductory course are basic skills in algebra and a knowledge of (a) exponential or scientific notation, (b) logarithms, and (c) quadratic equations. This appendix reviews each of the last three topics.

B.1
ELECTRONIC CALCULATORS

The directions for calculator use in this section are given for calculators using "algebraic" logic. Such calculators are the most common type used by students in introductory courses. For calculators using RPN logic (such as those made by Hewlett-Packard), the procedures will differ slightly.

The advent of inexpensive electronic calculators has made calculations in introductory chemistry much more straightforward. You are well advised to purchase a calculator that has the capability of performing calculations in scientific notation, has both base-10 and natural logarithms, and is capable of raising any number to any power and of finding any root of any number. In the discussion below, we shall point out in general how these functions of your calculator can be used.

Although electronic calculators have greatly simplified calculations, they have also forced us to focus again on significant figures. A calculator easily handles 8 or more significant figures, but real laboratory data are never known to this accuracy. Therefore, you are urged to review Appendix A on handling numbers.

B.2
EXPONENTIAL OR SCIENTIFIC NOTATION

In exponential or scientific notation, a number is expressed as a product of two numbers: $N \times 10^n$. The first number, N, is the so-called *digit term* and is a number between 1 and 10. The second number, 10^n, the *exponential term,* is some integer power of 10. For example, 1234 would be written in scientific notation as 1.234×10^3 or 1.234 multiplied by 10 three times.

$$1234 = 1.234 \times 10^1 \times 10^1 \times 10^1 = 1.234 \times 10^3$$

Conversely, a number less than 1, such as 0.01234, would be written as 1.234×10^{-2}. This notation tells us that 1.234 should be divided twice by 10 in order to obtain 0.01234.

$$0.01234 = \frac{1.234}{10^1 \times 10^1} = 1.234 \times 10^{-1} \times 10^{-1} = 1.234 \times 10^{-2}$$

Some other examples of scientific notation are

$10000 = 1 \times 10^4$ $12345 = 1.2345 \times 10^4$
$1000 = 1 \times 10^3$ $1234 = 1.234 \times 10^3$
$100 = 1 \times 10^2$ $123 = 1.23 \times 10^2$
$10 = 1 \times 10^1$ $12 = 1.2 \times 10^1$
$1 = 1 \times 10^0$ (any number to the zero power = 1)
$1/10 = 1 \times 10^{-1}$ $0.12 = 1.2 \times 10^{-1}$
$1/100 = 1 \times 10^{-2}$ $0.012 = 1.2 \times 10^{-2}$
$1/1000 = 1 \times 10^{-3}$ $0.0012 = 1.2 \times 10^{-3}$
$1/10000 = 1 \times 10^{-4}$ $0.00012 = 1.2 \times 10^{-4}$

When converting a number to scientific notation, notice that the exponent n is positive if the number is greater than 1 and negative if the number is less than 1. The value of n is the number of places by which the decimal was shifted to obtain the number in scientific notation.

$$1\ 2\ 3\ 4\ 5. = 1.2345 \times 10^4$$

Decimal shifted 4 places to the left. Therefore, n is positive and equal to 4.

$$0.0\ 0\ 1\ 2 = 1.2 \times 10^{-3}$$

Decimal shifted 3 places to the right. Therefore, n is negative and equal to 3.

If you wish to convert a number in scientific notation to the usual form, the procedure above is simply reversed.

$$6\ .\ 2\ 7\ 3 \times 10^2 = 627.3$$

Decimal point moved 2 places to the right, since n is positive and equal to 2.

$$0\ 0\ 6.273 \times 10^{-3} = 0.006273$$

Decimal point shifted 3 places to the left, since n is negative and equal to 3.

There are two final points to be made concerning scientific notation. First, if you are used to working on a computer you may be in the habit of writing a number such as 1.23×10^3 as 1.23E3 or 6.45×10^{-5} as 6.45E-5. Second, some electronic calculators allow you to convert numbers readily to the scientific notation. If you have such a calculator, you can change a number shown in the usual form to scientific notation simply by pressing the EE or EXP key and then the "=" key.

1. Adding and Subtracting Numbers

When adding or subtracting two numbers, they must first be converted to the same powers of ten. The digit terms are then added or subtracted as appropriate.

$$(1.234 \times 10^{-3}) + (5.623 \times 10^{-2}) = (0.1234 \times 10^{-2}) + (5.623 \times 10^{-2})$$
$$= 5.746 \times 10^{-2}$$

$$(6.52 \times 10^{2}) - (1.56 \times 10^{3}) = (6.52 \times 10^{2}) - (15.6 \times 10^{2})$$
$$= -9.1 \times 10^{2}$$

2. Multiplication

The digit terms are multiplied in the usual manner, and the exponents are added algebraically. The result is expressed with a digit term with only one nonzero digit to the left of the decimal.

$$(1.23 \times 10^{3})(7.60 \times 10^{2}) = (1.23)(7.60) \times 10^{3+2}$$
$$= 9.35 \times 10^{5}$$

$$(6.02 \times 10^{23})(2.32 \times 10^{-2}) = (6.02)(2.32) \times 10^{23-2}$$
$$= 13.966 \times 10^{21}$$
$$= 1.40 \times 10^{22} \text{ (answer in 3 significant figures)}$$

3. Division

The digit terms are divided in the usual manner, and the exponents are subtracted algebraically. The quotient is written with one nonzero digit to the left of the decimal in the digit term.

$$\frac{7.60 \times 10^{3}}{1.23 \times 10^{2}} = \frac{7.60}{1.23} \times 10^{3-2} = 6.18 \times 10^{1}$$

$$\frac{6.02 \times 10^{23}}{9.10 \times 10^{-2}} = \frac{6.02}{9.10} \times 10^{(23)-(-2)} = 0.662 \times 10^{25} = 6.62 \times 10^{24}$$

4. Powers of Exponentials

When raising a number in exponential notation to a power, treat the digit term in the usual manner. The exponent is then multiplied by the number indicating the power.

$$(1.25 \times 10^{3})^{2} = (1.25)^{2} \times 10^{3\times 2}$$
$$= 1.5625 \times 10^{6} = 1.56 \times 10^{6}$$

$$(5.6 \times 10^{-10})^{3} = (5.6)^{3} \times 10^{(-10)\times 3}$$
$$= 175.6 \times 10^{-30} = 1.8 \times 10^{-28}$$

Electronic calculators usually have two methods of raising a number to a power. To square a number, enter the number and then press the "x^2" key. To raise a number to any power, use the "y^x" key. For example, to raise 1.42×10^2 to the 4th power,

(a) enter 1.42×10^2

(b) press "y^x"

(c) enter 4 (this should appear on the display)

(d) press "=" and $4.0658 \ldots \times 10^8$ will appear on the display. (The number of digits will depend upon the calculator in use.)

As a final step, express the number in the correct number of significant figures (4.07×10^8 in this case).

5. Roots of Exponentials

Unless you use an electronic calculator, the number must first be put into a form where the exponential is exactly divisible by the root. The root of the digit term is found in the usual way, and the exponent is divided by the desired root.

$$\sqrt{3.6 \times 10^7} = \sqrt{36 \times 10^6} = \sqrt{36} \times \sqrt{10^6} = 6.0 \times 10^3$$

$$\sqrt[3]{2.1 \times 10^{-7}} = \sqrt[3]{210 \times 10^{-9}} = \sqrt[3]{210} \times \sqrt[3]{10^{-9}} = 5.9 \times 10^{-3}$$

To take a square root on an electronic calculator, enter the number and then press the "$\sqrt{x}$" key. To find a higher root of a number, such as the 4th root of 5.6×10^{-10},

(a) enter the number

(b) press the "$\sqrt[x]{y}$" key (On most calculators, the sequence you actually use is to press "2ndF" and then "$\sqrt[x]{y}$." Alternatively, you press "INV" and then "y^x.")

(c) enter the desired root, 4 in this case

(d) press "=". The answer here is 4.8646×10^{-3} or 4.9×10^{-3}.

A general procedure for finding any root is to use the "y^x" key. For a square root, x is 0.5 (or ½), whereas it is 0.33 (or ⅓) for a cube root, 0.25 (or ¼) for a 4th root, and so on.

B.3 LOGARITHMS

There are two types of logarithms used in this text: (a) common logarithms (abbreviated log) whose base is 10 and (b) natural logarithms (abbreviated ln) whose base is e (= 2.71828).

$$\log x = n \qquad \text{where } x = 10^n$$

$$\ln x = m \qquad \text{where } x = e^m$$

Most equations in chemistry and physics were developed in natural or base e logarithms and this practice is followed in this text. The relation between log and ln is

$$\ln x = 2.303 \log x$$

Aside from the different bases of the two logarithms, they are used in the same manner. What follows is largely a description of the use of common logarithms.

A common logarithm is the power to which you must raise 10 to obtain the number. For example, the log of 100 is 2, since you must raise 10 to the second power to obtain 100. Other examples are

$$\begin{aligned} \log 1000 &= \log (10^3) = 3 \\ \log 10 &= \log (10^1) = 1 \\ \log 1 &= \log (10^0) = 0 \\ \log 1/10 &= \log (10^{-1}) = -1 \\ \log 1/10000 &= \log (10^{-4}) = -4 \end{aligned}$$

To obtain the common logarithm of a number other than a simple power of 10, you must resort to a log table or an electronic calculator. For example,

$$\begin{aligned} \log 2.10 &= 0.3222, \text{ which means that } 10^{0.3222} = 2.10 \\ \log 5.16 &= 0.7126, \text{ which means that } 10^{0.7126} = 5.16 \\ \log 3.125 &= 0.49485, \text{ which means that } 10^{0.49485} = 3.125 \end{aligned}$$

To check this on your calculator, enter the number and then press the "log" key. When using a log table, the logs of the first two numbers above can be read directly from the table. The log of the third number (3.125), however, must be interpolated. That is, 3.125 is midway between 3.12 and 3.13, so the log is midway between 0.4942 and 0.4955.

To obtain the natural logarithm, ln, of the numbers above, use a calculator having this function. Enter each number and press "ln."

$$\begin{aligned} \ln 2.10 &= 0.7419, \text{ which means that } e^{0.7419} = 2.10 \\ \ln 5.16 &= 1.6409, \text{ which means that } e^{1.6409} = 5.16 \end{aligned}$$

To find the common logarithm of a number greater than 10 or less than 1 with a log table, first express the number in scientific notation. Then find the log of each part of the number and add the logs. For example,

$$\begin{aligned} \log 241 &= \log (2.41 \times 10^2) = \log 2.41 + \log 10^2 \\ &= 0.382 + 2 = 2.382 \\ \log 0.00573 &= \log (5.73 \times 10^{-3}) = \log 5.73 + \log 10^{-3} \\ &= 0.758 + (-3) = -2.242 \end{aligned}$$

LOGARITHMS AND NOMENCLATURE: The number to the left of the decimal in a logarithm is called the ***characteristic,*** *and the number to the right of the decimal is the* ***mantissa.***

Significant Figures and Logarithms Notice that the mantissa has as many significant figures as the number whose log was found. (So that you could more clearly see the result obtained with a calculator or a table, this rule was not strictly followed until the last two examples.)

Obtaining Antilogarithms If you are given the logarithm of a number and find the number from it, you have obtained the "antilogarithm" or "antilog" of the number. There are two common procedures used by electronic calculators to do this:

Procedure A	*Procedure B*
(a) enter the log or ln	(a) enter the log or ln
(b) press 2ndF	(b) press INV
(c) press 10^x or e^x	(c) press log or ln x

Test one or the other of these procedures with the following examples:

1. Find the number whose log is 5.234.
   Recall that $\log x = n$ where $x = 10^n$. In this case $n = 5.234$. Enter that number in your calculator and find the value of 10^n, the antilog. In this case,

$$10^{5.234} = 10^{0.234} \times 10^5 = 1.71 \times 10^5$$

   Notice that the characteristic (5) sets the decimal point; it is the power of 10 in the exponential form. The mantissa (0.234) gives the value of the number x. Thus, if you use a log table to find x, you need only look up 0.234 in the table and see that it corresponds to 1.71.

2. Find the number whose log is -3.456.

$$10^{-3.456} = 10^{0.544} \times 10^{-4} = 3.50 \times 10^{-4}$$

   Notice here that -3.456 must be expressed as the sum of -4 and $+0.544$.

Mathematical Operations Using Logarithms Because logarithms are exponents, operations involving them follow the same rules as the use of exponents. Thus, multiplying two numbers can be done by adding logarithms.

$$\log xy = \log x + \log y$$

For example, we multiply 563 by 125 by adding their logarithms and finding the anti-logarithm of the result.

$$\begin{aligned} \log 563 &= 2.751 \\ \log 125 &= \underline{2.097} \\ \log xy &= 4.848 \end{aligned}$$

$$xy = 10^{4.848} = 10^4 \times 10^{0.848} = 7.05 \times 10^4$$

One number (x) can be divided by another (y) by subtraction of their logarithms.

$$\log \frac{x}{y} = \log x - \log y$$

For example, to divide 125 by 742,

$$\begin{aligned} \log 125 &= \quad 2.097 \\ -\log 742 &= \quad \underline{2.870} \\ \log x/y &= -0.773 \end{aligned}$$

$$x/y = 10^{-0.773} = 10^{0.227} \times 10^{-1} = 1.69 \times 10^{-1}$$

Similarly, powers and roots of numbers can be found using logarithms.

$$\log x^y = y(\log x)$$

$$\log \sqrt[y]{x} = \log x^{1/y} = \frac{1}{y} \log x$$

As an example, find the fourth power of 5.23. We first find the log of 5.23 and then multiply it by 4. The result, 2.874, is the log of the answer. Therefore, we find the antilog of 2.874.

$$(5.23)^4 = ?$$
$$\log (5.23)^4 = 4 \log 5.23 = 4\ (0.719) = 2.874$$
$$(5.23)^4 = 10^{2.874} = 748$$

As another example, find the fifth root of 1.89×10^{-9}.

$$\sqrt[5]{1.89 \times 10^{-9}} = (1.89 \times 10^{-9})^{1/5} = ?$$

$$\log (1.89 \times 10^{-9})^{1/5} = \frac{1}{5} \log (1.89 \times 10^{-9}) = \frac{1}{5}(-8.724) = -1.745$$

The answer is the antilog of -1.745.

$$(1.89 \times 10^{-9})^{1/5} = 10^{-1.745} = 1.80 \times 10^{-2}$$

B.4 QUADRATIC EQUATIONS

Algebraic equations of the form $ax^2 + bx + c = 0$ are called **quadratic equations.** The coefficients a, b, and c may be either positive or negative. The two roots of the equation may be found using the *quadratic formula.*

$$x = \frac{-b \pm \sqrt{b^2 - 4ac}}{2a}$$

As an example, solve the equation $5x^2 - 3x - 2 = 0$. Here $a = 5$, $b = -3$, and $c = -2$. Therefore,

$$x = \frac{3 \pm \sqrt{(-3)^2 - 4(5)(-2)}}{2(5)}$$

$$= \frac{3 \pm \sqrt{9 - (-40)}}{10} = \frac{3 \pm \sqrt{49}}{10} = \frac{3 \pm 7}{10}$$

$$x = 1 \text{ and } -0.4$$

How do you know which of the two roots is the correct answer? You have to decide in each case which root has physical significance. However, it is *usually* true in this course that negative values are not significant.

When you have solved a quadratic expression, you should always check your values by substitution into the original equation. In the example above, we find that $5(1)^2 - 3(1) - 2 = 0$ and that $5(-0.4)^2 - 3(-0.4) - 2 = 0$.

The most likely place you will encounter quadratic equations is in the chapters on chemical equilibria, particularly in Chapters 15 and 16. Here you may be faced with solving an equation such as

$$1.8 \times 10^{-4} = \frac{x^2}{0.0010 - x}$$

This equation can certainly be solved by using the quadratic equation (to give $x = 3.4 \times 10^{-4}$). However, you may find the *method of successive approximations* to be especially convenient. Here you begin by making a reasonable approximation of x. This approximate value is substituted into the original equation, and this is solved to give what is hoped to be a more correct value of x. This process is repeated until the answer converges on a particular value of x, that is, until the value of x derived from two successive approximations is the same.

Step 1 First assume that x is so small that $(0.0010 - x) \approx 0.0010$. This means that

$$x^2 = 1.8 \times 10^{-4}(0.0010)$$
$$x = 4.2 \times 10^{-4} \text{ (to 2 significant figures)}$$

Step 2 Substitute the value of x from Step 1 into the denominator (but *not* the numerator) of the original equation and again solve for x.

$$x^2 = (1.8 \times 10^{-4})(0.0010 - 0.00042)$$
$$x = 3.2 \times 10^{-4}$$

Step 3 Repeat Step 2 using the value of x found in that step.

$$x = \sqrt{1.8 \times 10^{-4}(0.0010 - 0.00032)} = 3.5 \times 10^{-4}$$

Step 4 Continue by repeating the calculation, using the value of x found in the previous step.

$$x = \sqrt{1.8 \times 10^{-4}(0.0010 - 0.00035)} = 3.4 \times 10^{-4}$$

Step 5 $x = \sqrt{1.8 \times 10^{-4}(0.0010 - 0.00034)} = 3.4 \times 10^{-4}$

Here we find that iterations after the fourth step give the same value for x, indicating that we have arrived at a valid answer (and the same one obtained from the quadratic formula).

There are several final thoughts on using the method of successive approximations. First, there are cases where the method does not work. Successive steps may give answers that are random or that diverge from the correct value. For quadratic equations of the form $K = x^2/(C - x)$, the method of approximations will work as long as $K < 4C$ (assuming one begins with $x = 0$ as the first guess, that is, $K \approx x^2/C$). This will always be true for weak acids and bases.

Second, values of K in the equation $K = x^2/(C - x)$ are usually known only to two significant figures. Therefore, we are justified in carrying out successive steps until two answers are the same to two significant figures.

Finally, we highly recommend this method of solving quadratic equations. If your calculator has a memory function, successive approximations can be carried out easily and rapidly.

APPENDIX

C

Common Units, Equivalences, and Conversion Factors

C.1 FUNDAMENTAL UNITS OF THE SI SYSTEM

The metric system was begun by the French National Assembly in 1790 and has undergone many modifications. The International System of Units or *Système International* (SI), which represents an extension of the metric system, was adopted by the 11th General Conference of Weights and Measures in 1960. It is constructed from seven base units, each of which represents a particular physical quantity (Table C.1).

The first five units listed in Table C.1 are particularly useful in general chemistry. They are defined as follows.

1. The *meter* is the length of the path travelled by light in vacuum during a time interval of 1/299792458 of a second.
2. The *kilogram* represents the mass of a platinum-iridium block kept at the International Bureau of Weights and Measures at Sèvres, France.
3. The *second* was redefined in 1967 as the duration of 9,192,631,770 periods of a certain line in the microwave spectrum of cesium-133.

Table C.1

SI Fundamental Units

Physical Quantity	Name of Unit	Symbol
length	meter	m
mass	kilogram	kg
time	second	s
temperature	kelvin	K
amount of substance	mole	mol
electric current	ampere	A
luminous intensity	candela	cd

4. The *kelvin* is 1/273.16 of the temperature interval between absolute zero and the triple point of water.
5. The *mole* is the amount of substance that contains as many entities as there are atoms in exactly 0.012 kg of carbon-12 (12 g of ^{12}C atoms).

C.2
PREFIXES USED WITH TRADITIONAL METRIC UNITS AND SI UNITS

Decimal fractions and multiples of metric and SI units are designated by using the prefixes listed in Table C.2. Those most commonly used in general chemistry are in italics.

C.3
DERIVED SI UNITS

In the International System of Units, all physical quantities are represented by appropriate combinations of the base units listed in Table C.1. A list of the derived units frequently used in general chemistry is given in Table C.3.

Table C.2

Traditional Metric and SI Prefixes

Factor	Prefix	Symbol	Factor	Prefix	Symbol
10^{12}	tera	T	10^{-1}	*deci*	d
10^{9}	giga	G	10^{-2}	*centi*	c
10^{6}	mega	M	10^{-3}	*milli*	m
10^{3}	*kilo*	k	10^{-6}	micro	μ
10^{2}	hecto	h	10^{-9}	*nano*	n
10^{1}	deka	da	10^{-12}	*pico*	p
			10^{-15}	femto	f
			10^{-18}	atto	a

Table C.3

Derived SI Units

Physical Quantity	Name of Unit	Symbol	Definition
area	square meter	m^2	
volume	cubic meter	m^3	
density	kilogram per cubic meter	kg/m^3	
force	newton	N	$kg \cdot m/s^2$
pressure	pascal	Pa	N/m^2
energy	joule	J	$kg \cdot m^2/s^2$
electric charge	coulomb	C	$A \cdot s$
electric potential difference	volt	V	$J/(A \cdot s)$

Common Units of Mass and Weight

1 pound = 453.59 grams = 0.45359 kilogram

1 kilogram = 1000 grams = 2.205 pounds

1 gram = 10 decigrams = 100 centigrams = 1000 milligrams

1 gram = 6.022×10^{23} atomic mass units

1 atomic mass unit = 1.6605×10^{-24} gram

1 short ton = 2000 pounds = 907.2 kilograms

1 long ton = 2240 pounds

1 metric tonne = 1000 kilograms = 2205 pounds

Common Units of Length

1 inch = 2.54 centimeters (exactly)

1 mile = 5280 feet = 1.609 kilometers

1 yard = 36 inches = 0.9144 meter

1 meter = 100 centimeters = 39.37 inches = 3.281 feet = 1.094 yards

1 kilometer = 1000 meters = 1094 yards = 0.6215 mile

1 Ångstrom = 1×10^{-8} centimeter = 0.1 nanometer = 100 picometers
= 1×10^{-10} meter = 3.937×10^{-9} inch

Table C.3

Common Units of Volume

1 quart = 0.9463 liter

1 liter = 1.0567 quarts

1 liter = 1 cubic decimeter = 1000 cubic centimeters = 0.001 cubic meter

1 milliliter = 1 cubic centimeter = 0.001 liter = 1.056×10^{-3} quart

1 cubic foot = 28.316 liters = 29.924 quarts = 7.481 gallons

Common Units of Force and Pressure*

1 atmosphere = 760 millimeters of mercury (exactly) = 1.013×10^{5} pascals

= 14.70 pounds per square inch

1 bar = 10^{5} pascals

1 torr = 1 millimeter of mercury

1 pascal = 1 kg/m s^2 = 1 N/m^2

*Force: 1 newton (N) = 1 kg · m/s^2, i.e., the force that when applied for 1 second gives a 1 kilogram mass a velocity of 1 meter per second.

Common Units of Energy

1 joule = 1×10^{7} ergs

1 thermochemical calorie* = 4.184 joules (exactly) = 4.184×10^{7} ergs

= 4.129×10^{-2} liter-atmospheres

= 2.612×10^{19} electron volts

1 erg = 1×10^{-7} joule = 2.3901×10^{-8} calorie

1 electron volt = 1.6022×10^{-19} joule = 1.6022×10^{-12} erg = 96.485 kJ/mol†

1 liter-atmosphere = 24.217 calories = 101.32 joules = 1.0132×10^{9} ergs

1 British thermal unit = 1055.06 joules = 1.05506×10^{10} ergs = 252.2 calories

*The amount of heat required to raise the temperature of one gram of water from 14.5 °C to 15.5 °C.
†Note that the other units in this line are per particle and must be multiplied by 6.022×10^{23} to be strictly comparable.

APPENDIX

D

Physical Constants

Quantity	Symbol	Traditional Units	SI Units
acceleration of gravity	g	980.7 cm/s^2	9.806 m/s^2
atomic mass unit (1/12 the mass of ^{12}C atom)	amu or u	1.6606×10^{-24} g	1.6606×10^{-27} kg
Avogadro's number	N	6.0221367×10^{23} particles/mol	6.0221367×10^{23} particles/mol
Bohr radius	a_0	0.52918 Å	5.2918×10^{-11} m
		5.2918×10^{-9} cm	
Boltzmann constant	k	1.3807×10^{-16} erg/K	1.3807×10^{-23} J/K
charge-to-mass ratio of electron	e/m	1.7588×10^{8} coulomb/g	1.7588×10^{11} C/kg
electronic charge	e	1.6022×10^{-19} coulomb	1.6022×10^{-19} C
		4.8033×10^{-10} esu	
electron rest mass	m_e	9.1094×10^{-28} g	9.1094×10^{-31} kg
		0.00054858 amu	
Faraday constant	F	96,485 coulombs/mol e$^-$	96,485 C/mol e$^-$
		23.06 kcal/volt mol e$^-$	96,485 J/V mol e$^-$
gas constant	R	$0.08206 \dfrac{\text{L} \cdot \text{atm}}{\text{mol} \cdot \text{K}}$	$8.3145 \dfrac{\text{Pa} \cdot \text{dm}^3}{\text{mol} \cdot \text{K}}$
		$1.987 \dfrac{\text{cal}}{\text{mol} \cdot \text{K}}$	8.3145 J/mol · K
molar volume (STP)	V_m	22.414 L/mol	22.414×10^{-3} m^3/mol
			22.414 dm^3/mol
neutron rest mass	m_n	1.67493×10^{-24} g	1.67493×10^{-27} kg
		1.008665 amu	
Planck's constant	h	6.6261×10^{-27} erg · s	$6.6260755 \times 10^{-34}$ J · s
proton rest mass	m_p	1.6726×10^{-24} g	1.6726×10^{-27} kg
		1.007276 amu	
Rydberg constant	R_∞	3.289×10^{15} cycles/s	1.0974×10^{7} m^{-1}
		2.1799×10^{-11} erg	2.1799×10^{-18} J
velocity of light (in a vacuum)	c	2.9979×10^{10} cm/s (186,282 miles/second)	2.9979×10^{8} m/s

$\pi = 3.1416$
e = 2.7183
$\ln X = 2.303 \log X$

$2.303\,R$ = 4.576 cal/mol · K = 19.15 J/mol · K
$2.303\,RT$ (at 25 °C) = 1364 cal/mol = 5709 J/mol

Naming Simple Organic Compounds and Coordination Compounds

The systematic nomenclature for organic compounds was proposed by the International Union of Pure and Applied Chemistry (IUPAC). The IUPAC set of rules provides different names for the more than 10 million known organic compounds, and allows names to be assigned to new compounds as they are synthesized. Many organic compounds also have *common* names. Usually the common name came first and is widely known. Many consumer products are labeled with the common name, and when only a few isomers are possible, the common name adequately identifies the product for the consumer. However, as illustrated in Section 11.4, a system of common names quickly fails when several structural isomers are possible.

E.1 HYDROCARBONS

The name of each member of the hydrocarbon classes has two parts. The first part, the prefix (*meth-*, *eth-*, *prop-*, *but-*, and so on) reflects the number of carbon atoms. When more than four carbons are present, the Greek or Latin number prefixes are used: *pent-*, *hex-*, *hept-*, *oct-*, *non-*, and *dec-*. The second part of the name, or the suffix, tells the class of hydrocarbon. Alkanes have carbon-carbon single bonds, alkenes have carbon-carbon double bonds, and alkynes have carbon-carbon triple bonds.

Unbranched Alkanes and Alkyl Groups

The names of the first twenty unbranched (straight-chain) alkanes are given in Table E.1.

Alkyl groups are named by dropping "-ane" from the parent alkane and adding "-yl" (see Table 11.5 for examples).

Branched-Chain Alkanes

The rules for naming branched-chain alkanes are as follows:

1. *Find the longest continuous chain of carbon atoms: this chain determines the parent name for the compound.* For example, the following compound has two methyl groups attached to a *heptane* parent.

Table E.1

Names of Unbranched Alkanes			
CH_4	methane	$C_{11}H_{24}$	undecane
C_2H_6	ethane	$C_{12}H_{26}$	dodecane
C_3H_8	propane	$C_{13}H_{28}$	tridecane
C_4H_{10}	butane	$C_{14}H_{30}$	tetradecane
C_5H_{12}	pentane	$C_{15}H_{32}$	pentadecane
C_6H_{14}	hexane	$C_{16}H_{34}$	hexadecane
C_7H_{16}	heptane	$C_{17}H_{36}$	heptadecane
C_8H_{18}	octane	$C_{18}H_{38}$	octadecane
C_9H_{20}	nonane	$C_{19}H_{40}$	nonadecane
$C_{10}H_{22}$	decane	$C_{20}H_{42}$	eicosane

$$CH_3CH_2CH_2\underset{\substack{|\\CH_3}}{C}HCH_2\underset{\substack{|\\CH_3}}{C}HCH_3$$

The longest continuous chain may not be obvious from the way the formula is written, especially for the straight-line format that is commonly used. For example, the longest continuous chain of carbon atoms in the following chain is *eight*, not *four* or *six*.

```
   |
  —C—
   |
  —C—
   |   |   |   |
  —C—C—C—C—   is equivalent to  —C   C   C   C
   |   |   |   |                  \ / \ / \ / \
          —C—                      C   C   C   C—
           |
          —C—
           |
```

2. *Number the longest chain beginning with the end of the chain nearest the branching. Use these numbers to designate the location of the attached group. When two or more groups are attached to the parent, give each group a number corresponding to its location on the parent chain.* For example, the name of

$$\overset{7}{C}H_3\overset{6}{C}H_2\overset{5}{C}H_2\underset{\substack{|\\CH_3}}{\overset{4}{C}}H\overset{3}{C}H_2\underset{\substack{|\\CH_3}}{\overset{2}{C}}H\overset{1}{C}H_3$$

is 2,4-dimethylheptane. The name of the compound below is 3-methylheptane, not 5-methylheptane or 2-ethylhexane.

$$\overset{7}{CH_3}-\overset{6}{CH_2}-\overset{5}{CH_2}-\overset{4}{CH_2}-\overset{3}{CH}-CH_3$$
$$|$$
$${}^{2}CH_2$$
$$|$$
$${}^{1}CH_3$$

3-methylheptane

3. *When two or more substituents are identical, indicate this by the use of the prefixes di-, tri-, tetra-, and so on. Positional numbers of the substituents should have the smallest possible sum.*

$$\begin{array}{c} CH_3 \quad CH_3 \\ 1 \quad 2 \quad 3| \; 4 \quad 5| \quad 6 \quad 7 \quad 8 \\ CH_3CH_2CCH_2CHCHCH_2CH_3 \\ | \qquad\quad | \\ CH_3 \qquad CH_3 \end{array}$$

The correct name of this compound is 3,3,5,6-tetramethyloctane.

4. *If there are two or more different groups, the groups are listed alphabetically.*

$$\begin{array}{c} CH_3 \\ 1 \quad 2| \; 3 \quad 4 \quad 5 \quad 6 \\ CH_3CCH_2CHCH_2CH_3 \\ | \qquad | \\ CH_3 \quad CH_2 \\ | \\ CH_3 \end{array}$$

The correct name of this compound is 4-ethyl-2,2-dimethylhexane. Note that the prefix "di" is ignored in determining alphabetical order.

Alkenes

Alkenes are named by using the prefix to indicate the number of carbon atoms and the suffix *-ene* to indicate one or more double bonds. The systematic names for the first two members of the alkene series are *ethene* and *propene.*

$$CH_2{=}CH_2 \qquad CH_3CH{=}CH_2$$

When groups, such as methyl or ethyl, are attached to carbon atoms in an alkene, the longest hydrocarbon chain is numbered from the end that will give the double bond the lowest number and then numbers are assigned to the attached groups. For example, the name of

$$\begin{array}{c} CH_3 \\ 5 \quad 4| \quad 3 \qquad 2 \quad 1 \\ CH_3CHCH{=}CHCH_3 \end{array}$$

is 4-methyl-2-pentene. See Section 11.5 for a discussion of *cis-trans* isomers of alkenes.

Alkynes

The naming of alkynes is similar to that of alkenes, with the lowest number possible being used to locate the triple bond. For example, the name of

$$\overset{1}{C}H_3\overset{2}{C}\equiv\overset{3}{C}\overset{\overset{\displaystyle CH_3}{|}}{\overset{4}{C}}H\overset{5}{C}H_3$$

is 4-methyl-2-pentyne.

Benzene Derivatives

Monosubstituted benzene derivatives are named by using a prefix for the substituent. Some examples are

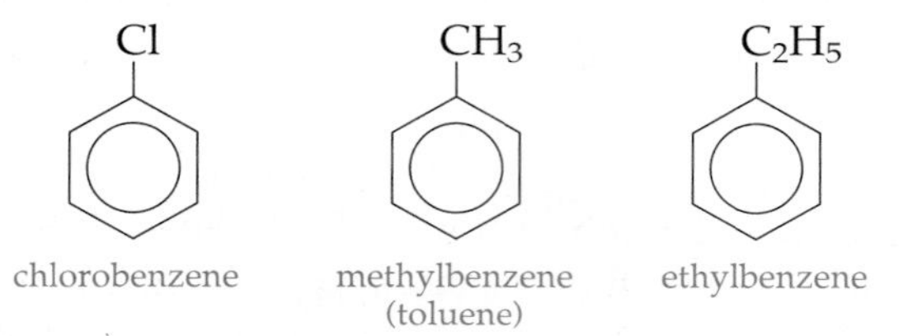

chlorobenzene   methylbenzene (toluene)   ethylbenzene

Three isomers are possible when two groups are substituted for hydrogen atoms on the benzene ring. The relative positions of the substituents are indicated either by the prefixes *ortho-*, *meta-*, and *para-* (abbreviated *o-*, *m-*, *p-*) or by numbers. For example,

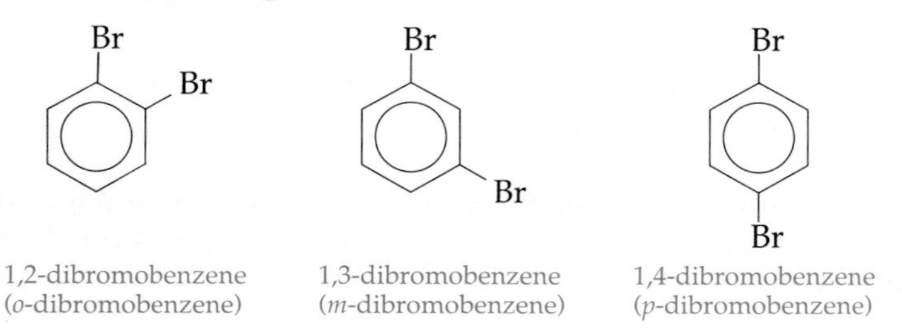

1,2-dibromobenzene (*o*-dibromobenzene)   1,3-dibromobenzene (*m*-dibromobenzene)   1,4-dibromobenzene (*p*-dibromobenzene)

The dimethylbenzenes are called *xylenes.*

If more than two groups are attached to the benzene ring, numbers must be used to identify the positions. The benzene ring is numbered to give the lowest possible numbers to the substituents.

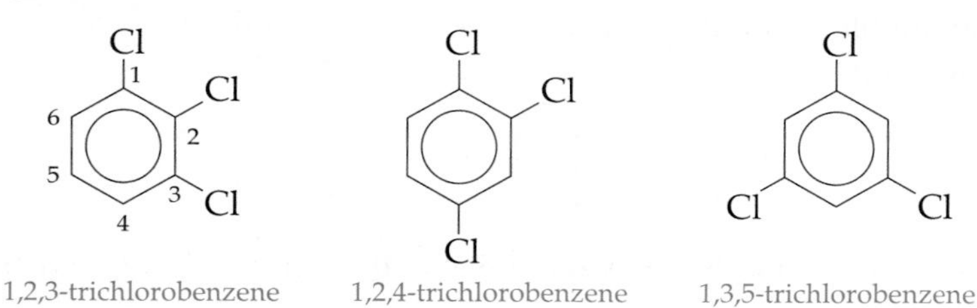

1,2,3-trichlorobenzene   1,2,4-trichlorobenzene   1,3,5-trichlorobenzene

E.2
FUNCTIONAL GROUPS

An atom or group of atoms that defines the structure of a specific class of organic compounds and determines their properties is called a **functional group.** The millions of organic compounds include classes of compounds that are obtained by replacing hydrogen atoms of hydrocarbons with functional groups (Chapter 13). The important functional groups are shown in Table E.2.

The "R" attached to the functional group represents the hydrocarbon framework with one hydrogen removed for each functional group added. The IUPAC system provides a systematic method for naming all members of a given class. For example, alcohols end in *-ol* (methan*ol*); aldehydes end in *-al* (methan*al*); carboxylic acids end in *-oic* (ethan*oic* acid); and ketones end in *-one* (propan*one*).

Alcohols

Isomers are also possible for molecules containing functional groups. For example, three different alcohols are obtained when a hydrogen atom in pentane is replaced by —OH, depending on which hydrogen atom is replaced. The rules for naming the "R" or hydrocarbon framework are the same as those for hydrocarbon compounds.

$CH_3CH_2CH_2CH_2CH_2OH$ 1-pentanol

$CH_3CH_2CH_2CH(OH)CH_3$ 2-pentanol

$CH_3CH_2CH(OH)CH_2CH_3$ 3-pentanol

Compounds with one or more functional groups and alkyl substituents are named so as to give the functional groups the lowest numbers. For example, the correct name of

$$\overset{1}{C}H_3\overset{2}{C}H(OH)\overset{3}{C}H_2\overset{4}{C}(CH_3)_2\overset{5}{C}H_3$$

is 4,4-dimethyl-2-pentanol.

Aldehydes and Ketones

The systematic names of the first three aldehydes are methanal, ethanal, and propanal.

Table E.2

Classes of Organic Compounds Based on Functional Groups*

General Formulas of Class Members	Class Name	Typical Compound	Compound Name	Common Use of Sample Compound
R—X	halide	$H—CH(Cl)—Cl$ (CH_2Cl_2)	dichloromethane (methylene chloride)	solvent
R—OH	alcohol	$H—CH_2—OH$	methanol (wood alcohol)	solvent
R—C(=O)—H	aldehyde	H—C(=O)—H	methanal (formaldehyde)	preservative
R—C(=O)—OH	carboxylic acid	$H—CH_2—C(=O)—OH$	ethanoic acid (acetic acid)	vinegar
R—C(=O)—R′	ketone	$H—CH_2—C(=O)—CH_2—H$	propanone (acetone)	solvent
R—O—R′	ether	$C_2H_5—O—C_2H_5$	diethyl ether (ethyl ether)	anesthetic
R—C(=O)—O—R′	ester	$CH_3—C(=O)—O—C_2H_5$	ethyl ethanoate (ethyl acetate)	solvent in fingernail polish
R—N(H)—H	amine	$H—CH_2—NH_2$	methylamine	tanning (foul odor)
R—C(=O)—N(H)—R′	amide	$CH_3—C(=O)—NH_2$	acetamide	plasticizer

*R stands for an H or a hydrocarbon group such as $—CH_3$ or $—C_2H_5$. R′ could be a different group from R.

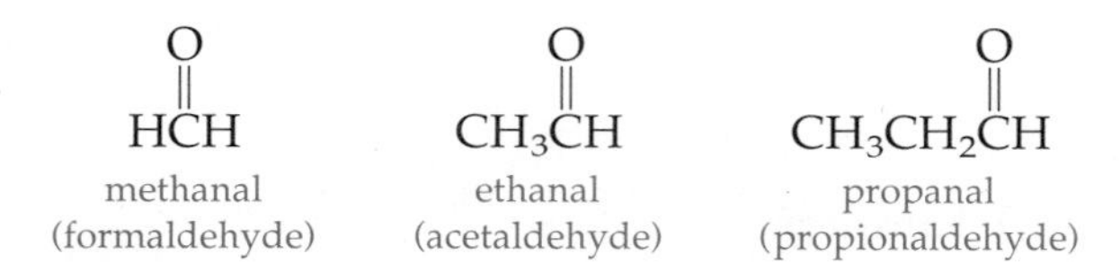

For ketones, a number is used to designate the position of the carbonyl group, and the chain is numbered in a way that gives the carbonyl carbon the smallest number.

$$\underset{\text{2-propanone (acetone)}}{CH_3\overset{O}{\overset{\|}{C}}CH_3} \qquad \underset{\text{2-butanone (methyl ethyl ketone)}}{CH_3CH_2\overset{O}{\overset{\|}{C}}CH_3} \qquad \underset{\text{4-penten-2-one}}{CH_3\overset{O}{\overset{\|}{C}}CH_2CH{=}CH_2}$$

Carboxylic Acids

The systematic names of carboxylic acids are obtained by dropping the final *e* of the name of the corresponding alkane and adding *oic acid*. For example, the name of

$$CH_3CH_2CH_2CH_2CH_2COOH$$

is hexanoic acid. The systematic names of the first five carboxylic acids are given in Table 13.7. Other examples are

$$\underset{\text{2-methylbutanoic acid}}{\overset{4}{C}H_3\overset{3}{C}H_2\overset{2}{\underset{}{C}}(CH_3)H\overset{1}{C}OOH} \qquad \underset{\text{2-butenoic acid}}{\overset{4}{C}H_3\overset{3}{C}H{=}\overset{2}{C}H\overset{1}{C}OOH}$$

Esters

The systematic names of esters are derived from the names of the alcohol and the acid used to prepare the ester. The general formula for esters is

$$R{-}\overset{O}{\overset{\|}{C}}{-}OR'$$

As shown in Section 13.5, the $R{-}\overset{O}{\overset{\|}{C}}$ comes from the acid and the R′O comes from the alcohol. The alcohol part is named first, followed by the name of the acid changed to end in *-ate*. For example,

$$CH_3CH_2\overset{O}{\overset{\|}{C}}{-}OCH_3$$

is named methyl propanoate and

$$CH_3\overset{O}{\overset{\|}{C}}{-}OCH{=}CH_2$$

is named ethenyl ethanoate.

E.3 NAMING COORDINATION COMPOUNDS

Just as there are rules for naming simple inorganic and organic compounds, coordination compounds are named according to an established system. The following compounds are named according to the rules outlined below.

Compound	*Systematic Name*
$[Ni(H_2O)_6]SO_4$	hexaaquanickel(II) sulfate
$[Cr(en)_2(CN)_2]Cl$	dicyanobis(ethylenediamine)chromium(III) chloride
$K[Pt(NH_3)Cl_3]$	potassium amminetrichloroplatinate(II)

As you read through the rules, notice how they apply to these examples.

1. In naming a coordination compound that is an ionic compound, name the cation first and then the anion, as is usually done.
2. When giving the name of the complex ion or molecule, the ligands are named first, in alphabetical order, followed by the name of the metal.
   a. If a ligand is an anion whose name ends in *-ite* or *-ate,* the final *e* is changed to *o* (as in sulfate → sulfato or nitrite → nitrito).
   b. If the ligand is an anion whose name ends in *-ide,* the ending is changed to *o* (as in chloride → chloro or cyanide → cyano).
   c. If the ligand is a neutral molecule, its common name is used. The important exceptions at this point are water, which is called *aqua,* ammonia, which is called *ammine,* and CO, called *carbonyl.*
   d. When there is more than one of a particular monodentate ligand with a simple name, the number of ligands is designated by the appropriate prefix: *di, tri, tetra, penta,* or *hexa.* If the ligand name is complicated (whether monodentate or bidentate), the prefix changes to *bis, tris, tetrakis, pentakis,* or *hexakis,* followed by the ligand name in parentheses.
   e. If the complex ion is an anion, the suffix *-ate* is added to the metal name.
3. Following the name of the metal, the charge (or oxidation number, Section 10.8) of the metal is given in Roman numerals.

Complex ions can become more complicated than those described in Section 10.10, and even more rules of nomenclature must be applied. However, the brief set just outlined is sufficient for the vast majority of complexes.

Ionization Constants for Weak Acids at 25 °C

Acid	Formula and Ionization Equation	K_a
acetic	$CH_3COOH \rightleftharpoons H^+ + CH_3COO^-$	1.8×10^{-5}
arsenic	$H_3AsO_4 \rightleftharpoons H^+ + H_2AsO_4^-$	$K_1 = 2.5 \times 10^{-4}$
	$H_2AsO_4^- \rightleftharpoons H^+ + HAsO_4^{2-}$	$K_2 = 5.6 \times 10^{-8}$
	$HAsO_4^{2-} \rightleftharpoons H^+ + AsO_4^{3-}$	$K_3 = 3.0 \times 10^{-13}$
arsenous	$H_3AsO_3 \rightleftharpoons H^+ + H_2AsO_3^-$	$K_1 = 6.0 \times 10^{-10}$
	$H_2AsO_3^- \rightleftharpoons H^+ + HAsO_3^{2-}$	$K_2 = 3.0 \times 10^{-14}$
benzoic	$C_6H_5COOH \rightleftharpoons H^+ + C_6H_5COO^-$	6.3×10^{-5}
boric	$H_3BO_3 \rightleftharpoons H^+ + H_2BO_3^-$	$K_1 = 7.3 \times 10^{-10}$
	$H_2BO_3^- \rightleftharpoons H^+ + HBO_3^{2-}$	$K_2 = 1.8 \times 10^{-13}$
	$HBO_3^{2-} \rightleftharpoons H^+ + BO_3^{3-}$	$K_3 = 1.6 \times 10^{-14}$
carbonic	$H_2CO_3 \rightleftharpoons H^+ + HCO_3^-$	$K_1 = 4.2 \times 10^{-7}$
	$HCO_3^- \rightleftharpoons H^+ + CO_3^{2-}$	$K_2 = 4.8 \times 10^{-11}$
citric	$H_3C_6H_5O_7 \rightleftharpoons H^+ + H_2C_6H_5O_7^-$	$K_1 = 7.4 \times 10^{-3}$
	$H_2C_6H_5O_7^- \rightleftharpoons H^+ + HC_6H_5O_7^{2-}$	$K_2 = 1.7 \times 10^{-5}$
	$HC_6H_5O_7^{2-} \rightleftharpoons H^+ + C_6H_5O_7^{3-}$	$K_3 = 4.0 \times 10^{-7}$
cyanic	$HOCN \rightleftharpoons H^+ + OCN^-$	3.5×10^{-4}
formic	$HCOOH \rightleftharpoons H^+ + HCOO^-$	1.8×10^{-4}
hydrazoic	$HN_3 \rightleftharpoons H^+ + N_3^-$	1.9×10^{-5}
hydrocyanic	$HCN \rightleftharpoons H^+ + CN^-$	4.0×10^{-10}
hydrofluoric	$HF \rightleftharpoons H^+ + F^-$	7.2×10^{-4}
hydrogen peroxide	$H_2O_2 \rightleftharpoons H^+ + HO_2^-$	2.4×10^{-12}
hydrosulfuric	$H_2S \rightleftharpoons H^+ + HS^-$	$K_1 = 1.0 \times 10^{-7}$
	$HS^- \rightleftharpoons H^+ + S^{2-}$	$K_2 = 1.3 \times 10^{-13}$
hypobromous	$HOBr \rightleftharpoons H^+ + OBr^-$	2.5×10^{-9}
hypochlorous	$HOCl \rightleftharpoons H^+ + OCl^-$	3.5×10^{-8}
nitrous	$HNO_2 \rightleftharpoons H^+ + NO_2^-$	4.5×10^{-4}
oxalic	$H_2C_2O_4 \rightleftharpoons H^+ + HC_2O_4^-$	$K_1 = 5.9 \times 10^{-2}$
	$HC_2O_4^- \rightleftharpoons H^+ + C_2O_4^{2-}$	$K_2 = 6.4 \times 10^{-5}$

Acid	Formula and Ionization Equation	K_a
phenol	$HC_6H_5O \rightleftharpoons H^+ + C_6H_5O^-$	1.3×10^{-10}
phosphoric	$H_3PO_4 \rightleftharpoons H^+ + H_2PO_4^-$	$K_1 = 7.5 \times 10^{-3}$
	$H_2PO_4^- \rightleftharpoons H^+ + HPO_4^{2-}$	$K_2 = 6.2 \times 10^{-8}$
	$HPO_4^{2-} \rightleftharpoons H^+ + PO_4^{3-}$	$K_3 = 3.6 \times 10^{-13}$
phosphorous	$H_3PO_3 \rightleftharpoons H^+ + H_2PO_3^-$	$K_1 = 1.6 \times 10^{-2}$
	$H_2PO_3^- \rightleftharpoons H^+ + HPO_3^{2-}$	$K_2 = 7.0 \times 10^{-7}$
selenic	$H_2SeO_4 \rightleftharpoons H^+ + HSeO_4^-$	K_1 = very large
	$HSeO_4^- \rightleftharpoons H^+ + SeO_4^{2-}$	$K_2 = 1.2 \times 10^{-2}$
selenous	$H_2SeO_3 \rightleftharpoons H^+ + HSeO_3^-$	$K_1 = 2.7 \times 10^{-3}$
	$HSeO_3^- \rightleftharpoons H^+ + SeO_3^{2-}$	$K_2 = 2.5 \times 10^{-7}$
sulfuric	$H_2SO_4 \rightleftharpoons H^+ + HSO_4^-$	K_1 = very large
	$HSO_4^- \rightleftharpoons H^+ + SO_4^{2-}$	$K_2 = 1.2 \times 10^{-2}$
sulfurous	$H_2SO_3 \rightleftharpoons H^+ + HSO_3^-$	$K_1 = 1.7 \times 10^{-2}$
	$HSO_3^- \rightleftharpoons H^+ + SO_3^{2-}$	$K_2 = 6.4 \times 10^{-8}$
tellurous	$H_2TeO_3 \rightleftharpoons H^+ + HTeO_3^-$	$K_1 = 2 \times 10^{-3}$
	$HTeO_3^- \rightleftharpoons H^+ + TeO_3^{2-}$	$K_2 = 1 \times 10^{-8}$

Ionization Constants for Weak Bases at 25 °C

Base	Formula and Ionization Equation	K_b
ammonia	$NH_3 + H_2O \rightleftharpoons NH_4^+ + OH^-$	1.8×10^{-5}
aniline	$C_6H_5NH_2 + H_2O \rightleftharpoons C_6H_5NH_3^+ + OH^-$	4.2×10^{-10}
dimethylamine	$(CH_3)_2NH + H_2O \rightleftharpoons (CH_3)_2NH_2^+ + OH^-$	7.4×10^{-4}
ethylenediamine	$(CH_2)_2(NH_2)_2 + H_2O \rightleftharpoons (CH_2)_2(NH_2)_2H^+ + OH^-$	$K_1 = 8.5 \times 10^{-5}$
	$(CH_2)_2(NH_2)_2H^+ + H_2O \rightleftharpoons (CH_2)_2(NH_2)_2H_2^{2+} + OH^-$	$K_2 = 2.7 \times 10^{-8}$
hydrazine	$N_2H_4 + H_2O \rightleftharpoons N_2H_5^+ + OH^-$	$K_1 = 8.5 \times 10^{-7}$
	$N_2H_5^+ + H_2O \rightleftharpoons N_2H_6^{2+} + OH^-$	$K_2 = 8.9 \times 10^{-16}$
hydroxylamine	$NH_2OH + H_2O \rightleftharpoons NH_3OH^+ + OH^-$	6.6×10^{-9}
methylamine	$CH_3NH_2 + H_2O \rightleftharpoons CH_3NH_3^+ + OH^-$	5.0×10^{-4}
pyridine	$C_5H_5N + H_2O \rightleftharpoons C_5H_5NH^+ + OH^-$	1.5×10^{-9}
trimethylamine	$(CH_3)_3N + H_2O \rightleftharpoons (CH_3)_3NH^+ + OH^-$	7.4×10^{-5}

APPENDIX

H

Solubility Product Constants for Some Inorganic Compounds at 25 °C

Substance	K_{sp}	Substance	K_{sp}
Aluminum compounds		Calcium compounds	
$AlAsO_4$	1.6×10^{-16}	$Ca_3(AsO_4)_2$	6.8×10^{-19}
$Al(OH)_3$	1.9×10^{-33}	$CaCO_3$	3.8×10^{-9}
$AlPO_4$	1.3×10^{-20}	$CaCrO_4$	7.1×10^{-4}
Antimony compounds		$CaC_2O_4 \cdot H_2O$*	2.3×10^{-9}
Sb_2S_3	1.6×10^{-93}	CaF_2	3.9×10^{-11}
Barium compounds		$Ca(OH)_2$	7.9×10^{-6}
$Ba_3(AsO_4)_2$	1.1×10^{-13}	$CaHPO_4$	2.7×10^{-7}
$BaCO_3$	8.1×10^{-9}	$Ca(H_2PO_4)_2$	1.0×10^{-3}
$BaC_2O_4 \cdot 2H_2O$*	1.1×10^{-7}	$Ca_3(PO_4)_2$	1.0×10^{-25}
$BaCrO_4$	2.0×10^{-10}	$CaSO_3 \cdot 2H_2O$*	1.3×10^{-8}
BaF_2	1.7×10^{-6}	$CaSO_4 \cdot 2H_2O$*	2.4×10^{-5}
$Ba(OH)_2 \cdot 8H_2O$*	5.0×10^{-3}	Chromium compounds	
$Ba_3(PO_4)_2$	1.3×10^{-29}	$CrAsO_4$	7.8×10^{-21}
$BaSeO_4$	2.8×10^{-11}	$Cr(OH)_3$	6.7×10^{-31}
$BaSO_3$	8.0×10^{-7}	$CrPO_4$	2.4×10^{-23}
$BaSO_4$	1.1×10^{-10}	Cobalt compounds	
Bismuth compounds		$Co_3(AsO_4)_2$	7.6×10^{-29}
$BiOCl$	7.0×10^{-9}	$CoCO_3$	8.0×10^{-13}
$BiO(OH)$	1.0×10^{-12}	$Co(OH)_2$	2.5×10^{-16}
$Bi(OH)_3$	3.2×10^{-40}	CoS (α)	5.9×10^{-21}
BiI_3	8.1×10^{-19}	$Co(OH)_3$	4.0×10^{-45}
$BiPO_4$	1.3×10^{-23}	Co_2S_3	2.6×10^{-124}
Bi_2S_3	1.6×10^{-72}	Copper compounds	
Cadmium compounds		$CuBr$	5.3×10^{-9}
$Cd_3(AsO_4)_2$	2.2×10^{-32}	$CuCl$	1.9×10^{-7}
$CdCO_3$	2.5×10^{-14}	$CuCN$	3.2×10^{-20}
$Cd(CN)_2$	1.0×10^{-8}	Cu_2O ($Cu^+ + OH^-$)†	1.0×10^{-14}
$Cd_2[Fe(CN)_6]$	3.2×10^{-17}	CuI	5.1×10^{-12}
$Cd(OH)_2$	1.2×10^{-14}	Cu_2S	1.6×10^{-48}
CdS	3.6×10^{-29}	$CuSCN$	1.6×10^{-11}

Substance	K_{sp}
Copper compounds *(continued)*	
$Cu_3(AsO_4)_2$	7.6×10^{-36}
$CuCO_3$	2.5×10^{-10}
$Cu_2[Fe(CN)_6]$	1.3×10^{-16}
$Cu(OH)_2$	1.6×10^{-19}
CuS	8.7×10^{-36}
Gold compounds	
$AuBr$	5.0×10^{-17}
$AuCl$	2.0×10^{-13}
AuI	1.6×10^{-23}
$AuBr_3$	4.0×10^{-36}
$AuCl_3$	3.2×10^{-25}
$Au(OH)_3$	1×10^{-53}
AuI_3	1.0×10^{-46}
Iron compounds	
$FeCO_3$	3.5×10^{-11}
$Fe(OH)_2$	7.9×10^{-15}
FeS	4.9×10^{-18}
$Fe_4[Fe(CN)_6]_3$	3.0×10^{-41}
$Fe(OH)_3$	6.3×10^{-38}
Fe_2S_3	1.4×10^{-88}
Lead compounds	
$Pb_3(AsO_4)_2$	4.1×10^{-36}
$PbBr_2$	6.3×10^{-6}
$PbCO_3$	1.5×10^{-13}
$PbCl_2$	1.7×10^{-5}
$PbCrO_4$	1.8×10^{-14}
PbF_2	3.7×10^{-8}
$Pb(OH)_2$	2.8×10^{-16}
PbI_2	8.7×10^{-9}
$Pb_3(PO_4)_2$	3.0×10^{-44}
$PbSeO_4$	1.5×10^{-7}
$PbSO_4$	1.8×10^{-8}
PbS	8.4×10^{-28}
Magnesium compounds	
$Mg_3(AsO_4)_2$	2.1×10^{-20}
$MgCO_3 \cdot 3H_2O$*	4.0×10^{-5}
MgC_2O_4	8.6×10^{-5}
MgF_2	6.4×10^{-9}
$Mg(OH)_2$	1.5×10^{-11}
$MgNH_4PO_4$	2.5×10^{-12}
Manganese compounds	
$Mn_3(AsO_4)_2$	1.9×10^{-11}
$MnCO_3$	1.8×10^{-11}
$Mn(OH)_2$	4.6×10^{-14}
MnS	5.1×10^{-15}
$Mn(OH)_3$	$\sim 1 \times 10^{-36}$

Substance	K_{sp}
Mercury compounds	
Hg_2Br_2	1.3×10^{-22}
Hg_2CO_3	8.9×10^{-17}
Hg_2Cl_2	1.1×10^{-18}
Hg_2CrO_4	5.0×10^{-9}
Hg_2I_2	4.5×10^{-29}
$Hg_2O \cdot H_2O(Hg_2^{2+} + 2OH^-)$*†	1.6×10^{-23}
Hg_2SO_4	6.8×10^{-7}
Hg_2S	5.8×10^{-44}
$Hg(CN)_2$	3.0×10^{-23}
$Hg(OH)_2$	2.5×10^{-26}
HgI_2	4.0×10^{-29}
HgS	3.0×10^{-53}
Nickel compounds	
$Ni_3(AsO_4)_2$	1.9×10^{-26}
$NiCO_3$	6.6×10^{-9}
$Ni(CN)_2$	3.0×10^{-23}
$Ni(OH)_2$	2.8×10^{-16}
NiS (α)	3.0×10^{-21}
NiS (β)	1.0×10^{-26}
NiS (γ)	2.0×10^{-28}
Silver compounds	
Ag_3AsO_4	1.1×10^{-20}
$AgBr$	3.3×10^{-13}
Ag_2CO_3	8.1×10^{-12}
$AgCl$	1.8×10^{-10}
Ag_2CrO_4	9.0×10^{-12}
$AgCN$	1.2×10^{-16}
$Ag_4[Fe(CN)_6]$	1.6×10^{-41}
Ag_2O ($Ag^+ + OH^-$)†	2.0×10^{-8}
AgI	1.5×10^{-16}
Ag_3PO_4	1.3×10^{-20}
Ag_2SO_3	1.5×10^{-14}
Ag_2SO_4	1.7×10^{-5}
Ag_2S	1.0×10^{-49}
$AgSCN$	1.0×10^{-12}
Strontium compounds	
$Sr_3(AsO_4)_2$	1.3×10^{-18}
$SrCO_3$	9.4×10^{-10}
$SrC_2O_4 \cdot 2H_2O$*	5.6×10^{-8}
$SrCrO_4$	3.6×10^{-5}
$Sr(OH)_2 \cdot 8H_2O$*	3.2×10^{-4}
$Sr_3(PO_4)_2$	1.0×10^{-31}
$SrSO_3$	4.0×10^{-8}
$SrSO_4$	2.8×10^{-7}
Tin compounds	
$Sn(OH)_2$	2.0×10^{-26}

Substance	K_{sp}	Substance	K_{sp}
Tin compounds *(continued)*		Zinc compounds	
SnI_2	1.0×10^{-4}	$ZnCO_3$	1.5×10^{-11}
SnS	1.0×10^{-28}	$Zn(CN)_2$	8.0×10^{-12}
$Sn(OH)_4$	1×10^{-57}	$Zn_3[Fe(CN)_6]$	4.1×10^{-16}
SnS_2	1×10^{-70}	$Zn(OH)_2$	4.5×10^{-17}
Zinc compounds		$Zn_3(PO_4)_2$	9.1×10^{-33}
$Zn_3(AsO_4)_2$	1.1×10^{-27}	ZnS	1.1×10^{-21}

*Since $[H_2O]$ does not appear in equilibrium constants for equilibria in aqueous solution in general, it does *not* appear in the K_{sp} expressions for hydrated solids.

†Very small amounts of oxides dissolve in water to give the ions indicated in parentheses. Solid hydroxides are unstable and decompose to oxides as rapidly as they are formed.

Standard Reduction Potentials in Aqueous Solution at 25 °C

Acidic Solution	Standard Reduction Potential, E^0 (volts)
$F_2(g) + 2\,e^- \longrightarrow 2\,F^-(aq)$	2.87
$Co^{3+}(aq) + e^- \longrightarrow Co^{2+}(aq)$	1.82
$Pb^{4+}(aq) + 2\,e^- \longrightarrow Pb^{2+}(aq)$	1.8
$H_2O_2(aq) + 2\,H^+(aq) + 2\,e^- \longrightarrow 2\,H_2O$	1.77
$NiO_2(s) + 4\,H^+(aq) + 2\,e^- \longrightarrow Ni^{2+}(aq) + 2\,H_2O$	1.7
$PbO_2(s) + SO_4{}^{2-}(aq) + 4\,H^+(aq) + 2\,e^- \longrightarrow PbSO_4(s) + 2\,H_2O$	1.685
$Au^+(aq) + e^- \longrightarrow Au(s)$	1.68
$2\,HClO(aq) + 2\,H^+(aq) + 2\,e^- \longrightarrow Cl_2(g) + 2\,H_2O$	1.63
$Ce^{4+}(aq) + e^- \longrightarrow Ce^{3+}(aq)$	1.61
$NaBiO_3(s) + 6\,H^+(aq) + 2\,e^- \longrightarrow Bi^{3+}(aq) + Na^+(aq) + 3\,H_2O$	~1.6
$MnO_4{}^-(aq) + 8\,H^+(aq) + 5\,e^- \longrightarrow Mn^{2+}(aq) + 4\,H_2O$	1.51
$Au^{3+}(aq) + 3\,e^- \longrightarrow Au(s)$	1.50
$ClO_3{}^-(aq) + 6\,H^+(aq) + 5\,e^- \longrightarrow \frac{1}{2}\,Cl_2(g) + 3\,H_2O$	1.47
$BrO_3{}^-(aq) + 6\,H^+(aq) + 6\,e^- \longrightarrow Br^-(aq) + 3\,H_2O$	1.44
$Cl_2(g) + 2\,e^- \longrightarrow 2\,Cl^-(aq)$	1.358
$Cr_2O_7{}^{2-}(aq) + 14\,H^+(aq) + 6\,e^- \longrightarrow 2\,Cr^{3+}(aq) + 7\,H_2O$	1.33
$N_2H_5{}^+(aq) + 3\,H^+(aq) + 2\,e^- \longrightarrow 2\,NH_4{}^+(aq)$	1.24
$MnO_2(s) + 4\,H^+(aq) + 2\,e^- \longrightarrow Mn^{2+}(aq) + 2\,H_2O$	1.23
$O_2(g) + 4\,H^+(aq) + 4\,e^- \longrightarrow 2\,H_2O$	1.229
$Pt^{2+}(aq) + 2\,e^- \longrightarrow Pt(s)$	1.2
$IO_3{}^-(aq) + 6\,H^+(aq) + 5\,e^- \longrightarrow \frac{1}{2}\,I_2(aq) + 3\,H_2O$	1.195
$ClO_4{}^-(aq) + 2\,H^+(aq) + 2\,e^- \longrightarrow ClO_3{}^-(aq) + H_2O$	1.19
$Br_2(\ell) + 2\,e^- \longrightarrow 2\,Br^-(aq)$	1.066
$AuCl_4{}^-(aq) + 3\,e^- \longrightarrow Au(s) + 4\,Cl^-(aq)$	1.00
$Pd^{2+}(aq) + 2\,e^- \longrightarrow Pd(s)$	0.987
$NO_3{}^-(aq) + 4\,H^+(aq) + 3\,e^- \longrightarrow NO(g) + 2\,H_2O$	0.96
$NO_3{}^-(aq) + 3\,H^+(aq) + 2\,e^- \longrightarrow HNO_2(aq) + H_2O$	0.94

Acidic Solution	Standard Reduction Potential, E^0 (volts)
$2\,Hg^{2+}(aq) + 2\,e^- \longrightarrow Hg_2^{2+}(aq)$	0.920
$Hg^{2+}(aq) + 2\,e^- \longrightarrow Hg(\ell)$	0.855
$Ag^+(aq) + e^- \longrightarrow Ag(s)$	0.7994
$Hg_2^{2+}(aq) + 2\,e^- \longrightarrow 2\,Hg(\ell)$	0.789
$Fe^{3+}(aq) + e^- \longrightarrow Fe^{2+}(aq)$	0.771
$SbCl_6^-(aq) + 2\,e^- \longrightarrow SbCl_4^-(aq) + 2\,Cl^-(aq)$	0.75
$[PtCl_4]^{2-}(aq) + 2\,e^- \longrightarrow Pt(s) + 4\,Cl^-(aq)$	0.73
$O_2(g) + 2\,H^+(aq) + 2\,e^- \longrightarrow H_2O_2(aq)$	0.682
$[PtCl_6]^{2-}(aq) + 2\,e^- \longrightarrow [PtCl_4]^{2-}(aq) + 2\,Cl^-(aq)$	0.68
$H_3AsO_4(aq) + 2\,H^+(aq) + 2\,e^- \longrightarrow H_3AsO_3(aq) + H_2O$	0.58
$I_2(s) + 2\,e^- \longrightarrow 2\,I^-(aq)$	0.535
$TeO_2(s) + 4\,H^+(aq) + 4\,e^- \longrightarrow Te(s) + 2\,H_2O$	0.529
$Cu^+(aq) + e^- \longrightarrow Cu(s)$	0.521
$[RhCl_6]^{3-}(aq) + 3\,e^- \longrightarrow Rh(s) + 6\,Cl^-(aq)$	0.44
$Cu^{2+}(aq) + 2\,e^- \longrightarrow Cu(s)$	0.337
$HgCl_2(s) + 2\,e^- \longrightarrow 2\,Hg(\ell) + 2\,Cl^-(aq)$	0.27
$AgCl(s) + e^- \longrightarrow Ag(s) + Cl^-(aq)$	0.222
$SO_4^{2-}(aq) + 4\,H^+(aq) + 2\,e^- \longrightarrow SO_2(g) + 2\,H_2O$	0.20
$SO_4^{2-}(aq) + 4\,H^+(aq) + 2\,e^- \longrightarrow H_2SO_3(aq) + H_2O$	0.17
$Cu^{2+}(aq) + e^- \longrightarrow Cu^+(aq)$	0.153
$Sn^{4+}(aq) + 2\,e^- \longrightarrow Sn^{2+}(aq)$	0.15
$S(s) + 2\,H^+(aq) + 2\,e^- \longrightarrow H_2S(aq)$	0.14
$AgBr(s) + e^- \longrightarrow Ag(s) + Br^-(aq)$	0.0713
$2\,H^+(aq) + 2\,e^- \longrightarrow H_2(g)$ (reference electrode)	0.0000
$N_2O(g) + 6\,H^+(aq) + H_2O + 4\,e^- \longrightarrow 2\,NH_3OH^+(aq)$	−0.05
$Pb^{2+}(aq) + 2\,e^- \longrightarrow Pb(s)$	−0.126
$Sn^{2+}(aq) + 2\,e^- \longrightarrow Sn(s)$	−0.14
$AgI(s) + e^- \longrightarrow Ag(s) + I^-(aq)$	−0.15
$[SnF_6]^{2-}(aq) + 4\,e^- \longrightarrow Sn(s) + 6\,F^-(aq)$	−0.25
$Ni^{2+}(aq) + 2\,e^- \longrightarrow Ni(s)$	−0.25
$Co^{2+}(aq) + 2\,e^- \longrightarrow Co(s)$	−0.28
$Tl^+(aq) + e^- \longrightarrow Tl(s)$	−0.34
$PbSO_4(s) + 2\,e^- \longrightarrow Pb(s) + SO_4^{2-}(aq)$	−0.356
$Se(s) + 2\,H^+(aq) + 2\,e^- \longrightarrow H_2Se(aq)$	−0.40
$Cd^{2+}(aq) + 2\,e^- \longrightarrow Cd(s)$	−0.403
$Cr^{3+}(aq) + e^- \longrightarrow Cr^{2+}(aq)$	−0.41
$Fe^{2+}(aq) + 2\,e^- \longrightarrow Fe(s)$	−0.44
$2\,CO_2(g) + 2\,H^+(aq) + 2\,e^- \longrightarrow (COOH)_2(aq)$	−0.49
$Ga^{3+}(aq) + 3\,e^- \longrightarrow Ga(s)$	−0.53

Acidic Solution	**Standard Reduction Potential, E^0 (volts)**
$HgS(s) + 2\,H^+(aq) + 2\,e^- \longrightarrow Hg(\ell) + H_2S(g)$	−0.72
$Cr^{3+}(aq) + 3\,e^- \longrightarrow Cr(s)$	−0.74
$Zn^{2+}(aq) + 2\,e^- \longrightarrow Zn(s)$	−0.763
$Cr^{2+}(aq) + 2\,e^- \longrightarrow Cr(s)$	−0.91
$FeS(s) + 2\,e^- \longrightarrow Fe(s) + S^{2-}(aq)$	−1.01
$Mn^{2+}(aq) + 2\,e^- \longrightarrow Mn(s)$	−1.18
$V^{2+}(aq) + 2\,e^- \longrightarrow V(s)$	−1.18
$CdS(s) + 2\,e^- \longrightarrow Cd(s) + S^{2-}(aq)$	−1.21
$ZnS(s) + 2\,e^- \longrightarrow Zn(s) + S^{2-}(aq)$	−1.44
$Zr^{4+}(aq) + 4\,e^- \longrightarrow Zr(s)$	−1.53
$Al^{3+}(aq) + 3\,e^- \longrightarrow Al(s)$	−1.66
$H_2(g) + 2\,e^- \longrightarrow 2\,H^-(aq)$	−2.25
$Mg^{2+}(aq) + 2\,e^- \longrightarrow Mg(s)$	−2.37
$Na^+(aq) + e^- \longrightarrow Na(s)$	−2.714
$Ca^{2+}(aq) + 2\,e^- \longrightarrow Ca(s)$	−2.87
$Sr^{2+}(aq) + 2\,e^- \longrightarrow Sr(s)$	−2.89
$Ba^{2+}(aq) + 2\,e^- \longrightarrow Ba(s)$	−2.90
$Rb^+(aq) + e^- \longrightarrow Rb(s)$	−2.925
$K^+(aq) + e^- \longrightarrow K(s)$	−2.925
$Li^+(aq) + e^- \longrightarrow Li(s)$	−3.045

Basic Solution	Standard Reduction Potential, E^0 (volts)
$ClO^-(aq) + H_2O + 2\,e^- \longrightarrow Cl^-(aq) + 2\,OH^-(aq)$	0.89
$OOH^-(aq) + H_2O + 2\,e^- \longrightarrow 3\,OH^-(aq)$	0.88
$2\,NH_2OH(aq) + 2\,e^- \longrightarrow N_2H_4(aq) + 2\,OH^-(aq)$	0.74
$ClO_3^-(aq) + 3\,H_2O + 6\,e^- \longrightarrow Cl^-(aq) + 6\,OH^-(aq)$	0.62
$MnO_4^-(aq) + 2\,H_2O + 3\,e^- \longrightarrow MnO_2(s) + 4\,OH^-(aq)$	0.588
$MnO_4^-(aq) + e^- \longrightarrow MnO_4^{2-}(aq)$	0.564
$NiO_2(s) + 2\,H_2O + 2\,e^- \longrightarrow Ni(OH)_2(s) + 2\,OH^-(aq)$	0.49
$Ag_2CrO_4(s) + 2\,e^- \longrightarrow 2\,Ag(s) + CrO_4^{2-}(aq)$	0.446
$O_2(g) + 2\,H_2O + 4\,e^- \longrightarrow 4\,OH^-(aq)$	0.40
$ClO_4^-(aq) + H_2O + 2\,e^- \longrightarrow ClO_3^-(aq) + 2\,OH^-(aq)$	0.36
$Ag_2O(s) + H_2O + 2\,e^- \longrightarrow 2\,Ag(s) + 2\,OH^-(aq)$	0.34
$2\,NO_2^-(aq) + 3\,H_2O + 4\,e^- \longrightarrow N_2O(g) + 6\,OH^-(aq)$	0.15
$N_2H_4(aq) + 2\,H_2O + 2\,e^- \longrightarrow 2\,NH_3(aq) + 2\,OH^-(aq)$	0.10
$[Co(NH_3)_6]^{3+}(aq) + e^- \longrightarrow [Co(NH_3)_6]^{2+}(aq)$	0.10
$HgO(s) + H_2O + 2\,e^- \longrightarrow Hg(\ell) + 2\,OH^-(aq)$	0.0984
$O_2(g) + H_2O + 2\,e^- \longrightarrow OOH^-(aq) + OH^-(aq)$	0.076
$NO_3^-(aq) + H_2O + 2\,e^- \longrightarrow NO_2^-(aq) + 2\,OH^-(aq)$	0.01
$MnO_2(s) + 2\,H_2O + 2\,e^- \longrightarrow Mn(OH)_2(s) + 2\,OH^-(aq)$	−0.05
$CrO_4^{2-}(aq) + 4\,H_2O + 3\,e^- \longrightarrow Cr(OH)_3(s) + 5\,OH^-(aq)$	−0.12
$Cu(OH)_2(s) + 2\,e^- \longrightarrow Cu(s) + 2\,OH^-(aq)$	−0.36
$S(s) + 2\,e^- \longrightarrow S^{2-}(aq)$	−0.48
$Fe(OH)_3(s) + e^- \longrightarrow Fe(OH)_2(s) + OH^-(aq)$	−0.56
$2\,H_2O + 2\,e^- \longrightarrow H_2(g) + 2\,OH^-(aq)$	−0.8277
$2\,NO_3^-(aq) + 2\,H_2O + 2\,e^- \longrightarrow N_2O_4(g) + 4\,OH^-(aq)$	−0.85
$Fe(OH)_2(s) + 2\,e^- \longrightarrow Fe(s) + 2\,OH^-(aq)$	−0.877
$SO_4^{2-}(aq) + H_2O + 2\,e^- \longrightarrow SO_3^{2-}(aq) + 2\,OH^-(aq)$	−0.93
$N_2(g) + 4\,H_2O + 4\,e^- \longrightarrow N_2H_4(aq) + 4\,OH^-(aq)$	−1.15
$[Zn(OH)_4]^{2-}(aq) + 2\,e^- \longrightarrow Zn(s) + 4\,OH^-(aq)$	−1.22
$Zn(OH)_2(s) + 2\,e^- \longrightarrow Zn(s) + 2\,OH^-(aq)$	−1.245
$[Zn(CN)_4]^{2-}(aq) + 2\,e^- \longrightarrow Zn(s) + 4\,CN^-(aq)$	−1.26
$Cr(OH)_3(s) + 3\,e^- \longrightarrow Cr(s) + 3\,OH^-(aq)$	−1.30
$SiO_3^{2-}(aq) + 3\,H_2O + 4\,e^- \longrightarrow Si(s) + 6\,OH^-(aq)$	−1.70

*Selected Thermodynamic Values**

Species	ΔH_f°(298.15K) kJ/mol	S°(298.15K) J/K · mol	ΔG_f°(298.15K) kJ/mol
Aluminum			
Al(s)	0	28.3	0
$AlCl_3$(s)	−704.2	110.67	−628.8
$Al_2O(_3$(s)	−1675.7	50.92	−1582.3
Barium			
$BaCl_2$(s)	−858.6	123.68	−810.4
BaO(s)	−553.5	70.42	−525.1
$BaSO_4$(s)	−1473.2	132.2	−1362.2
Beryllium			
Be(s)	0	9.5	0
$Be(OH)_2$	−902.5	51.9	−815.0
Bromine			
Br(g)	111.884	175.022	82.396
$Br_2(\ell)$	0	152.2	0
Br_2(g)	30.907	245.463	3.110
BrF_3(g)	−255.60	292.53	−229.43
HBr(g)	−36.40	198.695	−53.45
Calcium			
Ca(s)	0	41.42	0
Ca(g)	178.2	158.884	144.3
Ca^{2+}(g)	1925.90	—	—
CaC_2(s)	−59.8	69.96	−64.9
$CaCO_3$(s; calcite)	−1206.92	92.9	−1128.79
$CaCl_2$(s)	−795.8	104.6	−748.1
CaF_2(s)	−1219.6	68.87	−1167.3
CaH_2(s)	−186.2	42.	−147.2
CaO(s)	−635.09	39.75	−604.03
CaS(s)	−482.4	56.5	−477.4
$Ca(OH)_2$(s)	−986.09	83.39	−898.49
$Ca(OH)_2$(aq)	−1002.82	−74.5	−868.07
$CaSO_4$(s)	−1434.11	106.7	−1321.79
Carbon			
C(s, graphite)	0	5.740	0
C(s, diamond)	1.895	2.377	2.900

Species	ΔH_f°(298.15K) kJ/mol	S°(298.15K) J/K · mol	ΔG_f°(298.15K) kJ/mol
C(g)	716.682	158.096	671.257
CCl_4(ℓ)	−135.44	216.40	−65.21
CCl_4(g)	−102.9	309.85	−60.59
$CHCl_3$(liq)	−134.47	201.7	−73.66
$CHCl_3$(g)	−103.14	295.71	−70.34
CH_4(g, methane)	−74.81	186.264	−50.72
C_2H_2(g, ethyne)	226.73	200.94	209.20
C_2H_4(g, ethene)	52.26	219.56	68.15
C_2H_6(g, ethane)	−84.68	229.60	−32.82
C_3H_8(g, propane)	−103.8	269.9	−23.49
C_6H_6(ℓ, benzene)	49.03	172.8	124.5
CH_3OH(ℓ, methanol)	−238.66	126.8	−166.27
CH_3OH(g, methanol)	−200.66	239.81	−161.96
C_2H_5OH(ℓ, ethanol)	−277.69	160.7	−174.78
C_2H_5OH(g, ethanol)	−235.10	282.70	−168.49
CO(g)	−110.525	197.674	−137.168
CO_2(g)	−393.509	213.74	−394.359
CS_2(g)	117.36	237.84	67.12
$COCl_2$(g)	−218.8	283.53	−204.6
Cesium			
Cs(s)	0	85.23	0
Cs^+(g)	457.964	—	—
CsCl(s)	−443.04	101.17	−414.53
Chlorine			
Cl(g)	121.679	165.198	105.680
Cl^-(g)	−233.13	—	—
Cl_2(g)	0	223.066	0
HCl(g)	−92.307	186.908	−95.299
HCl(aq)	−167.159	56.5	−131.228
Chromium			
Cr(s)	0	23.77	0
Cr_2O_3(s)	−1139.7	81.2	−1058.1
$CrCl_3$(s)	−556.5	123.0	−486.1
Copper			
Cu(s)	0	33.150	0
CuO(s)	−157.3	42.63	−129.7
$CuCl_2$(s)	−220.1	108.07	−175.7
Fluorine			
F_2(g)	0	202.78	0
F(g)	78.99	158.754	61.91
F^-(g)	−255.39	—	—
F^-(aq)	−332.63	−13.8	−278.79
HF(g)	−271.1	173.779	−273.2
HF(aq)	−332.63	−13.8	−278.79
Hydrogen			
H_2(g)	0	130.684	0
H(g)	217.965	114.713	203.247

Species	ΔH_f°(298.15K) kJ/mol	S°(298.15K) J/K · mol	ΔG_f°(298.15K) kJ/mol
$H^+(g)$	1536.202	—	—
$H_2O(\ell)$	−285.830	69.91	−237.129
$H_2O(g)$	−241.818	188.825	−228.572
$H_2O_2(\ell)$	−187.78	109.6	−120.35
Iodine			
$I_2(s)$	0	116.135	0
$I_2(g)$	62.438	260.69	19.327
$I(g)$	106.838	180.791	70.250
$I^-(g)$	−197.	—	—
$ICl(g)$	17.78	247.551	−5.46
Iron			
$Fe(s)$	0	27.78	0
$FeO(s)$	−272.	—	—
Fe_2O_3(s, hematite)	−824.2	87.40	−742.2
Fe_3O_4(s, magnetite)	−1118.4	146.4	−1015.4
$FeCl_2(s)$	−341.79	117.95	−302.30
$FeCl_3(s)$	−399.49	142.3	−344.00
FeS_2(s, pyrite)	−178.2	52.93	−166.9
$Fe(CO)_5(\ell)$	−774.0	338.1	−705.3
Lead			
$Pb(s)$	0	64.81	0
$PbCl_2(s)$	−359.41	136.0	−314.10
PbO(s, yellow)	−217.32	68.70	−187.89
$PbS(s)$	−100.4	91.2	−98.7
Lithium			
$Li(s)$	0	29.12	0
$Li^+(g)$	685.783	—	—
$LiOH(s)$	−484.93	42.80	−438.95
$LiOH(aq)$	−508.48	2.80	−450.58
$LiCl(s)$	−408.701	59.33	−384.37
Magnesium			
$Mg(s)$	0	32.68	0
$MgCl_2(g)$	−641.32	89.62	−591.79
$MgO(s)$	−601.70	26.94	−569.43
$Mg(OH)_2(s)$	−924.54	63.18	−833.51
$MgS(s)$	−346.0	50.33	−341.8
Mercury			
$Hg(\ell)$	0	76.02	0
$HgCl_2(s)$	−224.3	146.0	−178.6
HgO(s, red)	−90.83	70.29	−58.539
HgS(s, red)	−58.2	82.4	−50.6
Nickel			
$Ni(s)$	0	29.87	0
$NiO(s)$	−239.7	37.99	−211.7
$NiCl_2(s)$	−305.332	97.65	−259.032
Nitrogen			
$N_2(g)$	0	191.61	0
$N(g)$	472.704	153.298	455.563

Species	ΔH_f°(298.15K) kJ/mol	S°(298.15K) J/K · mol	ΔG_f°(298.15K) kJ/mol
NH_3(g)	−46.11	192.45	−16.45
N_2H_4(ℓ)	50.63	121.21	149.34
NH_4Cl(s)	−314.43	94.6	−202.87
NH_4Cl(aq)	−299.66	169.9	−210.52
NH_4NO_3(s)	−365.56	151.08	−183.87
NH_4NO_3(aq)	−399.87	259.8	−190.56
NO(g)	90.25	210.76	86.55
NO_2(g)	33.18	240.06	51.31
N_2O(g)	82.05	219.85	104.20
N_2O_4(g)	9.16	304.29	97.89
NOCl(g)	51.71	261.69	66.08
HNO_3(ℓ)	−174.10	155.60	−80.71
HNO_3(g)	−135.06	266.38	−74.72
HNO_3(aq)	−207.36	146.4	−111.25
Oxygen			
O_2(g)	0	205.138	0
O(g)	249.170	161.055	231.731
O_3(g)	142.7	238.93	163.2
Phosphorus			
P_4(s, white)	0	164.36	0
P_4(s, red)	−70.4	91.2	−48.4
P(g)	314.64	163.193	278.25
PH_3(g)	5.4	310.23	13.4
PCl_3(g)	−287.0	311.78	−267.8
P_4O_{10}(s)	−2984.0	228.86	−2697.7
H_3PO_4(s)	−1279.0	110.5	−1119.1
Potassium			
K(s)	0	64.18	0
KCl(s)	−436.747	82.59	−409.14
$KClO_3$(s)	−397.73	143.1	−296.25
KI(s)	−327.90	106.32	−324.892
KOH(s)	−424.764	78.9	−379.08
KOH(aq)	−482.37	91.6	−440.50
Silicon			
Si(s)	0	18.83	0
$SiBr_4$(ℓ)	−457.3	277.8	−443.8
SiC(s)	−65.3	16.61	−62.8
$SiCl_4$(g)	−657.01	330.73	−616.98
SiH_4(g)	34.3	204.62	56.9
SiF_4(g)	−1614.94	282.49	−1572.65
SiO_2(s, quartz)	−910.94	41.84	−856.64
Silver			
Ag(s)	0	42.55	0
Ag_2O(s)	−31.05	121.3	−11.20
AgCl(s)	−127.068	96.2	−109.789
$AgNO_3$(s)	−124.39	140.92	−33.41

Species	**ΔH_f°(298.15K) kJ/mol**	**S°(298.15K) J/K · mol**	**ΔG_f°(298.15K) kJ/mol**
Sodium			
$Na(s)$	0	51.21	0
$Na(g)$	107.32	153.712	76.761
$Na^+(g)$	609.358	—	—
$NaBr(s)$	−361.062	86.82	−348.983
$NaCl(s)$	−411.153	72.13	−384.138
$NaCl(g)$	−176.65	229.81	−196.66
$NaCl(aq)$	−407.27	115.5	−393.133
$NaOH(s)$	−425.609	64.455	−379.494
$NaOH(aq)$	−470.114	48.1	−419.150
$Na_2CO_3(s)$	−1130.68	134.98	−1044.44
Sulfur			
S(s, rhombic)	0	31.80	0
$S(g)$	278.805	167.821	238.250
$S_2Cl_2(g)$	−18.4	331.5	−31.8
$SF_6(g)$	1209.	291.82	−1105.3
$H_2S(g)$	−20.63	205.79	−33.56
$SO_2(g)$	−296.830	248.22	−300.194
$SO_3(g)$	−395.72	256.76	−371.06
$SOCl_2(g)$	−212.5	309.77	−198.3
$H_2SO_4(\ell)$	−813.989	156.904	−690.003
$H_2SO_4(aq)$	−909.27	20.1	−744.53
Tin			
Sn(s, white)	0	51.55	0
Sn(s, gray)	−2.09	44.14	0.13
$SnCl_4(\ell)$	−511.3	248.6	−440.1
$SnCl_4(g)$	−471.5	365.8	−432.2
$SnO_2(s)$	−580.7	52.3	−519.6
Titanium			
$Ti(s)$	0	30.63	0
$TiCl_4(\ell)$	−804.2	252.34	−737.2
$TiCl_4(g)$	−763.2	354.9	−726.7
TiO_2	−939.7	49.92	−884.5
Zinc			
$Zn(s)$	0	41.63	0
$ZnCl_2(s)$	−415.05	111.46	−369.398
$ZnO(s)$	−348.28	43.64	−318.30
ZnS(s, sphalerite)	−205.98	57.7	−201.29

*Taken from "The NBS Tables of Chemical Thermodynamic Properties," 1982.

APPENDIX

K

The Valence Bond Hybridization Theory of Covalent Bonding

by John M. DeKorte
Northern Arizona University

The purpose of this appendix is to provide a summary of the valence bond hybridization theory of covalent bonding for anyone wishing to cover this topic in lecture.

Electrostatic attractions between covalently bonded atoms were discussed in Section 10.1.

According to the **valence bond theory of covalent bonding,** shared pairs of electrons in covalent bonds are located in overlapping orbitals. Orbital overlaps are the result of electrostatic attractions between the bonded atoms, and each shared pair of electrons occupies an overlap region that is formed by overlap of one orbital from each bonded atom. The orientation of these overlapping orbitals in space must be compatible with the observed shapes of molecules. In addition, the number of covalent bonds formed must be compatible with the number of orbitals available for bonding and the number of valence electrons that each atom can provide for bonding.

Valence bond theory differs from the VSEPR model by accounting for orbitals occupied by bonding electrons.

The atomic orbitals that were introduced in Section 9.2 can be used to describe the bonding in diatomic molecules with a single bond between atoms (H_2 and Cl_2, for example) since all diatomic molecules are necessarily linear (Figure 1). However, hybrid atomic orbitals are commonly used to describe the bonding in molecules containing multiple bonds (CO, for example) or with three or more atoms. **Hybrid atomic orbitals** are formed by mathematical combinations of the *s* and *p* or *s*, *p*, and *d* atomic orbitals of the same atom. Hybrid atomic orbitals have different shapes and orientations in space than unhybridized atomic orbitals. However, the total number of hybrid atomic orbitals formed is always equal to the number of atomic orbitals combined.

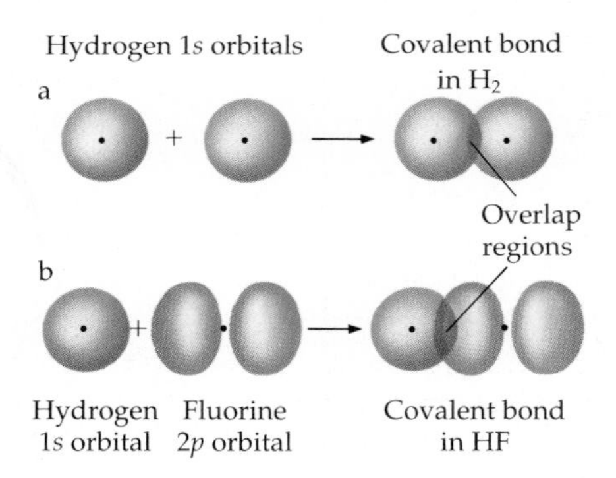

Figure 1

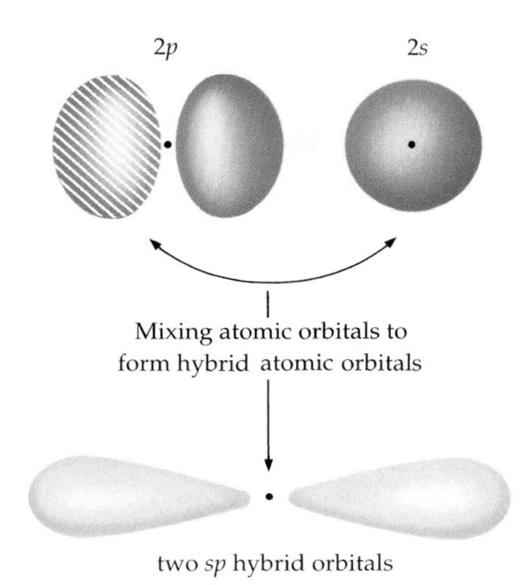

Figure 2

The mathematical combination of one *s* orbital and one *p* orbital of the same atom results in two *sp* hybrid atomic orbitals that lie 180° apart along the axis of the *p* orbital involved in the combination (Figure 2). The mathematical combination of one *s* orbital and two *p* orbitals leads to the formation of three sp^2 (read "s-p-two") hybrid atomic orbitals. The sp^2 orbitals lie in the plane of the *p* orbitals involved in the combination and point towards the corners of an equilateral triangle; they are 120° apart. Similarly, the mathematical combination of one *s* orbital and three *p* orbitals leads to four hybrid atomic orbitals pointing towards the corners of a tetrahedron. The mathematical combination of one *s* orbital, three *p* orbitals, and one *d* orbital (the d_{z^2} orbital) leads to five hybrid atomic orbitals pointing towards the corners of a triangular bipyramid, and the mathematical combination of one *s* orbital, three *p* orbitals, and two *d* orbitals (the d_{z^2} and $d_{x^2-y^2}$ orbitals) leads to six hybrid atomic orbitals pointing towards the corners of an octahedron. Hence, these orientations correspond to the electron-pair geometries that were predicted for two to six sets of electrons using the VSEPR model, as shown in Figure 11.5 on page 456.

The label sp^2 *indicates the hybrid orbitals were formed by the mathematical combination of one* s *orbital and two* p *orbitals from the same atom.*

Let's see how hybrid atomic orbitals can account for the numbers of bonds formed by atoms. Gaseous $BeCl_2$ is known to exist as a nonpolar linear molecule whereas the $2s^2$ valence electron configuration of Be suggests that there are no unpaired electrons available for bonding with Cl. According to the valence bond hybridization theory, *sp* hybridization for Be provides two unpaired electrons in *sp* hybrid atomic orbitals 180° apart for bonding with Cl (Figure 3). The Be-Cl bonds can be described as resulting from the overlap of an *sp* hybrid orbital of Be with a 3*p* orbital of Cl, each containing one electron. Hybrid atomic orbitals are not postulated for the Cl atoms, since there is no experimental evidence for the arrangement of the electron pairs about the Cl atoms.

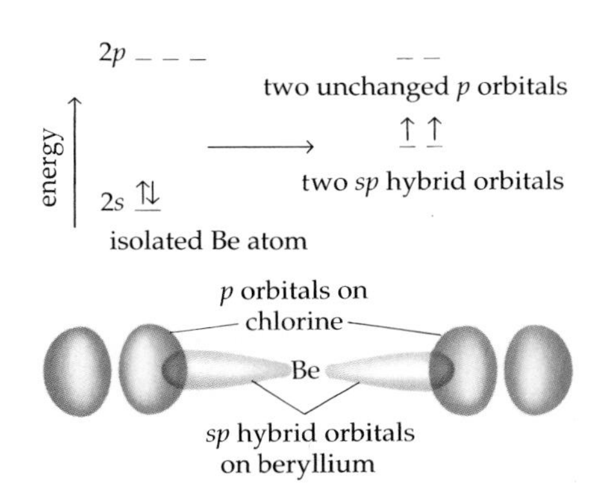

Figure 3

Similarly, sp^3 hybridization accounts for the formation of four sp^3 hybrid orbitals and the four bonds between C and H in tetrahedral methane, CH_4, the simplest carbon-hydrogen compound, whereas the $2s^2 2p^1 2p^1$ valence electron configuration of C suggests that the simplest carbon-hydrogen

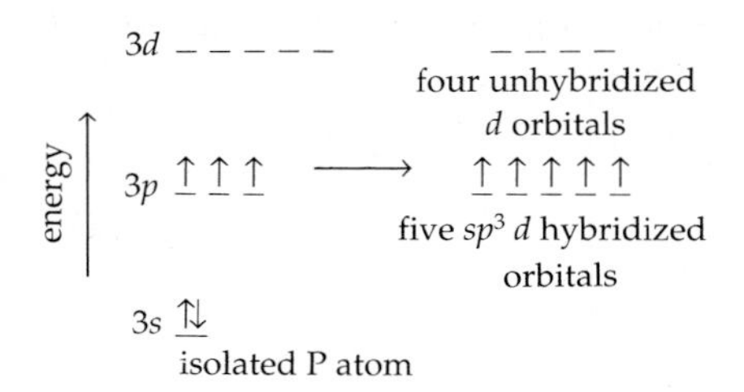

Figure 4

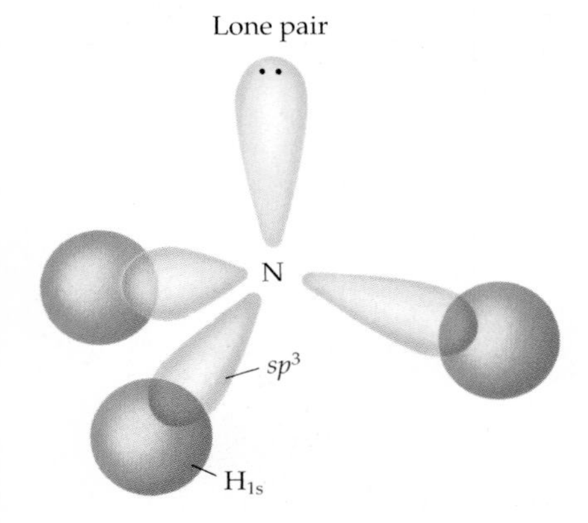

Figure 5

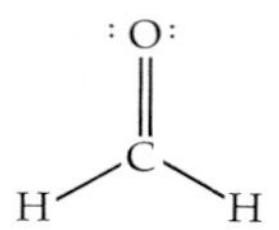

Figure 6

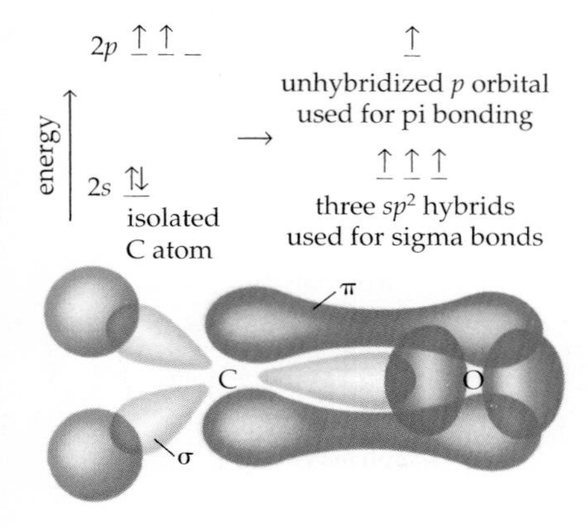

Figure 7

compound should be CH_2. In addition, sp^3d hybridization accounts for the formation of triangular bipyramidal PF_5 (Figure 4), whereas the $3s^23p^13p^13p^1$ valence electron configuration of P suggests that the only compound that P could form with F would be PF_3.

Since the VSEPR model and valence bond theory can both account for the electron-pair geometries of molecules and ions, we can use Lewis diagrams to predict the types of hybrid atomic orbitals used by central atoms (Table 1). This procedure can be used with species having lone pairs of electrons about the central atoms or for those without them. For example, the Lewis diagram for NH_3 suggests that the electron-pair geometry about the central nitrogen atom will be approximately tetrahedral, and this is compatible with the use of four sp^3 hybrid atomic orbitals by nitrogen (Figure 5).

Table 1

Electron-pair Geometries and Hybrid Orbital Sets for Two to Six Electron Pairs

Number of single bond and lone electron pairs	Electron-pair geometry	Hybrid orbital set	Example
2	Linear	sp	Be in $BeCl_2$
3	Triangular planar	sp^2	B in BF_3
4	Tetrahedral	sp^3	C in CH_4
5	Triangular bipyramidal	sp^3d	P in PF_5
6	Octahedral	sp^3d^2	S in SF_6

Hybrid atomic orbitals can also be used to describe the bonding in molecules with multiple bonds. Consider formaldehyde, HCHO, which has two carbon-hydrogen single bonds and one carbon-oxygen double bond (Figure 6). The triangular planar electron-pair geometry about the carbon atom suggests sp^2 hybridization for carbon. Two of the sp^2 hybrid orbitals of carbon overlap with 1*s* orbitals of the hydrogen atoms to form carbon-hydrogen bonds, and the third sp^2 orbital of carbon overlaps with a 2*p* orbital of oxygen to form a carbon-oxygen bond (Figure 7). Bonds of this type, which place electron density along the line joining the nuclei of the bonded atoms, are called **sigma bonds (σ bonds).**

The second bond in the carbon-oxygen double bond in formaldehyde involves pi (π) bonding. **Pi bonds** result from side-by-side overlap of *p* orbitals and place electron density above and below the line joining the bonded atoms. One-half of the pi overlap is above the line joining the bonded atoms, and the other half is below it. The unhybridized *p* orbital of carbon in formaldehyde overlaps with a *p* orbital of oxygen to form the pi bond (Figure 7). It is important to note that atoms that are involved in just sigma bonds are able to rotate freely about the bond axis, whereas atoms involved in double bonds are unable to rotate freely about the bond axis, since this would require breaking the pi overlap.

Hybridized orbitals can be used to describe the bonding in compounds having more than one central atom. Consider the series of carbon-hydrogen compounds ethane (C_2H_6), ethylene (C_2H_4), and acetylene (C_2H_2) having the Lewis structures shown in Figure 8. The geometry about each carbon atom in ethane is tetrahedral and is compatible with sp^3 hybridization for carbon. Similarly, the geometries about the carbon atoms in ethylene and acetylene are triangular planar and linear, respectively, and are compatible with sp^2 and sp hybridizations for the respective carbon atoms.

Since the electron densities of sigma and pi bonds between two atoms occupy the same general region of space, multiple bonds can be counted as single bonds when using the VSEPR model to predict electron-pair geometries (see Section 11.1).

```
  H H         H H
H:C:C:H     H:C::C:H     H:C:::C:H
  H H
```

Figure 8

Hybridization is especially useful for describing the geometry and bonding in carbon compounds.

There are only single bonds formed by sigma overlaps in ethane. Hence, there is free rotation about the carbon-carbon bond, and the lowest energy configuration is actually one in which the H atoms on the second carbon atom are exactly offset from those on the first carbon atom. The double bond in ethylene suggests the presence of one sigma overlap and one pi overlap between the carbon atoms, and the triple bond in acetylene suggests the presence of one sigma overlap and two pi overlaps (at right angles to each other) between these carbon atoms (Figure 9). The pi overlap prevents free rotation about the carbon-carbon bonds in ethylene. Furthermore, since pi overlaps are generally weaker than sigma overlaps, a carbon-carbon double bond is stronger than a carbon-carbon single bond but not twice as strong. Likewise, a carbon-carbon triple bond is stronger than a carbon-carbon single or double bond but not three times as strong as a single bond.

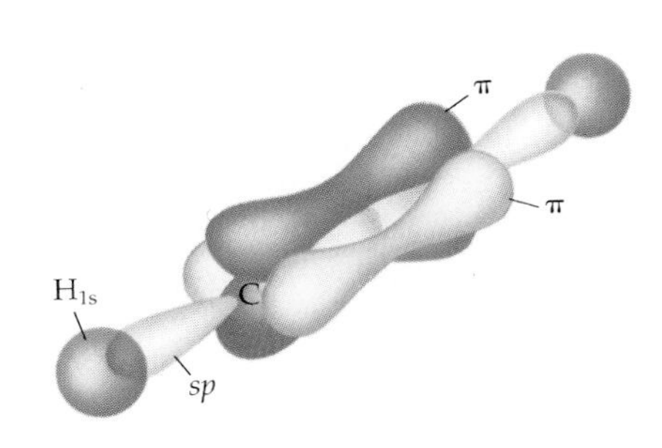

Figure 9

The lack of free rotation about double bonds in alkenes is actually the reason for the existence of cis *and* trans *isomers of alkenes (Section 11.5) and for the activation energy for the transformation of* cis-2-*butene to* trans-2-*butene (Section 8.1).*

Although valence bond hybridization theory is able to account for the number of covalent bonds formed by atoms and the electron-pair geometries of molecules and ions, it has some limitations. For example, it predicts that all the electrons will be paired in O_2 and is therefore unable to account for the observed paramagnetic behavior of O_2. An alternative theory of covalent bonding called the **molecular orbital theory** is widely used and is able to account for the presence of two unpaired electrons in O_2. Hence, we should always remember that we are using a theoretical construct of our own making to account for the behavior of atoms and realize that our theoretical models might fail to account for each and every case.

A P P E N D I X

L

Answers to Exercises

CHAPTER 2

2.1 The thin sheets of mica suggest that, at the atomic level, the atoms are arranged in sheets. This is indeed the case. The elements that make up mica—silicon, oxygen, aluminum, and magnesium—are bound tightly to one another in layers several atoms thick. These layers are then bound to one another through weaker forces, as indicated by the fact that one layer can be peeled away from another.

2.2 This photo is one of a soft drink being poured into a glass; gas bubbles are clearly formed. The solids in the photo are the glass and the soft drink bottle, and the liquid is a homogeneous mixture (a solution) of the ingredients of the soft drink. The gas, carbon dioxide, is in the bubbles.

2.3 1.00×10^3 ~~g~~ air ($1\ cm^3/1.12 \times 10^{-3}$ ~~g~~) =
$8.93 \times 10^5\ cm^3$ (or 893 L)

2.4 77 K = ? °C + 273.15 °C
? °C = −196 °C

2.5 The sand/iron filings mixture is heterogeneous, while the blue *solution* (of a compound called copper sulfate in water) is a homogeneous mixture.

2.6 (a) Na is sodium, Cl is chlorine, Cr is chromium
(b) zinc = Zn, aluminum = Al, silicon = Si

2.7 Chemical changes: burning of the wick and wax
Physical changes: melting and vaporization of wax
Energy: energy is evolved in the form of light and heat

2.8 10.0 ~~inches~~ (2.54 cm/~~inch~~) = 25.4 cm
25.4 ~~cm~~ (1 m/100 ~~cm~~) = 0.254 m
25.4 ~~cm~~ (10 mm/~~cm~~) = 254 mm
In the same way, we find that 8.00 inches = 20.3 cm = 0.203 m = 203 mm

2.9 (a) 750 ~~mL~~ (1 L/1000. ~~mL~~) = 0.75 L
(b) 2.0 ~~quart~~ (1 gallon/4 ~~quarts~~) = 0.50 gallon
0.50 ~~gallon~~ (3.7865 L/1 ~~gallon~~) = 1.9 L

2.10 (a) 500. ~~mg~~ (1 g/1000 ~~mg~~) = 0.500 g
(b) 2.00 ~~pound~~ (453.6 g/1 ~~pound~~)(1 kg/1000 g) = 0.907 kg

2.11 area of platinum sheet = $(2.50\ cm)^2 = 6.25\ cm^2$
volume of platinum sheet = 1.656 ~~g~~ ($1\ cm^3/21.45$ ~~g~~) = $0.07720\ cm^3$
Since the volume of the sheet = area × thickness, this means that
thickness = volume/area = $0.07720\ cm^3/6.25\ cm^2$ = 0.0124 cm
thickness in millimeters = 0.0124 cm (10 mm/1 cm) = 0.124 mm

2.12 15.0 ~~g earring~~ (0.58 g gold/1 ~~g earring~~) = 8.7 g gold

CHAPTER 3

3.1 (a) When clothes dry, the molecules of water escape from the clothes into the air. (b) When moisture appears on the outside of a glass of ice water, molecules of water that were in the air come together on the surface of the glass to form liquid water. (c) When crystals of solid sugar dissolve in water, the sugar molecules separate from other sugar molecules and become mixed with water molecules in the liquid water. (d) Sugar dissolves faster in hot water because at a higher temperature the molecules move faster, and it is easier for sugar and water molecules to push past one another to get mixed; the mixing takes less time.

3.2 The radius of a nucleus is 1/100,000 (or 10^{-5}) times the radius of an atom. $100\ m \times 10^{-5} = 10^{-3}\ m$ = 0.001 m = 1 mm. A crystal of salt in a salt shaker or the tip of a dull pencil is about this size.

3.3 (a) Mass number for copper with 34 neutrons = 29 protons + 34 neutrons = 63
(b) All Ni atoms have 28 protons and electrons. Therefore, the number of neutrons for ^{59}Ni is given by:

$$\text{mass number} = 59 = \text{number of protons} + \text{number of neutrons}$$
$$= 28 + \text{number of neutrons}$$
$$\text{number of neutrons} = 59 - 28 = 31$$

3.4 Silicon atoms have an atomic number of 14. The mass number of silicon with 14 neutrons is, therefore,

$$\text{mass number} = 14 \text{ protons} + 14 \text{ neutrons} = 28$$

The symbol for this isotope is $^{28}_{14}Si$. In the same manner, the mass numbers for the other isotopes are 29 and 30, and their symbols are $^{29}_{14}Si$ and $^{30}_{14}Si$, respectively.

3.5 Atomic weight of chlorine = (0.7577)(34.96885 amu) + (0.2423)(36.96590 amu) = 35.45 amu

3.6 There are 8 elements in the third period. Metals: sodium (Na), magnesium (Mg), and aluminum (Al). Metalloid: silicon (Si). Nonmetals: phosphorus (P), sulfur (S), chlorine (Cl), and argon (Ar).

3.7 The formula for a compound of indium and fluorine should be similar to $AlCl_3$; the formula is InF_3. The formula for sodium oxide is Na_2O; for sodium chloride, NaCl. There are half as many Na atoms in the chloride. The general formula for alkaline earth oxides is XO; for alkaline earth chlorides there should be half as many metal atoms. Since there is only one metal atom in XO, $Mg_{0.5}Cl$ would be predicted, but half an atom of Mg makes no sense. If we double the number of atoms of each kind, there would still be twice as many Cl atoms as Mg atoms in the formula: $MgCl_2$.

CHAPTER 4

4.1 Graphite has a layered structure with weak binding between the layers (Figure 2.8). The slippery feel of graphite and its use as a lubricant or in pencil "lead" come from the fact that these layers break away from one another when a little force is applied.

4.2 Oxalic acid: $C_2H_2O_4$
DDT: $C_{14}H_9Cl_5$

4.3 The molecular formula of glycine is $C_2H_5NO_2$. There are 9 bonds in the molecule. A condensed formula would be NH_2CH_2COOH.

4.4 (a) K^+ (d) V^{2+}
(b) Se^{2-} (e) Co^{2+}
(c) Be^{2+} (f) Cs^+

4.5 1. (a) 1 Na^+ and 1 F^-
(b) 1 Cu^{2+} and 2 NO_3^-
(c) 1 Na^+ and 1 $CH_3CO_2^-$
2. $FeCl_2$ and $FeCl_3$
3. Na_2S and Na_3PO_4. BaS and $Ba_3(PO_4)_2$

4.6 Br_2 is a diatomic molecule, K is a metal, and KBr is an ionic compound.

4.7 1. (a) NH_4NO_3
(b) $CoSO_4$
(c) $Ni(CN)_2$
(d) V_2O_3
(e) BaO
(f) $Ca(ClO)_2$
2. (a) magnesium bromide
(b) lithium carbonate
(c) potassium hydrogen sulfite
(d) potassium permanganate
(e) ammonium sulfide
(f) copper(I) chloride and copper(II) chloride

4.8 1. (a) CO_2 (e) BF_3
(b) PI_3 (f) O_2F_2
(c) SCl_2 (g) C_9H_{20}
(d) XeO_3
2. (a) dinitrogen tetrafluoride
(b) hydrogen bromide
(c) sulfur tetrafluoride
(d) chlorine trifluoride
(e) boron trichloride
(f) diphosphorus pentaoxide
(g) heptane

4.9 1. $C_{12}H_{26}$ and $C_{24}H_{50}$
2. hexadecane = $C_{16}H_{34}$

4.10 1. 2.5 mol Al (27.0 g/1 mol) = 68 g Al
2. 1.00 pound (454 g/pound)(1 mol/207.2 g) = 2.19 mol Pb

4.11 1. (a) $CaCO_3$ molar mass = 100.09 g/mol
(b) $C_8H_{10}N_4O_2$ molar mass = 194.19 g/mol
2. 454 g $CaCO_3$ (1 mol/100.09 g) = 4.54 mol $CaCO_3$
3. 2.50×10^{-3} mol caffeine (194.19 g/mol) = 0.485 g caffeine

4.12 (a) In 1.00 mol of NaCl (molar mass = 58.5 g/mol) there is 23.0 g of Na and 35.5 g of Cl. Na is 39.3% and Cl is 60.7%. For example,

$$\%\text{Na} = (23.0 \text{ g Na}/58.44 \text{ g})100\% = 39.4\%$$

(b) 1.00 mol of octane (molar mass = 114.2 g/mol) contains

$$8.00 \text{ mol C } (12.01 \text{ g C/mol}) = 96.1 \text{ g C}$$
$$18.0 \text{ mol H } (1.008 \text{ g H/mol}) = 18.1 \text{ g H}$$
$$\%\text{C} = (96.1 \text{ g C}/114.2 \text{ g})100 = 84.2\% \text{ C}$$

and so %H = 15.8%

(c) 1.00 mol of $(NH_4)_2SO_4$ (molar mass = 132.2 g/mol) contains

2 mol N (14.01 g/mol) = 28.0 g N
8 mol H (1.008 g/mol) = 8.06 g H
1 mol S (32.07 g/mol) = 32.1 g S
4 mol O (16.00 g/mol) = 64.0 g O
%N = (28.0 g N/132.2 g)100 = 21.2% N

and in the same manner %H is 6.10%, %S is 24.3%, and %O is 48.4%.

4.13 78.14 g B (1 mol/10.811 g) = 7.228 mol B
21.86 g H (1 mol/1.0079 g) = 21.69 mol H
mole ratio = 21.69 mol H/7.228 mol B = 3.000 mol H/1.000 mol B
empirical formula = BH_3
(27.7 g/mol)/(13.83 g/BH_3 unit) = 2 BH_3 units/mol
Therefore, the molecular formula is $(BH_3)_2$ or B_2H_6.

4.14 0.532 g Ti (1 mol Ti/47.88 g) = 0.0111 mol Ti
2.108 g Ti_xCl_y − 0.532 g Ti = 1.576 g Cl combined with the Ti
1.576 g Cl (1 mol Cl/35.453 g) = 0.04445 mol Cl
mole ratio = 0.04445 mol Cl/0.0111 mol Ti = 4/1
empirical formula = $TiCl_4$

4.15 1.056 g hydrated compound − 0.838 g $RuCl_3$ = 0.218 g H_2O
0.838 g $RuCl_3$ (1 mol/207.4 g) = 4.04×10^{-3} mol $RuCl_3$
0.218 g H_2O (1 mol/18.02 g) = 1.21×10^{-2} mol H_2O
mole ratio = 1.21×10^{-2} mol H_2O/4.04×10^{-3} mol $RuCl_3$ = 2.99/1.00 = 3/1
formula of the hydrate = $RuCl_3 \cdot 3\,H_2O$

CHAPTER 5

5.1 (a) Stoichiometric coefficients: 4, 3, and 2
(b) Law of conservation of matter states that the total mass of products must equal the total mass of reactants. Therefore, a maximum of 1.56 g of products can be obtained.
(c) 8000 atoms Fe (3 molecules O_2/4 atoms Fe) = 6000 molecules O_2

5.2 $C_5H_{12}(g) + 8\,O_2(g) \longrightarrow 5\,CO_2(g) + 6\,H_2O(\ell)$

5.3 Epsom salt, $MgSO_4 \cdot 7\,H_2O$, is a water-soluble ionic salt and so is a strong electrolyte. CH_3OH does not form ions in water (much like the closely related ethyl alcohol) and so is a nonelectrolyte.

5.4 (a) KNO_3 contains two ions (K^+ and NO_3^-) that lead to water-soluble compounds.
(b) $CaCl_2$ contains at least one ion (Cl^-) that leads to water-soluble compounds.
(c) CuO contains the Cu^{2+} and O^{2-} ions. Oxides are not usually water soluble.
(d) Sodium acetate contains two ions (Na^+ and $CH_3CO_2^-$) that usually lead to water-soluble compounds.

5.5 (a) $HClO_4(aq) \longrightarrow H^+(aq) + ClO_4^-(aq)$
(b) $Ca(OH)_2(s) \longrightarrow Ca^{2+}(aq) + 2\,OH^-(aq)$

5.6 (a) SeO_2 is an acidic oxide, much like SO_2.
(b) MgO is a basic metal oxide.
(c) P_4O_{10} is a nonmetal oxide and so is an acidic oxide.

5.7 (a) $BaCl_2(aq) + Na_2SO_4(aq) \longrightarrow BaSO_4(s) + 2\,NaCl(aq)$

net ionic equation:
$Ba^{2+}(aq) + SO_4^{2-}(aq) \longrightarrow BaSO_4(s)$

(b) As a first step, we write the complete, balanced equation including information on the water solubility of the compounds.

complete equation:
$Pb(NO_3)_2(aq) + 2\,KCl(aq) \longrightarrow PbCl_2(s) + 2\,KNO_3(aq)$

This can then be followed by the net ionic equation.

net ionic equation:
$Pb^{2+}(aq) + 2\,Cl^-(aq) \longrightarrow PbCl_2(s)$

5.8 (a) $2\,Cu(s) + O_2(g) \longrightarrow 2\,CuO(s)$
(b) $S_8(s) + 24\,F_2(g) \longrightarrow 8\,SF_6(g)$
(c) $Ba(s) + Br_2(\ell) \longrightarrow BaBr_2(s)$

5.9 $PbCO_3(s) \longrightarrow PbO(s) + CO_2(g)$

5.10 $2\,AgNO_3(aq) + K_2CrO_4(aq) \longrightarrow Ag_2CrO_4(s) + 2\,KNO_3(aq)$
$2Ag^+(aq) + CrO_4^{2-}(aq) \longrightarrow Ag_2CrO_4(s)$

5.11 $Mg(OH)_2(s) + 2\,HCl(aq) \longrightarrow MgCl_2(aq) + 2\,H_2O(\ell)$
$H_3O^+ + OH^- \longrightarrow H_2O$

5.12 $PbCO_3(s) + 2\,HNO_3(aq) \longrightarrow Pb(NO_3)_2(aq) + H_2O(\ell) + CO_2(g)$
lead (II) nitrate, water, carbon dioxide

5.13 (a) $CuCO_3(s) + H_2SO_4(aq) \longrightarrow CuSO_4(aq) + H_2O(\ell) + CO_2(g)$
Gas-forming reaction.
(b) $Ba(OH)_2(s) + 2\,HNO_3(aq) \longrightarrow Ba(NO_3)_2(aq) + 2\,H_2O(\ell)$
Acid-base reaction.
(c) $ZnCl_2(aq) + (NH_4)_2S(aq) \longrightarrow ZnS(s) + 2\,NH_4Cl(aq)$
Precipitation reaction.

5.14 (a) $Ba(OH)_2(aq) + H_2SO_4(aq) \longrightarrow BaSO_4(s) + 2\,H_2O(\ell)$
(b) $BaCO_3(s) + H_2SO_4(aq) \longrightarrow BaSO_4(s) + H_2O(\ell) + CO_2(g)$

5.15 (a) $NaOH(aq) + HNO_3(aq) \longrightarrow NaNO_3(aq) + H_2O(\ell)$
Acid-base reaction.
(b) $4\,Cr(s) + 3\,O_2(g) \longrightarrow 2\,Cr_2O_3(s)$
Oxidation-reduction reaction. Oxygen is an oxidizing agent and metals (Cr) are reducing agents.
(c) $NiCO_3(s) + 2HCl(aq) \longrightarrow NiCl_2(s) + CO_2(g) + H_2O(\ell)$
Decomposition reaction.
(d) $Cu(s) + Cl_2(g) \longrightarrow CuCl_2(s)$
Oxidation-reduction reaction. Chlorine is an oxidizing agent and metals (Cu) are reducing agents.

5.16 $C_{10}H_{16}(\ell)$ (reducing agent) $+ 14\,O_2(g)$ (oxidizing agent) $\longrightarrow 10\,CO_2(g) + 8\,H_2O(g)$

CHAPTER 6

6.1 454 g O_2(1 mol O_2/32.00 g O_2) = 14.2 mol O_2
14.2 mol O_2 (2 mol C consumed/1 mol O_2 used) = 28.4 mol C consumed
28.4 mol C(12.01 g/mol) = 341 g C
14.2 mol O_2 consumed (2 mol CO produced/1 mol O_2 consumed) = 28.4 mol CO produced
28.4 mol CO (28.01 g/mol) = 795 g CO produced
(Notice that the 795 of CO produced is the sum of the mass of C and O_2 used.)

6.2 (a) Balanced equation: $S_8(s) + 8\,O_2(g) \longrightarrow 8\,SO_2(g)$
(b) Find the limiting reagent
20.0 g S_8 (1 mol/256.5 g) = 0.0780 mol S_8
160. g O_2 (1 mol/32.00 g) = 5.00 mol O_2
5.00 mol O_2 (1 mol S_8 required/8 mol O_2 available) = 0.625 mol S_8 required
The quantity of S_8 available (0.0780 mol) is much less than the quantity required to consume all the oxygen. Therefore, sulfur is the limiting reagent.
(c) Reagent left after reaction.
Since S_8 is the limiting reagent, O_2 must remain after reaction. To find the excess O_2, first calculate the quantity of O_2 required to consume all the available S_8.

0.0780 mol S_8 (8 mol O_2 required/1 mol S_8 available) = 0.624 mol O_2
O_2 remaining = 5.00 mol available − 0.624 mol required = 4.38 mol O_2
4.38 mol O_2 (32.00 g O_2/1 mol) = 140. g O_2 remain

(d) Mass of SO_2 formed. This calculation must be based on the quantity of limiting reagent.

0.0780 mol S_8 (8 mol SO_2/1 mol S_8) = 0.624 mol SO_2 produced
0.624 mol SO_2 (64.06 g/1 mol) = 40.0 g SO_2 produced

6.3 (18.9 g $NaBH_4$)(1 mol/37.83 g) = 0.500 mol $NaBH_4$
0.500 mol $NaBH_4$ (2 mol B_2H_6/3 mol $NaBH_4$) = 0.333 mol B_2H_6
0.333 mol B_2H_6 (27.67 g/mol) = 9.21 g
% yield of B_2H_6 = (7.50 g/9.21 g)100 = 81.4%

6.4 2.357 g before − 2.108 g after = 0.249 g H_2O
0.249 g H_2O (1 mol H_2O/18.02 g) = 0.0138 mol H_2O
0.0138 mol H_2O (1 mol $BaCl_2 \cdot 2\,H_2O$/2 mol H_2O) = 0.00690 mol $BaCl_2 \cdot 2\,H_2O$
0.00690 mol $BaCl_2 \cdot 2\,H_2O$ (244.3 g/mol) = 1.69 g $BaCl_2 \cdot 2\,H_2O$
weight % = (1.69 g $BaCl_2 \cdot 2\,H_2O$/2.357 g)100% = 71.7%

6.5 0.600 g CO_2 (1 mol/44.01 g)(1 mol C/1 mol CO_2) = 0.0136 mol C
0.0136 mol C (12.01 g/mol) = 0.163 g C
0.163 g H_2O (1 mol/18.02 g)(2 mol H/1 mol H_2O) = 0.0181 mol H
0.0181 mol H (1.008 g/mol) = 0.0182 g H
0.400 g solid − (0.163 g C + 0.0182 g H) = 0.219 g O
0.219 g O (1 mol/16.00 g) = 0.0137 mol O
moles C/moles O = 0.0136 mol C/0.0137 mol O = 1 C to 1 O
moles H/moles O = 0.0181 mol H/0.0136 mol O = 1.32 H to 1 O (or 1.32 H to 1 C)
This gives a formula of $C_1H_{1.32}O_1$ or an empirical formula of $C_3H_4O_3$.

6.6 26.3 g $NaHCO_3$ (1 mol/84.01 g) = 0.313 mol $NaHCO_3$
0.313 mol $NaHCO_3$/0.200 L = 1.57 M

6.7 For 1 M HCl, $[H^+]$ = 1 M and $[Cl^-]$ = 1 M
$HCl(aq) \longrightarrow H^+(aq) + Cl^-(aq)$
$NaSO_4(s) \longrightarrow 2Na^+(aq) + SO_4^{2-}(aq)$
For 0.5 M Na_2SO_4, $[Na^+]$ = 1 M and $[SO_4^{2-}]$ = 0.5 M

6.8 moles of $KMnO_4$ required = 0.500 L (0.0200 mol/L) = 0.0100 mol $KMnO_4$
0.0100 mol $KMnO_4$ (158.0 g/mol) = 1.58 g $KMnO_4$
Weigh out 1.58 g of solid $KMnO_4$, place it in the volumetric flask, and then fill the flask to the mark with distilled water. After it is shaken thoroughly, the resulting solution has a $KMnO_4$ concentration of 0.0200 M.

6.9 moles of NaOH required = 0.250 L (1.00 mol/L) = 0.250 mol NaOH
volume of 2.00 M solution required = 0.250 mol NaOH (1 L/2.00 mol) = 0.125 L
Measure out, using a pipet or other volumetric glassware, exactly 125 mL of 2.00 M NaOH, place it in a 250. mL volumetric flask, and dilute with distilled water to a total volume of 250. mL.

6.10 (0.15 M)(0.0060 L) = C_d(0.010 L)
C_d = molarity of diluted solution = 0.090 M

6.11 moles of HCl used = 0.0500 L (0.450 mol/L) = 0.0225 mol HCl
0.0225 mol HCl (2 mol NaCl/2 mol HCl) = 0.0225 mol NaCl
0.0225 mol NaCl (58.44 g/mol) = 1.31 g NaCl

6.12 moles of NaOH used = 0.02833 L (0.953 mol/L) = 0.0270 mol NaOH
0.0270 mol NaOH (1 mol acid/1 mol NaOH) = 0.0270 mol acid
0.0270 mol CH_3CO_2H (60.05 g/mol) = 1.62 g CH_3CO_2H
0.0270 mol acid/0.0250 L = 1.08 M

6.13 reaction involved: HCl(aq) + NaOH(aq) $\longrightarrow$ NaCl(aq) + $H_2O(\ell)$
moles of HCl used = (0.02967 L)(0.100 mol/L) = 0.00297 mol
moles of NaOH used = 0.00297 mol
concentration of NaOH = (0.00297 mol/0.02500 L) = 0.119 M

CHAPTER 7

7.1 Energy in the form of heat is transferred to the egg and causes chemical and physical changes. Some of the grease in the pan acquires enough energy that it can spatter out of the pan.

7.2 (a) 160 Calories (1000 calories/1 Calorie)(4.184 J/calorie) = 6.7×10^5 J
(b) 3.0 hours (3600 s/hour) = 1.1×10^4 s
1.1×10^4 s (75 watts) = 8.1×10^5 joules
(c) 16 kJ (1000 J/kJ)(1 cal/4.184 J)(1 kcal/1000 cal) = 3.8 kcal

7.3 24.1 kJ (1000 J/kJ) = 2.41×10^4 J
2.41×10^4 J = (0.902 J/g · K)(250. g)(T_{final} − 5.0 °C)
T_{final} = 112 °C

7.4 A given quantity of heat will raise the temperature more for an object with a smaller specific heat. Therefore, the rock, with the smaller specific heat, should be heated to a higher temperature.

7.5 (400. g)(0.451 J/g · K)(32.8 °C − $T_{initial}$) = −(1000. g)(4.184 J/g · K)(32.8 °C − 20.0 °C)
$T_{initial}$ = 330 °C

7.6 237 g ice (333 J/1 g ice) = 7.89×10^4 J required to melt the ice
(4.184 J/g · K)(237 g)(25.0 °C − 0.0 °C) = 2.48×10^4 J required to warm the water
total heat required = 1.037×10^5 J (or 103.7 kJ)

7.7 (a) 10.0 g I_2 (1 mol/253.8 g) = 0.0394 mol
0.0394 mol I_2 (62.4 kJ/mol) = 2.46 kJ
(b) The condensation of I_2 vapor is exothermic.

3.45 g (1 mol/253.8 g)(62.4 kJ/mol) = 0.848 kJ

The amount of iodine being condensed is approximately 1/3 of the amount being vaporized in part (a). Therfore, the heat evolved on condensation is about 1/3 the amount of heat required for the vaporization of the iodine.

7.8 12.6 g H_2O (1 mol/18.02 g) = 0.699 mol H_2O
0.699 mol H_2O (285.8 kJ/mol) = 2.00×10^2 kJ

7.9 PbS(s) + $\frac{3}{2}O_2(g)$ $\longrightarrow$ PbO(s) + $SO_2(g)$ $\quad \Delta H_1 = -413.7$ kJ
PbO(s) + C(graphite) $\longrightarrow$ Pb(s) + CO(g) $\quad \Delta H_2 = +106.8$ kJ

PbS(s) + C(graphite) + $\frac{3}{2}O_2(g)$ $\longrightarrow$ Pb(s) + CO(g) + $SO_2(g)$
$\Delta H_{overall} = \Delta H_1 + \Delta H_2 = -413.7$ kJ + 106.8 kJ = −306.9 kJ
The reaction is exothermic.
454 g PbS (1 mol of PbS/239.3 g)(306.9 kJ/mol) = 582 kJ
582 kJ (1000 J/kJ) = 5.82×10^5 J

7.10 ΔH°_{rxn} = (6 mol)$\Delta H^\circ_f[CO_2(g)]$ + (3 mol)$\Delta H^\circ_f[H_2O(\ell)]$ − {(1 mol)$\Delta H^\circ_f[C_6H_6(\ell)]$}
= (6 mol)(−393.5 kJ/mol) + (3 mol)(−285.8 kJ/mol) − (1 mol)(49.0 kJ/mol)
= −3267.4 kJ
Note that the enthalpy of formation of O_2 is zero.

7.11 change in temperature = ΔT = 2.32 deg
heat transferred to the calorimeter water = (1.50×10^3 g)(4.184 J/g · K)(2.32 deg) = 14.6×10^3 J
heat transferred to the calorimeter bomb = (837 J/deg)(2.32 deg) = 1.94×10^3 J
total heat transferred by 1.00 g of sucrose = −16.5 kJ
heat of combustion per mole = (−16.5 kJ/g)(342.3 g/mol) = −5650 kJ/mol

7.12 ΔS for liquid $\longrightarrow$ vapor = q/T = 30,800 J/353.3 K = 87.2 J/K
ΔS for vapor $\longrightarrow$ liquid = −87.2 J/K

7.13 (a) Gas changes to a solid, so the entropy decreases.
(b) The entropy increases as solid salt dissolves.
(c) $MgCO_3$ decomposes into another solid and a gas. Therefore, entropy increases.

7.14 (a) A negative enthalpy change and a negative entropy change indicate the reaction is generally product-favored at low temperature but not necessarily at high temperature.
(b) A positive enthalpy change and a positive entropy change indicate the reaction is not product-favored at low temperature but may be product-favored at a sufficiently high temperature.

(c) A negative enthalpy change and a positive entropy change invariably lead to a product-favored reaction.

(d) A positive enthalpy change and a negative entropy change mean the reaction is never product-favored under any conditions of temperature.

7.15 The enthalpy change for this reaction, $\Delta H^\circ_{rxn} = 2\,\Delta H^\circ_f$ for HCl(g) = (2 mol)(−92.3 kJ/mol). That is, the reaction is strongly exothermic. Since two gas molecules give two molecules of HCl(g), the entropy change is small.

$$\begin{aligned}\Delta S^\circ_{system} &= 2\,S^\circ[\text{HCl(g)}] - S^\circ[\text{H}_2\text{(g)}] - S^\circ[\text{Cl}_2\text{(g)}]\\ &= 2\text{ mol}\,(186.908\text{ J/K}\cdot\text{mol})\\ &\quad - 1\text{ mol}\,(130.684\text{ J/K}\cdot\text{mol})\\ &\quad - 1\text{ mol}\,(223.036\text{ J/K}\cdot\text{mol})\\ &= 19.886\text{ J/K}\end{aligned}$$

$$\begin{aligned}\Delta S^\circ_{surroundings} &= -\Delta H^\circ/T = -(185{,}000\text{ J})/298\text{ K}\\ &= 619\text{ J/K}\end{aligned}$$

$$\Delta S^\circ_{universe} = 619\text{ J/K} + 20.096\text{ J/K} = 639\text{ J/K}$$

Since the entropy change for the universe is positive, the reaction is predicted to be product-favored. The combination of a negative enthalpy change (ΔH°_{rxn}) and a positive entropy change (ΔS°_{rxn}) for a reaction means, in general, that the reaction is product-favored.

7.16 $$\begin{aligned}\Delta H^\circ_{system} &= \Delta H^\circ[\text{CH}_3\text{OH}(\ell)]\\ &\quad - \Delta H^\circ[\text{CO(g)}] - 2\,\Delta H^\circ[\text{H}_2\text{(g)}]\\ &= (1\text{ mol})(-238.66\text{ kJ/mol})\\ &\quad - (1\text{ mol})(-110.525\text{ kJ/mol}) - (2\text{ mol})(0)\\ &= -128.14\text{ kJ}\end{aligned}$$

$$\begin{aligned}\Delta S^\circ_{system} &= S^\circ[\text{CH}_3\text{OH}(\ell)] - S^\circ[\text{CO(g)}] - 2\,S^\circ[\text{H}_2\text{(g)}]\\ &= (1\text{ mol})(126.8\text{ J/K}\cdot\text{mol})\\ &\quad - (1\text{ mol})(197.674\text{ J/K}\cdot\text{mol})\\ &\quad - (2\text{ mol})(130.684\text{ J/K}\cdot\text{mol})\\ &= -332.2\text{ J/K}\end{aligned}$$

$$\begin{aligned}\Delta G^\circ &= \Delta H^\circ - T\Delta S^\circ\\ &= -128.14\text{ kJ} - (298.15\text{ K})(-0.3322\text{ kJ/K})\\ &= -29.09\text{ kJ}\end{aligned}$$

The sign of ΔG° is negative, indicating the reaction is product-favored.

7.17 (a) Step (i): $\Delta H^\circ = +824.2$ kJ, $\Delta S^\circ = +0.276$ kJ/K, $\Delta G^\circ = +742.2$ kJ

The reaction is reactant-favored, owing to the positive free energy change.

Step (ii): $\Delta H^\circ = -1675.7$ kJ, $\Delta S^\circ = -0.313$ kJ/mol, $\Delta G^\circ = -1582.3$ kJ

The reaction is product-favored, as indicated by the negative free energy change. Note that both the enthalpy and entropy changes are negative. However, the large negative value for ΔH° outweighs the small entropy change.

(b) Net reaction: $Fe_2O_3(s) + 2\,Al(s) \longrightarrow 2\,Fe(s) + Al_2O_3(s)$

$\Delta H^\circ_{rxn} = -851.5$ kJ, $\Delta S^\circ_{rxn} = -0.037$ kJ/K, $\Delta G^\circ_{rxn} = -840.1$ kJ

This reaction is product-favored. Even though the entropy change is negative, it is small and is outweighed by the large, negative value of ΔH°_{rxn}.

(c) We have coupled a strongly product-favored reaction (the formation of aluminum oxide) to the reactant-favored decomposition of iron(III) oxide. The large negative free energy change for the formation of aluminum oxide allows the net process to be product-favored.

CHAPTER 8

8.1 $C_0 = 0.050$ M and $C_2 = 0.033$ M; $\Delta C = -0.017$ M

$\Delta C/\Delta t = -0.017\text{ M}/2\text{ hr} = -8.5 \times 10^{-3}$ M/hr

$C_6 = 0.015$ M and $C_8 = 0.010$ M; $\Delta C = -5 \times 10^{-3}$ M

$\Delta C/\Delta t = -5 \times 10^{-3}\text{ M}/2\text{ hr} = -3 \times 10^{-3}$ M/hr

Notice that the reaction becomes slower with time.

8.2 (a) The order is 2 with respect to NO and 1 with respect to H_2.

(b) The reaction rate increases by 4.

(c) The reaction rate is also halved.

8.3 Rate = $k[Pt(NH_3)_2Cl_2] = (0.090\text{ hr}^{-1})(0.020\text{ M}) = 1.8 \times 10^{-3}$ M/hr

Since $Pt(NH_3)_2Cl_2$ is a reactant and is being used up,

$$\frac{\Delta[\text{Pt(NH}_3)_2\text{Cl}_2]}{\Delta t} = -1.8 \times 10^{-3}\text{ M/hr}$$

Since Cl^- is formed at the same rate that $Pt(NH_3)_2Cl_2$ is used up, $\Delta[Cl^-]/\Delta t = 1.8 \times 10^{-3}$ M/hr.

8.4 The hill has to be 24 kJ/mol high from left to right and 36 kJ/mol high from right to left. Since B is below A on the energy scale, the reaction is exothermic.

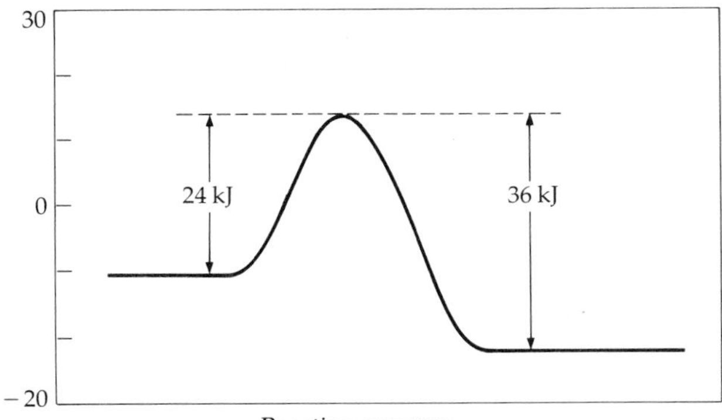

8.5 The overall, balanced stoichiometric equation is

$$2\,NH_3(aq) + OCl^-(aq) \longrightarrow N_2H_4(aq) + Cl^-(aq) + H_2O(\ell)$$

The second step is slow and is therefore the rate limiting step. The intermediates are NH_2Cl, OH^-, and $N_2H_5^+$.

8.6 (a) True. The concentration of a catalyst *must* appear in the rate law, since the catalyst affects the reaction rate.

(b) False. A catalyst is never consumed in the *overall* reaction. (A catalyst is always consumed in one step of the reaction and then regenerated in another step.)

(c) False. A catalyst may be in a different phase than the reactants.

8.7 (a) $K = \dfrac{[H_2S]}{[H_2]}$

(b) $K = \dfrac{[H_2]}{[HCl]}$

(c) $K = [Ag^+][Cl^-]$

(d) $K = \dfrac{[Cu^{2+}][NH_3]^4}{\{[Cu(NH_3)_4]^{2+}\}}$

8.8 (a) Multiplying the stoichiometric coefficients in the reaction by 2 means the value of K is raised to the second power. Therefore,

$$K_{new} = K^2{}_{old} = (2.5 \times 10^{-29})^2 = 6.3 \times 10^{-58}$$

(b) The reaction in (a) has been reversed, so the value of K for this new reaction is the reciprocal of the value for the reaction in (a).

$$K_{new} = 1/K_{old} = 1/6.3 \times 10^{-58} = 1.6 \times 10^{57}$$

8.9 Since K for AgCl is larger than for AgI, the concentration of Ag^+ in the AgCl beaker will be larger than the concentration of Ag^+ in the AgI beaker.

8.10 The equilibrium constant for the reverse reaction is the reciprocal of $K_{forward}$.

$$K_{reverse} = 1/K_{forward} = 1/1.8 \times 10^{-5} = 5.6 \times 10^4$$

This is large, indicating that the reaction of acetate ions with hydronium ions is strongly product-favored.

8.11

	NO_2	N_2O_4
Initial conc. (M)	0.90	0.00
Change in conc. (M)	−0.64	+0.32
Equilibrium conc. (M)	0.26	0.32

$$K = \frac{[N_2O_4]}{[NO_2]^2} = \frac{0.32}{(0.26)^2} = 4.7$$

The reaction is product-favored at this temperature.

8.12 At the lower temperature K is larger, which means the concentration of N_2O_4 is larger (and that of NO_2 is smaller) than at a higher temperature. This is predicted by Le Chatelier's principle, since the reaction is exothermic. As the temperature is lowered the reaction shifts left-to-right, which releases heat and thereby reduces the lowering of the temperature. The figure shows that the gas is less brown at the lower temperature, indicating that the concentration of NO_2 decreases as the temperature is lowered.

CHAPTER 9

9.1 $104.5\text{ MHz} = 104.5 \times 10^6\ s^{-1} = 1.045 \times 10^8\ s^{-1}$

$$\lambda = \frac{c}{\nu} = \frac{2.998 \times 10^8\ m\ s^{-1}}{1.045 \times 10^8\ s^{-1}} = 2.869\ m$$

9.2 Frequency decreases when wavelength increases.

9.3 (a) One photon of ultraviolet radiation has more energy than one proton of microwave radiation.

(b) One photon of blue light has more energy than one photon of green light.

9.4 $\nu = \dfrac{c}{\lambda} = \dfrac{2.998 \times 10^8\ m\ s^{-1}}{2.36 \times 10^{-9}\ m} = 1.27 \times 10^{17}\ s^{-1}$

x-ray photon energy $= h\nu$
$= (6.626 \times 10^{-34}\ J \cdot s)(1.27 \times 10^{17}\ s^{-1})$
$= 8.42 \times 10^{-17}\ J$

The energy of one photon of orange light (3.18×10^{-19} J) is less than that of an x-ray photon.

9.5 There are many energy levels within the hydrogen atom. In a collection of hydrogen atoms, many different transitions may occur simultaneously in the different atoms.

9.6 The maximum number of s orbitals that may be found in a given electron shell is 1. The maximum number is 3 for p orbitals, 5 for d orbitals, and 7 for f orbitals.

9.7 The total number of electrons in the $n = 3$ level is $2(3^2) = 18$. The $n = 3$ energy level has two electrons in an s orbital, two electrons in each of three p orbitals, and two electrons in each of five d orbitals.

9.8 S: $1s^22s^22p^63s^23p^4$ Noble gas notation: $[Ne]3s^23p^4$

1s	2s	$2p_x$	$2p_y$	$2p_z$	3s	$3p_x$	$3p_y$	$3p_z$
↑↓	↑↓	↑↓	↑↓	↑↓	↑↓	↑↓	↑	↑

A sulfur atom has two unpaired electrons.

9.9 Cl: Noble gas notation: $[Ne]3s^23p^5$ Chlorine has 7 electrons in the $n = 3$ level.

S: Noble gas notation: $[Ne]3s^23p^4$ Sulfur has 6 electrons in the $n = 3$ level.

9.10 Se: $[Ar]3d^{10}4s^24p^4$ Te: $[Kr]4d^{10}5s^25p^4$:S̈e· :T̈e·

9.11 Rb· ·Ṡi· :B̤r· ·Ba·

9.12 (a) ns^2np^1

(b) Ga^{3+}

(c) $[Ar]3d^{10}$

9.13 (a) Ni: Noble gas notation: $[Ar]3d^84s^2$

(b)

3d	3d	3d	3d	3d	4s
↑↓	↑↓	↑↓	↑	↑	↑↓

A nickel atom has two unpaired electrons.

(c) A Ni^{2+} ion is formed by removing two 4s electrons from a Ni atom. The orbital box diagram for Ni^{2+} is

3d	3d	3d	3d	3d
↑↓	↑↓	↑↓	↑	↑

A Ni^{2+} ion has two unpaired electrons.

9.14 C < Si < Al

9.15 Ion sizes: $N^{3-} > O^{2-} > F^-$. All ions have 10 electrons in this isoelectronic series but the number of protons attracting the electrons is 7, 8, and 9, respectively. This increasing nuclear charge attracts the 10 electrons more strongly and causes a corresponding decrease in the size of the isoelectronic ions.

9.16 Ba < Al < Mg < P < F

9.17 Since phosphorus has five valence electrons, the largest increase would be expected between the 5th and 6th ionization energy.

9.18 (a) $Mg(s) + H_2(g) \longrightarrow MgH_2(s)$

(b) Pt would be predicted to form interstitial hydrides.

9.19 $2\ Rb(s) + 2\ H_2O(\ell) \longrightarrow 2\ RbOH(aq) + H_2(g)$

Rb would be more reactive than K. Since K reacts explosively with water, the Rb reaction would be extremely violent.

9.20 (a) Sr_3N_2 (b) $MgBr_2$ (c) Ca_3P_2 (d) BaO

9.21 The primary reason is the lower ionization energies of the heavier alkali and alkaline earth metals.

9.22 (a) $GeO_2(s) + C(s) \longrightarrow Ge(s) + CO_2(g)$ or

$GeO_2(s) + 2\ C(s) \longrightarrow Ge(s) + 2\ CO(g)$

(b) $2\ PbS(s) + 3\ O_2(g) \longrightarrow 2\ PbO(s) + 2\ SO_2(g)$

CHAPTER 10

10.1 (a) Nitrogen (Group 5A) has five valence electrons and needs to share three electrons to achieve an octet. Each fluorine (Group 7A) has seven valence electrons and needs to share one electron to achieve an octet.

:F̤—N̈—F̤:
 |
:F̤:

(b) Sulfur (Group 6A) has six valence electrons and needs two for an octet. Each hydrogen can provide one for sharing.

H—S̤̈—H

(c) The total number of valence electrons is 14 (5 from each N and 1 from each H). Since H atoms form only one bond, the two N atoms must be bonded to each other.

H—N̈—N̈—H
 | |
 H H

10.2

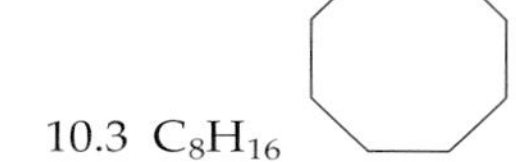

10.3 C_8H_{16}

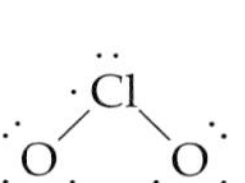

10.4 (b) and (c) are correct. (a) is incorrect because it uses 20 electrons when only 16 electrons are available (4 from C and 12 from 2 O). (d) has too many electrons around the fluorine on the left. The correct structures for (a) and (d) are shown below.

:Ö=C=Ö: :F̤—N̈—C̤l:
     |
     :C̤l:

10.5 NO^+ Total valence electrons = 5 from N + 6 from O − 1 (positive charge) = 10

$[:N{\equiv}O:]^+$

HCN Total valence electrons = 1 from H + 4 from C + 5 from N = 10

$H{-}C{\equiv}N:$

10.6 BeF_2 ClO_2 PCl_5

:F̤—Be—F̤:

·C̤l
/ \
:Ö: :Ö:

:C̤l:
 | :C̤l:
:C̤l—P<
 | :C̤l:
:C̤l:

10.7 (a) C—S is shorter than C—Si because the S atom is smaller than the Si atom.

(b) C—Cl is shorter than C—Br because the Cl atom is smaller than the Br atom.

(c) N≡O is shorter than N=O because triple bonds are shorter than double bonds.

10.8 The decreasing order of bond distance is C—N > C═N > C≡N. Decreasing bond energy is in the reverse order. The shorter multiple bond has more electron density in the bonding region, and, as a result, has a higher bond energy.

10.9 $C_5H_{12} + 8\,O_2 \longrightarrow 5\,CO_2 + 6\,H_2O$

```
   H  H  H  H  H
   |  |  |  |  |
H—C—C—C—C—C—H + 8 O═O ⟶
   |  |  |  |  |
   H  H  H  H  H

                    5 O═C═O + 6 H—O—H
```

Bonds Broken

4 C—C	4 × 347 =	1388 kJ
12 C—H	12 × 414 =	4968 kJ
8 O═O	8 × 498 =	3984 kJ
Total:		10,340 kJ mol^{-1}

Bonds Formed

10 C═O	10 × 803 =	8030 kJ
12 O—H	12 × 463 =	5556 kJ
Total:		13,586 kJ mol^{-1}

Thus ΔH°_{rxn} = 10,340 kJ mol^{-1} − 13,586 kJ mol^{-1} = −3246 kJ mol pentane^{-1}

10.10 NO_2^-

$[\,\ddot{O}—\ddot{N}═\ddot{O}\,]^- \longleftrightarrow [\,\ddot{O}═\ddot{N}—\ddot{O}\,]^-$ $[\,O\cdots\ddot{N}\cdots O\,]^-$

resonance structures   resonance hybrid structure

According to the resonance hybrid given, the bond length should be halfway between N—O (136 pm) and N═O (115 pm). The predicted value is 126 pm, which is in good agreement with the experimentally measured value of 124 pm.

10.11 (a) B—Cl is more polar than B—C because Cl is more electronegative than C. The positive end of the dipole in both bonds is B.
(b) O—H is more polar than N—H because O is more electronegative than N. The positive end of the dipole in both bonds is H.

10.12 (a) +4 (b) +4 (c) +3 (d) +7 (e) +4 (f) +5

10.13 In A all atoms have a formal charge of 0. However, in B, the B atom has a formal charge of −1 and the F of the F═B double bond has a +1 formal charge. A formal charge of +1 on F is not reasonable, since F is the most electronegative element. Therefore, structure A is much more important.

10.14 (a) The complex ion is $[Co(NH_3)_4(CO_3)]^+$.
(b) The ligands are NH_3 and CO_3^{2-}.
(c) The metal ion is Co^{3+}. It has a +3 charge because CO_3^{2-} is −2 and the total charge of the complex ion is +1.

CHAPTER 11

11.1 (a) Cl_2O has two bond pairs and two lone pairs on O. The electron-pair geometry is tetrahedral. (b) SO_3^{2-} has three bond pairs and one lone pair on S. The electron-pair geometry is tetrahedral. (c) SiO_4^{4-} has four bond pairs on Si. The electron-pair geometry is tetrahedral. The molecular geometries are (a) angular, (b) triangular pyramidal and (c) tetrahedral.

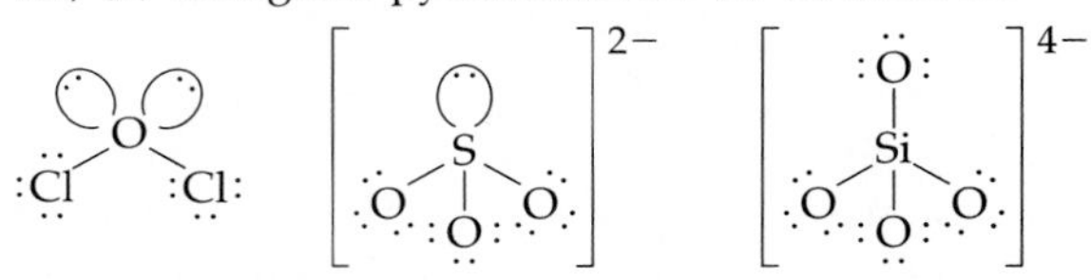

11.2 ICl_2^- has two bond pairs and three lone pairs on I. The electron-pair geometry is triangular bipyramidal and the molecular geometry is linear.

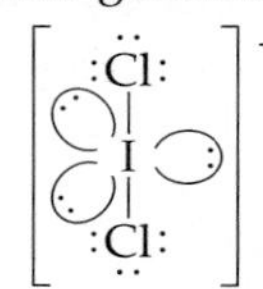

11.3 SCl_2 has two bond pairs and two lone pairs on S. Since lone pairs occupy more volume than bond pairs, the ClSCl angle would be expected to be less than 109.5°. (The experimental value is 102.7°.) PCl_3 has three bond pairs and one lone pair on the P. The ClPCl angle would be expected to be less than 109.5°. (The experimental value is 100.3°.)

$:\ddot{Cl}—S—\ddot{Cl}:$ (angular, two lone pairs on S)   P with three :Cl: and one lone pair

11.4 $BFCl_2$ is polar, with the B—F side more negative. NH_2Cl, with a triangular pyramidal molecular geometry, is polar with the Cl atom the more negative side. SCl_2, which has an angular molecular geometry, is polar with the negative end of the dipole toward the Cl atoms.

:F: δ− over B with :Cl δ+ Cl: :Cl: δ− over N with H δ+ H   δ+ S with :Cl δ− Cl:

11.5 *EXPANDED STRUCTURAL FORMULAS:*

```
   H  H  H  H  H
   |  |  |  |  |
H—C—C—C—C—C—H
   |  |  |  |  |
   H  H  H  H  H
```

```
    H  H     H     H
    |  |     |     |
H—C—C——C——C—H
    |  |     |     |
    H  H  H—C—H  H
             |
             H

           H
           |
   H  H—C—H  H
   |       |     |
H—C————C————C—H
   |       |     |
   H  H—C—H  H
           |
           H
```

CONDENSED FORMULAS:

$CH_3CH_2CH_2CH_2CH_3$ — pentane

$CH_3CH_2CHCH_3$ with CH_3 on the third carbon — 2-methylbutane

CH_3CCH_3 with CH_3 above and below the central carbon — 2,2-dimethylpropane

11.6 (a) $CH_3CHCH_2CH_2CH_3$ with CH_3 on C-2 — 2-methylpentane

(b) $CH_3CH_2CHCH_2CH_3$ with CH_3 on C-3 — 3-methylpentane

(c) $CH_3CCH_2CH_3$ with two CH_3 on C-2 — 2,2-dimethylbutane

(d) $CH_3CHCHCH_3$ with CH_3 on C-2 and C-3 — 2,3-dimethylbutane

11.7 $CH_3C{=}CCH_2CH_3$ with CH_3 on C-2 and C-3 — 2,3-dimethyl-2-pentene

11.8 Benzene ring with CH_3 groups in positions 1, 2 and 4 — 1,2,4-trimethylbenzene

11.9 Only *c* can have *cis* and *trans* isomers.

cis: CH_2Br and Cl on one carbon of $C{=}C$, CH_3 and H on the other (CH_2Br and CH_3 on the same side)

trans: Cl and CH_2Br on one carbon of $C{=}C$, CH_3 and H on the other (CH_2Br and CH_3 on opposite sides)

11.10 The molecule in *b* is chiral because there are four different groups attached to the carbon atom.

Left: C with OH, H, Cl, CH_3 | mirror | Right: C with OH, H_3C, Cl, H

CHAPTER 12

12.1 $$\text{pressure} = 29.1 \text{ in Hg}\left(\frac{2.54 \text{ cm}}{1 \text{ in}}\right)\left(\frac{10 \text{ mm}}{1 \text{ cm}}\right)$$
$$= 739 \text{ mm Hg}$$
$$\text{pressure} = 739 \text{ mm Hg}\left(\frac{1 \text{ atm}}{760 \text{ mm Hg}}\right) = 0.972 \text{ atm}$$

12.2 We have P_1, V_1, and P_2. We can solve for V_2 by using

$$V_2 = \frac{P_1V_1}{P_2} = \frac{(1.00 \text{ atm})(400. \text{ mL})}{0.750 \text{ atm}} = 533 \text{ mL}$$

12.3 We have V_1, T_1, and V_2. We can solve for T_2 by using

$$T_2 = \frac{T_1V_2}{V_1} = \frac{(25 + 273 \text{ K})(175 \text{ mL})}{100. \text{ mL}} = 522 \text{ K}$$
$$= 522 \text{ K} - 273 \text{ K} = 249\ ^\circ\text{C}$$

12.4 Since the coefficients in the equation are 1 for CH_4 and 2 for O_2, the volume of oxygen needed will be twice the volume of methane. Since only 1/5 of the air admitted is oxygen, 5 L of air will be required to supply 1 L of oxygen. Thus the volume of air supplied needs to be $2 \times 5 = 10$ times the volume of methane, and the tube supplying air needs to be ten times larger. To supply a slight excess of oxygen, the tube might be made slightly larger than this. However, supplying too much air will cool the flame, since the nitrogen and any unused oxygen will be heated by the exothermic combustion reaction.

12.5 Using $PV = nRT$, Boyle's law is concerned with a fixed amount of gas (n) at a constant temperature (T). Therefore, the term nRT is a constant and the equation becomes PV = constant, which is Boyle's law. In Charles's law, n and V are constants, so upon rearranging the ideal gas law equation we get $P =$

constant × T, where the constant is nR/V. This is the expression for Charles's law.

12.6 The radius of the balloon is 10. cm, so its volume is

$$V = \tfrac{4}{3}\pi r^3 = \tfrac{4}{3}(3.14159)(10.\text{ cm})^3 = 4.2 \times 10^3 \text{ cm}^3$$
$$= 4.2 \text{ L}$$

From volume, temperature (assuming 20 °C is room temperature), and pressure, calculate the amount of gas:

$$n_{CO_2} = \frac{PV}{RT} = \frac{(2.0 \text{ atm})(4.2 \text{ L})}{(0.0821 \text{ L} \cdot \text{atm/mol} \cdot \text{K})(293 \text{ K})}$$
$$= 0.33 \text{ mol}$$

Then use the stoichiometric ratio and molar mass to calculate the mass of $NaHCO_3$:

$$m_{NaHCO_3} = 0.33 \text{ mol } CO_2\left(\frac{1 \text{ mol } NaHCO_3}{1 \text{ mol } CO_2}\right) \cdot \left(\frac{84.0 \text{ g}}{\text{mol } NaHCO_3}\right) = 30 \text{ g}$$

12.7 Using $PV = nRT$ and recognizing that n (moles) = (mass/molar mass), substitute for n in $PV = nRT$

$P = 0.850$ atm; $V = 1.00$ L; $T = 293.15$ K; mass = 1.13 g; M = ? g/mol

$$M = \frac{(1.13 \text{ g})(0.0821 \text{ L} \cdot \text{atm/K} \cdot \text{mol})(293.15 \text{ K})}{(0.850 \text{ atm})(1.00 \text{ L})}$$

$$M = 32.0 \text{ g/mol}$$

The gas is probably oxygen.

12.8 First, calculate the number of moles of each gas. For nitrogen, $n_N = 7.0 \text{ g}/(28.0 \text{ g/mol}) = 0.25$ mol. For hydrogen, $n_H = 6.0 \text{ g}/2.02 \text{ g/mol}) = 3.0$ mol. The total number of moles is 3.2 mol. Then

$$X_H = \frac{3.0 \text{ mol}}{3.3 \text{ mol}} = 0.91$$

$$X_N = \frac{0.25 \text{ mol}}{3.3 \text{ mol}} = 0.076$$

Use the ideal gas law to solve for P:

$$P = \frac{nRT}{V} =$$
$$\frac{(3.3 \text{ mol})(0.0821 \text{ L} \cdot \text{atm/mol} \cdot \text{K})(500. + 273 \text{ K})}{5.0 \text{ L}} = 42 \text{ atm}$$

Now the partial pressures are given by

$$P_H = P_{total}X_H = (42 \text{ atm})(0.91) = 38 \text{ atm}$$
$$P_N = P_{total}X_N = (42 \text{ atm})(0.076) = 3.2 \text{ atm}$$

12.9 The pressure increases in a container of fixed volume and constant temperature when more gas is added because more molecules are available to collide with the walls of the container.

12.10 The ratio of the rates of diffusion of N_2 and H_2 is given by

$$\frac{\text{rate}_N}{\text{rate}_H} = \sqrt{\frac{2.0}{28}} = 0.27$$

12.11 The H---F bond is the strongest hydrogen bond because fluorine has the largest electronegativity of all elements.

12.12 (a) Dispersion forces, because the N_2 molecules are nonpolar.
(b) Induced dipoles in CO_2 molecules are attracted to polar water molecules. Also dispersion forces.
(c) Hydrogen bonding exists between NH_3 molecules and CH_3OH molecules. Also induced dipole and dispersion forces.

12.13 The $CHBr_3$ molecule has the larger molar mass and hence more electrons. This causes the dispersion forces to be stronger and the surface tension larger.

12.14 By making plots of ΔH_{vap} and boiling point vs. molar mass for Ar, Kr, and Xe, you can estimate that ΔH_{vap} for Kr is about 9.5 kJ/mol and the b.p. is about 122 K. The actual values are 9.1 kJ/mol and 119.8 K. For NO_2, you should use the values from Table 12.5 for O_2 and SO_2. Plots of ΔH_{vap} and boiling point vs. molar mass for these compounds show that ΔH_{vap} for NO_2 is about 20 kJ/mol and the b.p. is about 290 K, or 17 °C. The actual values are 19 kJ/mol and 21.15 °C.

12.15 Using Figure 12.24, the estimated boiling point of water at 400 mm Hg is about 83 °C. At 1000 mm Hg, the boiling point would be about 109 °C.

CHAPTER 13

13.1 $C_7H_{16} \longrightarrow C_6H_5CH_3 + 4\ H_2$

13.2 Although the reaction for the production of synthesis gas is endothermic (130 kJ/mol), combustion of the components of synthesis gas, CO and H_2, releases heat. $\Delta H_{net} = -395$ kJ.

$$C + H_2O \longrightarrow CO + H_2 \quad \Delta H = 130 \text{ kJ/mol C}$$
$$\tfrac{1}{2}[2\ CO + O_2 \longrightarrow 2\ CO_2] \quad \Delta H = -566 \text{ kJ } (-283 \text{ kJ/mol CO})$$
$$\tfrac{1}{2}[2\ H_2 + O_2 \longrightarrow 2\ H_2O] \quad \Delta H = -484 \text{ kJ } (-242 \text{ kJ/mol } H_2)$$

Net: $C + O_2 \longrightarrow CO_2$
$+130$ kJ/mol C + (-283 kJ/mol CO) +
(-242 kJ/mol H_2) = -395 kJ for overall reaction

13.3 Substitution reactions involve the replacement of hydrogen in alkanes or benzene derivatives with other atoms or groups of atoms. Addition reactions take place at the double or triple bond in alkenes and alkynes.

$$C_2H_6(g) + 2\ Cl_2(g) \xrightarrow[\text{or heat}]{\text{UV light}} C_2H_4Cl_2(g) + 2\ HCl(g)$$

$$C_2H_4(g) + Cl_2(g) \xrightarrow{\text{dark, 25 °C}} C_2H_4Cl_2(g)$$

13.4 The reaction products are

(a) $CH_3CH_2C(=O)-H$ (b) $CH_3C(=O)CH_3$

13.5 (a) 2-propanol or isopropyl alcohol
(b) ethylene glycol
(c) ethanol
(d) glycerol (glycerin)
(e) methanol

13.6 (a) $CH_3C(=O)OH + CH_3OH \longrightarrow CH_3C(=O)OCH_3$

(b) $CH_3CH_2C(=O)OH + HOCH_2CH_2OH \longrightarrow$
$CH_3CH_2C(=O)OCH_2CH_2OC(=O)CH_2CH_3$

13.7 (a) $H_2C=CH_2$ (H, H / H, H) (b) H, H / H, Cl on C=C (c) H, H / H, C_6H_5 on C=C

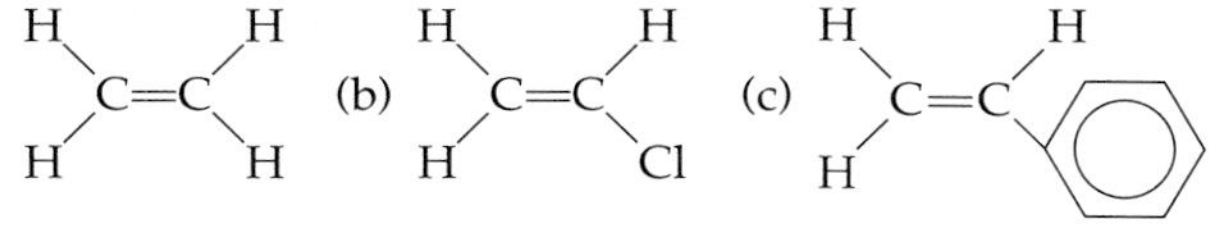

13.8 $n\,HO-C(=O)-C_6H_4-C(=O)-OH$ +
Terephthalic acid

$n\,HO-CH_2-CH_2-OH \longrightarrow$
Ethylene glycol

$$\left(-C(=O)-C_6H_4-C(=O)-O-CH_2-CH_2-O-\right)_n$$

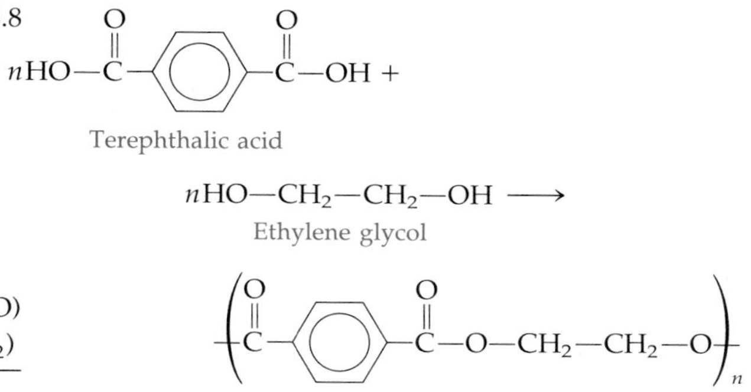

CHAPTER 14

14.1 The enthalpy of fusion of NaCl is 30.21 kJ/mol.
mol NaCl = (100.0 g) (1 mol/58.443 g) = 1.711 mol
heat = 30.21 kJ/mol (1.711 mol) = 51.69 kJ

14.2 1 Po atom per unit cell, 2 Li atoms per unit cell, 4 Ca atoms per unit cell.

14.3 We can calculate the edge of the unit cell, s, using $s\sqrt{3} = 4 \times 144$ ppm, where $s\sqrt{3}$ is the diagonal of the body-centered unit cell (see Figure 14.11).

$$s = 333 \text{ pm} = 3.33 \times 10^{-8} \text{ cm}$$

So, the volume of the unit cell is $(3.33 \times 10^{-8} \text{ cm})^3 = 3.69 \times 10^{-23} \text{ cm}^3$
To calculate the density, we need the mass of one gold atom. The molar mass of gold is 196.97 g/mol. The mass of one gold atom is

mass Au atom = (196.97 g/mol) ·
(1 mol/6.022×10^{23} atoms) = 3.304×10^{-22} g/atom

Since the bcc unit cell contains two atoms, the density is

$$\text{Density} = \frac{2 \text{ atoms } (3.304 \times 10^{-22} \text{ g/atom})}{3.66 \times 10^{-23} \text{ cm}^3} = 18.05 \text{ g/cm}^3$$

14.4 Since the center atom is the Cs atom and each Cl atom at each of the eight corners of the cell contributes $\frac{1}{8}$, the formula is Cs + 8($\frac{1}{8}$) Cl = CsCl.

14.5 The unit cell edge is 167 pm + 2(152 pm) + 167 pm = 638 pm.
Therefore KCl has the larger unit cell. (NaCl unit cell has an edge of 566 pm.)
The volume of the unit cell is $(638 \text{ pm})^3 = 2.60 \times 10^8 \text{ pm}^3$
V(in cm^3) = $2.60 \times 10^8 \text{ pm}^3 (10^{-10} \text{ cm/pm})^3 = 2.60 \times 10^{-22} \text{ cm}^3$

The mass of KCl per unit cell is

$= 4$ KCl formula units (74.55 g/mol) $\cdot$
(1 mol/6.022×10^{23} formula units)
$= 4.952 \times 10^{-22}$ g

This means the density of KCl is

$$\text{density KCl} = 4.95 \times 10^{-22}\ \text{g}/2.60 \times 10^{-22}\ \text{cm}^3 = 1.91\ \text{g/cm}^3$$

14.6 Using the densities of Al and Hg of 2.70 g/cm^3 and 13.54 g/cm^3 respectively, we can calculate the mass of 1 cm^3 of each. The heat required to melt each sample can then be calculated from the enthalpies of fusion and molar masses.

mass Hg $= 1.00\ \text{cm}^3(13.54\ \text{g/cm}^3) = 13.54$ g
heat required to melt 1.00 cm^3 Hg $=$
13.54 g Hg $\cdot$ (1 mol Hg/200.59 g Hg) (2.3 kJ/mol) $= 0.16$ kJ

mass Al $= 1.00\ \text{cm}^3(2.70\ \text{g/cm}^3) = 2.70$ g
heat required to melt 1.00 cm^3 Al $=$
2.70 g Al $\cdot$ (1 mol Al/26.98 g Al)(10.7 kJ/mol) $= 1.07$ kJ

14.7 Since diamond is more dense than graphite, high pressure would help bring about the change. A high temperature would help cause the change to occur more quickly. In practice, both high pressures and high temperatures are used.

14.8 Student drawings should look like Figures 14.23, 14.24, and 14.25.

CHAPTER 15

15.1 The long hydrocarbon chain in octanol is hydrophobic. Neighboring octanol molecules are attracted to one another strongly by dispersion forces. Therefore this alcohol has little solubility in water. Methanol, on the other hand, contains a much smaller hydrocarbon part and has much smaller dispersion forces. Methanol molecules can readily hydrogen bond with water molecules, and do, giving rise to solubility in water. Solubility in gasoline would be the opposite because the hydrocarbon molecules in gasoline are attracted to one another by dispersion forces.

15.2 At a partial pressure of 1520 mm Hg,

$$S_{N2} = k_H P_{N_2} = 8.4 \times 10^{-7}\ \text{M/mm Hg} \cdot 1520.\ \text{mm Hg} = 1.3 \times 10^{-3}\ \text{M}$$

At a partial pressure of 20. mm Hg,

$$S_{N2} = k_H P_{N_2} = 8.4 \times 10^{-7}\ \text{M/mm Hg} \cdot 20.\ \text{mm Hg} = 1.7 \times 10^{-5}\ \text{M}$$

15.3 Endothermic. The solubility is greater at higher temperature, which means that the solubility equilibrium shifts toward the right as temperature increases. According to Le Chatelier's principle, the equilibrium will shift to counteract the increase in temperature by absorbing heat. If a shift from left to right absorbs heat, the solution process must be endothermic.

15.4

$$\text{mass fraction Hg} = \frac{8.7 \times 10^{-9}\ \text{g}}{1000\ \text{g}} \cdot 10^9\ \text{ppb} = 0.0087\ \text{ppb}$$

$$\text{mass fraction Hg} = \frac{8.7 \times 10^{-9}\ \text{g}}{1000\ \text{g}} \cdot 10^{12}\ \text{ppt} = 8.7\ \text{ppt}$$

15.5 Note: The following summation of lead intake represents an extreme upper limit that is never reached in most people.

Lead from food $= 2$ kg/day $\cdot$ 200 μg/kg $= 400$ μg/day
Lead from water and beverages $=$
3 L/day $\cdot$ 100 μg/L $= 300$ μg/day
Lead from air $=$
11,500 L/day $\cdot$ 2.5 μg/m^3 $\cdot$ m^3/1000 L $= 29$ μg/day

Totaling these values gives 700 μg/day, or about three times what the body can easily get rid of per day. If we ingested this much lead every day we would probably suffer from some form of lead toxicity.

15.6 The mass fraction expressed as ppm gives the grams of lead per million grams of solution. Use this factor to calculate the mass of solution that contains 100.0 μg Pb:

$$\text{mass of solution} = 100.0\ \mu\text{g Pb} \cdot \frac{10^6\ \text{g soln.}}{0.025\ \text{g Pb}} \cdot \frac{1\ \text{g Pb}}{10^6\ \mu\text{g Pb}} = 4000\ \text{g soln.}$$

Assume that, since the solution is very dilute, its density is 1.0 g/mL. Then 4000 g corresponds to 4000 mL or 4.0 L.

15.7 $\Delta T_b = K_b\, m_{solute} i_{solute} = 2.53\ °\text{C}/m \cdot 0.10\ \text{m} \cdot 1 = 0.25\ °\text{C}$

$T_b = 80.10\ °\text{C} + 0.25\ °\text{C} = 80.35\ °\text{C}$

15.8 $\Delta T_b = -30.\ °\text{C}$; $i_{solute} = 1$; therefore

$$\text{m}_{solute} = \frac{\Delta T_b}{K_f \cdot i_{solute}} = \frac{-30.\ °\text{C}}{-1.86\ °\text{C/m} \cdot 1} = 16.\ \text{mol ethylene glycol/kg water}$$

Since there is 4.0 kg water (4.0 L),

$$4.0 \text{ kg water} \cdot \frac{16 \text{ mol ethylene glycol}}{\text{kg water}} = 64 \text{ mol ethylene glycol}$$

$$64 \text{ mol ethylene glycol} \cdot \frac{62.1 \text{ g}}{\text{mol}} = 4.0 \times 10^3 \text{ g}$$

$$4.0 \times 10^3 \text{ g} \cdot \frac{\text{mL}}{1.113 \text{ g}} = 3.6 \times 10^3 \text{ mL} = 3.6 \text{ L}$$

15.9 From the osmotic pressure equation calculate the molarity of hemoglobin

$$C = \frac{\pi}{RT} = \frac{1.8 \times 10^{-3} \text{ atm}}{0.0821 \text{ L atm/mol K} \cdot 298 \text{ K}} = 7.4 \times 10^{-5} \frac{\text{mol}}{\text{L}}$$

There are 5.0 g of horse hemoglobin in 1.0 L of water. If we assume that the solution volume is also 1.0 L, then the 5.0 g corresponds to 7.4×10^{-5} mol. The molar mass is

$$M = \frac{5.0 \text{ g}}{7.4 \times 10^{-5} \text{ mol}} = 6.8 \times 10^4 \text{ g/mol}$$

15.10 $K_{sp} = [Ag^+][Cl^-] = S^2$; $S = \sqrt{1.8 \times 10^{-10}} = 1.3 \times 10^{-5}$ mol/L

$$\frac{1.3 \times 10^{-5} \text{ mol}}{\text{L}} \cdot \frac{143 \text{ g}}{\text{mol}} = 1.9 \times 10^{-3} \text{ g/L}$$

15.11 Possible compounds are $PbBr_2$, $PbCl_2$, HgI_2, and PbI_2. HgI_2 has the smallest K_{sp} and is therefore least soluble.

15.12 Let S be the solubility of AgCl. Then $S = [Ag^+]$. (We cannot express S in terms of $[Cl^-]$ because extra Cl^- is available from the NaCl.) Since S will be small, assume that all the Cl^- in the solution comes from the NaCl. Then $[Cl^-] = 0.010$ M. Using the K_{sp} expression,

$$K_{sp} = 1.8 \times 10^{-10} = [Ag^+][Cl^-] = S \cdot 0.010$$

$$S = \frac{1.8 \times 10^{-10}}{0.010} = 1.8 \times 10^{-8}$$

Since $AgCl(s) \rightleftharpoons Ag^+(aq) + Cl^-(aq)$ the concentration of Cl^- produced by dissolving AgCl is 1.8×10^{-8} M, which is smaller than the 0.010 M from the NaCl; our assumption is correct.

15.13 First, calculate the mass of C_2H_6O in 1000. mL of water.

$$\text{mass of } C_2H_6O = 1000. \text{ mL} \cdot \frac{1 \text{ g } H_2O}{1 \text{ mL } H_2O} \cdot \frac{2.5 \text{ g } C_2H_6O}{10^6 \text{ g } H_2O} = 2.5 \times 10^{-3} \text{ g } C_2H_6O$$

Write a balanced equation for the oxidation of C_2H_6O.

$$C_2H_6O(aq) + 3\ O_2(g) \longrightarrow 2\ CO_2(g) + 3\ H_2O(\ell)$$

Calculate the amount (moles) of C_2H_6O.

$$\text{amount of } C_2H_6O = 2.5 \times 10^{-3} \text{ g } C_2H_6O \cdot \frac{1.0 \text{ mol } C_2H_6O}{46.1 \text{ g } C_2H_6O} = 5.4 \times 10^{-5} \text{ mol } C_2H_6O$$

From the balanced equation, the stoichiometric factor is $\frac{3 \text{ mol } O_2}{1 \text{ mol } C_2H_6O}$

So, the amount of O_2 required is

$$\text{amount of } O_2 = 5.4 \times 10^{-5} \text{ mol } C_2H_6O \cdot \frac{3 \text{ mol } O_2}{1 \text{ mol } C_2H_6O} = 1.6 \times 10^{-4} \text{ mol } O_2$$

And the mass of O_2 is

$$\text{mass } O_2 = 1.6 \times 10^{-4} \text{ mol } O_2 \cdot \frac{32 \text{ g } O_2}{1.0 \text{ mol } O_2} = 5.2 \times 10^{-3} \text{ g } O_2$$

According to Table 15.2, this value is about one-half the solubility of O_2 in water at 10 °C.

CHAPTER 16

16.1 (a) HBr, strong acid; (b) Br^-, extremely weak base; (c) HNO_2, weak acid; (d) PH_3, weak base (by analogy with NH_3); (e) H_2SO_3, weak acid; (f) $HClO_4$, strong acid; (g) CN^-, weak base.

16.2 (a) H_2CO_3 conjugate base is HCO_3^-; HNO_3 conjugate base is NO_3^-; (b) NH_3 conjugate acid is NH_4^+; CN^- conjugate acid is HCN.

16.3 (a) $PO_4^{3-} + H_3O^+ \rightleftharpoons HPO_4^{2-} + H_2O$
$HPO_4^{2-} + H_3O^+ \rightleftharpoons H_2PO_4^- + H_2O$
$H_2PO_4^- + H_3O^+ \rightleftharpoons H_3PO_4 + H_2O$
(b) $H_2C_2O_4 + H_2O \rightleftharpoons HC_2O_4^- + H_3O^+$
$HC_2O_4^- + H_2O \rightleftharpoons C_2O_4^{2-} + H_3O^+$

16.4 In 0.040 M NaOH the $[OH^-] = 0.040$ M
$pOH = -\log[OH^-] = -\log(0.040) = 1.40$
$pH + pOH = 14.00$; $pH = 14.00 - 1.40 = 12.60$

16.5 $pH = -\log[H_3O^+] = -\log(1.5 \times 10^{-10}) = 9.82$
If $[H_3O^+]$ is multiplied by 3000, the new

$[H_3O^+] = 3000 \cdot 1.5 \times 10^{-10} = 4.5 \times 10^{-7}$
$pH = -\log (4.5 \times 10^{-7}) = 6.35$, which is acidic.

16.6 $H_2SO_3(aq) + H_2O(\ell) \rightleftharpoons H_3O^+(aq) + HSO_3^-(aq)$

for H_2SO_3, $K_a = \dfrac{[H_3O^+][HSO_3^-]}{[H_2SO_3]}$

$HF(aq) + H_2O(\ell) \rightleftharpoons H_3O^+(aq) + F^-(aq)$

$K_a = \dfrac{[H_3O^+][F^-]}{[HF]}$

$HSO_3^-(aq) + H_2O(\ell) \rightleftharpoons H_3O^+(aq) + SO_3^{2-}(aq)$

$K_a = \dfrac{[H_3O^+][SO_3^{2-}]}{[HSO_3^-]}$

$HCN(aq) + H_2O(\ell) \rightleftharpoons H_3O^+(aq) + CN^-(aq)$

$K_a = \dfrac{[H_3O^+][CN^-]}{[HCN]}$

$NO_2^-(aq) + H_2O(\ell) \rightleftharpoons HNO_2(aq) + OH^-(aq)$

$K_b = \dfrac{[HNO_2][OH^-]}{[NO_2^-]}$

$HS^-(aq) + H_2O(\ell) \rightleftharpoons H_2S(aq) + OH^-(aq)$

$K_b = \dfrac{[H_2S][OH^-]}{[HS^-]}$

$CN^-(aq) + H_2O(\ell) \rightleftharpoons HCN(aq) + OH^-(aq)$

$K_b = \dfrac{[HCN][OH^-]}{[CN^-]}$

$PO_4^{3-}(aq) + H_2O(\ell) \rightleftharpoons HPO_4^{2-}(aq) + OH^-(aq)$

$K_b = \dfrac{[HPO_4^{2-}][OH^-]}{[PO_4^{3-}]}$

16.7 The acid ionizes according to the balanced equation

$$HC_3H_5O_3(aq) + H_2O(\ell) \rightleftharpoons H_3O^+(aq) + C_3H_5O_3^-(aq)$$

The pH gives us the equilibrium concentration of H_3O^+.

$$[H_3O^+] = 10^{-pH} = 10^{-2.43} = 3.7 \times 10^{-3}$$

The equilibrium concentrations of the other species can be seen using a table.

	$[HC_3H_5O_3]$	$[H_3O^+]$	$[C_3H_5O_3^-]$
initial concentration (mol/L)	0.10	1.0×10^{-7}	0
change in concentration on reaction (mol/L)	$-x$	$+x$	$+x$
concentration at equilibrium (mol/L)	$0.10 - x$	x	x

The quantity x is given by the pH measurement, so $x = 3.7 \times 10^{-3}$. Using these values in the equilibrium constant expression for the acid,

$$K_a = \frac{[H_3O^+][C_3H_5O_3^-]}{[HC_3H_5O_3]} = \frac{(3.7 \times 10^{-3})(3.7 \times 10^{-3})}{(0.10 - 3.7 \times 10^{-3})} = 1.4 \times 10^{-4}$$

Since K_a for lactic acid is larger than K_a for propanoic acid, lactic acid is stronger.

16.8 $K_b(CH_3COO^-) = 5.6 \times 10^{-10}$ (from Table 16.2)
$K_b(HS^-) = 1 \times 10^{-7}$ (from Table 16.2)
Since K_b (HS^-) is larger, HS^- is the stronger base.
Since $\dfrac{K_b(HS^-)}{K_b(CH_3COO^-)} = \dfrac{1 \times 10^{-7}}{5.6 \times 10^{-10}} = 200$, HS^- is 200 times stronger.

16.9 $K_b(C_6H_5O^-) = \dfrac{K_w}{K_a(C_6H_5OH)} = \dfrac{1.0 \times 10^{-14}}{1.3 \times 10^{-10}} = 7.7 \times 10^{-5}$

The species in Table 16.2 with K_b closest to 7.7×10^{-5} is $Fe(H_2O)_5OH^+$.

16.10 $HCl + H_2O \longrightarrow H_3O^+ + Cl^-$
HCl is stronger acid than H_3O^+
H_2O is stronger base than Cl^-

$H^- + H_2O \longrightarrow OH^- + H_2$
H^- is stronger base than OH^-
H_2O is stronger acid than H_2

$O^{2-} + H_2O \longrightarrow OH^- + OH^-$
O^{2-} is stronger base than OH^-
H_2O is stronger acid than OH^-

$CO_3^{2-} + H_2O \rightleftharpoons HCO_3^- + OH^-$
OH^- is stronger base than CO_3^{2-}
HCO_3^- is stronger acid than H_2O

$NH_3 + H_2O \rightleftharpoons NH_4^+ + OH^-$
OH^- is stronger base than NH_3
NH_4^+ is stronger acid than H_2O
Generalization: Stronger acid reacts with stronger base to give weaker base and weaker acid.

16.11 (a) Will occur. S^{2-} is a strong base.

$$\underset{\text{base}}{S^{2-}} + \underset{\text{acid}}{H_2O} \rightleftharpoons \underset{\text{acid}}{HS^-} + \underset{\text{base}}{OH^-}$$

(b) Will occur. OH^- is a strong base.

$$\underset{\text{acid}}{HPO_4^{2-}} + \underset{\text{base}}{OH^-} \rightleftharpoons \underset{\text{base}}{PO_4^{3-}} + \underset{\text{acid}}{H_2O}$$

(c) Will occur. CO_3^{2-} is a moderately strong base and HSO_4^- is a moderately strong acid.

$$\underset{\text{base}}{CO_3^{2-}} + \underset{\text{acid}}{HSO_4^-} \rightleftharpoons \underset{\text{acid}}{HCO_3^-} + \underset{\text{base}}{SO_4^{2-}}$$

(d) $H_2PO_4^-$ is a stronger acid than HCO_3^-, so HCO_3^- must act as a base. The reaction is

$$H_2PO_4^- + HCO_3^- \rightleftharpoons HPO_4^- + H_2CO_3$$

Because H_2CO_3 is a stronger acid than $H_2PO_4^-$ and HPO_4^{2-} is a stronger base than HCO_3^-, the equilibrium lies toward the reactants.

16.12 The reaction is

$$B(OH)_3(H_2O) + H_2O \rightleftharpoons B(OH)_4^- + H_3O^+$$

Set up a table

	$B(OH)_3(H_2O)$	$B(OH)_4^-$	H_3O^+
initial concentration (mol/L)	0.10	0	1.0×10^{-7}
change in concentration on reaction (mol/L)	$-x$	$+x$	$+x$
concentration at equilibrium (mol/L)	$0.10 - x$	x	x

$$K_a = \frac{x \cdot x}{0.10 - x} = 7.3 \times 10^{-10}$$

Assume that x is much smaller than 0.10. Then the equation simplifies to

$x^2 = 7.3 \times 10^{-10} \cdot 0.10 = 7.3 \times 10^{-11}$

$x = \sqrt{7.3 \times 10^{-11}} = 8.5 \times 10^{-6}$ (which is indeed much smaller than 0.10)

$[H_3O^+] = x = 8.5 \times 10^{-6}$ mol/L

$pH = -\log(8.5 \times 10^{-6}) = 5.07$

16.13 First, set up a table

	CO_3^{2-}	HCO_3^-	OH^-
Initial concentration (mol/L)	1.0	0	1.0×10^{-7}
Change in concentration on reaction (mol/L)	$-x$	$+x$	$+x$
Concentration at equilibrium (mol/L)	$1.0-x$	x	x

$$K_b = \frac{x \cdot x}{1.0 - x} = 2.1 \times 10^{-4}$$

Assume that x is much smaller than 1.0. Then

$x^2 = 2.1 \times 10^{-4}$; $x = \sqrt{2.1 \times 10^{-4}} = 1.4 \times 10^{-2}$ mol/L (which is much smaller than 1.0)

$pOH = -\log(1.4 \times 10^{-2}) = 1.84$

$pH = 14.00 - 1.84 = 12.15$

16.14 The equilibrium is

$$Al(H_2O)_6^{3+} + H_2O \rightleftharpoons Al(H_2O)_5OH^{2+} + H_3O^+$$

$$K_a = \frac{[Al(H_2O)_5OH^{2+}][H_3O^+]}{[Al(H_2O)_6^{3+}]} = 7.9 \times 10^{-6}$$

Set up a table

	$Al(H_2O)_6^{3+}$	$Al(H_2O)_5OH^+$	H_3O^+
Initial concentration (mol/L)	0.20	0	1.0×10^{-7}
Change in concentration on reaction (mol/L)	$-x$	$+x$	$+x$
Concentration at equilibrium (mol/L)	$0.20 - x$	x	x

$$7.9 \times 10^{-6} = \frac{x \cdot x}{0.20 - x}$$

Assume that $x \ll 0.20$; then

$x^2 = 0.20 \cdot 7.9 \times 10^{-6}$; $x = \sqrt{1.6 \times 10^{-6}} = 1.3 \times 10^{-3}$

$[H_3O^+] = x = 1.3 \times 10^{-3}$ mol/L

$pH = -\log[H_3O^+] = -\log(1.3 \times 10^{-3}) = 2.90$

16.15 An NH_4NO_3 solution will be acidic because of the hydrolysis of NH_4^+ ion. The NO_3^- ion, being the conjugate base of a strong acid, does not hydrolyze.

$$NH_4^+(aq) + H_2O(\ell) \rightleftharpoons H_3O^+(aq) + NH_3(aq)$$

A NaBr solution will be neutral because Na^+ does not undergo hydrolysis in water and the Br^- ion is the conjugate base of a strong acid.

A K_3PO_4 solution will be basic because of the hydrolysis of the PO_4^{3-} ion. The K^+ ions do not undergo hydrolysis.

$$PO_4^{3-}(aq) + H_2O(\ell) \rightleftharpoons HPO_4^{2-}(aq) + OH^-(aq)$$

A $Fe(NO_3)_2$ solution will be slightly acidic because of the hydrolysis of $Fe(H_2O)_6^{2+}$ ions. The NO_3^- ion does not hydrolyze because it is the conjugate base of a strong acid.

$$Fe(H_2O)_6^{2+}(aq) + H_2O(\ell) \rightleftharpoons H_3O^+(aq) + Fe(H_2O)_3(OH)^+(aq)$$

For NH_4HCO_3 both ions hydrolyze, and so we need to look up K_a and K_b values. From Table 16.2, $K_a(NH_4^+) = 5.6 \times 10^{-10}$ and $K_b(HCO_3^-) = 2.4 \times 10^{-8}$. This indicates that HCO_3^- is a stronger base than NH_4^+ is an acid, and so the solution will be slightly basic.

16.16 (a) $ZnCO_3$ soluble because H_2CO_3 forms and decomposes to CO_2 gas.
(b) FeS is soluble because S^{2-} is the conjugate base of a very weak acid. H_2S will form.
(c) AgBr is insoluble because Br^- is the very weak conjugate base of a strong acid.

(d) $BaCO_3$ soluble because H_2CO_3 forms and decomposes to CO_2 gas.
(e) $Ca(OH)_2$ is soluble because OH^- reacts with H_3O^+ and as long as solution remains acidic, insufficient OH^- is available to reform precipitate.

16.17 Look for acids whose K_a values are within a factor of 10 of the desired $[H_3O^+]$.
(a) formic acid, $K_a = 1.8 \times 10^{-4}$, and formate ion
(b) benzoic acid, $K_a = 6.3 \times 10^{-5}$, and benzoate ion
(c) dihydrogen phosphate ion, $K_a = 6.2 \times 10^{-8}$, and hydrogen phosphate ion or hydrogen sulfite ion, $K_a = 6.2 \times 10^{-8}$, and sulfite ion
(d) hydrogen carbonate ion, $K_a = 4.8 \times 10^{-11}$, and carbonate ion

16.18 In all cases, $\log\left(\frac{\text{base}}{\text{acid}}\right) = \log(0.25/0.10) = .040$

H_2CO_3/HCO_3^- $\quad pH = -\log(4.2 \times 10^{-7}) + 0.40 = 6.38 + .040 = 6.78$

$H_2PO_4^-/HPO_3^{3-}$ $\quad pH = -\log(6.2 \times 10^{-8}) + 0.40 = 7.20 + 0.40 = 7.61$

NH_4^+/NH_3 $\quad pH = -\log(5.6 \times 10^{-10}) + 0.40 = 9.25 + 0.40 = 9.65$

16.19 (a) Equivalence point is at 50.0 mL and solution is above pH = 7; pH in buffer region (at 25.0 mL added) is 7.46.
(b) Since base is in flask, titration curve starts at high pH (13, since NaOH is a strong base and concentration is 0.100 M). Since HNO_3 is a strong acid, no buffer is formed; the pH remains high until 50.0 mL HNO_3 has been added, which is the equivalence point.
(c) Since the NaOH is twice as concentrated as the HNO_2, only half the volume of NaOH (12.5 mL) is required to reach the equivalence point. Buffer region (6.25 mL) has pH = $-\log(4.5 \times 10^{-4})$ = 3.35.

16.20 The equivalence point is at pH = 8.87. Therefore the chosen indicator should change color near that pH. Possible indicators are phenol red, m-nitrophenol, o-cresolphthalein, and phenolphthalein.

16.21

Antacid	*Molar Mass*	*Mole ratio**	*Moles acid/gram antacid*
$Mg(OH)_2$	53.320 g/mol	2 mol acid/mol antacid	0.03429 mol acid
$CaCO_3$	100.09 g/mol	2 mol acid/mol antacid	0.01998 mol acid
$NaHCO_3$	84.007 g/mol	1 mol acid/mol antacid	0.01190 mol acid
$Al(OH)_3$	78.003 g/mol	3 mol acid/mol antacid	0.03846 mol acid
$NaAl(OH)_2CO_3$	144.00 g/mol	4 mol acid/mol antacid	0.02778 mol acid

**from balanced neutralization equation*

On a per gram basis, $Al(OH)_3$ is best.

16.22 (a) PH_3, Lewis base (b) BCl_3, Lewis acid
(c) H_2S, Lewis base (d) NO_2, Lewis acid
(e) Ni^{2+}, Lewis acid (f) CO, Lewis base

CHAPTER 17

17.1 (a) $\overset{0}{2\,Fe(s)} + \overset{0}{3\,Cl_2(g)} \longrightarrow 2\,\overset{+3}{Fe}\overset{-1}{Cl_3}(s)$
Fe: reducing agent; Cl_2: oxidizing agent

(b) $\overset{0}{2\,H_2(g)} + \overset{0}{O_2(g)} \longrightarrow 2\,\overset{+1}{H_2}\overset{-2}{O}(\ell)$
H_2: reducing agent; O_2: oxidizing agent

(c) $\overset{0}{Cu(s)} + 2\,\overset{+5}{N}\overset{-2}{O_3}^-(aq) + 4\,\overset{+1}{H^+}(aq) \longrightarrow \overset{+2}{Cu^{2+}}(aq) + \overset{+4}{N}\overset{-2}{O_2}(g) + 6\,\overset{+1}{H_2}\overset{-2}{O}(\ell)$
Cu: reducing agent; NO_3^-: oxidizing agent

(d) $\overset{0}{C(s)} + \overset{0}{O_2(g)} \longrightarrow \overset{+4}{C}\overset{-2}{O_2}(g)$
C: reducing agent; O_2: oxidizing agent

(e) $6\,\overset{+2}{Fe^{2+}}(aq) + \overset{+6}{Cr_2}\overset{-2}{O_7}^{2-}(aq) + 14\,\overset{+1}{H^+}(aq) \longrightarrow 6\,\overset{+3}{Fe^{3+}}(aq) + 2\,\overset{+3}{Cr^{3+}}(aq) + 7\,H_2O(\ell)$
Fe^{2+}: reducing agent; $Cr_2O_7^{2-}$: oxidizing agent

17.2 Oxidation half-reaction: $Zn(s) \longrightarrow Zn^{2+}(aq) + 2\,e^-$
Reduction half-reaction: $2\,H^+(aq) + 2\,e^- \longrightarrow H_2(g)$

$Zn(s) + 2\,H^+(aq) \longrightarrow H_2(g) + Zn^{2+}(aq)$

17.3 **Step 1. Recognize the reaction is an oxidation-reduction process.** The oxidation number of Zn changes from 0 in Zn(s) to +2 in $Zn^{2+}(aq)$, and the oxidation number of Cr changes from +6 in $Cr_2O_7^{2-}(aq)$ to +3 in $Cr^{3+}(aq)$.

Step 2. Break the overall reaction into half-reactions.

$$Cr_2O_7^{2-}(aq) \longrightarrow Cr^{3+}(aq)$$
$$Zn(s) \longrightarrow Zn^{2+}(aq)$$

Step 3. Balance each half-reaction for mass. Begin by balancing all atoms except for O and H; these are always the last to be balanced because they often appear in more than one reactant or product.

$Cr_2O_7^{2-}$ half-reaction: The two Cr atoms on the left require two Cr on the right, and an oxygen-rich species must be added to the right side in order to balance the oxygens.

$$Cr_2O_7^{2-}(aq) \longrightarrow 2\,Cr^{3+}(aq) + (7 \text{ oxygen atoms})$$

In aqueous solutions, we can always add H_2O to the side of the reaction requiring oxygens. So the partially completed half-reaction becomes

$$Cr_2O_7^{2-}(aq) \longrightarrow 2\,Cr^{3+}(aq) + 7\,H_2O(\ell)$$

Note that now there are 14 H atoms on the right and none on the left. Since the solution is acidic, H^+ ions can always be added to either side of the reaction as needed, so 14 H^+ are placed on the left side of the half-reaction.

$$14\,H^+(aq) + Cr_2O_7^{2-}(aq) \longrightarrow 2\,Cr^{3+}(aq) + 7\,H_2O(\ell)$$

The half-reaction is now balanced for mass.

Zinc half reaction: This reaction is balanced as written.

$$Zn(s) \longrightarrow Zn^{2+}(aq)$$

Now both half-reactions are balanced for mass.

Step 4. Balance the half-reactions for charge. The mass-balanced zinc half-reaction has a net charge of 0 on the left side and 2+ on the right. Therefore 2 e^- are added to the positive side.

$$Zn(s) \longrightarrow Zn^{2+}(aq) + 2\,e^-$$

This means Zn is the reducing agent (it loses electrons and gets oxidized). The two electrons lost are also in keeping with the earlier observation in Step 1 that each Zn atom loses two electrons in changing oxidation number from 0 to +2.

The mass balanced $Cr_2O_7^{2-}$ half-reaction has a charge of 12+ on the left and 6+ on the right. Therefore, 6 e^- are added to the left. This means that $Cr_2O_7^{2-}$ is the oxidizing agent.

$$14\,H^+(aq) + Cr_2O_7^{2-}(aq) + 6\,e^- \longrightarrow 2\,Cr^{3+}(aq) + 7\,H_2O(\ell)$$

Step 5. Multiply the half-reactions by appropriate factors so that the reducing agent donates as many electrons as the oxidizing agent consumes. The zinc half-reaction should be multiplied by 3 so that each half-reaction involves 6 electrons.

$$3\,[Zn(s) \longrightarrow Zn^{2+} + 2\,e^-]$$
$$14\,H^+(aq) + Cr_2O_7^{2-}(aq) + 6\,e^- \longrightarrow 2\,Cr^{3+}(aq) + 7\,H_2O(\ell)$$

Step 6. Add the half-reactions to give the overall reaction.

$$3\,Zn(s) \longrightarrow 3\,Zn^{2+} + 6\,e^-$$
$$14\,H^+(aq) + Cr_2O_7^{2-}(aq) + 6\,e^- \longrightarrow 2\,Cr^{3+}(aq) + 7\,H_2O(\ell)$$

$$3\,Zn(s) + 14\,H^+(aq) + Cr_2O_7^{2-}(aq) + 6\,e^- \longrightarrow 2\,Cr^{3+}(aq) + 3\,Zn^{2+}(aq) + 7\,H_2O(\ell) + 6\,e^-$$

Step 7. Cancel common reactants and products. Here, 6 e^- appear on the left and on the right.

$$3\,Zn(s) + 14\,H^+(aq) + Cr_2O_7^{2-}(aq) + \longrightarrow 2\,Cr^{3+}(aq) + 3\,Zn^{2+}(aq) + 7\,H_2O(\ell)$$

Step 8. Check the final results to make sure there is mass and charge balance.

Charge balance: H2 on both sides.
Mass balance: Both sides of the equation have 2 Zn, 7 O, 2 Cr, and 14 H atoms.

17.4 The half-reactions are:

$Ni(s) \longrightarrow Ni^{2+}(aq) + 2\,e^-$ oxidation at the anode
$Ag^+(aq) + e^- \longrightarrow Ag(s)$ reduction at the cathode

Electrons will flow from the nickel anode to the silver cathode. Nitrate ions will flow through the salt bridge in the direction of the nickel anode.

17.5 The two half-cell reactions are

$Fe(s) \longrightarrow Fe^{2+}(aq) + 2\,e^-$	$E° = ?$ V
$Cu^{2+}(aq) + 2\,e^- \longrightarrow Cu(s)$	$E° = +0.34$ V
$Fe(s) + Cu^{2+}(aq) \longrightarrow Fe^{2+}(aq) + Cu$	$E° = +0.78$ V

E° for the Fe/Fe^{2+} half-cell oxidation reaction must be +0.44 V. Table 17.1 gives −0.44 V for the standard reduction reaction.

17.6 Writing the two half-reactions and their E° values:

$Zn^{2+}(aq) + 2\,e^- \longrightarrow Zn(s)$	$E° = -0.763$ V
$H_2(g) \longrightarrow 2\,H^+(aq) + 2\,e^-$	$E° = 0.000$ V

Therefore $E°_{net} = -0.763$ V, so the reaction is not product favored. The value of $\Delta G°$ is given by $-nFE°$

$$\Delta G^\circ = -2 \text{ mole electrons transferred} \cdot \frac{9.65 \times 10^4 \text{ J}}{\text{V} \cdot \text{mol e}^-} \cdot -0.763 \text{ V} \cdot \frac{1 \text{ kJ}}{10^3 \text{ J}}$$
$$= 147 \text{ kJ}$$

17.7 The anode reaction is:

$$Cd(s) + 2\,OH^-(aq) \longrightarrow Cd(OH)_2(s) + 2\,e^-$$

The cathode reaction is:

$$2\,NiO(OH)(s) + 2\,H_2O(\ell) + e^- \longrightarrow 2\,Ni(OH)_2(s) + OH^-(aq)$$

17.8 Sand by the seashore (b) would rank first because it is moist and contains a relative large amount of dissolved salts to act as an electrolyte. Oxygen is also present in the air. Moist clay (a) would rank next because it is moist and will contain some dissolved salts, although not usually as much as moist sea sand. Oxygen is also present in the air. The dry desert environment (c) would contain less moisture, although plenty of oxygen, so it would promote slower oxidation of iron than (a) or (b). Finally, the moon (d) would be the least likely to promote oxidation of a piece of iron because there is little or no moisture or oxygen in the atmosphere.

17.9 (a) Molten NaBr would contain Na^+ and Br^- ions. Passing an electric current through molten NaBr would produce the following reactions.

At the anode (oxidation): $2\,Br^- \longrightarrow Br_2 + 2e^-$
At the cathode (reduction): $Na^+ + e^- \longrightarrow Na$

(b) Aqueous NaBr contains the following species. Na^+ and Br^- ions, and H_2O molecules. Water may be either oxidized or reduced. To determine which species will be reduced, we need to consider the reduction potentials for H_2O and Na^+.

$$Na^+(aq) + e^- \longrightarrow Na(s) \qquad E^\circ = -2.7 \text{ V}$$
$$2\,H_2O(\ell) + 2e^- \longrightarrow H_2(g) + 2\,OH^-(aq) \qquad E^\circ = -0.83 \text{ V}$$

The reduction of water would be less negative than the reduction of Na^+ ions. The reaction at the cathode will be the reduction of water to produce hydrogen gas and OH^- ions.

At the anode, the oxidation of Br^- or the oxidation of water will take place.

$$2\,Br^-(aq) \longrightarrow Br_2(\ell) + 2\,e^- \qquad E^\circ = -1.08 \text{ V}$$
$$6\,H_2O(\ell) \longrightarrow O_2(g) + 4\,H_3O^+(aq) + 4\,e^- \qquad E^\circ = -1.229$$

Having the less negative E° value, the oxidation of Br^- will be favored.

(c) In aqueous $SnCl_2$, the ions present are Sn^{2+} and Cl^-. Water is also present at both electrodes and may be either oxidized or reduced.

At the cathode, either Sn^{2+} or H_2O will be reduced.

$$Sn^{2+}(aq) + 2\,e^- \longrightarrow Sn(s) \qquad E^\circ = -0.14 \text{ V}$$
$$2\,H_2O(\ell) + 2\,e^- \longrightarrow H_2(g) + 2\,OH^-(aq) \qquad E^\circ = -0.83 \text{ V}$$

Having a less negative reduction potential, Sn^{2+} ions will be reduced. At the anode, either Cl^- or H_2O will be oxidized. Using the arguments in the text H_2O will be oxidized in preference to Cl^-.

17.10 The cathode reaction is $Na^+ + e^- \longrightarrow Na$
The current of 25×10^3 amperes for 1.0 hour will give the number of coulombs of electrical charge that flows through the cell.

$$\text{charge (coulombs)} = \text{current} \cdot \text{time}$$
$$= 25 \times 10^3 \text{ amperes} \cdot 1 \text{ hr} \cdot \frac{3600 \text{ sec}}{1 \text{ hr}} \cdot \frac{1 \text{ coulomb}}{\text{ampere} \cdot \text{sec}}$$
$$= 9.0 \times 10^7 \text{ coulombs}$$

The moles of Na, and grams of Na are given by using the Faraday constant and the half-reaction that tells us one mole of electrons produce on emole of Na.

$$\text{moles Na} = 9.0 \times 10^7 \text{ coulombs} \cdot \frac{1 \text{ mol Na}}{9.65 \times 10^4 \text{ coulombs}}$$
$$= 930 \text{ moles Na}$$

$$\text{grams Na} = 930 \text{ mol Na} \cdot \frac{23.0 \text{ g Na}}{1 \text{ mol Na}}$$
$$= 21{,}000 \text{ g Na or } 21 \text{ kg Na}$$

17.11 First, write the reduction reaction for Al^{3+} ions.

$$Al^{3+} + 3\,e^- \longrightarrow Al$$

This means 3 mole of electrons are required to produce a mole of Al metal. Next, calculate the moles of aluminum from the 2000. tons given.

$$\text{mol Al} = 2000.\text{ ton Al} \cdot \frac{2000 \text{ lb}}{1 \text{ ton}} \cdot \frac{453.6 \text{ g}}{1 \text{ lb}} \cdot \frac{1 \text{ mol Al}}{26.982 \text{ g Al}}$$
$$= 6.724 \times 10^7 \text{ mol Al}$$

Next, calculate the coulombs of charge required to produce this amount of Al.

$$\text{charge} = 6.724 \times 10^7 \text{ mol Al} \cdot \frac{3 \text{ mol e}^-}{1 \text{ mol Al}} \cdot \frac{9.65 \times 10^4 \text{ coulombs}}{1 \text{ mol e}^-}$$

$$= 1.946 \times 10^{13}$$

The joules of energy represented by this many coulombs of charge flowing through the cell at a potential of 4 V is given by

$$\text{energy} = 1.95 \times 10^{13}\text{ C} \cdot 4.0\text{ V} = 7.8 \times 10^{13}\text{ J}$$

Finally, using the conversion factor of 1 kwh = 3.60×10^6 J, we get

$$\text{kwh} = 7.8 \times 10^{13}\text{ J} \cdot \frac{1\text{ kwh}}{3.60 \times 10^6\text{ J}} = 2.2 \times 10^7\text{ kwh}$$

17.12 The cathode reaction tells us that 3 moles of electrons are required to plate one mole of Au.

$$Au^{3+}(aq) + 3\,e^- \longrightarrow Au(s)$$

The quantity (moles) of electrons passed by the current of 0.50 amperes for 20. minutes can be calculated using the faraday.

$$\text{moles of e}^- = 0.50 \text{ ampere} \cdot 20 \text{ min} \cdot \frac{60 \text{ sec}}{1 \text{ min}} \cdot \frac{1 \text{ coulomb}}{1 \text{ ampere} \cdot \text{sec}} \cdot \frac{1 \text{ mol e}^-}{9.65 \times 10^4 \text{ coulombs}} = 6.2 \times 10^{-3} \text{ mole e}^-$$

$$\text{mass Au} = 6.2 \times 10^{-3} \text{ mole e}^- \cdot \frac{197 \text{ g Au}}{3 \text{ mol e}^-} = 0.41 \text{ g Au}$$

CHAPTER 18

18.1 $2\,NaCl(\ell) \longrightarrow 2\,Na(\ell) + Cl_2(g)$

Coulombs (C) = (A)(s) =

$$2.0 \times 10^4 \text{ A} \cdot 24 \text{ hr} \cdot \frac{3600 \text{ s}}{\text{hr}} = 1.7 \times 10^9 \text{ C}$$

$$\text{moles e}^- = 1.7 \times 10^9 \text{ C} \cdot \frac{1 \text{ mole e}^-}{9.65 \times 10^4 \text{ C}} = 1.8 \times 10^4 \text{ moles e}^-$$

$$\text{tons Na} = 1.8 \times 10^4 \text{ moles e}^- \cdot \frac{1 \text{ mol Na}}{1 \text{ mol e}^-} \cdot \frac{23.0 \text{ g Na}}{1 \text{ mol Na}} \cdot \frac{1 \text{ lb}}{454 \text{ g}} \cdot \frac{1 \text{ ton}}{2000 \text{ lb}} = 0.46 \text{ tons Na}$$

$$\text{tons } Cl_2 = 1.8 \times 10^4 \text{ moles e}^- \cdot \frac{1 \text{ mol } Cl_2}{2 \text{ mol e}^-} \cdot \frac{70.9 \text{ g } Cl_2}{1 \text{ mol } Cl_2} \cdot \frac{1 \text{ lb}}{454 \text{ g}} \cdot \frac{1 \text{ ton}}{2000 \text{ lb}} = 0.70 \text{ tons } Cl_2$$

18.2 $C = 2.00 \times 10^4 \text{ A} \cdot 100. \text{ hr} \cdot \frac{3600 \text{ s}}{\text{hr}} = 7.20 \times 10^9 \text{ C}$

$$\text{tons NaOH} = 7.20 \times 10^9 \text{ C} \cdot \frac{1 \text{ mol e}^-}{9.65 \times 10^4 \text{ C}} \cdot \frac{1 \text{ mol NaOH}}{1 \text{ mol e}^-} \cdot \frac{40.00 \text{ g NaOH}}{1 \text{ mol NaOH}} \cdot \frac{1 \text{ lb}}{454 \text{ g}} \cdot \frac{1 \text{ ton}}{2000 \text{ lb}} = 3.29 \text{ tons NaOH}$$

18.3 $2\,HF \longrightarrow H_2 + F_2$

$$\text{Mass } F_2 = 10. \text{ Faradays} \cdot \frac{1 \text{ mol e}^-}{\text{Faraday}} \cdot \frac{1 \text{ mol } F_2}{2 \text{ mol e}^-} \cdot \frac{38.0 \text{ g } F_2}{1 \text{ mol } F_2} = 190 \text{ g } F_2$$

$$\text{Mass } H_2 = 190 \text{ g } F_2 \cdot \frac{2.00 \text{ g } H_2}{38.0 \text{ g } F_2} = 10. \text{ g } H_2$$

$$\text{Mass HF} = 190 \text{ g } F_2 \cdot \frac{1 \text{ mol } F_2}{38.0 \text{ g } F_2} \cdot \frac{2 \text{ mol HF}}{1 \text{ mol } F_2} \cdot \frac{20.0 \text{ g HF}}{1 \text{ mol HF}} = 2.0 \times 10^2 \text{ g HF}$$

18.4 $\text{moles Mg} = 2.00 \text{ tons Mg} \cdot \frac{2000 \text{ lb}}{\text{ton}} \cdot \frac{454 \text{ g}}{\text{lb}} \cdot \frac{1 \text{ mol Mg}}{24.3 \text{ g Mg}} = 7.47 \times 10^4 \text{ mol Mg}$

$$C = 7.47 \times 10^4 \text{ moles Mg} \cdot \frac{2 \text{ mol e}^-}{1 \text{ mol Mg}} \cdot \frac{9.65 \times 10^4 \text{ C}}{1 \text{ mol e}^-} = 1.44 \times 10^{10} \text{ C}$$

$$\text{time} = \frac{1.44 \times 10^{10} \text{ C}}{5000. \text{ A}} = 2.88 \times 10^6 \text{ s}$$

$$\text{hr} = 2.88 \times 10^6 \text{ s} \times \frac{1 \text{ hr}}{3600 \text{ s}} = 800. \text{ hr}$$

18.5 $C = (A)(s) = 200. \text{ A} \cdot 365 \text{ days} \cdot \frac{24 \text{ hr}}{\text{day}} \cdot \frac{3600 \text{ s}}{\text{hr}} = 6.31 \times 10^9 \text{ C}$

$$\text{moles e}^- = 6.31 \times 10^9 \text{ C} \cdot \frac{1 \text{ mol e}^-}{9.65 \times 10^4 \text{ C}} = 6.54 \times 10^4 \text{ moles e}^-$$

$$\text{kg Cu} = 6.54 \times 10^4 \text{ moles e}^- \cdot \frac{1 \text{ mol Cu}}{2 \text{ moles e}^-} \cdot \frac{63.5 \text{ g Cu}}{1 \text{ mol Cu}} \cdot \frac{1 \text{ k}}{1000 \text{ g}} \cdot 0.900 \text{ efficient} = 1.87 \times 10^3 \text{ kg Cu}$$

CHAPTER 19

19.1 (a) $^{237}_{93}Np \longrightarrow {}^{4}_{2}\alpha + {}^{233}_{91}Pa$
(b) $^{35}_{16}S \longrightarrow {}^{0}_{-1}\beta + {}^{35}_{17}Cl$

19.2 $^{235}_{92}U \longrightarrow {}^{4}_{2}\alpha + {}^{231}_{90}Th$
$^{231}_{90}Th \longrightarrow {}^{0}_{-1}\beta + {}^{231}_{91}Pa$
$^{231}_{91}Pa \longrightarrow {}^{4}_{2}\alpha + {}^{227}_{89}Ac$
$^{227}_{89}Ac \longrightarrow {}^{4}_{2}\alpha + {}^{223}_{87}Fr$
$^{223}_{87}Fr \longrightarrow {}^{0}_{-1}\beta + {}^{223}_{88}Ra$

19.3 (a) $^{13}_{7}N \longrightarrow {}^{13}_{6}C + {}^{0}_{+1}e$
(b) $^{41}_{20}Ca + {}^{0}_{-1}\beta \longrightarrow {}^{41}_{19}K$
(c) $^{90}_{38}Sr \longrightarrow {}^{90}_{39}Y + {}^{0}_{-1}\beta$
(d) $^{11}_{6}C \longrightarrow {}^{11}_{5}B + {}^{0}_{+1}e$
(e) $^{43}_{21}Sc \longrightarrow {}^{42}_{20}Ca + {}^{1}_{1}H$

19.4 (a) β-decay: $^{32}_{14}Si \longrightarrow {}^{0}_{-1}\beta + {}^{32}_{15}P$
(b) positron emission: $^{45}_{22}Ti \longrightarrow {}^{0}_{+1}e + {}^{45}_{21}Sc$
(c) α-emission: $^{239}_{94}Pu \longrightarrow {}^{4}_{2}\alpha + {}^{235}_{92}U$

19.5 Mass defect = $\Delta m = -0.03438$ g/mol
$\Delta E = (-4.68 \times 10^{-6}\ \text{kg/mol})(2.998 \times 10^{8}\ \text{m/s})^2 = -3.090 \times 10^{12}\ \text{J/mol}$
E_b per nucleon = 5.150×10^{8} kJ/nucleon
E_b for ^{6}Li is smaller than E_b for ^{4}He.

19.6 (a) $^{90}_{38}Sr \longrightarrow {}^{0}_{-1}\beta + {}^{90}_{39}Y$
(b) Disintegrations are reduced to 1000 after 1 half-life, to 500 after 2 half-lives, to 250 after 3 half-lives, and to 125 after 4 half-lives. Thus, a total of $4 \times 28 = 112$ years is required.

19.7 $k = 0.693/t_{1/2} = 8.90 \times 10^{-3}\ \text{hr}^{-1}$
$\ln(0.10/1.0) = -kt = -(8.90 \times 10^{-3}\ \text{hr}^{-1})t$
$t = 259$ hr

19.8 $\ln(12.9/14.0) = -(1.21 \times 10^{-4}/\text{yr})t$
$t = 676$ years
$1990 - 676 = 1314$ AD

19.9 (a) $^{13}_{6}C + {}^{1}_{0}n \longrightarrow {}^{4}_{2}\alpha + {}^{10}_{4}Be$
(b) $^{14}_{7}N + {}^{4}_{2}\alpha \longrightarrow {}^{1}_{0}n + {}^{17}_{9}F$
(c) $^{253}_{99}Es + {}^{4}_{2}\alpha \longrightarrow {}^{1}_{0}n + {}^{256}_{101}Md$

CHAPTER 20

20.1 To convert percent to ppm, multiply by 10,000.

	ppm
Nitrogen	780,840
Oxygen	209,480
Argon	9,340
Carbon dioxide	330
Neon	18.2
Hydrogen	10
Helium	5.2
Methane	2
Krypton	1
Xenon	0.08

20.2 (a) $H_2O + h\nu \longrightarrow HO + H$
(b) $CH_4 + HO \longrightarrow H_2O + CH_3$
(c) $H + O_2 \longrightarrow HO + O$

20.3 Industrial smog: SO_2, NO_2, smoke and ash where
SO_2 comes from burning coal and hydrocarbon fuels containing sulfur.
NO_2 comes from combustion reactions that form NO, which gets oxidized to NO_2.
Smoke and ash come from burning processes such as those in industrial boilers and foundries.
Photochemical smog: NO_2, O_3, hydrocarbons.
NO_2 comes from combustion processes that form NO, which gets oxidized to NO_2.
O_3 comes from O, produced by the photodecomposition of NO_2, which reacts with O_2.
Hydrocarbons come from unburned or spilled fuel and also from trees and other natural sources.

20.4 The balanced reactions are:

$$N_2(g) + O_2(g) \longrightarrow 2\,NO(g)$$
$$NO(g) + \tfrac{1}{2}O_2(g) \longrightarrow NO_2(g)$$
$$2\,NO_2 + H_2O \longrightarrow HNO_2(g) + HNO_3(g)$$
$$4\,NO_2(g) + 2\,H_2O(g) + O_2(g) \longrightarrow 4\,HNO_3(g)$$

However, ultimately all the nitrogen atoms in the quantity of N_2 will be converted to HNO_3, so the overall reaction can be written as

$$N_2(g) + 5/2\,O_2(g) + H_2O(g) \longrightarrow 2\,HNO_3(g)$$

Using the molar masses for N_2 and HNO_3,

$$\text{mass HNO}_3 = 400\text{t N}_2 \cdot \frac{2000\ \text{kgN}_2}{\text{tN}_2} \cdot \frac{1000\ \text{gN}_2}{\text{kgN}_2} \cdot \frac{1\ \text{mol N}_2}{28.02\ \text{gN}_2} \cdot \frac{2\ \text{mol HNO}_3}{1\ \text{mol N}_2} \cdot \frac{63.02\ \text{gHNO}_3}{1\ \text{mol HNO}_3} \cdot \frac{1\ \text{kg HNO}_3}{1000\ \text{g HNO}_3} \cdot \frac{1 + \text{HNO}_3}{2000\ \text{kg HNO}_3}$$
$$= 2000\ \text{t HNO}_3$$

20.5 $NO_2 + h\nu \longrightarrow NO + O$
$O_3 + h\nu \longrightarrow O_2 + O$
$O_2 + O \longrightarrow O_3$

20.6 The ratio is 750 t BAP for every 10^6 t coal burned in a power plant.

$$\text{mass of BAP formed} = 700\,\frac{\text{t coal}}{\text{hr}} \cdot 24\ \text{hr} \cdot \frac{750\ \text{t BAP}}{10^6\ \text{t coal}} \cdot \frac{2000\ \text{lb}}{\text{t}} \cdot \frac{1\ \text{kg}}{2.2\ \text{lb}}$$
$$= 11{,}000\ \text{kg BAP}$$

20.7 The reaction is $CaO(s) + SO_2(g) \longrightarrow CaSO_3(s)$

$$\text{mass } CaSO_3 \text{ per day} = 56 \text{ t } SO_2/\text{hr} \cdot 24 \text{ hr} \cdot \frac{120 \text{ t } CaSO_3}{64 \text{ t } SO_2}$$

$$= 2520 \text{ t } CaSO_3 \cdot 2000 \text{ lb/ton}$$

$$= 5.0 \times 10^6 \text{ lb (or } 2.3 \times 10^6 \text{ kg)}$$

$$\text{volume of } CaSO_3 = 2520 \text{ t } CaSO_3 \cdot \frac{1 \text{ cm}^3}{2.3 \text{ g}} \cdot \frac{1000 \text{ g}}{2.2 \text{ lb}} \cdot \frac{2000 \text{ lb}}{1 \text{ t}} \cdot \frac{(1 \text{ in})^3}{(2.54 \text{ cm})^3} \cdot \frac{(1 \text{ ft})^3}{(12 \text{ in})^3}$$

$$= 3.5 \times 10^4 \text{ ft}^3 \, CaSO_3 \text{ (or } 990 \text{ m}^3)$$

20.8 A pH 3.0 solution caused by HNO_3, a strong acid, has a $[H^+] = 1 \times 10^{-3}$ mol/L. The moles of acid are given by the concentration times the volume.

$$\text{moles } HNO_3 = 1 \times 10^{-3} \text{ mol/L}(2.0 \times 10^6 \text{ L}) = 2 \times 10^3 \text{ mol}$$

The neutralization reaction is

$$CaO(s) + 2\,H^+(aq) \longrightarrow H_2O(\ell) + Ca^{2+}(aq)$$

so, the pounds of CaO required is given by

$$\text{pounds CaO} = 2 \times 10^3 \text{ mol acid} \cdot \frac{1 \text{ mol CaO}}{2 \text{ mol acid}} \cdot \frac{56.1 \text{ g CaO}}{\text{mol CaO}} \cdot \frac{1 \text{ lb CaO}}{454 \text{ g CaO}}$$

$$= 100 \text{ lbs CaO}$$

20.9 The molar mass of CF_3Cl is 104.5 g/mol.

$$\text{molecules of } CF_3Cl = 100 \text{ t } CF_3Cl \cdot \frac{2000 \text{ lb}}{\text{t}} \cdot \frac{454 \text{ g}}{\text{lb}} \cdot \frac{6.02 \times 10^{23} \text{ molecules}}{104.5 \text{ g}}$$

$$= 5 \times 10^{29} \text{ molecules}$$

$$O_3 \text{ molecules destroyed} = 5 \times 10^{29} \text{ molecules } CF_3Cl \cdot (100{,}000 \; O_3 \text{ destroyed}/CF_3Cl \text{ molecule})$$

$$= 5 \times 10^{34} \text{ molecules } O_3$$

20.10 The sum of the emissions in Table 20.7 is 1246×10^6 pounds.

The surface area of the earth is 5.1×10^8 km^2 or 1.97×10^8 mi^2, and the U.S. surface area is 3.7×10^6 mi^2.

$$\text{atmospheric mass over the U.S.} = 5.1 \times 10^{15} \text{ metric tons} \cdot \frac{3.7 \times 10^6 \text{ mi}^2}{1.97 \times 10^8 \text{ mi}^2}$$

$$= 9.6 \times 10^{13} \text{ metric tons}$$

Converting the mass of the emissions in pounds to metric tons,

$$\text{emissions} = 1246 \times 10^6 \text{ lb} \cdot \frac{1 \text{ kg}}{2.2 \text{ lb}} \cdot \frac{1 \text{ metric ton}}{1000 \text{ kg}}$$

$$= 5.7 \times 10^5 \text{ metric tons}$$

As a percent of the mass of the atmosphere:

$$= \frac{5.7 \times 10^5 \text{ metric tons}}{9.6 \times 10^{13} \text{ metric tons}} \times 100\% = 5.9 \times 10^{-7}\%$$

CHAPTER 21

21.1 Since amino acid molecules contain both the acidic—COOH group and the basic $-NH_2$ group, an acid-base reaction takes place within the molecule to give the zwitterion structure.

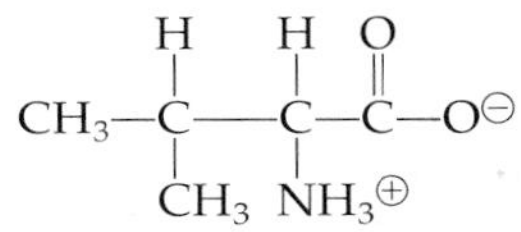

$$CH_3-\underset{CH_3}{\overset{H}{C}}-\underset{NH_3^{\oplus}}{\overset{H}{C}}-\overset{O}{\overset{\|}{C}}-O^{\ominus}$$

zwitterion of valine

21.2 $$H_2N-\underset{\underset{SH}{CH_2}}{\overset{H}{C}}-\overset{O}{\overset{\|}{C}}-\underset{H}{N}-\underset{\underset{C_6H_5}{CH_2}}{\overset{H}{C}}-\overset{O}{\overset{\|}{C}}-\underset{H}{N}-\underset{\underset{OH}{CH_2}}{\overset{H}{C}}-\overset{O}{\overset{\|}{C}}-\underset{H}{N}-\underset{CH_3}{\overset{H}{C}}-COOH$$

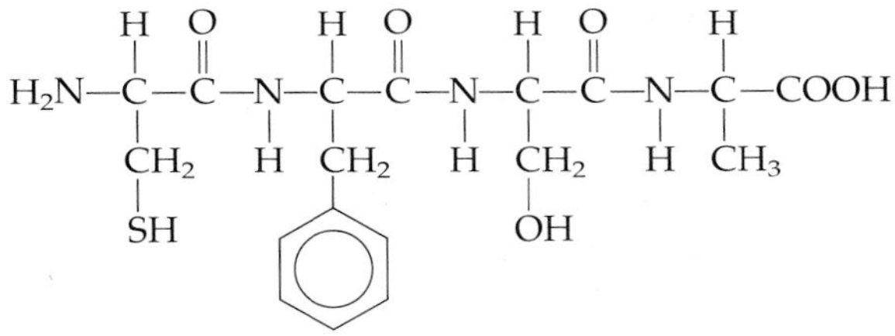

Structure of tetrapeptide Cys-Phe-Ser-Ala

21.3 The effectiveness of an enzyme is related to the match of the shape of the substrate, in this case a dipeptide, to the active site of the enzyme. A right-handed D-dipeptide would not fit the active site of an enzyme that accommodates a left-handed L-dipeptide. Consider the analogy of trying to put a shoe for your left foot on your right foot.

21.4 Humans lack the enzyme that catalyzes the hydrolysis of the β-ring glucose units in cellulose.

21.5 The amount of cholesterol in the average diet is more than is needed in the synthesis of other molecules, such as bile salts and hormones. Cholesterol not needed by the body deposits on the inner walls of arteries along with fatty acids to form atherosclerotic plaque, which reduces blood flow to the heart.

21.6 ...AGGCTA...

APPENDIX

M

Answers to Selected Study Questions

CHAPTER 2

14. (a) carbon
    (b) sodium
    (c) chlorine
    (d) phosphorus
    (e) magnesium
    (f) calcium
16. (a) Li
    (b) Ti
    (c) Fe
    (d) Si
    (e) Co
    (f) Zn
18. 557. g
20. 499. g, 1.10 lb
22. 0.911 g/mL
24. The most likely metal would be aluminum, with a density of 2.70 g/cm^3.
26. 298 K
28. (a) 289 K (b) 310. K (c) 230 K
30. Your body temperature is around 37 °C; therefore gallium would be in a liquid state.
32. 190 mm, 0.19 m, 7.5 inches
34. 1.0×10^2 km/hr
36. 22 cm 28 cm area = 620 cm^2
38. 800. cm^3, 0.800 L, 8.00×10^{-4} m^3
40. 4.10×10^3 cm^3, 4.10 L
42. 5.63×10^{-3} kg, 5.63×10^3 mg
44.

Milligrams	*Grams*	*Kilograms*
693	0.693	0.000693
156	0.156	0.000156
2.23×10^6	2230	2.23

46. 80.1% silver, 19.9% copper
48. 244 g of sulfuric acid
50. 0.197 nm, 197 pm
52. 0.178 nm^3, 1.78×10^{-22} cm^3
54. The troy ounce is larger (31.103 g).
56. 0.995 g of platinum
58. 15.2% lost, 3630 kernels in a pound
60. The water in the cylinder will rise 18.0 mL.
62. The foil is 1.8×10^{-2} mm thick.
64. 293 feet
66. The H_2O occupies a larger volume.
68. 2.5×10^{-9} m; the layer is about one molecule thick.
70. (a) Mg or Gg (b) km or Mm (c) pg, fg, or ag (d) g (e) μm
72. See Chemistry You Can Do; Separation of Dyes

CHAPTER 3

20. (a) 9 (# protons + # neutrons) (b) 48 (c) 70
22. (a) $^{23}_{11}Na$ (b) $^{39}_{18}Ar$ (c) $^{60}_{31}Ga$
24. (a) 20 electrons, 20 protons, 20 neutrons
    (b) 50 electrons, 50 protons, 69 neutrons
    (c) 94 electrons, 94 protons, 150 neutrons
26.

Symbol	^{45}Sc	^{33}S	^{17}O	^{56}Mn
# of protons	21	16	8	25
# of neutrons	24	17	9	31
# of e^-	21	16	8	25

28. ^{241}Am, americium, has an atomic number of 95. Therefore, there are 95 protons and electrons, 146 neutrons.
30. $^{10}_{9}X$, $^{13}_{9}X$, and $^{9}_{9}X$; *not* $^{12}_{10}X$
32. (^6Li mass)(% abundance) + (^7Li mass)(% abundance) = atomic weight of Li
    (6.015121 amu)(0.075) + (7.016003 amu)(0.925) = 6.94 amu.
34. ^{69}Ga abundance is 60.119%, ^{71}Ga abundance is 39.881%
36. Group 4A has five elements in it:

C	carbon	nonmetal
Si	silicon	metalloid
Ge	germanium	metalloid
Sn	tin	metal
Pb	lead	metal

38. Period 7, actinides, all are radioactive, and most are artificial.
40. (a) magnesium and iron
    (b) hydrogen
    (c) silicon
    (d) iron
    (e) three halogens are considered, chlorine is most abundant
42. BeO, MgO, CaO, SrO, BaO, RaO

51.

S	N
B	I

CHAPTER 4

14. (a) $C_{12}H_{22}O_{11}$ has more O atoms and more total atoms
16. (a) C_6H_6 (b) $C_6H_8O_6$ (c) $BaSO_4$
18. (a) calcium—1, carbon—2, oxygen—4
    (b) carbon—8, hydrogen—8
    (c) sulfur—1, nitrogen—2, oxygen—4, hydrogen—8
    (d) platinum—1, nitrogen—2, chlorine—2, hydrogen—6
    (e) iron—1, potassium—4, carbon—6, nitrogen—6
20. (a) $C_3H_6O_3$ (b) $C_6H_8O_7$
22. aluminum: Al^{3+}, selenium: Se^{2-}
24. (a) Mg^{2+} (b) Zn^{2+} (c) Fe^{2+} or Fe^{3+} (d) Ga^{3+}
26. (a) 2 K^+, 1 S^{2-} (b) 1 Ni^{2+}, 1 $SO_4{}^{2-}$ (c) 3 $NH_4{}^+$, 1 $PO_4{}^{3-}$
28. CoO, Co_2O_3
30. (a) $AlCl_3$ (b) NaF (c) correct (d) correct
32. MgO; charges of the ions are greater and ionic radii are smaller; this results in stronger attractions among ions.
34. (a) potassium sulfide (b) nickel(II) sulfate
    (c) ammonium phosphate
36. (a) $(NH_4)_2CO_3$ (b) CaI_2 (c) $CuBr_2$ (d) $AlPO_4$
38. (a) nitrogen trifluoride (b) hydrogen iodide
    (c) boron tribromide (d) hexane
40. (a) C_4H_{10} (b) N_2O_5 (c) C_9H_{20} (d) $SiCl_4$ (e) B_2O_3
42. (a) 27 g (b) 0.48 g (c) 0.698 g (d) 2610 g
44. (a) 1.9998 mol Cu (d) 3.1×10^{-4} mol K
    (b) 0.499 mol Ca (e) 2.1×10^{-5} mol Am
    (c) 0.6208 mol Al
46. 2.19 mol Na
48. 8.0 mL
50. 4.131×10^{23} atoms Cr
52. 1.055×10^{-22} g
54. (a) Fe_2O_3—molar mass: Fe = 55.85 g/mol, O = 16.00 g/mol
    (2)(55.85 g/mol) + (3)(16.00 g/mol) = 159.70 g/mol
    (b) (1)(10.81 g/mol) + (3)(19.00 g/mol) = 67.81 g/mol
    (c) (2)(14.01 g/mol) + (1)(16.00 g/mol) = 44.02 g/mol
    (d) (1)(54.94 g/mol) + (2)(35.45 g/mol) + (4)(18.02 g/mol) = 197.92 g/mol
    (e) (6)(12.01 g/mol) + (8)(1.01 g/mol) + (6)(16.00 g/mol) = 176.14 g/mol
56. (a) 3.12×10^{-2} mol (d) 4.06×10^{-3} mol
    (b) 0.0101 mol (e) 5.99×10^{-3} mol
    (c) 0.0125 mol
58. 40.0 mol
60. (a) 1.80×10^{-3} mol $C_9H_8O_4$
    2.266×10^{-2} mol $NaHCO_3$
    5.205×10^{-3} mol $C_6H_8O_7$
    (b) 1.08×10^{21} molecules
62. 5.67 mol, 3.41×10^{24} molecules, 3.41×10^{24} sulfur atoms, 1.02×10^{25} oxygen atoms ($3 \times 3.41 \times 10^{24}$)
64. (a) molar mass = 239.3 g/mol, 86.60% Pb, 13.40% S
    (b) molar mass = 30.08 g/mol, 79.89% C, 20.11% H
    (c) molar mass = 60.06 g/mol, 40.00% C, 53.28% O, 6.71% H
    (d) molar mass = 80.06 g/mol, 35.00% N, 59.97% O, 5.04% H
66. (a) molar mass = 53.07 g/mol
    (b) 67.91% carbon, 26.40% nitrogen, 5.70% hydrogen
68. The molar mass of the empirical formula = 29 g/mol, therefore the molecular formula is $C_4H_4O_4$.
70. The empirical formula is CH, molecular formula = C_2H_2.
72. The empirical formula is N_2O_3.
74. The empirical formula is CH_2O, molecular formula is $C_2H_4O_2$.
76. Empirical formula is C_2H_6As, molecular formula is $C_4H_{12}As_2$.
78. There are 7 molecules of H_2O per $MgSO_4$ formula unit.
80. There are 6 F atoms for every S atom.

CHAPTER 5

16. (a) $4\,Al(s) + 3\,O_2(g) \longrightarrow 2\,Al_2O_3(s)$
    (b) $N_2(g) + 3\,H_2(g) \longrightarrow 2\,NH_3(g)$
    (c) $2\,C_6H_6(\ell) + 15\,O_2(g) \longrightarrow 6\,H_2O(\ell) + 12\,CO_2(g)$
18. (a) $UO_2(s) + 4\,HF(\ell) \longrightarrow UF_4(s) + 2\,H_2O(\ell)$

(b) $B_2O_3(s) + 6\,HF(\ell) \longrightarrow 2\,BF_3(g) + 3\,H_2O(\ell)$
(c) $BF_3(g) + 3\,H_2O(\ell) \longrightarrow 3\,HF(\ell) + H_3BO_3(s)$

20. (a) $H_2NCl(aq) + 2\,NH_3(g) \longrightarrow NH_4Cl(aq) + N_2H_4(aq)$
(b) $(CH_3)_2N_2H_2(\ell) + 2\,N_2O_4(\ell) \longrightarrow 3\,N_2(g) + 4\,H_2O(g) + 2\,CO_2(g)$
(c) $CaC_2(s) + 2\,H_2O(\ell) \longrightarrow Ca(OH)_2(s) + C_2H_2(g)$

22. (a) $CuCl_2$ (b) $AgNO_3$ (c) KCl, K_2CO_3 and $KMnO_4$

24. (a) $NaCH_3CO_2$ (b) BaS (c) NaOH (d) AgCl (many more correct answers)

26. (a) Na^+, I^- (b) K^+, SO_4^{2-} (c) K^+, HSO_4^- (d) Na^+, CN^-

28. (a) soluble, Ba^{2+} and Cl^-
(b) soluble, Cr^{2+} and NO_3^-
(c) soluble, Pb^{2+} and NO_3^-
(d) insoluble

30.

Soluble	*Insoluble*	
Br^-	CO_3^{2-}	(many more correct answers)
NO_3^-	OH^-	

32. $HNO_3(aq) + H_2O(\ell) \longrightarrow H_3O^+(aq) + NO_3^-(aq)$

34. (a) $Zn(s) + 2\,HCl(aq) \longrightarrow H_2(g) + ZnCl_2(aq)$
$Zn(s) + 2\,H^+(aq) \longrightarrow Zn^{2+}(aq) + H_2(g)$
(b) $Mg(OH)_2(s) + 2\,HCl(aq) \longrightarrow MgCl_2(aq) + 2\,H_2O(\ell)$
$Mg(OH)_2(s) + 2\,H^+(aq) \longrightarrow Mg^{2+}(aq) + 2\,H_2O(\ell)$
(c) $2\,HNO_3(aq) + CaCO_3(s) \longrightarrow Ca(NO_3)_2(aq) + H_2O(\ell) + CO_2(g)$
$2\,H^+(aq) + CaCO_3(s) \longrightarrow Ca^{2+}(aq) + H_2O(\ell) + CO_2(g)$
(d) $4\,HCl(aq) + MnO_2(s) \longrightarrow MnCl_2(aq) + Cl_2(g) + 2\,H_2O(\ell)$
$4\,H^+(aq) + 2\,Cl^-(aq) + MnO_2(s) \longrightarrow Mn^{2+}(aq) + Cl_2(g) + 2\,H_2O(\ell)$

36. (a) $Ba(OH)_2(aq) + 2\,HNO_3(aq) \longrightarrow Ba(NO_3)_2(aq) + 2\,H_2O(\ell)$
not soluble
$H^+(aq) + OH^-(aq) \longrightarrow H_2O(\ell)$
(b) $BaCl_2(aq) + Na_2CO_3(aq) \longrightarrow BaCO_3(s) + 2\,NaCl(aq)$
$Ba^{2+}(aq) + CO_3^{2-}(aq) \longrightarrow BaCO_3(s)$
(c) $2\,Na_3PO_4(aq) + 3\,Ni(NO_3)_2(aq) \longrightarrow Ni_3(PO_4)_2(s) + 6\,NaNO_3(aq)$
$3\,Ni^{2+}(aq) + 2\,PO_4^{3-}(aq) \longrightarrow Ni_3(PO_4)_2(s)$

38. (a) $2\,Mg(s) + O_2(g) \longrightarrow 2\,MgO(s)$ magnesium oxide
(b) $2\,Ca(s) + O_2(g) \longrightarrow 2\,CaO(s)$ calcium oxide
(c) $4\,In(s) + 3\,O_2(g) \longrightarrow 2\,In_2O_3(s)$ indium oxide

40. (a) $2\,K(s) + Cl_2(g) \longrightarrow 2\,KCl(s)$ potassium chloride
(b) $Mg(s) + Br_2(g) \longrightarrow MgBr_2(s)$ magnesium bromide
(c) $2\,Al(s) + 3\,F_2(g) \longrightarrow 2\,AlF_3(s)$ aluminum fluoride

42. (a) $2\,C(s) + O_2(g) \longrightarrow 2\,CO(g)$
(b) $2\,Ni(s) + O_2(g) \longrightarrow 2\,NiO(s)$
(c) $4\,Cr(s) + 3\,O_2(g) \longrightarrow 2\,Cr_2O_3(s)$

44. (a) $BeCO_3(s)$ + heat $\longrightarrow BeO(s) + CO_2(g)$ beryllium oxide
(b) $NiCO_3(s)$ + heat $\longrightarrow NiO(s) + CO_2(g)$ nickel(II) oxide
(c) $Al_2(CO_3)_3(s)$ + heat $\longrightarrow Al_2O_3(s) + 3\,CO_2(g)$ aluminum oxide

46. $Ba(OH)_2(s) + 2\,HNO_3(aq) \longrightarrow Ba(NO_3)_2(aq) + 2\,H_2O(\ell)$

48. $CdCl_2(aq) + 2\,NaOH(aq) \longrightarrow Cd(OH)_2(s) + 2\,NaCl(aq)$
$Cd^{2+}(aq) + 2\,OH(aq) \longrightarrow Cd(OH)_2(s)$

50. $Pb(NO_3)_2(aq) + 2\,KCl(aq) \longrightarrow PbCl_2(s) + 2\,KNO_3(aq)$
lead(II) nitrate + potassium chloride $\longrightarrow$ lead(II) chloride + potassium nitrate

52. $MnCO_3(s) + 2\,HCl(aq) \longrightarrow MnCl_2(s) + H_2O(g) + CO_2(g)$
manganese(II) carbonate + hydrochloric acid $\longrightarrow$ manganese(II) chloride + water + carbon dioxide

54. (a) $MnCl_2(aq) + Na_2S(aq) \longrightarrow 2\,NaCl(aq) + MnS(s)$ precipitation reaction
(b) $Na_2CO_3(aq) + ZnCl_2(aq) \longrightarrow ZnCO_3(s) + 2\,NaCl(aq)$ precipitation reaction
(c) $K_2CO_3(aq) + 2\,HClO_4(aq) \longrightarrow 2\,KClO_4(aq) + H_2O(g) + CO_2(g)$ gas forming

56. (a) $NaOH(aq) + HNO_3(aq) \longrightarrow NaNO_3(aq) + H_2O(\ell)$
(b) $KOH(aq) + HCl(aq) \longrightarrow KCl(aq) + H_2O(\ell)$
(c) $3\,Ca(OH)_2(aq) + 2\,H_3PO_4(aq) \longrightarrow Ca_3(PO_4)_2(s) + 6\,H_2O(\ell)$
(d) $2\,CsOH(aq) + H_2SO_4 \longrightarrow Cs_2SO_4(aq) + 2\,H_2O(\ell)$

58. (a) $NiCl_2(aq) + 2\,NaOH(aq) \longrightarrow Ni(OH)_2(s) + 2\,NaCl(aq)$
(b) $Na_2CO_3(aq) + SrCl_2(aq) \longrightarrow SrCO_3(s) + 2\,NaCl(aq)$
(c) $K_2S(aq) + NiSO_4(aq) \longrightarrow NiS(s) + K_2SO_4(aq)$
(d) $BaCl_2(aq) + CaSO_4(aq) \longrightarrow BaSO_4(s) + CaCl_2(aq)$

60. (a) $K_2CO_3(aq) + 2\,HNO_3(aq) \longrightarrow 2\,KNO_3(aq) + H_2O(g) + CO_2(g)$

(b) $CaCO_3(s) + 2\ HCl(aq) \longrightarrow CaCl_2(aq) + CO_2(g) + H_2O(g)$
(c) $FeCO_3(s) + 2\ HNO_3(aq) \longrightarrow Fe(NO_3)_2(aq) + CO_2(g) + H_2O(g)$

62. (a) $NaOH(aq) + HNO_3(aq) \longrightarrow NaNO_3(aq) + H_2O(\ell)$
(b) $Sr(OH)_2(s) + H_2SO_4(aq) \longrightarrow SrSO_4(s) + 2\ H_2O(\ell)$
(c) $Zn(OH)_2(s) + H_2S(aq) \longrightarrow ZnS(s) + 2\ H_2O(\ell)$
64. (a) precipitation exchange reaction
(b) an oxidation-reduction reaction. Ca is losing electrons and is being oxidized ($Ca(s) \longrightarrow Ca^{2+} + 2\ e^-$). O is gaining electrons and is being reduced ($O_2(g) + 4\ e^- \longrightarrow 2\ O^{2-}$).
(c) an acid-base reaction
66. Common oxidizing agents: O_2, HNO_3, MnO_4^-, and H^+; see Table 5.3.

CHAPTER 6

3. 0.699 g of Ga needed, 0.751 g of As needed
5. 4.5 mol O_2, 310 g Al_2O_3
7. 22.7 g Br_2, 25.3 g Al_2Br_6 formed
9. (a) 318 g Fe
(b) 239 g CO
11. (a) sulfur dioxide, calcium carbonate, oxygen, calcium sulfate, carbon dioxide
(b) 234 g $CaCO_3$ needed
(c) 318 g $CaSO_4$ produced
13. (a) $NH_4NO_3(s) \longrightarrow N_2O(g) + 2\ H_2O(g)$
(b) 5.50 g N_2O produced, 4.50 g H_2O produced
15. (a) Cl_2 is the limiting reactant
(b) 5.07 g $AlCl_3$ can be produced
(c) 1.67 g of Al left over
17. CO is limiting reactant, 85.2 g CH_3OH can be obtained in theory, 1.3 g H_2 left in excess
19. 68.1 g of NH_3 can be formed
21. 73.5% yield of NH_3
23. 88.1% yield
25. 91.6% of weight is the hydrate
27. 84.0% $Al(C_6H_5)_3$ in the sample
29. The empirical formula is CH.
31. The empirical formula is $C_3H_6O_2$.
33. 0.254 M; sodium-ion concentration is 0.508 M, carbonate-ion concentration is 0.254 M.
35. 0.495 g
37. 5080 mL
39. concentration = 0.0150 M
41. Method b would work.
43. 0.12 M $BaCl_2$ splits up into Ba^{2+} ions and Cl^- ions. Ba^{2+} ion concentration is 0.12 M. Cl^- ion concentration is 2 times the Ba^{2+} ion concentration; therefore, it is 0.24 M.
45. 0.206 g of Na_2CO_3 required
47. 60 g of NaOH is formed
49. 193 mL of $Na_2S_2O_3$
51. 40.9 mL
53. 42.5 mL
55. 0.0219 g $C_3H_5O(CO_2H)_3$
57. citric acid
59. 0.500 g
61. 1.50×10^3 mL
63. 0.179 g AgCl formed, NaCl is in excess, 8.33×10^{-3} M
65. 21.6 g $N_2(g)$
67. 86.3 g of Al_2Br_6
69. 402 g
71. (c) $FeCO_5$
73. The concentrations of H^+ and Cl^- are 0.102 M.
75. 0.567 g
77. 31.8% Pb
79. 67.3% Cu
81. $FeBr_3$, $2\ Fe(s) + 3\ Br_2(g) \longrightarrow 2\ FeBr_3$
83. (3) represents the product mixture. (b) is true.
85. (a) 2/1
(b) Carbon
89. (a) The reactions are oxidation-reduction reactions. In the first reaction Au is oxidized and O_2 is reduced. In the second reaction Au is reduced and Zn is oxidized.
(b) 30 L
(c) 0.06 kg
(d) 0.9 cm

CHAPTER 7

16. 400. Calories
18. 1.12×10^4 J
20. Warming water would require more energy.
22. 101 kJ
24. 413 kJ
26. 2.18×10^4 J (required to raise temp. to melting point)
1.12×10^4 J (required to melt lead)
3.30×10^4 J (total)
28. 1.81×10^5 J
30. 329.9 °C
32. 37.3 °C
34. The reaction has the 38 kJ written with the reactants. This means $\Delta H = +38$ kJ; if ΔH is positive, the reaction is endothermic.
36. −1450 kJ/mol

38. 35.6 kJ produced
40. $\Delta H = -1220$ kJ
42. $\Delta H = -434.6$ kJ, 260 kJ evolved from 250 g Pb.
44. $Ag(s) + \frac{1}{2} Cl_2(g) \longrightarrow AgCl(s)$
46. (a) $2 Al(s) + \frac{3}{2} O_2(g) \longrightarrow Al_2O_3(s)$
    $\Delta H_f^\circ = -1675.7$ kJ/mol
    (b) $Ti(s) + 2 Cl_2(g) \longrightarrow TiCl_4(\ell)$
    $\Delta H_f^\circ = -804.2$ kJ/mol
    (c) $N_2(g) + 2 H_2(g) + \frac{3}{2} O_2(g) \longrightarrow NH_4NO_3(s)$
    $\Delta H_f^\circ = -365.56$ kJ/mol
48. $\Delta H^\circ_{rxn} = -98.9$ kJ
50. -905.2 kJ, exothermic
52. -228 kJ
54. 41.2 kJ of heat liberated ($\Delta H^\circ = -41.2$ kJ)
56. -6.62 kJ
58. -394 kJ/mol
60. (a) CO_2 vapor has higher entropy (b) dissolved sugar has higher entropy (c) a beaker with a mixture has higher entropy
62. (a) $AlCl_3(s)$ (b) $CH_3CH_2I(\ell)$ (c) $NH_4Cl(aq)$
64. $\Delta S = 112$ J/K · mol
66. $2 H_2O(\ell) \longrightarrow 2 H_2(g) + O_2(g)$
    $\Delta H > 0$, endothermic; $\Delta S > 0$; reaction is reactant favored at low temperatures, but product favored at high temperatures.
68. (a) product favored at low temperature; reactant favored at high temperature
    (b) reactant favored
70. 4835.0 J/K, product favored
72. $\Delta G^\circ = 28.62$ kJ, reactant favored
74. HF: ΔG° is negative (-80.3 kJ), reaction is product favored; HCl: ΔG° is positive (309.4 kJ), reaction is reactant favored.
76. Because the specific heat is lower for Au, it would heat up faster; it takes less heat to reach a given temperature with a given mass for a substance with a low heat capacity.
78. 75.4 g of ice melted
80. (a) endothermic (b) 2.66 kJ absorbed
82. -165.2 kJ/mol
84. $+15.3$ kJ
86. 2.67×10^8 kJ of heat is evolved in the overall process.
88. -11 kJ
90. 4.4 mol of CH_4 needed, or 70. g
92. -16.67 kJ/g for N_2H_4; -30.0 kJ/g for $N_2H_2(CH_3)_2$; $N_2H_2(CH_3)_2$ is more exothermic per gram.
93. The reaction is reactant favored at 25 °C. be raised to 460K. To become product favored, the temperature must be raised to 460K.

CHAPTER 8

18.

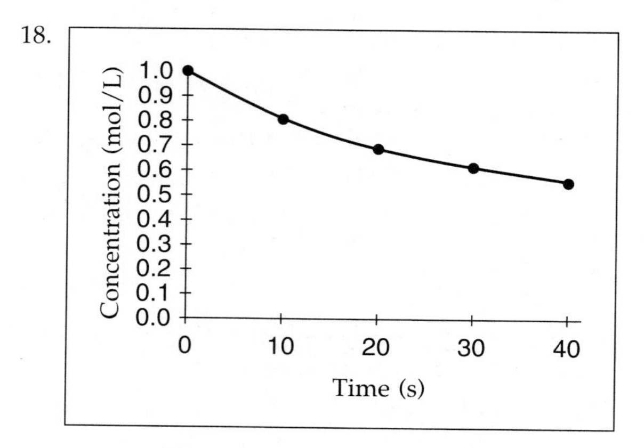

Rate of Change	*Time Interval*
-1.67×10^{-2} mol/L · s	0–10 s
-1.19×10^{-2} mol/L · s	10–20 s
-8.9×10^{-3} mol/L · s	20–30 s
-7.0×10^{-3} mol/L · s	30–40 s

(a) Rate of change decreases because concentration of reactants is decreasing.
(b) Rate of change of [B] is double that of [A] because there are two molecules of B being formed for every molecule of A used up. The rate of change of [B] for the interval from 10 to 20 s is 2.38×10^{-2} mol/L · s.

20. (a) rate = $k[NO_2]^2$ (b) the rate would decrease by $\frac{1}{4}$
    (c) the rate would remain unaffected
22. (a) first order in A, second order in B
    (b) first order in A and B
    (c) first order in A
    (d) third order in A, first order in B
24. (a) rate = -9.0×10^{-4} mol/L · hr
    (b) rate = -1.8×10^{-3} mol/L · hr
    (c) rate = -3.6×10^{-3} mol/L · hr
    As the initial concentration increases, the rate of disappearance of $Pt(NH_3)_2Cl_2$ increases; as the rate of disappearance of $Pt(NH_3)_2Cl_2$ increases, the rate of formation of Cl^- increases equally.
26. (a) When [CO] is doubled and $[NO_2]$ is held constant, the initial rate doubles; first order in CO.
    When $[NO_2]$ is doubled and [CO] is held constant, the initial rate doubles; first order in NO_2.
    rate = $k[CO][NO_2]$
    (b) first order in both reactants
    (c) $k = 1.9$ L/mol · hr
28. (a) When $[CH_3COCH_3]$ is increased by 1.3 times, and both $[Br_2]$ and $[H^+]$ are held constant, the rate in-

creases by 1.3 times. Therefore, the reaction is first order in CH_3COCH_3.
When $[Br_2]$ is doubled, and both $[CH_3COCH_3]$ and $[H^+]$ are held constant, the rate remains the same. Therefore, the reaction is zero order in Br_2.
When $[H^+]$ is doubled, and both $[CH_3COCH_3]$ and $[Br_2]$ are held constant, the rate doubles. Therefore, the reaction is first order in H^+.
rate = $k[H^+][CH_3COCH_3]$
(b) $k = 0.004$ L/mol · s
(c) rate = 2×10^{-5} mol/L · s

30. Activation energy is smaller when $A + B \longrightarrow C + D$. Therefore reaction is exothermic as written.

32. Step 1 $NO_2(g) + F_2(g) \longrightarrow FNO_2(g) + F(g)$
Step 2 $NO_2(g) + F(g) \longrightarrow FNO_2(g)$

$2\ NO_2(g) + F_2(g) \longrightarrow 2\ FNO_2(g)$

Step 1 is rate-determining because it is the slower step.

34. (a) true (b) false (c) false (d) false

36. (a) $H_3O^+(aq)$ is the catalyst, and it is homogeneous
(b) no catalyst
(c) Pt(s) is the catalyst, and it is heterogeneous
(d) no catalyst

38. (a) $K = \dfrac{[O_2][H_2O]^2}{[H_2O_2]^2}$ (c) $K = [CO]^2$
(b) $K = \dfrac{[PCl_5]}{[PCl_3][Cl_2]}$ (d) $K = \dfrac{[H_2S]}{[H_2]}$

40. (a) $K = [Cl_2]$ (c) $K = \dfrac{[CO_2][H_2]}{[H_2O][CO]}$
(b) $K = \dfrac{[Cl^-]^4[Cu^{2+}]}{[CuCl_4{}^{2-}]}$ (d) $K = \dfrac{[Cl_2][Mn^{2+}]}{[Cl^-]^2[H_3O^+]^4}$

42. The relationship would be $K_2 = 1/K_1{}^2$.

44. 1.1×10^{47}

46. [isobutane] = 2.5 mol/L

48. $K = 67$

50. (a) $K = 1.6$ at 986 °C
(b) 6.4×10^{-2} mol of CO and H_2O

52.

Change	*[Br_2]*	*[HBr]*	*K*
Some H_2 added to container	decrease	increase	no change
Temp. of gases inside container is increased	increase	decrease	decrease
The pressure of HBr is increased	increase	increase	no change

54. (a) adding more sulfur(g) shift right
(b) adding more H_2(g) shift right
(c) raising the temperature shift left

56. (b) equilibrium conc. of lead(II) ion will decrease, NaCl is soluble

58. rate = $k[H_2][NO]^2$

60. (c) The $[Ba^{2+}]$ will not change, $BaSO_4$ is insoluble.

62. K = 0.075 at 50 °C.

64. [isobutane] = 0.024 M, [butane] = 0.010 M

66. (a) [isobutane] = 2.9 M, [butane] = 1.1 M
(b) [isobutane] = 2.9 M, [butane] = 1.1 M

68. (a) The slow step is unimolecular; the fast step is bimolecular.
(b) $Ni(CO)_4 \Leftrightarrow \cancel{Ni(CO)_3} + CO$
$\cancel{Ni(CO)_3} + L \longrightarrow Ni(CO)_3L$

$Ni(CO)_4 + L \longrightarrow Ni(CO)_3L + CO$

(c) $Ni(CO)_3$ is an intermediate.
(d) rate = $k[Ni(CO)_4]$
(e) 0.364 g
(f) $\Delta H_{rxn} = +160.81$ kJ

CHAPTER 9

19. (a) orange and red
(b) violet
(c) blue

21. 3.9×10^6 m, 2.4×10^3 miles

23. $\xrightarrow[\text{increasing energy}]{}$
a < b < d < e < c

25. (a) less energy needed than $n = 1 \rightarrow n = 4$ (λ is sufficient)
(b) less energy needed (λ is sufficient)
(c) more energy needed (shorter λ needed)
(d) less energy needed (λ is sufficient)

27. When $n = 4$, there are four subshells: $4s$, $4p$, $4d$, $4f$

29. Mg: $1s^2\ 2s^2\ 2p^6\ 3s^2$ or [Ne] $3s^2$
Cl: $1s^2\ 2s^2\ 2p^6\ 3s^2\ 3p^5$ or [Ne] $3s^2\ 3p^5$

31. (a) [Kr] $5s^2$
(b) [Kr] $4d^{10}\ 5s^2\ 5p^2$

33. (a) ·F꞉ (with 7 dots) (b) ·In꞉ (c) ꞉Te· (d) Cs·

35. (a) $1s^2\ 2s^2\ 2p^6$ Na^+
(b) $1s^2\ 2s^2\ 2p^6$ Al^{3+}
(c) $1s^2\ 2s^2\ 2p^6\ 3s^2\ 3p^6$ Cl^-
Na^+ and Al^{3+} are isoelectronic.

37. Mn [Ar] $3d^5\ 4s^2$

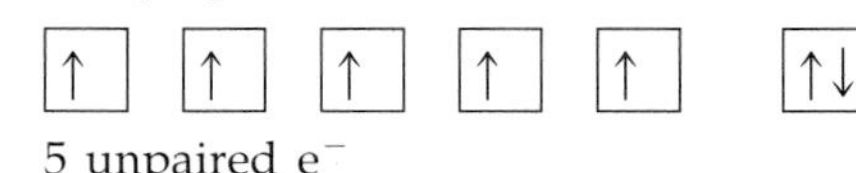

5 unpaired e^-
Mn^{2+} [Ar] $3d^5$

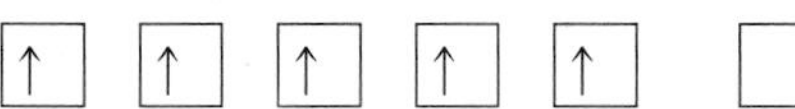

5 unpaired e^-

Mn^{3+} [Ar] $3d^4$

[↑] [↑] [↑] [↑] [ ] [ ]

4 unpaired e^-

39. V: [Ar] $3d^3\ 4s^2$
41. (a) [Xe] $4f^7\ 6s^2$, Eu
    (b) [Xe] $4f^{14}\ 6s^2$, Yb
43. K < Li < C < N
    increase in IE →
45. Smallest IE → Largest IE
    First IE K—First IE Li—First IE Be—2nd IE Na—2nd IE Be
    First IEs increase across a period and decrease down a group. Second IEs are larger than first IEs.
47. C < B < Al < Na < K
49. (a) Cl^- is larger (b) Al is larger (c) In is larger
51. (a) K has the largest atomic radius.
    (b) K < Li < C < N
53. (a) S < O < F
    (b) O
55. $3\ Mg + N_2 \longrightarrow Mg_3N_2$
57. (a) $SiO_2(s) + 2\ C(s) \longrightarrow Si(\ell) + 2\ CO(g)$
    (b) $Si(s) + 2\ Cl_2(g) \longrightarrow SiCl_4(g)$
    (c) $2\ PbS(s) + 3O_2(g) \longrightarrow 2\ PbO(s) + 2\ SO_2(g)$
    (d) $PbO(s) + CO(g) \longrightarrow Pb(s) + CO_2(g)$
    (e) $4\ Ga(s) + 3O_2 \longrightarrow 2\ Ga_2O_3(s)$
    (f) $2\ In(s) + 3\ Br_2(\ell) \longrightarrow 2\ InBr_3(s)$
59. (a) $HNO_3(aq) + KOH(aq) \longrightarrow KNO_3(aq) + H_2O(\ell)$
    (b) $NH_4NO_3(s) + \text{heat} \longrightarrow N_2O(g) + 2\ H_2O(g)$
    (c) $K(s) + O_2(g) \longrightarrow KO_2(s)$
    (d) $UO_2(s) + 4\ HF(aq) \longrightarrow UF_4(s) + 2\ H_2O(\ell)$
61. (a) 19, potassium (b) 7 (c) 12 (d) 0
63. argon
65. (a) sulfur (b) radium (c) nitrogen (d) ruthenium
    (e) copper
67. (a) metal (b) non-metal (c) B (d) B
69. In^{4+}, unlikely because it does not have a noble-gas configuration.
    Fe^{6+}, unlikely because the energy required to remove 6 e^- is too large; Fe^{2+} or Fe^{3+} form much more easily.
    Sn^{5+}, unlikely because the energy required to remove 5 e^- is too large.
71. 200,000 tons
73. (a) S is located in the third period
    (b) $1s^2\ 2s^2\ 2p^6\ 3s^2\ 3p^4$
    (c) Sulfur has the smallest ionization energy; oxygen has the smallest radius.
    (d) 804 g of Cl_2 is needed.
    (e) Less $SOCl_2$ is produced with 20.0 g Cl_2, so Cl_2 is the limiting reactant. The theoretical yield is 16.8 g.

CHAPTER 10

17. (a) :Cl—Si—Cl: with :Cl: above and :Cl: below Si (c) H—O—Cl
    (b) [:O—Cl—O: with :O: below Cl]$^-$ (d) [:O—S—O: with :O: below S]$^{2-}$
19. (a) :Cl—C—F: with H above and :F: below C (b) H—O—C—H with H, H on C (c) H—C—H with H above C and N below C, N bonded to H and H
21. (a) O=C—O: with H on C and H on O (c) H(Cl)C=C(H)H
    (b) H—C—C≡N: with H above and H below the first C
23. (a) :F—Br—F: with :F: below Br (c) :F—Xe—F: with :F: above and :F: below Xe
    (b) [:I—I—I:]$^-$
25. (a) can't (b) can (c) can't (d) can't (e) can
    (f) can't (g) can (h) can
27. (a) B—Cl is shorter
    (b) C—O is shorter
    (c) P—O is shorter
    (d) C=O is shorter
29. The carbon-oxygen bond in carbon monoxide is shorter.
31. The average O—F bond energy is 192 kJ/mol in OF_2.
33. $\Delta H^\circ_{rxn} = -87$ kJ
35. $\Delta H^\circ_{rxn} = -128$ kJ
37. (a) SO_3 resonance structures: O—S(=O)—O ⟷ O—S—O (S=O) ⟷ O=S—O (S—O)
    (b) [:S—C≡N:]$^-$ ⟷ [S=C=N]$^-$ ⟷ [:S≡C—N:]$^-$

39. The average N—O bond length would be longer for NO_3^-. The bonds look more like single bonds, which are longer than double bonds.
41. (a) $\overset{}{\underset{\delta+\ \delta-}{\text{C—O}}}$ more polar than $\underset{\delta+\ \delta-}{\text{C—N}}$
    (b) $\underset{\delta+\ \delta-}{\text{B—O}}$ more polar than $\underset{\delta+\ \delta-}{\text{P—S}}$
    (c) $\underset{\delta+\ \delta-}{\text{P—N}}$ more polar than $\underset{\delta+\ \delta-}{\text{P—H}}$
    (d) $\underset{\delta+\ \delta-}{\text{B—I}}$ more polar than $\underset{\delta+\ \delta-}{\text{B—H}}$
43. (a) The C—H and C═O bonds are polar; the C═C and C—C bonds are nonpolar.
    (b) The most polar bond in the molecule is the C═O bond. The O atom is the negative end of the dipole.
45. (a) H = 1, O = −2 (e) O = −2, Cl = 1
    (b) H = 1, O = −1 (f) H = 1, N = 5, O = −2
    (c) S = 4, O = −2 (g) Bi = 3, O = −2, Cl = −1
    (d) N = 1, O = −2
47. (a) $2\,NaHCO_3 \longrightarrow Na_2CO_3 + H_2O + CO_2$ not redox
    (b) $Ti + 2\,Cl_2 \longrightarrow TiCl_4$ redox
    (c) $2\,H_2O \longrightarrow 2\,H_2 + O_2$ redox
49. (a) oxidation numbers: O = −2, H = +1
    formal charge: O = 6 − 4 − 2 = 0
    H = 1 − 0 − 1 = 0
    (b) oxidation numbers: C = −4, H = +1
    formal charge: C = 4 − 0 − 4 = 0
    H = 1 − 0 − 1 = 0
    (c) oxidation numbers: O = −2, N = +5
    formal charge: O = 6 − 4 − 2 = 0
    N = 5 − 0 − 4 = +1
    (d) oxidation numbers: H = +1, O = 0, F = −1
    formal charge: H = 1 − 0 − 1 = 0
    O = 6 − 4 − 2 = 0
    F = 7 − 6 − 1 = 0
51. (a) :N≡N—Ö: ⟷ N═N═O ⟷ :N—N≡O:
    #1 #2 #3
    (b)

Formal Charges for #1	*Formal Charges for #2*
N_{left} = 5 − 2 − 3 = 0	N_{left} = 5 − 4 − 2 = −1
N_{right} = 5 − 0 − 4 = +1	N_{right} = 5 − 0 − 4 = +1
O = 6 − 6 − 1 = −1	O = 6 − 4 − 2 = 0

Formal Charges for #3
N_{left} = 5 − 6 − 1 = −2
N_{right} = 5 − 0 − 4 = +1
O = 6 − 2 − 3 = +1

    (c) Structure #1 is most reasonable because the negative charge is on the most electronegative atom, oxygen.
53. Silicon atoms can accommodate five or six pairs of valence electrons; carbon atoms cannot.
55. (a) −4 (b) −2 (c) −2 (d) 0 (e) +2 (f) +4
57. Estimated radius of C is 76.8 pm, close to half of 154 pm for C—C which would be twice the radius. Radii transfer pretty well from one molecule to another, but not exactly.
59. (a) $\Delta H^\circ = -870$ kJ
    (b) $\Delta H^\circ = 1310$ kJ
    (c) $\Delta H^\circ = -266$ kJ
61. (a) The shorter carbon-carbon bond is C═C
    (b) The stronger carbon-carbon bond is C═C
    (c) The most polar bond is the C≡N bond, and the negative end is the nitrogen.
63. CH_4 provides the greater amount of heat per gram: 50.2 kJ/g

CHAPTER 11

23. (a) N bonded to H, Cl, H (with lone pair on N) — The central N atom is surrounded by one lone pair and 3 bond pairs. The electron-pair geometry is tetrahedral and the molecular geometry is triangular pyramidal.
    (b) :Cl—O—Cl: — The central O atom is surrounded by 2 lone pairs and 2 bond pairs. The electron-pair geometry is tetrahedral and the molecular geometry is angular.
    (c) [:S═C═N:]⁻ — The central C atom is surrounded by 2 double bonds or 4 bond pairs. We treat double bonds as single bonds when using VSEPR. The ion geometry is linear.
    (d) H—O—F: — There are 2 bond pairs and 2 lone pairs around the central atom. The electron-pair geometry is tetrahedral and the molecular geometry is angular.
25. (a) O═C═O — electron-pair geometry = linear; molecular geometry = linear
    (b) [O—N═O]⁻ ⟷ — electron-pair geometry = triangular planar; molecular geometry = angular
    (c) O═S—O ⟷ — electron-pair geometry = triangular planar; molecular geometry = angular

(d) 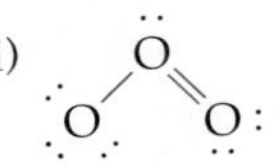 [Lewis structure of O_3] electron-pair geometry = triangular planar; molecular geometry = angular

(e) 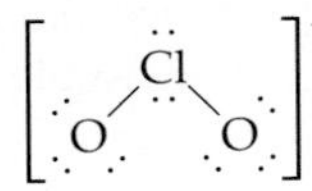 [Lewis structure of ClO_2^-] electron-pair geometry = tetrahedral; molecular geometry = angular

comment: all but CO_2 have angular geometry

27. (a) 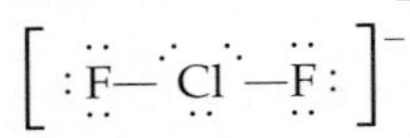 $[:F—Cl—F:]^-$ electron-pair geometry = triangular bipyramidal ($5\ e^-$ pairs); molecular geometry = linear (2 bonding pairs, 3 lone pairs)

(b) 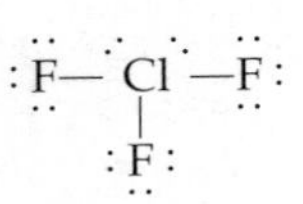 [Lewis structure of ClF_3] electron-pair geometry = triangular bipyramidal ($5\ e^-$ pairs); molecular geometry = T-shaped (3 bonding pairs, 2 lone pairs)

(c) 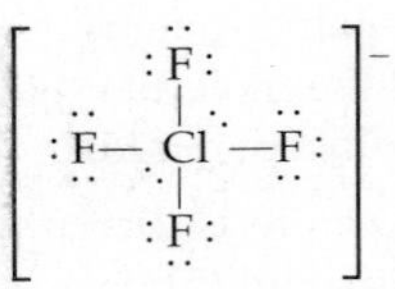 [Lewis structure of ClF_4^-] electron-pair geometry = octahedral ($6\ e^-$ pairs); molecular geometry = square planar (4 bonding, 2 lone pairs)

(d) 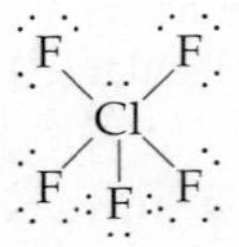 [Lewis structure of ClF_5] electron-pair geometry = octahedral ($6\ e^-$ pairs); molecular geometry = square pyramidal (5 bonding, 1 lone pair)

29. (a) 120° (b) 120° (c) 1 = 109.5°, 2 = 120° (d) 1 = 120°, 2 = 180°

31. (a) SeF_4 forms a seesaw-shaped structure. One F—Se—F bond angle is 120° and the other two are 90°.
(b) 90° and 120°
(c) 90°

33. NO_2^- [Lewis structure] NO_2^+ $[:O=N=O:]^+$

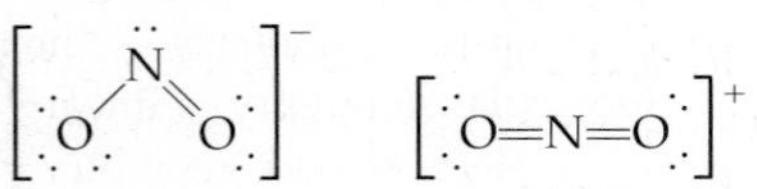

NO_2^- is angular (109.5°), NO_2^+ is linear (180°). Therefore, NO_2^+ would have a greater bond angle.

35. (a) H_2O has the most polar bonds because of the large electronegativity difference (1.4) between O and H.
(b) CO_2 and CCl_4 are not polar.
(c) The F atom is more negatively charged. F is more electronegative than Cl.

37. (a) not polar
(b) 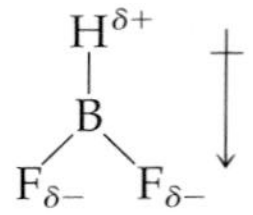 [BF_2H structure: $H^{\delta+}$ on B, $F_{\delta-}$ $F_{\delta-}$; dipole arrow]
(c) 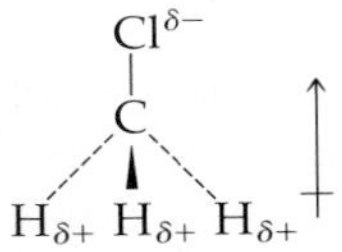 [CH_3Cl structure: $Cl^{\delta-}$ on C, $H_{\delta+}$ $H_{\delta+}$ $H_{\delta+}$; dipole arrow]
(d) not polar

39. $CH_3CH_2CH_2CH_2CH_2CH_3$; $CH_3CH(CH_3)CH_2CH_2CH_3$; $CH_3CH_2CH(CH_3)CH_2CH_3$; $CH_3C(CH_3)_2CH_2CH_3$; $CH_3CH(CH_3)CH(CH_3)CH_3$

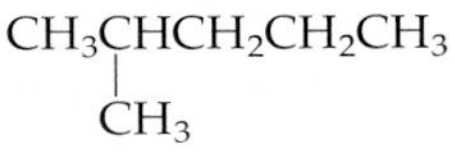

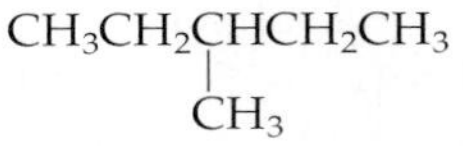

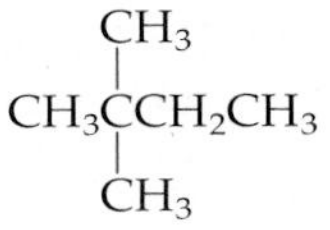

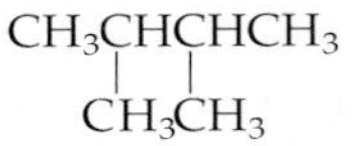

41. (a) $CH_3CH=CHCH_2CH_3$ (b) 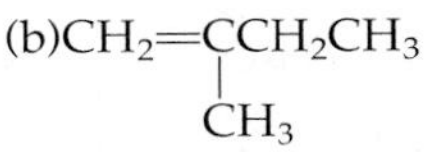 $CH_2=C(CH_3)CH_2CH_3$
(c) $CH_3C(CH_3)=CHCH_3$ (d) 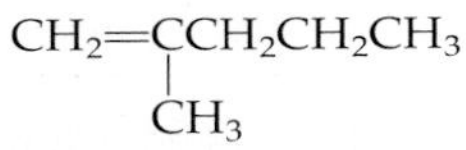 $CH_2=C(CH_3)CH_2CH_2CH_3$
(a), (b), and (c) are all constitutional isomers of each other; (d) isn't

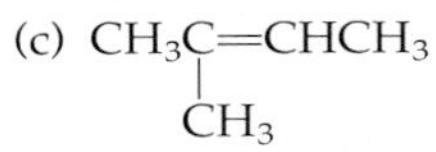

43. (a) 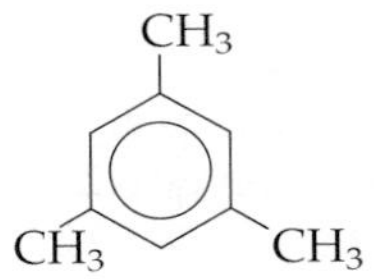 [benzene ring with CH_3, CH_3, CH_3 at 1,3,5 positions]
(b) 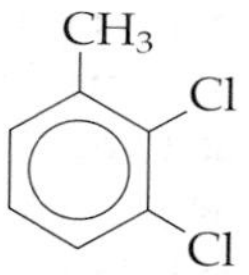 [benzene ring with CH_3, Cl, Cl at 1,2,3 positions]
(c) [benzene ring with Br, Br, Br at 1,2,4 positions]
(d) 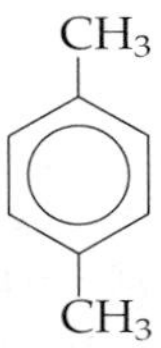 [benzene ring with CH_3 and CH_3 at 1,4 positions]

45. *cis*-2-pentene: [H_3C and CH_2CH_3 on same side of C=C, H and H]
trans-2-pentene: [H_3C and H on top, H and CH_2CH_3 on bottom of C=C]

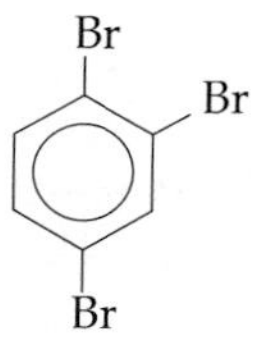

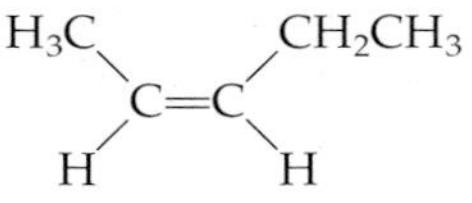

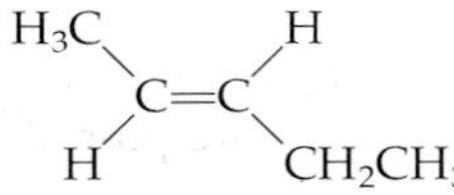

47. (a) The second and third carbons going from left to right are chiral.
(b) None are chiral.
(c) The third carbon going from left to right is chiral.

49. (a) 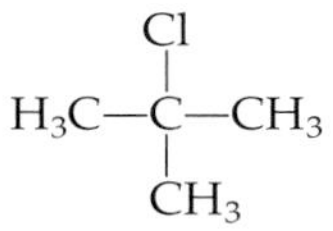 $H_3C—C(Cl)(CH_3)—CH_3$ (b) $H_2C(Cl)—CH(CH_3)—CH_3$

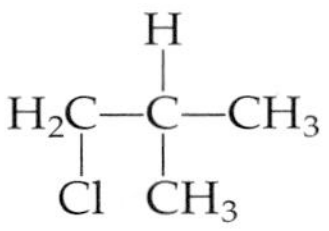

No chiral atoms. No chiral atoms.

(c) —C—C(Br)—C—C—C—

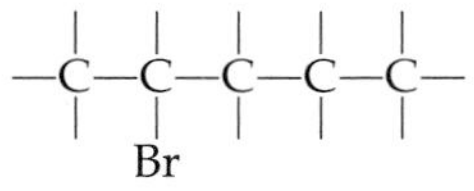

(left to right)
carbon 2 is chiral

(d) 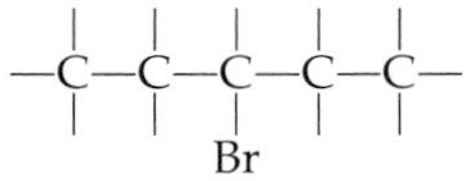 —C—C—C(Br)—C—C— No chiral atoms.

51. 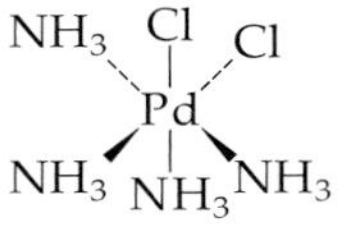 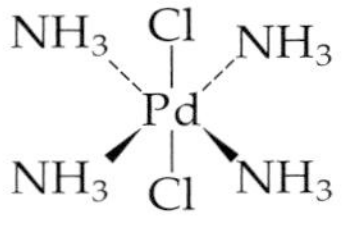 NH_3, Cl, Cl, NH_3, NH_3, NH_3 around Pd — *cis*; NH_3, Cl, NH_3, NH_3, Cl, NH_3 around Pd — *trans*

53. 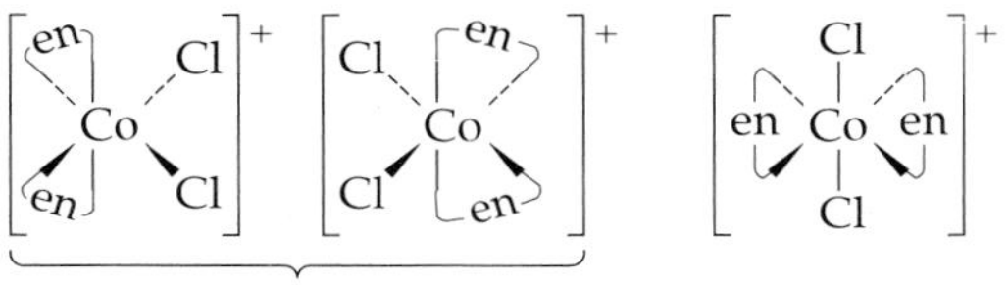 $[Co(en)_2Cl_2]^+$ — optical *cis* isomers; $[Co(en)_2Cl_2]^+$ — *trans* isomer

55. 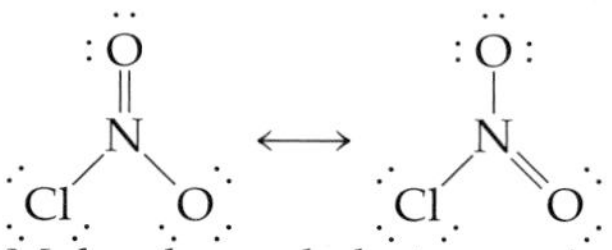 $:\ddot{O}=N(—\ddot{Cl}:)—\ddot{O}:$ ⟷ $:\ddot{O}—N(—\ddot{Cl}:)=\ddot{O}$

Molecular and electron pair geometries are both triangular planar; bond angles are all about 120°.

57. 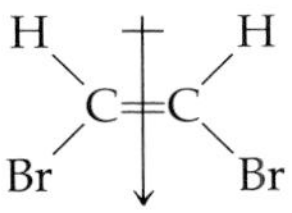 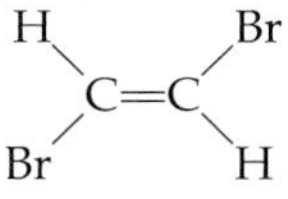 H(Br)C=C(H)(Br) (cis, with dipole arrow) H(Br)C=C(Br)(H) (trans) *trans* has no dipole moment; *cis* has dipole moment as illustrated.

59. (a) angle 1, 120°; angles 2 and 3, 109.5°.
(b) most polar, C=O; N—H is close behind.

61. 1.32×10^{-19} C compared to 1.602×10^{-19} C shows that KF is not completely ionic, but about 82% ionic.

63. 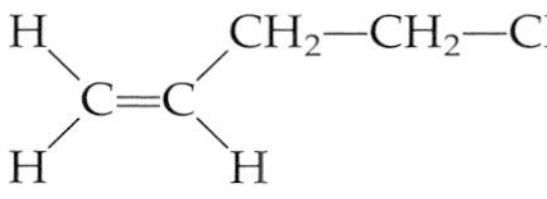 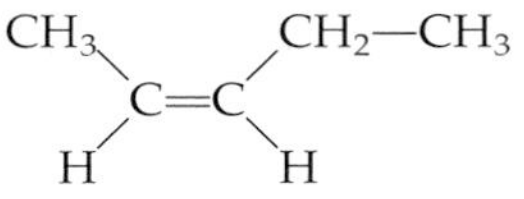 $H_2C=CH—CH_2—CH_2—CH_3$ $CH_3—CH=CH—CH_2—CH_3$

The second structure has *cis/trans* isomers, but otherwise only these two are possible.

65. 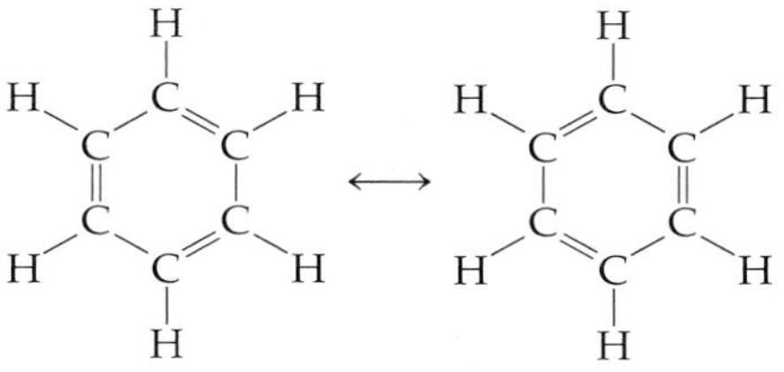 C_6H_6 (benzene resonance structures) ⟷ ; cyclohexane C_6H_{12}

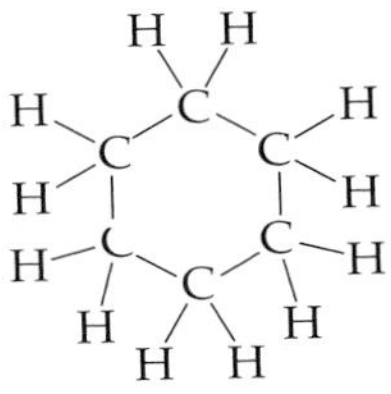

Benzene is planar because each carbon atom is triangular planar. In cyclohexane each carbon atom has four bonds at 109.5° angles, which cause the ring to be puckered.

CHAPTER 12

23. (a) 0.95 atm (b) 950. mm Hg (c) 542 torr (d) 98.7 kPa (e) 7 atm
25. Height of the oil would be 14 m.
27. 62.5 mm Hg
29. 210 mL
32. 26 mL
34. −100° C
36. 4.00 atm
38. 501 mL
40. 0.51 atm
42. −209 °C
44. 0.90 g
45. C_6H_6
46. 3.7×10^{-7} g/mL
48. *n* is proportional to *V*

$$2\,H_2 + O_2 \longrightarrow 2\,H_2O$$

Because it would take half as much O_2 as H_2 and we have 1.5 L of H_2, so we would need 0.75 L of O_2.

50. 31 g of H_2 can be produced.
53. 148 g/mol, the metal is most likely Sr.
54. 1.44 g of $Ni(CO)_4$

55. (a) The average kinetic energy per molecule of CO_2 would be equal to that of H_2 because the temperatures of the flasks are equal.
    (b) The average molecular velocity should be smaller for CO_2 since the CO_2 molecules are heavier than the H_2 molecules.
    (c) The number of molecules of CO_2 must be greater because the pressure is greater, yet the temperatures and volumes are the same.

58. lowest rate of effusion $\xrightarrow{Xe < C_2H_6 < CO < He}$ highest rate of effusion
59. molar mass = 36 g/mol.
63. (a) 1.75×10^{-7} atm (b) 1.75×10^{-7} (c) 500 μg
64. Acetone has a dipole moment because of the electronegativity of the oxygen. Water also has a dipole moment. Dipole-dipole attractions, some hydrogen bonding, and dispersion forces are involved in the interaction.

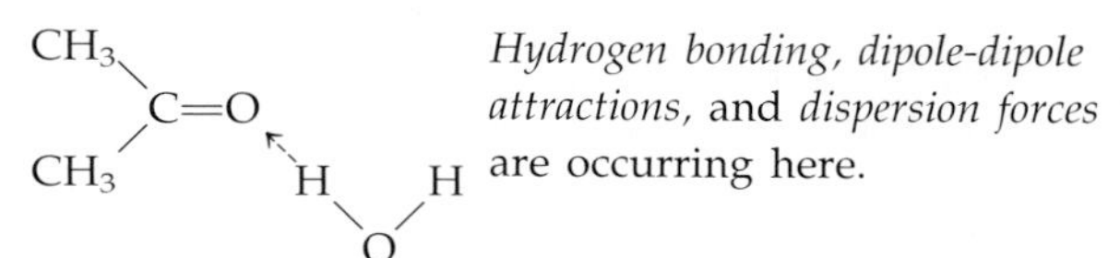

67. 9.46×10^{-2} mm Hg
69. 180. kJ
70. Methyl alcohol has a large boiling point because of hydrogen bonding. Formaldehyde has a lower boiling point because it doesn't hydrogen bond. For these reasons, the same vapor pressure can be achieved at different temperatures.
72. (a) 29 g/mol
    (b) mole fraction of N_2 = 0.83 or 83% N_2
    mole fraction of O_2 = 0.17 or 17% O_2
74. % yield = 73.6%

CHAPTER 13

22. (a) $R-NH_2$ (b) $R-\overset{\overset{O}{\|}}{C}-OH$

    (c) $R-OH$ (d) $R-\overset{\overset{O}{\|}}{C}-R'$

    (e) $R-\overset{\overset{O}{\|}}{C}-H'$ (f) $R-\overset{\overset{O}{\|}}{C}-O-R'$

24. (a) carboxylic acid and alcohol
    (b) ester and carboxylic acid
    (c) ketone and alcohol
    (d) amine and carboxylic acid
    (e) ester and carboxolic acid
    (f) amine, alcohol, and aldehyde

26. $CH_3-\underset{\underset{CH_3}{|}}{\overset{\overset{CH_3}{|}}{C}}-O-CH_3$

28. $2Na(s) + 2CH_3OH(\ell) \longrightarrow H_2(g) + 2Na^+(aq) + 2CH_3O(aq)$

30. (a) $CH_3-CH_2OH \longrightarrow CH_3-\overset{\overset{O}{\|}}{C}-H$

    $CH_3-\overset{\overset{O}{\|}}{C}-H \longrightarrow CH_3-\overset{\overset{O}{\|}}{C}-OH$

    (b) $CH_3-CH_2-CH_2-CH_2-OH \longrightarrow$

    $CH_3-CH_2-CH_2-\overset{\overset{O}{\|}}{C}H$

    $CH_3-CH_2-CH_2-\overset{\overset{O}{\|}}{C}H \longrightarrow$

    $CH_3-CH_2-CH_2-\overset{\overset{O}{\|}}{C}-OH$

32. 90 proof vodka is 45% ethanol

34. $\overset{H\quad H}{\underset{\underset{O}{\|}}{C}}$ aldehyde $CH_3-\underset{\underset{O}{\|}}{C}-CH_3$ ketone

36. (a) $CH_3-CH_2-\overset{\overset{O}{\|}}{C}H$ (b) $CH_3-\overset{\overset{O}{\|}}{C}-CH_3$

38. (a) $H\overset{\overset{O}{\|}}{C}-O-CH_3$ (b) $CH_3-(CH_2)_2-\overset{\overset{O}{\|}}{C}-O-C_2H_5$

    (c) $CH_3-\overset{\overset{O}{\|}}{C}-O-CH_2-CH_2-CH_2-CH_3$

    (d) $CH_3-CH_2-\overset{\overset{O}{\|}}{C}-O-C-\overset{\overset{H}{|}}{C}(CH_3)_2$

42. $\left(\begin{array}{ll} H & CH_3 \\ | & | \\ C- & C- \\ | & | \\ H & C-O-CH_3 \\ & \| \\ & O \end{array}\right)$

44. (a) $\overset{H}{\underset{H}{}}\!>C=C<\!\overset{H}{\underset{H}{}}$ (b) $\overset{H}{\underset{H}{}}\!>C=C<\!\overset{H}{\underset{CH_3}{}}$ (c) $\overset{H}{\underset{H}{}}\!>C=C<\!\overset{H}{\underset{C_6H_5}{}}$ (benzene ring)

    (d) $\overset{H}{\underset{H}{}}\!>C=C<\!\overset{CH_3}{\underset{Cl}{}}$ (e) $\overset{H}{\underset{H}{}}\!>C=C<\!\overset{H}{\underset{C(=O)-OC_2H_5}{}}$

46. $\overset{H}{\underset{H}{}}\!>C=C<\!\overset{H}{\underset{CN}{}}$

48. $\left(-NH-\bigcirc-NH-\overset{\overset{O}{\|}}{C}-\bigcirc-\underset{\underset{O}{\|}}{C}-\right)_n$ (benzene rings)

50. *Compound A* Reacts with sodium to give H_2 (g)—proof of an alcohol. The boiling point being fairly high compared with the other compound is proof of hydrogen bonding.

    $H-\underset{\underset{H}{|}}{\overset{\overset{H}{|}}{C}}-\underset{\underset{H}{|}}{\overset{\overset{H}{|}}{C}}-OH$ Ethanol

    Compound B No hydrogen bonding

    CH_3-O-CH_3 Dimethyl ether

52. $\left(-CH_2\overset{}{\underset{H}{}}C=C\underset{H}{}-CH_2-CH_2\underset{H}{}C=C\underset{H}{}-CH_2-CH_2-\underset{H\quad CN}{C}-CH_2-\right)_n$

54. 1.2 kg of chloroform and 0.40 kg of hydrogen fluoride would be needed.

CHAPTER 14

25. r = 0.219 nm for Xe.
27. 1 Cs^+ ion per unit cell.
    1 Cl^- ion per unit cell.
29. Length of diagonal = 0.700 nm
    Length of side = .404 nm
31. No, the ratio of ions in the unit cell must reflect the empirical formula.
34. The specific heat capacity of ice is smaller than that of liquid water. The heat of fusion of ice would be equal. Since it requires less heat to raise the temperature of ice, b) would require more energy.

35. Heat required = 27 kJ
36. 1970 J given off during freezing. The same amount of heat would have to be added to change from solid to liquid.
38. Total heat evolved = 2180 J
39. Total heat required = 1.6×10^4 kJ
40. The product of the ionic charges is proportional to the stength of attraction of ions in the solid. Therefore, magnesium oxide has a higher melting point than sodium fluoride.
44. The % of empty space = 47.6%
45. (a) ionic
    (b) molecular
    (c) network (quartz)
        amorphous (vitreous silica)
    (d) amorphous
47. highest MP: (b) CaO
    lowest MP: (c) CO
49. greatest electrical conductivity: Ag
    smallest electrical conductivity: P_4
50. greatest electrical conductivity: Rb
    least electrical conductivity: diamond
51. (a) network
    (b) metallic
    (c) ionic
    (d) molecular
58. Using *KC? Discoverer* to graph electrical conductivity vs. thermal conductivity shows that most elements fall on a straight line. The larger the electrical conductivity is, the larger the thermal conductivity is. In general metals are good conductors of electricity and thermal energy. A few elements fall off the line and are exceptions. Farthest off the line are Si, Ge, B, and Te, all semimetals. They have lower electrical conductivity than would be predicted on the basis of their thermal conductivity.

CHAPTER 15

29. 8.50×10^{-4} mol/L
31. 2.5×10^{-4}%
33. 11 grams Pb
35. molality = 1.45 m
    weight % = 8.26%
39. molality = 16.2 m
    weight % = 37.1%
41. weight % = 30.0%
    molarity = 10.0 M
43. 1.9×10^{-3} g/L
45. 85.16 °C
47. Sugar solution < NaCl solution < $CaCl_2$ solution
49. 106.85°C
51. molar mass = 360 g/mol
    molecular formula is $C_{20}H_{16}Fe_2$
53. $CaCl_2 \approx NaCl < Na_2SO_4 \approx$ sugar
55. 981 g ethylene glycol required
57. 0.296 M
59. Solubility = 4.5×10^{-12} M at 25°C
61. Solubility = 1.2×10^{-2} M
63. 3.5×10^{-7} mg/L
65. 170 mg
67. (a) 1×10^{-7} M
    (b) 2×10^{-14} M
    (c) 2×10^{-10} M
69. In pure water: 1.0×10^{-6} M
    In NaSCN solution: 1.0×10^{-10} M

CHAPTER 16

18. (a) $HBr + H_2O \longrightarrow H_3O^+ + Br^-$
    (b) $CF_3COOH + H_2O \rightleftharpoons CF_3COO^- + H_3O^+$
    (c) $HSO_4^- + H_2O \rightleftharpoons SO_4^{2-} + H_3O^+$
    (d) $HNO_2 + H_2O \rightleftharpoons NO_2^- + H_3O^+$
20. (a) HCN, hydrogen cyanide
    (b) HSO_4^-, hydrogen sulfate ion
    (c) H_2S or S^{2-}, hydrogen sulfide or sulfide ion
    (d) HS^-, hydrogen sulfide ion
    (e) H_2SO_4 or SO_4^{2-}, sulfuric acid or sulfate ion
    (f) $HCOO^-$, formate ion
22. (a) reactants: HI(aq) is the acid, $H_2O(\ell)$ is the base
        products: I^-, is the conjugate base, H_3O is the conjugate acid
    (b) reactants: NH_4^+ is the acid, OH^- is the base
        products: NH_3 is the conjugate base, H_2O is the conjugate acid
    (c) reactants: H_2CO_3 is the acid, NH_3 is the base
        products: NH_4^+, is the conjugate acid, HCO_3^- is the conjugate base
24. (a) $H_2SO_3(aq) + H_2O(\ell) \longrightarrow H_3O^+(aq) + HSO_3^-(aq)$
        $HSO_3^-(aq) + H_2O(\ell) \longrightarrow H_3O^+(aq) + SO_3^{2-}(aq)$
    (b) $S^{2-}(aq) + H_2O(\ell) \longrightarrow HS^-(aq) + OH^-(aq)$
        $HS^-(aq) + H_2O(\ell) \longrightarrow H_2S(aq) + OH^-(aq)$
    (c) $NH_3CH_2COOH^+(aq) + H_2O(\ell) \longrightarrow$
        $NH_3CH_2COO(aq) + H_3O^+(aq)$
        $NH_3CH_2COO(aq) + H_2O(\ell) \longrightarrow$
        $NH_2CH_2COO^-(aq) + H_3O^+(aq)$
26. $[H^+] = 5.0 \times 10^{-4}$ M. It is acidic.
28. pH = 2.89, pOH = 11.11
30. $[OH^-] = 4.5 \times 10^{-4}$ M, 9.6×10^{-3} g $Ba(OH)_2$ needed

28. pH = 2.89, pOH = 11.11
30. $[OH^-] = 4.5 \times 10^{-4}$ M, 9.6×10^{-3} g $Ba(OH)_2$ needed
32.

	pH	$[H_3O^+]$(M)	$[OH]^-$(M)	
(a)	1.00	0.10	1.0×10^{-13}	acidic
(b)	10.5	3×10^{-11}	3×10^{-4}	basic
(c)	3.74	1.8×10^{-4}	5.6×10^{-11}	acidic
(d)	9.25	5.6×10^{-10}	1.8×10^{-5}	basic
(e)	9.37	4.3×10^{-10}	2.3×10^{-5}	basic

34. (a) strong acid (b) weak acid (c) amphiprotic (d) weak base (e) CH_4 could be considered a very weak acid. (f) weak base
36. (a) HNO_3 is stronger. (b) HCl is stronger. (c) $HClO_4$ is stronger. (d) H_2SO_4 is stronger.
38. (a) Valeric acid is weaker than glutaric acid, which is weaker than hypobromous acid.
    (b) The pH of hypobromous acid is lower than that of glutaric acid, which is lower than that of valeric acid.
41. (a) Barbituric acid is the stronger acid; it has a higher K_a.
    (b) Nicotinic acid
43. (a) NH_3 will be more basic. (b) K_2S will be more basic.
44. (a) between 2 and 6 (b) between 2 and 6
    (c) between 8 and 12 (d) between 8 and 12
    (e) between 6 and 8 (f) less than 2
    (g) between 2 and 6 (h) between 6 and 8
46. (a) $[H_3O^+] = 2.1 \times 10^{-3}$ M (b) $K_a = 3.4 \times 10^{-4}$
48. $[H_3O^+] = 1.3 \times 10^{-5}$ M $[A^-] = 1.3 \times 10^{-5}$ M
    $[HA] = 0.040$ M
50. (a) $pK_a = 9.25$
    $K_a = 5.6 \times 10^{-10}$
    (b) $pK_a = 7.96$
    $K_a = 1.1 \times 10^{-8}$
    (c) $pK_a = 9.32$
    $K_a = 4.8 \times 10^{-10}$
    (d) $pK_a = 4.70$
    $K_a = 2.0 \times 10^{-5}$
52. pH = 8.85
54. pH = 3.28
56. (a) $OH^- > Na^+ \gg H^+ \gg NaOH$
    (b) $Cl^- > Mg^{2+} > Mg(OH)^+ \approx H^+ \gg OH^- \gg MgCl_2$
    (c) $HCN \gg H^+ > CN^- \gg OH^-$
    (d) $Na^+ > CH_3CH_2COO^- > CH_3CH_2COOH \approx OH^- \gg H^+$
57. (a) Reactant favored, both NH_3 (base) and HBr (acid) are stronger than Br^- (base) and NH_4^+ (acid).
    (b) Product favored, both CH_3COOH (acid) and CN^- (base) are stronger than HCN (acid) and CH_3COO^- (base).
    (c) Product favored, both NH_2^- (base) and H_2O (acid) are stronger than NH_3 (acid) and OH^- (base).
59. (a) HF is the strongest acid, HS^- is the weakest.
    (b) HF has the weakest conjugate base.
    (c) HS^- has the strongest conjugate base.
61. (a) reactant favored: $NH_4^+(aq) + HPO_4^{2-}(aq) \rightleftharpoons NH_3(aq) + H_2PO_4^-(aq)$
    (b) product favored: $CH_3COOH(aq) + OH^-(aq) \rightleftharpoons H_2O(\ell) + CH_3COO^-(aq)$
    (c) reactant favored: $HSO_4^-(aq) + H_2PO_4^-(aq) \rightleftharpoons H_2SO_4(aq) + HPO_4^{2-}(aq)$
    product favored: $HSO_4^-(aq) + H_2PO_4^-(aq) \rightleftharpoons H_3PO_4(aq) + SO_4^{2-}(aq)$
    (d) reactant favored: $CH_3COOH(aq) + F^-(aq) \rightleftharpoons CH_3COO^-(aq) + HF(aq)$
63. (a) pH < 7 (b) pH > 7 (c) pH = 7 (d) pH > 7
65. In general, insoluble, inorganic salts containing anions derived from weak acids tend to be soluble in solutions of strong acids. CO_3^{2-} would be considered an anion derived from a weak acid. SO_4^{2-} would be considered an anion derived from a strong acid.
66. (a) pH = 2, $Cu(OH)_2(s) + 2\,H_3O^+(aq) \rightleftharpoons Cu^{2+}(aq) + 4\,H_2O(\ell)$
    (b) pH = 2, $CuSO_4(s) + H_3O^+(aq) \rightleftharpoons Cu^{2+}(aq) + HSO_4^-(aq) + H_2O(\ell)$
    (c) pH = 2, $CuCO_3(s) + H_3O^+(aq) \rightleftharpoons Cu^{2+}(aq) + HCO_3^-(aq) + H_2O(\ell) \longrightarrow CO_2(g)$
    (d) pH = 2, $CuS(s) + H_3O^+(aq) \rightleftharpoons Cu^{2+}(aq) + HS^-(aq) + H_2O(\ell) \longrightarrow H_2S(g)$
    (e) pH = 2, $Cu_3(PO_4)_2(s) + 2H_3O^+(aq) \rightleftharpoons 3\,Cu^{2+}(aq) + 2\,HPO_4^{2-} + 2\,H_2O(\ell)$
68. First step that would occur for all, at a low pH:
    $MgCO_3(s) \rightleftharpoons Mg^{2+}(aq) + CO_3^{2-}(aq)$
    $CO_3^{2-}(aq) + H_3O^+(aq) \rightleftharpoons HCO_3^-(aq) + H_2O(\ell)$
    $HCO_3^-(aq) + H_3O^+(aq) \rightleftharpoons H_2CO_3(aq) + H_2O(\ell)$
    $H_2CO_3(aq) \longrightarrow CO_2(g) + H_2O(\ell)$
    then:
    (a) add H_2SO_4, $Mg^{2+}(aq) + H_2SO_4(aq) + 2H_2O(\ell) \longrightarrow MgSO_4(aq) + 2H_3O^+(aq)$
    (b) add HBr, $Mg^{2+}(aq) + 2HBr(aq) + 2H_2O(\ell) \longrightarrow MgBr_2(aq) + 2H_3O^+(aq)$
    (c) add $HClO_4$, $Mg^{2+}(aq) + 2HClO_4(aq) + 2H_2O(\ell) \longrightarrow MgClO_4(aq) + 2H_3O^+(aq)$
    (d) add CH_3COOH and NaOH,
    $Mg^{2+}(aq) + 2\,CH_3COOH(aq) + 2\,NaOH(aq) \longrightarrow Mg(CH_3COO)_2(aq) + 2\,Na^+(aq) + 2\,H_2O(\ell)$

69. 26 g benzoic acid needed
71. (b) NaH_2PO_4/Na_2HPO_4 ($pK_a = 7.21$)
73. (a) $H_3PO_4/H_2PO_4^-$
    (b) $H_2PO_4^-/H_2PO_4^{2-}$
    (c) $HC_3H_5O_2/C_3H_5O_2^-$
    (d) HCO_3^-/CO_3^{2-}
75. 4.1 g of NaC_2H_3O needed
77. pH of buffer is 9.55.
    pH of solution is 9.50.
79. (a) 51.3 mL
    (b) 34.8 mL
    (c) 49.3 mL
    (d) 0.0651 mL
81.

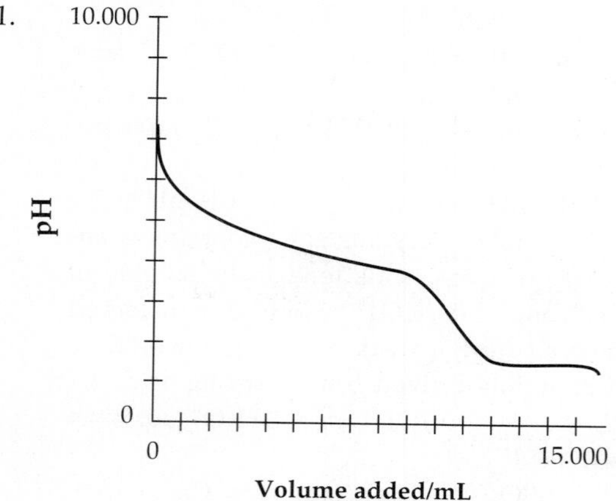

83. (a) A
    (b) D
    (c) 0.10 M
    (d) A and B
    (e) C and D
    (f) approximate pK_a values: A—9
    B—7
    C—5
    D—3
    (g) C
    (h) ammonium ion (any weak acid with a $pK_a \approx 9$)
85. pH = 2.0
87. (a) pH = 11.11
    (b) 5.28
    (c) 9.25
    (d) bromocresol green
89. (a) base
    (b) acid
    (c) acid
    (d) acid
    (e) base

91.

	Lewis acid	*Lewis base*
(a)	$SO_2(aq)$	$H_2O(\ell)$
(b)	$H_3BO_3(aq)$	$OH^-(aq)$
(c)	$Cu^{2+}(aq)$	$NH_3(aq)$
(d)	$SnCl_2(aq)$	$Cl^-(aq)$

93. ICl_3 :Cl: :Cl: I: :Cl: T-shaped molecule

acts as a Lewis acid when it reacts with the chloride ion.

ICl_4^- [:Cl: :Cl: :I: :Cl: :Cl:]$^-$ square planar ion

94. $Na^+ \approx Cl^- \gg H^+ \approx OH^-$
    $K^+ > HCO_3^- \gg H^+ > CO_3^{2-} \gg OH^-$
    $K^+ > HSO_4^- \gg H^+ > SO_4^{2-} \gg OH^-$
    neglecting H_2O molecules for all three solutions
96. pH = 2.82
98. (a) because NaOH is a strong base, full dissociation will occur: pH = 11.70
    (b) pink
    (c) red
100. (a) increase
    (b) stay the same
    (c) stay the same
102. initial pH = 2.25, after NaF pH = 3.62
103. $[CH_3COO^-] > [CH_3COOH] > [H_3O^+] \gg [OH^-]$; ($H_2O(\ell)$ and $Na^+(aq)$ ions are also present.)
    pH of the buffer is 4.95.
    pH of the buffer solution after adding 80 mg of NaOH is 5.04.
    $OH^-(aq) + CH_3COOH(aq) \rightleftharpoons CH_3COO^-(aq) + H_2O(\ell)$
    $CH_3COO^-(aq) + H_2O(\ell) \rightleftharpoons CH_3COOH(aq) + OH^-(aq)$
104. Choose an indicator that changes color in the pH range 6–8. Bromphenol blue will work. Make several standards of known pH to be certain pH changes you measure are accurate. It should be noted the EPA hazardous-waste regulations may not allow adjusting pH to render a waste nonhazardous.
106. reaction by which blood pH can be raised:

$$H_2CO_3(aq) \longrightarrow CO_2(g) + H_2O(\ell)$$
$$H_3O^+(aq) + HCO_3^-(aq) \longrightarrow H_2CO_3(aq)$$

Since $[H_3O^+]$ decreases, $[OH^-]$ increases. When breathing from a paper bag, you breathe mainly your

CHAPTER 17

17. (a) $Cu(s) \longrightarrow Cu^{2+}(aq) + 2e^-$
    $2\ Ag^+(aq) + 2e^- \longrightarrow 2\ Ag(s)$
    (b) $Cu(s) \longrightarrow Cu^{2+}(aq) + 2e^-$ (oxidation) occurs at anode.
    $2\ Ag^+(aq) + 2e^- \longrightarrow 2\ Ag(s)$ (reduction) occurs at cathode.
19. (a) $2\ Li(s) \longrightarrow 2\ Li^+(aq) + 2e^-$ (oxidation) $E° = -(-3.045\ V)$
    $F_2(g) + 2e^- \longrightarrow 2F^-(aq)$ (reduction) $E° = +2.87\ V$
    (b) $E° = 5.92\ V$
    $\Delta G° = -nFE° = -1.14 \times 10^6\ J$
21. (a) $V^{2+}(aq)$ (b) $Cl_2(g)$ (c) $V(s)$
    (d) $Cl^-(aq)$ (e) No (f) No
    (g) $I_2(s) + Cl_2(g)$
23. (a) $Cd(s) + Ni^{2+}(aq) \longrightarrow Cd^{2+}(aq) + Ni(s)$.
    (b) Cadmium is oxidized, nickel is reduced. Cadmium is the reducing agent, and nickel is the oxidizing agent.
    (c) Cd is the anode, Ni is the cathode. The Cd electrode is negative.
    (d) $E°_{total}$ 0.15 V
    (e) from right to left
    (f) toward the Cd compartment
25. $\Delta G = -6.66 \times 10^5\ J$
27. (a) $2\ Al(s) + 3\ Cl_2(g) \longrightarrow 2\ Al^{3+}(aq) + 6\ Cl^-(aq)$
    $2\ Al(s) \longrightarrow 2\ Al^{3+}(aq) + 6e^-$, $E° = 1.66\ V$
    $3\ Cl_2(g) + 6e^- \longrightarrow 6\ Cl^-(aq)$, $E° = 1.360\ V$
    (b) The $Al(s)/Al^{3+}(aq)$ half reaction occurs at the anode, negative polarity.
    The $Cl_2(g)/Cl^-(aq)$ half reaction occurs at the cathode, positive polarity.
    (c) $E°_{net} = 1.66\ V + 1.360\ V = 3.02\ V$
29. 0.16 g Ag
31. 5.93 g Cu
33. 3×10^5 g Al(s)
35. 190 g Pb
37. 0.1 g Zn is consumed.
39. 0.04 g Li(s) used.
    0.6 g Pb(s) used.
41. (a) As$[Cd^{2+}]$ is reduced, the cell emf will be larger.
    (b) 470 hr
43. (a) 9×10^6 g HF is consumed
    (b) 1700 kwh
45. 6.9×10^{-3} g Cu plates out.
    2.0×10^{-3} g Al plates out.

CHAPTER 18

15. (a) $2Na^+(aq) + 2Cl^-(aq) + 2H_2O(\ell) \longrightarrow H_2(g) + Cl_2(g) + 2NaOH(aq)$
    (b) Yes. No; electrolysis is not the only source of those chemicals.
17. 27 hr
19. (a) 1700 kwh
    (b) 1.1×10^5 g HF is consumed
    5.4×10^3 g H_2 is produced
    1.0×10^5 g F_2 is produced
21. 3.50×10^4
23. 4.1×10^7 g
25. (a) $Ca^{2+}(\ell) + 2e^- \longrightarrow Ca(\ell)/2Cl^-(\ell) \longrightarrow Cl_2(g) + 2e$
    (b) 28 g
27. (a) 2.06×10^5 g
    (b) 7.2×10^3 L.
29. approximately 10 hr; (3.62×10^4 s)

CHAPTER 19

11. (a) $^{56}_{26}Fe$ (b) $^{1}_{0}n$ (c) $^{32}_{15}P$ (d) $^{97}_{43}Tc$ (e) $^{0}_{-1}e$
13. (a) $^{0}_{-1}e$ (b) $^{87}_{37}Rb$ (c) $^{4}_{2}He$ (d) $^{226}_{88}Ra$ (e) $^{0}_{-1}e$
    (f) $^{24}_{11}Na$
15. 1st step: $^{235}_{92}U \longrightarrow {}^{4}_{2}He + {}^{231}_{90}Th$
    2nd step: $^{231}_{90}Th \longrightarrow {}^{0}_{-1}\beta + {}^{231}_{91}Pa$
    3rd step: $^{231}_{91}Pa \longrightarrow {}^{4}_{2}He + {}^{227}_{89}Ac$
    4th step: $^{227}_{89}Ac \longrightarrow {}^{0}_{-1}\beta + {}^{227}_{90}Th$
    5th step: $^{227}_{90}Th \longrightarrow {}^{4}_{2}He\ {}^{223}_{88}Ra$
17. ^{10}B binding energy $= 6.26 \times 10^{11}$ J/nucleon
    ^{11}B binding energy $= 6.70 \times 10^{11}$ J/nucleon
    ^{11}B is more stable and is therefore more abundant
19. 0.469 mg remain.
21. (a) $^{131}_{53}I \longrightarrow {}^{0}_{-1}\beta + {}^{131}_{54}Xe$
    (b) 1.56 mg remain.
23. (a) 0.625 mg remain after 21.2 yr.
    (b) 1.91×10^{-5} mg remain after 100 yr.
25. 34.8 days
27. The object is about 1850 years old, and so was made in approximately 150 AD.
29. $^{239}_{94}Pu + 2\ {}^{1}_{0}n \longrightarrow {}^{0}_{-1}\beta + {}^{241}_{95}Am$
31. $^{12}_{6}C$
33. 1600 tons of coal
35. 130 mL

CHAPTER 20

33. 0.2 metric tons of $CaCO_3$ needed.
35. 2×10^4 metric tons of coal need to be burned.
37. 1.0×10^{-3} mol CO

CHAPTER 20

33. 0.2 metric tons of $CaCO_3$ needed.
35. 2×10^4 metric tons of coal need to be burned.
37. 4.2×10^{-5} mol CO
39. (a) $C_8H_{18}(\ell) + 25/2\ O_2(g) \longrightarrow 8\ CO_2(g) + 9\ H_2O(g)$
    (b) $1. \times 10^3$ L CO_2 would be produced.
41. (a) $H_2S + O_3 \longrightarrow SO_2 + H_2O$
    (b) $\approx 3 \times 10^8$ metric tons of H_2SO_4 would be produced.
43. 100 ppm corresponds to 0.0001 atm at STP.
    20,000 ppm corresponds to 0.02 atm at STP.

CHAPTER 21

19. The L-form, common in nature, is most likely the form that this peptide follows in synthesis.

$$NH_2-\underset{\displaystyle CH_2CH_2C(=O)OH}{CH}-\overset{O}{\overset{\|}{C}}-NH-\underset{\displaystyle CH_2SH}{CH}-\overset{O}{\overset{\|}{C}}-NH-\underset{\displaystyle H}{CH}-\overset{O}{\overset{\|}{C}}-OH$$

21. Phe-ser-val
    Phe-val-ser
    Ser-val-phe
    Ser-phe-val
    Val-ser-phe
    Val-phe-ser

23. $$NH_2-\underset{\displaystyle CH_3}{CH}-\underset{\displaystyle O}{\underset{\|}{C}}-NH-\underset{\displaystyle H}{CH}-\underset{\displaystyle O}{\underset{\|}{C}}-NH-\underset{\displaystyle CH_2C_6H_5}{CH}-\overset{O}{\overset{\|}{C}}-OH$$

25. When a protein is denatured, a physical or chemical process occurs that changes the protein's secondary, tertiary, or quaternary structure and makes it incapable of performing its normal function.
27. Both proteins and nylon are polyamides. *If* nylon occurred naturally, it wouldn't be a protein. Proteins have one carbon with an "R" group sandwiched between two peptide bonds. Nylon has a larger number of carbons between the amide bonds (peptide bonds in protein).
29. Amylopectin.
31. While both amylose and amylopectin have α-D-glucose as monomers, amylose is straight-chained and soluble in hot water, whereas amylopectin is arranged in a branched chain. Amylose appears blue/black when given the I_2 test. Amylopectin appears red during same test.
33. (a) deoxyribose (b) ribose
35. Canola oil is very low in saturated fat, the most harmful fat for humans.
37. The codon is the sequence message (in a three-base code) carried by the m-RNA (later fits the anti-codon). The anti-codon is the trinucleotide base sequence at one end of t-RNA that fit the bases of the m-RNA (the codon).
39. Complementary bases are the bases found in DNA and m-RNA. The m-RNA bases are structured so they "fit" the DNA bases.
41. (a) ACAGUCACCCGGCGA
    (b) UGUCAGUGGGCCGCU
    (c) Thr-val-thr-arg-arg
43. An enzyme increases the rate of a biological reaction by acting as a catalyst for a *specific* reaction.
45. (a) amylopectin (b) α-2-deoxy-D-ribose (c) sucrose
    (d) glucose
47. $6CO_2 + 6H_2O + 2880\ kJ_{(sunlight)} \longrightarrow C_6H_{12}O_6 + 6O_2$
49. (a) ATP (adenosine triphosphate) stores energy in the form of phosphorous/oxygen bonds. ADP (adenosine diphosphate) has one less P—O bond. When ATP becomes ADP, a P—O bond is broken, and energy is released (30 kJ/mol).
    (b) The conversion of ATP to ADP provides the energy required to form glucose and other carbohydrates from CO_2 and H_2O in the process called photosynthesis.
51. (a) 80 L (b) 80 L (c) 400 food calories

Index/Glossary

Note: Page numbers in *italics* refer to illustrations; page numbers followed by t refer to tables. Glossary terms, printed in **boldface**, are defined here as well as in the text. Some terms used generally in the text are defined here without giving specific page references.

The Elements

Name	Symbol	Atomic Number	Atomic Weight*
Actinium	Ac	89	(227)
Aluminum	Al	13	26.981539
Americium	Am	95	(243)
Antimony	Sb	51	121.75
Argon	Ar	18	39.948
Arsenic	As	33	74.92159
Astatine	At	85	(210)
Barium	Ba	56	137.327
Berkelium	Bk	97	(247)
Beryllium	Be	4	9.012182
Bismuth	Bi	83	208.98037
Boron	B	5	10.811
Bromine	Br	35	79.904
Cadmium	Cd	48	112.411
Calcium	Ca	20	40.078
Californium	Cf	98	(251)
Carbon	C	6	12.011
Cerium	Ce	58	140.115
Cesium	Cs	55	132.90543
Chlorine	Cl	17	35.4527
Chromium	Cr	24	51.9961
Cobalt	Co	27	58.93320
Copper	Cu	29	63.546
Curium	Cm	96	(247)
Dysprosium	Dy	66	162.50
Einsteinium	Es	99	(252)
Erbium	Er	68	167.26
Europium	Eu	63	151.965
Fermium	Fm	100	(257)
Fluorine	F	9	18.9984032
Francium	Fr	87	(223)
Gadolinium	Gd	64	157.25
Gallium	Ga	31	69.723
Germanium	Ge	32	72.61
Gold	Au	79	196.96654
Hafnium	Hf	72	178.49
Helium	He	2	4.002602
Holmium	Ho	67	164.93032
Hydrogen	H	1	1.00794
Indium	In	49	114.82
Iodine	I	53	126.90447
Iridium	Ir	77	192.22
Iron	Fe	26	55.847
Krypton	Kr	36	83.80
Lanthanum	La	57	138.9055
Lawrencium	Lr	103	(260)
Lead	Pb	82	207.2
Lithium	Li	3	6.941
Lutetium	Lu	71	174.967
Magnesium	Mg	12	24.3050
Manganese	Mn	25	54.93805
Mendelevium	Md	101	(258)
Mercury	Hg	80	200.59
Molybdenum	Mo	42	95.94
Neodymium	Nd	60	144.24
Neon	Ne	10	20.1797
Neptunium	Np	93	(237)
Nickel	Ni	28	58.69
Niobium	Nb	41	92.90638
Nitrogen	N	7	14.00674
Nobelium	No	102	(259)
Osmium	Os	76	190.2
Oxygen	O	8	15.9994
Palladium	Pd	46	106.42
Phosphorus	P	15	30.973762
Platinum	Pt	78	195.08
Plutonium	Pu	94	(244)
Polonium	Po	84	(209)
Potassium (Kalium)	K	19	39.0983
Praseodymium	Pr	59	140.90765
Promethium	Pm	61	(145)
Protactinium	Pa	91	231.03588
Radium	Ra	88	(226)
Radon	Rn	86	(222)
Rhenium	Re	75	186.207
Rhodium	Rh	45	102.90550
Rubidium	Rb	37	85.4678
Ruthenium	Ru	44	101.07
Samarium	Sm	62	150.36
Scandium	Sc	21	44.955910
Selenium	Se	34	78.96
Silicon	Si	14	28.0855
Silver	Ag	47	107.8682
Sodium	Na	11	22.989768
Strontium	Sr	38	87.62
Sulfur	S	16	32.066
Tantalum	Ta	73	180.9479
Technetium	Tc	43	(98)
Tellurium	Te	52	127.60
Terbium	Tb	65	158.92534
Thallium	Tl	81	204.3833
Thorium	Th	90	232.0381
Thulium	Tm	69	168.93421
Tin	Sn	50	118.710
Titanium	Ti	22	47.88
Tungsten	W	74	183.85
Unnilennium†	Une	109	(266)
Unnilhexium†	Unh	106	(263)
Unniloctium†	Uno	108	(265)
Unnilpentium†	Unp	105	(262)
Unnilquadium†	Unq	104	(261)
Unnilseptium†	Uns	107	(262)
Uranium	U	92	238.0289
Vanadium	V	23	50.9415
Xenon	Xe	54	131.29
Ytterbium	Yb	70	173.04
Yttrium	Y	39	88.90585
Zinc	Zn	30	65.39
Zirconium	Zr	40	91.224

* Based on relative atomic mass of ^{12}C = 12; 1987 IUPAC values. Values in parentheses are the mass numbers of the isotopes of the longest half-life.

† In 1992 the following names were proposed for elements 107 to 109: 107, nielsbohrium (Ns); 108, hassium (Hs); and 109, meiterium (Mt). These will not become official until they are approved by the International Union of Pure and Applied Chemistry (IUPAC). Previously, the names rutherfordium and hahnium had been proposed for elements 104 and 105. No name has been proposed for element 106.

Top 25 Chemicals in the United States, 1992

Rank	Name	Production (billions of pounds)	How Made	End Uses
1.	Sulfuric acid	88.8	Burning sulfur to SO_2, oxidation of SO_2 to SO_3, reaction with water. Also recovered from metal smelting.	Fertilizers, petroleum refining, manufacture of metals and chemicals.
2.	Nitrogen	58.7	Separated from liquid air.	Blanketing atmospheres for metals, electronics, etc., freezing agent for foods, ammonia production.
3.	Oxygen	42.4	Separated from liquid air.	Steel production, metal fabricating and chemical processing.
4.	Ethylene	40.4	Cracking hydrocarbons from oil and natural gas.	Plastics, antifreeze production, fibers and solvents.
5.	Ammonia	36.0	Catalytic reaction of nitrogen, air, and hydrogen.	Fertilizers, plastics, fibers and resins.
6.	Calcium oxide (lime)	34.7	Heating limestone ($CaCO_3$).	Steel production, water treating, refractions, pulp and paper.
7.	Phosphoric acid	25.4	Sulfuric acid reacted with phosphate rock; elemental phosphorus burned and dissolved in water.	Fertilizers, detergents, and water-treating compounds.
8.	Sodium hydroxide	24.0	Electrolysis of NaCl solution.	Chemicals, pulp and paper, aluminum, textiles, oil refining.
9.	Propylene	22.6	Cracking oil and oil products.	Plastics, fibers, solvents.
10.	Chlorine	22.3	Electrolysis of NaCl, recovery from HCl users.	Chemical production, plastics, solvents, pulp and paper.
11.	Sodium carbonate	20.9	Trona ore, made from NaCl and limestone with ammonia.	Glass, chemicals, paper and pulp.
12.	Urea	16.8	React NH_3 and CO_2 under pressure.	Fertilizers, animal feeds, adhesives and plastics.
13.	Nitric acid	16.1	Oxidation of ammonia to nitrogen dioxide and dissolved in water.	Ammonium nitrate and phosphate fertilizers, nitro explosives, plastics, dyes and lacquers.
14.	Ethylene dichloride	15.9	Chlorination of ethylene.	Production of vinyl chloride.
15.	Ammonium nitrate	15.3	Reaction of ammonia and nitric acid.	Explosives, fertilizers, source of laughing gas, matches.
16.	Vinyl chloride	11.7	Dehydrochlorination of ethylene dichloride.	Polymers, films, coatings and moldings.
17.	Benzene	12.0	From oil and coal tar.	Polystyrene, other resins, nylon, and rubber.
18.	Ethylbenzene	11.0	Alkylation of benzene.	Production of styrene.
19.	Carbon dioxide	10.9	Burn hydrocarbons, heat limestone.	Production of urea, sodium carbonate, beverages; fire extinguishers.
20.	Methyl *tert*-butyl ether	10.9	Acid-catalyzed reaction of methanol with isobutene.	Gasoline additive.
21.	Styrene	8.9	Dehydrogenation of ethylbenzene.	Polymers, rubber, polyesters.
22.	Methanol	8.7	From natural gas. Methane oxidized to CO and H_2; catalytic conversion to alcohol.	Polymers, adhesives.
23.	Formaldehyde (37%)	7.0	Oxidation of methyl alcohol.	Adhesives, plastics.
24.	Xylene	6.4	From oil and oil cracking products.	Production of terephthalic acid.
25.	Toluene	6.0	Catalytic reforming of petroleum hydrocarbons.	High octane gasolines, urethane resins, solvent, intermediate to make TNT, benzaldehyde and benzoic acid.

TOP 25 CHEMICAL PRODUCERS IN THE UNITED STATES, 1992
(Chemical Sales in Billions of Dollars)*

Du Pont (15.5), Dow Chemical (12.9), Exxon (10.6), Hoechst Celanese (6.50), Monsanto (5.38), Union Carbide (4.87), General Electric (4.85), Occidental Petroleum (4.23), BASF (4.05), Eastman Kodak (3.93), Mobil (3.71), Amoco (3.69), W. R. Grace (3.48), Shell Oil (3.28), Arco Chemical (3.10), Rohm & Haas (3.06), Chevron (2.87), Air Products (2.81), Miles (2.65), Praxair (2.60), Allied-Signal (2.60), Ciba-Geigy (2.52), Ashland Oil (2.49), American Cyanamid (2.37), Rhone-Poulenc (2.30).

*Chemical sales are only a fraction of total sales. For example, the chemical sales for DuPont were 41.1% of the total sales. In contrast, the chemical sales for Dow Chemical were 68.1% of the total sales. Data from *Chemical and Engineering News,* June 28, 1993.